Modern Automotive Technology

James E. Duffy
Automotive Writer

Publisher
THE GOODHEART-WILLCOX COMPANY, INC.
Tinley Park, Illinois

Library of Congress Cataloging in Publication Data
Duffy, James E.
  Modern automotive technology / James E. Duffy.

    p. cm.
  Includes index.
  ISBN 1-56637-610-6
    1. Automobiles—Design and construction.   I. Title.
  TL146.D84 2000
  629.2'3—dc21   99-10562
                                                                    CIP

Introduction

Welcome to the exciting world of automotive technology! **Modern Automotive Technology** is an easy-to-understand, up-to-date book summarizing the operation and repair of all makes and models of vehicles. The text uses a building-block approach that starts with the simple and progresses gradually to the more complex. Short sentences, concise definitions, and thousands of color illustrations will help you learn quickly and easily.

No longer can the untrained person hope to fix the modern automobile. Multiple on-board computers are now used to monitor and control the engine, transmission, suspension, braking, emission control, and other systems. Although computer systems are discussed in almost every chapter, three chapters—Chapter 17, *Computer System Fundamentals;* Chapter 18, *On-Board Diagnostics and Scan Tools;* and Chapter 19, *Computer System Service*—explain the operation and service of these important systems in detail.

Additionally, Chapter 46, *Advanced Diagnostics,* emphasizes the use of the latest diagnostic equipment and techniques to locate engine performance problems. A few of the other topics discussed in the text include anti-lock brakes, four-wheel steering, four-valve cylinders, active suspension systems, passive restraint systems, security systems, and navigation systems.

The 2000 edition of **Modern Automotive Technology** is organized around the eight ASE automobile test areas and is correlated to the NATEF Task List. Terminology has been updated throughout the text to reflect the SAE J1930 standard.

Each automotive system is presented in two or more chapters. The first chapter explains the construction and operation of a specific system. The following chapter expands on this by detailing the troubleshooting and repair of the same system.

Modern Automotive Technology is a valuable reference for anyone interested in the operation, construction, and repair of automobiles and light trucks. Vehicle owners who need a general guide to automotive service will find the book both interesting and informative. Those who are preparing for a career in automotive technology will find the text a "must." Experienced technicians can use it when preparing for the ASE certification tests.

Type Styles Used in This Text

Various type styles are used throughout this text to emphasize words, identify important terms, and highlight figure references.

Italic type is used to emphasize words and terms. For example, the word *not* is often printed in italic type when it is imperative that an operation be avoided.

Important terms appear in ***bold-italic type.*** These terms are defined when introduced and most are listed in the Important Terms list at the end of the chapter, as well as in the Glossary at the back of the text. Study the ***bold-italic terms*** carefully.

Figure references in the body of the text and in the captions always appear in **bold type.** This makes them easy to identify.

Chapter Components

Each chapter opens with a list of *learning objectives.* These objectives identify the topics covered and goals to be achieved in the chapter. Review the objectives before reading the chapter to determine what you can expect to learn. After completing the chapter, read the objectives once more and make sure you have met each objective.

A *summary* is found at the end of each chapter. The summary highlights the material covered in the chapter. Review the summary after completing a chapter.

A list of *important terms* is also included at the end of each chapter. The terms in the list appear in the order in which they are presented in the chapter. After completing a chapter, review each term. If a term cannot be defined, review the related section in the chapter.

Conventional *review questions,* as well as a separate section of *ASE-type questions,* are presented at the end of each chapter. After completing a chapter, answer all the questions on a separate sheet of paper. This is a great way of reviewing the material presented in the chapter. It will also help prepare you for the types of questions encountered on the ASE certification tests.

Each chapter closes with a number of *activities*. These activities are automotive-related exercises that emphasize math and communication skills, as well as improve performance on the job.

Special Notices

There are a variety of special notices used throughout this text. These notices contain technical information, cautions, warnings, and references to pertinent material in other parts of the text. The notices are identified by color and by an icon.

A *note* may contain a reference to another section of the text that relates to the subject at hand. It may also highlight important technical information. For example:

Note
For more information on on-board diagnostics, refer to Chapter 18, *Computer System Troubleshooting.*

A *caution* identifies a situation that may cause damage to a vehicle, equipment, or tools if the proper procedures are not followed. For example:

Caution
Tighten an alternator belt only enough to prevent slippage. Overtightening is a common mistake that can quickly ruin alternator bearings.

A *warning* identifies repair operations that can result in personal injury if the proper procedures or safety measures are not followed. For example:

Warning
Never remove a radiator cap while the engine or radiator is hot. Boiling coolant can spray from the radiator, causing serious burns.

Tech tips provide supplemental technical information and service hints related to the procedure or system being explained. For example:

Tech Tip
Oxygen sensors should be replaced at periodic intervals. After prolonged service, they become coated with exhaust byproducts. As this happens, fuel economy and emissions will be adversely affected. If gas mileage is 10% to 15% lower than normal, suspect the oxygen sensor of slow response.

Procedures present common service and repair operations in an easy-to-follow, step-by-step format. For example:

To perform an injector balance test:
1. Connect a pressure gauge to the test fitting on the fuel rail.
2. Close off the valve for measuring fuel volume if provided on the fuel gauge assembly.
3. Connect the balance tester wiring to the injectors or injector in question.
4. Turn the ignition key on to pressurize the system. Then, turn the ignition key off.
5. Press the injector balance tester button while watching the pressure gauge drop.
6. Record the pressure drop reading.
7. Repeat this on the other fuel injectors. This will allow you to measure how much fuel each injector is feeding into the engine when energized.

Troubleshooting Charts

Troubleshooting charts have been added to the end of each service chapter. These charts will help the reader diagnose and repair common problems.

Color Use

Color is used extensively throughout this text to enhance understanding and highlight important information. In illustrations, dark yellow is used for primary emphasis in illustrations and blue is used for secondary emphasis. Other colors are used as needed to help clarify the illustrations. For example, red arrows are often used to show motion. Color is also used to represent different pressures, states of matter, temperatures, etc.

Enhancing the Text

To aid in the learning process, a comprehensive workbook and a series of instructional videos have been created. These items are designed to be used with this text.

The workbook contains a variety of questions that correlate with the text and a number of jobs that are related to specific service or repair procedures.

The automotive videos also complement the material presented in this text. There are 49 videos, which can be purchased individually or as a package.

Both the workbook and videos can be purchased directly from Goodheart-Willcox Company, Inc.

Contents

Section 6

Cooling and Lubrication Systems

Section 7

Emission Control Systems

Section 8
Engine Performance

Section 9
Engine Service and Repair

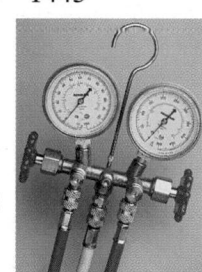

Introduction to Automotive Technology

The number of vehicles in the United States has increased by 40% over the last 25 years. Economists predict a strong demand for skilled automotive technicians and other automotive-related professionals for the foreseeable future. We are truly a "nation on wheels."

Section 1 will introduce you to the "basics" of automotive technology. It contains information on automobile construction and operation, ASE certification, safety, tools, service information, electricity, and vehicle maintenance

This section will also give you the knowledge needed to secure an entry-level job. You will learn to perform hands-on tasks, such as changing oil, checking vehicle fluids, replacing engine belts, and doing a "grease job." The information in this section will lay the groundwork for later chapters, which provide in-depth coverage of automotive technology.

Chapter 1

The Automobile

After studying this chapter, you will be able to:

- Identify and locate the most important parts of a vehicle.
- Describe the purpose of the fundamental automotive systems.
- Explain the interaction of automotive systems.
- Describe major automobile design variations.
- Comprehend later text chapters with a minimum amount of difficulty.
- Correctly answer ASE certification test questions that require a knowledge of the major parts and systems of a vehicle.

The term *automobile* is derived from the Greek word *autos,* which means self, and the French word *mobile,* which means moving. Today's "self-moving" vehicles are engineering marvels of safety and dependability. Over the last century, engineers and skilled workers the world over have used all facets of *technology* (the application of math, science, physics, and other subjects) to steadily give us a better means of transportation.

You are about to begin your study of the design, construction, service, maintenance, and repair of the modern automobile. This chapter provides a "quick look" at the major automotive systems. By knowing a little about each of these systems, you will be better prepared to learn the more detailed information presented later in this text.

Today, failure of one system can affect the operation of a seemingly unrelated system. This makes a thorough understanding of how the whole automobile works especially important.

Tech Tip

Try to learn something new about your profession every day! By studying just five minutes a day after graduation, you will constantly increase your knowledge and improve your skills. This will enable you to become a better technician.

Parts, Assemblies, and Systems

A *part* is the smallest removable item on a car. A part is not normally disassembled. The word *component* is frequently used when referring to an electrical or electronic part. For example, a spark plug is an ignition system component that ignites the fuel in the engine.

An *assembly* is a set of fitted parts designed to complete a function. For example, the engine is an assembly that converts fuel into useable power to move the vehicle. Technicians must sometimes take assemblies apart and put them back together during maintenance, service, and repair operations.

An automotive *system* is a group of related parts and assemblies that performs a specific job. For example, a vehicle's steering system is comprised of the steering wheel, gears, swivel joints, and other parts. This system allows the driver to turn the front wheels (and sometimes all four wheels) when maneuvering (turning) the vehicle. The brake system is another example of a basic automotive system. This system uses a variety of parts to slow and stop a vehicle.

Figure 1-1 shows the major systems of a vehicle. Memorize the name and general location of each system. Automotive parts and systems can be organized into ten major categories:

- *Body and frame*—support and enclose the vehicle.
- *Engine*—provides dependable, efficient power for the vehicle.
- *Computer systems*—monitor and control various vehicle systems.
- *Fuel system*—provides a combustible air-fuel mixture to power the engine.
- *Electrical system*—generates and/or distributes the power needed to operate the vehicle's electrical and electronic components.
- *Cooling and lubrication systems*—prevent engine damage and wear by regulating engine operating

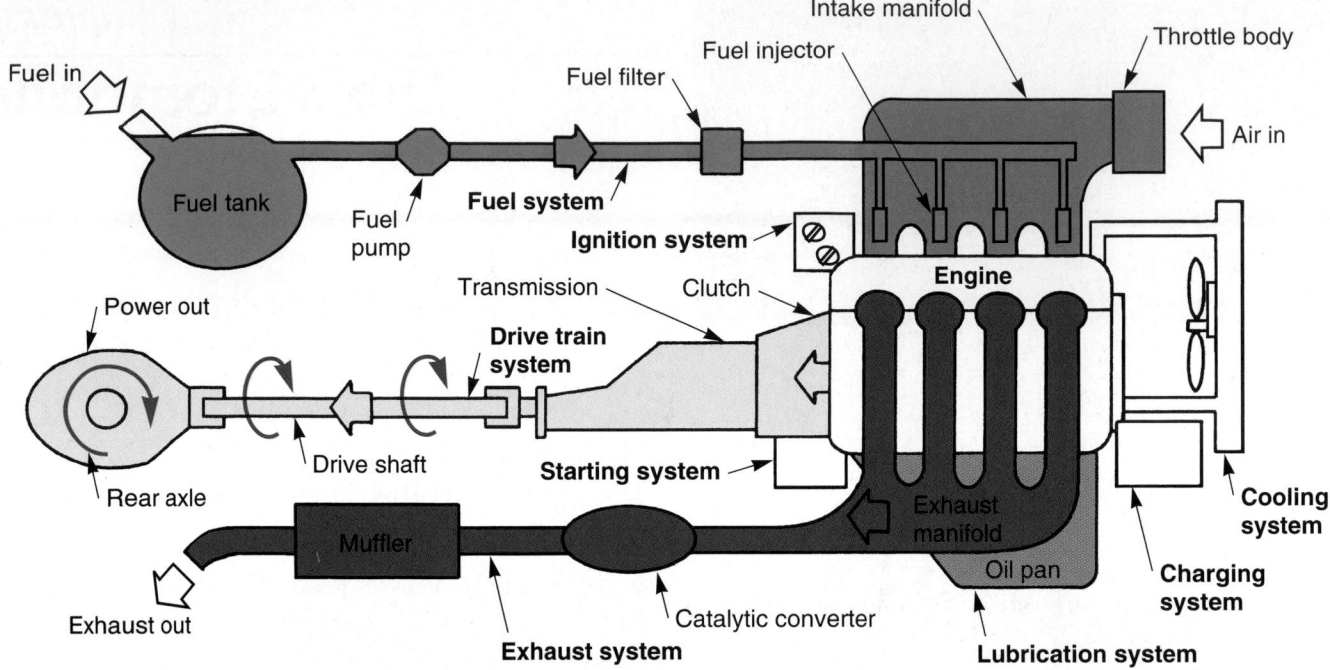

Figure 1-1. Note the general location of the major vehicle systems. Study the flow of fuel, air, exhaust, and power.

temperature and reducing friction between internal engine parts.

- *Exhaust and emission control systems*—quiet engine noise and reduce toxic substances emitted by the vehicle.

- *Drive train systems*—transfer power from the engine to the drive wheels.

- *Suspension, steering, and brake systems*—support and control the vehicle.

- *Accessory and safety systems*—increase occupant comfort, safety, security, and convenience.

Body and Frame

The body and frame are the two largest sections of a vehicle. The *body* includes the sheet metal, plastic, or fiberglass parts that form the passenger compartment and serve as an attractive covering for the chassis. The *chassis* generally includes everything but the body—engine, suspension, steering, brakes, wheels, tires, etc.

The *frame* is a very strong metal structure that supports various vehicle components. Some full-size cars and most pickup trucks have a metal frame. The body bolts onto the frame. This configuration is referred to as *body-over-frame construction.* Most late-model cars do not use a separate frame. Instead, the frame is an integral part of the body structure. This is called *unitized construction*, or *unibody construction*. With unibody con-

struction, the inner body sections are strengthened so that they can support the engine, suspension, and other major parts of the vehicle. See **Figure 1-2.**

Body Types

Automobiles are available in several body types, including the sedan, hardtop, convertible, hatchback, and station wagon. In addition, the minivan and the sport-utility vehicle have become increasingly popular.

A *sedan* is a car that has front and back seats and will carry four to six people. It has center body pillars, or "B" pillars, between the front and rear doors, **Figure 1-3A.** Both two-door and four-door sedans are available.

A *hardtop* is similar to the sedan, but it has no "B" pillars. Hardtop vehicles are also available in both two- and four-door models.

A *convertible* has a vinyl or cloth top that can be raised and lowered. A convertible has no door pillars, and its strength is designed into the frame or floor pan. Although most convertibles are two-door models, **Figure 1-3B,** a few four-door convertibles have been produced.

A *hatchback,* or *liftback,* has a large rear door for easy access when hauling items. This style car is available in three- and five-door models, **Figure 1-3C.**

A *station wagon* has a long, straight roof that extends to the rear of the vehicle. Station wagons have large rear interior compartments and come in two- and four-door

Figure 1-2. Compare unibody and body-over-frame construction. A—This car has unibody construction. The frame is an integral part of the body. B—This truck has a strong perimeter frame that is separate from the body. The body bolts onto the thick steel frame. (Ford, Chrysler)

A

B

C

D

E

F

Figure 1-3. Note the various vehicle body styles. A—Sedan. B—Convertible. C—Hatchback. D—Station wagon. E—Minivan. F—Sport-utility vehicle. (Buick, Pontiac, Honda)

models. Some station wagons have space for up to nine passengers, **Figure 1-3D.**

The *minivan* is similar to the station wagon, but it has a higher roofline for more headroom and cargo space. Most minivans are designed to carry seven passengers. See **Figure 1-3E.**

Sport-utility vehicles are often equipped with four-wheel-drive systems and have a tall body design. They provide the comfort of a passenger car, the interior space of a station wagon, and the durability of a truck, **Figure 1-3F.**

Common names for various automobile body parts are shown in **Figure 1-4.** Note that a vehicle's right and left sides are denoted as if you were sitting in the car looking forward.

Engine

The *engine* provides the energy to propel (move) the vehicle and operate the other systems. Most engines consume gasoline or diesel fuel. The fuel burns in the engine to produce heat. This heat causes gas expansion, creating

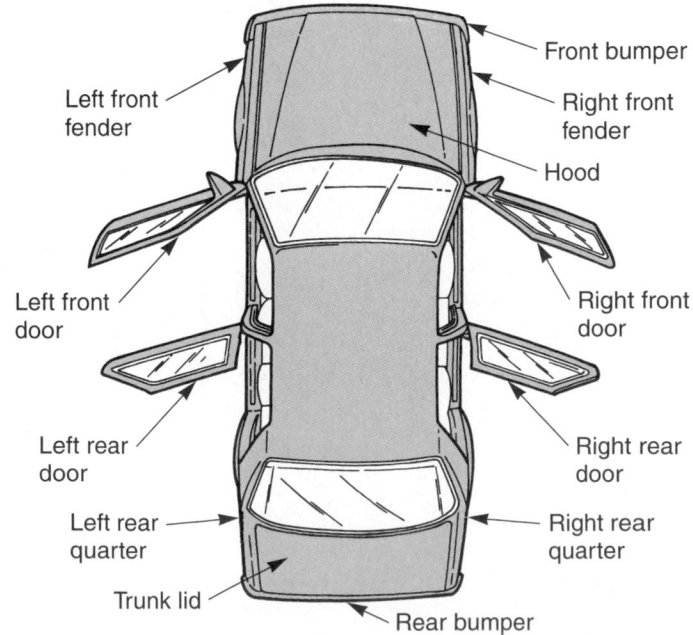

Figure 1-4. The right and left sides of a vehicle are denoted as if you were sitting forward inside passenger compartment. (Cadilliac)

pressure inside the engine. The pressure moves internal engine parts to produce power. See **Figure 1-5.**

The engine is usually located in the front portion of the body. Placing the heavy engine in this position makes the vehicle safer in the event of a head-on collision. In a few vehicles, the engine is mounted in the rear to improve handling. Refer to **Figure 1-6.**

Basic Engine Parts

The basic parts of a simplified one-cylinder engine are shown in **Figure 1-7.** Refer to this illustration as each part is introduced.

- The *block* is metal casting that holds all the other engine parts in place.
- The *cylinder* is a round hole bored (machined) in the block. It guides piston movement.
- The *piston* is a cylindrical component that transfers the energy of combustion (burning of air-fuel mixture) to the connecting rod.
- The *rings* seal the small gap around the sides of the piston. They keep combustion pressure and oil from leaking between the piston and the cylinder wall (cylinder surface).
- The *connecting rod* links the piston to the crankshaft.

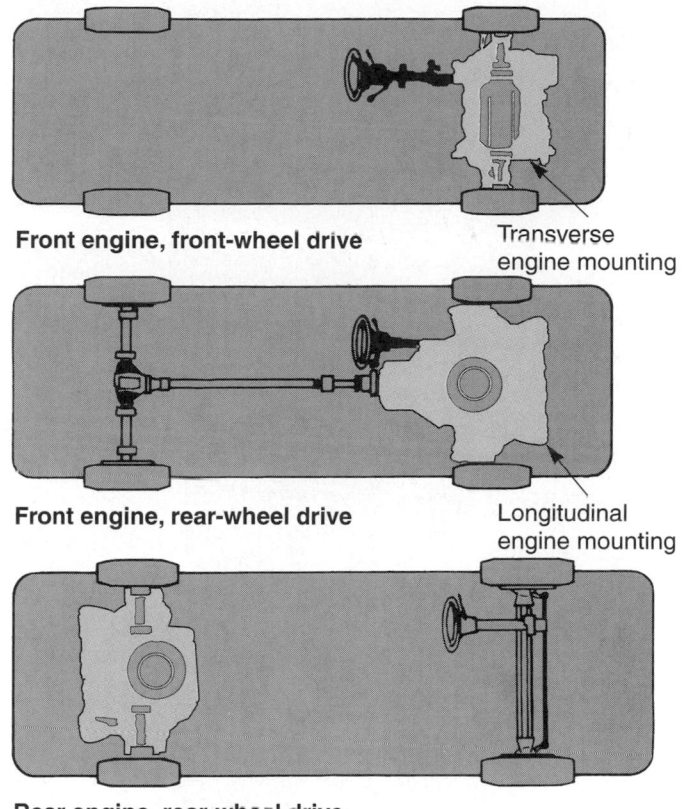

Front engine, front-wheel drive

Transverse engine mounting

Front engine, rear-wheel drive

Longitudinal engine mounting

Rear engine, rear-wheel drive

Figure 1-6. The engine can be located in the front or rear of the vehicle. (Dana Corp.)

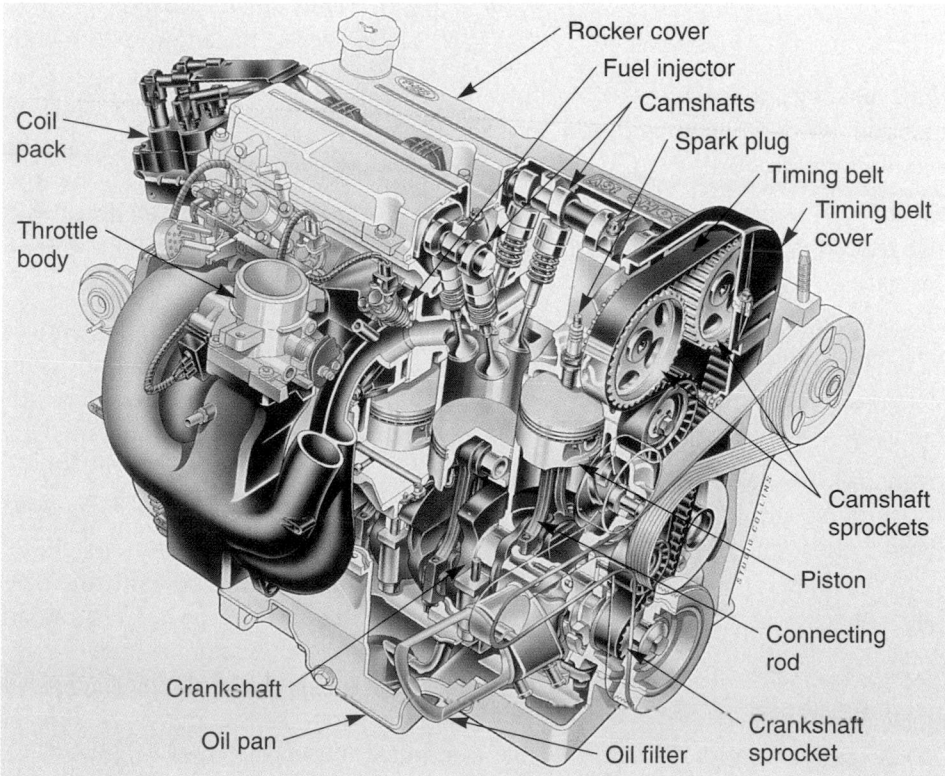

Coil pack

Throttle body

Rocker cover

Fuel injector

Camshafts

Spark plug

Timing belt

Timing belt cover

Camshaft sprockets

Piston

Connecting rod

Crankshaft

Oil pan

Oil filter

Crankshaft sprocket

Figure 1-5. An automotive engine commonly burns gasoline or diesel fuel to produce power. Note the part names. (Ford)

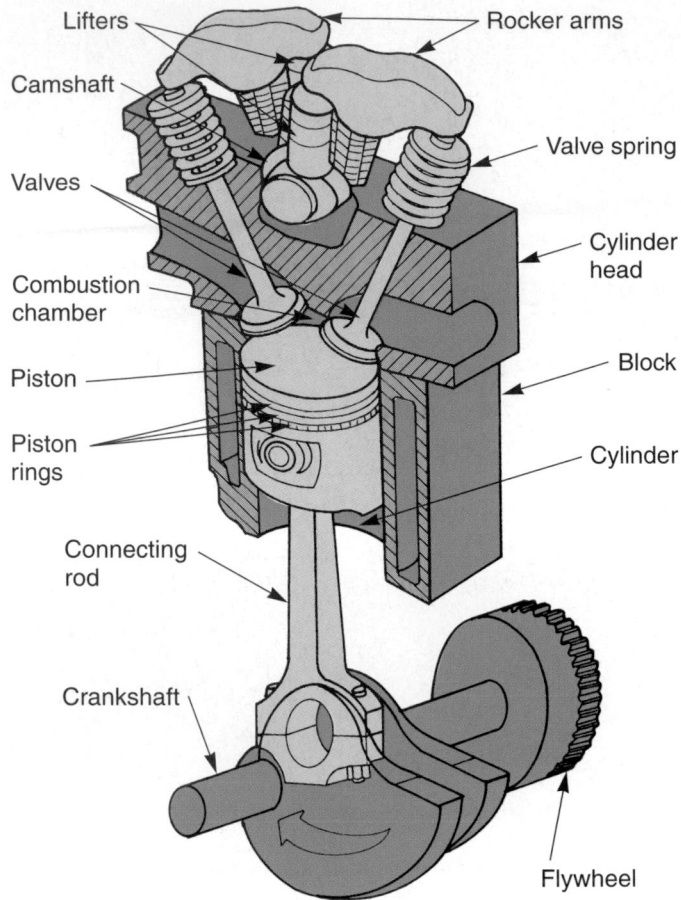

Figure 1-7. Memorize the basic parts of this one-cylinder engine.

- The *crankshaft* changes the *reciprocating* (up and down) motion of the piston and rod into useful *rotary* (spinning) motion.

- The *cylinder head* covers and seals the top of the cylinder. It also holds the valves, rocker arms, and often, the camshaft.

- The *combustion chamber* is a small cavity (hollow area) between the top of the piston and the bottom of the cylinder head. The burning of the air-fuel mixture occurs in the combustion chamber.

- The *valves* open and close to control the flow of the air-fuel mixture into the combustion chamber and the exhaust gases out of the combustion chamber.

- The *camshaft* controls the opening of the valves.

- The *valve springs* keep the valves closed when they do not need to be open.

- The *rocker arms* transfer camshaft action to the valves.

- The *lifters,* or *followers,* ride on the camshaft and transfer motion to the other parts of the valve train.

- The *flywheel* helps keep the crankshaft turning smoothly. It also provides a large gear for the starting motor.

Four-Stroke Cycle

Automobile engines normally use a *four-stroke cycle.* Four separate piston *strokes* (up or down movements) are needed to produce one *cycle* (complete series of events). The piston must slide down, up, down, and up again to complete one cycle.

As the four strokes are described below, study the simple drawings in **Figure 1-8.**

1. The *intake stroke* draws the air-fuel mixture into the engine's combustion chamber. The piston slides down while the intake valve is open and the exhaust valve is closed. This produces a vacuum (low-pressure area) in the cylinder. Atmospheric pressure (outside air pressure) can then force air and fuel into the combustion chamber.

2. The *compression stroke* prepares the air-fuel mixture for combustion. With both valves closed, the piston slides upward and compresses (squeezes) the trapped air-fuel mixture.

3. The *power stroke* produces the energy to operate the engine. With both valves still closed, the spark plug arcs (sparks) and ignites the compressed air-fuel mixture. The burning fuel expands and develops pressure in the combustion chamber and on the top of the piston. This pushes the piston down with enough force to keep the crankshaft spinning until the next power stroke.

4. The *exhaust stroke* removes the burned gases from the combustion chamber. During this stroke, the piston slides up while the exhaust valve is open and the intake valve is closed. The burned fuel mixture is pushed out of the engine and into the exhaust system.

During engine operation, these four strokes are repeated over and over. With the help of the heavy flywheel, this action produces smooth, rotating power output at the engine crankshaft.

Obviously, other devices are needed to lubricate the engine parts, operate the spark plug, cool the engine, and provide the correct fuel mixture. These devices will be discussed shortly.

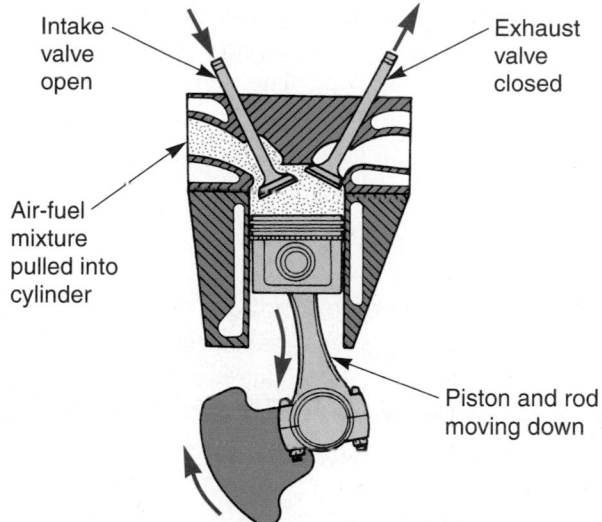

1—Intake stroke. Intake valve open. Exhaust valve closed. Piston slides down, forming vacuum in cylinder. Atmospheric pressure pushes air and fuel into combustion chamber.

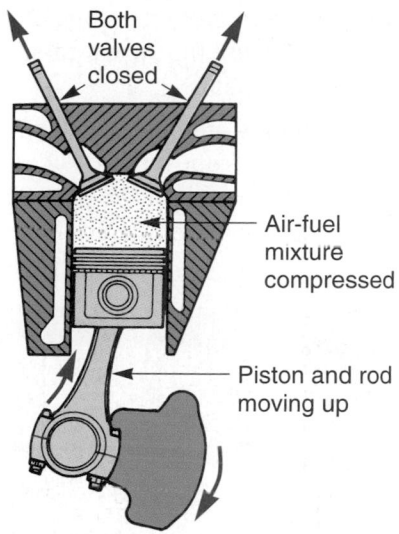

2—Compression stroke. Both valves are closed. Piston slides up and pressurizes air-fuel mixture. This readies mixture for combustion.

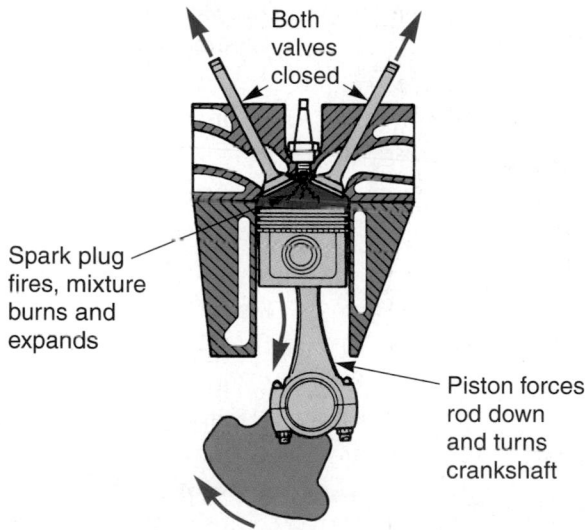

3—Power stroke. Spark plug sparks. Air-fuel mixture burns. High pressure forces piston down with tremendous force. Crankshaft rotates under power.

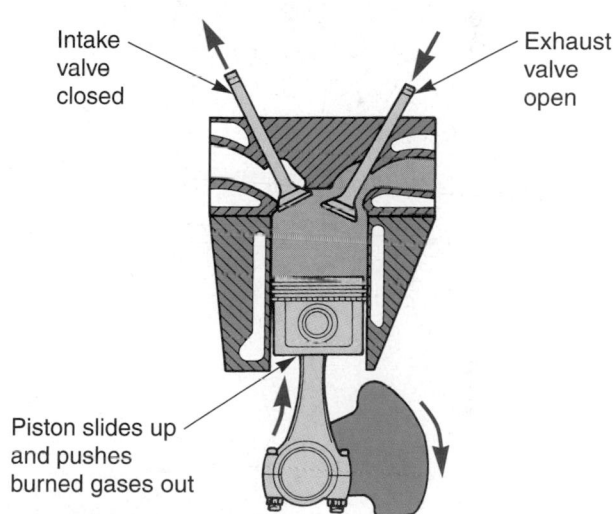

4—Exhaust stroke. Exhaust valve opens. Intake valve remains closed. Piston slides up, pushing burned gases out of cylinder. This prepares combustion chamber for another intake stroke.

Figure 1-8. A gasoline engine normally operates on a four-stroke cycle. Study the series of events.

Automotive Engines

Unlike the basic one-cylinder engine just discussed, automotive engines are **multi-cylinder engines,** which means they have more than one piston and cylinder. Vehicles commonly have 4-, 6-, 8-, or 10-cylinder engines. The additional cylinders smooth engine operation because there is less time (degrees of crank rotation) between power strokes. Additional cylinders also increase power output.

An actual automotive engine is pictured in **Figure 1-9.** Study the shape, location, and relationship of the major parts.

Computer System

The *computer system* uses electronic and electrical devices to monitor and control various systems in the vehicle, including the fuel, ignition, drive train, safety, and security systems. See **Figure 1-10.** The use of computer systems has improved vehicle efficiency and dependability. Additionally, most of these systems have self-diagnostic capabilities. There are three major parts of an automotive computer system:

- *Sensors*—input devices that can produce or modify electrical signals with a change in a condition, such as motion, temperature, pressure, etc. The sensors are the "eyes, ears, and nose" of the computer system.

- *Control module*—computer (electronic circuit) that uses signals from input devices (sensors) to control various output devices. The control module is the "brain" of the computer system.

- *Actuators*—output devices, such as small electric motors, that can move parts when energized by the control module. The actuators serve as the "hands and arms" of the computer system.

A modern car can have several control modules and dozens of sensors and actuators. These components will be detailed throughout this book.

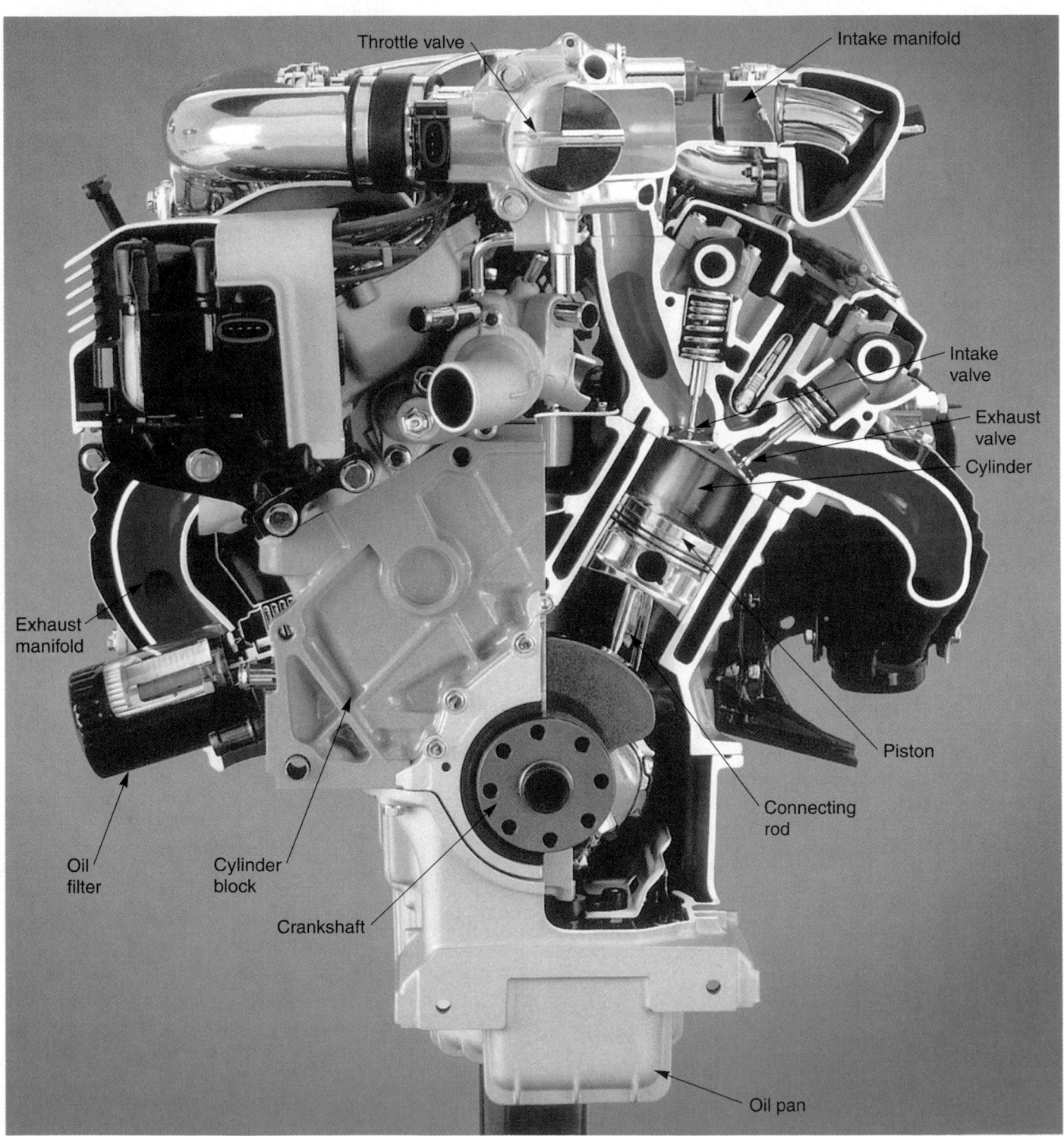

Figure 1-9. Automotive engines are multi-cylinder engines. Locate the major parts and visualize their operation. (Ford)

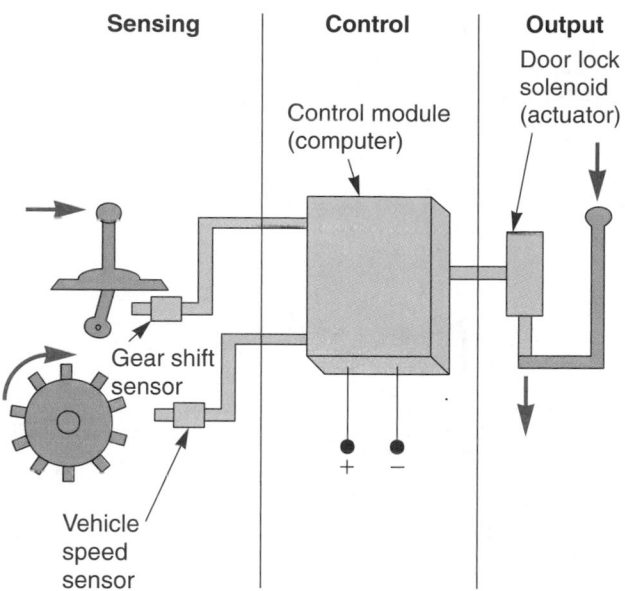

Sensing   **Control**   **Output**

Door lock
solenoid
(actuator)

Control module
(computer)

Gear shift
sensor

Vehicle
speed
sensor

Figure 1-10. This computer-controlled lock system automatically locks the doors as soon as the vehicle starts moving after being shifted into drive or reverse. When the gear shift sensor and the vehicle speed sensor send the correct signals to the control module, the module energizes the solenoid (actuator). The solenoid then converts the electrical signal from the control module to a linear motion, locking the doors.

Tech Tip

Learn all you can about electricity and electronics. Nearly every automotive system is now monitored or controlled by a computer. It is almost impossible to service any system of a car without handling some type of electric or electronic component. This book covers electronics in almost every chapter.

Fuel System

The *fuel system* must provide the correct mixture of air and fuel for efficient *combustion* (burning). This system must add the right amount of fuel to the air entering the cylinder. This ensures that a very *volatile* (burnable) mixture enters the combustion chambers.

The fuel system must also alter the *air-fuel ratio* (percentage of air and fuel) with changes in operating conditions (engine temperature, speed, load, and other variables).

There are three basic types of automotive fuel systems: gasoline injection systems, diesel injection systems, and carburetor systems. Look at the three illustrations in **Figure 1-11**.

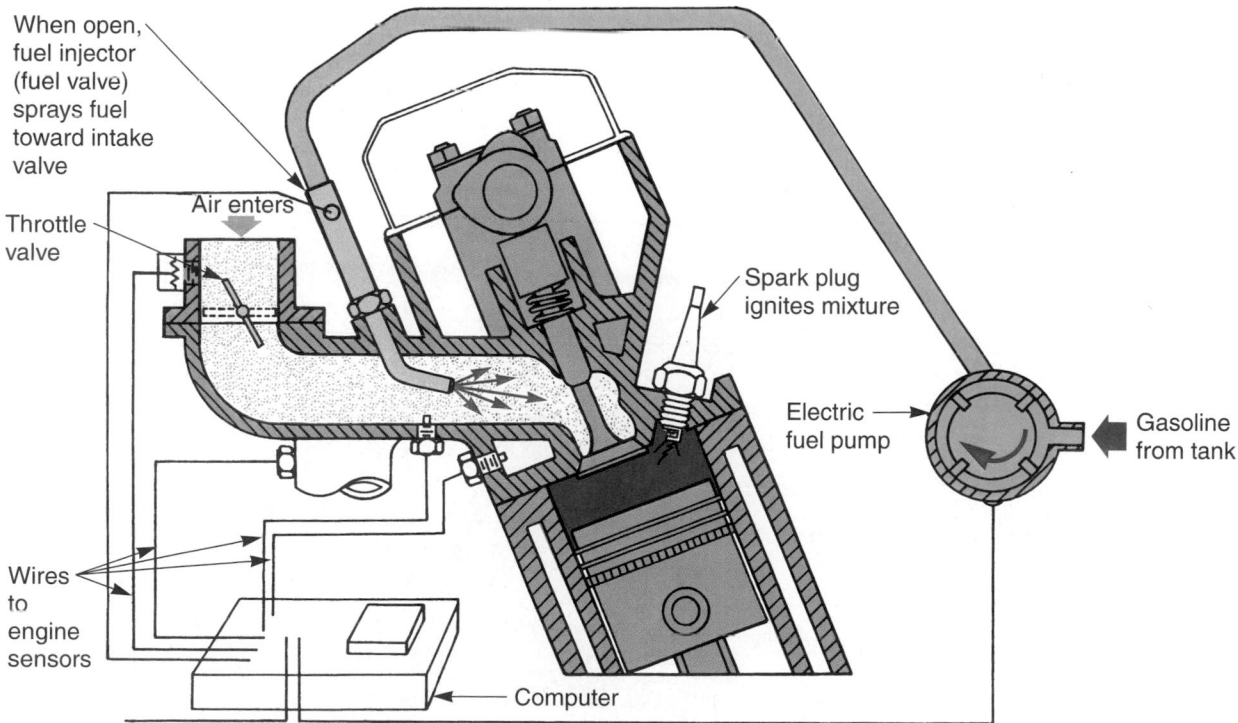

When open,
fuel injector
(fuel valve)
sprays fuel
toward intake
valve

Throttle
valve

Air enters

Spark plug
ignites mixture

Electric
fuel pump

Gasoline
from tank

Wires
to
engine
sensors

Computer

A—Gasoline injection system. Engine sensors feed information (electrical signals) to computer about engine conditions. Computer can then open injector for right amount of time. This maintains correct air-fuel ratio. Spark plug ignites fuel.

Figure 1-11. Note the three basic types of fuel systems. Compare differences.

(Continued)

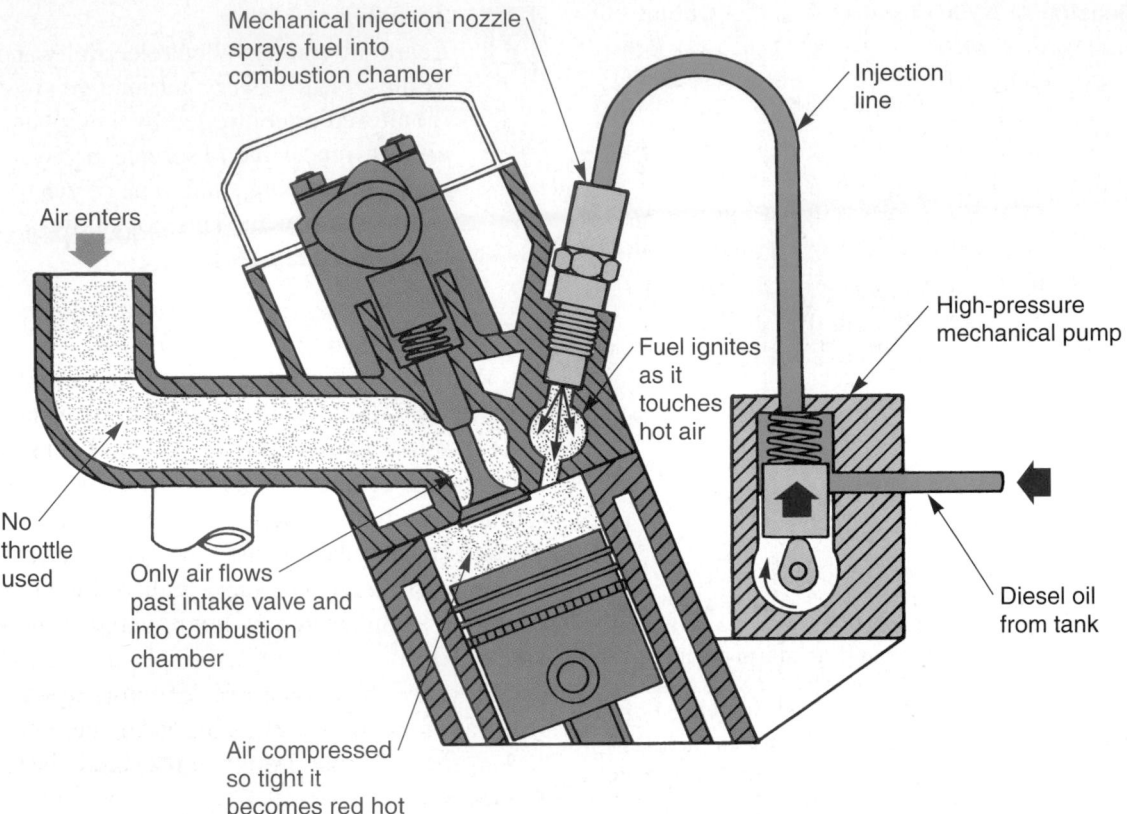

B—Diesel injection system. High-pressure mechanical pump sprays fuel directly into combustion chamber. Piston squeezes and heats air enough to ignite diesel fuel. Fuel begins to burn as soon as it touches heated air. Note that no throttle valve or spark plug is used. Amount of fuel injected into chamber controls diesel engine power and speed.

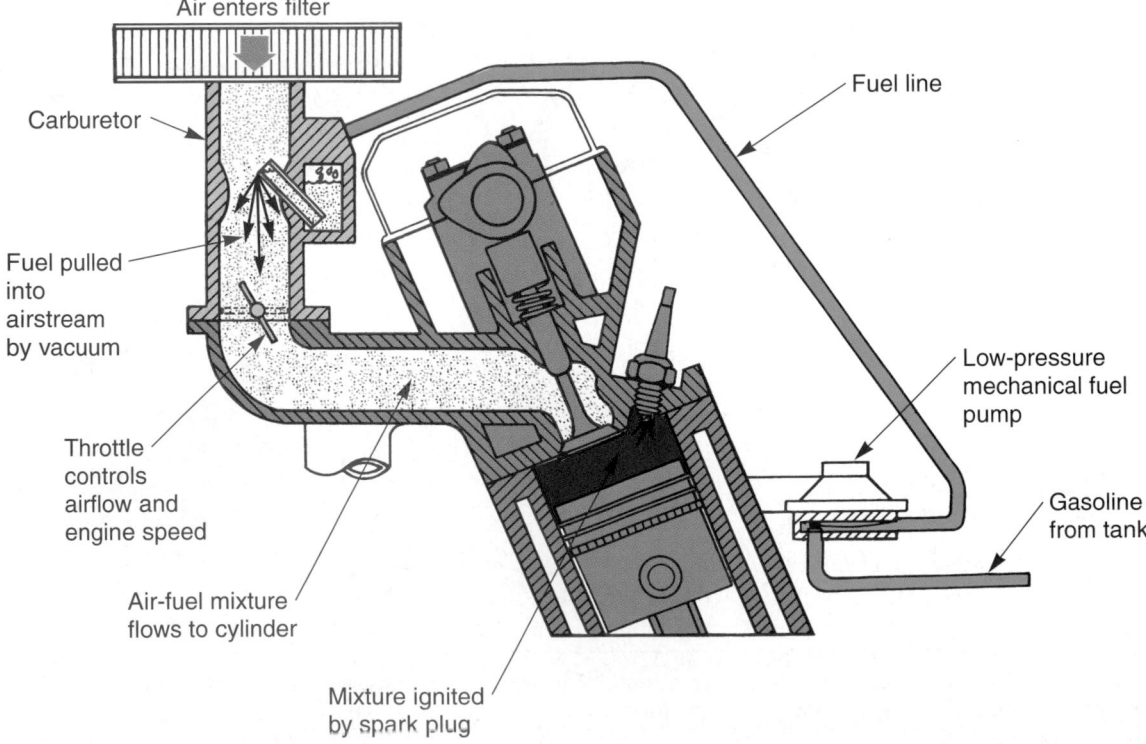

C—Carburetor fuel system. Fuel pump fills carburetor with fuel. When air flows through carburetor, fuel is pulled into engine in correct proportions. Throttle valve controls airflow and engine power output.

Figure 1-11. *(Continued)*

Gasoline Injection System

Modern *gasoline injection systems* use a control module, sensors, and electrically operated *fuel injectors* (fuel valves) to meter fuel into the engine. This is the most common type of fuel system on gasoline, or spark ignition, engines. See **Figure 1-11A.**

An electric *fuel pump* forces fuel from the fuel tank to the engine. The control module, reacting to electrical data it receives from the sensors, opens the injectors for the correct amount of time. Fuel sprays from the open injectors, mixing with the air entering the combustion chambers.

A *throttle valve* controls airflow, engine speed, and engine power. When the throttle valve is open for more engine power output, the computer holds the injectors open longer, allowing more fuel to spray out. When the throttle valve is closed, the computer opens the injectors for only a short period of time, reducing power output.

The throttle valve (air valve) is connected to the accelerator pedal. When the pedal is pressed, the throttle valve opens to increase engine power output.

Diesel Injection System

A *diesel fuel system* is primarily a mechanical system that forces diesel fuel (not gasoline) directly into the combustion chambers. Unlike the gasoline engine, the diesel engine does *not* use spark plugs to ignite the air-fuel mixture. Instead, it uses the extremely high pressure produced during the compression stroke to heat the air in the combustion chamber. The air is squeezed until it is hot enough to ignite the fuel. Refer to **Figure 1-11B.**

When the mechanical pump sprays the diesel fuel into a combustion chamber, the hot air in the chamber causes the fuel to begin to burn. The burning fuel expands and forces the piston down on the power stroke. Electronic devices are commonly used to monitor and help control the operation of today's diesel injection systems.

Carburetor Fuel System

The *carburetor fuel system* uses engine *vacuum* (suction) to draw fuel into the engine. The amount of airflow through the carburetor determines the amount of fuel used. This automatically maintains the correct air-fuel ratio. Refer to **Figure 1-11C.**

Either a mechanical or an electric fuel pump draws fuel out of the tank and delivers it to the carburetor. The engine's intake strokes form a vacuum inside the intake manifold and carburetor. This causes gasoline to be drawn from the carburetor and into the air entering the engine.

Electrical System

The vehicle's *electrical system* consists of several subsystems (smaller circuits): ignition system, starting system, charging system, and lighting system. Each subsystem is designed to perform a specific function, such as "fire" the spark plugs to ignite the engine's fuel mixture, rotate the crankshaft to start the engine, illuminate the highway for safe night driving, etc.

Ignition System

An *ignition system* is needed on gasoline engines to ignite the air-fuel mixture. It produces an extremely high voltage surge, which operates the *spark plugs.* A very hot electric arc jumps across the tip of each spark plug at the correct time. This causes the air-fuel mixture to burn, expand, and produce power. Study **Figure 1-12.**

With the ignition switch on and the engine running, the system uses sensors to monitor engine speed and other operating variables. Sensor signals are fed to the control module. The control module then modifies and *amplifies* (increases) these signals into on-off current pulses that trigger the ignition coil. When triggered, the *ignition coil* produces a high voltage output to "fire" the spark plugs. When the ignition key is turned off, the coil stops functioning and the spark-ignition engine stops running.

Starting System

The *starting system* has a powerful electric *starting motor* that rotates the engine crankshaft until the engine "fires" and runs on its own power. The major parts of the starting system are shown in **Figure 1-13A.**

A *battery* provides the electricity for the starting system. When the key is turned to the *start* position, current flows through the starting system circuit. The starting motor is energized, and the starting motor pinion gear engages a gear on the engine flywheel. This spins the crankshaft. As soon as the engine starts, the driver must shut off the starting system by releasing the ignition key.

Charging System

The *charging system* is needed to replace electrical energy drawn from the battery during starting system operation. To re-energize the battery, the charging system forces electric current back into the battery. The fundamental parts of the charging system are shown in **Figure 1-13B.** Study them!

When the engine is running, a *drive belt* spins the alternator pulley. The *alternator* (generator) can then produce electricity to recharge the battery and operate other electrical needs of the vehicle. A *voltage regulator,* usually built into the alternator, controls the voltage and current output of the alternator.

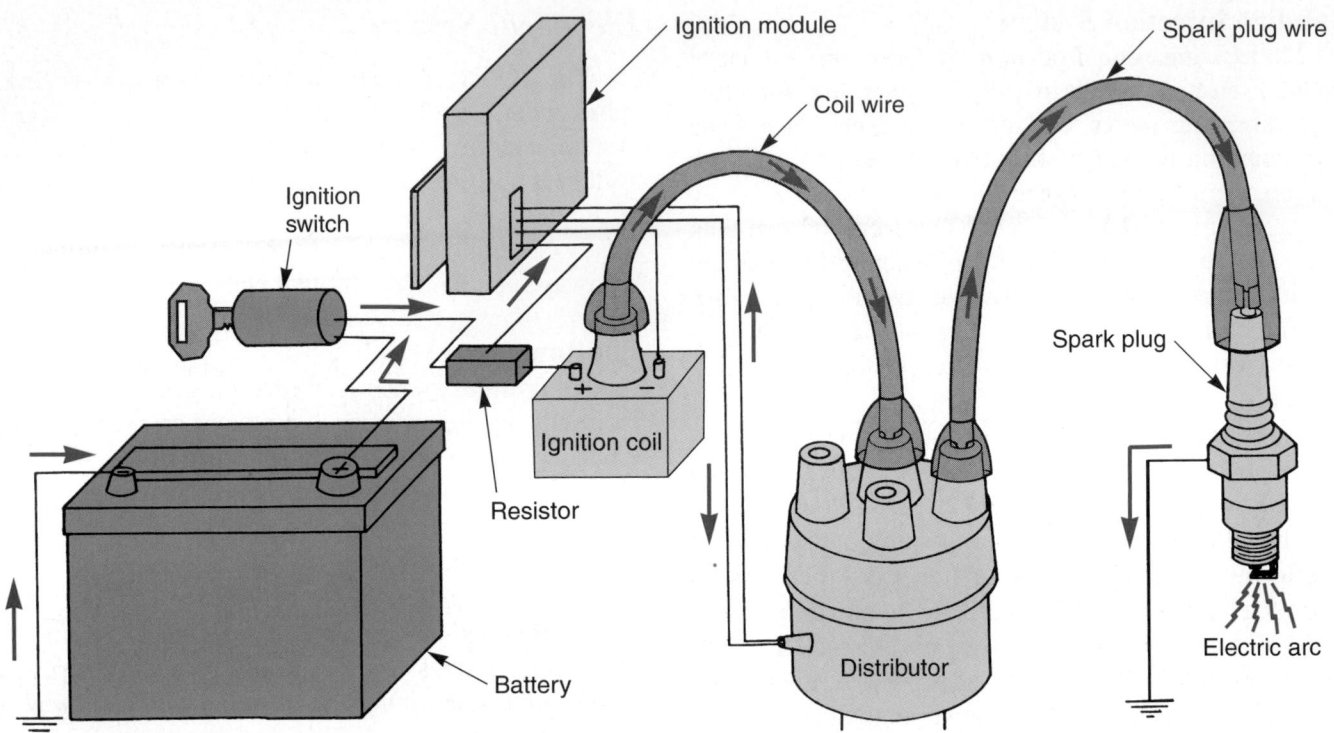

Figure 1-12. The ignition system is used on gasoline engines to start combustion. The spark plug must fire at the correct time during the compression stroke. A distributor or crankshaft sensor operates the ignition module. The module operates the ignition coil. The coil produces high voltage for the spark plugs.

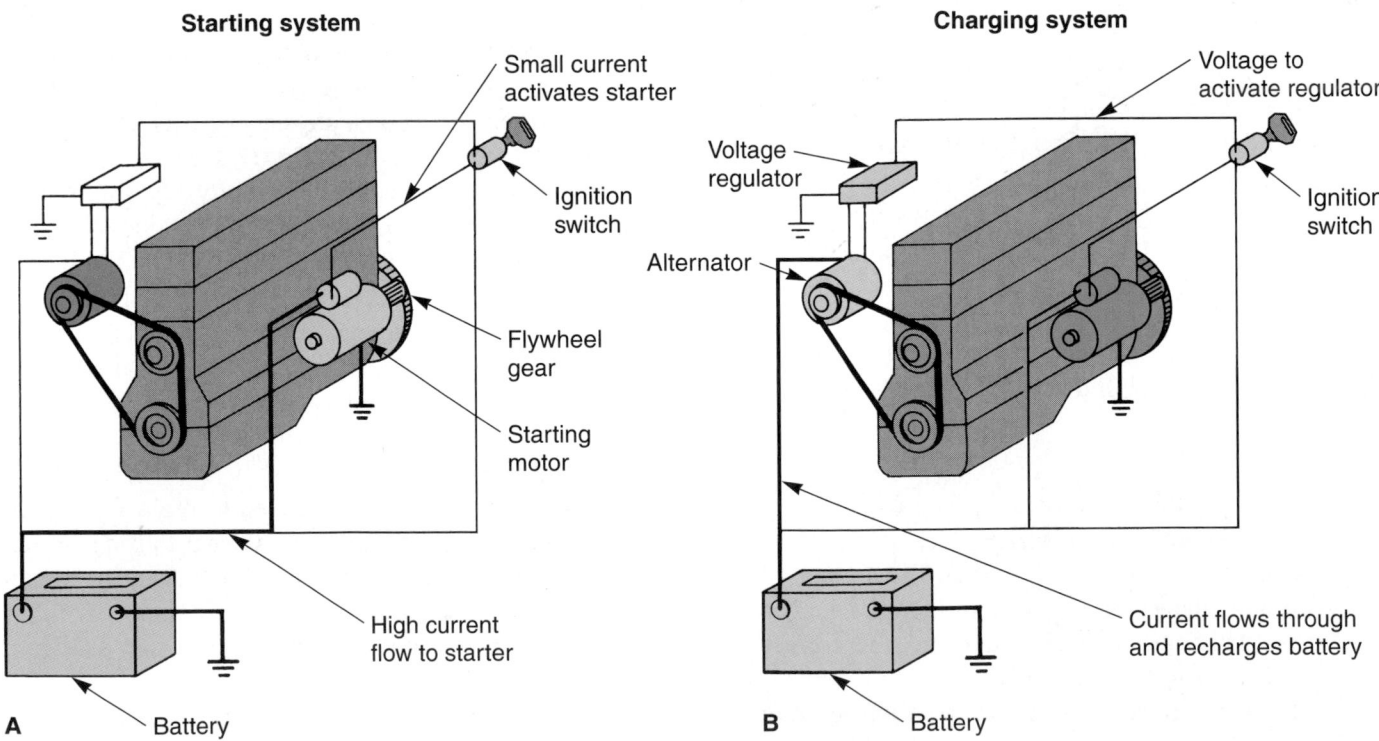

Figure 1-13. Note the basic actions and components of the starting and charging systems.

Lighting System

The *lighting system* consists of the components that operate a vehicle's interior and exterior lights (fuses, wires, switches, relays, etc.). The exact circuit and part configurations will vary from one model to another.

The *exterior lights* typically include the headlights, turn signals, brake lights, parking lights, backup lights, and side marker lights. The *interior lights* include the dome light, trunk light, instrument panel lights, and other courtesy lights.

Cooling and Lubrication Systems

The cooling and lubrication systems are designed to prevent engine damage and wear. They are important systems that prevent the engine from self-destructing.

The *cooling system* maintains a constant engine operating temperature. It removes excess combustion heat to prevent engine damage and also speeds engine warm-up. Look at **Figure 1-14.**

The *water pump* forces *coolant* (water and antifreeze solution) through the inside of the engine, hoses, and radiator. The coolant collects heat from the hot engine parts and carries it back to the radiator.

The *radiator* allows the coolant heat to transfer into the outside air. An *engine fan* draws cool air through the radiator. The *thermostat* controls coolant flow and engine temperature. It is usually located where the top radiator hose connects to the engine.

The *lubrication system* reduces friction and wear between internal engine parts by circulating filtered engine oil to high-friction points in the engine. The lubrication system also helps cool the engine by carrying heat away from internal engine parts.

Study the parts and operation of the lubrication system shown in **Figure 1-15.** Note how the *oil pump* pulls oil out of the pan and pushes it to various moving parts of the engine.

Exhaust and Emission Control Systems

The *exhaust system* quiets the noise produced during engine operation and routes engine exhaust gases to the rear of the vehicle body. **Figure 1-16** illustrates the basic parts of an exhaust system. Trace the flow of exhaust gases from the engine exhaust manifold to the tailpipe. Learn the names of the parts.

Various *emission control systems* are used to reduce the amount of *toxic* (poisonous) substances produced by an engine. Some systems prevent fuel vapors from entering the *atmosphere* (air surrounding the earth). Other emission control systems remove unburned and partially burned fuel from the engine exhaust. Later chapters cover these systems in detail.

Drive Train Systems

The *drive train* transfers turning force from the engine crankshaft to the drive wheels. Drive train configurations vary, depending on vehicle design. See **Figure 1-17.**

The drive train parts commonly found on a front-engine, rear-wheel-drive vehicle include the clutch, transmission, drive shaft, and rear axle assembly. The drive train parts used on most front-engine, front-wheel-drive vehicles include the clutch, transaxle, and drive axles. Refer to **Figure 1-16** as these components and assemblies are discussed.

Clutch

The *clutch* allows the driver to engage or disengage the engine and manual transmission or transaxle. When the clutch pedal is in the released position, the clutch locks the engine flywheel and the transmission input shaft together. This causes engine power to rotate the transmission gears and other parts of the drive train to

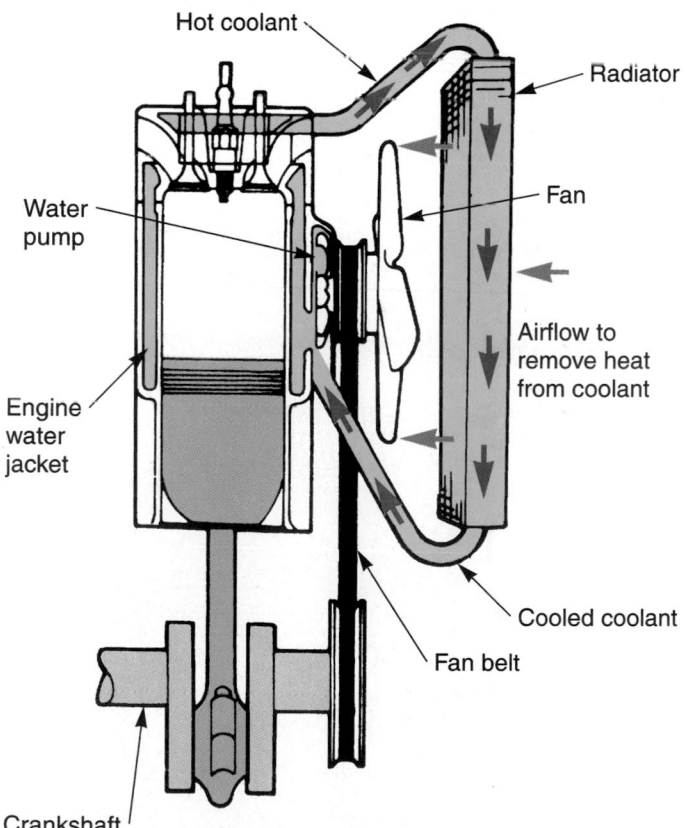

Figure 1-14. The cooling system must protect the engine from the heat of combustion. Combustion heat could melt and ruin engine parts. The system must also speed warm-up and maintain a constant operating temperature. Study the part names.

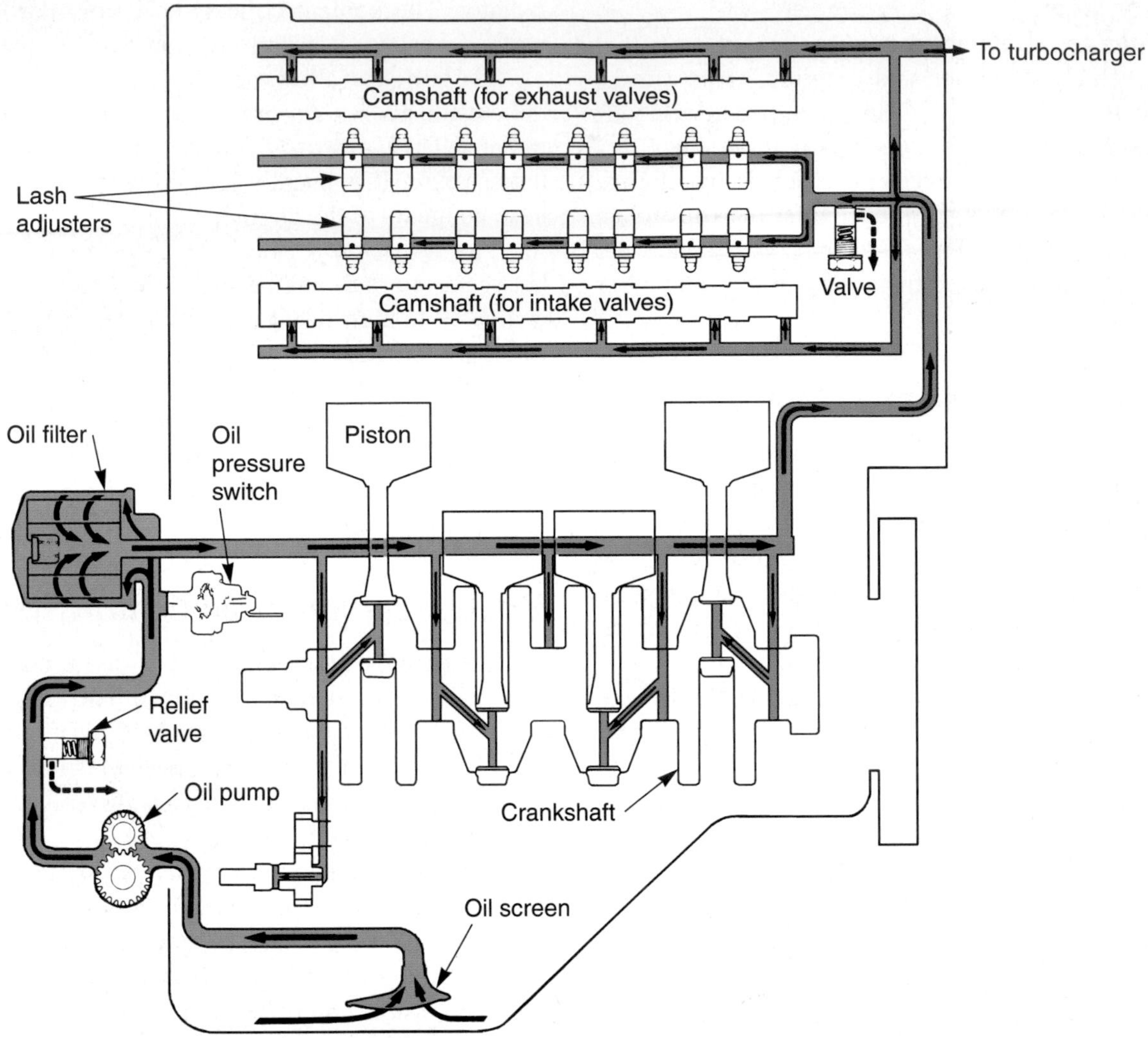

To turbocharger

Camshaft (for exhaust valves)

Lash
adjusters

Valve

Camshaft (for intake valves)

Oil filter

Oil
pressure
switch

Piston

Relief
valve

Oil pump

Crankshaft

Oil screen

Figure 1-15. The lubrication system uses oil to reduce friction and wear. The pump forces oil to high-friction points. (Renault)

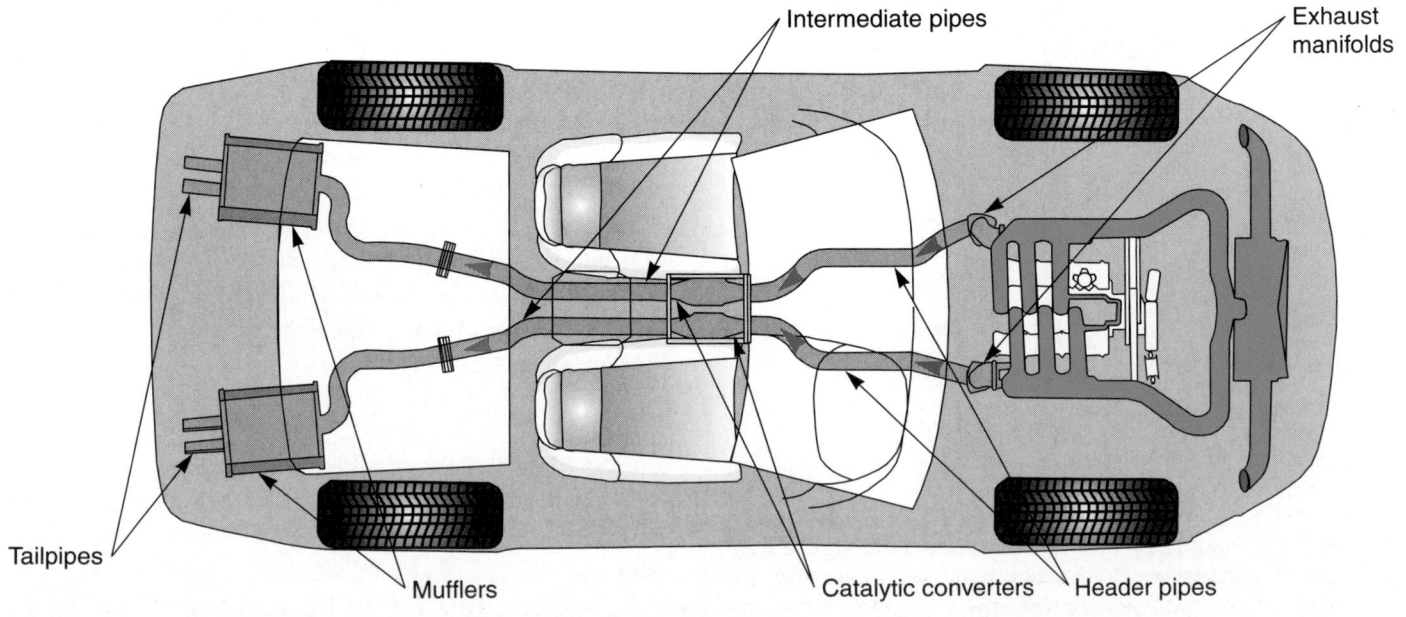

Intermediate pipes

Exhaust
manifolds

Tailpipes

Mufflers

Catalytic converters

Header pipes

Figure 1-16. The exhaust system carries burned gases to the rear of the vehicle. It also reduces engine noise. (Nissan)

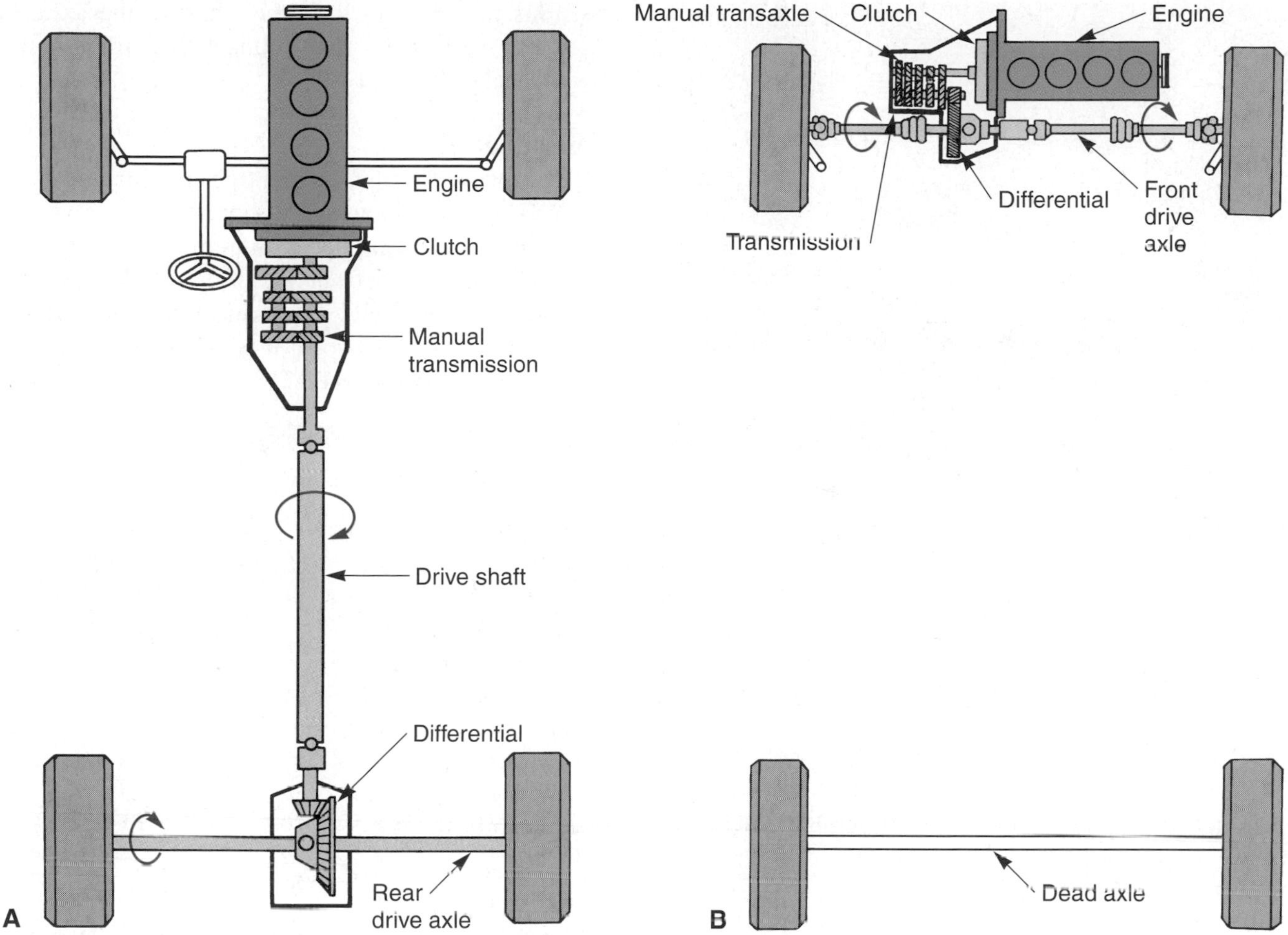

Figure 1-17. The drive train transfers engine power to the drive wheels. Study the differences between the two common types of drive trains. A—Front-engine, rear-wheel-drive vehicle. B— Front-engine, front-wheel-drive vehicle.

propel the vehicle. When the driver presses the clutch pedal, the clutch disengages power flow and the engine no longer turns the transmission input shaft and gears.

Transmission

The *transmission* uses various gear combinations, or ratios, to multiply engine speed and torque to accommodate driving conditions. Low gear ratios allow the vehicle to accelerate quickly. High gear ratios permit lower engine speed, providing good gas mileage

A *manual transmission* lets the driver change gear ratios to better accommodate driving conditions, **Figure 1-18.** An *automatic transmission,* on the other hand, does not have to be shifted by the driver. It uses an internal hydraulic system and, sometimes, electronic controls to shift gears. The input shaft of an automatic transmission is connected to the engine crankshaft through a *torque converter* (fluid coupling) instead of a clutch. The elementary parts of an automatic transmission are pictured in **Figure 1-19.**

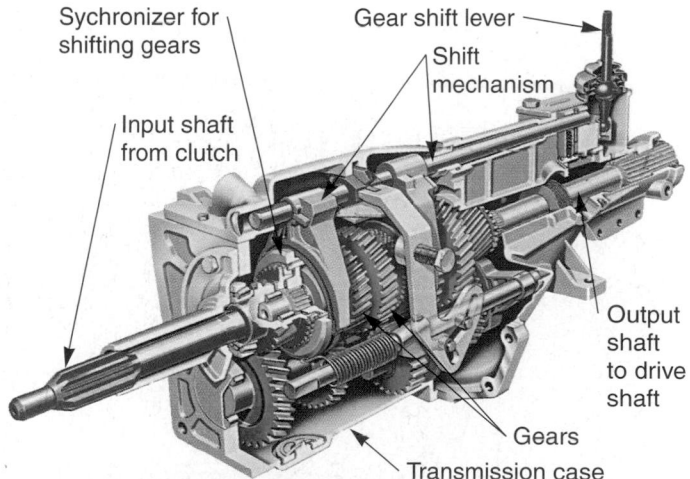

Figure 1-18. A manual transmission uses gears and shafts to achieve various gear ratios. The speed of the output shaft compared to the speed of the input shaft varies in each gear position. This allows the driver to change the amount of torque going to the drive wheels. In lower gears, the car accelerates quickly. When in high gear, engine speed drops while vehicle speed stays high for good fuel economy. (Ford)

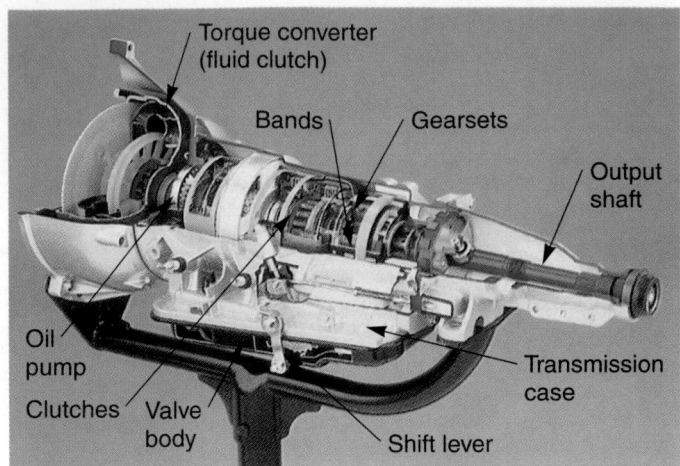

Figure 1-19. An automatic transmission serves the same function as a manual transmission. However, it uses a hydraulic pressure system to shift gears. (Ford)

Drive Shaft

The *drive shaft,* or propeller shaft, transfers power from the transmission to the rear axle assembly. Look at **Figure 1-20.** It is a hollow metal tube with two or more universal (swivel) joints. The universal joints allow the rear suspension to move up and down without damaging the drive shaft.

Rear Axle Assembly

The *rear axle assembly* contains a differential and two axles. The *differential* is a set of gears and shafts that transmits power from the drive shaft to the axles. The *axles* are steel shafts that connect the differential and drive wheels, **Figure 1-20.**

Transaxle

The *transaxle* consists of a transmission and a differential in a single housing. Although a few rear-wheel-drive vehicles are equipped with transaxles, they are most commonly used with front-wheel-drive vehicles, **Figure 1-21.** Both manual and automatic transaxles are available. The internal parts of a modern transaxle assembly are illustrated in **Figure 1-22.**

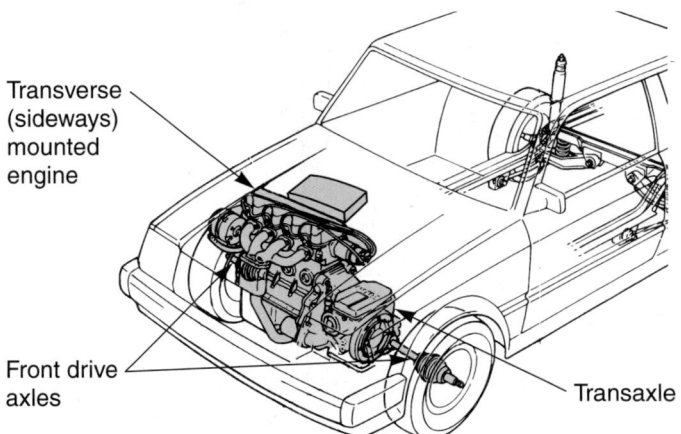

Figure 1-21. Front-wheel-drive vehicles do not have a drive shaft or a rear drive axle assembly. The complete drive train is in the front of the vehicle. (Ford)

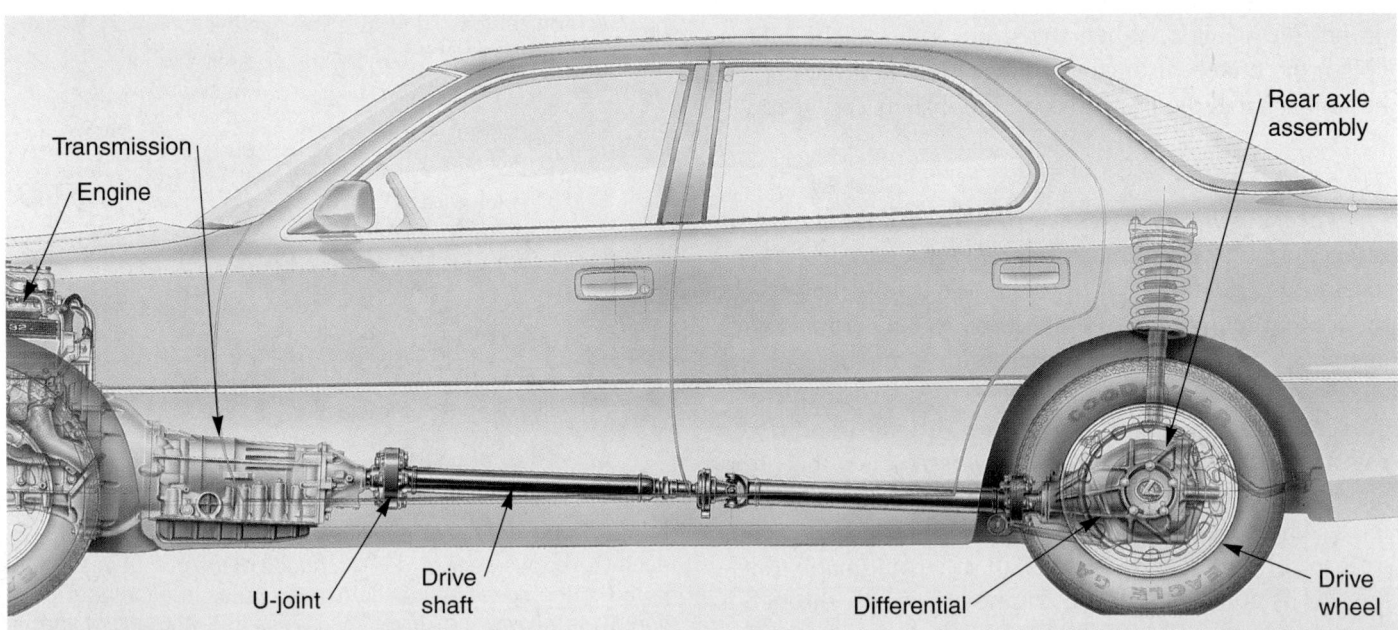

Figure 1-20. The drive shaft sends power to the rear axle assembly. The rear axle assembly contains the differential and two axles that turn the rear drive wheels. (Lexus)

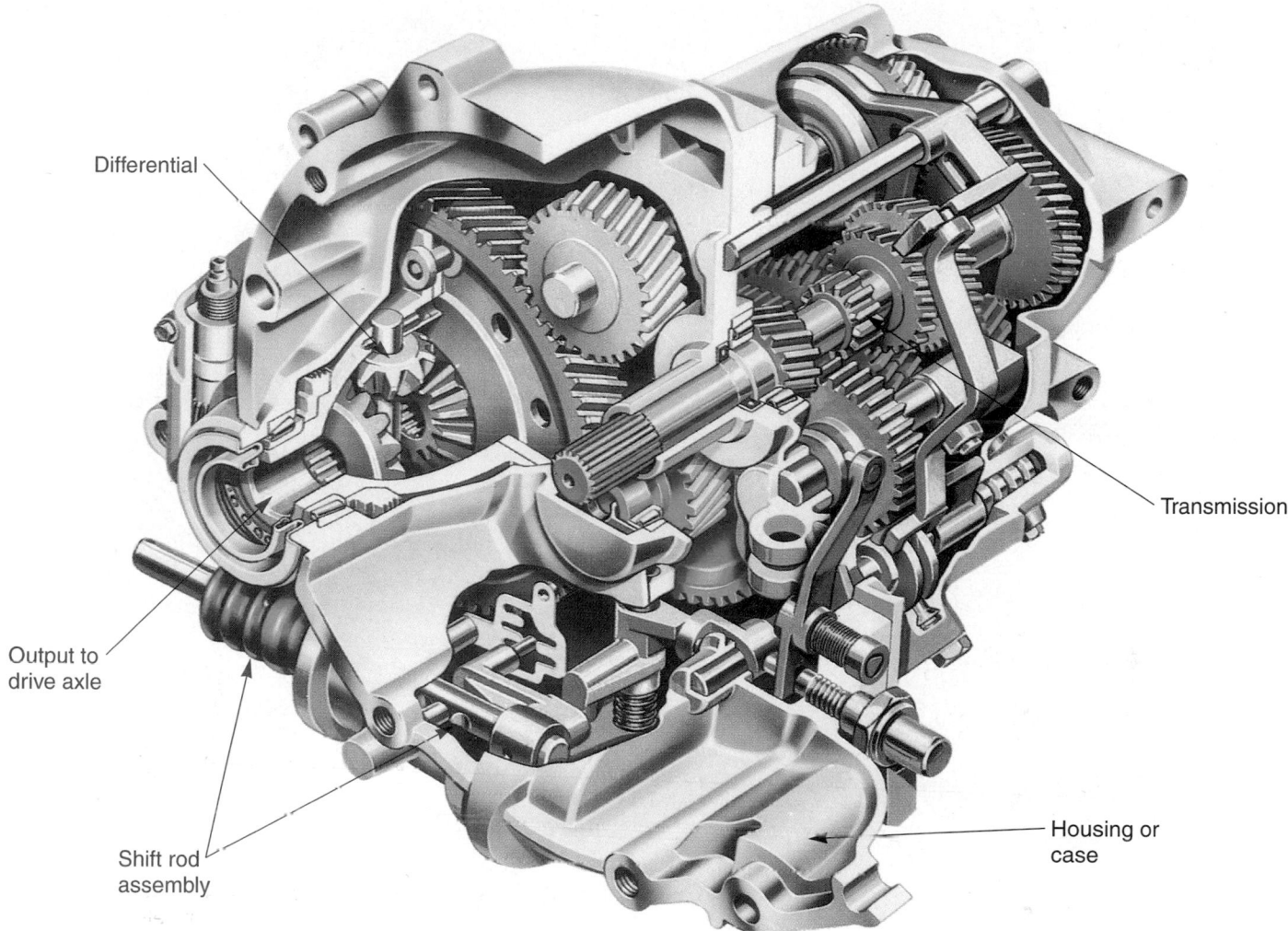

Differential

Transmission

Output to
drive axle

Housing or
case

Shift rod
assembly

Figure 1-22. A transaxle contains a transmission and a differential in one housing. (Ford)

Front Drive Axles

The *front drive axles* connect the transaxle differential to the hubs and wheels of the vehicle. These axles are equipped with constant-velocity joints, which allow the front wheels to be turned to the left or right and to move up and down.

Suspension, Steering, and Brake Systems

The suspension, steering, and brake systems are the movable parts of the chassis. They bolt or anchor to the frame and provide important functions that will be explained in the following sections.

Suspension System

The *suspension system* allows the vehicle's wheels and tires to move up and down with little effect on body movement. This makes the vehicle's ride smooth and safe. The suspension system also prevents excessive body lean when turning corners quickly.

As you can see in **Figure 1-23,** various springs, bars, swivel joints, and arms make up the suspension system.

Steering System

The *steering system* allows the driver to control vehicle direction by turning the wheels right or left. It uses a series of gears, swivel joints, and rods to do this. Study the names of the parts in **Figure 1-23.**

Brake System

The *brake system* produces friction to slow or stop the vehicle. When the driver presses the brake pedal, fluid pressure actuates a brake mechanism at each wheel. These mechanisms force friction material (brake pads or shoes) against metal discs or drums to slow wheel rotation. **Figure 1-24** shows the fundamental parts of a brake system.

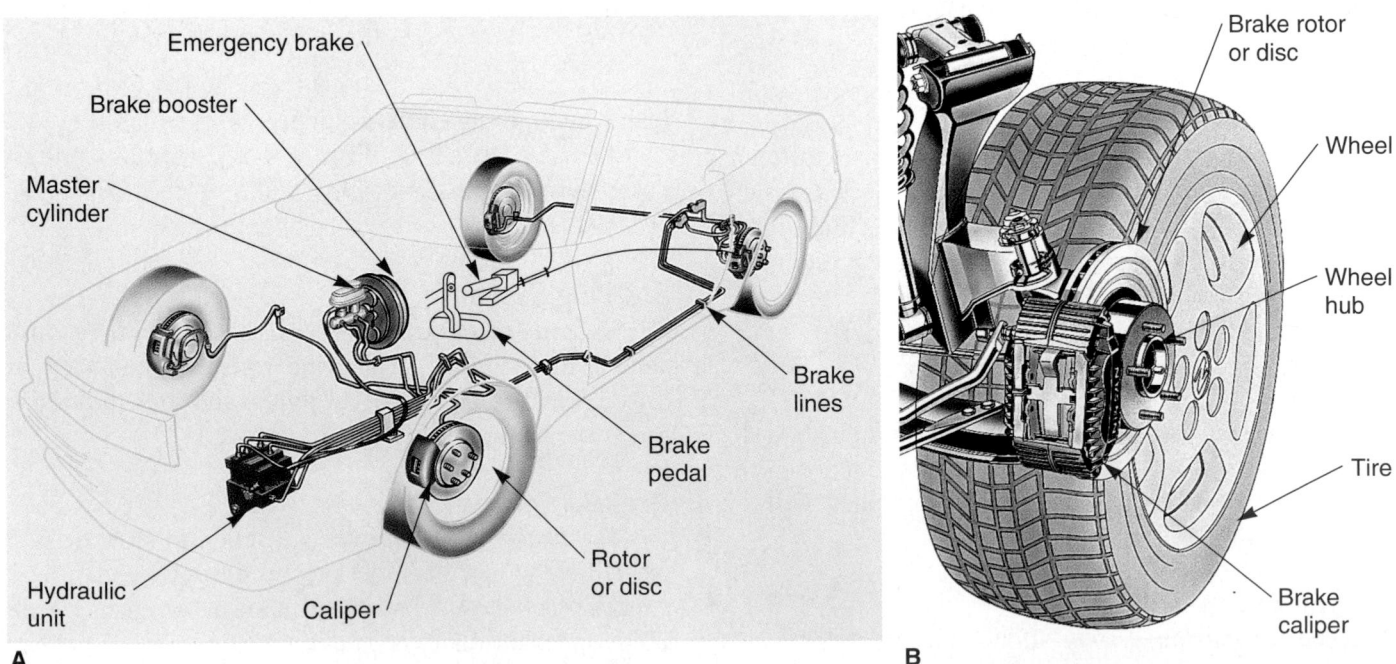

Figure 1-23. The suspension and steering systems mount on the frame. Study the part names. (Saab-Scania)

Figure 1-24. When the brake pedal is pressed, pressure is placed on a confined fluid. The fluid pressure transfers through the system to operate the brakes. An emergency brake is a mechanical system that applies the rear wheel brakes. A—Complete system. B—Close-up. (Cadillac, Nissan)

Accessory and Safety Systems

Common *accessory systems* include the air conditioner, sound system, power seats, power windows, and rear window defogger. Common *safety systems* include seat belts, air bags, and security systems. See **Figure 1-25.**

Summary

- The body and frame support, stop, and enclose the vehicle.
- Engines provide dependable, efficient power for the vehicle.
- The intake stroke draws the air-fuel mixture into the engine combustion chamber.
- The compression stroke prepares the fuel mixture for combustion.
- The power stroke produces the energy to operate the engine.
- The exhaust stroke must remove the burned gases from the engine cylinders
- The computer system uses electronic and electrical devices to monitor and control various systems in the vehicle.
- The fuel system provides the correct mixture of air and fuel for efficient combustion.

- Electrical systems operate the electrical-electronic devices.
- The cooling system maintains a constant engine operating temperature.
- The lubrication system reduces friction between internal engine parts.
- Emission control systems reduce air pollution produced by the vehicle.
- Drive train systems transfer turning force from the engine crankshaft to the drive wheels.
- Suspension, steering, and brake systems support and control the vehicle.
- Accessory and safety systems increase passenger comfort, safety, security and convenience.

Important Terms

Automobile	Body-over-frame
Technology	construction
Part	Unitized construction
Component	Unibody construction
Assembly	Engine
System	Four-stroke cycle
Body	Intake stroke
Chassis	Compression stroke
Frame	Power stroke

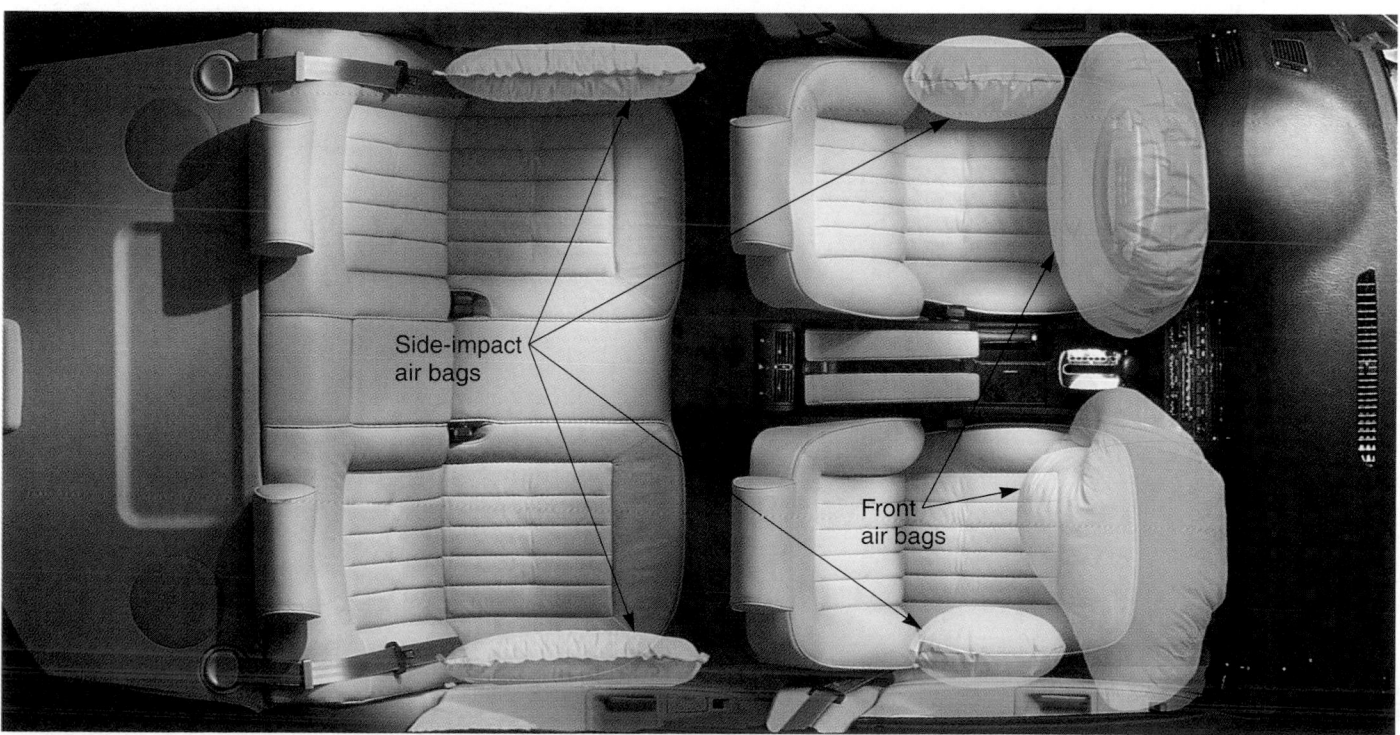

Side-impact air bags

Front air bags

Figure 1-25. Various safety systems are used on modern vehicles to protect both the driver and the passengers. This vehicle is equipped with both front and side-impact air bags. (Audi)

Exhaust stroke
Multi-cylinder engines
Computer system
Fuel system
Air-fuel ratio
Gasoline injection
　systems
Diesel fuel system
Carburetor fuel system
Ignition system
Starting system
Charging system

Lighting System
Cooling system
Lubrication system
Exhaust system
Emission control
　systems
Drive train
Suspension system
Steering system
Brake system
Accessory systems
Safety systems

Review Questions—Chapter 1

Please do not write in this text. Place your answers on a separate sheet of paper.

1. What is an automotive system?

2. Automotive parts and systems can be grouped into ten categories. Name them.

3. Which of the following is *not* part of an engine?
    (A) Block.
    (B) Piston.
    (C) Muffler.
    (D) Crankshaft.

4. Explain the engine's four-stroke cycle.

5. Most car engines are multi-cylinder engines. True or False?

6. List and describe the three common types of fuel systems.

7. A diesel engine does *not* use spark plugs. True or False?

8. The car's electrical system consists of the:
    (A) ignition, starting, lubrication, and lighting systems.
    (B) ignition, charging, lighting, and hydraulic systems.
    (C) lighting, charging, starting, and ignition systems.
    (D) None of the above.

9. The _____ _____ system reduces the amount of toxic substances released by the vehicle.

10. What is the difference between a manual transmission and an automatic transmission?

11. A one-piece drive shaft rotates the drive wheels on most front-wheel drive cars. True or False?

12. A rear axle assembly contains two _____ and a(n) _____.

13. Explain the term "transaxle."

14. The suspension system mounts the car's wheels solidly on the frame. True or False?

15. List four accessory systems.

16–25. Identify the parts and systems illustrated below. Write the numbers 16–25 on your paper. Then write the correct letter and words next to each number.

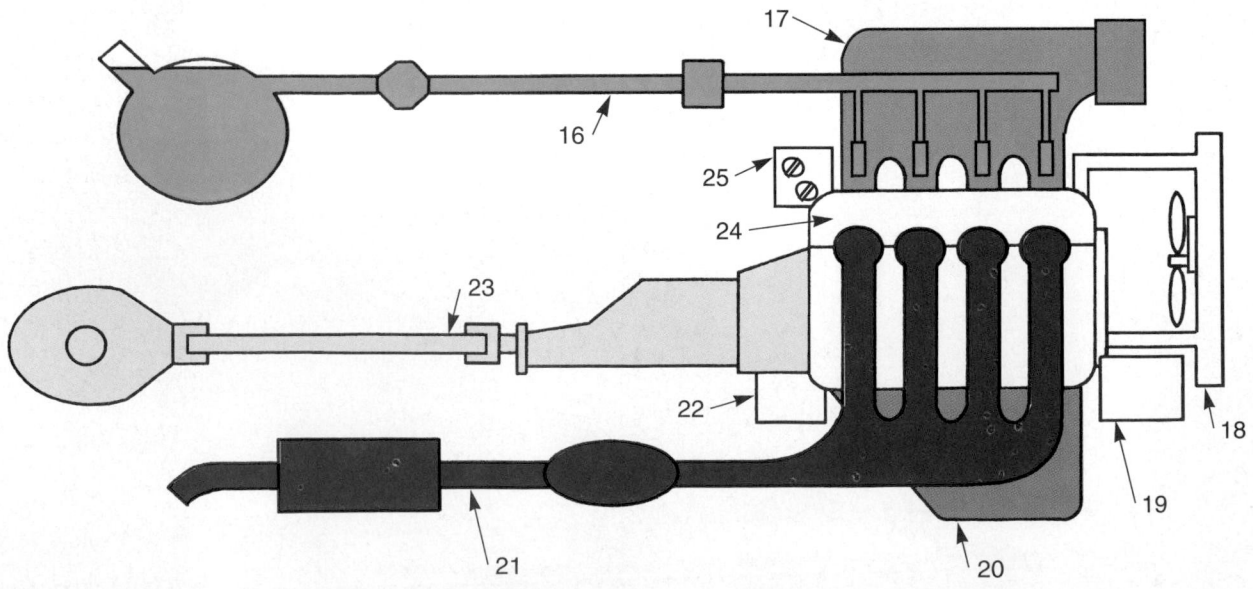

Can you identify the following parts and systems?
(A) Starting system. (B) Charging system. (C) Drive train. (D) Fuel system. (E) Cooling system. (F) Engine.
(G) Ignition system. (H) Lubrication system. (I) Exhaust system. (J) Intake manifold.

✹ ASE-Type Questions

1. A vehicle is brought into the shop with a slipping clutch. Technician A says that the clutch is part of the drive train system. Technician B says that the clutch is part of the suspension system. Who is correct?
   (A) *A only.*
   (B) *B only.*
   (C) *Both A and B.*
   (D) *Neither A nor B.*

2. When the internal body structure of a vehicle is used as its frame, it is called:
   (A) *unibody construction.*
   (B) *body-frame construction.*
   (C) *integral construction.*
   (D) *body-over-frame construction.*

3. The _____ controls the opening of engine's valves.
   (A) *camshaft*
   (B) *crankshaft*
   (C) *valve springs*
   (D) *combustion chamber*

4. Which piston stroke of the four-stroke cycle prepares the fuel mixture for combustion?
   (A) *Power.*
   (B) *Intake.*
   (C) *Exhaust.*
   (D) *Compression.*

5. All of the following are major components in the computer system *except:*
   (A) *regulators.*
   (B) *sensors.*
   (C) *actuators.*
   (D) *computer.*

6. Each of the following is a basic type of automotive fuel system *except*:
   (A) *carburetor.*
   (B) *auto injection.*
   (C) *diesel injection.*
   (D) *gasoline injection.*

7. Tests show that an engine is not getting spark at the spark plugs. Technician A says it could be due to the diesel injection system. Technician B says to test the ignition coil. Who is right?
   (A) *A only.*
   (B) *B only.*
   (C) *Both A and B.*
   (D) *Neither A nor B.*

8. A car with a dead battery is brought into the shop. Technician A says to check the output of the alternator. Technician B says to check the condition of the spark plugs. Who is right?
   (A) *A only.*
   (B) *B only.*
   (C) *Both A and B.*
   (D) *Neither A nor B.*

9. Since an automatic transmission does not have to be shifted by hand, Technician A believes it uses a hydraulic system to shift gears. Technician B believes it uses oil pressure to shift gears. Who is right?
   (A) *A only.*
   (B) *B only.*
   (C) *Both A and B.*
   (D) *Neither A nor B.*

10. A transaxle case contains both the:
   (A) *carburetor and drive shaft.*
   (B) *transmission and differential.*
   (C) *multi-cylinder engine and clutch.*
   (D) *suspension components and brakes.*

Activities—Chapter 1

1. Draw an automotive engine and drive train and label the parts. Then describe how the power is transferred from the engine to the drive wheels.

2. Using illustrations from the text, produce overhead transparencies of the four-stroke cycle and demonstrate the cycle to your class.

3. Arrange a field trip to tour an automobile assembly plant or to an auto shop.

Chapter 2

Automotive Careers and ASE Certification

After studying this chapter, you will be able to:

- List the most common automotive careers.
- Describe the type of skills needed to be an auto technician.
- Explain the tasks completed by each type of auto technician.
- Summarize the ASE certification program.

Over the last 25 years, the number of vehicles in the United States has increased by 40%. Today, there are well over one-hundred million vehicles on the road. In a single year, Americans spend approximately four-hundred billion dollars to own and operate their vehicles. Amazingly, there are about fourteen million people employed in the automotive field.

Economists predict a continued demand for skilled automotive technicians and other automotive-related professions for many years. Our country is, and will continue to be, a "nation on wheels."

The Automotive Technician

An *automotive technician* makes a living diagnosing, servicing, and repairing cars, vans, and light trucks. The technician must be highly skilled and well trained. He or she must be a "jack of all trades," being able to perform a wide variety of tasks. For example, an experienced master automotive technician is usually capable of performing operations common to the following occupations:

- *Machinist* (precision measurements, brake part machining).
- *Plumber* (working with fuel lines and power steering lines).
- *Welder* (gas and arc welding on exhaust systems, parts repair).
- *Electrician* (charging, starting, lighting system service).

- *Electronic technician* (servicing a vehicle's electronic parts).
- *Air conditioning technician* (repairing and recharging auto air conditioning).
- *TV-radio technician* (installing and repairing vehicle stereo, cellular phones, and radios).
- *Computer technician* (servicing a vehicle's on-board computers).
- *Bookkeeper* (business-type tasks, such as filling out repair orders, calculating hours on a job, ordering parts, totaling work order costs, etc.).

As this list demonstrates, an automobile technician's job can be very challenging. The technician is called on to perform a variety of repair tasks, which prevents boredom on the job. If you like to use your mind and your hands, automotive service can be a rewarding and interesting profession.

General Job Classifications

A wide variety of jobs are available in the automotive field. Many of these jobs involve troubleshooting, service, and repair.

Service Station Attendant

A job as a *service station attendant* requires little mechanical experience, yet provides an excellent learning experience. A "gas station" with a repair area provides better training than a station without a repair area. You could learn to make simple repairs and work your way into a position as a "light mechanic." As a service station attendant, you might do oil changes, grease jobs, and similar service tasks.

Apprentice

Another way to get started in automotive service is to become an apprentice. The *apprentice*, or helper mechanic, works under the direction of an experienced technician.

This is a good way to get paid for an education. As an apprentice, you would learn about automotive technology by running for parts, cleaning parts, maintaining tools, and helping with repairs. See **Figure 2-1.**

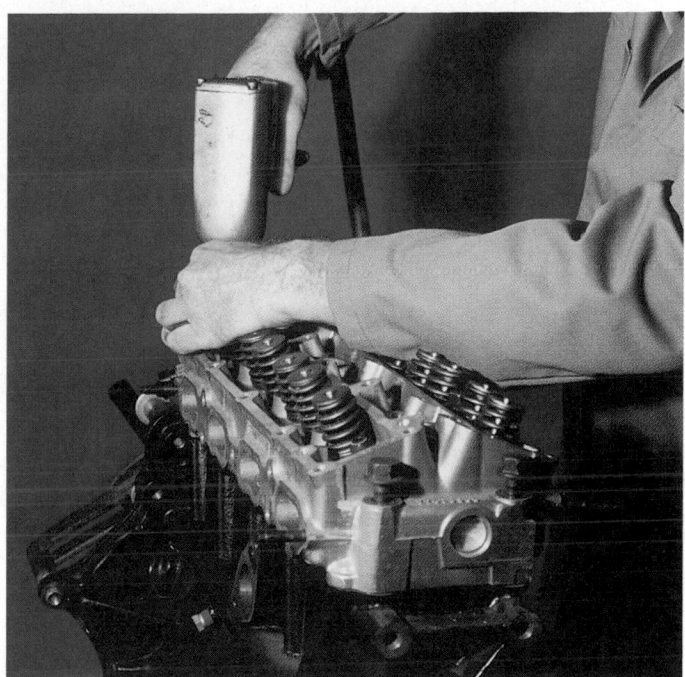

Figure 2-2. The engine technician must be highly skilled. Late-model engines are very complex. One mistake during assembly can cause major engine damage. (Oldsmobile)

Figure 2-1. An apprentice works under the direction of an experienced auto technician. This is an excellent way to learn the trade. (Fluke)

Specialized Technician

A *specialized technician* is an expert on one automotive system. Because of the increasingly complex nature of today's vehicles, the trend is toward specialization. It is much easier to learn to repair one system than all systems. After specializing in one area, you can expand your abilities to include other areas of repair.

Common areas of specialization include engines, transmissions, steering and suspension, brake, electrical, heating and cooling, driveability and performance, and lubrication.

An *engine technician* troubleshoots, services, and repairs automobile engines. Refer to **Figure 2-2.** This requires a knowledge of all types of engines: gasoline, diesel, 4-cylinder, 6-cylinder, 10-cylinder, etc. The engine technician has one of the most physically demanding automotive jobs. It requires a fairly strong individual who can lift heavy parts and easily torque large fasteners.

A *transmission technician* works on automatic and manual transmissions, transaxles, clutches, and, sometimes, rear axle assemblies. Because transmissions are so complex, the transmission technician must receive very specialized training and must frequently retrain. Some large service facilities have a rear axle specialist, who works on nothing but differentials, axle shafts, and drive shafts.

A *steering and suspension technician* is responsible for checking, replacing, and adjusting steering and suspension components. This technician must use specialized equipment, such as the wheel alignment rack, to line up the wheels. A steering and suspension technician may also take care of tire and wheel problems.

A *brake technician* specializes in brake system service and repair, **Figure 2-3.** This individual must be capable of rapidly diagnosing problems and making adjustments or repairs. A brake technician's job is one of the easiest to master. Jobs are available in both small and large shops, service stations, and tire outlets.

The *electrical system technician* must be able to test and repair lighting systems, charging systems, computer control systems, starting systems, and other electrical systems. Compared to other specialties, this area of repair might be desirable because it requires less physical strength than other areas. See **Figure 2-4.**

A *heating and air conditioning technician* must troubleshoot, service, and repair heaters, vents, and air conditioning systems. In some instances, this technician will install new air conditioning systems in vehicles. This requires considerable skill.

Figure 2-3. This brake technician is machining a brake drum. Brake repairs must be done correctly, since the safety of the customer and passengers is dependent on the operation of the brake system.

Figure 2-4. This electrical system technician is using a digital voltmeter to measure battery voltage. (Fluke)

The *driveability and performance technician* must test and service engine fuel, ignition, and emission systems. As pictured in **Figure 2-5,** this involves the use of special test equipment to keep engines in top running condition. The tune-up expert must change spark plugs, as well as adjust and repair carburetors, fuel injection systems, and ignition system components.

Figure 2-5. The driveability and performance technician must use state-of-the-art equipment, such as this diagnostic analyzer, to find the source of engine performance problems. (ASE)

The *lubrication specialist* changes engine oil, filters, and transmission fluid. He or she checks various fluid levels and performs "grease jobs" (lubricate pivot points on suspension and steering systems).

Master Technician

A *master technician,* or *general technician,* is an experienced professional who has mastered *all* the specialized areas of automotive technology and is capable of working on almost any part of a vehicle. This person can service and repair engines, brakes, transmissions, axles, heaters, air conditioners, and electrical systems. A master technician generally has enough experience to advance to a position as a shop foreman, a supervisor, or an instructor.

Shop Supervisor

The *shop supervisor* is in charge of all the other technicians in the service facility. The supervisor must be able to help others troubleshoot problems in all automotive areas. The shop supervisor must also communicate with the service manager, parts manager, and technicians.

Service Manager

The *service manager* is responsible for the complete service and repair operation of a large repair facility. This person must use a wide range of abilities to coordinate the efforts of the shop supervisor, parts specialist, service writer, service dispatcher, and other shop personnel. The service manager must also handle customer complaints, answer questions, and ensure that the technicians are providing quality service for their customers.

Other Automotive Careers

There are numerous other automotive careers that do not require extensive mechanical ability. They do, however, require a sound knowledge of automotive technology. A few of these careers are discussed below.

An *auto parts specialist* must have a general knowledge of the components and systems of a vehicle. This person must be able to use customer requests, parts catalogs, price lists, and parts interchange sheets to quickly and accurately find needed parts.

A *service writer,* or *service advisor,* prepares work orders for vehicles entering the shop for repair or service. This person greets customers and listens to descriptions of their problems. The service writer must then fill out the repair order, describing what might be wrong.

A *service dispatcher* must select, organize, and assign technicians to perform each auto repair. The dispatcher must also keep track of all the repairs taking place in the shop.

An *auto salesperson* informs potential buyers of the features and equipment on a vehicle while trying to make a sale. By understanding how a vehicle works, the salesperson will be better prepared to answer questions. Positions are also available selling automotive-related parts and equipment, **Figure 2-6.**

An *automotive designer* has art training and can make sketches or models of new body and part designs. This person is employed by automobile or automotive part manufacturers, **Figure 2-7.**

An *automotive engineer* designs new and improved automotive systems and parts. This person must use math, physics, and other advanced technologies to improve automotive designs. An engineer is a highly

Figure 2-7. Automotive designers must have extensive engineering knowledge and a strong art background to design buildable yet aesthetically pleasing vehicles. (Pontiac)

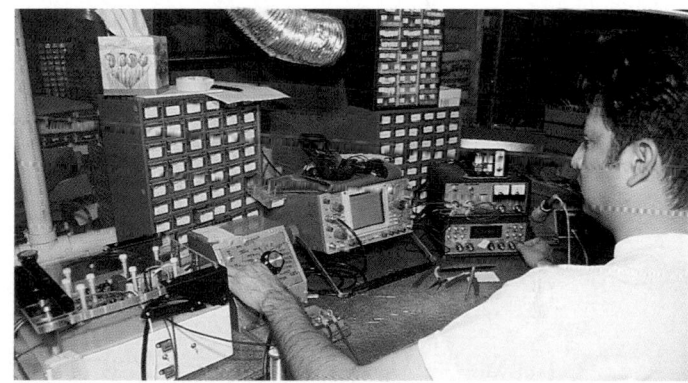

Figure 2-8. Automotive engineers normally have a bachelor's degree in a related engineering discipline. A working knowledge of automotive technology is a must for the engineer. (MSD Ignitions)

Figure 2-6. A convention allows manufacturers of parts, tools, and equipment to display new and established products. People well versed in automotive technology and sales are needed to work at these events.

Figure 2-9. An automotive instructor must have a strong background in automotive technology and must also be able to communicate well.

paid, college-trained individual with a working knowledge of the entire automobile, **Figure 2-8.**

Automotive instructors are experienced technicians capable of sharing their knowledge effectively. In addition to on-the-job experience, most instructors are required to have a college degree, **Figure 2-9.**

There are dozens of other job titles in the automotive field. Check with your school guidance counselor for more information. The chart in **Figure 2-10** shows automotive job opportunities. Trace the flow from manufacturer to service technician.

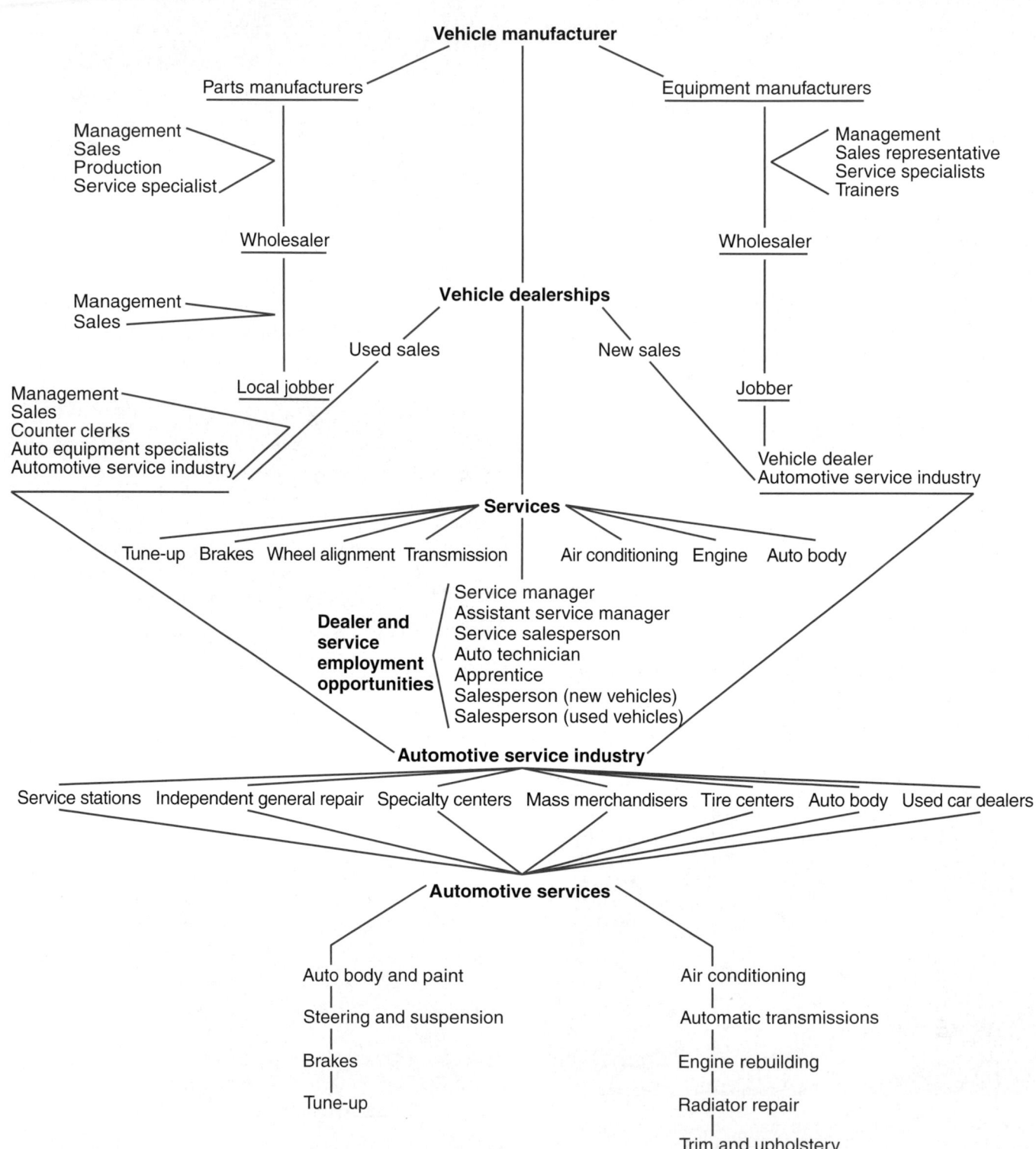

Figure 2-10. Note the many positions available in the automotive field. (Florida Dept. of Voc. Ed.)

Tech Tip
Many people start their careers as automotive technicians and then move to other related job areas. Never limit your sights on the future. Always try to improve your skills and potential for a new and better job.

Preparing for a Career in Automotive Technology

A career as an automobile technician can be quite challenging. Late-model cars and light trucks are very complex. They are constantly being updated and redesigned. This makes it difficult to keep up with the latest technology.

Many of the basic skills needed to succeed in automotive technology are learned in the classroom. Automobile technicians must have above-average math and English skills. They must be proficient at working with numbers to calculate part clearances and fill out work orders. They must also have good verbal skills to effectively communicate with customers and other technicians.

Many schools offer *cooperative training programs.* These programs allow you to earn school credit and a wage by working in a commercial repair shop. The employer gets a tax credit while helping the student technician learn the trade. Ask your guidance counselor or automotive instructor about a possible cooperative training program in your school.

On-the-Job Traits

Although automotive technicians are valued for their technical skills, employers look for other traits, as well. When you accept employment, you take on responsibilities beyond basic technical skills. These responsibilities include:

- *Reliability*, which means attending work regularly and arriving on time. Unreliable workers hurt themselves and other workers. Customers will be unhappy when their car or truck is not done on time because of unreliable workers.

- Maintaining a *positive attitude,* which means maintaining a cheerful outlook concerning other coworkers and about the job. If you are happy and think positive, you will do better work.

- *Productivity* means that you are able to get a lot of work done while maintaining quality. You are paid to work; use your time effectively. If you are not productive, you and your employer will make less money.

- *Following orders* means doing what the boss tells you to do. Someone must be in charge. You must assume that management knows how the job should be done. If you fail to follow orders, you will not have the job very long.

- *Pride* means looking and working like a professional. Come to the job in clean clothes and be well groomed. Also keep your tools and work area clean. This instills confidence in the customer. If you look sloppy, it implies your work is sloppy!

- *Congeniality* means getting along with co-workers. In a way, you live with your co-workers eight hours a day. If you get along with them, they will help you with tough jobs and make your life easier. If you argue with co-workers, everyone will suffer.

ASE Certification

The ***National Institute for Automotive Service Excellence,*** or ASE, is a nonprofit, *nonaffiliated* (no ties to industry) organization formed to help ensure the highest standards in automotive service.

ASE directs an organized testing and certification program under the guidance of a 40-member board of directors. The members of this board represent all aspects of the automotive industry—educators, shop owners, consumer groups, government agencies, aftermarket parts distributors, auto manufacturers, etc. This broad group of experts guides the ASE certification program and helps ASE stay in touch with the needs of the industry.

ASE certification is a program in which persons take written tests to prove their knowledge of automotive technology. ASE certification tests are voluntary. They do not have to be taken, and they do not license technicians.

Some countries have made technician certification a requirement. In the United States, however, technicians take the tests for personal benefit and to show their employers and customers they are fully qualified to work on a system of a car or an engine.

Over 500,000 technicians have passed ASE certification tests. Thousands of these technicians have been retested and recertified after five years to maintain their credentials.

Test Categories

In automotive technology, there are eight test categories: Engine Repair, Automatic Transmission/Transaxle, Manual Drive Train and Axles, Suspension and Steering, Brakes, Electrical Systems, Heating and Air Conditioning, and Engine Performance.

You can take any one or all of these tests. However, only four tests (200 questions maximum) can be taken at one testing session.

There are also seven medium/heavy truck tests, five collision repair and refinish tests, three engine machinist tests, and one advanced engine performance test. **Figure 2-11** provides a breakdown of each automotive test.

Certification Status

ASE-certified automobile technician status is granted for each ASE test passed. *ASE-certified master automobile technician* status, previously called certified general technician status, is granted when all eight test areas have been passed.

Sleeve or shirt patches and certificates are awarded for each test passed. This will let your customers know that you are fully capable of repairing their vehicles. See **Figure 2-12.**

Tech Tip

Become a certified automobile technician! If you master the material in this textbook, you should have no trouble passing all the ASE certification tests. As an ASE-certified technician, you will gain added respect from your employer and your customers.

Figure 2-12. If you become certified, your customers and your employer will know that you are prepared to properly repair cars and light trucks. (ASE)

Test title:		Test content:
Engine repair (80 questions)	TEST A1	Valve train, cylinder head, and block assemblies; lubricating, cooling, ignition, fuel, exhaust, and battery and starting systems
Automatic transmission/transaxle (50 questions)	TEST A2	Controls and linkages; hydraulic and mechanical systems
Manual drive train and axles (40 questions)	TEST A3	Manual transmissions, clutches, front and rear drive systems
Suspension and steering (40 questions)	TEST A4	Manual and power steering, suspension systems, alignment, and wheels and tires
Brakes (55 questions)	TEST A5	Drum, disc, combination, and parking brake systems; power assist and hydraulic systems
Electrical/electronic systems (50 questions)	TEST A6	Batteries; starting, charging, lighting, and signaling systems; electrical instruments and accessories
Heating and air conditioning (50 questions)	TEST A7	Refrigeration, heating and ventilating, AC controls
Engine performance (70 questions)	TEST A8	Oscilloscopes and exhaust analyzers; emission control and charging systems; cooling, ignition, fuel and carburetion, exhaust, and battery and starting systems

Figure 2-11. To become certified in a given automotive service area, you must pass the certification test for that area and have two years of related work experience. To become a master technician, you must pass all eight tests. (ASE)

Applying for ASE Tests

Anyone may take the ASE tests; however, a passing grade does not lead to automatic certification. To receive certification, the applicant must also have at least two years of experience as an automobile or truck technician. This experience does not have to be in any specific area of automotive service. In some cases, training programs, apprenticeship programs, or time spent performing similar work can be substituted for all or part of the work experience.

In some cases, formal training can be substituted for all or part of the experience requirement:

- High school training for three full years in automotive technology can be substituted for one year of work experience.
- Post-high school training for two years in a public or private facility can be substituted for one year of work experience.
- Two months of short training courses can be substituted for one month of work experience.
- Three years in an apprenticeship program can be substituted for two years of work experience.

To have schooling substituted for work experience, you must include a copy of your transcript (list of courses taken) or a certificate verifying your training or apprenticeship with your registration form and fee payment. Each should give your length of training and subject area.

To apply to take ASE tests, begin by acquiring a registration booklet. The registration booklet contains the proper registration form and all the information needed to complete the form. When you receive the form, fill it out carefully. For more information on auto technician certification or to obtain a registration booklet, send your name and address to:

ASE Registration Booklet
P.O. Box 591
Herndon, VA 20172-0591

ASE tests are given in the spring and the fall of each year. The tests are usually held during a two-week period and are given on weeknights and Saturdays. The tests are given at designated test centers at over 300 locations in the United States.

Consult the ASE Information Bulletin for test locations. Be sure to determine the closest test center and record its number in the appropriate space on the application form. Most test centers are located at local colleges, high schools, or vocational schools.

When submitting the application, you must include a check or money order to cover all necessary fees. A fee is charged to register for the test series, and a separate fee is charged for each test taken. Refer to the latest ASE Information Bulletin for the current fee structure. In some cases, employers pay the registration and test fees. Check with your employer before submitting your application.

After receiving your application and fees, ASE will send you an admission ticket to the test center. You should receive the ticket by mail about two weeks after submitting your application.

Test-Taking Techniques

Follow all instructions given by the test administrators. During the test, read each question carefully before deciding on a proper answer. You must select the *most correct* response. Sometimes more than one response is correct. However, one answer will always be more correct than the others.

You will not be required to recall exact specifications unless they are general and apply to most makes and models of cars. For example, compression test pressure readings and engine clearances are typically about the same for all gasoline engines. This type of general information might be needed to answer some questions.

After completing all the questions in a particular test, recheck your answers to ensure that you did not make a careless error. In most cases, rechecking your answers more than once is unnecessary and may lead you to change correct answers to incorrect ones. The time allowed for each test is usually about four hours. However, you may leave after completing your last scheduled test and handing in all test material.

A few tips that might help you pass ASE certification tests include:

- Read the statements or questions slowly. You might want to read through them twice to make sure you fully understand the question.
- Analyze the statement or question. Look for hints that make some of the possible answers wrong.
- Analyze the question as if you were the technician trying to fix the car. Think of all possible situations and use common sense to pick the most correct response.
- When two technicians give statements concerning a problem, try to decide if either is incorrect. If both are valid statements about a situation, choose the answer stating that both technicians are correct. If only one is correct or neither is correct, mark the answer accordingly. This is one of the most difficult types of questions.

- If the statement only gives limited information, make sure you do not pick one answer as correct because it is a more common condition. If the statement does not let you conclude one answer is better than another, both answers are equally correct.

- Your first thought about which answer is correct is usually the correct response. If you think about a question too much, you may read something into the question that is not there. Read the question carefully and make a decision.

- Do not waste time on any one question. Make sure you have enough time to answer all the questions on the test.

- Visualize yourself performing a test or repair when trying to answer a question. This will help you solve the problem more accurately.

Types of ASE Test Questions

ASE tests are designed to measure your knowledge of three things:

- The operation of various automotive systems and components.

- The diagnosis and testing of various automotive systems and components.

- The repair of automotive systems and components.

All test questions are multiple choice and contain four possible answers. Sample questions are given below. The answer to each question is explained in detail.

One-Part Questions

In a one-part question, you must choose the best answer out of all the possibilities.

1. Which of the following components ignites the fuel in a gasoline engine?
   (A) Injector.
   (B) Valve.
   (C) Spark plug.
   (D) Glow plug.

The spark plug produces the electric arc to start the fuel burning. Therefore, the correct answer is *(C) Spark plug*. The injector simply sprays fuel into the engine. The valve allows the air-fuel mixture to flow into the engine. The glow plug is only used in a diesel engine to warm the combustion chamber to aid combustion.

Two-Part Questions

Two-part questions require you to read two statements and decide if they are true. Both statements can be true or both can be false. In some cases, only one of the statements is true.

1. Technician A says a locking rear differential assembly can be refilled with regular gear oil. Technician B says the differential assembly allows the vehicle to turn corners without wheel hop. Who is right?
   (A) A only.
   (B) B only.
   (C) Both A & B.
   (D) Neither A nor B.

In this question, the statement made by Technician A is wrong. A locking rear wheel differential assembly must be refilled with special nonslip oil. Technician B's statement is correct, since the purpose of the differential assembly is to allow the vehicle to turn corners without wheel hop. Therefore, the correct answer is *(B) B only*. Note that Technician A and Technician B appear in many ASE test questions. You must carefully evaluate the statements of each technician before deciding which answer is correct.

Negative Questions

Some questions are called negative questions. These questions require you to identify the *incorrect* answer. Negative questions will usually contain the word *"except."*

1. An engine contains all of the following bearings *except:*
   (A) *connecting rod bearings.*
   (B) *main crankshaft bearings.*
   (C) *camshaft bearings.*
   (D) *reverse idler bearings.*

Since reverse idler bearings are used in transmissions and there is no bearing with that name used in the engine, the correct answer is *(D) reverse idler bearings.*

A variation of the negative question contains the word *"least."*

1. An automatic transmission installed in a late-model vehicle slips during acceleration. Which of these defects is *least* likely to be the cause?
   (A) *Clogged transmission oil filter.*
   (B) *Defective transmission oil pump.*
   (C) *Maladjusted throttle linkage.*
   (D) *Low fluid level.*

In this case, the least likely cause of transmission slippage is a maladjusted throttle linkage, which is much more likely to cause shifting problems than slippage. Therefore, the correct answer is *(C) Maladjusted throttle linkage.*

Completion Questions

Some test questions are simply sentences that must be completed. One of the four possible answers correctly completes the sentence.

1. A torque wrench is used to measure:
   (A) *twisting force on fasteners.*
   (B) *shear applied to fasteners.*
   (C) *horsepower applied to fasteners.*
   (D) *transmission slip yoke angles.*

Once again, the question calls for the best answer. The torque wrench is used to measure twisting force, or torque, so *(A) twisting force on fasteners* is correct.

Test Results

The results of your test will be mailed to your home. Only you will find out how you did on the tests. You can then inform your employer if you like.

Test scores will be mailed out a few weeks after you have completed the test. If you pass a test, you can consider taking more tests. If you fail, you will know that more study is needed before retaking the test.

Recertification Tests

Once you are certified in any area, you must take a *recertification test* every five years to maintain your certification. Recertification test questions concentrate on recent developments. This ensures that certified technicians will keep up with changes in technology.

Applying to take recertification tests is similar to applying for original certification tests. Use the same form and enclose the proper recertification test fees. If you allow your certification to lapse, you must take the regular certification test(s) to regain your certification.

Text Organization and ASE Certification

This textbook will help you prepare for ASE certification. The content, scope, and organization were developed with the ASE certification tests in mind. Additionally, each chapter contains a section of ASE-type questions, which will help you pass the ASE certification tests.

Entrepreneurship

An *entrepreneur* is someone who starts a business, such as a muffler shop, tune-up shop, parts house, or similar facility. To be a good entrepreneur, you must be able to organize all aspects of the business: bookkeeping, payroll, facility planning, hiring, etc. After gaining several years of on-the-job experience, you might want to consider starting your own business.

Most successful entrepreneurs have a quality known as leadership. Effective leaders have the courage to set a course of action and get the cooperation of others in meeting goals. Leaders are willing to accept responsibility for their actions. If their decisions show signs of failure, they take action to correct mistakes. If their decisions are good, they are willing to share the "glory." Good leaders readily credit the work of others who have contributed to the success of the business venture.

Summary

- Economists are predicting a continued demand for auto technicians for many years.
- An automotive technician makes a living by diagnosing, servicing, and repairing cars and light trucks.
- A service station attendant requires little mechanical experience, yet provides an excellent learning experience. It is a very common entry job.
- One way to get started in automotive service is to become an apprentice or helper mechanic, working under an experienced technician.
- A specialized technician is an expert on one system of a car. This type specialist may work only on brakes, transmissions, engines, tune-ups, electrical systems, or air conditioning.
- An engine technician must be able to troubleshoot, service, and repair automobile engines.
- A transmission technician usually works on automatic and manual transmissions, transaxles, clutches, and, sometimes, rear axle assemblies.
- A steering and suspension technician is responsible for checking, replacing, and adjusting steering and suspension components.
- A brake technician specializes in brake system service and repair.
- The electrical system technician must be able to test and repair lighting systems, charging systems, computer control systems, and starting systems.
- The driveability and performance technician must test and adjust engine fuel, ignition, and emission systems.
- A master technician is an experienced professional who has mastered *all* specialized areas and is capable of working on almost any part of a vehicle.

- A shop supervisor is in charge of other technicians in a large garage.

- An entrepreneur is someone who starts a business. This might be a muffler shop, tune-up shop, parts house, or similar facility.

- Auto technician certification is a program where persons voluntarily take written tests to prove their knowledge as an auto technician.

Important Terms

Automotive technician	Cooperative training
Service station attendant	programs
Apprentice	National Institute for
Specialized technician	Automotive Service
Master technician	Excellence (ASE)
Shop supervisor	Automotive technician
Service manager	certification
Auto parts specialist	ASE-certified
Service writer	automobile technician
Service dispatcher	ASE-certified
Auto salesperson	master automobile
Automotive designer	technician
Automotive engineer	Entrepreneur
Automotive instructors	

Review Questions—Chapter 2

Please do not write in this text. Place your answers on a separate sheet of paper.

1. List four skills that may be needed when working as an automotive technician.

2. Which of the following is *not* a typical specialized technician?
   (A) Engine technician.
   (B) Steering and suspension technician.
   (C) Brake technician.
   (D) Drive shaft technician.

3. Describe some of the responsibilities of a driveability and performance technician.

4. A(n) _____ specialist may have to do grease jobs.

5. What is a master technician?

6. Explain the job of a service manager.

7. What is a cooperative training program?

8. List the eight test categories of ASE certification.

9. You will receive a(n) _____ and a(n) _____ for each ASE certification exam passed.

10. A(n) _____ is an individual who starts a business.

ASE-Type Questions

1. Technician A says that an automotive technician must be a "jack of all trades." Technician B says that the technician's job is seldom boring. Who is right?
   (A) A only.
   (B) B only.
   (C) Both A and B.
   (D) Neither A nor B.

2. All the following are automatic benefits of ASE certification *except:*
   (A) recognition.
   (B) a sleeve patch.
   (C) a pay raise.
   (D) a wall certificate.

3. A vehicle comes into the shop for repairs. Technician A says to check with the service writer for the details of the customer complaint. Technician B says to check with the shop supervisor to get information on the complaint. Who is correct?
   (A) A only.
   (B) B only.
   (C) Both A and B.
   (D) Neither A nor B.

4. A used car needs a wheel alignment. Which technician would usually complete this operation?
   (A) Engine technician.
   (B) Steering and suspension technician.
   (C) Wheel technician.
   (D) Brake technician.

5. Technician A says some ASE questions cover the operation of various automotive systems and components. Technician B says some ASE questions cover the repair of various automotive systems and components. Who is right?
   (A) A only.
   (B) B only.
   (C) Both A and B.
   (D) Neither A nor B.

Activities—Chapter 2

1. Research an automotive career of your choice; using a computer or a typewriter, prepare a written report covering such topics as: duties, working conditions, pay range, and opportunities for advancement.

2. Interview a manager of a parts department for a local garage. Report on the duties performed.

Chapter 3

Basic Hand Tools

After studying this chapter, you will be able to:

- Identify common automotive hand tools.
- List safety rules for hand tools.
- Select the right tool for a given job.
- Maintain and store tools properly.
- Use hand tools safely.
- Correctly answer ASE certification test questions referring to hand tools.

Professional auto technicians invest thousands of dollars on tools, and for good reason. It is almost impossible to do even the simplest auto repair without using some type of tool. Tools serve as extensions to parts of the body. They increase the physical abilities of fingers, hands, arms, legs, eyes, ears, and back. A well-selected set of tools speeds up repairs, improves work quality, and increases profits.

This chapter will cover the basic hand tools commonly used in the shop. Specialized hand tools are covered in later chapters. Use the index to locate these tools as needed.

 **Tech Tip**
It is very frustrating trying to fix a vehicle without the right tools. It can be like trying to "fight a forest fire with a squirt gun"—*impossible!* Invest in a complete set of quality tools.

Tool Rules

There are several basic tool rules that should be remembered. These are as follows.

- Purchase quality tools—With tools, you usually get what you pay for. Quality tools are lighter, stronger, easier to use, and more dependable than off-brand, bargain tools. Many manufacturers of quality tools provide guarantees. Some are for the lifetime of the tool. If the tool fails, the manufacturer will replace it free of charge. This can save money in the long run.

- Keep tools organized—A technician has hundreds of different tools. For each tool to be used quickly, the tools should be neatly arranged. There should be a place for every tool, and every tool should be in its place. If just thrown into the toolbox, time and effort are wasted digging for tools.

- Keep tools clean—Wipe tools clean and dry after each use. A greasy or oily tool can be dangerous! It is very easy to lose a grip on a dirty tool, cutting or breaking a finger or hand.

- Use the right tool for the job—Even though several different tools may be used to loosen a bolt, usually one will do a better job. It may be faster, grip the bolt better, be less likely to break, or require less physical effort. A good technician knows when, where, and why a particular tool will work better than another. Keep this in mind as you study automotive tools.

 Tech Tip
The time spent maintaining your tools and toolbox is time well spent. Well-organized tools will save time on each job and help you get more work done. Unorganized or poorly maintained tools will hurt your on-the-job performance.

Tool Storage

A *toolbox* stores and protects a technician's tools when not in use. There are three basic parts to a typical toolbox, **Figure 3-1.** These include:

- A lower roll-around bottom cabinet.
- An upper tool chest that sits on the roll-around cabinet.

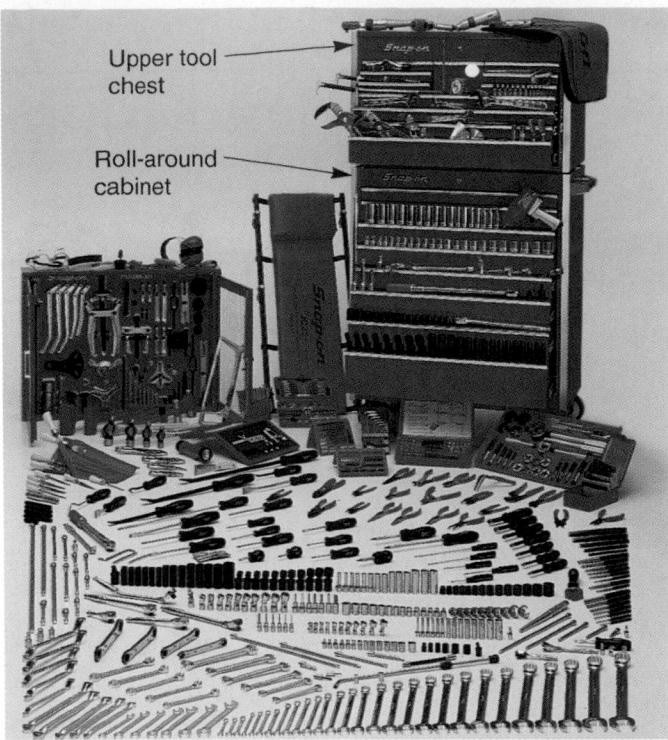

Figure 3-1. A toolbox is used to organize the wide variety of tools a technician needs. (Snap-on Tool Corp.)

- A small carrying tray. This is usually placed in the upper tool chest.

Commonly used tools are normally placed in the **upper tool chest.** Being near eye level, tools can be easily seen and reached without bending. This saves time and energy, and increases production.

The **lower roll-around cabinet** holds the bulky, heavy tools. Large power tools are normally kept in this part of the box. Extra storage compartments can be bolted to the sides of the roll-around cabinet.

The small **carrying (tote) tray** is for holding frequently used tools. For example, if a technician frequently does brake repairs, all the special brake tools can be kept in the tray and taken to the vehicle more easily.

Toolbox Organization

Related tools are normally kept in the same toolbox drawer. For example, various types of hammers may be stored in one drawer and all screwdrivers in another. Small or delicate tools should *not* be kept with large, heavy tools to prevent damage. **Tool holders** help organize small tools. These include small clip or magnetic racks, cloth or plastic pouches, or socket trays. They are often used to protect tools and to keep them organized by size. Holders also allow a full set of tools to be taken to the job.

Warning
Never open more than two toolbox drawers at one time. If you do, the heavy toolbox might flip over. Serious injury can result since a toolbox can weigh up to 1000 pounds. Close each drawer before opening the next.

Wrenches

Wrenches are used to install and remove nuts and bolts. **Wrench size** is determined by measuring across the wrench jaws. Refer to **Figures 3-2** and **3-3.** Wrenches come in both conventional (inch) and metric (millimeter) sizes. The size is stamped on the side of the wrench. Here are a few wrench rules to follow.

- Always select the right size wrench. It must fit the bolt head snugly. A loose fitting wrench will round-off the corners of the bolt head.

- Never hammer on a standard wrench to break loose a bolt. Use a longer wrench with more leverage or a special **slug wrench.** A slug wrench is designed to be used with a hammer.

- When possible, pull on the wrench. Then, if the wrench slips, you are less likely to hurt your hand. When you must push, use the palm of your hand and keep your fingers open.

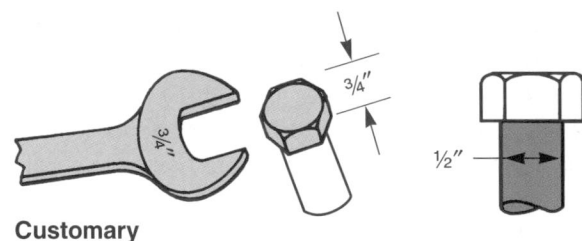

Customary

Figure 3-2. Customary tool sizes are given in fractions of an inch. The measurement is the width of the jaw opening. As shown here, these sizes are *not* the same as bolt sizes. (Deere & Co.)

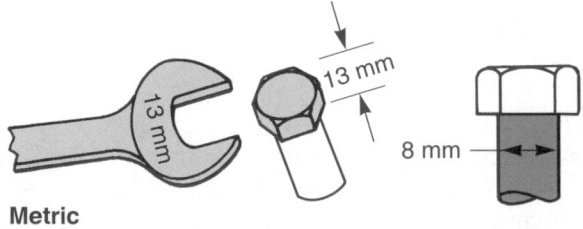

Metric

Figure 3-3. Metric wrench sizes given in millimeters. The measurement is the width of the jaw opening. The wrench size is *not* the same as the bolt size. (Deere & Co.)

- Never use a steel bar or pipe to increase the length of a wrench for leverage. Excess force can bend or break the wrench.

To be able to select the right wrench for the job, you must learn the advantages and disadvantages of each type. These advantages and disadvantages are covered in the next sections. Study the material carefully.

Open-End Wrenches

An *open-end wrench* has an open jaw on both ends. Each end is a different size and set at an angle, **Figure 3-4A.** This angle allows the open-end wrench to turn bolts and nuts with little wrench swing space. The wrench can be turned over between each swing to get a new "bite" on the bolt head. An open-end wrench has weak jaws. It should *not* be used on extremely tight nuts or bolts. Its jaws will flex outward and round-off the bolt head.

Box-End Wrenches

Box-end wrenches are completely closed on both ends. They fully surround and grip the head of a bolt or nut, **Figure 3-4B.** A box-end wrench will not round off bolt heads as easily as an open-end wrench. Box-end wrenches are available with either 6- or 12-point openings. A 6-point opening is the strongest configuration. It should be used on extremely tight, rusted, or partially rounded bolt or nut heads.

Combination Wrenches

A *combination wrench* has a box-end jaw on one end and an open end on the other, **Figure 3-4C.** Both ends are usually the same size. A combination wrench provides the advantage of two types of wrenches for the price of one.

Line Wrenches

A *line wrench,* also called a *tubing wrench* or *flare nut wrench,* is a box-end wrench with a small opening or split in the jaw, **Figure 3-4D.** The opening allows the wrench to be slipped over fuel lines, brake lines, or power steering lines and onto the fitting nut. A line wrench prevents damage to soft fittings.

Socket Wrenches

A *socket* is a cylinder-shaped, box-end tool for removing or installing bolts and nuts. See **Figure 3-5.** One end fits over the fastener. The other end has a square hole that fits on a handle used for turning.

A socket's *drive size* is the size of the square opening for the handle. As pictured in **Figure 3-6,** sockets come

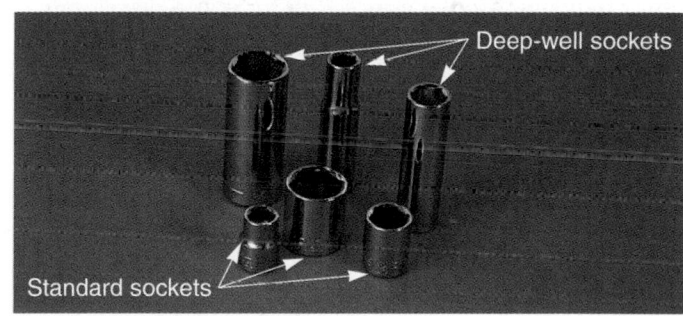

Figure 3-5. Different socket types. Note that both standard and deep-well sockets are shown.

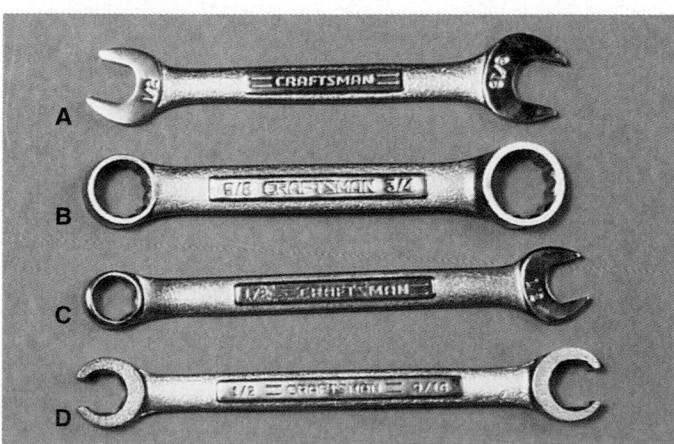

Figure 3-4. The four basic types of hand wrenches. A—Open-end wrench. B—Box-end wrench. C—Combination wrench. D—Tubing or line wrench.

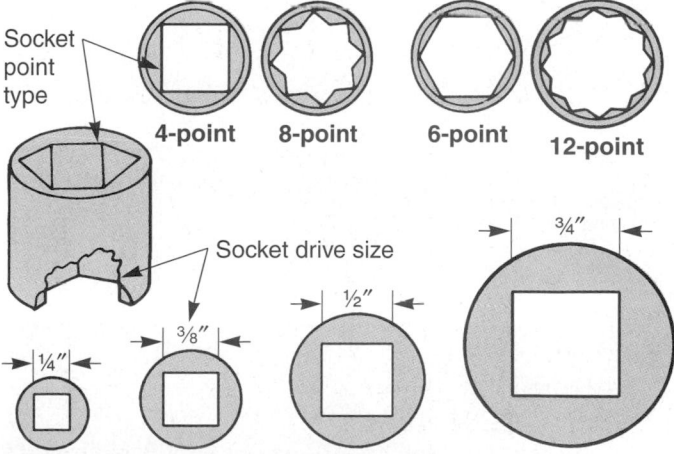

Figure 3-6. Sockets come in four drive sizes and four point types. The most commonly used drives are 3/8″ and 1/2″. The most common point types are 6-point and 12-point.

in four drive sizes: 1/4″, 3/8″, 1/2″, and 3/4″. They also come in four different points: 4-point, 6-point, 8-point, and 12-point. *Points* are the configuration of the box for the bolt head.

If a small drive size is used on very large or tight fasteners, the socket or handle can be broken. A large drive may be too awkward for small nuts and bolt heads. Generally, a 1/4″ drive socket and handle should be used on bolt and nut heads 1/4″ and smaller. A 3/8″ socket set is adequate on bolt head and nut sizes between 1/4″ and 5/8″. A 1/2″ drive is strong enough to handle bolt or nut heads from 5/8″ to 1″. A 3/4″ drive is for bolts and nuts with a head size larger than 1″.

Socket Handles

Socket handles fit into the square opening in the top of the socket. Several types are shown in **Figure 3-7.** A *ratchet* is the most commonly used and versatile socket handle. It has a small lever that can be moved for either loosening or tightening bolts. A *flex bar,* or *breaker bar,* is the most powerful and strongest socket handle. It should be used when breaking loose large or extremely tight bolts and nuts. A *speed handle* is the fastest hand-operated socket handle. After a bolt is loosened, a speed handle will rapidly spin out the bolt.

Extensions are used between a socket and its handle. See **Figure 3-8A.** They allow the handle to be placed farther from the workpiece, giving you room to swing the handle and turn the fastener. A *universal joint* is a swivel that lets the socket wrench reach around obstructions, **Figure 3-8B.** It is used between the socket and drive handle, with or without an extension. Avoid putting too much bend into a universal joint, or it may bind and break.

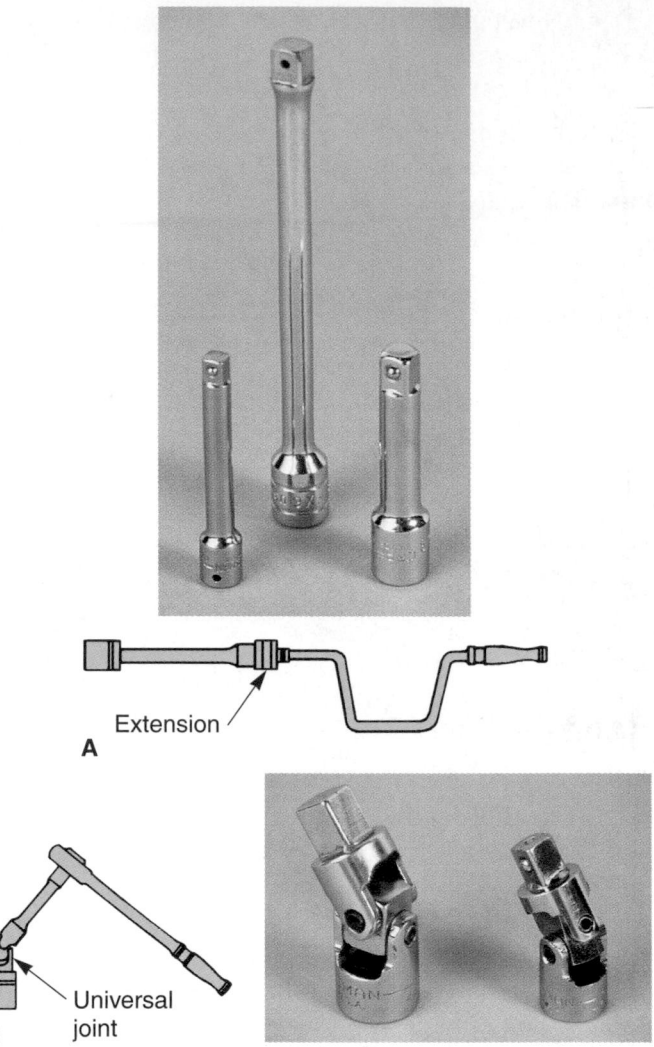

Figure 3-8. A—An extension moves the socket away from the handle for more clearance. B—A universal joint allows the socket to be turned from an angle.

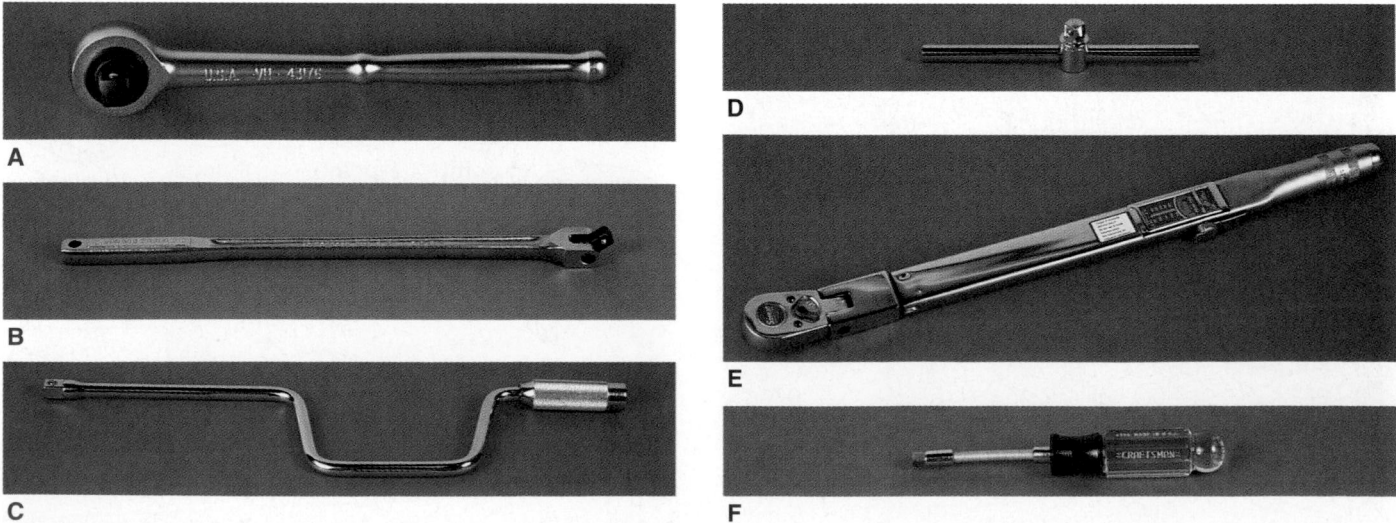

Figure 3-7. Various socket handles. A—Ratchet. B—Breaker bar or flex handle. C—Speed handle. D—T-handle. E—Torque wrench. F—Flexible driver.

Other Wrench Types

An *adjustable wrench,* or *Crescent wrench,* has jaws that can be adjusted to fit different size bolt and nut heads, **Figure 3-9A.** It should be used only when other type wrenches will *not* fit. An adjustable wrench is a handy tool to carry for emergencies. It is like having a full set of open-end wrenches.

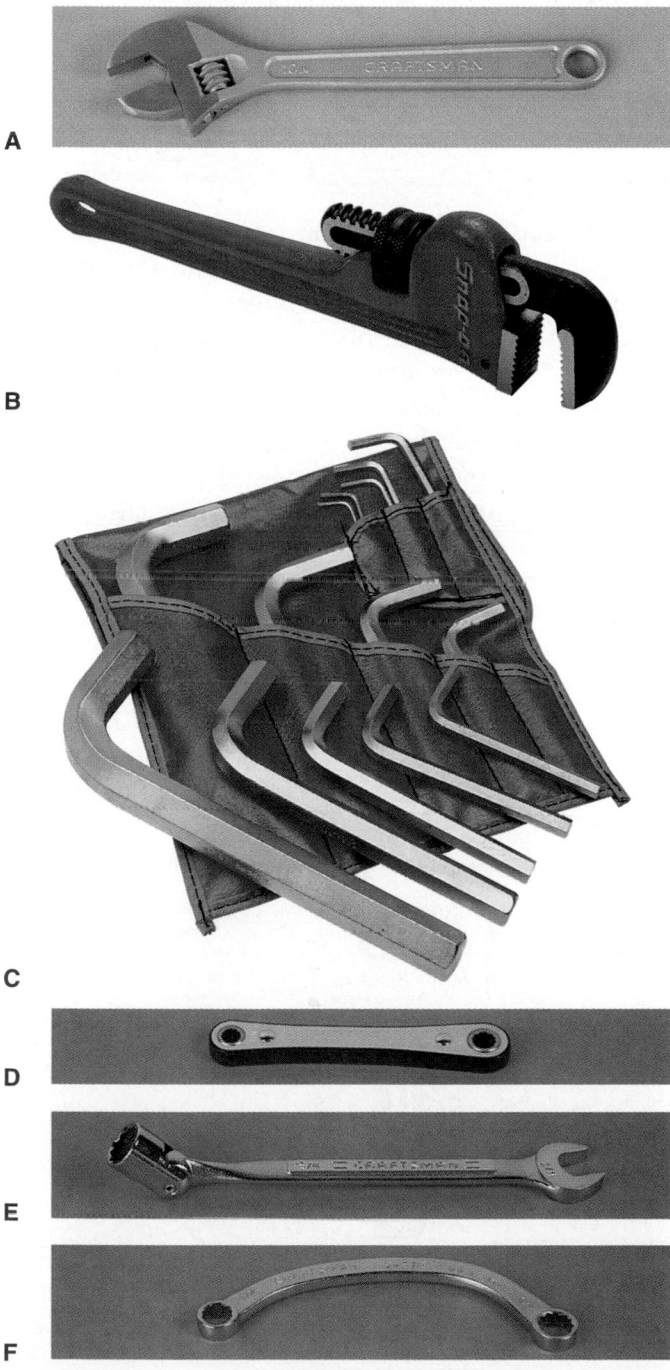

Figure 3-9. A—Adjustable, or Crescent, wrench. B—Pipe wrench. C—Allen, or hex, wrenches. D—Ratchet wrench. E—Flex-combination wrench. F—Half-moon, 12-point wrench for tight quarters. (Snap-on Tool Corp.)

A *pipe wrench* is an adjustable wrench used to grasp cylindrical objects. See **Figure 3-9B.** The toothed jaws actually dig into the object. For this reason, never use it on parts that will be ruined by marks or nicks.

An *Allen wrench* is a hexagonal (six sided) shaft-type wrench, **Figure 3-9C.** It is used to turn set screws on pulleys, gears, and knobs. To prevent damage, make sure the Allen wrench is fully inserted in the fastener before turning.

There are several other types of wrenches. These are shown in **Figure 3-9.**

Screwdrivers

Screwdrivers are used to remove or install screws. They come in many shapes and sizes. A *standard screwdriver* has a single blade that fits into a slot in the screw head. See **Figure 3-10A.** A *Phillips screwdriver* has two crossing blades that fit into a star-shaped screw slot, **Figure 3-10B.** A *Reed and Prince* screwdriver is similar to a Phillips, but it has a slightly different tip shape,

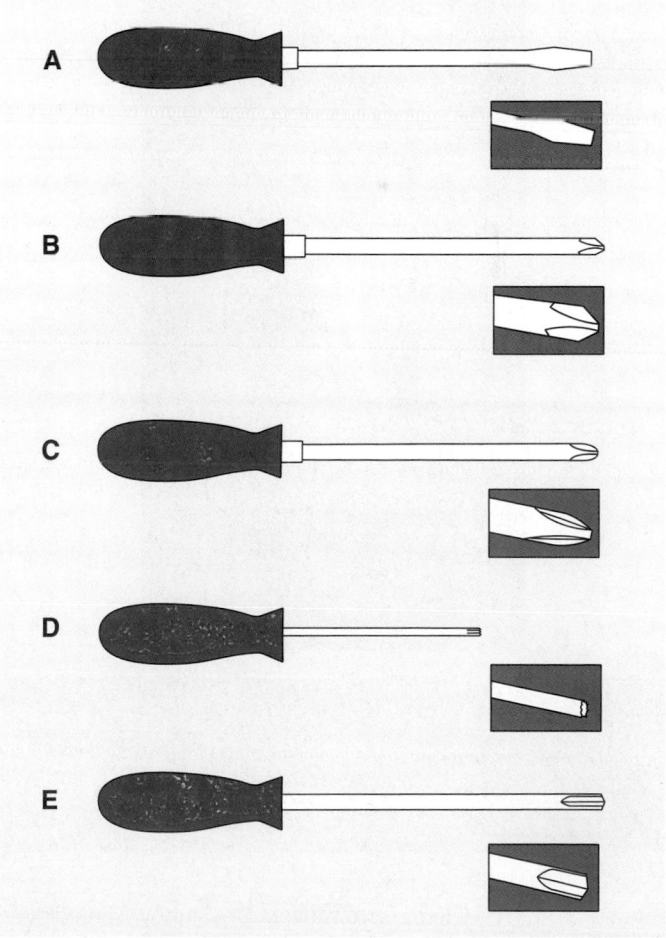

Figure 3-10. Screwdriver types. A—Standard. B—Phillips. C—Reed and Prince. D—Torx. E—Clutch. (Snap-on Tool Corp.)

Figure 3-10C. *Torx* and *clutch head* are special types of screwdrivers and are shown in **Figures 3-10D** and **E.**

Offset and *stubby screwdrivers* are good in tight places, **Figures 3-11A** and **B.** For example, a stubby screwdriver is needed for loosening screws inside a glove box. *Starting screwdrivers* hold the screw securely until started in its hole, **Figure 3-11C.** They prevent the screw from being dropped and lost. A *scratch awl* looks like a

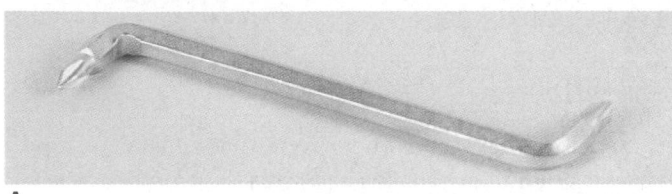

A

B

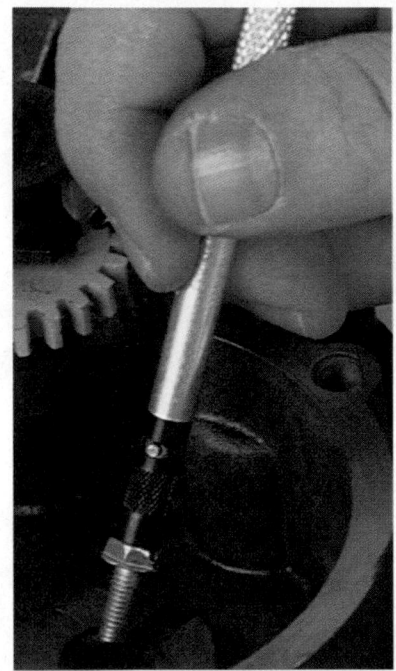

C

D

Figure 3-11. A—Offset screwdriver. B—Stubby screwdriver. C—Starting screwdriver. D—Scratch awl. It is similar to a screwdriver, but has a pointed tip for marking sheet metal and other parts.

screwdriver, but it has a sharp, pointed tip, **Figure 3-12D.** It is used for marking sheet metal and other parts. An *impact driver* can be used to loosen extremely tight screws. When struck with a hammer, the driver exerts powerful turning and downward forces. This is shown in **Figure 3-12.**

When selecting a screwdriver, pick one that is wide and thick enough to *completely fill* the screw slot, **Figure 3-13.** If too large or too small, damage to the screwdriver or screw may occur. Most screwdrivers are *not* designed to be hammered on or pried with. Only heavy-duty screwdrivers with a full shank can withstand *light* hammering and prying.

Pliers

Pliers are used to grip, cut, crimp, hold, and bend various parts. Different pliers are helpful for different situations. Several types of pliers are pictured in **Figure 3-14.** Never use pliers when another type tool will work. Pliers can nick and scar a part.

Combination pliers, or *slip-joint pliers,* are the most common pliers used by an automotive technician. The slip joint allows the jaws to be adjusted to grasp different size parts, **Figure 3-14A.** *Rib joint pliers,* also called *channel lock pliers* or *water pump pliers,* open extra wide for holding very large objects, **Figure 3-14B.**

Needle nose pliers are excellent for handling extremely small parts or reaching into highly restricted areas, **Figure 3-14C.** Do *not* twist too hard on needle nose pliers, or the long thin jaws can be bent. *Diagonal cutting pliers* are the most commonly used cutting pliers, **Figure 3-14D.** The jaw shape allows cutting flush with a surface.

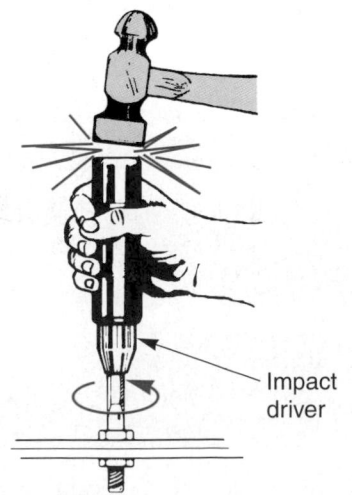

Impact driver

Figure 3-12. An impact driver loosens stubborn fasteners. Hit it with a hammer to free and turn the screw. (Lisle)

Correct tip size fills screw slot

Too small a tip will damage screwdriver

Too large a tip will strip screwhead

Figure 3-13. The screwdriver tip must fit in the slot perfectly. If not, the screwdriver and screw can be ruined.

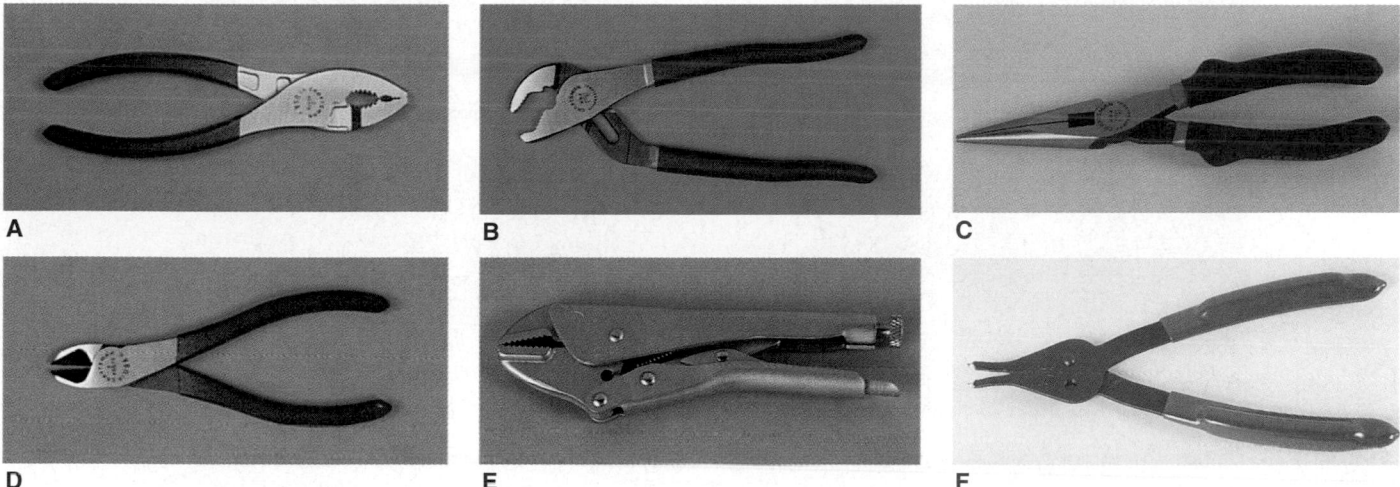

Figure 3-14. Different types of pliers. A—Slip joint. B—Rib joint. C—Needle nose. D—Diagonal cutting. E—Vise grips. F—Snap ring.

Locking pliers, or *vise grips,* clamp onto and hold a part, **Figure 3-14E.** This frees both hands to do other tasks. Because of their clamping power, vise grips can sometimes be used to unscrew fasteners with stripped or rounded heads. However, *never* use them on undamaged nuts or bolts. *Snap ring pliers* have sharp, pointed tips for installing and removing special clips called snap rings. A pair of snap ring pliers is shown in **Figure 3-14F.**

Hammers

Various types of hammers are used for operations that involve striking a tool or part. It is important to use the right hammer and to use it properly. Several hammers are shown in **Figure 3-15.** The following are some general rules governing hammers.

- Select the right size hammer. If a large part is struck by a small hammer, the hammer can fly backwards dangerously. If the hammer is too large, however, it may damage the part.

- Always check that the hammer head is tight on the handle. If not, the head may fly off and cause injury or damage.

- Never hit a hardened part with a steel hammer. Metal chips may fly off. Use a brass or lead hammer.

- Grasp the hammer near the end of the handle and strike the part or tool squarely.

A *ball peen hammer* is the most common type of hammer used in automotive work, **Figure 3-15A.** It has a flat face for general striking. It also has a round end for shaping metal parts, such as sheet metal or rivet heads.

A *sledge hammer* has a very large head, **Figure 3-15B.** It is usually the heaviest hammer and produces powerful blows. A sledge hammer is sometimes used to free frozen parts.

The *brass* or *lead hammer* has a soft, heavy head and is useful when scarring the surface of a part must be avoided, **Figure 3-15C.** The relatively soft head deforms to protect the part surface from damage.

A *plastic* or *rawhide hammer* is light and has a soft head, **Figure 3-15D.** It is used where light blows are needed to prevent part breakage or damage to surfaces on small and delicate parts.

A *rubber mallet* has a head made of solid rubber, **Figure 3-15E.** It will rebound, or bounce, upon striking

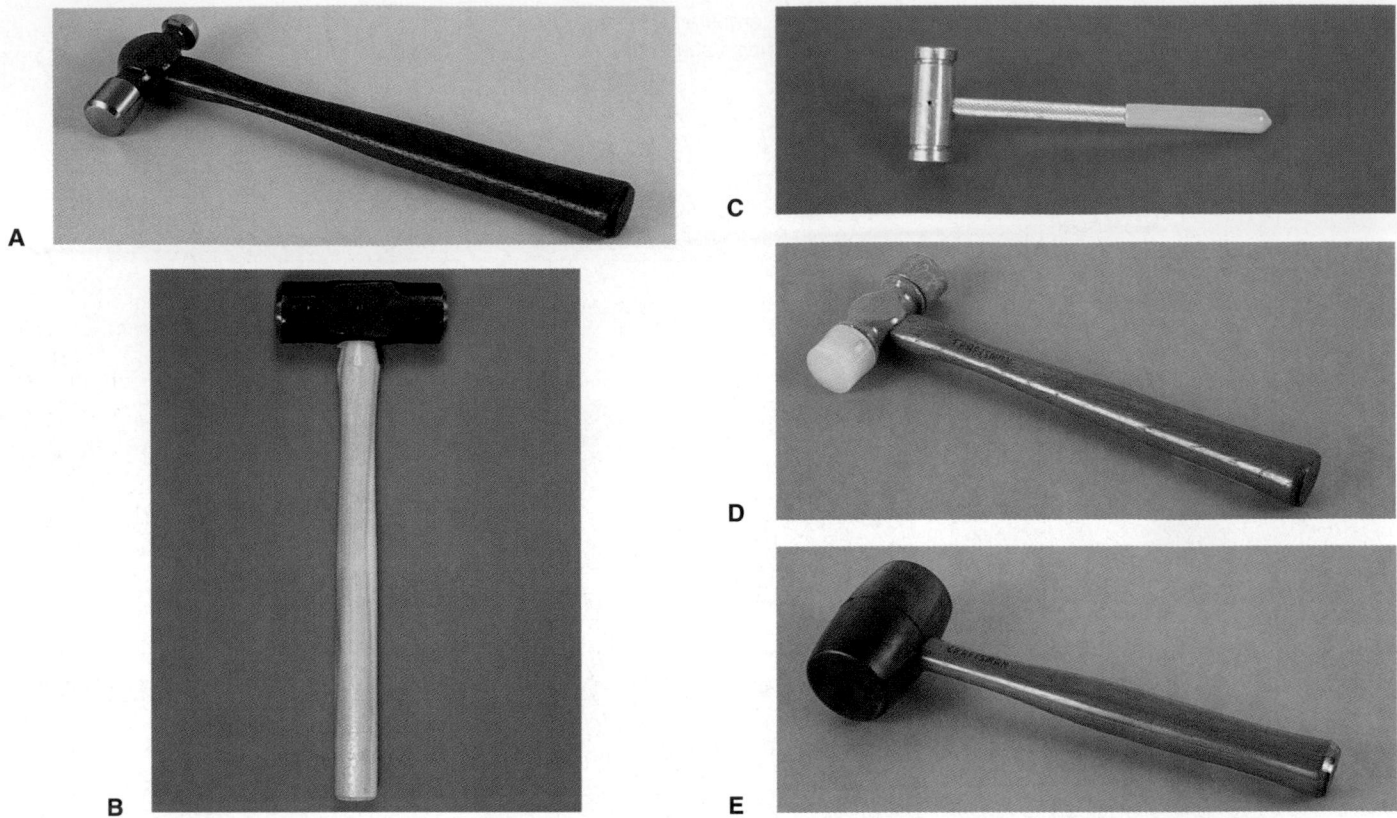

Figure 3-15. Different hammer types. A—Ball peen. B—Sledge. C—Brass. D—Plastic tipped. E—Rubber mallet.

and is not effective on solid metal parts. It is recommended on many sheet metal or plastic parts, such as garnish molding and wheel covers.

A ***dead blow hammer*** has a plastic-coated, metal face and is filled with small metal balls called lead shot. The extra weight prevents a rebound of the hammer when striking. The plastic coating avoids surface damage.

Chisels and Punches

Chisels are for cutting off damaged or badly rusted nuts, bolts, and rivet heads. There are various chisel shapes, **Figure 3-16.** Use common sense when selecting a chisel shape.

Punches also come in several configurations. See **Figure 3-16.** A *center punch* is frequently used to mark parts for reassembly and to start a hole before drilling. Look at **Figure 3-17.** The indentation made by a center punch will keep a drill bit from moving when first starting to drill.

A ***starting punch,*** or ***drift punch,*** has a shank tapered all the way to the end. It is strong and can withstand moderate blows. It is used to drive pins, shafts, and metal rods part of the way out of a hole. A ***pin punch*** has a straight shank and is lighter than a starting punch. It is used *after*

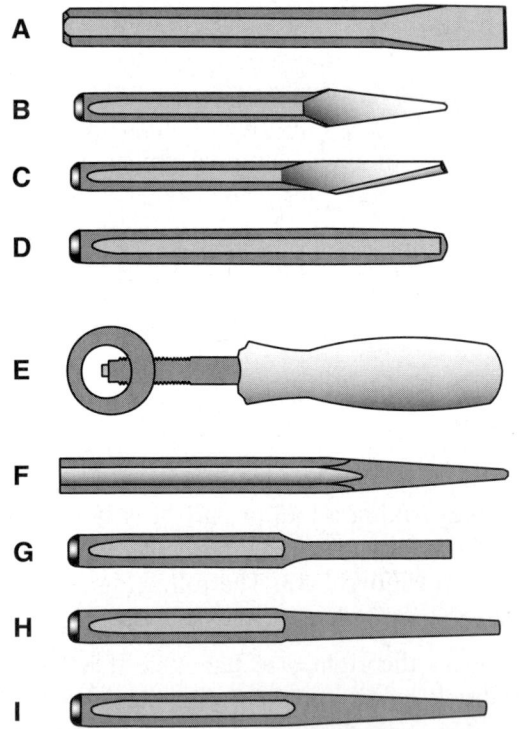

Figure 3-16. Different chisels and punches. A—Flat chisel B—Cape chisel. C—Round-nose cape chisel. D—Diamond-point chisel. E—Chisel or punch holder. F—Center punch. G—Pin punch. H—Long, tapered punch. I—Starting punch. (Snap-on Tool Corp. and Proto Tools)

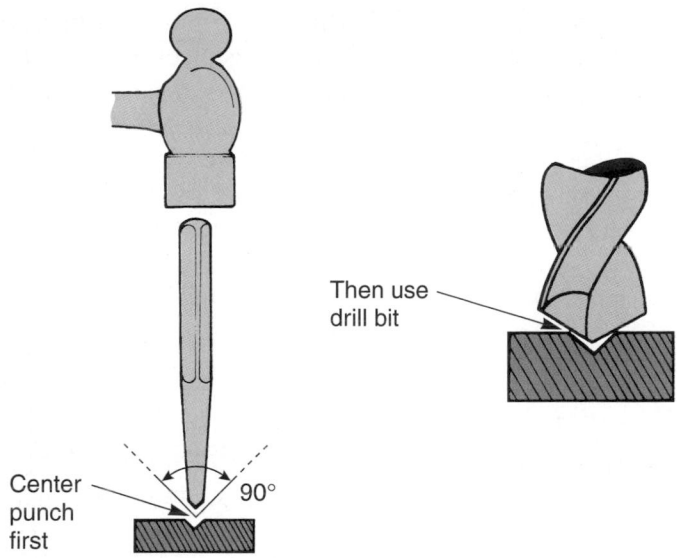

Figure 3-17. A center punch makes a small indentation in metal parts. This can then be used to start a drill bit. (Florida Dept. of Voc. Ed.)

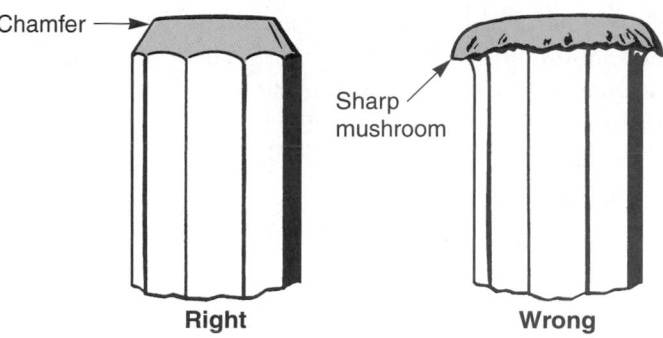

Figure 3-18. Always keep the top of a chisel or punch ground to a chamfer. A sharp, mushroomed end is dangerous. (Deere & Co.)

a starting punch to push a shaft or rod the rest of the way out of a hole.

An ***aligning punch*** is long and tapered. It is handy for lining up parts during assembly. An aligning punch can be inserted into holes in mating parts and then wiggled to match up the holes. Never use an aligning punch as a center punch. Its tip is too soft and would be ruined.

Remember these chisel and punch rules:

- Use the largest punch or chisel that will work. If a small punch is used on a very large part, the punch can rebound and fly out with tremendous force. The same is true for chisels.

- Keep both ends of a chisel or punch properly ground and shaped. A chisel's cutting edge should be sharp and square. A starting punch or a pin punch should also be ground flat and square. A center punch should have a sharp point.

- After prolonged hammering, the top of a chisel or punch can become deformed and enlarged. This is called mushrooming. A mushroomed chisel or punch is dangerous! Grind off the mushroom and form a chamfer, as shown in **Figure 3-18.**

- When grinding a chisel or punch, grind slowly to avoid overheating the tool. Excessive heat will cause the tool to turn blue, lose its temper, and become soft.

- Make sure to wear eye protection when using or grinding a chisel or punch.

Files

Files remove burrs, nicks, and sharp edges and perform other smoothing operations. They are useful when only a small amount of material must be removed. The basic parts of a file are shown in **Figure 3-19.**

A file is classified by its length, shape, and cutting surface. Generally, a *coarse file* with large cutting edges

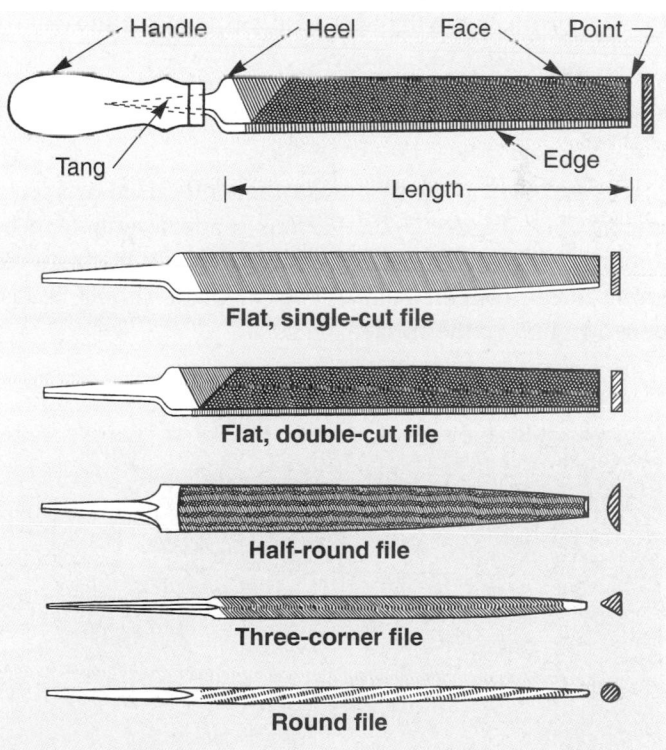

Figure 3-19. A file is used for smoothing metal. Note the different parts of a file and the different types of files available. (Starrett)

should be used on soft materials, such as plastic, brass, and aluminum. A *fine file* with small cutting edges is needed to produce a smoother surface and to cut harder materials, like cast iron or steel.

There are several file safety rules that should be remembered. These are:

- Never use a file without a handle securely attached. If the file's pointed tang is not covered by a handle, it can puncture your hand or wrist.

- To prevent undue file wear, apply pressure only on the forward stroke. Lift the file on the backstroke.

- When filing, place one hand on the handle and the other on the file tip. Hold the file firmly but *do not* press too hard.

- *Do not* file too rapidly. One file stroke every second is fast enough. Count to yourself: one thousand one, one thousand two, one thousand three, one thousand four. This will time your strokes properly at about 50–60 strokes per minute.

- If a file becomes clogged, clean it with a file card or a stiff wire brush.

- Never hammer on or pry with a file. A file is very brittle and will break easily. Bits of the file can fly into your face or eyes.

Saws

A *hacksaw* is the saw most frequently used by a technician. See **Figure 3-20.** Various blade lengths can be mounted in its adjustable frame. The blade teeth should point *away* from the handle, and the blade should be fastened tightly in the frame.

Select the appropriate blade for the job. As a rule of thumb, at least two saw teeth should contact the material being cut at any given time. If not, the teeth can catch and break.

When cutting, place one hand on the hacksaw handle and the other on the end of the frame. Press down lightly on the forward stroke and release pressure on the backstroke. As with a file, use 50–60 strokes per minute. If cuts are made faster than this, the blade will quickly overheat, soften, and become dull.

Holding Tools

There are several different types of tools used for holding objects in the automotive shop. These tools are covered in the next sections.

Vise

A *vise* is used to hold parts during cutting, drilling, hammering, and pressing operations. See **Figure 3-21.** It is mounted on a workbench. Avoid clamping a smooth, machined part in the uncovered jaws of a vise. If a machined surface is scarred, the part may be ruined.

Vise caps or wood blocks should be used when mounting precision parts in a vise. Vise caps are soft metal jaw covers. They will not only protect the part, but will provide a more secure grip on the part.

A few vise rules include:

- Never hammer on a vise handle to tighten or loosen the vise. Use the weight of your body.

- Keep the moving parts of the vise clean and oiled.

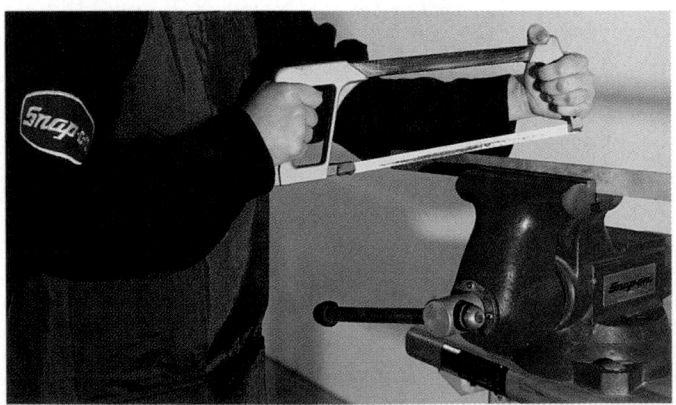

Figure 3-20. Hacksaws are used to cut metal. Hold the saw as shown and only push down on the forward stroke.

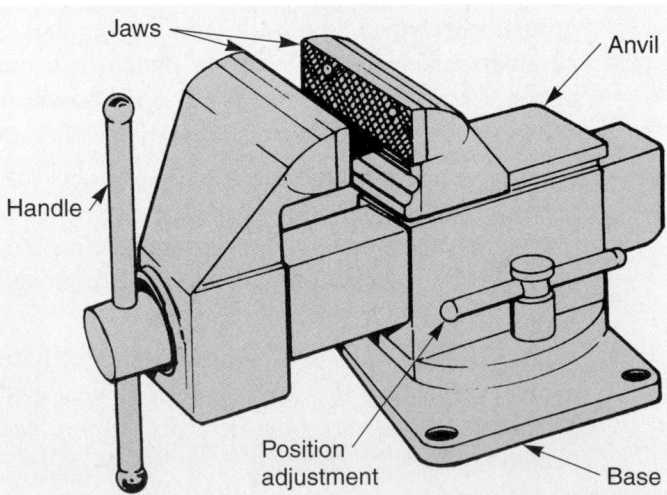

Figure 3-21. A vise mounts on the workbench. It holds parts securely. (Snap-on Tool Corp.)

- Wear safety glasses when using a vise. Tremendous clamping force can be exerted and parts may break and fly out.
- Be careful not to damage parts in the jaws of a vise.
- Use vise caps when a precision part is held in a vise. This will prevent part damage.

C-clamp

A *C-clamp* holds parts on a work surface when drilling, filing, cutting, welding, or doing other operations. Being portable, it can be taken to the job. Refer to **Figure 3-22.** C-clamps come in many different sizes.

Stands and Holding Fixtures

Stands and *holding fixtures* can be used to help secure heavy or clumsy parts while working, **Figure 3-23.** Cylinder head stands, transmission fixtures, rear axle holding stands, and others all make your work safe and easier. Always use them when available.

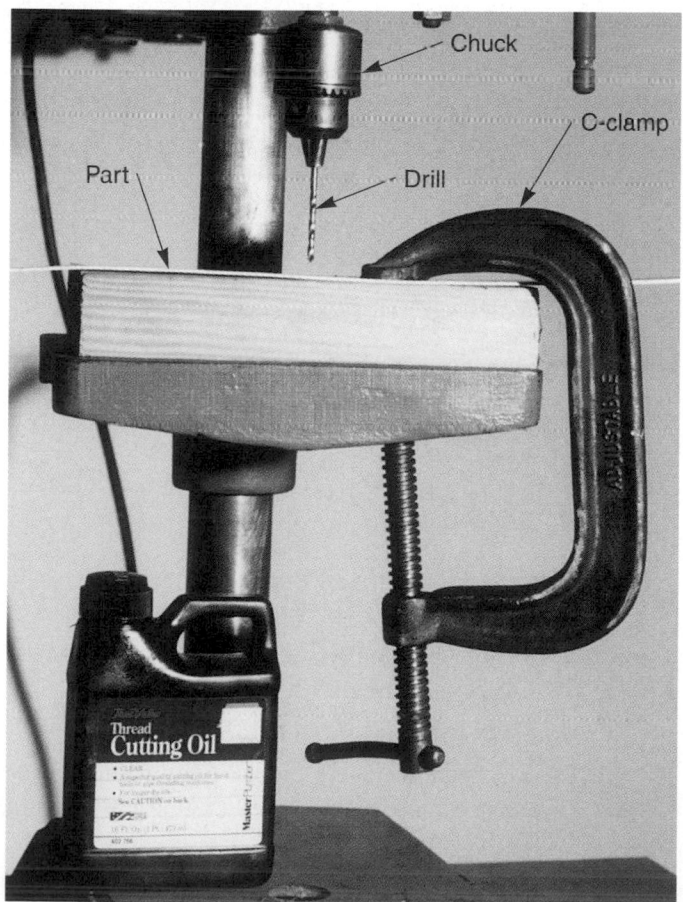

Figure 3-22. A C-clamp is a portable means of securing parts. It can also be used for light pressing operations.

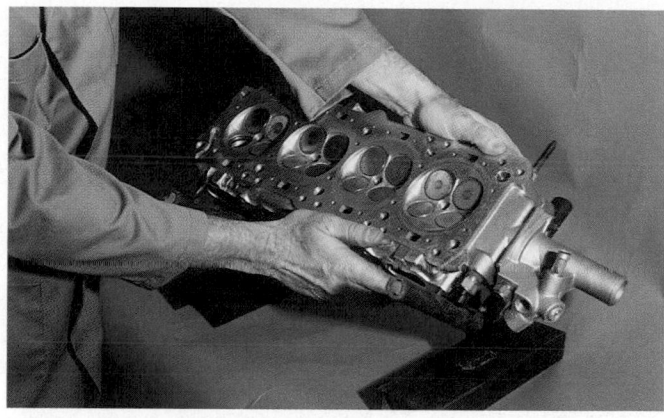

Figure 3-23. A cylinder head stand holds the head in position during valve and seat work. Other specialized stands are available for transmissions, differentials, and other parts.

Cleaning Tools

There is an old saying, "if you do the job right, you will spend most of your time cleaning parts." Dirt is a major enemy of a vehicle. One grain of sand can cause a major breakdown by clogging a passage or scarring a part. *Cleaning tools,* such as scrapers and brushes, help the technician remove carbon, rust, dirt, grease, old gaskets, and dried oil from parts.

Scrapers remove grease, gaskets, sludge, dried oil, and carbon on parts. They are used on flat surfaces. Never scrape toward your body. Keep your other hand out of the way. *Brushes* are used to remove light rust and dirt on parts. They are slow and only used when necessary.

Probe and Pickup Tools

Pickup and *probing tools* are needed when bolts, nuts, or other small parts are dropped and cannot be reached by hand. A *magnetic pickup tool* is a magnet hinged to the end of a rod. It can usually be shortened or lengthened and swiveled to reach into any area. If a ferrous (iron) metal part is dropped, it will be attracted and stick to the magnet, **Figure 3-24A.**

A *finger pickup tool* grasps nonmagnetic parts (aluminum, plastic, or rubber), which will *not* stick to a magnet, **Figure 3-24B.** A *mirror probe* allows you to look around corners or behind parts, **Figure 3-24C.** For example, a mirror probe will allow you to see an oil leak behind the engine.

Pry Bars

Pry bars are strong steel bars. They are helpful during numerous assembly, disassembly, and adjustment operations. For example, they are commonly used when

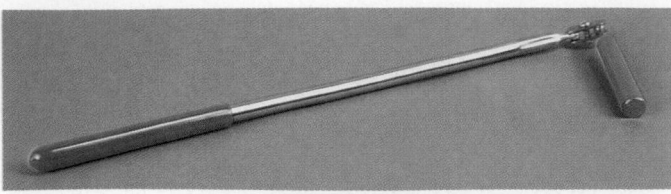

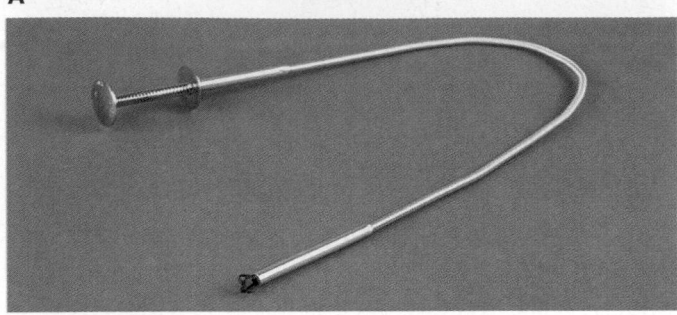

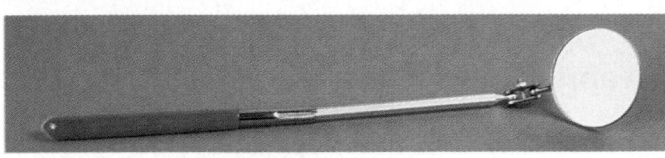

Figure 3-24. Different probe and pickup tools. A—Magnetic pickup tool. B—Finger pickup tool. C—Mirror probe.

adjusting the tension of engine belts. They are also used to align heavy parts. When prying, always be careful to *not* damage any part of the vehicle.

Summary

- It is almost impossible to do even the simplest auto repair without using some type of tool.
- Professional auto technicians invest thousands of dollars in tools. A well-selected set of tools will speed up repairs, improve work quality, and increase profits.
- Purchase quality tools. Quality tools are lighter, stronger, easier to use, and more dependable than off-brand, bargain tools.
- Keep tools organized. There should be a place for every tool and every tool should be in its place.
- Use the right tool for the job. A good technician will know when, where, and why a particular tool will work better than another.
- A toolbox stores and protects a technician's tools when not in use.
- A 6-point wrench is the strongest wrench configuration.
- A socket is a cylinder-shaped, box-end tool for removing or installing bolts and nuts.

- Socket handles fit into the square opening in the top of the socket.
- A ratchet is the most commonly used and versatile socket handle. It can either loosen or tighten bolts.
- Extensions are used between a socket and its handle.
- Pliers are used to grip, cut, crimp, hold, and bend various parts.
- A hacksaw is the saw most frequently used by the technician.
- A vise is used to hold parts during cutting, drilling, hammering, and pressing operations.
- Cleaning tools, such as scrapers and brushes, help a technician remove carbon, rust, dirt, grease, old gaskets, and dried oil from parts.
- Pry bars are strong steel bars that are helpful during numerous assembly, disassembly, and adjustment operations.

Important Terms

Toolbox	Sledge hammer
Tool holders	Lead hammer
Wrenches	Rawhide hammer
Wrench size	Rubber mallet
Socket	Dead blow hammer
Drive size	Chisels
Socket handles	Punches
Extensions	Files
Universal joint	Hacksaw
Adjustable wrench	Vise
Pipe wrench	Vise caps
Allen wrench	C-clamp
Screwdrivers	Stands
Scratch awl	Holding fixtures
Impact driver	Cleaning tools
Pliers	Probing tools
Ball peen hammer	Pry bars

Review Questions—Chapter 3

Please do not write in this text. Place your answers on a separate sheet of paper.

1. List and explain four general tool rules.
2. A bolt head is rusted and partially rounded off. Which wrench would work best for removing the bolt?
   (A) Open-end wrench.
   (B) 6-point box-end wrench.
   (C) 12-point box-end wrench.
   (D) None of the above.

3. What are the four socket drive sizes? Explain when each should be used.

4. _____ or _____ screwdrivers are useful in very tight places, such as inside a glove box.

5. Describe four rules to follow when using hammers.

6. What is the difference between a center punch, a starting punch, and an aligning punch?

7. A coarse file should be used on _____ materials. A fine cut file should be used on _____ materials.

8. When should you use vise caps?

9. List and explain four vise rules.

10. Which of the following tools are *not* cleaning tools?
    (A) Hand scraper.
    (B) Chisel.
    (C) Probe.
    (D) Hand brush.

● ASE-Type Questions

1. A bolt head is badly rusted and is difficult to loosen. Technician A says to use a pair of pliers to loosen the rusted bolt. Technician B recommends a six-point wrench or socket. Who is correct?
   (A) A only.
   (B) B only.
   (C) Both A and B.
   (D) Neither A nor B.

2. Which of the following are *not* common socket drive sizes?
   (A) 1/4"
   (B) 3/8"
   (C) 5/8"
   (D) 1/2"

3. The most commonly used and versatile socket handle is the:
   (A) ratchet.
   (B) flex bar.
   (C) breaker bar.
   (D) speed handle.

4. A(n) _____ is a swivel that lets the socket wrench reach around obstructions.
   (A) extension
   (B) hand spinner
   (C) flexible driver
   (D) universal joint

5. When working with sockets, Technician A states that a socket's "point" is the size of the square opening for the handle. Technician B states the point of a socket is the box configuration for the bolt head. Who is right?
   (A) A only.
   (B) B only.
   (C) Both A and B.
   (D) Neither A nor B.

6. Which type of screwdriver is especially good to use in tight spaces?
   (A) Torx.
   (B) Offset.
   (C) Phillips.
   (D) Clutch head.

7. Pliers are used on various parts to do each of these *except:*
   (A) cut.
   (B) grip.
   (C) bend.
   (D) screw.

8. Which type of pliers open extra wide to hold very large objects?
   (A) Rib joint.
   (B) Channel lock.
   (C) Water pump.
   (D) All the above.

9. The heaviest kind of hammer is:
   (A) lead.
   (B) brass.
   (C) sledge.
   (D) dead blow.

10. Which punch configuration is used to mark parts for reassembly or to start a drilled hole?
    (A) Center.
    (B) Aligning.
    (C) Tapered.
    (D) Diamond.

Activities—Chapter 3

1. Collect automotive catalogs and create a list of hand tools needed to equip an automotive shop. Provide an estimate of what it will cost to purchase the tools.

2. Discuss tool safety with your instructor. Prepare a list of safety regulations for your shop area.

Chapter 4

Power Tools and Equipment

After studying this chapter, you will be able to:

- List the most commonly used power tools and equipment.
- Describe the uses for power tools and equipment.
- Explain the advantages of one type of tool over another.
- Explain safety rules that pertain to power tools and equipment.
- Correctly answer ASE certification test questions that require a knowledge of power tools and equipment.

To be a productive technician in today's automotive service facility, you must know when and how to use power tools and equipment. *Power tools* are tools driven by compressed air, electricity, or pressurized liquid. They make many repair operations easier and quicker. Large shop tools, such as floor jacks, parts cleaning tanks, and steam cleaners, are called *shop equipment.*

This chapter discusses properly selecting and using common power tools and shop equipment. They can be very dangerous if misused. Always follow the operating instructions for the particular tool or piece of equipment before use. If in doubt, ask your instructor for a demonstration. Specialized power tools and equipment are covered in later chapters. Refer to the index to find more information on them as needed.

Compressed-Air System

The components of a *compressed-air system* include an air compressor, air lines, and air tools. In addition, a pressure regulator, filter, and lubricator may be attached to the system. Air tools are driven by the compressed-air system. Air-powered tools can be found in nearly every service facility.

Air Compressor

An *air compressor* is the source of compressed air for an automotive service facility. Look at **Figure 4-1.** An air compressor normally has an electric motor that spins an air pump. The air pump forces air into a large, metal storage tank. The air compressor turns on and off automatically to maintain a preset pressure in the system. Metal air lines feed out from the tank to several locations in the shop. A technician can then connect flexible air hoses to the metal lines.

Warning

Shop air pressure is usually around 100–150 psi (689–1034 kPa). This is enough pressure to kill or severely injure a person. Respect shop air pressure!

Air Hoses

Flexible, high-pressure *air hoses* are connected to the metal lines from the air compressor. These hoses allow the technician to take a source of air pressure to the vehicle being repaired. *Quick-disconnect connectors* are used on air hoses. These allow a technician to connect or disconnect hoses or tools without using a wrench. They have an outer sleeve that slides back with finger pressure as the technician pushes or pulls on the hose.

Other Components

A *pressure regulator* is used to set a specific pressure in the compressed-air system. This pressure is often called *shop pressure.* In most cases, shop pressure is between 100 and 150 pounds per square inch (psi). A filter may be connected to the system. The *filter* traps water so that it can be removed. This increases the life of air tools. In addition, a lubricator may also be connected to the system. The *lubricator* introduces oil into the airstream. This also increases the life of air tools.

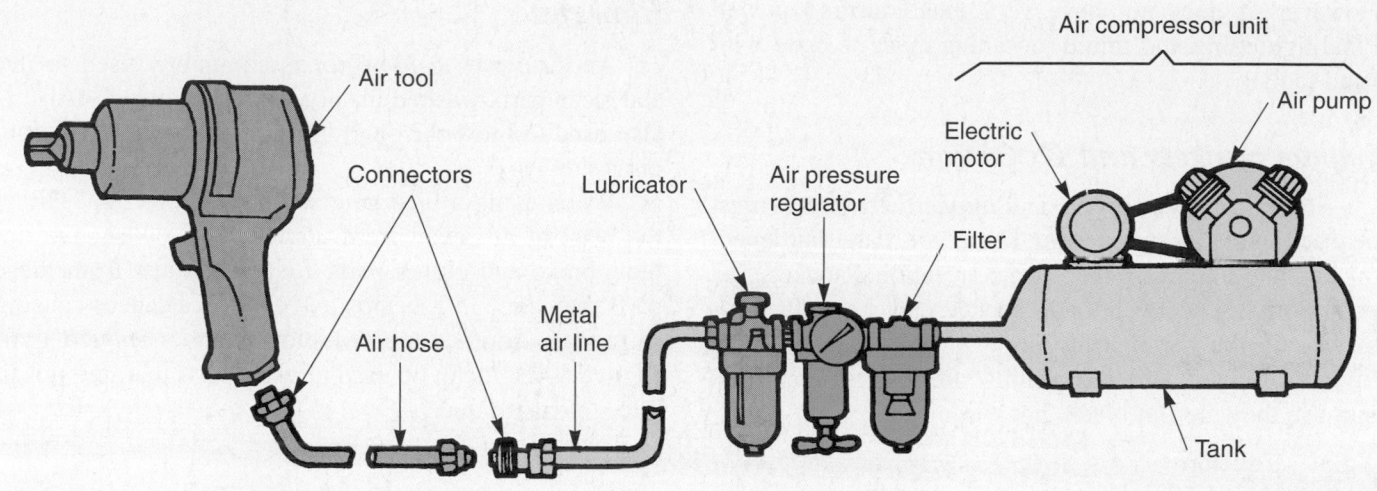

Figure 4-1. The basic parts of a typical shop air pressure system. The air compressor unit develops air pressure. The filter removes moisture. The regulator allows the technician to control system pressure. Metal lines and flexible hoses carry pressurized air to the tool. (Florida Dept. of Voc. Ed.)

Air Tools

Air tools use air pressure for operation. They are also called *pneumatic tools.* Air tools are labor-saving devices and well worth their cost. Always lubricate an air tool before use. Squirt a few drops of air tool oil into the tool's air inlet fitting. This protects the internal parts of the tool, increasing the life and power of the tool.

Air Wrenches

Air wrenches, or *impact wrenches,* provide a very fast means of installing or removing threaded fasteners. Look at **Figure 4-2A.** An impact wrench uses compressed air to rotate a driving head. The driving head holds a special impact socket.

Impact wrenches come in 3/8″, 1/2″, and 3/4″ drive sizes. A 3/8″ drive impact is ideal for small fasteners, such as 1/4″–9/16″ bolts. A 1/2″ drive is for general purpose use with medium to large fasteners, such as 1/2″–1″ bolts. The 3/4″ drive impact is for extremely large fasteners. It is *not* commonly used in automotive service. A button or switch on the impact wrench controls the direction of rotation. In one position, the impact wrench tightens the fastener. With the switch in the other position, the wrench loosens the fastener.

 Caution

Until you become familiar with the operation of an air wrench, be careful not to overtighten bolts and nuts or leave them too loose. It is easy to strip or break fasteners with an air tool.

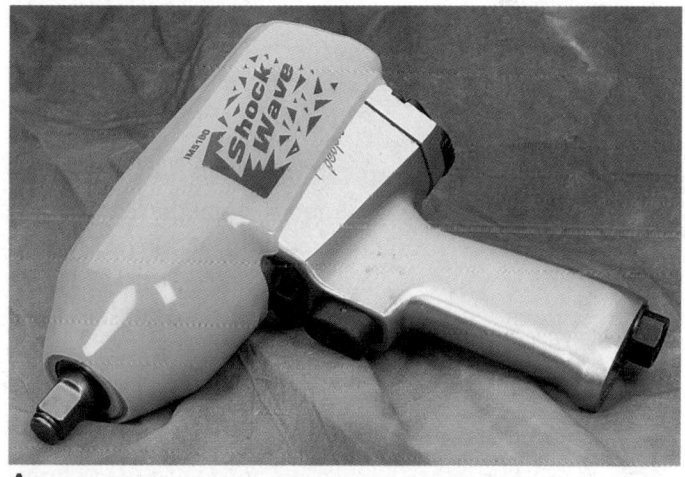

A

B

Figure 4-2. A—1/2″ drive impact wrench. B—3/8″ drive air ratchet. (Snap-on Tool Corp.)

Air Ratchet

An *air ratchet* is a special impact wrench designed for working in limited space. Look at **Figure 4-2B.** For instance, an air ratchet is commonly used when removing water pumps. It will fit between the radiator and engine easily. It works in much the same way as a hand-tool ratchet. An air ratchet normally has a 3/8″ drive.

However, it does not have very much turning power. Final tightening and initial loosening must be done with hand tools.

Impact Sockets and Extensions

Special *impact sockets* and *impact extensions* must be used with air wrenches. These are case hardened, thicker, and much stronger than conventional sockets and extensions. A conventional socket can be ruined or broken by the hammering blows of an impact wrench. Impact sockets and extensions are easily identified because they are flat black, not chrome.

Caution

Know when and when not to use power tools. In most situations, power tools will speed up your work. However, there are many times when they should not be used. For example, never use an impact wrench in place of a torque wrench. An impact wrench will not torque critical fasteners to their correct specification. Problems and comebacks will result.

Air Hammer

An *air hammer,* or *air chisel,* is useful during various driving and cutting operations. Look at **Figure 4-3.** An air hammer is capable of producing about 1000–4000 impacts per minute. Several different cutting or hammering attachments are available. Be sure to select the correct one for the job.

Warning

Never turn an air hammer on unless the tool is pressed tightly against the workpiece. If you do, the tool head can fly out of the hammer with great force, as if shot from a gun!

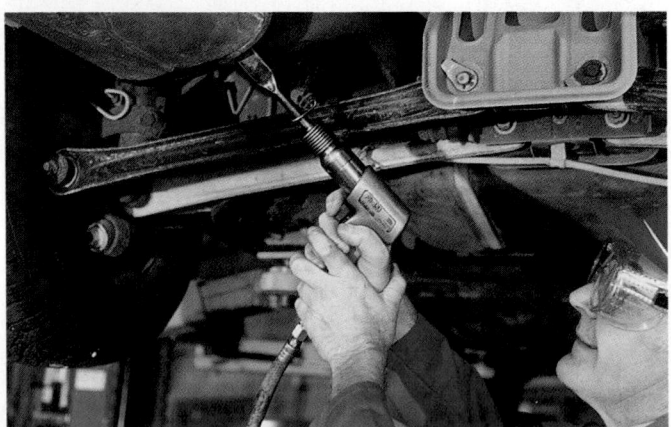

Figure 4-3. An air hammer is useful for tasks such as quickly cutting off a rusted exhaust system. *Always* wear safety glasses.

Blowgun

An air-powered *blowgun* is commonly used to dry and clean parts washed in solvent. See **Figure 4-4A.** It is also used to blow dust and loose dirt from a part before disassembly.

When using a blowgun, wear eye protection. Direct the blast of air away from yourself and others. Do not blow brake and clutch parts clean. The dust from these parts may contain asbestos. Asbestos is a cancer-causing substance. Another type of blowgun is a *solvent gun,* **Figure 4-4B.** It can be used to wash parts that will not fit into a cleaning tank.

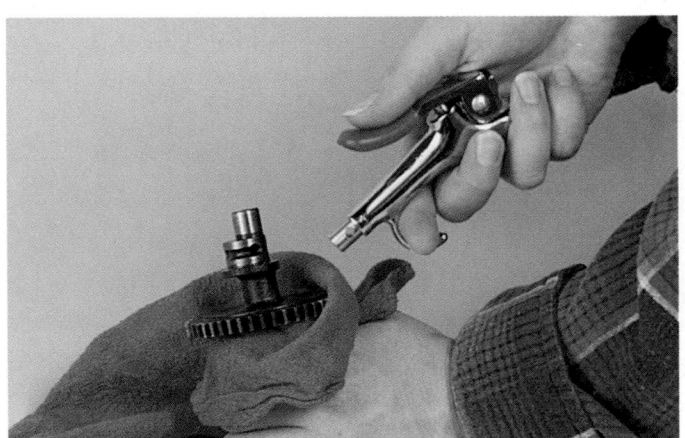

A

B

Figure 4-4. A—A blowgun is commonly used to blow parts clean and to dry parts after washing them in solvent. B—A solvent gun can be used to wash parts.

Air Drill

An *air drill* is excellent for many repairs because of its power output and speed adjustment capabilities. Its power and rotating speed can be set to match the job at hand. Look at **Figure 4-5.** With the right attachments, air drills can drill holes, grind, polish, and clean parts.

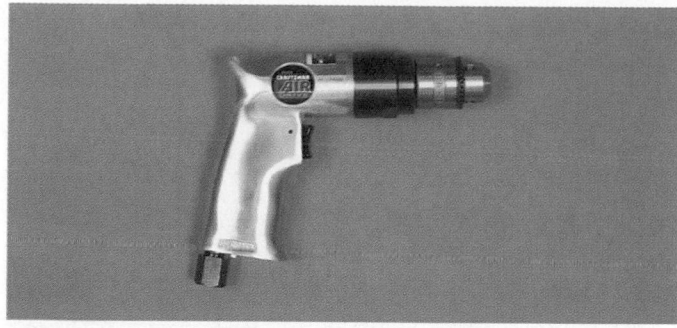

Figure 4-5. An air drill. The speed of the air drill can be adjusted. The air drill is capable of very high turning force.

A *rotary brush* is used in an air or electric drill for rapid cleaning of parts, **Figure 4-6.** It can quickly remove old gasket material, carbon deposits, and rust with a minimum amount of effort.

An *abrasive pad* is another type of cleaning tool that can be used in an air drill. It is used for removing old gasket material. It has the advantage of not scratching aluminum like a rotary brush can.

A *rotary file,* or *stone,* can be used in an air drill, electric drill, or air (die) grinder, **Figure 4-7.** It is handy for removing metal burrs and nicks. Make sure the stone is not turned too fast by the air tool. Normally, the maximum speed is printed on the file or stone container.

Warning

Use a high-speed rotary brush in an air drill. A brush designed for an electric drill may fly apart. To be safe, always adjust an air drill to the slowest acceptable speed when using a rotary brush. Also, always wear eye protection.

Electric Tools

There are many *electric tools* that can be useful to a technician. Some of these tools, such as a drill press or

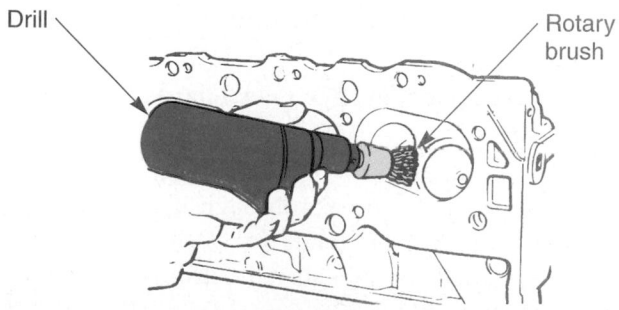

Drill

Rotary brush

Figure 4-6. A rotary brush is commonly used in a drill for cleaning off carbon deposits or old gaskets. *Always* wear eye protection.

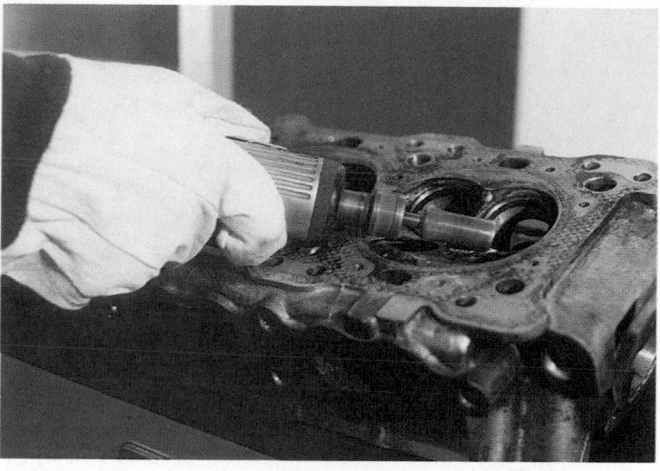

Figure 4-7. This die grinder is equipped with a high-speed stone. This tool is used for removing burrs and for other smoothing operations. Here, the technician is making minor repairs to a damaged cylinder head combustion chamber.

grinder, may be fixed to the floor or a bench. Other tools, such as a drill, are portable and can be taken to the job site.

Bench Grinder

A *bench grinder* can be used for grinding, cleaning, or polishing operations, **Figure 4-8.** A bench grinder usually has two wheels— a grinding wheel and a wire wheel. The hard, abrasive *grinding wheel* is used for sharpening and deburring. The soft *wire wheel* is used for cleaning and polishing. A few bench grinder rules to follow are:

- Always wear eye protection and keep your hands away from the wheel.

- Make sure the grinder shields are in place.

- Keep the tool rest adjusted close to the wheel. If the rest is *not* close to the wheel, the part being ground can catch in the grinder.

- Do *not* use a wire wheel to clean soft metal parts, such as aluminum pistons or brass bushings. The abrasive action of the wheel can remove metal or scuff the part and ruin it. Instead, use a solvent and a dull hand scraper on soft metal parts that could be damaged.

Drills

Drills are used to create holes in metal and plastic parts. Some drills are portable; others are mounted on a workbench or the floor. Drills use different-size bits to create the size of hole needed.

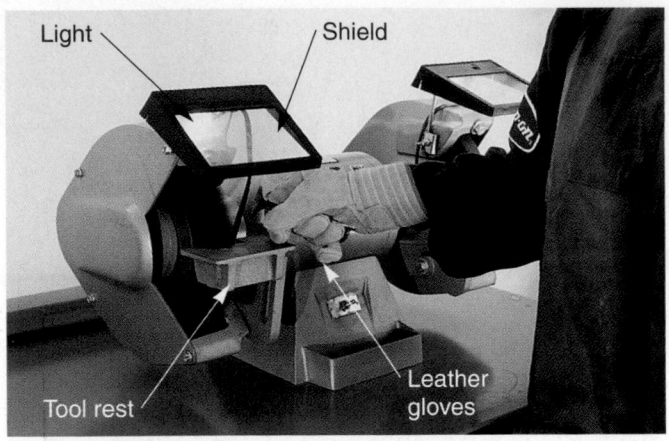

Figure 4-8. The stone on a bench grinder is used to sharpen tools. The brush can be used to clean and polish small parts. *Always* keep shields, tool rests, and guards in place.

Bits

Drills use *drill bits,* or *twist drills,* to drill holes in metal and plastic parts, **Figure 4-9.** A drill bit is *chucked,* or *mounted,* in a drill. A special key, called a *chuck key,* is used to tighten the drill bit in the drill chuck, **Figure 4-10.** Drill bits are commonly made of either carbon steel or high-speed steel. High-speed steel is better because of its resistance to heat. It will not lose its hardness when slightly overheated.

Portable Electric Drill

Portable electric drills are hand-held drills. They come in different sizes. The size of a drill is an indication of the capacity of its chuck. Commonly used sizes are 1/4″, 3/8″, and 1/2″, **Figure 4-11.** Portable electric drills work fine on most small drilling operations.

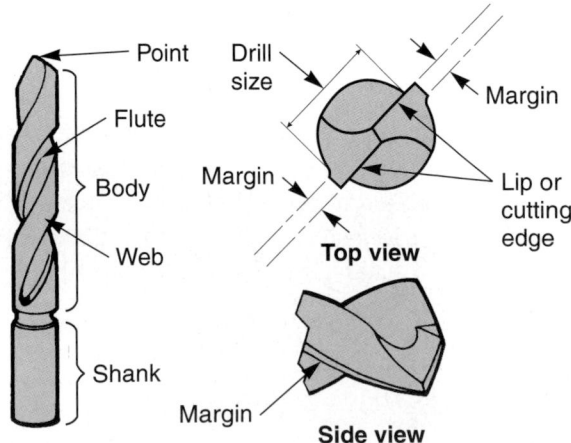

Figure 4-9. The basic parts of a drill bit. (Florida Dept. of Voc. Ed.)

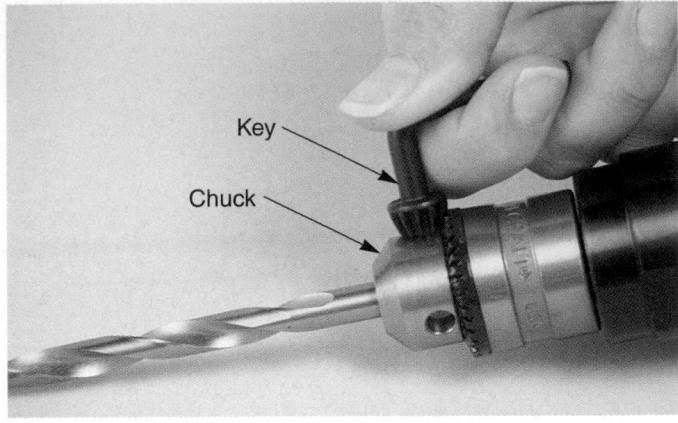

Figure 4-10. A key is used to tighten a bit in the chuck.

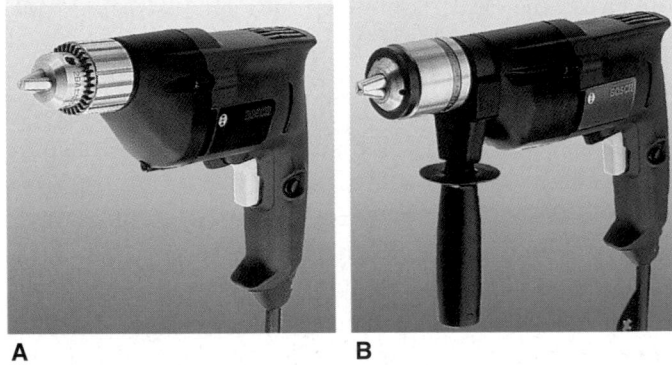

Figure 4-11. Portable electric drills. A—3/8″ drill. B—1/2″ drill. (Robert Bosch)

Drill Press

A *drill press* is a large, floor- or bench-mounted drill needed for drilling large holes, deep holes, or a great number of holes in several parts, **Figure 4-12.** The drill press handle allows the bit to be pressed into the work with increased force. Also, very large bits can be used. A few drill press rules to follow include:

- Remove the key from the chuck before turning on the drill press.
- Secure the part to be drilled in a vise or with C-clamps.
- Use a center punch to indent the part and start the hole.
- To prevent injury, release drilling pressure right before the bit breaks through the bottom of the part. A drill bit tends to catch when breaking through. This can cause the drill or part to rotate dangerously.
- Oil the bit as needed.

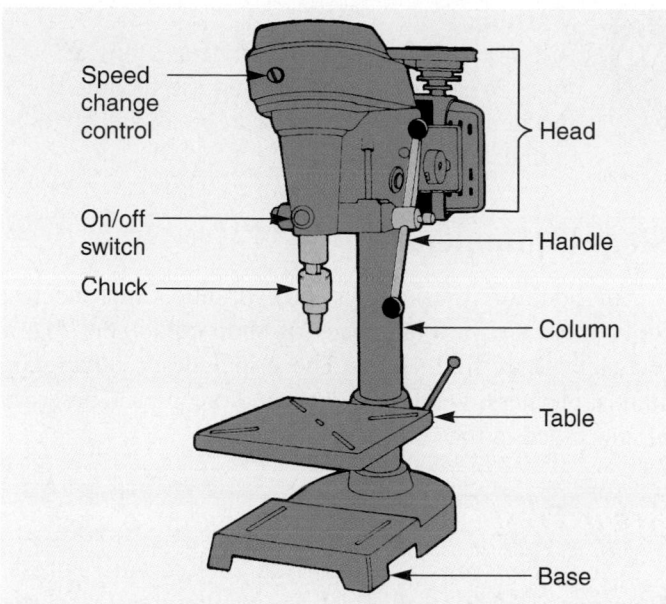

Figure 4-12. The parts of a drill press. A drill press is for drilling deep or large holes when a part will fit on the table. (Florida Dept. of Voc. Ed.)

Hydraulic Tools

Hydraulic tools are powered by pressurized liquid. The hydraulic tools typically used in the automotive shop include jacks, cranes, and presses. These tools are discussed in the next sections.

Floor Jack

A *floor jack* is used to raise either the front, sides, or rear of a vehicle. Look at **Figure 4-13.** To avoid vehicle damage, place the jack saddle under a solid part of the car such as the frame, suspension arm, or axle housing. If the saddle is not properly located, it is very easy to smash the oil pan, muffler, floor pan, or another part of the vehicle. To raise the vehicle, turn the jack handle or knob clockwise and pump the handle. To lower the vehicle, turn the handle or knob counterclockwise slowly to release the pressure-relief valve.

When raising the front of a vehicle, place the transmission in neutral and release the parking brake. This lets the vehicle roll, preventing it from pulling off the jack. After raising, secure the vehicle on jack stands. Place an automatic transmission in park and a manual transmission in gear. Apply the emergency brake and block the wheels. It is then safe to work under the vehicle.

> ⚠️ **Caution**
> Most floor jack handles tend to stick when the pressure-relief valve is released. This makes it easy for you to lower the vehicle too quickly. When releasing the valve, turn it very slowly. This will prevent the car or truck from slamming to the ground violently!

Transmission Jack

Transmission jacks are designed to hold transmissions during removal or installation. One type is similar to a floor jack. However, the saddle is enlarged to fit the bottom of a transmission. Another type of transmission jack is designed to be used when the vehicle is raised on a lift, **Figure 4-14.** It has a long post that can reach high into the air to support the transmission.

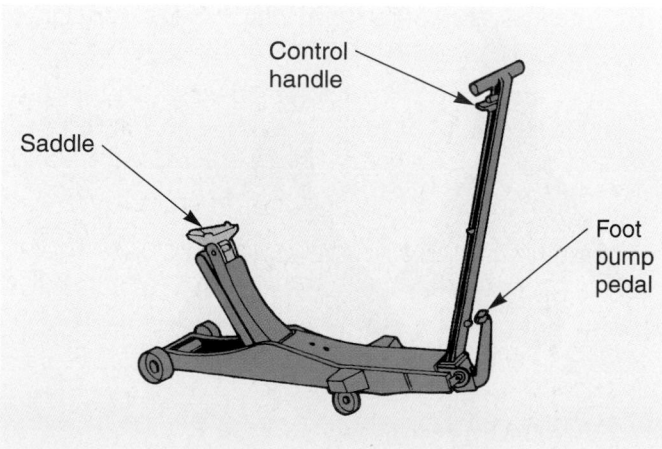

Figure 4-13. A floor jack is for raising the car only.

Figure 4-14. A transmission jack is designed for holding transmissions during removal, transporting, and installation. This foot-operated, hydraulic stand can be extended to a height of 72″. (OTC Div. of SPX Corp.)

Engine Crane

A portable *engine crane* is used to remove and install engines, **Figure 4-15.** It has a hydraulic hand jack for raising engines and a pressure-release valve for lowering engines. The engine crane is also handy for lifting intake manifolds, cylinder heads, engine blocks, transmissions, transaxles, and other heavy parts.

Hydraulic Press

A *hydraulic press* is used to install or remove gears, pulleys, bearings, seals, and other parts requiring a high pushing force. One is shown in **Figure 4-16.** A hydraulic ram extends as the pump handle is worked. The ram presses the parts against a table.

Figure 4-15. A hydraulic engine crane can be used to lift heavy objects, such as engines, transmissions, transaxles, and rear axle assemblies. This technician has used a crane to mount an engine on a stand.

Figure 4-16. A hydraulic press is needed for numerous pressing operations. It is commonly used to remove and install bearings, bushings, seals, and other pressed-on parts.

 **Warning**
A hydraulic press can literally exert tons of force. Wear eye protection and use recommended procedures. Parts can break and fly out with deadly force!

Shop Equipment

In addition to pneumatic, hydraulic, and electric tools, there are various pieces of shop equipment that a technician may find useful. These include tire changers, stands, cleaners, welders, lights, and creepers. These are all discussed in the following sections.

Arbor Press

An *arbor press* works like a hydraulic press. However, it is all mechanical. Hydraulic pressure is not used; therefore, the operating pressure is much lower. An arbor press is suited for smaller jobs.

Tire Changer

A *tire changer* is used to remove and replace tires on wheels. It is a common piece of shop equipment. Some tire changers are pneumatic; others are hand operated. Do not attempt to operate a tire changer without proper supervision. Follow the directions provided with the changer.

Jack Stands

Jack stands support a vehicle during repair. After raising the vehicle with a jack, place stands under the vehicle, **Figure 4-17.** Be sure the stands are placed in secure positions. For example, place jack stands under the frame, axle housing, or suspension arm.

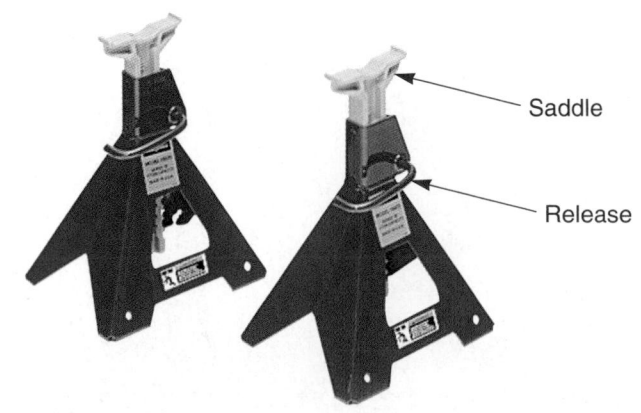

Figure 4-17. Jack stands are needed to safely support the weight of a vehicle.

Warning

It is *not safe* to work under a vehicle held up by only a jack. Secure the vehicle on jack stands before placing any part of your body under the vehicle. Even a small car can weigh well over a ton. The next chapter details the safe use of lifts, jacks, and jack stands.

Engine Stand

An *engine stand* is used to hold an engine once it is removed from the vehicle for rebuilding or repair. The engine bolts to the stand just as it would to the transmission or transaxle. The engine can usually be rotated and held in different positions on the stand, making it easy to work on different parts.

Cold Solvent Tank

A *cold solvent tank* contains a pump, reservoir, and solvent. It is used to remove grease and oil from parts, **Figure 4-18.** After removing all old gaskets and scraping off excess grease, you can scrub the parts clean in the solvent. A blow gun is then normally used to dry the solvent.

Figure 4-18. A cold-solvent tank is used to remove oil and light grease from parts. The unit sprays filtered solvent onto parts. Rub the parts with a brush for rapid cleaning.

Steam Cleaner and High-Pressure Washer

A *steam cleaner* or *high-pressure washer* is used to remove heavy deposits of dirt, grease, and oil from the outside of large assemblies, such as engines, transmissions, and transaxles. Look at **Figure 4-19.** To help keep the environment clean, wire brush the item to be cleaned and collect oil-soaked dirt before steaming or washing. Then, dispose of the oil-soaked material properly.

Figure 4-19. A high-pressure washer will remove greasy buildup from the outside of assemblies before teardown.

Warning

A steam cleaner operates at high pressures and temperatures. Follow the manufacturer's safety rules and specific operating instructions.

Oxyacetylene Torch

An *oxyacetylene torch* can be used to cut, bend, weld, or braze metal parts, **Figure 4-20.** The oxyacetylene setup consists of an oxygen tank, an acetylene tank, pressure regulators, hoses, and a hand-held oxyacetylene torch. Tremendous heat is produced by the burning acetylene gas and oxygen. The rapid cutting action of this torch is extremely beneficial. For example, an oxyacetylene torch is often used to remove old, rusted exhaust systems.

Welder

A *welder* uses high electric current to create an electric spark, or arc. This arc is hot enough to melt and fuse metal parts together, **Figure 4-20.** Be sure to complete proper training before attempting to weld. Using a welder improperly can result in personal injury or damage to parts.

Warning

Whenever working with a torch or welder, there is always a chance of fire. Always observe standard safety practices.

Soldering Gun

A *soldering gun* or *soldering iron* is used to join wires, **Figure 4-21.** An electric current heats the tip of the

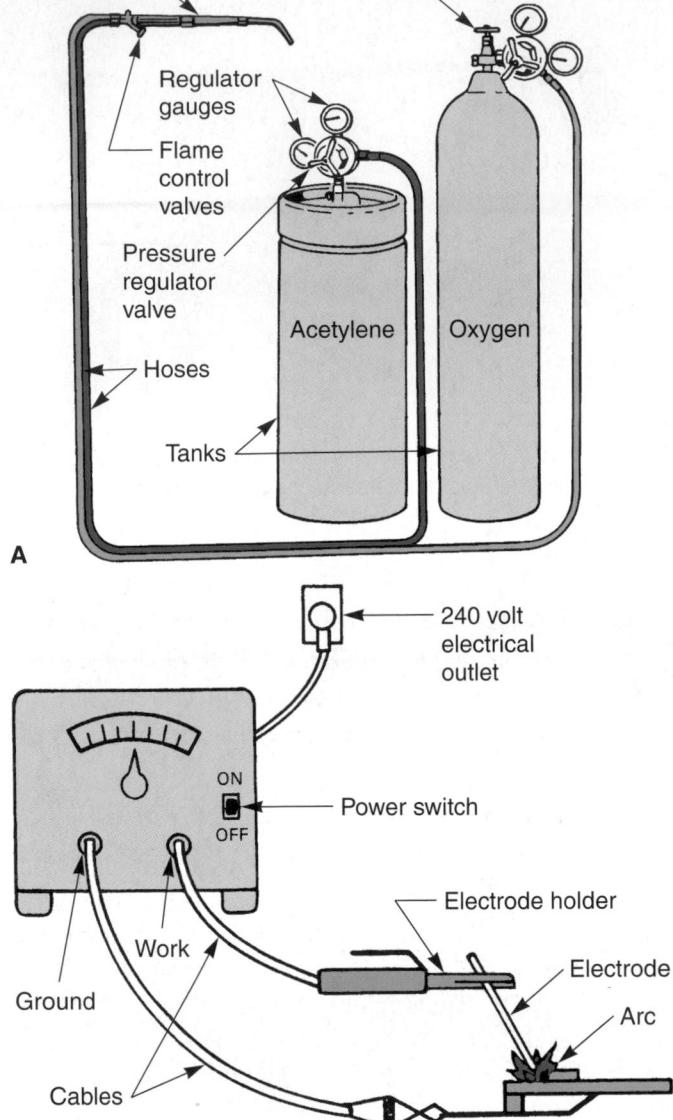

A

B

Figure 4-20. A—An oxyacetylene outfit can be used for cutting or welding metal. B—A basic arc welder. (Sun)

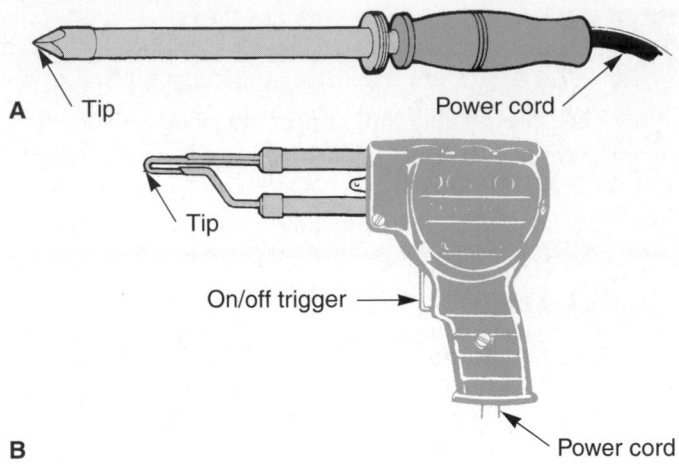

A

B

Figure 4-21. Soldering irons and guns produce enough heat to melt solder for joining wires and small metal terminals. A—Soldering iron. B—Soldering gun. (Florida Dept. of Voc. Ed.)

Warning

The gases around the top of a battery can explode. Always connect the battery charger leads to the battery *before* turning on the charger. This prevents sparks, which could ignite the battery gases.

Droplight

A *droplight* provides a portable source of light. The light can be taken to the repair area under the vehicle or in the engine compartment. Several types of droplights are shown in **Figure 4-22.**

gun. The hot gun tip is used to heat the wires. Solder is then applied to the hot wires and it melts. *Solder* is a lead-tin alloy. When the solder cools, it hardens into a strong, solid connection.

Battery Charger

A *battery charger* is used to recharge a dead car battery. It forces current through the battery and realigns the ions. The red charger lead connects to the positive (+) battery terminal. The black charger lead connects to the negative (−) battery terminal.

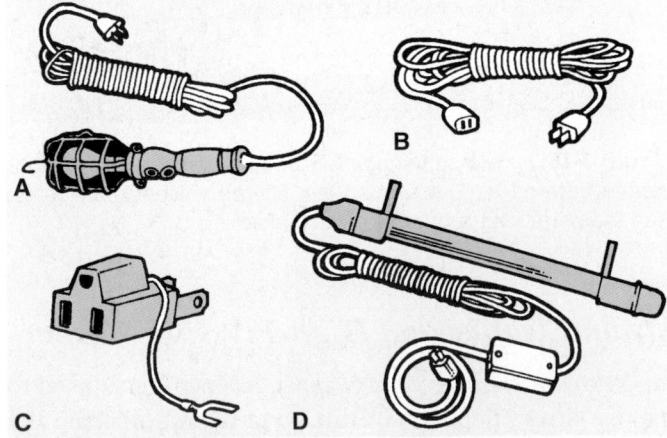

A   **B**

C   **D**

Figure 4-22. A—Droplight. B—Drop or extension cord. C—Three-prong adapter with ground terminal and ground wire for safety. D—Fluorescent droplight. (Florida Dept. of Voc. Ed.)

Pullers

Wheel pullers are used to remove seals, gears, pulleys, steering wheels, axles, and other pressed-on parts. A few puller types are pictured in **Figure 4-23.** Special pulling operations are covered in later chapters.

Warning
Pullers can exert tons of force. They must be used properly to prevent personal injury or part damage. Always wear eye protection!

Jumper Cables

Jumper cables are used to start a vehicle that has a dead battery. The cables are connected between the dead battery and another battery. The second battery is often in a running vehicle. Once the cables are connected, the car with the dead battery can be started. See **Figure 4-24.**

Figure 4-24. Jumper cables are for emergency starting. Connect the red lead to the positive terminal of both batteries. Black is for negative and ground. Never connect the terminals backwards or the batteries may explode.

When connecting jumper cables, connect positive to positive and negative to negative. Also, keep sparks away from the dead battery.

There is a specific sequence that should be followed when connecting the cables.
1. Connect the red jumper to the positive terminal on the dead battery.
2. Connect the red jumper to the positive terminal on the good battery.
3. Connect the black jumper to the negative terminal of the good battery or to a good ground on the vehicle with the good battery.
4. Connect the black jumper to a ground on the vehicle with the dead battery so sparks will not occur near the battery.

Warning
As soon as the car starts, remove the black jumper from ground *immediately*. If you leave the cables connected, both batteries may explode! Carefully remove the remaining jumpers in reverse order.

Creepers

A *creeper* is useful when working under a car supported on jack stands, **Figure 4-25A.** It lets the technician easily roll under vehicles without getting dirty. A *stool creeper* allows the technician to sit while working on parts that are near the ground. See **Figure 4-25B.** For example, a stool creeper is often used during brake system repairs. The brake parts and tools can be placed on the creeper. The service technician can sit and still be at eye level with the brake assembly.

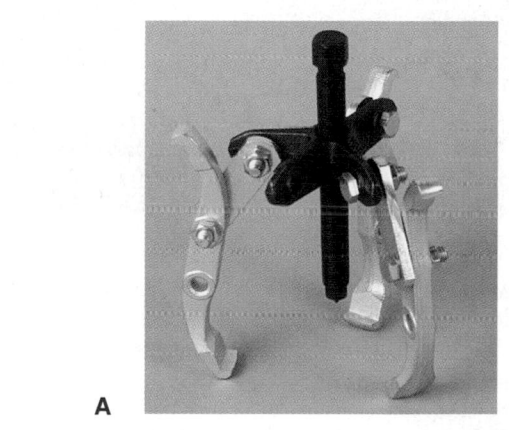

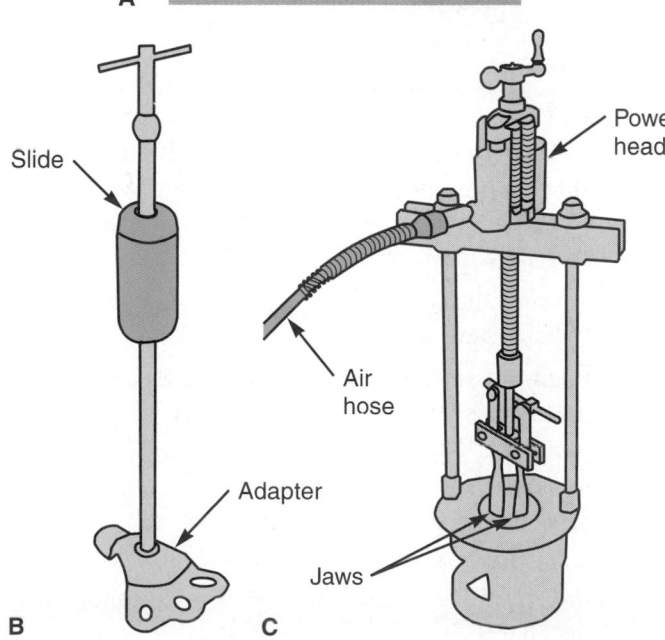

Figure 4-23. A—Three-way puller. B—Slide hammer puller. C—Power puller.

Slide

Power head

Air hose

Adapter

Jaws

A

B

C

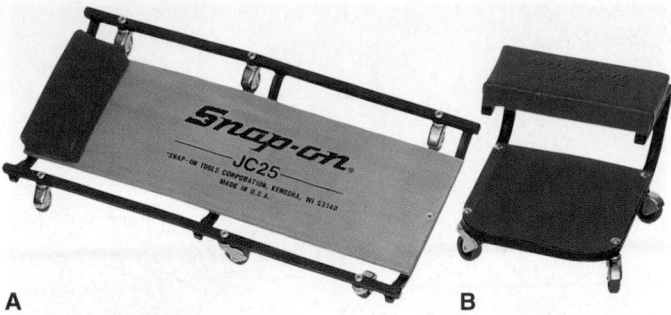

A   **B**

Figure 4-25. A—A creeper is for working under a vehicle. B—The stool creeper is commonly used during brake and suspension repairs. You can sit on the stool and store tools on the bottom. (Snap-on Tool Corp.)

Roll-Around Cart

A large *roll-around cart* or table is handy for taking a number of tools to the job. One is pictured in **Figure 4-26.** A technician can quickly place all needed tools in the cart and take them to the vehicle. The cart places the tools within hand's reach. This saves time and effort before, during, and after the job.

Covers

Fender covers are placed over fenders, upper grille, or other body sections. They protect the paint or finish from nicks, scratches, and grease. See **Figure 4-27A.** Never lay tools on a painted surface. Costly scratches may result.

Seat covers are placed over seats to protect them from dirt, oil, and grease that might be on your work clothes. These covers should be used while driving the vehicle or while working in the passenger compartment. Look at **Figure 4-27B.**

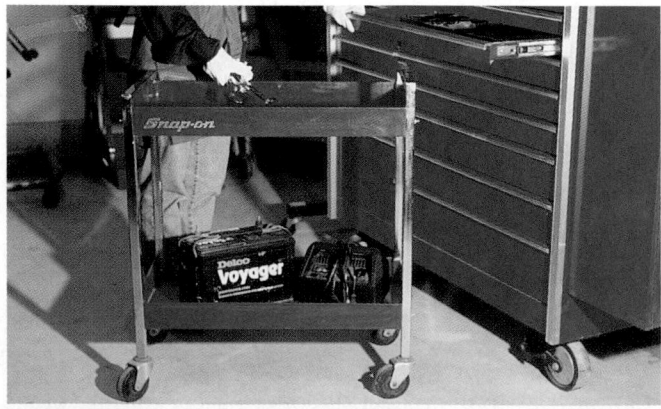

Figure 4-26. A roll-around cart allows you to take many tools to the vehicle. This saves several trips to the toolbox. It also saves time during cleanup at the end of the day. (Snap-on Tool Corp.)

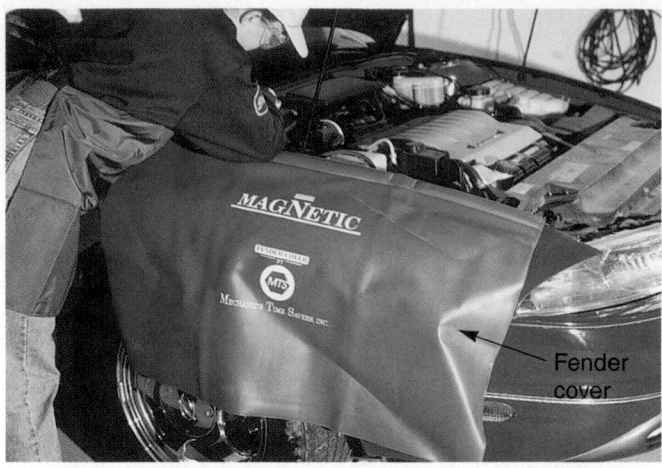

A

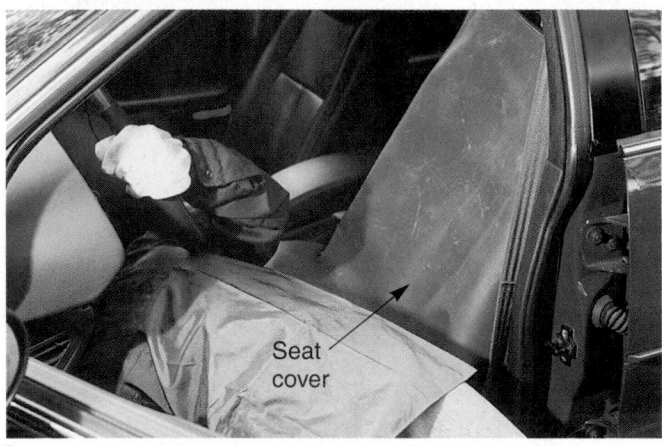

B

Figure 4-27. Always take good care of your customer's vehicle. A—Fender covers protect the paint from nicks and dents. B—Seat cover protects the upholstery from dirty work clothes.

Summary

- Power tools use electricity, compressed air, or hydraulic pressure (liquid confined under pressure). Large shop tools such as floor jacks, parts cleaning tanks, and steam cleaners, are classified as shop equipment.

- An air compressor is the source of compressed (pressurized) air for the auto shop.

- High-pressure air hoses are connected to the metal lines from the air compressor.

- Air tools, also called pneumatic tools, use air pressure for operation. They are labor-saving tools and are well worth their cost.

- Air wrenches, or impact wrenches, provide a very fast means of installing or removing threaded fasteners.

- Special impact sockets and impact extensions must be used with air wrenches.

- A blowgun is commonly used to dry and clean parts washed in solvent.

- An air drill is excellent for many repairs because of its power output and speed adjustment capabilities.

- A bench grinder can be used for grinding, cleaning, or polishing operations. The hard grinding wheel is used for sharpening or deburring. The soft wire wheel is for cleaning and polishing.

- Drill bits, or twist drills, are used to drill holes in metal and plastic parts.

- A drill bit is chucked in and rotated by a drill. A special key is commonly used to tighten the drill bit in the drill chuck.

- A floor jack is used to raise either the front, sides, or rear of a vehicle.

- A tire changer is a common piece of shop equipment used to remove and replace tires on wheels.

- Jack stands support a vehicle during repair. After raising a vehicle with a jack, place stands under it.

- A portable engine crane is used to remove and install engines.

- An engine stand is used to hold an engine while it is overhauled (rebuilt) or repaired.

- A cold solvent tank can be used to remove grease and oil from parts.

- An oxyacetylene torch outfit can be used to cut, bend, weld, or braze metal parts.

- A welder is used to melt and fuse metal parts together.

- A soldering gun or iron is used to solder wires.

- A battery charger is used to recharge a discharged car battery.

- A droplight provides a portable source of light.

- Pullers are needed to remove seals, gears, pulleys, steering wheels, axles, and other pressed-on parts.

- Jumper cables are used to start engines that have a dead (discharged) battery.

- Fender covers are placed over the fenders, the upper grille, or other body sections to protect them.

Important Terms

Power tools	Quick-disconnect
Shop equipment	connectors
Air compressor	Pressure regulator
Air hoses	Filter

Lubricator	Transmission jacks
Air tools	Engine crane
Air wrenches	Hydraulic press
Impact wrenches	Arbor press
Air ratchet	Tire changer
Impact sockets	Jack stands
Impact extensions	Engine stand
Air hammer	Cold solvent tank
Air chisel	Steam cleaner
Blowgun	High-pressure washer
Solvent gun	Oxyacetylene torch
Air drill	Welder
Rotary brush	Soldering gun
Abrasive pad	Soldering iron
Rotary file	Battery charger
Stone	Droplight
Electric tools	Pullers
Bench grinder	Jumper cables
Drill bits	Creeper
Portable electric drills	Stool creeper
Drill press	Roll-around cart
Hydraulic tools	Fender covers
Floor jack	Seat covers

Review Questions—Chapter 4

Please do not write in this text. Place your answers on a separate sheet of paper.

1. Power tools use _____, _____ _____, or _____ _____ as sources of energy.

2. Which of the following is *not* a commonly used air tool?
   (A) Impact wrench.
   (B) Air ratchet.
   (C) Air chisel.
   (D) Air saw.

3. A(n) _____ is used to blow dirt off parts and to dry parts after cleaning.

4. A rotary file is frequently used to remove _____.
   (A) old gasket materials
   (B) carbon deposits
   (C) metal burrs and nicks
   (D) None of the above.

5. List four important rules for a bench grinder.

6. List five important rules for a drill press.

7. Use this tool to support the car while working under the car.
   (A) Floor jack.
   (B) Jack stands.
   (C) Transmission jack.
   (D) Bumper jack.

8. Explain the use of a solvent tank.

9. What are wheel pullers for?

10. A 1/2″ drive impact wrench is used for fasteners with head sizes between 1/2″ to 1″. Which of the following sockets can be used by this particular tool?
    (A) 5/8″ chrome plated socket.
    (B) 9/16″ flat black socket.
    (C) 7/16″ flat black socket.
    (D) None of the above.

ASE-Type Questions

1. Power tools are tools that use:
   (A) electricity.
   (B) hydraulics.
   (C) compressed air.
   (D) All the above.

2. Technician A says that shop air pressure is usually around 100 to 150 psi. Technician B says shop air pressure is much higher, around 300 psi. Who is correct?
   (A) A only.
   (B) B only.
   (C) Both A and B.
   (D) Neither nor B.

3. This is *not* a common impact wrench drive size:
   (A) 1/4″.
   (B) 3/8″.
   (C) 1/2″.
   (D) 3/4″.

4. Technician B says to use a 1/4″ drive on sockets from 1/4″ to 9/16″. Technician B says to use a 3/8″ drive on these socket sizes. Who is correct?
   (A) A only.
   (B) B only.
   (C) Both A and B.
   (D) Neither A nor B.

5. Special impact sockets and extensions are easily identified because they are:
   (A) chrome.
   (B) aluminum.
   (C) flat black.
   (D) hard rubber.

6. Each of these can be used to clean parts *except:*
   (A) air drill.
   (B) blowgun.
   (C) air ratchet.
   (D) bench grinder.

7. Which of the following is *not* a rule to follow when using a bench grinder?
   (A) Wear eye protection.
   (B) Make sure shields are in place.
   (C) Use the wire wheel to clean soft metal parts.
   (D) Keep the tool rest adjusted close to the stone and brush.

8. When using a drill press, Technician A believes drilling pressure should not be released until the bit breaks completely through the bottom of the part. Technician B believes pressure release should occur just before the bit breaks through. Who is right?
   (A) A only.
   (B) B only.
   (C) Both A and B.
   (D) Neither A nor B.

9. After using a floor jack to raise the front of a car:
   (A) place the car in park.
   (B) block the car's wheels.
   (C) secure the car on jack stands.
   (D) All of the above.

10. The _____ press performs the same function as a hydraulic press, but at lower pressures.
    (A) arbor
    (B) steam
    (C) rotary
    (D) ratchet

Activities for Chapter 4

1. Using an automotive tool catalog, develop a list of power tools needed to equip your school's automotive repair shop. Find prices and add up the cost.

2. Research safety literature on power equipment used in an automotive repair facility.
   (A) Develop a bibliography of resources for safe use of power equipment.
   (B) Develop a list of safety rules for their use.

Chapter 5

The Auto Shop and Safety

After studying this chapter, you will be able to:

- Describe the typical layout and sections of an auto shop.
- List the types of accidents that can occur in an auto shop.
- Explain how to prevent auto shop accidents.
- Describe general safety rules for the auto shop.

An auto shop can be a safe and enjoyable place to work. Most shops are clean, well lighted, and relatively safe, **Figure 5-1**. However, if basic safety rules are *not* followed, an auto shop can be very dangerous. In this chapter, the layout of a typical automotive service facility will be discussed and the most important safety rules will be emphasized.

Auto Shop Layout

There are several different areas in an auto shop. You must know their names and the basic rules that apply to each. It is important that you learn your shop layout and organization to improve work efficiency and safety. The auto shop includes the following work areas:

- Repair area (includes the shop stall, lift, alignment rack, and outside work area).
- Toolroom.
- Classroom.
- Locker room (dressing room).

Repair Area

The *repair area* includes any location in the shop where repair operations are performed. It normally includes every area *except* the classroom, locker room, and toolroom.

Shop Stall

A *shop stall* is a small work area where a car can be parked for repairs. Sometimes, each stall is numbered and marked off with lines painted on the floor.

Lift

The *lift* is used to raise a vehicle into the air. Refer to **Figure 5-2.** It is handy for working under the car

Figure 5-1. A well-maintained automotive shop can be an enjoyable place to work. Always do your part to keep the shop clean and well organized.

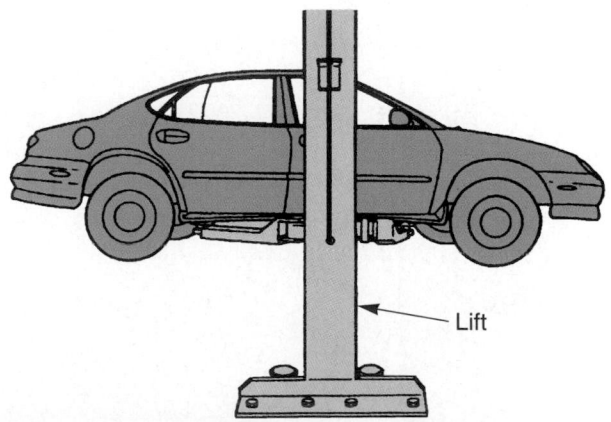

Figure 5-2. A lift is handy for repairs on parts located under the vehicle. It is commonly used when changing oil, greasing the chassis, and repairing the exhaust system. (Ford)

(draining oil, greasing front end parts, or repairing exhaust system).

Remember these *lift safety rules:*

- Ask your instructor for a demonstration and get permission before using the lift.

- Center the vehicle on the lift as described in a service manual, **Figure 5-3.** Raise vehicle slowly.

- Check ceiling clearance before raising trucks and campers. Make sure the vehicle roof does not hit overhead pipes, lights, or the ceiling.

- Make sure the lift's safety catch is engaged. Do not walk under the lift without the catch locked into position, **Figure 5-4.**

Alignment Rack

The *alignment rack,* or *front end rack,* is another specialized stall used when working on a car's steering and suspension systems. One is shown in **Figure 5-5.** It

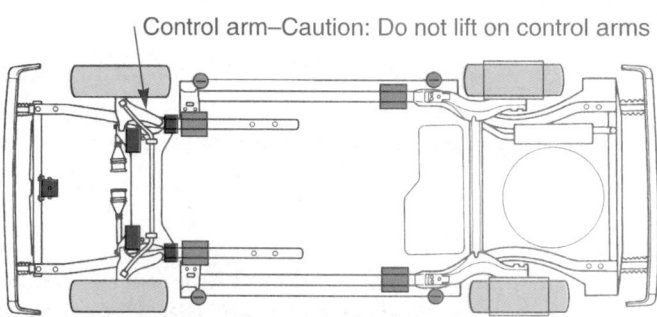

Control arm–Caution: Do not lift on control arms

Lift points on frame rails
- ◼ Twin post lift points
- ◼ Frame contact or floor jack
- ◻ Drive on hoist
- ◼ Scissors jack (emergency) locations

Figure 5-3. Follow the service manual instructions when raising a car on a lift. Note the specific lifting instructions and lift points for this vehicle. (Chrysler)

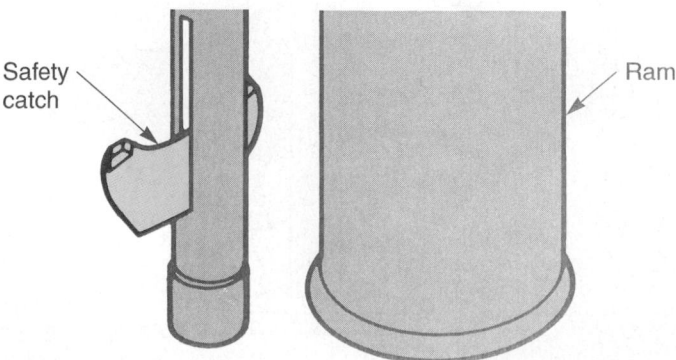

Safety catch

Ram

Figure 5-4. Most lifts have a safety catch. It must be engaged before working under the vehicle. (Ford)

Figure 5-5. An alignment rack is used in most shops. It is often needed when servicing steering and suspension systems. (Hunter)

may contain a special tool board and equipment used when replacing worn suspension parts, steering parts, and for adjusting wheel alignment.

When using an alignment rack, the car should be pulled on the rack slowly and carefully. Someone should *guide* the driver and help keep the tires centered on the rack. As with other complicated and potentially dangerous equipment, obtain a full demonstration before using the alignment rack.

Outside Work Area

Some auto shop facilities have an *outside work area* adjacent to the garage overhead doors. In good weather, this area can be used for auto repairs.

Always raise the shop doors all the way and pull cars through the doors very slowly. Check the height of trucks and campers to make sure they will clear (top of vehicle will not hit doors).

Toolroom

The *toolroom* is a shop area normally adjacent to (next to) the main shop or classroom. It is used to store shop tools, small equipment, and supplies (nuts, bolts, oil).

When working in the toolroom, you will be responsible for keeping track of shop tools. Every tool checked out of the toolroom must be recorded and called in before the end of the class period.

Normally, the tools hang on the walls of the toolroom for easy access. Each tool may have a painted silhouette, which indicates where it should be kept, **Figure 5-6.** Your instructor will detail specific toolroom policies and procedures.

Classroom

The classroom is often located adjacent to the repair area. It is used for seminars, demonstrations, and other

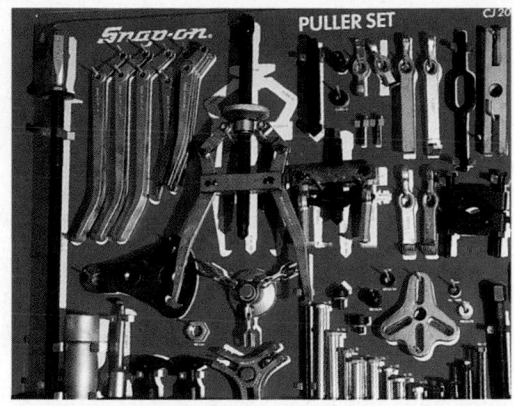

Figure 5-6. Keep all shop tools clean and organized. Make sure you return every tool to its correct location. (Snap-on Tools)

technician training activities. The classroom may also be used for employee meetings.

Locker Room

The **locker room** or **dressing room** is usually located adjacent to the main shop. It provides an area for changing into your work clothes. Always do your part to keep the locker room *clean* and *orderly.*

Shop Safety

Every year, thousands of technicians are accidentally injured or killed on the job. Most of these accidents resulted from a broken safety rule. The injured persons learned to respect safety rules the hard way–by experiencing a painful injury. You must learn to respect safety rules the easy way–by studying and following the safety rules given in this book.

Note
Specific safety rules on hand tools, power tools, shop equipment, and special operations are given elsewhere in this text. It is much easier to understand and remember these rules when they are covered fully.

While working, constantly think of safety. Look for unsafe work habits, unsafe equipment, and other potentials for accidents. See **Figure 5-7.**

When working in an auto shop, you must always remember that you are surrounded by other technicians. This makes it even more important that you concentrate on safety to prevent injury to yourself and to others in the shop.

Types of Accidents

Basically, you should be aware of and try to prevent six kinds of accidents:

- Fires.
- Explosions.
- Asphyxiation (airborne poisons).
- Chemical burns.
- Electric shock.
- Physical injuries.

If an accident or injury occurs in the shop, notify your instructor immediately. Use common sense when deciding whether to get a fire extinguisher or to take other actions.

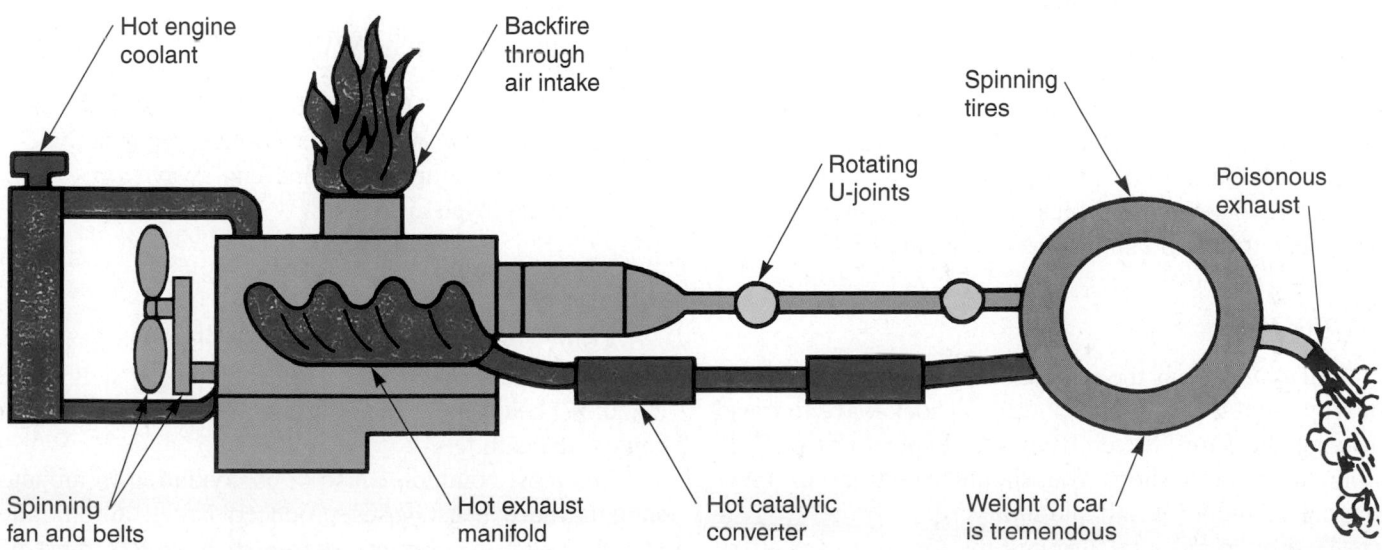

Figure 5-7. An automotive shop has the potential to be very dangerous. Just a few of the dangers present around an automobile are shown.

Fires

Fires are terrible accidents capable of causing severe injury and permanent scar tissue. Therefore, every precaution must be taken to prevent fires in the automotive shop.

There are numerous combustible substances (gasoline, oily rags, paints, thinners) found in an auto shop. *Gasoline* is by far the most dangerous and underestimated flammable in an auto shop. Gasoline has astonishing potential for causing a tremendous fire. Just a cupful of gasoline can instantly engulf a car in flames.

A few gasoline safety rules include:

- Store gasoline and other flammables in approved, sealed containers.

- When disconnecting a vehicle's fuel line or hose, wrap a shop rag around the fitting to keep fuel from squirting or leaking.

- Disconnect the battery before working on a fuel system.

- Wipe up gasoline spills immediately. Do not place oil absorbent (oil-dry) on gasoline because the absorbent will become highly flammable.

- Keep any source of heat away from fuel system parts.

- Never use gasoline as a cleaning solvent.

Oily rags can also start fires. Soiled rags should be stored in an approved safety can (can with lid).

Paints, thinners, and other combustible materials should be stored in a fire cabinet. Also, never set flammables near a source of sparks (grinder), flames (welder or water heater), or heat (furnace for example).

Note the location of all fire extinguishers in your shop. A few seconds can be a "lifetime" during a fire.

Electrical fires can result when a "hot wire" (wire carrying current to component) touches ground (vehicle frame or body). The wire can heat up, melt the insulation, and burn. Then, other wires can do the same. Dozens of wires could burn up in a matter of seconds.

To prevent electrical fires, always disconnect the battery when told to do so in a service manual.

Explosions

An *explosion* is the rapid, almost instant, combustion of a material that causes a powerful shock wave to travel through the shop. Several types of explosions are possible in an auto shop. You should be aware of these sources of sudden death and injury.

Hydrogen gas can surround the top of a car battery that is being charged or discharged (used). This gas is highly explosive. The slightest spark or flame can ignite

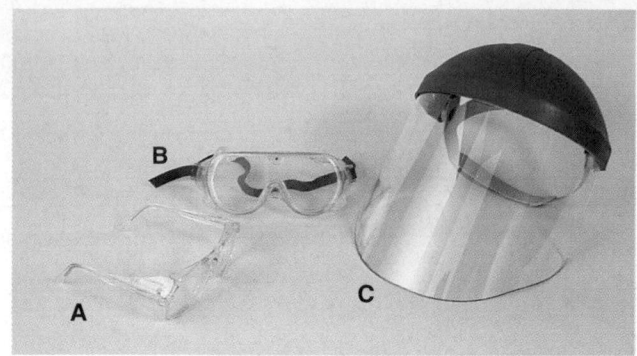

Figure 5-8. Wear approved eye and face protection when needed. A—Safety glasses. B—Safety goggles. C—Face shield.

and cause the battery to explode. Battery acid and pieces of the battery case can blow into your eyes and face. Blindness, facial cuts, acid burns, and scars can result. Always wear eye and face protection when working around a battery. See **Figure 5-8.**

Fuel tanks can explode, even seemingly empty ones. A drained fuel tank can still contain fuel gum and varnish. When this gum is heated and melts, it can emit vapors that may ignite.

Keep sparks and heat away from fuel tanks. When a fuel tank explodes, one side will usually blow out. Then, the tank will shoot across the shop as if shot out of a cannon. You or other workers could be killed or seriously injured.

Various other sources can cause shop explosions. For example, special sodium-filled engine valves, welding tanks, and propane-filled bottles can explode if mishandled. These hazards will be discussed in later chapters.

 Warning

Air bags should be handled with extreme care. If accidentally deployed, they can break bones or even kill. Carry them with the metal housing facing downward and away from your body. Keep all sources of electricity away from undeployed air bags.

Asphyxiation

Asphyxiation is caused by breathing toxic or poisonous substances. Mild cases of asphyxiation will cause dizziness, headaches, and vomiting. Severe asphyxiation can cause death.

The most common cause of asphyxiation in an auto shop is the exhaust gases produced by an automobile engine. *Exhaust gases are poison.* If a vehicle must be operated in an enclosed shop, connect the vehicle's tailpipe to the shop's exhaust ventilation system as shown in **Figure 5-9.** Also, make sure the exhaust ventilation system is turned *on.*

Discussed in related chapters, other shop substances are harmful if inhaled. A few of these harmful substances include asbestos (brake lining dust, clutch disc dust), parts cleaners, and paint spray. ***Respirators*** (filter masks) should be worn when working around any airborne impurities. Refer to **Figure 5-10**

Chemical Burns

Solvents (parts cleaners), battery acid, and various other shop substances can cause ***chemical burns*** to the skin. Always read the directions on all chemical containers. Also, be sure to wear proper protective gear when handling solvents and other caustic materials. See **Figure 5-11.**

Carburetor cleaner (decarbonizing types), for example, is very powerful and can severely burn your skin in a matter of seconds. Wear rubber gloves when using carburetor cleaner. If a skin burn occurs, follow label directions.

Warning

If your eyes are chemically burned, the material safety warning label may recommend flushing them with water. An ***eye flushing station*** is sometimes used to wash chemicals from your eyes after an accident.

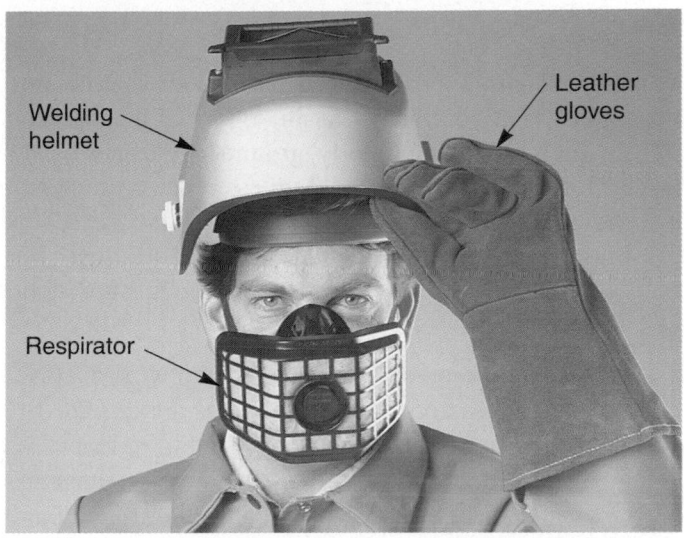

Figure 5-10. This technician is wearing a respirator to protect himself from toxic welding fumes. The welding helmet will shield the technician's face and eyes from hot sparks and the bright welding arc. (Lab Safety)

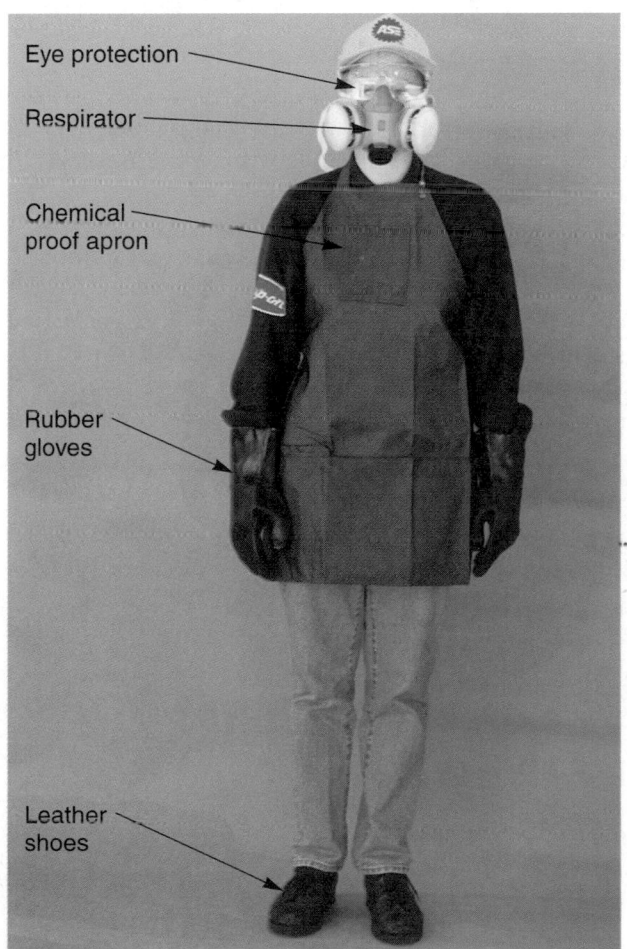

Figure 5-11. Always dress properly when handling substances that can cause chemical burns. Note that this technician is wearing rubber gloves, a chemical-proof apron, a respirator, and safety goggles.

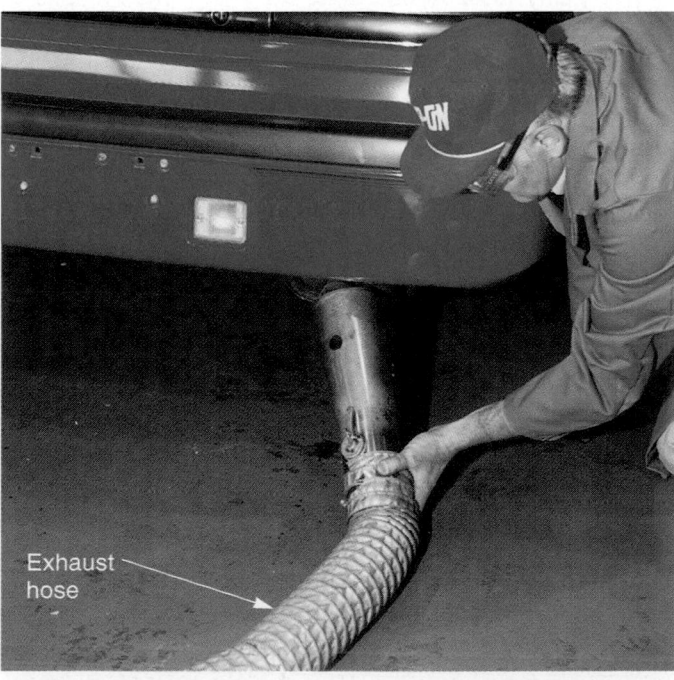

Figure 5-9. Place an exhaust hose over the tailpipe of any car running in an enclosed shop. This will prevent the shop from filling with deadly fumes. (Kent-Moore)

Electric Shock

Electric shock is a result of electric current passing through parts of your body, causing injury or death. It can occur when using improperly grounded electric power tools. Never use an electric tool unless it has a functional *ground prong* (third, round prong on plug socket). This prevents current from accidentally passing through your body. Also, never use an electric tool on a wet shop floor.

Warning
Some late-model cars have heated windshields. The alternators on these vehicles are designed to put out more than 100 volts ac to quickly warm the windshield. This is enough voltage to cause electrocution. Work carefully around this high voltage.

Physical Injury

Physical injuries (cuts, broken bones, strained backs) can result from hundreds of different accidents. As a technician, you must evaluate every repair technique. Decide whether a particular operation is safe and take action as required.

For instance, if you are pulling on a hand wrench as hard as you can and the bolt will not turn, *stop!* Find another wrench that is larger. A larger tool has more leverage and is, therefore, safer. This approach will help prevent injuries and improve your mechanical abilities.

Warning
Never overexert your back by improperly lifting heavy assemblies. Once you injure your back, it can take months to recover. Sometimes, surgery is needed to repair the damage. When lifting a heavy object, keep your back straight. Bend your knees and lift the item with your legs. If necessary, use power equipment to move heavy objects.

General Safety Rules

Listed are several general safety rules that should be followed at all times.

- Wear eye protection during any operation that could endanger your eyes. This would include operating power tools, working around a running car engine, carrying batteries, etc.

- Avoid anyone who does not take shop work seriously. Remember, a joker is "an accident just waiting to happen."

- Keep your shop organized. Return all tools and equipment to their proper storage areas. Never lay tools, creepers, or parts on the floor.

- Dress in an appropriate manner. Never wear loose clothing, neckties, shorts, or open-toed shoes when working in the shop. Remove rings, bracelets, necklaces, watches, and other jewelry. They can get caught in engine fans, belts, drive shafts, and other rotating parts, causing serious injury. Also, button or roll up long sleeves and secure long hair; they too can get caught in spinning parts.

- Never carry sharp tools or parts in your pockets. They can puncture the skin.

- Wear full face protection when grinding, welding, and performing other operations where severe hazards are present.

- Work like a professional. When learning auto repair, it is easy to get excited about your work. However, avoid working too fast. You could overlook a repair procedure or safety rule.

- Use the right tool for the job. There is usually a "best tool" for each repair task. Always be thinking about whether a different tool will work better than another, especially when you run into difficulty.

- Keep guards or shields in place. If a power tool has a safety guard, use it.

- Lift with your legs, not your back. There are many assemblies that are very heavy. When lifting, bend at your knees while keeping your back straight. On extremely heavy assemblies (transmissions, engine blocks, rear axles, transaxles), use a portable crane.

- Use adequate lighting. A portable shop light not only increases working safety, but it increases working speed and precision.

- Ventilate when needed. Turn on the shop ventilation fan anytime fumes are present in the shop.

- Never stir up asbestos dust. Asbestos dust (particles found in brake and clutch assemblies) is a powerful *cancer-causing agent.* Do *not* use compressed air to blow the dust from brake and clutch parts. Use an enclosed vacuum system to remove asbestos dust from brake assemblies, **Figure 5-12.**

- Jack up a vehicle slowly and safely. A car can weigh between one and two tons. Never work under a vehicle not supported by jack stands. It is *not* safe to work under a floor jack if it is the only support. See **Figure 5-13.**

Figure 5-12. Asbestos dust from brakes and clutches can cause lung cancer. Always use a vacuum system like this one to remove asbestos dust from parts. Never blow asbestos dust into the shop area. (Nilfisk)

- Drive slowly when in the shop area. With all the other students and vehicles in the shop, it is very easy to have an accident.

- Report unsafe conditions to your instructor. If you notice any type of hazard, let your instructor know about it.

- Stay away from engine fans. The fan on an engine is like a spinning knife. It can inflict serious injuries. Also, if a part or tool is dropped into the fan, it can fly out and hit someone. Electric fans can turn on even with the ignition key off!

- Respect running engines. When an engine is running, make sure the transmission is in park or neutral, the emergency brake is set, and the wheels are blocked. If these steps are not taken and the car is accidentally knocked into gear, it could run over you or a friend.

- Do not smoke in the auto shop. Smoking is a serious fire hazard, considering fuel lines, cleaning solvents, paints, and other flammables may be exposed.

- Read material safety data sheets when in doubt about any dangers. The material safety data sheet contains all the information needed to work safely with the hazardous material. See **Figure 5-14.**

- Obtain instructor permission before using any new or unfamiliar power tool, lift, or other shop equipment. If necessary, your instructor will give a demonstration.

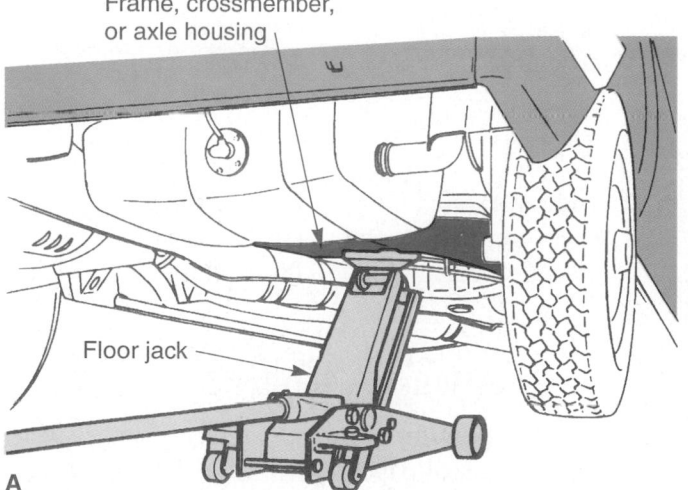

Figure 5-13. Never work under a car supported only by a floor jack. A—A jack must only be used for initial lifting. B—Jack stands are used to secure the car before working under it. Place the stands under the recommended lift points. (Subaru and Tech-Cor)

MATERIAL SAFETY DATA SHEET		DATE PREPARED 3/94
HI-TEMP PRODUCTS CO. 14936 GROVER ST. OMAHA, NEBRASKA 68144	EMERGENCY PHONE NUMBER 1-402-333-8323 1-402-359-5796	TELEPHONE NUMBER FOR INFORMATION 1-402-330-3344

1 | PRODUCT NAME:
HT 0801 AKRYA Cut

2 | PRECAUTIONARY STATEMENT FROM PRODUCT LABEL

WARNING

WEAR SAFETY GLASSES. MAY CAUSE EYE IRRITATION. IF EYE IRRITATION OCCURS, FLUSH EYES FOR 15 MINUTES WITH WATER. IF SWALLOWED, DO NOT INDUCE VOMITING. DRINK WATER OR MILK. IN EITHER CASE, CALL A PHYSICIAN IMMEDIATELY. REFER TO MSDS FOR CHEMICAL HAZARDOUS INFORMATION.

KEEP OUT OF THE REACH OF CHILDREN.

+ HEALTH	1
^ FIRE	0
* REACTIVE	0

HAZARD RATING
0 – LEAST
1 – SLIGHT
2 – MODERATE
3 – SERIOUS
4 – SEVERE

3 | HAZARDOUS COMPONENTS OSHA –>

INGREDIENTS	%	CAS NO.	PEL/TLV ppm	mg/m³	TWA ppm	mg/m³	STEL ppm	mg/m³	CEILING ppm	mg/m³
ISOPROPYL ALCOHOL	<3	67-63-0			400	983	500	1230		

4 | PHYSICAL/CHEMICAL CHARACTERISTICS

Boiling Point	212° F	Specific Gravity (H2O=1)	N/A
Vapor Pressure (mm Hg.)	N/A	Melting Point	N/A
Vapor Density (Air = 1)	N/A	Evaporation Rate (Butyl Acetate = 1)	N/A
Solubility in water	MISCIBLE		
Appearance and Odor	PURPLE THICK EMULSION, BANANA ODOR		

5 | Fire and Explosion Hazard Data

Flash Point	N/A	Flammable Limits	N/A	LEL	N/A	UEL	N/A
Extinguishing Media	WATER TO COOL DOWN CONTAINERS						
Special Fire Fighting Procedures	N/A	DIRECT FLAME FROM BUTANE TORCH					
WILL NOT IGNITE MATERIAL.							
Unusual Fire and Explosion Hazards	N/A						

6 | HEALTH HAZARD INFORMATION – SIGNS AND SYMPTOMS OF EXPOSURE

SKIN: MAY IRRITATE SKIN, CAUSE LIGHT REDDENING IN PATIENTS WITH PREEXISTING SKIN DISORDERS.
EYES: MAY CAUSE EYE IRRITATION, BURNING, AND REDNESS.

SWALLOWING: MAY CAUSE MOUTH AND THROAT IRRITATION, BURNING, AND POSSIBLE ABDOMINAL DISCOMFORT OR NAUSEA.
BREATHING: PROLONGED BREATHING OF MIST MAY IRRITATE NASAL PASSAGES.

7 | EMERGENCY AND FIRST AID PROCEDURES

ON SKIN: WASH WITH SOAP AND WATER. IF IRRITATION PERSISTS, SEEK MEDICAL ATTENTION.

IN EYES: FLUSH EYES WITH PLENTY OF RUNNING WATER FOR 15 MINUTES, LIFTING EYELIDS OCCASIONALLY. GET IMMEDIATE MEDICAL ATTENTION.

SWALLOWED: CALL A PHYSICIAN IMMEDIATELY. DO NOT INDUCE VOMITING. IF CONSCIOUS, DRINK PLENTY OF WATER OR MILK.

BREATHED: REMOVE TO FRESH AIR. GIVE ARTIFICIAL RESPIRATION IF NOT BREATHING. SEEK IMMEDIATE MEDICAL ATTENTION.

8 | TOXICITY DATA

ORAL: N/A	DERMAL: N/A

INHALATION: N/A

CARCINOGENICITY: NOT CONSIDERED TO BE A CARCINOGEN BY IARC.

NTP? N/A   IARC MONOGRAPHS? N/A   OSHA REGULATED? N/A

9 | PERSONAL PROTECTION

VENTILATION: LOCAL OR MECHANICAL EXHAUST.
RESPIRATORY PROTECTION: IF NECESSARY, WEAR A PARTICLE MASK OR AN OSHA APPROVED MASK FOR MIST CONCENTRATIONS.
EYE PROTECTION: WEAR GOGGLES OR SAFETY GLASSES.
SKIN AND PROTECTIVE CLOTHING: WEAR RUBBER GLOVES.

10 | HAZARDOUS REACTIVITY

STABILITY:	UNSTABLE?		CONDITIONS TO AVOID: N/A
	STABLE?	X	

INCOMPATIBILITY: N/A

HAZARDOUS BYPRODUCTS: N/A

HAZARDOUS POLYMERIZATION:	MAY OCCUR		CONDITIONS TO AVOID: N/A
	WILL NOT OCCUR	X	

Figure 5-14. Study the types of information given on a material safety data sheet. (Hi-Temp Products Co.)

Summary

- An auto shop can be a very safe and enjoyable place to work. However, if basic safety rules are *not* followed, an auto shop can be very dangerous.

- The shop repair area includes any location in the shop where repair operations are performed.

- The toolroom is a shop area normally adjacent (next to) the main shop or classroom. It is used to store shop tools, small equipment, and supplies (nuts, bolts, oil).

- Every year, thousands of technicians are injured or killed on the job. Most of these accidents resulted from a broken safety rule.

- Fires are capable of causing instant and permanent scar tissue. There are numerous combustible substances found in an auto shop.

- Gasoline is by far the most dangerous and often underestimated flammable in an auto shop.

- Electrical fires can result when a "hot wire" (wire carrying current to component) touches ground (vehicle frame or body).

- An explosion is the rapid, almost instant combustion of a material which causes a powerful shock wave to travel out through the shop.

- Asphyxiation is caused by breathing toxic or poisonous substances in the air.

- Respirators should be worn when working around any kind of airborne impurities.

- Electric shock results when electric current passes through your body, causing injury or death.

- Physical injuries (cuts, broken bones, strained backs) can result from a variety of accidents.

Important Terms

Repair area	Explosion
Shop stall	Asphyxiation
Lift	Respirators
Alignment rack	Chemical burns
Toolroom	Eye flushing station
Outside work area	Electric shock
Locker room	Ground prong
Dressing room	Physical injuries
Fires	

Review Questions—Chapter 5

Please do not write in this text. Place your answers on a separate sheet of paper.

1. List four safety rules to follow when using a vehicle lift.

2. A(n) _____ _____ is used when working on a car's steering and suspension systems. It has special equipment for aligning the vehicle's wheels.

3. _____ is the most common and dangerous flammable found in an auto shop.

4. What causes an electrical fire in an automobile?

5. Car batteries can explode. True or False?

6. Which of the following cannot cause electric shock?
   (A) Missing ground prong on cord.
   (B) Using electric drill on wet floor.
   (C) Using electric tools with a ground prong.
   (D) None of the above.

7. Explain what must be done to prevent physical injuries.

8. If you are pulling on a wrench as hard as you can and the fastener does *not* turn, what should you do to prevent injury?

9. When lifting heavy objects, always lift with your _____.
   (A) arms
   (B) legs
   (C) back
   (D) None of the above.

10. List 20 general safety rules.

⚙ ASE-Type Questions

1. In which auto shop area would an exhaust system repair most likely be done?
   (A) *Shop stall.*
   (B) *Grease rack.*
   (C) *Alignment rack.*
   (D) *Outside work area.*

2. Rules to remember when using gasoline include each of the following *except:*
   (A) *Store gas in approved containers.*
   (B) *Keep gas away from sources of heat.*
   (C) *Use quick dry to absorb any gas spills.*
   (D) *Never use gasoline as a cleaning solvent.*

3. Which of the following is a possible source of explosions in an auto shop?
   (A) *Fuel tanks.*
   (B) *Car batteries.*
   (C) *Welding tanks.*
   (D) *All the above.*

4. Asphyxiation can be caused by:
   (A) *touching a current-carrying wire.*
   (B) *improper lifting techniques.*
   (C) *breathing toxic substances.*
   (D) *None of the above.*

5. Asbestos dust, which can cause cancer, is found in:
   (A) *fuel tanks.*
   (B) *transmissions.*
   (C) *propane-filled bottles.*
   (D) *brake and clutch assemblies.*

6. A respirator is a(n):
   (A) *filter mask.*
   (B) *type of chemical burn.*
   (C) *machine guard.*
   (D) *device to put out small fires.*

7. Eye protection should be worn when:
   (A) *carrying batteries.*
   (B) *operating power tools.*
   (C) *working by a running engine.*
   (D) *All of the above.*

8. Which of the following is *not* a good tip when dressing for work?
   (A) *Secure long hair.*
   (B) *Roll up long shirt sleeves.*
   (C) *Make sure all jewelry fits well.*
   (D) *Don't carry sharp tools in pocket.*

9. A engine needs to be moved. Technician A says two people can slide the engine out of the way. Technician B says that an engine crane should be used to move the engine. Who is right?
   (A) *A only.*
   (B) *B only.*
   (C) *Both A and B.*
   (D) *Neither A nor B.*

10. When removing asbestos dust from parts, Technician A believes a vacuum system should be used. Technician B believes dust should be blown away using compressed air. Who is right?
    (A) *A only.*
    (B) *B only.*
    (C) *Both A and B.*
    (D) *Neither A nor B.*

Activities for Chapter 5

1. Sketch out a floor plan of your shop and label the different areas. Study the safety cautions in this chapter and determine if there are any safety hazards. Mark their location on the floor plan.

2. On the same floor plan, mark the location of fire extinguishers, exits, and water fountains.

3. Examine a fire extinguisher in the shop area; read the instructions carefully. Demonstrate its use.

Chapter 6

Automotive Measurement and Math

After studying this chapter, you will be able to:

- Describe both customary and metric measuring systems.
- Identify basic measuring tools.
- Describe the use of common measuring tools.
- Use conversion charts.
- List safety rules relating to measurement.
- Summarize basic math facts.
- Correctly answer ASE certification test questions that require a basic understanding of measurement and math.

As a vehicle is driven, many of its moving parts slowly wear out. With enough part wear, mechanical failures and performance problems result. Manufacturers give **specifications,** or **"specs,"** which are maximum wear limits and dimensions of important parts. If the measurements are *not* within these specifications, the part must be adjusted, repaired, or replaced. Therefore, a technician must be able to make accurate measurements.

This chapter introduces the most important types of measurements performed by a service technician. General measuring tools and methods are explained using both customary and metric systems. Work-related math skills are also covered. Study this chapter very carefully. It prepares you for other textbook chapters and for hundreds of in-shop tasks.

Measuring Systems

Two **measuring systems** are commonly used when working in an auto shop: the U.S. Customary Units system and the SI Metric system. The U.S. Customary Units system is also called the customary system or the English system. Most countries use the metric system. The customary system is mainly used in the United States. However, the United States uses the metric system

as well. All vehicles manufactured in the U.S. by foreign companies and many vehicles made by U.S. companies use metric bolts, nuts, and other parts. Manufacturer specifications are often given in both customary and metric values for U.S. made vehicles. **Figure 6-1** summarizes and compares the two measuring systems.

Customary Measuring System

The **customary measuring system** originated from sizes taken from parts of the human body. For example, the width of the human thumb was used to standardize

Quantity	Customary (abbreviation)	Metric (abbreviation)
Length	Inch (in) Foot (ft) Mile (mi)	Meter (m)
Weight (mass)	Ounce (oz) Pound (lb)	Kilogram (kg)
Area	Square inch (sq-in)	Square meter (m²)
Dry volume	Cubic inch (cu-in)	Cubic meter (m³) Cubic centimeter (cc)
Liquid volume	Ounce (oz) Pint (pt) Quart (qt) Gallon (gal)	Liter (L) Cubic centimeter (cc)
Road speed	Miles per hour (mph)	Kilometer per hour (km/h)
Torque	Foot-pounds (ft-lb)	Newton meter (N·m)
Power	Horsepower (hp)	Kilowatt (kW)
Pressure	Pounds per square inch (psi)	Kilopascal (kPa)
Temperature	Degrees fahrenheit (°F)	Degrees celsius (°C)

Figure 6-1. This chart compares U.S. customary and metric values. Study them carefully.

the inch, **Figure 6-2.** The length of the human foot was used to standardize the foot as 12 inches. The distance between the tip of a finger and nose was used to set the standard for the yard as 3 feet. Obviously, these are not very scientific standards since these distances vary from person to person.

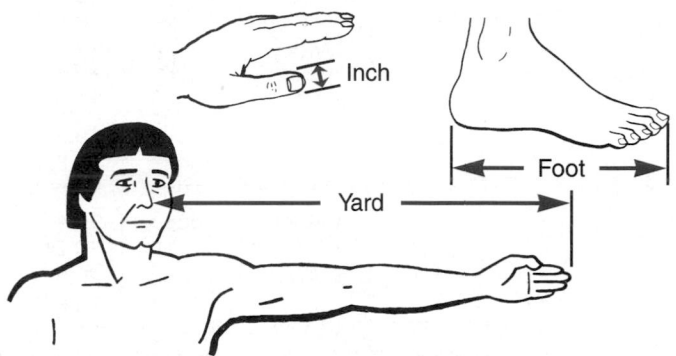

Figure 6-2. The customary system was originally based on parts of the human body. (Starrett)

Metric (SI) Measuring System

The *metric (SI) measuring system* uses a power of 10 for all basic units. It is a simpler and more logical system than the customary system. Computation often requires nothing more than moving the decimal point. For instance, one meter equals 10 decimeters, 100 centimeters, or 1000 millimeters.

Conversion Charts

A measuring system *conversion chart* is needed when changing from one measuring system to another, such as when changing from inches to centimeters, gallons to liters, or liters to gallons. A conversion chart lets the technician quickly convert customary values to equivalent metric values, or vice versa. One is shown in **Figure 6-3.**

A *decimal conversion chart* is commonly used to find equivalent values for fractions, decimals, and millimeters. See **Figure 6-4.** Fractions are only accurate to about 1/64 of an inch. For smaller measurements, either

Measurement		When you know:	You can find:	If you multiply by:
	Length	inch (in)	millimeter (mm)	25.4
		foot (ft)	meter (m)	.3
		yard (yd)	meter (m)	.9
		mile (mi)	kilometer (km)	1.6
		millimeter (mm)	inch (in)	.04
		centimeter (cm)	inch (in)	.39
		meter (m)	yard (yd)	1.09
		kilometer (km)	mile (mi)	.6
	Pressure	pounds per square inch (psi)	kilopascal (kPa)	6.89
		kilopascal (kPa)	pounds per square inch (psi)	.145
	Power	horsepower (hp)	kilowatt (kW)	.746
		kilowatt (kW)	horsepower (hp)	1.34
	Torque	foot-pounds (ft-lb)	Newton-meter (N·m)	1.36
		Newton-meter (N·m)	foot-pounds (ft-lb)	.74
	Volume	quart (qt)	liter (L)	.95
		liter (L)	quart (qt)	1.06
		cubic inch (cu-in)	liter (L)	.016
		liter (L)	cubic inch (cu-in)	61.02
	Mass	ounce (oz)	gram (g)	28.35
		gram (g)	ounce (oz)	.035
		pound (lb)	kilogram (kg)	.45
		kilogram (kg)	pound (lb)	2.20
	Speed	miles per hour (mph)	kilometers per hour (km/h)	1.61
		kilometers per hour (km/h)	miles per hour (mph)	.62

Figure 6-3. To convert from one system to another, multiply the known value by the number in right column. This will give an approximately equal value.

Fraction				Inches	mm
			1/64	.01563	.397
		1/32		.03125	.794
			3/64	.04688	1.191
	1/16			.06250	1.588
			5/64	.07813	1.984
		3/32		.09375	2.381
			7/64	.10938	2.778
1/8				.12500	3.175
			9/64	.14063	3.572
		5/32		.15625	3.969
			11/64	.17188	4.366
	3/16			.18750	4.763
			13/64	.20313	5.159
		7/32		.21875	5.556
			15/64	.23438	5.953
1/4				.25000	6.350
			17/64	.26563	6.747
		9/32		.28125	7.144
			19/64	.29688	7.541
	5/16			.31250	7.938
			21/64	.32813	8.334
		11/32		.34375	8.731
			23/64	.35938	9.128
3/8				.37500	9.525
			25/64	.39063	9.922
		13/32		.40625	10.319
			27/64	.42188	10.716
	7/16			.43750	11.113
			29/64	.45313	11.509
		15/32		.46875	11.906
			31/64	.48438	12.303
1/2				.50000	12.700

Fraction				Inches	mm
			33/64	.51563	13.097
		17/32		.53125	13.494
			35/64	.54688	13.891
	9/16			.56250	14.288
			37/64	.57813	14.684
		19/32		.59375	15.081
			39/64	.60938	15.478
5/8				.62500	15.875
			41/64	.64063	16.272
		21/32		.65625	16.669
			43/64	.67188	17.066
	11/16			.68750	17.463
			45/64	.70313	17.859
		23/32		.71875	18.256
			47/64	.73438	18.653
3/4				.75000	19.050
			49/64	.76563	19.447
		25/32		.78125	19.844
			51/64	.79688	20.241
	13/16			.81250	20.638
			53/64	.82813	21.034
		27/32		.84375	21.431
			55/64	.85938	21.828
7/8				.87500	22.225
			57/64	.89063	22.622
		29/32		.90625	23.019
			59/64	.92188	23.416
	15/16			.93750	23.813
			61/64	.95313	24.209
		31/32		.96875	24.606
			63/64	.98438	25.003
1				1.00000	25.400

Figure 6-4. A decimal conversion chart is commonly used in the auto shop. This chart lets you interchange fractions, decimals, and millimeters. What are equal decimal and millimeter values for 1/4″, 5/32″, 43/64″, and 7/8″? (Parker Hannifin Corp.)

decimals or millimeters should be used. A decimal conversion chart may be needed to change a fractional measurement to a decimal measurement.

Measuring Tools

There are various tools used by a technician to make accurate measurements. Common measuring tools include the steel rule, caliper, micrometer, and dial indicator. Most of these are available in both customary and metric units. These and other tools are covered in the next sections.

Steel Rule

A *steel rule*, or *scale*, is frequently used to make low-precision linear measurements. It is accurate to about 1/64″ (0.4 mm) in most instances. A *customary rule* has number labels that represent full inches, **Figure 6-5.** The smaller, unnumbered lines, or graduations, represent fractions of an inch, such as 1/2″, 1/4″, 1/8″, and 1/16″.

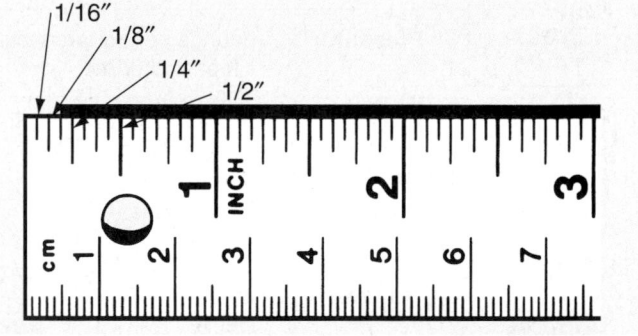

Figure 6-5. Compare inches to centimeters. Ten millimeters equals one centimeter. Twenty-five millimeters is a little less than one inch. The customary rule is divided into 1/16″ fractions. (Fairgate)

The shortest graduation lines represent the smallest fractions. In **Figure 6-5,** this is 1/16″. A *metric rule* normally has lines or divisions representing millimeters (mm). Each numbered line usually equals 10 mm, or 1 cm (centimeter). This is also shown in **Figure 6-5.**

A *pocket rule,* or *pocket scale,* is typically 6″ long. It is small enough to fit in your shirt pocket. A *combination square* is a sliding square that is mounted on a steel rule. It is needed when the rule must be held perfectly square against the part being measured. See **Figure 6-6.**

A *tape measure,* or *tape rule,* extends to several feet or meters in length. It is sometimes needed for large distance measurements during body, suspension, and exhaust system repairs. Look at **Figure 6-7.** A *yardstick* or *meterstick* is a rigid measuring device used for large lineal measurements up to one yard or one meter.

Dividers

Dividers look like a drafting compass, but have straight, sharply pointed tips, **Figure 6-8A.** They are commonly used for layout work on sheet metal parts. The sharp points can scribe circles and lines on sheet metal and plastic. Dividers can also be used to transfer and make surface measurements.

Calipers

An *outside caliper* is used to make external measurements when 1/64″ (approximately 0.40 mm) accuracy is sufficient. See **Figure 6-8B.** The caliper is fitted over the outside of parts and adjusted so that each tip just touches the part. Then, the caliper is held against a rule and the distance between the tips is measured to determine part size.

An *inside caliper* is designed for internal measurements in holes and other openings, **Figure 6-8C.** It is placed inside a hole and adjusted until the tips just touch

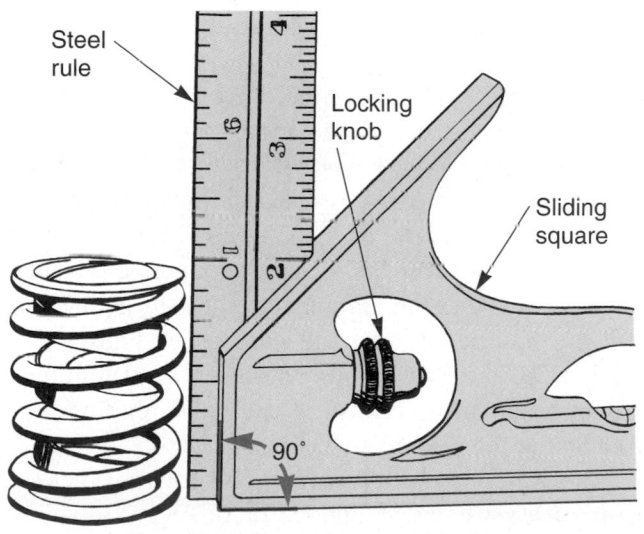

Figure 6-6. A combination square is needed when the scale must be held perfectly parallel to the part. (Cadillac)

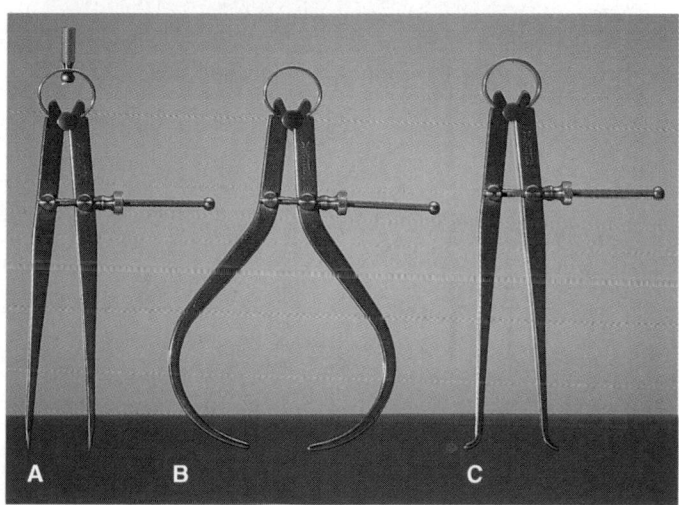

Figure 6-8. A—Dividers have sharp points for measuring or marking on metal parts. B—An outside caliper for measurements on the outside of a part. C—An inside caliper for internal measurements. (Starrett)

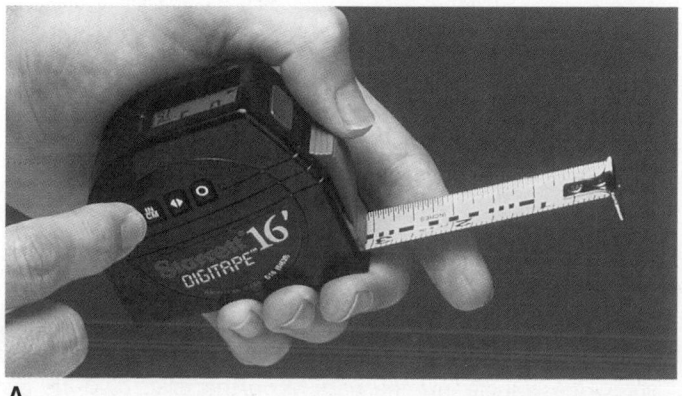

Figure 6-7. A—A digital-reading tape measure. B—A tape measure is used to make large straight-line measurements.

the part. Then, it is held against a rule and the distance between the tips is measured.

A *vernier caliper* is a sliding measuring device that can make inside, outside, and depth measurements with considerable accuracy. One is pictured in **Figure 6-9.** Many vernier calipers can take measurements as small as 0.001″ (0.025 mm). Some vernier calipers have a dial gauge attached. The dial makes the "thousandths" part of a measurement easier to read. A vernier caliper is fast and easy to use, making it a very useful tool for the automotive technician to have.

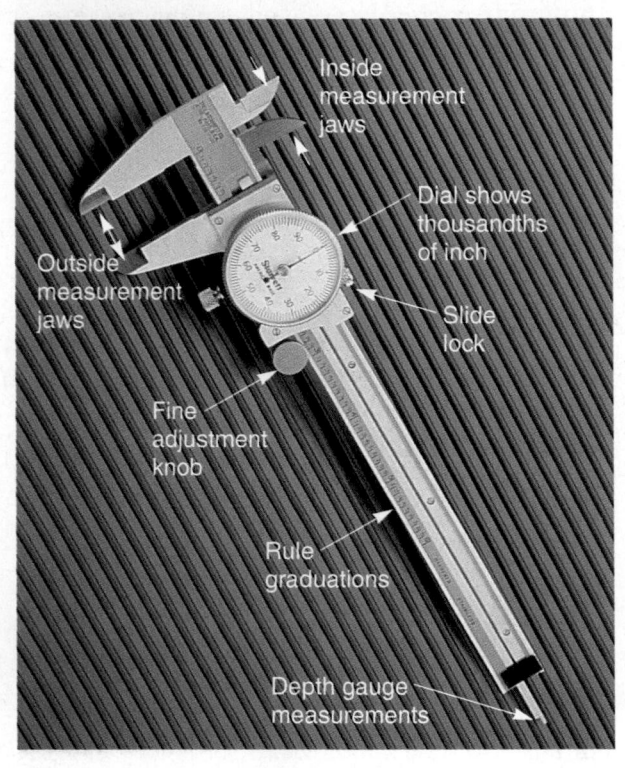

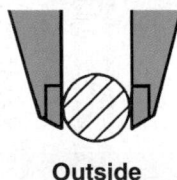 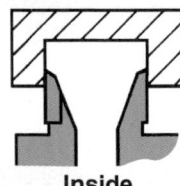 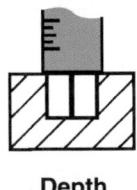

Outside   **Inside**   **Depth**

Figure 6-9. A vernier caliper can be used to quickly check inside, outside, and depth measurements. (Starrett and K-D Tools)

Micrometers

A *micrometer,* nicknamed a *mike,* is used to make very accurate measurements. It can measure to one ten-thousandth of an inch (.0001″) or one thousandth of a millimeter (0.001 mm). There are several types of mikes used in automotive service and repair. These include

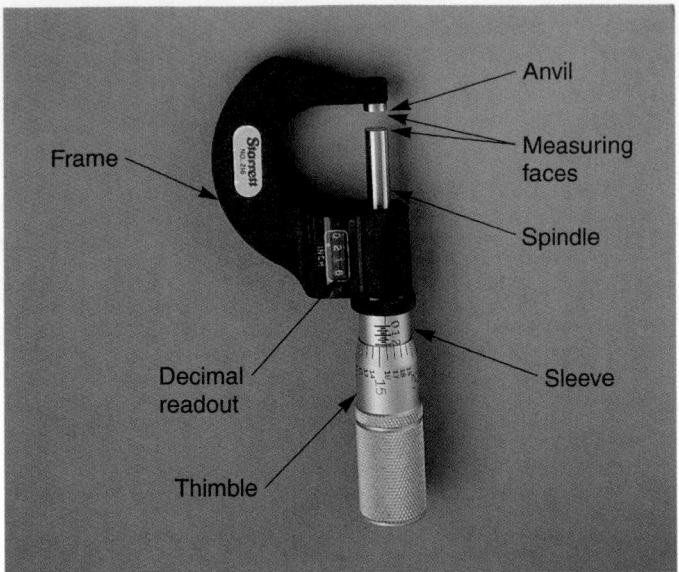

Figure 6-10. A micrometer is the precision measuring tool most commonly used by a technician. This one is easy to use because it has a digital readout. (Starrett)

outside, inside, and depth micrometers. In addition, a telescoping gauge, or a hole gauge, may be used with a micrometer.

An *outside micrometer* is used for measuring external dimensions, diameters, or thicknesses, **Figure 6-10.** To use an outside micrometer, place it around the outside of the part. Then, turn the thimble until both the spindle and anvil are lightly touching the part, as in **Figure 6-11.** Finally, read the graduations on the hub and thimble to determine the measurement. Reading a micrometer is discussed in the next section.

An *inside micrometer* is used for internal measurements of large holes, cylinders, or other part openings, **Figure 6-12A.** To use an inside micrometer, place it

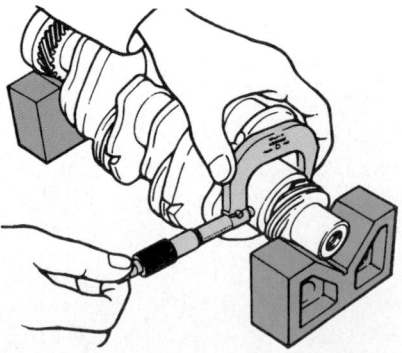

Figure 6-11. To use a micrometer, gently rotate the thimble to screw the spindle into the part. Move the mike over the part while holding it squarely. When you feel a slight drag, remove the mike and read the measurement. (Subaru)

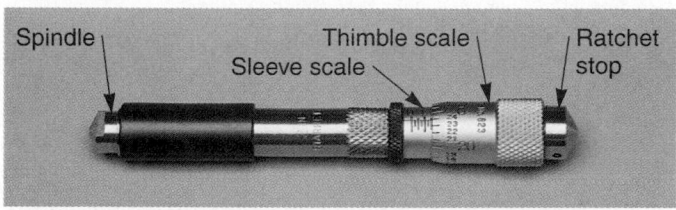

Spindle | Thimble scale | Ratchet stop | Sleeve scale

A

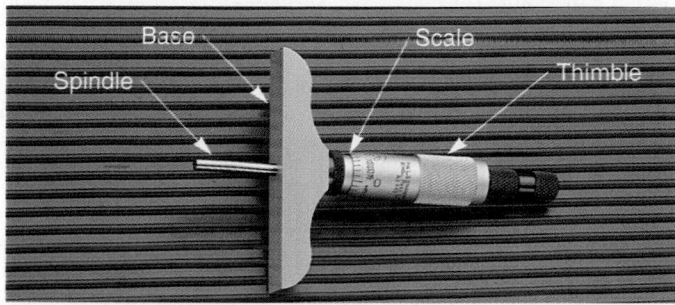

Base | Scale | Spindle | Thimble

B

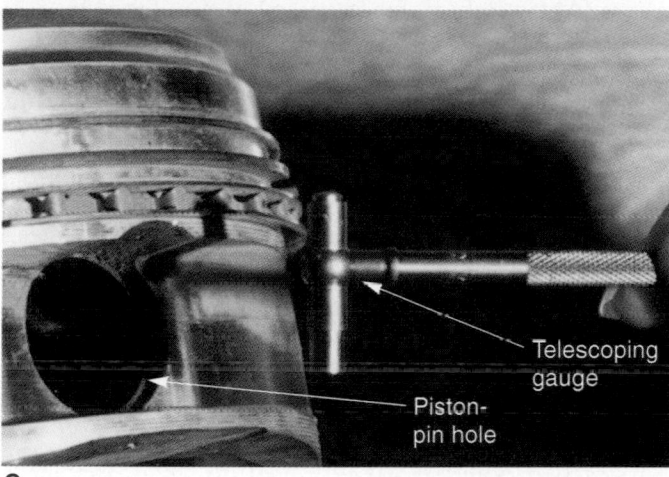

Telescoping gauge | Piston-pin hole

C

Figure 6-12. A—Inside micrometer. B—Depth micrometer. C—Telescoping gauge. (Starrett and Snap-on Tools)

inside the hole. Then, adjust it until the part is just being touched by the micrometer. Finally, remove the micrometer from the hole and read the measurement. The inside micrometer is read in the same manner as an outside mike.

A ***depth micrometer*** is helpful when precisely measuring the depth of an opening. Look at **Figure 6-12B.** The base of the micrometer is positioned squarely on the part. Then, the thimble is turned until the spindle contacts the bottom of the opening. The depth micrometer is read in the same way as an outside micrometer. However, the hub markings are *reversed.*

A ***telescoping gauge*** is used to measure internal part bores or openings, **Figure 6-12C.** To use the gauge, compress the spring-loaded extensions and lock them with the thumb wheel. Then, insert the gauge into the hole and release the thumb wheel. The extensions "snap" to the edges of the hole. Use the thumb wheel to lock the

extensions to the size. Finally, use an outside micrometer to measure the distance across the extensions.

A ***hole gauge*** is used for measuring very small holes in parts. To use a hole gauge, first loosen the thumb wheel. Then, insert the gauge into the hole and tighten the thumb wheel until the gauge just touches the part. Finally, remove the gauge and measure it with an outside micrometer.

Reading a Customary Micrometer

To read a customary micrometer, follow the four steps listed below. Refer to **Figure 6-13.**

1. Note the *largest* number visible on the micrometer sleeve. Each number equals 0.100″ (2 = 0.200″, 3 = 0.300″, 4 = 0.400″).

2. Count the number of graduation *lines* to the right of the sleeve number. Each full sleeve graduation equals 0.025″ (2 full lines = 0.050″, 3 = 0.075″).

3. Note the *thimble graduation* aligned with the horizontal sleeve line. Each thimble graduation equals 0.001″ (2 thimble graduations = 0.002″, 3 = 0.003″). Round off when the sleeve line is not directly aligned with a thimble graduation.

4. Add the decimal values from steps 1, 2, and 3. Also, add any full inches. This sum is the micrometer reading in inches.

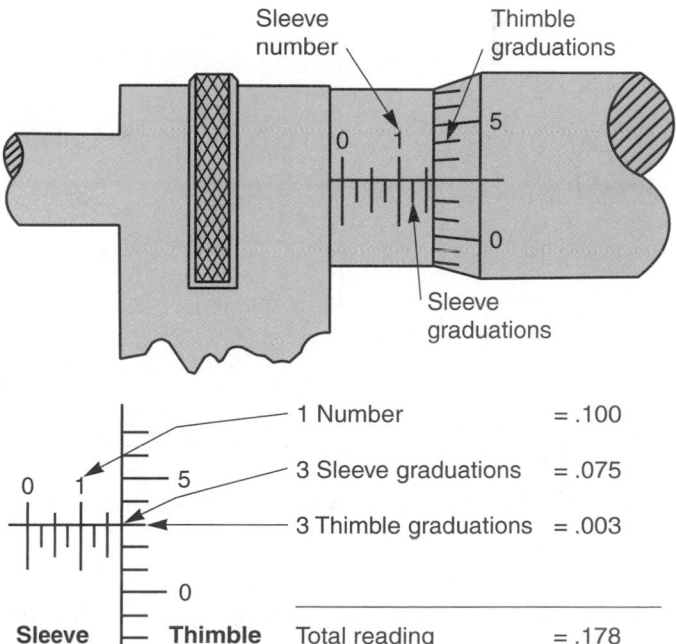

Sleeve number | Thimble graduations | Sleeve graduations

1 Number	= .100
3 Sleeve graduations	= .075
3 Thimble graduations	= .003
Total reading	= .178

Sleeve | Thimble

Figure 6-13. To read a micrometer, read the sleeve number first. Then, read the sleeve graduations. Finally, read the thimble number. Add these three values to obtain the reading. (Starrett)

Reading a Metric Micrometer

Micrometers are also available in metric units. They are similar to customary micrometers but have graduations and numbers in metric values. For example, one revolution of the thimble equals 0.500 mm. To read a metric micrometer, follow the four steps given below. Refer to **Figure 6-14.**

1. Read the *largest* number visible on the micrometer sleeve. Each number equals 1.00 mm (2 = 2.00 mm, 3 = 3.00 mm).

2. Count the number of graduation *lines* to the right of the sleeve number. Each full sleeve graduation equals 0.50 mm (2 = 1.00 mm, 3 = 1.50 mm).

3. Read the *thimble graduation* aligned with the horizontal sleeve line. Each thimble graduation equals 0.01 mm (2 = 0.02 mm, 3 = 0.03 mm).

4. Add the values from the steps 1, 2, and 3. This sum is the metric micrometer reading in millimeters.

Micrometer Rules

A few important micrometer rules to remember include:

- Never drop or overtighten a micrometer. It is very delicate, and its accuracy can be thrown off easily.

- Store micrometers where they cannot be damaged. Keep them in wooden or plastic storage boxes.

- Grasp the micrometer frame in your palm and turn the thimble with your thumb and finger. The measuring faces should just drag on the part being measured.

- Hold the micrometer squarely with the work or false readings can result. Closely watch how the spindle is contacting the part.

- Rock or swivel the micrometer as it is touched on round parts. This will ensure that the most accurate diameter measurement is obtained.

- Place a thin film of oil on the micrometer during storage. This will keep the tool from rusting.

- Always check the accuracy of a micrometer if it is dropped, struck, or after a long period of use. Standardized gauge blocks are used for checking micrometer accuracy.

Tech Tip

An easy way to practice using a micrometer is to measure the thickness of feeler gauge blades. Since the thickness is printed on the blades, you can read the mike and compare your results to the actual thickness of each blade.

Feeler Gauges

A *feeler gauge* is used to measure small clearances or gaps between parts. There are two basic types of feeler gauges: flat feeler gauges and wire feeler gauges. Both types are available in customary and metric versions.

A *flat feeler gauge* has precision-ground steel blades of various thicknesses, **Figure 6-15A.** Thickness is written on each blade in thousandths of an inch and/or in hundredths of a millimeter. A flat feeler gauge is normally used to measure distances between *parallel surfaces.*

A *wire feeler gauge* has precise-size wires labeled by diameter or thickness, **Figure 6-15B.** It is normally used to measure slightly larger spaces or gaps than a flat feeler gauge. A wire gauge is also used for measuring the distance between *unparallel* or *curved surfaces.*

Using a Feeler Gauge

To measure with either type feeler gauge, find the gauge thickness that just fits between the two parts being measured. The gauge should drag slightly when pulled between the two surfaces. The size given on the gauge is the clearance between the two components.

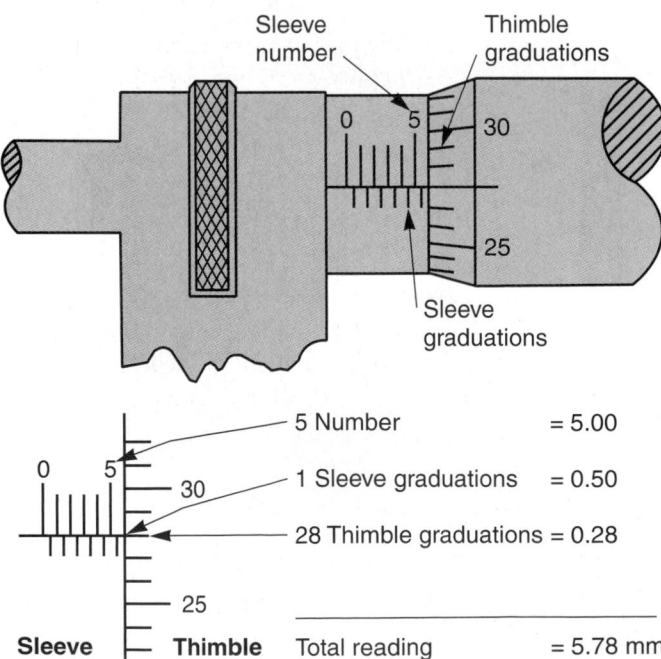

Figure 6-14. A metric micrometer is read like a customary micrometer. However, metric values are used for the sleeve and thimble. (Starrett)

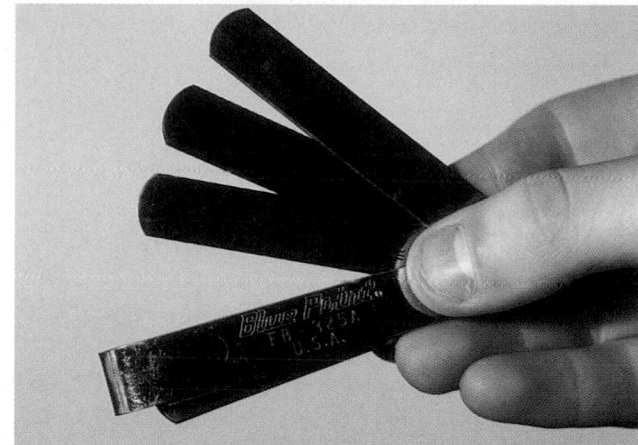

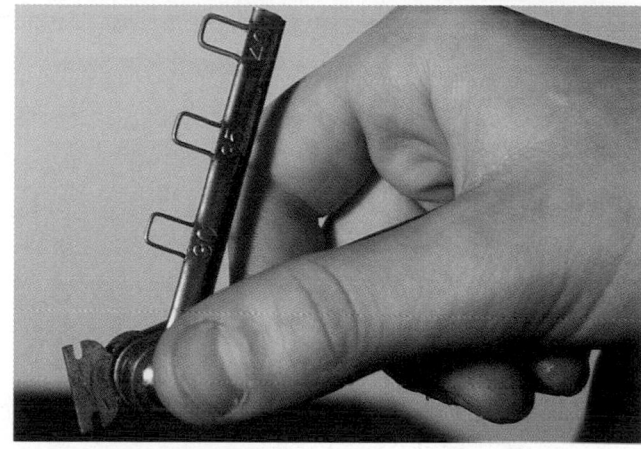

Figure 6-15. A—Flat, or blade, feeler gauge set. B—Wire feeler gauge set.

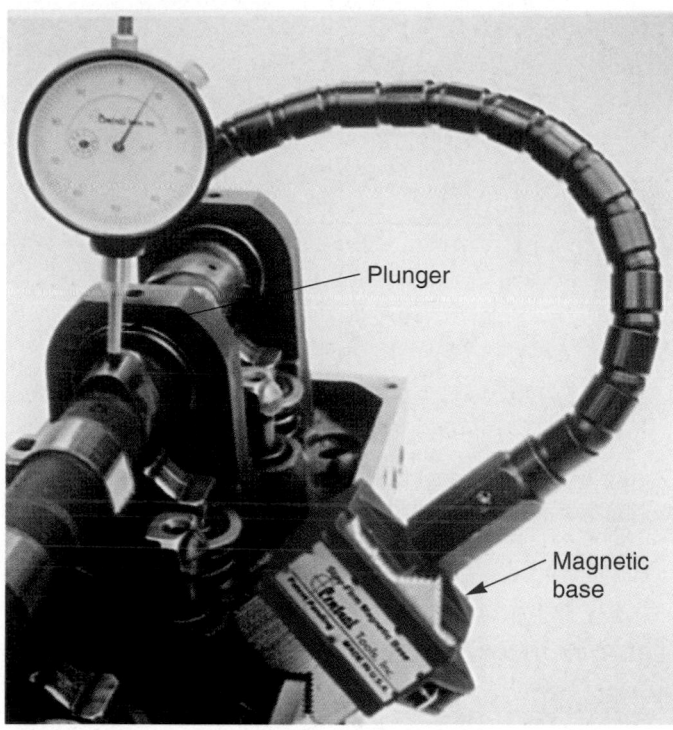

Figure 6-16. A dial indicator is used when measuring part movement. In this example, the tool is set up to check height and wear of a camshaft lobe. The cam is rotated and the indicator reading is compared to specs to find wear. (Central Tool Co.)

Dial Indicator

A *dial indicator* is used to measure part movement in thousandths of an inch (hundredths of a millimeter). See **Figure 6-16.** The needle on the indicator face registers the amount of plunger movement. A dial indicator is frequently used to check gear teeth backlash (clearance), shaft end play, cam lobe lift, and similar kinds of part movements. A magnetic mounting base or clamp mechanism is normally used to secure the dial indicator to or near the work. Be careful not to damage a dial indicator. It is very delicate.

Using a Dial Indicator

To measure with a dial indicator, follow these basic rules:

1. Mount the indicator securely and position the dial plunger parallel with the movement to be measured.

2. Partially compress the indicator plunger before locking the indicator into place. This allows part movement in either direction to be measured.

3. Move the part back and forth or rotate the part while reading the indicator.

4. Subtract the lowest reading from the highest reading. The result equals the distance the part moved, the clearance, or the runout.

Other Measurements and Measuring Tools

A service technician may make other types of measurements and use other types of measuring tools from those discussed to the point. Some of these are discussed in the next sections. More specialized measurements are covered in later chapters.

Angle Measurement

A circle is divided into 360 equal parts, called *degrees,* **Figure 6-17.** Degrees are abbreviated with "deg." or the degree symbol (°). One-half of a circle equals 180°, one-quarter of a circle equals 90°, and one-eighth of a circle equals 45°. Specifications are normally given in degrees when you are measuring rotation of a part or an angle formed by a part. Later text chapters discuss this.

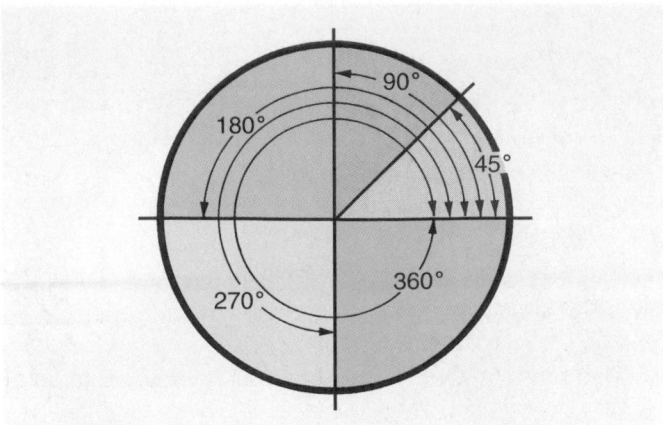

Figure 6-17. The amount of rotation and angles are measured in degrees. Note how many degrees are in a full circle and in fractions of a circle.

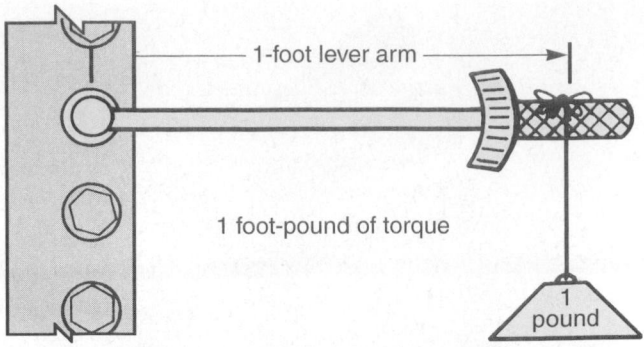

Figure 6-19. One foot-pound equals one pound of pull on a one-foot-long lever arm. This provides a means of measuring torque, or twisting motion.

Temperature Measurement

Temperature gauges, or *thermometers,* are used to measure temperature. For example, air conditioning output temperature or radiator temperature may need to be determined. The temperature obtained with the gauge can be compared to specifications. Then, if the temperature is too low or too high, you know that a repair or adjustment is needed. Temperature gauges are available that can read in either customary *Fahrenheit* (F) or metric *Celsius* (C), **Figure 6-18.**

Torque Wrenches

A *torque wrench* is not used for taking measurements. Rather, it is used to apply a specific amount of turning force to a fastener, such as a bolt or nut. A torque wrench uses the principle illustrated in **Figure 6-19.** Torque wrench scales usually read in foot-pounds (ft-lb) and Newton-meters (N•m). The three general types of torque wrenches are the *flex bar, dial indicator,* and *ratcheting* types. These are shown in **Figure 6-20.**

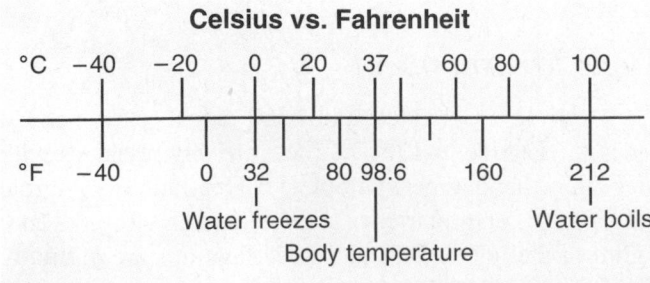

Celsius vs. Fahrenheit

°C	−40	−20	0	20	37	60	80	100

| °F | −40 | 0 | 32 | 80 | 98.6 | 160 | 212 |

Water freezes
Body temperature
Water boils

Figure 6-18. Study the relationship between customary Fahrenheit (°F) and metric Celsius (°C) temperature values.

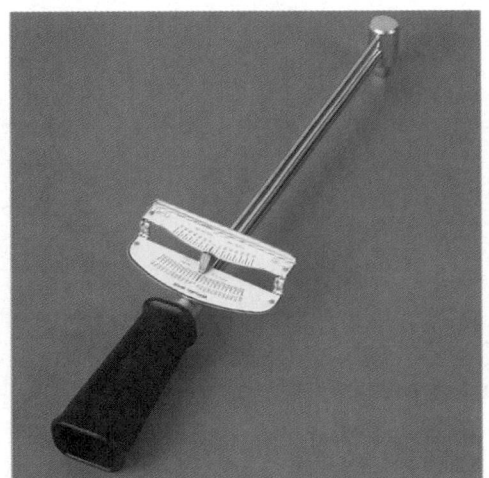

A

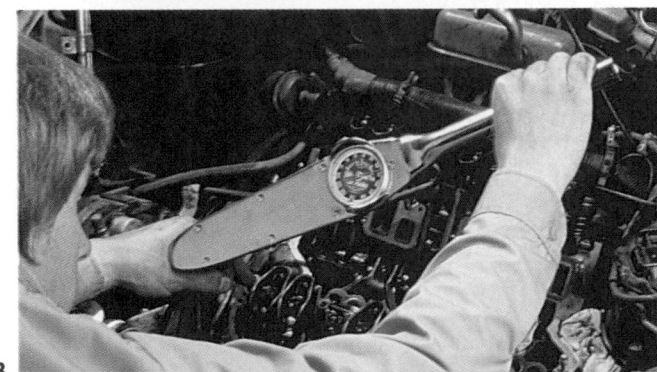

B

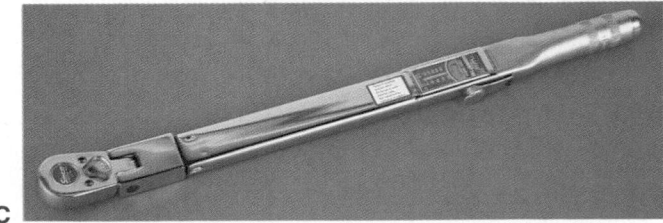

C

Figure 6-20. Different torque wrench types. A—A flex bar torque wrench uses a bending metal beam to make the pointer read torque on the scale. B—A dial indicator torque wrench is very accurate. C—A ratcheting, or snap-type, torque wrench is fast. The torque value is set by turning the handle. Then, the fastener is tightened until a click or popping sound is heard.

Pressure Gauge

A *pressure gauge* is used to measure air and fluid pressure in various systems and components. For example, pressure gauges may be used to check tire air pressure, fuel pump pressure, air conditioning system pressure, or engine compression stroke pressure. Look at **Figure 6-21.** A pressure gauge normally reads in pounds per square inch (psi), kilograms per square centimeter (kg/cm^2), or kilopascals (kPa).

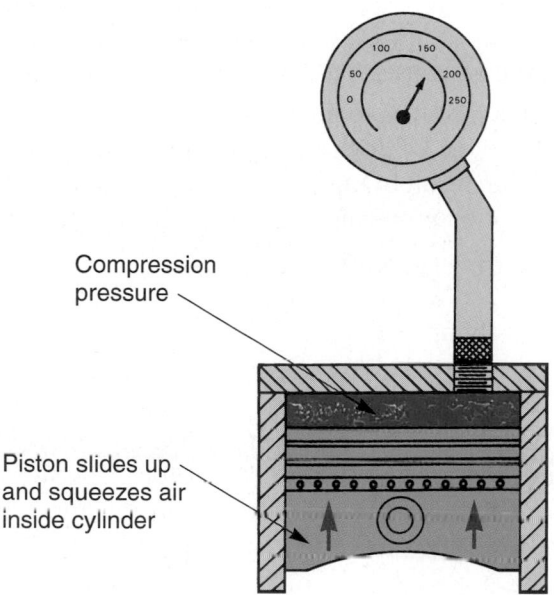

Compression pressure

Piston slides up and squeezes air inside cylinder

Figure 6-21. A technician must frequently measure pressure. This example shows the pressure developed during the engine compression stroke. If the pressure is not high enough, engine mechanical problems are indicated.

Vacuum Gauge

A *vacuum gauge* is used to measure negative pressure, or vacuum. It is similar to a pressure gauge. However, the gauge reads in inches of mercury (in./hg.) or metric kilograms per square centimeter (kg/cm^2). For example, a vacuum gauge is used to measure the vacuum in an engine's intake manifold. If the reading is low or fluctuating, it may indicate an engine problem.

Using Basic Mathematics

Automotive technicians often use mathematics during the service and repair of vehicles. Technicians must be able to do the four basic math operations: addition, subtraction, multiplication, and division. They must also be able to work with fractions and decimals.

Math skills are needed when working with specifications. For example, if you are working on an engine, you may need to measure piston diameter and cylinder diameter. By subtracting piston diameter from cylinder diameter, you would find piston clearance. This math must be done properly if the engine is to run normally after repairs.

You will also use math when filling out work orders. You must calculate part prices, labor charges, tax percentage, and total charges.

The following section will quickly review basic math calculations.

Addition

Addition is the combining of two or more numbers to find the total quantity or number of something. The result of the addition process is called the *sum* or the *total*. A plus sign (+) is used to indicate that the numbers are to be added. Numbers to be added may be written two ways:

in a string: 5 + 3 + 4 = 12 *(sum)*
or
in a column: 5
                  3
                 + 4
                 12 *(sum)*

When there are large numbers or a long series of numbers, it is best to write them in a column so sums of 10 and over can be carried to the next column. Always start adding from the right-hand column so that sums exceeding 9 can be carried from that column to the next column to the left:

122	- - 2	*First, add the right-*
804	- - 4	*hand column.*
644	- - 4	
+829	- - 9	
	19	

$\quad$ 1 $\quad$ *Since the total is 19,*
- 2 - *place the "9" under-*
- 0 - *neath the right-hand* $\quad$ 1 - - $\quad$ *Now, add up the next*
- 4 - *column and add the* $\quad$ 8 - - $\quad$ *column to the left and*
- 2 - *"1" to the next* $\quad$ 6 - - $\quad$ *place the sum below*
99 *column. Add up the* $\quad$ 8 - - $\quad$ *that column.*
$\quad$ *2nd column.* $\quad$ **23**99 *(Answer is 2399)*

Addition is used in adding up the cost of parts and labor when preparing a customer's bill. If, for example, parts total $125, labor charges are $95, and tax is $8, the total bill would be $228.

Subtraction

Subtraction is taking away a certain quantity from another. The amount that is left after subtracting is called the *remainder* or the *difference.* The minus sign (–) indicates that the number to the right of it is to be subtracted. Subtraction problems might be written in two ways:

in a string: 495 – 125 = 370 *(remainder)*

or

in a column: 495
              – 125
               370 *(remainder)*

Subtraction might be used in determining a customer's bill. It might also be used by the technician to check the deductions made on a paycheck for taxes and social security. Suppose that a customer's bill totaled $253, but there had been a $25 deposit before the work was done. To determine the amount due, you would subtract $25 from $253 ($253 – $25).

$253   *Notice that we had to borrow 10 from the second*
– 25   *column, since 5 cannot be subtracted from 3.*
$228

Division

To find out how many times one number is contained in another, we use *division.* The division sign (÷) indicates that one number is to be divided by another. The number being divided is called the *dividend.* The number a dividend is divided by is called the *divisor.* The answer is called the *quotient.* A division problem can be written one of three ways:

in a string: 860 ÷ 10 = 86 *(quotient)*

$$\begin{array}{r} 86 \text{ (quotient)} \\ 10 \overline{)\,860} \end{array}$$

or: $\dfrac{860}{10}$ = 86 *(quotient)*

The technician must use division frequently in an automotive repair facility. For example, suppose that 10 fuel pumps had been ordered and placed in stock. The total bill for the pumps came to $860. What is the cost of each fuel pump?

The cost of each fuel pump is found by dividing $860 (total cost) by 10 (number of pumps). The cost of each pump is $86. This information would be used to determine what the customer would be charged for the pump.

When dividing, not all answers come out to full numbers. In such cases, a decimal point is placed to the right of the last number of the dividend. A decimal point is also placed in the answer directly above the decimal point in the dividend. One or more zeros may be added to the dividend, depending on how many places the decimal number must carried out. For example, suppose that the cost of the fuel pumps in the previous example came to $865 instead of $860.

$$\begin{array}{r} 86.50 \text{ (quotient—each pump cost \$86.50)} \\ 10 \overline{)\,865.00} \\ \underline{80} \\ 65 \\ \underline{60} \\ 50 \\ \underline{50} \\ 0 \end{array}$$

Multiplication

Multiplication is a shortcut for adding the same number over and over. Suppose that the number 15 were to be added 12 times: One could set up the problem as 15 + 15 + 15 + 15 + 15..., until there were 12 additions. It is faster, however, to multiply 15 by 12 once. The multiplication sign (×) indicates that numbers are to be multiplied. The result of multiplication is called the *product.* Multiplication problems can be written two ways:

in a string: 15 × 12 = 180 *(product)*

or

in a column: 15
              × 12
               180 *(product)*

Multiplication is often useful in the automotive field. Suppose that a customer purchased four new tires. The tires cost $104 each. Rather than adding the price of each tire individually, it is easier to multiply $104 by 4.

    104
  × 4
    416

The price for four tires would be $416.

Numbers of more than one digit used as multipliers are multiplied one digit at a time. The products for each multiplication are stacked and then added together. Suppose that the customer in the previous example purchased 41 tires at $104 each.

    104
  × 41
    104
   416
   4264 *(product)*

Note that the second product (416) is shifted one column to the left. This is done because the multiplier is actually 40, not 4. To help make this clear, mentally place a 0 after the 6 in the second product (104 × 40 = 4160).

Fractions and Decimal Fractions

Fractions and decimal fractions are used to represent a portion of a whole number.

Fractions are written as two numbers, one over the other or one beside the other:

$$\frac{4}{5} \text{ or } 4/5 \text{ } (\textit{The fraction is read as "four-fifths."})$$

The number below the line or after the slash is called the denominator. This number tells how many parts the whole is divided into; the number above the line or ahead of the slash tells how many parts are present in the fraction. This number is called the numerator. When reading a fraction, the top or first number is always read first; thus, read 12/32 as "twelve thirty seconds."

Decimal fractions also have a numerator and denominator. The denominator is always a multiple of 10. However, it is never written. A dot or period, called a decimal point, is used in its place. For example, 9/10 is written as 0.9 in decimal notation. The number of digits to the right of the decimal point tell what multiple of 10 the denominator is. Thus:

0.9 is 9/10 (*nine-tenths*)
0.09 is 9/100 (*nine-hundredths*)
0.009 is 9/1000 (*nine-thousandths*)
0.0009 is 9/10,000 (*nine ten-thousandths*)

Since decimal fractions are easier to work with than fractions, it is common to convert fractions to decimal fractions. This is especially true in the automotive service field. Very small measurements are given in thousandths of an inch. However, wrenches are still sized in fractions.

Decimal fractions are used for fine measurements, such as the exact size of machined engine parts. Often, the technician must use a micrometer to check a dimension, such as the diameter of a crankshaft journal or the runout on a brake rotor. Decimal fractions can be added, subtracted, multiplied, and divided in the same manner as whole numbers.

There are rules that must be remembered when working with decimal numbers. The first set of rules has to do with placement of zeros.

- A zero placed between a number and a decimal point changes the value of the number (.45 is not the same as .045).

- A zero placed to the right of a decimal number does not change the value of the number (.45 is the same as .450).

- A zero placed to the left of the decimal point does not change the value of the decimal number (.45 is the same as 0.45).

Addition and Subtraction of Decimals

The rules for addition and subtraction of decimal fractions are:

- Line up the decimal points in a column.

- The decimal point in the answer must be in the same position as the decimal point in the column.

- Since some decimal fractions will have more numbers to the right of the decimal point than others, you may fill in with zeros on the shorter numbers. This is optional.

Example:			
1.5	*could also be*		1.500
9.356	*written with*		9.356
3.62	*zeros in*		3.620
.96	*the blanks*		0.960
15.436			15.436

Multiplication and Division of Decimals

Multiplying decimal numbers is not much different than multiplying whole numbers. The rules explain how to deal with the decimal point.

- In setting up the problem, the decimal points do not need to be aligned.

- Multiply the two numbers, ignoring the decimal points.

- Count the total number of digits (places) to the right of the decimal points of both numbers. Starting from the right-hand digit, count to the left the same number of digits in the answer. Place the decimal point to the left of the last digit counted. Dividing decimals is also similar to dividing whole numbers. Several steps are involved.

- If neither the dividend nor divisor contain decimal points, but the division does not come out even:

  Place a decimal point to the right of the last number of the dividend. Add one or more zeros after the decimal and continue dividing to the number of decimal places necessary.

For example:
```
        7.71      Division carried out
    7 ) 54.00     two decimal places
        49
        50
        49
        10
         7
         3
```

- When the dividend has a decimal and the divisor does not:

    Divide as usual.

    Place a decimal point in the answer directly above the decimal point in the dividend. It will occur at the time that the division process moves past (to the right) of the decimal point.

    For example:

$$
\begin{array}{r}
2.01 \\
25\,\overline{)\,50.25} \\
\underline{50} \\
02 \\
\underline{\;0} \\
25 \\
\underline{25} \\
0
\end{array}
$$

- When the divisor has a decimal point:

    If the dividend does not have a decimal point, add one at the far right.

    If the dividend has a decimal point, move it one place to the right for each decimal place in the divisor. Move the decimal point in the divisor accordingly to the right. Use zeros as place holders, if necessary.

    Divide as usual.

    Place a decimal point in the answer directly above the relocated decimal point in the dividend. It will occur when the division process moves to the right past the decimal point.

    For example: $2.5\,\overline{)\,50.25}$

$$
\begin{array}{r}
20.1 \\
25.\,\overline{)\,502.5} \\
\underline{50} \\
02 \\
\underline{\;0} \\
25 \\
\underline{25} \\
0
\end{array}
$$

Summary

- As a vehicle is driven, its moving parts slowly wear out. With enough part wear, mechanical failures and performance problems result.
- Auto manufacturers give "specs" or specifications (measurements) for maximum wear limits and dimensions of specific parts.
- Our customary measuring system originated from sizes taken from parts of the human body.
- The metric (SI) measuring system uses a power of 10 for all basic units.

- A steel rule, also called scale, is frequently used to make low precision linear (straight-line) measurements.
- A dial caliper is a sliding caliper with a dial gauge attached.
- A micrometer, nicknamed a "mike," is commonly used when making very accurate measurements.
- Never drop or overtighten a micrometer. It is very delicate and its accuracy can be thrown off easily.
- A feeler gauge is used to measure small clearances or gaps between parts.
- A dial indicator will measure part movement in thousandths of an inch (hundredths of a millimeter).
- A torque wrench measures the amount of turning force applied to a fastener (bolt or nut).
- A pressure gauge is frequently used in the auto shop to measure air and fluid pressure in various systems and components.
- Automotive technicians use mathematics during servicing and repair of vehicles.

Important Terms

Measuring systems	Outside caliper
Customary measuring system	Inside caliper
Metric (SI) measuring system	Vernier caliper
	Micrometer
Conversion chart	Telescoping gauge
Decimal conversion chart	Hole gauge
Steel rule	Hole gauge
Scale	Feeler gauge
Customary rule	Dial indicator
Metric rule	Thermometers
Pocket rule	Torque wrench
Combination square	Pressure gauge
Tape measure	Vacuum gauge
Yardstick	Addition
Meterstick	Subtraction
Dividers	Division
	Multiplication

Review Questions—Chapter 6

Please do not write in this text. Place your answers on a separate sheet of paper.

1. The two measuring systems are the _____ _____ _____, also called the _____ system and the _____ _____.

2. Parts of the human body are used as the basis for the customary measuring system. True or False?

3. The metric system uses a power of 16 for all basic units. True or False?

4. Which of the following is *not* a metric value?
   (A) Decimeter.
   (B) Octimeter.
   (C) Millimeter.
   (D) Meter.

5. What is a measuring system conversion chart?

6. A decimal system conversion chart is used to _____ and find equal values for _____,_____, and _____.

7. Describe the four steps for reading a customary outside micrometer.

8. Describe the four steps for reading a metric outside micrometer.

9. Which of the following is *not* a special micrometer used in auto technology?
   (A) Inside micrometer.
   (B) Depth micrometer.
   (C) Width micrometer.
   (D) All of the above.

10. Describe the differences between a flat feeler gauge and a wire feeler gauge.

11. A dial indicator will measure part _____ in thousandths of an inch or _____ of a millimeter.

12. List the four basic rules for measuring with a dial indicator.

13. The three types of torque wrenches are the _____ _____, _____ _____, and _____ _____.

14. For measuring purposes, the circle is divided into 720 parts, called degrees. True or False?

15. Explain the use of a vacuum gauge.

16. Note the inch rule below and give the measurement by full inches and fractions of an inch.

17. Give the micrometer reading for the micrometer scale shown below.

18. An automotive piston's connecting rod is 8″ long. What is the approximate length of this connecting rod in millimeters?
    (A) 42 mm.
    (B) 173.6 mm.
    (C) 203.2 mm.
    (D) 307.8 mm.

19. Which of the following instruments measures in pounds per square inch (psi)?
    (A) Pressure gauge.
    (B) Flat feeler gauge.
    (C) Depth micrometer.
    (D) Combination square.

20. When using a vacuum gauge, Technician A states that the gauge reads in inches of mercury. Technician B says that the gauge reads in kilograms per square centimeter. Who is right?
    (A) A only.
    (B) B only.
    (C) Both A and B.
    (D) Neither A nor B.

⬤ ASE-Type Questions

1. "Specs" are:
   (A) *dimensions of parts.*
   (B) *maximum wear limits of parts.*
   (C) *measurements of specific parts.*
   (D) *All of the above.*

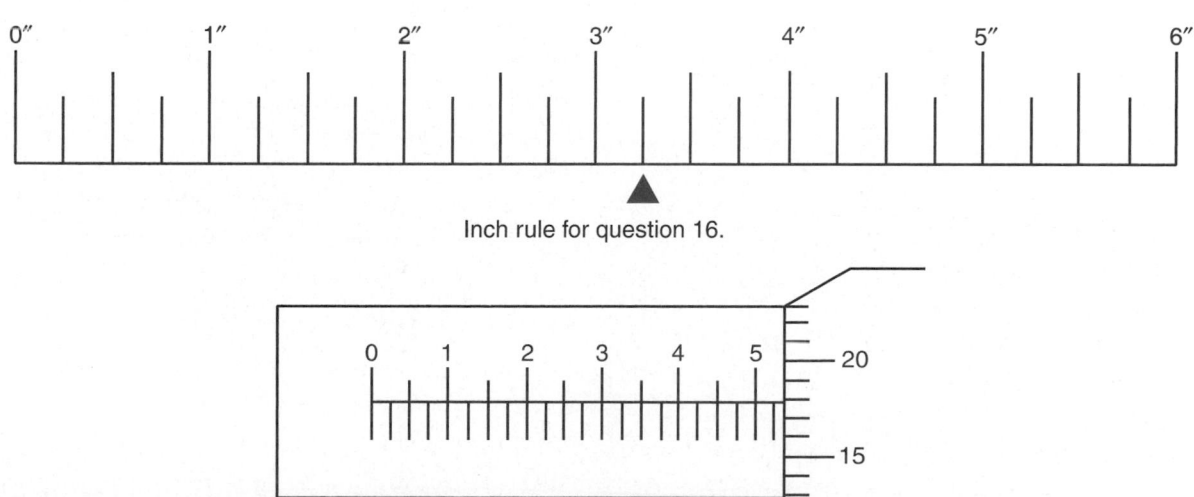

Inch rule for question 16.

Micrometer reading for question 17.

2. Technician A says the customary measuring system originated from sizes taken from parts of the human body. Technician B says customary units were derived from the landscape. Who is right?
   (A) A only.
   (B) B only.
   (C) Both A and B.
   (D) Neither A nor B.

3. Which of the following is used to make low-precision linear measurements?
   (A) Steel rule.
   (B) Pocket scale.
   (C) Inside caliper.
   (D) Combination square.

4. Which of the following measuring devices has a gauge attached which makes measurements easier to read?
   (A) Divider.
   (B) Dial caliper.
   (C) Sliding caliper.
   (D) Combination square.

5. Volume can be measured in each of these except:
   (A) liters.
   (B) quarts.
   (C) kilopascals.
   (D) cubic inches.

6. When using a decimal conversion chart, fractions are only accurate to about:
   (A) 1/16 of an inch.
   (B) 1/64 of an inch.
   (C) 1 centimeter.
   (D) 0.025 millimeters.

7. Each number on a customary micrometer equals:
   (A) 0.100"
   (B) 0.200"
   (C) 0.300"
   (D) 0.400"

8. Which of the following is not a good rule to remember when using a micrometer?
   (A) Never drop or overtighten a micrometer.
   (B) Hold the micrometer squarely with the work.
   (C) Grasp frame with fingers and turn with palm.
   (D) Put a little oil on a micrometer when in storage.

9. This type of gauge is used to measure the distance between unparallel or curved surfaces:
   (A) flat feeler gauge.
   (B) wire feeler gauge.
   (C) telescoping gauge.
   (D) spring gauge.

10. A dial indicator measures part movement in:
    (A) kilopascals.
    (B) millimeters.
    (C) hundredths of an inch.
    (D) thousandths of an inch.

Activities—Chapter 6

1. Demonstrate the proper reading of a dial indicator.

2. Use the micrometer to measure a part supplied by your instructor.

3. Demonstrate to the class the use of a feeler gauge.

Chapter 7

Using Service Information

After studying this chapter, you will be able to:

- Describe the different types of service manuals.
- Find and use the service manual index and contents sections.
- Explain the different kinds of information and illustrations used in a service manual.
- Describe the three basic types of troubleshooting charts found in service manuals.
- Explain how to use computer-based service information.
- Correctly answer ASE certification test questions concerning service information.

Vehicles contain thousands of parts. Many of these parts are assembled to close tolerances and require precise assembly and adjustment. Sometimes a technician needs specific technical information to properly repair a vehicle. In these cases, the technician must refer to a service manual, CD-ROM, computer network, or other service information.

Figure 7-1. Service manuals will answer almost any repair question. The manuals are essential reference tools of automotive technicians. (Deere & Co.)

Service Manuals

Service manuals, also called *shop manuals,* are books with detailed information on how to repair a vehicle. They have step-by-step procedures, specifications, diagrams, part illustrations, and other data for each vehicle model. Every service facility normally has a set of service manuals. They help technicians with difficult repairs. Refer to **Figure 7-1.**

Service manuals are written in very concise, technical language. They are designed to be used by well-trained technicians. A service manual is one of a technician's *most important tools.*

Service Manual Types

There are various types of service manuals. These include manufacturer's manuals, specialized manuals, and general repair manuals. It is important to understand the differences between each type.

Manufacturer's manuals, also called *factory manuals,* are published by a vehicle's manufacturer. Each manual covers a specific vehicle produced by that company during a given model year.

Specialized manuals cover only specific repair areas. They usually come in several volumes, each covering one section of the vehicle. One may cover engines and another body components or electrical systems. Specialized manuals are published by vehicle manufacturers or aftermarket companies. Aftermarket companies are suppliers other than a vehicle manufacturer.

General repair manuals are published by companies other than the major vehicle makers. Some of these companies include Mitchell Manuals, Motor Manuals, and

Chilton Manuals. General repair manuals are like manufacturer's manuals, but are usually *not* as detailed. They may include data on all American cars produced for several years. Other general repair manuals only cover foreign cars, light trucks, or large trucks. It is often too costly for a service facility to buy service manuals from every vehicle manufacturer. Instead, they may buy two or three general repair manuals for all types of vehicles. These manuals summarize the most important and most needed information.

Service Manual Sections

A service manual is divided into sections, such as general information, engine, transmission, and electrical. See **Figure 7-2.** To effectively use a service manual, you need to understand these sections and how they are organized.

The **general information section** of a service manual helps you with a vehicle's identification, basic maintenance, lubrication, and other general subjects. An important topic in this section is the **vehicle identification number (VIN).** The VIN provides data about the car. It is commonly used when ordering parts. The number, which is usually found on a plate located on the vehicles dashboard, contains a code. The manual explains what each part of this number code means. Look at **Figure 7-3.** The VIN tells you engine type, transmission type, and other useful information.

The **repair sections** of a service manual cover the vehicle's major systems. These sections explain how to recognize and diagnose problems and inspect, test, and repair each system. One page may describe how to remove the engine. Another page might explain how to disassemble the engine. Specifications such as bolt tightening limits, capacities, clearances, and operating temperatures are given in the repair sections. They are commonly used during service and repair operations. The repair section also refers to special tools that are needed for a limited number of repair tasks. These tools may be pictured at the end of the manual section. Refer to **Figure 7-4.**

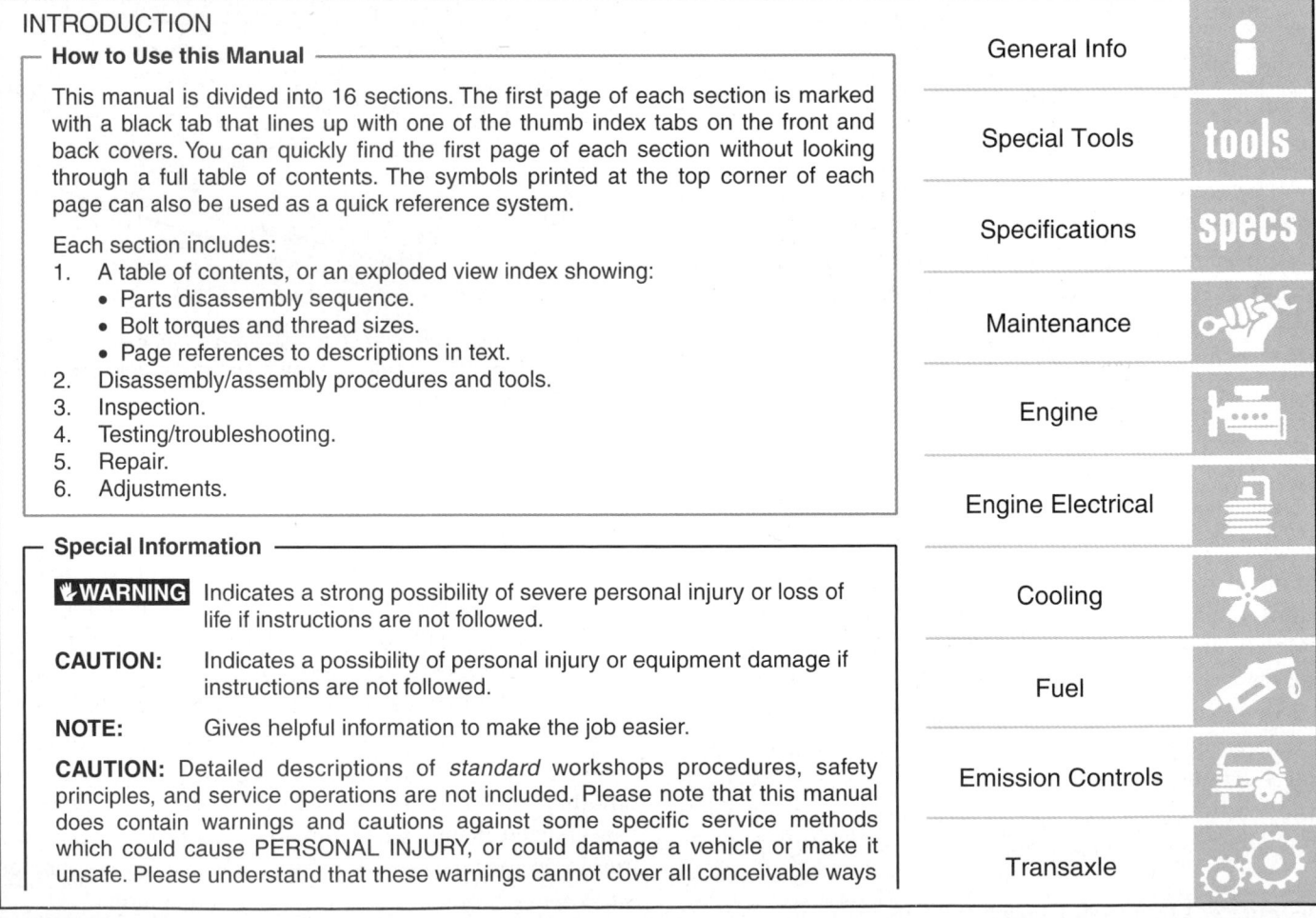

Figure 7-2. A service manual is divided into several repair sections. Be sure to read the introduction and any special information. (Honda)

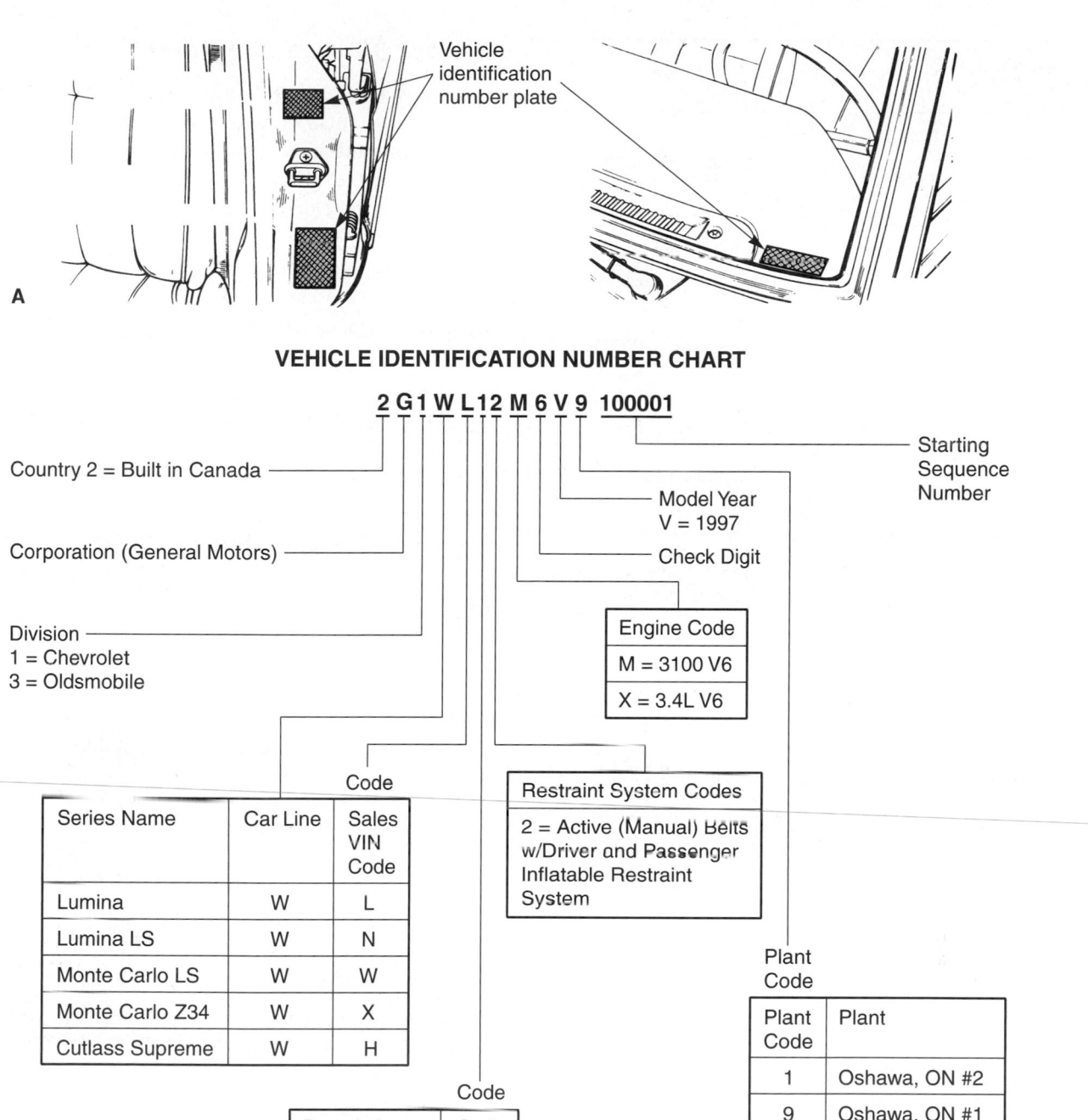

Figure 7-3. A—The vehicle identification number (VIN) can be located on the door, on the dashboard, or in the engine compartment. B—The VIN is a code. A service manual will explain the code, as shown. (Subaru and General Motors)

Service Manual Illustrations

Various types of *service illustrations* are used to supplement the written information in a service manual. Some show how to measure part wear, while others show how to install a part. Others show an exploded, or disassembled, view of parts. When using a service manual, you will find the illustrations essential for a full understanding of the procedures and specifications. They may show you what parts look like, how they fit together, where leaks might occur, or how a part works. **Figure 7-5** shows some common types of service manual illustrations.

Tool Number & Description	Illustration
49-0813-310 Centering tool, clutch disc	
49-0500-330 Installer, transition bearing	
49-0259-440 Turning holder, mainshaft	
49-0862-350 Guide, shift fork assembly	

Figure 7-4. The service manual will explain any special tool numbers. Note these special tools for clutch and transmission repairs. (Mazda)

Service Manual Diagrams

Diagrams are drawings used when working with electrical circuits, vacuum hoses, and hydraulic circuits. They represent how wires, hoses, passages, and parts connect together. *Wiring diagrams* show how the wiring connects to the electrical components. See **Figure 7-5C.** This subject is covered later in the text.

Vacuum diagrams help the technician determine how vacuum hoses connect to the engine and vacuum-operated devices, **Figure 7-5D.** *Hydraulic diagrams* show how a fluid, usually oil, flows in a circuit or part. They are helpful in understanding how a component operates or how to troubleshoot problems. Hydraulic diagrams are commonly given for automatic transmissions and power steering systems.

Service Manual Abbreviations

Abbreviations are letters that stand for an entire word. They are often used in service manuals. Sometimes,

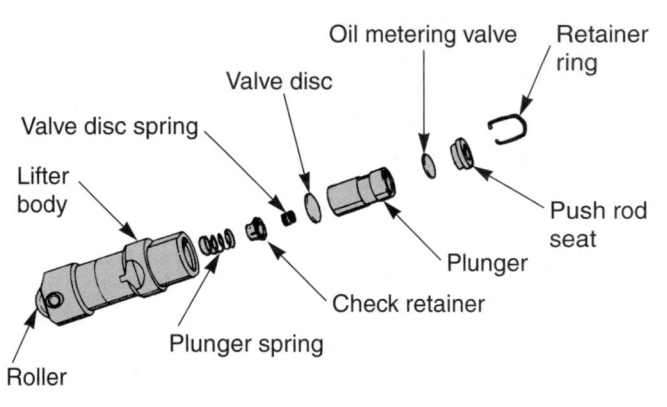

A

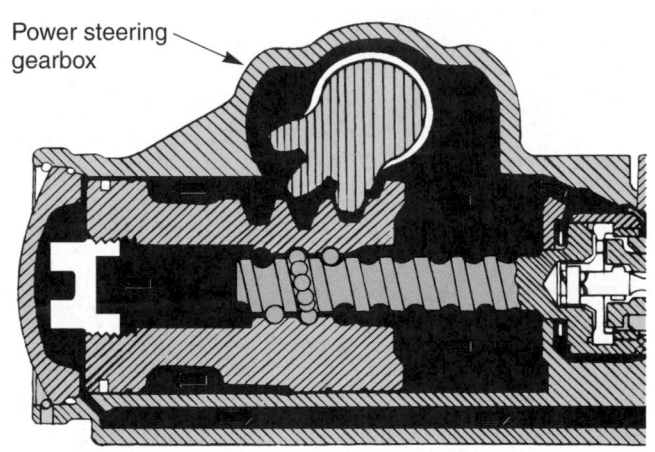

B

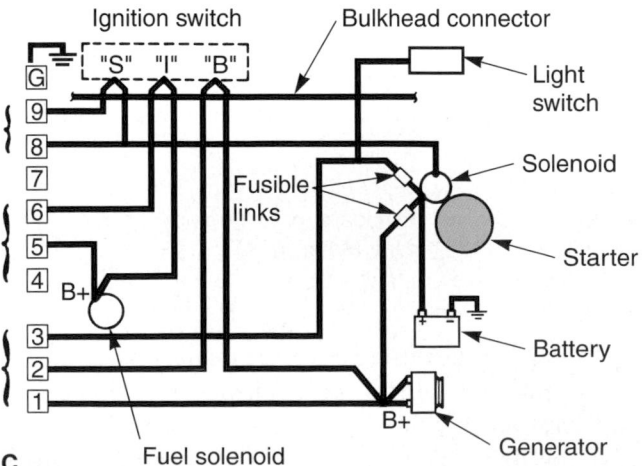

C

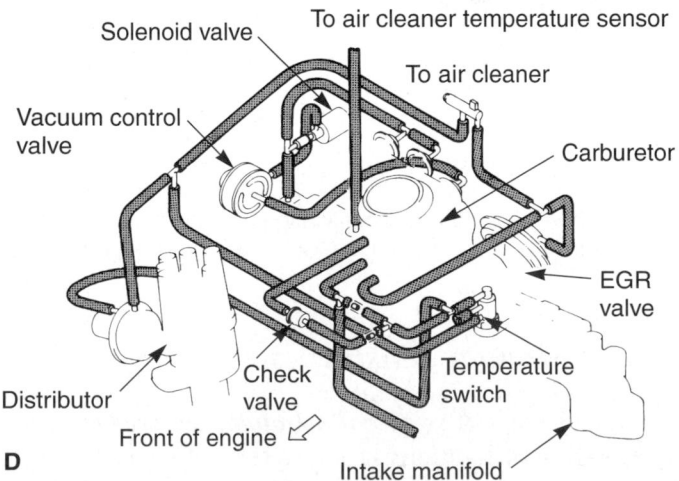

D

Figure 7-5. Typical service manual illustrations. A—An exploded view shows how parts fit together. B—An operational illustration shows how parts function. C—A wiring diagram shows how wires connect to components. D—A vacuum diagram shows how hoses connect to components. (Oldsmobile, Buick, and Subaru)

abbreviations are explained as soon as they are used. They may also be explained at the front or rear of the manual in a chart. This textbook uses only universally accepted abbreviations. It does *not* use abbreviations that only apply to one manufacturer. **Figure 7-6** gives some of the abbreviations recommended by SAE International.

Term	Abbrevations	Term	Abbrevations
Accelerator pedal	AP	Ignition control	IC
Air cleaner	ACL	Ignition control module	ICM
Air conditioning	A/C	Inertia fuel shutoff	IFS
Automatic transaxle	A/T	Intake air	IA
Automatic transmission	A/T	Intake air temperature	IAT
Barometric pressure	BARO	Knock sensor	KS
Battery positive voltage	B+	Malfunction indicator lamp	MIL
Camshaft position	CMP	Manifold absolute pressure	MAP
Carburetor	CARB	Manifold differential pressure	MDP
Charge air cooler	CAC	Manifold surface temperature	MST
Closed loop	CL	Manifold vacuum zone	MVZ
Closed throttle position	CTP	Mass airflow	MAF
Clutch pedal position	CPP	Mixture control	MC
Continuous fuel injection	CFI	Multiport fuel injection	MFI
Continuous trap oxider	CTOX	Nonvolatile random access memory	NVRAM
Crankshaft position	CKP	On-board diagnostic	OBD
Data link connector	DLC	Open loop	OL
Diagnostic test mode	DTM	Oxidation catalytic converter	OC
Diagnostic trouble code	DTC	Oxygen sensor	O2S
Direct fuel injection	DFI	Park/neutral position	PNP
Distributor ignition	DI	Periodic trap oxidizer	PTOX
Early fuel evaporation	EFE	Positive crankcase ventilation	PCV
EGR temperature	EGRT	Power steering pressure	PSP
Electrically erasable programmable read only memory	EEPROM	Powertrain control module	PCM
		Programmable road only memory	PROM
Electronic ignition	EI	Pulsed secondary air injection	PAIR
Engine control	EC	Random access memory	RAM
Engine control module	ECM	Read only memory	ROM
Engine coolant level	ECL	Relay module	RM
Engine coolant temperature	ECT	Scan tool	ST
Engine modification	EM	Secondary air injection	AIR
Engine speed	RPM	Sequential multiport fuel injection	SFI
Erasable programmable read only memory	EPROM	Service reminder indicator	SRI
		Smoke puff limiter	SPL
Evaporative emission	EVAP	Supercharger	SC
Exhaust gas recirculation	EGR	Supercharger bypass	SCB
Fan control	FC	System readiness test	SRT
Flash electrically erasable programmable read only memory	FEEPROM	Thermal vacuum valve	TVV
		Third gear	3GR
Flash erasable programmable read only memory	FEPROM	Three way + oxidation catalytic converter	TWC+OC
		Three way catalytic converter	TWC
Flexible fuel	FF	Throttle body	TB
Fourth gear	4GR	Throttle body fuel injection	TBI
Fuel level sensor	-----	Throttle position	TP
Fuel pressure	-----	Torque converter clutch	TCC
Fuel pump	FP	Transmisson control module	TCM
Fuel trim	FT	Transmission range	TR
Generator	GEN	Turbocharger	TC
Governor	-----	Vehicle speed sensor	VSS
Governor control module	GCM	Voltage regulator	VR
Ground	GND	Volume airflow	VAF
Heated oxygen sensor	HO2S	Warm up oxidation catalytic converter	WU-OC
Idle air control	IAC	Warm up three way catalytic converter	WU-TWC
Idle speed control	ISC	Wide open throttle	WOT

Figure 7-6. SAE-recommended abbreviations.

Diagnostic Charts

Diagnostic charts, or *troubleshooting charts,* give steps for finding and correcting problems in an automobile. These steps may include inspection, testing, measurement, and repair. If the source of the problem is hard to find, a diagnostic chart should be used. It will guide you to the most common causes of specific problems. There are four basic types of diagnostic charts. These are tree, block, illustrated, and component location.

A *tree diagnosis chart* provides a logical sequence for what should be inspected or tested when trying to solve a repair problem, **Figure 7-7.** For instance, if a horn will not work, the top of the tree chart may tell you to check the horn's fuse. Then, if the fuse is good, it may have you measure the voltage going to the horn. You can work your way down the "tree" until the problem is fixed.

A *block diagnosis chart* lists conditions, causes, and corrections in columns, **Figure 7-8.** The most common cause is the top listing. For example, if an engine overheats, loss of coolant appears at the top of the "causes" column. Check the coolant level. If this is the problem, fill the radiator and check for leaks. If the coolant level is OK, go to the next listing.

An *illustrated diagnosis chart* uses pictures, symbols, and words to guide the technician through a

sequence of tests. This type of troubleshooting chart is illustrated in **Figure 7-9.** If an engine oil pressure gauge shows low oil pressure, for example, the chart shows you exactly what to do, step by step, until the problem is corrected. This type of diagnosis chart not only tells you what to do, but it shows you how to do it.

A *component location chart* shows where various parts are located on the vehicle. This type of chart is often used for the engine compartment. The chart helps locate the numerous sensors, relays, fuses, and other components housed in this area. See **Figure 7-10.**

Using a Service Manual

To use a service manual, follow these basic steps:

1. Locate the proper service manual. Some manuals come in sets or volumes that cover different repair areas. Others cover all subjects and all car makes. If you are working on engines, find the manual that gives the most information for your type of engine.

2. Turn to the table of contents or the index to quickly find the needed information. *Never* thumb through a manual looking for a subject.

3. Use the page listings given at the beginning of each repair section. Most manuals have a small

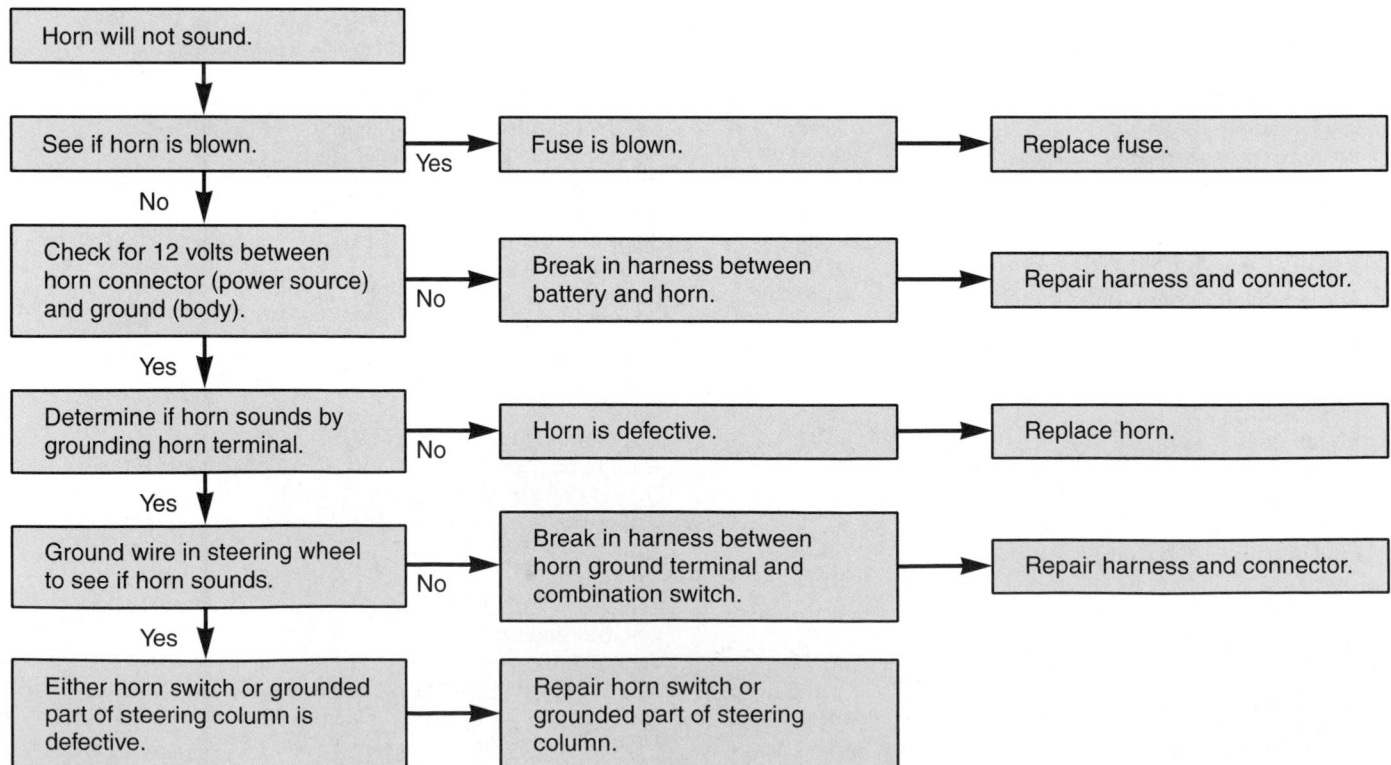

Figure 7-7. A tree diagnosis chart starts at the top and guides you through repair operations.

Condition	Possible Cause	Correction
• Loss of coolant.	• Pressure cap and gasket. • Exhaust leakage. • Internal leakage.	• Inspect, wash gasket, and test. Replace only if cap will not hold pressure test specifications. • Pressure test system. • Inspect hose, hose connections, radiator, edges of cooling system gaskets, core plugs, drain plugs, transmission oil cooler lines, water pump, heater system components. Repair or replace as required. • Check for obvious restrictions. • Check torque of head bolts. Retorque if necessary. • Disassemble engine as necessary— check for: cracked intake manifold, blown head gaskets, warped head or block gasket surfaces, cracked cylinder head, or engine block.
• Engine overheats.	• Low coolant level. • Loose fan belt. • Pressure cap. • Radiator or A/C condenser obstruction. • Closed thermostat. • Fan drive clutch. • Ignition. • Temperature gauge or cold light. • Engine. • Exhaust system.	• Fill as required. Check for coolant loss. • Adjust. • Test. Replace if necessary. • Remove bugs and leaves. • Test. Replace if necessary. • Test. Replace if necessary. • Check timing and advance. Adjust as required. • Check electrical circuits and repair as required. • Check water pump and block for blockage. • Check for restrictions.
• Engine fails to reach normal operating temperature.	• Thermostat stuck open. • Temperature gauge or cold light inoperative.	• Test. Replace if necessary. • Check electrical circuits and repair as required. Refer to electrical section.

Figure 7-8. A block diagnosis chart lists conditions, causes, and corrections in columns. Read to the right to match a condition with possible causes and corrections. (Ford)

table of contents at the beginning of each section.

4. Read the procedures carefully. A service manual gives highly detailed instructions. You must *not* overlook any step or the repair may fail.

5. Study the manual illustrations closely. The pictures in a service manual contain essential information. They cover special tools, procedures, torque values, and other data essential to the repair.

Tech Tip

Service manual layouts and features vary. The best way to learn their use is to practice finding information in them. For example, on your lunch hour, look up specific service information in several types of manuals. Compare the organization and differences in each manual. This will help you find data more quickly when on the job.

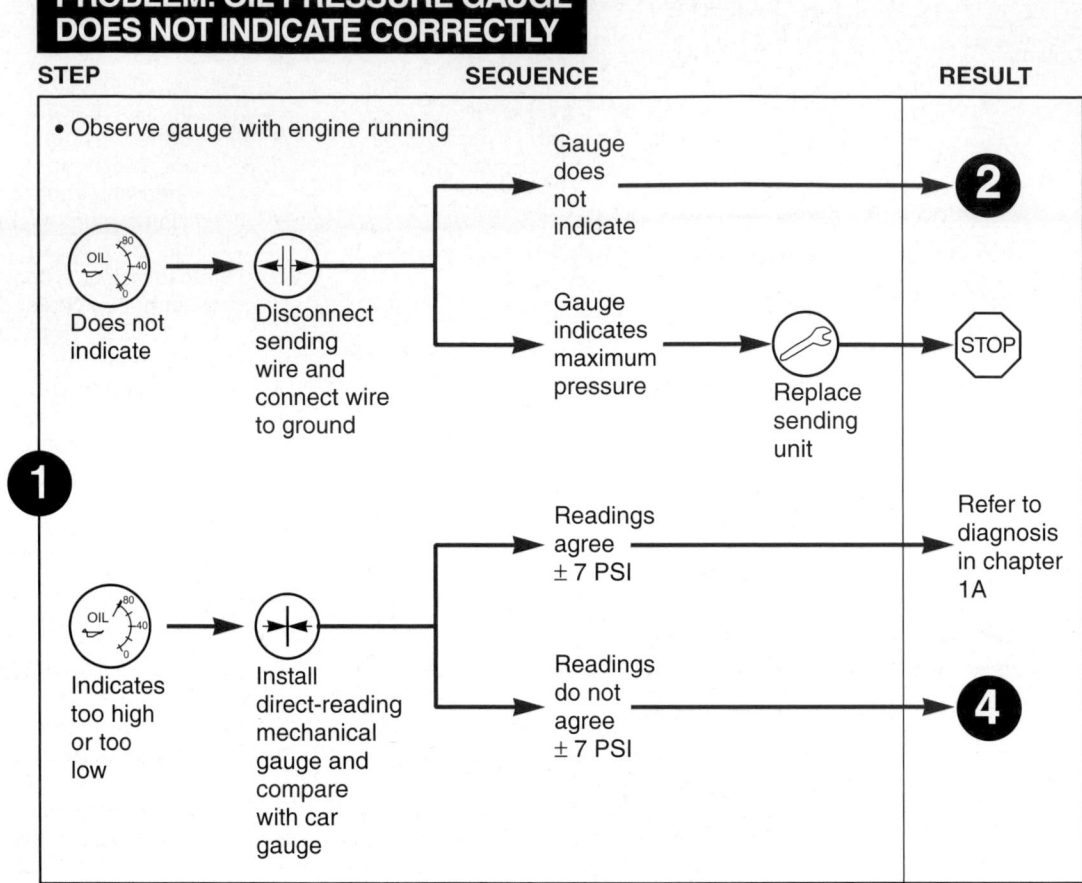

PROBLEM: OIL PRESSURE GAUGE DOES NOT INDICATE CORRECTLY

STEP   SEQUENCE   RESULT

- Observe gauge with engine running

Does not indicate → Disconnect sending wire and connect wire to ground → Gauge does not indicate → **2**

Gauge indicates maximum pressure → Replace sending unit → STOP

Indicates too high or too low → Install direct-reading mechanical gauge and compare with car gauge → Readings agree ± 7 PSI → Refer to diagnosis in chapter 1A

Readings do not agree ± 7 PSI → **4**

Figure 7-9. An illustrated diagnosis chart uses small illustrations and symbols to show how to find and correct a problem.

Service Publications

A service manual is just one kind of book that contains technical information on a vehicle. Other types, called *service publications,* include owner's manuals, flat rate manuals, and technical bulletins.

Owner's Manual

An *owner's manual* is a small booklet given to the purchaser of a new vehicle. It contains basic information on starting the engine, maintaining the car, and operating vehicle accessories.

Flat Rate Manual

A *flat rate manual* is used to calculate how much labor to charge the customer for a repair. It contains an estimate of how much time a specific repair should take. This time can then be multiplied by the shop's hourly labor rate to find the labor charge in dollars. Using the flat rate manual, you will be able to give the customer a cost estimate before beginning the actual repair.

Technical Bulletins

Technical bulletins help the technician stay up-to-date with recent technical changes, repair problems, and other service-related information. Usually only a few pages long, these publications are mailed to the service manager, who passes them along to the technicians. Technical bulletins are published by auto manufacturers and equipment suppliers.

 **Tech Tip**
Technical bulletins often describe common troubles with certain makes of vehicles. This allows you to check these common problems first.

Computerized Service Data

Computerized service data is information stored or retrieved electronically using a personal computer, **Figure 7-11.** Information may be stored on a floppy disk, hard drive, CD-ROM, or a computer network. A computer can find and retrieve this information quicker than a technician using a book. Modern repair shops are using more and more computerized service data.

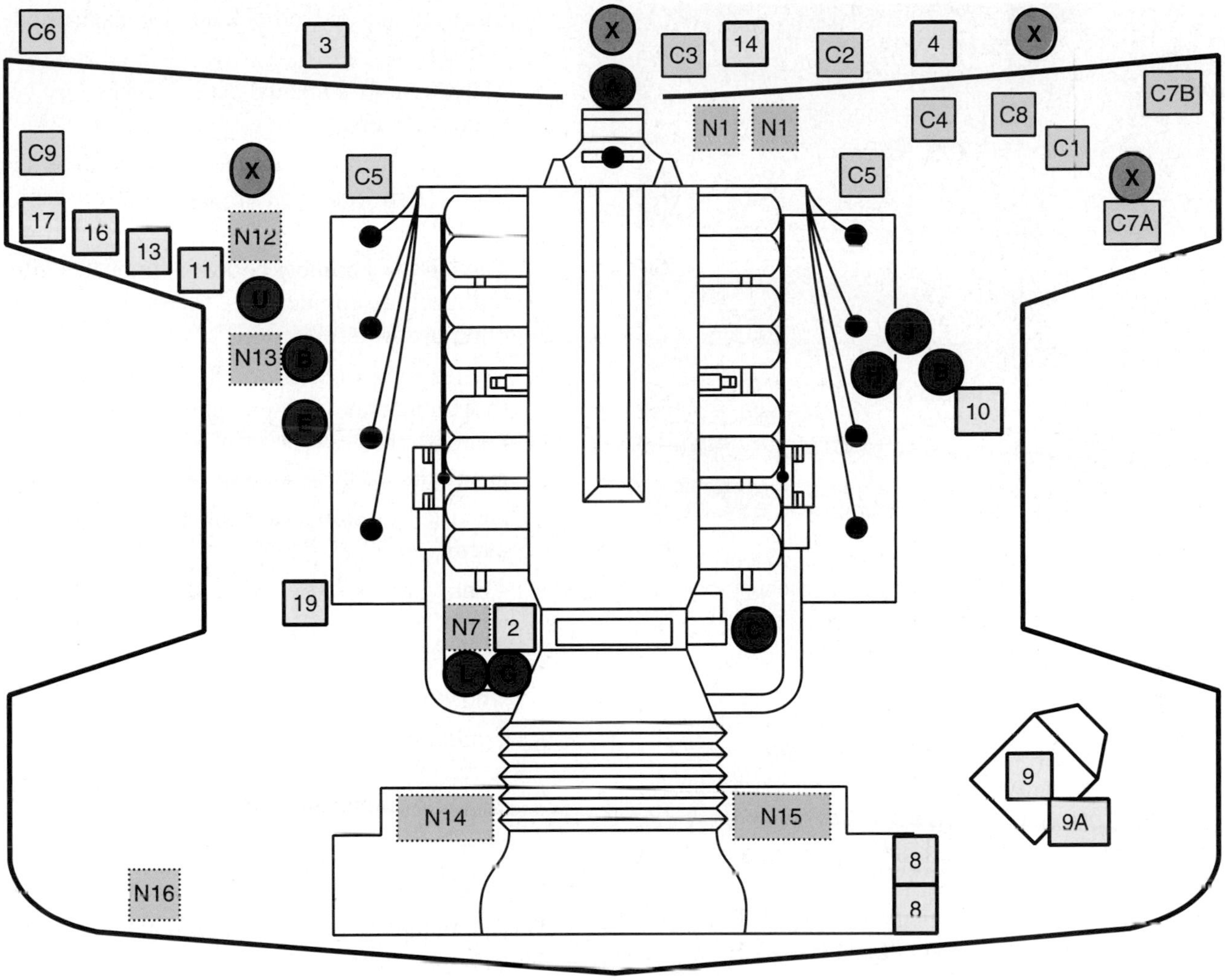

Computer Harness

C1   Engine Control Module (ECM)
C2   Data Link Connector (DLC)
C3   Malfunction Indicator Lamp (MIL)
C4   Electronic Ignition diagnostic
       connector
C5   ECM harness grounds
C6   I/P fuse panel
C7A Underhood fuse block
C7B Underhood fuse block
C8   Fuel pump test connector
C9   TP sensor interface module

Not ECM Connected

N1   Crankcase ventilation valve
N7   Oil pressure sensor gauge
N12 A/C pressure cycling switch
N13 A/C high pressure cycling switch
N14 Secondary cooling fan (FAN 2)
N15 Primary cooling fan (FAN 1)
N16 Secondary air inlet valve electric
       vacuum pump

Controlled Devices

2   Idle Air Control (IAC) valve
3   Fuel Pump (FP) relay (primary)
4   Fuel Pump (FP) relay (secondary)
8   Cooling fan relay(s)
9   Secondary Air Injection (AIR) pump
9A   Air pump relay
10   Air bypass valve
11   2nd & 3rd gear block out solenoid
13   A/C clutch relay
14   2nd & 3rd gear block out solenoid
16   Secondary SFI control module #1
17   Secondary SFI control module #2
19   Linear EGR valve

Information Sensor

A   Manifold Absolute Pressure (MAP)
       sensor
B   Heated Oxygen Sensor (HO2S)
C   Throttle Position (TP) sensor
E   Crankshaft Position sensor
F   Vehicle Speed Sensor (VSS)
       (mounted on transmission, not
       shown)
G   Intake Air Temperature (IAT) sensor
H   Camshaft Position Sensor
J   Knock Sensor (KS)
L   Engine oil temperature sensor
U   A/C cooling fan switch

X   SIR System Components. Refer to
       section 9J of the Service Manual,
       for "Cautions" and information on
       SIR System Components.

Figure 7-10. Study the location of the parts shown in a component location chart. The service manual will often give this type of chart for the exact make and model car being serviced. (Chevrolet)

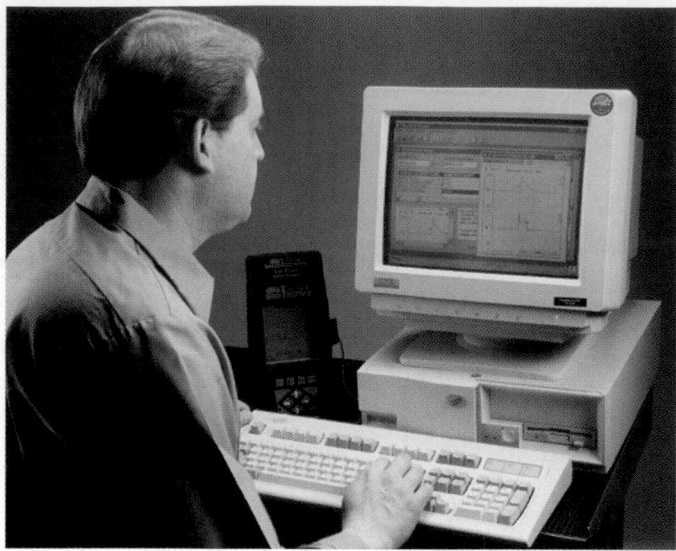

Figure 7-11. A computer is now a commonly used tool of many technicians. It can help with everything from ordering parts to troubleshooting hard-to-find problems. (OTC)

A floppy disk is an inexpensive way to hold computer information. However, the storage capacity is limited. A **CD-ROM,** or **compact disc,** is a computer storage device with tremendous capacity for holding large amounts of data. One CD can hold the information from a whole set of service manuals.

The **Internet** is a computer network that allows you to communicate with other computers all over the world. A **modem** is connected to the computer and used to access the Internet or other on-line service through a phone line. Various Internet sites are available for finding common vehicle troubles, manufacturer recalls, and new tools. On-line networks allow a service facility to find service information that is not available in the manuals or electronic media they have on hand.

Service facility computer systems can also store information on various items, such as:

- Repair procedures.
- Repair illustrations.
- Part prices.
- Labor times.
- Safety rules.
- Troubleshooting charts.
- Trouble code charts.

Summary

- Service manuals are books with detailed information on how to repair a vehicle. They have step-by-step procedures, specifications, diagrams, part

illustrations, and other data for each model of vehicle.
- Manufacturer's manuals are published by vehicle manufacturers.
- A service manual is divided into sections, such as general information, engine, transmission, and electrical.
- The VIN is a number code that indicates information such as engine type, transmission type, and other useful information.
- Various types of service illustrations are used to supplement the written information in a service manual.
- Diagrams are drawings used when working with electrical circuits, vacuum hoses, and hydraulic circuits.
- Abbreviations are letters that stand for entire words.
- A flat rate manual is used to calculate how much labor to charge the customer for a repair.
- Computerized service data is information stored or retrieved using a personal computer instead of a book or publication.

Important Terms

Service manuals	Component location
Manufacturer's manuals	chart
Specialized manuals	Service publications
General repair manuals	Owner's manual
Vehicle identification	Flat rate manual
number (VIN)	Technical bulletins
Service illustrations	Computerized service
Abbreviations	data
Diagnostic charts	

Review Questions—Chapter 7

Please do not write in this text. Place your answers on a separate sheet of paper.

1. What is a service manual?
2. Which of the following is *not* a service manual containing information on car repairs?
   (A) Manufacturer's manual.
   (B) Owner's manual.
   (C) General repair manual.
   (D) All of the above.
3. Explain the purpose of the following.
   (A) Wiring diagrams.
   (B) Vacuum diagrams.
   (C) Hydraulic diagrams.

4. List ten common abbreviations and explain them.

5. The following is *not* a common type of diagnostic chart.
   (A) Track diagnosis chart.
   (B) Tree diagnosis chart.
   (C) Block diagnosis chart.
   (D) Illustrated diagnosis chart.

6. Write the five basic steps for using a service manual.

7. A _____ _____ manual is used to calculate how much labor to charge for a repair.

8. _____ _____ help the mechanic stay up-to-date with recent technical changes, repair problems, and other service related information.

9. Why has computerized service data become more common?

10. The computer network that allows you to communicate with computers all over the world is called the _____.
    (A) Modem
    (B) CD-ROM
    (C) Internet
    (D) None of the above

ASE-Type Questions

1. Each of the following is a type of service manual *except:*
   (A) general repair manuals.
   (B) owner's manuals.
   (C) factory manuals.
   (D) specialized manuals.

2. Technician A says VIN information is generally found in the repair section of a service manual. Technician B says VIN information is generally located in the general information section of a service manual. Who is right?
   (A) A only.
   (B) B only.
   (C) Both A and B.
   (D) Neither A nor B.

3. A technician who is having trouble finding the cause of a vehicle problem should refer to the following first.
   (A) Service bulletin.
   (B) Troubleshooting chart.
   (C) Spec sheet.
   (D) Repair procedures.

4. Which of the following should be used when calculating the cost of a repair?
   (A) Troubleshooting chart.
   (B) Service bulletin.
   (C) Flat rate manual.
   (D) Cold rate manual.

5. Technician A says technical bulletins describe common problems encountered on specific vehicles. Technician B says technical bulletins are often published by independent repair facilities. Who is right?
   (A) A only.
   (B) B only.
   (C) Both A and B.
   (D) Neither A nor B.

Activities—Chapter 7

1. Obtain a flat rate manual and a parts catalog from your instructor and prepare a bill for replacement of a fuel pump on a vehicle of your choice. Check the manual for the amount of labor involved. Then, consult the parts catalog for the cost of the part. Add up the costs plus the state tax for your state. (Figure labor cost at $48/hour.)

2. Using a service manual, demonstrate to your class how to find the procedure for removal and replacement of a part chosen by your instructor.

Chapter 8

Basic Electricity and Electronics

After studying this chapter, you will be able to:

- Explain the principles of electricity.
- Describe the action of basic electric circuits.
- Compare voltage, current, and resistance.
- Describe the principles of magnetism and magnetic fields.
- Identify basic electric and electronic terms and components.
- Explain different kinds of automotive wiring.
- Perform fundamental electrical tests.
- Correctly answer ASE certification test questions that require a basic understanding of electricity and electronics.

Almost every system in a modern vehicle uses some type of electric or electronic part, or **component.** Electronic ignition systems, electronic fuel injection, computerized engine management systems, anti-lock brakes, and other advanced systems require technicians skilled in electricity and electronics. Even specialized technicians need this background to fix today's vehicles. This chapter covers the most important aspects of automotive electricity and electronics.

Electricity

Everything is made from fewer than 100 types of atoms. These atoms can be gold, carbon, lead, hydrogen, oxygen, or other elements. You are made of atoms. This book is made of atoms, so are your chair, your desk, and air—everything!

An **atom** basically consists of small particles called protons, neutrons, and electrons. Negatively charged **electrons** circle around a center of neutrons and positively-charged **protons.** See **Figure 8-1.** The makeup of atoms varies in different substances or elements.

Electricity is the movement of electrons from atom to

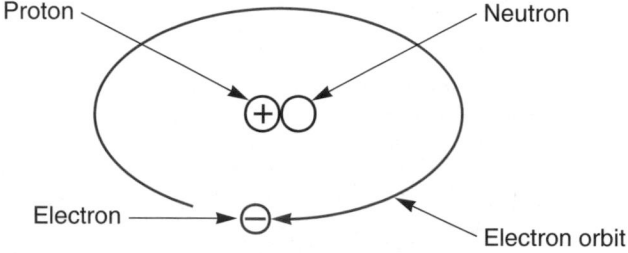

Figure 8-1. Atoms are made up of electrons, protons, and neutrons. Electrons are negatively charged. Protons are positively charged.

atom. Some substances have atoms that allow electrical flow; others do not. **Conductors,** such as wires and other metal objects, have atoms that allow the flow of electricity. The atoms contain free electrons. **Free electrons** are extra electrons in orbit of the atom that are not locked to protons.

Insulators, such as plastic, rubber, and ceramics, do *not* contain free electrons. They resist the flow of electricity. Wire conductors are usually covered with plastic or rubber as an insulating material.

The maximum number of electrons in the atom's **valence,** or outer, band is eight. An atom with eight valence electrons is normally a good insulator. Since the valence band is filled, it is in a state of equilibrium and is reluctant to give or take an electron from the valence band of a neighboring atom. Atoms with one to three electrons in their valence band are usually good conductors. These atoms contribute electrons to, and receive electrons from, neighboring atoms. Metals such as gold, silver, and copper have a single electron in their valence band and are excellent conductors.

Current, Voltage, and Resistance

The three basic elements of electricity are current, voltage, and resistance. Current is measured in amps.

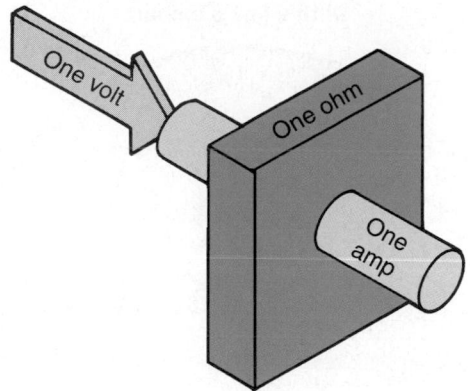

Figure 8-2. Voltage is a pressure, or pushing force. Current is the flow of electrons. Resistance opposes current flow. One volt can push one amp of current through one ohm of resistance.

Voltage is measured in volts. Resistance is measured in ohms. See **Figure 8-2.**

Current is the *flow* of electrons through a conductor. Just as water flows through a garden hose, electrons flow through a wire in a circuit. When current flows through a lightbulb, the electrons rub against the atoms in the bulb filament. This produces an "electrical friction." The friction heats the filament, making it glow. Current is abbreviated *I.*

Two theories have been used to explain how current flows through a circuit: the conventional theory and the electron theory. The *conventional theory* states that current flows from positive to negative. The *electron theory,* on the other hand, states that current flows from negative to positive. The electron theory is more widely accepted than the conventional theory. Therefore, it will be used in this text.

Voltage is the force, or *electrical pressure,* that causes electron flow. This is similar to how water pressure causes water to squirt out the end of a garden hose. An increase in voltage causes an increase in current. A decrease in voltage causes a decrease in current. Automobiles normally use a 12V electrical system. Voltage is abbreviated *V* or *E.*

Resistance is the *opposition* to current flow. Resistance is needed to control the flow of current in a circuit. Just as the on/off valve on a garden hose can be opened or closed to control water flow, circuit resistance can be increased or decreased to control the flow of electricity. High resistance reduces current. Low resistance increases current. Resistance is abbreviated *R.*

Simple Circuit

Figure 8-3 shows a simple circuit. A simple circuit consists of:

- A *power source,* such as a battery, alternator, or generator, which supplies electricity for circuit.
- A *load* which is an electrical device that uses electricity.
- *Conductors,* such as wires or metal parts, that carry current between power source and load.

The power source feeds electricity to the conductors and load. The conductors carry the electricity out to the load and back to the source. The load changes the electricity into another form of energy, such as light, heat, or movement.

Types of Circuits

A *series circuit* has more than one load connected in a single electrical path, **Figure 8-4A.** For example, inexpensive Christmas tree lights are often wired in series. With only one electrical path, if one bulb burns out, all

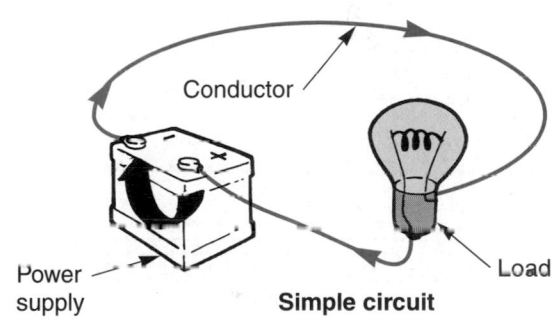

Figure 8-3. A simple electric circuit consists of a power source, a load, and a conductor. (British Leyland)

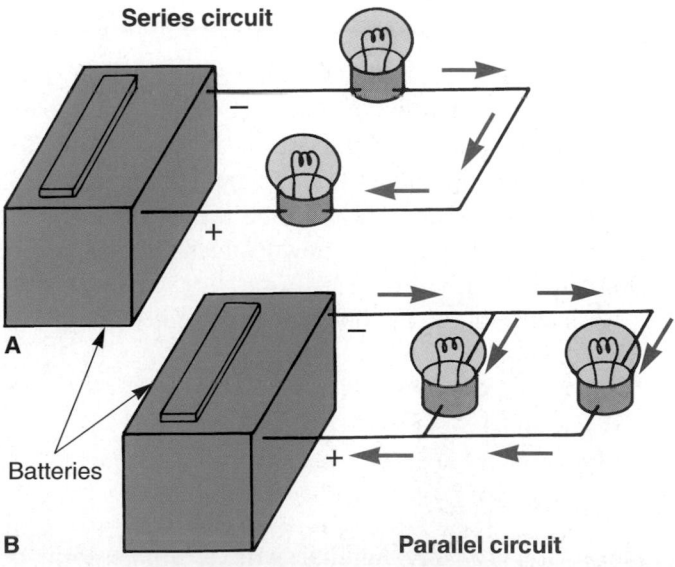

Figure 8-4. A—Series circuit only has one path for current. B—Parallel circuit has a separate path for each load.

the bulbs stop glowing. The circuit path is open, or broken, and current stops.

A *parallel circuit* has more than one electrical path, **Figure 8-4B.** When Christmas tree lights are wired in parallel, one bulb can burn out without affecting the others. The other bulbs have their own leg, or path, to receive current. A *series-parallel circuit* contains both a series circuit and a parallel circuit.

In a *one-wire circuit,* or *frame-ground circuit,* the vehicle's frame or body serves as an electrical conductor, **Figure 8-5.** A cable is used to connect the negative battery terminal to the frame. Insulated wires are used to connect the frame to the load's ground wire and the positive battery terminal to the load's hot wire. Current from the negative battery terminal travels through the vehicle's frame. From the frame, current travels through the load and into the positive battery terminal.

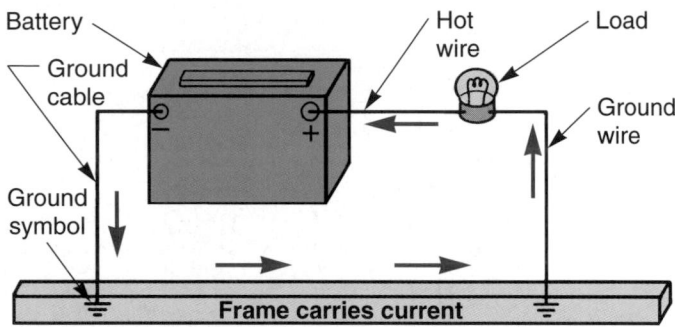

Figure 8-5. Automobile wiring commonly uses a frame ground. Metal parts of the vehicle carry current in the circuit. This reduces the number of wires needed.

Ohm's Law

Ohm's law is a simple formula for calculating volts, amps, or ohms when two of the three values are known. This is illustrated in **Figure 8-6.** If you know, for example, that a circuit has 12 volts applied and a current flow of 6 amps, Ohm's law can be used to find circuit resistance. Simply plug the known values into the correct formula.

Resistance = voltage divided by current

$$R = \frac{12 \text{ Volts}}{6 \text{ amps}}$$

R = 2 ohms

Magnetic Field

You are probably familiar with a simple form of magnetism—a refrigerator magnet. This is a permanent

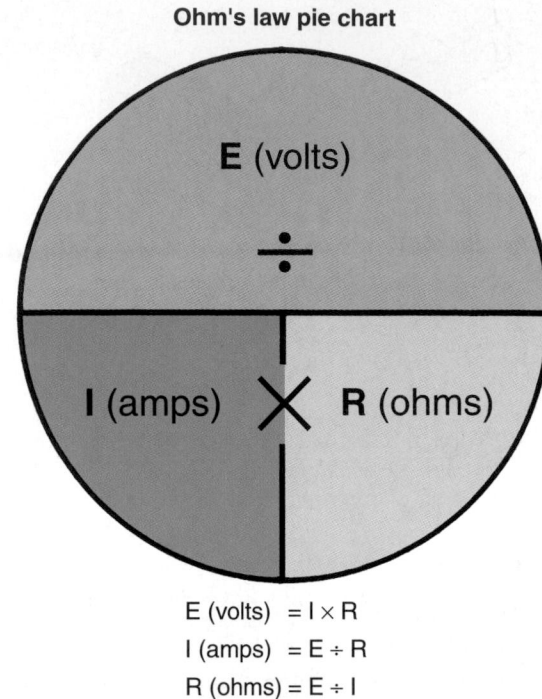

Ohm's law pie chart

E (volts) = I × R
I (amps) = E ÷ R
R (ohms) = E ÷ I

Figure 8-6. Ohm's law pie chart. Use your finger to cover one of the letters in the chart. This will show you the Ohm's law formula.

magnet. It produces an invisible *magnetic field* that attracts ferrous, or iron-containing, objects.

A magnetic field can also be created using electricity, **Figure 8-7A.** A long piece of wire can be wound into a coil. The ends of the wire can be connected to a battery or other source. Then, when current passes through the wire, a magnetic field is produced. To make the field, or lines of force, stronger, a soft iron bar can be inserted into the center of the coil. The iron core will become magnetized, making an electromagnet.

Magnetism can also create electricity. If a magnetic field is passed over a wire, an electric current will be generated in the wire. The wire cutting the lines of force causes a tiny amount of electricity to flow through the wire. Look at **Figure 8-7B.** This action is called *induction.*

Many automotive components use the characteristics of magnetism and magnetic fields. Electronic fuel injection, electric motors, relays, ignition systems, and on-board control modules are just a few examples.

Electrical Terms and Components

There are several electrical terms and components that auto technicians must know. Some of the most important ones are discussed here.

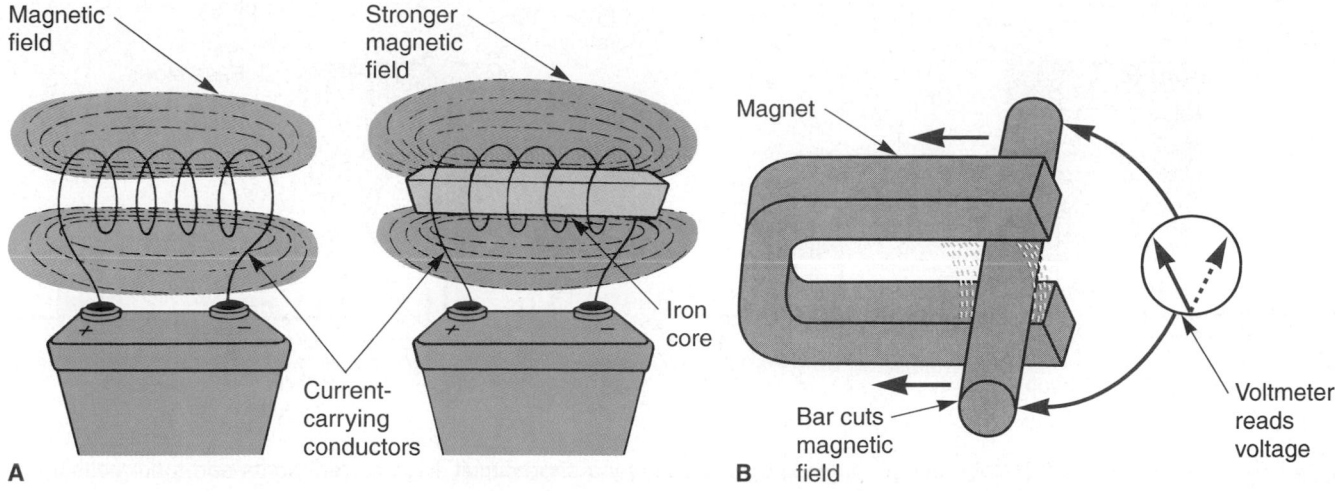

Figure 8-7. A—When current flows through a wire, a magnetic field forms around the wire. The wire can be wound into a coil to strengthen the field. An iron core strengthens the field even more. B—A magnetic field can be used to produce electricity. When an iron bar is moved through a magnet's field, current is induced in the bar and wire. (British Leyland and Deere & Co.)

A *switch* allows an electric circuit to be turned on or off. When the switch is *closed,* or on, the circuit is complete and operates. When the switch is *open,* or off, the circuit is broken and does not function. See **Figure 8-8.**

A *short circuit,* or "short," is an accidental low-resistance connection that results in excessive current flow. See **Figure 8-9.** If a short to ground exists between the battery and load, ***unlimited current flow*** can melt and burn the wire insulation.

A *fuse* protects a circuit against damage caused by a short circuit. The link in the fuse will melt and burn in half to stop excess current and further circuit damage, **Figure 8-10.** A *fuse box* is often located under the dashboard. It contains fuses for the various circuits, **Figure 8-11.**

A *fusible link,* or *fuse link,* is a small section of wire designed to burn in half when excess current is present in the circuit. It is often used as protection between the battery and main fuse box. If a major wire, like the starting

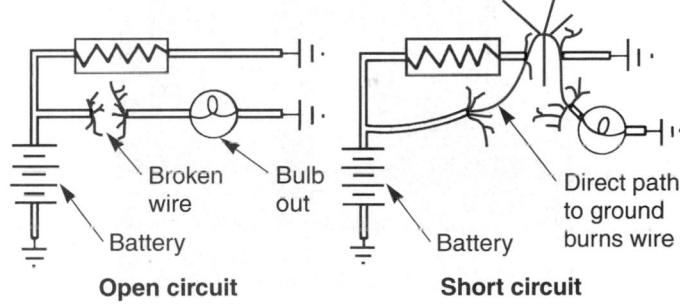

Figure 8-9. An open circuit has a break in a wire or an electric component. Current stops flowing through the circuit. A short circuit has a wire touching ground. A high amount of current flows through the short.

motor wire, is shorted, the fusible link will burn in half to prevent an electrical fire and further damage.

A *circuit breaker* performs the same function as a fuse. It disconnects the power source from the circuit when current becomes too high. See **Figure 8-12.** Normally, a circuit breaker will reset itself when current returns to a normal level.

A *relay* is an electrically operated switch. It allows a small dash switch to control another circuit from a distant point. It also allows very small wires to be used behind the dash, while large wires may be needed in the relay-operated circuit. Look at **Figure 8-13.**

Automotive Electronics

Some electrical components, such as relays and circuit breakers, use moving mechanical contacts. These

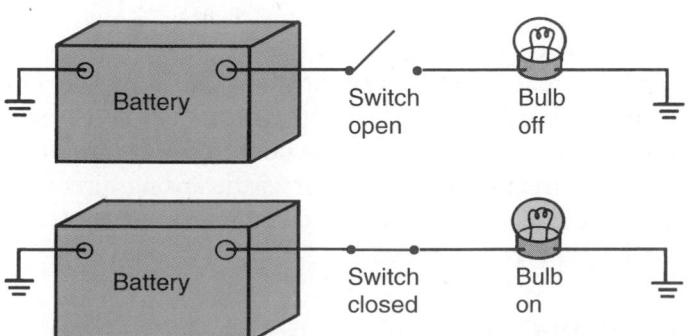

Figure 8-8. A switch is used to break (open) and complete (close) a circuit.

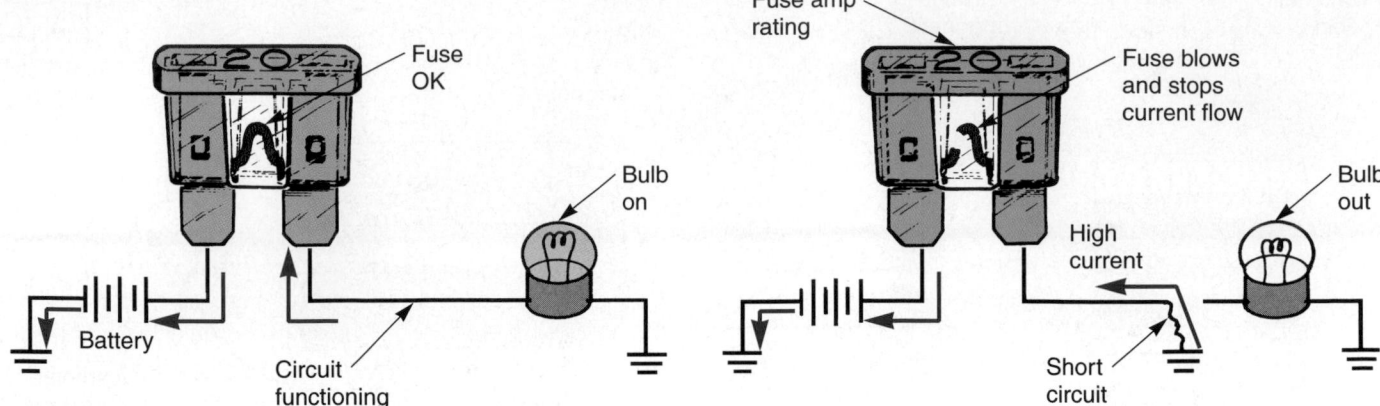

Figure 8-10. A fuse protects against damage that would be caused by a short circuit. High current heats and then melts the link, creating an open circuit. This stops the current flow in the circuit.

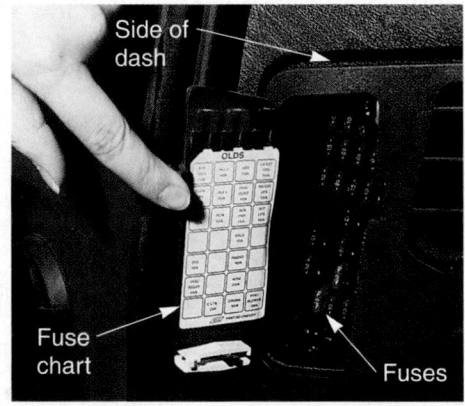

Figure 8-11. A fuse box is sometimes located in the side of the dash. Fuses are normally labeled with a name or circuit.

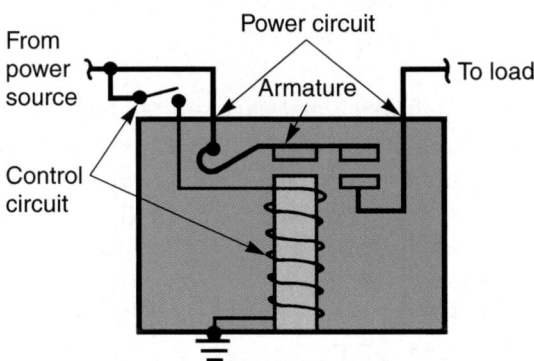

Figure 8-13. A relay is a remote-control switch. When current enters the control circuit, it creates a magnetic field that pulls the points closed. This completes the main circuit to a load. (Ford)

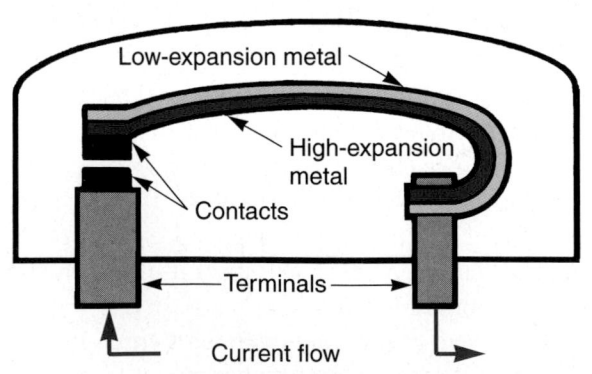

Figure 8-12. A circuit breaker performs the same function as a fuse. High current heats a bimetal strip, causing it to deform and open the contacts. This stops the current flow in the circuit. (Ford)

contact points can wear, burn, or pit. Also, mechanical parts are relatively slow in electrical components. In electronic systems, the components are solid state and do *not* have moving parts. Solid state circuits use semiconductors. A ***semiconductor*** is a special substance capable of acting as both a conductor and an insulator. This characteristic enables solid state devices to control current without mechanical points.

Diode

A ***diode*** is an "electronic check valve" that only allows current to flow in one direction. See **Figure 8-14.** When ***forward biased,*** current is entering in correct direction and the diode acts as a conductor. When ***reverse biased,*** current is trying to enter in the wrong direction and the diode acts as an insulator, stopping current from passing through the circuit.

Transistor

A ***transistor*** performs the same basic function as a relay. It acts as a remote control switch or current amplifier. However, it is much more efficient than a relay.

A transistor can sometimes turn on and off more than 200 times a second. It does this without using moving parts, which can wear and deteriorate. Look at **Figure 8-15.** A transistor **amplifies,** or increases, a small control or base current. The small base current energizes the semiconductor material, changing it from an insulator to a conductor. This allows the much larger circuit current to pass through the transistor.

Other Electronic Devices

Capacitors are devices used to absorb unwanted electrical pulses, such as voltage fluctuations, in a circuit. They are used in various types of electrical and electronic circuits. A capacitor is often connected to the supply wires going to a car radio. It absorbs any electrical voltage pulses from alternator or ignition system that may be heard in the radio speakers as buzzing.

An *integrated circuit (IC)* contains microscopic diodes, transistors, resistors, and capacitors in a wafer-like chip. This chip is a small plastic housing with metal terminals. See **Figure 8-16.** Integrated circuits are used in very complex electronic circuits.

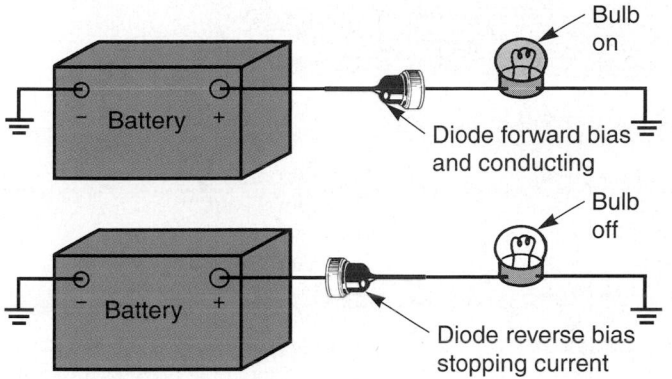

Figure 8-14. A diode allows current flow in only one direction. Diodes are used in wide range of electric and electronic circuits.

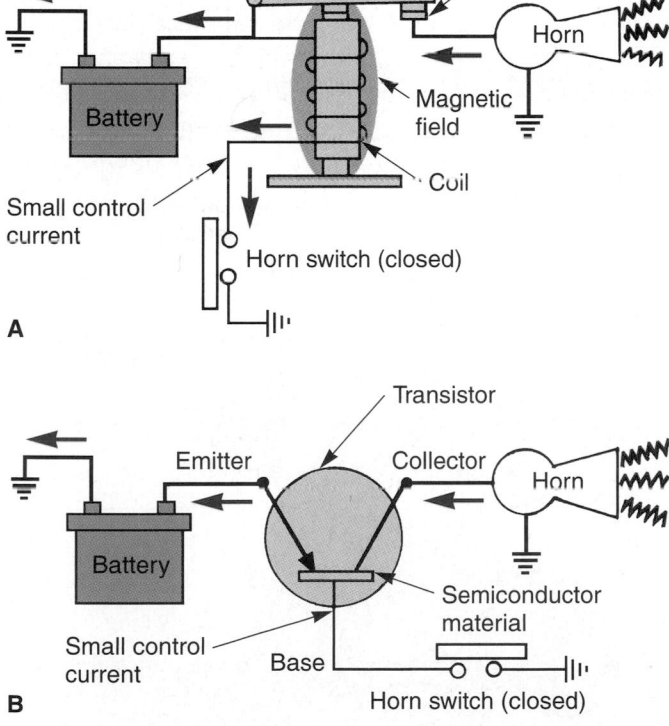

Figure 8-15. Basically, relays and transistors perform the same function. They allow a small control current to operate a larger current to a load. A—In a relay system, when the horn button is pressed, a small current enters the relay coil. The coil field attracts the point arm. Then, battery current can reach and operate the horn through the relay. B—In a transistor system, when the horn button is pressed, a small base current enters the transistor. This changes the semiconductor material in the transistor from an insulator to a conductor. Then, current can flow through the transistor and to the horn. (Echlin)

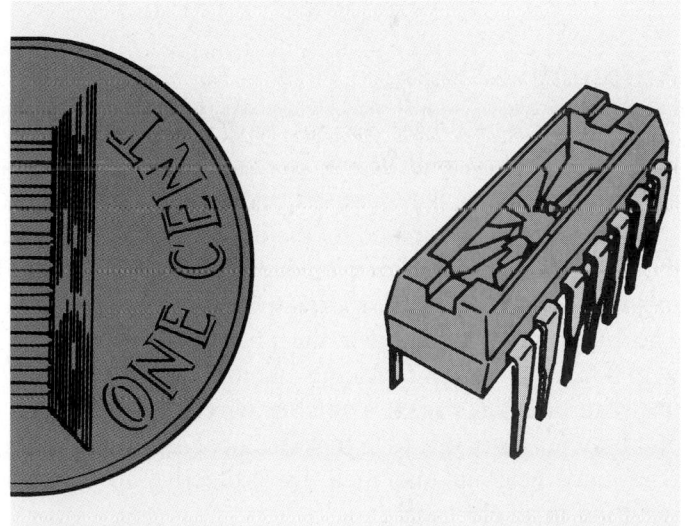

Figure 8-16. An integrated circuit is a tiny chip containing microscopic components such as transistors, diodes, resistors, and conductors. ICs are used in modern electronic circuits.

Printed circuits use flat conductor strips mounted on an insulating board. This is pictured in **Figure 8-17.** Printed circuits are normally used instead of wires on the back of the instrument panel. This eliminates the need for a bundle of wires going to the indicators, gauges, and instrument bulbs.

An *amplifier* is an electronic circuit designed to use a very small current to control a very large current. A good example of an amplifier is an ignition control module, which was introduced in Chapter 1. It uses small

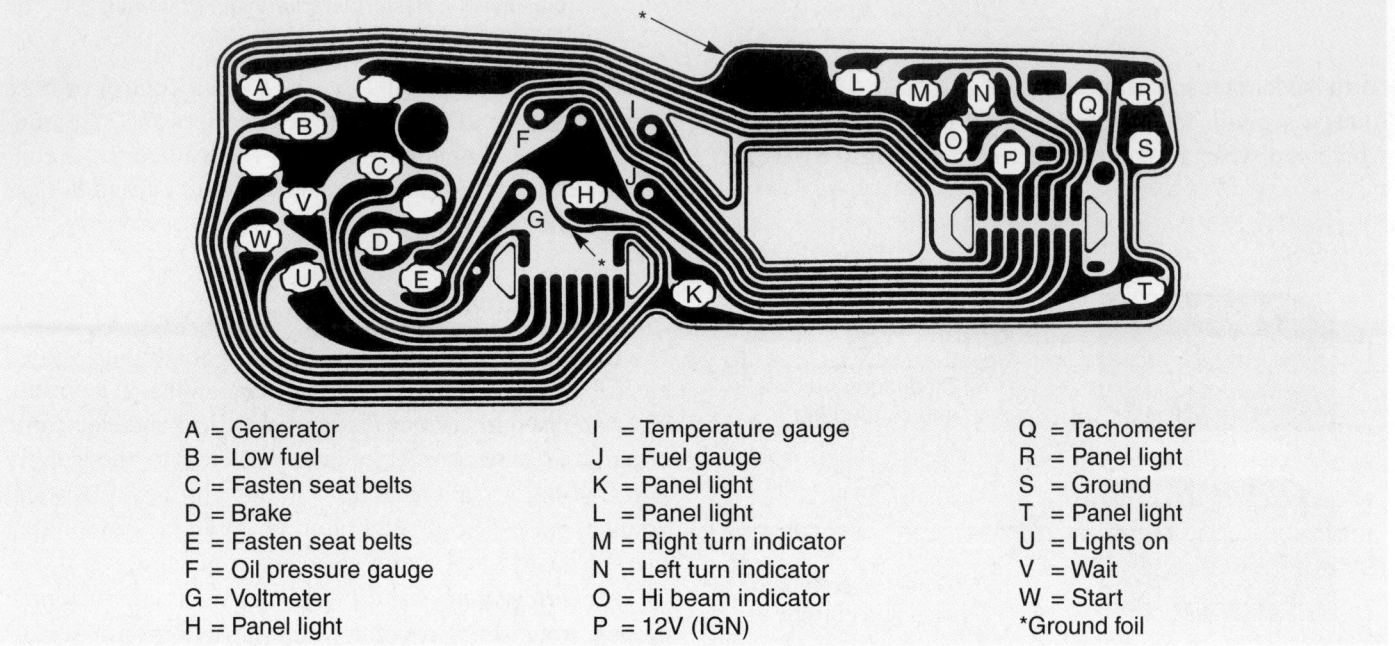

A = Generator
B = Low fuel
C = Fasten seat belts
D = Brake
E = Fasten seat belts
F = Oil pressure gauge
G = Voltmeter
H = Panel light

I = Temperature gauge
J = Fuel gauge
K = Panel light
L = Panel light
M = Right turn indicator
N = Left turn indicator
O = Hi beam indicator
P = 12V (IGN)

Q = Tachometer
R = Panel light
S = Ground
T = Panel light
U = Lights on
V = Wait
W = Start
*Ground foil

Figure 8-17. A printed circuit has flat conductor strips mounted on an insulating board. (Oldsmobile)

electrical pulses from the distributor to produce strong on/off cycles to operate the ignition coil.

Automotive Wiring

An automobile uses various types of wiring in its many electrical systems. It is important that you learn the different types, how they are used, and how to repair them.

Wire size is determined by the diameter of the wire's metal conductor. The wire's conductor diameter is stated in *gauge size.* Gauge size is a relative numbering system. The larger the gauge number, the smaller the diameter of wire conductor. When replacing a section of wire, always use wire of equal size. If a smaller wire is used, the circuit may not work due to high resistance. Also, undersize wire may heat up and melt its protective insulation, resulting in an electrical fire.

Wire Types

A *primary wire* is small and carries battery or alternator voltage. A primary wire normally has plastic insulation to prevent shorting. The insulation is usually color coded for easy troubleshooting, **Figure 8-18.** As shown in **Figure 8-19,** primary wires are often grouped in a wiring harness. A *wiring harness* is a group of wires with

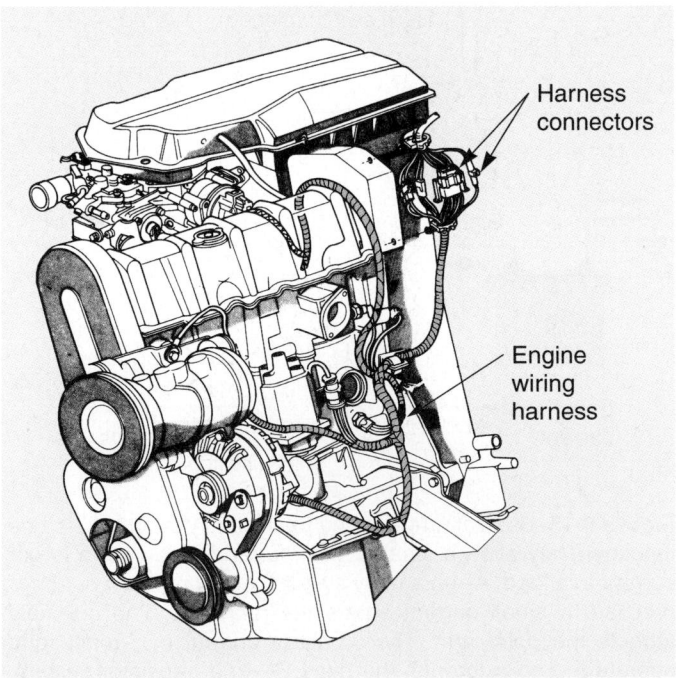

Figure 8-19. A wiring harness usually consists of a plastic covering and a group of primary wires. The covering organizes and protects the wires. Also note the harness connectors. (Chrysler)

Code	Color
B	Black
Br	Brown
G	Green
Gy	Gray
L	Blue
Lb	Light blue
Lg	Light green
O	Orange
R	Red
W	White
Y	Yellow

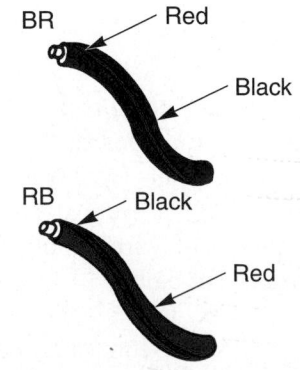

Figure 8-18. Primary wires are color-coded. This lets you trace a wire through the vehicle.

a plastic or tape covering that helps protect and organize the wires.

A *secondary wire,* also called *high tension cable,* is only used in a vehicle's ignition system for spark plug or coil wires. It has extra thick insulation to prevent the high voltage from short circuiting. The conductor is designed for very low current.

Battery cable is an extremely large-gauge wire capable of carrying high currents from the battery to the starting motor. Look at **Figure 8-20.** Usually, a starting motor draws more current than all of the other electrical components combined, normally over 100 amps. For this reason, a very large conductor is required.

Ground wires or *ground straps* connect electrical components to the chassis or ground of the car. Insulation is not used on these wires.

Wiring Repairs

Crimp connectors and *terminals* can be used to quickly repair automotive wiring. See **Figure 8-21.** Connectors, or splicers, allow a wire to be connected to another wire. Terminals allow a wire to be connected to an electrical component.

Harness connectors are multi-wire terminals that connect several wires together. They usually have a two-part plastic housing that snaps together. To free, or disconnect, a harness connector, you must usually disengage a tab or plastic clip that secures the two halves. Designs

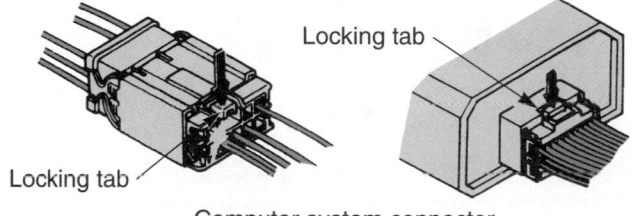

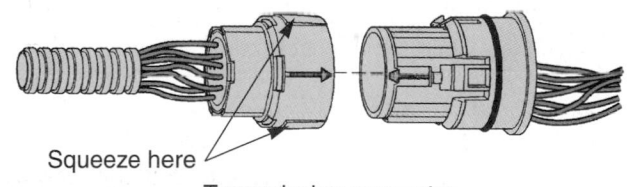

Figure 8-21. Various types of wire terminals and connectors. (Belden, General Motors, Honda)

vary, so look at the connector carefully before attempting to disengage it to prevent damage.

Crimping pliers are used to deform the connector or terminal around the wire. **Figure 8-22** shows a technician installing a crimp terminal.

A soldering gun or iron can also be used to permanently fasten wires to terminals or to other wires, **Figure 8-23.** The *soldering gun* produces enough heat to melt solder. The soldering gun is first touched to the wire and other component to preheat them. Then, the solder is touched to the joint and melts. When cooled, the solder makes a solid connection between the electrical components.

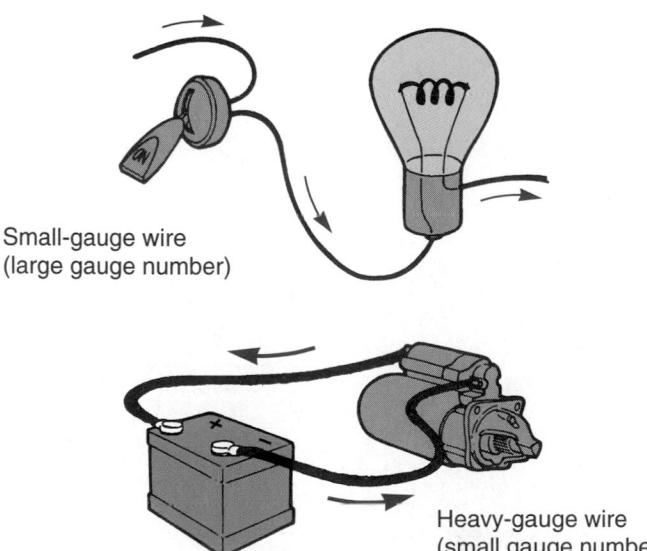

Figure 8-20. Wire gauge size is matched to current draw. Small gauge will only handle small current. Larger gauge is needed for high current draws, such as for a starting motor. (British Leyland)

A

B

Figure 8-22. Installing crimp-type connectors and terminals. A—Strip off a short section of insulation. B—Use the correct-size crimping jaw to form the terminal or connector around the wire. Tug on the wire lightly to check the connection. (Vaco Tool)

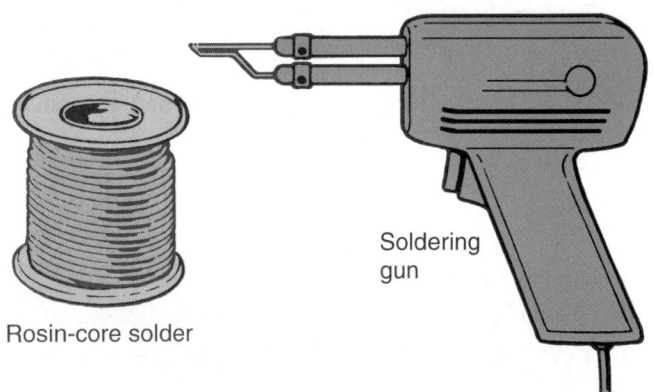

Soldering gun

Rosin-core solder

Figure 8-23. Rosin-core solder and a soldering gun make permanent connections between wires and components.

Rosin-core solder should be used on all electrical repairs. It is usually purchased in a roll for easy use and handling. *Acid-core solder* can cause corrosion of electrical components. It is recommended for nonelectrical repairs, such as radiator and heater core repairs.

 Caution
Most late-model electrical connectors require special methods or a special tool during disconnection. If you simply force the connector apart, you may create a new problem—a high-resistance connection. There are many connector designs and special tools available. Always use the right tool and methods when disconnecting wiring harnesses.

Basic Electrical Tests

Various electrical tests and testing devices are used by an auto technician. To be prepared for many later chapters, you should have a general understanding of these tools and how to use them.

Jumpers and Test Lights

A *jumper wire* is handy for testing switches, relays, solenoids, wires, and other components. The jumper can be substituted for the component, as shown in **Figure 8-24.** If the circuit begins to function with the jumper in place, then the component being bypassed is defective.

A *test light* is used to quickly check a circuit for power, or voltage. It has an alligator clip that connects to ground. Then, the pointed tip can be touched to the circuit to check for voltage. If there is voltage, the light will glow. If it does not glow, there is an open or break between the power source and the test point. See **Figure 8-25.**

A *self-powered test light* is similar to a flashlight with test leads attached. It contains batteries and is used to check to see if there is a complete electrical path. To use this type test light, the normal source of power must be disconnected. If the light glows, the circuit or part has continuity. If it does *not* glow, there is an open or break between the two test points.

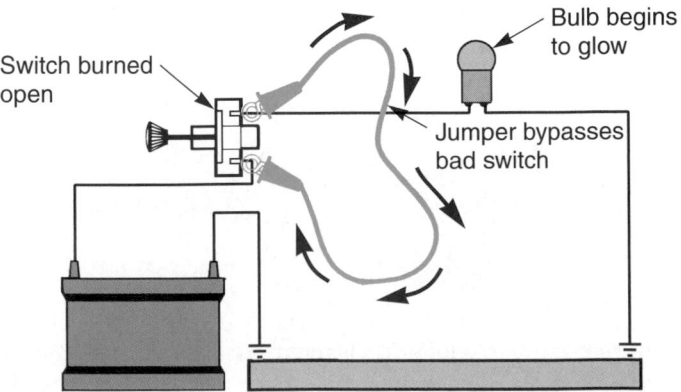

Switch burned open

Bulb begins to glow

Jumper bypasses bad switch

Figure 8-24. A jumper wire is handy for bypassing electrical components. It can also be used to supply power to a section of a circuit. (Ford)

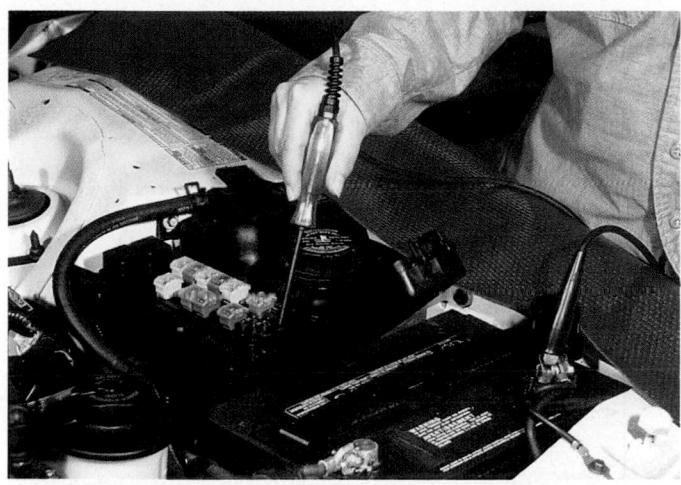

Figure 8-25. A test light will quickly check for power in a circuit. Connect the alligator clip to ground and touch the tip to the circuit. The light will glow if there is power in the circuit.

Tech Tip

Some test lights have a built-in flashlight. This is handy when probing circuits and wires in poorly lit areas, like under the dash. They will save time and pay for themselves in a short period.

Voltmeter, Ammeter, Ohmmeter

A *voltmeter* is used to measure the amount of voltage in a circuit, **Figure 8-26A.** It is normally connected across, or in parallel with, the circuit. The voltmeter reading can be compared to specifications to determine whether an electrical problem exists.

An *ammeter* measures the amount of current in a circuit, **Figure 8-26B.** Conventional types must be connected in *series* with the circuit. All the current in the circuit must pass through the ammeter. A modern *inductive,* or *clip-on, ammeter* is simply slipped over the outside of the wire insulation. It uses the magnetic field around the outside of the wire to determine the amount of current in the wire. An inductive ammeter is very fast and easy to use.

An *ohmmeter* will measure the amount of resistance in ohms in a circuit or component. To prevent damage, an ohmmeter must *never* be connected to a source of voltage. The wire or part being tested must be disconnected from the vehicle's battery. As shown in **Figure 8-26C,** the ohmmeter is connected across the wire or component being tested. Then, the ohmmeter reading can be compared to specifications. If the resistance is too high or too low, the part is defective.

A *multimeter,* also called a *VOM,* is an ohmmeter, ammeter, and voltmeter combined in one case. As pictured in **Figure 8-27,** a function knob can be turned to

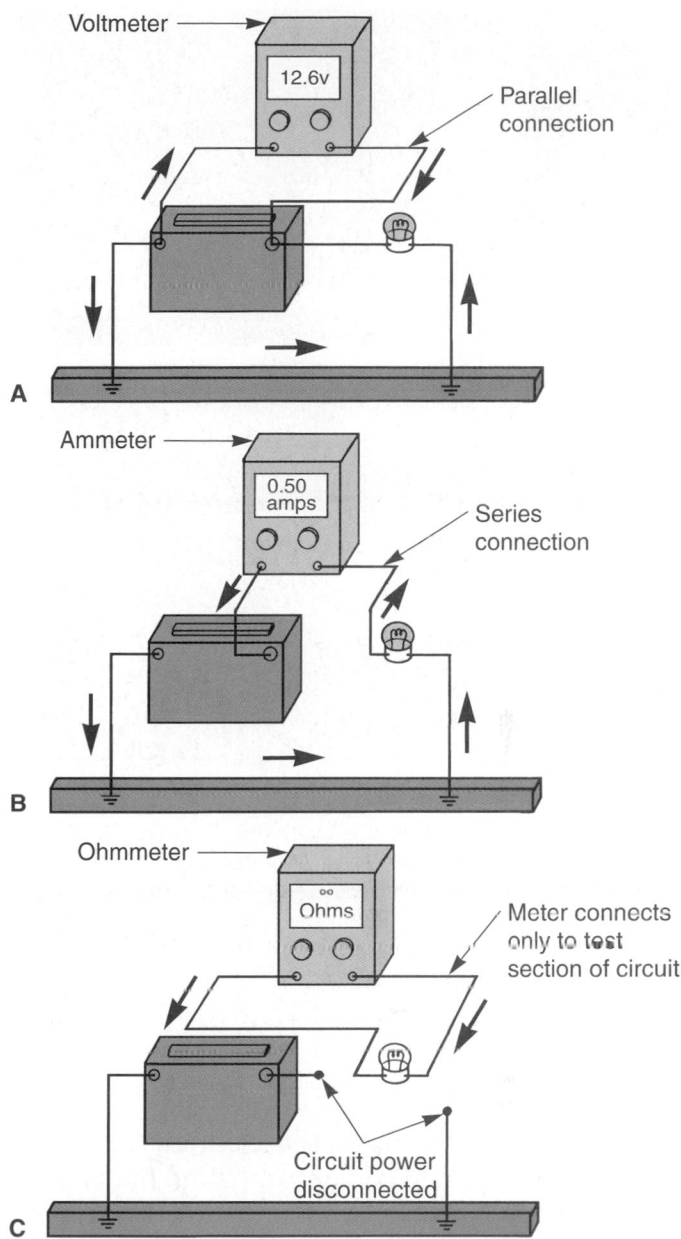

Figure 8-26. Three basic meter connections. A—A voltmeter connects in parallel. It measures the amount of electrical pressure, or potential, in a circuit. B—An ammeter connects in series with the circuit. Current flows through the meter and circuit. C—An ohmmeter is connected to the circuit with the power disconnected. Voltage can damage some ohmmeters.

select the type measurement to be made, such as volts, amps, or ohms. It must be connected to the circuit as described for each individual meter.

Caution

Only use a high-resistance test light or meter when probing today's vehicle circuits. If you use an older low-resistance, high-current-draw test light or meter, you can damage delicate electronic components.

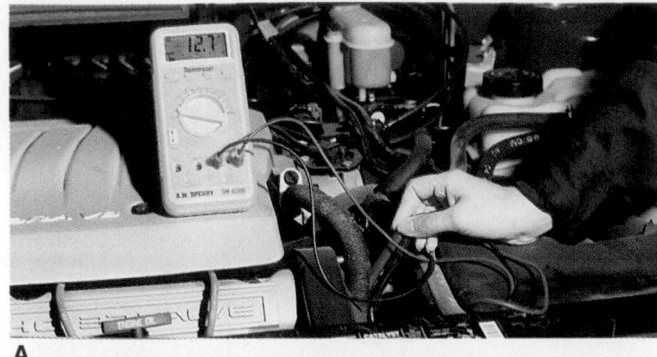

A

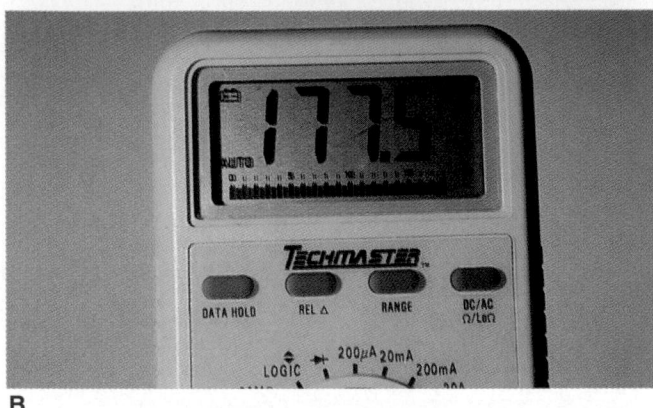

B

Figure 8-27. A multimeter is a voltmeter, ammeter, and ohm-meter combined. A—Multimeter is commonly used to check the condition of numerous electrical and electronic parts. B—This is a digital-analog meter face. The number display gives a digital readout. The bar across the bottom is an analog readout for interpreting fluctuating values.

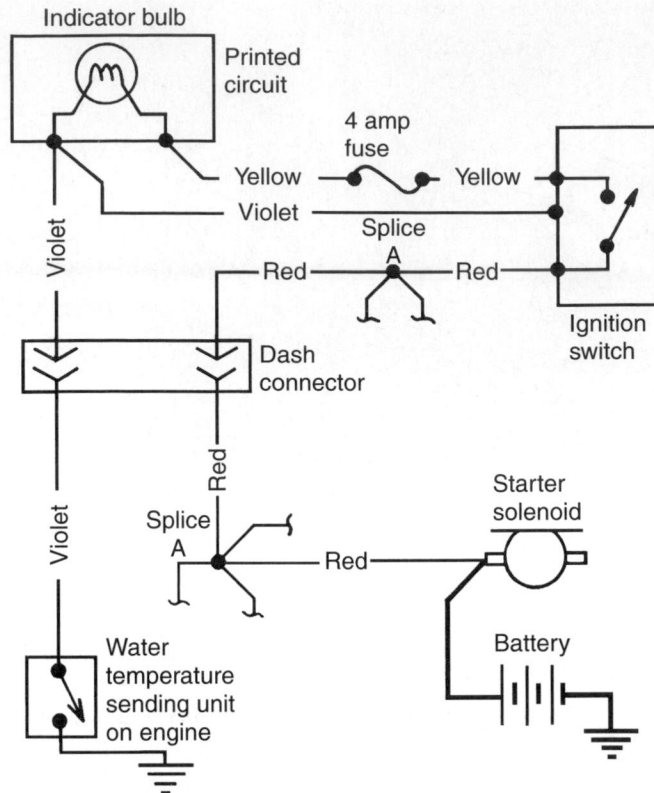

Figure 8-28. Note how a wiring diagram uses symbols to represent the parts of an electrical circuit.

Wiring Diagrams

A *wiring diagram* shows how electrical components are connected by wires. It serves as an "electrical map" to help the technician with difficult electrical repairs. Look at **Figure 8-28.**

Wiring diagrams use *symbols* to represent the electrical components in a circuit. The lines on the diagram represent the wires. In this way, each wire can be traced to see how it connects to each component.

Oscilloscope

An *oscilloscope,* or *scope,* is an electronic measuring instrument. It obtains the same type of information as a voltmeter. However, a scope displays voltage readings as a *trace,* or white line, on a display screen. This allows a scope to show voltage variations very accurately as a *waveform.*

Voltage is shown as an up or down movement of the trace. Time is shown horizontally. A scope is usually built as part of an engine analyzer that can do other functions besides voltage measurement. See **Figure 8-29.**

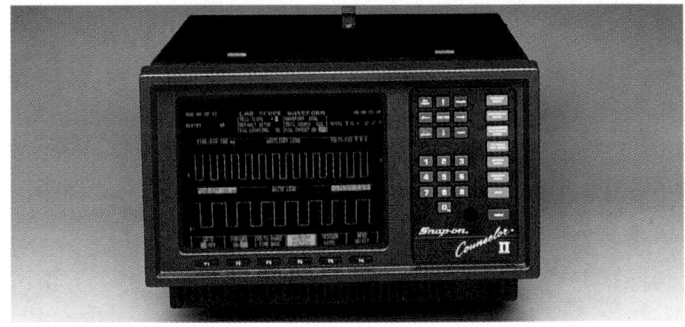

Figure 8-29. An oscilloscope is a graphic voltmeter. It displays voltage and time as a line on the display screen. It is needed when analyzing rapidly changing voltage values, such as ignition or computer system voltages. This is a dual trace scope that is showing two square waves simultaneously. (Snap-On Tools)

A *dual-trace scope* can read and show two separate waveforms or traces. This is handy if you need to compare the timing, or amplitude, of two voltage levels simultaneously when doing advanced troubleshooting.

Scan Tools

A *scan tool* is a diagnostic tool that helps find problems with onboard computer systems, **Figure 8-30.** It is an electronic unit that plugs into the vehicle's diagnostic connector, **Figure 8-31.** It can then communicate with the vehicle's control module to tell you which part might be at fault. For more information on scan tools, scopes, and other advanced electronic tools, refer to the index. Many other sections of this book discuss their use in more detail.

Note

The principles of electronics are explained in numerous other chapters. The chapters on starting systems, charging systems, fuel systems, and computer systems all cover this topic. Refer to the index as needed for added information.

Figure 8-30. A scan tool is used to help find computer system troubles. It exchanges data with the vehicle computer and tells which part of the computer circuit might be at fault. (Snap-On Tools)

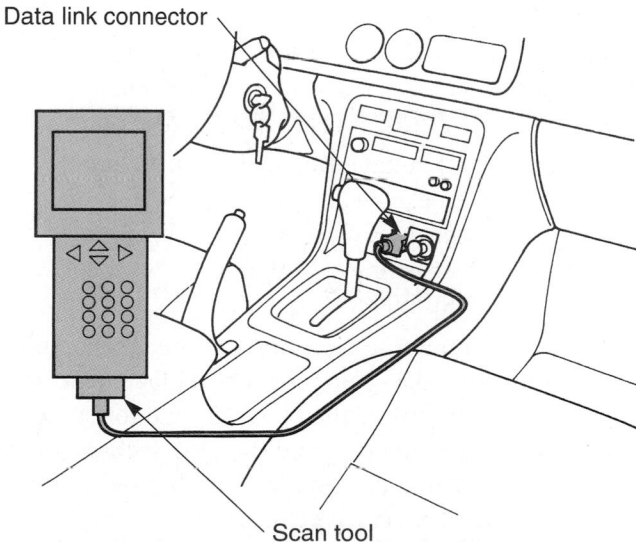

Figure 8-31. A scan tool plugs into the vehicle's data link connector, or diagnostic connector. Then, the scan tool can give instructions on testing the computer system for faults. Most scan tools convert trouble code numbers into a brief description of what might be wrong with the computer system. (Honda)

Summary

- Almost every system in a vehicle uses some type of electric or electronic component.
- An atom consists of small particles called protons, neutrons, and electrons.
- Negatively charged electrons circle around neutrons and protons.
- Current (abbreviated I) is the *flow* of electrons through a conductor.
- Voltage (abbreviated V or E) is the force that causes electron flow.
- Resistance (abbreviated R) is the *opposition* to current flow.
- Ohm's law is a simple formula for calculating volts, amps, or ohms when two of the three values are given.
- If a magnetic field is passed over a wire, an electric current is generated in the wire.
- When a switch is closed, a circuit is complete. When the switch is open, a circuit is broken and does not function.
- A fuse protects a circuit against damage caused by a short circuit.
- A relay is an electrically operated switch.
- A semiconductor is made of a special substance capable of acting as both a conductor and an insulator.
- A diode is an "electronic check valve" that only allows current to flow in one direction.
- A transistor performs the same basic function as a relay. It acts as a remote control switch or current amplifier.
- A condenser or capacitor is a device used to absorb unwanted electrical pulses in a circuit.
- An integrated circuit (IC) is a wafer-like chip that contains almost microscopic diodes, transistors, resistors, and capacitors.
- A jumper wire is handy for testing switches, relays, solenoids, wires, and other components.
- A multimeter, also called a VOM, is an ohmmeter, ammeter, and voltmeter combined in one case.

- A wiring diagram shows how electrical components are connected by wires.

Important Terms

Component	Semiconductor
Atom	Diode
Electrons	Transistor
Protons	Capacitors
Electricity	Integrated circuit(IC)
Conductors	Printed circuits
Free electrons	Amplifier
Insulators	Wire size
Current	Gauge size
Voltage	Primary wire
Resistance	Wiring harness
Power source	Secondary wire
Load	Ground wires
Conductors	Ground straps
Series circuit	Crimp connectors
Parallel circuit	Terminals
Series-parallel circuit	Harness connectors
Frame-ground circuit	Soldering gun
Ohm's law	Jumper wire
Magnetic field	Test light
Induction	Voltmeter
Switch	Ammeter
Short circuit	Ohmmeter
Unlimited current flow	Multimeter
Fuse	Wiring diagram
Fuse box	Oscilloscope
Fusible link	Trace
Circuit breaker	Waveform
Relay	Scanner

Review Questions—Chapter 8

Please do not write in this text. Place your answers on a separate sheet of paper.

1. What is electricity?
2. Explain the difference between a conductor and an insulator.
3. Which of the following is *not* part of a simple circuit?
   (A) Wheatstone bridge.
   (B) Load.
   (C) Power source.
   (D) Conductors.
4. List and explain the three basic elements of electricity.
5. A(n) _____ circuit has more than one load connected in a single electrical path.
6. A(n) _____ circuit has more than one electrical path or leg.
7. What is a one-wire circuit?
8. Using Ohm's law, find the resistance in a circuit with 12 volts and 3 amps.
9. Define the term "short circuit."
10. Explain the functions of fuses and circuit breakers.
11. A(n) _____ is an electrical, not electronic, device that allows a small current to control a larger current.
12. Explain the difference between an electric component and an electronic component.
13. Which of the following is not an electronic component.
    (A) Diode.
    (B) Transistor.
    (C) Circuit breaker.
    (D) Integrated circuit.
14. A(n) _____ is an electronic circuit that uses a very small current to control a very large current.
15. A(n) _____ wire is used to carry battery voltage and has plastic insulation.
16. Why are wires color coded?
17. Which of the following should *not* be used for electrical repairs?
    (A) Acid-core solder.
    (B) Rosin-core solder.
    (C) Crimp connectors.
    (D) Soldering gun.
18. Explain the use of a test light, voltmeter, ohmmeter, ammeter, and wiring diagrams.
19. Which of the following devices is used to absorb unwanted voltage fluctuations in a circuit?
    (A) Diode.
    (B) Capacitor.
    (C) Transistor.
    (D) Multimeter.
20. While repairing a section of wire, Technician A believes the segment of replacement wire should be larger than the old wire for smoother flow. Technician B believes the section of new wire should be smaller for improved performance. Who is right?
    (A) A only.
    (B) B only.
    (C) Both A and B.
    (D) Neither A nor B.

● ASE-Type Questions

1. Technician A says that all things are made of atoms. Technician B says that only some atoms allow current flow. Who is correct?
   (A) A only.
   (B) B only.
   (C) Both A and B.
   (D) Neither A nor B.

2. The movement of electrons from atom to atom is called:
   (A) induction.
   (B) electricity.
   (C) atomic flow.
   (D) conductivity.

3. Each of these is an insulator *except:*
   (A) metal.
   (B) plastic.
   (C) rubber.
   (D) ceramics.

4. The three basic elements of electricity are:
   (A) watts, volts, and ohms.
   (B) current, voltage, and resistance.
   (C) Both A and B.
   (D) Neither A nor B.

5. The electron theory of flow states that current flows from:
   (A) electrons to volts.
   (B) atoms to a current.
   (C) positive to negative.
   (D) negative to positive.

6. Which of the following is abbreviated by an I?
   (A) Current.
   (B) Voltage.
   (C) Resistance.
   (D) None of the above.

7. Technician A says a simple circuit has only one path for current. Technician B says a series circuit has more than one path for current. Who is right?
   (A) A only.
   (B) B only.
   (C) Both A and B.
   (D) Neither A nor B.

8. Which type of circuit uses a vehicle body as a return wire to the power source?
   (A) Series.
   (B) Parallel.
   (C) One-wire.
   (D) Series-parallel.

9. Which of the following is *not* a correct example of Ohm's law?
   (A) $E = I \times R$.
   (B) $R = I \times E$.
   (C) $R = E \div I$.
   (D) $I = E \div R$.

10. Which of the following is an electronic check valve that will only allow current to flow in one direction?
    (A) Diode.
    (B) Relay.
    (C) Transistor.
    (D) Condenser.

Activities—Chapter 8

1. Using sketches and principles you have learned about basic electricity, prepare a presentation showing how electricity can be created through magnetism.

2. Demonstrate the use of a continuity tester and explain what it tells you about a circuit.

3. Using the pie chart in **Figure 8-6,** solve the following problem: If a circuit with 12 volts has a current of 2 amperes, what is the resistance? Explain how you got the answer.

Chapter 9

Fasteners, Gaskets, Seals, and Sealants

After studying this chapter, you will be able to:

- Identify commonly used automotive fasteners.
- Select and use fasteners properly.
- Remove, select, and install gaskets, seals, and sealants correctly.
- Summarize safety rules relating to fasteners, gaskets, seals, and sealants.
- Correctly answer ASE certification test questions that require a knowledge of fasteners, gaskets, seals, and sealants.

Many different fasteners, gaskets, seals, and sealants are used by the automotive technician. It is almost impossible to connect any two parts of a vehicle without them.

This chapter covers fasteners, gaskets, and seals. This knowledge is important and will prepare you for many repair operations and other text chapters.

Fasteners

Fasteners are devices that hold the parts of a car together. Thousands of fasteners are used in vehicles. **Figure 9-1** shows some of the most common types.

Bolts and Nuts

A *bolt* is a metal rod with external threads on one end and a head on the other. When a high-quality bolt is threaded into a part other than a nut, it can also be called a *cap screw*. A *nut* has internal threads and usually a

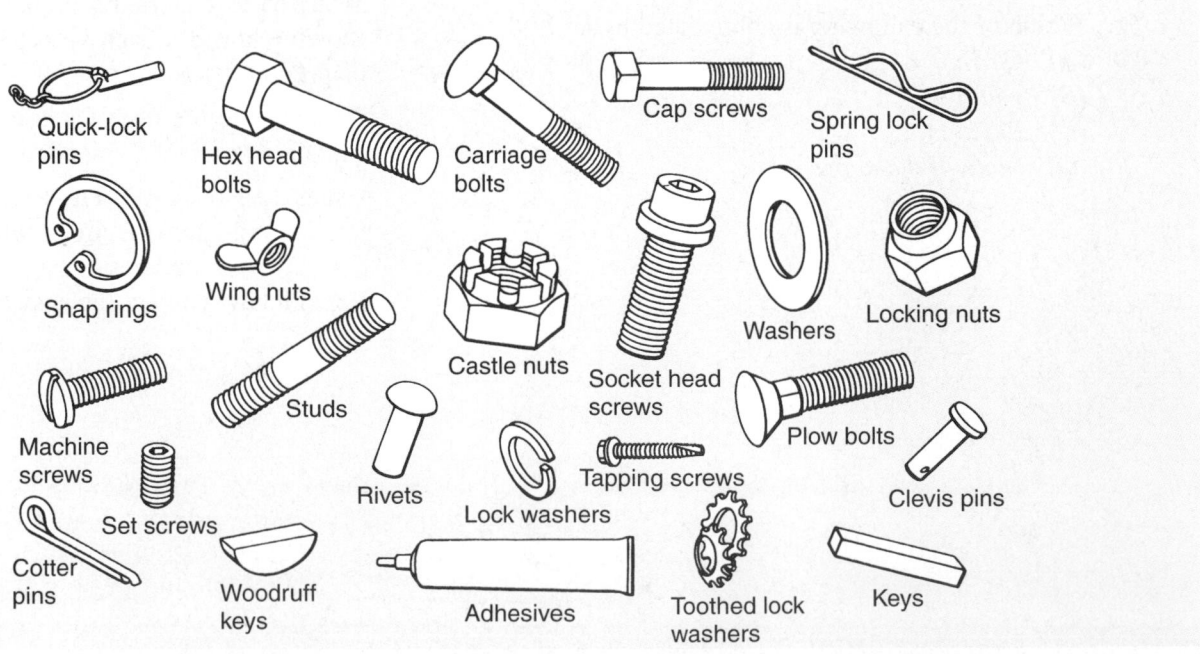

Figure 9-1. A fastener is any device or adhesive used to hold parts together. Study these basic automotive fasteners. (Deere & Co.)

six-sided outer shape. When a nut is screwed onto a bolt, a powerful clamping force is produced, **Figure 9-2.**

In automotive technology, bolts and nuts are named after the parts they hold. For instance, the bolts holding the cylinder head on the block are called cylinder head bolts. The bolts on an engine connecting rod are called connecting rod bolts.

Bolt and Nut Terminology

Bolts and nuts come in various sizes, grades (strengths), and thread types. It is important to be familiar with these differences. The most important bolt dimensions are:

- *Bolt size*—the measurement of the outside diameter of the bolt threads. See **Figure 9-3.**
- *Bolt head size*—the distance across the flats or outer sides of the bolt head. It is the same as the wrench size.
- *Bolt length*—measured from the bottom of the bolt head to the threaded end of the bolt.
- *Thread pitch*—the same as thread coarseness. With U.S. conventional system fasteners, thread pitch is the number of threads per inch. With metric fasteners, it is the distance between each thread in millimeters. Refer again to **Figure 9-3.**

Thread Types

There are three basic types of threads used on fasteners:

- *Coarse threads* (UNC-Unified National Coarse).
- *Fine threads* (UNF-Unified National Fine).
- *Metric threads* (SI).

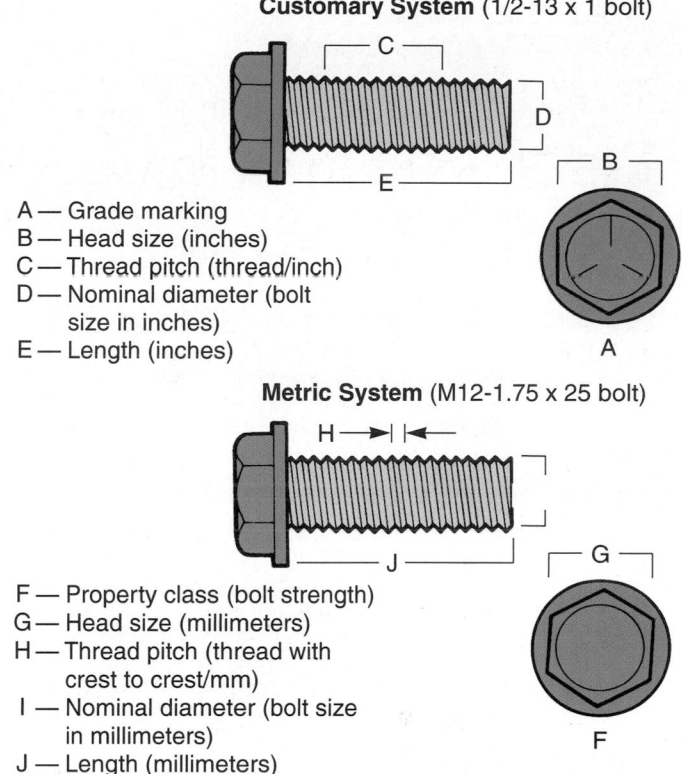

Customary System (1/2-13 x 1 bolt)

A — Grade marking
B — Head size (inches)
C — Thread pitch (thread/inch)
D — Nominal diameter (bolt size in inches)
E — Length (inches)

Metric System (M12-1.75 x 25 bolt)

F — Property class (bolt strength)
G — Head size (millimeters)
H — Thread pitch (thread with crest to crest/mm)
I — Nominal diameter (bolt size in millimeters)
J — Length (millimeters)

Figure 9-3. The naming systems for both customary and metric bolts. (Ford)

Never interchange thread types or thread damage will result. As shown in **Figure 9-4,** metric threads *look* like customary threads. If a metric bolt is forced into a hole with fine threads, either the bolt or part threads will be ruined.

Bolts and nuts also come in right- and left-hand threads. With *right-hand threads,* the fastener must be turned clockwise to tighten. This is the most common style of thread. With *left-hand threads,* turn the fastener in a counterclockwise direction to tighten. Left-hand threads are not very common. The letter L may be stamped on fasteners with left-hand threads.

Bolt Grade

Tensile strength, or *grade,* refers to the amount of pull a fastener can withstand before breaking. Bolts are made of different metals, some stronger than others. Tensile strengths can vary. *Bolt head markings,* also called *grade markings,* specify the tensile strength of the bolt. U.S. customary bolts are marked with lines or slash marks. The more lines, the stronger the bolt. A metric bolt is marked with a numbering system. The larger the number, the stronger the bolt. Look back at **Figure 9-3.**

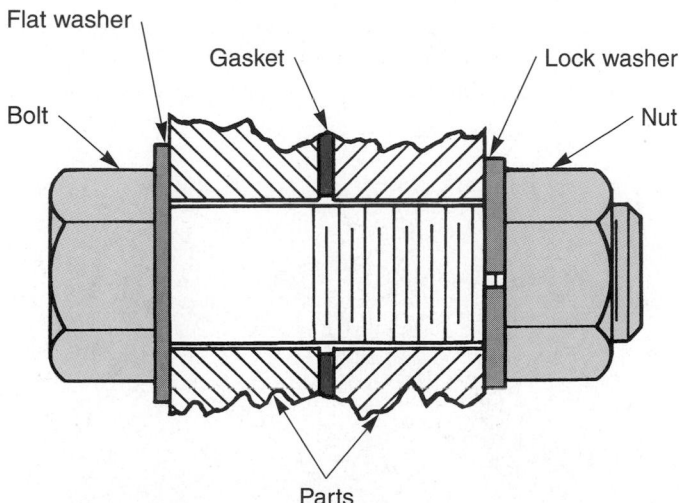

Flat washer
Gasket
Bolt
Lock washer
Nut
Parts

Figure 9-2. A bolt and nut exert a powerful clamping force on parts. Notice the washers and gasket used.

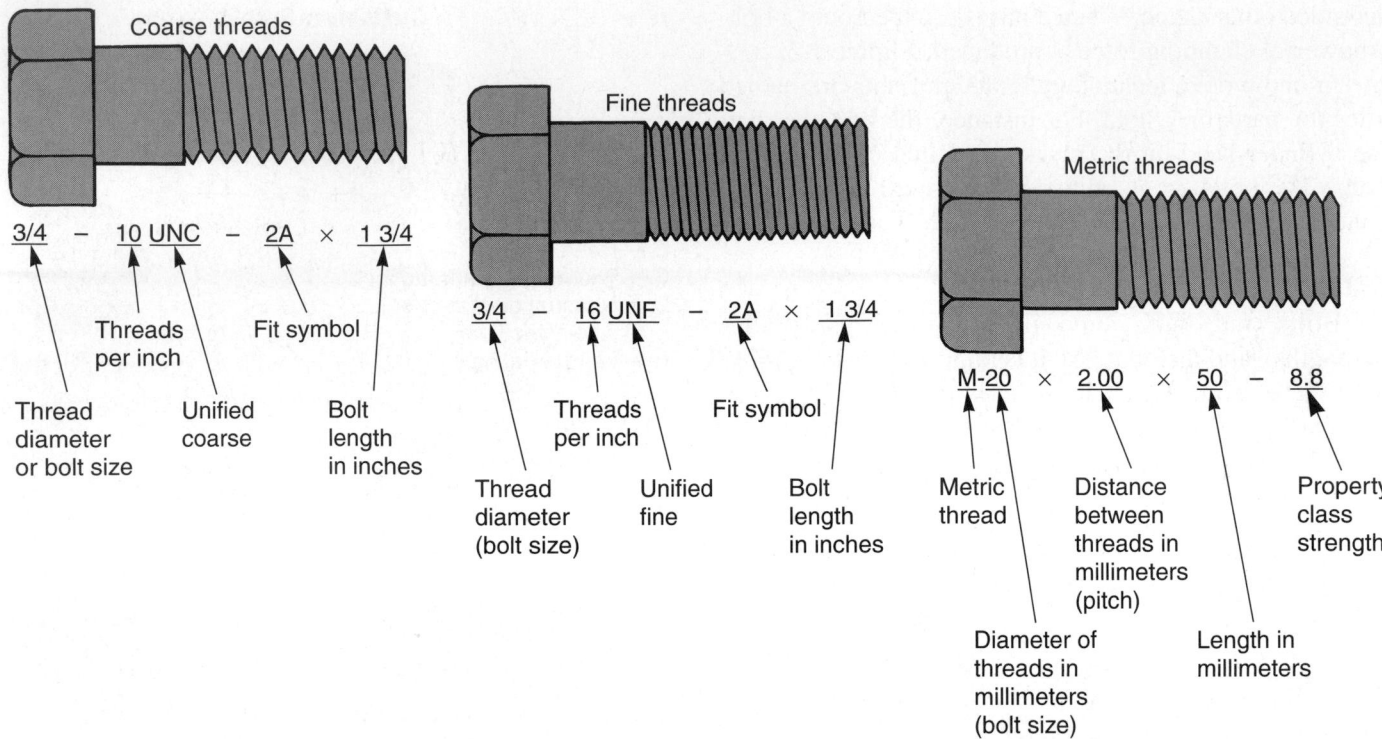

Figure 9-4. The bolt designation number gives information about the bolt. This number is commonly used when purchasing new bolts.

Warning

Never replace a high-grade bolt with a lower grade bolt. The weaker bolt could easily snap, possibly causing part failure and a dangerous situation.

Bolt Description

A *bolt description* is a series of numbers and letters that describe the bolt, as shown in **Figure 9-4.** When purchasing new bolts, the bolt description information is needed.

Nut Types

Many types of nuts are used in a vehicle. The most common ones are pictured in **Figure 9-5.** Study their names closely. A slotted nut uses a safety device called a *cotter pin.* It fits through a hole in the bolt or part. This keeps the nut from turning and possibly coming off. Look at **Figure 9-6.**

Washers

Washers are used under bolt heads and nuts. The two basic types are flat washers and lock washers. Look at

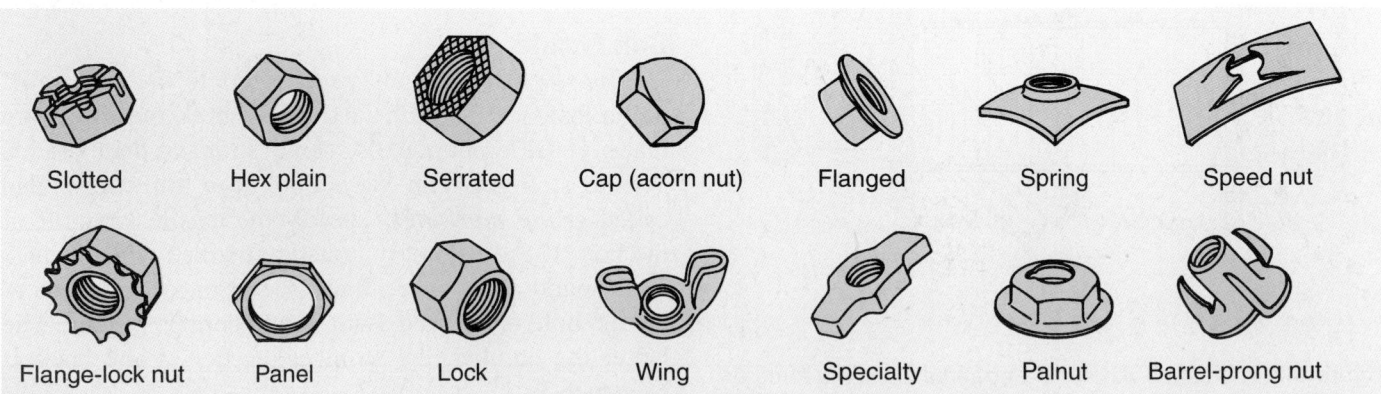

| Slotted | Hex plain | Serrated | Cap (acorn nut) | Flanged | Spring | Speed nut |
| Flange-lock nut | Panel | Lock | Wing | Specialty | Palnut | Barrel-prong nut |

Figure 9-5. Common types of nuts used in vehicles. (Deere & Co.)

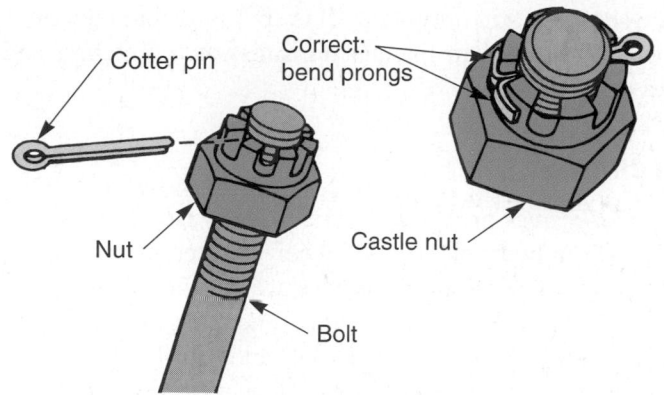

Figure 9-6. A cotter pin slides through a castle nut and a hole in the bolt. This makes sure the nut cannot turn and come off. (Deere & Co.)

Figure 9-7. A *flat washer* increases the clamping surface under the fastener. It prevents the bolt or nut from digging into the part.

A *lock washer* prevents the bolt or nut from becoming loose under stress and vibration. *Lock tabs,* or *lock plates,* perform the functions of both flat washers and lock washers. They increase clamping surface area and secure the fastener.

Machine Screws

Machine screws are similar to bolts, but they normally have screw-type heads. They are threaded along their full length and are relatively small. Refer back to **Figure 9-1**. Machine screws are used to secure parts when clamping loads are light. They come in various head shapes.

Sheet Metal Screws

Sheet metal screws, or *tapping screws,* are commonly used on plastic and sheet metal parts, such as body

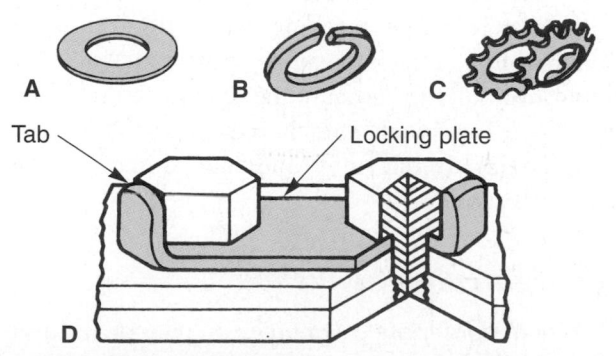

Figure 9-7. Basic washer types. A—Plain flat washer. B—Split lock washer. C—Toothed lock washer. D—Lock plate.

trim, dashboard panels, and grills. Several are shown in **Figure 9-8.** Sheet metal screws have tapering threads that are very widely spaced. They come in a wide range of head configurations and sizes.

Nonthreaded Fasteners

Numerous types of *nonthreaded fasteners,* such as snap rings, clips, and adhesives, are used in the assembly of a vehicle. It is essential to learn the most common types.

Snap Rings

A *snap ring* fits into a groove in a part and commonly holds shafts, bearings, gears, pins, and other similar components in place. **Figure 9-9** shows several types of snap rings. *Snap ring pliers* are needed to remove and install snap rings. As pictured in **Figure 9-9,** these pliers have special jaws that fit and grasp the snap ring.

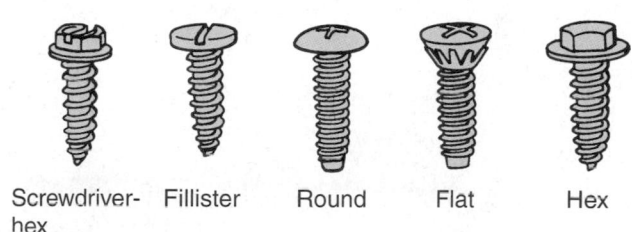

Screwdriver-hex   Fillister   Round   Flat   Hex

Figure 9-8. Basic types of tapping screws. (Deere & Co.)

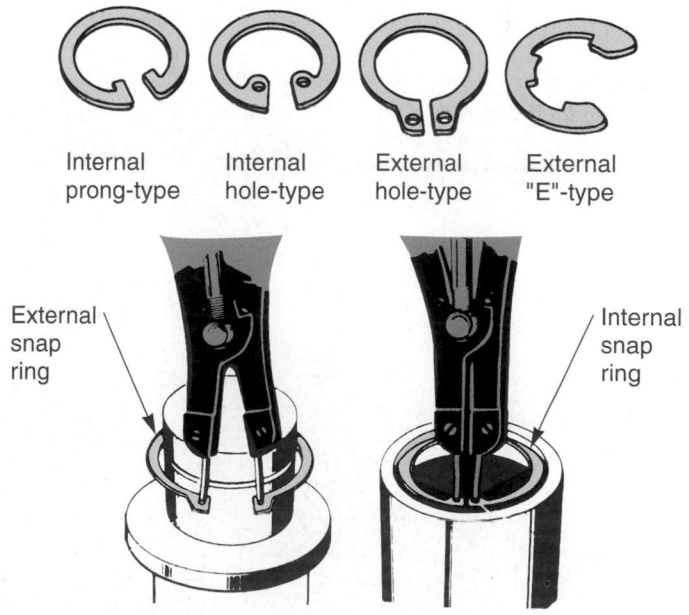

Internal prong-type   Internal hole-type   External hole-type   External "E"-type

External snap ring

Internal snap ring

Figure 9-9. Different snap ring types. External snap rings fit into a groove on a shaft. Internal snap rings fit into a groove inside a hole.

Warning

Wear eye protection when working with a snap ring. When flexed, the ring can shoot into your face with considerable force.

Keys and Set Screws

A metal *key* fits into a *keyseat,* or slot, cut into a shaft and a *keyway* cut into the mating part, such as a gear, pulley, or collar. The key prevents the part from turning on its shaft. Refer to **Figure 9-10A.** *Set screws* are normally used to lock a part onto a shaft. See **Figure 9-10B.** They can be used with or without a key and keyway. A set screw is a headless fastener normally designed to accept a hex (Allen) wrench or screwdriver.

Splines

Splines are a series of slots cut into a shaft and a mating part. See **Figure 9-11.** Splines have an advantage over keys in that they allow the gear or collar to slide on

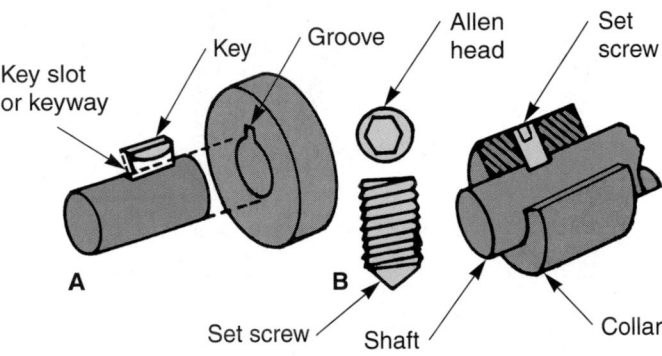

Figure 9-10. A—A key fits into the keyseat, or slot, in a shaft and keyway in the mating part. This keeps the part from turning on the shaft. B—A set screw also locks a part to a shaft, but with less strength. (Florida Dept. of Voc. Ed.)

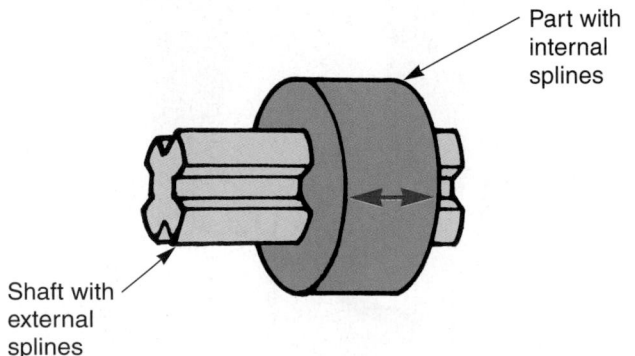

Figure 9-11. Splines allow a part to slide on a shaft, but not turn on the shaft.

the shaft but still *not* rotate. This sliding locking action is commonly used in manual transmissions, clutches, and drive shaft yokes.

Adhesives

Adhesives are special glues widely used in vehicles. They hold body moldings, rubber weather stripping, and body emblems. Some adhesives are designed to stay soft and pliable; others dry hard. Some take hours to dry, while others dry in seconds. Observe all directions and safety precautions when using adhesives.

Torquing Bolts and Nuts

It is very important that bolts and nuts are tightened properly. This is called *torquing.* If overtightened, a bolt will stretch and possibly break. The threads could also fail. If undertightened, a bolt may loosen and fall out. Part movement could also shear the fastener or break a gasket, causing leakage.

Torque specifications are tightening values given by the auto manufacturer. Torque specifications are normally given for all precision assemblies, such as engines, transmission, and differentials. **Figure 9-12** shows a general torque specification chart that gives average bolt tightening values. It can be used when factory specifications are not available. Notice how bolt torque increases with bolt size and grade.

Service manuals sometimes recommend *new* bolts because of a torque-to-yield process. Discard the old bolts in such cases. *Torque-to-yield* is a bolt tightening method that requires a specific bolt torque, followed by turning the bolt a specific number of degrees. After using a torque wrench, a degree wheel adapter is placed between the wrench and socket. The fastener is then turned until the degree wheel reads as specified by the manufacturer. This stretches the bolt to its correct yield point and preloads the fastener for better clamping under varying conditions.

Torque stretch is determined by measuring bolt length change while torquing the bolt. For example, when building a racing engine, you can "mike" connecting rod bolts to measure the length before and after tightening. Too much stretch indicates bolt weakness. Not enough stretch may indicate thread problems affecting torque.

Bolt Tightening Sequence

A *bolt tightening sequence,* or pattern, is used to make sure that parts are fastened evenly. An incorrect sequence or uneven tightening can cause breaks,

> **Caution**
> The torque specifications listed below are approximate guidelines only and may vary depending on conditions when used such as amount and type of lubricant, type of plating on bolt, etc.

SAE Standard / Foot-pounds

Grade of bolt	SAE 1 & 2	SAE 5	SAE 6	SAE 8		
Min. tensile strength	64,000 P.S.I.	105,000 P.S.I.	133,000 P.S.I.	150,000 P.S.I.		
Markings on head	⬢	✦	✦	✳	Size of socket or wrench opening	
U.S. standard					U.S. regular	
Bolt diameter	Foot-pounds				Bolt head	Nut
1/4	5	7	10	10.5	3/8	7/16
5/16	9	14	19	22	1/2	9/16
3/8	15	25	34	37	9/16	5/8
7/16	24	40	55	60	5/8	3/4
1/2	37	60	85	92	3/4	13/16
9/16	53	88	120	132	7/8	7/8
5/8	74	120	167	180	15/16	1.
3/4	120	200	280	296	1-1/8	1-1/8

Metric Standard

Grade of bolt	5D	8G	10K	12K		
Min. tensile strength	71,160 P.S.I.	113,800 P.S.I.	142,200 P.S.I.	170,679 P.S.I.		
Grade markings on head	5D	8G	10K	12K	Size of socket or wrench opening	
Metric					Metric	
Bolt dia.	U.S. dec equiv.	Foot-pounds			Bolt head	
6mm	.2362	5	6	8	10	10mm
8mm	.3150	10	16	22	27	14mm
10mm	.3937	19	31	40	49	17mm
12mm	.4720	34	54	70	86	19mm
14mm	.5512	55	89	117	137	22mm
16mm	.6299	83	132	175	208	24mm
18mm	.709	111	182	236	283	27mm
22mm	.8661	182	284	394	464	32mm

Figure 9-12. A general bolt torque chart. Note how torque values increase as the bolt size and grade increase.

warping, gasket leaks, and other problems. Generally, a group of fasteners on a part are tightened in a *crisscross pattern.* This creates an even, gradual clamping force along the entire mating surface of the parts. A service manual will illustrate the proper sequence when a torque pattern is critical. Refer to **Figure 9-13.**

Tech Tip
When tightening engine covers and covers on other large assemblies, remember that "less is usually better." The most common mistake is to overtighten covers and damage or split the gasket or seal. It usually takes much less torque than you may think to make a new gasket seal properly.

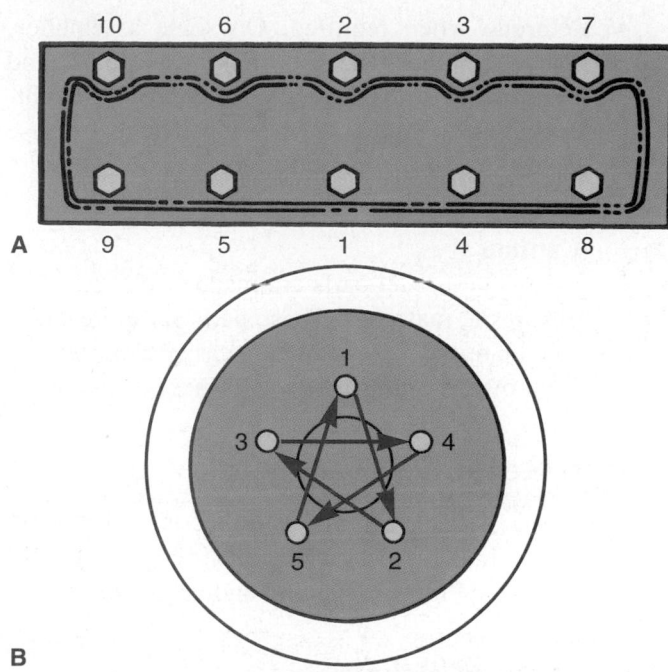

Figure 9-13 A crisscross pattern is recommended when multiple fasteners hold a part. A—A service manual pattern for engine cylinder head. B—A service manual pattern for wheel lug nuts.

Using a Torque Wrench

A *torque wrench* is used to apply the proper amount of torque when tightening a threaded fastener. Use the following basic rules.

- Keep a steady pull on the wrench. Do *not* use short, jerky motions.
- Clean fastener threads.
- When possible, avoid using swivel joints. They can affect torque wrench accuracy.
- When reading a torque wrench, look straight down at the scale. Viewing the scale from an angle can give a false reading.
- A general torque value chart should only be used when the manufacturer's specifications are *not* available.
- When the manufacturer's torque patterns are not available, use a general crisscross pattern for tightening fasteners.
- Pull only on the handle of the torque wrench. *Do not* allow beam of a beam-type torque wrench to touch anything.
- Tighten bolts and nuts in four steps: one-half recommended torque, three-fourths torque, full torque, and full torque a second time.

- Retorque when required. On some assemblies, such as cylinder heads, intake manifolds, and exhaust manifolds, bolts may have to be retightened after operation and heating. This is because expansion and contraction due to temperature changes can cause the fasteners to loosen.

 Caution
Many late-model parts are made of plastic or composite materials. These parts are more delicate than metal parts and are easily damaged from overtorquing.

Thread Repairs

Threaded holes in parts can become damaged, requiring repairs. A technician must be capable of repairing damaged threads quickly and properly.

Minor Thread Repairs

Minor thread damage includes nicks, partial flattening, and other less serious problems. Minor thread damage can usually be repaired with a thread chaser or thread file, **Figure 9-14**. A *thread chaser* "cleans up" slightly damaged internal and external threads. The chaser is run through or over the threads to restore them.

Major Thread Repairs

Major thread damage generally includes badly smashed or stripped threads. Sometimes, major thread damage is repaired with either a tap or die, **Figure 9-15**. A *tap* is a threaded tool for cutting internal threads in holes. Various tap shapes are available. Some are for starting the threads. Others are for cutting the threads all the way to the bottom of a hole. A *die* cuts external

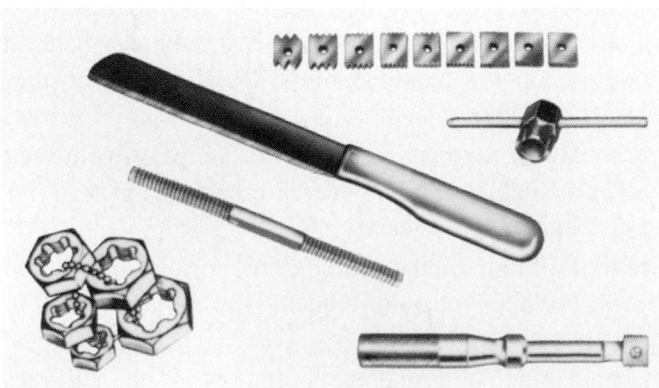

Figure 9-14. Thread chasers or files are used to repair threads. (Snap-On Tools)

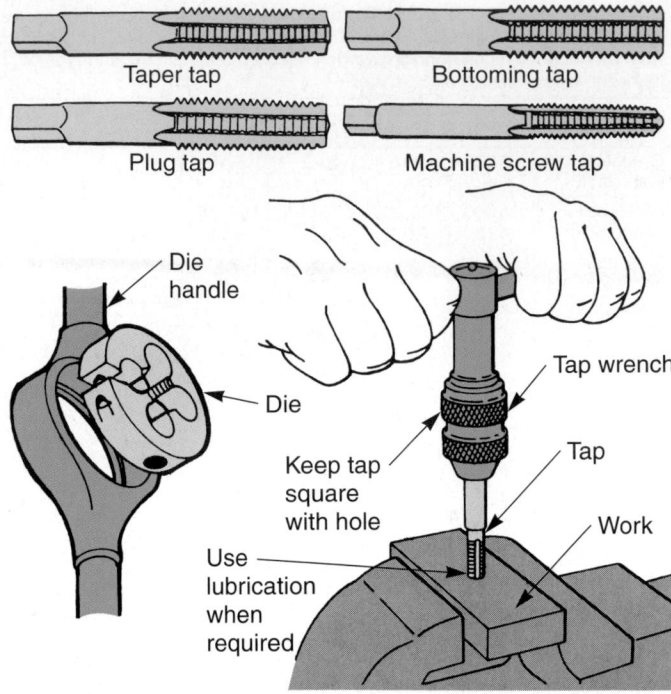

Figure 9-15. A tap or die fits into a special handle. The handle is held square as the bit is turned into the work. Back the handle off to clean metal out of the threads. A taper tap is used to start threads in a hole. Then, a plug tap and a bottoming tap are used to cut the threads to the bottom of the hole.

threads. It can be used to cut threads on metal rods, bolts, shafts, and pins.

Taps and dies are mounted in special handles, called *tap handles* or *die handles*. See **Figure 9-15**. The tool must be held squarely while being rotated into the work. As soon as the tap or die begins to bind, back the tool off about a quarter turn. This clears away the metal cuttings. Then, the cut can be made another half turn deeper. Keep rotating a half turn in and a quarter turn out until the cut is complete.

Tap and Die Rules

- Never force a tap handle or the tool may break. Back off the handle to clean out metal shavings.
- Keep the tap and die well oiled to ease cutting.
- Always use the right size tap in the correct size hole.
- Use coarse threads when threading or tapping into soft metal, like aluminum. Coarse threads hold better than fine threads.

Tapping Oversize

When a thread chaser or tap cannot be used to clean up damaged threads, the hole can be drilled and tapped

American National Screw Thread Pitches

Coarse Standard Thread (N.C.) Formerly U.S. Standard Thread				
Bolt or tap size	Threads per inch	Outside diameter at screw	Drill sizes	Decimal equivalent of drill
1	64	.073	53	0.0595
2	56	.086	50	0.0700
3	48	.099	47	0.0785
4	40	.112	43	0.0890
5	40	.125	38	0.1015
6	32	.138	36	0.1065
8	32	.164	29	0.1360
10	24	.190	25	0.1495
12	24	.216	16	0.1770
¼	20	.250	7	0.2010
⁵⁄₁₆	18	.3125	F	0.2570
⅜	16	.375	⁵⁄₁₆	0.3125
⁷⁄₁₆	14	.4375	U	0.3680
½	13	.500	²⁷⁄₆₄	0.4219
⁹⁄₁₆	12	.5625	³¹⁄₆₄	0.4843
⅝	11	.625	¹⁷⁄₃₂	0.5312
¾	10	.750	²¹⁄₃₂	0.6562
⅞	9	.875	⁴⁹⁄₆₄	0.7656
1	8	1.000	⅞	0.875
1 ⅛	7	1.125	⁶³⁄₆₄	0.9843
1 ¼	7	1.250	1 ⁷⁄₆₄	1.1093

Fine Standard Thread (N.F.) Formerly S.A.E. Thread				
Bolt or tap size	Threads per Inch	Outside diameter at screw	Drill sizes	Decimal equivalent of drill
0	80	.060	³⁄₆₄	0.0469
1	72	.073	53	0.0595
2	64	.086	50	0.0700
3	56	.099	45	0.0820
4	48	.112	42	0.0935
5	44	.125	37	0.1040
6	40	.138	33	0.1130
8	36	.164	29	0.1360
10	32	.190	21	0.1590
12	28	.216	14	0.1820
¼	28	.250	3	0.2130
⁵⁄₁₆	24	.3125	I	0.2720
⅜	24	.375	Q	0.3320
⁷⁄₁₆	20	.4375	²⁵⁄₆₄	0.3906
½	20	.500	²⁹⁄₆₄	0.4531
⁹⁄₁₆	18	.5625	0.5062	0.5062
⅝	18	.625	0.5687	0.5687
¾	16	.750	¹¹⁄₁₆	0.6875
⅞	14	.875	0.8020	0.8020
1	14	1.000	0.9274	0.9274
1 ⅛	12	1.125	1 ³⁄₆₄	1.0468
1 ¼	12	1.250	1 ¹¹⁄₆₄	1.1718

Figure 9-16. A tap drill chart tells you what size hole should be drilled for different taps. The drill bit size is in two right columns. The tap or bolt size is in left column.

oversize. A ***drill and tap size chart*** is used to determine right size drill bit and tap, **Figure 9-16.** For example, a 27/64″ drilled hole should have a 1/2″ coarse tap used in it. First, drill the hole one diameter or size larger than the original hole. Then, cut new threads in the drilled hole with the correct size tap. A larger bolt can then be installed in the threaded hole.

Thread Repair Insert

A ***thread repair insert*** should be used when the use of an oversize hole and fastener is not acceptable. An insert takes the place of damaged internal threads and allows the use of the original-size bolt. Look at **Figure 9-17.**

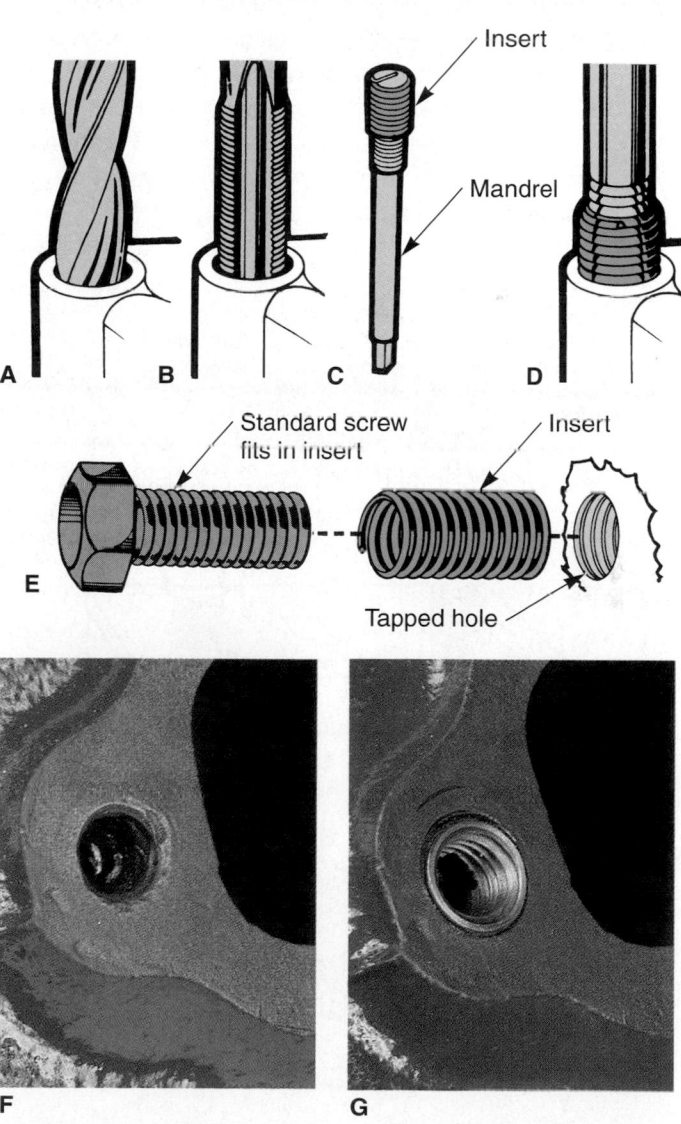

Figure 9-17. Using an insert to repair stripped threads. A—First, drill the hole oversize. B—Next, tap the hole oversize. C—Mount the insert on a mandrel. D—Thread the insert into the hole. E—The insert allows the use of the original-size bolt. F—Damaged threads before a repair. G—An installed insert. (Buick, Chrysler, and The Eastwood Company)

To use a thread repair insert, drill the hole oversize as described in the insert manufacturer's instructions. Then, tap the hole. Finally, screw the insert into the threaded hole. The inside of the insert contains threads that are the same size as those in the original hole.

Removing Damaged Fasteners

An automotive technician must be able to remove broken bolts, screws, studs, and fasteners having rusted or rounded-off heads. Certain tools and methods are needed for removing problem fasteners. Refer to **Figure 9-18.**

- *Locking pliers* can sometimes be used to remove fasteners with heads that are badly rusted and rounded off. Lock the pliers tightly on the bolt or nut for removal.

- A *stud puller,* or *stud wrench,* can remove studs and bolts broken off above the surface of the part. This tool is also used to install studs. Position the stud puller so that it will not clamp onto and damage the threads.

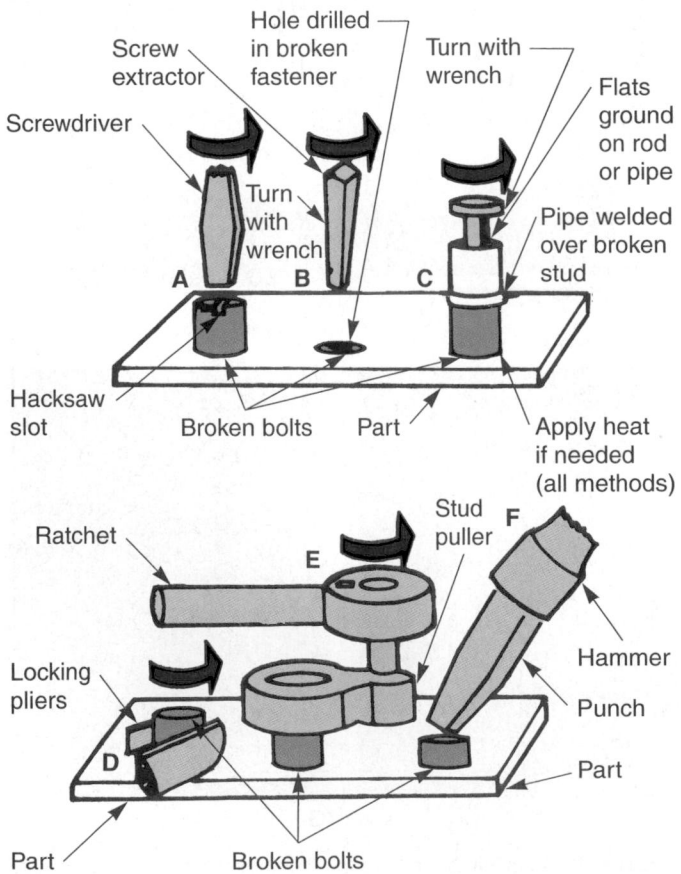

Figure 9-18. Various ways to remove broken bolts. A—Hacksaw a slot and use a screwdriver. B—Drill a hole and use a screw extractor. C—Weld a pipe or shaft onto the bolt and use a wrench. D—Use locking pliers. E—Use a stud extractor and ratchet. F—Use a hammer and punch.

- In some cases, broken fasteners are too short to grasp with any tool. Either cut a slot in the fastener with a hacksaw or weld on another bolt head. Then use a screwdriver or wrench to unscrew the broken bolt.

- When the fastener is broken *flush* with the part surface, a *hammer* and *punch* can sometimes be used to remove it. Angle the punch so that blows from the hammer can drive out the broken fastener.

- A *screw extractor,* or "easy-out," can also be used to remove bolts that are broken flush or below the part surface. See **Figure 9-19.** To use a screw extractor, drill a hole in the center of the broken fastener. Then, lightly tap the extractor into the hole using a hammer. Finally, unscrew the broken bolt by turning the extractor with a wrench.

- On some broken bolts, you may have to drill a hole almost as large as the inside diameter of the threads. Then, use a tap or punch to remove the thread shell. The thread shell is the thin layer of threads remaining in the hole.

 Caution
Be extremely careful not to break a tap or a screw extractor. They are hardened steel and cannot be easily drilled out of a hole. You will compound your problems if you over twist and break one of these tools.

Rust Penetrant

Rust penetrant is a chemical that dissolves rust or corrosion. It is often applied to rusted fasteners to aid in their removal. The penetrant is sprayed or squirted on the rusty fastener and allowed to soak for a few minutes. This often helps the fastener free up without breakage. See **Figure 9-20.**

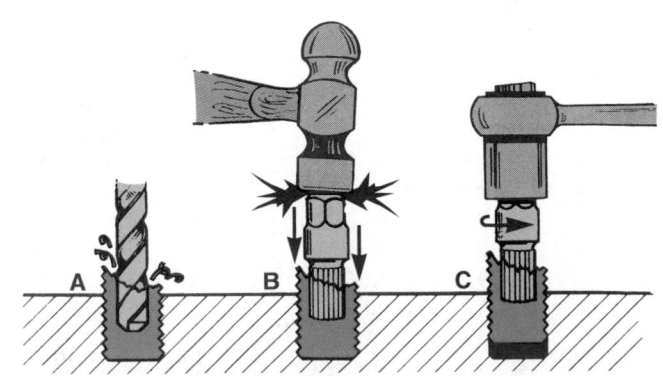

Figure 9-19. Using a screw extractor. A—First, drill a hole in the center of the broken bolt. B—Next, tap the extractor into the hole. C—Finally, unscrew the extractor and the broken bolt with a wrench. (Lisle Tools)

Figure 9-20. On rusted threads, such as those on this brake bleed screw, use rust penetrant to help ease turning and prevent breakage of the screw. (Fel-Pro Gaskets)

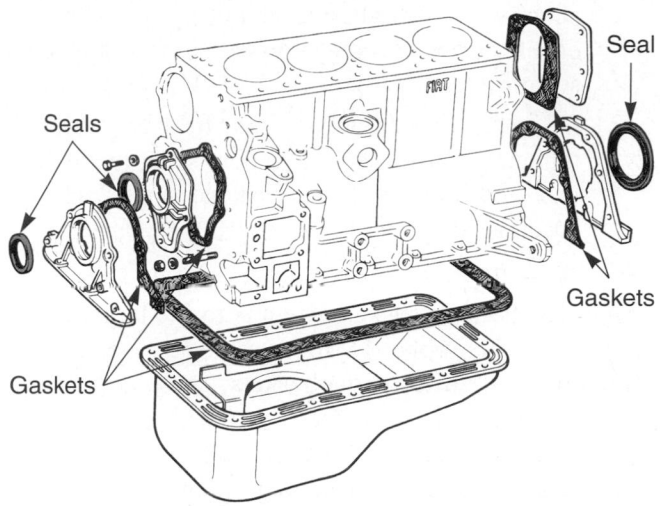

Figure 9-21. Gaskets prevent leakage between stationary parts. Seals prevent leakage between a moving part and a stationary part. (Fiat)

Gaskets and Seals

Gaskets and seals are used between parts to prevent leakage of engine oil, coolant, transmission oil, and other fluids. It is important to understand a few principles about gaskets and seals. If they are serviced improperly, customer complaints and serious mechanical failures can result.

Gaskets

A *gasket* is a soft, flexible material placed between parts to prevent leakage. See **Figure 9-21.** It can be made of fiber materials, rubber, neoprene (synthetic rubber), cork, treated paper, or thin steel.

When the parts are fastened tightly together, the gasket is compressed and deformed. This forces the gasket material to fill small gaps, scratches, dents, or other imperfections in the mating surfaces. A leakproof seal is produced.

Gasket Rules

When working with gaskets, remember the following:

- Inspect for leaks before disassembly. If the two parts are leaking, the part surfaces should be inspected closely for problems.

- Be careful not to nick, gouge, or dent mating surfaces while removing parts. The slightest unevenness can cause leakage.

- Clean off old gaskets carefully. All the old gasket material must be scraped or wire brushed from the parts. Use care, especially on aluminum and brass. These soft metals are easily damaged. Use a dull scraper and wire brush lightly.

- Wash and dry parts thoroughly using solvent. Blow them dry with compressed air. Then, wipe mating surfaces with a clean shop towel.

- Compare the new gasket to the shape of the mating surface. All holes and sealing surfaces must match perfectly.

- Use sealer if needed! Some gaskets require sealer. Sealer is normally used where two different gaskets come together. It will prevent leakage where gaskets overlap. Check a service manual for details. However, use sealer sparingly. Too much sealer could clog internal passages in the assembly.

- After fitting the gasket and parts in place, screw all bolts in by hand. This will assure proper part alignment and threading of fasteners. It also lets you check bolt lengths.

- Tighten fasteners in steps! When more than one bolt is used to hold a part, tighten each bolt a little at a time. First, tighten all the bolts to about half of their torque specification. Next, tighten them to three-fourths torque. Then, tighten the fasteners to full torque. As a final precaution, *retorque* each fastener.

- Use a crisscross tightening pattern. Either a basic crisscross or factory-recommended torque pattern should be used when tightening parts. This will ensure even gasket compression and proper sealing.

- Do not overtighten fasteners. It is very easy to tighten the bolts enough to dent sheet metal parts and smash or break the gaskets. Apply only the specified torque.

Tech Tip

The trend in gasket design is to use large, synthetic-rubber O-ring-type seals instead of fiber or treated paper gaskets. These new seals can be round, D-shaped, or odd-shaped to help seal with the cover design. Follow the manufacturer's instructions when installing these types of gaskets or seals.

Sealers

A gasket is commonly coated with a *sealer* to help prevent leakage and to hold the gasket during assembly, **Figure 9-22.** There are numerous kinds of sealers. They have different properties and are designed for different uses, **Figure 9-23.** Always read the manufacturer's label and the service manual before selecting a sealer.

Hardening sealers are used on permanent assemblies, such as fittings and threads, and for filling uneven surfaces. They are usually resistant to heat and most chemicals. *Nonhardening sealers* are for semipermanent assemblies, such as cover plates, flanges, threads, and hose connections. They are also resistant to most chemicals and moderate heat. *Shellac* is a nonhardening sealer.

It is a sticky substance that remains pliable. It is frequently used on fiber gaskets as a sealer and to hold the gasket in place during assembly.

Form-in-Place Gaskets

Form-in-place gaskets are special sealers used instead of conventional fiber or rubber gaskets. Two common types of form-in-place gaskets are *room temperature vulcanizing (RTV) sealer* and *anaerobic sealer.*

Figure 9-22. Always use the recommended type of sealer or adhesive on gaskets. Some adhesives are oil and fuel soluble and can cause leakage. Both spray adhesive and brush-on sealer are available. They will hold the gasket in place during the assembly of parts. (Fel-Pro)

Type	Temperature range	Use	Resistant to	Characteristics
shellac	−65° to 350° F (−54° to 177° C)	general assembly: gaskets of paper, felt, cardboard, rubber, and metal	gasoline, kerosene, grease, water, oil, and antifreeze mixtures	dries slowly sets pliable alcohol soluble
hardening gasket sealant	−65° to 400° F (−54° to 205° C)	permanent assemblies: fittings, threaded connections, and for filling uneven surfaces	water, kerosene, steam, oil, grease, gasoline, alkali, salt solutions, mild acids, and antifreeze mixture	dries quickly sets hard alcohol soluble
nonhardening gasket sealant	−65° to 400° F (−54° to 205° C)	semipermanent assemblies: cover plates, flanges, threaded assemblies, hose connections, and metal-to-metal assemblies	water, kerosene, steam, oil, grease, gasoline, alkali, salt solutions, mild acids, and antifreeze solutions	dries slowly nonhardening alcohol soluble

Figure 9-23. Different uses and characteristics of various types of sealers. (Fel-Pro)

When selecting a form-in-place gasket, refer to a manufacturer's service manual. Scrape or wire brush all gasket surfaces to remove all loose material. Check that all gasket rails are flat. Using a shop towel and solvent, wipe off oil and grease. The sealing surfaces must be *clean* and *dry* before using a form-in-place gasket.

RTV sealer, also called **silicone sealer,** dries in contact with air. It is used to form a rubber-like gasket on thin, flexible flanges. RTV sealer normally comes in a tube, as shown in **Figure 9-24.** Depending upon the brand, it can have a shelf life from one year to two years. Always inspect the package for the expiration date before using it. If too old, RTV sealer will *not* cure and seal properly.

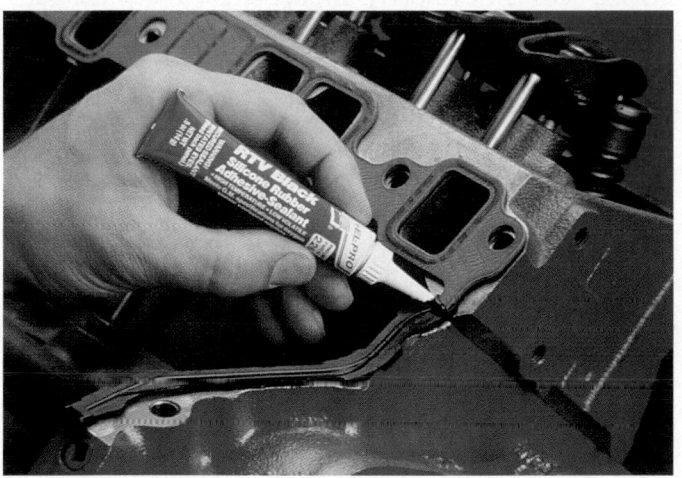

Figure 9-24. RTV sealer is commonly recommended where two different gaskets join. This sealer prevents leakage between the two gaskets.

RTV sealer should be applied in a continuous bead that is approximately 1/8″ (3 mm) wide. All mounting holes must be circled. Locating dowels are often used to prevent the sealing bead from being smeared. If the continuous bead is broken, a leak may result. Uncured RTV can be removed with a rag. Components should be torqued in place while the RTV is still wet to the touch, usually within about 10 minutes of application.

Anaerobic sealer cures to a plastic-like substance in the absence of air and is designed for tightly fitting, thick parts. It is used between two smooth, true surfaces, *not* on thin, flexible flanges. **Figure 9-25** shows the use of both anaerobic sealer and RTV sealer on a water pump. The RTV-sealed section contacts the flexible engine oil pan. The anaerobic-sealed section touches the strong, machined engine block.

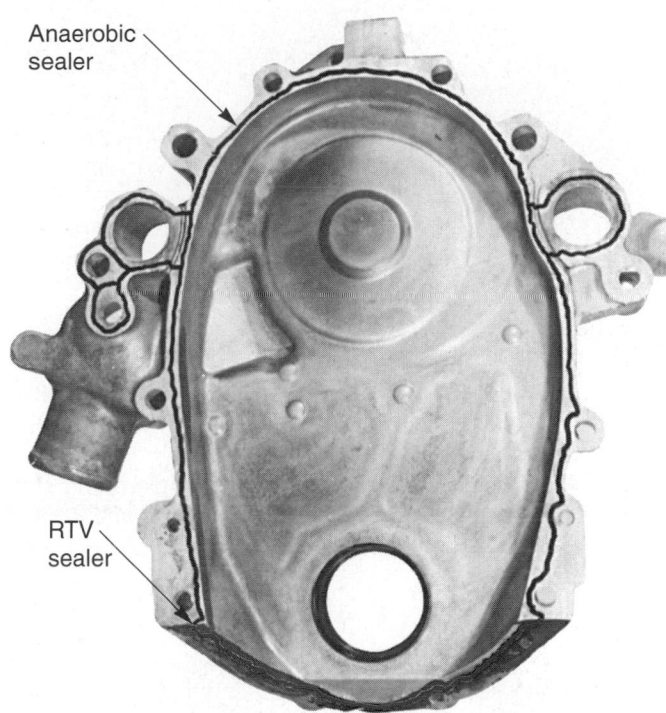

Anaerobic sealer

RTV sealer

Figure 9-25. Notice the use of both RTV and anaerobic sealers. RTV is for flexible flanges. Anaerobic sealer is for solid, tight-fitting castings. Use the recommended bead size and form a continuous bead to avoid leakage. (Pontiac)

Anaerobic sealer should be applied sparingly. Use a 1/16″–3/32″ (1.5 mm–2 mm) wide bead on one gasket surface. Be certain that the sealer surrounds each mounting hole. Typically, bolts should be torqued within 15 minutes of sealer application.

Tech Tip

A few gasket manufacturers sell precut gaskets designed to replace form-in-place gaskets. When working on an engine installed in a vehicle, it can be difficult to properly clean the sealing surfaces. It may also be impossible to fit a part on the engine without hitting and breaking the bead of sealant. When this is the case, a precut gasket might work better than a form-in-place gasket.

Seals

Seals prevent leakage between a stationary part and a moving part. They can be found in engines, transmissions, power steering units, and almost any part containing fluid and moving parts. A seal allows a shaft to spin or slide inside a nonmoving part without fluid leakage. Seals are normally made of synthetic rubber molded onto a metal body, **Figure 9-26.**

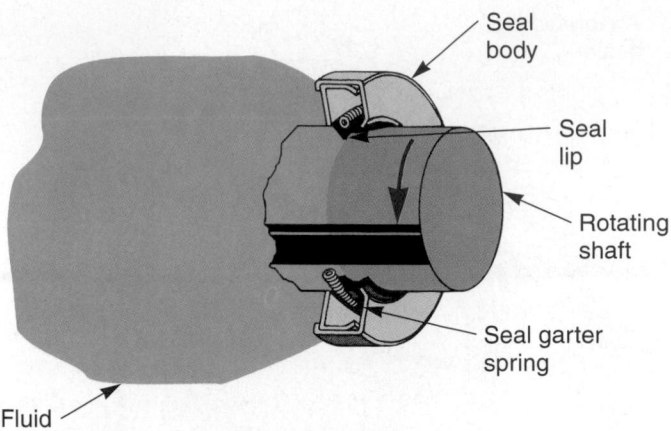

Figure 9-26. A seal mounted in a stationary part. The shaft spins inside the seal. The seal lip faces the fluid and keeps it inside the part. (Caterpillar Tractor)

Seal Rules

There are several important procedures to remember when working with seals.

- Inspect the seal for leakage before disassembly. If a seal is leaking, there may be other problems besides a defective seal. Look for a bent shaft, misaligned seal housing, or damaged parts. Leakage requires close inspection after disassembly.

- Remove the old seal carefully, without scratching the seal housing. Sometimes, a special puller is required for seal removal. This is discussed in later chapters.

- Inspect shafts for wear and burrs, **Figure 9-27.** Look at the shaft closely where it contacts the seal. It should be smooth and flat. File off any burrs that may cut the new seal. A badly worn shaft will require polishing, a shaft sleeve repair kit, or replacement.

- Compare the old seal to the new seal. Hold them next to each other. Both the inside diameter (ID) and the outside diameter (OD) must be the same. To double-check the ID, slip the seal over the shaft. It should fit snugly to prevent leakage.

- Install new seal correctly. Coat the outside of the seal housing with an approved sealer. Coat the inner lip of the seal with system fluid. Install the seal with the sealing lip facing the *inside* of the part. If installed backwards, a tremendous leak will result. Also, check that the seal is squarely and fully seated in its bore.

O-Ring Seals

An *O-ring seal* is a stationary seal that fits into a groove between two parts, **Figure 9-28.** When the parts are assembled, the synthetic rubber seal is partially compressed to form a leakproof joint. Normally, O-ring seals should be coated with system fluid, such as engine oil, diesel fuel, or transmission fluid before installation. This helps the parts slide together without scuffing or cutting the seal. Usually, sealants are *not* used on O-ring type seals. When in doubt about any seal installation, refer to a shop manual.

Figure 9-29 shows a special engine seal that fits into an odd-shaped groove in an engine part. When installing this type seal, make sure the part groove is perfectly clean. During assembly, make sure the seal does not fall out of its groove, or leakage will result.

Other Information

Special gaskets and seals sometimes require other installation techniques. These special situations are discussed in later chapters.

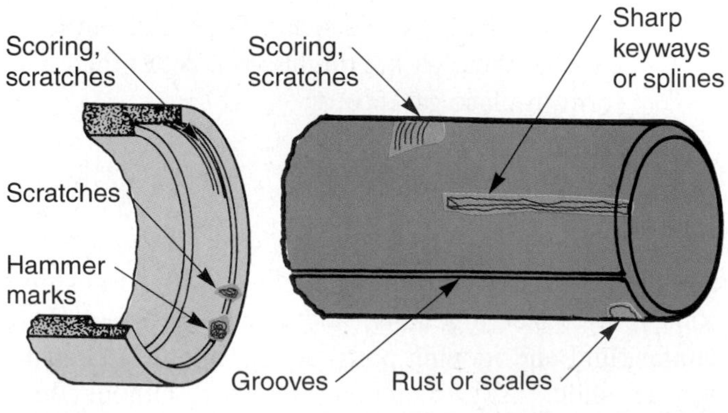

Figure 9-27. Always inspect seals and shafts for damage. Slight nicks or scratches could cause leakage. (Federal Mogul)

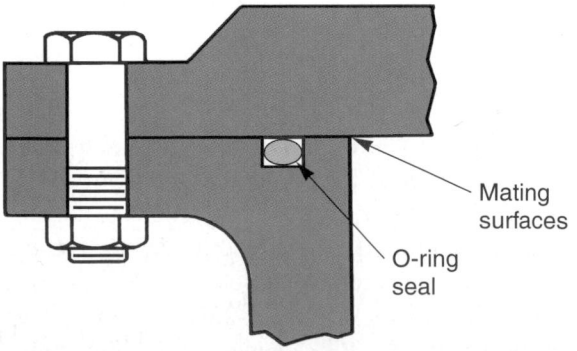

Figure 9-28. O-ring seals prevent leakage by applying pressure to multiple surfaces, as shown. When the parts are bolted together, they partially compress the O-ring. (Deere & Co.)

Figure 9-29. Many modern engines use synthetic seals instead of gaskets. This specially-shaped seal fits down into a groove formed in the part.

- RTV sealer, also called silicone sealer, cures in the presence of air. It is used to form a rubber-like gasket on thin, flexible flanges.
- Seals prevent leakage between a stationary part and a moving part.

Important Terms

Fasteners	Torquing
Bolt	Torque specifications
Cap screw	Torque-to-yield
Nut	Bolt tightening sequence
Tensile strength	Crisscross pattern
Bolt head markings	Torque wrench
Cotter pin	Minor thread damage
Washers	Major thread damage
Machine screws	Tap
Sheet metal screws	Die
Tapping screws	Thread repair insert
Nonthreaded fasteners	Locking pliers
Snap ring	Stud puller
Key	Screw extractor
Keyseat	Rust penetrant
Keyway	Gasket
Set screws	Sealer
Adhesives	Seals

Summary

- Fasteners are devices that hold the parts of a vehicle together.
- When a high-quality bolt is threaded into a part without a nut, it can also be called a cap screw.
- Bolt size is a measurement of the outside diameter of the bolt threads.
- Tensile strength, or grade, refers to the amount of pull or stretch a fastener can withstand before breaking.
- A flat washer increases the clamping surface under the fastener.
- A lock washer prevents the bolt or nut from becoming loose under stress and vibration.
- It is very important that bolts and nuts are torqued properly.
- Torque specifications are tightening values given by the vehicle manufacturer.
- Generally, tightening a group of fasteners on a part follows a crisscross pattern.
- A torque wrench measures the twisting force applied when tightening a threaded fastener.
- A tap is a threaded tool for cutting internal threads in holes. A die cuts external threads.
- Machine screws are similar to bolts, but they normally have screwdriver type heads.
- Numerous types of nonthreaded fasteners such as snap rings, clips, and adhesives are utilized in the assembly of a vehicle.
- Adhesives are widely used on most vehicles.
- A gasket is a soft, flexible material placed between parts to prevent leakage.

Review Questions—Chapter 9

Please do not write in this text. Place your answers on a separate sheet of paper.

1. Define the terms *bolt* and *nut*.
2. How are bolts and nuts usually named in automotive terminology?
3. List and explain the four basic dimensions of a bolt.
4. Customary bolt heads are marked with _____ or _____ marks to indicate bolt strength. Metric bolts use a _____ system to indicate bolt strength.
5. A(n) _____ _____ is a safety device commonly used with a slotted nut.
6. Describe the difference between a flat washer and a lock washer.
7. What are torque specifications?
8. What is a bolt or nut tightening sequence?
9. Which of the following is *not* used for thread repair?
   (A) Tap.
   (B) Die.
   (C) Chaser.
   (D) Chisel.

10. A _____ and _____ _____ _____ is needed to select the right size drill and tap.

11. How do you use a thread repair insert?

12. Describe six ways to remove broken fasteners.

13. Which of the following group of sealers is used on permanent assemblies?
    (A) Hardening.
    (B) Nonhardening.
    (C) Form-in-place.
    (D) All the above.

14. Explain when RTV and anaerobic sealers are recommended.

15. Describe five rules for working with seals.

⬥ ASE-Type Questions

1. When rebuilding a steering system, a can of bolts and nuts was somehow lost. Technician A says to simply go to the hardware store and buy the same size fasteners. Technician B says to order the fasteners from the manufacturer. Who is correct?
   (A) A only.
   (B) B only.
   (C) Both A and B.
   (D) Neither A nor B.

2. Which of the following is a safety device that keeps a slotted nut from turning or coming off?
   (A) Lock tab.
   (B) Snap ring.
   (C) Cotter pin.
   (D) Thread pitch.

3. Which of the following is determined by measuring bolt length change while torquing a bolt?
   (A) Torque stretch.
   (B) Torque-to-yield.
   (C) Torquing sequence.
   (D) Torque specification.

4. Fastener tightening generally follows a:
   (A) clockwise pattern.
   (B) crisscross pattern.
   (C) left-to-right pattern.
   (D) counterclockwise pattern.

5. Which of the following is *not* a basic rule to follow when using a torque wrench?
   (A) Avoid using swivel joints.
   (B) Pull only on the wrench handle.
   (C) Clean fastener threads.
   (D) Use short, jerky pull motions for accuracy.

6. Which threaded tool is used for cutting internal threads in holes?
   (A) Tap.
   (B) Helicoil.
   (C) Thread chaser.
   (D) Thread repair insert.

7. Each of the following is a tool which can be used to help remove problem fasteners *except:*
   (A) hacksaw.
   (B) stud puller.
   (C) die.
   (D) screwdriver.

8. Snap rings hold and fit into grooves in:
   (A) pins.
   (B) gears.
   (C) shafts.
   (D) All the above.

9. Set screws are headless fasteners designed to accept a(n):
   (A) hex wrench.
   (B) screwdriver.
   (C) Allen wrench.
   (D) All the above.

10. While installing gaskets, Technician A believes gaskets alone will produce leakproof seals. Technician B believes that some gaskets require sealer to further prevent leakage where gaskets overlap. Who is right?
    (A) A only.
    (B) B only.
    (C) Both A and B.
    (D) Neither A nor B.

Activities for Chapter 9

1. Prepare a large chart for the shop showing how to read information given on a bolt.

2. Demonstrate the proper methods for repairing thread damage.

Chapter 10

Vehicle Maintenance, Fluid Service, and Recycling

After studying this chapter, you will be able to:

- Check a car's fluid levels.
- Explain the importance of vehicle maintenance.
- Locate fluid leaks.
- Replace engine oil and filter.
- Change automatic transmission fluid and filter.
- Perform a grease job.
- Inspect for general problems with hoses, belts, and other components.
- Demonstrate safe practices while working with vehicle fluids.
- Correctly answer ASE certification test questions on fluid service and vehicle maintenance.

Vehicle fluids include engine oil, coolant, brake fluid, transmission fluid, power steering fluid, and other liquids. All automotive technicians will, at some time, service vehicle fluids. Service station attendants, apprentice mechanics, and even experienced technicians must check, add, or replace fluids.

Many technicians' first job is as a service station attendant. They "cut their teeth" doing fluid service, grease jobs, and light mechanical repairs. Therefore, this chapter is extremely important. It is your chance to learn fluid service and prepare for what may be your first job. Study this material carefully.

The last section of the chapter discusses recycling. To help save our environment, you should recycle as many automotive parts and materials as possible. Plastic bumpers, batteries, tires, and used fluids can all be recycled into new products. Find recyclers in your area who can take your used parts and make them into new products. This will help prevent larger landfills. It will also save energy because less energy is often needed to manufacture new products from recycled ones.

Lubrication Service

Lubrication service is vital to keeping a vehicle in good working order. A technician must be familiar with all aspects of lubrication service, which include:

- Checking fluid levels and conditions.
- Adding fluids as needed.
- Changing engine oil and filter.
- Changing automatic transmission fluid.
- Lubricating (greasing) certain chassis parts.
- Locating fluid leaks and other obvious problems.
- Following state regulations for recycling and disposal of fluids.

Vehicle Maintenance

Vehicle maintenance includes any operation that will keep a vehicle in good operating condition. Without proper care, the life of an automobile may be reduced by thousands of miles. For example, fluids can become contaminated and change chemically after prolonged use. This can cause wear, corrosion, and mechanical failure of parts.

Tech Tip

A wise saying goes, "You can pay now or you can pay later." In the automotive field, this means that the customers can pay a little now for maintenance or pay much more later for repairs. A poorly serviced vehicle will wear out and break down sooner than a well-maintained vehicle. In the long run, vehicle maintenance saves the customer money.

Fluid Service

A service manual contains detailed information on how to check fluid levels. The manual will usually describe:

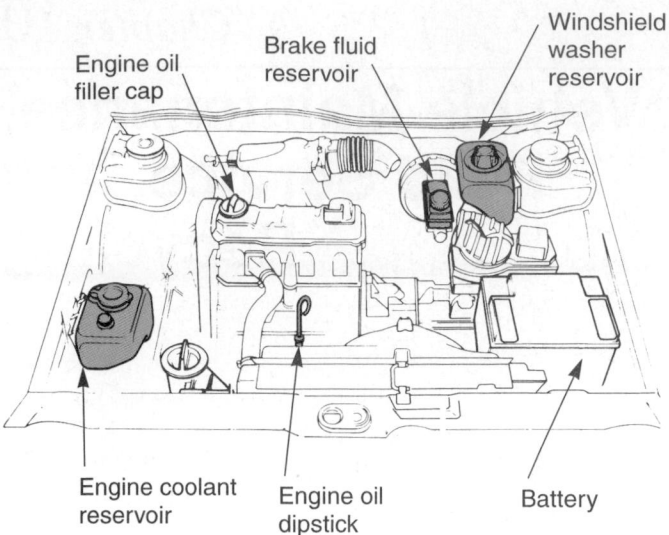

Engine oil filler cap

Brake fluid reservoir

Windshield washer reservoir

Engine coolant reservoir

Engine oil dipstick

Battery

Figure 10-1. A service manual gives the locations of all fluid checkpoints. This manual illustration shows engine compartment fluid check points. (VW)

- Location of fluid check points, **Figure 10-1.**
- Location of fluid fill points.
- Correct interval (time or mileage) between fluid checks and changes.
- Correct type and quantity of fluid to be used.

This information varies from vehicle to vehicle. For example, a diesel engine may require more frequent oil changes than a gasoline engine. Automatic transmission or transaxle fluids, differential lubricants, and other fluids can vary in chemical content.

Tech Tip
A car's warranty may become void if incompatible fluids, incorrect service procedures, or improper intervals are used. For this reason, refer to the manufacturer's recommendations when servicing fluids.

Checking Engine Oil

To check engine oil, warm the engine to operating temperature. Shut off the engine and allow it to sit for a few minutes. Locate and remove the engine oil dipstick. Wipe off the dipstick and replace it all the way into its tube. Pull the dipstick back out and hold it over your shop rag. See **Figure 10-2.**

As shown in **Figure 10-3,** the oil level should be between the ADD (LOW) and FULL marks on the dipstick. Before reinstalling the dipstick, inspect the condition of the oil. The oil should not be too thick or thin, smell like gasoline, or be too dirty.

A

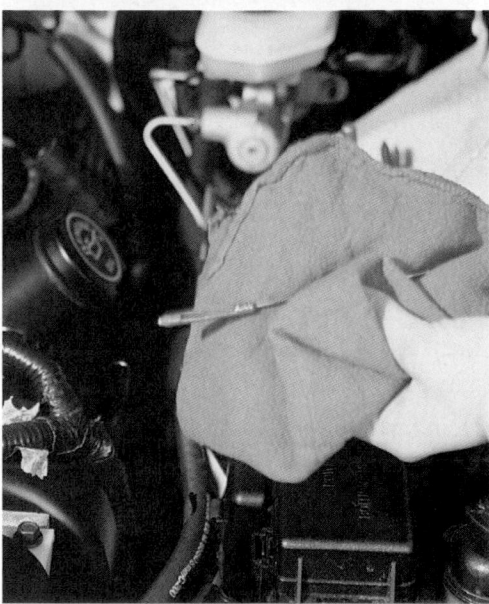

B

C

Figure 10-2. Check the amount of oil in the engine using the dipstick. A—Locate the dipstick and remove it from the engine. B—Wipe the oil from the dipstick; then reinsert the dipstick in its tube. C—Remove the dipstick and check the oil level. Hold the dipstick over a shop rag to prevent oil from dripping onto the engine or the shop floor.

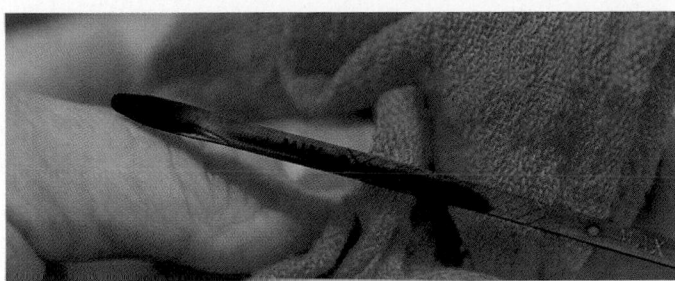

Figure 10-3. Oil should be between the ADD and FULL marks. Be sure to check the condition of the oil before reinserting the dipstick.

 **Warning**
Be careful when checking or changing vehicle fluids. At operating temperature, oil and other fluids can be hot enough to burn your hand.

If the oil level is low, you must add the correct amount and type of oil. If the oil level is down to the ADD mark, typically one quart is needed. If the dipstick reads halfway between ADD and FULL, you would need to add only one-half quart.

Never add too much oil to an engine. Pour in only enough oil to reach the FULL mark. Overfilling can cause *oil foaming* (the oil absorbs air bubbles), which reduces the oil's lubricating ability.

Adding engine oil

To add engine oil, obtain the right kind of oil. Look for a lubrication sticker in the engine compartment or on the driver's door. Use the same type of oil that was installed during the last oil change, if possible.

To add oil to the engine, remove the oil cap, which is usually on the valve cover. Install a small, clean funnel into the engine opening. Without spilling, pour the oil slowly into the funnel and engine filler tube or opening.

Changing Engine Oil and Filter

When changing engine oil, make sure the engine is warm and the vehicle is sitting level. This will ensure that more of the oil contaminants are suspended in the oil and are drained out of the engine. If the oil is cold, the oil will drain more slowly and debris will settle to the bottom of the oil pan.

 To change the engine oil:
  1. Warm up the engine.
  2. Raise the car on a lift or place it on jack stands in a level position.
  3. Place a catch pan under the oil drain plug.

4. Unscrew the plug and allow enough time for the oil to drain completely. See **Figure 10-4.**
5. Reinstall the drain plug. Be careful. The threads in the pan and on the plug can strip easily. Apply only enough torque to draw the plug tight and prevent leakage. A *stripped oil drain plug* can damage the oil pan threads.
6. Position your catch pan under the oil filter.
7. Using an oil filter wrench, as in **Figure 10-5,** unscrew the filter.
8. Obtain the correct replacement filter. Compare the old and new filters. Make sure the rubber O-ring on the new filter has the same diameter as the O-ring on the old filter.

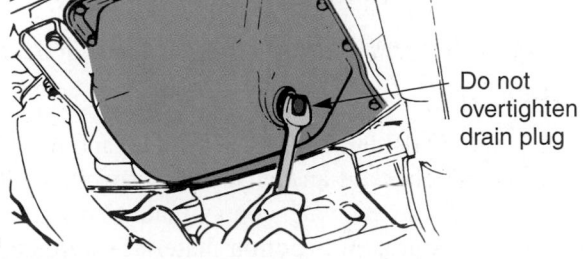

Figure 10-4. To drain the engine oil, remove the oil pan drain plug. Allow the oil to pour into a catch pan. Be careful not to overtighten the drain plug. Its threads will strip out easily. (Subaru)

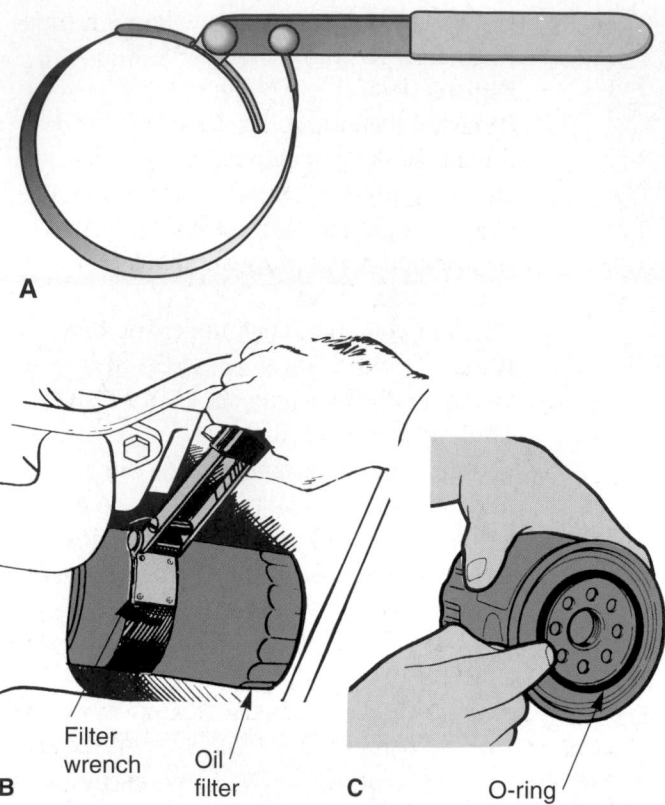

Figure 10-5. Changing the engine oil filter. A—An oil filter wrench is needed to unscrew the oil filter. B—Turn in a counterclockwise direction. C—When installing a new filter, coat the O-ring seal with clean oil and only hand tighten the filter. (Lisle)

9. Wipe some clean oil on the O-ring and install the new filter.
10. Tighten the filter by hand only, *not* with the filter wrench. Overtightening will smash the O-ring and cause leakage.
11. Lower the car to the ground and add the correct amount and type of oil.
12. Start the engine and make sure the oil pressure light *goes out.*
13. Let the engine run while checking for leaks under the engine.

Automatic Transmission Fluid and Filter Service

Like engine oil, automatic transmission and transaxle fluid should be checked and changed at specified intervals. The fluid can become contaminated (filled) with metal, dirt, moisture, and friction material (nonmetallic, heat-resistant fibrous substance) from internal parts. This can cause rapid part wear and premature transmission failure.

To check the fluid in an automatic transmission, warm up the engine and move the gear selector through all positions. Apply the parking brake. Place the transmission in park and block the wheels.

With the engine still running, locate the transmission dipstick. See **Figure 10-6.** It is normally behind the engine, near the front of the transmission.

Tech Tip

Don't go crazy looking for the transmission dipstick on some new cars and trucks. You won't find one. Some transmissions and transaxles are sealed at the factory. They are designed to not require fluid changes and periodic checking of fluid level.

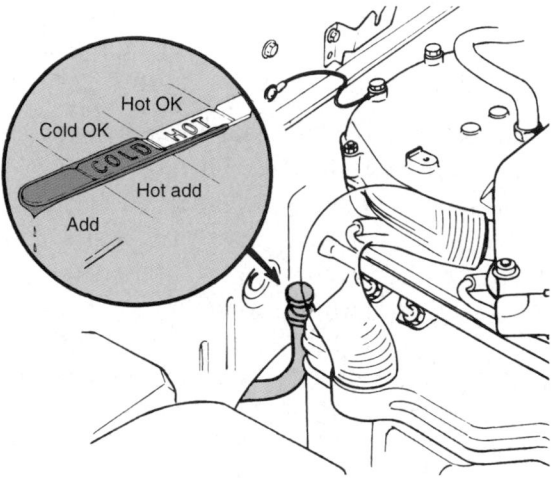

Figure 10-6. An automatic transmission fluid dipstick is normally behind engine, on the side. Check it with the engine running and the transmission in park. If needed, add the correct fluid. (Saab)

Pull out the dipstick. Wipe it off and reinsert it into the tube. Remove the dipstick again and hold it over a shop rag. The fluid should read between the ADD and FULL marks. Also, inspect the fluid for discoloration and odor. If it smells burned or looks dirty, the fluid should be changed.

It is very easy to overfill an automatic transmission. Seldom do you have to add a full quart. Normally, if the dipstick reads ADD, only a fraction of a quart is needed to fill the transmission. Sometimes, instructions are written on the dipstick. If in doubt, check a shop manual.

 To change the fluid and filter in an automatic transmission:

1. Warm the engine and transmission.
2. Raise the vehicle.
3. Remove the bolts securing the transmission pan, **Figure 10-7.** Be careful not to spill the hot transmission fluid. It can cause painful burns!

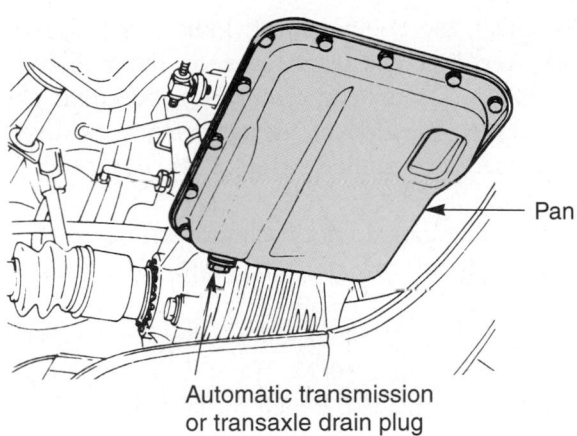

Pan

Automatic transmission
or transaxle drain plug

Figure 10-7. Usually, the transmission pan must be removed to drain the fluid. A few pans, however, have a drain plug. Do not spill hot fluid on yourself. (Subaru)

4.   Unscrew the last pan bolt while holding the transmission pan with a shop rag.
5.   Let the fluid pour into a catch pan.
6.   If needed, replace or clean the transmission filter, **Figure 10-8.**
7.   Scrape the old gasket off the transmission pan and housing.
8.   Position the new pan gasket using an approved sealer. Use the sealer sparingly. You do not want any to squeeze out of the gasket and into the transmission housing.
9.   Start all the pan bolts with your fingers.
10. Tighten the pan bolts in a crisscross pattern to their recommended torque specification. Overtightening can split the gasket or distort the transmission pan.

11. If recommended, drain the torque converter (fluid coupling in front of the transmission). A drain plug may be located in the converter. It is usually under a rock shield on the front of the transmission housing.
12. Refill the transmission with the correct type and amount of transmission fluid. If required, check a service manual for details. In most cases, you must pour fresh fluid into the dipstick tube.
13. Start the engine and shift through the gears.
14. Check under the car for leaks.

Manual Transmission Fluid Service

To check the fluid in a manual transmission, locate and remove the transmission fill plug, **Figure 10-9.** It is normally on one side of the transmission. Generally, warm fluid should be even with the fill hole. With the transmission cold, the fluid can be slightly below the fill hole.

Some manufacturers suggest that manual transmission fluid be changed periodically; others do not. If a fluid change is needed, remove the transmission drain plug on the bottom of the transmission. Install the right kind and quantity of fluid. Lubricate the gear shift mechanism and clutch release as described in a service manual.

Some new transmissions and transaxles are designed to *never* need fluid replacement with normal use. They do not have a fluid dipstick and are permanently sealed. The chemical makeup of the fluid, improved filtering, and less part wear have allowed this technological advance.

Differential Fluid Service

To measure the fluid level in a differential (rear axle assembly), remove the fill plug. It will normally be on the

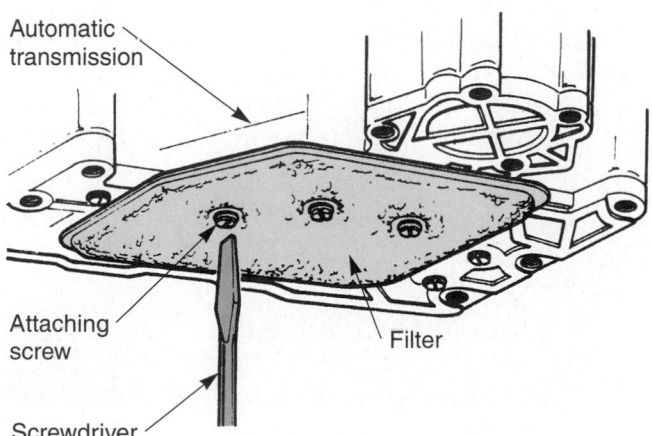

Automatic transmission

Attaching screw

Filter

Screwdriver

Figure 10-8. Some manufacturers recommend periodic replacement of the automatic transmission filter. It is located inside the transmission pan. Tighten all fasteners to specs when assembling.

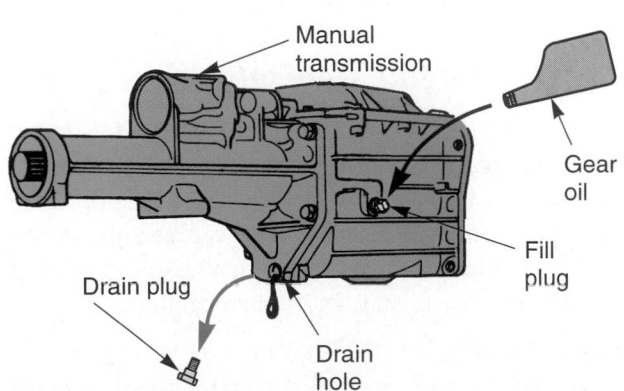

Manual transmission

Gear oil

Fill plug

Drain plug

Drain hole

Figure 10-9. A manual transmission will have a fill plug for checking the fluid level. The fluid should be almost even with the hole when the fluid is warm. Check the service manual for details. (Chrysler)

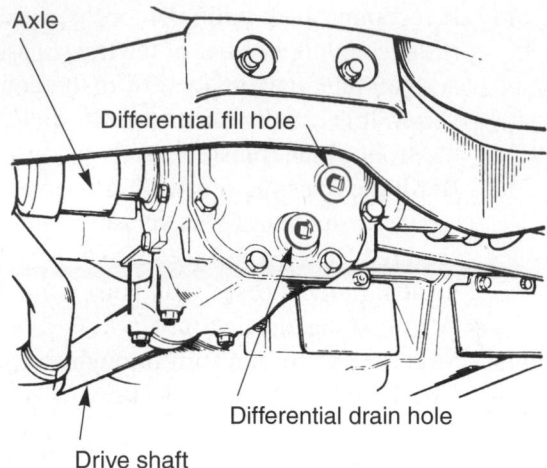

Axle

Differential fill hole

Differential drain hole

Drive shaft

Figure 10-10. The differential fill hole allows for a check of the lubricant level. Do not accidentally remove the drain plug. (Subaru)

front, back, or side of the differential. See **Figure 10-10.** The lubricant should be even with the fill hole when hot. When cold, it should be slightly below the hole.

At the manufacturer's recommended change interval, remove the drain plug. It will be on the bottom of the differential. After draining, reinstall the plug and fill with the proper lubricant. If a drain plug is not provided, a special siphon (suction) gun can be used to draw out the old fluid.

Caution
Positive-traction, or limited-slip, differentials (both wheels turn for added traction) require a special lubricant. Refer to the vehicle identification number, a service manual, and Chapter 62, *Differential and Rear-Drive Axle Diagnosis and Repair,* for details. If you install the wrong lubricant, differential action and traction can be adversely affected.

Checking Engine Coolant

Engine coolant (mixture of water and antifreeze) is used in an engine's cooling system. Engine coolant must be changed at least once every two years. After prolonged use, the coolant will deteriorate. It can become very corrosive and filled with rust. This may result in premature water pump, thermostat, and radiator failure.

Warning
Never remove a radiator cap while the engine or radiator is hot. Boiling coolant can spray out of the radiator, causing serious burns.

To check the coolant level, look at the side of the plastic overflow tank connected to the radiator. See **Figure 10-11.** The coolant should be between the hot and cold marks on the side of the tank. When an overflow tank is used, the radiator cap does *not* need to be removed.

Some older cars do not use an overflow tank. In this case, the radiator cap must be removed to check the coolant level. The coolant should be about an inch (25 mm) down in the radiator. Also, inspect the condition of the coolant, **Figure 10-12.** If rusty, the coolant should be drained and replaced. Watch for system leaks.

Figure 10-11. Modern cooling systems have a reservoir tank. You can check the coolant level without removing the radiator cap. On older systems, remove the cap only after the engine has cooled.

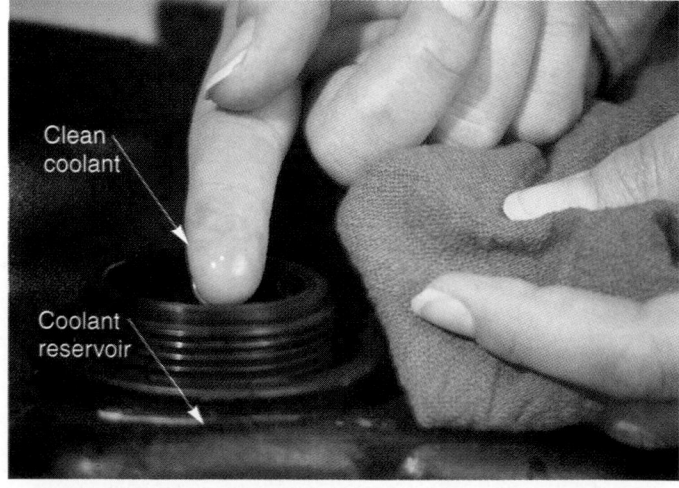

Clean coolant

Coolant reservoir

Figure 10-12. Check the condition of the coolant in the reservoir or the radiator. If the coolant is rusty, it should be drained and replaced.

Checking Power Steering Fluid

Power steering fluid level should be checked regularly. The engine should not be running when checking power steering fluid. If the fluid is contained in a clear plastic reservoir, simply compare the fluid level to the markings on the side of the reservoir. See **Figure 10-13A.** In some vehicles, the level is checked by removing a dipstick from the power steering pump, **Figure 10-13B.** Check the fluid level on the dipstick. If low, inspect for leaks and add the correct type and amount of power steering fluid.

Checking Brake Fluid

The amount of brake fluid in a master cylinder should be inspected at least twice a year. Look at **Figure 10-14.** The master cylinder is normally mounted on the

A

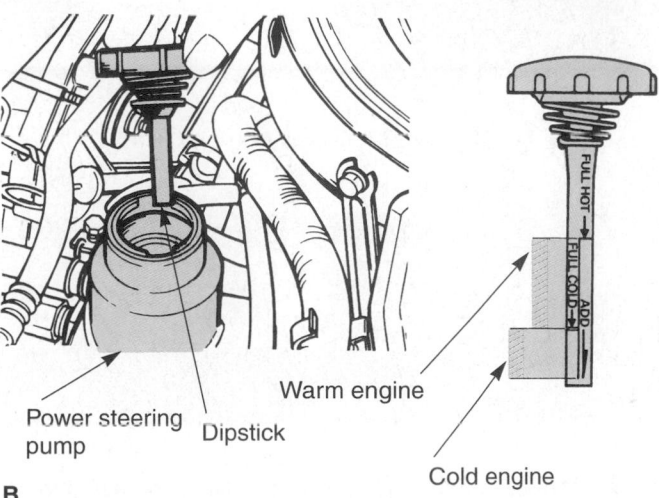

Warm engine

Power steering pump   Dipstick

Cold engine

B

Figure 10-13. Checking power steering fluid. A—Comparing the fluid level to markings on the side of the reservoir. B—On some vehicles, the power steering pump has a cap with a dipstick. Check the fluid with the engine off. Compare the fluid level to markings on the dipstick. (Subaru)

firewall (body section between the engine and the passenger compartment).

When the master cylinder reservoir is clear plastic, simply compare the fluid level to the markings on the reservoir. The fluid should be between the ADD and FULL marks.

With many master cylinders, you must remove the reservoir cover to check the fluid. Generally, the fluid should be about 1/4″ (about 6 mm) down from the top of the master cylinder. Add the recommended type of brake fluid as needed.

 Caution
Never let anything (oil, grease, dirt) contaminate the brake fluid. Oil and grease, for example, will attack the rubber parts in the brake system. Major repairs would be needed and the vehicle could lose braking ability.

Checking Clutch Fluid

Some manual transmission clutches do *not* use mechanical linkage rods or cables. Instead, they use a hydraulic system to disengage the clutch. A clutch master cylinder, similar to a brake master cylinder, produces hydraulic pressure to activate the clutch release. The fluid in the clutch master cylinder should be checked. If low, add brake fluid to fill the reservoir almost full. Always watch for leaks.

Checking Manual Steering Box Fluid

Manual steering box fluid is checked by removing either a fill plug or designated bolt from the top of the

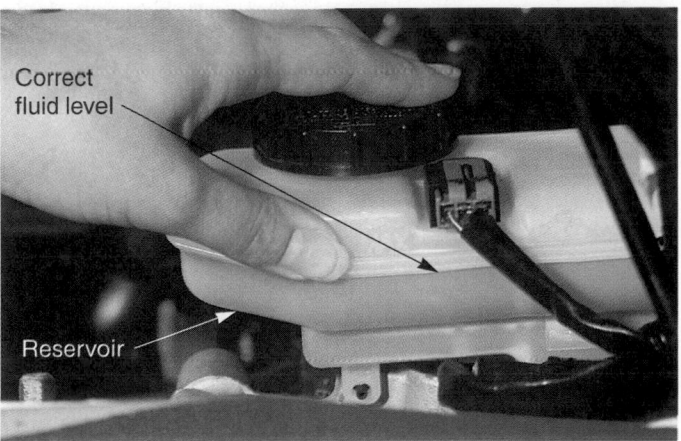

Correct fluid level

Reservoir

Figure 10-14. Check the brake fluid at the master cylinder reservoir. The master cylinder is mounted on the firewall, in front of the driver. The maximum fluid level is often indicated on the side of the reservoir. If not, fluid should be slightly below the the top of the reservoir.

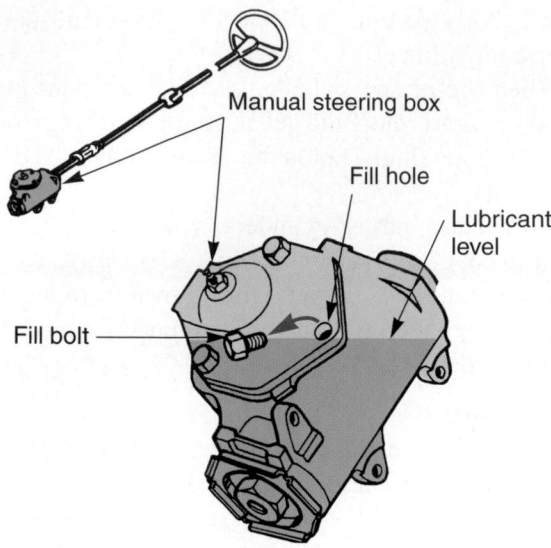

Figure 10-15. The manual steering box will have a bolt or plug for checking the lubricant. If needed, fill with recommended type of fluid up to the fill hole. (Chrysler)

box. See **Figure 10-15.** The lubricant should be almost even with the plug or bolt opening. If not, add the recommended type fluid.

Checking Windshield Washer Solvent

The windshield washer solution is normally visible through the side of the plastic storage tank. Refer back to **Figure 10-1.** If low, add an approved washer solution. The solution will aid windshield cleaning and also prevent ice formation in cold weather.

Checking Battery

New cars use maintenance-free batteries, which do not require an electrolyte (acid) check. However, make sure that the battery terminals and case top are clean. A battery post and cable cleaning tool can be used on corroded connections, **Figure 10-16.**

Filter Service

Quite often, various filters used in a vehicle are replaced during lubrication service. Besides the engine oil and transmission filters, the technician may need to change or clean the air and fuel filters.

If an air filter is extremely dirty, it is normally replaced. However, some manufacturers permit light dirt and dust to be blown from the filter. Special foam or oil-bath (oil-filled) air filters can be cleaned as described in a service manual.

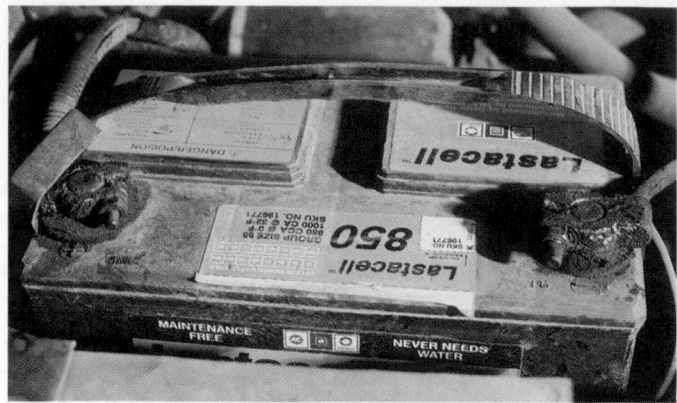

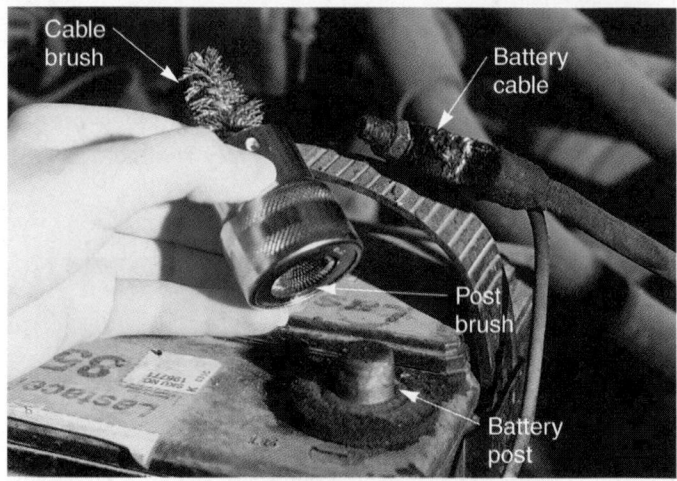

Figure 10-16. Checking battery condition is important to vehicle maintenance. Battery problems are the number one cause of engine "no start" problems. A—A dirty battery top will drain the battery. Corroded terminals prevent charging and starting. B—A post cleaning tool will remove corrosion from the surfaces of posts and cable ends. C—Wash the top of the battery with baking soda and secure terminals.

Fuel filters can be located almost anywhere in the fuel circuit. Modern fuel systems often use inline fuel filters between the fuel tank and the engine. In older systems, the fuel filter can be located at the inlet to the throttle body injector or in the carburetor. Most fuel systems also have a fuel strainer on the pickup tube in the fuel tank. Refer to the service manual for exact filter locations.

 **Warning**
Hold a shop rag around fuel line fittings when loosening. This will keep fuel from spraying out, preventing a possible explosion and fire.

Chassis Lubrication

Chassis lubrication generally involves greasing high-friction points in the suspension, steering, and drive train systems. It may also involve lubricating locks, hinges, latches, and other body parts.

Grease Job

During a *grease job*, you must lubricate high-friction pivot points on the suspension, steering, and drive train systems. Most service manuals illustrate which parts must be lubricated, **Figure 10-17.**

A *grease gun,* **Figure 10-18,** is used to force lubricant (chassis grease) into small fittings. Inject only enough grease to fill the cavity in the part. Overgreasing can sometimes rupture the rubber boot surrounding the joint.

Figure 10-18. This technician is using a power grease gun to lubricate fittings on a suspension system.

Body Lubrication

When performing a complete chassis lubrication job, you should also lubricate high-friction points on the body (hinges and latches on doors, hood, and trunk). See **Figure 10-19.** This will help prevent squeaking doors, sticking hinges, and wear problems.

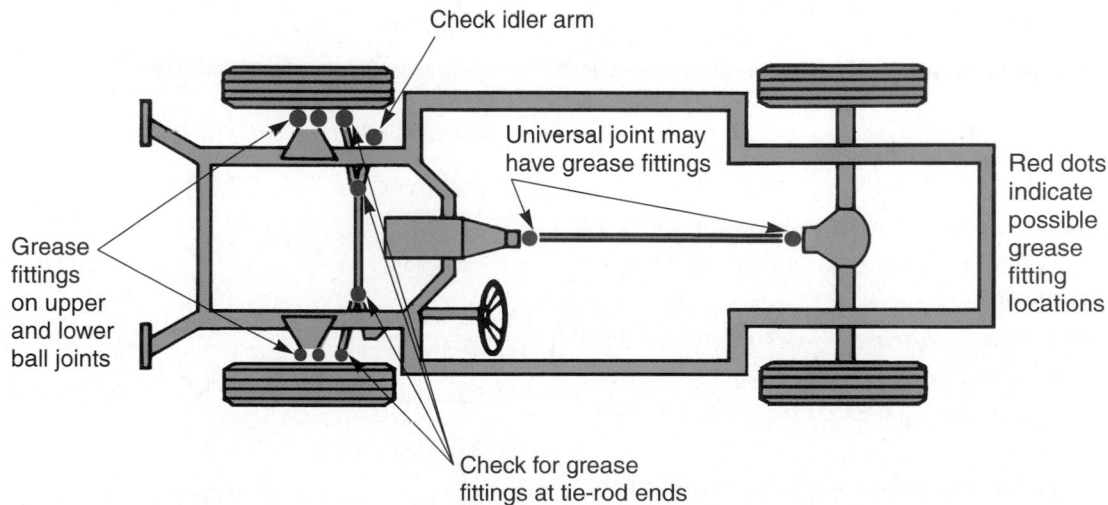

Figure 10-17. A grease job involves lubricating the pivot points shown. Some cars have more grease fittings than others. Check closely.

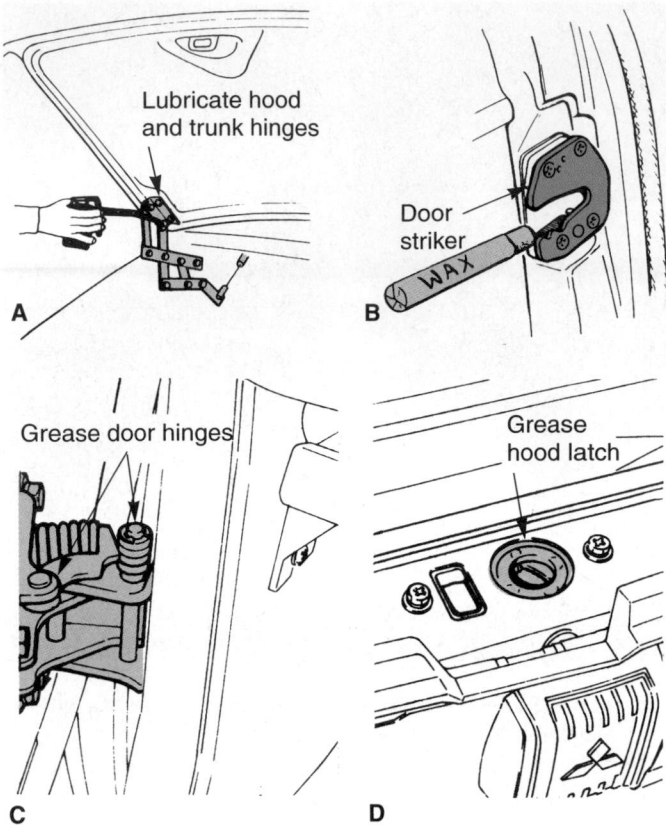

Fig. 10-19. During lubrication service, lubricate body components to prevent squeaks and wear. A—Lubricating hood and trunk hinges. B—Using wax on door strikers. C—Place a dab of grease on door hinges. D—The hood latch is a common rust problem. (Volvo)

Be careful to always use the prescribed lubricant. Normally, rubber and plastic parts will deteriorate if exposed to petroleum-based lubricants (oils and grease). Silicone lubricant should be used on plastic and rubber components. The most common types of body lubricants are listed below:

- *Engine oil*—used on hard-to-reach high-friction points.
- *Graphite*—excellent for door and trunk locks. It will not collect dust and dirt, which could upset lock operation.
- *Dry stick (wax) lubricant*—desirable on door latches and strikers (post that engages the door latch). See **Figure 10-19.** It will not stain clothing.
- *Chassis grease*—good all-around body lubricant. It can be used on easy-to-reach hinges and latches.
- *Silicone lubricant*—often comes in a spray can. It is especially suited for rubber door weather stripping and windows. It is a dry lubricant that will not soil windows and clothing.

Service Intervals

A *service interval* is the amount of time (in months) or the number of miles between recommended service checks or maintenance operations. The factory service manual will give exact intervals for the particular make, model, and year of vehicle. New vehicles tend to have longer intervals before service is required.

Figure 10-20 shows the service manual recommendations for chassis maintenance on one vehicle. Note the intervals for each service operation. They are typical.

Note
Chapter 47, *Engine Tune-Up,* gives general engine maintenance intervals. Refer to this chapter if needed.

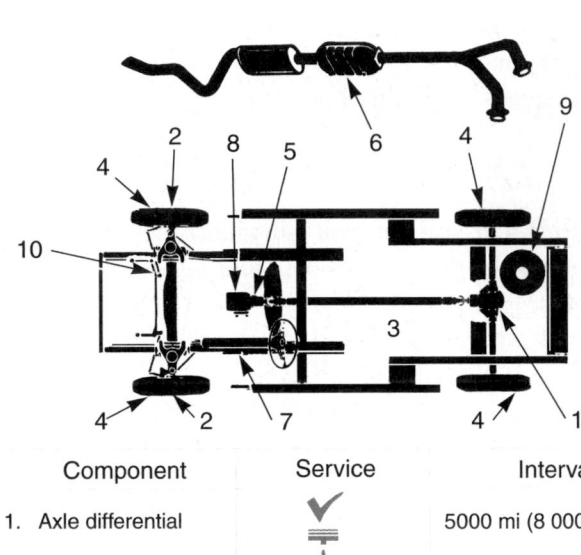

Component	Service	Interval
1. Axle differential	✓	5000 mi (8 000 km)
		30,000 mi (48 000 km)
2. Bearings, front wheel	a.	30,000 mi (48 000 km)
3. Body lubrication	b.	15,000 mi (24 000 km)
4. Brake inspection	✓ c.	15,000 mi (24 000 km)
5. Clutch lever		30,000 mi (48 000 km)
6. Exhaust system inspection	✓ d.	15,000 mi (24 000 km)
7. Manual steering gear	✓ e.	5000 mi (8 000 km)
8. Manual transmission	✓ e.	5000 mi (8 000 km)
9. Spare tire	✓ f.	7500 mi (12 000 km)
10. Steering, suspension, and chassis	✓ g.	15,000 mi (24 000 km)
	h.	30,000 mi (48 000 km)

Figure 10-20. Study the chassis maintenance information from a service manual. Recommendations for other parts of the car are also given in the manual. (Chrysler)

General Inspection and Problem Location

As you perform lubrication service or any kind of auto repair, always watch for mechanical problems. Visually inspect the vehicle for any signs of wear, deterioration, loose parts, or leaks. Check the condition of fan belts, water hoses, fuel hoses, vacuum hoses, and wiring. This can be done as you are working.

- *Hose inspection* includes checking for hardening, softening, cracking, splitting, or other signs of impending failure. See **Figure 10-21.** Squeeze all the hoses. If deteriorating (hard or soft), inform the customer or shop supervisor of the problem.

- *Drive belt inspection* includes looking for splitting, tears, cuts, and wear. If worn or loose, the belt may slip and squeal. Refer to **Figure 10-22.**

- *Wiring inspection* involves looking for improper routing, cracked or brittle insulation, or other obvious problem signs. Make sure wires are away from all moving or hot parts.

- *Tire inspection* is done by looking for excessive wear, improper inflation, or physical damage. This is very important from a safety standpoint.

- *Steering system inspection* includes checking for excessive wear and play in moving parts. The steering wheel should *not* move more than about an inch (25 mm) without causing front wheel movement. If it does, wear in the steering mechanism is indicated.

- *Exhaust system inspection* involves looking for damaged, rusted, or leaking parts. The exhaust system should be inspected any time a vehicle is raised on a lift. Poisonous exhaust fumes make a

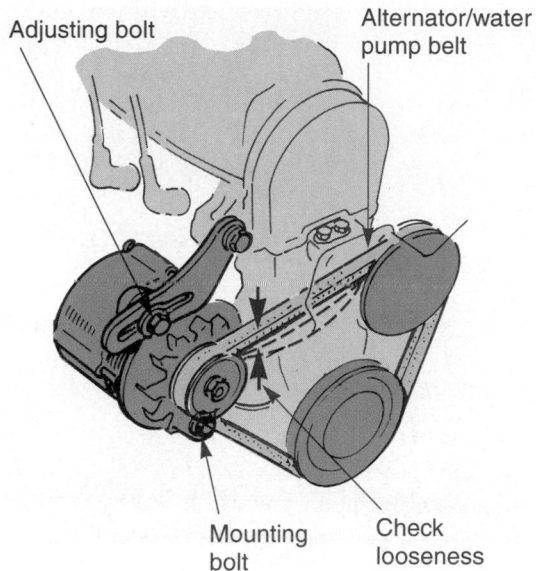

Figure 10-22. Belts should not be too loose or too tight. To adjust, loosen correct mounting bolts and the adjustment bolt. Using directions in service manual, pry the component outward and tighten the adjusting bolt. Then, tighten the mounting bolts. Recheck belt tightness. (Honda)

leaking exhaust system very dangerous. Look for rust holes in the pipes, muffler, and other parts.

When working on a vehicle, be alert for these kinds of problems. This will show the shop supervisor and the customer that you are a concerned, competent technician.

Fluid Leaks

Fluid leaks result from bad gaskets, seals, or hoses; cracks in parts; and similar troubles. Leaks are very common problems that should be corrected. See **Figure 10-23.** To become good at leak detection and correction, you should:

- Become familiar with the *color*, *smell*, and *feel* (texture) of the different fluids. Then, you will be able to quickly identify a fluid leak. Does the fluid feel more like water or oil? Dab a white paper towel into a puddle of fluid to show its color more accurately. Oil will be slippery to the touch and will be dark brown or black if used for an extended period of time. If the fluid is clear or brown and feels "squeaky" when rubbed between your fingers, it is probably brake fluid or hydraulic clutch fluid. Antifreeze can be green, orange, or rust colored and will feel slick. Automatic transmission fluid can be dark brown, red, or dark green with some friction material feel (gritty). Power steering fluid can be amber, red, or clear and will feel like transmission fluid.

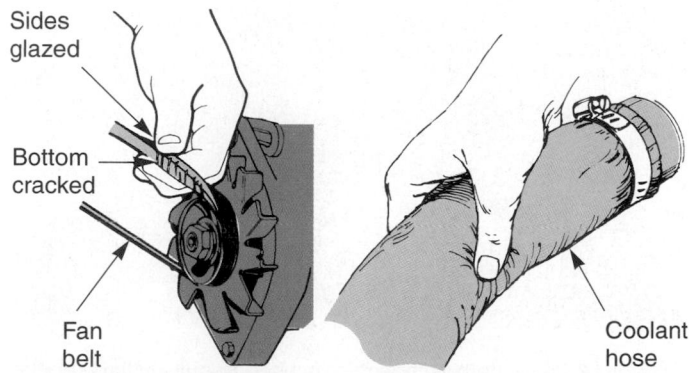

Figure 10-21. Check the condition of all hoses and belts. Inspect belts for glazing, cracking, and fraying. Feel hoses for hardening or softening. Look for leaks. (Gates Rubber Co.)

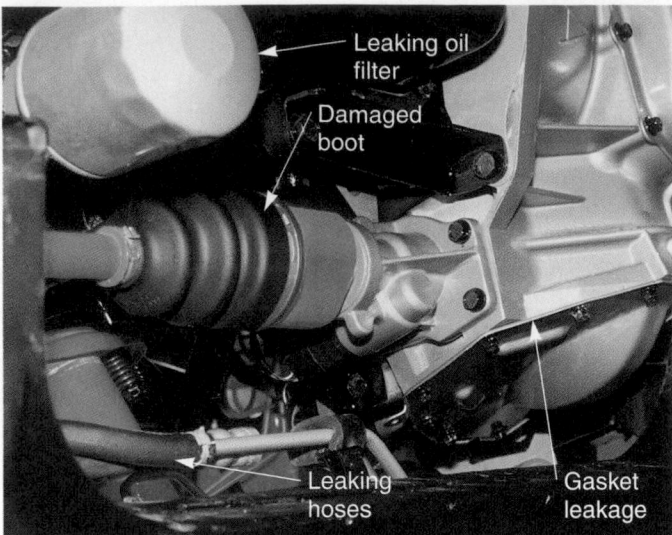

Figure 10-23. When working under a vehicle, always look for signs of fluid leakage and other obvious problems. Inspect rubber boots and hoses for signs of damage.

- Fluid leaks tend to flow downward and to the rear of the vehicle. For this reason, look for leaks above and in front of where you find fluid dripping off the vehicle.
- If multiple leaks are indicated, fix the leak located the highest and farthest forward on the vehicle. Then, repair other leaks.
- If the leaking part is badly soiled, clean the area thoroughly. Then you will be able to see fresh fluid leaking out of the part more easily.
- The most frequent cause of fluid leakage is broken gaskets and worn seals. Replacement will usually correct the problem. However, you should always check the parts for warpage, cracks, and dents.

Some *stop-leak products* are designed to recondition the leaking seal. If the seal has hardened and shrunk in size, stop-leak chemicals can fix the leak with little time and effort. If the seal is torn or the part is warped, stop-leak products will not work.

Other stop-leak products (for the cooling system, for example) contain small particles (fibrous or metallic materials) that collect at and fill the opening causing the leak. These materials can also act as a conditioner or rust preventative.

Tech Tip

Stop-leak products will not work on hose leaks and large leaks from metal parts. These products are designed to work on small seepage problems. Refer to the manufacturer's recommendations on stop-leak products.

Noise Detection and Location

Abnormal noises are unwanted sounds that indicate part wear or other mechanical problems. They are common to almost all systems of a vehicle. When inspecting a vehicle, listen for unusual sounds (knocks, clunks, rattles, clicks, and hisses). As you work, always listen for abnormal noises.

A *stethoscope* (similar to that used by a doctor to listen to a patient's heart) is commonly used by an auto technician. It will help the technician pinpoint the source of internal part noises. To use the stethoscope, touch the probe on the component near the unwanted sound, **Figure 10-24.** Move the stethoscope around until the sound is the loudest.

A long screwdriver can also be used in place of a stethoscope. Place the tip of the screwdriver on the part. Place the handle next to your ear. Sound will travel through the screwdriver and permit noise diagnosis. Make sure you keep the screwdriver away from moving parts or you could be injured.

A section of vacuum hose, **Figure 10-25,** is a handy device for finding sounds *not* coming from internal parts. The hose is useful for locating hissing sounds, rattles, whines, and squeaks. Place one end of the hose to your ear. Then, move the other end around the area of the sound. When the noise becomes the *loudest*, you have

Figure 10-24. A stethoscope can be used to quickly find knocks and rattles inside components. Move the tip around on parts. When noise becomes the loudest, you have found the source of the problem.

Figure 10-25. A piece of vacuum hose can be used like a stethoscope to find external noises. It will find vacuum leaks, squeaks, wind noise, and other abnormal sounds.

pinpointed the problem. Again, keep the hose away from moving or hot parts. By removing the metal end from a stethoscope, you can also listen for these kinds of noises.

Recycling and Disposal of Auto Shop Wastes

Recycling and disposal of auto shop wastes are needed to help save our planet's natural resources and to reduce the amount of materials being sent to landfills. Laws have been passed that require specific procedures when handling and discarding potentially harmful materials. The following sections summarize this important information.

Auto Shop Wastes

Automotive maintenance may generate hazardous wastes that come under the requirements of the Resource Conservation and Recovery Act. This federal act covers businesses that generate, transport, and manage hazardous wastes. Any business that maintains or repairs vehicles, heavy equipment, or farm equipment is classified as a vehicle maintenance facility by this act.

Vehicle maintenance fluid and solid wastes include:

- Used motor oil (combustible and may contain toxic chemicals).

- Other discarded lubricants, such as transmission and differential fluids (like motor oil, may contain toxic chemicals).
- Used parts.
- Cleaners and degreasers that are contaminated from parts-cleaning operations.
- Carburetor and fuel injection system cleaners (contain flammable or combustible liquids).
- Rust removers (may contain strong acidic or alkaline solutions).
- Paint thinners or reducers (may be ignitable or contain toxic additives).
- Worn out batteries (lead and toxic chemicals).
- Tires and catalytic converters.

Repair and maintenance facilities (such as service stations, automotive dealerships, and independent auto repair shops) that generate 220 lb. (100 kg) of hazardous waste monthly must file a Uniform Hazardous Waste Manifest before removing the wastes. The manifest must list the proper Department of Transportation (DOT) shipping descriptions for a number of wastes. Tables listing these descriptions are available from each state's hazardous waste management agency or a regional EPA office.

However, EPA regulations also state that no manifest is needed for used oil or lead-acid batteries if they are sent for recycling. In such cases, the material is not regarded as hazardous. Your state may have its own requirements; check with your state hazardous waste management agency.

Unless recycled for scrap metal, used oil filters are considered hazardous waste. If not recycled, they must be listed on the monthly manifest as hazardous. Before disposal, filters should be gravity drained so they do not contain free-flowing oil. Store them uncrushed in a closed, labeled container for pickup by a recycler.

Recycling Motor Oil

Used motor oil is considered hazardous waste unless it is destined to be recycled. The old oil should be stored in an approved container for recycling. One gallon of used motor oil can be refined into two and one-half quarts of high-quality motor oil. It takes about 40 gallons of crude oil to produce this much motor oil. Recycling old oil not only saves our environment from pollution, but it also helps save our natural resources. Always send used motor oil to a recycling center! Put the oil in an approved container. Some recycling companies provide a pickup service, while others require you to take the old oil to their facility.

Recycling Coolants

Antifreeze has been classified as a hazardous waste due to heavy metal and chlorinated solvents that it picks up circulating through cooling systems. It should never be mixed with used oil. The entire mixture would then be classified as a hazardous waste, even though the used oil may not be, under federal regulations.

Regulations require that spent antifreeze solutions be collected by a registered hazardous waste hauler. Several major companies offer pickup and recycling services.

Recycling Refrigerants

Refrigerants, such as R-12 and R-134a, removed from automotive air conditioning systems during servicing, should not be vented to the atmosphere. State regulations require that refrigerants be recovered and recycled. See **Figure 10-26.**

As you will learn in later chapters, systems are now available for recovering, cleaning, and recycling air conditioning refrigerants, which cannot be exhausted into the atmosphere.

Other Automotive Recyclables

Other recyclable materials that are commonly removed from service during maintenance and repair of vehicles include:

- Catalytic converters, which contain platinum.
- Worn tires, which can be sold to a retreader (if the carcass is sound) or to a shredder. Shredded rubber is an ingredient in road resurfacing materials and other products that give the rubber a second use.
- Batteries can be recycled and used to make new batteries. This saves lead, acid, and other materials from adding to our waste disposal problems.
- Brake shoes can be recycled and sold as cores for making reconditioned brake shoes.
- Many small assemblies (alternators, starters, master cylinders, etc.) can be recycled and made into rebuilt parts.
- Plastic bumpers and other body parts can be recycled into a variety of new products.

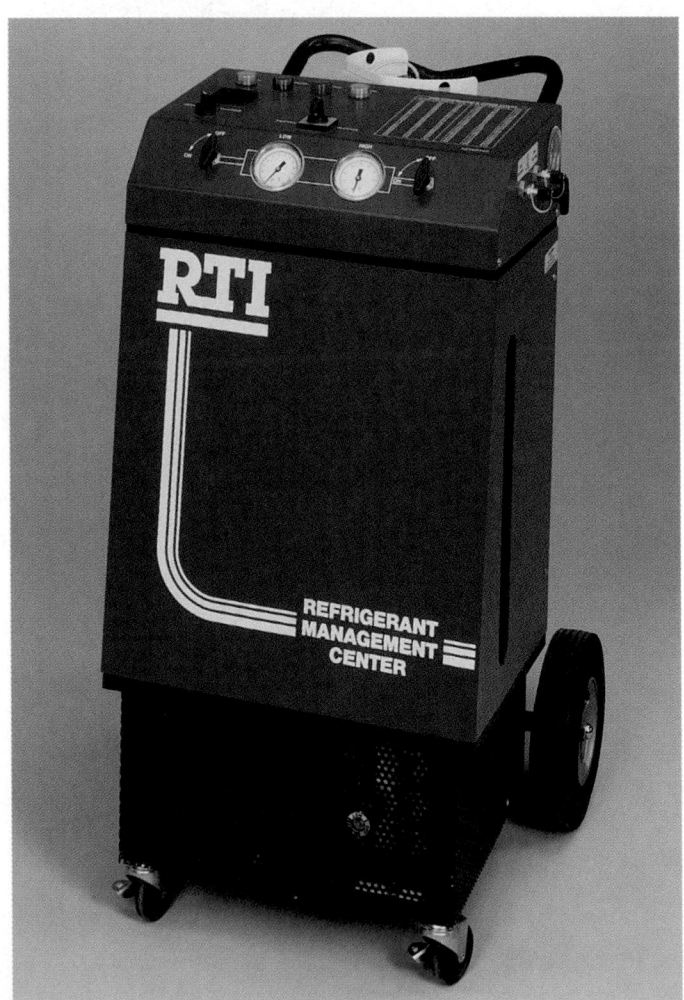

Figure 10-26. A recovery station will pull old refrigerant out of the air conditioning system. The machine will then treat used refrigerant for reuse in the vehicle. (RTI)

Summary

- Vehicle fluids include engine oil, coolant, brake fluid, transmission fluid, power steering fluid, and other liquids.
- Vehicle maintenance includes any operation that will keep a vehicle in good operating condition.
- A car's warranty can become void if improper fluids or incorrect service procedures or intervals are used.
- Be careful! Oil and other fluids at operating temperature can be hot enough to burn.
- EPA (Environmental Protection Agency) guidelines and state regulations affect how you must handle and dispose of used fluids, solvents, and other shop chemicals.
- Never remove a radiator cap while the engine or radiator is hot. Boiling coolant can spray out of the radiator, causing serious burns.
- Hold a shop rag around fuel line fittings when loosening. This will keep fuel from spraying or leaking out, preventing a possible explosion and fire.
- During a grease job, you must lubricate high-friction pivot points on the suspension, steering, and drive train systems.

- Wheel bearings are usually packed (filled) with grease during lubrication service.
- As you perform a lubrication job or any kind of auto repair, always watch for mechanical problems. Visually inspect the vehicle for any signs of wear, deterioration, loose parts, or leaks.
- A stethoscope (similar to that used by a doctor to listen to a patient's heart) is commonly used by a technician.
- A section of vacuum hose can be used as a handy device for finding sounds not coming from inside parts.

Important Terms

Vehicle fluids	Packed
Lubrication service	Service interval
Vehicle maintenance	Hose inspection
Oil foaming	Drive belt inspection
Stripped oil drain plug	Wiring inspection
Grease job	Tire inspection
Grease gun	Steering system
Engine oil	inspection
Graphite	Exhaust system
Dry stick (wax)	inspection
lubricant	Fluid leaks
Chassis grease	Stop-leak products
Silicone lubricant	

Review Questions—Chapter 10

Please do not write in this text. Place your answers on a separate sheet of paper.

1. What seven steps does lubrication service typically involve?
2. _____ _____ includes any operation that will keep the car in good operating condition.
3. When checking engine oil, allow the engine to cool completely. True or False?
4. What can happen when too much oil is added in an engine?
5. Which of the following should *not* be done when changing an engine's oil and filter?
   (A) Torque the drain plug only enough to prevent leaking and loosening.
   (B) Use an oil filter wrench to remove the old filter.
   (C) Wipe clean engine oil on the new filter O-ring seal.
   (D) Use a filter wrench to tighten the filter.
6. The automatic transmission dipstick is normally located in front of the engine. True or False?
7. Check automatic transmission or transaxle fluid with the engine running. Check engine oil with the engine off. True or False?
8. Explain how to check the following:
   (A) Engine coolant level and condition.
   (B) Power steering fluid.
   (C) Brake fluid.
   (D) Manual steering fluid.
   (E) Battery condition.
9. A(n) _____ _____ involves lubricating the steering, suspension, and drive train of a vehicle.
10. List and explain the use of five lubricants.
11. A(n) _____ _____ is the amount of time between recommended service or maintenance operations.
12. Describe six general inspection points that should be checked during vehicle maintenance.
13. Which of the following should be done to help with leak detection and troubleshooting skills?
    (A) Become familiar with the color and smell of different fluids.
    (B) Trace the problem to the highest point of wetness or leakage.
    (C) Clean the area around the leak if the leak is difficult to isolate.
    (D) All of the above.
14. A(n) _____ is commonly used to find the source of noises inside parts.
15. List four automotive items that should be recycled.

✹ ASE-Type Questions

1. During a complete lubrication service, all of the following are done *except*:
   (A) Change engine oil.
   (B) Change oil filter.
   (C) Check all fluid levels.
   (D) Check ride height.
2. During an oil change, Technician A says to use an oil filter wrench to remove the old oil filter. Technician B says to apply a little oil to the seal on the new filter to aid installation. Who is correct?
   (A) A only.
   (B) B only.
   (C) Both A and B.
   (D) Neither A nor B.

3. When changing engine oil, Technician A says the engine oil should be cool. Technician B says the oil should be warm. Who is correct?
   (A) *A only.*
   (B) *B only.*
   (C) *Both A and B.*
   (D) *Neither A nor B.*

4. When checking automatic transmission fluid, the following should be done.
   (A) *Engine off, transmission in park.*
   (B) *Engine running, transmission in park.*
   (C) *Engine off, transmission in neutral.*
   (D) *Engine on, transmission in neutral.*

5. Engine coolant consists of:
   (A) *water and antifreeze.*
   (B) *oil and antifreeze.*
   (C) *water and stop leak.*
   (D) *water only.*

Activities—Chapter 10

1. On a vehicle chosen by your instructor, determine the capacity of the cooling system. Then determine the amounts of water and antifreeze needed to produce a 50/50 mixture.

2. Survey the motor oils offered for sale at a local outlet. Determine the cost to change the motor oil in a vehicle named by your instructor. Add labor cost at $36/hour.

3. Change the oil and filter on a vehicle designated by your instructor.

Cutaway of a late-model V-6 engine. With proper maintenance, this engine will provide years of trouble free operation. (Lexus)

Engines

The engine produces the energy needed to propel the vehicle and operate other systems. If you know how an engine is constructed and how it is designed to operate, you will be better prepared to diagnose problems when it malfunctions.

Section 2 details the construction and operation of late-model automotive engines. It reviews the four-stroke cycle, explains the names and locations of major parts, describes design variations, and summarizes engine size and performance measurements.

The information presented in this section will help you pass ASE Test A1, *Engine Repair.* It will also help you prepare for later chapters on engine service and repair.

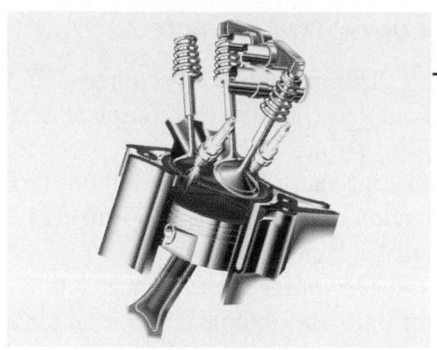

Chapter 11

Engine Fundamentals

After studying this chapter, you will be able to:

- Identify the major parts of a typical automotive engine.
- Describe the four-stroke cycle.
- Define common engine terms.
- Explain the basic function of the major parts of an automotive engine.
- Cite and demonstrate safe working practices related to engines.
- Correctly answer ASE certification test questions that require knowledge of the basic operation of piston engines.

In the first chapter, you learned a little about how an engine operates. This chapter will build upon that information by explaining each engine part in more detail. What you learn here will help you prepare for other coverage of engine types, engine construction, engine diagnostics, and engine service.

Note
If needed, quickly review the material on engines and engine systems in Chapter 1, *The Automobile*. A sound understanding of engine theory and engine support systems is very important.

Engine Operation

The *engine* is the source of power for the vehicle. For this reason, it is also called a power plant. An energy source, or *fuel* (usually gasoline or diesel oil), is burned inside the engine's *combustion chamber* (hollow area between the top of the piston and the bottom of the cylinder head) to produce heat. The heat causes *expansion* (enlargement) of the gases in the engine.

The burning and expansion in the combustion chamber produces pressure. The engine piston, connecting rod, and crankshaft convert this pressure into motion for moving the car and operating its other systems.

Figure 11-1 shows how an engine converts fuel into a useful form of energy. Combustion pressure forces the piston down. By linking the piston to the crankshaft, an engine can produce a powerful spinning motion. The rotating crankshaft can be used to drive gears, chains and sprockets; belts and sprockets; and other drive mechanisms.

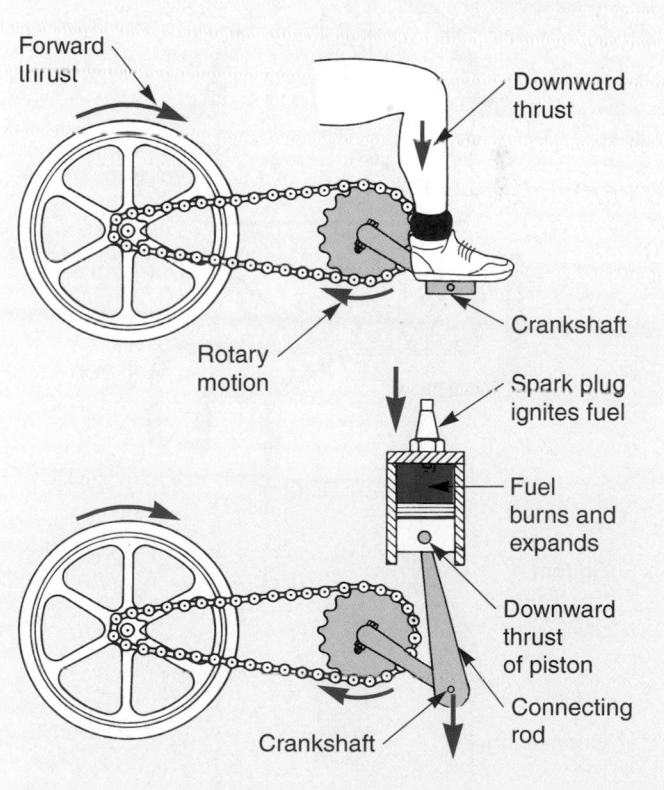

Figure 11-1. A crankshaft converts the downward thrust of the piston into useful rotating motion. The rotating motion can be used to operate the drive mechanism.

Piston Travel (TDC, BDC)

The distance the piston can travel up or down in the cylinder is limited by the crankshaft. When the piston is at the *highest point* in the cylinder, it is at **top dead center (TDC)**. When the piston slides to its *lowest point* in the cylinder, it is at **bottom dead center (BDC)**. See **Figure 11-2**.

Piston Stroke

Piston stroke is the distance the piston slides up or down from TDC to BDC. This takes one-half turn of the crankshaft. The crank rotates 180° during one piston stroke. Refer to **Figure 11-2**.

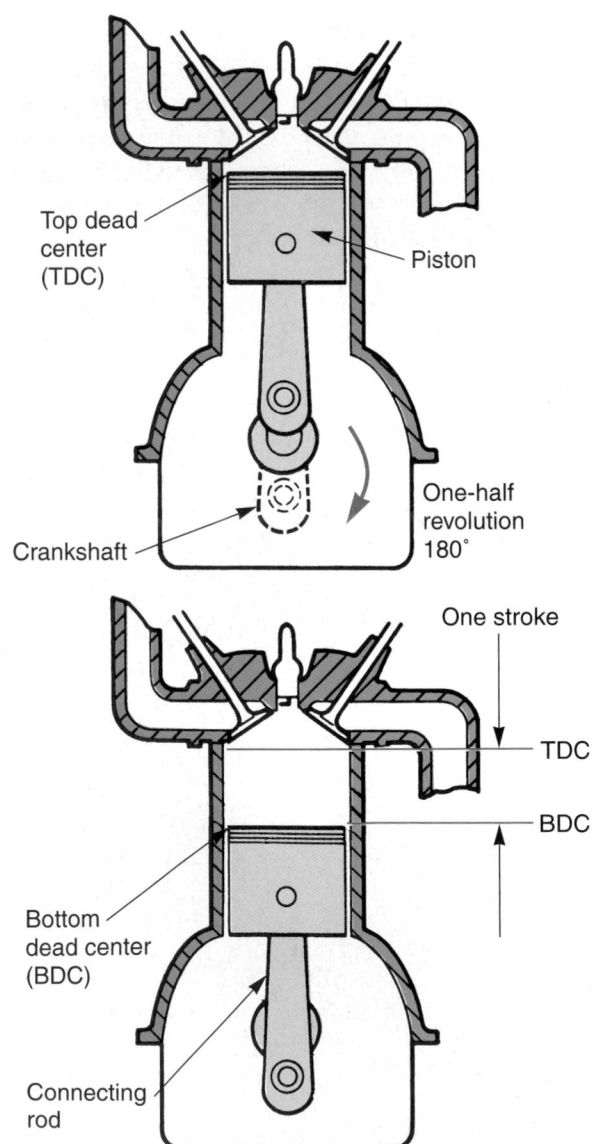

Figure 11-2. TDC means the piston is at the top of its stroke. BDC means the piston is at the bottom of its stroke. A stroke is one piston movement. (Ford)

Four-Stroke Cycle

The **four-stroke cycle** requires four piston strokes to complete one cycle (complete series of events). Every four strokes, the engine produces one *power stroke* (useful energy). Almost all automobiles use four-stroke-cycle engines. Look at **Figure 11-3** to review the four-stroke cycle.

The **intake stroke** of a gasoline engine draws fuel and air into the engine. The intake valve is open and the exhaust valve is closed. The piston slides down and forms a low-pressure area, or vacuum, in the cylinder. Outside air pressure then pushes the air-fuel mixture into the engine.

The **compression stroke** squeezes the air-fuel mixture to prepare it for combustion (burning). The mixture is more combustible when pressurized. During this stroke, the piston slides up with both valves closed.

The **power stroke** burns the air-fuel mixture and pushes the piston down with tremendous force. This is the only stroke that does not consume energy—it produces energy. When the spark plug fires (gasoline engine), it ignites the air-fuel mixture. Since both valves are still closed, pressure forms on the top of the piston. The piston is forced down, spinning the crankshaft.

The **exhaust stroke** removes the burned gases from the engine and readies the cylinder for a fresh charge of air and fuel. During this stroke, the piston moves up. The intake valve is closed and the exhaust valve is open. The burned gases are pushed out the exhaust port and into the exhaust system.

The crankshaft must rotate *two* complete revolutions to complete the four-stroke cycle. With the engine running, this series of events happens over and over very quickly.

Engine Bottom End

The term **engine bottom end** generally refers to the block, crankshaft, connecting rods, pistons, and related components. Another name for engine bottom end is the **short block**. It is an assembled engine block with the cylinder heads, intake manifold, exhaust manifold, and other external parts *removed*.

Engine Block

The **engine block,** also called **cylinder block,** forms the main body of the engine. Other parts bolt to or fit inside the block. **Figure 11-4** shows a cutaway view of a basic block with parts installed.

The **cylinders,** also known as the **cylinder bores,** are large, round holes machined through the block from top

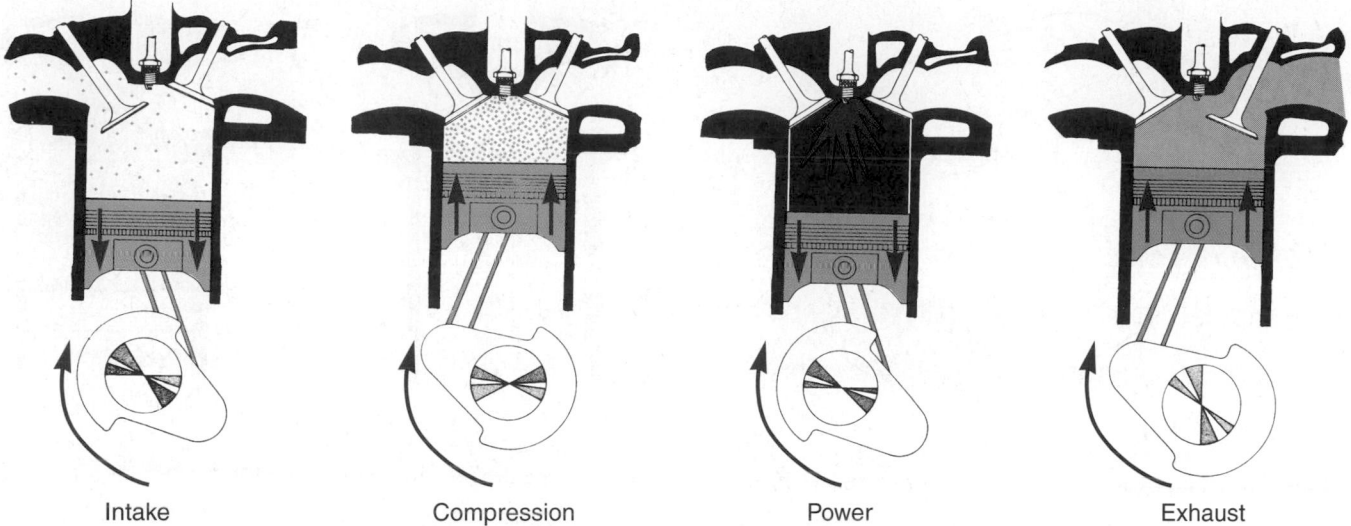

Intake   Compression   Power   Exhaust

Figure 11-3. Restudy the basic four-stroke cycle. (TRW)

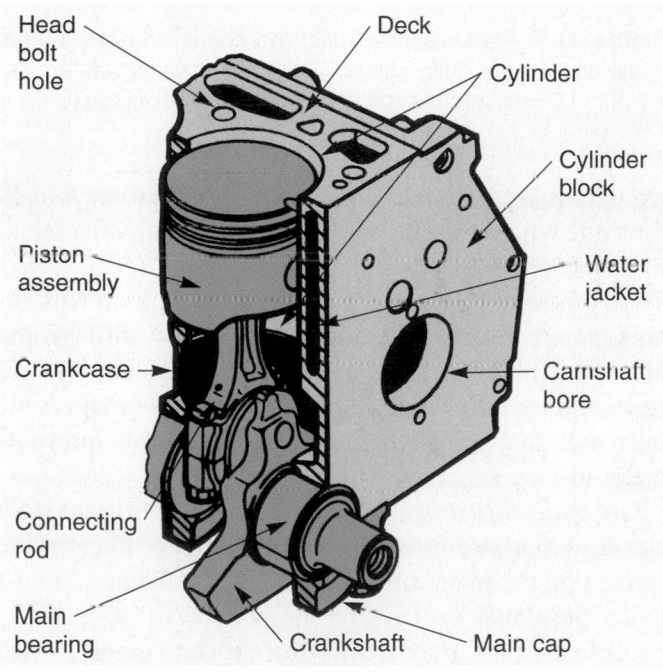

Head bolt hole
Deck
Cylinder
Cylinder block
Piston assembly
Water jacket
Crankcase
Camshaft bore
Connecting rod
Main bearing
Crankshaft
Main cap

Figure 11-4. The block is the main supporting member of the engine. Note how the other parts fit into the block. (Ford)

to bottom. The *pistons* fit into the cylinders. The cylinders are slightly larger than the pistons. This lets the pistons slide up and down freely. The *deck,* or *deck surface,* is the top of the block surrounding the cylinders. It is machined perfectly flat. The cylinder head bolts to the deck. Oil and coolant passages through the deck surface allow lubrication and cooling of the cylinder head parts.

Water jackets are coolant passages through the block. They allow a solution of water and antifreeze to cool the cylinders.

Core plugs, or *freeze plugs,* are round, metal plugs on the outside of the block. They seal holes left in the block after *casting* (manufacturing in a foundry). The plugs prevent coolant leakage out of the water jackets. Some new engines do *not* have freeze plugs.

The *main bearing bores* are holes machined in the bottom of the block to hold the crankshaft. Removable *bearing inserts* fit into these bores.

Main caps bolt to the bottom of the block and hold the crankshaft and main bearing inserts in place. Two or four large bolts normally secure each cap to the block. The caps and the block together form the main bearing bore.

The crankcase is the lowest portion of the block. The crankshaft rotates inside the crankcase.

Crankshaft

The *crankshaft* harnesses the tremendous force produced by the downward thrust of the pistons. It changes the up-and-down motion of the pistons into a rotating motion. The crankshaft fits into the bottom of the engine block, **Figure 11-5. Figure 11-6** pictures an engine crankshaft. Refer to this illustration as it is explained.

The *crankshaft main journals* are surfaces that are precisely machined and polished. They fit into the block's main bearings.

The *crankshaft rod journals* are also machined and polished surfaces, but they are offset from the main journals. The connecting rods bolt to the rod journals. With the engine running, the rod journals circle around the centerline of the crankshaft.

Counterweights are formed on the crankshaft to prevent vibration. The weights counteract the weight of the

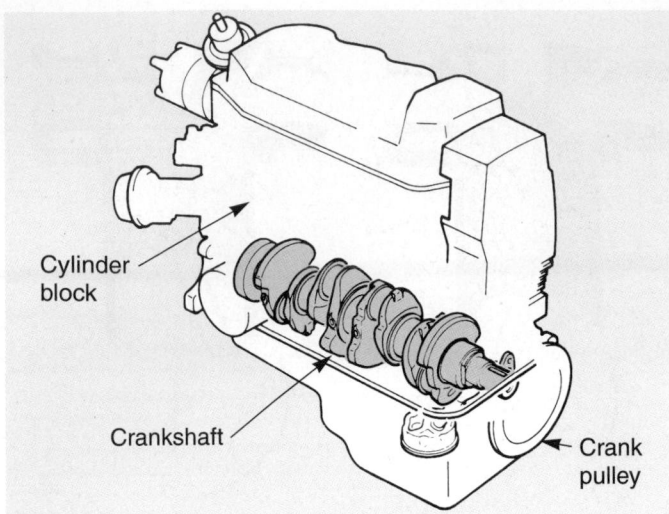

Figure 11-5. The crankshaft fits into the bottom of the block. (Ford)

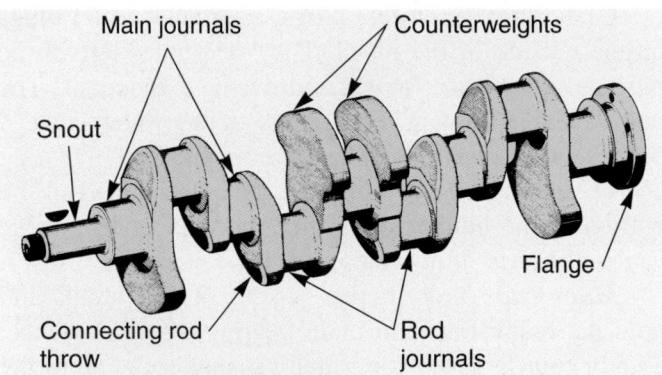

Figure 11-6. Study the basic parts of a crankshaft. Journals are very smooth surfaces for the bearings. (Peugeot)

connecting rods, pistons, rings, and rod journal offset. See **Figure 11-7.**

The *crankshaft snout* sticks through the front of the block. It provides a mounting place for the camshaft drive mechanism, front damper, and fan belt pulleys.

A *crankshaft flange* holds the flywheel. The flywheel bolts to this flange. The center of the flange has a pilot hole or bushing for the transmission torque converter or input shaft.

Automobile engines normally have 4, 6, or 8 cylinders. The crankshaft rod journals are arranged so there is always at least one cylinder on a power stroke. As a result, force is always being transmitted to the crankshaft to smooth engine operation.

Engine Main Bearings

The *engine main bearings* are removable inserts that fit between the block main bore and the crankshaft

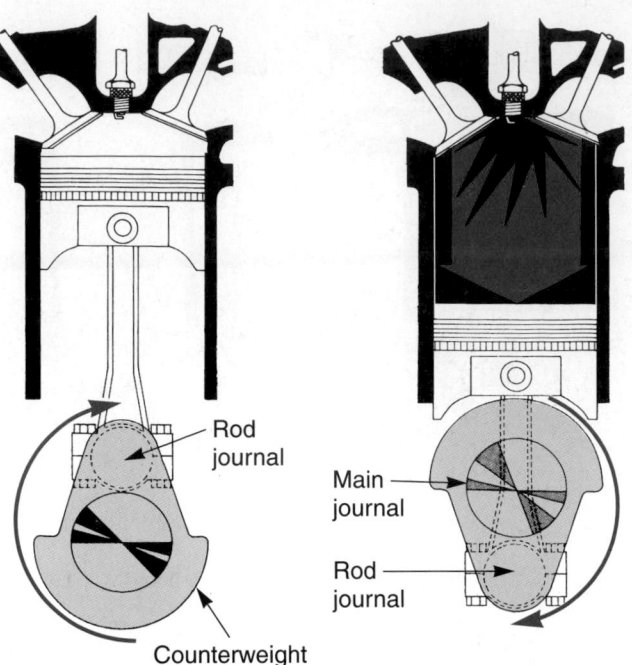

Figure 11-7. As an engine runs, the connecting rod journal spins around the main journal. The counterweight offsets the weight of the piston and rod to prevent vibration. (TRW)

main journals. One-half of each insert fits into the block. The other half fits into the block main caps. Refer to **Figure 11-8** and study the parts.

Oil holes in the upper bearing insert line up with oil holes in the block. This allows oil to flow through the block and main bearings, and into the crankshaft. The oil flows through the crankshaft to lubricate the main bearings and the connecting rod bearings. This prevents metal-to-metal contact.

A *main thrust bearing* limits how far the crankshaft can slide forward or rearward in the block. Flanges are formed on the main thrust bearing. These flanges almost touch the thrust surfaces on the crankshaft. This limits crankshaft end play (front-to-rear movement). See **Figure 11-9.** Normally, only one of the main bearings serves as a thrust bearing.

Main bearing clearance is the space between the crankshaft main journal and the main bearing insert. The clearance allows lubricating oil to enter and separate the journal and bearing. This allows the two to rotate without rubbing on each other and wearing.

Crankshaft Oil Seals

Crankshaft oil seals keep oil from leaking out the front and rear of the engine. The oil pump forces oil into the main and rod bearings. This causes oil to spray out of the bearings. Seals are placed around the front and rear of the crankshaft to contain this oil.

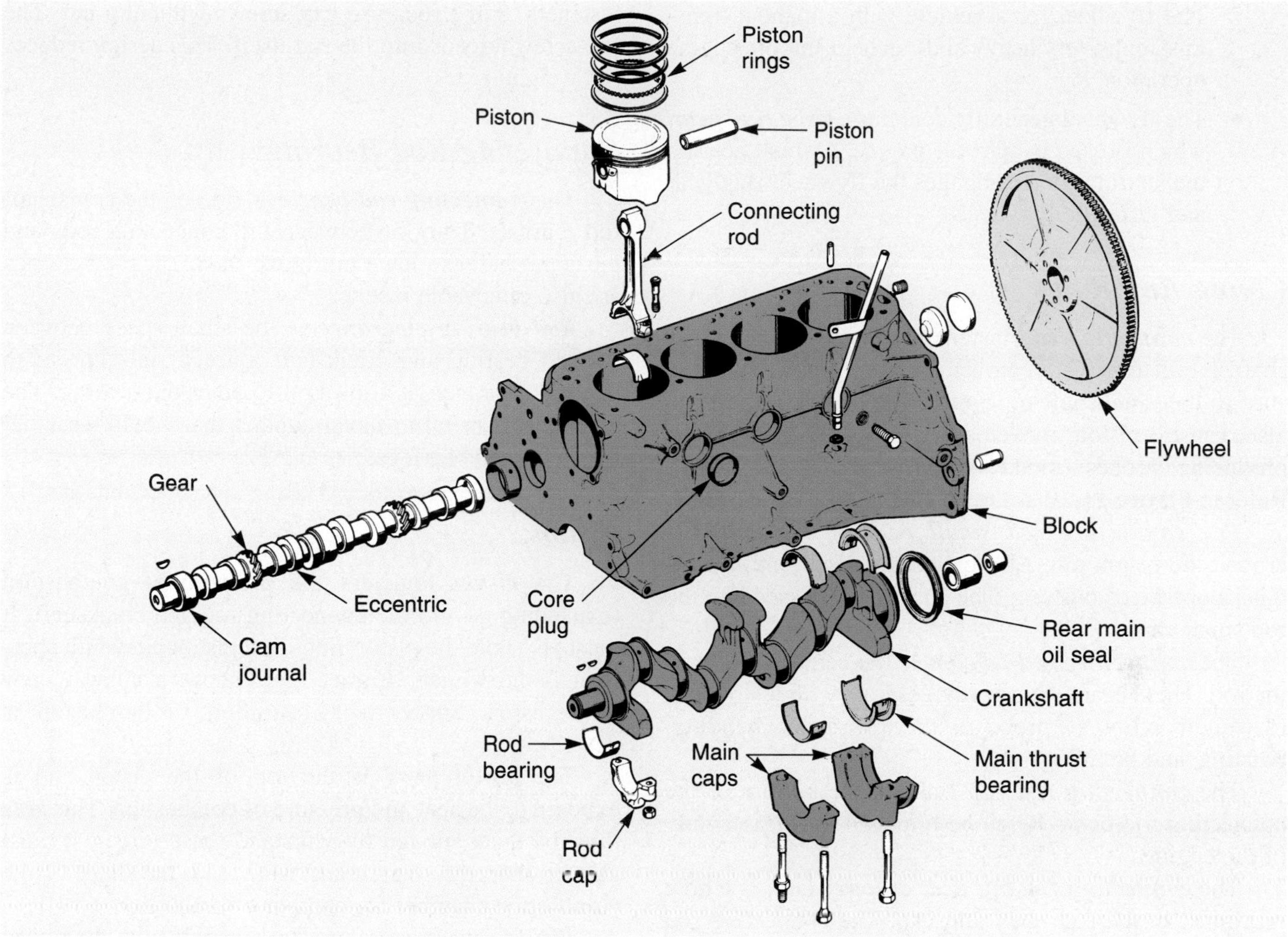

Piston rings

Piston

Piston pin

Connecting rod

Flywheel

Gear

Block

Eccentric

Core plug

Cam journal

Rear main oil seal

Crankshaft

Rod bearing

Main caps

Main thrust bearing

Rod cap

Figure 11-8. The engine bottom end consists of these basic parts. Note the crankshaft bearings and block main caps.

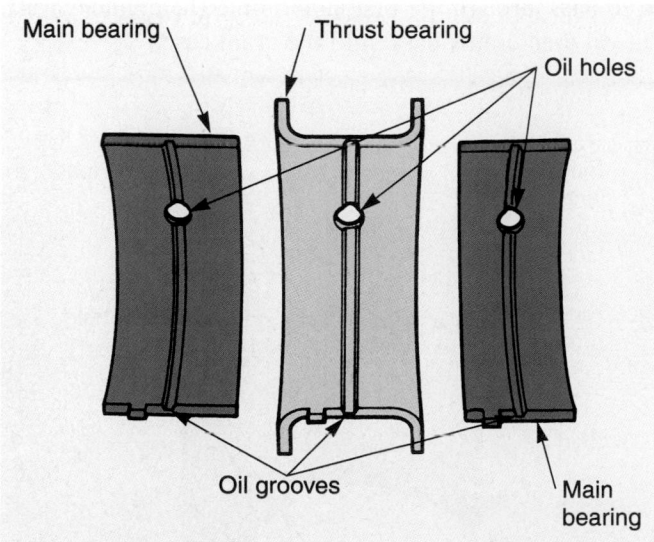

Figure 11-9. Main bearing inserts fit between the crankshaft main journals and the block. One bearing has thrust surfaces to control crankshaft end play. Oil holes and grooves allow oil to lubricate the bearings. (Federal Mogul)

The *rear main oil seal* fits around the rear of the crankshaft to prevent oil leakage, as pictured in **Figure 11-8.** It can be a one- or two-piece seal. The seal lip rides on a machined and polished surface on the crankshaft.

The front seal prevents oil leakage around the crankshaft snout. It is normally a one-piece seal that is pressed into the engine's front cover (metal housing that bolts to the front of the engine). The seal lip may contact the crankshaft directly, or it may contact a sleeve that fits over the crankshaft snout.

Flywheel

A *flywheel* is a large wheel mounted on the rear of the crankshaft. Look at **Figure 11-8.** A flywheel can have several functions:

- The flywheel connects the engine crankshaft to the transmission or transaxle. Either the manual clutch or the automatic transmission torque converter bolts to the flywheel.

- The flywheel for a vehicle with a manual transmission is very heavy and can help smooth engine operation.

- The flywheel generally contains a large *ring gear,* which is used to start the engine. A small gear on the starting motor engages the flywheel ring gear and turns the flywheel.

Connecting Rod

The *connecting rod* fastens the piston to the crankshaft. It transfers piston movement and combustion pressure to the crankshaft rod journals. The connecting rod also causes piston movement during the non-power-producing strokes (intake, compression, and exhaust). Refer to **Figure 11-10** as the connecting rod is discussed.

The *connecting rod small end,* or *top end,* fits around the piston pin. Also called the upper end, it contains a one-piece bushing. The bushing is pressed into the rod small end.

The *connecting rod I-beam* is the center section of the rod. The I-beam shape provides a very high strength-to-weight ratio. It prevents the rod from bending, twisting, and breaking.

The *connecting rod cap* bolts to the bottom of the connecting rod body. It can be removed for disassembly of the engine.

The *connecting rod big end,* or *lower end,* is a hole machined in the rod body and cap. The connecting rod bearing fits into the big end.

Connecting rod bolts and *nuts* clamp the rod cap and rod together. They are special high tensile strength fasteners. Some rods use cap screws without a nut. The cap screw threads into the rod itself. This design reduces rod weight.

Connecting Rod Bearings

The *connecting rod bearings* ride on the crankshaft rod journals. They fit between the connecting rods and the crankshaft as shown in **Figure 11-8.** The rod bearings are also removable inserts.

Rod bearing clearance is the small space between the rod bearing and crankshaft journal. As with main bearing clearance, it allows oil to enter the bearing. The oil prevents metal-to-metal contact that would wear out the crank and bearings.

Piston

The *piston* transfers the pressure of combustion (expanding gas) to the connecting rod and crankshaft. It must also hold the piston rings and piston pin while operating in the cylinder. **Figure 11-11** shows a cutaway view of a piston. Study this illustration as the piston is described.

The *piston head* is the top of the piston and is exposed to the heat and pressure of combustion. This area must be thick enough to withstand these forces. It must also be shaped to match and work with the shape of the combustion chamber for complete combustion.

Piston ring grooves are slots machined in the piston for the piston rings. The upper two grooves hold the compression rings. The lower piston groove holds the oil ring.

Piston oil holes in the bottom ring groove allow the oil to pass through the piston and onto the cylinder wall. The oil then drains back into the crankcase.

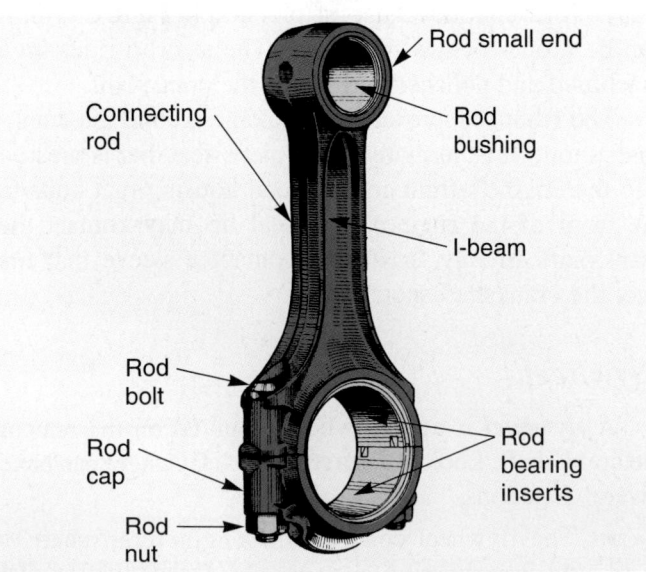

Figure 11-10. The connecting rod is the link between the piston and crankshaft. (Peugeot)

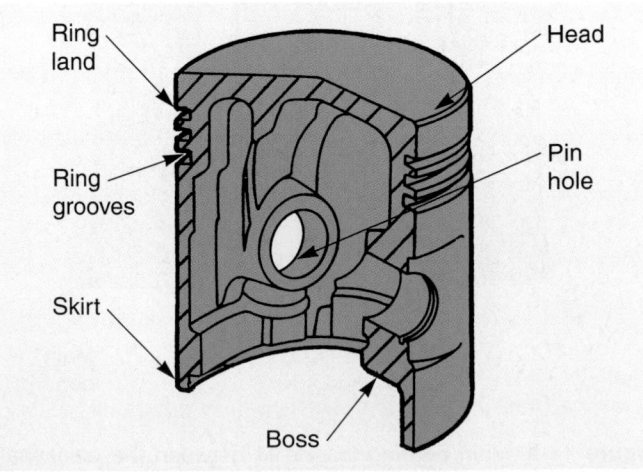

Figure 11-11. The piston rides in the cylinder and is exposed to the combustion flame. It must be light and strong. (Deere & Co.)

The *piston ring lands* are the areas between and above the ring grooves. They separate and support the piston rings as they slide on the cylinder.

The *piston skirt* is the side of the piston below the last ring. It keeps the piston from tipping in its cylinder. Without a skirt, the piston could cock and jam in the cylinder.

The *piston boss* is a reinforced area around the piston pin hole. It must be strong enough to support the piston pin under severe loads.

A *piston pin hole* is machined through the pin boss for the piston pin. It is slightly larger than the piston pin.

Piston Pin

The *piston pin,* also called *wrist pin,* allows the piston to swing on the connecting rod. The pin fits through the hole in the piston and the connecting rod small end. This is pictured in **Figure 11-12.**

Piston Clearance

Piston clearance is the amount of space between the sides of the piston and the cylinder wall. Clearance allows a lubricating film of oil to form between the piston and the cylinder. It also allows for expansion when the piston heats up. The piston must always be free to slide up and down in the cylinder block.

Piston Rings

The *piston rings* seal the clearance between the outside of the piston and the cylinder wall. They must keep combustion pressure from entering the crankcase. They must also keep oil from entering the combustion chambers.

Most pistons use three rings: two upper compression rings and one lower oil ring. This is shown in **Figure 11-13.** Note ring locations.

The *compression rings* prevent blowby (combustion pressure leaking into engine crankcase). **Figure 11-14** shows how compression rings function in an engine.

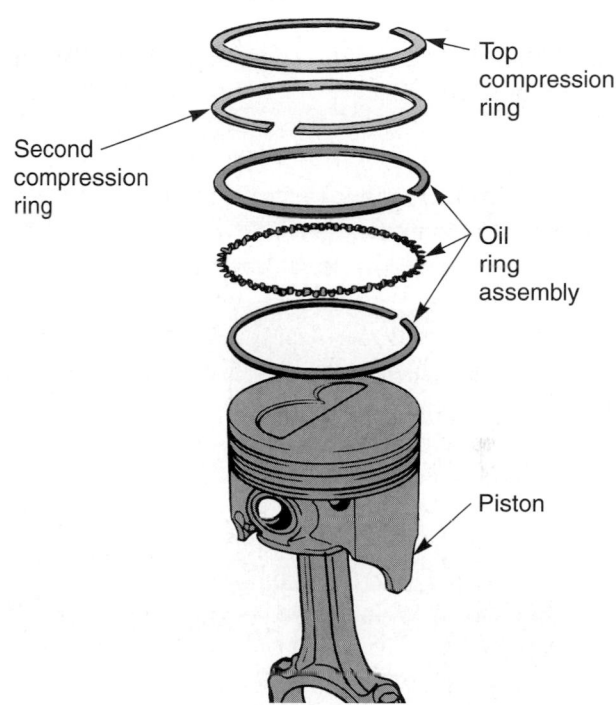

Figure 11-13. The two top piston rings are compression rings. The bottom ring is an oil ring. They fit into grooves cut in the piston. (Oldsmobile)

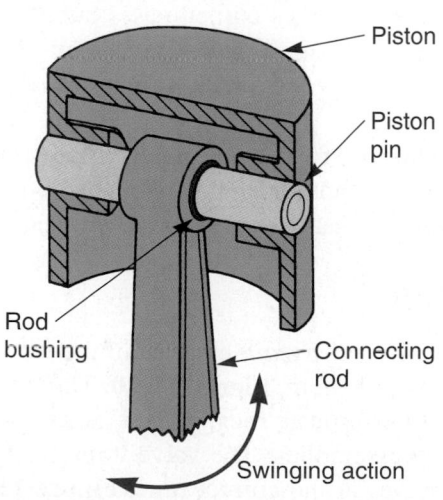

Figure 11-12. The piston pin allows the connecting rod to swing in the piston. This allows crankshaft and rod bottom end movement.

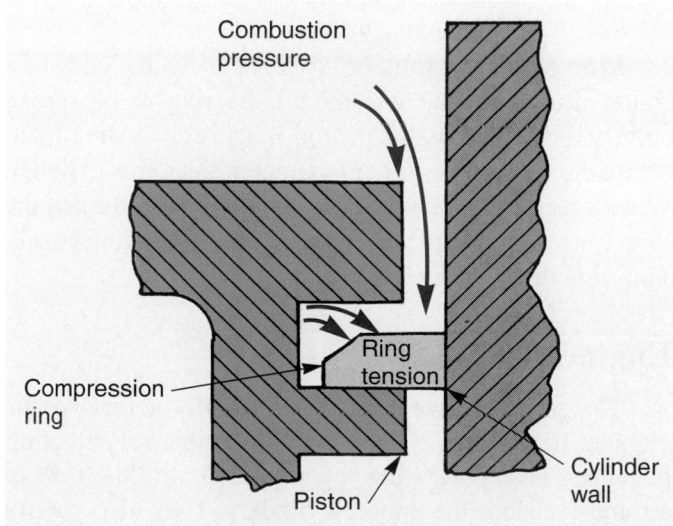

Figure 11-14. The compression ring must prevent combustion pressure from leaking between the piston and cylinder wall. Pressure actually helps push the ring against the cylinder to aid sealing.

On the compression stroke, pressure is trapped between the cylinder and piston grooves by the compression rings. Combustion pressure pushes the compression rings down in their grooves and out against the cylinder wall. This produces an almost leakproof seal.

The main job of *oil rings* is to prevent engine oil from entering the combustion chamber. They scrape excess oil off the cylinder wall, **Figure 11-15.** If too much oil got into the combustion chamber and was burned, blue smoke would come out of the vehicle's exhaust pipe.

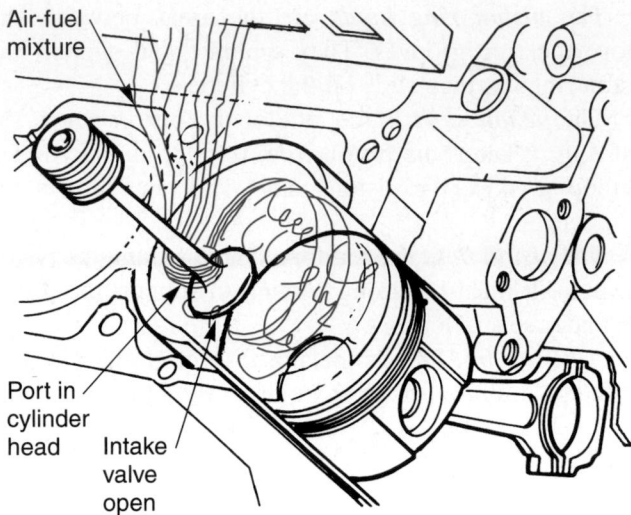

Figure 11-16. The engine top end controls the flow of the mixture into the cylinder. It also controls the flow of exhaust out of the cylinder. (Ford)

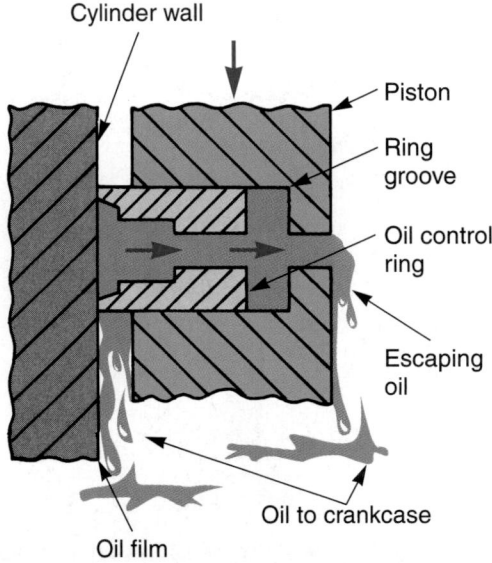

Figure 11-15. The oil control ring scrapes excess oil off the cylinder wall. If this oil entered the combustion chamber, the engine would emit blue smoke.

Ring gap is the split, or space, between the ends of a piston ring. The ring gap allows the ring to be spread open and installed on the piston. It also allows the ring to be made slightly larger in diameter than the cylinder. When squeezed together and installed in the cylinder, the ring spreads outward and presses on the cylinder wall. This aids ring sealing.

Engine Top End

The term *engine top end* generally refers to the cylinder heads, valves, camshaft, and other related components. These parts work together to control the flow of air and fuel into the engine cylinders. They also control the flow of exhaust out of the engine. **Figure 11-16** shows the fuel charge entering engine.

Cylinder Head

The *cylinder head* bolts to the deck of the cylinder block. It covers and encloses the top of the cylinders. Refer to **Figure 11-17.**

Combustion chambers are small pockets formed in the cylinder heads. The combustion chambers are located directly over the pistons. Combustion occurs in these areas of the cylinder head. Spark plugs (gasoline engine) or injectors (diesel engine) protrude through holes and into the combustion chambers. **Figure 11-18** shows a combustion chamber.

Intake and exhaust ports are cast into the cylinder head. The *intake ports* route air (diesel engine) or air and fuel (gasoline engine) into the combustion chambers. The *exhaust port* routes burned gases out of the engine. *Valve guides* are small holes machined through the cylinder head for the valves. The valves fit into and slide in these guides.

Valve seats are round, machined surfaces in the combustion chamber port openings, **Figure 11-19.** When a valve is closed, it seals against the valve seat.

Valve Train

The *engine valve train* consists of the valves and the parts that operate them, **Figure 11-20.** These include the camshaft, lifters, push rods, rocker arms, valves, and valve spring assemblies. The valve train must open and close the valves at the correct time. **Figure 11-21** illustrates basic valve train action.

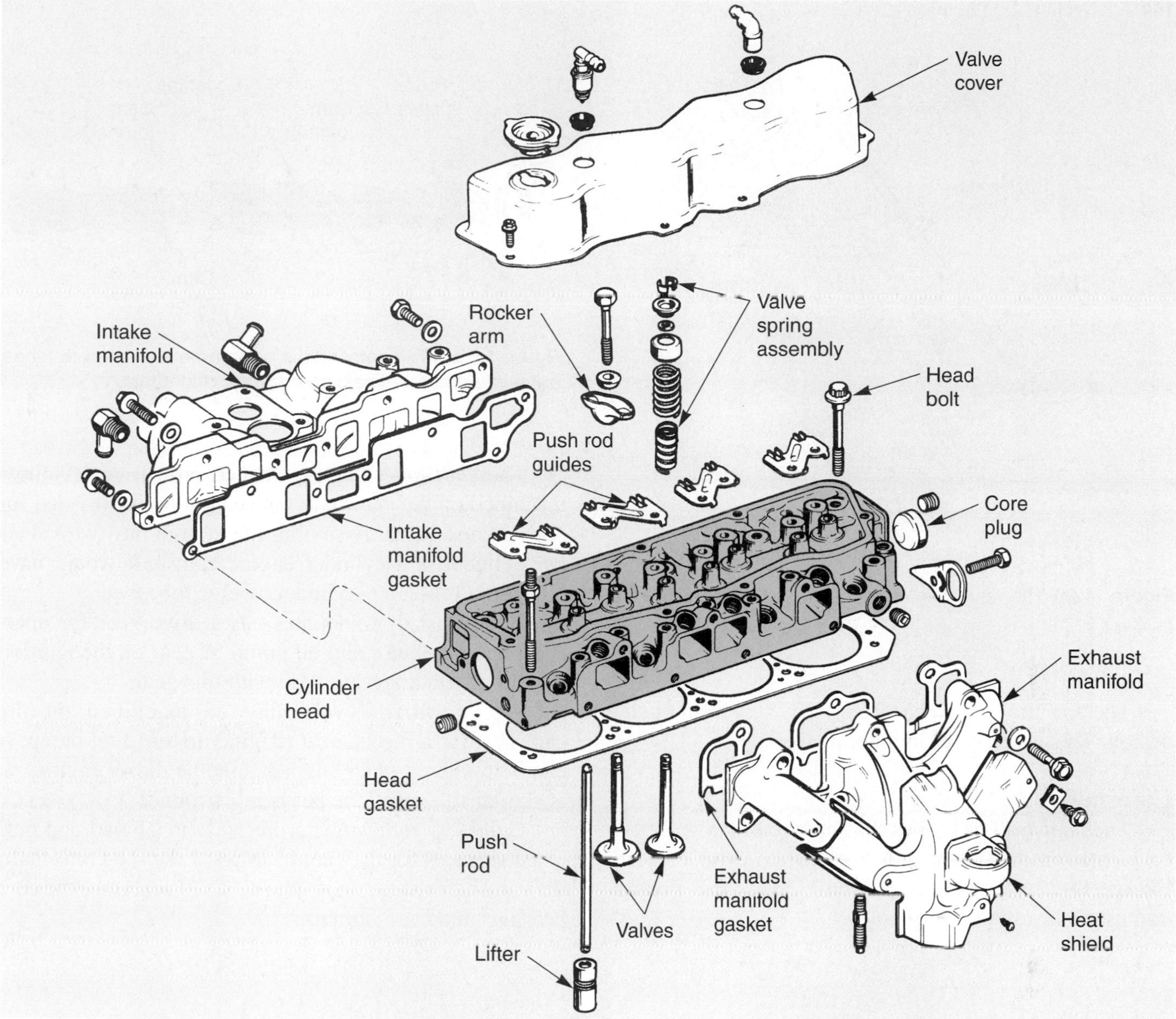

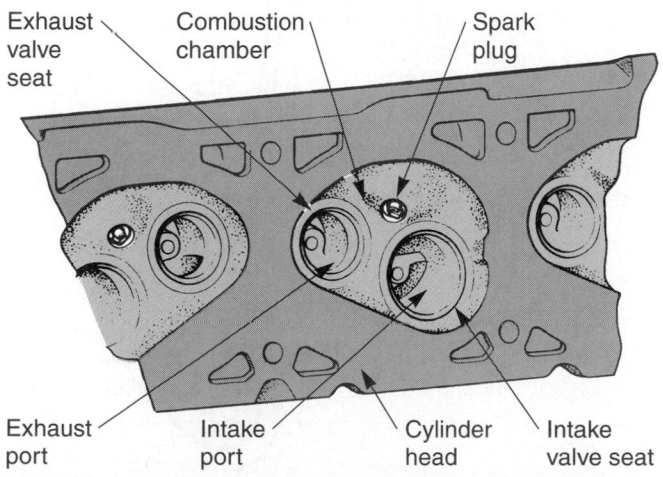

Figure 11-17. Study the basic engine top end components. The cylinder head is the foundation for these parts. (Chrysler)

Figure 11-18. The combustion chamber is formed in the cylinder head. Valve ports enter the chamber. Also, note the spark plug tip and valve seats. (Cadillac)

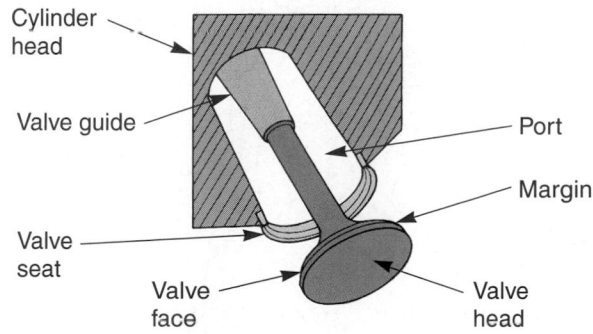

Figure 11-19. A valve slides up and down in the guide during operation. When the valve is closed, it seals against the valve seat to close off the port.

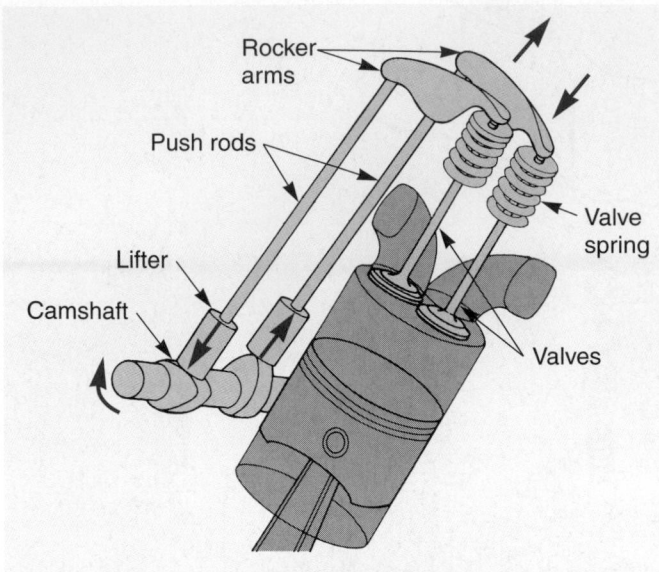

Figure 11-20. The valve train operates engine valves. Study the parts.

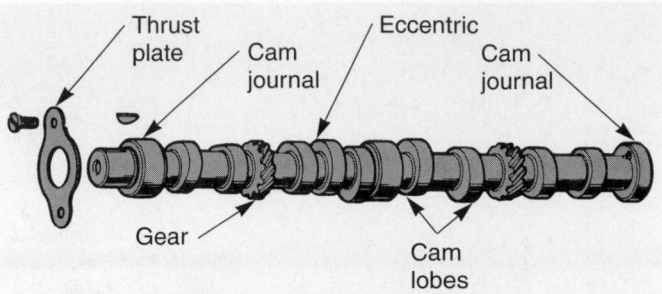

Figure 11-22. The camshaft is a long metal shaft with lobes, journals, and sometimes, an eccentric and gears.

The specific parts of a valve train vary with engine design. This is discussed in later chapters.

Camshaft

The *camshaft* has lobes that open each valve. It can be located in the engine block or in the cylinder head. **Figure 11-22** illustrates a camshaft. Study this illustration as the camshaft is explained.

The *cam lobes* are egg-shaped protrusions (bumps) machined on the camshaft. One cam lobe is provided for each engine valve. Assuming the engine has two valves per cylinder, a 4-cylinder engine camshaft would have eight cam lobes; a 6-cylinder, twelve lobes; etc.

The camshaft sometimes has a drive gear for operating the distributor and oil pump. A gear on the ignition system distributor may mesh with this gear.

An *eccentric* (oval) may be machined on the camshaft for a mechanical (engine driven) fuel pump. It can be found on older engines or some diesel engines. It is similar to a cam lobe but is more round.

Camshaft journals are precisely machined and polished surfaces for the cam bearings. Like the crankshaft, the camshaft rotates on its journals. Oil separates the cam bearings and cam journals.

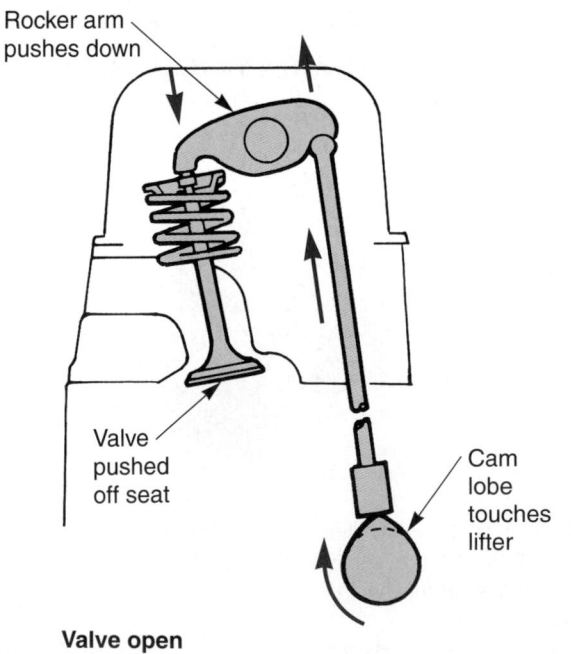

Valve open

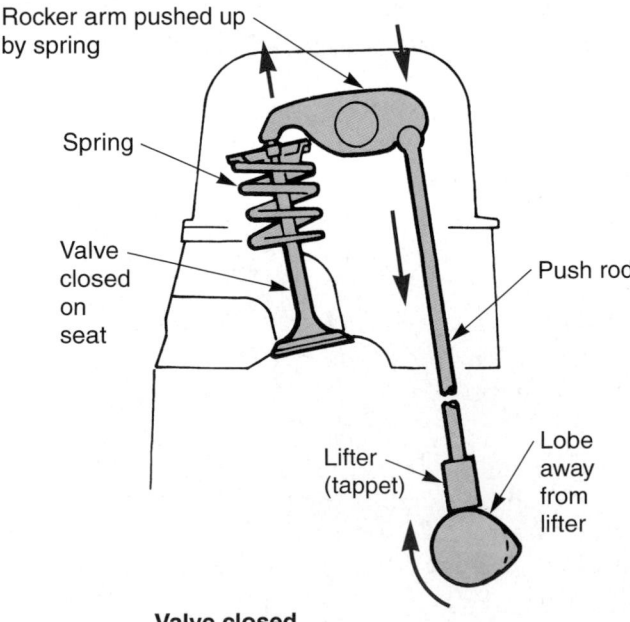

Valve closed

Figure 11-21. When the camshaft lobe turns into the lifter, the valve is pushed open. When the lobe rotates away from the lifter, the valve spring pushes the valve closed. (Ford)

Valve Lifters

A *valve lifter,* also called a *tappet,* usually rides on the cam lobes and transfers motion to the rest of the valve train. Refer back to **Figure 11-20.** The lifters can be located in the engine block or cylinder head. They fit into lifter bores, which are machined holes in the block or head.

When the cam lobe moves toward the lifter, the lifter is pushed up in its bore. This opens the valve. Then, when the lobe rotates away from the lifter, the lifter is pushed down in its bore by the valve spring. This keeps the lifter in constant contact with the camshaft.

Push Rods

Push rods transfer motion between the lifters and the rocker arms, **Figure 11-21.** They are needed when the camshaft is located in the cylinder block. They are *not* needed when the camshaft is in the cylinder head.

A push rod is a hollow metal tube with a ball or socket formed on each end. One end of the push rod fits into the lifter. The other end fits against the rocker arm. In this way, when the lifter slides up, the push rod moves the rocker arm.

Rocker Arms

Rocker arms can be used to transfer motion to the valves. They mount on top of the cylinder head. A pivot mechanism allows the rockers to rock back and forth, opening and closing the valves. See **Figure 11-21.**

Valves

Engine *valves* open and close the ports in the cylinder head. Until recently, only two valves were used per cylinder: one intake valve and one exhaust valve. To improve efficiency, many late-model engines are equipped with four valves per cylinder: two intake valves and two exhaust valves.

The *intake valve* is the larger valve. It controls the flow of the fuel mixture (gasoline engine) or air (diesel) into the combustion chamber. The intake valve fits into the port leading from the intake manifold.

The *exhaust valve* controls the flow of exhaust gases out of the cylinder. It is smaller than the intake valve. The exhaust valve fits into the port leading to the exhaust manifold.

Valve Parts

Look at **Figure 11-23** as the basic parts of a valve are introduced.

The *valve head* is the large, disc-shaped surface exposed to the combustion chamber. Its outside diameter determines the size of the valve.

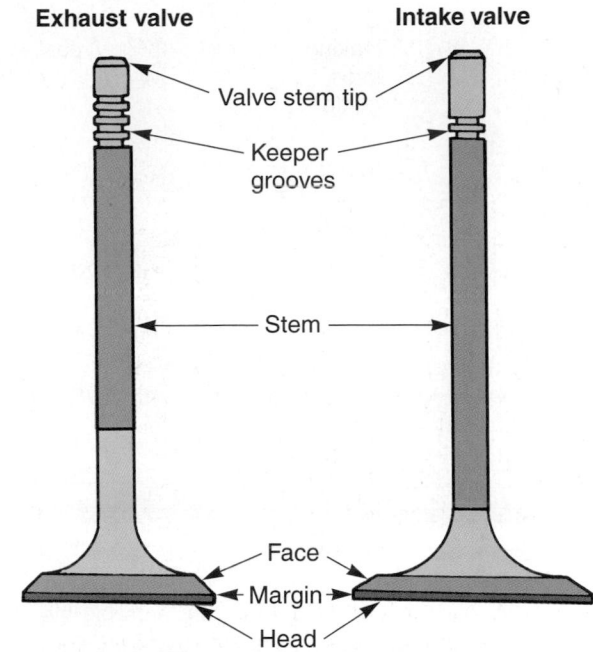

Figure 11-23. The intake valve is larger than the exhaust valve. Note the parts of each valve.

The *valve face* is a machined surface on the back of the valve head. It touches and seals against the seat in the cylinder head.

The *valve margin* is the flat surface on the outer edge of the valve head. It is located between the valve head and face. The margin is needed to allow the valve to withstand the high temperatures of combustion. Without a margin, the valve head would melt and burn.

The *valve stem* is a long shaft extending out of the valve head. The stem is machined and polished. It fits into the guide machined through the cylinder head.

Keeper grooves, or *lock grooves,* are machined into the top of the valve stem. They accept small keepers or locks that hold the spring on the valve.

Valve Seals

Valve seals prevent oil from entering the combustion chambers through the valve guides. This is illustrated in **Figure 11-24.**

The valve seals fit over the valve stems and keep oil from entering through the clearance between the stems and guides.

Without valve seals, oil could be drawn into the engine cylinders and burned during combustion. Oil consumption and engine smoking could result.

Valve Spring Assembly

The *valve spring assembly* is used to close the valve. It basically consists of a valve spring, a retainer, and two

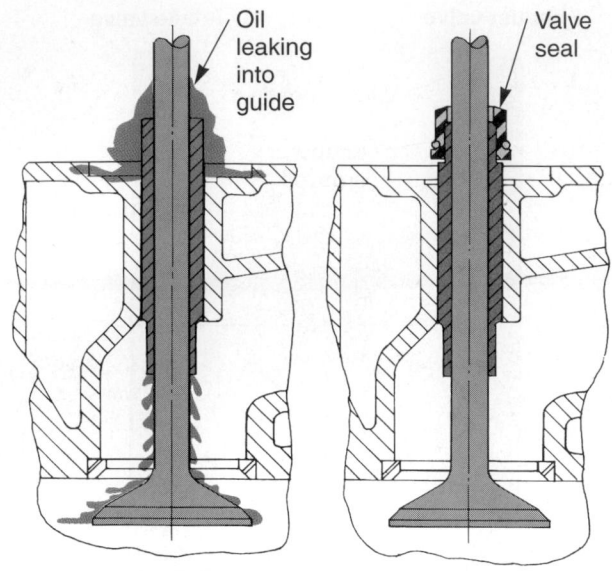

Figure 11-24. A valve seal keeps oil from entering the guide and combustion chamber. (American Hammered Piston Rings)

keepers. The *keepers* fit into the grooves cut in the valve stem. This locks the retainer and spring on the valve. See **Figure 11-25.**

Intake Manifold

The *intake manifold* bolts to the side of the cylinder head or heads. On late-model engines, the fuel injectors and the throttle body mount on the intake manifold. On older engines, the carburetor is mounted on the top of the

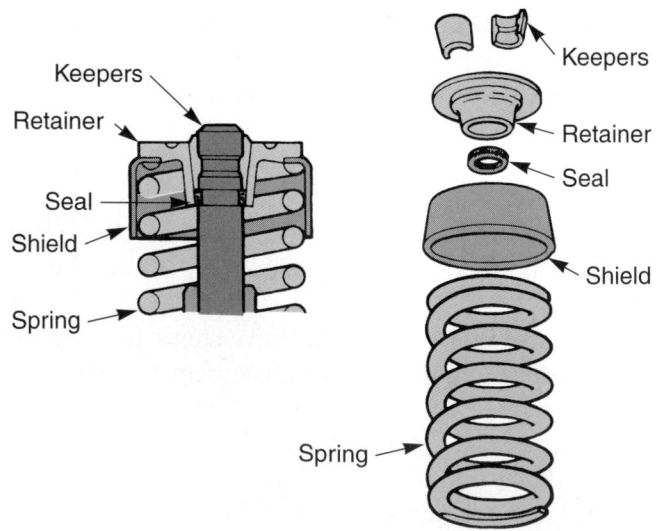

Figure 11-25. The valve spring assembly basically consists of a spring, retainer, keepers, and sometimes, a shield. Note how this type of seal fits on the valve stem. (Buick)

manifold. The intake manifold contains runners (passages) going to each cylinder head port. Air and fuel (gasoline engine) are routed through these runners, **Figure 11-26.**

Exhaust Manifold

The *exhaust manifold* also bolts to the cylinder head; however, it fastens over the exhaust ports to carry burned gases into the exhaust system, **Figure 11-26.** During the exhaust strokes, hot gases blow into this manifold before entering the rest of the exhaust system. An engine exhaust manifold can be made of heavy cast iron or lightweight aluminum or stainless steel tubing. The trend is toward lighter designs.

Valve Cover

The *valve cover,* also called the *rocker cover,* is a thin metal or plastic cover over the top of the cylinder head. It simply keeps valve train oil spray from leaking out of the engine. Look at **Figure 11-26.**

Engine Front End

The *engine front end* operates the engine camshaft and sometimes the oil pump, distributor, engine sensors, and diesel injection pump. Basically, the engine front end consists of a drive mechanism for the camshaft and other devices, a front cover, an oil seal, and a crankshaft damper.

> **Tech Tip**
> Engine front end assemblies are much more complicated than in the past. With today's dual overhead cam engines, the cam drive mechanism can be difficult to comprehend. Carefully study all the illustrations in this book to compare front end design differences. This will prepare you to work on various configurations.

Camshaft Drive

A *camshaft drive* is needed to turn the camshaft at one-half engine speed. A belt and sprockets, gears, or a chain and sprockets can be used to turn the camshaft. Look at **Figure 11-27.**

These parts are often called the timing belt, timing gears, or timing chain because they time the camshaft with the crankshaft. See **Figures 11-28, 11-29,** and **11-30.**

The camshaft is designed to turn at one-half engine speed. As a result, each valve will open only once for every two crankshaft revolutions.

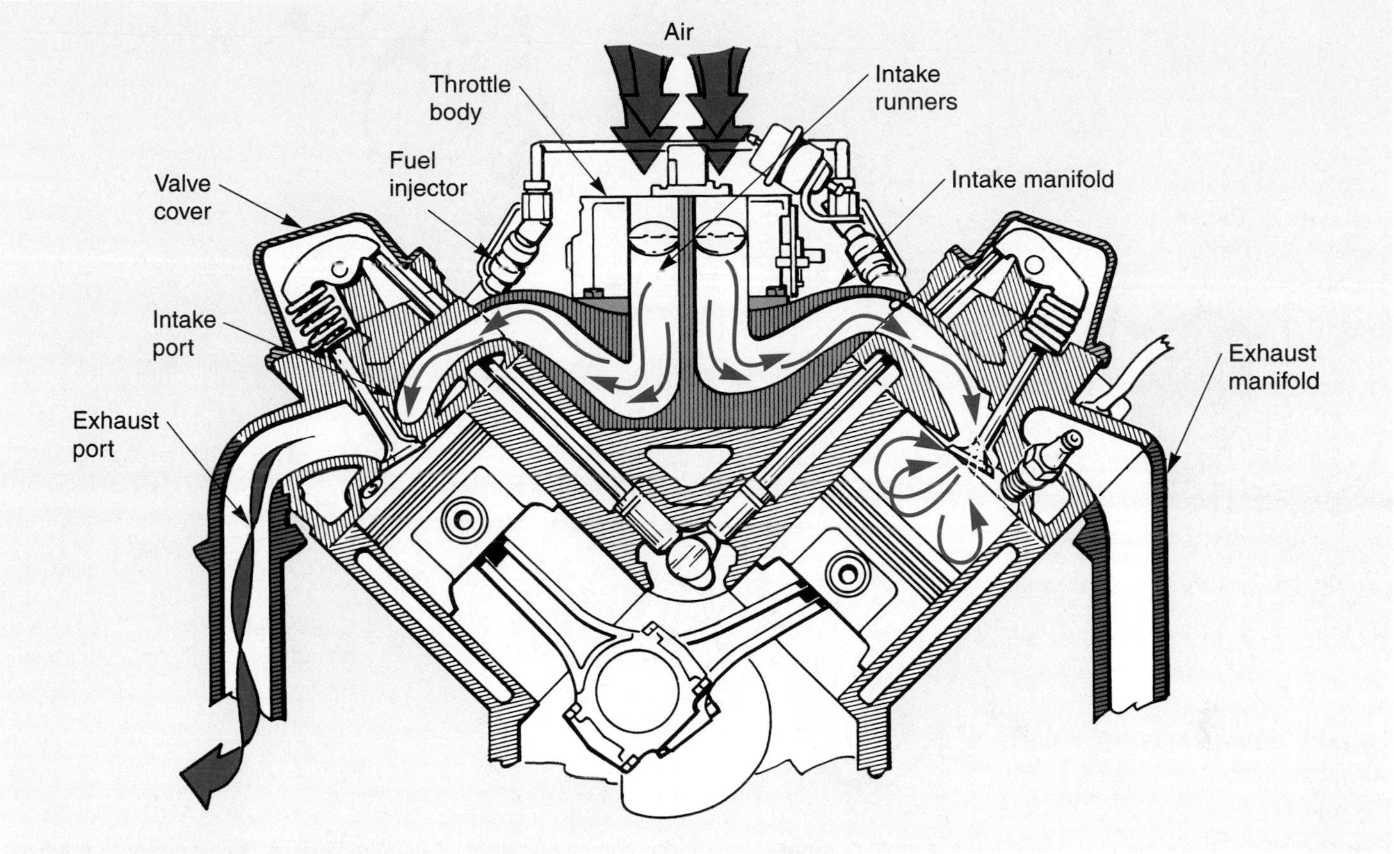

Air

Throttle
body

Intake
runners

Valve
cover

Fuel
injector

Intake manifold

Intake
port

Exhaust
manifold

Exhaust
port

Figure 11-26. Intake and exhaust manifolds bolt to the cylinder head. The intake manifold contains runners that route the fuel mixture into the cylinder heads. The exhaust manifolds route burned gases into the exhaust system.

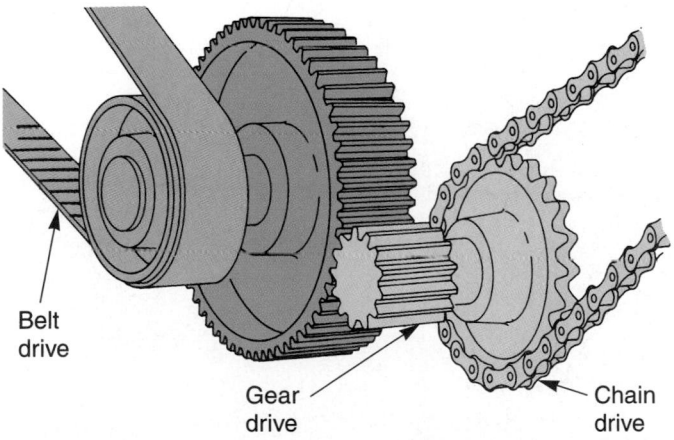

Belt
drive

Gear
drive

Chain
drive

Figure 11-27. The camshaft can be turned by one of three drive mechanisms: a belt drive, a gear drive, or a chain drive.

For instance, the intake valve must open only on the intake stroke, not the compression, power, or exhaust strokes. To do this, the camshaft gear or sprocket is twice as big as the gear or sprocket on the crankshaft.

Front Cover

The *front cover* bolts over the crankshaft snout. It holds an oil seal that seals the front of the crankshaft.

When the engine uses a gear- or chain-type camshaft drive, the front cover can also be called the timing cover, **Figure 11-28.** With a belt drive, this cover does not enclose the cam drive or timing mechanism. A second cover is installed over the belt, **Figure 11-29.**

Crank Damper

A *crank damper* is a heavy wheel on the crankshaft snout. It is mounted in rubber and helps prevent crankshaft vibration and damage. This damper is also called the harmonic balancer or vibration damper.

Summary

- The engine is the source of power for the vehicle. For this reason, it is also called a power plant.

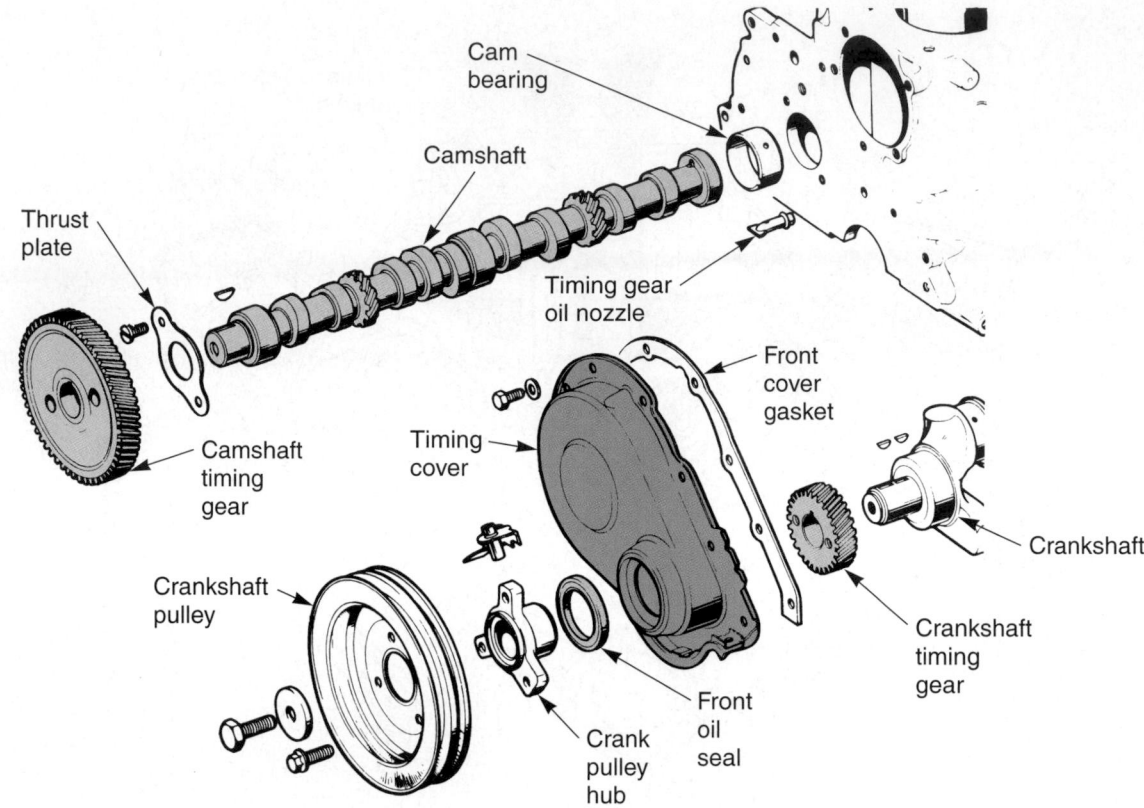

Figure 11-28. The engine front end components primarily operate the engine camshaft. This engine uses timing gears to drive the camshaft at one-half engine speed. The front cover encloses the gears. The front seal prevents leakage around the crankshaft snout. (Chrysler)

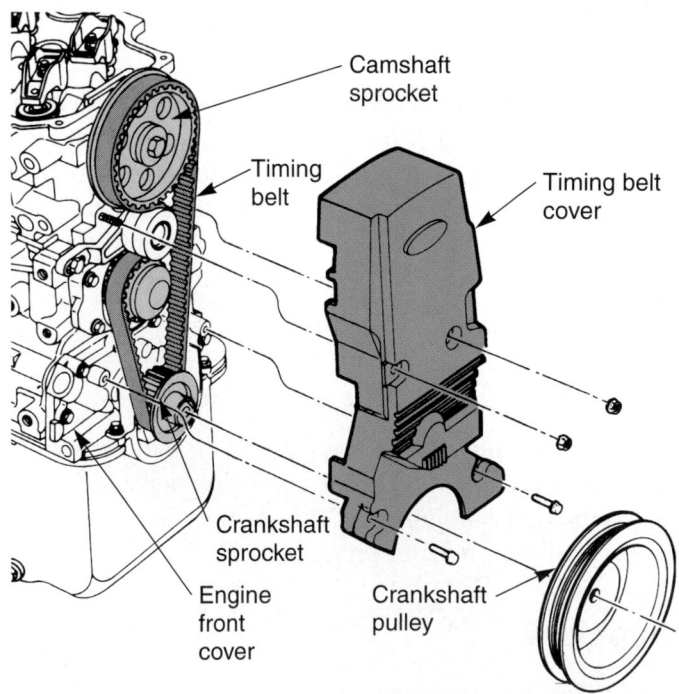

Figure 11-29. The crankshaft sprocket turns the timing belt. The timing belt turns the camshaft sprocket and camshaft. The front cover simply houses the front oil seal. The timing cover fits over the belt. (Ford)

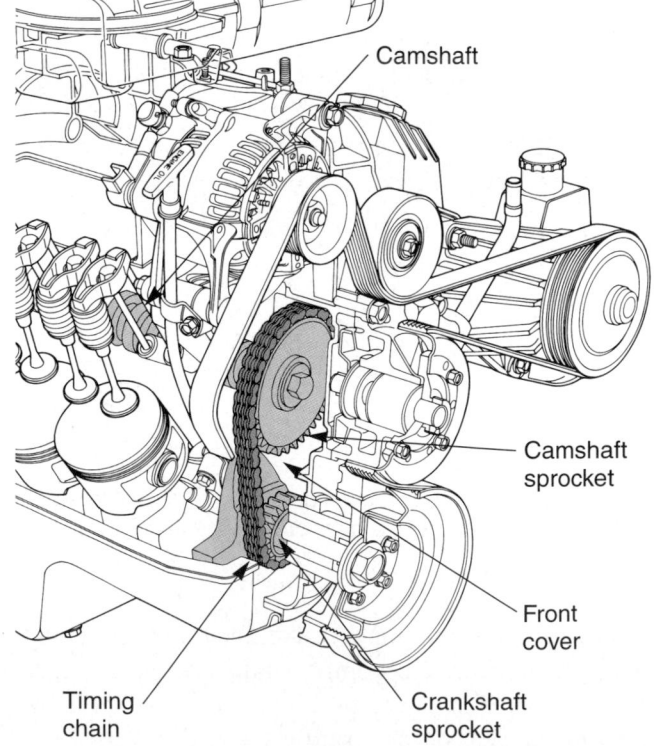

Figure 11-30. The timing chain and sprockets operate the camshaft in this engine. (Chrysler)

- When the piston is at its *highest point* in the cylinder, it is at TDC (top dead center). When the piston slides to its *lowest point* in the cylinder, it is at BDC (bottom dead center).

- The term engine bottom end generally refers to the block, crankshaft, connecting rods, pistons, and related components.

- The crankshaft harnesses the tremendous force produced by the downward thrust of the pistons.

- The engine main bearings are removable inserts that fit between the block main bore and crankshaft main journals.

- Main bearing clearance is the space between the crankshaft main journal and the main bearing insert.

- Crankshaft oil seals keep oil from leaking out the front and rear of the engine.

- The connecting rod fastens the piston to the crankshaft. It transfers piston movement and combustion pressure to the crankshaft rod journals.

- The engine piston transfers the pressure of combustion (expanding gas) to the connecting rod and crankshaft.

- Piston clearance is the amount of space between the sides of the piston and the cylinder wall.

- The piston rings seal the clearance between the outside of the piston and the cylinder wall.

- The compression rings prevent blowby (combustion pressure leaking into engine crankcase).

- The main job of oil rings is to prevent engine oil from entering the combustion chamber.

- The term engine top end generally refers to the cylinder heads, valves, camshaft, and other related components.

- The camshaft has lobes that open the valves.

- A valve lifter usually rides on the cam lobes and transfers motion to the rest of the valve train.

- Engine valves open and close the ports in the cylinder head.

- The valve spring assembly is used to close the valve.

- The engine front end operates the engine camshaft and, sometimes, the oil pump, distributor, engine sensors, and diesel injection pump.

- A crank damper is a heavy wheel on the crankshaft snout. It is mounted in rubber and helps prevent crankshaft vibration and damage.

Important Terms

Engine	Piston
Fuel	Wrist pin
Combustion chamber	Piston rings
Expansion	Compression rings
Top dead center (TDC)	Oil rings
Bottom dead center (BDC)	Engine top end
	Cylinder head
Piston stroke	Combustion chambers
Four-stroke cycle	Intake ports
Intake stroke	Exhaust port
Compression stroke	Valve guides
Power stroke	Valve seats
Exhaust stroke	Engine valve train
Engine bottom end	Camshaft
Engine block	Valve lifter
Cylinders	Tappet
Pistons	Push rods
Deck	Rocker arms
Water jackets	Intake valve
Core plugs	Exhaust valve
Main bearing bores	Valve seals
Bearing inserts	Valve spring assembly
Main caps	Keepers
Crankshaft	Intake manifold
Crankshaft oil seals	Exhaust manifold
Rear main oil seal	Valve cover
Flywheel	Engine front end
Ring gear	Camshaft drive
Connecting rod	Front cover
Connecting rod bearings	Crank damper

Review Questions—Chapter 11

Please do not write in this text. Place your answers on a separate sheet of paper.

1. Usually, _____ or _____ _____ is burned inside the engine to produce heat, expansion, and resulting pressure.

2. What do TDC and BDC mean?

3. Every four strokes, the engine produces two power or energy producing strokes. True or False?

4. Explain the intake stroke.

5. Explain the compression stroke.

6. Explain the power stroke.

7. Explain the exhaust stroke.

8. _____ _____ bolt to the bottom of the block and hold the crankshaft in place.

9. The crankcase is the highest portion in the block. True or False?

10. The crankshaft changes the up-and-down motion of the piston into a useful _____ motion.

11. What is the function of crankshaft counter-weights?

12. Describe the function of the main thrust bearing.

13. The _____ _____ transfers piston movement to the crankshaft.

14. The distance from the centerline of the crankshaft to the centerline of a rod journal is 3″ (76 mm). What is the total vertical distance that the piston will travel in the cylinder (from TDC to BDC)?

15. Which of the following is *not* part of a connecting rod?
    (A) I-beam.
    (B) Lobe.
    (C) Cap.
    (D) Bushing.

16. Why is rod bearing clearance needed?

17. Explain the function of compression and oil rings.

18. Which of the following is part of the cylinder head?
    (A) Combustion chambers.
    (B) Intake and exhaust ports.
    (C) Valve guides.
    (D) All of the above.

19. List and explain the basic parts of a camshaft.

20. _____ open and close the ports in the cylinder head.

21. The intake valves are larger than the exhaust valves. True or False?

22. Describe the five basic parts of an engine valve.

23. What do valve seals do and what would happen without valve seals?

24. Explain the function of the following parts.
    (A) Intake manifold.
    (B) Exhaust manifold.
    (C) Valve cover.

25. Identify the engine parts in the illustration below. Write A through N on your paper. Then write the name of the part next to each letter.

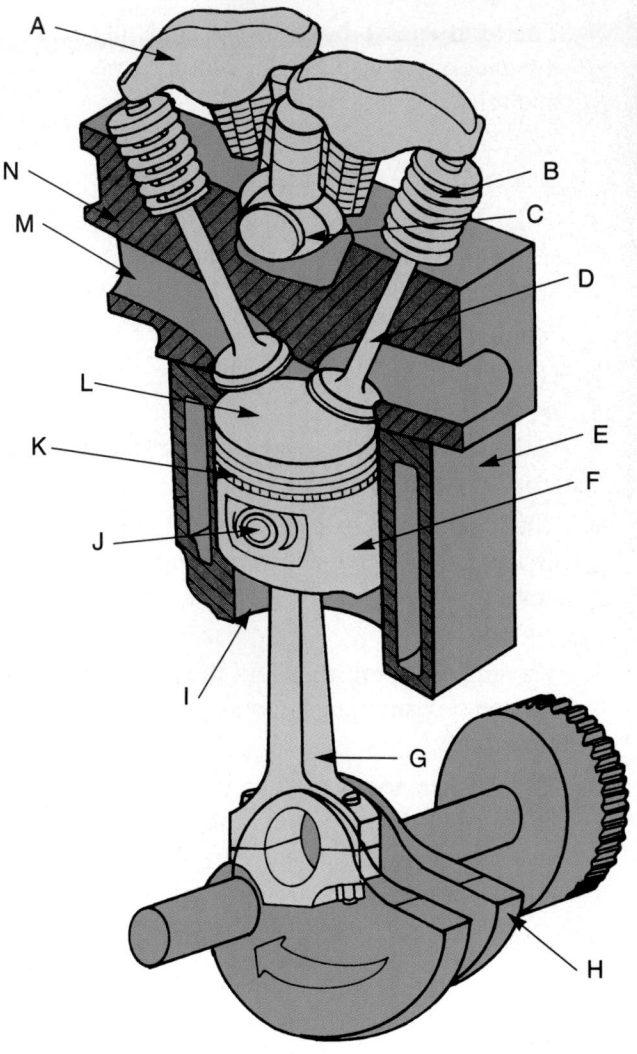

ASE-Type Questions

1. Technician A says the engine camshaft converts reciprocating motion of the piston into rotary motion for power output. Technician B disagrees and says the crankshaft performs this function. Who is right?
   (A) *A only.*
   (B) *B only.*
   (C) *Both A and B.*
   (D) *Neither A nor B.*

2. The distance a piston travels up or down in a cylinder is limited by the:
   (A) *flywheel.*
   (B) *camshaft.*
   (C) *crankshaft.*
   (D) *valve train.*

3. The air-fuel mixture is actually burned during the:
   (A) *power stroke.*
   (B) *intake stroke.*
   (C) *exhaust stroke.*
   (D) *compression stroke.*

4. How many crankshaft rotations are needed to complete the four-stroke cycle?
   (A) *One.*
   (B) *Two.*
   (C) *Three.*
   (D) *Four.*

5. Automobile engines normally have:
   (A) *four cylinders.*
   (B) *six cylinders.*
   (C) *eight cylinders.*
   (D) *All of the above.*

6. A flywheel performs each of these functions *except:*
   (A) *contains a gear used for engine starting.*
   (B) *smoothes engine operation.*
   (C) *provides lubrication to parts.*
   (D) *connects crankshaft to transmission.*

7. Which of the following causes piston movement during non-power-producing strokes?
   (A) *Flywheel.*
   (B) *Connecting rod.*
   (C) *Counterweights.*
   (D) *Main thrust bearing.*

8. Which of the following components transfers combustion pressure to the crankshaft and connecting rods?
   (A) *Piston.*
   (B) *Camshaft.*
   (C) *Valve train.*
   (D) *Thrust bearing.*

9. Which of the following keeps a piston from tipping in its cylinder?
   (A) *Piston pin.*
   (B) *Piston boss.*
   (C) *Piston skirt.*
   (D) *Piston head.*

10. An engine comes into the shop with blowby. Technician A says the engine could have a cracked flywheel. Technician B says the rings could be worn. Who is right?
    (A) *A only.*
    (B) *B only.*
    (C) *Both A and B.*
    (D) *Neither A nor B.*

11. Which of the following is *not* a valve train component?
    (A) *Camshaft.*
    (B) *Crankshaft.*
    (C) *Rocker arms.*
    (D) *Valve springs.*

12. While discussing camshaft operation, Technician A states that cam lobes open each valve in the engine. Technician B states that the camshaft journals operate engine valves. Who is right?
    (A) *A only.*
    (B) *B only.*
    (C) *Both A and B.*
    (D) *Neither A nor B.*

13. The valve train parts that transfer motion to the rest of the valve train are the:
    (A) *lifters.*
    (B) *tappets.*
    (C) *Both A and B.*
    (D) *Neither A nor B.*

14. Which of the following is used to turn the camshaft at one-half engine speed?
    (A) *Timing belt.*
    (B) *Timing chain.*
    (C) *Timing gears.*
    (D) *All of the above.*

15. Which of the following helps prevent crankshaft vibration and damage?
    (A) *Crank damper.*
    (B) *Vibration damper.*
    (C) *Harmonic balancer.*
    (D) *All of the above.*

Activities—Chapter 11

1. A camshaft changes rotary motion to up-and-down motion. Find and describe or sketch at least three other examples of ways that a mechanical system changes the type, direction, or force of a motion.

2. If possible, obtain from a repair shop a bearing or other engine part that shows severe wear. Clean the part so the wear can be seen easily and pass it around to your classmates. Discuss how the wear was caused.

3. Make a chart showing the position (open or closed) of the intake valve and the exhaust valve during each stroke of the four-stroke cycle.

Chapter 12

Engine Design Classifications

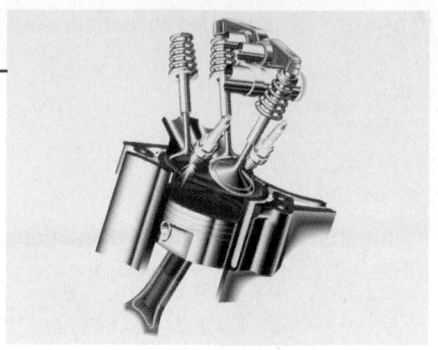

After studying this chapter, you will be able to:

- Describe basic automotive engine classifications.
- Compare gasoline and diesel engines.
- Contrast combustion chamber designs.
- Discuss alternative engine types.
- Compare two- and four-stroke cycle engines.
- Correctly answer ASE certification test questions that require a knowledge of engine classifications and design differences.

Now that you have learned about the basic parts of an engine, you should become familiar with the various engine types used in automobiles and light trucks. An experienced engine technician can glance into an engine compartment and "rattle off" dozens of engine facts.

For example, you might hear a technician say, "This is an inline, 4-cylinder, 16-valve, overhead-cam engine." You must understand these kinds of terms when troubleshooting and repairing vehicle engines. Study this chapter carefully. It will help you learn the "language" of an automotive technician.

Engine Classifications

Even though basic parts are the same, design differences can change the way engines operate and how they are repaired. For this reason, you must be able to classify engines.

Some common engine classifications include:

- Cylinder arrangement.
- Number of cylinders.
- Cooling system type.
- Valve location.
- Camshaft location.
- Combustion chamber design.
- Type of fuel burned.

- Type of ignition.
- Number of strokes per cycle.
- Number of valves per cylinder.

Cylinder Arrangement

Cylinder arrangement refers to the position of the cylinders in relation to the crankshaft. There are four basic cylinder arrangements: inline, V-type, slant, and opposed. These are shown in **Figure 12-1.**

In an *inline engine,* the cylinders are lined up in a single row. Each cylinder is located in a straight line that is parallel to the crankshaft. Four-, five-, and six-cylinder engines are commonly inline engines.

Viewed from either end, a *V-type engine* looks like the letter "V." The two banks of cylinders lie at an angle to each other. This arrangement reduces the length and the height of the engine.

A *slant engine* has only one bank of cylinders. This bank leans to one side. As with V-type engines, this saves space. It allows the body and hood of the vehicle to be much lower. A relatively large engine can fit into a small engine compartment.

Cylinders of an *opposed engine* lie flat on either side of the crankshaft. Because of its appearance, this type of engine is sometimes called a "pancake engine" or "boxer engine." An opposed engine may be found in older Volkswagens and a few late-model foreign sports cars (Porsche and Subaru, for example).

One advantage of an "opposed engine" is that there is less frictional horsepower loss at the main bearings. Inline and V-type engines produce a downward thrust on the crankshaft that causes a small but unwanted power loss. In an opposed engine, the pistons move right and left against each other to prevent a downward thrust on the main bearings. This type of engine also has a much lower center of gravity than other car and truck engines. This can be an advantage with a sports car or racing car, where a low center of gravity increases cornering ability.

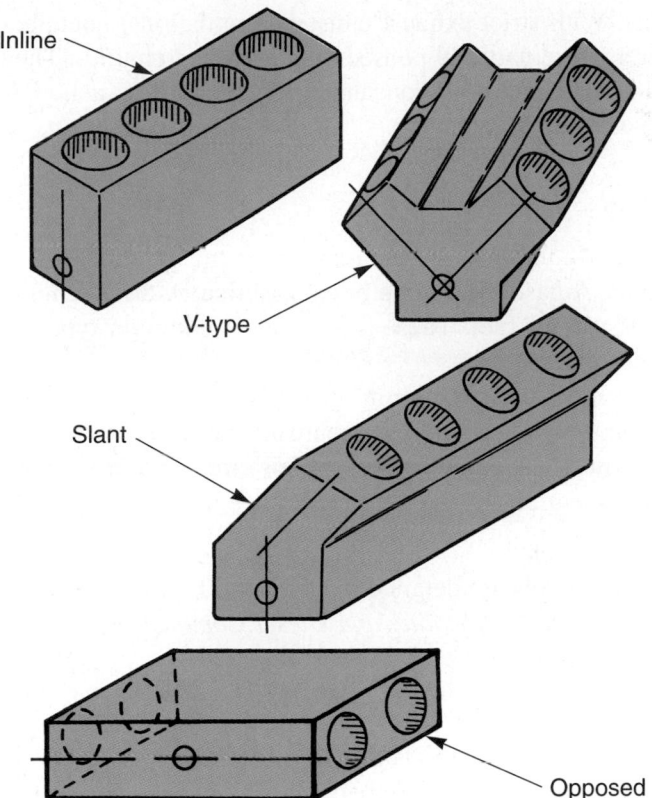

Figure 12-1. Engines can be classified by the way the cylinders are arranged in the block. The four basic cylinder arrangements used in automobiles are shown.

Number of Cylinders

Normally, car and truck engines have either 4, 6, or 8 cylinders. A few engines have 3, 5, 10, 12, or 16 cylinders.

A greater number of cylinders generally increases engine smoothness and power. For instance, an 8-cylinder engine produces twice as many power strokes per crank revolution as a 4-cylinder engine. This reduces power pulsations and roughness (vibrations) at idle. Four-cylinder engines usually have inline, slant, or opposed cylinder arrangements. Six-cylinder engines can have inline, slant, or V-type arrangements. Five-cylinder engines are normally inline engines. Eight-, 10-, and 12-cylinder engines are commonly V-type engines.

Cylinder Numbering and Firing Order

Engine manufacturers number each engine cylinder to help technicians make repairs. The service manual will provide an illustration showing the number of each cylinder, as shown in **Figure 12-2.**

Cylinder numbers are normally stamped on the connecting rods. Sometimes, they are cast into the intake manifold. Cylinder numbering varies from one manufacturer to another. You should keep this in mind when referring to engine classifications. Two V-6 engines, for

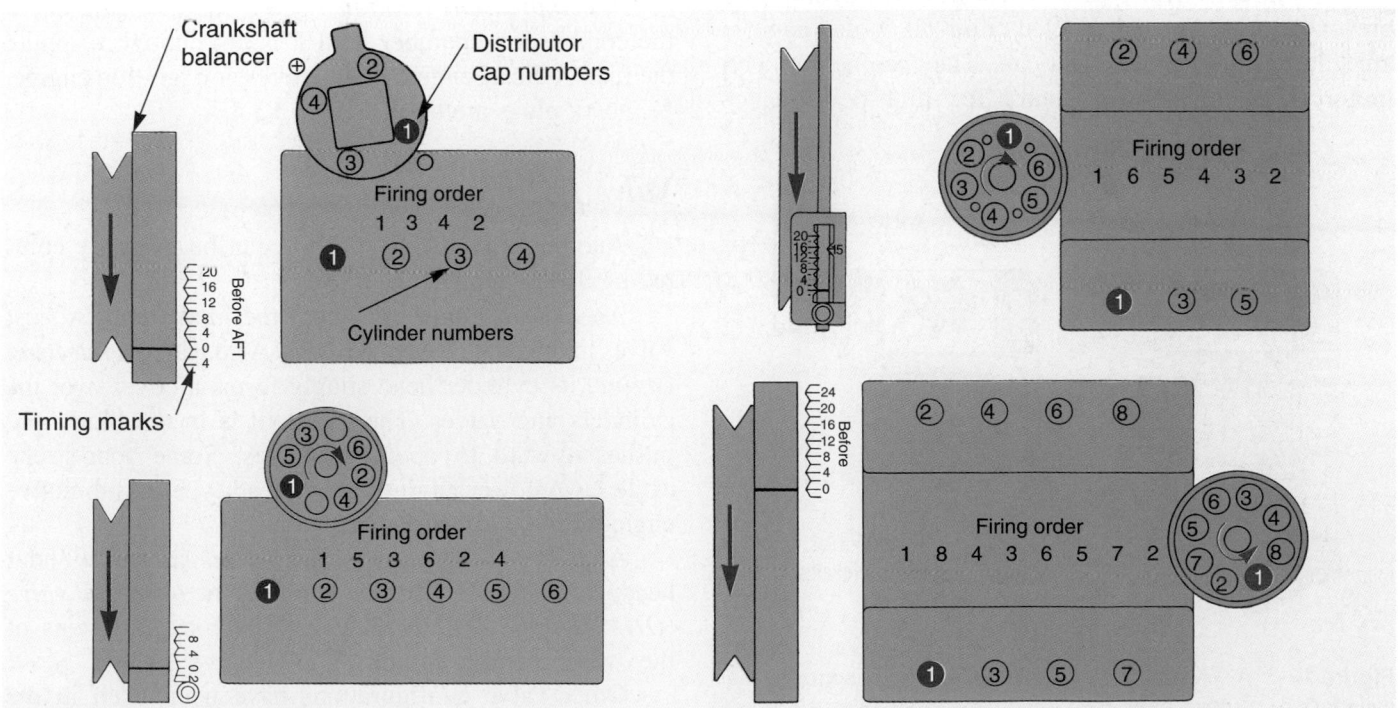

Figure 12-2. Engine cylinder numbers, distributor cap numbers, and firing order numbers can be found in the service manual. Cylinder numbers are usually stamped on the connecting rods and may be cast into the intake manifold. These numbers vary from engine to engine. (Mitchell Manuals)

example, can have completely different cylinder numbering systems.

Firing order refers to the sequence in which the cylinders fire. The position of the crankshaft rod journals in relation to each other determines engine firing order. The service manual will have a drawing showing the firing order for the engine. This information may also be given on the engine intake manifold.

It is important to note that two similar engines can have completely different firing orders. For example, a 4-cylinder, inline engine may have a firing order of 1-3-4-2 or 1-2-4-3. Firing orders for 6- and 8-cylinder engines also vary.

You must know an engine's firing order when working on the ignition system. It can be used when replacing spark plug wires, installing a distributor, or performing other tune-up related operations.

Cooling System Type

There are two types of cooling systems: liquid cooling systems and air cooling systems. The liquid cooling system is the most common. See **Figure 12-3.**

The *liquid cooling system* surrounds the cylinder with coolant (water and antifreeze solution). The coolant carries combustion heat out of the cylinder head and engine block to prevent engine damage.

An *air cooling system* circulates air over cooling fins on the cylinders. This removes heat from the cylinders to prevent overheating. Air-cooled or air-oil cooled engines are seldom used in passenger cars. They can be found on motorcycles, lawnmowers, and a few high-performance

cars. With strict exhaust emission regulations, manufacturers have partially phased out air-cooled engines. They cannot maintain as constant a temperature as a liquid-type cooling system.

Fuel Type

An engine is also classified by the type of fuel it burns. A gasoline engine burns gasoline. A diesel engine burns diesel fuel. These are the most common types of fuel for vehicles.

Liquefied petroleum gas (LPG), gasohol (10% alcohol, 90% gasoline), and pure alcohol can also be used to power an engine. These fuels are used in limited quantities.

 Note
Fuels are detailed in Chapter 20, *Automotive Fuels, Gasoline and Diesel Combustion.*

Ignition Type

Two basic methods are used to ignite the fuel in an engine combustion chamber: spark ignition (spark plug) and compression ignition (compressed air).

A *spark ignition engine* uses an electric arc at the spark plug to ignite the fuel. The arc produces enough heat to start the fuel burning. Gasoline engines use spark ignition, **Figure 12-4A.**

A *compression ignition engine* squeezes the air in the combustion chamber until it is hot enough to ignite the fuel. A diesel engine is a compression ignition engine. No spark plugs are used, **Figure 12-4B.**

Valve Location

Another engine classification can be made by comparing the location of the valves.

An *L-head engine* has both the intake and exhaust valves in the block, **Figure 12-5A.** Also called a *flat head engine*, its cylinder head simply forms a cover over the cylinders and valves. The camshaft is in the block and pushes upward to open the valves. Some four-stroke cycle lawnmower engines are L-head types. Automotive engines are no longer L-head types.

An *I-head engine* has both valves in the cylinder head. Another name for this design is the *overhead valve (OHV) engine,* **Figure 12-5B.** Numerous variations of the overhead valve engine are now in use.

Other valve configurations have been used in the past. However, they are so rare that their mention is not important.

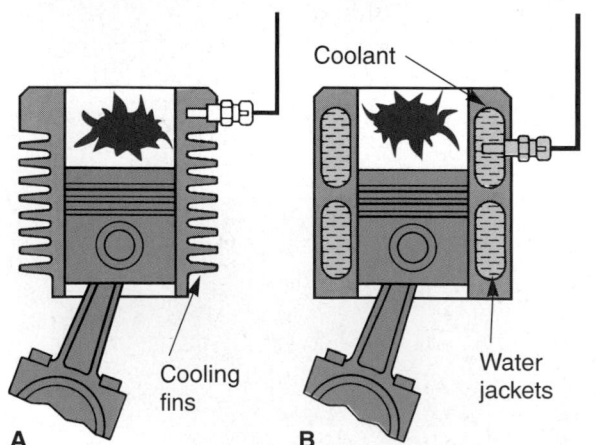

Figure 12-3. A—Air cooling systems uses large fins around the cylinders to remove heat. B—Liquid cooling systems surround the cylinders with coolant. Liquid cooling systems are commonly used in automobiles. Motorcycles and lawnmowers use air cooling systems.

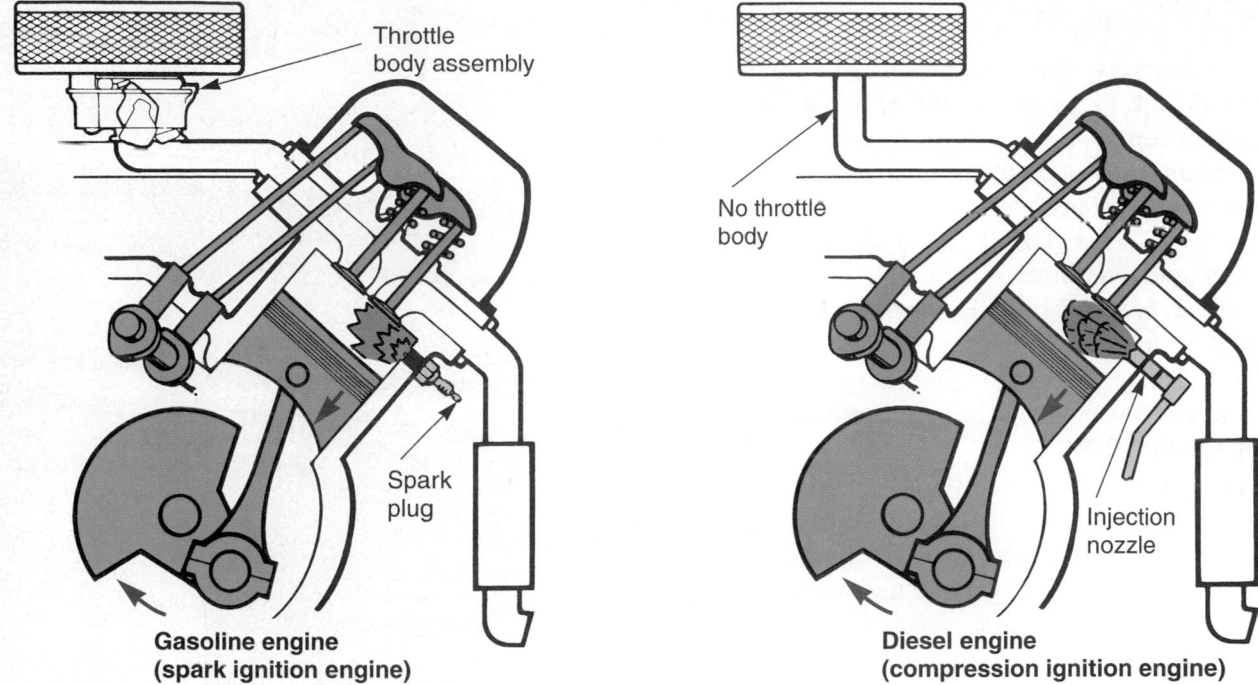

Figure 12-4. Gasoline and diesel engines use different means to ignite fuel. A—A gasoline engine uses a spark plug to start the power stroke. B—A diesel engine compresses air in the cylinder until it is hot. When fuel is injected into the cylinder, the hot air makes the fuel burn.

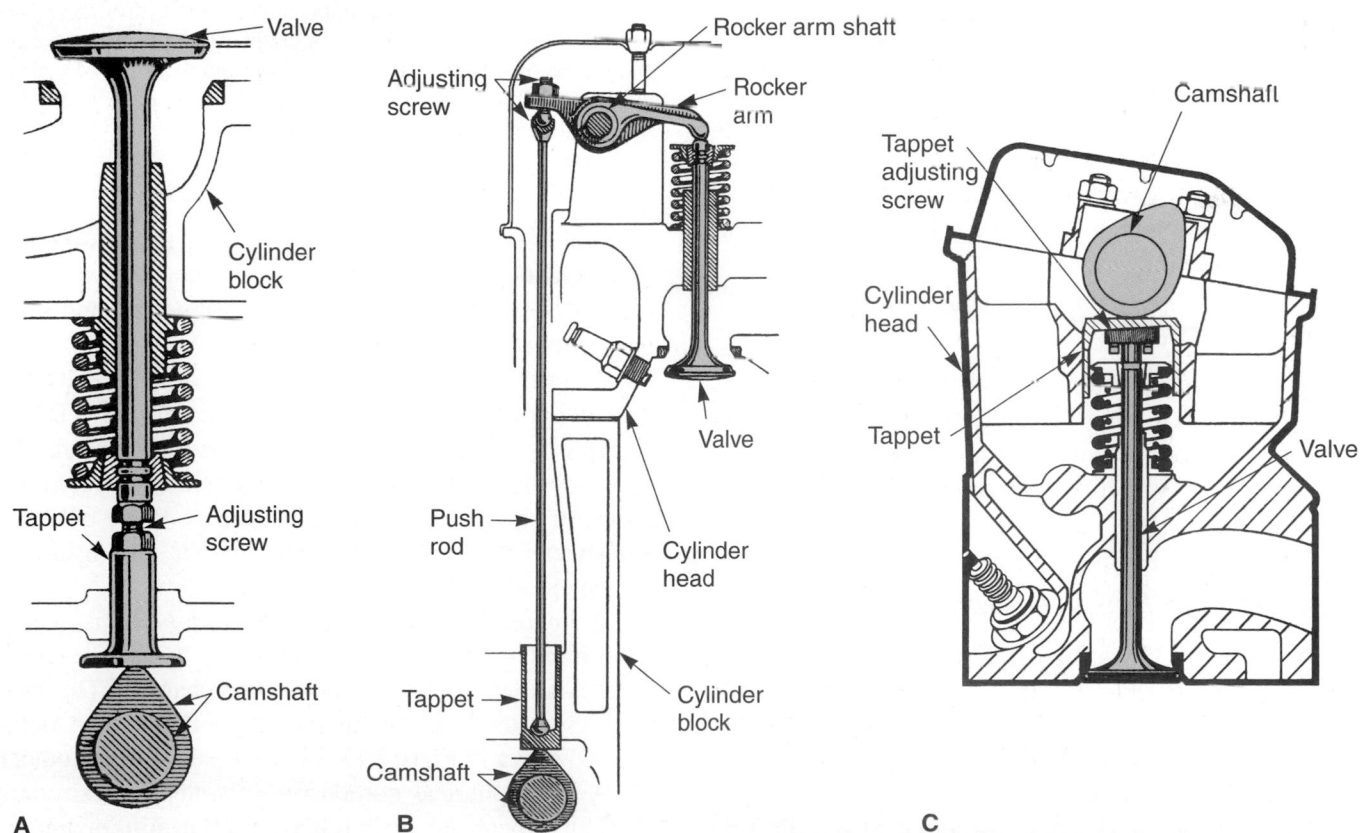

Figure 12-5. Three common valve-camshaft locations. A—The valve-in-block, or L-head, engine is no longer used in automobiles. This arrangement is still used in small gas engines, such as lawnmower engines. B—The cam-in-block engine, also called the overhead valve or I-head engine, is common. C—The overhead cam engine is another form of the I-head engine. It is also very common in today's vehicles. (Black & Decker, Chrysler)

Camshaft Location

There are two basic locations for the engine camshaft: in the block and in the cylinder head. Both locations are common.

A *cam-in-block engine* uses push rods to transfer motion to the rocker arms and valves, **Figure 12-5B.** The term overhead valve (OHV) is sometimes used when referring to a cam-in-block engine.

In an *overhead cam (OHC) engine,* the camshaft is located in the top of the cylinder head. Push rods are *not* needed to operate the rockers and valves. This type of engine is a refinement of the overhead valve engine. Refer to **Figure 12-5C.**

With the cam in the head, the number of valve train parts is reduced. This cuts the weight of the valve train. Also, the valves can be placed at an angle to improve breathing (airflow through cylinder head ports).

OHC engines were first used in racing cars to improve their high rpm (revolutions per minute) efficiency. Now they are commonly used in small, high-rpm, car engines. With lower valve train weight, improved valve positioning, and no push rods to flex, the OHC engine is becoming very popular.

A *single overhead cam (SOHC) engine* has only one camshaft per cylinder head. The cam may act directly on the valves, or rocker arms may be used to transfer motion to the valves.

A *dual overhead cam (DOHC) engine* has two camshafts per cylinder head. One cam operates the intake valves; the other operates the exhaust valves. The dual overhead cam arrangement is frequently used in engines equipped with four-valve combustion chambers. A DOHC engine will be shown later in the chapter.

Tech Tip

Subscribe to and read a variety of automotive magazines. They often contain technical information on the latest engine designs. This will give you valuable information when working on new vehicles.

Combustion Chamber Shape

The shape of the combustion chamber provides still another method of classifying an engine. The three basic combustion chamber shapes for gasoline engines are pancake, wedge, and hemispherical. These are pictured in **Figure 12-6.**

The *pancake combustion chamber,* also called the *bath tub chamber,* has valve heads that are almost parallel to the top of the piston. The chamber forms a flat pocket over the piston head, **Figure 12-6A.**

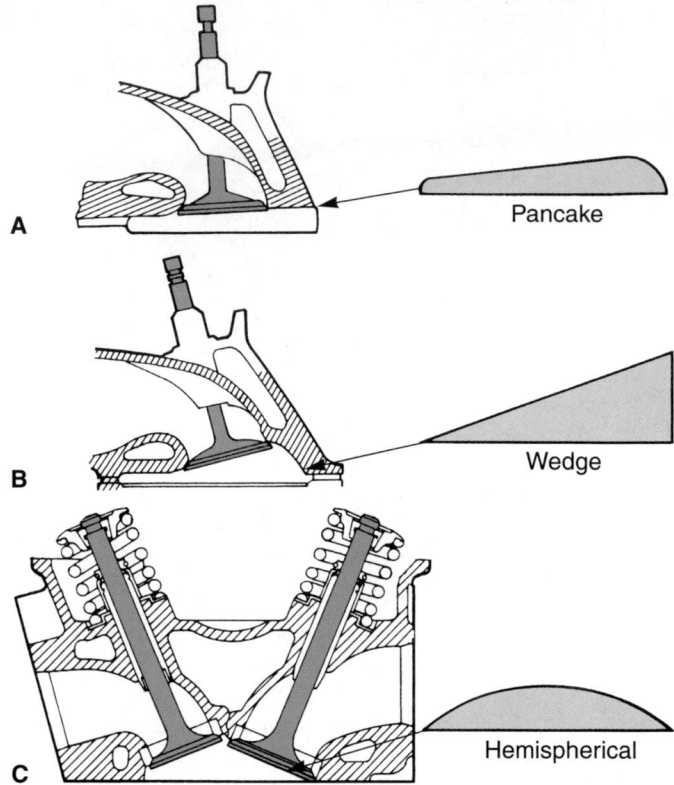

Figure 12-6. There are three basic combustion chamber shapes. A—In a pancake combustion chamber, the valves are almost parallel with the top of the piston. B—In the wedge combustion chamber, the valves are at an angle to the top of the piston. C—The valves are at an angle to each other in a hemispherical combustion chamber. (Chrysler)

A *wedge combustion chamber,* or *wedge head,* is shaped like a triangle or a wedge when viewed as in **Figure 12-6B.** Valves are placed side-by-side, and spark plug is located next to the valves.

A *squish area* is commonly formed inside a wedge-type cylinder head. When the piston reaches TDC, it comes very close to the bottom of the cylinder head. This squeezes the air-fuel mixture in that area and causes it to squirt, or squish, out into the main part of the chamber. Squish can be used to improve air-fuel mixing at low engine speeds.

A *hemispherical combustion chamber*, nicknamed *hemi-head*, is shaped like a dome. The valves are canted (tilted) on each side of the combustion chamber. The spark plug is located near the center of the chamber. A hemi-head is shown in **Figure 12-6C.** Compare it to the others.

A hemispherical combustion chamber is extremely efficient. There are no hidden pockets, minimizing the chances of incomplete combustion. The surface area is very small, reducing heat loss from the chamber. The centrally located spark plug produces a very short flame path for combustion. The canted valves help increase breathing ability.

The hemi-head was first used in high-horsepower racing engines. It is now used in many OHC passenger car engines. It allows the engine to operate at high rpm and makes it very fuel efficient. It also produces complete burning of the fuel to reduce emissions.

Combustion Chamber Types

Besides the three shapes just covered, there are several other combustion chamber classifications. Each type is designed to increase combustion efficiency, gas mileage, and power while reducing exhaust pollution.

A *swirl combustion chamber* is designed to cause the air-fuel mixture to swirl, or spin, as it enters from the intake port. Look at **Figure 12-7.** This causes the air and fuel to mix into a finer mist that burns better.

A *four-valve combustion chamber* uses two exhaust valves and two intake valves. This is illustrated in **Figure 12-8.** The extra valves increase flow in and out of the combustion chamber. This setup is now used on many modern passenger car engines.

In some four-valve combustion chambers, one of the intake valves is opened more than the other. This produces

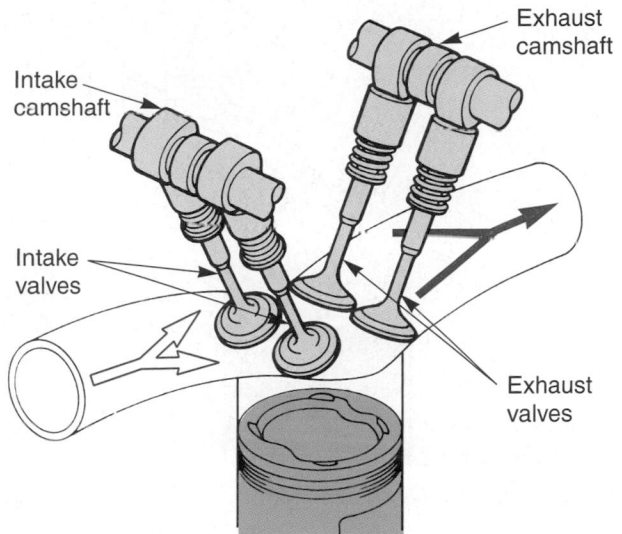

Figure 12-8. Four-valve combustion chambers are used in many late-model engines. Extra valves increase flow and engine power. (Toyota)

a swirling action on the air-fuel mixture entering the combustion chamber. The swirling action helps reduce low-speed emissions.

A *three-valve combustion chamber* has two intake valves and one exhaust valve. The two intake valves allow ample airflow into the combustion chamber on the intake stroke. The single exhaust valve provides enough surface area to handle exhaust flow. This is a relatively new design and is used by only a couple manufacturers. The three-valve combustion chamber provides almost as much performance gain as a four-valve chamber, but it has fewer moving parts. This reduces cost and increases reliability.

A *stratified charge combustion chamber* uses a small combustion chamber flame to ignite and burn the fuel in the main, large combustion chamber. A very *lean mixture* (high ratio of air to fuel) is admitted into the main combustion chamber. The mixture is so lean that it will not ignite and burn easily. A *richer mixture* (higher ratio of fuel to air) is admitted into the small chamber by an extra valve. When the fuel mixture in the small chamber is ignited, flames blow into the main chamber and ignite the lean mixture.

The stratified charge chamber allows the engine to operate on a lean, high-efficiency air-fuel ratio. Fuel economy is increased, and exhaust emission output is reduced.

An *air jet combustion chamber* has a single combustion chamber fitted with an extra air valve. Shown in **Figure 12-9,** a passage runs from the carburetor to the combustion chamber and jet valve.

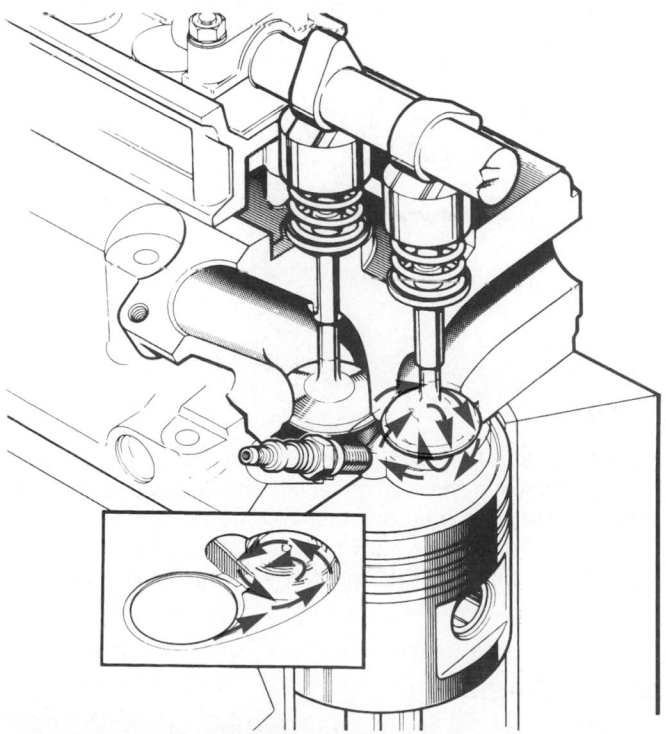

Figure 12-7. The swirl combustion chamber has a port entry designed to cause air-fuel mixture swirling. This helps stir the air and fuel into a finer mist for improved combustion. Many chambers use this principle. (Jaguar)

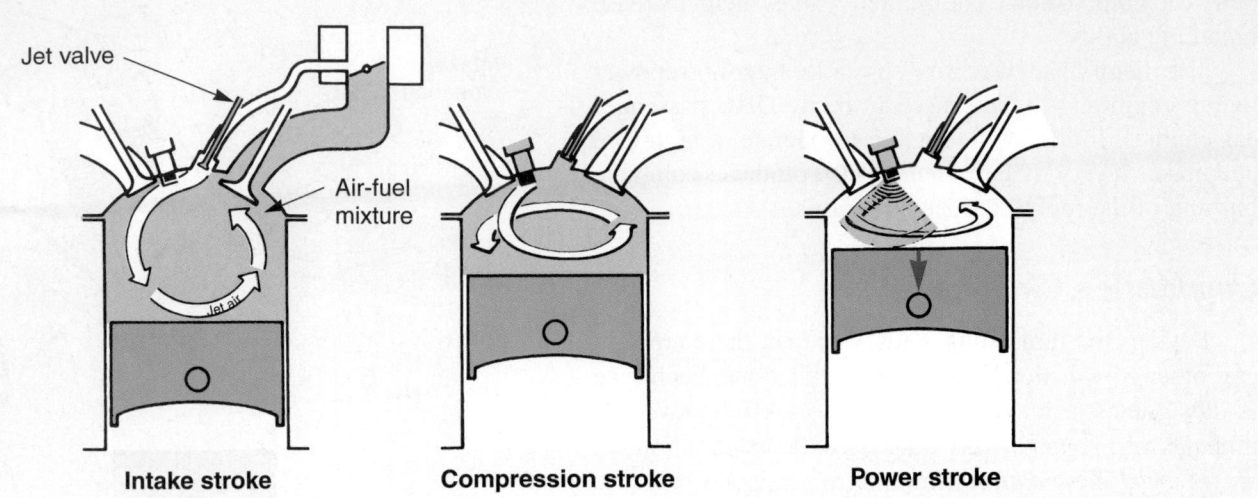

Figure 12-9. An air jet injects a stream of air into the chamber at idle to improve fuel mixing and combustion. (Chrysler)

During the intake stroke, the engine camshaft opens both the conventional intake valve and the air jet valve. This allows fuel mixture to flow past the conventional intake valve and into the cylinder. At the same time, a stream of air flows into the cylinder through the jet valve.

The *jet valve* action causes the fuel mixture in the cylinder to swirl and mix. This increases combustion efficiency by causing more of the fuel to burn during the power stroke. The jet valve only works at idle and low engine speeds. At higher rpm, normal air-fuel mixing is adequate for efficient combustion.

A *precombustion chamber* is commonly used in automotive diesel engines. It is similar in shape to the stratified charge chamber used in gasoline engines. Also called a *diesel prechamber,* the precombustion chamber is used to quiet engine operation and to allow the use of a glow plug (heating element) to aid cold weather starting. **Figure 12-10** shows a cutaway view of a diesel prechamber.

During combustion, diesel fuel is injected into the prechamber. If the engine is cold, the glow plug heats the air in the prechamber. This heat, along with the heat produced by compression, causes the fuel to ignite and burn. As it burns, the flame expands and moves into the main chamber to burn the remaining diesel fuel.

Alternative Engines

As you have learned, vehicles generally use internal combustion, 4-stroke cycle, reciprocating piston engines. *Alternative engines* include all other engine types that may be used to power a vehicle. Various engine types have been developed, but few have been placed into production.

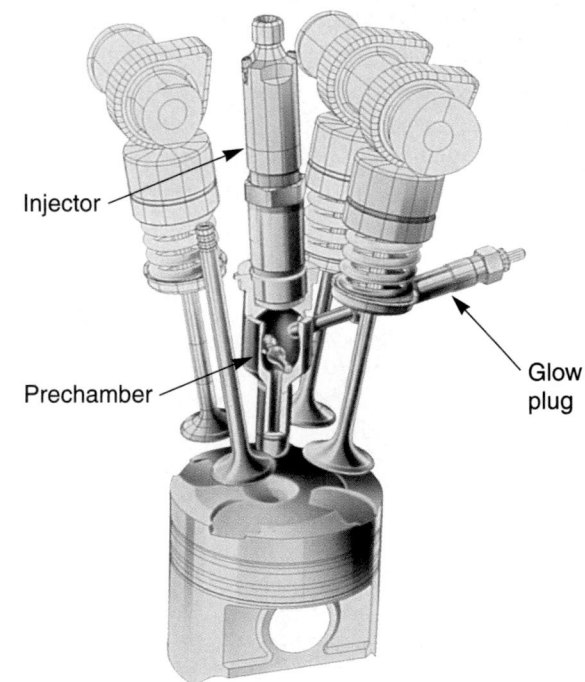

Figure 12-10. A diesel engine precombustion chamber should not be confused with a gasoline engine stratified charge chamber. The precombustion chamber quiets engine operation and allows the use of a glow plug. The glow plug is a heating element that improves cold weather starting. (Oldsmobile)

Rotary Engine

A *rotary engine,* also known as a *Wankel engine,* uses a triangular rotor instead of conventional pistons. The rotor turns inside a specially shaped chamber, as shown in **Figure 12-11.**

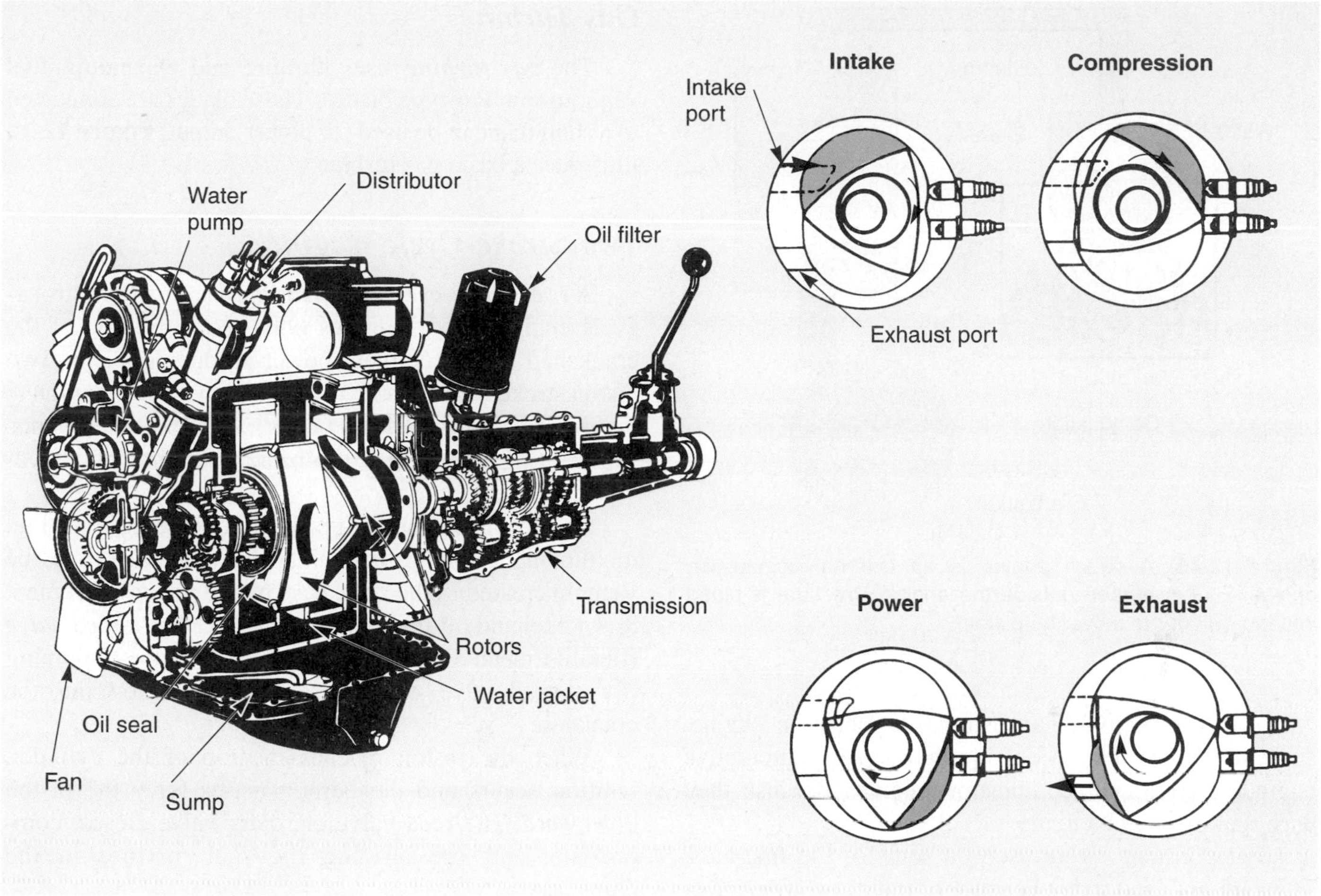

Intake

Intake port

Exhaust port

Compression

Power

Exhaust

Figure 12-11. The rotary engine does not use conventional pistons, which move up and down in the cylinder. Instead, a rotor spins in circular motion. Note the engine parts. (Mazda)

Water pump

Distributor

Oil filter

Transmission

Rotors

Water jacket

Oil seal

Fan

Sump

While turning in the chamber, the rotor orbits around a mainshaft. This eliminates the reciprocating (up-and-down) motion found in piston engines.

Three complete power-producing cycles take place during every revolution of the rotor: three rotor faces produce three intake, compression, power, and exhaust events per revolution. **Figure 12-12** illustrates the basic operation of a rotary engine.

A rotary engine is very powerful for its size. Also, because it spins—rather than moving up and down—engine operation is very smooth and vibration free.

A complicated emission control system is needed to make the rotary engine pass emission standards. This limited its use. The rotary engine was one of the few alternate engines to be mass-produced.

Steam Engine

A *steam engine* heats water to produce steam (hot water vapor). The steam pressure operates the engine pistons. Look at **Figure 12-13.**

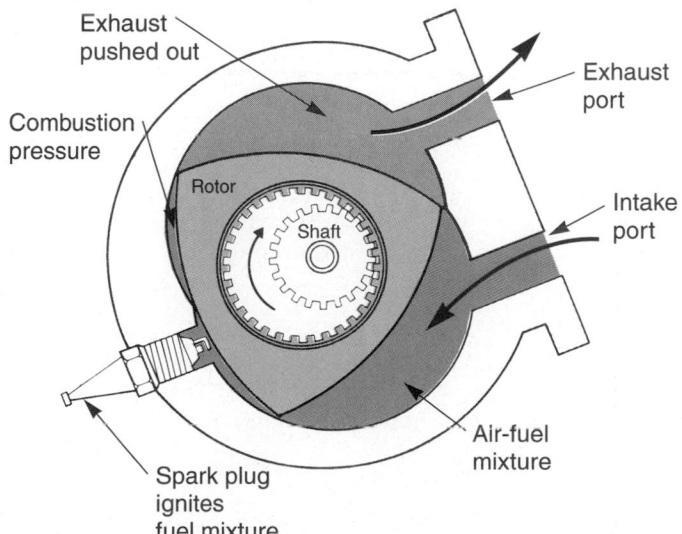

Exhaust pushed out

Combustion pressure

Rotor

Shaft

Exhaust port

Intake port

Spark plug ignites fuel mixture

Air-fuel mixture

Figure 12-12. In a rotary engine, rotor movement produces a low-pressure area, pulling the air-fuel mixture into the engine. As the rotor turns, the mixture is compressed and ignited. As the fuel burns, it expands and pushes on the rotor. The rotor continues to turn, and burned gases are pushed out of the engine.

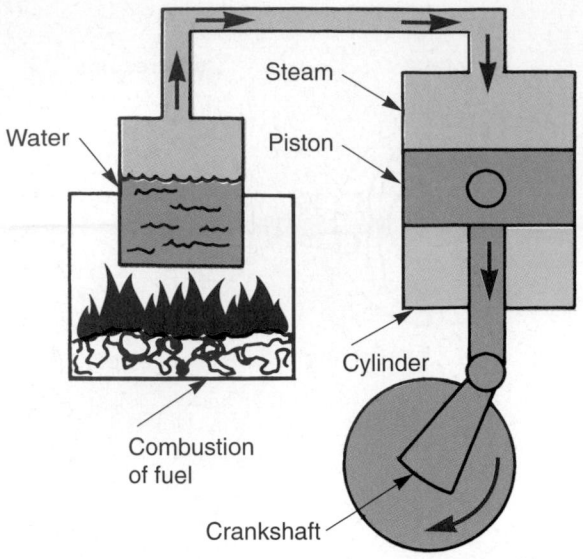

Figure 12-13. A steam engine is an external combustion engine. Fuel is burned outside the engine. Pressure is piped into the cylinder to move the piston.

A steam engine is an external combustion engine, since its fuel is burned outside the engine. Automotive engines are internal combustion engines because they burn fuel inside the engine.

Steam engines were used on some of the first cars. They are not used today because of their low efficiency.

Gas Turbine

The *gas turbine* uses burning and expanding fuel vapor to spin fan-type blades. These blades are connected to a shaft that can be used for power output. **Figure 12-14** illustrates a basic gas turbine.

Two-Stroke-Cycle Engine

A *two-stroke-cycle engine* is similar to a four-stroke-cycle engine, but it requires only one revolution of the crankshaft for a complete power-producing cycle. Two piston strokes (one upward and one downward) complete the intake, compression, power, and exhaust events. **Figure 12-15** illustrates the basic operation of a two-stroke engine.

As the piston moves up, it compresses the air-fuel mixture in the combustion chamber. At the same time, the vacuum created in the crankcase by the piston movement draws fuel and oil into the crankcase. Either a *reed valve* (flexible metal flap valve) or a *rotary valve* (spinning disc-shaped valve) can be used to control flow into the crankcase.

When the piston reaches the top of the cylinder, ignition occurs and the burning gases force the piston downward. The reed valve or rotary valve closes, compressing and pressurizing the fuel mixture in the crankcase.

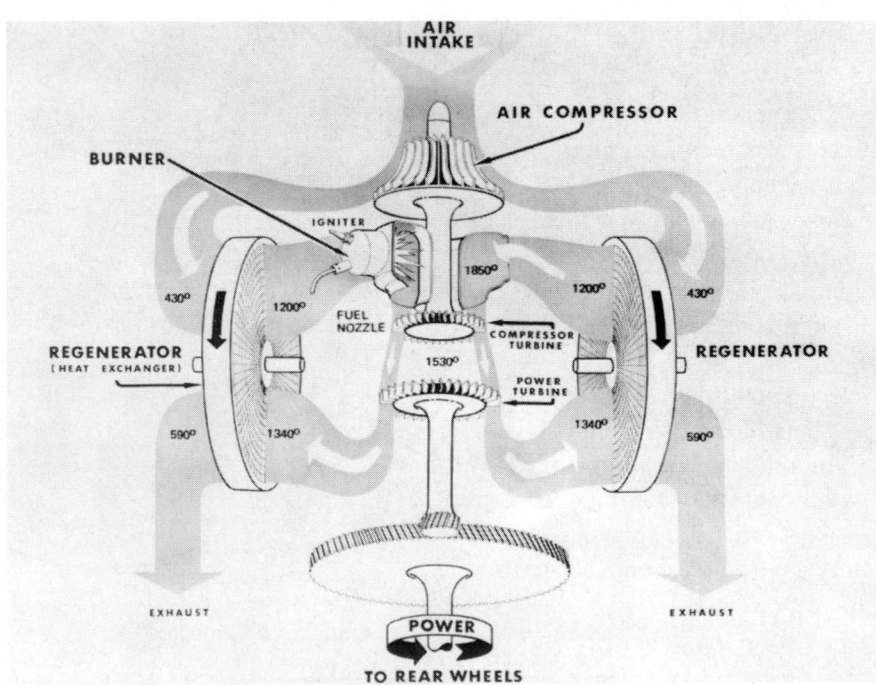

Figure 12-14. A gas turbine uses pressure from burning fuel to spin fan-like blades. This motion is used to rotate the shaft and gears for power output. The gas turbine is not commonly used because of its high manufacturing costs. It requires special metals, ceramics, and precision machining and balancing. Someday, the gas turbine may be a very common automotive engine.

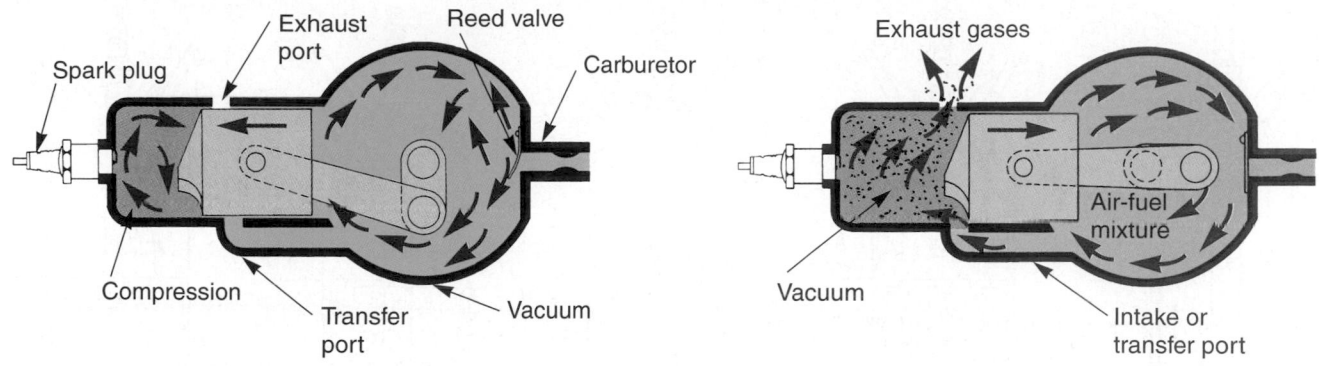

Figure 12-15. The two-stroke-cycle engine completes all four events in two piston movements. (Ethyl Corporation)

As the piston moves far enough down in the cylinder, it uncovers an exhaust port in the cylinder wall. Burned gases leave the engine through the exhaust port.

As the piston continues downward, it uncovers the transfer port. Pressure in the crankcase causes a fresh fuel charge to flow through the transfer port and into the cylinder. Upward movement of the piston again covers the transfer and exhaust ports, compression begins, and the cycle is repeated.

Since the crankcase is used as a storage chamber for each successive fuel charge, the fuel and lubricating oil are pre-mixed and introduced into the engine through the carburetor.

Inside the crankcase, some of the oil separates from the gasoline. The oil mist lubricates and protects the moving parts inside the engine.

Generally speaking, two-stroke-cycle engines are *not* used in vehicles because they:

- Produce too much exhaust pollution.
- Have poor power output at low speeds.
- Require more service than a four-stroke engine.
- Must have motor oil mixed into the fuel.
- Are not as fuel efficient as a four-stroke engine.

Miller-Cycle Engine

A ***Miller-cycle engine*** uses a modified four-stroke cycle. This engine is designed with a shorter compression stroke and a longer power stroke to increase efficiency. The intake valve remains open longer to delay compression. Because the intake valve remains open for a relatively long time, the air-fuel charge wants to be pushed back out the intake port. To compensate for this reverse flow, a supercharger is normally used to pressurize the intake manifold and block this flow, **Figures 12-16** and **17.**

Theoretically, a Miller-cycle engine can produce more power and is more economical than a conventional

Figure 12-16. Cutaway view of a Miller-cycle engine. It is similar to a conventional four-stroke piston engine. However, the opening and closing of the intake valves is modified to increase the duration of power stroke. Note the supercharger, which is needed to prevent backflow of the fuel charge into the intake manifold. (Mazda)

four-stroke-cycle engine of equal size. However, the engine is more complex to produce and maintain because of the supercharger. One auto manufacturer is now successfully using a Miller-cycle engine in its passenger cars.

Typical Automotive Engines

Figures 12-18 through **12-28** illustrate typical automotive engines. Study each of these carefully. Note the design variations between each type. Also, study the names of all the parts. This will help you in later chapters.

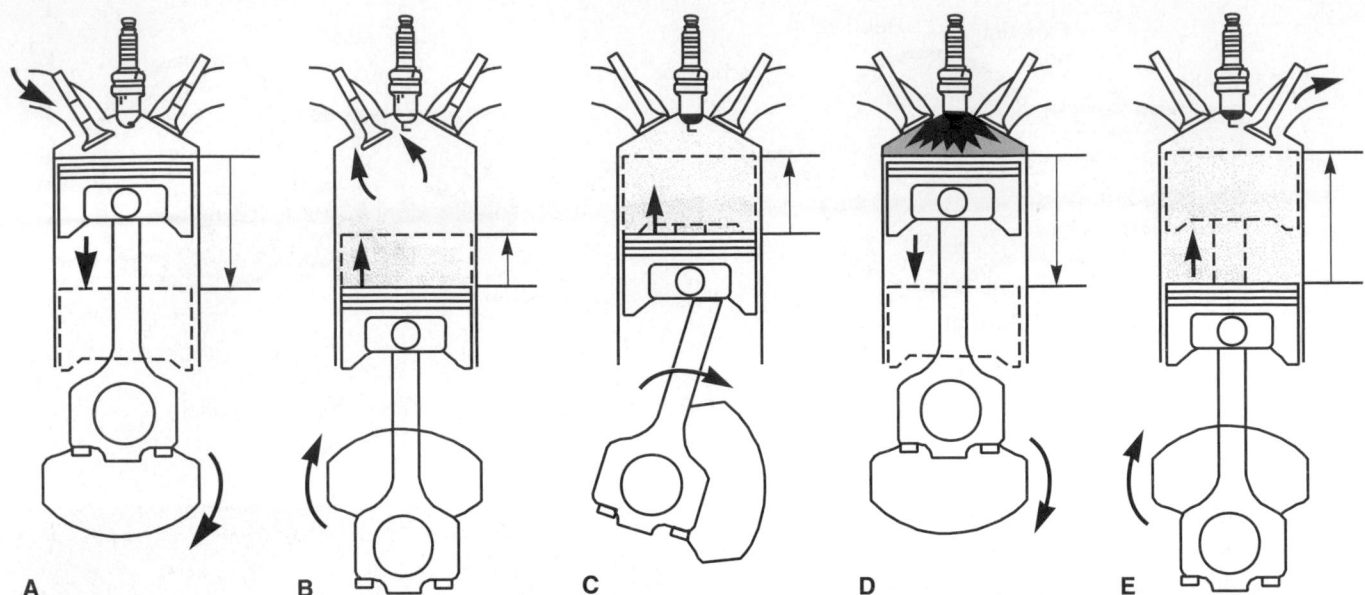

Figure 12-17. Illustrations show Miller-cycle engine operation. A—The piston slides down with the intake valve open. B—The intake valve remains open as the piston slides up. C—Supercharger pressure prevents backflow into the intake port. D—Power stroke. E—Exhaust stroke.

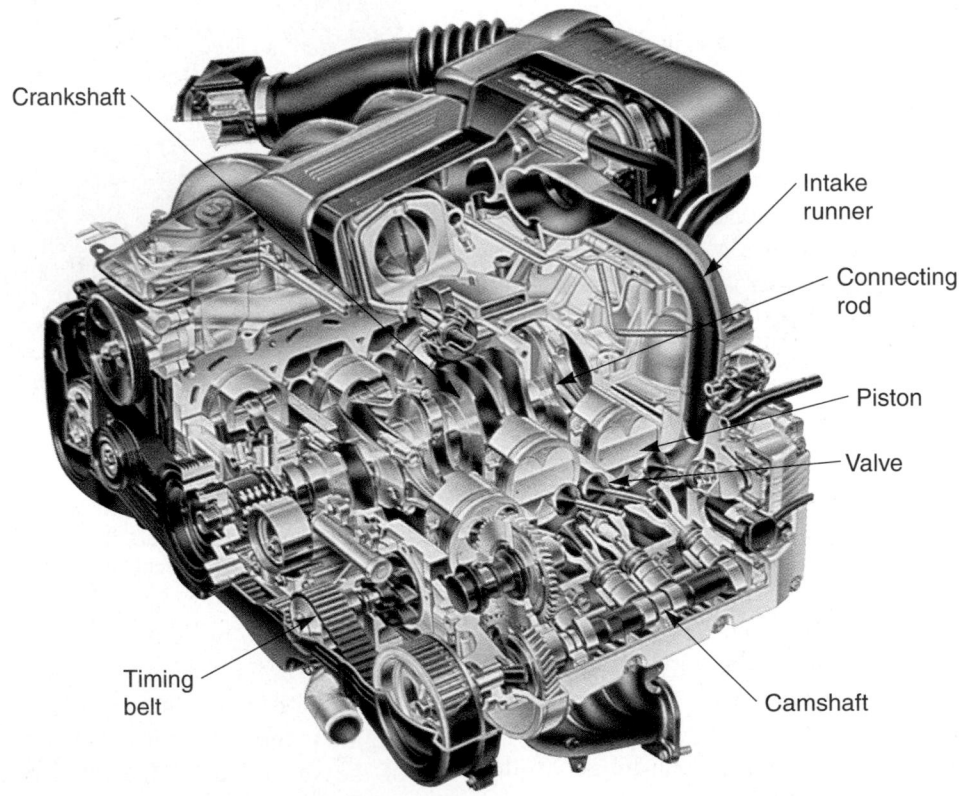

Figure 12-18. Cutaway shows a horizontally opposed, 24-valve, six-cylinder engine. This fuel-injected engine is liquid cooled, has dual overhead camshafts, and provides the lowest center of gravity of any piston engine. (Subaru)

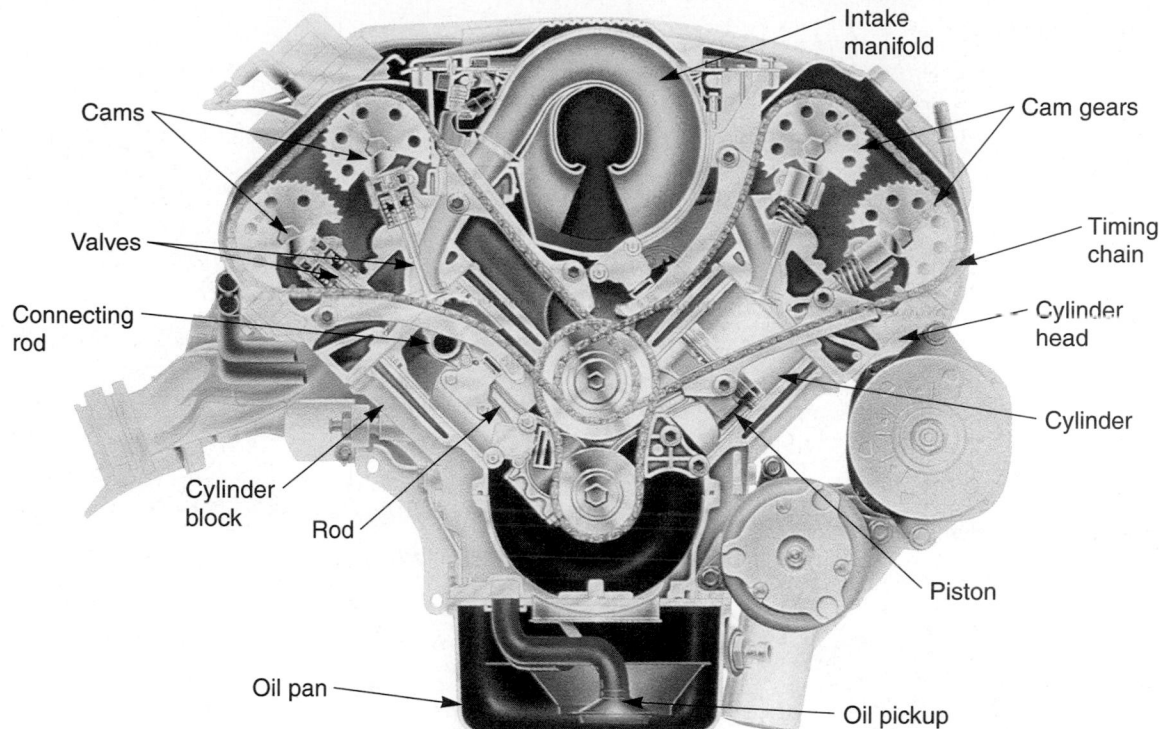

Figure 12-19. Cutaway view of an overhead cam V-8 engine that has four camshafts and 32 valves. The camshafts are chain driven. (Cadillac)

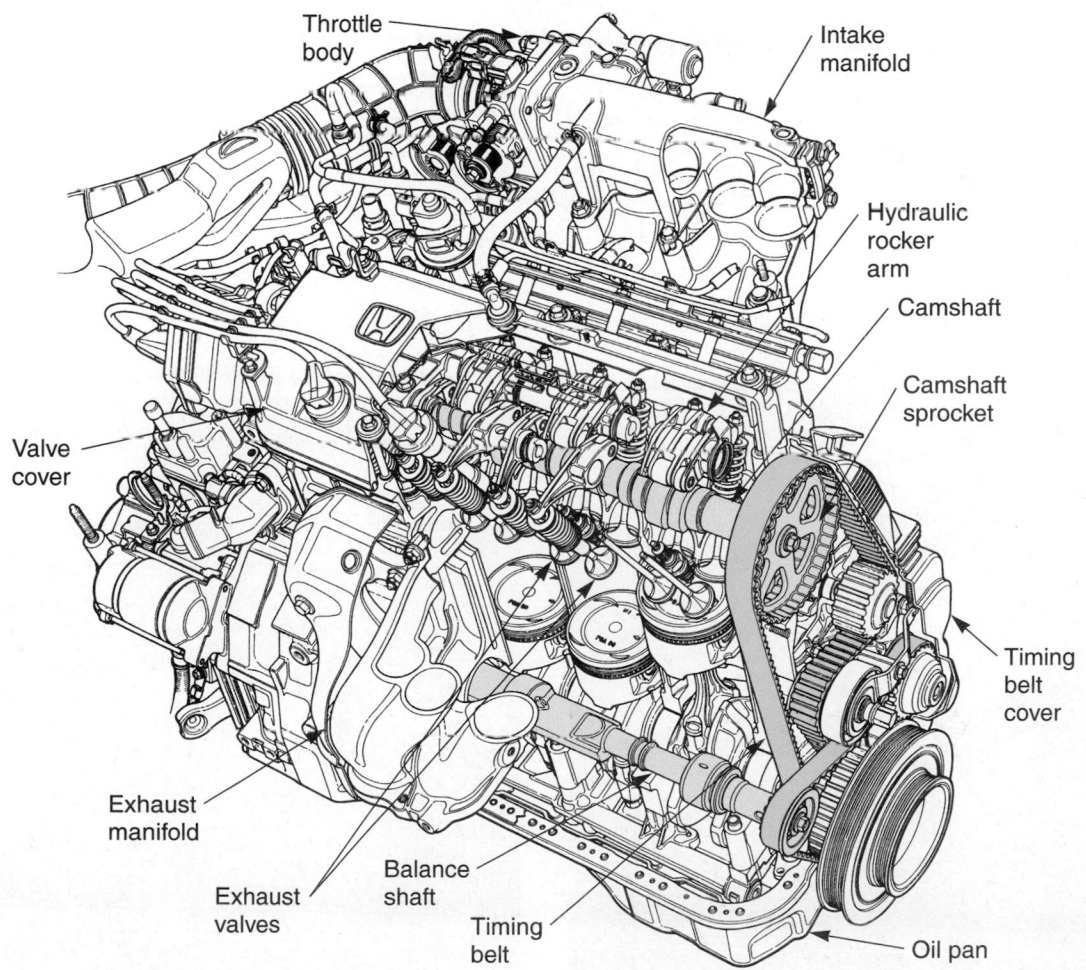

Figure 12-20. Cutaway of a SOHC, 16-valve, four-cylinder engine. Note that both the camshaft and the balance shaft are belt driven. (Honda)

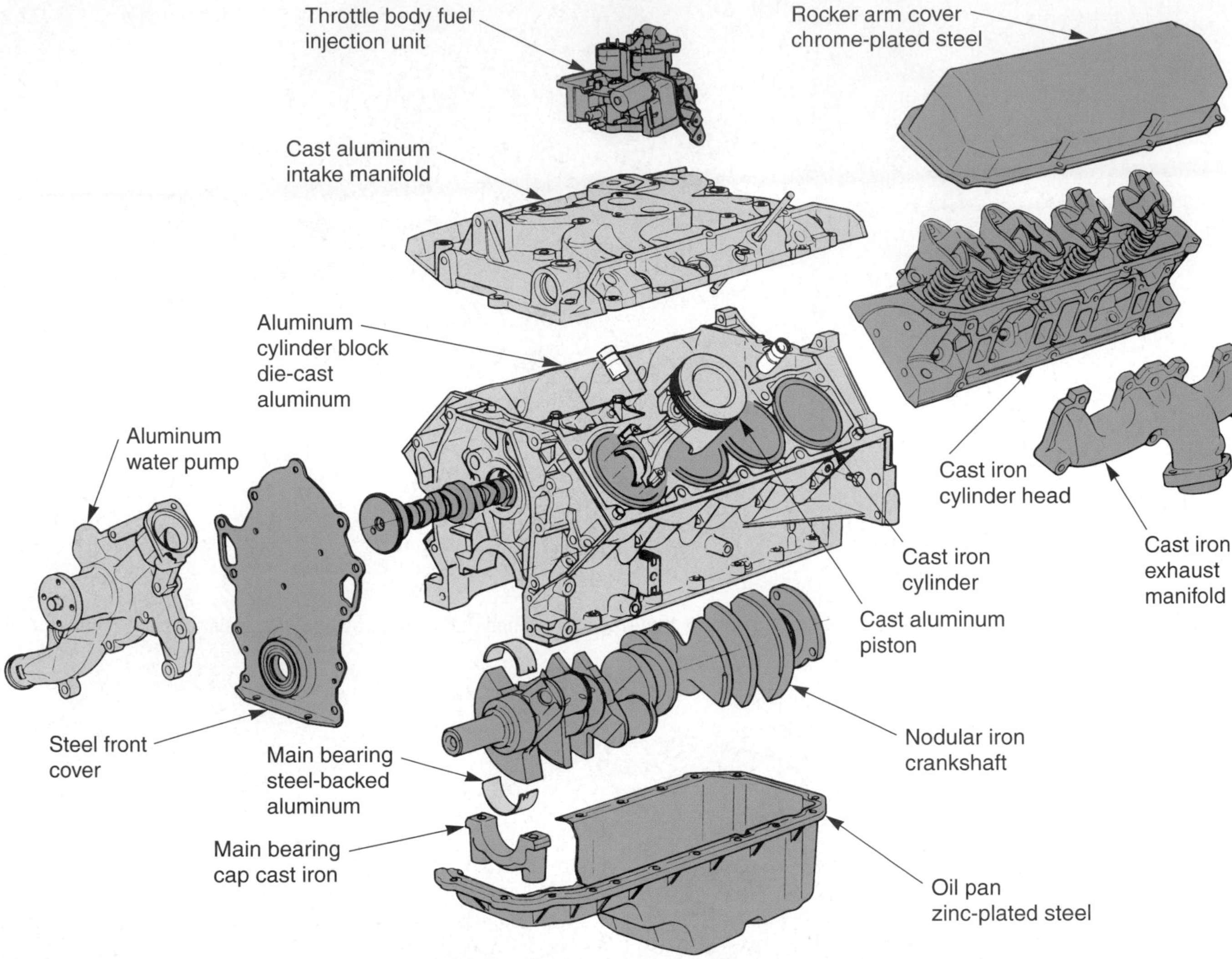

Throttle body fuel
injection unit

Rocker arm cover
chrome-plated steel

Cast aluminum
intake manifold

Aluminum
cylinder block
die-cast
aluminum

Aluminum
water pump

Cast iron
cylinder head

Cast iron
exhaust
manifold

Cast iron
cylinder

Cast aluminum
piston

Steel front
cover

Nodular iron
crankshaft

Main bearing
steel-backed
aluminum

Main bearing
cap cast iron

Oil pan
zinc-plated steel

Figure 12-21. This fuel-injected V-8 engine uses many aluminum parts. (Cadillac)

Figure 12-22. This OHC engine develops tremendous power for its small size. (Buick)

Figure 12-23. Cutaway of a modern DOHC V-8 engine. Can you find the starting motor in the center valley of the engine? (Cadillac)

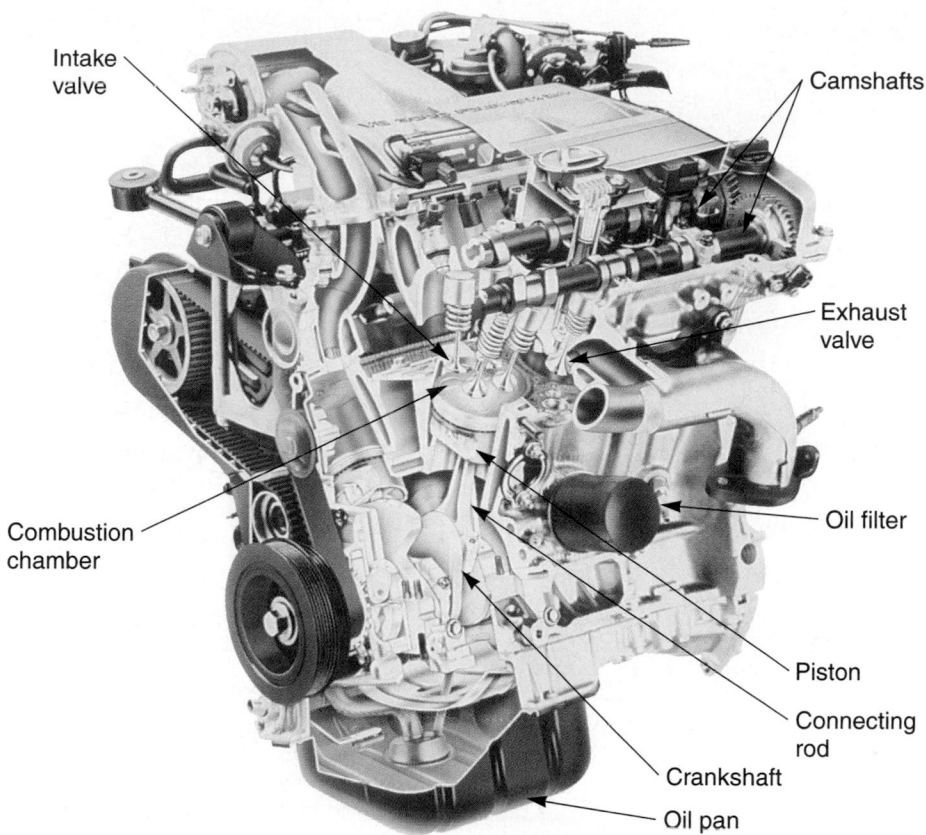

Figure 12-24. DOHC V-6 engine. Each cylinder head contains two camshafts— one to operate the intake valves and another to operate the exhaust valves. (Lexus)

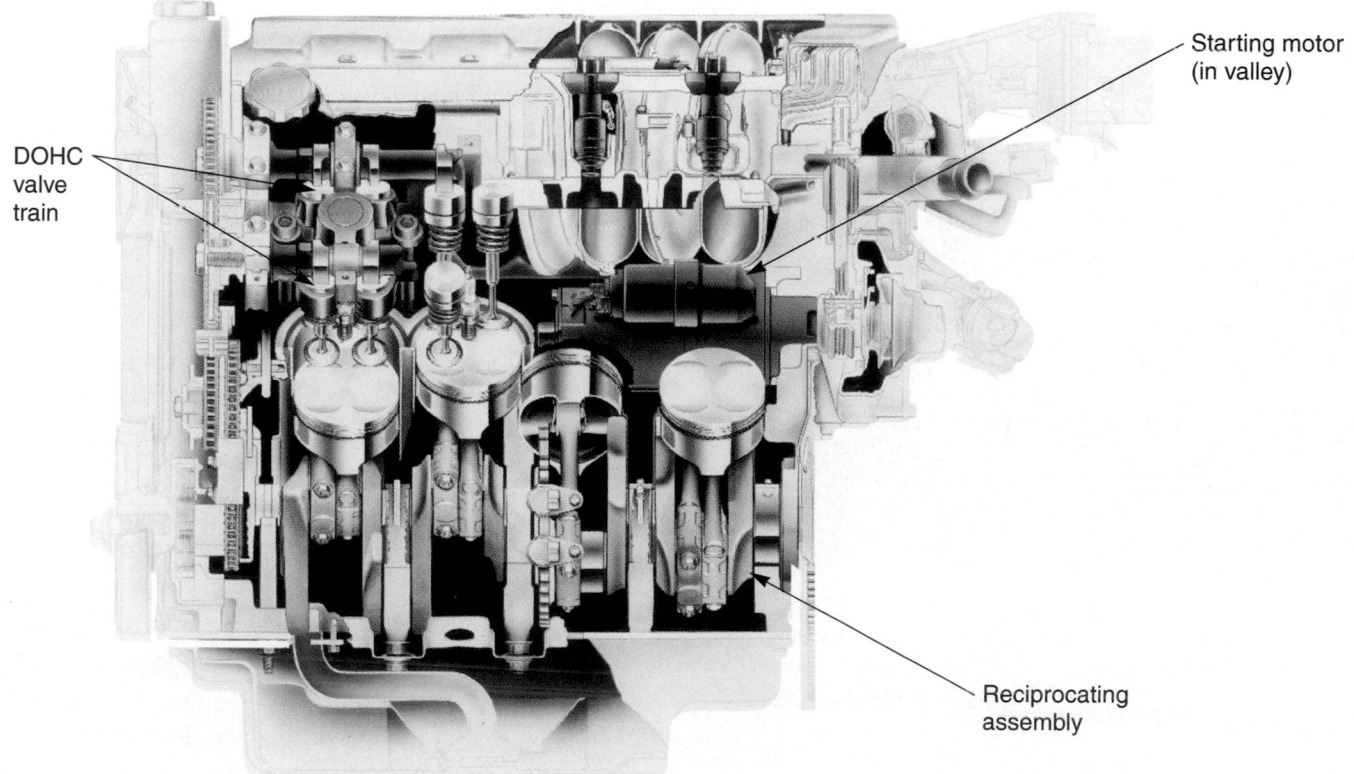

Figure 12-25. This side view of modern V-8 engine clearly shows the reciprocating assembly and the valve train. (Cadillac)

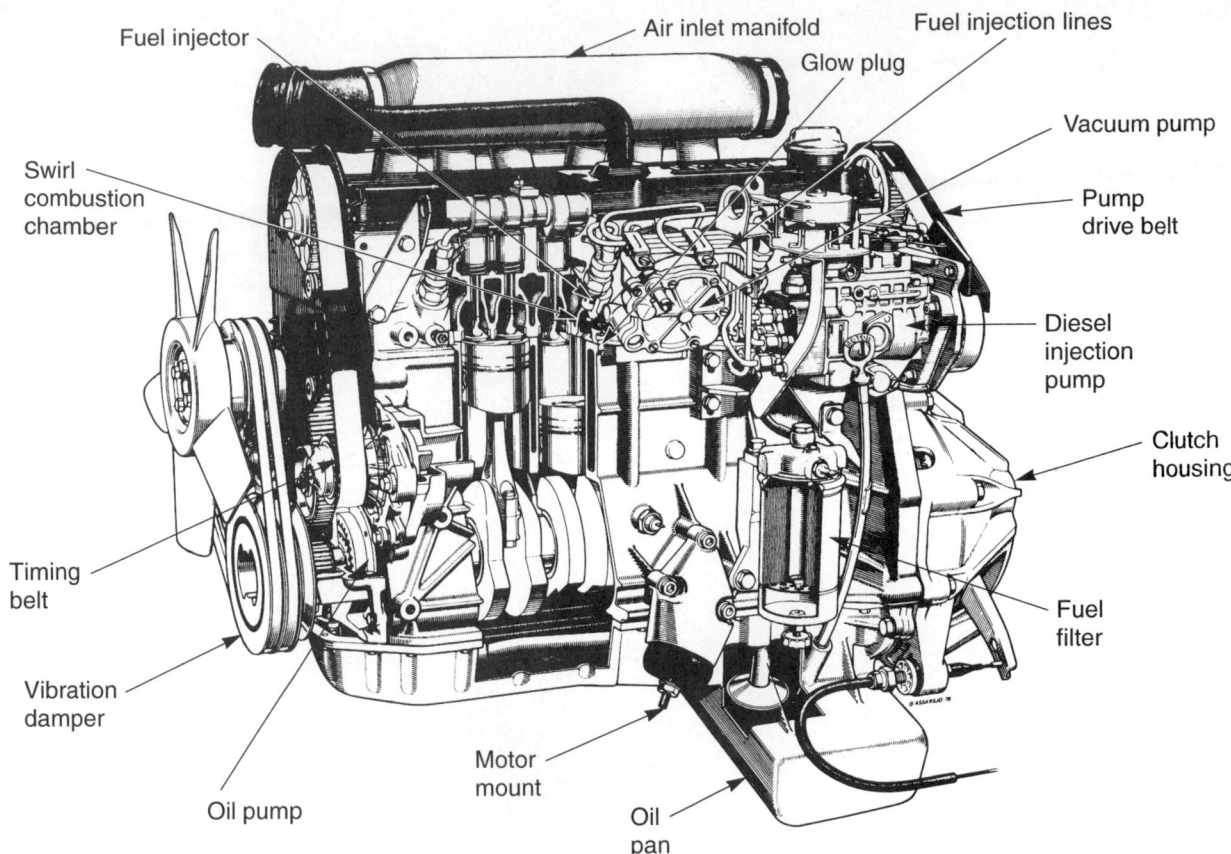

Fuel injector — Air inlet manifold — Fuel injection lines
Glow plug
Vacuum pump
Swirl combustion chamber
Pump drive belt
Diesel injection pump
Clutch housing
Timing belt
Fuel filter
Vibration damper
Motor mount
Oil pump
Oil pan

Figure 12-26. Inline, six-cylinder diesel engine. Note the names of the parts. Also, note the rear drive belt for the injection pump. (Volvo)

Figure 12-27. This fuel-injected V-6 engine has a distributorless ignition system. Each of the three coil packs (on top of the engine, next to the alternator) serves two cylinders. (Buick)

Summary

- Although basic engine parts are the same, design differences can change how engines operate and how they are repaired.
- Cylinder arrangement refers to the position of the cylinders in relation to the crankshaft.
- Firing order refers to the sequence in which combustion occurs in each engine cylinder.
- The liquid cooling system surrounds the cylinder with coolant (water and antifreeze solution).
- A spark ignition engine uses an electric arc at the spark plug to ignite the fuel.
- A compression ignition engine squeezes the air in the combustion chamber until it is hot enough to ignite the fuel.
- A cam-in-block engine uses push rods to transfer motion to the rocker arms and valves.
- In an overhead cam (OHC) engine, the camshaft is located in the top of the cylinder head.
- A four-valve combustion chamber uses two exhaust valves and two intake valves per cylinder.

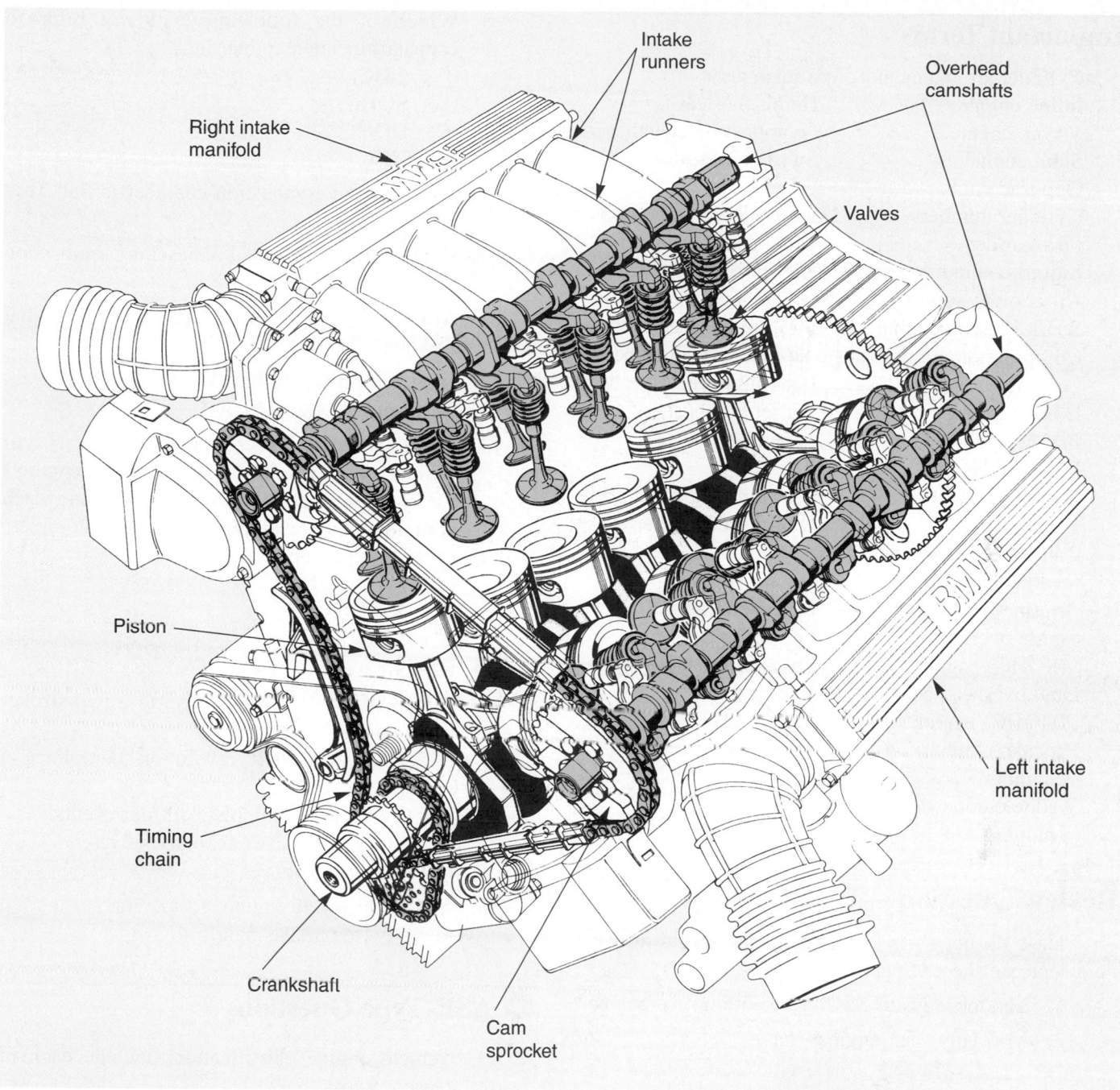

Figure 12-28. Study the construction of this high-performance V-12 engine. Two roller chains are used to drive the overhead camshafts. Twelve cylinders make this a very smooth-running engine because of the relatively high number of power strokes per crankshaft revolution. (BMW)

- A stratified charge combustion chamber uses a small combustion chamber flame to ignite and burn the fuel in the main combustion chamber.

- A precombustion chamber is commonly used in automotive diesel engines.

- Alternative engines include all engine types that can be used to power a vehicle except four-stroke-cycle, reciprocating piston engines.

- The gas turbine uses burning and expanding fuel vapor to spin fan-type blades.

- A two-stroke-cycle engine is similar to a four-stroke-cycle engine, but it requires only one revolution of the crankshaft for a complete power-producing cycle.

Important Terms

Cylinder arrangement
Inline engine
V-type engine
Slant engine
Opposed engine
Cylinder numbers
Firing order
Liquid cooling system
Air cooling system
Spark ignition engine
Compression ignition
  engine
L-head engine
Flat head engine
I-head engine
Overhead valve (OHV)
  engine
Cam-in-block engine
Overhead cam (OHC)
  engine
Single overhead cam
  (SOHC) engine
Dual overhead cam
  (DOHC) engine
Pancake combustion
  chamber
Wedge combustion
  chamber

Squish area
Hemispherical
  combustion chamber
Swirl combustion
  chamber
Four-valve combustion
  chamber
Three-valve combustion
  chamber
Stratified charge
  combustion chamber
Lean mixture
Air jet combustion
  chamber
Jet valve
Precombustion chamber
Diesel prechamber
Alternative engines
Rotary engine
Steam engine
Gas turbine
Two-stroke-cycle engine
Reed valve
Rotary valve
Miller-cycle engine

Review Questions—Chapter 12

Please do not write in this text. Place your answers on a separate sheet of paper.

1. List ten ways to classify an automotive engine.
2. Normally, car engines have _____, _____, or _____ cylinders.
3. Why are cylinders numbered?
4. Firing order refers to the _____ in which the spark plugs fire in the combustion chambers.
5. Most car engines have air cooling systems. True or False?
6. Explain the difference between spark ignition and compression ignition.
7. Where are the two typical locations for the engine camshaft?

8. Which of the following does *not* refer to camshaft location and design?
   (A) OHC.
   (B) SOHC.
   (C) DOHC.
   (D) UHC.
9. A hemi-head combustion chamber is flat. True or False?
10. Explain the operation of a diesel precombustion chamber.
11. A(n) _____ engine does *not* use pistons that slide up and down.
12. How does a gas turbine differ from a conventional piston engine?
13. Three complete power-producing cycles occur during every revolution of a rotary engine's rotor. If an engine has three rotors and each completes one revolution, how many complete cycles are produced?
    (A) 4.
    (B) 3.
    (C) 6.
    (D) 9.
14. Which of the following refers to a two-stroke-cycle engine?
    (A) One crankshaft revolution completes a power stroke.
    (B) Two strokes complete all four events.
    (C) Uses reed valves or rotary valves.
    (D) All of the above.
15. In a Miller-cycle engine, a supercharger is used to _____ the intake manifold.

● ASE-Type Questions

1. Vehicle engine classifications include each of these *except*:
   (A) *valve location.*
   (B) *type of ignition.*
   (C) *number of cylinders.*
   (D) *type of drive mechanism.*
2. Which of the following is *not* a basic cylinder arrangement?
   (A) *Slant.*
   (B) *Inline.*
   (C) *U-type.*
   (D) *V-type.*

3. A 4-cylinder, inline engine's firing order will be:
   (A) 1-3-4-2.
   (B) 1-2-4-3.
   (C) 1-2-3-4.
   (D) Either A or B.

4. The type of engine that has both an intake and exhaust valve in its cylinder head is called a(n):
   (A) I-head engine.
   (B) L-head engine.
   (C) flat head engine.
   (D) All of the above.

5. In an overhead cam engine, push rods are not needed to operate the:
   (A) rockers and valves.
   (B) camshaft and valves.
   (C) camshaft and rockers.
   (D) All the above.

6. The combustion chamber shape in which the valve heads are almost parallel with the top of the piston is called the:
   (A) swirl chamber.
   (B) wedge chamber.
   (C) pancake chamber.
   (D) stratified chamber.

7. Which of the following combustion chambers allows the use of a glow plug in automotive diesel engines?
   (A) Precombustion chamber.
   (B) Swirl chamber.
   (C) Air jet chamber.
   (D) Stratified charge chamber.

8. The engine type that uses a triangular rotor instead of conventional pistons is called a:
   (A) rotary engine.
   (B) opposed engine.
   (C) cam-in-block engine.
   (D) spark ignition engine.

9. A two-stroke-cycle engine requires how many crankshaft revolutions per power cycle?
   (A) One.
   (B) Two.
   (C) Four.
   (D) None of the above.

10. While discussing engines, Technician A says two-stroke-cycle engines are not used in vehicles because they have poor power output at low speeds. Technician B says two-stroke engines are not used in vehicles because they produce too much exhaust pollution. Who is right?
    (A) A only.
    (B) B only.
    (C) Both A and B.
    (D) Neither A nor B.

Activities for Chapter 12

1. Make a poster that shows the four types of engine cylinder arrangements. Label the arrangements.

2. Visit an auto dealership and identify the different engine types offered as standard across the range of automobile models under a single brand name (Ford, Honda, Chrysler, Volvo, etc.). Write a short report on your findings.

3. Find out about the Stanley Steamer or another steam-driven automobile. Describe to the class how its engine worked. Show a drawing or photograph, if possible.

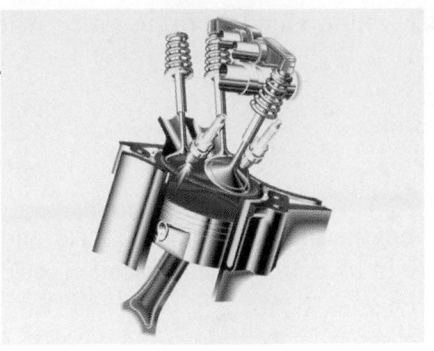

Chapter 13

Engine Top End Construction

After studying this chapter, you will be able to:

- Describe the construction of an engine cylinder head.
- Explain umbrella and O-ring type oil seals.
- Explain the purpose of valve spring shims, rotators, stem caps, and spring shields.
- Describe the construction and operation of a camshaft.
- Explain hydraulic and mechanical lifters.
- Describe different types of rocker arm assemblies.
- Explain the construction and design of intake and exhaust manifolds.
- Describe safety practices used when working on engine top end components.

- Answer ASE certification test questions that require a knowledge of engine top end construction.

An *engine top end* basically includes the cylinder head, valve train, valve cover, and intake and exhaust manifolds. This is illustrated in **Figure 13-1.** Understanding the construction of an engine top end helps improve your ability to troubleshoot and repair an engine. This chapter explains the engine top end and each of its components.

Cylinder Head Construction

A *bare cylinder head* is a cylinder head with all of its parts removed. These parts include the valves, keepers,

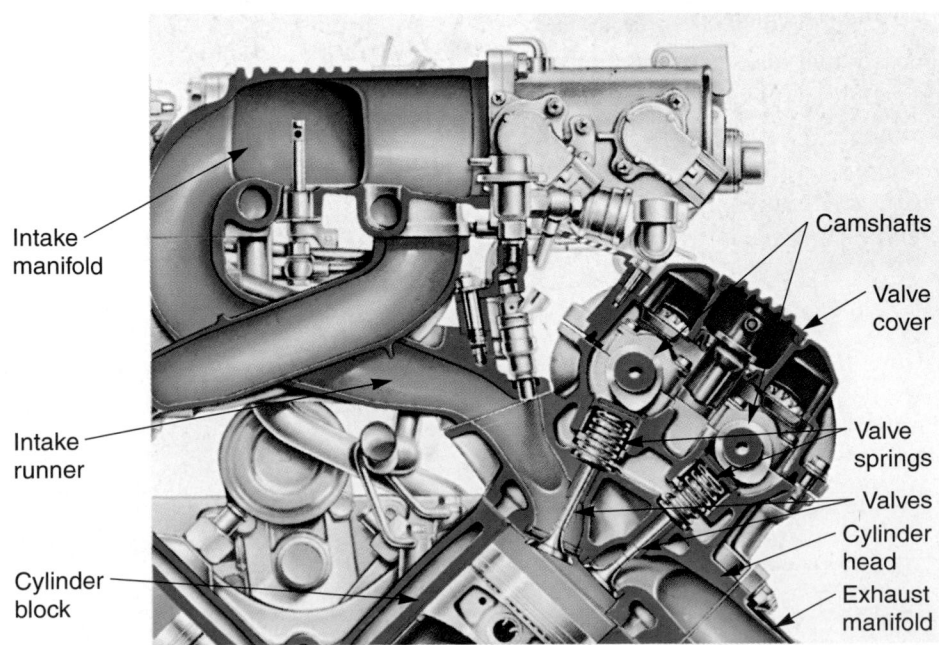

Figure 13-1. Generally, an engine top end includes parts that fasten to top of cylinder block. These include the head, valve train, intake manifold, exhaust manifold, and valve cover. (Honda)

retainers, springs, seals, and rocker arms. A bare cylinder head is commonly made of cast iron or aluminum. The parts that fit in or on a bare cylinder head are pictured in **Figure 13-2.** If a cylinder head becomes badly damaged, the technician may need to install a new bare head. All of the old, reusable parts (valves, keepers, retainers, etc.) can be removed and installed in the new head.

The construction of a cylinder head varies with engine design and type. It is critical that you understand the most important cylinder head variations. See **Figures 13-3** and **13-4.** For example, some engines have a drain hole at the rear of the cylinder head. This allows the oil from the valve train to flow back down through the engine and into the sump. With this design, the engine may be mounted in the vehicle at a slight angle.

Valve Guide Construction

There are two basic types of valve guides: integral guides and pressed-in guides. Both types are used in modern engines.

An *integral valve guide* is part of the cylinder head casting. One is shown in **Figure 13-5.** An integral guide is simply a hole machined through the cylinder head.

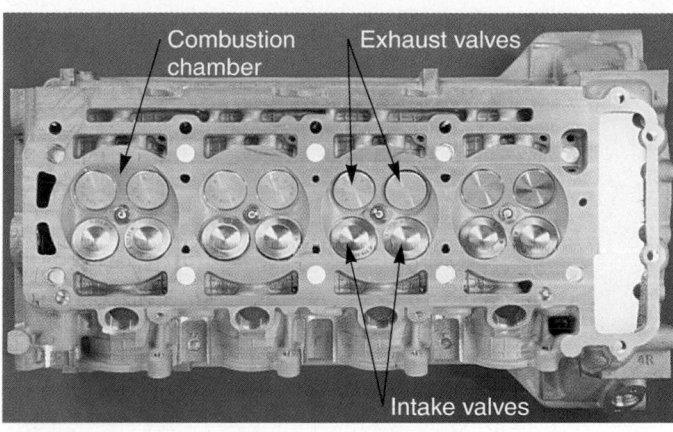

Figure 13-3. Cylinder heads can be cast from iron or aluminum. This cast aluminum cylinder head has four-valve combustion chambers. (Mercedes-Benz AG)

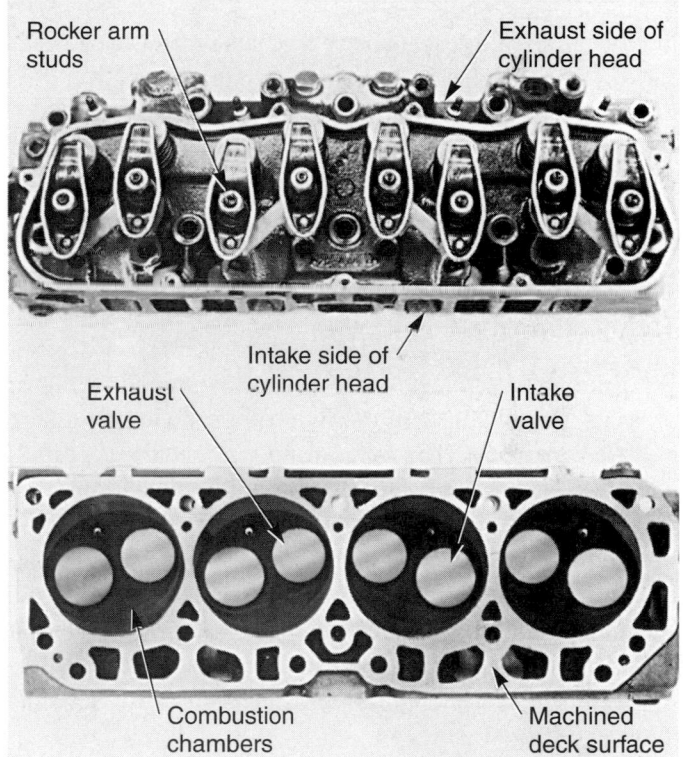

Figure 13-4. This cast iron cylinder head has a two-valve combustion chamber. (Chevrolet)

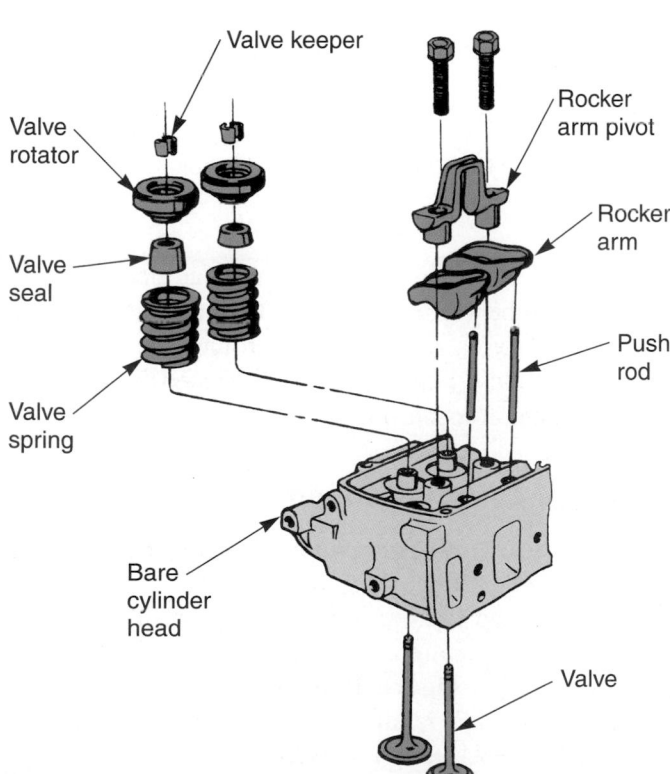

Figure 13-2. A bare cylinder head is a head casting with the components removed. Note the location and names of the parts. (GMC)

This type of guide is very common because of its low production cost.

A *pressed-in valve guide* is a separate sleeve forced into an oversize hole machined in the cylinder head. It can be made of cast iron or bronze. Look at **Figure 13-5.** Friction from the press fit holds the valve guide in the cylinder head. A pressed-in valve guide simplifies guide repair. A worn guide can be pressed out and a new guide

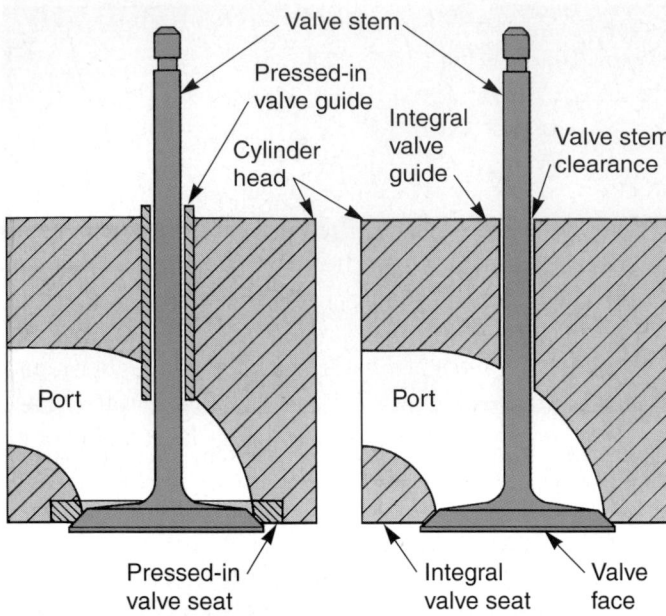

Figure 13-5. Valve guides and seats can be made as separate inserts or as part of the head.

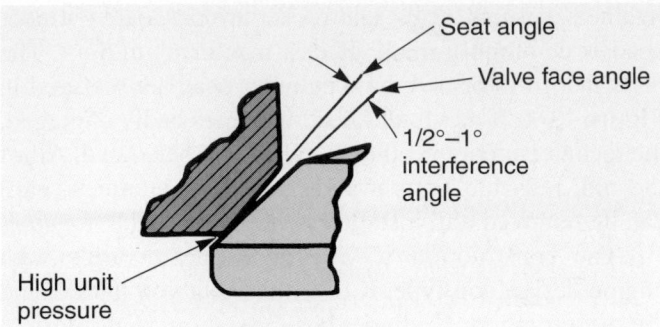

Figure 13-6. Common valve and seat angles are 30° and 45°. The interference angle is the 1° difference between the valve and seat angles to increase sealing pressure and to speed seating, or break-in. (TRW)

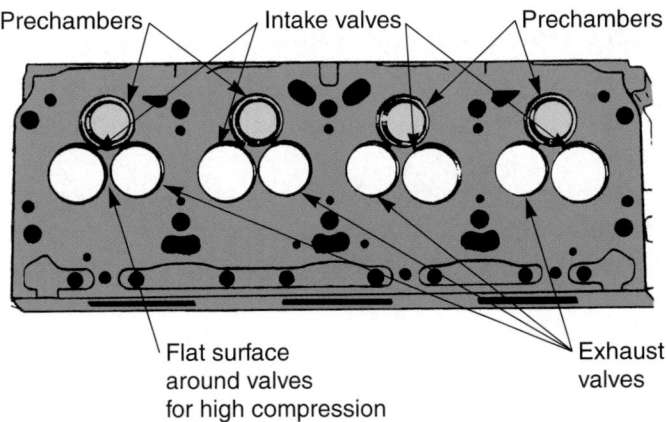

Figure 13-7. A diesel cylinder head normally has pressed-in prechamber cups. They form the precombustion chambers for the tips of glow plugs and injectors. (Oldsmobile)

can be quickly pressed in. This eliminates some of the machining needed when replacing a badly worn integral-type guide.

Valve Seat Construction

Like valve guides, valve seats can be integral or pressed-in. Both are commonly used in modern engines.

An *integral valve seat* is simply a machined portion of the cylinder head casting. Different cutters are used to machine a precise face on the valve port opening. A *pressed-in valve seat,* also called a *seat insert,* is a separate part that is forced into a recess cut into the head. Steel inserts are commonly used in aluminum cylinder heads. Steel is needed to withstand the extreme heat produced during engine operation. A pressed-in valve seat is shown in **Figure 13-5.**

The *valve seat angle* is the angle formed by the face of the seat. A 45° angle is commonly used on passenger car engines. However, some high-performance engines have seat angles of 30°. An *interference angle* is a 1/2–1° difference between the valve seat face angle and the angle of the valve face, see **Figure 13-6.** The interference angle reduces the contact area between the seat and valve. This increases pressure between the two and speeds valve seating (sealing) during engine operation.

Diesel Prechamber Cup

A *prechamber cup* is pressed into the cylinder head of some diesel engines. Refer to **Figure 13-7.** Holes are

precisely machined into the cylinder head deck surface. The prechambers are force-fit into these holes. Each prechamber forms an enclosure around the tip of an injector and glow plug. This area is heated by the glow plug for better cold starting.

Stratified Charge Chamber

A *stratified charge chamber* also fits into the cylinder head casting. It is found in gasoline engines that are designed to use a rich fuel mixture in the auxiliary chamber to ignite a lean mixture in the main combustion chamber. **Figure 13-8** shows how the stratified charge chamber and related components are assembled in the cylinder head. Study it closely.

 Note
For more information on stratified charge chambers, refer to Chapter 12, *Engine Design Classifications.*

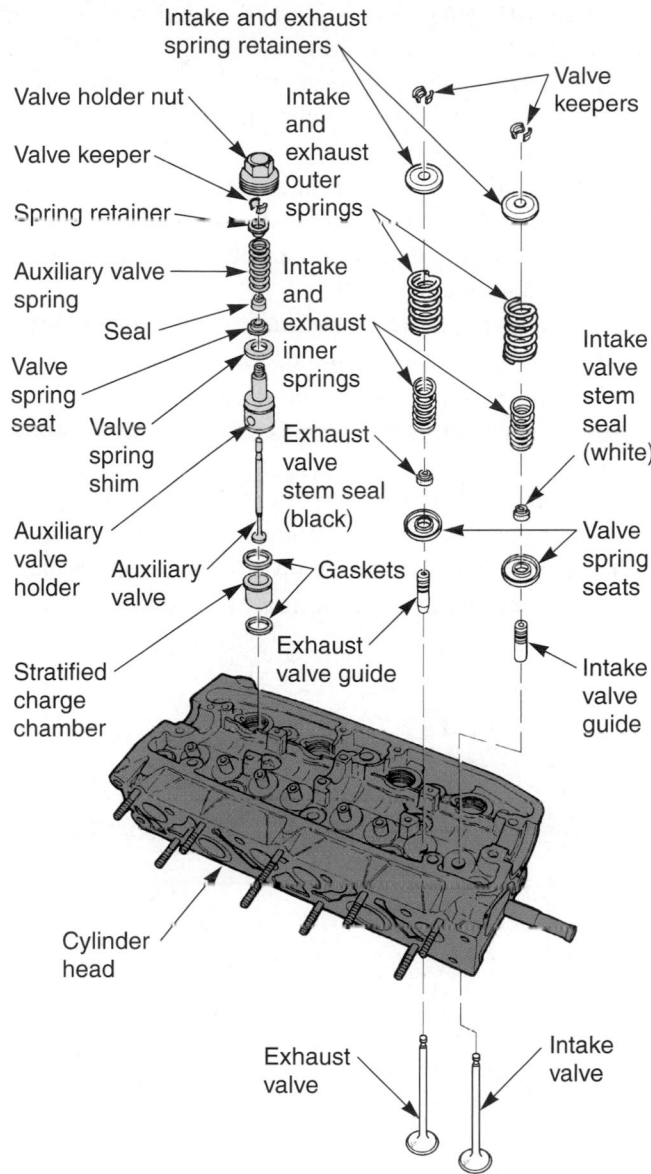

Figure 13-8. This is a gasoline engine cylinder head with stratified charge auxiliary chambers. Note how the chamber and extra intake valve are assembled. (Honda)

Valve Train Construction

The valve train controls the opening and closing of the cylinder head ports. Although the basic function of all valve trains is the same, valve train construction will vary with engine design. To be able to work on any type of valve train, you must understand these differences.

Valve Construction

Automotive engines commonly use *poppet valves,* or *mushroom valves.* These terms come from the valve's shape and action. The valve is shaped something like a

mushroom and it "pops" open. Several engine valves are shown in **Figure 13-9.**

Almost every surface of a valve is machined. The stem must accurately fit the guide. Some engine valve stems are chrome plated to better resist wear. The face must contact the seat perfectly. The margin must be thick enough to prevent valve burning. Grooves are also cut into the valve stem tops for the keepers.

The *valve face angle* is the angle formed between the valve face and valve head. Normal valve face angles are 45° and 30°. The exhaust valve is exposed to higher temperatures than the intake valve—the intake valve has cooler outside air flowing over it, while hot combustion gases flow over the exhaust valve. If the exhaust valve does not transfer heat into the cylinder head, it will burn.

Hollow-stem, *sodium-filled valves* are used when extra valve cooling action is needed. Refer to **Figure 13-9C.** During engine operation, the sodium inside the hollow valve melts. When the valve is pressed down, or opened, the sodium splashes down into the valve head and collects heat. Then, when the valve is released, or closed, the sodium splashes up into the valve stem. Heat transfers out of the sodium and into the stem, valve guide, and engine coolant. In this way, the valve is cooled.

A *stellite valve* has a stellite (very hard metal alloy) coating on its face. A stellite coating is often used in engines designed to burn unleaded fuel. Older engines

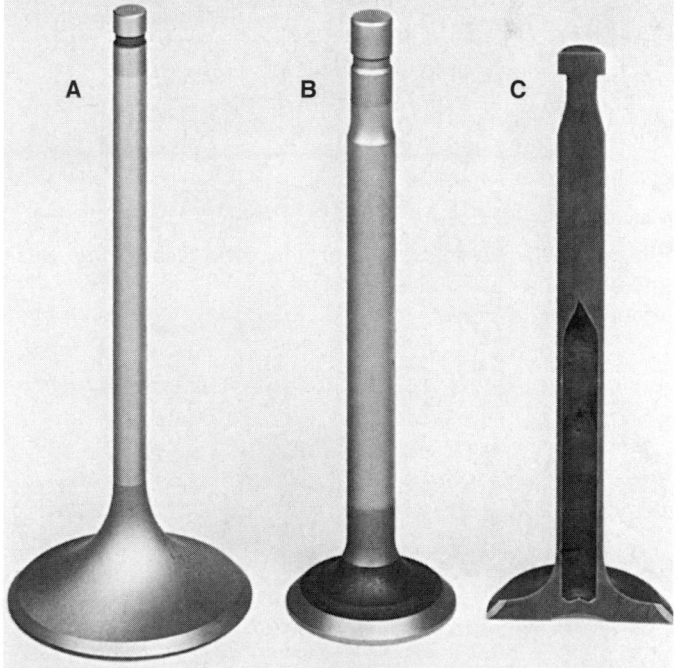

Figure 13-9. Three typical engine valves. A—Polished intake valve. B—Stock exhaust valve. C—Hollow exhaust valve designed to hold sodium for cooling. (Dana)

did not require the use of stellite valves, because leaded fuel (now phased out) helped lubricate the valve and seat faces. Look at **Figure 13-10.**

Warning
Sodium-filled valves are used in a few high-performance engines. They are very light and allow high engine rpm for prolonged periods. However, they can explode if placed in a fire and, therefore, must be disposed of properly.

Valve Seal Construction

Valve seals come in two basic types. These types are umbrella and O-ring. Both are common on modern engines.

An *umbrella valve seal* is shaped like a cup and can be made of neoprene rubber or plastic. Three are shown in **Figure 13-11.** An umbrella valve seal slides down over the valve stem before the spring and retainer. It covers the small clearance between the valve stem and guide. This keeps oil from being drawn into the cylinder head port and combustion chamber.

An *O-ring valve seal* is a small round seal that fits into an extra groove cut in the valve stem. Look at **Figure 13-12.** Unlike the umbrella type, it seals the gap between the retainer and valve stem, not the guide and stem. It stops oil from flowing through the retainer, down the stem, and into the guide. An O-ring valve seal fits onto the valve stem *after* the spring and retainer. It is made of soft synthetic rubber that allows it to be stretched over the valve stem and into place.

A *valve spring shield* is normally used with an O-ring type oil seal. The shield surrounds the top and upper sides of the spring and helps keep oil off the valve stem. See **Figure 13-12.** A *nylon shedder* can also be used to limit the amount of oil that splashes on the valve stem and guide opening. One is illustrated in **Figure 13-13.** The shedder is a cross between a conventional oil seal and a valve spring shield. It seals against the valve stem like a seal and encircles the upper spring like a shield.

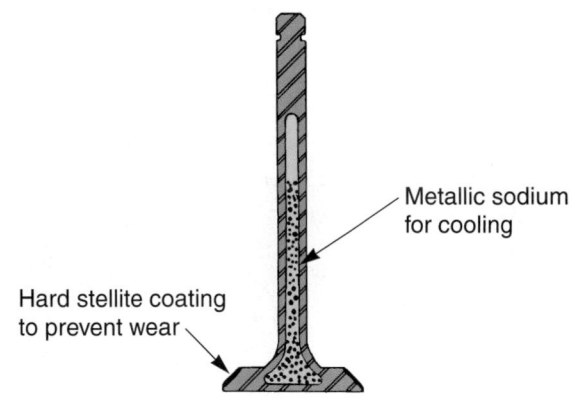

Figure 13-10. Stellite coating on the valve face retards wear and allows the use of unleaded fuel.

Figure 13-11. Umbrella valve seals form a covering over the opening at the top of valve guides. A—Synthetic rubber seal with a plastic shedder insert. B—All synthetic rubber seal. C—Plastic valve seal.

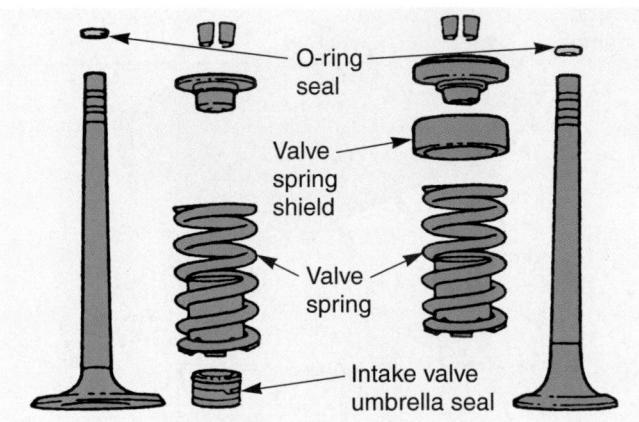

Figure 13-12. O-ring valve seal fits into an extra groove cut into the valve stem. To prevent seal damage, the spring and retainer must be installed before the seal. This valve spring assembly uses both an O-ring seal and an umbrella seal on the intake valve. (Buick)

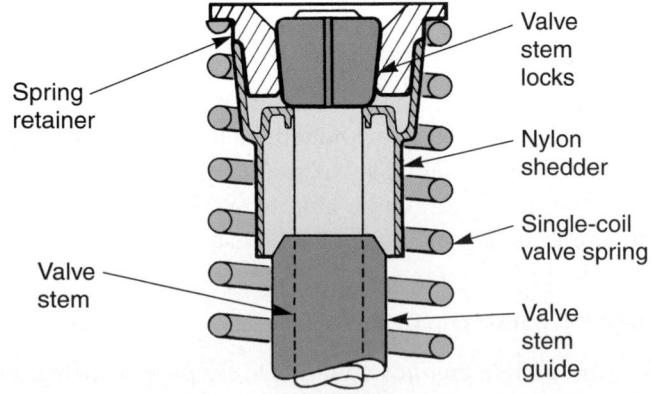

Figure 13-13. A nylon shedder can be used as both an O-ring seal and a shield to keep oil out of the valve guide. (Cadillac)

Valve Spring Construction

Valve spring construction is basically the same for all engines. However, the number and type of coils can vary. **Figure 13-13** shows single coil valve springs. **Figure 13-14** shows a valve spring with an inner and outer coil. The second coil increases the amount of pressure holding the valve closed.

Spring tension refers to the stiffness of a valve spring measured in psi or kPa. Spring tension is usually stated for both opened and closed valve positions. The service manual will give the tension in pounds or kilograms for specific compressed lengths.

Spring free length is the length of the valve spring when removed from the engine. *Spring open length* is the spring's length when installed on the engine with the valve fully open. *Spring closed length* is the length of the valve spring when installed on the engine with the valve closed. Both open length and closed length are measured from the bottom of the spring to the bottom of the spring retainer.

Spring specifications are important. They affect valve action. Low spring tension can cause valve float. *Valve float* occurs when the valve fails to close entirely at high rpms because the spring is too weak.

Valve Spring Shim

A *valve spring shim* is a very thin, accurately machined washer used to increase spring tension. When a shim is placed under a spring, the open and closed lengths of the valve spring are reduced. This compresses the spring to increase closing pressure. A used valve spring may weaken and lose some of its tension. Valve spring shims provide a means of restoring full spring tension and pressure without spring replacement.

 Note
Selection and installation of valve spring shims is covered in Chapter 51, *Engine Top End Service.*

Valve Retainers and Keepers

Valve retainers and *keepers* lock the valve spring onto the valve. The retainer is a specially-shaped washer that fits over the top of the valve spring. The keepers fit into the valve stem grooves. This holds the retainer and spring in place. Refer to **Figures 13-14** and **13-15.**

Valve Spring Seat

A *valve spring seat* is a cup-shaped washer installed between the cylinder head and the bottom of the valve spring. It provides a pocket to hold the bottom of the valve spring, as shown in **Figure 13-15.**

Valve Rotators

A *valve rotator* turns the valve to prevent carbon buildup and hot spots on the valve face. A valve rotator may be located under or on top of the valve spring. If under the seat, it is called a *seat-type rotator.* If on top of

Figure 13-14. A dual-coil valve spring increases the valve closing pressure. Also notice the other parts. (Ford)

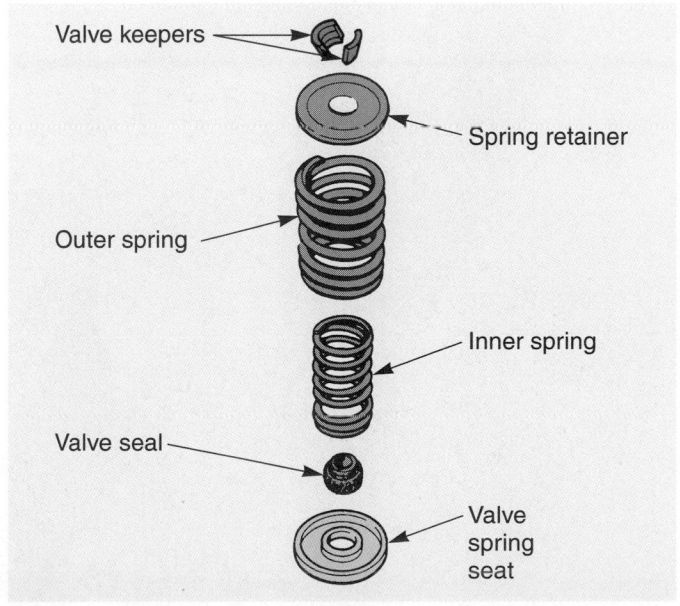
Figure 13-15. A valve spring assembly using a spring seat. The seat keeps the bottom of the spring in alignment with the stem. (Honda)

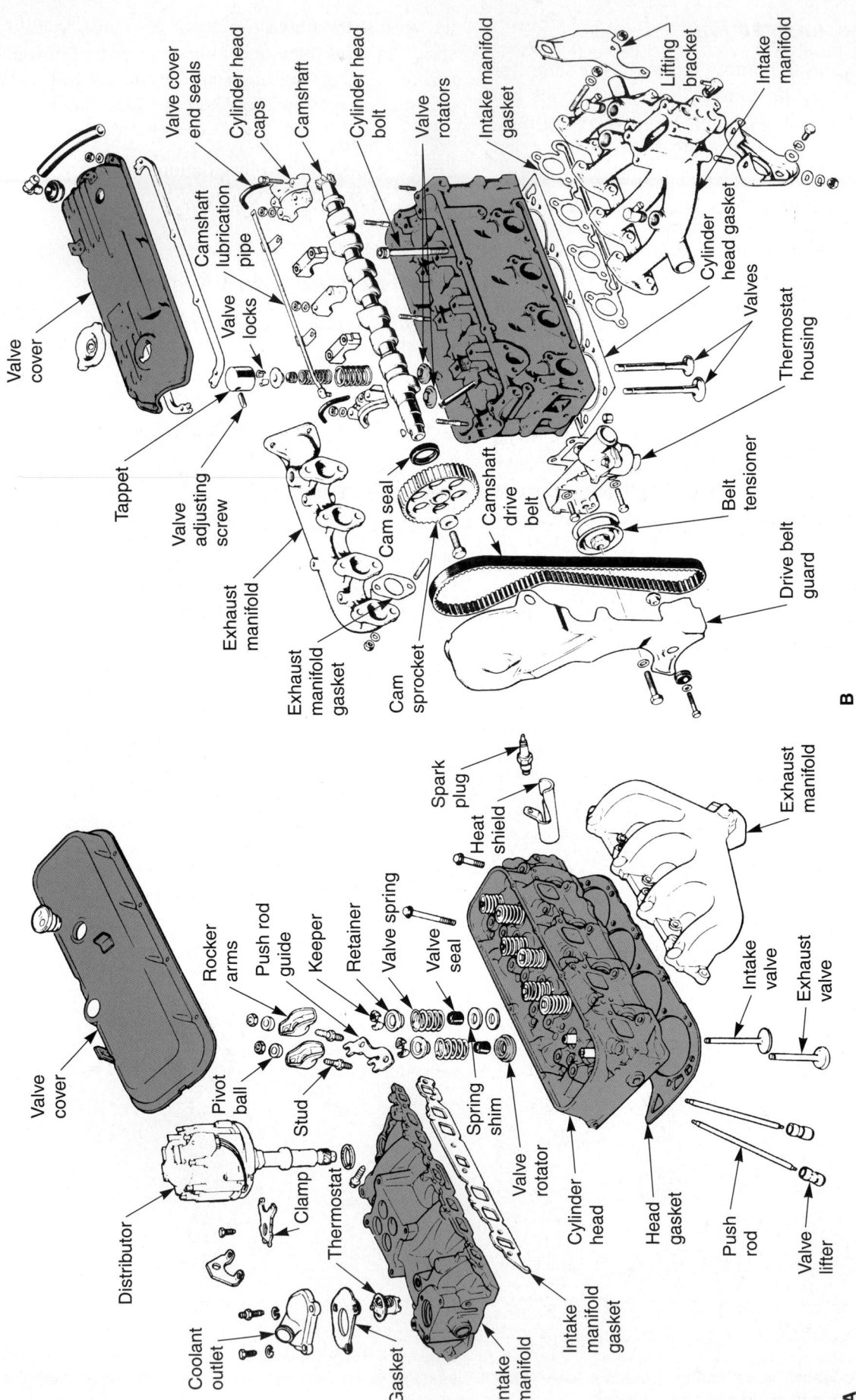

Valve cover

Valve cover end seals

Camshaft lubrication pipe

Cylinder head caps

Camshaft

Cylinder head bolt

Valve rotators

Intake manifold gasket

Valve locks

Cylinder head gasket

Lifting bracket

Intake manifold

Valves

Thermostat housing

Tappet

Valve adjusting screw

Cam seal

Camshaft drive belt

Belt tensioner

Exhaust manifold

Exhaust manifold gasket

Cam sprocket

Drive belt guard

B

Spark plug

Heat shield

Exhaust manifold

Rocker arms

Push rod guide

Keeper

Retainer

Valve spring

Valve seal

Intake valve

Exhaust valve

Valve cover

Pivot ball

Stud

Clamp

Thermostat

Spring shim

Valve rotator

Cylinder head

Head gasket

Push rod

Valve lifter

Distributor

Coolant outlet

Gasket

Intake manifold

Intake manifold gasket

A

Figure 13-16. These basic engine top end assemblies use valve rotators. A—This cam-in-block V-type engine has a valve rotator for each exhaust valve. B—This overhead cam engine uses a valve rotator for each valve. (GMC)

the valve spring, it is called a ***retainer-type rotator.*** Rotators are commonly used on engine exhaust valves, which are exposed to more heat than intake valves. See **Figure 13-16.**

Valve Stem Cap

A ***valve stem cap*** may be placed over the end of the valve stem. It helps prevent stem and rocker arm wear. A valve stem cap is free to turn on the valve stem. This serves as a bearing that reduces friction. Some valve stem caps are used to adjust the clearance in the valve train. Different cap thicknesses are available. Caps can be changed to alter the clearance between the rocker arm and the valve stem.

Camshaft Construction

A ***camshaft*** controls when the valves open and close. It can be driven by gears, a chain, or a belt. The number of lobes on a camshaft depends on the number of valves in the combustion chambers and the number of cylinders in the engine. Many engines only use one camshaft. However, others use two or more, **Figure 13-17.**

With dual overhead cam engines, there are two camshafts. One is the intake camshaft and the other is the exhaust camshaft. The ***intake camshaft*** operates all the intake valves in the cylinder head. The ***exhaust camshaft*** operates all the exhaust valves.

Cam Lobes

The ***cam lobes*** are precision-machined and polished surfaces on the camshaft. Each cam lobe consists of a nose, flank, heel, and base diameter, as shown in **Figure 13-18.** Variations in lobe shape control:

- When each valve opens in relation to piston position.
- How long each valve stays open.
- How far each valve opens.

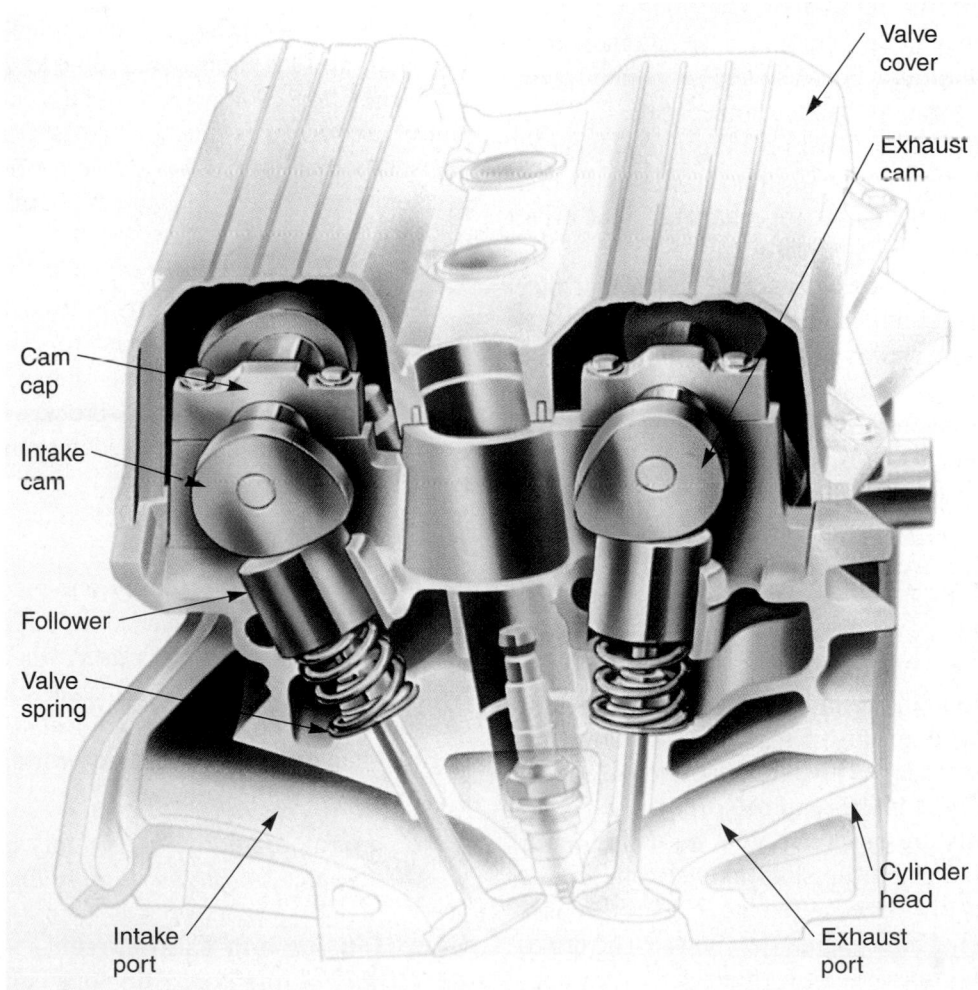

Figure 13-17. In a dual overhead cam engine, separate camshafts control the intake and exhaust valves. The shape of the cam lobes determines when the valves open, how far they open, and how long they remain open.

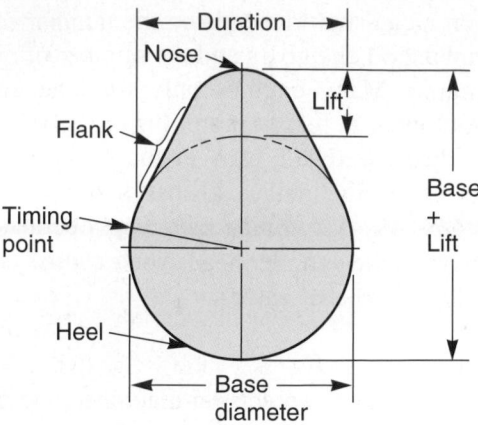

Figure 13-18. The basic parts and shape of a cam lobe. The height of the nose compared to base circle determines the amount of valve opening. The width or roundness of the nose determines how long valve stays open. (TRW)

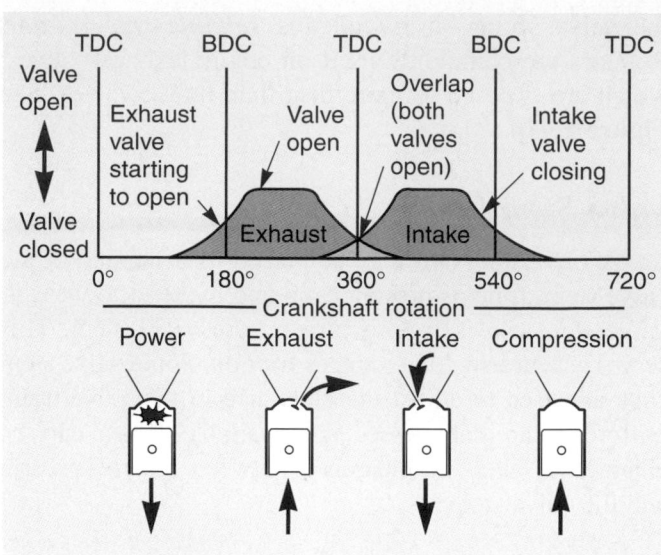

Figure 13-19. Valve timing is the measurement of when valves open in relation to the piston position in the cylinder. Valve overlap is the period where both the intake and exhaust valves in the same cylinder are open. (VW)

Camshaft lift is how far the valve opens. See **Figure 13-18.** Camshaft lift is found by subtracting the cam base diameter from the height of the cam lobe.

Camshaft duration determines how long the valve stays open. The shape of the cam lobe nose and flank regulates camshaft duration. For instance, a pointed cam lobe has a shorter duration than a rounded lobe.

Valve timing refers to valve opening and closing in relation to the position of the pistons in the cylinders. Valve timing is designed into the camshaft and drive sprockets or gears by the manufacturer.

Variable valve timing means that the engine can alter *when* the valves open with engine speed. Various mechanisms are used to provide variable valve timing. Some engines use an extra cam lobe that functions at high speeds. Others use electromechanical or hydraulic devices on the camshaft sprockets to advance or retard the cams.

Valve overlap is the time when both the intake and exhaust valves in the same cylinder are open. Look at **Figure 13-19.** Valve overlap is used to help draw burned gases out of the cylinder. It also helps pull a fresh fuel charge into the cylinder. With both the intake and exhaust valves open, the movement of the gases through one cylinder head port and the cylinder itself acts on the gases in the other port. This results in slightly more flow into and out of the cylinder. Valve overlap helps engine breathing, especially at higher engine speeds.

Some camshafts are machined with *dual cam lobes* that have two different profile shapes. One cam lobe is designed for good low-speed efficiency. The other is a high-speed profile for high engine rpm power. The driveline control module operates a solenoid valve that controls oil flow to shift the rocker arms from one lobe profile to the other.

Hollow camshafts can have their lobes pressed onto the shaft. To lock the splined lobes in place, an oversize steel ball is forced down through the center of the hollow shaft. Some new engines use hollow composite camshafts to lighten the valve train for quicker engine acceleration.

Camshaft Thrust Plate

A *camshaft thrust plate* is used to limit camshaft end-play. End-play is the front to rear movement. The thrust plate bolts to the front of the block or cylinder head. When the drive gear or sprocket is bolted in place, the thrust plate sets up a predetermined camshaft end play.

Cam Bearings

Cam bearings are usually one-piece inserts pressed into the block or cylinder head. See **Figure 13-20.** Two-piece inserts are sometimes used when the camshaft is mounted in the head. The camshaft journals ride in the cam bearings. Cam bearings are usually constructed like engine main and connecting rod bearings.

 Note

For more information on this subject, refer to Chapter 14, *Engine Bottom End Construction.*

Cam Housing and Cam Cover

A *cam housing* is a casting that bolts to the top of the cylinder head to hold the engine camshaft. It is used in some overhead cam engine designs. A *cam cover* is a lid

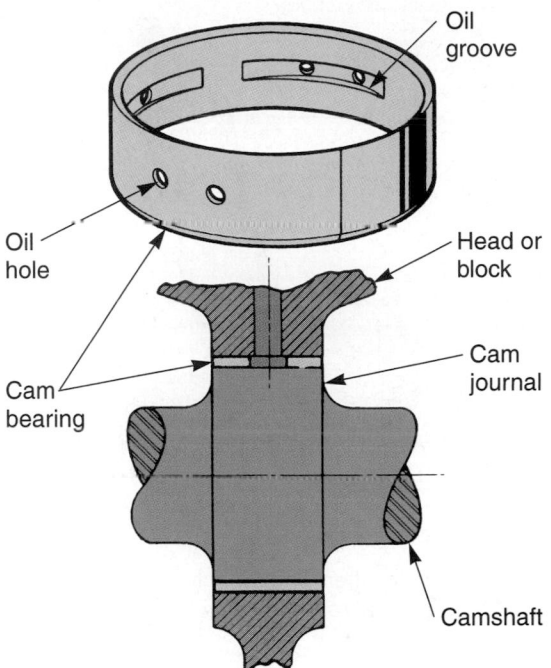

Figure 13-20. Camshaft journals ride in cam bearings. Cam bearings may be pressed into the block or cylinder head bore. (Federal Mogul)

over the top of the cam housing. It serves the same purpose as a valve cover on an overhead valve engine. Acoustically dampened construction means the part is designed to help reduce the transmission of sound. For example, some head or valve covers are acoustically dampened by making them out of two sheets of steel formed around a center layer of sound-dampening material.

Valve Lifters

Valve lifters, or *tappets,* ride on the camshaft lobes and transfer motion to the other parts of the valve train. Look at **Figure 13-21.** The bottom of a lifter is crowned,

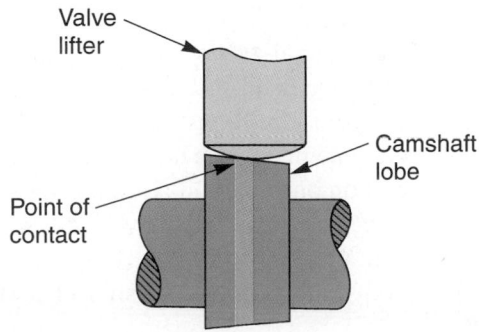

Figure 13-21. The bottom of a lifter is crowned and the cam lobe is tapered. This causes the rotating cam lobe to turn the lifter in its bore, reducing wear. (Dana)

or slightly curved. The camshaft lobe is also not perfectly square with the lifter base. As a result, one side of the lifter touches the cam lobe. This tends to rotate the lifter in its bore to reduce wear. There are four basic types of lifters. These are hydraulic, mechanical, roller, and OHC follower.

Hydraulic valve lifters are common because they operate *quietly* by maintaining zero valve clearance. Zero valve clearance means that there is no space between valve train parts. With zero clearance, the valve train does not clatter when the engine is running. The hydraulic lifter adjusts automatically with temperature changes and part wear. Look at **Figure 13-22.** During engine operation, oil

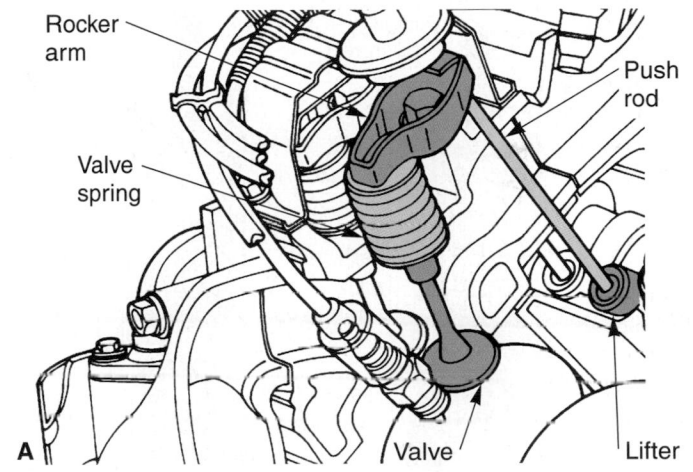

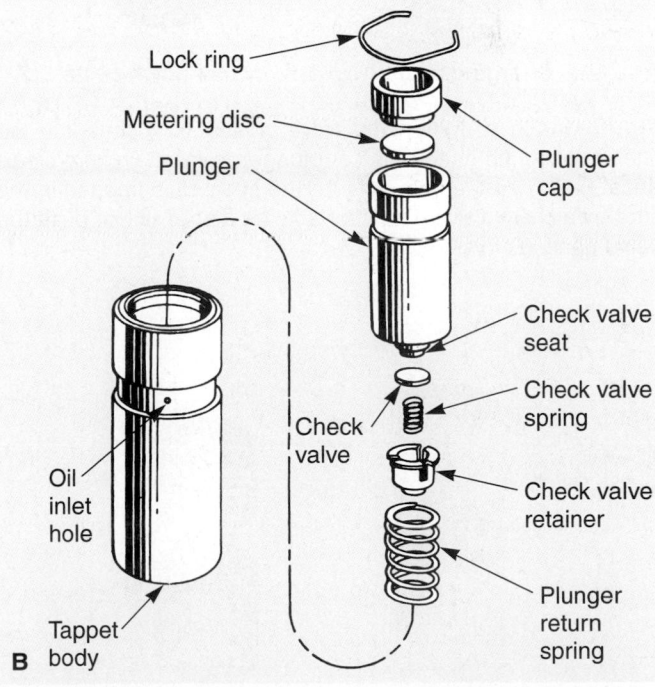

Figure 13-22. A—In a push rod engine, lifter bores are in the block. Push rods run up to the rocker arms. Rocker arms then change the upward movement into downward movement to open the valves. B—The parts of a hydraulic lifter. (Chrysler)

pressure fills the inside of the hydraulic lifter with motor oil, **Figure 13-23.** The pressure pushes the lifter plunger up in its bore until all the play is out of the valve train. As the camshaft pushes on the lifter, the lifter check valve closes to seal oil inside the lifter. Since oil is *not* compressible, the lifter acts as a solid unit to open the valve.

Mechanical lifters, also called *solid lifters,* do not contain oil. They simply transfer cam lobe action to the push rod. Mechanical lifters are *not* self-adjusting and require periodic setting. A screw adjustment is normally provided at the rocker arm when solid lifters are used. Turning the adjustment screw down reduces any "play"

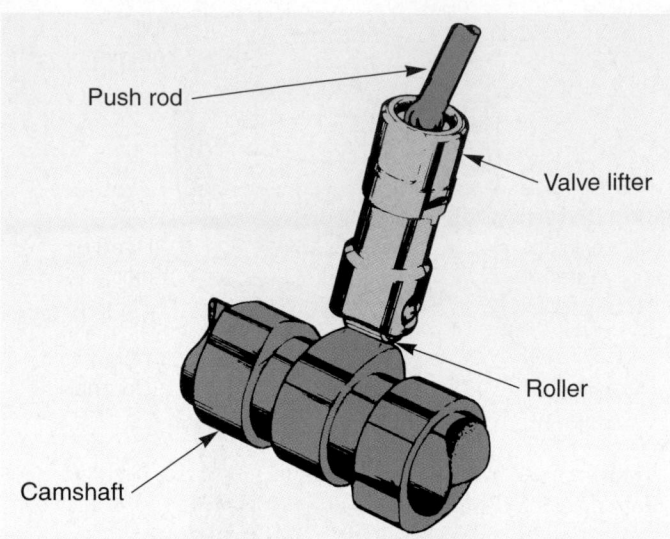

Figure 13-25. A roller lifter is commonly used in diesel engines to reduce friction. The roller spins as the camshaft rotates. (Oldsmobile)

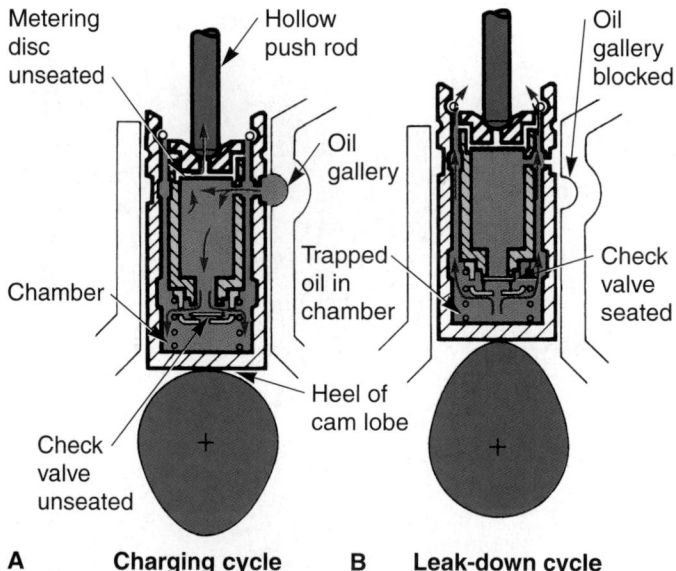

A   Charging cycle   B   Leak-down cycle

Figure 13-23. A—When the valve closes, oil rushes into the lifter, pushing up on the disc or plunger to maintain zero clearance. B—When the cam acts on the lifter, oil is trapped in the lifter by a check valve. The lifter acts as a solid unit to push the valve open. (Chrysler)

Figure 13-24. Typical mechanical, or solid, lifters.

in the valve train. Unscrewing, or backing off, the rocker arm adjustment increases clearance. The small clearance needed with solid lifters causes valve train noise. A clattering or clicking noise is produced as the valves open and close. See **Figure 13-24.**

A *roller lifter* has a small roller that rides on the camshaft lobe, **Figure 13-25.** These can be either mechanical or hydraulic. The point where the lifter touches the camshaft is one of the highest friction points in the engine. The roller helps reduce this friction and wear. A roller lifter is also used to reduce frictional losses of power.

An *OHC follower* fits between the camshaft and valve, **Figure 13-26.** The follower slides up and down in a bore machined in the head. Either an adjusting screw in the follower or different thickness shims can be used to adjust valve clearance.

Push Rod Construction

Push rods are metal tubes or rods with specially formed ends. They are used in cam-in-block engines to transfer motion from the lifters to the rocker arms, **Figure 13-27A.** Some push rods have a ball on each end. Others have a ball on one end and a female socket on the other end. Some hollow push rods have holes in the ends to feed oil from the lifter to the rocker arms. This prevents wear on the tip of the push rod and on the rocker arm.

In some engines, *push rod guide plates* are used to limit side movement of the push rods. See **Figure 13-27B.** The guides hold the push rods in alignment with the

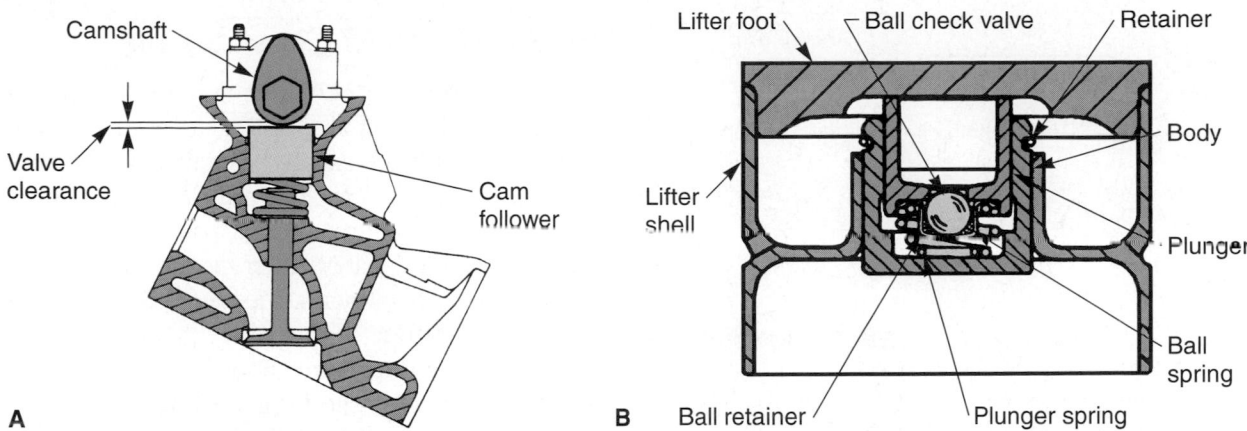

Figure 13-26. A—An overhead cam follower fits directly between the camshaft and valve stem. The follower slides up and down in a bore machined in the head. Either a spacer washer or a screw is used to adjust valve clearance. B—A cutaway of a cam follower. (GM)

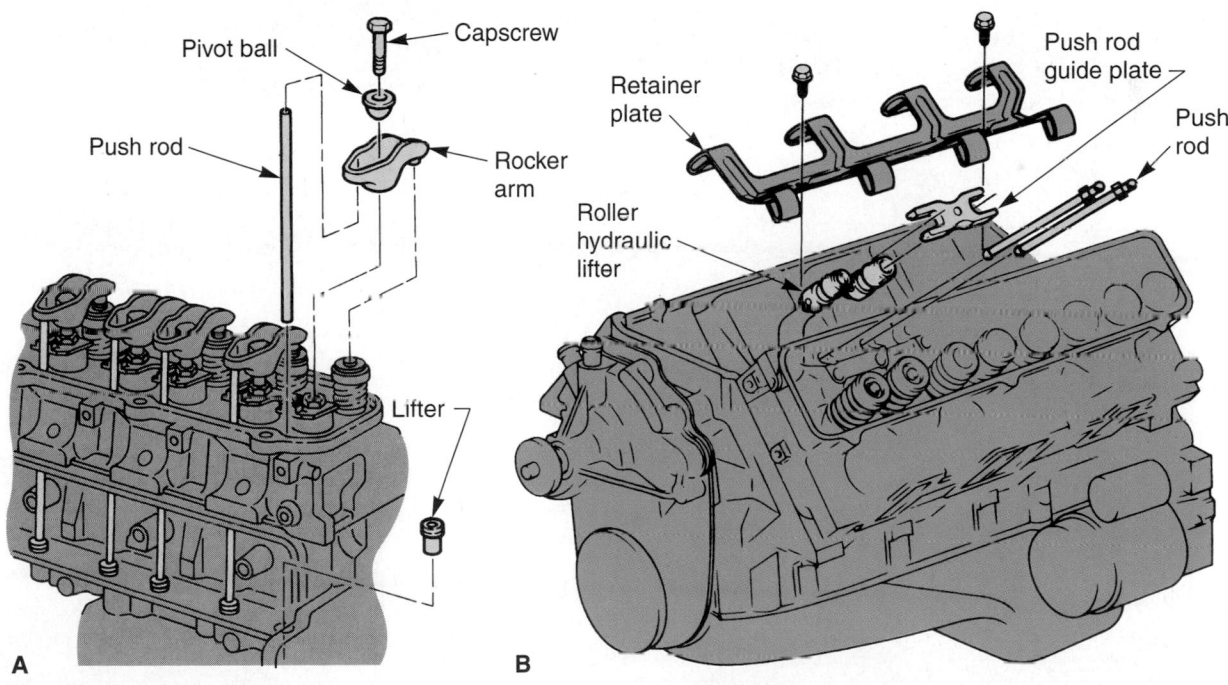

Figure 13-27. Two different push rod rocker arrangements. A—An in-line engine using a pivot ball to hold the rocker arm. B—A V-type engine using push rod guide plates. Note the location of the lifters in both engines. (Chrysler, Oldsmobile)

rocker arms. When the push rods pass through holes in the cylinder head or intake manifold, guide plates are *not* needed.

Rocker Arm Construction

Rocker arms transfer valve train motion to the valve stem tips. In OHC engines, the camshaft may act directly on the rocker arm. Then, the rocker acts on the valve. Some rocker arms are forked so that they can actuate two valves at once. In a push rod engine, the push rod acts on the rocker. Then, the rocker transfers the motion to the valve. See **Figure 13-28.** Rocker arms are usually made of either cast iron or steel.

Various methods are used to support the rocker arms on the cylinder head. Individual *pivot balls* or *stands* can be used to hold the rocker arm in place over the valve. **Figure 13-29** shows one such arrangement.

Adjustable rocker arms are used to change the valve train clearance. Either a screw is provided on the rocker arm or the rocker arm pivot point can be changed. Adjustable rocker arms *must* be used with mechanical lifters. *Nonadjustable rocker arms* have no means of changing the valve clearance. They are used only with

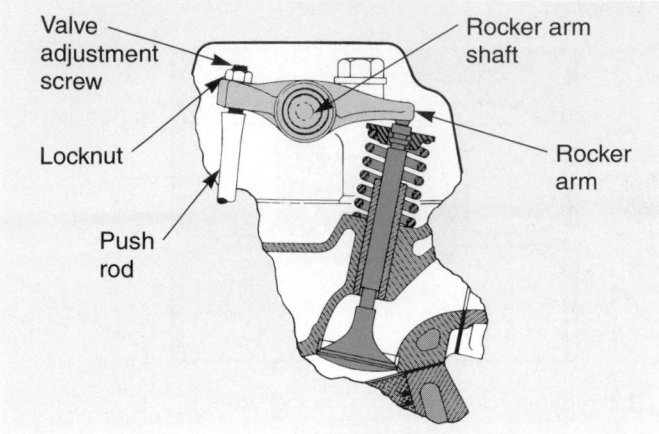

Figure 13-28. A rocker arm shaft holding the rocker arms in position. This rocker arm has an adjustment screw and lock nut for changing valve clearance. The rocker is cast iron. (Federal Mogul)

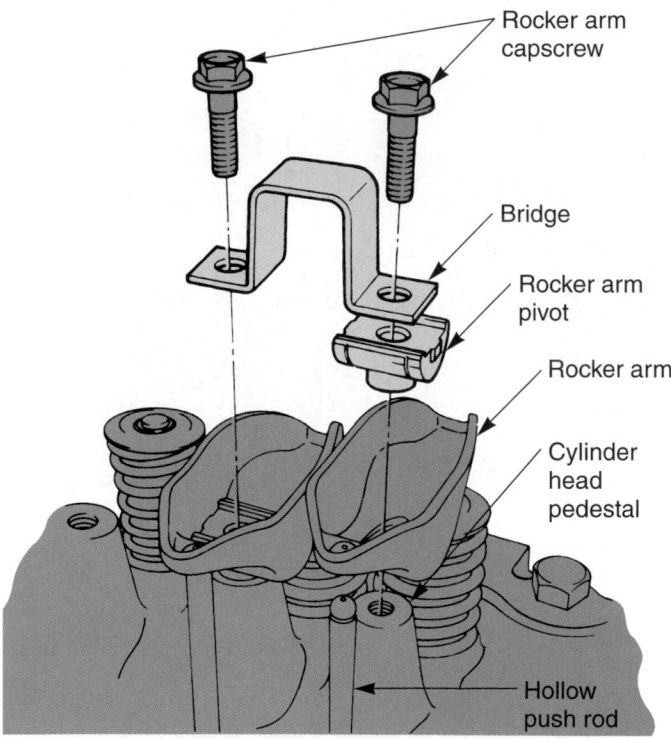

Figure 13-29. This engine uses rectangular rocker arm pivots. A bridge helps hold the rockers in position over the valve stems. These rockers are made of stamped steel. (Chrysler)

some hydraulic lifters. The rocker arm assembly is tightened to a specific torque. This presets the lifter plunger halfway in its travel. Then, during engine operation, the hydraulic lifter automatically maintains zero clearance. Push rod length can be changed for clearance adjustments when using nonadjustable rocker arms.

Intake Manifold Construction

An *intake manifold* holds the throttle body and has passages, called runners, going to each cylinder head port. The manifold usually contains water jackets for cooling. It can be cast of iron, aluminum, or plastic. *Manifold runners* are internal passages formed in the intake manifold to carry either the air-fuel mixture or air to the cylinder head ports. **Figure 13-30** is a cutaway of an intake manifold. Notice the exhaust passage, which warms the manifold, and the exhaust gas recirculation (EGR) system. Also see **Figure 13-31.**

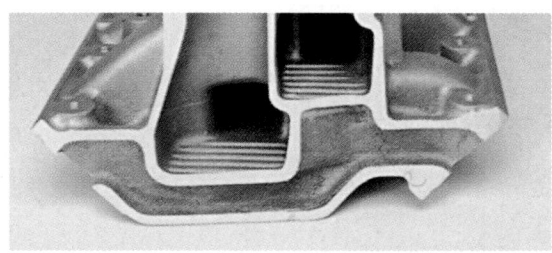

Figure 13-30. A cutaway of an engine intake manifold. Note the intake runners for fuel and the exhaust passages. This intake is made of aluminum to reduce weight. (Edelbrock)

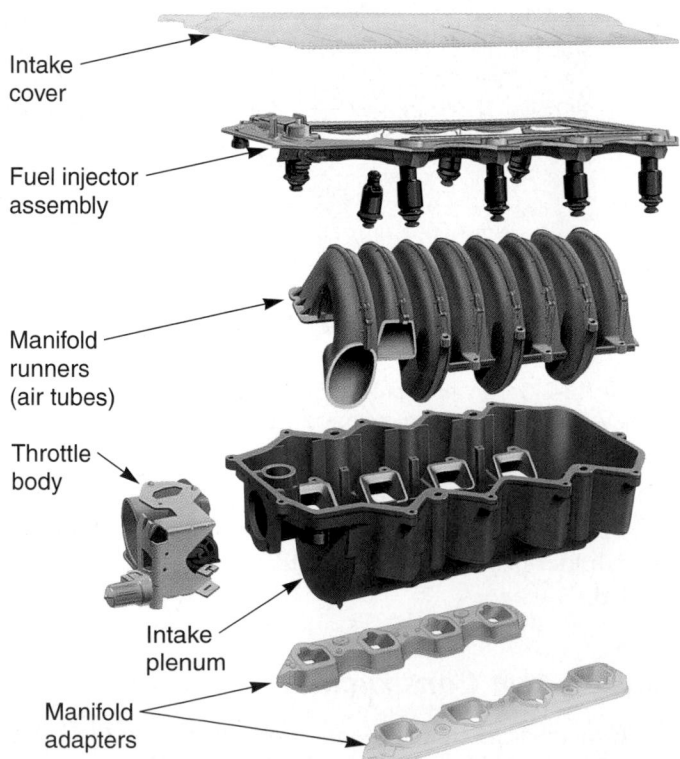

Figure 13-31. This "fluid induction" assembly takes the place of a traditional intake manifold in some types of vehicles. Individual tubes supply air to each cylinder. (Cadillac)

Intake manifolds can also be made of glass-filled nylon or plastic. These can weigh half as much as aluminum and several times less than iron. The inside of the plastic runners is also smoother than metal to improve airflow into engine for more power. With plastic manifolds, brass thread inserts are pressed into the plastic. This allows additional parts, such as sensors and vacuum fittings, to be bolted to the plastic part without thread damage. *Compression limiters* are metal inserts around bolt holes that limit the deformation of plastic parts. They allow enough bolt torque to provide good sealing without cracking or deforming the soft plastic.

A *flame arrester* is sometimes located before the engine intake manifold to prevent backfire damage to the air filter. It is made of metal mesh to prevent the flame of an engine backfire from entering the air filter housing. It should periodically be removed and cleaned in solvent for best engine performance.

Variable Induction System

A *variable induction system* has two sets of intake runners controlled by butterfly valves to aid engine efficiency and performance. At low engine speeds, the system uses the longer intake runners to "ram" more air and fuel into the cylinders for increased torque. Then, when a specific engine speed is reached, the computer opens the butterfly valves over the shorter intake runners. The short runners increase airflow at high rpm for added power. A variable induction system is like having two intake manifolds in one. One manifold is for low-speed torque and the other is for high-speed power.

Exhaust Manifold Construction

An *exhaust manifold* routes burned exhaust gases from the cylinder head exhaust port to the exhaust pipe. Because of the high operating temperatures, the exhaust manifold is usually made of cast iron. A few high-performance or sports car engines use lightweight, free-flowing steel tubing exhaust manifolds called *headers*. **Figure 13-32** shows both intake and exhaust manifolds for one particular engine.

Summary

- An engine top end basically includes the cylinder head, valve train, valve cover, and the intake and exhaust manifolds.
- A bare cylinder head is a head with all of its parts removed.

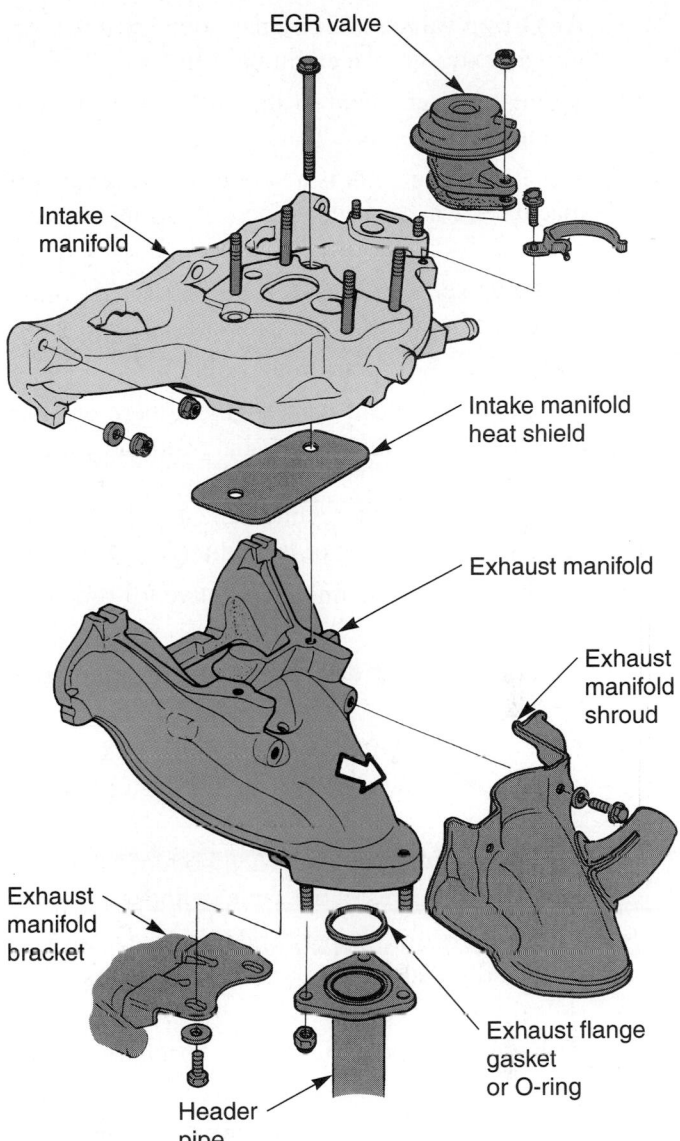

Figure 13-32. Intake and exhaust manifolds shown with their related components. (Honda)

- An integral valve guide is part of the cylinder head casting. A pressed-in valve guide is a separate sleeve forced into an oversize hole machined in the head.
- The valve seat angle is the angle formed by the face of the seat.
- An interference angle is a 1/2° to 1° difference between the valve seat face angle and the angle of the valve face.
- The valve face angle is the angle formed between the valve face and valve head.
- An umbrella valve seal is shaped like a cup and can be made of neoprene rubber or plastic.

- An O-ring valve seal is a small round seal that fits into an extra groove cut in the valve stem.
- Spring tension refers to the stiffness of a valve spring.
- A valve spring shim is a very thin and accurately machined washer used to increase spring tension.
- Valve retainers and keepers lock the valve spring on the valve.
- A valve rotator turns the valve to prevent carbon buildup and hot spots on the valve face.
- With dual overhead cam engines, there are two camshafts—the intake camshaft and the exhaust camshaft. The intake camshaft operates all of the intake valves in the cylinder head. The exhaust camshaft operates all of the exhaust valves.
- Camshaft lift is the amount of valve lift produced by the cam lobe.
- Camshaft duration determines how long the valve stays open.
- Valve timing refers to valve opening and closing in relation to the position of the pistons in the cylinders.
- Valve overlap is the time when both of the intake and exhaust valves in the same cylinder are open.
- Valve lifters, or tappets, ride on the camshaft lobes and transfer motion to the other parts of the valve train.
- Push rods are metal tubes with specially formed ends.
- Rocker arms transfer valve train motion to the valve stem tips.

Important Terms

Engine top end	Spring tension
Bare cylinder head	Spring free length
Integral valve guide	Spring open length
Pressed-in valve guide	Spring closed length
Integral valve seat	Valve float
Pressed-in valve seat	Valve spring shim
Seat insert	Valve retainers
Valve seat angle	Keepers
Interference angle	Valve spring seat
Prechamber cup	Valve rotator
Poppet valves	Valve stem cap
Mushroom valves	Camshaft
Valve face angle	Cam lobes
Umbrella valve seal	Camshaft lift
O-ring valve seal	Camshaft duration
Valve spring shield	Valve timing

Variable valve timing	Valve overlap
Push rods	Push rod guide plates
Camshaft thrust plate	Rocker arms
Cam bearings	Intake manifold
Cam housing	Manifold runners
Cam cover	Compression limiters
Valve lifters	Flame arrester
Hydraulic valve lifters	Variable induction
Mechanical lifters	system
Roller lifter	Exhaust manifold
OHC follower	

Review Questions—Chapter 13

Please do not write in this text. Place your answers on a separate sheet of paper.

1. What is a bare cylinder head?
2. A(n) _____ type valve guide is part of the cylinder head.
3. A(n) _____ valve guide is a separate sleeve forced into an oversize hole in the cylinder head.
4. Cylinder heads use _____ or _____ valve seats.
5. What is the function of an interference angle on the valve face and seat?
6. This part is pressed into the cylinder head on most automotive diesel engines.
   (A) Stratified charge chamber.
   (B) Glow plug.
   (C) Injector nozzle.
   (D) Prechamber cup.
7. Engines use poppet valves, also called _____ valves.
8. Define the term "valve face angle."
9. Why are some valves filled with sodium?
10. Stellite valves are used in engines designed to burn _____ _____.
11. Describe the two basic types of valve seals.
12. Which of the following does *not* pertain to valve springs?
    (A) Free length.
    (B) Tension.
    (C) Tensile strength.
    (D) Closed length.
13. What is the function of a valve spring shim?
14. What is used to lock the valve spring on the valve?
15. A(n) _____ _____ can be used to turn the valve and prevent hot spots on the valve face.

16. Variations in camshaft lobe shape control:
    (A) When each valve opens.
    (B) How long each valve stays open.
    (C) How far or wide each valve opens.
    (D) All of the above are correct.

17. Explain the following terms: Camshaft lift, Camshaft duration, Valve timing, Valve overlap.

18. Why are hydraulic valve lifters used more than solid or mechanical lifters?

19. _____ lifters reduce friction and are frequently used in diesel or high performance applications.

20. The cam base circle diameter of a camshaft is 3/4″. The camshaft's cam lobe height is 1 1/2″. What is the approximate camshaft lift of this engine?
    (A) 3/8″.
    (B) 1/4″.
    (C) 1/2″.
    (D) None of the above.

ASE-Type Questions

1. An engine top end typically consists of each of these *except:*
   (A) *valve train.*
   (B) *connecting rod.*
   (C) *cylinder head.*
   (D) *intake manifold.*

2. Which of the following is a hole machined through the cylinder head?
   (A) *Stellite face.*
   (B) *Valve seat insert.*
   (C) *Integral valve guide.*
   (D) *Pressed-in valve guide.*

3. Common passenger car engines have a valve seat angle of:
   (A) *10°.*
   (B) *20°.*
   (C) *25°.*
   (D) *45°.*

4. Which of the following is a cross between an oil seal and a valve spring shield?
   (A) *Nylon shedder.*
   (B) *O-ring valve seal.*
   (C) *Valve spring shim.*
   (D) *Umbrella valve seal.*

5. Low valve spring tension can cause:
   (A) *valve float.*
   (B) *valve rotation.*
   (C) *valve spring shim.*
   (D) *mushroomed valves.*

6. Which of the following is used to lock a valve spring on a valve?
   (A) *Retainer.*
   (B) *Keeper.*
   (C) *Seal.*
   (D) *Both A and B.*

7. Which of the following is *not* a method by which a camshaft can be driven?
   (A) *Belt.*
   (B) *Chain.*
   (C) *Gears.*
   (D) *Rotator.*

8. A cam lobe consists of each of these *except:*
   (A) *heel.*
   (B) *nose.*
   (C) *head.*
   (D) *flank.*

9. The device used to limit camshaft end-play is the:
   (A) *cam cover.*
   (B) *thrust plate.*
   (C) *cam bearings.*
   (D) *valve overlap.*

10. Which type of valve lifter is filled with oil?
    (A) *Solid.*
    (B) *Hydraulic.*
    (C) *Mechanical.*
    (D) *All of the above.*

Activities for Chapter 13

1. The rocker arm is a form of lever. Do research to find out what class of lever it is. Construct simple models of the three classes of levers and describe for the class how they work.

2. Make a sketch of a cam lobe, like the one shown in **Figure 13-18.** Assume that TDC is 0°. Use a protractor to mark points for a 90°, 180°, and 270° rotation. Using pencils or pens of different colors, draw the cam lobe in each of these positions.

3. If a disassembled engine is available, examine the head to determine whether integral or pressed-in valve guides and valve seats are used.

Chapter 14

Engine Bottom End Construction

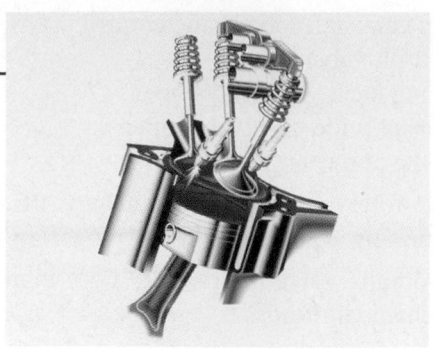

After studying this chapter, you will be able to:

- Compare the construction of different types of cylinder blocks.
- Explain how piston construction affects engine operation.
- Describe piston ring variations.
- Explain the construction of engine bearings.
- Compare design variations of different engine bottom end components.
- Explain safe practices when working with engine bottom end components.

- Correctly answer ASE certification test questions on engine bottom end construction.

The basic parts of an *engine bottom end* are the block, crankshaft, connecting rods, and piston assemblies. These components are shown in **Figure 14-1.** This chapter details construction techniques commonly used in an engine bottom end assembly. This information will help you understand how to repair engine bottom end components.

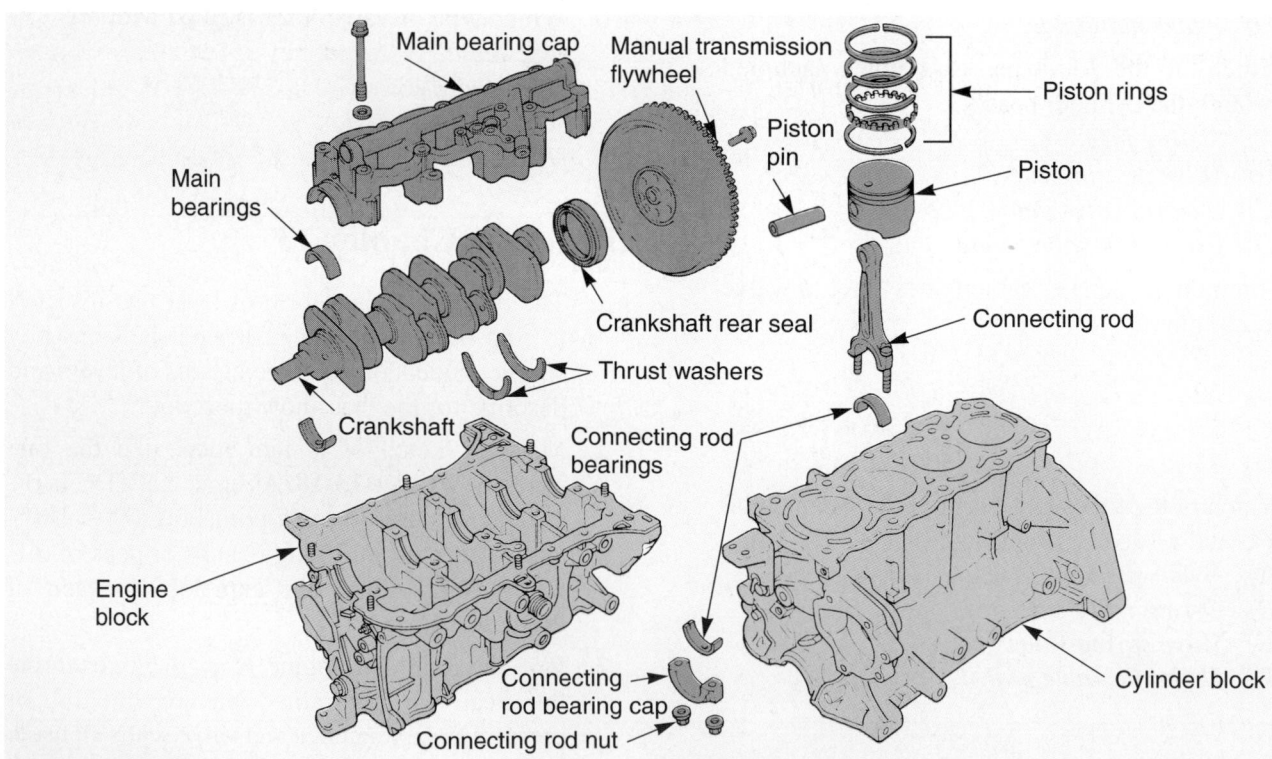

Figure 14-1. The engine bottom end assembly consists of the block, crank, rods, pistons, and rings. Understanding the construction of these components is very important to service and repair operations. (Honda)

Cylinder Block Construction

Engine cylinder blocks are normally made of cast iron or aluminum, **Figure 14-2.** A *cast iron cylinder block* is very heavy and strong. Nickel is sometimes added to the iron to improve strength and wear resistance. An *aluminum cylinder block* is relatively light. An aluminum block also dissipates heat better than a cast iron block. Many vehicles use aluminum blocks to reduce weight and increase fuel economy.

Cylinder Sleeves

Cylinder sleeves, or *liners,* are metal, pipe-shaped inserts that fit into the cylinder block. They act as cylinder walls for the piston to slide up-and-down on, **Figure 14-2. *Cast iron sleeves*** are commonly used in aluminum cylinder blocks. Sleeves can also be installed to repair badly damaged cylinder walls in cast iron blocks. There are two basic types of cylinder sleeves—dry sleeves and wet sleeves. These are shown in **Figure 14-3.**

A *dry sleeve* presses into a cylinder that has been bored, or machined, oversize. Look at **Figure 14-4A.** A dry sleeve is relatively thin and is not exposed to engine coolant. The outside of a dry sleeve touches the walls of the cylinder block. This provides support for the sleeve. When a cylinder becomes badly worn or is damaged, a dry sleeve can be installed. The original cylinder must be bored almost as large as the outside diameter of the sleeve. Then, the sleeve is pressed into the oversized hole. Next, the inside of the sleeve is machined to the original bore diameter. This allows the use of the original piston size.

Figure 14-3. Notice the difference between the two wet sleeves and the one dry sleeve. (Dana Corp.)

A *wet sleeve* is exposed to the engine coolant. It must withstand combustion pressure and heat without the added support of the cylinder block. Therefore, it must be thicker than a dry sleeve. Refer to **Figure 14-4B.** A wet sleeve will generally have a flange at the top. When the head is installed, the clamping action pushes down on the sleeve and holds it in position. The cylinder head gasket keeps the top of the sleeve from leaking. A rubber or copper O-ring is used at the bottom of a wet sleeve to prevent coolant leakage into the crankcase. The O-ring seal is pinched between the block and the liner to form a leakproof joint. Many vehicles use aluminum cylinder

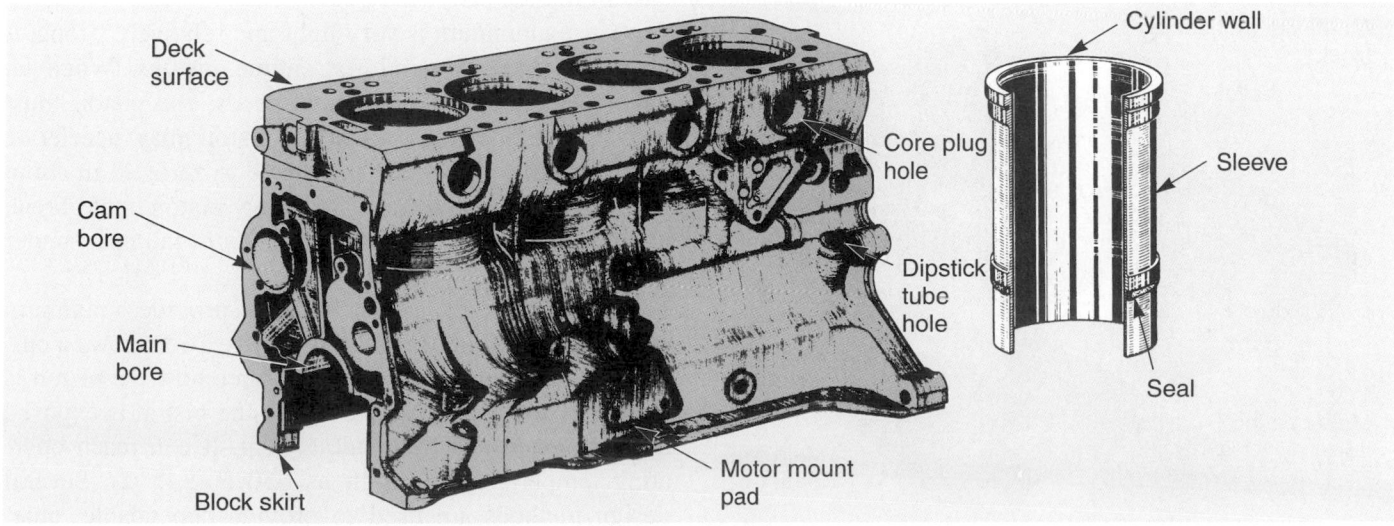

Figure 14-2. A cylinder block may be cast from iron or aluminum. The cylinder may be an integral part of the block or formed by a pressed-in liner. (Peugeot)

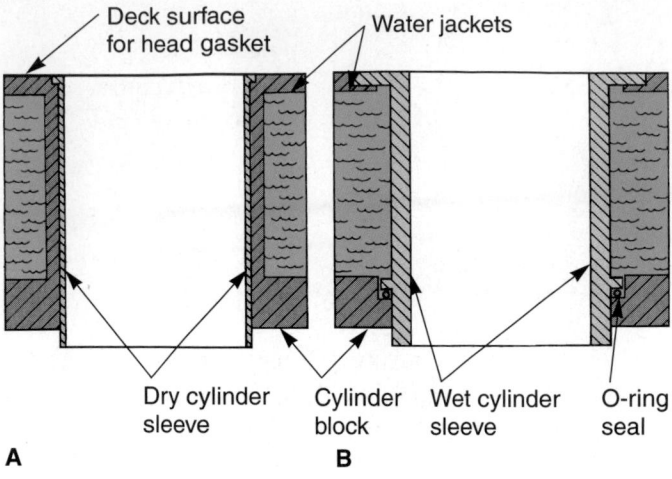

Figure 14-4. A—A dry sleeve presses into existing cylinder bore. It is not exposed to engine coolant. B—A wet sleeve is thicker to withstand combustion pressure and heat. It is also exposed to coolant.

blocks with cast iron, wet sleeves. The light aluminum block reduces weight for increased fuel economy. The cast iron sleeves wear very well, increasing engine service life. Refer to **Figure 14-5.**

Line Boring

The term *line boring* refers to a machining operation that cuts a series of holes through the block for the crankshaft bearings. A block may also be line bored for the camshaft bearings. The holes must be in perfect alignment for the crankshaft or camshaft to turn freely. Line boring can also be done to an OHC cylinder head.

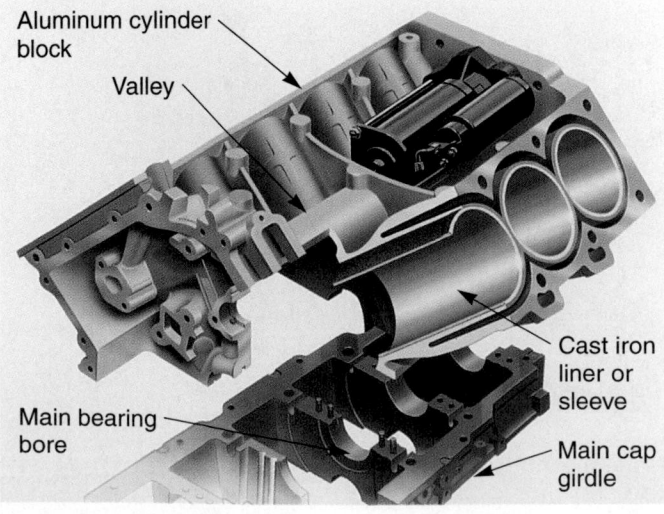

Figure 14-5. Modern cylinder blocks are frequently made of aluminum with pressed-in, cast iron wet sleeves. (Cadillac)

Two- and Four-Bolt Mains

A *two-bolt main block* only uses two cap screws to secure each main bearing cap to the cylinder block. A *four-bolt main block* has four cap screws holding each main cap. Four-bolt mains are used on high-performance engines. A few engines use six-bolt main caps. With extra bolts, the block can withstand more crankshaft downward pressure without part failure.

Crossbolted Block

A *crossbolted block* has extra cap screws going in through the sides of the block and main caps for added strength. This design is often used on high-performance engines.

Block Girdle

A *block girdle,* or *main bearing bedplate,* is a large one-piece cap that fits over the entire bottom of the block. Also called a *unit main cap,* it secures the main bearings. All the main caps are formed as one piece to increase strength and block stiffness. The bedplate can also hold any balancer shafts. Gears on the crankshaft are used to drive the balancer shafts.

Piston Construction

Engine pistons are normally cast or forged from an aluminum alloy. Cast pistons are relatively soft and are used in slow-speed, low-performance engines. Forged pistons are commonly used in today's fuel-injected, turbocharged, and diesel engines. These engines expose the pistons to much higher stress loads, which could break cast aluminum pistons.

Since aluminum is very light and relatively strong, it is an excellent material for engine pistons. When an engine is running at highway speeds, the piston must withstand tremendous loads. A piston may accelerate from zero to 60 mph and then back to zero, all in about four inches of piston travel. A heavy piston could break the connecting rod. A weak piston could fall apart under these loads.

The design of a piston must provide maximum strength and minimum weight. **Figure 14-6** shows a cutaway view of an engine piston. Notice how the piston is thicker at stress points. The top of the piston is exposed to combustion and tremendous heat. It can reach operating temperatures as high as 650°F (345°C). Several design methods are used to provide dependable, quiet piston operation.

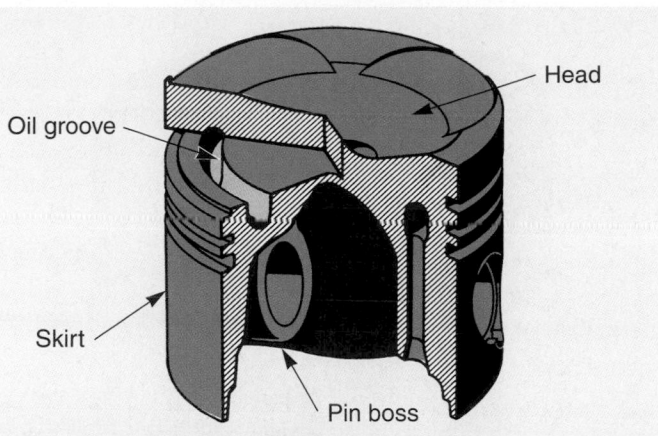

Figure 14-6. Piston construction is critical because of extreme loads. This piston is for a diesel engine and has a groove that allows an oil spray to help cool the piston. Diesel pistons must be made much thicker and heavier than pistons for gasoline engines. Pressure and temperatures in a diesel are higher. (Mercedes Benz)

Piston Dimensions

Figure 14-7 illustrates several piston dimensions. These dimensions affect how the piston functions in the cylinder. These dimensions are explained as follows.

- *Piston diameter*—the distance measured across the sides of the piston.
- *Pin hole diameter*—the distance measured across the inside of the piston pin hole.
- *Ring groove width*—the distance measured from the top to the bottom of the ring groove.

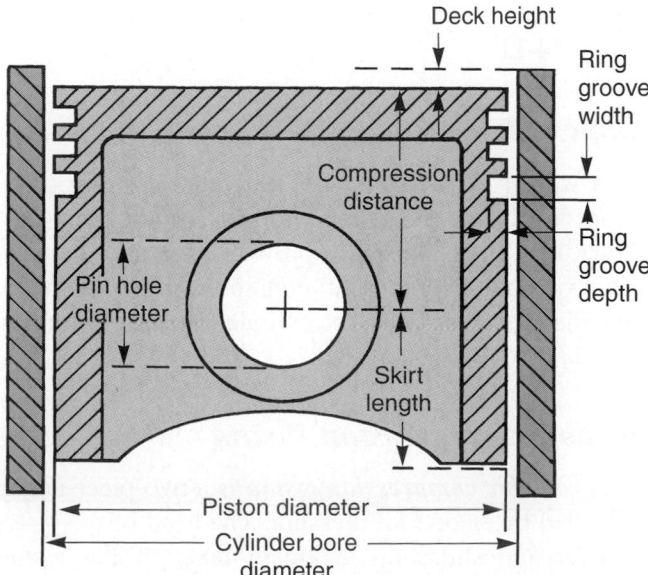

Figure 14-7. The basic dimensions of a piston.

- *Ring groove depth*—the distance measured from the ring land to the back of the ring groove.
- *Skirt length*—the distance from the bottom of the skirt to the centerline of the pin hole.
- *Compression distance*—the distance from the centerline of the pin hole to the top of the piston.

Cam-Ground Piston

A *cam-ground piston* is machined slightly out of round when viewed from the top. The piston is a few thousandths of an inch larger in diameter perpendicular to the piston pin centerline. See **Figure 14-8.**

Cam grinding is done to compensate for different rates of piston expansion due to differences in metal wall thickness. As the piston is heated by combustion, the thicker area around the pin boss causes the piston to expand more parallel to the piston pin. The oval-shaped piston becomes round when hot, and there is still enough clearance parallel to the piston pin.

A cold cam-ground piston has the correct piston-to-cylinder clearance. The unexpanded piston will *not* slap, flop sideways, and knock in the cylinder because of too much clearance. However, the cam-ground piston will

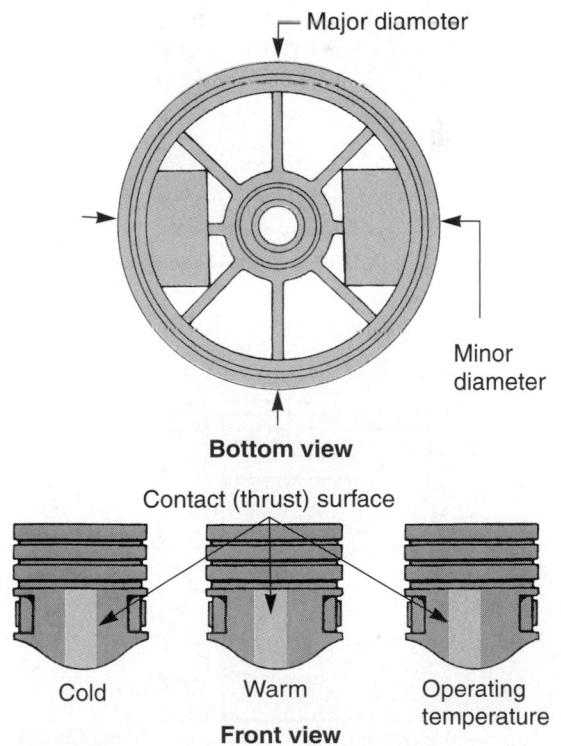

Figure 14-8. A cam-ground piston compensates for different rates of expansion. The piston is larger across the major diameter. The piston expands more across the minor diameter when heated. This causes the piston to become round when at full operating temperature. (Ford)

not become too tight in the cylinder when heated to full operating temperature.

Piston Taper

Piston taper is also used to maintain the correct piston-to-cylinder clearance. The top of the piston is machined slightly smaller than the bottom, **Figure 14-9.** Since the piston head gets hotter than the skirt, it expands more. The piston taper makes the piston almost equal in size at the top and bottom at operating temperature.

Piston Shape

Piston shape generally refers to the contour of the piston head. Usually, a piston head is shaped to match and work with the shape of the cylinder head combustion chamber. See **Figure 14-10.** The head of a *flat top piston*

Figure 14-10. A piston's head shape is designed to work with the shape of the cylinder head combustion chamber. This is a piston for a diesel engine having a direct injection nozzle. (Dana Corp.)

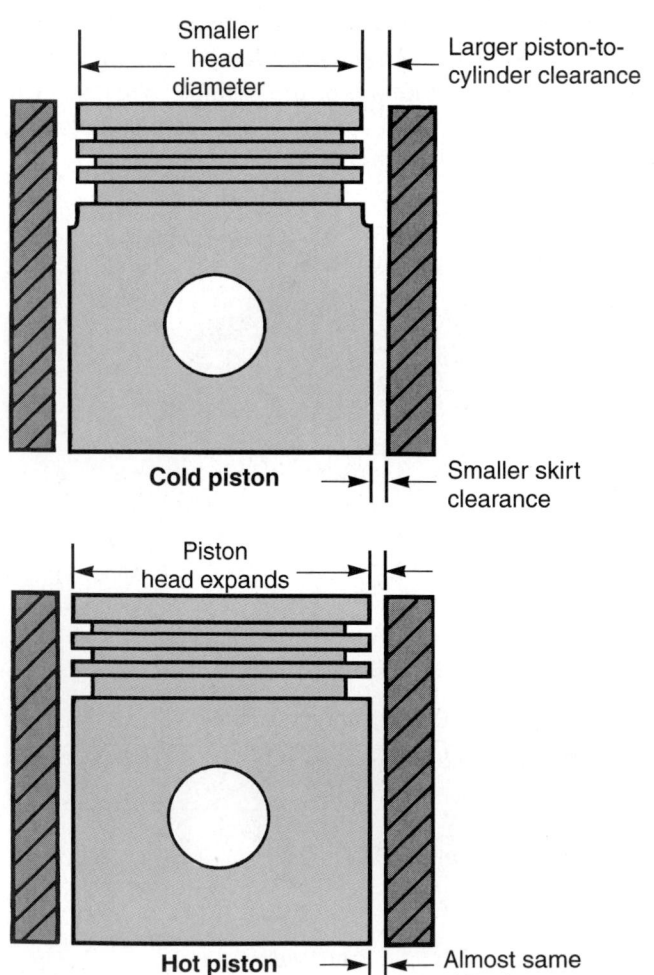

Figure 14-9. Piston taper compensates for more expansion around the piston head. The head becomes hotter than the skirt and expands more. By machining the head smaller, the piston diameter becomes almost equal at the top and bottom when hot.

is almost flat and is parallel with the block's deck surface. Refer to **Figure 14-11.** A flat top piston is commonly used with a wedge- or pancake-type cylinder head.

A *domed,* or *pop-up, piston* has a head that is convex, or curved upward. This type is normally used with a hemi-type cylinder head and some four-valve heads. The piston crown must be enlarged to fill the domed combustion chamber and produce enough compression pressure.

Valve reliefs are small indentations either cast or machined into the piston crown to provide ample piston-to-valve clearance. Without valve reliefs, the valve heads could strike the pistons. Valve reliefs are shown in **Figure 14-11.**

Slipper Skirt

A *slipper skirt* is produced when the portions of the piston skirt below the piston pin ends are removed, as in **Figure 14-11.** A slipper skirt provides clearance between the piston and the crankshaft counterweights. The piston can slide farther down in the cylinder without hitting the crankshaft.

Variable Compression Piston

A *variable compression piston* is a two-piece design controlled by engine oil pressure. The head of the piston fits over and slides on the main body of the piston. Engine oil pressure is fed between the two halves to form a hydraulic cushion. With normal driving, the oil pressure

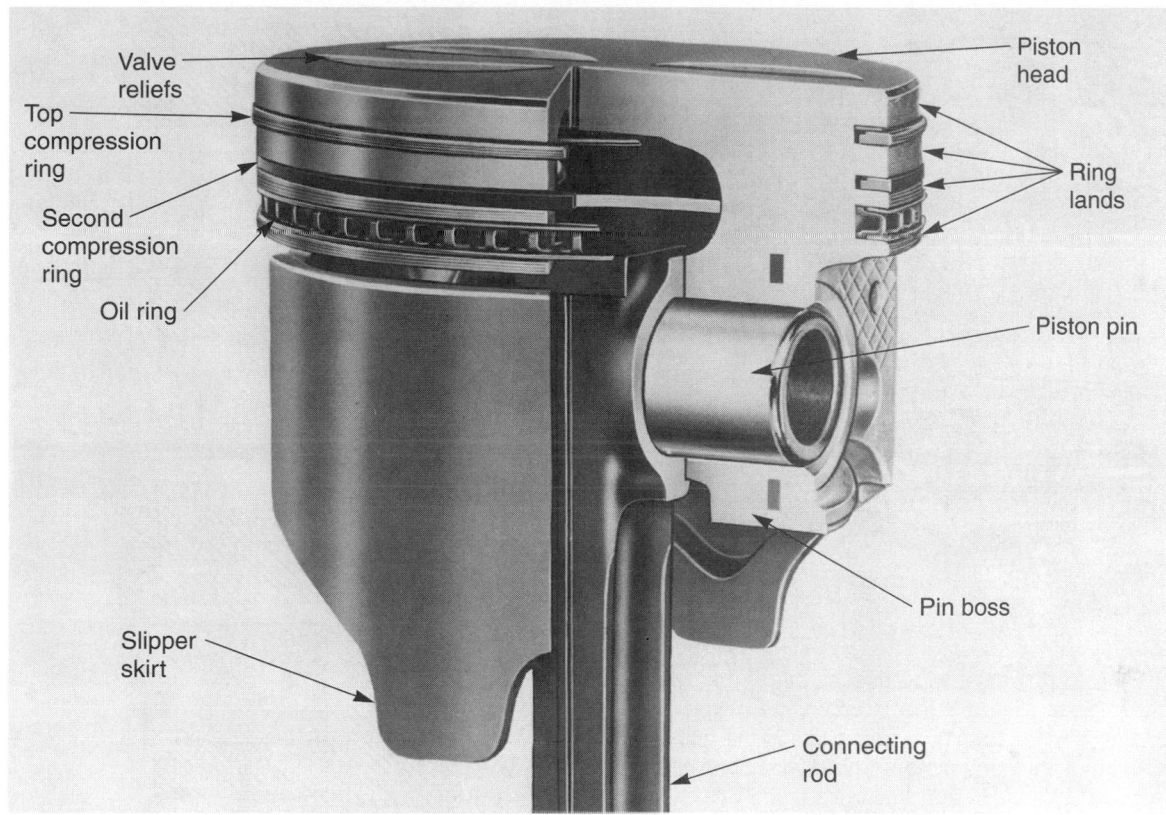

Figure 14-11. The basic parts of a typical piston and rod assembly. This piston pin is press fit in the rod. (Deere & Co.)

extends the top of the piston out for maximum compression ratio and power. When engine speed increases, the added combustion pressure pushes the head of the piston down to lower the compression ratio. This prevents engine knocking or pinging.

Piston Ring Construction

Automotive pistons normally use three rings—two compression rings and one oil ring. Refer to **Figure 14-11.** It is important for you to understand how variations in ring construction provide different operating characteristics.

The **compression rings** prevent pressure leakage into the crankcase. They also wipe some of the oil from the cylinder walls. To accomplish these functions, ring shapes vary, as shown in **Figure 14-12.** These shapes help the ring seal and remove oil from the cylinder. Compression rings are usually made of cast iron. An outer layer of chrome or other metal may be used to increase wear resistance. The face of compression rings may also be grooved to speed ring seating. **Ring seating** is the initial ring wear that makes the ring perfectly match surface of cylinder.

Oil rings are available in two basic designs: rail-spacer type and one-piece type, **Figure 14-13.** The

primary function of oil rings is to keep crankcase oil out of the combustion chambers. A three-piece oil ring consisting of two rails and a spacer is the most common. A **ring expander-spacer** is part of a three-piece oil ring. It holds the two steel oil ring scrapers apart and helps push

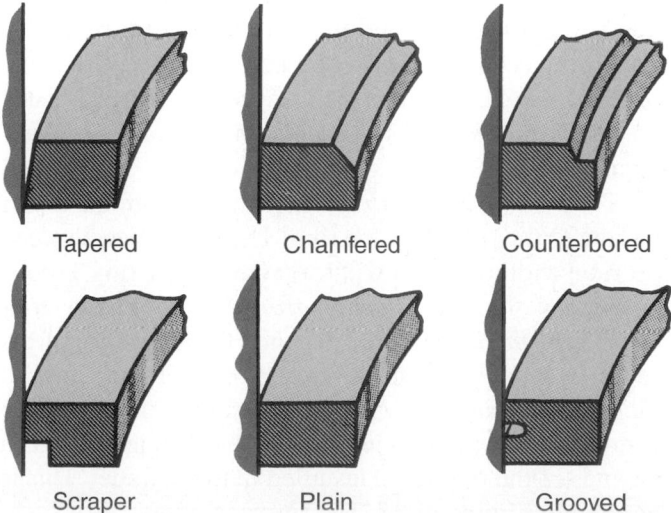

Figure 14-12. Various compression ring shapes are available. Each type is designed to help the ring prevent combustion pressure from leaking into the crankcase. (Ford)

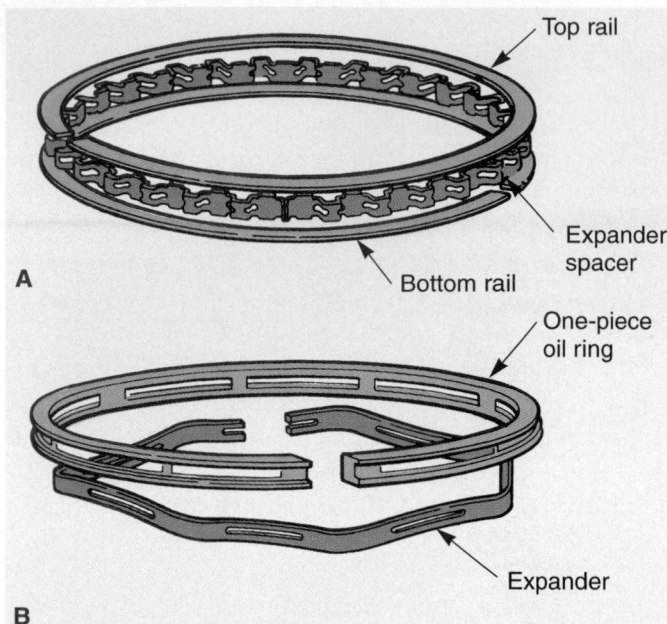

Figure 14-13. An oil ring must wipe excess oil off the cylinder wall. A—A three-piece oil ring is the most common type. B—A one-piece oil ring is made from cast iron. Slots in the ring allow oil to flow through holes in the piston groove and back into the oil pan. (Ford)

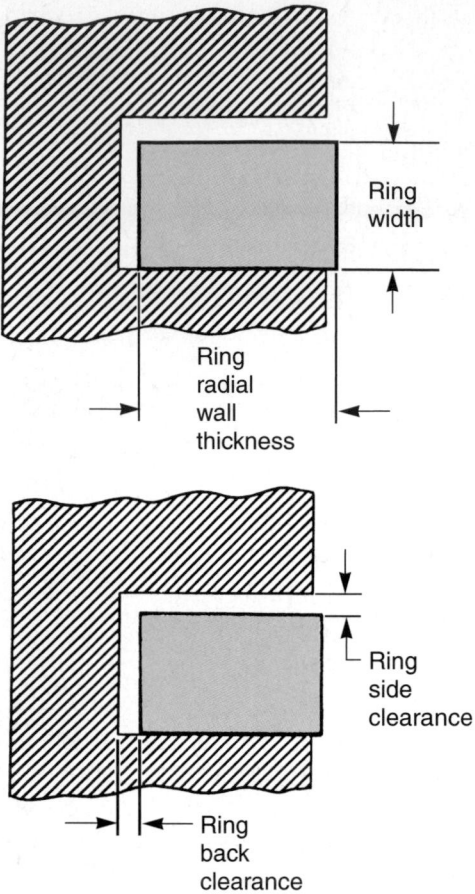

Figure 14-14. Ring width, ring groove depth, ring radial wall thickness, and ring groove height determine the ring back and side clearances. (Perfect Circle)

them outward, **Figure 14-13A.** A *ring expander* can be placed behind a one-piece oil ring to increase ring tension, **Figure 14-13B.** It can also be used behind the second compression ring. The expander helps push the ring out against the cylinder wall, increasing the ring's sealing action.

Piston Ring Dimensions

Basic piston ring dimensions include the width, radial wall thickness, and gap. Refer to **Figure 14-14.** These dimensions affect the operation of the engine. Ring side and back clearances are also very important. They must be within factory specifications or poor engine performance could result.

The *piston ring width* is the distance from the top of the ring to the bottom of the ring. The difference between the ring width and the width of the piston ring groove determines the *ring side clearance.* The *piston ring radial wall thickness* is the distance from the face of the ring to its inner wall. The difference between the ring wall thickness and ring groove depth determines the *ring back clearance.* The *piston ring gap* is the distance between the ends of the ring when installed in the cylinder. This is pictured in **Figure 14-15.** The ring gap allows the ring to be installed on the piston and to "spring outward" in its cylinder. The gap also allows the ring to conform to any variation in the cylinder diameter due to wear.

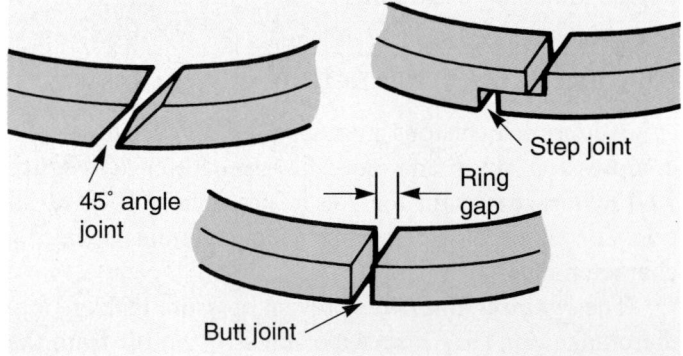

Figure 14-15. The ring gap is the small space between the ends of the ring when it is installed on the piston and the piston is in the cylinder. Most modern piston rings use a butt joint. (Ford)

Piston Ring Coatings

Soft ring coatings of porous metal, usually iron, help the ring wear in quickly, forming a good seal. The soft outer surface will wear away rapidly so the ring conforms to the shape of the cylinder. Rings with these coatings, often called *quick seal rings,* are commonly recommended for used cylinders that are slightly worn.

Hard ring coatings, such as chrome or moly, are used to increase ring life and reduce friction. Rings with these coatings are used in new or freshly machined cylinders that are perfectly round and *not* worn. To aid break-in, chrome-plated rings usually have ribbed faces. The ribs hold oil and wear quickly to produce a good seal.

Tech Tip

Avoid using rings with hard coatings when rebuilding a passenger car engine. The hard rings are more difficult to seat than softer rings. Soft rings are more forgiving of cylinder wall imperfections. Engine smoking problems are more common in rebuilt engines when hard rings are used, because the rings take much longer to seat.

Piston Pin Construction

Piston pins are normally made of case-hardened steel. *Case-hardening* is a heating and cooling process that increases the wear resistance of the piston pin. It hardens the outer layer of metal on the pin. The inner metal remains unhardened so the pin is not too brittle. The hollow piston pin is also machined and polished to a very precise finish. Piston pins are held in the piston by one of two means—snap rings or a press-fit.

A *full-floating piston pin* is secured by snap rings and is free to rotate in both the rod and piston. Look at **Figure 14-16A.** The pin is free to "float" in both the piston pin bore and the connecting rod small end. Full-floating piston pins are better than press-fit pins because they reduce friction and wear. A bronze bushing is usually used in the connecting rod. The piston pin hole serves as the other bearing surface for the pin. The snap rings fit into grooves machined inside the piston pin hole.

A *press-fit piston pin* is forced tightly into the connecting rod's small end. It can rotate freely in the piston pin hole. However, the pin is *not* free to move in the connecting rod, **Figure 14-16B.** This holds the pin inside the piston and prevents it from sliding out and rubbing on the cylinder wall. The press-fit is a very dependable piston pin design. It is also inexpensive to manufacture.

Piston Pin Offset

A *piston pin offset* locates the piston pin hole slightly to one side of the piston centerline. This helps quiet the piston during use. The pin hole is moved toward the piston's major thrust surface. This is the surface of the piston that is pushed tightly against the cylinder wall during the power stroke. If the pin hole is centered in the piston, the piston could slap or knock in the cylinder. As the piston moves up in the cylinder, it could be positioned opposite the major thrust surface. Then, during combustion, the piston could be rapidly pushed to the opposite side of the cylinder, producing a knocking sound. With

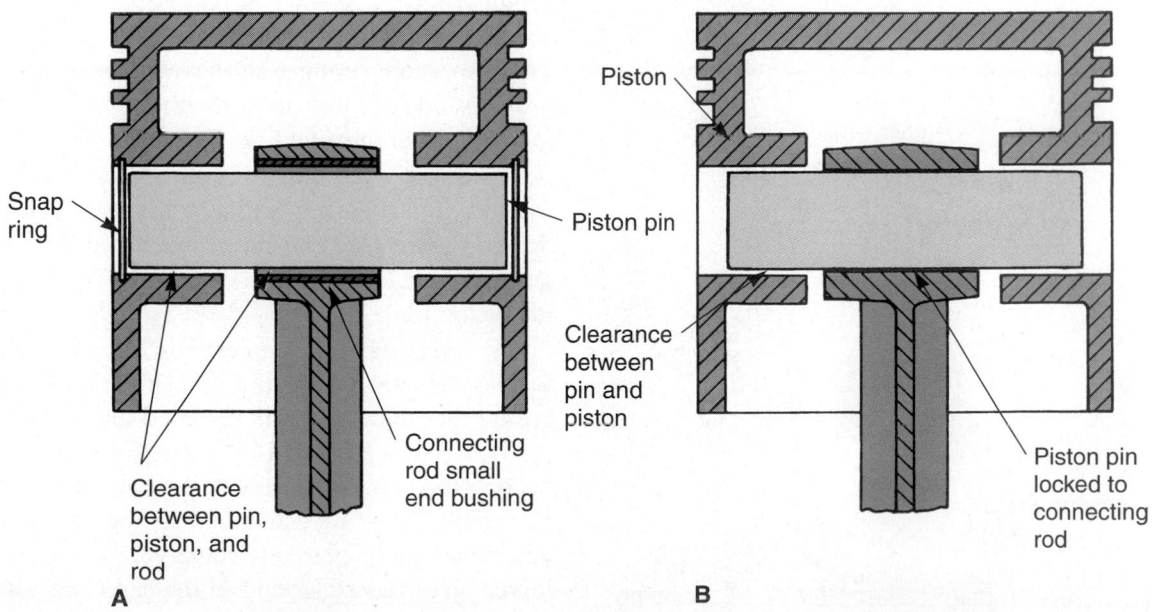

Figure 14-16. A—A full-floating piston pin has snap rings in grooves on the piston. This holds the pin in place. B—A press-fit piston pin is forced into the connecting rod. The side of the rod hits the piston boss before the pin can protrude out of the piston to strike the cylinder wall.

the pin offset, the piston tends to be pushed against its major thrust surface. This reduces its tendency to slap sideways in the cylinder.

A *piston notch* or other marking on the head of the piston is frequently used to indicate piston pin offset and the front of the piston. Look at **Figure 14-17.** The piston may also have the word "front" or an arrow stamped on it. This information lets you know how to position the piston in the block for correct location of the piston pin offset.

Figure 14-18 shows an exploded view of a piston and connecting rod assembly. Note how the parts are assembled.

Connecting Rod Construction

Most connecting rods are made of steel. The rod must withstand tons of force as the piston moves up and down in the cylinder. Connecting rods normally have an I-beam shape. This shape has a high strength-to-weight ratio.

Connecting rods must be very strong, but as light as possible for low inertia forces from their changes in

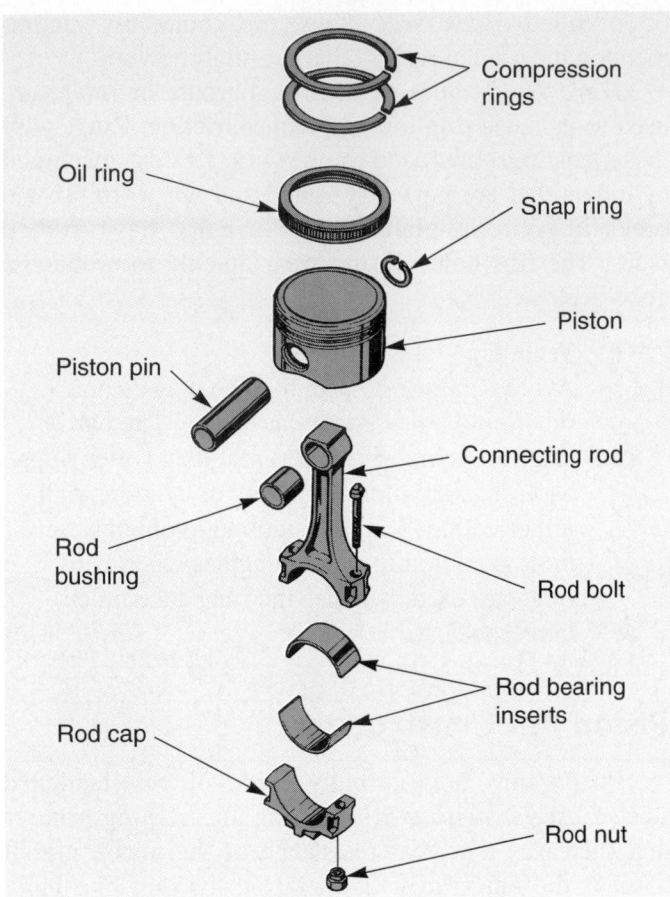

Figure 14-18. An exploded view of a piston and rod assembly. This piston has a full-floating piston pin.

direction at top dead center (TDC) and bottom dead center (BDC). *Low-inertia parts* are light parts that will accelerate quickly.

Some connecting rods have an *oil spurt hole* that provides added lubrication for the cylinder walls, piston pin, and other surrounding parts. See **Figure 14-19.** Oil pressure forces oil out when the holes in the crankshaft journal and bearing align with the spurt hole. A *drilled connecting rod* has a machined hole through its entire length. The purpose of this hole is to supply oil to the piston pin.

Connecting rod numbers are used to ensure proper location of each connecting rod in the engine. They also ensure that the rod cap is installed on the rod body correctly. Look at **Figure 14-20.** During manufacture, connecting rod caps are bolted to the connecting rods. Then, the crank holes are machined in the rods. Since these holes may not be perfectly centered, rod caps must *not* be mixed up or turned around. If the cap is installed without the rod numbers in alignment, the bore will *not* be perfectly round. Severe rod, crankshaft, and bearing damage will result.

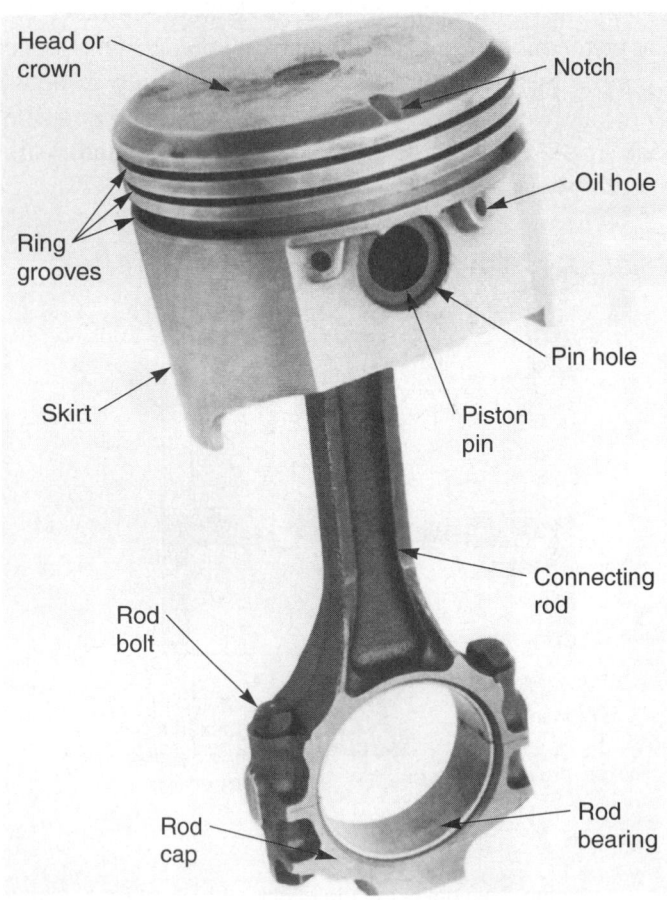

Figure 14-17. A piston assembly. (Chevrolet)

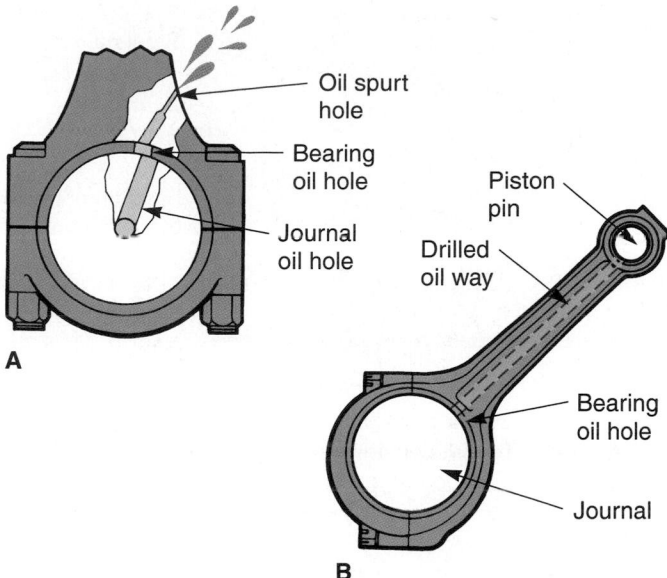

Figure 14-19. A—Oil spurt holes provide added lubrication for the piston pin, cam lobes, cylinder walls, and other surrounding parts. B—A drilled rod allows oil to enter the clearance between the pin and bushing. (Federal Mogul)

A *broken-surface rod* is scribed and broken off when manufactured to produce a rough, irregular mating surface between the rod body and cap. This is done to help lock the rod and cap into alignment. The broken, irregular surfaces match perfectly to prevent the rod and cap from shifting during engine operation. This type of rod cannot be rebuilt. However, oversize rod bearings can still be installed during an engine rebuild.

Powdered metal forging refers to a process that forms the rough shape of the part out of metal powder before final shaping in a powerful forge. Some connecting rods are powered metal forged to help control the shape and weight while reducing machining.

Machined block forging involves initial turning in a lathe to bring the blank of metal to size before forming it in a drop or press forge. This process helps eliminate flashing. *Flashing* is the small lip of rough metal produced when the two halves of the forge come together to "smash" the metal into shape. By reducing flashing, a step is removed from the manufacturing process.

Crankshaft Construction

Engine crankshafts are usually made of cast iron or forged steel. Forged steel crankshafts are needed for heavy-duty applications, such as turbocharged or diesel engines. A steel crankshaft is stiffer and stronger than a cast iron crankshaft. It will withstand greater forces without flexing, twisting, or breaking.

Oil passages leading to the rod and main bearings are either cast or drilled in the crankshaft, **Figure 14-21.** Oil

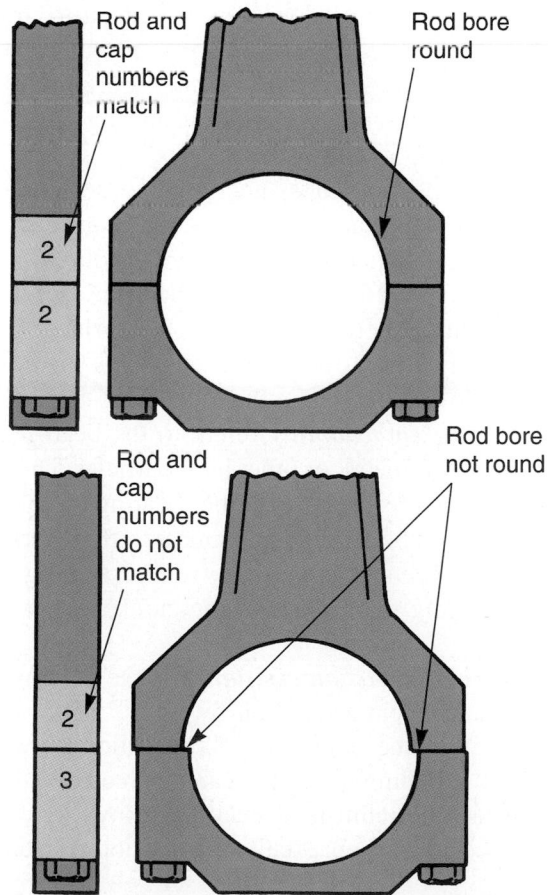

Figure 14-20. A rod cap must be installed on the rod correctly. If rod caps are mixed up or turned, the bore for the bearing may not be round. The bearing could be crushed into the crankshaft journal, damaging both.

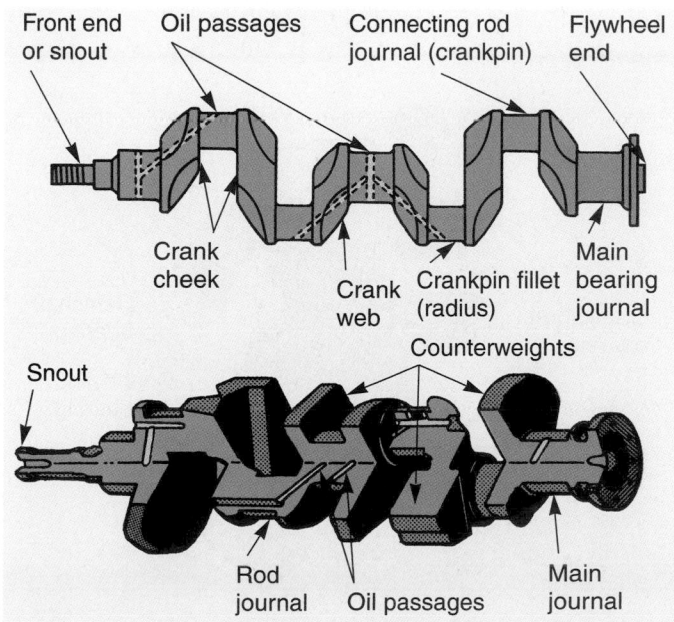

Figure 14-21. A crankshaft has internal passages to supply oil to the connecting rod bearings. (Ford)

enters the crankshaft at the main bearings and passes through holes in the main bearing journals. It then flows through passages in the crankshaft and out to the connecting rod bearings.

With an inline engine, only *one* connecting rod fastens to each rod journal. With a V-type engine, *two* connecting rods bolt to each rod journal. The amount of rod journal offset controls the stroke of the piston. The journal surfaces are precision machined and polished to very accurate tolerances. It is common to have reduced journal, or *crankpin,* diameters in order to reduce friction in the bearings.

A *fully counterweighted crankshaft* has weights formed opposite every crankpin. A *partially counterweighted crankshaft* only has weights formed on the center areas. A fully counterweighted crankshaft will operate with less vibration than a partially counterweighted crankshaft.

Engine Bearing Construction

There are three basic types of engine bearings: connecting rod bearings, crankshaft main bearings, and camshaft bearings. This is illustrated in **Figure 14-22.**

Steel is normally used for the bearing body, or backing, which contacts the stationary part of the engine. Softer alloys are bonded over this backing to form the bearing surface. Any one of three metal alloys can be plated over the steel backing. These are babbitt, copper, or aluminum. *Babbitt* is a lead-tin alloy. These three metals may be used in different combinations to design bearings for use in light-, medium-, heavy-, or extra-heavy-duty applications. See **Figure 14-23.**

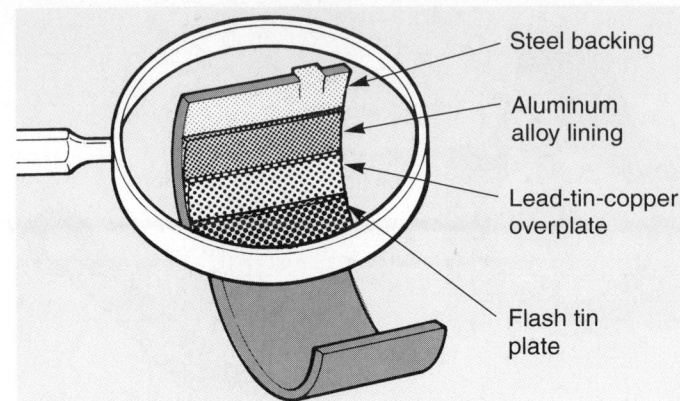

Figure 14-23. Typical construction of an engine bearing. Steel backing forms the body of the bearing. Other alloys are plated over the backing to provide better operating characteristics. (Federal Mogul)

Bearing Characteristics

Engine bearings must operate under tremendous loads, severe temperature variations, abrasive conditions, and corrosive surroundings. Essential bearing characteristics are described below. It is important to understand each of these.

Bearing load strength is the bearing's ability to withstand pounding and crushing during engine operation. The piston and rod can produce several tons of force. The bearing must not fail under these loads. If bearing load resistance is too low, the bearing can smash, spin, or split. This can ruin the bore or journal.

Bearing conformability is the bearing's ability to adjust to imperfections in the journal surface. Usually, a soft metal is plated over a hard steel. This lets the bearing conform to any defects in the journal.

Bearing embedability refers to the bearing's ability to absorb dirt, metal, or other hard particles. Dirt or metal is sometimes carried into the bearings. The bearing should allow the particles to sink into the bearing material. This prevents the particles from scratching, wearing, and damaging the surface of the crankshaft or camshaft journal.

Bearing corrosion resistance is the bearing's ability to withstand being acted on by acids, water, and other impurities in the engine oil. Combustion blowby gases cause oil contamination that can also corrode the engine bearings. Aluminum-lead and other alloys are commonly used because of their excellent corrosion resistance.

Bearing Crush

Bearing crush is used to help prevent the bearing from spinning inside its bore during engine operation.

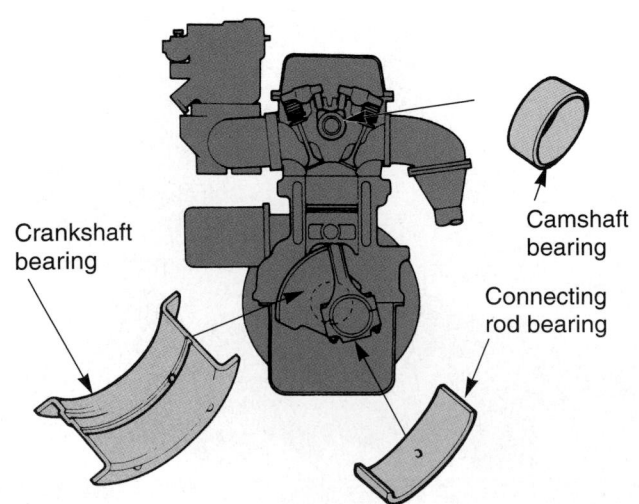

Figure 14-22. The three basic types of engine bearings are crankshaft main bearings, connecting rod bearings, and camshaft bearings. (Federal Mogul)

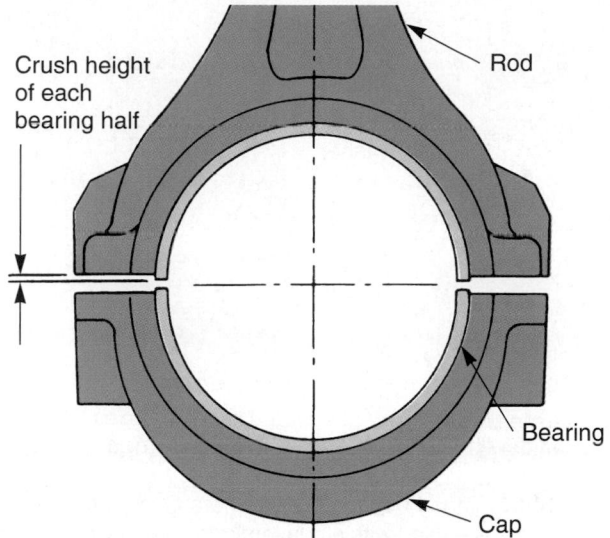

Figure 14-24. Bearing crush is produced when bearings are made slightly larger than the bearing bore. When the cap is bolted down, the bearing is forced into the bore. This keeps the bearing from turning with the crankshaft. (Deere & Co.)

Pictured in **Figure 14-24,** the bearing is made slightly larger in diameter than the bearing bore. The end of the bearing is slightly above the bore. When the rod or main cap is tightened, the bearing ends press against each other. This jams the back sides of the bearing inserts tightly against the bore, locking them in place.

Bearing Spread

Bearing spread is used on split-type engine bearings to hold the bearing in place during assembly. The distance across the parting line of the bearing is slightly wider than the bearing bore. This causes the bearing insert to stick in its bore when pushed into place with your fingers. Tension from bearing spread keeps the bearing from falling out of its bore as you assemble the engine.

Standard and Undersize Bearings

A *standard bearing* has the original dimensions specified by the engine manufacturer for a new, unworn, or unmachined crankshaft. A standard bearing may have the abbreviation "STD" stamped on its back. An *undersize bearing* is designed to be used on a crankshaft journal that has been machined to a smaller diameter. If the crankshaft journals have been worn or damaged, they can be ground undersize by a machine shop. Then, undersize bearings are needed. Connecting rod and main bearings are available in undersizes of 0.010″, 0.020″, 0.030″, and sometimes 0.040″. The undersize is normally

stamped on the back of the bearing, as in **Figure 14-25.** The crankshaft may also have an undersize number stamped on it by the machine shop.

Bearing Locating Lugs and Dowels

Bearing locating lugs and *dowels* position split bearings in their bores. Look at **Figure 14-26.** The bearing usually has a lug that fits into a recess machined in the bearing bore or cap. Sometimes a dowel in the cap or

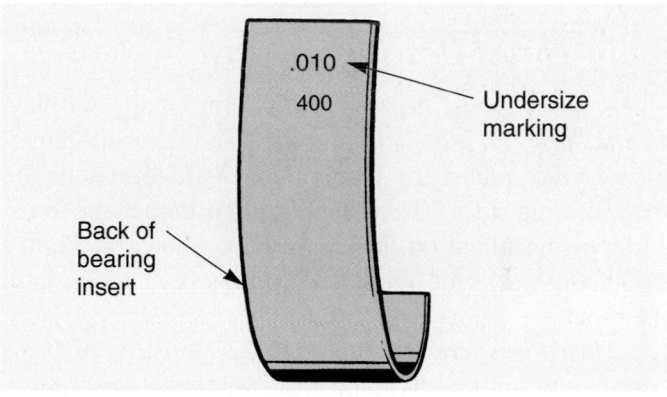

Figure 14-25. The bearing undersize is stamped on the back of the bearing. An undersize marking of 0.010″, for example, means that crankshaft journal has been ground 0.010″ smaller in diameter. Undersize bearings are needed to maintain correct bearing clearance. (Buick)

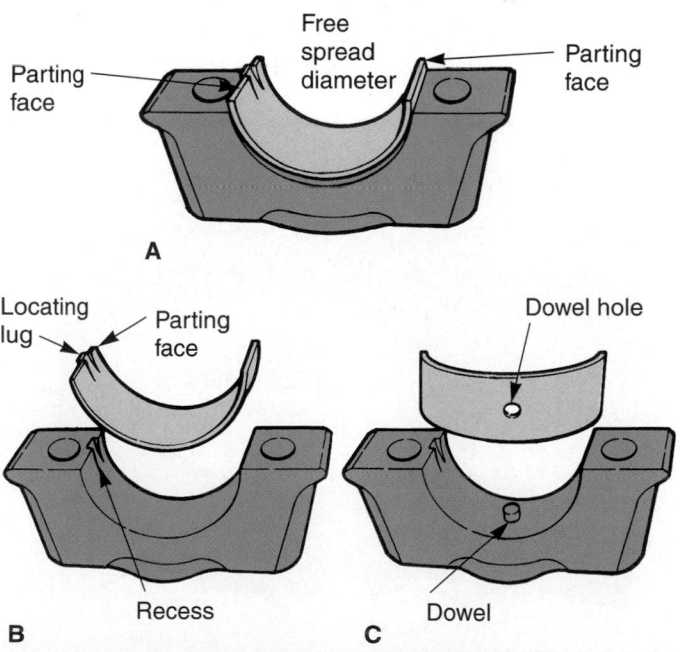

Figure 14-26. Bearing locating lugs or dowels can be used to position engine bearings in their bore. A—Spread. B—Lug. C—Dowel. (Federal Mogul)

bore fits in a hole in the bearing insert. Either method helps keep the insert from shifting or turning during crankshaft rotation.

Bearing Oil Holes and Grooves

Bearing oil holes and *grooves* in the engine bearings permit bearing lubrication. The holes allow oil to flow through the block and into the clearance between the bearing and the journal. The grooves provide a channel so oil can completely encircle the bearing before flowing over and out of it. See **Figure 14-27.**

Main Thrust Bearings, Thrust Washers

A *main thrust bearing* limits crankshaft end play. *Crankshaft end play* is the forward and rearward movement of the crankshaft. Thrust flanges are formed on the main bearing sides. These flanges almost touch the thrust surfaces machined on the crankshaft. Shown in **Figure 14-28,** this keeps the crank from sliding back and forth in the block.

Thrust washers are sometimes used instead of thrust bearings to limit crank end play. The thrust washers are separate parts from the main bearing. They slide down into the space between the crankshaft and block, as pictured in **Figure 14-28.**

Figure 14-29 shows how a main bearing, thrust washers, main cap, and related components assemble to the block.

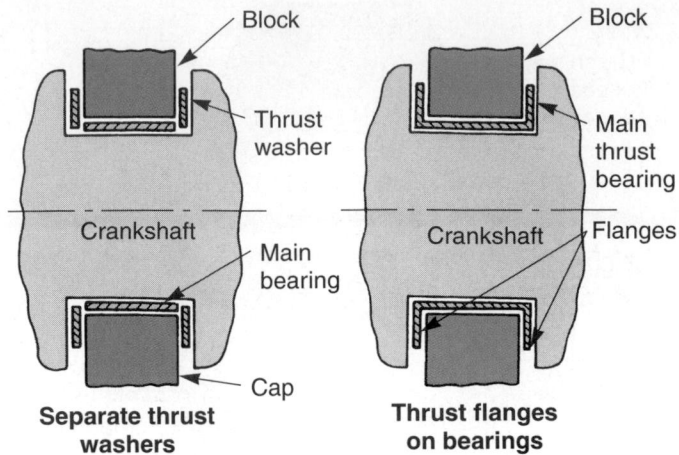

Figure 14-28. Thrust bearings or washers limit crankshaft end play. Washers are used with a conventional main bearing insert. A main thrust bearing has thrust flanges formed as part of the main bearing. (Deere & Co.)

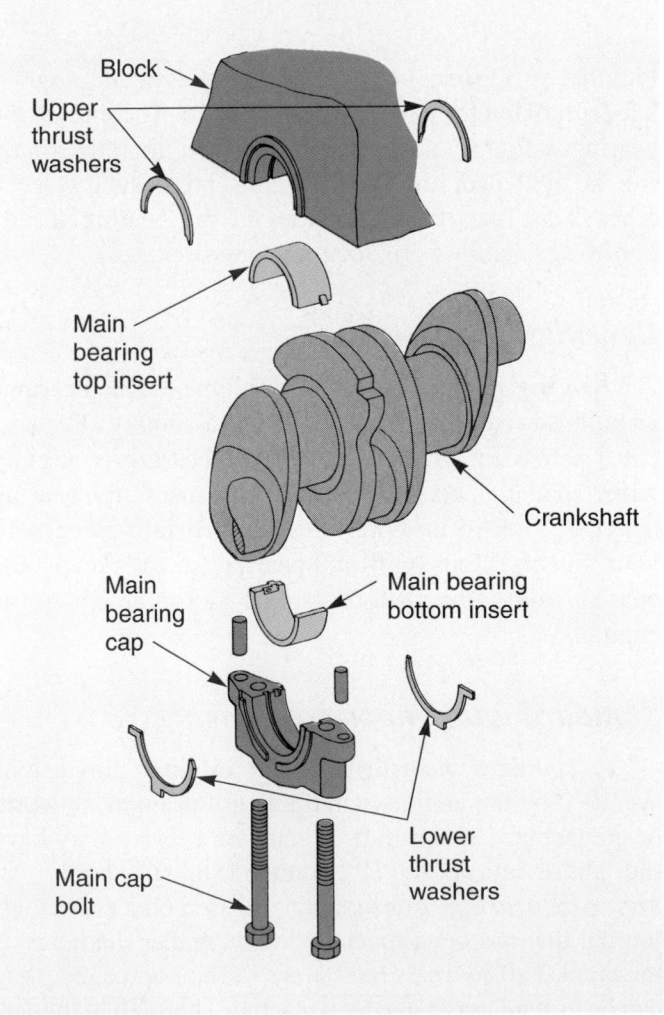

Figure 14-29. Note how the main bearing, thrust washers, crankshaft, and cap fit together. (Mercedes Benz)

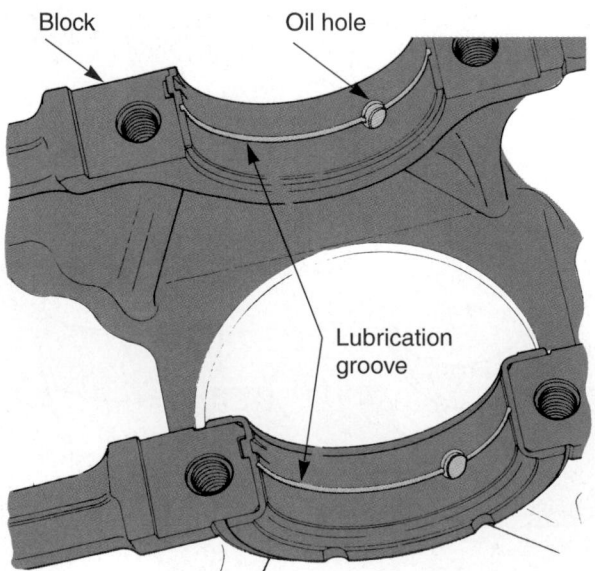

Figure 14-27. Main bearings have holes that let oil enter the bearing clearance. Grooves allow oil to circle the bearing to evenly distribute lubricating oil. (Chrysler)

Rear Main Bearing Oil Seal Construction

The engine *rear main bearing oil seal* prevents oil leakage around the back of the crankshaft. There are several different types. These include a two-piece neoprene seal, a one-piece neoprene seal, and a wick or rope seal.

A *two-piece neoprene rear oil seal* usually fits into a groove cut into the block and rear main cap. Look at **Figures 14-30** and **14-31**. The seal has a lip that traps oil and another lip that keeps dust and dirt out of the engine. The sealing lips ride on a machined surface of the crankshaft. Spiral grooves may be used on the crank surface to help throw oil inward, preventing leakage.

A *one-piece neoprene rear oil* seal fits around the rear flange on the crankshaft. **Figure 14-32** shows how a retainer may be used to hold the seal on the rear of the engine. The one-piece rear seal has sealing lips that are similar to those on a two-piece neoprene seal.

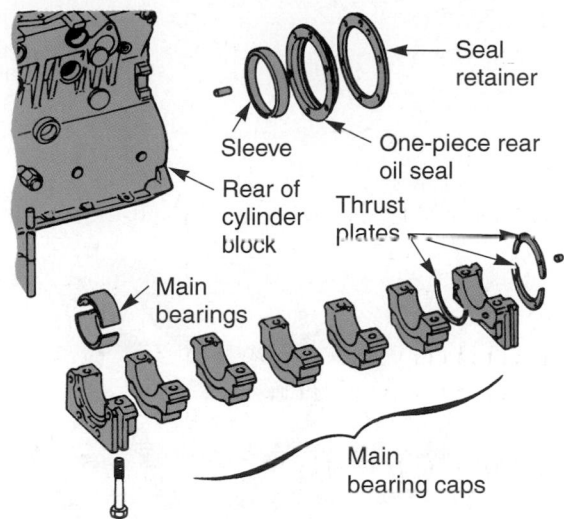

Figure 14-32. Some rear oil seals are one-piece. They assemble into the rear of the block around the crankshaft flange. (Chrysler)

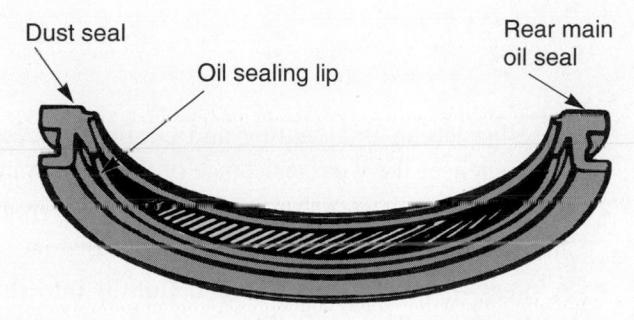

Figure 14-30. A rear main oil seal has a sealing lip that must face the inside of the engine. The dust seal faces the outside of the engine. (GMC)

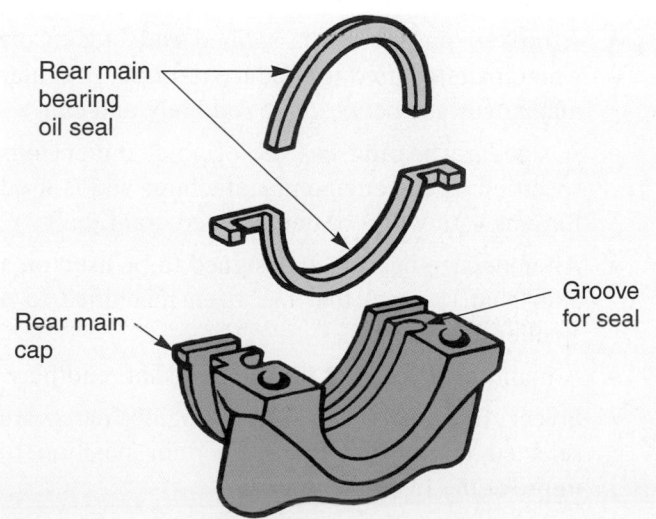

Figure 14-31. A two-piece seal fits into a groove machined in the block and rear main cap.

A *rope,* or *wick, rear oil seal* is simply a woven rope filled with graphite. One piece of the rope seal fits into a groove in the block. Another piece fits in a groove in the main cap. This type seal is not as common as one- and two-piece neoprene seals.

Select-Fit Parts

Select-fit means that some engine parts are selected and installed in a certain position to improve the fit or clearance between parts. For example, pistons are commonly selected to fit precisely into their cylinders. To do so, the engine manufacturer measures the diameter of the cylinders. If one cylinder is machined slightly larger than another, a slightly larger piston is installed. This is select fit. Because of select-fit parts, it is important that you reinstall parts in their original locations whenever possible.

Balancer Shafts

Balancer shafts are used in some engines to cancel the vibrating forces produced by crankshaft, piston, and rod movement. They are usually found on 4- and 6-cylinder engines. **Figure 14-33** shows the balancer shafts used in one particular engine.

The balancer shafts, also called *silencer shafts*, are installed in the right and left sides of the cylinder block. Usually, a chain is used to spin the shafts at twice crankshaft rpm. The shafts are supported on bearings fitted in a bore machined in the block. Oil is pressure-fed to these bearings to provide lubrication.

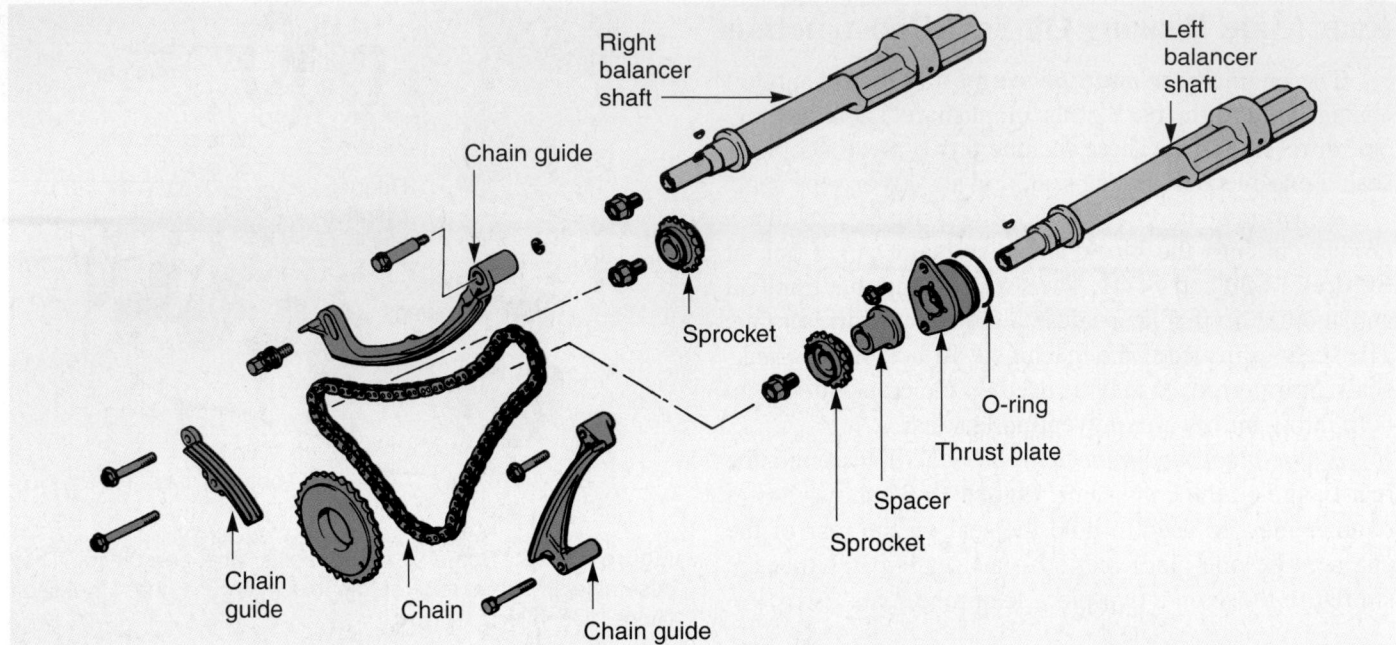

Figure 14-33. Balancer shafts are turned by a chain, sprockets, and the engine crankshaft. They counteract any vibration produced by the crankshaft, pistons, and rods.

Summary

- The basic parts of an engine bottom end are the block, crankshaft, connecting rods, and piston assemblies.

- Cylinder sleeves or liners are metal, pipe-shaped inserts that fit into the cylinder block.

- Line boring refers to a machining operation that cuts a series of holes through the block for the crankshaft bearings.

- A two-bolt main block only uses two bolts to secure each main bearing cap to the cylinder block. A four-bolt main block has four bolts holding each main cap.

- A main bearing bedplate is a large part that fits over the whole bottom of the block.

- A cam-ground piston is machined slightly out-of-round when viewed from the top.

- Piston taper is normally used to maintain the correct piston-to-cylinder clearance.

- A slipper skirt is produced when the portions of the piston skirt below the piston pin ends are removed.

- Piston ring width is the distance from the top of the ring to the bottom of the ring.

- Piston ring gap is the distance between the ends of the ring when it is installed in the cylinder.

- Case-hardening is a heating and cooling process that increases the wear resistance of the piston pin.

- A full-floating piston pin is secured by snap rings and is free to rotate in both the rod and piston.

- A press-fit piston pin is forced tightly into the connecting rod's small end.

- Piston pin offset locates the piston pin hole slightly to one side of the piston centerline to quiet piston operation.

- Connecting rod numbers are used to ensure proper location of each connecting rod in the engine.

- A broken surface rod is scribed and broken off when manufactured to produce a rough, irregular mating surface between the rod body and cap.

- A standard bearing has the original dimensions specified by the engine manufacturer and is used for a new, unworn, or unmachined crankshaft.

- An undersize bearing is designed to be used on a crankshaft journal that has been machined to a smaller diameter.

- A main thrust bearing limits crankshaft end play.

- Select fit means that some engine parts are selected and installed in a certain position to improve the fit between parts.

- Balancer shafts are used in some engines to cancel the vibrating forces produced by crankshaft, piston, and rod movement.

Important Terms

Engine bottom end
Cast iron cylinder block
Aluminum cylinder block
Cylinder sleeves
Dry sleeve
Wet sleeve
Line boring
Two-bolt main block
Four-bolt main block
Crossbolted block
Block girdle
Piston diameter
Pin hole diameter
Ring groove width
Ring groove depth
Skirt length
Compression distance
Cam-ground piston
Piston taper
Piston shape
Valve reliefs
Slipper skirt
Variable compression
  piston
Compression rings
Ring seating
Ring expander-spacer
Ring expander
Piston ring width
Ring side clearance
Piston ring radial wall
  thickness

Ring back clearance
Piston ring gap
Soft ring coatings
Hard ring coatings
Full-floating piston pin
Press-fit piston pin
Piston pin offset
Oil spurt hole
Drilled connecting rod
Connecting rod numbers
Broken-surface rod
Crankpin
Fully counterweighted
  crankshaft
Partially counterweighted
  crankshaft
Bearing crush
Bearing spread
Standard bearing
Undersize bearing
Bearing locating lugs
Bearing oil holes
Dowels
Main thrust bearing
Crankshaft end play
Thrust washers
Rear main bearing oil
  seal
Select-fit
Balancer shafts

Review Questions—Chapter 14

Please do not write in this text. Place your answers on a separate sheet of paper.

1. Cylinder blocks can be made of _____ _____ or _____.

2. Explain the difference between wet sleeves and dry sleeves.

3. Aluminum blocks commonly use _____ _____, _____ sleeves.

4. List and define six piston dimensions.

5. A(n) _____ _____ piston is machined slightly out-of-round when viewed from the top to compensate for different rates of piston _____.

6. When the top of the piston outside diameter is machined slightly smaller than the bottom, it is called a:
   (A) pop-up piston.
   (B) dished piston.
   (C) cam-ground piston.
   (D) piston taper.

7. _____ _____ are small indentations either cast or machined in the piston top.

8. What is a ring expander-spacer?

9. The difference between the ring width and the width of the piston ring groove determines _____ _____ _____.

10. The difference between the ring wall thickness and ring groove depth determines _____ _____ _____.

11. Define ring gap.

12. When should you use piston rings with soft coatings?

13. Explain why some parts are case hardened.

14. Which of the following does *not* relate to modern piston assemblies:
    (A) Full-floating pin.
    (B) Press-fit pin.
    (C) Bolted piston pin.
    (D) Offset piston pin.

15. Describe piston slap and piston pin offset.

16. A(n) _____ on the head of a piston is used to indicate piston pin offset.

17. Name two reasons connecting rod numbers are used.

18. Engine crankshafts are usually made of _____ _____ or _____ _____.

19. How does lubricating oil get to the connecting rod bearings?

20. Which of the following does *not* pertain to engine bearings?
    (A) Load strength.
    (B) Lubrication absorption.
    (C) Conformability.
    (D) Embedability.

21. Explain bearing crush.

22. Explain bearing spread.

23. A(n) _____ bearing has the original dimensions specified by the manufacturer for a new, unworn, or unmachined crankshaft.

24. A(n) _____ bearing is designed to be used on a crankshaft journal that has been machined to a smaller diameter.

25. Describe three rear main oil seal variations.

⬤ ASE-Type Questions

1. Which of the following is *not* a basic part of an engine bottom end?
   (A) *Block.*
   (B) *Valve train.*
   (C) *Crankshaft.*
   (D) *Connecting rods.*

2. A piston head can reach operating temperatures as high as:
   (A) *450°F.*
   (B) *500°F.*
   (C) *650°F.*
   (D) *700°F.*

3. The distance from the piston pin hole centerline to the piston top is called:
   (A) *skirt length.*
   (B) *piston diameter.*
   (C) *pin hole diameter.*
   (D) *compression distance.*

4. The piston type commonly used with a wedged cylinder head is:
   (A) *concave.*
   (B) *pop-up.*
   (C) *flat top.*
   (D) *All of the above.*

5. Which of the following provides clearance between the piston and crankshaft counterweights?
   (A) *Slipper skirt.*
   (B) *Piston taper.*
   (C) *Valve reliefs.*
   (D) *Compression rings.*

6. The most common oil rings consist of:
   (A) *two rails and a spacer.*
   (B) *ten slots and a spacer.*
   (C) *three rails and an expander.*
   (D) *seven slots and an expander.*

7. The difference between piston ring width and ring groove width determines:
   (A) *ring gap.*
   (B) *ring side clearance.*
   (C) *ring wall thickness.*
   (D) *ring back clearance.*

8. The purpose of a drilled connecting rod is to:
   (A) *rotate the piston pin.*
   (B) *create piston pin offset.*
   (C) *prevent pressure leakage.*
   (D) *supply oil to the piston pin.*

9. How many connecting rods bolt to each crankshaft journal in a V-type engine?
   (A) *One.*
   (B) *Two.*
   (C) *Three.*
   (D) *Four.*

10. Basic types of engine bearings include each of these *except:*
    (A) *piston bearings.*
    (B) *camshaft bearings.*
    (C) *crankshaft bearings.*
    (D) *connecting rod bearings.*

Activities for Chapter 14

1. Use inside and outside micrometers to measure the diameter of a piston and the engine cylinder bore. Then, find the amount of piston clearance by subtracting the piston diameter from the bore measurement.

2. If a number of pistons are available, determine whether any are cam ground. Do this by measuring the diameter parallel to the piston pin, and then measuring the diameter perpendicular (at a 90° angle) to the piston pin. If the piston is cam ground, the measurement perpendicular to the pin will be larger by a few thousandths of an inch.

3. Disassemble a rod cap from its connecting rod, turn it end-for-end so the numbers no longer align, and bolt it back into place. Can you visually detect an "out-of-round" rod bore? If not, try to think of a way to determine whether the bore is perfectly round. When finished with this activity, reassemble the rod and cap properly.

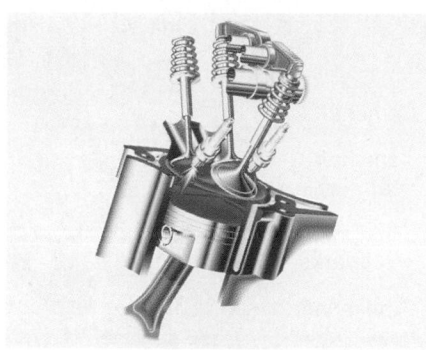

Chapter 15

Engine Front End Construction

After studying this chapter, you will be able to:

- Explain the function and construction of a vibration damper.

- Compare the three types of camshaft drives.

- Explain the construction of a timing gear, timing chain, and timing belt assembly.

- Summarize the construction of engine front covers, oil slingers, and other related components.

- Describe safety practices related to working on engine front end components.

- Correctly answer ASE certification test questions on engine front end construction.

The typical *engine front end* assembly consists of the parts that attach to the front of the engine. These parts include the camshaft drive mechanism, front cover-mounted oil pumps, water pumps, auxiliary shafts, etc. This chapter explains the construction and operation of engine front end assemblies. Study it carefully. It will help you understand later chapters on service and repair.

Note
Oil pumps, water pumps, and other front end-related parts are detailed in other textbook chapters. Refer to the index for more information on these components.

Vibration Damper Construction

Harmonic vibration is a high-frequency movement resulting from twisting and untwisting of the crankshaft. Each piston and rod assembly can exert over a ton of downward force on its journal. This can actually flex, or bend, the crank throws in relation to each other. If harmonic vibration is not controlled, the crankshaft could vibrate like a musician's tuning fork or a string-type musical instrument. Serious engine damage, usually a broken crankshaft, can result. A *vibration damper,* or *harmonic balancer,* is used to control harmonic vibration. The vibration dampener also cuts load variation on the engine timing belt, chain, or gears, so these parts last longer.

The vibration damper is a heavy wheel mounted in rubber to control harmonic vibration. Look at **Figure 15-1.** The balancer is keyed to the crankshaft snout. This makes the damper spin with the crankshaft. **Figure 15-2** illustrates the basic construction of a typical vibration damper. Notice how a rubber ring separates the outer *inertia ring* and the inner sleeve. The inertia ring and the rubber ring set up a damping action on the crankshaft as it tries to twist and untwist. This deadens vibrating action. A *dual-mass harmonic balancer* has one weight mounted on the outside of the crankshaft pulley and another on the inside. The extra rubber-mounted weight helps reduce vibration at high engine speeds.

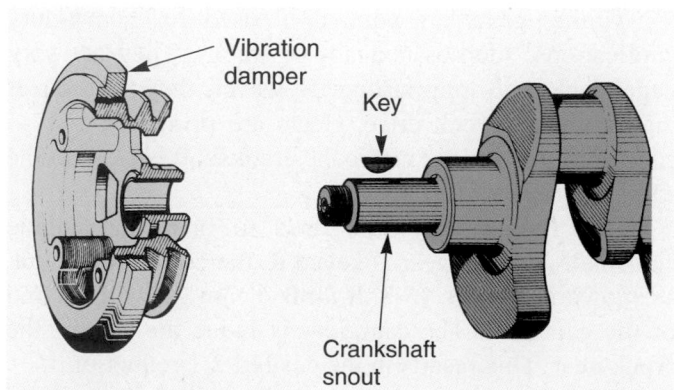

Figure 15-1. The vibration damper, or harmonic balancer, installs on the front of the crank. A key locks the damper to the crank snout. A large bolt is commonly used to hold the damper in place. (Peugeot)

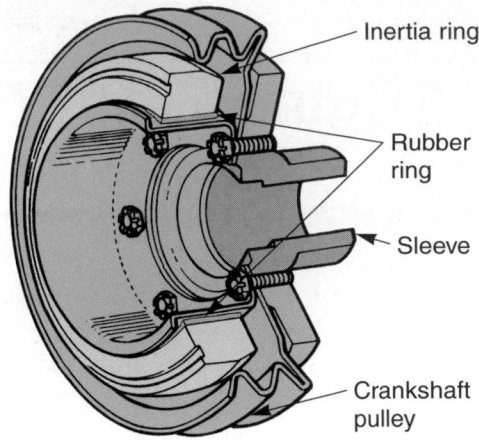

Figure 15-2. Study the construction of a vibration damper. A rubber ring separates the inertia ring and inner sleeve. This acts to deaden, or cancel, crankshaft harmonic vibration. (Chrysler)

Crankshaft Pulley

The *crankshaft pulley* operates belts for the alternator, water pump, and other units. As shown in **Figure 15-2,** it is often part of the harmonic balancer. It may also simply be bolted to the front of the balancer. The pulley has either V or ribbed grooves for the belts.

Camshaft Drive Construction

A *camshaft drive* must turn the camshaft at one-half of the crankshaft speed. It must do this smoothly and dependably. There are three basic types of camshaft drives: timing gears, a timing chain and sprockets, or a timing belt and sprockets.

Timing Gears

Timing gears are commonly used for heavy-duty applications, such as taxicabs or trucks. They are very dependable and long lasting. However, they are noisier than a chain or belt drive. Gears are primarily used in cam-in-block engines where the crankshaft is close to the camshaft.

Two *timing gears* are used to drive the engine camshaft. A crank gear is keyed to the crankshaft snout, as shown in **Figure 15-3.** It turns a cam gear on the end of the camshaft. The cam gear is twice the size of the crank gear. This results in the desired 2:1 reduction.

Timing marks on the two gears show the technician how to install the gears properly. Refer to **Figure 15-3.** The marks may be circles, indentations, or lines on the gears. The timing marks must line up for the camshaft to be in time with the crankshaft.

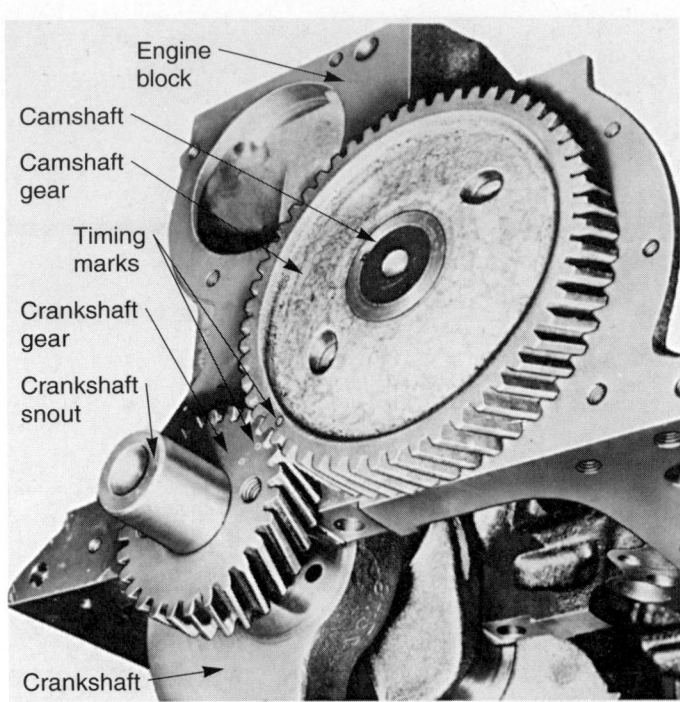

Figure 15-3. A timing gear setup is very dependable. The crank gear turns the cam gear at one-half of the crank speed. Timing marks allow for the proper assembly of the gears. (Chevrolet)

Timing Chain and Sprockets

A *timing chain* and *sprockets* can also be used to turn the camshaft. See **Figure 15-4.** This is the most common type of camshaft drive arrangement on cam-in-block engines. It may also be used in OHC engines. A

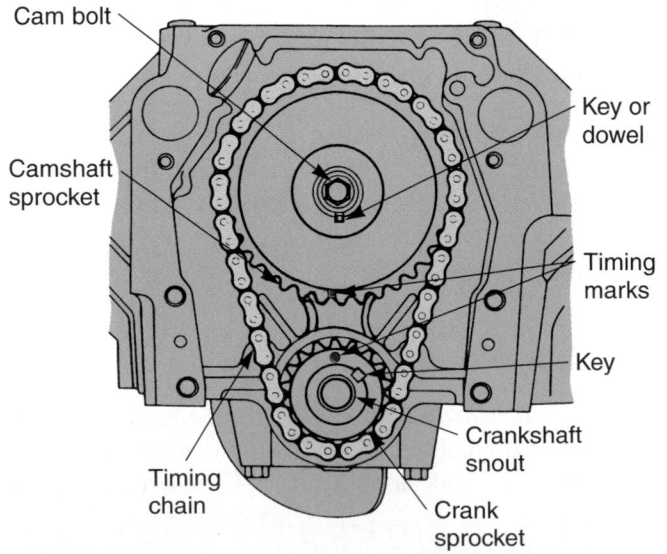

Figure 15-4. A timing chain setup is very common. Study the part names. (Buick)

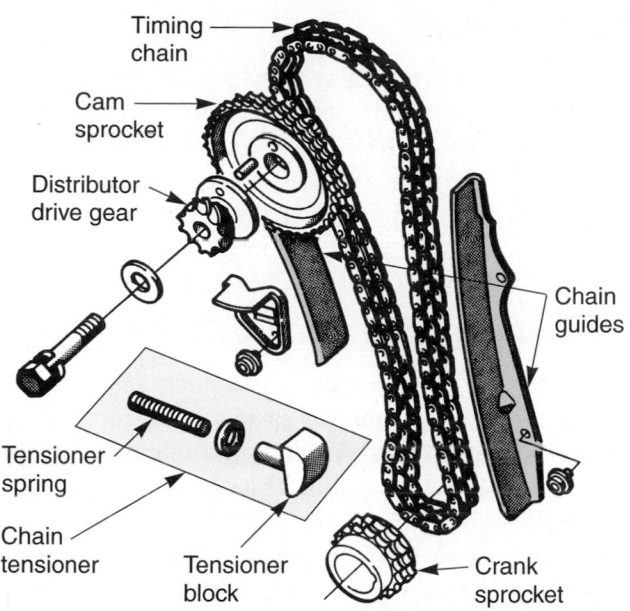

Figure 15-5. An OHC timing chain assembly normally uses chain guides and sometimes a tensioner. (Chrysler)

crank sprocket is keyed to the crankshaft snout. A larger cam sprocket, with either metal or plastic teeth, bolts to the camshaft. The timing chain transfers power from the crank sprocket to the cam sprocket. Like timing gears, the chain sprockets have timing marks. The marks must line up to properly time the camshaft with the crankshaft.

A *chain guide* may be needed to prevent chain slap. *Chain slap* is when the chain flaps back and forth because of excessive slack in the chain. **Figure 15-5** shows a timing chain using chain guides. The guides have a metal body with either a plastic or Teflon® face. This allows the chain to slide on the guide with minimum friction and wear. A *chain tensioner* may be used to take up the slack in the chain as it and the sprockets wear. One is shown in **Figure 15-5**. It is usually a spring-loaded plastic or fiber block. The spring pushes the block outward, keeping a constant tension on the chain. A chain-type cam drive for an OHC engine commonly uses guides and sometimes a tensioner. Refer to **Figure 15-6** for a front view of this setup.

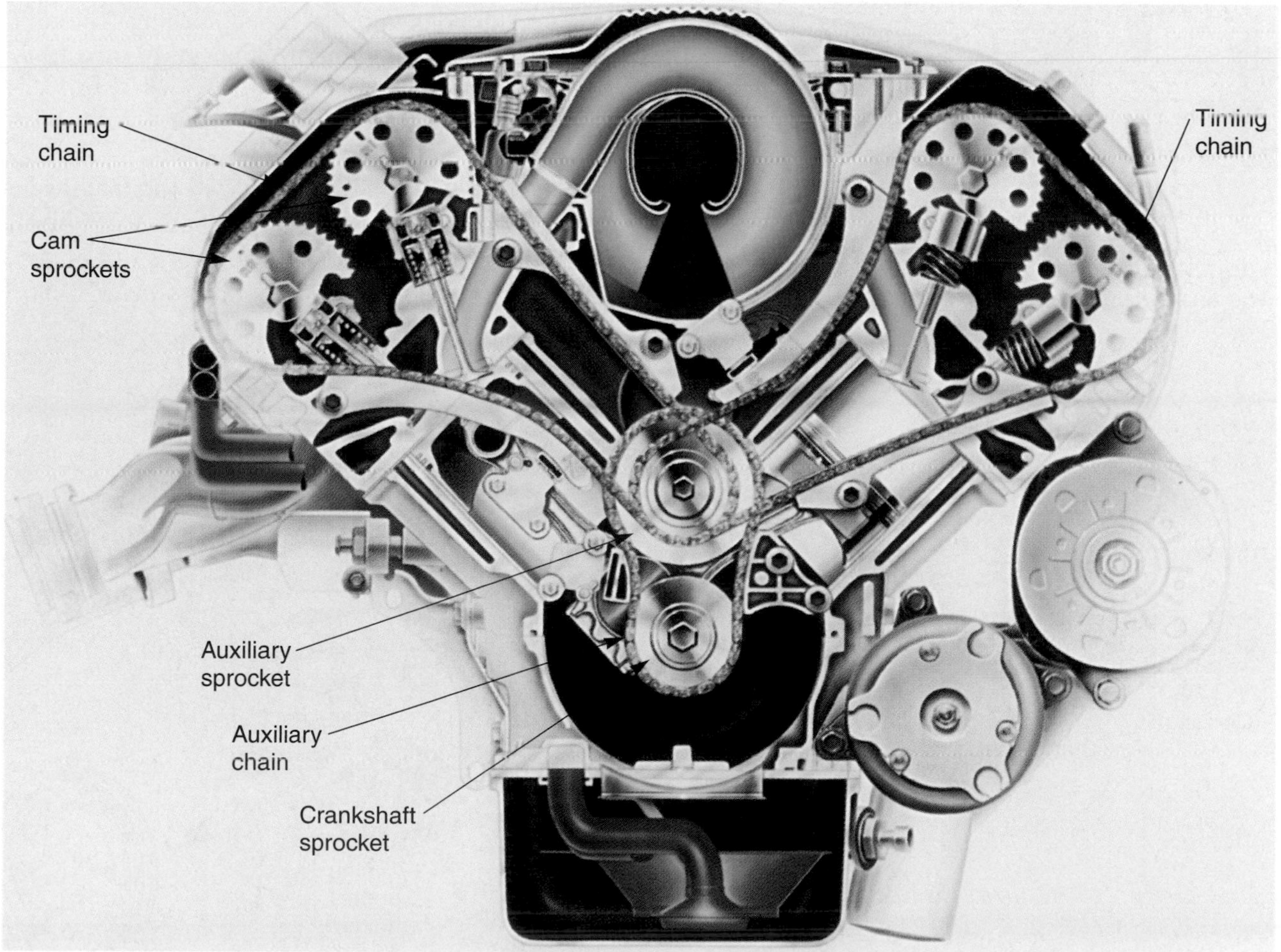

Figure 15-6. Note the dual timing chains and the auxiliary chain in this OHC engine. Chain guides are used on the timing chains, and a tensioner is used on the auxiliary chain. (Cadillac)

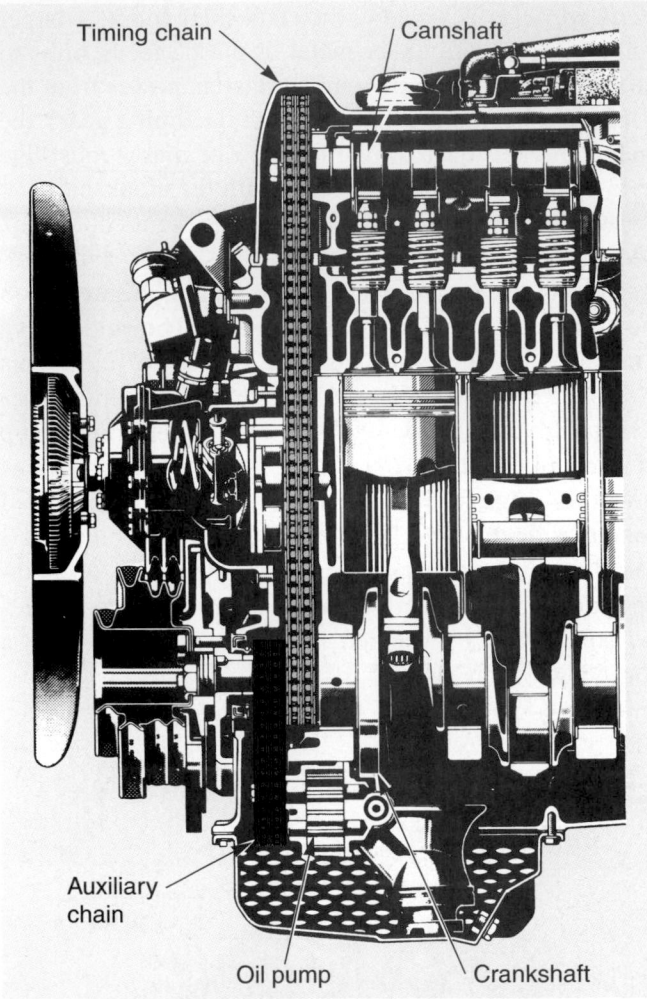

Figure 15-7. A side view of a diesel engine using both a timing chain and an auxiliary drive chain. (Mercedes Benz)

Auxiliary Chain

An *auxiliary chain* may be used to drive the engine oil pump, balancer shafts, and other units on the engine. **Figures 15-6** and **15-7** show two examples. An auxiliary chain is driven by an extra sprocket, usually placed in front of the crankshaft timing chain sprocket. Do *not* confuse an auxiliary chain with a timing chain.

Engine Front Cover Construction

An *engine front cover,* also called *timing cover,* encloses the timing chain or gear mechanism and prevents oil leakage from the front of the engine. As pictured in **Figure 15-8,** it bolts to the front of the engine block. The cover holds the front oil seal, timing pointer, probe holder, and other parts. A *front oil seal* prevents oil leakage between the crankshaft and cover. The seal is press-fit in the cover.

An engine front cover is commonly made of thin, stamped steel or cast aluminum. Sometimes the water pump bolts onto the front cover. Some water pump housings are cast into the front of the engine block. The core of the pump bolts onto the block housing. The oil pump may also be housed in the front cover, **Figure 15-9.**

Oil Slinger

An *oil slinger* is a washer-shaped part that fits in front of the crankshaft sprocket. Refer back to **Figure 15-8.** When the engine is running, oil squirts out of the front main bearing and strikes the spinning slinger. Centrifugal

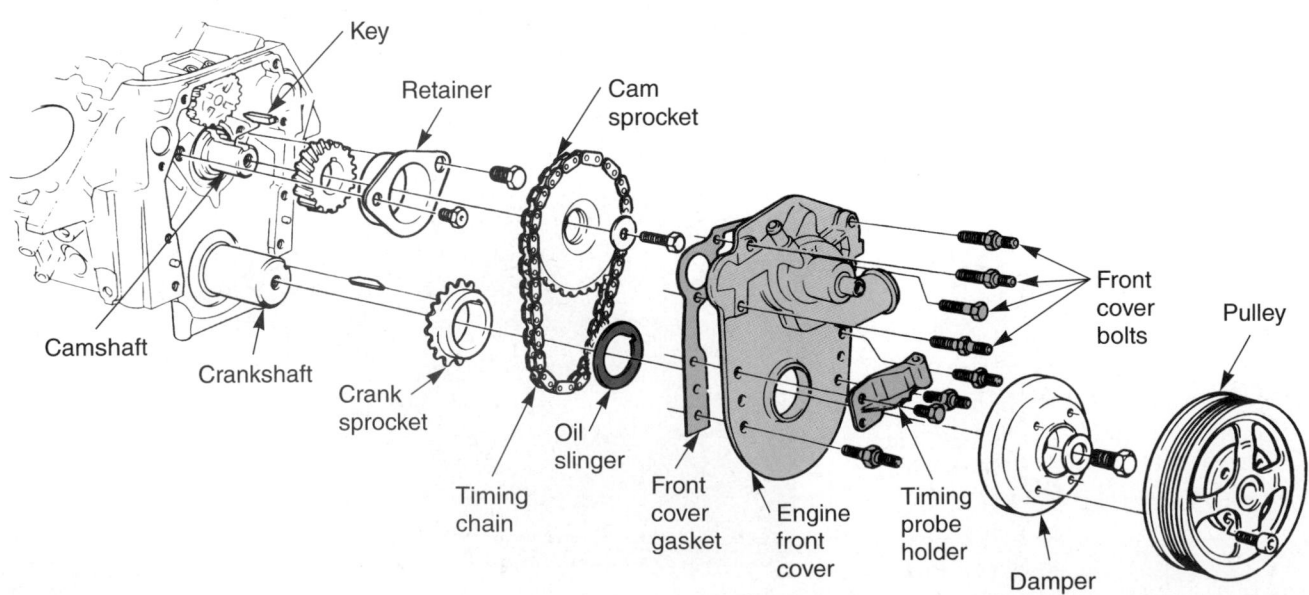

Figure 15-8. Study the construction of the front cover and related parts. In particular, note the oil slinger that installs in front of the crank sprocket. (Buick)

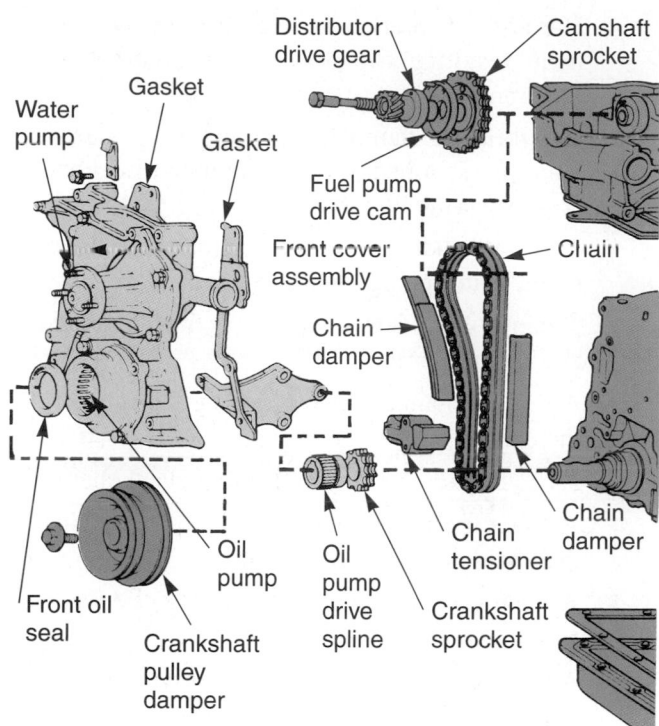

Figure 15-9. This front cover houses the engine oil pump and water pump. Drive splines in front of the crank sprocket power the oil pump. A V-belt drives the water pump. (Honda)

force then throws the oil outward and onto the timing chain or gears. The oil slinger not only helps lubricate the timing mechanism, but it blocks the oil from striking the front seal, preventing leakage.

Timing Belt and Sprockets

A timing belt drive mechanism basically consists of a crank sprocket, cam sprocket, timing belt, and belt tensioner. It is used to drive the camshaft on OHC engines. Look at **Figure 15-10.**

A *timing belt* provides a very smooth and accurate method of turning the camshaft. The timing belt and the sprockets have cogged, or square, teeth that prevent belt slippage. Some timing belts are made of fiberglass-reinforced nitrile rubber for increased strength and durability. Some are designed for a service life of over 100,000 miles (160 000 km) if not damaged by road debris or oil.

The *timing belt sprockets* are usually made of cast iron or aluminum, **Figure 15-10.** The crank sprocket is keyed to the crankshaft snout. The cam sprocket normally bolts to the front of the camshaft. A dowel pin may be used to position the cam sprocket correctly. Belt sprockets have timing marks, just like timing gears and chain sprockets. The marks must be aligned with specific points on the engine. This will properly time the opening of the engine valves.

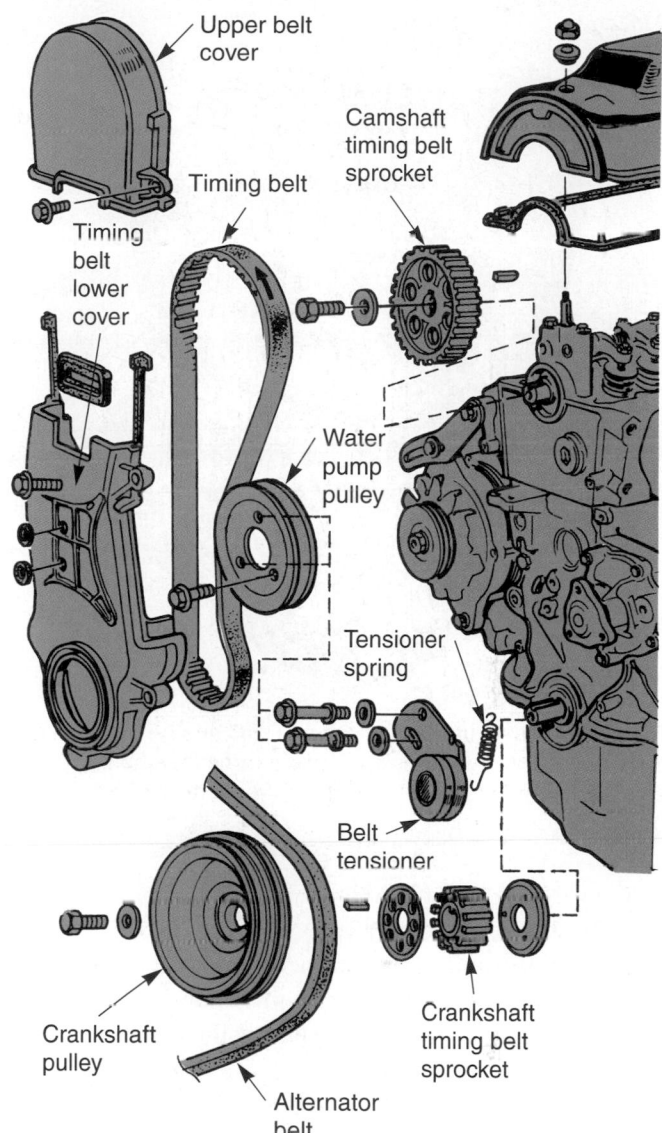

Figure 15-10. Study the parts that install on the front of this OHC engine. Cogged sprockets are used with the cam timing belt. A V-belt operates in the crank pulley. A belt cover surrounds the timing belt and sprockets. (Honda)

A *belt tensioner* is a spring-loaded wheel that keeps the timing belt firmly seated on its sprockets. Illustrated in **Figure 15-11,** the tensioner pushes inward on the back of the belt. This prevents the belt teeth from slipping on the sprocket teeth. The tensioner wheel is mounted on an antifriction bearing. This bearing is filled with grease and permanently sealed at the factory.

Some belt tensioners use both spring tension and hydraulic pressure to maintain belt tightness on its sprockets. The spring tensioner keeps the belt tight when the engine is shut off. The hydraulic tensioner adjusts the belt tension with engine speed. At higher rpms, belt tension is increased to keep the belt from slipping or flying off.

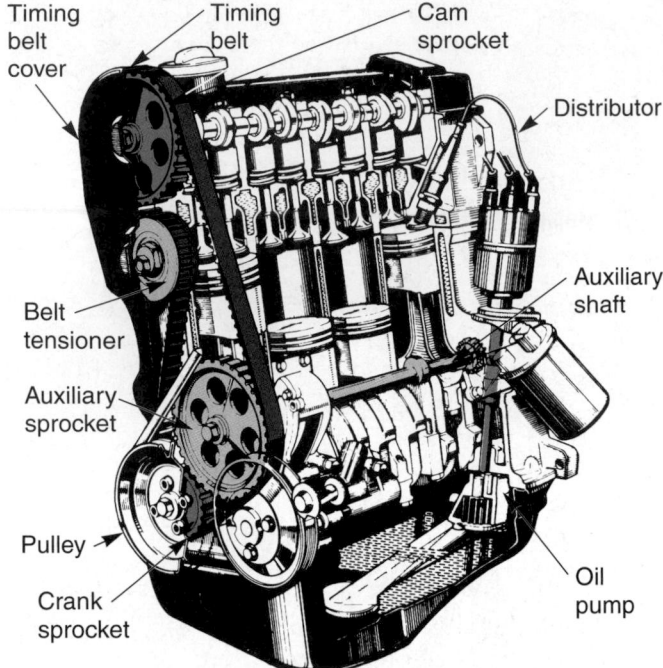

Figure 15-11. A cutaway view of an OHC engine using an auxiliary drive shaft. An auxiliary sprocket turns the shaft. The shaft also has a small gear for operating the ignition distributor and oil pump. (Chrysler)

The timing belt may also be used to drive the oil pump, diesel injection pump, or ignition distributor. **Figure 15-11** shows an engine in which the timing belt drives an auxiliary shaft. The shaft is then used to operate the distributor and oil pump.

Timing Belt Cover

A timing belt cover simply protects the belt from damage and the technician from injury. It is made from sheet metal or plastic. See **Figures 15-10** and **15-11**. A gasket may seal the mating surfaces between the timing belt cover and the block. Timing belt covers are tightly sealed at the bottom to keep road debris and water off the rubber belt. If ice forms on the belt sprocket, it can cause the belt to jump off its sprockets, upsetting valve timing. This can lead to severe engine damage.

Tech Tip
A timing belt cover should not be confused with an engine front cover. A timing belt cover does *not* contain oil or an oil seal.

Summary

- The typical engine front end assembly consists of the parts that attach to the front of the engine.

- Harmonic vibration is a high-frequency movement resulting from twisting and untwisting of the crankshaft.
- A vibration damper, also called a harmonic balancer, is a heavy wheel mounted in rubber to control harmonic vibration.
- The crankshaft pulley operates belts for the alternator, water pump, and other units.
- Timing gears are two gears that operate the engine camshaft.
- A timing chain and sprockets can be used to turn the camshaft.
- A chain guide may be needed to prevent chain slap.
- A chain tensioner may be used to take up slack as the chain and sprockets wear.
- A front oil seal prevents oil leakage between the crankshaft and cover.
- A timing belt mechanism basically consists of a crank sprocket, cam sprocket, timing belt, and a belt tensioner.

Important Terms

Engine front end	Chain guide
Harmonic vibration	Chain slap
Vibration damper	Chain tensioner
Harmonic balancer	Auxiliary chain
Inertia ring	Engine front cover
Dual-mass harmonic	Timing cover
balancer	Front oil seal
Crankshaft pulley	Oil slinger
Camshaft drive	Timing belt drive
Timing gears	mechanism
Timing marks	Timing belt
Timing chain	Timing belt sprockets
Sprockets	Belt tensioner
Crank sprocket	

Review Questions—Chapter 15

Please do not write in this text. Place your answers on a separate sheet of paper.

1. Define the term "crankshaft harmonic vibration."
2. A(n) _____ _____ is a heavy wheel mounted in rubber to control harmonic vibration.
3. The _____ _____ operates belts for the alternator, water pump, and other units.
4. A camshaft drive must turn the camshaft at _____ crankshaft speed.

5. Which of the following does *not* refer to timing gears?
   (A) Used on heavy-duty applications.
   (B) Dependable and long lasting.
   (C) Very quiet.
   (D) All of the above.

6. The most common camshaft drive setup for cam-in-block engines consists of a(n) _____ _____ and set of _____ .

7. Why is a chain guide used?

8. Explain the function of a chain tensioner.

9. What does an auxiliary chain commonly drive?

10. Is an engine front cover the same as a timing belt cover?

ASE-Type Questions

1. High-frequency movement resulting from the twisting and untwisting of the crankshaft is called:
   (A) engine knock.
   (B) after-running.
   (C) oscillating vibration.
   (D) harmonic vibration.

2. Which of the following is used to control the high-frequency movement that results from crankshaft twisting?
   (A) Vibration damper.
   (B) Crankshaft pulley.
   (C) Harmonic balancer.
   (D) Both A and C.

3. Which of the following normally operates belts for the alternator, water pump, and other units.
   (A) Auxiliary chain.
   (B) Timing sprockets.
   (C) Vibration damper.
   (D) Crankshaft pulley.

4. All of the following are basic types of camshaft drives *except:*
   (A) rod drive.
   (B) belt drive.
   (C) gear drive.
   (D) chain drive.

5. A camshaft drive must turn the camshaft at:
   (A) twice crankshaft speed.
   (B) crankshaft speed.
   (C) one-half crankshaft speed.
   (D) None of the above.

6. The two gears that operate the camshaft are the:
   (A) helical and spur gears.
   (B) crank and camshaft gears.
   (C) countershaft and crank gears.
   (D) cam sprocket and pulley gears.

7. Which of the following transfers power from the crank sprocket to the cam sprocket?
   (A) Spur gear.
   (B) Timing belt.
   (C) Timing chain.
   (D) Timing gears.

8. Which of the following devices is used to prevent chain slap?
   (A) Chain guide.
   (B) Auxiliary chain.
   (C) Timing sprocket.
   (D) Chain link.

9. A timing belt mechanism contains each of the following *except:*
   (A) timing belt.
   (B) belt damper.
   (C) cam sprocket.
   (D) belt tensioner.

10. The washer-shaped part that fits directly in front of the crankshaft sprocket is the:
    (A) pulley.
    (B) damper.
    (C) retainer.
    (D) oil slinger.

Activities for Chapter 15

1. On an engine with the timing cover removed, locate the timing marks on the crankshaft and camshaft pulleys, gears, or sprockets. If the engine can be rotated by hand, determine whether the timing marks align properly.

2. Sketch a front-end view of an engine with a timing chain or timing belt, showing the belt or chain path and labeling the sprockets that it engages.

Chapter 16

Engine Size and Performance Measurements

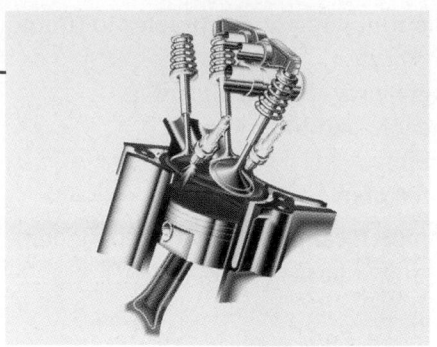

After studying this chapter, you will be able to:

- ■ Describe engine size measurements based on bore, stroke, displacement, and number of cylinders.
- ■ Explain engine compression ratio and how it affects engine performance.
- ■ Explain engine torque and horsepower ratings.
- ■ Describe the different methods used to measure and rate engine performance.
- ■ Explain volumetric efficiency, thermal efficiency, mechanical efficiency, and total engine efficiency.
- ■ Follow safe practices when making engine performance measurements.
- ■ Correctly answer ASE certification test questions on engine size and performance measurements.

Engine size and performance measurements are important to the technician. Shop manuals list many engine size and performance values for specific engines. You must be able to understand this information to properly communicate with others.

Engine Size Measurement

Engine size is determined by the number of cylinders, the cylinder diameter, and the amount of piston travel per stroke. Any of these variables can be changed to alter engine size. Engine size information is used when ordering parts and when measuring wear during major engine repairs.

Bore and Stroke

Cylinder bore is the diameter of the engine cylinder. See **Figure 16-1.** It is measured across the cylinder, parallel with the top of the block. Cylinder bores vary in size from 3–4″ (75–100mm).

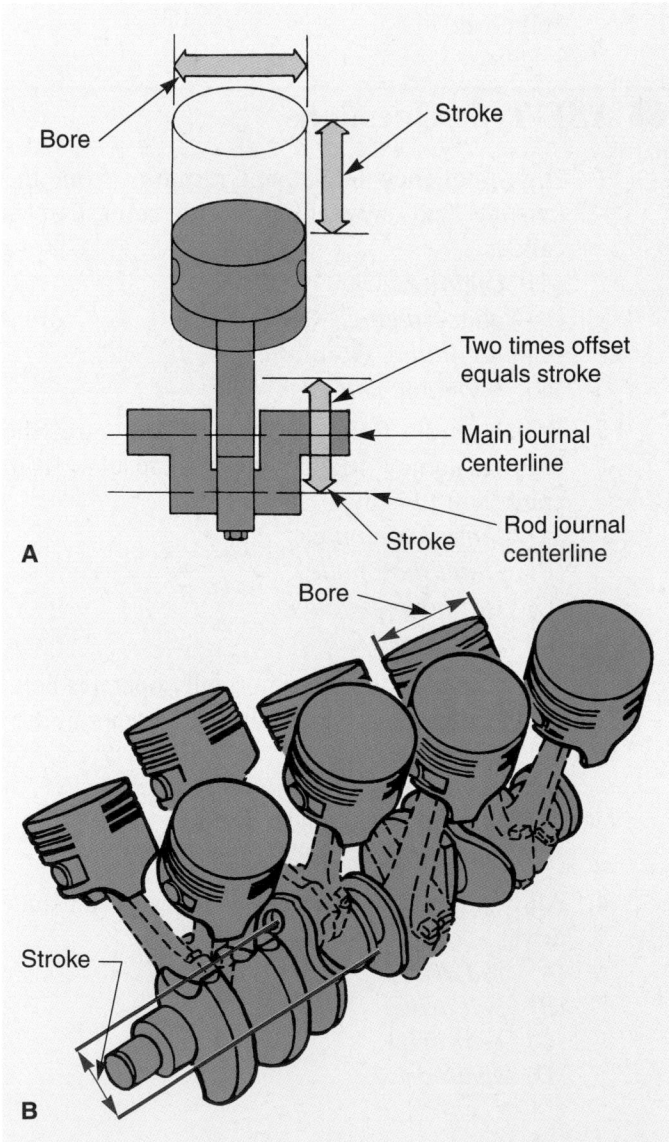

Figure 16-1. Bore and stroke measurements. A—Cylinder bore is measured across the cylinder, parallel with the deck of the engine block. Piston stroke is the distance the piston moves from BDC to TDC. Piston displacement is the amount of volume the piston moves in one upward stroke. B—Engine displacement is the displacement for all the pistons. (Ford)

Piston stroke is the distance the piston moves from top dead center (TDC) to bottom dead center (BDC), **Figure 16-1.** The amount of offset built into the crank journal (throw) controls the piston stroke. The stroke also varies from about 3–4″ (75–100mm).

A shop manual normally gives bore and stroke specifications. For instance, suppose a specification for bore and stroke is given as 4.00″ × 3.00″. This means that the engine cylinder is 4″ in diameter and the piston stroke is 3″. Bore is always the first value given and stroke is the second.

Generally, a larger bore and a longer stroke make an engine more powerful. It can pull in more fuel and air on each intake stroke. As a result, more pressure is exerted on the head of the piston during the power stroke.

Piston Displacement

Piston displacement is the volume the piston displaces (moves) as it travels from BDC to TDC. It is found by comparing cylinder diameter and piston stroke. A large cylinder diameter and large piston stroke produce a large piston displacement.

Piston displacement is calculated using a simple formula:

$$\text{piston displacement} = \frac{\text{bore squared} \times 3.14 \times \text{stroke}}{4}$$

Example: If an engine has a bore of 4″ and a stroke of 3″, what is its piston displacement?

Solution: $\text{piston displacement} = \dfrac{(4''^2) \times 3.14 \times 3''}{4}$

$$= 37.68 \text{ cu. in.}$$

Engine Displacement

Engine displacement, or *engine size,* is the volume displaced by all the pistons in an engine (piston displacement multiplied by the number of engine cylinders). For example, if one piston displaces 25 cu. in. and the engine has four cylinders, the engine displacement would be 100 cu. in. (25 × 4 = 100).

Cubic inch displacement (CID), cubic centimeters (cc), and *liters (L)* are units of engine displacement. For example, a V-8 engine might have a 350 CID. A V-6 could be a 3.3 L engine. A four cylinder engine might have a displacement of 2300 cc. Since one liter equals 1000 cc, a 2 liter engine would have 2000 cc.

Engine displacement is usually matched to the weight of the car. A heavier car, truck, or van needs a larger engine, which produces more power. A light, economy car only needs a small, low-power engine for adequate acceleration.

Force, Work, and Power

Force is a pushing or pulling action. When a spring is compressed, an outward movement and force is produced. Force is measured in pounds or newtons.

Work is done when force causes movement. If a compressed spring moves another engine part, work has been done. If the spring does not cause movement, no work has been done. Work is measured in foot-pounds, or joules. The formula for work is:

work = distance moved × force applied

Example: If you use a hoist to lift a 400 lb engine 3′ in the air, how much work has been done?

Solution: work = 3′ × 400 lb
$$= 1200 \text{ foot pounds (ft lb)}$$

Power is the rate, or speed, at which work is done. It is measured in foot-pounds per second or per minute. The metric unit for power is the watt (kilowatt).

High power output can do a large amount of work. Lower power can only do a small amount of work. The formula for power is:

$$\text{power} = \frac{\text{distance (feet)} \times \text{force (pounds)}}{\text{time (minutes)}}$$

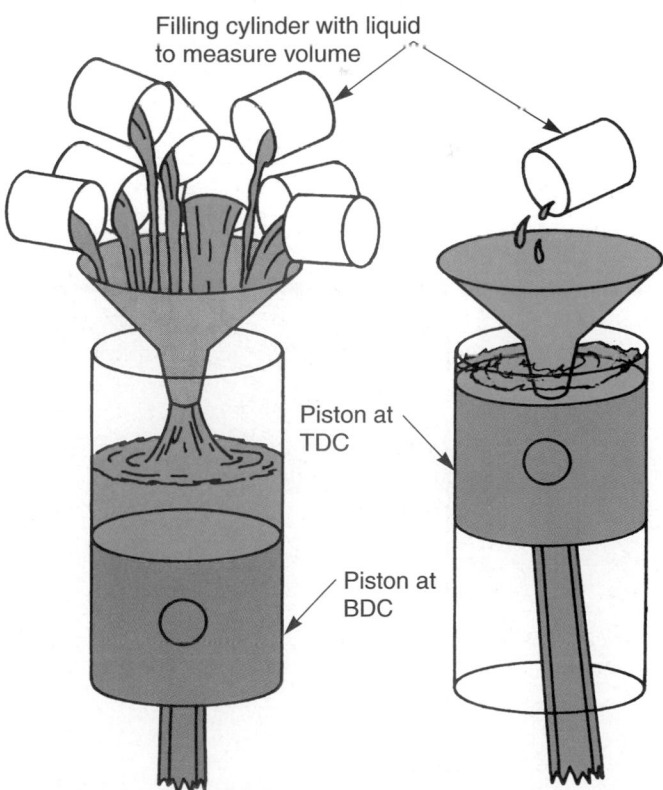

Filling cylinder with liquid to measure volume

Piston at TDC

Piston at BDC

Figure 16-2. Compression ratio is a comparison of cylinder volumes with the piston at TDC and BDC. This engine has eight times the volume at BDC, producing an 8:1 compression ratio.

Example: If an engine moves a 3000 lb car 1000′ in one minute, how much power is needed?

Solution: power = $\dfrac{1000 \text{ lb} \times 3000'}{1 \text{ min.}}$

$= \dfrac{3{,}000{,}000 \text{ ft lb}}{\text{min.}}$

Compression Ratio

Engine compression ratio compares the cylinder volume with the piston at TDC to the cylinder volume with the piston at BDC. Look at **Figure 16-2.** An engine's compression ratio illustrates how much the air-fuel mixture is compressed on the compression stroke.

A compression ratio is given as two numbers. For example, an engine may have a compression ratio of 9:1 (9 to 1). This means the maximum cylinder volume is nine times as large as the minimum cylinder volume. At BDC a cylinder has maximum volume. Minimum cylinder volume occurs at TDC.

Figure 16-3 illustrates two examples of compression ratios. When the gasoline engine piston is at BDC, the

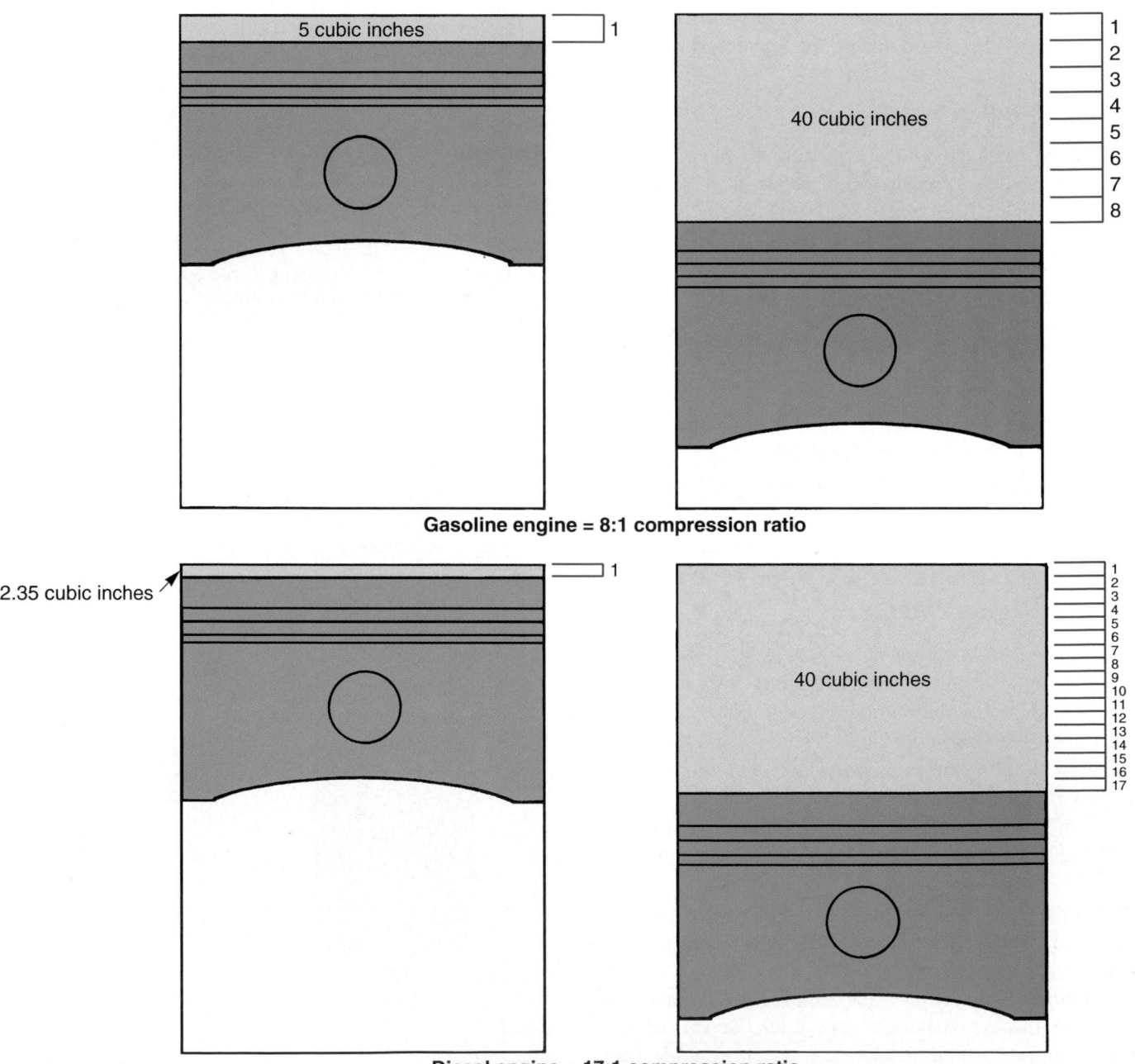

Figure 16-3. A diesel engine has a much higher compression ratio than a gasoline engine. A diesel must squeeze the fuel mixture very tight to cause combustion.

cylinder volume is 40 cu. in. (0.65 L). When the piston slides to TDC, the volume is reduced to 5 cu. in. (0.08 L). Dividing 40 by 5 (0.65 by 0.08), the compression ratio for this engine would be 8:1.

In a diesel (compression ignition) engine, BDC cylinder volume is 17 times as large as the TDC cylinder volume. The compression ratio is 17:1.

Older engines designed for leaded gasoline had higher compression ratios. A high compression ratio increases engine fuel efficiency and power. However, it also causes higher exhaust emission levels.

Today's gasoline engines use a lower compression ratio (8:1 or 9:1). This allows the use of cleaner-burning, unleaded fuel. There is a slight reduction in engine power and efficiency, however.

Diesel engines have a very high compression ratio (17:1 to 25:1, typically). The piston must compress the air in the cylinder enough for it to heat up and ignite the diesel fuel. Automotive fuels and combustion will be explained in detail in the next chapter.

Compression Pressure

Compression pressure is the amount of pressure in the cylinder on the compression stroke. Compression pressure is normally measured in pounds per square inch (psi) or kilopascals (kPa).

A gasoline engine has compression pressure from 130–180 psi (900–1200 kPa). A diesel engine has a much *higher* compression pressure of about 250–400 psi (1700–2800 kPa).

A *compression gauge* is used to measure compression pressure. The gauge is screwed into the spark plug, injector nozzle, or glow plug hole. The ignition or injection system (whichever contains the removed part) is disabled. Then, the engine is cranked over. The gauge measures the compression pressure.

Compression pressure is an indicator of engine condition. If it is low, something is allowing air to leak out of the cylinder. The engine may have bad rings, burned valves, or a blown head gasket.

Engine Torque

Torque is a turning, or twisting, force. When you turn a steering wheel, you have applied torque to the steering wheel.

Engine torque is a rating of the turning force at the engine crankshaft. When combustion pressure pushes the piston down, a strong rotating force is applied to the crankshaft. This turning force is sent to the transmission or transaxle, to drive shaft or drive axles, and to wheels, propelling the car.

Engine torque specifications are given in a shop manual. An example of torque is 78 ft lb @ 3000 rpm. This engine would be capable of producing a maximum of 78 ft lb of torque at an operating speed of 3000 revolutions per minute. In metrics, engine torque is often stated in newton-meters (N•m).

Horsepower

Horsepower (hp) is a measure of an engine's ability to perform work (power). At one time, one horsepower was the approximate strength of a horse, **Figure 16-4.** A 300 hp engine could theoretically do the work of 300 horses.

One horsepower equals 33,000 ft lb of work per minute. To find engine horsepower, use the following formula:

$$hp = \frac{work\ (ft\ lb)}{33,000}$$

or

$$hp = \frac{distance\ (ft) \times weight\ (lb)}{33,000}$$

Example: For a small engine to lift 500 lb a distance of 700′ in one minute, about how much horsepower would be needed?

Solution: $hp = \dfrac{500\ lb \times 700'}{33,000}$

$$= 10.6\ hp$$

Factory Horsepower Ratings

Automobile makers rate engine power at a specific engine speed. For instance, a high performance, turbocharged engine might be rated at 300 hp @ 5000 rpm. This engine power rating is normally stated in a service manual. There are several different methods of calculating engine horsepower.

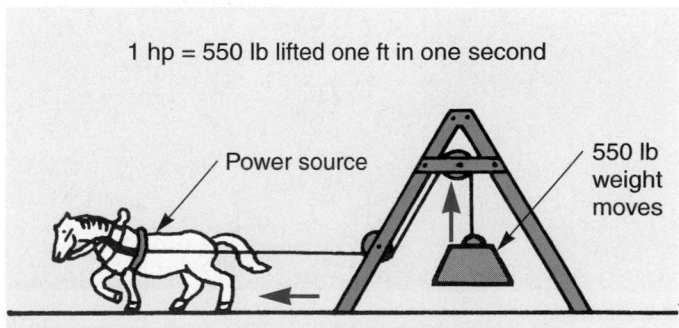

Figure 16-4. This represents one horsepower. In automotive work, one horsepower equals 33,000 lb moved one foot in one minute.

Brake horsepower (bhp) measures the usable power at the engine crankshaft. Shown in **Figure 16-5,** a prony brake was first used to measure brake horsepower. The engine turns the prony brake when the braking mechanism is applied, producing a force on the scale. The resulting amount of pointer deflection is then used to find brake horsepower.

An *engine dynamometer (dyno)* is used to measure the brake horsepower of modern car engines. Refer to **Figure 16-6.** It functions in much the same way as a prony brake. Either an electric motor or a fluid coupling is used to place a drag on the engine crankshaft. Then, power output can be determined.

A *chassis dynamometer* measures the horsepower delivered to the drive wheels. See **Figure 16-7.** It indicates the amount of horsepower available to propel the car.

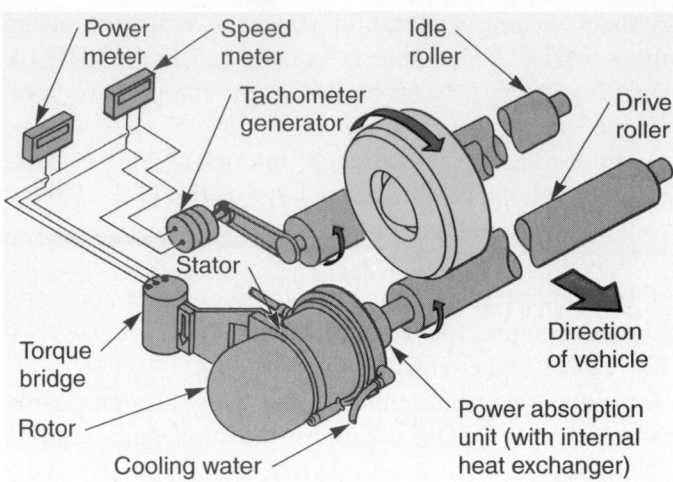

Figure 16-7. A chassis dynamometer measures turning power at the drive wheels. This accounts for any power consumed by the drive train. (Clayton)

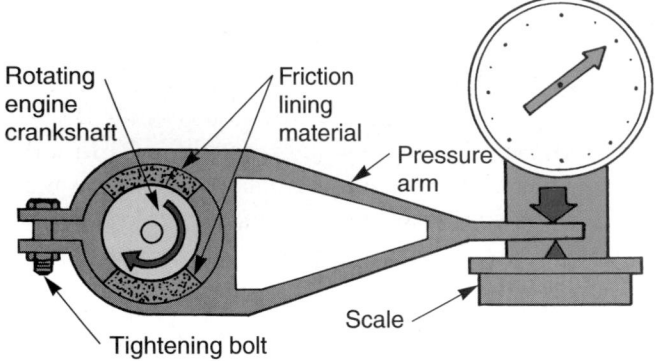

Figure 16-5. A prony brake measures engine brake horsepower. A brake is applied to the engine crankshaft. The amount of needle deflection can be used to find horsepower.

Indicated horsepower (ihp) refers to the amount of power formed in the engine combustion chambers. A special pressure-sensing device is placed in the cylinder. The pressure readings are used to determine the indicated horsepower.

Frictional horsepower (fhp) is the power needed to overcome engine friction. It is a measure of the resistance to movement between engine parts, or power lost to friction. It reduces the amount of power left to propel the car.

Net horsepower, or SAE net horsepower, is the maximum power developed when an engine is loaded by all accessories (alternator, water pump, fuel pump, air injection pump, air conditioning, and power steering pump). Net horsepower indicates the amount of power available to move the car. See **Figure 16-8.**

Figure 16-6. A technician is using an engine dynamometer to measure engine performance. The dyno loads the engine to simulate driving conditions while monitoring horsepower output, fuel efficiency, emissions, and many other engine functions. (MSD Ignition Systems)

Figure 16-8. Net horsepower is the available horsepower with the engine operating all accessories. (Chevrolet)

Gross horsepower (ghp) is similar to net horsepower, but it is the engine power available with only basic accessories installed (alternator, fuel pump, and water pump). Gross horsepower does *not* include the power lost to the power steering pump, air injection pump, air-conditioning compressor, or other extra units. Look at **Figure 16-9.**

Taxable horsepower is simply a general rating of engine size. In many states, it is used to find the tax placed on a car. The formula for taxable horsepower (thp) is:

thp = bore squared × number of cylinders × 0.4

Engine Efficiency

Engine efficiency is the ratio of usable power at the engine crankshaft (brake horsepower) to the power supplied to the engine (heat content of fuel). By comparing fuel consumption to engine power output, you can find engine efficiency.

If all the heat energy in the fuel were converted into useful power, the engine would be 100% efficient. Modern piston engines are only about 20% efficient.

Figure 16-10 illustrates how the heat energy of the fuel is used by a piston engine. About 70% of the fuel's energy is used by the cooling and exhaust systems. This leaves a small portion to drive the pistons.

Volumetric Efficiency

Volumetric efficiency is the ratio of air drawn into the cylinder and the maximum possible amount of air that

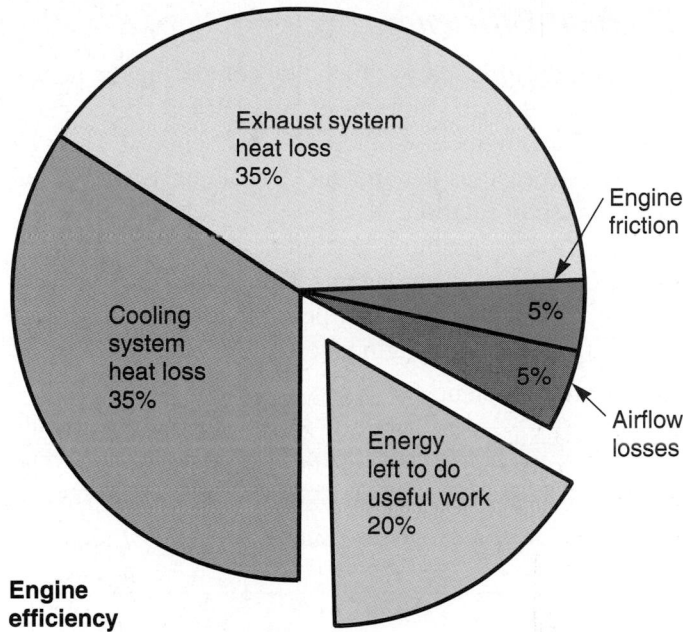

Engine efficiency

Figure 16-10. Pie chart shows how fuel's heat energy is used by a piston engine. Note that most of the heat energy is wasted. Aerodynamic body designs increase vehicle fuel economy. The vehicle must pass through air, which is actually a fluid by nature. The aerodynamic design reduces drag. (Buick)

could enter the cylinder. It refers to how well an engine can "breathe" on its intake stroke.

If volumetric efficiency was 100%, the cylinder would completely fill with air on the intake stroke. Engines are capable of only 80–90% volumetric efficiency. Restrictions in the ports and around the valves limit airflow.

High volumetric efficiency increases engine power because more fuel and air can be burned in the combustion chambers. The formula for volumetric efficiency is:

$$\text{volumetric efficiency} = \frac{\text{actual volume of air taken into cylinder}}{\text{volume of cylinder}}$$

Mechanical Efficiency

Mechanical efficiency compares brake horsepower and indicated horsepower. It is a measurement of mechanical friction. Indicated horsepower is the theoretical power produced by combustion. Brake horsepower is the actual power at the engine crankshaft. The difference between the two is due to friction losses.

Mechanical efficiency of 70–80% is normal. This means that 20–30% of the engine's power is lost to friction (frictional hp loss). The friction between the piston rings and cylinder walls accounts for most of this loss.

Figure 16-9. Gross horsepower is similar to net horsepower, but it does not include power lost to unneeded accessories. (Chevrolet)

Thermal Efficiency

Thermal efficiency is found by comparing the horsepower output to the amount of fuel burned. It indicates how well an engine uses the fuel's heat energy. Thermal efficiency measures the amount of heat energy converted into crankshaft rotation.

A gallon of gasoline has about 19,000 Btu (British thermal units) of heat energy. One horsepower equals about 42.4 Btu of heat energy per minute. With this information, you can find engine thermal efficiency:

$$\text{thermal efficiency} = \frac{\text{bhp}}{450 \times \text{gallons of fuel per minute}}$$

Generally, engine thermal efficiency is 20–30%. The rest of the heat energy is absorbed by the metal parts of the engine or blown out the exhaust.

Summary

- Engine size is determined by cylinder diameter, amount of piston travel on each stroke, and number of cylinders.
- Cylinder bore is the diameter of the engine cylinder.
- Piston stroke is the distance the piston moves from TDC to BDC.
- Piston displacement is the volume the piston displaces (moves) from BDC to TDC.
- Engine displacement or engine size equals piston displacement times the number of engine cylinders.
- Force is a pushing or pulling action.
- Work is done when force causes movement.
- Power is the rate or speed at which work is done.
- Engine compression ratio compares cylinder volumes with the piston at TDC and BDC.
- Compression pressure is the amount of pressure produced in the engine cylinder on the compression stroke.
- Torque is a turning or twisting force.
- Engine torque is a rating of the turning force at the engine crankshaft.
- Horsepower (hp) is a measure of an engine's ability to perform work.
- Brake horsepower (bhp) measures the usable power at the engine crankshaft.
- An engine dynamometer (dyno) is used to measure the brake horsepower of modern car engines.
- A chassis dynamometer measures the horsepower delivered to the drive wheels.
- Engine efficiency is the ratio of power produced by the engine (brake horsepower) and the power supplied to the engine (heat content of fuel).
- Volumetric efficiency is the ratio of actual air drawn into the cylinder and the maximum possible amount of air that could enter the cylinder.
- Mechanical efficiency compares brake and indicated horsepowers. It is a measurement of mechanical friction.
- Thermal efficiency is heat efficiency found by comparing fuel burned and horsepower output.

Important Terms

Cylinder bore	Engine torque
Piston stroke	Brake horsepower (bhp)
Piston displacement	Engine dynamometer
Engine displacement	(dyno)
Engine size	Chassis dynamometer
Cubic inch displacement	Indicated horsepower
(CID)	(ihp)
Cubic centimeters (cc)	Frictional horsepower
Liters (L)	(fhp)
Force	Net horsepower
Work	Gross horsepower (ghp)
Power	Taxable horsepower
Engine compression ratio	Engine efficiency
Compression pressure	Volumetric efficiency
Compression gauge	Mechanical efficiency
Torque	Thermal efficiency

Review Questions—Chapter 16

Please do not write in this text. Place your answers on a separate sheet of paper.

1. What three factors determine engine size?
2. Cylinder bore is measured across the cylinder, parallel with the top of the block. True or False?
3. Piston stroke is the distance the piston moves during a complete four-stroke cycle. True or False?
4. _____ _____ is the volume the piston moves from BDC to TDC.
5. If an engine has a bore of 3.5″ and a stroke of 3″, what is the piston displacement?
6. Define the term "engine displacement."
7. Explain the difference between force, work, and power.

8. When a gasoline engine's piston is at BDC, the cylinder volume is 45 cu. in. When the piston slides to TDC, the cylinder volume is 9 cu. in. What is the compression ratio for this engine?
   (A) 8:1.
   (B) 5:1.
   (C) 7:1.
   (D) 9:1.

9. Which of the following would *not* be a compression ratio for a car engine?
   (A) 2:1.
   (B) 8:1.
   (C) 17:1.
   (D) All of the above.

10. A gasoline engine may produce a compression pressure of _____ to _____ psi (_____ to _____ kPa).

11. A diesel engine can produce a compression pressure of about _____ to _____ psi (_____ to _____ kPa).

12. What is engine torque?

13. Explain the term "horsepower."

14. The ratio of actual air drawn into an engine to the maximum possible air that could enter the engine is called _____ _____.

15. A piston engine's thermal efficiency is approximately _____.
   (A) 80-100%
   (B) 70-80%
   (C) 40-50%
   (D) 20-30%

⚙ ASE-Type Questions

1. Engine size is determined by:
   (A) cylinder diameter.
   (B) number of cylinders.
   (C) piston travel per stroke.
   (D) All of the above.

2. The distance a piston moves from TDC to BDC is called:
   (A) piston taper.
   (B) piston stroke.
   (C) piston clearance.
   (D) piston displacement.

3. Piston displacement times the number of engine cylinders equals:
   (A) engine size.
   (B) cylinder bore.
   (C) engine torque.
   (D) compression ratio.

4. Work can be measured in each of these *except:*
   (A) watts.
   (B) joules.
   (C) kilograms.
   (D) foot pounds.

5. Modern gasoline engines have a compression ratio of about:
   (A) 8 or 9:1.
   (B) 2 or 13:1.
   (C) 15 or 16:1.
   (D) 20 or 21:1.

6. Compression pressure is measured in:
   (A) kilopascals.
   (B) pounds per square inch.
   (C) Both A and B are correct.
   (D) Neither A nor B are correct.

7. Diesel engines have compression pressure of about:
   (A) 130 to 180 psi.
   (B) 250 to 400 psi.
   (C) 425 to 500 psi.
   (D) 510 to 525 psi.

8. One horsepower equals:
   (A) 28,000 kW of work per minute.
   (B) 25,000 ft lb of work per hour.
   (C) 30,000 in lb of work per second.
   (D) 33,000 ft lb of work per minute.

9. Which of the following is used to measure brake hp of an engine?
   (A) Brake tachometer.
   (B) Compression gauge.
   (C) Engine dynamometer.
   (D) Chassis dynamometer.

10. Which abbreviation indicates the amount of power or pressure formed in engine combustion chambers?
   (A) ihp.
   (B) fhp.
   (C) ghp.
   (D) thp.

Activities—Chapter 16

1. Use a compression gauge to measure the compression of each cylinder of an engine. Check your findings against the specifications for that engine. Are all the readings "in spec," or do one or more cylinders read high or low?

2. Convert your compression readings from pounds per square inch to kilopascals, or vice versa. Multiply psi readings by 6.895 to find kPa; divide kPa readings by 6.895 to find psi.

Computer Systems

Today's vehicles are as much electronic as they are mechanical. Computers control and monitor every major system in a modern car or light truck. It is virtually impossible to work on almost any vehicle without a basic knowledge of computers.

Section 3 details the theory and operation of computer systems, explains how to use a scan tool to find troubles in computerized systems, and describes the use of pinpoint tests to verify the basic electrical/electronic faults indicated by the scan tool.

This important section will prepare you for almost every other chapter in this textbook. Study it carefully! This section will also help you pass several ASE certification tests requiring a knowledge of computers, sensors, actuators, scan tools, and related topics, including Test A6, *Electrical/Electronic Systems,* and Test A8, *Engine Performance.*

Chapter 17

Computer System Fundamentals

After studying this chapter, you will be able to:

- Compare computer systems to the human body's nervous system.
- Describe the input, processing, and output sections of a basic computer system.
- Explain input sensor and output device classifications and operation.
- Summarize computer system signal classifications.
- Sketch a block diagram for a computer system.
- Summarize where computers, control modules, sensors, and actuators are typically located.
- Summarize the flow of data through a computer.
- Explain how a computer uses sensor inputs to determine correct outputs.
- Correctly answer ASE certification test questions that require a knowledge of automotive computer system fundamentals.

A *computer* is a complex electronic device that will produce programmed electrical output signals after receiving specific electrical input signals. Computers now monitor and control all major systems of a modern vehicle. In fact, today's cars and trucks are as much electronic as they are mechanical.

It is impossible to work on almost any system without having a basic knowledge of vehicle computers. In the past, automotive systems (ignition, cooling, emissions, and fuel, for example) worked independently of each other. Today, automotive systems work together in a tightly integrated package and share many of the same sensors and actuators. This makes it important for you to fully understand how automotive computer systems operate.

This chapter will give you a basic picture of automotive computer systems. It will also provide you with the background information needed to understand the material presented in the next two chapters—*On-board Diagnostics, Scan Tools* and *Computer System Service.* It will also help prepare you for almost all the following chapters in this book.

Note
Before starting this chapter, you may want to review Chapter 1, *The Automobile,* and Chapter 8, *Basic Electricity and Electronics.*

Cybernetics

The term *cybernetics* is often used to refer to the study of how electrical-mechanical devices can duplicate the action of the human body. Comparing the human body to a computer system is an easy way to explain this subject. Just as your brain communicates with and controls parts of your body, an automotive computer system can communicate with and control parts of a vehicle. See **Figure 17-1.**

The Nervous System (Computer Input)

The human nervous system uses chemical-electrical signals to control the body. If you touch a sharp needle, nerve cells in your finger "fire" and send a signal through a strand of nerve cells in your hand, up your arm, and into your brain. The strand of nerve cells forms a "wire" that connects your finger and brain. The nerve cells in the tip of your finger are comparable to a sensor, or input device, in a computer system.

The Brain (Computer Processing)

The brain uses billions of cells interconnected by linking cells called neurons. Each neuron chemically-electrically connects tens of thousands of neighboring cells. When the brain "thinks," minute electrical impulses

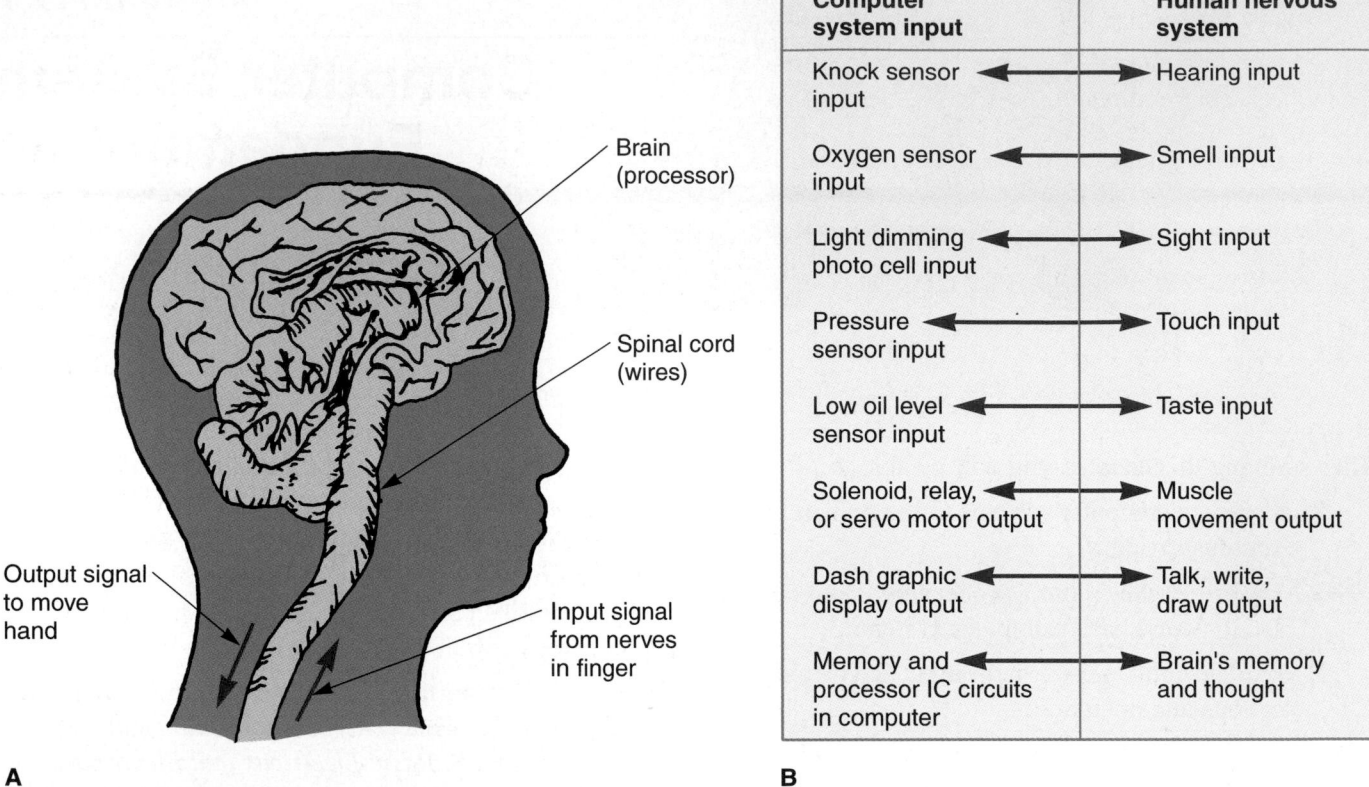

Figure 17-1. A—Cybernetics compares the human body to a computer system. B—Study this chart comparing the human nervous system and an automotive computer system.

travel from neuron to neuron. One simple thought might involve electrical impulses between hundreds of thousands of cells in a specific pattern. This is not unlike the electrical action inside the circuits of a computer.

The brain is comparable to a super powerful computer. It can process the inputs from the nervous system and determine what actions should be taken.

In the preceding example, the chemical-electrical signal of a sharp needle prick would tell your brain, "my finger is being injured." Your brain would then take corrective action to protect your finger.

The brain makes decisions much like computer chips produce logical outputs. The cells in the brain can be either chemically-electrically charged (on) or not be charged (off). By connecting all the brain cells, the brain can decide what to do in each situation.

The Reflex Action (Computer Output)

The needle prick signal activates specific brain cells, and a reflex output is produced. The brain sends a signal back into your arm. This chemical-electrical output signal stimulates the muscles in your arm to pull back, protecting your finger from the sharp object. The reflex

action of your muscles is similar to the action of an actuator, or output device, in a car's computer system.

Computer Advantages

There are several reasons that computers are being used in modern vehicles. Computers provide several advantages.

- Computer systems can compensate for mechanical wear. Also, they do not have as many mechanical parts to wear and go out of calibration.

- Computers are very fast and can alter outputs in milliseconds (thousandths of a second). This lets a computer alter outputs almost instantly as inputs and operating conditions change.

- Computers reduce fuel consumption and lower emissions by more precise control of fuel metering into the engine. The computer "sniffs" the exhaust gases to find out if too much or too little fuel is entering the engine.

- Computers can increase engine power by more accurate control of ignition timing, fuel injection, emission control system operation, etc.

- Computers reduce vehicle weight because they are much lighter than mechanical control mechanisms.
- Most computers have on-board diagnostics that can detect and record system problems. The on-board computer can produce and store an output code that tells the technician where a fault might be located.
- Computers can increase driver convenience by better control of the passenger compartment environment and dash displays.
- Computers can improve passenger safety by controlling the anti-lock brakes, air bags, suspension, and other systems.
- Computers can compensate for component wear and failure to keep the car driveable.

Digital Electronics

Digital electronics is a field of study dealing with the ways a computer uses on-off signals to produce "artificial intelligence."

A computer analyzes input signals from sensors. It has memory in which fixed and semi-fixed data is stored. The computer can make logical choices on how to control specific outputs by comparing the input signals to the data stored in memory.

Binary Numbering System

The **binary numbering system** uses only two numbers, zero and one, and is the key to how computers operate. The numbers zero (0) and one (1) can be arranged in different sequences to represent other numbers, letters, words, a computer input, a computer output, or a condition.

To use the binary system, a computer turns switches (transistors, for example) either on or off. *Off* would represent a zero (0) and *on* would represent one (1) in the binary system, **Figure 17-2.**

Note that a 0011 in binary would equal a three (3) in the base 10 decimal system. A 0110 in binary would equal a six (6) in the base 10 system, **Figure 17-3.**

In binary language, a zero or a one is called a *bit*. A pattern of four bits is a *nibble*. A pattern of eight bits (zeros or ones) is called a *byte,* (pronounced "bite."). One or more bytes are referred to as a *word*. You are more likely to hear this terminology when you work with home or personal computers. Look at **Figure 17-4.**

Gating Circuits

A *gate* is an electronic circuit that produces a specific output voltage for given input voltages. Just as a diode will pass current when forward biased (output lead would have voltage representing one) and stop current when

Decimal number	Binary number code 8 4 2 1	Binary to decimal conversion
0	0 0 0 0	= 0 + 0 + 0 + 0 = 0
1	0 0 0 1	= 0 + 0 + 0 + 1 = 1
2	0 0 1 0	= 0 + 0 + 2 + 0 = 2
3	0 0 1 1	= 0 + 0 + 2 + 1 = 3
4	0 1 0 0	= 0 + 4 + 0 + 0 = 4
5	0 1 0 1	= 0 + 4 + 0 + 1 = 5
6	0 1 1 0	= 0 + 4 + 2 + 0 = 6
7	0 1 1 1	= 0 + 4 + 2 + 1 = 7
8	1 0 0 0	= 8 + 0 + 0 + 0 = 8

Figure 17-3. Binary numbers can be converted into decimal (base ten) numbers. Note how the right-hand binary number equals one and the left-hand number equals eight.

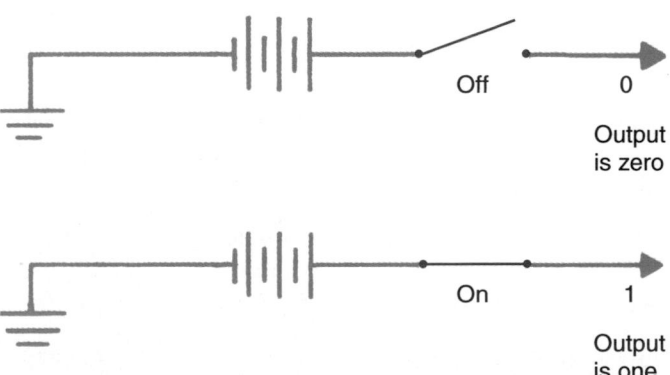

Figure 17-2. Since electronic components can be either on or off, the binary numbering system is ideal for digital logic and computer circuits. The binary system has only two numbers, zero and one, which represent *off* and *on* conditions.

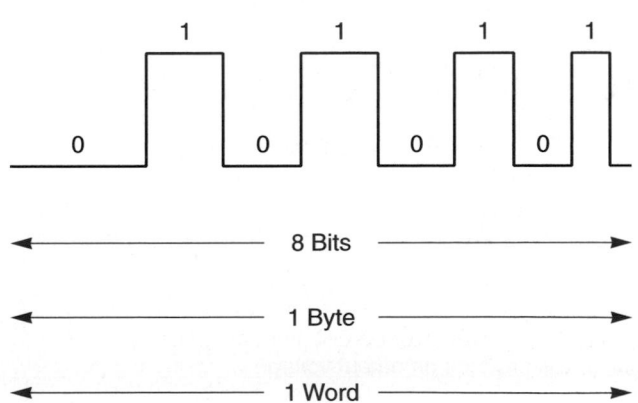

Figure 17-4. Each *one* (voltage on) or *zero* (voltage off) signal equals a bit of computer data. Eight bits equal a byte, or a word.

reverse biased (output would be zero or no voltage), gates have *programmed* (previously known) outputs. Shown in **Figure 17-5,** the most common computer gates are:

- *NOT gate*—also called an *inverter gate*, this gate will reverse its input. If the input has voltage applied (1), the output terminals will *not* have voltage (0) or vice versa. A NOT gate can be used to make other gates.

- *AND gate*—an AND gate requires voltage (1) at *both inputs* to produce a voltage (1) at the output. If pins A and B are *both one* (1), the output will be *1*. If only pin A is one, the output will be 0.

- *NAND gate*—is an inverted AND gate. Its output will be opposite that of an AND gate. Note the small circle or dot on the output lead of the gate. The small circle represents an inverter.

- *OR gate*—will produce an output (one or on) if *either* input is energized (1). A or B input voltage (1) will result in voltage (1) at the output lead.

- *NOR gate*—is an inverted OR gate. Note the small circle that represents an inverter. The output is inverted to produce an output opposite that of an OR gate.

Truth Tables

A truth table is a chart that shows what the output of a gate will be with different inputs. Look at the truth table for an OR gate shown in **Figure 17-6.** Note that if two switches are wired in parallel, either switch A or switch B (OR gate) will turn the light *on* (output 1). The output will be *on* (1) with A or B or both energized. Only when neither input is *on* will the output be *off* (0). A truth table graphically shows how a gate functions.

Gates are often called *logic devices* because they make logical decisions (outputs) based on specific inputs (facts). If an AND gate was compared with two switches

Not (inverter)

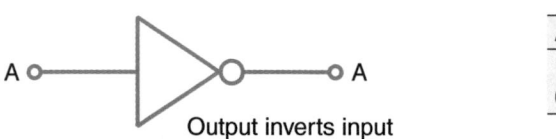

A	A
1	0
0	1

Output inverts input

A—Inverter simply reverses its input.

AND

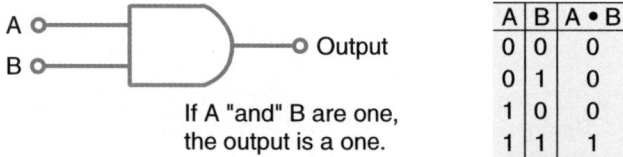

If A "and" B are one, the output is a one.

A	B	A•B
0	0	0
0	1	0
1	0	0
1	1	1

B—AND gate requires both inputs to be on for an output.

NAND

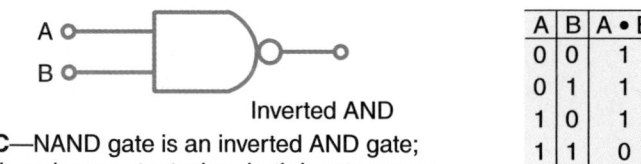

Inverted AND

A	B	A•B
0	0	1
0	1	1
1	0	1
1	1	0

C—NAND gate is an inverted AND gate; there is no output when both inputs are on.

OR

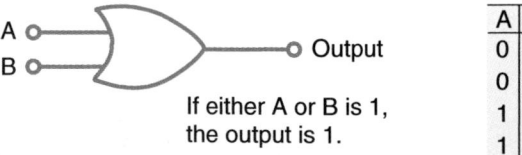

If either A or B is 1, the output is 1.

A	B	A+B
0	0	0
0	1	1
1	0	1
1	1	1

D—OR gate only needs input voltage on either terminal to get an output voltage.

NOR

A	B	A+B
0	0	1
0	1	0
1	0	0
1	1	0

Inverted OR

E—NOR gate is an inverted OR gate; no input voltage will produce an output voltage.

Figure 17-5. Study the basic types of computer gates. Truth tables on the right show what the output of each gate will be with different inputs.

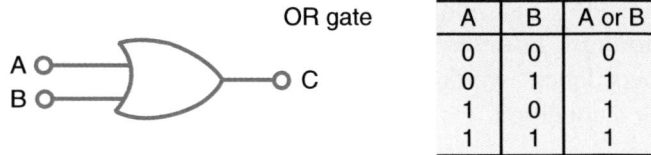

OR gate

A	B	A or B
0	0	0
0	1	1
1	0	1
1	1	1

A or B on = C on

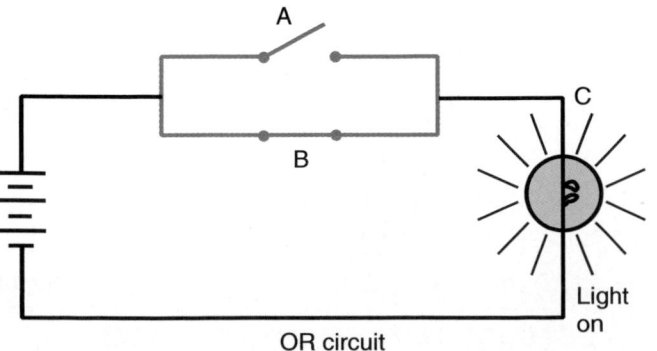

OR circuit

Figure 17-6. Compare the OR gate and the switch setup in this simple circuit. Either A or B will turn on the light and produce a one, or on, output.

wired in series, both switches (A and B) must be *on* to activate the starter motor, **Figure 17-7.** Compare these circuits to the action of the gates and the truth tables.

Using Gates

Logic gates can be connected together to form super-complex circuits. Millions of gates can be interconnected to produce thousands of preprogrammed outputs (decisions) from numerous inputs (facts). This is how a computer works, or thinks. Its circuitry and software are designed to make the correct output signals based on various input signals. In this way, a computer knows what to do to keep the vehicle operating efficiently with many different variables.

Integrated Circuits

Discussed briefly in Chapter 8, an *integrated circuit (IC)* is an electronic circuit that has been reduced in size and etched on the surface of a tiny semiconductor chip, **Figure 17-8. Figure 17-9** shows the basic construction of an integrated circuit. Note how different semiconductor substances are deposited on the silicone chip and then etched to produce resistors, diodes, and transistors.

Metal conductors on the top of the chip connect these various electronic components to form the circuit. Wire leads allow for input and output connections to and from the IC chip.

Computer Signals

A *computer signal* is a voltage variation over short periods of time. It is a specific arrangement of pulses or

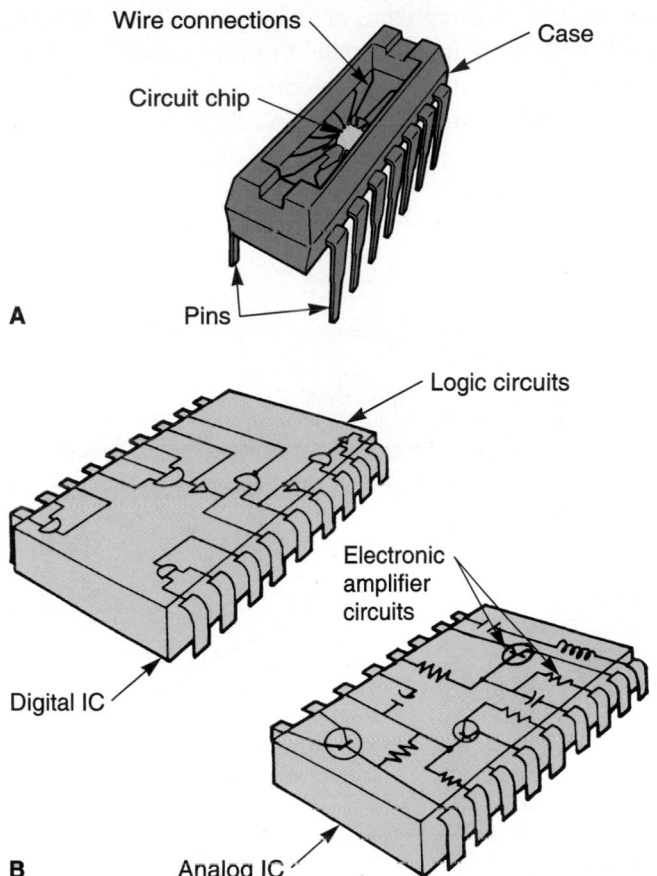

Figure 17-8. A—Note how a tiny chip is installed inside this plastic case. Tiny wires connect the chip to each metal pin. The pins then plug in or are soldered to other parts of the circuit. B—These are two broad classifications of integrated circuits. Digital circuit uses gates to produce logic circuits for computers. An analog IC is a small amplifier circuit for increasing output strength or altering output.

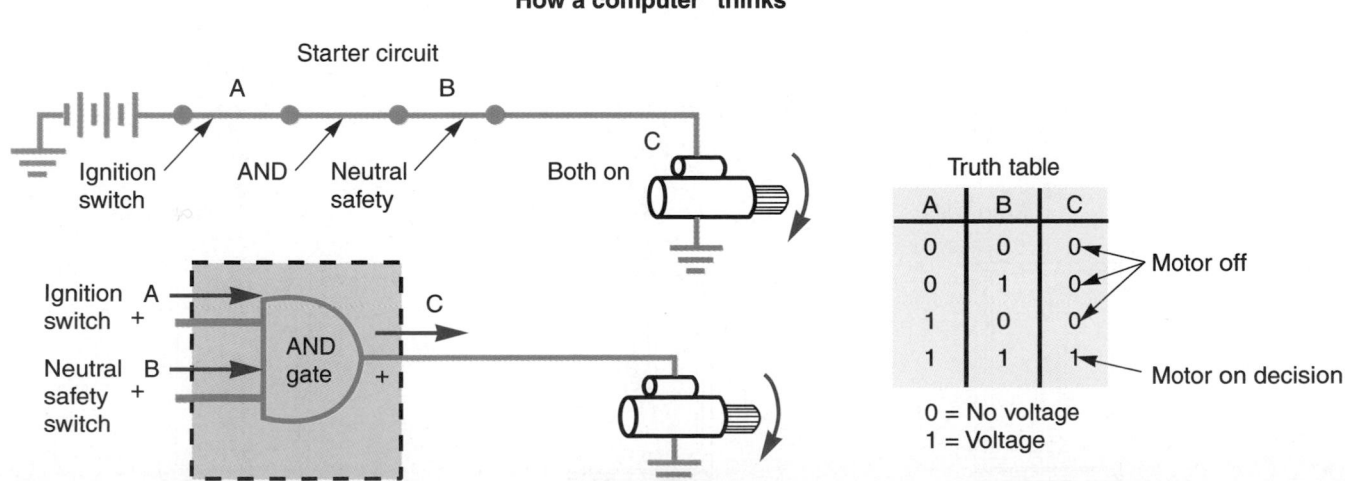

Figure 17-7. Compare this starting circuit with the computer gate circuit and truth table. It shows how a computer gate can make a decision. Two input conditions (A *and* B) must be satisfied in both circuits to produce an output to the starting motor. The same principle, only with thousands of gates and dozens of inputs and outputs, is used inside an automotive computer.

waves used to carry data, or information. You must be familiar with computer signals to understand the operation and service computer systems.

Computer signals can be digital or analog. **Figure 17-10** compares analog and digital signals.

As discussed earlier, *digital signals* are on-off signals, like those produced by a rapidly flipping switch. An example of a sensor providing a digital signal is the crankshaft sensor that shows engine rpm. Voltage output goes from maximum to minimum. A digital signal basically produces a square wave. See **Figure 17-11A.**

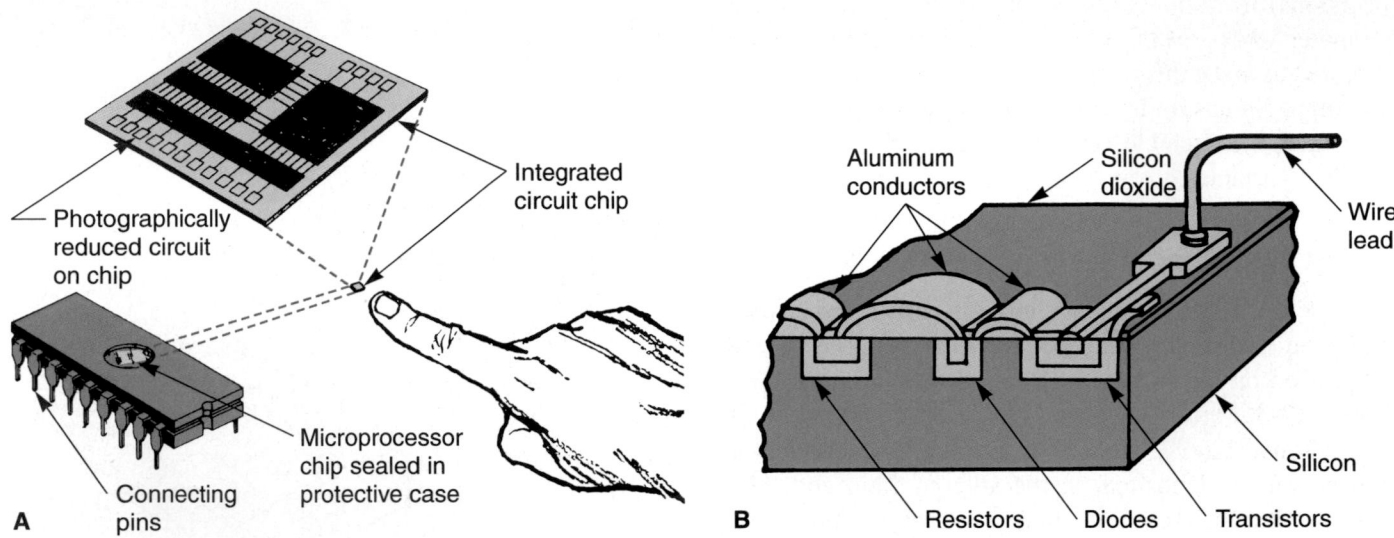

Figure 17-9. A—An integrated circuit has been photographically reduced in size, etched on a silicon chip, and placed inside a protective case. B—Note how components are deposited on doped silicone in an IC. (Ford)

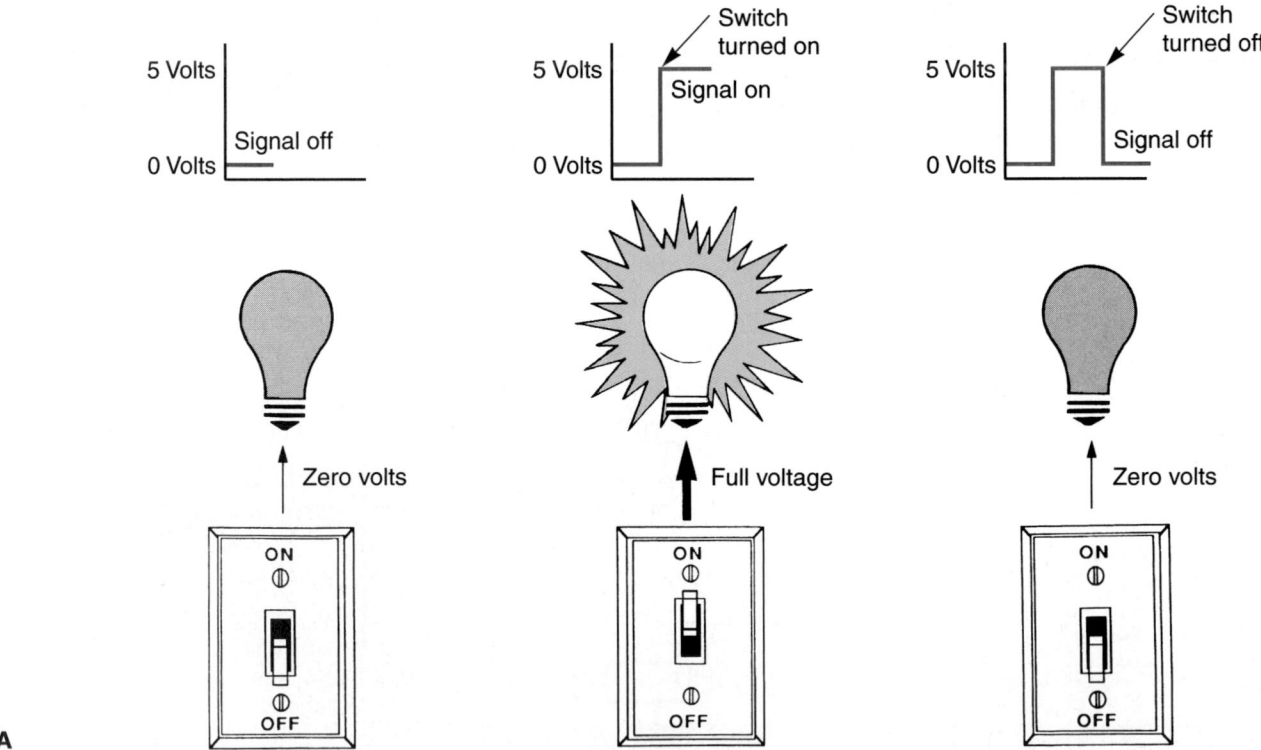

Figure 17-10. Compare digital and analog signals. A—A digital signal is similar to the output from a light switch, which would produce a square wave. B—An analog signal would be similar to the output from a light dimming switch. Circuit voltage would progressively go up or down as the variable resistor is rotated. (Ford) *(Continued)*

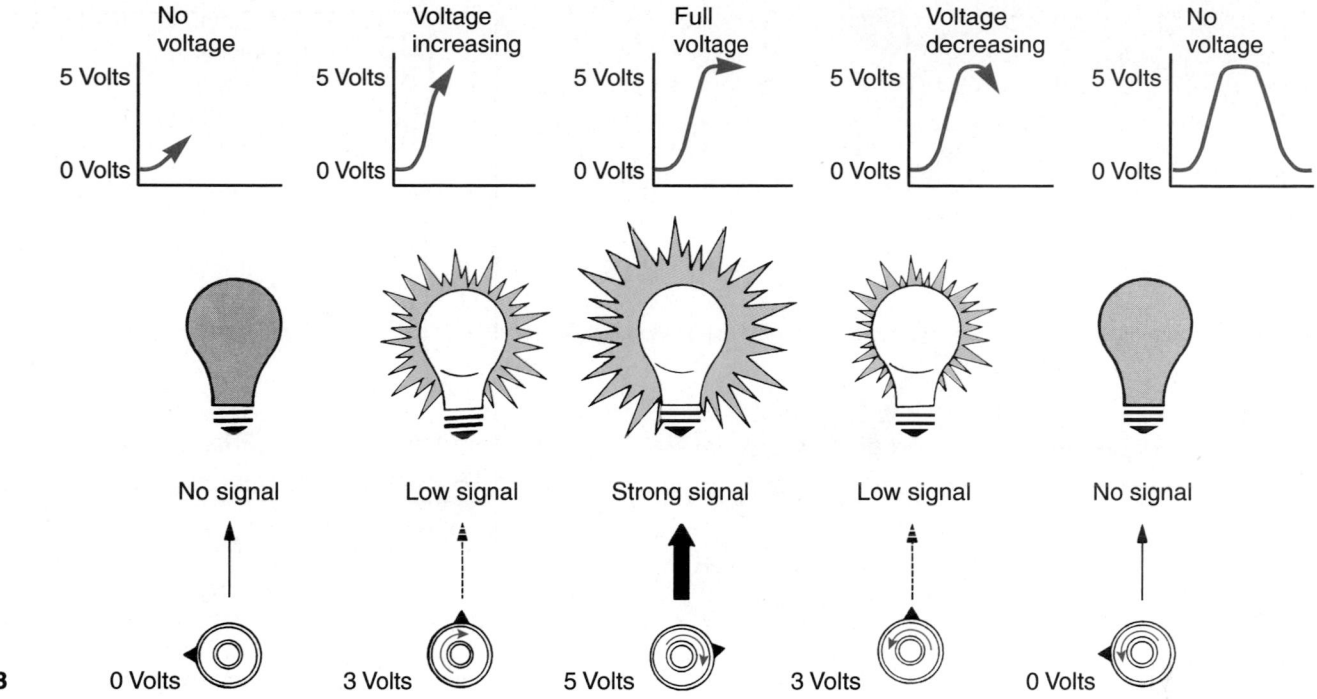

Figure 17-10. *(Continued)*

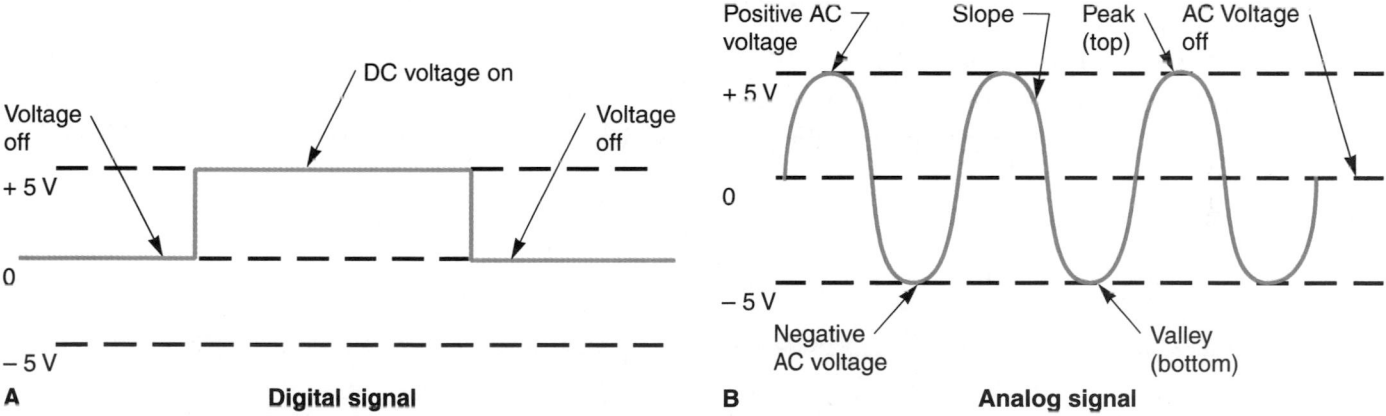

Figure 17-11. These are the two basic signals found in computer system circuits. Voltage is shown vertically and time is shown horizontally. An oscilloscope can be used to measure computer signals. A—Digital signal is an on-off voltage change shown by a vertical line going between zero and the applied voltage. When at zero volts, the signal would be off. When at 5 volts, for this example, the signal would be on. B—An analog signal progressively changes voltage, as shown by the smooth waveform curve. Many sensors produce analog signals.

An *analog signal* gradually changes in strength. For example, a sensor's internal resistance (and resulting current or voltage signal) may smoothly increase or decrease with changes in temperature, pressure, or part position. An analog signal will produce a curved wave or an irregular wave. See **Figure 17-11B.**

Computer signals can be measured with an oscilloscope (introduced in Chapter 4 and detailed in Chapter 46).

If a scope is connected to a sensor that generates a signal voltage, a trace of the signal, or **waveform,** can be viewed on the scope. **Figure 17-12** shows how a scope will display a waveform for a magnetic speed sensor. There are many types of electrical waveforms within the analog-digital categories, **Figure 17-13.** A few of the most common include:

- *AC sine wave*—a curving signal that smoothly fluctuates above and below zero volts. The electricity in your home or inside an alternator would produce an ac sine wave.

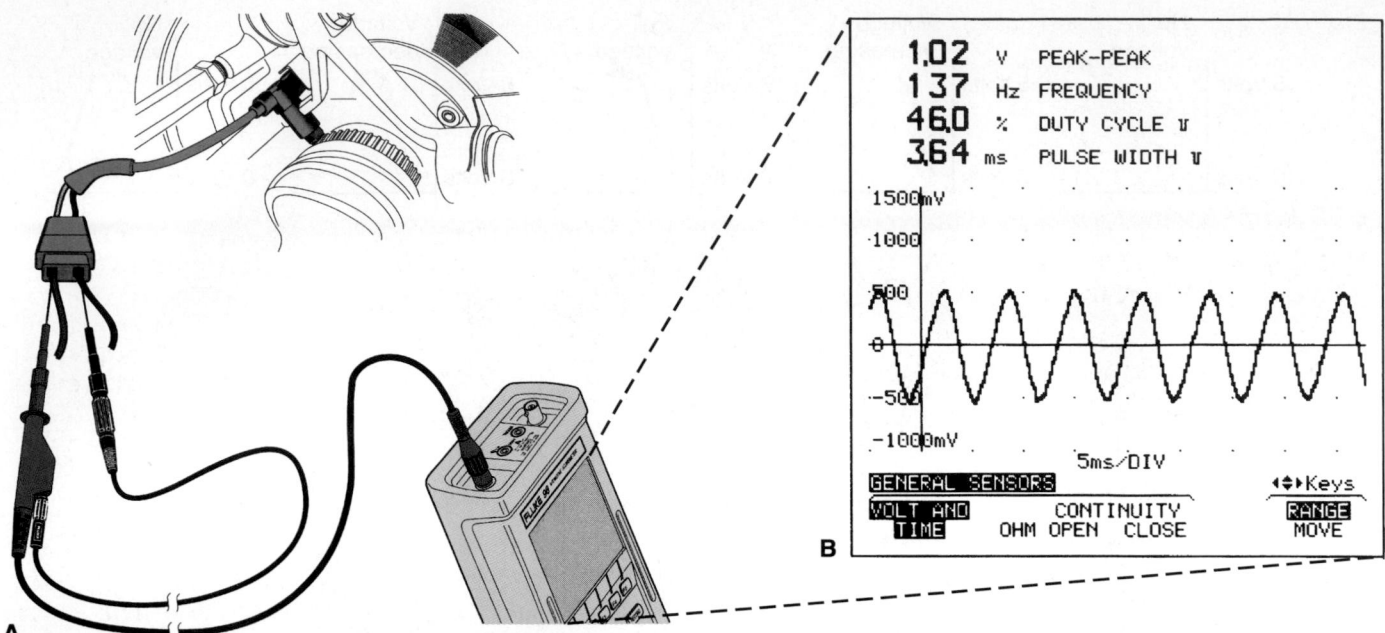

Figure 17-12. A scope is commonly used to check the output from sensors, computer, or output actuators. This is a waveform for a wheel speed sensor in an anti-lock brake system. Its signal is similar to other magnetic speed sensors found on the engine, transmission, etc. A—A hand-held scope is connected to measure sensor output and waveform. B—The top part of this screen gives a digital readout of electrical values. The waveform allows you to check for other problems, such as interference from other circuits. (Fluke)

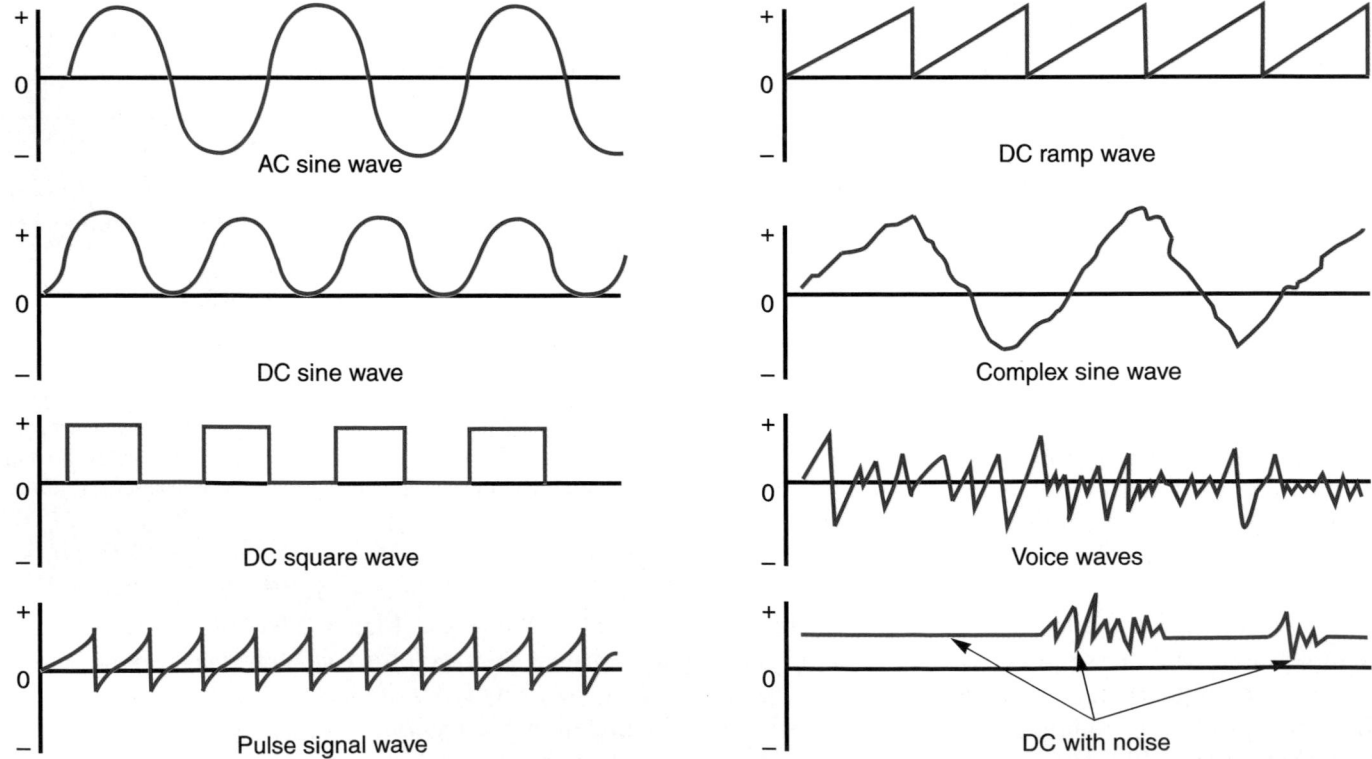

Figure 17-13. These are some of the waveforms, or signals, produced by various automotive systems.

- *DC sine wave*—a curving, fluctuating signal that stays at or above zero volts.
- *DC square wave*—digital, on-off signal that stays at or above zero volts.
- *DC ramp wave*—moves steadily up from zero and then switches off almost instantly.
- *Pulse wave*—signal that rises progressively from zero and then almost instantly returns to zero.
- *Complex sine wave*—signal voltage that moves up and down erratically but gradually.
- *Voice wave*—an irregular wave that corresponds to the frequency of the human voice.
- *DC noise wave*—a straight line (dc) with some voltage fluctuations (noise). Noise is usually unwanted voltage spikes in a circuit and is caused by induced voltage from an outside source.

Computer Signal Terminology

There are several terms used to describe computer signals. Signal frequency, or *pulse width,* generally refers to how fast a voltage signal changes over time. The signal frequency from a sensor might be measured in milliseconds, or thousandths of a second.

A high-frequency signal would be very short (narrow), or have a short pulse width. A low-frequency signal would be longer (wider), or would have a long pulse width, **Figure 17-14.**

Signal amplitude refers to how much voltage is present in the wave. A *high-amplitude signal* would have more voltage than a *low-amplitude signal.* Most automotive computer systems use a signal amplitude of about 3-5 volts. See **Figure 17-15.**

Duty cycle is the percentage of *on* time compared to

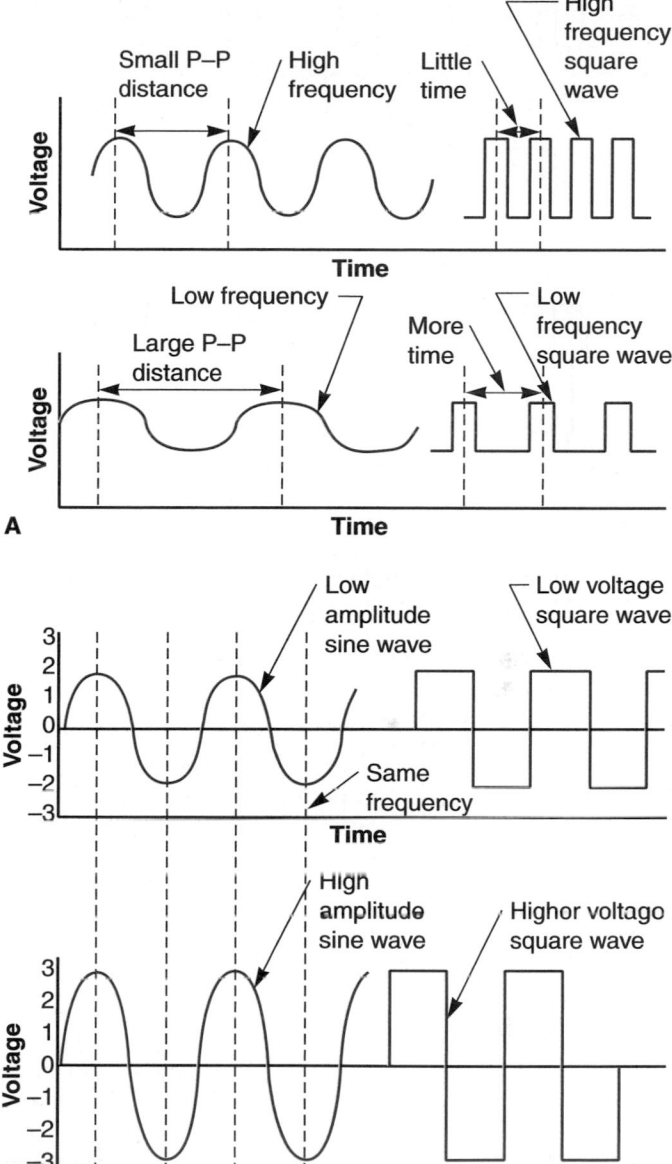

Figure 17-15. A—Compare the differences in waveform frequencies. B—Note how the waveform amplitude, or voltage level, is denoted.

total cycle time. Along with pulse width, it can be used to measure a signal or pulse as a percentage. For example, if a solenoid is sent a 50% duty cycle, it will be turned on half the time and off half the time, **Figure 17-16.**

Computer System Operation

There are three stages of computer system operation:

- *Input*—vehicle sensors convert a condition into an electrical signal that can be used by the computer.
- *Processing* and *Storage*—the computer compares sensor inputs to data and operating parameters in

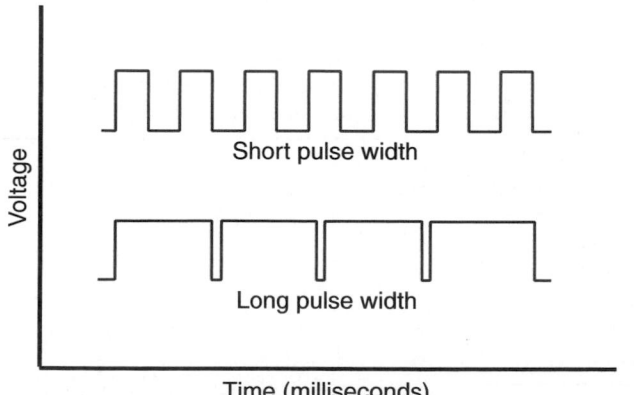

Figure 17-14. Signal frequency can also be denoted as pulse width. A short pulse width, or high-frequency signal, is not very wide. A long pulse width, or low-frequency signal, would appear wider on a scope.

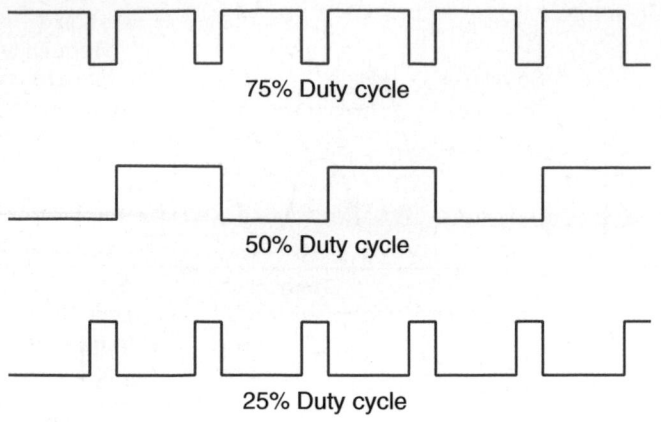

Figure 17-16. Duty cycle is another way of denoting a circuit's on-time. It is expressed as a percentage.

its memory to determine what action should be taken to control the vehicle's systems. The computer can remember programming, as well as store trouble codes in its memory chips.

- *Output*—the computer produces an electrical output signal so actuators can perform physical actions to alter component operation.

Figure 17-17 shows the stages of computer system operation. You must be able to visualize this flow of electrical-electronic data and the changes in operation that result.

Computer System Block Diagram

A *computer system block diagram* is a simple service manual drawing that shows how the sensors, the actuators, and the computer interact. It uses basic squares or rectangles to represent components and lines to represent wires. A computer block diagram is handy when trying to find out what types of sensors are used and what conditions are controlled by a specific computer system. A block diagram for one computer control system is shown in **Figure 17-18.** Study the various inputs and outputs.

Sensors

Most *vehicle sensors,* or transducers, change a physical condition into an electrical signal. *Transduce* means to change from one form to another. Just as our eyes, ears, nose, fingers, etc., can sense conditions, vehicle sensors can detect the operating conditions of a car or truck.

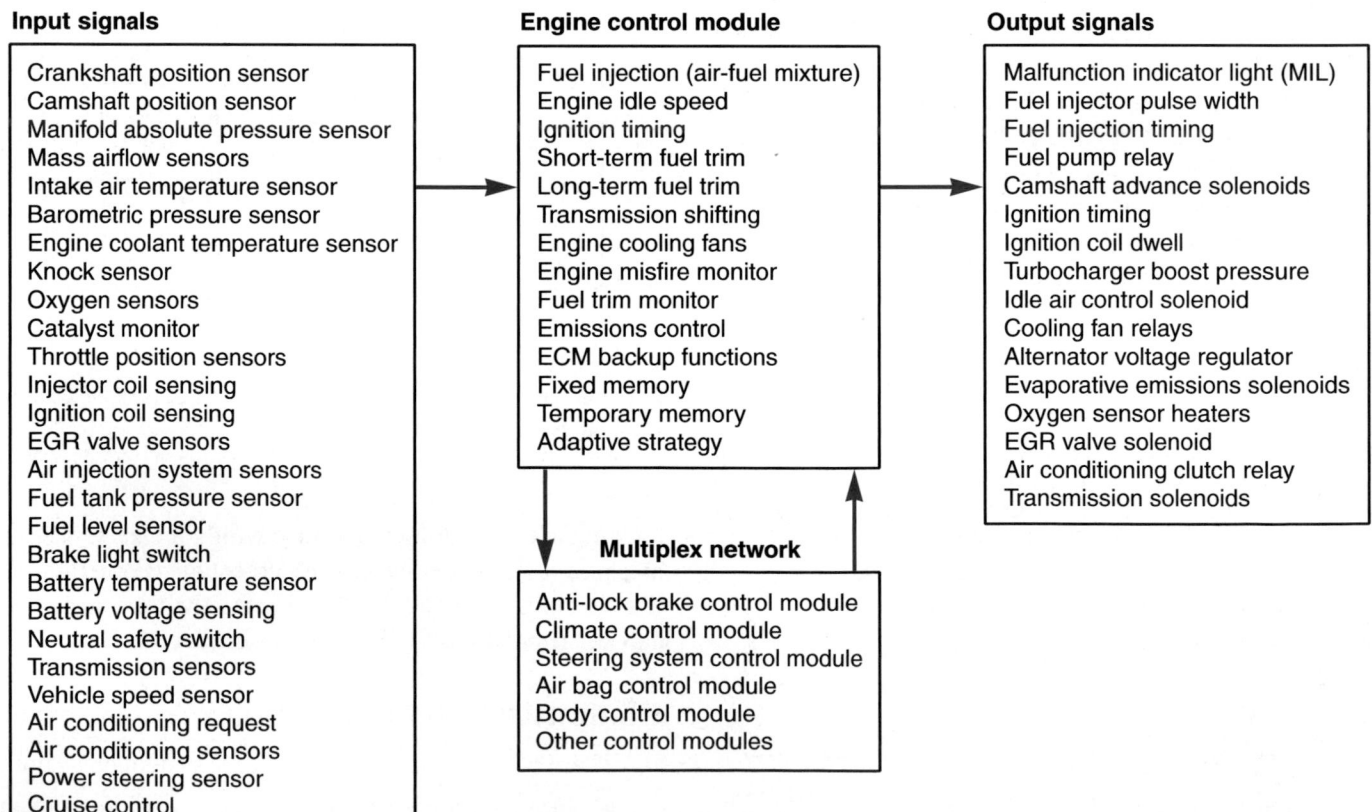

Figure 17-17. Modern computer systems can have numerous inputs and outputs. Study these to prepare for later chapters.

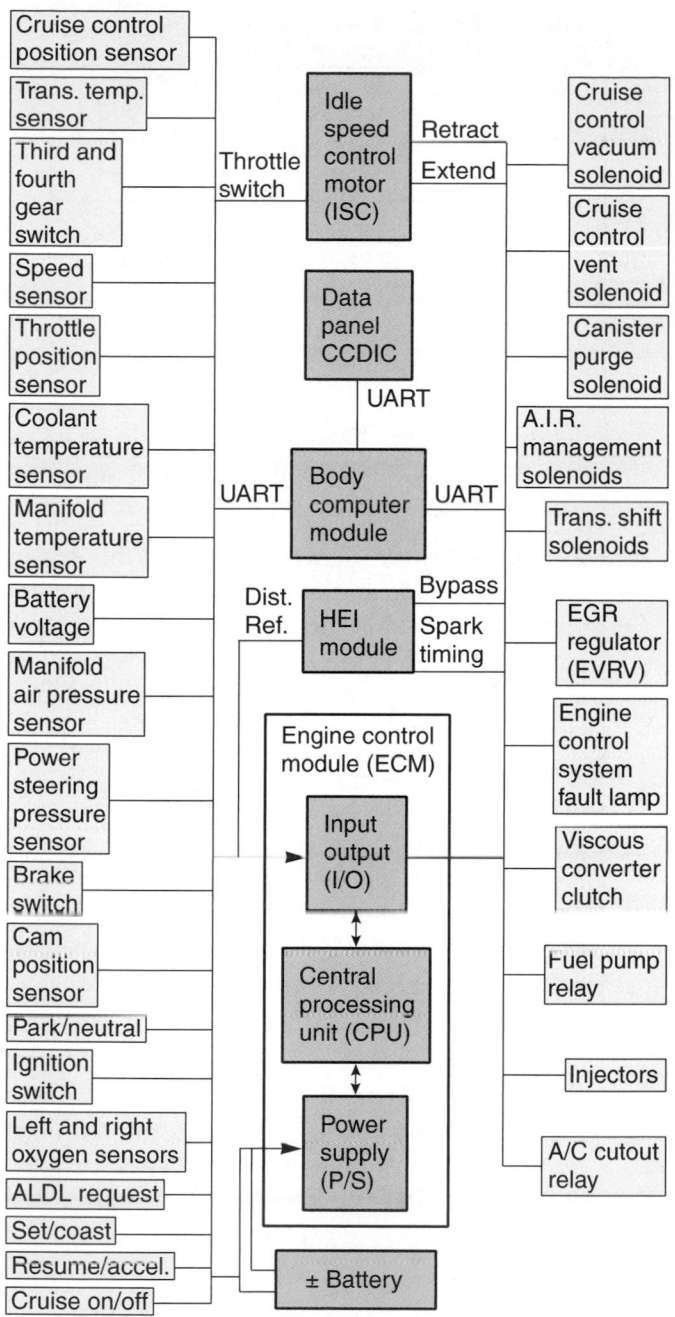

Figure 17-18. Study the block diagram for this computer system. Note the different electronic control modules and the various input and output devices. (Cadillac)

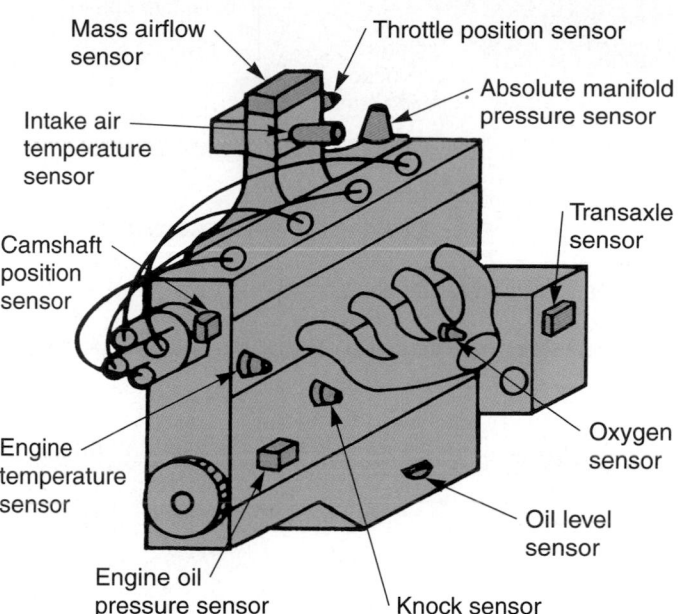

Figure 17-19. Here are a few of the many sensors that can be found on the engine and transaxle assemblies. Other sensors are used elsewhere on the car.

The computers use the voltage signals from sensors to control the actuators.

Sensor Locations

Sensors can be found almost anywhere on a vehicle. Many are mounted on the engine, **Figure 17-19.** Others can be located on or in the transmission or transaxle, in the exhaust system, on the wheel hubs, on and in the fuel tank, on the suspension, or even in the trunk (impact

sensor). If in doubt, refer to the service manual for the specific vehicle to find sensor types and locations.

Sensor Classifications

An automobile uses several types of sensors to provide electrical data to the computer. There are dozens of specific names for vehicle sensors. However, they can all be classified into two general categories: active sensors and passive sensors.

An **active sensor** produces its own voltage signal internally. This very weak signal is fed back to the computer for analysis. **Shielded wire,** which has a flexible metal tube around the conductor to block induced voltage and interference, is often used with active sensors. Look at **Figure 17-20A.**

A **passive sensor** is a variable resistance sensor. Voltage is fed to the sensor from the computer. The passive sensor's resistance varies with changes in a condition (temperature, pressure, motion, etc.). The computer can detect the resulting change in voltage caused by the change in resistance. See **Figure 17-20B.** Within these two categories are several sensor types:

- **Variable resistor sensor**—changes its internal resistance with a change in a condition; its ohms value may change with temperature, pressure, etc. It is an analog sensor. Examples of variable resistance sensors are throttle position and temperature sensors. A **potentiometer** is a variation of a variable resistor sensor. It has three external

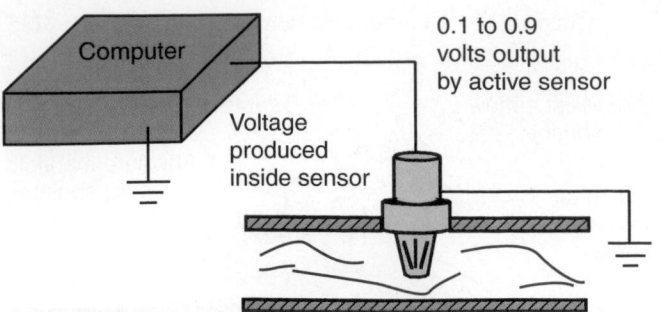

0.1 to 0.9
volts output
by active sensor

Computer

Voltage
produced
inside sensor

A—Active sensor produces its own voltage and current.
The weak signal is sent to the computer.

Computer

5 volts fed to
passive sensor

Sensor
resistance
changes
with condition

Voltage or
current signal
flows back to
computer

B—Passive sensor cannot produce its own voltage and
current. A reference voltage must be fed to the sensor by
the computer.

Figure 17-20. Compare active and passive sensor operation.

connections instead of two. One lead connects to each end of the resistor, and the remaining lead connects to a wiping arm that slides over the fixed

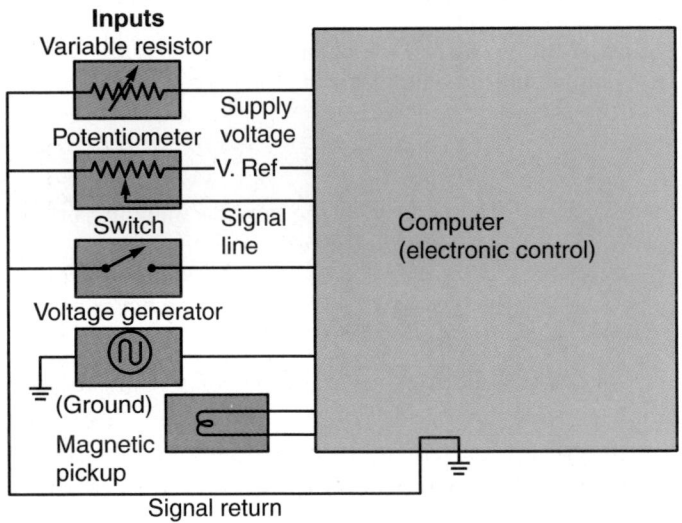

Inputs

Variable resistor

Potentiometer

Switch

Voltage generator

(Ground)

Magnetic
pickup

Supply
voltage

V. Ref

Signal
line

Computer
(electronic control)

Signal return

Figure 17-21. These are the basic classifications of input sensors. Each sensor feeds a different type of signal to the computer.

resistor. This arrangement more precisely controls the current throughput then a two-connection variable resistor sensor.

- **Switching sensor**—opens or closes the sensor circuit to provide an electrical signal for the computer. Transmission pressure switches are examples of this type of sensor. A switching sensor can detect almost any condition and produces a digital signal, **Figure 17-21.**

- **Magnetic sensor**—uses part movement and induced current to produce a signal for the computer. This type of sensor is commonly used to monitor speed (vehicle speed sensor) or part rotation. It usually produces an analog signal. Magnetic sensors are also referred to as permanent magnet (PM) generators.

- **Hall-effect sensor**—uses a special semiconductor chip to sense part movement and speed, **Figure 17-22.** It produces a digital signal. These sensors are used as crankshaft and camshaft position sensors, as well as pickup devices in some electronic ignition distributors.

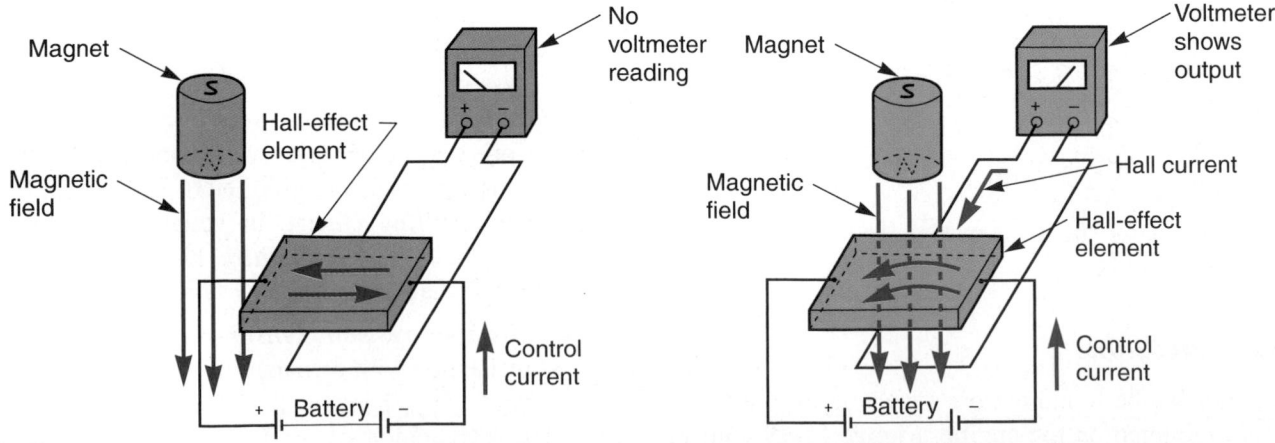

Figure 17-22. Hall-effect sensors use a semiconductor chip that reacts to magnetic fields. A—Battery current flows lengthwise through the Hall-effect element. No current flow is indicated by the meter. B—When a magnetic field acts on the element, some current is diverted into the sensor, as indicated by the meter. This signal is used by the computer to detect part rotation.

- *Optical sensor*—uses light-emitting diodes and photo diodes to produce a digital signal. It may be used to sense part rotation and speed. See **Figure 17-23.** These sensors are used in some distributors and as speed sensors mounted outside the speedometer.

- *Piezoelectric sensor*—generates voltage from a physical shock or motion. It is sometimes used to listen to a condition, such as engine knocking (knock sensor), **Figure 17-24.**

- *Solar sensor*—converts sunlight directly into an electrical signal. It is made of a semiconductor material that converts photons (light energy) directly into direct current, **Figure 17-25.**

- *Direction sensor*—detects the polarity of a moving magnet. This is a variation of a magnetic

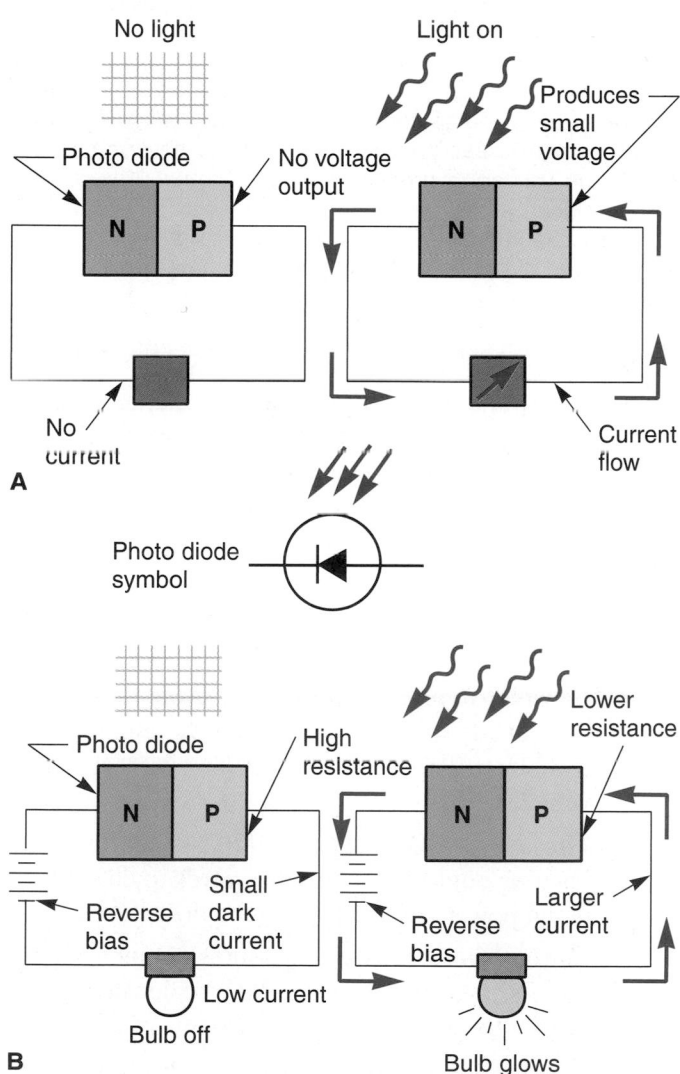

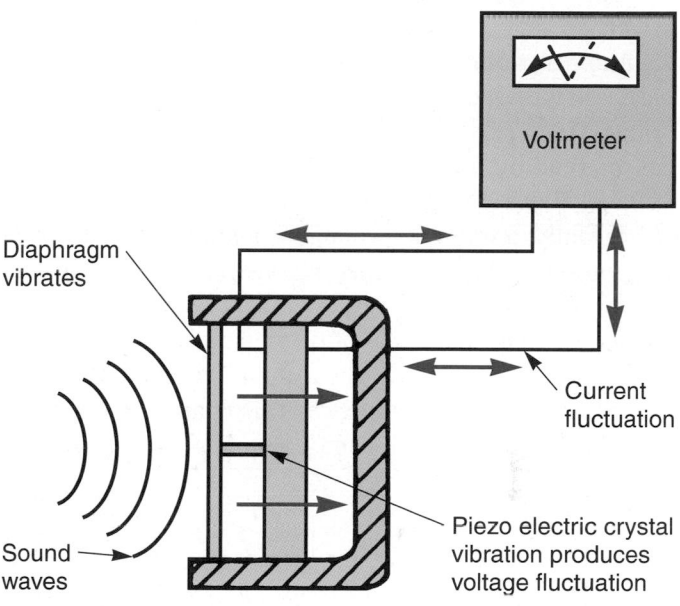

Figure 17-24. A piezoelectric crystal converts physical pressure into a tiny electrical signal. This is ideal for microphones and listening sensors. The crystal produces a voltage signal proportional to the amount of sound or vibration detected.

Figure 17-23. An optical sensor uses light to produce a signal. A—When in a dark area, the photodiode has no voltage output. When it is exposed to a light source, the photodiode produces a voltage signal. B—A reverse-biased photodiode is used to switch current on and off. Without light, it acts as an insulator. When in light, the photodiode conducts current. This operating principle is used in some automatic headlight systems.

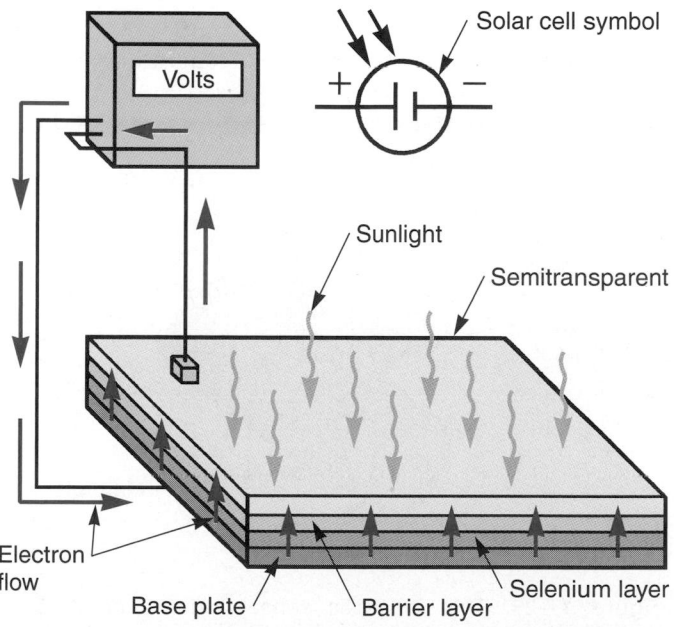

Figure 17-25. A solar sensor converts sunlight directly into electrical current. It is another type of voltage-generating sensor. The semiconductor surface feeds the electrons in sunlight to the output leads.

sensor. Unlike a magnetic sensor, it can signal which direction a part is rotating. It is used in some computer-controlled steering systems, **Figure 17-26.**

Reference Voltage

The computer sends a reference voltage to a passive sensor so that a signal can be fed back for processing. *Reference voltage,* sometimes abbreviated *Vref,* is the base voltage used to carry a computer signal. A reference voltage is needed so that a change in sensor resistance can be read by the computer as a change in current and voltage, **Figure 17-27.**

Usually, a reference voltage is applied to a sensor by the computer. The reference voltage is typically around 5 volts. The computer steps down battery voltage so that a smooth, constant supply of dc voltage is fed to the passive sensors. The sensor alters the reference voltage by changing its internal resistance. This change in reference voltage is interpreted by the computer as a change in condition or state.

Sensor Types

The most common sensors used in late-model vehicles include the following:

- *Intake air temperature (IAT) sensor*—measures the temperature of intake air as it enters the intake manifold. The IAT is also called a manifold air temperature (MAT) or an inlet air temperature sensor.

- *Engine coolant temperature (ECT) sensor*—measures the temperature of engine coolant. In most cases, it is simply referred to as a coolant temperature sensor.

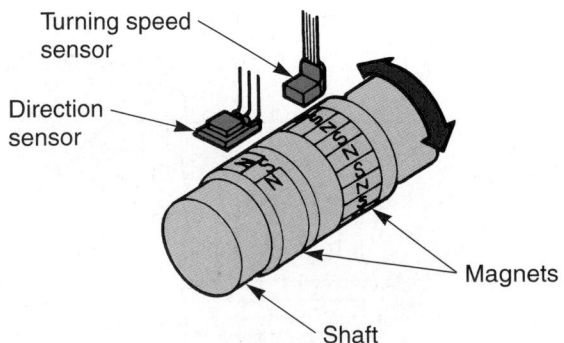

Figure 17-26. This magnetic sensor setup can measure rotational speed and direction. Permanent magnets on the rotating part induce voltage into both sensors. The speed sensor only detects how fast the magnets pass by. The direction sensor, using two pickups, can also detect magnet polarity to determine the direction of rotation. (Honda)

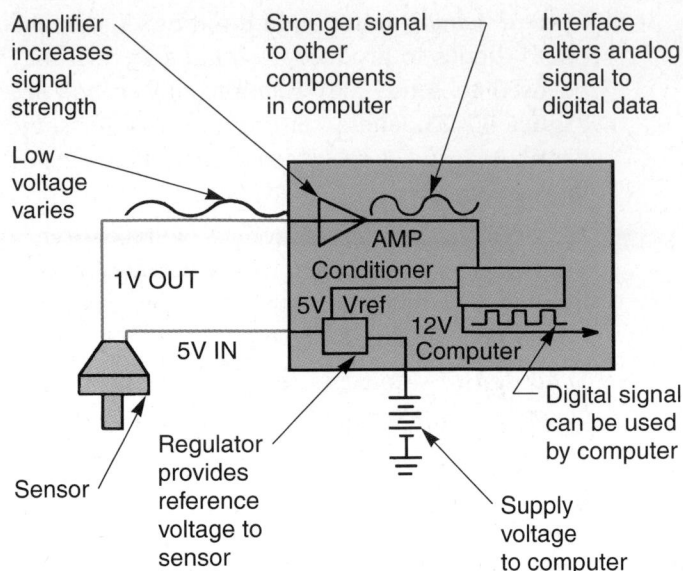

Figure 17-27. A computer power supply or voltage regulator sends voltage to each passive sensor. Most passive sensors modify the voltage to produce an analog signal. If weak, the analog signal is first amplified, or increased in strength. A conditioner, or interface, converts the analog signal into a digital signal. The digital signal can then be processed by the computer.

- *Oxygen sensors (O_2)*—measure the amount of oxygen in the engine's exhaust gases before and after the catalytic converter. Most newer sensors contain heaters and are referred to as heated oxygen sensors (HO_2S).

- *Manifold absolute pressure (MAP) sensor*—measures outside air pressure in relation to vacuum inside the engine intake manifold.

- *Barometric pressure (BARO) sensor*—measures the outside air pressure around the engine. It is combined with the MAP sensor on most newer vehicles.

- *Throttle position sensor (TPS)*—measures the opening angle of the throttle valves to detect how much power is requested by the driver.

- *Engine speed sensor*—measures engine rpm for ignition system operation. May be located in the distributor or next to the crankshaft or camshaft.

- *Crankshaft position sensor*—measures crankshaft position, rotation, and speed.

- *Camshaft sensor*—checks the engine camshaft position and rotation.

- *Mass airflow (MAF) sensor*—measures the amount of intake air flowing into the engine. It is often referred to as an airflow sensor or an airflow meter.

- *Knock sensor (KS)*—piezoelectric sensor that detects engine pinging, preignition, or detonation so the computer can retard ignition timing or reduce turbocharger boost pressure.

- *Transaxle/transmission sensor*—checks transaxle or transmission gear selection. It is usually a part of the neutral safety switch.

- *Brake switch*—detects brake pedal application.

- *Wheel speed sensor*—measures wheel rotational speed for anti-lock brake and traction control systems.

- *Oil level sensor*—measures the amount of oil in the engine oil pan.

- *EGR sensor*—measures the position of the exhaust gas recirculation valve pintle.

- *Impact sensors*—detect a collision to shut off the fuel pump or the engine. Another type of impact sensor is used to deploy the air bag system.

- *Vehicle speed sensor*—measures the vehicle's road speed so the computer can adjust fuel, ignition, transmission, suspension, and other system operations. It is located in the transmission/transaxle housing.

- *Fuel tank pressure sensor*—measures fuel tank and system pressure as part of some evaporative emission control systems. Similar construction and operation to a MAP sensor.

- *Battery temperature sensor*—monitors battery temperature so the computer can adjust vehicle operation to affect charging system output as needed.

Older cars use none or just a few of these sensors. Newer cars might use these and many other sensors.

Circuit Sensing

Circuit sensing involves using the computer itself instead of dedicated sensors to monitor component and circuit operation. The computer monitors current flow through various components and circuits. For example, some circuit sensing systems can monitor fuel injector operation (injector coil winding current), ignition coil action (current through ignition coil windings), and computer operation (current through computer circuits). The windings and the wires in the circuits serve as the sensors.

Computers

The term computer is a general term that refers to any electronic circuit configuration that can use multiple inputs to find outputs. Automobile manufacturers use many names for their computers, including:

- Central processing unit (CPU)
- Electronic control unit (ECU)
- Electronic control module (ECM)
- Engine control module (ECM)
- Electronic control assembly (ECA)
- Powertrain control module (PCM)
- Vehicle control module (VCM)
- Microprocessor
- Logic module

Keep these names in mind when reading service manuals. To prevent confusion, this textbook will use the terms *computer* and *control module* when referring to computers in general. When discussing a computer used to control one or more systems, the text will use the terms recommended by the Society of Automotive Engineers (J1930). Additionally, the term "module" will be used for an electronic circuit used to amplify and/or modify a single signal or control a single system, such as the module used to control the operation of the anti-lock brake (ABS) system.

Computer Types

Several types of computers can be used in a car, **Figure 17-28.** The number and types of computers will vary with the manufacturer. The most common types are:

- *Vehicle control module*—large, powerful computer that processes data from sensors and other,

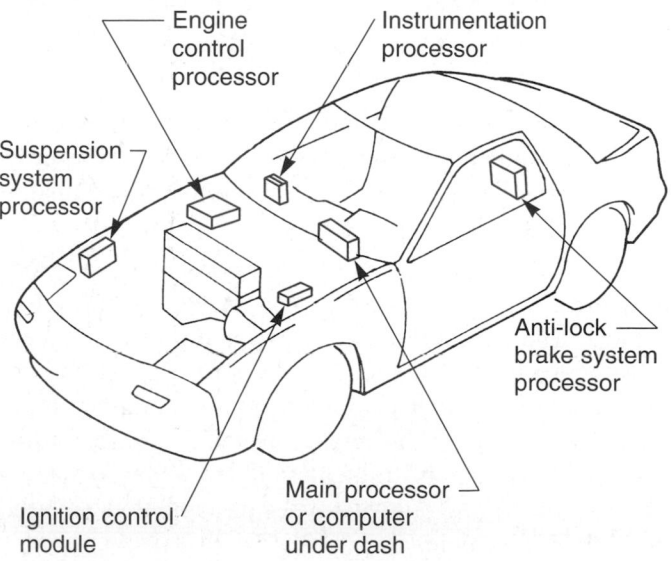

Figure 17-28. One or more computers, or control modules, can be used on the same car. Note the potential locations.

less powerful control units. It coordinates engine, transmission, traction control, and anti-lock brake functions, for example.

- *Powertrain control module*—powerful computer used to monitor and control the engine, transmission, and other systems.

- *Engine control module*—computer that uses sensor inputs to control engine idle speed, fuel injection, ignition timing, emission control devices, and other operating parameters.

- *Anti-lock brake module*—small computer that uses wheel speed sensor inputs and other inputs to control anti-lock brake application.

- *Instrumentation module*—small computer that uses sensor inputs to operate a digital dash display.

- *Ignition module*—small computer that uses sensor inputs to control ignition timing, spark plug firing, or ignition coil pack operation.

- *Suspension system module*—small computer that uses vehicle speed, suspension, and steering sensor inputs to control ride stiffness or shock absorber action.

- *Climate control module*—small computer used to control the operation of the heating, ventilation, and air conditioning systems.

- *Air bag module*—small computer that controls the vehicle's air bag system. This module also stores power to deploy the air bags in the event that battery power is lost in a collision.

- *High-power control module*—computer used to control or increase current flow or process output signals from a few sensors and the main computer. It is also called a *power control module*.

- *Body module*—computer that provides memory and other functions for the radio, driver's information center, electronic compass, trip computer, cellular telephone, etc.

Again, the number and types of computers and modules used will vary with the year, make, and model of vehicle. The trend is to use a main computer to process most of the input data and control most outputs. In conjunction with the main computer, smaller computer modules are used for the brake system, ignition coils, suspension system, instrumentation, etc. Refer to the service manual to find out what types of computers are used on a specific vehicle.

Computer Locations

Automotive computers may be located under the dash on the passenger side of the vehicle, **Figure 17-28.** This protects the delicate circuits and components in the computer from engine heat, vibration, and moisture. However, computers can also be located in the engine compartment, in the trunk, under the seats, etc.

When located in the engine compartment, the computer is closer to most sensors and actuators. Less wiring and fewer connectors are needed to tie the system together. Modules used to operate the anti-lock brakes and other systems are often found in the engine compartment. Engine and powertrain control modules are located under the dash, in the engine compartment, or in another central location. Modules used with specific systems, such as the ignition or air conditioning systems, are close to, in, or under the system components.

Computer Operation

A computer is sometimes nicknamed the "*black box.*" This is because it is enclosed in a box-shaped housing, sometimes made of black plastic or metal, and contains circuits that can do complex operations.

This section will summarize the operation of a computer. Even though you will probably never repair a computer, it will be helpful to you as a technician to understand how a computer uses sensor inputs and produces actuator outputs.

Figure 17-29 shows the inside of an automotive computer. It is composed of printed circuit boards,

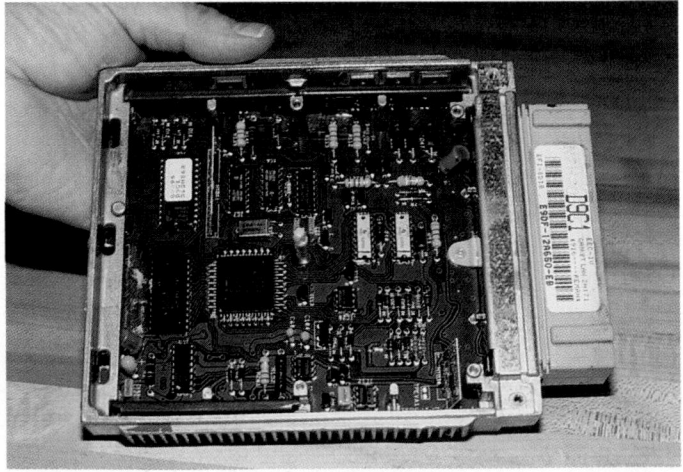

Figure 17-29. This photo shows the inside of a computer. Note that it uses integrated circuits attached to a printed circuit board.

integrated circuits, capacitors, resistors, power transistors, and many other basic electronic components. You learned about these in Chapter 8.

Parts of a Computer

All computers can be divided into sections. Each section has a specific function. Basically, a computer can be divided into eleven parts. These include:

- *Voltage regulator*—converts battery and other voltages into lower voltages that can be used by the computer and its sensors.
- *Amplifiers*—increase strength of signals from input devices, **Figure 17-30.**
- *Conditioners*—alter signals for use by the computer and its actuators.
- *Buffer*—serves as a temporary storage area for data.
- *Microprocessor*—integrated circuit that makes decisions or calculations for the computer.
- *Memory*—integrated circuit that stores data for the microprocessor, **Figure 17-30.**

- *Clock*—integrated circuit that produces a constant pulse rate to coordinate computer operations.
- *Output drivers*—power transistors that step up current or provide a ground path to operate actuators or modules.
- *Circuit board*—fiberboard with flat metal conductors printed on its surface that connects and holds components.
- *Harness connector*—multipin terminal that attaches to the vehicle's wiring harness.
- *Computer housing*—metal or plastic enclosure that protects electronic components from induced currents and physical damage.

Computer Voltage Regulator Operation

A *computer voltage regulator* provides a reduced voltage for the electronic components in the computer. This must be a very smooth dc voltage that does not vary and that does not have *spikes* (abrupt changes in voltage).

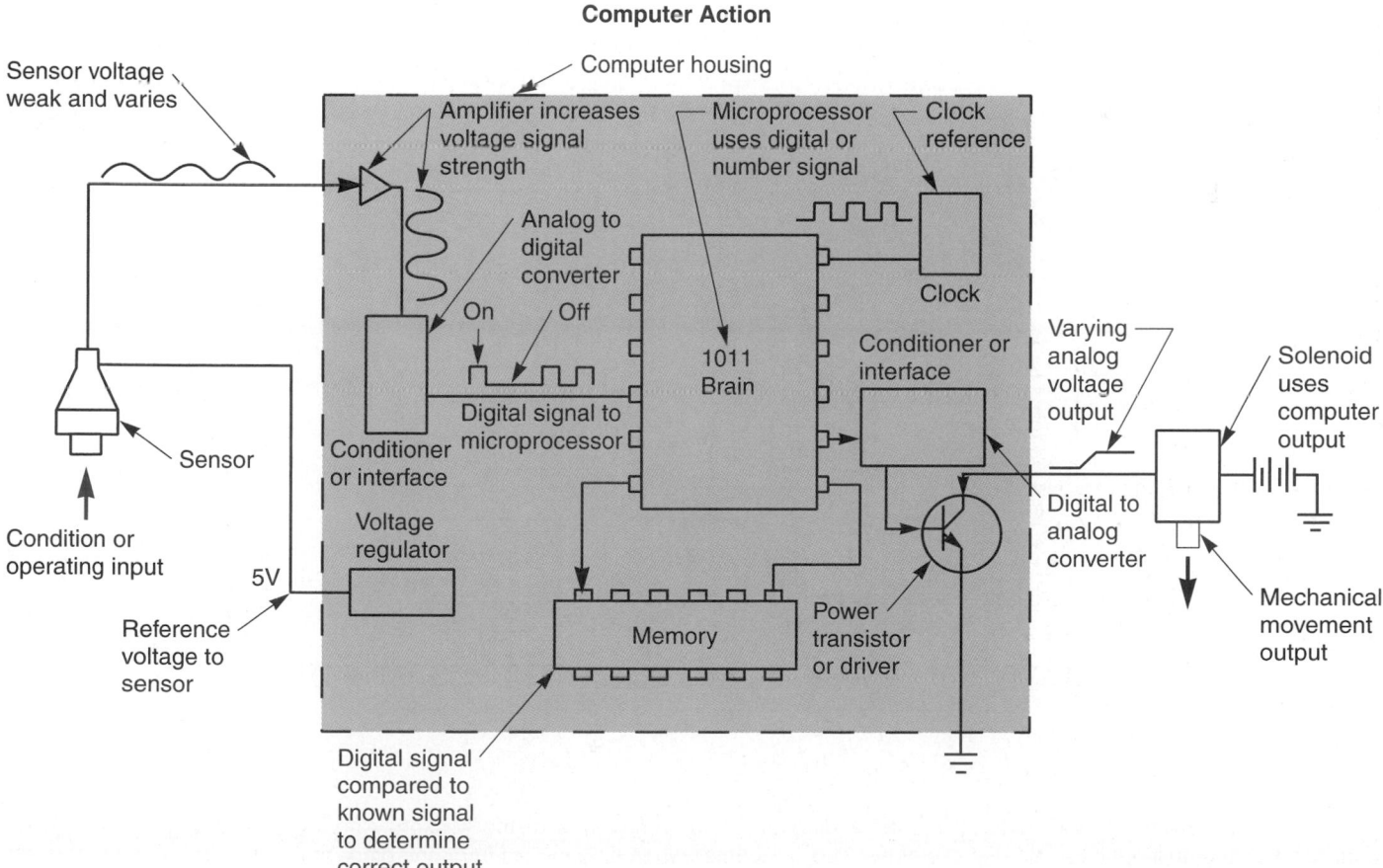

Figure 17-30. Trace the flow of data through this simplified computer system. Can you explain the purpose of each component?

Computer Amplifier Operation

A *computer amplifier* simply strengthens various signals when inside the computer. An amplifier might increase the voltage signal from the oxygen sensor, which is less than one volt. Then, the signal is strong enough to be used by circuits in the computer.

Computer Conditioner Operation

There are two basic types of conditioners in a computer: input conditioners and output conditioners. A conditioner can be called a *converter* or an *interface*. An *input conditioner* alters the input signals from some sensors. It treats incoming data (voltage and current) so it can be utilized by the computer. For example, many sensor signals are analog. The conditioner can convert the analog signal into a comparable digital signal that the computer can understand. An *output conditioner* is needed to change digital signals back into analog signals. The output of the computer must usually be analog to operate the actuators and other control modules.

Both input and output conditioners protect the computer from shorted or grounded circuits. If a sensor, output device, or related circuit is shorted or grounded, the conditioner will not accept the signal and will open the circuit to protect the ECM from damage. This is similar to the operation of a circuit breaker in a shorted electrical circuit.

Buffer

A buffer is a computer device that can serve as a temporary storage area for data. It can also protect internal computer chips from improper data. For example, if data is fed into the computer too quickly, the buffer can hold the data and then feed it into the other devices at a controlled rate. In many cases, the buffer is built into the input conditioners.

Computer Microprocessor Operation

The term *microprocessor* means *small* (micro) *computer* (processor). A microprocessor is a small computer chip or an integrated circuit capable of analyzing data and calculating appropriate outputs. It is the brain of a computer, **Figure 17-31.**

A microprocessor uses the binary number system to make decisions, comparisons, or calculations. Digital pulses from the conditioners are fed into the microprocessor. Since these inputs are off (0) or on (1) voltages, they can be used by the logic gates in the processor.

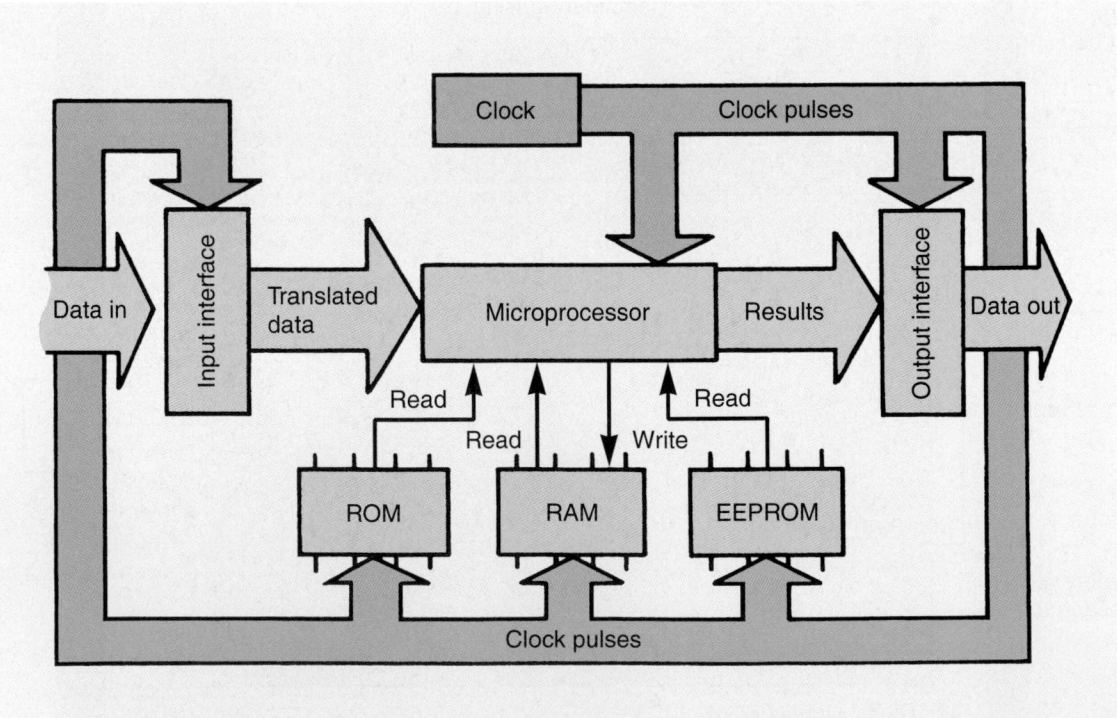

Figure 17-31. Note the flow of data in this block diagram. Data comes in from the sensors. The input interface, or conditioner, changes the analog input data into digital signals. The clock times when data moves from one place to another. The microprocessor compares the input information to permanent memory and writes into temporary memory. The microprocessor decides on the appropriate outputs using logic gates. The results are sent from the microprocessor to the output interface, which changes digital data back into analog signals that can be used by the actuators. (General Motors)

A microprocessor also uses data stored in the computer's memory. It compares input signals to memory data to decide what the outputs should be for maximum efficiency.

Tech Tip
The computers used on late-model vehicles have more computing power and memory than their predecessors. For example, computers used in late-model vehicles have 16- or 32-bit processors and large read only memory chips. The computers found in older vehicles contain 8-bit processors and smaller memory chips and are not powerful enough to monitor and control present-day vehicle systems.

Computer Memory Operation

Computer memory uses gates that are capable of storing data as voltage charges. The integrated circuits inside the memory chips will hold the data (*on* or *off* charges) until needed by the microprocessor. There are several basic types of computer memories:

- *Random access memory (RAM)*—is a memory chip used by the computer to store information or data *temporarily*. This data is erased if battery power to the computer is removed.

- *Read only memory (ROM)*—stores permanent data that cannot be removed from memory. This memory chip contains calibration tables and look-up tables for the general make and model car.

- *Programmable read only memory (PROM)*—is a memory chip containing permanent data that is more specific than the data stored in ROM. The microprocessor can read from the PROM, but it cannot write to the PROM.

The PROM contains specific information about the vehicle's engine (number of cylinders, valve sizes, compression ratio, fuel system type), transaxle (shift points, gear ratios, etc.), weight, tire size, optional accessories, and any unique features. For example, a car with a manual transaxle will have a different PROM than one with an automatic transaxle.

The PROM is the only part of some computers that is commonly serviced. During computer replacement on some vehicles, the PROM chip can be removed from the old computer and reused in the new computer. PROMs are also replaced to correct a performance or operation problem. The PROM seldom fails, and it is programmed for the specific make and model car. Data is retained in the PROM even when the chip is removed from the computer. The ECMs used on late-model vehicles use a different type of read only memory that is permanently fixed to the circuit board but can be erased and reprogrammed.

- *Erasable programmable read only memory (EPROM)*—can be changed. However, in most cases, the changes can only be made by the manufacturer using special equipment. It is also responsible for storing semipermanent data, such as odometer or mileage readings for an electronic dash display.

- *Electrically erasable programmable read only memory (EEPROM)*—can be altered by the technician in the field. This allows the manufacturer to easily change operating parameters if a performance or driveability problem is discovered.

- *Flash erasable programmable read only memory (FEPROM)*—is similar to EEPROMs in all respects. FEPROM is simply another name for an EEPROM.

- *Keep alive memory (KAM)*—is a memory chip that allows the computer to have an adaptive strategy.

An *adaptive strategy* is needed as parts wear and components deteriorate. The information stored in KAM allows the computer to maintain normal vehicle performance with abnormal inputs from sensors. It gives the computer the ability to also ignore false inputs to maintain good driveability.

Tech Tip
Different computer designs will use different memory chips. Also, the names of these chips can vary.

Output Drivers

Output drivers, sometimes referred to as *quad drivers* or *power transistors,* control current flow through the actuators. When energized by the computer, the drivers ground the actuator circuits. The actuators can then produce movement, such as turn on the electric cooling fans.

Processor-Memory Bus

The processor-memory bus is the pathway by which sections of a computer communicate, or exchange data. When the computer is operating, data rapidly shuffles between the memory chips and the microprocessor chip. The microprocessor chip controls this flow of data. Sometimes, it writes data about vehicle operation into memory or it may read out data about how the vehicle should operate.

Multiple Sensor Inputs

The computer uses multiple inputs (inputs from more than one sensor) to determine the needed output. As shown in **Figure 17-32,** the computer uses signals from the engine speed sensor to determine when to fire the fuel injectors. However, if a temperature sensor signals a cold engine, the computer would know to increase injector pulse width to enrich the fuel mixture for good cold engine operation.

Multiplexing (Integrated Computer System)

A *computer network* is a series of computers that control different systems but work together to improve overall vehicle efficiency. The computers in the network share common parts, such as wires, input signals, and

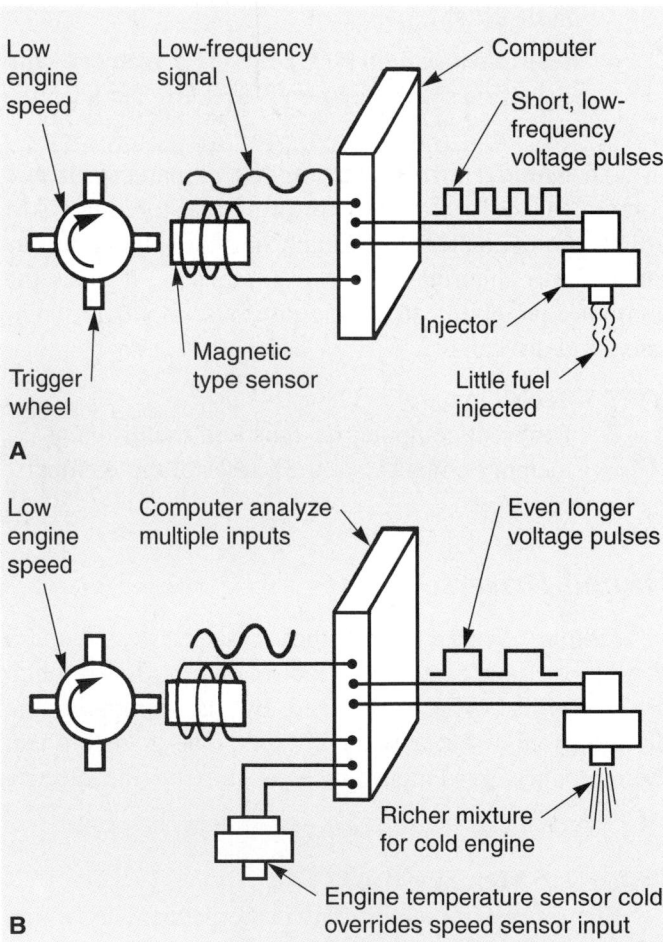

output signals. They exchange data from sensors to prevent duplication of parts and to reduce wiring. The term used to describe this interaction in vehicles is called *multiplexing.* Also termed an *integrated computer system,* the computer network allows all the on-board computers to use feedback data from several systems (engine transmission, braking, suspension, traction control for example) to better control all vehicle systems. The wiring used for the multiplex network is often referred to as a *data bus.*

Some systems are controlled from a central computer instead of individual ones. This allows a single computer to monitor all functions to better decide how to control braking, throttle action, ride stiffness, engine management, and differential action for optimum efficiency. The signals between each computer are coded to indicate which computer the signal is being sent to, and in what order the signals should be processed. This prevents the signal from being processed by the wrong computer or in the wrong order.

Each computer and system needed for vehicle operation has its own default operating or limp-in mode, so the vehicle can be driven home or to a shop for service.

In the near future, fiber optics, which uses light to transmit data, may form the network connections. This will allow faster and simultaneous transmission of data from each computer.

Actuators

Mentioned earlier, *actuators* are the "hands and arms" of a computer system. They allow computers to do work and alter the operation of other components.

Actuator Locations

Actuators can be found almost anywhere on a vehicle. They can be located on the engine, in the doors, under the hood, and in the trunk. Check an appropriate service manual for exact locations.

Actuator Classifications

A computer system uses several types of output devices, or *actuators,* to control part operation. All actuators can be grouped into the following categories:

- *Solenoid*—current through the solenoid winding forms a magnetic field that moves a metal core. The core acts upon other components.

- *Relay*—current from the computer energizes a relay to control high current flow to another electrical component.

Figure 17-32. A computer system uses more than one sensor input to make most control decisions. A—Low engine speed along with normal engine operating temperature causes the computer to produce a short pulse width to inject a small amount of fuel into the engine. B—If the engine coolant temperature sensor signals a cold engine, the computer would use both speed and temperature signals to increase injector pulse width to enrich the mixture for cold engine operation.

- *Servo motor*—current is sent to a small dc motor that can produce an output by moving parts.
- *Display devices*—current is sent to a vacuum-fluorescent or liquid-crystal display to provide output data to the instrument cluster.
- *Control module*—the computer sends an electrical signal to a control module. The module then amplifies and/or modifies the signal to operate a device.

Look at **Figure 17-33.** Imagine how these devices could be used to control engine, transmission, and other vehicle functions.

Actuator Operation

When the computer turns on an actuator, it normally provides the device with a ground circuit. Then, current can operate the actuator, **Figure 17-34.** The following sections detail the operation of the three most common actuator types: solenoids, relays, and servo motors.

Solenoid Operation

Figure 17-35 shows how the computer can use a solenoid to lock the vehicle's doors once a certain road speed is achieved. As one example, input from the vehicle speed sensor enters the computer. When the computer detects vehicle travel or forward motion, it can use the solenoids to lock all the vehicle's doors.

The computer grounds the solenoid circuits and current flows through the solenoid windings. This produces a magnetic field in the windings. The magnetic field moves the plunger mounted in the solenoid windings. This plunger movement locks the doors.

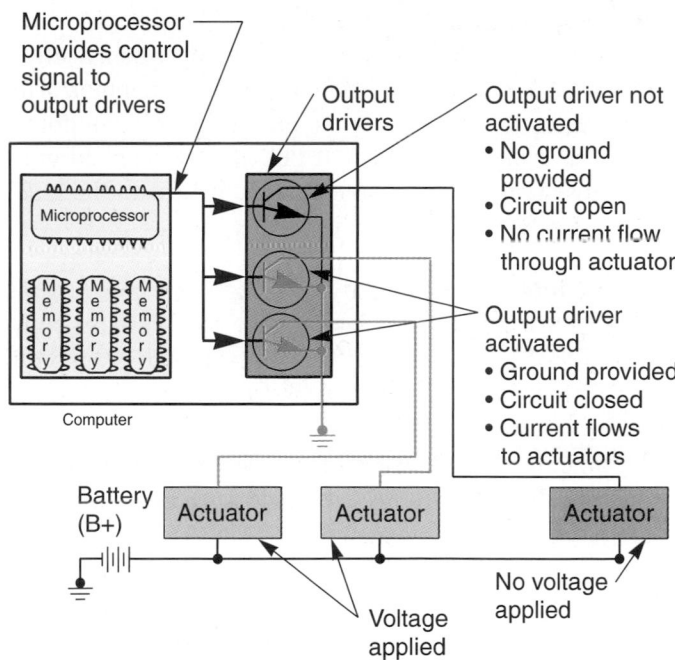

Figure 17-34. A computer will normally activate each output by grounding its circuit. Power is present at the output devices at all times, but there is not a complete circuit path. When a computer's output drivers, or power transistors, are turned on, they will conduct current through ground to energize the output devices. (Ford)

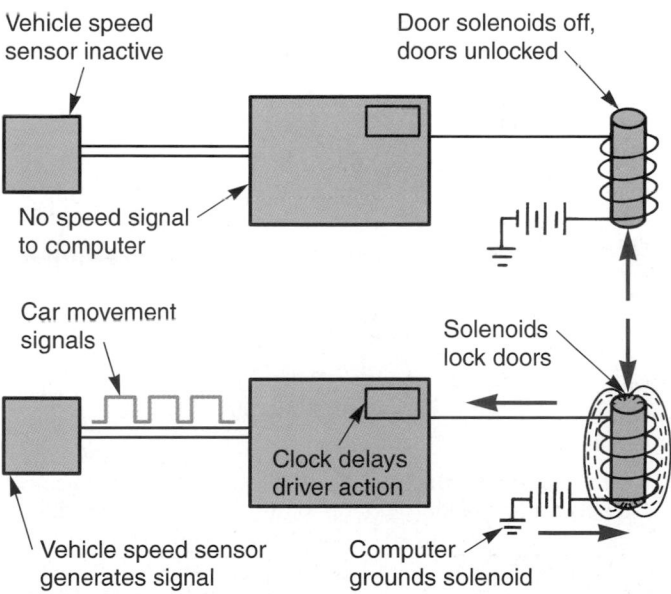

Figure 17-35. Simplified illustration of how sensor inputs can be used by the computer to operate solenoids. When the vehicle speed sensor produces a signal, the computer grounds the door lock solenoid. The solenoid movement is used to lock the car doors automatically.

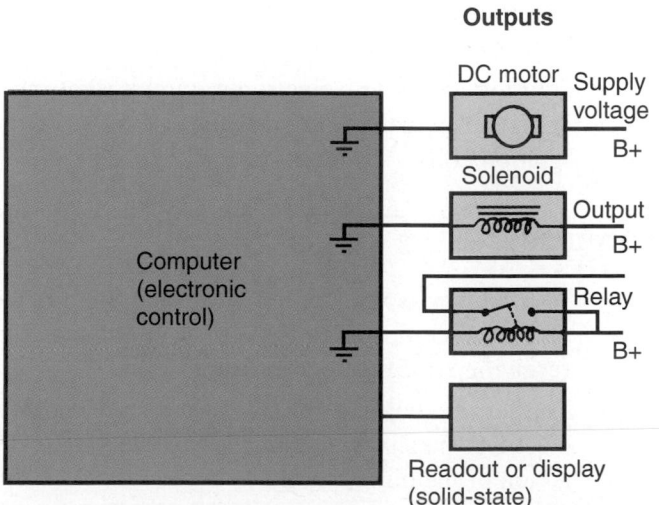

Figure 17-33. These are the basic output actuator classifications. They serve as the "hands and arms" of a computer system.

Relay Operation

A relay is often used as an actuator when a high-current load must be controlled by the computer. The computer will simply ground the relay coil windings. Then, the relay coil field will pull the mechanical contacts closed, allowing a large current to flow to the load, **Figure 17-36.**

Servo Motor Operation

A servo motor provides another way for the computer to produce an output. The computer can ground the motor circuit, turning the motor on or off or reversing motor rotation as needed.

Sometimes, the servo motor is simply a reversible dc motor. The motor will turn a threaded mechanism to produce a controlled movement of a part. A good example would be an idle speed motor. Look at **Figure 17-37.** Note that some motors also serve as sensors for the computer. They can inform the computer as to their positions.

Specific Actuators

Some common actuators used on late-model vehicles include the following:

- **Fuel injector**—solenoid valve that controls fuel flow.
- **Fuel pump**—electric motor that drives a pumping mechanism to force fuel out of the tank and to the engine.

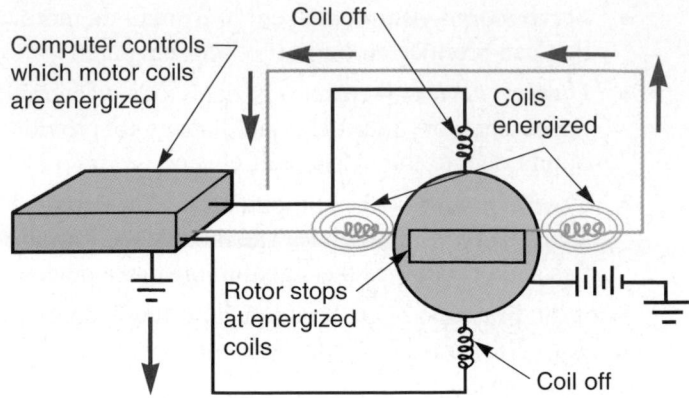

Figure 17-37. A servo, or stepper, motor can be stopped in an exact position. The computer can energize specific coils to attract and stop the armature next to the energized coils.

- **Idle air solenoid**—controls airflow into the engine to help control idle speed.
- **Idle speed motor**—small reversible dc motor that opens and closes the engine throttle valve to control idle speed.
- **EGR solenoids**—solenoids that open and close small ports to control exhaust gas flow back into engine to control emissions.
- **Canister purge solenoids**—solenoids that control vacuum flow to draw fuel vapors into the engine for burning to reduce emissions.
- **Door lock motors**—Solenoids that move latch mechanisms to lock or unlock the doors
- **Electric seat motors**—reversible dc motors that move the seat into the desired position.
- **Ignition coil**—uses high current flow to change low voltage into the high voltage that operates the spark plugs.
- **Ignition module**—electronic circuit that uses computer signals to control the operation of the ignition coils.

Summary

- A computer is a complex electronic circuit that will produce programmed electrical outputs after receiving specific electrical input signals.
- The term cybernetics refers to the study of how electrical-mechanical devices can duplicate the action of the human body.
- The binary numbering system only uses two numbers, zero and one, and is the key to how computers operate.

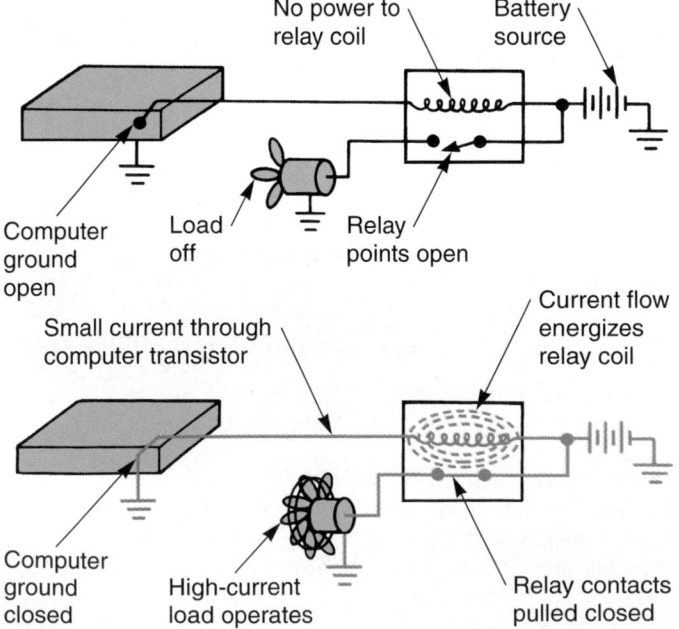

Figure 17-36. A relay can also serve as an actuator to control high current. Low-voltage computer control current is used to close the relay points. Then, high current flows to the lead.

- A gate is an electronic circuit that produces a specific output voltage for given input voltages.
- Digital signals are on-off signals, like from a rapidly flipping switch.
- An analog signal changes in strength or has a variable voltage.
- A computer signal, waveform, or wave is a voltage variation over short periods of time.
- An integrated circuit is an electronic circuit that has been reduced in size and placed on the surface of a tiny semiconductor chip.
- A vehicle sensor is a transducer that changes a condition into an electrical signal.
- Circuit sensing has the computer monitor component and circuit current flow to monitor other conditions without using a sensor.
- The computer sends a reference voltage to passive sensors so that a signal can be fed back for processing.
- Microprocessor means *small* (micro) *computer* (processor).
- Computer memory uses gates that are capable of storing data as voltage charges.
- The three basic types of computer memory are RAM, ROM, and PROM.
- Output drivers control current flow through the actuators.
- A computer network is a series of computers that control different systems but work together to improve overall vehicle efficiency.
- Actuators allow the computer to alter the operation of other components.

Important Terms

Computer
Cybernetics
Digital electronics
Binary numbering system
Gate
Integrated circuit (IC)
Computer signal
Digital signals
Analog signal
Waveform
Pulse width
Signal amplitude
Duty cycle
Input
Processing
Storage
Output
Computer system block diagram
Vehicle sensors
Transduce
Active sensor
Shielded wire
Passive sensor
Potentiometer
Reference voltage
Vref
Circuit sensing

Vehicle control module
Powertrain control module
Engine control module
Anti-lock brake module
Instrumentation module
Ignition module
Suspension system module
Climate control module
Airbag module
High-power control module
Body module
Voltage regulator
Amplifiers
Conditioners
Buffer
Microprocessor
Memory
Clock
Output drivers
Circuit boards
Harness connector
Computer housing
Random access memory (RAM)
Read only memory (ROM)
Programmable read only memory (PROM)
Erasable programmable read only memory (EPROM)
Electrically erasable programmable read only memory (EEPROM)
Flash erasable programmable read only memory (FEPROM)
Keep alive memory (KAM)
Adaptive strategy
Computer network
Multiplexing
Data bus.
Actuators
Solenoid
Relay
Servo motor
Display devices
Control module

Review Questions—Chapter 17

Please do not write in this text. Place your answers on a separate sheet of paper.

1. Define the term "cybernetics."
2. The binary numbering system only uses _____ and _____ and is the key to _____ _____.
3. Describe the five basic types of computer gates.
4. Name and describe the two general sensor classifications.
5. An active sensor requires a reference voltage. True or False?
6. Typically, a reference voltage is about _____ .
7. Explain seven types of computers used on a car.
8. What is an EPROM?
9. What is meant by the term "adaptive strategy"?
10. _____ are the "hand and arms" of a computer network.

● ASE—Type Questions

1. Technician A says the term "cybernetics" refers to the study of how chemical-mechanical devices can duplicate the action of the human body. Technician B says the term "cybernetics" refers to the study of how electrical-mechanical devices can duplicate the action of the human body. Who is right?
   (A) A only.
   (B) B only.
   (C) Both A and B.
   (D) Neither A nor B.

2. Which of the following are known as the "linking cells" of the human brain?
   (A) Neutrons.
   (B) Neurons.
   (C) Protons.
   (D) None of the above.

3. Technician A says digital electronics deals with the way an alternator charges a battery. Technician B says digital electronics deals with the way a computer system uses discrete on and off signals to produce artificial intelligence. Who is right?
   (A) A only.
   (B) B only.
   (C) Both A and B.
   (D) Neither A nor B.

4. Which of the following uses only two numbers, zero and one, and is the key to computer operation?
   (A) Electronic system.
   (B) Base 10 system.
   (C) Decimal system.
   (D) Binary System.

5. Technician A says the numbers 1 and 2 are used in the binary numbering system. Technician B says the numbers 0 and 1 are used in the binary numbering system. Who is right?
   (A) A only.
   (B) B only.
   (C) Both A and B.
   (D) Neither A nor B.

6. Technician A says in computer language a 0 or 1 is called a byte. Technician B says in computer language a 0 or 1 is called a bit. Who is right?
   (A) A only.
   (B) B only.
   (C) Both A and B.
   (D) Neither A nor B.

7. Technician A says the term "gate" refers to an electronic circuit that stops current from flowing in a particular direction. Technician B says the term "gate" refers to an electronic circuit that produces a specific output voltage for given input voltages. Who is right?
   (A) A only.
   (B) B only.
   (C) Both A and B.
   (D) Neither A nor B.

8. Technician A says an AND gate requires voltage at both inputs to produce a voltage at the output. Technician B says an AND gate requires voltage at only one input to produce a voltage at the output. Who is right?
   (A) A only.
   (B) B only.
   (C) Both A and B.
   (D) Neither A nor B.

9. Technician A says both inputs of a NAND gate must be *on* to produce an output. Technician B says both inputs of a NAND gate must be *off* to produce an output. Who is right?
   (A) A only.
   (B) B only.
   (C) Both A and B.
   (D) Neither A nor B.

10. Technician A says an IC is an electronic circuit that has been reduced in size and etched on the surface or a semiconductor chip. Technician B says an IC is a small amplifier circuit for increasing output strength. Who is right?
    (A) A only.
    (B) B only.
    (C) Both A and B.
    (D) Neither A nor B.

11. Technician A says a variable resistor sensor opens or closes the sensor circuit to provide an electrical signal for the computer. Technician B says a variable resistor sensor changes its internal resistance with a change in a condition. Who is right?
    (A) A only.
    (B) B only.
    (C) Both A and B.
    (D) Neither A nor B.

12. Which of the following is another name for an automotive computer?
    (A) Electronic Control Unit.
    (B) Electronic Control Module.
    (C) Electronic Control Assembly.
    (D) All of the above.

13. Technician A says an automotive computer can be located under the dash. Technician B says an automotive computer can be located in the automobile's engine compartment. Who is right?
    (A) A only.
    (B) B only.
    (C) Both A and B.
    (D) Neither A nor B.

14. Which of the following memories can be altered in the field to adjust for operating parameters?
    (A) EEPROM.
    (B) XPROM.
    (C) KAM.
    (D) RAM.

15. Technician A says a servo motor is one type of output device used in an automotive computer system. Technician B says a solenoid is one type of output device used in an automotive computer system. Who is right?
    (A) A only.
    (B) B only.
    (C) Both A and B.
    (D) Neither A nor B.

Activities—Chapter 17

1. Convert your ZIP Code or your telephone number into binary number form. Make a chart that shows how the decimal numbers were converted to binary.

2. Visit several automobile dealerships and gather literature on the anti-lock brake systems offered on their cars. If possible, compare the number of models that offer ABS as standard equipment. Also compare the cost of ABS systems as an option from different manufacturers.

Chapter 18

On-Board Diagnostics and Scan Tools

After studying this chapter, you will be able to:

- Discuss the purpose and operation of on-board diagnostic systems.
- Explain the use of scan tools to simplify reading of trouble codes.
- Compare OBD I and OBD II system capabilities and procedures.
- Locate the data link connector on most makes and models of cars.
- Activate on-board diagnostics and read trouble codes with and without a scan tool.
- Use a trouble code chart in a service manual or code conversion by a scan tool.
- Erase diagnostic trouble codes.
- Correctly answer ASE certification test questions concerning late-model on-board diagnostics and scan tool use.

On-board diagnostics refers to a vehicle computer's ability to analyze the operation of its circuits and to output data showing any problems. All new cars and light trucks have this self-test feature. It is critical that you know how to use this vital troubleshooting aid.

Today, the first thing a technician often does when diagnosing a problem in a computerized system is check for vehicle diagnostic trouble codes with a scan tool. Introduced in Chapter 4, a *scan tool* is used to communicate with the vehicle's computers to retrieve trouble codes, display circuit and sensor electrical values, run tests, and give helpful hints for finding problems. This can all be done quickly and easily, without disconnecting wires or removing parts.

This chapter will summarize recent changes in on-board diagnostic capabilities and explain the fundamental use of scan tools. It will prepare you for other text chapters on troubleshooting and servicing vehicle systems.

Note

This chapter provides the basics of using scan tools and reading trouble codes. More advanced scan tool functions are explained at the beginning of most service chapters and are covered in detail in Chapter 46, *Advanced Diagnostics.*

On-Board Diagnostic Systems

Modern automotive computer systems are designed to detect problems and indicate where they might be located. The computer is programmed to detect abnormal operating conditions. It actually scans its input and output circuits to detect an incorrect voltage, resistance, or current.

Today's on-board diagnostics will check the operation of almost every electrical–electronic part in every major vehicle system. A vehicle's engine control module can detect engine misfiring and air-fuel mixture problems. It monitors the operation of the fuel injectors, ignition coils, fuel pump, emissions system parts, and other major components that affect vehicle performance and emissions control. You can scan for problems in the engine and its support systems, the transmission, the suspension system, the anti-lock brake system, and other vehicle systems. This has greatly simplified the troubleshooting of complex automotive systems.

If the on-board computer finds any abnormal values, it will store a trouble code and light a malfunction indicator light on the instrument panel. This will inform the driver and the technician that something is wrong and must be fixed.

Since some vehicles have six or more computers, on-board diagnostics can be a time-saver when trying to narrow down possible problems. The computers can interact with dozens of sensors and actuators and, in some cases, with each other. No longer can the untrained "shade tree mechanic" hope to repair modern vehicles. It takes the skill of a well-trained technician versed in on-board diagnostics to troubleshoot and repair today's vehicles.

Early On-Board Diagnostic Systems

Most early on-board diagnostic systems could only check a limited number of items. Although these systems were able to detect a problem in the circuit, they were unable to determine what type of problem the circuit, sensor, or system had. Technicians who were unfamiliar with a particular manufacturer's line of vehicles found it difficult to accurately diagnose problems caused by a computer system failure.

Also, there was little or no standardization among these early systems. A wide range of connectors and methods were used to retrieve stored trouble codes. This made it confusing for the tool manufacturers and necessary for the shop or technician to purchase a variety of harness adapters, program cartridges, and service literature. Even the names of the systems and their components were different, making part identification difficult.

Early diagnostic systems are often referred to as **on-board diagnostics generation one** or **OBD I.** See **Figure 18-1.** There are millions of vehicles on the road that use OBD I systems.

On-Board Diagnostics II (OBD II)

A poorly tuned or malfunctioning automobile is a serious source of air pollution. It can produce several times the normal amount of emissions. For this reason, the California Air Resources Board (CARB), along with the Environmental Protection Agency (EPA), recommended and passed regulations that require on-board diagnostic systems to detect problems *before* they could result in the production of harmful exhaust emissions. These regulations also require auto manufacturers to standardize the performance monitoring systems on their cars and light trucks.

As mentioned, OBD I diagnostic systems simply stored a code and illuminated a dash light once a sensor or circuit stopped working completely. The new standard requires on-board diagnostic systems to go a step further by monitoring how efficiently each part of the system is operating. **On-board diagnostics generation two,** abbreviated **OBD II,** is designed to more efficiently monitor the condition of hardware and software that affect emissions. OBD II diagnostics detect part deterioration (changes in performance), not just complete part failure. For example, if a sensor becomes lazy or remains in the low end of its normal operating range, this potential problem is stored as a trouble code in the computer memory for retrieval at a later date. Refer to **Figure 18-1.**

OBD II is designed to keep the vehicle running efficiently for at least 100,000 miles (160,000 km).

The on-board computers used in OBD II systems have greater processing speed, more memory, and more complex programming than computers used in OBD I systems. New vehicles now monitor more functions and can warn the driver and technician of more possible problems that affect driveability and emissions.

The OBD II diagnostic system can produce *over 500* engine performance-related trouble codes. It checks the operating parameters of switches, sensors, actuators, in-system components, their related wiring, and the computer itself.

For example, with OBD II, engine misfire (engine not completely burning the fuel mixture) and fuel system malfunctions will cause the malfunction indicator light in the dash to flash on and off. This is to warn that the misfire or poor mixture could damage the catalytic converter. This warns the driver that the vehicle could be damaged and should be serviced immediately.

OBD II also standardized data link connections, trouble codes, sensor and output device terminology, and scan tool capabilities. In the past, one manufacturer required over a dozen different connectors for the ECMs used in their vehicles. This made it very expensive for the small repair shop to purchase all of the necessary adapters. To solve this problem, the federal government and the Society of Automotive Engineers (SAE) have set standards for all automakers to use.

Malfunction Indicator Light (MIL)

If an unusual condition or electrical value is detected, the computer will turn on an amber-colored indicator light in the dash instrument panel or driver information center. In OBD II systems, this light is referred to as a **malfunction indicator light (MIL).** OBD I systems use a red- or amber-colored **check engine light, service engine soon light,** or other lights with a similar name. Some vehicles use the silhouette of an engine as its MIL light.

The malfunction indicator light will notify the driver that the vehicle needs service. If the MIL has a *continuous glow,* the trouble is not critical but should be repaired. If the MIL light comes on and then goes out, this means that the problem may be intermittent. However, a *flashing* MIL in an OBD II equipped vehicle means that the trouble could damage the catalytic converter and is, therefore, considered critical. The MIL will flash on and off every second conditions that could damage the catalytic converter exist. Whenever the MIL is illuminated, drivers should be advised to bring the vehicle in for service as soon as possible. The technician can then use a scan tool to retrieve information about the problem.

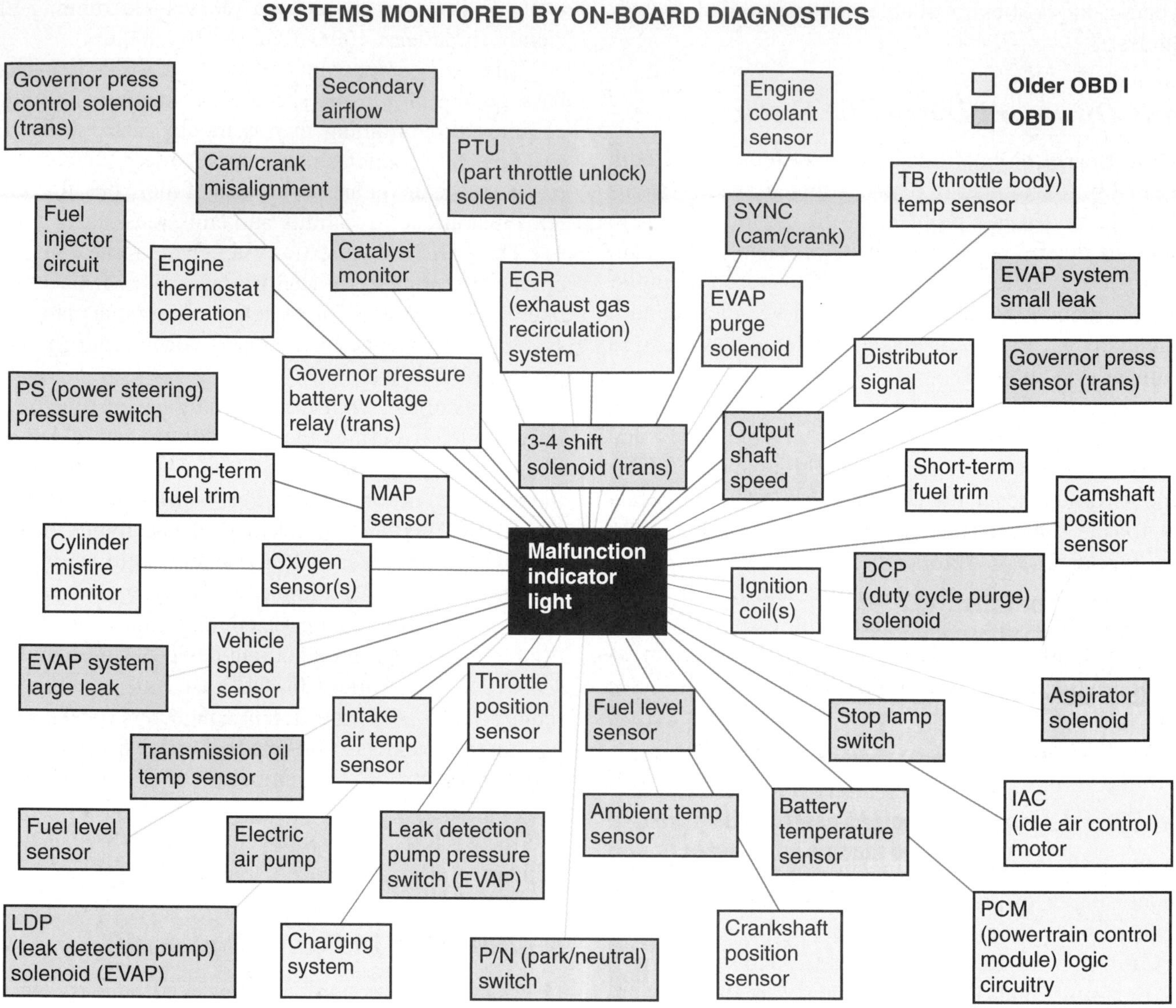

SYSTEMS MONITORED BY ON-BOARD DIAGNOSTICS

Figure 18-1. Compare the OBD I diagnostic system's capabilities with the OBD II system's abilities.

A *trouble code chart* in the service manual will state what each number code represents. Most scan tools have the capability to perform trouble code conversion. *Trouble code conversion* means the scan tool is programmed to automatically convert the number code into abbreviated words that explain the potential problem. The technician can then use the service manual to further isolate the problem.

Diagnostic Trouble Codes

Diagnostic trouble codes (DTC) are digital signals produced and stored by the computer when an operating parameter is exceeded. An *operating parameter* is an acceptable minimum and maximum value. The parameter might be an acceptable voltage range from the oxygen

sensor, a resistance range for a temperature sensor, current draw from a fuel injector coil winding, or an operational state from a monitored device. In any case, the computer "knows" the operating parameters for most inputs and outputs. This information is stored in its permanent memory chips.

Computer System Problems

Common problems that can affect vehicle performance and cause the computer system to set a code and light the MIL in the dash are shown in **Figure 18-2.** These problems include:

- *Loose electrical connection*—input signal from a sensor not reaching the computer properly or an actuator not responding to the computer's output.

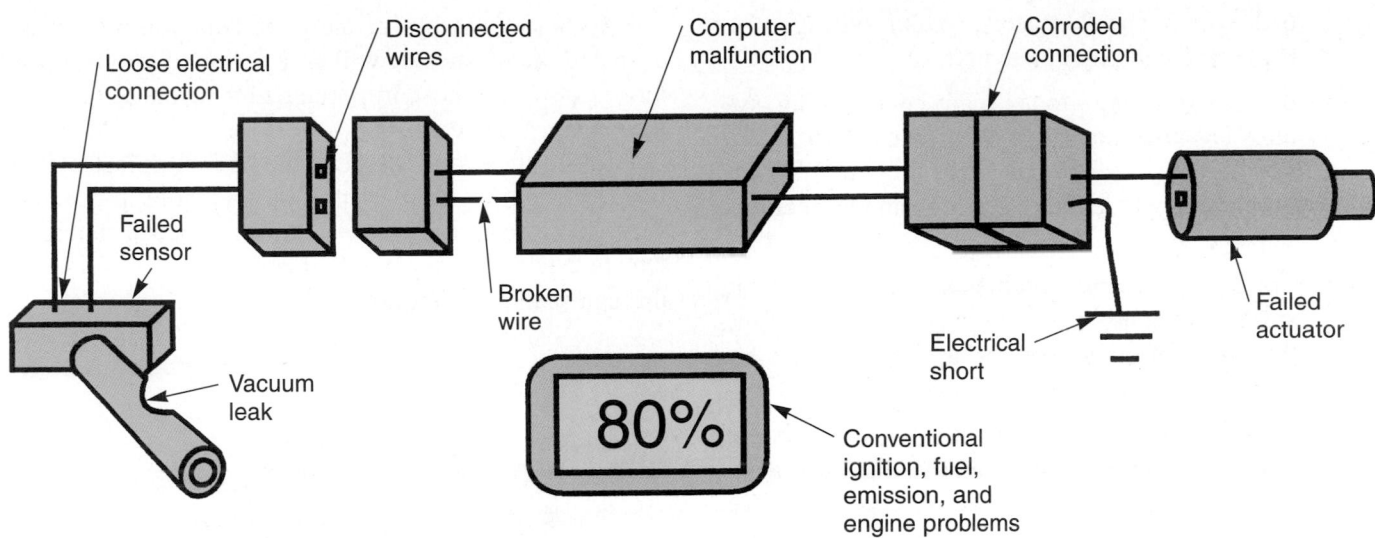

Figure 18-2. Always remember that about 80% of all performance problems are *not* caused by the computer, its sensor, or its actuators. Most problems are the result of conventional problems, such as loose connections, broken wires, vacuum leaks, mechanical problems, etc.

- *Corroded electrical connection*—high resistance in a wiring connector, upsetting sensor input or actuator output.

- *Failed sensor*—opened or shorted sensor or other sensor malfunction preventing normal computer system operation.

- *Failed actuator*—solenoid, servo motor, relay, or display shorted, open, or does not react to computer signals.

- *Leaking vacuum hose*—vacuum leak or poor operation of engine or vacuum-operated actuator that reduces engine or system performance.

- *Electrical short*—wires touching ground or each other to cause a current increase or incorrect current path.

- *Ignition system problems*—open spark plug wires, fouled spark plugs, weak ignition coil voltage, bad crankshaft sensor, etc. For example, a spark plug misfire causing unburned fuel to enter the exhaust can trick the oxygen sensor into trying to create a leaner mixture. The misfire upsets computer system operation and can be detected in OBD II systems by variations in the crankshaft sensor signal.

- *Fuel system problems*—leaking or clogged injectors, bad pressure regulator, faulty electric fuel pump or other problems.

- *Emission system problems*—problems with the catalytic converter, EGR valve, vapor storage system, etc. Many emission components are monitored electronically and will set a trouble code if a malfunction occurs.

- *Engine problems*—mechanical problem that cannot be compensated for by the computer modifying system operation. Engine misfire due to mechanical wear will also trip a trouble code on OBD II systems.

- *Computer malfunction*—an incorrect PROM, wrong internal programming, internal failure of integrated circuit, or failure of other components can disable the computer and alter the operation of related systems.

- *Weak or lazy component*—sensor, actuator, or computer is not outputting normal operating values. In some cases, a sensor's current, voltage, or resistance values are within specs, but the component is sending signals to the ECM at a reduced rate of speed. A lazy sensor can trick the computer system into compensating for an artificial lean or rich condition; it may trip codes on OBD II-equipped vehicles.

- *Transmission problems*—electronically controlled transmissions are monitored and will trip trouble codes if there is a mechanical problem. Transmission problems include a bad vehicle speed sensor, a faulty shift sensor or solenoid, or faulty wiring.

- *Anti-lock brake system problems*—modern anti-lock brakes are monitored by an on-board computer. Anti-lock brake system problems include bad wheel speed sensors, faulty wiring, or a malfunctioning hydraulic unit.

- *Air conditioning*—today's air conditioning systems are also monitored electronically for operational state, leaks, and high pressure. Typical problems include faulty pressure and temperature sensors.

- *Air bag problems*—problems with the air bag system, such as faulty impact sensors, a malfunctioning arming sensor, or a damaged air bag module, will trip trouble codes.

- *Other system faults*—most other vehicle systems have some monitored functions that will trip a trouble code.

Tech Tip
Most computer system problems are conventional (loose electrical connection, mechanical problem, etc.). Only about 20% of all performance problems are caused by an actual fault in the computer or one of its sensors. For this reason, always check for the most common problems before testing more complex computer-controlled components.

Scanning Computer Problems

A scan tool is an electronic test instrument designed to retrieve trouble codes from the computer's memory and to display the codes as a number and words explaining the problem. Also called a *diagnostic readout tool*, the scan tool makes it easier to read diagnostic trouble codes. In some cases, it is the only way to access the computer's diagnostic system. Refer to **Figure 18-3.**

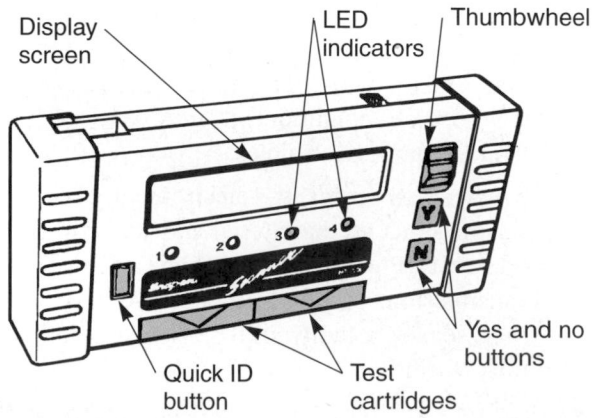

Figure labels: Display screen, LED indicators, Thumbwheel, Quick ID button, Test cartridges, Yes and no buttons

Figure 18-3. A scan tool is now the most important tool of the automotive technician. It will tell you where problems are located. (Snap-on Tools)

A scan tool is by far the most common way to use on-board diagnostics. It will save time and effort. A scan tool is now the most important tool of the automobile technician, **Figure 18-4.**

To use the scan tool, read the operating instructions for the specific type of scan tool. Operating procedures vary. Some scan tools have buttons to control functions. Others have a rotary knob that lets you scroll down through scan tool functions.

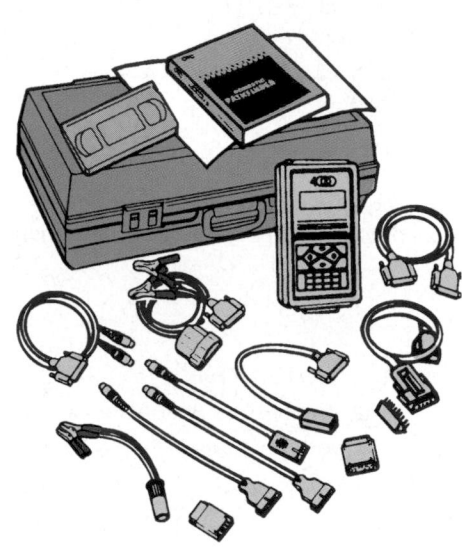

Figure 18-4. Scan tool designs vary. Always read the owner's manual that comes with the tool before use. (OTC)

Scan Tool Program Cartridges

Most scan tools come equipped with several *program cartridges.* These removable cartridges house one or more computer chips that contain specific information about the vehicle to be scanned, **Figure 18-5.**

One type of scan tool cartridge is a *vehicle program cartridge.* This type of program cartridge provides data for one or more vehicle manufacturers (GM, Ford, Chrysler, Asian, European, etc.). Scan tool cartridges must match the model year (vehicle manufacturing date). Program cartridges are also available for certain systems, such as anti-lock brakes, automatic transmissions, etc. New cartridges must be purchased as the on-board diagnostic systems are modified. Some scan tool manufacturers now offer generic storage cartridges that can be updated by downloading the up-to-date specifications to the scan tool from a computer.

Caution
Avoid touching the cartridge or scan tool terminals. Static electricity can destroy the delicate electronics in these units. See **Figure 18-6.**

Program cartridge

Figure 18-5. Scan tool cartridges contain stored information for troubleshooting a specific make, model, and year of vehicle. Troubleshooting cartridges sometimes give added instructions to help solve a problem. (OTC)

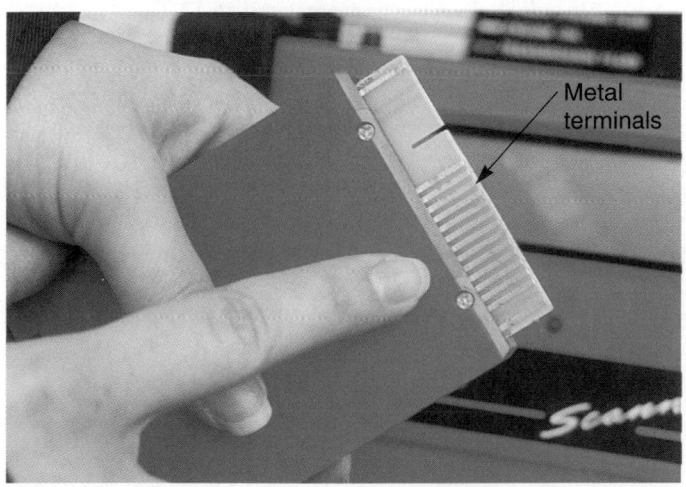

Metal terminals

Figure 18-6. When installing a cartridge into a scan tool, do not touch the metal terminals. Static electricity could damage the cartridge's internal chips or electronics. (Snap-on Tools)

A few scan tools also come with a ***troubleshooting cartridge,*** which can give additional information on how to verify the source of various trouble codes. This is a handy device that can help guide you to the most common sources of trouble. Look at **Figure 18-7.** However, the troubleshooting cartridge must be used in conjunction with the vehicle cartridge. This makes it

Figure 18-7. Install the correct cartridge(s) into the scan tool. Make sure it is fully seated. (Snap-on Tools)

necessary to have a scan tool that can access two cartridges at the same time.

Many scan tools will hold two cartridges, one for the vehicle being tested and another for added convenience. However, most scan tools can access the information from one cartridge at a time. A few scan tools can access both cartridges at the same time. This capability allows for the use of the troubleshooting cartridge discussed earlier. Install the right cartridge(s) into the scan tool. Slide the cartridge straight into the tool to prevent damage.

Data Link Connector (Diagnostic Connector)

The ***data link connector (DLC)*** is a multipin terminal used to link the scan tool to the computer. In the past, this connector was identified by a variety of names, including ***diagnostic connector*** and ***assembly line diagnostic link (ALDL).***

OBD I diagnostic connectors came in various shapes and sizes, and were equipped with a varying number of pins or terminals. With OBD II, the DLC is a *standardized* 16-pin connector. The female half of the connector is on the vehicle, and the male half is on the scan tool cable.

Some of the most common locations for the diagnostic connector include:

- Under the dash, within arm's reach when sitting in the driver's seat, **Figures 18-8A** and **18-8B.** This is the standard OBD II location.
- Near the firewall in the engine compartment, **Figure 18-8C.**
- Near or on the side of the fuse box, **Figure 18-8D.**
- Near the inner fender panel in the engine compartment, **Figure 18-8E.**
- Under the center console, **Figure 18-8F.**

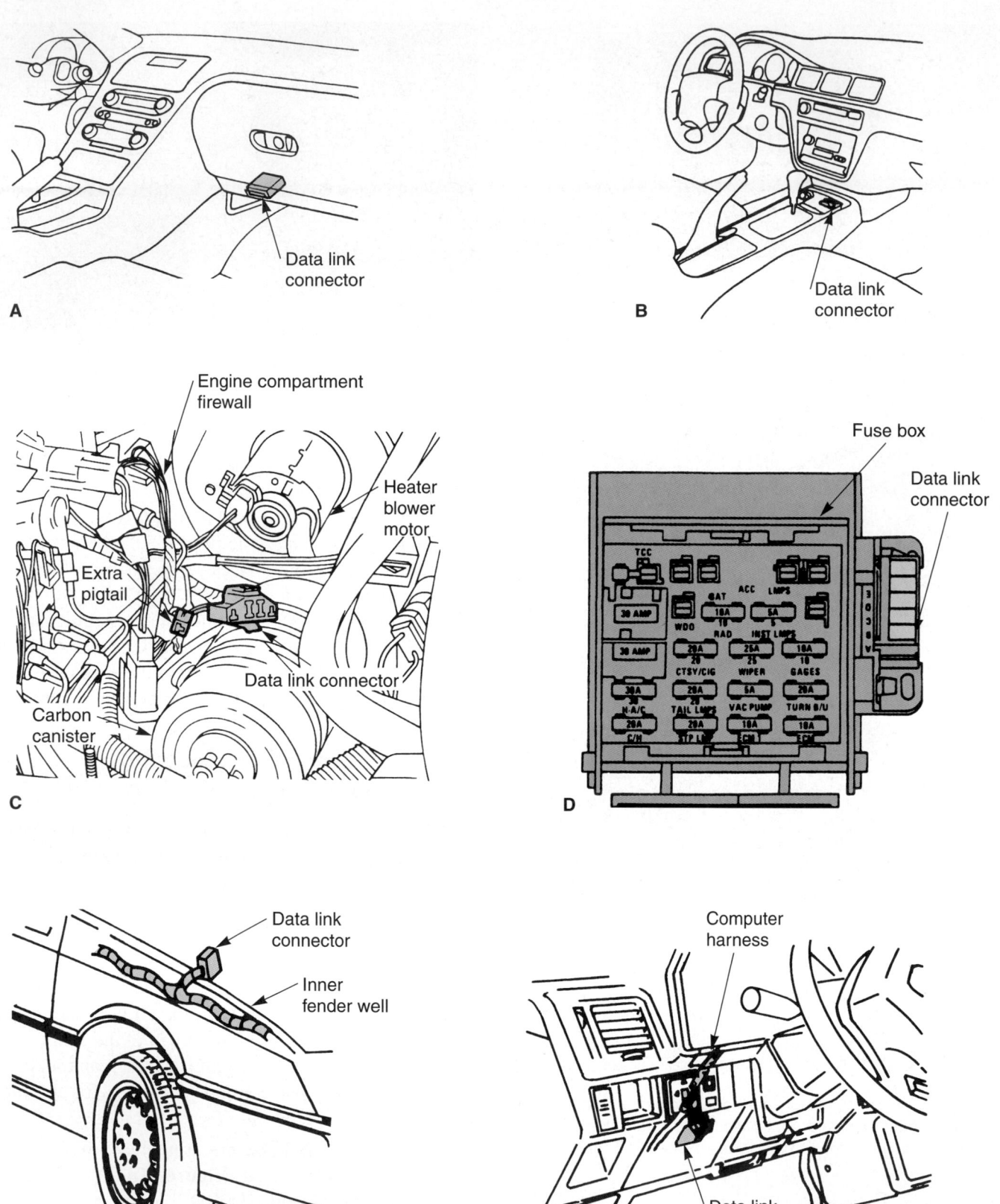

Figure 18-8. Data link connector locations vary. A—OBD II vehicles have their data link connector below the dash, within easy reach of the driver's seat. B—The OBD II connector is sometimes located in the center console. C—Some Ford diagnostic connectors are on the firewall, near the back of the engine. D—Some early General Motors connectors are under the right side of the dash or next to the fuse box. E—Early Chrysler diagnostic connectors are located in the engine compartment. F—Other data link connectors may be located under the dash, in or behind the glove box, under the center console, etc. Refer to the service manual if needed. (General Motors, Ford, and Snap-on)

With OBD II diagnostic systems, you should be able to connect a scan tool to the vehicle's data link connector with one hand while sitting in the driver's seat or kneeling outside the vehicle.

Tech Tip
Some OBD I vehicles are equipped with a 16-pin, OBD II–style data link connector. Do not assume that a vehicle equipped with a 16-pin connector is OBD II compliant.

Connecting the Scan Tool

The scan tool cable should slide easily into the vehicle's data link connector. If not, something is wrong. Never force the two together or you could damage the pins on the tool cable or the data link connector. You may have to use an *adapter* so the scan tool connector will fit the vehicle's DLC or communicate with different pin configurations, **Figure 18-9.**

If not powered through the DLC, connect the scan tool to battery power. In most cases, you can use a cigarette lighter adapter to connect power to the scan tool. See **Figure 18-10.** You can also use alligator clips to connect the tool to the battery.

Caution
Make sure you are connecting the scan tool to the vehicle properly. Some technicians have mistakenly connected scan tools to the wrong connector (tach connector, for example), which can damage the scan tool.

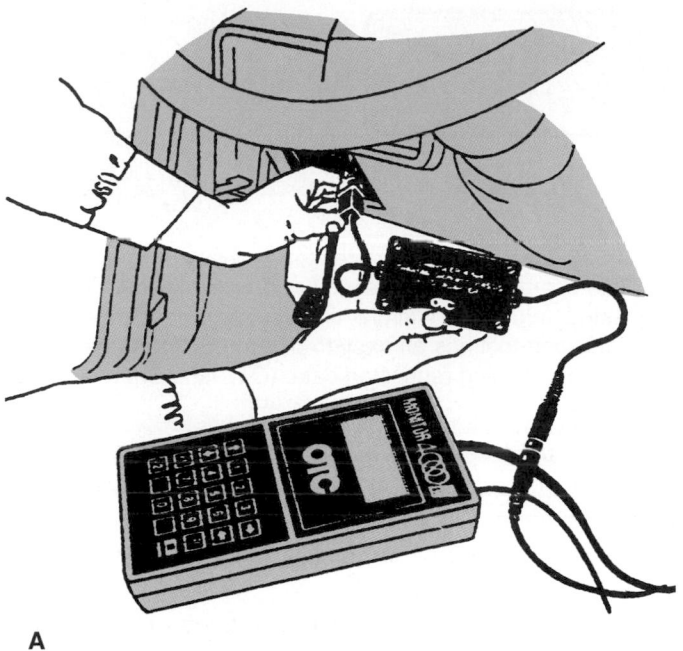

A

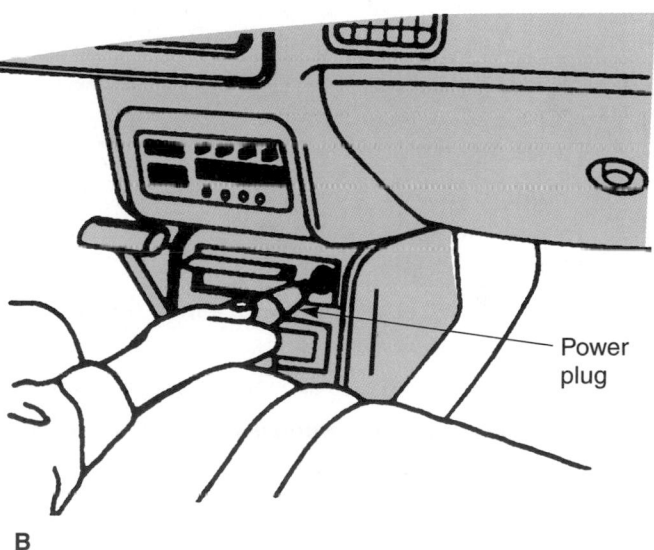

Power plug

B

Figure 18-10. A—Connect the scan tool cable to the vehicle's data link connector. Make sure the pins match up. Do not force them together. B—Scan tools can be powered by a cigarette lighter plug, or they can be connected directly to the battery. (OTC)

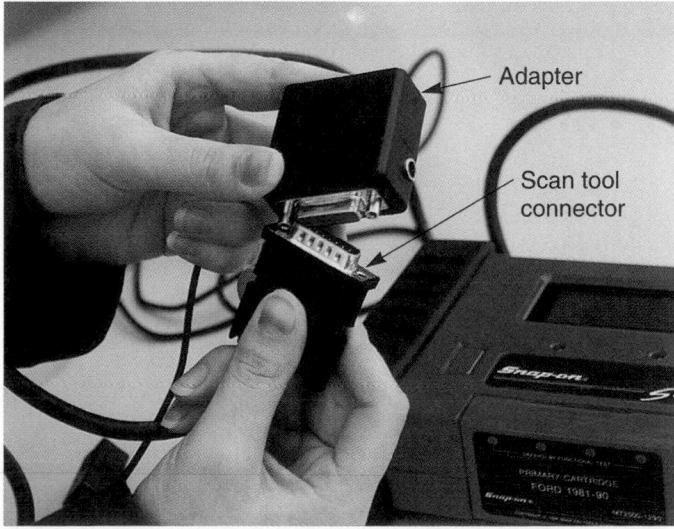

Adapter

Scan tool connector

Figure 18-9. An adapter is sometimes needed between the scan tool cable and the vehicle's data link connector. As OBD II–equipped vehicles become more commonplace, these adapters will be needed less. (Snap-On Tools)

Using Scan Tools

Modern scan tools will give *prompts,* or step-by-step instructions in their display windows. The prompts tell you how to input specific vehicle information and run diagnostic tests. See **Figure 18-11.** Scan tool instructions are procedures and specifications programmed into the cartridge.

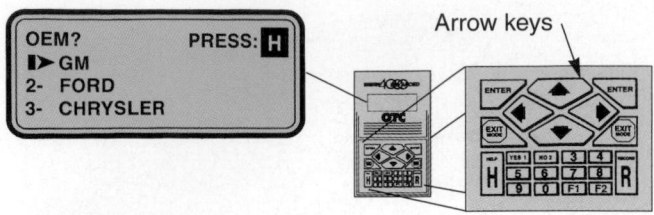

Figure 18-11. Controls on a scan tool will vary by manufacturer. A—This scan tool uses a keypad and arrow keys for inputting requested information about the vehicle and desired tests. B—This scan tool has an easy-to-use rotary knob and yes–no buttons for inputting requested data. (OTC and Snap-on Tools)

First, the scan tool may ask you to input VIN information from the plate on the top of the dash. You may be asked to input specific numbers and letters from the VIN. Refer to **Figure 18-12.**

The VIN data lets the scan tool know which engine, transmission, and options are installed on the vehicle. With some makes, however, the on-board computer will contain this data and will automatically download it into the scan tool. Then, you will be able to select the information that you would like the scan tool to give you. Some of the information you can request includes:

- *Stored diagnostic trouble codes*—gives trouble code number.

- *Fault description*—explains what each stored diagnostic code means. This information is given with the trouble code number on most scan tools.

- *Datastream information*—displays the operating values of all monitored circuits and sensors.

- *Run tests*—performs sensor and actuator tests.

- *Oxygen sensor monitoring*—performs detailed tests of the O$_2$ sensor signal.

- *Failure record*—lists the number of times a particular trouble code has occurred by keystarts or warm-ups.

Figure 18-12. The scan tool may ask you to input VIN information. This lets the tool know how the vehicle is equipped— engine type, fuel system type, computer configuration, etc. (Snap-on Tools)

- *Freeze frame*—takes a snapshot of sensor and actuator values when a problem occurs.

- *Troubleshooting*—Provides help and instructions for diagnosing faults.

Figure 18-13 gives the general sequence for using a scan tool. You can ask the scan tool to give more information on a trouble code, and the tool will display words that give sensor resistance values, common problems, and other useful information. **Figures 18-14** and **18-15** show some examples of scan tool troubleshooting tips.

As stated earlier, some scan tools are able to take a snapshot, or *freeze frame,* of sensor and actuator values when a problem occurs. The tool records the values from all monitored components so they can be further evaluated. This helps you locate and correct intermittent problems more easily.

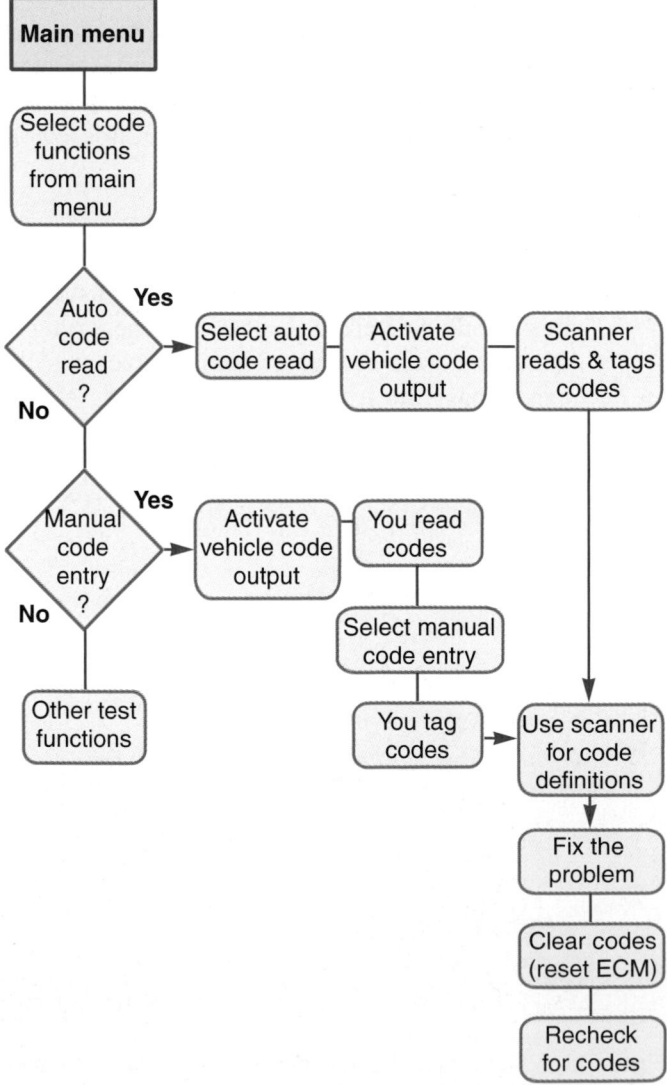

Figure 18-13. Flowchart showing the basic steps for using a scan tool. (Snap-on Tools)

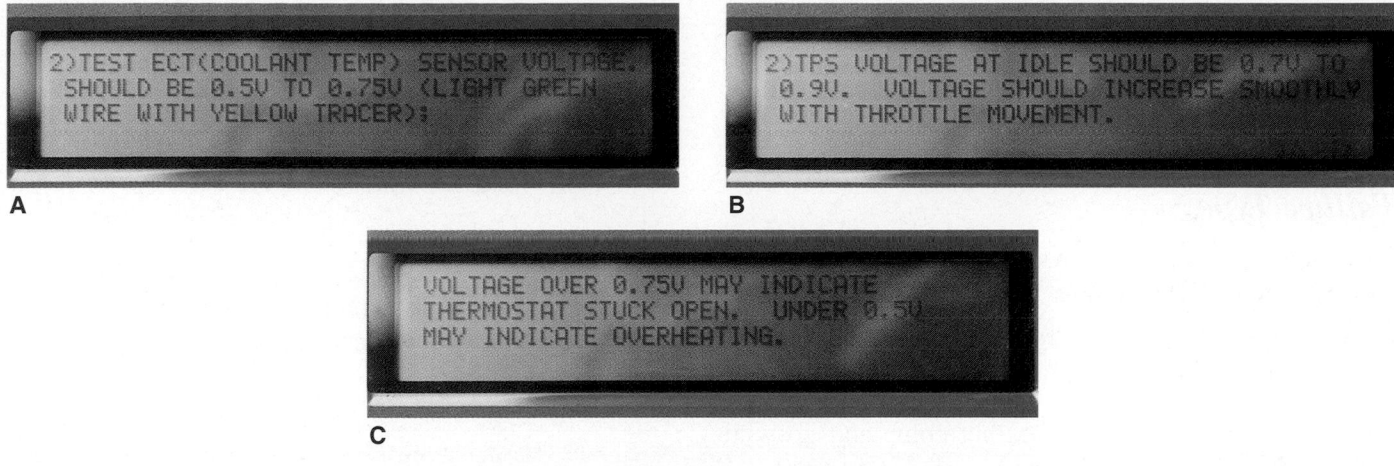

Figure 18-14. Here are examples of how scan tools will give specifications and tips for finding the source of a problem. A—The scan tool is showing normal sensor voltage and which wire to probe when measuring actual voltage. B—This scan tool is giving more information for testing the throttle position sensor. C—Scan tools can also give hints on how engine overheating or overcooling can fool the computer into signaling a problem with the engine coolant temperature sensor.

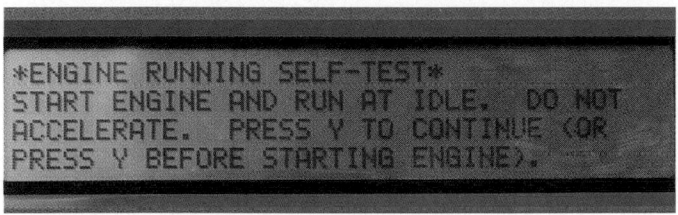

Figure 18-15. A key-on/engine-on test is sometimes needed to further diagnose problems. You must allow the engine to reach operating temperature first so all sensors are operating normally.

Tech Tip

Most technicians check the ECM for stored diagnostic trouble codes before performing tests on specific components. This is a quick way to pinpoint any hard failures, so they can be repaired first.

Always correct the cause of the *lowest number* diagnostic trouble code first. Sometimes, fixing the cause of the lowest code will clear other codes because of component interaction. If not, you can use other scan tool features to find and solve more complex problems. A trouble code does *not* mean that a certain component is bad. It simply indicates that a possible problem has been detected in that particular device or circuit.

Most ECMs count the number of times a trouble code has occurred. This information is stored in a *failure record,* or *failure recorder,* that also indicates the number of keystarts since the last time the trouble code occurred. The failure recorder in OBD II systems counts the number of times the engine reached operating temperature, rather than keystarts, since the last time the code was set. If one

code has occurred more frequently than the others, investigate this code first. In many cases, the lowest number code and the most frequently stored code are the same.

Diagnostic Trouble Code Identification

As mentioned, early on-board diagnostic systems were not standardized. Often, technicians would have to refer to the service manual to find out what a particular code number meant. OBD I and earlier codes were different for each manufacturer.

To simplify troubleshooting, OBD II requires all auto manufacturers to use a set of *standardized alpha-numeric trouble codes.* Each trouble code identifies the same problem in all vehicles, regardless of the manufacturer.

OBD II codes contain a letter and a four-digit number. The letter in all OBD II codes indicates the *general function* of the affected system (power train, chassis, etc.).

The first digit of the number indicates whether the code is a standard trouble code or a nonuniform code. *Standard trouble codes,* or SAE codes, are indicated by the number *0.* *Nonuniform codes* (nonstandard OBD II codes that are assigned by the auto manufacturers) have the number *1* after the system letter. The second digit of the OBD II code number indicates the *specific function* of the system where the fault is located, such as fuel, computer, etc.

The code's last two digits refer to the specific fault designation. The *fault designation* pinpoints exactly which component or circuit of the system might be at fault, as well as the type of problem. Regardless of the type of vehicle being serviced, the core trouble code numbers will be the same. The scan tool must explain the code and, in some cases, may describe how to fix the problem.

Figure 18-16 gives a breakdown of the OBD II diagnostic code. Study it carefully. **Figure 18-17** shows the display window of a scan tool that has found a stored trouble code.

Failure Types

Computer system failures can be grouped into two general types: hard failures and soft failures. A ***hard failure*** is a problem that is always present in a computer system. An example of a hard failure is a disconnected wire or another problem that would cause a general circuit failure. A hard failure does not come and go with varying conditions. After the computer memory is cleared, any hard failures will usually reset as soon as the engine is started or the affected system is energized.

A ***soft failure,*** or *intermittent failure,* is a problem that only occurs when certain conditions are present. It might be present one minute and gone the next. Soft failures will usually be stored in memory for 30-50 key starts or engine warm-ups. An example of a soft failure is a loose terminal that connects and disconnects as the vehicle travels over bumps in the road. Low-input, high-input, and improper range failures are usually classified as soft failures.

Computer system failure types can be further broken down into four general categories:

- A ***general circuit failure*** means the circuit or component has a fixed value, no output, or an output that is out of specifications. This is the most severe fault, but it is the easiest to locate. It

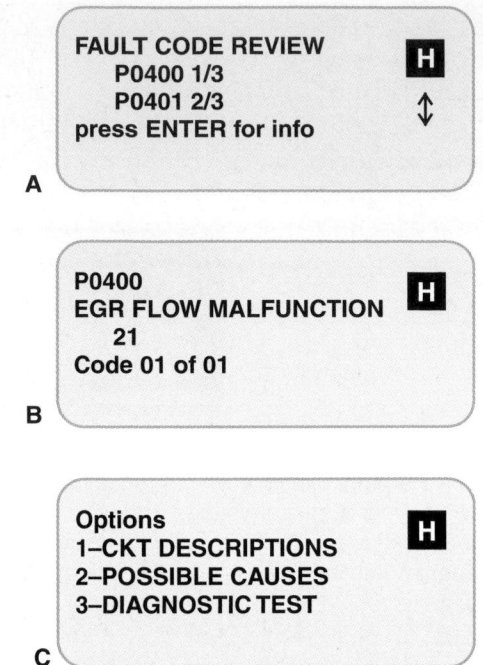

Figure 18-17. This is an example of what you might see on the display of a scan tool. A—The scan tool will give you the trouble code numbers. B—If you request information on the stored trouble codes, the tool will explain what each code means. C—Options will allow you to use the scan tool to get detailed descriptions of each code, list possible causes, or perform diagnostic tests. (OTC)

is caused by disconnected wires, high-resistance connections, shorts, or a component constantly operating out of parameter.

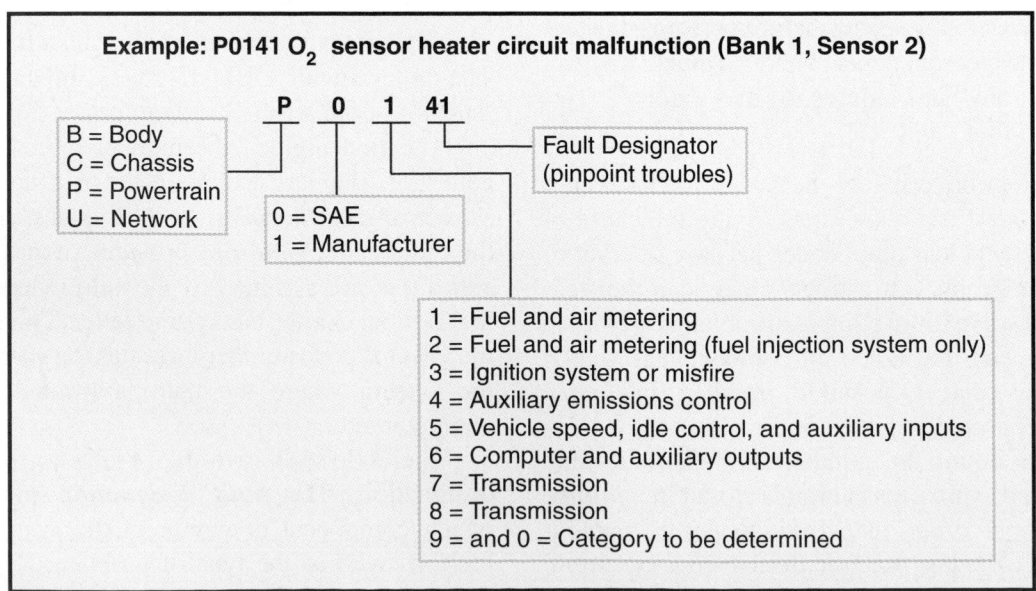

Figure 18-16. OBD II trouble codes are alpha-numeric. Note what each part of the trouble code means. The first part of the code tells you which system is having problems. The last part of the code identifies the specific problem circuit or component.

- A *low-input failure* is one that produces a voltage, current, or signal frequency below normal operating parameters. A weak or abnormally low signal is being sent to the on-board computer. This type of failure is often caused by high circuit resistance, a poor electrical connection, a contaminated or failed sensor, or a similar problem.

- A *high-input failure* results when the signal reaching the on-board computer has more voltage, more current, or a higher frequency than normal. This type of failure is often caused by a faulty sensor or a mechanical fault that is "fooling" the computer system.

- An *improper range/performance failure* occurs when a sensor or actuator is producing values slightly lower or higher than normal. The circuit is still functioning, but not as well as it should under normal conditions. This type of failure can be caused by a contaminated sensor, a partial sensor failure, a poor electrical connection, and similar problems. Improper range/performance failures were not detected in OBD I systems and were often difficult to find. OBD II systems can detect improper range/performance failures.

Datastream Values

Datastream values, or diagnostic scan values, produced by the vehicle's computer give electrical operating values of sensors, actuators, and circuits. These values can be read on a scan tool's digital display and compared to known normal values in the service manual. Datastream values give additional troubleshooting information when trying to locate a problem.

Note
For more information on scan tools and datastream values, see Chapter 46, *Advanced Diagnostics.*

Engine-Off/Key-On Diagnostics

In order to access the ECM data on most vehicles, it is necessary to turn the ignition key on. *Engine-off/key-on diagnostics* are performed by triggering the ECM's on-board diagnostic system with the ignition key in the *on* position but *without* the engine running. This allows you to access any stored trouble codes in the computer's memory chips. Engine-off/key-on diagnostics are usually performed *before* engine-on/key-on diagnostics. Look at **Figure 18-18.**

If you anticipate working in the engine-off/key-on diagnostic mode for over 30 minutes, connect a battery

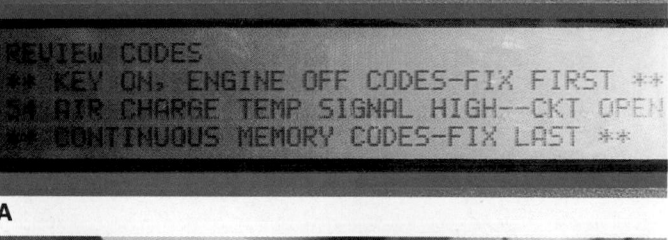

A

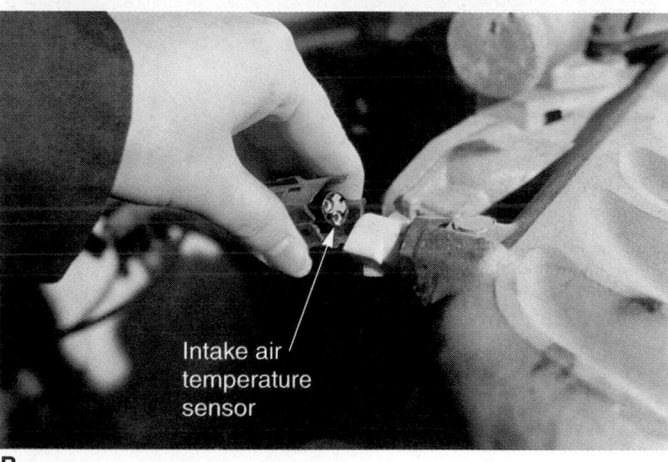

Intake air temperature sensor

B

Figure 18-18. Most technicians check for stored trouble codes first. A—This scan tool readout shows a problem with the intake air (air charge) temperature sensor. B—You would then know to check that sensor and its wiring for problems. (Snap-on Tools)

charger to the vehicle. This will prevent the extended current draw from draining the battery and upsetting the operation of the computer while in the diagnostic mode. False trouble codes could result from a partially drained battery.

Wiggle Test

Many computer system failures, especially intermittent failures, are caused by loose, dirty, or corroded connections. These can be found by performing a *wiggle test,* or "flex" test.

To perform a wiggle test:
1. Connect a scan tool to the vehicle and choose the appropriate test options. Refer to the instructions provided with the scan tool.
2. Place the vehicle in the engine-off/key-on diagnostic mode.
3. Flex wires and wiggle harness connectors while scanning for problems.
4. If wiggling a wire or connector produces a new diagnostic trouble code, check the wire or electrical connection more closely. It may be loose, corroded, or damaged, **Figure 18-19.**

Figure 18-19. The wiggle test involves moving wires and connectors while scanning for trouble codes. If wiggling a wire trips a code, you found the location of the problem.

Some technicians perform a wiggle test while the engine is running. If engine operation changes suddenly (stalls or idles high, for example) when a connector or wire is flexed, the problem is located at or near that point. Be careful when performing a wiggle test on a running vehicle.

You might also want to use a heat gun to heat potentially faulty components during a wiggle test. For example, electronic amplifiers and modules tend to malfunction when hot. This could help find an intermittent problem.

Caution
Exercise care when using a heat gun. The heat generated by the gun can easily melt most plastics and damage electronic components.

Engine-On/Key-On Diagnostics

Engine-on/key-on diagnostics are performed with the engine running at full operating temperature. These tests check the condition of the sensors, actuators, computer, and wiring while they are operating under normal conditions.

Switch Diagnostic Test

A *switch diagnostic test* involves activating various switches while using a scan tool. The scan tool will tell you which switch to move and will monitor its operation. The scan tool will quickly indicate if the switch is working normally, **Figure 18-20.**

Figure 18-20. Most scan tools will also perform switch and actuator tests. This is sometimes done automatically. You may be prompted to close different switches to make sure each one is working. The scan tool may also be able to perform additional actuator tests. (Snap-on Tools)

For example, you might be told to shift the transmission shift lever through the gears, press on the brake pedal, and turn the air conditioning on and off. As each step is performed, the scan tool will indicate if the affected switch is *OK and* whether or not the ECM is reading the switch input. Refer to the service manual for details of the switch diagnostic.

Actuator Diagnostic Tests

An *actuator diagnostic test* uses the scan tool to order the vehicle's computer to energize specific output devices with the engine on or off. This will let you find out if the actuators are working. Most actuator diagnostic tests are considered intrusive tests, since engine or vehicle operation will be drastically affected while the device is being tested. Actuator diagnostic tests might:

- Fire or prevent the firing of the ignition coil.
- Open and close the fuel injectors.
- Cycle the idle speed motor or solenoid.
- Energize the digital EGR valve solenoids.

You can then watch or listen to make sure these actuators are working. With OBD II, the scan tool will give readouts showing whether there is trouble with any actuators.

Not all vehicle manufacturers provide switch and actuator diagnostic tests. Refer to the service manual or scan operating manual for details.

Scanning During a Test Drive

With a modern scan tool, you can also check for problems while driving the vehicle to simulate the conditions present when the trouble happens, **Figure 18-21.** For example, if the problem occurs only while driving at a specific speed when the engine is cold, you can scan under these conditions. Start the cold engine and drive at the specified speed while scanning for problems. You can

Figure 18-21. Scan tools are sometimes used while test driving vehicles. This will allow you to check engine and vehicle operating parameters while duplicating the conditions present when the problem occurred. (OTC)

then take a snapshot, or freeze-frame, (if the scan tool has this feature) when the problems occur. For more information on advanced diagnostics, refer to Chapter 46.

Energizing OBD I Systems Without a Scan Tool

If you do not have a scan tool and are working on an OBD I-equipped vehicle, there are several ways to activate the computer's on-board diagnostics and to retrieve trouble codes, **Figure 18-22.** The most common methods include:

- Using a jumper wire to ground one of the data link connector terminals and then reading the flashing code on the dash-mounted check engine light, **Figure 18-23.**

- Connecting an analog voltmeter to vehicle ground and to one terminal on the data link connector while jumping from the pigtail (extra wire) to the data link connector. The code is produced by the meter's needle movement.

- Turning the ignition key on and off several times within a few seconds and reading the flashing code on the dash-mounted check engine light. See **Figure 18-23.**

- Pushing two dash-mounted climate control buttons at the same time and read the dash display, **Figure 18-24.**

Always refer to the service manual for detailed instructions. Procedures vary from model to model as well as from year to year. These methods will not work on vehicles equipped with OBD II.

Tech Tip
Some older vehicles with on-board computers do *not* have on-board diagnostics. You must use conventional testing methods to pinpoint problems on these vehicles.

Reading Trouble Codes Without a Scan Tool

Reading trouble codes manually involves noting the computer output after the on-board diagnostics have been energized. There are several different ways that trouble codes can be read on older vehicles. The most typical methods include:

- Observing the check engine light as it flashes on and off.

- Noting an analog voltmeter's needle as it deflects back and forth.

- Watching a test light connected to the data link connector flash on and off.

- Reading the digital display in a climate control panel or driver's information center.

- Observing the LED display on the side of the ECM.

Reading the Flashing Check Engine Light

The check engine light indicates each code as it blinks on and off with long and short pauses between each flash. Some codes are single-digit numbers and others are two-digit numbers. The number of flashes between each pause designates a single-digit code or one-half of a two-digit code, **Figure 18-25.** With a single-digit code, count the number of pulses and this equals the code number. After a pause, the next code number would be given.

Reading a two-digit code is a bit more involved. If, for example, the check engine light blinks twice, pauses momentarily, and then blinks two more times, the trouble code would be 22 (2 pause 2). If the light flashes once, pauses, and flashes three more times, the trouble code would be 13 (1 pause 3). When there are multiple two-digit codes, there will be a relatively long pause between codes.

Reading Voltmeter Needle Deflections

An analog voltmeter code is read by counting the number of needle deflections between each pause. This is similar to the dash light flashes. However, the computer usually produces 5 volt pulses for the test meter.

RETRIEVING OBD I TROUBLE CODES (WITHOUT SCAN TOOL)

The three major auto makers use different procedures to make their car's on-board computer spit out trouble codes. The computer can actually detect if a sensor or actuator has failed or if a bad electrical connection has developed. This will help you know where to test for possible computer system problems.

The following is a summary of typical methods used to make an on-board computer produce these codes:

General Motors Corporation Trouble Codes

1. Locate diagnostic connector. It is usually under dash near fuse panel or steering column.
2. Use a jumper wire or paper clip to short across designated terminals in connector.
3. Watch engine light flash on and off in a Morse type code. Count number of flashes between each pause and note them. Three flashes, a pause, and two flashes would equal code 32.
4. Refer to trouble code chart in service manual for a explanation of code number.
5. Test suspected component or circuit with a digital VOM. Compare your test readings to factory specs.
6. Note that some GM cars require you to press two climate control buttons on dash at same time to enter self-diagnosis. Then, trouble code number will appear in dash. You would then need to find the same number in service manual trouble code chart.

Ford Motor Company Trouble Codes

1. Locate diagnostic connector. It is usually in engine compartment on firewall, fenderwall, or near engine intake manifold.
2. Connect an analog or needle type VOM to designated terminals in diagnostic connector.
3. Use a jumper wire to connect extra pigtail near connector to service-manual-designated terminal in connector.
4. Observe needle fluctuations on voltmeter as you did when watching engine light for a GM car. Count needle movements between each pause. Two needle movements, a pause and then six needle movements would equal code 26.
5. Refer to service manual trouble code chart to find out what number code means.
6. Use conventional testing methods and your VOM to pinpoint cause for problem.

Chrysler Corporation Trouble Codes

1. Chrysler provides a diagnostic connector in engine compartment on late model cars. However, connector is NOT needed to energize self-diagnosis. It is provided so a scanner-tester can be connected to system.
2. To trip trouble codes, simply turn ignition key on and off three times within five seconds. Turn key on, off, on, off, and then leave it ON.
3. Observe engine light flashing on and off. Count number of flashes between each pause. Three flashes, a pause, and then one flash would equal a trouble code of 31.
4. Refer to the service manual trouble code chart to find out which component or circuit is indicated by the trouble code.
5. Use conventional VOM tests to find the source of the trouble. Test the sensor or actuator and the wiring between the device and the computer.

Note! computer self-diagnosis systems and procedures can vary from the methods just described. Always refer to a factory service manual when in doubt!

Figure 18-22. Study the basic methods for reading OBD I trouble codes without a scan tool. (TIF Instruments)

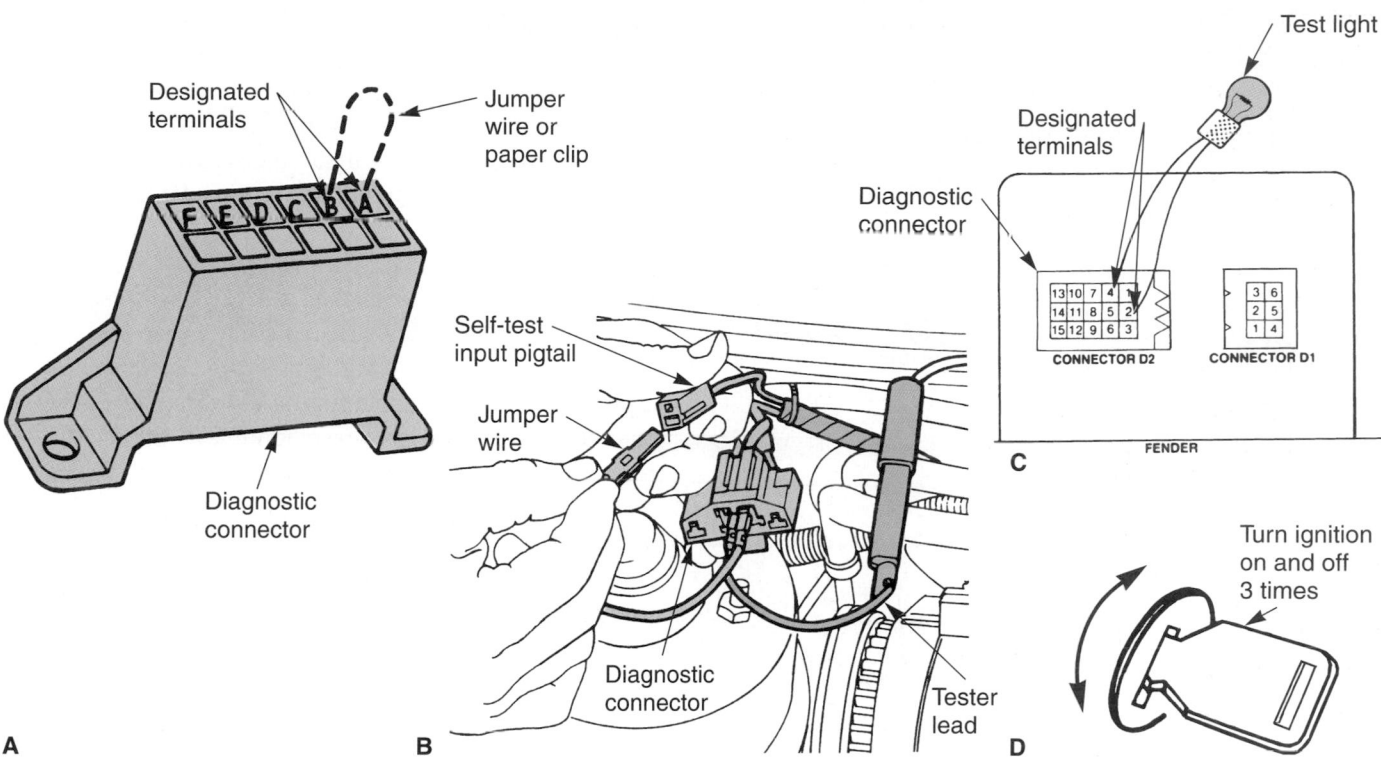

Figure 18-23. Various methods are used to energize OBD I systems. A—Use a jumper wire or paper clip to ground specified terminals in most GM connectors. B—Jump from the extra pigtail to a specified terminal in many Ford connectors. C—Connect a test light across specified terminals in this connector. D—Turning the ignition key on, off, on, off, and then on within five seconds will energize on-board diagnostics with most Chrysler cars. (Chrysler, GM, and Ford)

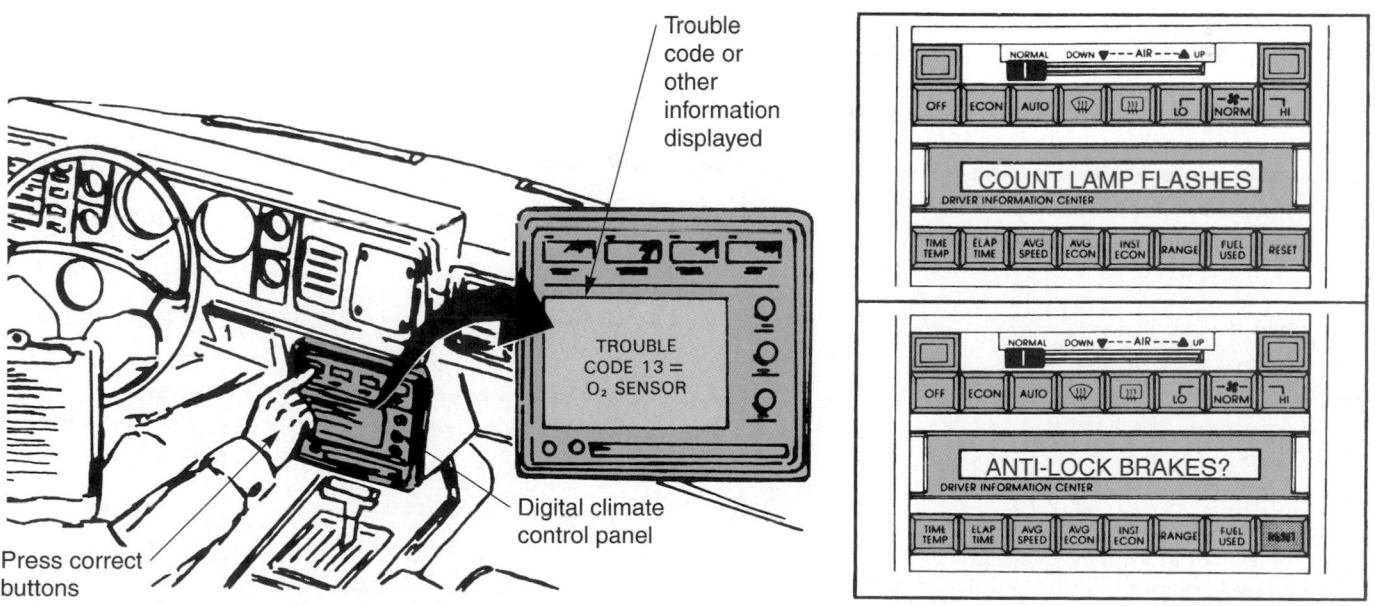

Figure 18-24. You can read trouble codes on a few vehicles in the dash climate control digital readout panel. A—By pressing two buttons at same time, the readout will give you any stored trouble code numbers. B—Some systems will actually describe each trouble code. (Ford and General Motors)

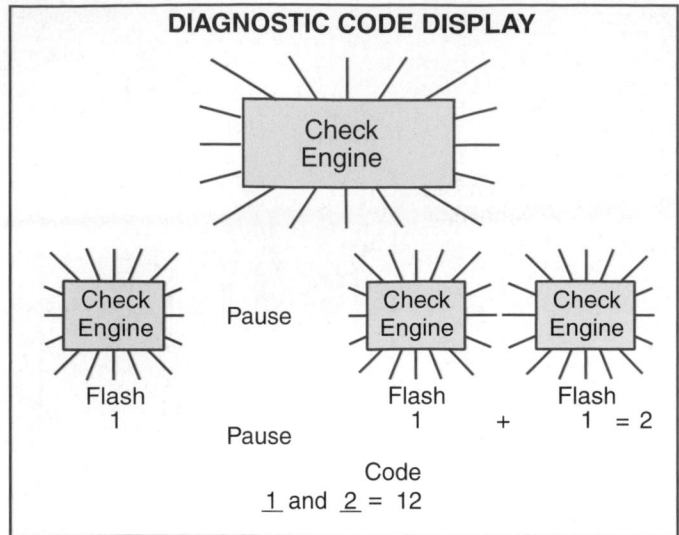

DIAGNOSTIC CODE DISPLAY

Figure 18-25. A dash indicator light will normally glow if the computer detects a potential fault. This tells the driver and technician something is wrong. After energizing the computer's diagnostic mode, the dash light may flash on and off to produce a number code. Note how this code is read. (General Motors)

An example of a two-digit code would be if the needle deflected once, paused, and then deflected four more times. The first digit would be one and the second would be four. This would be a trouble code 14. A few digital test meters have a bar graph that will show trouble code pulses. Look at **Figure 18-26.** However, most digital meters cannot read trouble codes. A test light code is read by noting the flashes of the tester bulb.

Caution

Use only a high-impedance (10 megohms or higher) test light or multimeter when testing computer circuits. A high current draw from a conventional test light could shorten electronic circuit life or even destroy delicate integrated circuits.

Reading Digital Codes and LED Displays

Dash digital codes are read like scan tool codes. These codes are retrieved after pressing two dash buttons, usually climate control buttons, at the same time. The climate control or temperature readout will then show any trouble codes.

An LED trouble code is produced by indicator lights on the side of the computer. This is a less common method to display computer trouble codes. **Figure 18-27** summarizes how to use the LED trouble code.

Trouble Code Charts

A trouble code chart in the service manual will explain what each trouble code number means, **Figure 18-28.** This will help you know where to start further tests on specific components.

Erasing Trouble Codes

Erasing trouble codes, also termed clearing diagnostic codes, removes the stored codes from computer memory after system repairs have been made. In most

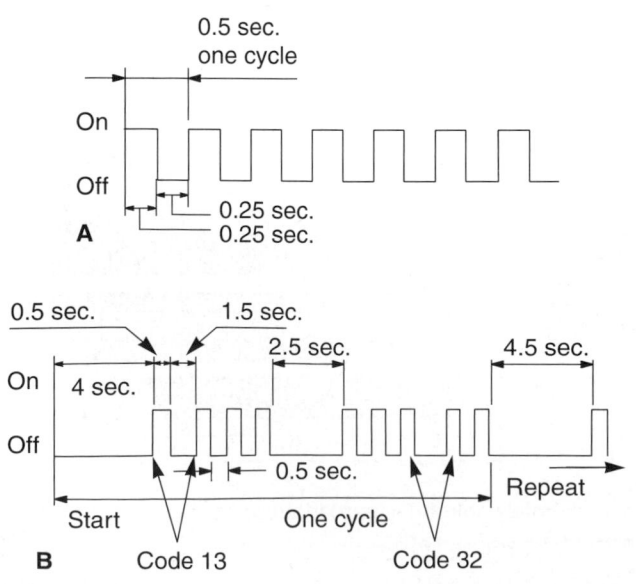

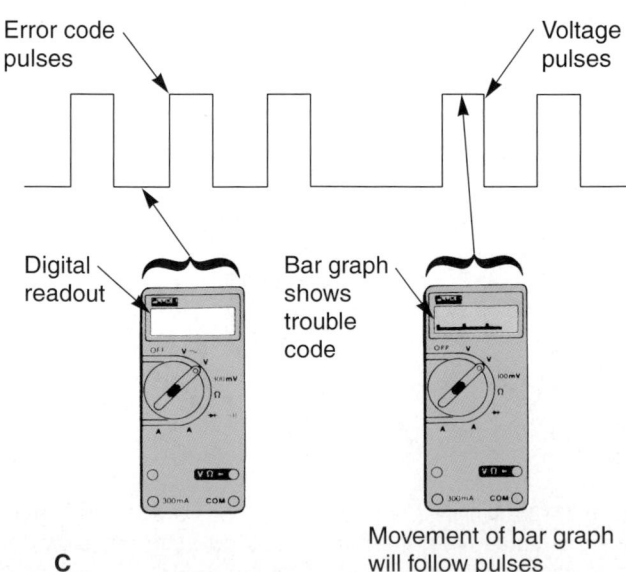

Figure 18-26. Another method of reading a computer code is by noting light flashes or pulses. A—If the computer system is normal, pulses or flashes will occur two times per second. B—However, if a malfunction is present, the pulses and pauses denote any stored trouble codes. C—This digital VOM has a bar graph for reading coded voltage pulses. (Fluke)

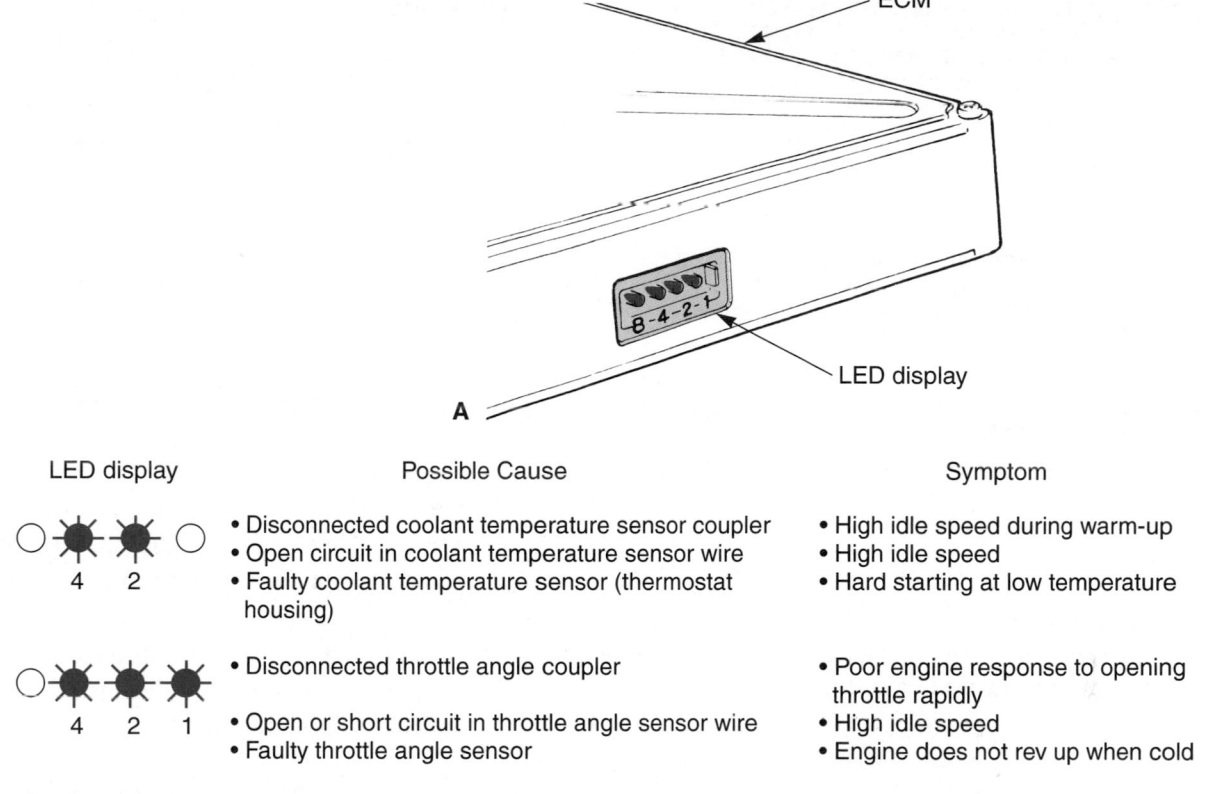

Figure 18-27. This computer has light emitting diodes on the side of its casing. They can be read to obtain trouble codes. Note how LED codes are read. A—LED display. B—Code chart. (Honda)

DTC. No.	DTC dectecting condition	Trouble area
P0171	When the air-fuel ratio feedback is stable after engine warming up, the fuel trim is considerably in error on the RICH side.	• Air intake (hose loose) • Fuel line presure • Injector blockage • Heated oxygen sensor malfunction • Mass airflow meter • Engine coolant temperature sensor
P0171	When the air-fuel ratio feedback is stable after engine warming up, the fuel trim is considerably in error on the LEAN side.	• Fuel line pressure • Injector leak, blockage • Heated oxygen sensor malfunction • Mass airflow meter • Engine coolant temperature sensor

Figure 18-28. This is a trouble code chart for one type of vehicle. Study how different code numbers show possible problems and causes. (General Motors)

cases, codes will be automatically erased after 30-50 engine starts or warm-ups. However, codes should be erased to prevent a possible misdiagnosis by the next technician who works on the vehicle. There are various methods used to erase trouble codes from the computer:

• Use a scan tool to remove stored diagnostic codes from the on-board computer. This is the best way to remove old codes after repairs. In OBD II systems, the ECM may retain stored codes for several days without battery power. See **Figure 18-29.**

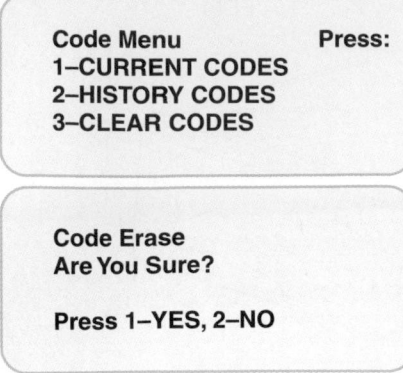

Figure 18-29. Using a scan tool is the fastest and easiest way to erase stored trouble codes. With most scan tools, simply choose the menu selection to clear codes and then press YES. (OTC)

- Disconnect the battery ground cable or strap. However, this will also erase the digital clock memory, all radio presets, and any adaptive strategy information from the computer.

- Unplug the fuse to the ECM. This will also erase all other information stored in the computer's temporary memory.

Caution

Some auto manufacturers warn against unplugging or plugging in the computer harness connector with the ignition key on or the engine running. This could cause a voltage spike that could damage the computer.

After erasing trouble codes, reenergize the on-board diagnostics and check for diagnostic trouble codes. If no trouble codes are displayed, you have corrected the problem.

Tech Tip

A memory saver, which consists of a small battery and alligator clips, can be connected across the battery terminals before disconnecting the battery. It will provide enough power to keep the clock, stereo, and computer from losing the information stored in their memories. When using a memory saver, turn off all accessories (radio, blower, etc.). The current drain from these devices, combined with even the smallest voltage drop, could cause electronic devices (computer, clock, radio, etc.) to lose their preprogrammed data.

Summary

- On-board diagnostics refers to a vehicle computer's ability to analyze the operation of its circuits and output data showing any problems.

- A scan tool is used to communicate with the vehicle's computers to retrieve trouble codes, display circuit and sensor electrical values, run tests, and give helpful hints for finding problem sources.

- OBD I and earlier on-board diagnostic systems could only check a limited number of items.

- On-Board Diagnostics II, abbreviated OBD II, is designed to more efficiently monitor the condition of hardware and software that affect emissions. New vehicle diagnostics detect part deterioration and not just complete part failure.

- If an unusual condition or electrical value is detected, the computer will turn on a malfunction indicator light (MIL) in the dash instrument panel or driver information center.

- Code conversion means the scan tool is programmed to automatically convert the number code into abbreviated words that explain what might be wrong without referring to a service manual.

- Diagnostic trouble codes (DTCs) are digital signals produced by the computer when an operating parameter is exceeded.

- Most scan tools come equipped with different program cartridges. A scan tool cartridge contains information about the specific make of vehicle to be scanned.

- The data link connector (DLC) is a multipin terminal for reading computer trouble codes or scanning problems.

- A scan tool snapshot or freeze frame is an instantaneous readout of operating parameters at the time of a malfunction.

- OBD II codes contain a letter and a four-digit number.

- The letter in all OBD II codes indicates the general function of the affected system.

- The first digit of the number in OBD II codes indicates whether the code is a standard trouble code or a nonuniform code.

- The second digit of the number in the OBD II code indicates the specific function of the system where the fault is located.

- The last two digits in the OBD II code refer to the specific fault designation.
- A wiggle test is done by moving wires and harness connectors while scanning to find soft failures.
- If you do not have a scan tool or are working on older computer-controlled vehicles, there are several other ways to activate computer on-board diagnostics to pull out trouble codes.

Important Terms

On-board diagnostics	Prompts
Scan tool	Freeze frame
On-board diagnostics generation one (OBD I)	Failure record
On-board diagnostics generation two (OBD II)	Standardized alpha-numeric trouble codes
Malfunction indicator light (MIL)	Standard trouble codes
Check engine light	Non-uniform codes
Service engine soon light	Fault designation
Trouble code chart	Hard failure
Trouble code conversion	Soft failure
Diagnostic trouble codes (DTC)	General circuit failure
Operating parameter	Low-input failure
Program cartridges	High-input failure
Troubleshooting cartridge	Improper range/performance failure
Data link connector (DLC)	Datastream values
Diagnostic connector	Engine-off/key-on diagnostics
Assembly line diagnostic link (ALDL)	Wiggle test
Adapter	Engine-on/key-on diagnostics
	Switch diagnostic test
	Actuator diagnostic test

Review Questions—Chapter 18

Please do not write in this text. Place your answers on a separate sheet of paper.

1. What is computer on-board diagnostics?
2. What is the first thing most technicians look at when diagnosing a computer system problem?
3. If an unusual condition or electrical value is detected, most computer systems will turn on a(n) _____ _____ _____.
4. List and summarize 17 types of problems that can affect computer system operation.
5. Only about _____ of all performance problems are caused by the computer, sensors, and actuators.
6. Give five locations for the data link connector.
7. Summarize the OBD II alpha-numeric diagnostic code.
8. Which of the following is a standardized OBD II code for a malfunction in the computer or auxiliary outputs?
   (A) P0605.
   (B) P1600.
   (C) P0141.
   (D) P0505.
9. Name the four general types of computer system failures.
10. A(n) _____ _____ is always present and a(n) _____ _____ is intermittent.
11. Engine-off diagnostics is done by triggering on-board diagnostics with the _____ on and the _____ off.
12. What is a wiggle test?
13. A non-OBD II car enters the shop with the computer warning light on. Technician A says to trigger on-board diagnostics, you may have to connect a jumper wire across specified terminals in the diagnostic connector. Technician B says to trigger on-board diagnostics, you may have to turn the ignition key on and off a specified number times. Who is right?
   (A) A only.
   (B) B only.
   (C) Both A and B.
   (D) Neither A nor B.
14. Explain several ways to read trouble codes without a scan tool.
15. What are trouble code charts?

ASE-Type Questions

1. Technician A says an automotive computer is able to scan its input and output circuits to detect incorrect voltage problems. Technician B says an automotive computer is able to scan its input and output circuits to detect an incorrect resistance problem. Who is right?
   (A) A only.
   (B) B only.
   (C) Both A and B.
   (D) Neither A nor B.

2. Technician A says if an automotive computer system detects an abnormal condition, the car's malfunction indicator light will normally be activated. Technician B says if an automotive computer system detects an abnormal condition, the car's low oil warning light will normally be activated. Who is right?
   (A) A only.
   (B) B only.
   (C) Both A and B.
   (D) Neither A nor B.

3. An automobile's computer malfunction indicator light is flashing. Technician A says this means the problem could be damaging the catalytic converter. Technician B says that this simply indicates a soft problem. Who is right?
   (A) A only.
   (B) B only.
   (C) Both A and B.
   (D) Neither A nor B.

4. Technician A says a faulty actuator can affect the operation of an automotive computer system. Technician B says a leaking vacuum hose can affect the operation of an automotive computer system. Who is right?
   (A) A only.
   (B) B only.
   (C) Both A and B.
   (D) Neither A nor B.

5. Technician A says a spark plug misfire can affect the operation of an automotive computer system. Technician B says an automotive computer system is not affected by a spark plug misfire. Who is right?
   (A) A only.
   (B) B only.
   (C) Both A and B.
   (D) Neither A nor B.

6. An automobile has a small fuel tank leak. Technician A says this problem may activate the car's computer system "malfunction indicator light." Technician B says this type of problem has no effect on the car's computer system. Who is right?
   (A) A only.
   (B) B only.
   (C) Both A and B.
   (D) Neither A nor B.

7. Technician A says about 50% of all automotive engine performance problems are not caused by the computer system. Technician B says about 80% of all automotive engine performance problems are not caused by the computer system. Who is right?
   (A) A only.
   (B) B only.
   (C) Both A and B.
   (D) Neither A nor B.

8. Which of the following is a possible location for an automotive computer system's data link connector?
   (A) Under right side of dash.
   (B) Near the firewall in the engine compartment.
   (C) Under the center console.
   (D) All of the above.

9. When a trouble code number is looked up in a service manual, the trouble code chart says oxygen sensor. Technician A says to test the sensor and its circuit. Technician B says to replace the oxygen sensor. Who is right?
   (A) A only.
   (B) B only.
   (C) Both A and B.
   (D) Neither A nor B.

10. Technician A says the term "hard failure" refers to an intermittent automotive computer problem. Technician B says the term "hard failure" refers to a constant automotive computer problem. Who is right?
    (A) A only.
    (B) B only.
    (C) Both A and B.
    (D) Neither A nor B.

11. A wiggle test is being performed on an automotive computer system. Technician A performs this test with the engine off and the ignition key off. Technician B performs this test with the engine off and the ignition key on. Who is right?
    (A) A only.
    (B) B only.
    (C) Both A and B.
    (D) Neither A nor B.

12. All of the following can normally be performed during an automotive computer system actuator self-test *except:*
    (A) open and close injectors.
    (B) fire the ignition coil.
    (C) operate the reed valve.
    (D) activate the idle speed motor.

13. Technician A says on certain models of automobiles, you can activate the computer's self-diagnostics by pushing two dash climate control buttons at the same time. Technician B says on certain models of automobiles, you can activate the computer's self-diagnostics by turning the ignition key on and off within a few seconds. Who is right?
    (A) A only.
    (B) B only.
    (C) Both A and B.
    (D) Neither A nor B.

14. Which of the following test instruments can be used to read automotive computer system trouble codes?
    (A) Test light.
    (B) Voltmeter.
    (C) Scan tool.
    (D) All of the above.

15. Trouble codes need to be erased from an OBD II computer system. Technician A wants to accomplish this by unplugging the ECM fuse. Technician B wants to accomplish this by using the shop's scan tool. Who is right?
    (A) A only.
    (B) B only.
    (C) Both A and B.
    (D) Neither A nor B.

Activities—Chapter 18

1. Demonstrate, on at least one vehicle, the proper method for using a scan tool.

2. Videotape a service technician using a scan tool to "check out" an engine. Ask the technician to explain each step as he or she performs works. Show the completed tape to the class.

Chapter 19

Computer System Service

After studying this chapter, you will be able to:

- Perform a visual inspection of the engine, its sensors, actuators, and the systems they monitor and control.
- Test sensors and their circuits.
- Remove and replace sensors.
- Test and replace actuators.
- Remove and replace a computer.
- Remove and replace a computer PROM.
- Program an EEPROM.
- Demonstrate safe working practices when servicing automotive computers.
- Correctly answer ASE certification test questions on servicing computer system components.

Since almost all vehicle systems now use computer components, you must have a basic knowledge of computer service before studying the remaining chapters in this text. Otherwise, you will not fully understand the chapters on fuel injection, ignition, and emissions systems. This chapter will briefly summarize how to test computer components and circuits and help you develop the skills needed to verify *where problems are.*

After you have checked the computer for trouble codes, you can find the exact source of the problem by doing pinpoint tests. *Pinpoint tests* are more specific tests of individual components. The service manual will normally have charts that explain how to do each pinpoint test. It will show specific tests, as well as provide component electrical values, and other critical information.

Remember that trouble codes only indicate the area of trouble and sometimes the type of problem, *not* what part or circuit is at fault. It is therefore imperative that you know how to do basic electrical tests on individual components.

Preliminary Visual Inspection

A *preliminary visual inspection* involves looking for signs of obvious trouble: loose wires, leaking vacuum hoses, part damage, etc. For example, if the trouble code says there is something wrong in the coolant temperature sensor circuit, you should check the sensor resistance and the wiring going to the sensor. You should also check the coolant level and the thermostat. A low coolant level or engine overheating could also set this code.

When there is a malfunction in a system, always remember that the cause is often something simple. It is easy for the untrained person to instantly think computer problem when an engine runs rough, fails to start properly, or exhibits some other performance problem.

For example, *contaminated engine oil* can trigger a computer trouble code, **Figure 19-1.** Fumes from the contaminated oil can be drawn into the engine's intake manifold from the crankcase. If these fumes are excessively strong, the oxygen sensor could be tricked into signaling a rich air-fuel mixture. The computer would then lean the mixture to compensate for the crankcase fumes. An oxygen sensor trouble code may be produced and, in some cases, an engine performance problem could result.

As this points out, it is critical that you check for conventional or simple problems *first.* Start checking for computer problems only after all of the conventional causes have been eliminated.

KISS Principle

KISS is an acronym that could help you find the source of performance problems on a computer-controlled vehicle. *KISS* stands for *keep it simple, stupid.* This means you should start your troubleshooting with the simple checks and tests. Then, as the common problems are eliminated, you will move to more complex tests.

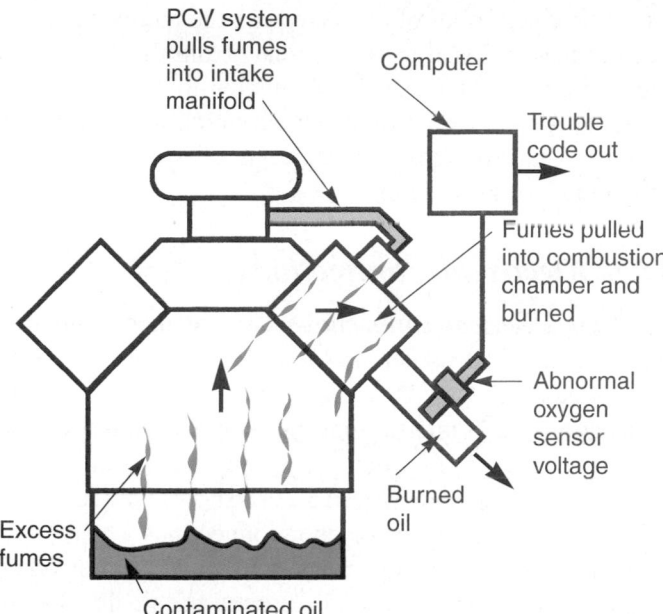

Figure 19-1. This example shows that a trouble code can be tripped by factors other than a computer system problem. Contaminated engine oil can cause excess fumes to be pulled into the PCV system. When burned in the combustion chambers, the fumes could trick the computer into sensing that something is wrong with the oxygen sensor circuit.

Electrical Component Damage

Semiconductor devices, such as transistors and integrated circuits, are very easy to damage. They can be damaged by static electricity, voltage spikes, heat, and impact shocks. Here are some things to remember when working with semiconductor devices and their wiring:

- Arc welding can damage the on-board computers. If welding on the vehicle is necessary, remove all of the on-board computers. If this is not possible or too time-consuming, unplug their connectors and make sure the welder's lead is securely grounded.

- Never disconnect the battery cables while the engine is running. In the past, some technicians would do this to see if the alternator was working. This can destroy or weaken electronic circuits, causing failure in a short period of time.

- Do not disconnect or connect wiring, especially the computer wiring, with the ignition key on. This can cause a current surge that can damage the computer.

- Make sure you do not reverse the battery cable connections. This can destroy electronic components.

- Wear an anti-static wrist strap whenever you handle static-sensitive components (removable PROM chips for example) to protect them from damage.

- Only use high impedance test lights and meters when checking electronic circuits or their wiring. A conventional test light or meter will draw too much current and destroy electrical components.

- Do not disconnect a scan tool from the data link connector while the ignition key is on. This could create a voltage spike that can damage the computer.

Computer System Circuit Problems

Almost all electrical-electronic problems are actually basic circuit problems. A *basic circuit problem* is caused by a problem in a circuit that increases or decreases current, resistance, or voltage. For example, a broken wire could stop current flow or a charging system problem could decrease output voltage and current flow.

Unfortunately, when a minor problem occurs in a complex circuit like a computer control system, it may *not* seem like a minor problem. For example, a poor electrical connection in a feed wire to a body ground may cause the computer system or one of the systems it controls to lose power. This can result in a shutdown of the fuel injection system, the emissions control system, the ignition system, or the entire vehicle.

You might think that any of these systems or the computer itself is at fault. Several systems could appear to have a problem. In reality, it is simply a poor electrical connection in one wire that is causing all of the problems. It is important for an automotive technician to remain calm when diagnosing electrical problems. If analyzed properly, problems can usually be found and corrected easily.

Locating Computer Problems

The most difficult aspect of making computer system repairs is finding the source of the problem. To find the source of computer problems, you must ask yourself the following types of questions:

- What could be causing the specific symptoms? Mentally picture the parts in the circuit and how they function. Trace through the circuit while referencing a wiring diagram to find out which wires, connections, and components are in the circuit leading to potential trouble source.

- How many components are affected? If several components are *not* working, something close to a common power source or ground point is at fault. If only one or two sections of the circuit are faulty, begin your tests at those sections of the circuit.

- Is the problem always present or is it ***intermittent*** (only occurs under some conditions)? If the problem is intermittent, the conditions causing the problem will have to be simulated. For example, a loose electrical connection could open and close with vibration or movement. You might simulate driving conditions by wiggling wires and connectors in the circuit to make the problem occur.

- Is the problem's occurrence related to heat or cold? If it occurs only on a hot day or when the engine is warmed to full operating temperature, heat is related to the problem's occurrence. Electronic circuits (transistors in particular) are greatly affected by heat. In fact, too much heat can ruin an electronic component. You can use a heat gun to simulate the heat in an engine compartment.

- Is the problem's occurrence affected by moisture? If the trouble occurs only on wet or humid days, you have information to use when analyzing the source of a problem. In most cases, moisture cannot enter a sealed electronic component, but it may enter and affect the wire connections and any components exposed to the environment.

Stress testing refers to the use of heat, cold, or moisture to simulate extreme operating conditions of components, like spark plug wires, explained in a later chapter under spark plug wire leakage.

Sensor and Actuator Problems

As with other electrical and electronic components, sensors, actuators, and their circuits can develop opens, shorts, or abnormal resistance or voltage values. When your pinpoint tests find a problem, the sensor or actuator should be replaced or the circuit repaired.

In most cases, a scan tool is used to find the problem circuit and a digital multimeter is used to measure the resistance in the circuit and the actual sensor output or actuator input. Then, this value (voltage, resistance, current, or action) is compared to factory specifications. If the test value is too high or too low, you would know that the sensor or actuator is faulty and must be replaced.

The shop manual will also have a wiring diagram, or ***schematic,*** for the computer system. The diagram will show the color codes of the wires and the number of connectors that are used to feed signals from the sensors to the computer and from the computer to the actuators. This can be very helpful when servicing any computer system. The following paragraphs discuss the most frequent circuit problems.

Poor Electrical Connections

Poor electrical connections are the most common cause of electrical-related problems in a computer system. Discussed in Chapter 18, a wiggle test will help find poor connections and intermittent problems. Always check electrical connections when diagnosing sensors and other electronic components. **Figure 19-2** shows how to test a wiring harness for opens or poor connections.

Poor electrical connections can be due to corroded terminals, loose terminal ends, burned terminals, chafed wires, and other problems. Dirt and moisture can get into connectors, causing high resistance. Any of these conditions can prevent a normal sensor signal from returning to the computer. They can also prevent the control current from reaching an actuator.

Vacuum Leaks

Vacuum leaks are frequently caused by deteriorated, broken, or loose vacuum hoses. Vacuum leaks often make a *hissing sound*. Some vacuum leaks can upset the operation of a computer system and cause a wide range of

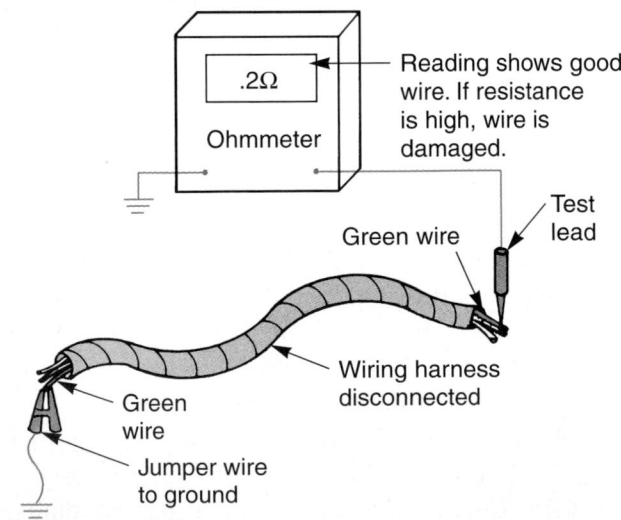

Figure 19-2. If you suspect a wire is broken inside the harness, this is an easy way to test the wire. Disconnect the wiring harness at both ends, and ground the suspect wire on one end. Then use an ohmmeter to check the wire's resistance. If the wire's resistance is high, the wire must be repaired, bypassed, or replaced.

symptoms. Also, some engine sensors and actuators rely on engine vacuum for operation.

Always check for vacuum leaks when they could be causing a performance problem, **Figure 19-3.** For example, if the trouble code indicates a problem with the MAP (manifold absolute pressure) sensor, check the vacuum lines leading to the sensor. If there is a vacuum leak, the sensor cannot function normally. Also check the intake manifold gasket area, as this is a common location for vacuum leaks.

Air leaks after a mass airflow sensor can also cause problems. The sensor cannot measure the actual amount of air being drawn into the engine, and an incorrect air-fuel mixture will result.

Figure 19-3. Always inspect the engine compartment for signs of trouble. A loose or corroded connection or a vacuum leak may be setting trouble codes or tricking the computer system.

Sensor Service

Sensor service involves testing and, if needed, replacing computer system sensors. Since sensor designs vary and some can be damaged by incorrect testing methods, it is important for you to know the most common ways of checking sensor values.

For testing purposes, you can classify sensors into one of two categories: passive sensors or active sensors. This will help you select a testing method.

As detailed in Chapter 17, a passive sensor varies its internal resistance as an operating condition changes. There are two common types of passive sensors: variable resistance sensors and switching sensors. The variable resistance sensor modifies a reference voltage that is sent

from the computer. A switching sensor acts as either a conductor or an insulator, switching on and off with condition changes. Common passive sensors include the following:

- Intake air temperature sensors.
- Coolant temperature sensors.
- Throttle position sensors.
- Transmission linkage position sensors.
- EGR pintle position sensors.
- Manifold absolute pressure sensors.
- Fuel tank pressure sensors.
- Mass airflow sensors.
- Oil level sensors.
- Brake fluid level sensors.
- Suspension height sensors.

Active sensors, or voltage-generating sensors, produce a very weak internal voltage, which is sent to the computer for analysis. Typical active sensors include:

- Oxygen sensors.
- Engine speed sensors.
- Camshaft sensors.
- Crankshaft sensors.
- Vehicle speed sensors.
- Knock sensors.
- Solar sensors.

Each type of sensor needs a slightly different testing method. Refer to the service manual for exact sensor types and locations.

Testing Passive Sensors

Since passive sensors do *not* generate their own voltage, the computer must feed them a reference voltage. A passive sensor can change its internal resistance with a change in system or vehicle condition or operation. This resistance change modifies the reference voltage, which is interpreted and used by the computer to control the engine's various systems.

To test a passive sensor, either measure its internal resistance with an ohmmeter or measure the voltage drop across the sensor with its reference voltage applied. Depending on the sensor or manufacturer, both tests may be performed. You can also read computer data stream values with your scan tool. However, the scan tool cannot isolate the sensor, wiring, or computer for individual tests. Any unusual sensor readings by a scan tool should be verified using a multimeter.

Testing Variable Resistance Sensors

To test a variable resistance sensor with an ohmmeter:
1. Disconnect the sensor wires.
2. Connect the ohmmeter's test leads to the sensor terminals, **Figure 19-4A.**
3. Compare the ohmmeter reading to the manufacturer's specifications.

The sensor's resistance must be within factory specifications. For example, if you are testing a temperature sensor, connect an ohmmeter to the sensor terminals and then measure coolant temperature with a digital thermometer. Note the resistance readings at various coolant temperatures and compare these readings to service manual specifications.

If the sensor's resistance is within specifications, check for opens or shorts in the wires going to the sensor. If the wires are good, suspect the ECM. The service manual will give detailed instructions for testing the control module.

To check a variable resistance sensor with a voltmeter:
1. Connect the voltmeter in parallel with the sensor. Be sure to leave the computer wires connected to the sensor, **Figure 19-4B.**
2. Measure the voltage drop across the passive sensor with the computer reference voltage applied.
3. Compare your measurements to specifications.

Testing Switching Sensors

With a switching sensor, such as a power steering pressure switch, you can use an ohmmeter to check that the switch is opening and closing. As in **Figure 19-4C,** connect your ohmmeter and move the switch opened and closed. Your meter should register infinite ohms and then zero ohms. Replace the sensor if it is defective. You could also use a high impedance test light to quickly check the operation of a switching sensor.

Note that some auto manufacturers do not give resistance specifications for passive sensors, they only give voltage drops. You may have to use a test harness to connect the meter in parallel with the sensor. You can make or purchase this type of test harness.

Testing Reference Voltage

A reference voltage (typically *5 volts*) is fed to passive sensors. Then, when conditions and sensor resistance change, the amount of voltage flowing back to the computer also changes. The reference voltage is needed so that a signal returns to the computer.

To measure reference voltage to a passive sensor:
1. Disconnect the wires leading to the sensor.
2. Connect a digital voltmeter to the wires.
3. Turn the ignition key on and note your readings.
4. Compare your voltage readings to specifications.

Typically, the open circuit voltage should be about 5 volts. Refer to **Figure 19-5.**

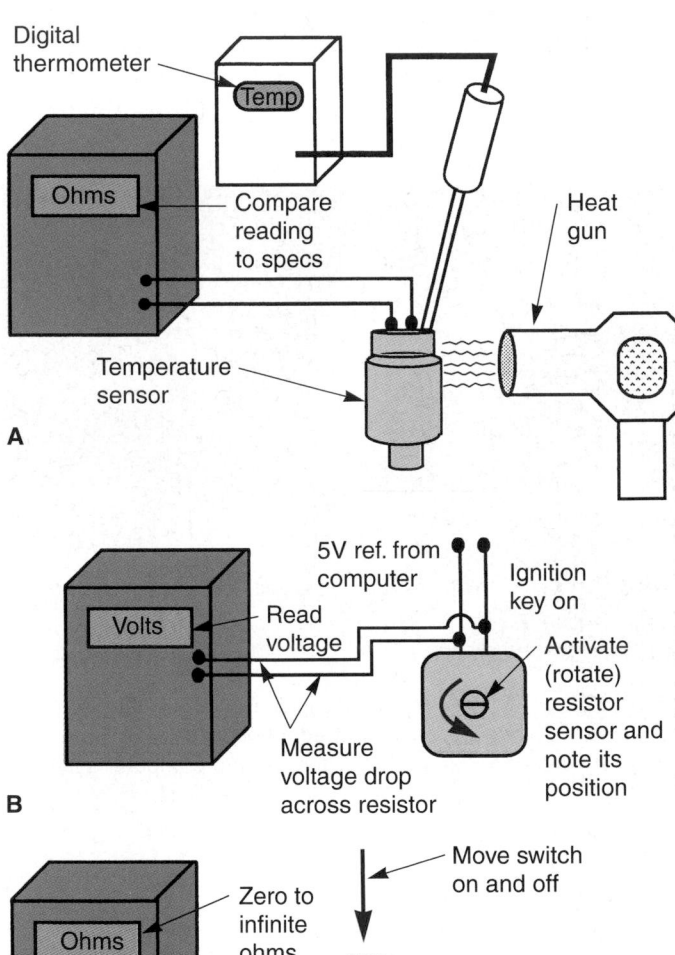

Figure 19-4. There are many ways to test passive sensors. A—An ohmmeter can be used to measure internal resistance. Some sensors can be heated with a heat gun while measuring resistance. A digital thermometer may also be used to compare actual temperature with sensor resistance. B—A voltmeter can be used if sensor specifications are only given as a voltage drop. The sensor must have reference voltage while testing occurs. C—Switching sensors may be tested with an ohmmeter or a high-impedance test light. The meter or light should show a change in condition when the switch is turned on and off.

If the reference voltage to the sensor is low, check the wiring harness for high resistance, as something is preventing the full reference voltage from reaching the sensor. Low reference voltage would cause a sensor to produce erroneous readings. It is possible for the reference voltage to be too high if a current-carrying wire is shorted into the circuit. High voltage may also be caused by a computer malfunction, but this is rare.

Testing Active Sensors

As mentioned, an active sensor produces its own voltage and sends it back to the computer. The voltage produced by an active sensor is very low, often under 1 volt. This makes sensor wiring harness continuity very critical. One poor electrical connection can keep the low voltage from returning to the computer.

Figure 19-6 shows several ways to test an active sensor. In **Figure 19-6A,** an ohmmeter is connected to a

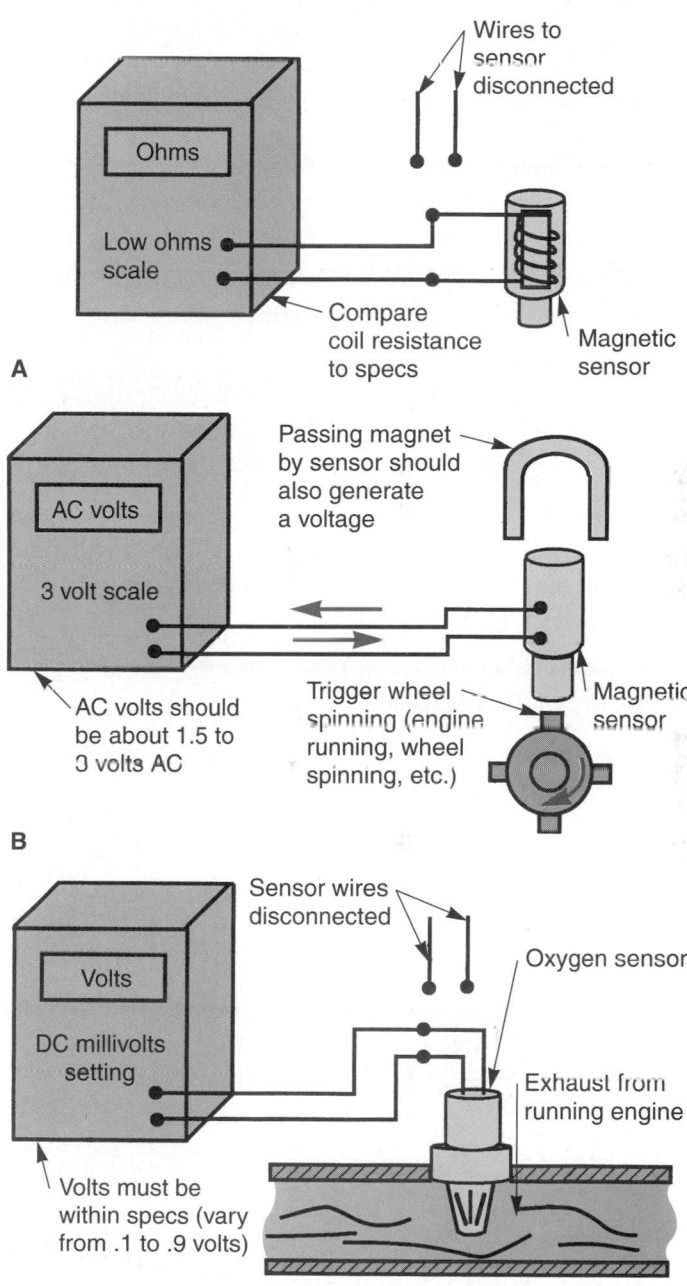

Figure 19-6. Since an active sensor generates its own output signal, testing methods for active sensors are slightly different than those for passive sensors. A—Magnetic sensor internal coil resistance can be measured with an ohmmeter. B—An ac voltmeter can be used to check a magnetic sensor while it is operating. With the triggering device moving, an ac voltage signal should be generated. C—Sometimes it is recommended to connect a meter directly to the sensor. You can compare the operating sensor's output to specifications. Be sure to test the harness leading to the sensor.

Figure 19-5. Passive sensors rely on a computer reference voltage for proper operation. A—Use a digital voltmeter to measure the amount of voltage supplied to the sensor. Typical reference voltage might be 5 volts, but refer to the service manual specifications. B—If the reference voltage is not correct, test for voltage at the computer. If the reference voltage at the computer is correct, a problem with the wiring harness is indicated.

common magnetic sensor. The ohmmeter will measure the resistance of the coil winding. Replace the sensor if the resistance is high or low.

In **Figure 19-6B,** an ac voltmeter is connected to a magnetic sensor. The trigger wheel must be rotated (engine cranked over, wheel or hub in an ABS system turned, etc.) to make the sensor generate voltage. A magnetic sensor should typically produce about 1.5 to 3 volts ac. A magnet can also be passed by a coil to make it produce a voltage.

In **Figure 19-6C,** a digital voltmeter is connected to an oxygen sensor. With the engine running in closed loop, the voltmeter should show the sensor's output voltage. If the output voltage from the sensor is low or high, the sensor may require replacement.

 Tech Tip
Whenever a sensor tests good, check the wiring leading to the sensor. Bad wiring may be blocking current flow back to the computer.

Figure 19-7 shows how to use small jumper wires to connect a meter to check a sensor while it is still functioning in the circuit. More advanced testing will be explained in later chapters.

Replacing Sensors

When replacing a sensor, there are several general rules you should remember:

- Always purchase an exact sensor replacement. Even though two sensors may look identical, their internal resistance or circuitry may be different. See **Figure 19-8.**
- Release the sensor connector properly. Most connectors have positive locks that must be released. If you damage the connector, intermittent problems may result from a loose connection.
- Use special tools as needed. Some sensors, such as oxygen sensors, require the use of a *sensor socket.* This socket has a deep pocket and a cutout that will fit over the sensor and any wires. Conventional deep sockets may not fit over the wiring or the sensor head.
- Use thread sealant sparingly. If the sensor seals a coolant, oil, or vacuum passage, do not use too much sealant.
- Use thread and engine sealants that are safe for oxygen sensors. Some sealants can poison the oxygen sensors.
- Tighten the sensor properly. Overtightening a sensor could damage it. Undertightening could cause leaks.
- Adjust the sensor, if needed. Some throttle position sensors require adjustment after installation.
- Scan for trouble codes after sensor replacement.

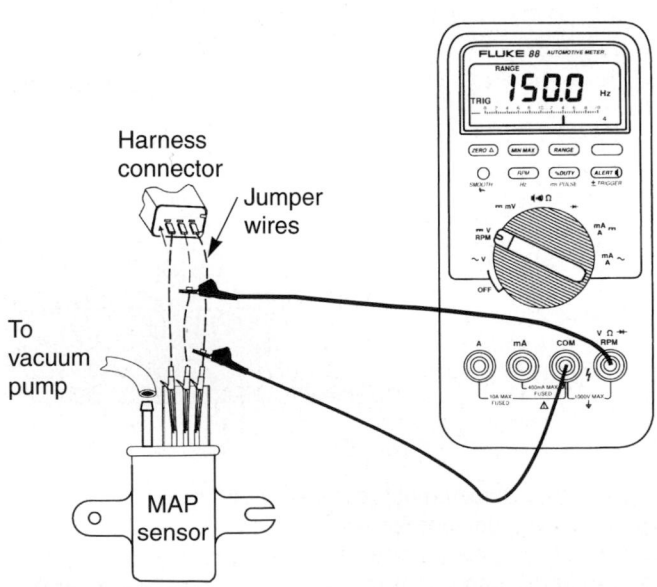

Figure 19-7. Small jumper wires are often used when measuring sensor circuit values. Make sure you do not short the jumper wires together or to ground. This test is used to measure the frequency signal in hertz from a manifold absolute pressure sensor. (Fluke)

Figure 19-8. Always make sure the replacement sensor is same as the old one. These two temperature sensors look the same but have different temperature and resistance values. If the wrong one is installed, the vehicle may not function correctly. (Snap-On Tools)

Actuator Service

Actuator service involves testing the actuators for possible electrical or mechanical problems and replacing them, if necessary. If an actuator fails, the computer cannot control the engine and vehicle systems properly.

Testing Actuators

Since actuators are simply relays, solenoids, and motors, they are fairly easy to test. **Figure 19-9** shows several ways to test actuators.

Testing Servo Motors

In **Figure 19-9A**, a voltage source has been connected to a servo motor. The wiring harness to the motor has been disconnected. Jumper wires feed current directly to the motor. This is a simple way to check the operation of an electric or servo motor. If the motor begins to function with an external voltage source, you should test the wire harness leading to the motor.

Testing Solenoids

In **Figure 19-9B,** a power supply is being used to check a solenoid. When jumper wires are connected to the vehicle's battery and the solenoid, the solenoid should operate. If the solenoid tests good, you should check the voltage coming to the solenoid through its harness.

An ohmmeter can also be used to test an actuator. You can use the meter to measure the internal resistance of the unit. By comparing resistance readings to specifications in the service manual, you can find out if the actuator must be replaced.

Testing Relays

Figure 19-9C shows how to test a relay. Check the voltage entering the relay and the voltage leaving the relay. It is possible that voltage is applied to the relay but the relay points are not sending voltage out to the controlled device.

Since relays contain movable contact points, they are a common source of computer system problems. The scan tool may indicate a problem with the circuit containing the relay. However, you must test the relay to pinpoint the problem source.

Relays can be located almost anywhere on a vehicle: in the engine compartment, under the dash, under the seat, or in the trunk. The service manual will give exact locations. See **Figure 19-10.**

An integrated *junction block* encloses most or all of the vehicle's mechanical relays in a single housing. Some of the relays found in this block are the fuel pump relay, cooling fan relay, wide open throttle relay, air conditioning relay, and a host of others. The junction block is usually mounted in the engine compartment.

When testing a relay, refer to the service manual wiring diagrams for pin numbers and wire color codes. Special *relay testers* can be used to quickly test relay operation. They plug into the relay and test the unit

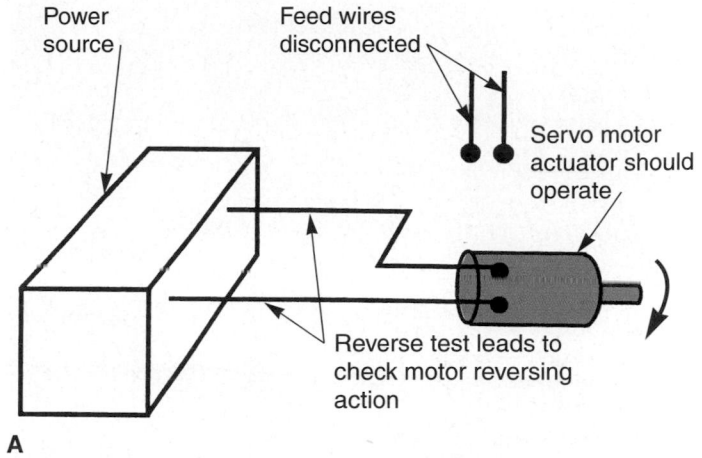
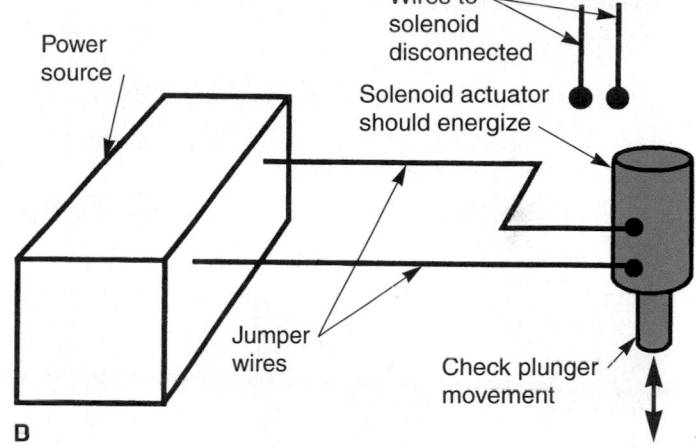
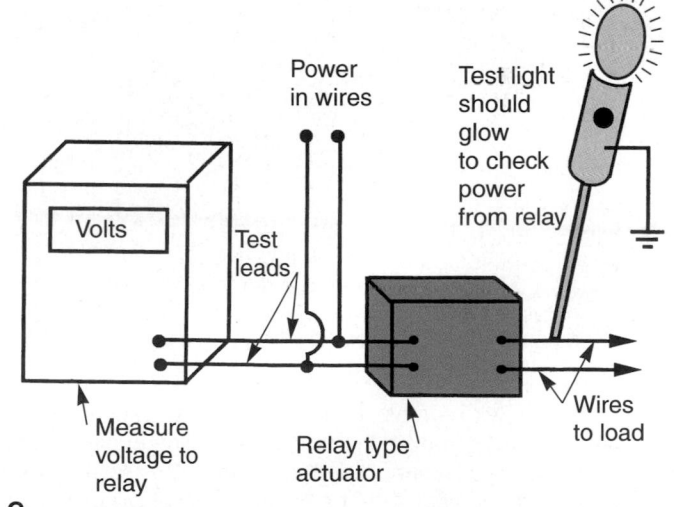

Figure 19-9. Actuator testing is straight-forward. It is similar to testing a conventional motor, relay, or solenoid. A—Voltage can be jumped to a servo motor. The motor should function when energized by the power source. B—A solenoid can also be tested in the same manner as a servo motor. C—A relay is slightly more complex to test. You must make sure there is an output when voltage is supplied to its input terminals. This can be done using a voltmeter or a high-impedance test light.

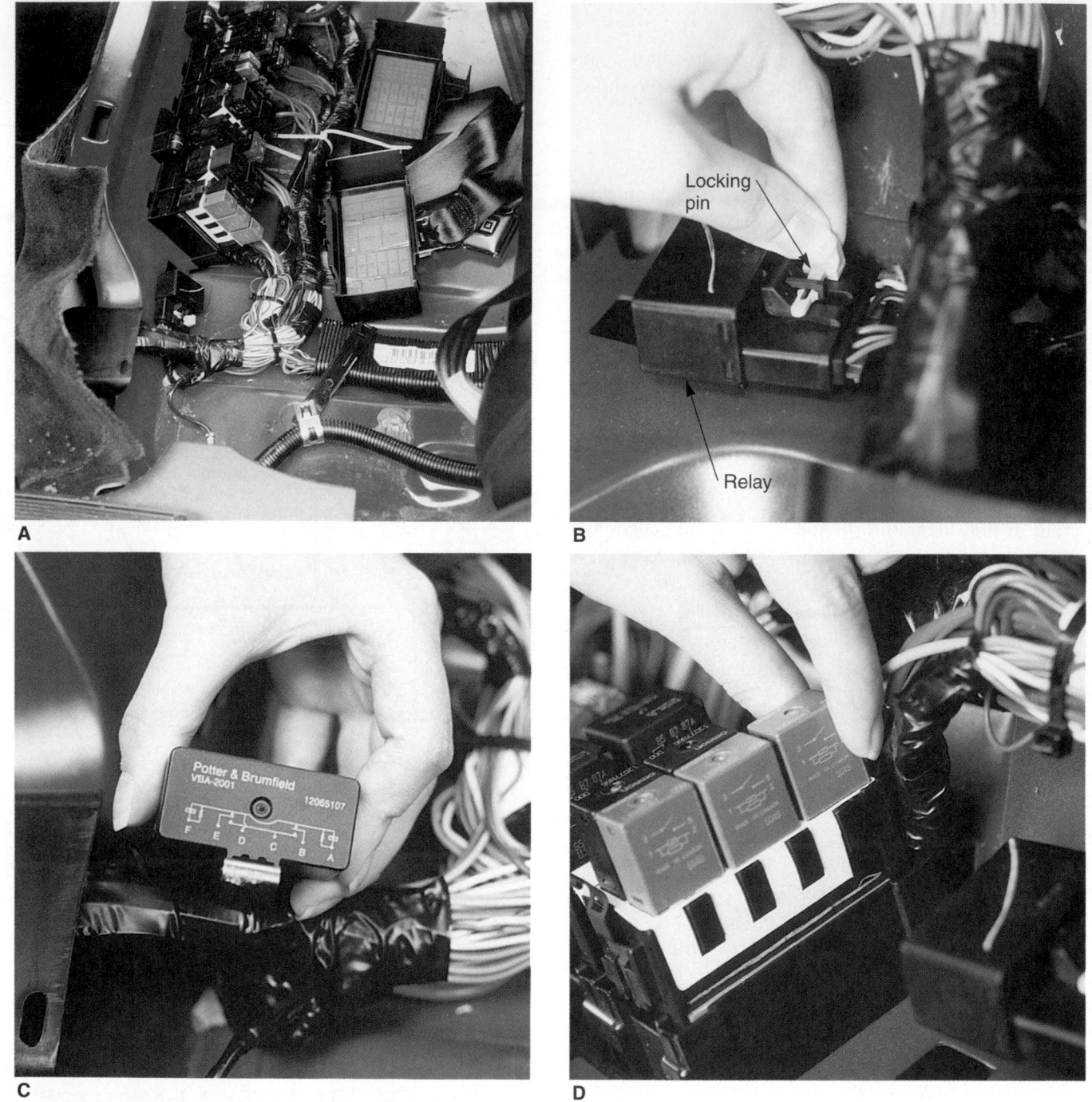

Figure 19-10. Relays have moving parts that can fail in service. Relays can be located almost anywhere on the vehicle. A—Technician has removed the rear seat cushion to gain access to the fuses and relays. B—This relay has a locking pin that must be removed before disconnecting the harness. C—Relays can have different internal construction. Make sure the replacement unit is identical to the original. D—The service manual will give relay locations for pinpoint testing and replacement.

automatically. For information on actuator service, refer to the index in this textbook.

You can find instructions on testing fuel injectors, glow plugs, and more specialized actuators.

Replacing Actuators

Actuator replacement will vary with the type and location of the unit. Here are a few rules to follow when replacing an actuator.

- Do not damage the wire connectors when releasing them. They are made of plastic and will break if forced open improperly.
- Do not drop the actuator mounting screws during removal or installation. If a screws falls into an engine, major problems can occur.
- Make sure you have the correct replacement actuator before attempting installation.
- Check that the actuator is fully seated before tightening the mounting screws.
- Double check actuator operation after replacement.

Tech Tip

Make sure you obtain the correct replacement relay. Two relays may look the same, but relays often have different internal construction.

Computer Service

Computer service involves replacing the computer. The computer is the last component to be suspected of being the problem source, only after all other potential sources of trouble have been eliminated. It is sometimes possible for an integrated circuit, transistor, or other electronic part in the computer to fail and upset system operation.

The diagnostic trouble code will tell you which computer or electronic control unit is having a problem. This information can help you find when the computer circuitry is at fault.

Measuring Computer Output

If the computer is not tested, in most cases, a defective computer is found through the process of eliminating the sensors, actuators, and related wiring as the cause of the problem. In the process of sensor and actuator testing, it was necessary for you to measure the computer's output. A *computer output* can be a reference voltage to a sensor or a supply voltage to an actuator.

Use a voltmeter to make sure the correct reference voltage is being sent to a sensor. Most computers produce a reference voltage of about 5 volts. If the reference voltage is not correct, check the wiring before condemning the computer. You can also measure voltage to make sure the correct voltage output is being fed to the actuators.

Each computer input and output passes through an individual metal terminal, or pin, which plugs into the vehicle wiring harness. *Pin numbers* identify the location and purpose (electrical value and internal connection) of each terminal in a computer wiring harness connector.

Discussed in Chapter 46, you may have to probe computer terminal pins to find the source of complex problems.

Again, always refer to service manual for exact procedures when testing a computer system. One wrong electrical connection can destroy the computer.

Caution

Never use an ohmmeter to check a computer, as it will damage the computer's internal circuitry. If it is necessary to use an ohmmeter to check the continuity of a wire or circuit in the computer harness, disconnect all wiring harnesses from the computer before testing.

Saving Memory

Saving memory can be done by connecting a small battery (such as a 9-volt battery) across the two battery cables *before* the vehicle's battery is disconnected. This will provide enough power to keep the clock, stereo, and computer from losing the information stored in their memories. When using a memory saver, turn off all accessories (radio, blower, etc.). The current drain from these devices, combined with even the smallest voltage drop, could cause electronic devices (computer, clock, radio, etc.) to lose their preprogrammed data. You are still disconnecting the vehicle battery for safety. The smaller battery cannot produce enough current to cause an electrical fire or operate the starting system.

Computer Replacement

Before disconnecting the battery and removing the computer from the vehicle, you should scan the computer and obtain the PROM identification number or the EEPROM calibration number. This information is needed to check for updated PROMs that should be installed and calibration programs that should be downloaded to the new computer. External identification numbers are not always placed on the PROM.

When removing a computer, the ignition key should be off and the vehicle's negative battery cable disconnected. This will prevent voltage spikes from damaging the computer when the harness connectors are removed. Remove any shields or components necessary to access the computer. Unbolt the brackets holding the computer in place and unplug the computer connectors.

Identification information is usually stamped or printed on the computer. Use this data and the year, make, and model of vehicle to order the correct replacement computer. The VIN (vehicle identification number) may be helpful, as well.

Caution

When handling computers, keep one hand on chassis ground and use the other to remove the component. This will prevent a static electrical charge from entering and damaging the electronic circuitry. If available, wear an anti-static wrist strap when working on computer circuits. Static electricity may not instantly ruin an electronic part, but it can reduce the part's useful service life from years to days.

PROM Service

Many computers use a PROM to store data for the specific make and model vehicle. In most cases, you must remove the PROM from the old computer and install it in the new computer.

Remove the cover over the PROM. Then, use a PROM tool to grasp and pull the PROM out of its socket. Most PROMs use a *carrier*, which is a plastic housing that surrounds the outside of the integrated circuit chip. Avoid touching the PROM terminals with your fingers because the body oils on your hand could adversely affect its operation.

Before installing the PROM in the new computer, check that the pins (terminals) are straight. Check for the presence of *reference marks* (indentations or other markings to show how to reinstall the unit) on the PROM or the carrier. If you install a PROM backwards, it will usually require replacement because of physical damage.

To install the PROM, install the carrier and PROM in the computer with the reference mark correctly positioned. First press down on the carrier only. Then, carefully press the PROM down into the computer. Press on each corner until the PROM is fully seated in its socket. If the PROM socket has locking tabs, make sure they are connected to the carrier housing, **Figure 19-11.**

After installing the PROM, install the access cover, connect the wiring harness to the computer, and install the computer into its mounts. Reconnect the battery, turn on the ignition, and activate self-diagnostics. As a final check of the computer and PROM, make sure no trouble codes are set. A code might be set if the PROM is not fully seated or a pin is bent over.

Updated PROMS

An *updated PROM* is a modified integrated circuit produced by the auto manufacturers to correct a driveability problem or improve a vehicle's performance. The old PROM is simply removed from the computer and replaced with the updated PROM.

Updated PROMs are produced to correct problems like surging, extended cranking periods, excessive

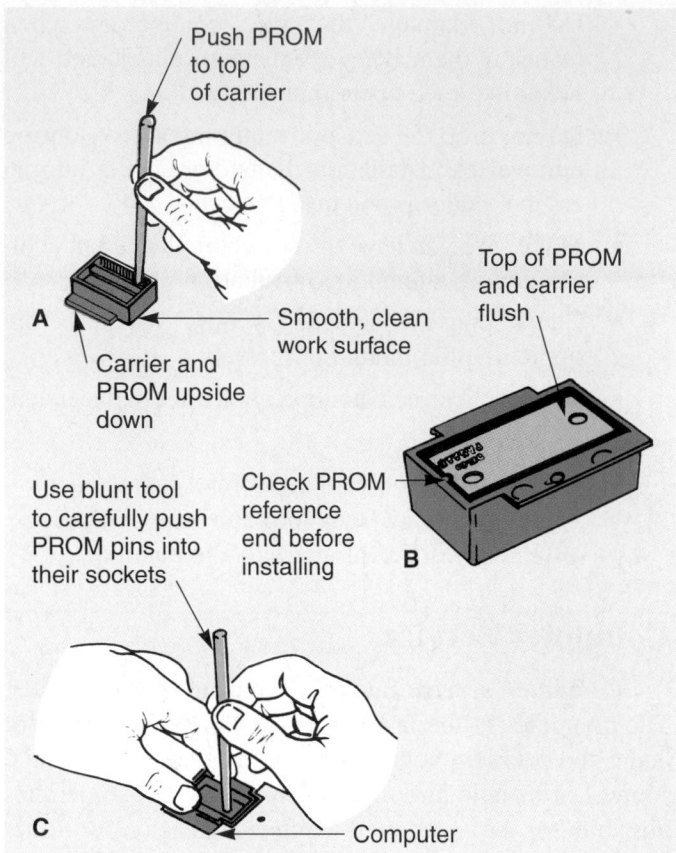

Figure 19-11. Most PROMs use a carrier, which is a plastic case around the IC chip. The PROM from the old computer may have to be installed in the new computer. A—Before installing the PROM in the new computer, use a blunt tool to push the top of the chip flush with the top of the carrier. The IC pins should be sticking up and straight. B—Make sure reference marks are positioned correctly before installation. If you install the PROM backwards, it will be ruined. C—Touching only the carrier, position the PROM pins into the socket in the computer. Use a blunt tool or a small wooden dowel to carefully push the PROM into its socket. Push lightly on all corners until the PROM is fully seated. Do not press too hard. (General Motors)

emissions, cold and hot start problem, and unusual driveability problems that cannot be isolated to one system. If you are faced with a problem and cannot find the cause, you should check with the local dealership to find out if there are any updated PROMs for the vehicle that address the problem.

Due to the popularity of aftermarket chips, or so-called "hot PROM" performance chips (PROM chips that enhance engine performance, and in many cases, increase exhaust emissions), federal regulations require the computers in all new vehicles to be equipped with fixed PROMs. With the frequent number of updated PROMs released in the past, electrically erasable PROMs, or EEPROMs, are now used. EEPROM programming is covered in the next section.

EEPROMs can be reprogrammed to correct drive-ability problems or improve performance. EEPROMs are permanently fixed to the circuit board, which makes the installation of a hot PROM almost impossible.

EEPROM Programming Using Computerized Equipment

Most newer computers use Electrically Erasable Programmable Read Only Memory (EEPROM) or Flash Erasable Programmable Read Only Memory (FEPROM) chips that are permanently soldered to the circuit board. These chips must be programmed using electronic equipment. They can also be reprogrammed in order to correct driveability and performance problems.

EEPROMs are programmed using a method referred to as flash programming. *Flash programming* may be performed by downloading the vehicle's information through a computer, a computerized diagnostic analyzer, or a scan tool. Actual programming details vary between manufacturers, but the basic procedure begins by placing the computer in the programming mode. One of three methods is used to program the computer:

- Direct programming using a service computer or computerized analyzer.
- Indirect programming using a scan tool and a computer or computerized analyzer.
- Remote programming with the computer off the vehicle.

Direct Programming

Direct programming is the fastest and simplest method. The new information is downloaded by attaching a shop recalibration device (usually a computer or computerized analyzer) directly to the data link connector. The erasure and programming is done by accessing the programming menu and following the instructions as prompted by the computer. Then, the vehicle's operating information and parameters are entered into the vehicle's computer through the connector.

Indirect Programming

To perform *indirect programming,* the proper scan tool must be available to connect to the programming computer and to the vehicle, as well as to reset some computer-controlled vehicle systems after programming. The programming computer may resemble the personal computer (PC) used in the home, or it may be a computerized analyzer like the one used for direct programming.

In this type of programming, the vehicle information is downloaded from the PC or computerized analyzer into the scan tool. It is then downloaded from the scan tool into the vehicle's computer. The scan tool menu is accessed using the keypad. Most scan tools will use a high-capacity generic program cartridge to store the information. Some newer scan tools have sufficient fixed memory to hold the information and, therefore, do not use a separate program cartridge. In either case, follow the manufacturer's procedure as prompted.

Remote Programming

Remote programming is done with the vehicle's computer removed from the vehicle. This procedure is used when changes need to be made through a direct modem connection to a manufacturer's database computer. It can also be done in cases where direct or indirect programming is not practical or possible. Since special connectors and tools are required for this type of programming, this procedure is done only at new vehicle dealerships.

EEPROM Programming Procedure

To begin programming the EEPROM-equipped computer, make sure that the battery is fully charged. Recharge the battery if necessary. Do not charge the battery during the programming procedure, as damage to the computer will result. Connect the service computer or scan tool to the computer data link connector. Make any other vehicle connections as needed before proceeding with the programming sequence.

 Caution
Do not disconnect the scan tool or service computer from the data link connector during the programming sequence. Doing so will damage the vehicle's computer.

To start the programming sequence, the analyzer or scan tool prompt may ask you to enter the engine type, vehicle type, and vehicle identification number (VIN), in a specific sequence. Once the vehicle information is entered, go to the programming software and follow the directions as prompted.

Depending on the manufacturer, it may be necessary to turn the ignition switch on or off during the connection and programming procedure. Double check any instructions on ignition switch position before making any connections or beginning computer erasure and programming. The next step is to determine the type of programming that is needed.

Programming a New Computer

If a new computer is being installed, only program that computer. In many cases, other on-board computers can be accessed and programmed using the data link connector. While most computers have internal circuitry that protects them from accidental programming, be careful not to program the wrong computer. Do not attempt to program a new computer with information from the old computer or a computer from another vehicle. Any attempt to do this will set a failed programming sequence code in the new computer's memory.

Tech Tip

In some cases, an erasure may need to be performed on a new computer before initial programming can take place.

Reprogramming Computers

Before reprogramming a vehicle's computer, determine the date that the current programming was downloaded or check the program's calibration number. If the program installed is the latest version, no further actions are required. If the latest program has not been downloaded, proceed with the reprogramming sequence.

Before reprogramming most computers, you must first erase the existing information. After this step is complete, select the updated calibration information from the reprogramming computer or scan tool menu. Then download the new information into the computer. On some systems, the erasure step is not necessary since the service computer or scan tool will erase or overwrite the old information as it loads the new program into the computer.

Tech Tip

If the computer does not accept the new program or cannot be programmed, check all connections first. Ensure that the correct computer is being reprogrammed with the proper information and that all procedures are being followed. If the computer still cannot be programmed, it may need to be replaced.

Allow sufficient time for the programming to take place. Monitor the computer or scan tool to determine when the programming sequence is complete. Do not touch any connections until you are sure the programming sequence is complete. After programming is completed, turn the ignition switch to the position called for and disconnect the computer or scan tool. Then, use a scan tool to check the computer and control system operation. While doing this, make sure you have installed the proper program into the computer by checking the program calibration number.

Computer Learn Procedures

Once the computer is connected and programmed, when necessary, it must adjust to, or "learn," the vehicle's sensor inputs and to control output actuators. The computer must receive inputs to formulate adaptive strategy to set some of its output parameters. This is usually done by driving the vehicle for a few minutes to allow the computer to learn the sensors and output actuators, and to adjust system operation according to the EEPROM programming. In a few cases, sensors or actuators may have to be manually adjusted using a scan tool.

Summary

- Pinpoint tests are more specific tests of individual components. They are completed after scanning for trouble codes.

- A preliminary inspection involves looking for signs of obvious trouble: loose wires, leaking vacuum hoses, part damage, etc.

- Semiconductors are very easy to damage. They can be damaged by static electricity, voltage spikes, heat, and impact shocks.

- Stress testing refers to the use of heat, cold, or moisture to simulate extreme operating conditions of components, like a electronic control unit.

- Poor electrical connections are the most common cause of electrical-related problems in a computer system.

- Vacuum leaks are frequently caused by deteriorated, broken, or loose vacuum hoses. Vacuum leaks often make a *hissing sound*.

- A reference voltage (typically *5 volts*) is fed to switching and variable resistance type sensors.

- Sensor sockets have a deep pocket and cutout from any sensor pigtail (wires).

- Relays, since they contain movable contact points, are a common source of computer system problems.

- Computer service usually involves a few tests and computer replacement if needed.

- Many computers use the old PROM (memory chip) during computer replacement, since it stores data for the specific make and model vehicle.

Important Terms

Pinpoint tests
Preliminary visual
  inspection
Basic circuit problem
Intermittent
Stress testing
Schematic
Poor electrical
  connections
Vacuum leaks
Sensor service

Sensor socket
Actuator service
Junction block
Relay testers
Computer service
Computer output
Pin numbers
Saving memory
Carrier
Reference marks

Review Questions—Chapter 19

Please do not write in this text. Place your answers on a separate sheet of paper.

1. What should you look for during a preliminary inspection of a computer system?

2. Explain five questions you should ask yourself when analyzing computer system problems.

3. How can a vacuum leak upset the operation of a computer system?

4. An engine runs poorly only when cold. Symptoms and a trouble code indicates a problem with the engine coolant temperature sensor. Technician A says to measure the engine coolant temperature sensor resistance with a digital ohmmeter while measuring coolant temperature with a digital thermometer. Technician B says to measure its ac voltage output while measuring its temperature with a digital thermometer. Who is right?
   (A) A only.
   (B) B only.
   (C) Both A and B.
   (D) Neither A nor B.

5. How do you check sensor reference voltage?

6. Explain three ways to test an active sensor.

7. Describe three ways to test an actuator.

8. Relays are a common source of problems in a modern computer system. True or false?

9. How do you save computer memory?

10. The computer must be removed from a late-model vehicle. Technician A says to keep one hand on chassis ground when handling the computer. Technician B says to wear an anti-static wrist strap when handling the computer. Who is right?
   (A) A only.
   (B) B only.
   (C) Both A and B.
   (D) Neither A nor B.

ASE-Type Questions

1. Technician A says automotive computer system trouble codes are used to detect faulty components in the system. Technician B says automotive computer system trouble codes are used to indicate the general area of trouble in a computer system. Who is right?
   (A) A only.
   (B) B only.
   (C) Both A and B.
   (D) Neither A nor B.

2. A trouble code is triggered in an automotive computer system. Technician A says fuel contaminated engine oil could possibly trigger this trouble code. Technician B says fuel contaminated engine oil has no effect on computer system operation. Who is right?
   (A) A only.
   (B) B only.
   (C) Both A and B.
   (D) Neither A nor B.

3. Technician A says a basic circuit problem is caused by something in the circuit that increases or decreases current. Technician B says a basic circuit problem is caused by something in the circuit that increases or decreases resistance. Who is right?
   (A) A only.
   (B) B only.
   (C) Both A and B.
   (D) Neither A nor B.

4. A scan tool indicates a problem with the MAP sensor. Technician A says you should check the operating condition of the MAP sensor. Technician B says you should check for poor electrical connections in the system. Who is right?
   (A) A only.
   (B) B only.
   (C) Both A and B.
   (D) Neither A nor B.

5. Technician A says a passive automotive sensor is capable of generating its own voltage. Technician B says a passive automotive sensor is not capable of generating its own voltage. Who is right?
   (A) A only.
   (B) B only.
   (C) Both A and B.
   (D) Neither A nor B.

6. Which of the following is not considered a passive automotive computer sensor?
 (A) Intake air temperature sensor.
 (B) EGR sensor.
 (C) Oxygen sensor.
 (D) Mass airflow sensor.

7. Technician A says a reference voltage of about 12 volts is normally required for an automotive variable resistance type sensor to operate properly. Technician B says a reference voltage of about 5 volts is normally required for an automotive variable resistance type sensor to operate properly. Who is right?
 (A) A only.
 (B) B only.
 (C) Both A and B.
 (D) Neither A nor B.

8. A trouble code shows a problem with an active sensor. Technician A says the voltage produced by an active sensor is very low, often under 1 volt. This makes sensor output and wiring harness continuity very critical. Technician B says that active sensor signals are just as strong as passive sensor signals. Who is right?
 (A) A only.
 (B) B only.
 (C) Both A and B.
 (D) Neither A nor B.

9. An automobile's oxygen sensor output needs to be checked. Technician A is going to use a digital voltmeter to check this sensor's output. Technician B is going to use an analog ohmmeter to check this sensor's output. Who is right?
 (A) A only.
 (B) B only.
 (C) Both A and B.
 (D) Neither A nor B.

10. An oxygen sensor tested faulty and must be replaced. Technician A says to avoid using sealer because it can contaminate the new sensor. Technician B says to use a sensor socket to prevent sensor damage. Who is right?
 (A) A only.
 (B) B only.
 (C) Both A and B.
 (D) Neither A nor B.

Activities—Chapter 19

1. Describe and sketch the procedure you would use to test a temperature sensor for proper operation.

2. Demonstrate two methods of checking a knock sensor—with a multimeter and by manual tapping.

3. Obtain an unserviceable vehicle computer with a removable PROM from a junkyard or other source and use it to practice removing and replacing the PROM.

Computer System Sensor and Actuator Diagnosis		
Condition	**Possible Causes**	**Correction**
Hard starting.	1. Faulty coolant temperature sensor. 2. Defective intake air temperature sensor. 3. Maladjusted or defective throttle position sensor. 4. Faulty crankshaft position sensor. 5. Faulty manifold absolute pressure sensor. 6. Bad mass airflow sensor. 7. Faulty idle speed motor. 8. Bad fuel injectors.	Test components and related circuitry. Service or replace as necessary.
Engine stalling.	1. Faulty coolant temperature sensor. 2. Defective intake air temperature sensor. 3. Maladjusted or defective throttle position sensor. 4. Faulty crankshaft position sensor. 5. Faulty manifold absolute pressure sensor. 6. Bad mass airflow sensor. 7. Faulty idle speed motor. 8. Bad fuel injectors.	Test components and related circuitry. Service or replace as necessary.
Rough idle or surging.	1. Faulty coolant temperature sensor. 2. Defective intake air temperature sensor. 3. Maladjusted or defective throttle position sensor. 4. Faulty crankshaft position sensor. 5. Faulty manifold absolute pressure sensor. 6. Bad mass airflow sensor. 7. Faulty idle speed motor. 8. Bad fuel injectors.	Test components and related circuitry. Service or replace as necessary.
Erratic idle speeds.	1. Maladjusted or defective throttle position sensor. 2. Faulty crankshaft position sensor. 3. Bad idle speed motor.	Test components and related circuitry. Service or replace as necessary.
Cold engine warm-up problems.	1. Faulty coolant temperature sensor. 2. Faulty intake air temperature sensor. 3. Faulty manifold absolute pressure sensor. 4. Bad mass airflow sensor. 5. Bad idle speed motor.	Test components and related circuitry. Service or replace as necessary.
Engine hesitation.	1. Faulty coolant temperature sensor. 2. Faulty air temperature sensor. 3. Maladjusted or defective throttle position sensor. 4. Faulty EGR position sensor. 5. Defective or contaminated oxygen sensor. 6. Maladjusted or faulty crankshaft position sensor. 7. Faulty manifold absolute pressure sensor. 8. Bad mass airflow sensor. 9. Bad fuel injectors.	Test components and related circuitry. Service or replace as necessary.

(Continued)

Computer System Sensor and Actuator Diagnosis		
Condition	**Possible Causes**	**Correction**
Poor performance and gas mileage.	1. Faulty coolant temperature sensor. 2. Faulty air temperature sensor. 3. Maladjusted or defective throttle position sensor. 4. Faulty EGR position sensor. 5. Defective or contaminated oxygen sensor. 6. Maladjusted or faulty crankshaft position sensor. 7. Faulty manifold absolute pressure sensor. 8. Bad mass airflow sensor. 9. Bad fuel injectors.	Test components and related circuitry. Service or replace as necessary.
Erratic acceleration.	Faulty EGR position sensor.	Test sensor and related circuitry. Service as necessary.
Pinging.	1. Bad coolant temperature sensor. 2. Maladjusted or defective throttle position sensor. 3. Defective EGR position sensor. 4. Faulty manifold absolute pressure sensor. 5. Bad mass airflow sensor. 6. Faulty knock sensor.	Test sensors and related circuitry. Service or replace as necessary.
Surging at highway speeds.	1. Bad coolant temperature sensor. 2. Maladjusted or defective throttle position sensor. 3. Defective EGR position sensor. 4. Faulty manifold absolute pressure sensor. 5. Bad mass airflow sensor.	Test sensors and related circuitry. Service or replace as necessary.
Backfiring.	1. Maladjusted or defective throttle position sensor. 2. Faulty manifold absolute pressure sensor. 3. Bad mass airflow sensor.	Test sensors and related circuitry. Service or replace as necessary.
Black exhaust smoke.	1. Bad coolant temperature sensor. 2. Faulty intake air temperature sensor. 3. Faulty oxygen sensor. 4. Faulty manifold absolute pressure sensor. 5. Bad mass airflow sensor. 6. Bad fuel injectors.	Test components and related circuitry. Service or replace as necessary.
No torque converter lockup.	1. Maladjusted or defective throttle position sensor. 2. Faulty EGR position sensor. 3. Faulty manifold absolute pressure sensor. 4. Bad mass airflow sensor. 5. Faulty torque converter lockup solenoid. 6. Faulty or maladjusted brake light switch.	Test components and related circuitry. Service or replace as necessary.
Run on.	Defective idle speed motor.	Test motor and related circuitry. Service or replace as necessary.

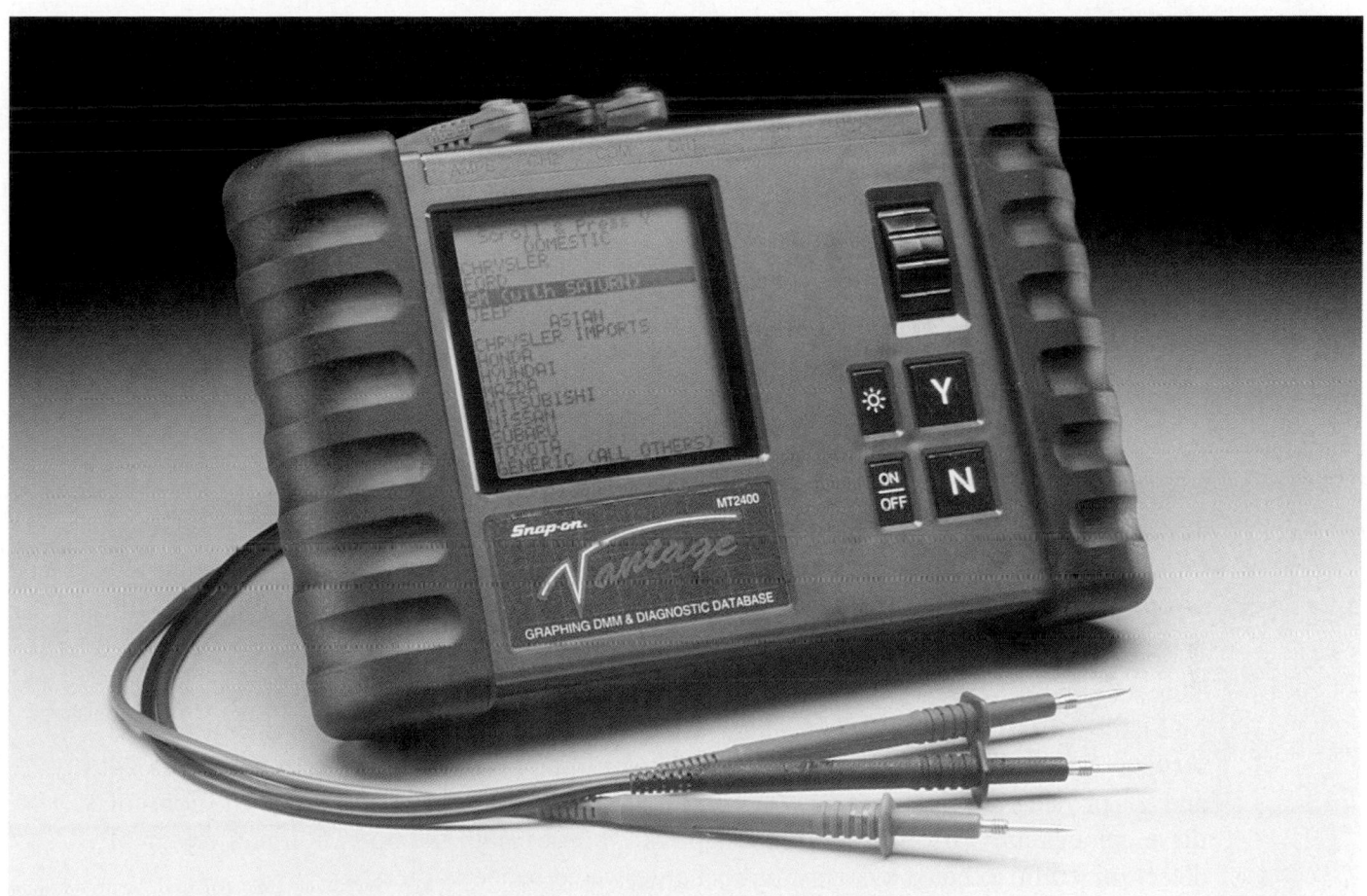

Hand-held scopes are often used when troubleshooting today's computer-controlled vehicles. This particular unit provides information on testing procedures, records dual waveforms, and performs flight record functions. For more information on using a scope, refer to Chapter 46, *Advanced Diagnostics.*

Fuel Systems

Today's fuel systems must meter a precise amount of fuel into the engine under a wide range of constantly changing operating conditions. The fuel system has the important job of optimizing engine performance while keeping fuel consumption to a minimum. This is no easy task!

This section will explain modern fuel systems in detail. Chapters 20 and 21 cover fuels, fuel tanks, fuel pumps, and fuel filters. Chapters 22 and 23 describe the operation and service of electronic fuel injection, the most common type of fuel system. Chapters 24 and 25 provide a brief overview of carburetor operation and repair, since millions of these devices are still in use today. Chapters 26-28 explain the operation and repair of diesel injection, exhaust systems, turbochargers, and superchargers.

The information in this section will help you pass ASE Test A6, *Electrical/Electronic Systems,* and Test A8, *Engine Performance.*

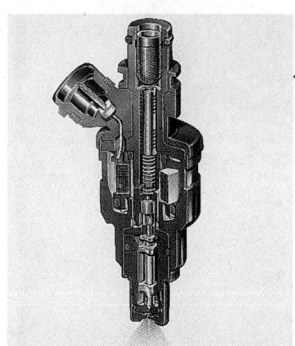

Chapter 20

Automotive Fuels, Gasoline and Diesel Combustion

After studying this chapter, you will be able to:

■ Summarize how crude oil is converted into gasoline, diesel fuel, liquefied petroleum gas, and other products.

■ Describe properties of gasoline and diesel fuel.

■ Explain octane and octane ratings.

■ Describe normal and abnormal combustion of gasoline and diesel fuel.

■ Summarize the properties of alternative fuels.

■ Correctly answer ASE certification test questions on automotive fuels and combustion.

An automotive engine burns a *fuel* as a source of energy. Various types of fuel will burn in an engine: gasoline, diesel fuel, gasohol, alcohol, liquefied petroleum gas, and other alternative fuels. An automotive technician must understand how fuel burns inside an engine. *Combustion* (burning) is a primary factor controlling fuel economy, power, emissions, and engine service life.

Petroleum (Crude Oil)

Petroleum, also called *crude oil*, is oil taken directly out of the ground. It is used to make gasoline, diesel fuel, liquefied petroleum gas, and many nonfuel materials (asphalt, motor oil). **Figure 20-1** shows some of the products made from petroleum.

Natural crude oil is a mixture of *semisolids* (neither solid nor liquid), liquids, and gases. Chemically, crude oil consists of highly flammable hydrocarbons. *Hydrocarbons* are chemical mixtures of about 12% hydrogen (flammable gas vapor) and 82% carbon (heavy, black solid). Crude oil also contains sulfur, nitrogen, metals, and saltwater. These elements must be removed.

Processing Crude Oil

Crude oil deposits are contained inside the earth. Oil companies perform *exploration tests* (seismic studies, surface mapping, test drilling) to find oil. After determining where oil might be located, a drill crew bores a hole thousands of feet into the ground. A huge steel derrick is used for the drilling operation. It has a cutting bit capable of passing through dirt, sand, and rock. See **Figure 20-2.**

Once the oil deposit has been reached, the oil is pumped to the surface. Then, oil is sent to the refinery. The *refinery* converts the crude oil into more useful substances.

Distillation is the first conversion process. During distillation, a *fractionating tower* is used to break the crude oil down into different parts, or fractions, (LPG, gasoline, kerosene, fuel oil, lubricating oils). Look at **Figure 20-3.** After distillation, other processes purify these products.

Gasoline

Gasoline is the most common type of automotive fuel. It is an abundant and highly flammable part of crude oil. Extra chemicals, called additives (detergents, antioxidants), are mixed into gasoline to improve its operating characteristics.

Antiknock additives slow down the burning of gasoline. This helps prevent *engine ping,* or *knock* (knocking sound produced by abnormal, rapid combustion).

Gasoline Octane Ratings

The *octane rating* of gasoline is a measurement of the fuel's ability to resist knock or ping. A high octane rating indicates the fuel will *not* knock or ping easily. High-octane gasoline should be used in high-compression engines and turbocharged engines. Low-octane gasoline is suitable for low-compression engines.

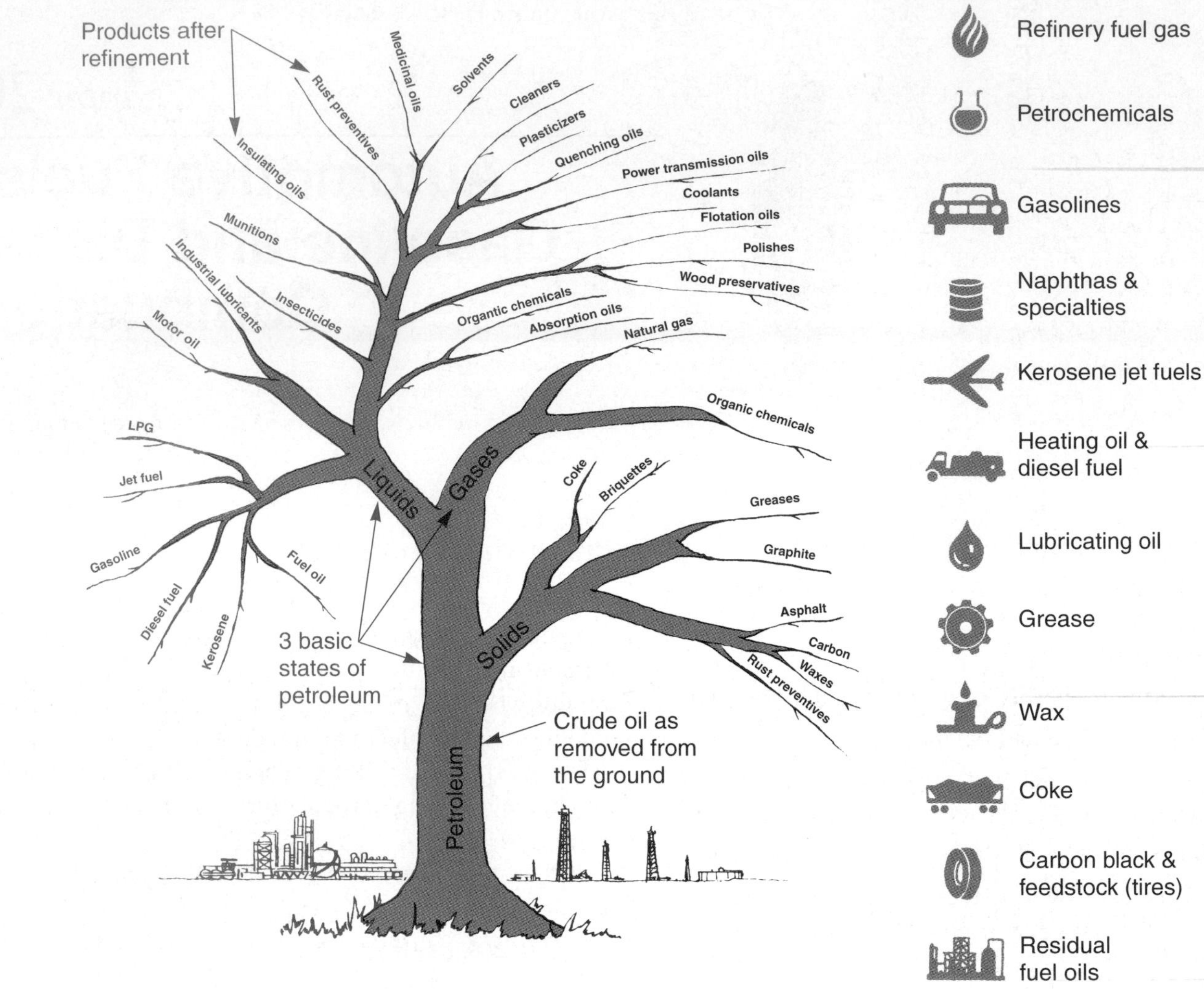

Products after refinement

Medicinal oils
Solvents
Rust preventives
Cleaners
Insulating oils
Plasticizers
Quenching oils
Munitions
Power transmission oils
Coolants
Industrial lubricants
Insecticides
Flotation oils
Polishes
Motor oil
Organic chemicals
Wood preservatives
Absorption oils
LPG
Natural gas
Jet fuel
Organic chemicals
Coke
Briquettes
Greases
Gasoline
Fuel oil
Graphite
Diesel fuel
Asphalt
Kerosene
Carbon
Waxes
Rust preventives

Liquids   Gases

Solids

3 basic states of petroleum

Crude oil as removed from the ground

Petroleum

Refinery fuel gas

Petrochemicals

Gasolines

Naphthas & specialties

Kerosene jet fuels

Heating oil & diesel fuel

Lubricating oil

Grease

Wax

Coke

Carbon black & feedstock (tires)

Residual fuel oils

Figure 20-1. Petroleum is used to make many products besides gasoline and diesel fuel. (Gulf Oil Corp. and Ethyl Corp.)

Figure 20-2. This large cutter can penetrate dirt and rock to drill holes deep into ground. When found, oil can be pumped to surface. (Texaco)

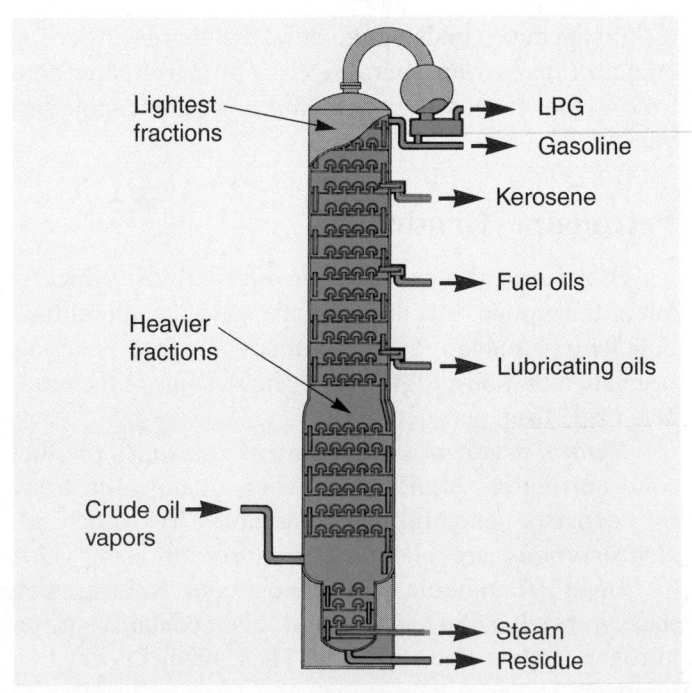

Lightest fractions

LPG
Gasoline
Kerosene
Fuel oils

Heavier fractions

Lubricating oils

Crude oil vapors

Steam
Residue

Figure 20-3. A fractionating tower allows crude oil vapors to condense and separate into trays. (Ford Motor Co.)

Octane numbers give the antiknock value of gasoline. Gasoline with a high octane number (91, for example) will resist knock and ping better than gasoline with a low octane number (87, for example). Automakers recommend octane ratings of fuel for their engines.

Octane numbers are given on the side of the gas station pump, **Figure 20-4.** The owner's manual will give the octane number of the fuel recommended for the car's engine. Use a fuel with an octane number as high as or higher than the automaker's recommendations. **Figure 20-5** summarizes several factors that control engine octane requirements.

In the past, tetraethyl lead (a heavy metal) was used to increase the octane rating of gasoline. It was phased out because it posed a health hazard.

Octane enhancers, or *oxygenates*, are now used as blending components in gasoline to increase octane levels and reduce engine knock. Oxygenates are alcohols that are made up of hydrogen, carbon, and oxygen. Examples of octane enhancers include toluene, ethanol, and MTBE.

Gasoline Combustion

For gasoline (or any other fuel) to burn properly, it must be mixed with the right amount of air. The mixture must then be compressed and ignited. The resulting combustion produces heat, expansion of gases, and pressure. The pressure pushes down on the piston to rotate the crankshaft. Refer to **Figure 20-6.**

Normal Gasoline Combustion

Normal gasoline combustion occurs when the spark plug ignites the fuel and burning progresses smoothly through the fuel mixture. Maximum cylinder pressure should be produced a few degrees of crank rotation after piston TDC on the power stroke. **Figure 20-7** illustrates normal gasoline combustion:

A— A spark at the spark plug starts the fuel burning. A small ball of flame forms around the tip of the plug. The piston is moving up in the cylinder, compressing the fuel mixture.

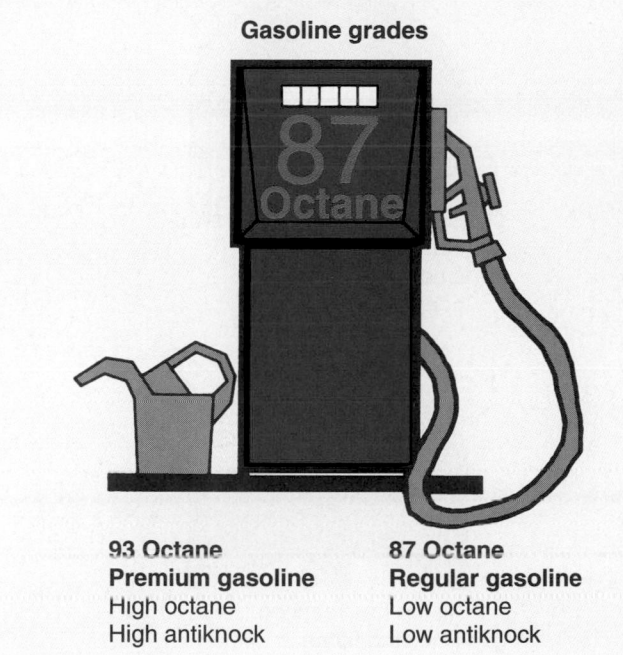

Gasoline grades

93 Octane
Premium gasoline
High octane
High antiknock

87 Octane
Regular gasoline
Low octane
Low antiknock

Figure 20-4. The gasoline grade indicates the antiknock value of gasoline.

Octane requirement factors	
Octane number requirement tends to go *up* when:	Octane number requirement tends to go *down* when:
1. Ignition timing is advanced.	1. Car is operated at higher altitudes (lower barometric pressure).
2. Air density rises due to supercharging, a larger throttle opening, or higher barometric pressures.	2. Fuel-air ratio is richer or leaner than that producing maxium knock.
3. Humidity or moisture content of air decreases.	3. Spark plug location in combustion chamber provides shortest path of flame travel.
4. Inlet air temperature goes up.	4. Combustion chamber design gives maximum turbulence of fuel-air charge.
5. Lean fuel-air ratios.	5. Compression ratio is lowered.
6. Compression ratio is increased.	6. Exhaust gas recycle system operates at part-throttle.
7. Coolant temperature is raised.	7. Ignition timing retard devices are used.
8. Antifreeze (glycol) engine coolant is used.	8. Humidity of the air increases.
9. Combustion chamber design provides little or no quench area.	9. Ignition timing is retarded.
10. Vehicle weight is increased.	10. Inlet air temperature is decreased.
11. Engine loading is increased, such as when climbing a grade, pulling a trailer, or increasing wind grade, or increasing wind resistance with a car-top carrier.	11. Reduced engine loads are employed.

Figure 20-5. Various factors that control octane requirements. (Ethyl Corp.)

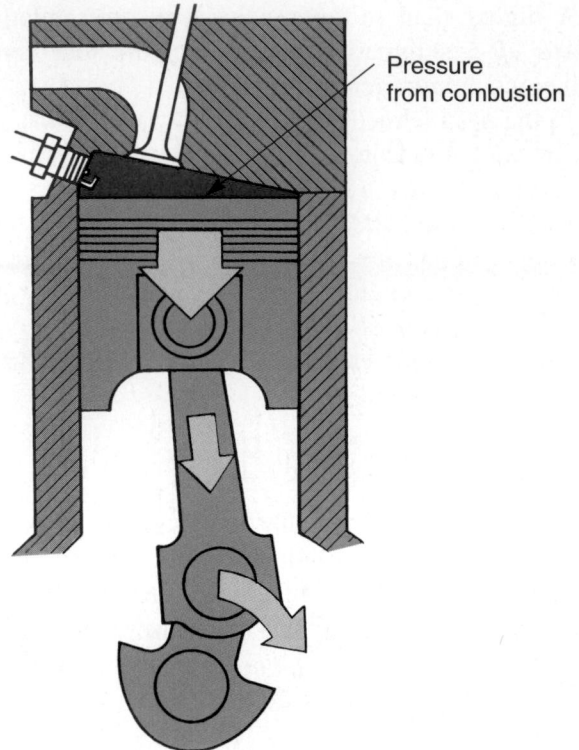

Pressure from combustion

Figure 20-6. Combustion produces heat. Heat causes gases to expand. Expansion causes pressure. Pressure pushes the piston down on power stroke. (Ford)

B— The flame spreads faster and moves about halfway through the mixture. Generally, the flame is moving evenly through the fuel mixture. The piston is nearing TDC, causing increased pressure.

C— The piston reaches TDC. The flame picks up more speed.

D— The flame shoots out to consume the rest of the fuel in the chamber. Combustion is complete with the piston only a short distance down in the cylinder.

Normal combustion only takes about 3/1000 of a second. This is much slower than an explosion. Dynamite explodes in about 1/50,000 of a second. Under some undesirable conditions, however, gasoline can burn too quickly, making part of combustion like an explosion. This is detailed later.

Air-Fuel Mixture

For proper combustion and engine performance, the correct amounts of air and fuel must be mixed. If too much fuel or too much air is used, engine power, fuel economy, and efficiency will suffer.

A *stoichiometric fuel mixture* is a chemically correct air-fuel mixture. For gasoline, it is a mixture ratio of

A

Spark occurs, ball of flame forms around electrodes.

B

Flame front spreads smoothly while piston nears TDC.

C

Piston has reached TDC and flame is "shooting" through chamber, forming heat.

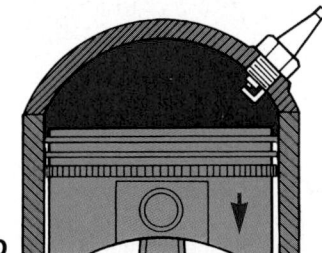

D

Combustion pressure is maximum and piston is a few degrees down in cylinder.

Figure 20-7. Normal combustion. A single flame moves smoothly through the air-fuel mixture.

14.7:1 (14.7 parts air to 1 part fuel, by weight). Under constant engine conditions, this ratio helps ensure that all the fuel is burned during combustion.

As you will learn in later chapters, the conditions in an engine are not always ideal. The fuel system must change the air-fuel ratio with changes in engine operating conditions.

Lean and Rich Air-Fuel Mixture

A *lean air-fuel mixture* contains a large amount of air compared to fuel. Look at **Figure 20-8.** For gasoline, 20:1 is a very lean mixture.

A *rich air-fuel mixture* is the opposite of a lean mixture; more fuel is mixed with the air. For gasoline, 8:1 (8 parts air to one part fuel) is a very rich fuel mixture. Refer to **Figure 20-8.**

A slightly lean mixture is desirable for high gas mileage and low exhaust emissions. Extra air in the cylinder ensures that all the fuel is burned. Too lean a mixture, however, can cause poor engine performance (lack of power, missing, and even engine damage).

A slightly rich mixture tends to increase engine power. However, it also increases fuel consumption and exhaust emissions. An over-rich mixture will reduce engine power, foul spark plugs, and cause incomplete burning (black smoke at engine exhaust).

Abnormal Combustion

Abnormal combustion occurs when the flame does not spread evenly and smoothly through the combustion chamber. The lean air-fuel mixtures, high operating temperatures, and low-octane fuels of today make abnormal combustion a problem.

Detonation

Detonation results when part of the unburned air-fuel mixture explodes violently. This is the most severe and engine-damaging type of abnormal combustion.

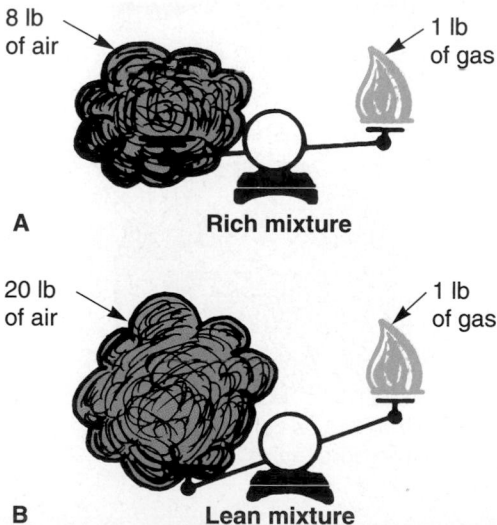

Figure 20-8. A—Rich fuel mixture has more fuel mixed into the air. B—Lean fuel mixture has less fuel mixed with the air.

Engine knock is a symptom of detonation. The combustion chamber pressure rises so quickly that parts of the engine vibrate. Detonation sounds like a hammer hitting the side of the engine. **Figure 20-9** shows what happens during detonation. Study the four phases.

Spark occurs, combustion is slow but normal.

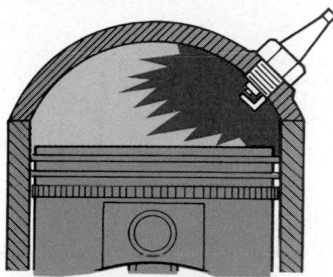

Normal combustion spreads very slowly.

End gas auto-ignites and two flame fronts spread rapidly.

Flames collide with pressure "spike" and knock.

Figure 20-9. Detonation is caused by combustion being too slow. End gas, or unburned air-fuel mixture, ignites and two flame fronts collide with a loud knock.

As you can see, the **end gas** (unburned portion of mixture) is heated and pressurized for an extended period. Combustion is too slow because of an incorrect air-fuel mixture, a lack of turbulence, or a fuel distribution problem. This causes the end gas to explode with a "bang" (knock).

Detonation can greatly increase the pressure and heat in the engine combustion chamber. **Detonation damage** includes cracked cylinder heads, blown head gaskets, burned pistons, and shattered spark plugs. See **Figures 20-10** and **20-11**.

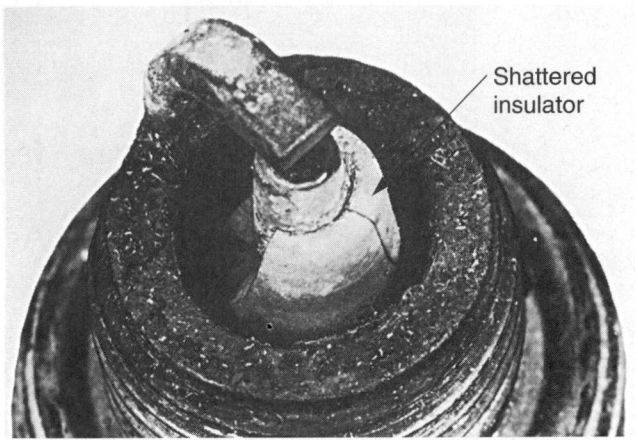

Figure 20-10. Detonation can shatter the insulator on a spark plug. (Champion Spark Plugs)

Figure 20-11. Detonation can blow a hole in the piston head. (Champion Spark Plugs)

Preignition

Preignition results when an overheated surface in the combustion chamber ignites the air-fuel mixture. Termed **surface ignition,** a "hot spot" (overheated bit of carbon, sharp metal edge, hot exhaust valve) causes the mixture to burn prematurely.

Hot carbon deposit ignites fuel mixture.

Spark plug "fires" and two flame fronts form.

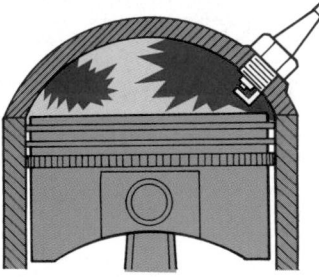

Both flame fronts shoot toward each other at high speed.

Two flames collide causing pressure "spike" and a knock.

Figure 20-12. Preignition is caused by early ignition of the fuel mixture. Abnormal and normal flames collide, producing a pinging noise.

A *ping,* or a mild knock, is a light tapping noise that can be heard during preignition. It is not as loud nor as harmful as detonation knock. Study **Figure 20-12.**

Preignition is similar to detonation, but the actions are reversed. Detonation begins *after* the start of normal combustion. Preignition begins *before* the start of normal combustion.

 Caution
Prolonged preignition can produce harmful detonation. If an engine pings or knocks excessively, serious engine damage can result. Correct the problem right away.

Dieseling

Dieseling, also called **after-running** or **run-on**, is a problem where the engine keeps running after the key is turned off. A knocking, coughing, or fluttering noise is heard as the fuel ignites and the crankshaft spins uncontrollably. When dieseling, the gasoline engine ignites the fuel from heat and pressure, somewhat like a diesel engine. With the ignition key off, the engine runs without voltage to the spark plugs.

The most common causes of dieseling are a high idle speed, carbon deposits in the combustion chambers, low-octane fuel, an overheated engine, or spark plugs that have too high a heat range. This problem will be discussed later in the text.

Spark Knock

Spark knock is an engine combustion problem caused by the spark plug firing too soon in relation to the position of the piston. Spark timing that is advanced too far causes combustion pressure to slam into the upward moving piston. This causes maximum cylinder pressure before TDC, not just after TDC as it should.

Figure 20-13 shows what happens during spark knock. Spark knock can also lead to preignition and more damaging detonation.

Spark knock and preignition produce about the same symptoms—pinging under load. To find the cause of pinging, first check the ignition timing. If timing is correct, check other possible causes.

Diesel Fuel

Diesel fuel is the second most popular type of automotive fuel. A gallon of diesel fuel contains more heat energy than a gallon of gasoline. It is a thicker fraction (part) of crude oil. Diesel fuel can produce more cylinder pressure and vehicle movement than an equal amount of gasoline. Diesel fuel now costs about the same as gasoline.

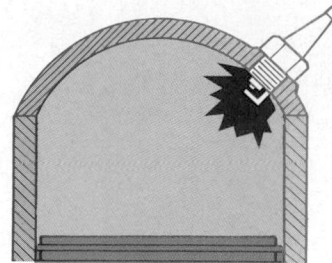
Spark plug "fires" too soon.

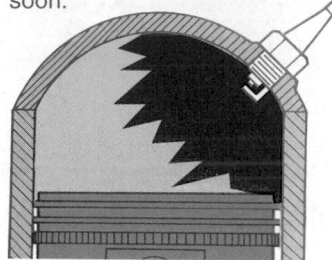
Piston moves toward flame front.

Pressure builds as piston slams into combustion flame.

Spark knock occurs because of excessive pressure in cylinder.

Figure 20-13. Spark knock is a ping or knock caused by an ignition timing problem.

Since diesel fuel is thicker and has different burning characteristics than gasoline, a high-pressure injection system must be used to spray the fuel directly into the combustion chambers. A low-pressure injection system or carburetor would not meter the thicker diesel fuel properly. Look at **Figure 20-14.**

Diesel fuel will not *vaporize* (change from a liquid to a gas) as easily as gasoline. If diesel fuel enters the intake

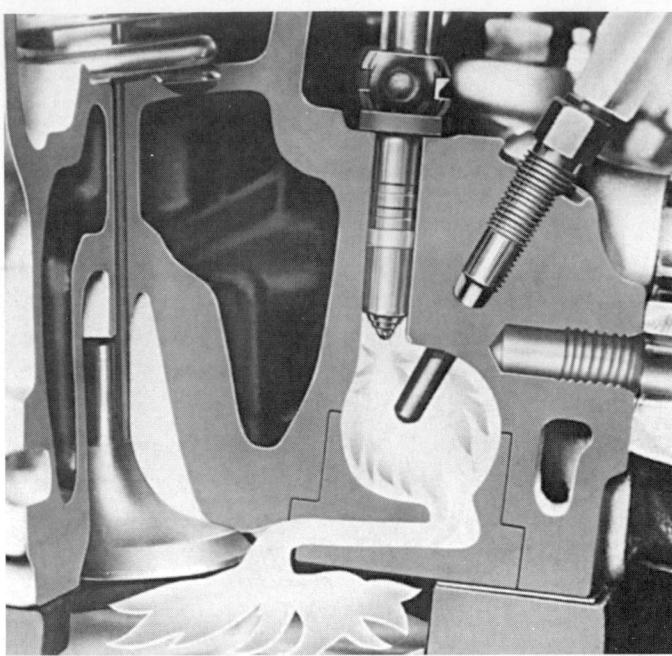

Figure 20-14. A diesel engine is a compression-ignition engine. High pressure heats air in the cylinder. When fuel is sprayed into the hot air, combustion begins.

manifold of an engine, it collects on the inner walls of the manifold. This upsets engine operation.

Combustion requires fuel to be in a vapor state. Diesel engines inject the diesel fuel directly into the combustion chamber. The compressed, hot air vaporizes and burns the fuel.

Diesel Fuel Grades

Diesel fuel grades ensure that diesel fuel sold all over the country has uniform standards. Diesel engine makers are then able to select a diesel fuel grade that meets the needs of their engines. There are three diesel fuel grades: No. 1 diesel, No. 2 diesel, and No. 4 diesel.

No. 2 diesel is normally recommended for use in diesel automotive engines. It is also the only grade of fuel available at many service stations. No. 2 diesel has a medium *viscosity* (thickness or weight), which provides proper operating traits for the widest range of conditions.

No. 1 diesel is thinner than No. 2 diesel and is sometimes recommended as a winter fuel. Low temperatures tend to thicken diesel fuel, causing performance problems (hard starting, poor fuel delivery).

Some auto manufacturers allow the use of only No. 2 diesel fuel, with special cold weather additives being used in the winter. In some engines, the thin No. 1 diesel fuel will not provide adequate lubrication. Metal-to-metal contact may occur, causing serious engine damage. When in doubt, refer to the vehicle's service manual.

 Caution
Diesel fuel should not be confused with fuel oil or home heating oil. Diesel fuel contains fewer impurities than fuel oil. Fuel oil should *never* be used in a diesel engine or damage will result.

Diesel Fuel Cloud Point

One of the substances in diesel fuel is *paraffin* (wax). At very cold temperatures, this wax can separate from the other parts of the diesel fuel. When this happens, the fuel will turn cloudy or milky. *Cloud point* is the temperature at which paraffin separates out of the fuel. At cloud point, the paraffin can clog the fuel filters and prevent diesel engine operation.

Diesel Fuel Water Contamination

Water contamination is a common problem with diesel engines. When mixed with diesel fuel, water can clog filters and corrode components. The parts in diesel injection pumps and nozzles are very precise. They can be easily damaged by water.

Many late-model diesel injection systems have *water separators* to prevent water damage. These will be covered in later chapters.

Diesel Fuel Cetane Rating

A *cetane rating* indicates the cold starting ability of diesel fuel. The higher the cetane number, the easier the engine will start and run in cold weather. Most automakers recommend a cetane rating of about 45. This is the average cetane value of No. 2 diesel fuel.

A cetane number, in some ways, is the opposite of a gasoline octane number. This is shown in **Figure 20-15.** A high cetane number means the fuel will ignite easily from heat and pressure. In a diesel, the fuel must ignite and burn as soon as it touches the hot air in the combustion chamber.

Diesel Combustion

A diesel engine is a *compression-ignition engine*. This engine compresses air until it is hot enough to ignite the fuel. A spark plug would not ignite diesel fuel properly. If gasoline were used in a diesel engine, it would detonate on the compression stroke and not produce useful energy.

Diesel fuel is thick and hard to ignite. A high-pressure mechanical pump and nozzle force the fuel into the engine combustion chamber. An extremely high compression ratio heats the air in the cylinder. Then, when fuel is

Cetane and octane comparison

Cetane rating

Octane rating

Fast burning

60 cetane

50 cetane

40 cetane

30 cetane

70 octane

80 octane

90 octane

100 octane

Slow burning

Diesel fuel

Gasoline

Figure 20-15. Diesel fuel cetane rating is the opposite of gasoline octane rating.

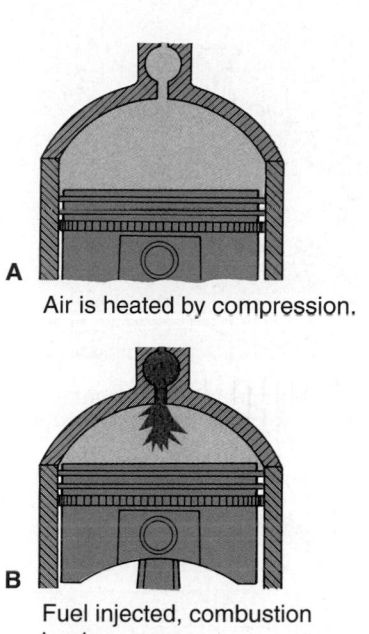

A Air is heated by compression.

B Fuel injected, combustion begins.

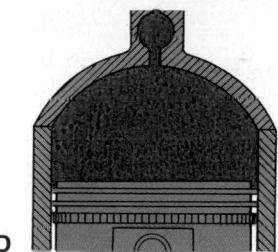

C More fuel sprays into chamber.

D Combustion continues and pressure is formed while piston moves down.

Figure 20-16. Normal diesel combustion.

sprayed into the hot air, it begins to burn. **Figure 20-16** shows the phases of normal diesel combustion:

A— The piston moves up to compress and heat the air in the cylinder. Note that this is different than a gasoline engine, which compresses both fuel and air.

B— Diesel fuel is injected directly into the combustion chamber. The hot air makes the fuel begin to burn and expand.

C— More fuel is sprayed into the chamber. More pressure is developed and the piston begins to move down in the cylinder.

D— The rest of the fuel is injected into the chamber. Pressure continues to form, pushing the piston down on the power stroke.

Note that fuel was injected into the engine for several degrees of crankshaft rotation. This caused a smooth, steady buildup of pressure for quiet diesel engine operation.

Diesel Combustion Knock

When compared to gasoline engines, diesel engines knock almost all the time. The engine clatters and rattles as the diesel fuel ignites in the combustion chambers.

Diesel knock occurs when too much fuel ignites at one time, producing a loud knocking noise. Excessive diesel knock can reduce engine power, fuel economy, and engine life.

Ignition lag is the time it takes diesel fuel to heat up, vaporize, and begin to burn. It is the time lapse between initial fuel injection and actual ignition (burning).

Ignition lag is a major controlling factor of diesel knock. If lag time is too long, a large amount of fuel can ignite, producing a louder-than-normal knock. A high cetane fuel, which has a short lag time, reduces the

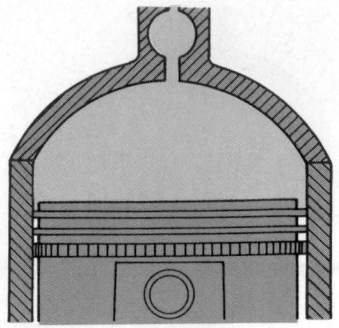

Air is compressed and heated.

Fuel is injected but fails to vaporize and burn.

More fuel injected, but still no combustion.

All of the fuel suddenly ignites with a "bang."

Figure 20-17. Diesel knock is caused by too much fuel igniting at one time. The fuel does not ignite quick enough when injection begins.

chances of diesel knock. **Figure 20-17** shows the basic phases of diesel knock. Study this illustration.

Diesel knock is caused by too much fuel burning at one time. This can be due to a cold engine, a low-cetane fuel, an improper fuel spray pattern, or incorrect injection timing. These will be discussed in later chapters.

Alternative Fuels

Alternative fuels include any fuel other than gasoline and diesel fuel. Liquefied petroleum gas, alcohol, and hydrogen are examples of alternative fuels.

Liquefied Petroleum Gas

Liquefied petroleum gas (LPG) is sometimes used as a fuel for automobiles and trucks. It is one of the lightest fractions of crude oil. Chemically, LPG is similar to gasoline. However, at room temperature and pressure, LPG is a vapor, *not* a liquid. Refer again to **Figure 20-3.**

A special fuel system is needed to meter the gaseous LPG into the engine. One is shown in **Figure 20-18.** Note the names and construction of the basic parts.

LPG is commonly used in industrial equipment, such as fork lifts. It is also used in some fleet trucks that always refuel at the same location. Being a gas, LPG burns cleanly, producing few exhaust emissions.

Alcohol

Alcohol has the potential to be an excellent alternative fuel for automotive engines. The two types of alcohol used in automobiles are ethyl alcohol and methyl alcohol.

Ethyl alcohol, also called **grain alcohol** or **ethanol**, is made from farm crops. Grain, wheat, sugarcane, potatoes, fruit, oats, soy beans, and other crops rich in carbohydrates can be made into ethyl alcohol. This type alcohol is a colorless, harsh tasting, toxic, and highly flammable liquid.

Methyl alcohol, also termed *"wood alcohol"* or **methanol**, can be made out of wood chips, petroleum, garbage, and animal manure. It has a strong odor, is colorless, poisonous, and very flammable.

Alcohol is a clean-burning fuel for automobiles. It is not commonly used because it is expensive to use and produce. Also, the vehicle's fuel system requires modification before it can burn straight alcohol. Almost twice as much alcohol must be burned, compared to gasoline. This reduces fuel economy by 50%.

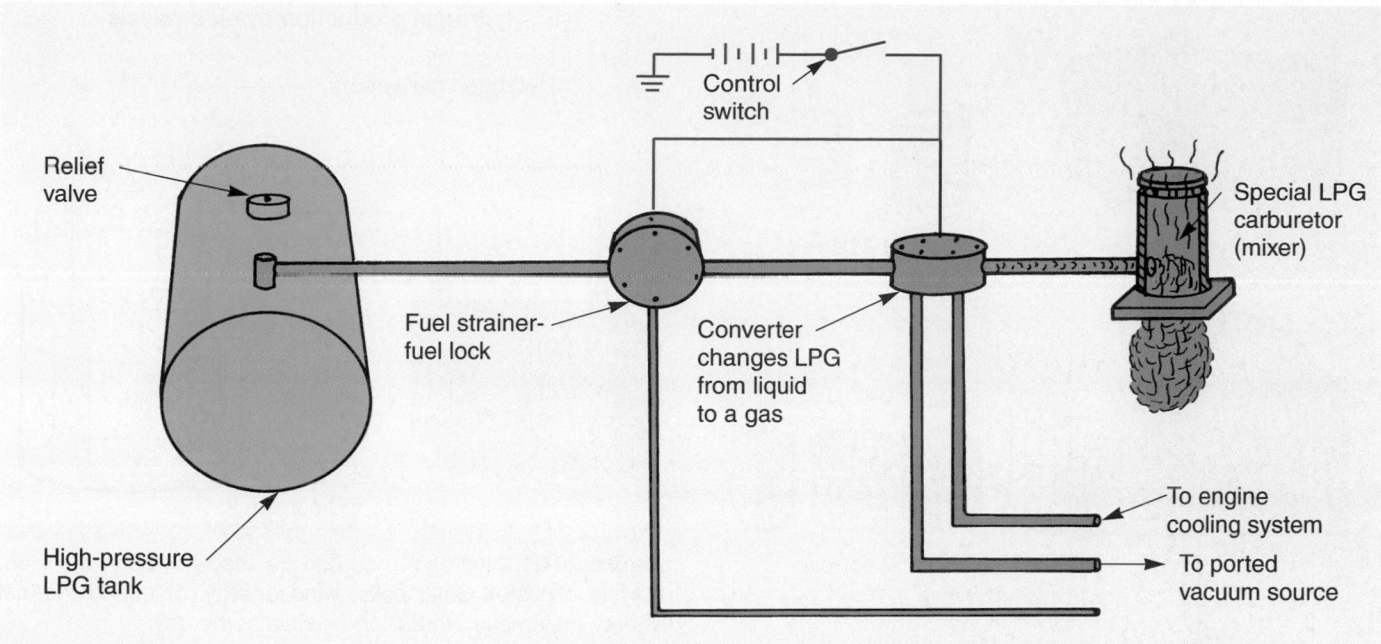

Figure 20-18. An LPG fuel system uses a high-pressure storage tank. A fuel strainer-fuel lock cleans fuel and prevents leakage when the engine is not running. A converter uses heat from the engine coolant to change the liquid LPG into a gas. A special carburetor meters LPG into the engine.

Tech Tip
Methanol is commonly used as a racing fuel. It burns very hot but does not produce a visible flame. This can be very dangerous because you cannot see the flames if there is a fire. For increased safety, racing organizations have required the use of additives in methanol racing fuel to make its flame visible.

Gasohol

Gasohol, as the name implies, is a mixture of gasoline (usually 87 octane gasoline) and alcohol (usually grain alcohol). The mixture can range from 2–20% alcohol. In most cases, gasohol is a blend of 10% alcohol and 90% gasoline. See **Figure 20-19.**

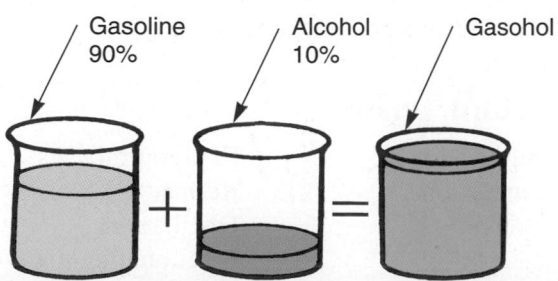

Figure 20-19. Gasohol is usually a mixture of 10% alcohol and 90% gasoline. (Ethyl Corp.)

Gasohol is commonly used as an alternative fuel in motor vehicles because fuel system and engine modifications are not needed. Many gas stations sell gasohol as a high-octane fuel. The alcohol tends to reduce the knocking tendencies of the gasoline. It acts like an anti-knock additive.

For example, 10% alcohol can increase 87 octane gasoline to 91 octane. Gasohol can be burned in a high-compression, high-horsepower engine without detonating and knocking.

Synthetic Fuels

Synthetic fuels are fuels made from coal, shale oil (rock filled with petroleum), and tar sand (sand filled with petroleum). See **Figure 20-20.**

Synthetic fuels are synthesized (changed) from a solid hydrocarbon state to a liquid or gaseous state. Synthetic fuels are being experimented with as a means of supplementing crude oil. As crude oil-based fuels become more expensive, synthetic fuels will become more practical.

Hydrogen

Hydrogen is a highly flammable gas that is a promising alternative fuel of the future. Hydrogen is one of the most abundant elements on our planet. It can be produced through the electrolysis of water (sending

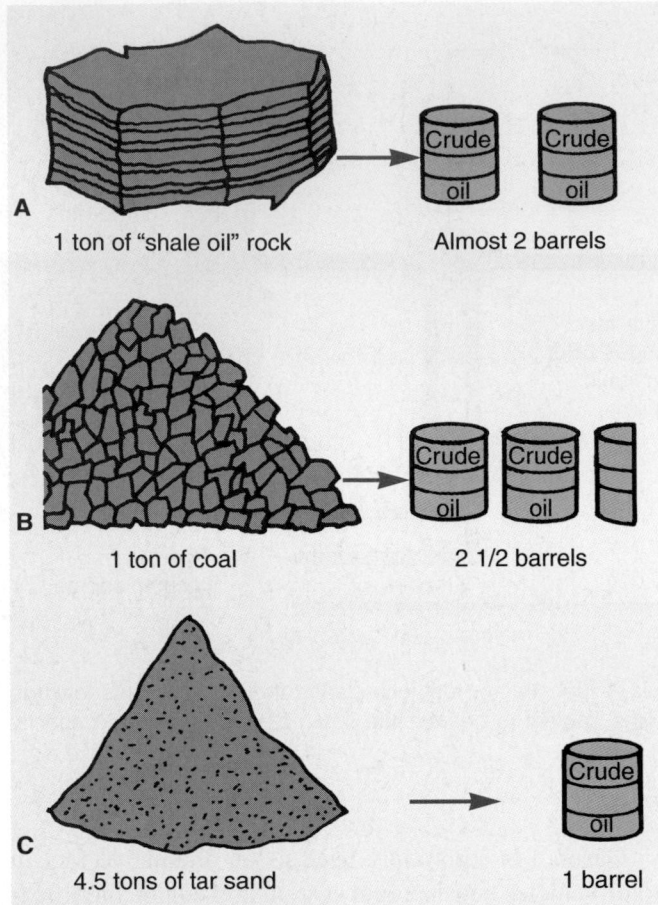

Figure 20-20. Synthetic sources of oil. A—Shale rock can be converted into oil. B—Coal can produce about two and one-half barrels of oil per ton. C—Four and one-half tons of tar can be changed into about one barrel of oil.

Hydrogen production by electrolysis

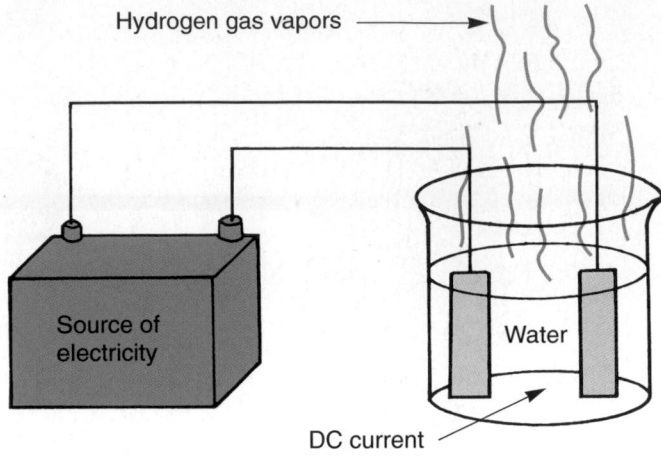

Figure 20-21. Hydrogen gas can be made through the electrolysis of water. Solar cells, wind energy, or ocean thermal energy may make production feasible some day.

electric current through saltwater). This is illustrated in **Figure 20-21.**

Hydrogen is an ideal fuel. Hydrogen burns almost perfectly, leaving only water and harmless carbon dioxide as by-products.

Presently, hydrogen is too expensive to make and store. However, as we use up our supply of crude oil, we may someday make hydrogen a major source of automotive fuel.

Summary

- Petroleum, also called crude oil, is oil taken directly out of the ground.
- Gasoline is the most common type of automotive fuel.
- The octane rating of gasoline is a measurement of the fuel's ability to resist knocking or pinging.

- Normal gasoline combustion occurs when the spark plug ignites the fuel and burning progresses smoothly through the air-fuel mixture.
- A stoichiometric fuel mixture is a chemically correct, or perfect, air-fuel mixture.
- Detonation results when part of the unburned air-fuel mixture explodes violently.
- Preignition results when an overheated surface in the combustion chamber ignites the fuel mixture.
- Dieseling, also called after-running or run-on, occurs when the engine keeps running after the key is turned off.
- Diesel fuel is the second most popular type of automotive fuel.
- No. 2 diesel fuel is normally recommended for use in automotive diesel engines.
- A cetane rating indicates the cold starting ability of diesel fuel.
- Alternative fuels include any fuel other than gasoline and diesel fuel. LPG, alcohol, and hydrogen are examples of alternative fuels.

Important Terms

Fuel	Exploration tests
Combustion	Refinery
Petroleum	Distillation
Crude oil	Fractionating tower
Natural crude oil	Gasoline
Hydrocarbons	Antiknock additives

Engine ping
Knock
Octane rating
Octane numbers
Octane enhancers
Oxygenates
Normal gasoline
  combustion
Stoichiometric fuel
  mixture
Lean air-fuel mixture
Rich air-fuel mixture
Abnormal combustion
Detonation
Engine knock
End gas
Detonation damage
Preignition
Surface ignition
Ping
Dieseling
After-running
Run-on
Spark knock
Diesel fuel

Vaporize
Diesel fuel grades
Viscosity
Paraffin
Cloud point
Water contamination
Water separators
Cetane rating
Compression ignition
  engine
Diesel knock
Ignition lag
Alternative fuels
Liquefied petroleum gas
  (LPG)
Alcohol
Ethyl alcohol
Grain alcohol
Ethanol
Methyl alcohol
"Wood alcohol"
Methanol
Gasohol
Synthetic fuels
Hydrogen

Review Questions—Chapter 20

Please do not write in this text. Place your answers on a separate sheet of paper.

1. _____, also called _____, is oil taken directly out of the ground.

2. What are hydrocarbons?

3. Explain the difference between leaded and unleaded gasoline.

4. The lead in leaded gasoline acts as a lubricant for the engine valves and valve seats. True or False?

5. The _____ of gasoline is a measurement of the fuel's ability to resist knock or ping.

6. If an automaker recommends gasoline with an octane number of 91, 87 octane gasoline is also acceptable. True or False?

7. Which of the following is *not* needed for proper combustion?
   (A) Air.
   (B) Compression.
   (C) Condensation.
   (D) Ignition.

8. Describe normal gasoline combustion.

9. Define the term "stoichiometric fuel mixture."

10. A(n) _____ air-fuel mixture ratio contains a large amount of air.

11. A(n) _____ air-fuel mixture ratio contains a large amount of fuel.

12. What are the results of lean and rich air-fuel mixtures?

13. Explain detonation, preignition, spark knock, and dieseling in a gasoline engine.

14. How does diesel fuel differ from gasoline?

15. Explain how ignition lag affects diesel combustion.

16. _____ is made from farm crops and _____ can be made out of wood chips, petroleum, garbage, and animal manure.

17. What is gasohol?

18. Gasohol can normally be used without major engine or fuel system modifications. True or False?

19. A certain gasohol mixture contains 5% alcohol. What is the approximate percentage of gasoline in this mixture?
   (A) 85%.
   (B) 92%.
   (C) 90%.
   (D) 95%.

20. Which of the following does *not* pertain to hydrogen as an alternative fuel?
   (A) Made from most abundant element.
   (B) Produced through electrolysis.
   (C) Burns without toxic emissions.
   (D) Economical or inexpensive to make.

● ASE-Type Questions

1. Petroleum is used to make each of these *except:*
   (A) *LPG.*
   (B) *gasoline.*
   (C) *hydrocarbons.*
   (D) *hydrogen.*

2. Crude oil is a mixture of:
   (A) *liquids.*
   (B) *semisolids.*
   (C) *hydrocarbons.*
   (D) *All of the above.*

3. Which of these is a commonly used antiknock additive?
   (A) *Naphthas.*
   (B) *Oxygenates.*
   (C) *Petrochemicals.*
   (D) *None of the above.*

4. Which of the following is the measurement of a fuel's ability to resist knock?
   (A) *Octane rating.*
   (B) *Cetane rating.*
   (C) *Clouding point.*
   (D) *All of the above.*

5. Normal combustion takes about:
   (A) *1/100 of a second.*
   (B) *3/1000 of a second.*
   (C) *3/10,000 of a second.*
   (D) *1/50,000 of a second.*

6. A chemically correct air-fuel ratio is called a:
   (A) *rich air-fuel mixture.*
   (B) *lean air-fuel mixture.*
   (C) *combustible fuel mixture.*
   (D) *stoichiometric fuel mixture.*

7. Each of these is a form or result of abnormal combustion *except:*
   (A) *detonation.*
   (B) *ignition lag.*
   (C) *engine knock.*
   (D) *after-running.*

8. A diesel fuel substance that can separate from other fuel parts and clog fuel filters is:
   (A) *lead.*
   (B) *water.*
   (C) *carbon.*
   (D) *paraffin.*

9. Automakers recommend a cetane rating of about:
   (A) *45.*
   (B) *50.*
   (C) *55.*
   (D) *60.*

10. When looking at cetane ratings, Technician A says that the higher a cetane number is, the easier an engine starts and runs in cold weather. Technician B says that the higher a cetane number is, the easier a fuel ignites from heat and pressure. Who is right?
    (A) *A only.*
    (B) *B only.*
    (C) *Both A and B.*
    (D) *Neither A nor B.*

Activities—Chapter 20

1. Obtain literature about automotive fuels and prepare a written report on additives and their properties.

2. Research magazines and newspapers for information about the manufacture of alcohol for use as an automotive fuel.

3. Research and discuss modifications of engines for use of LPG as a fuel.

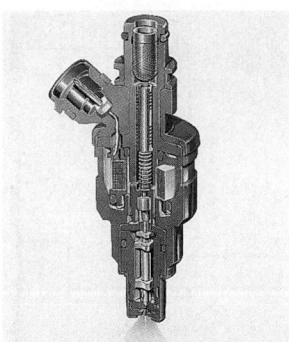

Chapter 21

Fuel Tanks, Pumps, Lines, and Filters

After studying this chapter, you will be able to:

- List the components of a fuel supply system.

- Describe the operation of mechanical and electric fuel pumps.

- Describe the construction and action of air filters.

- Explain the tests used to diagnose problems with fuel pumps, fuel filters, and fuel lines.

- Repair a fuel line or replace a fuel hose.

- Locate and replace fuel filters in both gasoline and diesel fuel systems.

- State safety rules for working on fuel supply systems.

- Correctly answer ASE certification test questions on fuel tanks, fuel pumps, fuel lines, fuel filters, and air filters.

A *fuel system* provides a combustible air-fuel mixture to power the engine. Introduced in Chapter 1, there are several types of fuel systems. Today's cars commonly use gasoline injection systems or diesel injection systems. Older vehicles use carburetors.

A modern fuel system has three subsystems:

- *Fuel supply system*—provides filtered fuel to the gasoline injectors, the diesel injection pump, or the carburetor.

- *Air supply system*—provides clean combustion air for engine.

- *Fuel metering system*—controls the amount of fuel that mixes with the air entering engine.

This chapter explains the construction, operation, and service of fuel tanks, fuel lines, fuel filters, air filters, and fuel pumps. These parts make up the fuel and air supply systems. This information will prepare you for later chapters on fuel metering systems. Study carefully!

Fuel Supply System

A fuel supply system draws fuel from the fuel tank and forces it into the fuel metering device (gasoline injectors, diesel injection pump, or carburetor), **Figure 21-1.** Modern vehicles use electric fuel pumps. Older engines use mechanical (engine-driven) fuel pumps. Some older vehicles use both types: an electric pump on the fuel tank and a mechanical pump on the engine.

The basic parts of a fuel supply system include:

- *Fuel tank*—stores gasoline, diesel fuel, gasohol, or LPG.

- *Fuel lines*—carry fuel between the tank, pump, and other parts.

- *Fuel pump*—draws fuel from the tank and forces it to the engine or fuel-metering device.

- *Fuel filters*—remove contaminants in the fuel.

Fuel Tanks

An automotive *fuel tank* must safely hold an adequate supply of fuel for prolonged engine operation. It is normally mounted in the rear of the car, under the trunk or the rear seat. See **Figure 21-1.**

The size of a fuel tank determines, in part, a car's *driving range* (distance a vehicle can be driven without stopping for fuel). *Fuel tank capacity* is the amount of fuel a fuel tank can hold. Average fuel tank capacity is 12–25 gallons (45–95 liters).

Fuel Tank Construction

Fuel tanks are usually made of thin sheet metal or plastic. The main body of a metal tank is made by soldering or welding two formed pieces of sheet metal together. As shown in **Figure 21-2,** other parts (filler neck, baffles, vent tubes, expansion chamber) are added to form the complete fuel tank assembly. A lead-tin alloy is normally plated to the sheet metal to keep the tank from rusting.

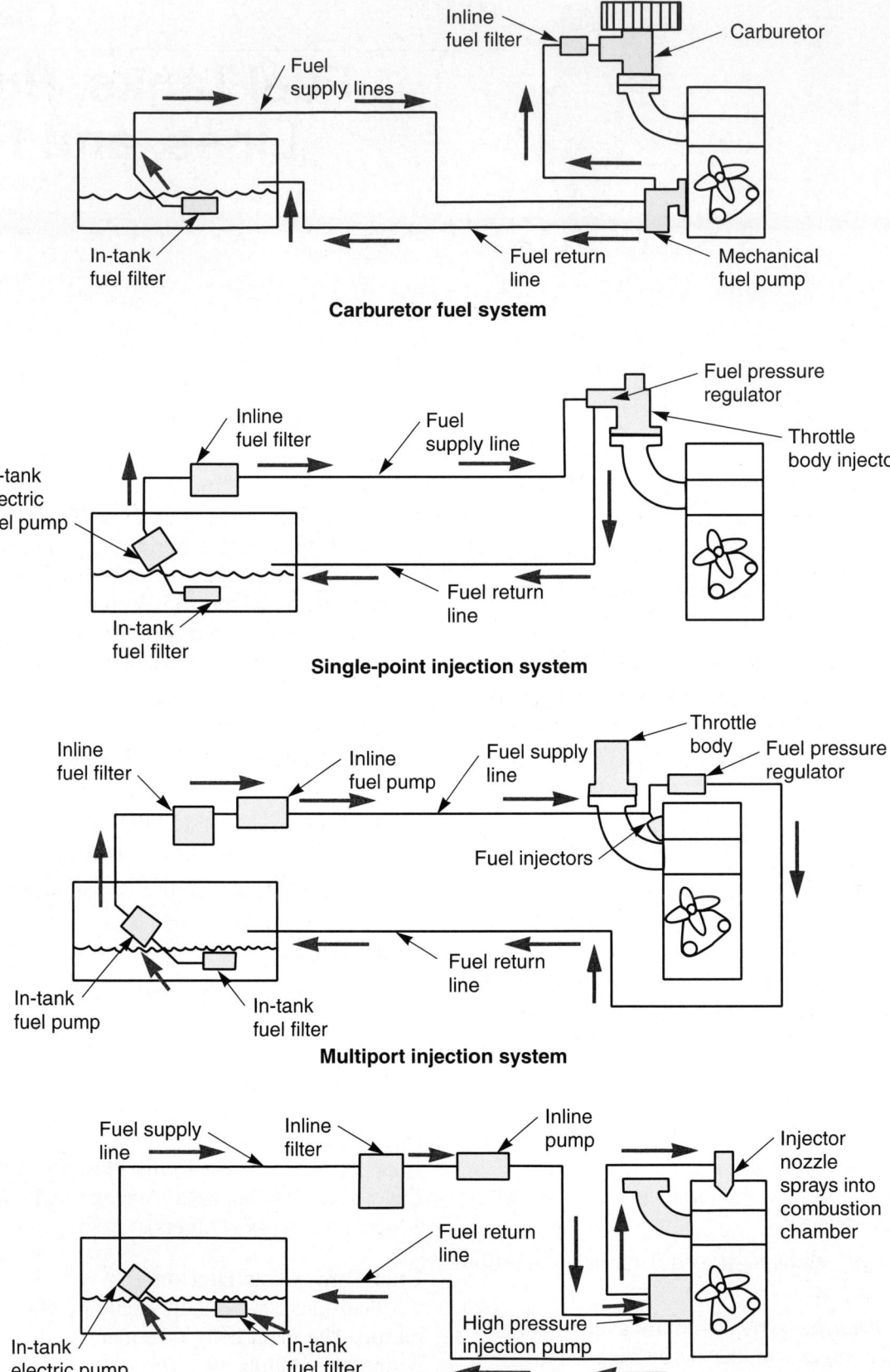

Figure 21-1. Basic types of fuel supply systems. Note the differences between carburetor, single-point, multiport, and diesel systems.

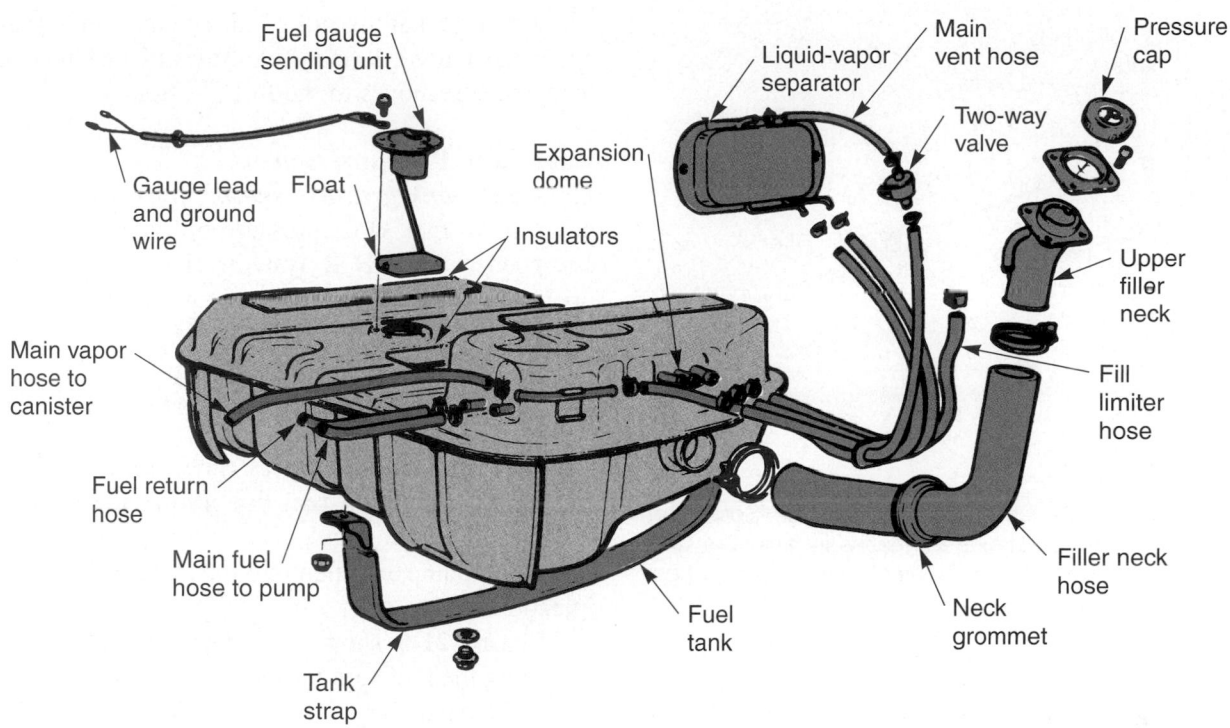

Figure 21-2. Basic parts of a fuel tank assembly. (Chrysler)

The *fuel tank filler neck* is the extension tube on the tank for filling the tank with fuel. The *filler cap* fits on the end of the filler neck, **Figure 21-2.** The neck extends from the tank through the body of the car. A flexible hose is normally used as part of the filler neck. It allows tank vibration without part breakage.

A *spillback ball* is a large "ping pong–type" ball in the fuel tank filler neck to prevent fuel from leaking out of the vehicle during fillups.

Modern *fuel tank caps* are sealed to prevent the escape of fuel and fuel vapors (emissions) from the tank. Normally, it is *not* vented to the atmosphere.

Fuel tank baffles are placed inside the fuel tank to keep fuel from sloshing, or splashing, around in the tank. The baffles are metal plates that restrict fuel movement when the car accelerates, decelerates, or turns a corner.

Fuel tank straps are used to secure the tank to the vehicle. They are thick steel bands that bolt around the fuel tank to hold it in place, **Figure 21-3.** A *fuel tank cover* is sometimes used to prevent damage from road debris.

Tank Pickup-Sending Unit

A *tank pickup-sending unit* extends down into the tank to draw out fuel and operate the fuel gauge. One is shown in **Figure 21-4.** A coarse filter is usually placed on the end of the pickup tube to strain out debris.

The tank sending unit is a variable resistor. Its resistance changes with changes in the fuel level. This causes

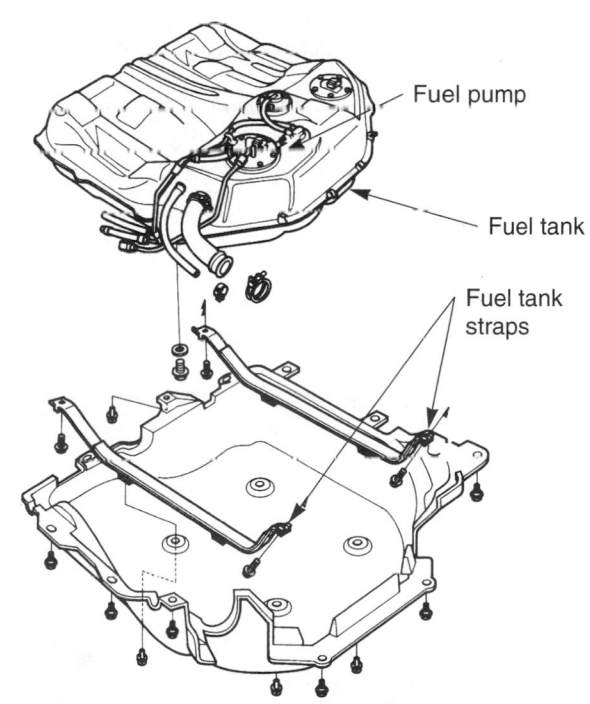

Figure 21-3. Large straps hold the fuel tank in the vehicle. Straps, wires, and the filler neck must be disconnected before tank removal. (Honda)

it to control the amount of current reaching the fuel gauge in the instrument panel. **Figure 21-5** shows the basic action of a fuel tank sending unit.

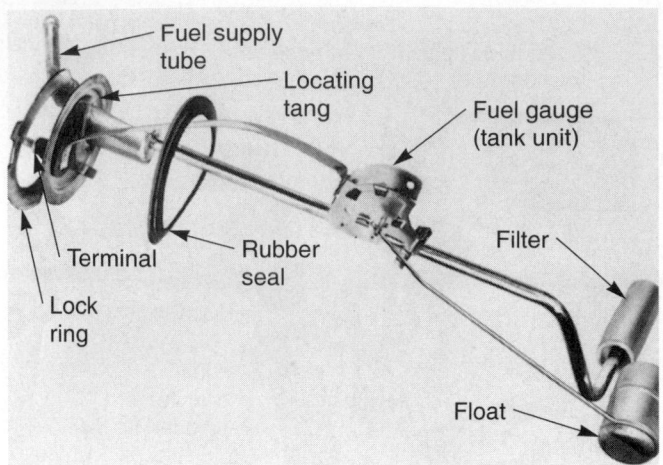

Figure 21-4. Typical fuel tank pickup-sending unit. "Sock" filter on the end of the pickup strains out debris in tank. The tank sending unit operates the instrument panel fuel gauge. (Chrysler)

When the fuel level in the tank is low (float down), the sending unit has a high resistance. Only a small amount of current flows to the gauge. The gauge shows a low fuel level.

When the tank is full (float up) the sending unit has a low resistance. More current flows to the gauge, and the gauge needle moves to the full position.

Fuel Tank Pressure Sensor

A *fuel tank pressure sensor* is used to monitor the buildup of fuel tank pressure in late-model vehicles equipped with OBD II systems. The sensor aids in the operation of the fuel vapor recovery emission control system.

Fuel Lines and Hoses

Fuel lines and *fuel hoses* carry fuel from the tank to the engine. The main fuel line allows the fuel pump to draw fuel out of the tank. The fuel is pulled through this line to the pump and then into the metering section of the injection system.

Figure 21-6 shows a complete set of fuel lines, including the fuel vapor lines for the evaporation control (emission control) system. Study the routing of the lines. Note how they connect to the fuel system components.

Fuel lines are normally made of strong, double-wall steel tubing. For fire safety reasons, a fuel line must be

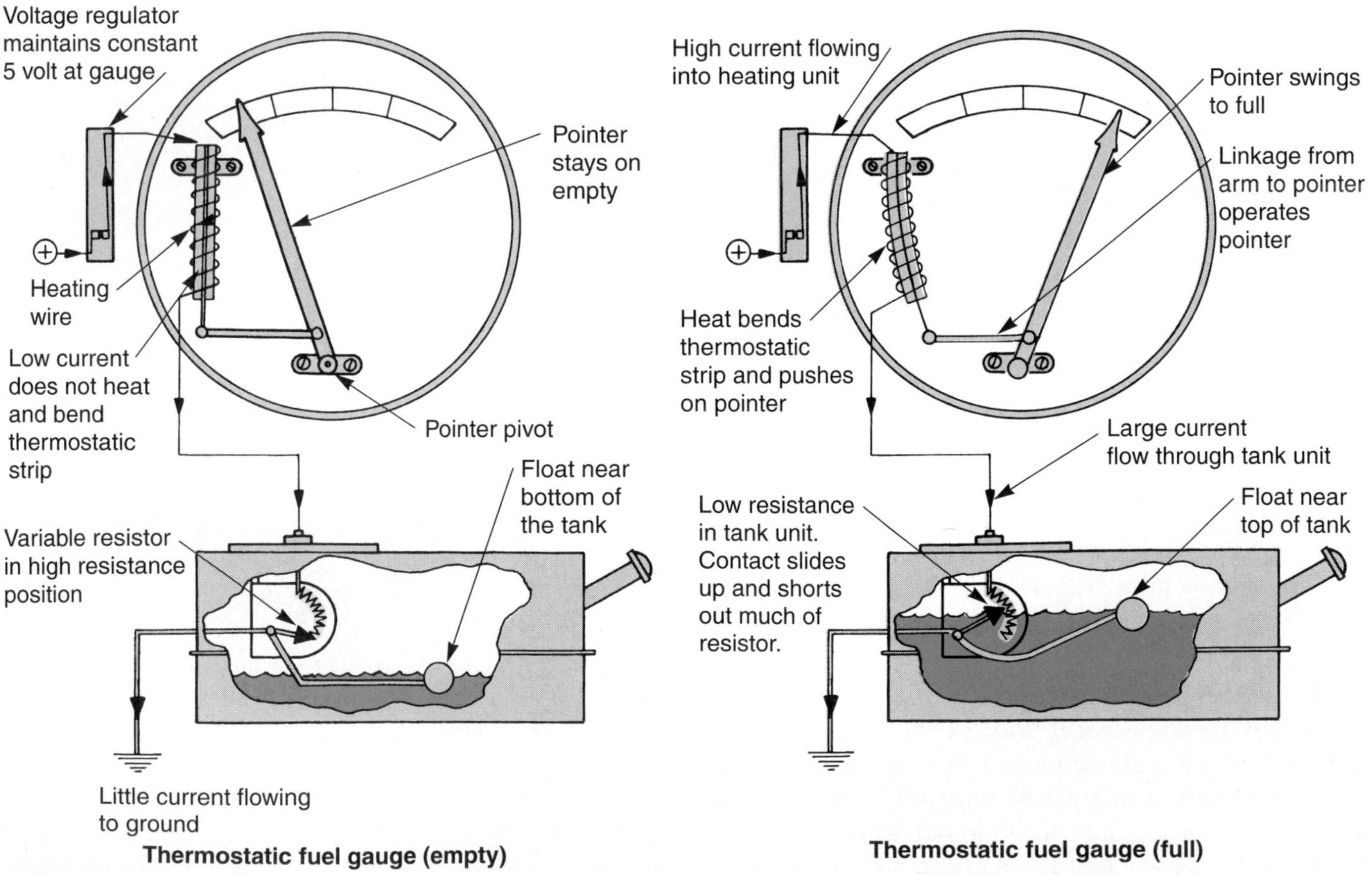

Thermostatic fuel gauge (empty)

Thermostatic fuel gauge (full)

Figure 21-5. Fuel tank sending unit and fuel gauge operation.

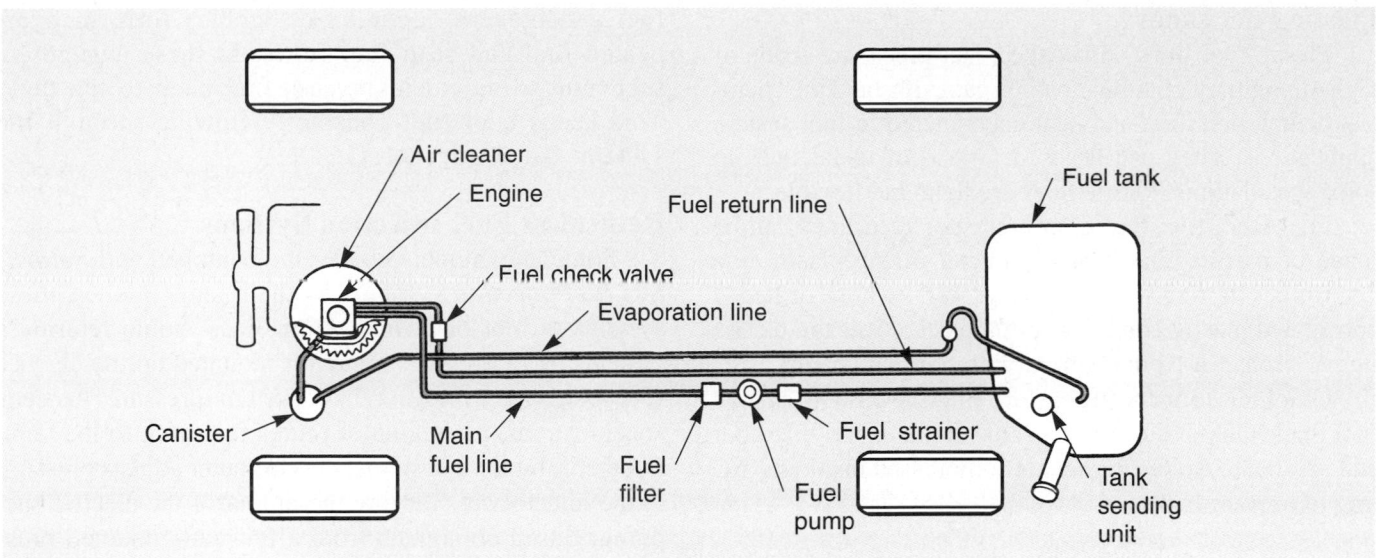

Figure 21-6. Study fuel and emission control lines. Note the location of the fuel pump, filters, and other devices. (Mazda)

able to withstand the constant and severe vibration produced by the engine and road surface.

A *fuel rail* is a large diameter fuel line that feeds fuel into multiport gasoline injectors. Is serves the same purpose as the fuel line but provides a greater volume of fuel right before the injectors. The fuel rail prevents any injector from starving for fuel under high demand conditions.

Fuel hoses made of synthetic rubber are needed where severe movement occurs between parts. For example, a fuel hose is used between the main fuel line and the engine. The engine is mounted on rubber motor mounts. The soft mounts allow the engine some movement in the car frame, **Figure 21-7.** A flexible hose can absorb this movement without breaking. *Hose clamps* secure fuel hoses to fuel lines or metal fittings.

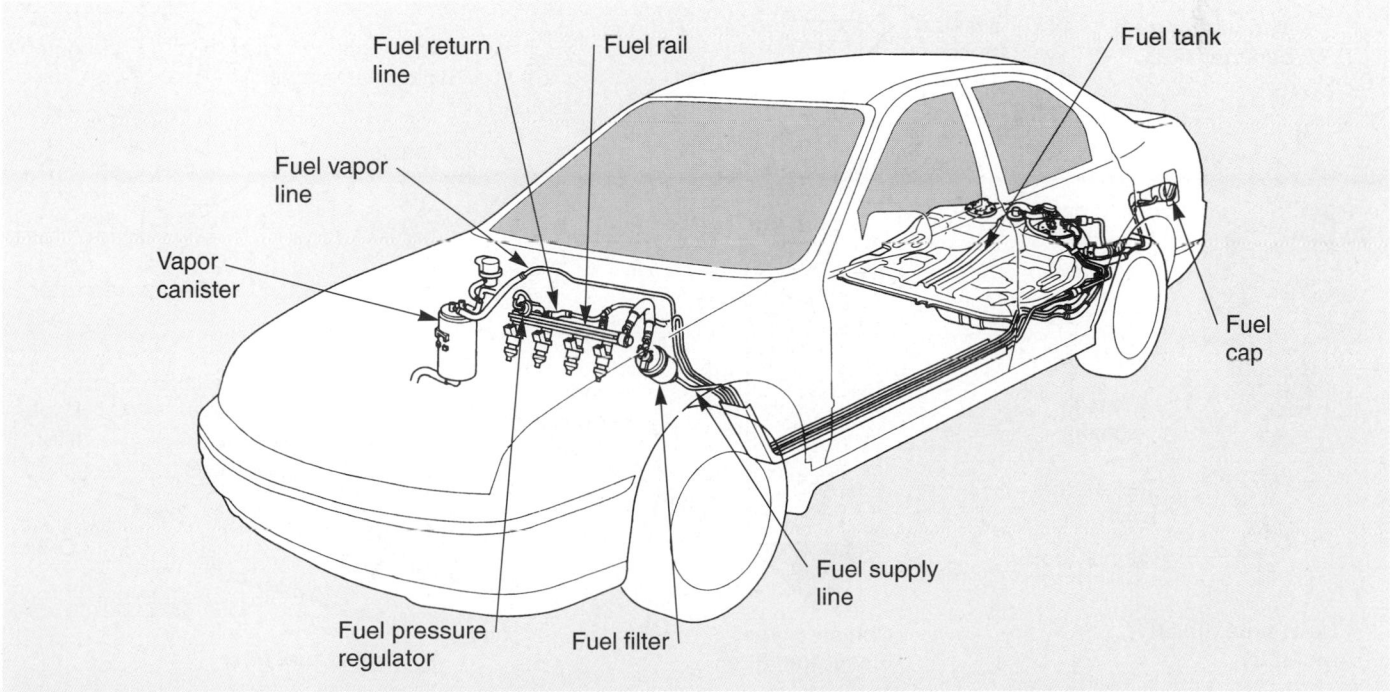

Figure 21-7. Three metal lines often run between the fuel tank and the engine: the main fuel supply line (tank to engine), the fuel return line (engine to tank), and the fuel vapor line (tank to storage canister). (Honda)

Plastic Fuel Lines

Plastic fuel lines, also called fuel pipes, are made of high-strength nylon tubing, which can withstand the chemical action, pressure, and heat encountered in fuel system applications. They can be used instead of metal lines in some installations. Plastic lines are light and flexible.

Fuel feed, fuel return, and injector feed lines can be made of plastic. The inside diameter of the plastic line determines its size when ordering. Rubber hose or heavy corrugated plastic conduit is often placed over the plastic line to protect it from chafing, vibration, and heat.

Quick-disconnect fittings are often used with plastic fuel lines. One end of the fuel line is male, and the other end is female. An O-ring seal is compressed inside the fitting to prevent leaks.

Fuel Return System

Most fuel injected vehicles, as well as some carbureted vehicles, use a *fuel return system* to cool the fuel and prevent vapor lock (bubbles form in overheated fuel and stop fuel flow). In these systems, a fuel return line carries excess fuel back to the tank. This keeps cool fuel constantly flowing through the system. See **Figure 21-1.**

Returnless Fuel Injection Systems

Some late–model vehicles are equipped with *returnless fuel injection systems.* As their name implies, these systems do not have fuel return lines. Some returnless systems use a pressure regulator mounted on the electric fuel pump assembly. When system pressure exceeds specifications, the regulator routes fuel back to the tank. In other returnless systems, the computer adjusts pressure at the injectors by varying the output of the electric fuel pump. Based on signals from a fuel rail–mounted pressure sensor, as well as signals from several other sensors, the computer increases or decreases the current sent to the pump motor.

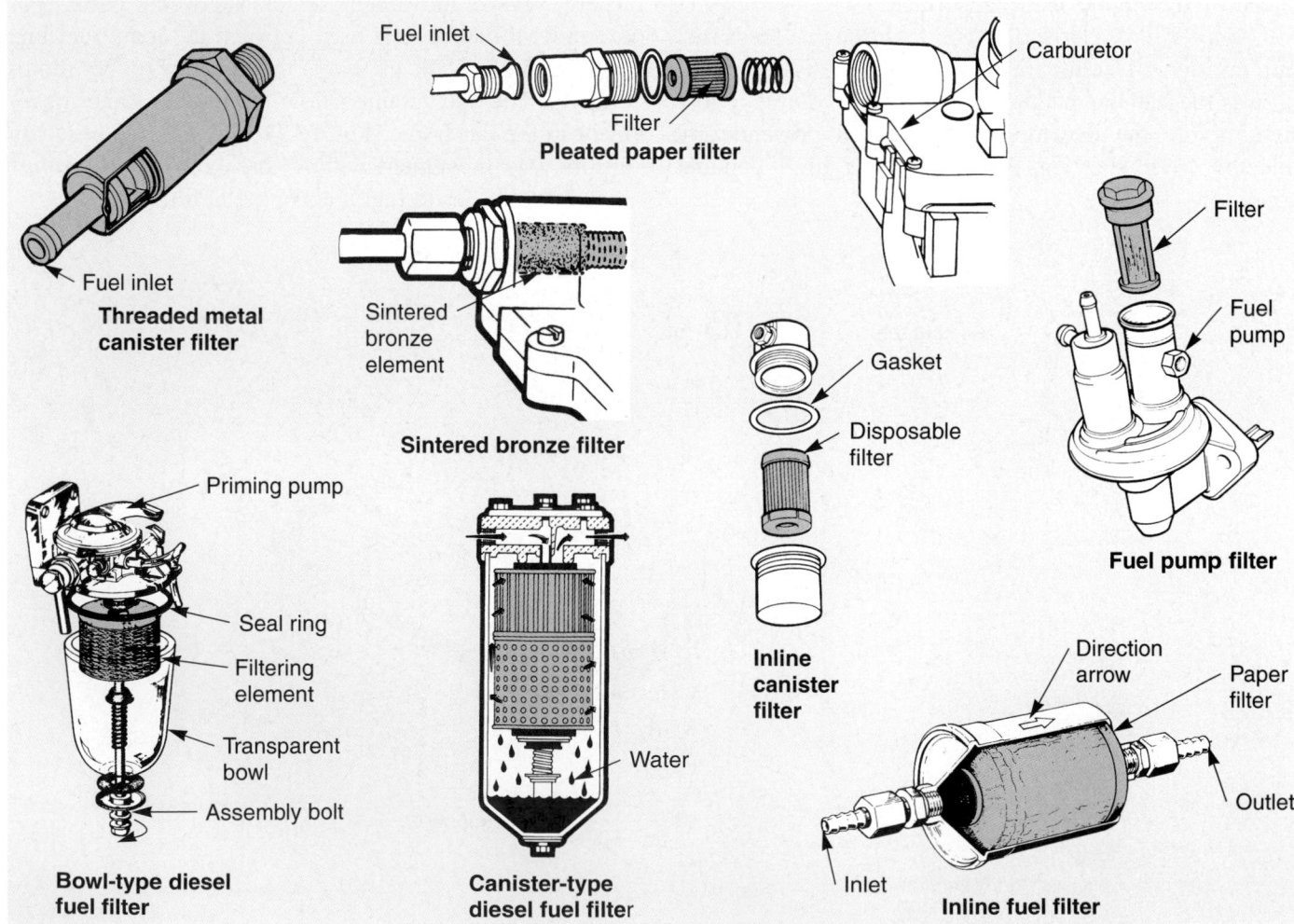

Figure 21-8. Variations of automotive fuel filters. (Fram, Saab, Ford, Chrysler)

Fuel Filters

Fuel filters stop contaminants (rust, water, corrosion, dirt) from entering the fuel lines, hoses, throttle body, injectors, pressure regulator, injection pump, and any other part that could be damaged by foreign matter. A fuel filter is normally located on the fuel tank pickup tube. A second fuel filter is commonly located in the main fuel line.

Fuel Filter Construction

Most fuel filters use *pleated paper elements* to trap contaminants present in the fuel, **Figure 21-8.** Others use a *sintered bronze* (porous metal) element. Both types are capable of stopping very small particles. Some filters are also capable of trapping water.

A *bowl fuel filter* is used on both gasoline and diesel engines. Look at **Figure 21-8.** Many types use a glass bowl that allows the technician to see the filter element. When the bowl and element become dirty, the technician can remove the bowl for service.

A *canister fuel filter* is sometimes used on diesel engines. Refer to **Figure 21-8.** The filter element is housed inside a metal container. When the filter requires replacement, a new element can be installed inside the container.

Fuel Pumps

A *fuel pump* forces fuel out of the tank and to the engine under pressure. There are two basic types of fuel pumps: mechanical and electric. Electric pumps are used on today's vehicles.

Mechanical Fuel Pumps

A *mechanical fuel pump* is usually powered by an *eccentric* (egg-shaped lobe) on the engine camshaft. See **Figure 21-9.** The mechanical fuel pump bolts to the side of the engine block. A gasket prevents oil leakage between the pump and engine block.

Mechanical fuel pumps are commonly used with carburetor-type fuel systems. They are still found on many vehicles. Since the mechanical pump uses a back-and-forth motion, it is a *reciprocating* pump.

Mechanical Fuel Pump Construction

The *rocker arm,* also called the actuating lever, is a metal arm that pivots in the middle, **Figure 21-9.** The outer end of the rocker arm rides on the camshaft eccentric. The inner end operates the diaphragm.

The fuel pump *return spring* keeps the rocker arm pressed against the eccentric. Without a return spring, the rocker arm would make a loud clattering sound as the eccentric lobe hit the rocker arm.

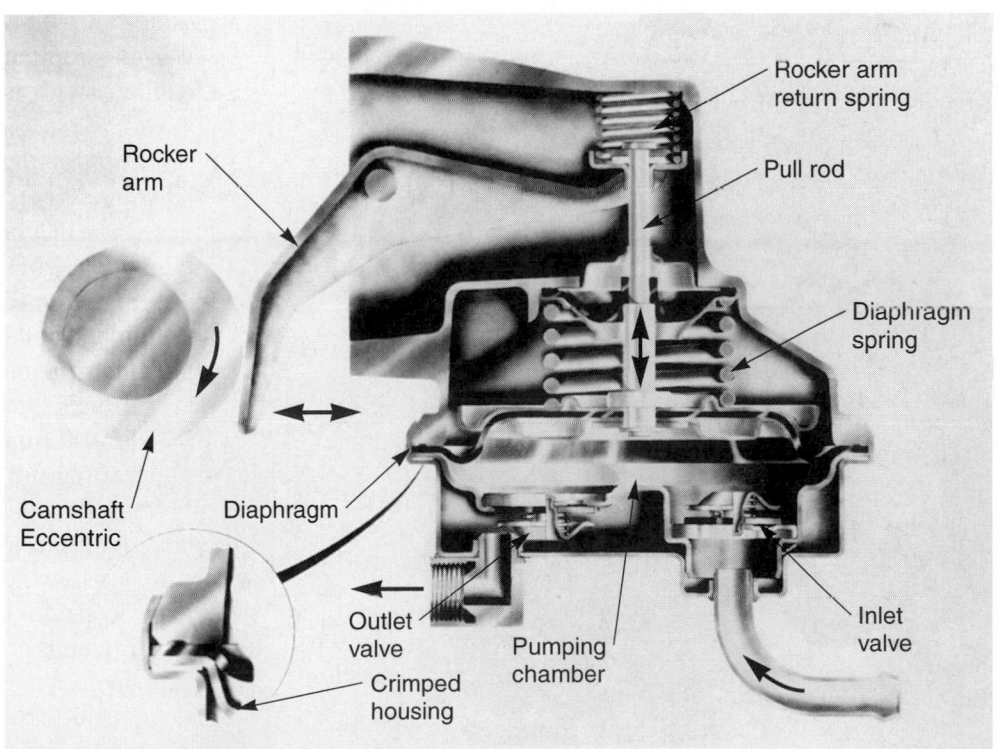

Figure 21-9. Cutaway view of a mechanical fuel pump. Note part names and locations. (AC-Delco)

The *diaphragm* is a synthetic rubber disc clamped between the halves of the pump body. A metal pull rod is mounted on the diaphragm to connect the diaphragm with the rocker arm. The *diaphragm spring,* when compressed, pushes on the diaphragm to produce fuel pressure and flow.

Two *check valves* are used in a mechanical fuel pump: the inlet check valve and the outlet check valve. Fuel flows easily through the valves in one direction but cannot flow through in the other direction. The two check valves are reversed. This causes the fuel to enter through one valve and exit through the other.

Mechanical Fuel Pump Operation

During the *intake stroke,* the eccentric lobe pushes on the rocker arm. See **Figure 21-10A.** This pulls the diaphragm up and compresses the diaphragm spring. The area in the pumping chamber increases, and the resulting vacuum pulls fuel through the inlet check valve. This fills the pump with fuel.

On the *output stroke,* the eccentric lobe rotates away from the rocker arm. This releases the diaphragm, **Figure 21-10B.** The diaphragm spring then pushes on the diaphragm and pressurizes the fuel in the pumping chamber. The amount of spring tension controls fuel pressure. The fuel flows out through the outlet check valve.

When an engine is running slowly, the fuel pump would produce more fuel than the engine consumes. For this reason, the fuel pump is made to idle when fuel is not needed. See **Figure 21-10C.** The diaphragm pull rod is free to slide through the rocker arm when fuel pressure compresses the pump spring. This lets the rocker arm move up and down while the diaphragm remains stationary. Diaphragm spring tension maintains fuel pressure.

Vapor Lock

Vapor lock is a problem created when bubbles in overheated fuel reduce or stop fuel flow. **Figure 21-11** shows vapor lock in a mechanical fuel pump.

During vapor lock, engine heat transfers through the metal parts of the pump and into the fuel. The fuel "boils," forming bubbles that displace fuel. This can reduce fuel pump output and cause engine performance problems.

Electric Fuel Pumps

An *electric fuel pump,* like a mechanical pump, produces fuel pressure and flow for the fuel metering section of a fuel system. Electric fuel pumps are commonly used on all types of modern engines.

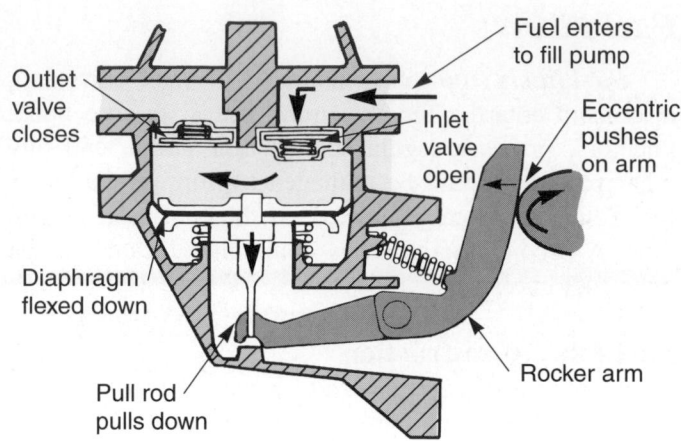

A — Intake stroke

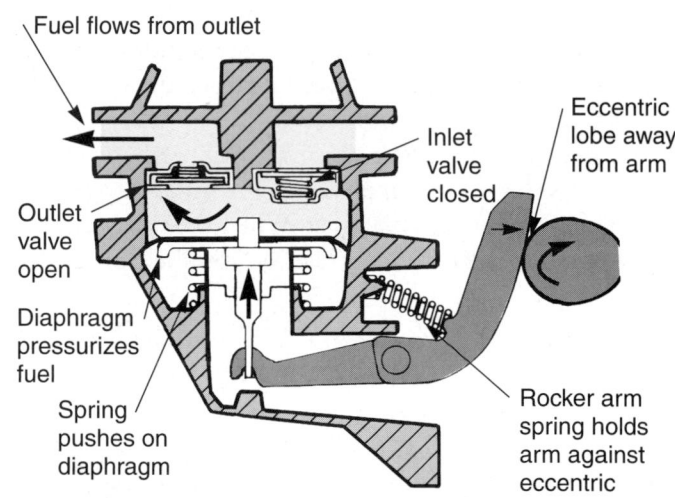

B — Output stroke

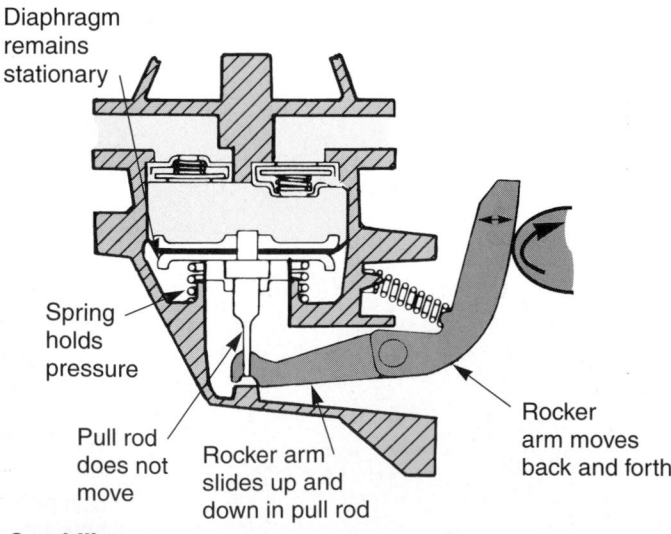

C — Idling

Figure 21-10. Mechanical fuel pump operation.

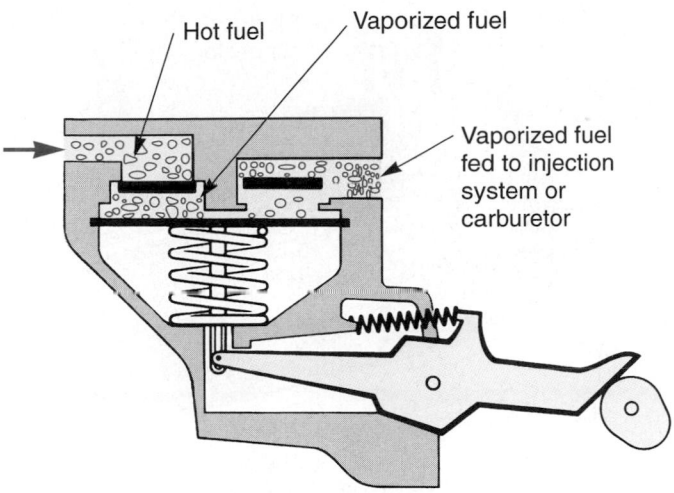

Figure 21-11. Vapor lock is caused when too much engine heat is transferred into the fuel. Bubbles form and displace fuel. This can prevent fuel from flowing through system. (Ford)

An electric fuel pump can be located inside the fuel tank or in the fuel line between the tank and the engine. A few vehicles have one pump in the fuel tank and a second pump in the fuel line. In many late-model vehicles, the tank sending unit is an integral part of the in-tank fuel pump assembly. See **Figure 21-12.**

An electric fuel pump has advantages over a mechanical fuel pump. An electric pump can produce almost instant fuel pressure. A mechanical pump slowly builds pressure as the engine is cranked for starting. Also, most electric fuel pumps are rotary-type pumps. They produce a smoother flow of fuel (fewer pressure pulsations) than reciprocating, mechanical pumps.

Since most electric fuel pumps are located away from the engine, vapor lock is less likely. An electric fuel pump pressurizes all the fuel lines that are near engine heat. This also helps prevent vapor lock because the pressure makes it more difficult for bubbles to form.

Rotary Fuel Pumps

Rotary fuel pumps include the impeller type, the roller vane type, and the sliding vane type. They all use a circular, or spinning, motion to produce pressure.

An **impeller electric fuel pump** is a centrifugal pump. Normally, it is located inside the fuel tank. See **Figure 21-13.** This pump uses a small dc motor to spin the impeller (fan blade). The impeller blades cause the fuel to move outward due to centrifugal force (spinning matter is propelled outward). This produces enough pressure to move the fuel through the fuel lines. **Figure 21-14** shows an in-tank impeller pump and its related hardware.

A **roller vane electric fuel pump** is a positive-displacement pump (each pump rotation moves a specific amount of fuel). It is normally located in the main fuel line. Look at **Figure 21-15.** Small rollers and an offset-mounted rotor disc produce fuel pressure.

When the rotor disc and rollers spin, they pull fuel in on one side of the pump. Then, the fuel is trapped and pushed to a smaller area on the opposite side of the pump housing. This squeezes the fuel between the rollers, and the fuel flows out under pressure.

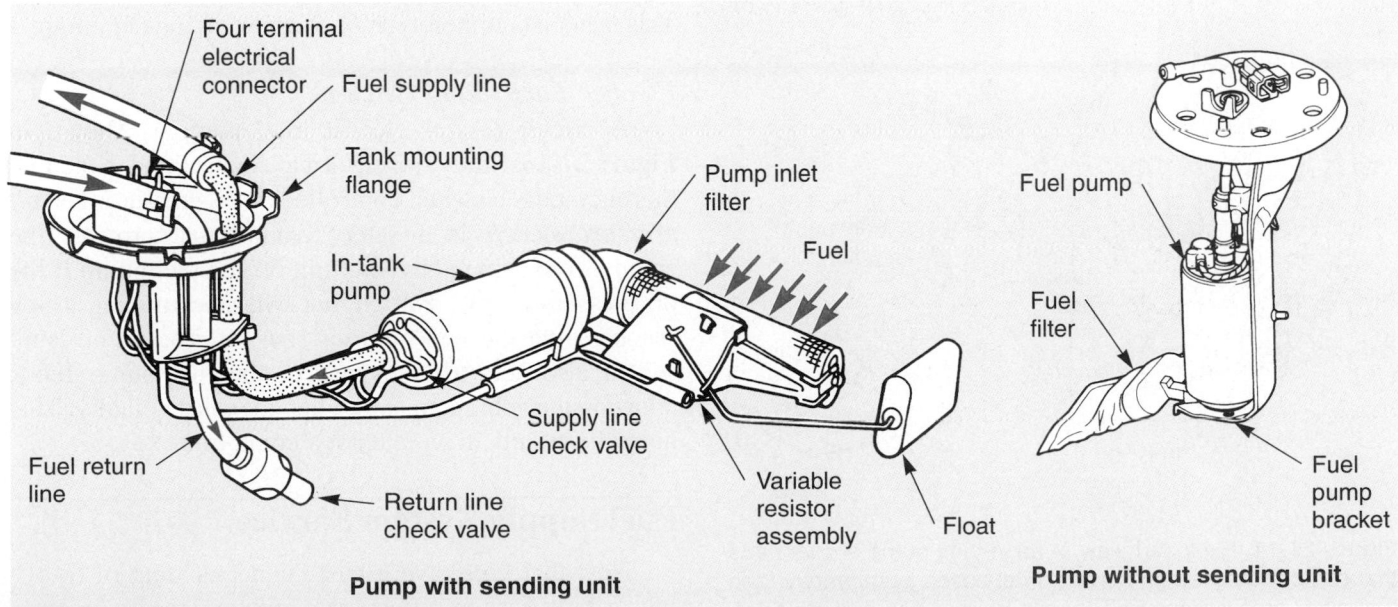

Pump with sending unit

Pump without sending unit

Figure 21-12. In-tank electric fuel pumps. (Chrysler and Honda)

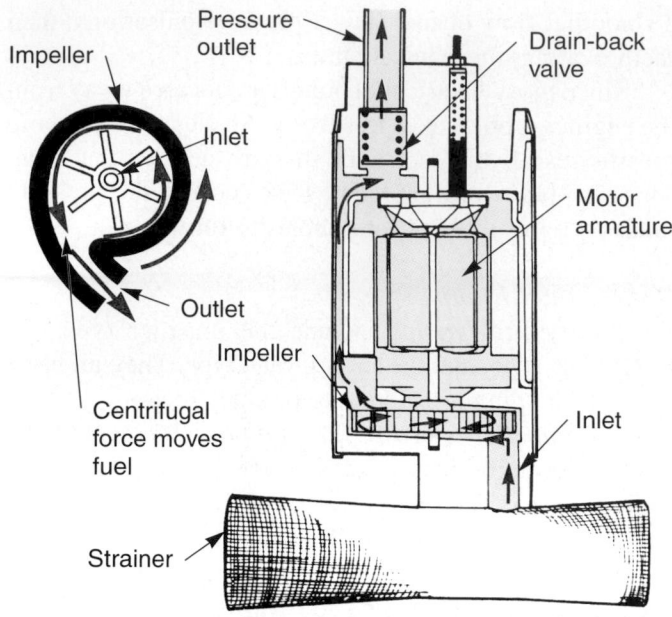

Figure 21-13. In-tank, impeller-type electric fuel pump. The impeller produces a very smooth flow of fuel through the system. A check valve prevents fuel from draining out of lines and back into the tank when the pump is not running. (Volvo, Ford)

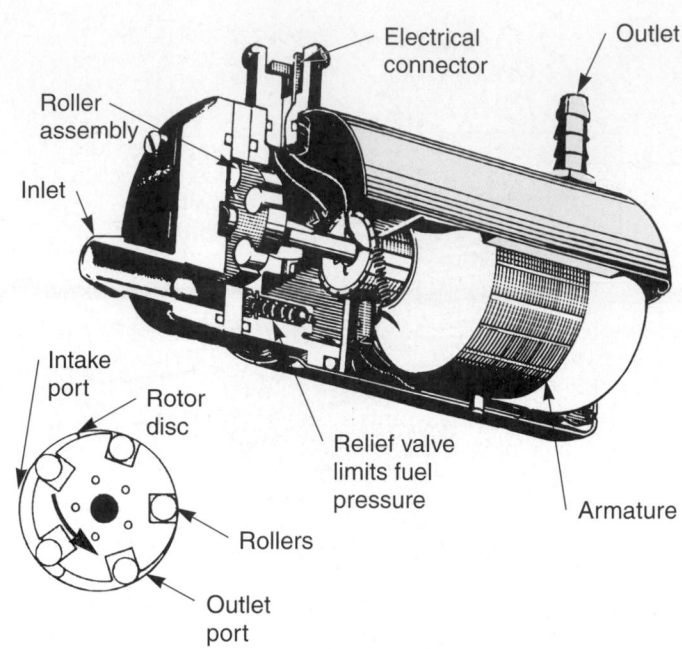

Figure 21-15. A roller vane electric fuel pump is usually capable of producing higher pressure and greater volume than an impeller type pump. A relief valve limits the fuel pressure. (Volvo)

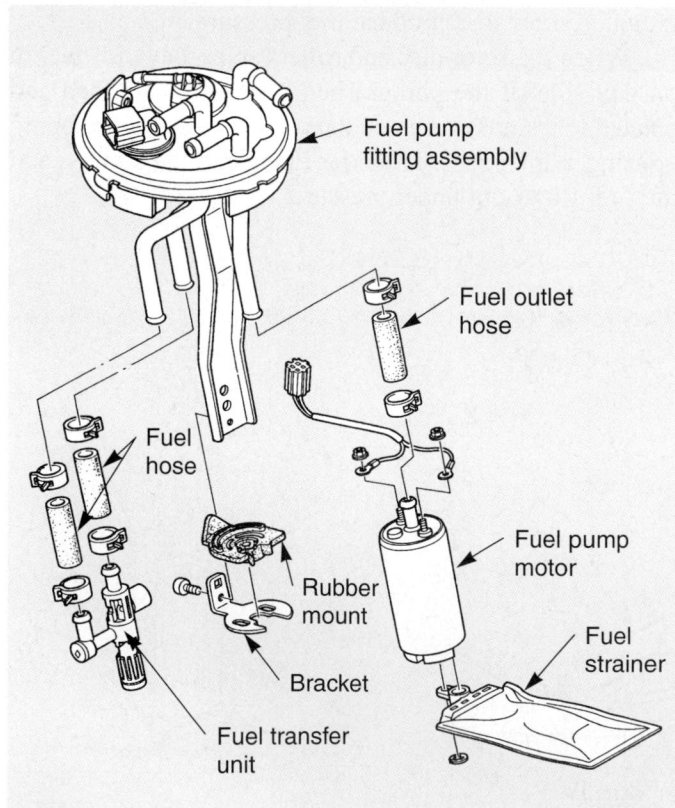

Figure 21-14. Exploded view of an in-tank pump shows parts that can be removed for service. Fuel hoses, pump motor, and a strainer or low-pressure filter can be removed and replaced. (Mazda)

A *sliding vane electric fuel pump* is similar to a roller vane pump. Vanes are used instead of rollers.

Most rotary electric fuel pumps also have check valves and relief valves, **Figure 21-13** and **21-15.** The *check valves* keep fuel from draining out of the fuel line when the pump is not running. A *relief valve* limits the maximum output pressure of the pump.

A *reciprocating electric fuel pump* has the same basic action as a mechanical fuel pump. However, it uses a solenoid instead of a rocker arm to produce a pumping motion. This is not a common type of original equipment pump.

Electric Fuel Pump Circuit

A circuit for an electric fuel pump is shown in **Figure 21-16.** Study the electrical connections. Note that this circuit has a switch controlled by oil pressure. The *oil pressure switch* is a safety feature that protects the engine from damage by shutting off the fuel pump if the oil pressure is low. The engine will stop running a few seconds after the fuel pump is shut off. The oil pressure switch also acts as a bypass to the fuel pump relay, allowing the vehicle to start when the relay is faulty. This normally results in an extended crank time.

Fuel Supply System Service

Now that you have a basic understanding of how a fuel supply system functions, you are ready to learn about problems, tests, and repairs common to the parts.

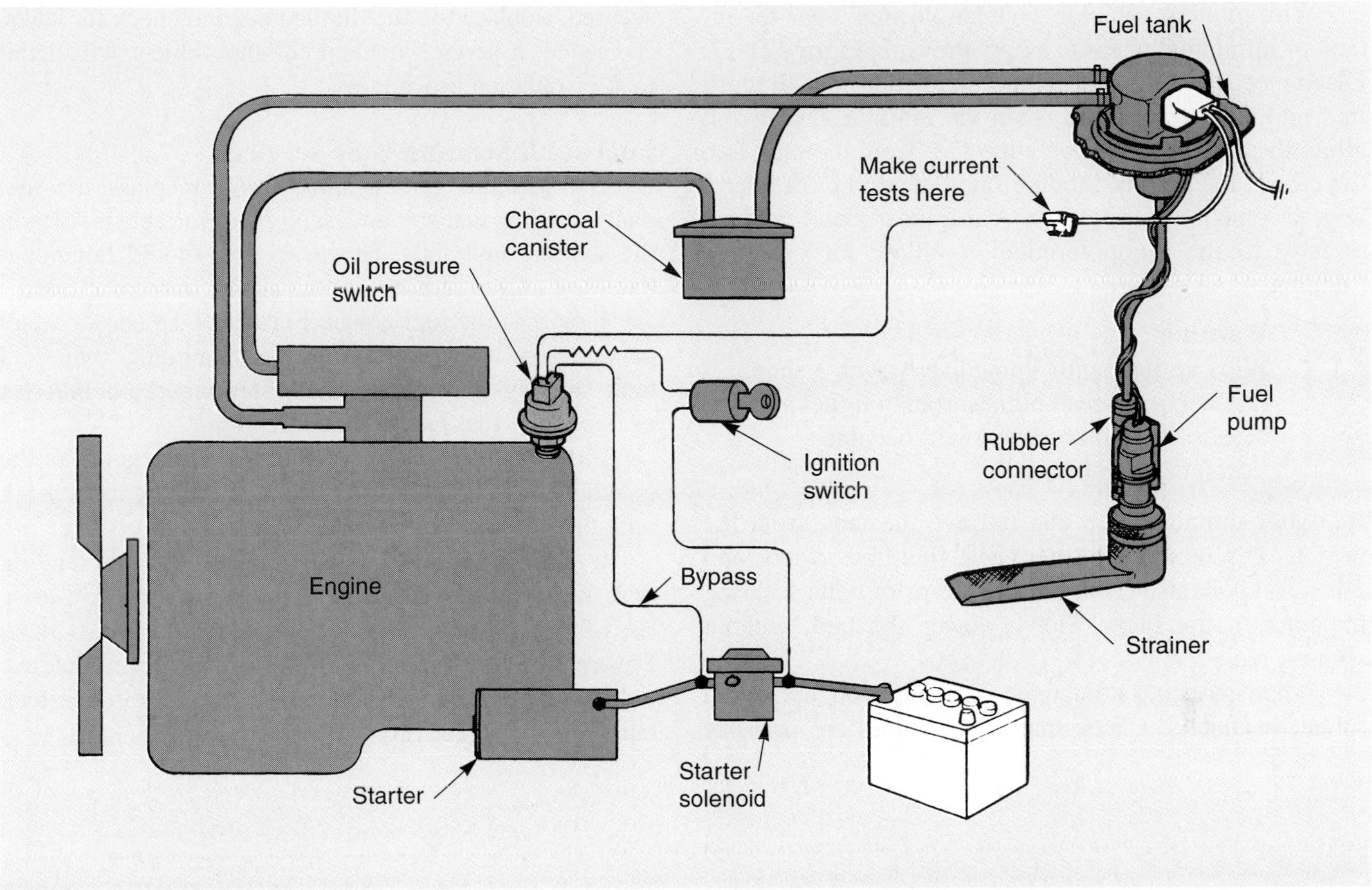

Fuel tank

Make current
tests here

Charcoal
canister

Oil pressure
switch

Fuel
pump

Ignition
switch

Rubber
connector

Engine

Bypass

Strainer

Starter

Starter
solenoid

Figure 21-16. A typical electric fuel pump circuit with oil pressure switch. If the engine oil pressure drops too low, the switch shuts off the fuel pump. Note other connections and parts of the circuit. (Ford)

Warning
Always keep a fire extinguisher handy when working on a car's fuel system. During a fire, a few minutes spent looking for an extinguisher can be a *lifetime!*
Never weld or solder a fuel tank. If not badly rusted, a leaking tank should be sent to a well-trained specialist. Even an empty tank can *explode.* The heat produced during soldering or welding operations can cause fuel gum to melt, vaporize, and ignite.

Fuel Supply Systems Scanning

Depending on the symptoms, you will often want to scan for diagnostic trouble codes before starting fuel supply system repairs. Vehicles equipped with OBD II systems monitor fuel tank pressure, fuel level, and other conditions. A scan tool will quickly let you find any stored codes. Modern scanners also display operating values (voltage, current, resistance) for the electrical circuit, which can speed troubleshooting.

Fuel Tank Service

Typical *fuel tank problems* include fuel leakage, physical damage (dents, holes, etc.), and contamination (rust, dirt, and water). Vibration or rusting can cause a fuel tank to develop pinhole leaks. Rocks can puncture the tank. Internal deterioration of the tank or foreign matter in the fuel can cause contamination problems.

Fuel Tank Removal and Replacement

A fuel tank can be located under the trunk, in a body panel, or under the rear seat. It can be held in the vehicle by large metal straps or by bolts passing through the tank flange. With rear-engine vehicles, the fuel tank can be located in the front.

Before servicing a fuel tank, empty the tank. A full tank is very heavy and can rupture if dropped.

To remove fuel from the tank, unscrew the drain plug and drain the fuel into an approved safety can. If a drain is *not* provided, use an approved pumping method to draw the fuel out of the tank.

With modern vehicles, you can normally use the in-tank or inline fuel pump to empty the tank, **Figure 21-17.** Disconnect the fuel hose at the main fuel line. Route it into an approved gas can. Turn the ignition key on and allow the fuel pump to force the fuel from the tank into the can. If the control module shuts the pump off after a few seconds, use a jumper wire to connect voltage directly to the pump terminal to allow for complete draining.

> ⚠ **Warning**
> Wipe up fuel spills immediately with a shop rag. Do *not* spread oil absorbent on fuel spills, because the oil absorbent will become extremely flammable.

After draining, you can remove the tank from the vehicle. Disconnect the filler neck, fuel lines, wires, and other components. Then, remove straps or bolts securing the tank to the body. Slowly lower the tank without spilling fuel.

When installing a fuel tank, make sure you replace the rubber insulators. Check that all fuel lines are properly secured. Replace the fuel in the tank and check for leaks. If needed, a service manual for the vehicle will detail tank installation procedures.

Fuel Tank Sending Unit Service

A *faulty fuel tank sending unit* can make the fuel gauge reading inaccurate. Usually, the variable resistor in the sending unit fails. However, you should remember that the fuel gauge or the gauge circuit may be at fault.

First, test the fuel gauge. **Figure 21-18** shows a fuel gauge tester. It is connected to the wire going to the fuel tank sending unit. When the tester is set on *full*, for example, the fuel gauge should read *full*.

If the gauge does not function, either the gauge or the gauge circuit is faulty. If the fuel gauge begins to work with the tester in place, the tank sending unit is bad.

If your tests indicate an inoperative tank sending unit, remove the unit after draining the tank. Unscrew the cam lock holding the sending unit in the fuel tank, **Figure 21-19.** If you do not have a special cam tool, use a drift punch and light hammer blows to rotate the lock tabs, **Figure 21-20.** Lift the unit out of the tank.

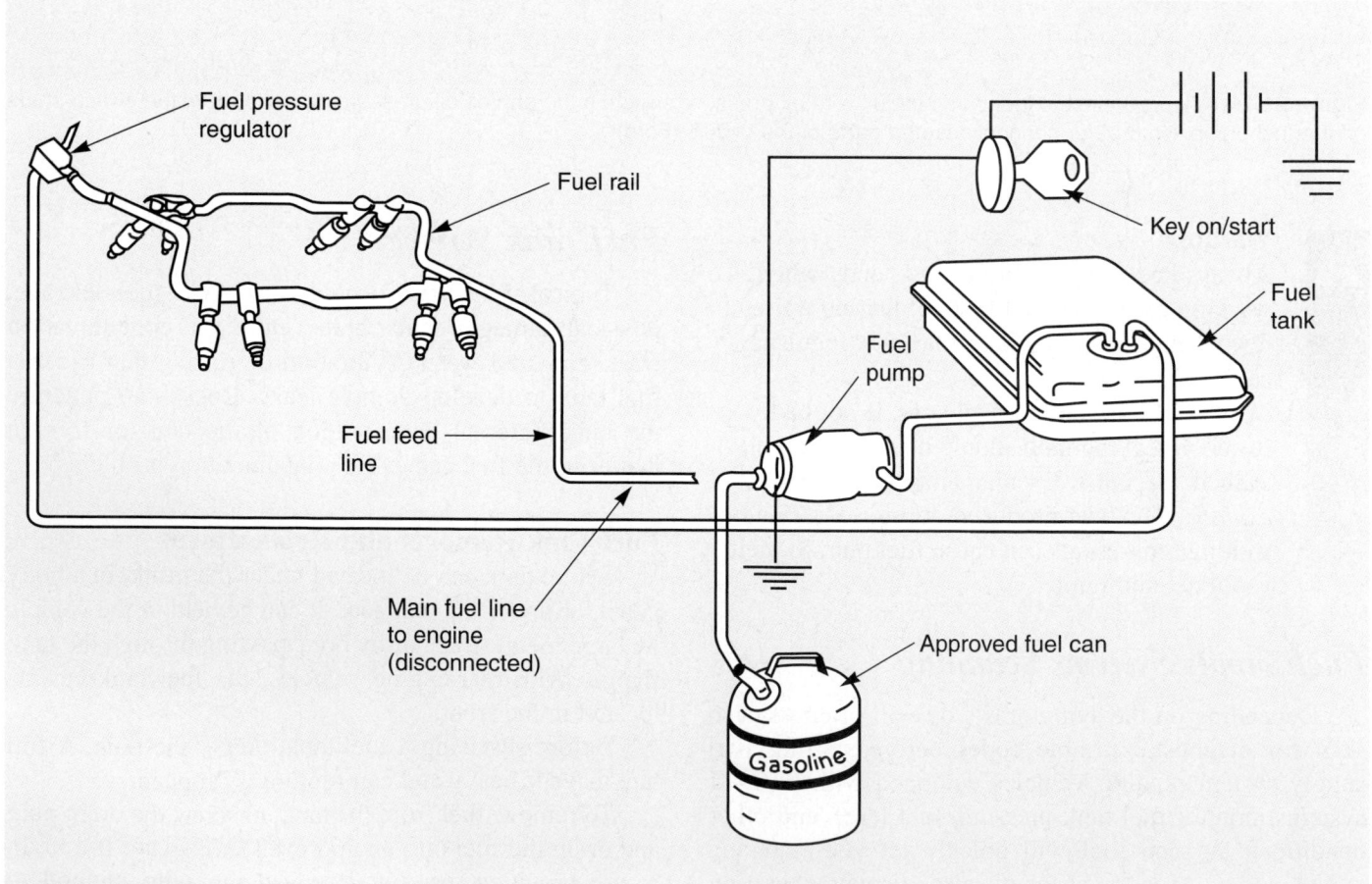

Figure 21-17. With electric fuel pumps, you can use the in-tank or inline pump to drain gasoline out of the tank. Connect a long hose from the pump outlet to an approved fuel can. Energize the pump and make sure fuel does not spill out.

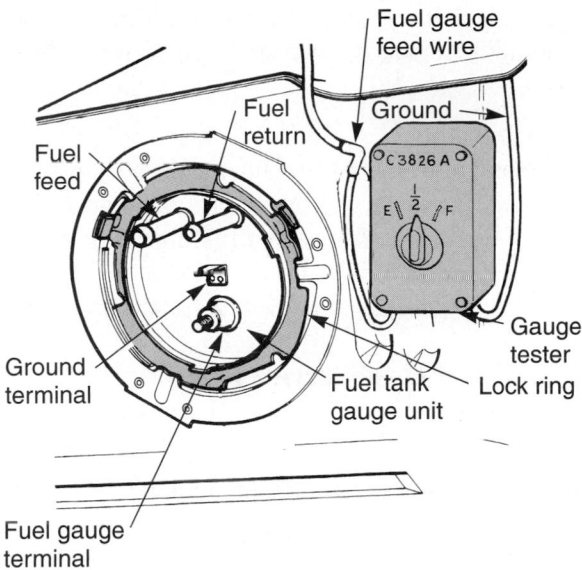

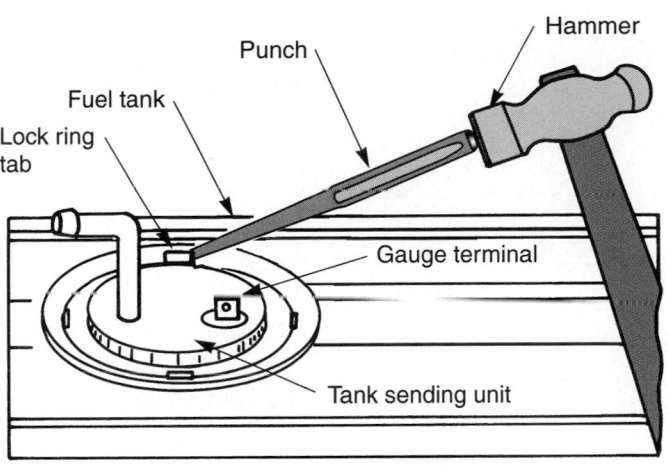

Figure 21-20. A long drift punch and light hammer blows will rotate and free the lock ring on a sending unit or fuel pump.

Figure 21-18. A tester will check the condition of the fuel gauge and the circuit. If the circuit and the gauge are working well, the problem may be in tank sending unit. Note how the sending unit is held in the fuel tank by a lock ring. (Chrysler)

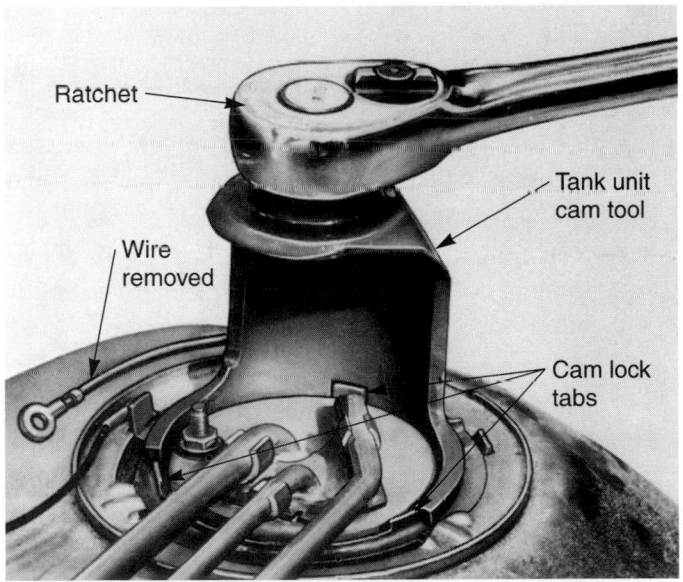

Figure 21-19. To remove the tank sending unit, turn the cam lock ring. A cam tool makes this easy. Light taps with a hammer on a drift punch will also work. (Cadillac)

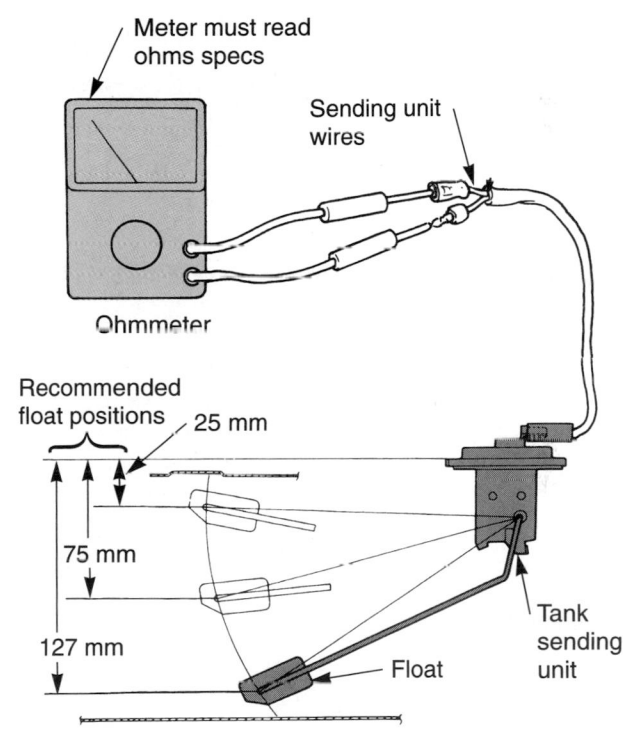

Figure 21-21. An ohmmeter can be used to check the condition of the tank sending unit. The reading should be within specifications with the float in prescribed positions. (Honda)

With the sending unit removed, measure its resistance with an ohmmeter, **Figure 21-21.** If the resistance is not within the factory specifications, install a new sending unit.

Also, check the float for leakage. Shake the float next to your ear. If you can hear liquid splashing, replace the float. If the tank unit resistance is correct, check the tank ground. A poor ground could prevent operation.

Fuel Line and Hose Service

Faulty fuel lines and hoses are common sources of fuel leaks. See **Figure 21-22.** Fuel hoses can become hard and brittle after being exposed to engine heat and the environment. Engine oil softens and swells hoses. Always inspect hoses closely and replace any in poor condition.

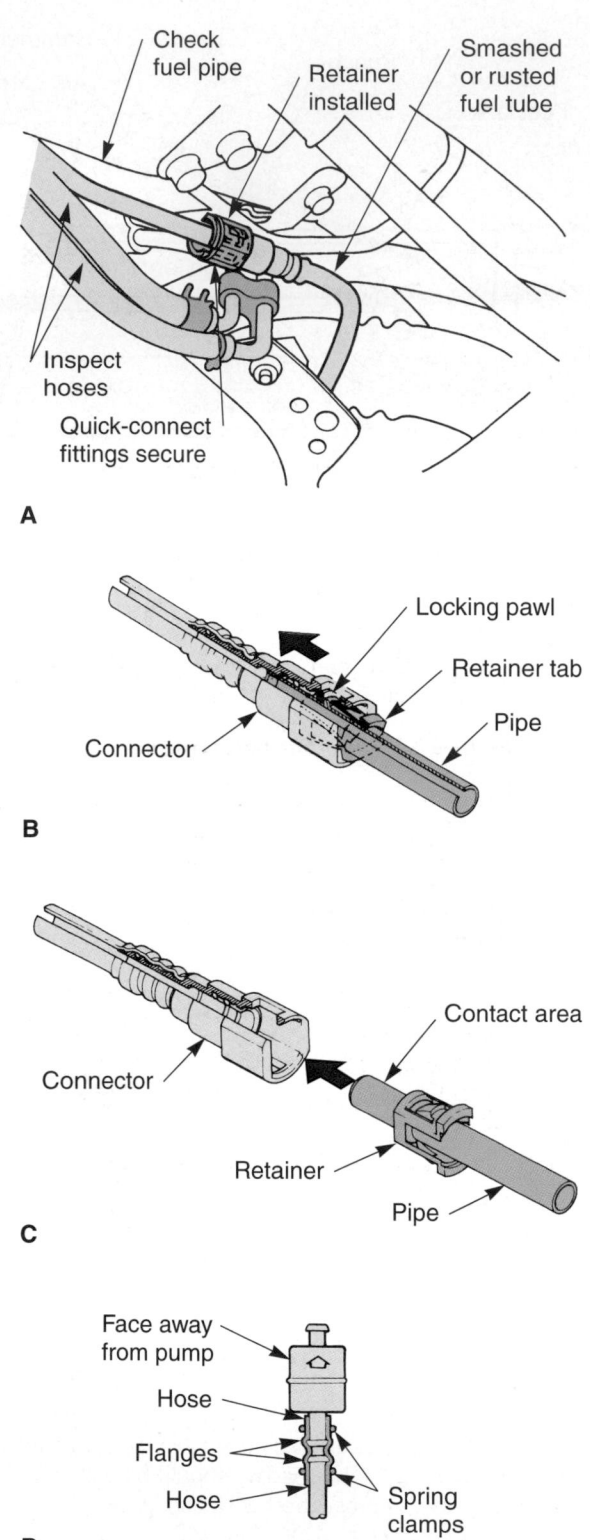

Figure 21-22. A—Always inspect fuel lines, hoses, and fittings for leakage and damage. Pay special attention to plastic lines. B—To disconnect this fuel line fitting, press down on the locking tab while pulling apart. C—To reconnect fitting, simply push the two halves of the fitting together. D—When installing fuel hoses, make sure they are completely over raised areas on the fuel line. Clamps should be on pressure side of flanges. (Honda and Chrysler)

Metal fuel lines seldom cause problems. However, they should be replaced when smashed, kinked, rusted, or leaking.

 Caution

Some late-model fuel lines have a snap-type fitting. Do not try to pry the fitting apart or it will be damaged. You need a fuel line fitting tool to release the fuel line fitting for service. This tool can be purchased at most auto part stores.

Fuel Line and Hose Service Rules

Remember these rules when working with fuel lines and hoses:

- Place a shop rag around the fuel line fitting during removal. This will keep fuel from spraying on you or on the hot engine. Use a flare nut or tubing wrench on fuel system fittings.
- Only use approved double-wall steel tubing for fuel lines. Never use copper or plastic tubing, unless it is OEM equipment.
- Make smooth bends when forming a new fuel line. Use a bending spring or bending tool.
- Form double-lap flares on the ends of the fuel line. A single-lap flare is not approved for fuel lines. Refer to **Figure 21-23.**
- Reinstall fuel line hold-down clamps and brackets. If not properly supported, the fuel line can vibrate and fail.
- Route all fuel lines and hoses away from hot or moving parts. Double-check clearance after installation.
- Route plastic fuel lines in their original locations. They are more prone to damage from engine heat and abrasion than metal lines.
- A plastic line will kink if bent too sharply. This will restrict fuel flow and may cause the line to fail in service.
- Use a factory-recommended tool to release clip-type fittings used to secure plastic fuel lines.
- If a plastic line or fitting is damaged, replace it with a new one. Do not attempt repairs, such as inline splices.
- Do not use non-OEM plastic tubing to repair fuel systems. The line pressure rating and the composition of the plastic may be different. The incorrect plastic line can rupture and cause a serious fire.
- Plastic lines tend to stiffen in service. Avoid bending or rerouting used plastic lines.
- Always cover plastic fuel lines with a wet towel when welding, torch cutting, or grinding near

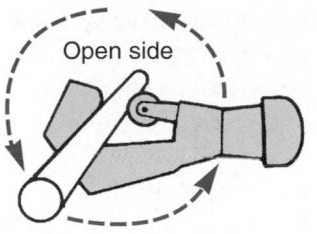

Step 1. Cut tubing with tube cutter.

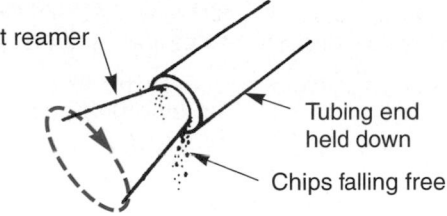

Step 2. Remove burrs.

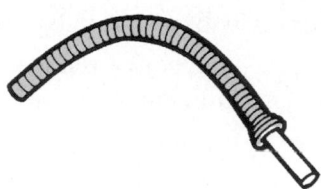

Step 3. Slip bender over tubing.

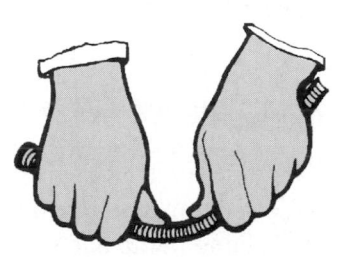

Step 4. Bend tubing to shape of old fuel line.

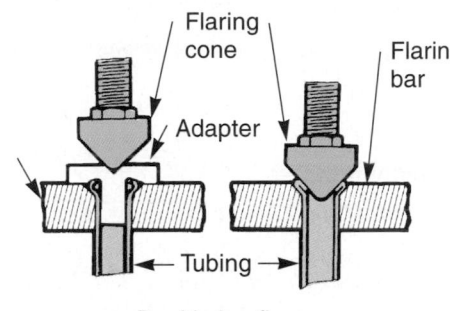

Step 5. Use a flaring tool adapter to fold the tube end inward.

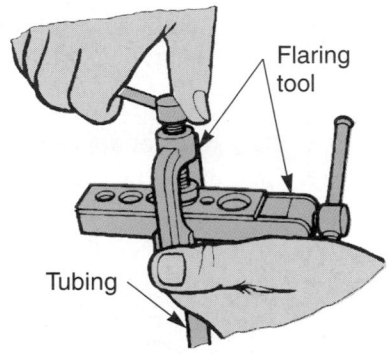

Step 6. Tighten flaring tool into tubing.

Figure 21-23. Fuel lines need double-lap flares. Study basic steps for making a new fuel line. This procedure also applies to other lines (brake lines, steel vacuum lines, etc.). (Florida Dept. of Voc. Ed. and Chrysler)

them. If a hot piece of metal melts the line, a serious fire can result.

- Never allow battery acid to contact a metal or plastic fuel line. Lines can be weakened or eaten away by acid.

- Only use approved synthetic rubber hoses in a fuel system. If vacuum-type rubber hose is accidentally used, the fuel can chemically attack and rapidly ruin the hose. A dangerous leak could result.

- Make sure a fuel hose fully covers its fitting or line before installing the clamps. Pressure in the fuel system could force the hose off if it is not installed properly.

- When servicing press-fit fuel line fittings, you may need to replace the O-ring seal to prevent dangerous fuel leakage. Make sure you purchase the correct seal. If you install the wrong seal, a fire can result.

- Double-check all fittings for leaks. Start the engine and inspect the connections closely.

Warning

Most fuel injection systems have very high fuel pressure. Follow recommended procedures for releasing pressure before disconnecting a fuel line or fitting. This will prevent fuel spray from possibly causing injury or a fire!

Fuel Filter Service

Fuel filter service involves periodic replacement or cleaning of system filters. It may also include locating clogged fuel filters that are upsetting fuel system operation. Replace paper-element filters when clogged or after prolonged service.

A *clogged fuel filter* can restrict the flow of fuel to the pressure regulator, gasoline injectors, diesel injection pump, or carburetor. Engine performance problems will usually show up at higher cruising speeds.

With a clogged fuel filter, the engine may temporarily lose power or stall when a specific engine speed is reached. This is due to the flow restriction caused by a partially clogged filter. A partially clogged filter may pass enough fuel at low engine speeds. However, when engine speed and fuel flow increase, the engine may "starve" for fuel.

On modern vehicles, a *clogged in-tank strainer* is a common and difficult-to-diagnose problem. When clogged, the in-tank filter can collapse and stop all fuel flow. Then, after the engine stalls, the strainer can open again, leaving no trace of a restriction.

Some fuel filters have a check valve that opens when the filter becomes clogged. This allows fuel contaminants to flow into the system. When contaminants are found in the filters and system, the tank, pump, and lines should be flushed with clean fuel.

Fuel Filter Locations

Fuel filters may be found in the following locations:

- In the fuel line before the fuel rail, fuel injectors, diesel injection pump, or carburetor.
- Inside the fuel pump.
- In the fuel line right after the electric fuel pump.
- Under the fuel line fitting in the carburetor.
- In the fuel tank on the end of the fuel pickup tube.

When in doubt about fuel filter locations, refer to a service manual. It will give information about service intervals, cleaning, and replacement of all system filters.

Tech Tip

If a fitting must be loosened when changing a fuel filter, use a flare nut wrench. Do not over-tighten and strip the fitting when replacing the filter. Double-check that the fuel hose, if used, is fully installed over the fitting barbs.

Fuel Pump Service

Fuel pump problems are usually low fuel pressure, inadequate fuel flow, abnormal pump noise, or fuel leakage from the pump. Both mechanical and electric fuel pumps can fail after prolonged operation.

Low fuel pump pressure can be caused by worn pump bearings, high resistance in electric pump circuit, leaking check valves, or physical wear of moving parts. Low fuel pump pressure can make the engine starve for fuel at higher engine speeds.

High fuel pump pressure, more frequent with electric pumps, indicates an inoperative pressure relief valve. If the relief valve fails to open, both pressure and volume can be above normal. A faulty fuel pressure regulator can also cause high fuel pressure. This can produce a rich fuel mixture or even flood the engine.

Mechanical fuel pump noise (clacking sound from inside pump) is commonly caused by a weak or broken rocker arm return spring, wear of the rocker arm pin, or wear of the arm itself. Mechanical fuel pump noise can be easily confused with valve or tappet clatter. They sound very similar. To verify mechanical fuel pump noise, place a stethoscope on the pump body. The point of the loudest noise is the source of the problem.

Most electric fuel pumps make some noise (buzz or whirl sound) when running. Only when the pump noise is abnormally loud should an electric fuel pump be considered faulty. A clogged tank strainer can also cause excessive pump noise. Pump speed can increase because fuel is not entering the pump properly.

Fuel pump leaks are caused by physical damage to the pump body or deterioration of the diaphragm or gaskets. Inline electric pump leaks are usually from the inlet and outlet hoses.

Most mechanical fuel pumps have a small vent hole in the pump body. When the diaphragm is ruptured, fuel will leak out of this hole.

It is possible for a ruptured mechanical fuel pump diaphragm to contaminate the engine oil with gasoline. Fuel can leak through the diaphragm, through the pump, into the side of the block, and down into the oil pan. When a gasoline smell is noticed in the oil, correct the pump problem; then change the oil and filter.

Fuel Pump Tests

Fuel pump testing commonly involves measuring fuel pump pressure and volume. Exact procedures vary, depending upon fuel system. Refer to a manual for exact testing methods. Sometimes, fuel pump vacuum is measured as another means of determining pump and supply line condition.

Always remember there are several other problems that can produce symptoms like those caused by a faulty fuel pump. Before testing a fuel pump, check for:

- Restricted fuel filters.
- Smashed or kinked fuel lines or hoses.
- Air leak into vacuum side of pump or line.
- Injection system or carburetor troubles.
- Ignition system problems.
- Low engine compression.

Measuring Fuel Pump Pressure

To measure fuel pump pressure:
1. Connect a pressure gauge to the output line of the fuel pump or to the appropriate test fitting, **Figure 21-24.** Modern fuel injected engines normally have a test fitting on the engine fuel rail.
2. To test a mechanical fuel pump, start the engine and allow it to idle at the specified rpm. To test an electric fuel pump, you may only need to activate (supply voltage to) the pump motor.
3. Compare your pressure readings to specifications.

Tech Tip

With fuel injection, you are also testing the fuel pressure regulator, not just the pump.

Fuel pressure for a carburetor-type fuel system should be about 4–6 psi (30–40 kPa). A gasoline injection system will usually have a higher pressure output. Fuel pressure can run from 15–40 psi (100–280 kPa). A diesel

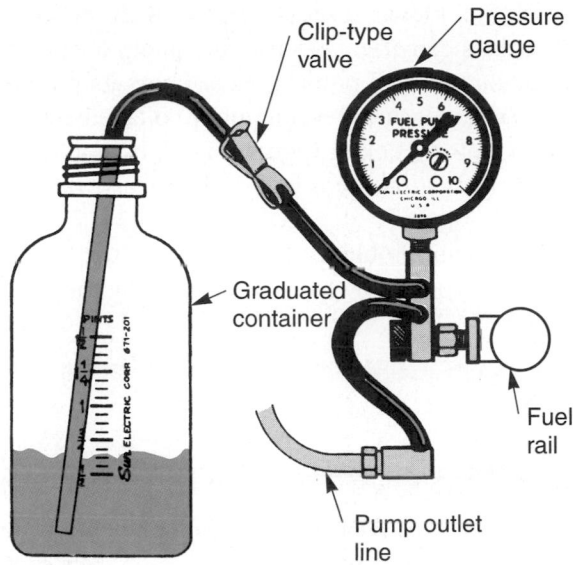

Figure 21-24. To test fuel pump pressure and volume, connect test equipment as shown. Connect a gauge to the fuel line before the fuel manifold, throttle body, or carburetor. Close the clip and start the engine to measure pressure. Open the clip to measure volume over a prescribed time span. (Chrysler)

supply pump should produce around 6–10 psi (40-70 kPa). It feeds fuel to the high-pressure injection pump.

Always remember to use factory values when determining fuel pump condition. Pressures vary from system to system.

If fuel pump pressure is *not* within specifications, check pump volume and the lines and filters before replacing the fuel pump. Also, isolate the fuel pressure regulator from the pump. This can be done by pinching the fuel hose going to the fuel return line or by taking the regulator out of the system.

If needed, you can also connect the pressure gauge directly to the output of the electric fuel pump. This will take the pressure regulator, feed lines, and other parts out of the system for isolating the problem.

Measuring Fuel Pump Volume

Fuel pump volume, also called **fuel pump capacity,** is the amount of fuel the pump can deliver in a specific amount of time. It is measured by allowing fuel to pour into a graduated (marked) container for a certain time period.

 To check fuel pump volume:
1. Route the output line from the fuel pump into a graduated container, **Figure 21-24.** For safety, a valve or clip should be used to control fuel flow into the container.

2. With the engine idling at a set speed, allow fuel to pour into the container for the prescribed amount of time (normally 30 seconds).
3. Close off the clip or valve.
4. Compare volume output (amount of fuel in the container) to specifications.

Fuel pump volume output should be a minimum of one pint (0.47 liters) in 30 seconds for carburetor systems. Fuel injection systems typically have a slightly higher volume output from the supply pump. Refer to factory service manual values for the particular fuel pump and automobile.

Measuring Fuel Pump Vacuum

Fuel pump vacuum should be checked when a fuel pump fails pressure and volume tests. A vacuum test will rule out possible problems in the fuel lines, hoses, filters, and pickup screen.

For example, a clogged fuel pickup screen could make the fuel pump fail the volume test. If the same pump passes a vacuum test, you need to check the lines and filters for problems.

 To measure fuel pump vacuum:
1. Connect a vacuum gauge to the inlet side of the pump, **Figure 21-25.**
2. Leave the fuel hose in your graduated container from the volume test.
3. Open the valve and activate the pump (start engine and allow to run on fuel in carburetor or connect voltage to electric pump).
4. Compare your vacuum reading to specifications.

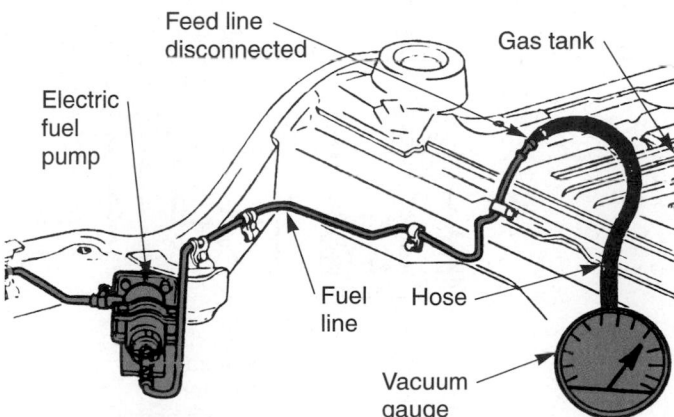

Figure 21-25. Fuel pump vacuum specifications are sometimes given. Connect vacuum gauge to inlet of inline pump and compare to specifications. (Cadillac)

Typically, *fuel pump vacuum* should be about 7–10 in. Hg. A good vacuum reading indicates a good fuel pump. If the pump failed the pressure or volume test but passed the vacuum test, the fuel supply lines or filters may be at fault.

Electric Fuel Pump Circuit Tests

Many electric fuel pump problems are caused by electrical circuit problems, **Figure 21-26.** Broken wires, bad relays, shorts, blown fuses, computer malfunctions, and other troubles can affect electric fuel pump operation.

If an electric fuel pump does *not* pass its pressure or volume tests, measure the amount of voltage being fed to the pump motor. Look at **Figure 21-27.** If supply voltage is low, there is a problem in the electrical circuit to the pump.

When circuit problems must be found, use the service manual and your knowledge of basic electrical testing

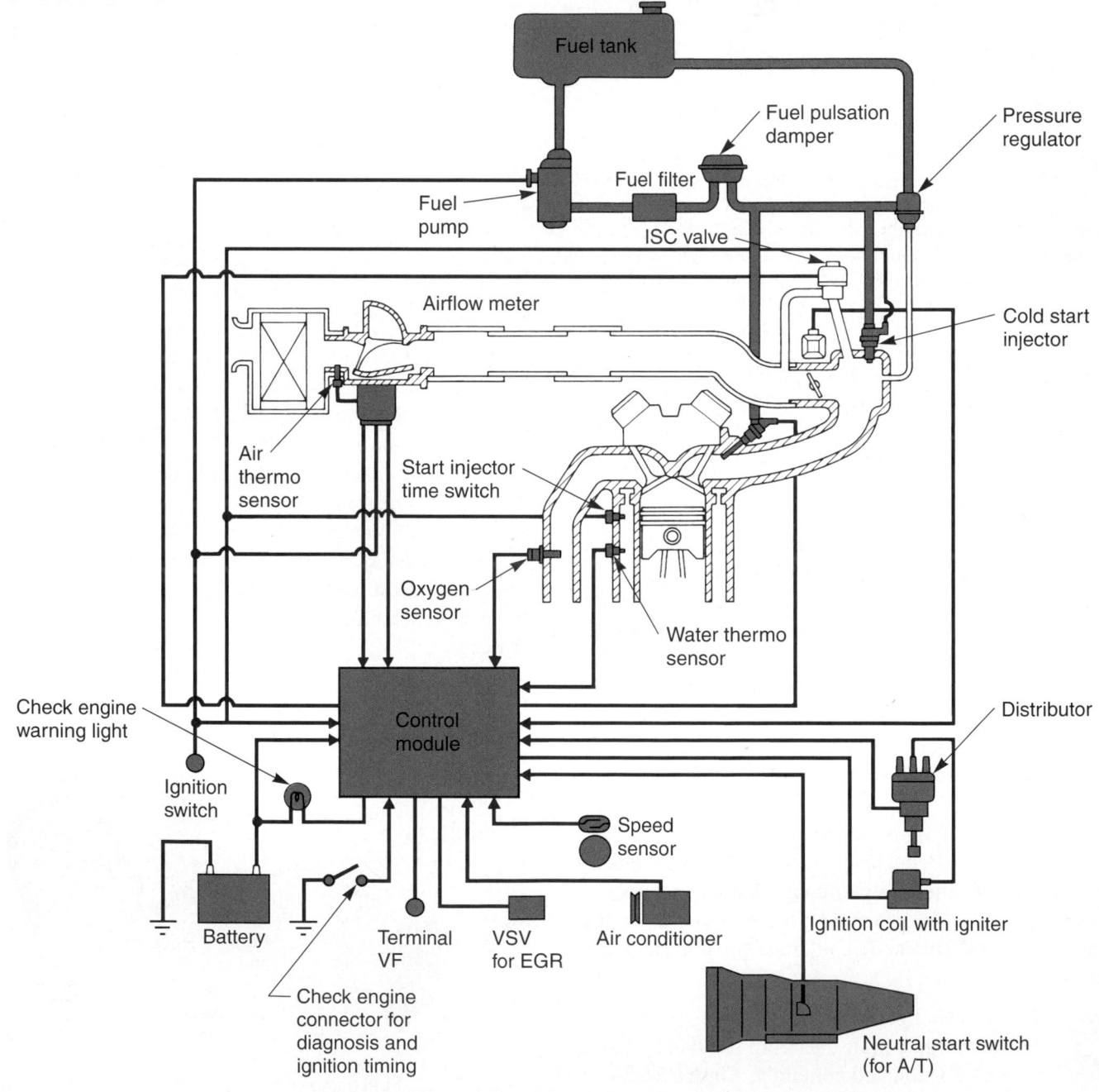

Figure 21-26. When problems are difficult to find, refer to the schematic in the service manual for the fuel pump circuit. It will help you trace and find problems. The diagram shows how the control module controls the fuel pump circuit. Trace wire from control module to fuel pump. (Toyota)

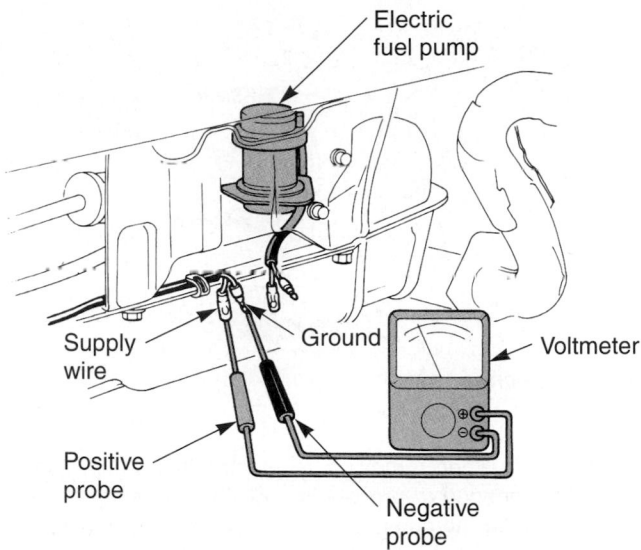

Figure 21-27. Before condemning an electric fuel pump, make sure the circuit is in good condition. Measure the amount of voltage being supplied to the pump and compare to specifications. If voltage is low, repair the circuit. (Honda)

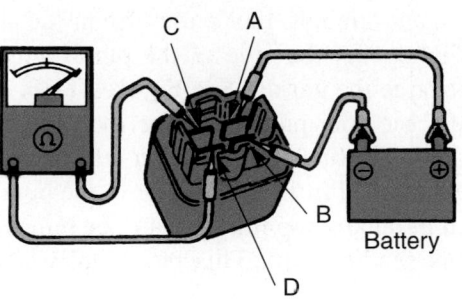

Specification	B+: Battery positive voltage
Terminal A — B	**Terminal C — D**
B+ applied	Continuity
B+ not applied	No continuity

Figure 21-29. The fuel pump relay can fail and prevent pump operation. Note that you can energize the relay with battery power and then check point closing with ohmmeter. (Mazda)

procedures. Generally, test the circuit at various points until the source of the trouble is found. **Figure 21-28** shows a wiring diagram for one electric fuel pump circuit. Note that there are only a few components and connections that could upset pump operation. Make sure to check the electric fuel pump relay if power is not reaching the pump, **Figure 21-29.**

Fuel Pump Shutoff Circuits

An *inertia switch* can be used to block current flow to the electrical fuel pump after a severe impact or collision.

It is a safety device that can prevent a serious fire after an auto accident. The inertia switch is usually located in the trunk or near the electric fuel pump. After a collision, you must press a button on the inertia switch before the electric pump will function again.

A *oil pressure switch* can be used to shut off the electric fuel pump if engine oil pressure drops too low. Its circuit is designed to protect the engine from major mechanical damage.

Fuel Pump Removal and Replacement

When a fuel pump does *not* pass its performance tests, it must be removed and replaced or rebuilt. To remove a mechanical fuel pump, simply disconnect the

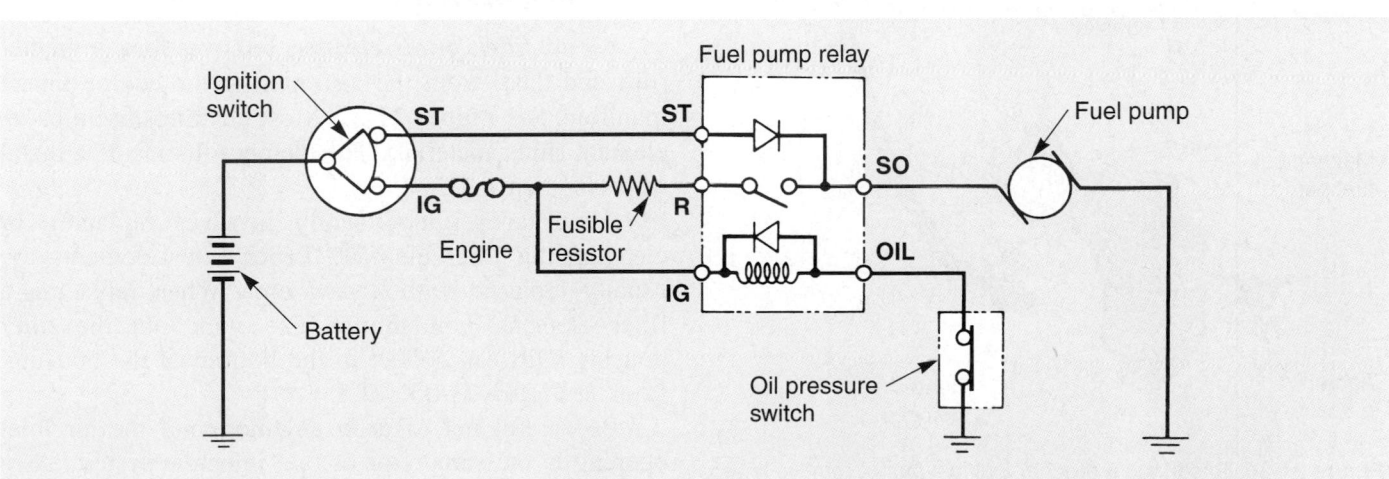

Figure 21-28. A fuel pump control circuit. Note fusible resistor, relay, oil pressure switch, and pump. Any defective part could upset the pump operation. (Toyota)

fuel lines and unbolt the pump from the engine. If needed, lightly tap the side of the pump with a plastic hammer to free the gasket. See **Figure 21-30.**

An electric fuel pump may be located in the main fuel line or in the fuel tank. With an inline pump, simply disconnect the fuel fittings and remove the pump, **Figure 21-31.** An in-tank pump must usually be removed as part of the tank sending unit. This procedure was described earlier in the chapter.

Some vehicles provide an in-tank *fuel pump access door* in the trunk so you do not have to remove the fuel tank to service the pump. By removing the small screws that hold the access door in place, you can service the in-tank pump more easily. If the electric pump is mounted on the top of the tank and no access door is provided, you will have to remove the tank to service the fuel pump.

Fuel Pump Rebuilding

Older mechanical fuel pumps are held together with screws and can be overhauled if parts are available.

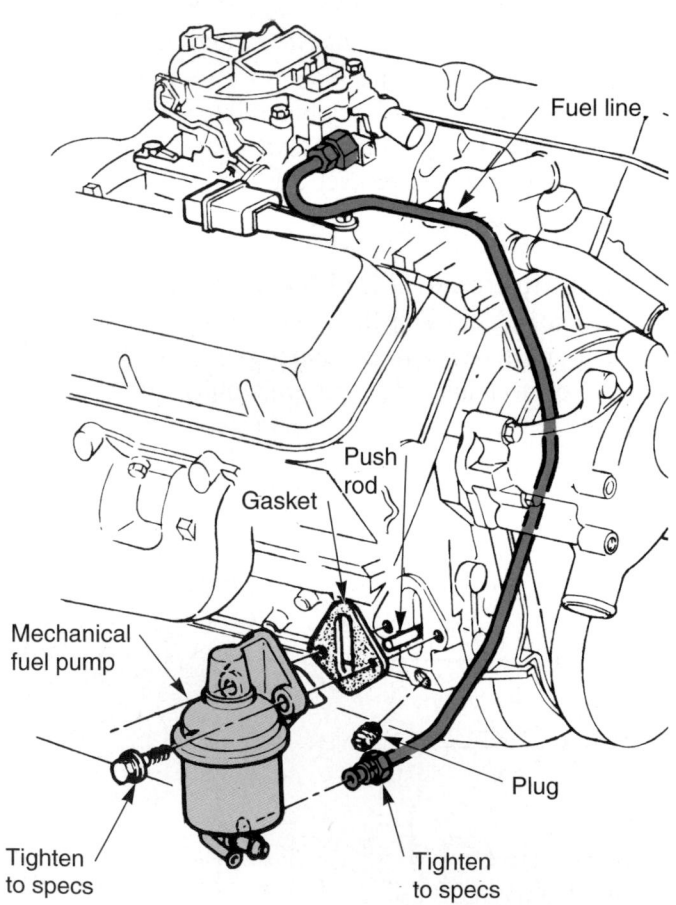

Figure 21-30. To service a mechanical fuel pump, remove lines and hoses. Then remove pump-to-block fasteners. Use a new gasket when installing. Torque fasteners to specifications. (Chevrolet)

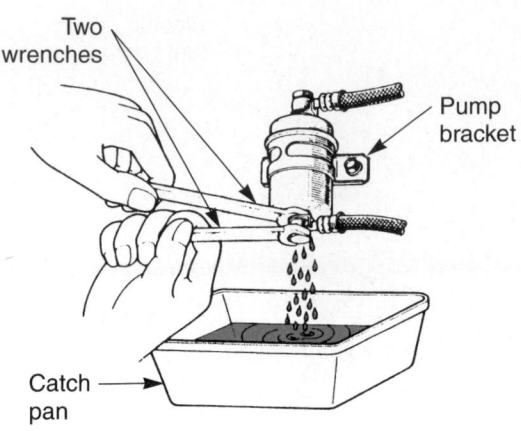

Figure 21-31. When removing an electric fuel pump, use two wrenches if needed to prevent line or fitting damage. Place a catch pan under the pump to prevent fuel spillage. (Toyota)

Normally, when either a mechanical or electric fuel pump is bad, it is replaced with a new pump.

To rebuild a mechanical pump, you must install the new parts provided in the rebuild kit. This normally includes a new diaphragm, check valves, springs, and other parts that suffer wear.

 Tech Tip

When installing a mechanical fuel pump, position the camshaft eccentric *away* from the pump rocker arm. This will make it much easier to hold the pump against the engine while starting the bolts. If a push rod is used, coat it with heavy grease. Then, push it up into place. This will hold the push rod out of the way while you install the fuel pump.

Air Filter Service

An *air filter,* or *air cleaner,* removes foreign matter (dirt and dust) from the air entering the engine intake manifold. See **Figure 21-32.** Most air filters use a paper element (filter material). The element fits inside a metal or plastic housing.

Air filter service usually involves replacing or cleaning the filter element. Paper filter elements are usually replaced with a new unit. When replacing a filter element, you should also wipe out the filter housing. Dirt can collect in the bottom of the housing. Look at **Figure 21-33.**

Be careful not to drop anything into the air inlet opening in the carburetor or fuel injection system. As a precaution, place a clean shop rag over the engine's air inlet.

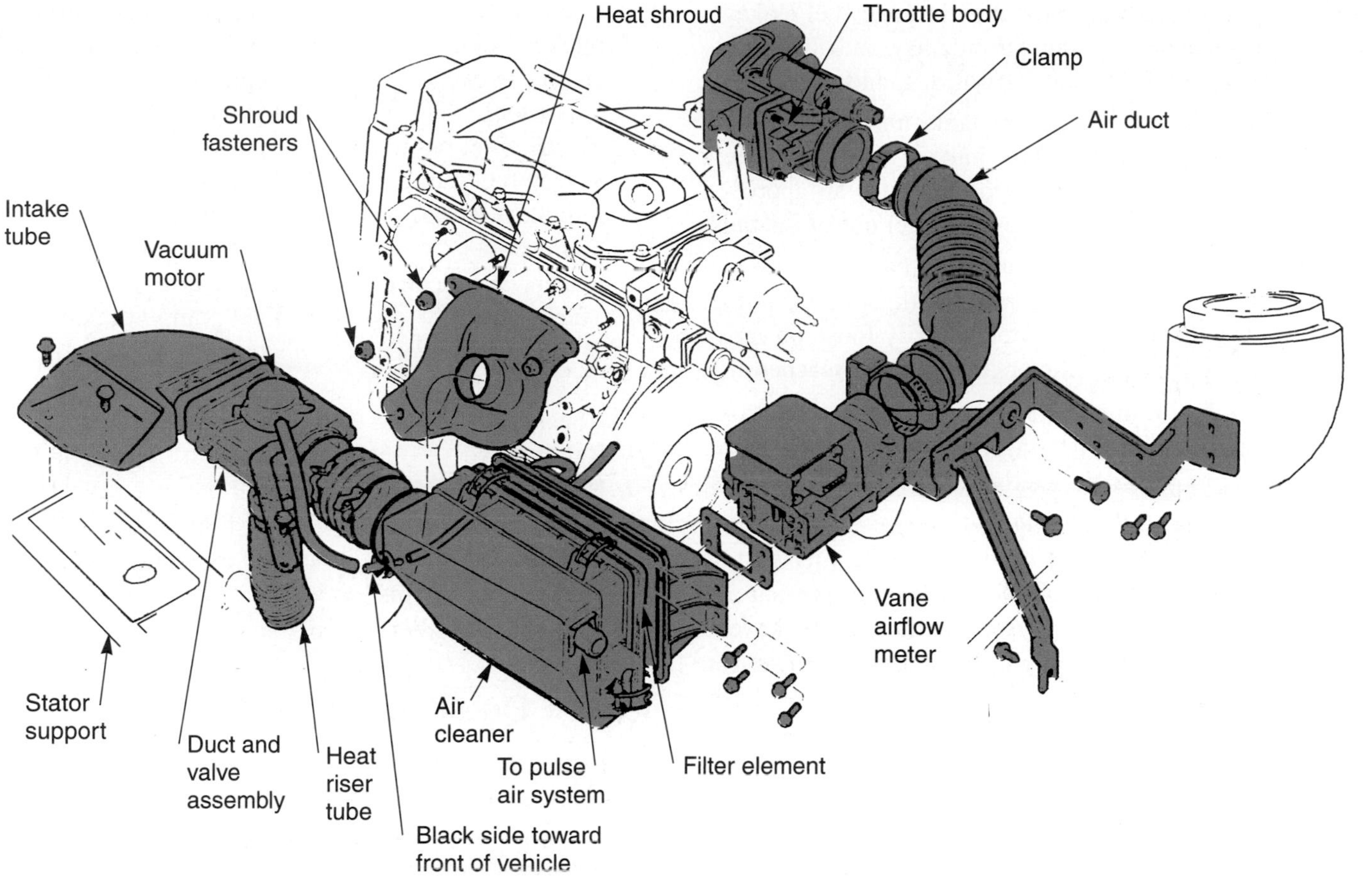

Heat shroud

Throttle body

Clamp

Air duct

Shroud
fasteners

Intake
tube

Vacuum
motor

Stator
support

Duct and
valve
assembly

Heat
riser
tube

Air
cleaner

To pulse
air system

Black side toward
front of vehicle

Filter element

Vane
airflow
meter

Figure 21-32. Parts of modern air filter assembly. (Toyota)

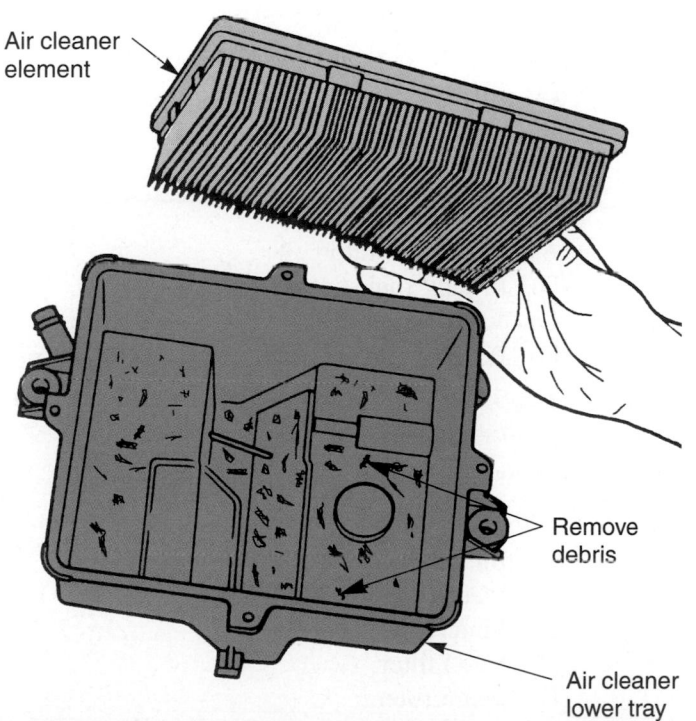

Air cleaner
element

Remove
debris

Air cleaner
lower tray

Figure 21-33. Paper filter elements are normally replaced when dirty. Make sure the housing is clean, secure, and not leaking. (Ford)

Summary

- A fuel supply system draws fuel from the fuel tank and forces it into the fuel metering device (gasoline injectors, diesel injection pump, or carburetor on older vehicles).

- Fuel tank capacity is the amount of fuel a fuel tank can hold. An average fuel tank capacity is 12–25 gallons (45–95 liters).

- A tank pickup-sending unit extends down into the tank to draw out fuel and to operate the fuel gauge.

- A fuel return system helps cool the fuel and prevents vapor lock (bubbles form in overheated fuel and stop fuel flow).

- An electric fuel pump, like a mechanical pump, produces fuel pressure and flow for the fuel metering section of a fuel system.

- The oil pressure switch is a safety feature that protects the engine from damage by shutting off the fuel pump with low oil pressure.

- Typical fuel tank problems include fuel leakage, physical damage (auto accidents), and contamination by foreign matter (rust, dirt, and water).

- To remove fuel from the tank, unscrew the drain plug and drain the fuel into an approved safety can. If a drain is *not* provided, use an approved pumping method to draw the fuel out of the tank.

- A faulty fuel tank sending unit can make the fuel gauge reading inaccurate. With a clogged fuel filter, the engine may temporarily loose power or stall when a specific engine speed is reached.

- Low fuel pump pressure can be caused by worn, dragging pump bearings, high resistance in electric pump circuit, leaking check valves, or physical wear of moving parts.

- Fuel pressure for a carburetor type fuel system should be about 4–6 psi (30–40 kPa) pressure. A gasoline injection system will usually have a higher pressure output. Fuel pressure can run from 15–40 psi (100 to 280 kPa). A diesel supply pump should produce around 6–10 psi (40–70 kPa).

- Fuel pump volume, also called capacity, is the amount of fuel the pump can deliver in a specific amount of time.

- An inertia switch can be used to block current flow to the electrical fuel pump after a severe impact or collision.

- Some vehicles provide an in-tank fuel pump access door in the trunk so you do not have to remove the fuel tank to service the pump.

Important Terms

Fuel system	Tank pickup-sending
Fuel supply system	unit
Air supply system	Fuel hoses
Fuel metering system	Fuel return system
Fuel tank	Fuel filters
Fuel lines	Pleated paper elements
Fuel pump	Sintered bronze
Fuel filters	Bowl fuel filter
Driving range	Canister fuel filter
Fuel tank capacity	Fuel pump
Fuel tank filler neck	Mechanical fuel pump
Filler cap	Eccentric
Spillback ball	Reciprocating
Fuel tank caps	Rocker arm
Fuel tank baffles	Return spring
Fuel tank straps	Diaphragm
Fuel tank cover	Diaphragm spring

Check valves	Clogged in-tank strainer
Intake stroke	Low fuel pump pressure
Output stroke	High fuel pump pressure
Vapor lock	Mechanical fuel pump
Electric fuel pump	noise
Rotary fuel pumps	Fuel pump leaks
Impeller electric fuel	Fuel pump testing
pump	Fuel pump volume
Roller vane electric	Capacity
fuel pump	Fuel pump volume
Sliding vane electric	output
fuel pump	Fuel pump vacuum
Check valves	Inertia switch
Relief valve	Oil pressure switch
Reciprocating electric	Fuel pump access door
fuel pump	Air filter
Oil pressure switch	Air cleaner
Fuel tank problems	
Clogged fuel filter	

Review Questions—Chapter 21

Please do not write in this text. Place your answers on a separate sheet of paper.

1. List and describe the three subsystems of a modern fuel system.

2. A(n) _____ _____ system draws fuel from the fuel tank and forces it to the fuel metering device.

3. An average fuel tank capacity is around 12–25 gallons (45–95 liters). True or False?

4. What is the purpose of a filler neck restrictor in a fuel tank assembly?

5. Which of the following is *not* part of a fuel tank pickup-sending unit?
   (A) Vapor separator.
   (B) In-tank fuel strainer or filter.
   (C) Variable resistor.
   (D) Pickup tube.

6. Fuel lines are normally made of single-wall steel tubing. True or False?

7. When are fuel hoses needed?

8. Explain the operation of a fuel return system.

9. The following could *not* be a fuel filter.
   (A) Pleated paper filter.
   (B) Sintered iron filter.
   (C) Bowl filter.
   (D) Foam filter.

10. The two basic types of fuel pumps are the _____ and _____ types.

11. Explain the difference between the two types of fuel pumps in Question 10.

12. The _____ _____ controls fuel pressure in a mechanical fuel pump.

13. During the intake stroke of a mechanical fuel pump, the camshaft's eccentric pushes on the pump rocker arm. True or False?

14. Can a mechanical fuel pump idle? Explain.

15. Define the term "vapor lock."

16. Where can electric fuel pumps be located?

17. Which of the following is an advantage of an electric fuel pump?
    (A) Instant fuel pressure.
    (B) Very little pressure pulsations.
    (C) Helps prevent vapor lock.
    (D) All of the above.

18. Rotary type electric fuel pumps include the _____, _____ _____, and _____ _____ types.

19. A(n) _____ _____ limits the maximum output pressure of an electric fuel pump.

20. Why is an oil pressure switch sometimes used in an electric fuel pump circuit?

21. List nine service rules for fuel lines and hoses.

22. Name five typical fuel filter locations.

23. Fuel pump pressure for a carburetor type fuel system should be approximately _____ to _____ psi or _____ to _____ kPa.

24. During an electric fuel pump test, where can you connect your pressure gauge to remove the pressure regulator from the circuit?
    (A) Pump inlet.
    (B) Pump outlet.
    (C) Fuel rail.
    (D) None of the above.

25. What is a fuel pump volume test?

ASE-Type Questions

1. Which of the following is *not* one of the three fuel system subsystems?
    (A) Cooling system.
    (B) Air supply system.
    (C) Fuel supply system.
    (D) Fuel metering system.

2. Which of these is a fuel metering device?
    (A) Carburetor.
    (B) Gasoline injector.
    (C) Diesel injection pump.
    (D) All of the above.

3. Which of the following are placed inside fuel tanks to prevent fuel splashing?
    (A) Caps.
    (B) Filters.
    (C) Baffles.
    (D) Restrictors.

4. When bubbles form in overheated fuel and prevent fuel flow, it is called:
    (A) blowby.
    (B) vapor lock.
    (C) fuel blocking.
    (D) condensation.

5. Which of the following are used to stop contaminants from entering parts that could be damaged by foreign matter?
    (A) Fuel filters.
    (B) Pleated paper elements.
    (C) Sintered bronze elements.
    (D) All of the above.

6. A mechanical fuel pump is powered by a(n):
    (A) pull rod.
    (B) eccentric.
    (C) rocker arm.
    (D) outlet valve.

7. Each of these is an electric rotary fuel pump *except:*
    (A) vacuum.
    (B) impeller.
    (C) roller vane.
    (D) sliding vane.

8. When performing fuel tank repairs, Technician A believes, for safety reasons, leaks should never be mended by welding or soldering. Technician B agrees, but says any fuel tank leak repair should be handled strictly by a specialist. Who is right?
    (A) A only.
    (B) B only.
    (C) Both A and B.
    (D) Neither A nor B.

9. Fuel filters are located:
    (A) in the fuel line.
    (B) in the fuel pump.
    (C) under the fuel line.
    (D) All of the above.

10. Fuel pressure for a typical fuel injection system should be about:
    (A) 4–6 psi.
    (B) 6–10 psi.
    (C) 15–40 psi.
    (D) 28–41 psi.

11. Which of the following can be used to block current flow to an electric fuel pump after a severe collision?
    (A) *Hose clamp.*
    (B) *Inertia switch.*
    (C) *Fusible resistor.*
    (D) *Fuel pump relay.*

12. To remove a mechanical fuel pump, do each of these *except:*
    (A) *disconnect fuel lines.*
    (B) *unbolt pump from engine.*
    (C) *remove diaphragm spring.*
    (D) *free gasket with plastic hammer.*

Activities—Chapter 21

1. Prepare a "tree diagnosis chart" for a fuel pump problem assigned by your instructor. Your resource will be a shop manual for the specific vehicle and your instructor. If your instructor requests it, review the procedure to the class using an overhead projector and sketches.

2. Review electrical principles on resistance in a circuit and explain with sketches how a thermostatic fuel gauge works.

Fuel Delivery System Diagnosis

Condition	Possible Cause	Correction
No fuel delivery.	1. No fuel. 2. Tank filter clogged. 3. Fuel lines kinked or clogged. 4. Vapor lock. 5. Fuel pump inoperative. 6. Fuel filter or filters clogged. 7. Frozen fuel line. 8. Air leak between fuel pump and tank. 9. Clogged injectors. 10. Inoperative injection pump (diesel).	1. Add fuel to the tank. 2. Replace tank filter. 3. Straighten or clean fuel lines. 4. Cool fuel lines. Change to less volatile fuel. Protect fuel lines from heat. 5. Replace pump. 6. Replace filters. 7. Thaw and remove water from the fuel system. 8. Repair leak. 9. Clean or replace injectors, locate source of deposits. 10. Repair or replace pump.
Insufficient fuel delivery.	1. Tank filter partially clogged. 2. Gas lines kinked or clogged. 3. Vapor lock. 4. Air leak between fuel pump and tank. 5. Fuel filter partially clogged. 6. Defective fuel pump. 7. Insufficient voltage reaching fuel pump. 8. Wrong pump. 9. Clogged injection pipes. 10. Defective injectors. 11. Faulty injection pump (diesel). 12. Fuel injection pulse width incorrect. 13. Mass airflow sensor malfunction. 14. Defective oxygen sensor.	1. Replace filter. 2. Straighten or clean fuel lines. 3. Cool fuel lines. Change to less volatile fuel. Protect fuel lines from heat. 4. Repair air leak. 5. Replace filter. 6. Replace pump. 7. Repair fuel pump circuit. 8. Install correct pump. 9. Clean or replace pipes. 10. Clean or replace injectors. 11. Rebuild or replace pump. 12. Check control module and sensors. Replace as needed. 13. Replace mass airflow sensor. 14. Replace oxygen sensor.
Excessive fuel delivery or pressure.	1. Defective fuel pump. 2. Incorrect pump. 3. Fuel injection pulse width excessive. 4. Fuel injection pressure regulator malfunctioning. 5. Engine control module defective. 6. Mass airflow sensor malfunction. 7. Defective oxygen sensor.	1. Replace pump. 2. Replace pump. 3. Check control module and sensors. Replace as needed. 4. Adjust or replace pressure regulator. 5. Replace control module. 6. Replace mass airflow sensor. 7. Replace oxygen sensor.

Chapter 22

Gasoline Injection Fundamentals

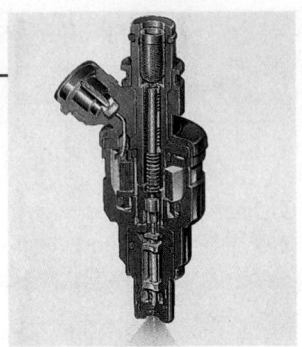

After studying this chapter, you will be able to:

- List some of the possible advantages of gasoline injection.
- Describe the classifications of gasoline injection.
- Explain the operation of electronic throttle body gasoline injection.
- Explain the operation of electronic multiport gasoline injection.
- Summarize the operation of electronic airflow sensing, hydraulic-mechanical (continuous), and manifold-pressure-sensing gasoline injection systems.
- Compare the various types of gasoline injection systems.
- Correctly answer ASE certification test questions on gasoline injection systems.

This chapter introduces the operating principles of gasoline injection systems. Specific systems vary, but many of the parts—such as sensors, fuel injectors, and control modules—are very similar. This chapter provides a broad background in the many gasoline injection systems found on today's vehicles. However, electronic multiport fuel injection is the main focus of the chapter, since it is the most common system.

Gasoline Injection Fundamentals

A *gasoline injection system* uses pressure from an electric fuel pump to spray fuel into the engine's intake manifold. See **Figure 22-1.** Like a carburetor, the gasoline injection system must provide the engine with the correct air-fuel mixture for specific operating conditions. Unlike a carburetor, however, *pressure,* not engine *vacuum,* is used to feed fuel into the engine in this type of system. This makes a gasoline injection system very efficient.

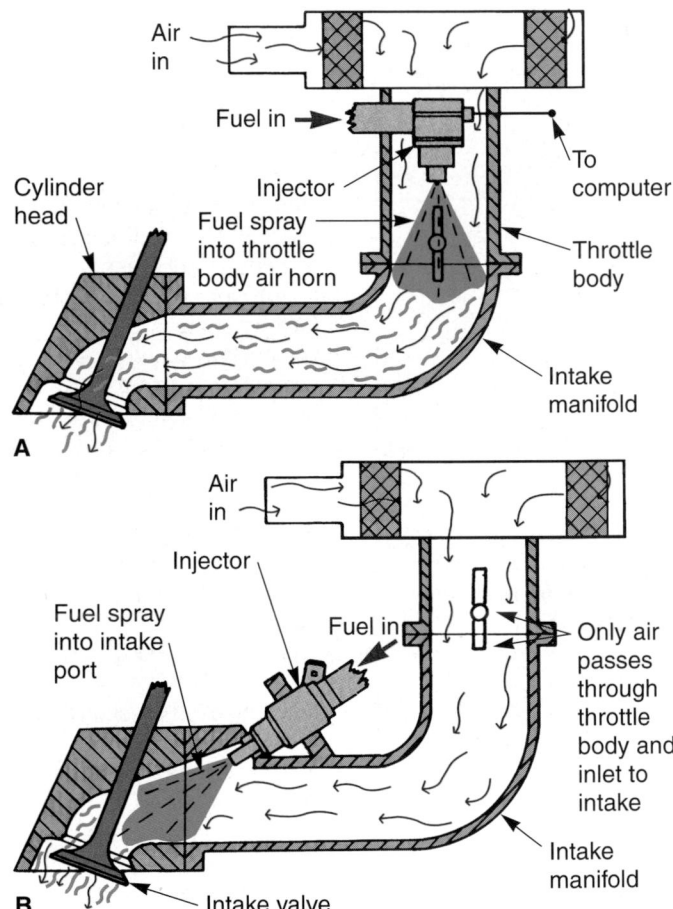

Figure 22-1. A—In a throttle body injection system, the fuel injector is located in the throttle body. B—In a multiport injection system, an injector is located right before each intake valve. This improves efficiency and fuel distribution.

Gasoline Injection Advantages

A gasoline injection system has several advantages over a carburetor-type fuel system. A few of these include:

- Improved atomization. Fuel is forced into the intake manifold under pressure, which helps break fuel droplets into a fine mist.

- Better fuel distribution. There is a more equal flow of fuel vapors into each cylinder.

- Smoother idle. A lean fuel mixture can be used without rough idle because of better fuel distribution and low-speed atomization.

- Improved fuel economy. Higher efficiency results from more precise fuel metering, atomization, and distribution.

- Lower emissions. A lean, efficient air-fuel mixture reduces exhaust pollution.

- Better cold-weather driveability. Injection provides better control of mixture enrichment than a carburetor choke.

- Increased engine power. Precise metering of fuel to each cylinder and increased airflow can result in more horsepower output.

- Simpler. Electronic fuel injection systems have fewer parts than computer-controlled carburetor systems.

Atmospheric Pressure

Atmospheric pressure is the pressure formed by the air surrounding the earth. At sea level, the atmosphere exerts 14.7 psi (103 kPa) of pressure on everything. This pressure is caused by the weight of the air, as shown in **Figure 22-2.**

Vacuum

A *vacuum* is lower than atmospheric pressure in an enclosed area. Suction is another word for vacuum. Any space with less than 14.7 psi (103 kPa) of pressure at sea level has a vacuum.

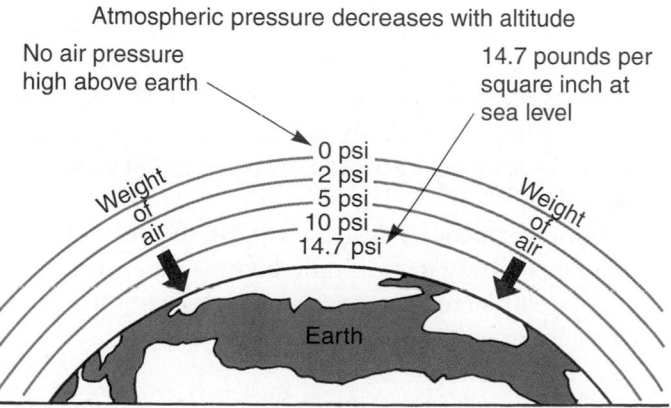

Figure 22-2. Atmospheric pressure is produced by the weight of air. Pressure changes with altitude. Atmospheric pressure is what moves air and fuel mixture into cylinders on intake strokes.

Differences in Pressure Cause Flow

A difference in pressure between two areas can be used to cause flow. For instance, when you suck on a straw, atmospheric pressure pushes down on the liquid in the glass. This causes the liquid to flow through the straw and into the vacuum in your mouth. An engine uses differences in pressure to force fuel and air into its cylinders. The engine acts as a vacuum pump, producing a low-pressure area, or vacuum, in the intake manifold. See **Figure 22-3.**

Engine Throttle Valve

The *engine throttle valve* controls airflow and gasoline engine power output. It is a "butterfly" or flap-type valve in the throttle body assembly. When closed, the throttle valve restricts the flow of air and the resulting flow of fuel into the engine.

When the throttle valve is opened, airflow, fuel flow, and engine power increase. This keeps engine speed and power low for *idling.*

When the driver presses on the accelerator, the throttle cable slides inside its housing. This swings the throttle valve open. Atmospheric pressure then pushes more air into the engine intake manifold. Engine sensors detect the resulting changes and increase fuel flow through the injectors. With more air and fuel entering the cylinders, the pressure produced on the power strokes is increased. Engine speed and horsepower output then increase to accelerate the vehicle. When the accelerator is released, a throttle return spring pulls the throttle valve closed. This returns the engine to idle speed. **Figure 22-4** shows how a vehicle's accelerator and throttle cable control the throttle valve.

Drive-by-Wire System

Instead of a mechanical cable or linkage, a *drive-by-wire system* uses a pedal sensor, a control module, and an actuator to operate the engine throttle valve on a gasoline engine (or the fuel injection pump on a diesel engine). When you press on the accelerator, the pedal moves a variable resistor. The control module measures current flow through the resistor and calculates how far the pedal is depressed. It then sends a signal to the actuator to control engine speed.

The benefit of drive-by-wire systems is that they can gradually increase power to prevent undue strain on the driveline and reduce fuel consumption. Even if the accelerator is tromped to the floor, the control module can slowly increase engine speed for better efficiency.

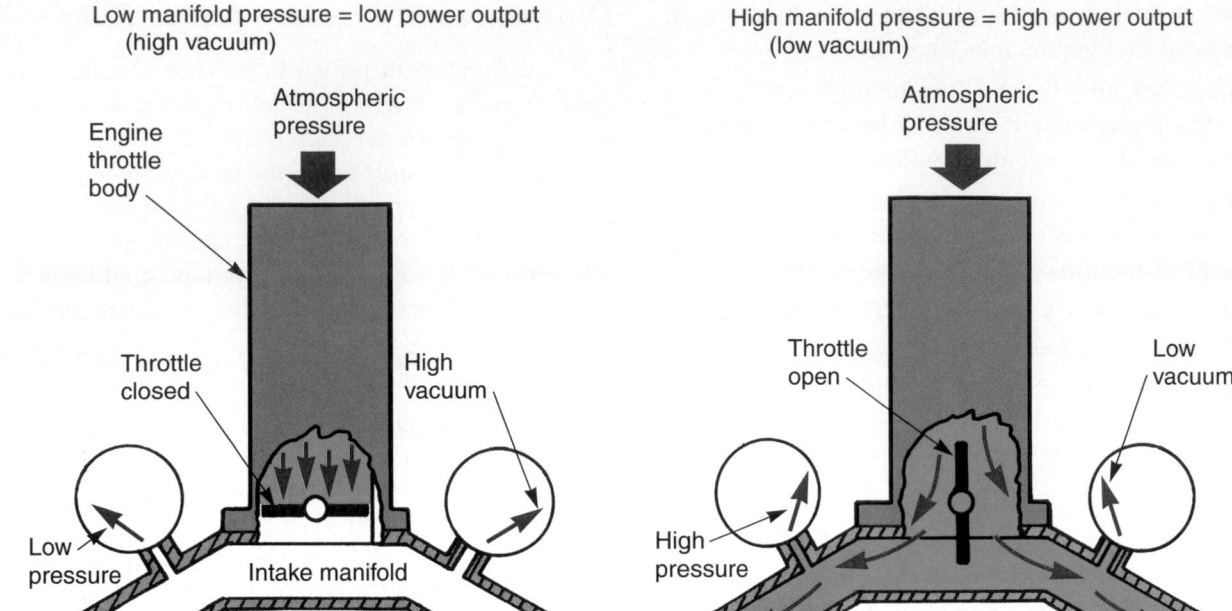

Low manifold pressure = low power output
(high vacuum)

High manifold pressure = high power output
(low vacuum)

Figure 22-3. Throttle valve position controls airflow and the amount of vacuum in the intake manifold. A—A closed throttle valve produces high vacuum in the manifold. The engine tries to draw air through the throttle body, but cannot. B—An open throttle allows airflow, reducing the vacuum in the intake manifold.

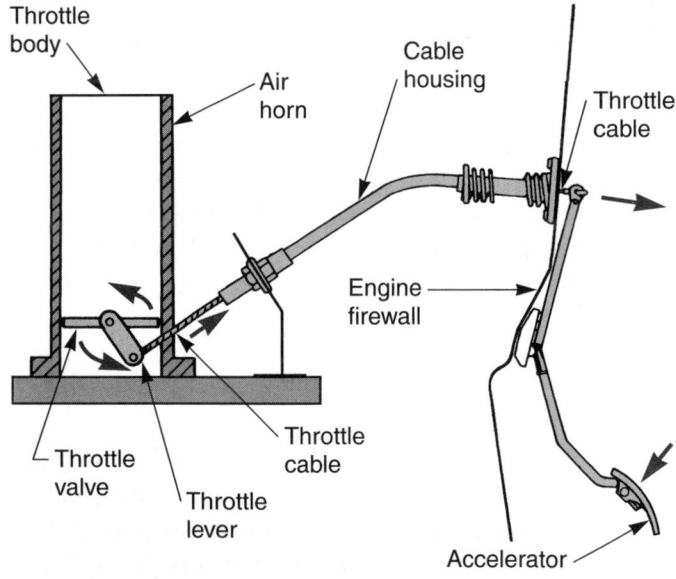

Figure 22-4. The accelerator is connected to the throttle valve in the throttle body. This valve controls airflow and engine power output. Pedal movement slides the cable inside the housing to transfer motion to the valve on the engine.

Gasoline Injection Classifications

There are many types of gasoline injection systems. Before studying the most common ones, you should have a basic knowledge of the different classifications of gasoline injection. This will help you understand the similarities and differences between systems.

A gasoline injection system is commonly called a *fuel injection system*. To avoid confusion, remember that a diesel injection system is also a fuel injection system. The two are quite different, however.

Throttle Body and Multiport Injection

The location, or point, of the fuel injectors is one way to classify a gasoline injection system. A *throttle body injection system* has the injector nozzles in a throttle body assembly on top of the engine. Fuel is sprayed into the top center of the intake manifold. Throttle body injection is illustrated in **Figure 22-1A.** A *multiport injection system* has fuel injectors in the *intake ports* (air-fuel runners or passages going to each cylinder). See **Figure 22-1B.** Gasoline is sprayed into each intake port, toward each intake valve.

Multiport injection systems control the air-fuel mixture more precisely than throttle body systems, lowering emissions and increasing power output. Therefore, these systems are used in all late-model vehicles.

Indirect and Direct Injection

An *indirect injection system* sprays fuel into the engine intake manifold. Most gasoline injection systems are indirect systems. A ***direct injection system*** forces fuel into the engine combustion chambers. All diesel injection systems are a direct type. Both indirect and direct injection systems are shown in **Figure 22-5.**

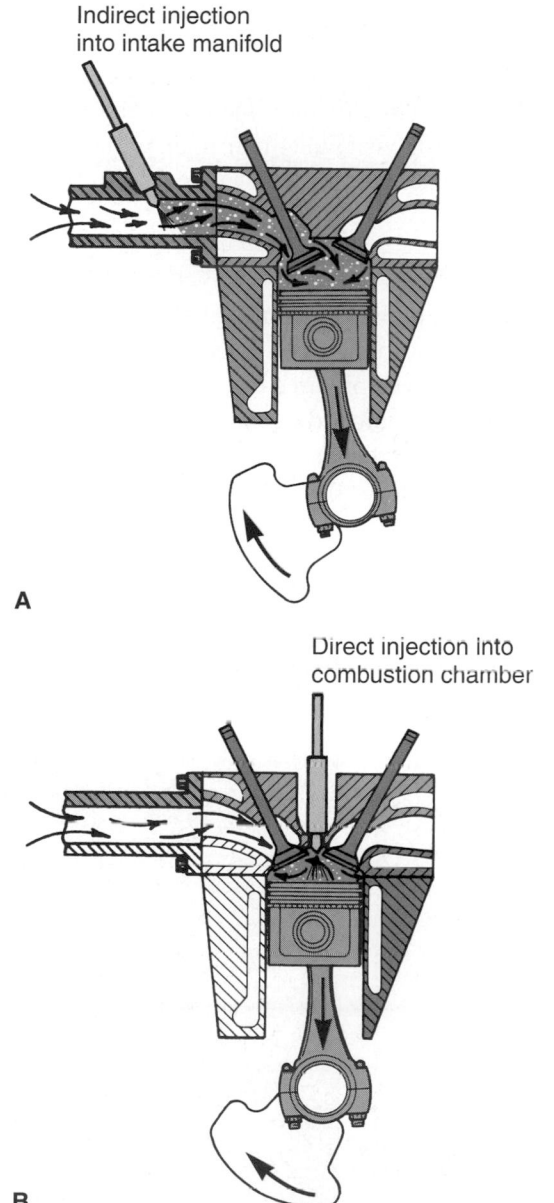

A

B

Figure 22-5. A—Indirect injection sprays fuel into the intake manifold. Gasoline injection systems are usually this type. B—Direct injection sprays fuel into the combustion chambers.

Gasoline Injection Controls

There are three common methods used to control the amount of gasoline injected into the engine. These are electronic controls, hydraulic controls, and mechanical

controls. Older gasoline injection systems use a combination of the three.

Electronic fuel injection uses various engine sensors and a control module to regulate the opening and closing of the injection valves. This is the most common type of gasoline injection system. It is covered in detail later in this chapter.

Hydraulic fuel injection utilizes hydraulic control devices moved by air or fuel pressure. The hydraulic control system uses an airflow sensor and a fuel distributor to meter gasoline into the engine. A ***fuel distributor*** is a hydraulic valve mechanism.

Mechanical fuel injection uses throttle linkage, a mechanical pump, and a governor speed device to control injection volume. This is a very old, seldom-used type injection system found mainly on high-performance and diesel engines. Diesel systems are covered in later chapters.

Gasoline Injection Timing

The ***timing*** of a gasoline injection system links the engine valve action to the time when fuel is sprayed into the intake manifold. There are three basic classifications of gasoline injection timing. These are intermittent, timed, and continuous.

An ***intermittent gasoline injection system*** opens and closes the injection valves independently of the engine intake valves. This type of injection system may spray fuel into the engine when the valves are open or when they are closed. Another name for an intermittent injection system is ***modulated injection system.***

A ***timed injection system*** squirts fuel into the engine right before or as the intake valves open. The best example of timed injection is a diesel injection system.

A ***continuous gasoline injection system*** sprays fuel into the intake manifold at all times. Anytime the engine is running, some fuel is forced out of the injector nozzles and into the engine. The air-fuel ratio is controlled by increasing or decreasing fuel pressure at the injectors. This increases or decreases fuel flow out of the injectors. This type is discussed near the end of this chapter.

Injector Opening Relationship

Simultaneous injection means that all the injectors open at the same time. The injectors are pulsed *on* and *off* together. ***Sequential injection*** means that the injectors open one after the other. ***Group injection*** has several, but not all, injectors opening at the same time. For example, a V-8 engine might have one group of four injectors that open at one time and another group of four that open at a different time.

Electronic Fuel Injection

An electronic fuel injection (EFI) system can be divided into four subsystems:

- Fuel delivery system.
- Air induction system.
- Sensor system.
- Computer control system.

These four subsystems are illustrated in **Figure 22-6. Figure 22-7** shows the major parts of a complete fuel injection system.

Fuel Delivery System

The *fuel delivery system* of an EFI system includes an electric fuel pump, a fuel filter, a pressure regulator, fuel injectors (injector valves), and connecting lines and hoses. See **Figure 22-8.**

The *electric fuel pump* draws gasoline out of the tank and forces it into the pressure regulator. The *fuel pressure regulator* controls the amount of pressure entering the injector valves. When sufficient pressure is attained, the regulator returns excess fuel to the tank. This maintains a preset amount of fuel pressure for injector operation.

Engine vacuum is ported into the fuel pressure regulator. This allows the pressure regulator to lower fuel pressure slightly at idle speed (low engine load) and increase it with higher engine speed (higher engine load). Look at **Figures 22-9 and 22-10.** A *fuel injector* for an EFI system is simply a coil- or solenoid-operated fuel valve, **Figure 22-11.** When not energized, spring pressure holds the injector closed, keeping fuel from entering the engine. When current flows through the injector coil, the magnetic field attracts the injector armature and the injector valve opens. Fuel then squirts into the intake manifold under pressure. See **Figure 22-12.**

Air Induction System

An *air induction system* for EFI typically consists of an air filter, throttle valves, sensors, and connecting ducts. **Figure 22-13** pictures the air induction system for one type of EFI-equipped engine.

Airflow enters the inlet duct and flows through the air filter. The air filter traps dust and debris and prevents it from entering the inside of the engine. Plastic ducts then route the clean air into the throttle body assembly. The throttle body assembly with multiport injection simply contains the throttle valve and idle air control

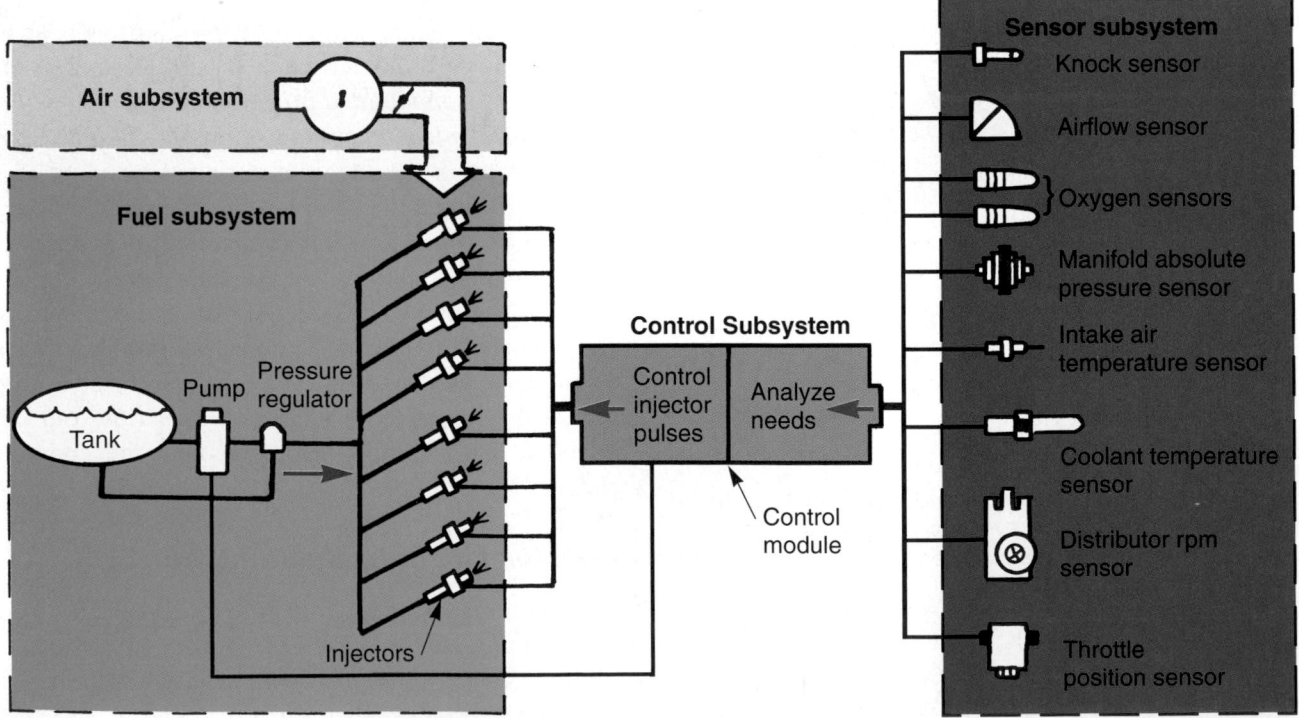

Figure 22-6. The four subsystems of an electronic gasoline injection system. The sensor system feeds data to the control module. The control module uses this data to operate the fuel delivery system. Parts of the air system can also be controlled by the control module.

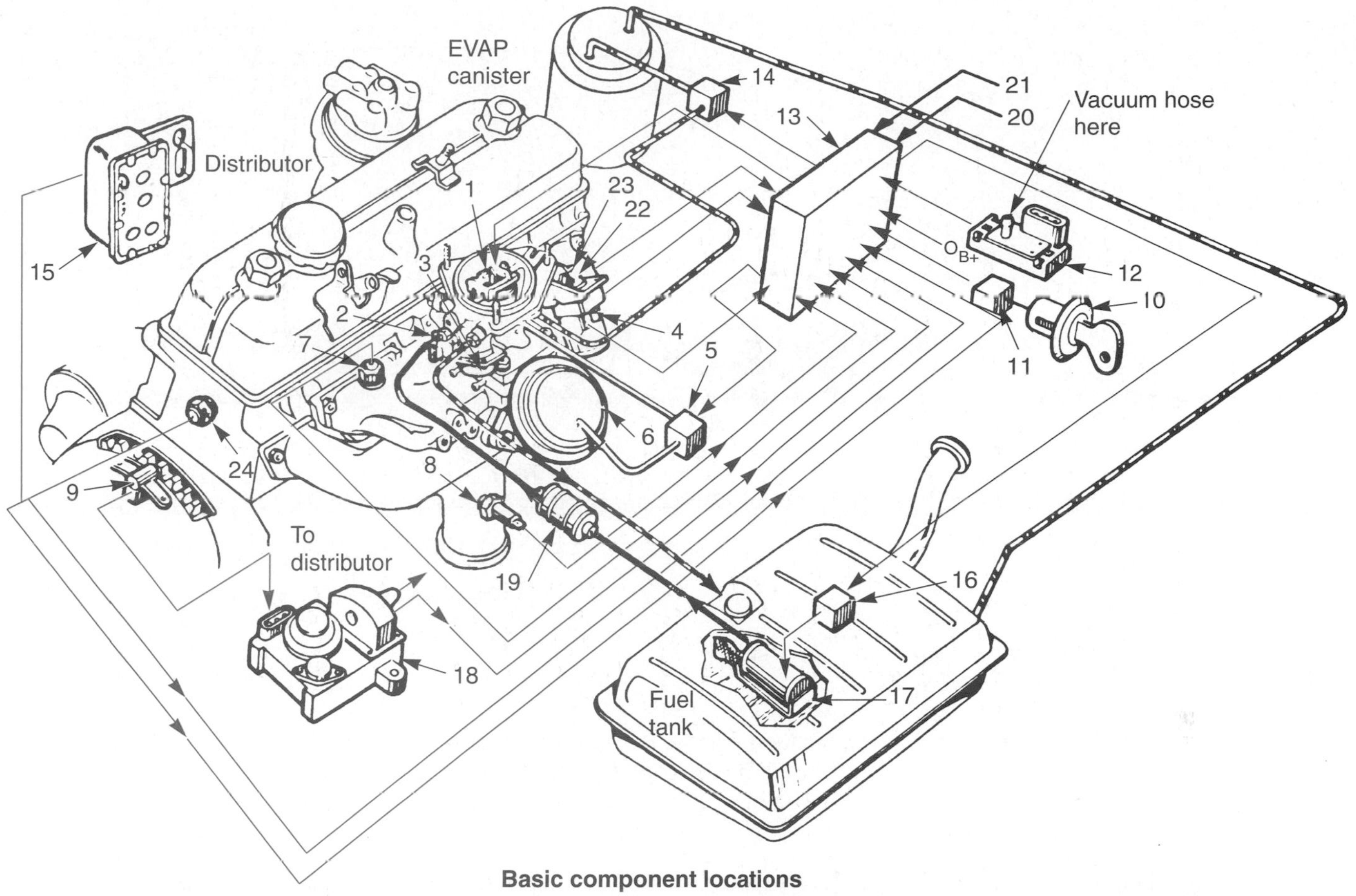

EVAP canister

Distributor

To distributor

Basic component locations

1. Injector
2. Throttle position sensor
3. Pressure regulator
4. Idle speed control motor
5. Solenoid-to-EGR valve
6. EGR valve
7. Manifold air-fuel temperature sensor
8. O_2 sensor
9. Speed sensor
10. Ignition switch
11. Power relay
12. Manifold pressure sensor
13. Engine control module
14. Solenoid-to-EVAP canister control
15. Starter motor relay
16. Fuel pump relay
17. Fuel pump
18. Ignition control module
19. In-line fuel filter
20. Air conditioner on
21. Transaxle neutral/park switch
22. Closed throttle (idle) switch
23. Wide open throttle (WOT) switch
24. Temperature sensor (coolant)

Figure 22-7. Note the general location of electronic fuel injection parts. This will help you grasp the details of each component.

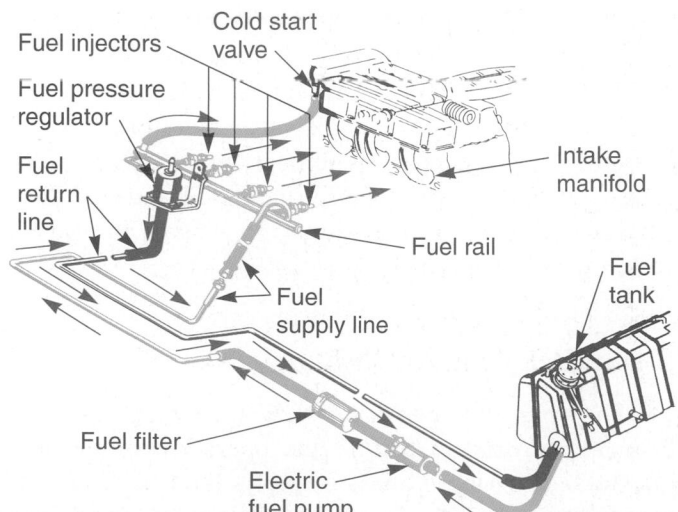

Figure 22-8. A fuel delivery system typically consists of the parts shown above. (Honda)

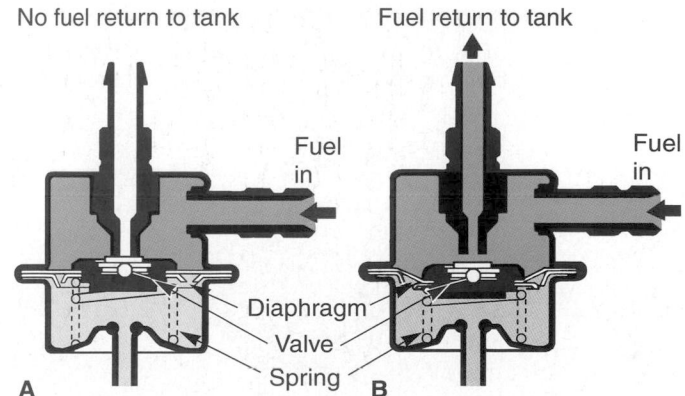

Figure 22-9. Fuel pressure regulator operation. A—Low engine vacuum indicates high engine load. The spring then holds the regulator return closed to increase fuel pressure for more power. B—High engine vacuum indicates low load. Vacuum acts on the diaphragm, opening the regulator return to the tank. This reduces or limits fuel pressure. (Lancia)

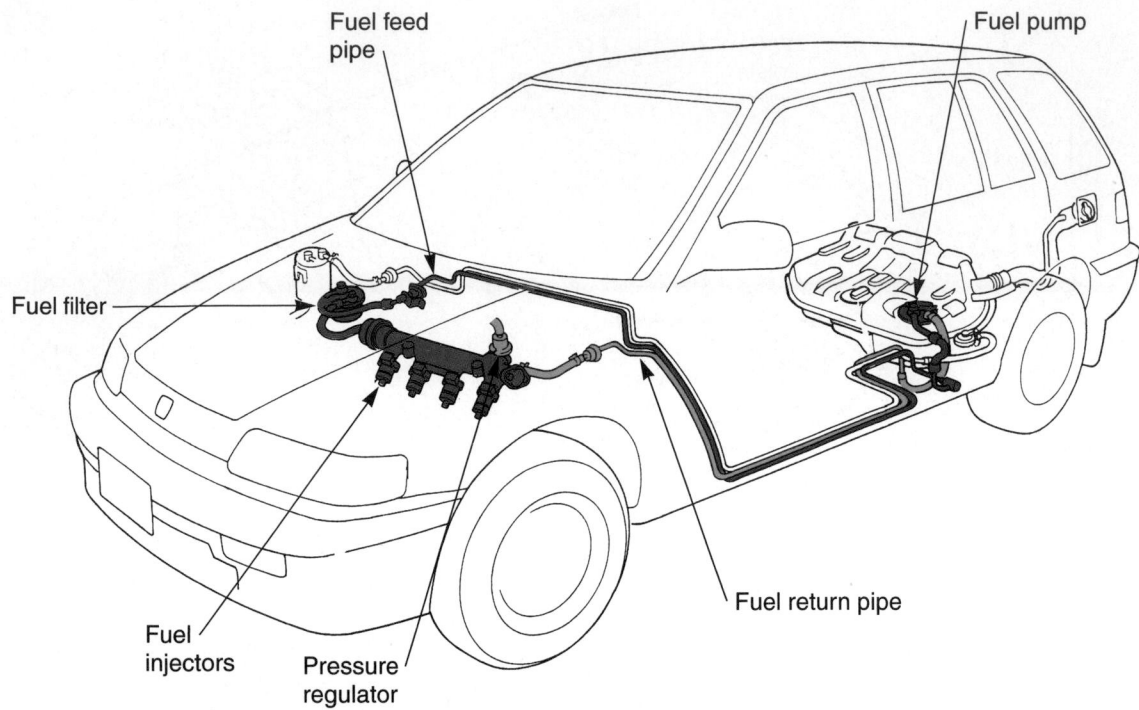

Fuel feed pipe

Fuel pump

Fuel filter

Fuel injectors

Pressure regulator

Fuel return pipe

Figure 22-10. Note how the electric fuel pump forces fuel up to the fuel rail, the injectors, and the fuel pressure regulator. The regulator bypasses some fuel back to tank through the return line to limit pressure and to keep the fuel cool, preventing vapor lock. (Honda)

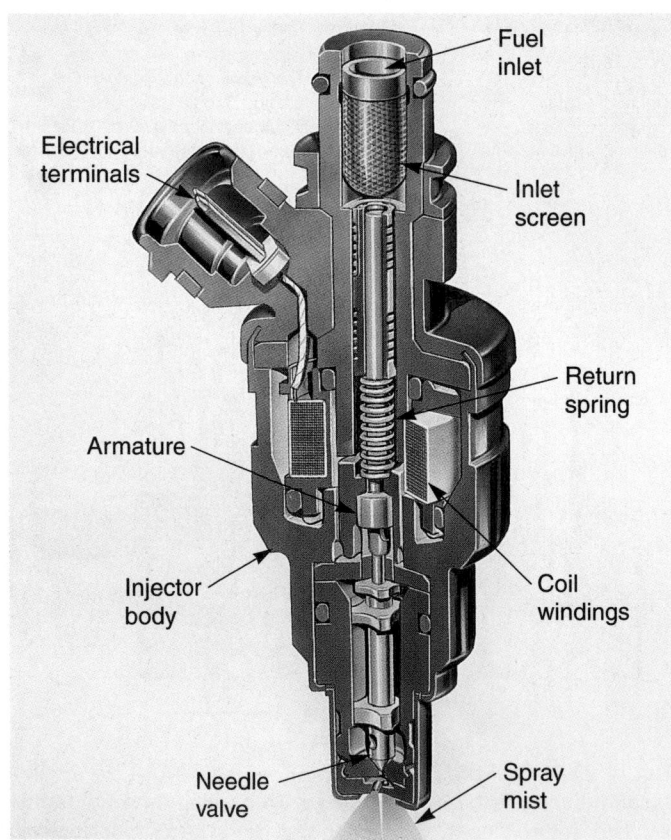

Fuel inlet

Electrical terminals

Inlet screen

Return spring

Armature

Injector body

Coil windings

Needle valve

Spray mist

Figure 22-11. This cutaway view shows the major parts of an electronic fuel injector. (Lexus)

device. After leaving the throttle body, air flows into the engine intake manifold. The manifold is divided into runners (passages) that route the air into each cylinder head intake port.

Sensor System

The *EFI sensor system* monitors engine operating conditions and reports this information to the control module. See **Figure 22-14.** An *engine sensor* is an electrical device that changes circuit resistance or voltage with a change in a condition, such as temperature, pressure, or position. For example, a temperature sensor's resistance may decrease as temperature increases. The control module analyzes the increased current flow through the sensor to determine if a change in injector valve opening is needed.

Computer Control System

The *computer control system* uses electrical data from the sensors to control the operation of the fuel injectors. A wiring harness connects the sensors to the input of the computer (control module). Another wiring harness connects the output of the computer to the fuel injectors. Refer to **Figure 22-15.**

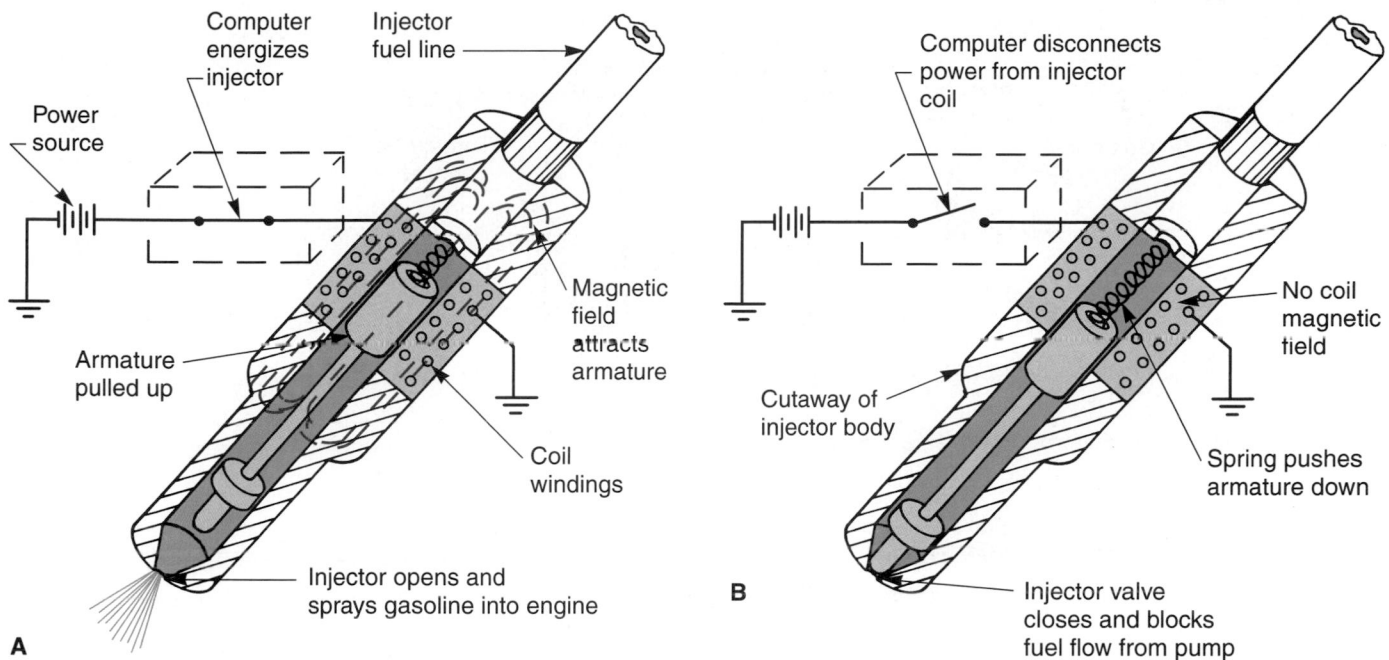

Figure 22-12. EFI injector operation. A—Current through the injector coil builds a magnetic field. The field attracts and pulls up on the armature to open the injector. Fuel then sprays out of the injector. B—When the control module breaks the circuit, the spring pushes the injector valve closed to stop the fuel spray.

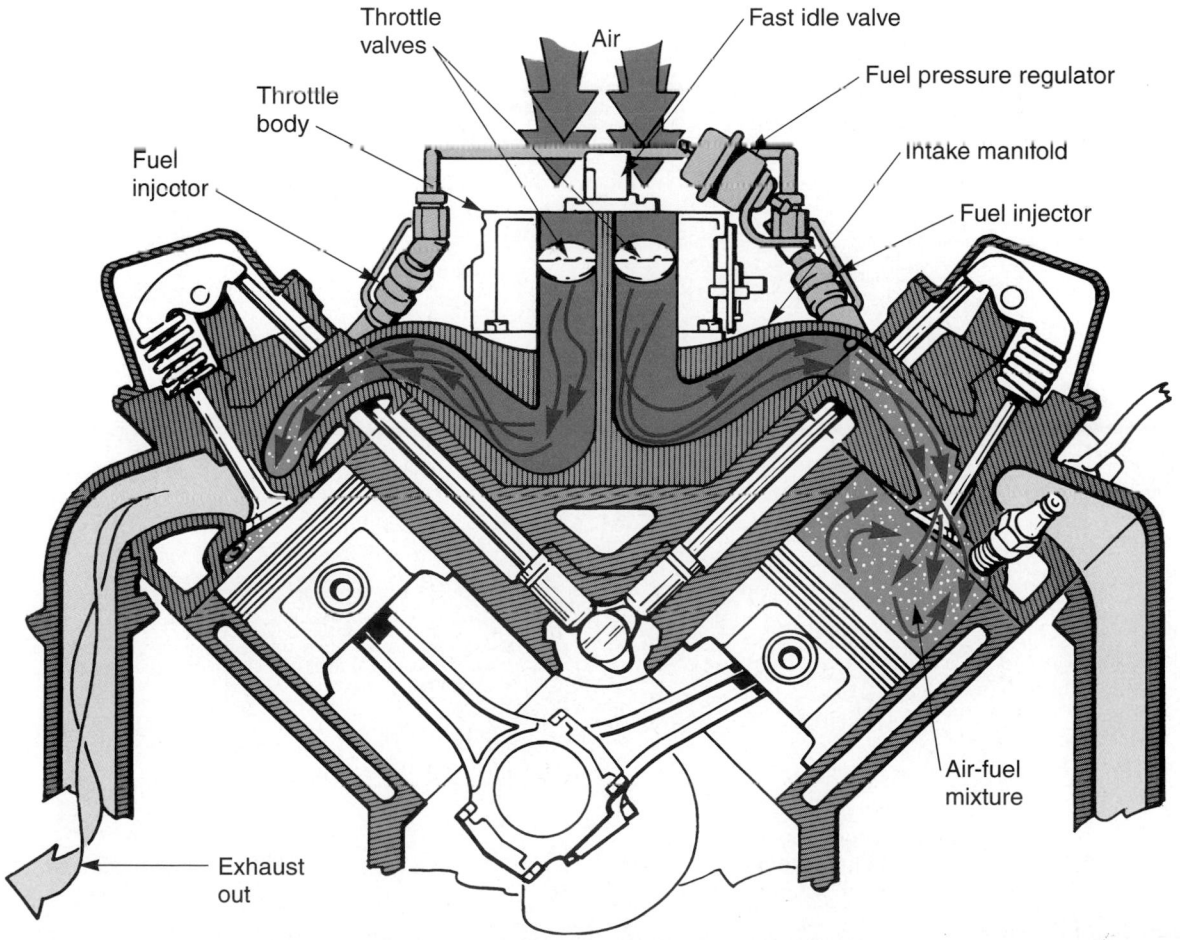

Figure 22-13. An air induction system consists mainly of a throttle body. The throttle body contains throttle plates, which control the airflow into the engine. Also, note the location of the injectors in this V-type engine. (Cadillac)

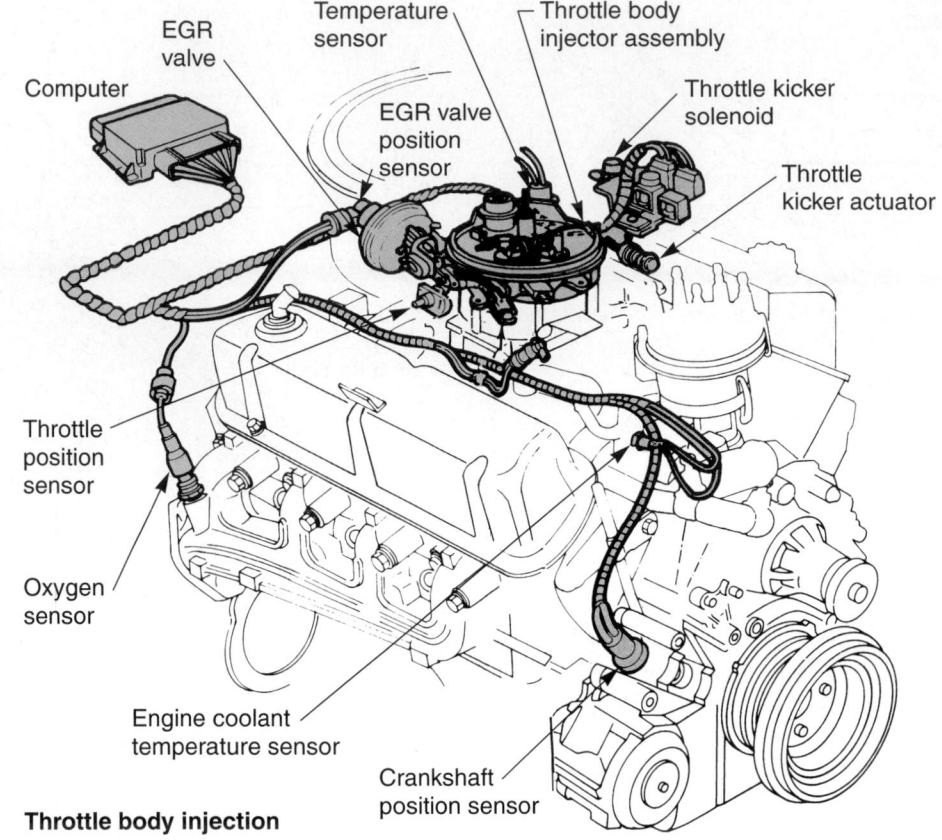

Throttle body injection

A

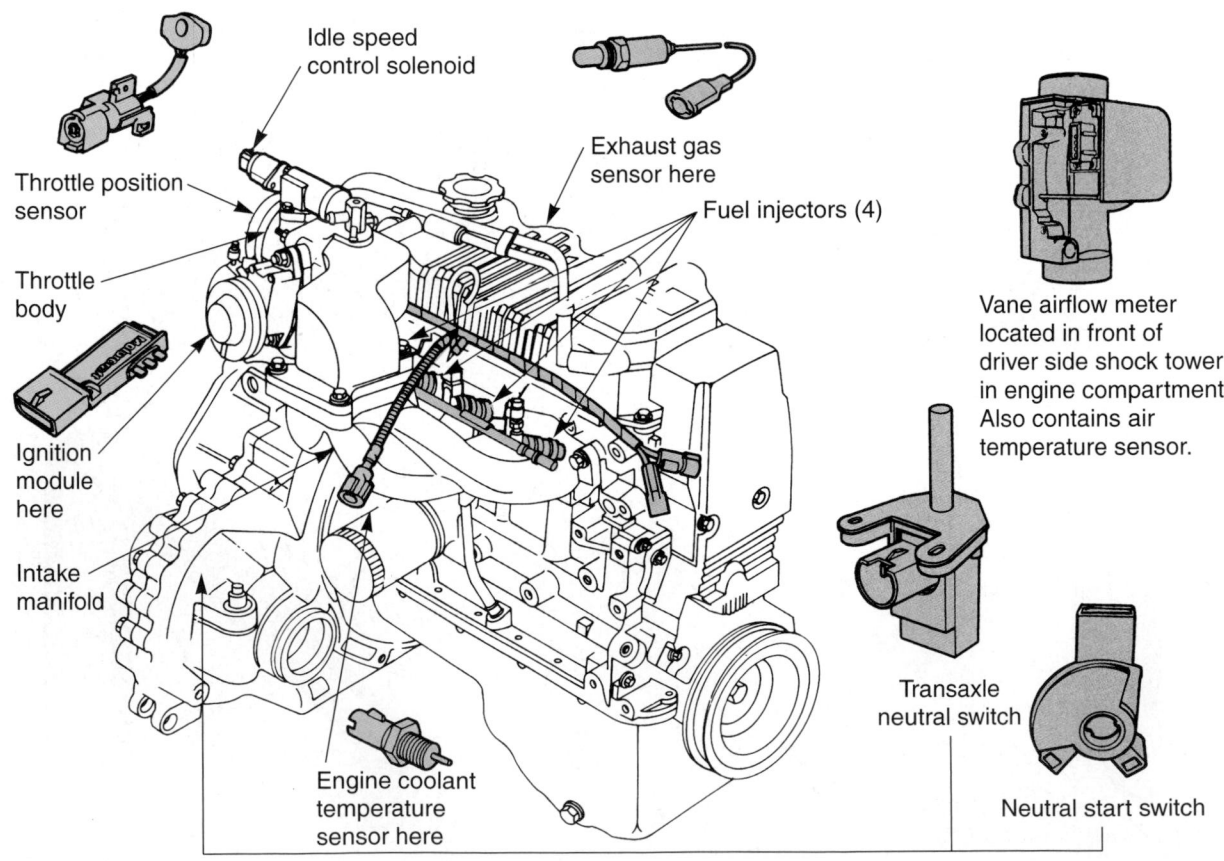

Multiport injection

B

Figure 22-14. Study the sensors for throttle body and multiport injection systems. (Ford)

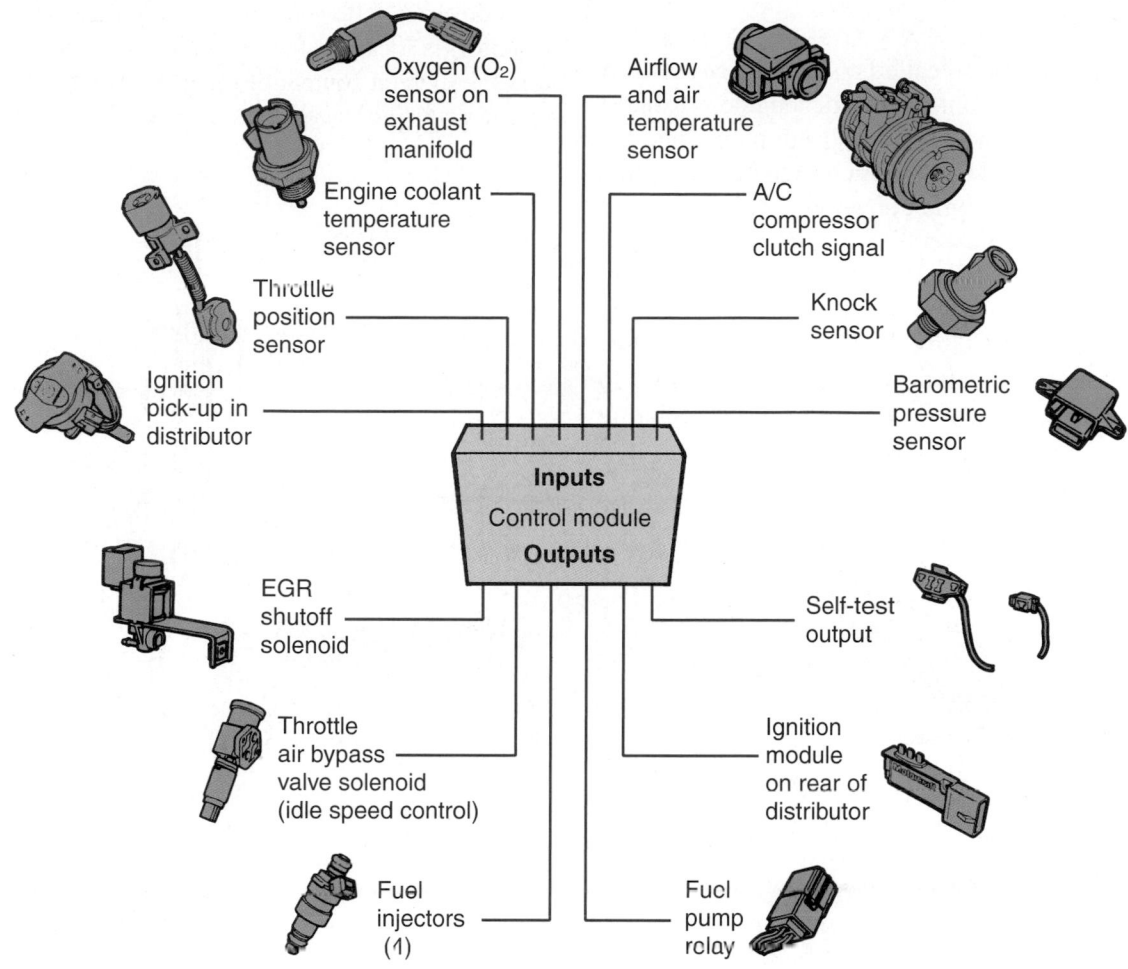

Figure 22-15. Sensors feed information to the control module. The control module uses this data to operate other system components. (Ford)

The *engine control module* is the "brain" of an electronic fuel injection system. Refer to **Figure 22-16.** It is a preprogrammed microcomputer. The control module uses sensor inputs to calculate when and how long to open the fuel injectors. To open an injector, the computer connects the injector coil to battery voltage. To close the injector, the computer opens the circuit between the battery and the injector coil.

Engine Sensors

Typical sensors for an EFI system include the:

- Exhaust gas or oxygen sensor.
- Manifold absolute pressure (MAP) sensor.
- Throttle position sensor.
- Engine temperature sensor.
- Airflow sensor.
- Inlet air temperature sensor.
- Crankshaft position sensor.

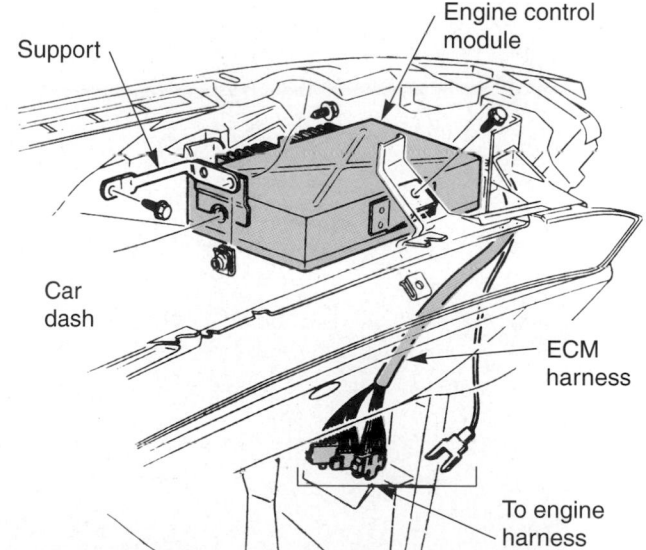

Figure 22-16. The control module is commonly mounted behind the instrument panel. This keeps it away from damaging engine heat and vibration. Some control modules are mounted on the air cleaner or elsewhere in the engine compartment. (Cadillac)

Oxygen Sensor

An *oxygen sensor,* also called an *exhaust gas sensor,* measures the oxygen content in the engine's exhaust system as a means of checking combustion efficiency. It often fits into the exhaust manifold and the exhaust pipe at a point before the catalytic converter. Look at **Figures 22-17** and **22-18.**

Vehicles with OBD II use at least two oxygen sensors—one before the catalytic converter and one after the catalytic converter. See **Figure 22-19.** The primary sensor is located before the catalytic converter and is used to monitor the oxygen content of the exhaust gases entering the converter. The secondary oxygen sensor, or catalyst monitor, is located after the catalytic converter. It monitors the oxygen content of the gases leaving the converter to determine how well the catalyst elements are working.

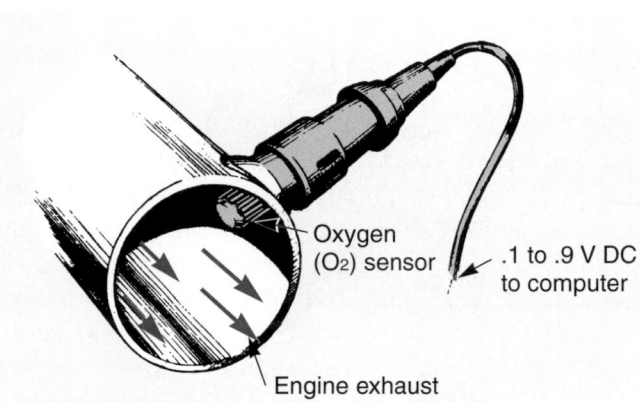

Figure 22-17. An exhaust gas sensor is a very important sensor commonly used in EFI systems. It allows the system to test the air-fuel mixture setting by measuring oxygen in the exhaust.

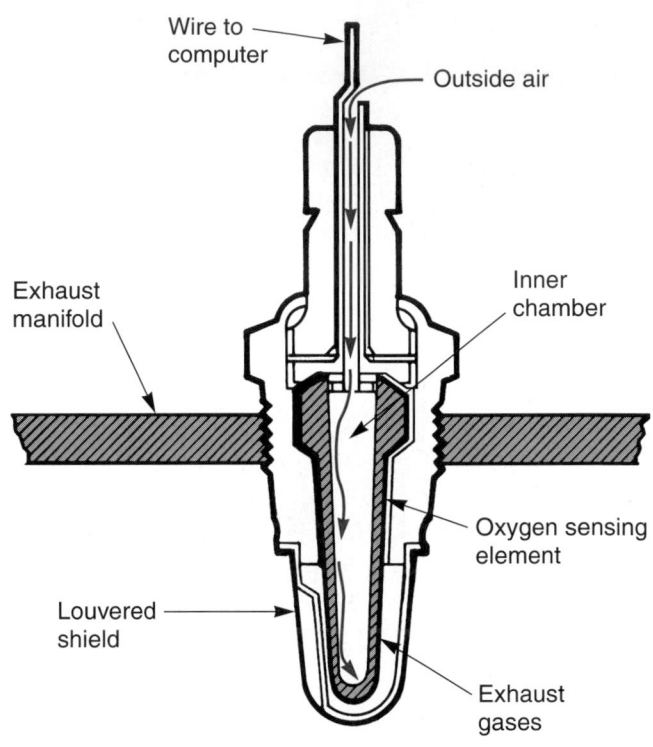

Figure 22-18. An exhaust gas sensor compares the amount of oxygen in the exhaust with the amount of oxygen in the outside air. (GMC)

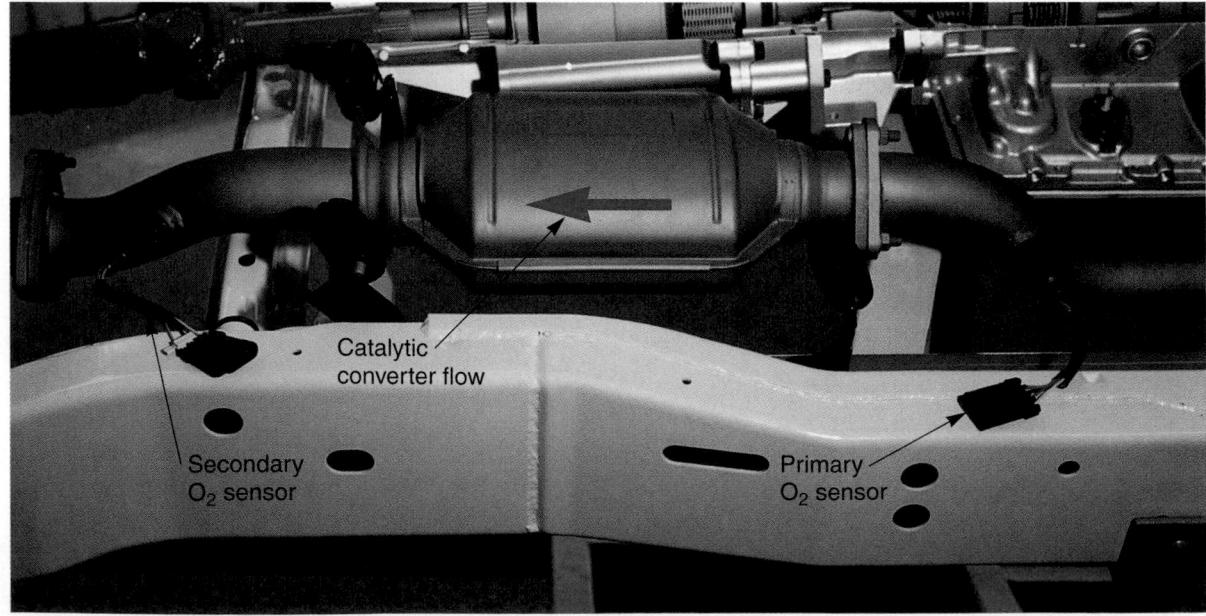

Figure 22-19. With OBD II systems, two oxygen sensors are used to more closely monitor exhaust emissions. The first oxygen sensor is called the primary oxygen sensor. The second oxygen sensor is called the secondary oxygen sensor.

The voltage output (or resistance) of the oxygen sensor varies with changes in the oxygen content of the exhaust. For example, an increase in oxygen from a lean mixture makes the sensor output voltage decrease. A decrease in oxygen from a rich mixture causes the sensor output to increase. In this way, the sensor supplies data on oxygen content to the computer. The computer can then alter the opening and closing of the injectors to adjust the air-fuel ratio for maximum efficiency.

Open Loop and Closed Loop

When in *open loop,* the electronic injection system does *not* use engine exhaust gas content as a main indicator of the air-fuel mixture. Instead, the system operates on information stored in the computer. See **Figure 22-20A.** For example, right after cold engine starting, the computer operates in open loop. Several sensors, especially the oxygen sensor, *cannot* provide accurate information before the engine reaches normal operating temperature. The computer is set to ignore these sensor inputs when the engine is cold.

Closed loop means that the computer is using information from the oxygen sensor and the other sensors. This information forms an imaginary loop, or circle, from the computer, through the fuel system, into the exhaust system, and back to the computer, **Figure 22-20B.**

Under most operating conditions, an electronic gasoline injection system functions in closed loop. This lets the computer double-check the fuel mixture it is providing to the engine.

Manifold Absolute Pressure Sensor

A *manifold absolute pressure (MAP) sensor* measures the pressure, or vacuum, inside the engine intake manifold. Manifold pressure is an excellent indicator of engine load. High pressure (low intake vacuum) indicates a high load, requiring a rich mixture. Low manifold pressure (high intake vacuum) indicates very little load, requiring a leaner mixture. The manifold pressure sensor varies resistance with changes in engine load, **Figure 22-21.** This data is used by the computer to alter the fuel mixture.

Throttle Position Sensor

A *throttle position sensor* is a variable resistor connected to the throttle plate shaft. Look at **Figure 22-22.** When the throttle opens or closes, the sensor changes resistance and signals the computer. The computer can then richen or lean the mixture as needed.

Engine Temperature Sensor

An *engine temperature sensor* monitors the operating temperature of the engine. It is mounted so that it is exposed to the engine coolant. When the engine is cold, the sensor might provide a low resistance (high current flow). The computer would then richen the air-fuel mixture for cold engine operation. When the engine warms, the sensor supplies information, such as high resistance, so the computer can make the mixture leaner.

Airflow Sensor

An *airflow sensor* is used in many EFI systems to measure the amount of outside air entering the engine. This helps the computer determine how much fuel is needed. Refer to **Figure 22-23.** The airflow sensor usually contains an air flap or door that operates a variable resistor, **Figure 22-24.** Increased airflow opens the flap more, changing the position of the variable resistor.

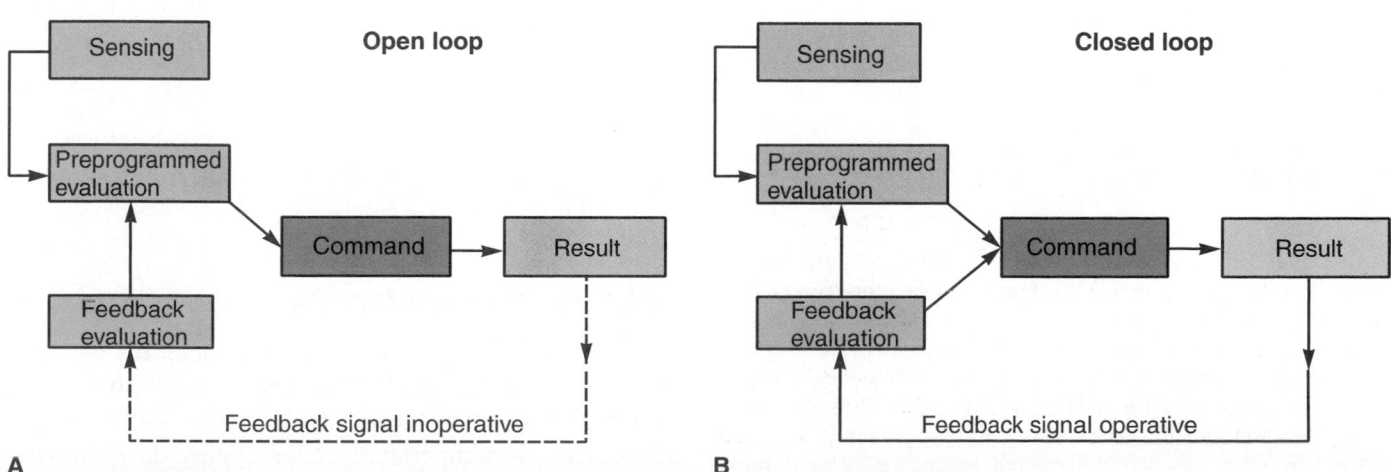

Figure 22-20. The basic flow of information in an EFI system. A—In open loop. B—In closed loop. (Chrysler)

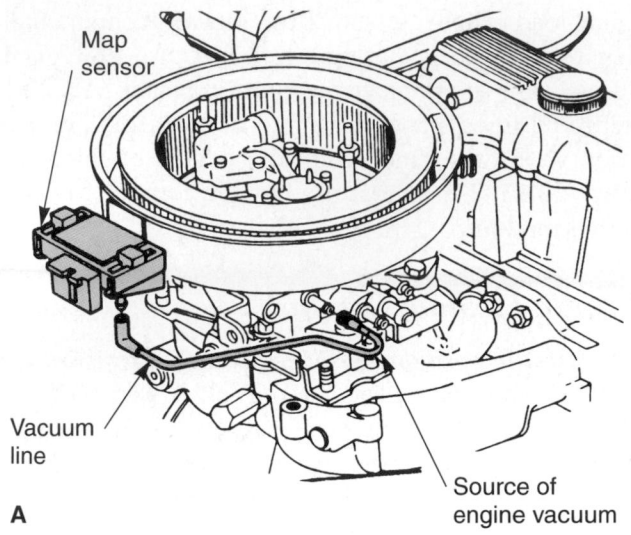

A

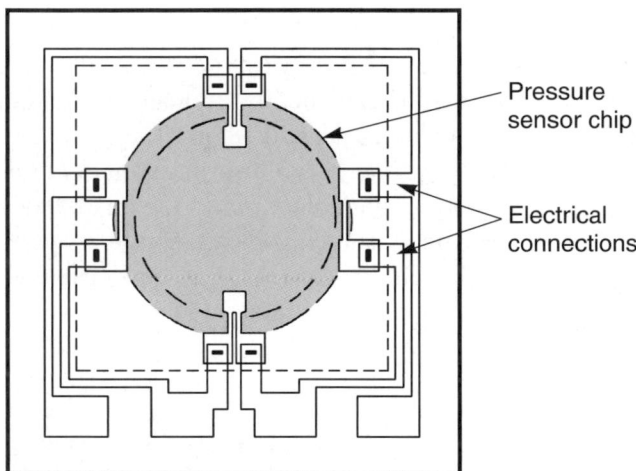

B

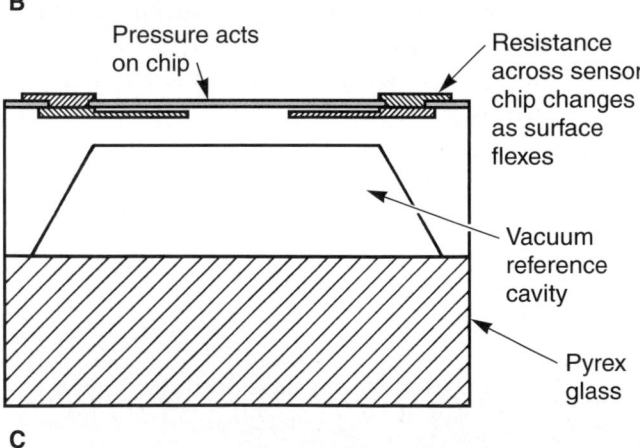

C

Figure 22-21. The engine manifold absolute pressure (MAP) sensor changes resistance with changes in vacuum (pressure). A—A MAP sensor often mounts on top of the engine or in the engine compartment. This one attaches to air cleaner housing. B—This MAP sensor uses a piezo-resistor chip that converts a change in pressure into a resistance change. C—This side view shows the vacuum reference cavity. Engine manifold vacuum flexes the chip, altering its internal resistance. The control module uses the resulting current flow change to determine the load on the engine. (General Motors)

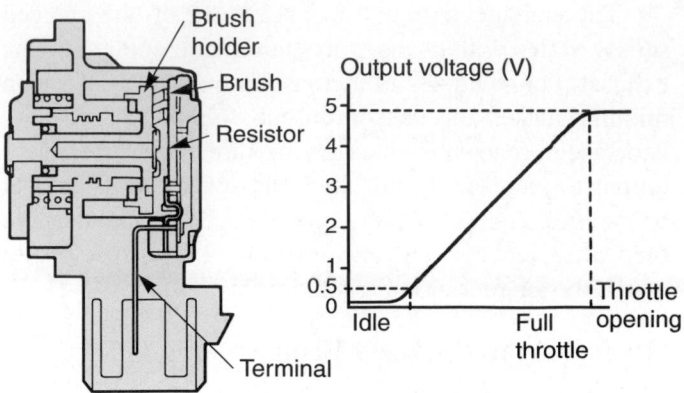

Figure 22-22. The throttle position sensor uses a variable resistor to report the amount of throttle opening to the control module. Throttle shaft rotation causes the arm contact to slide on the resistor. In this way, different current levels are produced for different throttle positions. The control module can then alter the fuel mixture for idle and wide open throttle positions. (Honda)

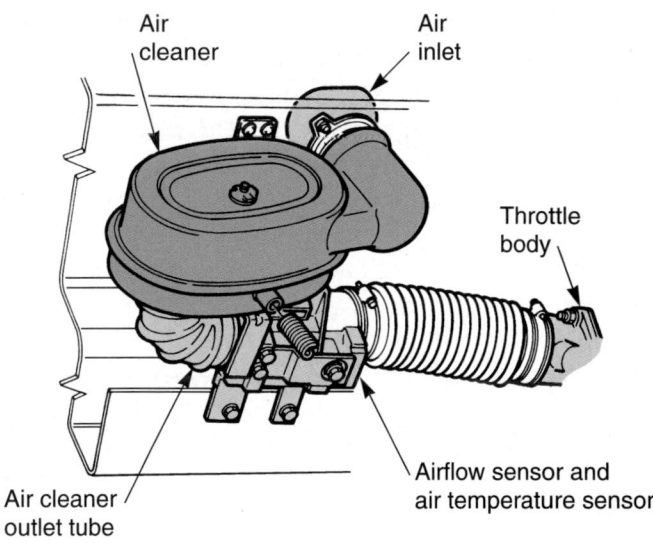

Figure 22-23. The airflow sensor and the inlet air temperature sensor are housed under the air cleaner on this engine. (Ford)

Information is then sent to the computer indicating air inlet volume.

Inlet Air Temperature Sensor

An *inlet air temperature sensor* measures the temperature of the air entering the engine. Cold air is more dense than warm air, requiring more fuel for the proper ratio. The air temperature sensor helps the computer compensate for changes in outside air temperature and maintain an almost perfect air-fuel mixture ratio.

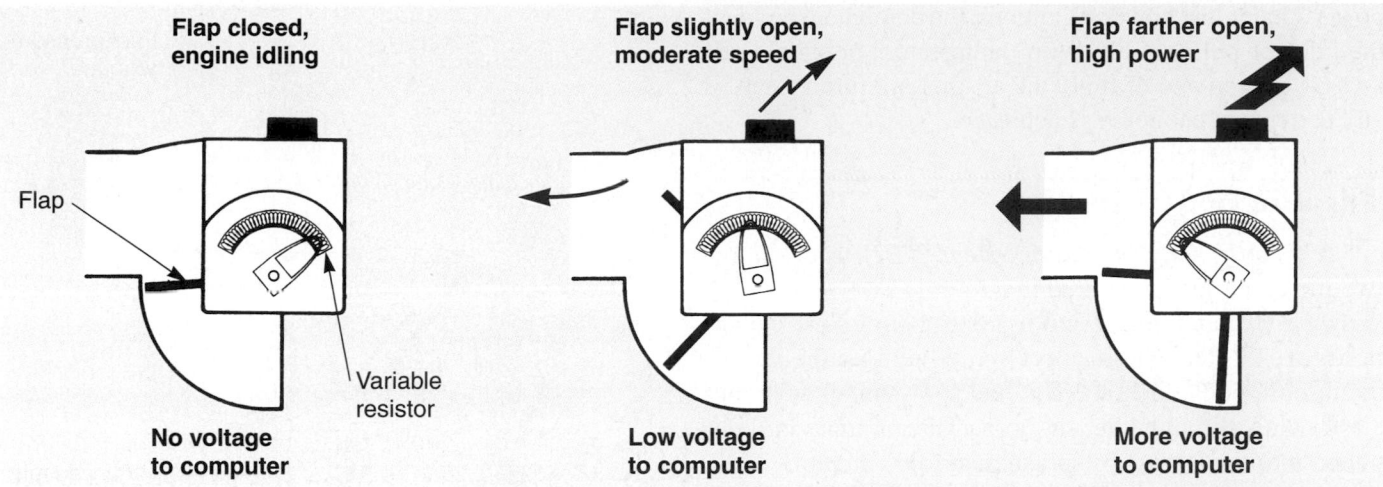

Figure 22-24. The airflow sensor operates a variable resistor. Low airflow at idle does not open the sensor flap, and resistance stays high. As airflow increases, the flap swings open, decreasing sensor resistance and increasing current flow to the control module. (VW)

Crankshaft Position Sensor

A *crankshaft position sensor* is used to detect engine speed. Refer to **Figure 22-14A.** It allows the computer to change injector timing and duration with changes in engine rpm. Higher engine speeds generally require more fuel.

Other Sensors

Other sensors can also be used to affect the operation of a fuel injection system. These include an A/C compressor sensor, transmission sensors, EGR sensor, vehicle speed sensor, and engine knock sensor. They provide additional data about operating conditions affecting engine fuel needs.

Analog and Digital Signals

The signal from the engine sensors can be either digital or analog. *Digital signals* are on-off signals. An example of a sensor providing a digital signal is the crankshaft position sensor, which monitors engine rpm. Voltage output or resistance goes from maximum to minimum, like a switch. An *analog signal* changes in strength to let the computer know about a change in a condition. For example, the internal resistance of a sensor may smoothly increase or decrease with changes in temperature, pressure, or part position. The sensor acts as a variable resistor.

Injector Pulse Width

The *injector pulse width* indicates the amount of time each injector is energized and kept open. The computer controls the injector pulse width. **Figure 22-25** shows

how pulse width controls injector output. Study the drawing carefully!

As an example, under full acceleration, the computer will sense a wide-open throttle, high intake manifold pressure, and high inlet airflow. The computer then lengthens the injector pulse width to richen the mixture for more

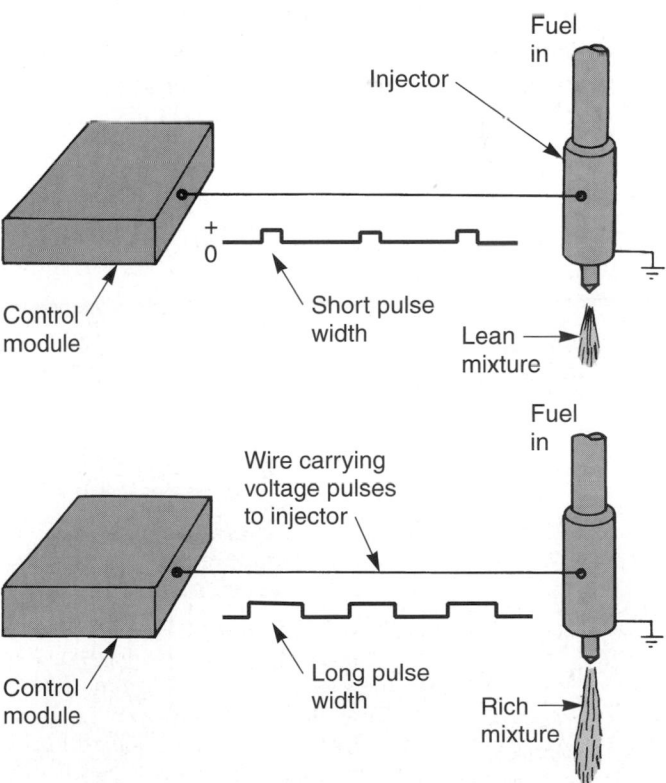

Figure 22-25. Pulse width is used to control the amount of fuel injected into the engine. A longer pulse width richens the mixture. A shorter pulse width leans the mixture.

power. Under low power conditions, the computer shortens the injector pulse width. With the injectors being closed a larger percentage of time, the air-fuel mixture is leaner and better fuel economy is achieved.

Throttle Body Injection

A *throttle body injection system (TBI)* uses one or two injector valves mounted in a throttle body assembly. A diagram of an injector and its control circuits is shown in **Figure 22-26.** The injectors spray fuel into the top of the throttle body air horn. The fuel spray mixes with the air flowing through the air horn. The mixture is then pulled into the engine by intake manifold vacuum.

TBI Assembly

The *TBI assembly*, **Figure 22-27,** typically consists of:

- *Throttle body housing*—this is the metal casting that holds the injectors, fuel pressure regulator, throttle valves, and other parts.
- *Fuel injector*—a solenoid-operated fuel valve mounted in the upper section of the throttle body assembly.
- *Fuel pressure regulator*—a spring-loaded bypass valve that maintains constant pressure at the injectors.
- *Throttle positioner*—a motor assembly that opens or closes the throttle plates to control engine idle speed.

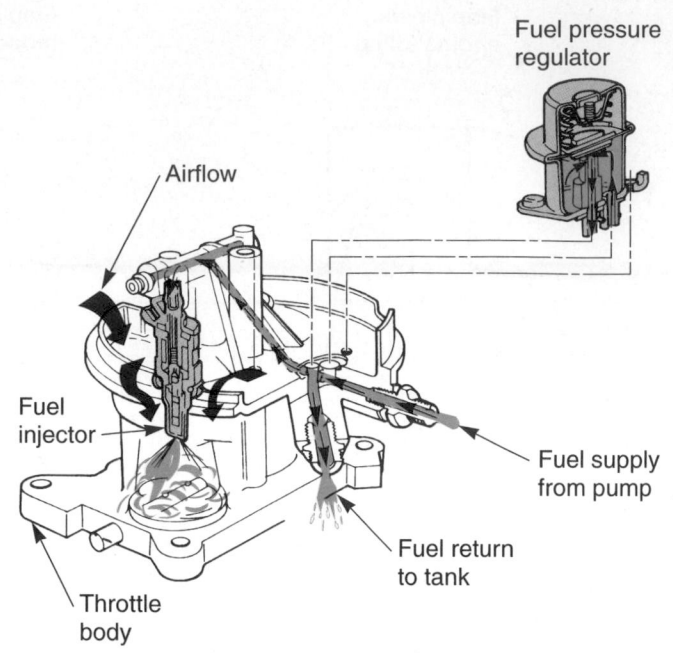

Figure 22-27. Fuel enters the throttle body from the pump. It then enters the pressure regulator before passing into the injector. Fuel sprays out of the injector and mixes with air entering the air horn. (Ford)

- *Throttle position sensor*—a variable resistor that senses opening or closing of the throttle plates.
- *Throttle plates*—butterfly valves that control air-flow through the throttle body.

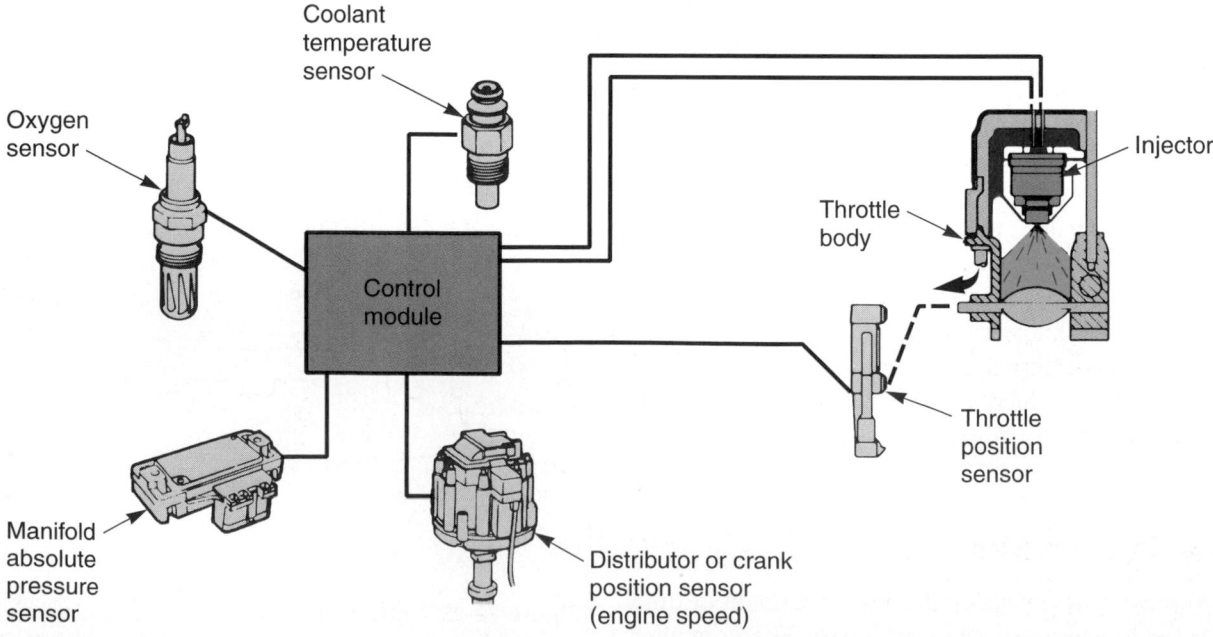

Figure 22-26. Throttle body injection has sensors and a control module–operated injector mounted inside a throttle body. (Buick)

TBI Throttle Body

The TBI *throttle body* bolts to the pad on the intake manifold. Throttle plates are mounted in the lower section of the throttle body. A linkage mechanism or cable connects the throttle plates with the accelerator. An inlet fuel line connects to one fitting on the throttle body. An outlet return line to the tank connects to another fitting on the throttle body.

TBI injector

A *throttle body injector* consists of the electric solenoid coil, armature or plunger, ball or needle valve, ball or needle seat, and injector spring. These parts are pictured in **Figure 22-28** and on the right in **Figure 22-29.** Wires from the control module connect to the terminals on the injectors. When the computer energizes the injectors, a magnetic field is produced in the injector coil. The magnetic field pulls the plunger and valve up to open the injector. Fuel can then squirt through the injector nozzle and into the engine.

TBI Pressure Regulator

The *throttle body pressure regulator* consists of a fuel valve, diaphragm, and spring. When fuel pressure is low,

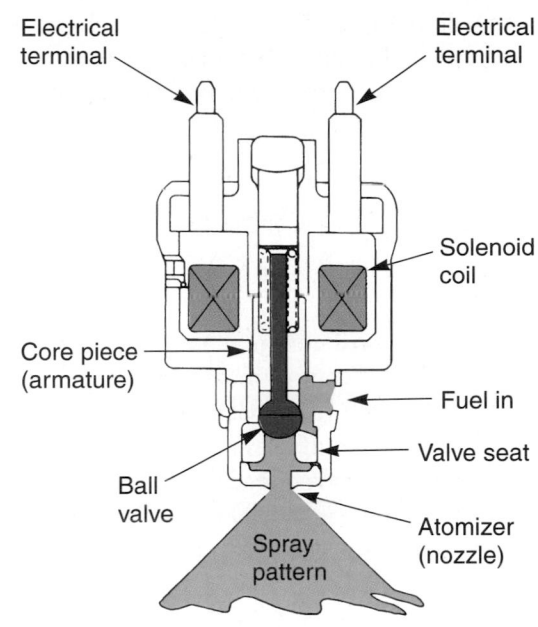

Figure 22-28. This TBI injector uses a ball-type valve instead of a pointed needle valve. Note the part names.

such as when starting the engine, the spring holds the fuel valve closed. This causes pressure to build as fuel flows into the regulator from the electric fuel pump. Refer to **Figure 22-29.** When a preset pressure is reached, pressure

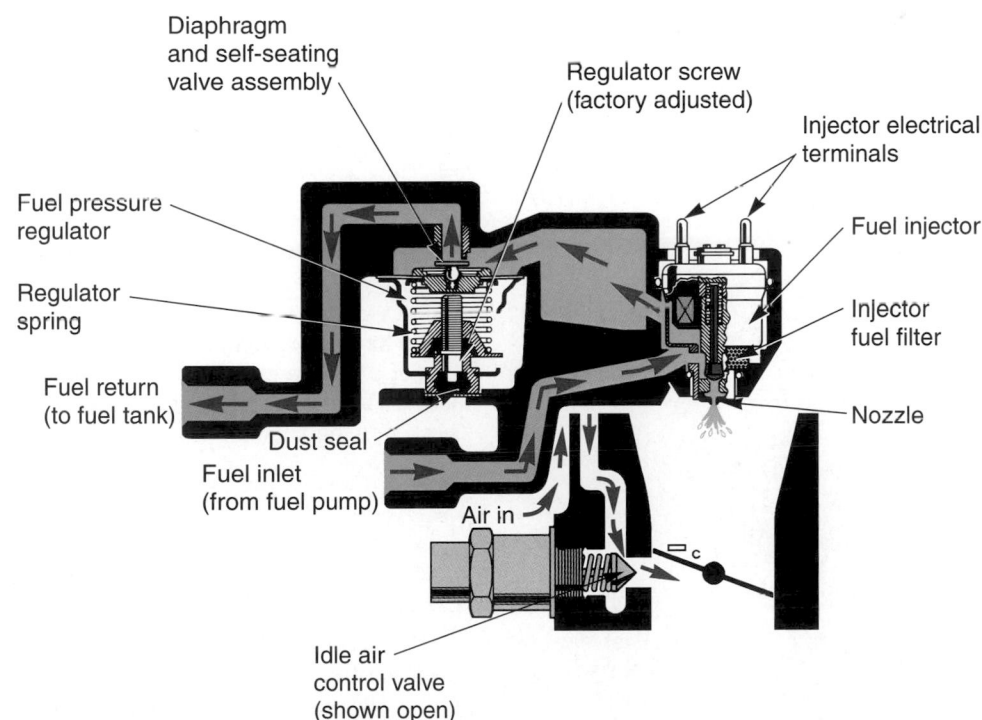

Figure 22-29. This cutaway illustration shows the basic action inside a typical throttle body assembly. The regulator limits the maximum pressure inside the injector to a preset level. Then, the pulse width accurately controls the air-fuel ratio. The idle air control valve is used to increase or decrease idle speed. (Pontiac)

acts on the diaphragm. The diaphragm compresses the spring and opens the fuel valve. Fuel can then flow back to the fuel tank. This limits the maximum fuel pressure at the injectors.

Engine Idle Speed Control

There are many different ways to provide a smooth engine idle under all operating conditions with fuel injection systems. The most common are the fast idle thermo valve and the idle air control motor or solenoid.

A *fast idle thermo valve* opens to increase idle speed when the engine is cold. It uses a thermowax plunger that expands when heated and contracts when cooled. Shown in **Figure 22-30,** the thermo valve opens when cold to allow air to bypass the throttle valve. This extra air increases engine idle speed to prevent cold engine stalling. As both the engine and thermo valve warm, the valve closes the bypass air off to return to a normal engine idle speed.

An *idle air control motor* may also be used to help control engine idle speed. It is a solenoid- or servo-motor-operated air bypass valve. It works something like the thermo valve, but it is computer controlled. See **Figure 22-31.** The system control module opens the idle air control valve when temperature sensors signal a cold engine. It then progressively closes the air bypass air as the engine warms. It is used on both multiport and throttle body systems, **Figure 22-32.** When open, the idle air control valve allows more air to enter the intake manifold. This tends to increase idle speed. When closed, the valve decreases bypass air and idle speed goes down. The

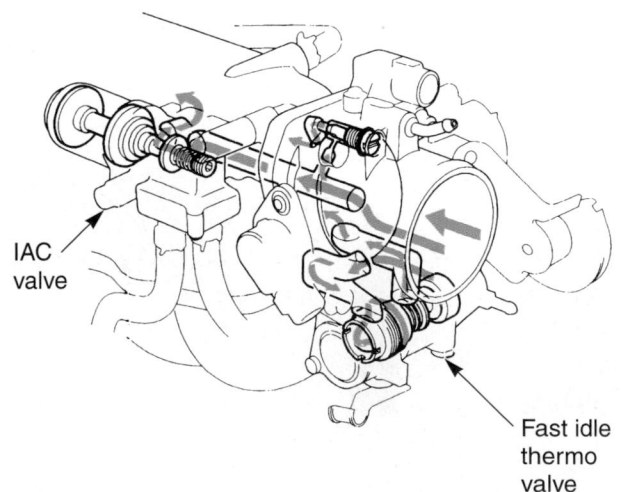

Figure 22-30. Note how the idle air control valve and the thermo valve can affect air that bypasses the throttle valve to alter engine idle speed. (Honda)

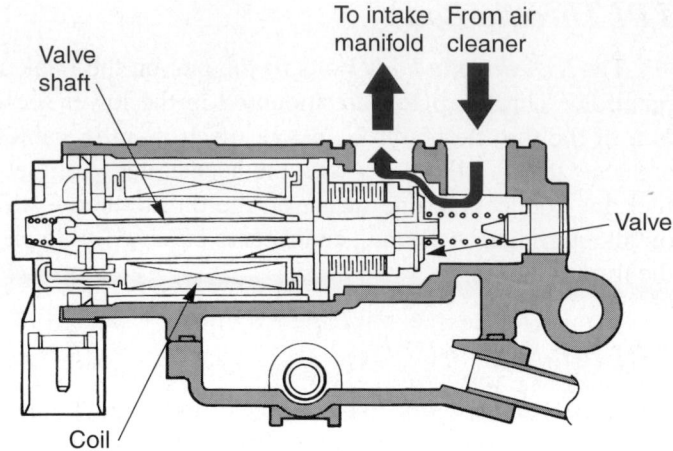

Figure 22-31. This cutaway shows the basic parts inside the idle air control motor. When the coil is energized, it attracts and pulls on the shaft to open the air valve. Spring tension pushes the valve closed when the coil is de-energized. (Honda)

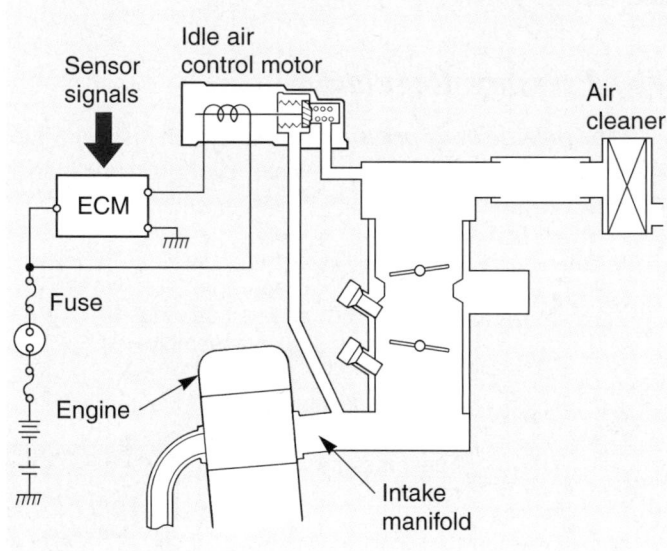

Figure 22-32. This diagram shows how the control module can open or close the idle air motor to affect engine idle speed. As more air bypasses the throttle valve, engine speed increases. This keeps the engine from stalling when cold. (Honda)

valve can be used to control both slow and fast idle speeds. It is comparable to a carburetor's fast idle cam.

In addition to the idle air control valve, a throttle positioner is often used on throttle body assemblies to control engine idle speed. The computer actuates the positioner to open or close the throttle plates. In this way, the computer can maintain a precise idle speed with changes in engine temperature, load, and other conditions.

Continuous Throttle Body Injection

A *continuous throttle body injection (CTBI)* system sprays a solid stream of fuel into the air horn. Unlike the more common modulated system just discussed, it does *not* pulse the injectors on and off to control the air-fuel mixture. To increase or decrease fuel flow, the CTBI system alters the pressure applied to the nozzles in the throttle body. The system measures fuel flow, airflow, and other engine conditions. The computer then increases or decreases the speed of the *control pump,* or fuel pump, to meet engine needs. The air cleaner and throttle body assembly of this type injection system contain the computer, airflow sensor, fuel control motor, *spray bar* or injector nozzles, and other components. This is not a commonly used system.

Electronic Multiport Injection

Electronic multiport injection systems use a computer, engine sensors, and one solenoid injector for each engine cylinder. This is a very common system on late model cars. Look at **Figure 22-33**. The operation of an electronic multiport type system is similar to a modulated throttle body injection system covered earlier. However, fuel is injected at each intake port instead of at the top center of the intake manifold.

A *multiport throttle body* assembly contains the throttle plates, throttle position sensor, but does *not* contain the injector valves. See **Figure 22-34**. Its main function is to control airflow into the engine.

A *multiport pressure regulator* is mounted in the fuel line before or after the injectors, **Figure 22-35**. It performs the same function as the pressure regulator covered earlier. It maintains a constant pressure at the inlet to the injector valves by acting as a bypass branch.

A *fuel rail* feeds fuel to several of the injectors. It is a tubing assembly that connects the main fuel line to the inlet of each injector, **Figure 22-36**.

EFI Multiport Injector

An EFI *multiport injector valve* is usually press fitted into the runner (port) in the intake manifold. Each injector is aimed to spray toward an engine intake valve. It is constructed something like an intermittent throttle body injector. **Figure 22-37** shows an EFI multiport injector. Study this illustration carefully. An EFI multiport injector typically consists of:

- *Electric terminals*—these are electrical connections for completing a circuit between the injector coil and control module.

- *Injector solenoid*—an armature and coil that opens and closes the needle valve.
- *Injector screen*—a screen filter for trapping debris before it can enter the injector nozzle.
- *Needle valve*—located on the end of the armature and seals on needle seat.
- *Needle seat*—a round hole in the end of the injector that seals against the needle valve tip.
- *Injector spring*—a small spring that returns the needle valve to the closed position.
- *O-ring seal*—a rubber seal that fits around the outside of the injector body and seals in the intake manifold.
- *Injector nozzle*—the outlet of the injector that produces the fuel spray pattern.

There are several variations of electronic multiport injection. It is important that you understand the primary differences between each type system.

Air-Fuel Emulsion Injector

An *air-fuel emulsion injector* mixes air with the fuel creating a slurry. The slurry is then injected into the intake manifold. This helps atomize the fuel and improve combustion. One is illustrated in **Figure 22-38**. Low pressure air from the air injection pump can be used to provide this air to the injector body.

Unitized Multiport Injection

Unitized multiport injection is a system that mounts all of the fuel injectors into a single assembly, **Figure 22-39**. For example, on a six-cylinder engine, all six injectors are housed in one enclosure on top of the engine plenum or intake manifold. One large multi-pin terminal connects each injector to its own control module or to the engine control module. The unitized injector unit is often mounted in rubber so it is isolated from engine vibration. Fuel lines connect to the unit to feed fuel pressure to each injector valve, **Figure 22-40**.

Plastic *fuel injection transfer lines* connect each injector to its poppet valve. One end of each line connects to the bottom of the fuel injector assembly. The other end connects to a poppet valve, **Figure 22-40**. A *fuel injection poppet valve* is simply a spring loaded valve that prevents fuel leakage between injector pulses. They are normally closed. When one of the fuel injectors opens, fuel pressure overcomes poppet valve spring pressure and fuel sprays out the poppet valve. The mist of fuel enters the intake port and is pulled into the combustion chamber.

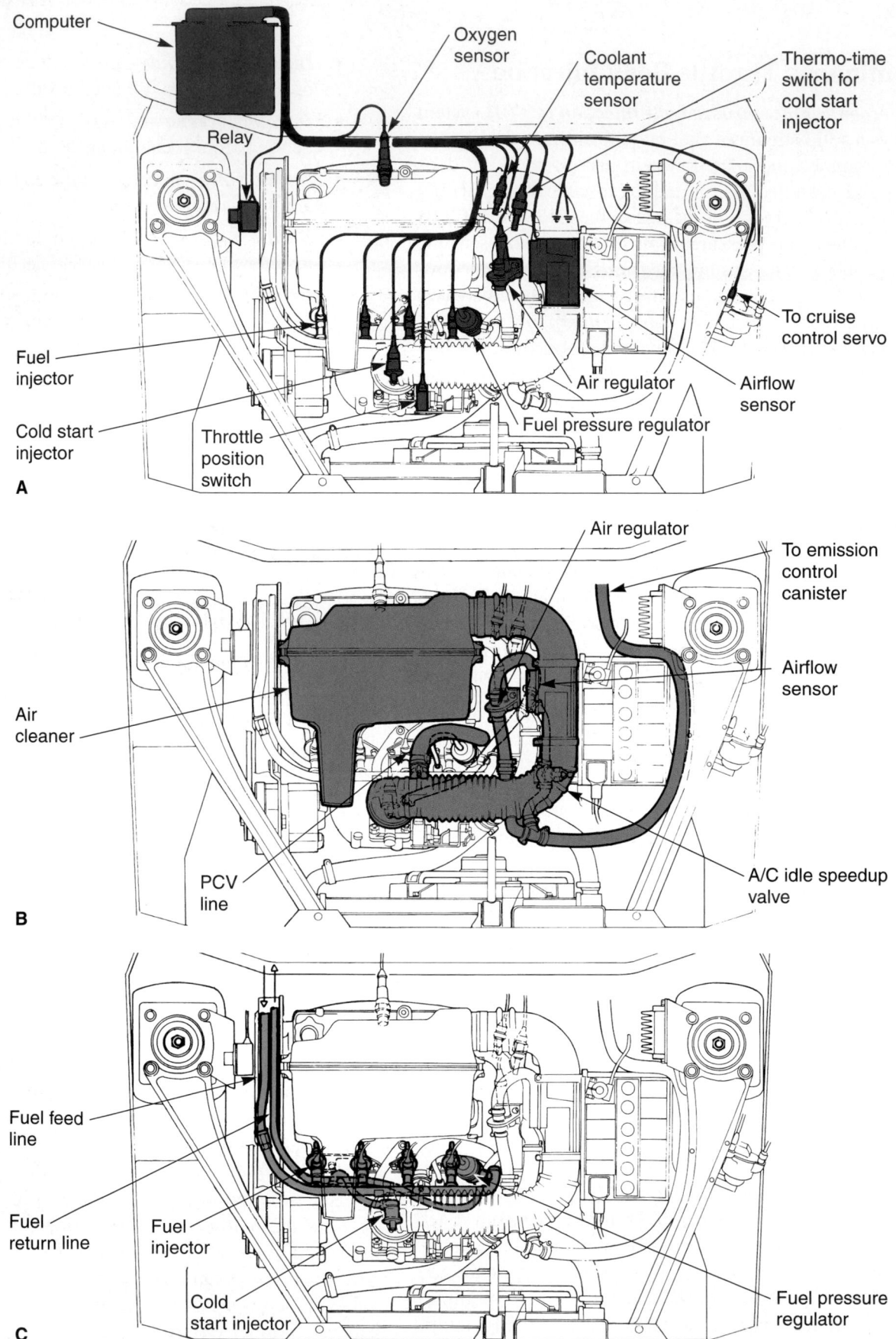

Figure 22-33. Note the systems of a multiport electronic fuel injection system. A—The sensor and control systems of EFI multiport system. Sensors feed data to the control module. The control module can then operate the injectors and other components for maximum efficiency. B—Air delivery system parts. Note the airflow sensor, which monitors air volume entering the engine. C—The fuel delivery system basically includes the injectors, the pressure regulator, the lines, and the hoses. Some systems use a cold start injector. (Lancia)

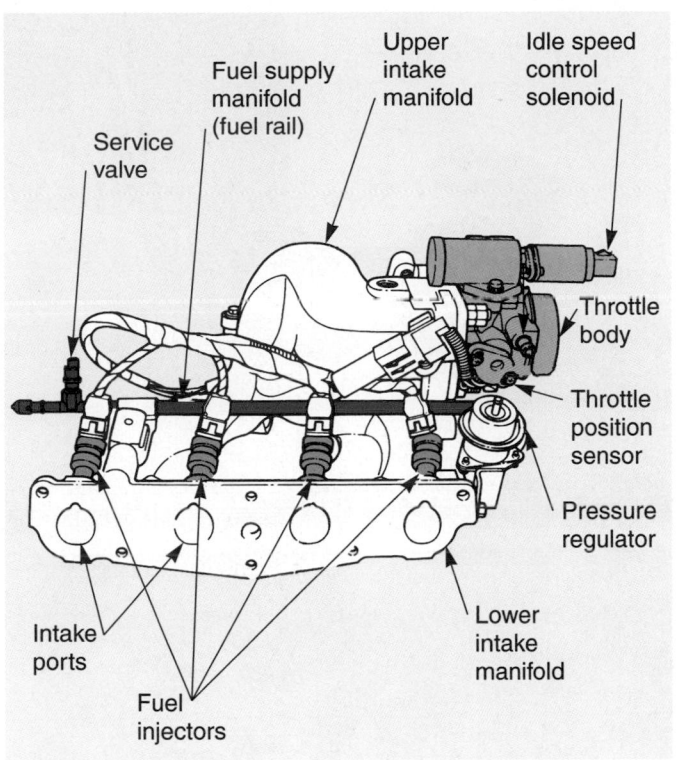

Figure 22-34. A throttle body assembly is part of this multiport injection system. Note how the injectors install in the intake runners. (Ford)

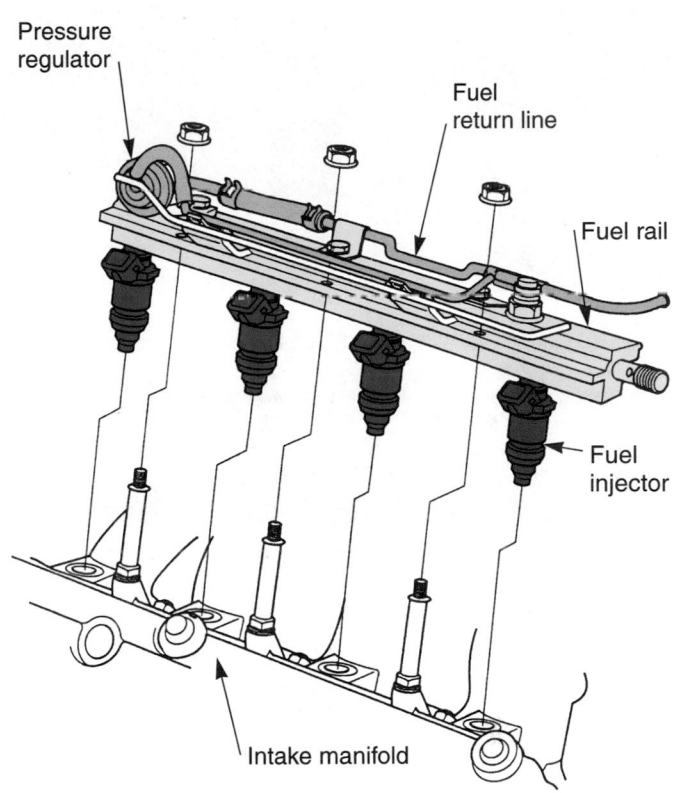

Figure 22-36. The fuel rail feeds fuel pressure to each injector. Note the fuel pressure regulator mounted on this rail. (Honda)

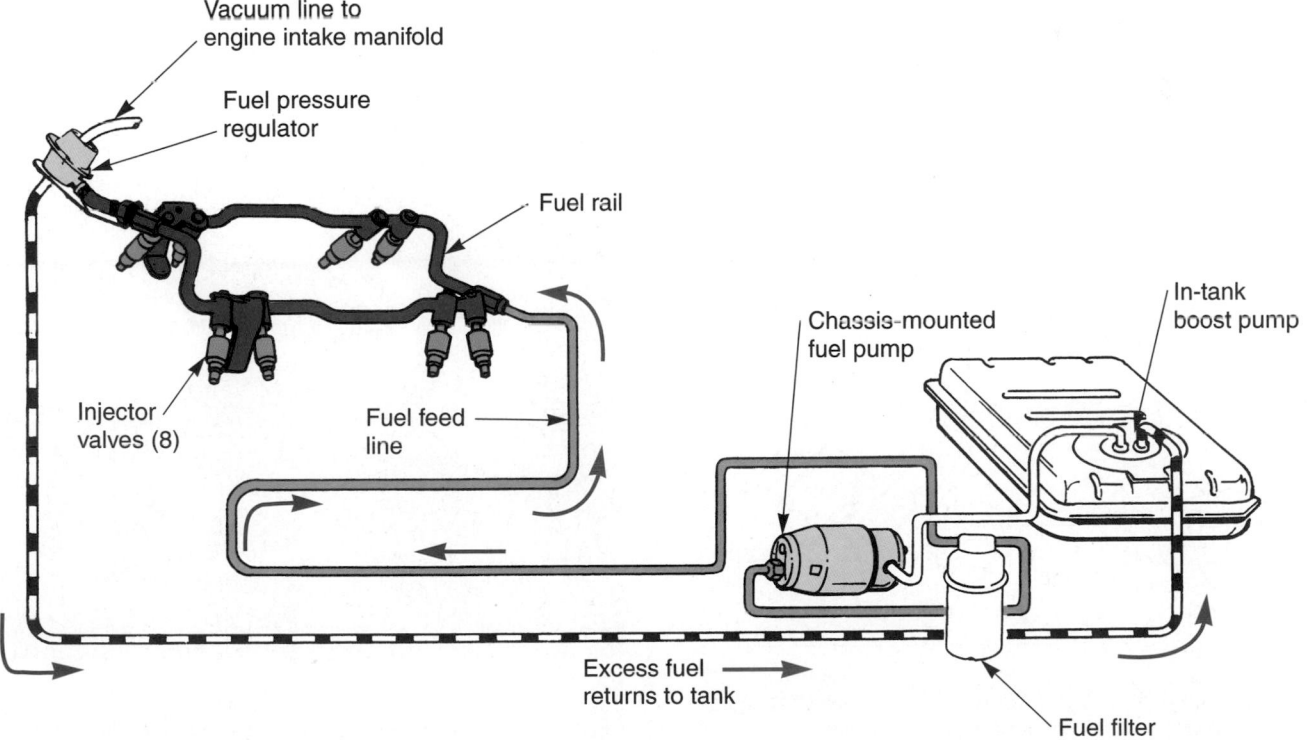

Figure 22-35. Pressure regulator action. The fuel pump forces fuel into the fuel rail, injectors, and regulator. The regulator allows excess fuel to flow back to the fuel tank. The vacuum supplied to the regulator causes fuel pressure to increase and decrease with changes in engine vacuum and load. (Cadillac)

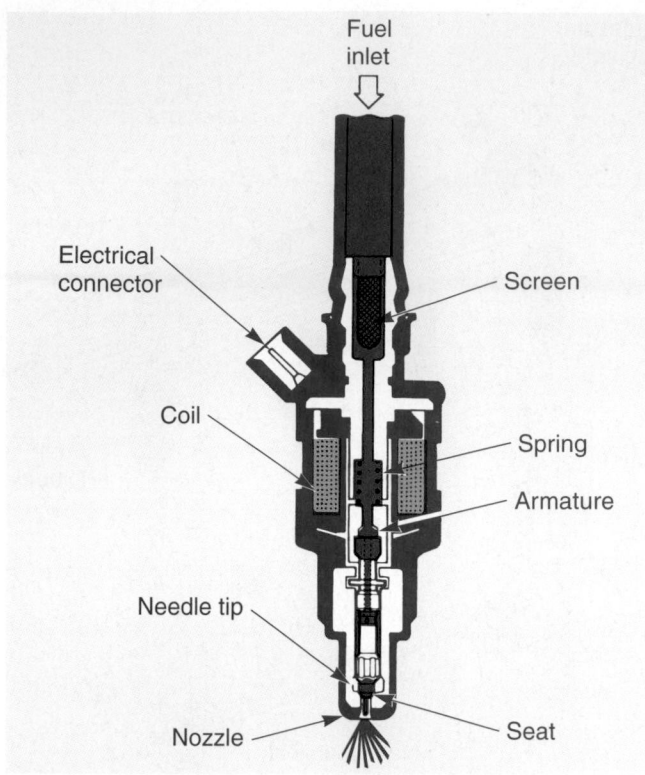

Figure 22-37. Study the basic parts of this electronic multiport fuel injector. The solenoid opens the injector when current flowing through coil builds a magnetic field that acts on the armature. (Lancia)

A *unitized fuel injector* is similar to an electronic multiport or port-mounted injector. However, it has a plastic hose fitting for the transfer line instead of a spray nozzle. It uses an electromagnetic coil to pull up on an armature to open the valve, allowing fuel injection. See **Figure 22-41.**

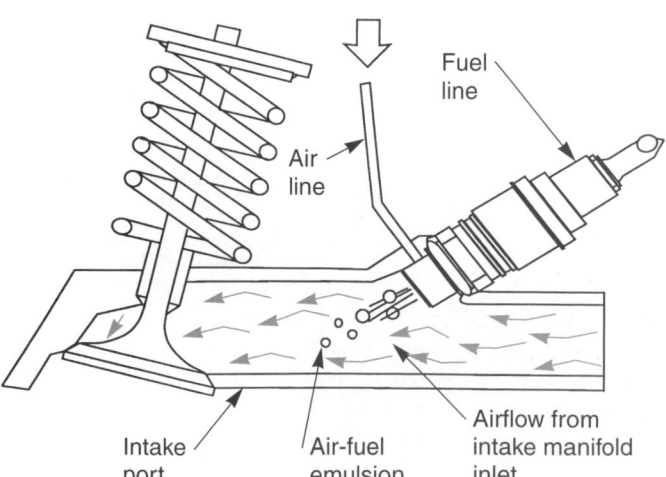

Figure 22-38. A few engines use low air pressure to help fuel atomization. Air is forced into the injector so the air bubbles help break the fuel up into a fine mist during injection.

A

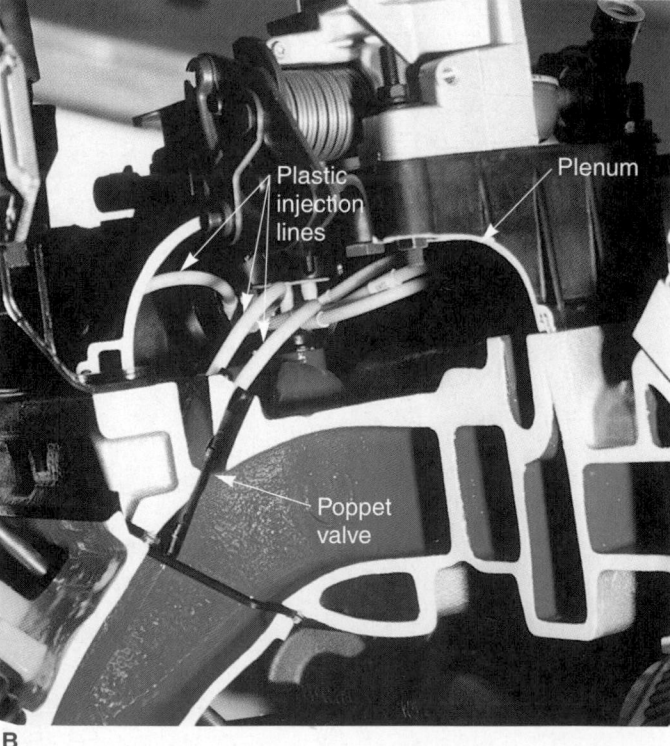

B

Figure 22-39. This engine has a unitized multiport injection system. A—The fuel injection assembly mounts on top of the engine intake manifold. B—Since injectors are mounted in a housing, plastic lines are needed to carry fuel to the poppet valves in the intake ports.

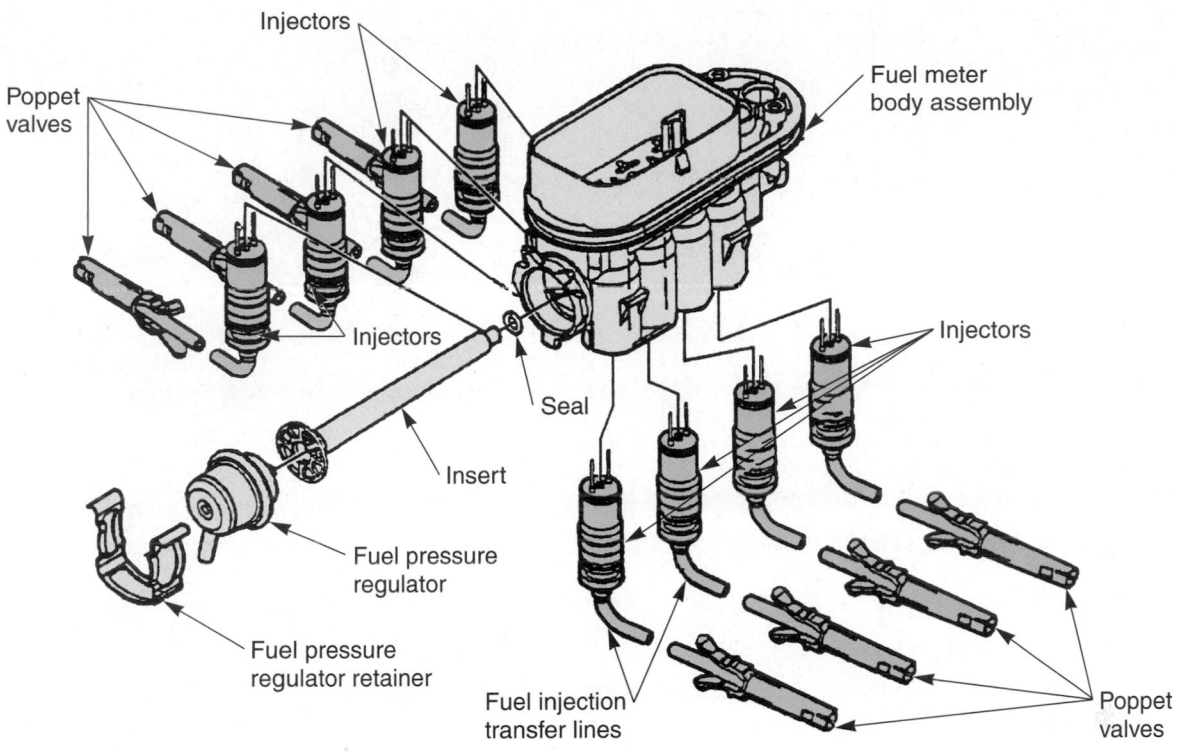

Figure 22-40. This exploded view shows a unitized injector assembly, plastic transfer lines, and poppet valves.

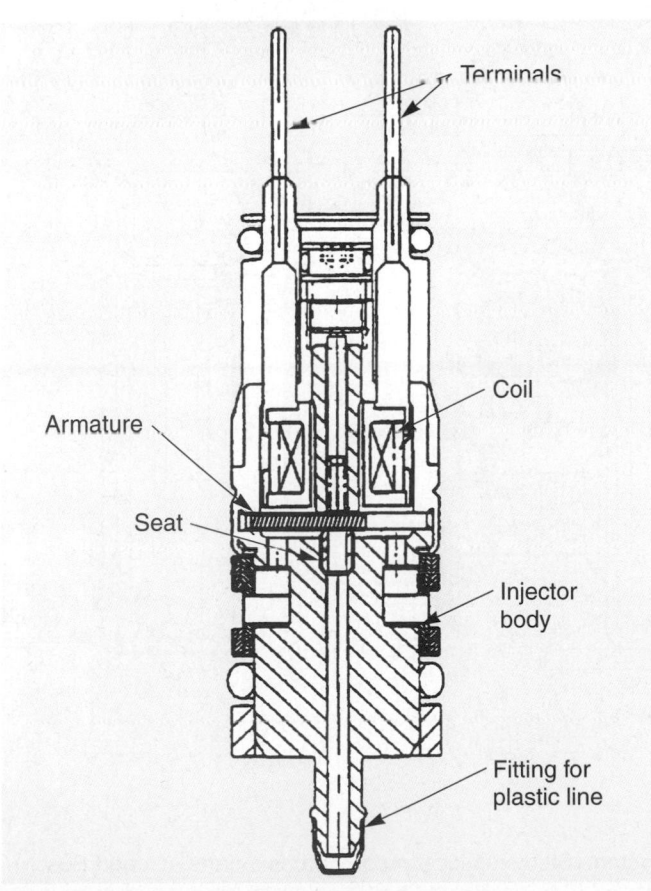

Figure 22-41. This cutaway shows an injector for a unitized system. Note the fitting for the plastic transfer line.

Injector Resistor Pack

An *injector resistor pack* is a set of low-ohm resistors that control current flow to each injector coil. They protect the injector windings from excess current. Each resistor is usually around 5–7 ohms. The injector resistor pack is often mounted in the engine compartment. **Figure 22-42** shows the basic circuit for a typical injector resistor pack.

Airflow Sensing Multiport EFI

An *airflow sensing multiport EFI* uses an airflow sensor as a main control of the system. As shown in **Figure 22-43,** an airflow sensor is placed at the inlet to the intake manifold. It and other engine sensors provide electrical data to the computer.

The airflow sensor is a flap-operated variable resistor, **Figure 22-44.** Airflow through the sensor causes an air door (flap) to swing to one side. Since the air door is connected to a variable resistor, the amount of airflow into the engine is converted into an electrical signal for the computer.

Airflow Sensor Operation

When the throttle valve is closed at idle, the airflow meter door remains almost closed. See **Figure 22-45.**

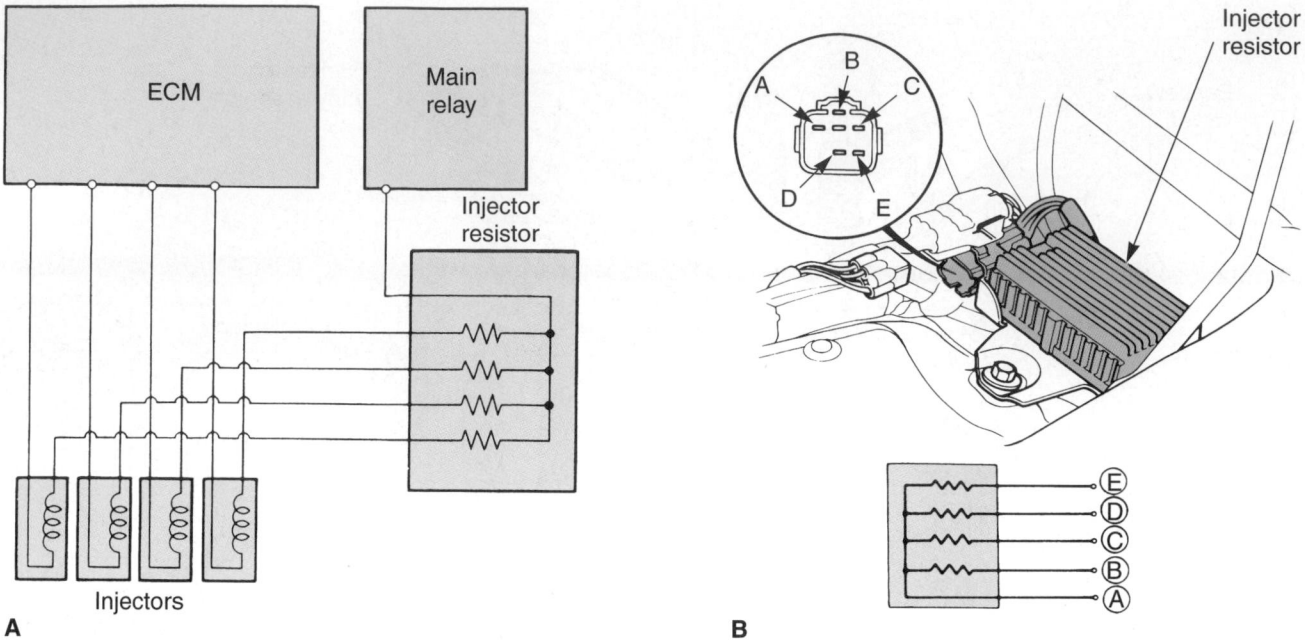

Figure 22-42. A—Schematic of an injector circuit containing an injector resistor pack. B—Each pin in the resistor pack terminal corresponds to a specific resistor.

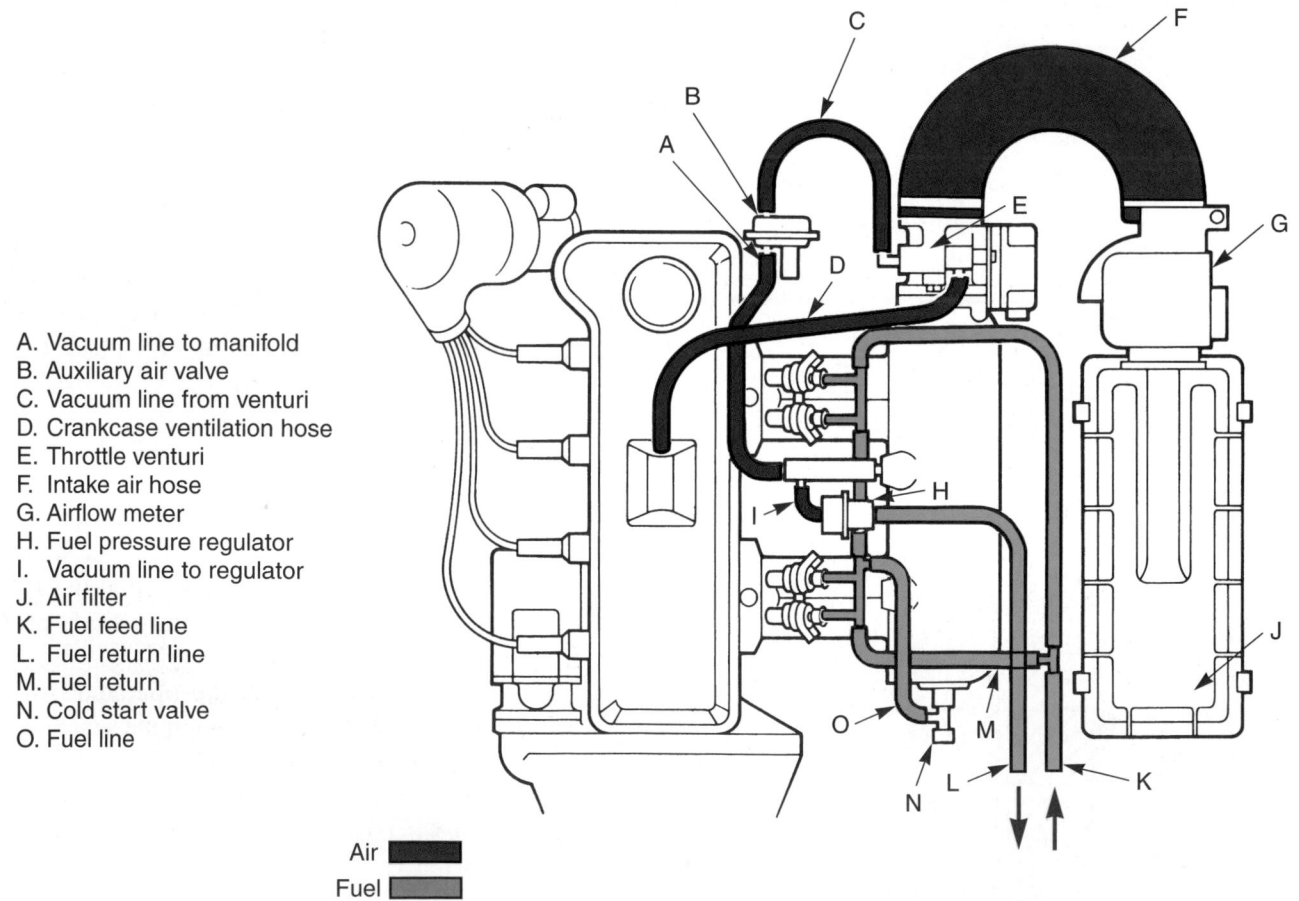

A. Vacuum line to manifold
B. Auxiliary air valve
C. Vacuum line from venturi
D. Crankcase ventilation hose
E. Throttle venturi
F. Intake air hose
G. Airflow meter
H. Fuel pressure regulator
I. Vacuum line to regulator
J. Air filter
K. Fuel feed line
L. Fuel return line
M. Fuel return
N. Cold start valve
O. Fuel line

Air
Fuel

Figure 22-43. The top view of an airflow-sensing multiport injection system. Study the location of the basic parts. (Robert Bosch)

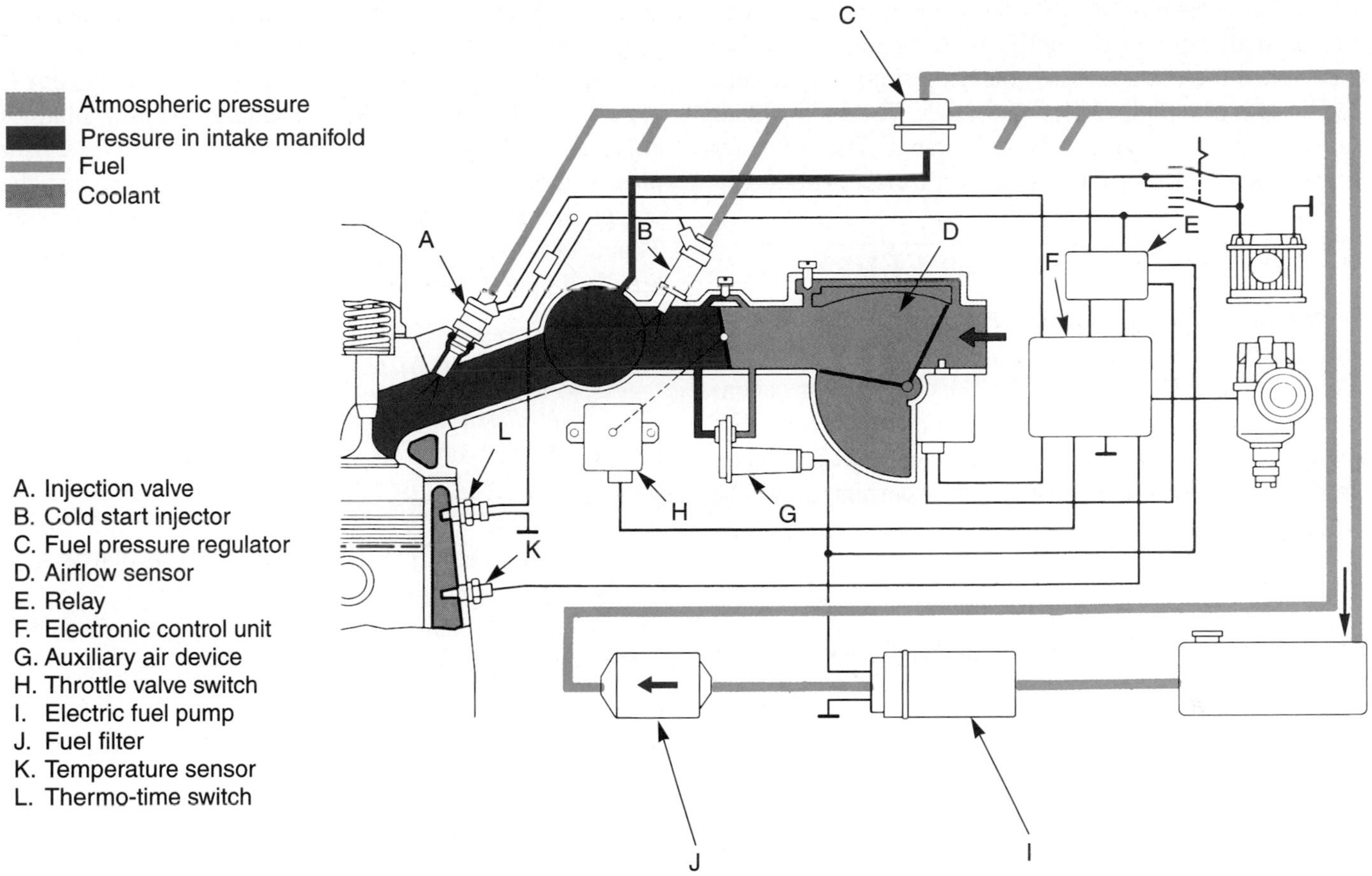

Atmospheric pressure
Pressure in intake manifold
Fuel
Coolant
A
B
C
D
E
F
G
H
I
J
K
L
A. Injection valve
B. Cold start injector
C. Fuel pressure regulator
D. Airflow sensor
E. Relay
F. Electronic control unit
G. Auxiliary air device
H. Throttle valve switch
I. Electric fuel pump
J. Fuel filter
K. Temperature sensor
L. Thermo-time switch

Figure 22-44. This diagram shows how each part is connected in an airflow-sensing EFI system. The airflow sensor is the primary sensor. Also, find the cold start injector. (Robert Bosch)

Idle speed adjusting screw
Bypass port
Throttle chamber
Throttle valve
Air bypass screw
Airflow meter
Flap
Flap bypass
Increased airflow swings flap open
A
B

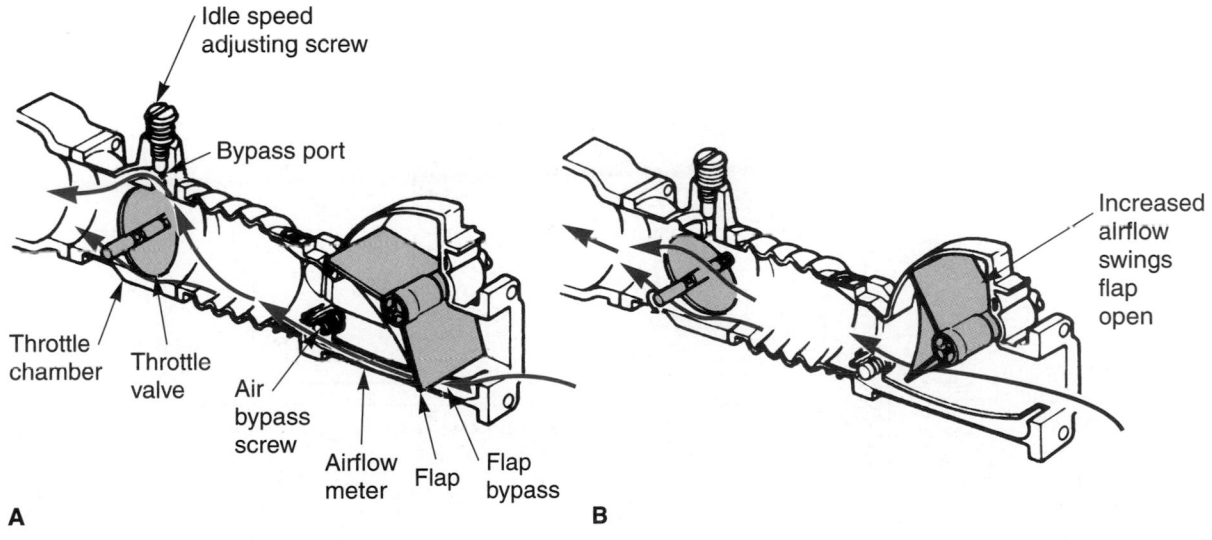

Figure 22-45. The throttle valve controls engine speed and power output. A—The throttle is almost closed. The engine is running slowly. The airflow sensor detects little airflow. The control module produces a short injection pulse width for a small injection quantity. B—The throttle is moved open for more power. Increased flow pushes the sensor flap open. The control module increases the pulse width for a richer mixture. (Nissan)

The computer then produces a short injector pulse width. Only a small amount of fuel is injected into the intake ports. When the driver presses the accelerator and swings open the throttle plate, airflow increases. The airflow door is pushed open, changing sensor resistance. The computer then increases injector pulse width for a richer mixture.

Pressure-Sensing Multiport EFI

Pressure-sensing multiport injection uses intake manifold pressure (vacuum) as a primary control of the system. Look at **Figure 22-46.** A pressure sensor is connected to a passage going into the intake manifold. The *pressure sensor* converts changes in manifold pressure into changes in electrical resistance or current flow. The computer uses this electrical data to calculate engine load and air-fuel ratio requirements.

Hydraulic-Mechanical Continuous Injection System

A *hydraulic-mechanical continuous injection system* uses a mixture control unit to operate the injectors. A *mixture control unit* is an airflow sensor-fuel distributor assembly. **Figure 22-47** shows a diagram of this system. This is not an electronic type system.

The airflow sensor for this type injection system is pictured in **Figure 22-48.** A round air sensor plate is attached to a hinge inside the sensor housing. The sensor plate is mounted on a sensor arm. When airflow into the engine increases, the airflow pushes up the sensor plate. This action also pulls up on the sensor arm. The sensor arm then operates the fuel distributor.

CIS Fuel Distributor

A *fuel distributor* is a hydraulically-operated valve mechanism that controls fuel flow or pressure to each CIS injector. See **Figure 22-48.** The fuel control plunger is located in the center of the distributor. Fuel is fed from the plunger to spring-loaded diaphragms. The diaphragms compensate for pressure differences in each injection line. They help ensure that the same amount of fuel is sent to each injector.

Tech Tip
A fuel distributor is only used in one type of CIS system, which is found on many foreign cars.

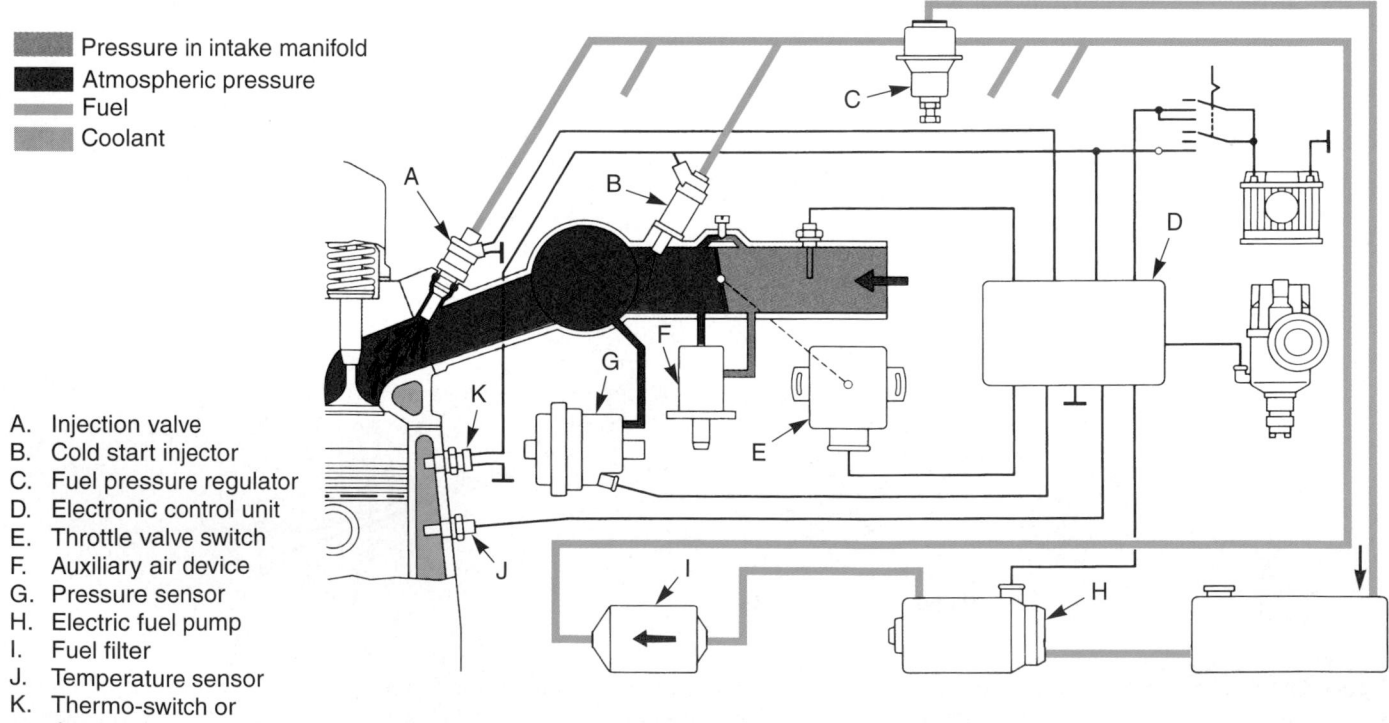

Pressure in intake manifold
Atmospheric pressure
Fuel
Coolant

A. Injection valve
B. Cold start injector
C. Fuel pressure regulator
D. Electronic control unit
E. Throttle valve switch
F. Auxiliary air device
G. Pressure sensor
H. Electric fuel pump
I. Fuel filter
J. Temperature sensor
K. Thermo-switch or thermo-time switch

Figure 22-46. A pressure-sensing gasoline injection system uses intake manifold vacuum as the main source of information. High intake manifold vacuum indicates a low-load condition that requires a lean air-fuel mixture. Low intake vacuum indicates a high-load condition, requiring a richer mixture.

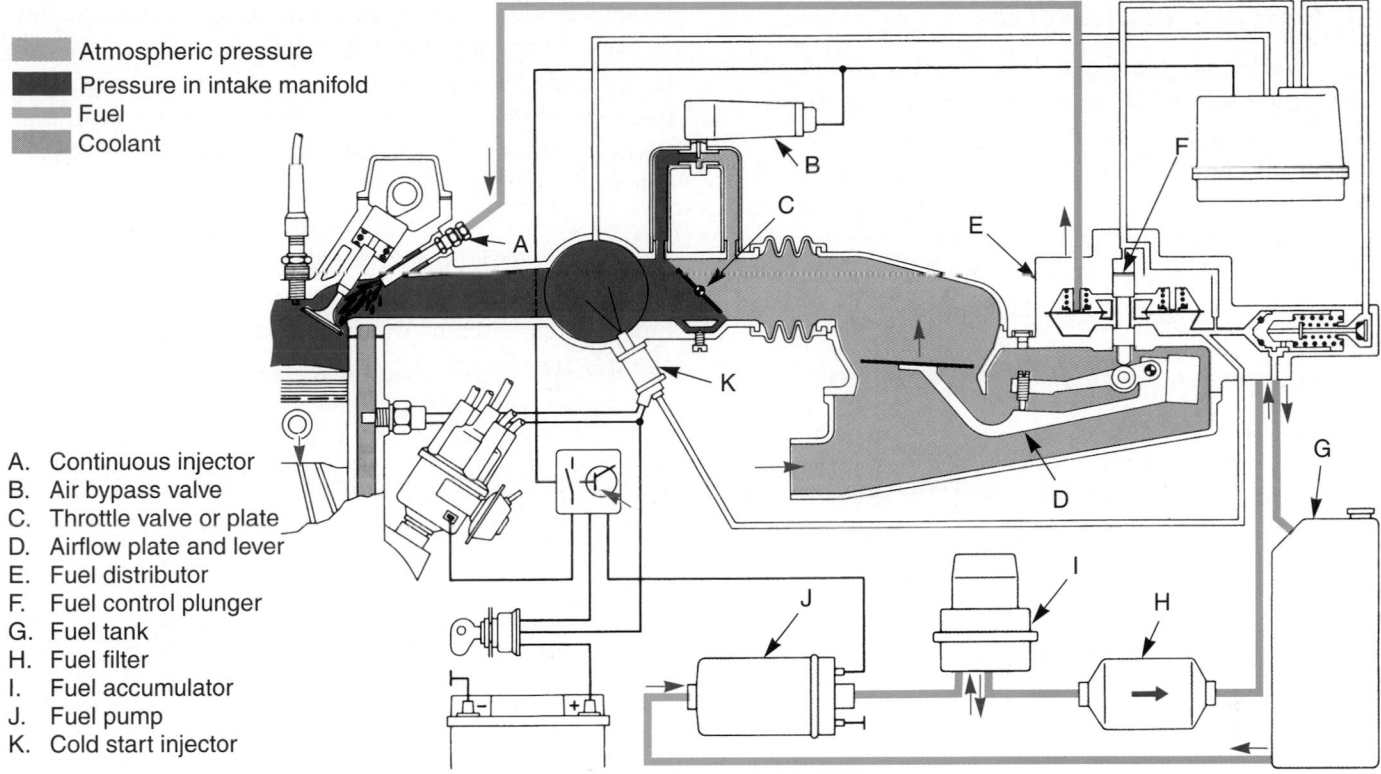

Atmospheric pressure
Pressure in intake manifold
Fuel
Coolant

A. Continuous injector
B. Air bypass valve
C. Throttle valve or plate
D. Airflow plate and lever
E. Fuel distributor
F. Fuel control plunger
G. Fuel tank
H. Fuel filter
I. Fuel accumulator
J. Fuel pump
K. Cold start injector

Figure 22-47. A hydraulic-mechanical injection system uses a mechanical airflow sensor to operate a hydraulic fuel distributor assembly. Note that a continuous injector is used to spray fuel into engine anytime the engine is running. (Robert Bosch)

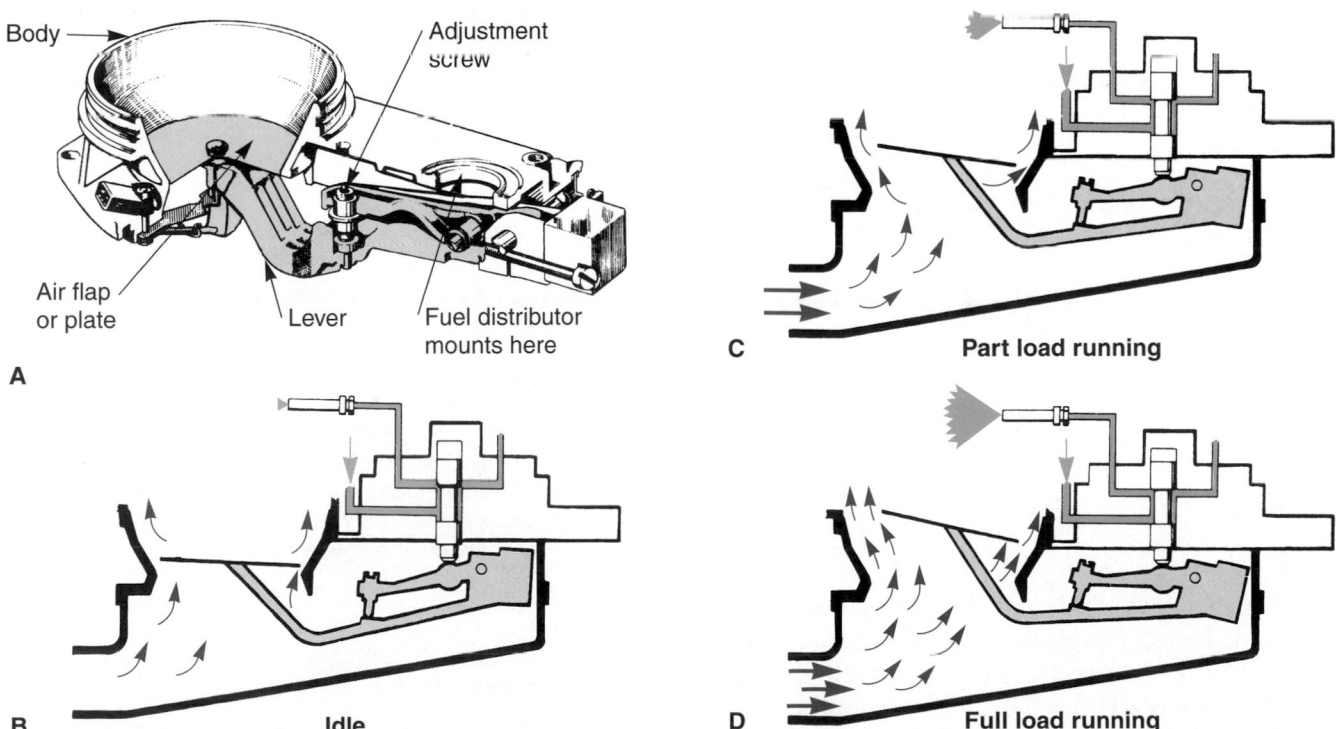

Figure 22-48. A—The airflow sensor has a large disc-shaped plate that is hinged in the air horn. The plate operates a lever arm. B—At idle, low airflow only moves the sensor plate a little. The lever arm pushes up lightly on the fuel control plunger for a small injection quantity. C—At partial load, more airflow moves the sensor and the control plunger up more. More fuel sprays out of the injectors. D—At full load, high airflow pushes the sensor plate up high. This opens the fuel control plunger fully for maximum injection pressure and volume. (Robert Bosch)

Continuous Fuel Injector

A *continuous fuel injector* is simply a spring-loaded valve. It injects fuel all the time when the engine is running. See **Figure 22-49.** A spring holds the valve in a normally closed position. A filter in the injector traps dirt. When the engine is cranked for starting, fuel pressure builds and pushes the injector valve open. A steady stream of gasoline then sprays toward each engine intake valve. The fuel is pulled into the engine when the intake valves open.

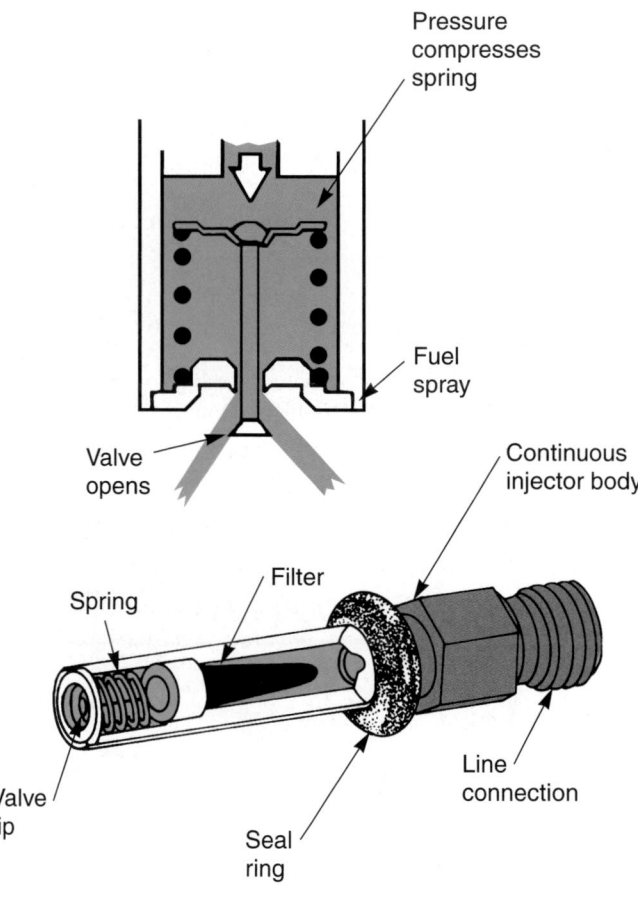

Figure 22-49. A continuous-type injector is simply a spring-loaded valve. With enough fuel pressure, the injector valve opens and fuel sprays into the intake port of the engine.

With CIS injectors, the quantity of air-fuel is controlled by increasing or decreasing fuel pressure to the injectors. The injector is usually push-fitted into plastic bushings in the cylinder head or intake manifold.

Cold Start Injector

A *cold start injector* is an additional fuel injector valve used to supply extra fuel for cold engine starting.

Either a thermo-time switch or the system control module is used to operate the cold start injector. A cold start injector can be used in electronic airflow–sensing, electronic pressure–sensing, and hydraulic–mechanical multiport systems. It is constructed like a conventional, solenoid type injector.

Cold Start Injector Operation

Figure 22-50 shows a basic cold start injector thermo-time switch circuit. When the sensor detects a cold engine, the switch closes to energize the cold start injector. The cold start injector and the other injectors all spray fuel into the intake manifold. Like a carburetor choke, this provides a very rich mixture to sustain cold engine operation.

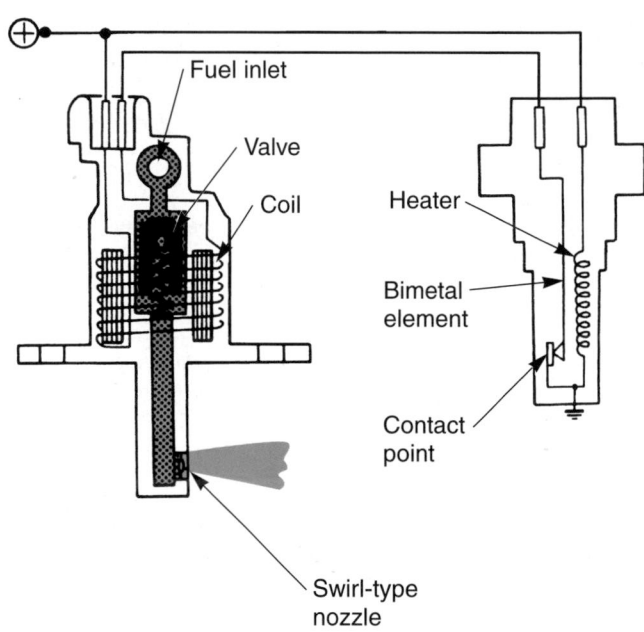

Figure 22-50. A basic cold start injector thermo-time switch circuit. The thermo-time switch energizes the cold start injector when engine temperature is low enough. The injector then sprays extra fuel into the engine to help keep the cold engine running smoothly. When the engine warms enough, the thermo-time switch opens to shut off the injector. A heating element in the switch ensures that the injector stays on for a short period of time, even if the engine is very cold.

Fuel Accumulator

A *fuel accumulator* can be used in an injection system to dampen pressure pulses. One is shown in **Figure 22-51.** The accumulator may also maintain pressure when the system is shut down. This aids engine restarting.

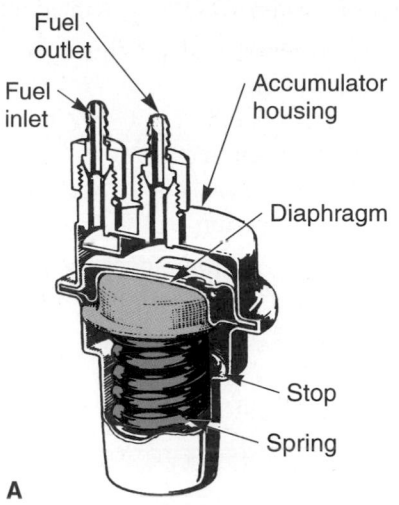
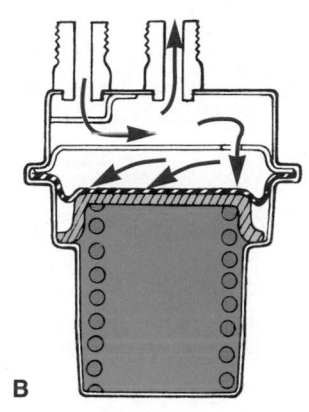
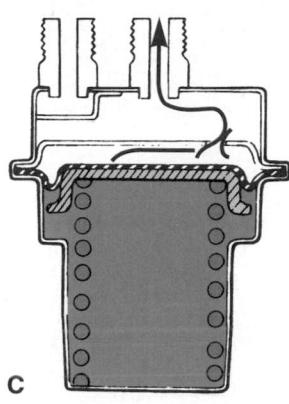

Figure 22-51. A fuel accumulator is simply a spring-loaded diaphragm. It dampens pressure pulsations in the system. It also maintains fuel pressure when the engine is shut off. A—Study the basic parts of a fuel accumulator. B—With the engine running, fuel pressure compresses the diaphragm spring. C—When the engine is shut off, the spring pushes up on the diaphragm to hold pressure in the system. (Volvo)

Summary

- A gasoline injection system uses pressure from an electric fuel pump to spray fuel into the engine intake manifold.

- Atmospheric pressure is the pressure formed by the air surrounding the earth. At sea level, the atmosphere exerts 14.7 psi (103 kPa) of pressure.

- The engine throttle valve controls airflow and engine power output.

- A throttle body injection system (TBI) has the injector nozzles in a throttle body assembly on top of the engine.

- A multiport injection system has fuel injectors in the air-fuel runners going into each cylinder.

- A timed injection system squirts fuel into the engine right before or as the intake valves open.

- A fuel injector for an EFI system is simply a coil or solenoid operated fuel valve.

- An engine sensor is an electrical device that changes circuit resistance or voltage with a change in a condition, such as temperature, pressure, or part position.

- The engine control module is the "brain" of the electronic fuel injection system.

- An exhaust gas sensor, also called an oxygen sensor, measures the oxygen content in the engine's exhaust system as a means of checking combustion efficiency.

- When in open loop, the electronic injection system does *not* use engine exhaust gas content as a main control of the air-fuel mixture.

- Closed loop means that the computer is using information from the oxygen sensor and other sensors.

- An idle air control motor or solenoid may be used to help control engine idle speed.

- Unitized multiport injection is a system that mounts all of the fuel injectors into a single assembly.

- An injector resistor pack is a set of low-ohm resistors that control current flow to each injector coil.

Important Terms

Gasoline injection system	Electronic fuel injection
Atmospheric pressure	Hydraulic fuel injection
Vacuum	Fuel distributor
Engine throttle valve	Mechanical fuel injection
Idling	Timing
Drive-by-wire system	Intermittent gasoline injection system
Fuel injection system	Modulated injection system
Throttle body injection system	Timed injection system
Multiport injection system	Continuous gasoline injection system
Intake ports	Simultaneous injection
Indirect injection system	Sequential injection
Direct injection system	Group injection
	Fuel delivery system

Electric fuel pump
Fuel pressure regulator
Fuel injector
Air induction system
EFI sensor system
Engine sensor
Computer control
  system
Engine control module
Oxygen sensor
Exhaust gas sensor
Open loop
Closed loop
Manifold absolute
  pressure (MAP)
  sensor
Throttle position
  sensor
Engine temperature
  sensor
Airflow sensor
Inlet air temperature
  sensor
Crankshaft position
  sensor
Digital signals
Analog signal
Injector pulse width
Throttle body injection
  system (TBI)
TBI assembly
Throttle body housing
Fuel injectors
Fuel pressure regulator
Throttle positioner
Throttle position
  sensor
Throttle plates
Throttle body
Throttle body injector

Throttle body pressure
  regulator
Fast idle thermo valve
Idle air control motor
Throttle positioner
Continuous throttle
  body injection (CTBI)
Control pump
Spray bar
Electronic multiport
  injection
Multiport throttle body
Multiport pressure
  regulator
Fuel rail
Multiport injector valve
Air-fuel emulsion
  injector
Unitized multiport
  injection
Fuel injection transfer
  lines
Fuel injection poppet
  valve
Unitized fuel injector
Injector resistor pack
Airflow sensing
  multiport EFI
Airflow sensor
Pressure-sensing
  multiport injection
Pressure sensor
Hydraulic-mechanical
  continuous injection
  system
Mixture control unit
Fuel distributor
Continuous fuel injector
Cold start injector
Fuel accumulator

Review Questions—Chapter 22

Please do not write in this text. Place your answers on a separate sheet of paper.

1. A gasoline injection system uses pressure from a(n) _____ _____ _____ to spray fuel into the engine _____ _____.

2. List seven possible advantages of a gasoline injection system over a carburetor system.

3. Explain the difference between throttle body and multiport injection systems.

4. Gasoline injection systems use _____ injection.

5. This is the most common and modern type of gasoline injection system.
   (A) Mechanical fuel injection.
   (B) Hydraulic fuel injection.
   (C) Electronic fuel injection.
   (D) Pneumatic fuel injection.

6. This type of gasoline injection pulses the injectors open and closed independently of the engine valve action.
   (A) Timed injection.
   (B) Intermittent injection.
   (C) Continuous injection.
   (D) Bank injection.

7. List the parts typically included in an EFI fuel delivery system.

8. How does an EFI injector open and close?

9. Explain the action of an EFI system throttle valve.

10. An engine _____ is an electrical device that changes circuit resistance or voltage with a change in a condition, such as temperature, pressure, or part position.

11. Define the term "EFI control module."

12. A(n) _____ _____ _____, also called a(n) _____ _____, measures the oxygen content in the engine's exhaust system as a means of checking _____ _____.

13. When the intake manifold absolute pressure sensor detects high pressure (low vacuum), the computer would know that a(n) _____ mixture is needed for load conditions.

14. A throttle position sensor is a(n) _____ _____ connected to the _____ _____ _____.

15. Which of these is *not* a typical EFI system sensor?
   (A) Exhaust back pressure sensor.
   (B) Throttle position sensor.
   (C) Engine temperature sensor.
   (D) Inlet air temperature sensor.

16. Explain the difference between sensor analog and digital signals.

17. When an EFI system is in _____ loop, the computer uses stored information to operate the system.

18. When an EFI system is in _____ loop, the computer uses engine sensor information to control the system.

19. Define the term "injector pulse width."

20. List and explain the six major parts of a TBI unit.

21. What are the main differences between the throttle body for multiport injection and throttle body injection?

22. An EFI multiport injector fits into the _____ or _____ in the _____ manifold.

23. List and explain the eight major parts of an EFI multiport injector.

24. Describe the mixture control unit in a hydraulic mechanical CIS.

25. A(n) _____ fuel injector is a spring-loaded fuel valve that does *not* use an electric coil.

✸ ASE-Type Questions

1. Gasoline injection systems feed fuel into an engine using:
   (A) *suction.*
   (B) *vacuum.*
   (C) *pressure.*
   (D) *All of the above.*

2. Gasoline injection systems have several advantages over carburetor-type fuel systems, including each of these *except:*
   (A) *lower emissions*
   (B) *uses direct injection.*
   (C) *improved atomization.*
   (D) *better fuel distribution.*

3. In which fuel injection system is gasoline sprayed into the top center of an intake manifold?
   (A) *Port.*
   (B) *Electronic.*
   (C) *Multiport.*
   (D) *Throttle body.*

4. Which of the following is *not* a common method used to control the amount of gasoline injected into an engine?
   (A) *Hydraulic controls.*
   (B) *Computer controls.*
   (C) *Electronic controls.*
   (D) *Thermal controls.*

5. Which injection system squirts fuel into the engine right before or as the intake valves open?
   (A) *Timed.*
   (B) *Modulated.*
   (C) *Continuous.*
   (D) *Intermittent.*

6. The sensor used to detect engine speed is the:
   (A) *throttle position sensor.*
   (B) *manifold pressure sensor.*
   (C) *crankshaft position sensor.*
   (D) *idle valve actuator.*

7. Which of these is *not* a throttle body injector part?
   (A) *Bypass valve.*
   (B) *Injector spring.*
   (C) *Ball or needle valve.*
   (D) *Armature or plunger.*

8. While discussing electronic injection systems, Technician A states that most multiport EFI uses two solenoid injectors for each engine cylinder. Technician B says such a system uses only one solenoid injector per engine cylinder. Who is right?
   (A) *A only.*
   (B) *B only.*
   (C) *Both A and B.*
   (D) *Neither A nor B.*

9. Which of the following is a flap-operated variable resistor?
   (A) *Venturi.*
   (B) *Fuel rail.*
   (C) *Airflow sensor.*
   (D) *Port fuel injector.*

10. A cold start injector is used on which type of multiport system?
    (A) *Hydraulic-mechanical.*
    (B) *Electronic airflow sensing.*
    (C) *Electronic pressure-sensing.*
    (D) *All of the above.*

Activities—Chapter 22

1. Using a service manual for a fuel-injected vehicle, study the section on fuel injector operation. Demonstrate to your instructor your understanding of how the injectors are controlled.

2. Develop a drawing of the fuel injection system for a gasoline engine and produce an overhead transparency from it. Using the transparency or handouts, explain to the class how the system works.

3. Prepare simple sketches to show the basic difference between a carbureted fuel system and a throttle body injection fuel system.

Chapter 23

Gasoline Injection Diagnosis and Repair

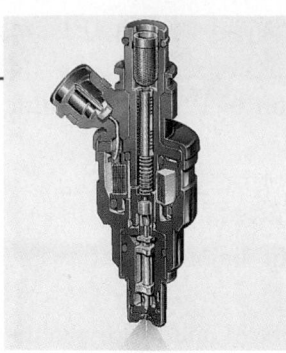

After studying this chapter, you will be able to:

- Diagnose typical gasoline injection system problems.
- Measure fuel pressure regulator output.
- Test both electronic and continuous fuel injectors.
- Explain OBD II testing features used on late-model fuel injection systems.
- Use a service manual when making basic adjustments on gasoline injection systems.
- Cite safety rules for injection system service.
- Correctly answer ASE certification test questions about fuel injection system diagnosis, service, and repair.

As you have learned, several types of gasoline injection systems are used by automakers. Most types have similar parts (injectors, control module, and sensors). This chapter describes the most common gasoline injection system problems, tests, adjustments, and repair procedures. It also explains the service methods used on vehicles equipped with OBD II systems.

Gasoline Injection Problem Diagnosis

To diagnose problems in a gasoline injection system, you must use:

- Your knowledge of system operation.
- Basic troubleshooting skills.
- A service manual.
- A scan tool on late-model vehicles.

As you try to locate problems, visualize the operation of the four subsystems: air, fuel, sensor, control. Relate the function of each subsystem component to the problem. This will let you eliminate several possible problem sources and concentrate on others.

Verify the Problem

Before you test the fuel injection system, verify the problem or complaint. Make sure that the customer or service writer has accurately described the symptoms.

Never attempt to repair a gasoline injection system until you have checked all possible problem sources. The ignition system, for example, normally causes more problems than the injection system.

Inspecting Injection System

A general inspection of the engine and related components will sometimes locate gasoline injection system troubles. Look at **Figure 23-1.** Check the condition of all hoses, wires, and other injection system components. Look for fuel leaks, vacuum leaks, kinked lines, loose electrical connections, and other problems. Thoroughly check the components most likely to cause the particular symptoms. See **Figure 23-2.**

Figure 23-1. Close inspection of the engine compartment is important. Look for loose wires, leaking hoses, and other obvious troubles.

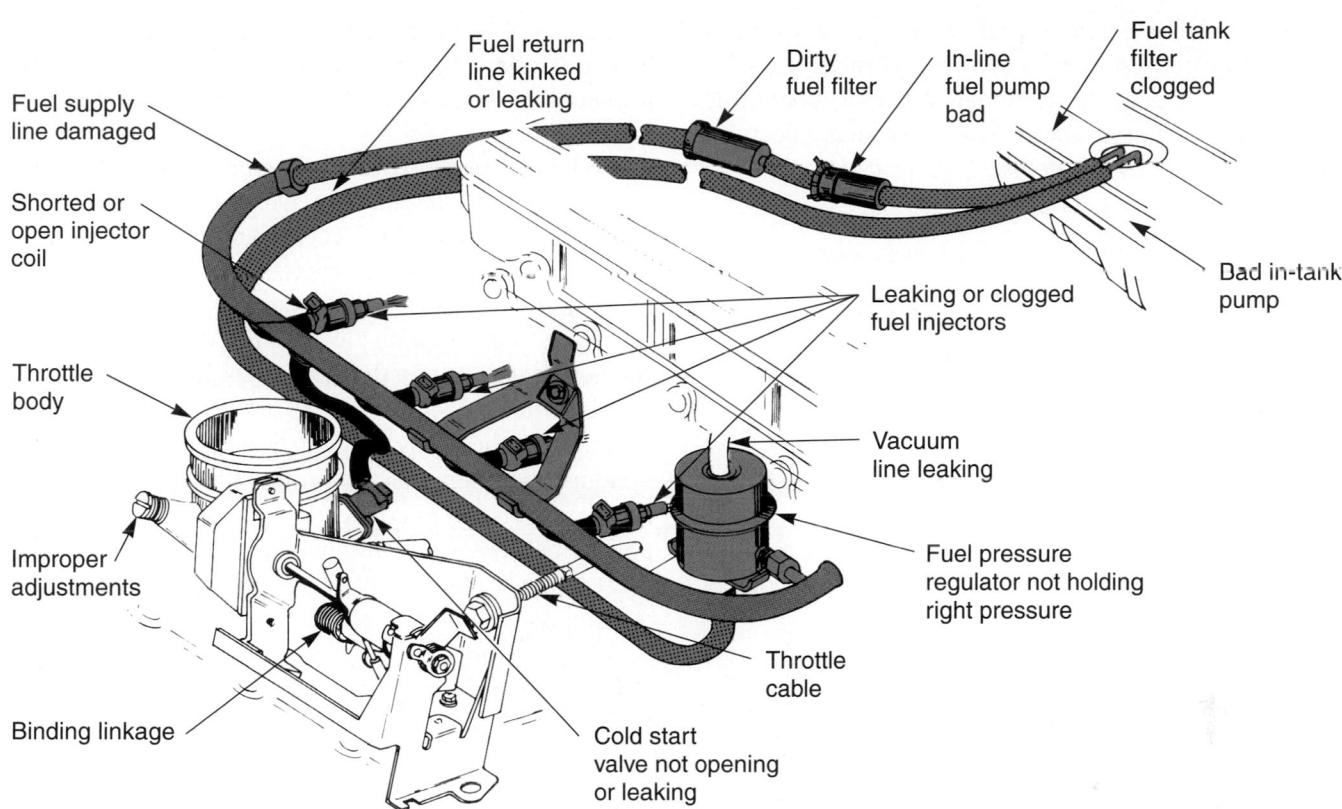

Fuel return line kinked or leaking

Dirty fuel filter

In-line fuel pump bad

Fuel tank filter clogged

Fuel supply line damaged

Shorted or open injector coil

Throttle body

Improper adjustments

Binding linkage

Leaking or clogged fuel injectors

Bad in-tank pump

Vacuum line leaking

Fuel pressure regulator not holding right pressure

Throttle cable

Cold start valve not opening or leaking

Figure 23-2. After evaluating symptoms, try to narrow the list of components that could be at fault. Visualize the operation of each part and do not overlook common problems, such as a clogged fuel filter. (Flat)

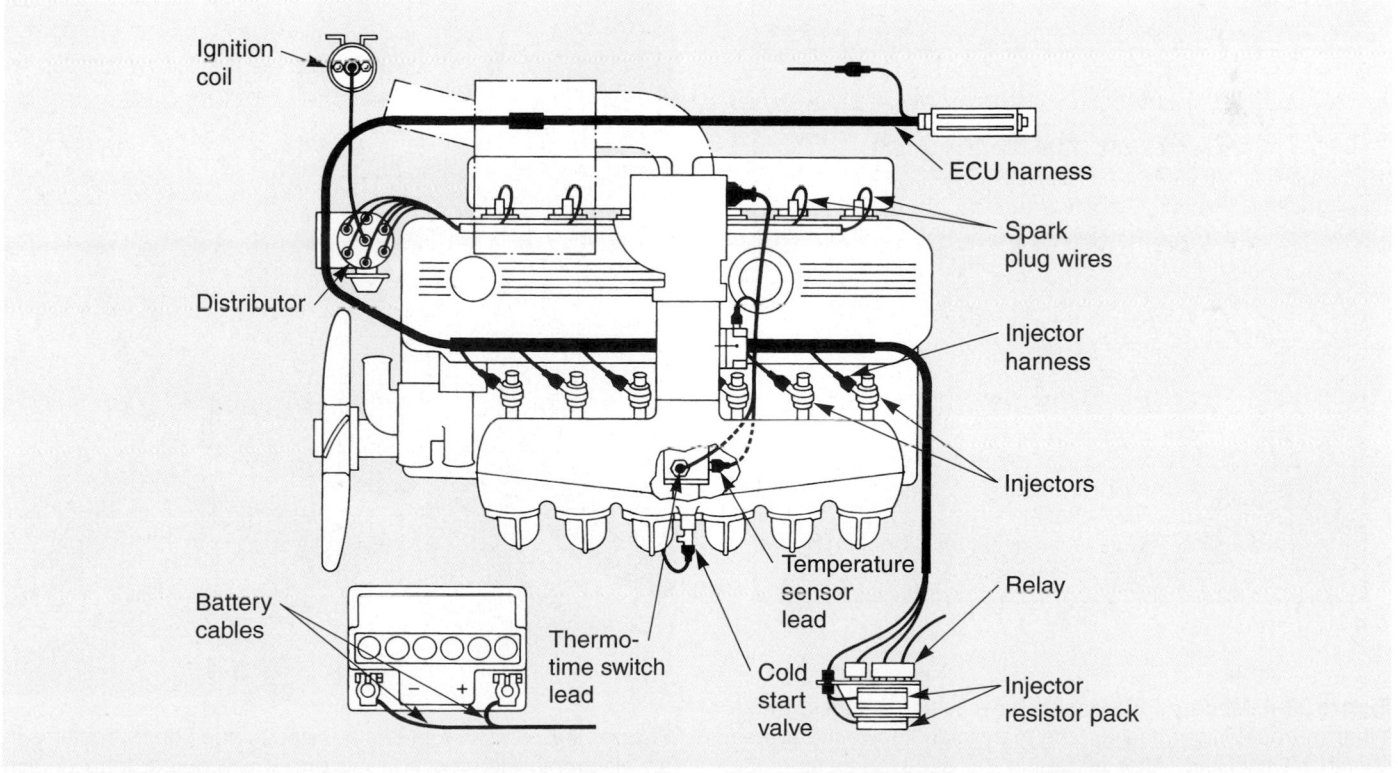

Ignition coil

ECU harness

Spark plug wires

Distributor

Injector harness

Injectors

Battery cables

Thermo-time switch lead

Temperature sensor lead

Cold start valve

Relay

Injector resistor pack

Figure 23-3. When diagnosing problems in EFI systems, check the electrical connections closely. One loose or disconnected wire could upset system operation.

With an EFI system, you may need to disconnect and check the terminals of the wiring harness. Look at **Figure 23-3.** Inspect the terminals for rust, corrosion, and burning. High resistance at the terminal connections is a frequent cause of problems.

 Caution
Do not disconnect an EFI harness terminal when the ignition switch is in the *on position.* This can damage the control module. Refer to a service manual for details. You may be instructed to disconnect the negative battery terminal during EFI service.

On-Board Diagnostics

Most EFI systems have on-board diagnostic abilities, which means the vehicle's control module can detect and record possible faults. The control module can detect a bad component and produce a *trouble code* pinpointing the problem. Specific systems vary depending on make, model, and year.

When a *malfunction indicator light* (MIL) in the dashboard glows, it tells the driver and the technician that something is wrong. The technician can then use service manual procedures to activate the self-diagnostic mode or to connect a scan tool to the system.

Some older diagnostic systems display the trouble codes as *numbers* in a digital display located on the dash panel. Other systems produce an *on-off* type code by flashing the check engine light. See **Figure 23-4.** In some systems, the diagnostic codes must be retrieved with an analog meter that is connected to a specified circuit test point. In all cases, the numbers have to be compared to a chart in the service manual to pinpoint the faulty section or component.

A *scan tool* will find and display many problems related to an electronic fuel injection system, **Figure 23-5.** It is connected to a test connector on the vehicle.

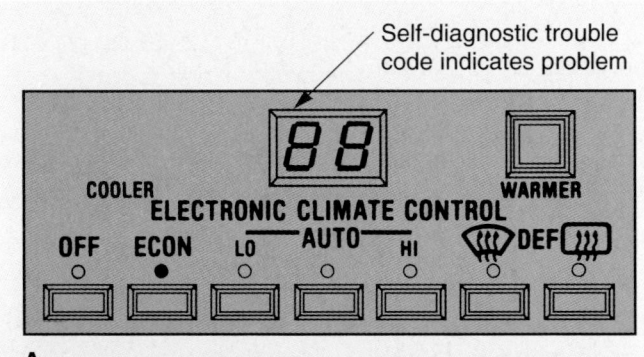

Programmed ECM Trouble Codes

Code	Circuit affected
12	No tack signal.
13	Oxygen sensor not ready.
14	Shorted coolant sensor circuit.
15	Open coolant sensor circuit.
16	Generator output voltage out of range.
17	Crank signal circuit high.
18	Open crank signal circuit.
19	Fuel pump circuit high.
20	Open fuel pump circuit.
21	Shorted TPS circuit.
22	Open TPS circuit.
23	EST/by-pass circuit shorted or open.
24	Speed sensor failure.

Figure 23-4. Modern EFI systems have a self-diagnostic mode. When a trouble light flashes, the technician knows the system should be checked. After activating the self-diagnostic mode, some older systems display the number code in the dash. A—Dash number indicates trouble code. B—Partial list of trouble codes from one service manual. (Cadillac)

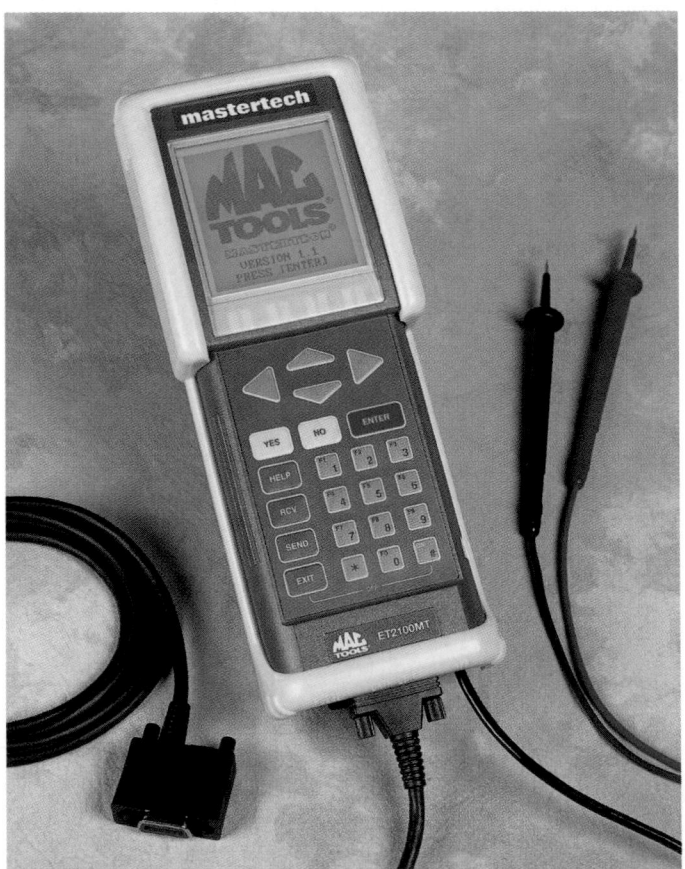

Figure 23-5. A scan tool can be used to find some problems in an electronic fuel injection system. It will detect lean or rich conditions, random misfiring, bad sensor circuits, inoperative actuators, etc. This particular scan tool is also equipped with a multimeter. (Mac Tools)

Always scan for trouble codes before attempting other diagnostic procedures. This may help you find the source of the problem more quickly. Modern scanners will automatically convert the trouble code number into an explanation of potential problems.

OBD II Fuel System Monitoring

With late-model on-board diagnostic (OBD II) systems, the vehicle's control module will record potential problems in its memory. The scanner can read this data and convert it into a brief description of the problem. The scanner will tell you which sensors, injectors, and other monitored components are not operating normally. The scan tool may also indicate what might be wrong with the component. This will save considerable time when troubleshooting.

Vehicles equipped with OBD II systems can set trouble codes that pinpoint injector problems. For example, if the scan tool readout shows a problem with the number three injector, you know to test this injector and its supply circuit for problems. With late-model vehicles, the scan tool will also help you find common problems with system sensors and other actuators. You would then perform pinpoint tests to validate the scanner readout.

Note
For more information on on-board diagnostics, refer to Chapter 18, *On-Board Diagnostics and Scan Tools.*

Fuel system monitoring involves checking whether a stoichiometric, or theoretically perfect, fuel mixture is being fed to the engine. Primarily, the feedback from the upstream oxygen sensor (located before catalytic converter) is used to determine fuel mixture content. If the oxygen sensor detects abnormal combustion resulting from a poor fuel mixture, it trips a trouble code.

Before a trouble code is tripped, however, the control module will try to adjust the fuel mixture as needed. It will alter the temporary fuel trim or injector pulse width to compensate for abnormal combustion by-products.

Short term fuel trim refers to the temporary adjustment of injector pulse width to correct the fuel mixture. *Long term fuel trim* is a permanent adjustment of injector pulse width to compensate for altered operating parameters. The control module uses this data to determine if the vehicle will pass an emission test.

Minor fuel trim adjustments are normal with part wear. However, if the control module determines the vehicle might flunk an emission test from fuel trim data, it will trip a trouble code to warn the driver and technician.

OBD II Oxygen Sensor Monitoring

Oxygen sensor monitoring is done to keep the normally closed loop mode of operation efficient. It usually checks the upstream oxygen sensor for abnormal voltage, slow response times, and similar problems. It also checks the downstream oxygen sensor for normal voltage output. If not, a trouble code is produced.

Note that presently oxygen sensors in OBD II systems are heated to speed warm-up. Quicker oxygen sensor warm-up allows the control module system to go from open loop (startup or cold mode) to closed loop (full operating temperature mode) in less time. In closed loop, the control module system make its own adjustments based on feedback information for sensors.

The oxygen sensor *heater monitor* checks the action of the heating element in the sensor. It does this by rapidly turning the heating element on and off while checking its response. If the control module detects an abnormal heater response, it trips a trouble code.

Note
For more information on scanning fuel system related problems, refer to the text index. Other chapters have related information.

EFI Testers

When the early EFI system does not have on-board diagnostics, an EFI tester can sometimes be used to locate system troubles. The tester, also called an EFI analyzer, is connected to the wiring harness of the system.

An *EFI tester* uses indicator lights and sometimes a digital meter (volt-ohmmeter-ammeter) to check system operation. The technician refers to the instructions with the tester and uses indicator light action to make various tests.

EFI testers are usually "make" specific. Each tester is designed to test only one make of vehicle. EFI testers are often used in large new car dealerships.

Oscilloscope Tests

An *oscilloscope* can sometimes be used to test or view the electrical waveforms (voltage values) at the EFI injectors. This provides a quick and easy way of diagnosing injector, wiring harness, and control module problems.

Note
For information on using oscilloscopes, see Chapter 46, *Advanced Diagnostics.*

Fuel Pressure Regulator Service

A faulty *fuel pressure regulator* can cause an extremely rich or lean mixture. If the output pressure is high, too much fuel will spray out of the injectors, causing a rich mixture. If the regulator bypasses too much fuel to the tank (low pressure), not enough fuel will spray out each injector. This causes a lean mixture.

Relieving EFI System Pressure

Warning

Always relieve fuel pressure before disconnecting any EFI fuel line. Many gasoline injection systems maintain fuel pressure (as high as 60 psi or 414 kPa), even when the engine is *not* running. At this pressure, fuel could spray out with great force, causing eye injury or a fire!

Some EFI systems have a special *relief valve* for bleeding pressure back to the fuel tank; or the pressure regulator may allow pressure relief. Look at **Figure 23-6.**

When a relief valve or a pressure regulator relief are *not* provided, remove the fuse for the fuel pump or disconnect its wires. Start and idle the engine. When the engine stalls from lack of fuel, the pressure has been removed from the system. It is then safe to work.

Testing Fuel Pressure Regulator

Although exact procedures vary, you can test a fuel pressure regulator and fuel pump pressure with a pressure gauge, **Figure 23-7.**

Connect the gauge to the *Schrader valve* (fuel pressure test fitting) on the fuel rail. Start the engine and note the pressure gauge reading. If the reading is higher or lower than specs, the pressure regulator may be bad.

With low fuel pressure, check the fuel pump and fuel filters *before* replacing the regulator. A bad electric fuel pump or partially clogged filter could be lowering fuel pressure. High pressure is commonly due to a bad pressure regulator.

To test maximum fuel pump pressure, place a rag over the fuel return hose. Use pliers to squeeze and block the hose. This will prevent fuel pressure regulator fuel return. The pressure gauge reading should then increase to maximum fuel pump pressure. If fuel pressure remains too low with the return line blocked, the fuel pump, not the pressure regulator, is at fault. See **Figure 23-8.**

Fuel Pressure Regulator Replacement

With throttle body injection, the fuel pressure regulator is normally located on the throttle body assembly. With multiport injection, the pressure regulator is normally located on a branch connected to the fuel manifold (rail) or injectors. Look at **Figure 23-9.** Replace

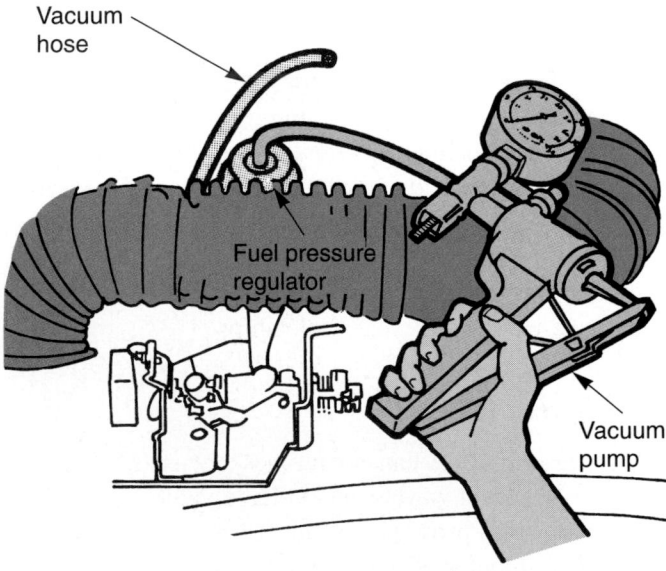

Figure 23-6. Bleed off system pressure before working on the system. When a vacuum-operated pressure regulator is used, applying vacuum to the regulator will relieve the fuel pressure. Fuel flows back to the fuel tank. (Fiat)

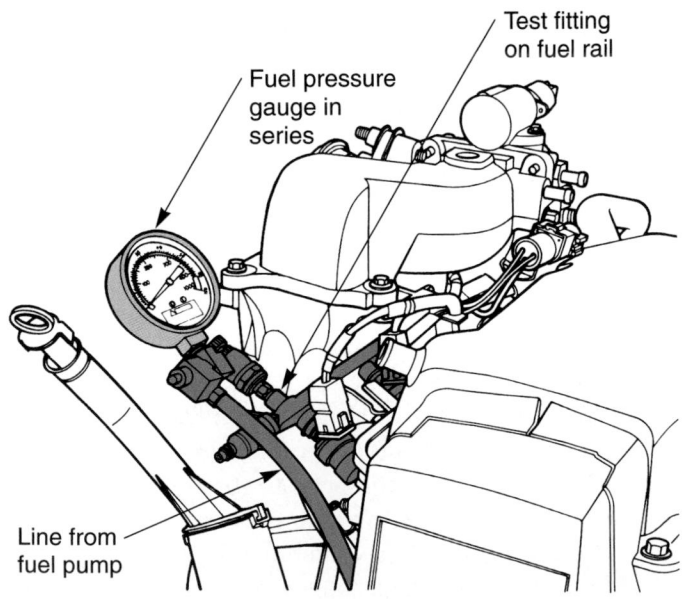

Figure 23-7. To check the pressure regulator, fuel pump, and filters, connect the pressure gauge before the injectors. Pressure should be within specs. If not, check the filters and pump before replacing the regulator. (Ford)

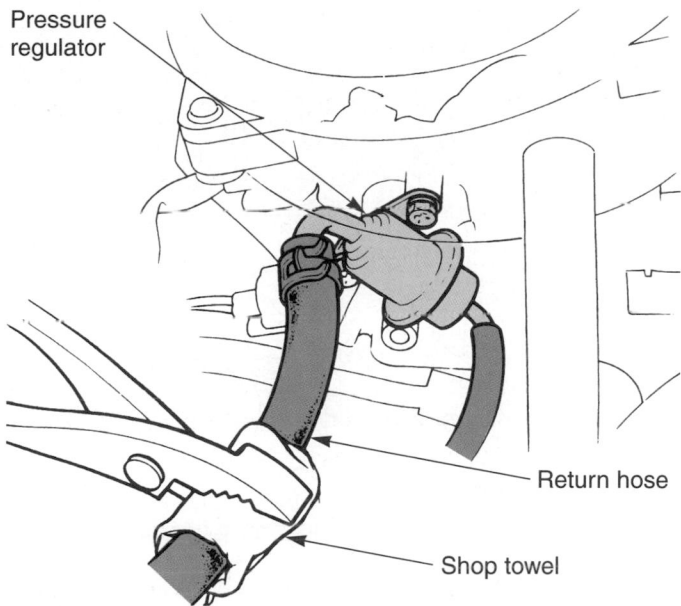

Figure 23-8. If fuel pressure is low, pinch the fuel return hose. This will isolate the trouble to either the fuel pump or the pressure regulator. If pressure increases to above normal with the hose pinched, the regulator is probably bad. (Honda)

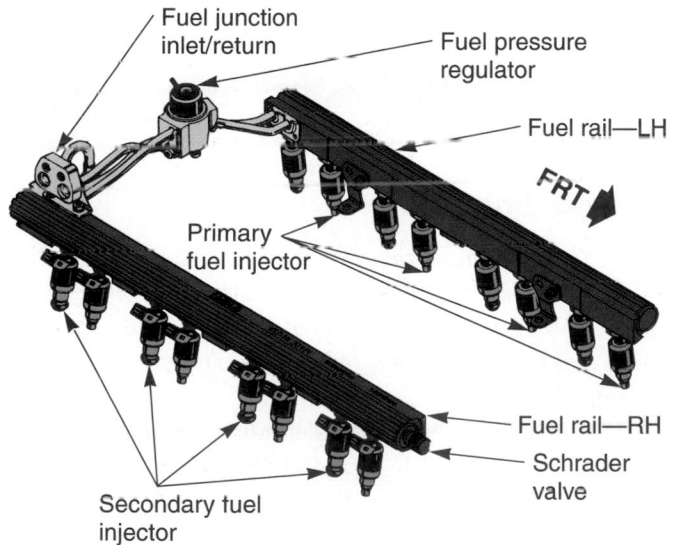

Figure 23-9. Note the location of the pressure regulator and service valve on this multiport injection fuel rail. (Chevrolet)

a fuel pressure regulator according to service manual directions. Use new seals or gaskets between the regulator and rail.

Injector Problems

A *bad injector* can cause a wide range of problems: rough idle, hard starting, poor fuel economy, engine miss.

It is very important that each fuel injector provide the correct fuel spray pattern.

A *leaking injector* richens the fuel mixture by allowing extra fuel to drip from the closed injector valve. The injector valve may be worn or dirty, or the return spring may be weak or broken.

A *dirty injector* can also restrict fuel flow and make the air-fuel ratio too lean. If foreign matter collects in the valve, a poor spray pattern and inadequate fuel delivery can result.

An *inoperative EFI injector* normally has shorted or opened coil windings. Current is reaching the injector, but since the coil is bad, a magnetic field cannot form and open the injector valve.

A continuous injector that does not use a solenoid will usually operate, but may have other problems. It may have a poor spray pattern or weak spring (incorrect opening pressure).

Throttle Body Injector Service

You can quickly check the operation of a throttle body injection system by watching the fuel spray pattern in the throttle body. Remove the air cleaner. With the engine cranking (no-start problem) or running, each injector should form a rapidly pulsing spray of fuel.

Testing TBI injectors

If the TBI injector does not spray fuel (already checked system pressure), follow service manual procedures to test for power to the injector solenoid. **Figure 23-10** illustrates some simple tests recommended by one automaker.

No power (current) to the injector would indicate a problem with the wiring harness or control module.

If you have power, but no fuel spray pattern, then the injector may be bad. Make sure you have adequate fuel pressure before condemning a fuel injector.

Replacing TBI Injectors

Although exact procedures vary with throttle body design, there are a few rules to follow when replacing a TBI injector. These rules include:

- Relieve system fuel pressure before removing any throttle body component.
- Disconnect the negative battery cable.
- Label all hoses and wires before removal, **Figure 23-11.**
- Avoid damage to the throttle body housing when pulling or prying out the injector, **Figure 23-12.**

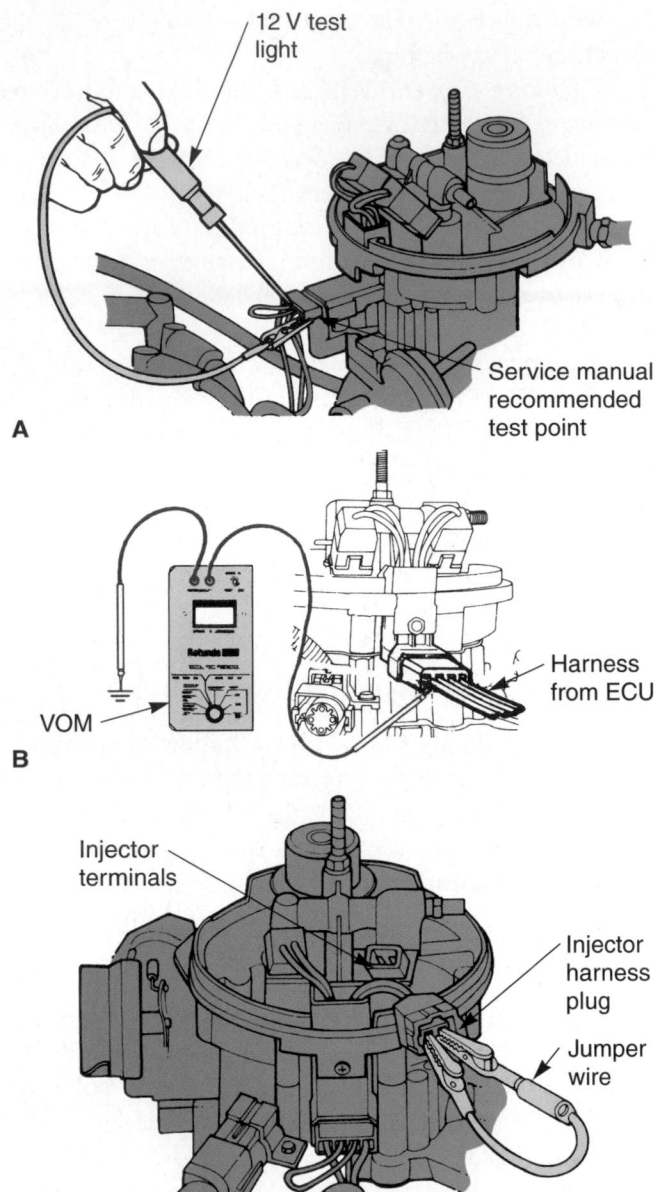

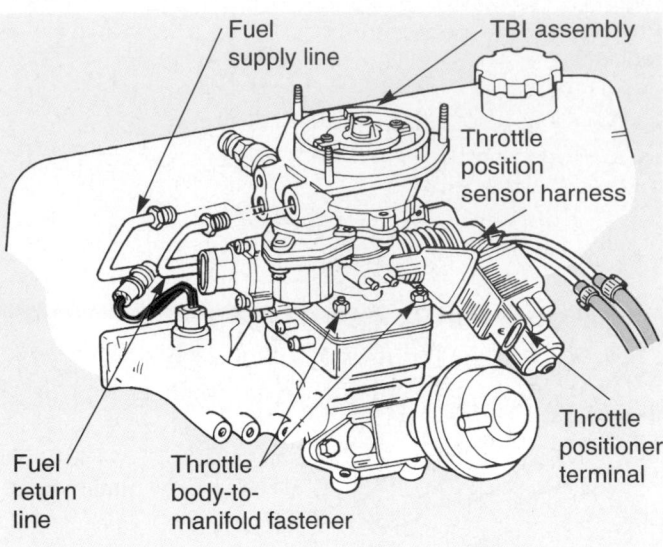

Figure 23-11. When removing the throttle body, label wires and hoses. Unscrew the four fasteners and lift off unit. When installing, double-check the connection of wires and hoses. Torque the fasteners properly using a crisscross pattern. (Renault)

- Make sure the injector is fully seated in the throttle body. Normally, a tab must fit into a notch, assuring correct alignment.

- Check for fuel leaks after TBI injector installation. Test drive the car.

Throttle Body Rebuild

A *throttle body rebuild*, like a carburetor rebuild, involves replacing all gaskets, seals, and other worn parts. You must remove and disassemble the throttle body, making sure all plastic, rubber, and electrical parts are removed.

The metal parts are soaked in carburetor cleaner, washed in cold soak cleaner, and blown dry.

 **Warning**
Carburetor cleaner is a very powerful solvent. Wear eye protection and rubber gloves when working this type of cleaner. Follow the directions on the solvent label in case of an accident.

Inspect each part carefully for signs of wear or damage. Then, reassemble the TBI unit following the instructions in a manual.

Install the throttle body on the engine as you would a carburetor. Use a new base plate gasket. Tighten the hold-down fasteners equally and to the proper torque. Make sure all vacuum hoses are connected to the correct vacuum port on the throttle body. Start the engine. Check for leaks and smooth system operation.

Figure 23-10. Follow service manual directions when checking EFI system voltages and resistances. An incorrect connection or meter setting could damage parts. A—Checking for supply voltage to an injector from the control module. B—Measuring exact supply voltage with a digital VOM. C—Jumper wire allows a check of circuit resistance. An ohmmeter can be used at the other end of the harness. (Ford)

- Install new rubber O-rings. You may need to lubricate the O-ring with approved lubricant to aid new injector installation.

- Double-check that you have installed all injector washers and O-rings correctly.

- Push injector into the throttle body by hand. Avoid using a hammer or pliers. You may damage the new injector. Look at **Figure 23-11.**

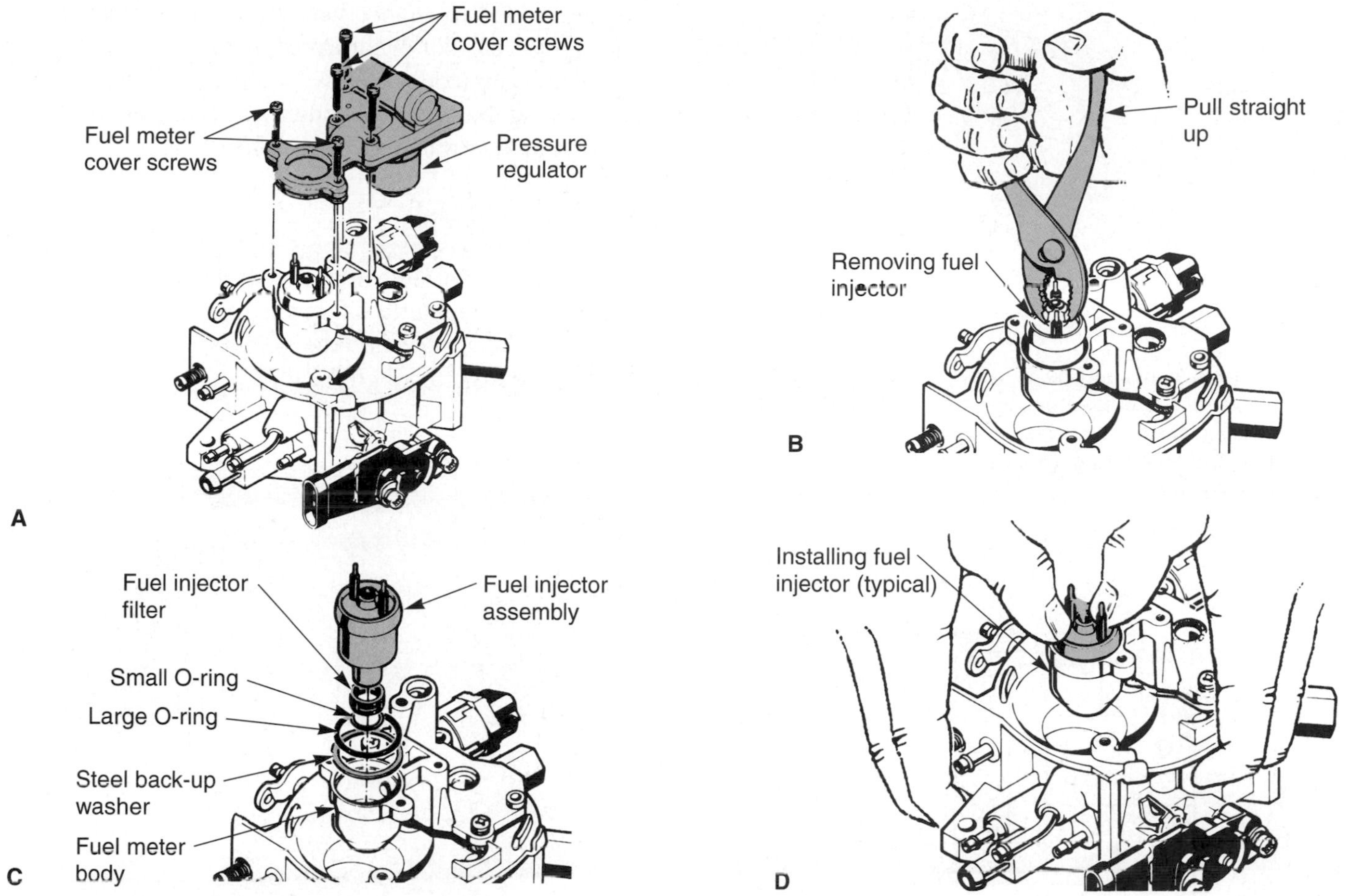

Fuel meter
cover screws

Fuel meter
cover screws

Pressure
regulator

A

Pull straight
up

Removing fuel
injector

B

Fuel injector
filter

Fuel injector
assembly

Small O-ring

Large O-ring

Steel back-up
washer

Fuel meter
body

C

Installing fuel
injector (typical)

D

Figure 23-12. Basic procedure for removing a TBI injector. A—Unscrew fasteners holding the fuel metering body. Lift off while keeping screw lengths organized. B—Use pliers to pull out old injector. C If needed, use new O-rings and washers. D—Carefully align and push the injector into the throttle body with your thumbs. (General Motors)

Note
Chapter 22, *Gasoline Injection Fundamentals,* covers subjects related to rebuilding a throttle body assembly.

Tech Tip
Icing or freeze up can occur in the throttle body of fuel injection systems. This can happen when the thermostatic flap in the air inlet system sticks in the open position. Only cool outside air, not engine preheated warm air, is flowing through the throttle body. Throttle body icing can occur even when the outside temperature is well above freezing.

Servicing EFI Multiport Injectors

The service of an injector for a multiport EFI system is like the service procedures for a TBI system. However, multiport injectors are located in the intake manifold runners, instead of the throttle body assembly.

Testing EFI Multiport Injector Operation

To quickly make sure each EFI injector is opening and closing, place a *stethoscope* (listening device) against each injector, **Figure 23-13.** A clicking sound means the injector is opening and closing. If you do *not* hear a clicking sound, the injector is not working. The injector solenoid, wiring harness, or control module control circuit may be bad.

With the engine off, you can check the condition of the coils on the inoperative injector. Use an ohmmeter. Measure the resistance across the injector coil and check for shorts to ground. If the coil is open (infinite resistance) or shorted (zero resistance to ground), you must replace the injector. Special injector coil testers are also available, **Figure 23-13.**

If the injector tests good, you may need to check the wiring going to that injector. Following the service manual, check supply voltage to the inoperative injector. You may also need to measure the resistance in the circuit between the injector solenoid and the control

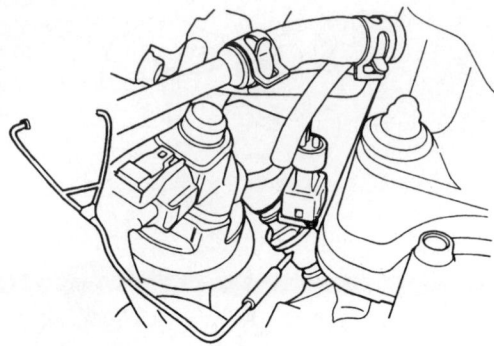

Figure 23-13. If the engine is missing and you suspect an injector, place a stethoscope on the side of the injector. A clicking sound shows that power is reaching the injector and the injector should be working. (Honda)

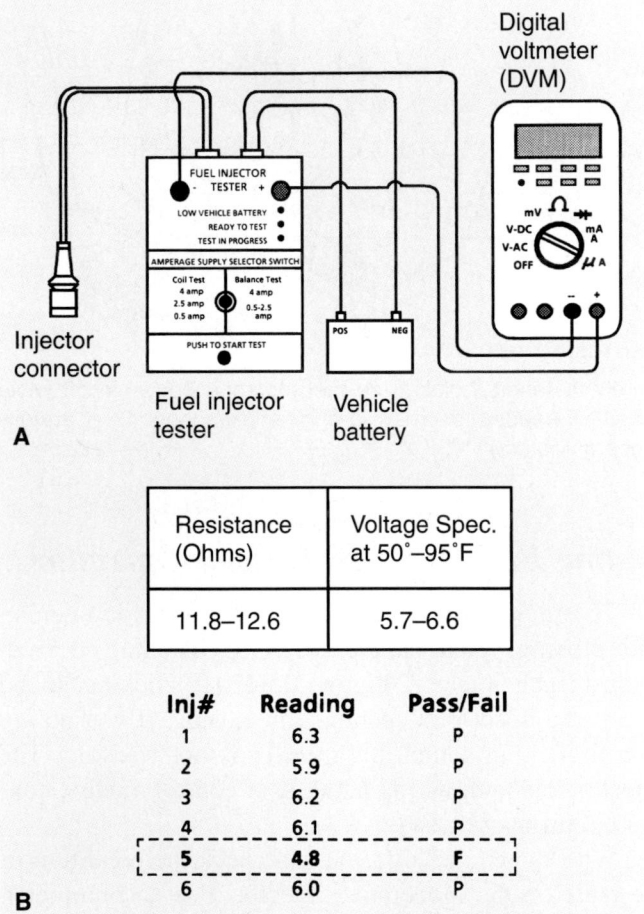

Resistance (Ohms)	Voltage Spec. at 50˚–95˚F
11.8–12.6	5.7–6.6

Inj#	Reading	Pass/Fail
1	6.3	P
2	5.9	P
3	6.2	P
4	6.1	P
5	4.8	F
6	6.0	P

Figure 23-14. This injector tester checks the injector coil for problems. It is faster and more accurate to use than a voltmeter or an ohmmeter alone, because it sends normal current through the injector. Since injector coil resistance and current flow drops as the injector warms, use the lowest reading obtained during the test. If the reading jumps up and down, suspect a partially open winding in the coil. A—Connections for using a fuel injector coil tester and digital multimeter. B—Compare readings from all injectors to determine if any are bad. Note that injector No. 5 is low and should be replaced. (Chevrolet)

module. A high resistance would indicate a frayed wire, broken wire, or poor electrical connection.

An injector coil tester is available for more accurately testing the electrical condition of the fuel injectors. Its use is demonstrated in **Figure 23-14.**

Injector Noid Light Test

A *noid light* is a special test light for checking electronic fuel injector feed circuits, **Figure 23-15.** Different noid lights are made to fit wiring harnesses from each auto manufacturer. Usually the make of vehicles the noid light will fit are printed on the tool.

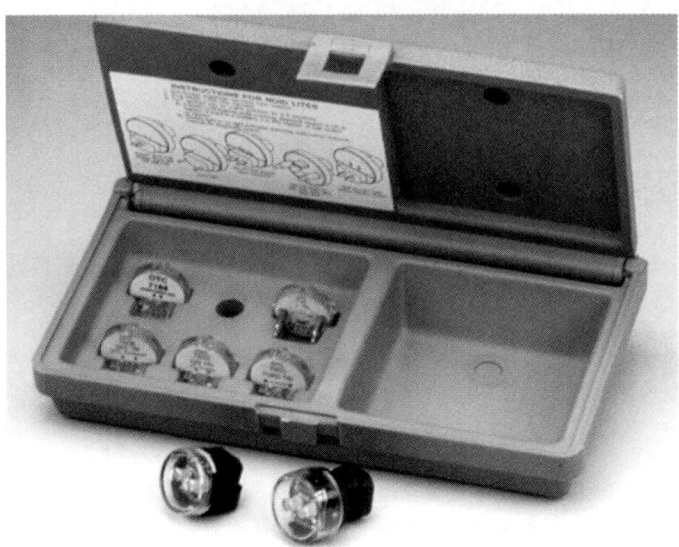

Figure 23-15. Noid lights are designed to check for normal current pulse being sent to each injector. Correct noid light must be used to prevent wiring connector damage. (OTC)

To use a noid light:
1. Disconnect the wiring harness from the fuel injector, **Figure 23-16A.** Make sure you release the connector properly to prevent part damage. Most harness connectors use a positive lock to keep the wiring from vibrating loose.
2. Fit the correct style noid light into the injector harness connector, **Figure 23-16B.**
3. Start the engine and check the light.
4. If the noid light flashes, you know that power is reaching the injector from the control module or control unit. If the noid light does *not* flash on and off, something is keeping current from reaching the noid light and injector. You could have an open in the wiring, a bad connection, open injector resistors, or control module troubles.

Figure 23-16. A—To use a noid test light, disconnect the harness from the injector. To avoid damage, release the locking mechanism on the connector. Do not force it off. B—Install the noid light into the injector connector. Start the engine. The noid light should flash on and off, showing electrical pulses for the injector. If not, check the wiring, connections, and control unit.

5. Repeat the noid light test on any injector that is not operating. Your stethoscope will quickly tell you which injector is "dead" and not clicking.

Refer to a wiring diagram when solving complex fuel injection electrical problems. The diagram will show all electrical connections and components that can upset the function of the injection system.

 Caution
Some EFI multiport systems use dropping resistors before the injectors. The resistors lower the supply voltage to the injectors. Do *not* connect direct battery voltage to this type of injector or coil damage may occur.

Injector balance test

An *injector balance test* uses a fuel injector tester to measure the amount of fuel flowing through each injector. This is a common test performed on modern multiport, electronic fuel injection systems. It will tell you if any injectors are clogged or not opening fully.

An electronic fuel injector tester is a tool that will fire an injector for a specific amount of time. Instead of the control module, the tool feeds current to the injector coil to make it open for a controlled time span. One is shown in **Figure 23-17.**

 To perform an injector balance test:
1. Connect a pressure gauge to the test fitting on the fuel rail. Make sure all fittings are tight and not leaking.
2. Close off the valve for measuring fuel volume if provided on the fuel gauge assembly.

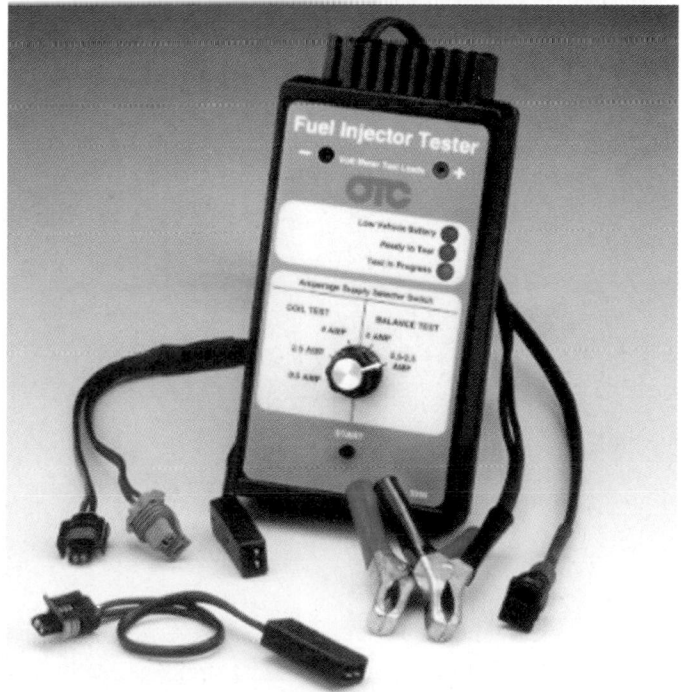

Figure 23-17. This is an injector flow balance tester. It can find clogged or partially failed fuel injectors. (OTC)

3. Connect the balance tester wiring to the injectors or injector in question. See **Figure 23-18.**
4. Turn the ignition key on to pressurize the system. Then, turn the ignition key off.
5. Press the injector balance tester button while watching the pressure gauge drop.
6. Record your pressure drop reading.
7. Repeat this on the other fuel injectors. This will allow you to measure how much fuel each injector is feeding into the engine when energized.

If one of the injectors shows a relatively low pressure drop, it is not injecting as much fuel. That injector could be clogged or worn. If an injector shows a higher-than-normal pressure drop, it could be sticking open and not closing properly. Refer to **Figure 23-19.**

Replacing EFI Multiport Injectors

An EFI multiport injector is easy to replace. After bleeding off fuel pressure, simply remove the hose from the injector and fuel manifold, **Figure 23-20.** Unplug the electrical connection and remove any fasteners holding the injector. Pull the injector out of the engine.

Inspect the boot and other rubber parts closely. Some manufacturers suggest that you replace the boot, seals, and hose if the injector is removed for service. Refer to **Figure 23-21.**

A

B

C

Figure 23-18. An injector flow balance test will find clogged or failing fuel injectors. A—Connect a pressure gauge to the fuel rail test fitting. Close the gauge valve when measuring fuel flow. B—Connect the injector balance tester to the injector terminals and to the battery. Pressurize the system by turning the ignition key on or starting the engine. Then, shut the engine off. C—Press the button on the tester to fire the injector. Note the pressure drop after the injector has been fired. Compare all injector readings to determine if any are bad.

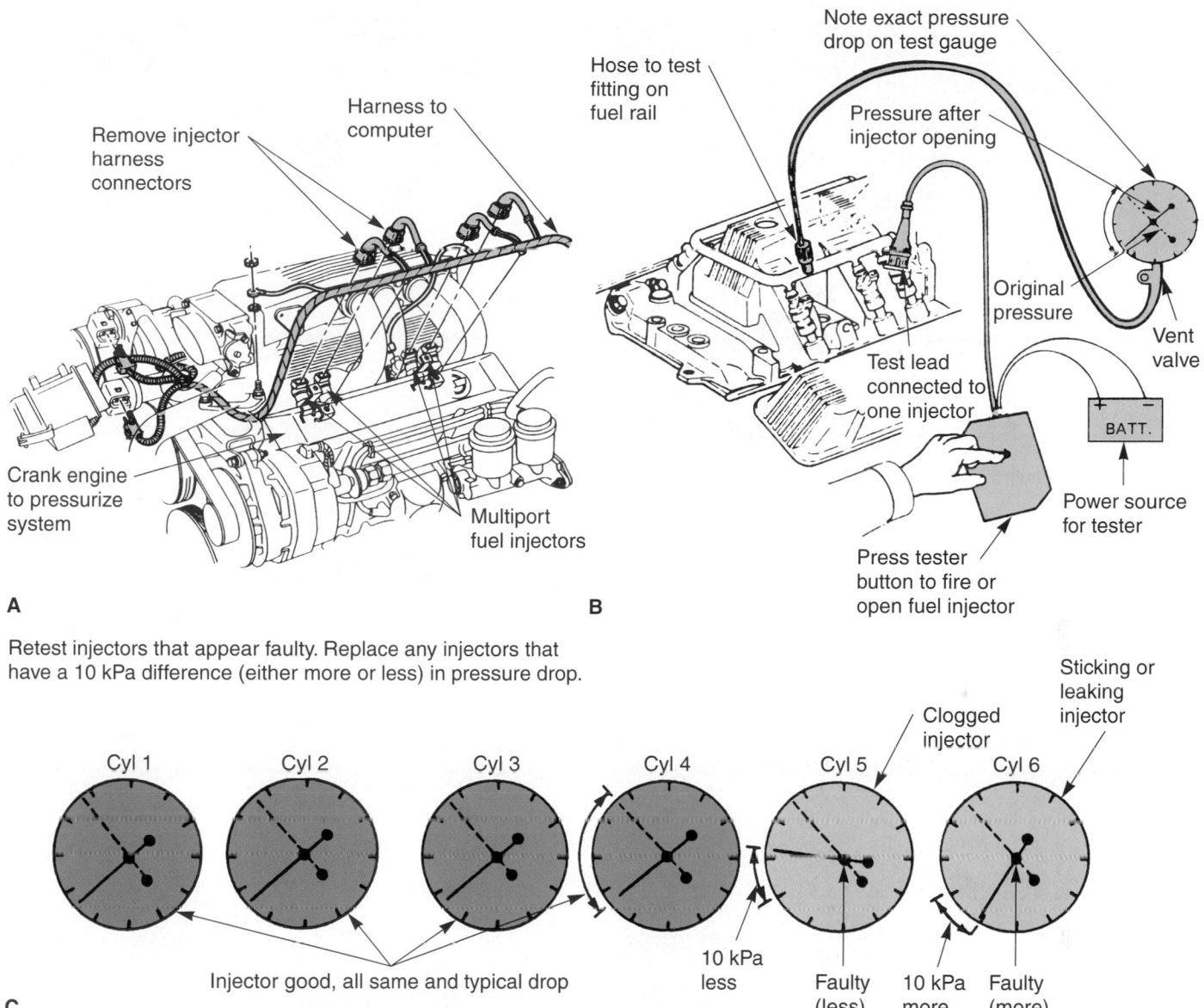

Retest injectors that appear faulty. Replace any injectors that have a 10 kPa difference (either more or less) in pressure drop.

Figure 23-19. Basic steps for doing a balance test on multiport injectors. A—Disconnect all wires from the injectors. Do not damage plastic connectors during removal. B—Connect the test harness to an injector and connect the pressure gauge to the test fitting on the fuel rail. Tester normally connects to the battery. Press tester button and note the pressure gauge reading. C—Pressure drop at the gauge should be the same for each injector. Turn the key on or activate the fuel pump between each test so the pressure is the same at each injector. If one injector has less pressure drop when fired, it might be clogged. If pressure drop is excessive, the injector might be worn or sticking. If pressure does not hold, the injector is leaking. (Chevrolet)

Install the new or serviced injector in reverse order, **Figure 23-22.** Use the directions in the shop manual for details. Exact procedures can vary.

Electronic Fuel Injector Cleaning

If the EFI injectors fail the balance test and show low drop, they may be clogged, and in need of cleaning, **Figure 23-23.**

A *fuel injector cleaning tool* uses shop air pressure to force a cleaning solution through the injectors to remove deposits. As shown in **Figure 23-24,** connect the

tool to the fuel rail. Following tool instructions, force the cleaning agent through the injectors. If engine performance improves, the injectors have been successfully cleaned. The cleaning agent is combustible and will burn as the engine operates.

⚠ Warning

Make sure a shop exhaust hose is connected to the tailpipe when cleaning injectors. Exhaust fumes that occur during testing can be more toxic than normal.

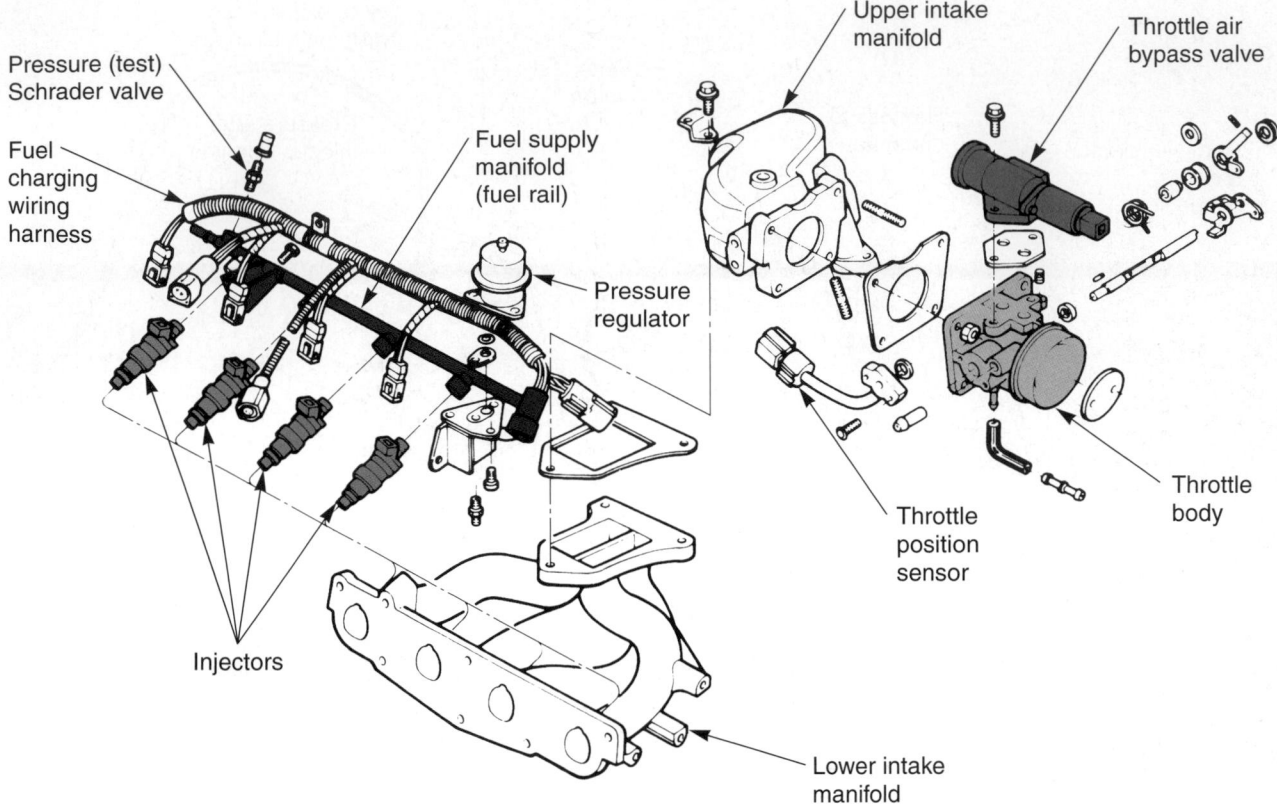

Figure 23-20. Exploded view shows multiport injectors, fuel rail, pressure regulator, air bypass valve, throttle body, throttle position sensor, and intake manifold assembly. Study part locations. (Ford)

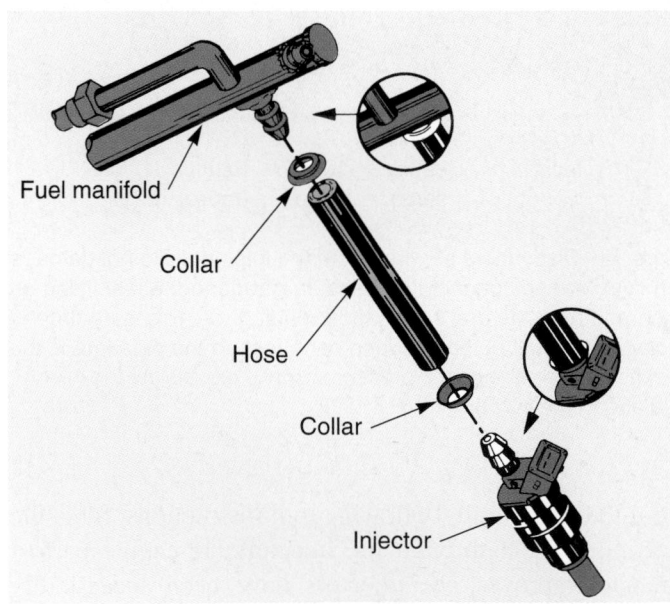

Figure 23-21. Multiport injector removal is simple. Remove the fuel hose and the electrical plug. Bolts may hold the injector in the manifold, or the injector may use a press-fit with a boot. (Fiat)

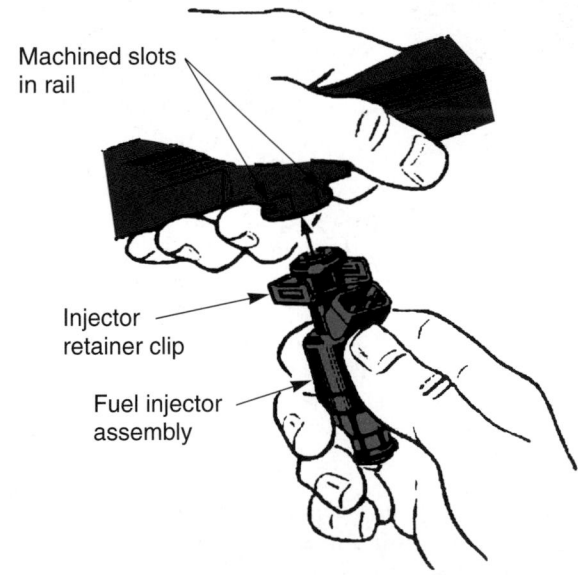

Figure 23-22. When installing a multiport injector, make sure the seals, washers, and boots are in perfect condition. Some automakers suggest using new ones. (Fiat)

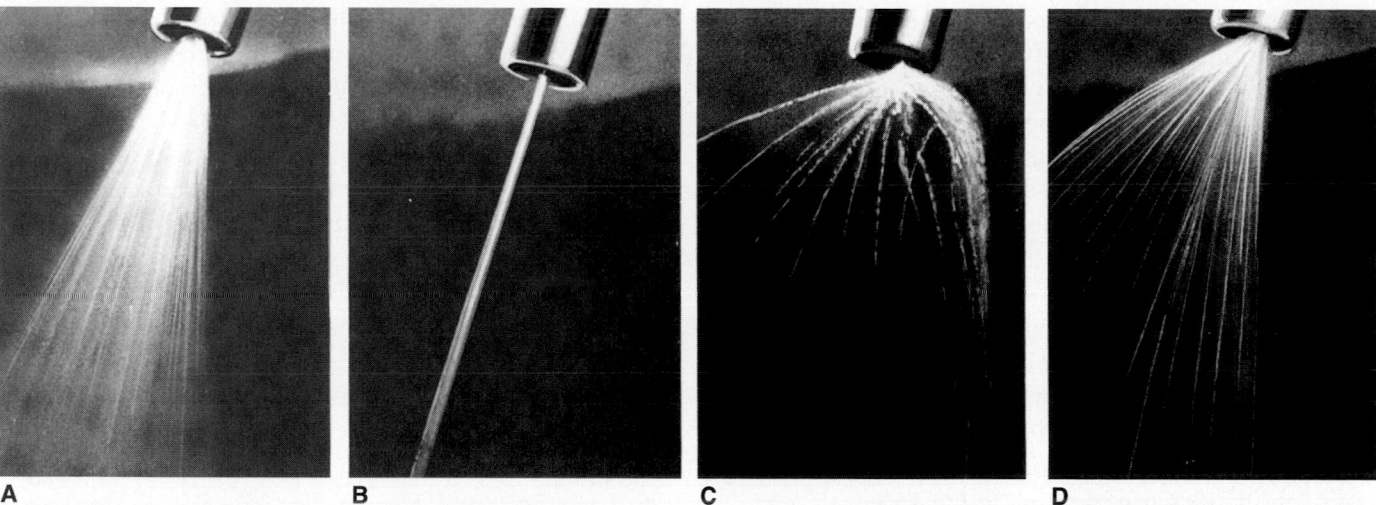

Figure 23-23. Gasoline injector spray patterns. A—Good, even, partially atomized pattern. B—Solid stream, poor spray pattern. C—Dirty nozzle causing poor spray pattern. D—Uneven spray pattern is unacceptable. (Saab)

Figure 23-24. Fuel injector cleaner will force solvent through clogged or restricted injectors. Sometimes, this will return them to normal operation without removal. (OTC)

Engine Sensor Service

Most EFI engine sensors can be checked with a scan tool. A modern scan tool will give operating values which can be compared to specs. However, an engine sensor can also be checked with a digital meter or a test light if not equipped with advanced on-board diagnostics.

 Caution
When a service manual directs you to measure voltage or resistance in an EFI sensor, use a high-impedance meter or a digital meter. An inexpensive analog meter could damage some sensors.

Throttle Position Sensor Service

A throttle position sensor should produce a given amount of resistance for different throttle openings. For example, the sensor might have high resistance with the throttle plates closed and a lower resistance when the throttles are open. Compare your ohmmeter readings to specs to determine the condition of a throttle position sensor.

Throttle Position Sensor Service

A *bad throttle position sensor* (TPS) can affect fuel metering, ignition timing, and other control module outputs. It can also trip several trouble codes on some systems and can be a frequent cause of problems. Many throttle position switches use contact points or variable resistors that can wear and fail.

A throttle position sensor is comparable to a carburetor accelerator pump and metering rod. It signals the control module when the gas pedal is depressed to different positions for acceleration, deceleration, idle, cruise, and full power. It can cause a wide range of performance problems. If shorted, it might make the fuel mixture too rich or, if open, too lean.

The throttle position sensor can sometimes be tested at a special tester terminal in the wiring harness. Some manuals say you should measure voltage drops across the sensor at specified throttle positions. A reference voltage is fed to the sensor by the control module.

Many manuals also recommend checking the resistance of the throttle position sensor at different throttle openings. The manual might have you measure ohms at idle, half throttle, and full throttle. If resistance is within specs, check the wiring leading to the TPS.

Throttle Position Sensor Removal

To remove many throttle position sensors, you must file or grind off stakes (small welds) on the sensor screws. You might also have to drill into the screws from the bottom of the throttle body assembly. This will let you turn and remove the screws and TPS. Refer to the service manual for details.

Throttle Position Sensor Adjustment

Some throttle position sensors must be adjusted; some cannot be adjusted. Many are mounted so they can be rotated on the carburetor or throttle body. Either a special tester or an ohmmeter is commonly used to adjust a throttle position sensor.

Basically, you must measure sensor resistance or note tester output with the throttle at specific positions. You may have to insert a feeler gauge under the throttle lever, or have the throttle plates at curb idle, for example. With the throttle plates at the correct angle, the TPS should trigger the tester or show a specified ohms value. If an adjustment is needed and possible, loosen the sensor mounting screws. Rotate the TPS until the correct ohms reading is obtained. Then, tighten the mounting screws and recheck the meter reading.

Manifold Absolute Pressure Sensor Service

A *bad manifold pressure sensor* can affect the air-fuel ratio when the engine accelerates and decelerates. It serves the same basic function as a power valve in a carburetor. It senses engine vacuum to signal when more fuel is needed under a load or when gaining speed. It might also have some effect on ignition timing and a few other control module outputs.

To test a manifold pressure sensor, use your scan tool or measure sensor circuit voltage or sensor resistance

while applying vacuum to the unit. Use a vacuum pump on the vacuum fitting of the sensor. Apply specified vacuum levels while measuring the output of the MAP sensor.

Some manuals instruct you to measure sensor voltage at a specified test terminal. Others might have you disconnect the wires from the sensor and compare ohmmeter readings to specs. In any case, sensor values must be within limits at the various vacuum levels. If testing at the test terminal, check the wiring harness before condemning the sensor. A poor connection or short could upset a reference voltage flowing to the sensor.

Oxygen Sensor Service

A *bad oxygen sensor* will primarily upset the fuel injection system or the computerized carburetor system. The voltage signal from the sensor represents air-fuel ratio. If the oxygen sensor produces a false output (incorrect voltage), the control module cannot precisely control how much fuel is metered into the engine. A rich mixture or lean misfire condition could result.

Tech Tip

Oxygen sensors should be replaced at periodic intervals. After prolonged service, they become coated or fouled with exhaust by-products. As this happens, fuel economy and emissions will be adversely affected. If gas mileage is 10% to 15% lower than normal, suspect the oxygen sensor of slow response. One- and two-wire sensors should be replaced at about 50,000–60,000 miles. Heated three-wire oxygen sensors should be replaced at about 100,000 miles.

Oxygen Sensor Contamination

Normally, an oxygen sensor is designed to last about 50,000 miles (81,000 km). However, its life can be shortened by contamination, blocked outside air, shorts, and poor electrical connections.

Oxygen sensor contamination can be caused by:

- *Leaded fuel.* Leaded fuel is the most common cause of oxygen sensor contamination. Lead coats the ceramic element and the sensor cannot produce enough voltage output for control module.

- *Silicone.* Sources are antifreeze, RTV silicone sealers, waterproofing sprays, and gasoline additives. Silicone forms a glassy coating.

- *Carbon.* Carbon contamination results from rich fuel mixtures. Carbon in fuel coats the sensor.

Carbon and moderate lead contamination can sometimes be reversed. Run the engine at high speeds with a large vacuum hose (PCV hose, for example) removed and with only unleaded fuel in the tank. This will sometimes burn off light lead and most carbon deposits. The sensor may start working normally again.

Oxygen Sensor Inspection

Also, check that the outside of the sensor and its electrical connection are free of oil, dirt, undercoating, and other deposits. If outside air cannot circulate through the oxygen sensor, the sensor will not function.

An oxygen sensor generates only a tiny voltage (an average of about 0.5 volts). A poor electrical connection can prevent this small voltage from reaching the control module. Always check the sensor's electrical connections.

Oxygen Sensor Testing

Discussed earlier, most control module systems will now produce a trouble code and give operating voltages indicating when the output from the oxygen sensor is *not* within normal parameters. This would tell you to do further tests on the oxygen sensor and its circuit.

Many control module systems have a limp-in mode. If the oxygen sensor or some other sensor fails and produces an incorrect output, the system will go into this emergency limp-in mode. A predetermined oxygen sensor voltage (0.5 volts for example) will be simulated by the control module and used to keep the engine running well enough to drive in for repairs.

A digital voltmeter can also be used to test the output of an oxygen sensor. Warm the engine to full operating temperature to shift the control module system into closed loop. The sensor must be hot (about 600°F or 315°C) to operate properly. You may have to warm the engine at fast idle for up to 15 minutes with some cars. Note that a few systems can drop out of closed loop at idle.

Warning
Only use a high-impedance digital meter to measure oxygen sensor voltage. A conventional analog or low-resistance meter can draw too much current and damage the oxygen sensor.

Oxygen Sensor Output Voltage

In most cases, *oxygen sensor output voltage* should cycle up and down from about 0.2–0.8 volts (200–800 mV). A 0.2 volt or low reading would show a lean air-fuel ratio and a 0.8 volt or high reading would show a richer condition. A high or low reading does not always mean the oxygen sensor is bad. Another problem (leaking or clogged fuel injector, for example) could make the sensor read high or low.

A quick test to see if the oxygen sensor reacts to a change in air-fuel mixture is to pull off a large vacuum hose, like the hose to the PCV valve. This extra air should make the oxygen sensor try to richen the fuel mixture and compensate for the air leak (lean condition). The output voltage should then go *down* (to about 0.2 or 0.3 volts), to signal a need for more fuel to adjust for the vacuum leak or extra air.

When the engine throttle is snapped open and closed, oxygen sensor output should also cycle up and down to show the change in air-fuel mixture.

If you block the air inlet at the air cleaner or inject propane gas into the air inlet (creating a rich mixture), the oxygen sensor voltage should *increase* (go up to about 0.8 to 0.9 volts). It should try to signal the control module that too much fuel, or not enough air, is entering the combustion chambers.

If the oxygen sensor voltage does not change properly as you simulate rich and lean air-fuel ratios, the oxygen sensor is faulty. You might try running the engine at high speed, with a large vacuum hose removed, to clean off light lead or carbon contamination.

A faulty oxygen sensor will usually be locked at one voltage level and will not cycle voltage up and down. It also may not produce enough voltage.

Testing Oxygen Sensor Circuit

If the oxygen sensor has normal voltage, you should check the circuit leading to the sensor. Measure the resistance of the wires leading to the oxygen sensor. You can use long test leads. You can also ground one end of the sensor wire and check it for continuity at the other end.

Oxygen Sensor Replacement

Disconnect the negative battery cable. Then, separate the sensor from the wiring harness by unplugging the connector. Never pull on the wires themselves, as damage may result.

The oxygen sensor may have a permanently attached pigtail. Never attempt to remove it. Use a wrench to unscrew the sensor. Inspect its condition. Some sensors may be difficult to remove at temperatures below 120°F. Use care to avoid thread damage.

Tech Tip
When attempting to remove an oxygen sensor with seized threads, you can unintentionally round off the hex nut with a slotted sensor socket. If this happens, cut off the sensor wires and use a six-point deep-well or spark plug socket. If the sensor's hex nut is not badly damaged, the socket should grasp and turn the "stubborn" oxygen sensor.

Follow these rules when installing an oxygen sensor:

- Do *not* touch the sensor element with anything (water, solvents, etc.)
- Coat oxygen sensor threads before installation with anti-seize compound to prevent seizure and thread damage.
- Do *not* use silicone-based sealers on or around exhaust system components. Use only low volatile silicone sealers sparingly on engine components. The PCV system can draw silicone fumes into engine intake manifold and over oxygen sensor.
- Hand start sensor to prevent cross threading.
- Do *not* overtighten the oxygen sensor. It could be damaged.
- Make sure outside vents are clear so that air can circulate through the sensor.
- Make sure wiring is reconnected securely to sensor.
- If sensor checks out good, check continuity of wiring between sensor and control module.
- Check oxygen sensor output and fuel system operation after installing sensor.
- Repair any engine oil leaks that might contaminate new oxygen sensor right away.

Coat the threads of the new sensor with anti-seize compound. Start the sensor by hand. Then screw in and tighten the sensor with a wrench. Do not overtighten or the sensor may be damaged. Check fuel injection system operation after sensor installation.

Reading Oxygen Sensor

Reading an oxygen sensor can indicate fuel system problems, silicone contamination, leaded fuel in the tank, and other troubles. To *read an oxygen sensor*, inspect the color of the sensor's tip. See **Figure 23-25.**

- A *light gray tip* is normal for an oxygen sensor.
- A *white sensor tip* might indicate a lean mixture or silicone contamination. Sensor must usually be replaced.
- A *tan sensor tip* could be lead contamination. It can sometimes be cleaned away but a new sensor is usually needed.
- A *black sensor tip* normally indicates a rich mixture and carbon contamination which can sometimes be cleaned after correcting the cause.

Note that some manufacturers recommend oxygen sensor replacement after only 12,000 miles (19,308 km) when the sensor is removed. Therefore, the oxygen sensor is normally replaced with a new one when unscrewed from the exhaust system.

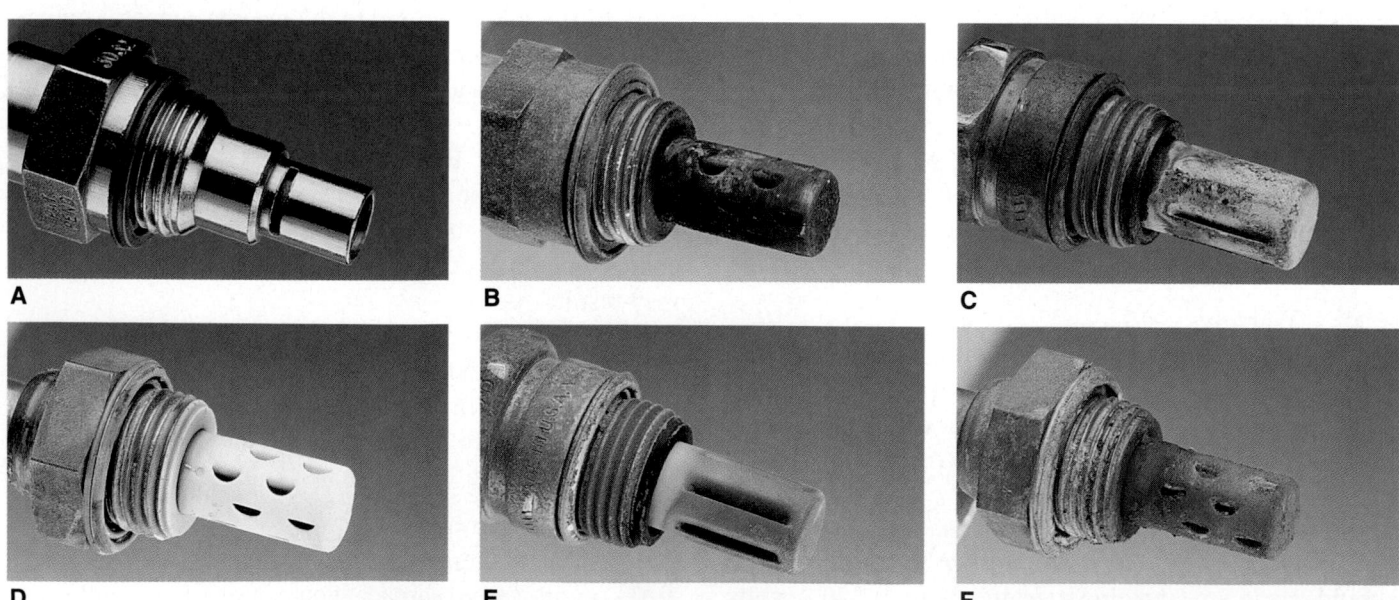

Figure 23-25. After removing an oxygen sensor, note the color of its tip. This can give you information about the condition of the engine and its support systems. A—The new oxygen sensor must be identical to and have the same part number as the old one. This is a high-quality platinum oxygen sensor that is designed to provide long service life. B—A rich mixture caused by a fuel system problem has coated this sensor with black soot. C—Antifreeze contamination was caused by a leaking head gasket or a crack in the engine. D—Silicone contamination resulted from the use of too much silicone sealer during an engine rebuild. E—Lead contamination was caused by using leaded racing fuel or a lead octane additive. Check the fuel tank filler neck for modification. F—Oil contamination caused by worn piston rings and valve seals. (Tomco Inc.)

Oxygen Sensor Signal Generator

An *oxygen sensor signal generator* is a tool for sending a false 200–800 mV signal to the control module for testing purposes. You can then alter air-fuel ratio (simulate vacuum leak for example) and perform other testing functions. Since the control module thinks it is closed loop, you can do tests without the control module tripping a trouble code or trying to compensate for your intentional modifications of engine operating conditions.

Temperature Sensor Service

Most EFI systems use both a coolant temperature sensor and an inlet air temperature sensor. If these sensors are bad, they will make the engine run either rich or lean. The internal resistance of these sensors changes with temperature.

Engine Coolant Temperature Sensor Service

A *bad coolant temperature sensor* can also affect air-fuel ratio and ignition timing by not accurately informing the control module of the engine operating temperature. The coolant sensor serves a similar function to a carburetor choke; it richens the mixture when cold and leans the mixture when hot. If open, the coolant sensor might affect cold engine driveability. If shorted, it might affect warm engine driveability.

If the older vehicle is not equipped with on-board diagnostics, a digital ohmmeter is often recommended for testing a temperature sensor. The service manual will give resistance values for various temperatures. If the ohmmeter test readings are not within specs for each temperature value, the sensor is bad and must be replaced.

Note that temperature sensor operation and ratings can vary. When purchasing a new temperature sensor, or any sensor, make sure you have the *right one*. If you install the wrong sensor, it will upset the operation of the control module system.

When replacing a coolant sensor, use a deep socket or six-point wrench to unscrew the old unit. You might want to coat the sensor threads with approved sealer. Then, hand start and tighten the sensor in the engine. Do not overtighten the coolant sensor or it could bottom out in the engine and be ruined.

Intake Air Temperature Sensor Service

A *bad intake air temperature sensor* will usually not have a pronounced effect on vehicle operation. It normally allows the control module system to make fine adjustments of air-fuel ratio and timing with changes in outside air temperature. If the sensor fails, it will normally trigger a self-diagnosis code and you would know to test the sensor and its circuit.

A scan tool or an ohmmeter is commonly used for checking an air temperature sensor. As with a coolant sensor, the unit is frequently a thermistor that changes internal resistance with temperature. The sensor should have spec ohms for certain temperatures.

Airflow Sensor Service

A *bad airflow sensor* will normally cause the system to go into limp-in mode, as will several other sensors. If shorted or opened, the control module will begin to operate on preprogrammed values. The car will perform poorly and get lower fuel economy.

Since there are various types of airflow sensors, you must refer to the shop manual for exact procedures. Many manufacturers have you use an ohmmeter to check the airflow sensor when it is a variable resistance type. Others have integrated circuits. Special testing methods are required to prevent sensor damage. On late-model vehicles, the scan tool will indicate a problem with an airflow sensor.

If faulty, you must remove and replace the airflow sensor. During replacement, make sure that you have the correct unit. Also, tighten all fittings carefully. An *air leak* after the airflow sensor will upset its operation, and can trigger trouble codes.

Servicing Other EFI Sensors

The other sensors in an electronic fuel injection system are tested using the same general procedures just discussed. You would use the self-diagnosis mode, a special analyzer, or a digital meter to check each sensor. Refer to a service manual for exact procedures.

 Note
For detailed information on sensor service, refer to index. Sensor testing and replacement is covered in numerous chapters.

Control Module Service

When diagnostic trouble codes and/or pinpoint tests indicate a faulty control module, the module must generally be replaced. Procedures for replacing a control module vary from manufacturer to manufacturer. Always refer to an appropriate service manual for specific instructions.

As discussed in Chapter 17, a *PROM* (programmable read only memory) is an integrated circuit chip found in all control modules. In many late-model vehicles, this chip cannot be removed from the control module. In older vehicles (pre-OBD II), however, the original PROM can be removed from the faulty control module and reused in the new unit. This will ready the new control module for a specific vehicle. See **Figure 23-26.**

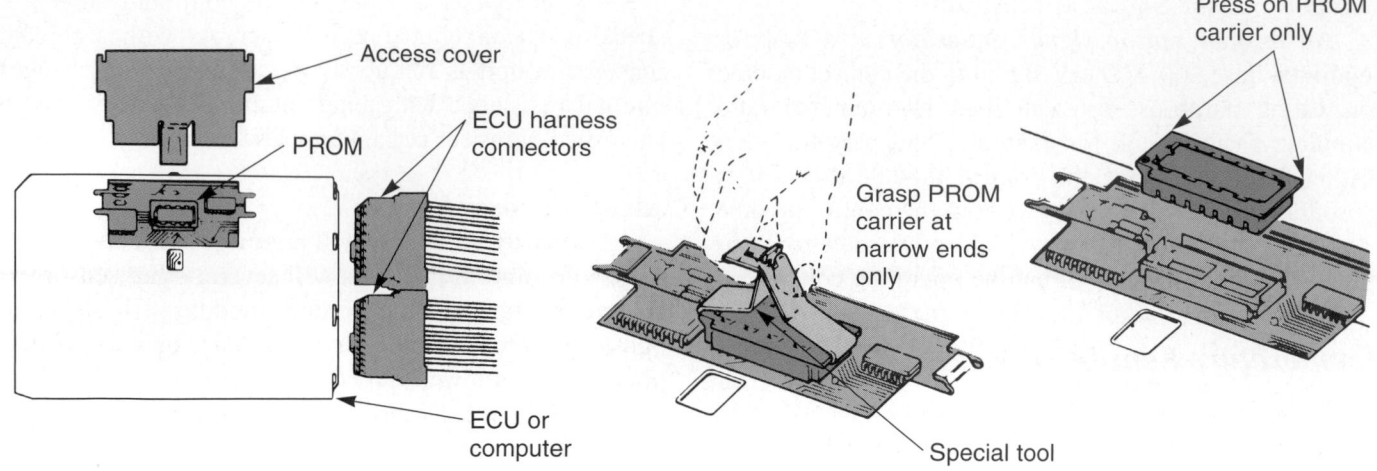

Figure 23-26. When tests find a faulty control module, you may need to replace the entire module. Sometimes, however, the PROM (computer chip calibrated for exact engine) must be reused. A—Unplug the control module and remove the access cover. B—Pull the PROM out using a special tool. C—When installing a new PROM, keep your fingers off the chip itself. Only press on the outer carrier. (GM)

The control module is normally mounted under the dash. This keeps the electronic circuits away from engine heat and vibration. In a few cars, however, the control module is in the engine compartment.

 Note
Control module system operation, diagnosis, and repair are detailed in Chapters 17, 18, and 19.

Gasoline Injection Adjustments

As with a carburetor, there are several tune-up adjustments needed on a gasoline injection system. These include:

- Engine idle speed adjustment, **Figure 23-27.**
- Throttle plate stop adjustment.
- Idle air-fuel mixture adjustment.
- Throttle cable adjustment.

Since there are so many types of systems, refer to a service manual for exact procedures. **Figure 23-28** shows a wiring diagram for a typical electronic fuel injection system.

 Tech Tip
Many automakers warn against making some adjustments on a gasoline injection system. For example, only make mixture adjustments when major problems exist or an exhaust analyzer shows high emission levels. If you must make large adjustments, there are usually other problems in the system.

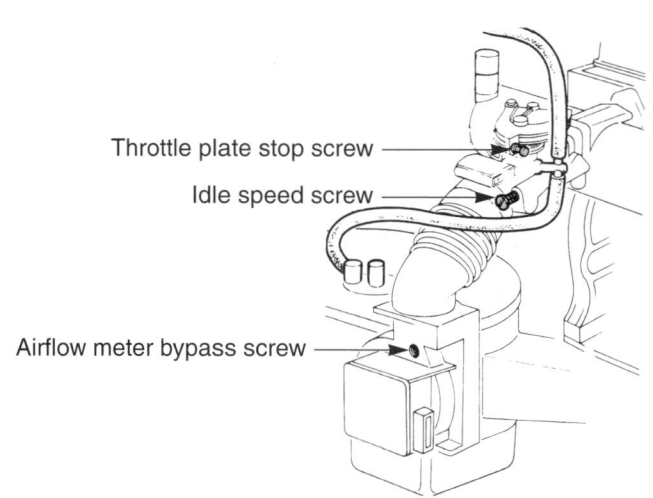

Figure 23-27. Adjustments and their access locations will vary with the particular model and type of gasoline injection system. These are typical of one system. (Renault)

Testing Idle Control Motor

A *bad idle speed motor* may not be able to maintain the correct engine idle speed. Engine idle speed may be too low or too high for conditions. The servo motor could have shorted windings, open windings, bad internal parts, or other problems.

To check an idle speed motor, jump battery voltage to specific terminals on the servo motor. This should make the idle speed motor plunger retract and extend as the connections are reversed. A faulty motor will usually not function.

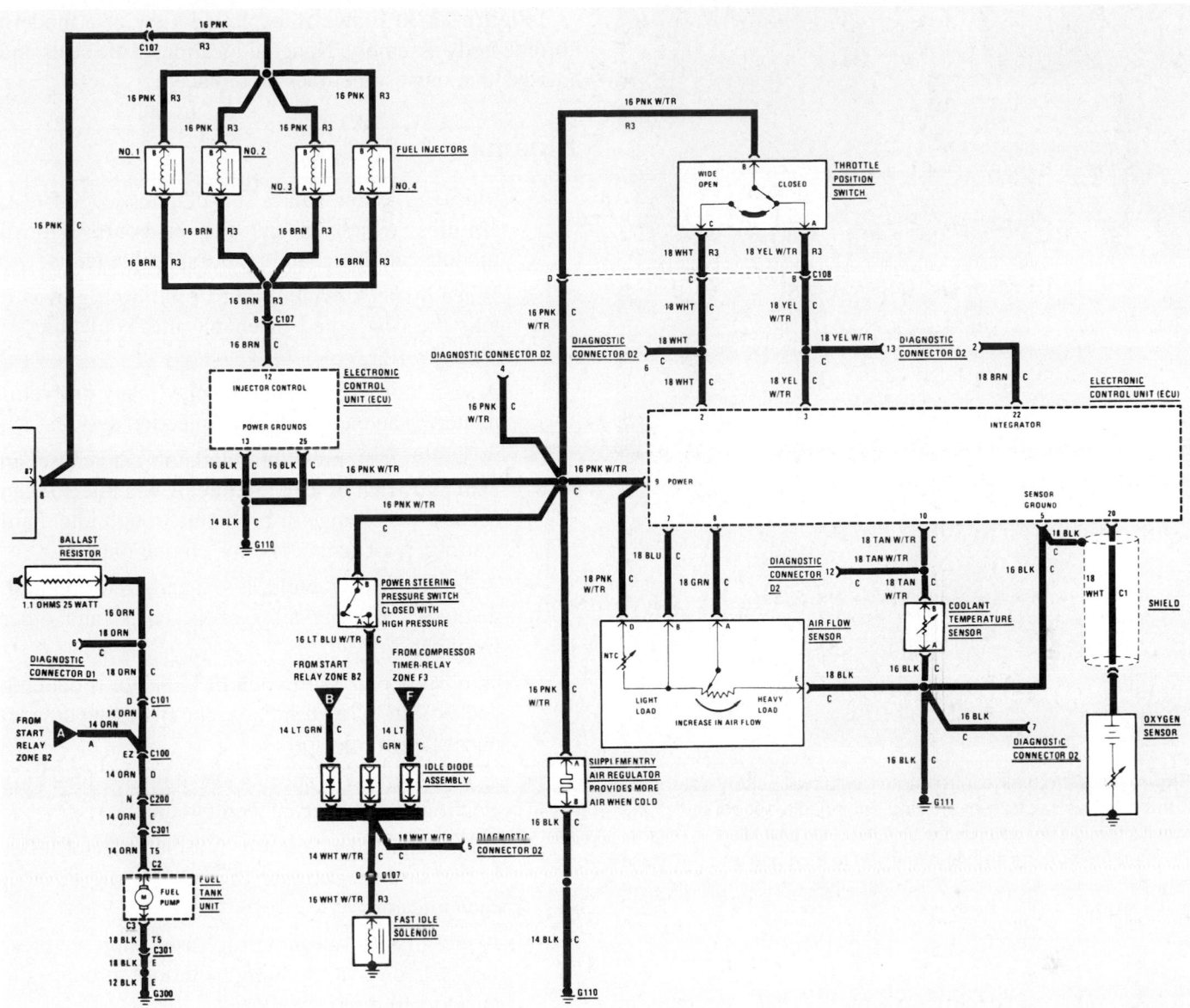

Figure 23-28. Study this partial wiring diagram for an EFI multiport injection system. Note the diagnostic (troubleshooting) connectors. Also note how injectors, oxygen sensor, airflow sensor, throttle position switch, and other components connect to ECU (control unit). (Renault)

If the idle speed motor works when jumped to battery voltage, check wiring leading to motor. Wiring harness could have an open or short. A control module or relay problem could also prevent motor operation.

Idle Air Control Valve Service

A *bad idle air control valve* will upset engine idle speed like a bad idle speed motor. It uses solenoid action to open and close an air passage bypassing the throttle plates. In this way, it can increase or decrease engine idle rpm. If it fails, engine rpm will be constant and may not increase with a cold engine or decrease as the engine warms to operating temperature.

To check the idle air control valve, jump battery voltage to the windings. This should trigger the solenoid and change engine speed. If the engine speed does not change, check for blockage in the passage at the idle air control valve before replacing the unit. An ohmmeter may also be recommended to check the windings.

Idle Air Control Motor Tester

An *idle air control motor tester* energizes the solenoid or servo motor to check its affect on engine idle speed. See **Figure 23-29.** Disconnect the wiring going to the idle air motor. Connect the tester harness to the motor.

Figure 23-29. Idle air control motor tester will quickly determine if the motor is capable of altering engine idle speed. Install the tool on the idle air control motor connector and idle the engine. Pushing the buttons should allow you to increase and decrease engine idle speed. If not, check for passage restrictions or a failed motor.

As you press the up arrow on the tester, engine idle speed should increase. As you press the down arrow, engine idle speed should decrease. If engine idle speed does not change, you may have a bad idle air control motor, clogged passage, or bad vacuum leak.

If the tester shows a functioning idle air control motor, check that the control module is sending the right amount of voltage to the motor. You could have an open in the wiring or control module problem.

Multiport Throttle Body Service

The service of a multiport throttle body usually involves replacing dried, deteriorated gaskets and seals. Basically, remove the throttle body from the engine. Disassemble the unit while keeping all parts organized. Inspect moving parts for wear. The throttle shaft and its bushings can wear with prolonged service. Install new gaskets and seals and reassemble the unit in reverse order of disassembly.

Figure 23-30 shows an exploded view of a modern throttle body assembly. Note the location of the seals and gaskets that must sometimes be replaced.

Summary

- Most EFI systems have self-diagnostic (self-test) abilities which means the on-board control module can detect and record possible faults.
- When a *check engine light* in the dash glows, it tells the driver and mechanic that something is wrong.
- A scan tool will find and display many problems related to an electronic fuel injection system.
- A faulty fuel pressure regulator can cause an extremely rich or lean mixture. A bad injector can cause a wide range of problems: rough idle, hard starting, poor fuel economy, engine miss, etc.
- A throttle body rebuild, like a carburetor rebuild, involves replacing all gaskets, seals, and other worn parts.
- To quickly make sure each EFI injector is opening and closing, place a stethoscope (listening device) against each injector.
- A noid light is a special test light for checking electronic fuel injector feed circuits.
- An injector balance test uses a fuel injector tester to measure the amount of fuel flowing through each injector.
- A fuel injector cleaning tool uses shop air pressure to force a cleaning solution through the injectors to remove deposits.

Important Terms

Trouble code	Noid light
Malfunction indicator light	Injector balance test
Scan tool	Fuel injector cleaning tool
Fuel system monitoring	Oxygen sensor contamination
Short term fuel trim	Leaded fuel
Long term fuel trim	Silicone
Oxygen sensor monitoring	Carbon
Heater monitor	Oxygen sensor output voltage
EFI tester	
Oscilloscope	Oxygen sensor signal generator
Fuel pressure regulator	
Relief valve	PROM
Schrader valve	Idle air control motor tester
Throttle body rebuild	
Stethoscope	

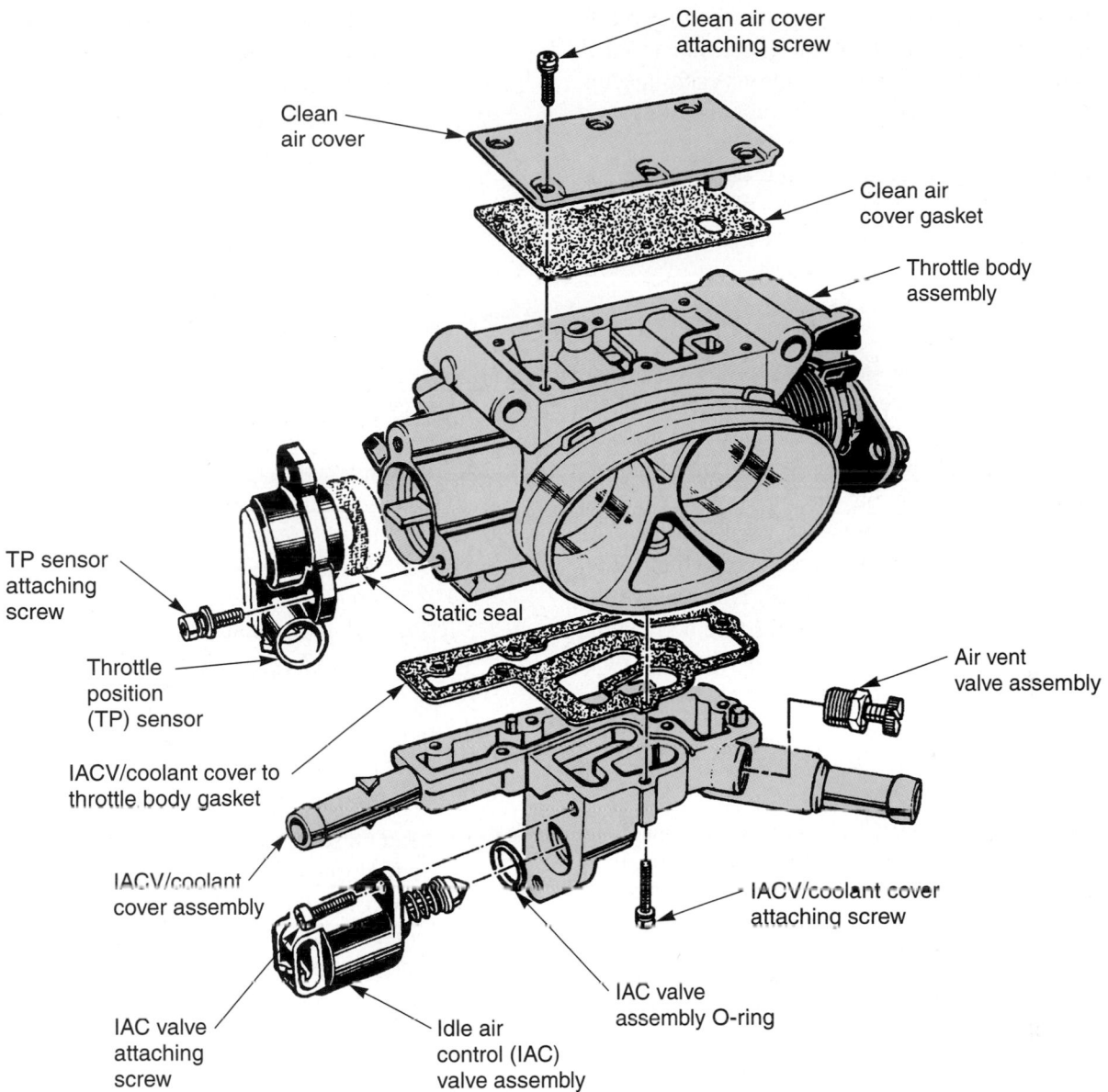

Clean air cover attaching screw

Clean air cover

Clean air cover gasket

Throttle body assembly

TP sensor attaching screw

Static seal

Throttle position (TP) sensor

Air vent valve assembly

IACV/coolant cover to throttle body gasket

IACV/coolant cover assembly

IAC valve attaching screw

Idle air control (IAC) valve assembly

IAC valve assembly O-ring

IACV/coolant cover attaching screw

Figure 23-30. Service of throttle body for multiport fuel injection is similar to if not easier than servicing a TBI unit. Gaskets and seals sometimes require replacement. Check the throttle shaft for wear. Reinstall the sensors and IAC motor properly. (Chevrolet)

Review Questions—Chapter 23

Please do not write in this text. Place your answers on a separate sheet of paper.

1. What four things must be used when diagnosing problems in a gasoline injection system?

2. Explain the EFI self-diagnosis mode found on many late-model cars.

3. How does an EFI tester work?

4. An oscilloscope can sometimes be used to view the electrical waveforms at the injectors for checking the control module, wiring, and injector coil. True or False?

5. Many EFI systems can maintain a fuel pressure as high as _____ psi or _____ kPa.

6. Which of the following does *not* pertain to a fuel pressure regulator?
   (A) Can cause a rich fuel mixture.
   (B) Can cause a lean fuel mixture.
   (C) Tested with a dwell meter.
   (D) Tested with a pressure gauge.

7. Where are the locations of the fuel pressure regulators with throttle body and multiport injection?

8. An engine with electronic multiport injection has a rough idle, as if one cylinder is dead (not

firing). The ignition system and engine are in good condition. Technician A says that a stethoscope should be used to listen to each injector. One of the injectors may not be operating (making a clicking sound). Technician B says that all of the injectors should be removed and checked for a proper spray pattern. Who is correct?
(A) A only.
(B) B only.
(C) Both A and B.
(D) Neither A nor B.

9. An engine with electronic throttle body injection will not run. The ignition is producing adequate spark and the engine is in good condition. Technician A says that the fuel pressure should be measured. The pressure regulator may be bad or there may be a clogged fuel filter. Technician B says that you should remove the air filter and look inside the throttle body while trying to start the engine. This will show you if the EFI system is working. Who is correct?
(A) A only.
(B) B only.
(C) Both A and B.
(D) Neither A nor B.

10. List nine rules to follow when replacing a throttle body injector.

11. An ohmmeter can be used to check an EFI injector coil for shorts or opens. True or False?

12. Why is injector fuel volume output important?

13. How do you do an electronic fuel injector balance test?

14. A digital meter is better than a conventional needle type meter when testing EFI sensors, because it draws less current. True or False?

15. A throttle position sensor should produce a given amount of _____ for different throttle openings.

16. An oxygen sensor is very sensitive and can be damaged if tested improperly with a VOM. True or False?

17. A(n) _____ _____ is often recommended for testing an EFI temperature sensor.

18. What is an EFI system PROM?

19. Where is the EFI system control module normally mounted?

20. List and explain four gasoline injection system adjustments.

21. Many automakers warn against making some gasoline injection system adjustments unless absolutely necessary. True or False?

22. Oxygen sensors are normally designed to last about _____ miles or _____ km.

23. A faulty oxygen sensor will usually be locked at _____ _____ _____ and will *not* _____.

24. How do you adjust a throttle position sensor?

25. A black oxygen sensor tip normally indicates a(n) _____ mixture and _____ contamination.

⬢ ASE-Type Questions

1. Which of the following can be used to test voltage values at EFI injectors?
(A) *Oscilloscope.*
(B) *Milliammeter.*
(C) *Ohmmeter.*
(D) *All of the above.*

2. A faulty fuel pressure regulator can cause:
(A) *an extremely rich fuel mixture.*
(B) *an extremely lean fuel mixture.*
(C) *Either A or B may result.*
(D) *Neither A nor B will result.*

3. After testing a fuel pressure regulator and finding pressure low, Technician A suggests checking the fuel pump before actually replacing the regulator. Technician B proposes checking the fuel filters before replacing the regulator. Who is right?
(A) *A only.*
(B) *B only.*
(C) *Both A and B.*
(D) *Neither A nor B.*

4. Which of the following is *not* a common fuel pressure regulator location?
(A) *On the fuel rail.*
(B) *On control solenoid.*
(C) *On throttle body assembly.*
(D) *Connected to fuel manifold.*

5. Which injector problem can create a lean air-fuel mixture?
(A) *Dirty injector.*
(B) *Leaking injector.*
(C) *Stuck open injector.*
(D) *All of the above.*

6. Electronic fuel injection pressure can reach:
   (A) 20 psi
   (B) 30 psi
   (C) 60 psi
   (D) 100 psi

7. Which of the following can be used to *quickly* check EFI multiport injector operation?
   (A) Ohmmeter.
   (B) Oscilloscope.
   (C) EFI analyzer.
   (D) Stethoscope.

8. If an EFI multiport injector is removed for service, manufacturers suggest replacing the:
   (A) boot.
   (B) hose.
   (C) seals.
   (D) All of the above.

9. An injector balance test shows a below normal pressure drop on one injector. Technician A says to try running cleaning solution through the injector. Technician B says that injector is bad and should be replaced. Who is right?
   (A) A only.
   (B) B only.
   (C) Both A and B.
   (D) Neither A nor B.

10. Typical oxygen sensor output should be about:
    (A) 0.5 volts.
    (B) 0.6 volts.
    (C) 0.8 volts.
    (D) 0.10 volts.

Activities—Chapter 23

1. Study the fuel injection section of any shop manual given you by your instructor. Develop a "diagnosis tree" for an injector that is not working.

2. Develop an overhead transparency showing various injector spray patterns. Using the transparency, explain the patterns to the class.

3. Demonstrate the proper procedure for replacement of a multiport injector.

Gasoline Injection System Diagnosis

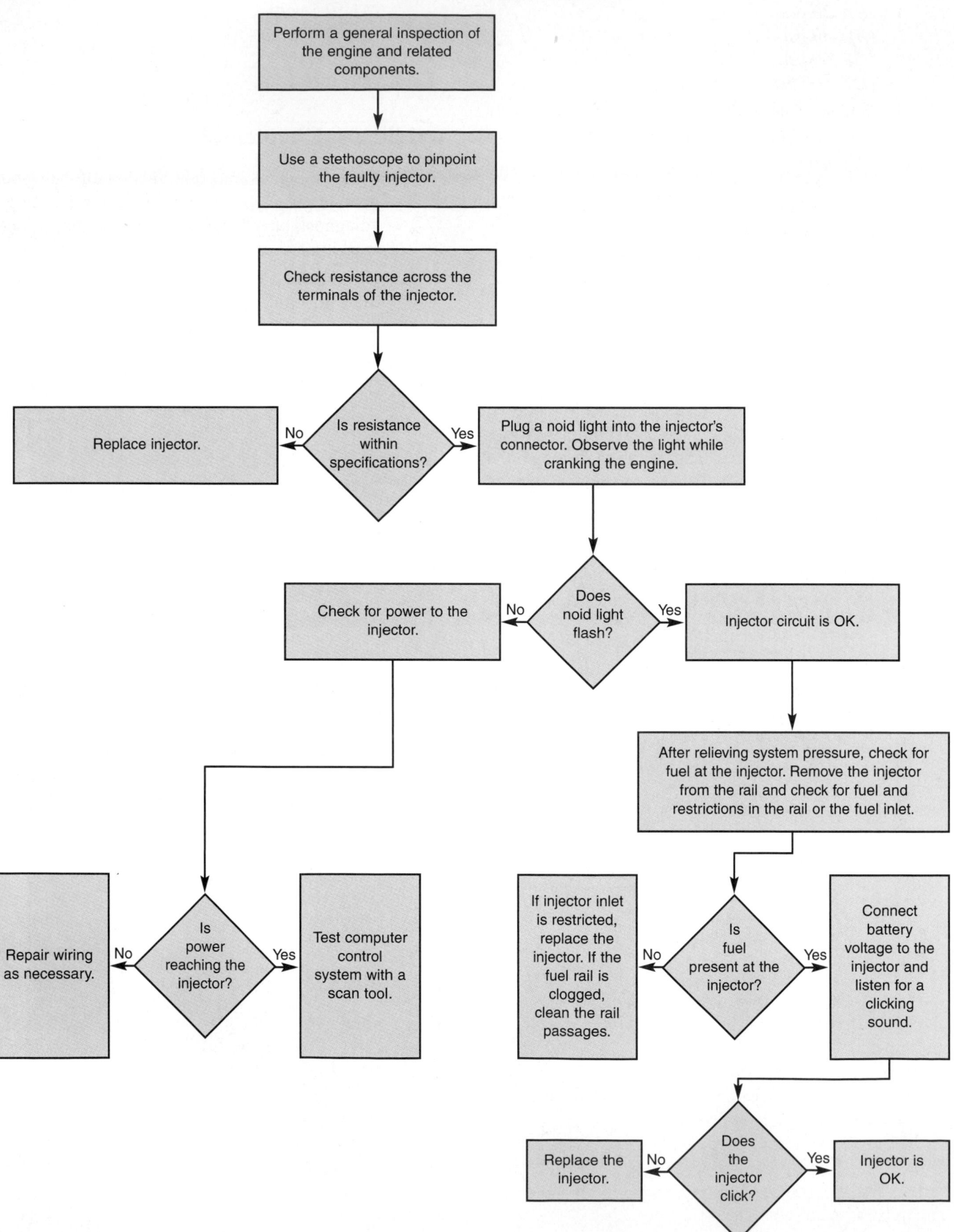

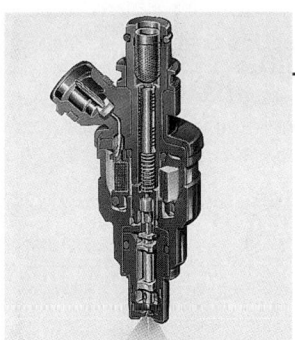

Chapter 24

Carburetor Fundamentals

After studying this chapter, you will be able to:

- Describe and identify the basic parts of a carburetor.
- Compare carburetor design differences.
- List and explain the fundamental carburetor systems.
- Explain special carburetor devices.
- Describe the operation of computer-controlled carburetors.
- Correctly answer ASE certification test questions that require a knowledge of the fundamentals of carburetors.

Even though gasoline engines now have electronic fuel injection, millions of older cars and trucks on the road today are equipped with carburetors. These vehicles commonly need service and repair. This makes a basic understanding of carburetor operation and repair important to today's technician.

This chapter introduces the fundamental principles of carburetion. It discusses carburetor systems, design differences, auxiliary control devices, and computer control. It will prepare you to study the service and repair of carburetors in the next chapter.

Basic Carburetor

A *carburetor* is basically a device that mixes air and fuel in the correct proportions (amounts) for efficient combustion.

The carburetor bolts to the engine's intake manifold. The air cleaner fits over the top of the carburetor to trap dust and dirt. See **Figure 24-1.**

When the engine is running, downward-moving pistons on their intake strokes produce a suction (low-pressure area) in the intake manifold. Air rushes through the carburetor and into the engine to fill this low-pressure

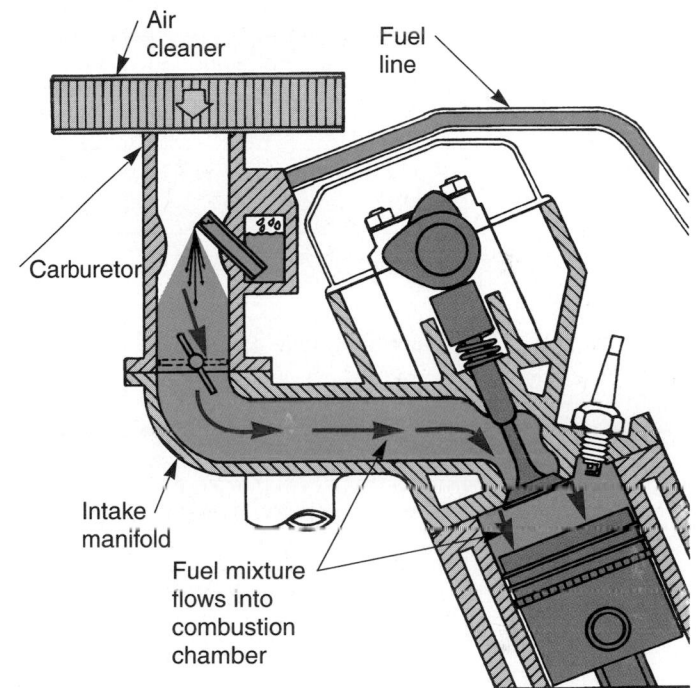

Figure 24-1. The carburetor bolts to the engine intake manifold. It meters and mixes fuel with incoming air.

area. The airflow through the carburetor is used to meter fuel and mix it with the air.

Basic Carburetor Parts

Many carburetor parts perform the same functions as corresponding gasoline injection system parts, **Figure 24-2.** The basic parts of a carburetor are shown in **Figure 24-3.** Refer to this illustration as these parts are introduced.

The *carburetor body* is a cast metal housing for the other carburetor components. It contains cast and drilled passages for air and fuel. The main discharge tube, venturi, and fuel bowl are normally integral parts of the

Carburetor System	Gasoline Injection System	Part Function
Choke	Engine temperature sensor, manifold temperature sensor, or cold start injector	Meters in more fuel for cold engine operation
Float	Fuel pressure regulator, electric fuel pump	Provides additional supply of fuel
Idle circuit	Air bypass valve, idle control motor, engine speed sensor	Meters air/fuel at low speeds
Accelerator pump	Throttle position switch, manifold pressure sensor	Enriches fuel mixture for acceleration
Fast idle cam	Throttle positioner or air bypass valve	Increases idle speed of cold engine
Power valve or metering rod	Manifold pressure sensor, throttle position switch, oxygen sensor	Enriches mixture for rapid acceleration
Fuel metering jets or mixture control solenoid	Injector valves, fuel pressure regulator	Meters fuel into right amount of air
Throttle valves	Throttle valves	Controls engine speed and power
Carburetor body	Throttle body	Houses throttle valve and idle speed device
Venturi and main discharge	Airflow sensor	Measures amount of air entering engine
System computer (if used)	System computer	Monitors and controls system
System sensors (if used)	System sensors	Measure operating conditions

Figure 24-2. A carburetor fuel system can be compared to a gasoline injection system. Study the parts and their functions. This will help you understand carburetor operation.

carburetor body. A flange on the bottom of the body allows the carburetor to be bolted to the engine.

The carburetor *air horn,* also called *throat* or *barrel,* routes outside air into the engine intake manifold, **Figure 24-3.** It contains the throttle valve, venturi, and outlet end of the main discharge tube.

The *throttle valve* is a butterfly valve located in the air horn. It controls the power output of a gasoline engine. When the throttle valve is closed, it restricts airflow through the carburetor. This reduces the amount of fuel flowing into the engine.

A *venturi* produces sufficient suction to pull fuel out of the main discharge tube. Venturi action is illustrated in **Figure 24-4.** Vacuum is highest inside the venturi. The narrowed airway increases air velocity, forming a low-pressure area in the air horn.

The *main discharge tube* uses venturi vacuum to feed fuel into the air horn and engine. Also called *main fuel nozzle,* it is a passage in the carburetor body and air horn that connects the fuel bowl to the center of the venturi. Refer to **Figure 24-3.**

The carburetor *fuel bowl* holds a supply of fuel that is *not* under fuel pump pressure. Several additional carburetor parts are mounted in the fuel bowl. These will be discussed later.

Basic Carburetor Systems

A *carburetor system* is a network of passages and related parts that helps control the air-fuel ratio under a specific engine operating condition. Also called a *carburetor circuit,* each system supplies a predetermined

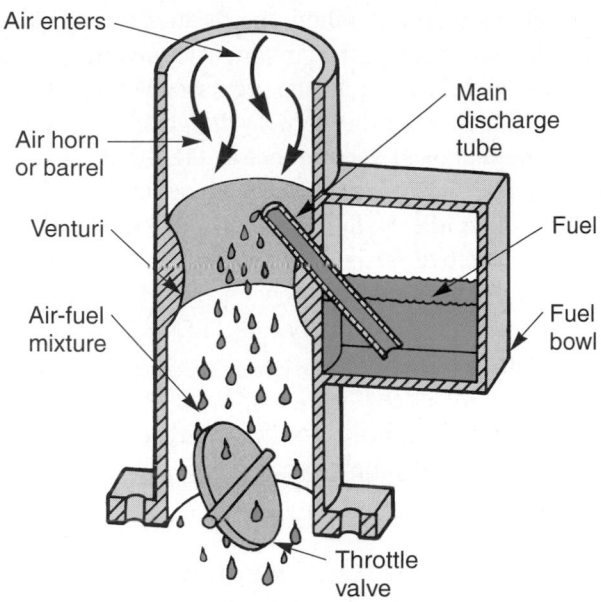

Figure 24-3. A carburetor controls the amount of air and fuel entering an engine. Study its basic parts.

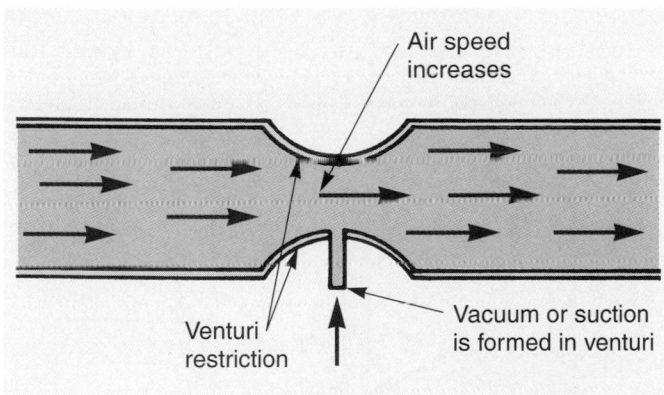

Figure 24-4. A venturi is used to produce vacuum from airflow. Note how the vacuum is highest inside venturi. (Ford)

air-fuel mixture as the temperature, speed, and engine load change.

For example, a gasoline engine's air-fuel mixture may vary from a rich 8:1 ratio to a lean 18:1 ratio, depending on operating conditions. An automotive carburetor, using its various systems, must be capable of providing varying air-fuel ratios:

- 8:1 for cold engine starting.
- 16:1 for idling.
- 15:1 for part throttle.
- 13:1 for full acceleration.
- 18:1 for normal cruising at highway speeds.

The seven basic carburetor systems are the float, idle, off-idle, acceleration, high-speed, full-power, and choke systems.

Float System

The *float system* must maintain the correct level of fuel in the carburetor bowl. The float system keeps the fuel pump from forcing too much gasoline into the carburetor bowl. Look at **Figure 24-5.**

The *carburetor float* rides on top of the fuel in the bowl to open and close the needle valve. It is normally made of thin brass or plastic. One end of the float is hinged to the side of the carburetor body. The other end is free to swing up and down.

The *needle valve* in the top of the fuel bowl regulates the amount of fuel passing through the fuel inlet and needle seat.

The carburetor float *needle seat* works with the needle valve and float to control fuel flow into the bowl. It is a brass fitting that threads into the carburetor body, **Figure 24-5.**

A *bowl vent* prevents a pressure or vacuum buildup in the carburetor fuel bowl. Without venting, pressure could form in the bowl as the fuel pump fills the carburetor. This could also cause vacuum to form in the bowl as fuel is drawn out of the carburetor and into the engine.

When engine speed or load increases, fuel is rapidly pulled out of the fuel bowl and into the venturi.

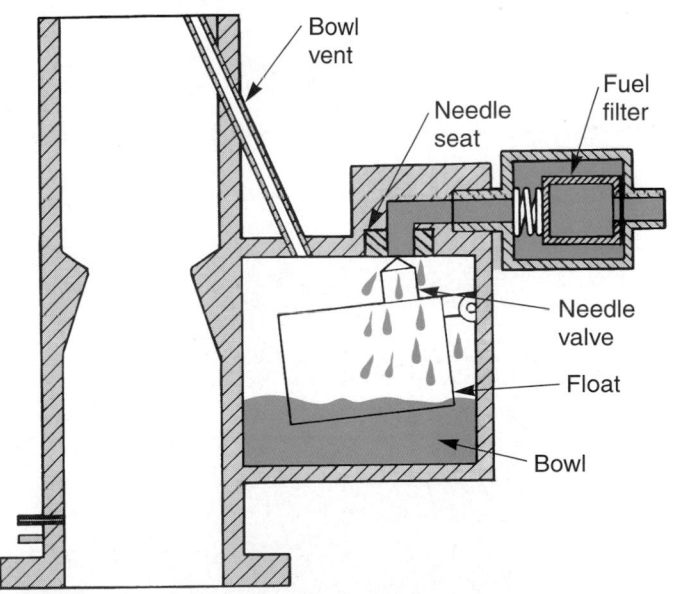

Figure 24-5. Basic parts of a float system. The float opens and closes a needle valve as the fuel level falls and rises. Study the part names.

Illustrated in **Figure 24-6,** this makes the fuel level and float drop in the bowl. The needle valve also drops away from its seat. The fuel pump can then force more fuel into the bowl.

As the fuel level in the bowl rises, the float pushes the needle valve back into the seat. When the fuel level is high enough, the float closes the opening between the needle valve and seat.

When the engine is running, the needle valve usually lets some fuel leak into the bowl. As a result, the float system maintains a stable quantity of fuel. This is very important because the fuel level in the bowl can affect the air-fuel ratio.

Idle System

A carburetor's **idle system** provides the engine's air-fuel mixture at speeds below approximately 800 rpm or 20 mph (32 km/h).

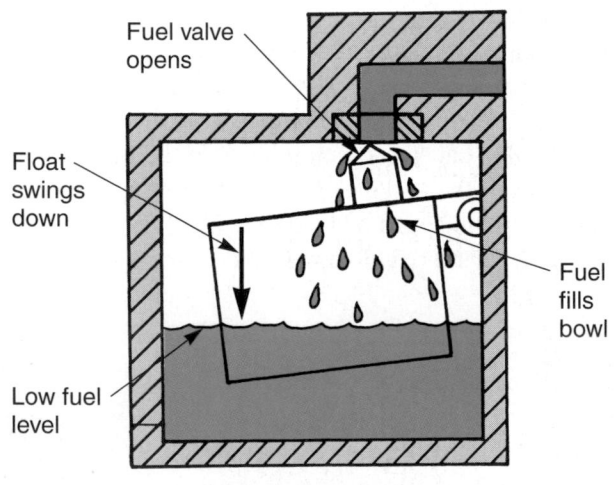

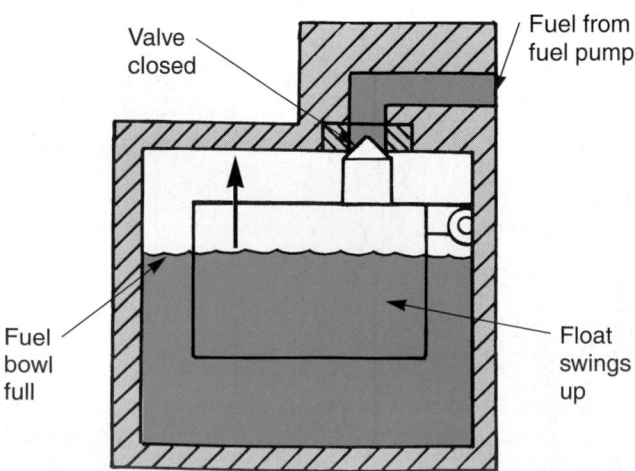

Figure 24-6. Basic float operation.

When an engine is idling, the throttle valve is almost closed. Airflow through the air horn is too restricted to produce enough vacuum in the venturi. Venturi vacuum cannot pull fuel out of the main discharge tube. Instead, the high intake manifold vacuum *below* the throttle plate and a separate idle circuit are used to feed fuel into the air horn. The parts of an idle system are shown in **Figure 24-7.**

The **low-speed jet** is a restriction in the idle passage that limits maximum fuel flow in the idle circuit. It is placed in the fuel passage before the idle air bleed and economizer.

The **idle air bleed** works with the **economizer** and **bypass** to add air bubbles to the fuel flowing to the idle port. The air bubbles help break up, or atomize, the fuel. This makes the air-fuel mixture burn more efficiently once in the engine.

The **idle passage** carries the air-fuel slurry (mixture of liquid fuel and air bubbles) to the idle screw port. The **idle screw port** is an opening into the air horn below the throttle valve.

The **idle mixture screw** allows adjustment of the size of the opening in the idle port. Turning the idle screw *in* reduces the size of the idle port. This reduces the amount of fuel entering the air horn, producing a leaner fuel

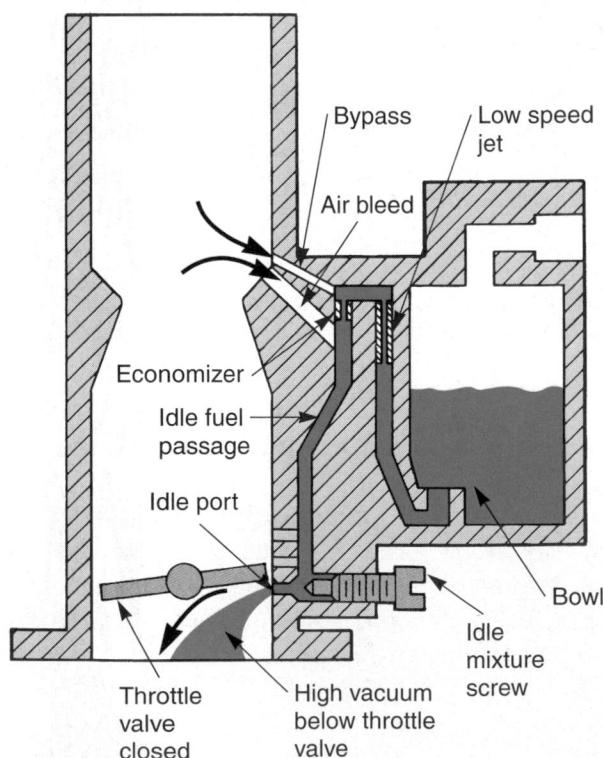

Figure 24-7. The idle system feeds fuel into the air horn when the throttle is closed for low engine speed operation. High vacuum below the throttle pulls fuel out of the idle port. A mixture screw allows adjustment of the mixture at idle. An air bleed helps premix the air and fuel.

mixture at idle. Turning the idle screw *out* increases the size of the idle port and the amount of fuel flowing into the air horn, enriching the fuel mixture at idle.

Tech Tip

Idle mixture screw adjustment is critical to exhaust emissions. Therefore, most late-model carburetors have *sealed idle mixture screws.* These idle mixture screws are covered with metal plugs to prevent tampering with the factory idle mixture setting.

Sometimes, plastic *limiter caps* are pressed over the heads of the idle mixture screws. The caps restrict how far the screws can be turned toward either rich or lean settings.

For the idle system to function, the throttle plate must be closed. Then, high intake manifold vacuum can pull fuel out of the idle circuit.

Off-Idle System

The *off-idle system,* often termed the *part throttle circuit,* feeds more fuel into the air horn when the throttle plates are partially open. Look at **Figure 24-8.** This system is an extension of the idle system. It functions *above* 800 rpm or 20 mph (32 km/h).

Without the off-idle system, the fuel mixture would become too lean slightly above idle. The off-idle system helps supply fuel during the transition (change) from idle circuit to high-speed circuit. The off-idle system begins to function when the driver presses lightly on the accelerator, cracking open the throttle plates.

Figure 24-9 shows the bottom of a carburetor air horn. Notice the idle screw port, idle screw tip, and off-idle ports. Note that the throttle plate exposes all the ports to vacuum when it is partially opened.

Acceleration System

The carburetor's *acceleration system,* like the off-idle system, provides extra fuel when changing from the idle circuit to the high-speed circuit (main discharge).

The acceleration system "squirts" a stream of extra fuel into the air horn whenever the accelerator is

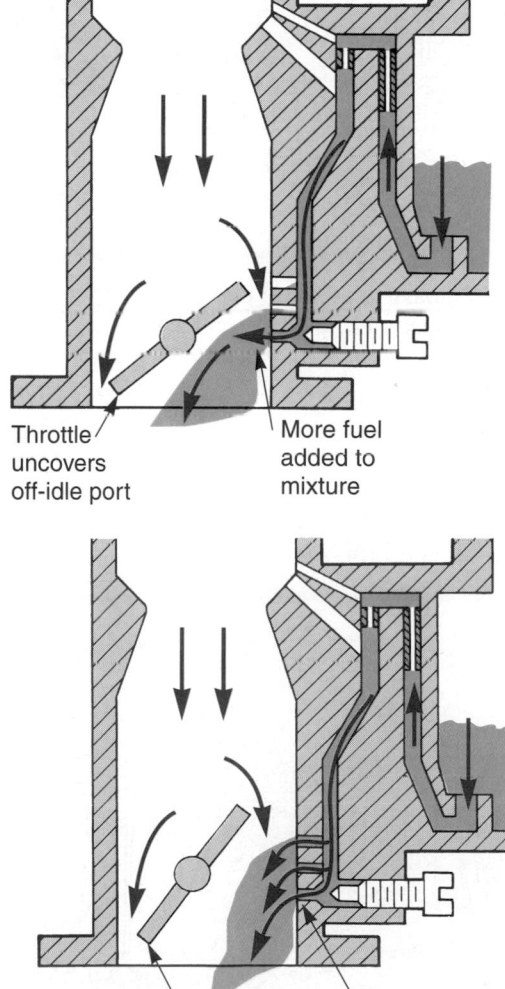

Figure 24-8. Off-idle system feeds fuel to the engine when the throttle is opened slightly. It adds a little extra fuel to the extra air flowing around throttle valve.

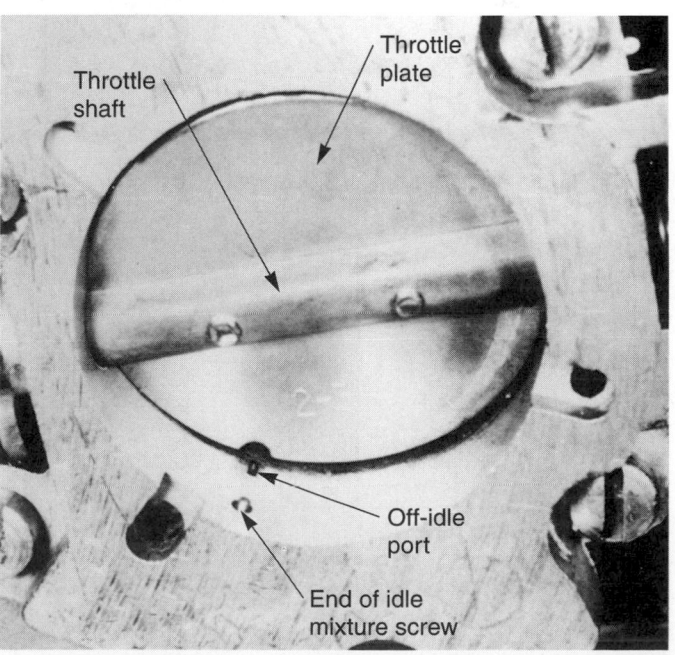

Figure 24-9. Bottom view of a carburetor shows the idle mixture screw tip, idle port opening, and off-idle port. (Carter Carburetor Div.)

pressed (throttle valves swing open). This is illustrated in **Figure 24-10.**

Without the acceleration system, too much air would rush into the engine as the throttle is quickly opened. The mixture would become too lean for combustion, and the engine would *hesitate* or *stall.* The acceleration system prevents a lean air-fuel mixture from upsetting a smooth increase in engine speed.

The *accelerator pump* develops the pressure to force fuel out of the pump nozzle and into the air horn. There are two types of accelerator pumps: piston pumps and diaphragm pumps. See **Figures 24-11** and **24-12.**

The *pump check ball* allows fuel to only flow into the pump reservoir. It stops fuel from flowing back into the fuel bowl when the pump is actuated.

The *pump check weight* prevents fuel from being pulled into the air horn by venturi vacuum. Its weight seals the passage to the pump nozzle and prevents fuel siphoning.

The *pump nozzle,* or *pump jet,* has a fixed orifice (opening) that helps control fuel flow out of the pump circuit. It also guides the fuel stream into the center of the air horn.

When the driver presses the accelerator, the throttle plates swing open. This causes the accelerator pump

piston or diaphragm to compress the fuel in the pump reservoir.

The accelerator pump pressure closes the pump check ball and the fuel flows toward the pump check weight. Pressure lifts the pump check weight off its seat, and fuel squirts into the carburetor air horn, as if from a *toy squirt gun.*

A spring is used on the accelerator pump assembly to produce a smooth, steady flow of fuel out of the pump

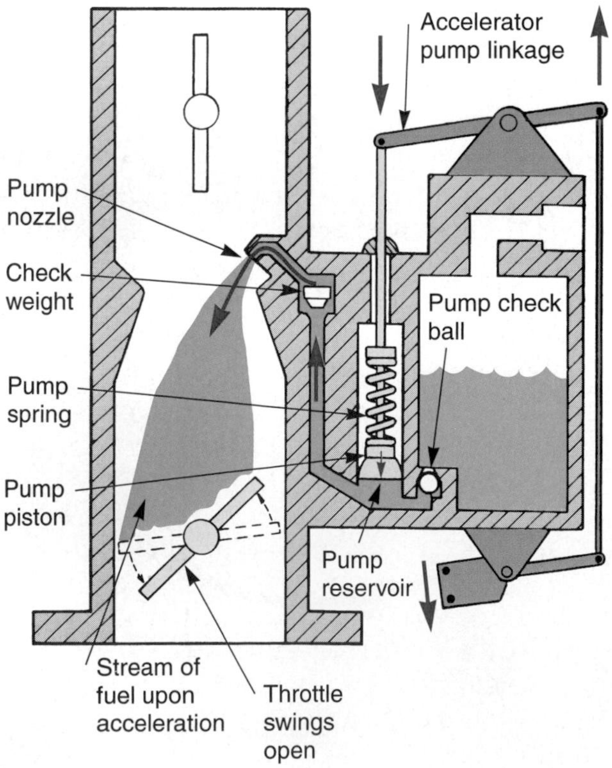

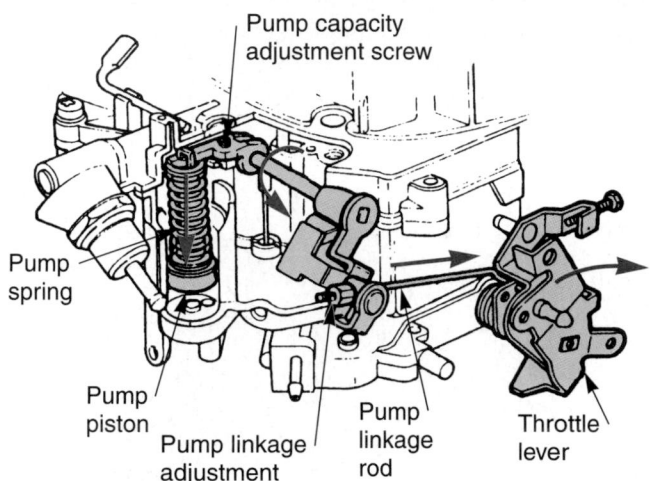

Figure 24-11. Most accelerator pump systems use mechanical linkage from the throttle lever. When the driver presses the gas pedal for acceleration, both the throttle valve and pump are actuated. (Ford)

Figure 24-10. The accelerator pump system squirts fuel into the air horn every time the throttle is opened. This adds fuel to the rush of air entering the engine and prevents a temporary lean mixture. Study the part names.

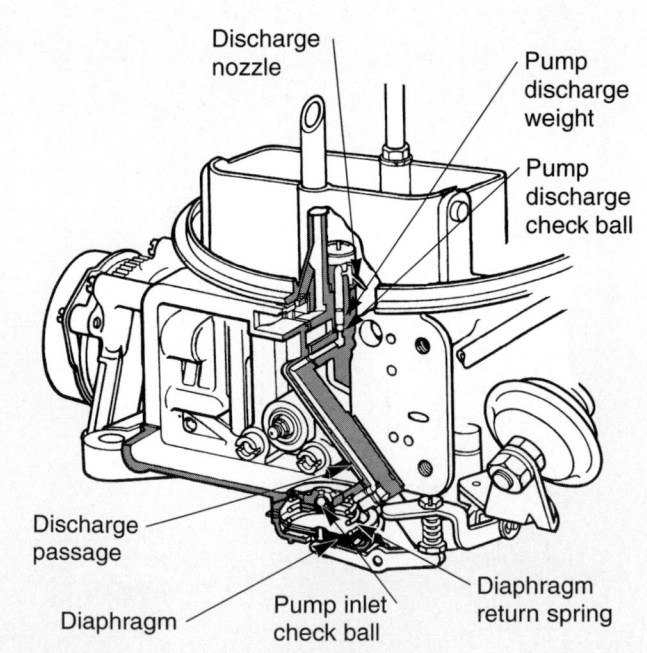

Figure 24-12. Cutaway view of a carburetor using a diaphragm-type accelerator pump. (Holley Carburetors)

nozzle. Throttle opening compresses the spring. Then the compressed spring pushes on the pump piston to pressurize the fuel and produce prolonged fuel flow.

As the accelerator is released, the pump piston or diaphragm retracts. This closes the check weight and opens the pump check. Fuel flows out of the bowl to refill the accelerator pump reservoir. The system is then ready to spray another stream of fuel into the air horn when the vehicle accelerates.

High-Speed System

The carburetor's *high-speed system,* also called *main metering system,* supplies the engine's air-fuel mixture at cruising speeds, **Figure 24-13.**

This circuit begins to function when the throttle plates are open wide enough for venturi action to occur. Airflow through the carburetor must be relatively high for venturi vacuum to draw fuel out of the main discharge tube.

The high-speed system provides the leanest, most fuel efficient air-fuel ratio. It functions from about 20–55 mph (32–90 km/h) or 2000–3000 rpm.

The *high-speed jet* is a fitting with a precision hole drilled in its center. This jet screws into a threaded hole in the fuel bowl. One jet is used for each air horn.

The hole size in the high-speed jet determines how much fuel flows through the circuit. A number is usually stamped on the jet to denote the diameter of the hole. Since jet numbering systems vary, refer to the carburetor manufacturer's manual for information on jet size.

The *emulsion tube* and the air bleed add air to the fuel flowing through the main discharge tube. The premixing of air with fuel helps the fuel atomize as it discharges into the air horn.

The *primary venturi* is the venturi formed in the side of the carburetor air horn. One or two *booster venturis* can be added inside the primary venturi to increase vacuum at lower engine speeds.

When engine speed is high enough, airflow through the carburetor forms a high vacuum in the venturi. The vacuum pulls fuel through the main metering system. Fuel flows through the main jet, which meters the amount of gasoline entering the circuit. Then, the fuel flows into the main discharge tube and emulsion tube.

The emulsion tube causes air from the air bleed to mix with the fuel. The air-fuel mixture is finally pulled out of the main nozzle and into the engine.

Full-Power System

The carburetor *full-power system* provides a means of enriching the fuel mixture for high-speed, high-power conditions. This circuit operates when the driver presses the gas pedal to pass another vehicle or to climb a steep hill. A simplified illustration of a full-power system is given in **Figure 24-14.**

The full-power system is usually an addition to the main metering system. Either a metering rod or a power valve (jet) can be used to provide a variable high-speed air-fuel ratio.

Metering Rod

A *metering rod* is a stepped rod that moves in and out of the main jet to alter fuel flow. As shown in **Figure 24-15,** when the metering rod is inside the jet, flow is restricted and a leaner fuel mixture results. When the metering rod is pulled out of the jet, more fuel can flow through the system to enrich the mixture for more power output.

Either mechanical linkage, engine vacuum, or an electric solenoid and computer (modified metering rod or valve assembly) can be used to operate a metering rod. The metering rod can be linked to the throttle lever. Then, whenever the throttle is opened wide, the linkage lifts the metering rod out of the jet.

A metering rod controlled by engine vacuum is connected to a diaphragm. When power demands are low, engine vacuum is high. As power demands increase,

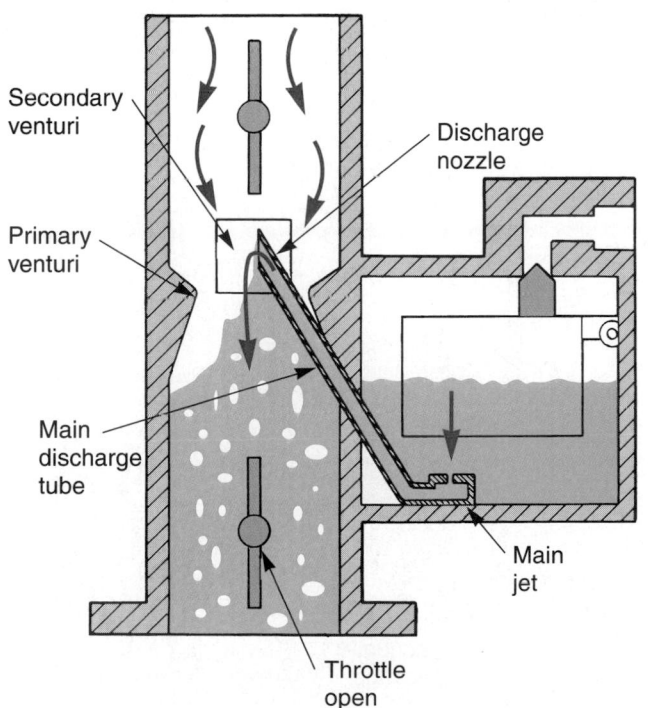

Figure 24-13. The high-speed system is simple. The main jet controls fuel flow and mixture. At higher engine speeds, there is enough airflow through the venturi to produce vacuum. This pulls fuel through the main discharge. Study the part names.

intake manifold vacuum drops. This vacuum-load relationship is ideal for controlling a metering rod or power valve.

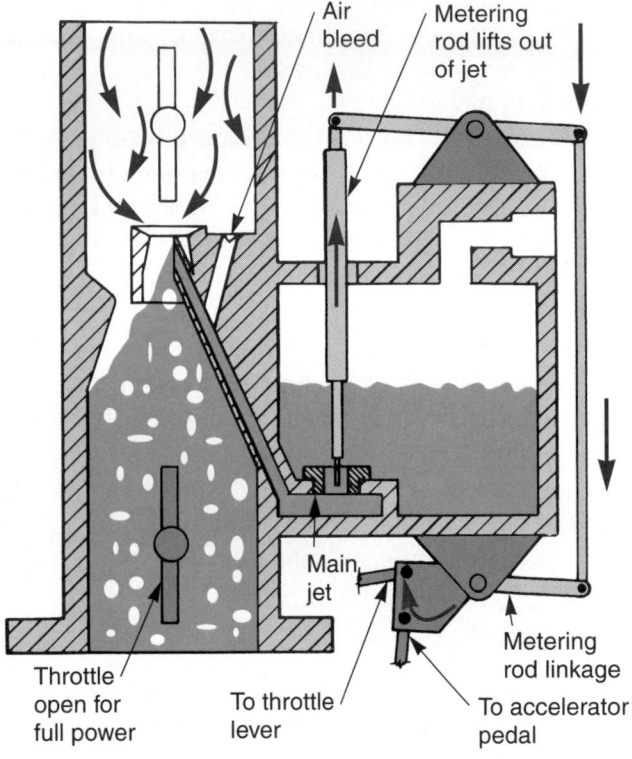

Figure 24-14. The full-power system enriches the high-speed circuit when needed. When the accelerator pedal is pushed down for full power, the throttle linkage acts on the metering rod linkage. The metering rod is lifted out of the main jet to add more fuel to the mixture.

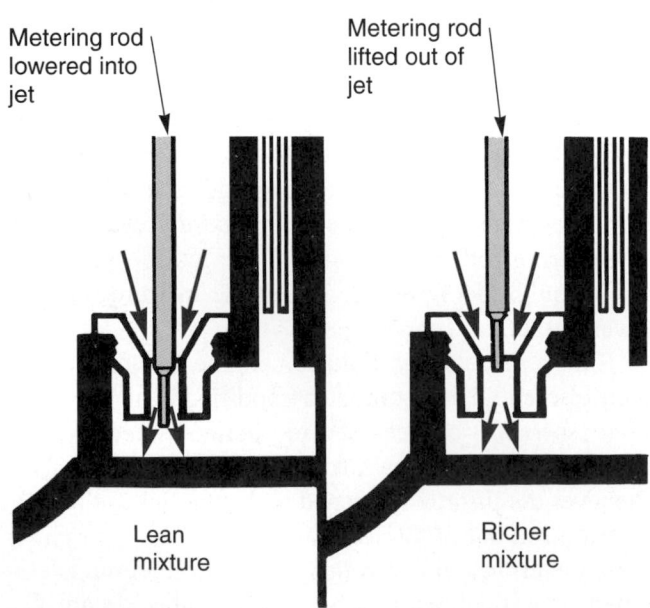

Figure 24-15. Metering rod action.

Power Valve

A *power valve,* also known as an *economizer valve,* performs the same function as a metering rod: it provides a variable high-speed fuel mixture. A power valve consists of a fuel valve, a vacuum diaphragm, and a spring.

Look at **Figure 24-16.** The spring holds the power valve in the normally open position. A vacuum passage runs to the power valve diaphragm. When the power valve is open, it serves as an extra jet that feeds fuel into the high-speed circuit.

When the engine is cruising at normal highway speeds, engine intake manifold vacuum is high. This vacuum acts on the power valve diaphragm and pulls the valve closed, **Figure 24-17.** No additional fuel is added to the main metering system under normal driving conditions.

However, when the throttle plates are swung open for passing another vehicle or climbing a hill, engine manifold vacuum drops. Then, the spring in the power valve can push the valve open. Fuel flows through the power valve and into the main metering system. This adds more fuel, increasing engine power.

Choke System

The *choke system* is designed to supply an extremely rich air-fuel ratio to aid cold engine starting. For the fuel mixture to burn properly, the fuel entering the intake must atomize and vaporize.

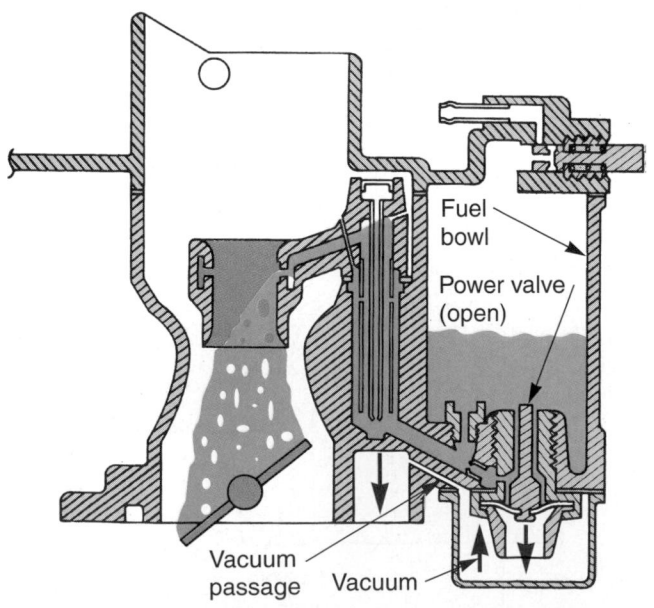

Figure 24-16. A power valve serves the same function as a metering rod. It enriches the air-fuel mixture under high load, low intake manifold vacuum conditions. When vacuum is low, a spring opens the power valve. Extra fuel can then flow through the valve and into the main discharge.

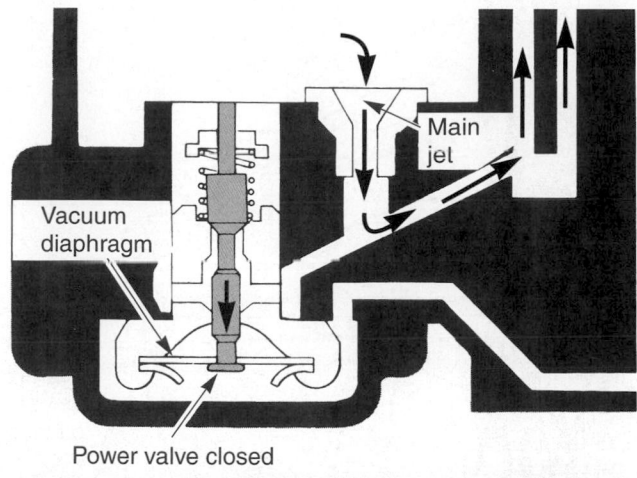

Low power, high vacuum

High power, low vacuum

Figure 24-17. Power valve action.

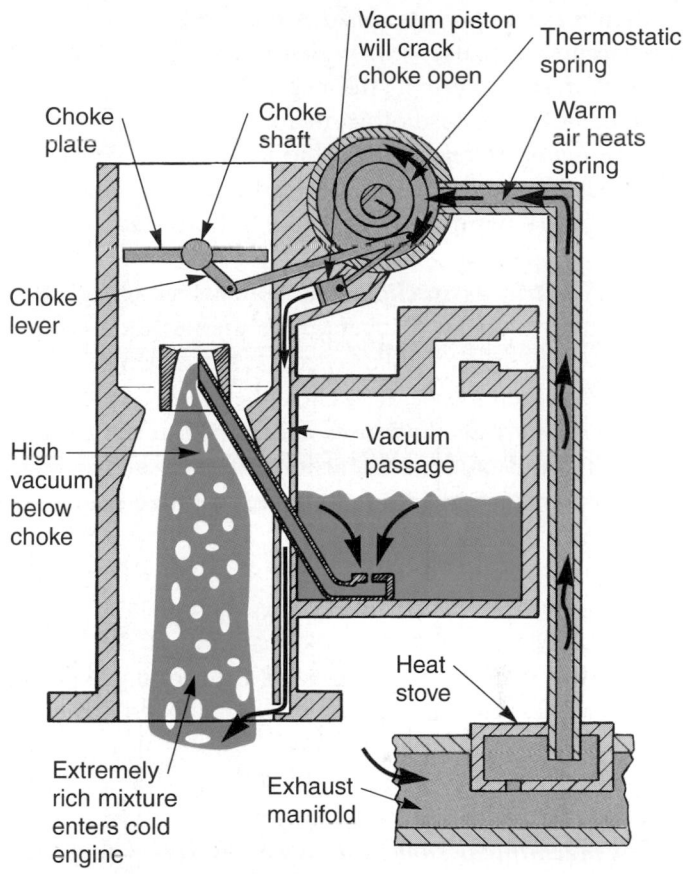

Figure 24-18. Basic choke system parts. The thermostatic spring is the main control of choke operation. When the engine is cold, the spring closes the choke. High vacuum below the choke pulls a large amount of fuel out of the main discharge. When the engine warms, hot air causes the spring to open the choke. The vacuum piston cracks the choke upon engine starting to prevent flooding.

When the engine is cold, the fuel entering the intake tends to condense into a liquid. As a result, not enough fuel vapors enter thc combustion chambers and the engine will miss or stall. A choke is used to prevent this lean condition. See **Figure 24-18.**

The *choke plate* is a butterfly (disc) valve located near the top of the carburetor air horn. When the choke plate is closed, it blocks normal airflow through the carburetor. This causes high intake manifold vacuum to form below the choke plate. Vacuum pulls on the main discharge tube, even though air is not flowing through the venturi. Fuel is pulled out to prime the engine with extra fuel.

A *thermostatic spring* may be used to open and close the choke. The thermostatic spring is a *bimetal spring* (spring made of two dissimilar metals). The two metals have different rates of expansion that make the spring coil tighter when cold. It uncoils when heated. This coiling-uncoiling action is used to operate the choke.

Before the engine starts, the cold thermostatic spring holds the choke closed. When the engine is started, the closed choke causes high vacuum in the carburetor air horn. This pulls a large amount of fuel out the main discharge. The rich mixture helps keep the cold engine running.

As the engine and thermostatic spring warm, the spring uncoils and opens the choke. This produces a leaner mixture. A warm engine would not run properly if the choke were to remain closed.

Manual and Automatic Chokes

A *manual choke* simply uses a cable mechanism that allows the driver to open or close the choke plate. Manual chokes were used on early model vehicles. They are still found on small gas engines and heavy equipment. *Automatic chokes,* which open and close the choke plate with changes in temperature, are more common.

An *integral hot air choke* is mounted on the side of the carburetor. It uses *warm air* from the engine to heat the thermostatic spring. An integral choke may also use engine coolant instead of warm air.

In a *nonintegral choke,* the thermostatic spring is mounted in the top of the intake manifold. Then, as the engine and manifold warm, the thermostatic spring uncoils to open the choke plate.

An *electric assist choke* uses both hot air and an electric heating element to operate the thermostatic spring. The electric assist choke system uses a temperature-sensitive switch to operate a choke heating element.

An *all electric choke* uses neither hot air nor coolant to aid thermostatic spring action. Instead, a two-stage heating element provides full control of choke operation. See **Figure 24-19.**

When the engine is cold, only the first stage of the heating element is activated. The thermostatic spring warms slowly to keep the choke partially closed. When the engine warms, both stages of the heating element operate. The element heats up very quickly to make the thermostatic spring open the choke.

Mechanical Choke Unloader

A *mechanical choke unloader* physically opens the choke plate whenever the throttle swings fully open. Look at **Figure 24-20.**

A mechanical choke unloader uses a metal lug on the throttle lever. When the throttle lever moves to the fully open position, the lug pushes on the choke linkage (fast idle linkage). This gives the driver a means of opening

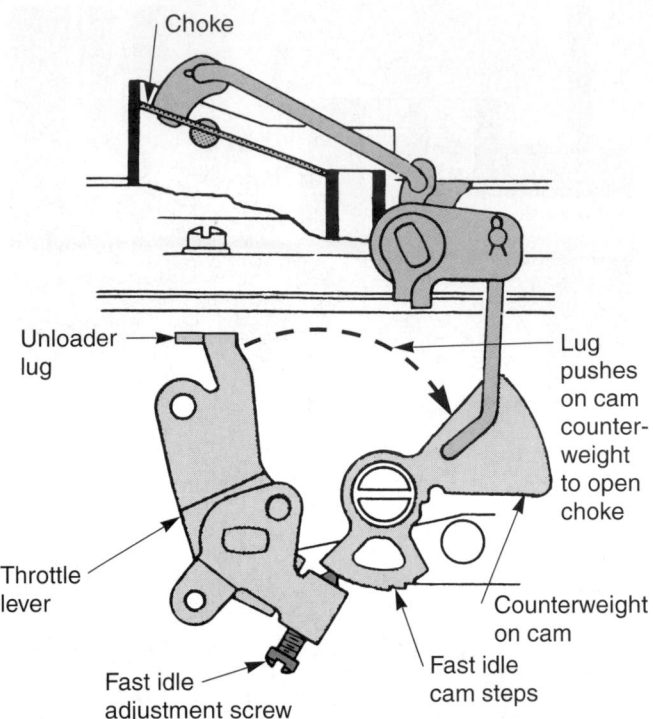

Figure 24-20. A choke unloader physically opens the choke when the accelerator pedal is pushed to the floor. The unloader lug moves the choke linkage. This lets the driver clean out a flooded engine. (Carter)

the choke. Air can then enter the air horn to help clear a flooded engine (engine with too much liquid fuel in the cylinders and intake manifold).

Vacuum Choke Unloader

A *vacuum choke unloader,* also called *choke break,* uses engine vacuum to crack open the choke plate as soon as the engine starts. It automatically prevents the engine from flooding with too much fuel.

A vacuum choke unloader consists of a manifold vacuum fitting, vacuum hose, vacuum diaphragm (choke break), and linkage connected to the choke lever.

Before the engine starts, the choke spring holds the choke plate almost completely closed. This primes the engine with enough fuel for starting. Then, as the engine starts, intake manifold vacuum acts on the choke break diaphragm. The diaphragm pulls on the choke linkage and lever to swing the choke plate open slightly. This helps prevent an overly rich mixture and improves cold engine driveability.

Carburetor Devices

Various carburetor devices have been used over the years to increase efficiency. A few of these devices will be summarized in the following sections.

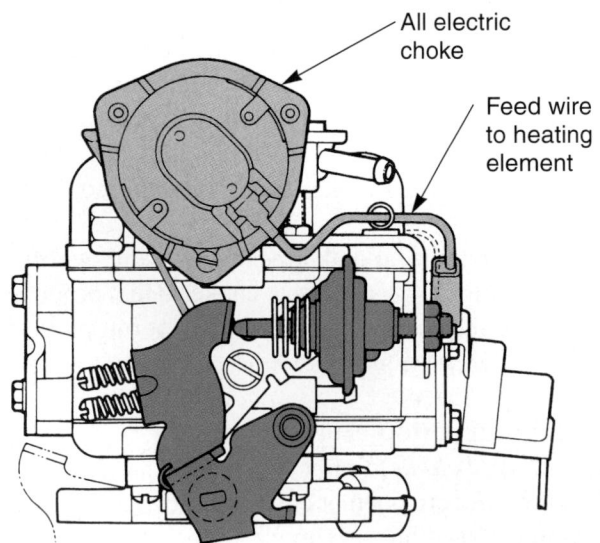

Figure 24-19. An all electric choke has a two-stage heating element. One stage warms the thermostatic spring when the engine is cold. When the engine is partially warm, both heating stages function. (Ford)

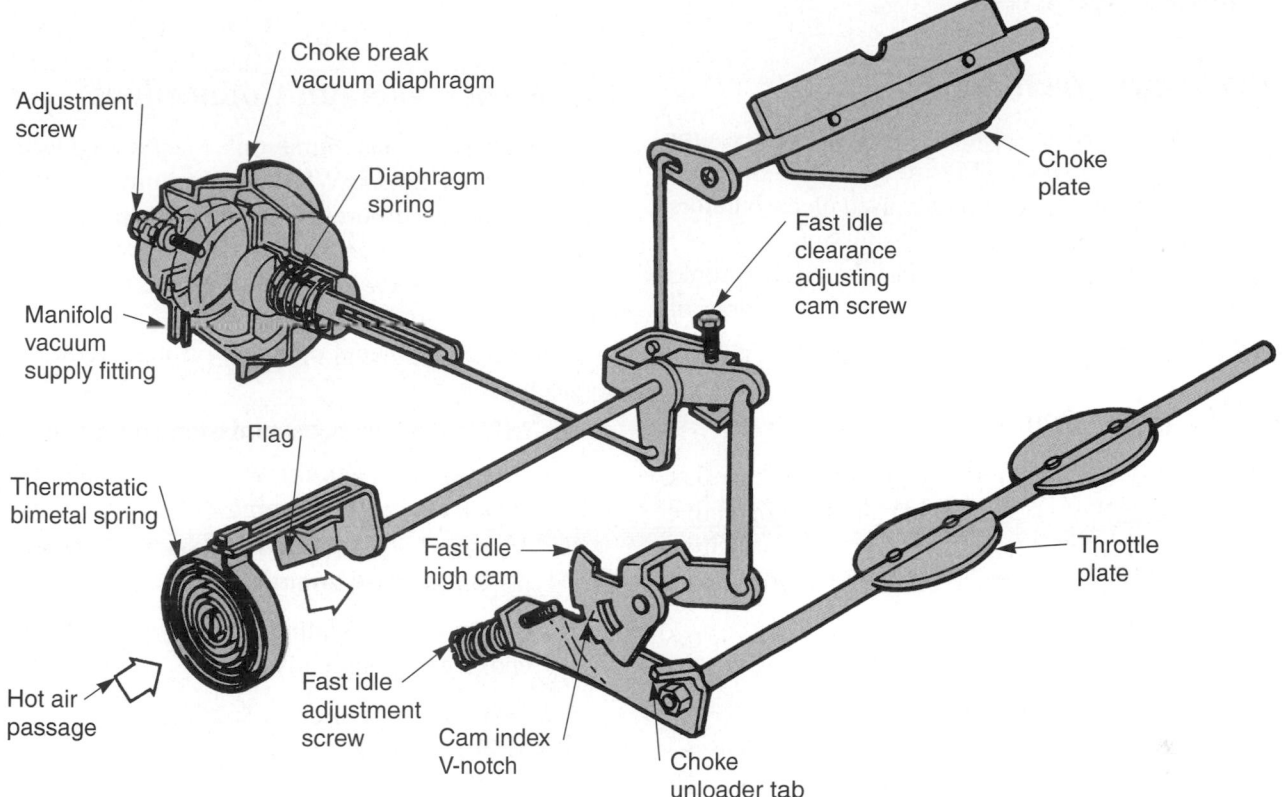

Figure 24-21. Note relationship of these parts. They work together to provide smooth engine operation when the engine is cold. (Ford)

Fast Idle Cam

A *fast idle cam* increases engine idle speed when the choke is closed. It is a stepped cam fastened to the choke linkage, **Figure 24-21.**

When the choke closes, the fast idle cam swings around in front of a fast idle screw. The fast idle screw is mounted on a throttle lever. As a result, the fast idle cam and fast idle screw prevent the throttle plates from closing. Engine idle speed is increased to smooth cold engine operation and prevent stalling.

As soon as the engine warms and the choke opens, the fast idle cam is deactivated. When the throttle is opened, the choke linkage swings the cam away from the fast idle screw and the engine returns to curb idle (normal, hot idle speed).

Fast Idle Solenoid

A fast idle solenoid opens the carburetor throttle plates during engine operation but allows the throttle plates to close as soon as the engine shuts off. In this way, a faster idle speed can be used while still avoiding dieseling (engine keeps running even though ignition key is turned off).

Sometimes termed an *anti-dieseling solenoid,* the fast idle solenoid is mounted on the carburetor so that it contacts the throttle lever, as in **Figure 24-22.**

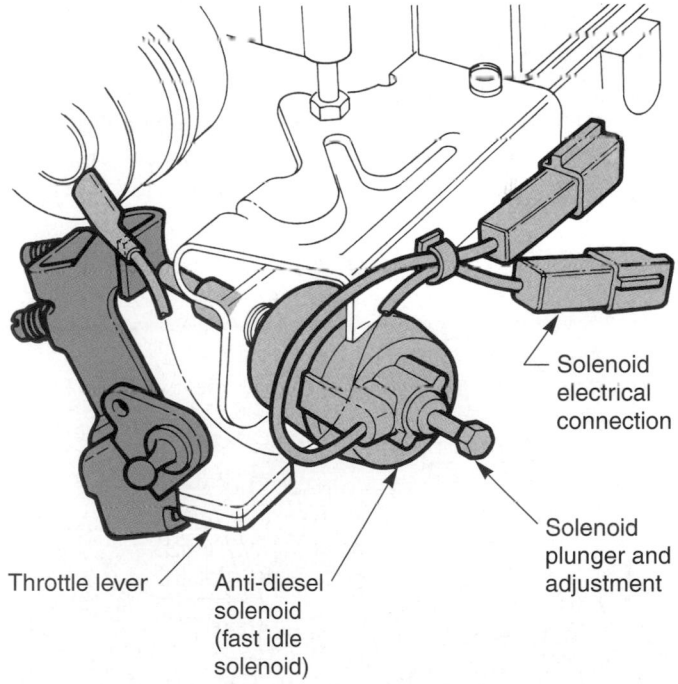

Figure 24-22. A fast idle solenoid holds the throttle open when the engine is running. When the ignition key is turned off, the solenoid allows the throttle to close more. This keeps engine from dieseling, or after-running. (Ford)

Throttle Return Dashpot

A *throttle return dashpot* causes the carburetor throttle plates to close slowly, **Figure 24-23.** Often called an *anti-stall dashpot,* it is commonly used on carburetors for cars equipped with automatic transmission.

Without a throttle return dashpot, the engine could stall when it returns too quickly to an idle. The drag of the automatic transmission could "kill" the engine.

Hot Idle Compensator

A *hot idle compensator* is a carburetor device that prevents engine stalling or a rough idle under high engine temperature conditions. It is a temperature-sensitive valve that admits extra air into the engine to increase idle speed and smoothness.

Altitude Compensator

An *altitude compensator* can be used to change the carburetor's air-fuel mixture with changes in the vehicle's altitude (distance above or below sea level). An altitude compensator normally has an *aneroid* (bellows device that expands and contracts with changes in atmospheric pressure).

When a car is driven up a mountain, for example, the density of the air around the car decreases. This tends to make the air-fuel mixture richer. The reduced air pressure causes the aneroid to expand, opening an air valve. Extra air flows into the air horn and the air-fuel mixture becomes leaner as needed.

Carburetor Vacuum Connections

A carburetor has numerous vacuum connections. Look at **Figure 24-24.** When the vacuum connection or port is *below* the carburetor throttle plate, the port *always* receives full intake manifold vacuum. However, when the vacuum port is *above* the throttle plate, vacuum is only present at the port when the throttle is opened.

Typical components operated off carburetor vacuum connections include:

- *EGR valve*—exhaust emission control device.
- *Distributor vacuum advance*—diaphragm for advancing ignition timing.
- *Charcoal canister*—emission control container for storing fuel vapors.
- *Choke break*—diaphragm for partially opening choke when engine is running.

Carburetor Barrels

Multiple barrel carburetors are used to provide increased air intake ("engine breathing"). The amount of fuel and air that enters the engine is a factor limiting engine horsepower output. Extra carburetor barrels allow more air and fuel into the engine at wide-open throttle. This allows the engine to develop more power. See **Figure 24-25.**

Primary and Secondary

Inline two-barrel carburetors and all four-barrel carburetors have two sections: the primary and the secondary.

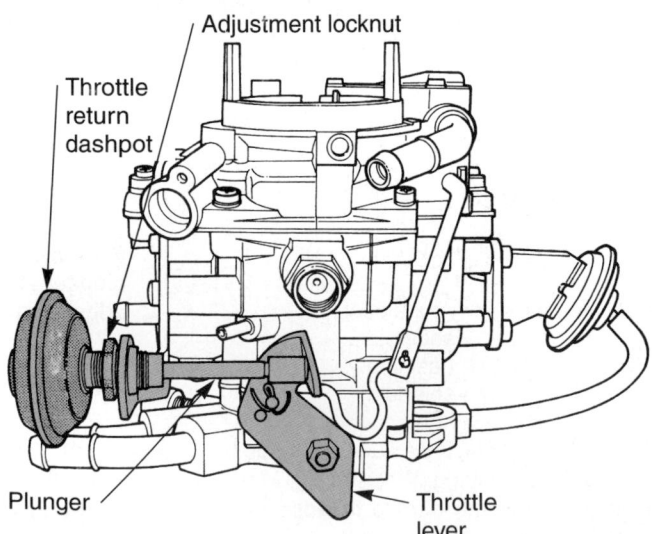

Figure 24-23. A throttle return dashpot keeps the engine from stalling when quickly returned to an idle. It is normally used on cars with an automatic transmission. The dashpot makes throttle plates close slowly. (Chrysler)

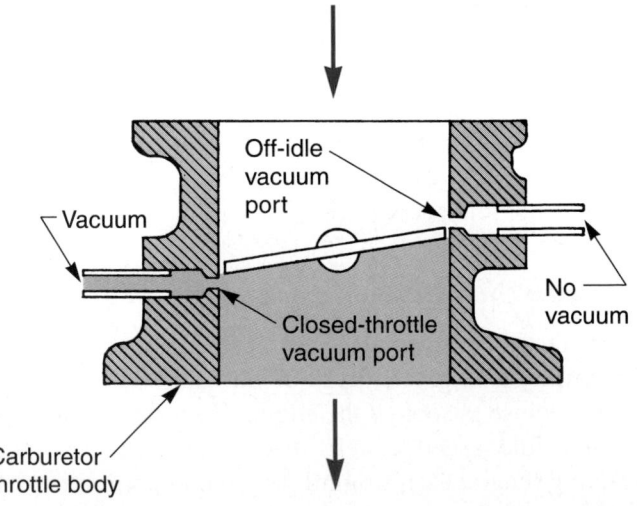

Figure 24-24. The position of the vacuum port in the carburetor determines when vacuum is present. When the port is below the throttle valve, vacuum is present at idle. When the port is above the throttle, vacuum is present above idle. (Ethyl Corp.)

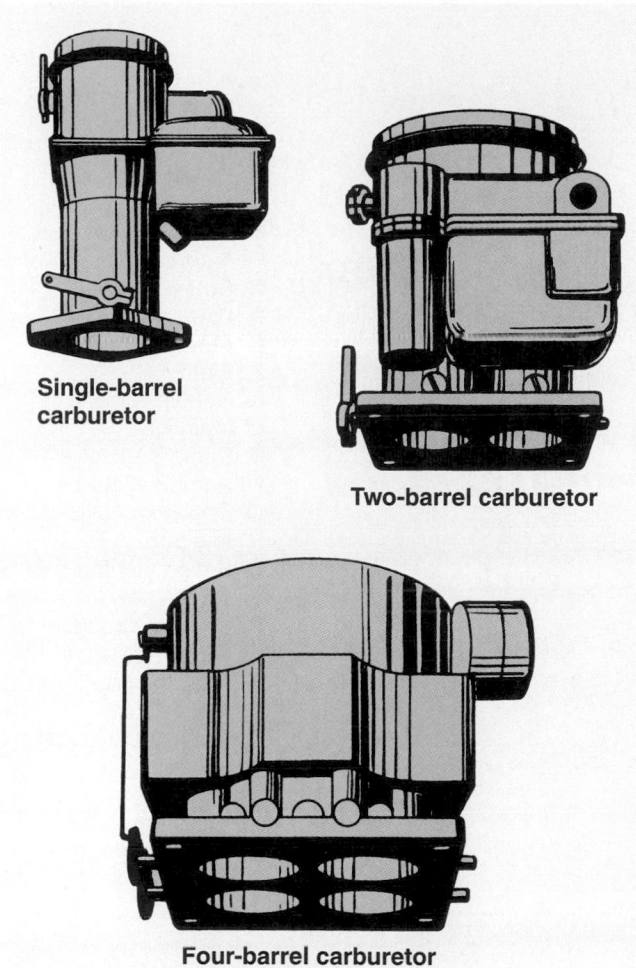

Single-barrel carburetor

Two-barrel carburetor

Four-barrel carburetor

Figure 24-25. Carburetors have one, two, or four barrels. More barrels, or air horns, are used with larger engines.

The *primary* of a carburetor includes the components that operate under normal driving conditions. In a four-barrel carburetor, the primary consists of the two front throttle plates and related components. Refer to **Figure 24-26.**

The *secondary* of a carburetor consists of the components or circuits that function under high engine power output conditions, **Figure 24-26.** In a four-barrel carburetor, secondary would include the two rear barrels. They only function when more fuel is needed for added power.

A *secondary diaphragm* is normally used to open the secondary throttle plates, causing the secondary circuits to function. As illustrated in **Figure 24-26,** a diaphragm is connected to the secondary throttle lever. A vacuum passage runs from this diaphragm to the venturi in the primary throttle bore.

Under normal driving conditions, vacuum in the primary is *not* high enough to actuate the secondary diaphragm and throttles. The engine will run using only the primary of the carburetor.

Increased airflow in the primary (if the driver passes another vehicle, for example) produces enough vacuum to actuate the secondary diaphragm. Vacuum pulls on the diaphragm and compresses the diaphragm spring. This opens the secondary throttle plates for increased engine horsepower.

Figure 24-27 shows many of the parts just discussed. Can you explain their function?

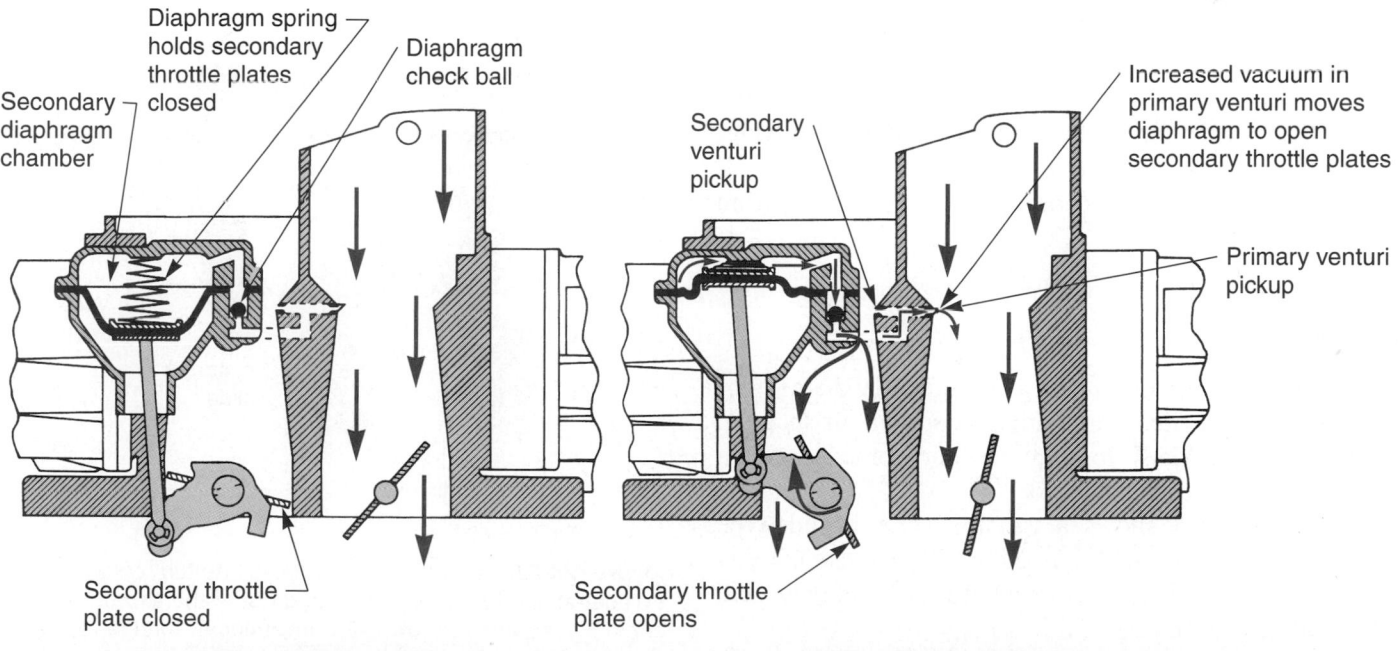

Low speed/light load

High speed/heavy load

Figure 24-26. The primary is the front barrel or barrels of the carburetor. The secondary is the rear barrels of the carburetor. Note how the secondary diaphragm opens the rear throttle plates when engine power output is high. (Holley Carburetors)

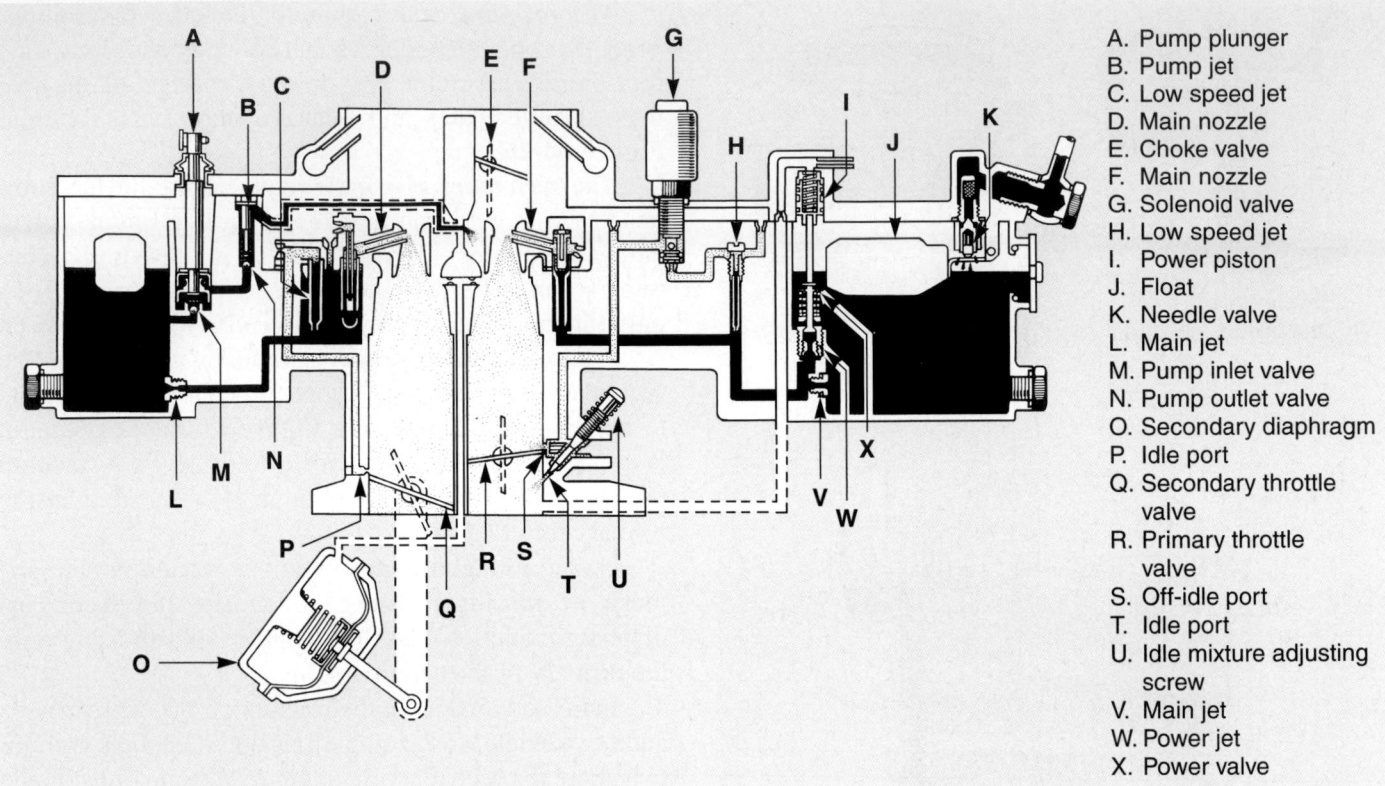

A. Pump plunger
B. Pump jet
C. Low speed jet
D. Main nozzle
E. Choke valve
F. Main nozzle
G. Solenoid valve
H. Low speed jet
I. Power piston
J. Float
K. Needle valve
L. Main jet
M. Pump inlet valve
N. Pump outlet valve
O. Secondary diaphragm
P. Idle port
Q. Secondary throttle valve
R. Primary throttle valve
S. Off-idle port
T. Idle port
U. Idle mixture adjusting screw
V. Main jet
W. Power jet
X. Power valve

Figure 24-27. Try to identify and explain carburetor parts before looking at the listed names. (Toyota)

Carburetor Size

Generally, *carburetor size* is stated in CFM (cubic feet of air per minute). This is the amount of air that can flow through the carburetor at wide-open throttle. CFM is an indication of maximum airflow capacity.

Usually, small-CFM carburetors are more fuel efficient than larger carburetors. Air velocity, fuel mixing, and atomization is better with small throttle bores. A larger CFM rating would be desirable for high engine power output.

Variable Venturi Carburetor

A *variable venturi carburetor* adjusts the diameter of the venturi to maintain a relatively constant air speed in the carburetor. Many foreign cars and some American cars use this type of carburetor.

Figure 24-28 shows a variable venturi, slide-type carburetor. A piston slides in and out to regulate the size of the venturi. This type of carburetor is commonly used on motorcycles and some foreign cars.

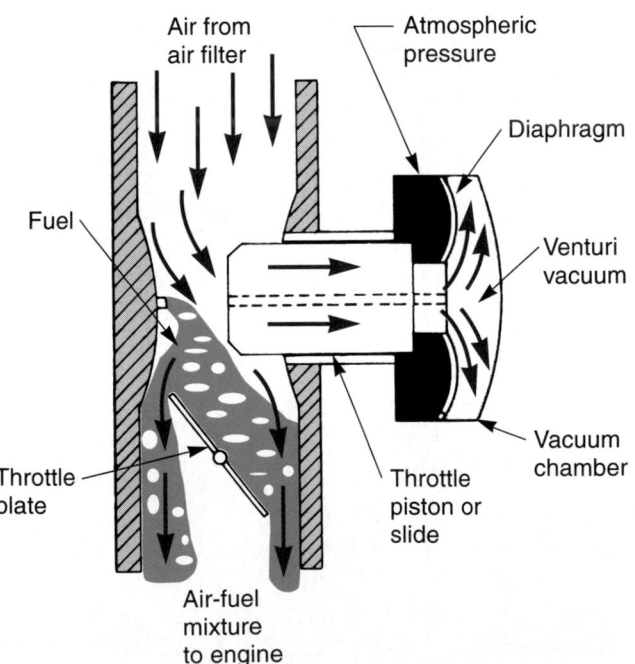

Figure 24-28. A slide-type, variable venturi carburetor. A cylinder-shaped slide moves in and out of the air horn to help control fuel and airflow. A conventional throttle valve is connected to the accelerator pedal.

Computer-Controlled Carburetors

A *computer-controlled carburetor* normally uses a solenoid-operated valve to respond to commands from a computer (control module). A typical system is shown in **Figure 24-29.**

The system uses various sensors to send information to the computer. Then, the computer calculates how rich or lean to set the carburetor's air-fuel mixture.

An *electromechanical carburetor* has both electrical and mechanical control devices. This type is shown in **Figure 24-30.** It is commonly used with a computer control system.

A *mixture-control solenoid* in the computer-controlled carburetor alters the air-fuel ratio. Electrical signals from the computer activate the solenoid to open and close air and fuel passages in the carburetor. See **Figure 24-31.**

An *idle speed actuator* may also be used to allow the computer to change engine idle speed. It is usually a tiny electric motor and gear mechanism that holds the carburetor throttle lever in the desired position.

Computer-Controlled Carburetor Operation

In a computer-controlled carburetor, the air-fuel ratio is maintained by cycling the mixture-control solenoid on and off several times a second. **Figure 24-32** shows how

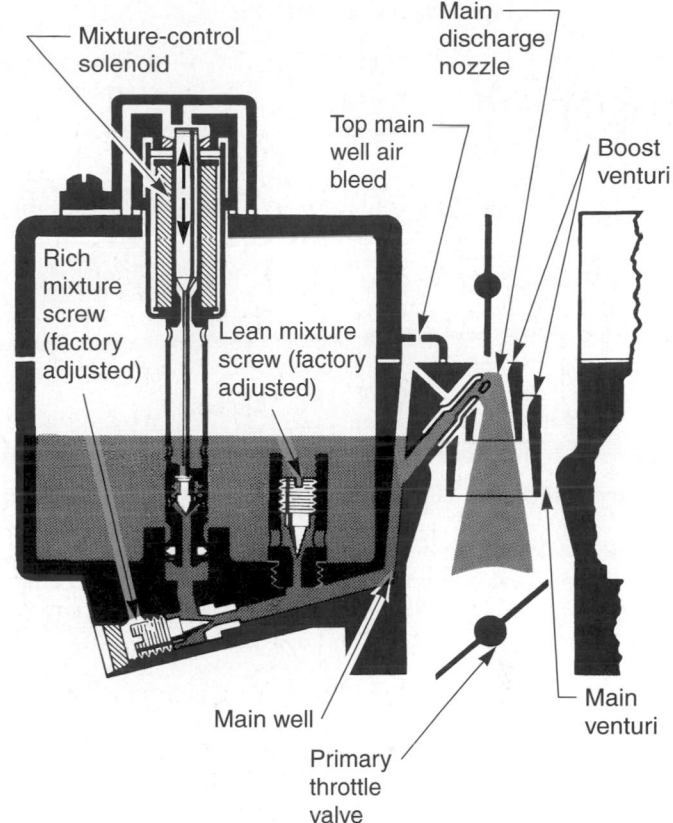

Figure 24-30. A mixture-control solenoid can quickly open or close a valve to change the carburetor's air-fuel ratio. The computer sends electrical pulses to the solenoid. The solenoid magnetic field acts on a metal plunger, operating the mixture valve. (Pontiac)

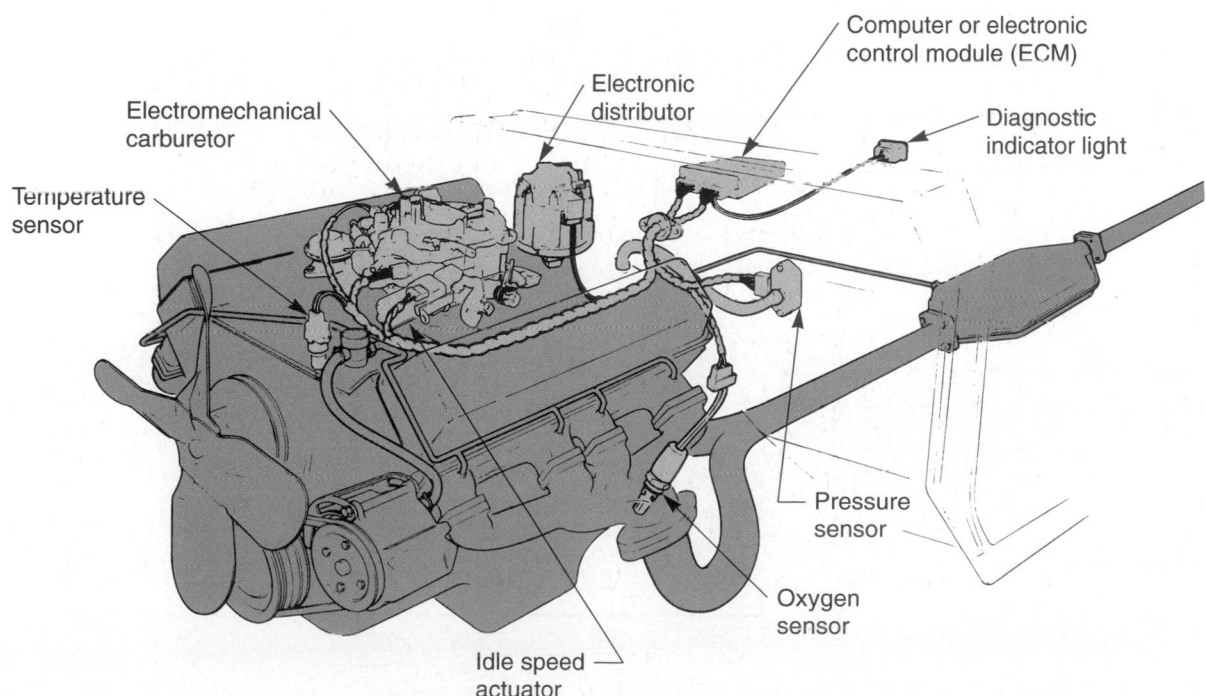

Figure 24-29. Typical computer-controlled carburetor system. Study the part names and locations. (AC-Delco)

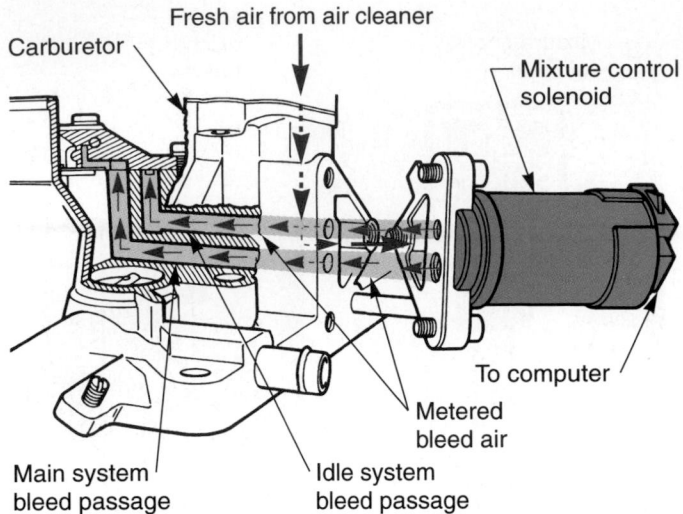

Carburetor

Fresh air from air cleaner

Mixture control solenoid

To computer

Metered bleed air

Main system bleed passage

Idle system bleed passage

Figure 24-31. Another design for a mixture-control solenoid. This solenoid valve opens and closes the air passage to control the air-fuel ratio. This also allows the computer to control the carburetor. (Ford)

the control signals from the computer can be used to meter different amounts of fuel out of the carburetor.

When the computer sends a rich command to the solenoid, the signal voltage to the mixture-control solenoid is *off* more than it is on. This causes the solenoid to stay open more. During a lean signal from the computer, the signal has more *on* time. This causes less fuel to pass through the solenoid valve. The mixture becomes leaner.

Summary

- A carburetor is basically a device for mixing air and fuel in the correct proportions (amounts) for efficient combustion.

- The float system must maintain the correct level of fuel in the carburetor bowl.

- The off-idle system, often termed the part throttle circuit, feeds more fuel into the air horn when the throttle plates are partially open.

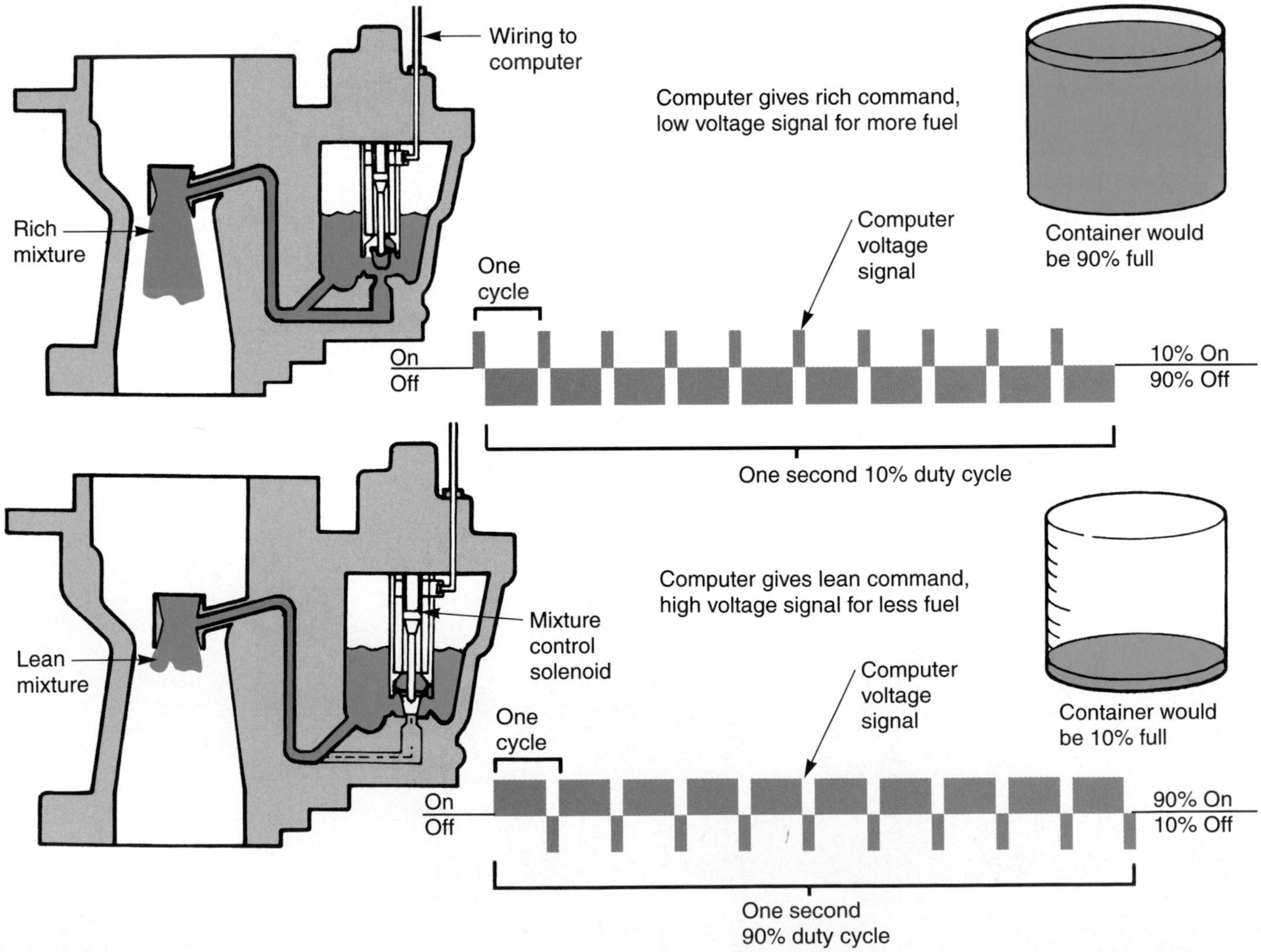

Wiring to computer

Rich mixture

Computer gives rich command, low voltage signal for more fuel

One cycle

Computer voltage signal

Container would be 90% full

On
Off

10% On
90% Off

One second 10% duty cycle

Lean mixture

Mixture control solenoid

Computer gives lean command, high voltage signal for less fuel

Container would be 10% full

One cycle

Computer voltage signal

On
Off

90% On
10% Off

One second 90% duty cycle

Figure 24-32. Computer rapidly pulses mixture control solenoid ON and OFF to control the air-fuel ratio. With this carburetor, note how longer OFF time enriches the mixture. Shorter OFF time makes the mixture lean. (Chrysler)

- The carburetor's high-speed system, also called main metering system, supplies the engine's air-fuel mixture at normal cruising speeds.

- The carburetor full-power system provides a means of enriching the fuel mixture for high-speed, high-power conditions.

- The choke system is designed to supply an extremely rich air-fuel ratio to aid cold engine starting.

- A computer-controlled carburetor normally uses a solenoid-operated valve to respond to commands from a microcomputer (control module).

Important Terms

Carburetor	Main metering system
Carburetor body	High-speed jet
Air horn	Emulsion tube
Throat	Primary venturi
Barrel	Booster venturis
Throttle valve	Full-power system
Venturi	Metering rod
Main discharge tube	Power valve
Main fuel nozzle	Economizer valve
Fuel bowl	Choke system
Carburetor system	Choke plate
Carburetor circuit	Thermostatic spring
Float system	Bimetal spring
Carburetor float	Manual choke
Needle valve	Automatic choke
Needle seat	Integral hot air choke
Bowl vent	Nonintegral choke
Idle system	Electric assist choke
Low-speed jet	All electric choke
Idle air bleed	Mechanical choke
Economizer	unloader
Bypass	Vacuum choke unloader
Idle passage	Choke break
Idle screw port	Fast idle cam
Idle mixture screw	Anti-dieseling solenoid
Sealed idle mixture	Throttle return dashpot
screws	Anti-stall dashpot
Limiter caps	Hot idle compensator
Off-idle system	Altitude compensator
Part throttle circuit	Aneroid
Acceleration system	EGR valve
Accelerator pump	Distributor vacuum
Pump check ball	advance
Pump check weight	Charcoal canister
Pump nozzle	Choke break
Pump jet	Primary
High-speed system	Secondary

Secondary diaphragm	Electromechanical
Carburetor size	carburetor
Variable venturi	Mixture-control
carburetor	solenoid
Computer-controlled	Idle speed actuator
carburetor	

Review Questions—Chapter 24

Please do not write in this text. Place your answers on a separate sheet of paper.

1. A carburetor is basically a device for _____ air and fuel in the correct _____ for efficient combustion.

2. List and explain the five major parts of a carburetor.

3. The air horn is the main fuel passage in the carburetor. True or False?

4. A carburetor _____ is a network of passages and related parts that controls the air-fuel ratio under specific operating conditions.

5. A venturi is used to produce vacuum in the carburetor air horn. True or False?

6. Which of the following is *not* a typical air-fuel ratio for a gasoline engine?
   (A) 8:1
   (B) 16:1
   (C) 3:1
   (D) 18:1

7. List and explain the seven major carburetor systems or circuits.

8. The carburetor _____ rides on top of the fuel to open and close the needle valve as needed.

9. A bowl vent is used to prevent a pressure or vacuum buildup in the bowl. True or False?

10. The carburetor _____ system provides the engine's air-fuel mixture at speeds below about _____ rpm or 20 mph (32 km/h).

11. What is the purpose of an idle air bleed?

12. What happens when you turn the idle mixture screw in and out?

13. Why are idle mixture screws sometimes covered with a metal plug?

14. A carburetor accelerator pump squirts fuel into the air horn when the throttle is opened for acceleration. True or False?

15. The _____ _____ carburetor system provides the leanest, most efficient air-fuel mixture at normal cruising speeds.

16. What is a high-speed jet?

17. Which of the following could *not* be included in the full-power system?
    (A) Power valve.
    (B) Metering rod.
    (C) Economizer valve.
    (D) Choke plate.

18. What is the function of a carburetor choke?

19. List and explain four types of automatic choke.

20. A mechanical choke unloader uses a metal lug on the throttle lever to physically open the choke plate at full throttle. True or False?

21. How does a vacuum choke break work?

22. What carburetor component is used to prevent dieseling?

23. Describe the difference between the primary and secondary of a carburetor.

24. How do you measure carburetor size?

25. Explain the purpose and operation of a mixture control solenoid used in a computer-controlled carburetor system.

✹ ASE-Type Questions

1. A basic carburetor consists of each of these *except:*
   (A) *venturi.*
   (B) *vacuum.*
   (C) *air horn.*
   (D) *fuel bowl.*

2. The device that routes outside air into the engine intake manifold is called the carburetor:
   (A) *barrel.*
   (B) *throat.*
   (C) *air horn.*
   (D) *All of the above.*

3. Which of the following creates suction (vacuum) that pulls fuel from main discharge tube?
   (A) *Venturi.*
   (B) *Main discharge tube.*
   (C) *Fuel bowl.*
   (D) *Throttle valve.*

4. For idling, carburetors must provide an air-fuel ratio of about:
   (A) *13:1.*
   (B) *15:1.*
   (C) *16:1.*
   (D) *18:1.*

5. Which of the following is *not* one of the seven basic carburetor systems?
   (A) *Float.*
   (B) *Injection.*
   (C) *Full-power.*
   (D) *High-speed.*

6. Which carburetor system provides air-fuel mixture at speeds below 20 mph?
   (A) *Idle.*
   (B) *Choke.*
   (C) *Off-idle.*
   (D) *Acceleration.*

7. The two types of accelerator pumps are the:
   (A) *throttle and spring.*
   (B) *thermal and vacuum.*
   (C) *piston and diaphragm.*
   (D) *primary and secondary.*

8. High-speed carburetor circuits consist of each of the following components *except:*
   (A) *primary venturi.*
   (B) *air bleed.*
   (C) *high speed jet.*
   (D) *emulsion tube.*

9. While figuring metering rod action, Technician A states that an engine vacuum is used to operate a metering rod. Technician B says that an electric solenoid and computer are used to operate a metering rod. Who is right?
   (A) *A only.*
   (B) *B only.*
   (C) *Both A and B.*
   (D) *Neither A nor B.*

10. Which of the following provides a variable high-speed fuel mixture?
    (A) *Power valve.*
    (B) *Metering rod.*
    (C) *Economizer valve.*
    (D) *All of the above.*

Activities—Chapter 24

1. On a carburetor of your instructor's choice, demonstrate your understanding of its systems by naming its parts and explaining their purpose.

2. Draw a simple carburetor cutaway and show how fuel and air are mixed.

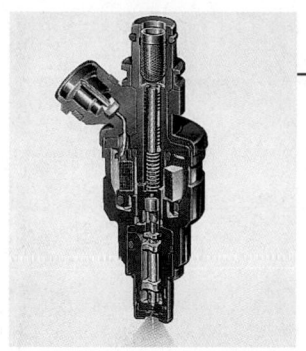

Chapter 25

Carburetor Diagnosis, Service, and Repair

After studying this chapter, you will be able to:

- Diagnose common carburetor problems.
- Describe the troubles that frequently occur in carburetor systems.
- Properly remove and replace a carburetor.
- Disassemble and reassemble a carburetor while following the procedures in a service manual.
- Perform basic carburetor adjustments.
- Describe and observe all safety rules and cautions relating to carburetor service.
- Correctly answer ASE certification test questions about carburetor diagnosis and repair.

The carburetor is very important to the efficiency and reliability of an engine. If a carburetor passage becomes clogged with dirt or a gasket or seal deteriorates, engine performance can suffer. Mechanical part wear of linkages and throttle shafts also affect performance.

Carburetor Problem Diagnosis

The first step in carburetor problem diagnosis is to determine whether the carburetor or another system is at fault. Troubles in the ignition system, emission control systems, and engine can produce the same general symptoms (missing, poor fuel economy, engine not starting) as a faulty carburetor.

A visual inspection of the carburetor may provide clues to the problem, **Figure 25-1.** Remove the air cleaner. Look for leaking fuel, a sticking choke, binding linkage, and missing or disconnected vacuum hoses. A heavy covering of road dirt usually indicates that the carburetor has been in service for a long time. Adjustments or repairs may be needed.

While inspecting the carburetor, check the rest of the engine compartment. Look for disconnected wires and hoses. Listen for the hissing sound of a vacuum leak.

Make sure the distributor cap is not cracked. Try to locate anything that could upset normal engine operation.

Exhaust Gas Analyzer

An *exhaust gas analyzer* is a testing device that measures the chemical content of the engine exhaust gases. It can be used to determine the air-fuel mixture of a fuel injection system or carburetor.

The exhaust gas analyzer draws a sample of the exhaust gases (combustion byproducts) from the vehicle's tailpipe. It measures the amount of carbon monoxide (CO), hydrocarbons (HC), carbon dioxide (CO_2), oxygen (O_2), and nitrogen oxides (NO_x) in the exhaust. This information indicates combustion efficiency (how efficiently the engine is burning its fuel).

Carburetor Problems

When diagnosing carburetor problems, try to determine which system is at fault. For example, if the engine only runs poorly when cold, think of things (mechanical parts, sensors, actuators) that would be affected by cold temperatures. Check the choke and fast idle systems or the engine temperature sensor on a computer-controlled fuel system. Use this type of logic to narrow down the possible sources of problems.

Carburetor float system troubles can cause a wide range of performance problems: flooding, rich fuel mixture, lean fuel mixture, fuel starvation (no fuel), stalling, low-speed engine miss, and high-speed engine miss. See **Figure 25-2.**

Float problems can affect the operation of the engine at all speeds. The float system provides fuel for all the other carburetor circuits.

Carburetor flooding occurs when fuel pours out the top of the carburetor (vent, air bleed, or main discharge) or leaks around the bowl gasket. This problem arises

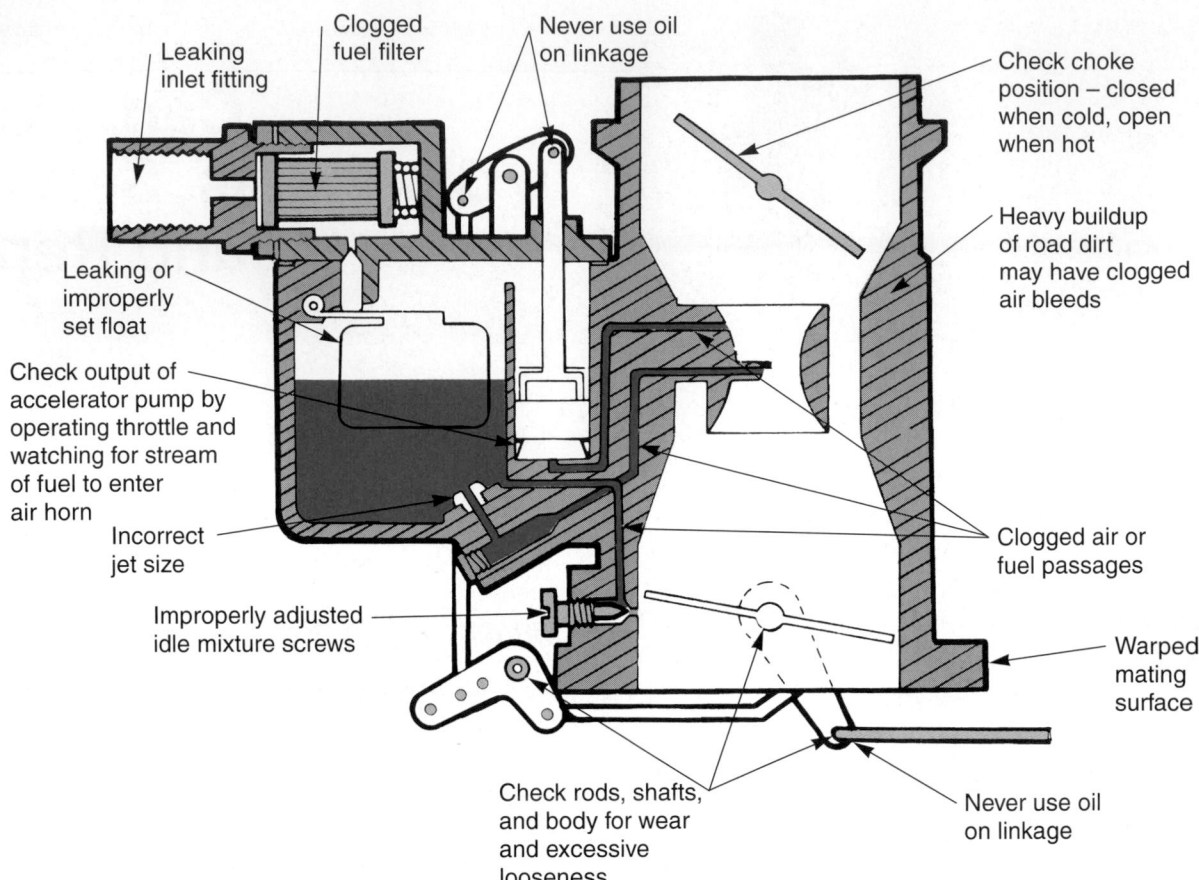

Figure 25-1. An external check of the carburetor can sometimes find the source of the problem. A few possible problems are illustrated here. (Chrysler)

when the float needle valve does not stop fuel flow from the fuel pump. The carburetor bowl overflows.

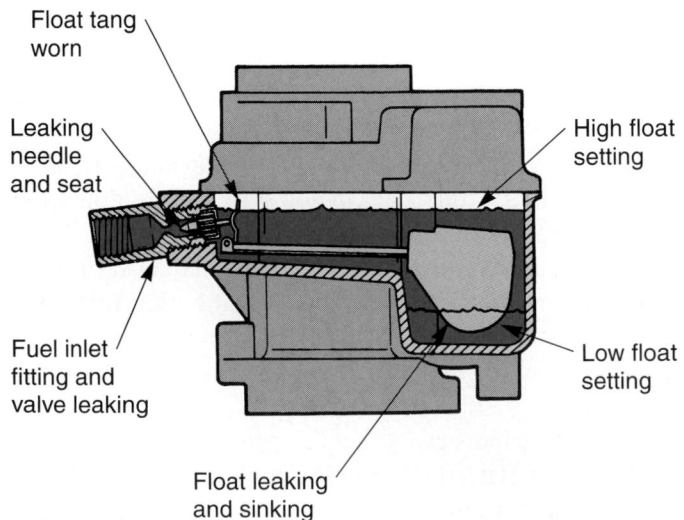

Figure 25-2. Float system problems can cause flooding, fuel starvation, and poor performance at all engine speeds.

Warning

If a carburetor floods, shut off the engine immediately. Gasoline leaking onto the intake manifold could be ignited by a spark or the hot exhaust manifold.

To correct carburetor flooding, disassemble the carburetor. Check the needle valve and seat for wear. They may need replacement. If fuel has leaked into the float, the float must be replaced. A carburetor bowl that is full of debris can also cause the needle valve to stick open. A complete carburetor overhaul is usually necessary to correct this problem.

Engine flooding is a problem resulting from too much fuel in the intake manifold and combustion chambers. Excess fuel could foul the spark plugs (fuel keeps sparks from jumping gaps), keeping the engine from running.

Engine flooding may be due to any number of reasons besides a float system problem. For example, if the driver pumps the gas pedal repeatedly, operating the accelerator pump, the engine could flood. A weak spark, stuck choke, or bad choke break diaphragm could also lead to engine flooding.

A *high float level* richens the air-fuel mixture. It can cause high fuel consumption, rough idling, and, possibly, black engine exhaust. To check the float setting, start the engine and let it idle. While wearing safety glasses, check for fuel dripping from the main discharge tube (venturi). If fuel leaks out the main nozzle at idle, the float setting may be too high.

A *low float level* can produce a lean air-fuel mixture. It can cause a high-speed miss, stalling when cornering, and other symptoms common to fuel starvation. Usually, you need to remove the air horn or bowl cover to check for a low fuel level. A few carburetors, however, have a glass inspection window or a fuel level screw on the side of the fuel bowl.

A high or low float level can usually be corrected by adjusting the float mechanism. This adjustment will be discussed later in the chapter.

Idle System Problems

Carburetor *idle system problems* normally show up as a rough idle, an incorrect idle speed, or stalling at low speeds. The engine may run fine at higher speeds, but poorly at idle. **Figure 25-3** points out common idle circuit problems.

A *clogged idle passage* can restrict fuel flow to the idle port. This causes a very lean mixture and poor idle. The jets in the idle circuit are very small. A tiny bit of dirt can easily upset idle circuit operation.

A *clogged idle air bleed,* can enrich the mixture by obstructing the premixing of air, upsetting engine operation at idle. Since most of the air bleed jets are at the top of the carburetor, inspect the air bleeds and clean them, if needed.

To check for idle system problems on older cars, adjust the idle mixture screws. If turning the screws has little or no effect on idle speed and smoothness, the idle circuit is not functioning properly. The carburetor probably needs an overhaul.

Tech Tip

Sometimes you can clear out idle passages by revving the engine to high rpm and placing a leather glove over the carburetor air inlet. This will create a powerful suction to pull debris out of the tiny passages.

Acceleration System Problems

Carburetor *acceleration system problems* usually cause the engine to hesitate, stall, pop, or backfire when the car first moves. If the accelerator pump does not squirt a strong stream of fuel into the air horn as the throttles open, the mixture will be too lean for proper engine acceleration. Look at **Figure 25-4.**

To check accelerator pump operation, shut the engine off and remove the air cleaner. Open and close the throttle while looking inside the carburetor air horn.

Each time the throttle is opened, a strong stream of fuel should spray into the air horn. The stream should start as soon as you move the throttle open. It should

Figure 25-3. Idle circuit problems show up at very low engine speeds. Jets and passages are so small they can be clogged by tiny dirt particles. Improper adjustment of mixture screw is another common trouble. (Chrysler)

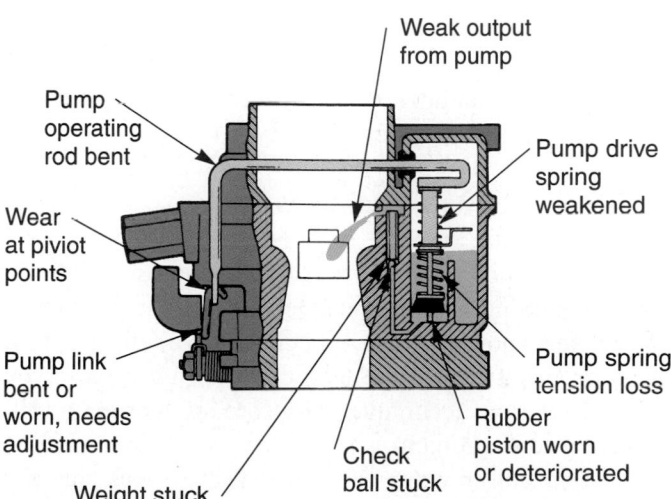

Figure 25-4. A bad accelerator pump will make the engine hesitate upon acceleration. Worn or misadjusted linkage, a deteriorated pump or diaphragm, or other problems may keep fuel from squirting out of the nozzle when the throttle opens. (Chrysler)

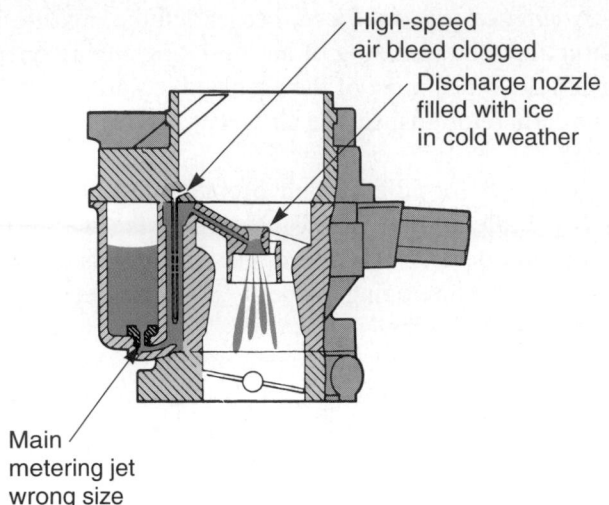

Figure 25-5. The main metering system is very simple and dependable.

continue for a short time after full throttle opening. If the stream is weak, adjust or repair the accelerator pump.

High-Speed System Problems

High-speed system problems normally show up as a lean air-fuel mixture (miss at cruising speeds) or a rich mixture (poor fuel economy). Refer to **Figure 25-5.** The high-speed system is the simplest and most dependable carburetor circuit.

Engine surge is a condition in which engine power at cruising speeds seems to alternately increase and decrease. It is frequently caused by an extremely lean air-fuel mixture. Surge is a common problem stemming from the extremely lean, high-efficiency mixtures of today's carburetors.

A *faulty mixture control solenoid* or computer-control circuit can upset high-speed, midrange, and low-speed carburetor operation.

Full-Power System Problems

Full-power system problems can reduce fuel economy or limit engine power in high-power situations, such as passing another car. Depending on the type of carburetor (power valve, mechanical metering rod, vacuum metering rod, solenoid metering valve), you must use various techniques to locate and correct a full-power systems problem.

A *bad power valve* can allow fuel to leak into the vacuum passage. A common problem, it causes a rough idle and poor gas mileage. When the power valve diaphragm ruptures, fuel may be pulled through the diaphragm and into the air horn by vacuum. This richens

the mixture. The ruptured diaphragm can also affect high-speed engine performance.

A *faulty metering rod* mechanism will generally upset only high-speed engine operation. If the metering rod is adjusted too far into the jet, the high-speed mixture will be too lean. If set too far out of the jet, the high-speed mixture will be too rich. A vacuum-operated metering rod can cause problems if the diaphragm ruptures or the vacuum piston sticks.

Choke System Problems

Carburetor *choke system problems* will usually make the engine perform poorly when cold or right after warm-up. The choke should operate only when the engine is cold. Problems usually occur during or right after starting. **Figure 25-6** shows some choke-related problems.

If the choke sticks closed, a super-rich air-fuel mixture will pour into the engine. The engine will run extremely rough, and black smoke will blow out the vehicle's tailpipe. The engine will lack power and may stall as it warms.

If the choke sticks open, the engine may not start when cold. Pumping the gas pedal several times will help. However, the engine will accelerate poorly and may stall several times after cold starting.

A slightly *rich choke setting* can cause the engine to run poorly after reaching operating temperature. By remaining partially closed, the choke pulls too much fuel into the engine. A slight engine miss or roll and a small amount of black exhaust smoke may result.

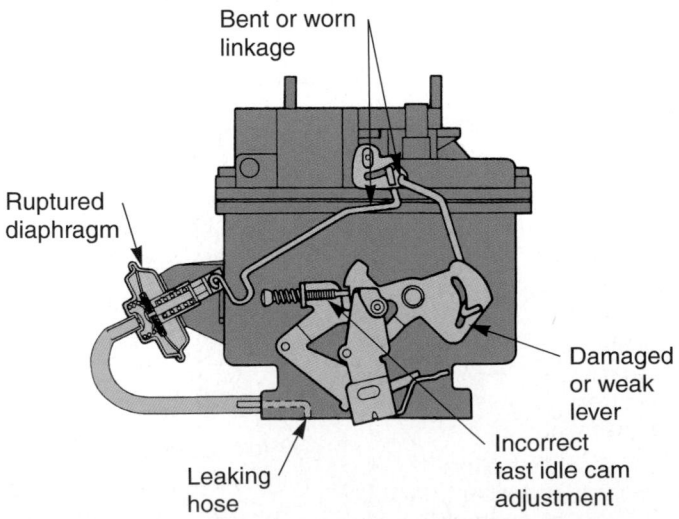

Figure 25-6. Choke malfunctions cause problems when the engine is started cold. The engine may not start easily, may stall, or may run roughly. (Chrysler)

A *lean choke setting,* as with an open choke, will make the engine hard to start in cold weather. No problems may be noticed in warm weather. A lean choke adjustment can cause the engine to stall when trying to accelerate with the engine cold.

Fast idle cam problems can make the engine idle too fast or too slow, usually when cold If the fast idle is too high, a cold engine may race for extended periods. If the fast idle is too slow, a cold engine will stall right after start-up. Fast idle cam problems may stem from choke problems or an incorrect linkage adjustment.

Factory Diagnosis Charts

Use a factory or service manual *diagnosis chart* when you have difficulty locating a carburetor problem. Such charts are designed for each type of carburetor.

Carburetor Removal

When diagnosis points to *internal carburetor problems* (dirt-clogged passages, leaking gaskets or seals, worn parts), carburetor removal is usually necessary. Begin by removing the air cleaner and all wires, hoses, and linkages that would prevent removal of the carburetor.

 **Warning**
Wrap a shop rag around the fitting when removing a fuel line. This will keep fuel from leaking on the engine, which would create a fire hazard.

You may want to label vacuum hoses and wires that connect to the carburetor. This will simplify reassembly. See **Figure 25-7.** If the hoses and wires are not labeled

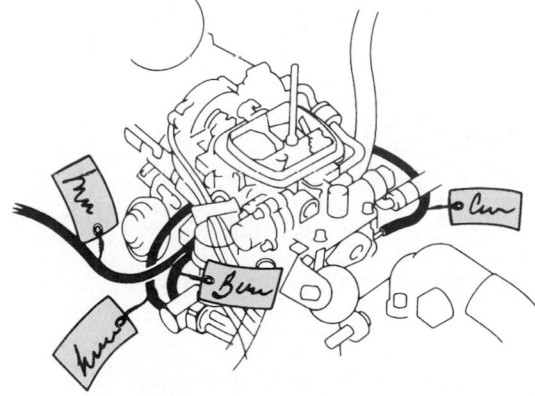

Figure 25-7. Always mark or tag wires and hoses when servicing a carburetor. This can save time and avoid confusion when reassembling. (Toyota)

before carburetor removal, you may have to trace color codes on a wiring diagram.

Remove the four nuts or bolts that secure the carburetor to the engine intake manifold. Carefully lift the carburetor off the engine while holding it in a level position. Do not splash the fuel in the bowl around. This would stir up any dirt in the bottom of the bowl.

 Tech Tip
To prevent foreign objects from entering the intake manifold, place a shop rag or towel in the manifold's inlet opening after the carburetor has been removed. If you accidentally drop a washer or nut into the manifold, you may be doing major engine disassembly operations for free! The washer or nut can bend valves, knock holes in a piston, or score cylinder walls.

Carburetor Body Sections

Automotive carburetors are usually constructed in three sections: air horn body, main body, and throttle body, **Figure 25-8.** Sometimes, the main body and throttle body are combined into a single casting.

The *air horn body,* or *upper body,* fits on the top of the main body and serves as a lid for the fuel bowl. Shown in **Figure 25-8,** it is held on the main body with screws. A gasket fits between the main body and the air horn body. The parts that normally fasten to the air horn body are the choke, hot idle compensator, fast idle linkage rod, choke vacuum break, and, sometimes, the float and pump mechanisms.

The *main body* is the largest section of the carburetor. It forms the air horn and fuel bowl. Small passages are cast or drilled in this section to carry fuel and air. Usually, the main body houses the fuel bowl, main jets, air bleeds, power valve, pump checks, diaphragm-type accelerator pump, venturis, circuit passages, and float mechanism.

The *throttle body* is the lower section of the carburetor and contains the throttle valves. It fastens to the bottom of the main body with screws. A throttle shaft passes through the sides of the throttle body to provide a hinge for the throttle valves. A gasket between the throttle body and the main body makes the joint leakproof. The main parts found on a throttle body, besides the throttle butterflies and throttle shaft, are the throttle levers, idle mixture screws, idle speed screw, fast idle cam, dashpot, and fast idle solenoid.

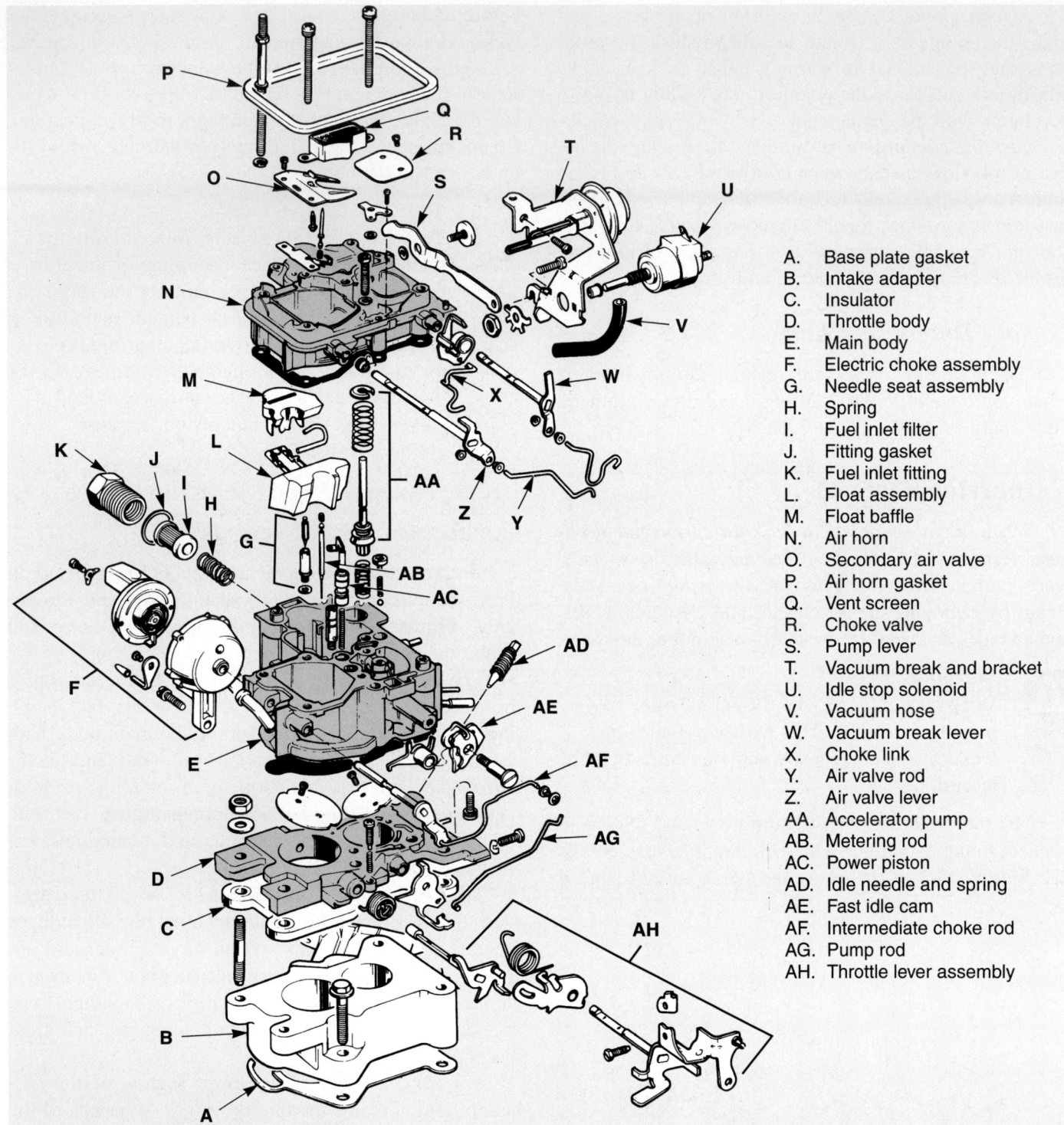

A. Base plate gasket
B. Intake adapter
C. Insulator
D. Throttle body
E. Main body
F. Electric choke assembly
G. Needle seat assembly
H. Spring
I. Fuel inlet filter
J. Fitting gasket
K. Fuel inlet fitting
L. Float assembly
M. Float baffle
N. Air horn
O. Secondary air valve
P. Air horn gasket
Q. Vent screen
R. Choke valve
S. Pump lever
T. Vacuum break and bracket
U. Idle stop solenoid
V. Vacuum hose
W. Vacuum break lever
X. Choke link
Y. Air valve rod
Z. Air valve lever
AA. Accelerator pump
AB. Metering rod
AC. Power piston
AD. Idle needle and spring
AE. Fast idle cam
AF. Intermediate choke rod
AG. Pump rod
AH. Throttle lever assembly

Figure 25-8. Exploded view of a carburetor shows the body sections and how each part fits into place. (Chrysler)

Carburetor Rebuild

A *carburetor rebuild,* or *overhaul,* is needed when carburetor passages become clogged, gaskets or seals leak, rubber parts deteriorate, or components wear and fail.

A carburetor rebuild generally involves:

1. Disassembly and cleaning of major parts.

2. Inspection for wear or damage.

3. Installation of a *carburetor rebuild kit* (new gaskets, seals, needle valve and seat, pump diaphragm or cup, and other nonmetal parts).

4. Reassembly of the carburetor.

5. Major adjustments.

When disassembling a carburetor, follow the kit directions or the detailed directions in a service manual. Generally, remove all plastic and rubber parts. Remove any part that will prevent thorough cleaning of the carburetor passages. If not worn or damaged, the throttle plates and shaft can be left in place.

To prevent part damage, always use proper disassembly techniques. When unscrewing jets, use a special jet tool or the correct size screwdriver.

During disassembly, keep parts organized. Note the location of each part. Jet sizes are sometimes different. Therefore, jets must be reinstalled in the same location.

Carburetor cleaner, also known as *decarbonizing cleaner,* removes deposits. A very powerful cleaning agent, it will remove gum, carbon, oil, grease, and other deposits from inside air and fuel passages and on external parts. Place metal components in a tray and lower them into the cleaner. Allow the carburetor parts to soak for the recommended amount of time. Do not allow carburetor cleaner to contact plastic and rubber parts. The cleaner will ruin nonmetallic parts.

Warning
Carburetor cleaner can cause serious chemical burns to the skin and eyes. Wear rubber gloves and eye protection when working with any chemical.

After soaking, rinse all carburetor parts with clean water, kerosene, or cold soak cleaning solution. Dry the parts with compressed air. Force air through all passages to make sure they are clear. Do *not* use wire or a drill bit to clean out carburetor passages. This could scratch and enlarge the passages, upsetting carburetor operation.

Inspect each cleaned carburetor part carefully. Check for:

- Wear and excessive play between the throttle shaft and throttle body.
- Binding of the choke plate and linkage.
- Warpage, cracks, and other problems with carburetor bodies.
- Float leakage or a bent hinge arm.
- Nicks, burrs, and dirt on gasket mating surfaces.
- Damaged or weakened springs and stripped fasteners.
- Damage to tips of idle mixture screws.

Replace all parts that show wear or damage.

Carburetor Reassembly

To reassemble a carburetor, follow the detailed directions in the service manual. Some car makers have as many as twenty different carburetors for a single model year. This results in thousands of different reassembly procedures and adjustments. Each is critical to proper performance.

Basically, the carburetor is assembled in reverse order of disassembly. An exploded view of the carburetor, **Figure 25-9,** shows how parts fit together.

As you assemble the carburetor, several adjustments must be completed. Check a service manual for details. Adjustments typically include:

- Float drop adjustment—The float drop is the distance the float can move down in the bowl. It must be properly set to keep the float from hitting and possibly sticking to bottom. The float drop is usually set by bending a tang on the float. See **Figure 25-10.**

- Float level adjustment—The float level is usually set by measuring the distance from the float to the air horn cover or the top of the bowl. To change the float level, bend the other metal tang on the float. Wet float level is set by checking how much fuel enters the bowl before the needle valve closes. See **Figure 25-11.**

- Idle mixture screw "rough" adjustment—Rough adjustment of idle mixture screws typically involves bottoming screw lightly, then backing out about 2 1/2 turns, **Figure 25-12.** Final adjustment is done with the engine running.

- Choke linkage and spring adjustments—Choke adjustment is done so that the choke is almost fully closed when the engine is cold and fully open when the engine is warm. Adjustment procedures vary, so refer to the service manual for specific instructions. You may have to rotate a choke spring housing, bend linkage rods, or perform similar adjustments.

- Antistall dashpot adjustment—The throttle return dashpot is adjusted by moving the unit closer to or farther from the throttle lever. To check adjustment, depress the plunger and measure the distance between the plunger and the throttle lever with a feeler gauge. See **Figure 25-13.**

- Accelerator pump adjustment—Basically, an accelerator pump should be adjusted so that fuel squirts into the air horn as soon as the throttle is moved off idle. Adjustments are done by bending linkage rods or arms and changing rods in their holes. Refer to the service manual for details.

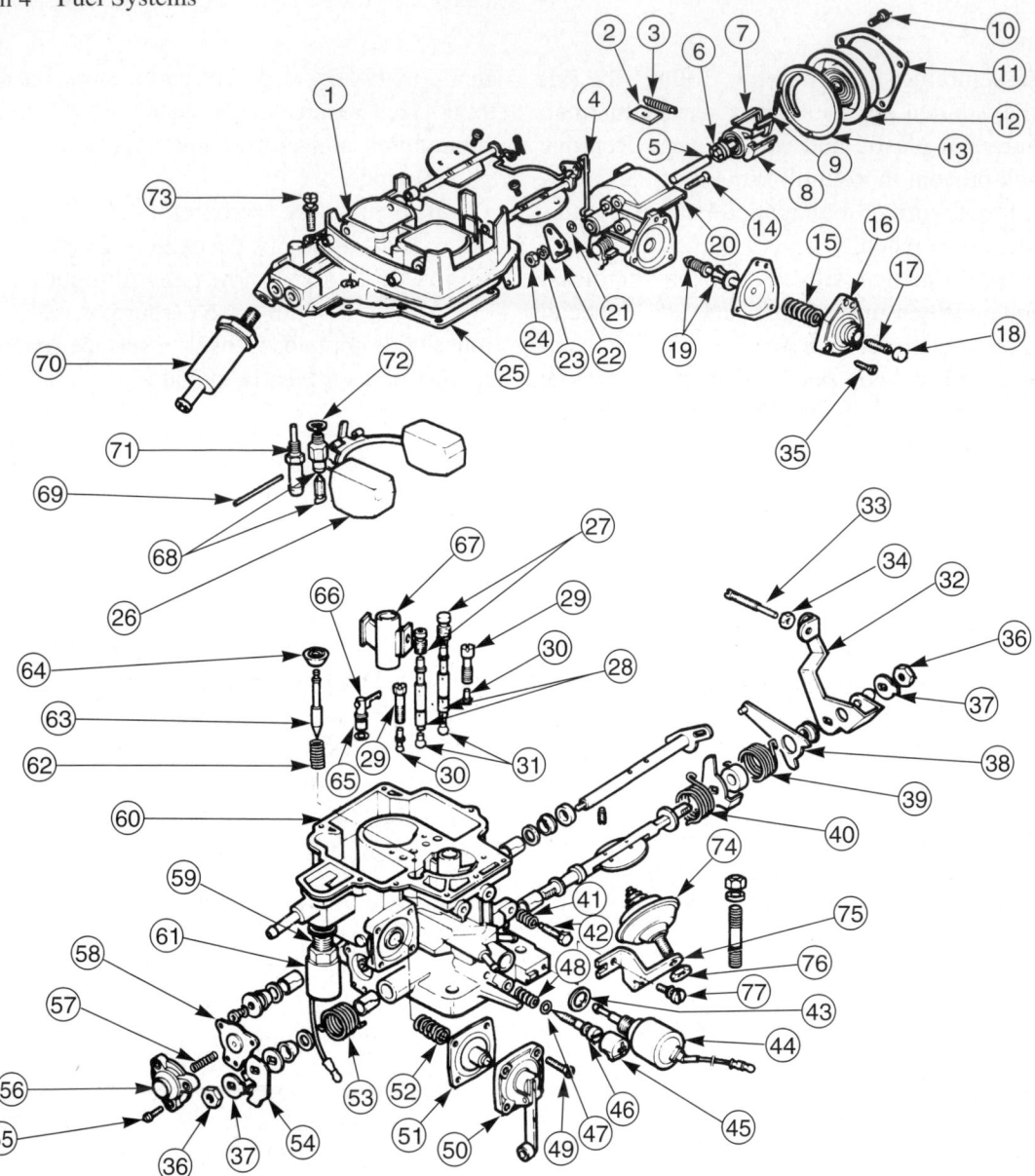

1. Upper body
2. Primary choke link dirt seal
3. Dirt seal retainer
4. Primary choke link
5. Choke bimetal shaft bushing
6. Fast idle cam spring
7. Choke bimetal lever
8. Choke bimetal shaft
9. Choke assist spring
10. Electric choke retaining rivets (2)
11. Electric choke retaining ring
12. Electric choke unit
13. Choke housing index shield
14. Choke housing screws
15. Choke pull-down spring
16. Cover
17. Choke pulldown adjusting screw
18. Choke pulldown adjusting seal
19. Choke pulldown diaphragm assembly
20. Choke housing assembly
21. Choke housing vacuum seal (O-ring)
22. Choke lever
23. Choke bimetal shaft lock washer
24. Choke bimetal shaft nut
25. Cover gasket
26. Fuel bowl float
27. High speed air bleeds
28. Well tubes
29. Idle jet holder
30. Idle jet
31. Main jet
32. Throttle lever
33. Fast idle speed adjusting screw
34. Fast idle speed adjusting screw locknut
35. Choke pulldown diaphragm cover rivet
36. Primary throttle shaft nut
37. Primary throttle shaft nut locking tab
38. Secondary throttle operating lever
39. Secondary throttle return spring
40. Primary throttle return spring
41. Idle speed screw spring
42. Idle speed screw
43. Idle fuel shutoff solenoid washer
44. Idle fuel shutoff solenoid
45. Idle mixture screw plug
46. Idle mixture screw
47. Idle mixture screw O-ring
48. Idle mixture screw spring
49. Accelerator pump cover screw
50. Accelerator pump cover assembly
51. Accelerator pump diaphragm
52. Accelerator pump spring
53. Primary throttle return spring
54. Accelerator pump cam
55. Power valve cover screw
56. Power valve cover
57. Power valve spring
58. Power valve diaphragm washer
59. Fuel bowl vent solenoid washer
60. Main body assembly
61. Fuel bowl vent solenoid
62. Bowl vent spring
63. Bowl vent arm
64. Viton bowl vent seal
65. O-ring seal for pump nozzles
66. Pump shooter
67. Fuel discharge nozzles
68. Fuel inlet seat and needle
69. Float hinge pin
70. Fuel filter
71. Fuel return line check valve and fitting
72. Fuel inlet seat gasket
73. Cover holddown screws
74. Dashpot
75. Dashpot mounting bracket
76. Dashpot adjusting locknut
77. Dashpot mounting bracket screw

Figure 25-9. Exploded view of a carburetor shows how parts fit together. (Ford)

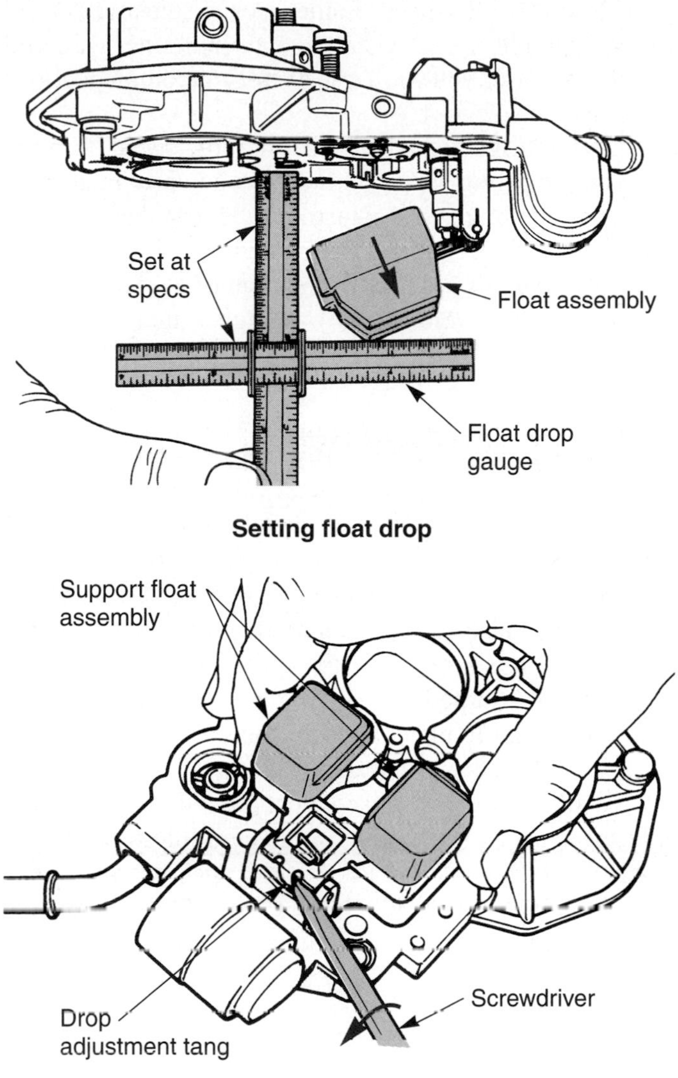

Setting float drop

Set at specs

Float assembly

Float drop gauge

Figure 25-10. Adjusting the float drop. (Chrysler)

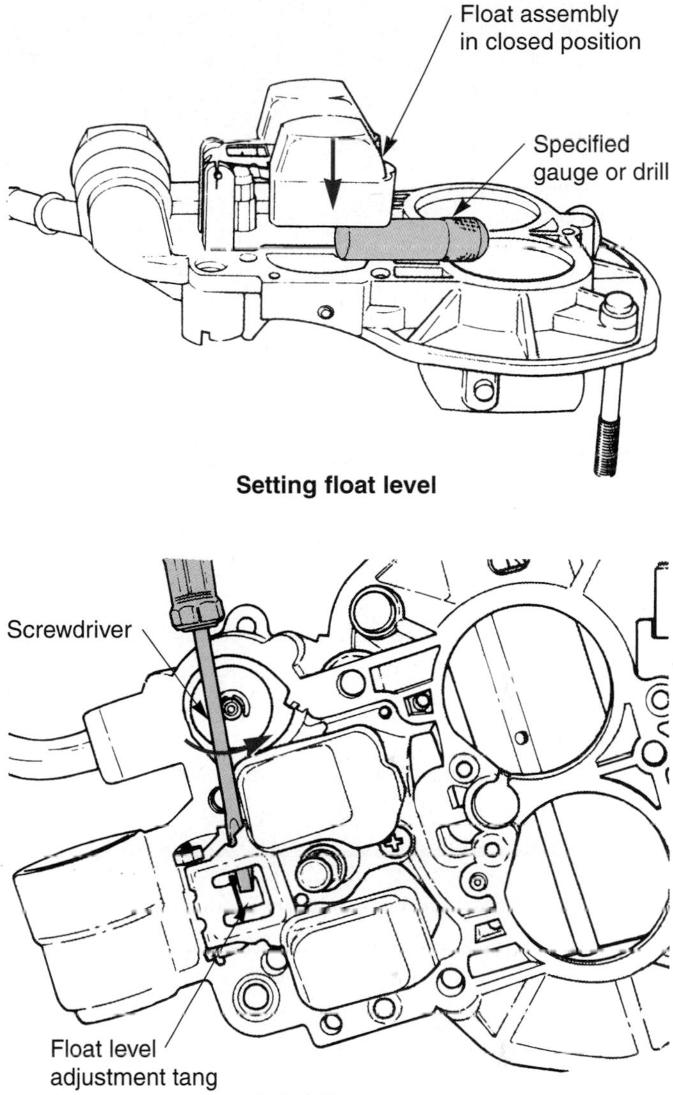

Float assembly in closed position

Specified gauge or drill

Setting float level

Support float assembly

Drop adjustment tang

Screwdriver

Adjusting tang

Screwdriver

Float level adjustment tang

Adjusting tang

Figure 25-11. Dry float level adjustment.

Carburetor Installation

To install a carburetor, place a new base plate gasket on the intake manifold. Then, fit the carburetor on the manifold. Install the fasteners that secure the carburetor to the intake manifold and torque them to specification.

 Caution
Do not overtighten the carburetor hold-down nuts or bolts. Overtightening can damage threads or warp the carburetor base plate.

Reinstall all the hoses, lines, wires, and linkages. Double-check that the vacuum hoses have been installed in their correct positions. Hand operate the throttle to make sure the throttle plates are not binding or hitting on the base plate gasket.

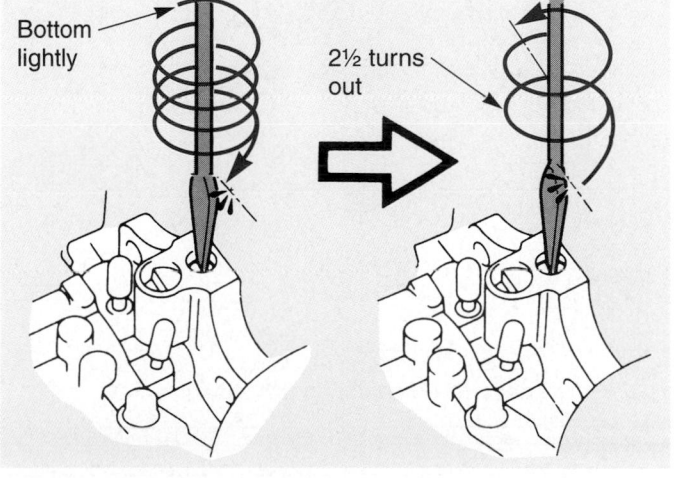

Bottom lightly

2½ turns out

Figure 25-12. Rough adjustment of the idle mixture screws. (Toyota)

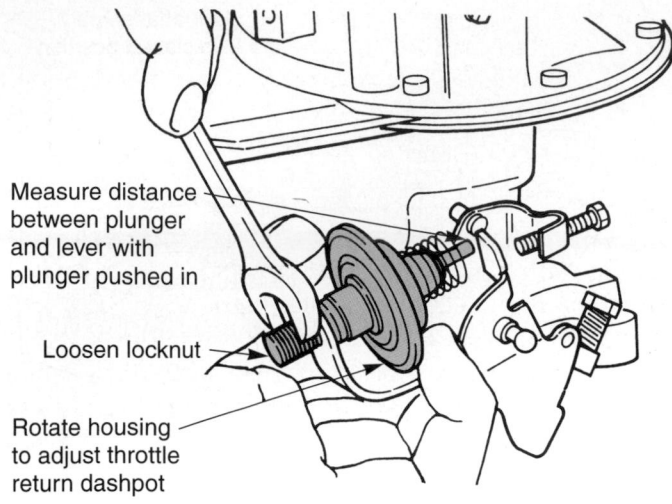

Figure 25-13. Adjusting the throttle return dashpot. (Ford)

To set the cold fast idle or fast idle cam, connect a tachometer to the engine. Warm the engine to operating temperature. Set the emergency brake and block the wheels of the car.

While following service manual instructions, set the fast idle cam to hold the throttle open for a fast idle. The fast idle screw must contact a specified step on the fast idle cam, **Figure 25-14.**

Turn the fast idle screw until the tachometer reads within specifications. Cold fast idle speeds vary from 750–950 rpm. Some specifications are given with the automatic transmission in drive.

Many carburetors use a solenoid to control the hot idle speed of the engine. When the engine is running, the solenoid acts on the throttle lever to hold the throttle open for the correct idle speed. Sometimes, a solenoid is used to maintain idle speed when the air conditioner is turned on. In any case, you must use factory recommendations to adjust the solenoid.

Look at **Figure 25-15.** To set hot idle speed, turn the solenoid adjusting screw until the tachometer reads correctly. Sometimes the adjustment is on the solenoid mounting bracket or on the solenoid plunger itself. Typically, hot idle speed is from about 650–850 rpm. It is lower than cold idle speed but higher than curb idle speed.

Curb Idle Speed Adjustment

Curb idle speed is the lowest idle speed setting. On older carburetors, it is the idle speed setting that controls engine speed under normal, warm engine conditions.

However, on carburetors using a fast idle solenoid, curb idle can also be termed *idle drop* or *low idle speed.* In this case, it is a very low idle speed that only occurs as the ignition is turned off. When the solenoid is de-energized, the throttle plates drop to the curb idle setting to keep the engine from dieseling.

To adjust curb idle, disconnect the wires going to the idle speed solenoid. Then, turn the curb idle speed screw until the tachometer reads within specifications. See **Figure 25-16.** Curb idle speed is usually very low (500–750 rpm).

Idle Mixture Adjustment

After setting the idle speed, you may also need to adjust the carburetor idle mixture. There are several

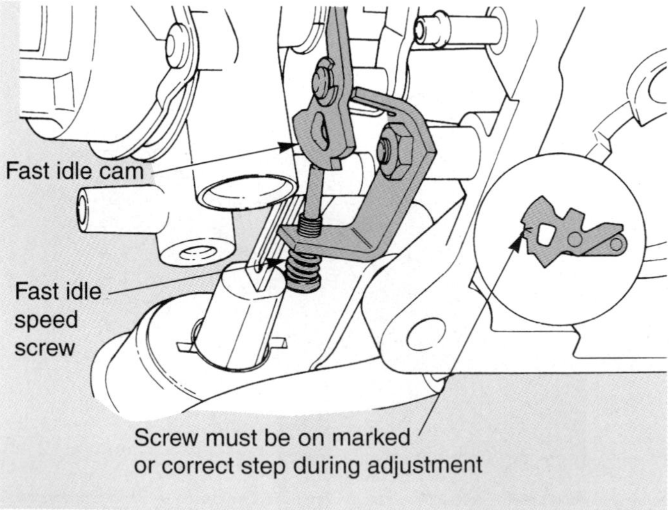

Figure 25-14. Fast idle speed adjustment involves turning the fast idle screw with the screw contacting the correct step on the cam. A tachometer is used to measure engine speed.

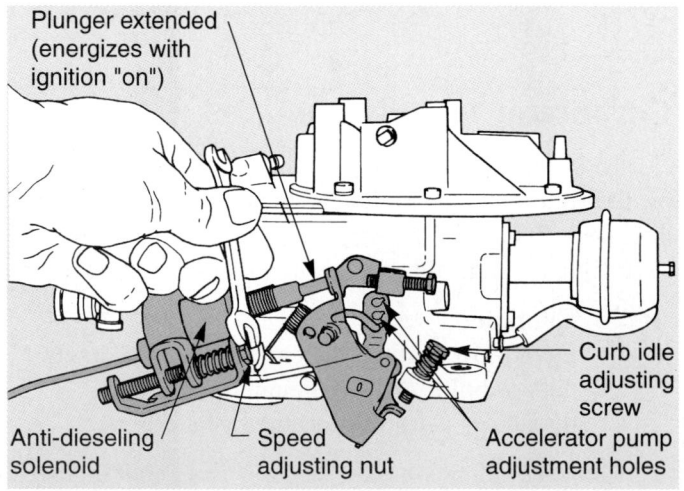

Figure 25-15. When an idle solenoid is used, energize the solenoid and adjust the engine speed to specifications by moving the solenoid or solenoid plunger. Also, note curb idle screw for lowest speed setting. (Ford)

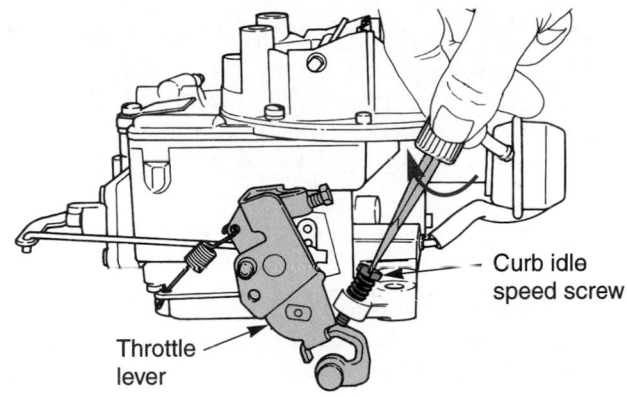

Figure 25-16. The curb idle speed screw simply moves the throttle when turned. Adjust to specifications using a tachometer. (Ford)

methods for doing this. With older, pre-emission carburetors, the idle mixture screw is adjusted until the smoothest idle mixture is obtained.

To adjust idle mixture:
1. Turn the mixture screw in until the engine misses *(lean miss)*.
2. Turn the same screw out until the engine rolls *(rich miss)*.
3. Set the mixture screw halfway between the lean miss and rich roll settings.

4. Perform steps 1-3 on the other mixture screw, if needed (two- and four-barrel carburetors).
5. Recheck the idle speed. If you must reset the idle speed, then the idle mixture must be readjusted.

Propane idle mixture adjustment is required on some carburetors. A bottle of propane is connected to the air cleaner or intake manifold. The propane enriches the fuel mixture during carburetor adjustment. Following service manual procedures, you must meter a certain amount of propane into the engine while adjusting the mixture screws. This will provide a leaner setting, producing less exhaust emissions.

Tech Tip

Most carburetors on late-model vehicles have sealed idle mixture screws that are preset at the factory. Unless major carburetor repairs have been made, do not tamper with these screws. To avoid a violation of federal law, follow the manufacturer's procedures for carburetor adjustments.

An *exhaust gas analyzer* can be used to adjust the idle mixture on some cars. The mixture screws are adjusted until the exhaust sampling is within specified limits. See **Figure 25-17.**

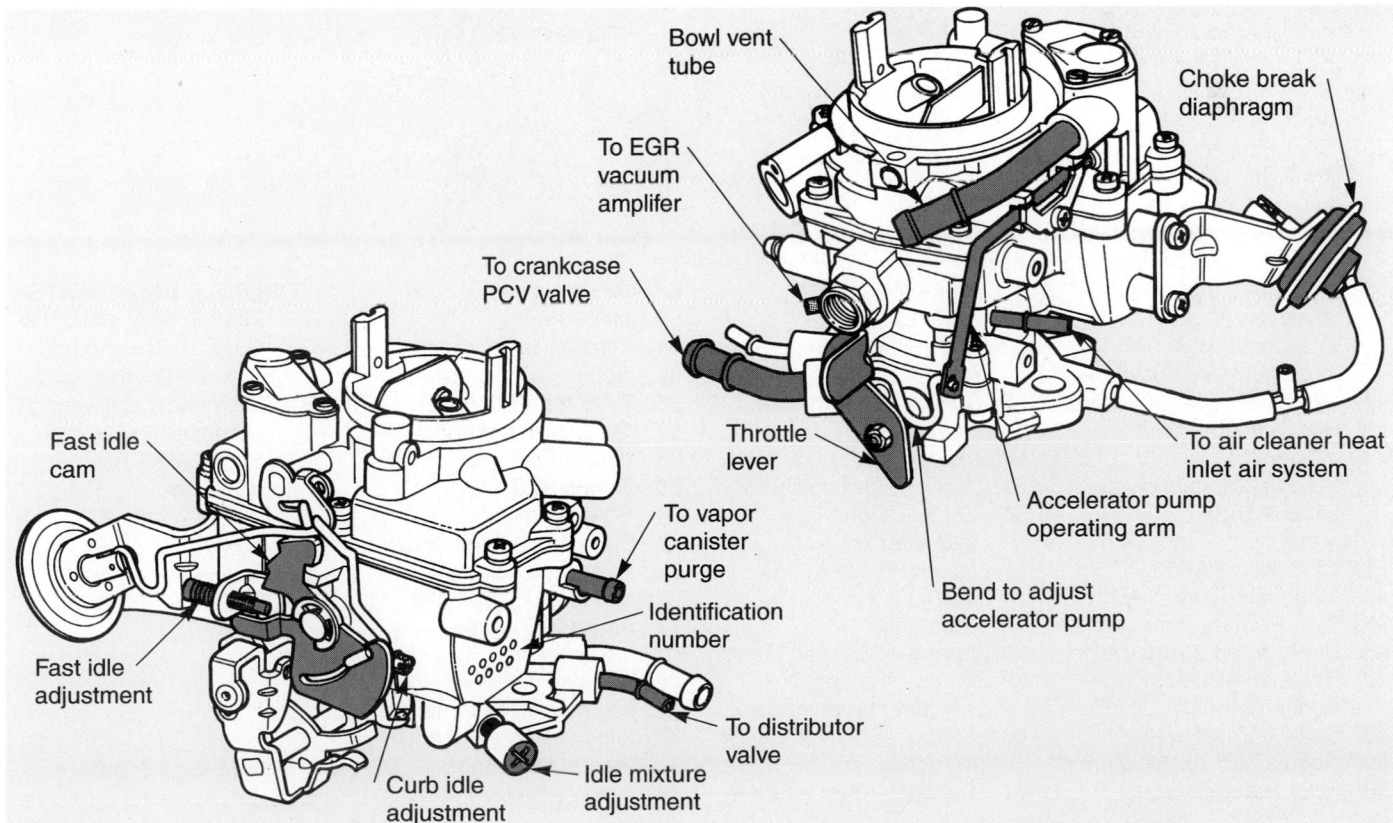

Figure 25-17. Two views show important carburetor parts and connections. Study them. This is a one-barrel carburetor.

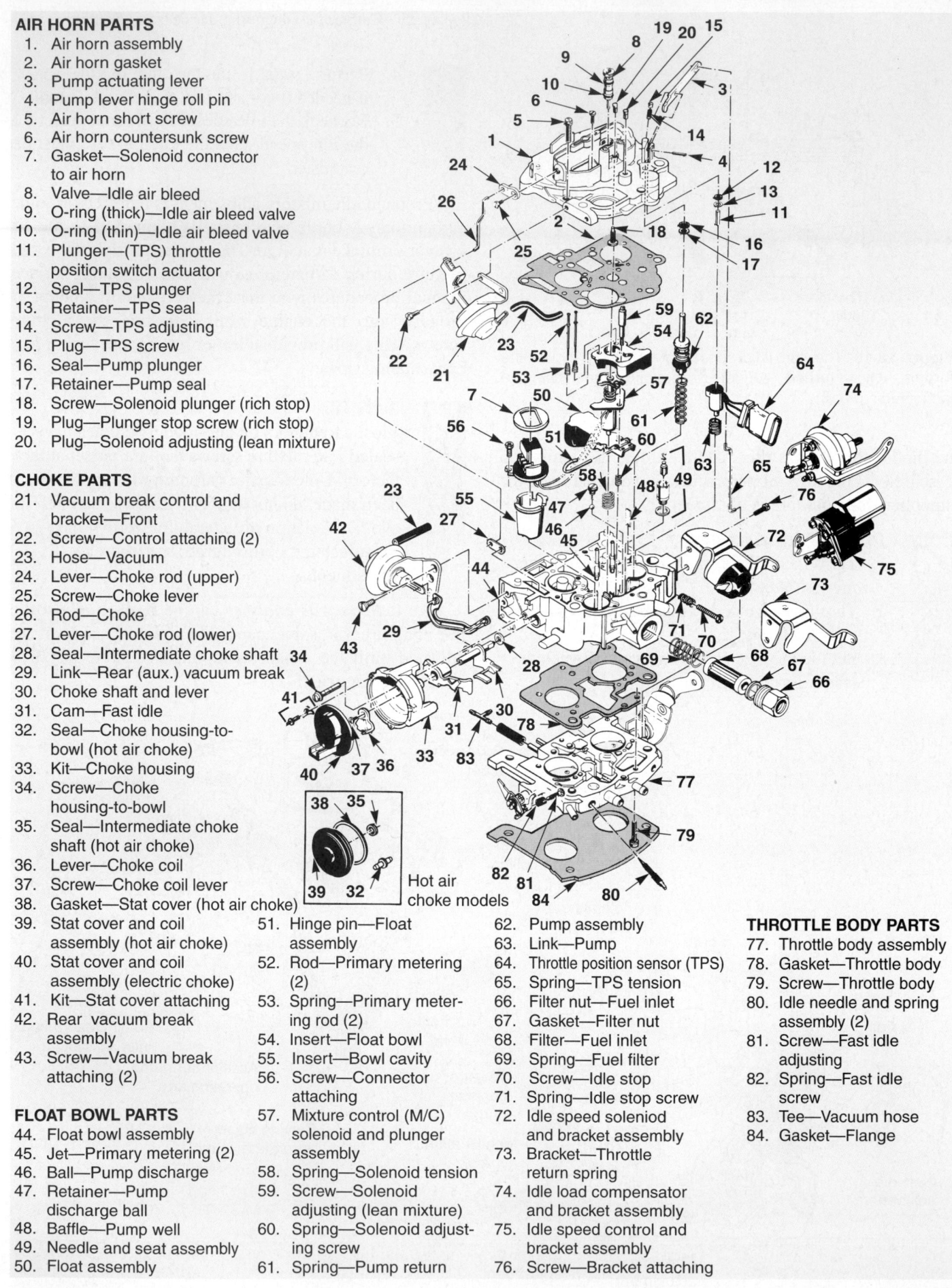

AIR HORN PARTS
1. Air horn assembly
2. Air horn gasket
3. Pump actuating lever
4. Pump lever hinge roll pin
5. Air horn short screw
6. Air horn countersunk screw
7. Gasket—Solenoid connector to air horn
8. Valve—Idle air bleed
9. O-ring (thick)—Idle air bleed valve
10. O-ring (thin)—Idle air bleed valve
11. Plunger—(TPS) throttle position switch actuator
12. Seal—TPS plunger
13. Retainer—TPS seal
14. Screw—TPS adjusting
15. Plug—TPS screw
16. Seal—Pump plunger
17. Retainer—Pump seal
18. Screw—Solenoid plunger (rich stop)
19. Plug—Plunger stop screw (rich stop)
20. Plug—Solenoid adjusting (lean mixture)

CHOKE PARTS
21. Vacuum break control and bracket—Front
22. Screw—Control attaching (2)
23. Hose—Vacuum
24. Lever—Choke rod (upper)
25. Screw—Choke lever
26. Rod—Choke
27. Lever—Choke rod (lower)
28. Seal—Intermediate choke shaft
29. Link—Rear (aux.) vacuum break
30. Choke shaft and lever
31. Cam—Fast idle
32. Seal—Choke housing-to-bowl (hot air choke)
33. Kit—Choke housing
34. Screw—Choke housing-to-bowl
35. Seal—Intermediate choke shaft (hot air choke)
36. Lever—Choke coil
37. Screw—Choke coil lever
38. Gasket—Stat cover (hot air choke)
39. Stat cover and coil assembly (hot air choke)
40. Stat cover and coil assembly (electric choke)
41. Kit—Stat cover attaching
42. Rear vacuum break assembly
43. Screw—Vacuum break attaching (2)

FLOAT BOWL PARTS
44. Float bowl assembly
45. Jet—Primary metering (2)
46. Ball—Pump discharge
47. Retainer—Pump discharge ball
48. Baffle—Pump well
49. Needle and seat assembly
50. Float assembly

Hot air choke models

51. Hinge pin—Float assembly
52. Rod—Primary metering (2)
53. Spring—Primary metering rod (2)
54. Insert—Float bowl
55. Insert—Bowl cavity
56. Screw—Connector attaching
57. Mixture control (M/C) solenoid and plunger assembly
58. Spring—Solenoid tension
59. Screw—Solenoid adjusting (lean mixture)
60. Spring—Solenoid adjusting screw
61. Spring—Pump return

62. Pump assembly
63. Link—Pump
64. Throttle position sensor (TPS)
65. Spring—TPS tension
66. Filter nut—Fuel inlet
67. Gasket—Filter nut
68. Filter—Fuel inlet
69. Spring—Fuel filter
70. Screw—Idle stop
71. Spring—Idle stop screw
72. Idle speed solenoid and bracket assembly
73. Bracket—Throttle return spring
74. Idle load compensator and bracket assembly
75. Idle speed control and bracket assembly
76. Screw—Bracket attaching

THROTTLE BODY PARTS
77. Throttle body assembly
78. Gasket—Throttle body
79. Screw—Throttle body
80. Idle needle and spring assembly (2)
81. Screw—Fast idle adjusting
82. Spring—Fast idle screw
83. Tee—Vacuum hose
84. Gasket—Flange

Figure 25-18. An exploded view of a computer-controlled carburetor. Note the mixture control solenoid and the idle speed control motor. (Buick)

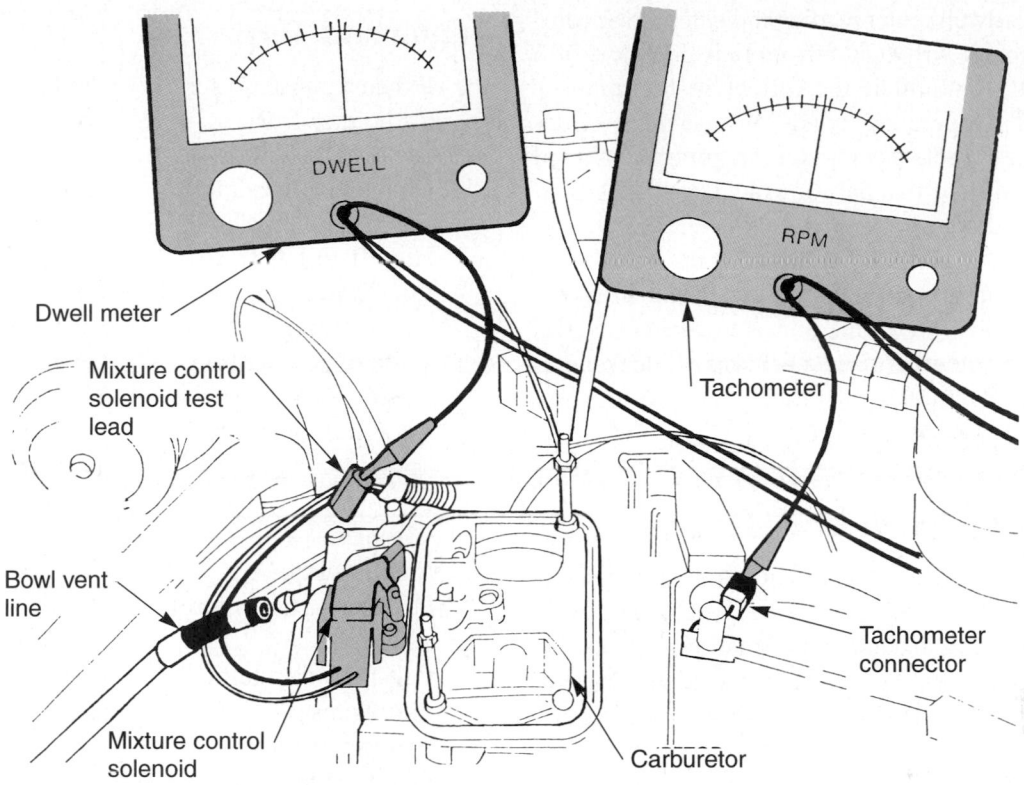

Figure 25-19. Some computer-controlled carburetor systems can be checked with a dwell meter. The dwell meter measures the duration of voltage pulses from computer. Dwell readings may also be used when adjusting the idle mixture. Follow the service manual directions; procedures vary for particular systems. (Chrysler)

Note

For information on using an exhaust gas analyzer, refer to Chapter 44, *Emission Control System Testing, Service, and Repair.*

Computer-Controlled Carburetor Service

A computer-controlled carburetor system requires specialized service techniques, **Figure 25-18.** Most systems have a self-diagnostic mode, which allows easy location of system problems. One computer system flashes a light on and off a number of times. The technician uses the code chart in the service manual to pinpoint the system trouble.

Figure 25-19 shows testing methods used on one computer-controlled carburetor system. A tachometer and a dwell meter are connected to the system. The tachometer measures engine speed. The dwell meter measures the duration of the computer's electrical signal going to the mixture-control solenoid in the carburetor. This allows the technician to determine whether the computer or the mixture-control solenoid is causing a problem.

Figure 25-20 shows the relationship between the dwell meter reading and the computer output. Note that with this system, a high dwell signal indicates a lean

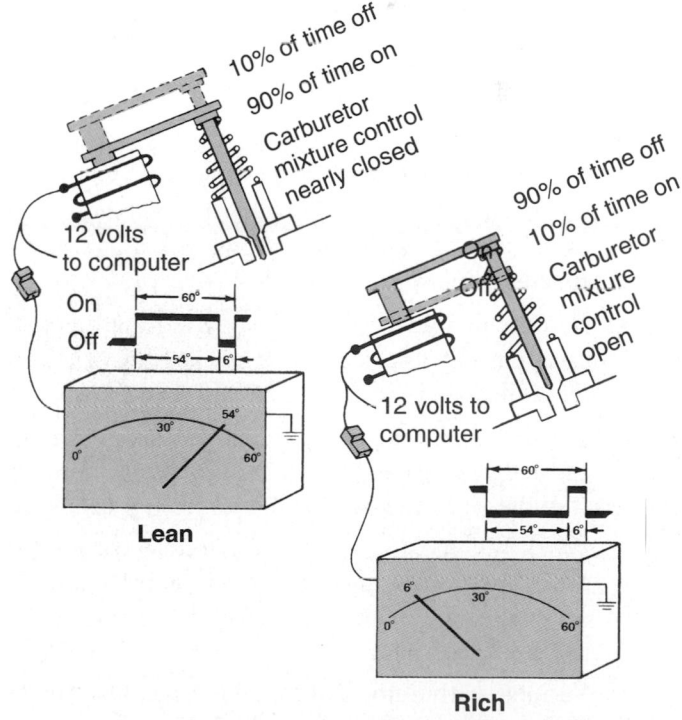

Figure 25-20. Dwell meter reading shows computer output to the carburetor mixture control solenoid. Study the connections and values. (General Motors)

command. A low dwell meter reading indicates a rich command. The computer will switch from rich (low dwell) to lean (high dwell) to maintain the correct air-fuel ratio.

Under certain operating conditions, such as wide-open throttle, the vehicle's computer may produce a fixed output. Refer to a shop manual for exact dwell specifications and test methods.

To test an older carburetor idle mixture solenoid, a hand vacuum pump is connected to the mixture solenoid. Then, the leak-down rate of the unit is measured in both the energized and de-energized positions. If leakage is not within manufacturer's specifications, the mixture control solenoid must be replaced.

Computer Self-Diagnosis

Some computer-controlled carburetor systems have a self-diagnostic, or on-board diagnostic feature. The computer will produce a trouble code that represents the possible problem source. This topic is discussed in several places in this textbook. Refer to the index as needed.

Summary

- Troubles in the ignition system, emission control systems, and engine can produce the same general symptoms (missing, poor fuel economy, engine not starting) as carburetor problems.

- An exhaust gas analyzer is a testing device that measures the chemical content of the engine exhaust gases.

- Carburetor float system troubles can cause a wide range of performance problems: flooding, rich fuel mixture, lean fuel mixture, fuel starvation (no fuel), stalling, low-speed engine miss, and high-speed engine miss.

- Carburetor idle system problems normally show up as a rough idle, stalling at low speeds, or incorrect idle speed.

- Carburetor acceleration system problems usually cause the engine to hesitate, stall, pop, backfire, or stumble when the car first moves from a standstill.

- When diagnosis points to internal carburetor problems (dirt clogged passages, leaking gaskets or seals, worn parts), carburetor removal is usually necessary.

- Various carburetor adjustments must be made during assembly and after installation.

Important Terms

Exhaust gas analyzer
Carburetor float
  system troubles
Carburetor flooding
Engine flooding
High float level
Low float level
Idle system problems
Clogged idle passage
Clogged idle air bleed
Acceleration system
  problems
High-speed system
  problems
Engine surge
Faulty mixture control
  solenoid
Full-power system
  problems
Bad power valve
Faulty metering rod
Choke system problems
Rich choke setting
Lean choke setting
Fast idle cam problems
Diagnosis chart
Internal carburetor
  problems
Air horn body
Upper body
Main body
Throttle body
Carburetor rebuild
Overhaul
Carburetor rebuild kit
Carburetor cleaner
Decarbonizing cleaner
Idle drop
Low idle speed
Lean miss
Rich miss

Review Questions—Chapter 25

Please do not write in this text. Place your answers on a separate sheet of paper.

1. The first step in diagnosis is to determine whether the _____ or _____ is at fault.

2. What should you look for when inspecting a carburetor?

3. Describe the symptoms of a lean air-fuel mixture.

4. Describe the symptoms of a rich air-fuel mixture.

5. List some causes for lean and rich air-fuel mixtures.

6. How can an exhaust gas analyzer be used when troubleshooting carburetor problems?

7. When diagnosing carburetor problems, try to determine which carburetor _____ is at fault.

8. Carburetor float system problems can cause flooding, a rich fuel mixture, a lean fuel mixture, stalling, missing, and other performance problems. True or False?

9. Define the terms "carburetor flooding" and "engine flooding."

10. A clogged idle air bleed usually tends to richen the air-fuel mixture. True or False?

11. A car tends to hesitate or lose power when accelerating from a dead stop with the engine both cold and warm. Technician A says that the choke is set too rich and that a choke adjustment may be needed. Technician B says that the accelerator pump may need adjustment or replacement because a lean mixture is indicated. Who is correct?
    (A) A only.
    (B) B only.
    (C) Both A and B.
    (D) Neither A nor B.

12. What is engine surge?

13. List and explain the three major body sections of a carburetor.

14. Describe the five major steps for a carburetor rebuild.

15. Explain a method for testing a mixture control solenoid in a computer-controlled carburetor system.

● ASE-Type Questions

1. A carburetor must adjust an air-fuel mixture as the engine's:
   (A) load changes.
   (B) speed changes.
   (C) temperature changes.
   (D) All of the above.

2. Which of these air-fuel mixtures will cause a high-speed miss?
   (A) Rich.
   (B) Lean.
   (C) Stoichiometric.
   (D) All of the above.

3. An exhaust gas analyzer measures the amount of each of these in a vehicle's exhaust except:
   (A) oxygen.
   (B) hydrocarbons.
   (C) helium.
   (D) carbon dioxide.

4. Which problem will not create a rich air-fuel mixture?
   (A) High float level.
   (B) Float system trouble.
   (C) Clogged fuel passage.
   (D) Clogged idle air bleed.

5. A faulty metering rod will generally upset:
   (A) low-speed engine operation.
   (B) high-speed engine operation.
   (C) midrange-speed engine operation.
   (D) All of the above.

6. While discussing choke system problems, Technician A states that when a choke sticks closed, black smoke blows from the car's tailpipe. Technician B says a closed choke will prevent the engine from starting when cold. Who is right?
   (A) A only.
   (B) B only.
   (C) Both A and B.
   (D) Neither A nor B.

7. When diagnosing problems, carburetor removal is usually necessary when there are:
   (A) worn parts.
   (B) dirt-clogged passages.
   (C) leaking gaskets or seals.
   (D) All of the above.

8. Which of the following fits on top of the main body and serves as a lid for the fuel bowl?
   (A) Air horn body.
   (B) Throttle body.
   (C) Vacuum hose.
   (D) Float assembly.

9. Each of these parts is found on a throttle body except:
   (A) dashpot.
   (B) butterflies.
   (C) pump lever.
   (D) fast idle solenoid.

10. To set cold idle speed, connect this to the engine:
    (A) tachometer.
    (B) dwell meter.
    (C) speed sensor.
    (D) hand vacuum pump.

Activities—Chapter 25

1. Diagnose carburetor problems from symptoms given to you by your instructor or by a customer of the automotive shop.

2. Disassemble and clean a carburetor in preparation for overhaul.

3. Determine specifications for a selected carburetor and set the float drop accurately using the appropriate measuring tool.

Carburetor Diagnosis		
Condition	**Possible Cause**	**Correction**
Flooding.	1. Defective inlet needle valve. 2. Float level too high. 3. Dirty inlet needle valve. 4. Excessive fuel pressure. 5. Float sunk. 6. Binding float. 7. Heavy fuel flow pulsations. 8. Internal circuit leakage.	1. Replace inlet needle valve and seat. 2. Adjust float level. 3. Clean needle valve, fuel lines, and filters. 4. Reduce pressure. 5. Reset float level or replace float. 6. Align float. 7. Install or replace pulsation damper. 8. Repair or replace carburetor.
Stalling and/or rough idling.	1. Idle speed too slow. 2. Inoperative throttle return dashpot. 3. Improperly adjusted throttle dashpot. 4. Improper fuel level in bowl. 5. Improperly adjusted idle mixture screw or screws. 6. Clogged idle system. 7. Inoperative or misadjusted choke. 8. Carburetor flooding. 9. Clogged air cleaner. 10. Vacuum leak. 11. Ruptured vacuum power jet diaphragm. 12. Idle mixture needles grooved. 13. Inoperative hot idle compensator (air valve). 14. Clogged bowl vent. 15. Clogged idle air bleeds. 16. Defective float needle valve. 17. Exhaust heat valve stuck open. 18. Incorrect distributor vacuum advance. 19. Restricted air cleaner and/or exhaust. 20. Secondary throttle sticking open. 21. Inoperative fast idle solenoid. 22. Contaminated fuel.	1. Increase idle speed. 2. Replace dashpot. 3. Adjust dashpot. 4. Adjust fuel level. 5. Adjust mixture screws. 6. Clean carburetor. 7. Repair or adjust choke. 8. See *Carburetor flooding.* 9. Clean or replace air cleaner. 10. Repair leak. 11. Replace power jet diaphragm. 12. Replace needles. 13. Replace air valve. 14. Open vent. 15. Clean carburetor. 16. Replace needle valve. 17. Free heat valve or replace. 18. Adjust to specifications. 19. Repair restriction or replace restricted part. 20. Free throttle. 21. Replace solenoid. 22. Flush system. Add fresh, clean fuel.
Idle speed varies.	1. Defective or improperly adjusted throttle return dashpot. 2. Loose or defective power valve. 3. Dirty throttle linkage. 4. Weak throttle return spring. 5. Sticking accelerator pedal. 6. Sticking fuel inlet valve. 7. Sticking fast idle cam. 8. Defective air cleaner vacuum motor.	1. Replace or adjust dashpot. 2. Tighten or replace power valve. 3. Clean linkage. 4. Replace with stronger spring. 5. Clean and lubricate accelerator cable and/or linkage. 6. Clean or replace valve. 7. Free cam linkage. 8. Replace motor.
Poor acceleration.	1. Disconnected accelerator pump linkage. 2. Improperly adjusted accelerator pump linkage. 3. Ruptured accelerator pump diaphragm. 4. Worn or cracked accelerator pump piston. 5. Dirty or defective accelerator pump check valve.	1. Connect linkage. 2. Adjust correctly. 3. Replace diaphragm. 4. Replace piston. 5. Clean or replace check valve.

(Continued)

Carburetor Diagnosis		
Condition	**Possible Cause**	**Correction**
Poor acceleration. *(Cont.)*	6. Clogged accelerator jets. 7. Fuel level in bowl too low. 8. Weak accelerator pump follow-through spring. 9. Exhaust manifold heat control valve stuck. 10. Inoperative power valve. 11. Clogged main fuel passage. 12. Air leaks. 13. Defective secondary diaphragm. 14. Secondary throttle plates stuck. 15. Clogged air cleaner. 16. Clogged fuel filters.	6. Clean jets. 7. Raise fuel level. 8. Replace spring. 9. Free heat control valve. 10. Clean or replace power valve. 11. Clean carburetor. 12. Repair air leaks. 13. Replace diaphragm. 14. Free throttle plates. 15. Clean or replace air filter. 16. Clean or replace filters.
Low top speed.	1. Incorrect throttle linkage adjustment. 2. Choke sticking on. 3. Low or high float level. 4. Air leak. 5. Inoperative secondary throttle valves. 6. Improper secondary throttle adjustment. 7. Main jet too small. 8. Main jet clogged. 9. Obstruction under accelerator pedal.	1. Adjust linkage correctly. 2. Clean and adjust choke. 3. Adjust float level. 4. Repair air leak. 5. Clean and rebuild carburetor. 6. Adjust throttle correctly. 7. Install proper size jet. 8. Clean carburetor. 9. Remove obstruction.
Hard starting when cold.	1. Inoperative or improperly adjusted choke. 2. Flooding from excessive use of accelerator pump or rich choke. 3. Failure to depress throttle to allow choke valve to close. 4. Air leak. 5. Incorrect fuel level in bowl. 6. Clogged bowl vent. 7. No fuel delivery. 8. Stale or contaminated fuel. 9. Gasoline not sufficiently volatile. 10. Vacuum leaks.	1. Clean and adjust choke. 2. Hold throttle in wide open position and crank engine. Advise driver. 3. Advise driver. 4. Repair air leak. 5. Adjust fuel level. 6. Clean bowl vent. 7. Check tank and delivery system. 8. Drain tank. Flush system. Fill with fresh fuel. 9. Change to more volatile fuel. 10. Locate and repair.
Hard starting when hot.	1. Flooding from excessive use of choke or accelerator pump. 2. Vapor lock. 3. Clogged bowl vent. 4. Air leak. 5. Clogged air cleaner. 6. Fuel level incorrect. 7. No fuel delivery. 8. Stale or contaminated fuel. 9. Improper choke unloader adjustment. 10. Vacuum hoses split, kinked, or loose. 11. Fuel vapor boil over. 12. Overheated carburetor.	1. Hold throttle wide open and crank until engine starts. 2. Cool lines. Change to less volatile fuel and protect lines from heat. 3. Clean bowl vent. 4. Repair air leak. 5. Clean or replace air filter. 6. Adjust fuel level. 7. Check delivery system. 8. Drain fuel system. Fill with fresh fuel. 9. Adjust properly. 10. Replace and secure connections. 11. Hold throttle full open while cranking. 12. Check for proper heat shield and/or insulator installation.

Chapter 26

Diesel Injection Fundamentals

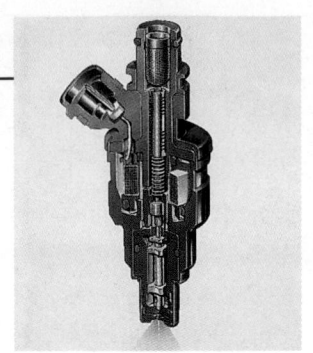

After studying this chapter, you will be able to:

- Explain the operating principles of a diesel injection system.
- Summarize the differences between gasoline and diesel engines.
- Describe the major parts of a diesel injection system.
- Compare variations in the design of diesel injection systems.
- Correctly answer ASE certification test questions that require a knowledge of the fundamentals of diesel injection.

A *diesel fuel injection* system is a high-pressure, mechanical system that delivers fuel directly into the engine combustion chambers. It is unlike a lower-pressure, gasoline system, which meters fuel into the engine intake manifold. Diesel fuel injection is relatively simple. It uses a mechanical, high-pressure pump with electronic controls.

Note
Several earlier chapters discuss information essential to this chapter. If needed, use the index to locate and review coverage of diesel engines, diesel fuel, fuel pumps, fuel filters, compression ratios, and diesel combustion. They are quite different from a gasoline, spark ignition engine.

Diesel and Gasoline Engine Differences

Before covering the parts of a diesel injection system in detail, you should review the major differences between a gasoline engine and a diesel engine. This will help you understand diesel injection.

- Diesel engines use a very high compression ratio (approximately 17:1 to 25:1). A gasoline engine's compression ratio is only 8:1 or 9:1.

- A diesel engine is a compression ignition engine because it uses the heat from compressed air to ignite the fuel. Gasoline engines are classified as spark ignition engines because they use an electric arc (from the spark plug) to ignite the fuel.

- A diesel engine has no throttle valve to control airflow into the engine. Gasoline engines use a throttle valve to control airflow and engine power. Diesel engines use an injection pump to control engine power output.

- A diesel engine compresses air on its compression stroke. A gasoline engine compresses an air-fuel mixture.

- A diesel engine injects fuel directly into the combustion chambers. A gasoline engine meters fuel into the intake manifold.

- In a diesel engine, speed and power are controlled by the amount of fuel injected into the engine. More fuel produces more power. A gasoline engine controls engine power by regulating air and fuel flow with a throttle valve.

Basic Diesel Injection System

Refer to **Figure 26-1** which shows a basic diesel injection system.

- *Injection pump*—high-pressure, mechanical pump that meters the correct amount of fuel and delivers it to each injector nozzle at the proper time.

- *Injection lines*—high-strength steel tubing that carries fuel to each injector nozzle.

- *Injector nozzles*—spring-loaded valves that spray fuel into each combustion chamber.

- *Glow plugs*—electric heating elements that warm air in precombustion chambers to aid starting of cold engine.

420

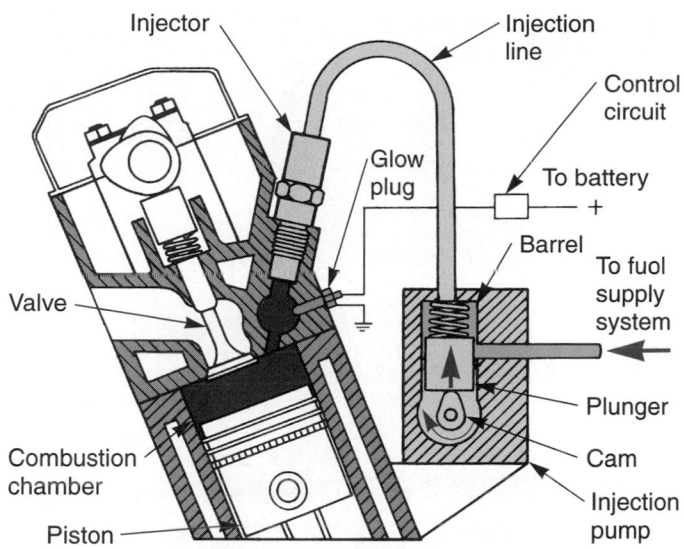

Figure 26-1. Basic parts of a simplified diesel injection system. The injection pump plunger produces very high pressure to open injector valve. A glow plug warms the air in the precombustion chamber when the engine is cold.

The *diesel fuel supply system* feeds fuel to the injection pump, normally using an inline electric pump, **Figure 26-2.** The engine-driven injection pump then controls how much fuel is forced to the injector nozzles. The high-pressure injection lines carry fuel to the injectors. The spring-loaded nozzles are normally closed. However, when the injection pump produces enough pressure, each nozzle opens and squirts a fuel charge into the engine *cylinders* to start combustion. A return line carries excess fuel back to the tank.

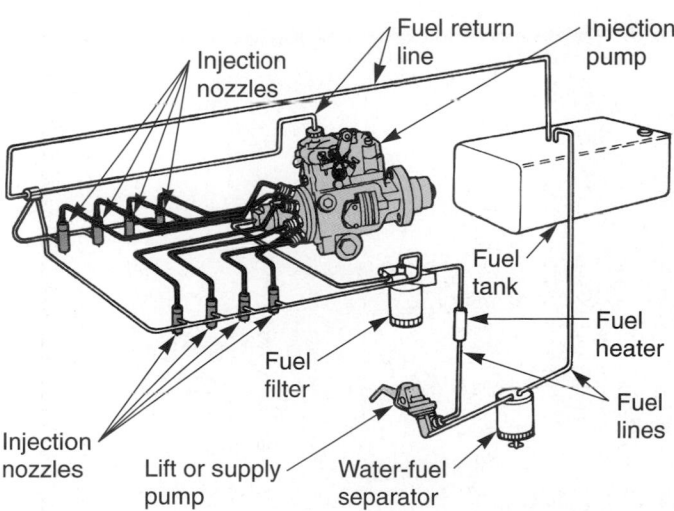

Figure 26-2. Schematic of complete diesel fuel injection system. The lift pump feeds fuel into the injection pump. The injection pump produces high pressure for injection into combustion chambers. (Ford)

Diesel Injection Pumps

A diesel injection pump has several important functions:

- Meters the correct amount of fuel to each injector.
- Circulates fuel through fuel lines and nozzles.
- Produces extremely high fuel pressure.
- Times fuel injection to meet the speed and load of engine.
- Provides a means for the driver to control engine power output.
- Controls engine idle speed and maximum engine speed.
- Helps close injector nozzles after injection.
- Provides a means of shutting off the engine.

The diesel injection pump is normally bolted to the side or top of the engine. See **Figure 26-3.** The fuel supply system (fuel tank, lines, filters, conventional fuel pump) pushes clean, filtered fuel under low pressure to the injection pump.

Figure 26-3. This diesel has an injection pump on top of the engine, at upper right of photo. Can you find it? Study other parts. (Audi)

The injection pump then pressurizes the fuel. A camshaft acts on a plunger. The plunger slides up in its barrel (cylinder), compressing and pressurizing the fuel. The fuel then flows through the injection line and out the injector nozzle.

A diesel injection pump is powered by the engine. Power may be transferred by a set of gears, a chain, or a toothed belt. There are two common types of automotive diesel injection pumps: inline pumps and distributor pumps.

Inline Diesel Injection Pumps

An *inline diesel injection pump* has one plunger (piston) for each engine cylinder. The plungers are lined up in a row, like the pistons in an inline engine.

The major parts of an inline injection pump are shown in **Figures 26-4** and **26-5**. Refer to these illustrations as the parts are introduced.

The inline *injection pump camshaft* operates the plungers. It has lobes like an engine camshaft. When the engine turns the pump camshaft, the lobes push on the roller tappets to move them up and down.

The *injection pump roller tappets* transfer camshaft action to the plungers. Like roller lifters in an engine, the roller tappets reduce friction and wear on the cam lobes.

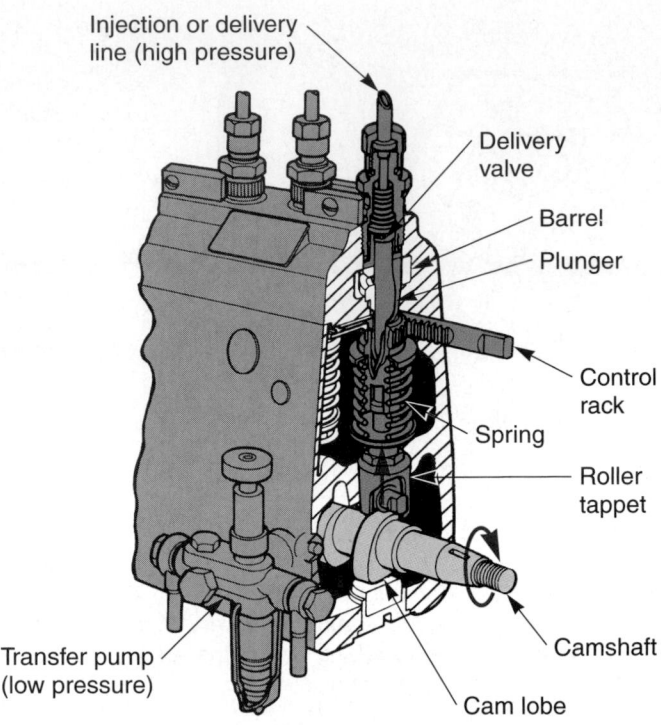

Figure 26-5. The camshaft acts on the roller tappet. The tappet pushes on the plunger to push fuel out the delivery valve and into the injection line. The control rack can be used to change injection quantity. (Chrysler)

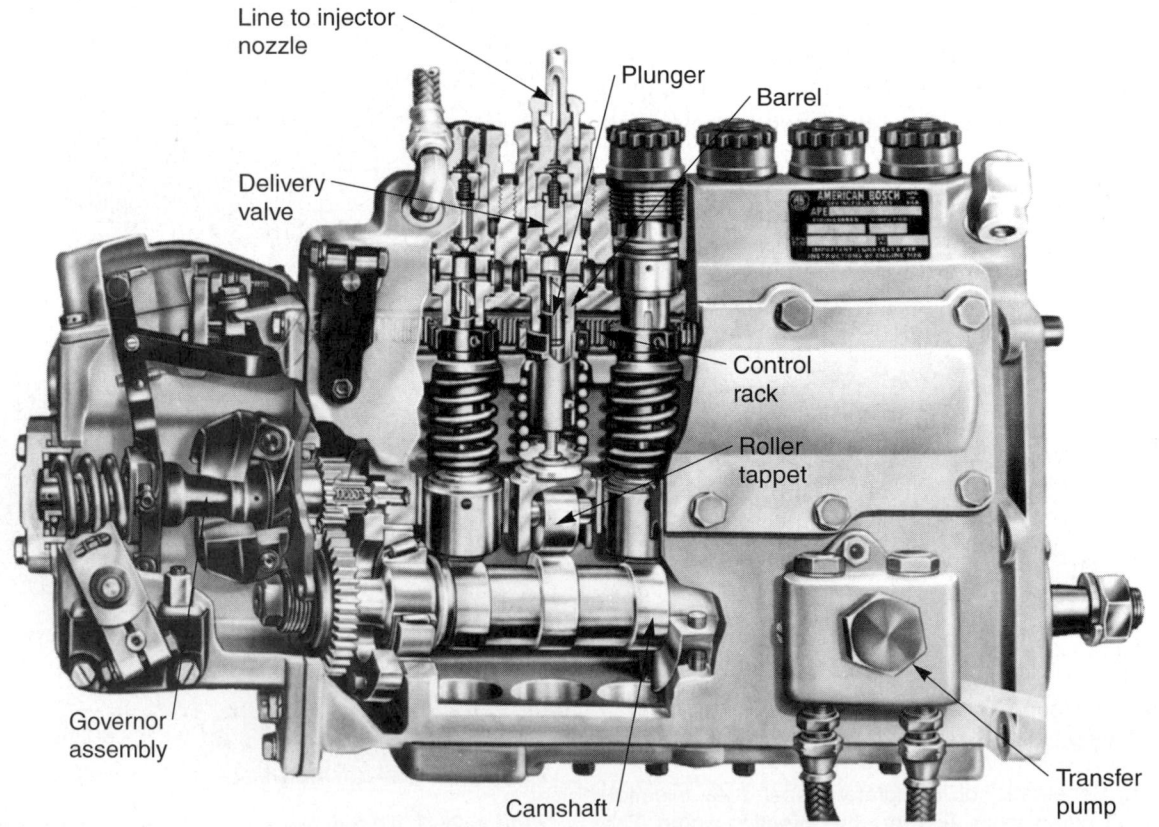

Figure 26-4. Cutaway shows basic parts of an inline diesel injection pump. (Robert Bosch)

Inline pump plungers are small pistons that push on and pressurize the diesel fuel. When the cam lobe acts on the roller tappet, both the tappet and the plunger are pushed upward.

The *barrels* are small cylinders that hold the plungers. When the plunger slides upward in its barrel, extremely high fuel pressure is formed.

The *plunger return springs* keep a downward pressure on the plungers and roller tappets. This holds the tappets against the camshaft when the lobes rotate away from the rollers.

Control sleeves alter the amount of fuel pushed to each injector nozzle. See **Figure 26-6** and locate the control sleeve.

A *control rod,* or *rack,* is a toothed shaft that acts as a throttle to control diesel engine speed and power. It rotates the control sleeves to increase or decrease injection pump output and engine power.

Delivery valves are spring-loaded valves in the outlet fittings to the injection lines. They help ensure quick, leak-free closing of the injector nozzles.

Inline Injection Pump Operation

When the engine is running, the injection pump camshaft rotates at one-half engine speed. When the cam lobe is not pushing on the roller tappet, the lift pump (fuel supply pump) fills the barrel with fuel. Then, as the cam lobe puts pressure on the roller tappet, the plunger is forced upward in its barrel. This forces fuel through the delivery valve, through the injection line, out of the nozzle, and into the engine. See **Figure 26-7.**

As the plunger reaches the end of its stroke, pressure drops and the delivery valve closes. The delivery valve action helps reduce injection line pressure rapidly, keeping fuel from dripping out of the injector nozzle.

When the cam lobe moves away from the roller tappet, the plunger return spring pushes the plunger down. Fuel can again fill the barrel. This readies the plunger to supply fuel for another power stroke.

Inline Injection Pump Fuel Metering

To control the amount of fuel injected into the engine, the control rod (or rack) slides across the control sleeves to rotate them. Look at **Figure 26-8.**

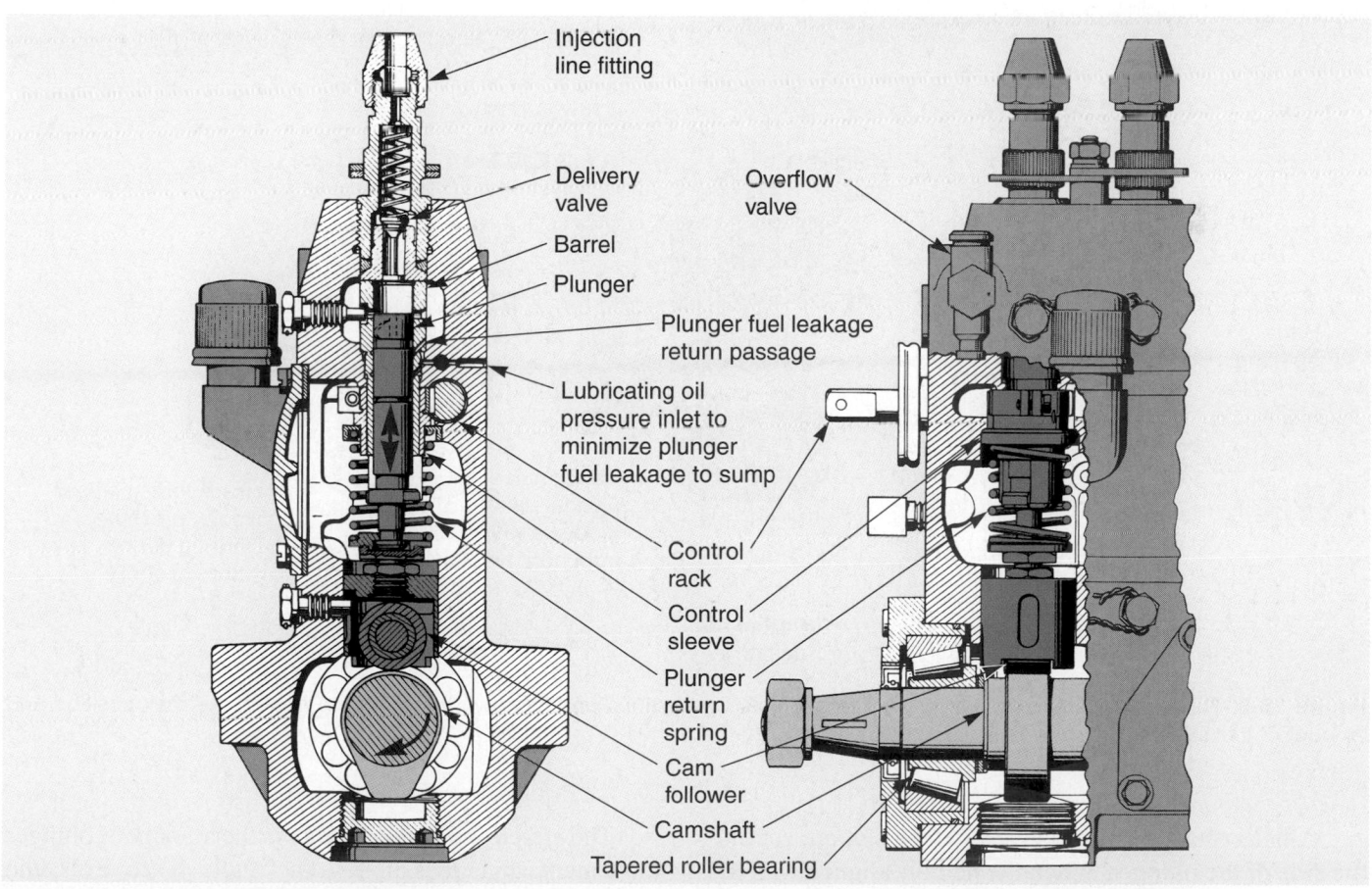

Figure 26-6. Another view of an inline diesel injection pump. Compare this illustration to previous ones. (Waukesha)

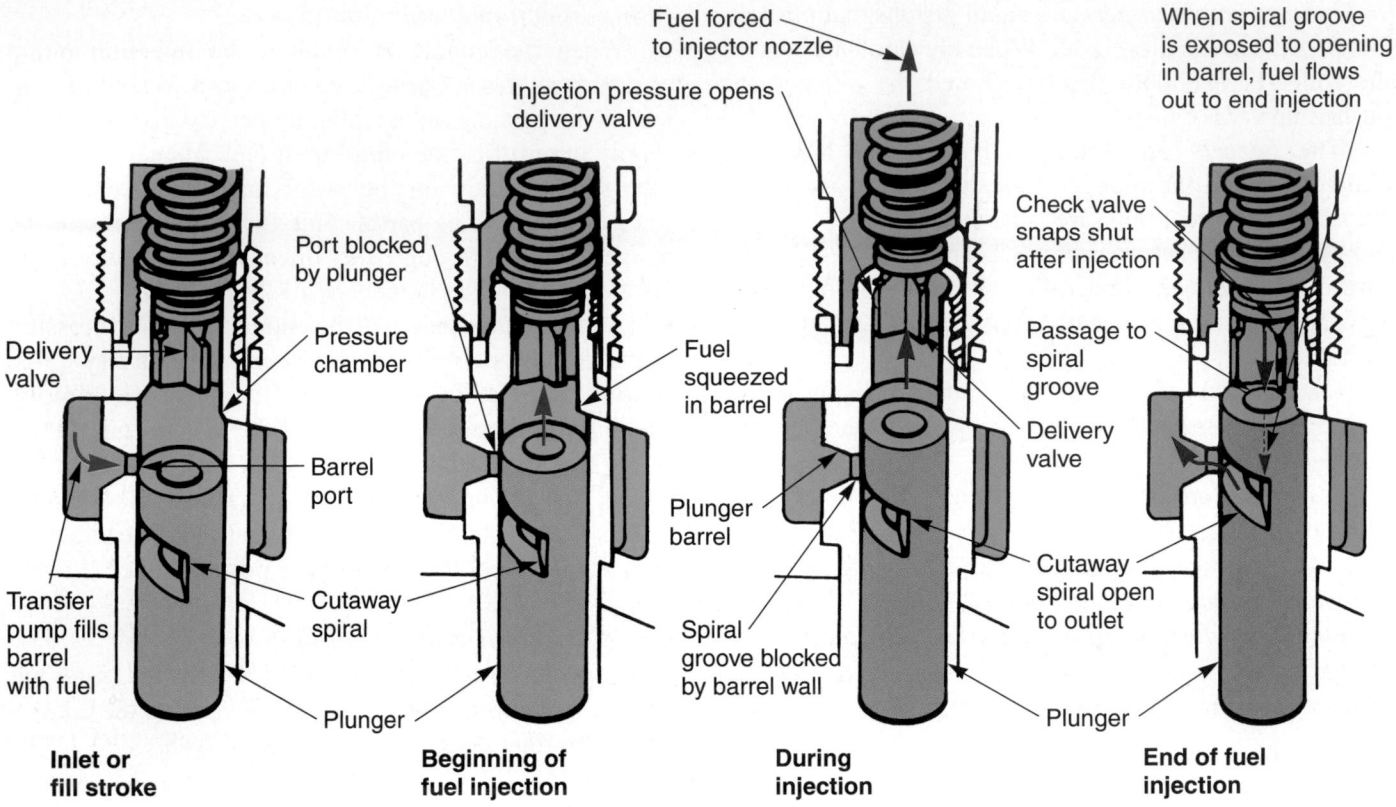

Figure 26-7. Plunger action in an inline injection pump. (Chrysler)

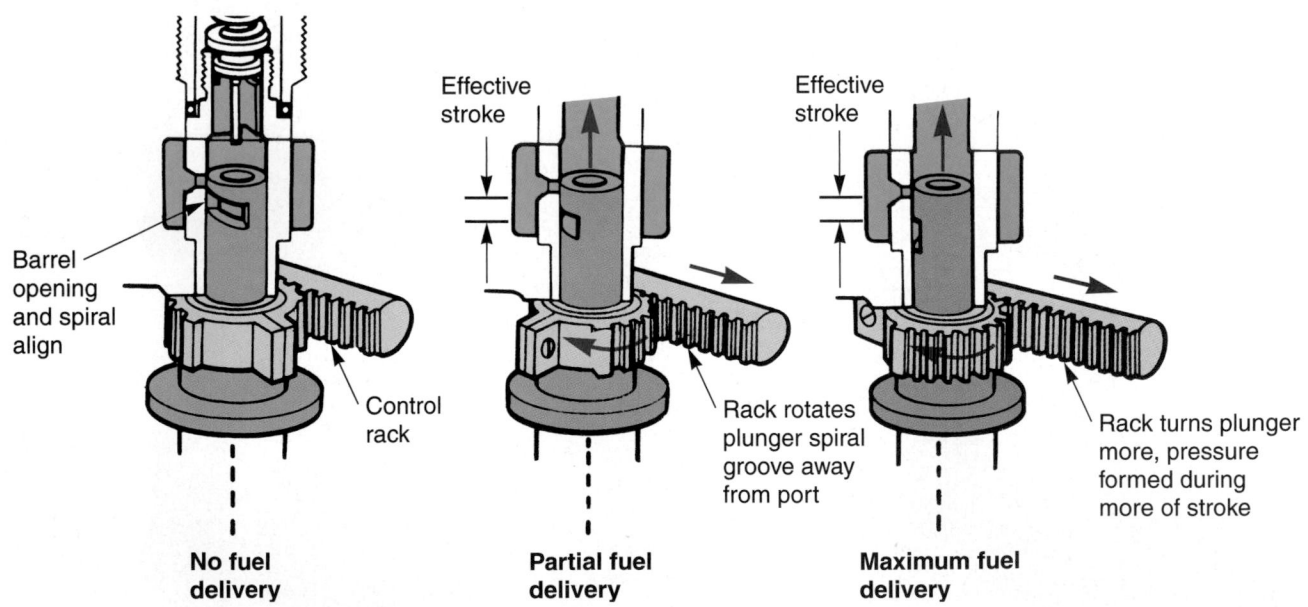

Figure 26-8. The control rack is linked to the gas pedal and governor. Two forces work together to control the rack position and amount of fuel injected. (Chrysler)

A helix-shaped (spiral) groove and a slot are cut into the side of the plunger. When the helix is aligned with the port (hole) in the side of the barrel, the plunger cannot develop pressure. Fuel will flow down the slot, through the helix groove, and out of the port.

The *effective plunger stroke* is the amount of plunger movement that pressurizes the fuel. It controls the amount of fuel delivered to the injectors.

When the plunger moves up and the helix is not aligned with the barrel port, fuel is trapped and

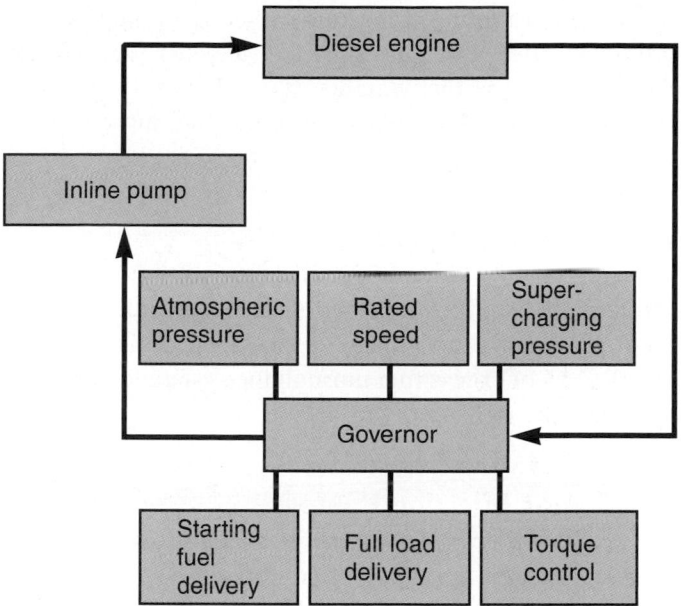

Figure 26-9. Diagram illustrates governor operation.

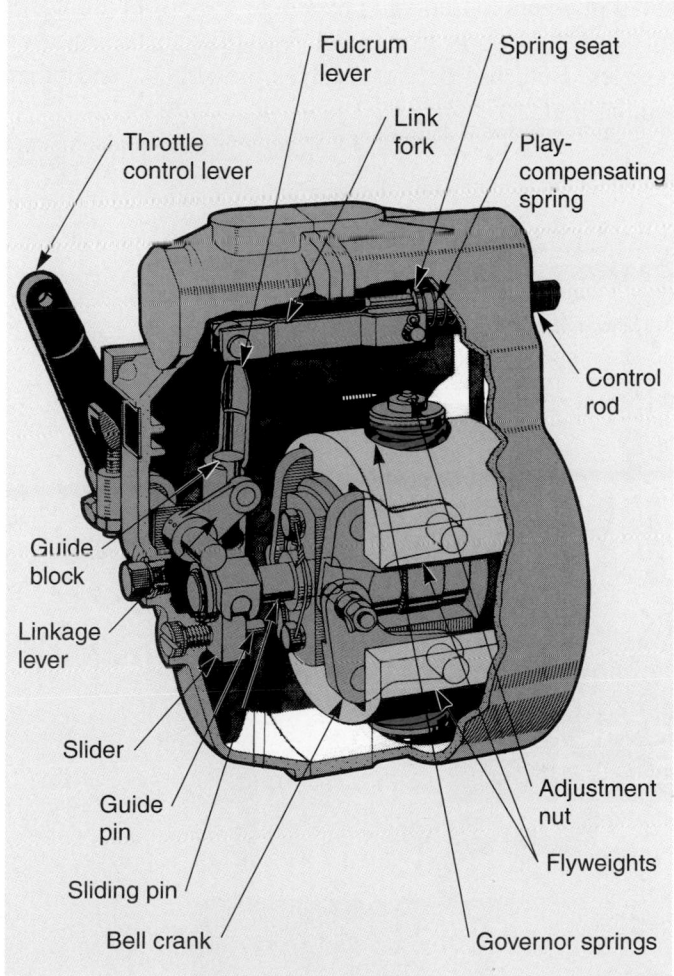

Figure 26-10. Governor for an inline injection pump. Note that the throttle control lever and the governor are both connected to the control rod. (Robert Bosch)

pressurized. In this way, rotation of the sleeve can be used to regulate how much fuel is injected into the engine's combustion chambers.

Inline Injection Pump Governor

A ***governor*** is used on an inline injection pump to control engine idle speed and limit maximum engine speed. Look at **Figure 26-9.** A diesel engine can be damaged if allowed to run too fast. **Figure 26-10** shows a cutaway view of an inline injection pump governor.

Notice how the governor uses ***centrifugal*** (spinning) weights, springs, and levers, **Figure 26-11.** The levers are connected to the control rack. If engine speed increases too much, the governor weights are thrown outward. This moves the levers and control rack to reduce the effective stroke of the plungers. Engine speed is limited.

When the driver presses the gas pedal for more power, it moves a throttle lever on the side of the injection pump governor. See **Figure 26-12.** This causes throttle lever spring pressure to overcome governor spring pressure. The control rack is moved to increase fuel delivery and engine power. Only when engine speed reaches a preset level does the governor overcome the full-throttle lever position.

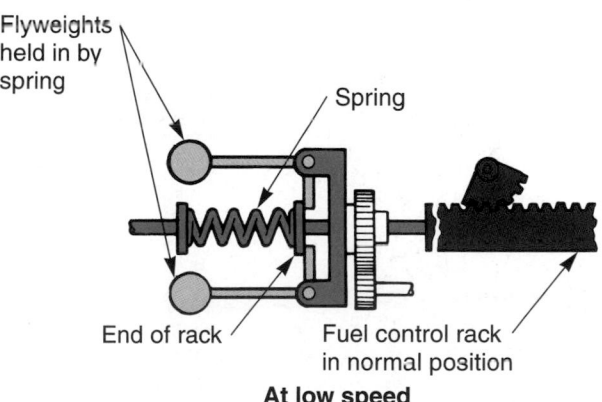

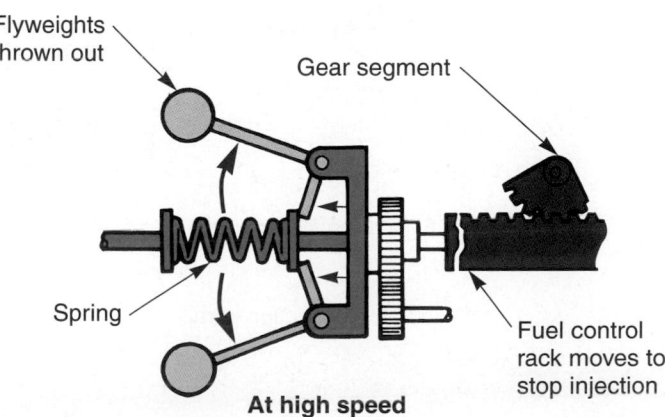

Figure 26-11. Basic diesel injection pump centrifugal governor operation. (Chrysler)

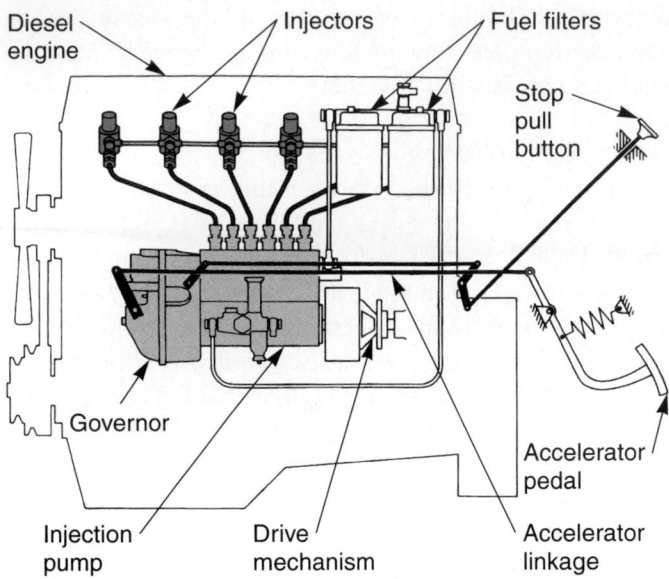

Figure 26-12. Basic accelerator-injection pump linkage arrangement. (Robert Bosch)

Injection Timing

Injection timing refers to when fuel is injected into the combustion chambers in relation to the engine's piston position. It is similar to spark timing in a gasoline engine.

Injection timing in an inline injection pump is usually controlled by spring-loaded weights. As engine speed increases, the weights fly outward to advance injection timing. This gives the diesel fuel more time to ignite and burn properly.

Inline Injection Pump Fuel Flow

Look at **Figure 26-13.** It illustrates the flow of fuel through a diesel injection system using an inline pump. Note how return fuel lines are provided. This allows a steady flow of excess fuel through the system to help cool and lubricate moving parts.

Distributor Injection Pumps

A **distributor injection pump** normally uses only one or two plungers to supply fuel for all of the engine's cylinders. It is the most common type of pump used on passenger cars, **Figure 26-14.**

In many ways, the operation of a distributor pump is similar to the action of an inline injection pump. Both use small plungers to trap and pressurize fuel. Both align and misalign fuel ports to control fuel flow to the injector nozzles. Both use delivery valves, governors, and other similar parts.

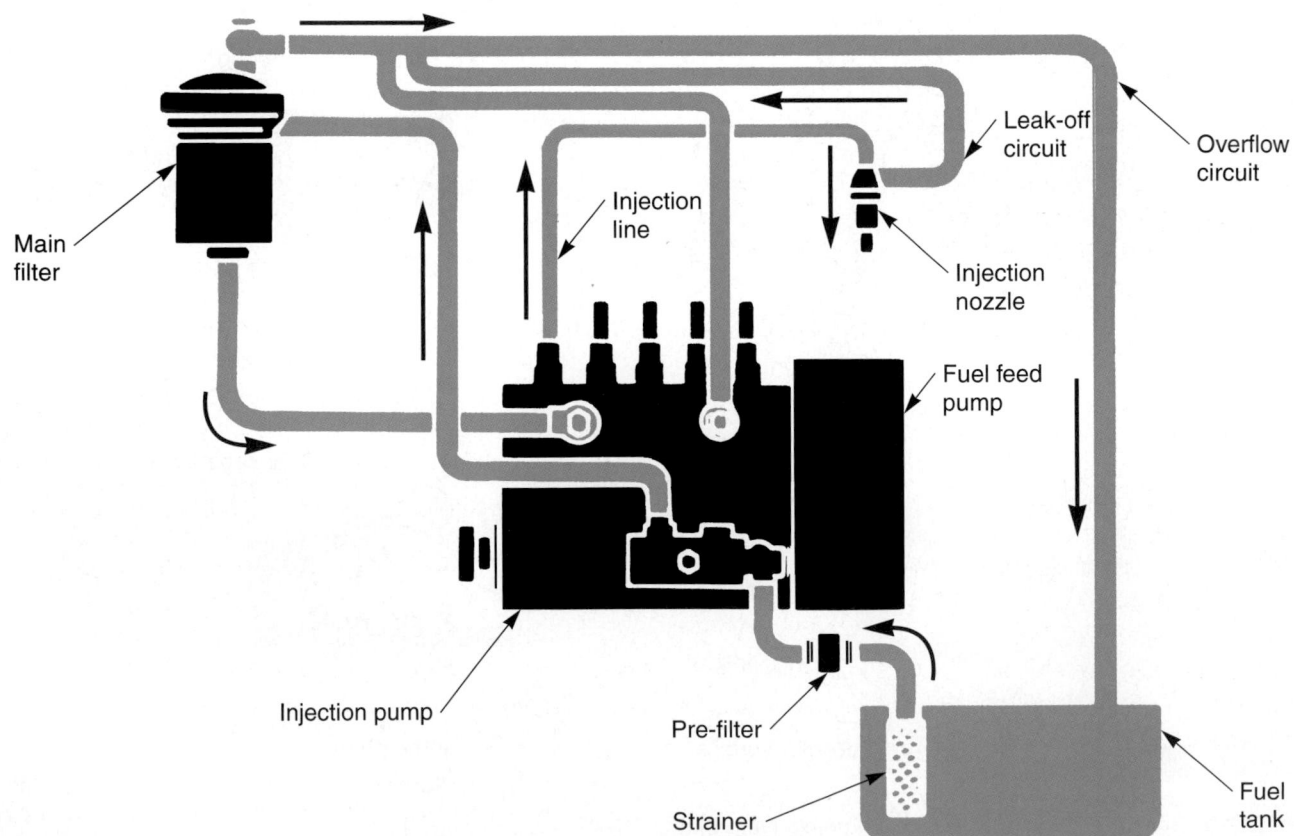

Figure 26-13. Trace the flow of fuel through the injection pump and lines. (Mercedes Benz)

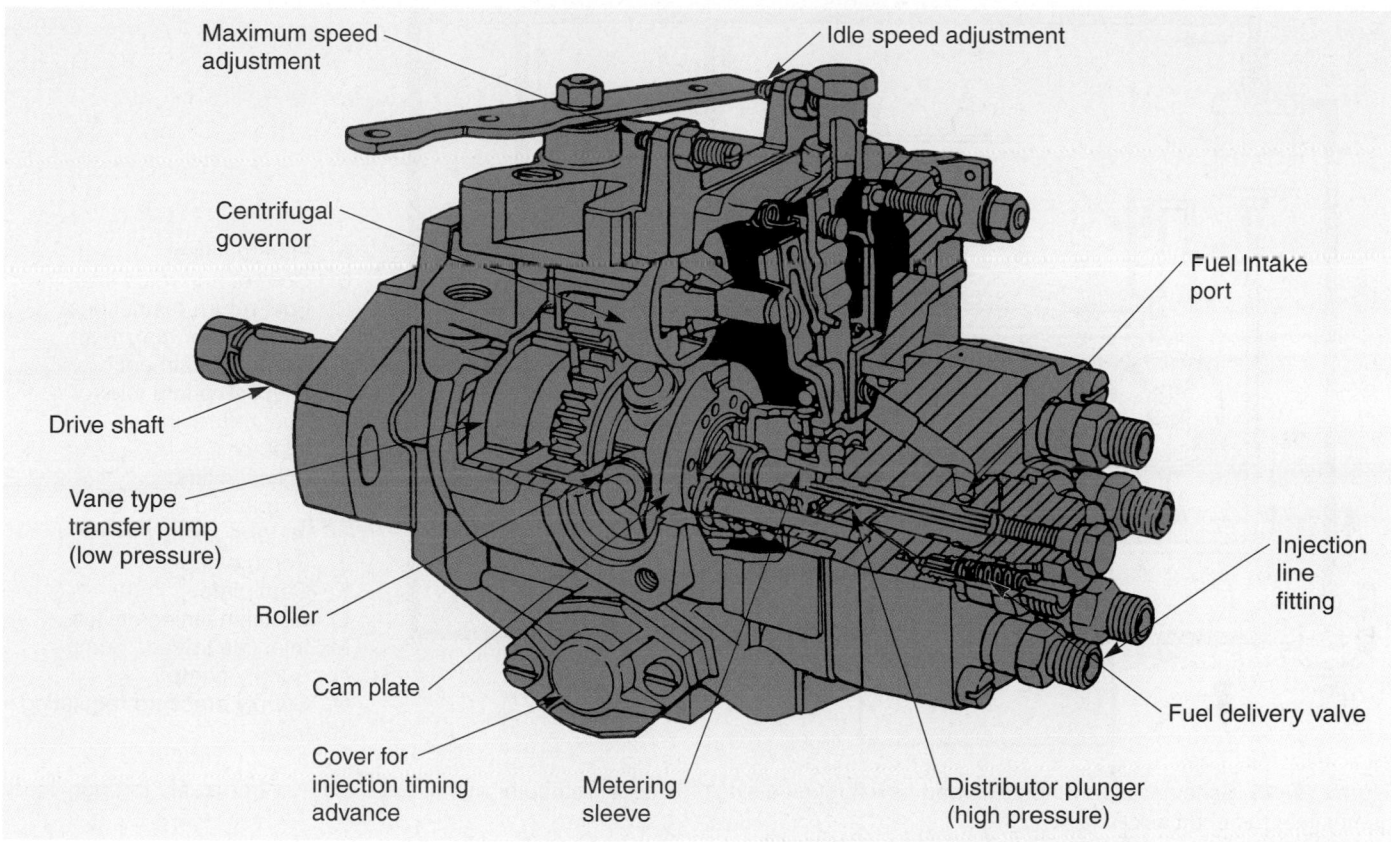

Figure 26-14. Single-plunger distributor diesel injection pump. Study the part names. (Chrysler)

There are two common variations of the distributor injection pump: single-plunger and two-plunger. Both will be discussed.

Single-Plunger Distributor Injection Pump

The major parts of a single-plunger distributor injection pump are shown in **Figures 26-15** and **26-16.** Refer to these illustrations as the parts are discussed.

The *drive shaft* uses engine power to operate the parts in the injection pump. The outer end of the shaft holds either a gear, chain sprocket, or a belt sprocket. This provides a drive mechanism for the pump.

A *transfer pump* is a small pump that forces diesel fuel into the injection pump. This lubricates the pump and fills the pumping chambers. Most transfer pumps for distributor pumps are a vane type.

A *distributor plunger* is a small piston that produces high fuel pressure. It is comparable to an inline plunger.

A *cam plate* is a rotating lobed disc that operates the plunger. Like an inline pump camshaft, the cam plate forces the plunger to move and develop injection pressure.

A *fuel metering sleeve* can be slid sideways on the plunger to change the effective plunger stroke (plunger movement that causes fuel pressure). It surrounds the plunger. The fuel metering sleeve performs the same function as the sleeves and control rack in an inline pump. The sleeve controls injection quantity, engine speed, and power output.

The *hydraulic head* is the housing around the plunger. It contains passages for filling the plunger barrel with fuel and for injecting fuel into the delivery valves.

A *centrifugal governor* helps control the amount of fuel injected and engine speed. Flyweight action moves the metering sleeve to limit top speed.

Single-Plunger Distributor Pump Operation. As the injection pump shaft rotates, the fill port in the hydraulic head lines up with the port in the plunger. At this point, the transfer pump can force fuel into the high-pressure chamber in front of the plunger. Refer to **Figure 26-17.**

With more shaft and plunger rotation, the fill port moves out of alignment and an injection port lines up. At this instant, the cam plate lobe pushes the plunger sideways. Fuel is forced out of the injection port to the correct injector nozzle.

This process is repeated several times during each rotation of the injection pump drive shaft. Fuel injection must be timed to occur at each nozzle as that engine piston nears TDC (top dead center) on its compression stroke.

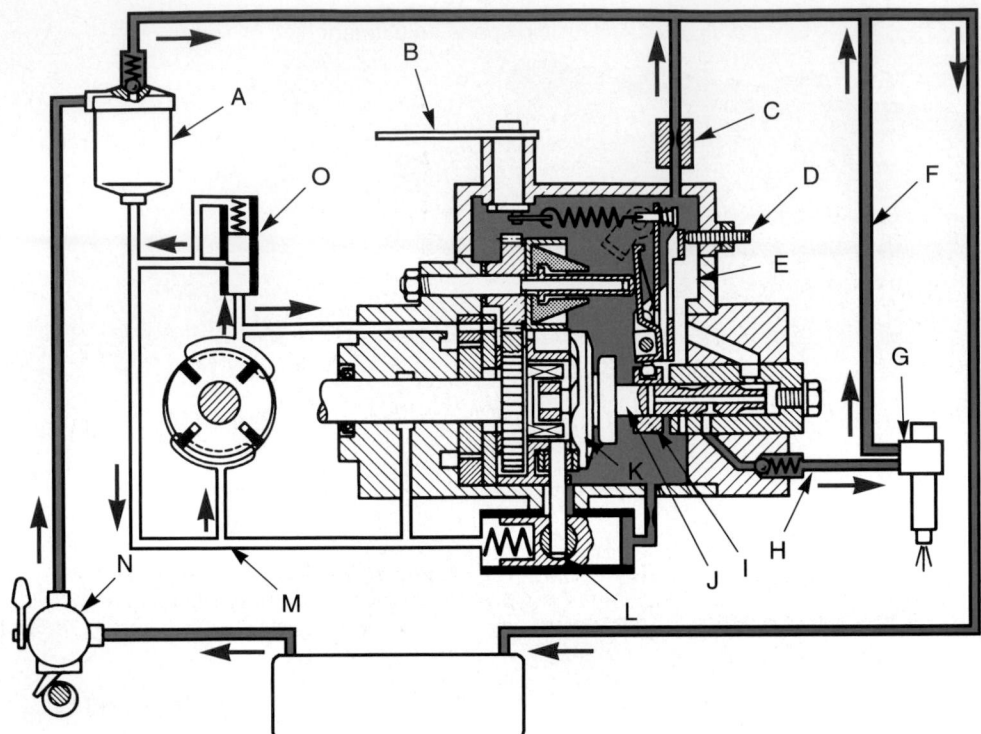

A. Fine fuel filter
B. Control or throttle lever
C. Fuel return restrictions
D. Fuel load or maximum
   speed adjustment
E. Injection pump interior
F. Fuel return line
G. Injector
H. Injection line
I. Regulating collar or
   metering sleeve
J. Pump plunger
K. Cam plate
L. Injection timing device
M. Inlet line to vane pump
N. Supply pump
O. Supply pressure regulating
   valve

Figure 26-15. Schematic showing parts and flow through a single-plunger distributor pump. Trace the flow from tank, through lines, pump, injector, and back to tank.

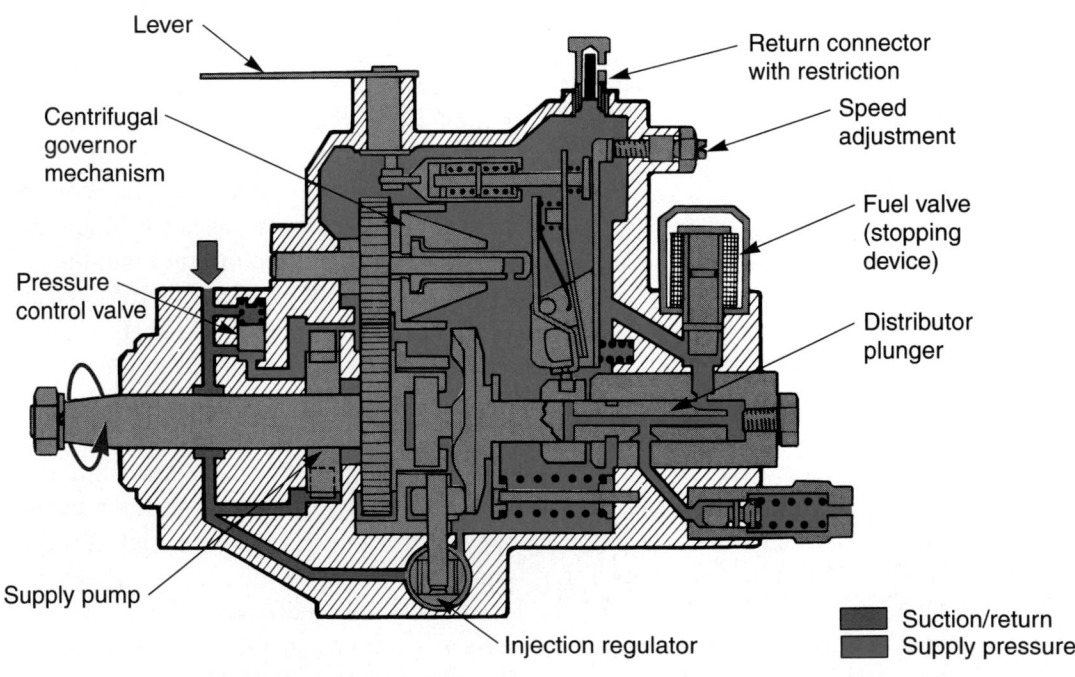

Figure 26-16. Main parts of single-plunger distributor pump.

Single-Plunger Distributor Pump Fuel Metering. In a single-plunger distributor injection pump, the amount of fuel injected is controlled by movement of the sleeve on the plunger. The sleeve slides one way to increase fuel delivery by covering the spill port (relief port). The sleeve moves the other way to reduce delivery by uncovering the spill port. This allows the fuel to flow into the return line.

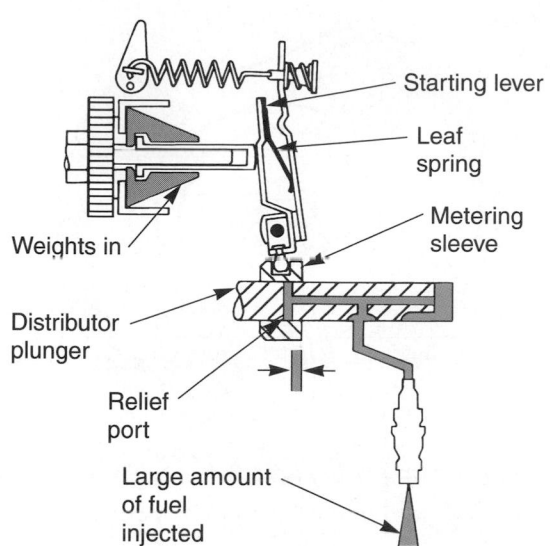

Starting—Leaf spring presses starting lever to left so metering sleeve moves to right. Distributor plunger moves farther before relief port is exposed. Injection lasts longer.

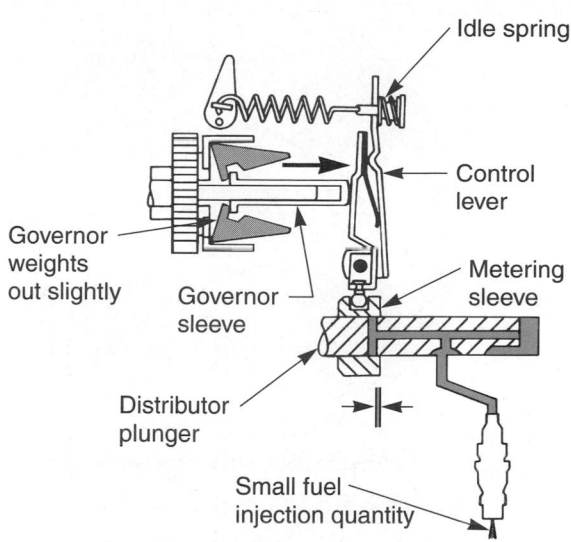

Idle—Weights in centrifugal governor are partly expanded so governor sleeve moves to right. Metering sleeve moves to left. Distributor plunger now moves a short distance before relief port is uncovered.

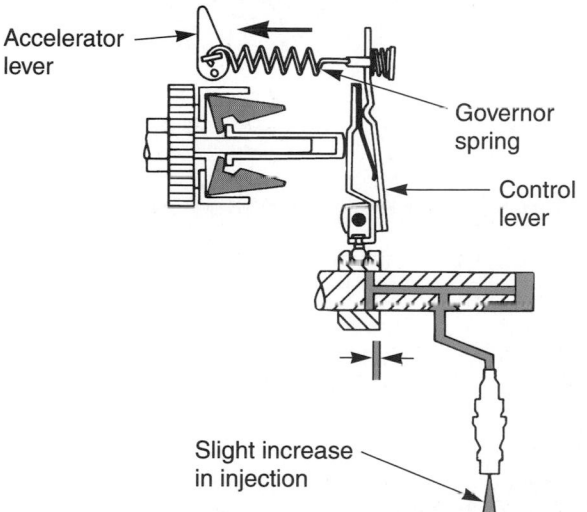

Acceleration—Control lever is pulled to left by linkage from accelerator pedal. Metering sleeve is moved to right. Engine speed increases until governor "neutralizes" effect of pedal linkage.

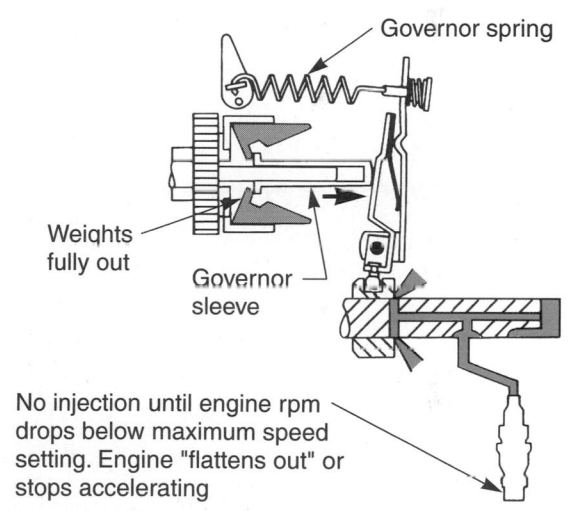

Maximum speed—Governor is spinning with enough centrifugal force for governor sleeve to stretch governor spring. Metering sleeve uncovers relief port at beginning of distributor plunger stroke.

Figure 26-17. These illustrations show the basic operation of a single-plunger distributor diesel injection pump.

Single-Plunger Distributor Pump Injection Timing. At the end of the engine compression stroke, diesel fuel must be injected directly into the precombustion chamber. Injection must continue past TDC to make sure all of the fuel burns and adequate power is developed.

As engine speed increases, injection must occur sooner to ensure peak combustion pressure right after TDC. Increased engine speed causes the transfer (vane) pump to spin faster. This increases the pressure output of the transfer pump. The pressure is used to move an injection advance piston. The piston, in turn, causes the cam plate ramps (lobes) to engage the plunger sooner, advancing the injection timing. Look at **Figure 26-18.**

Two-Plunger Distributor Pump

A two-plunger distributor injection pump is pictured in **Figure 26-19.** Besides many of the basic parts already covered, this injection pump consists of:

- *Two plungers*—two small pistons that move in and out to force fuel to each injector nozzle.
- *Distributor rotor*—slotted shaft that controls fuel flow to each injector nozzle.

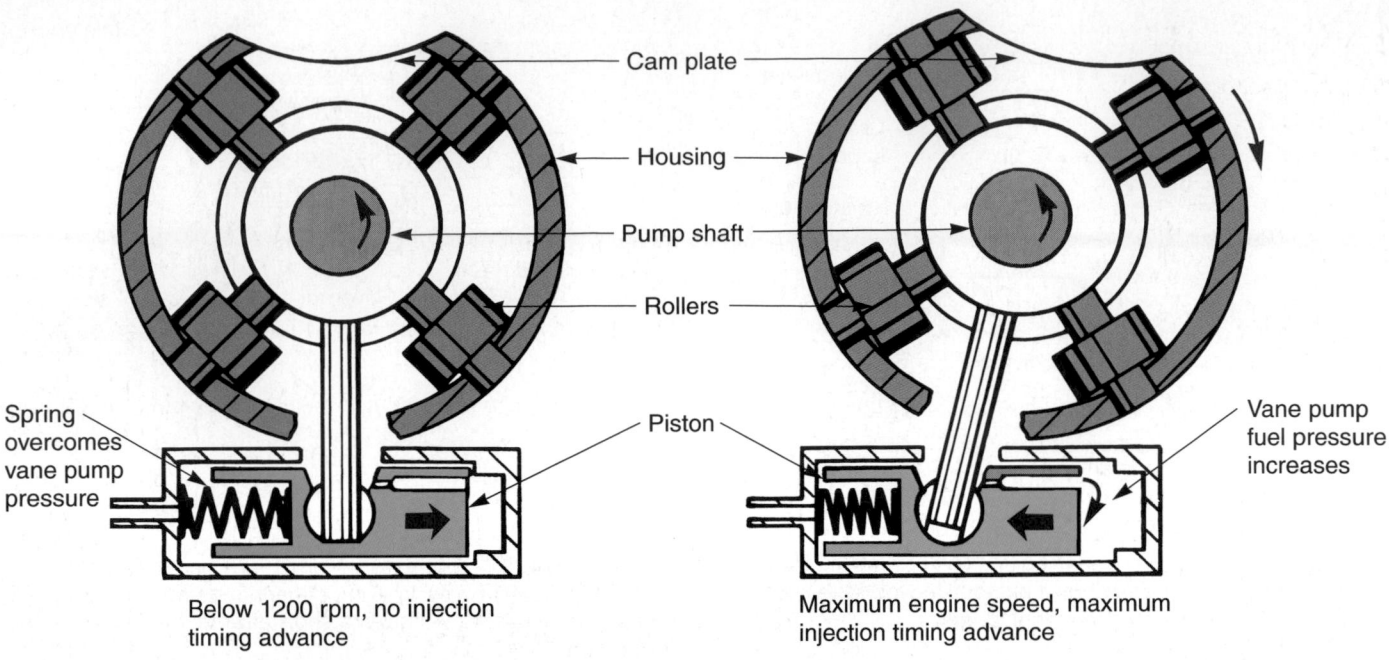

Cam plate

Housing

Pump shaft

Rollers

Piston

Spring overcomes vane pump pressure

Vane pump fuel pressure increases

Below 1200 rpm, no injection timing advance

Maximum engine speed, maximum injection timing advance

Figure 26-18. Vane pump pressure can be used to control injection timing. (VW)

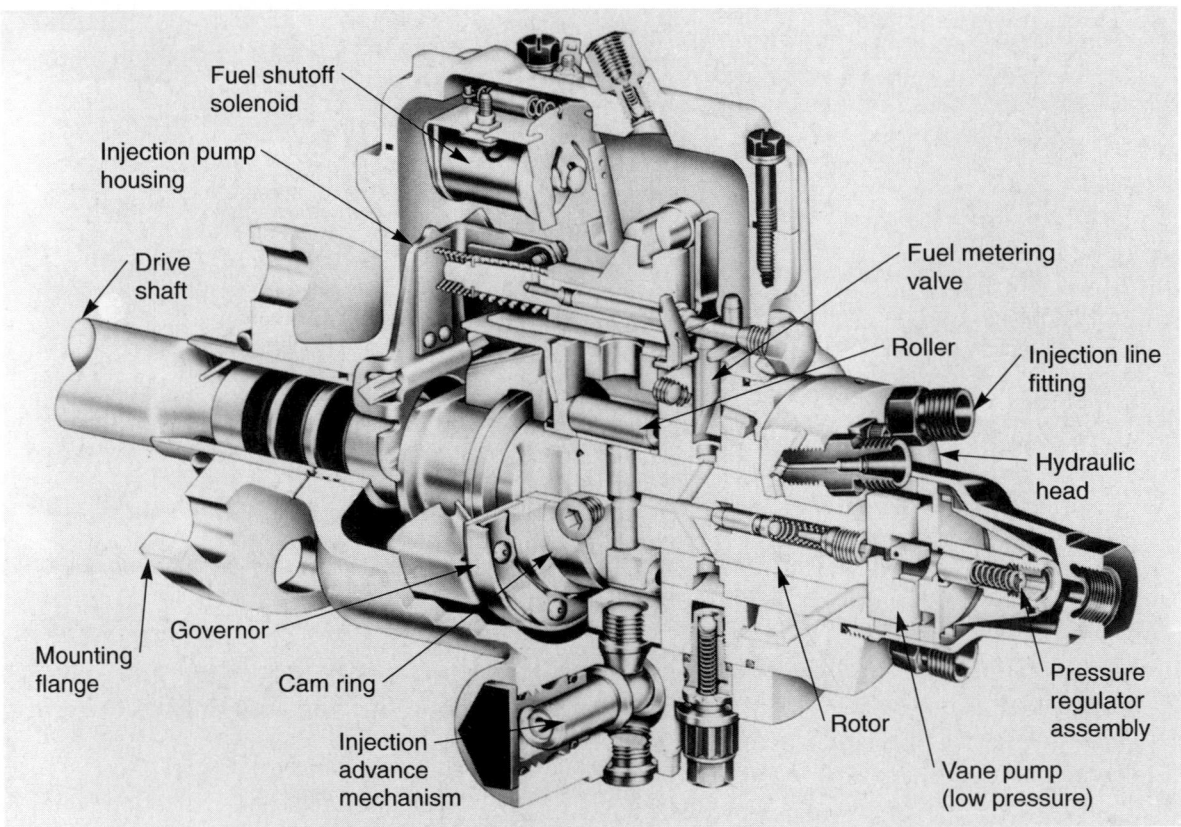

Fuel shutoff solenoid

Injection pump housing

Drive shaft

Fuel metering valve

Roller

Injection line fitting

Hydraulic head

Governor

Mounting flange

Cam ring

Rotor

Pressure regulator assembly

Vane pump (low pressure)

Injection advance mechanism

Figure 26-19. A two-plunger distributor pump is similar to a single-plunger pump. (Oldsmobile)

- *Internal cam ring*—lobed collar that acts as a cam to force plungers inward for injection of more fuel.

- *Fuel metering valve*—rotary valve that regulates fuel injection quantity by controlling how far the two plungers move apart on the pump's fill stroke.

The other parts of a two-plunger distributor pump (transfer pump, hydraulic head, delivery valve) are almost the same as a single-plunger pump. **Figure 26-20** gives a circuit diagram for a two-plunger distributor pump. Study this illustration carefully.

Two-Plunger Distributor Pump Operation. When the engine is running, the drive shaft turns the transfer (vane) pump. This pulls fuel into the injection pump. When the charging ports line up, fuel fills the high-pressure pumping chamber.

As the shaft continues to turn, the charging ports move out of alignment and the discharge ports line up. At this instant, the plungers are forced inward by the cam ring. This pressurizes the fuel and pushes it out of the hydraulic head to the injector nozzles. Look at **Figure 26-19** to locate the plungers.

Other Diesel Injection Pump Features

There are numerous design variations of injection pumps. For details of each pump design, refer to the pump manufacturer's service manual. It will explain the construction and operation of the particular pump.

A few additional features that many automotive diesel injection pumps have include:

- *Electric fuel shutoff*—a solenoid that stops fuel injection when the ignition key is turned off. See **Figure 26-19.**
- *Viscosity compensating valve*—this allows for different fuel weights or thicknesses and temperatures.
- *Injection pump vent*—a small passage that allows fuel to return to the fuel tank. This helps bleed air out of system, **Figure 26-20.**

Diesel Injector Nozzles

Diesel injector nozzles are spring-loaded valves that spray fuel directly into the engine precombustion chambers. See **Figure 26-21.**

The injector nozzles are threaded into the cylinder head. One injector is provided for each cylinder. The inner tip of the injector nozzle is exposed to the heat of combustion, like a spark plug in a gasoline engine.

Diesel Injector Parts

The basic parts of a diesel injector are pictured in **Figures 26-22** and **26-23.** Look at these illustrations as the parts are explained.

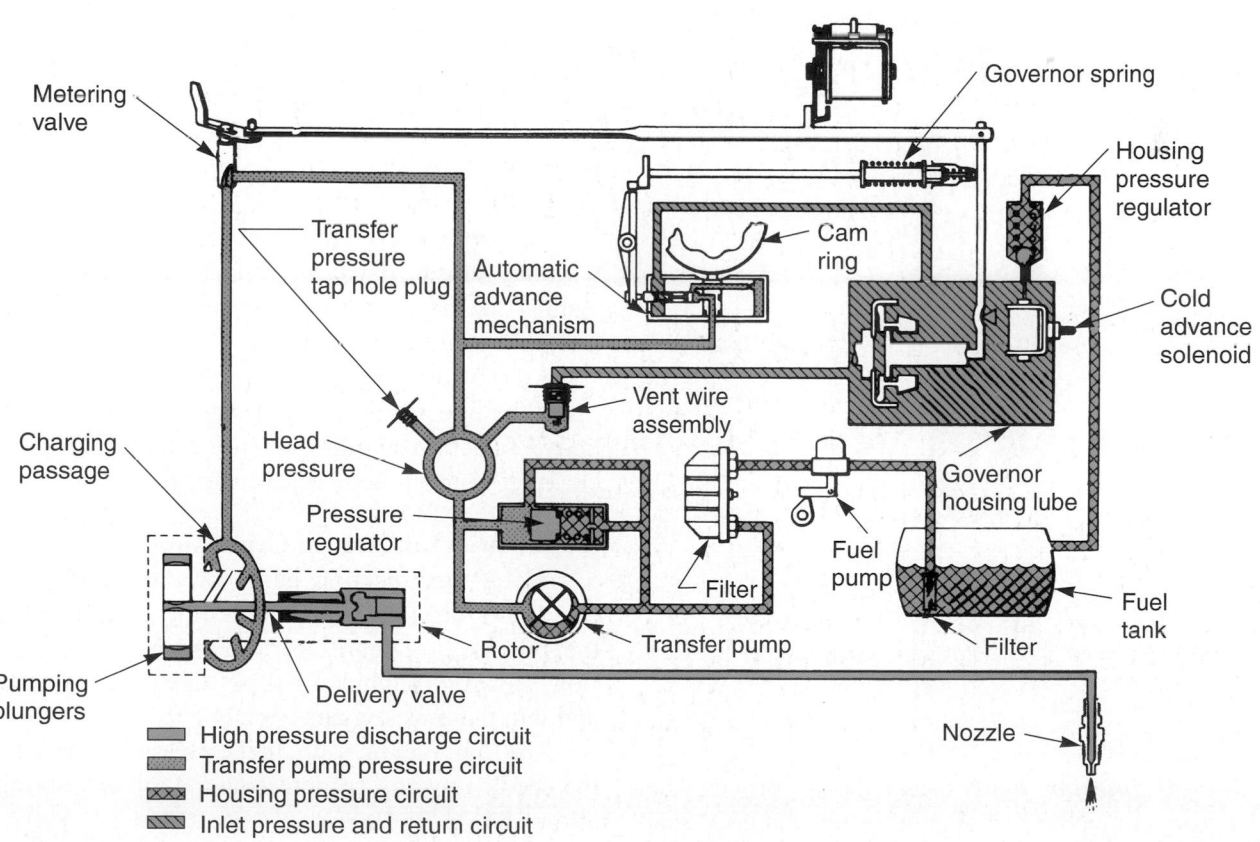

Figure 26-20. Fuel circuit diagram for a two-plunger distributor injection pump. Can you trace fuel through system? (Oldsmobile)

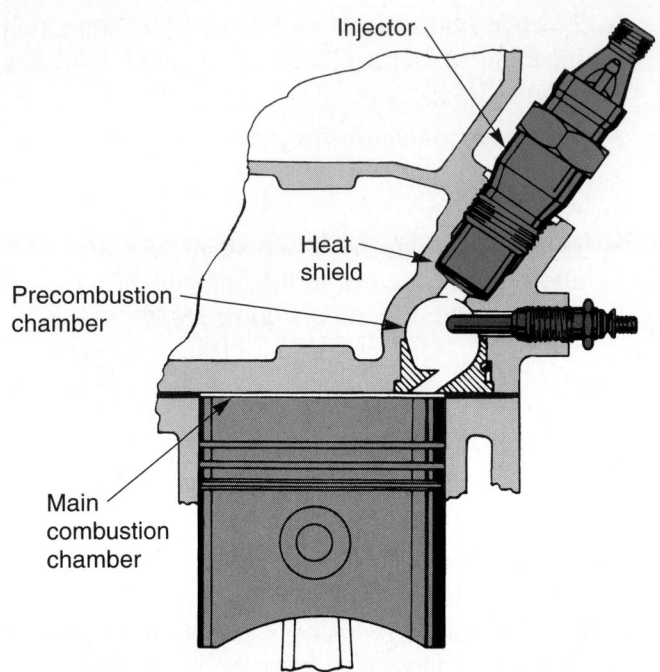

Figure 26-21. The diesel injector screws into the cylinder head. A heat shield fits between the injector body and the cylinder head. Fuel is injected into precombustion chamber. (VW)

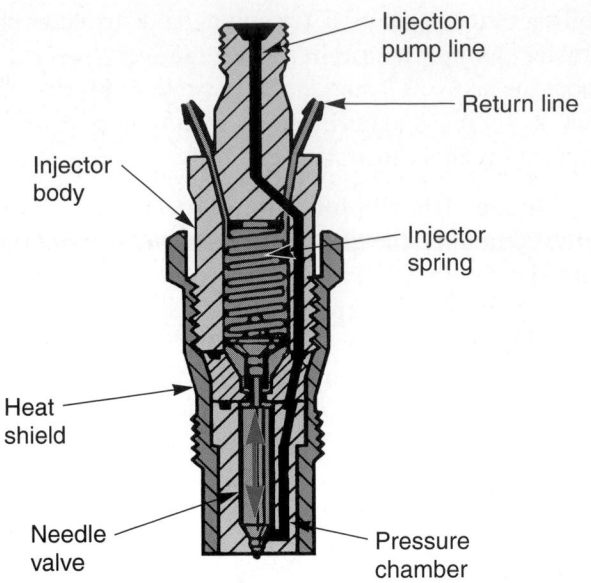

Figure 26-23. High-pressure fuel from the injection pump enters the top of the injector valve. Fuel flows through the body passage to the pressure chamber. With enough injection pressure, the needle valve is pushed up and fuel sprays into the precombustion chamber. (VW)

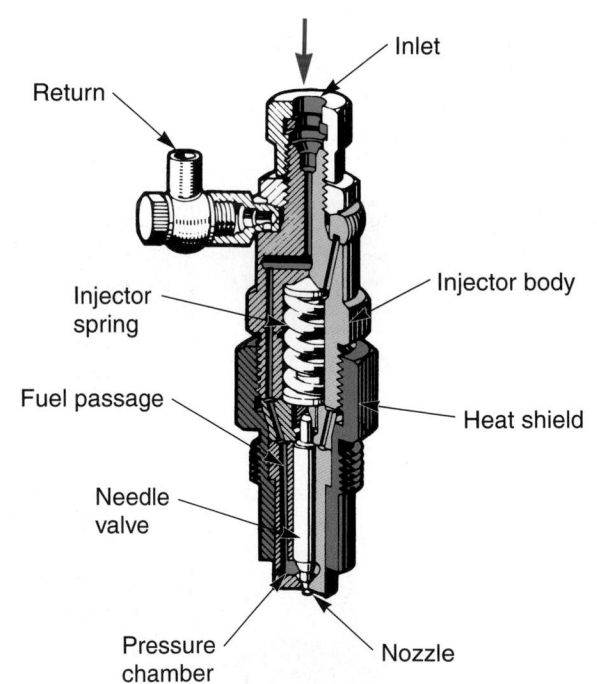

Figure 26-22. Basic diesel injector construction. Note the part names, shapes, and locations. (Robert Bosch)

A *diesel injector heat shield* helps protect the injector from engine heat. It also helps make a good seal between the injector and the cylinder head.

The *injector body* is the main section of the injector, which holds the other parts. The body threads into the heat shield. The heat shield threads into the cylinder head. Fuel passages are provided in the injector body. A needle seat is formed by the lower opening in the injector body.

The *diesel injector needle valve* opens and closes the nozzle (fuel opening). It is a precisely machined rod with a specially shaped tip. The tip of the needle seals against the injector body when closed.

The *injector spring* holds the injector needle in the normally closed position. It fits around the needle and against the injector body. Spring tension helps control injector opening pressure.

An *injector pressure chamber* is formed around the tip of the injector needle and inner cavity in the injector body. Injection pump pressure forces fuel into this chamber to push the needle valve open.

Diesel Injector Nozzle Operation

When the injection pump produces high pressure, fuel flows through the injection line and into the inlet of the injector nozzle. Look at **Figure 26-23.** Fuel then flows down through the fuel passage in the injector body and into the pressure chamber.

The high fuel pressure in the pressure chamber forces the needle upward, compressing the injector spring. This allows diesel fuel to spray out, forming a cone-shaped spray pattern. Some fuel leaks past the injector needle and returns to the fuel tank through the return lines.

Diesel Injector Nozzle Types

Several types of injector nozzles are used in diesel engines:

- Inward opening injector nozzle, **Figure 26-24.**
- Outward opening injector nozzle.
- Pintle injector nozzle.
- Hole injector nozzle.

Most automotive diesels use an inward opening, pintle injector nozzle. **Figure 26-25** shows some typical spray patterns.

Glow Plugs

Glow plugs are heating elements that warm the air in precombustion chambers to help start a cold diesel engine. Refer to **Figure 26-26** and **Figure 26-27.**

The glow plugs are threaded into holes in the cylinder head. The inner tip of the glow plug extends into the precombustion chamber.

Glow Plug Control Circuit

A *glow plug control circuit* automatically disconnects the glow plugs after a few seconds of operation. **Figure 26-28** shows a typical glow plug circuit.

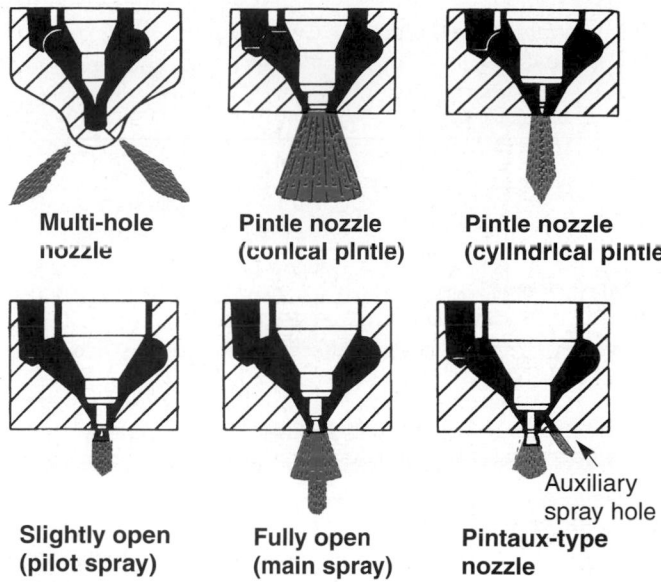

Figure 26-25. Study spray patterns for different types of injectors. (Chrysler)

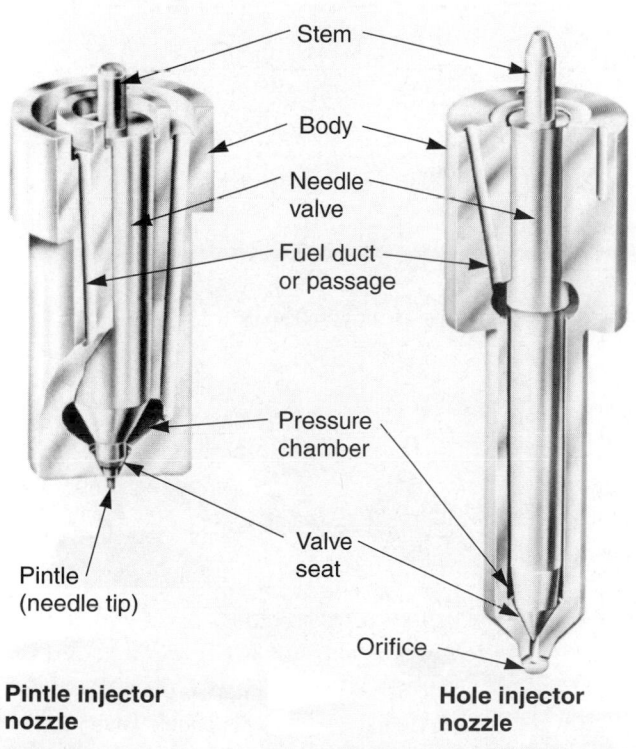

Figure 26-24. Inward opening injector nozzles. (Robert Bosch)

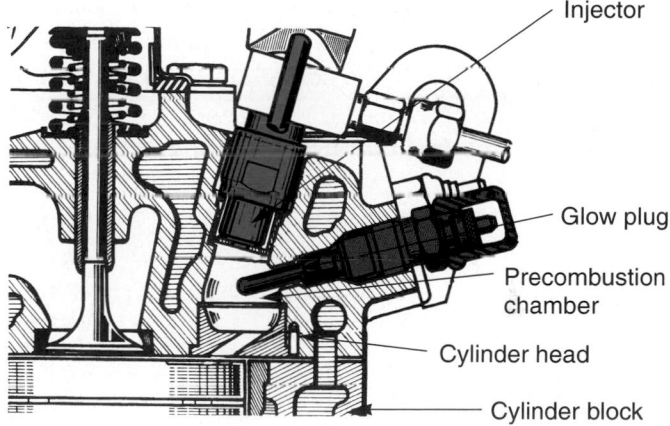

Figure 26-26. Glow plug screws into the cylinder head next to the injector. Its tip protrudes into precombustion chamber.

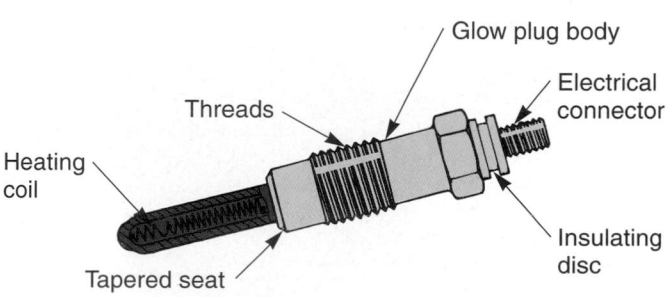

Figure 26-27. A glow plug is simply an electric heating element. Current flow through plug heats the element to warm air in precombustion chamber. This aids the starting of a cold engine. (Mercedes Benz)

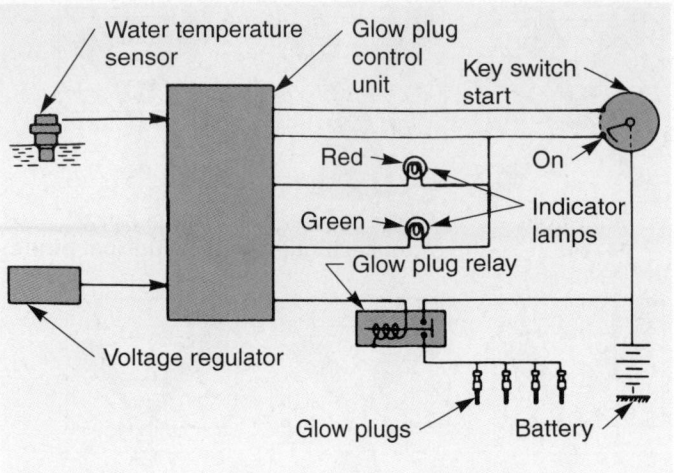

Figure 26-28. Basic circuit for glow plugs. The control unit monitors engine temperature and informs the driver whether glow plugs have been activated long enough for the engine to start. (Chrysler)

The *engine temperature sensor,* or *water temperature sensor,* checks the temperature of the engine coolant. It feeds this electrical data to a control unit. Thus, if the engine is already warm, the control unit will not turn on the glow plugs.

Indicator lights, also operated by the control unit, inform the driver whether or not the engine is ready to start. The glow plugs need a few seconds to heat up.

Glow Plug Operation

When the engine is cold and the driver turns the ignition switch to run, a large current flows from the battery to the glow plugs. In a few seconds, the glow plug tips will heat to a dull red glow.

When the glow plug indicator light goes out, the driver can start the engine. The compression stroke pressure and heat, along with the heat from the glow plugs, helps the engine to start easier.

Diesel Injection System Accessories

Various diesel injection system accessories are used to aid in the dependability of a diesel engine. These accessories are needed because diesel fuel is susceptible to problems caused by moisture and low temperatures.

Water Detector

A *water detector* may be used to warn the driver of water in the diesel fuel. Such contamination is very harmful to a diesel fuel system. The water mixes with the fuel and can cause corrosion of the precision parts in the injection pump and injectors. **Figure 26-29** shows a circuit using an in-tank water detector.

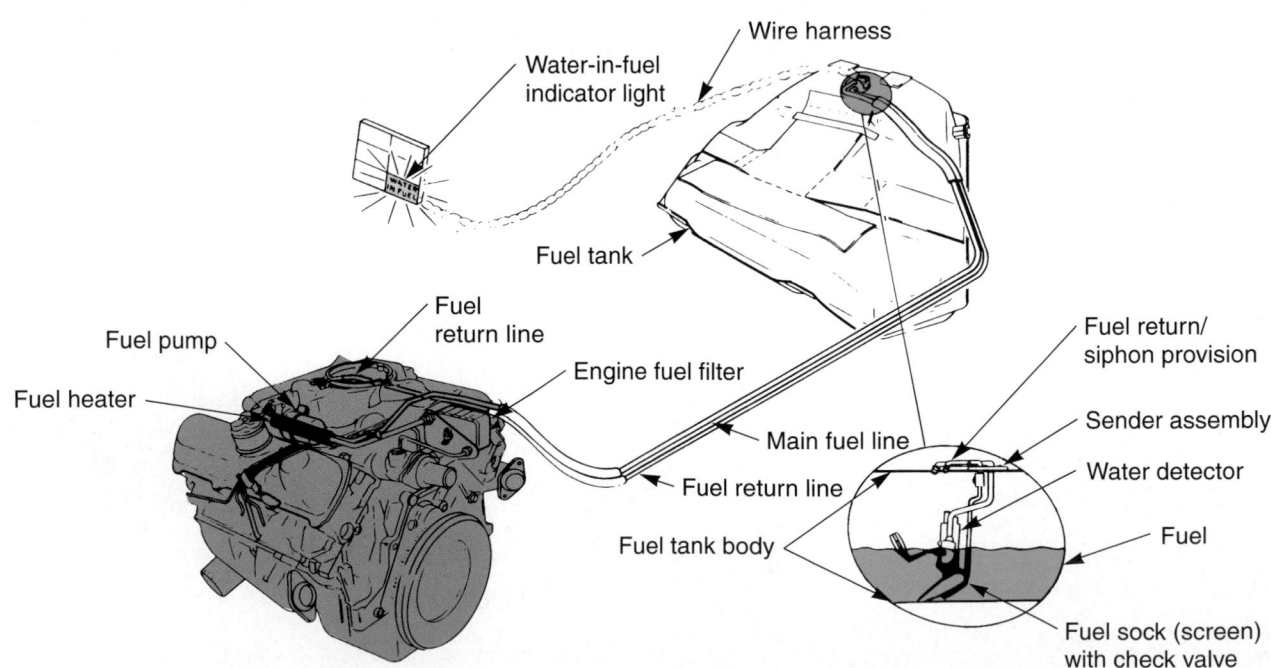

Figure 26-29. This diesel injection system has a water detector-warning light system and a fuel heater. The indicator light glows if there is an excessive amount of water in the tank. The fuel heater warms the fuel in the line before it enters injection pump. Heat is only needed in cold weather. (Oldsmobile)

Fuel Heater

A *fuel heater* is sometimes used to warm the diesel fuel, preventing the fuel from jelling (turning into a semisolid). An optional device, it is needed in very cold climates. The heater is simply an electric heating element in the fuel line ahead of the injection pump. See **Figures 26-29** and **26-30.**

Block Heater

A *block heater* may be used to warm the engine block in cold weather. It is a heating device that plugs into a home wall electric outlet. It keeps the engine warm overnight to make the diesel engine easier to start on cold mornings.

Vacuum Pump

A *vacuum pump* is frequently used on a diesel engine to provide a source of vacuum (suction) for the vehicle. Vacuum is needed for the power brakes, A/C ventilation system, and emission control devices.

A gasoline engine has a natural source of vacuum in the intake manifold. A diesel engine has very little vacuum in the intake manifold because it does not have a throttle valve restriction. For this reason, a vacuum pump is needed.

A vacuum pump on a diesel engine may be a reciprocating diaphragm type bolted to the front or rear of the engine, **Figure 26-31.** It may also be a rotary-type pump at the rear of the alternator, **Figure 26-32.**

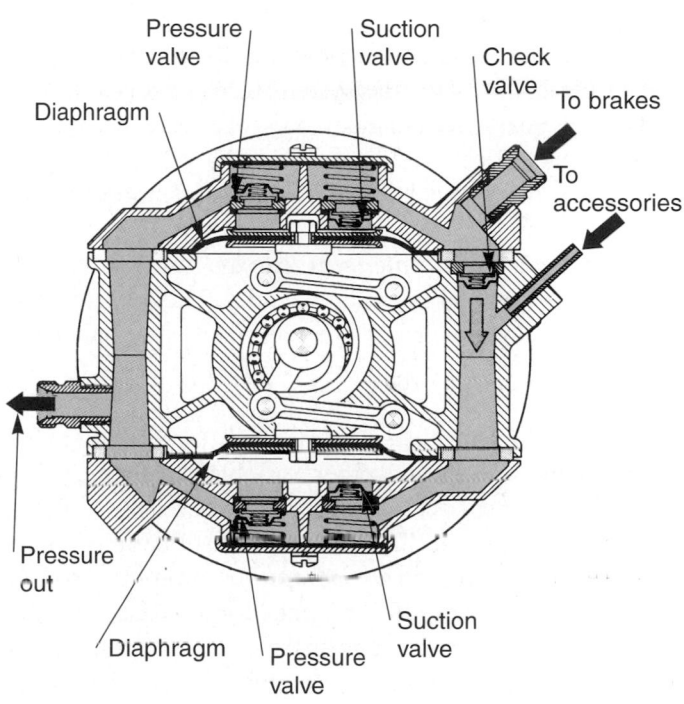

Figure 26-31. Construction of a diaphragm vacuum pump. (Mercedes Benz)

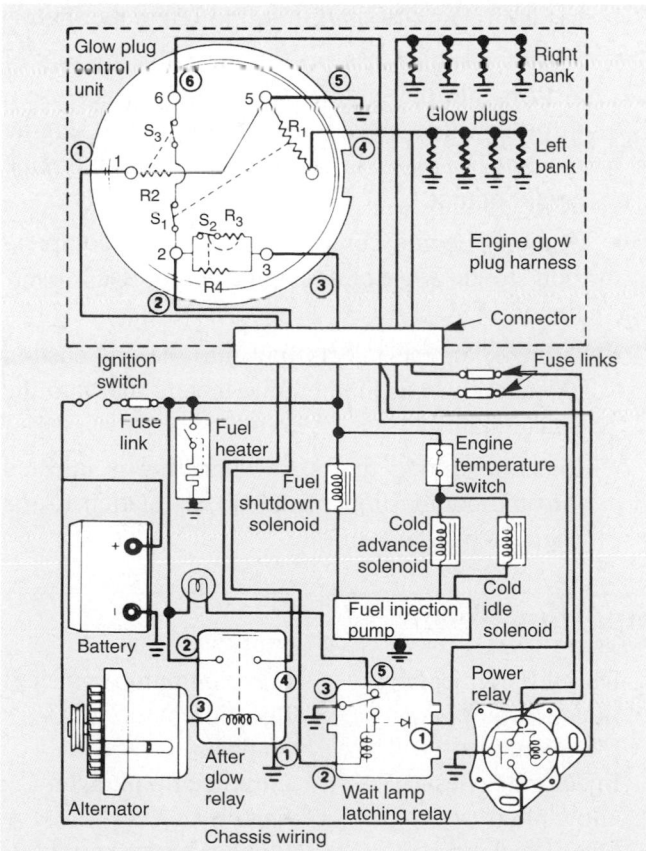

Figure 26-30. Complete electrical circuit for a typical diesel injection system. Note the parts and electrical connections. (Ford)

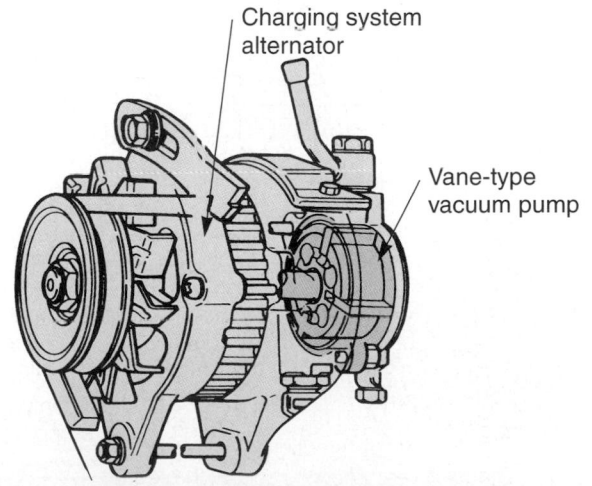

Figure 26-32. A vane-type vacuum pump is mounted in back of the charging system alternator. It produces vacuum for power brakes and other vacuum-operated devices. (Chrysler)

A vacuum pump uses the same principle as a fuel pump (see Chapter 21). However, it pumps air, not liquid fuel, out of an enclosed area to reduce pressure and form a vacuum. The diesel engine provides power for the vacuum pump, usually through a belt.

Computer-Controlled Diesel System

A *computer-controlled diesel system* uses a computer, sensors, and actuators to increase the efficiency of mechanical diesel injection. A simplified illustration is given in **Figure 26-33.**

In the past, diesel injection was totally mechanical. All of the controls for the system were linkages, levers, rods, and gears. Mechanical controls were heavy and slow to react. Electronic components have been designed to replace many mechanical devices and improve efficiency.

Injection pump solenoids can be used to control injection timing and injection quantity. They react to signals from the computer. When energized or de-energized electrically, they can open or block fuel passages to alter pump operation.

Diesel injection sensors are used to monitor engine temperature, speed, and other variables that affect fuel needs. The sensor feeds signals to the computer. These sensors are similar to those used for gasoline injection.

The *diesel injection computer* monitors sensor inputs and calculates the outputs for the actuators. In this way, the computer can more precisely control the mechanical, high-pressure plungers inside the diesel injection pump.

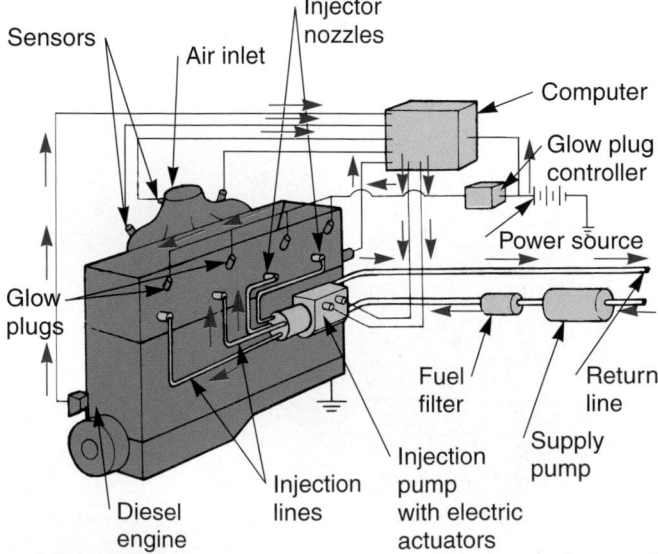

Figure 26-33. Study the basic parts used in an electronic diesel injection system.

Summary

- A diesel fuel injection system is a high-pressure, mechanical system that delivers fuel directly into the engine combustion chambers.

- The injection pump is a high-pressure, mechanical pump that meters the correct amount of fuel and delivers it to each injector nozzle at the proper time.

- Injection lines are high-strength steel tubing that carries fuel to each injector nozzle.

- Injector nozzles are spring-loaded valves that spray fuel into each combustion chamber.

- Glow plugs are electric heating elements that warm air in the precombustion chambers to aid starting of a cold engine.

- Diesel engines use a very high compression ratio (17:1 to 25:1). A gasoline engine's compression ratio is only 8:1 or 9:1.

- A diesel engine is a compression ignition engine because it uses the heat from compressed air to ignite the fuel. Gasoline engines are classified as spark ignition engines because they use an electric arc (from the spark plug) to ignite the fuel.

- A diesel engine has no throttle valve to control airflow into the engine. (Gasoline engines use a throttle valve to control airflow and engine power.) The diesel injection pump controls engine power output.

- A diesel engine compresses air on its compression stroke. A gasoline engine compresses an air-fuel mixture.

- A diesel engine injects fuel into the combustion chambers. A gasoline engine meters fuel into the intake manifold.

- Injection timing refers to when fuel is injected into the combustion chambers in relation to the engine's piston position.

Important Terms

Diesel fuel injection	Injection pump camshaft
Injection pump	Injection pump roller
Injection lines	tappets
Injector nozzles	Inline pump plungers
Glow plugs	Barrels
Diesel fuel supply	Plunger return springs
system	Control sleeves
Inline diesel injection	Control rod
pump	Rack

Delivery valves
Effective plunger
  stroke
Governor
Centrifugal
Injection timing
Distributor injection
  pump
Drive shaft
Transfer pump
Distributor plunger
Cam plate
Fuel metering sleeve
Hydraulic head
Centrifugal governor
Distributor rotor
Internal cam ring
Fuel metering valve
Electric fuel shutoff
Viscosity
  compensating valve
Injection pump vent
Diesel injector nozzles
Diesel injector heat
  shield

Injector body
Diesel injector needle
  valve
Injector spring
Injector pressure
  chamber
Glow plugs
Glow plug control
  circuit
Engine temperature
  sensor
Indicator lights
Water detector
Fuel heater
Block heater
Vacuum pump
Computer-controlled
  diesel system
Injection pump
  solenoids
Diesel injection sensors
Diesel injection
  computer

Review Questions—Chapter 26

Please do not write in this text. Place your answers on a separate sheet of paper.

1. A diesel fuel injection system is a super-high pressure, mechanical system that delivers fuel directly to the engine combustion chambers. True or False?

2. Which of the following is *not* related to a diesel engine?
   (A) Spark ignition.
   (B) No throttle valves for air control.
   (C) Fuel quantity controls engine speed.
   (D) All of the above are correct.

3. List and explain the four major components of a diesel injection system.

4. There are two common types of diesel injection pumps: _____ and _____ types.

5. What is the function of a plunger in an injection pump?

6. Roller tappets are commonly used in an inline injection pump. True or False?

7. Explain how an inline injection pump alters the amount of fuel forced to the injector nozzles.

8. _____ _____ _____ are spring-loaded valves in the outlet fittings to the injection lines for assuring quick, leak-free closing of the injector nozzles.

9. Define the term "effective plunger stroke."

10. Why is a governor needed on a diesel engine?

11. _____ _____ refers to when fuel is injected into the combustion chambers in relation to piston position.

12. A(n) _____ _____ _____ normally uses only one or two plungers and is the most common type diesel injection pump.

13. Which of the following is *not* part of a single-plunger distributor injection pump?
    (A) Drive shaft.
    (B) Fuel metering sleeve.
    (C) Hydraulic head.
    (D) Roller tappet.

14. How does a single-plunger distributor pump develop injection pressure?

15. How does a single-plunger distributor pump control injection quantity?

16. Which of the following is *not* part of a two-plunger distributor injection pump?
    (A) External camshaft.
    (B) Internal cam ring.
    (C) Distributor rotor.
    (D) Fuel metering valve.

17. A vane type transfer pump is commonly used to pull fuel into the two-plunger distributor injection pump. True or False?

18. An electric fuel shutoff is only used on inline injection pumps. True or False?

19. A diesel _____ _____ is a spring-loaded valve that sprays fuel into the engine precombustion chamber.

20. List and explain the five major parts of a diesel injector.

21. Why are glow plugs needed in a diesel engine?

22. Explain the purpose of a water detector.

23. A(n) _____ _____ can be used to help keep the diesel fuel from jelling in cold weather.

24. A block heater is a heating device operated by the car battery. True or False?

25. Explain why a diesel engine, unlike a gasoline engine, needs a vacuum pump.

ASE-Type Questions

1. Diesel engine compression ratio is about:
   (A) 8:1 or 9:1.
   (B) 10:1 to 12:1.
   (C) 14:1 to 16:1.
   (D) 17:1 to 25:1.

2. Which of the following engine types does *not* use a throttle valve to control airflow into the engine?
   (A) Diesel.
   (B) Gasoline.
   (C) Carburetor.
   (D) None of the above.

3. The spring-loaded valves that spray fuel into each combustion chamber are called:
   (A) injection lines.
   (B) injector pumps.
   (C) injector nozzles.
   (D) supply or lift pumps.

4. Each of the following is a diesel injection pump function *except:*
   (A) produce extremely high fuel pressure.
   (B) open injector nozzles before injection.
   (C) circulate fuel through lines and nozzles.
   (D) control engine idle and maximum speeds.

5. Power is transferred from engine to injection pump using a:
   (A) chain.
   (B) set of gears.
   (C) toothed belt.
   (D) Any of the above.

6. With a diesel engine running, Technician A believes the inline injection pump camshaft is rotating at one-half engine speed. Technician B says the camshaft is rotating equal to engine speed. Who is right?
   (A) Technician A only.
   (B) Technician B only.
   (C) Both A and B are correct.
   (D) Neither A nor B are correct.

7. Which of the following is used on an inline injection pump to control engine idle speed and limit maximum engine speed?
   (A) Fulcrum.
   (B) Governor.
   (C) Camshaft.
   (D) Control rod.

8. How many plungers does a distributor injection pump use to supply fuel to the engine cylinders?
   (A) One for each cylinder.
   (B) Two for each cylinder.
   (C) One or two for all cylinders.
   (D) None of the above.

9. *All* distributor injection pumps have each of these basic parts *except* a(n):
   (A) delivery valve.
   (B) transfer pump.
   (C) hydraulic head.
   (D) internal cam ring.

10. Which of the following is *not* a common diesel injector nozzle type?
    (A) Hole.
    (B) Pilot.
    (C) Pintle.
    (D) Inward opening.

Activities—Chapter 26

1. Prepare an overhead transparency showing how the various parts of a diesel injection system are interrelated.

2. Show in a sketch the relationship of a compression ratio of 17:1.

3. Research centrifugal force as a scientific principle; relate this principle to the operation of a governor; devise a way of demonstrating the principle to the shop class.

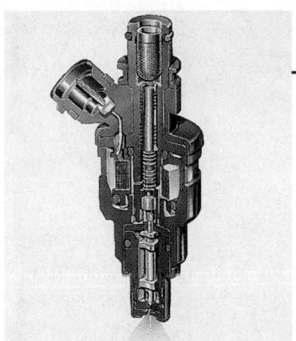

Chapter 27

Diesel Injection Diagnosis, Service, and Repair

After studying this chapter, you will be able to:

- Diagnose typical diesel injection problems.
- List safety precautions pertaining to diesel injection service.
- Test, rebuild, and replace diesel injectors.
- Test and replace glow plugs.
- Perform basic maintenance operations on a diesel injection system.
- Describe basic adjustments common to diesel injection systems.
- Demonstrate safe work habits in diesel injection service.
- Correctly answer ASE certification test questions relating to diesel injection system diagnosis, service, and repair.

It is important for you to understand how to work on a diesel injection system because almost all auto manufacturers offer a diesel engine. While diesel injection service procedures are different from those for a gasoline injection system, they are still relatively simple.

Diesel Injection Maintenance

Refer to a service manual for details on periodic maintenance of a diesel injection system. You must change or clean filters periodically. Maintenance also involves inspecting the system for signs of trouble.

Fuel filters are normally located in the fuel tank (sock filter), in the fuel line (main filter), and, sometimes, in the injector assemblies (inlet connection screens or final filter screens), **Figure 27-1.** To ensure proper engine performance, these filters must be kept clean.

The main fuel filter may have a drain. The drain can be used to bleed off trapped water. When mixed with diesel oil, water causes rapid corrosion and pitting of injection system components.

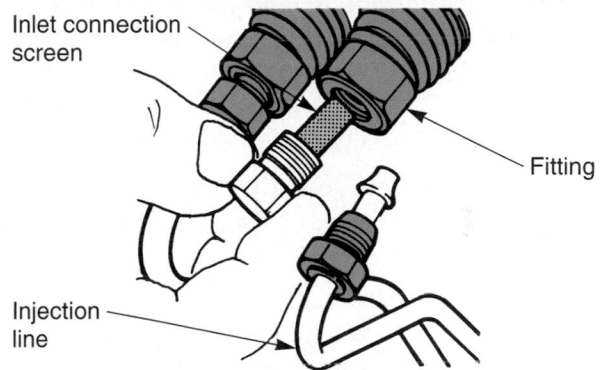

Figure 27-1. Inlet connection screens at injectors are frequently used. Check them if problems point to the injection system. (Chrysler)

If you detect signs of fuel leakage, use a piece of cardboard to find the leak. See **Figure 27-2.** Move the cardboard around each fuel fitting. If there is a serious leak, fuel will strike the cardboard and not your hand.

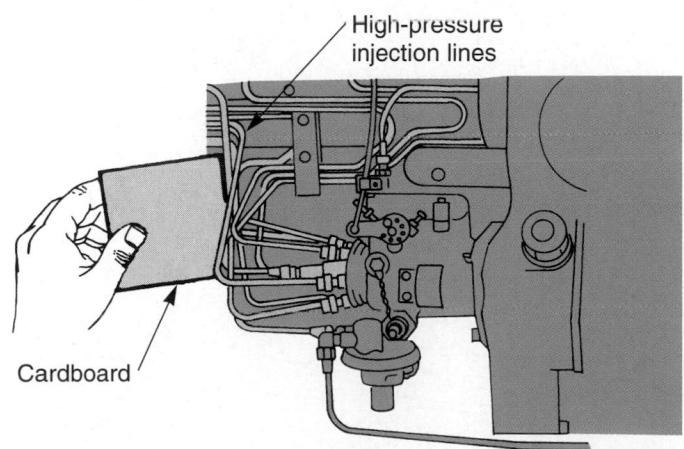

Figure 27-2. Use a piece of cardboard to locate injection system leaks. Remember that injection pressure is high enough to make the fuel spray puncture skin. (Chrysler)

Replace any injection line or return hose that is not in perfect condition, **Figure 27-3.**

Warning

Never attempt to stop a diesel engine by covering the air inlet opening. Since there is no throttle valve, there is enough suction to cause a hand injury or to suck rags and other objects into the intake manifold.

Diesel Injection Cleanliness

When servicing a diesel engine, cleanliness is very important. A small bit of foreign material can upset the operation of the injection pump or the injectors. These parts are machined to precision tolerances measured in microns (millionths of a meter). Keep this in mind during service.

A few rules of cleanliness to remember are:

- Always cap and seal any injection system fitting that is disconnected. This will prevent dust from entering the system.
- Use clean, lint-free shop rags when wiping off components. Even a piece of lint could upset the operation of the injection system.
- Use compressed air and clean shop rags to remove dirt from around fittings before disassembly. Look at **Figure 27-4.**
- Do not spray water on a hot diesel engine. This could crack or warp the injection pump.

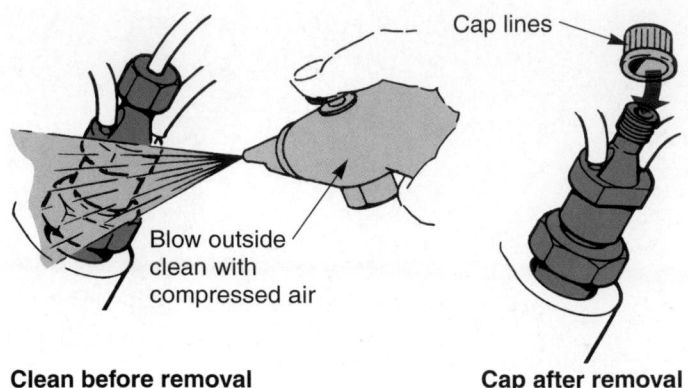

Clean before removal   **Cap after removal**

Figure 27-4. Before removing an injection line, blow away dirt with compressed air. Cap the line after it is removed. A small amount of dirt can upset injector operation.

Diesel Injection Diagnosis

Diesel injection diagnosis requires you to use your knowledge of engine and injection system operation and your basic troubleshooting skills. Begin diagnosis by checking the operation of the engine. Check for:

- Abnormal exhaust smoke.
- Excessive knock.
- Engine miss.
- A "no start" condition.
- Lack of power.
- Poor fuel economy.

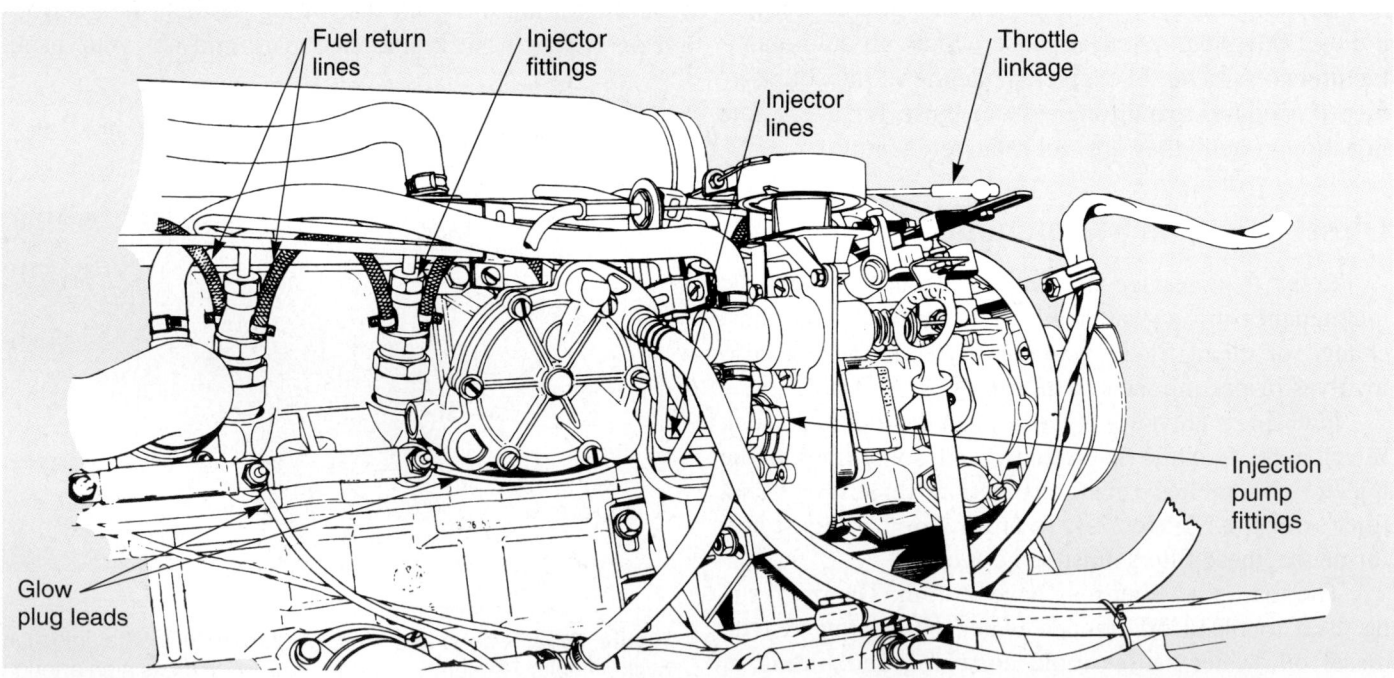

Figure 27-3. A visual check of hoses, lines, filters, and linkages will sometimes reveal the source of the problem. (Volvo)

Refer to the troubleshooting chart in the vehicle's service manual. It will list the possible causes for these and other conditions. The service manual chart will be accurate because it is designed for one type of diesel injection system.

Abnormal Exhaust Smoke

Excessive diesel exhaust smoke is normally due to incomplete combustion caused by injection system or engine troubles. A small amount of exhaust smoke is normal during initial start-up, cold engine operation, or rapid acceleration. Abnormal exhaust smoke may be black, white, or blue. See **Figure 27-5.**

The main cause of excessive **black smoke** is too much fuel. A rich air-fuel mixture allows carbon (ash) to blow out of the exhaust system. Black smoke may be due to problems with the injection pump, injection timing, air cleaner, injectors, fuel, or the engine itself.

White smoke occurs mainly during cold starts. The smoke usually consists of condensed fuel particles. The cold engine parts cause the fuel to condense into a liquid, which will not burn. The most common reasons for white exhaust smoke are inoperative glow plugs, low engine compression, thermostat stuck open, bad injector spray pattern, late injection timing, and cold start (injection pump) problems.

Tech Tip

White smoke can also be caused by coolant leakage into the combustion chambers. The engine may have a leaking head gasket, cracked cylinder head, or cracked block.

Excessive **blue smoke** may be due to oil consumption from worn piston rings, scored cylinder walls, or leaking valve stem seals. White-blue smoke, however, is normally caused by incomplete combustion or injection system problems.

Smoke Meter

A **smoke meter** is a testing device used to measure the amount of smoke (ash or soot) in diesel exhaust. See **Figure 27-6.** The smoke meter measures the amount of light that can shine through an exhaust sample. If the

A

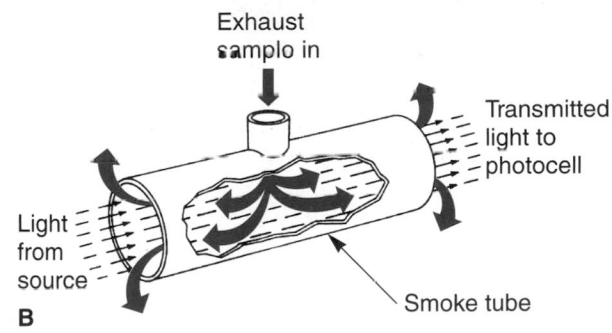

B

Figure 27-6. A smoke meter can be used to check the amount of smoke in diesel exhaust. A—Smoke meter. B—Principle of smoke meter operation. (Hartridge)

Black smoke—rich fuel mixture

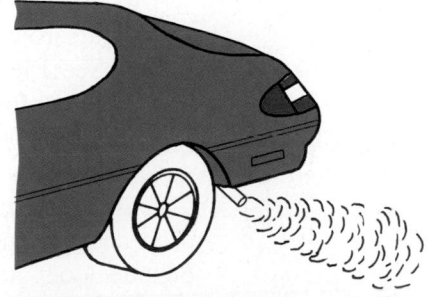

White smoke—partially burned fuel

Blue smoke—burning oil

Figure 27-5. Exhaust smoke will tell you much about diesel engine operation.

exhaust smoke blocks too much light, engine or injection system repairs or adjustments are needed.

Excessive Knock

All diesel engines produce a knocking sound when running. This occurs because the fuel ignites spontaneously and burns rapidly. Very high pressures produce a rumble or a dull clattering sound.

Ignition lag is the time span between the injection of the diesel fuel and the ignition of the fuel. It is a controlling factor affecting diesel knock. If ignition lag is too long, too much fuel will ignite at once and a mild explosion (loud knock) will result.

Abnormally loud diesel knock can be due to low operating temperatures (thermostat stuck open), early injection timing, low compression, fuel contamination, and oil consumption.

Engine Miss

A diesel engine *miss* results from one or more cylinders not firing (burning fuel) properly. Just as a gasoline engine will miss if a spark plug does not produce a spark, a diesel engine can also run roughly because of injection system problems.

A miss in a diesel engine can be due to faulty injectors, clogged fuel filters, incorrect injection timing, low cylinder compression, injection system leaks, air leaks, or a faulty injection pump.

Engine No Start Condition

A *no start condition* may be due to inoperative glow plugs, restricted air or fuel flow, a bad fuel shutoff solenoid, contaminated fuel, or a defective injection pump.

A slow cranking speed is a common cause of a diesel no start condition. Being a compression-ignition engine, a diesel must crank fast enough to produce sufficient heat for combustion.

Lack of Power

When a diesel engine lacks power, check the throttle cable adjustment, governor setting, fuel filters, air filter, engine compression, and other factors affecting combustion. Keep in mind, however, that a diesel engine does not produce as much power as a gasoline engine of equal size.

Poor Fuel Economy

Poor fuel economy may be due to a fuel leak, a clogged air filter, incorrect injection timing, or leaking injectors. Normally, a diesel engine will get better fuel economy than a gasoline engine.

Check the EPA fuel economy values for the vehicle before measuring actual fuel economy. If the measured fuel economy is much lower than the EPA values, adjustments or repairs should be made.

Scanning Diesel Problems

Many late-model diesel engines are computer controlled. You can usually connect a scan tool to the system to aid troubleshooting. An OBD II system will check the injection pump solenoids, electric feed pumps, fuel tank pressure, water sensors, and other devices.

For more information on scanning computer control systems, refer to the index. This subject is explained in several other chapters.

Testing Diesel Injection Operation

There are several ways to check the operation of a diesel injection system. We will briefly explain the most common testing methods. Always refer to your service manual for exact instructions. Recommended testing procedures vary with the design of the injection pump, return lines, and injection nozzles.

Cylinder Balance Test

A diesel *cylinder balance test* involves disabling one injector at a time to check the firing of that cylinder. Just as you can remove a spark plug wire on a gasoline engine to check for combustion, you can loosen the injection line to disable the injector.

 Warning
When loosening an injection line, only unscrew the fitting enough to allow fuel to drip from the connection. Wear safety glasses and leather gloves, and obtain instructor approval before completing this test. Also, refer to the service manual because it may describe a safer testing method.

To perform a cylinder balance test on a diesel engine, wrap a rag around the injector and loosen the injector line fitting. See **Figure 27-7.** When the fitting is loosened, fuel should slowly leak out of the connection. Fuel leakage will prevent the injector from opening and spraying fuel into the combustion chamber.

As the injector line is loosened, engine speed should drop and the engine should idle roughly. If "killing" a cylinder does not affect engine operation, that cylinder has not been firing. There may be a bad injector, low compression, or an injection pump problem. Further tests will be needed. Check all injectors.

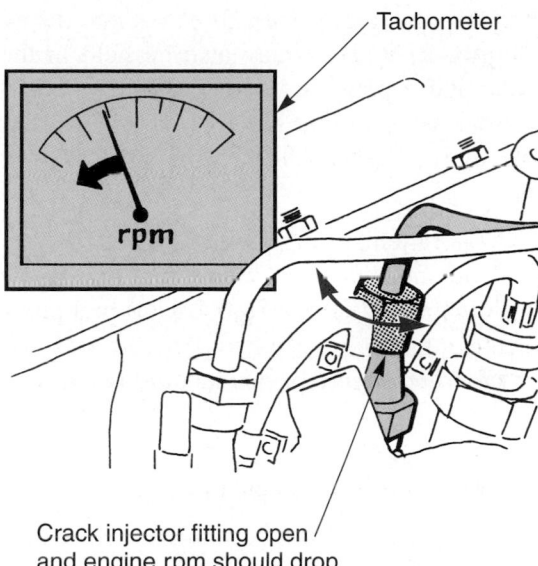

Crack injector fitting open
and engine rpm should drop

Figure 27-7. To check for a dead cylinder or cylinder not firing, crack open each injector line one at a time. Let fuel drip from fitting with the engine running. This will keep the injector from opening. If engine speed drops, that injector and cylinder are functioning. If engine speed does not drop with injector disabled, that cylinder is not firing. (Volvo)

Diesel Engine Compression Test

A diesel engine *compression test* is similar to a compression test for a gasoline engine. However, do not use a compression gauge intended for a gasoline engine. It can be damaged by the high pressures. A diesel compression gauge must read up to approximately 600 psi (4134 kPa).

To perform a diesel compression test:
1. Remove either the injectors or the glow plugs. Refer to a service manual for instructions.
2. Install the compression gauge in the recommended hole. Usually, a heat shield must be used to seal the gauge when it is installed in place of an injector.
3. Disconnect the fuel shutoff solenoid to disable the injection pump.
4. Crank the engine and note the highest reading on the gauge.
5. Compare your compression gauge readings to specifications.

Typical compression pressure for an automotive diesel engine is 400–500 psi (2800–3400 kPa). Readings should be within about 50–75 psi (350–500 kPa) of each other. If not within this range, engine repairs are needed.

For more information on compression testing, refer to the text index. Compression testing is covered in several other textbook locations.

Caution

Some automakers warn against performing a wet compression test on a diesel engine. If too much oil is squirted into the cylinder, hydraulic lock and part damage could result because the oil will not compress.

Glow Plug Resistance-Balance Test

A *glow plug resistance-balance test* provides a safe way of finding out if each cylinder is firing. When combustion is occurring in a cylinder, it will raise the temperature and internal resistance of the glow plug in that cylinder.

To perform this test, unplug the wires to all the glow plugs. Connect a digital ohmmeter across each glow plug and ground. Write down the ohms reading of each glow plug. Then, start the engine and let it run for a few minutes. Shut the engine off and recheck glow plug resistance. If a cylinder is not firing, the resistance of its glow plug will not increase as much as the resistance of the others.

Digital Pyrometer Balance Test

A *digital pyrometer* is an electronic device that is used to take very accurate temperature measurements. It can sometimes be used to check the operation of a diesel engine.

Touch the digital pyrometer on the exhaust manifold at each exhaust port. With the engine running, the temperature of the exhaust manifold should be almost the same at each port. If the manifold reading is cooler next to one exhaust port, that cylinder is not firing.

Injection-Pressure Test

An *injection-pressure tester* uses special valves and a high-pressure gauge to measure the amount of pressure in the injection lines. See **Figure 27-8.**

Connect the injection-pressure tester between the injection pump and the injectors. Follow the instructions provided with the particular tester. Some injection-pressure testers allow you to check the following.

- Injector opening pressure.
- Injector nozzle leakage.
- Injection line pressure balance.
- Injection pump condition.

An injection-pressure tester is a very informative test instrument. It enables you to pinpoint problems without removing major parts from the engine. For example, you can quickly locate a leaking injector nozzle, a clogged injector filter, or a bad injection pump. A leaking nozzle will show up as a pressure drop. A clogged filter will cause higher-than-normal pressure readings. A bad injector pump will produce lower-than-normal pressure readings.

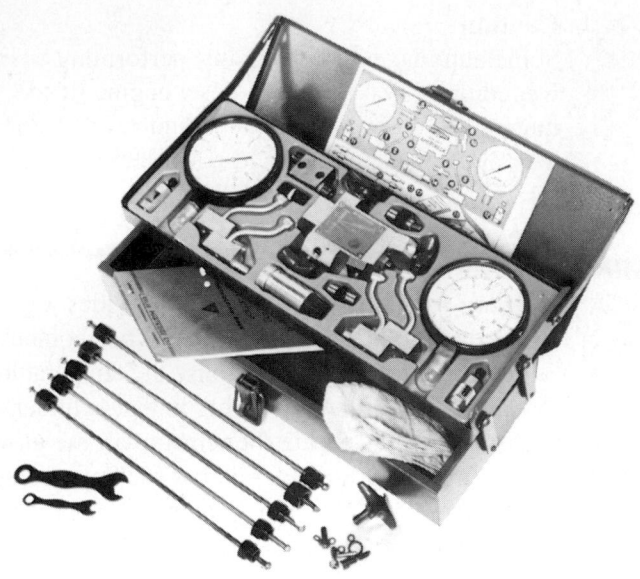

Figure 27-8. Test setup for checking injection system operation while still on engine. This tester makes diagnosis quick and easy. (Hartridge)

Diesel Injection Service

As each diesel injection component is discussed in the following sections, possible symptoms, problems, and corrections for that component will be explained.

Injector Nozzle Service

A bad diesel injector usually causes the engine to miss. It may also reduce engine power or cause smoking and knocking. The injector nozzles are exposed to the direct heat and by-products of combustion. They can wear, become clogged with carbon, or be damaged. This can result in an incorrect opening pressure, incorrect spray pattern, fuel leakage, and other problems.

Injector Substitution

Injector substitution is an easy way to verify an injector problem. It involves installing a good injector for the one being tested. If the cylinder fires with the good injector, then the old injector is faulty. If the cylinder still misses with the good injector, then other engine or injection problems exist.

Injector Removal

If your tests indicate a faulty injector, you should remove the injector for service. Following the directions in the service manual, disconnect the battery to prevent engine cranking. Using the appropriate tools, remove the injection line. Be careful not to bend or kink the high-pressure line.

Normally, the injectors are threaded into the cylinder head, **Figure 27-9.** They may also be held in the head with bolts and a press fit. With press-fit injectors, you may need to use a special impact tool to force them out of the head. See **Figure 27-10.**

⚠️ **Warning**

Never attempt to remove an injection system component with the engine running. With 6000-8000 psi (42,000 to 56,000 kPa) fuel pressure, fuel could squirt out and puncture your skin. *Blood poisoning* or *death* could result.

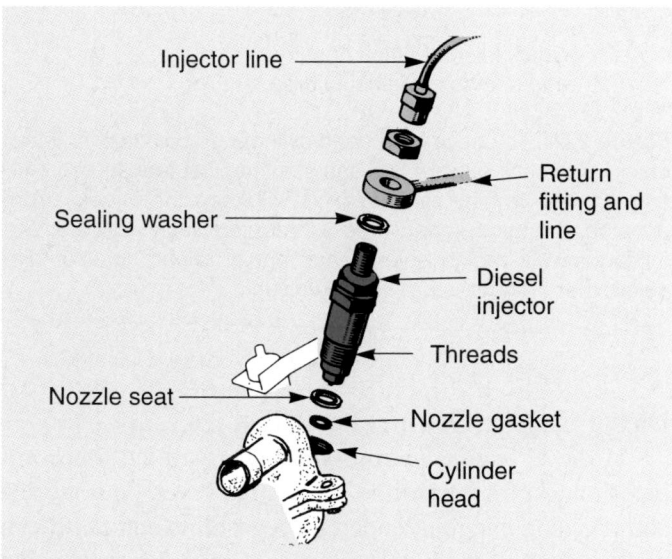

Figure 27-9. When removing an injector, note the position of all parts. This injector screws into the cylinder head. Some are clamped and press-fit into the head. (Toyota)

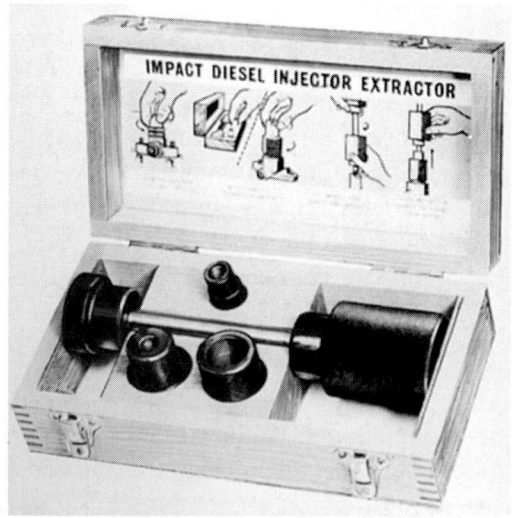

Figure 27-10. An impact or slide hammer puller may be needed to remove press-fit injectors. (Hartridge)

Pop Testing Diesel Injector Nozzles

A *pop tester* is a device for checking a diesel injection nozzle while it is out of the engine. One type of pop tester is pictured in **Figure 27-11.**

To use a pop tester:
1. Fill the tester reservoir with the correct calibration fluid (test liquid). Do not use diesel fuel because it is too flammable and test results may not be reliable.
2. Open the tester valve and pump the tester handle.
3. As soon as solid fluid (no air bubbles) sprays out of the tester, close the valve.
4. Connect the injection nozzle to the tester as shown in **Figure 27-12.**
5. To check injector opening pressure, purge any remaining air from the nozzle by pumping the tester lever up and down. Then, pump the handle slowly while watching the pressure gauge.
6. Note the pressure reading when the injection nozzle opens.
7. Repeat the test until you are sure you have an accurate reading.

Warning

Extremely high pressures are developed when pop testing a diesel injector nozzle. Wear eye protection and keep your hands away from the fuel spraying out of the nozzle. It can puncture the skin.

Typical diesel injector opening pressure is approximately 1700–2200 psi (12,000–15,000 kPa). If opening pressure is not within the service manual specifications, rebuild or replace the injector.

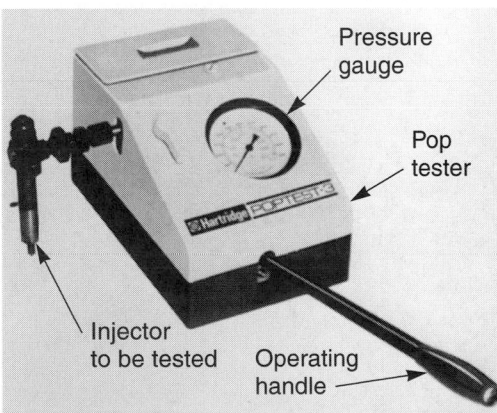

Figure 27-11. This pop tester checks diesel injector operation. It will test the injector spray pattern, opening pressure, and leakage. (Hartridge)

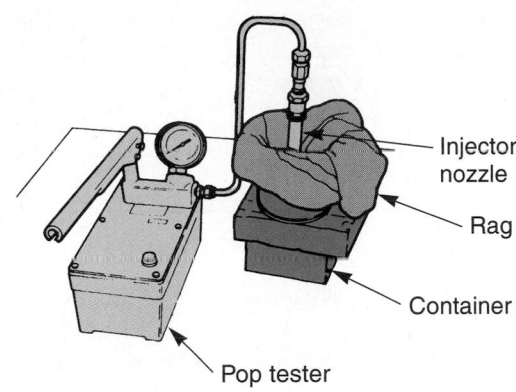

Figure 27-12. To use a pop tester, mount the injector on tester. Point the injector into an approved catch container. Surround the container with rags to prevent splashing. Pump the tester handle while observing the spray pattern, opening pressure, and leakage. Follow directions for the specific tester. (Ford)

To check the injector's *spray pattern,* operate the pop tester handle while watching the fluid spray out of the injector nozzle. As shown in **Figure 27-13,** there should be a narrow, cone-shaped mist of fluid. A solid stream of fuel, an uneven spray, an excessively wide spray, or a spray filled with liquid droplets indicates that the injector requires service or replacement.

To check *injector leakage,* slowly operate the tester handle to maintain a pressure that is lower than the nozzle opening pressure. Many auto manufacturers recommend a pressure about 300 psi (2000 kPa) below opening pressure. With this pressure, the diesel injector should not leak or drip for 10 seconds. Leakage would point to a dirty injector nozzle or worn components.

Tech Tip

Some diesel injectors make a chattering sound during operation; others do not. All nozzles, however, should make a swishing or pinging sound when spraying fuel.

Diesel Injector Nozzle Rebuild

An injector nozzle rebuild involves disassembling, cleaning, inspecting, replacing bad parts, reassembling, and testing the injector. Since injector designs vary, always refer to the service manual for detailed instructions. It will give specific assembly methods, torque values, and other critical information.

To disassemble the injector, unscrew the body using a six-point wrench or socket. As illustrated in **Figure 27-14,** remove and inspect the injector needle and nozzle opening.

Clean the nozzle parts in solvent. Special cleaning tools (brass scrapers, brass brushes, etc.) may be needed

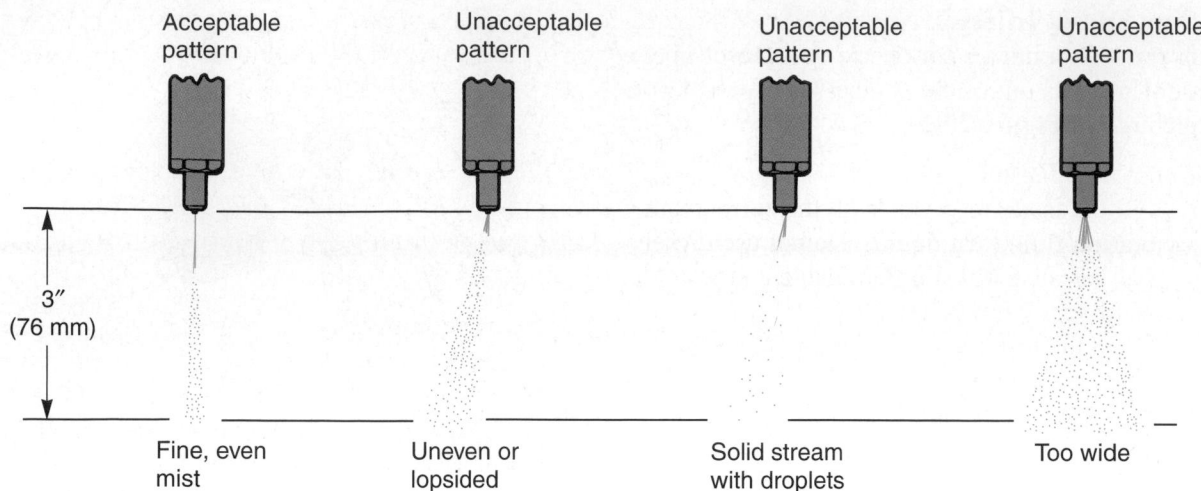

Figure 27-13. Typical diesel injector spray patterns. An acceptable spray produces uniform mist of fuel. An unacceptable spray is uneven and does not have proper mist density. (Chrysler)

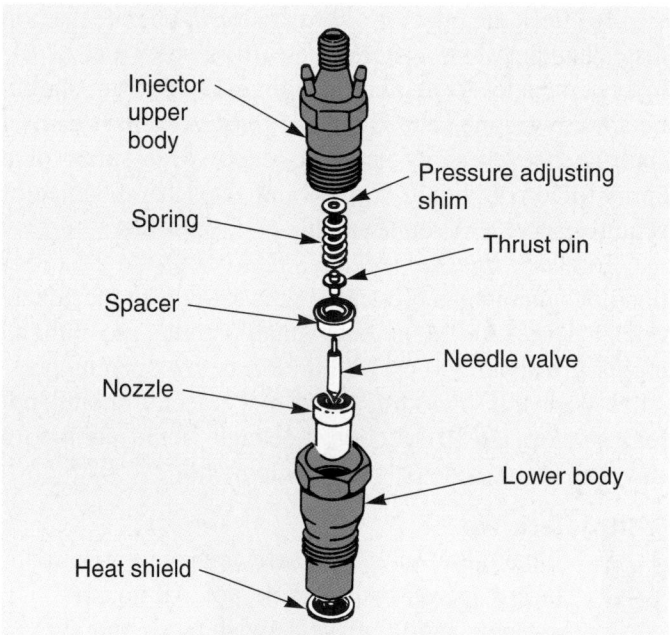

Figure 27-14. When rebuilding a diesel injector, be careful not to damage it. Keep all parts together. Clamp the injector lightly in a vise to unscrew the body. Inspect each part closely for carbon buildup, wear, or damage. (VW)

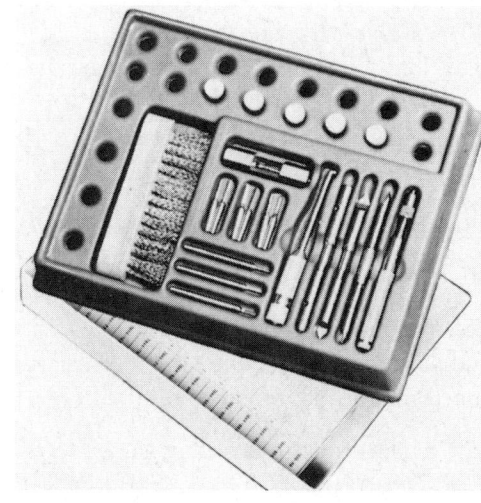

Figure 27-15. Special tool kit for servicing diesel injectors. It contains a soft brush, gauges, and other devices for rebuilding injectors. (Hartridge)

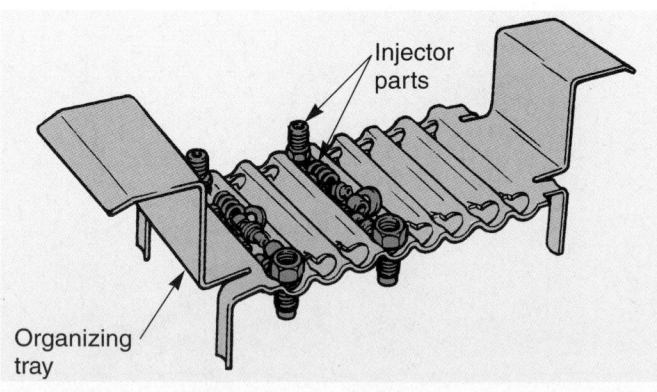

Figure 27-16. An organizing tray is handy when servicing injectors. It will help prevent a mix-up of parts. If even one part is installed in wrong injector, it may not operate properly. (Buick)

to remove hard carbon deposits, **Figure 27-15.** Use extreme care not to scratch the needle and nozzle. The smallest scratch can upset injector operation. Replace any component that shows signs of wear or damage.

If more than one injector is disassembled, make sure you do not mix parts. Use an organizing tray to keep the parts from each injector separate, **Figure 27-16.** The components in each injector may have been select-fit at the factory and must not be interchanged.

An *injector shim* is frequently used to adjust spring tension and valve-opening pressure. If the pop test shows a low opening pressure, install a thicker injector shim. This increases spring tension and raises the opening pressure. The manufacturer will give details on how shim thickness affects pressure.

Reassemble the injector as described in the service manual. Lubricate all parts with diesel fuel. Make sure all parts are positioned properly. Torque the injector body to specifications.

Installing Diesel Injectors

When installing a diesel injector, coat the threads with antiseize compound. Use a new heat shield or seal to prevent compression leakage. Screw the injector into the cylinder head by hand. Then, torque it to the recommended specification. Look at **Figure 27-17.** Reconnect the injector line without bending it. Tighten the connection properly.

Reconnect the wire on the fuel shutoff solenoid and start the engine. Check the system for leaks and proper operation. A few systems may require you to bleed air out of the system. Refer to a manual for exact details.

 Warning
Always double-check that all fittings have been torqued before starting the engine.

Glow Plug Service

Inoperative glow plugs will make a diesel engine hard to start when cold. There will not be enough heat from compression alone to ignite the fuel. If only one or two glow plugs are inoperative, the engine may miss while the engine is cold.

If glow plug problems are indicated, use a test light to check for voltage to the glow plugs. Touch the test light on the feed wires to the glow plugs. A clip-on ammeter can also be used to check current to the glow plugs. An incorrect current to the glow plugs can be caused by the supply circuit or by the glow plugs themselves.

 Caution
Refer to a service manual when testing a glow plug circuit. It is possible to damage some glow plugs (6-volt type) by connecting them to direct battery voltage.

An ohmmeter can be used to determine the condition of the glow plugs. Connect the ohmmeter between each glow plug terminal and ground. The resistance should be within specifications. If the resistance is too high or low, replace the glow plug. See **Figure 27-18.**

Glow Plug Replacement

To remove a glow plug disconnect the wires going to it. Then, use a deep-well socket and ratchet to unscrew the glow plug.

 **Warning**
A glow plug can reach temperatures above 1000°F (550°C). This can cause serious burns to your hand. Use extreme caution when removing glow plugs from the engine.

When the glow plugs are removed, inspect each one closely. Look for damage or a heavy coating of carbon. A carbon buildup can insulate the plug, making the engine hard to start. Clean any glow plugs that are to be reused. Replace all plugs that are faulty.

When installing the glow plugs, coat their threads with antiseize compound. Start the plugs by hand and then tighten them to specifications. Overtightening can easily damage a glow plug.

Some glow plugs operate on 12 volts, and some operate on 6 volts. Make sure you have the type recommended for the vehicle. Reconnect the wires and check glow plug operation.

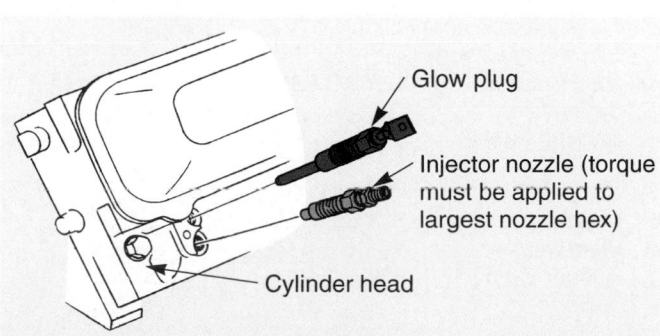

Figure 27-17. When installing injectors or glow plugs, coat the threads with antiseize compound. Torque to specifications. (Oldsmobile)

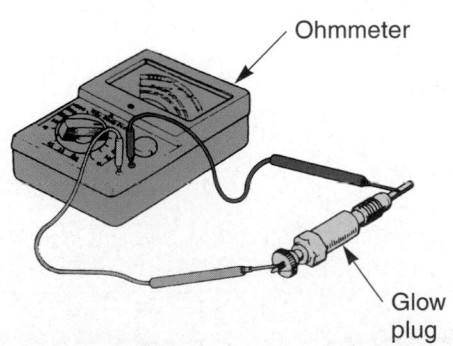

Figure 27-18. Glow plug condition can be checked with a common ohmmeter. Plug resistance must be within specifications. (Toyota)

Injection Pump Service

A bad injection pump can prevent an engine from running and cause engine missing, smoking, and other problems. The injection pump is usually a very trouble-free unit. However, water contamination, prolonged service, leaking seals, and physical damage may require replacement, repairs, or adjustment.

Most shops install a new or factory-rebuilt diesel injection pump when internal parts are faulty. The injection pump is a very precise mechanism. Specialized tools and equipment are needed for repair work.

Figure 27-19 shows an injection pump test stand, which is used to check pump performance. Its use is normally limited to specialized shops, not automotive shops.

It is sometimes possible to replace leaking external gaskets and seals without major pump disassembly. On most pumps, other external parts (idle stop solenoid, vacuum valve, cold advance solenoid, fuel shut-off solenoid) can also be replaced in the shop. Refer to your service manual for details. See **Figure 27-20.**

To replace a diesel injection pump:
1. Crank the engine until the No. 1 piston is at TDC.
2. Study the timing marks on the pump and engine. This will make reinstallation easier.
3. Disconnect the battery cable to prevent accidental cranking and fuel injection.
4. Remove the injection lines, wires, linkages, and fasteners to allow pump removal, **Figure 27-21.**

5. Cap all lines to keep out dust and dirt.
6. Carefully lift off the old pump.
7. Transfer parts (solenoids, vacuum valves, drive sprocket, etc.) from the old pump to the new pump. See **Figure 27-22.**
8. Install the new pump on the engine.
9. Align the timing marks noted during disassembly.
10. Torque all fasteners and lines properly.
11. Before starting the engine, adjust injection timing as described in the next section.

Caution
Never hammer or pry on an injection pump housing. Also, be careful not to drop an injection pump. It can be damaged easily and is very expensive to replace.

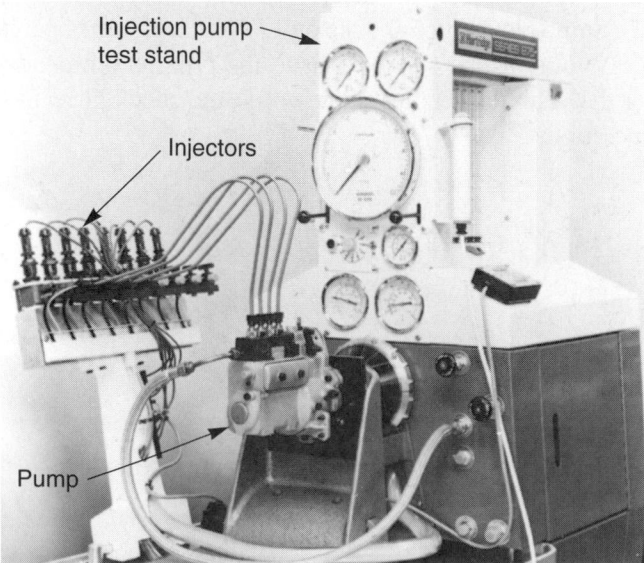

Figure 27-19. An injection pump test stand is needed to check the operation of an injection pump. It is found in shops specializing in injection pump service. (Hartridge)

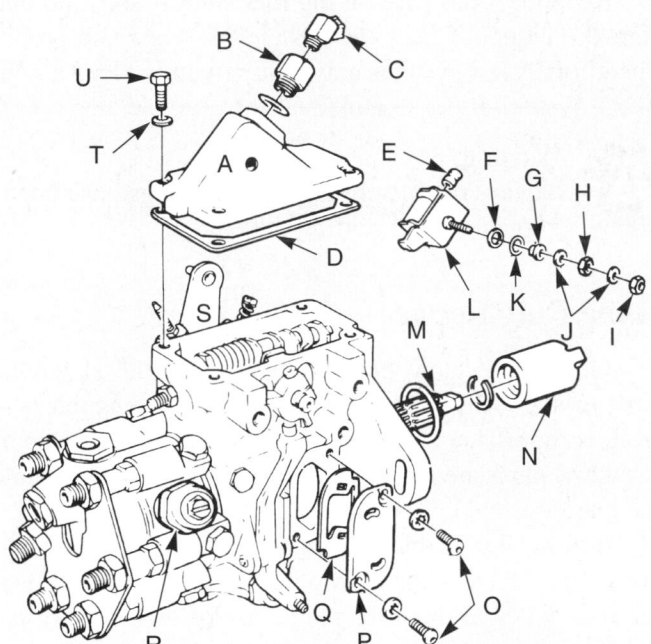

A. Governor control cover
B. Black leakage connection
C. Elbow connection
D. Joint (gasket)
E. Spring filter
F. Backing washer
G. Insulating washer
H. Nut
I. Locknut
J. Plain washer
K. Rubber O-ring
L. Cold advance solenoid
M. Rubber O-ring
N. Tang drive hub
O. Screw—dome headed
P. Inspection cover
Q. Rubber sealing joint (gasket)
R. Stop solenoid
S. Throttle lever
T. Shakeproof washer
U. Hexagon socket screw

Figure 27-20. External repairs can sometimes be made on injection pumps. Leaking seals or gaskets, bad fittings, and bad solenoids can be replaced without pump teardown. Internal repairs cannot be done in an average garage. (Buick)

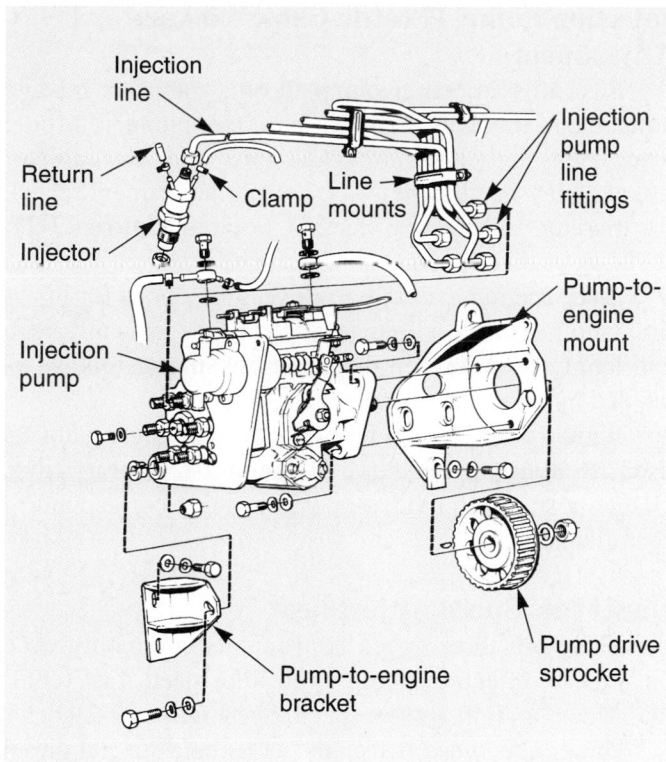

Figure 27-21. Note the parts that must be disconnected to remove injection pump. When installing, torque all fasteners and fittings to specifications. Make sure dirt does not enter system. (Volvo)

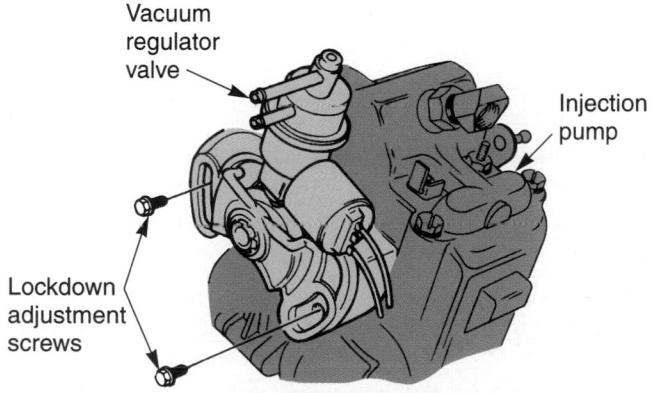

Figure 27-22. Normally, the vacuum regulator valve must be reused when installing a new or rebuilt injection pump. Adjustment involves a check of vacuum action while rotating the valve on its mount. (Oldsmobile)

Injection Pump Timing

Diesel injection pump timing is adjusted by rotating the injection pump on its mounting. To advance injection pump timing, turn the pump opposite the direction of pump drive shaft rotation. To retard injection pump timing, turn the pump housing in the same direction as the drive shaft rotation.

Injection pump timing must be set whenever the pump is removed from the engine or when an incorrect adjustment is discovered. There are several methods for adjusting injection pump timing. Procedures vary with the particular type of engine and injection system. A diesel does not have an electrically operated ignition system for triggering a timing light. For this reason, other methods are needed to determine when the No. 1 cylinder fires.

Injection pump timing can be adjusted by several methods:

- Align timing marks on the engine and injection pump with the No. 1 piston at TDC (rough adjustment), **Figure 27-23.**

- Use a dial indicator to measure injection pump stroke in relation to engine piston position. See **Figure 27-24.**

- Use a luminosity (light) device to detect combustion flame in the No. 1 cylinder. Look at **Figure 27-25.**

- Use a fuel pressure (injection) detector to detect the minute flex and ping that occurs as high fuel pressure forces the diesel injection nozzle open.

- Use a scope tool on the crankshaft position sensor and a transducer pickup tool around the No. 1 injection line to see a graphical representation of injection timing. This setup is shown in **Figure 27-26.**

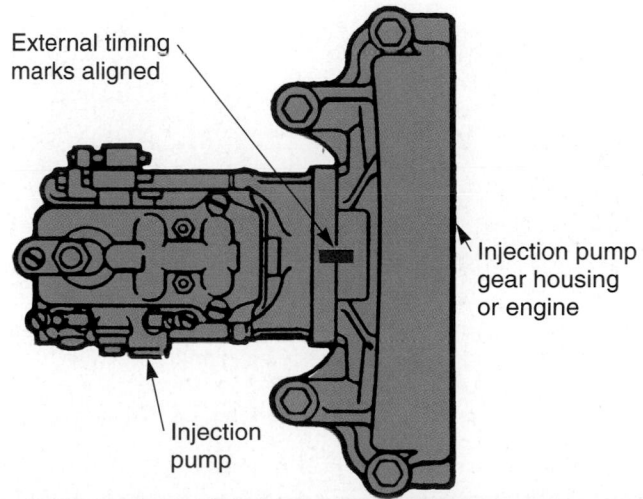

Figure 27-23. A simple method of setting injection timing involves aligning marks on pump and engine. When the two marks align, the pump should be timed properly. (Ford)

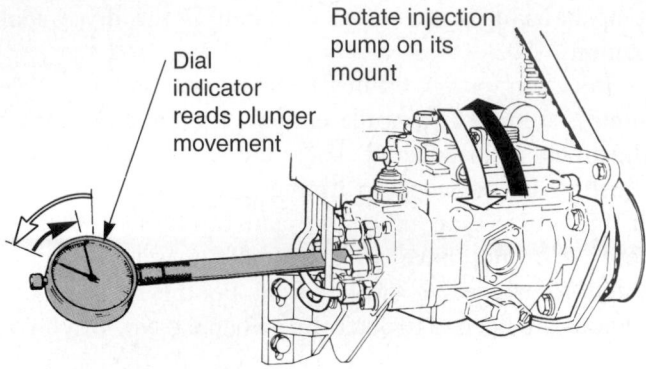

Figure 27-24. Some automakers recommend using a dial indicator when setting injection timing. After removing a plug in the end of the pump, the dial indicator stem is positioned against the plunger. The indicator will then register plunger movement. To change injection timing, rotate the injection pump on the engine. Tighten pump mounting fasteners to lock timing in place. (Volvo)

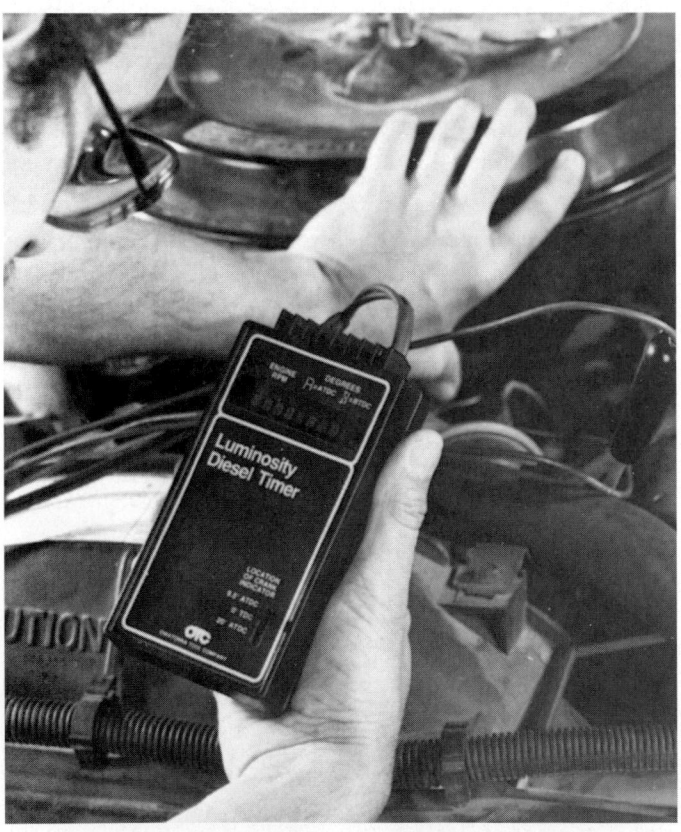

Figure 27-25. A mechanic using a luminosity meter to detect combustion timing for adjustment of injection timing. (OTC Tools)

Refer to a factory shop manual for details. The manual will describe which adjustment method should be used. Also, follow the operating instructions for the specific type of timing device. Injection timing is very critical to diesel engine performance.

Injection Pump Throttle Cable/Linkage Adjustment

Basically, injection pump throttle cable or linkage adjustment involves making sure the pump's throttle lever opens and closes freely. As parts wear, the cable or linkage may need to be reset. During adjustment, check for binding. Lubricate if needed. Look at **Figure 27-27.** It shows basic adjustments.

A service manual will give exact directions for injection pump cable or linkage adjustment. Typically, you can lengthen or shorten the cable housing or linkage as needed by simply loosening a locknut.

Figure 27-28 shows how a carburetor angle gauge is used to adjust a transmission vacuum regulator valve. Gauge position and valve operation must correspond to specifications.

Diesel Idle Speed Adjustment

There are three speed adjustments commonly used on a diesel injection pump: curb idle speed, fast (cold) idle speed, and maximum speed. See **Figure 27-29.**

Diesel *idle speed* is usually set using a special diesel tachometer. An idle stop screw on the injection pump is turned to vary engine speed.

A diesel tachometer may be similar to or part of the injection timing device. It may also be an instrument like the one shown in **Figure 27-30.** With all designs, the diesel tachometer will read engine speed in rpm (revolutions per minute).

Diesel Curb Idle Speed

To adjust diesel **curb idle speed,** place the transmission in neutral or park. Start the engine and allow it to run until it reaches full operating temperature. Then, connect the tachometer to the engine.

Compare the readings with the specifications. Curb idle speed will be given on the engine compartment emission control decal or in the service manual.

If needed, turn the curb idle speed adjusting screw on the injection pump to raise or lower idle rpm. Usually, you must loosen a locknut before the screw will turn. Hold the screw and retighten the locknut when the tachometer reads correctly.

Diesel Cold Idle Speed Adjustment

To set diesel **cold idle speed**, connect a jumper wire from the battery positive terminal to the fast idle solenoid terminal. This will activate the fast idle, even when the engine is warm.

Raise the engine speed momentarily to release the solenoid plunger. Again, compare your tachometer readings to specifications. Adjust the solenoid if needed. Refer to **Figure 27-31.**

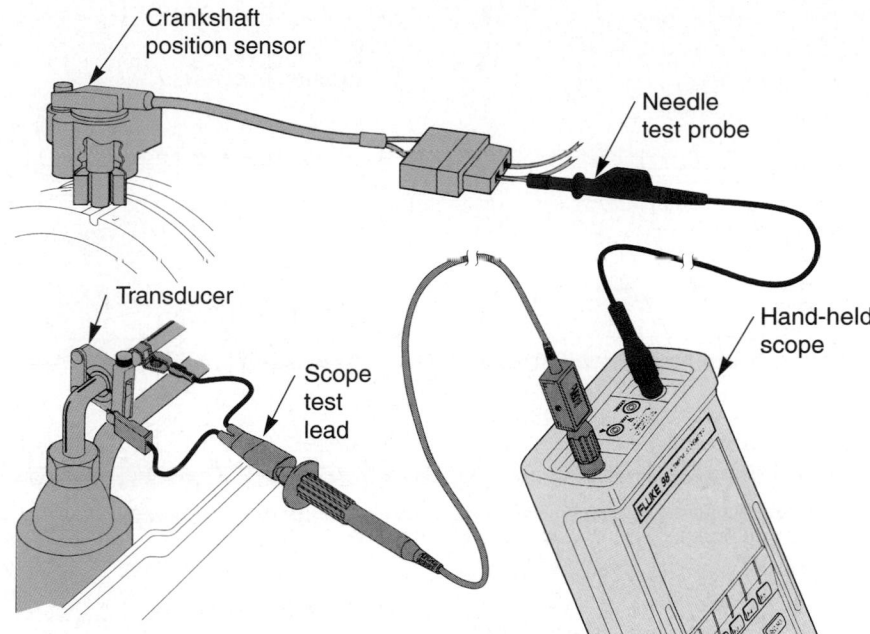

A—Note the connections for injection timing measurement if the engine has a crankshaft position sensor. A transducer pickup clamps around the injection line to detect fuel pressure and flow increase from the high-pressure pump.

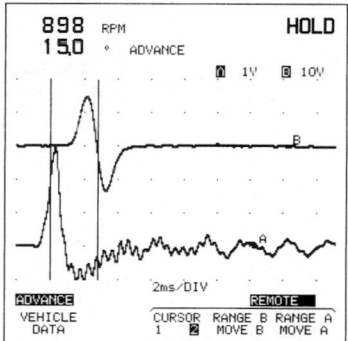

B—The scope traces and digital readouts show the alignment and amplitude at low engine idle speed.

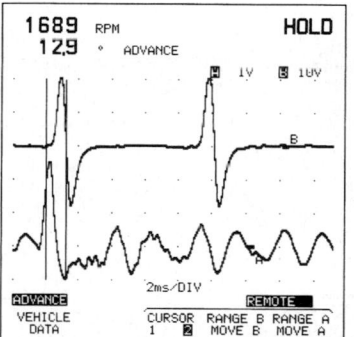

C—With engine speed increased to 1700 rpm, the scope shows advance of injection timing. Injection occurs sooner, or to left of the scope screen. This gives diesel fuel enough time to burn at higher speeds.

Figure 27-26. If the diesel has crankshaft position sensor, connect a scope to the sensor and a transducer pickup tool around the No. 1 injection line. The tool will then allow you to observe advance-retard (trace movement from side to side) electronically. Rapid, jerky movement of the waveform could mean worn injection pump parts. You can also find weak sensor signal. This scope also gives numerical readout of injection advance. (Fluke)

Sometimes a cold start lever replaces a cold idle solenoid. Refer to manufacturer manuals for details.

Diesel Maximum Speed Adjustment

A diesel ***maximum speed adjustment*** is used to limit the highest attainable engine rpm. If maximum governor rpm is too high, it may damage internal engine compo-nents. If maximum rpm is too low, the engine will not produce enough power.

To adjust the maximum speed of a diesel, position your tachometer so that it can be read from the driver's compartment. With the transmission in neutral or park, hold your left foot on the brake. Press the accelerator pedal slowly to the floor.

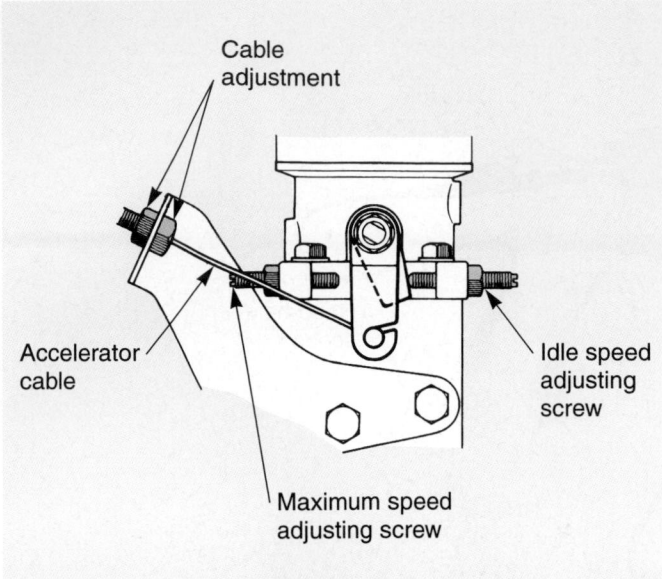

Figure 27-27. Note basic injection pump adjustments: throttle cable, idle speed, and maximum speed. (Toyota)

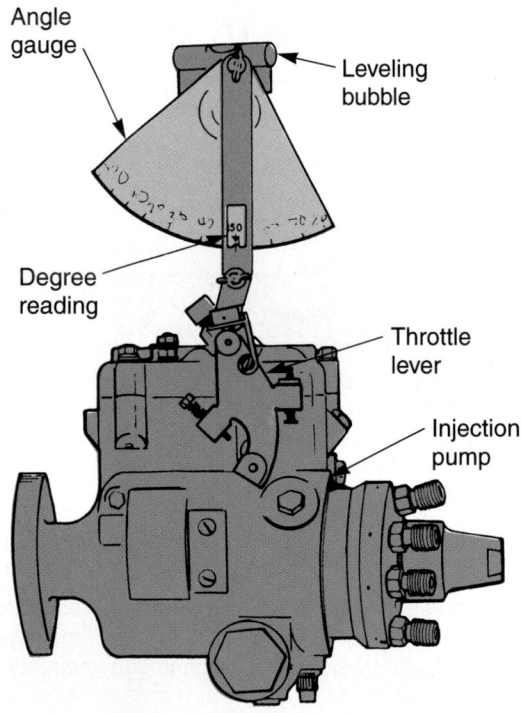

Figure 27-28. A carburetor angle gauge is needed to set vacuum valve on this particular injection pump. Follow service manual directions for details. (GM)

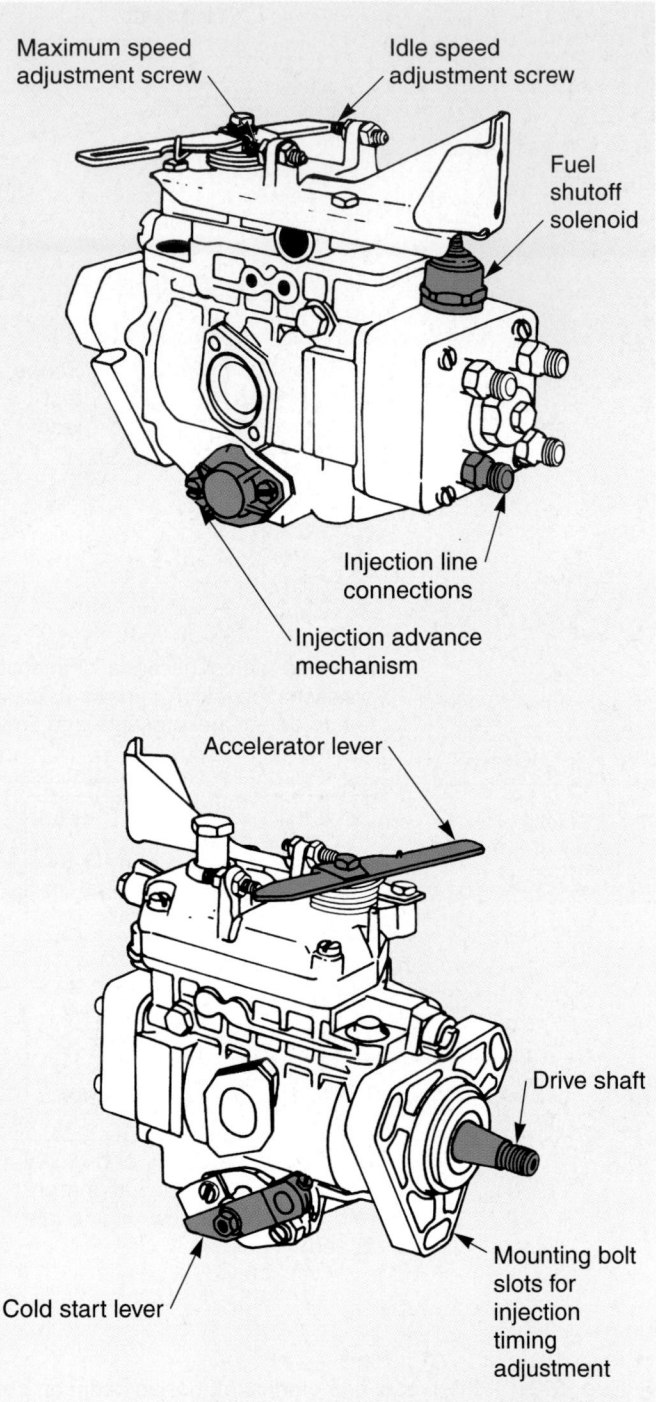

Figure 27-29. Two views of this injection pump show external adjustments and other parts. (VW)

Figure 27-32. Tighten the locknut on the screw after adjustment. Recheck maximum engine speed.

 Caution
When performing a maximum speed test, be ready to release the accelerator pedal at any time. If the engine rpm goes above specifications, engine damage could result.

Once the tachometer reads the maximum speed specification, engine speed should no longer increase. If the maximum speed is not within specifications, turn the maximum speed adjusting screw on the injection pump,

Figure 27-30. A diesel engine does not have electrical ignition system. Therefore, a special tachometer is needed. It senses crankshaft damper position, rather than gasoline engine ignition system operation. (Kent-Moore Corp.)

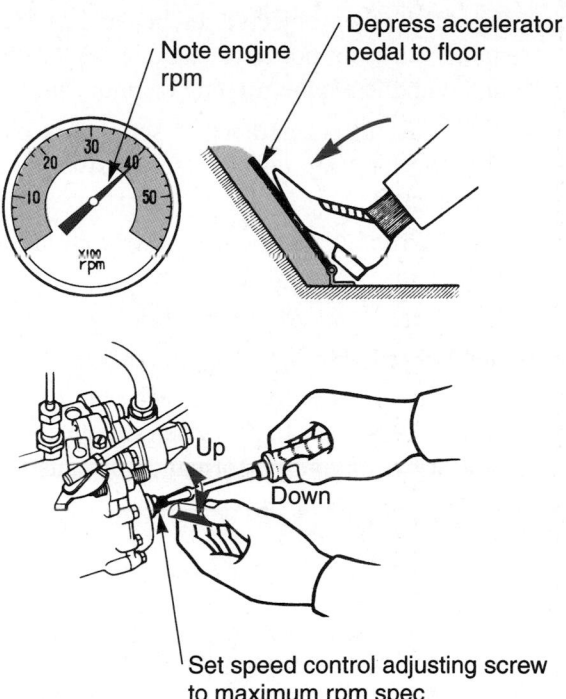

Figure 27-32. To set diesel engine maximum speed, press the gas pedal to the floor while watching the tachometer. If maximum rpm is too high or too low, turn the maximum speed adjusting screw on the pump. Be careful not to let engine speed go above the maximum specified. (Toyota)

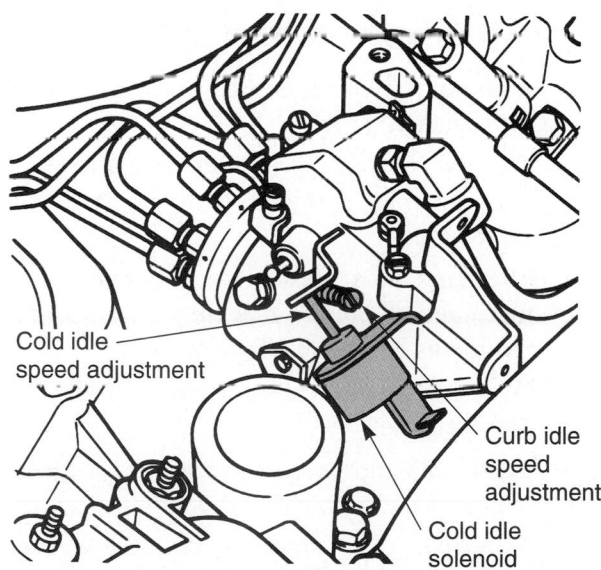

Figure 27-31. Cold idle speed solenoid is adjusted like a carburetor cold idle speed solenoid. Note the location of the curb idle speed adjustment on injection pump. (Ford)

Computerized Diesel Injection Service

Diagnosing and repairing a diesel system that is controlled by a computer is similar to working on other computer-controlled systems. Some systems have self-diagnostics, which will help indicate the location of prob-

lems. You can use a digital meter to check for opens and shorts in wiring and to test the sensors and actuators.

For more information on servicing computer systems, refer Chapters 17-19.

Diesel Injection Service Rules

When servicing a diesel injection system, remember these basic rules:

- Always cap lines and fittings to prevent the entry of foreign matter.
- Never pry on or drop a diesel injector or injection pump. They can be damaged.
- Remember that high pressure inside a diesel injection system can cause serious injury.
- Clean around fittings before they are disconnected.
- Adhere to all torque specifications. This is extremely important on a diesel engine.
- Never use a bent, frayed, or kinked injection line.
- Some diesel injection systems must be bled (air removed) after repairs.
- Place a piece of screen mesh over the air inlet when the engine is to be operated without the air cleaner.

Rags and other objects can be sucked into the engine. Also, do not cover the air inlet with your hand with the engine running or injury may result.

- Check fuel filters and water separators periodically. Water can corrode expensive injection system parts.
- Follow all safety rules concerning ventilation and fire hazards.
- Wear safety glasses when working on a diesel injection system.
- When in doubt, refer to a service manual for vehicle being serviced. The slightest mistake could upset engine performance or cause engine damage.

Summary

- Refer to a service manual for details on periodic maintenance of a diesel injection system.
- Diesel injection diagnosis requires you to use your knowledge of engine and injection system operation, as well as your basic troubleshooting skills.
- Excessive diesel exhaust smoke is normally due to incomplete combustion caused by injection system problems or engine troubles.
- Ignition lag is the time span between the injection of the diesel fuel and the ignition of the fuel.
- A diesel engine miss results from one or more cylinders not firing properly.
- Never attempt to remove an injection system component when the engine is running. With 6000-8000 psi (42,000 to 56,000 kPa) fuel pressure, fuel could squirt out and puncture your skin. Blood poisoning or death could result.
- A diesel cylinder balance test involves disabling one injector at a time to check the firing of the cylinders.
- A glow plug resistance-balance test provides a safe way to find out if each cylinder is firing.
- A bad diesel injector usually causes the engine to miss. It may also reduce engine power or cause smoking and knocking.
- Injector substitution is an easy way to verify an injector problem.
- A pop tester is a device used to check injector spray pattern, opening pressure, and leakage.
- Inoperative glow plugs will make a cold diesel engine hard to start.

- A bad injection pump can keep the engine from running or cause engine missing, smoking, and other problems.
- Diesel injection pump timing is adjusted by rotating the injection pump on its mounting.
- Diesel idle speed is usually set while using a special diesel tachometer. An idle stop screw on the injection pump is turned to vary engine speed.

Important Terms

Black smoke
White smoke
Blue smoke
Smoke meter
Ignition lag
Miss
No start condition
Cylinder balance test
Compression test
Glow plug resistance-balance test
Digital pyrometer
Injection-pressure tester
Injector substitution
Pop tester
Spray pattern
Injector shim
Idle speed
Curb idle speed
Cold idle speed
Maximum speed adjustment

Review Questions—Chapter 27

Please do not write in this text. Place your answers on a separate sheet of paper.

1. The main fuel filter will often have a water drain that must be serviced periodically. True or False?
2. The parts in a diesel injection pump can be machined so precise that they are measured in millionths of a meter. True or False?
3. What is the main cause of black smoke from a diesel engine?
4. Explain the most common reasons for excessive white smoke from a diesel engine.
5. Excessive _____ smoke from a diesel may be due to oil consumption from worn _____ _____, scored _____ _____, or leaking _____ _____.
6. A smoke meter is for measuring the amount of smoke in diesel exhaust. True or False?
7. Which of the following is *not* a normal cause of excessive diesel knock?
   (A) Long ignition lag time.
   (B) Low engine-operating temperature.
   (C) Early injection timing.
   (D) All of the above are normal causes of excessive knock.

8. Explain a diesel cylinder balance test.

9. A compression gauge reading up to 250 psi (1,723 kPa) can be used to measure compression pressure in a diesel engine. True or False?

10. What are typical compression test readings with a diesel engine in good condition?

11. Why do many service manuals warn against doing a wet compression test on a diesel?

12. A glow plug _____-_____ test relies on the fact that glow plug electrical resistance increases with temperature.

13. How can a digital pyrometer be used to check the operation of a diesel engine?

14. What four values do some on-car injection-pressure testers allow you to measure?

15. Injector nozzles are exposed to combustion and can become clogged with carbon. True or False?

16. Which of the following should *not* be done when removing a diesel injector?
    (A) Disconnect battery.
    (B) Bend injection line out of the way.
    (C) Use impact tool if needed.
    (D) All of the above are correct.

17. A diesel engine has a dead miss (one cylinder not firing). Engine analyzer cranking tests indicate good engine compression. Technician A says all of the injectors should be removed and rebuilt. They must be at fault. Technician B says you should perform a cylinder balance test or on-car pressure tests. Who is correct?
    (A) A only.
    (B) B only.
    (C) Both A and B.
    (D) Neither A nor B.

18. Explain the use of a pop tester.

19. Injector nozzle parts can be interchanged. True or False?

20. Inoperative _____ _____ will make a diesel engine hard to start when cold.

21. Which of the following can be used to test diesel glow plug system operation?
    (A) Test light.
    (B) Ammeter.
    (C) Ohmmeter.
    (D) All of the above are correct.

22. Glow plugs can reach temperatures above _____ °F or _____ °C.

23. Tests show that a diesel injection pump has an internal problem. Technician A says the pump should be disassembled and repaired. Technician B says the pump should be replaced with a new or factory-rebuilt unit. Who is correct?
    (A) A only.
    (B) B only.
    (C) Both A and B.
    (D) Neither A nor B.

24. What is an injection pump test stand?

25. Explain four methods of adjusting diesel injection timing.

⚙ ASE-Type Questions

1. This is sometimes located in the injector assemblies:
   (A) *Sock filter.*
   (B) *Main filter.*
   (C) *Final filter screens.*
   (D) *All of the above.*

2. Diesel injection parts are machined to precision tolerances measured in:
   (A) *Inches.*
   (B) *Microns.*
   (C) *Millimeters.*
   (D) *Centimeters.*

3. Which smoke color indicates engine oil is entering the combustion chamber?
   (A) *Blue.*
   (B) *Black.*
   (C) *White.*
   (D) *All of the above.*

4. Abnormally loud diesel knock can be caused by each of these *except:*
   (A) *Oil consumption.*
   (B) *Rich fuel mixture.*
   (C) *Fuel contamination.*
   (D) *Early injection timing.*

5. This is *not* a cause of diesel engine miss:
   (A) *Air leak.*
   (B) *Faulty injectors.*
   (C) *Clogged fuel filters.*
   (D) *Shorted fuel cutoff solenoid.*

6. Which of these tests involves disabling one injector at a time?
   (A) *Cylinder balance test.*
   (B) *Injection-pressure test.*
   (C) *Diesel compression test.*
   (D) *Digital pyrometer balance test.*

7. Compression pressure for an automotive diesel engine is:
   (A) 50–75 psi.
   (B) 100–200 psi.
   (C) 300–375 psi.
   (D) 400–500 psi.

8. When checking diesel injection nozzles with a pop tester, Technician A believes the tester reservoir should be filled with the correct diesel oil. Technician B feels the tester reservoir should be filled with the correct calibration fluid. Who is right?
   (A) A only.
   (B) B only.
   (C) Both A and B.
   (D) Neither A nor B.

9. When replacing an injection pump, these need to be transferred from the old pump to the new pump:
   (A) Solenoids.
   (B) Vacuum valves.
   (C) Drive sprockets.
   (D) All of the above.

10. Injection pump timing can be adjusted by using any of these *except:*
    (A) light device.
    (B) dial indicator.
    (C) digital pyrometer.
    (D) fuel pressure detector.

Activities—Chapter 27

1. Prepare an overhead transparency on the steps for troubleshooting the injectors in a vehicle that has abnormal exhaust smoke. Use it to demonstrate the steps to your class.

2. Perform a cylinder balance test on the engine's injectors.

3. Using an ohmmeter, demonstrate the proper method for checking the condition of a glow plug.

Diesel Engine Diagnosis		
Condition	**Possible Cause**	**Correction**
Low power output.	1. Low engine rpm. 2. Restriction in air inlet system. 3. Water in fuel. 4. Fuel with low specific gravity. 5. Low fuel pressure. 6. Fuel rack setting incorrect. 7. Fuel too hot due to blocked constant bleed valve.	1. Adjust governor linkage to rest against the high idle stop when accelerator pedal is depressed. Adjust high idle rpm with adjustment screw. If proper setting cannot be made, disassemble, inspect, and service governor. 2. Inspect air cleaner element and ductwork for damage and excessive bends and turns. Install new air cleaner element. 3. Test for fuel in water. Drain fuel tank as needed to remove water. Install new fuel filters and fill tanks with clean fuel. 4. Test fuel API gravity. Replace fuel if API gravity is greater than 38. 5. Check fuel pressure. 6. Check rack setting and adjust if needed. 7. Check by removing fuel return line and checking for sufficient fuel flow from valve. Replace valve if blocked.
Misfiring and rough running.	1. Air in fuel system. 2. Fuel injection timing incorrect. 3. Automatic timing advance malfunction. 4. Defective fuel nozzles. 5. Incorrect camshaft timing. 6. Fuel leakage at nozzle nut or adapter.	1. Bleed air from lines. 2. Check and make necessary adjustments. 3. Check for correct timing. Replace timing mechanism if defective. 4. Locate misfiring injector; clean and service as required. 5. Check timing. 6. Tighten nut to specifications.
Black or gray exhaust smoke (engine runs smooth).	1. High altitude operation at 2500 ft. (762 m) or greater. 2. Dirty air cleaner. 3. Air inlet system restriction. 4. Fuel injection timing incorrect. 5. Fuel rack setting incorrect. 6. Fuel with low specific gravity. 7. Leaking fuel nozzles.	1. Adjust fuel rack position. 2. Replace dirty air cleaner element. 3. Inspect ductwork for damage and/or excessive turns and bends. Adjust or repair as needed. 4. Check and make necessary adjustments. 5. Check rack setting and adjust to specifications as needed. 6. Test fuel API gravity. 7. Locate misfiring injector and service as required.
Black or gray exhaust smoke (engine runs rough)	1. Automatic timing advance malfunction. 2. Air in fuel system.	1. Check for correct timing. If light is not available, check for smooth acceleration from low to high idle. 2. Check for leaks and bleed air from system.
White exhaust smoke.	1. Cold outside temperatures. 2. Long idle periods. 3. Fuel with low specific gravity. 4. Air in fuel system. 5. Fuel injection timing incorrect. 6. Automatic timing advance malfunction. 7. Bad fuel nozzles.	1. Normal condition until engine warms to operating temperature. 2. Reduce unnecessary idling or use fuel heating system. 3. Test fuel API gravity. 4. Check for leaks and bleed air from system. 5. Check and make necessary adjustments. 6. Check for correct timing. 7. Locate misfiring injector. Clean and service as required.

Diesel Engine Diagnosis		
Condition	**Possible Cause**	**Correction**
Excessive fuel consumption.	1. Air inlet system restriction. 2. External fuel system leakage. 3. Fuel injection timing incorrect. 4. Leaking fuel nozzles. 5. Fuel injection pump calibration incorrect.	1. Inspect ductwork for damage and/or excessive turns and bends. 2. Check fuel system external piping and tubing for signs of leaks. Repair as needed. 3. Check and make necessary adjustments. 4. Locate leaking injector and service as required. 5. Remove injection pump and nozzle assemblies from engine. Check calibration and adjust.
Erratic engine speed.	1. Air leaks in fuel system. 2. Throttle linkage loose or out of adjustment. 3. Injection pump governor failure.	1. Check for air leaks and make needed repairs. 2. Check and adjust linkage. 3. Check injection pump for damaged or broken springs or other components. Check fuel rack for free travel. Check for correct governor spring. Install new parts as needed and recalibrate injection pump.
Engine stalls at low speeds.	1. Idle speed too low. 2. Fuel tank vent plugged. 3. Low fuel supply. 4. Defective fuel injection nozzle or pump.	1. Adjust idle as required. 2. Check vent arrangement and make needed repairs. 3. Check tank for fuel. Check fuel lines for sharp bends and restrictions. Check fuel pressure. If pressure is low, replace fuel filters. If pressure is still low, replace transfer pump. Bleed fuel system. 4. Inspect, test, and replace nozzle parts or injection pump.
Engine does not reach no-load governed speed.	1. Air in fuel system. 2. Accelerator linkage loose or misadjusted. 3. Restricted fuel lines/stuck overflow valve. 4. High idle adjustment set too low. 5. Fuel injection pump calibrated incorrectly. 6. Internal fuel pump governor wear.	1. Check for leaks and bleed air from lines. 2. Check linkage and adjust as needed. 3. Check for restrictions and for defective spring, poor valve setting, or sticking. Make all necessary repairs. 4. Check setting and adjust as needed. 5. Remove injection pump and nozzle assemblies from engine and recalibrate. 6. Remove injection pump from engine and make all necessary repairs.
Difficult starting (crankshaft turns).	1. Cold outside temperatures. 2. Air in fuel system. 3. Water in fuel. 4. Low fuel pressure. 5. Fuel injection timing incorrect. 6. Bad fuel nozzles. 7. No exhaust smoke visible while starting.	1. Use starting aids. 2. Bleed air from lines. 3. Test for water in fuel. Drain fuel tank as needed to remove water. Install new fuel filter and fill tanks with clean fuel. 4. Check fuel pressure at fuel pump housing. 5. Check and make necessary adjustments. 6. Locate misfiring injector and clean and service as required. 7. No fuel in tank. Tank valves inadvertently closed. No fuel from injection pump due to cold weather waxing or fuel line restriction. Faulty fuel injection pump shut-off solenoid. Repair as needed.

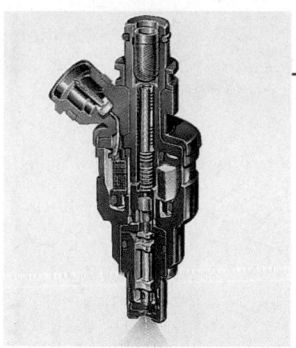

Chapter 28

Exhaust Systems, Turbochargers, and Superchargers

After studying this chapter, you will be able to:

- Describe the basic parts of an exhaust system.
- Compare exhaust system design differences.
- Perform exhaust system repairs.
- Explain the fundamental parts of a turbocharging system.
- Describe the construction and operation of a turbocharger and waste gate.
- Remove and replace a turbocharger and waste gate
- Summarize the construction and operation of a supercharging system.
- Demonstrate an understanding of safety procedures for working on exhaust systems, turbochargers, and superchargers.
- Correctly answer ASE certification test questions on exhaust system, turbocharger, and supercharger operation and service.

This chapter begins by covering the basic parts of an exhaust system. It then explains how to make common repairs by replacing rusted or damaged components. The second section of the chapter covers the theory and repair of turbocharging and supercharging systems.

Exhaust Systems

An *exhaust system* quiets engine operation and carries exhaust fumes to the rear of the vehicle. Typical parts, which are shown in **Figure 28-1,** include the following:

- *Exhaust manifold*—connects the cylinder head exhaust ports to header pipe.
- *Header pipe*—steel tubing that carries exhaust gases from the exhaust manifold to the catalytic converter or muffler.

- *Catalytic converter*—device that removes pollutants from engine exhaust.
- *Intermediate pipe*—tubing sometimes used between the header pipe and muffler or catalytic converter and muffler.
- *Muffler*—metal chamber for damping pressure pulsations to reduce exhaust noise.
- *Tailpipe*—tubing that carries exhaust from the muffler to the rear of car body.
- *Hangers*—devices for securing the exhaust system to the underside of the car body.
- *Heat shields*—metal plates that prevent exhaust heat from transferring into another object.
- *Exhaust system clamps*—U-bolts for connecting parts of the exhaust system together.

When an engine is running, extremely hot gases blow out of the cylinder head exhaust ports. The gases enter the exhaust manifold. They flow through the header pipe, catalytic converter, intermediate pipe, muffler, and tailpipe.

Many header pipes are connected to the exhaust manifold by a spring-loaded coupling. The coupling compresses and holds the "doughnut" seal between the exhaust manifold and header pipe. This design allows the engine to move on its motor mounts without moving the exhaust system and damaging the seal.

Exhaust Back Pressure

Exhaust back pressure is the pressure developed in the exhaust system when the engine is running. High back pressure reduces engine power. A well-designed exhaust system should have low back pressure.

The size of the exhaust pipes, catalytic converter, and muffler contributes to exhaust back pressure. Larger pipes and a free-flowing muffler, for example, would reduce back pressure.

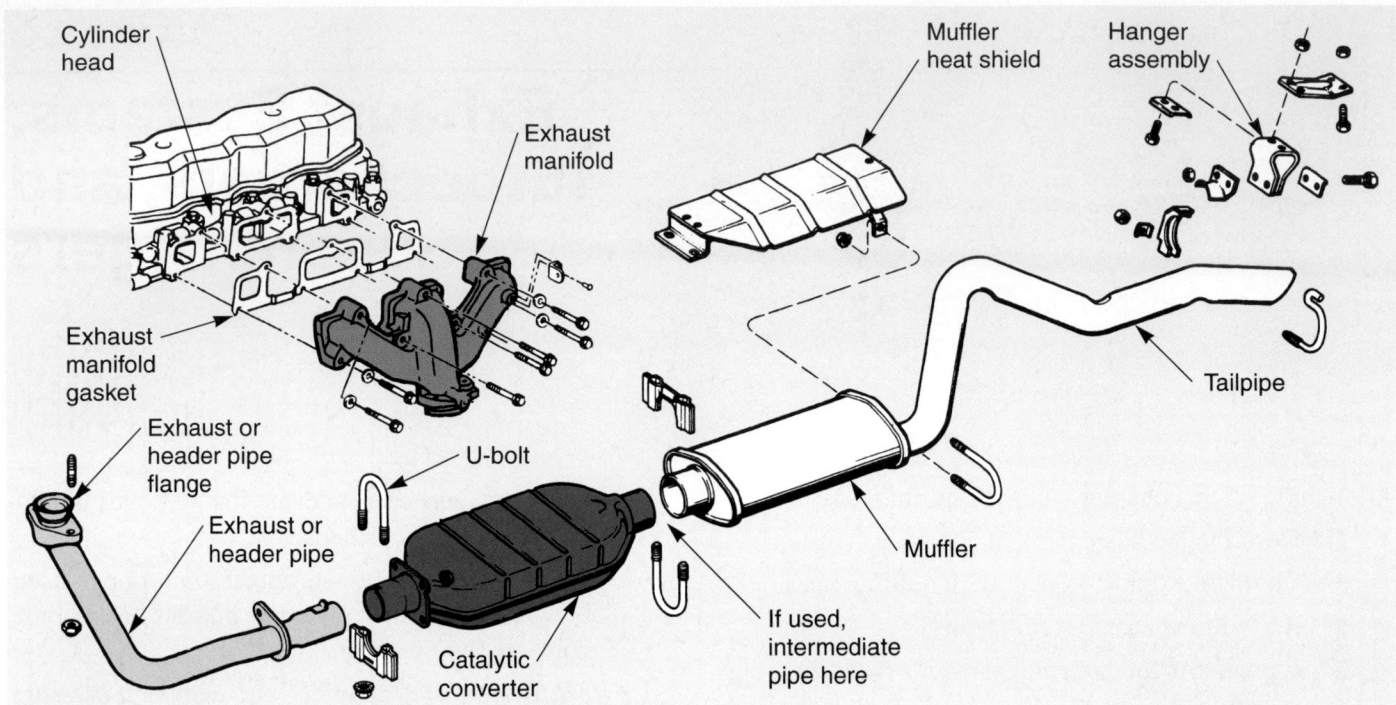

Figure 28-1. Note the parts of a typical exhaust system. Exhaust comes out of the cylinder head, into the manifold, and then through system. (Chrysler)

Single and Dual Exhaust Systems

A *single exhaust system* has one path for exhaust flow through the system. Typically, it has only one header pipe, a main catalytic converter, a muffler, and a tailpipe. The most common type of exhaust system, it is used from the smallest four-cylinder engines on up to large V-8 engines.

A *dual exhaust system* has two separate exhaust paths to reduce back pressure. It is essentially two single exhaust systems used on one engine. A dual exhaust system is sometimes used on high-performance cars with large V-6 or V-8 engines. It lets the engine "breath" better at high rpm.

A *crossover pipe* normally connects the right and left header pipes to equalize back pressure in a dual system. This also increases engine power slightly.

Exhaust Manifold

An exhaust manifold bolts to the cylinder head to enclose the exhaust port openings, **Figure 28-2.** The manifold is usually made of cast iron. High-flow, high-performance manifolds, which are commonly called headers, are sometimes made of stainless steel or light-weight steel tubing.

The cylinder head mating surface is machined smooth and flat. An exhaust manifold gasket is commonly used between the cylinder head and manifold to help prevent leakage. The outlet end of the exhaust manifold has a round opening with holes for stud bolts or cap screws. A gasket or an O-ring (*exhaust manifold doughnut*) seals the connection between the exhaust manifold outlet and header pipe to prevent leakage.

Exhaust Manifold Heat Valve

The *exhaust manifold heat valve,* also called the heat control valve or the heat riser, forces hot exhaust gas to flow into the intake manifold to aid cold weather starting. Look at **Figure 28-3.**

An exhaust manifold heat valve may be located in the outlet of the exhaust manifold. A heat-sensitive spring or a vacuum diaphragm and temperature sensing vacuum switch operate the butterfly valve.

When the engine is cold, the valve is closed. This increases exhaust system back pressure. Hot exhaust gases flow into an exhaust passage in the intake manifold, **Figure 28-4.** This warms the floor of the intake manifold to hasten fuel vaporization. The manifold heat valve opens as the engine warms up.

Exhaust Pipes

The *exhaust pipes* (header pipe, intermediate pipe, and tailpipe) are usually made of rust-resistant steel tubing. One end of each pipe may be enlarged to fit over the end of the next pipe. The inlet end of the header pipe

Oxygen
sensor

Exhaust
manifold

Figure 28-2. The exhaust manifold bolts over the exhaust ports on the side of the cylinder head. Note the oxygen sensor at the end of the manifold. (Chrysler)

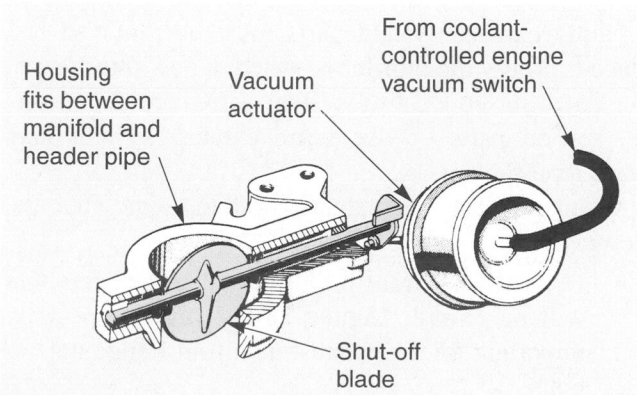

Housing
fits between
manifold and
header pipe

Vacuum
actuator

From coolant-
controlled engine
vacuum switch

Shut-off
blade

Figure 28-3. The heat control valve, sometimes called the heat riser, forces hot exhaust gas into the intake manifold. This helps the engine run smoothly. The valve opens as the engine warms up. (Chrysler)

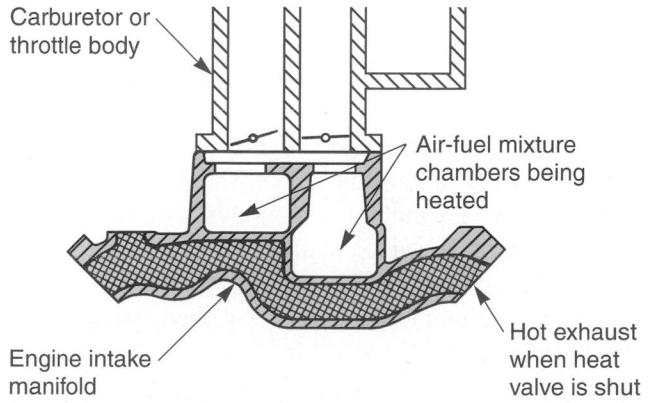

Carburetor or
throttle body

Air-fuel mixture
chambers being
heated

Engine intake
manifold

Hot exhaust
when heat
valve is shut

Figure 28-4. The heat control valve increases back pressure in exhaust system. This directs a large amount of hot exhaust into a chamber in the bottom of intake manifold. This action warms and helps vaporize fuel. (Pontiac)

has a flange for securing the pipe to the exhaust manifold studs, **Figure 28-1.**

Heat Shields

Heat shields are located in areas where the exhaust system components (especially the catalytic converter and muffler) are close to the car body or near the ground. The shields reduce the amount of heat transferred into the car's body and protect items under the vehicle. Refer to **Figure 28-1.**

Warning
Always reinstall all exhaust system heat shields. If the heat shields are not installed, undercoating, carpeting, dry leaves on the ground, and other flammable materials could catch on fire!

Catalytic Converter

A catalytic converter is used to reduce the amount of exhaust pollutants entering the atmosphere. One or more catalytic converters can be located in the exhaust system, **Figure 28-1.** For details of catalytic converters, refer to Chapter 43, *Emission Control Systems,* and Chapter 44, *Emission Control System Testing, Service, and Repair.*

Muffler

A muffler reduces the pressure pulses and resulting noise produced by the engine exhaust. When an engine is running, the exhaust valves are rapidly opening and closing. Each time an exhaust valve opens, a blast of hot gas shoots out of the engine. Without a muffler, these exhaust gas pulsations are very loud.

Figure 28-5 shows the inside of a muffler. Note how chambers, tubes, holes, and baffles are arranged to cancel out the pressure pulsations in the exhaust.

Exhaust System Service

Exhaust system service is usually needed when a component in the system rusts and begins to leak. Engine combustion produces water and acids. Therefore, an exhaust system can corrode and fail in a relatively short time.

A leaking exhaust can allow toxic gases to flow through any opening in the car's body and into the passenger compartment. People have died from engine exhaust fumes!

Exhaust System Inspection

To inspect an exhaust system, raise the car on a lift. Using a droplight, closely inspect the system for problems

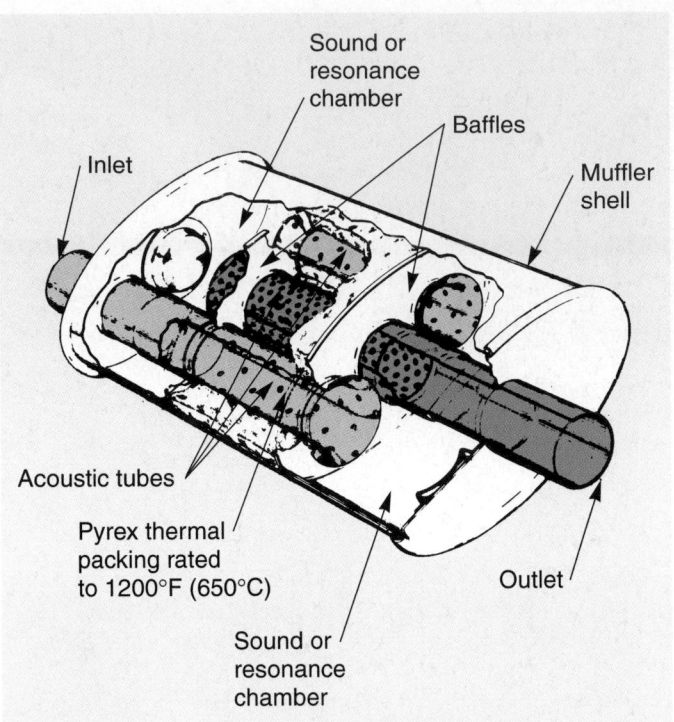

Figure 28-5. Mufflers contain baffles, resonance chambers, and acoustic tubes to reduce exhaust noise. (American Exhaust Industries)

(rusting, loose connections, leaks). Pay particular attention to the muffler and all pipe connections, gaskets, and pipe bends. *Exhaust leaks* will often show up as gray or white carbon lines coming from openings.

Warning
Parts of the exhaust system, especially the catalytic converter, can be very hot. Do not touch any hot part until after it has cooled.

Exhaust System Repairs

Faulty exhaust system parts must be removed and replaced. If only the muffler is rusted, a new muffler can be installed in the existing system. After prolonged service, several parts or the entire exhaust system may require replacement.

When repairing an exhaust system, remember the following:

- Use rust penetrant on all threaded fasteners that will be reused, **Figure 28-6.** This is especially important on the exhaust manifold flange nuts or bolts.
- Use an air chisel, cut-off tool, cutting torch, or hacksaw to remove faulty parts. Make sure you do not damage parts that will be reused. See **Figure 28-7** for some examples.

- A six-point socket and ratchet or an impact wrench will usually allow quick fastener removal without rounding off the fastener heads. Refer to **Figure 28-8.**

- Wear safety glasses or goggles to keep rust and dirt from entering your eyes.

- Obtain the correct replacement parts.

- A pipe expander should be used to enlarge pipe ends as needed, **Figure 28-9.** A pipe shaper can be used to straighten dented pipe ends, **Figure 28-10.**

- Position all clamps properly, **Figure 28-11.**

- Install any necessary adapters. See **Figure 28-12.**

- Make sure all pipes are fully inserted.

- Double-check the routing of the exhaust system. Keep adequate clearance between the exhaust system components and the vehicle's body and chassis. See **Figure 28-13.**

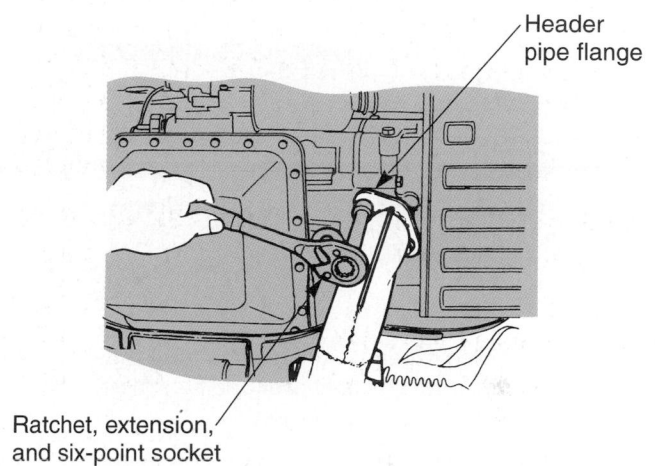

Figure 28-8. Header pipe fasteners can be difficult to remove. Use rust penetrant, six-point socket, extension, and ratchet. This will usually remove fasteners. (Subaru)

Figure 28-6. Rust penetrant or solvent will ease removal of badly rusted fasteners. (AP Parts)

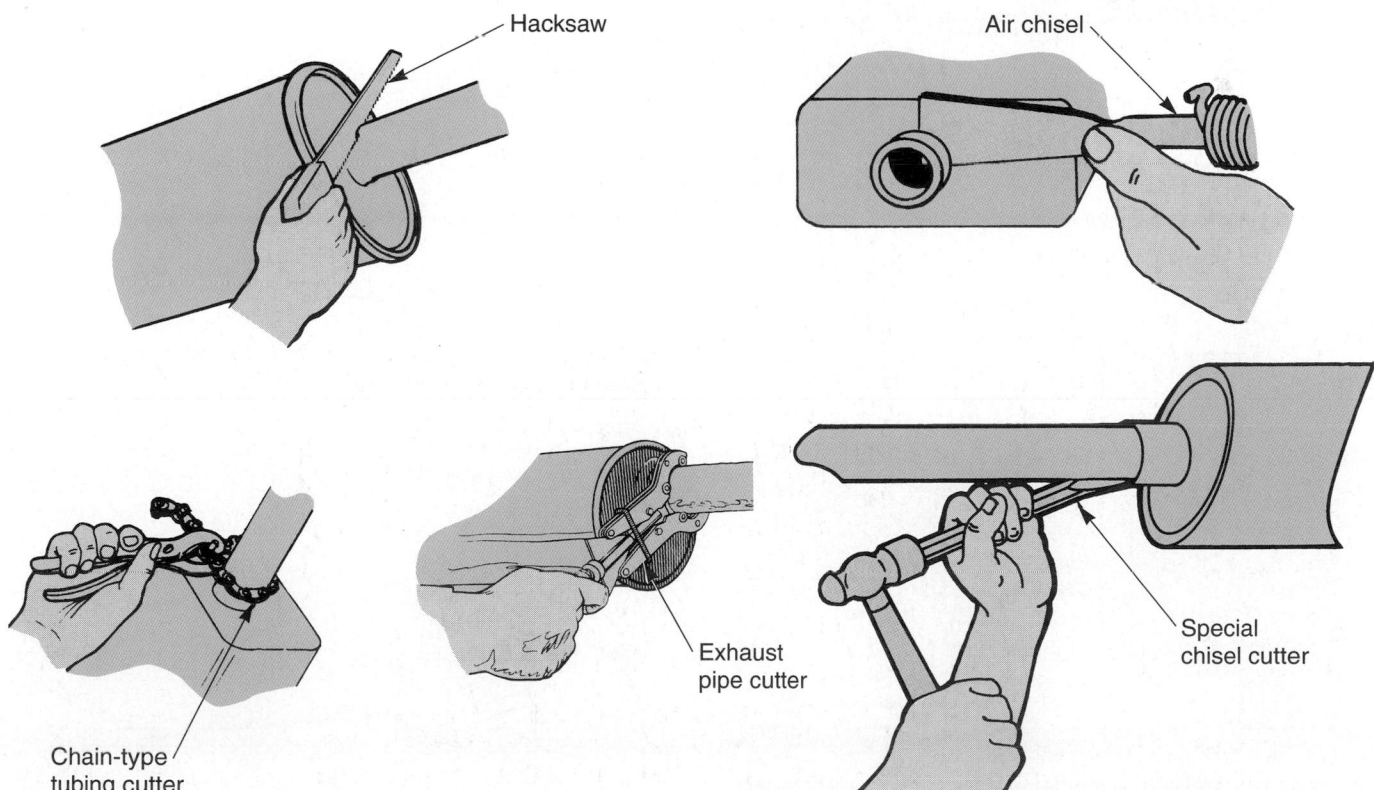

Figure 28-7. Several methods of removing old exhaust system parts. (AP Parts, Lisle, and Florida Dept. of Voc. Ed.)

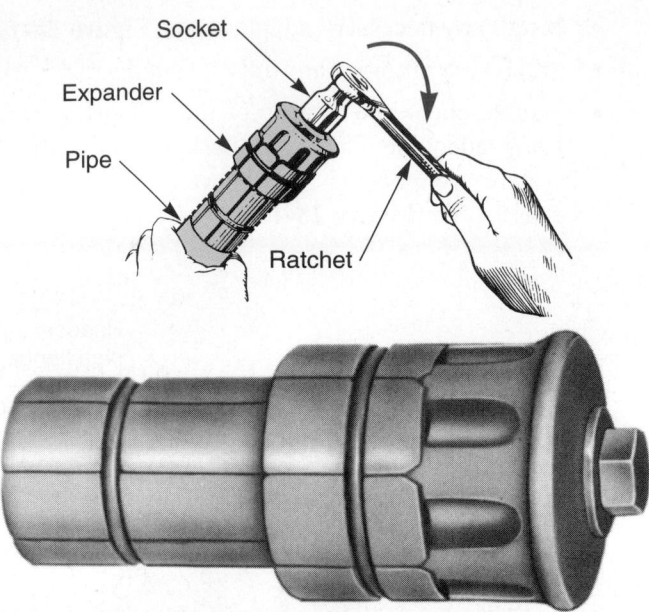

Figure 28-9. A pipe expander will enlarge the inside diameter (ID) of pipes. One pipe should fit over another. (Lisle)

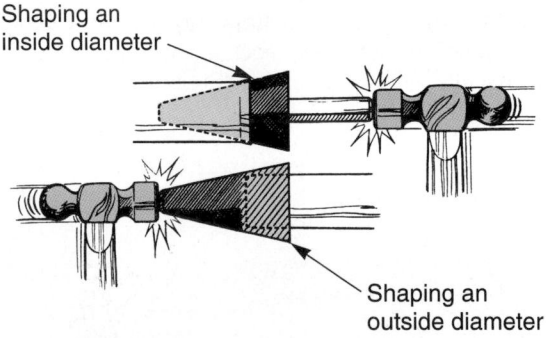

Figure 28-10. A pipe shaper will round dented pipe ends. (Lisle)

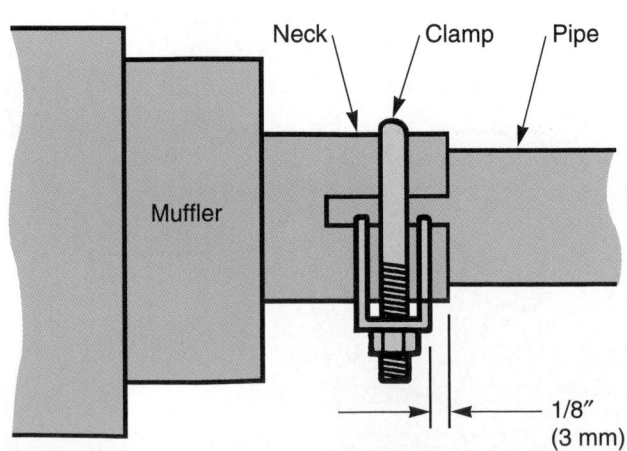

Figure 28-11. Make sure muffler clamps are installed correctly. The clamp must be positioned around both pipes. If not, one pipe can pull out of the other. (Florida Dept. of Voc. Ed.)

- Tighten all clamps and hangers evenly. Torque the fasteners only enough to hold the parts. Overtightening will smash and deform the pipes, possibly causing leakage, **Figure 28-14.**

- When replacing an exhaust manifold, use a gasket and check sealing surface flatness. If the manifold

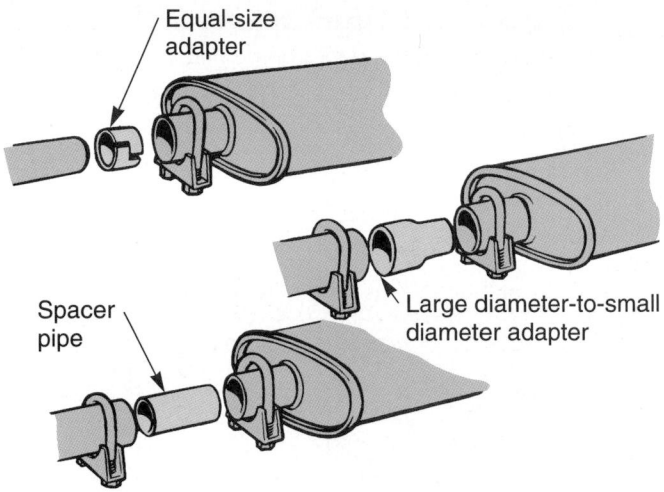

Figure 28-12. Adapters are sometimes needed to make a muffler work on an existing system. (AP Parts)

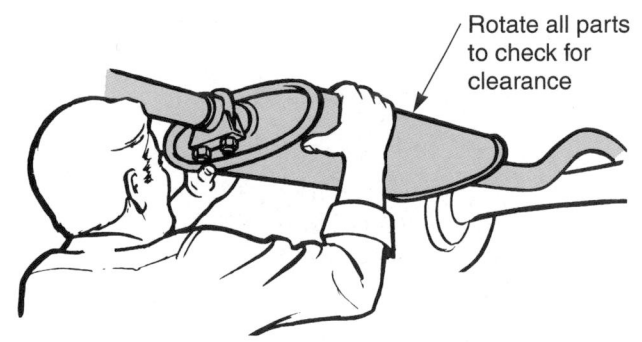

Figure 28-13. Double-check the exhaust system-to-car clearance carefully. (AP Parts)

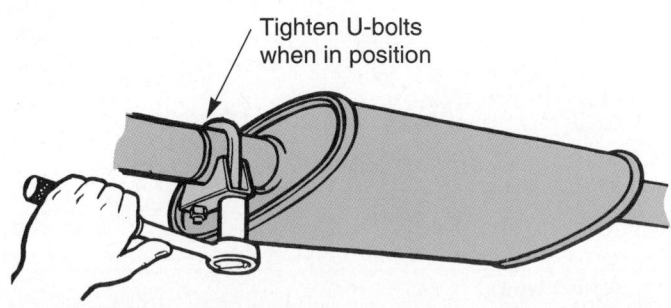

Figure 28-14. After checking clearance, tighten all clamps evenly and properly. (AP Parts)

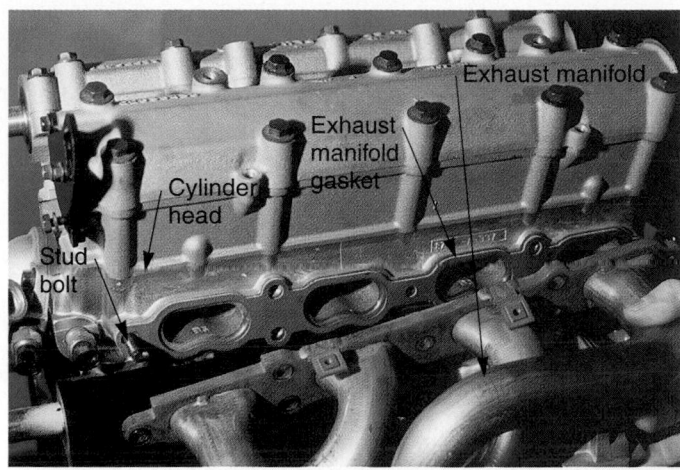

Figure 28-15. An exhaust gasket is often recommended. The gasket is held in position as all fasteners are started by hand. Torque fasteners to recommended value in a crisscross pattern. (Oldsmobile)

is warped, it must be machined flat. Torque the exhaust manifold bolts to specification, **Figure 28-15.**

- Always use new gaskets and O-rings.
- Check heat valve operation using the information in a service manual.
- Install all heat shields.
- Check the system for leaks and rattles after repairs.

Stainless Steel Exhaust System Repairs

Many new vehicle exhaust systems are made of stainless steel. Stainless steel will not rust and will provide much longer service life than ordinary steel systems.

When servicing stainless steel exhaust systems, use heavy-duty clamps designed for this type of system. Conventional muffler clamps are not strong enough and may allow dangerous exhaust gas leakage. They will not compress the stiff stainless steel pipe enough to make a good connection.

When cutting or welding stainless steel, use the correct rod or wire material. Keep in mind that stainless steel does not react in the same manner as carbon steel when heated near its melting point. Stainless steel can be "red hot" when it looks cool.

Superchargers and Turbochargers

A *normally aspirated engine*, also called an *atmospheric engine,* uses atmospheric pressure (14.7 psi or 100 kPa at sea level) to push air into the engine. With outside air pressure as a moving force, only a limited amount of fuel can be burned on each power stroke.

The term *supercharger* originally meant that some type of air pump was used to increase engine power by pushing more air and fuel into the combustion chambers. With a greater amount of air and fuel (a denser air-fuel mixture), combustion can generate more heat energy and pressure to push the pistons down in their cylinders.

A *supercharger* is a blower driven by a belt, gears, or a chain. Superchargers are used on large diesel truck engines and racing engines. They are sometimes found on high-performance passenger cars.

The *turbocharger,* or *"turbo,"* is a blower driven by engine exhaust gases. Turbochargers are commonly used on passenger cars, trucks, and competition engines.

Sometimes called a *blower,* the supercharger or turbocharger raises the air pressure in the engine intake manifold. Then, when the intake valves open, more air-fuel mixture (gasoline engine) or air (diesel engine) can flow into the cylinders. Engine horsepower can be doubled in racing applications with engine modifications and a blower.

Turbochargers

A turbocharger is an exhaust-driven fan or blower that forces air into the engine under pressure. Turbochargers are frequently used on small gasoline and diesel engines to increase power output. By harnessing exhaust energy, a turbocharger can also improve engine efficiency (fuel economy and emission levels). This is especially true with diesel engines.

The basic parts of a turbocharger are shown in **Figure 28-16**:

- *Turbine wheel*—exhaust-driven fan that turns the turbo shaft and compressor wheel.
- *Turbine housing*—outer enclosure that routes exhaust gases around the turbine wheel.
- *Turbo shaft*—steel shaft that connects the turbine and compressor wheels. It passes through center of the bearing housing.
- *Compressor wheel*—driven fan that forces air into the engine intake manifold under pressure.
- *Compressor housing*—part of the turbo housing that surrounds the compressor wheel. Its shape helps pump air into the engine.
- *Bearing housing*—enclosure around the turbo shaft that contains bearings, seals, and oil passages.

 **Tech Tip**

Some late-model turbocharger impellers are made of carbon fiber reinforced plastic. This can cut impeller weight in half and reduce the problem of turbo lag considerably.

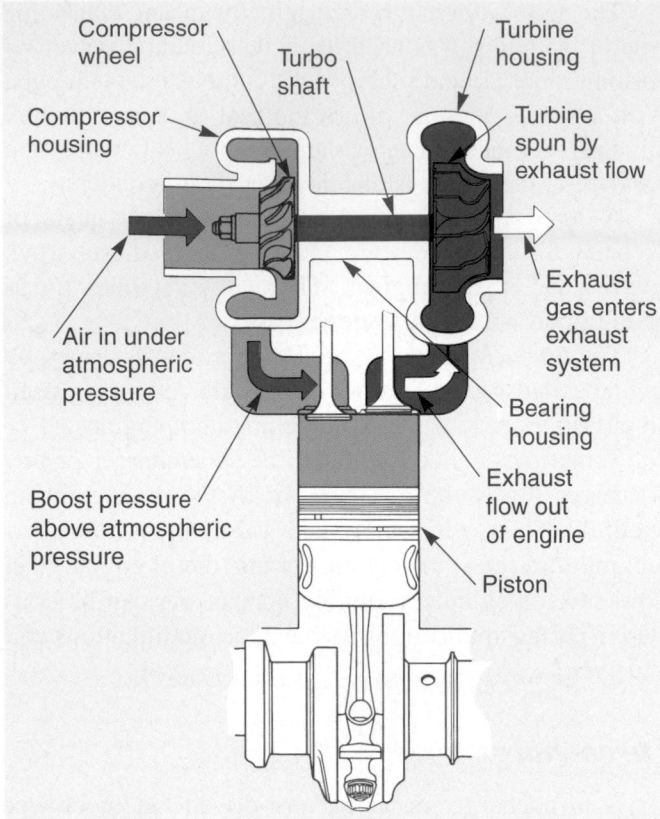

Figure 28-16. A turbocharger uses exhaust gas flow to spin a turbine wheel. The turbine wheel spins a shaft and a compressor wheel. The compressor wheel then pressurizes the air entering the engine for more power output. (Mercedes Benz)

Turbocharger Operation

When the engine is running, hot exhaust gases blow out through the open exhaust valve ports and into the exhaust manifold. The exhaust manifold and connecting tubing route these gases into the turbine housing. Refer to **Figure 28-17**.

As the gases pass through the turbine housing, they strike the fins on the turbine wheel. When engine load is high enough, there is enough exhaust gas flow to rapidly spin the turbine wheel, **Figure 28-17**.

Since the turbine wheel is connected to the compressor wheel by the turbo shaft, the compressor wheel rotates with the turbine. Compressor wheel rotation pulls air into the compressor housing. Centrifugal force throws the spinning air outward. This causes air to flow out of the turbocharger and into the engine cylinder under pressure.

Turbocharger Location

A turbocharger is usually located on one side of the engine. An exhaust pipe connects the engine exhaust manifold to the turbine housing. The exhaust system header pipe connects to the outlet of the turbine housing. See **Figure 28-18**.

A *blow-through turbo system* has the turbocharger located before the carburetor or throttle body. The turbo compressor wheel only pressurizes air. Fuel is mixed with the air after air leaves the compressor.

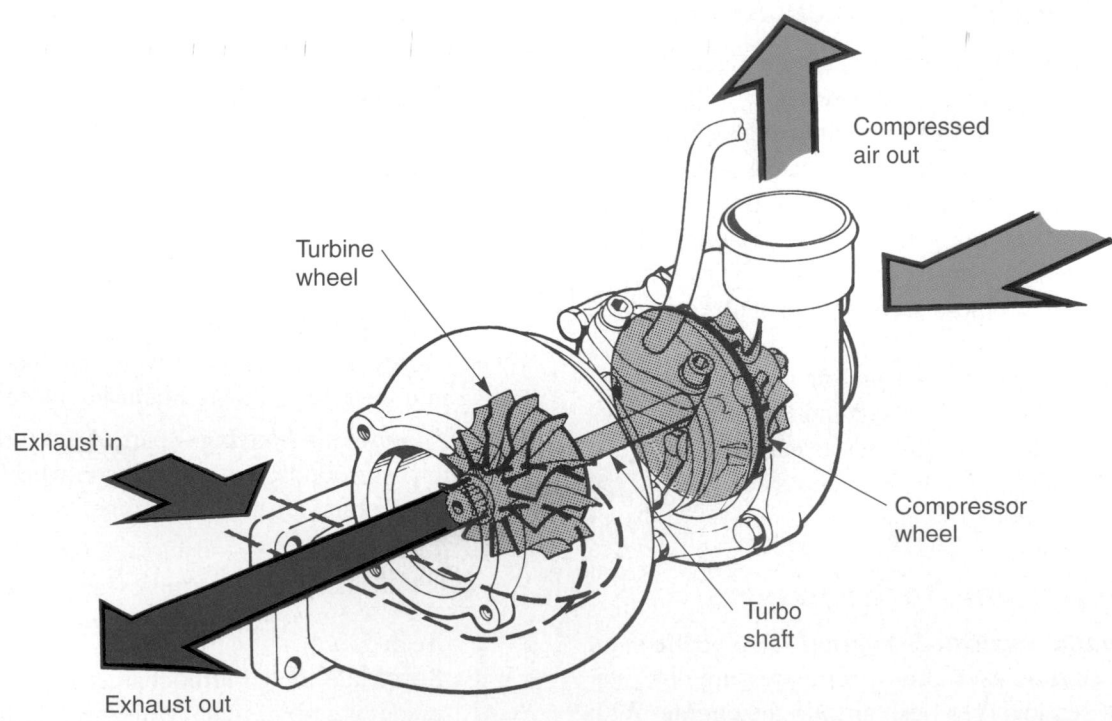

Figure 28-17. Exhaust flow spins the turbine wheel shaft and the compressor wheel. Normally, wasted energy in exhaust is used to increase compression stroke pressure in cylinders for more violent combustion. (Saab)

A *draw-through turbo system* locates the turbocharger after the carburetor or throttle body assembly. As a result, both air and fuel (gasoline engine) pass through the compressor housing in carburetor systems and throttle body fuel injection system. In port fuel injection systems, only air passes through the housing.

Theoretically, the turbocharger should be located as close to the engine exhaust manifold as possible. Then, a maximum amount of exhaust heat will enter the turbine housing. This ensures that the gases are still expanding as they enter the turbocharger. This expansion will help spin the turbocharger, increasing boost pressure and engine power.

Turbocharger Lubrication

Turbocharger lubrication is needed to protect the turbo shaft and bearings from damage. A turbocharger can operate at speeds up to 100,000 rpm. For this reason, the engine lubrication system forces motor oil into the turbo shaft bearings, **Figure 28-19.**

Oil passages are provided in the turbo housing and bearings. An oil supply line runs from the engine to the turbo. With the engine running, oil enters the turbo under pressure.

Sealing rings (piston-type rings) are placed around the turbo shaft at each end of the turbo housing. They prevent oil leakage into the compressor and turbine housings, **Figure 28-19.** A *drain passage* and *drain line* allow oil to return to the oil pan after passing through the turbo bearings.

Turbo Lag

Turbo lag refers to a short delay before the turbo develops sufficient boost (pressure above atmospheric pressure) to meet engine demands. See **Figure 28-20.**

When the car's accelerator pedal is pressed down for rapid acceleration, the engine may lack power for a few seconds. This is caused by the compressor and turbine wheels not spinning fast enough. It takes time for the exhaust gas to bring the turbo up to operating speed.

Modern turbo systems suffer very little from turbo lag. Their turbine and compressor wheels are very light so that they can accelerate up to speed quickly.

Turbocharger Intercooler

A *turbocharger intercooler* is an air-to-air heat exchanger that cools the air entering the engine. It is a

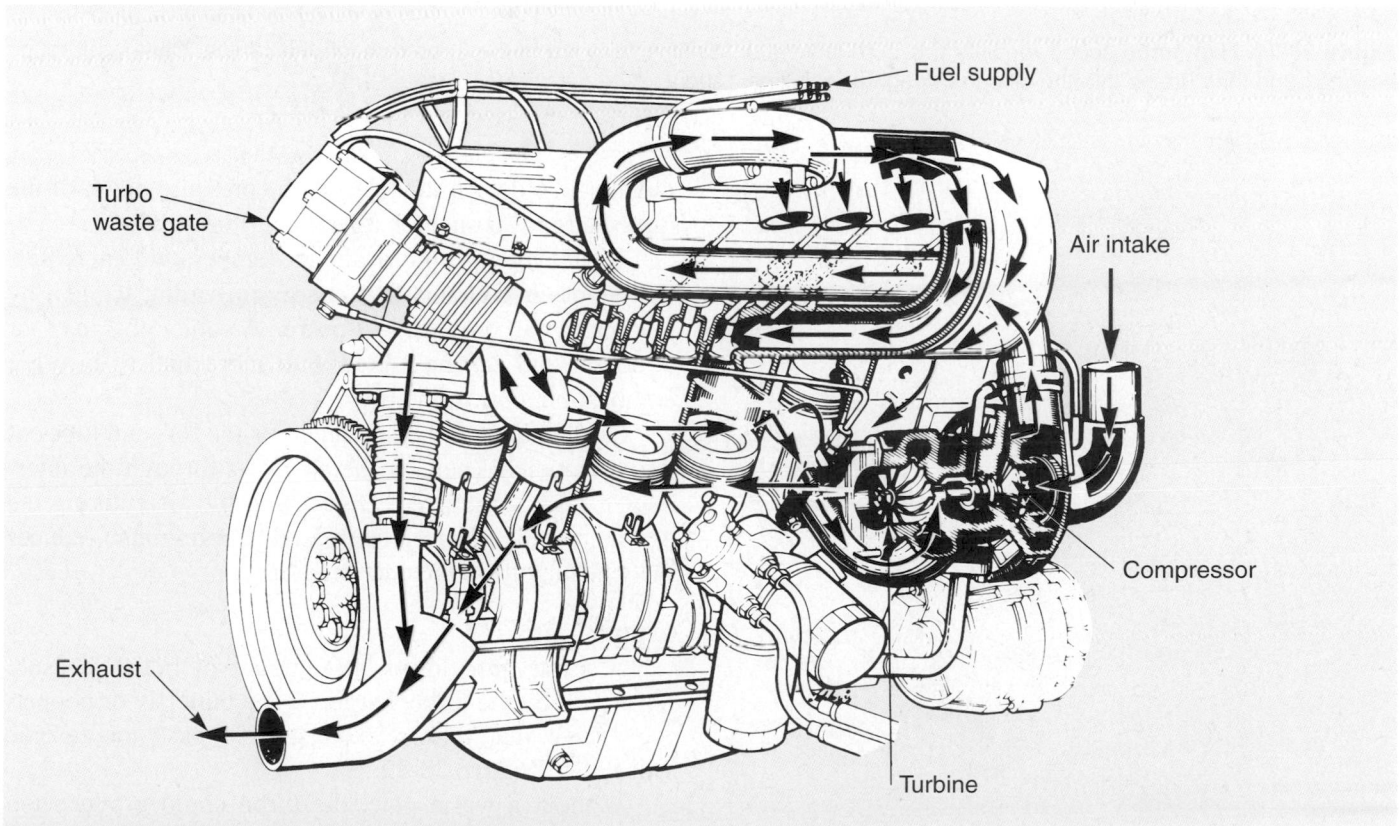

Figure 28-18. Turbocharger normally bolts to one side of engine. Pipes route exhaust through turbine housing. Compressed air leaves turbo and enters intake tract and engine. (Audi)

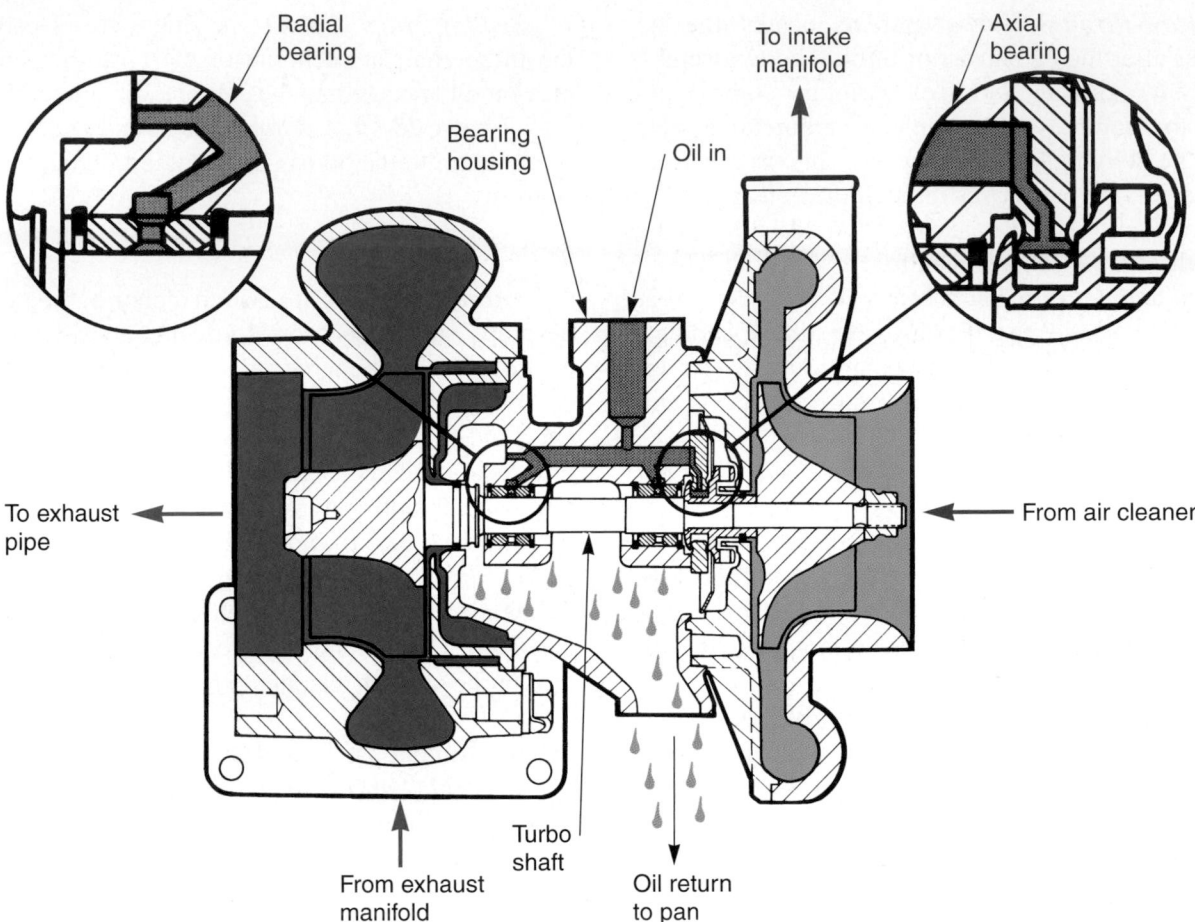

Figure 28-19. High turbo speed requires pressure lubrication. Engine oil is fed to the turbo through the oil line. Oil flows through bearings and then drains into the oil pan through a drain line. (Audi)

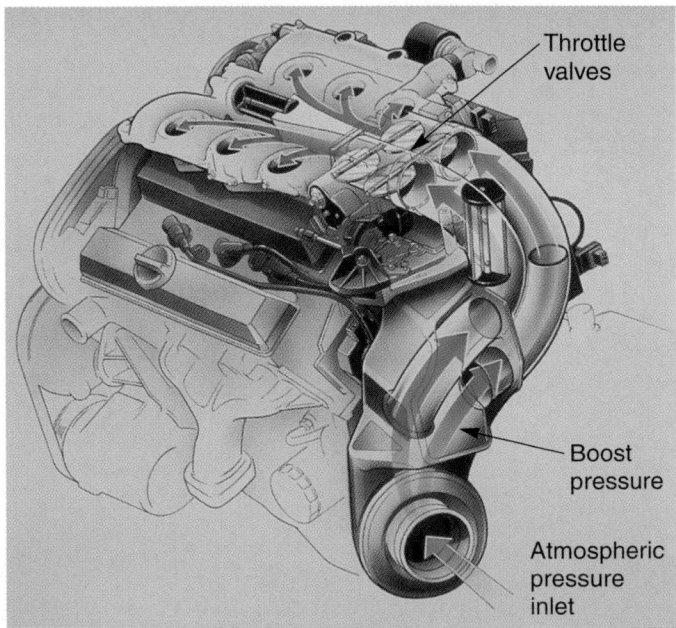

Figure 28-20. Modern turbo systems suffer from very little turbo lag. Turbo wheels are smaller and lighter and spin up to speed quickly. (Saab)

radiator-like device mounted at the pressure outlet of the turbocharger (or supercharger). See **Figure 28-21.**

When you compress air, its temperature increases. Since hot air contains less energy-providing oxygen by volume, it will produce less power. A cooler charge of air is denser and can be mixed with more fuel to increase combustion and engine power.

Outside air flows over and cools the fins and tubes of the intercooler. Then, as the air flows through the intercooler, heat is removed. By cooling the air entering the engine, engine power is increased. Cooling also reduces the tendency for engine detonation.

Waste Gate

A *waste gate* limits the amount of boost pressure developed by the turbocharger. It is a butterfly or poppet-type valve that allows exhaust to bypass the turbine wheel. See **Figure 28-22.**

Without a waste gate, the turbo could produce too much pressure in the combustion chambers. This could lead to detonation (spontaneous combustion) and engine damage.

Figure 28-21. Intercooler is an air-to-air heat exchanger or radiator. It cools the air charge entering the engine for increased horsepower output. (Saab)

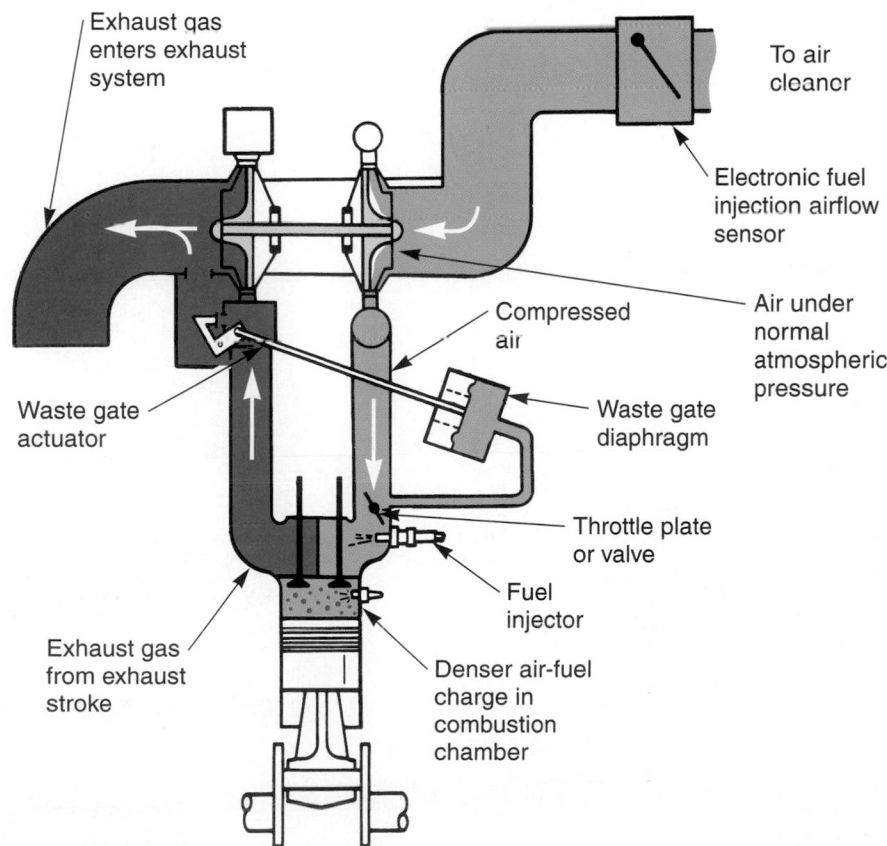

Figure 28-22. Study basic exhaust and inlet airflow through complete turbo system. (Ford)

Basically, a waste gate is a valve operated by a diaphragm assembly, **Figure 28-23.** Intake manifold pressure acts on the diaphragm to control waste gate valve action. The valve controls the opening and closing of a passage around the turbine housing.

Waste Gate Operation

Figure 28-24 illustrates the basic operation of a turbocharger waste gate. Under partial load, the system routes all the exhaust gases through the turbine housing. The waste gate is closed by the diaphragm spring. This

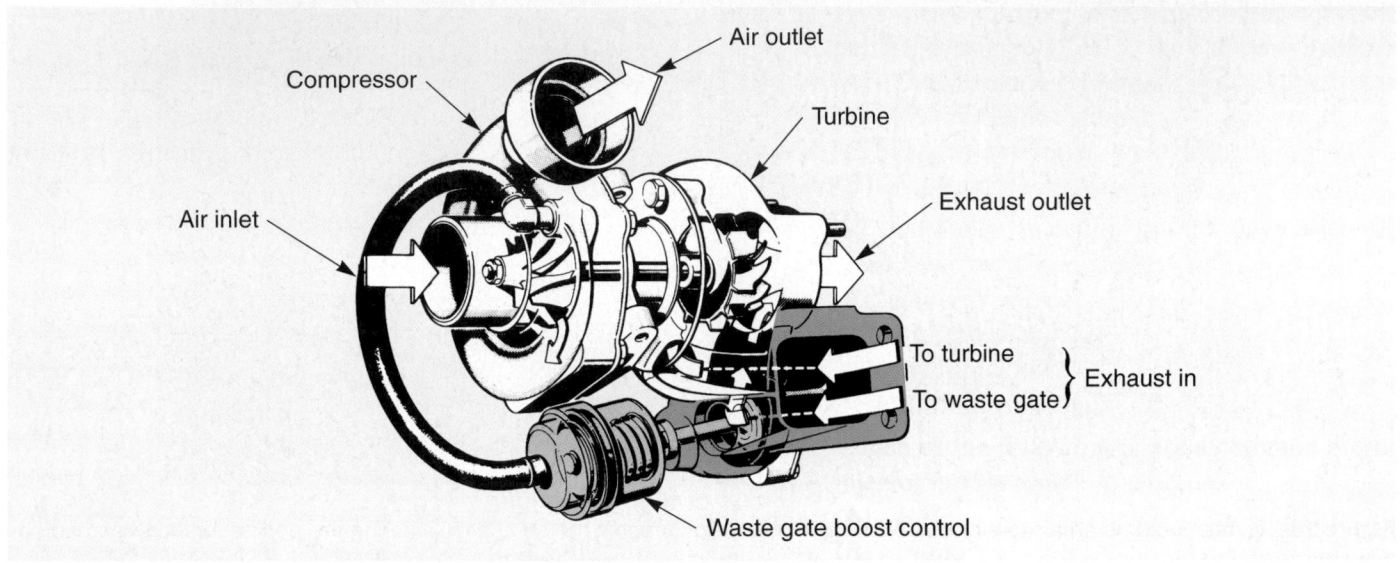

Figure 28-23. The waste gate or boost control is a valve in the turbine housing. When needed, it can open to limit boost pressure by reducing the amount of exhaust acting on the turbine wheel. (Mercedes Benz)

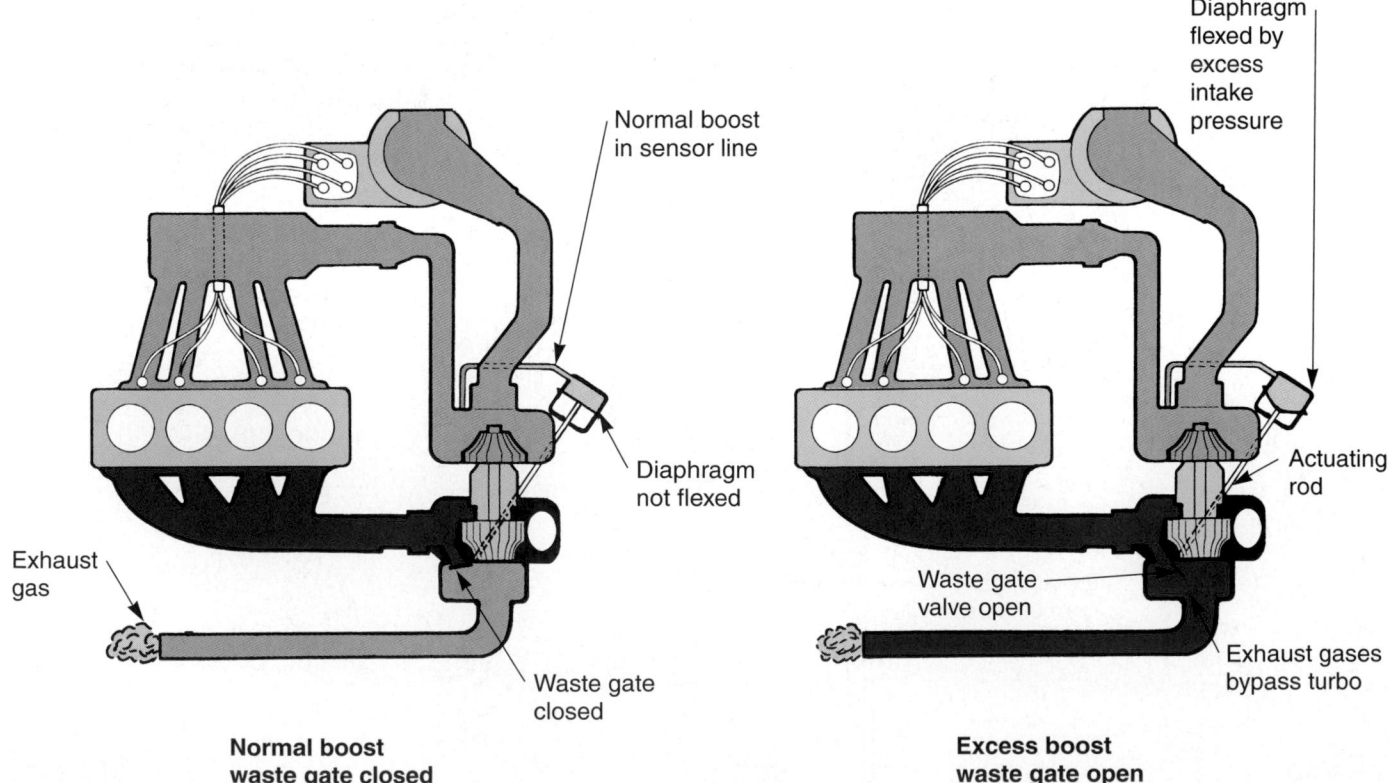

**Normal boost
waste gate closed**

**Excess boost
waste gate open**

Figure 28-24. Study operation of complete turbo system. (Saab)

ensures that there is adequate boost to increase engine power.

Under full load, boost may become high enough to overcome the diaphragm spring pressure. Manifold pressure compresses the spring and opens the waste gate valve. This permits some of the exhaust gases to flow through the waste gate passage and into the exhaust system. Less exhaust is left to spin the turbine. Boost pressure is limited to a preset value.

Turbocharged Engine Modifications

A turbocharged engine normally has several modifications to make it withstand the increased horsepower. A few of these are shown in **Figure 28-25** and include:

- Lower compression ratio.
- Stronger rods, pistons, and crankshaft.
- Higher volume oil pump and an oil cooler.
- Larger cooling system radiator.

- O-ring type head gasket.
- Heat resistant valves.
- Knock sensor (ignition retard system).

Turbo Computer Control

As with other systems, the vehicle computer often controls the turbocharger by operating the waste gate and by retarding the ignition timing when needed. As shown in **Figure 28-26,** several sensor input signals are sent to the computer. These commonly include inputs from the manifold pressure sensor (boost pressure), manifold air temperature sensor, knock sensor, throttle position sensor, and other sensors.

The computer uses preprogrammed data to determine if boost pressure or ignition timing should be altered. The computer must limit boost and timing advance to prevent knock and possible engine damage. It can then produce outputs to open the waste gate or retard timing if needed.

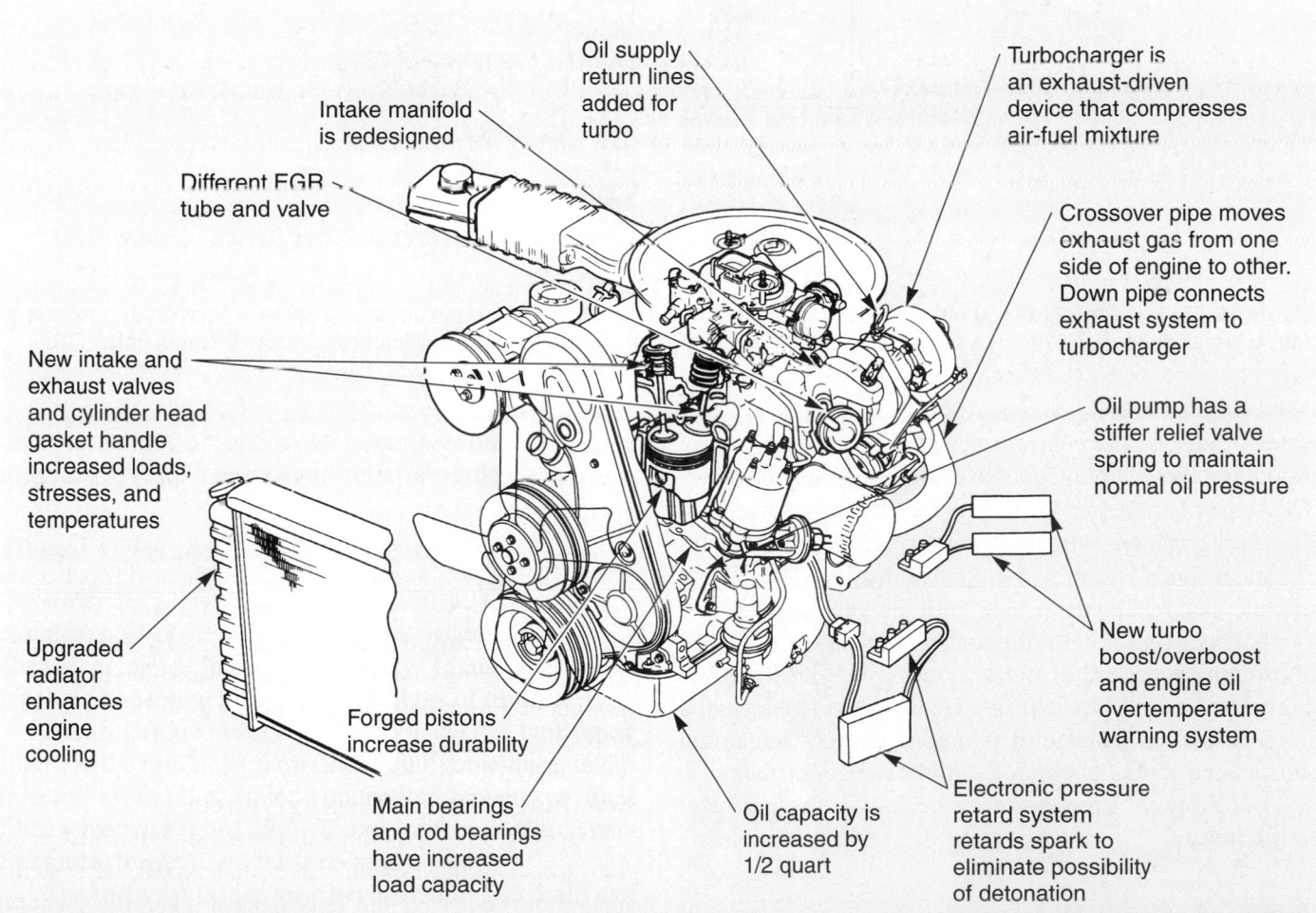

Figure 28-25. Note many engine modifications are commonly used with turbocharging. Turbocharging increases demands on engine. (Ford)

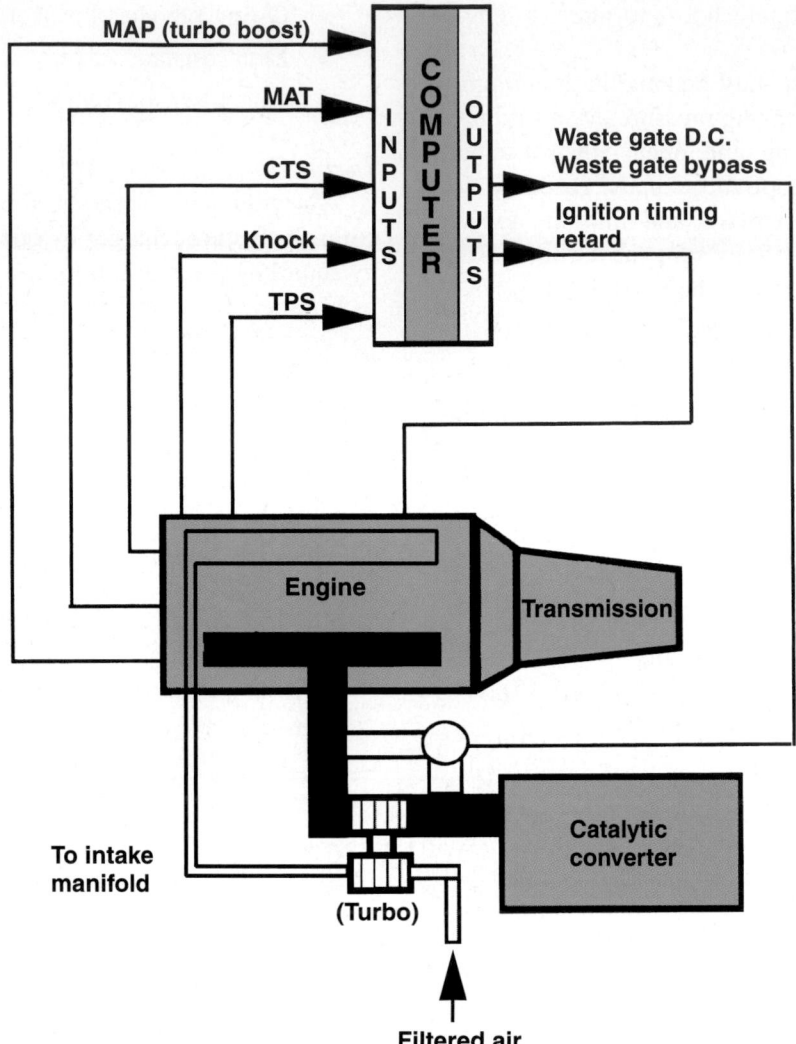

Figure 28-26. The on-board computer is commonly used to monitor and control the turbocharging system. Note how sensors feed data to the computer and how the computer can then output current signals to retard ignition timing or open waste gate. (OTC)

Knock Sensor. A *knock sensor* signals the engine control module if the engine begins to knock (detonate or ping). The sensor is mounted on the engine. It works something like a microphone. When it "hears" a knocking sound, an electrical signal is sent to the control module. The control module then retards the timing until the knock stops.

A knock sensor helps the control module keep the ignition timing advanced as much as possible. This improves engine power and gas mileage. It also protects the engine from detonation damage. It is one of the most important sensors in a computer-controlled turbocharger system.

For more information on turbocharger service, refer to the index.

Turbocharging System Service

Turbocharging system problems usually show up as inadequate boost pressure (lack of engine power), leaking shaft seals (oil consumption), damaged turbine or compressor wheels (vibration and noise), or excess boost (detonation).

To protect a turbocharger from damage, most automakers recommend that the oil in a turbocharged engine be changed more frequently (about every 3000 miles or 4827 km). Because of the high rotating speeds, the turbo bearings and shaft are very sensitive to oil contaminants. Engine oil must be kept clean to ensure long turbocharger life.

Scanning a Turbocharging System

Use a scan tool to check for trouble codes relating to the turbocharging system. Late-model OBD II systems may show codes for the knock sensor, throttle position sensor, manifold pressure sensor, manifold temperature sensor, and other sensors. These codes can all relate to the turbo system.

Tech Tip

Remember that on-board diagnostic systems and scan tools can sometimes be "confused" by mechanical problems. For example, a worn piston pin can knock and "fool" the knock sensor into thinking the engine is detonating. It will then mistakenly retard ignition timing and lower boost pressure. A trouble code might be tripped when something else is causing the problem. Keep this in mind when using a scan tool.

Refer to a factory service manual for a detailed troubleshooting chart. It will list the common troubles for the particular turbocharging system.

Checking a Turbocharging System

There are several checks that can be made to determine turbocharging system condition:

- Check connection of all vacuum lines to the waste gate and oil lines to the turbocharger, **Figure 28-27.**

- Use a regulated low-pressure air hose to check for waste gate diaphragm leakage and operation.

- Use the dash gauge or a test gauge to measure boost pressure (pressure developed by turbo under a load). If needed, connect the pressure gauge to an intake manifold fitting. Compare the gauge readings to specifications.

- Use a stethoscope to listen for bad turbocharger bearings.

Checking a Turbocharger

To check the internal condition of a turbocharger, remove the unit from the engine, as in **Figure 28-28.** Unbolt the connections at the turbo. Remove the oil lines and take the unit to your workbench.

Inspect the turbocharger wheels for physical damage. The slightest nick or dent will throw the unit out of balance, causing vibration. **Figure 28-29** shows how to measure turbo bearing and shaft wear.

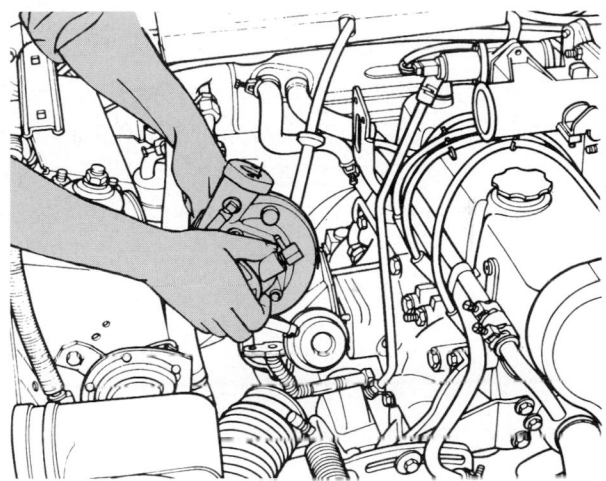

Figure 28-28. After removing mounting fasteners and any other parts, a turbocharger can be lifted off for replacement. A turbo cannot be repaired in the shop. A new unit is normally installed. (Ford)

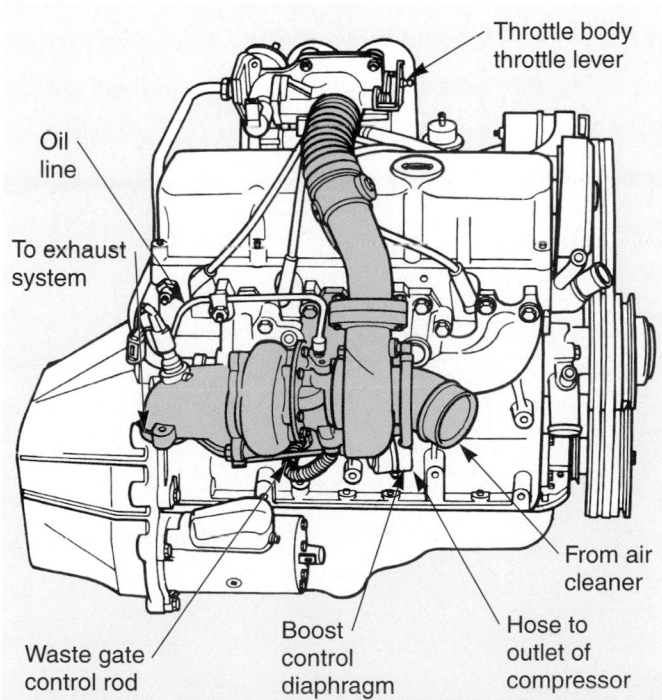

Figure 28-27. Side view of engine shows location and mounting of turbocharger. (Ford)

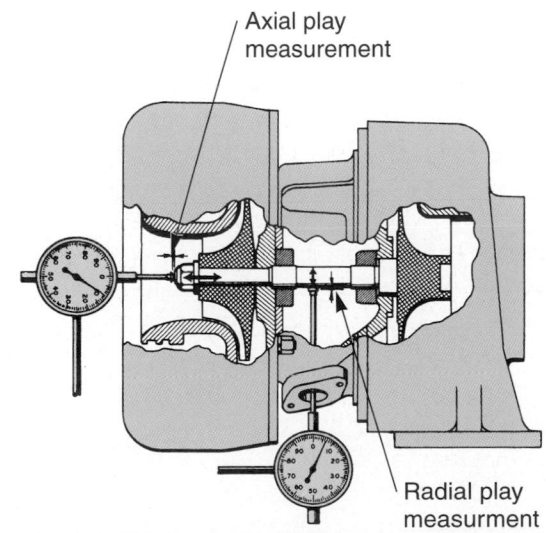

Figure 28-29. A dial indicator can be used to check radial and axial play of turbo shaft. If not within specifications, replace the unit. (Waukesha)

Caution
Never use a hard metal object or sandpaper to remove carbon deposits from the turbine wheel. If you gouge the wheel, it can vibrate and destroy the turbo when its spins up to speed. Only use a soft wire brush and solvent to clean the turbo wheels.

Installing a New Turbocharger

Many turbocharger problems are not repaired in the field, **Figure 28-30.** Most technicians install a new or rebuilt unit. When installing a turbocharger, you should:

- Make sure the new turbo is the correct type. Compare part numbers.
- Use new gaskets and seals.
- Torque all fasteners to specifications.
- If needed, change the engine oil and flush the oil lines before starting the engine.
- If the failure was oil related, check the oil supply pressure in the feed line to the turbocharger.

Waste Gate Service

An inoperative waste gate can either cause too much or too little boost pressure. If the waste gate is stuck open, the turbocharger will not produce boost pressure and the engine will lack power. If the gate is stuck closed, detonation and engine damage can result from excessive boost.

Before replacing the waste gate, always check other parts. Check the knock sensor and the ignition timing. Make sure the vacuum pressure lines are all connected properly.

Follow service manual instructions when testing or replacing a waste gate. As shown in **Figure 28-31,** waste gate removal is relatively easy. Simply unbolt the fasteners, remove the lines, and lift the unit off the engine. Many manuals recommend waste gate replacement, rather than in-shop repairs.

Superchargers

A *supercharger* is a compressor or blower driven by a belt, chain, or gears. Unlike a turbocharger, it is not

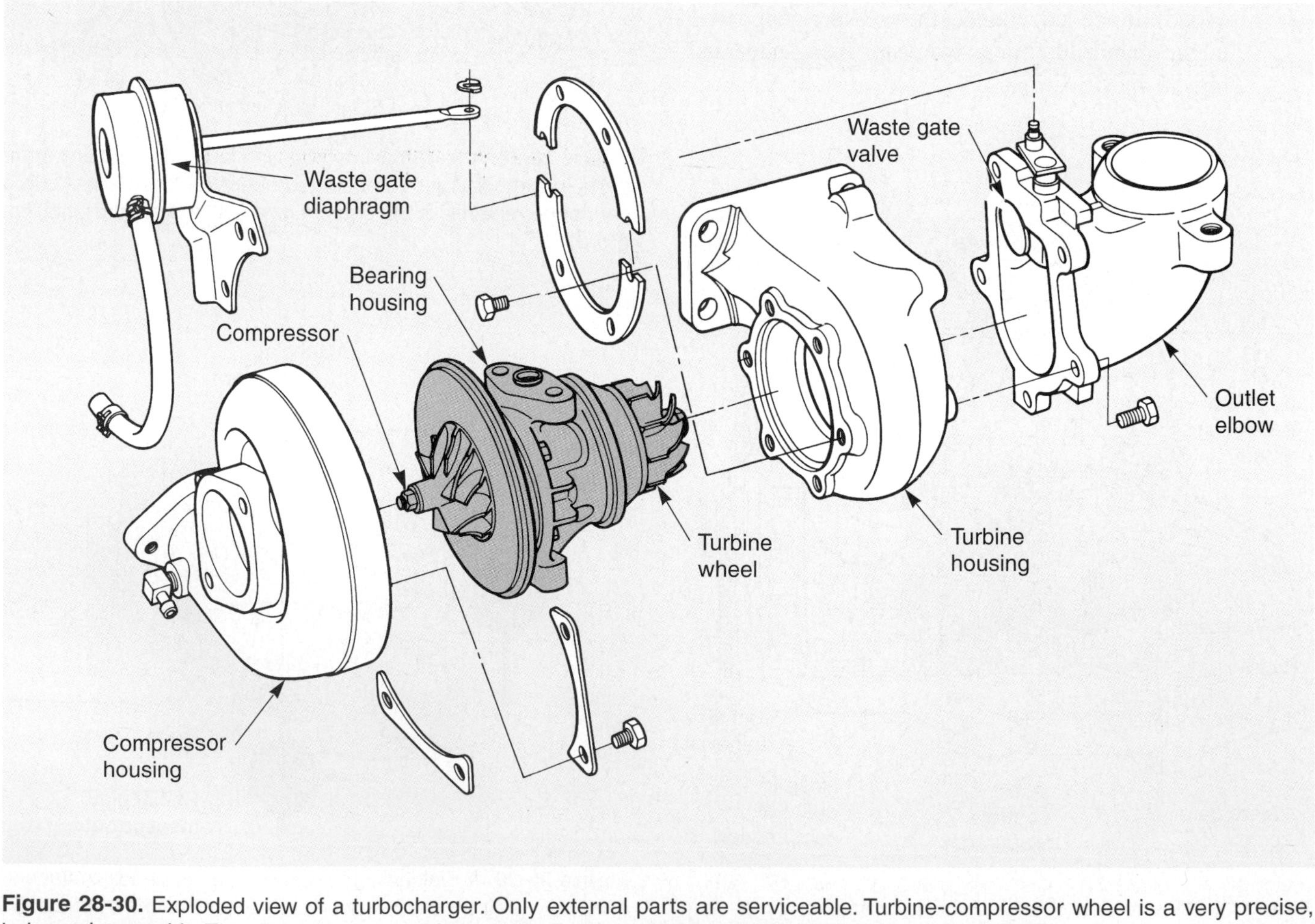

Figure 28-30. Exploded view of a turbocharger. Only external parts are serviceable. Turbine-compressor wheel is a very precise, balanced assembly. The slightest nick or chip on a blade can cause the unit to explode in service. (Ford)

driven by engine exhaust gases. Most passenger car superchargers are driven by a belt on the front of the engine. See **Figure 28-32.**

The *supercharger belt* drives the rotors inside the supercharger. As the rotors turn, they compress the air inside the housing and force the air, under pressure, into the engine intake manifold. See **Figure 28-33.**

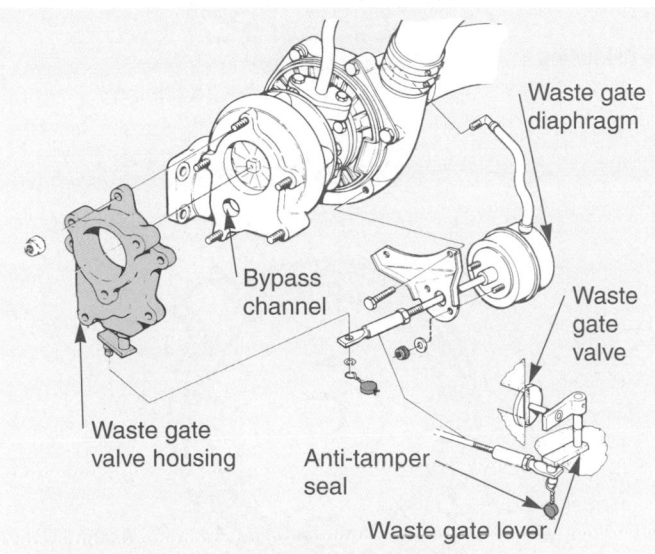

Figure 28-31. The waste gate normally bolts to the side of the turbo housing. A linkage rod connects the diaphragm with the valve mechanism. Also note the seal that prevents tampering with boost setting. Although overboost will increase power, it can also cause engine damage. (Saab)

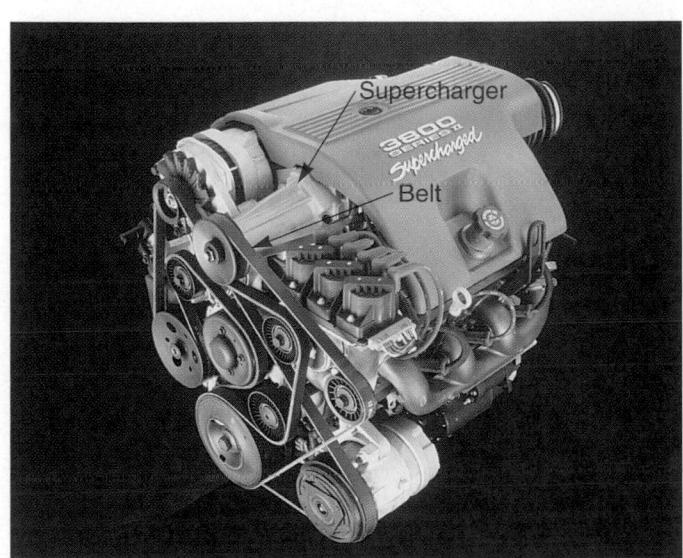

Figure 28-32. A supercharged engine provides added acceleration for entering highway ramps and passing other vehicles. The engine must be modified with a stronger reciprocating assembly (crankshaft, connecting rods, and pistons) to withstand the power increase without part failure. (General Motors Corp.)

An *electromagnetic clutch* is sometimes used to disengage the drive belt from the blower. It works like an air-conditioning compressor clutch to save energy when additional power is not needed. See **Figure 28-34.** An

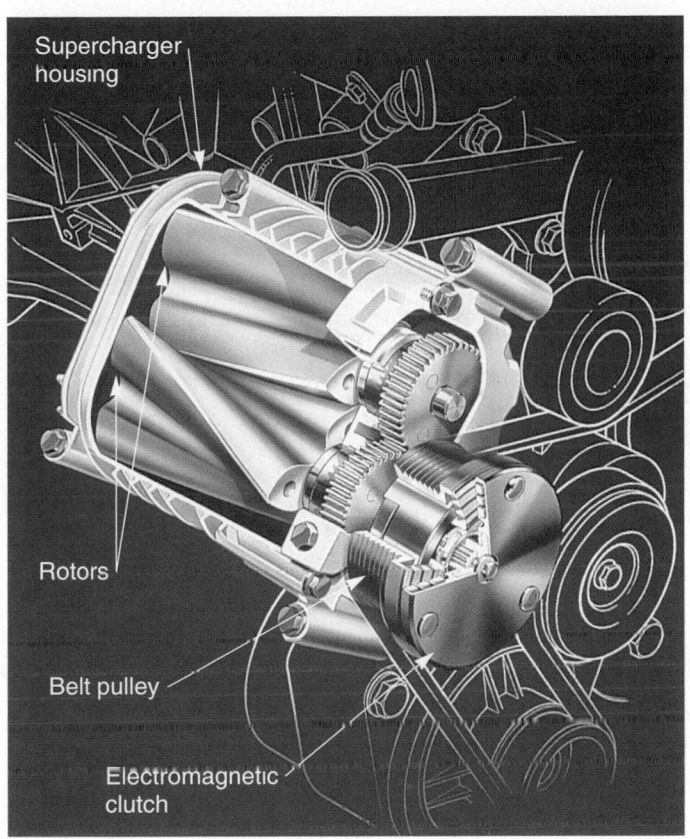

Figure 28-33. Cutaway of a supercharger. The rotors in this unit spin at speeds up to 12,000 rpm to pressurize incoming air. (Mercedes-Benz)

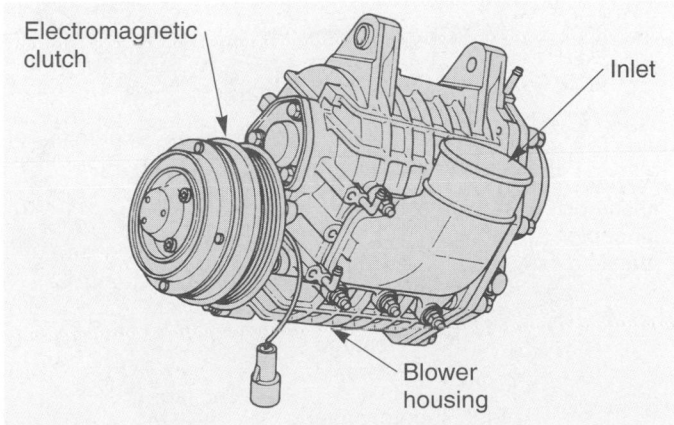

Figure 28-34. The supercharger is normally driven by belt running up from the engine crankshaft pulley. An electric clutch can be used to turn the blower on and off as needed. For example, the blower might only engage at full throttle when there is demand for more engine power. (Toyota)

intercooler, explained earlier, is commonly used between the supercharger outlet and the engine to cool the air and increase power (cool charge of air carries more oxygen needed for combustion).

Superchargers have the advantage of not suffering from turbo lag. A supercharger will instantly produce boost pressure at low engine speeds because it is mechanically linked to the engine crankshaft. This low-speed power and instant throttle response is desirable in a passenger car when passing other vehicles and when entering interstate highways.

Figure 28-35 shows the major parts included in a supercharger system.

Supercharger Types

There are three basic types of superchargers:

- Centrifugal supercharger.
- Rotor (Rootes) supercharger.
- Vane supercharger.

These types are shown in **Figure 28-36.** Note the differences in construction and operation.

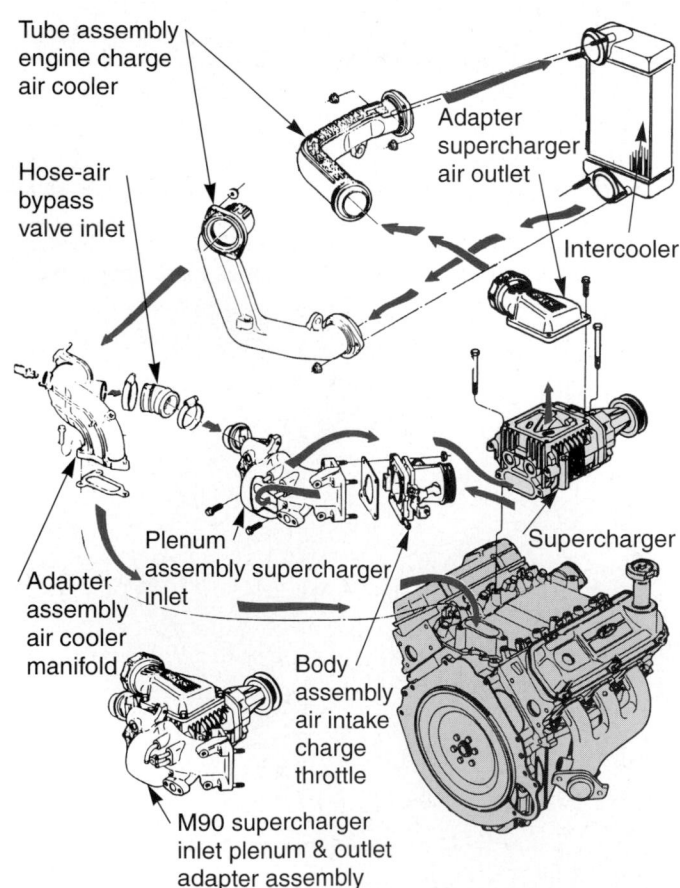

Figure 28-35. Basic components in a supercharging system. This supercharger bolts to top of the engine. Trace the flow of air through the components. (Ford)

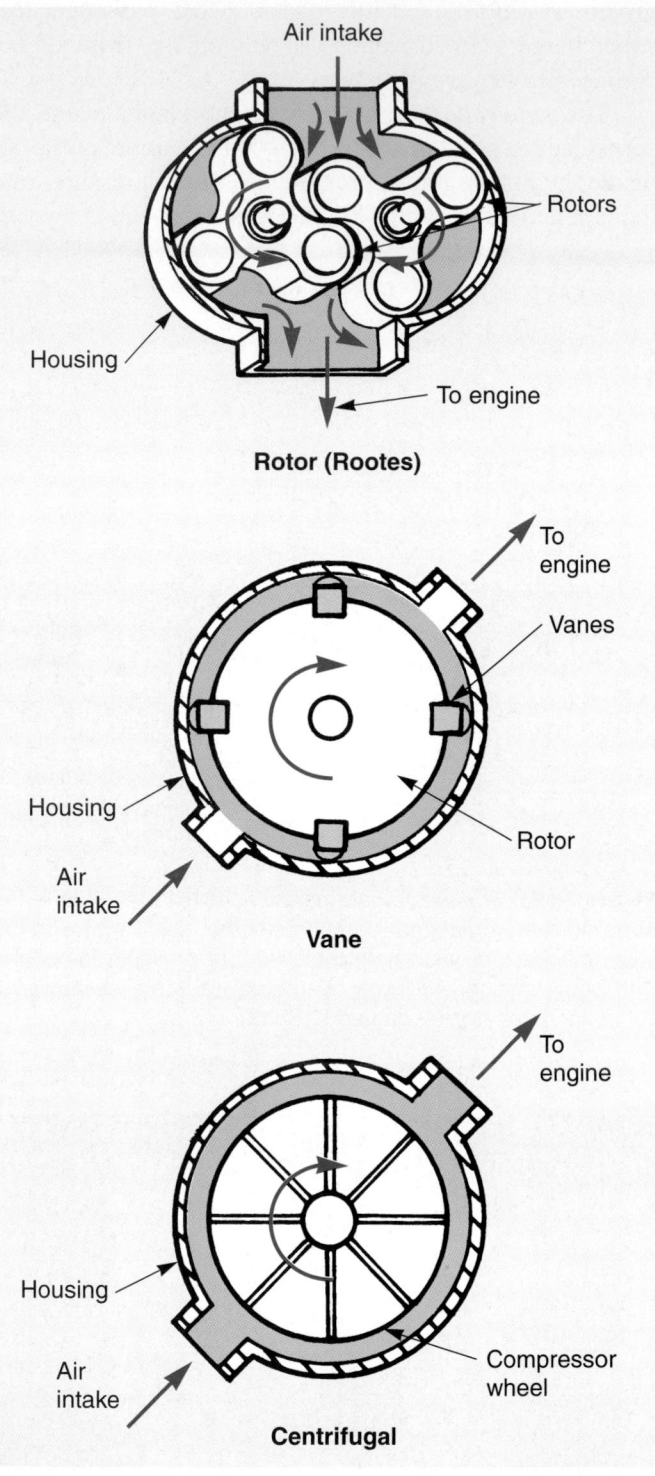

Figure 28-36. Three basic types of superchargers. (Chrysler)

Supercharger Service

A faulty supercharger will exhibit many of the same symptoms described for a faulty turbocharger: lack of power, blue engine smoke, abnormal noises, etc.

If the engine lacks power, measure boost pressure by connecting a pressure gauge to a fitting on the intake manifold. If boost is low, check the ***bypass actuator,*** which controls the supercharger boost pressure, **Figure 28-37.**

A bypass actuator that is stuck open can lower boost pressure and power. If the bypass is held closed and the boost will still not reach its maximum, suspect internal rotor or housing wear in the supercharger.

To isolate noises, place a stethoscope on the supercharger. Check for bearing noise at each end of the housing. Internal noises usually require supercharger removal and rebuilding, **Figure 28-38.**

A supercharger rebuild generally involves replacing all bearings and seals. You must also measure rotor lobe wear and internal housing wear. The clearances between the rotors and housing must be within specifications. If parts are worn, most shops simply install a new or rebuilt supercharger.

Scanning a Supercharging System

With vehicles equipped with onboard diagnostic systems, use a scan tool to help find supercharging system problems. The scan tool will show any diagnostic trouble codes and will also output system electrical values. You can check the operation of the knock sensor, manifold pressure sensor, and other devices that control supercharging.

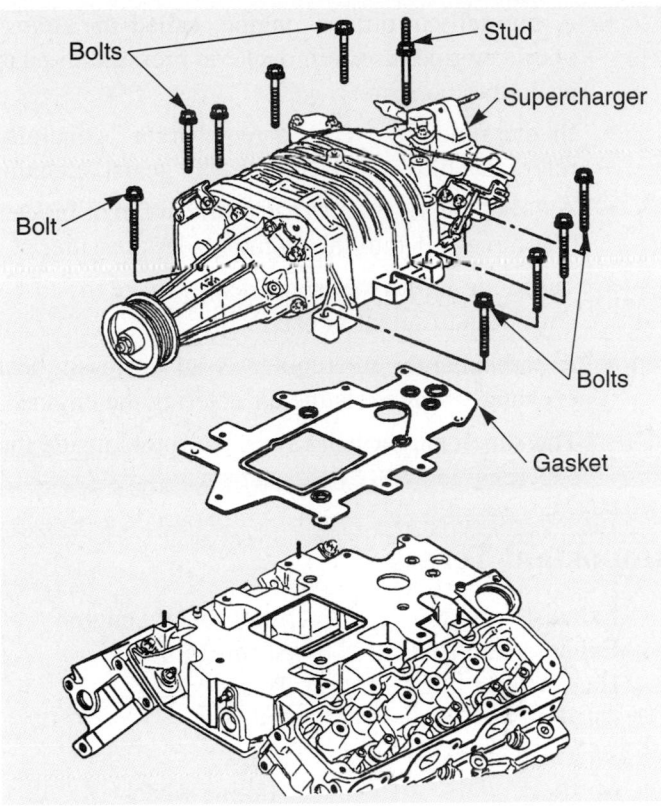

Figure 28-38. If your tests find a supercharger problem, remove it for replacement or repairs. Simply remove the belt and any parts (hoses, brackets, wires) that prevent supercharger removal. After removal of all bolts, light prying will free the supercharger from its gasket. (Oldsmobile)

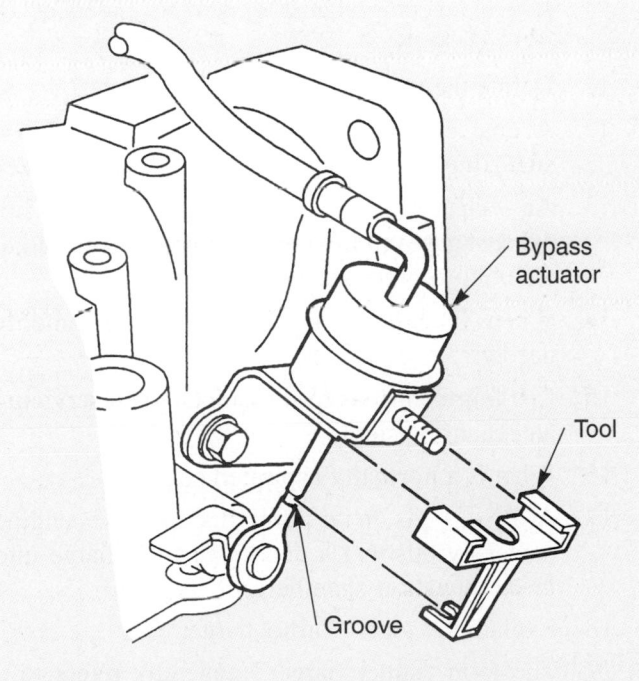

Figure 28-37. Supercharger bypass actuator serves the same purpose as a waste gate on a turbocharging system. When supercharger boost pressure reaches a maximum value, air pressure overcomes spring tension to bypass extra air, limit boost, and prevent engine damage. (Oldsmobile)

Summary

- An exhaust system quiets engine operation and carries exhaust fumes to the rear of the vehicle.

- An exhaust manifold connects cylinder head exhaust ports to header pipe.

- Header pipe steel tubing that carries exhaust gases from exhaust manifold to catalytic converter or muffler.

- A catalytic converter is a device for removing pollutants from engine exhaust.

- Intermediate pipe is tubing sometimes used between header pipe and muffler or catalytic converter and muffler.

- A muffler is a metal chamber for damping pressure pulsations to reduce exhaust noise.

- Heat shields are metal plates that prevent exhaust heat from transferring into another object.

- A supercharger is an air pump that increases engine power by pushing a denser air-fuel charge into the combustion chambers.

- A normally aspirated engine, called an atmospheric engine, uses atmospheric pressure to push air into the engine.

- In the field, the term "supercharger" generally refers to a blower driven by a belt, gears, or chain.

- A turbocharger is an exhaust-driven fan or blower that forces air into the engine under pressure.

- Turbo lag refers to a short delay before the turbo develops sufficient boost.

- A turbocharger intercooler is an air-to-air heat exchanger that cools the air entering the engine.

- The supercharger belt drives the rotors inside the supercharger.

Important Terms

Exhaust system	Atmospheric engine
Exhaust manifold	Supercharger
Header pipe	Turbocharger
Catalytic converter	Turbo
Intermediate pipe	Blower
Muffler	Turbine wheel
Tailpipe	Turbine housing
Hangers	Turbo shaft
Heat shields	Compressor wheel
Exhaust system clamps	Compressor housing
	Bearing housing
Exhaust back pressure	Blow-through turbo system
Single exhaust system	
Dual exhaust system	Draw-through turbo system
Crossover pipe	
Exhaust manifold doughnut	Turbocharger lubrication
	Sealing rings
Exhaust manifold heat valve	Drain passage
	Drain line
Exhaust pipes	Turbo lag
Heat shields	Turbocharger intercooler
Catalytic converter	Waste gate
Muffler	Knock sensor
Exhaust system service	Supercharger belt
Exhaust leaks	Electromagnetic clutch
Normally aspirated engine	Bypass actuator

Review Questions—Chapter 28

Please do not write in this text. Place your answers on a separate sheet of paper.

1. An exhaust system _____ engine operation and carries _____ _____ to the rear of the car.

Match the following exhaust system parts with their definition.

2. _____ Heat shield.
3. _____ Catalytic converter.
4. _____ Intermediate pipe.
5. _____ Hangers.
6. _____ Header pipe.
7. _____ Tailpipe.
8. _____ Muffler clamp.
9. _____ Exhaust manifold.
10. _____ Muffler.

A. U-bolt for connecting parts of exhaust system.
B. Tubing that connects exhaust manifold to rest of system.
C. Chamber for damping out pressure pulsations.
D. Carries exhaust from muffler to rear of car body.
E. Connects cylinder head exhaust ports to header pipe.
F. Prevents heat from transferring into other objects.
G. Connects exhaust manifold to tailpipe.
H. Device for removing pollutants from exhaust.
I. Pipe between catalytic converter and muffler.
J. Connects exhaust system to underside of car body.

11. Define the term "exhaust back pressure."

12. A dual exhaust system is commonly used on small, high fuel economy engines. True or False?

13. An exhaust _____ _____ _____ forces hot exhaust gases to flow into the intake manifold to aid cold weather starting.

14. When is exhaust system service commonly needed?

15. List fourteen rules to remember when servicing an exhaust system.

16. What is a normally aspirated engine?

17. A(n) _____ is an air pump that increases engine power by pushing a denser air-fuel charge into the combustion chambers.

18. Explain the term "Turbocharger."

19. The term "supercharger" generally refers to a blower driven by a(n) _____, _____, or _____.

20. List and explain the six basic parts of a turbocharger.

21. A(n) _____-_____ turbocharger has the turbo located before the carburetor or throttle body.

22. A turbocharger can operate at speeds up to _____ rpm.

23. _____ _____ refers to the short delay before the turbo develops sufficient _____ (pressure above atmospheric pressure).

24. A waste gate limits the minimum amount of boost produced by the turbocharger. True or False?

25. What could happen if a waste gate did not open?

26. List seven engine modifications commonly found on a turbocharged engine.

27. Which of the following is *not* a recommended practice when servicing a turbocharging system?
    (A) Inspect vacuum and oil lines to turbo and waste gate.
    (B) Remove carbon from turbo compressor with gasket scraper.
    (C) Use regulated, low-pressure air hose to check waste gate diaphragm leakage and operation.
    (D) Use pressure gauge to measure boost pressure.
    (E) Use stethoscope to listen for bad turbo bearings.

28. A turbocharger was badly damaged because of excess bearing and shaft wear. Technician A says that a new unit should be installed and that oil pressure to the turbo should be checked. Engine oil should be changed after a short break-in period. Technician B says that the oil should be drained and all lines should be flushed before installing the new turbocharger. Oil pressure to the unit should also be checked. Who is correct?
    (A) A only.
    (B) B only.
    (C) Both A and B.
    (D) Neither A nor B.

29. An inoperative turbocharger waste gate can cause:
    (A) high boost pressure.
    (B) low boost pressure.
    (C) detonation.
    (D) All of the above.

30. Which of the following is *not* a type of super-charger?
    (A) Vane.
    (B) Gear.
    (C) Rotor.
    (D) Centrifugal.

ASE-Type Questions

1. Each of these is a major part of an exhaust system *except:*
   (A) muffler.
   (B) hangers.
   (C) flyweights.
   (D) header pipe.

2. Which of the following connects the right and left side header pipes?
   (A) Tailpipe.
   (B) Main pipe.
   (C) Intermediate pipe.
   (D) Crossover pipe.

3. Which of these is used to reduce the amount of exhaust pollutants entering the atmosphere?
   (A) Heat shields.
   (B) Catalytic converter.
   (C) Pyrex thermal packing.
   (D) Cylinder head exhaust ports.

4. Which of the following turbocharger parts passes through the center of the turbo housing?
   (A) Turbo shaft.
   (B) Turbine wheel.
   (C) Bearing housing.
   (D) Compressor wheel.

5. The delay before a turbo develops sufficient pressure is known as:
   (A) turbo lag.
   (B) turbo boost.
   (C) turbo bearing.
   (D) boost pressure.

6. Which of the following limits the maximum amount of boost pressure developed by the turbocharger?
   (A) Diaphragm.
   (B) Waste gate.
   (C) Compressor.
   (D) Sealing rings.

7. Technician A believes a turbocharged engine will need a higher compression ratio to withstand an increase in horsepower. Technician B feels the engine will no longer need a knock sensor if horsepower is increased. Who is right?
   (A) A only.
   (B) B only.
   (C) Both A and B.
   (D) Neither A nor B.

8. Which of the following helps an on-board computer keep the ignition timing advanced as much as possible?
    (A) Diaphragm.
    (B) Control rod.
    (C) Knock sensor.
    (D) All of the above.

9. Vibration and noise in a turbocharging system is a result of:
    (A) damaged turbine.
    (B) leaking shaft seals.
    (C) inadequate boost pressure.
    (D) Any of the above.

10. When installing a turbo, each of these is a rule to remember *except:*
    (A) check and flush oil lines.
    (B) check and change engine oil.
    (C) torque all fasteners to specifications.
    (D) reuse good gaskets and seals.

Activities—Chapter 28

1. Research the internal combustion process and produce a cutaway sketch of an exhaust system. Use it to explain why high back pressure will reduce engine power.

2. Using knowledge of atmospheric pressure, demonstrate to the shop class why a turbocharger or a supercharger will increase engine power.

3. Collect literature from dealerships on turbocharged and non-turbocharged engines. Develop a table of horsepower-to-displacement ratios. (This can be done by dividing the rated horsepower by the number of cubic inches or liters of displacement for each engine.) Draw up a report of your findings and present them to the class.

Exhaust System Diagnosis		
Condition	**Possible cause**	**Correction**

Excessive exhaust noise (under hood).	1. Damaged exhaust manifold.
	2. Manifold to cylinder head leak.
	3. EGR valve leakage:
	  (a) EGR valve to manifold gasket.
	  (b) EGR valve to EGR tube gasket.
	  (c) EGR tube to manifold tube nut.
	4. Exhaust flex joint.
	  (a) Spring height, installed not correct.
	
	  (b) Exhaust sealing ring defective.
	
	5. Pipe and shell noise from front exhaust pipe.
		1. Replace manifold.
		2. Tighten manifold and/or replace gasket.
		3.
		  (a) Tighten nuts or replace gasket.
		  (b) Tighten nuts or replace gasket.
		  (c) Tighten tube nut.
		4.
		  (a) Check spring height, both sides (specification is 32.5 mm, 1.28 inch) look for source of spring height variation if out of specification.
		  (b) Inspect seal for damage on round spherical surface. If no damage is evident, check for exhaust obstruction causing high back pressure on heavy acceleration.
		5. Characteristic of single wall pipes.

Excessive exhaust noise.	1. Leaks at pipe joints. 2. Burned or blown or rusted out muffler, tailpipe of exhaust pipe. 3. Restriction in muffler or tailpipe. 4. Converter material in muffler.	1. Tighten clamps at leaking joints. 2. Replace muffler or muffler tailpipe or exhaust pipe. 3. Remove restriction, if possible or replace as necessary. 4. Replace muffler and converter assemblies. Check fuel injection and ignition systems for proper operation.
Exhaust system mechanical noise.	1. Improperly aligned exhaust system. 2. Support brackets loose, bent, or broken. 3. Incorrect muffler or pipes. 4. Baffle loose in muffler. 5. Manifold heat control valve rattles. 6. Worn engine mounts. 7. Damaged or defective catalytic converter.	1. Align system. 2. Tighten or replace brackets. 3. Install correct muffler or pipes. 4. Replace muffler. 5. Replace thermostatic spring. 6. Replace engine mounts. 7. Repair or replace converter.
Engine lacks power.	1. Clogged muffler. 2. Clogged or kinked exhaust or tailpipe. 3. Muffler or pipes too small for vehicle. 4. Catalytic converter clogged or crushed shut.	1. Replace muffler. 2. Replace pipe. 3. Install muffler and pipes of the correct size and type. 4. Replace with new converter.

Electrical Systems

Almost every system in a modern vehicle uses some type of electric or electronic component. Electronic ignition, electronic fuel injection, computerized engine management systems, anti-lock brakes, and other advanced systems require technicians skilled in electricity and electronics. Even specialized technicians need a background in electricity and electronics to fix today's vehicles.

This section details the operation, diagnosis, and repair of the major electrical systems. It starts with the battery, the source of initial electrical energy, and progresses through the starting, charging, ignition, and other systems.

Study this information carefully. A technician who is well versed in electricity and electronics is in high demand. The information in this section will help you pass ASE Test A6, *Electrical/Electronic Systems.*

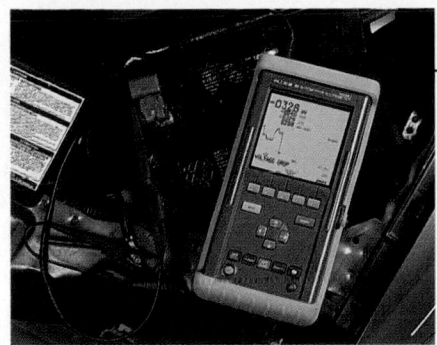

Chapter 29

Automotive Batteries

After studying this chapter, you will be able to:

- Explain the operating principles of a lead-acid battery.

- Describe the basic parts of an automotive battery.

- Compare conventional and maintenance-free batteries.

- Explain how temperature and other factors affect battery performance.

- Describe safety practices that should be followed when working with batteries.

- Correctly answer ASE certification test questions that require a basic knowledge of automotive batteries.

Earlier textbook chapters briefly introduced battery operation and electrical fundamentals. This chapter will build on this knowledge by discussing automotive batteries in more detail. It will prepare you to learn about battery service, as well as charging and starting systems.

Battery Principles

An *automotive battery* is an electrochemical device that produces and stores electricity. A cutaway of such a battery is shown in **Figure 29-1.** A battery produces direct current (*dc*) electricity, which flows in only one direction.

When *discharging* (current flowing out of the battery), the battery changes chemical energy into electrical energy. In this way, it releases stored energy.

During *charging* (current flowing into the battery from the charging system), electrical energy is converted to chemical energy. The battery can then store this energy until needed.

Battery cycling refers to repeated charging-discharging events. In extreme cases, the battery is almost fully discharged by sitting or cranking-current draw and is recharged once the engine is started.

Repeated *deep cycling* (going from a very low charge to a full charge) can shorten battery service life.

Basic Battery Cell

A simple *battery cell* consists of a negative plate, positive plate, container, and electrolyte (battery acid). Look at **Figure 29-2.**

The *battery plates* are made of lead and lead oxide. These act as dissimilar (unlike) metals. The container is usually plastic to resist corrosion. The electrolyte is a mixture of sulfuric acid and water.

If a *load* (current-using device) is connected to our simple battery cell, current will flow through the load. If

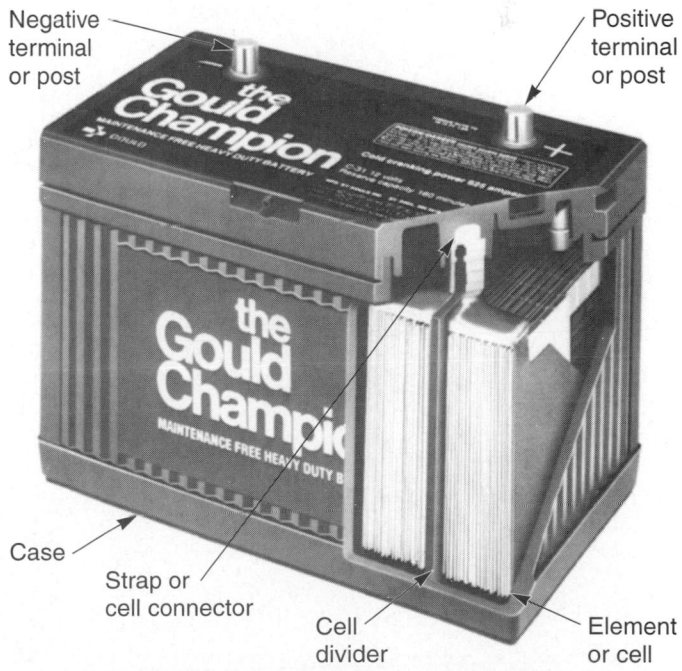

Figure 29-1. Before learning how a battery works, study in detail its basic parts. Note part names and locations. (Gould)

the load is a lightbulb, as in **Figure 29-2,** the bulb will glow because of electron movement.

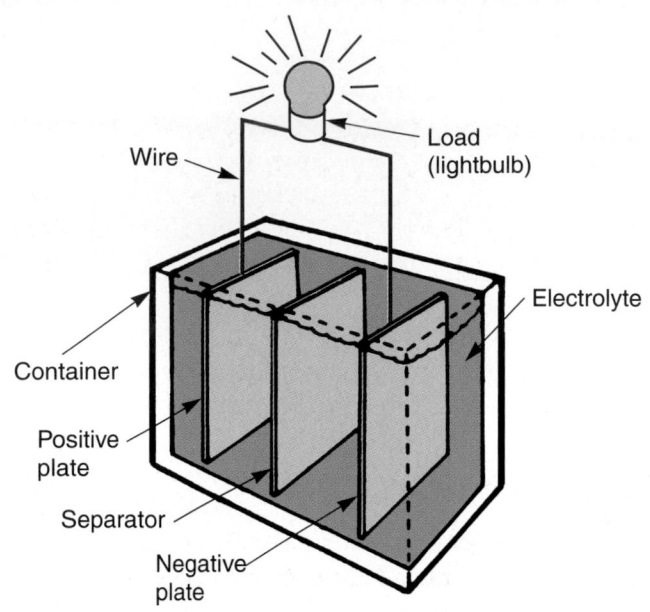

Figure 29-2. A simple lead-acid battery cell. The positive and negative plates are kept apart by a separator. An electrolyte causes a chemical reaction between plates, producing current flow through the circuit. One cell like this would produce 2.1 volts.

Battery Cell Action

Figure 29-3 shows the basic chemical-electrical action inside a battery cell. When the cell is being charged, the alternator causes free electrons (negative charges) to be deposited on the negative (–) plate. This causes the plates to have a difference in potential (electrical pressure, or voltage).

When a load is connected across the terminals, there is a current (flow of electrons) to equalize the difference in charges on the plates. The excess electrons move from the negative plate to the positive plate.

Battery Functions

A vehicle battery has several important functions. It must:

- Operate the starting motor, ignition system, electronic fuel injection system, and other electrical devices during engine cranking and starting.

- Supply all the electrical power for the vehicle whenever the engine is not running.

- Help the charging system provide electricity when current demands exceed the output limit of the charging system.

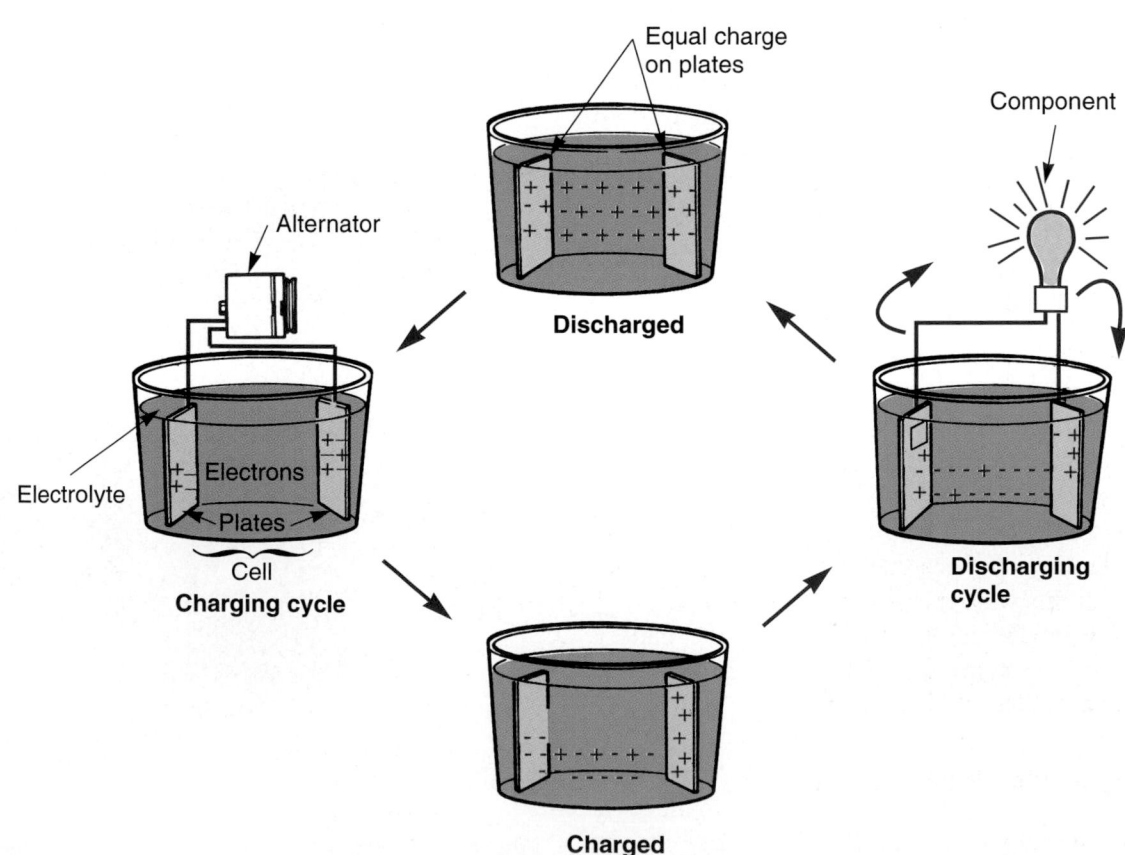

Figure 29-3. Basic battery cycling. (Chrysler)

- Act as a capacitor (voltage stabilizer) that smoothes current flow through the electrical system.
- Store energy (electricity) for extended periods.

Imagine the following sequence of events: You are sitting in your car with the radio on, but the engine is not running. The battery is supplying the electricity to operate the radio and any indicator lights. It is slowly discharging.

When you start the engine, the battery provides a tremendous amount of current. This energy operates the starting motor and essential engine systems. This, too, drains current out of the battery.

As soon as the engine starts, the charging system takes over. It recharges the battery while feeding current to the electrical units in the car.

If the load becomes too much for the charging system (engine idling slowly and all accessories on, for example), the battery may also feed current into the electrical system.

Battery Construction

Automobile batteries are built to withstand severe vibration, extreme temperatures, corrosive chemicals, high current discharge, and prolonged periods without use. To properly test and service batteries, you must understand battery construction.

Battery Element

A *battery element* is made up of positive plates, negative plates, straps, and separators. The element fits into a cell compartment in the battery case. Refer to **Figure 29-4.**

The *battery plates* are made of a grid (stiff mesh framework) coated with porous lead. Shown in **Figure 29-4,** several plates are needed in each cell to provide enough battery power.

A *lead strap* connects several negative plates to form a negative plate group. Look at **Figure 29-4.** Another lead strap connects the positive plates for the positive plate group.

The chemically active material in the negative plates is porous lead, **Figure 29-4.** The active material on the positive plates is lead peroxide. Calcium or antimony is normally added to the lead to increase battery performance and to decrease *gassing* (acid fumes forming during chemical reaction).

Since the lead on the plates is porous, like a sponge, the battery acid easily penetrates the material. This aids the chemical reaction and the production of electricity.

Lead battery straps, or connectors, run along the upper portion of the case to connect the plates. The battery terminals (top posts or side terminals) are constructed as part of one end of each strap.

Separators fit between the battery plates to keep them from touching each other and shorting. The separators are made of insulating material. They have openings that allow free circulation of the electrolyte around the battery plates.

Battery Case, Cover, and Caps

The *battery case,* usually made of high-quality plastic, holds the elements and electrolyte, **Figure 29-5.** The case must withstand extreme vibration, temperature change, and the corrosive action of the battery acid. Dividers in the case form individual containers for each element. A container and its element make up one cell.

The *battery cover* is bonded to the top of the battery case. It seals the top of the case. There is an opening above each battery cell for battery caps or a cell cover. Refer to **Figure 29-5.**

Battery caps snap into the holes in the battery cover. They keep electrolyte from splashing out of the battery. The caps also serve as spark arrestors (keep sparks or flames from igniting gases inside battery). Maintenance-free batteries have a large cover that is not removed during normal service.

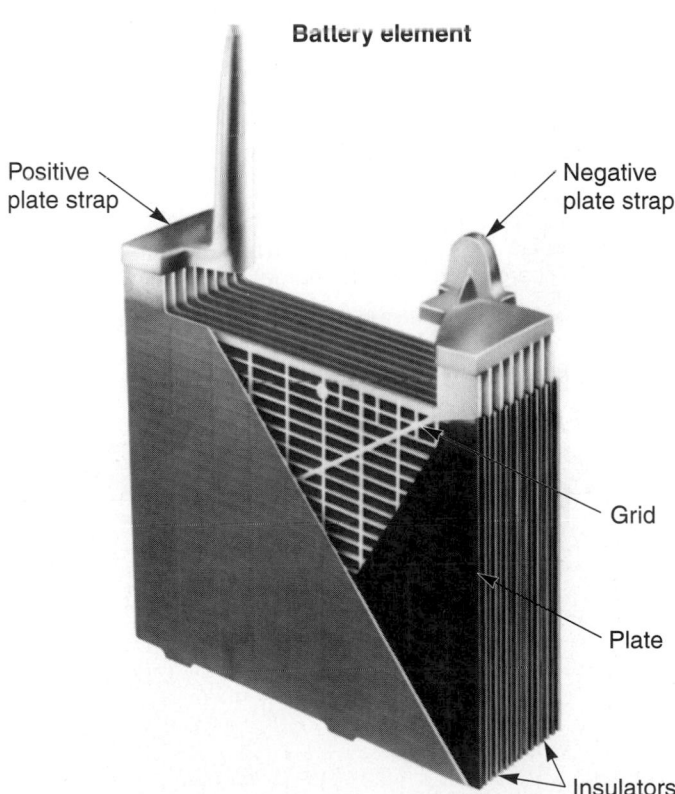

Battery element

Positive plate strap

Negative plate strap

Grid

Plate

Insulators

Figure 29-4. A battery element is made up of a positive plate group, negative plate group, separators, and straps. Most auto batteries have six elements. The elements fit into the battery case. (Gould)

Warning

Hydrogen gas can collect at the top of a battery. If this gas is exposed to a flame or spark, it can explode!

Electrolyte (Battery Acid)

Electrolyte, often called *battery acid,* is a mixture of sulfuric acid and distilled water, **Figure 29-6.** Electrolyte is poured into each cell until plates are covered.

Distilled water is used in batteries because it does not contain many of the impurities found in tap water. These impurities can contaminate the battery plates and reduce efficiency.

Warning

Avoid having electrolyte come in contact with your skin or eyes. The sulfuric acid in the electrolyte can cause serious skin burns or even blindness.

Battery Charge Indicator

A *battery charge indicator,* also called a *battery eye* or *test indicator,* shows the general charge condition of the battery. One is pictured in **Figure 29-7.**

The charge indicator changes color as the level of battery charge changes. For example, the indicator may be green with the battery fully charged. It may turn black

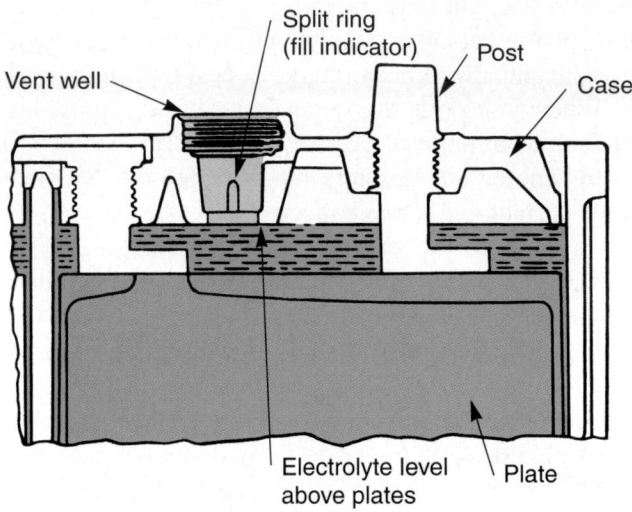

Figure 29-6. Electrolyte or battery acid covers plates. The acid should just touch the split ring in the top of the case. A vent allows gases to leave the case. (GMC)

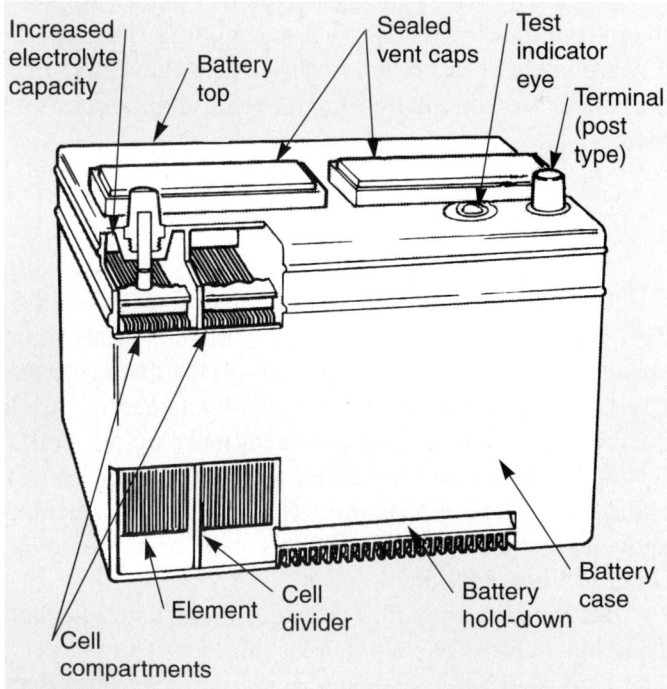

Figure 29-5. The battery case holds the elements and electrolyte. Note the part names. (Chrysler)

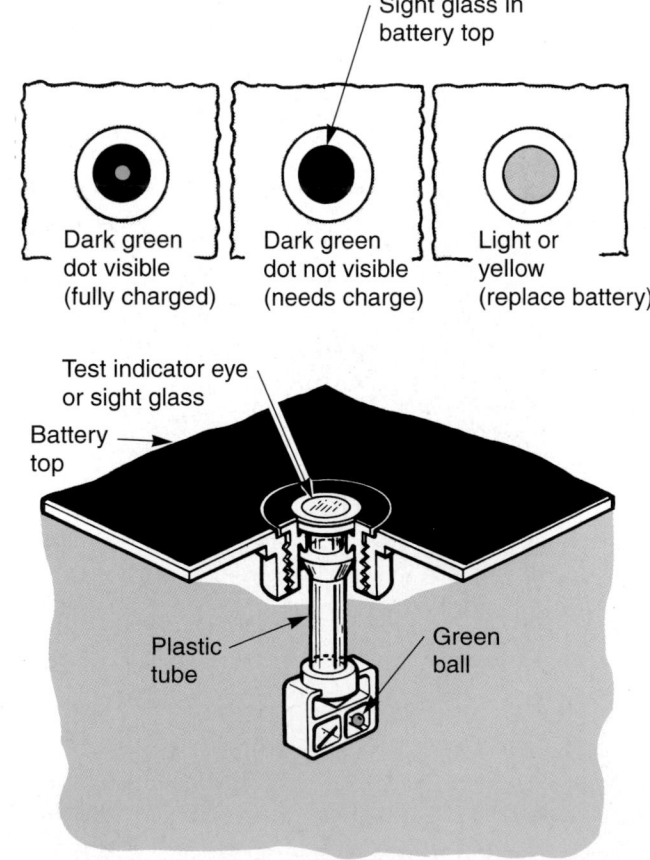

Figure 29-7. A charge indicator provides an easy way of checking battery condition. (Chrysler)

when discharged or yellow when the battery needs replacement.

Battery Terminals

Battery terminals provide a means of connecting the battery plates to the car's electrical system. Either two round posts or two side terminals can be used, as shown in **Figure 29-8.**

Battery posts are round metal terminals that extend through the top of the battery cover. They serve as male connections for female battery cable ends.

The *positive post* is larger than the negative post. It may be marked with red paint and a positive (+) symbol. The *negative post* is smaller and may be black or green in color. It normally has a negative (–) symbol on or near it.

Side terminals are electrical connections located on the side of the battery. They have female threads that accept a special bolt on the battery cable end. Side terminal polarity is identified by positive and negative symbols on the case.

Battery Voltage

Battery *open circuit* (no load) *cell voltage* is 2.1 volts, often rounded to 2 volts. Since the cells in a battery are connected in series, battery voltage depends on the number of cells. Refer to **Figure 29-9.**

A 12-volt battery has six cells, which produce an open circuit voltage of 12.6 volts. Modern vehicles use a 12-volt battery and 12-volt electrical system. Older vehicles are designed to use 6-volt batteries. See **Figure 29-9.**

Some cars with diesel engines use two 12-volt batteries connected in parallel. When two batteries are in parallel, their output voltage stays the same, but their current output increases. Dual batteries may be needed to crank and start a compression-ignition diesel engine.

 Caution
When two batteries are connected in *series,* their output voltage doubles. Keep this in mind when working with batteries. Two 12-volt batteries connected in series produces 24 volts, which could damage electrical devices.

Battery Cables

Battery cables are large wires that connect the battery terminals to the vehicle's electrical system.

The *positive cable* is normally red and fastens to the starter solenoid (introduced in Chapter 1). The *negative battery cable* is usually black and connects to ground on

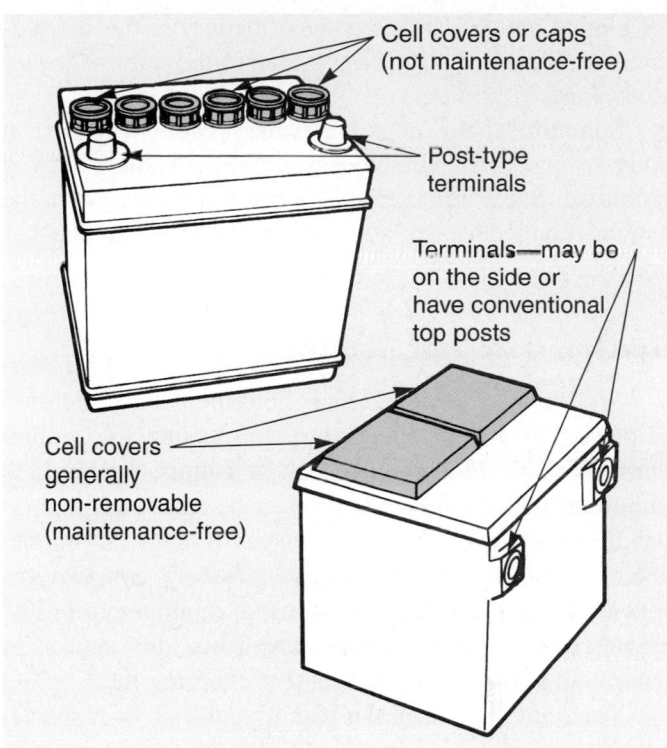

Figure 29-8. Study the differences between batteries. Maintenance-free batteries do not have vent caps. Post or side terminals may be used on conventional or maintenance-free batteries. (Chrysler)

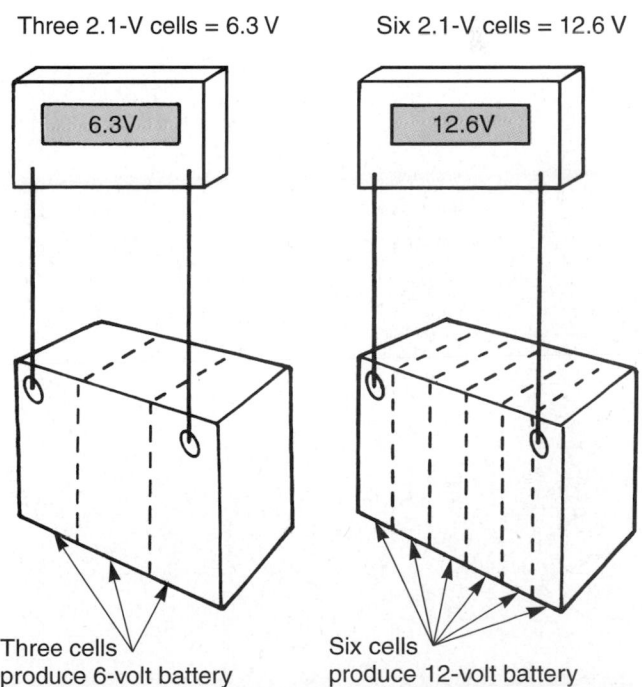

Figure 29-9. Three cells connected in series produce 6.3 volts. The three cells are rated as, or called, a 6-volt battery. A more common unit of six cells produces 12.6 volts, or a 12-volt battery.

the engine block. Various types of battery cables are pictured in **Figure 29-10.** Note that ground cables do not always use insulation.

Sometimes, the negative battery cable will have a body ground wire, which ensures that the vehicle body is grounded. See **Figure 29-11.** If this wire does not make a good connection, a component grounded to the car body may not operate properly.

Battery Tray and Retainer

A *battery tray* and *retainer* hold the battery securely in place. They keep the battery from bouncing around during vehicle movement. Look at **Figure 29-12.** It is important that the tray and retainer be tight and in good condition to prevent battery damage. In many late-model vehicles, the battery tray houses a *battery temperature sensor.* This sensor monitors battery temperature The engine control module uses data from this sensor to determine when to vary the battery charging rate.

The *battery hold-down* usually consists of a special bracket and fasteners that secure the battery to its tray. The hold-down must be tight enough to prevent battery movement.

Battery heat shields are designed to keep excess engine heat from damaging the battery. They usually fit over the battery or between the engine and battery case. Always reinstall battery heat shields to ensure normal battery life. See **Figure 29-13.**

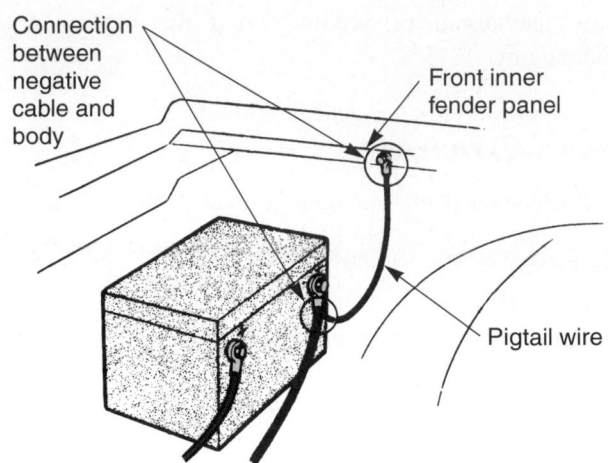

Figure 29-11. Note the cable connections to the battery. The negative cable grounds on the engine block. The positive cable connects to the electrical system. A pigtail grounds the car body to battery negative. (Sun)

Wet- and Dry-Charged Batteries

There is no difference in the materials used in wet- and dry-charged batteries. The difference is in how the batteries are prepared for service.

With a *wet-charged battery,* the battery is filled with electrolyte and charged at the factory. The battery is then tested and placed in stock, ready for service.

A *dry-charged battery* contains fully charged elements but does not contain electrolyte. It leaves the

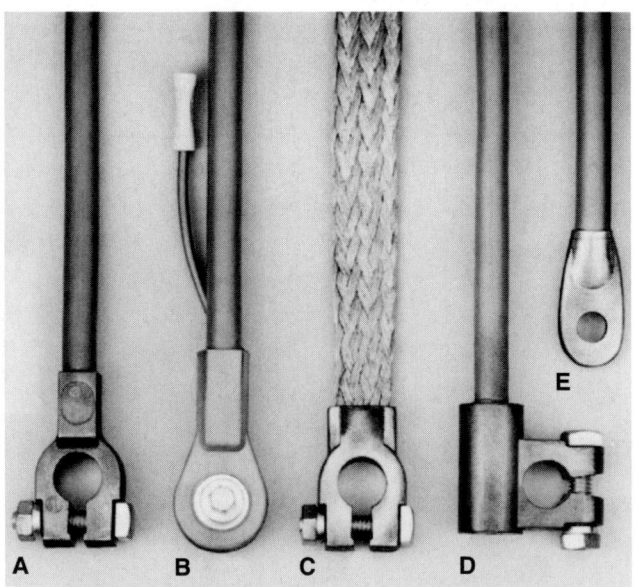

Figure 29-10. Battery cable types. A—Post-type battery cable. B—Side terminal battery cable with pigtail for ground or accessory connection. C—Braided ground cable. D—90° post-type cable. E—Solenoid-to-starter cable. Note the large conductor size for carrying a large amount of current to the starter. (Belden)

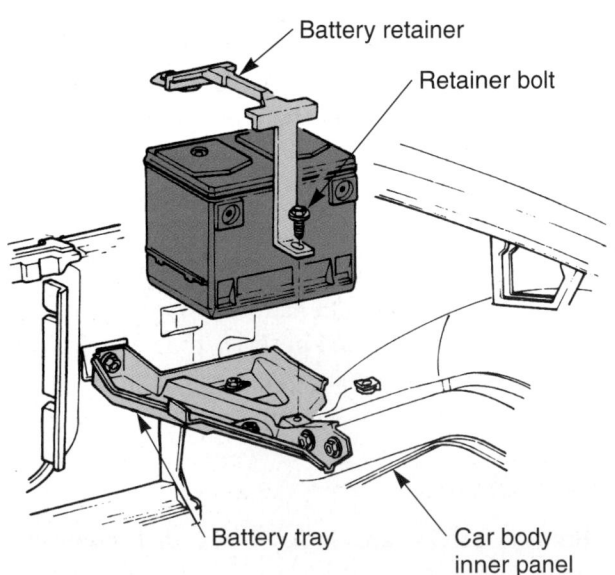

Figure 29-12. Battery tray and retainer hold the battery securely in place. The tray usually mounts on the inner body panel. (Cadillac)

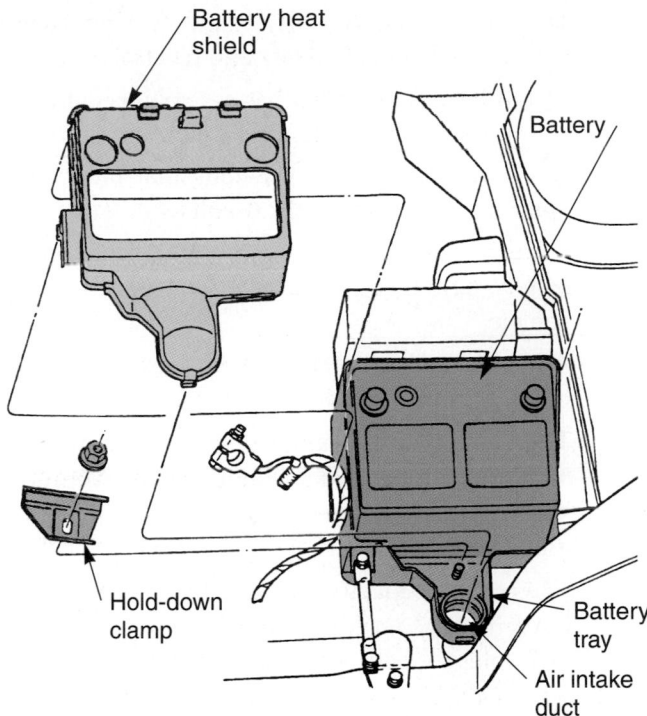

Figure 29-13. A battery heat shield is used to protect the battery from excess engine heat. Note the battery tray in this particular vehicle contains an air intake duct, which routes outside air to the area between the heat shield and the battery case. (Chrysler)

factory in a dry state. Before use, this type of battery is filled with electrolyte. A dry-charged battery is commonly used because it has a much longer shelf life than a wet-charged battery.

Note

Activation of a dry-charged battery is covered in Chapter 30, *Battery Testing and Service.*

Maintenance-Free Battery

A *maintenance-free battery* is easily identified because it does not use removable filler caps. The tops of the battery cells are covered with a large, snap-in cover. Calcium is used to make the battery plates, so water does not have to be periodically added to the electrolyte.

The calcium in the plates reduces the production of battery gases. As a result, battery gas does not carry as much of the chemicals out of the battery. This increases battery service life and decreases service requirements.

Battery Ratings

Battery ratings are set according to national test standards for battery performance. They let the technician and consumer compare the cranking power of one battery to another.

The two most common methods of rating lead-acid storage batteries are the cold cranking rating and reserve capacity rating. These ratings were developed by the Society of Automotive Engineers (SAE) and the Battery Council International (BCI).

Battery Cold Cranking Rating

The *cold cranking rating* determines how much current (in amperes) the battery can deliver for 30 seconds at 0°F (–18°C) while maintaining terminal voltage of 7.2 volts, or 1.2 volts per cell. This rating indicates the battery's ability to crank a specific engine (based on starter current draw) at a specified temperature.

For example, one auto manufacturer recommends a battery with 305 cold cranking amps for a small 4-cylinder engine but a battery with 450 cold cranking amps for a larger V-8 engine. A more powerful battery is needed to handle the heavier starter current draw of the larger engine.

A *battery watt rating* is another battery rating. It is the equivalent of the cold cranking rating. Remember that watts are a measure of power equal to voltage multiplied by current.

Battery Reserve Capacity Rating

The *reserve capacity rating* is the time needed to lower battery terminal voltage below 10.2 volts (1.7 volts per cell) at a discharge rate of 25 amps. This is with the battery fully charged and at 80°F (27°C).

Reserve capacity will be given as a time interval in minutes. For example, if a battery is rated at 90 minutes and the charging system fails, the driver has approximately 90 minutes of driving time under minimum electrical load before the battery goes completely dead.

Battery Amp-Hour Rating

The *amp-hour rating* was once used to indicate battery power. It was a measurement of how much current the battery could produce for 20 hours at 80°F (27°C) with battery voltage staying above 10.5 volts.

Battery Temperature and Efficiency

As battery temperature drops, battery power is reduced. At low temperatures, the chemical action inside the battery is slowed. When cold, a battery will not produce as much current as when warm. This affects a battery's ability to start an engine in extremely cold weather. Also, when an engine is cold, the motor oil is very thick. This increases the amount of current needed to crank the engine with the starting motor.

Figure 29-14 shows a chart comparing battery efficiency and required starting power. Note that at 0°F (–18°C), a battery has only 40% of its normal cranking power. In addition, starter current draw will be up approximately 200% of its normal value. The engine could be very difficult to start on a cold morning. The battery, starter, and electrical connections must be in almost perfect condition.

Parasitic Loads

A *parasitic load* includes any current draw present when all electrical devices are shut off. This load would include anything that requires a small current when the engine and ignition key are turned off. On-board computers and the dashboard clock need a small amount of current to retain memory. This results in a parasitic load on the battery.

If a vehicle sits unused for prolonged periods, the parasitic load can drain the battery enough to prevent starting. A battery must be in good condition to withstand prolonged parasitic load.

Summary

- An automotive battery is an electrochemical device for producing and storing electricity.

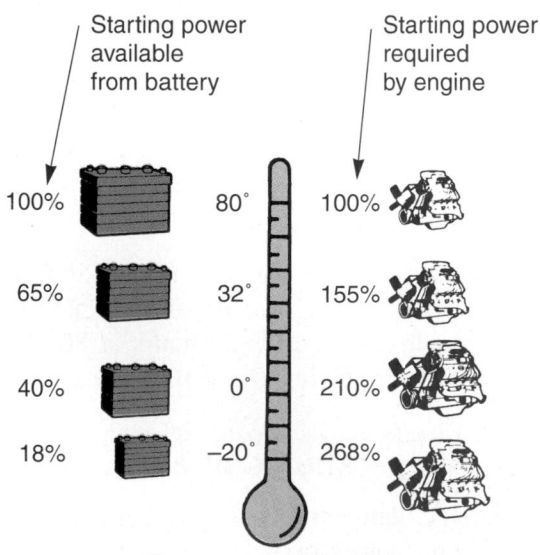

Figure 29-14. Temperature affects battery power and starter current draw. This is why engines crank slowly in very cold weather. (Champion Spark Plugs)

- The battery case, usually made of high quality plastic, holds the elements and electrolyte.
- Electrolyte, often called battery acid, is a mixture of sulfuric acid and distilled water.
- A 12-volt battery has six cells that produce an open circuit voltage of 12.6 volts.
- Battery cables are large wires that connect the battery terminals to the electrical system of the vehicle.
- A maintenance-free battery is easily identified because it does not use removable filler caps.
- The cold cranking rating determines how much current (in amperes) the battery can deliver for 30 seconds at 0°F (–18°C) while maintaining terminal voltage of 7.2 volts or 1.2 volts per cell.
- The reserve capacity rating is the time needed to lower battery terminal voltage below 10.2 volts (1.7 volts per cell) at a discharge rate of 25 amps.
- A parasitic load includes any current draw present when all electrical devices are shut off.

Important Terms

Automotive battery	Test indicator
DC	Battery terminals
Discharging	Battery posts
Charging	Side terminals
Battery cycling	Open circuit cell voltage
Battery cell	Battery cables
Battery plates	Battery tray
Load	Retainer
Battery element	Battery hold-down
Lead strap	Battery heat shields
Gassing	Wet-charged battery
Separators	Dry-charged battery
Battery case	Maintenance-free
Battery cover	battery
Battery caps	Cold cranking rating
Electrolyte	Battery watt rating
Battery acid	Reserve capacity rating
Battery charge	Amp-hour rating
indicator	Parasitic load
Battery eye	

Review Questions—Chapter 29

Please do not write in this text. Place your answers on a separate sheet of paper.

   1. A(n) _____ _____ is an electrochemical device for producing and storing electricity.

2. Define the terms battery "discharging" and "charging."

3. Which of the following is *not* part of a basic battery cell?
   (A) Positive plate.
   (B) Negative plate.
   (C) Electrolyte.
   (D) Spark arrestorsArrestors.

4. List five functions of a car battery.

5. Explain the function of a spark arrestor in an automotive battery.

6. _____ gas can collect around the top of batteries. If this gas is exposed to a flame or spark, it can _____ !

7. Electrolyte, also called battery acid, is a mixture of sulfuric acid and distilled water. True or False?

8. What is the purpose of a charge indicator or eye?

9. Describe the difference between battery posts and side terminals.

10. A 12-volt battery has _____ cells that produce an open circuit voltage of _____ volts.

11. Most modern vehicle batteries are 6-volt. True or False?

12. The battery positive cable normally connects to the _____ _____ and the negative cable connects to _____ on the engine.

13. Explain the difference between a wet- and a dry-charged battery.

14. A(n) _____ battery is easily identified because it usually does *not* have removable filler caps or covers.

15. Which of the following is *not* a conventional battery rating?
    (A) Hot cranking amps.
    (B) Reserve capacity rating.
    (C) Watt rating.
    (D) Cold cranking amps.

⚙ ASE-Type Questions

1. An automotive battery produces:
   (A) *hydrogen gas.*
   (B) *voltage.*
   (C) *direct current.*
   (D) *All of the above.*

2. Battery charging converts:
   (A) *atoms into electrolytes.*
   (B) *electrical energy into chemical energy.*
   (C) *chemical energy into electrical energy.*
   (D) *lead oxide into sulfuric acid and water.*

3. Which of the following is *not* a battery cell component?
   (A) *Eye.*
   (B) *Battery acid.*
   (C) *Positive plate.*
   (D) *Negative plate.*

4. Battery plates are grids which are coated with:
   (A) *plastic.*
   (B) *copper.*
   (C) *thin steel.*
   (D) *porous lead.*

5. Which of the following connect a battery's plates?
   (A) *Posts.*
   (B) *Straps.*
   (C) *Separators.*
   (D) *Side terminals.*

6. The active material on positive plates is:
   (A) *antimony.*
   (B) *sponge lead.*
   (C) *lead peroxide.*
   (D) *None of the above.*

7. Which of the following serve as spark arrestors?
   (A) *Battery caps.*
   (B) *Battery plates.*
   (C) *Battery separators.*
   (D) *All of the above.*

8. Which of the following shows the general charge condition of a battery?
   (A) *Indicator eye.*
   (B) *Test indicator.*
   (C) *Charge indicator.*
   (D) *All of the above.*

9. When testing a battery's charge condition, the indicator shows a visible dark green dot. Technician A interprets this as meaning the battery needs to be charged. Technician B feels the dot indicates the battery is fully charged. Who is right?
   (A) *A only.*
   (B) *B only.*
   (C) *Both A and B.*
   (D) *Neither A nor B.*

10. Each of the following provides a means of connecting battery plates to a car's electrical system *except:*
    (A) *battery caps.*
    (B) *battery posts.*
    (C) *side terminals.*
    (D) *battery terminals.*

Activities—Chapter 29

1. Survey the service managers of at least three businesses that sell and install vehicle batteries. Find out how they dispose of batteries they remove from vehicles.

2. Find the best battery value. Gather information from catalogs or advertisements, or visit stores that sell automotive batteries. Find the cold cranking ratings and prices of comparable-size batteries from different stores, and make a chart of your findings. Which will give you the most for your money?

3. Construct a model of a vehicle battery element (cell) for classroom display. Clearly label all components.

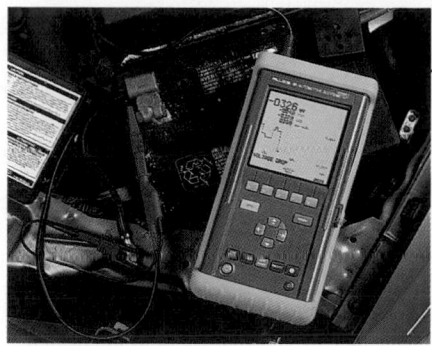

Chapter 30

Battery Testing and Service

After studying this chapter, you will be able to:

- Visually inspect a battery for obvious problems.
- Perform common battery tests.
- Clean a battery case and terminals.
- Charge a battery.
- Jump start a car using a second battery.
- Replace a defective battery.
- Describe safety practices to follow when testing and servicing batteries.
- Correctly answer ASE certification test questions on battery service.

A "*dead battery*" (discharged battery) is a very common problem. The engine will usually fail to crank and start. Even though the lights and horn may work, there is not enough "juice" (current) in the battery to operate the starting motor.

Since this is a common trouble, it is important for you to know how to inspect, test, and service vehicle batteries. This chapter covers the most common tasks relating to battery service. Study it carefully and you will be prepared for later chapters on starting and charging systems.

Battery Maintenance

If a battery is not maintained properly, its service life will be reduced. Battery maintenance should be done periodically—during tune-ups, grease jobs, or anytime symptoms indicate battery problems. *Battery maintenance* typically includes:

- Checking the electrolyte level or the indicator eye.
- Cleaning battery terminal connections.
- Cleaning the battery top.
- Checking the battery hold-down and tray.
- Inspecting for physical damage to the case and terminals.

Inspecting Battery Condition

Inspect the battery anytime the hood is opened. Check for the types of problems shown in **Figure 30-1.** Look for a dirt buildup on the battery top. Look for case damage, loose or corroded connections, or any other trouble that could upset battery operation. If a problem is found, correct it before it gets worse.

⚠ **Warning**
Wear eye protection when working around batteries. Batteries contain sulfuric acid, which can cause blindness. Even the film buildup on a battery can contain acid.

Battery Leakage Test

A *battery leakage test* will find out if current is discharging across the top of the battery case. A dirty battery

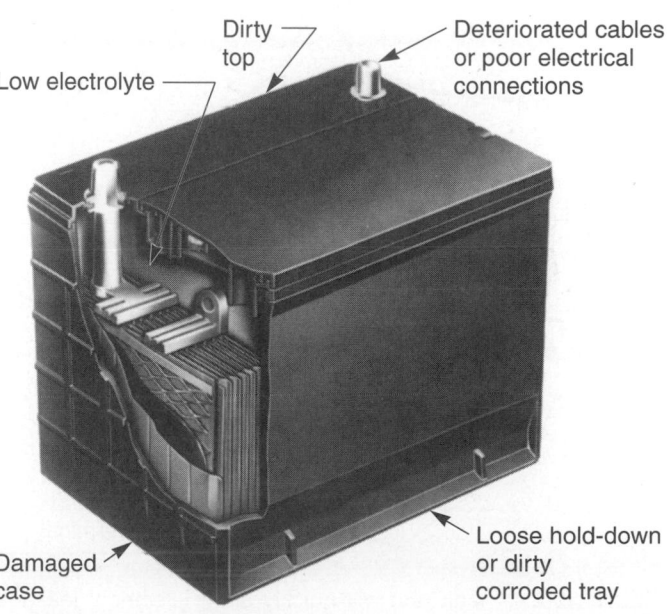

Figure 30-1. Visually inspect batteries for these kinds of problems. If any problems are found, correct them. (GMC)

can run down (discharge) when not in use. This can shorten battery life and cause starting problems.

To do a battery leakage test, set a voltmeter on a low setting. Touch the acid-resistant probes on the battery as shown in **Figure 30-2.** If the meter registers voltage, current is leaking out of the battery cells. You need to clean the battery top.

Cleaning the Battery Case

If the top of the battery is dirty, wash it with baking soda and water. See **Figure 30-3.** This will neutralize and remove the acid-dirt mixture. If not a maintenance-free battery, be careful not to let debris enter the filler openings.

Battery Terminal Test

A **battery terminal test** quickly checks for a poor electrical connection between the battery cables and terminals. A voltmeter is used to measure voltage drop across the cables and terminals, as in **Figure 30-4.**

To perform a battery terminal test:
1. Connect the negative voltmeter lead to the cable end.

2. Touch the positive voltmeter lead on the battery terminal.
3. Disable the ignition or injection system so the engine will not start.
4. Crank the engine while watching the volt-meter reading.

If the voltmeter shows over 0.5 volt, there is a high resistance at the cable connection. This would tell you to clean the battery connections. A clean, good electrical connection would have less than 0.5 volt drop.

Cleaning Battery Terminals

To clean the terminals, remove the battery cables. See **Figure 30-5.** Use a six-point wrench if the bolt or nut is extremely tight. Use pliers only on a spring-type cable

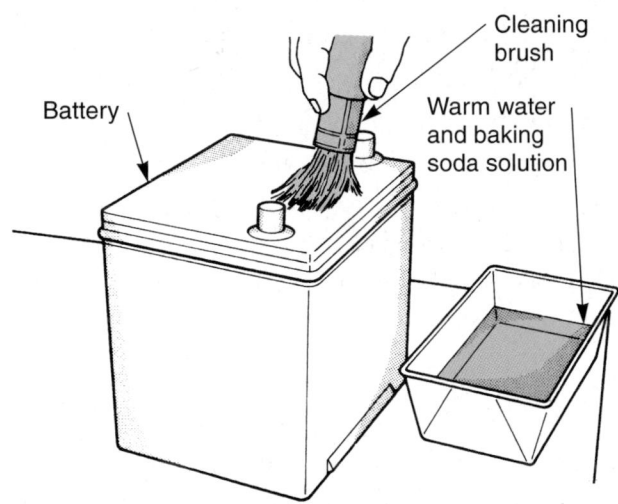

Figure 30-3. Clean battery with a baking soda–water solution and a brush. Keep dirt out of filler openings. (Chrysler)

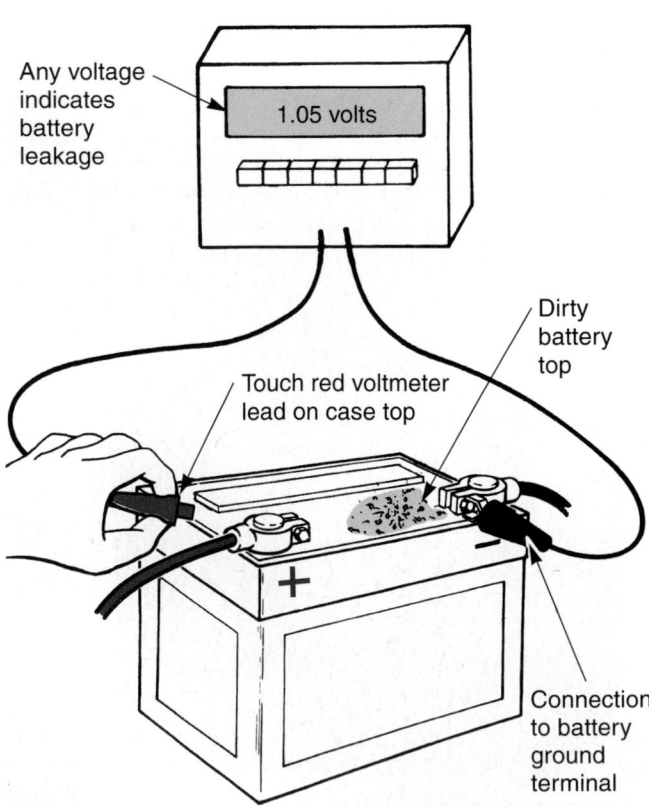

Figure 30-2. A leak test will quickly show electrical leakage across the top of a battery. If the voltmeter registers, clean the battery. (Sun Electric)

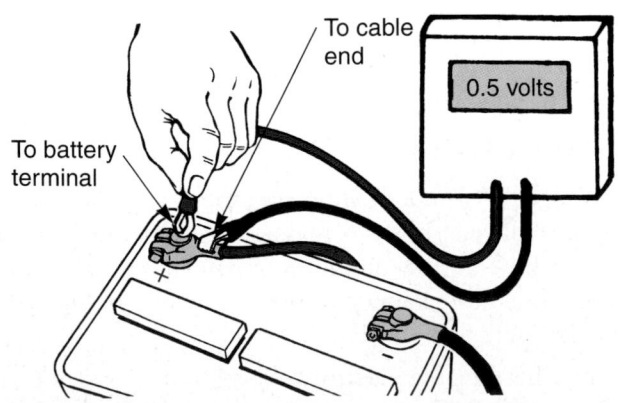

Figure 30-4. To quickly find out if battery terminals need cleaning, measure the voltage drop across the cable-to-terminal connection. Crank the engine with the ignition disabled. A reading of over 0.5 volt would require terminal and cable end cleaning. (NAPA)

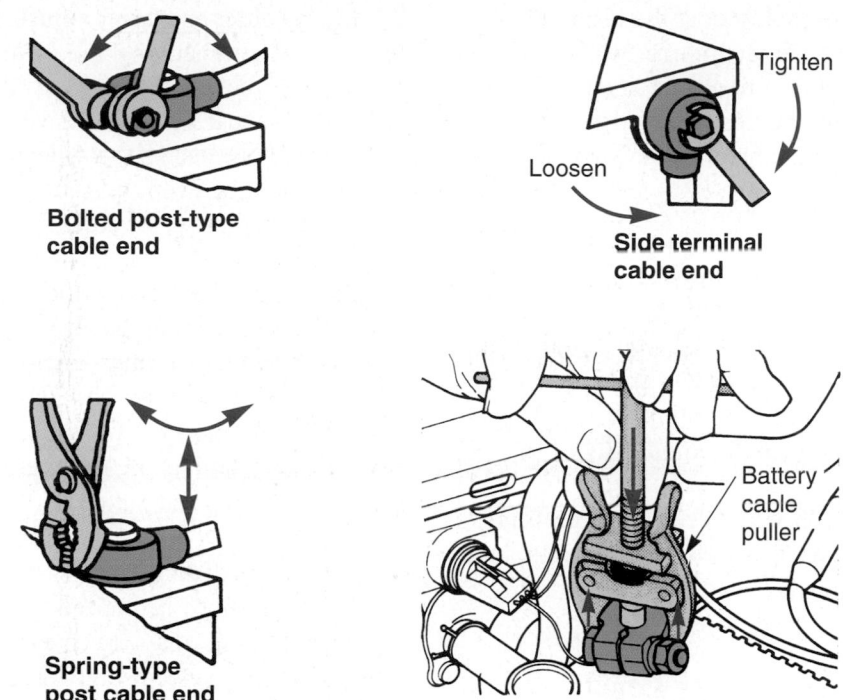

Figure 30-5. Note methods of removing a battery cable from a battery terminal. Be careful not to damage the post or side terminal. (Chrysler)

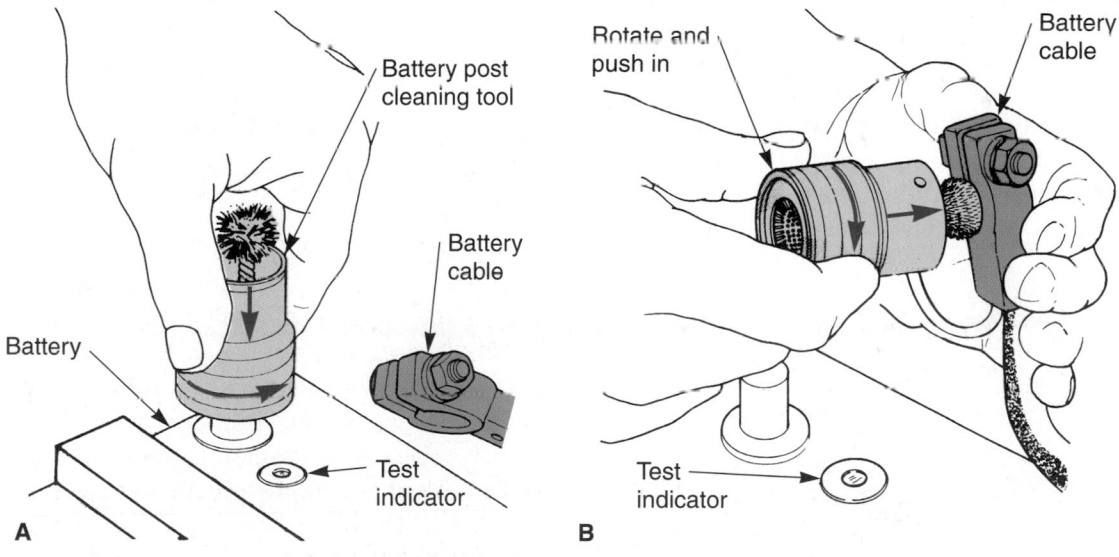

Figure 30-6. Cleaning battery posts and cable ends. Rotate the female end of the cleaner on the post. Use the male end of the cleaner on the cable end. Turn it until all corrosion is gone. (Chrysler)

end or when the fastener head is badly corroded and rounded off. Be careful not to damage the post or side terminal with excess side force.

To clean post-type terminals, use a cleaning tool like the one in **Figure 30-6.** Use the female end to clean the post. Use the male end on the terminal. Twist the tool to remove the oxidized outer surface on the connections. To clean side terminals, use a small wire brush. Polish both the cable end and the mating surface on the battery terminal.

Caution
Do not use a knife or scraper to clean battery terminals. This removes too much metal and can ruin the terminal connection.

When reinstalling the cables, coat the terminals with petroleum jelly or white grease, **Figure 30-7.** This will keep acid fumes off the connections and keep them from corroding again. Tighten the fasteners just enough to secure the connection. Overtightening can strip the cable bolt threads.

Tech Tip
When disconnecting the car battery, consider using a memory saver to keep power to the computer. This will keep all driver programmed information intact (clock, radio stations, etc.). It will also avoid having the computer relearn information, which could temporarily upset driveability. Some systems require over a hundred miles of driving to relearn everything (ideal ignition timing, injector pulse width, etc.) and operate normally.

Checking Battery Electrolyte Level

Unlike older batteries, maintenance-free batteries do not need periodic electrolyte service under normal conditions. Maintenance-free batteries are designed to operate for long periods without loss of electrolyte. Older batteries with removable vent caps, however, must have their electrolyte level checked.

Warning
The invisible hydrogen gas produced by the chemical reaction in a battery is flammable. Keep all sparks and flames away from the top of a battery. Batteries can explode if the gas is ignited!

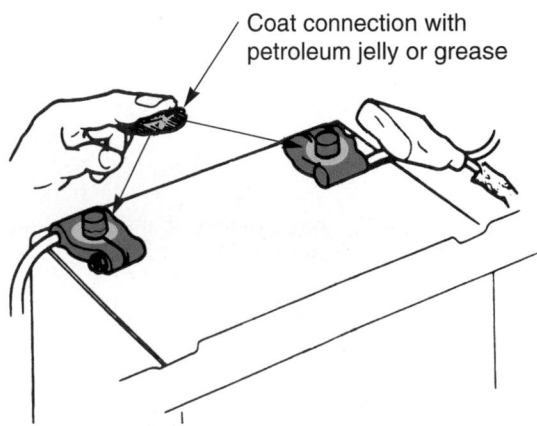

Coat connection with petroleum jelly or grease

Figure 30-7. Before reconnecting battery cables, coat the connection with petroleum jelly or white grease. This will help prevent corrosion from battery gases. Do not overtighten cable fasteners or damage may result. Most terminals are made of very soft lead. (Honda)

Many older batteries must have their vent caps removed when checking the electrolyte. The electrolyte should just cover the top of the battery plates and separators. Most batteries have a fill ring (electrolyte level indicator) inside the filler cap opening. The electrolyte should be even with the fill ring.

If the electrolyte is low, fill the cells to the correct level with *distilled water* (purified water). Distilled water should be used because it does not contain many of the impurities found in tap water. Water taken directly out of a water faucet can contain chemicals that reduce battery life. The chemicals can contaminate the electrolyte and collect in the bottom of the battery case. If enough contaminants collect in the battery, the cell plates can short out, ruining the battery.

Battery Overcharging
If water must be added at frequent intervals, the charging system may be overcharging the battery. A faulty charging system can force excessive current into the battery. Battery gassing can then remove water from the battery.

Note
Refer Chapter 34, *Charging System Diagnosis, Testing, and Repair,* for more information on this subject.

Checking Battery Charge

When measuring battery charge, you are checking the condition of the battery electrolyte and battery plates. For example, if lights are left on without the engine running, the battery will run down (discharge). Current flow out of the battery will steadily reduce available battery power. There are several ways to measure battery charge.

Some modern batteries use a charge indicator eye that shows battery charge. You simply look at the eye in the battery cover to determine battery charge, **Figure 30-8.** This was covered in the previous chapter.

Hydrometer Check
A *hydrometer* measures the specific gravity (weight or density) of a liquid. A battery hydrometer measures the specific gravity (and the state of the charge) for battery electrolyte. Look at **Figure 30-9.**

Water has a specific gravity standard of one (1.0). Fully charged electrolyte has a specific gravity of between 1.265 and 1.299. The larger number denotes that electrolyte is more dense (heavier) than water.

As a battery becomes discharged, its electrolyte has a larger percentage of water. Thus, a discharged battery's electrolyte will have a lower specific gravity than a fully

Figure 30-8. The battery is vital to starting a vehicle. Always inspect the battery and its cables during service. They can be located in the engine compartment, under the back seat, or in the trunk. (GMC)

charged battery. This rise and drop in specific gravity can be used to check the charge in a battery.

To use a float-type hydrometer, squeeze the hydrometer bulb. Immerse the end of the hydrometer in the electrolyte. Then release the bulb, **Figure 30-9.** Compare the numbers on the hydrometer float with the top of the

electrolyte. Hold the hydrometer even with your line of sight. Wear safety glasses and do not drip electrolyte on anything. See **Figure 30-10.**

Most float-type hydrometers are not temperature correcting. However, some models have a built-in thermometer and a conversion chart. This will let you compensate for battery temperature.

To use a ball-type hydrometer, draw electrolyte into the hydrometer with the rubber bulb. Then note the number of balls floating in the electrolyte. Instructions on the hydrometer will tell you whether the battery is fully charged or discharged.

A needle-type hydrometer uses the same principle as the ball-type. When the electrolyte is drawn into the hydrometer, it causes a plastic needle to register specific gravity.

Hydrometer Readings

A fully charged battery should have a hydrometer reading of at least 1.265. If the reading is below 1.265, the battery needs recharging or it may be defective.

A discharged battery could have several causes:

- Defective battery.
- Charging system problem (loose alternator belt, for example).
- Starting system problem.
- Poor cable connections.
- Engine performance problem requiring excessive cranking time.

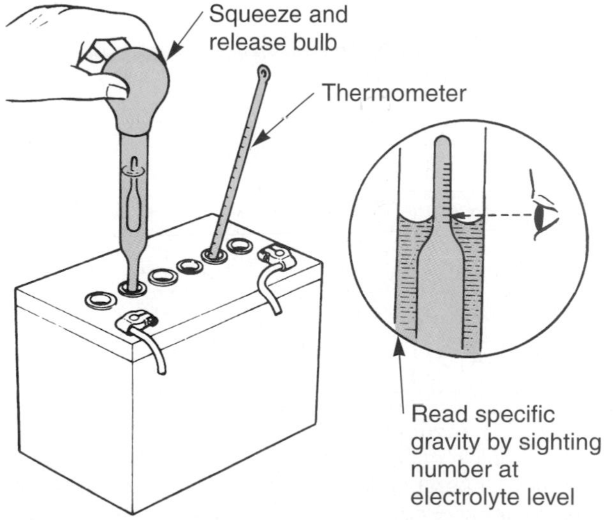

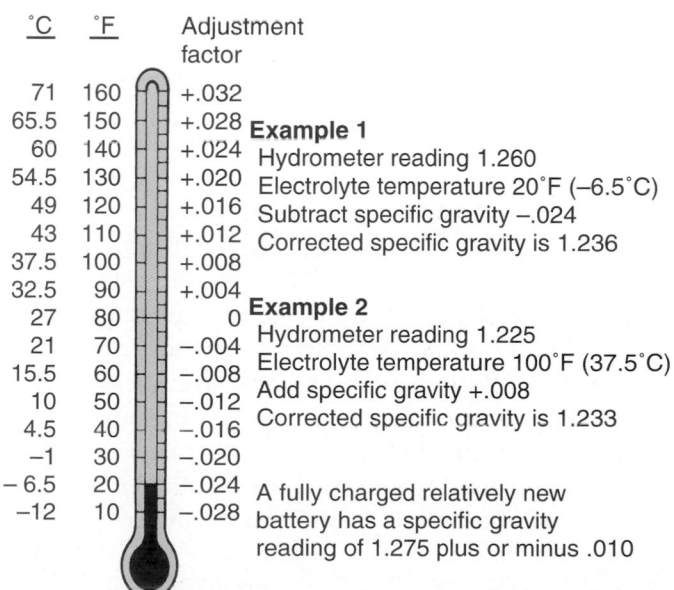

Figure 30-9. To check battery charge, draw electrolyte into a hydrometer by squeezing and releasing the bulb. Read specific gravity on the float at the top level of electrolyte. The temperature of the electrolyte will affect the reading. (Mazda)

Figure 30-10. Study examples of using a hydrometer correction chart. (Chrysler)

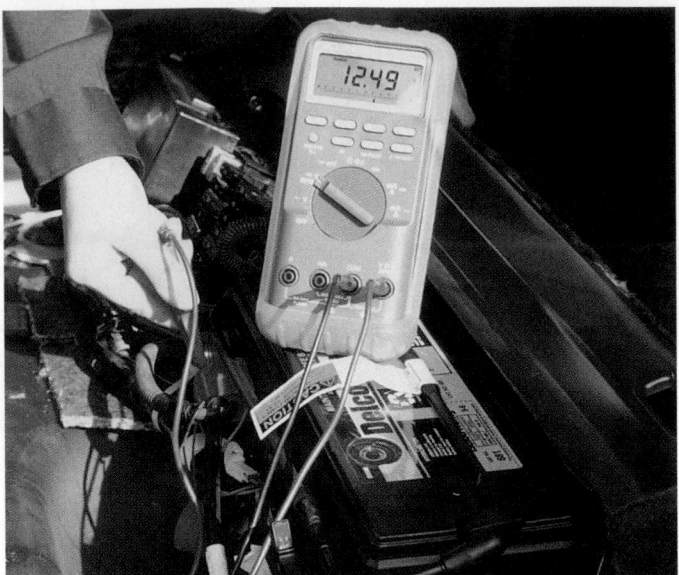

Figure 30-11. A digital voltmeter will check the general charge of a battery. Turn on headlights for a light load, then read the meter. Generally, voltage below 12.5 V indicates the amount of battery discharge. This battery shows 12.49 V, which is about 75% of a full charge. (Fluke)

- Electrical problem drawing current out of battery with ignition key off.

A defective battery can be found with a hydrometer by checking the electrolyte in every cell. If the specific gravity in any cell varies excessively from other cells (25 to 50 points), the battery is usually ruined. Cells with low readings may be shorted. When all of the cells have an equal specific gravity, the battery can usually be regenerated by recharging.

With maintenance-free batteries, the hydrometer is not commonly used. A voltmeter and ammeter or a load tester (covered later in the chapter) can be used to quickly determine battery condition.

Battery Voltage Test

A *battery voltage test* is done by measuring total battery voltage with a voltmeter or special tester. It will determine the general state of charge and battery condition quickly. A battery voltage test is used on maintenance-free batteries. These batteries do not have filler caps that can be easily removed for testing with a hydrometer.

Look at **Figure 30-11.** Connect the meter across the battery terminals. Turn on the car's headlights or heater blower to provide a light load. Read the meter.

A fully charged battery should have 12.6 volts. The lower the voltage, the less charge on the battery.

Other tests are needed to find the actual problem when a battery fails a voltage test. The chart in **Figure 30-12**

compares specific gravity (hydrometer readings) and battery voltage. Note the relationship.

Cell Voltage Test

A *cell voltage test* will let you know if the battery is defective or just discharged. Just like a hydrometer cell test, if the cell voltages vary by more than 0.2 volts, the battery must be replaced.

To do a cell voltage test, insert the special cadmium (acid-resistant metal) tips of a low voltage reading voltmeter into each cell. Start at one end of the battery. Work your way down, testing each cell carefully. Some manufacturers recommend battery fast-charging during this test. Refer to a service manual for details.

If cell voltages are low, but equal, recharging will usually restore the battery. If cell voltage readings vary more than 0.2 volts, the battery is faulty.

 **Warning**
Make sure you do not drip electrolyte on the vehicle or your skin. The acid in the electrolyte will eat the paint or burn your skin.

Battery Drain Test

A *battery drain test* will check for an abnormal current draw with the ignition key off. When a battery goes dead without being used, you may need to check for a current drain. It is possible that there is a short or other problem constantly discharging the battery.

A battery can be discharged if an electrical accessory remains on when the ignition switch is shut off. For example, a short in a switch could cause a glove box light

Load on battery (Amps)	**Battery Voltage**				
	Specific Gravity (Percent charge)				
	1.265 (100%)	1.250 (95%)	1.230 (75%)	1.200 (50%)	1.175 (25%)
0	12.7	12.6	12.5	12.4	12.2
5	12.5	12.4	12.3	12.1	11.8
15	12.3	12.2	12.0	11.7	11.3
25	12.1	11.9	11.6	11.2	10.7

This is the range in which most vehicle batteries normally operate in customer service.

At 1.180 and below, starting will be unreliable and function of other circuits may be erratic.

Figure 30-12. Chart shows how battery voltage relates to specific gravity (hydrometer readings). A voltage test is needed on maintenance-free batteries, which do not have removable filler caps. (Chrysler)

to always stay on. This could slowly drain the battery and cause a starting problem.

To perform a battery current drain test:
1. Make the ammeter connections shown in **Figure 30-13.**
2. Pull the fuse for the dash clock.
3. Close the doors and trunk.
4. Unscrew the underhood lightbulb, if needed.
5. Read the ammeter and compare your reading to specifications.

Caution

To prevent meter damage, do not operate starting motor or any high current draw device (blower motor) with the meter connected in series for measuring current drain. High current draw will blow the meter fuse or damage the meter.

If everything is off (good condition), the ammeter should read almost zero or only a few milliamps (10 mA, typically). However, an ammeter reading above this would point to a drain and a problem. To help pinpoint a drain, pull fuses one at a time. When the ammeter reads zero, the problem is in the circuit on that fuse.

Remember that normal parasitic current drain for the clock and computers can discharge a battery if the vehicle sits unused for an extended period. Also account for this small current draw when checking for a battery drain.

Battery Chargers

When tests show that a battery is discharged, a **battery charger** may be used to re-energize it. The battery charger will force current back into the battery to restore the charge on the plates and in the electrolyte. It contains a step-down transformer that changes wall outlet voltage (around

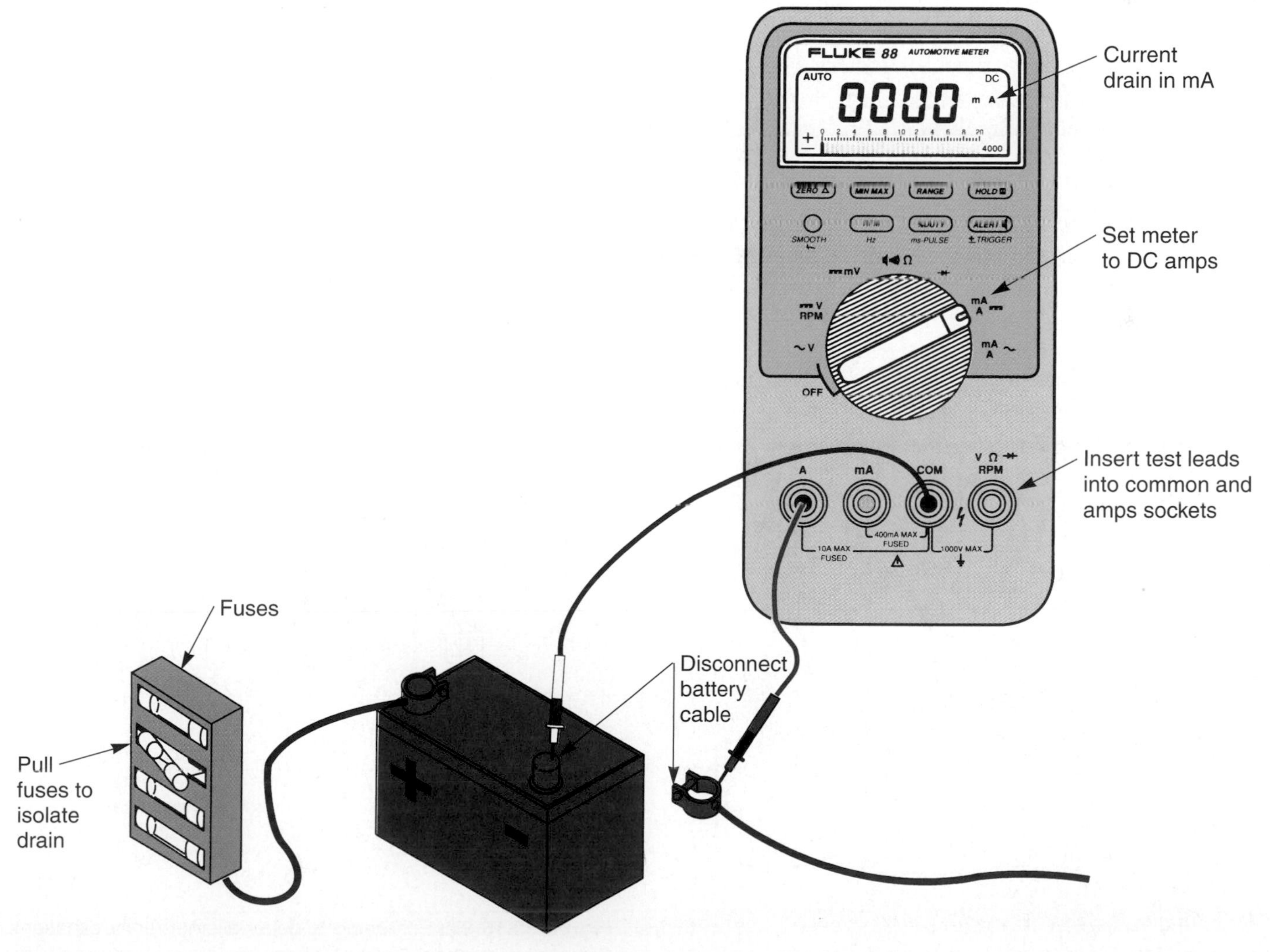

Current drain in mA

Set meter to DC amps

Insert test leads into common and amps sockets

Disconnect battery cable

Fuses

Pull fuses to isolate drain

Figure 30-13. If the battery discharges while not being used, perform a battery drain test. Connect an ammeter in series with the negative cable. If current is flowing out of the battery with everything turned off, an electrical problem is discharging the battery. By pulling fuses, you can isolate the problem circuit. If the drain stops flowing with the fuse pulled, that circuit is at fault. (Fluke)

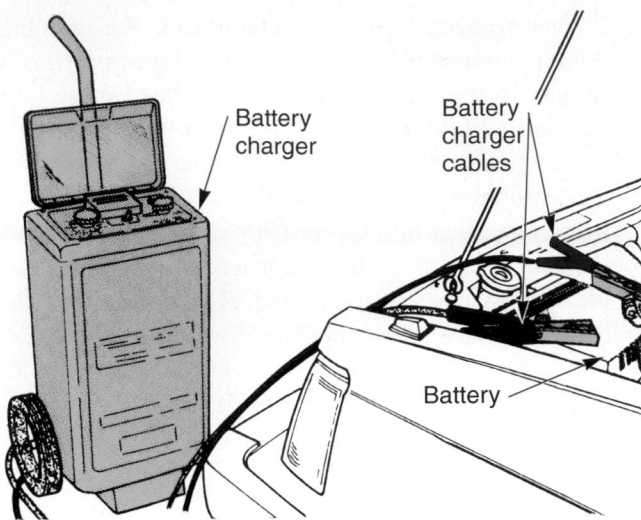

Figure 30-14. A battery charger forces current through the battery to restore the charge on the plates. (Chrysler)

120 volts) to slightly above battery voltage (14–15 volts). Refer to **Figure 30-14.** It shows a battery charger.

There are two basic types of battery chargers: the slow charger and the fast charger. Sometimes both are incorporated into one unit.

Slow Charger

A *slow charger,* also called a *trickle charger,* feeds a small amount of current into the battery. Charging time is

longer (about 12 hours at 10 amps). However, the chemical action inside the battery is improved. The active materials are plated back on the battery plates better. When time allows, use a slow charger. Look at **Figure 30-15.**

Fast (Quick) Charger

A *fast charger,* also called a *quick* or *boost charger,* forces a high current flow into the battery for rapid recharging. A fast charger is shown in **Figure 30-16.** It is commonly used in auto shops. When the customer needs the car, time may not allow the use of a slow charger.

Fast charging will usually allow engine starting in a matter of minutes. If possible, slow charging is recommended after fast charging.

Warning
Before connecting a battery charger to a battery, make sure the charger is turned off. Also, check that the work area is well ventilated. If a spark ignites any battery gas, the battery could explode. Wear eye protection!

Charging a Battery

To use a battery charger, connect the *red* charger lead to the positive terminal of the battery. Connect the *black* charger lead to the negative terminal of the battery. With side terminal batteries, use adapters like those shown in **Figure 30-17.**

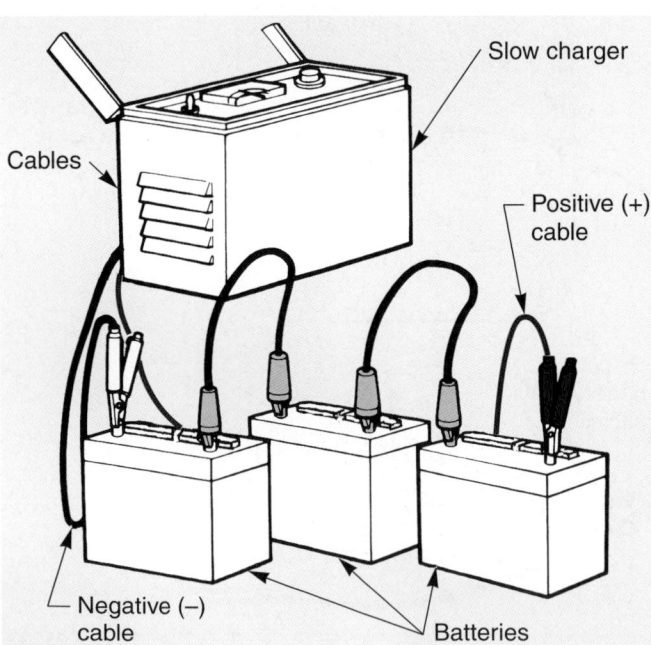

Figure 30-15. A slow charger forces only a small amount of current through the battery. Since slow charging requires several hours, several batteries may be connected to get more done.

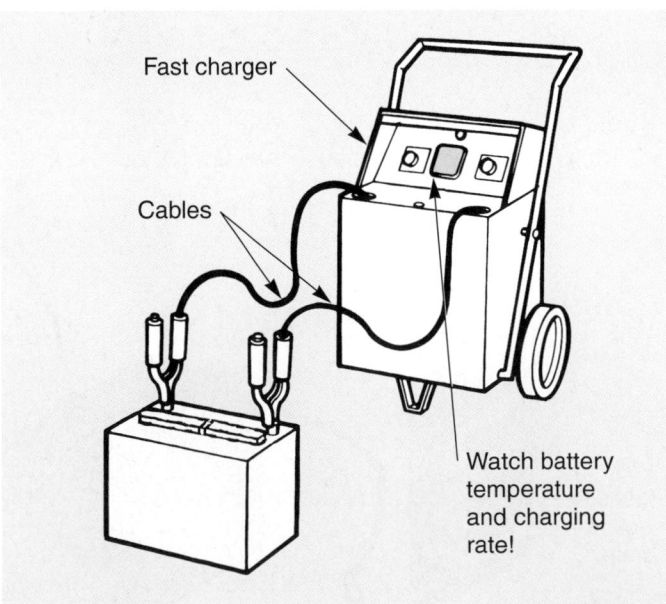

Figure 30-16. Fast charging is done in emergency situations. Slow charging should follow fast charging to restore the battery properly. Do not let battery temperature go above about 125°F (52°C) or battery damage may occur.

Caution
Make sure you do not reverse the charger connections on the vehicle's charging system or electronics could be damaged. Set the charger controls and turn on the power.

When fast charging, do not exceed a charge rate of 35 amps. Also, battery temperature must not exceed 125°F (52°C). Exceeding either charge rate or temperature could damage the battery.

If you suspect a battery is frozen or has ice in it from extreme cold weather, do not charge it. Charging a frozen battery can rupture the case and cause an explosion. Allow the battery to thaw in the warm shop before charging.

Jump Starting

In emergency situations, it may be necessary to *jump start* a vehicle by connecting another battery to the discharged battery. Look at **Figure 30-18.** The two batteries are connected *positive to positive* and *negative to negative*. Connect the red jumper cable to the positive terminal of both batteries. Then, connect one end of the black jumper cable to the negative terminal of the good battery or to a good ground on the vehicle with the good battery. Finally, connect the other end of the black jumper cable to a good ground on the vehicle with the bad battery. See **Figure 30-19.**

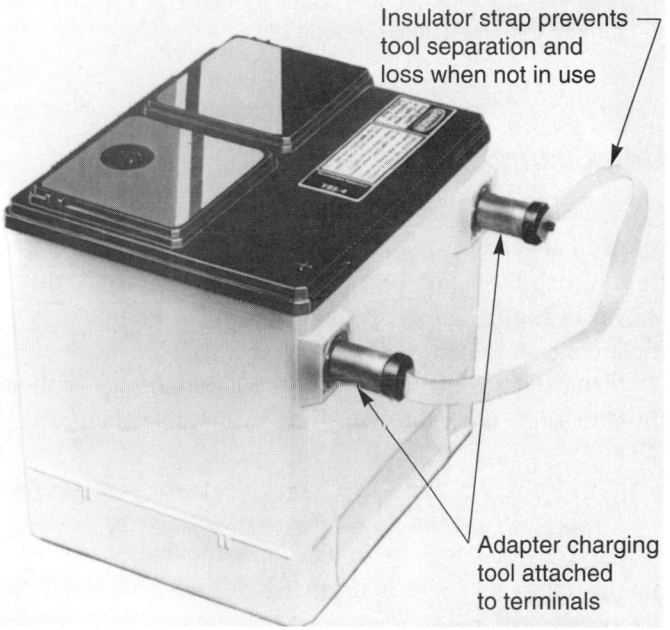
Figure 30-17. When charging a side terminal battery, use adapters. They will let you connect the charger clamps to the terminals. (Chevrolet)

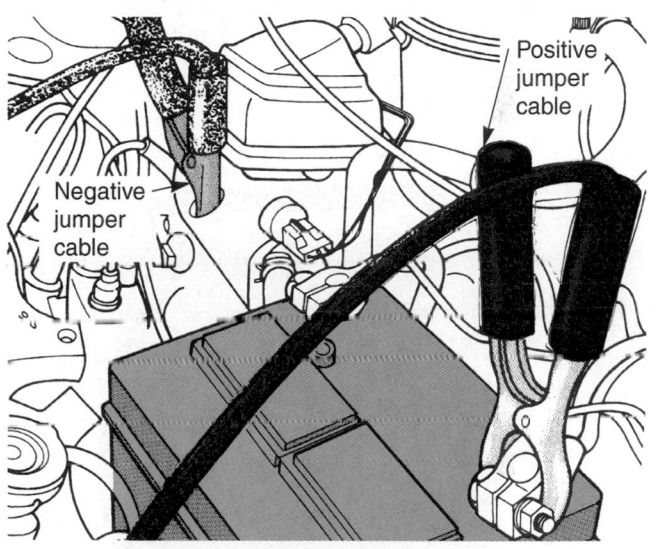
Figure 30-19. Close-up of jumper connections shows the connection of the negative (black) jumper to chassis ground. A spark near the dead battery could make battery gases explode. (Chrysler)

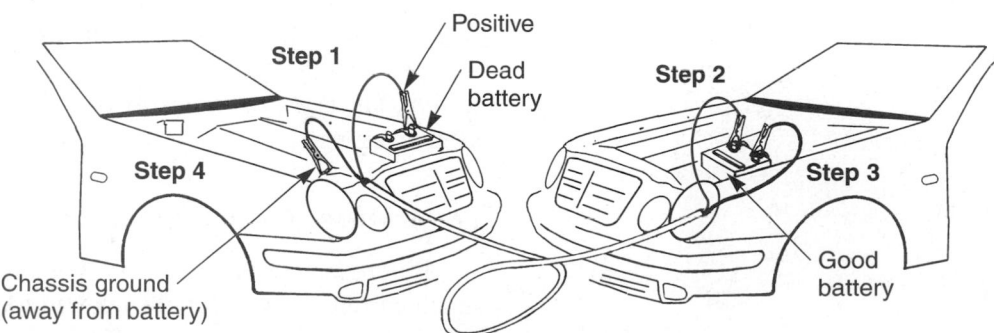
Figure 30-18. Jumper cables can be used to start a car with a dead battery. First, connect the red jumper to the positive terminal of the dead battery. Then connect the other end of the red jumper to the positive terminal of the good battery. Connect the black jumper to the negative cable of the good battery. Finally, connect the other end of the black jumper to a good ground, away from the dead battery. This will keep sparks away from battery gases. Run the engine in the car with the good battery while starting the vehicle with the dead battery. (Belden)

Warning
Do not short jumper cables together or connect them backwards. This could cause serious damage to the charging or computer systems. By connecting the last negative jumper cable to a ground away from the battery, any spark will be further away from the explosive battery gas.

Battery Load Test

A battery load test, also termed a battery capacity test, is one of the *best* methods of checking battery condition. It tests the battery under full current load.

The hydrometer and voltage tests were general indicators of battery condition. The battery load test, however, actually measures the current output and performance of the battery. It is one of the most common and informative battery tests used in automotive garages. Refer to **Figure 30-20.**

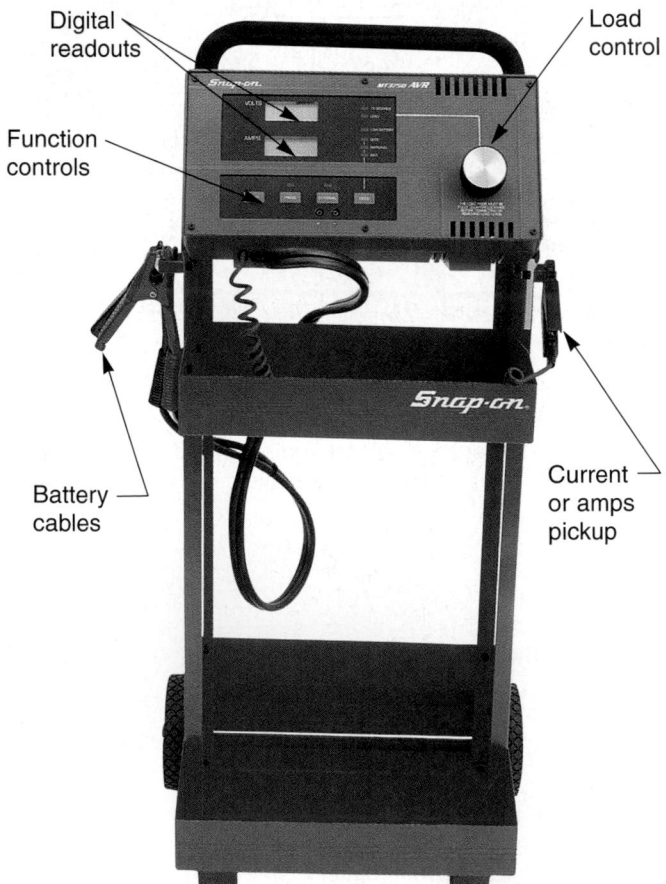

Figure 30-20. A battery load tester is the most accurate method of determining battery condition. It is a commonly used testing device that measures actual battery performance. (Snap-On Tools)

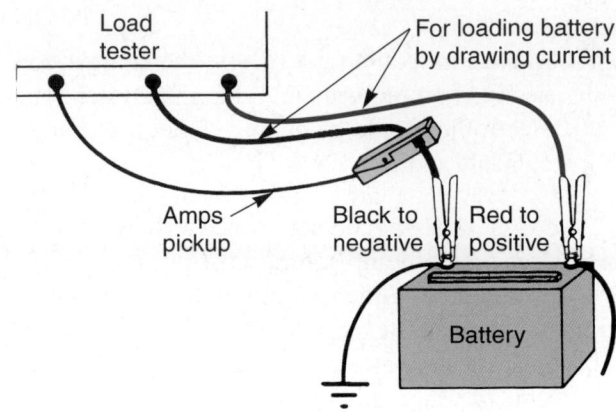

Figure 30-21. Load testers are connected as shown. Clamp the large cables to the battery. Clip the inductive amps pickup around the negative tester cable. Large cables load the battery by drawing current through the tester. Inductive pickup operates the ammeter in the tester. (Marquette)

Connecting a Load Tester

Connect the load tester to the battery terminals. If the tester is an inductive type (clip-on ammeter lead senses field around outside of cables), use the connections shown in **Figure 30-21.** If the tester is *not* inductive, you must connect the ammeter in series.

Control settings and exact procedures vary. Follow the directions provided with the testing equipment.

Double-Check Battery Charge

Before load testing, make sure the battery is adequately charged. Use a hydrometer or digital voltmeter as discussed earlier. The load tester can be used to check the battery charge. Adjust the load control to draw 50 amps for 10 seconds. This will remove any surface charge. Then, check no-load battery voltage (also called open circuit voltage, or OCV), by reading the voltmeter.

A *fully charged battery* should have an OCV (no-load, open circuit voltage) of 12.4 volts or higher. If battery voltage is below 12.4 volts, charge the battery before load testing. The battery is probably bad if it fails a second test after charging.

Determine Battery Load

Before load testing a battery, you must calculate how much current draw should be applied to the battery.

If the amp-hour rating is given, load the battery to three times its amp-hour rating. For example, if the battery is rated at 60 amp-hours, test the battery at 180 amps ($60 \times 3 = 180$).

Battery ratings			Load test amps
Cold cranking current	Amp-hour (approx.)	Watts	
200	35–40	1800	100 amps
250	41–48	2100	125 amps
300	49–62	2500	150 amps
350	63–70	2900	175 amps
400	71–76	3250	200 amps
450	77–86	3600	225 amps
500	87–92	3900	250 amps
550	93–110	4200	275 amps

Figure 30-22. This chart shows different battery ratings and calculated current values for load testing. (Marquette)

Approximate electrolyte temperature	Minimum acceptable voltage under load for good battery
60°F (16°C)	9.5
50°F (10°C)	9.4
40°F (4°C)	9.3
30°F (−1°C)	9.1
20°F (−7°C)	8.9
10°F (−12°C)	8.7
0°F (−18°C)	8.5

Figure 30-23. To load test a battery, turn the load control knob until the ammeter reads calculated test current. Hold the load or current for 15 seconds and read the voltmeter. If the reading is below voltages in the chart for the specific temperature, the battery is probably bad. (Snap-On Tools)

Many batteries are now rated in SAE cold cranking amps, rather than amp-hours. To determine the load test for these batteries, divide the cold crank rating by two. For instance, a battery with a 400-cold cranking amps rating should be loaded to 200 amps (400 ÷ 2 = 200). The watt is another battery performance rating.

Figure 30-22 gives a chart which compares battery ratings. A *load conversion chart* will normally be provided with the load testing equipment. Refer to this material when in doubt.

Loading the Battery

After checking the battery charge and finding the amp load value, you are ready to test the battery output. Double-check that the tester is connected properly. Then, turn the load control knob until the ammeter reads the correct load for your battery, **Figure 30-22.** Hold the load for 15 seconds. Next, read the voltmeter while the load is applied. Then, turn the load control completely off so the battery will not be discharged.

Load Test Results

If the voltmeter reads 9.5 volts or more at room temperature, the battery is good. Six-volt batteries should maintain 4.8 volts. These voltages are based on a battery temperature above 70°F (21°C).

A cold battery may show a lower voltage. You will need a temperature compensation chart, like the one in **Figure 30-23.** It allows for any reduced battery performance caused by a low temperature.

If the voltmeter reads below 9.5 volts at room temperature, battery performance is poor. This would show that the battery is not producing enough current to properly run the starting motor. Before replacing the battery, however, a quick charge test should be completed.

Quick (3 Minute) Charge Test

A *quick charge test,* also called a *3 minute charge test,* will determine if the battery is **sulfated** and the plates ruined. If the battery load test results are poor, fast charge the battery. Charge for three minutes at 30–40 amps. Test the voltage while charging, as shown in **Figure 30-24.**

If the voltage goes above 15.5 volts (12-volt battery) or 7.8 volts (6-volt battery), the battery plates are sulfated and ruined. A new battery should be installed in the vehicle.

Other Battery-Related Problems

If the battery passes all of its tests, but does not perform properly (starting motor does not crank, for example), the following are some likely problems to check out.

- Defective charging system.
- Battery drain (light or other accessory on).

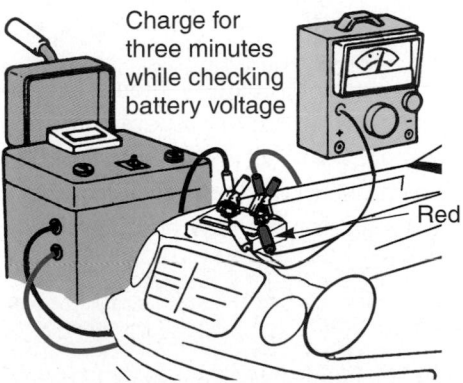

Charge for three minutes while checking battery voltage

Red

Figure 30-24. The 3 minute charge test will double-check the load test. Charge the battery at 30–40 amps while measuring the battery voltage for three minutes. If voltage increases above 15.5 volts, replace the battery. (GMC)

- Loose alternator belt.
- Corroded, loose, or defective battery cables.
- Defective starting system.

Note
Later textbook chapters cover the starting and charging systems. They will give you more information relating to battery service.

Activating Dry-Charged Battery

A new, dry-charged battery must be activated (readied for service) before installation. Put on safety glasses and rubber gloves. Remove the cell caps or covers. Using a plastic funnel (not a metal funnel) pour electrolyte into each cell. Pour in enough electrolyte to just cover the plates and separators.

Replace the caps. Charge the battery as recommended by the manufacturer. After charging, recheck the electrolyte level. Install the battery.

Removing and Replacing a Battery

To remove a battery, first disconnect the cables. Then loosen the battery hold-down. Using a *battery strap* or tool, **Figure 30-25,** carefully lift the battery out of the vehicle.

Warning
Always wear safety glasses when carrying a battery. If you drop a battery, electrolyte (acid) can squirt out of the vent caps or a broken case and into your face and eyes.

To install a battery, gently place the battery into its clean tray or box. Check that the battery fits properly. The tray edge must not cut through and rupture the plastic case. Bolt on the hold-down and install the cables. See **Figure 30-26.**

Tech Tip
The replacement battery should have a power rating *equal* to factory recommendations. If an *undersize battery* (lower watt rating) is installed, starting motor performance and battery service life may be reduced.

Summary

- A "dead battery" (discharged battery) is a very common problem. The engine will usually fail to crank and start.
- A battery leakage test will find out if current is discharging across the top of the battery case.

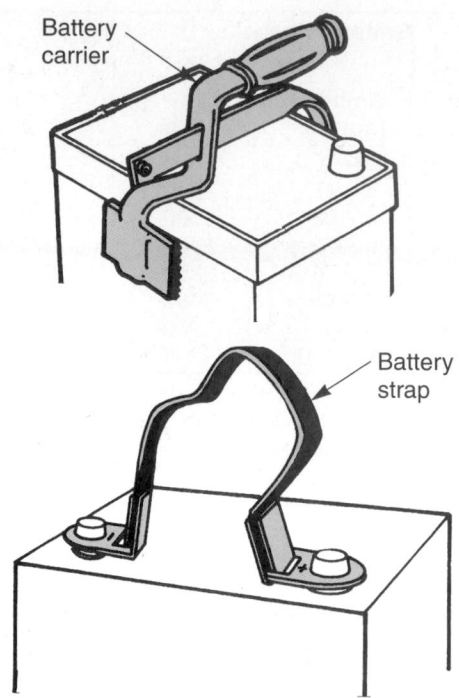

Figure 30-25. If you lose your grip and drop a battery, acid could splash out, causing eye or skin injury. Always use a battery strap or carrier and wear eye protection for safety. (Florida Dept. of Voc. Ed.)

Figure 30-26. When installing a battery, do not overtighten cable terminals or you could strip threads. Make sure the hold-down is secure. This battery is located under the rear seat in the passenger compartment.

- If the top of the battery is dirty, wash it down with baking soda and water.
- A battery terminal test quickly checks for a poor electrical connection between the battery cables and terminals.

- When reinstalling the cables, coat the terminals with petroleum jelly or white grease.
- Invisible hydrogen gas produced by the chemical reaction in a battery is very flammable. Keep all sparks and flames away from the top of a battery. Batteries can explode if the gas is ignited!
- A battery voltage test is done by measuring total battery voltage with an accurate voltmeter or special tester.
- A battery drain test will check for an abnormal current draw with the ignition key off.
- When tests show that a battery is discharged, a battery charger may be used to re-energize it.
- In emergency situations, it may be necessary to jump start a vehicle by connecting another battery to the discharged battery.
- A fully charged battery should have an OCV (no-load, open circuit voltage) of 12.4 volts or higher.
- The replacement battery should have a power rating equal to factory recommendations. If an undersize battery is installed, starting motor performance and battery service life will be reduced.

Important Terms

Dead battery	Fast charger
Battery maintenance	Quick
Battery leakage test	Boost charger
Battery terminal test	Jump start
Distilled water	Fully charged battery
Hydrometer	Load conversion chart
Battery voltage test	Quick charge test
Cell voltage test	3 minute charge test
Battery drain test	Sulfated
Battery charger	Battery strap
Slow charger	Undersize battery
Trickle charger	

Review Questions—Chapter 30

Please do not write in this text. Place your answers on a separate sheet of paper.

1. What five tasks does battery maintenance typically include?
2. A(n) _____ _____ _____ will find out if current is discharging across the top of the battery case.
3. If a voltmeter shows over 0.5 volt drop across the battery post-to-cable connection, what should be done?
4. A knife or scraper is a good method of cleaning corroded battery terminals. True or False?
5. When you measure _____ _____, you check the condition of the battery electrolyte and plates.
6. List six reasons for a discharged battery.
7. A battery _____ _____ is done by measuring total battery voltage with an accurate voltmeter. A(n) _____ _____ battery should have over 12 volts.
8. A customer complains that her battery goes dead when the car is not driven for an extended period. A new battery has been installed and tested by another shop. Technician A says that starting motor current draw should be measured. A shorted starting motor could be draining the battery. Technician B says that a battery drain test should be done. An electrical short could be discharging the battery, even with the ignition key off. Who is correct?
   (A) A only.
   (B) B only.
   (C) Both A and B.
   (D) Neither A nor B.
9. When using a battery charger, connect the red lead to positive and the black lead to negative. True or False?
10. When fast charging a battery, you should never exceed a charge rate of _____.
11. How do you connect jumper cables safely?
12. A(n) _____ _____ _____ , also termed a(n) _____ _____ _____ , is one of the best methods of checking battery condition.
13. What is an inductive-type ammeter lead?
14. Explain how to do a battery load test.
15. If a 12-volt battery shows below _____ volts during a load test at 70°F (21°C), the battery is bad.
    (A) 9.9 volts.
    (B) 9.5 volts.
    (C) 9.7 volts.
    (D) 10.0 volts.

ASE-Type Questions

1. Battery maintenance and inspection should be conducted:
   (A) *during tune-ups.*
   (B) *during grease jobs.*
   (C) *anytime the hood is opened.*
   (D) *All of the above.*

2. Each of these can be performed using a volt-meter *except:*
   (A) *battery plate test.*
   (B) *battery voltage test.*
   (C) *battery leakage test.*
   (D) *battery terminal test.*

3. Which of the following combinations will neutralize and remove an acid-dirt mixture from a battery case?
   (A) *Sulfur and water.*
   (B) *Baking soda and water.*
   (C) *White grease and a solvent.*
   (D) *Distilled water and gasoline.*

4. Fully charged electrolyte has a specific gravity of between:
   (A) *0.724 and 0.999.*
   (B) *1.000 and 2.000.*
   (C) *1.265 and 1.299.*
   (D) *2.435 and 2.922.*

5. Which of the following is *not* a type of hand-held battery hydrometer?
   (A) *Ball.*
   (B) *Nose.*
   (C) *Float.*
   (D) *Needle.*

6. A discharged battery can be caused by a(n):
   (A) *defective battery.*
   (B) *starting system problem.*
   (C) *engine performance problem.*
   (D) *All of the above.*

7. Each of these can be used to quickly determine battery condition on a maintenance-free battery *except:*
   (A) *ammeter.*
   (B) *voltmeter.*
   (C) *load tester.*
   (D) *hydrometer.*

8. Charging time for this device is about 12 hours.
   (A) *Fast charger.*
   (B) *Slow charger.*
   (C) *Quick charger.*
   (D) *Boost charger.*

9. Technician A feels the best test to determine battery current output and performance is the battery load test. Technician B feels hydrometer testing is the best choice. Who is right?
   (A) *A only.*
   (B) *B only.*
   (C) *Both A and B.*
   (D) *Neither A nor B.*

10. A fully charged battery should have an OCV of:
    (A) *5.5 volts.*
    (B) *12.4 volts or higher.*
    (C) *25.0 volts or higher.*
    (D) *None of the above.*

Activities—Chapter 30

1. Get permission from several classmates to conduct a battery maintenance inspection on their vehicles. Check for the points listed on page 493. Make and complete a simple checklist, or write a short report, for each vehicle inspected. Give the form or report to the vehicle owner.

2. Demonstrate to the class the proper way to make a hydrometer reading of a battery that has removable cell caps. Observe all safety precautions, and note them for the class during your demonstration.

3. Show the effects of splashed electrolyte on clothing. Obtain small pieces of different clothing fabrics, attach them to a thick sheet of cardboard, and label them. Wearing gloves and eye protection, carefully apply two drops of electrolyte from a battery to each sample. Ask class members to note the effect of the electrolyte on each fabric.

Battery Diagnosis		
Procedure	**Possible results**	**Correction**
Clean the battery and check for damage to the battery case, posts, etc.	1. Loose battery post, cracked case, leaks, or any other physical damage. 2. Battery OK.	1. Replace battery. 2. Check *state of charge*.
Determine state of charge by checking the color of the battery charge indicator.	1. Green. 2. Black. 3. Clear.	1. Battery is charged. Perform *battery voltage test*. 2. Charge battery. 3. Replace battery.
Perform battery voltage test.	1. Battery voltage is above 12.40 volts. 2. Battery voltage is below 12.40 volts.	1. Perform a *load test*. 2. Charge battery.
Charge battery.	1. Battery accepted charge. 2. Battery will not accept charge.	1. Ensure that the indicator eye is green and perform *battery voltage test*. 2. Replace battery.
Perform load test.	1. Acceptable minimum voltage. 2. Unacceptable minimum voltage.	1. Battery is OK. Perform *battery drain test*. 2. Replace battery and perform *battery drain test*.
Perform battery drain test.	1. Battery drain is within specifications. 2. Battery drain exceeds specifications.	1. Vehicle is normal. 2. Eliminate excess draw.

Chapter 31

Starting System Fundamentals

After studying this chapter, you will be able to:

- Explain the principles of an electric motor.
- Describe the construction and operation of a starting motor.
- Sketch a simple starting system circuit.
- Explain the operation of solenoids.
- List the functions of the main starter drive parts.
- Describe starter drive operation.
- Compare different types of starting motors.
- Describe starting system safety features.
- Correctly answer ASE Certification test questions that require a knowledge of starting system fundamentals.

The starting system has helped make the modern automobile a reliable and convenient means of transportation. It provides an easy method of starting the engine. Early "Model Ts" had to be cranked by hand to get the engine running. This took considerable strength and patience.

Few car owners would be physically capable of starting a modern, multi-cylinder engine using a hand crank. The electric starting motor is designed to crank the engine with the simple turn of a key.

Note
If any of the principles in this chapter are unclear, use the index to locate earlier text material or refer to Chapter 8, *Basic Electricity and Electronics.*

Starting System Principles

The ***starting system*** uses battery power and an electric motor to turn the engine crankshaft for engine starting. **Figure 31-1** shows the basic parts and action of a starting system. The main parts of a starting system include:

- ***Battery***—source of energy for the starting system.

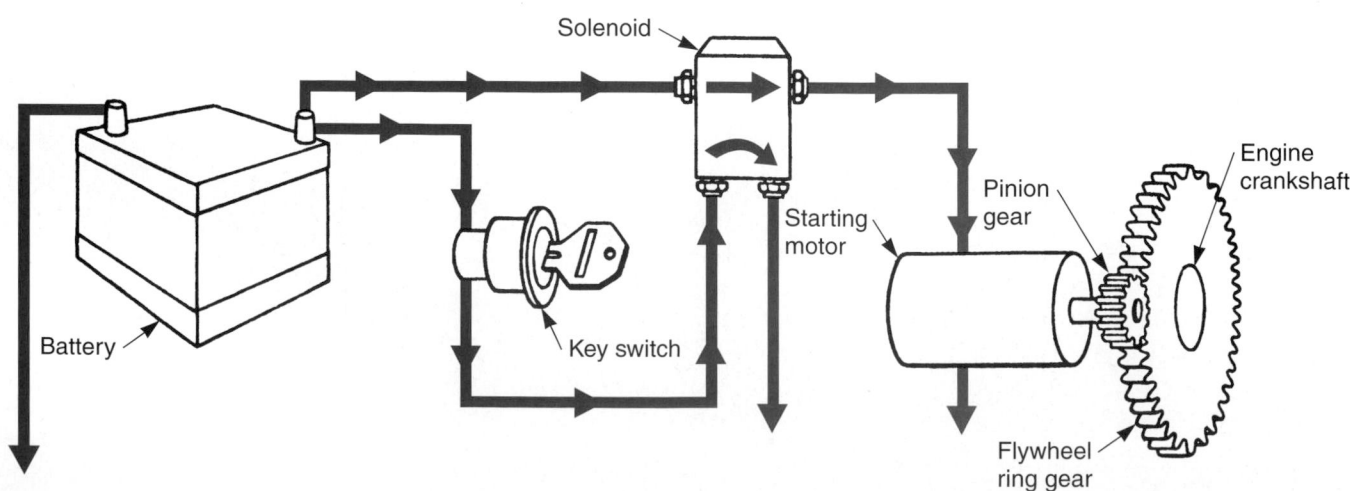

Figure 31-1. Basic starting system operation. The ignition switch energizes the solenoid. The solenoid then energizes the starting motor. The motor turns the flywheel gear for engine starting. (GMC)

- *Ignition switch*—allows the driver to control starting system operation.
- *Solenoid*—high-current relay (switch) for connecting the battery to the starting motor.
- *Starting motor*—high-torque electric motor for turning the gear on the engine flywheel.

Starting System Action

When you turn the ignition key to start an engine, current flows through the solenoid coil. This closes the solenoid contacts, connecting battery current to the starting motor. The motor turns the flywheel ring gear until the engine starts and runs on its own power.

When the engine starts, you release the ignition key. This breaks the current flow to the solenoid and starter. The starter stops turning, and the starter gear moves away from the flywheel gear.

Tech Tip

Some late-model starting systems are programmed to open the circuit between the ignition key and the starting solenoid when the engine is running. This prevents the driver from accidentally engaging the starting motor and grinding its pinion gear into the flywheel gear when the engine is running.

Starting Motor Fundamentals

The *starting motor* converts electrical energy from the battery into mechanical energy to crank the engine. It is similar to other electric motors. All electric motors (wiper motors, fan motors, fuel pump motors, etc.) produce a turning force through the interaction of magnetic fields inside the motor.

Magnetic Field Action

As explained in earlier text chapters, a *magnetic field* is made up of invisible magnetic lines of force. The lines of force flow between the poles of a permanent magnet. They also flow around the outside of a wire that is carrying current.

Since **like charges** (fields) *repel* each other and **unlike charges** *attract each other,* magnetic fields can be used to produce motion. Look at **Figure 31-2.** Note how the lines of force in the magnet and the lines of force around the conductor act upon each other. This principle is used in electric motors.

Simple Electric Motor

To build a simple electric motor, you would start by bending a piece of wire into a loop. When current is

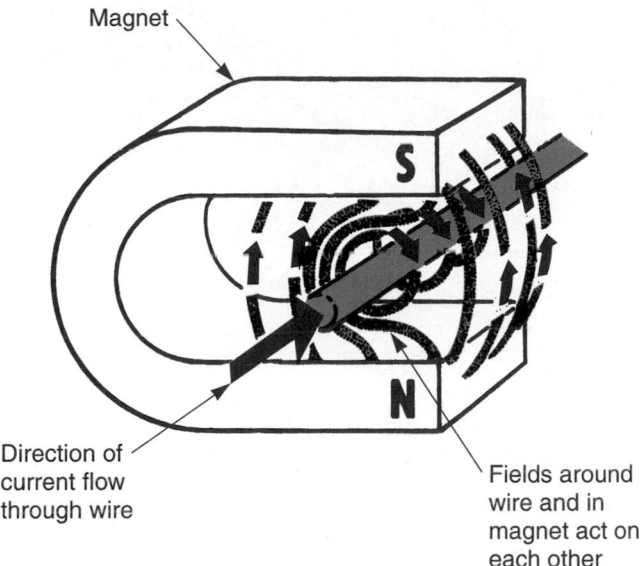

Figure 31-2. Magnetic fields from magnets and the fields around a current-carrying conductor can be used to produce motion. Note how the fields interact. (GMC)

passed through the wire loop (winding), a magnetic field forms around the wire. Refer to **Figure 31-3.**

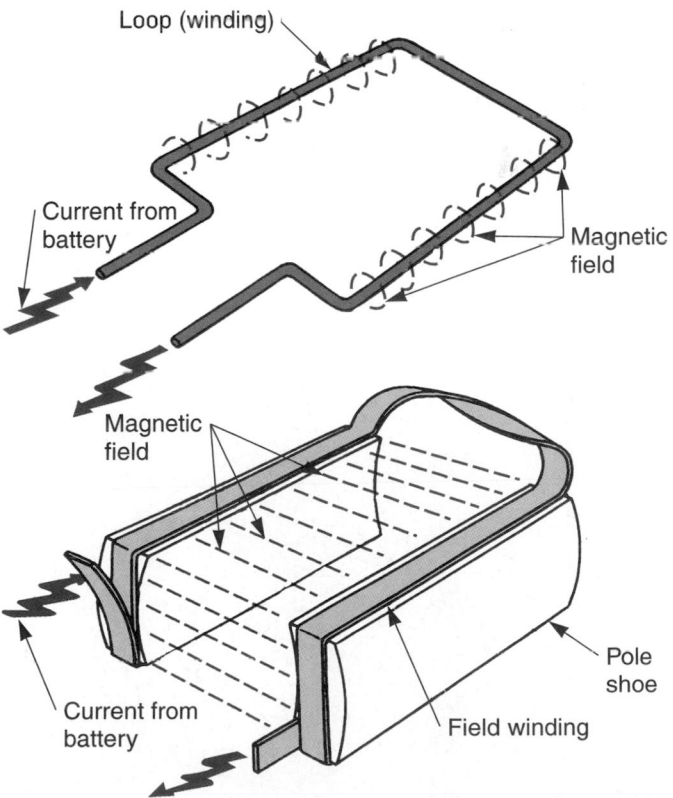

Figure 31-3. A simple electric motor. Wire bent into a loop forms the winding. When battery current flows through the winding, a field develops around the wire. Wires wrapped around pole shoes form a strong magnetic field. (Deere & Co.)

Figure 31-4. If a winding is placed inside a magnetic field, the winding rotates away from the pole shoes. (Deere & Co.)

A *magnet* or *pole piece* is needed to make the loop of wire move. A magnetic field is set up between the pole pieces, also called *pole shoes*. See **Figure 31-3.**

Changing Electricity into Motion

Electric current can be changed into a strong rotating motion by placing the winding inside the pole shoes. When current passes through the loop, the magnetic field around the loop and the field between the pole shoes act upon each other. The winding is moved toward a vertical position, **Figure 31-4.**

Commutator and Brushes

A commutator and several brushes are used to keep the electric motor spinning by controlling the current passing through the windings. Look at the simplified motor shown in **Figure 31-5.**

The *commutator* serves as a sliding electrical connection between the motor windings and the brushes. The commutator has many segments, which are insulated from each other.

The *motor brushes* ride on top of the commutator. They slide on the commutator to carry battery current to the spinning windings.

Figure 31-5 illustrates commutator and brush action. As the winding rotates away from the pole shoes, the commutator segments change the electrical connection between the brushes and the winding. This reverses the magnetic field around the winding. Then the winding is again pulled around and passes the other pole shoe. The constantly changing electrical connection keeps the motor spinning. A push-pull action is set up as each loop moves around inside the pole shoe area.

Increasing Motor Power

Several loops of wire and a commutator with many segments are used to increase motor power and smoothness. Shown in **Figure 31-6,** each winding is connected to its own segment on the commutator. This provides current flow through each winding as the brushes contact each segment. As the motor spins, many windings contribute to the motion. This produces a constant and smooth turning force.

Armature

A starting motor, unlike our simple motor, must produce very high *torque* (turning power) and relatively high speed. Therefore, a system to support the windings and

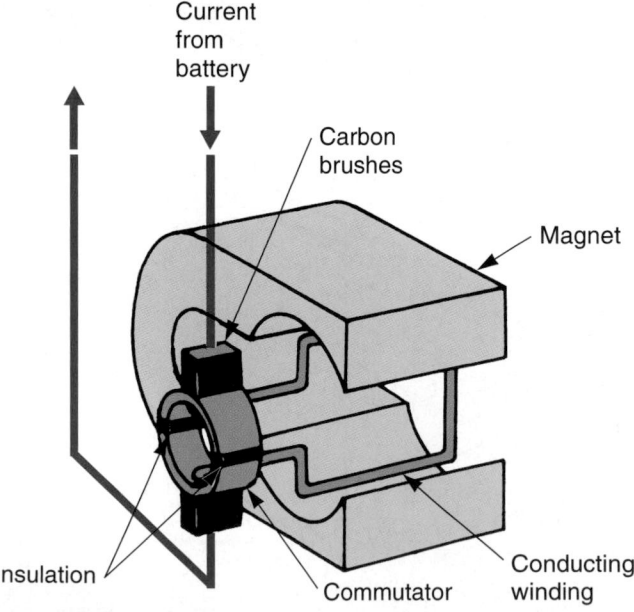

Figure 31-5. Brushes and a commutator are used to keep windings spinning. Note how the commutator reverses the electrical connection when the loop rotates around. (Robert Bosch)

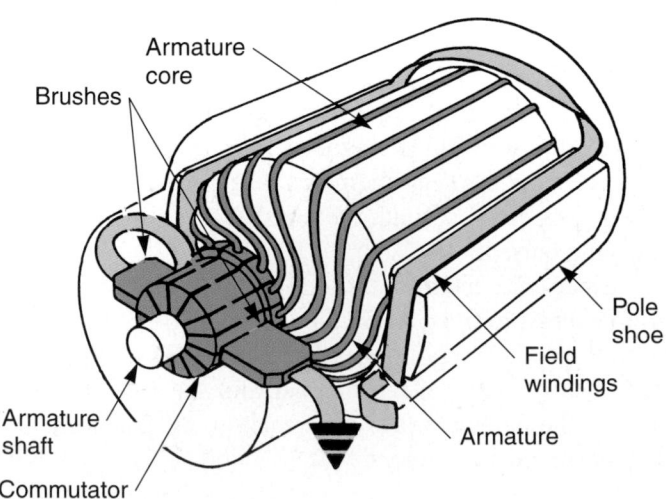

Figure 31-6. An actual starting motor has multiple commutator segments and windings to increase motor power and smoothness. (Deere & Co.)

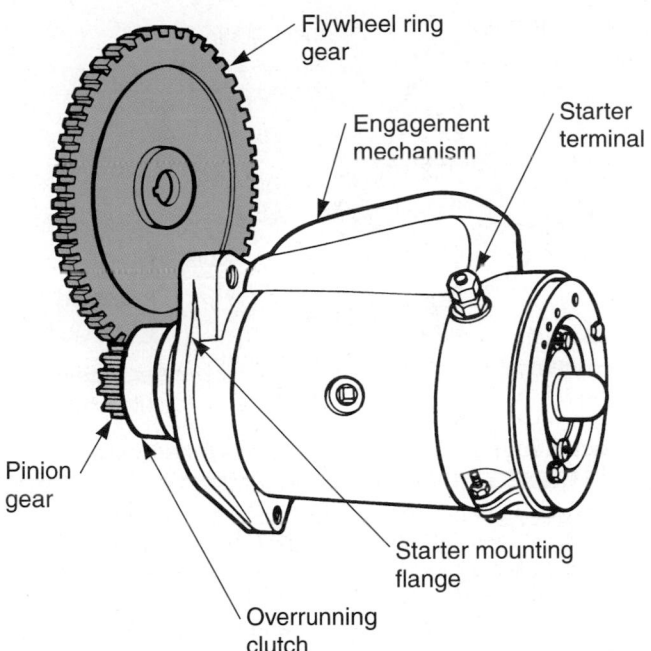

Figure 31-7. The starter pinion gear meshes with the large ring gear on the engine flywheel.

increase the strength of each winding's magnetic field is needed.

A *starter armature* consists of the armature shaft, armature core, commutator, and armature windings. The *starting motor shaft* supports the armature as it spins inside the starter housing. The commutator is mounted on one end of the armature shaft.

The *armature core* holds the windings in place. The armature core is made of iron to increase the strength of the magnetic field produced by the windings. Look at **Figure 31-6.**

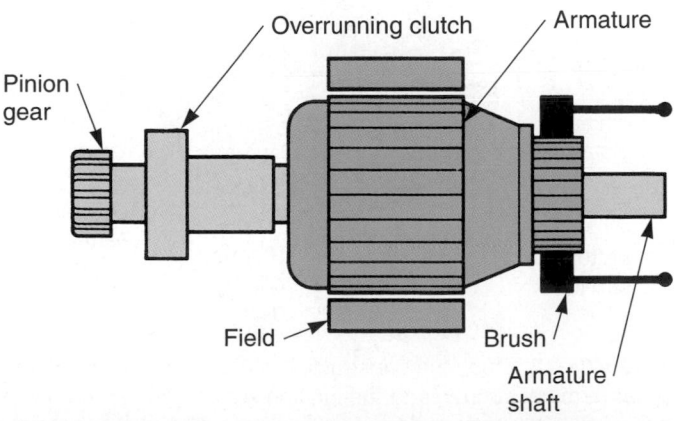

Figure 31-8. The pinion gear clutch assembly slides onto one end of the armature shaft. Note the other parts. (Robert Bosch)

Field Windings

A *field winding* is a stationary insulated wire wrapped in a circular shape. It creates a strong magnetic field around the motor armature. **Figure 31-6** shows a basic set of field windings.

When current flows through the field winding, the magnetic field between the pole shoes becomes very large. It can be 5–10 times that of a permanent magnet. As the magnetic field between the pole shoes acts against the field developed by the armature, the motor spins with extra power.

Starter Pinion Gear

A *starter pinion gear* is a small gear on the armature shaft that engages a large ring gear on the engine flywheel. It moves into and meshes with the *flywheel ring gear* anytime the starter is energized. In **Figure 31-7,** note the relationship of the gears.

Most starter pinion gears are made as part of a pinion drive mechanism (pinion gear, clutch, and housing). The pinion drive unit slides over one end of the armature shaft, **Figure 31-8.**

Overrunning Clutch

The *starter overrunning clutch* locks the pinion gear in one direction and releases it in the other. This allows the pinion gear to turn the flywheel ring gear for starting. It also lets the pinion gear freewheel when the engine begins to run.

Figure 31-9 shows a cutaway view of a pinion gear overrunning clutch assembly. Without the overrunning clutch, the starter could be driven by the engine flywheel. The flywheel gear could spin the starter too fast and cause armature damage.

Overrunning Clutch Operation

Figure 31-10 shows the basic operation of a starter overrunning clutch. Small spring-loaded rollers are located between the pinion gear collar and the clutch shell. The rollers wedge into the notches in the shell in the driving direction. They slide back and release when driven in the other direction (freewheeling direction).

There are spiral grooves on the armature shaft and the inside diameter of the clutch. The grooves force the shaft and pinion gear assembly to turn together. They also let the gear assembly slide on the armature shaft, **Figure 31-11.**

Starter Solenoid

The *starter solenoid* is a high-current relay. It makes an electrical connection between the battery and the starting

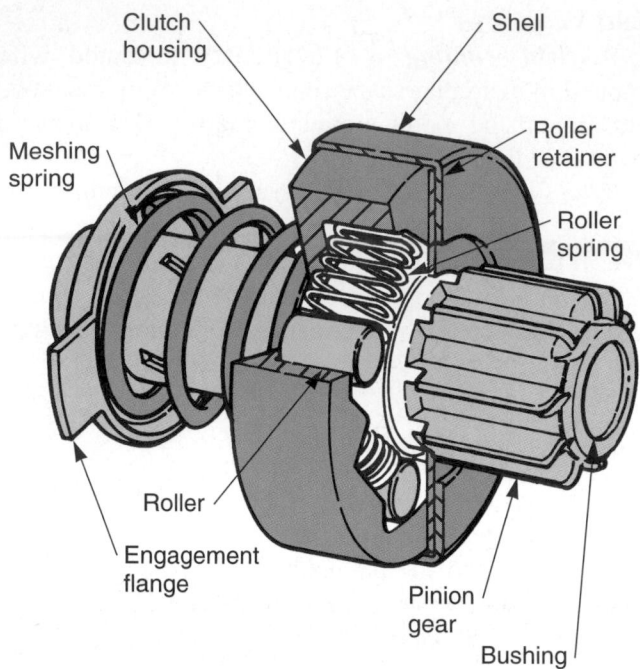

Figure 31-9. Study the construction of the starter overrunning clutch. It is simply a one-way clutch. It locks the flange to the pinion gear in one direction and releases in other direction. (Ford)

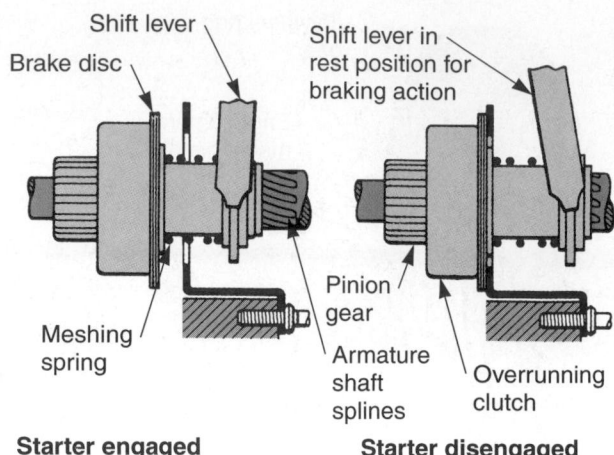

Starter engaged   **Starter disengaged**

Figure 31-11. The pinion gear assembly is splined to the armature shaft. This makes the pinion gear assembly turn with the shaft. It also lets pinion gear slide on the shaft for engagement with the flywheel gear. Note how this unit also has a disc brake for stopping the pinion after disengagement. (Robert Bosch)

motor. The starter solenoid is an ***electromagnetic switch*** (switch using electricity and magnetism for operation). It is similar to other relays but is capable of handling much higher current levels.

A cutaway view of a starter solenoid is given in **Figure 31-12.** Note the solenoid windings, contact disc, terminals, plunger, and other parts.

Starter Solenoid Operation

With the ignition key in start position, a small amount of current flows through the ***solenoid windings.*** This produces a magnetic field that pulls the solenoid plunger and disc into the coil windings. This causes the

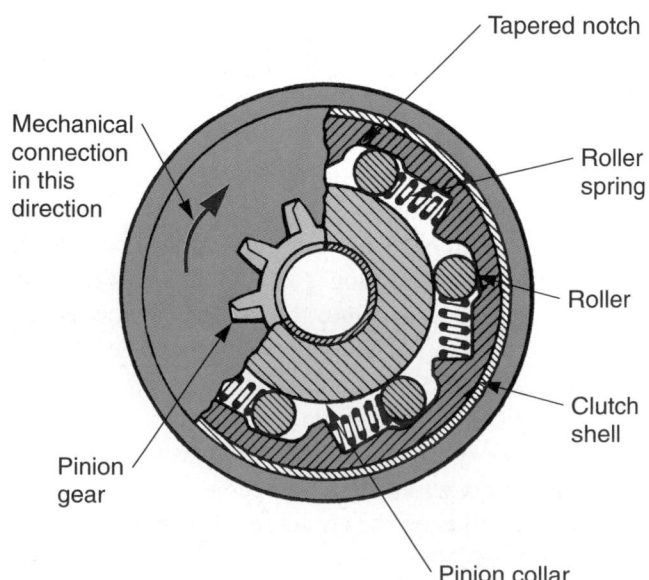

Figure 31-10. This cutaway shows how rollers fit in the clutch. Rollers jam and lock units together one way. Going the other way, they release, allowing the pinion to freewheel. (Robert Bosch)

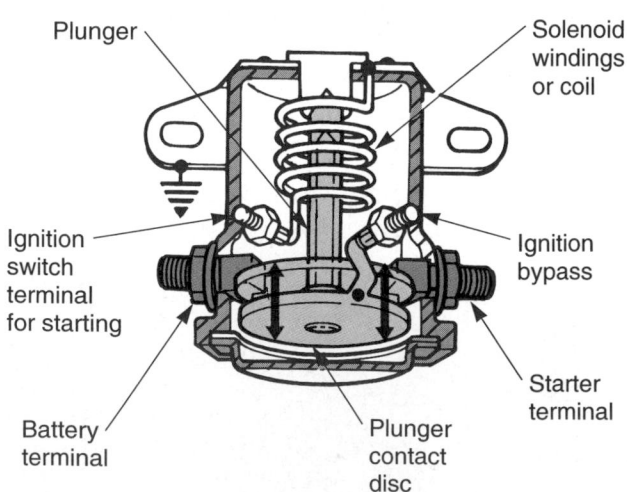

Figure 31-12. Study the construction of a starter solenoid. One small terminal connects to the ignition switch. Larger terminals connect to the battery and starting motor. The plunger movement pulls the disc into contact with two battery terminals to activate the starter. (Ford)

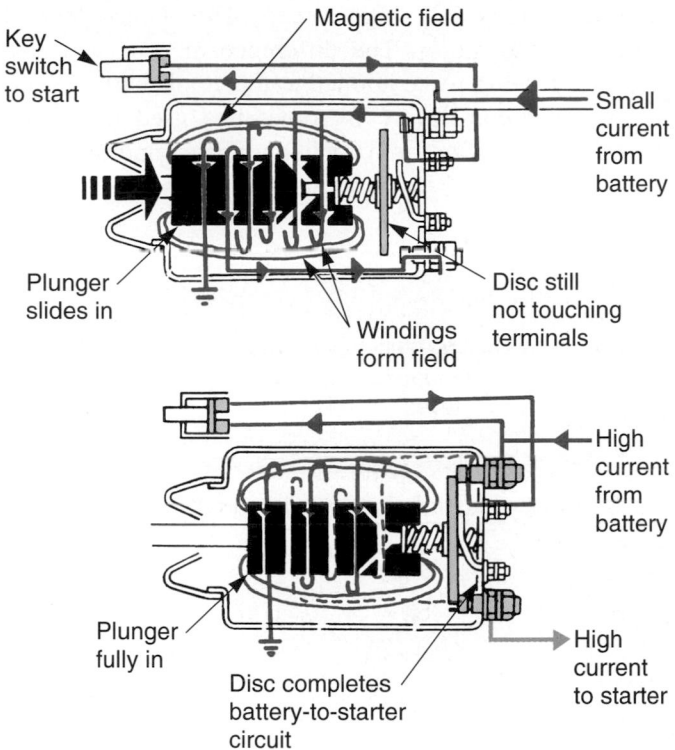

Figure 31-13. Solenoid operation. With the ignition key turned to start, current flows through the solenoid coil. This produces a field that pulls the plunger inward. As soon as the disc touches the terminals, a large amount of current flows to the starter.

solenoid disc to touch both of the high-current terminals, completing the battery-to-starter circuit. A current of 150–200 amps flows through the solenoid to the starter. Look at **Figure 31-13.**

When the ignition key is released, current is disconnected from the solenoid windings. The magnetic field collapses, and the plunger is free to slide out of the windings. This opens the disc-to-terminal connection. The open connection stops current to the starter, and the starter motor shuts off.

Starter Solenoid Functions

A starter solenoid may have three functions, depending on starter design:

- Close battery-to-starter circuit.
- Push the starter pinion gear into mesh with the flywheel ring gear.
- Bypass the resistance wire in the ignition circuit. (See Chapter 35.)

The starter solenoid may be located on the starting motor or away from it. When mounted on a body panel (away from starter), the solenoid simply makes and breaks electrical connections. When mounted on the starter, it also slides the pinion gear into the flywheel ring gear.

Starting Motor Construction

The construction of all starting motors is very similar. There are slight variations in design, however. As pictured in **Figure 31-14,** the main parts of a starting motor are:

- *Armature*—windings, core, starter shaft, and commutator assembly that spins inside a stationary field.
- *Pinion drive assembly*—pinion gear, overrunning clutch, and sometimes a shift lever and solenoid.
- *Commutator end frame*—end housing for brushes, brush springs, and shaft bushing.
- *Field frame*—center housing that holds field coils and shoes.
- *Drive end frame*—end housing around pinion gear; has bushing for armature shaft.

Starting Motor Types

Starting motors are classified by type of pinion gear engagement. There are two main starter types: moveable pole shoe starting motors and solenoid starting motors. Both types are pictured in **Figure 31-14.**

A *movable pole shoe starting motor* uses a yoke lever to move the pinion gear into contact with the flywheel gear, **Figure 31-15.** The shoe is hinged on the starter frame with a drive yoke. The drive yoke links the pole shoe and pinion gear.

When the starter is activated, the magnetic field of the motor pulls the pole shoe downward. Then, lever action of the yoke pushes the pinion gear outward on its shaft. This causes gear engagement as the armature begins to spin. When the motor is shut off, a spring moves the pinion gear and pole shoe into the released position.

A *starter-mounted solenoid* has a plunger that moves a shift lever to engage the pinion gear. The solenoid is mounted on the side of the starter field frame, as in **Figure 31-16.** With this starter design, the solenoid completes the battery-to-starter circuit and also operates the pinion gear. **Figure 31-17** shows the basic operation of a starter-mounted solenoid.

Permanent-Magnet Starter

A *permanent-magnet starter* uses special high-strength magnets in place of conventional field windings. The magnets produce a strong magnetic field capable of rotating the armature with enough torque to crank the engine. A permanent-magnet starting motor is shown in **Figure 31-18.**

Starting Motor Torque

A starting motor must produce high torque to start an engine. The pinion gear is much smaller than the flywheel ring gear. Therefore, the starting motor armature turns at a relatively high speed. This helps prevent stalling of the motor. The difference in gear size also increases turning force applied to the crankshaft.

A *reduction starter* is sometimes used to further increase the rotating force applied to the engine flywheel.

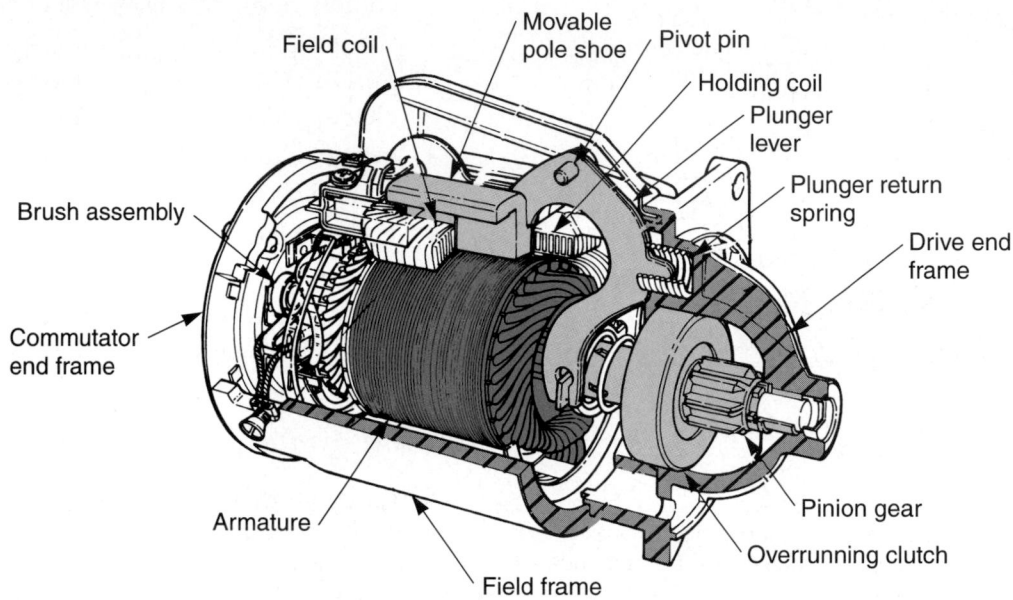

Starting motor with movable pole shoe

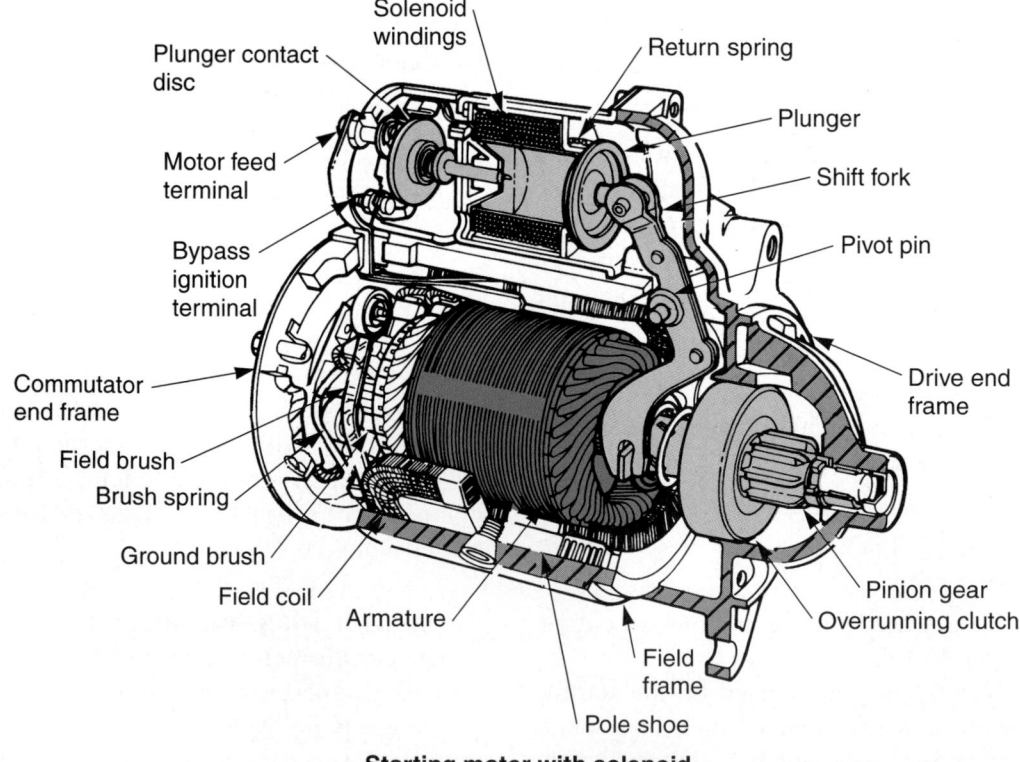

Starting motor with solenoid

Figure 31-14. Two common types of starting motors. (Ford)

It uses an extra set of gears to improve the gear reduction, as in **Figure 31-19.**

The starter pinion gear drives an idler gear. The idler gear drives a larger gear on the overrunning clutch assembly. This allows higher armature speed and higher torque output to the flywheel. More constant engine cranking speeds are produced by a reduction starter.

Internal Motor Circuits

Direct-current electric motors have three common types of internal connections: series, shunt, and compound. Refer to **Figure 31-20.**

Generally, *series-wound motors* develop maximum torque at initial start-up. Torque decreases as motor speed increases. *Shunt-wound motors* have less starting torque but more constant torque at varying speeds. *Compound-wound motors* have both series and shunt windings. They have good starting power with fairly constant operating speed.

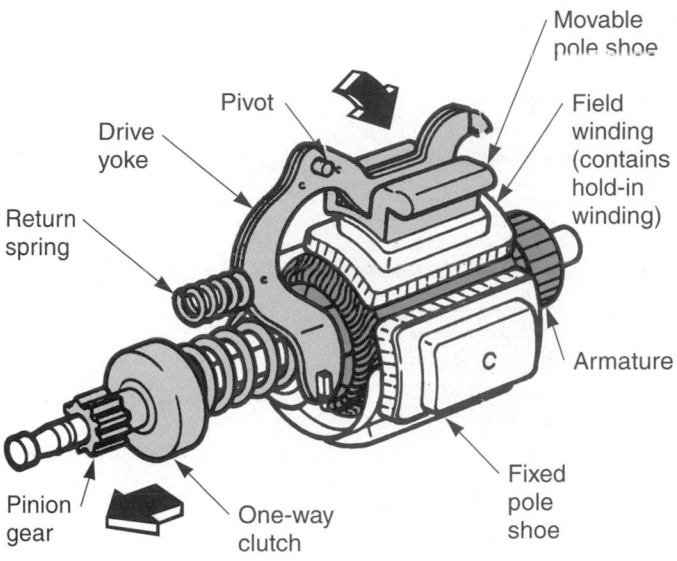

Figure 31-15. When the starter is powered, a magnetic field pulls the movable shoe toward the armature. The yoke then slides the pinion into mesh with the ring gear. When the starter is deactivated, a small spring pushes the pinion away from the flywheel. (Chrysler)

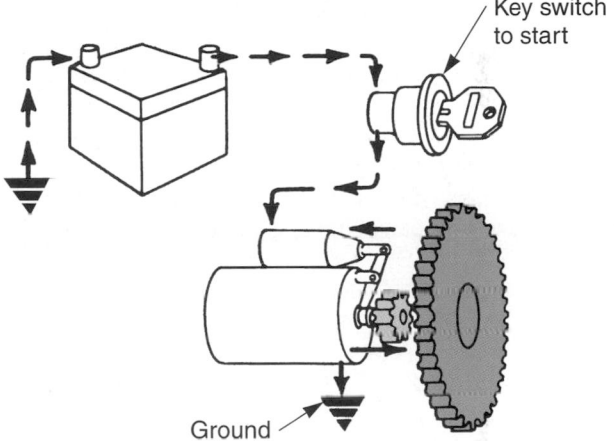

Solenoid moves pinion into ring gear

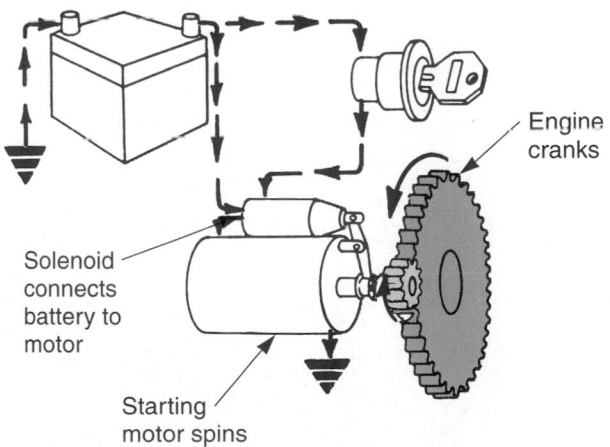

Once pinion gear is in place, circuit to starting motor is completed

Figure 31-17. When the ignition switch is turned to start, current activates the solenoid, which slides the pinion to the flywheel gear. At the end of the solenoid plunger travel, the disc closes the starter-to-battery circuit and the starting motor turns. (Deere & Co.)

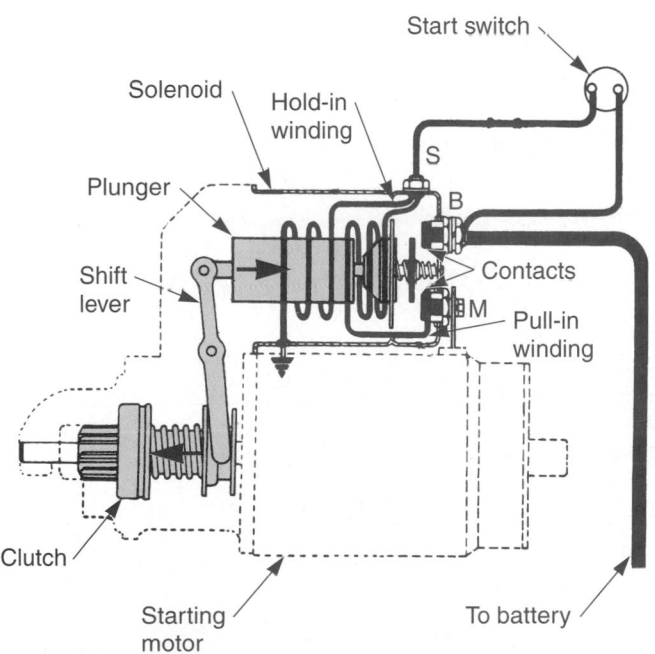

Figure 31-16. With a starter-mounted solenoid, a lever is connected to the solenoid plunger. When the plunger moves into the windings, the lever slides the pinion into the mesh. Spring action moves the pinion away from the ring gear when the engine starts. (GM Trucks)

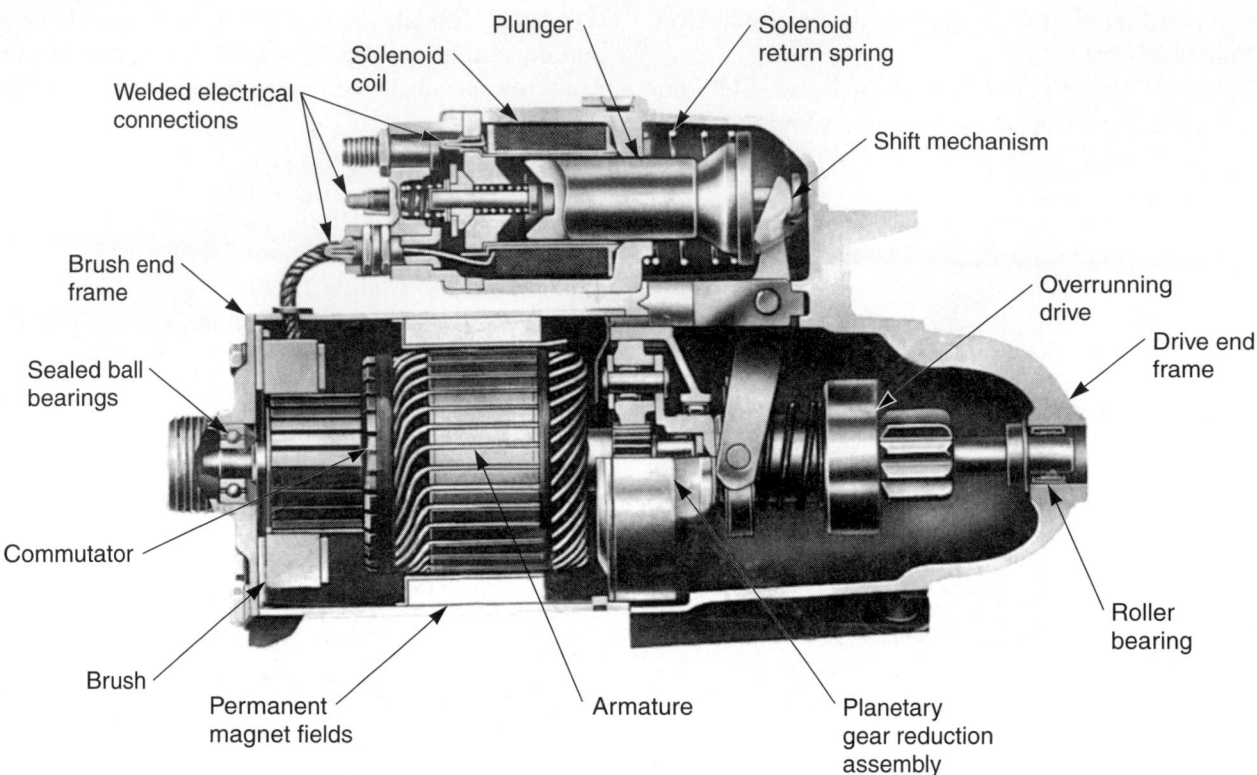

Figure 31-18. This is a permanent-magnet starting motor. Instead of coils of wire, it uses special magnets to produce a stationary magnetic field to act against the armature. (General Motors)

Neutral Safety Switch

A *neutral safety switch* prevents the engine from cranking unless the shift selector is in neutral or park. It keeps the starting system from working when the transmission is in gear. See **Figure 31-21.** The switch provides a safety feature to prevent the car from starting while in gear.

Cars with automatic transmissions commonly have a neutral safety switch. The switch may be mounted on the shift lever mechanism or on the transmission.

Figure 31-19. A reduction starter uses extra gears to increase motor torque. A gear on the armature shaft turns the idler gear. The idler turns a larger gear on the clutch for gear reduction. The exposed gear turns the flywheel gear. (Honda)

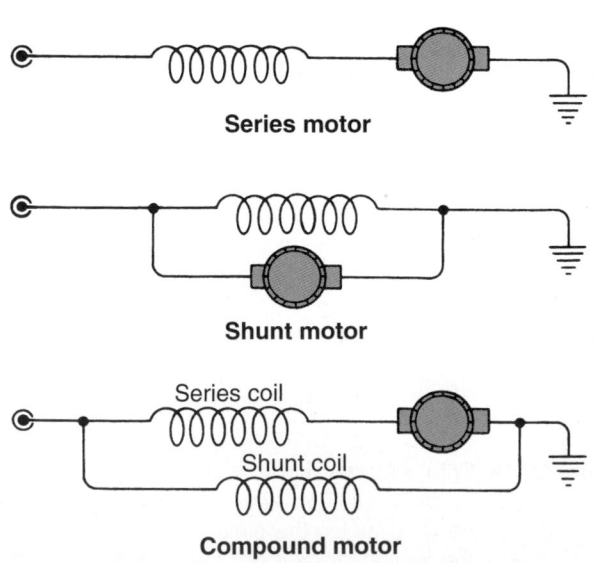

Figure 31-20. Three types of dc motor circuits. (Sun Electric)

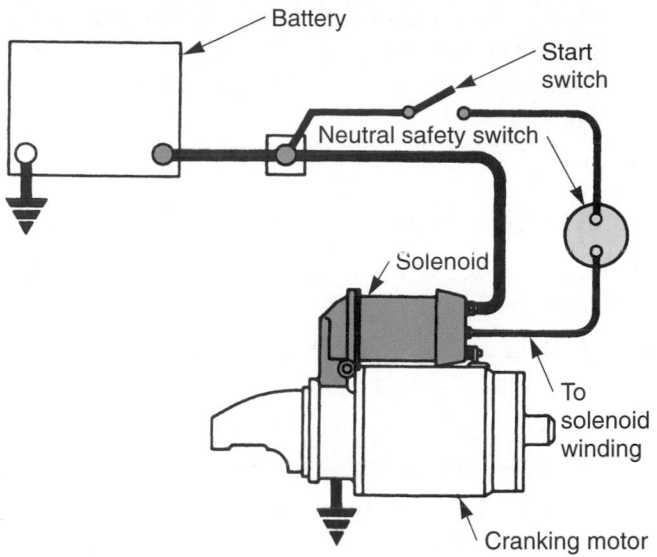

Figure 31-21. The neutral safety switch is in series with the starter solenoid. If it is open, the starter will not work. The shift mechanism on an automatic transmission operates the switch. (GMC)

Neutral Safety Switch Operation

The neutral safety is wired into the circuit going to the starter solenoid. When the transmission is in drive or reverse, the neutral safety switch is open (disconnected). This prevents current from activating the solenoid and starter when the ignition switch is turned to start.

With the transmission in neutral, the neutral safety switch is closed (connected). Current can flow to the starter when the ignition switch is turned.

Tech Tip

In most late-model cars, the brake light switch is wired into the same control circuit as the neutral safety switch. You must have the transmission in park (or neutral) and press the brake pedal for the starting motor to energize.

Starter Relay

As you learned in the chapter on electricity and electronics, a *relay* is a device that opens or closes one circuit by responding to an electrical signal from another circuit. Some starting systems use a relay between the ignition switch and the starter solenoid.

A *starter relay* uses a small current flow from the ignition switch to control a slightly larger current flow to the starter solenoid. This further reduces the load on the ignition key switch.

Starter Relay Operation

A starter relay circuit is shown in **Figure 31-22.** When the ignition switch is turned to start, current flows into the relay. This closes the relay contacts. The contacts complete the circuit to the solenoid windings and the starting system operates.

When the key is released, the relay opens. This stops the solenoid current to disengage the starting motor.

Summary

- The starting system uses battery power and an electric motor to turn the engine crankshaft for engine starting.

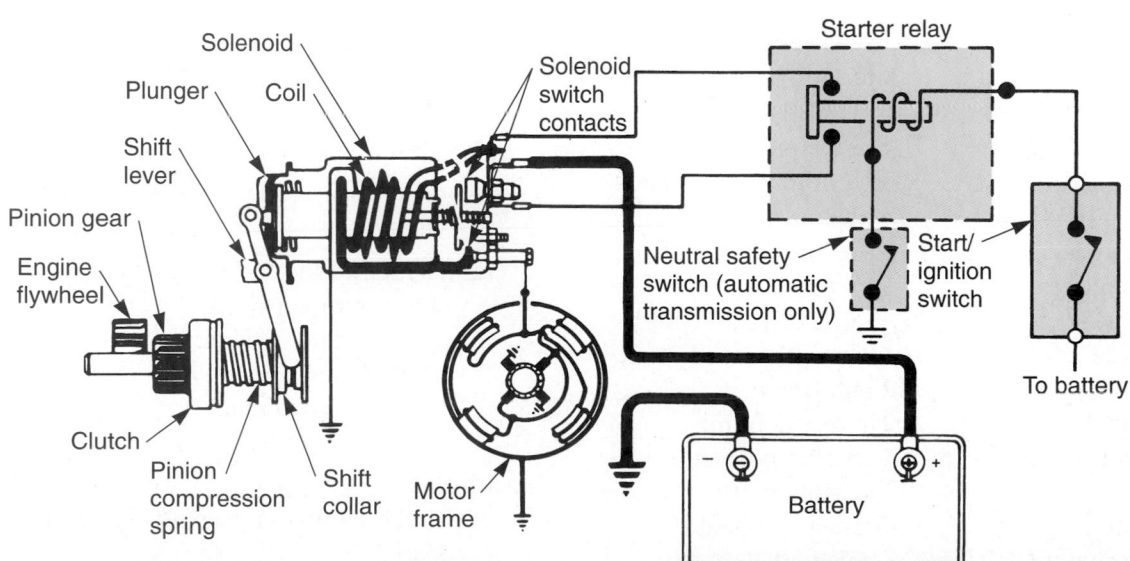

Figure 31-22. A complete starting system circuit. This circuit has a starter relay. The relay further decreases the amount of current flowing through the ignition switch. Also note how the relay winding is wired to the neutral safety switch. (Chrysler)

- The starting motor converts electrical energy from the battery into mechanical or rotating energy to crank the engine.
- The commutator serves as a sliding electrical connection between the motor windings and the brushes.
- A field winding is a stationary set of windings (insulated wire wrapped in circular shape).
- A starter pinion gear is a small gear on the armature shaft that engages a large ring gear on the engine flywheel.
- The starter overrunning clutch locks the pinion gear in one direction and releases it in the other.
- The starter solenoid is a high-current relay. It makes an electrical connection between the battery and starting motor.
- A permanent-magnet starter uses special high-strength magnets in place of conventional field windings.
- A reduction starter is sometimes used to further increase the rotating force applied to the engine flywheel.
- A neutral safety switch prevents the engine from cranking unless the shift selector is in neutral or park.
- A starter relay uses a small current flow from the ignition switch to control a slightly larger current flow to the starter solenoid.

Important Terms

Starting system	Flywheel ring gear
Battery	Starter overrunning
Ignition switch	clutch
Solenoid	Starter solenoid
Starting motor	Electromagnetic switch
Magnetic field	Solenoid windings
Like charges	Solenoid disc
Unlike charges	Armature
Magnet	Pinion drive assembly
Pole piece	Commutator end frame
Pole shoes	Field frame
Commutator	Drive end frame
Motor brushes	Movable pole shoe
Torque	starting motor
Starter armature	Starter-mounted
Starting motor shaft	solenoid
Armature core	Permanent-magnet
Field winding	starter
Starter pinion gear	Reduction starter

Series-wound motors	Neutral safety switch
Shunt-wound motors	Relay
Compound-wound motors	Starter relay

Review Questions—Chapter 31

Please do not write in this text. Place your answers on a separate sheet of paper.

1. List and describe the four major parts of a starting system.

2. The _____ _____ converts electrical energy from the battery into mechanical or rotating energy to crank the engine.

3. Like charges (fields) attract each other and unlike charges (fields) repel each other. True or False?

4. The _____ serves as a sliding electrical connection between the motor windings and the brushes.

5. Which of the following is *not* part of a starting motor?
   (A) Armature.
   (B) Field winding.
   (C) Commutator.
   (D) Slip ring.

6. What is the function of the starter pinion gear?

7. What would happen if the starting motor did *not* have an overrunning clutch (gear locked to armature shaft)?

8. The _____ _____ is a high current relay that completes the circuit between the battery and the starting motor.

9. List three functions of a starter solenoid.

10. List and explain the five major parts of a starting motor.

11. Describe the two main types of starting motors.

12. Why is a reduction starting motor sometimes used?

13. Which of the following is *not* a type of internal starting motor circuit?
    (A) Parallel wound.
    (B) Series wound.
    (C) Shunt wound.
    (D) Compound wound.

14. The _____ _____ _____ keeps the engine from cranking unless the shift selector is in neutral or park.

15. A starter relay uses a small current flow from the ignition switch to control a slightly larger current flow to the _____ _____.

ASE-Type Questions

1. Which of the following converts electrical energy from the battery into mechanical or rotating energy to crank the engine?
   (A) Solenoid.
   (B) Pinion gear.
   (C) Starting motor.
   (D) All of the above.

2. Magnetic fields that are *alike* will:
   (A) repel each other.
   (B) cancel each other.
   (C) attract each other.
   (D) None of the above.

3. Which of the following serves as a sliding electrical connection between motor windings and brushes?
   (A) Solenoid.
   (B) Armature.
   (C) Pole piece.
   (D) Commutator.

4. A starter armature consists of each of these *except:*
   (A) shaft.
   (B) flange.
   (C) windings.
   (D) commutator.

5. Which device locks a pinion gear in one direction and releases it in the other?
   (A) Roller retainer.
   (B) Starter solenoid.
   (C) Flywheel ring gear.
   (D) Overrunning clutch.

6. A starter solenoid is this type of relay switch.
   (A) Electric.
   (B) Magnetic.
   (C) Electromagnetic.
   (D) None of the above.

7. Which of the following is *not* a starter solenoid function?
   (A) Control electric motor spinning.
   (B) Close battery-to-starter circuit.
   (C) Push pinion gear into flywheel gear.
   (D) Bypass ignition circuit resistance wire.

8. When reviewing starter motor classifications, Technician A says starters are grouped by type of pinion gear engagement. Technician B feels starters are grouped by type of internal connections they have. Who is right?
   (A) A only.
   (B) B only.
   (C) Both A and B.
   (D) Neither A nor B.

9. Which of the following uses a plunger to move a shift lever which engages the pinion gear?
   (A) Reduction starter.
   (B) Movable pole shoe.
   (C) Starter-mounted solenoid.
   (D) Permanent-magnet starter.

10. Which of the following uses a small current flow from the ignition switch to control a larger current flow to the starter solenoid?
    (A) Clutch.
    (B) Starter relay.
    (C) Reduction starter.
    (D) None of the above.

Activities—Chapter 31

1. If possible, obtain an unserviceable starter motor from a shop or junkyard. Carefully disassemble and clean the components. Mount and label them for a classroom display.

2. Find out the advantages and disadvantages of installing a rebuilt starter (rather than a new one) in a vehicle. Report to the class.

Chapter 32

Starting System Testing and Repair

After studying this chapter, you will be able to:

- Diagnose common starting system troubles.
- Make orderly starting system tests.
- Remove and replace a starting motor.
- Explain typical procedures for a starting motor rebuild.
- Adjust a neutral safety switch.
- Describe the safety practices that should be followed when testing or repairing a starting system.
- Correctly answer ASE certification test questions on starting system diagnosis, service, and repair.

This chapter introduces the steps for diagnosing, testing, and repairing common starting system problems. It begins by explaining on-car diagnosis. Then, the service of each component is detailed. You will learn about the symptoms of faulty starting system components and how to check each component. Finally, procedures for part removal, repair, and replacement are summarized.

Starting System Diagnosis

The starting system is easier to work on than a car's other electrical systems. It has only a few major components that can cause problems, **Figure 32-1.** If any of these parts have high resistance, lower-than-normal resistance, damage, or wear, the engine may not crank normally.

Common Starting System Problems

In a *no-crank problem,* the engine crankshaft does not rotate properly when the ignition key is turned to the start position. The most common causes are a dead battery, poor electrical connection, or faulty system component.

A *slow-cranking condition* occurs when the engine crankshaft rotates at speeds that are lower than normal. This condition is usually caused by the same kinds of faults that produce a no-crank problem.

If a *solenoid buzzing* or *clicking* sound (with the engine not cranking) occurs when the ignition is activated, the battery may be discharged or the battery cable connections may be poor. Low current flow causes the solenoid plunger to rapidly kick in and out, making a clicking sound. A single solenoid click sound without engine cranking may point to a bad starting motor, burned solenoid contacts, a dead battery, or engine mechanical problems. The click is usually the solenoid closing or the pinion gear contacting the flywheel gear.

A *humming sound* that occurs after momentary engine cranking may be due to a bad overrunning clutch or a worn pinion gear unit. Pinion gear wear can make the gear disengage from the flywheel gear too soon. This can let the motor armature spin rapidly, producing a humming sound.

A *metallic grinding noise* may be caused by broken flywheel teeth or worn pinion gear teeth. The grinding may be caused as the gears clash against each other.

Normal cranking without starting is usually not caused by the starting system. There may be trouble in the fuel or ignition systems. With a diesel engine, check engine cranking speed. If cranking speed is low, the diesel may not start.

 Tech Tip

Sometimes the starting solenoid feeds current to the ignition system after the engine has started. If the engine starts and then dies (stops running) as the ignition key is released, check the voltage from the solenoid to the ignition system. You could have an open wire or an open connection in the solenoid circuit. A defective ignition switch or a wiring problem are other possible causes.

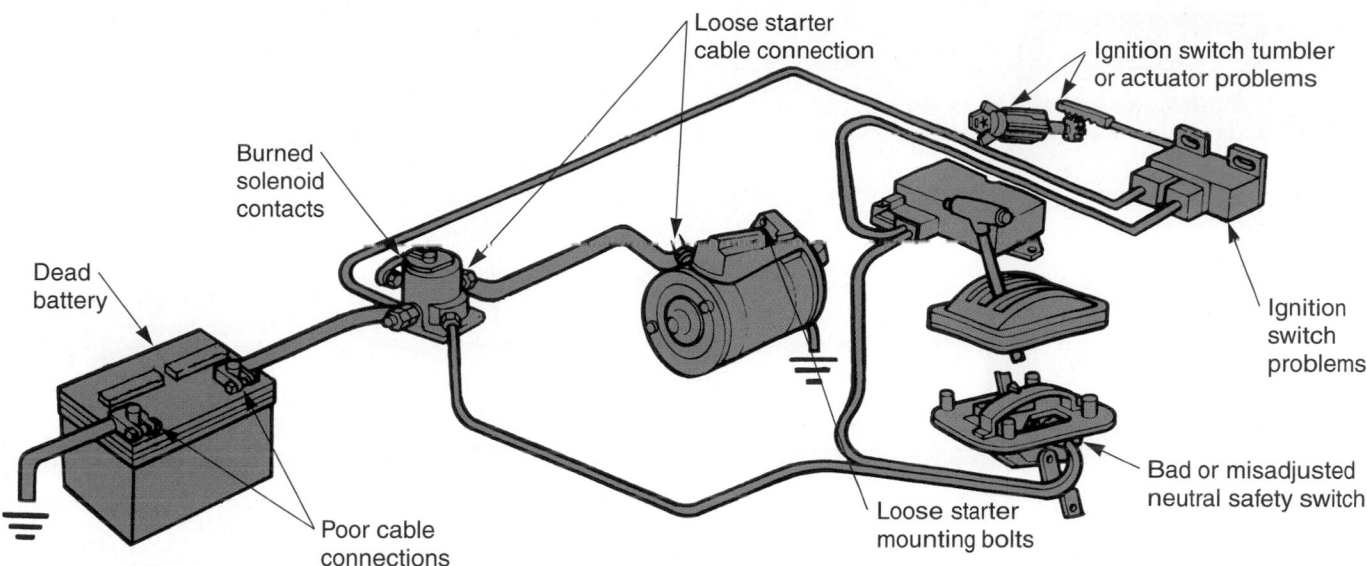

Figure 32-1. During initial inspection, check for these kinds of starting system troubles. A loose connection or discharged battery could keep engine from cranking.

Starting Headlight Test

A *starting headlight test* will quickly indicate the causes of trouble in a starting system. Turn on the headlights and try to start the engine. Note any sounds and watch the brightness of the headlights.

No cranking with no headlights points to a dead battery or an open in the electrical system. The battery connections may be bad. A burned fusible link will also kill power to most systems, **Figure 32-2.** A main feed wire to the fuse box could also be broken or disconnected.

If the headlights go out when cranking, the battery may be weak. The starting motor may be shorted. The engine could also be at fault. Dimming headlights indicate heavy current draw or poor current supply from the battery.

If the lights stay bright but the engine does not crank properly, a high resistance or an open in the starting circuit is likely. The problem could be in the ignition switch, wiring, solenoid, starter cable connections, or relay.

Depending on what the headlights and starter do when testing, you can decide what further tests are needed.

Using Service Manual Troubleshooting Charts

Service manual troubleshooting charts should be used when the causes of problems are hard to find. These charts are designed for the exact circuit and will often reduce the list of causes.

Checking the Battery

A dead or discharged battery is one of the most common reasons the starting system fails to crank the

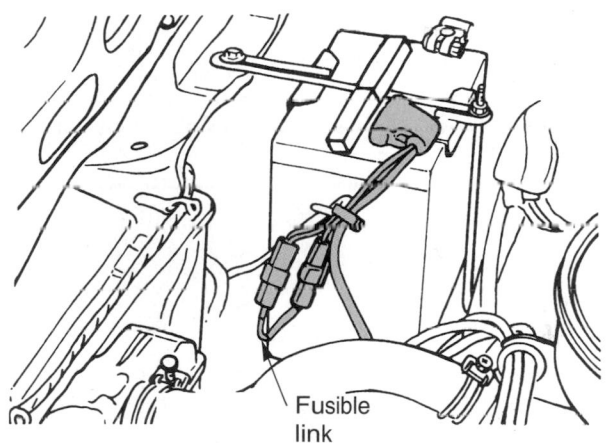

Figure 32-2. The battery and its cables are the most common cause of improper cranking. Note the fusible link attached to the positive battery cable. It can burn during a short to protect the electrical system. If major components (lights, horn, starting system) are dead, check the fusible link. (Mazda)

engine properly. The starting motor draws much more current (over 200 amps) than any other electrical component. A discharged or poorly connected battery can operate the lights, but may not have enough power to operate the starting motor.

If needed, load-test the battery as described in the chapter on battery service. Make sure the battery is in good condition and is fully charged. A starting motor will not function without a fully charged and well-connected battery.

Starter Current Draw Test

A *starter current draw test* measures the current used by the starting system. It will quickly tell you about

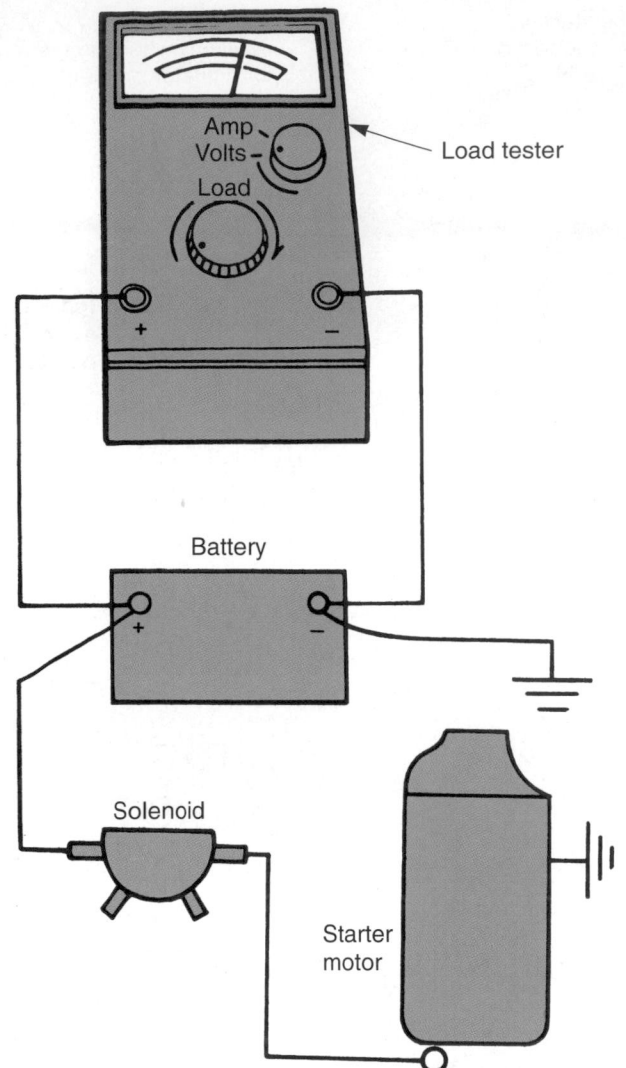

Figure 32-3. A battery load tester can be used to check the starter current draw. Crank the engine and note the voltage reading. Then, load the battery to obtain the same voltage. This will equal the current draw by starting motor. (Chrysler)

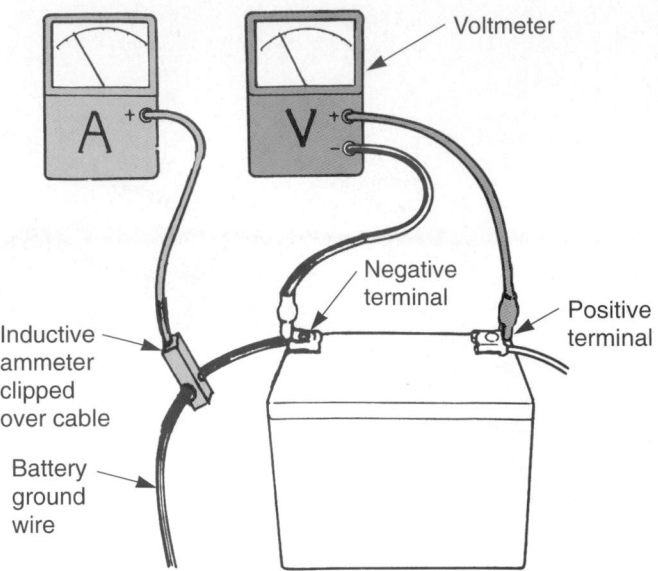

Figure 32-4. A voltmeter and an ammeter can also be used to measure starter current draw. A voltmeter reading is needed to compare different battery conditions. If current draw is not within specifications, there are starting system troubles. (Honda)

the condition of the starting motor and other system parts. If current draw is higher or lower than specifications, there is a problem.

To perform a current draw test, connect meters to measure the battery voltage and the current flow out of the battery. Two testing methods are shown in **Figures 32-3** and **32-4.** A load tester may also be used.

To keep the engine from starting during the test, disconnect the coil primary supply wire or ground the coil wire. You can also pull the fuse for the electric fuel pump if this is easier (direct ignition systems). Look at **Figure 32-5.**

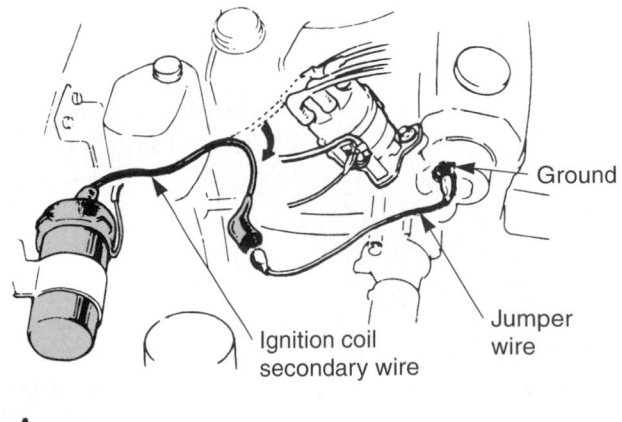

A

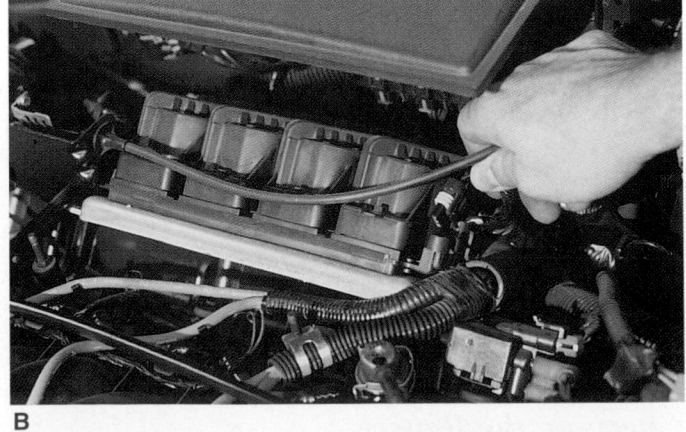

B

Figure 32-5. To keep the engine from running during a starter current draw test, disable the ignition system. (Honda) A—If a distributor is used, you can ground the coil wire. B—With a coil pack, disconnect the primary wires to disable the system.

With a diesel engine, you must disable the injection system. You may have to unhook the fuel shutoff solenoid. Check a shop manual for details.

Caution

Do not crank the engine for more than 15–30 seconds or starter damage may result. If cranked too long, the starter could overheat. Allow the starter to cool for a few minutes if more cranking time is needed.

Crank the engine and note the voltage and current readings. If they are not within specifications, something is wrong in the starting system or engine. Further tests are needed. **Figure 32-6** contains a troubleshooting chart and gives the average current draw values for various engine sizes.

Starting System Voltage Drop Tests

Voltage drop tests will quickly locate a part with higher-than-normal resistance. These tests provide an easy way of checking circuit condition. You do not need to disconnect wires and components to check internal resistance (voltage drops).

Insulated Circuit Resistance Test

An *insulated circuit resistance test* checks all parts between the positive battery terminal and the starting

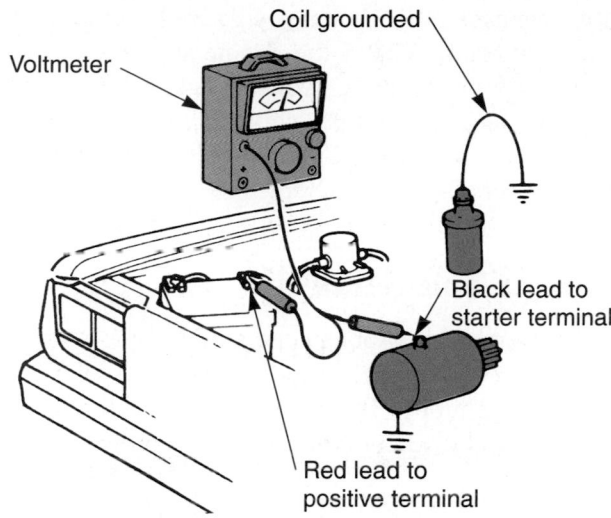

Insulated circuit resistance test

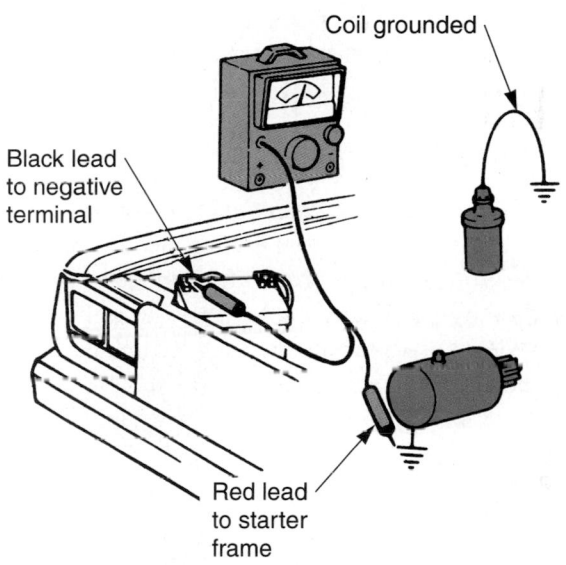

Starter ground circuit test

Figure 32-7. Voltage drop tests are a quick way of checking circuit resistance without disconnecting wires. (Chrysler)

ENGINE DISPLACEMENT	12-VOLT SYSTEM MAX. CURRENT
Most 4–6 Cylinders	125–175 Amps Max.
Under 300 C.I.D.	150–200 Amps Max.
Over 300 C.I.D.	175–250 Amps Max.

CRANKING CIRCUIT TROUBLESHOOTING CHART		
Cranking Voltage	Cranking Amps	Possible Problem
Voltage OK	Current OK	System OK
Voltage OK	Current Low Engine Cranks Slowly	Starter Circuit Connections Faulty
Voltage Low	Current Low Engine Cranks Slowly	Battery Low
Voltage Low	Current High	Starter Motor Faulty

Figure 32-6. This chart shows typical current draw values for different engine sizes. Ignore initial readings until the engine has cranked for a few seconds. Then current draw will stabilize. Study meter readings and trouble chart. (Marquette)

motor for excess resistance. **Figure 32-7** shows the basic connections for this test. Touch your voltmeter probes on the positive battery terminal and the starting motor input terminal.

Disable the ignition or injection system; then crank the engine. The voltmeter should not read over 0.5 volts. If voltage drop is greater, there is excess resistance in the circuit. There may be loose electrical connections, burned or pitted solenoid contacts, or other problems. Test each part individually.

Starter Ground Circuit Test

A *starter ground circuit test* checks the circuit between the starting motor ground and the negative battery terminal. Look at **Figure 32-7.** Touch the voltmeter leads to the battery negative terminal and the starter end frame. Crank the engine and note the meter reading. If higher than 0.5 volts, check the voltage drop across the negative battery cable. The engine may not be grounded properly. Clean, tighten, or replace the cable as needed.

Battery Cable Service

A battery cable problem can produce symptoms similar to those caused by a dead battery, bad solenoid, or weak starting motor. If the cables do not allow enough current flow, the starter will turn slowly or not at all.

Warning
Never remove starting system parts without disconnecting the battery. The engine could be cranked over, causing injury. Also, an electrical fire could result if wires are shorted.

Testing Battery Cables

To do a *battery cable connection test,* connect a voltmeter to the battery post and to the cable. See **Figure 32-8.** Measure the voltage drop across the connections while cranking the engine. If any cable shows a high voltage drop (above 0.3 volts) during cranking, clean and tighten its connections. Then retest the cable. If the voltage drop is still high, replace the cable.

Replacing Battery Cables

When replacing battery cables, make sure the new cables are the same as the old ones. Compare cable length and diameter. Lead terminals are better than steel terminals. The soft lead will conform to the shape of the battery terminal easier.

Caution
When tightening the connections on the ends of battery or starter cables, only snug down the fasteners. Many of the threaded studs, bolts, and nuts are made of soft lead or brass. They can strip and break easily.

Starter Solenoid Service

A bad starter solenoid can cause a variety of problems, including no cranking (with or without a click) or slow cranking. It can also keep the engine and starting motor running or prevent the starting motor from disengaging from the engine.

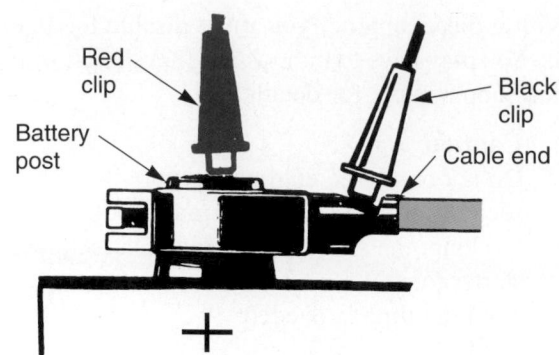
Figure 32-8. Checking battery terminals for corrosion and high resistance.

Usually, the large disc-shaped contact will burn and pit. The disc can develop high resistance that reduces current flow to the starter. An open or shorted solenoid winding can keep the contact disc from closing. No click will occur and the engine will not crank.

Testing the Starter Solenoid

To test the solenoid, connect a voltmeter as shown in **Figure 32-9.** This will measure the voltage drop and resistance of the solenoid lugs and contacts. Crank the

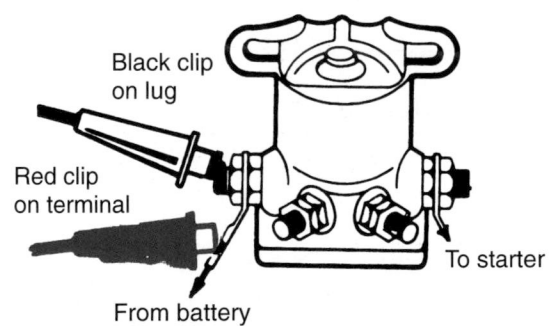
Check solenoid lug-to-cable connection

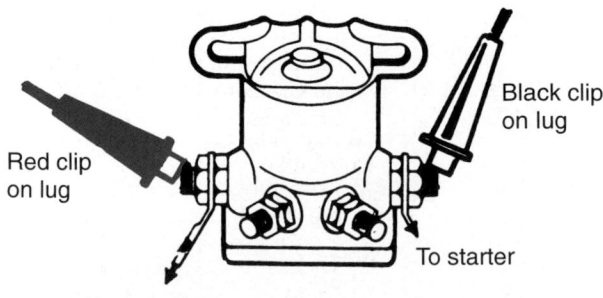
Check disc contact and terminal condition

Figure 32-9. Checking the starter solenoid. (Marquette)

engine and note the voltmeter reading. If either voltage measurement is above 0.3 volts, tighten the cable connections or replace the solenoid.

Replacing the Starter Solenoid

To replace a solenoid that is mounted away from the starter, simply remove the cables and wires. Unbolt the solenoid from the fender well and install the new one.

If the solenoid is a starter-mounted unit, the starter must be removed from the engine. Then, the solenoid is unscrewed from the starting motor and replaced. The procedures for removing, assembling, and installing a starter are covered later.

Warning
When disconnecting the positive cable from a starter solenoid, wrap electrical tape around its terminal. If you short this cable to ground, a powerful electric short will result. This could damage parts or cause burns.

Ignition Switch Service

A bad ignition switch (starter feed section of switch) can keep the starter solenoid from working normally. It can also keep the solenoid from releasing. The contacts in the ignition switch can wear or burn, causing either an open (no-cranking problem) or a short (engine cranks all the time).

Testing the Ignition Switch

To test an ignition switch, touch a test light to the starter solenoid start (S) terminal. If the ignition switch is good, the test light will glow when the key is turned to start the engine. The test light should go out when the key is released.

If the test light on the solenoid does not glow, either a wire in the ignition switch circuit is open or the ignition switch is defective. Test the wires coming out of the ignition switch to eliminate the possibility of a bad wire.

If the test light on the solenoid glows in both the start and run positions, the ignition switch is probably shorted. The engine would crank all the time.

These are simplified tests. You will need to use your own judgment and a service manual to perform more detailed tests.

Note
For more information on ignition switch service, refer to Chapter 36, *Ignition System Problems, Testing, and Repair.*

Starter Relay Service

A bad starter relay will keep power from the starter solenoid. This will prevent engine cranking. The winding or the contact points in the relay could be faulty. If the relay is bad, you will not hear the solenoid click and engage.

To test a starter relay, use a test light to check for voltage going into and coming out of the relay terminals. Refer to a wiring diagram for test points. Replace the relay if needed.

Neutral Safety Switch Service

A misadjusted or bad neutral safety switch can also keep the engine from cranking when the key is turned to start. If the neutral safety switch is open, current cannot flow from the ignition switch to the starter solenoid.

Checking the Neutral Safety Switch

Before testing the switch, move the transmission gear shift lever into various positions while trying to start the engine. The switch may close, letting the starter operate. If the starter begins to work, the neutral safety switch may only need adjustment.

Adjusting the Neutral Safety Switch

The procedure for adjusting a neutral safety switch is fairly simple:

1. Loosen the fasteners holding the switch. The switch may be located on the steering column, shift lever, or transmission.
2. With the switch loosened, place the shift selector into park.
3. While holding the ignition switch to start, slide the neutral switch on its mount until the engine cranks.
4. Without moving the switch, tighten its hold-down screws. This should allow the engine start only with the shift lever in park or neutral.
5. Check operation after adjustment.

Testing the Neutral Safety Switch

To test a neutral safety switch, touch a 12-volt test light to the switch output wire connection while moving the transmission shift lever. The light should glow as the shift lever is slid into park and neutral. The test light should not glow when the transmission is in other positions.

Check the mechanism that operates the neutral safety switch. There should be a prong or other device that

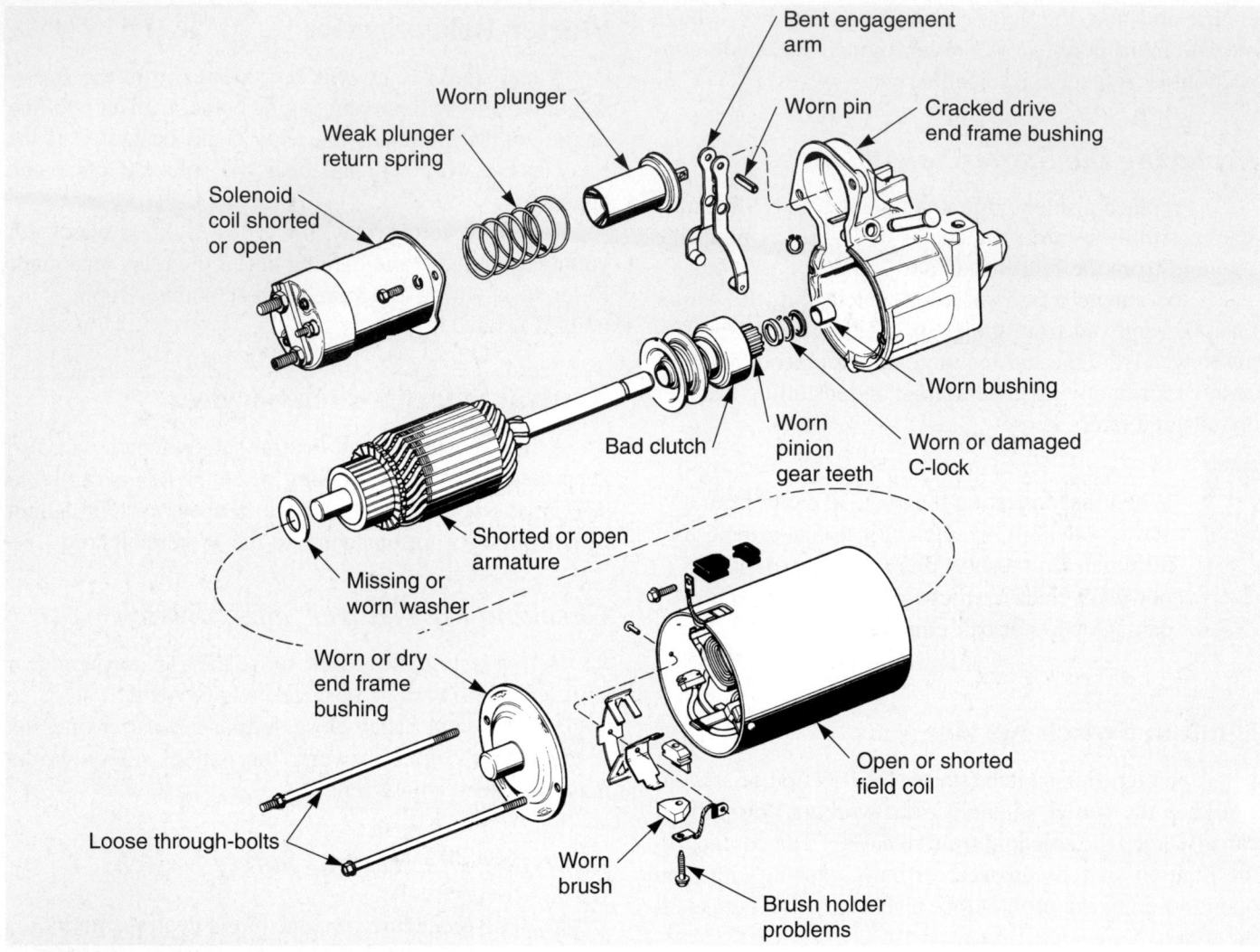

Figure 32-10. Study the types of problems requiring starting motor repairs.

actuates the neutral safety switch. If the problem is in the switch, remove, replace, and adjust it.

Starter Service

A faulty starting motor can cause a wide range of symptoms: slow cranking, no cranking, overheating of starter cables, and abnormal noises while cranking. If the battery, cables, solenoid, and other starting system parts are good but the engine does not crank properly, the starter may be bad.

A current draw test and other tests will suggest when the starting motor should be removed for further inspection and testing.

Figure 32-10 shows an exploded view of a typical starter. Study the types of problems that can occur and require a motor rebuild.

Starting Motor Rebuild

A *starting motor rebuild* typically involves:

1. Removing and disassembling the starting motor.
2. Cleaning parts and inspecting them for wear or damage.
3. Replacing brushes, bushings, and other worn or damaged parts.
4. Polishing or turning the commutator.
5. Lubricating, reassembling, testing, and installing the starting motor.

Many shops do not rebuild starting motors. They purchase and install new or factory-rebuilt units. The cost of labor is too high to make in-shop rebuilding economical. Also, the factory-rebuilt units will have a limited warranty.

Tech Tip

When the starter must be repaired, you may only need to disassemble a section of the starter. For example, a worn pinion gear can make the starter disengage before engine start-up. The clutch will freewheel before the engine has cranked. The pinion gear assembly can be replaced without complete starting motor disassembly. By removing only the drive-end frame and a C-lock, the brushes and the other end of the motor can usually be left together.

Starting Motor Removal

Before deciding to remove the starting motor, inspect it closely for problems. Check that the starter-to-engine bolts are tight. Loose starter bolts can upset motor operation by causing a poor ground or incorrect pinion gear meshing. Make sure all wires on the motor and solenoid are tight.

If no obvious problems are found, the starter should be removed. To remove a starter:

1. Disconnect the battery.
2. Unbolt the battery cable, solenoid wires (starter-mounted solenoid type), and any braces on the motor, **Figure 32-11.**
3. Unscrew the bolts while holding the motor. Look at **Figure 32-12.**

Starter shims may be used to adjust the space between the pinion gear and the flywheel ring gear. During starter removal, always check for shims. They must be returned to the same place during reassembly. If the shims are not replaced, the pinion and flywheel gears will not mesh properly. A grinding noise will result, and the pinion and flywheel gears may be damaged.

Tech Tip

In a few late-model engines, the starting motor is located under the engine intake manifold. See **Figure 32-13.** You must remove the manifold to service the starting motor on these engines. Usually, only a few easy-to-reach bolts secure the intake manifold.

Starter Disassembly

After the starter is removed, it must be disassembled. Refer to **Figure 32-14.**

1. If the starter has a solenoid, remove the fasteners that hold it in place.
2. Pull the solenoid off the motor using care not to lose the plunger spring. You may need to rotate the solenoid slightly.
3. After punch-marking the end frames, remove the through-bolts.
4. Tap off the end frames with a plastic hammer while noting the positions of the internal parts.
5. Slide the armature out of the field frame after removing the end frames.
6. Remove the C-clip holding the pinion gear on the armature shaft. Then slide the pinion gear off the shaft.

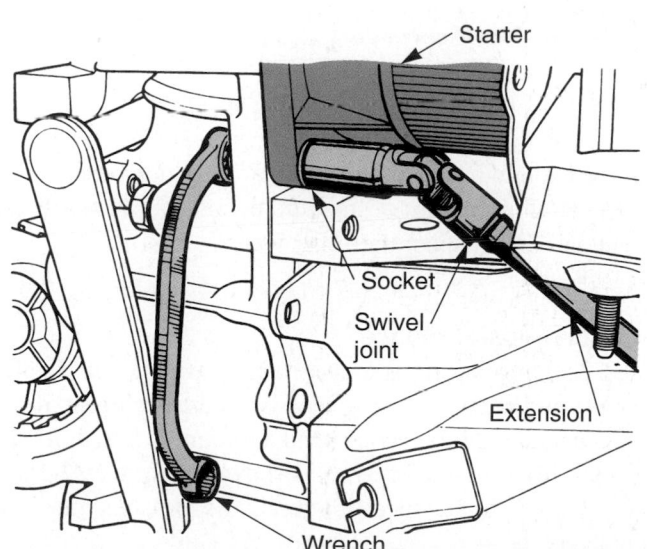

Figure 32-11. When starting motor fasteners are hard to reach, a swivel, extension, and ratchet may help. (Renault)

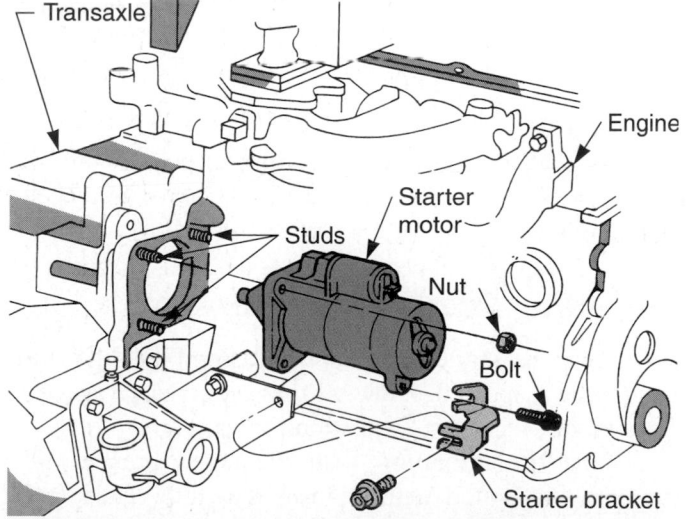

Figure 32-12. To repair a starter, remove fasteners, wires, and brackets or heat shields. Hold the starter firmly because it is fairly heavy. (Chrysler)

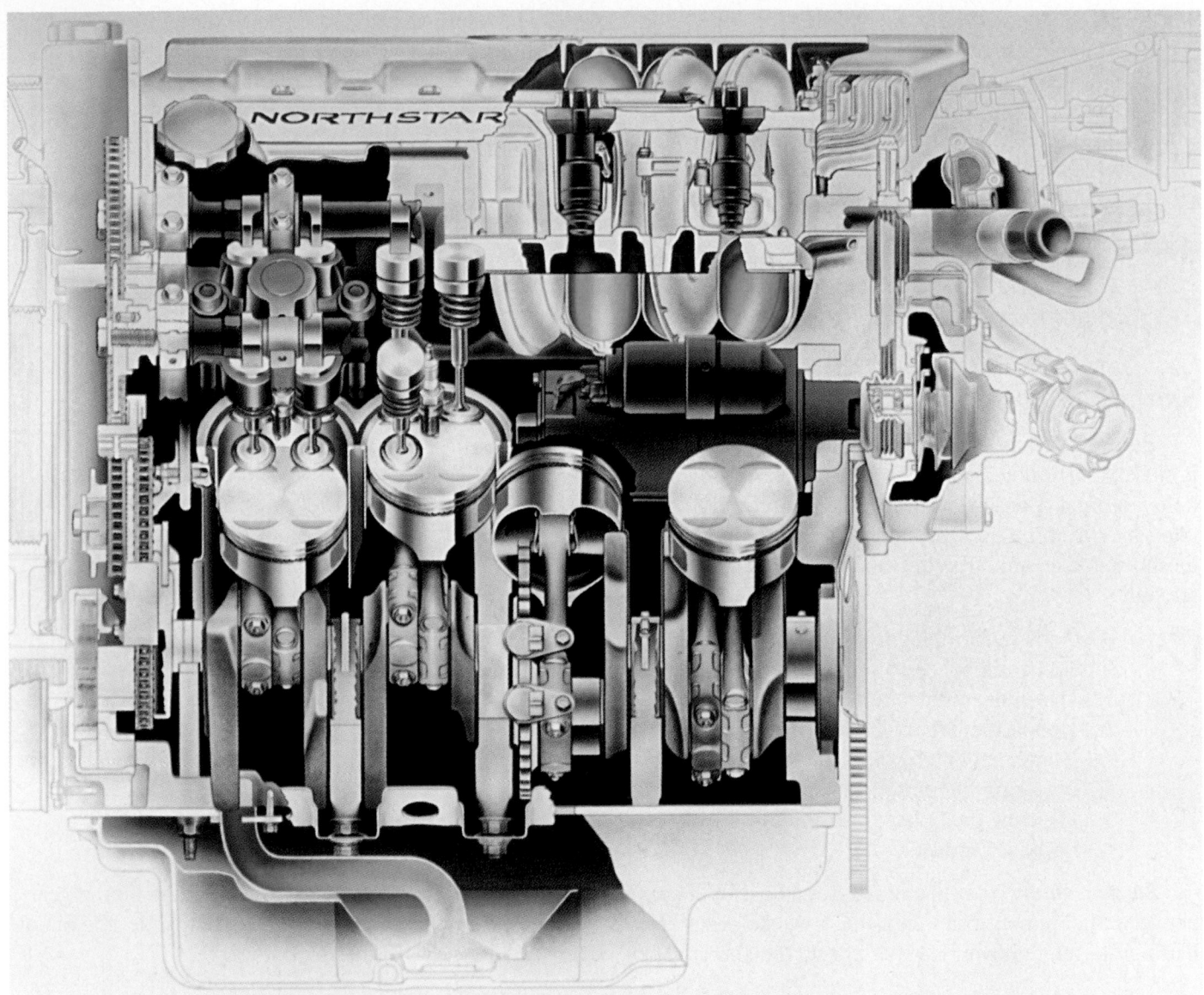

Figure 32-13. A starting motor mounted inside the valley of a V-8 engine. To get at the starting motor, you must remove the intake manifold. You can then gain easy access to the starting motor. (Cadillac)

Place all the parts in an organized pattern on your workbench. This will help you if you forget how to put something together.

Inspecting Starter Motor Parts

After the starter has been disassembled, the parts should be inspected. Blow all the parts clean with compressed air. Wear eye protection.

Wipe the armature, field windings, brushes, and overrunning clutch with a clean, dry cloth. These parts should *not* be cleaned using solvent. The solvent can

damage the wire insulation, soak into the brushes, or wash the lubricant out of the clutch. Once all the parts are clean and dry, inspect them for wear.

Starter Brush Service

Check for worn starter brushes, which can reduce starter torque and cause excessive starter current draw. Some manufacturers recommend a minimum length for the brushes. Measure the brushes as shown in **Figure 32-15.** If worn shorter than specifications, the brushes must be replaced.

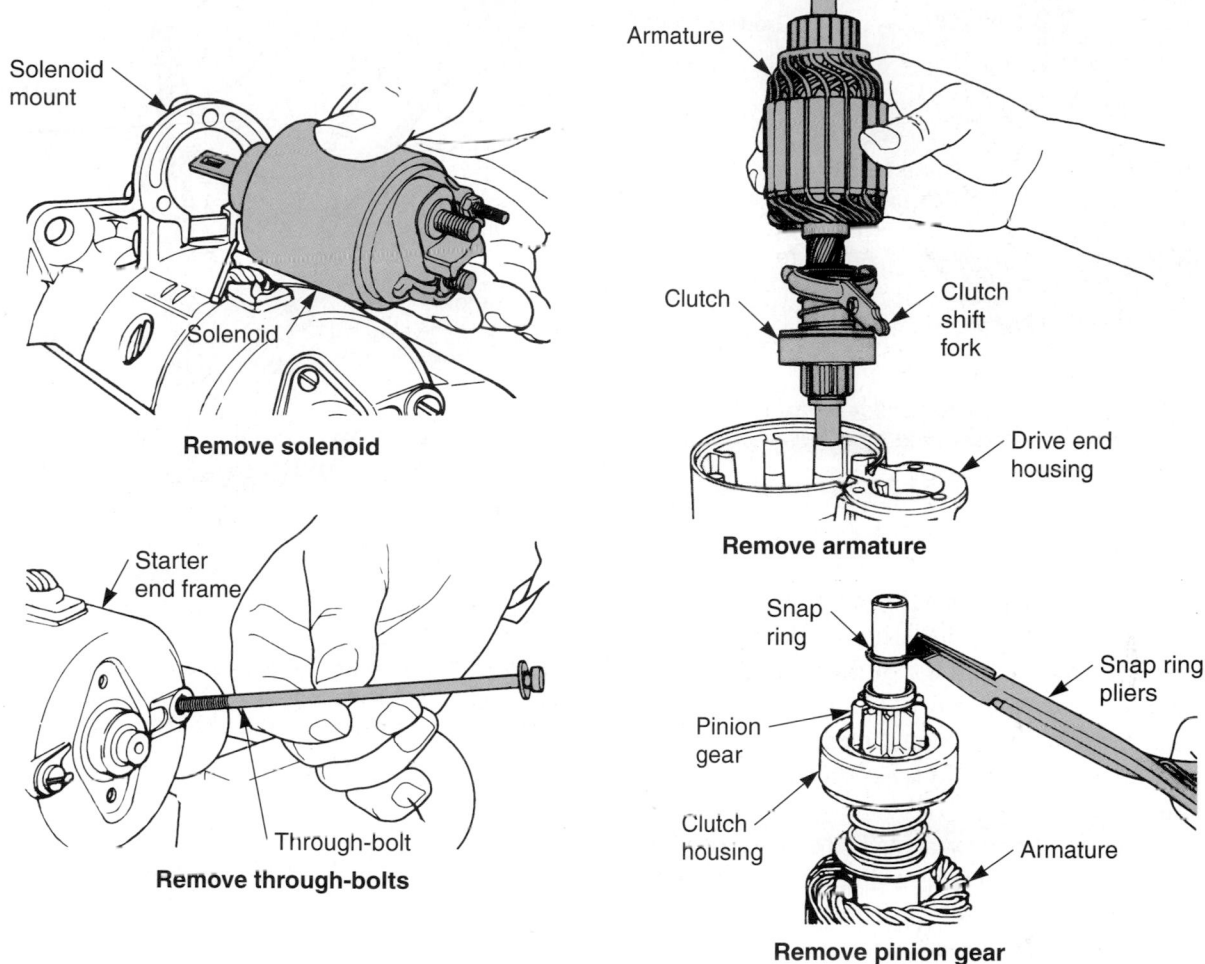

Figure 32-14. Basic steps for starter disassembly. (Chrysler)

The brush wire leads are usually soldered in place. A soldering gun and rosin-core (not acid-core) solder must be used when replacing the brushes. Also, check the brush holder for shorts to ground, **Figure 32-15.**

Armature Service

Inspect the armature for wear and damage. Look for signs of burning or overheating on the windings and com-mutator. If the armature has been rubbing on a field pole shoe, the shaft may be bent. Also, check the ends of the shaft for wear and burrs.

To check for an armature short circuit, mount the armature on a *growler* (armature tester). This is shown in **Figure 32-16.**

After reading the instructions for the growler, turn on the power. Hold a thin strip of metal or hacksaw blade

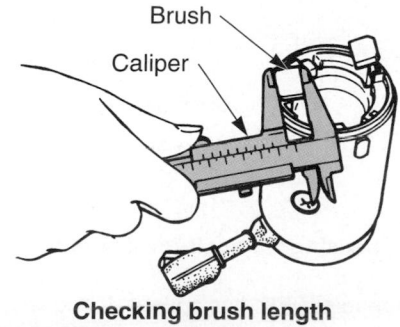

Checking brush length

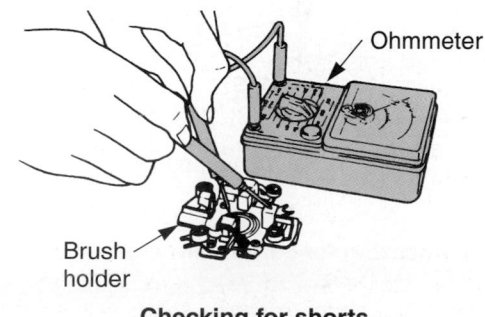

Checking for shorts

Figure 32-15. Measure brush length using a caliper and check for shorts using an ohmmeter. (Subaru)

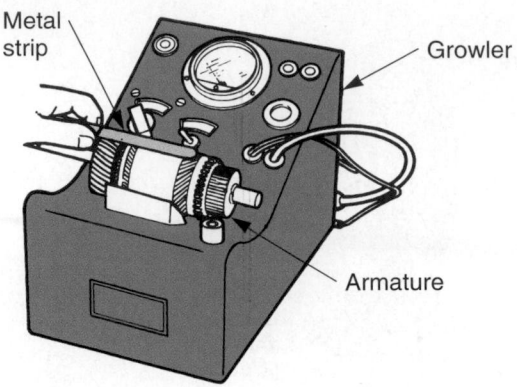

Figure 32-16. A growler will quickly check for armature shorts. The metal strip vibrates when moved over a shorted winding. (Mazda)

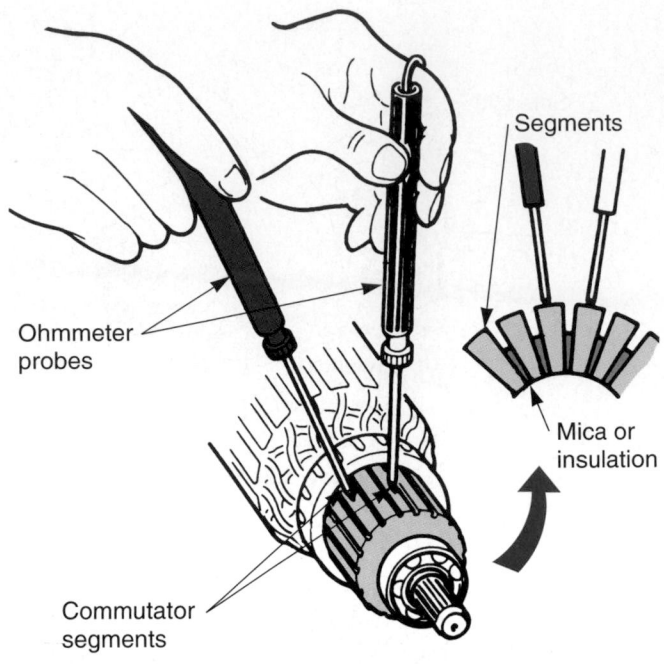

Figure 32-17. An ohmmeter is used to check armature continuity. If open or infinite resistance exists between any commutator segment, replace the armature. (Honda)

next to the armature while rotating the armature in the growler. The metal strip or hacksaw blade will vibrate when passed over the shorted leg of the windings.

To check armature continuity, do an open-circuit test. You can use a growler with an integral meter or an ohmmeter. Follow the directions provided with the test equipment.

When using an ohmmeter, touch the meter leads to each commutator segment, as in **Figure 32-17.** If the meter reads infinite resistance on any segment, that segment winding is open. The armature must be replaced.

You should also check for an armature ground (short from winding to shaft or core). This test is illustrated in **Figure 32-18.**

Touch the ohmmeter leads on the armature coil core and the commutator segments. Repeat this test on the commutator segments and armature shaft. If there is continuity (low resistance), then the armature is grounded and must be replaced.

If the windings are in good condition, the commutator should be cleaned using very fine sandpaper, not

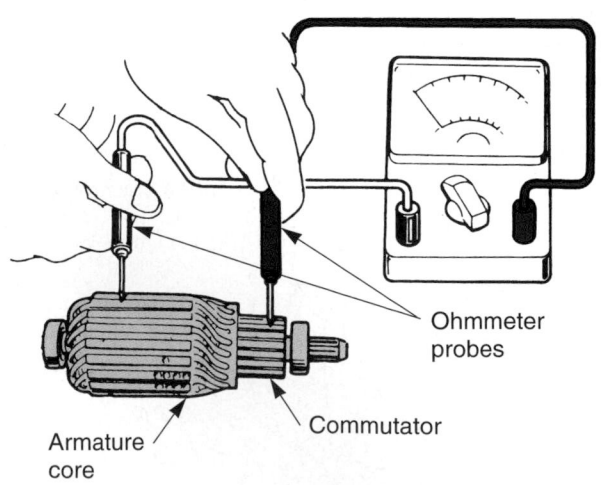

Checking for core-to-armature shorts

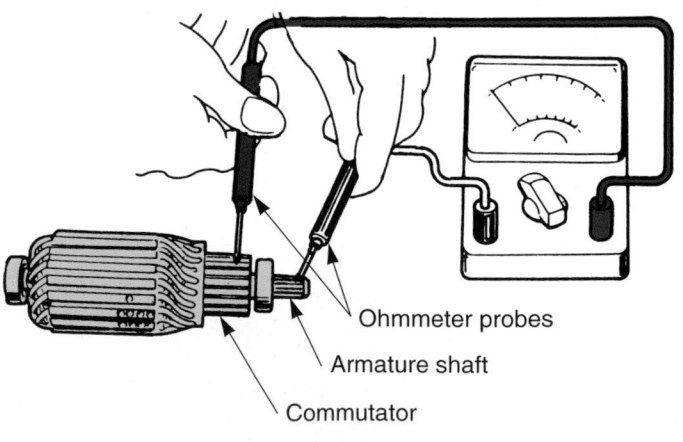

Checking for shaft-to-commutator shorts

Figure 32-18. Checking for armature shorts with an ohmmeter. A low resistance reading indicates a short. If shorted, the armature should be replaced. (Honda)

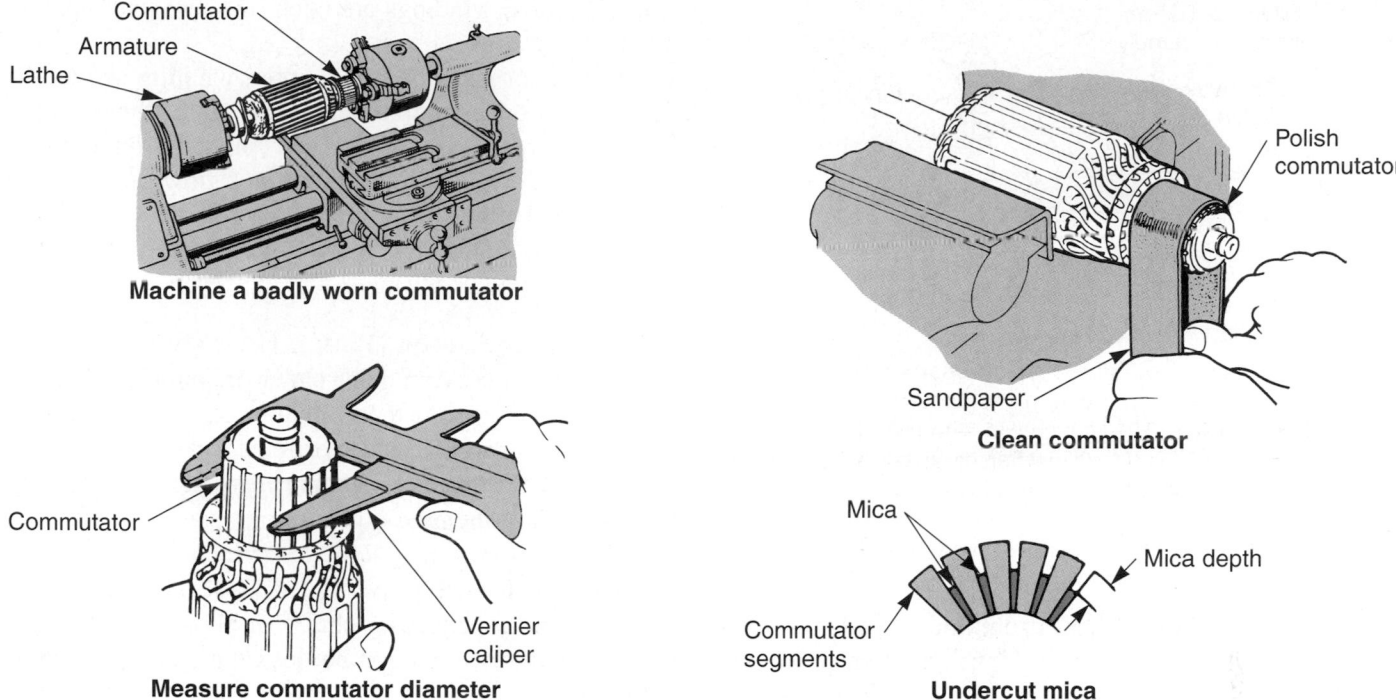

Figure 32-19. Armature service. Check a service manual for details. (Honda)

emery cloth. If the commutator is badly worn, it should be turned (machined) on a lathe. See **Figure 32-19.** Then, the mica (insulation) between each commutator segment may need to be undercut. A special tool or a hacksaw blade can be used to cut the mica lower than the surface of the segments.

Also check the armature shaft. If there are any burrs at the lock ring groove, file them off.

Field Coil Service

Inspect the field windings inside the starter frame. Look for signs of physical damage or burning. To test for open field coils, use a battery-powered test light or an ohmmeter. Look at **Figure 32-20.** Touch the test leads to wires or brushes that connect to the field windings. This connection may vary with some starters, so refer to a manual.

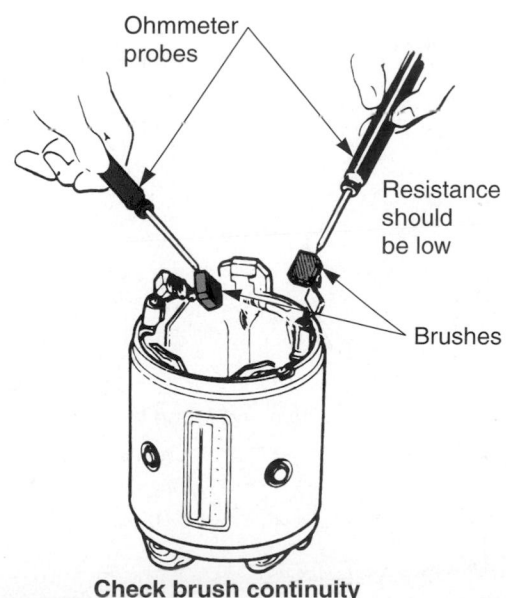

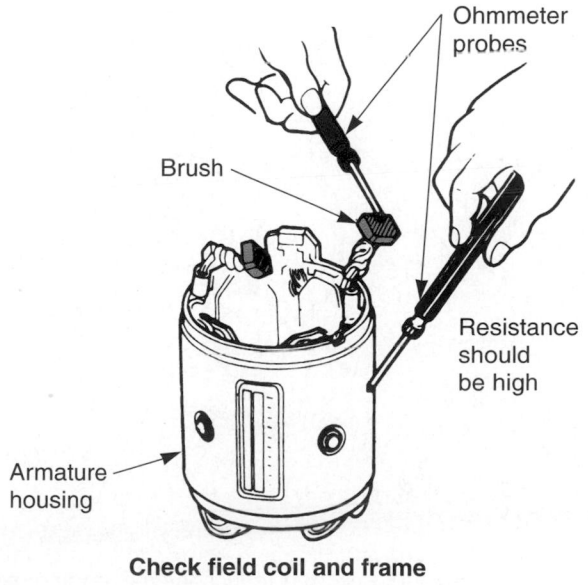

Figure 32-20. An ohmmeter is used to check for field problems. (Honda)

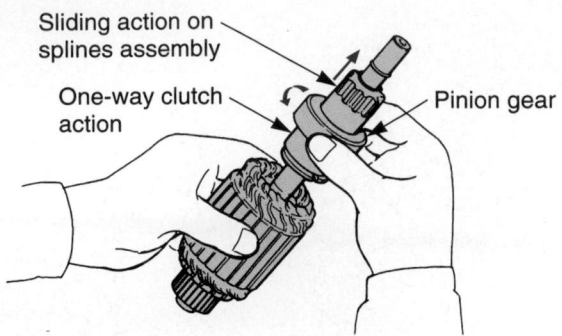

Figure 32-21. Checking starter pinion gear action. The over-running clutch should turn freely in one direction. It should lock in the other direction. The unit should also slide freely on the armature shaft splines. Most mechanics install a new pinion gear during starter service. (Honda)

If the test light glows or the ohmmeter reads zero, the field windings are not open. If the test light does not glow or the meter indicates infinite (maximum) resis-tance, the field windings are open (broken) and must be replaced.

To test for grounded field coils (winding shorted to the frame or other starter component), touch the test light or ohmmeter leads across the field coil and ground. Look at **Figure 32-20.** The light should not glow or the meter should read infinite resistance. If the starter has a shunt winding, disconnect it before making the test. If a winding is grounded, it must be replaced.

Overrunning Clutch (Pinion Gear) Service

Normally, the overrunning clutch or pinion gear assem-bly is replaced anytime the starting motor is disassembled. The pinion gear is subjected to extreme wear when engaged and disengaged from the engine flywheel. It is usually wise to replace the pinion gear during starter service.

If the pinion gear is to be reused, check the ends of the gear teeth for wear. See **Figure 32-21.** Also, check the action of the overrunning clutch. It should let the gear turn freely in one direction but lock the gear in the other direction.

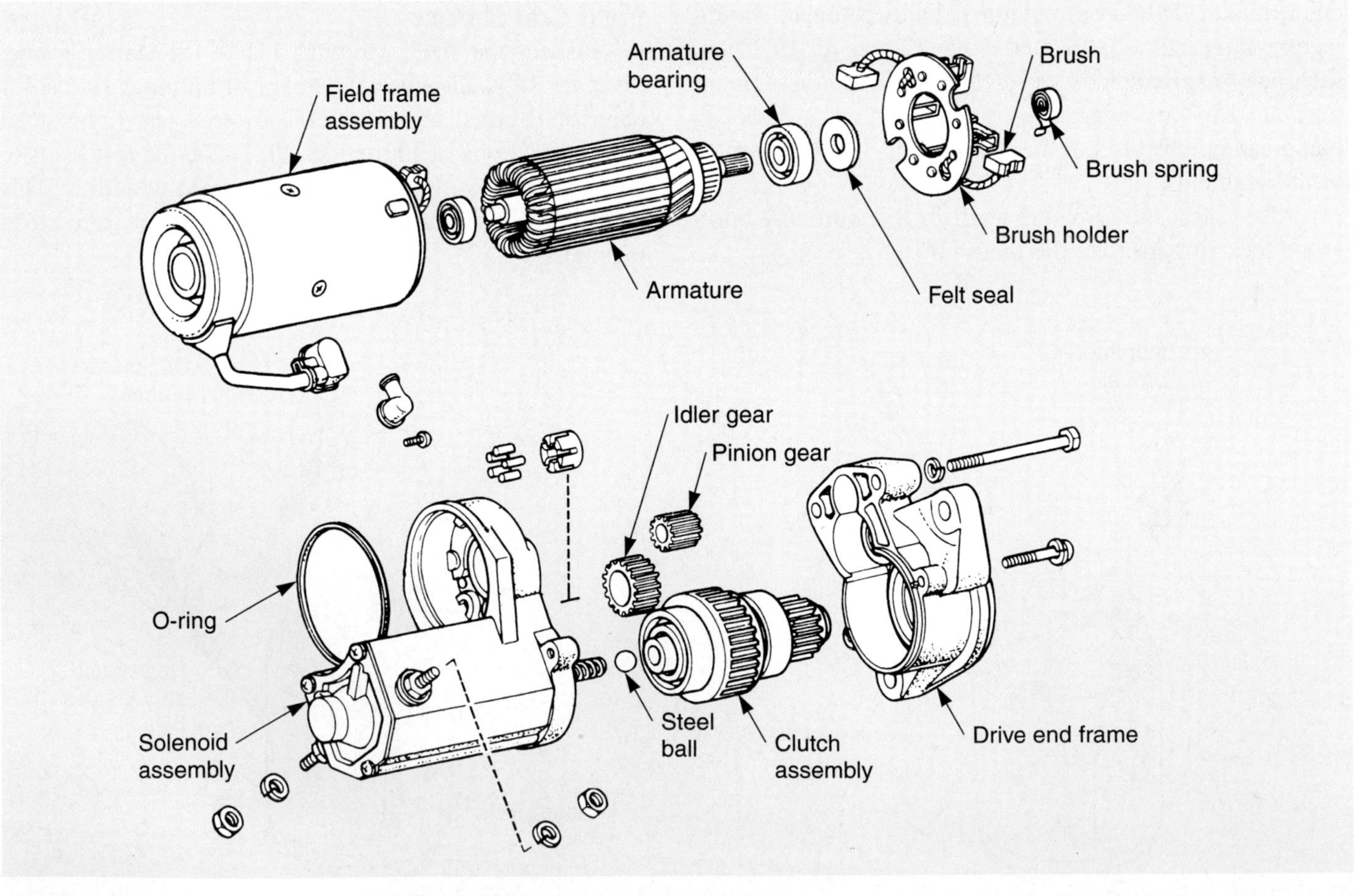

Figure 32-22. Exploded view of a reduction-type starting motor. A service manual will give a similar illustration for the starter you are servicing. It can help during reassembly. (Toyota)

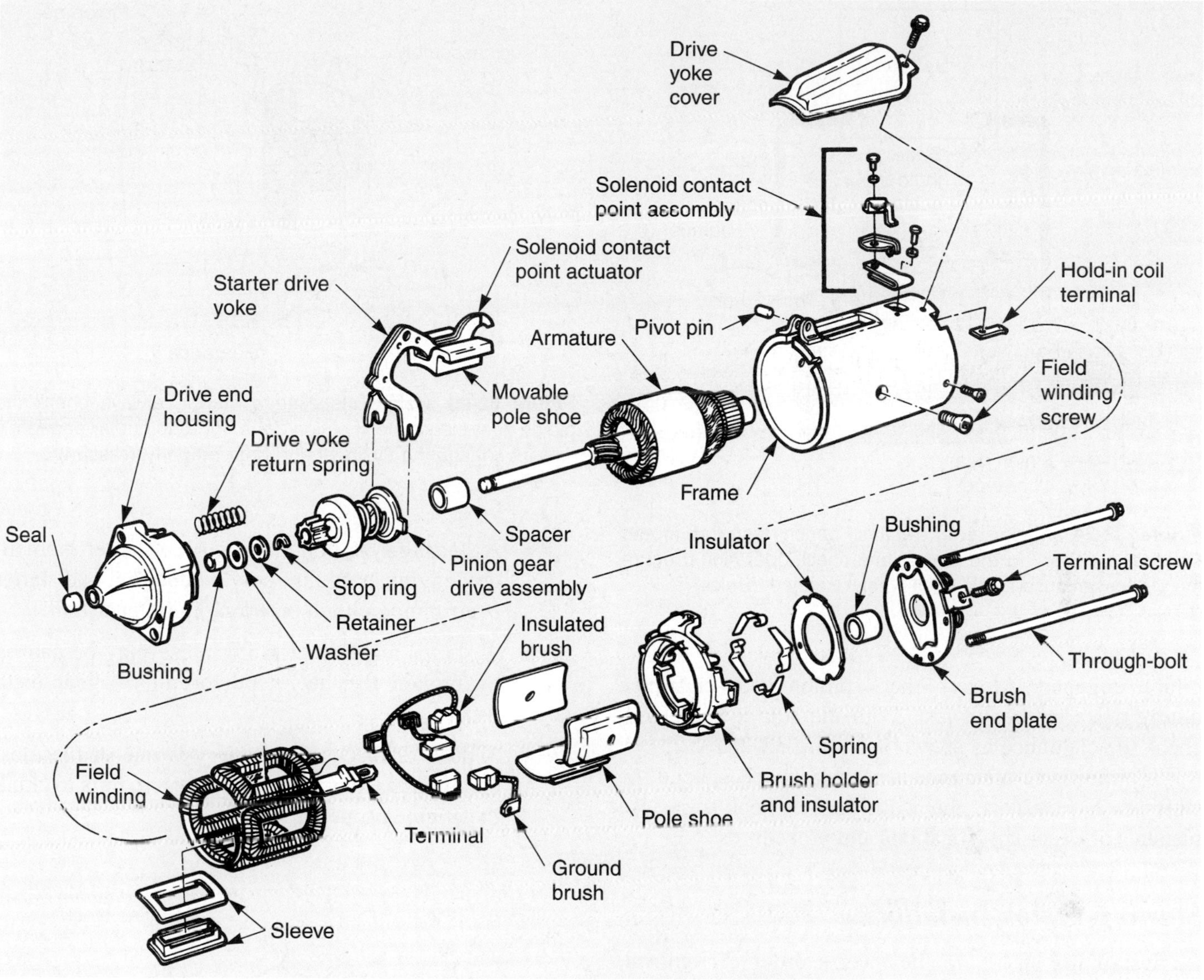

Drive
yoke
cover

Solenoid contact
point assembly

Solenoid contact
point actuator

Starter drive
yoke

Hold-in coil
terminal

Armature

Pivot pin

Field
winding
screw

Drive end
housing

Movable
pole shoe

Drive yoke
return spring

Seal

Frame

Spacer

Insulator

Bushing

Terminal screw

Pinion gear
drive assembly

Stop ring

Through-bolt

Retainer

Brush
end plate

Washer

Insulated
brush

Bushing

Spring

Field
winding

Brush holder
and insulator

Terminal

Pole shoe

Ground
brush

Sleeve

Figure 32-23. Exploded view of another typical starting motor. Study the part relationships. (Chrysler)

Starter Reassembly

Reassemble the starter using the reverse order of disassembly. Lubricate the armature shaft bushings, pinion gear splines, and other parts as recommended by the manufacturer.

 Caution
Do not use too much oil or grease to lubricate the bushings and other parts of a starting motor. If lubricant gets on the brushes and commutator, starter motor power and service life will be reduced.

Brush installation is the only part of starter reassembly that may be difficult. With many starters, the brushes can be locked out of the way using the brush springs. The springs are wedged on the sides of the brushes. They will hold the brushes up so you can slide the armature and commutator into place. Then, the brushes can be pushed down and snapped into place on the commutator.

Study **Figures 32-22** and **32-23.** They show exploded views of modern starting motors.

After reassembly, test the starter before mounting it on the engine. As pictured in **Figure 32-24,** connect the starter to a battery using jumper cables. Connect the positive cable first; then connect the negative cable. Hold or clamp the starter firmly because it will lurch and rotate when energized. Make sure the motor spins at the correct speed and that the pinion gear moves into the correct position.

Starter pinion gear clearance is the distance between the pinion and the drive end frame with the

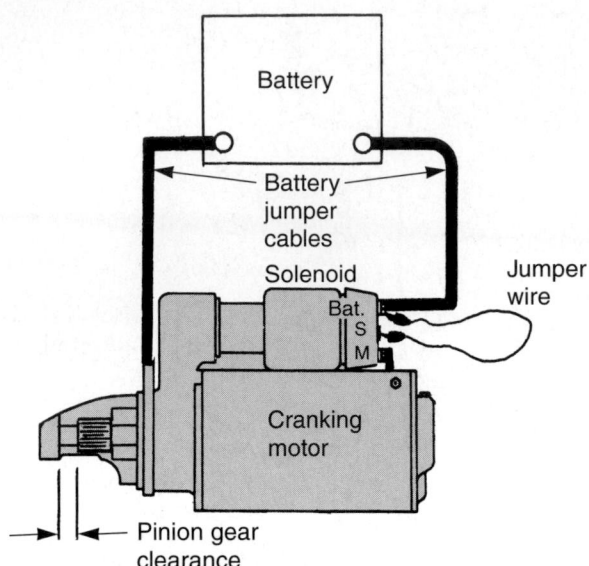

Figure 32-24. After starter repairs, bench test the motor. Connect a battery to the starter and check operation. Secure the starter because it can lurch when engaged. (Buick)

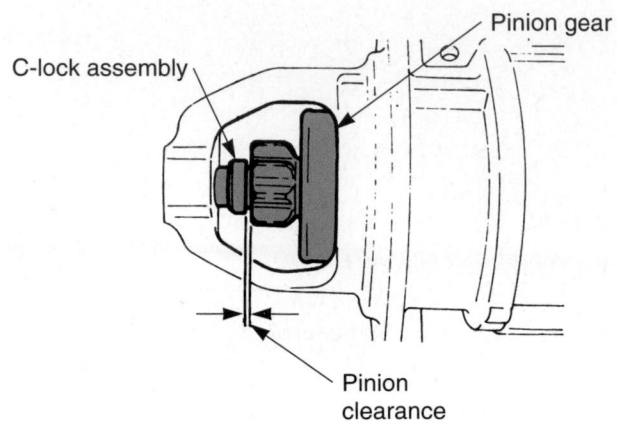

Figure 32-25. Before installing a starter, always check the pinion gear clearance. If pinion does not slide out far enough, it will not engage the flywheel ring gear properly. (Nissan)

pinion engaged. Always check pinion gear clearance during a starter bench test. With the starter energized, check the clearance as shown in **Figure 32-25.**

If pinion gear clearance is not within specs, bend the shift lever or replace the worn parts. Check a service manual for exact specifications and procedures.

Starting Motor Installation

Install the starter in the reverse order of removal. Make sure that any spacer shims are replaced between the motor and the engine block. If these shims are left out, the pinion gear may not mesh with the flywheel gear properly. Refer to **Figure 32-26.**

If the starter has a solenoid on it, connect the wires on the solenoid before bolting the starter to the engine. Torque the starter bolts to specifications. Replace any brackets or shields and reconnect the battery. Crank the engine several times to check starting motor operation.

Summary

- In a no-crank problem, the engine crankshaft does not rotate properly with the ignition key at start. The most common causes are a dead battery, poor electrical connection, or faulty system component.

- A solenoid buzzing or clicking sound, without cranking, is commonly due to a discharged battery or poor battery cable connections.

- A starting system humming sound, after momentary engine cranking, may be due to a bad starter overrunning clutch or worn pinion gear unit.

- A starter metallic grinding noise may be caused by broken flywheel teeth or pinion gear teeth wear.

- A dead or discharged battery is one of the most common reasons the starting system fails to crank the engine properly.

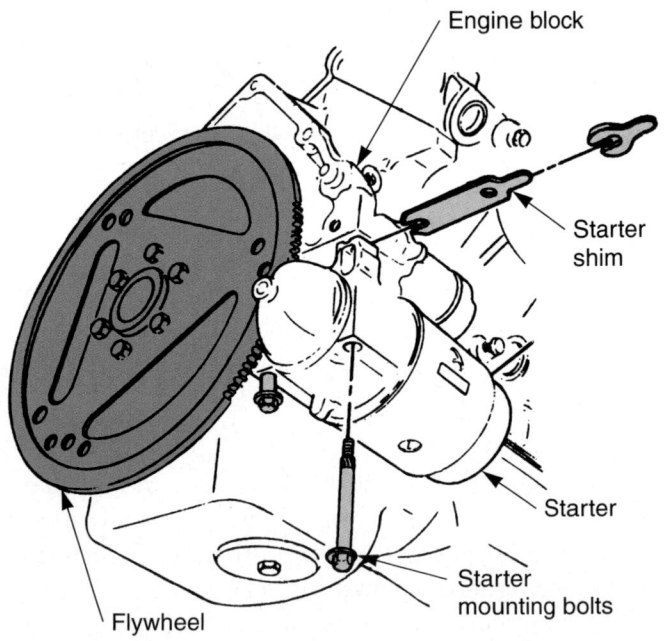

Figure 32-26. When installing the starter, replace any shims. Install wires without overtightening and stripping the terminal threads. Torque the mounting bolts or nuts to specifications. (GMC)

- A starter current draw test measures the current used by the starting system. It will quickly tell you about the condition of the starting motor and other system parts.

- Voltage drop tests will quickly locate a part with higher than normal resistance.

- A starter ground circuit test checks the circuit between the starting motor ground and the negative battery terminal.

- A battery cable problem can produce symptoms similar to a dead battery, bad solenoid, or weak starting motor.

- A bad starter solenoid can cause a range of symptoms: click with no cranking, no cranking with no click, or slow cranking. It can also keep the engine or starting motor from shutting off.

- A starting motor rebuild typically involves removal and disassembly of starting motor; cleaning and inspection for part wear or damage; replacement of brushes, bushings, and any other worn or damaged parts; polishing or turning of commutator; lubrication, reassembly, testing, and installation of starting motor.

- Starter shims may be used to adjust the space between the pinion gear and the flywheel ring gear.

- Starter pinion gear clearance is the distance between the pinion and the drive end frame with the pinion engaged.

Important Terms

No-crank problem
Slow cranking
  condition
Starting headlight test
Starter current draw
  test
Voltage drop tests
Insulated circuit
  resistance test
Starter ground circuit
  test
Battery cable connection
  test
Starting motor rebuild
Starter shims
Growler
Starter pinion gear
  clearance

Review Questions—Chapter 32

Please do not write in this text. Place your answers on a separate sheet of paper.

1. What are the most common causes of a no-crank problem?

2. A buzz or a click sound from the starter solenoid is normally due to a(n) _____ _____ or poor _____ _____ connections.

3. A humming sound, after momentary engine cranking, may be due to a bad _____ _____ or a bad _____ _____ unit.

4. What commonly causes a metallic grinding noise during starting?

5. Summarize how a starting headlight test is performed.

6. Why is a starter current draw test commonly used?

7. A high voltage drop in the starting system could indicate:
   (A) High resistance.
   (B) Loose electrical connection.
   (C) Corroded or burned terminal.
   (D) All of the above are correct.

8. How do you test the battery cables?

9. Explain some of the symptoms of a bad starter solenoid.

10. A bad ignition switch can keep the starting motor from working and can also make the engine crank all of the time. True or False?

11. How do you adjust a neutral safety switch?

12. List the five major steps for a starting motor rebuild.

13. _____ _____ may be used to adjust the space between the pinion gear and the flywheel ring gear.

14. Define the term "armature growler."

15. Always keep oil and grease away from starting motor brushes and commutator. True or False?

● ASE-Type Questions

1. A metallic grinding noise can be caused by each of the following *except:*
   (A) *clashing gears.*
   (B) *broken flywheel teeth.*
   (C) *pinion gear tooth wear.*
   (D) *broken main feed wire.*

2. After performing a starting headlight test, Technician A says the resulting bright lights indicate starting circuit high resistance. Technician B says the test results suggest an open starting circuit. Who is right?
   (A) *A only.*
   (B) *B only.*
   (C) *Both A and B.*
   (D) *Neither A nor B.*

3. Which of the following tests measures the number of amps used by a starting system?
   (A) Starter current draw test.
   (B) Starter ground circuit test.
   (C) Insulated circuit resistance test.
   (D) All of the above.

4. A battery cable problem produces symptoms similar to each of the following *except:*
   (A) bad solenoid.
   (B) dead battery.
   (C) weak starter motor.
   (D) brush holder problem.

5. Which of the following is not an indication of a bad starter solenoid?
   (A) Slow cranking.
   (B) Clicking without cranking.
   (C) Cranking without starting.
   (D) No cranking and no clicking.

6. When removing a starter motor, which of the following should be done first?
   (A) Disconnect the battery.
   (B) Mark the end frames.
   (C) Unbolt the battery cable.
   (D) Remove any solenoid fasteners.

7. Which of the following can be used to adjust the space between the pinion gear and the flywheel ring gear?
   (A) Ratchet.
   (B) Growler.
   (C) Starter shims.
   (D) Swivel wrench.

8. Which of the following is used to quickly check for armature shorts?
   (A) Growler.
   (B) Ohmmeter.
   (C) Tachometer.
   (D) Starter shims.

9. If an ohmmeter reads zero when testing a field coil, the windings:
   (A) are open.
   (B) are not open.
   (C) are shorted to the frame.
   (D) are shorted to a starter component.

10. With the pinion engaged, pinion gear clearance is the distance between the:
    (A) pinion and retainer.
    (B) pinion and flywheel.
    (C) pinion and drive end frame.
    (D) pinion and commutator segments.

Activities—Chapter 32

1. Demonstrate the use of a voltmeter in making voltage drop tests to detect excessive resistance in starting system components.

2. Make a videotape to show a procedure that would be hard to demonstrate for a class, such as starting motor removal and installation. Narrate the finished tape and play it for the class.

Starting System Diagnosis		
Condition	**Possible cause**	**Correction**
Starter fails to engage.	1. Battery discharged or faulty. 2. Faulty starting circuit wiring. 3. Defective starter relay. 4. Faulty starter solenoid. 5. Faulty starter. 6. Defective ignition switch. 7. Park/Neutral position switch faulty or misadjusted. 8. Clutch pedal position switch faulty or misadjusted.	1. Charge or replace battery. 2. Test and repair wiring. 3. Replace relay. 4. Replace solenoid or starter. 5. Replace starter. 6. Replace switch. 7. Replace or adjust switch. 8. Replace or adjust switch.
Starter engages but fails to turn engine.	1. Battery discharged or faulty. 2. Starting circuit wiring faulty. 3. Faulty starter. 4. Engine seized.	1. Charge or replace battery. 2. Test and repair wiring. 3. Replace starter assembly. 4. Repair engine.
Starter engages but spins out before engine starts.	1. Broken teeth on starter ring gear. 2. Faulty starter.	1. Replace ring gear. 2. Replace starter.
Starter fails to disengage.	1. Starter improperly installed. 2. Faulty starter relay. 3. Faulty ignition switch. 4. Faulty starter.	1. Install starter properly. 2. Replace relay. 3. Replace switch. 4. Replace starter.
Starter makes excessive noise.	1. Starter-to-flywheel housing mounting fasteners loose. 2. Dragging armature. 3. Dragging field pole shoes. 4. Dry bushings. 5. Chipped pinion teeth. 6. Chipped flywheel ring gear teeth. 7. Bent armature shaft. 8. Worn drive unit. 9. Loose starter through-bolts. Loose end frame bolts. 10. Flywheel ring gear misaligned.	1. Tighten mounting fasteners. 2. Replace armature and/or bushings. 3. Tighten pole shoes. 4. Lubricate bushings. 5. Replace pinion. 6. Replace ring gear. 7. Replace armature. 8. Replace starter drive unit. 9. Tighten all starter and frame cap bolts. 10. Install new ring gear.

Chapter 33

Charging System Fundamentals

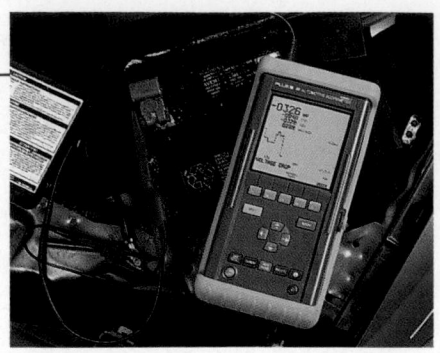

After studying this chapter, you will be able to:

- List the basic parts of a charging system.
- Explain charging system operation.
- Describe the construction of major charging system components.
- Compare alternator and voltage regulator design differences.
- Explain charging system indicators.
- Describe safety practices to follow when working with charging systems.
- Correctly answer ASE certification test questions that require a knowledge of charging system fundamentals.

In the two previous chapters, you learned how the starting motor uses battery power to crank the engine. The starter consumed electricity and discharged the battery as the motor was drawing current.

In this chapter, you will learn how the charging system recharges the battery and supplies electricity for all of the car's electrical systems. This chapter will give you a background in charging system terminology and prepare you for the next chapter on testing and repair.

Basic Charging System Parts

Figure 33-1 shows the major parts of a typical charging system. Study this illustration as each part is introduced.

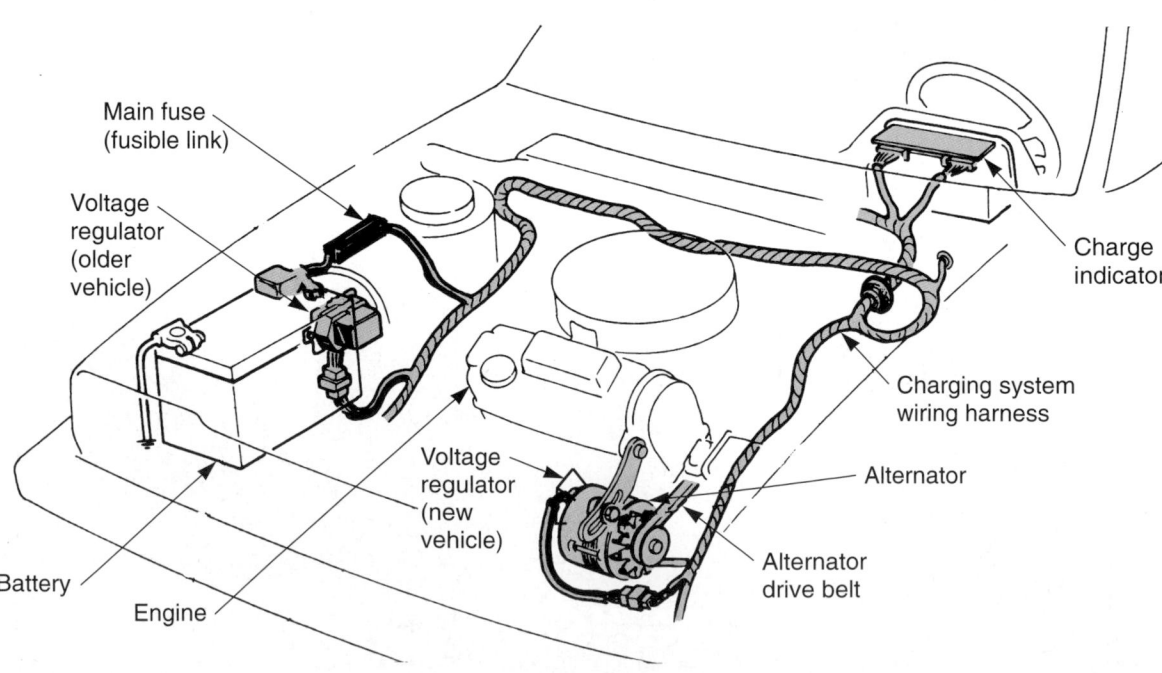

Figure 33-1. Review the names and locations of the basic parts of a charging system. (Honda)

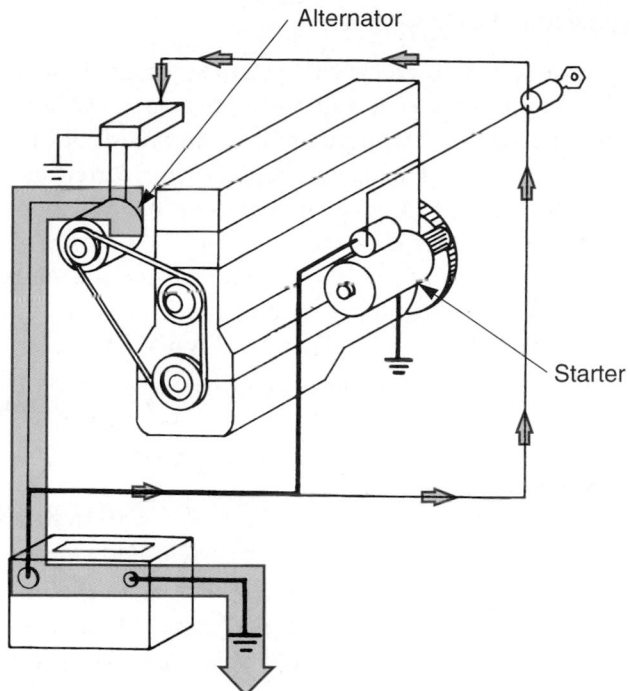

Figure 33-2. The charging system recharges the battery and supplies electricity when the engine is running.

- *Alternator*—generator that uses mechanical (engine) power to produce electricity.

- *Voltage regulator*—electronic device that controls the output voltage and current of the alternator. The engine control module may serve as the voltage regulator.

- *Alternator drive belt*—links the engine crankshaft pulley with the alternator pulley to drive the alternator.

- *Charge indicator*—ammeter, voltmeter, or warning light that informs the driver of the charging system condition.

- *Charging system harness*—wiring that connects the parts of the system.

- *Battery*—provides current to initially energize the alternator. Also helps to stabilize alternator output.

During engine cranking, the battery supplies all the electricity used by the automobile. Then, when the engine starts running, the charging system takes over to provide current to the vehicle's electrical systems.

The engine crankshaft pulley and alternator belt spin the alternator pulley. This powers the alternator and causes it to produce electricity, **Figure 33-2.**

The voltage regulator keeps alternator output at a preset *charging voltage* (13–15 volts). Since this is higher

than battery voltage (12.6 volts), current flows back into the battery and recharges it. Current also flows to the ignition system, electronic fuel injection system, on-board computer, radio, or any other device using electricity.

Charging System Functions

The *charging system* performs several functions:

- It recharges the battery after engine cranking or after the use of electrical accessories with the engine shut off.

- It supplies all the car's electricity when the engine is running.

- It provides a voltage output that is slightly higher than battery voltage.

- It changes output to meet different electrical loads.

Types of Charging Systems

There are two basic charging systems: alternator (ac generator) systems and dc generator systems. Alternator charging systems have replaced the older dc generator systems. Keep in mind, however, that an alternator is sometimes called a generator, meaning ac generator. **Figure 33-3** shows the fundamental differences between an alternator and a dc generator.

DC Generator

A *dc generator* is similar to an electric motor. It has a stationary magnetic field. The output conductor unit (armature) spins inside this field. This induces current output from within the armature.

The dc generator was fine for producing electricity on early model cars. However, today's cars have more electrical components requiring high-current output. A dc generator would not be able to supply enough current for a modern car at idle speeds. The efficiency of dc generators is poor at low engine speeds.

AC Generator (Alternator)

An *alternator,* or *ac generator,* has replaced the dc generator because of its improved efficiency. It is smaller, lighter, and more dependable than a dc generator. The alternator will also produce more output at idle than a dc generator. This makes it ideal for use in late-model cars.

The alternator has a spinning magnetic field, as illustrated in **Figure 33-3.** The output conductors (windings) are stationary. As the field rotates, it induces current in the output windings. In a way, alternator construction is the reverse of dc generator construction.

Alternator Operation

The two main parts of a simplified alternator are the rotor and stator, **Figure 33-4.** The alternator rotor is located in the center of the alternator housing. It creates a rotating magnetic field. The alternator drive belt turns the rotor, making the field rotate.

The alternator stator is a stationary set of windings. The stator surrounds the rotor. It is the output winding in the alternator.

When the rotor spins, its strong magnetic field cuts across the stator windings. This induces current in the stator windings. If the stator windings are connected to a load (a lightbulb, for example), the load would operate (glow).

AC Output

Alternating current (*AC*) flows one way and then the other. The simple alternator in **Figure 33-4** produces an ac output. As the rotor turns into one stator winding, current is induced in one direction. Then, when the same rotor pole moves into the other stator winding, current reverses and flows out in the other direction.

Rectified AC Current

An automobile electrical system is designed to use direct current (dc), which flows in only one direction. It cannot use the alternating current as it comes out of the alternator's stator. Therefore, the alternator current must be *rectified* (changed) into direct current before entering the vehicle's electrical system.

A *diode* (covered in Chapter 8) is an electronic device that allows current flow in only one direction. It serves as an "electrical check valve." Look at **Figure 33-5.**

When a diode is connected to a voltage source in such a way that current passes through the diode, the diode is said to be *forward biased.* A forward-biased diode acts as a conductor.

When *reverse biased,* the diode is connected to a voltage source in such a way that current does not pass through. A reverse-biased diode acts like an insulator.

If a diode were placed on the stator output of our simple alternator, current would only flow out through

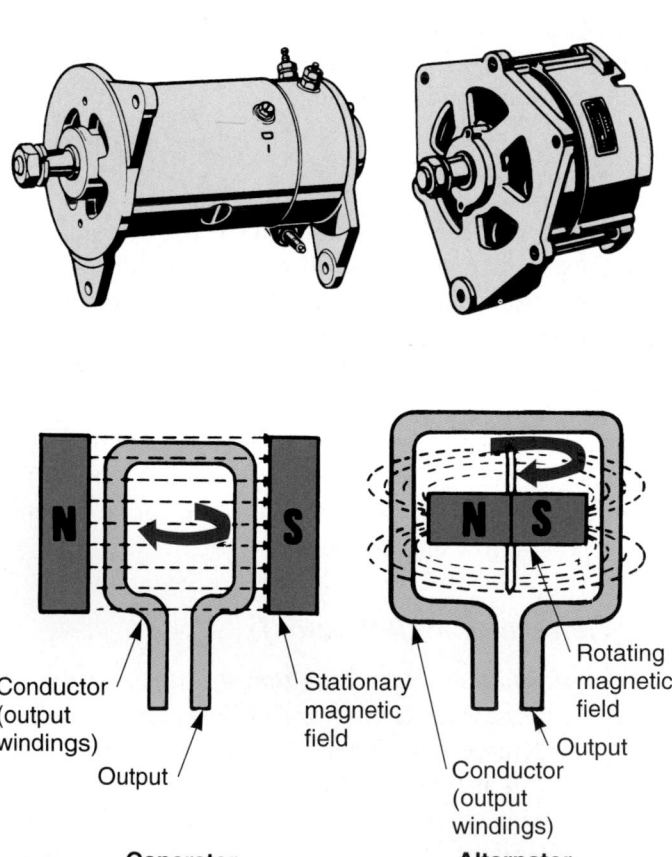

Figure 33-3. Comparison of an old dc generator and modern ac generator, or alternator. Note how fields and windings are in opposite locations. (Ford and Bosch)

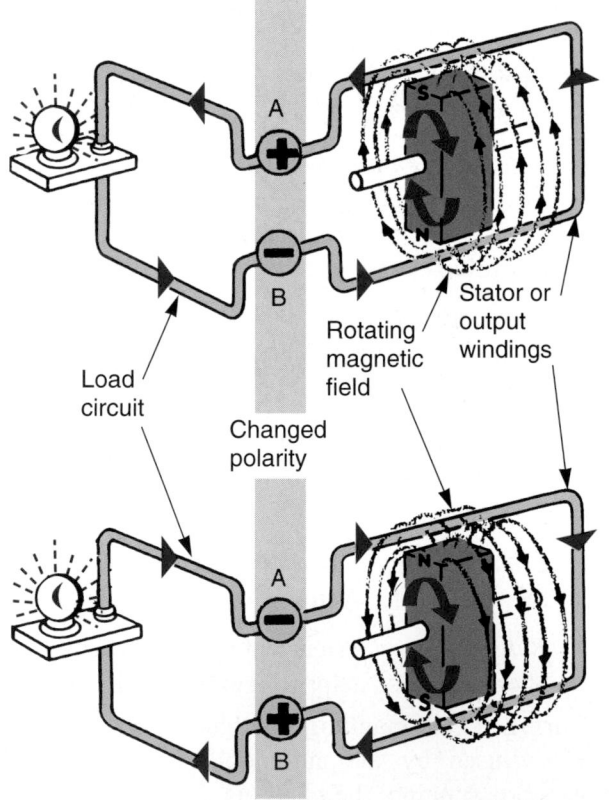

Figure 33-4. Basic alternator operation. A rotating magnetic field moves across the stationary windings. Current is induced in the windings and out to the load. When the rotating field turns one-half turn, the polarity of the windings is reversed. This causes current to flow out to the load in the opposite direction. Alternating current is produced. (Deere & Co.)

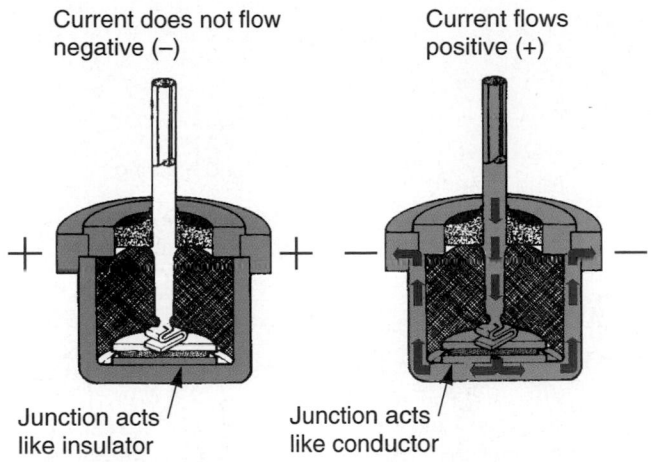

Figure 33-5. A diode is a one-way valve. When polarity is connected one way, current flows. When polarity is reversed, current is blocked. (Motorola)

the circuit in one direction. This is illustrated in **Figure 33-6.** Study diode action.

A single diode would not use all the alternator's output, however. Also, it would result in pulsing direct current, not smooth current flow. Therefore, an alternator uses several diodes connected into a rectifier circuit. This produces more efficient alternator output.

Alternator Construction

It is very important that you understand alternator construction. This information is essential if your are to properly test and repair alternators.

Pictured in **Figure 33-7,** the main components of a typical alternator are:

- *Rotor assembly*—field windings, claw poles, rotor shaft, and slip rings.
- *Brush assembly*—brush housing, brushes, brush springs, and brush wires.
- *Rectifier assembly*—diodes, heat sink or diode plate, and electrical terminals.
- *Stator assembly*—three stator windings or coils, stator core, and output wires.
- *Housing*—drive end frame, slip ring end frame, and end frame bolts.
- *Fan and pulley assembly*—fan, spacer, pulley, lock washer, and pulley nut.
- *Voltage regulator*—an electronic voltage regulator may be mounted in or on a modern alternator. In some cases, the engine control module serves as the voltage regulator.

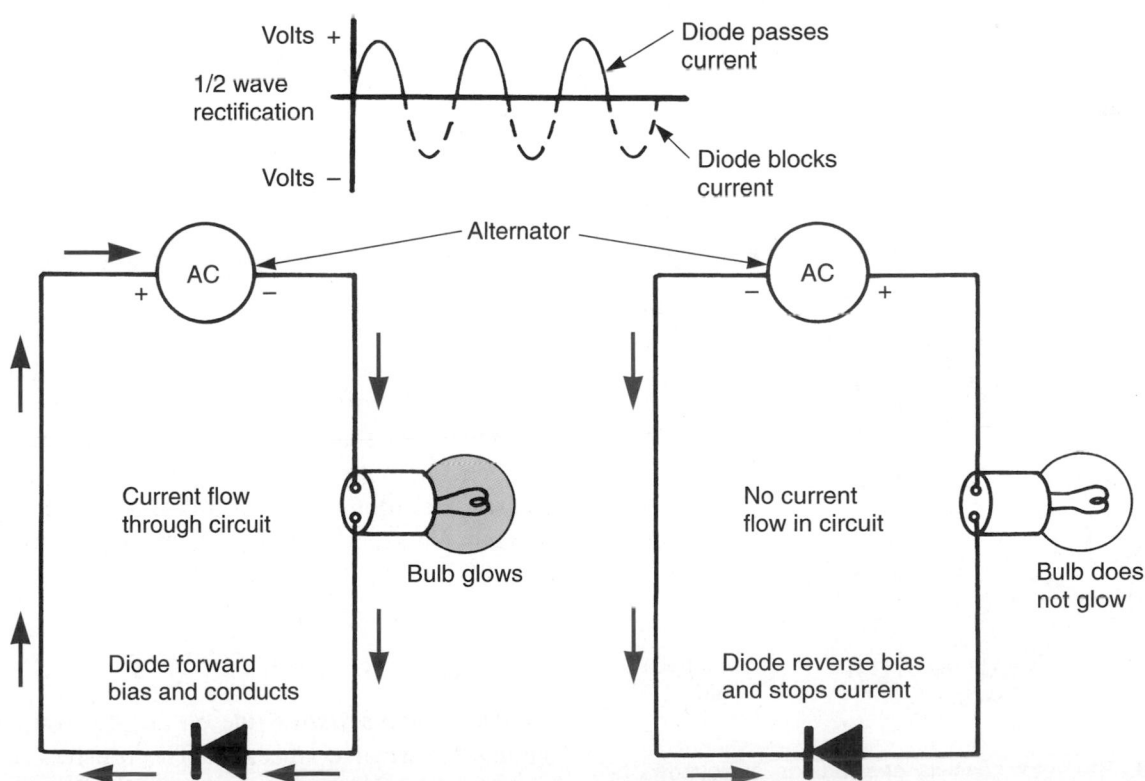

Figure 33-6. Simple circuit shows action of a diode. When the alternator output places forward bias on diode, current passes through circuit. With reverse bias, the diode prevents current flow. The bulb would only glow on the positive output wave from the alternator.

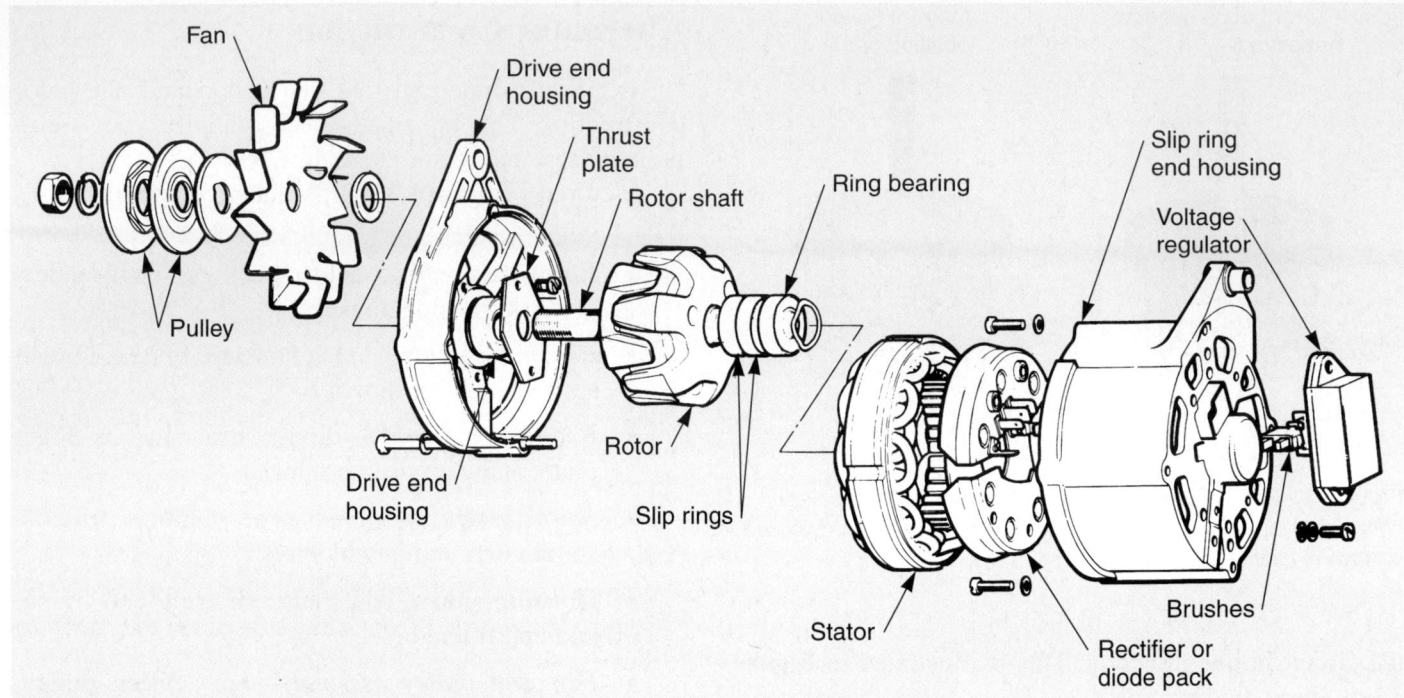

Figure 33-7. This exploded view shows the major parts of an alternator. Study the names of the parts and the general construction. (Ford)

Alternator Rotor

An *alternator rotor* consists of field coil windings mounted on a shaft. Two claw-shaped pole pieces surround the field windings to increase magnetic field strength. This is illustrated in **Figure 33-8.**

The fingers on one of the pole pieces produce S (south) poles. The fingers on the other pole piece form N (north) poles. As the rotor spins inside the alternator, the alternating polarity (North-South-North-South) produces an alternating current.

Alternator Slip Rings

Alternator slip rings are mounted on the rotor shaft to provide current to the rotor windings. Each end of the field coil connects to one of the slip rings. An external source of electricity is needed to excite the field. See **Figure 33-9.**

Alternator Bearings

Alternator bearings (needle or ball types) are commonly used to produce a low-friction surface for the rotor shaft. These bearings support the rotor and shaft as they spin inside the stator.

The alternator bearings are normally packed with grease. The front bearing is frequently held in place with a small plate and screws. The rear bearing is usually press-fit into place.

Alternator Brushes

Alternator brushes ride on the slip rings to make a sliding electrical connection. The brushes feed battery current into the slip rings and rotor windings.

Small *brush springs* hold the brushes in contact with the slip rings. Current flow into the rotor windings is low.

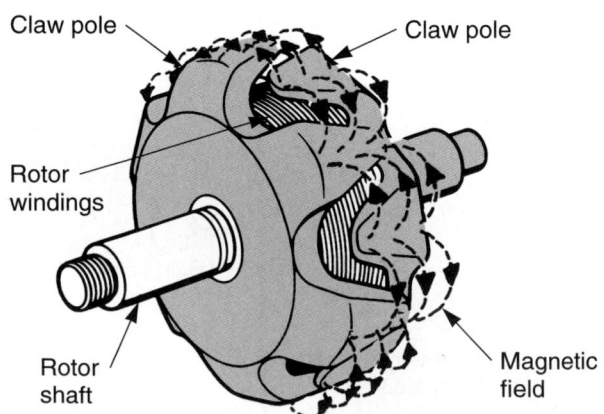

Figure 33-8. The rotor consists of windings surrounded by claw-shaped poles. The poles strengthen the magnetic field around the windings. The shaft supports the poles and windings. (Bosch)

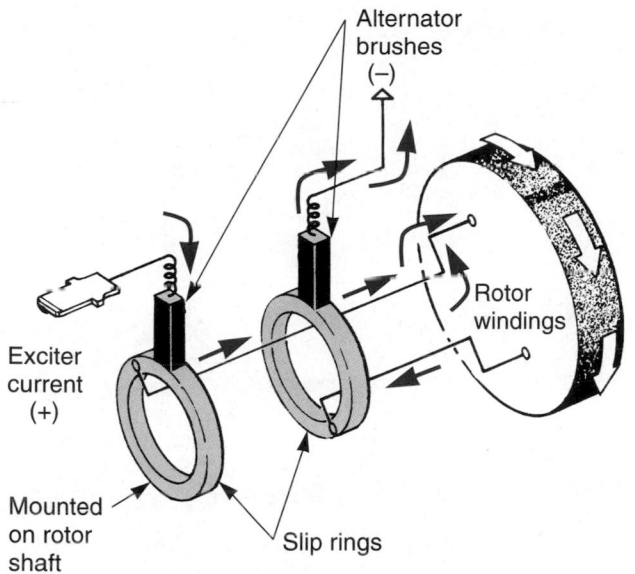

Figure 33-9. Brushes and slip rings allow current to be fed into rotor windings. This excites windings to produce a rotating magnetic field. (Motorola)

Therefore, the brushes are small compared to motor brushes. Look at **Figure 33-9** again.

Alternator Rectifier

An *alternator rectifier* assembly, also called a *diode assembly,* commonly uses six diodes to convert stator output (alternating current) to direct current. The diodes are usually wired as shown in **Figure 33-10.** The rectifier provides *full-wave rectification* (changes both positive and negative outputs into direct current) as the different polarity rotor claws pass the stator windings.

A *diode trio* may be used to supply current to the rotor field windings. In a diode trio, three diodes are connected to the field through a connection in the voltage regulator. The stator output feeds the diode trio. **Figure 33-11** shows how the rectifier assembly is connected to the stator assembly.

The *rectifier diodes* are mounted in a *diode frame* or *heat sink* (metal mount for removing excess heat from electronic parts). Three positive diodes are press-fit in an insulated frame, and three negative diodes are mounted in an uninsulated or grounded frame.

Alternator Stator

The *alternator stator* consists of three groups of windings. The coils are wrapped around a soft, laminated

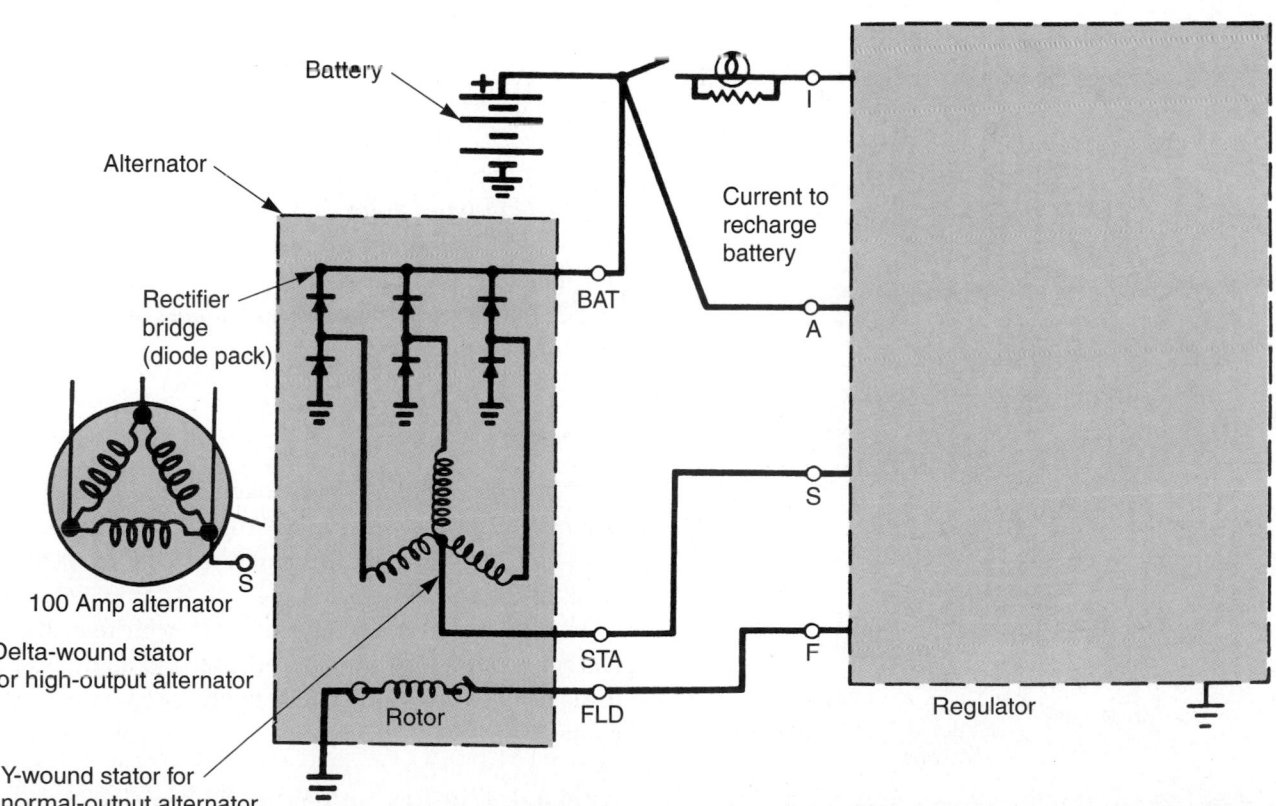

Figure 33-10. Wiring diagram shows the relationship between stator windings, rotor windings, diodes, and electrical connections. The diodes are organized so that current flows to the battery in only one direction. (Ford)

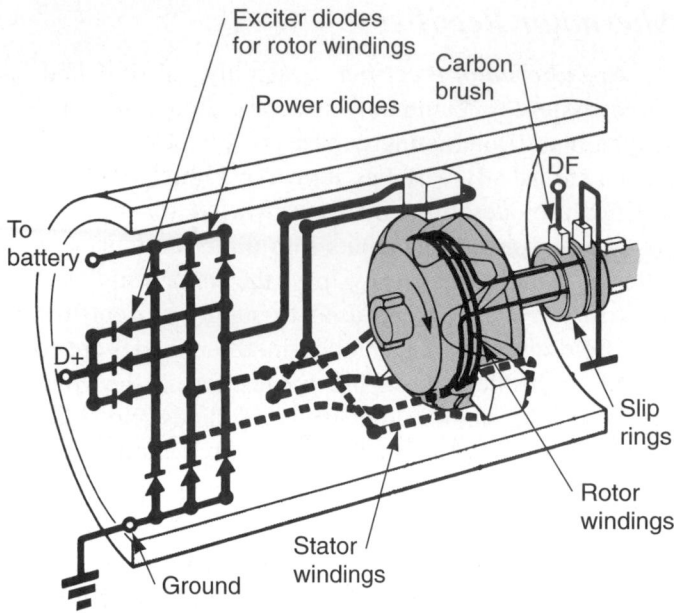

Figure 33-11. Stator windings surround the rotor windings. The rotor field cuts through stator windings to produce current output. Diodes change stator ac output into dc output before current leaves the alternator. Also note slip rings and brushes. (Robert Bosch)

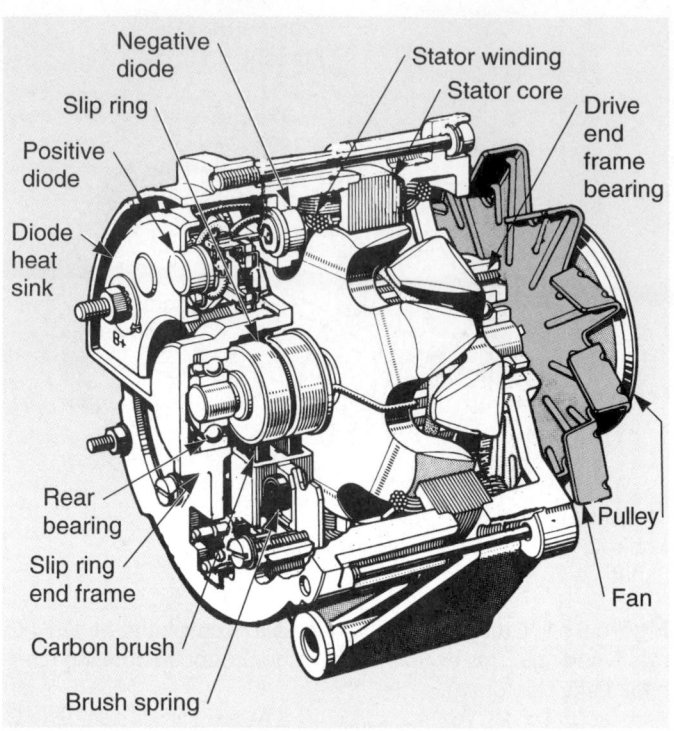

Figure 33-13. Cutaway view shows many of the parts already discussed. Note construction. (Bosch)

iron core or ring. A stator with an attached rectifier is shown in **Figure 33-12.**

The stator produces the electrical output of the alternator. As you can see, the stator windings connect to the

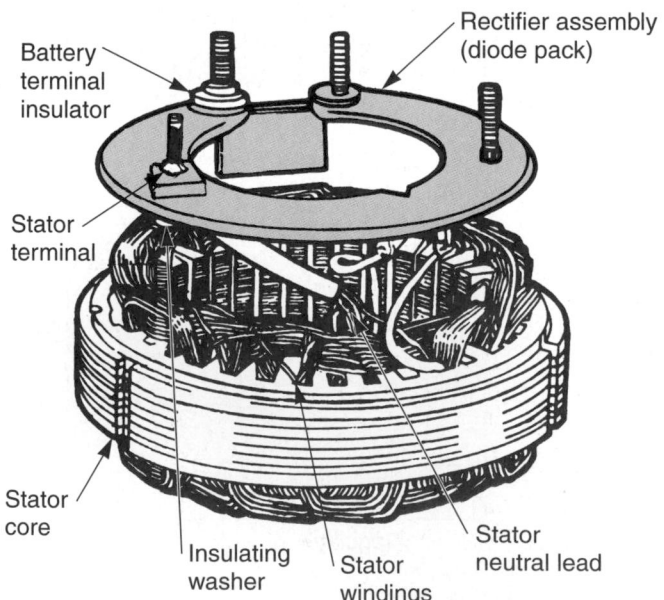

Figure 33-12. The rectifier assembly consists of six diodes in heat sink. Wires from the stator windings connect to the rectifier. An iron core around the stator windings increases induction. (Mercury)

diode assembly. The iron core is used to concentrate and strengthen the field around the stator windings.

A **Y-type stator** has the wire ends from the stator windings connected to a neutral junction. The circuit looks like the letter "Y," as shown in **Figure 33-10.** A Y-type stator provides good current output at low engine speeds.

A **delta-type stator** has the stator wires connected end to end. This is shown in **Figure 33-10.** With no neutral junction, two circuit paths are formed between the diodes during each phase. A delta-type stator is used in high-output alternators.

Alternator Fan

To provide cooling for the alternator, an **alternator fan** is mounted on the front of the rotor shaft. Normally, the fan is located between the pulley and the front bearing. Look at **Figure 33-13.**

As the rotor and shaft spin, the whirling alternator fan helps draw air through and over the alternator. This cools the windings and diodes to prevent overheating and damage.

Alternator Pulley and Belt

An **alternator pulley** provides a means of spinning the rotor through the use of a belt. The pulley is secured to the front of the rotor shaft by a large nut. See **Figure 33-14.**

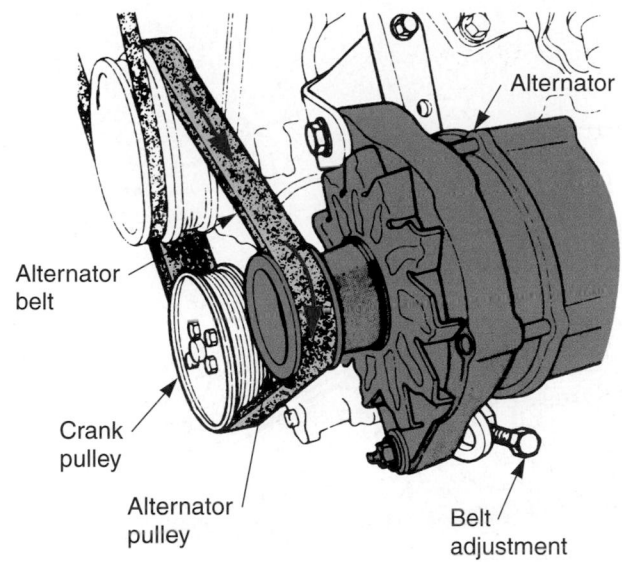

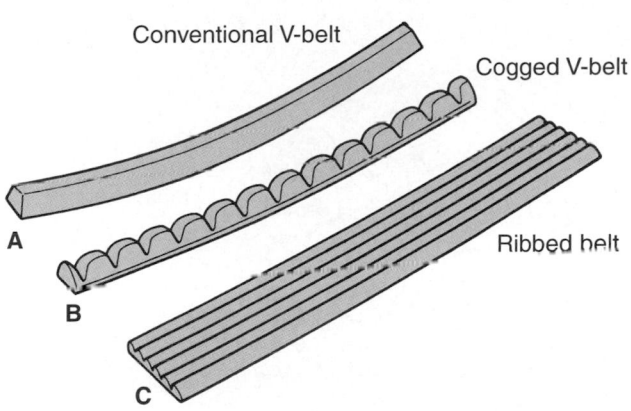

Figure 33-15. Three types of belts. A—Conventional V-belt. B—Clogged V-belt. C—Ribbed belt. (Ford)

Figure 33-14. The crankshaft pulley turns the alternator belt. The belt powers the alternator. This is a ribbed belt. (Sun Electric)

An *alternator belt,* powered by the crankshaft pulley, turns the alternator pulley and rotor. One of three belt types may be used: V-belt, cogged V-belt, and ribbed belt. These are pictured in **Figure 33-15.** Covered in other chapters, these types of belts are also used to drive the power steering pump, air-conditioning compressor, water pump, and other units.

Voltage Regulator

A *voltage regulator* controls alternator output by changing the amount of current flowing through the rotor windings. Any change in rotor winding current changes the field strength acting on the stator (output) windings. In this way, the voltage regulator can maintain a preset charging voltage.

Figure 33-16 shows a common location for a voltage regulator and its terminals. In many cases, the engine control module serves as the voltage regulator. However,

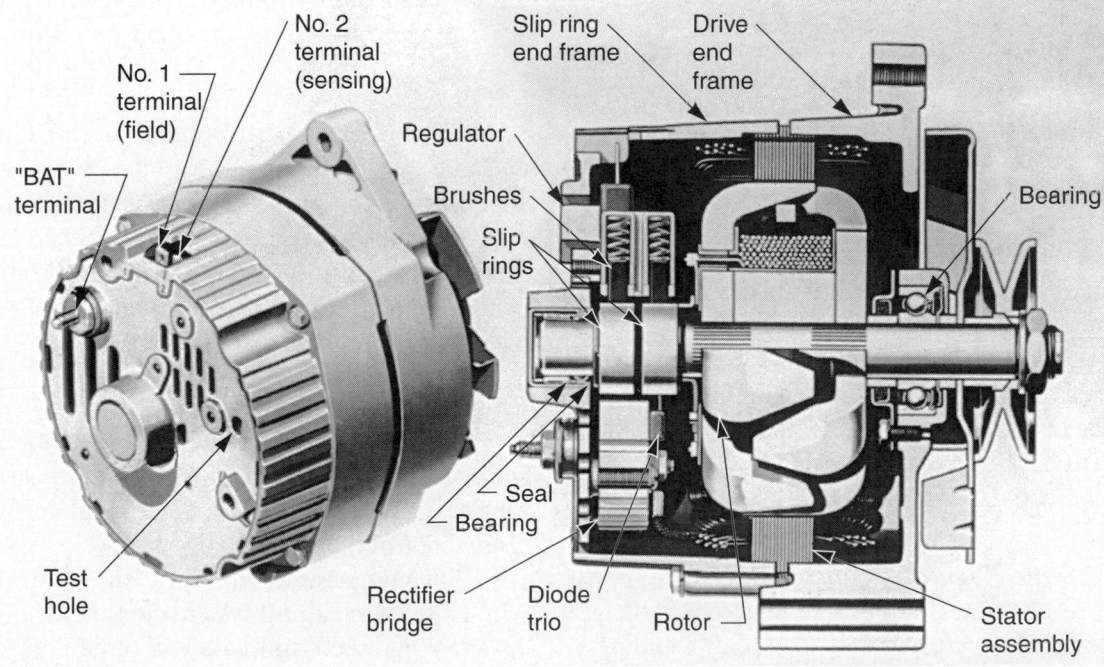

Figure 33-16. Another type of modern alternator. Study the part locations and terminal connections on the rear of the alternator. (Chevrolet)

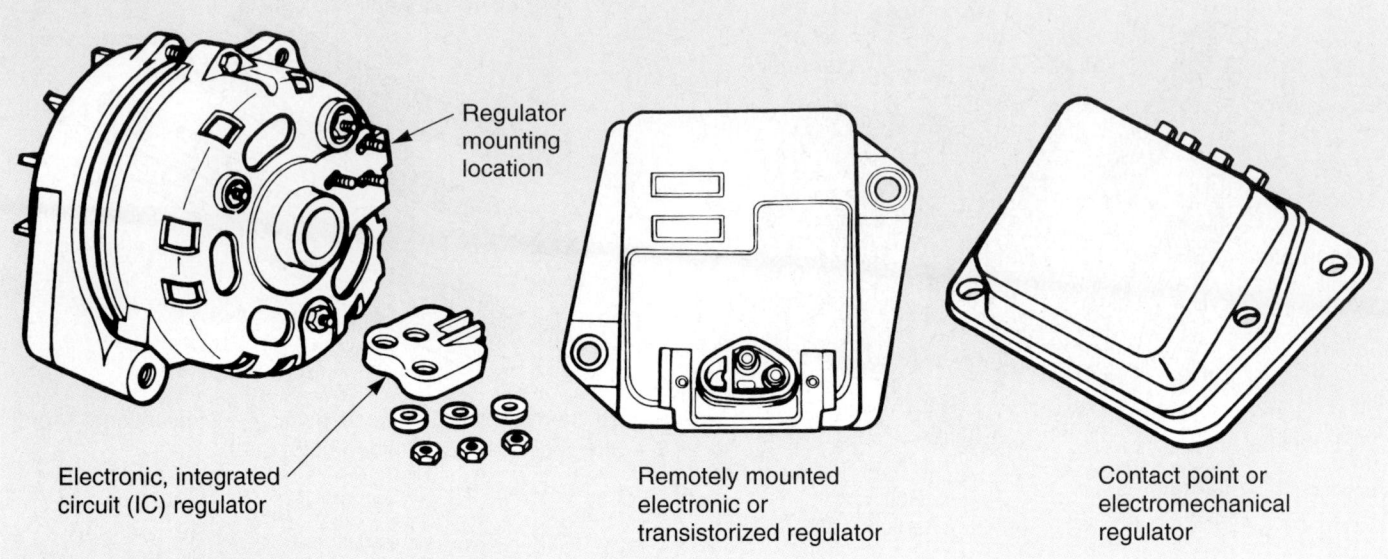

Figure 33-17. The three basic voltage regulators. (Plymouth)

if the vehicle is equipped with a conventional regulator, it is one of three basic types:

- Electronic regulator mounted inside or on the back of the alternator (integral voltage regulator).
- Electronic regulator mounted away from alternator in engine compartment.
- Contact-point regulator mounted away from alternator in engine compartment. See **Figure 33-17.**

Electronic Voltage Regulators

An *electronic voltage regulator* uses an electronic circuit (transistors, diodes, resistors, and capacitors) to control rotor field strength and alternator output, **Figure 33-18.**

An electronic voltage regulator is a sealed unit and cannot be repaired. The electronic circuit must be sealed because it can be damaged by moisture, excessive heat, and vibration. Usually the circuit is surrounded by a rubber-like gel for protection.

An *integral voltage regulator* is an electronic regulator comprised of integrated circuits and mounted inside or on the rear of the alternator. It is the most common type used on today's vehicles. The integral regulator is small, efficient, and dependable.

Electronic Voltage Regulator Operation

To increase alternator output, the electronic voltage regulator allows more current into the rotor windings. This strengthens the magnetic field around the rotor. More current is then induced in the stator windings, increasing alternator output. Look at **Figure 33-18.**

To reduce alternator output, the electronic regulator places more resistance between the battery and the rotor windings. Field strength drops and less current is induced in the stator windings.

Alternator speed and electrical load determine whether the regulator increases or decreases charging output. If the load is high or the rotor speed is low (engine idling), the regulator will sense a drop in system voltage. The regulator then increases rotor field current until a preset output voltage is obtained. If the load drops or rotor speed increases, the opposite occurs.

Battery Thermistor

In some systems, a *battery thermistor* is used to measure battery temperature so the charging system can alter charging output as needed. Generally, a cold battery requires more voltage for recharging than a hot battery. The battery thermistor is often mounted on the positive battery cable. It rests against the battery so it can measure battery temperature.

Computer Monitoring and Control

The *engine control module* or the *powertrain control module* is sometimes used to supplement or replace the conventional voltage regulator to more precisely control the charging circuit.

With computer monitoring, the control module is wired into the charging system circuit as shown in **Figure 33-19.** When the ignition key is turned on, power is fed to the alternator and regulator from the module.

The control module can monitor charging system output and react to changing operating conditions. For example, the module can shut the alternator off at wide-open throttle for better vehicle acceleration, saving several horsepower. It can more accurately control charge rate, allowing the use of a smaller, lighter battery. The module can also monitor system operation to simplify diagnosis and repair by producing trouble codes and problem descriptions.

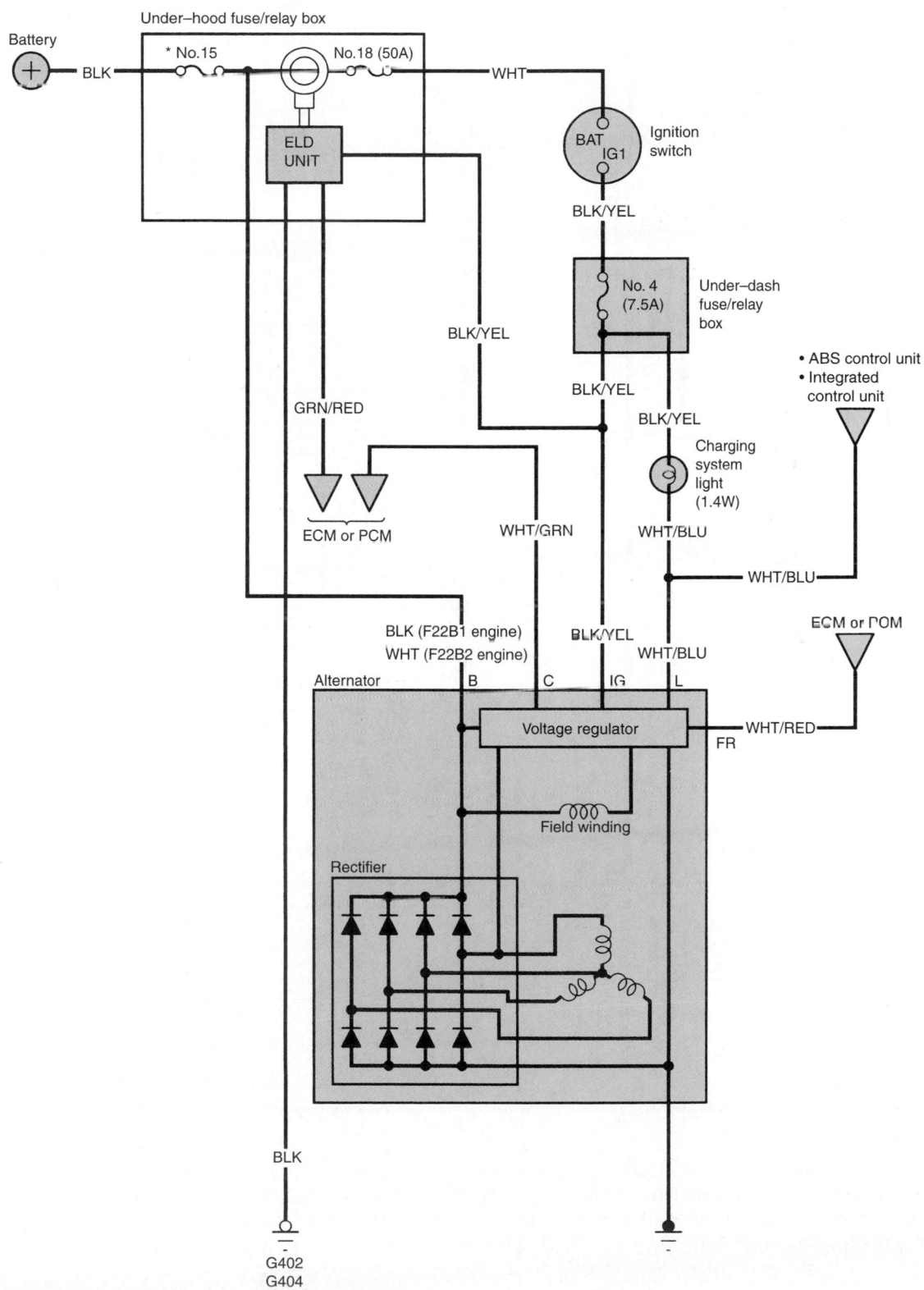

Figure 33-18. Charging system schematic diagram. The voltage regulator is an unserviceable miniature electronic circuit. Note other connections. (Honda)

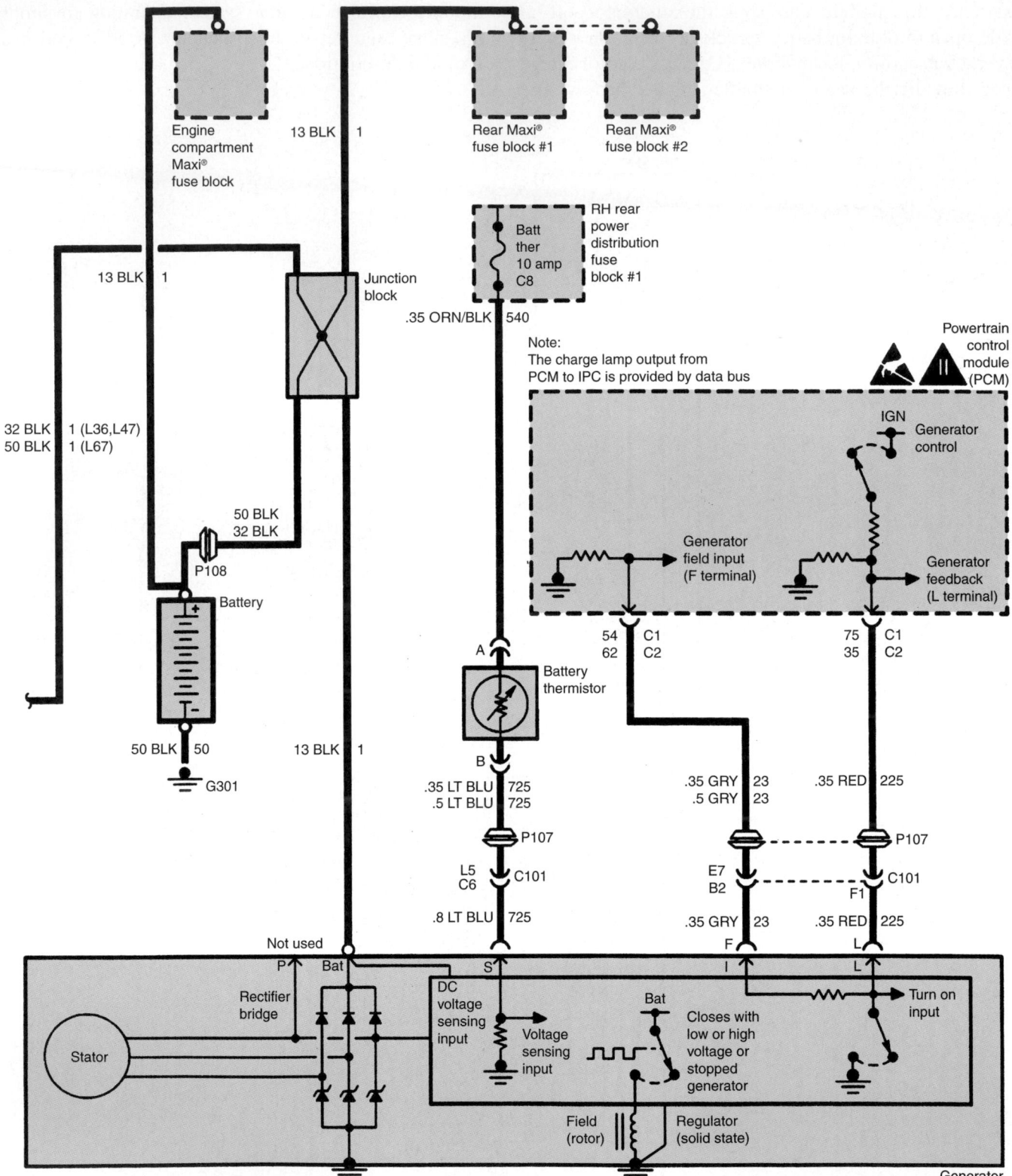

Figure 33-19. Study the circuit for a late-model charging system. Note how battery thermistor is wired to report battery temperature to integral voltage regulator. This allows the system to adjust the charging voltage higher for a cold battery or lower for a warm battery. Also note how the control module feeds power to the alternator and regulator. By adjusting charge voltage for vehicle needs, conservation of energy results in better fuel economy. (Oldsmobile)

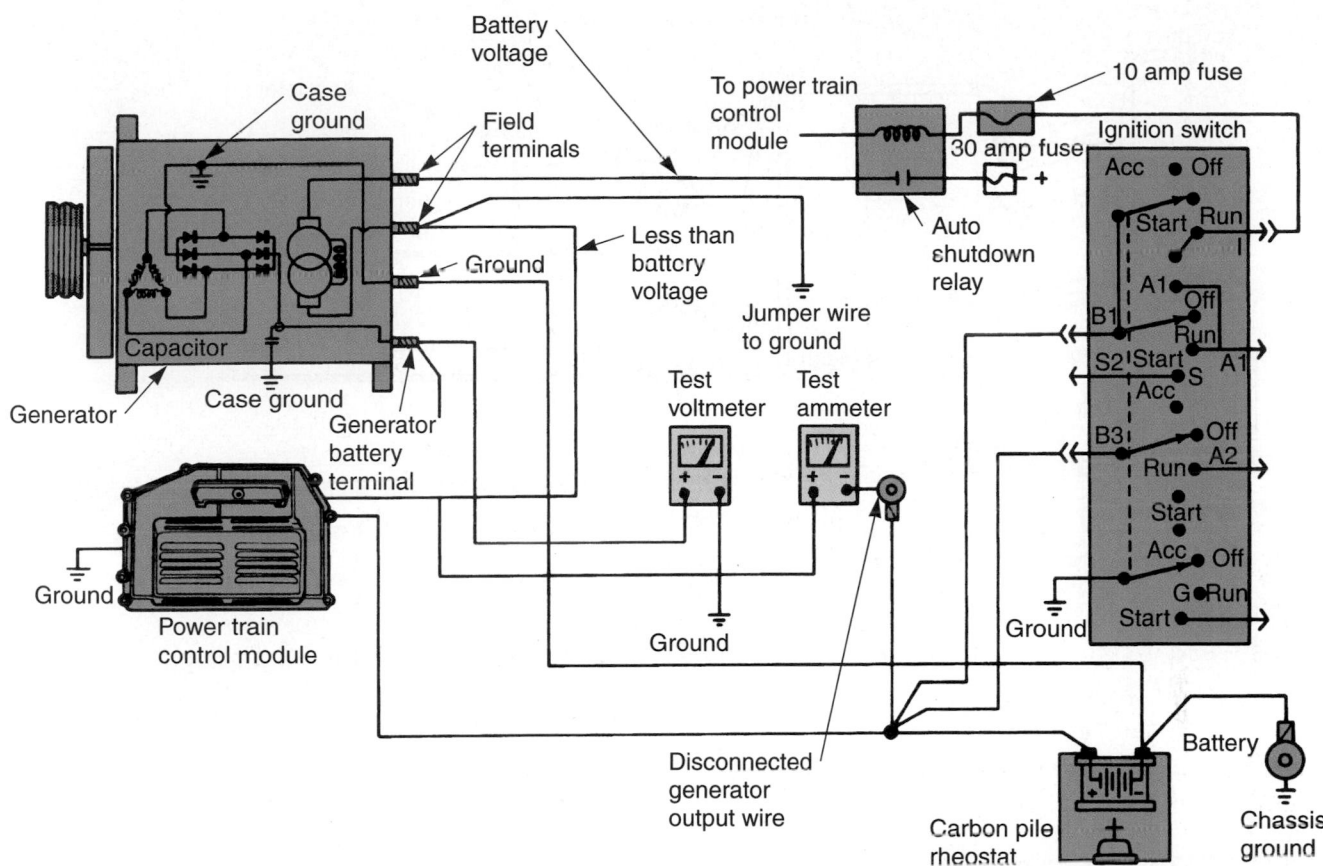

Figure 33-20. Note how this charging system does not have a voltage regulator inside or on alternator. Instead, a regulator circuit is included in the power train control module. This allows more precise control of the charging voltage for improved vehicle efficiency. Also note how meters can be connected for testing the current output of system. (Chrysler)

Voltage regulator switching is very fast (about 400 cycles per second) to help prevent radio noise. At low engine and alternator speeds and with a high load, the regulator may use a 90% on-time to produce an output to meet the high load.

Modern electronic voltage regulators, either integral or in-computer types, can also progressively switch on charging voltage. If the full load of the alternator were to come on instantly, it could lug the idling engine down.

Some modern charging systems do not have a conventional regulator. Instead, *computer voltage regulator circuit* is inside the control module. This is shown in **Figure 33-20.**

Fail-safe Circuit

Some late-model charging systems have *fail-safe circuits,* which disconnect alternator output if voltage levels become too high. This protects all the on-board electronics from being damaged by high voltage. The fail-safe circuit is designed into the voltage regulator.

Contact-Point Voltage Regulator

Contact-point voltage regulators use a coil, a set of points, and resistors to control alternator output. This is an older type of regulator that has been replaced by electronic or solid-state units.

Charge Indicators

A *charge indicator* informs the driver of the operating condition or output of the charging system. There are three basic types of charge indicators:

- *Warning light*—glows if alternator output is below a specified level.
- *Voltmeter indicator*—measures voltage output of alternator.
- *Ammeter indicator*—measures current output of alternator.

These indicators are mounted in the dash of the car. Normally, all cars have an indicator light. A voltmeter or

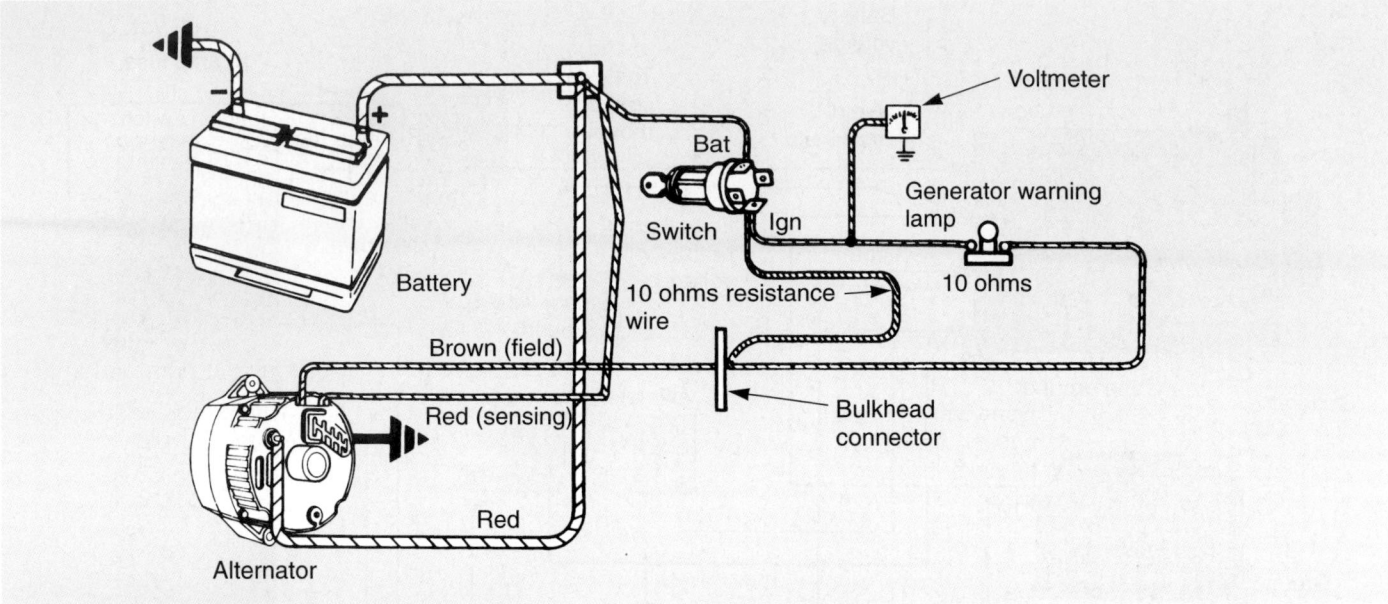

Figure 33-21. The charging circuit shows connections for a generator warning lamp and voltmeter. (Oldsmobile)

ammeter may be added for more precise monitoring of charging system action.

Alternator Warning Light

An **alternator warning light** is wired into the charging system so that it will glow when alternator output drops to a specified level. **Figure 33-21** shows one circuit using an alternator warning light.

If problems develop in the charging system, the field current, trying to increase alternator output, will increase enough to light the indicator bulb. If the charging system is in good operating condition, current flow through the field will be too low to light the bulb.

Voltmeter Indicator

A **voltmeter indicator** can also be used to warn the driver of charging system problems. Refer to **Figure 33-21.** The voltmeter reads the system voltage when the engine is running.

A battery has 12.6 volts when fully charged. To recharge the battery, alternator output must be higher than the battery voltage. Alternator output is normally 13–15 volts.

The voltmeter indicator simply shows voltage, which is an indicator of current output and charging system condition. If the voltmeter reading drops to battery voltage or below, the charging system has problems. If the voltmeter reading is too high, overcharging can occur. **Overcharging** results when excess current is forced through the battery. Battery overheating damage can

result after extended overcharging. This problem is often due to a faulty voltage regulator.

Ammeter Indicator

An **ammeter indicator** simply shows the current output of the alternator in amps. A simplified circuit containing an ammeter is shown in **Figure 33-22.** Study it carefully.

Basically, if the ammeter reads to the right side of the scale (positive dial mark), the battery is being charged. If the ammeter reads to the left side of the scale (negative dial mark), the charging system is not working properly.

Summary

- An alternator is a generator that uses mechanical (engine) power to produce electricity.
- The voltage regulator controls the alternator output voltage and current.
- The alternator belt links the engine crankshaft pulley with alternator pulley to drive the alternator.
- A charge indicator can be an ammeter, voltmeter, or warning light. Indicators inform the driver of charging system condition.
- The battery provides current to initially energize alternator and also helps stabilize alternator output.
- The voltage regulator keeps alternator output at a preset charging voltage (13–15 volts).

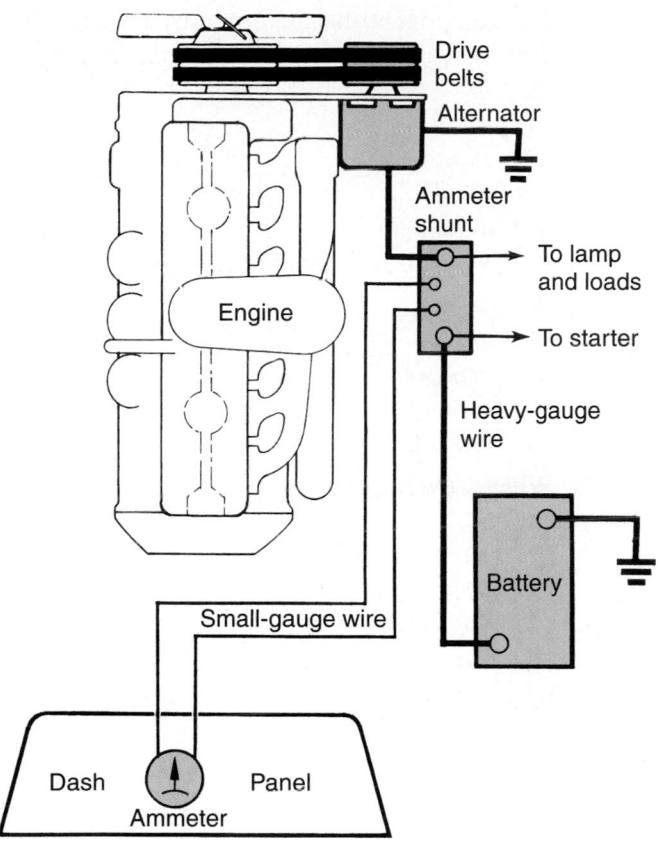

Figure 33-22. Circuit showing basic connections for ammeter indicator. (Motorola)

- The alternator rotor is a rotating magnetic field. It fits in the center of the alternator housing. The alternator drive belt turns the rotor, making the field spin.

- The alternator stator is a stationary set of windings. Alternator current must be rectified (changed) into direct current before entering the electrical system.

Important Terms

Alternator	Alternator stator
Voltage regulator	AC
Alternator drive belt	Rectified
Charge indicator	Diode
Charging system	Forward biased
harness	Reverse biased
Battery	Rotor assembly
Charging voltage	Brush assembly
Charging system	Rectifier assembly
DC generator	Housing
AC generator	Fan and pulley assembly
Alternator rotor	Alternator rotor

Alternator slip rings	Electronic voltage
Alternator brushes	regulator
Brush springs	Integral voltage
Alternator rectifier	regulator
Diode assembly	Battery thermistor
Full-wave rectification	Power train control
Diode trio	module
Rectifier diodes	Computer voltage
Diode frame	regulator
Heat sink	Fail-safe circuits
Alternator stator	Contact point voltage
Y-type stator	regulator
Delta type stator	Warning light
Alternator fan	Voltmeter indicator
Alternator pulley	Ammeter indicator

Review Questions—Chapter 33

Please do not write in this text. Place your answers on a separate sheet of paper.

1. List and explain the six major parts of a charging system.

2. The voltage regulator keeps alternator output at a preset charging voltage of approximately _____ to _____ volts.

3. What are four functions of a charging system?

4. The dc generator is the most common type of automotive charging system. True or False?

5. The ac generator, normally called alternator, has replaced the dc generator. True or False?

6. The alternator is a device for changing mechanical energy into _____ energy.

7. Explain the alternator's rotor and stator.

8. Define the term "rectified current."

9. This alternator device can be used to change alternating current into direct current.
   (A) Transistor.
   (B) Capacitor.
   (C) Stator.
   (D) Diode.

10. What do the terms forward and reverse bias mean?

11. List and explain the seven major parts of an alternator.

12. An alternator _____ consists of field coil windings mounted on a shaft.

13. Alternator brushes ride on _____ _____ to feed current to the rotor windings.

14. An alternator rectifier assembly provides full-wave rectification. True or False?

15. An alternator _____ consists of three groups of windings or coils that surround the spinning rotor.

16. A voltage regulator controls alternator output by changing the amount of current flowing through the _____ _____.

17. Which of the following is *not* a typical type of voltage regulator?
    (A) Contact point type mounted away from alternator.
    (B) Electronic type mounted away from the alternator.
    (C) Integral electronic type.
    (D) Integral contact point type.

18. A(n) _____ voltage regulator is an electronic unit mounted inside or on rear of alternator.

19. Alternator _____ and electrical _____ determine whether the regulator increases or decreases charging output.

20. List and explain three types of charging system indicators.

● ASE-Type Questions

1. A charge indicator can be any of these *except* a(n):
   (A) ammeter.
   (B) voltmeter.
   (C) ohmmeter.
   (D) warning light.

2. Technician A believes that once an engine starts running, the engine crankshaft pulley powers the alternator to produce electricity to a car's electrical systems. Technician B feels a battery is a car's sole electrical supplier. Who is right?
   (A) A only.
   (B) B only.
   (C) Both A and B.
   (D) Neither A nor B.

3. Charging voltage is approximately:
   (A) 12.6 volts.
   (B) 13–15 volts.
   (C) 16–19 volts.
   (D) 20–25 volts.

4. Which of the following is *not* a type of basic charging system?
   (A) Alternator.
   (B) AC generator.
   (C) DC generator.
   (D) Stationary diode.

5. The component that is turned by the alternator drive belt and creates a rotating magnetic field is the:
   (A) rotor.
   (B) stator.
   (C) rectifier.
   (D) diode trio.

6. Each of these is a rectifier assembly component *except:*
   (A) diodes.
   (B) heat sink.
   (C) stator.
   (D) electrical terminals.

7. Which device has stator wires connected end to end?
   (A) Y-type stator.
   (B) Delta-type stator.
   (C) Voltage regulator.
   (D) Stator neutral lead.

8. Which of the following is *not* a type of alternator belt?
   (A) V.
   (B) Ribbed.
   (C) Cogged V.
   (D) Integrated.

9. Electronic voltage regulators use these to control rotor field strength and alternator output.
   (A) Diodes.
   (B) Capacitors.
   (C) Transistors.
   (D) All of the above.

10. When fully charged, an automotive battery has:
    (A) 12.6 volts.
    (B) 13–15 volts.
    (C) 16–19 volts.
    (D) 20–25 volts.

Activities—Chapter 33

1. Vehicles from different manufacturers use various belt arrangements to drive the alternator. Identify at least three different belt arrangements and sketch them. Label each sketch with vehicle make and engine type.

2. A diode is sometimes called "an electrical check valve." Make sketches to show why it received that name.

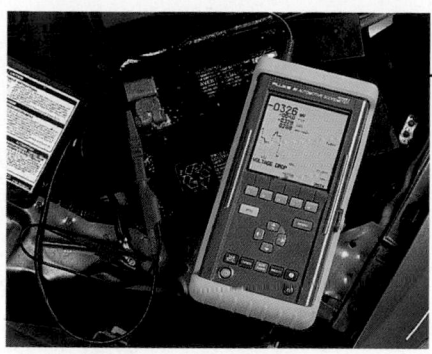

Chapter 34

Charging System Diagnosis, Testing, and Repair

After studying this chapter, you will be able to:

- Diagnose charging system troubles.
- Inspect a charging system.
- Test charging system output with a voltmeter or a load tester.
- Remove, test, repair, and replace an alternator.
- Adjust an alternator belt.
- Remove and replace a voltage regulator.
- Describe safety practices to follow when testing or repairing a charging system.
- Correctly answer ASE certification test questions on charging system diagnosis and repair.

This chapter begins by summarizing the types of troubles that occur in a modern charging system. You will learn how to perform fundamental tests that help locate problems. Then, the chapter covers parts removal, disassembly, repair, and reassembly. This should give you a sound background in the most common types of charging system service and repair jobs.

Charging System Diagnosis

Although a charging system has only two major parts (alternator and regulator), troubleshooting can be difficult. Sometimes, another system's fault (bad starting motor, defective battery, computer failure, or faulty wiring) will appear to be caused by problems in the charging system.

There are four common symptoms of charging system problems:

- Dead battery—slow or no cranking.
- Overcharged battery—water must be added frequently, or the battery suffers overheating damage.
- Abnormal noises—grinding, squealing, and buzzing.
- Indicator shows problem—light glows all the time or there is an incorrect indicator reading.

Verify these problems by starting or trying to start the engine. It is possible that the symptoms have been mistaken by the service writer or customer. For example, a problem described as a no-charge condition may really be a shorted starting motor, battery drain, or other trouble.

Warning

Some late-model cars are equipped with a heated windshield. When the windshield is being heated, the alternator produces 110 volts ac. This is enough voltage to cause electrocution. Keep this in mind when working on this type system.

Visual Inspection

Open the hood and visually inspect the parts of the charging system. Check for obvious troubles, like the ones given in **Figure 34-1.**

Check for battery problems, such as loose battery cables, a discharged battery, corroded terminals, a low water level, or a damaged battery case.

Check for alternator belt problems. Make sure the belt is adjusted properly. A loose alternator drive belt may squeal or flap and prevent normal charging, **Figure 34-2.** Also, check the condition of the belt. Look for cracks, glazing (hard, shiny surface), grease or oil contamination, and deterioration. These conditions are shown in **Figure 34-3.**

If needed, alternator belt tension can be adjusted as follows:
1. Loosen the alternator mounting and adjusting bolts.
2. Pry on a strong surface of the end frame. Pull hard enough to produce proper belt tension. Refer to **Figure 34-4.**

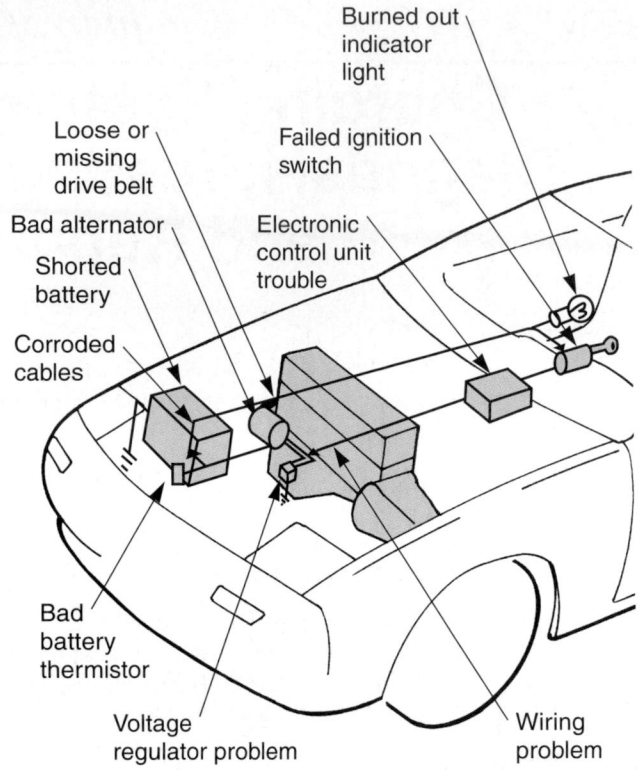

Figure 34-1. Note common problems in a charging system. Some of these could be found during inspection.

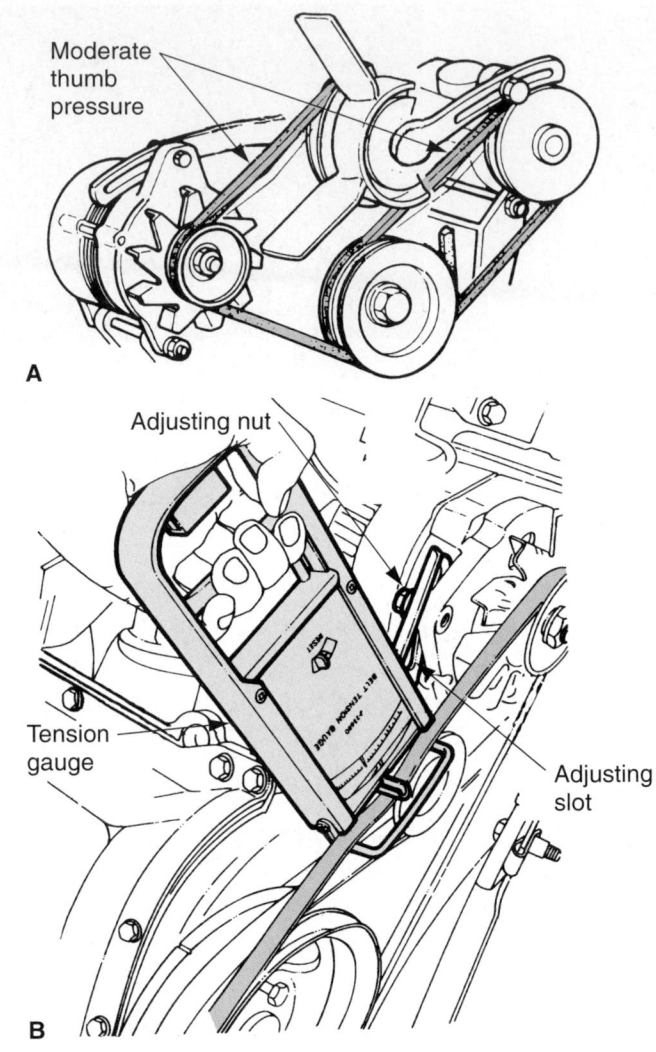

A

B

Figure 34-2. A loose alternator belt can cause low charging system output and a dead battery. The belt should deflect 1/2" when pressed with thumb. A belt tension gauge provides a slower but more accurate way of checking tension. (Mazda and Chrysler)

3. Tighten the adjusting bolt while holding the pry bar.
4. Tighten any other mounting bolts.
5. Recheck tension.

Caution

Tighten an alternator belt only enough to prevent belt slippage or flap. Overtightening is a common mistake that quickly ruins alternator bearings.

Figure 34-3. Inspect belts closely for these kinds of problems. Replace belt if needed. (Snap-On Tools and Chrysler Corp.)

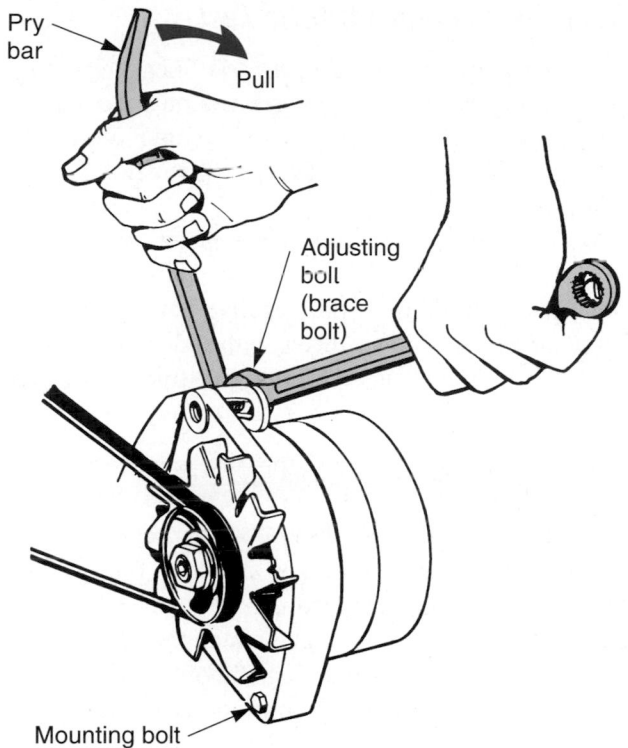

Figure 34-4. To adjust belt, loosen mounting and adjusting bolts. Pry on the thick area of end frame. Hold tension while tightening adjustment bolt. Check tension and tighten any other bolts. (Florida Dept. of Voc. Ed.)

Check for wiring problems in the charging system, **Figure 34-1.** Check for loose or corroded electrical connections, shorted wires (missing insulation), and other problems.

In particular, check the connections on the back of the alternator, on the regulator (type mounted away from alternator), and on the engine control module (computer-controlled voltage regulation). Wiggle the wires while the engine is running. If the indicator light goes out or the indicator gauge begins to read properly when a wire is moved, the problem is in that area of the wiring.

Tech Tip

Connect a voltmeter across the battery terminals while wiggling charging system wires. If the meter fluctuates to a higher voltage when a wire is wiggled, you have found the bad electrical connection. Otherwise, you would need a helper to watch the charge indicator light on the instrument panel.

Listen closely for abnormal noises in the alternator. If necessary, use a stethoscope to listen to the alternator. Try to detect internal grinding (worn, dry bearings), whining (leaking diodes or overcharging), or any other unusual sounds.

Scanning Charging System

On most vehicles with self-diagnostic systems, you can connect a scan tool to the vehicle to aid in troubleshooting. The scan tool will display charging voltage, battery temperature, and other data useful in finding problems. Capabilities vary with each model; refer to the service manual or scanner instructions for details.

Charging System Precautions

Observe the following rules when working on a charging system. They will help prevent possible damage to electrical-electronic systems.

- *Disconnect the battery before removing any charging system component.* If a **hot wire** (current-carrying wire) touches ground, parts could be ruined. Disconnect the battery before connecting it to a battery charger. A voltage surge or high voltage can damage electronic components in the alternator, regulator, and other computerized systems.
- *Never reverse polarity.* If the battery or jumper cables are connected backwards, serious electrical damage can occur. Reversing polarity can damage the diodes in the alternator, ruin the circuit in the regulator, and burn electronic components in other computer systems.
- *Do not operate the alternator with the output disconnected.* If the alternator is operated with the output wire disconnected, alternator voltage can increase above normal levels. Alternator or charging circuit damage could result.
- *Never short or ground any charging system terminal unless instructed to do so by a shop manual.* Some circuits can be grounded or shorted without damage; others will be seriously damaged. Refer to a service manual when in doubt.
- *Do not attempt to polarize an alternator.* DC generator systems had to be polarized (voltage connected to generator field) after repairs. This must *not* be done with an alternator.

Charging System Tests

Charging system tests should be done when symptoms point to low alternator voltage and current. These tests will quickly determine the operating condition of the charging system. There are five common charging system tests:

- Charging system output test—measures current and voltage output of the charging system under a load.

- Regulator voltage test—measures charging system voltage under low-output, low-load conditions.
- Regulator bypass test—connects full battery voltage to the alternator field.
- Scope testing—analyzes voltage waveform.
- Circuit resistance tests—measures resistance in insulated and ground circuits of the system.

Charging system tests are performed in three ways: using a load tester (same tester used to check the battery and starting system in earlier chapters), a scope tool, or a common VOM (volt-ohm-milliammeter).

A *load tester* provides the most accurate method of checking a charging system. See **Figure 34-5.** It will measure both system current and voltage while applying a load to the system.

Before testing the charging system, it is common practice to check the condition of the battery. Although charging system problems often show up as a dead battery, you must not forget that the battery itself may be bad. Measure the battery's state-of-charge and perform a battery load test (see Chapter 30). Then, you will be sure that the battery is not affecting your charging system tests.

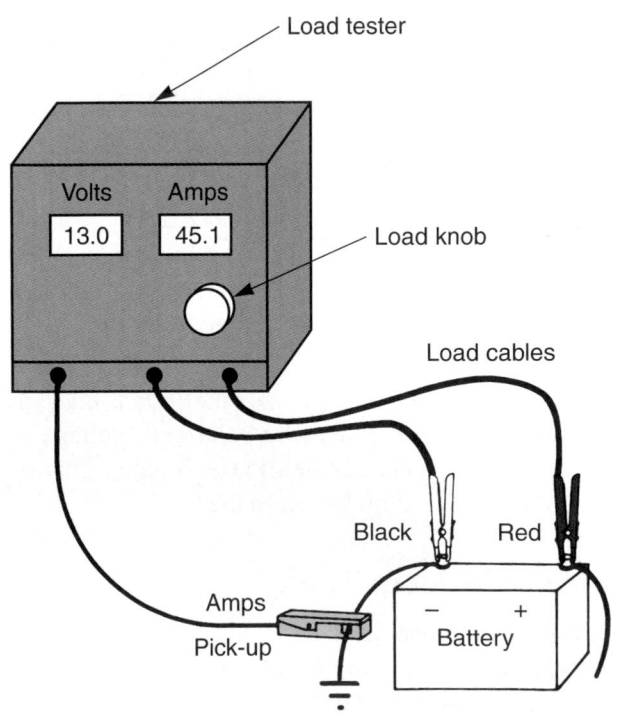

Figure 34-5. Load testers, as used during battery and starting system tests, will also check charging output. Modern testers have an inductive, clip-on current clamp. Older testers must have an ammeter connected in series. (Marquette)

Charging System Output Test

A *charging system output test* measures system current and voltage under maximum load (current output). To check charging system output with a load tester, connect the tester leads as described by the manufacturer.

With modern testers, two leads fasten to the battery terminals. The inductive (clip-on) amps pickup fits around the insulation on the negative battery cable. See **Figure 34-5.** Procedures differ with older, noninductive testers. Check the tester's operating instructions when in doubt.

 A charging system output test is performed as follows:

1. With the load tester controls set properly, turn the ignition key switch to run. Record this ammeter reading.
2. Start the engine and adjust the idle speed to the test specifications (about 2000 rpm).
3. Rotate the load control on the tester until the ammeter reads the specified current output, but do not let the voltage drop below specifications (about 12 volts). Record this ammeter reading.
4. Turn load control off. Evaluate the readings.

Charging System Output Test Results

To calculate charging system output, add your two ammeter readings (current with ignition switch at run plus current with engine running). This will give you the total charging system output in amps. Compare this figure to specifications.

 **Tech Tip**

Alternator current specifications are sometimes stamped on the alternator housing. If not, look up the alternator specifications in the service manual.

Current output specifications for charging systems depend on the size (rating) of the alternator. For instance, a car with few electrical accessories may have an alternator rated as low as 35 amps. A luxury car with many accessories (air conditioning, speed control, power windows, etc.) might have an alternator with a much higher rating (40–80 amps). Always look up exact factory values when evaluating charging system operation.

If the charging system output current is low, perform regulator voltage and regulator bypass tests. They will let you determine whether the alternator, regulator, or circuit wiring is at fault. Even if the output test is within 10% of specifications, perform a regulator voltage test.

Regulator Voltage Test

A *regulator voltage test* checks the calibration of the voltage regulator and detects a high or low voltage setting.

 To perform this test, use the following procedure:
1. Set the meter selector to the correct test position.
2. With the load control off, run the engine at 2000 rpm or at the specified test speed.
3. Note the voltmeter reading and compare it to specifications.

Most voltage regulators are designed to operate with a voltage of 13.5–14.5 volts. This range applies to a fully charged battery at normal temperature.

If the meter reading is steady and within recommended values, then the regulator setting is correct. If the voltage is steady but too high or too low, then the regulator may need adjustment or replacement. If the reading is not steady, there may be a bad wiring connection, an alternator problem, or a defective regulator.

If a charging system fails either the output test or the regulator voltage test, a regulator bypass test should be performed.

Regulator Bypass Test

A *regulator bypass test* is a quick and easy way of finding if the alternator, regulator, or circuit is faulty. Procedures for a regulator bypass test are similar to the those for the output test already explained. However, the regulator is taken out of the circuit.

Depending on system design, there are several ways to bypass the voltage regulator. If the rear of the alternator has a test tab, short it on the end frame, **Figure 34-6**. When the tab is shorted, the alternator should produce maximum output. With many vehicles, a jumper wire must be used to connect battery and field terminals, **Figure 34-7**. When the battery voltage (unregulated voltage) excites the rotor field, the alternator should produce maximum output.

 Caution
Follow manufacturer's directions to avoid damage when bypassing the voltage regulator. You must not short or connect voltage to the wrong wire, or the diodes or regulator may be ruined.

Regulator Bypass Test Results

If charging voltage and current increase to normal levels when the regulator is bypassed, you usually have a

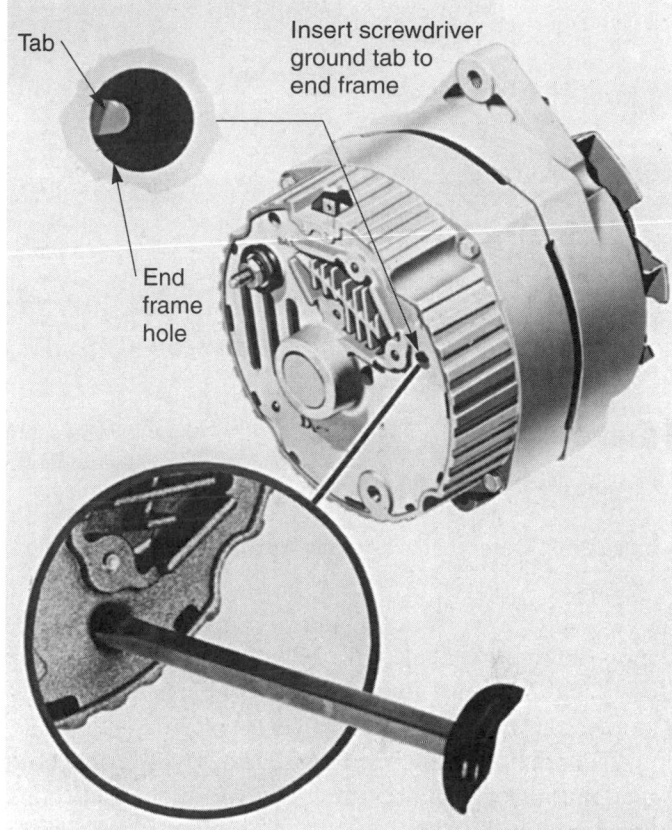

Figure 34-6. If charging system fails the current output test, bypass the regulator to pinpoint the problem. The shorting tab on this alternator should make the alternator produce maximum output. If it does, the regulator is bad. If it does not, then the alternator may be bad. (Pontiac)

bad regulator. If system output remains the same when the regulator is bypassed, you normally have a bad alternator.

Specialized charging system testers may have other test modes (positions). For example, they may be able to check the alternator stator, diodes, or regulator. Read the tester instructions.

Charging System Scope Testing

The vehicle's computer can also be used to control and/or monitor the charging system alternator by replacing or supplementing the voltage regulator. Basically, the computer controls the duty cycle of the alternator's field.

The *duty cycle* is the percentage of time that current is fed to the alternator's field windings. With a large electrical load, the duty cycle would be high (long) to make the alternator produce high output. A duty cycle of 50% is normal.

In some charging systems with internal voltage regulators or computer-controlled regulation, the regulators

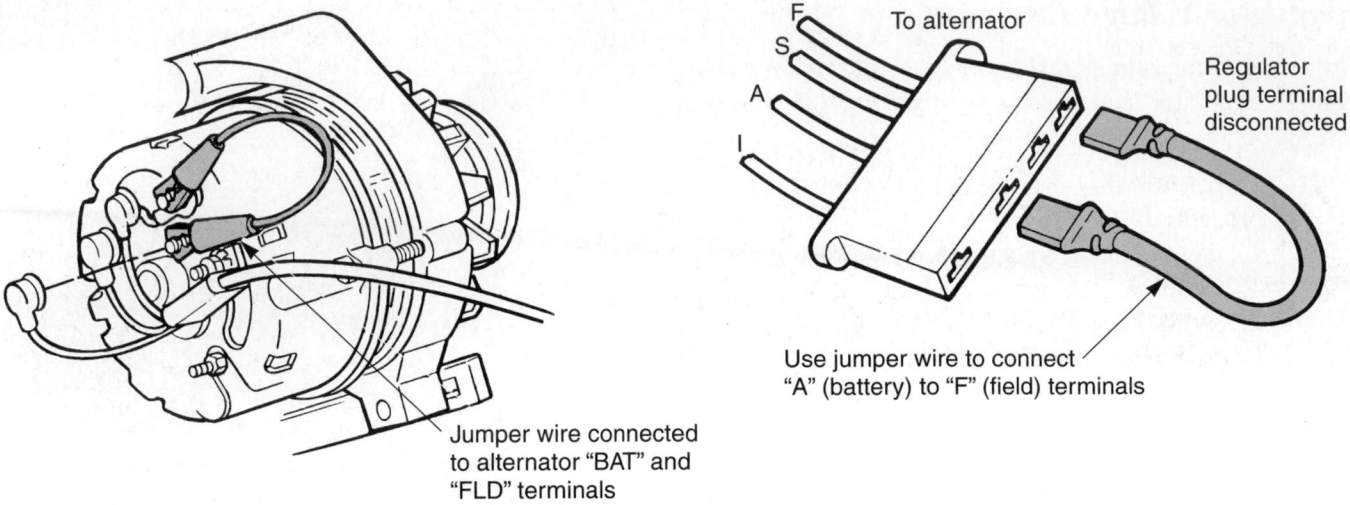

Jumper wire connected
to alternator "BAT" and
"FLD" terminals

To alternator

Regulator
plug terminal
disconnected

Use jumper wire to connect
"A" (battery) to "F" (field) terminals

Figure 34-7. Other methods of bypassing the regulator to find out if the alternator or the regulator is faulty. (Mercury)

cannot be bypassed and full-fielded. This makes it more difficult to isolate a low-charging problem to the alternator or the voltage regulator. When testing these systems, you must use a scope or load tester with a diode test function to check for abnormal voltage ripple (indicating bad alternator diodes).

A charging system *scope test* involves analyzing the alternator voltage waveform for signs of abnormal ripple.

 Use the following procedure when performing a scope test:

1. Connect the scope to the alternator output terminal.
2. Ground the other scope lead.
3. Start the engine and observe the scope trace (waveform).

As shown in **Figure 34-8,** the charging system should produce a relatively smooth dc voltage trace with minimum ripple. **Figure 34-9** shows both good and bad scope waveforms for typical charging systems.

 Note
For more information on these tests, refer to Chapter 46, *Advanced Diagnostics.*

Circuit Resistance Tests

Circuit resistance tests are used to locate wiring problems in a charging system: loose connections, corroded terminals, partially burned wires, and similar troubles. Resistance tests should be performed when symptoms point to problems other than the alternator or regulator. Two common circuit resistance tests are the insulated-circuit resistance test and ground-circuit resistance test.

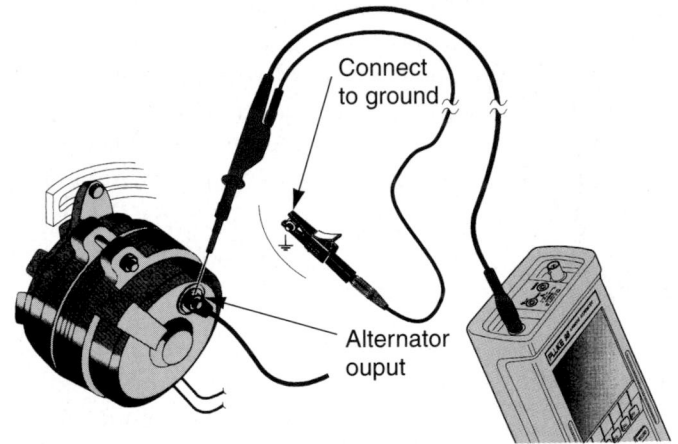

Connect
to ground

Alternator
ouput

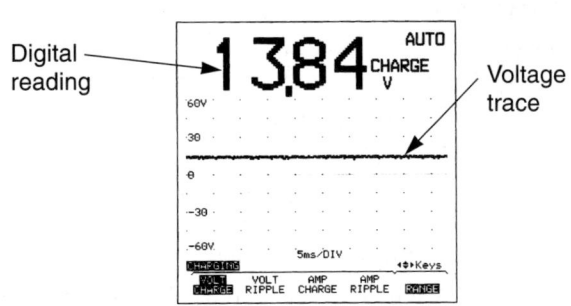

Digital
reading

Voltage
trace

Figure 34-8. A hand-held oscilloscope can be connected to the alternator to check for smooth dc output. This scope will show a digital readout of the voltage and a voltage trace. With a good charging system, there should be very little ripple in the trace. (Fluke)

To do an *insulated-circuit resistance test* on a charging system, connect the tester as described by the manufacturer. One type of connection is shown in **Figure 34-10.** Note how the voltmeter leads are

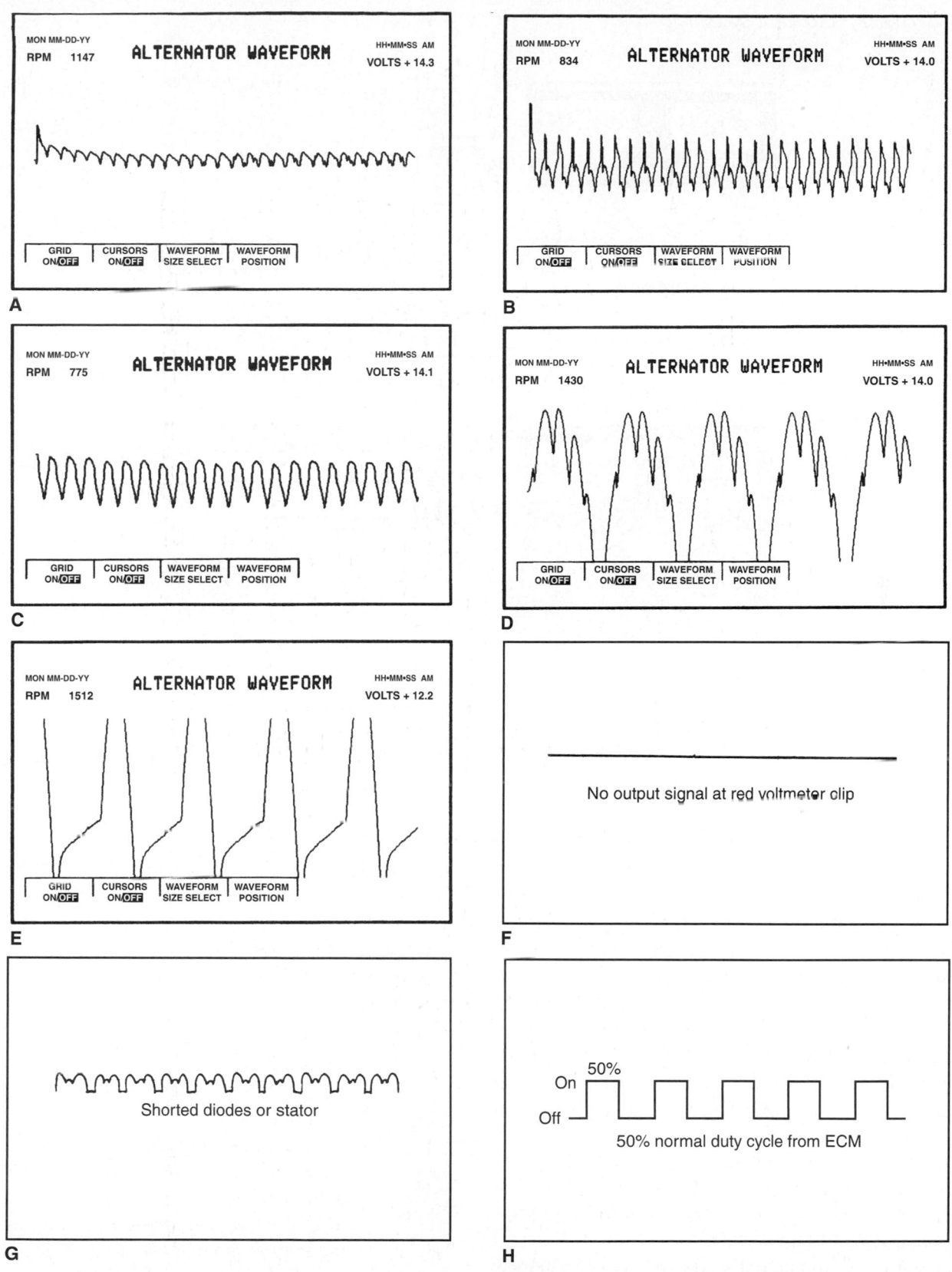

Figure 34-9. Oscilloscope will show faulty diodes and other problems before alternator removal and tear down. A—Normal alternator waveform with typical output. Note small, regularly spaced, even ripples. B—This is normal light load alternator waveform with some electronic regulators using duty cycles to control output. Slightly higher inductive peaks and spike can be normal with some systems. C—Normal alternator waveform with heavy load from a dead battery or many accessories on. It is similar to *A* but has a higher amplitude ripple from increased current output. D—This alternator waveform shows an open diode causing high spikes. E—This waveform is due to one open diode and one shorted diode. F—A solid straight line on scope indicates no output from the alternator. G—This waveform indicates shorted diodes or shorted stator windings. H—Duty cycle from ECU to alternator field windings can be checked with a scope. High load, as applied from load tester, should make duty cycle show more On time. If not, suspect the computer. (Snap-On Tools)

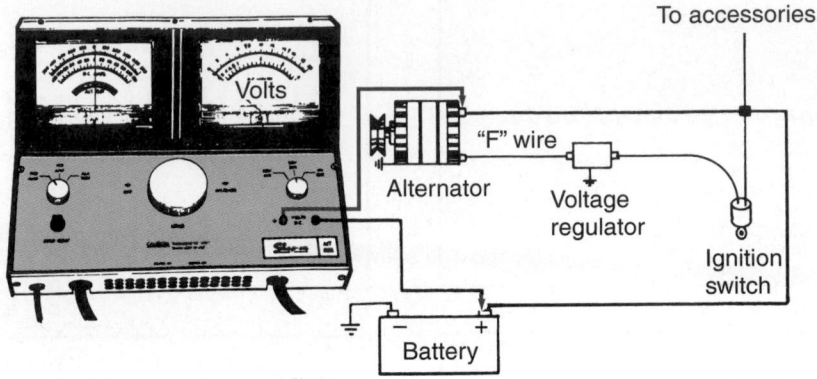

Insulated-circuit resistance test

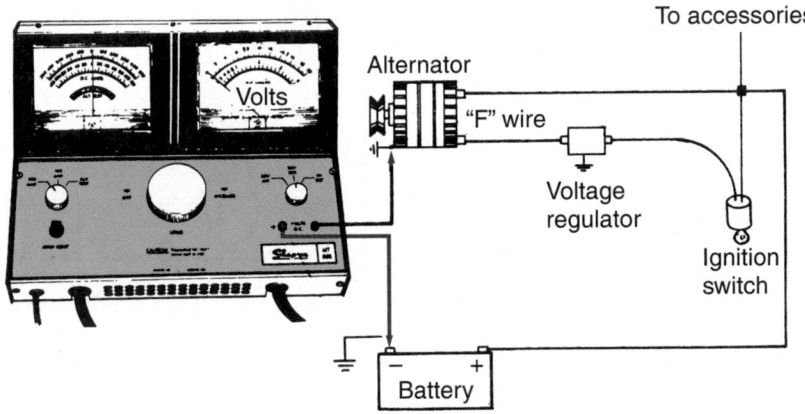

Ground-circuit resistance test

Figure 34-10. Voltage drop or resistance tests will find problems in wiring. (Snap-On Tools)

connected across the alternator output terminal and positive battery terminal.

With the vehicle running at a fast idle, turn the load control to obtain a 20 amp current flow at 15 volts or less. All lights and accessories should be off. Read the voltmeter.

If the circuit is in good condition, the voltmeter should not read over about 0.7 volts (0.1 volt per electrical connection). If the voltage drop is higher than 0.7 volts, circuit resistance is high. A poor connection exists in that section of the charging circuit.

A *ground-circuit resistance test* is similar to the insulated-circuit resistance test. However, the voltmeter is placed across the negative battery terminal and alternator housing. See **Figure 34-10.**

The voltmeter should not read over 0.1 volt per electrical connection. If the voltmeter reading is higher, look for loose connections, a burned plug socket, or similar problems.

Voltmeter Test of Charging System

A voltmeter can also be used to test the output of a charging system when a load tester is not available. It will measure charging system voltage with an accessory load (all electrical accessories on). The voltage reading will be an indicator of current output and charging system condition. If charging system output voltage is not above battery voltage, the battery cannot be recharged and a problem exists.

 A voltage test involves four steps. Refer to **Figure 34-11.**

1. Measure *base voltage*—battery voltage with the engine off.
2. Measure *no-load voltage*—battery voltage with the engine running and electrical accessories off.
3. Measure *load voltage*—battery voltage with the engine running and all electrical accessories on.
4. Calculate *charge voltage*—load voltage - battery voltage = charge voltage. Charge voltage should be 0.5 volts higher than battery voltage.

The load voltage must be higher than the base voltage for battery charging. No-load voltage must be within specifications.

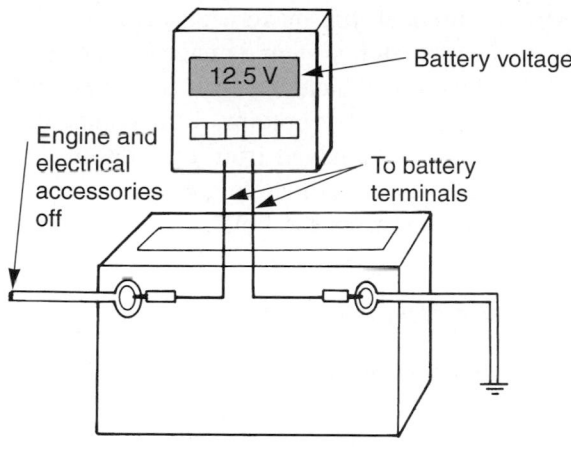

Base voltage

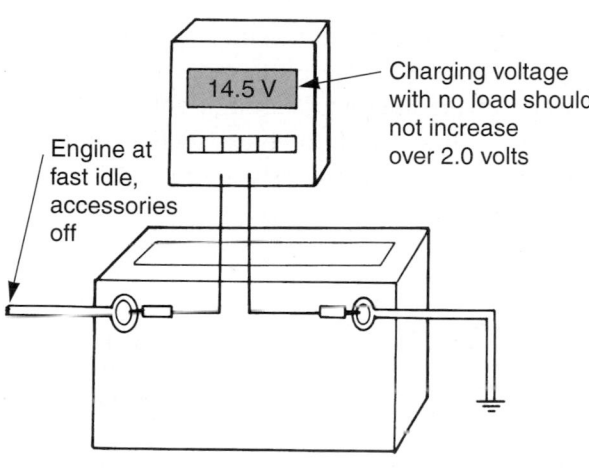

No-load voltage

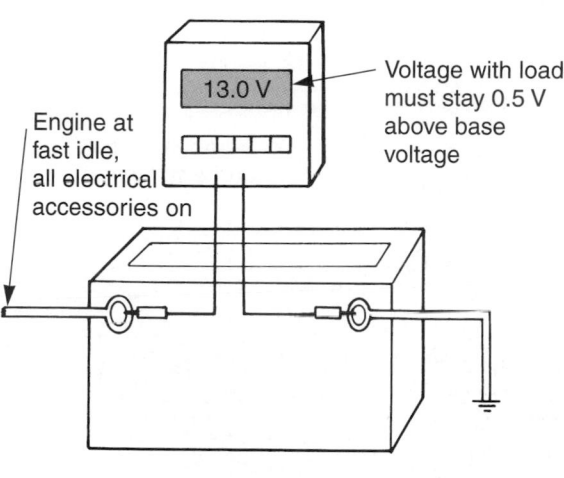

Load voltage

Figure 34-11. If a load tester is not available, a voltmeter can be used to test the charging system output. The electrical accessories are turned on to make sure the voltage stays high enough to charge the battery.

Base Voltage

When testing the charging system with a voltmeter, connect the meter probes across the terminals of the battery to measure the base voltage. This reading will compensate for any variation in the condition of the battery and accuracy of your meter. With a good, fully charged battery, the base voltage reading should be around 12.6 volts.

No-Load Voltage

A no-load test measures the charging system voltage with no current draw. Start and run the engine at about 1500 rpm with all electrical accessories off. The voltmeter reading should increase from the base reading, but not by more than two volts.

A no-load voltage that is 0.5–2 volts higher than the base voltage is normal. For example, if your base voltage is 12.5 volts, your no-load reading should be 13.0–14.5 volts. If the no-load voltage is more than 2–3 volts higher than base voltage, the alternator is overcharging the battery. Either the voltage regulator or wiring is bad.

If the no-load voltage reading is not higher than the base voltage, the charging system is not working. The alternator, regulator, or wiring may be bad. Bypass the regulator or perform resistance tests to isolate the problem.

Load Voltage

If the system passes the no-load voltage test, you should also complete a load test with your voltmeter. This will check the charging system output under high-current draw conditions. The load test shows if the charging system is providing current for all the electrical units and still has enough current to recharge the battery.

Start the engine and run it at about 2000 rpm. To load the charging system, turn on all electrical accessories (headlights, wipers, blower motor, air conditioning, etc.). The voltmeter should read at least 0.5 volts higher than the base voltage. If the load voltage is not 0.5 volts above the base voltage, bypass the regulator to determine which component (alternator or regulator) is faulty.

Alternator Service

A bad alternator will show up during your tests as a low voltage and current output problem. Even when the regulator is bypassed and full voltage is applied to the alternator field, charging voltage and current will not be up to specifications.

Alternator Removal

Before unbolting the alternator, disconnect the battery to prevent damage to parts if wires are shorted. As shown in **Figure 34-12,** most alternators are attached to

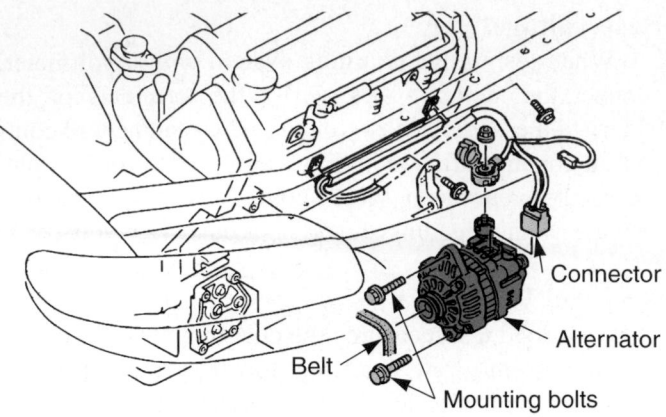

Figure 34-12. To remove the alternator, first remove parts that block access to the alternator mounting bolts. Then loosen bolts and slide the belt off the pulley. Disconnect wires and remove the bolts. (Mazda)

the front of the engine with two or three bolts. Loosen the bolts and remove the belt. Then remove the alternator.

When removing the wires from the back of the alternator, note their location and whether special insulating washers are used. If you make a mistake in reattaching wires to the alternator, system damage can occur.

Depending on vehicle design, the alternator may come out from the top or bottom of the engine compartment. If it must be removed from the bottom, you may need to remove a splash shield to gain access to the alternator mounting brackets and bolts.

Alternator Disassembly

To disassemble an alternator, first scribe marks on the outside of the housing. This will aid you in reassembly. When clamping the alternator in a vise, be careful not to damage the housing or bend the fan.

Use the directions in a shop manual to disassemble the alternator. **Figure 34-13** illustrates the basic disassembly steps. An Allen wrench may be needed to hold the shaft. Use a puller, if needed, to remove the pulley.

Remove the alternator through-bolts. Tap the drive end frame with a plastic or brass mallet. Slide the end frame from the rotor shaft. As you remove the remaining alternator parts, watch how everything fits together.

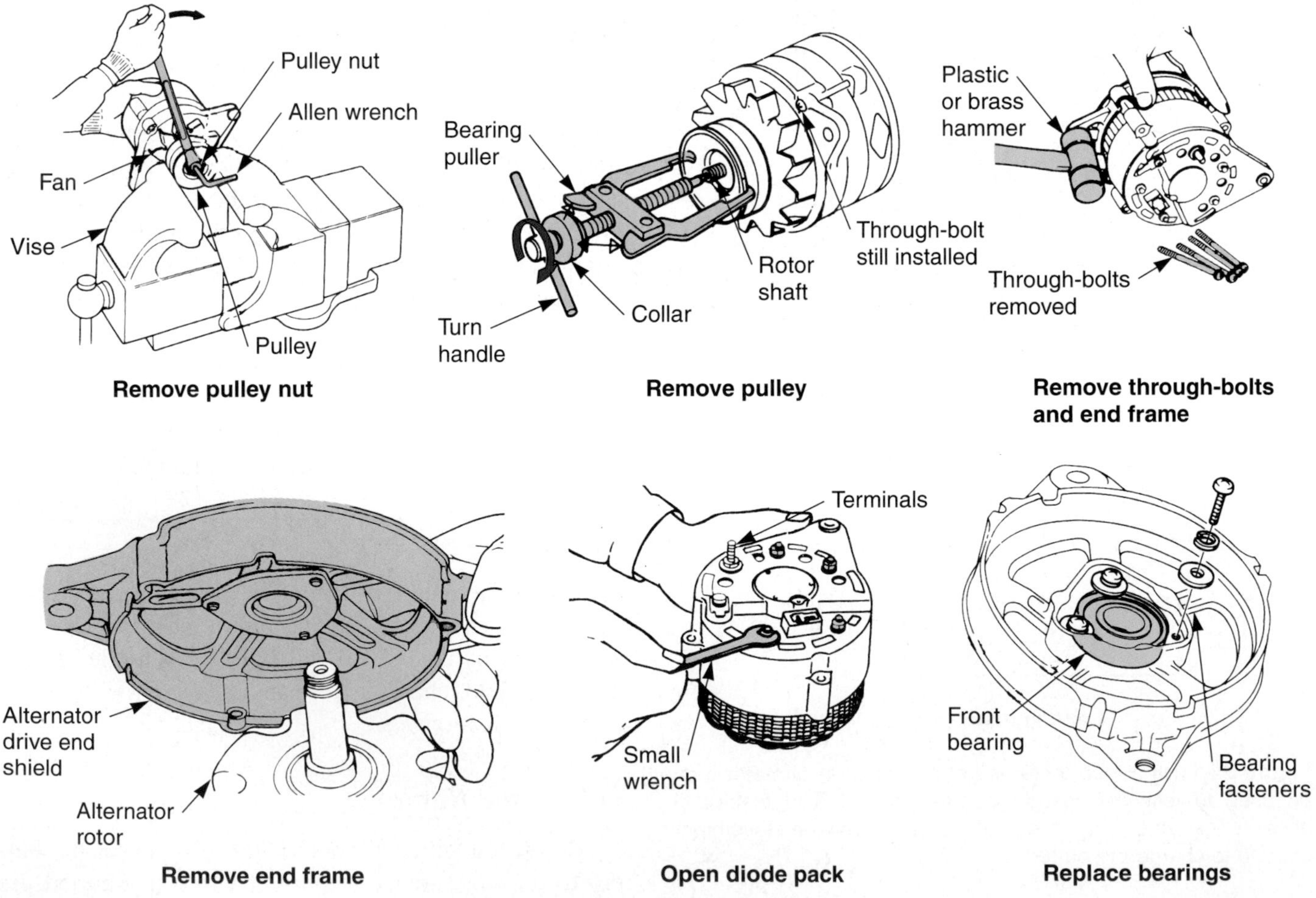

Remove pulley nut

Remove pulley

Remove through-bolts and end frame

Remove end frame

Open diode pack

Replace bearings

Figure 34-13. Basic steps for alternator disassembly. (Honda, Subaru, Chrysler)

Tech Tip

Depending on the type of repair, you may not need to completely disassemble the alternator. For example, when replacing worn alternator bearings, you do not have to remove the diodes, built-in regulator, or other unrelated parts.

Keep all your parts organized and clean, **Figure 34-14.** If you get grease on the brushes, replace them. Grease or oil will ruin the brushes.

Caution

Do not soak the rotor, stator, diode pack, regulator, or other electrical components in solvent. Solvent could ruin these components.

Alternator Rotor Service

A bad alternator rotor can have a bent shaft, scored slip rings, open windings, or shorted windings. Visually inspect the rotor closely. Make sure the rotor is in good condition before assembling the alternator. There are several tests designed to check an alternator rotor. Refer to **Figure 34-15.**

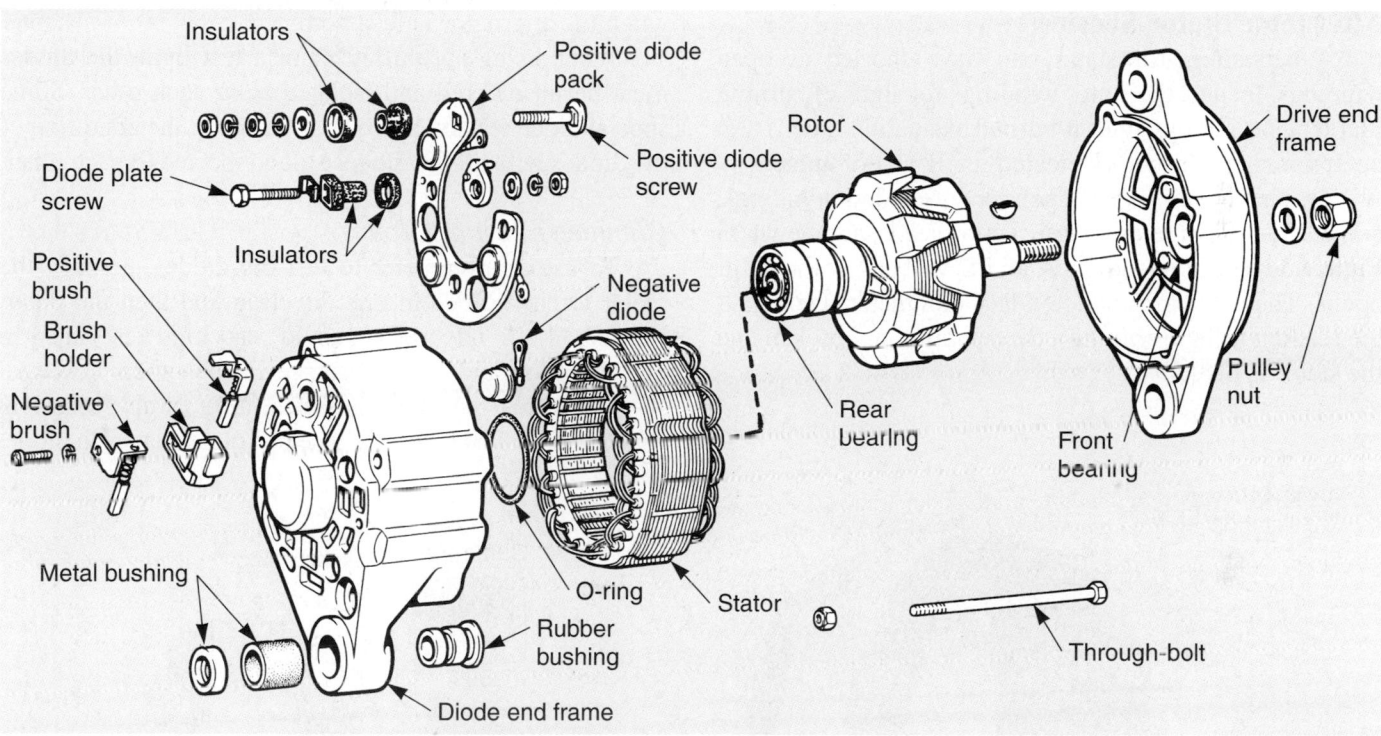

Figure 34-14. Keep all parts organized on the workbench during disassembly. Do not clean electrical parts in solvent. (Fiat)

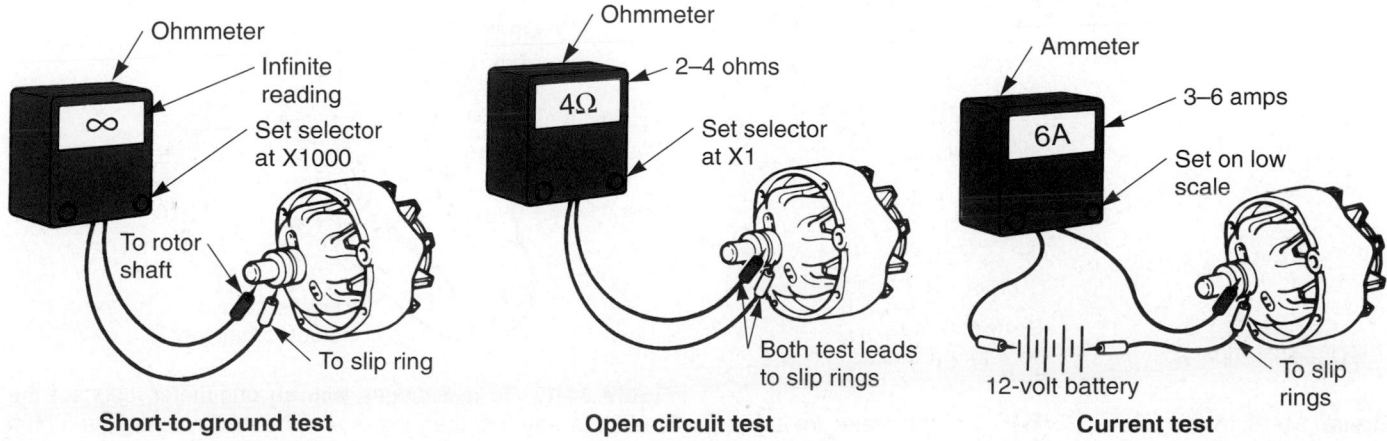

Short-to-ground test | **Open circuit test** | **Current test**

Figure 34-15. Alternator rotor tests. (Chrysler)

A *rotor winding short-to-ground test* measures resistance between the rotor shaft and the windings. The ohmmeter should read infinite (maximum) resistance to show no short to ground.

A *rotor winding open circuit test* measures the resistance between the two slip rings. The meter should read low resistance (2–4 ohms). This would indicate that the windings are not broken.

A *rotor current test* checks the windings for internal shorts. Connect a 12-volt battery and an ammeter to the slip rings. Measure the current and compare it to specifications. Typical rotor current should be 3–6 amps. Replace the alternator rotor if it fails any of these three tests.

Alternator Stator Service

A bad alternator stator can have shorted or open windings. Inspect the stator windings for signs of burning (darkened windings with a burned insulation smell). An open winding is usually detected using an ohmmeter.

To test a stator for open or grounded windings, connect an ohmmeter to the stator leads as shown in **Figure 34-16.** Connections A and B will check for stator opens. They should produce a low ohmmeter reading. If the reading is high (infinite), the windings are broken and the stator is defective.

Connection C tests for a grounded winding. An infinite reading is desirable. If the reading is low, the stator is grounded and should be replaced.

Alternator Diode Service

Bad alternator diodes reduce alternator output current and voltage, and may also cause voltage ripple that can upset computer system operation. Faulty diodes are a frequent cause for alternator failure. It is important to check the condition of the diodes when rebuilding an alternator.

There are various methods used to test alternator diodes: ohmmeter, test light, diode tester, and scope test. The ohmmeter is the most common testing tool used when the alternator is disassembled.

When using an ohmmeter or a test light, the diodes must be unsoldered and isolated from each other. Some special diode testers, however, will check the condition of the diodes with all the diodes still connected to each other.

Ohmmeter Test of Diodes

To use an ohmmeter to test the diodes, connect the meter to each diode in one direction and then the other, **Figure 34-17.** The meter should read high resistance in one direction and low resistance in the other. This will show you that the diode is functioning as an "electrical check valve." The test should be performed on each diode.

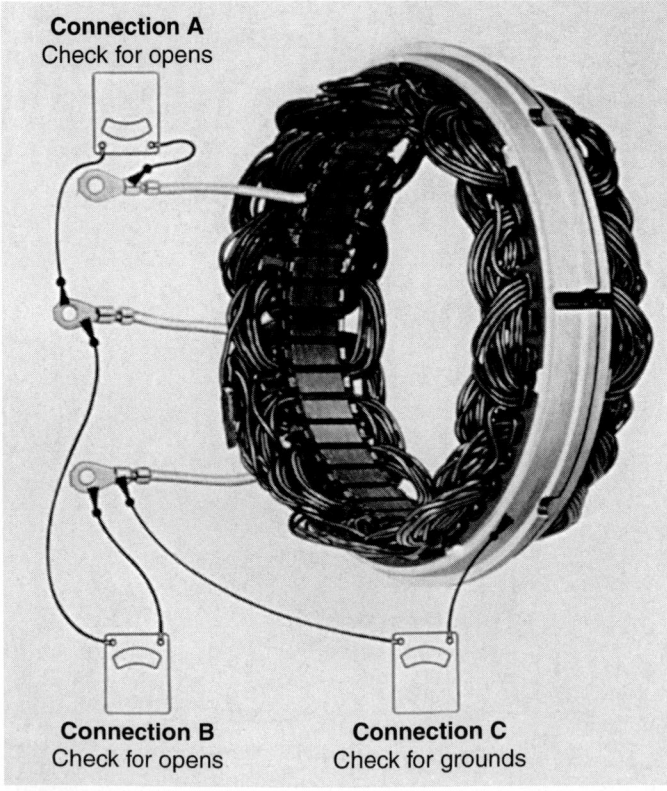

Figure 34-16. Stator tests. Connect an ohmmeter for three tests to check windings. (Chevrolet)

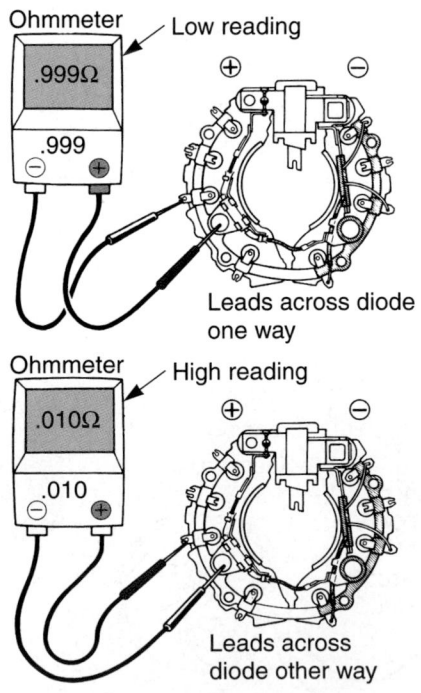

Figure 34-17. To test diodes with an ohmmeter, connect the leads one way and then the other. The meter should read high ohms in one direction and low in the other. (Toyota)

A bad diode can either be shorted or opened. An *open diode* will have a high (infinite) resistance in both directions. A *shorted diode* will have a low (zero) resistance in both directions. In either case, the diode must be replaced.

If diodes were unsoldered for testing, they must be resoldered during installation. After pressing in the new diode or obtaining a new diode pack, use a soldering gun and rosin-core solder to attach the diode leads, **Figure 34-18.** Heat the wires quickly to avoid overheating the diodes. Excess heat can ruin a diode.

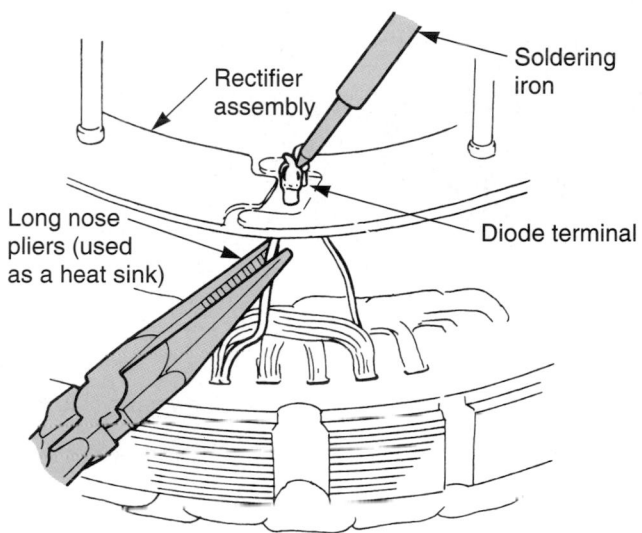

Figure 34-18. Use rosin-core solder to join stator-to-rectifier wires.

Alternator Bearing Service

Worn alternator bearings and *dry alternator bearings* produce a rumbling or grinding noise during operation. They can become loose enough to upset alternator output by allowing too much rotor shaft movement. When rebuilding an alternator, it is common practice to replace the bearings.

The front alternator bearing, also called the drive end bearing, is usually held in place with a cover plate and small screws. To replace the bearing, remove the screws and plate and lift out the old bearing.

The rear alternator bearing, also referred to as the diode end bearing or slip ring end bearing, is normally pressed into the rear end housing. It may be pressed or carefully driven out of the alternator housing for replacement.

If the bearings are relatively new and you do not replace them, make sure you put a moderate amount of grease into the rear bearing. The front bearing is usually sealed and cannot be greased.

To check the action of the front bearing, rotate it with your finger while feeling for roughness or dryness. Replace the bearing if there is any sign of failure.

Alternator Brush Service

Worn brushes can affect the output voltage and current of an alternator. As the brushes wear, spring tension and brush pressure on the slip rings will be reduced.

Inspect the brushes and measure their length. When the brushes are worn beyond specifications or soaked with oil or grease, replace them. Many technicians simply replace the brushes whenever the alternator is disassembled.

Alternator Assembly

After you have inspected and tested the components of an alternator, you are ready for reassembly. Alternator construction varies; refer to a service manual describing the particular unit. In general, assemble the alternator in the reverse order of disassembly. Study **Figure 34-19.**

 The following procedure is typical for alternator assembly:

1. Install all of the components in the rear end frame: electronic regulator, diode pack, rear bearing, terminals, and nuts.
2. If the brushes are not mounted on the outside rear of the end frame, you may need to use a piece of *stiff wire* or a *small Allen wrench* to install the brushes. Push the brush spring and brush into place, **Figure 34-20.** Then, slide your wire or Allen wrench into a hole in the rear end frame. Push the next spring and brush into place. Slide the wire the rest of the way through the hole. The wire or Allen wrench will hold the brushes out of the way as you slide the rotor into the housing.
3. Fit the front end frame into position and check the alignment pins or marks. Install and tighten the through-bolts.
4. Pull out the piece of wire or Allen wrench. You should hear the brushes click into place on the slip rings.
5. Install any spacer, the fan, front pulley, lock washer, and nut. Torque the pulley nut to specifications, **Figure 34-21.** Then, spin the rotor shaft and pulley to check for free movement. The rotor should spin freely without making unusual noises.

Test alternator output on a *bench tester* (unit for off-car output test of alternator) if one is available. If not, test the charging system output after alternator installation.

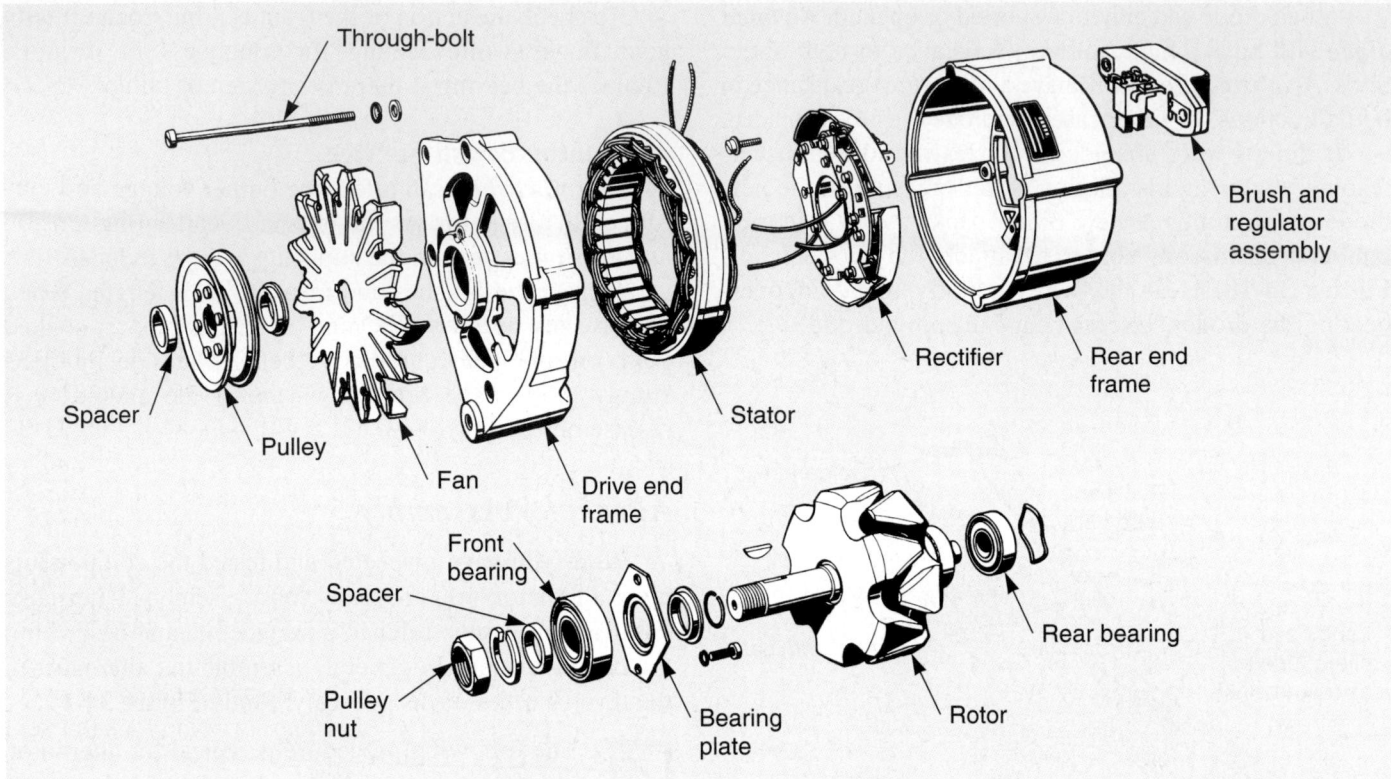

Figure 34-19. Refer to a service manual for an illustration like this one during assembly. Exploded view shows how parts fit together. (Bosch)

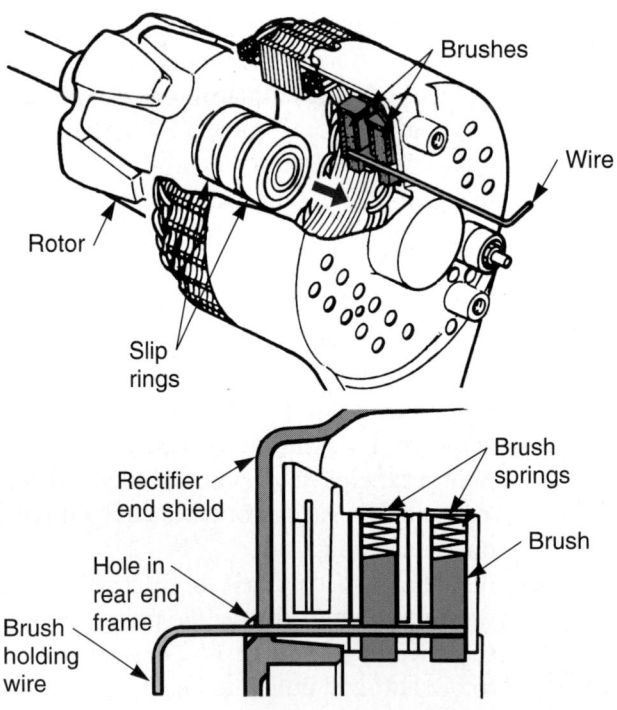

Figure 34-20. Top. With internal brushes, wire is often used to install the brushes. Push the brushes and springs up into place. Then slide the wire through the hole in the back of the alternator. Bottom. The wire will hold brushes as you fit the rotor into position. Slide the wire out and the brushes will snap into contact with the slip rings.

Alternator Installation

With the battery still disconnected, fit the alternator onto the front of the engine. If needed, install the wires on the back of the alternator first. Hand start the bolts and screw them in without tightening.

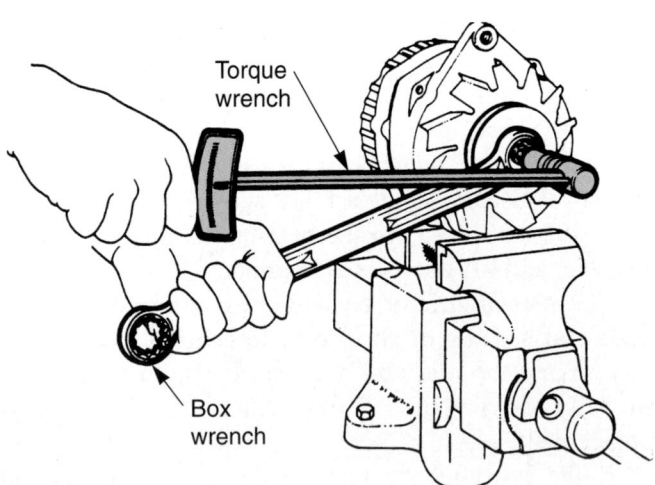

Figure 34-21. Use a torque wrench to properly tighten the pulley nut. Be careful not to damage the alternator housing or fan in the vise. (Chrysler)

Check the condition of the alternator belt. Replace it if needed. Slip the belt over the engine and alternator pulley. Make sure the belt is aligned properly on each pulley. Adjust belt tension and tighten the bolts. Reconnect the battery.

Regulator Service

When a modern electronic regulator fails a voltage test, it must be replaced. However, with older contact-point regulators, you can often simply adjust the regulator setting.

Electronic Regulator Service

It is a simple task to replace a faulty electronic regulator. The electronic regulator is normally located on the back of the alternator or inside the alternator. If not, it may also be located on a fender well.

Obtain the correct replacement regulator and install it. Then recheck charging system output.

Some electronic voltage regulators are adjustable. They can have a small adjusting screw inside a hole in the outer case. Turning the screw will adjust the voltage setting of the regulator. Check a shop manual for details.

Normally, the engine is run at idle with a voltmeter connected to measure no-load charging voltage. The regulator screw is rotated until the voltmeter reads within specifications.

Contact-Point Regulator Service

Contact-point regulators were used on dc generator and early alternator charging systems. The regulator, due to point wear and pitting, is a common cause of problems.

If you work on a car equipped with a contact-point regulator, refer to the service manual. Regulator operation and circuits vary. You will need a manual covering the exact type of regulator circuit. It will tell you how to file, test, and adjust the regulator points. If the regulator cannot be adjusted to obtain the proper voltage, replace it.

Computer Voltage Regulator Service

As discussed in the previous chapter, some alternators use an on-board computer or power train control module to control charging system output. To check this type system, use a scope to check the duty cycle and voltage being applied to the alternator field windings. You may also find poor electrical connections between the electronic control unit and the alternator.

With computer control of the alternator, charging voltage can be slightly lower than normal. However, charging voltage must still stay above battery voltage.

If an abnormal signal or no signal is being sent out of the control unit to the alternator, the regulator should be replaced. Refer to the service manual for more specific testing and servicing methods.

Summary

- Sometimes, another system fault (bad starting motor, battery, computer failure, or wiring) will appear to be caused by problems in the charging system.
- A loose drive belt may squeal or flap and may prevent normal charging.
- Listen closely for abnormal noises in the alternator. Use a stethoscope to listen to the alternator if necessary.
- A charging system output test measures system current and voltage under maximum load (current output).
- A regulator voltage test will check the calibration of the voltage regulator and detect an improper voltage setting.
- A regulator bypass test is a quick and easy way of finding if the alternator, regulator, or circuit is faulty.
- Duty cycle is the percentage of time that current is fed to the alternator's field windings.
- Circuit resistance tests are used to locate wiring problems in a charging system.
- A bad alternator will show up during your tests as a low voltage and current output problem.
- Bad alternator diodes reduce alternator output current, output voltage, and may also cause voltage ripple that can upset computer system operation.
- With a computer-controlled alternator, charging voltage can be slightly lower than normal.

Important Terms

Hot wire	Scope test
Charging system tests	Circuit resistance tests
Load tester	Insulated-circuit
Charging system	resistance test
output test	Ground-circuit
Regulator voltage test	resistance test
Regulator bypass test	Base voltage
Duty cycle	No-load voltage

Load voltage

Charge voltage

Rotor winding short-to-ground test

Rotor winding open circuit test

Rotor current test

Open diode

Shorted diode

Worn alternator bearings

Dry alternator bearings

Worn brushes

Bench tester

Review Questions—Chapter 34

Please do not write in this text. Place your answers on a separate sheet of paper.

1. Name four common symptoms caused by charging system problems.

2. Overtightening an alternator belt is a common mistake that ruins the alternator bearings. True or False?

3. List and explain five charging system service precautions.

4. List and explain four common charging system tests.

5. A(n) _____ _____ is an instrument that provides the most accurate method of checking the condition of a charging system.

6. A charging system output test measures system _____ and _____ under maximum conditions.

7. Depending upon alternator rating, charging system output can range from 35–80 amps. True or False?

8. Most voltage regulators are designed to maintain between 11 and 12 volts. True or False?

9. Explain the regulator bypass test.

10. A charging system has failed the output and regulator voltage tests. Voltage and current were too low. All belts, electrical connections, and other visible checks were OK. Technician A says that a regulator bypass test should be done next to help isolate the problem to the alternator, regulator, or circuit wiring. Technician B says that the alternator should be removed and tested on a bench. This will eliminate any wiring or circuit troubles. Who is right?
    (A) A only.
    (B) B only.
    (C) Both A and B.
    (D) Neither A nor B.

11. What is the purpose of charging system circuit resistance tests?

12. Charging system testers and an oscilloscope can be used to help diagnose charging system troubles. True or False?

13. A bad alternator rotor can have:
    (A) a bent shaft.
    (B) scored slip rings.
    (C) open windings.
    (D) All of the above.

14. Explain three rotor tests.

15. How do you test a stator?

⬤ ASE-Type Questions

1. When working on charging systems:
    (A) *never reverse polarity.*
    (B) *always polarize the alternator.*
    (C) *never short system terminals.*
    (D) *detach the battery before part removal.*

2. Each of these is a common charging system test *except:*
    (A) *regulator bypass test.*
    (B) *circuit resistance tests.*
    (C) *alternator mounting test.*
    (D) *charging system output test.*

3. In which test(s) is the regulator left out of circuit?
    (A) *Regulator bypass test.*
    (B) *Regulator voltage test.*
    (C) *Circuit resistance tests.*
    (D) *None of the above.*

4. Charging system tests are performed using a(n):
    (A) *VOM.*
    (B) *scope.*
    (C) *load tester.*
    (D) *All of the above.*

5. Most voltage regulators operate in this volt range:
    (A) *10–12 volts.*
    (B) *13.5–14.5 volts.*
    (C) *30.2–35.2 volts.*
    (D) *40.5–80.5 volts.*

6. Technician A believes that if charging voltage and current increase to normal levels during a regulator bypass test, the regulator is bad. Technician B feels such results will indicate a bad alternator. Who is right?
    (A) *A only.*
    (B) *B only.*
    (C) *Both A and B.*
    (D) *Neither A nor B.*

7. The percentage of time that current is fed to the alternator's field windings is called the:
   (A) feed time.
   (B) duty cycle.
   (C) circuit term.
   (D) charge ratio.

8. During a charging system voltmeter test, all electrical accessories are *on* when checking:
   (A) base voltage.
   (B) load voltage.
   (C) no-load voltage.
   (D) All of the above.

9. Which part should *not* be soaked in solvent during alternator disassembly?
   (A) Rotor.
   (B) Stator.
   (C) Diode pack.
   (D) All of the above.

10. Which of the following inspections checks rotor windings for internal shorts?
    (A) Rotor current test.
    (B) Rotor winding open circuit test.
    (C) Rotor winding short-to-ground test.
    (D) All of the above.

Activities—Chapter 34

1. Make a poster for classroom display showing the steps involved in checking a charging system with a voltmeter.

2. Demonstrate the use of a stiff wire to hold brushes in place when reassembling an alternator.

Charging System Diagnosis

Condition	Possible cause	Correction
No charge.	1. Alternator drive belt loose or broken. 2. Voltage regulator fusible link blown. 3. Sticking or worn commutator brushes. 4. Loose or corroded connection. 5. Rectifiers open. 6. Charging circuit open. 7. Open circuit in stator winding. 8. Field circuit open. 9. Defective field relay. 10. Defective voltage regulator. 11. Open isolation diode. 12. Open resistor wire. 13. Slipping drive pulley. 14. Oil soaked brushes. 15. Corroded or loose brush connections. 16. Seized bearings.	1. Tighten or replace belt. 2. Install new fusible link. 3. Free or replace brushes. 4. Clean and solder connections. 5. Correct cause and replace rectifiers. 6. Correct as needed. 7. Replace stator. 8. Test and correct as required. 9. Replace relay. 10. Replace voltage regulator. 11. Replace diode. 12. Replace resistor wire. 13. Install new key and tighten. 14. Replace brushes. 15. Clean and tighten connections. 16. Replace bearings. Check shaft for damage.
Low or erratic rate of charge.	1. Loose drive belt. 2. Open stator; grounded or shorted turns in stator windings. 3. High resistance in battery terminals. 4. High resistance in charging circuit. 5. Engine ground strap loose or broken. 6. Loose connections. 7. Defective rectifier. 8. Dirty, burned slip rings. 9. Grounded or shorted turns in rotor. 10. Brushes worn. Brush springs weak.	1. Tighten belt. 2. Replace stator. 3. Clean and tighten terminals. 4. Repair cause of high resistance. 5. Tighten or replace strap. 6. Tighten connections. 7. Replace rectifier. 8. Turn slip rings. 9. Replace rotor. 10. Replace brushes and/or springs.
Excessive rate of charge.	1. Faulty voltage regulator ground. 2. Defective voltage regulator. 3. Alternator field winding grounded. 4. Open rectifier. 5. Loose connections.	1. Ground regulator properly. 2. Replace regulator. 3. Repair grounded field winding. 4. Replace rectifier. 5. Tighten connections.
Noise.	1. Drive belt slipping. 2. Drive pulley loose or misaligned. 3. Mounting bolts loose. 4. Worn bearings. 5. Dry bearing. 6. Open or shorted rectifier. 7. Sprung rotor shaft. 8. Open or shorted stator winding. 9. Alternator fan dragging. 10. Excessive rotor end play. 11. Out-of-round or rough slip rings. 12. Hardened brushes.	1. Tighten belt. 2. Tighten or align pulley. 3. Tighten mounting bolts. 4. Replace bearings. 5. Lubricate or replace as required. 6. Replace rectifier. 7. Install new rotor. 8. Test. Replace stator as needed. 9. Adjust fan clearance. 10. Adjust for correct end play. 11. Turn slip rings. 12. Replace brushes.

(Continued)

Charging System Diagnosis		
Condition	**Possible cause**	**Correction**
Undercharged battery.	1. No charge or low charge rate.	1. See *No charge* or *Low or erratic rate of charge.*
	2. Excessive use of starter.	2. Recharge battery. Advise owner.
	3. Defective battery.	3. Replace battery.
	4. Excessive resistance in charging circuit.	4. Test circuit and remove resistance.
	5. Defective alternator.	5. Rebuild or replace alternator.
	6. Defective regulator.	6. Replace regulator.
	7. Electrical load exceeds alternator rating.	7. Reduce load or install higher capacity alternator.
	8. Electrical draw in system.	8. Test. Remove source of electrical draw.
	9. Excessive starter motor draw.	9. Rebuild or replace starter motor.
Overcharged battery.	1. Excessive resistance in voltage regulator circuit.	1. Clean and tighten connections.
	2. Regulator-alternator ground wire loose or open.	2. Tighten or replace wire.
	3. Defective battery.	3. Replace battery.
	4. Defective regulator.	4. Replace regulator.

Chapter 35

Ignition System Fundamentals

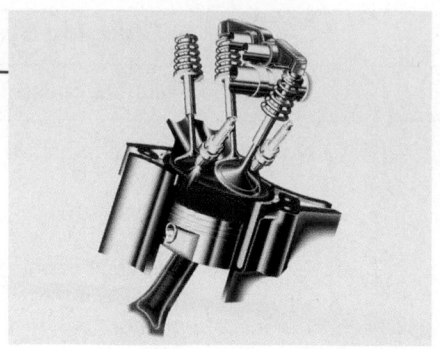

After studying this chapter, you will be able to:

- ▪ Explain the operating principles of an automotive ignition system.
- ▪ Compare contact point, electronic, and computer-controlled ignition systems.
- ▪ Describe the function of major ignition system components.
- ▪ Explain vacuum, centrifugal, and electronic ignition timing advance.
- ▪ Sketch the primary and secondary sections of an ignition system.
- ▪ Compare ignition coil, spark plug, and distributor design variations.
- ▪ Describe the safety practices that must be followed when working with ignition systems.
- ▪ Correctly answer ASE certification test questions that require a knowledge of ignition system fundamentals.

The ignition system of a spark-ignition engine produces the high voltage needed to ignite the fuel charges in the cylinders. The system must create an electric arc across the gaps at the spark plugs. These arcs must be timed so they happen exactly as each piston nears the top of its compression stroke. The heat of the arcs starts combustion and produces the engine's power stroke.

In recent years, different types of ignition systems have been developed to reduce emissions and improve engine performance, fuel economy, and dependability. This chapter explains electronic, computer-coil (distributorless), and direct ignition systems by comparing them to older, simple contact point systems. This should give you a sound knowledge of all automotive ignition system types.

Note

Several other sections of this book discuss information relevant to ignition systems. Chapter 8 summarized basic electrical-electronic components. Chapters 17–19 describe computer system operation, diagnosis, and repair. Refer to these chapters as needed.

Functions of an Ignition System

An ignition system performs the following six functions:

- • Provides a method of turning a spark ignition engine *on* and *off.*
- • Operates on various supply voltages (battery or alternator voltage).
- • Produces high voltage arcs at the spark plug electrodes to start combustion.
- • Distributes high-voltage pulses to each spark plug in the correct sequence.
- • Times the spark so that it occurs as the piston nears TDC on the compression stroke.
- • Varies spark timing as engine speed, load, and other conditions change.

Basic Ignition System

An *ignition system* changes battery voltage to a very high voltage and then sends the high voltage to the spark plugs, **Figure 35-1.** The parts needed to do this include the following:

- • *Battery*—provides power for system.

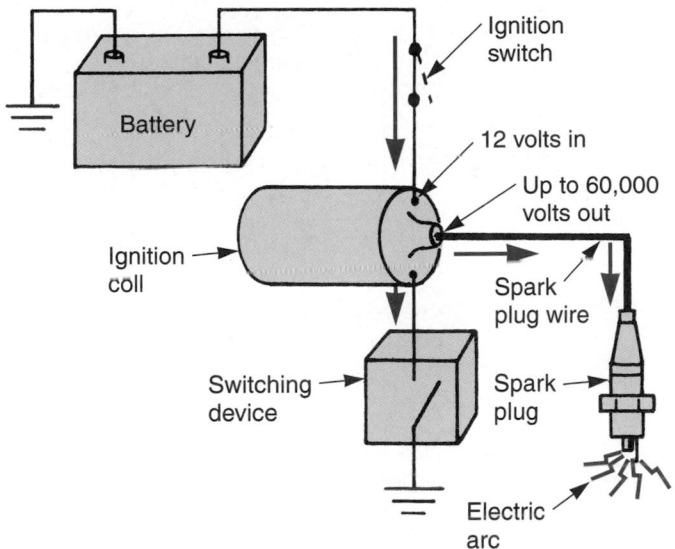

Figure 35-1. Study the operation of a basic ignition system for a one-cylinder engine. Battery voltage is stepped up to about 15,000 volts by the coil before it is sent to the spark plug. The switching device can be a set of breaker points or, more commonly, an electronic switching circuit.

- *Ignition switch*—allows driver to turn ignition and engine on and off.

- *Ignition coil*—changes battery voltage into 30,000 volts or more during normal operation. It has the potential to produce up to 60,000 volts.

- *Switching device*—mechanical or electronic switch that operates the ignition coil.

- *Spark plug*—uses high voltage from the ignition coil to produce an arc, igniting the fuel charge.

- *Ignition system wires*—connect ignition system components.

When the ignition switch is *on* and the switching device is closed (conducting current), current flows through the ignition coil. When the piston is near TDC on its compression stroke, the switching device opens to stop current flow through the ignition coil. This causes the coil to produce a high-voltage surge, which flows through the spark plug wire and arcs across the spark plug's electrodes.

The electric arc, or spark, at the spark plug ignites the fuel mixture. The mixture begins to burn, forming pressure in the cylinder for the engine's power stroke.

When the ignition key is turned to *off*, the battery-to-coil circuit is broken. Without current to the ignition coil, an arc *cannot* be produced at the spark plugs and the engine stops running.

An actual ignition system is much more complex than the one just discussed. Vehicles have multiple-cylinder engines, and the timing of the spark must vary with operating conditions.

Ignition System Supply Voltage

The *ignition system supply voltage* is fed to the ignition system by the battery and alternator. On late-model vehicles, these sources feed voltage through the engine control module or the power train control module. The battery provides electricity when starting the engine. After the engine is running, the alternator supplies a slightly higher voltage to the battery and ignition system.

Ignition Switch

The *ignition switch* is a key-operated switch in the driver's compartment. A hot wire (voltage supply wire) connects the switch to the battery. Other terminals on the switch are connected to the ignition system, starter solenoid, and other electrical devices.

Bypass and Resistance Circuits

An ignition system *bypass circuit* is sometimes used to supply direct battery voltage to the ignition system during starting motor operation. When an engine is being started, the ignition switch is held in the start position (fully clockwise). This connects the battery to the starting motor and to the ignition system, **Figure 35-2A.** The electric motor spins the engine until the engine begins to run.

The starting motor draws high current and causes battery voltage to drop below 12.6 volts. The bypass circuit ensures that there is still enough voltage and current for proper ignition coil operation and easy engine starting.

To protect the ignition system from damage, a *resistance circuit* is sometimes placed between the ignition switch and the coil to limit supply voltage to the ignition system during alternator operation. Look at **Figure 35-2B.** After the engine starts, the ignition key switch is released. A spring inside the switch causes it to return to the run position.

Either a special *resistance wire* (wire having internal resistance) or a *ballast resistor* (heat sensitive resistor that can regulate voltage to the ignition coil) is used in the resistance circuit. This circuit ensures that a relatively steady voltage of about 9.5–10.5 volts is applied to the ignition system. With new vehicles, the *computer voltage regulator* supplies current to the ignition circuit.

 Tech Tip
Many electronic and computer controlled ignition systems do *not* use bypass or resistance circuits. Their internal circuitry is capable of controlling voltage and current for optimum ignition system operation.

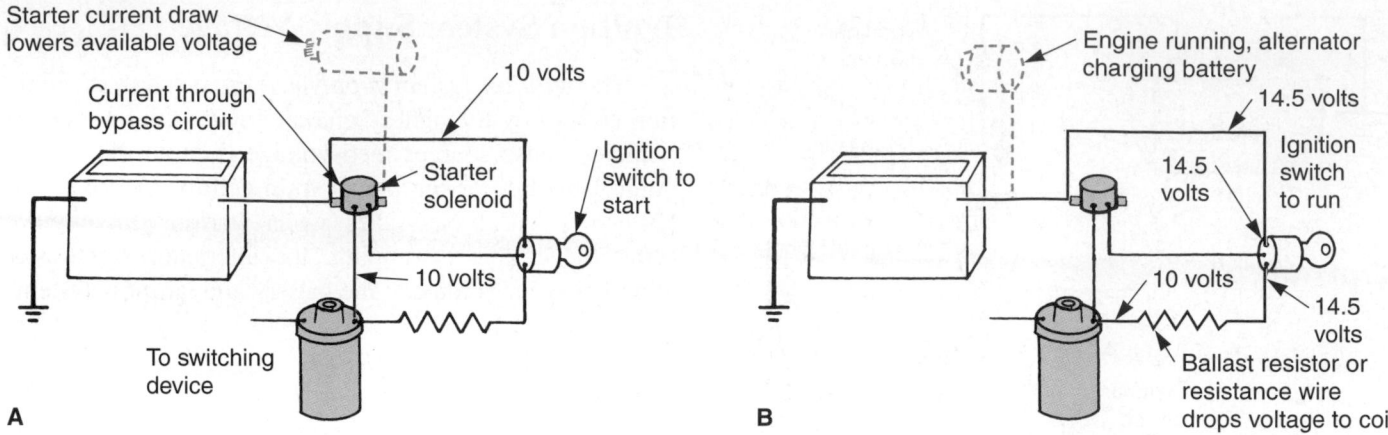

Figure 35-2. Some ignition systems use bypass and resistance circuits to feed current to the ignition coil. A—When cranking, the bypass circuit feeds direct battery voltage to the coil. B—After starting, the resistance circuit feeds controlled voltage to the coil. (Echlin)

Primary and Secondary Circuits

There are two main sections, or circuits, of an ignition system: the primary circuits and secondary circuits. The *primary circuit* of the ignition system includes all the components and wires operating on low voltage (battery or alternator voltage). *Ignition system primary voltage* is normally below battery voltage (12.6 volts). See **Figure 35-3A.**

The *secondary circuit* of the ignition system is the high-voltage section. The *ignition system secondary operating voltage* can normally range from 4000–30,000 volts, depending upon system design. The secondary consists of the wires and parts between the coil output and the spark plug ground, **Figure 35-3B.**

The primary circuit of the ignition system uses conventional wire, similar to the wire used in the other electrical systems of the car. The secondary wiring, however, must have much *thicker insulation* to prevent *electrical leakage* (electricity traveling through insulation or from one component to another).

Ignition Coil

An *ignition coil* is a pulse-type transformer that is capable of producing short bursts of high voltage to start combustion. The coil produces a low current, high voltage surge that jumps the rotor gap, passes through the secondary resistance wire, and jumps the spark plug gap.

Older vehicles with relatively small spark plug gaps operate at about 4000–8000 volts (4–8 kV). With today's wider spark plug gaps and leaner fuel mixtures, more voltage is needed to make the electricity surge through all the parts of the ignition system. Late-model vehicles operate at about 15,000 volts (15 kV) on average but can

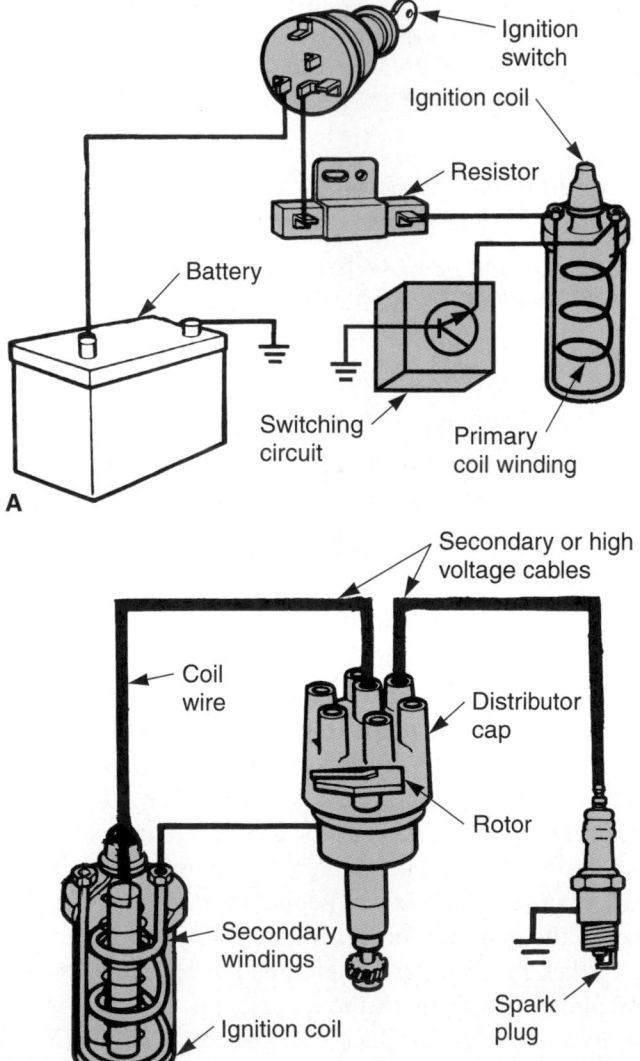

Figure 35-3. Two major sections of an ignition system. A—Primary circuit includes all the parts working on battery voltage. B—Secondary circuit consists of all the parts carrying coil output voltage (high voltage). (Mopar)

vary from 4000–30,000 volts, depending on ignition system design.

If everything is in good condition, the ignition coil operates below its potential maximum output voltage. This keeps some voltage in reserve. In most late-model systems, the *ignition coil's open-circuit voltage* (maximum voltage potential) is 40,000–60,000 volts (40–60 kV).

Ignition Coil Windings

Shown in **Figure 35-4,** the ignition coil consists of two sets of *windings* (insulated wire wrapped in circular pattern). It also contains two *primary terminals* (low-voltage connections), an *iron core* (long piece of iron inside windings), and a *high-voltage terminal* (output or coil wire connection).

The *primary windings* are made up of several hundred turns of heavy wire wrapped around or near the secondary windings. The *secondary windings* consist of several thousand turns of very fine wire located inside or near the primary windings, **Figure 35-5.** Both windings are wrapped around an iron core and are housed inside the coil case.

Ignition Coil Operation

When battery current flows through primary windings of the ignition coil, a strong magnetic field is generated. The action of the iron core helps concentrate and strengthen the field. Look at **Figure 35-6A.**

When the current flowing through the coil is interrupted, the magnetic field *collapses* across the secondary windings. Since the secondary windings have more turns

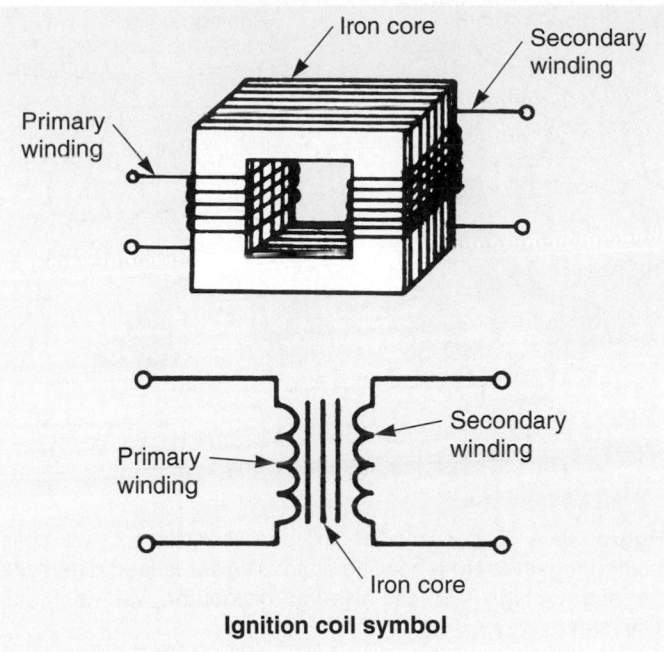

Figure 35-5. Ignition coil is a pulse-type step-up transformer. Note how the schematic symbol for a coil relates to its construction.

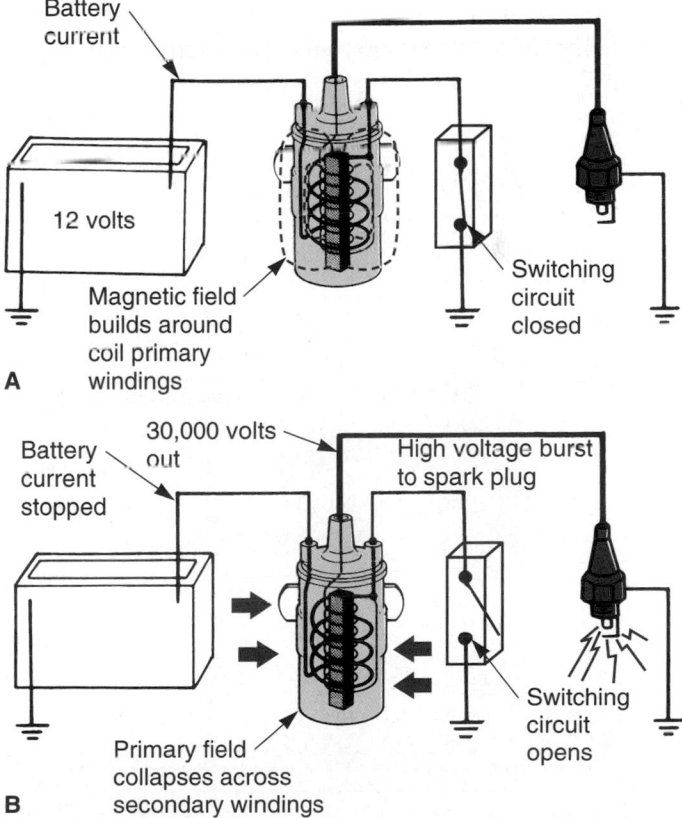

Figure 35-6. Ignition coil operation. A—With the switching device (points or electronic circuit) closed, current flows through the ignition coil primary windings. A strong magnetic field builds in the coil. B—When the switching device opens, current flow stops and the magnetic field collapses across the secondary windings. This induces high voltage in the secondary windings of the coil. The spark plug fires. (Saab)

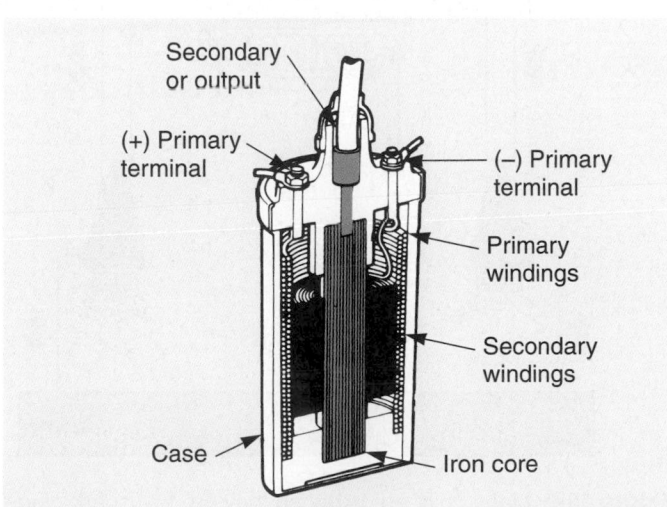

Figure 35-4. Cutaway of an ignition coil shows the basic parts. The primary windings surround the secondary windings. An iron core is mounted in the center of the windings. (Dodge)

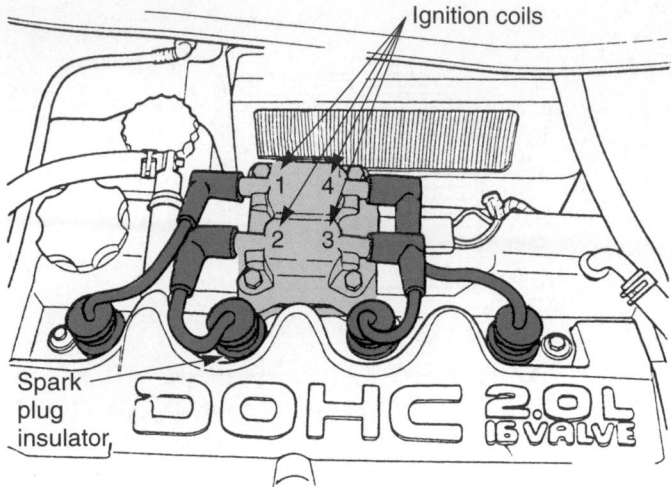

Ignition coils

Spark plug insulator

Figure 35-7. This particular coil pack consists of two coils molded together. Note how the spark plug wires feed directly to the spark plugs. No coil wire or distributor wire is used. (Chrysler)

than the primary windings, up to 60,000 volts is induced in the secondary windings. High voltage shoots out of the high-voltage terminal of the coil, traveling through the secondary circuit to a spark plug. See **Figure 35-6B.**

There are two methods used to break current flow and fire the coil: older mechanical breaker points or more common electronic switching circuit.

Ignition Coil Designs

With today's computer-controlled ignition systems, ignition coils can be more complex than those used in the past. Depending on ignition system design, an engine can have one or more ignition coils. A conventional system has one coil that serves all engine cylinders. A distributorless ignition system has one coil for every two cylinders. A direct ignition system has one coil for every engine cylinder.

A *coil pack* consists of several ignition coils combined into one assembly, **Figure 35-7.** As you will learn, this type of coil is used in distributorless ignition systems. No coil wire is needed with a coil pack. The spark plug wires are connected from the ignition coil pack to the spark plugs.

Figure 35-8 shows a top view of a coil pack. Note how the secondary connections are numbered so you can connect the right terminal to the correct spark plug. Also note the primary connections.

Wasted-Spark Ignition Coil

A *wasted-spark ignition coil* is wired so that it fires two spark plugs at the same time. As illustrated in **Figure 35-9,** each end of the coil's secondary winding is

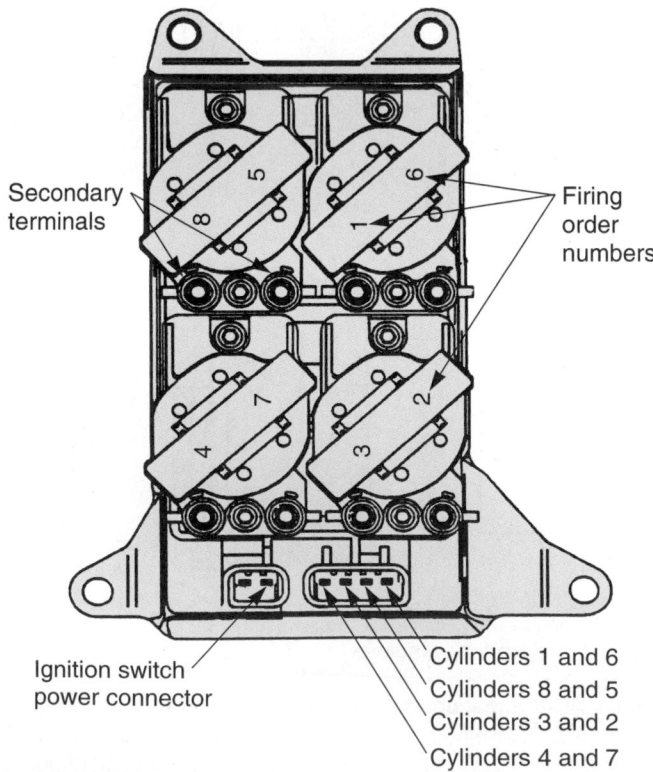

Secondary terminals

Firing order numbers

Ignition switch power connector

Cylinders 1 and 6
Cylinders 8 and 5
Cylinders 3 and 2
Cylinders 4 and 7

Figure 35-8. Top view of a coil pack shows the numbers for the spark plug wire connections and the pins for the low-voltage wire connections. (Chevrolet)

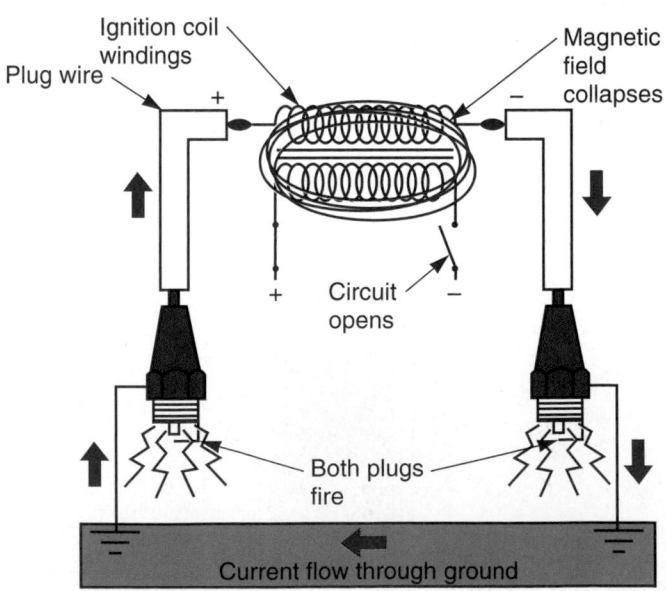

Ignition coil windings
Plug wire
Magnetic field collapses
Circuit opens
Both plugs fire
Current flow through ground

Figure 35-9. Many coils are wired so they fire two spark plugs simultaneously. One spark plug starts combustion on the power stroke. The other spark plug produces a wasted spark, which occurs during the engine exhaust stroke. Note current flow through the coil, plugs, and ground.

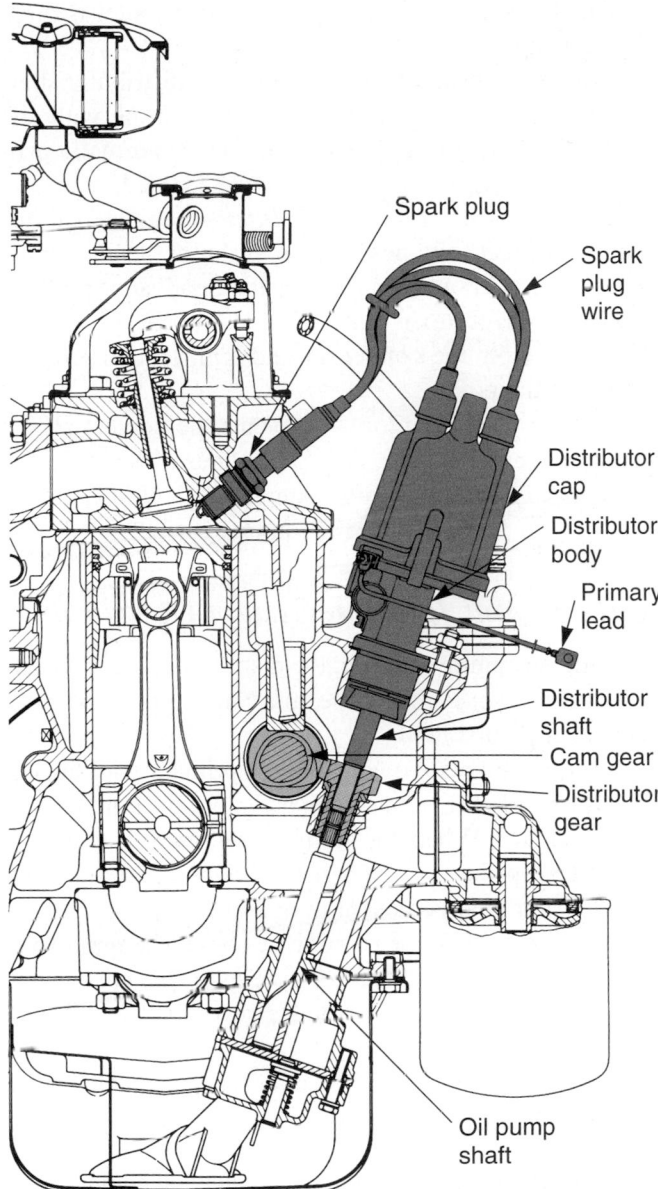

Figure 35-10. The ignition distributor is usually driven by the engine camshaft. A small gear on the cam drives the gear on the distributor at one-half engine rpm. The main purpose of the distributor is to feed coil voltage to the spark plugs. (Fiat)

connected to a spark plug wire. Then, when the primary magnetic field collapses across the secondary winding, a high-voltage, low-current surge is induced in both spark plug wires. Current flows through each spark plug gap with the opposite polarity. The engine ground connects the two spark plugs. Wasted-spark coils and distributorless ignition systems will be discussed in more detail later in this chapter.

Ignition Distributors

Typically, an *ignition distributor*, **Figure 35-10**, has several functions:

- It actuates the *on/off* cycles of current flow through the ignition coil primary windings.

- It distributes the coil's high-voltage pulses to the spark plug wire.

- It causes the spark to occur at each plug earlier in the compression stroke as engine speed increases and vice versa.

- It changes spark timing with changes in engine load. As more load is placed on the engine, the spark timing must occur later in the compression stroke to prevent spark knock (abnormal combustion).

- Sometimes, the bottom of the distributor shaft powers the engine oil pump.

- Some distributors (unitized distributors) house the ignition coil and electronic switching circuit in one assembly. Refer to **Figure 35-11.**

Distributor Types

An ignition distributor can be a:

- Contact point distributor (older, mechanical type).

- Pickup coil distributor (magnetic-sensing coil or winding type).

- Hall-effect distributor (magnetic-sensing, solid-state chip type).

- Optical distributor (LED and light-sensing type).

A contact point distributor is found on older cars. The pickup coil–type distributor is used on many modern automobiles and is the most common. Hall-effect and optical distributors can be found on a few makes of vehicles.

Distributor Drives

The *distributor drive* refers to the method used to rotate the distributor shaft to match engine rpm. On some engines, the *camshaft gear* is used to drive the distributor at one-half engine speed. A gear on the bottom of the distributor shaft meshes with the camshaft gear. Anytime the engine camshaft turns, the distributor shaft also turns. See **Figure 35-10.** With this drive arrangement, the distributor can be mounted on the rear of the engine, directly behind the camshaft centerline, or on the front of the engine, under the water pump. Both locations allow *direct distributor drive* off the camshaft. This eliminates the problem of backlash, or play, in a distributor drive gear, which can affect ignition timing.

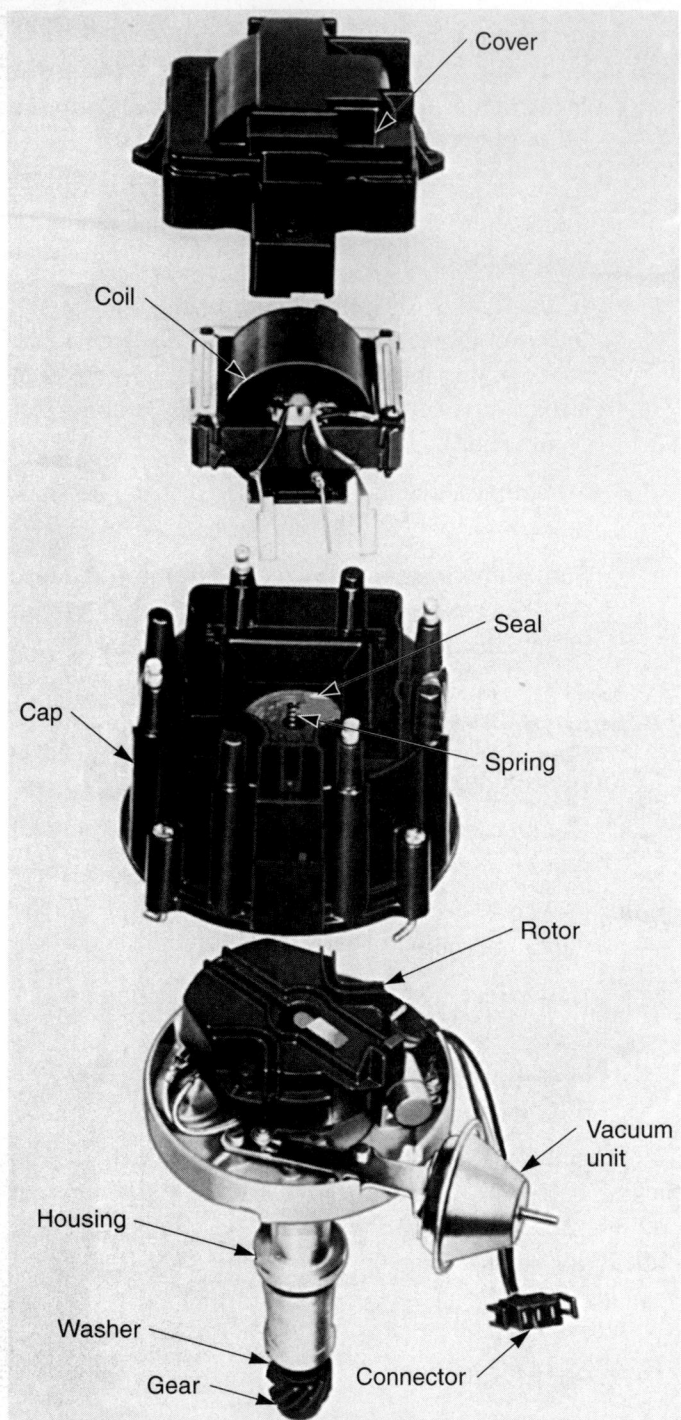

Cover

Coil

Seal

Cap

Spring

Rotor

Vacuum unit

Housing

Washer

Gear

Connector

Figure 35-11. Unitized distributor has an ignition coil and an ignition control module (electronic switching circuit) mounted inside its housing. Note the part names and locations. (Chevrolet)

Contact Point Ignition System

Before studying today's electronic ignition systems, you should have a basic understanding of contact point systems. The two systems are similar in many ways, **Figure 35-12.** Also, a contact point system is easier to understand.

The major parts of a contact point distributor include the following:

Distributor cam—Lobed part on the distributor shaft that opens the contact points. The cam turns with the shaft at one-half engine speed. One lobe is normally provided for each spark plug. See **Figure 35-12.**

Contact points (breaker points)—Act like spring-loaded electrical switches in the distributor. Small screws hold the contact points on the distributor advance plate. A rubbing block of fiber material rides on the distributor cam. Wires from the condenser and ignition coil primary connect to the points.

Condenser (capacitor)—Prevents the contact points from arcing and burning. It also provides a storage place for electricity as the points open. This electricity is fed back into the primary when the points close.

Contact Point Ignition System Operation

With the engine running, the distributor shaft and distributor cam rotate. This causes the cam to open and close the contact points.

Since the points are wired to the primary windings of the ignition coil, the points open and close the ignition coil primary circuit. When the points are closed, a magnetic field builds in the coil. When the points open, the field collapses and voltage is sent to one of the spark plugs.

With the distributor rotating at one-half engine rpm, and with one cam lobe per engine cylinder, each spark plug fires once during a complete revolution of the distributor camshaft.

As you will learn, other distributor and electronic devices are used to determine when the spark plugs fire.

Point Dwell (Cam Angle)

Point dwell, or *cam angle,* is the amount of time, given in degrees of distributor rotation, that the points remain closed between each opening, **Figure 35-13.** A dwell period is needed to ensure that the coil has enough time to build up a strong magnetic field.

Without enough point dwell, a weak spark would be produced. With too much dwell, the *point gap* (distance between fully open points) would be too narrow. Point arcing and burning could result.

Discussed in the next chapter, modern systems use several methods of controlling dwell electronically. Computer-controlled systems can use fixed dwell, variable dwell, and current-limiting dwell.

Electronic Ignition System

An *electronic ignition system,* also called a *solid state* or *transistor ignition system,* uses an ignition control

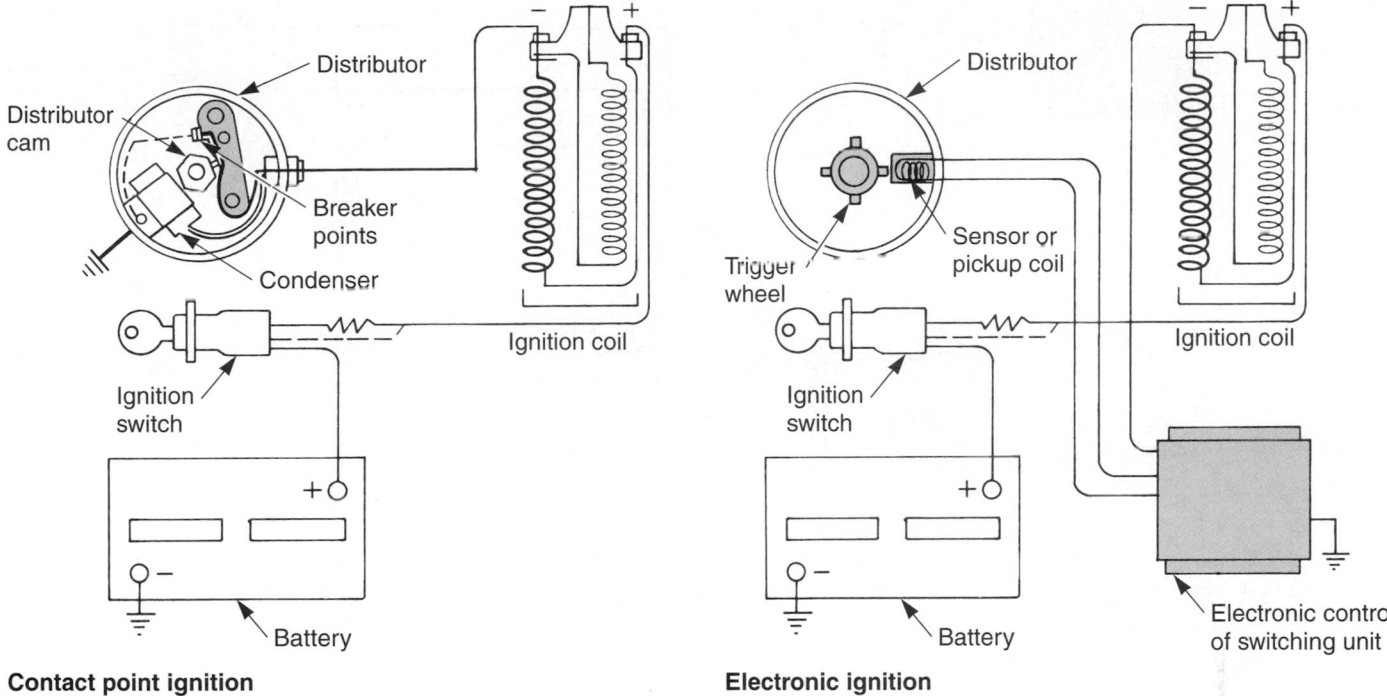

Contact point ignition

Electronic ignition

Figure 35-12. Compare a contact point ignition system and an electronic ignition system. Note that the pickup coil and the control unit replace the contact points in a modern system. (Deere & Co.)

module and a distributor pickup coil to operate the ignition coil. The pickup coil and the module replace the contact points, **Figure 35-14.**

An electronic ignition system is more dependable than a contact point system. There are no mechanical breakers to wear or burn. This helps prevent trouble with ignition timing and dwell.

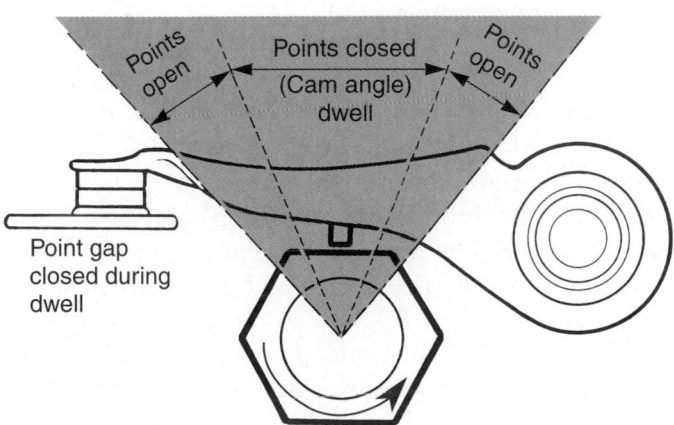

Figure 35-13. Dwell is the amount of time the points remain closed in degrees of distributor rotation. Point gap is the distance between the two points in the fully open position. Dwell affects point gap and vice versa. With modern electronic systems, the engine control module controls dwell time. (Echlin)

An electronic ignition is also capable of producing much higher secondary voltages than a contact point ignition. This is an advantage because wider spark plug gaps and higher voltages are needed to ignite the lean air-fuel mixtures now used for reduced exhaust emissions and fuel consumption.

Trigger Wheel

The **trigger wheel,** also called the **reluctor** or **pole piece,** is fastened to the upper end of the distributor shaft. See **Figure 35-15.** The trigger wheel replaces the distributor cam used in a contact point distributor. One tooth is normally provided on the wheel for each engine cylinder.

Pickup Coil

The **pickup coil,** also termed **sensor assembly** or **sensor coil,** produces tiny voltage pulses that are sent to the ignition control module. Look at **Figure 35-15.** The pickup coil is a small set of windings that form a coil.

As a trigger wheel tooth passes the pickup coil, it strengthens the magnetic field around the coil. This causes a change in the current flow through the coil. As a result, an electrical pulse (voltage or current change) is sent to the ignition control module as the trigger wheel teeth pass the pickup unit.

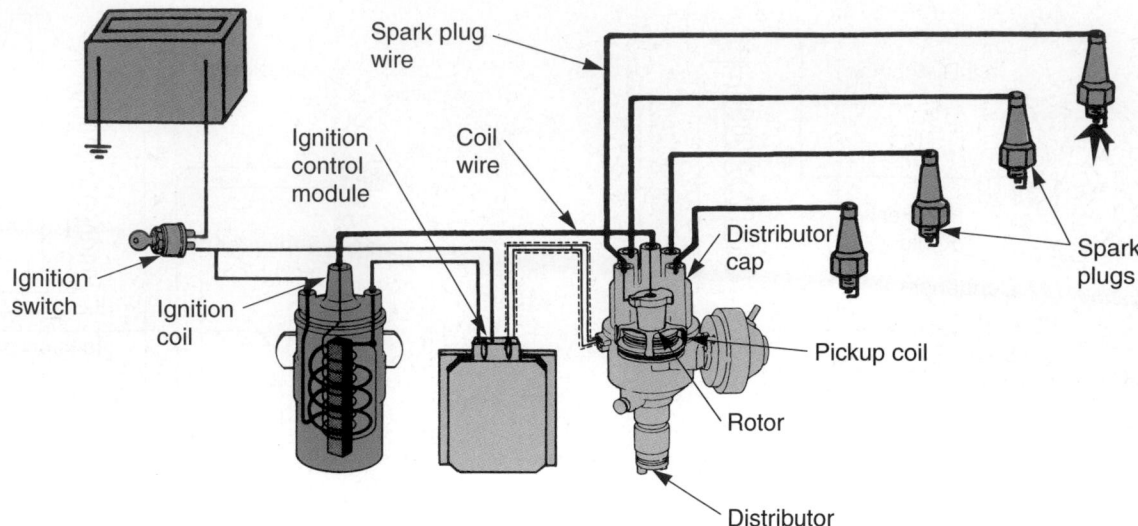

Figure 35-14. This is a simplified electronic ignition system. Note how the trigger wheel and pickup coil feed a signal to the ignition control module. The ignition control module can then alter and increase signal strength to turn the ignition coil on and off as needed.

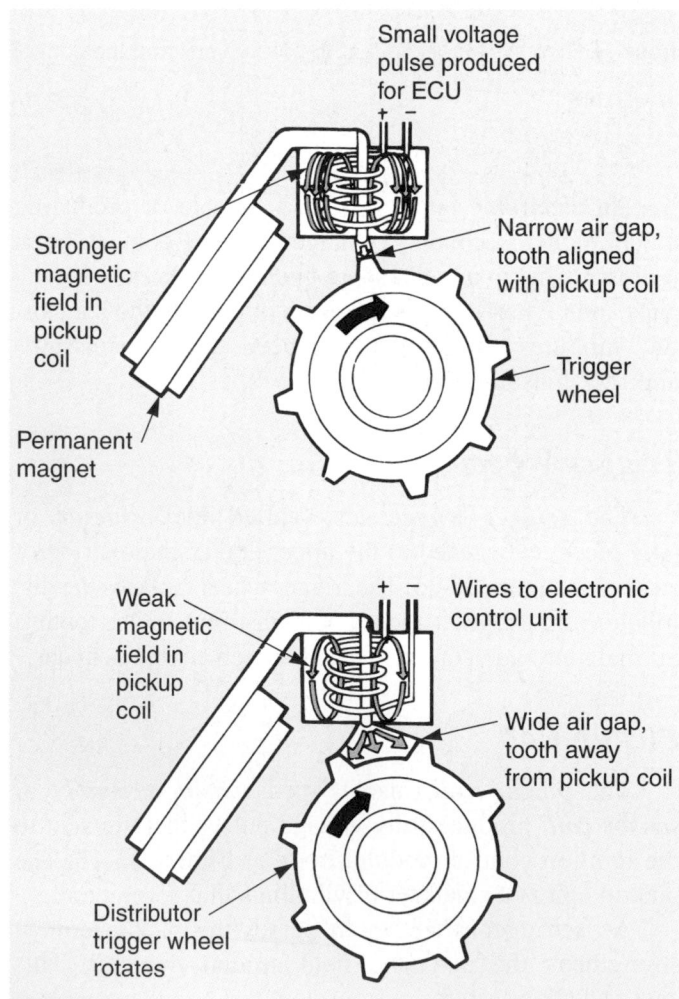

Figure 35-15. Study magnetic pickup coil operation. This principle applies to both distributor pickups and other magnetic coil–type sensors. (Chrysler)

Hall-Effect Pickup

A *Hall-effect pickup* is a solid-state chip or module that produces an electrical signal when triggered by a slotted wheel. A constant amount of current is sent through the device. A permanent magnet is located next to the Hall-effect chip. See **Figure 35-16.**

When the slotted wheel passes between the permanent magnet and the Hall-effect chip, the magnetic field is blocked, decreasing the chip's output voltage (sensor or switch *off*). When the slotted wheel's tooth moves out from between the magnet and chip, magnetic field action increases the chip's voltage output (sensor or switch *on*). This *on/off* action operates the ignition control module.

Optical Pickup

An *optical pickup* uses *LEDs* (light-emitting diodes) and *photo diodes* (light sensors) to produce an engine speed signal for the ignition system. As shown in **Figure 35-17,** a slotted rotor plate rotates between the light-emitting diodes and the photo diodes. When a slot, or window, passes between the two diodes, light from the LEDs strikes the photo diodes and an electrical signal is generated.

An optical pickup is seldom used because its operation is adversely affected by a dirt buildup on the LEDs and photo diodes.

Ignition Control Module

The *ignition control module,* or *electronic control unit amplifier,* is an "electronic switch" that turns the

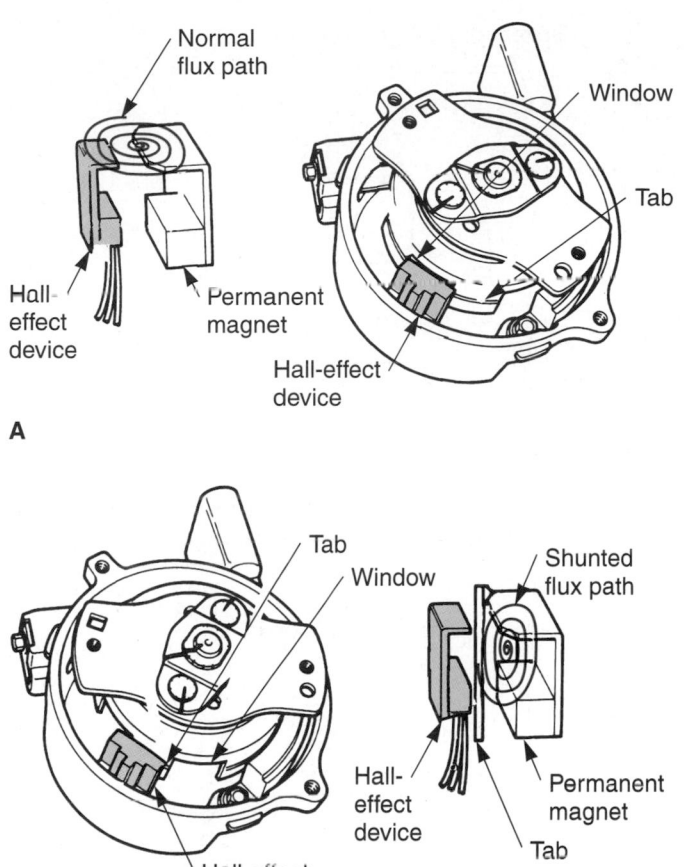

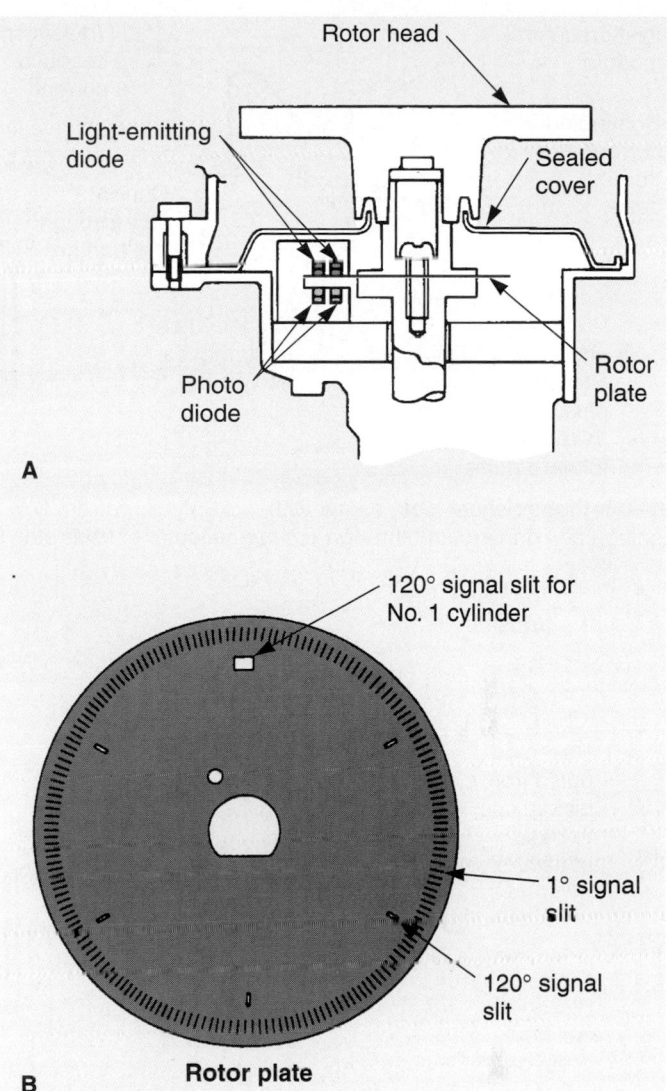

Figure 35-16. A Hall-effect pickup chip is similar to a magnetic pickup. A—The trigger wheel window (opening) allows a strong magnetic field to develop around the pickup. B—As the trigger wheel rotates, a tab or tooth moves between the pickup and a permanent magnet. This decreases pickup field strength and voltage. (Ford)

Figure 35-17. A—This distributor uses an optical pickup unit to sense engine speed. Light-emitting diodes shine light on photo diodes. When the rotor plate slits pass by the diodes, a speed signal is produced. B—Note the arrangement of the slits on the rotor plate. (Nissan)

ignition coil primary current on and off. The module does the same thing as contact points. See **Figure 35-18.**

An ignition control module is a network of transistors, resistors, capacitors, and other electronic components. The circuit is sealed in a plastic or metal housing. A typical module and related wiring is shown in **Figure 35-19.** The ignition control module can be located in the engine compartment, on the side of the distributor, inside the distributor, or under the vehicle's dash. See **Figure 35-20.**

Electronic Ignition System Operation

With the engine running, the trigger wheel spins inside the distributor. As the teeth pass the pickup, a change in the magnetic field causes a change in output voltage or current. This output voltage, which represents engine rpm, is sent to the ignition control module, **Figure 35-19.**

The ignition control module increases these tiny pulses into *on/off* current cycles for the ignition coil.

When the module is *on*, current flows through the primary windings of the ignition coil, developing a magnetic field. Then, when the trigger wheel and pickup turn *off* the module, the ignition coil field collapses and fires a spark plug.

Dwell time (number of degrees of camshaft rotation that the circuit conducts current to ignition coil) is designed into the ignition control module's electronic circuit. It is not adjustable.

Distributor Cap and Rotor

The *distributor cap* is an insulating plastic component that fits over the top of the distributor housing. Its

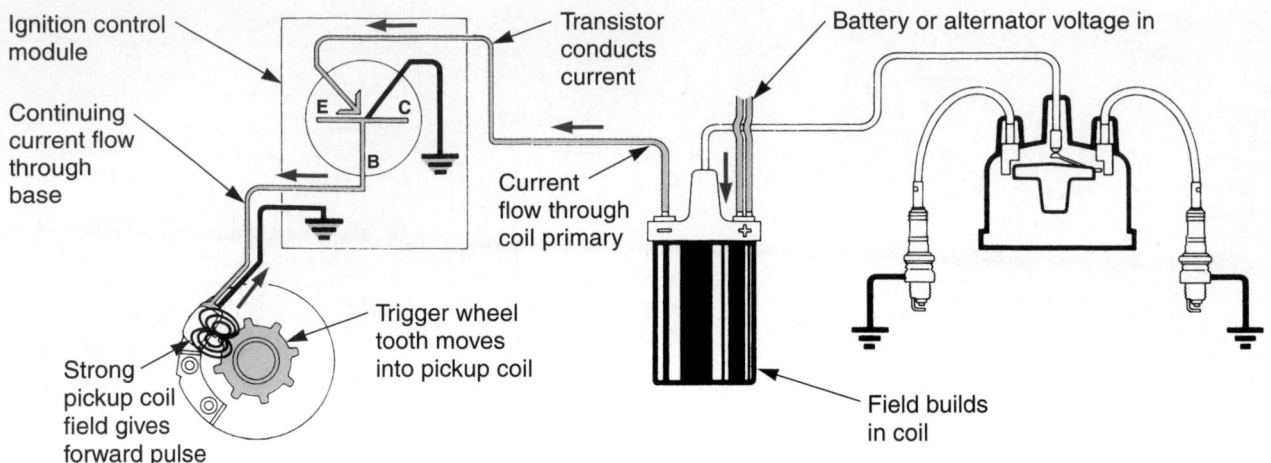

A—As trigger wheel tooth aligns with pickup coil, current flow through base of transistor turns transistor on. Current flows through ignition coil primary and through emitter-collector of transistor. Strong field builds in ignition coil primary windings.

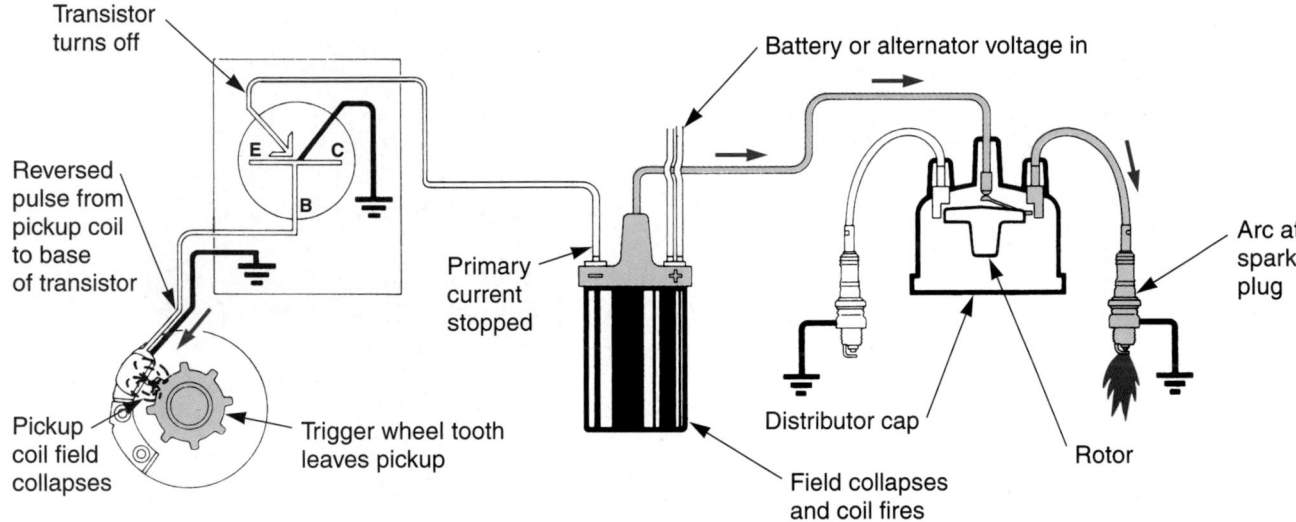

B—Just as trigger wheel passes pickup coil, current pulse flows out of pickup and to base of transistor. Electrical pulse is opposite the polarity of emitter-base voltage. This turns transistor off. Without current flow through emitter and collector, field collapses in ignition coil and 30,000 volts is induced into coil secondary windings. Spark plug fires.

Figure 35-18. Study this simplified illustration of how the pickup coil and ignition control module operate the ignition coil. (Echlin)

center terminal transfers voltage from the coil wire to the rotor, **Figure 35-21.**

The distributor cap also has outer terminals, or side terminals, that transfer electric arcs from the rotor to the spark plug wires. Metal terminals are molded into the plastic cap to make the electrical connections.

The *rotor* transfers voltage from the distributor cap center terminal (the coil wire) to the distributor cap outer terminals (spark plug wires). Look at **Figure 35-21.**

The rotor is mounted on top of the distributor shaft. It is a spinning electrical switch that feeds voltage to each spark plug wire in turn.

A metal terminal on the rotor touches the center terminal of the distributor cap. The outer end of the rotor terminal *almost* touches the outer cap terminals.

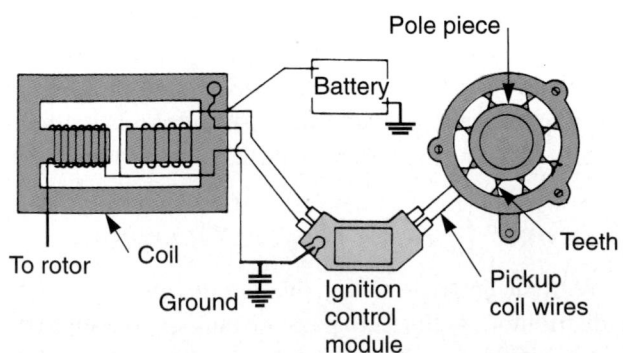

Figure 35-19. Note the electrical connections to a typical ignition control module. The module contains an electronic circuit, with several transistors and other miniaturized electronic devices. (Echlin)

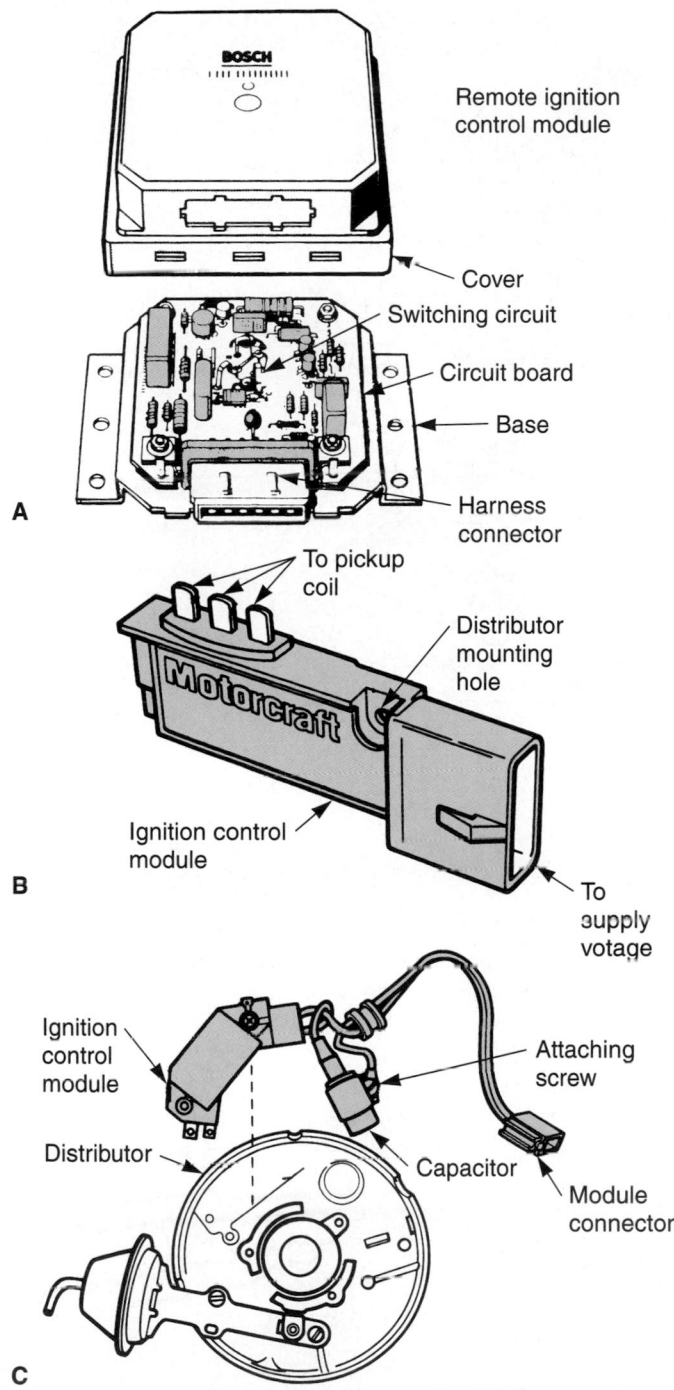

A

B

To pickup coil

Distributor mounting hole

Ignition control module

To supply votage

Ignition control module

Distributor

Capacitor

Attaching screw

Module connector

C

Figure 35-20. Study ignition control unit variations. A—Remote-mounted ignition control module. B—Ignition control module mounted on the outside of the distributor body. C—Ignition control module mounted inside the distributor. (Bosch, Ford, Mopar)

Remote ignition control module

Cover
Switching circuit
Circuit board
Base
Harness connector

Voltage is high enough that it can jump the air space between the rotor and cap. About 2000–6000 volts is needed for the spark to jump the rotor-to-cap gap in today's ignition systems.

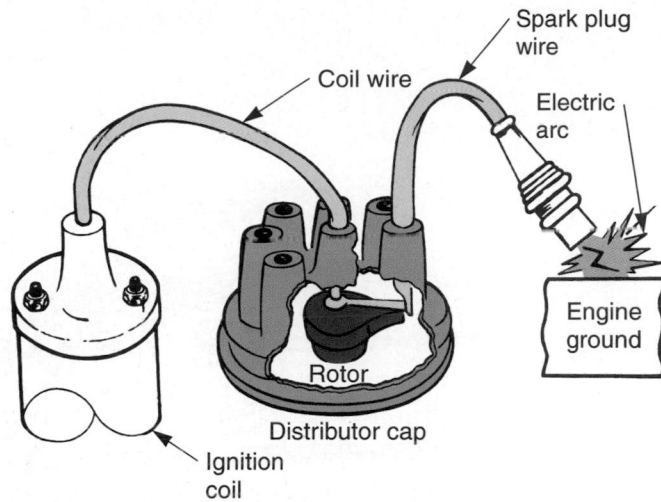

Figure 35-21. With a distributor, ignition coil output is fed through the coil wire to the distributor cap center terminal. The spinning rotor then feeds high voltage to each spark plug wire. (Echlin)

Coil wire
Spark plug wire
Electric arc
Engine ground
Rotor
Distributor cap
Ignition coil

Secondary Wires

Secondary wires carry the high voltage produced by the ignition coil. The *coil wire* carries voltage from the high voltage (high tension) terminal of the ignition coil to the distributor cap, **Figure 35-21.** *Spark plug wires* carry coil voltage from the side terminals of the distributor cap to the spark plugs. See **Figure 35-22.**

 Tech Tip
With a unitized distributor or distributorless ignition, a coil wire is *not* needed. Secondary wires carry high voltage from the ignition coil directly to the spark plugs. With a direct ignition system, explained shortly, no secondary wires are needed.

Solid secondary wires are used on racing engines and very old automobiles. The wire conductor is simply a stranded metal wire. Solid secondary wires are no longer used because they cause radio interference (noise or static in speakers).

Resistance wires are now used because they contain internal resistance, which helps prevent radio noise. They consist of carbon-impregnated strands and rayon braids. Look at **Figure 35-23.** Also called *radio suppression wires,* resistance wires have about 10,000 ohms of resistance per foot.

Insulated *boots* usually fit over both ends of the secondary wires to protect the metal connectors from corrosion, oil, and moisture.

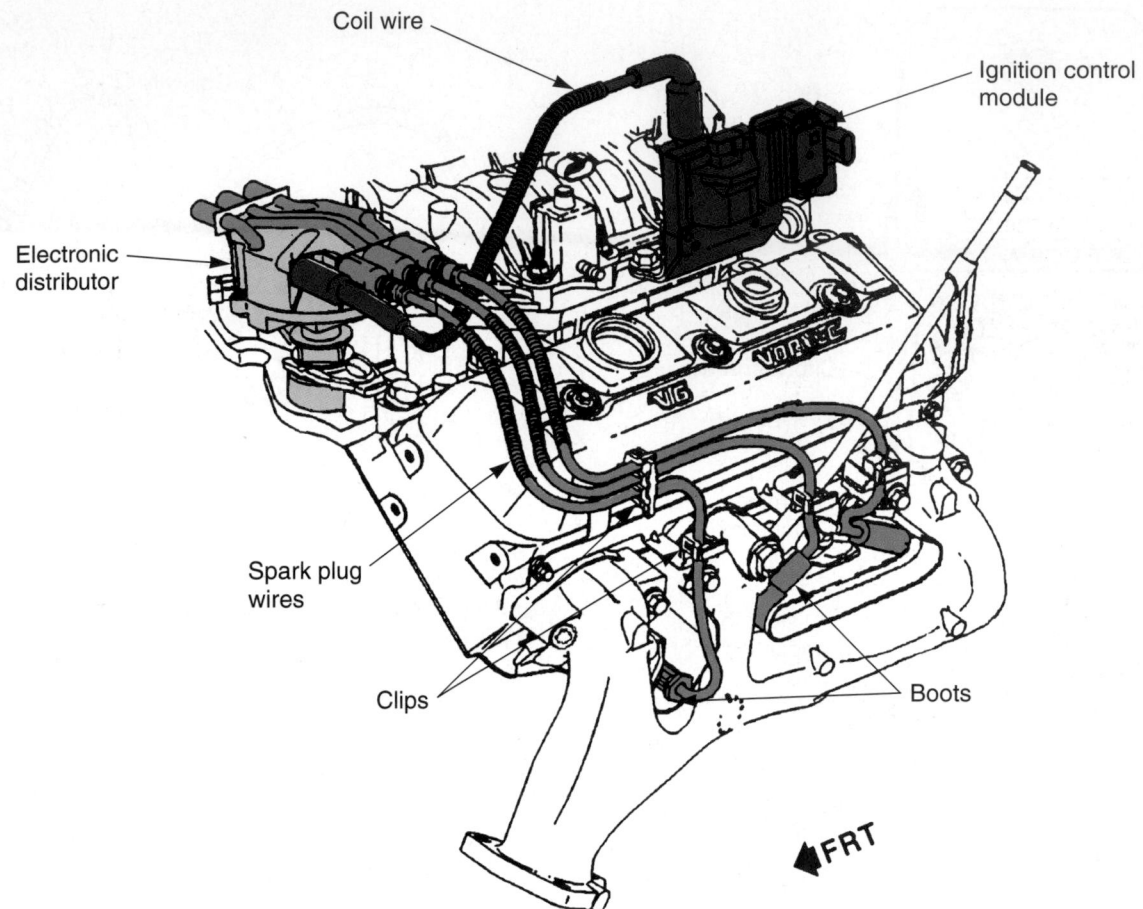

Figure 35-22. Secondary wires transfer high voltage between the ignition system components. Clips secure the wires so they do not get burned by hot exhaust manifolds or induce voltage into the computer system wires. (Chevrolet)

Spark Plugs

The *spark plugs* use ignition coil high voltage to ignite the fuel mixture. Somewhere between 4000 and 28,000 volts is needed to make current jump the gap at the spark plug electrodes. This is much lower than the coil's output voltage potential discussed earlier.

As shown in **Figure 35-24,** the basic parts of a spark plug include the:

- *Center terminal (center electrode)*—Conducts electricity into the combustion chamber.

- *Grounded side electrode*—Causes the electricity to jump the gap and return to the battery through frame ground.

- *Ceramic Insulator*—Keeps the high voltage at the plug wire from shorting to ground before producing a spark in the engine cylinder.

- *Steel shell*—Supports the other parts of the plug and has threads for screwing the plug into the engine cylinder head.

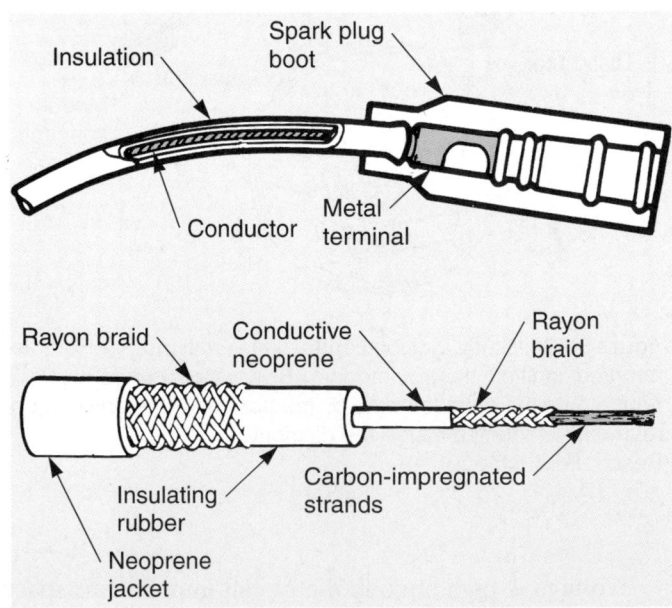

Figure 35-23. Secondary wire has very thick insulation. The secondary wire used on late-model vehicles contains carbon-impregnated strands that increase the resistance of the wire, preventing radio interference. (Champion Spark Plugs)

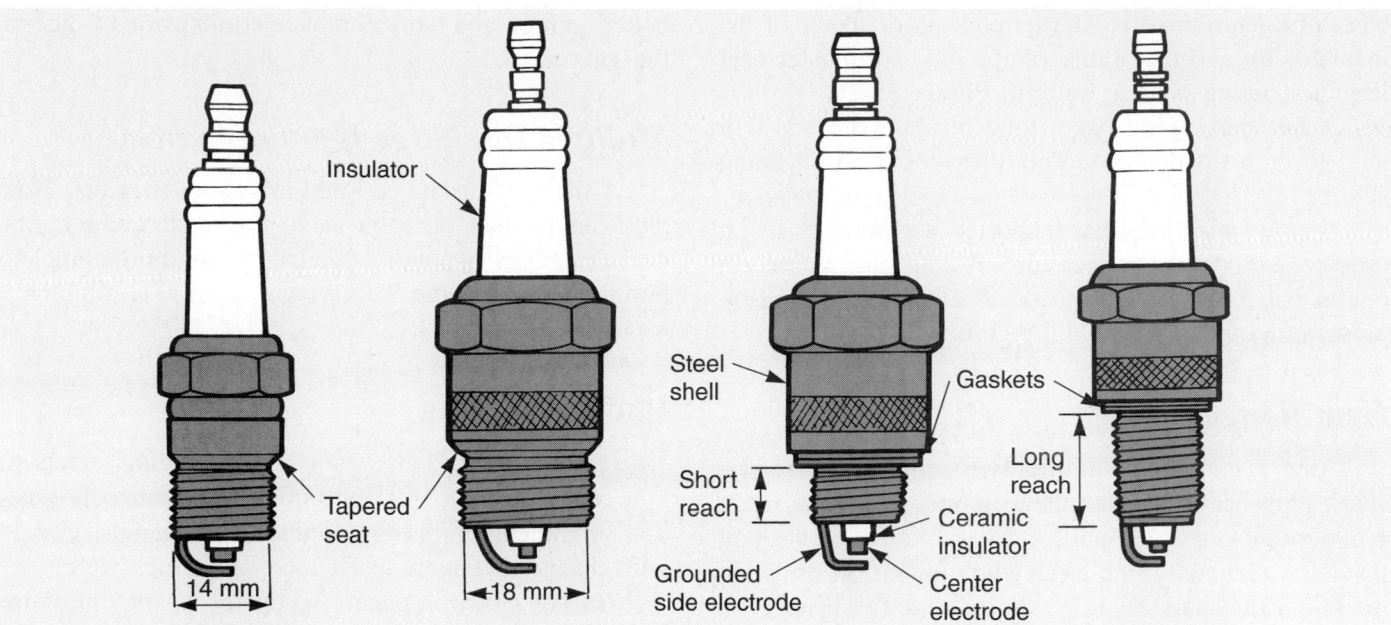

Figure 35-24. Note spark plug variations. Small 14 mm plugs are commonly used in today's engines. Larger 18 mm plugs are for older engines. Spark plug reach is the length of the plug threads. (Mopar)

Spark Plug Reach

Spark plug reach is the distance between the end of the plug threads and the seat or sealing surface on the plug shell. Refer to **Figure 35-24.** Plug reach determines how far the plug extends through the cylinder head.

If spark plug reach is too long, the plug electrode may be struck by the piston at TDC. If reach is too short, the plug electrodes may not extend far enough into the chamber and combustion efficiency may be reduced.

Resistor and Non-Resistor Spark Plugs

A *resistor spark plug,* like a resistor plug wire, has internal resistance (around 10,000 ohms) designed to reduce static in radios and television sets. Most newer vehicles require resistor plugs.

A *non-resistor spark plug* has a solid metal rod forming the center electrode. This type of plug is *not* commonly used, but it is found in some racing and off-road applications. **Figure 35-25** illustrates both resistor and non-resistor spark plugs.

Spark Plug Gap

Spark plug gap is the distance between the center and side electrodes, **Figure 35-25.** Normal gap specifications range from 0.030″–0.080″ (0.76 mm–2.0 mm).

Smaller spark plug gaps are used on older cars. Larger spark plug gaps are now used with modern electronic and computerized ignition systems.

Spark Plug Heat Range

Spark plug heat range is a rating of the operating temperature of the spark plug tip. Plug heat range is

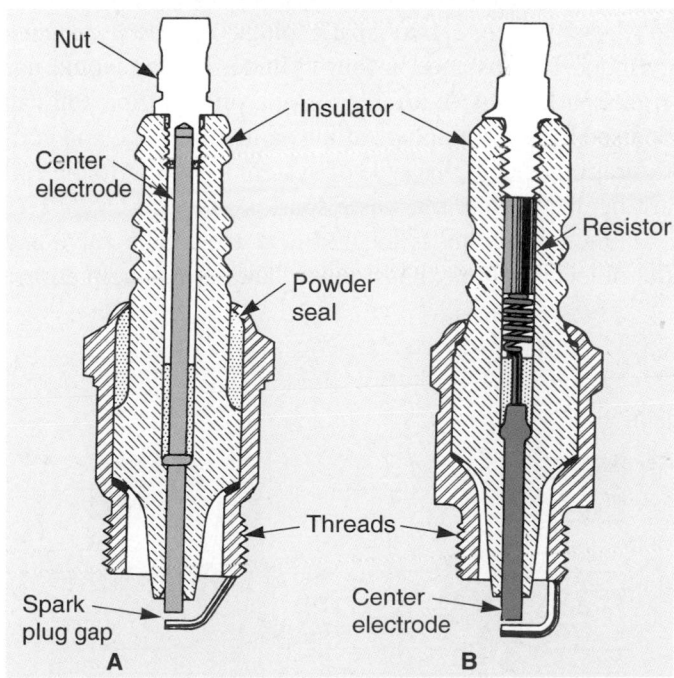

Figure 35-25. Cutaway shows the internal parts of a plug. A—Non-resistor plug has a solid metal center electrode. B—Resistor plug has a small resistor between the two-piece center electrode. Plug gap is the space between the side and center electrodes. (Ethyl Corp.)

basically determined by the length and diameter of the insulator tip and the ability of the plug to transfer heat into the cooling system. Refer to **Figure 35-26.**

A *hot spark plug* has a long insulator tip and will tend to burn off deposits. This provides a self-cleaning action.

A *cold spark plug* has a shorter insulator tip; its tip operates at a cooler temperature. A cold plug is used in engines operated at high speeds. The cooler tip will help prevent tip overheating and preignition.

Heat Range Ratings

Auto manufacturers normally recommend a specific spark plug heat range for their engines. The heat range will normally be coded and given as a number on the plug insulator. Generally, the larger the number on the plug, the hotter the spark plug tip will operate. For instance, a 52 plug would be hotter than a 42.

The only time you should deviate from plug heat range specifications is when abnormal engine or driving conditions are encountered. For example, a hotter plug may be installed in a worn out, oil-burning engine. The hotter plug will help burn off oil deposits. This will prevent spark plug oil fouling.

Dual Spark Plug Ignition System

A few engines have a *dual spark plug ignition system.* In these systems, two spark plugs are used in each cylinder. For instance, a four-cylinder engine would use eight spark plugs. With this design, one ignition coil can be used to fire both plugs at the same time. The coil configuration used in this type of system is similar to the coil arrangement used in a waste spark system.

One spark plug is located near the intake valve and the other near the exhaust valve. Dual plugs help ensure

better ignition and more complete combustion of the air-fuel mixture.

Multiple Discharge Ignition System

A multiple discharge ignition system fires the spark plugs more than once on each power stroke. By producing a series of sparks, it helps ensure more complete burning of fuel charge. This type of ignition is commonly used on racing engines.

Ignition Timing

Ignition timing, also called *spark timing,* refers to how early or late the spark plugs fire in relation to the position of the engine pistons. Ignition timing must change with changes in engine speed, load, and temperature.

Timing advance occurs when the spark plugs fire sooner on the engine's compression strokes. The timing is set several degrees before top dead center (BTDC). Even more timing advance is needed at higher engine speeds to give combustion enough time to develop pressure on the power stroke. See **Figure 35-27.**

Timing retard occurs when the spark plugs fire later on the compression strokes. It is the opposite of timing advance. Timing retard is needed at lower engine speeds and under high load conditions. Timing retard prevents the fuel from burning too much on the compression stroke, causing a spark knock or ping (abnormal combustion).

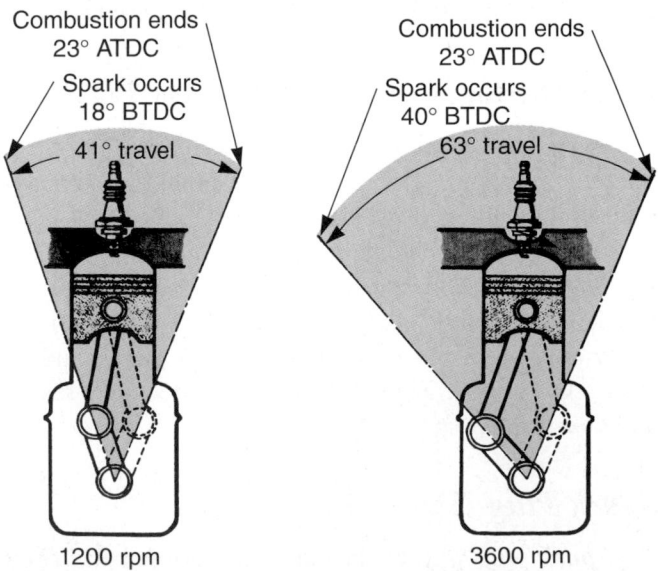

Figure 35-27. Since each combustion period takes about the same amount of time, the spark must start combustion sooner as engine speed increases. This will ensure that all the fuel is burned on the power stroke and that sufficient pressure acts upon the piston. (Sun Electric Corp.)

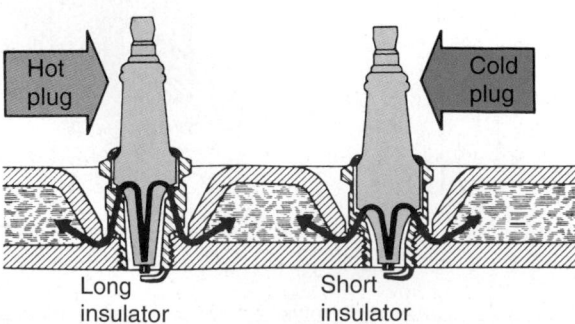

Figure 35-26. A hot plug has a long insulator that prevents heat transfer into the water jackets. It will burn off oil deposits. A cold plug has a shorter insulator. (Ethyl Corp.)

There are three basic methods used to control ignition system spark timing:

- *Distributor centrifugal advance*—Controlled by engine speed, **Figure 35-28**.

- *Distributor vacuum advance*—Controlled by engine intake manifold vacuum and engine load, **Figure 35-28**.

- *Electronic advance*—Controlled by a computer.

Distributor Centrifugal Advance

The *distributor centrifugal advance* makes the ignition coil and spark plugs fire sooner as engine speed increases. It uses spring-loaded weights, centrifugal force, and lever action to rotate the distributor cam or trigger wheel on the distributor shaft. By rotating the cam or trigger wheel against distributor shaft rotation, spark timing is advanced. Centrifugal advance helps maintain

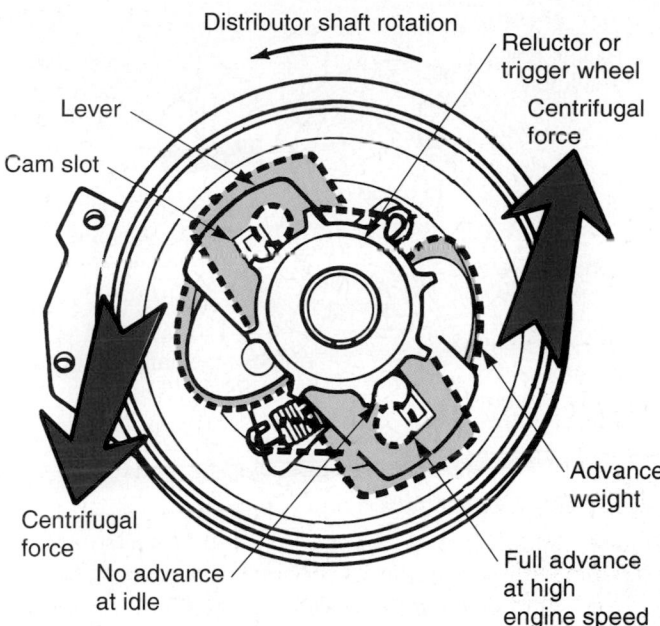

Figure 35-29. Springs hold the weights in at low engine speeds, and no centrifugal advance is produced. When engine speed increases, the weights swing outward. The weights push on and rotate the cam or the trigger wheel lever. This advances the ignition timing. (Dodge)

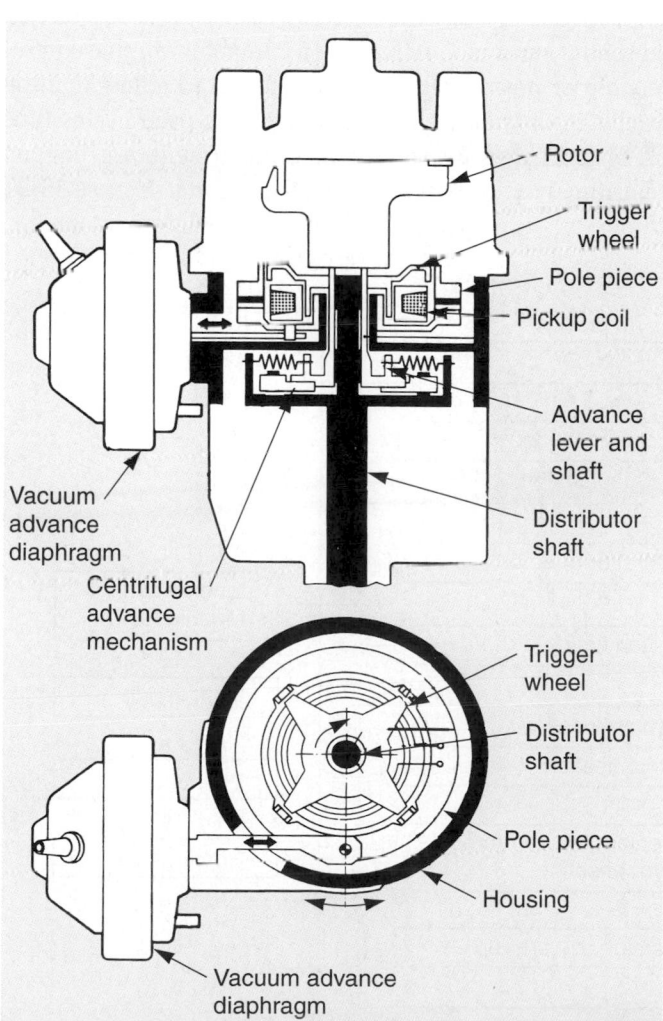

Figure 35-28. Study the parts of vacuum and centrifugal advance mechanisms. (Robert Bosch)

correct ignition timing for maximum engine power. See **Figure 35-29**.

At low engine speeds, small springs hold the advance weights inward to keep timing retarded. As engine speed increases, centrifugal force overcomes spring tension. The weights are thrown outward, and they act on the cam or trigger wheel lever. This rotates the distributor cam or trigger wheel. As a result, the points open sooner or the trigger wheel and pickup coil turn off the ignition control module sooner. This causes the ignition coil to fire with the engine pistons farther down in their cylinders.

As engine speed keeps increasing, the weights swing out farther and timing is advanced a greater amount. At a preset engine rpm, the lever strikes a stop, preventing the centrifugal advance mechanism from further advancing the ignition timing.

Distributor Vacuum Advance

The *distributor vacuum advance* provides additional spark advance at part (medium) throttle positions when the engine load is low. It is a method of matching ignition timing to *engine load*. The vacuum advance mechanism increases fuel economy because it helps maintain ideal spark advance at all times.

A distributor vacuum advance mechanism consists of a vacuum advance diaphragm, a link, a movable distributor plate, and a vacuum supply hose. Look at **Figure 35-30**.

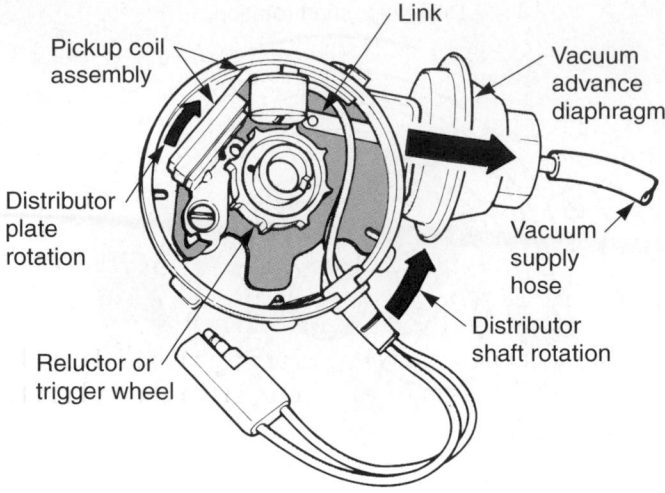

Figure 35-30. Vacuum advance uses a vacuum advance diaphragm to rotate the pickup coil or contact points against the direction of distributor shaft rotation. (Plymouth)

At idle, the vacuum port to the distributor advance is covered. Vacuum (suction) is *not* applied to the vacuum diaphragm. Spark timing is *not* advanced.

At part throttle, the throttle valve uncovers the vacuum port and the port is exposed to engine vacuum. This causes the distributor diaphragm to be pulled toward the vacuum. The distributor plate (points or pickup coil) is rotated against distributor shaft rotation and spark timing is advanced.

During acceleration and full throttle, engine vacuum drops. Thus, vacuum is *not* applied to the distributor diaphragm and the vacuum advance does *not* operate.

A *dual-diaphragm vacuum advance,* which was used on some distributors to aid early emission-control systems, contains two separate vacuum chambers: an advance chamber and a retard chamber.

A *vacuum delay valve* restricts the flow of air to slow down the vacuum action on a vacuum device. It keeps the vacuum advance from working too quickly, preventing possible knock or ping. The vacuum delay valve is usually located in the vacuum line going to the distributor diaphragm. A check valve allows the free release of vacuum from the diaphragm when returning to the retard position.

Electronic (Computer) Spark Advance

An *electronic spark advance system,* or *computer-controlled spark advance system,* uses engine sensors, an ignition control module, and/or a computer (engine control module or power train control module) to adjust ignition timing. A distributor may or may not be used in this type of system. If a distributor is used, it will *not* contain centrifugal or vacuum advance mechanisms. **Figure 35-31**

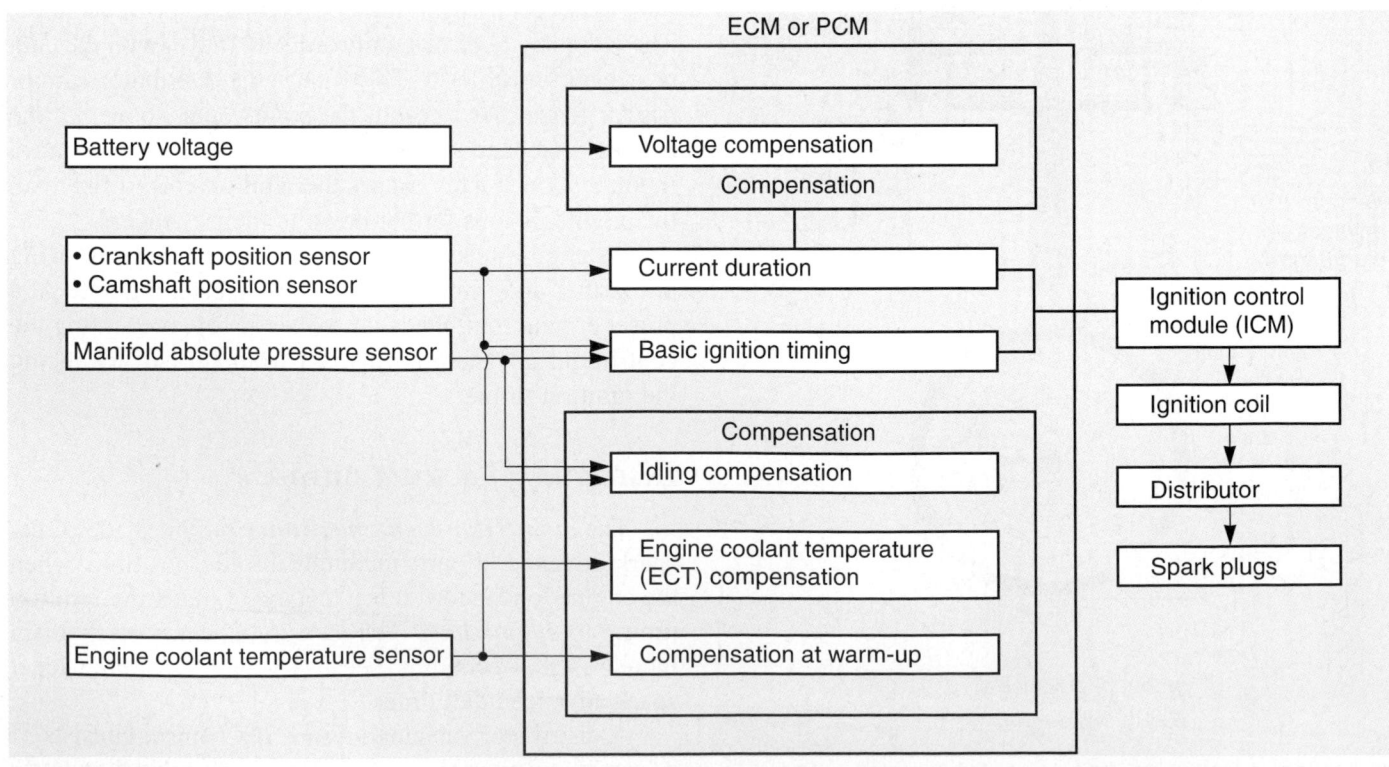

Figure 35-31. Chart shows flow of data for computer control of ignition timing. Designs can vary, however. (Honda)

shows a block diagram of how information is used to control engine ignition timing electronically.

The *engine sensors* check various operating conditions and send electrical data representing these conditions to the computer. The computer can then analyze the data and change ignition timing for maximum engine efficiency.

Sensors that influence ignition timing include:

- *Crankshaft position sensor*—Reports engine rpm to the computer.
- *Camshaft position sensor*—Tells the computer which cylinder is on its power stroke.
- *Manifold absolute pressure sensor*—Measures engine intake manifold vacuum, an indicator of load.
- *Intake air temperature sensor*—Checks temperature of air entering the engine.
- *Engine coolant temperature sensor*—Measures the operating temperature of the engine.
- *Knock sensor*—Allows the computer to retard timing when the engine pings or knocks.
- *Throttle position sensor*—Notes the position of the throttle.

The computer receives input signals (different current or voltage levels) from these sensors. It is programmed (preset) to adjust ignition timing to meet different engine conditions. The computer may be mounted on the air cleaner, on the fender inner panel, under the dash, or under a seat.

The computer can also measure battery voltage to compensate for voltage variations due to battery state of charge, accessory loads, etc.

Electronic Spark Advance Operation

For an example of electronic spark advance, imagine a car or light truck traveling down the highway at 55 mph (88 km/h). The crankshaft sensor would detect moderate engine rpm. The throttle position sensor would detect part throttle. The intake air and coolant temperature sensors would report normal operating temperatures. The manifold absolute pressure sensor would send high vacuum signals to the computer.

The computer could then calculate that the engine would need almost maximum spark advance. The timing would occur several degrees before TDC on the compression stroke. This would ensure that the engine attained high fuel economy on the highway. There would be enough time for all the fuel to burn and produce maximum pressure on the downward motion of the pistons.

If the driver began to pass a car, engine intake manifold vacuum would drop to a very low level. The manifold absolute pressure sensor signal would be fed to the computer. The throttle position sensor would detect wide open throttle (WOT). Other sensor outputs would stay about the same. Based on the signals from the manifold absolute pressure sensor and the throttle position sensor, the computer could then retard ignition timing to prevent spark knock or ping.

Since computer systems vary, refer to a service manual for more information. The manual will detail the operation of the specific system for that make and model vehicle.

Base Timing

Most new cars use the computer to control ignition timing advance. To check timing, you may have to trigger the computer to go to base timing. *Base timing* is the ignition timing without computer-controlled advance. This will be explained in more detail in the next chapter.

Crankshaft-Triggered Ignition

A *crankshaft-triggered ignition* can maintain more precise ignition timing than a system with a distributor-mounted pickup coil. There is no backlash or play in the distributor drive gear, timing chain, or gears to upset ignition timing. Crankshaft and piston position is "read" right off the crankshaft. **Figure 35-32** shows a simplified illustration of a crankshaft-triggered ignition.

A *crankshaft trigger wheel,* also called the *pulse ring,* or *reluctor ring,* is mounted on the crankshaft damper or the crankshaft itself to provide engine speed and crankshaft position data for the sensors and computer. In some cases, the wheel is cast or machined as an integral part of the shaft. The crankshaft trigger wheel performs the same function as the trigger wheel used in the distributor for an electronic ignition. The teeth on the trigger wheel correspond to the number of engine cylinders and position of the crank journals (pistons), **Figure 35-33.**

The *crankshaft position sensor* is mounted next to the crankshaft trigger wheel and sends electrical pulses to the system computer. It does the same thing as a distributor pickup, **Figure 35-33.**

If used, the distributor for a crankshaft-triggered ignition simply transfers high voltage to each spark plug wire through a spinning rotor and a stationary cap. No advance mechanisms, pickup coil, or related parts are inside the distributor in a crankshaft-triggered ignition system.

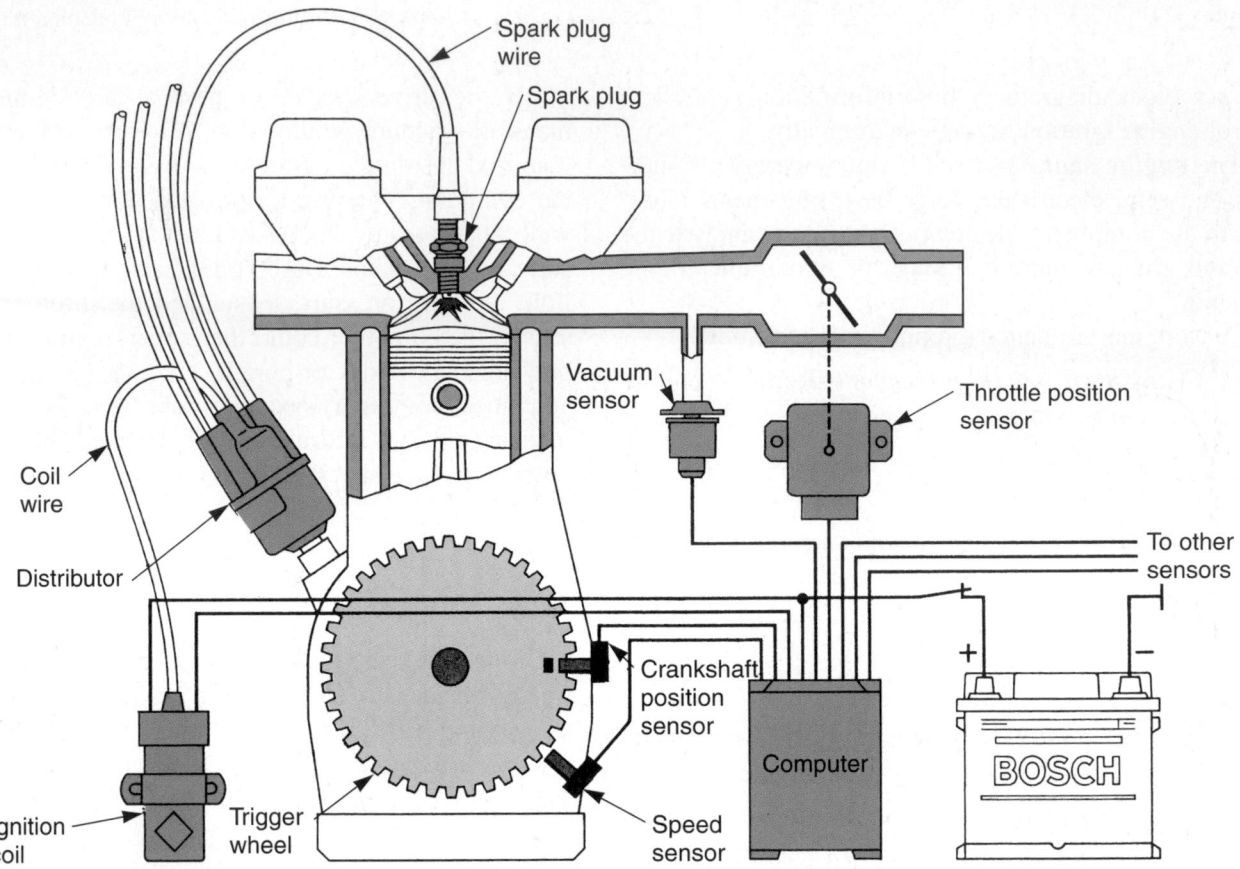

Figure 35-32. This simplified crankshaft-triggered ignition system places the pickup coil or coils next to the engine crankshaft damper. Teeth on the damper act as a trigger wheel to send electrical pulses to the computer. The computer can then operate the ignition coil and control spark advance or retard. Study the parts and wiring. (Robert Bosch)

Crankshaft-Triggered Ignition Operation

The operation of a crankshaft-triggered ignition is similar to that of the electronic systems already covered. Some designs use one crankshaft position sensor; others use two.

Figure 35-34 shows a crank fired ignition system that uses two crankshaft position sensors. Both sensors extend through the cylinder block and are sealed with rubber O-rings. Other engines can have the crankshaft position sensor mounted on the front of the engine in a small bracket.

As the crankshaft spins, the crankshaft trigger wheel and the crankshaft position sensor or sensors produce *on-off-on-off* voltage signals representing crankshaft speed and position. Depending on design, a different number of teeth can be used on the crankshaft trigger wheel. Some of the teeth are offset so that the sensors can detect crankshaft position.

A camshaft position sensor and a knock sensor can also be used to feed data to the engine control module. The **knock sensor** monitors preignition or knock so the computer can retard timing or reduce turbocharger boost pressure as necessary, **Figure 35-34.**

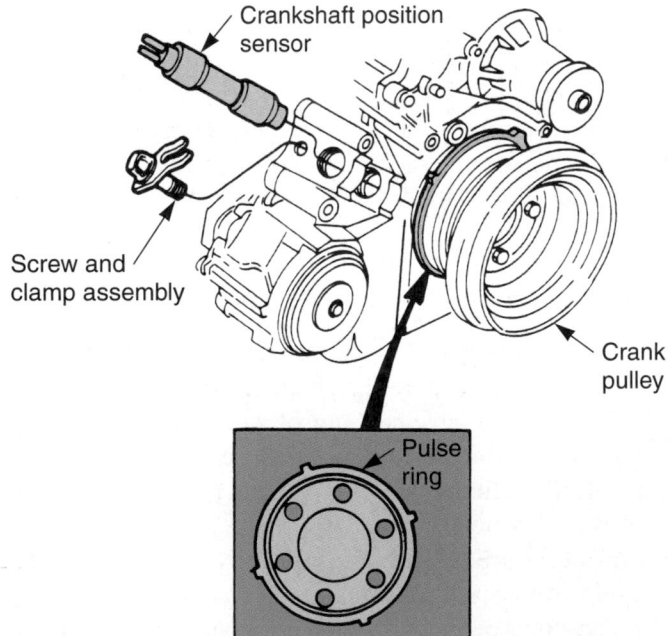

Figure 35-33. Pulse ring for this crankshaft-triggered ignition mounts behind the crank pulley or on the crankshaft inside the engine. The crankshaft position sensor fits in a hole in the front cover or slides through the side of the block. Wires from the sensor connect to the on-board computer. Teeth on the pulse ring change the magnetic field around the sensor to produce an electric pulse. (Ford)

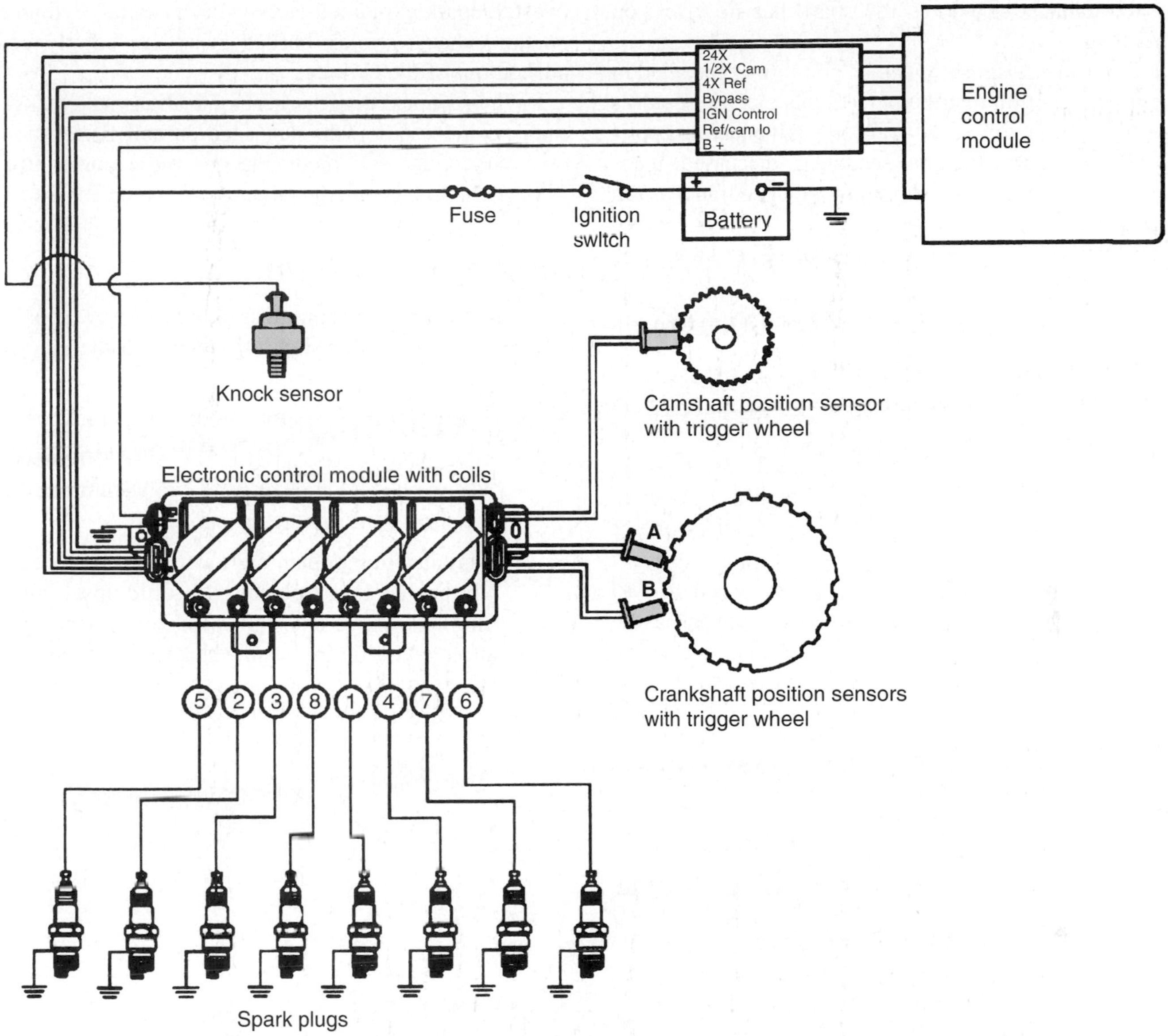

Figure 35-34 diagram with labels:

Engine control module

24X
1/2X Cam
4X Ref
Bypass
IGN Control
Ref/cam lo
B +

Fuse   Ignition switch   Battery

Knock sensor

Camshaft position sensor with trigger wheel

Electronic control module with coils

A
B

Crankshaft position sensors with trigger wheel

5 2 3 8 1 4 7 6

Spark plugs

Figure 35-34. Newer ignition system designs do not use a distributor. The on-board computer and an electronic coil control module operate multiple ignition coils. A crankshaft position sensor and a camshaft position sensor send electrical signals to help determine the time when the coils fire. Trace the connections from each component. (Oldsmobile)

Figure 35-35 shows a close-up of a crankshaft trigger wheel and a crankshaft position sensor. Note how the sensor mounts through the engine block. Also note the spacing of the notches on the trigger wheel.

Tech Tip
Most crankshaft trigger wheels are missing one tooth. The missing tooth represents top dead center or a reference for measuring crankshaft position.

Distributorless Ignition System

A *distributorless ignition system* (no distributor), also called a *computer-coil ignition,* uses multiple ignition coils, a coil control unit, engine sensors, and a computer (engine control module) to operate the spark plugs. A distributor is *not* needed. See **Figure 35-34.**

An *electronic coil module* consists of several ignition coils and an electronic circuit for operating the coils. The module's electronic circuit performs about the same function as the ignition control module in an electronic ignition. It is more complex, however, because it must analyze data from engine sensors and the system computer.

A four-cylinder engine would need an electronic coil module with two ignition coils. A six-cylinder engine would need a module with three ignition coils.

The coils are wired so that they fire *two spark plugs* at once. One spark plug is on the power stroke. The other is on the exhaust stroke, so the spark has no effect on engine operation.

A *camshaft position sensor* is commonly installed in place of the ignition distributor. It sends electrical pulses to the coil module giving data on camshaft and valve position. The crankshaft position sensor, as discussed, feeds pulses to the module that show engine speed and piston position. A knock sensor may be used to allow the system to retard timing if the engine begins to ping or knock.

Distributorless Ignition System Operation

Figure 35-36 illustrates how a distributorless ignition system works. The on-board computer monitors engine operating conditions and controls ignition timing. Some sensor data is also fed to the electronic coil module.

When the computer and sensors send correct electrical pulses to the coil module, the module fires one of the ignition coils.

Since each coil secondary output is wired to two spark plugs, both spark plugs fire. One produces the power stroke. The other spark plug arc does nothing

(wasted spark explained earlier) because that cylinder is on the exhaust stroke: burned gases are simply being pushed out of the cylinder.

When the next trigger wheel tooth aligns with the crank shaft position sensor, the next ignition coil fires. Another two spark plugs arc, producing one more power stroke. This process is repeated over and over as the engine runs.

Advantages of a Distributorless Ignition

A distributorless ignition system has several possible advantages over other ignition types. Some of these include:

- No rotor or distributor cap to burn, crack, or fail.

- Computer-controlled spark advance. No mechanical weights to stick or wear. No vacuum advance diaphragm to rupture or leak.

- Play in timing chain and distributor drive gear is eliminated as a problem that could upset ignition timing. The crankshaft position sensor is not affected by timing chain slack or gear play.

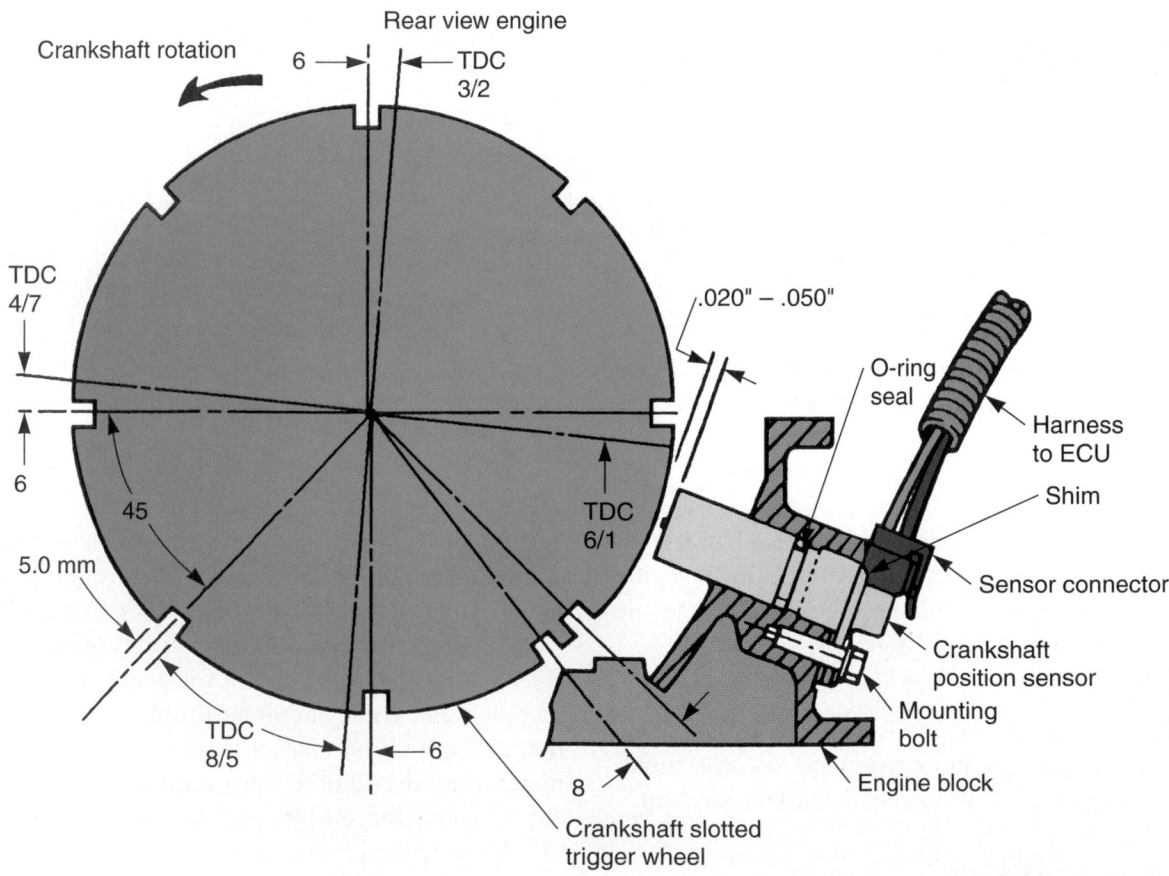

Figure 35-35. This crankshaft trigger wheel is cast and machined as an integral part of the crankshaft. Note how the crankshaft position sensor fits through the cylinder block. A rubber O-ring seal prevents oil leakage. An offset notch on the wheel allows the computer to detect crankshaft position, as well as speed.

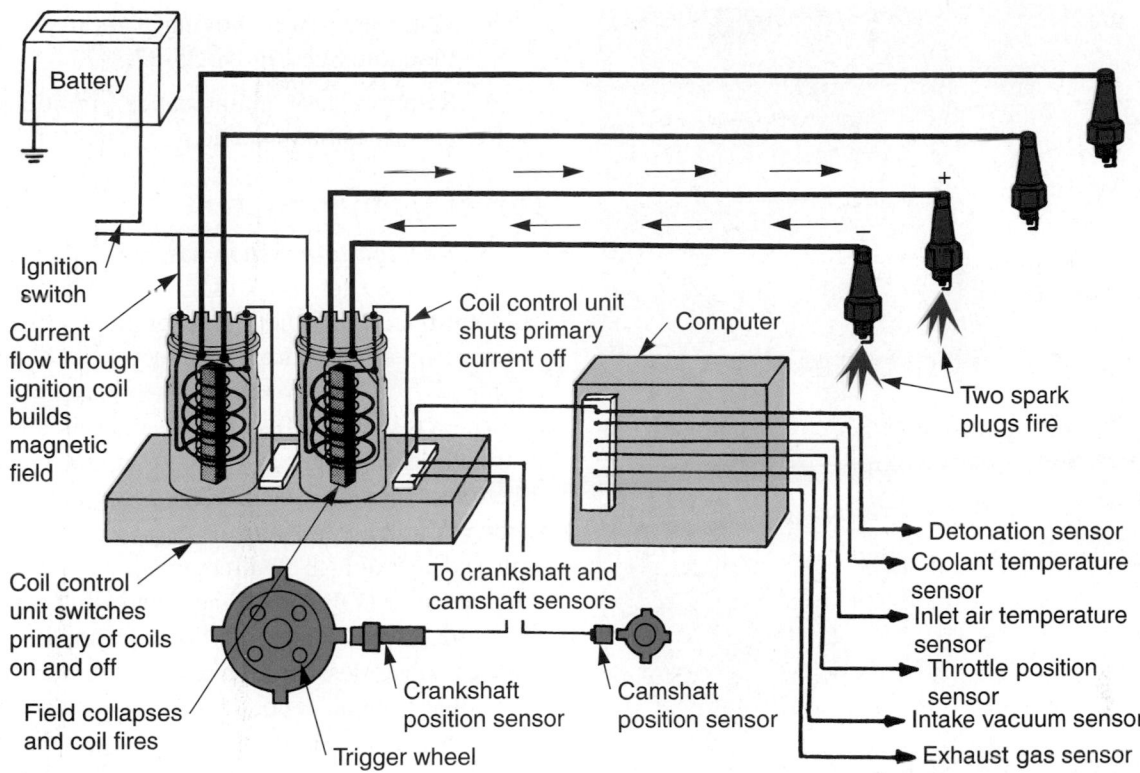

Figure 35-36. This simplified illustration shows the operation of an ignition system that does not use a distributor. Inputs to coil module include signals from the crankshaft position sensor, the camshaft position sensor, and the computer. With the correct inputs, the coil module fires one of the ignition coils, which, in turn, fires two spark plugs. One plug produces a power stroke. The other plug sparks as burned exhaust leaves the cylinder. Two coils would operate the ignition for a four-cylinder engine. Three coils would be needed for a six-cylinder engine.

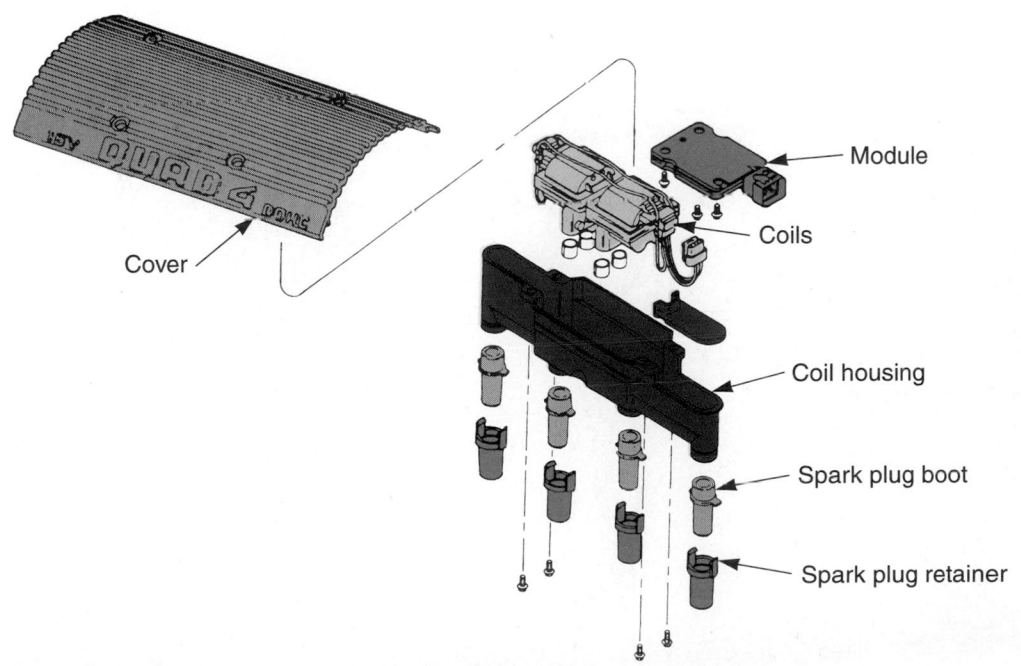

Figure 35-37. Note the location of the ignition coil and the coil module for this distributorless ignition. Two coils are needed for a four-cylinder engine. (Delco-Remy)

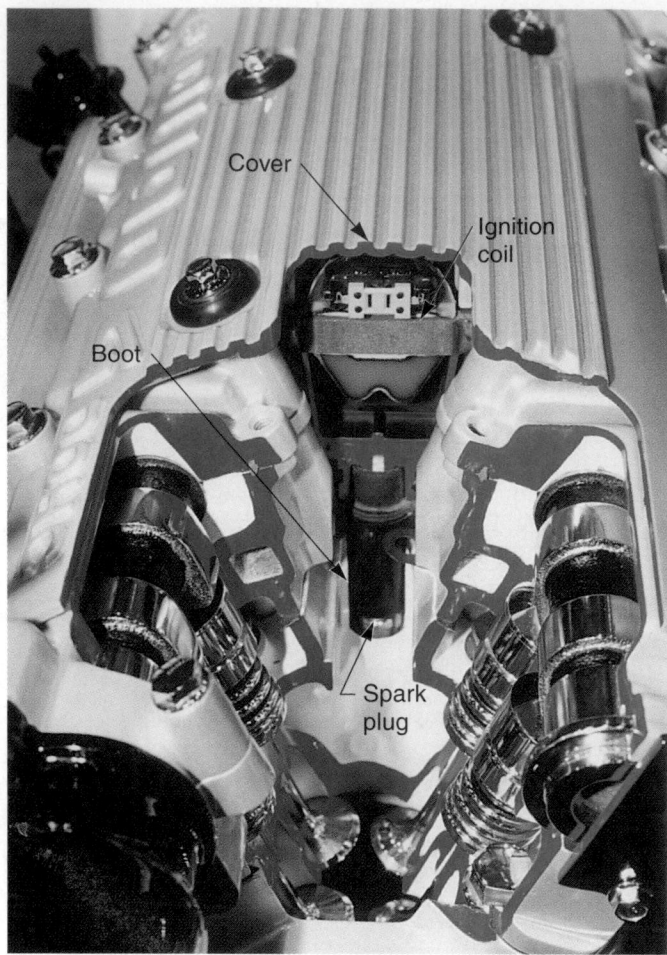

Figure 35-38. Engine cutaway shows a coil mounted over a spark plug.

- There are fewer moving parts to wear and malfunction. See **Figure 35-37.**

- Requires less maintenance. Ignition timing is usually *not* adjustable.

Direct Ignition System

A *direct ignition system* has an ignition coil mounted over the top of each spark plug. This system is newer and different than a distributorless ignition. There is no conductor strip or plug wire to connect the coil and the spark plug.

There is some confusion in the industry about the terms distributorless ignition and direct ignition systems. A distributorless ignition system may or may not use spark plug wires, but it will have some form of conductor between the coils and spark plugs. A direct ignition system mounts one ignition coil over the top of each spark plug without a spark plug wire or conductor strip. Some technicians mistakenly call a distributorless ignition a direct ignition system when high voltage output is not directly fed to each spark plug.

With a direct ignition system, one coil assembly is needed for each spark plug and engine cylinder, **Figure 35-38.** This allows the use of smaller ignition coils and fewer parts.

The other components in a direct ignition system (computer, sensors, etc.) are the same as those used in a distributorless system. The direct ignition coils only fire on the power stroke. They do *not* fire on the exhaust strokes like many distributorless ignition systems.

Figure 35-39 shows a cutaway of the direct ignition coils for a four-cylinder engine.

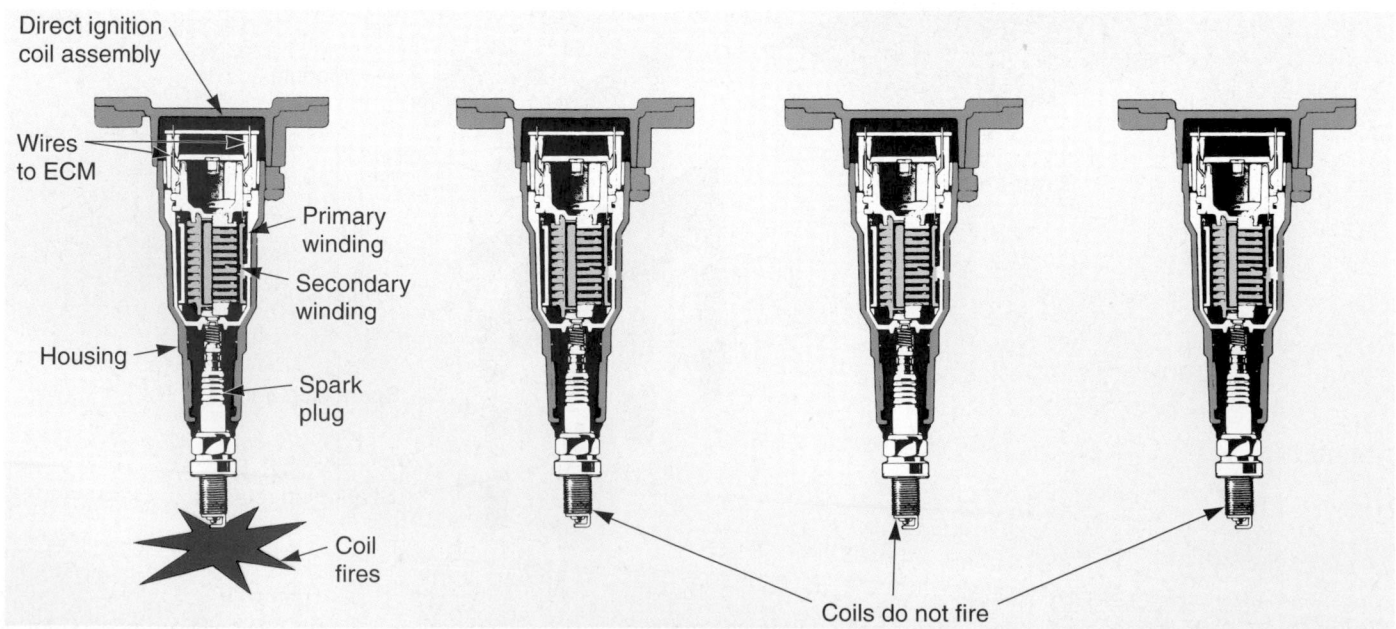

Figure 35-39. Cutaway of an ignition coil for a direct ignition system. With four-cylinder engine, four coils would be needed. (Saab)

Tech Tip

In some multi-coil ignition systems, the engine control computer measures the time needed to charge each ignition coil. Charge time varies with each coil's resistance/inductance and with the output voltage from the alternator. The computer uses this charge time to determine the ignition coil on-time for optimum spark plug arc duration and spark timing.

Ionization Knock Sensing

In some cases, an *ionization knock sensing* system is used to detect abnormal combustion and knocking. In these systems, the computer prompts the ignition coil to send a low-voltage discharge across the spark plug gap immediately following combustion. The quality of combustion affects the resistance across the plug gap by varying the degree of ionization (the process by which atoms gain or lose electrons) in the gases between the plug's electrodes. Consequently, the quality of combustion influences the strength of the discharge across the gap. The computer uses feedback from the discharge to determine if the spark advance or the turbocharger boost should be modified to reduce knocking. A conventional knock sensor is not needed in an ionization knock sensing system.

Engine Firing Order

Engine firing order refers to the sequence in which the spark plugs fire to cause combustion in each cylinder. A four-cylinder engine may have one of two firing orders: 1-3-4-2 or 1-2-4-3. The cylinders are numbered 1-2-3-4, starting at the front of the engine. In this way, you can tell which cylinders will fire in sequence. Firing orders and cylinder numbers for V-6 and V-8 engines vary. **Figure 35-40** shows typical firing orders for 4-, 6-, and 8-cylinder engines.

The engine firing order is sometimes cast into the top of the intake manifold. When not on the manifold, the firing order can be found in a service manual. See **Figure 35-41.**

Discussed in later chapters, the engine firing order is commonly used when installing spark plug wires and when doing other tune-up tasks.

Summary

- The vehicle's ignition system produces the high voltage needed to ignite the fuel charges in the cylinders of a spark ignition engine.
- The ignition switch is a key-operated switch in the driver's compartment.

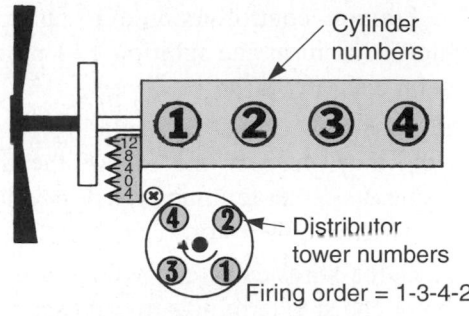

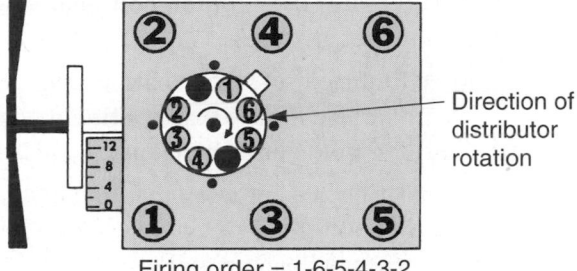

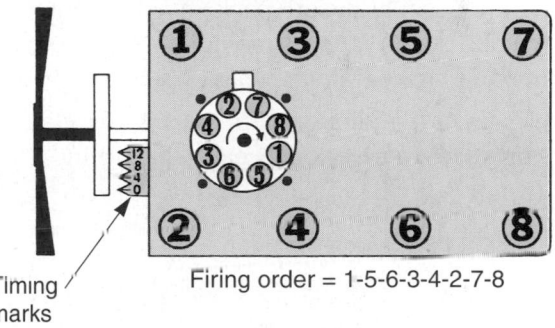

Figure 35-40. Firing order is the sequence in which the spark plugs fire. Firing order information is used when installing plug wires, installing a distributor, setting ignition timing, and other operations covered in the next chapter. (Mitchell Manuals and Echlin)

- The primary circuit of the ignition system includes all of the components and wires operating on low voltage (battery or alternator voltage).
- The secondary circuit of the ignition system is the high voltage section.
- The ignition system secondary operating voltage can normally range from 4000–30,000 volts depending upon system design.
- An ignition coil produces the high voltage (between 4000 and 30,000 volts under normal operating conditions) needed to make high voltage, low current flow through the system.
- The trigger wheel, also called reluctor or pole piece, is fastened to the upper end of the distributor shaft.

- The ignition control module is an "electronic switch" that turns the ignition coil primary current on and off.

- The distributor cap is an insulating plastic component that fits over the top of the distributor housing. Its center terminal transfers voltage from the coil wire to the rotor.

- Spark plug wires carry coil voltage from the distributor cap side terminals to each spark plug.

- The spark plugs use ignition coil high voltage to ignite the fuel mixture.

- Ignition timing, also called spark timing, refers to how early or late the spark plugs fire in relation to the position of the engine pistons.

- A crank-triggered ignition can maintain more precise ignition timing than a system with a distributor mounted pickup coil.

- A distributorless ignition system uses multiple ignition coils, a coil pack, a coil control unit, engine sensors, and a computer (engine control module) to operate the spark plugs.

- A direct ignition system has an ignition coil mounted over the top of each spark plug.

Important Terms

Ignition system
Battery
Ignition switch
Ignition coil
Switching device
Spark plug
Ignition system wires
Ignition system supply
  voltage
Bypass circuit
Resistance circuit
Resistance wire
Ballast resistor
Computer voltage
  regulator
Primary circuit
Ignition system
  primary voltage
Secondary circuit
Ignition system
  secondary operating
  voltage

Ignition coil open-
  circuit voltage
Coil pack
Wasted-spark ignition
  coil
Ignition distributor
Distributor drive
Distributor cam
Contact points
Condenser
Point dwell
Point gap
Electronic ignition
  system
Trigger wheel
Pickup coil
Hall-effect pickup
Optical pickup
Ignition control module
Distributor cap
Rotor

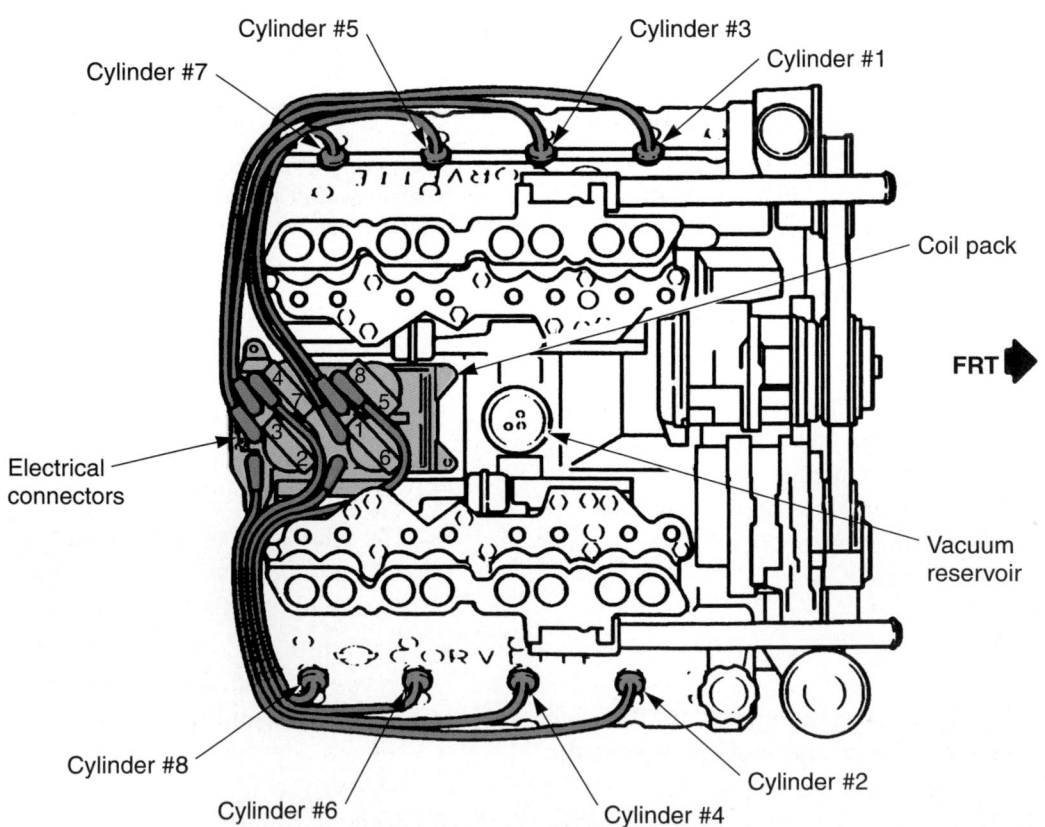

Figure 35-41. Firing order information for a distributorless ignition system. (Chevrolet)

Secondary wires
Boots
Spark plugs
Spark plug reach
Resistor spark plug
Non-resistor spark
  plug
Spark plug gap
Spark plug heat range
Dual spark plug
  ignition system
Ignition timing
Timing advance
Timing retard
Distributor centrifugal
  advance
Distributor vacuum
  advance
Dual-diaphragm
  vacuum advance

Electronic spark
  advance system
Engine sensors
Base timing
Crankshaft-triggered
  ignition
Crankshaft trigger wheel
Crankshaft position
  sensor
Knock sensor
Distributorless ignition
  system
Electronic coil module
Camshaft position
  sensor
Direct ignition system
Ionization knock
  sensing
Engine firing order

Review Questions—Chapter 35

Please do not write in this text. Place your answers on a separate sheet of paper.

1. What are the six basic functions of an ignition system?

2. List and explain the six major parts of an ignition system.

3. An ignition system _____ circuit is used to supply direct battery voltage to the system during starting motor operation.

4. Define the terms "primary circuit" and "secondary circuit."

5. An ignition coil is capable of producing an open-circuit voltage of:
   (A) 70–90 kV.
   (B) 80–100 kV.
   (C) 40–60 kV.
   (D) None of the above.

6. The primary windings of an ignition coil are several hundred turns of heavy wire. True or False?

7. When the current flowing through the ignition coil is broken, the magnetic field _____ and induces high _____ into the secondary.

8. Explain the differences between a contact point distributor and a pickup coil (electronic) distributor.

9. What is dwell?

10. An electronic ignition system uses a distributor pickup coil and an ignition control module to operate the _____ _____.

11. How does the pickup coil produce signals for the ignition control module?

12. The ignition control module is *not* normally located in the distributor. True or False?

13. Dwell time is not _____ in an electronic ignition system.

14. What is electrical leakage?

15. The _____ _____ is an insulating plastic component that fits over the distributor housing and sends voltage to each spark plug wire.

16. Explain the function of the distributor rotor.

17. Why do spark plug wires need internal resistance?

18. It normally takes about _____ to _____ volts to operate a spark plug.

19. Spark plug gap is the distance between the center electrode and side electrode. True or False?

20. A cold spark plug might be beneficial in an older engine that burns some oil. True or False? Explain.

21. Describe the difference between timing advance and timing retard.

22. List and explain the three methods of controlling ignition timing.

23. The distributor centrifugal advance depends upon engine _____ and the vacuum advance depends upon intake manifold pressure (vacuum), an indicator of engine _____.

24. Electronic spark advance uses engine _____ and a(n) _____ _____ _____ to control ignition timing.

25. How does a crankshaft triggered electronic ignition system work?

26. Which of the following does *not* relate to a distributorless ignition system?
    (A) No rotor.
    (B) No centrifugal or vacuum advance.
    (C) Two ignition coils per cylinder.
    (D) No spark plug wires.

27. With a distributorless ignition, two spark plugs fire at once. True or False?

28. List five possible advantages of a distributorless ignition.

29. What is a direct ignition system?

30. Engine _____ _____ refers to the sequence in which the spark plugs operate to cause combustion in each cylinder.

⬥ ASE-Type Questions

1. An ignition system uses this to change battery voltage into high secondary voltage:
   (A) spark plug.
   (B) bypass circuit.
   (C) coil.
   (D) ballast resistor.

2. Which of the following provides voltage to the battery and ignition system *after* an engine is running?
   (A) Alternator.
   (B) Condenser.
   (C) Primary circuit.
   (D) Secondary circuit.
   (E) None of the above.

3. Ignition system wires operating at battery voltage are considered part of the:
   (A) ignition coil.
   (B) primary circuit.
   (C) secondary circuit.
   (D) primary windings.

4. Technician A says that with modern ignition coils, open-circuit voltage can reach 60 kV. Technician B says that average operating voltages for an ignition system are about 15 kV. Who is right?
   (A) A only.
   (B) B only.
   (C) Both A and B.
   (D) Neither A nor B.

5. Which of the following is *not* a possible function of the ignition distributor?
   (A) House the ignition coil.
   (B) Power the engine oil pump.
   (C) Transfer high voltage to the coil.
   (D) Distribute coil pulses to each plug wire.

6. Technician A says an older contact point system produced more secondary voltage than modern electronic or computer controlled systems. Technician B says a contact point system was also more dependable than modern systems. Who is right?
   (A) A only.
   (B) B only.
   (C) Both A and B.
   (D) Neither A nor B.

7. An ignition control module can be located at each of these spots *except:*
   (A) in the vehicle's trunk.
   (B) inside the distributor.
   (C) under the vehicle dash.
   (D) in the engine compartment.

8. Which of the following spark plug components causes electricity to jump the gap and return to the battery through frame ground?
   (A) Shell.
   (B) Insulator.
   (C) Side electrode.
   (D) Center terminal.

9. Which of the following is *not* a typical electronic spark advance ignition system sensor?
   (A) Vehicle speed sensor.
   (B) Knock sensor.
   (C) Manifold absolute pressure sensor.
   (D) Crankshaft position sensor.

10. A direct ignition system uses:
    (A) spark plug wires.
    (B) secondary conductors strips.
    (C) one coil per spark plug.
    (D) All of the above.

Activities—Chapter 35

1. If you have access to a computer with a CAD or drawing program, use it to make a cross-sectional drawing of a resistor-type spark plug. Label the components, then output the drawing to a printer, if available.

2. Make a list of the various engine sensors used with an electronic spark advance system. Describe the purpose of each sensor.

3. Using shop manuals, identify the firing order of the spark plugs for two different makes of four–, six–, and eight–cylinder engines.

4. Make a wall chart comparing contact point, electronic, distributorless, and direct ignition system types.

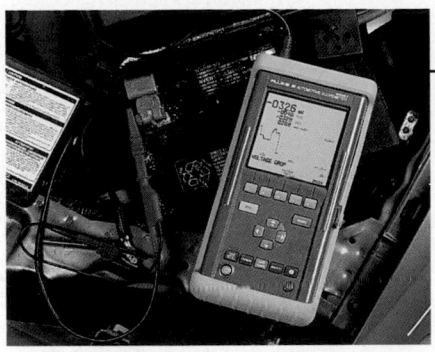

Chapter 36

Ignition System Problems, Testing, and Repair

After studying this chapter, you will be able to:

- Diagnose typical ignition system problems.
- List the symptoms produced by faulty ignition system components.
- Describe common tests used to find ignition system troubles.
- Explain how to replace or repair ignition system parts.
- Summarize contact point and pickup coil adjustments.
- Adjust ignition timing.
- Describe safety practices to follow when testing or repairing an ignition system.
- Correctly answer ASE certification test questions on the diagnosis and repair of ignition systems.

An ignition system is one of the most important systems on a vehicle. If a problem exists in the ignition, engine performance will suffer. As a service technician, you must be able to quickly and accurately correct ignition system troubles.

The previous chapter covered the fundamentals of ignition system operation. This chapter will let you use this information to diagnose, test, and repair modern ignition systems. Study carefully!

Note
Chapters 45–47 summarize driveability problems, oscilloscopes, and tune-up procedures. Refer to these chapters for more information on ignition system diagnosis and repair as needed.

Ignition System Problem Diagnosis

Diagnosis of ignition system problems can be very challenging. Because of computer networking, the ignition system and several other systems (fuel, emission, electrical, etc.) all work together. A problem in one system may affect, or appear to affect, the operation of another system.

For example, an inoperative gasoline injector can cause an engine miss or a rough idle. An oil-fouled spark plug can also cause an engine miss. The symptoms for each of these problems will be almost identical. Only proper testing methods will find the faulty component.

Preliminary Checks of Ignition System

Visually inspect the ignition system with and without the engine running. Look for obvious problems: loose primary connections, disconnected spark plug wires, deteriorated secondary wire insulation, cracked distributor cap, and other troubles. See **Figures 36-1** and **36-2**. At the same time, look over other engine systems. Try to find anything that could upset engine operation.

Scanning for Ignition System Problems

With late-model vehicles, especially those equipped with OBD II diagnostic systems, you can connect a scan tool to the vehicle computer to find many ignition related problems. See **Figure 36-3**. The scan tool can find troubles in the following ignition system circuits and components:

- Crankshaft position sensor.
- Crankshaft speed sensor.
- Camshaft position sensor.
- Knock sensor.
- Ignition coil(s) primary circuit.
- Ignition coil(s) secondary circuit.
- Timing reference signal.
- Other devices.

If your scan tool readout shows problems with any of these components or circuits, start your pinpoint tests at that part or circuit. You might test the component first and then its circuit. This will depend on the specific circumstances of the symptoms and vehicle design.

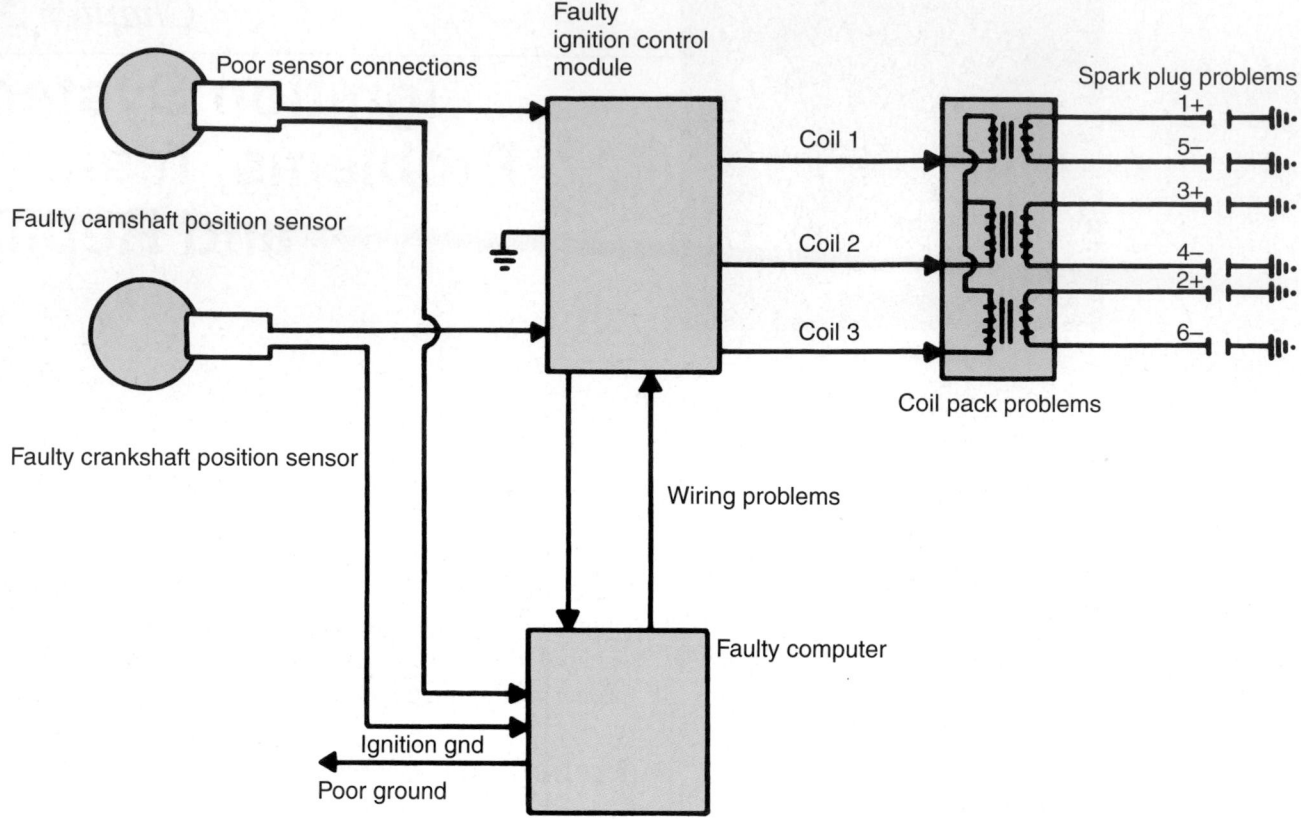

Figure 36-1. Note some of the problems you will find in a distributorless ignition system. (Snap-On Tools)

If your scanner readout indicates a ***misfire*** condition, the engine has failed to ignite and burn its air-fuel mixture properly. This increases exhaust emissions, lowers engine power, and increases fuel consumption. Misfire can be caused by problems in the ignition system, the fuel injection system, the emission control system, or the engine itself. Additional tests and the use of more specific scan tool data would be needed to pinpoint the problem causing the misfire condition.

Note
For more information on on-board diagnostics and misfire conditions, refer to Chapter 18, *On-Board Diagnostics and Scan tools.*

If you are working on a vehicle without on-board diagnostics or a vehicle equipped with an early self-diagnostic system, you will have to use basic testing methods to check logical problem sources.

Performing a Spark Intensity Test

A ***spark intensity test***, or ***spark test***, measures the brightness and length of the electric arc (spark) produced by the ignition system. It is a quick and easy way to check the general condition of the ignition system.

The spark intensity test is often used when an engine cranks but will not start. The test will tell you whether the trouble is in the fuel system ("no fuel" problem) or the ignition system ("no spark" problem). It may also be used to check the condition of the spark plug wires, distributor cap, and other secondary components.

Spark Test Procedures

A ***spark tester*** is a device that is used to check ignition system output voltage. It looks like a spark plug with a very large air gap and a ground wire. Look at **Figure 36-4.**

To perform a spark intensity test:
1. Remove one of the secondary wires from a spark plug.
2. Insert the spark tester into the wire.
3. Ground the tester on the engine.
4. Crank or start the engine and observe the spark at the tester's air gap, **Figure 36-5.**

Caution
Only run the engine for a short period of time with a spark plug wire off. Unburned fuel from the dead cylinder could foul and ruin the catalytic converter.

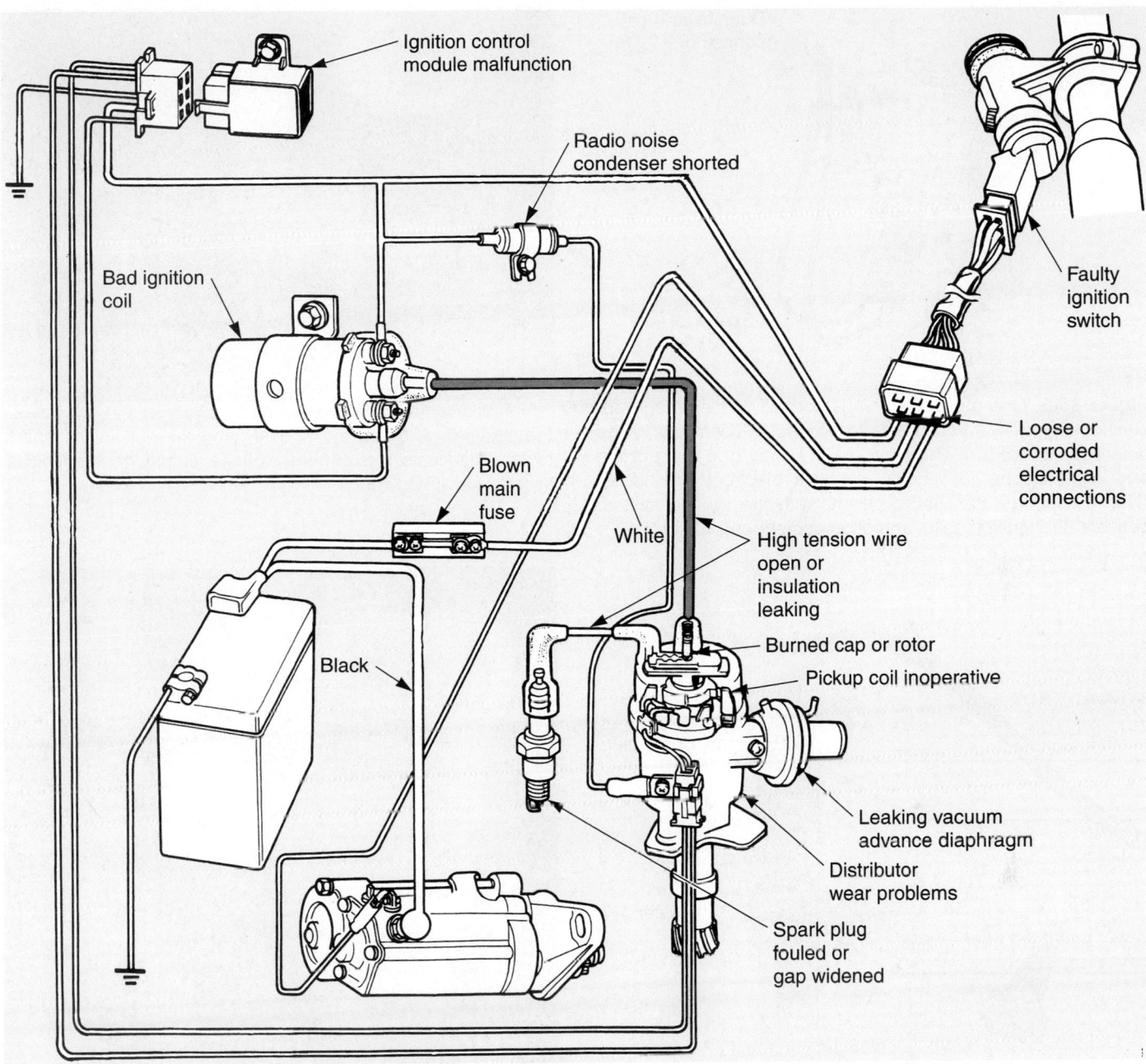

Figure 36-2. These problems are common to an electronic ignition system with a distributor. (Honda)

Spark Test Results

A strong spark (wide, bright, snapping electric arc) shows that ignition system voltage is good and that the ignition coil, pickup coil, electronic control unit, and other ignition system parts are functioning. If a strong spark is present, the engine no-start problem may be due to fouled spark plugs, fuel system problems, or engine trouble.

A weak spark or no spark indicates that something is wrong in the ignition system. If the spark is weak at all the spark plug wires, the problem is common to all the cylinders (bad ignition coil, rotor, coil wire). Other tests (to be covered shortly) are needed to pinpoint the trouble.

Depending upon the results of your scan tests and spark tests, you may want to check the coil supply voltage if you are not getting a spark. If a coil is not getting voltage or if the coil is faulty, the ignition system will not produce a secondary arc or the arc will be too weak to start combustion. See **Figure 36-6.**

In late-model systems, a bad crankshaft position sensor can upset ignition system operation and keep the engine from starting. Other reasons and tests for a no-spark condition will be discussed shortly.

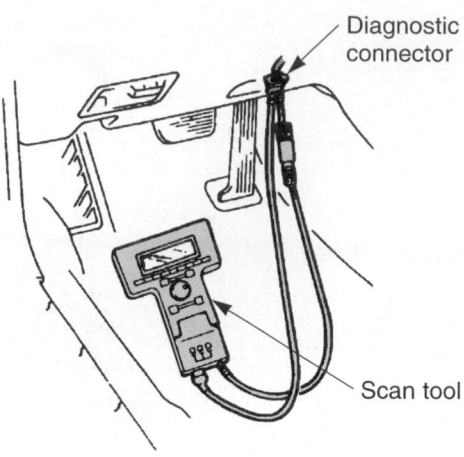

Figure 36-3. One of the first things normally done when troubleshooting today's vehicles is to connect a scan tool to the diagnostic connector. The scan tool will then give instructions and explanations for finding electrical-electronic problems. OBD II scanners can find problems with the ignition system sensors, the ignition coils, and other components. (Mazda)

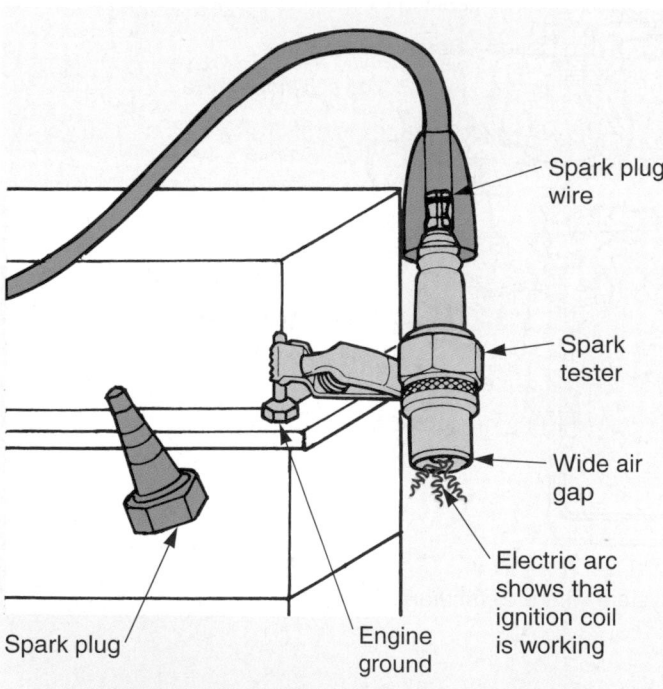

Figure 36-4. Connect the spark tester to the spark plug wire. Ground the tester on the engine. Crank the engine while watching the arc in the tester. No spark or a weak spark indicates ignition system troubles.

Checking for a Dead Cylinder

A *dead cylinder* is a cylinder (combustion chamber) that is not burning fuel on the power stroke. There may be ignition system troubles or problems in the engine, fuel system, or another system. A very rough idle and a

Figure 36-5. A bright arc should jump the gap in the spark tester. This tests the general voltage output of the ignition system. (Snap-On Tools)

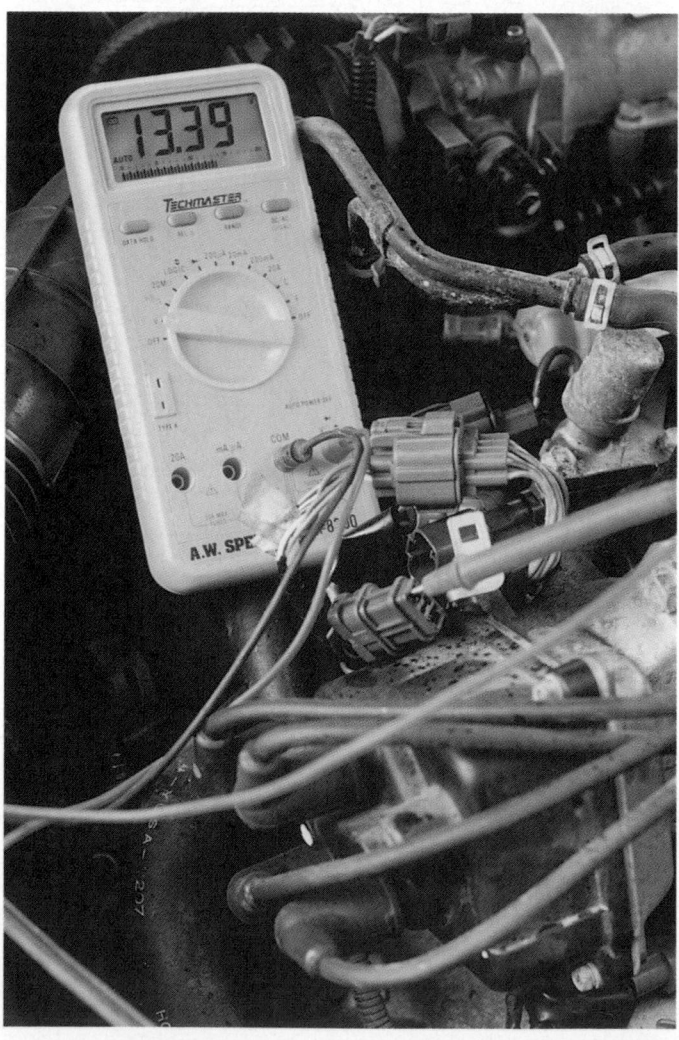

Figure 36-6. If you have a no-spark problem, you might want to check for voltage being fed to the ignition coils. Refer to the service manual to find the wire color codes and test points in the circuit. (A. W. Sperry)

puffing noise in the engine exhaust may indicate a dead cylinder. This is an extreme misfire condition because none of the fuel is "firing" and burning.

To check for a dead cylinder, pull off one spark plug wire at a time. On a "live," or firing, cylinder, pulling the wire off will cause engine rpm to drop and idle to become rougher. On a dead cylinder, idle smoothness and rpm *will not change* when the plug wire is removed. After locating a dead cylinder, check for spark and study the spark plug's condition. Also check for low cylinder compression. Some modern engine analyzers automatically check for dead cylinders.

Evaluating the Symptoms

After performing preliminary ignition system tests, you must evaluate the test results and narrow down the possible causes. Use your knowledge of system operation, a service manual troubleshooting chart, basic testing methods, and common sense to locate the trouble. In some cases, the following equipment can be used to help pinpoint the problem.

Using a Hand-Held Scope (Oscilloscope)

A **hand-held scope** is usually a VOM combined with an oscilloscope in one housing. It is a handy tool for advanced troubleshooting of electrical-electronic problems. Like a VOM, the hand-held scope can be used to take voltage, current, and resistance measurements. It can also show the instantaneous voltage variations of a scope. See **Figure 36-7.**

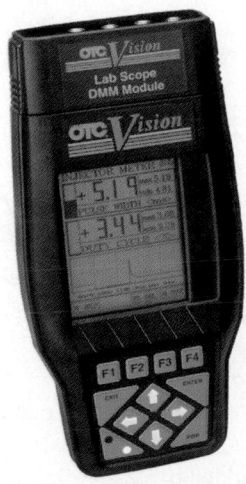

Figure 36-7. A hand-held scope is handy for advanced ignition system testing. It will give voltage, current, and resistance readings. It will also show scope waveforms for advanced troubleshooting of the ignition system and other systems. (OTC)

A scope is sometimes used to check ignition system operating voltages over time. It will also check the output signals from sensors and the signals leaving electronic control units. Voltage is shown horizontally on the oscilloscope screen, while time is shown vertically.

Using an Engine Analyzer

An *engine analyzer* contains several types of test equipment (oscilloscope, dwell meter, tachometer, VOM). They are housed in one large, roll-around cabinet. The analyzer is often used to check the operation of an ignition system when problems are hard to find. One type of analyzer is shown in **Figure 36-8.**

Introduced in earlier chapters, an *oscilloscope,* or "scope," will precisely measure the operating voltages of an ignition system. It uses a television-type display screen to show voltage changes in relation to time (or degrees of distributor or crankshaft rotation). To determine the condition of the ignition system, the technician can compare the scope test patterns with known good patterns.

 Note
For more information on using a scope or an engine analyzer to test ignition, computer, and other system components, refer to Chapter 46, *Advanced Diagnostics.*

Using an Electronic Ignition Tester

An *electronic ignition tester* is an instrument that speeds ignition system diagnosis by indicating specific problems. The ignition tester is connected to the ignition circuit or to a special test plug on the side of the engine compartment. Indicator lights on the tester show whether

Figure 36-8. An engine analyzer is commonly used to diagnose and locate ignition system problems. The oscilloscope shows ignition system voltages. (Snap-On Tools)

problems exist. Ignition testers are usually make and model specific and are found in large dealerships that service one make of vehicle. They are being replaced by diagnostic scan tools.

Spark Plug Service

Bad spark plugs can cause a wide range of problems, including misfiring, lack of power, poor fuel economy, and hard starting. After prolonged use, the spark plug tip can become coated with ash, oil, and other substances. Also, the plug electrodes can burn, widening the gap. This can make it more difficult for the ignition system to produce an arc between the electrodes. Wide spark plug gaps require an increased secondary voltage, which can burn other ignition system parts (distributor cap, rotor, ignition module, etc.).

To test the spark plugs, use an oscilloscope or your OBD II scanner misfire readings. Bad plugs will show up as abnormal scope waveforms (trace or pattern) or as a misfire code. If a scope is not available or the vehicle does not have OBD II diagnostics, remove the plugs and inspect their condition. Refer to **Figure 36-9.**

Spark Plug Removal

Before removing the spark plugs, check that the spark plug wires are numbered or located correctly in their clips. Then, grasp the spark plug wire boot and pull the wire off the plug, **Figure 36-10.** Twist the boot back and forth if it sticks.

Spark plug wire pliers are long pliers that have insulation on the handles and jaws. They can sometimes be helpful when removing spark plug wires that are difficult to reach.

 Caution

Never remove a spark plug wire by pulling on the wire. Always grasp and pull on the boot. If you pull on the wire, you can break the conductor in the wire and ruin it.

If the ignition coils are mounted over the spark plugs, the coils must be removed before the plugs can be serviced. Normally, a few small bolts secure the coils. After removing these bolts, you can lift off the ignition coils to gain access to the spark plugs. Look at **Figure 32-11.**

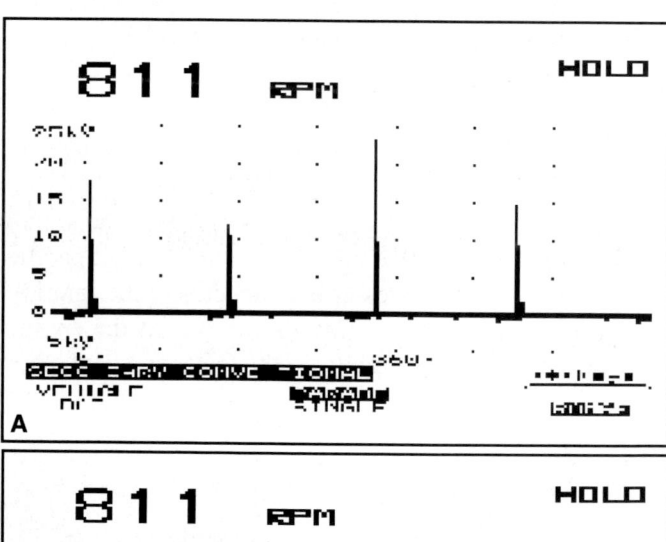

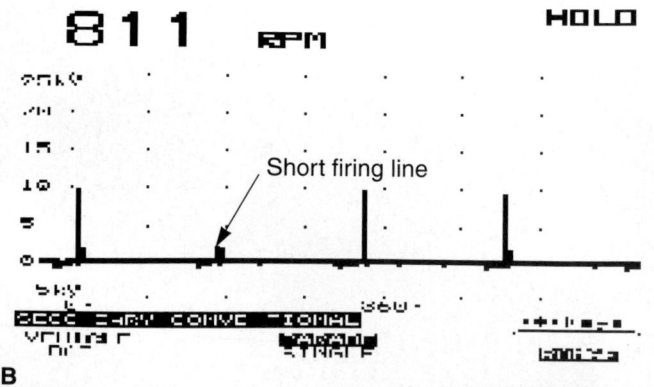

Figure 36-9. Secondary ignition system patterns can provide information on spark plug condition. A—Uneven firing lines are commonly caused by worn spark plugs. B—A short firing line is often caused by a fouled spark plug. (Fluke)

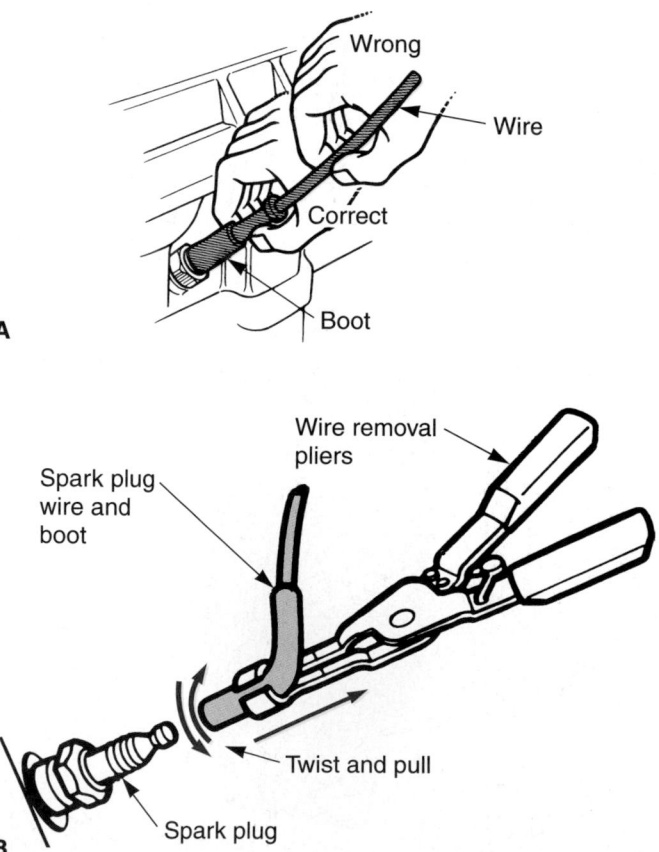

Figure 36-10. When removing a spark plug wire, pull on the boot, never on the wire. A—Using hands to remove a spark plug wire. B—Using special pliers to grasp and pull on the boot. (Toyota and Mopar)

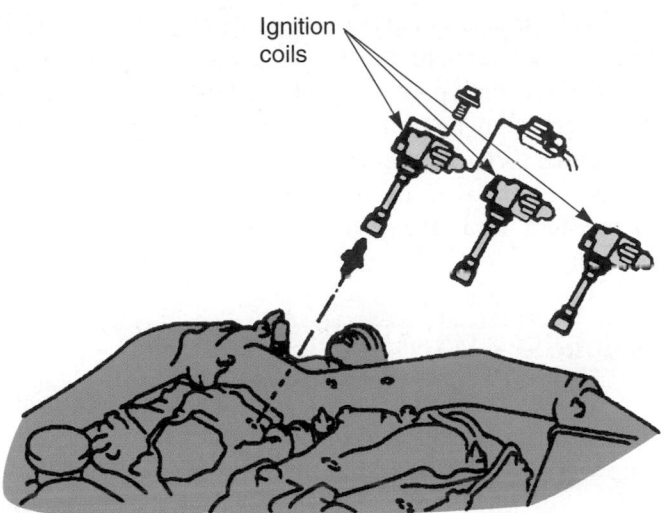

Figure 36-11. In a direct ignition system, the ignition coils are mounted over the top of the spark plugs. You must remove the coils before removing the plugs. Be careful not to drop the small coil mounting screws into the engine. (Mazda)

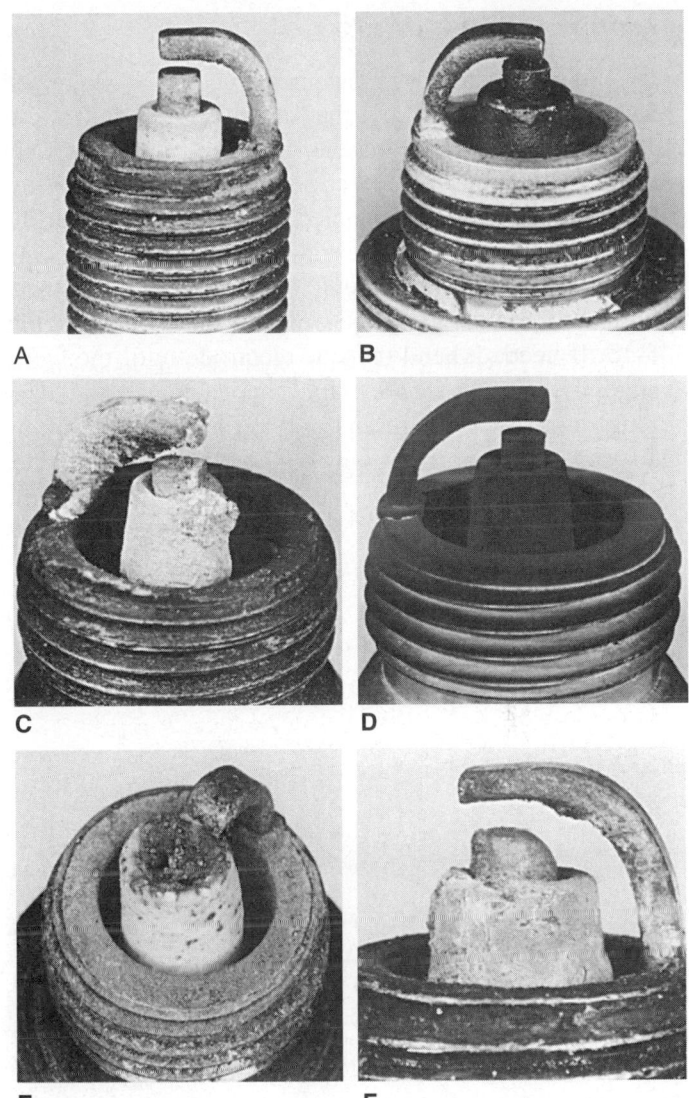

Figure 36-12. Read the spark plugs by inspecting the condition of the insulators and electrodes. Study these typical examples. A—Normal used plug (engine and ignition in good condition). B—Oil-fouled plug (worn rings, scored cylinder, or leaking valve seals). C—Ash-fouled plug (poor fuel quality or some oil entering cylinder). D—Carbon-fouled plug (slow-speed driving, plug heat range too cold, weak ignition, or rich mixture). E—Preignition damage (timing too far advanced, low-octane fuel, or plug heat range too high). F—Normal electrode erosion (old plug with prolonged use). (Champion Spark Plugs)

Before removing the plugs, use compressed air to blow debris away from the plug holes. This will prevent particles from falling into the cylinders when the plugs are removed.

Using a spark plug socket, an extension, and a ratchet, as needed, unscrew each spark plug. As you remove the plugs, lay them in order on the fender cover or workbench. Do not mix up the plugs. After all the plugs have been removed, inspect them to diagnose the condition of the engine.

Reading Spark Plugs

To read a spark plug, closely inspect and analyze the condition of its tip and insulator. This will give you information on the condition of the engine, fuel system, and ignition system.

For example, a properly burning plug should have a brown to grayish-tan color.

A black or wet plug indicates that the plug is not firing or that there is an engine problem (worn piston rings and cylinders, leaking valve stem seals, low engine compression, or rich fuel mixture).

Carefully study the spark plug conditions shown in **Figure 36-12**. Learn to read used spark plugs properly. They can provide valuable information when troubleshooting problems.

Cleaning Spark Plugs

Most manufacturers do not recommend spark plug cleaning. Blasting can roughen the insulator and lead to

fouling, misfiring, and loss of performance. However, if new spark plugs are not available, used plugs can be cleaned with a *spark plug cleaner* (air-powered device that blasts the plug tip with abrasive).

Be very careful that the abrasive (sand) does not wedge up inside the insulator. If this occurs, the sand could fall into the cylinder and cause cylinder scoring. Most shops install new plugs rather than taking the time to clean used spark plugs.

Gapping Spark Plugs

Obtain the correct replacement plug (heat range and reach) recommended by the manufacturer. Then, set spark plug gap by spacing the side electrode the correct distance from the center electrode. If the new spark plugs have been dropped, mishandled, or incorrectly set at the factory, the gap may not be within specifications.

A wire feeler gauge should be used to measure spark plug gap. Slide the gauge between the electrodes, **Figure 36-13.** If needed, bend the side electrode until the feeler gauge fits snugly in the gap. The gauge should drag lightly as it is pulled in and out of the gap. Spark plug gaps vary from approximately 0.030″ (0.76 mm) on older ignition systems to over 0.080″ (2.03 mm) on electronic systems.

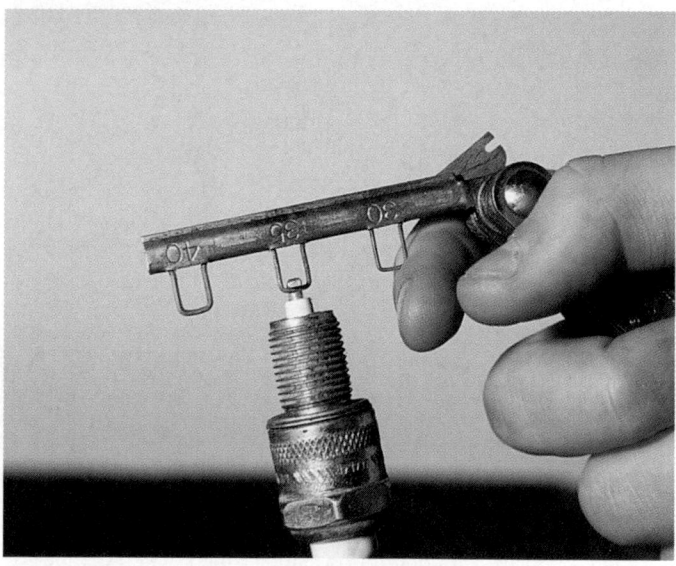

Figure 36-13. To gap a spark plug, use a wire feeler gauge to measure the gap. Both electrodes should just touch the gauge. If needed, bend the side electrode with the gapping tool to achieve the correct electrode spacing. (Snap-On Tools)

Installing Spark Plugs

Spark plugs set to the correct gap are ready for installation. Use your fingers, a spark plug socket, or a short piece of vacuum hose to start the plugs in their holes.

 Caution
Never use the ratchet to start spark plug threads because the plug and cylinder head threads could be crossthreaded and damaged. Hand start the spark plugs to make sure they go in smoothly.

With the spark plugs threaded into the head a few turns by hand, turn them in the rest of the way with your ratchet.

Tighten the spark plugs according to the manufacturer's recommendations. Some manufacturers specify a spark plug torque. Others recommend bottoming the plug on its seat and then turning it an additional one-quarter to one-half turn. Refer to a service manual for exact procedures.

Secondary Wire Service

A faulty spark plug wire can have a burned or broken conductor or deteriorated insulation. An open secondary wire will keep voltage and current from reaching the spark plug. Most spark plug wires have a resistance conductor that is easily damaged.

Secondary wire leakage occurs when the insulation is faulty and sparks jump through the insulation to ground or to another wire. This problem can keep the spark from reaching the spark plugs.

Removing Spark Plug Wires

Before removing spark plug wires, make sure they are numbered. This will simplify installation.

Remove each wire from its plug and from the distributor cap or coil pack. **Figure 36-14** shows several ways spark plug wires are removed from a distributor cap. **Figure 36-15** shows how the spark plug wires are installed in a distributorless ignition system.

Secondary Wire Resistance Test

A *secondary wire resistance test* can be used to check the condition of a secondary wire conductor (spark plug wire or coil wire). It is commonly used to check for a bad wire when an oscilloscope is not available or handy.

To perform a wire resistance test, connect an ohmmeter to each end of the secondary wire, **Figure 36-16.** The meter will measure the wire's internal resistance in ohms. Compare your reading to specifications.

Typically, secondary wire resistance should not be over approximately 12,000 ohms per foot or about 50,000 ohms maximum for long wires. Since specifications vary, always check an appropriate service manual for the exact value. A bad spark plug wire will often have almost infinite (maximum) resistance.

Secondary Wire Insulation Test

A *secondary wire insulation test* is used to check for sparks arcing through the insulation to ground. An ohmmeter test will not detect bad insulation.

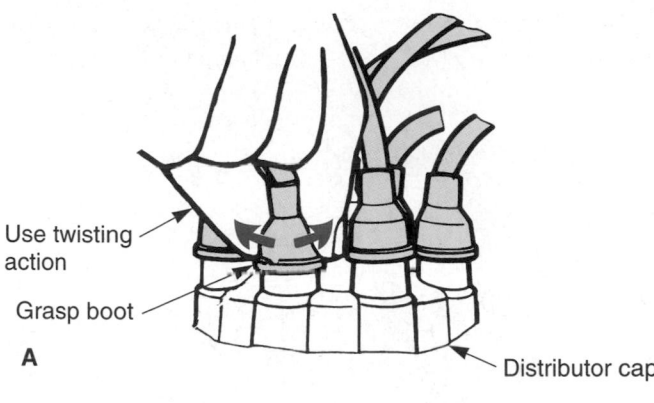

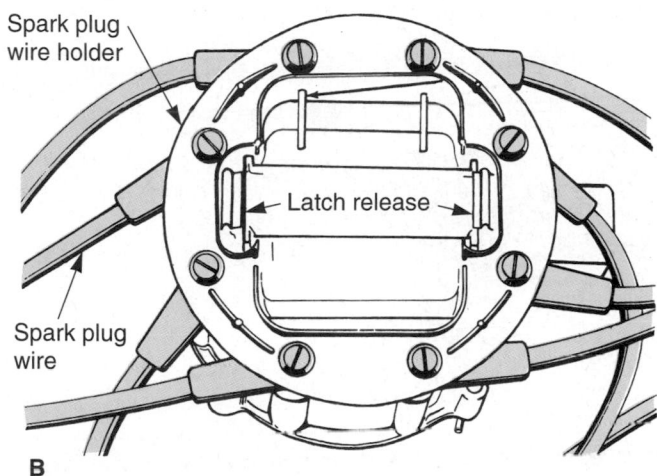

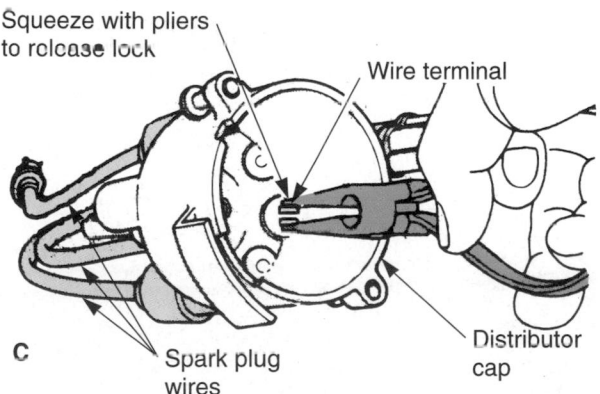

Figure 36-14. Three typical methods of securing spark plug wires in a distributor cap. A—Press fit. Twist and pull on the boot to remove the wire. B—Large ring holds all wires in the cap. Release the latches and lift the ring to release the plug wires. C—Snap locks hold the wire terminals in the cap. Use pliers to free the locks before pulling on the wire boots. (Mopar, Oldsmobile, Chrysler)

 To check secondary wire insulation:
1. Darken the shop or place fender covers over the sides of the vehicle's hood to block out light.
2. Start the engine and visually inspect each wire for arcing.

 3. If arcing is not evident during the visual inspection, move a high-impedance test light or a grounded screwdriver along the length of each wire. Hold the tip of the tool about 1″ from the wire insulation.
4. If an arc jumps through the insulation and onto the test light probe or the screwdriver, the wire is bad and must be replaced.

 Tech Tip
Secondary wire insulation leakage will usually show up when it rains or in damp weather. The moisture will cause excessive leakage, and the engine may miss or not start.

Replacing Spark Plug Wires

Installing new spark plug wires is simple, especially if one wire is replaced at a time. Compare the length of the old wire with the length of the new wire. Replace each wire with one of equal length. Make sure the new wire is fully attached on the plug and in the distributor or coil module.

Spark plug wire replacement is more complicated if all the wires are removed at once. Then, you must use the engine firing order and cylinder numbers to route each wire correctly.

Figure 36-17 shows a typical service manual firing order illustration. It can be used to trace the spark plug wires from each coil pack tower to the correct spark plug.

 Tech Tip
Some sensor circuits use lower-than-normal reference voltages (1.5V DC). This can make them more susceptible to interference from outside magnetic fields. To prevent induced voltage from entering these sensor wires, always keep spark plug wires in their clips.

Distributor Service

A distributor is critical to the proper operation of an ignition system. It distributes high voltage to each spark plug wire. It may also alter ignition timing and sense engine speed. If any part of the distributor is faulty, engine performance suffers.

Distributor Cap and Rotor Service

A bad distributor cap or rotor can cause engine missing, ***backfiring*** (popping noise from the throttle body or carburetor), and other engine performance problems.

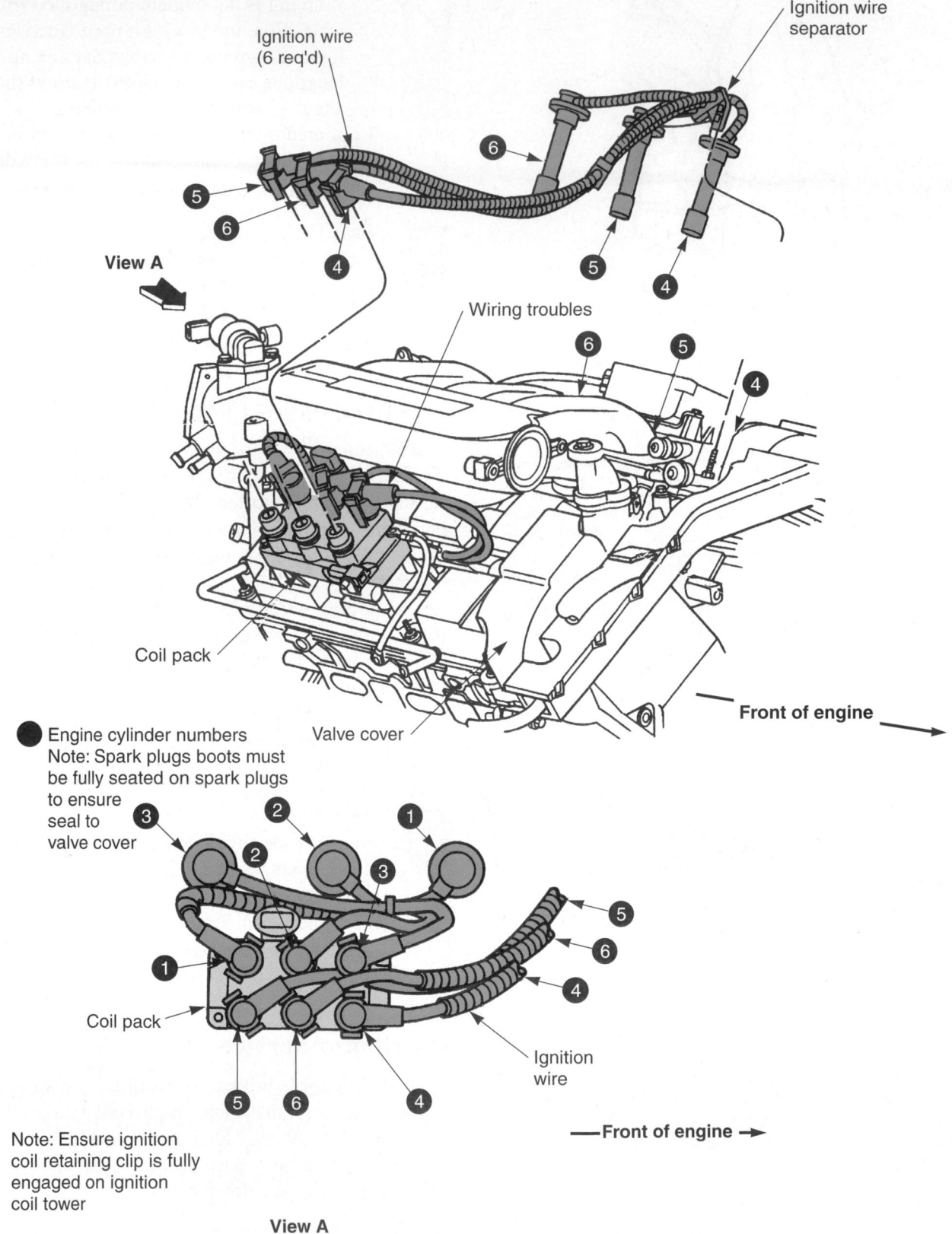

Figure 36-15. Service manual illustration gives information for removing and installing spark plug wires in a distributorless ignition system. (Ford)

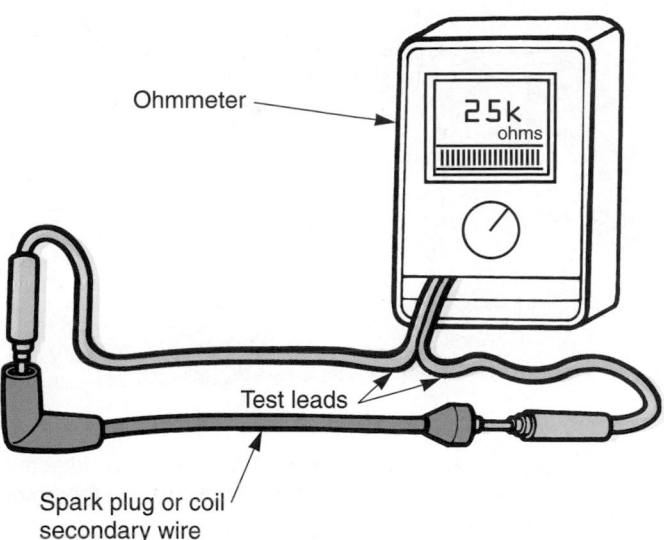

Figure 36-16. To check the conductor in a secondary wire, measure the wire's resistance with an ohmmeter. If the resistance is too high, the conductor or terminal is broken and the wire must be replaced. (Mopar)

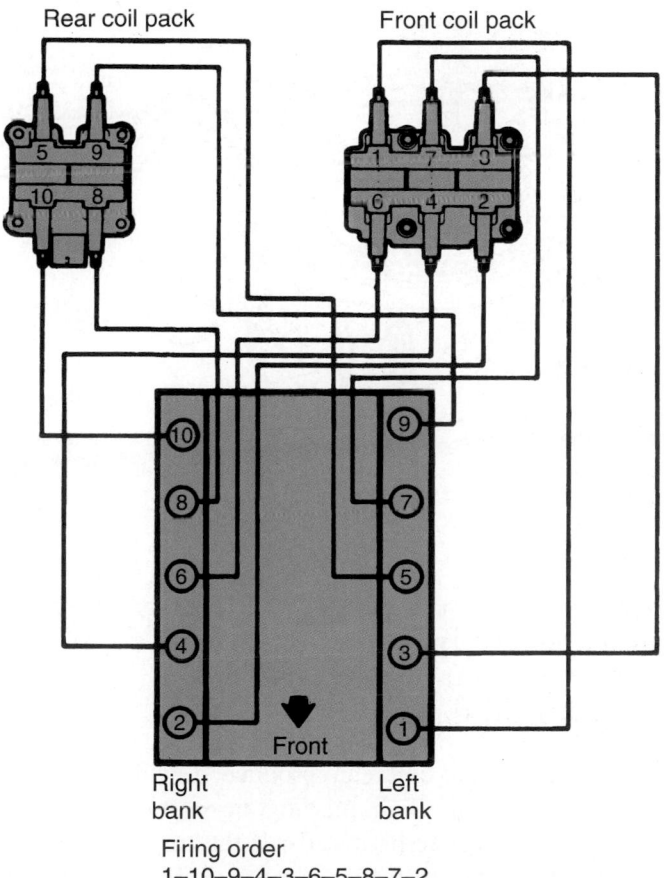

Firing order
1–10–9–4–3–6–5–8–7–2

Figure 36-17. If all the spark plug wires have been removed, locate a service manual illustration like this one for your engine. Then, trace and install the wires from the correct distributor cap or coil tower to the proper spark plug. (Chrysler)

Trouble often arises when a *carbon trace* (small line of carbon-like substance that conducts electricity) forms on the distributor cap, the rotor, the ignition coil, or the coil pack. The carbon trace will short coil voltage to ground or to the wrong plug wire. A carbon trace can cause the spark plugs to fire poorly, out of sequence, or not at all. See **Figure 36-18.**

When problems point to possible distributor cap or rotor troubles, remove and inspect them. Using a droplight, check the inside of the cap for cracks and carbon traces. A carbon trace is black, making it difficult to see on a black distributor cap or rotor.

If a crack or carbon trace is found, replace the cap and rotor as detailed in the following section. Also check the rotor tip for excessive burning, damage, or looseness. Make sure the rotor fits snugly on the distributor shaft.

Replacing the Distributor Cap and Rotor

Distributor caps can be secured by either screws or spring-type metal clips. Normally, turn the screws counterclockwise for removal. With spring clips, pry on the top of spring clips, being careful not to crack the cap. The clips should pop free. Wiggle and pull upward on the cap to remove it from the distributor body.

Tech Tip

In unitized distributor arrangements, the center terminal in the distributor cap commonly burns. If the spark plugs have not been changed and their gaps are too wide, high coil voltage will burn the center terminal and cap. Keep this in mind during unitized distributor service.

Rotors may be held by screws, or they may be press fit onto the distributor shaft. Pulling by hand will usually free a press-fit rotor. Refer to **Figure 36-19.** If stuck, carefully pry under the rotor.

To install a rotor, properly line up the rotor on the distributor shaft. With a press-fit rotor, a tab inside the rotor fits into a groove or slot in the shaft. With a screw-held rotor, the rotor may have round or square dowels that fit into holes in the distributor.

When installing a distributor cap, a notch or tab on the cap must line up with a tab or notch on the distributor housing. This ensures that the cap is correctly aligned with the rotor. Before securing the spring clips or screws, push down on the cap while twisting it from side to side. Make sure the cap does not wobble on the distributor.

In some ignition systems, the ignition coil is housed inside the distributor cap. In this case, the coil must be taken out of the old cap and installed in the new one.

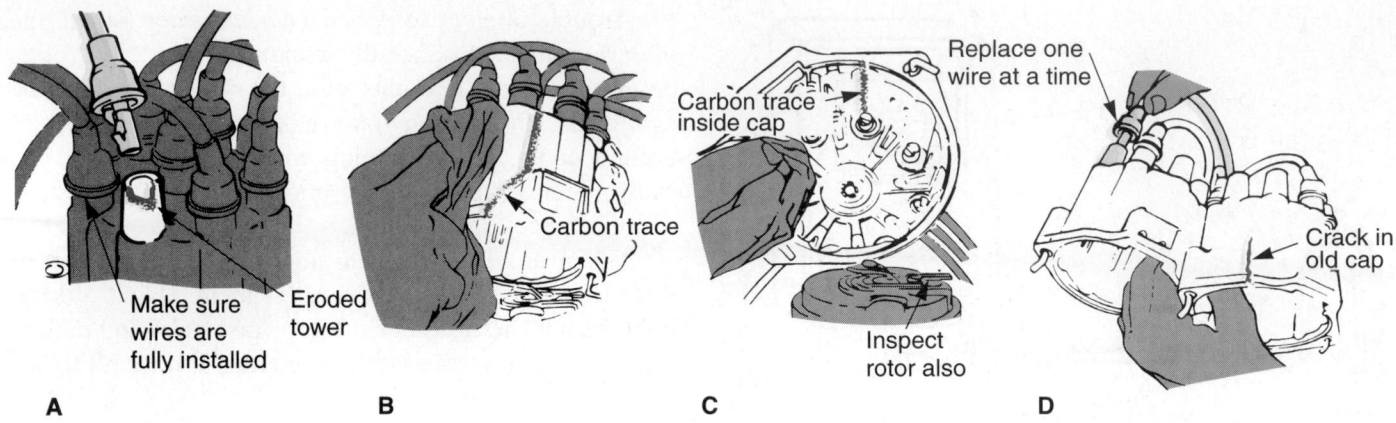

Figure 36-18. Inspect the distributor cap and rotor for burning, carbon traces, and cracks. Replace them if problems are detected. A—Burned, eroded cap tower. B—Carbon trace on the outside of the cap. C—Carbon trace in the cap and on the rotor. D—Crack in the cap. When replacing a damaged cap, replace the secondary wires one at a time. Install the wire in the proper tower on the new cap. (GMC)

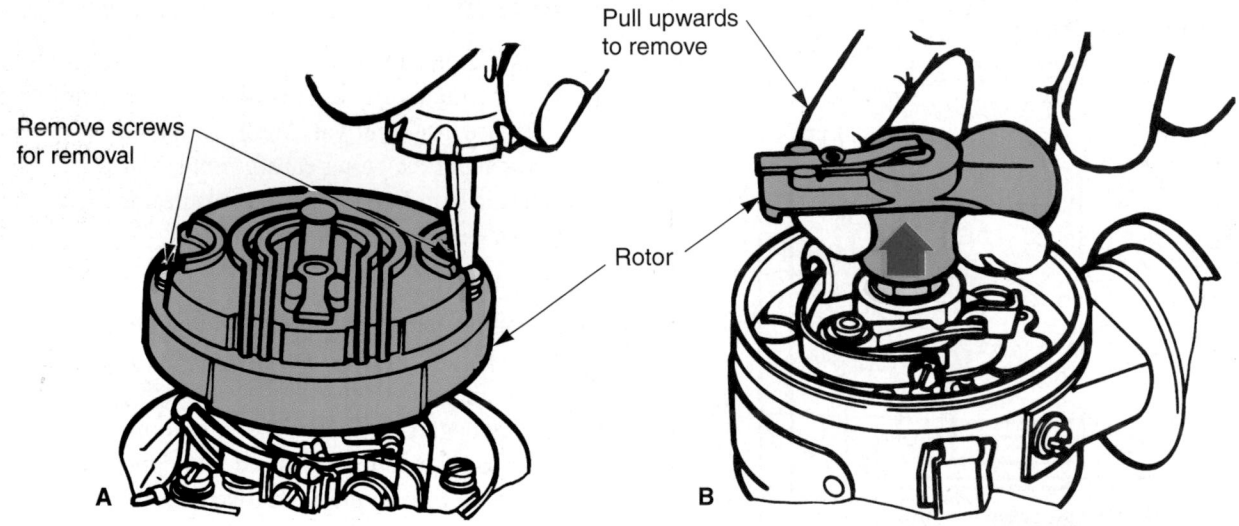

Figure 36-19. Two types of rotor mountings. A—Screws hold the rotor on the distributor. The rotor has dowels that must line up in the distributor. B—Press-fit rotor is pulled off and pushed on the distributor shaft. It has a lug that fits into a notch in the distributor shaft. (Mopar)

 Caution
If a distributor cap is not installed correctly, rotor and distributor damage can result. The spinning rotor can strike the sides of the distributor cap, breaking the plastic rotor and cap.

Electronic Ignition Distributor Service

As you have learned, most electronic ignition distributors use a pickup coil to sense trigger wheel (distributor shaft) rotation. The pickup coil sends small electrical impulses to the ignition control module.

If the distributor fails to operate properly, the complete ignition system can stop functioning. It is important to know how to make several basic tests on electronic ignition distributors.

Pickup Coil Service

A bad pickup coil can produce a wide range of engine problems: stalling, missing, no-start troubles, and loss of power at specific speeds. If the tiny windings in the pickup coil break, they can cause problems that only occur under certain conditions. Also, because of vibration and movement, the thin wire leads going to the pickup coil can break. Though the insulation may look fine, the conductor could be damaged inside the insulation. When this happens, the engine may lope, miss, or not run.

Testing a Magnetic Pickup Coil

A magnetic pickup coil or a speed sensor can be located in the distributor or on the engine block (crankshaft speed or position sensor). Tests for either type are similar.

Your scan tool may show a readout of primary circuit problems with a bad pickup coil or crankshaft sensor. Refer to the service manual for details.

A *magnetic pickup coil test* compares actual sensor coil resistance or voltage output with specifications. If resistance or voltage output is too high or low, the unit is bad.

 To perform a pickup coil test:
1. Connect an ohmmeter or low-reading ac voltmeter across the pickup coil output leads, **Figure 36-20.**
2. Observe the meter reading. Ohmmeter readings will usually vary between 250 and 1500 ohms. If an ac voltmeter is used, a small ac voltage (3-8 V) should be produced by the magnetic pickup coil when the engine is cranked. Check a service manual for exact specifications.
3. Wiggle the wires to the pickup coil while watching the meter. This will help locate a break in the leads to the pickup. Also, lightly tap on the coil with the handle of a screwdriver. This could uncover any break in the coil windings.
4. If the meter reading is not within specifications, or if the reading changes when the leads are moved or the coil is tapped, replace the pickup coil.

Refer to specifications for exact voltage values and procedures.

Testing Hall-Effect and Optical Sensors

Hall-effect and optical sensors are tested in much the same way as the more common magnetic sensor. You can check their output signals and compare them to specifications. However, they are often tested with a scope to more accurately analyze their output signals.

 Note
For information on scope testing magnetic sensors, Hall-effect sensors, optical sensors, etc., refer to Chapter 46, *Advanced Diagnostics.* Earlier chapters on computer system troubleshooting and service also include useful information on testing these sensors.

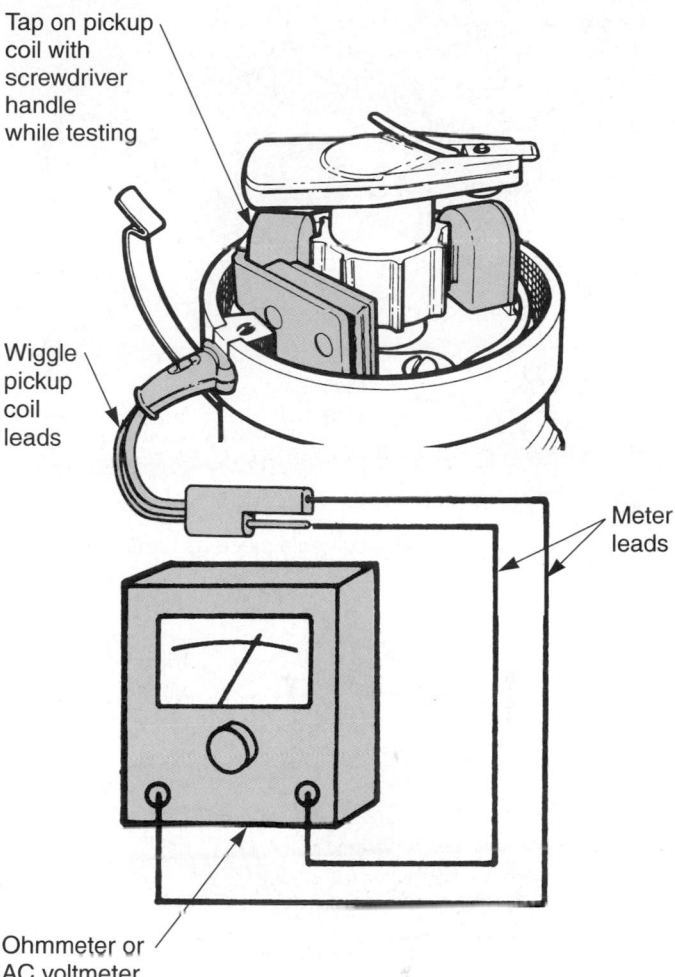

Figure 36-20. Connect an ohmmeter to the pickup coil leads. Wiggle the wires and tap on the coil with a screwdriver handle. If the ohmmeter reading does not remain steady and within specifications, replace the pickup coil. (Chrysler)

Replacing the Pickup Coil

A distributor pickup coil can usually be replaced simply by removing the distributor cap, the rotor, and the screws holding the coil to the advance plate. Sometimes the pickup coil is mounted around the distributor shaft. Since procedures vary, find detailed directions for the particular distributor in a service manual. See **Figure 36-21A.**

The *pickup coil air gap* is the space between the pickup coil and a trigger wheel tooth. In some designs, it must be set after installing the pickup coil.

To set the air gap, position the trigger wheel so that one of its teeth points at the pickup coil. Slide the correct size nonmagnetic feeler gauge (plastic or brass gauge) between the pickup coil and the trigger wheel tooth. Move the pickup coil in or out until the gauge fits snugly in the gap. Tighten the pickup screws and double-check the air gap setting. Look at **Figure 36-21B.**

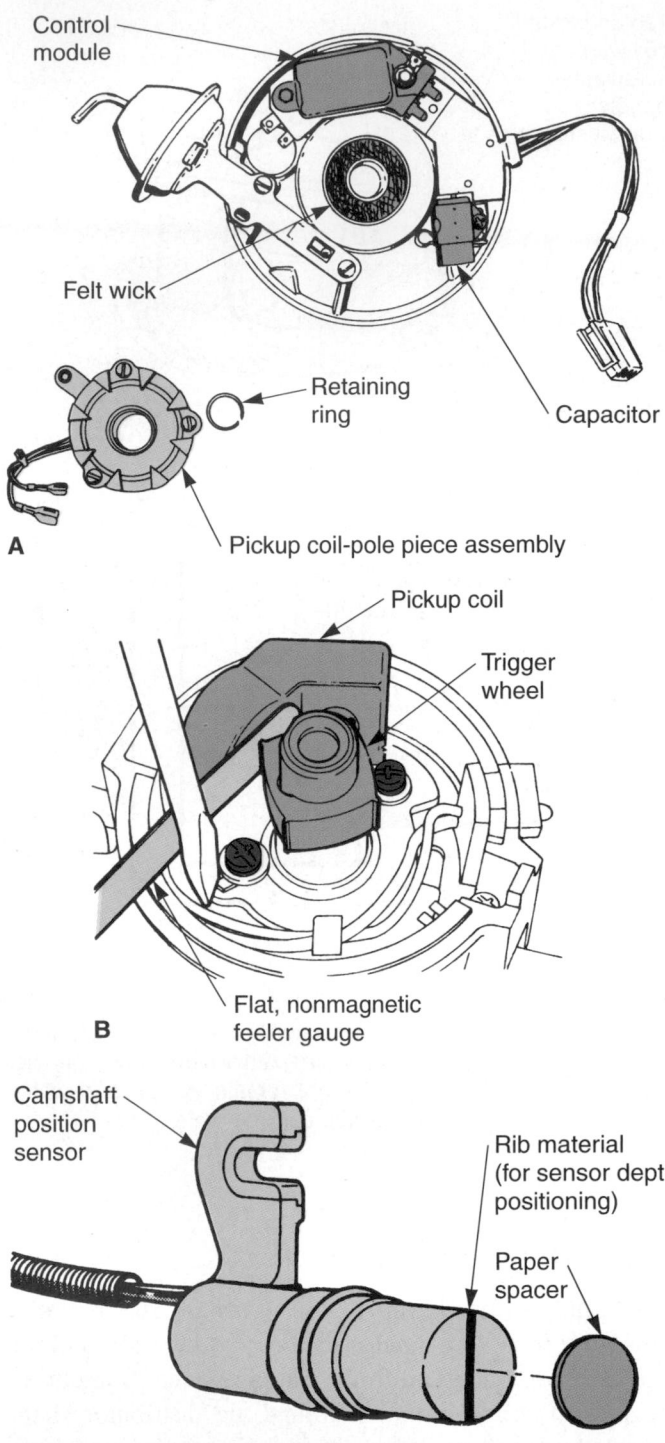

Figure 36-21. A—To remove this pickup coil, you must remove the distributor shaft and the retaining ring. Many other pickup coils can be removed by loosening two screws. B—In some distributors, pickup coil air gap must be measured and adjusted. With a trigger wheel tooth pointing at the pickup, slide the correct nonmagnetic feeler gauge into the gap. Adjust the pickup until the gauge just fits in the gap. C—This crankshaft position sensor must be adjusted. A special paper spacer is stuck on the end of the sensor before the sensor is installed in the block. The spacer will set the correct air gap. The paper spacer must be removed after the sensor mounting screws have been tightened. (Buick and Chrysler)

Replacing a Crankshaft Position Sensor

The crankshaft position sensor, which is used to monitor engine speed in many modern ignition systems, can be mounted on the front, side, or rear of the engine.

When mounted on the front of the engine, the crankshaft position sensor normally fits into a bracket hole and is secured with a small bolt. When mounted on the side of the engine block, the sensor normally extends into the crankcase and a rubber O-ring seal prevents oil leakage. When installing this type of sensor, always install a new O-ring seal and make sure there is no debris under the sensor. This could hold the sensor too far away from its trigger wheel and upset normal operation.

Most crankshaft position sensors simply lock into place and are not adjustable. However, a few require adjustment. Crankshaft position sensor air gap is the distance from the tip of the sensor to an unnotched area on the trigger wheel. This gap must be correct for proper sensor signal output. One method of crankshaft sensor adjustment is shown in **Figure 36-21C.**

Contact Point Distributor Service

Bad contact points (points having burned, pitted, misaligned contacts, or worn rubbing block) cause a wide range of engine performance problems. These problems include high-speed missing, no-start problems, and many other ignition-related troubles.

A faulty condenser may leak (allow some DC current to flow to ground), be shorted (direct electrical connection to ground), or be open (broken lead wire to condenser foils).A leaking or open condenser will cause point arcing and burning. If the condenser is shorted, primary current will flow to ground and the engine will not start.

Testing Distributor Points

Many technicians visually inspect the surfaces of the contact points to determine their condition. Points with burned and pitted contacts or a worn rubbing block must be replaced. If the points look good, point resistance should be measured. Many dwell-tachometers, have a scale for measuring point resistance.

To measure contact point resistance, crank the engine until the points are closed. Connect the meter to the primary point lead and to ground. If the resistance is too high (out of scale markings for condenser test), the points are burned and must be replaced.

Testing a Distributor Condenser

Most technicians simply replace a condenser anytime symptoms point to a condenser problem. However, an ohmmeter can be used to test a condenser.

The ohmmeter is connected to the condenser (capacitor) as shown in **Figure 36-22.** The meter should register momentarily and then return to infinity (maximum resistance). Any continuous reading other than infinity means that the condenser is leaking and must be replaced.

 Tech Tip
An ohmmeter condenser test can also be used to check the noise suppression condensers found in various circuits to prevent radio noise.

Replacing the Points and Condenser

Normally, the distributor points and condenser are held in place by small screws. To prevent dropping the screws, use a magnetic or clip-type screwdriver that firmly holds the screws. If you drop one of the screws under the distributor advance plate (breaker plate), the distributor may have to be removed from the engine to retrieve the screw. Use a small wrench to disconnect the primary wires from the points.

Wipe the distributor cam and advance plate clean before installing the new parts. If recommended, apply a small amount of oil to lubrication points on the distributor (wick in the center of shaft, cam wick, or oil hole in the side of the distributor housing). Do not apply too much oil because it could get on the points.

To prolong service life, place a small amount of grease on the side of the breaker arm rubbing block. This will reduce friction between the distributor cam and the fiber block.

Fit the points into the distributor. Install the point screws, the condenser, and the primary wires. If the distributor cap has a window (square metal plate), fully tighten the point hold-down screws. If the distributor cap does not have a window, only partially tighten the screws so the points can be adjusted.

Adjusting Distributor Points

Distributor points can be adjusted using either a feeler gauge (metal blade ground to precise thickness) or a dwell meter.

To use a feeler gauge to gap (set) distributor points, crank the engine until the points are fully open. The point rubbing block should be on top of a distributor cam lobe. This is illustrated in **Figure 36-23.**

Distributor point gap is the recommended distance between the contacts in the fully open position. Look up this specification in the service manual. It may also be given on the emission sticker in the engine compartment. Typical point gap settings average around 0.015″ (0.38 mm) for eight-cylinder engines and 0.025″ (0.53 mm) for six- and four-cylinder engines.

With the distributor points open, slide the specified thickness feeler gauge between the points, **Figure 36-23.** Adjust the points so that there is a slight drag on the blade of the gauge. Use a screwdriver or an Allen wrench, depending on point design, to open or close the points. If needed, tighten the hold-down screws and recheck point gap.

Make sure your feeler gauge is clean before inserting it between the points. Oil or grease will reduce the service life of the points.

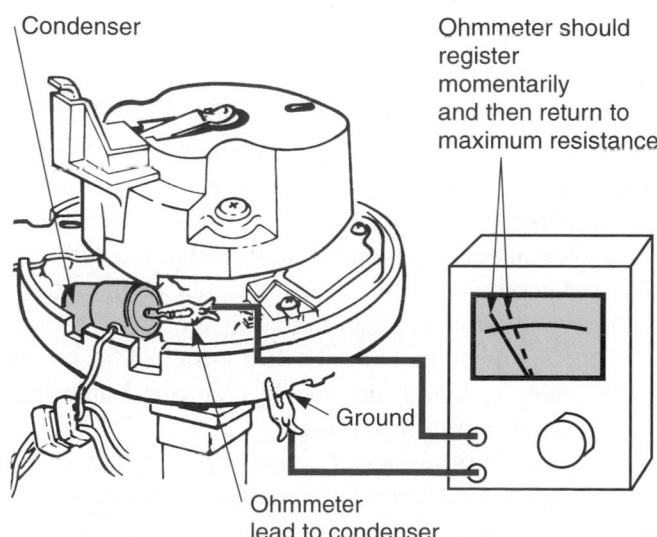

Figure 36-22. An ohmmeter can be used to check a condenser for shorts. When the meter is connected, it should register momentarily. Then, as the capacitor charges from the meter current, the reading should return to infinity. A continuous reading indicates a shorted condenser. (Echlin)

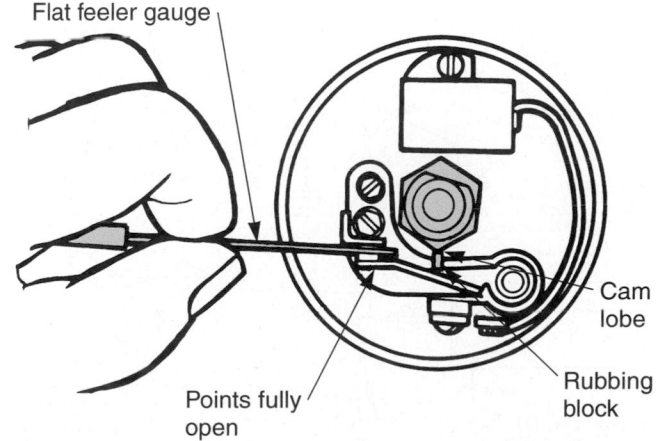

Figure 36-23. To set the points, "bump" the engine until the cam lobe is pushing on the rubbing block. The points must be fully open. Slide the correct thickness feeler gauge between the points. Open or close the points until the gauge just touches both contacts. (Mopar)

Using a Tach-Dwell Meter

The ***tach-dwell meter*** is composed of a tachometer and a dwell meter. It can be used to measure engine speed, dwell, and sometimes, resistance.

The ***tachometer,*** or ***tach,*** section of the tach-dwell meter measures engine speed by checking how often the ignition coil(s) fires. The tach is often used to adjust engine idle speed or to check engine speed during ignition timing adjustment.

The ***dwell meter*** portion of the tach-dwell meter measures current flow as compared to degrees of engine crankshaft or distributor rotation. On contact point ignition systems, it is used to measure point dwell, which is the time the points are closed and conducting primary current through the ignition coil windings. On electronic ignition systems, the dwell meter is used to measure the amount of time the ignition control module conducts current through the ignition coil(s).

When using a dwell meter to adjust distributor points, follow the directions provided with the meter, **Figure 36-24.** Typically, connect the red lead to the distributor side of the coil (wire going to contact points). Connect the black lead to ground (any metal part on engine).

If an opening is provided in the distributor cap, the points should be set with the engine running. Install the distributor cap and rotor. Start the engine. With the meter

controls set properly, adjust the points using an Allen wrench or a special screwdriver-type tool. Turn the point adjustment screw until the dwell meter reads within specs.

If the distributor cap does not have an adjustment window, set the points with the cap removed. Instead of starting the engine, ground the coil wire (connect output end on engine). Crank the engine with the starting motor. This will simulate engine operation and allow point adjustment with the dwell meter.

Dwell

Dwell specifications vary with the number of cylinders in the engine and the type of ignition system (point, electronic, or computerized). An 8-cylinder engine with contact points will usually require 30° of dwell. An engine with fewer cylinders will normally require more dwell time.

An electronic or computer-controlled system can have different amounts of dwell time to energize the ignition coil(s). Always obtain exact dwell values from a tune-up chart or shop manual.

Fixed and Variable Dwell

With modern electronic or computer-controlled ignition systems, the dwell can vary with system design. There are a few terms you should understand concerning ignition dwell.

Fixed dwell means that the dwell time should remain the same at all engine speeds. Fixed dwell is found on contact point systems and older electronic systems. If the dwell varies when it should not, it is usually due to distributor shaft or bushing wear or an ignition control module failure.

Variable dwell means the engine control module alters ignition coil dwell time with engine speed. This is common in most late-model vehicles. At low engine speeds, the control module can use a shorter dwell period to build an adequate magnetic field around the coil windings for good spark. However, at higher engine speeds, the module increases dwell time to make sure the coil(s) fires the spark plugs properly. If a system fails to alter dwell when it should, the engine control module has a problem and should be replaced.

Current-limiting dwell means the engine control module sends high current through the ignition coil windings until a strong magnetic field is developed around the coil windings. Once the module senses a ***saturated ignition coil*** (coil's magnetic field is fully formed), it reduces the amount of current sent through the coil windings. Only a small primary current is needed to maintain the strong magnetic field in the coil. At high engine

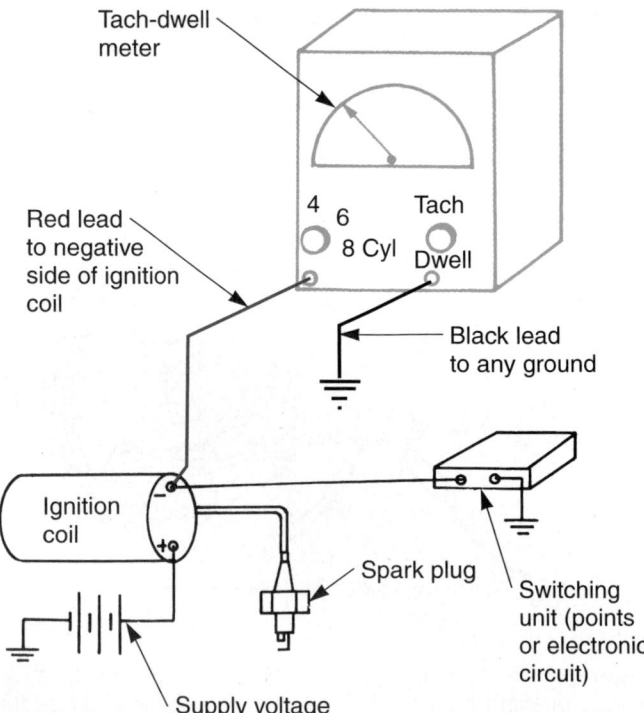

Figure 36-24. A tach-dwell meter is connected to contact point and some electronic ignition systems as shown. (TIF)

speeds, the current-limiting feature may not be needed. Full control module current output may be needed to fully charge the ignition coil to ready it to fire the spark plugs at high engine rpm.

Ignition Timing Adjustment

Initial ignition timing is the spark timing set by the technician with the engine idling (no advance). It must be adjusted whenever the distributor has been removed and reinstalled in an engine. During a tune-up, initial timing must be checked and then adjusted if not within specifications. Ignition timing is sometimes checked with a timing light, **Figure 36-25.**

Initial ignition timing is sometimes adjusted by turning the distributor housing in the engine. This makes the pickup coil and electronic control unit (or the contact points) fire the ignition coil earlier or later.

Many computer-controlled ignition systems have no provision for timing adjustment. A few, however, have a tiny screw or lever on the computer for minor ignition timing changes.

When the ignition timing is *too advanced,* the engine may suffer from spark knock or ping. A light tapping sound may result when the engine accelerates or is under a load. The ping (abnormal combustion) will sound like a small hammer tapping on the engine.

When ignition timing is *too retarded,* the engine will have poor fuel economy and will lack power. It will also be very sluggish during acceleration. If timing is extremely retarded, combustion flames blowing out of the opened exhaust valve can overheat the engine and crack the exhaust manifolds.

Figure 36-25. A timing light is sometimes used to check ignition timing. (OTC)

Energizing Base Timing

Base timing is the ignition timing without computer-controlled advance. Base timing is checked by disconnecting a wire connector in the computer wiring harness or by jumping across specific pins on a service connector, **Figure 36-26.** The connector may be on the engine, next to the distributor, or in the passenger compartment. Refer to a manual for the exact location. When in the base timing mode, you can use a conventional timing light to measure ignition timing.

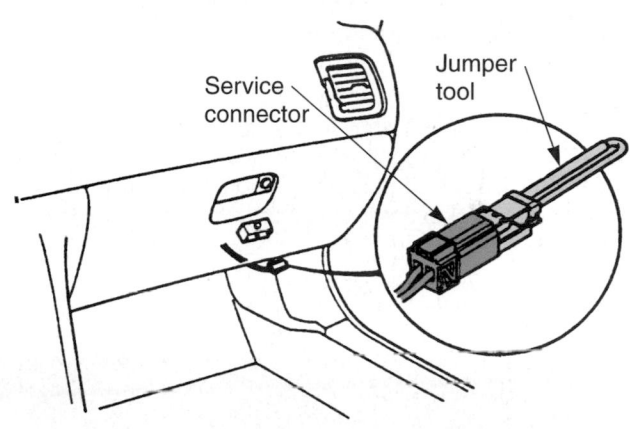

Figure 36-26. With many computer controlled ignition systems, you must trigger base timing to shut off the electronic advance. You may have to jump across special connector terminals or disconnect a timing connector. (Honda)

Measuring Ignition Timing

A *timing light* is sometimes used to measure ignition timing. As shown in **Figure 36-27,** a timing light normally has three leads. The two small leads connect to the battery. The larger lead generally connects to the *number one* spark plug wire.

Depending on the type of timing light, the large lead may be connected directly to the metal terminal of the plug wire (conventional type) or it may clip around the spark plug wire (inductive type).

When the engine is running, the timing light will flash *on* and *off* like a strobe light. The flashing action can make a moving object appear stationary (stand still), so it can be viewed.

Before measuring engine timing, disconnect and plug the vacuum advance hose going to the distributor (if used).

Start the engine and aim the timing light on the timing marks, **Figure 36-28.** The timing marks may be

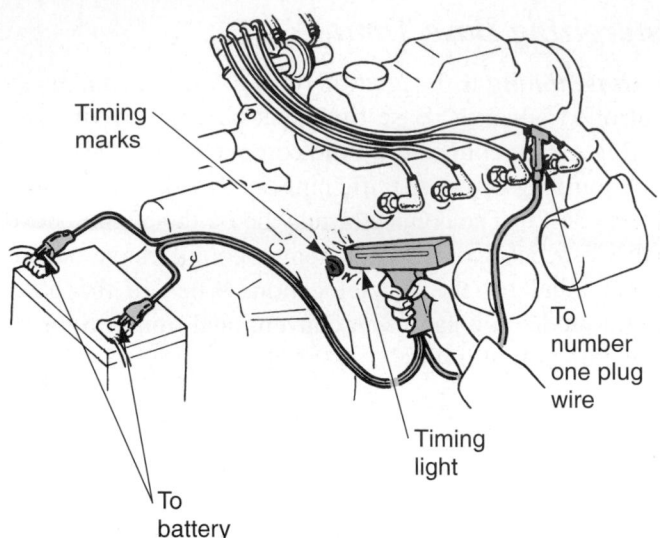

Figure 36-27. A timing light is connected to the system as shown. The large lead connects to the number one spark plug wire. The small leads connect to the battery. The light is aimed at the timing marks on the flywheel or at the front engine cover and damper. (Honda)

Figure 36-28. Timing marks can be difficult to see on modern vehicles. Find the correct angle for shining the light and viewing the marks when adjusting ignition timing. (Snap-On Tools)

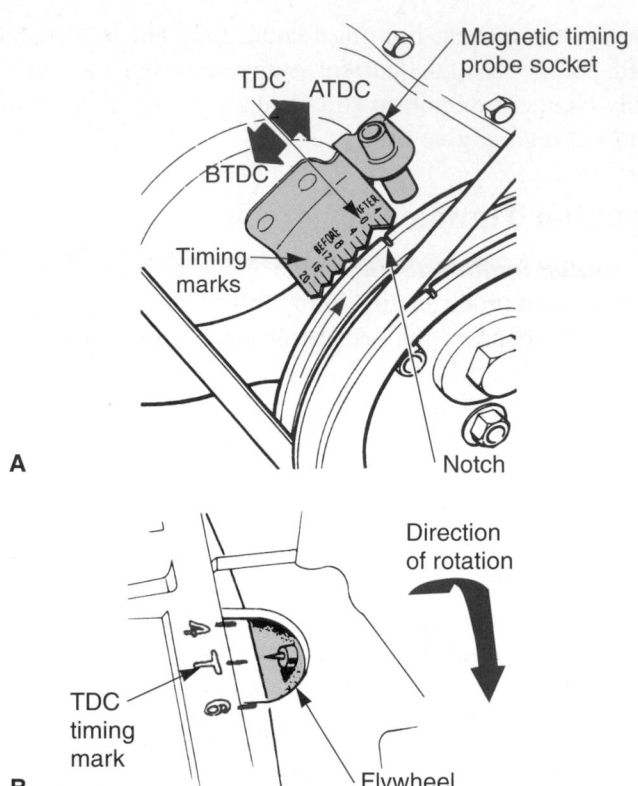

Figure 36-29. A—Timing marks on the front of the engine. B—Timing marks on the engine flywheel and clutch housing. (Chrysler and Honda)

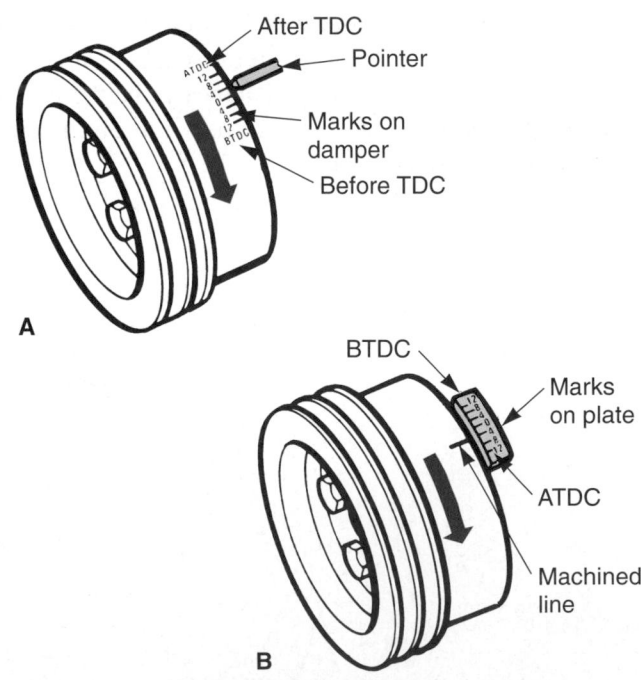

Figure 36-30. A—When degree markings are on a moving damper, read the marking lined up with the pointer. B—With stationary degree markings, read the marking lined up with the line machined in the damper. When the timing light is aimed at the markings, the rotating marks will appear to be stationary. (Chrysler)

on the front cover and harmonic balancer of the engine. The timing marks may also be on the engine flywheel. **Figure 36-29** shows ignition timing marks.

The flashing timing light will make the mark or marks on the harmonic balancer or flywheel appear to stand still. This will let you determine whether or not the engine is timed properly. For example, if there is only one reference line on the harmonic balancer, simply read initial timing by noting the degree mark that is aligned with the reference line, as in **Figure 36-30.**

Tech Tip
Sometimes a dual trace scope is needed to check ignition timing. One scope lead is connected to the ignition coil primary and another is connected to the crankshaft position sensor. The two waveforms can be compared to measure ignition timing advance or retard.

Ignition Timing Specifications

Ignition timing is very critical to the performance of an engine. If the timing is off by even 2 or 3 degrees, fuel economy and power can drop considerably. Always obtain the exact timing specifications from the vehicle's emission control sticker or an appropriate service manual.

Adjusting Ignition Timing

If the timing marks do not line up correctly, the ignition timing must be adjusted. As previously mentioned, timing can sometimes be adjusted by rotating the distributor or by moving the mounting for the engine speed sensor or crankshaft position sensor. In a few systems, a tiny screw or lever is provided for minor timing adjustments. Refer to an appropriate service manual for exact procedures

Tech Tip
In some systems, ignition timing cannot be adjusted. If timing is incorrect in these systems, the ECM or another component affecting timing must be replaced. Refer to the service manual for more information on specific vehicles.

To adjust timing in systems that require rotating the distributor:
1. Loosen the distributor hold-down bolt. A *distributor wrench* (long, specially shaped wrench for reaching under distributor housing) is handy for this purpose, **Figure 36-31.** Only loosen the distributor bolt enough to allow distributor rotation. Do *not* remove the bolt.
2. Shine the timing light on the engine timing marks.
3. Turn the distributor one way or the other until the timing marks line up correctly, **Figure 36-32.**
4. Tighten the hold-down bolt and double check the timing.
5. Reconnect the distributor vacuum hose and disconnect the timing light.

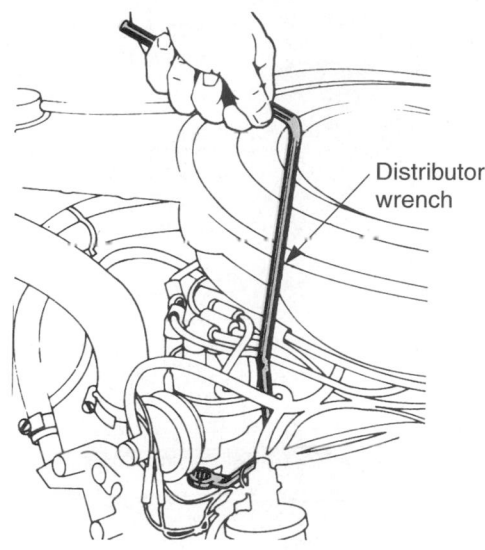
Figure 36-31. A distributor wrench may be needed to loosen the hold-down bolt on the distributor. The bolt can be hard to reach. (K-D Tools)

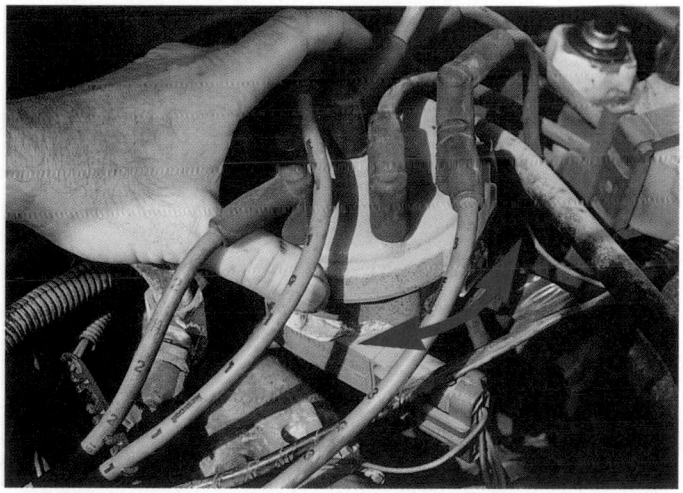
Figure 36-32. To change ignition timing, turn the distributor housing in the engine. Turning the distributor in the direction of rotor rotation will retard timing. Turning the distributor against rotor rotation will advance timing.

Warning
Keep your hands and the timing light leads away from the engine fan and belts. The spinning fan and belts can damage the light or cause serious injury!

Turning the distributor housing against distributor shaft rotation advances the timing. Turning the housing with shaft rotation retards the timing, **Figure 36-33.** With modern computer advance distributors, an arrow is sometimes stamped on the cap to show direction of shaft rotation.

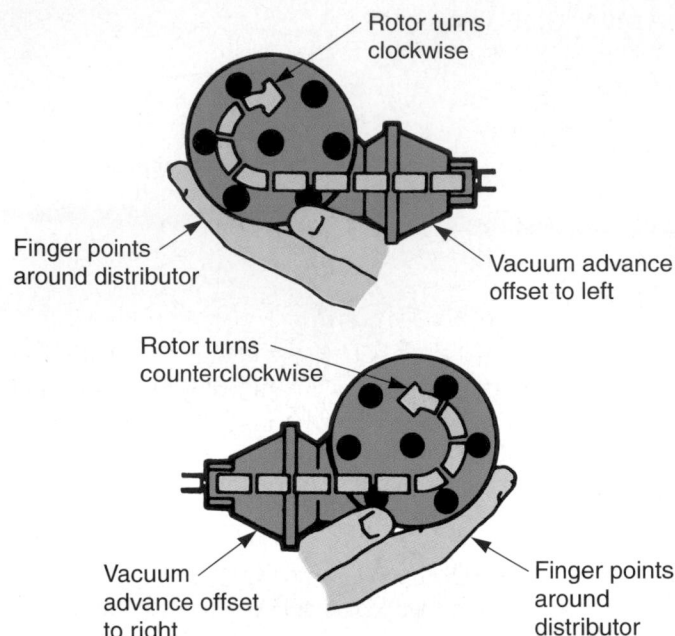

Rotor turns
clockwise

Finger points
around distributor

Vacuum advance
offset to left

Rotor turns
counterclockwise

Vacuum
advance offset
to right

Finger points
around
distributor

Figure 36-33. Use the finger rule to determine the direction of rotor rotation. Point around the distributor as shown. This will tell you how the rotor turns without removing the cap and watching the rotor. The direction of distributor rotation is needed when installing spark plug wires and setting ignition timing. (Florida Dept. of Voc. Ed.)

Testing Centrifugal and Vacuum Distributor Advance Systems

Both contact point–type distributors and older electronic distributors use similar advance mechanisms. A faulty advance mechanism will reduce engine performance and fuel economy. A timing light can be used to test the general operation of a centrifugal advance

mechanism. Connect the timing light. Remove the vacuum hose going to the distributor. Start and idle the engine.

While shining the timing light on the engine timing marks, slowly increase engine speed to approximately 3500 rpm. Note the movement of the timing mark. If the centrifugal advance is working, the mark should steadily move to a more advanced position with the increase in speed. See **Figures 36-34A** and **36-34B.** If the timing mark jumps around or does not move, the centrifugal advance mechanism is faulty. It may be worn or rusted. It may also have weak springs or other mechanical problems.

To test the distributor vacuum advance, connect a timing light. Remove the vacuum advance hose from the distributor. Start the engine and increase its speed to approximately 1500 rpm. Note the location of the timing marks. Then, reconnect the vacuum hose on the distributor diaphragm.

As soon as vacuum is reconnected, engine speed should increase and the timing mark should advance. See **Figure 36-34C.** If the vacuum advance is not working, check the vacuum advance diaphragm and the supply vacuum to the distributor.

Testing Vacuum Advance Diaphragm

To test the vacuum advance diaphragm, apply a vacuum to the unit using a vacuum pump, **Figure 36-35.** When suction is applied to the diaphragm, the advance plate in the distributor should swing around. When vacuum is released, the advance plate should snap back into its normal position.

If the advance diaphragm leaks and will not hold vacuum, it must be replaced. If the advance mechanism is stuck, check components for binding, rust, or other problems.

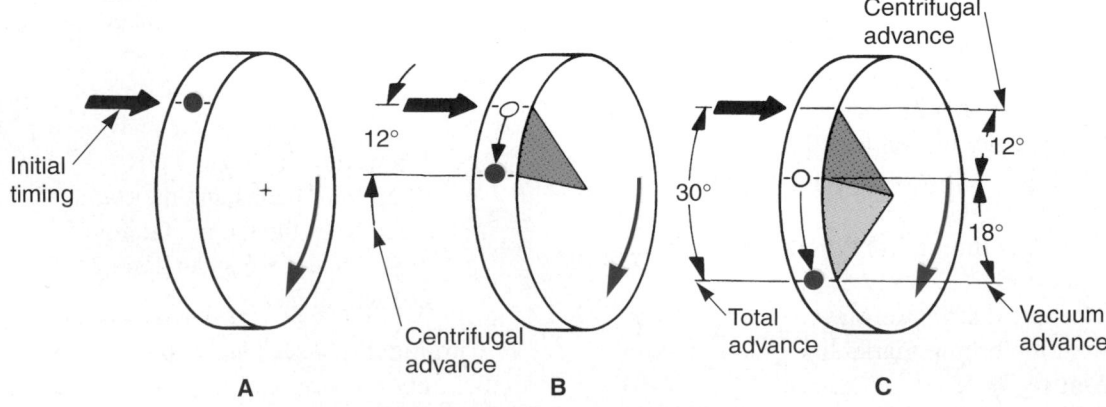

Initial
timing

12°

Centrifugal
advance

30°

Total
advance

Centrifugal
advance

12°

18°

Vacuum
advance

A   B   C

Figure 36-34. Timing advance mechanisms can be tested with a timing light. A—With the engine idling and the vacuum advance hose disconnected, initial timing marks should line up. B—When you increase engine speed, the timing marks should move to an advanced position. C—With vacuum applied to the distributor advance diaphragm, timing should advance more and be within specifications. (Snap-On Tools)

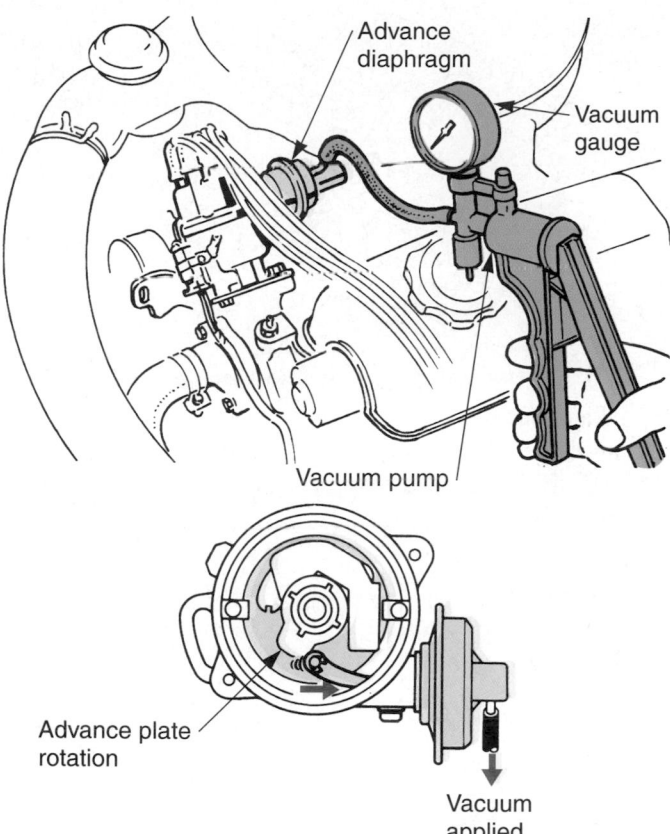

Figure 36-35. A leaking vacuum advance diaphragm is a common problem that will greatly reduce fuel economy. Use a vacuum pump to check for leakage. (Honda)

To check the supply vacuum going to the distributor, disconnect the diaphragm vacuum hose. Connect it to a vacuum gauge. Measure the amount of vacuum with the engine running at a specified rpm. If the vacuum is not sufficient, check all components controlling vacuum to the distributor (carburetor, thermal vacuum switch, delay valve, vacuum hose).

Tech Tip

The methods for checking a vacuum advance diaphragm can be applied to the vacuum diaphragms found in other systems (transmission, speed control, emission control, etc.).

Measuring Distributor Advance

A special timing light is available that can measure the exact distributor advance with the distributor installed in the engine. This timing light has a degree meter built into the back of its case. The meter will register advance quickly and accurately.

A *distributor tester* may also be used to check distributor operation. The distributor is removed and mounted in the tester. The tester will check all primary

distributor functions (pickup coil output, point dwell, centrifugal and vacuum advance).

Removing the Ignition Distributor

Before removing a distributor, carefully mark the position of the rotor tip on distributor housing and the engine with a scribe or marking pen. Look at **Figure 36-36.** Then, if the engine crankshaft is not rotated, you will be able to reinstall the distributor by simply lining up your marks.

To remove the distributor, take off the distributor cap, rotor, primary wires, and distributor hold-down bolt. Pull the distributor upward while rotating it back and forth. If stuck, use a slide hammer puller with a two-prong fork.

Keep in mind that distributor locations and service procedures can vary. For example, the distributor on one high-performance V-8 engine is mounted on the front of the engine, under the water pump, **Figure 36-37.**

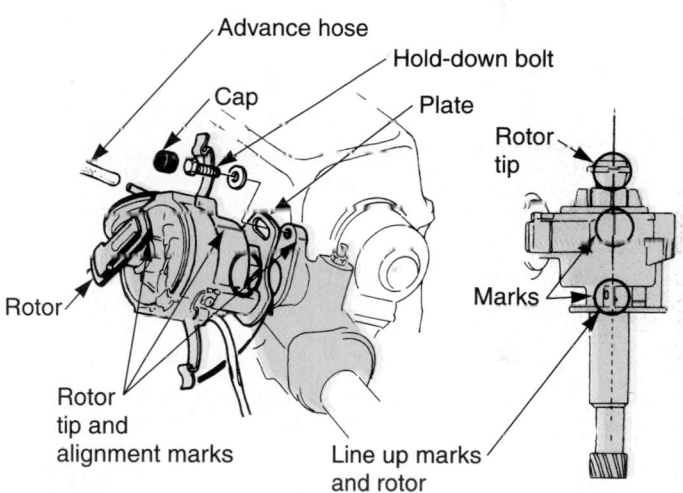

Figure 36-36. Before removing a distributor, mark the distributor housing and the engine below the rotor tip. This will let you quickly reinstall the distributor so the engine will at least start and run. (Honda)

Rebuilding a Distributor

A distributor rebuild involves disassembly, cleaning, inspection, worn part replacement, and reassembly. Depending on the distributor type, exact procedures vary, **Figure 36-38.** Always refer to a shop manual for detailed directions and specifications. The trend is to install a new or factory rebuilt unit. Many late-model distributors cannot be rebuilt in the shop.

The major steps for a distributor rebuild are illustrated in **Figure 36-39.** Assemble the distributor in reverse order of disassembly. See **Figure 36-40.**

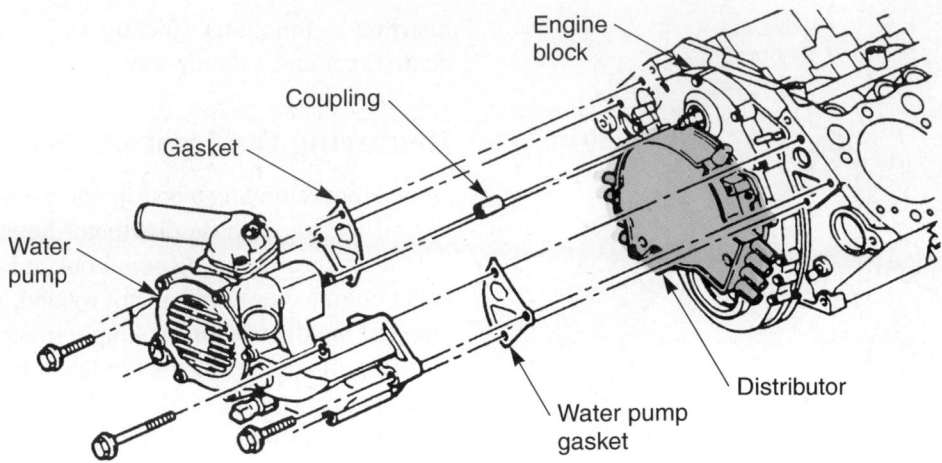

Figure 36-37. The distributor on this engine is hidden under the water pump. Water pump removal is needed to service the distributor. Keep this in mind when quoting repair prices. (Chevrolet)

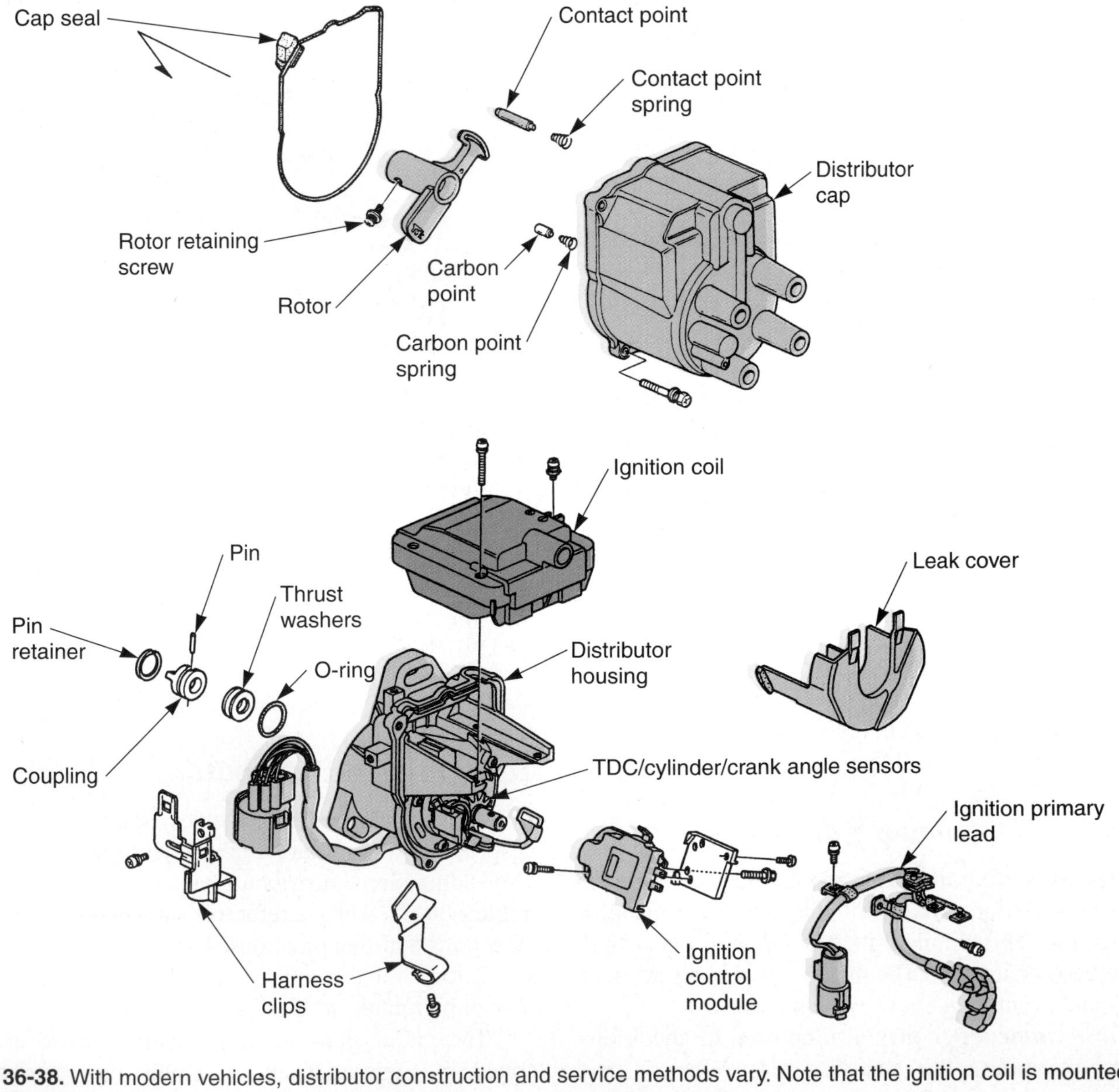

Figure 36-38. With modern vehicles, distributor construction and service methods vary. Note that the ignition coil is mounted under the distributor cap in the system shown above. (Honda)

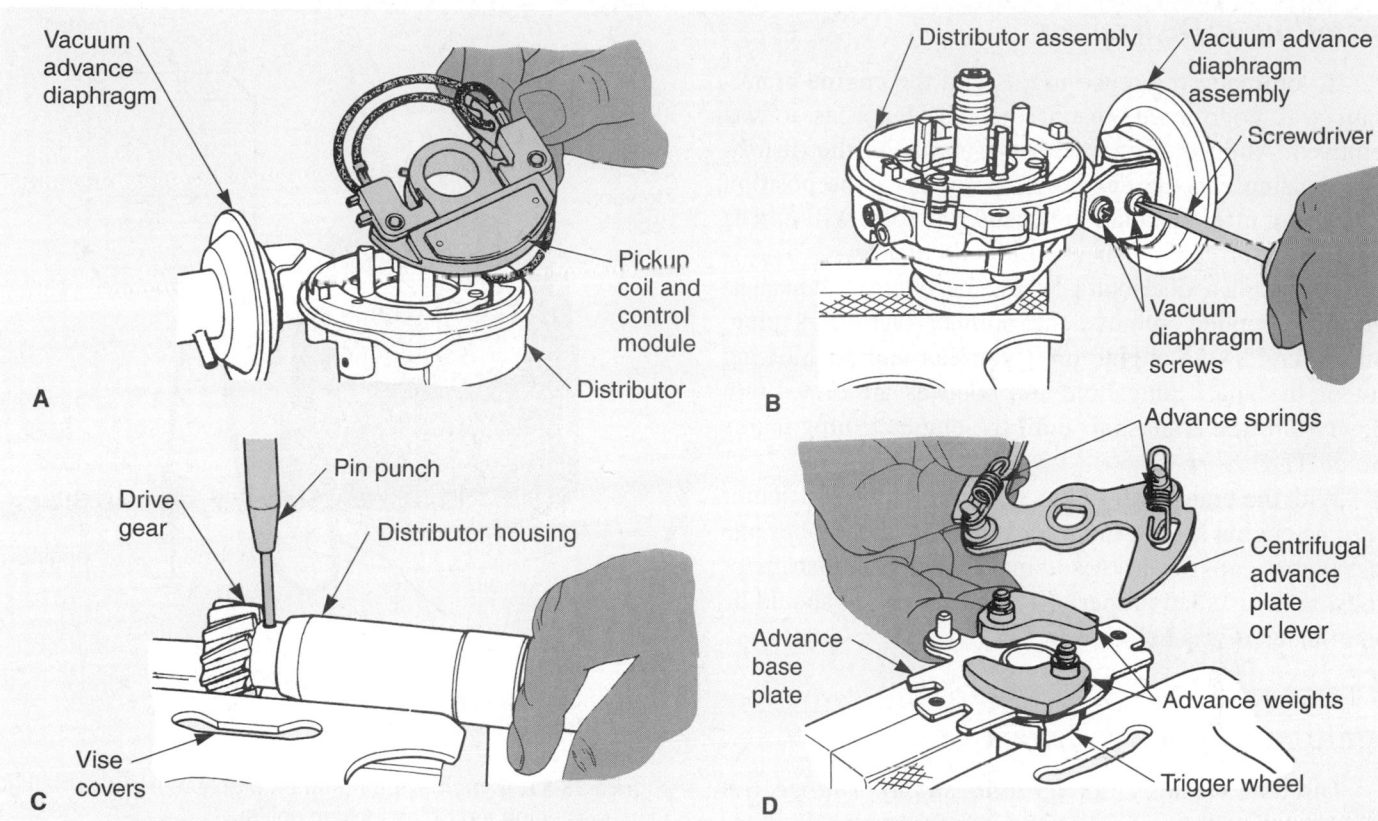

Figure 36-39. Major steps for distributor disassembly. A—Remove cap, pickup coil, and electronic control unit, if used. B—Remove vacuum diaphragm. A small snap ring may hold the advance lever on the advance plate. C—Drive the roll pin out of the shaft to free the drive gear. Slide the gear off the shaft and slide the shaft out of the housing. D—Disassemble the centrifugal advance mechanism. Check for wear, rust, and other problems. (Chrysler)

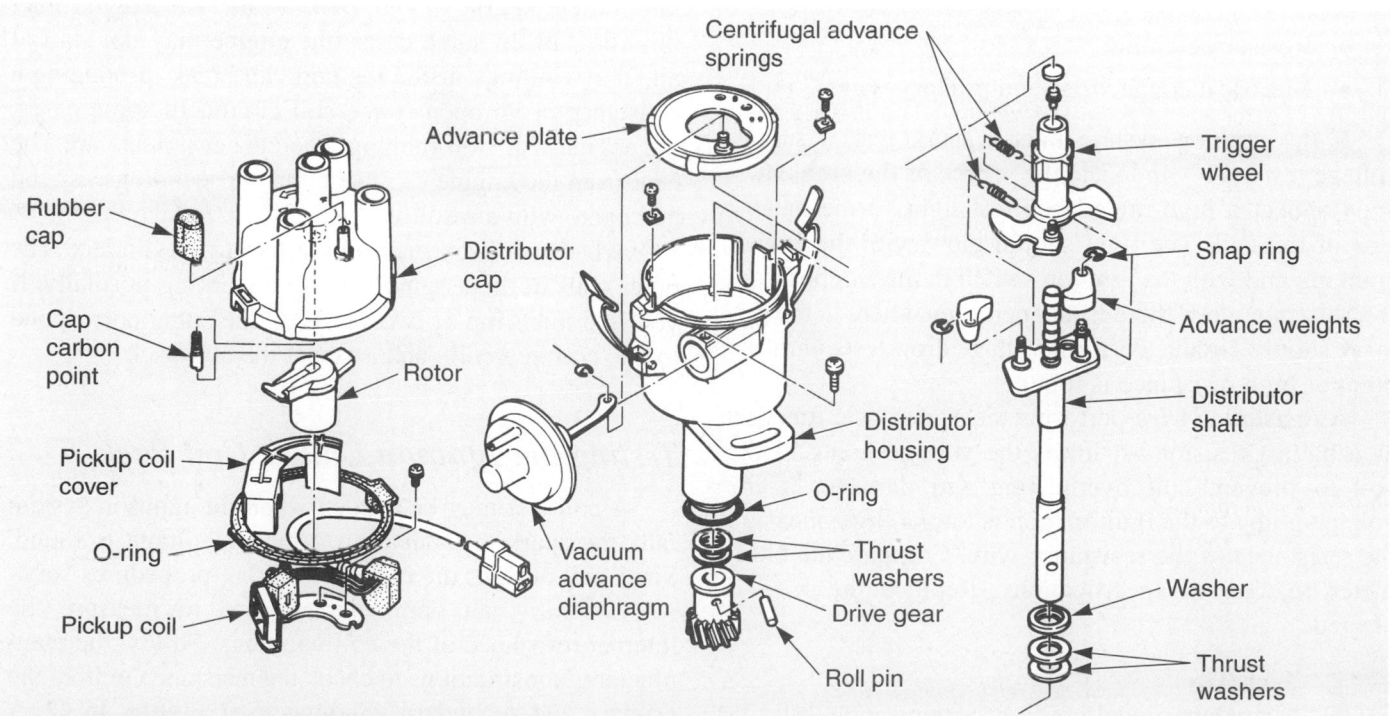

Figure 36-40. Exploded view shows how a distributor fits together. A service manual illustration may help with specific distributor assembly troubles. (Subaru)

Installing a Distributor

If you made reference marks and the engine crankshaft was not turned, install the distributor as it was removed. Align the rotor with the marks on the distributor housing and the engine. Double-check the position of the rotor after installation because the rotor will turn as the distributor gear meshes with its drive gear.

To install a distributor when the engine crankshaft has been rotated, remove the number one spark plug. Bump (crank) the engine until you can feel air blowing out of the spark plug hole. As soon as air blows out, slowly turn the crankshaft until the engine timing marks are on TDC.

With the crankshaft in this position, fit the distributor into the engine so that the rotor points at the number one distributor cap tower. Also, make sure the distributor housing is installed properly. The advance unit should be pointing as it was before removal.

Ignition Supply Voltage Test

Discussed briefly, an *ignition supply voltage test* checks the voltage going to the positive terminal of the ignition coil. It checks the circuit between the battery feed wire and the coil. An ignition supply voltage test will help locate troubles in the:

- Ignition switch.
- Bypass circuit.
- Resistance circuit.
- Electrical connections and primary wires.

If the ignition system fails a spark test, a supply voltage test may help locate the source of the problem.

Connect a high-impedance test light to the battery side of the coil. The light should glow with the engine cranking and with the ignition switch in the *run* position. If it does not glow, there is an open somewhere in the primary supply circuit. Perform voltage drop tests until the point of high resistance is found.

A *resistance wire* performs the same basic function as a ballast resistor—it limits the voltage going to the coil to prevent coil overheating and damage. If the voltage going to the ignition coil is low or high, measure the resistance of the resistance wire. Compare the ohmmeter reading to specifications. Replace the wire if needed.

Tech Tip

Many late-model ignition systems apply full battery voltage to the ignition coil primary terminal. See **Figure 36-41.**

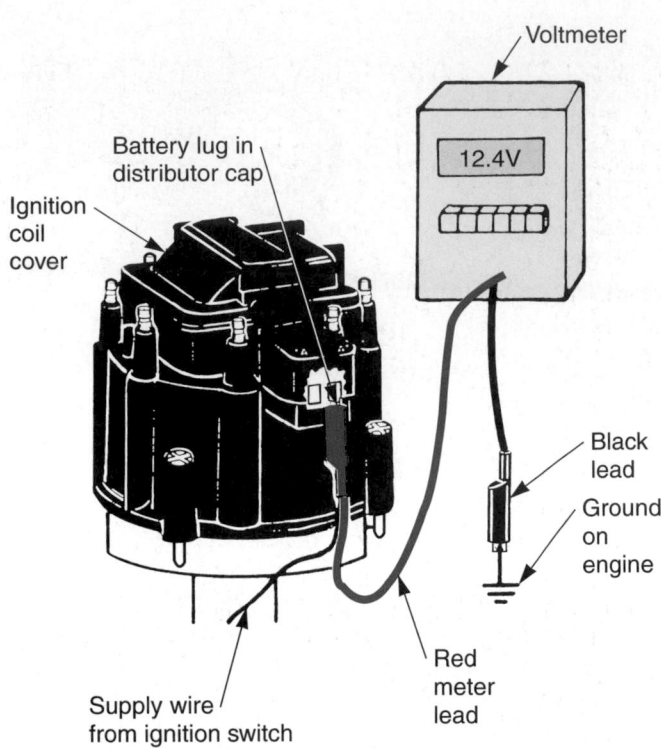

Figure 36-41. If you have a no-spark problem, make sure that voltage is being fed to the ignition coil(s).

Ignition Coil (Coil Pack) Service

A faulty ignition coil may result in a weak spark, an intermittent spark, or no spark at all. The engine may miss or stall. In some cases, the engine may not start at all. The windings inside the coil can break, producing a resistance or an open in the coil circuit. In some cases, the engine will stop running when the coil heats up. The heat from the engine can make the coil windings expand and open, with a resulting loss of high voltage output.

A bad coil pack may only affect two cylinders. The other coils in the assembly may be working normally. If you find misfiring or two dead cylinders that correspond to one coil in a coil pack, suspect the coil pack.

Testing the Ignition Coil or Coil Pack

A coil test may be needed when the ignition system fails the spark test but proper supply voltage is found. Since coil designs are different, testing procedures vary.

Generally, an ohmmeter is used to measure the internal resistance of the coil windings. Follow the manufacturer's instructions to check the resistance of both the primary and secondary windings. See **Figure 36-42.** A reading that is out of specifications indicates a faulty ignition coil.

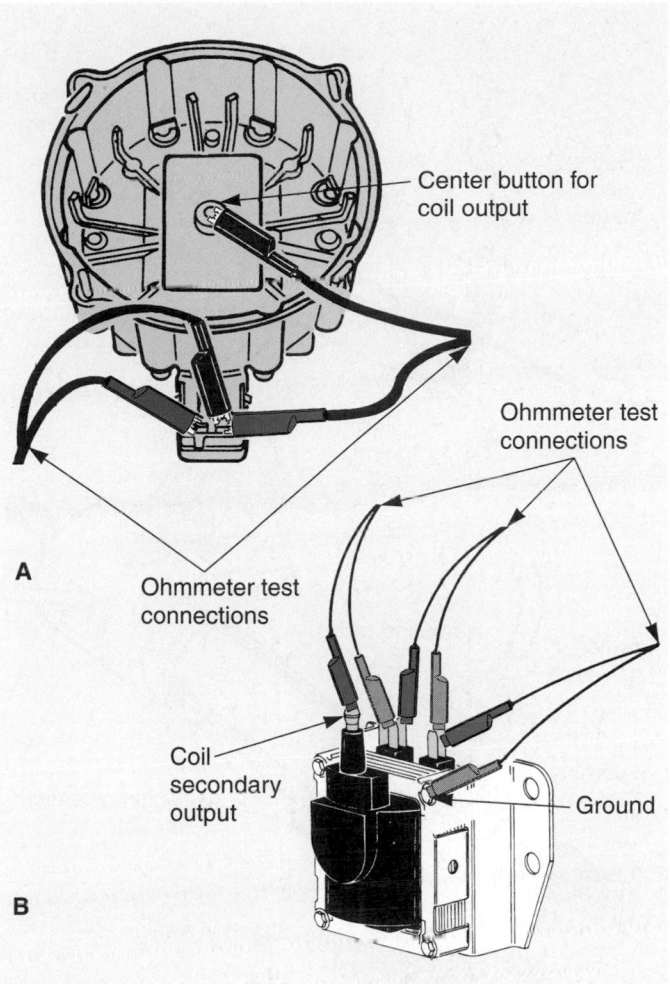

Figure 36-42. With a weak spark, you may need to test the ignition coil. Follow the service manual instructions to measure the resistance of the windings. A—Checking a coil mounted in the distributor cap. B—Checking an externally mounted ignition coil. (Peerless and Echlin)

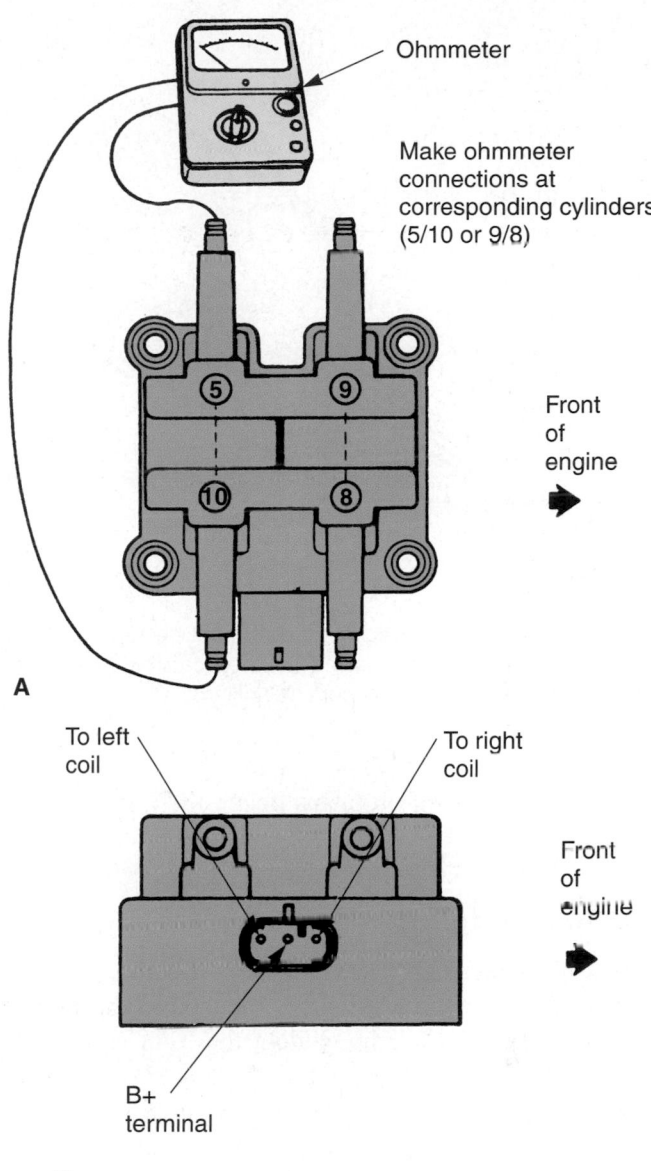

Figure 36-43. This service manual illustration shows how to test a specific coil pack used on a V-10 engine. Each coil in the pack must be tested individually. A—Checking secondary resistance. B—When checking primary resistance, the ohmmeter leads should be connected between the B+ terminal and the pin corresponding to the coil in question. Resistance readings must be within specifications. (Chrysler)

With a coil pack, test the windings with the spark problem. If, for example, the number two cylinder is not firing, the coil for that cylinder should be checked first. Normally, a bad coil pack winding will show infinite resistance, or open. **Figure 36-43** shows how to test one make of coil pack.

Replacing an Ignition Coil

When the ignition coil or coil pack is mounted on the engine, coil replacement involves removing the wires and bolts securing the old coil and lifting the coil from the engine. Then, simply bolt on the new coil in place and reconnect all wires, **Figure 36-44.**

Be careful not to connect the coil in *reverse polarity* (primary wires accidentally connected backwards). This would reduce secondary voltage output.

When the coil is mounted inside the distributor cap (unitized type ignition), the distributor cap must be disassembled to install the new coil.

Ignition Switch Service

A bad ignition switch can cause several problems: the engine may not crank or start, the engine may not shut off when the ignition key is turned off, or the starter may not disengage when the ignition key is returned to *run*.

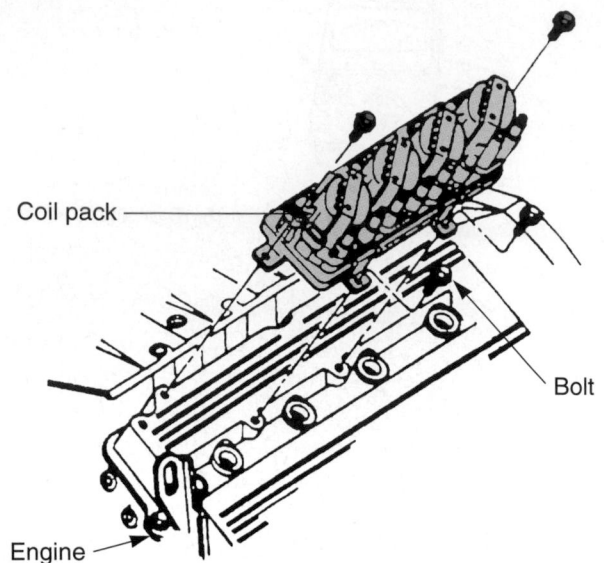

Figure 36-44. If a coil pack is bad, it must be replaced. Keep the spark plug wires organized so they can be quickly reinstalled on the correct coil towers. (Cadillac)

When these types of problems occur, the ignition switch should be tested.

Testing an Ignition Switch

A test light can be used to check the action of an ignition switch. When a test light is touched to the *start* terminal on the back of the switch, the light should glow only when the key has been turned to *start*. It should not glow when the key is released to the *run* position.

In the *run* position, the test light should glow when touched on the *run* terminal of the switch. With the ignition key in the *off* position, neither terminal (start or run) should make the test light glow.

Replacing an Ignition Switch

If the ignition switch is defective, it must be replaced. Before removing a dashboard-mounted switch, the **tumbler** (lock mechanism) must be removed from the switch. Normally, a small piece of wire is inserted into a hole in the front of the switch and the key is turned. This will release the tumbler from the switch so the tumbler can be pulled free. Then, unplug the wires and remove the old switch.

A steering column–mounted ignition switch is separate from the tumbler, **Figure 36-45.** It is normally about halfway down and on top of the steering column. To remove this type of switch, remove the fasteners holding the steering column to the dash. This will let the column drop down so you can replace the ignition switch.

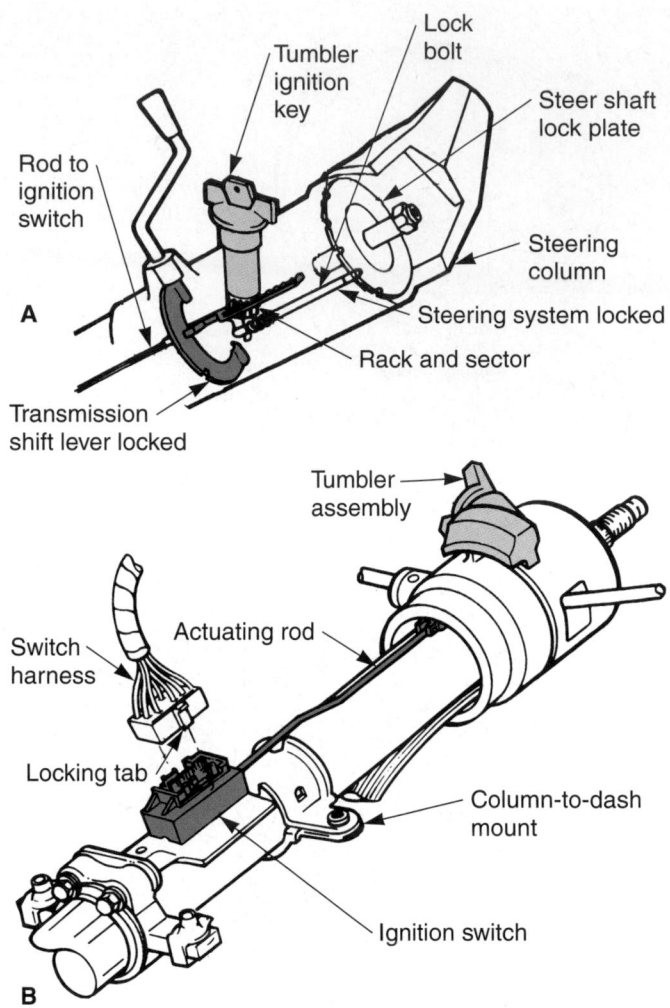

Figure 36-45. Late-model cars have the ignition switch and key tumbler in the steering column. A—Construction of a typical steering lock mechanism. B—A small rod runs down the column to the ignition switch. To remove the switch, drop the column by removing the column-to-dash fasteners. Then, remove the small nuts holding the switch to the top of the column. (Florida Dept. of Voc. Ed. and Ford)

Ignition Control Module Service

A faulty ignition control module will produce a wide range of problems: engine stalls when hot, engine cranks but fails to start, engine misses at high or low speeds, etc.

Quite often, an ignition control module problem will show up after a period of engine operation. Engine heat will soak into the module, raising its temperature. The heat will upset the operation of the electronic components in the unit.

Testing an Ignition Control Module

Many shop manuals list the ignition control module as one of the last components to test when troubleshooting an ignition system. If all the other components are in good working order, then the problem might be in the ignition control module.

If a specialized tester is available, it may be used to quickly determine the condition of the ignition control module. The wires going to the module are unplugged and the tester is connected to the module. The tester will then indicate whether an ignition control module fault exists.

Heating an Ignition Control Module

The microscopic components (transistors, diodes, capacitors, resistors) inside the ignition control module are very sensitive to high temperatures and vibration.

When testing the control module, many technicians use a heat gun or lightbulb to warm the unit. This will simulate the temperature in the engine compartment after the engine has been running. The heat may make the control module act up and allow you to find an intermittent problem. Refer to **Figure 36-46.**

Do not apply too much heat to an ignition control module or it may be ruined. Only heat the unit to a temperature equal to its normal operating temperature.

Replacing an Ignition Control Module

Replacing an ignition control module is a simple task.

 If the control module is mounted in the engine compartment or under the dash:
1. Carefully unplug the wiring harness.
2. Unbolt and remove the old unit.
3. Install the new unit.
4. Reconnect the wiring harness.

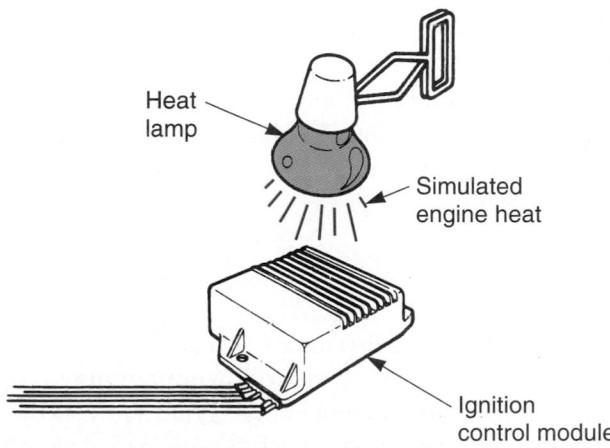

Figure 36-46. When testing an ignition control module, heat may help you locate an intermittent problem. Do not overheat the unit, or it could be damaged. (Ford)

 If the ignition control unit is mounted inside the distributor:
1. Remove the distributor cap.
2. Disconnect the wires leading to the module.
3. Remove the screws holding the module in place.
4. Remove the old control module.
5. Install the new unit according to service manual instructions.

In many cases, the bottom of the ignition control module must be coated with a special grease (silicone grease, dielectric heat transfer compound, or heat sink compound) before installation. This grease helps heat transfer into the distributor housing, protecting the module from overheating and circuit damage. See **Figure 36-47.**

Make sure you have the correct ignition control module. The new control module may look identical to the old one, but it may have internal circuit differences. Even cars of the same year and make can require different ignition control modules.

Distributorless Ignition System Service

As discussed, many of the components used in a computer-controlled ignition system are similar to those found in older electronic or contact point systems. Testing procedures for these parts (spark plugs, secondary wires, ignition coils, etc.) have been covered earlier in this chapter. However, the computerized ignition system contains engine sensors and a computer, which make it more difficult to troubleshoot and service these systems.

In systems containing a coil pack that fires two spark plugs at the same time, a bad ignition coil will kill two cylinders. For example, if a four-cylinder engine has two ignition coils, one bad coil will make the engine run on two cylinders, producing a very rough idle. If two dead cylinders correspond to a specific coil, test that ignition coil.

 Caution
A computerized ignition system can be seriously damaged if the wrong wire is shorted to ground or if a meter is connected improperly. Always follow manufacturer's testing procedures.

 Note
Refer to Chapter 19, *Computer System Service,* for more information on diagnosing computer-controlled ignition system problems. Normally, the computer system sensors (crankshaft position sensor, camshaft sensor, knock sensor, etc.) are common to both fuel and ignition systems.

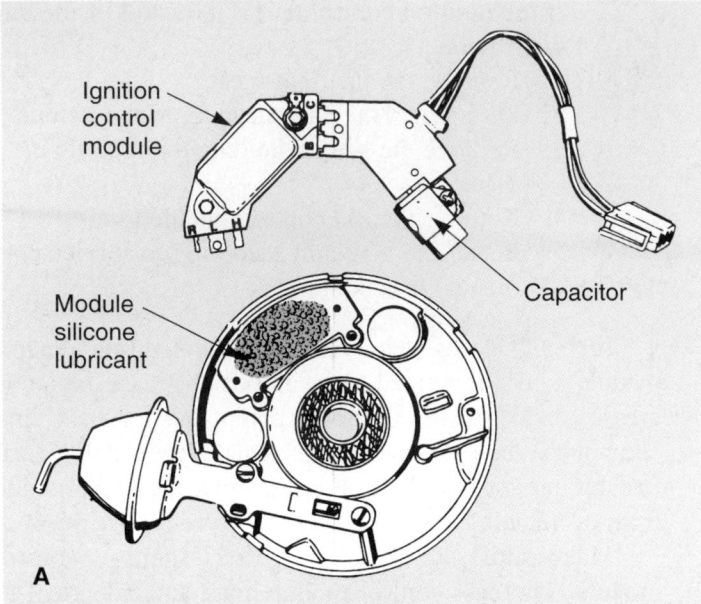

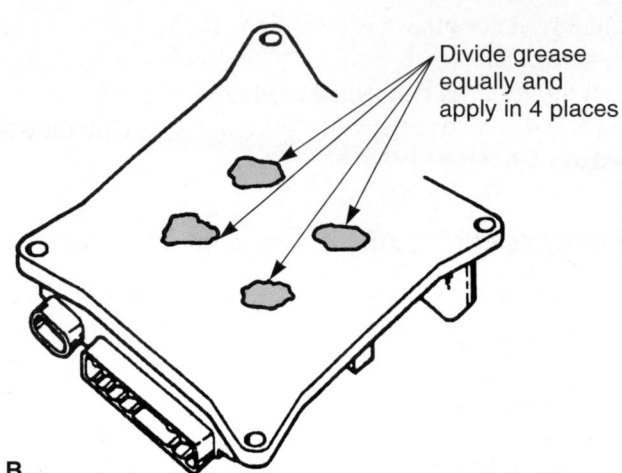

Figure 36-47. To prevent heat damage, silicone grease is sometimes applied to the ignition control module. A—When the ignition control module is in the distributor, apply silicone grease to the bottom of the module. Lubricant will help transfer heat out of the control module, ensuring proper operation. B—With larger control modules that are mounted on a heat sink, more grease may be needed. Follow the service manual instructions for the specific control module. (Oldsmobile and Chevrolet)

Knock Sensor Service

A knock sensor is used to detect abnormal combustion or ping. When it "hears" pinging or knocking, it will retard ignition timing or lower turbo boost with the turbo waste gate. A bad knock sensor can upset ignition timing and affect turbocharger boost pressure. Many computers will store a trouble code if there is a potential problem with the knock sensor.

To check a knock sensor operation, start the engine and allow it to reach operating temperature. Lightly tap on the engine block or a bracket with a wrench or small hammer. This will simulate pinging or knocking and should make the computer retard the ignition timing. The light taps should make the engine speed drop slightly. You might need to prop open the throttle to increase engine speed slightly so timing is advanced and will retard.

If tapping on the engine has no effect on timing and engine speed, you can check the sensor with a VOM or scope. Refer to the manual for recommendations. Remember to check the wiring leading to the knock sensor before removing or replacing the sensor!

Direct Ignition System Service

The procedures for servicing a direct ignition system are similar to the procedures described for other types of ignition systems. The main difference is that you must test the ignition coil of the affected cylinder. For example, if the #1 spark plug is not firing, tests should be performed on the #1 ignition coil.

A direct ignition coil is tested like other ignition coils. Measure both primary and secondary winding resistance. Also make sure you are getting primary voltage to the coil.

Figure 36-48 shows some tricks for working on a direct ignition system. Remove the coil cover and connect conventional spark plug wires between the coil output terminals and the spark plugs. This will let you connect a timing light, an inductive tachometer, a spark tester, etc. to the system.

 Note
Several other textbook chapters contain information that will be useful when troubleshooting and testing an ignition system. Refer to the index as needed.

Summary

- With late-model vehicles, a scan tool can be used to find troubles in the following ignition system circuits and components: crankshaft position sensor, crankshaft speed sensor, camshaft position sensor, knock sensor, ignition coil(s) primary circuit, ignition coil(s) secondary circuit, timing reference signal, other devices.

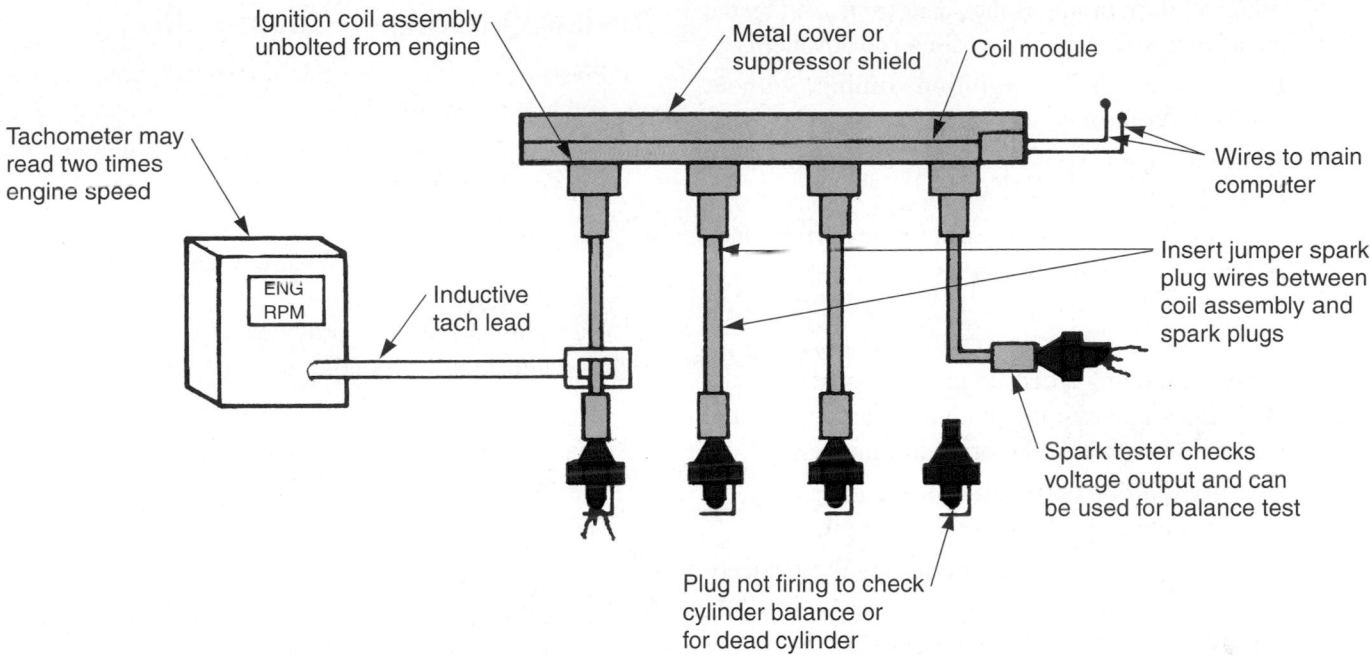

Figure 36-48. Study these direct ignition system tests. The coil module can be unbolted and removed from the engine. Then, spark plug wires can be used to jump from the coil assembly to the spark plugs. This will let you connect an inductive tachometer to the engine. The tachometer may read two times actual engine speed if two plugs fire at once. You can also use a spark tester to check high-voltage output and short out each plug to make sure each cylinder is firing and lowering engine rpm. A scope or timing light can be connected to the engine with this setup.

- If your scanner readout indicates a misfire, the engine has failed to properly ignite and burn its air fuel mixture properly.

- A spark intensity test, also called a spark test, measures the brightness and length of the electric arc (spark) produced by the ignition system.

- A dead cylinder is a cylinder (combustion chamber) that is not burning fuel on the power stroke.

- A hand-held scope is usually a VOM combined with an oscilloscope in one housing. It is a handy tool for advanced troubleshooting of electrical-electronic problems.

- An engine analyzer contains several types of test equipment (oscilloscope, dwell meter, tachometer, VOM, etc.).

- An oscilloscope will precisely measure the operating voltages of an ignition system. It uses a television-type display screen to show voltage changes in relation to degrees of distributor or crankshaft rotation.

- Bad spark plugs can cause a wide range of problems, including misfiring, lack of power, poor fuel economy, and hard starting.

- Never remove a spark plug wire by pulling on the wire. Always grasp and pull on the boot.

- To read spark plugs, closely inspect and analyze the condition of each used spark plug tip and insulator.

- A wire feeler gauge should be used to measure spark plug gap.

- A faulty spark plug wire can have either a damaged conductor or deteriorated insulation.

- Problems often arise when a carbon trace (small line of carbon-like substance that conducts electricity) forms on the distributor cap, the rotor, or the ignition coil (coil pack).

- A bad pickup coil can produce a wide range of engine problems, including stalling, missing, no-start troubles, and loss of power at specific speeds.

- Dwell specifications vary with the number of cylinders in the engine and type of ignition system (contact point, electronic, or computerized).

- Fixed dwell means that the dwell time should remain the same at all engine speeds.

- Variable dwell means that the ignition module or engine control module alters ignition coil dwell time with engine speed.

- Current-limiting dwell means that the ECU sends high current through the ignition coil windings until a strong enough magnetic field is developed around the coil windings.

- Initial ignition timing is the spark timing set by the technician with the engine idling (no advance).

- Base timing is the ignition timing without computer-controlled advance.

- A timing light is sometimes used to measure ignition timing.

- Ignition timing is very critical to the performance of an engine. If the timing is off by even 2 or 3 degrees, fuel economy and power can drop considerably.

- Before removing a distributor, carefully mark the position of the rotor tip on distributor housing and the engine with a scribe or a marking pen.

- A faulty ignition coil may result in a weak spark, an intermittent spark, or no spark at all.

- A faulty ignition control module will produce a wide range of problems: engine stalls when hot, engine cranks but fails to start, engine misses at high or low speeds, etc.

- In systems containing a coil pack that fires two spark plugs at the same time, a bad ignition coil will kill two cylinders.

- A bad knock sensor can upset ignition timing and affect turbocharger boost pressure.

Important Terms

Misfire	Pickup coil air gap
Spark intensity test	Distributor point gap
Spark tester	Tachometer
Dead cylinder	Dwell meter
Hand-held scope	Fixed dwell
Engine analyzer	Variable dwell
Oscilloscope	Current-limiting dwell
Electronic ignition tester	Saturated ignition coil
	Initial ignition timing
Spark plug wire pliers	Base timing
Spark plug cleaner	Timing light
Secondary wire resistance test	Distributor wrench
	Distributor tester
Secondary wire insulation test	Ignition supply voltage test
Backfiring	Resistance wire
Carbon trace	Reverse polarity
Magnetic pickup coil test	Tumbler

Review Questions—Chapter 36

Please do not write in this text. Place your answers on a separate sheet of paper.

1. Explain how to scan an ignition system and how it affects troubleshooting.

2. A(n) _____ _____ _____, also called a(n) _____ _____, measures the brightness and length of the electric arc produced by the ignition system. It provides a(n) _____ way of checking the condition of the ignition system.

3. Which of the following is commonly used to check the secondary output of an ignition system?
   (A) Voltmeter.
   (B) Ohmmeter.
   (C) Spark tester.
   (D) Ammeter.

4. Define the term "dead cylinder."

5. How do you find a dead cylinder without using special equipment?

6. What kinds of test equipment are usually found in an engine analyzer?

7. What does an oscilloscope measure?

8. Why should a technician "read" spark plugs?

9. Which of the following does *not* pertain to setting spark plug gap?
   (A) Wire feeler gauge.
   (B) Bend side electrode.
   (C) Space between center and side electrode.
   (D) All of the above.

10. When screwing a spark plug into the cylinder head, a socket, an extension, and a ratchet should be used to start the plug threads. True or False?

11. A faulty spark plug wire can have either a burned or broken _____ or deteriorated _____.

12. Explain both a resistance test and an insulation test of a secondary wire.

13. Define the term "carbon trace."

14. Describe how either an ohmmeter or a voltmeter can be used to test most distributor pickup coils.

15. Why is a nonmagnetic feeler gauge needed to set a pickup coil air gap?

ASE-Type Questions

1. An engine analyzer consists of all the following *except:*
   - (A) *VOM.*
   - (B) *tachometer.*
   - (C) *oscilloscope.*
   - (D) *spark tester.*

2. Which device will properly measure ignition system operating voltages over time?
   - (A) *Oscilloscope.*
   - (B) *Dwell meter.*
   - (C) *Spark tester.*
   - (D) *All of the above.*

3. Which of the following is *not* a problem that may be caused by bad spark plugs?
   - (A) *Misfiring.*
   - (B) *High idle speed.*
   - (C) *Hard starting.*
   - (D) *Lack of power.*

4. A properly burning spark plug should be:
   - (A) *blue.*
   - (B) *black.*
   - (C) *white.*
   - (D) *brown.*

5. Each of the following can be used to start spark plugs in their holes *except:*
   - (A) *a ratchet.*
   - (B) *your fingers.*
   - (C) *a spark plug socket.*
   - (D) *a short piece of vacuum hose.*

6. While conducting a secondary wire resistance test, Technician A states that wire resistance should be approximately 12,000 ohms per foot. Technician B says that resistance should be about 50k ohms (50,000 ohms) maximum for long spark plug cables. Who is right?
   - (A) *A only.*
   - (B) *B only.*
   - (C) *Both A and B.*
   - (D) *Neither A nor B.*

7. Which of the following is *not* a function of the distributor?
   - (A) *Sense engine speed.*
   - (B) *Change ignition timing.*
   - (C) *Maintain spark plug gap.*
   - (D) *Distribute voltage to plug wires.*

8. Most electronic ignition systems use this to sense trigger wheel rotation:
   - (A) *optical sensor.*
   - (B) *magnetic pickup coil.*
   - (C) *Hall-effect sensor.*
   - (D) *photo diode sensor.*

9. Which of the following is the ignition switch locking mechanism?
   - (A) *Tumbler.*
   - (B) *Steering column.*
   - (C) *Actuating rod.*
   - (D) *Suppresser shield.*

10. A car equipped with distributorless ignition enters the shop with two dead cylinders. Technician A says to check to see if the dead cylinders are operated by the same ignition coil in the coil pack. Technician B says to check for a bad crankshaft sensor. Who is right?
    - (A) *A only.*
    - (B) *B only.*
    - (C) *Both A and B.*
    - (D) *Neither A nor B.*

Activities—Chapter 36

1. Arrange to observe a technician at a tune-up shop or an auto dealership who is using an engine analyzer to diagnose ignition system condition. Ask the technician to explain what is being shown on the oscilloscope and other displays.

2. Demonstrate the use of a wire feeler gauge to measure the gap of a spark plug, then demonstrate the method used to set the gap to specifications.

3. Using a timing light to check the timing marks on two or three engines. Check your findings with the appropriate service manual to determine whether timing is "within spec."

Ignition System Diagnosis		
Condition	**Possible cause**	**Correction**
No spark.	1. Breaker points defective or misadjusted (point ignition).	1. Install new points.
	2. Distributor pickup defective or misadjusted (electronic ignition).	2. Replace or adjust as needed.
	3. Defective condenser.	3. Replace condenser.
	4. Discharged battery.	4. Charge battery.
	5. Faulty coil or primary circuit resistor.	5. Replace coil or resistor.
	6. No primary current to points.	6. Check ignition switch, coil, resistor, wiring.
	7. Defective coil high tension lead.	7. Replace lead.
	8. Defective rotor and/or distributor cap.	8. Replace cap and rotor.
	9. Defective plug wires.	9. Replace plug wires.
	10. Moisture in distributor cap and on points.	10. Dry cap and points.
	11. Breaker plate not grounded.	11. Replace or tighten ground wire.
	12. Defective ignition control module.	12. Replace ignition module.
	13. Loose, corroded, or open electronic control module ground lead.	13. Tighten, clean, or connect as needed.
	14. Loose, corroded, or disconnected primary connections.	14. Clean, cover with special, protective grease and shove firmly into distributor.
	15. Defective distributor electronic pickup.	15. Replace pickup unit.
	16. Trigger wheel positioned too high.	16. Reposition correctly.
	17. Incorrect trigger wheel-to-pickup air gap.	17. Set correctly. Use nonmagnetic feeler gauge.
	18. Defective cam or crankshaft sensor.	18. Replace defective sensor.
Weak or intermittent spark.	1. Breaker points defective (where used).	1. Install new points.
	2. Defective condenser (where used).	2. Install new condenser.
	3. Point dwell set incorrectly.	3. Set dwell correctly.
	4. Discharged battery.	4. Charge or replace battery.
	5. Loose or dirty primary wiring connections.	5. Clean and tighten connections.
	6. Weak coil.	6. Replace coil.
	7. Defective primary circuit resistor.	7. Replace resistor.
	8. Burned rotor and cap contacts.	8. Replace cap and rotor.
	9. Defective resistance spark plug wires.	9. Replace resistance wires.
	10. Insufficient system voltage.	10. Adjust regulator.
	11. Weak breaker spring pressure (point ignition only).	11. Increase spring pressure.
	12. Worn distributor bushings or bent shaft.	12. Replace bushings or shaft.
	13. Worn distributor cam (point ignition only).	13. Replace cam.
	14. Breaker arm sticking (point ignition only).	14. Free and lubricate bushing.
	15. Loose spark plug wires.	15. Clean and tighten connections.
	16. Defective distributor electronic pickup.	16. Replace pickup.
	17. Trigger wheel pin sheared.	17. Replace pin.
	18. Shorted primary wiring.	18. Replace wire and relocate.
	19. Loose wiring harness connectors.	19. Tighten connections.
Missing at idle or low speed.	1. Weak or intermittent spark at plugs.	1. See *Weak or intermittent spark*.
	2. Fouled spark plugs.	2. Clean or replace plugs.
	3. Spark plug gaps too narrow.	3. Adjust gaps to specifications.
	4. Improper plug heat range.	4. Install proper heat range.
	5. Damaged plugs.	5. Replace plug or plugs.
	6. Defective distributor electronic pickup.	6. Replace pickup unit.
	7. Loose harness connections.	7. Clean and tighten connections.
	8. Discharged battery.	8. Charge or replace battery.
	9. Coil polarity incorrect.	9. Reverse coil primary leads.

(Continued)

Ignition System Diagnosis		
Condition	**Possible cause**	**Correction**
Missing during acceleration.	1. Weak spark. 2. Plugs damp. 3. Fouled plugs. 4. Plug gap too wide. 5. Damaged plug. 6. Incorrect trigger wheel-to-pickup air gap. 7. Defective vacuum advance. 8. Incorrect coil polarity. 9. Crossfiring. 10. Weak ignition coil.	1. See *Weak or intermittent spark.* 2. Dry plugs. 3. Clean or replace plugs. 4. Gap as specified. 5. Replace plug. 6. Set gap to specifications. 7. Repair or replace vacuum advance. 8. Reverse coil primary leads. 9. Rearrange ignition secondary wires. 10. Replace ignition coil.
Missing during cruising and high-speed operation.	1. Weak spark. 2. Improper heat range plug (too hot). 3. Crossfiring. 4. Fouled plugs. 5. Plug gap incorrect. 6. Damaged plug. 7. Ignition timing incorrect. 8. Defective distributor. 9. Loose or corroded wire connections. 10. Defective spark plug wires. 11. Weak ignition coil.	1. See *Weak or intermittent spark.* 2. Install proper heat range plug. 3. Arrange wires properly. If needed, install new wires. 4. Clean or replace plugs. 5. Gap to specifications. 6. Replace plug. 7. Reset timing. 8. Repair or replace distributor. 9. Clean and tighten connections. 10. Replace wires. 11. Replace ignition coil.
Missing at all speeds.	1. Weak spark. 2. Fouled spark plugs. 3. Damaged plug. 4. Crossfiring. 5. Plug gap too wide or too narrow. 6. Plugs and/or distributor damp. 7. Improper plug heat range. 8. Defective distributor. 9. Defective spark plug wires. 10. Distributor trigger wheel pin sheared or missing. 11. Burned or corroded breaker points. 12. Defective condenser.	1. See *Weak or intermittent spark.* 2. Clean or replace plugs. 3. Replace plug. 4. Arrange wiring correctly and if needed, install new wires. 5. Adjust gap as needed. 6. Dry distributor and plugs. 7. Change to correct heat range. 8. Repair or replace distributor. 9. Replace plug wires. 10. Install new pin. 11. Replace and gap points. 12. Replace condenser.
Coil failure.	1. Carbon tracking on tower. 2. Excessive system voltage. 3. Oil leak in coil. 4. Engine heat damage. 5. Physical damage.	1. Replace coil and wire nipple. 2. Adjust voltage regulator. 3. Replace coil. 4. Replace coil. Relocate or baffle against heat. 5. Replace coil.
Short spark plug life.	1. Incorrect plug heat range (too hot—burns). 2. Incorrect plug heat range (too cold—fouls). 3. Mechanical damage during installation. 4. Loose spark plug (overheats and burns). 5. Incorrect plug reach (too short—fouls). 6. Incorrect plug reach (too long—strikes piston).	1. Install correct (cooler) spark plug. 2. Install correct (hotter) spark plug. 3. Install correctly. 4. Tighten plugs to proper torque. 5. Install plugs with correct reach. 6. Install plugs with correct reach.

(Continued)

Ignition System Diagnosis		
Condition	**Possible cause**	**Correction**
Short spark plug life. *(Continued)*	7. Worn engine—oil fouling. 8. Bending center electrode. 9. Detonation. 10. Preignition. 11. Lean mixture.	7. Switch to hotter plugs or overhaul engine. 8. Bend side electrode only. 9. Adjust timing. Change to higher octane gas and/or remove carbon buildup. 10. Remove carbon buildup, install valves with full margin, install cooler plugs. 11. Adjust air-fuel ratio.
Preignition.	1. Overheated engine. 2. Glowing pieces of carbon. 3. Spark plugs overheating. 4. Sharp valve edges. 5. Glowing exhaust valve.	1. Check cooling system. 2. Remove carbon. 3. Change to cooler plugs. 4. Install valves with full margin. 5. Check for proper tappet clearance, for sticking, and air leaks.
Detonation.	1. Ignition timing advanced. 2. Engine temperature too high. 3. Carbon buildup is raising compression ratio. 4. Low octane fuel. 5. Exhaust heat control valve stuck. 6. Excessive block or head metal removed to increase compression.	1. Retard timing. 2. Check cooling system. 3. Remove carbon. 4. Switch to high octane fuel. 5. Free valve. 6. Use thicker gasket, change head, or true warped head or block surface.

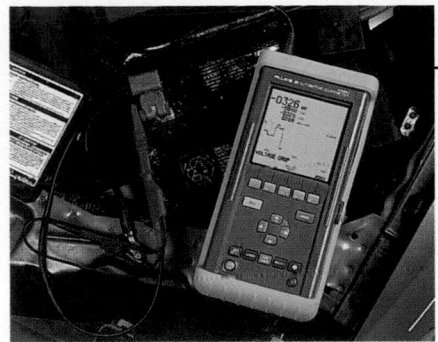

Chapter 37

Lights, Instrumentation, Wipers, and Horns— Operation and Service

After studying this chapter, you will be able to:

- Explain the operating principles of automotive light, wiper, and horn systems.
- Diagnose problems in light, wiper, and horn systems.
- Summarize automatic light and wiper systems.
- Replace burned-out bulbs.
- Explain how to aim headlights.
- Describe the safety practices to follow when working with light, wiper, and horn systems.

- Explain both analog and digital instrumentation.
- Summarize how to remove and service an instrument cluster.
- Correctly answer ASE certification test questions on light, instrumentation, wiper, and horn systems.

In this chapter, you will learn the theory of operation, basic testing, and repair methods for lighting, instrumentation wiper, and horn circuits, **Figure 37-1.** These are essential systems that perform important safety functions. They are installed on all passenger vehicles.

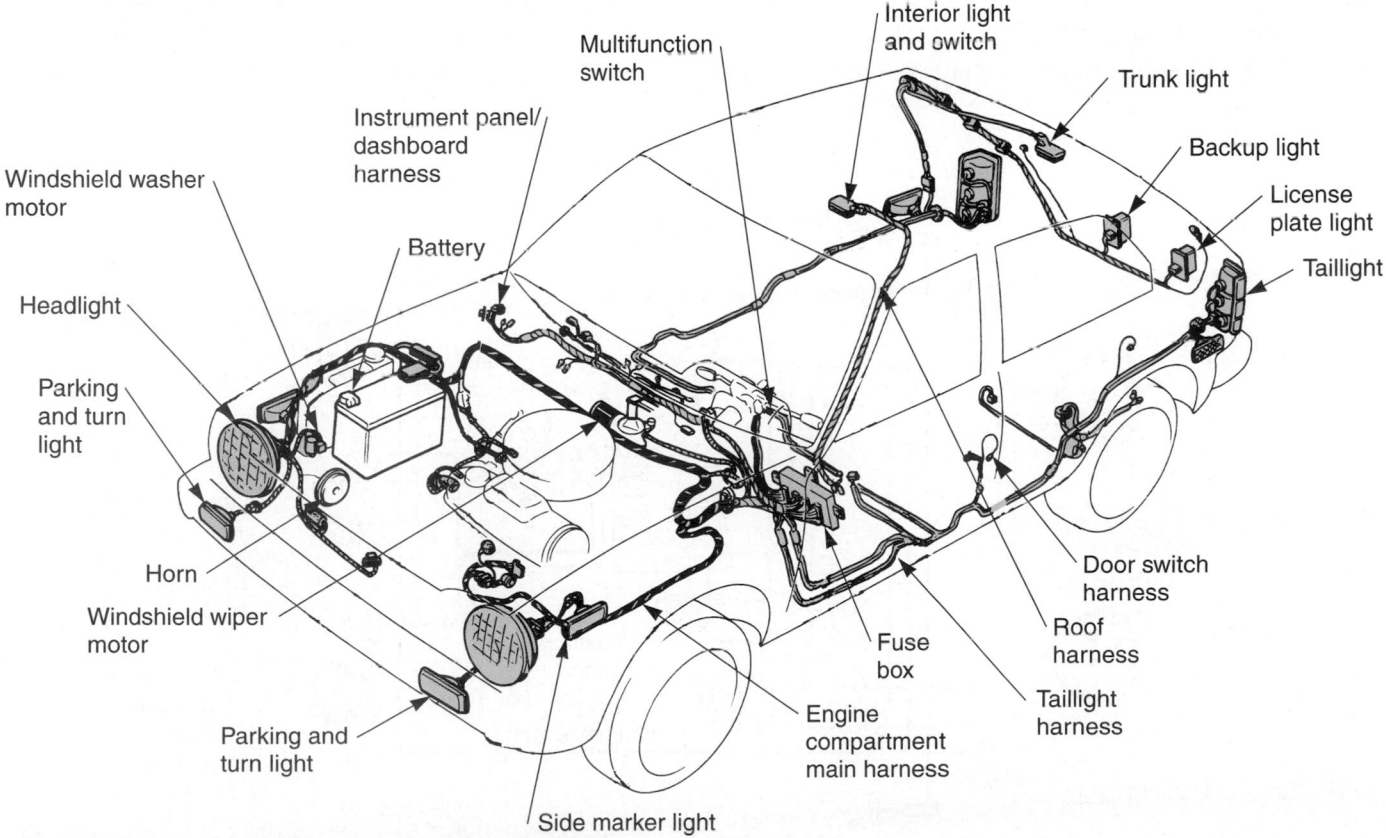

Figure 37-1. Light, wiper, and other major electrical systems are simple when studied separately. Note the location of the major components. (Honda)

After prolonged use, the lighting, instrument, wiper, and horn circuits may require repairs. Bulbs, switches, relays, motors, fuses, and wiring can fail. If you understand how the major parts function and follow basic testing techniques, each system is relatively easy to fix.

Lighting Systems

The *lighting system* consists of the components (fuses, wires, switches, and relays) that operate the interior and exterior lights on a vehicle. The exact circuit and part configurations will vary with each make and model of vehicle. However, the systems operate similarly.

Figure 37-2 shows a simplified drawing of a lighting system. As you will learn, actual vehicle light circuits are more complex than this.

The *exterior lights* typically include the headlights, turn signal lights, stoplights, parking lights, backup lights, and side marker lights.

The *interior lights* normally include the dome light, instrument lights in the dash, trunk light, and other courtesy lights.

Headlamp System

The *headlamp system* includes the battery, headlamp wiring, fuse panel, light switch, dimmer switch, headlamps, taillights, marker lights, and instrument lights. If the headlamps are concealed, the system also has either a vacuum or electric motor mechanism to operate the doors (flaps) over the headlamps.

The *headlamp switch* is an on/off switch in the dash panel or on the steering column. It controls current flow to the bulbs in the headlamps. It may also contain a *rheostat* (variable resistor) for adjusting the brightness of the instrument panel lights. The rheostat can also be mounted in the dash. A vacuum switch may be added to the headlamp switch or in the circuit when vacuum doors are used over the headlamps. **Figure 37-3** shows a dash-mounted headlamp switch.

A *multifunction switch* is capable of controlling several circuits—headlights (bright and dim circuits) and turn lights simultaneously for example. Most late-model vehicles use multifunction switches. Look at **Figure 37-4.**

Most multifunction switches have a *flash-to-pass feature,* which energizes the high beams when the switch arm is pulled back. This lets the driver quickly warn traffic that he or she is going to increase speed and pass. As soon as the multifunction switch arm is released, the headlamps return to low beams.

The *headlamp bulbs* are high-intensity bulbs that illuminate the road during night driving or bad weather conditions. Headlight bulbs can have one or two elements. Two elements are needed to provide low and high beams. *Low beams* are used for driving in traffic and *high beams* for greater visibility when there is no oncoming traffic.

When current flows through the *bulb element* or *filament,* it gets white-hot and glows to produce light energy. The reflector and lens direct this light forward. See **Figure 37-5.**

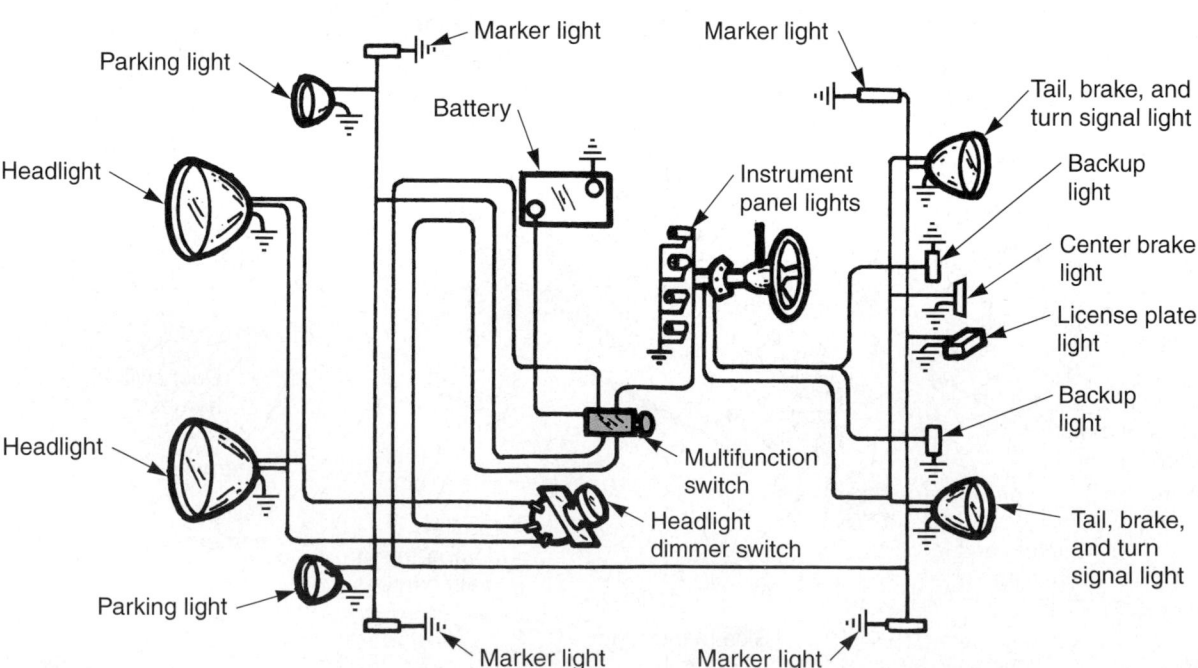

Figure 37-2. The light switch is the heart of the lighting system. It feeds current to circuits. (Florida Dept. of Voc. Ed.)

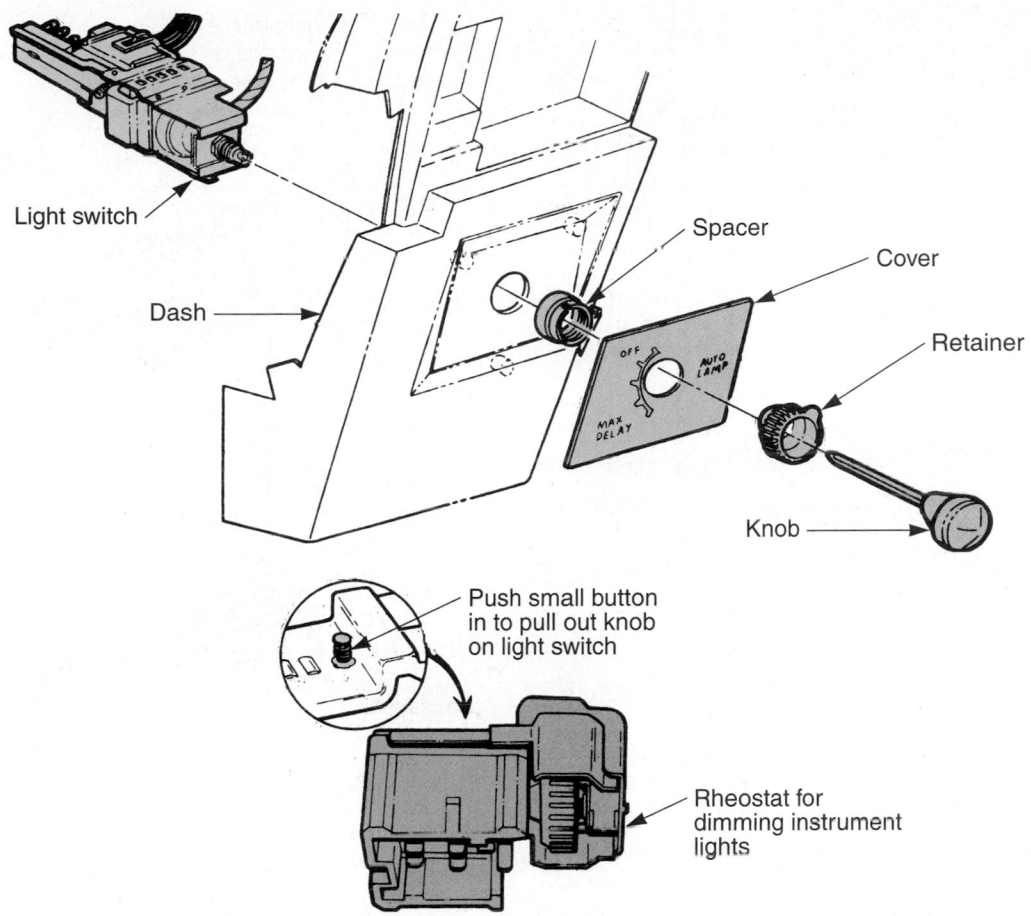

Figure 37-3. The light switch is complex and can cause problems after prolonged use. The light switch may mount in dash. A small button on the switch may be provided to release the knob from the switch. (Ford)

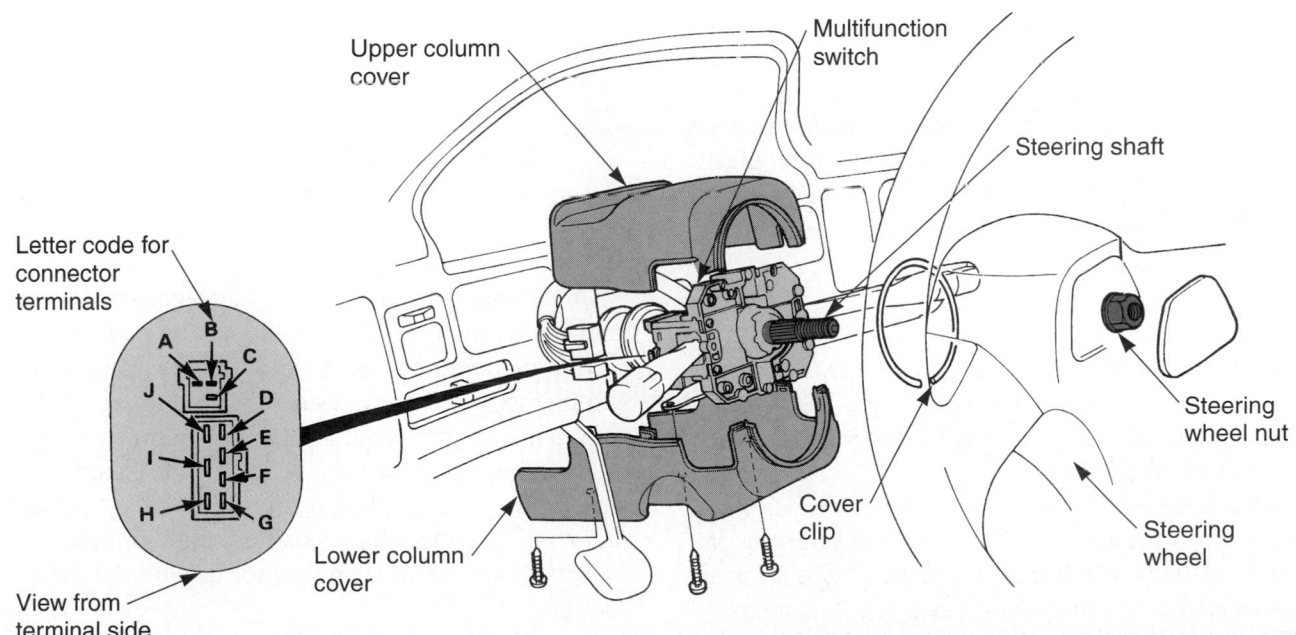

Figure 37-4. A multifunction switch on the steering column is common. It can be tested after removing the covers around the steering column. Letters on the connector can be compared to service manual charts and diagrams for probing voltage in different switch positions. (Honda)

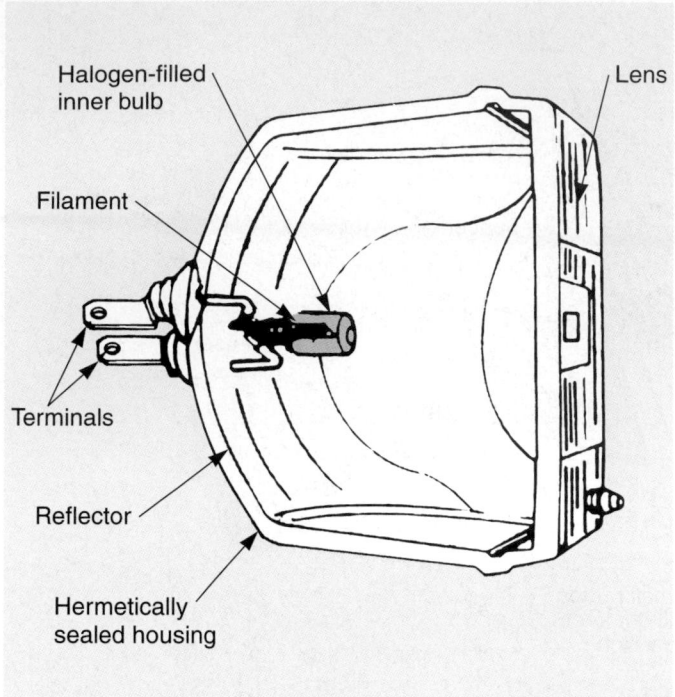

Figure 37-5. Sealed-beam lamp has an airtight glass housing around the filaments. This sealed-beam lamp has a small halogen inner bulb inside a glass lens. (Chrysler)

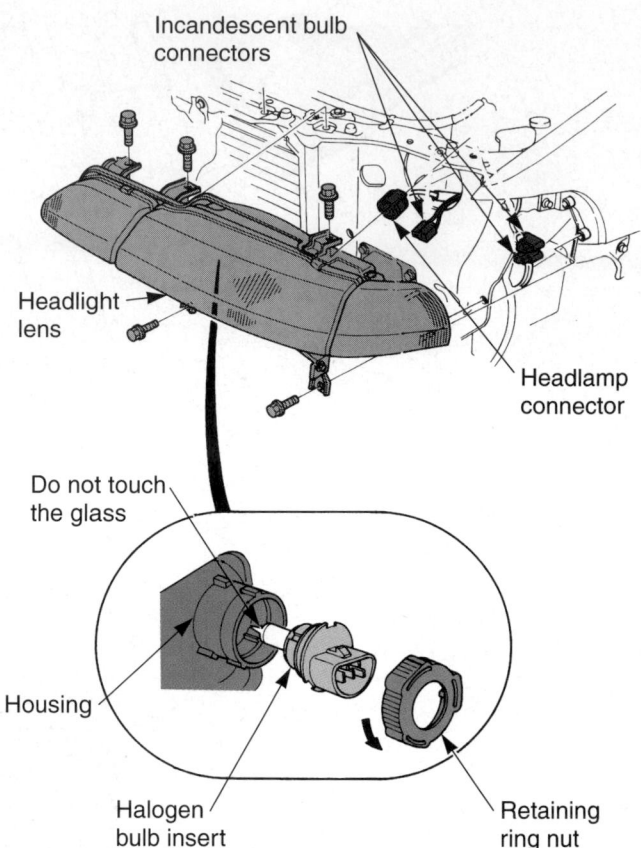

Figure 37-6. New vehicles commonly have halogen headlamp bulb inserts. The bulb insert installs in a plastic housing behind the lens. The lens and housing are not vacuum sealed. Note how the retaining ring nut locks the insert bulb into the housing. (Honda)

Sealed-beam headlamp bulbs have filament enclosed in a vacuum-sealed glass housing. There are two different types of sealed-beam bulbs: number 1 and number 2. A *number 1 lamp* has only one lighting element. The *number 2 lamp* has two lighting elements (high and low beam). Vehicles with only two headlamps use two number 2 lamps to provide high and low beams. These have been phased out for smaller halogen headlamp bulb inserts.

Most late-model cars use *halogen headlamp bulb inserts,* which are small bulbs that fit into a larger plastic housing. The bulb is made of heat-resistant quartz that is filled with halogen gas to protect the filament from damage. See **Figure 37-6.**

Compared to a conventional headlamp, a halogen headlamp increases light output by about 25% with no increase in current draw. The halogen bulb is also whiter, which increases visibility. Halogen bulbs are rated in watts. A typical low-beam bulb is 45 watts and a typical high-beam bulb is 65 watts.

Federal regulations limit how bright low-beam headlamps can be on public roads. The maximum brightness for low beams is 20,000 candle power. Brighter bulbs are available but should not be installed for use on public roads.

With a headlamp insert, a *headlight lens* disperses the light beam in front of the vehicle and protects the bulb. This lens can be made of glass or plastic. Look at **Figure 37-6.**

Tech Tip

Always make sure you have the correct replacement lamp. Normally, a lamp number is stamped on the housing. If needed, take the old lamp to the parts house so it can be matched up.

Headlamp Dimmer Switch

A *dimmer switch* controls the high and low headlamp beam function. This switch may be mounted on the steering column or floorboard. It simply controls which bulb filaments receive power. Refer to **Figure 37-7.**

When the driver activates the dimmer switch, it changes the electrical connection to the headlamps. In one position, the headlamp high beams, or brights, are turned on. In the other position, the dimmer switch changes to low beam filaments for driving in traffic.

Automatic Headlamp Systems

An *automatic headlight dimmer system* uses a light sensor, amplifier, and relay to control the high and low beams. The system automatically dims the lights when

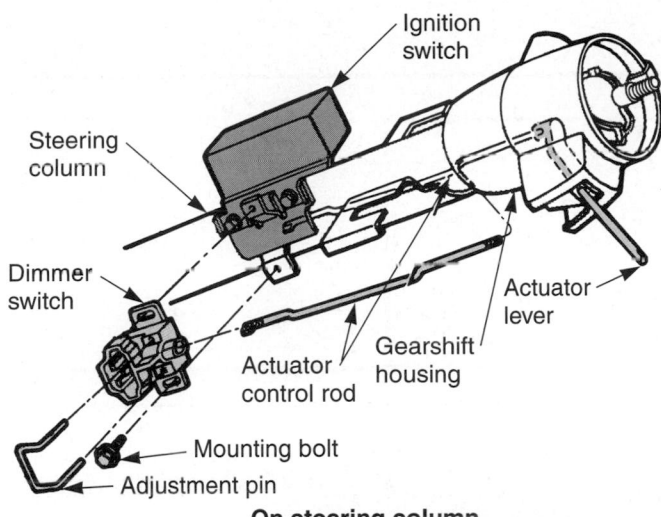

On steering column

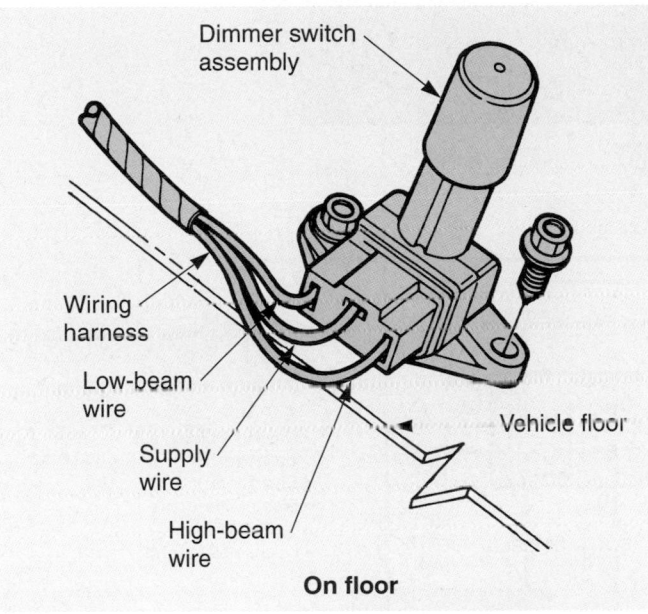

On floor

Figure 37-7. Dimmer switch locations. (Chrysler and Ford)

detecting light from oncoming traffic. It switches to high beams when no oncoming light is detected.

The light sensor, or photosensor, may be mounted in the grille area or on the dash. The sensor produces a small amount of electric current when exposed to light. The amplifier uses this current to operate a dimmer relay. The relay changes the electrical connection from high beam to low beam automatically to keep the bright lights from blinding oncoming traffic.

Other variations of automatic control of the lighting system are available. Some turn the headlamps on when dark. Others (*auto-off headlamp system* or *delayed exit system*) automatically turn the headlamps off after a short delay period. By remaining lit, the headlamps increase visibility and security when walking away from the vehicle.

Daylight Running Lights

With *daylight running lights* (DRL), the headlamps are lit anytime the engine is running. This is a safety feature that makes the vehicle more visible. **Figure 37-8** shows a lighting system circuit diagram that includes DRL.

Turn Signal, Emergency, and Brake Light Systems

The turn signal, emergency, and brake light systems are normally considered separate circuits. However, they commonly use some of the same wiring, electrical connections, and lightbulbs. This is illustrated in **Figure 37-9.**

The lights in these systems are small incandescent bulbs. They can contain either one or two filaments. **Figure 37-10** shows two common types of bulbs for turn, emergency, stop, and backup lights.

Turn Signal System

The *turn signal system* consists of a fuse, turn signal switch, flasher unit, turn signal bulbs, indicator bulbs, and related wiring. When the steering column–mounted switch is activated, it causes the right or left side turn lamps to flash. Turn indicator lights in the instrument panel or on the fenders also flash.

The turn signal switch may be mounted in the center of the steering column, behind the steering wheel. A multifunction switch can also be used to control turn lights, horn, and dimmer switch. Look at **Figure 37-11.**

The *turn signal flasher* automatically opens and closes the turn signal circuit, causing the bulbs to flash. The flasher unit contains a temperature-sensitive bimetallic strip and a heating element. The bimetal strip is connected to a set of contact points and to the fuse panel.

When current flows through the turn signal flasher, the bimetallic strip is heated and bends. This opens the contact points and breaks the circuit. As the bimetal strip rapidly cools, it closes the points and again completes the circuit. This heating and cooling cycle takes place in about a second. The turn lights flash as the points open and close.

The turn signal flasher is frequently mounted on the fuse panel, **Figure 37-12.** However, it may be located somewhere else, under the dash for example. A shop manual will give the flasher location.

 Tech Tip

A turn signal flasher is designed to operate with a specific current draw. If you alter the load (add trailer lights, for instance), it can affect the operation of the flasher. Generally, if you increase current draw through the circuit, the flasher will blink more quickly. If you decrease current draw (one bulb burns out), it will slow or even stop the blinking action of the flasher.

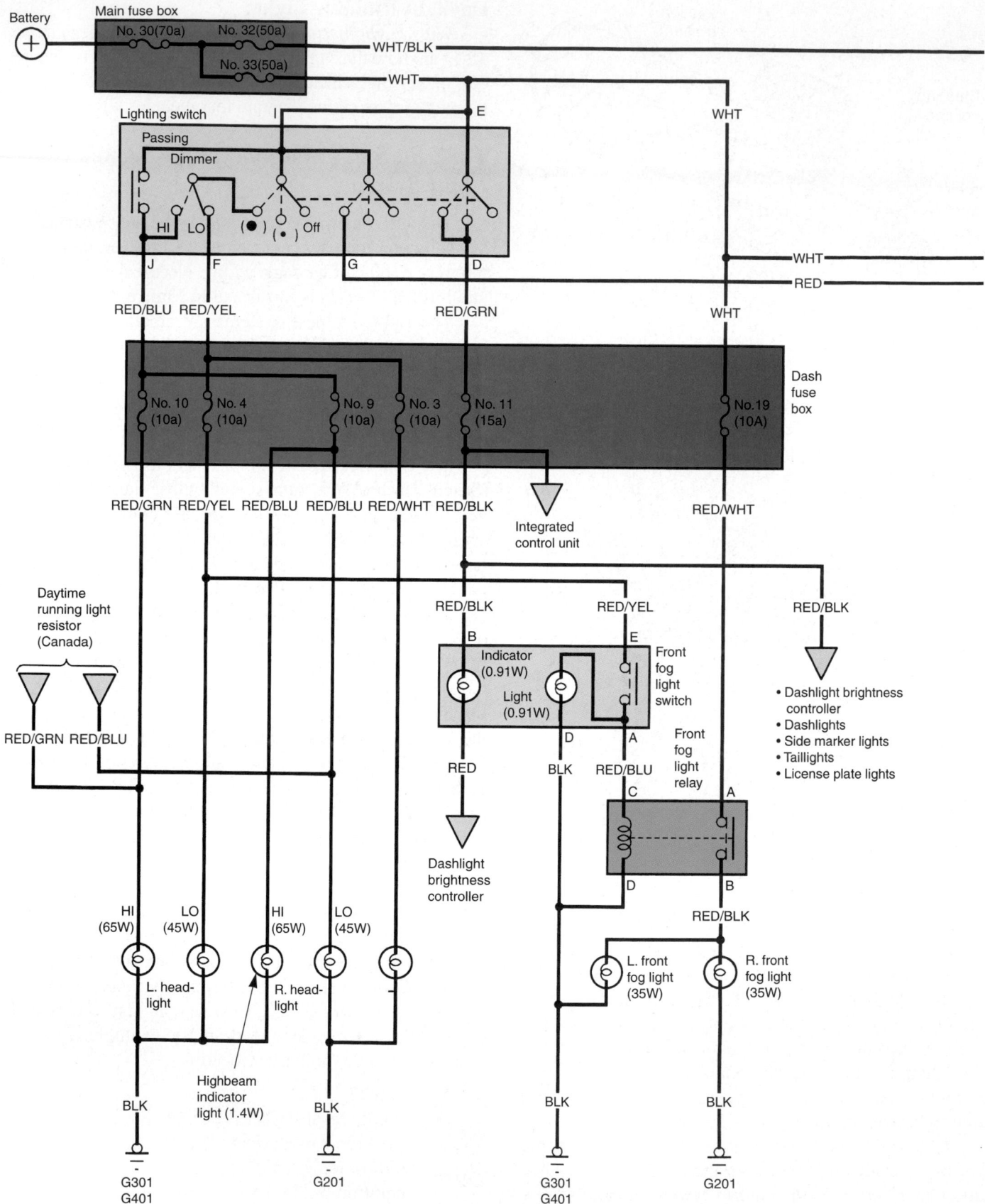

Figure 37-8. Typical wiring diagram for a lighting system. Trace electrical connections from the main fuse box (top left) to headlights (lower left) and daytime running lights (lower right). Color codes on wires are needed when testing for circuit problems. (Honda)

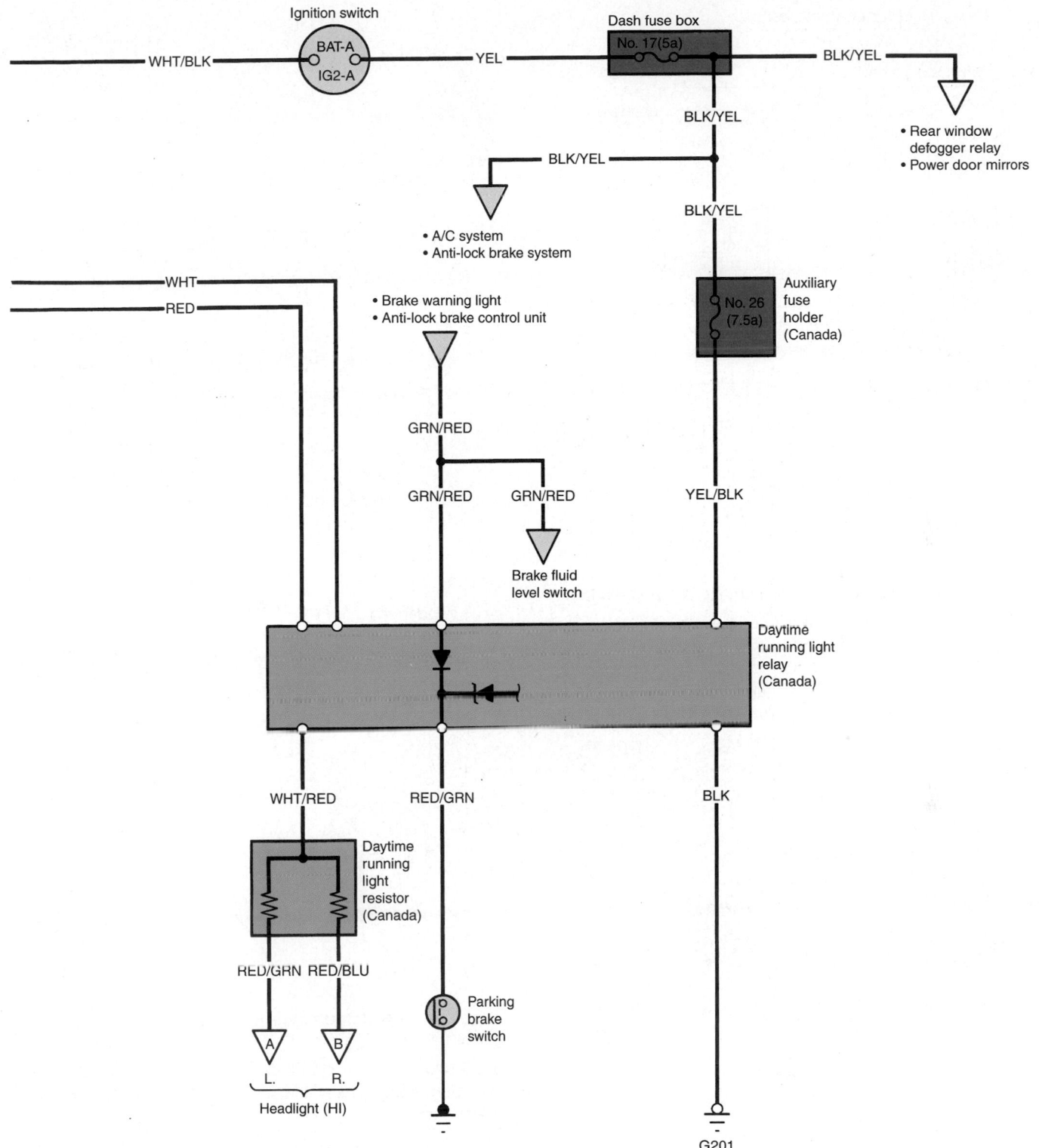

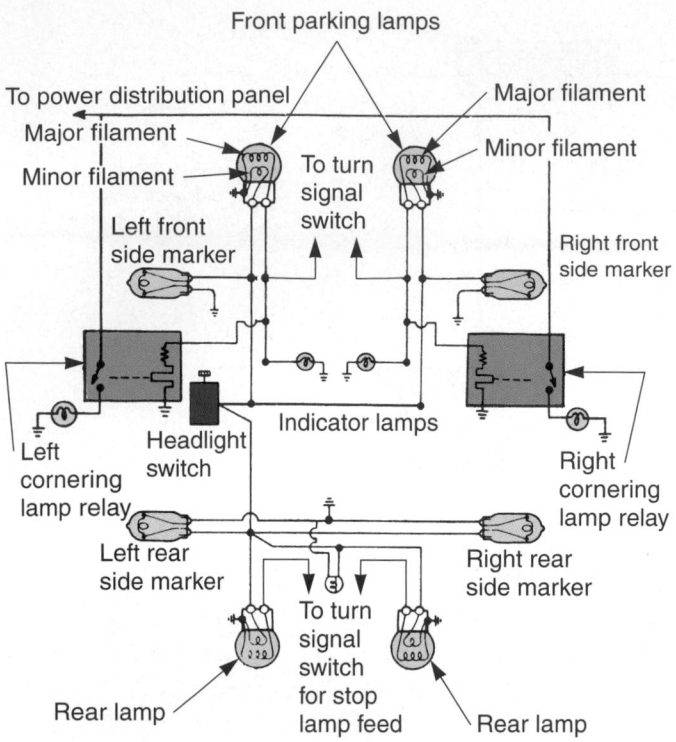

Figure 37-9. Study the basic circuit for turn signals and marker lights.

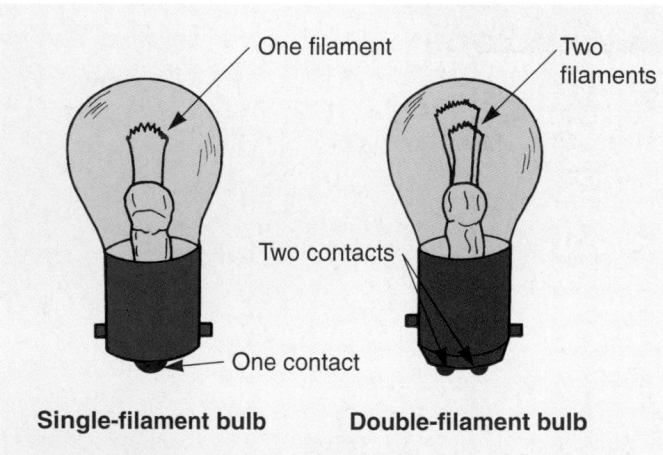

Figure 37-10. Single-filament bulbs are commonly used as backup and marker lights. Double-filament bulbs serve as parking, turning, or braking lights.

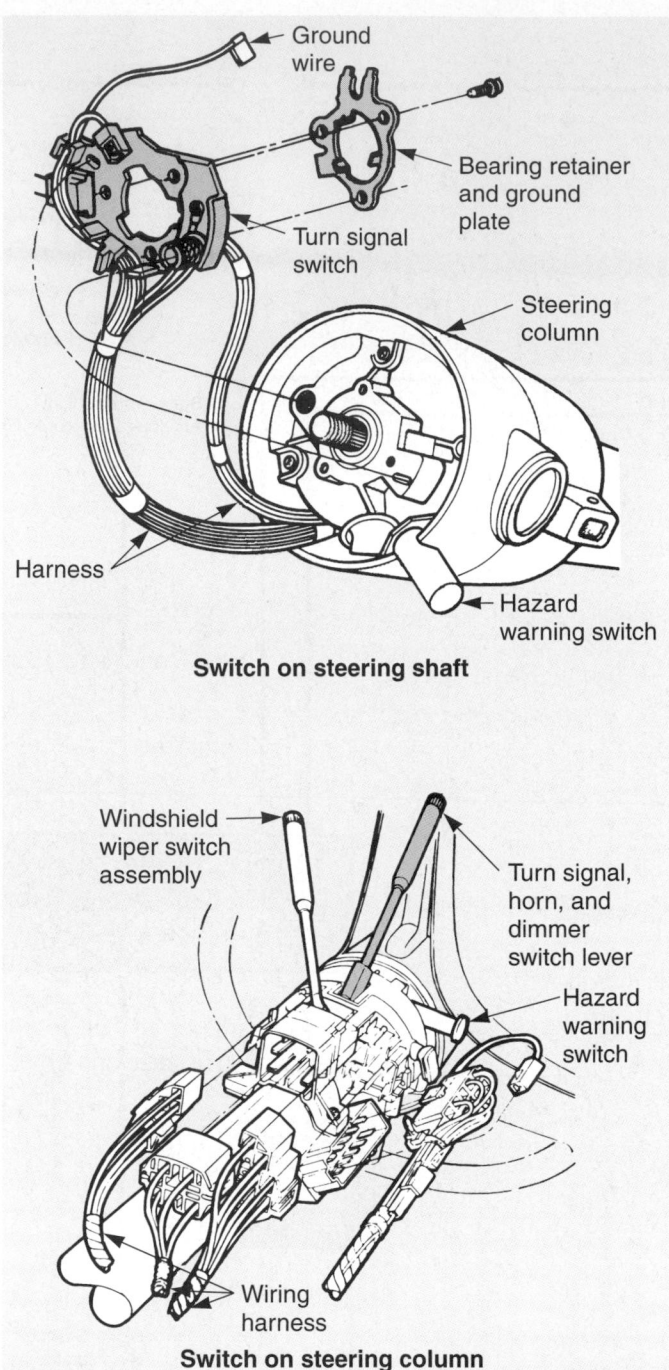

Figure 37-11. The turn signal switch can be mounted around the steering shaft or on the side of the steering column. (Chrysler and Ford)

Emergency Light System

The *emergency light system,* also termed *hazard warning system,* consists of a switch, a flasher unit, turn signal lamps, and related wiring. The emergency light switch is normally mounted on the steering column, **Figure 37-11.** It is usually a push-pull switch.

When the switch is closed, current flows through the emergency flasher. Like a turn signal flasher, the *emergency flasher* opens and closes the circuit to the lights. This causes the turn signals to flash. Oncoming traffic is warned of a possible emergency or hazard.

Brake Light System

The *brake light system* is commonly made up of a fuse, brake light switch, rear lamps, and related wiring. The brake light switch is normally mounted on the brake

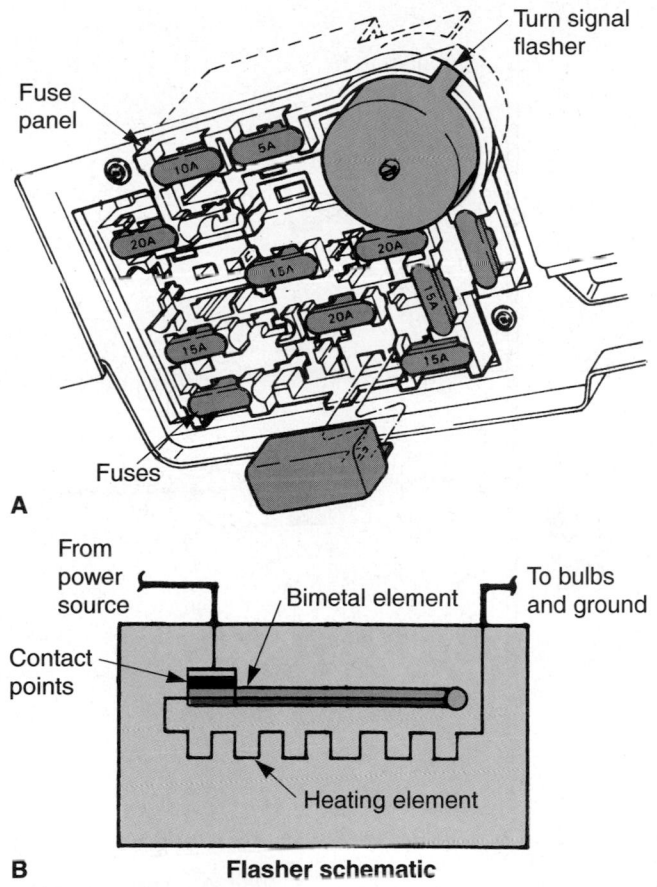

Figure 37-12. A—The turn signal flasher is often located on the fuse panel or under the dash. B—The flasher contains a bimetallic strip, contact points, and a heating element. Current heats the bimetallic strip to open the points. As the unit cools, the contacts close to complete the circuit. Flashing lights result. (Ford)

pedal, as in **Figure 37-13.** Battery power is fed to the brake light switch from the ignition switch. When the brake pedal is pressed, it closes the switch and current flows through the wiring to the brake lights. This turns on the rear brake lights to warn traffic of the stop.

The brake light switch can also be located on the master cylinder. In these switches, hydraulic pressure from the brake system closes the switch to turn on the brake lights.

> **Note**
> For more information on brake light systems, refer to Chapter 71, *Brake System Fundamentals,* and Chapter 72, *Brake System Diagnosis and Repair.*

Backup Light System

A *backup light system* typically has a fuse, gear shift- or transmission-mounted switch, two backup lamps, and wiring to connect these components.

The *backup lamp switch* closes the light circuit when the transmission is shifted into reverse. This illuminates the area behind the car. See **Figure 37-14.**

Illuminated Entry System

The *illuminated entry system* turns on the interior lights when you move the door handle or open the door. It allows you to see the key cylinders and the interior of the passenger compartment at night. Illuminated entry systems vary from simple to complex.

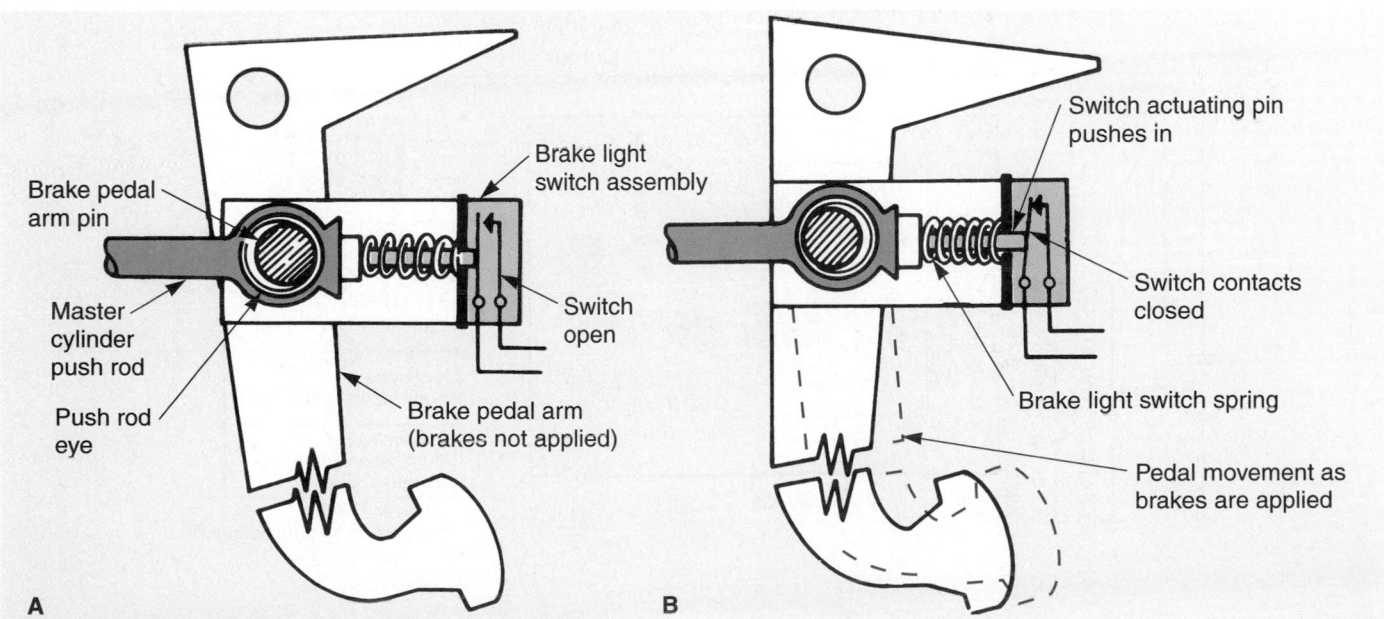

Figure 37-13. Study brake light switch action. A—Brake pedal released, contacts open, brake lights are off. B—Pedal depressed, contacts close, brake lights function. Note the part names. (Ford)

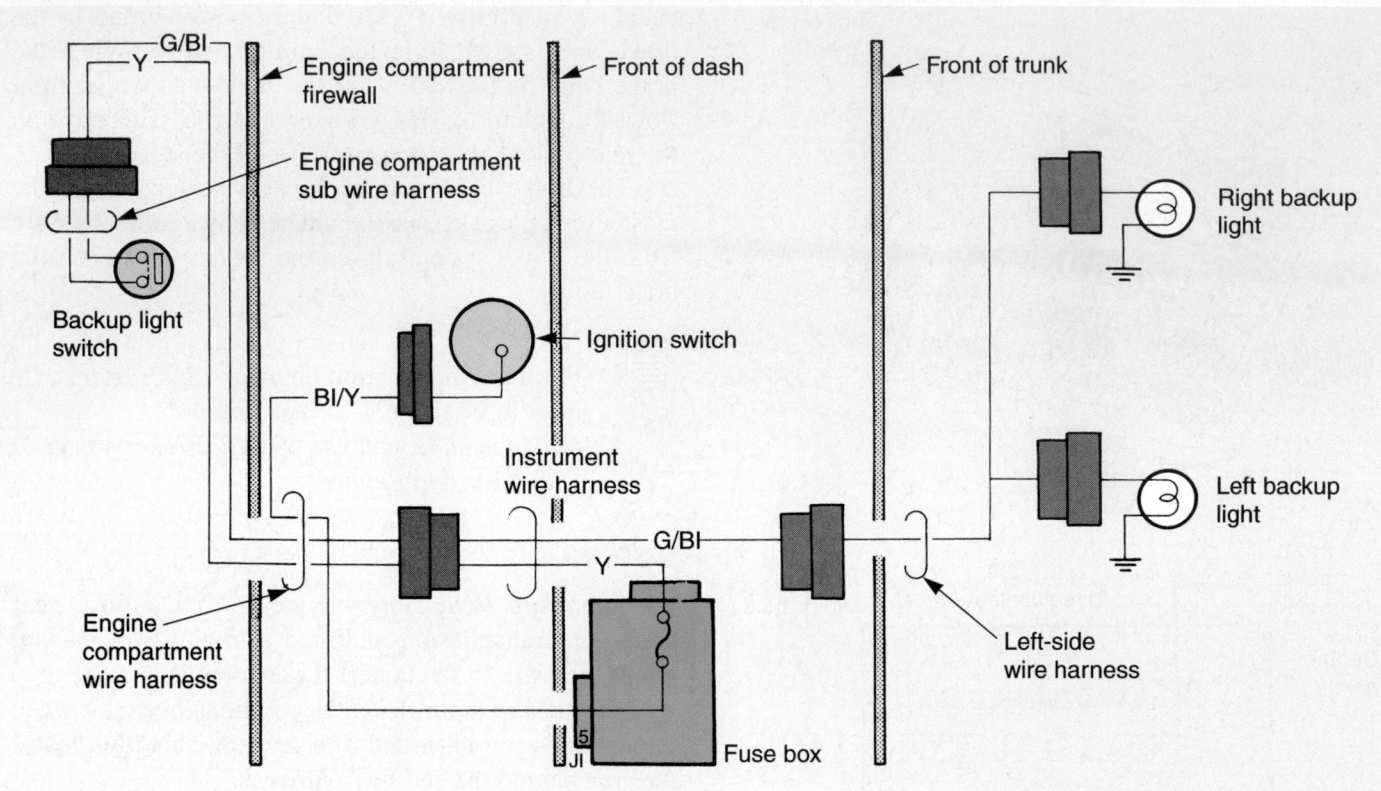

Figure 37-14. Note basic configuration of the backup light circuit. Vertical lines represent the firewall, dash, and trunk divider. (Honda)

A simple illuminated entry system uses door-jam switches to turn on the interior lights. When a door is opened, it closes that switch to energize the lights.

More complex illuminated entry systems use an electronic control unit, various switches, and several interior lights. As shown in **Figure 37-15,** both door switches and

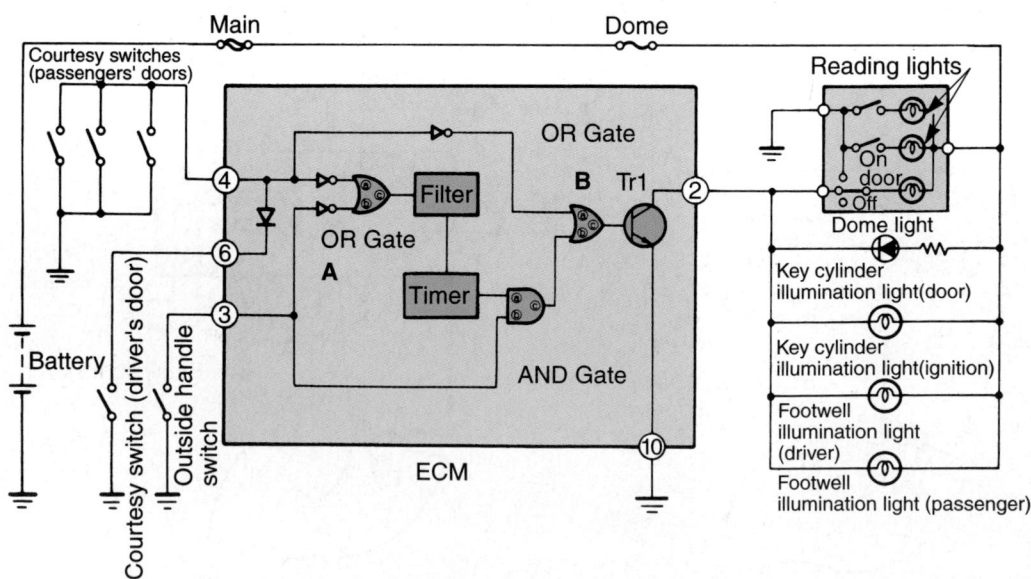

Figure 37-15. This illuminated entry system uses several switches (left) to signal the ECM (center), which operates the lights (right). When you pull up on the driver's door handle, it energizes the system. The lights fade in about ten seconds if no other switch is closed. (Lexus)

a driver's door handle switch energize the system. When one of the switches is closed, it signals the ECM. The ECM can then turn on the lights and time when they go off. Note that this circuit also illuminates the key cylinders and footwell in the passenger compartment.

Light System Service

Light system service involves changing burned-out bulbs, testing bulb sockets, checking fuses, finding shorted and open circuits, aiming headlights, and similar tasks. This section of the chapter will summarize the most important types of electrical tests and repairs performed on a car's lighting system.

Lamp Replacement

A *burned-out bulb* has the filament melted in half. Headlamps are usually held in place with small screws and a retaining ring. Halogen insert bulbs normally fit into the rear of the bulb housing. You must push and twist to install some headlamp bulb inserts. Sometimes a small ring screws over and secures the halogen insert. Spring clips can secure halogen fog lamp bulbs.

Figure 37-16 shows how several kinds of light assemblies fit together.

Most incandescent bulbs are housed in a lens. They are normally held in the socket by a spring and small dowels or a press fit. You may have to access the bulb by removing the lens or by reaching behind the housing. This is illustrated in **Figure 37-17.**

Always make sure you have the correct replacement bulb.

Caution
Never touch the glass surface of a high intensity bulb. The oil on your skin and the high operating temperature can make the glass shatter or shorten the bulb's service life.

No-Light Problem

A *no-light problem* is a total failure of the light circuit or bulb. First, check to see if the bulb is burned out. Close inspection of the filament will show whether it has burned. If the bulb is good, check for power to the bulb socket. See **Figure 37-18.**

Figure 37-19 shows how to check for current in a bulb socket. With the light turned on, there is a socket or circuit problem if the test light does not glow. If you do not have power to the socket, trace back through the circuit to find the open preventing current flow.

Flickering Light Problem

Flickering lights (lights go on and off) point to a loose electrical connection or a circuit breaker that is kicking out because of a short.

If all or several of the lights flicker, the problem is in a section of the circuit common to those lights. Check to see if the lights flicker only with the light switch in one position. For example, if the lights flicker only when the headlights are on high beam, you should check the components and wiring in the high beam section of the circuit.

If only one light flickers, the problem is in that section of the circuit. Check the bulb socket for corrosion. Clean the socket. Also, make sure the bulb terminals are

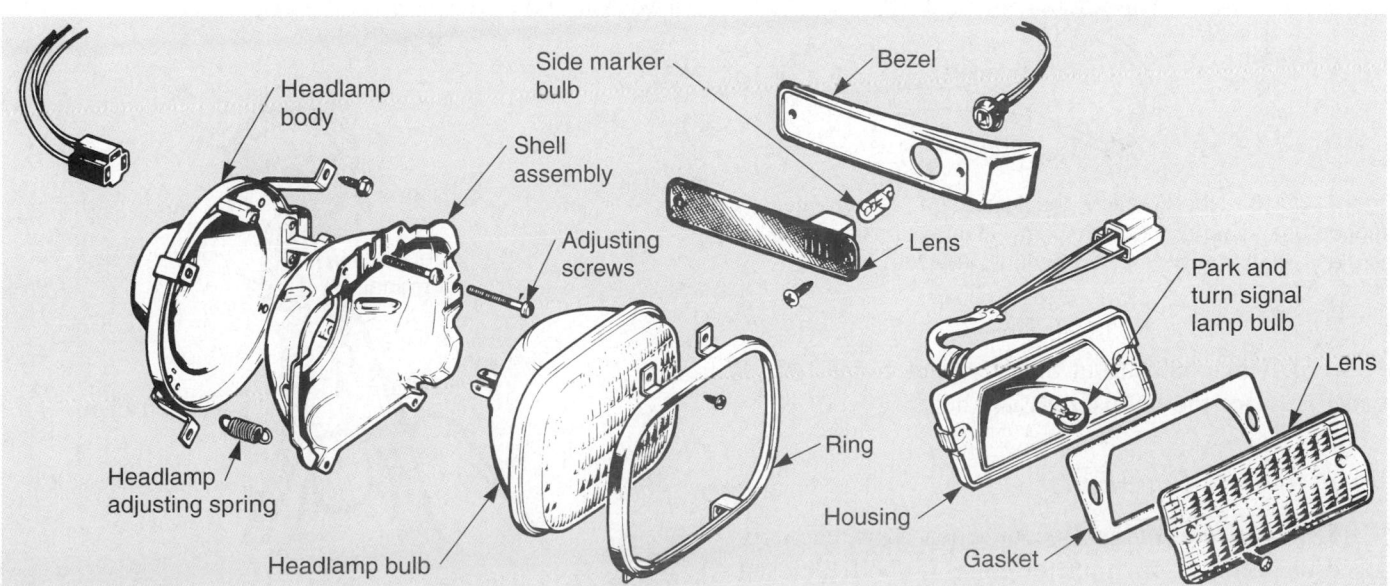

Figure 37-16. Note the construction of headlamp, turn signal, and side marker light assemblies. (Chrysler)

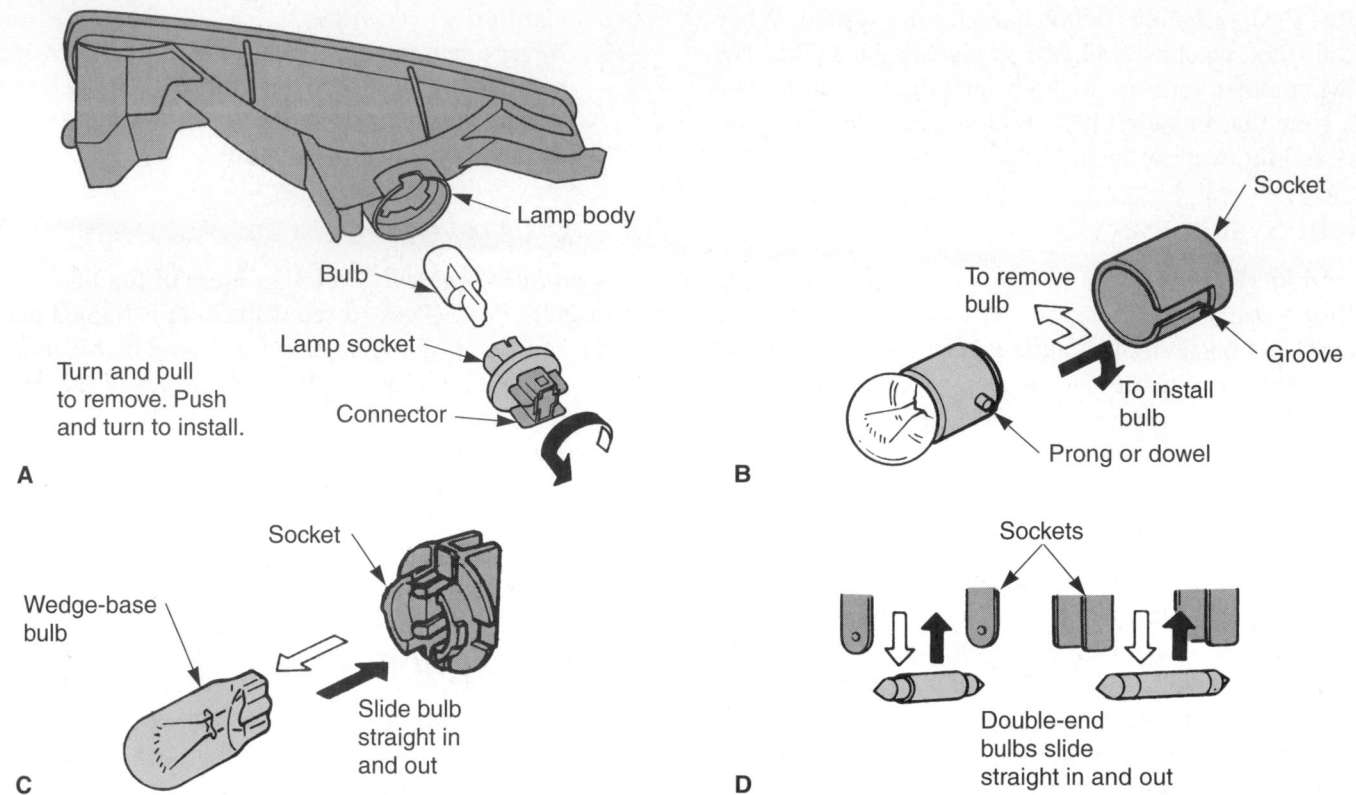

Turn and pull to remove. Push and turn to install.

A

Lamp body

Bulb

Lamp socket

Connector

To remove bulb

To install bulb

Socket

Groove

Prong or dowel

B

Wedge-base bulb

Socket

Slide bulb straight in and out

C

Sockets

Double-end bulbs slide straight in and out

D

Figure 37-17. Study bulb configurations. A—Socket has lugs that lock into the lamp body. Partial turn will free or install the bulb socket. B—Push in and turn to install or remove a bulb with prongs. Socket has spring-loaded terminals. C—Just pull in or out on wedge-base bulb. Be careful not to break the bulb and cut your fingers. D—Double-end bulbs are snapped in and out as shown. (Honda and Nissan)

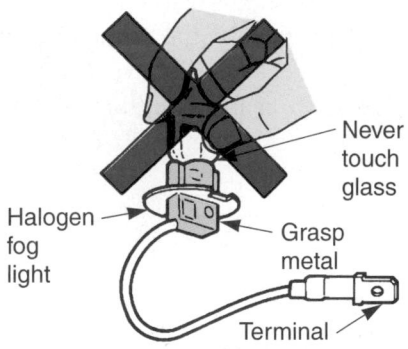

Never touch glass

Halogen fog light

Grasp metal

Terminal

Figure 37-18. Never touch the glass on a high-intensity halogen headlamp or fog lamp bulb. Oil on your skin can make the glass shatter and ruin the new bulb. (Mazda)

not worn. This could upset the electrical connection. If needed, replace the bulb socket and bulb.

Light Switch Problems

Light switch problems occur when a switch shorts closed or becomes open all the time. If shorted closed, the lights will always glow. If always open, the lights will not come on.

Figure 37-20 shows several of the types of switches used to operate lights and instrumentation systems.

To test a switch, use a test light or voltmeter to check for power going to and from the switch in the closed, or on, position. You can also disconnect the switch and test it with an ohmmeter, **Figure 37-21.**

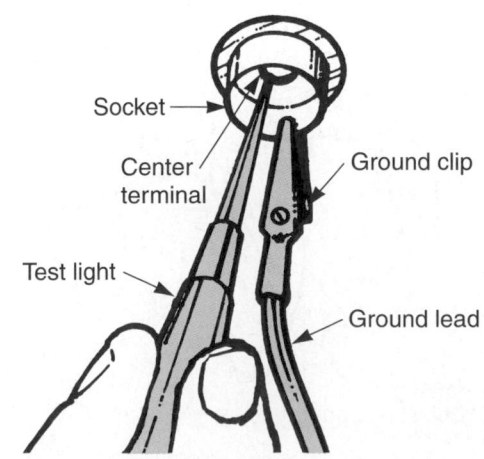

Socket

Center terminal

Ground clip

Test light

Ground lead

Figure 37-19. If the new bulb does not work, check for voltage to the socket with a test light. You may need to clean rust and corrosion out of the socket. (Florida Dept. of Voc. Ed.)

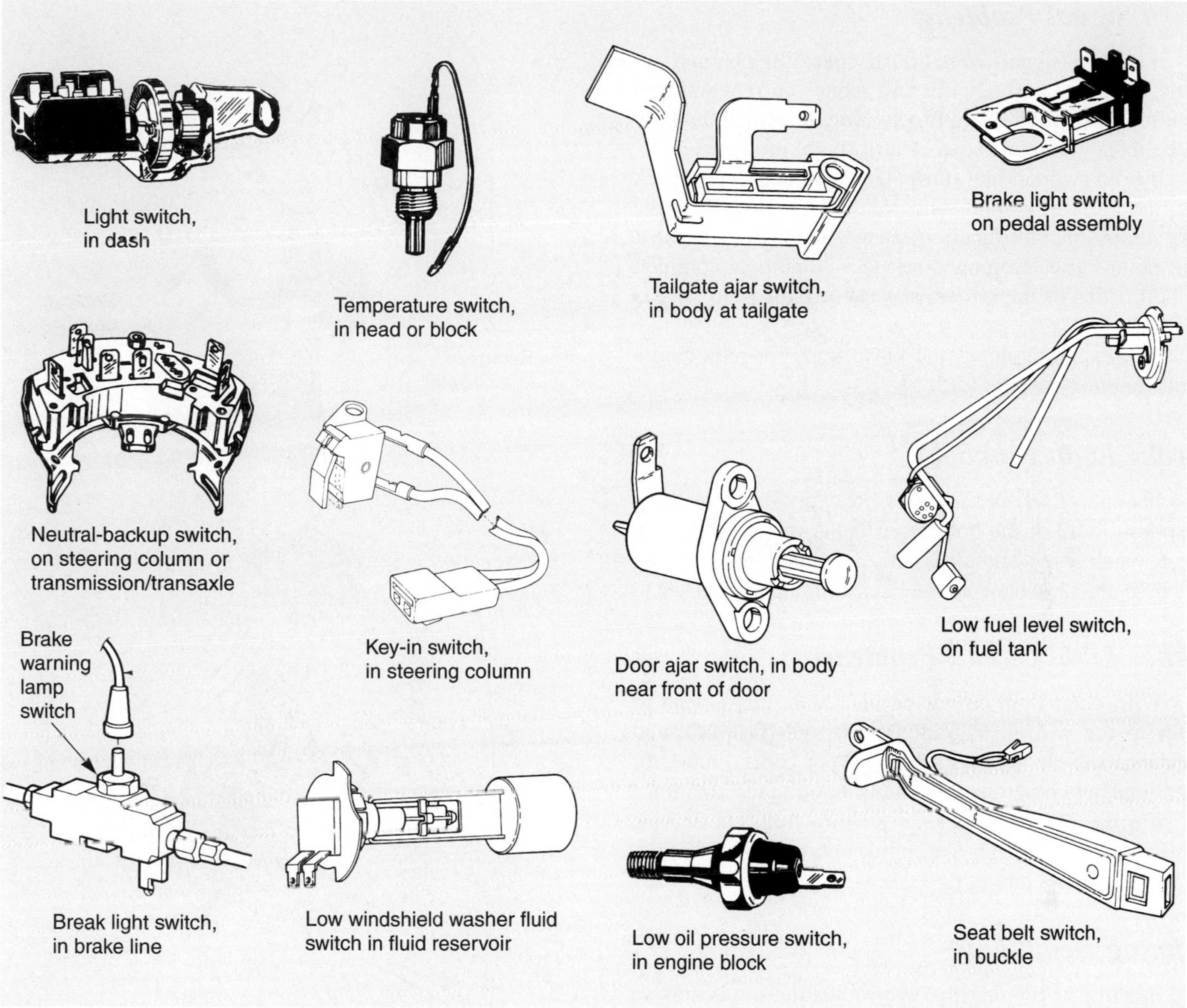

Light switch,
in dash

Temperature switch,
in head or block

Tailgate ajar switch,
in body at tailgate

Brake light switch,
on pedal assembly

Neutral-backup switch,
on steering column or
transmission/transaxle

Key-in switch,
in steering column

Door ajar switch, in body
near front of door

Low fuel level switch,
on fuel tank

Brake
warning
lamp
switch

Break light switch,
in brake line

Low windshield washer fluid
switch in fluid reservoir

Low oil pressure switch,
in engine block

Seat belt switch,
in buckle

Figure 37-20. Study the various switches used in light and instrumentation circuits. (GP Parts and Chrysler)

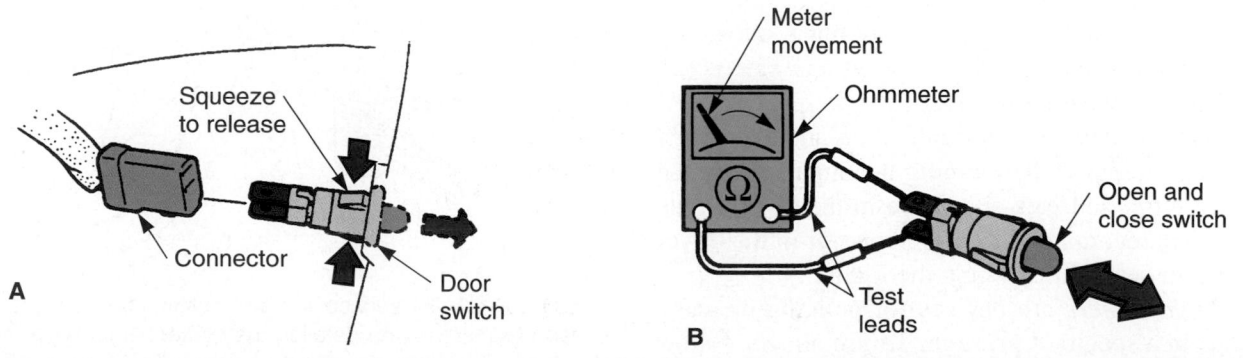

Squeeze
to release

Connector

Door
switch

A

Meter
movement

Ohmmeter

Open and
close switch

Test
leads

B

Figure 37-21. Switches, because of their moving parts, are a common source of trouble. They can short closed or become broken and remain opened. A—To test a switch, remove the wiring connector. This is door switch that must be squeezed lightly for removal from the door pillar. B—Connect an ohmmeter across the switch terminals. Activate the switch and the meter reading should go from almost zero to infinity. (Mazda)

Turn Signal Problems

If the turn signals do not flash, check for a burned-out bulb. Even one burned bulb will reduce current and prevent the flasher unit from functioning. A burned-out bulb is the most common cause of turn signal problems.

If both right and left turn signals do not work, check the fuse and flasher unit. The problem may also be in the turn signal or multifunction switch. It is prone to wear and failure after prolonged service. Something common to both sides of the circuit may be at fault if no bulb is coming on.

Figure 37-22 shows the basic steps for removing a multifunction switch.

Brake Light Problem

If none of the brake lights are working, something common to all of the bulbs is at fault, such as the brake light switch or feed circuit. If only one bulb is not working, the bulb and its section of the circuit should be checked.

Other Light Circuit Problems

With other light circuit problems, you may want to refer to the wiring diagram for the specific make and model of vehicle. It will give wire color codes, connector locations, and other useful troubleshooting information.

Figure 37-23 is a wiring diagram from one service manual. Trace through the circuit and note the information given in the diagram.

Aiming Headlights

Aiming headlights involves adjusting the beams so they are directed to light the area in front of the vehicle. Headlights can be aimed using mechanical aimers, a wall screen, or leveling bubbles on the headlight housings. These methods ensure that the headlight beams point in the direction specified by the vehicle manufacturer.

Headlights aimed too high could blind oncoming traffic. Headlights aimed too low or to one side could reduce visibility for the driver.

The car should have a half tank of gas and the correct tire inflation pressure when aiming headlights. Only the normal spare tire and jack should be in the trunk. Some manufacturers recommend that someone sit in the driver and passenger seat while aiming the lights.

Headlight aimers are devices for pointing the car's headlamps in a specified position. To use aimers, follow the instructions for the specific type of equipment. Some require a level floor. Others have internally leveling mechanisms to allow for an uneven shop floor. **Figure 37-24** shows the use of headlight aimers.

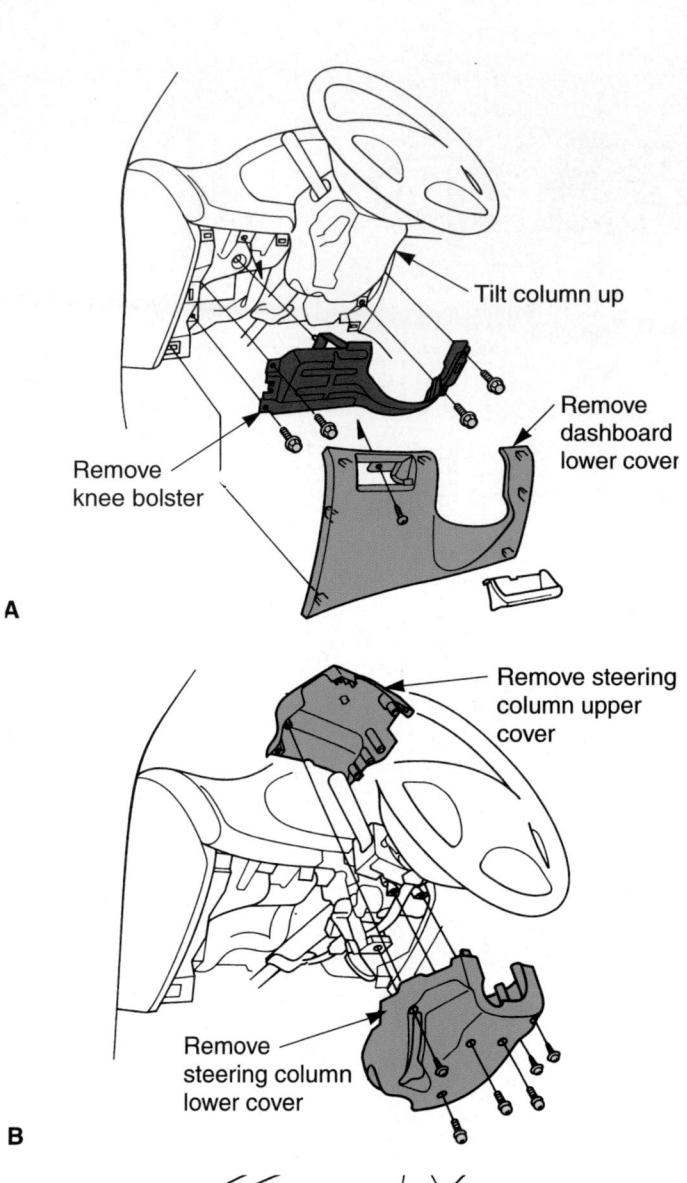

A

Tilt column up

Remove dashboard lower cover

Remove knee bolster

Remove steering column upper cover

Remove steering column lower cover

B

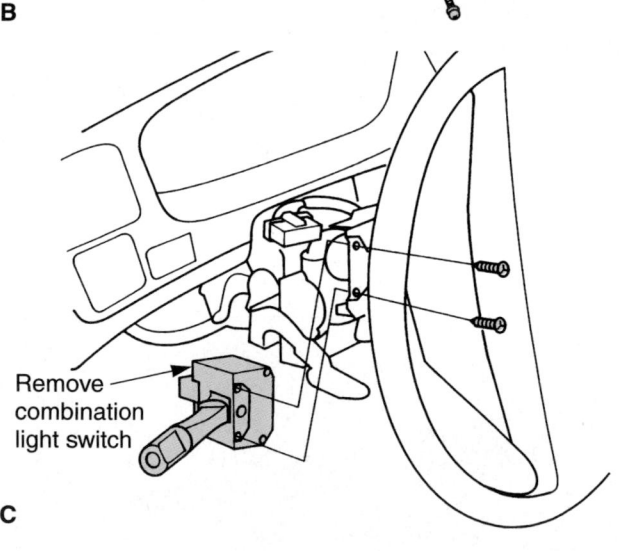

Remove combination light switch

C

Figure 37-22. To service a bad multifunction switch, follow these basic steps and detailed instructions in service manual. A—Remove electrical connections or test switch using service manual diagrams and instructions. B—If bad, replace the multifunction switch. This one can be replaced without steering wheel removal. Others require steering wheel removal, however. (Honda)

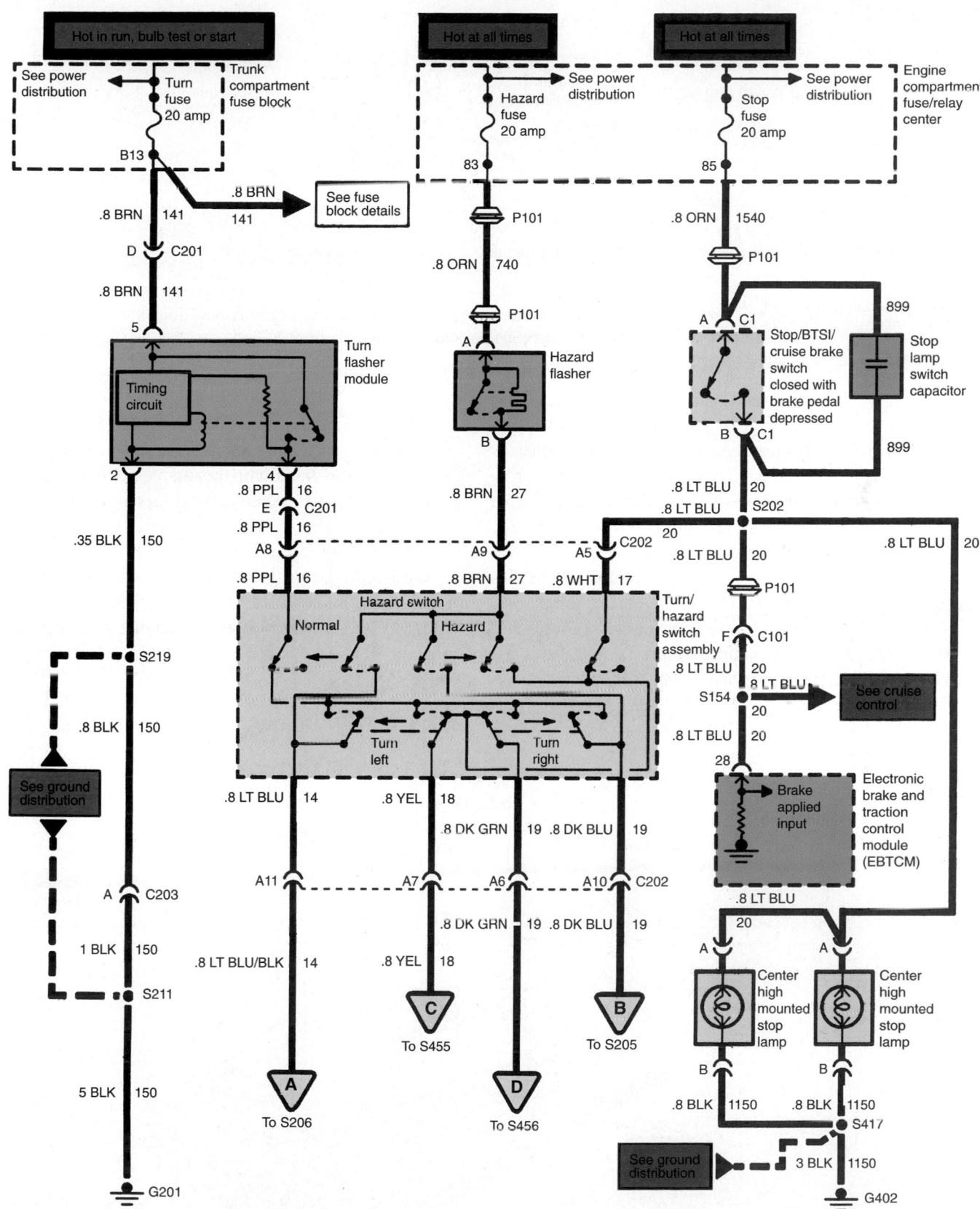

Figure 37-23. Carefully study the components of this exterior lamp circuit diagram. Note fuses, electrical connections, grommets, connector markings, and references to other service manual pages. Diagrams like this are essential when tracing troubles. (General Motors)

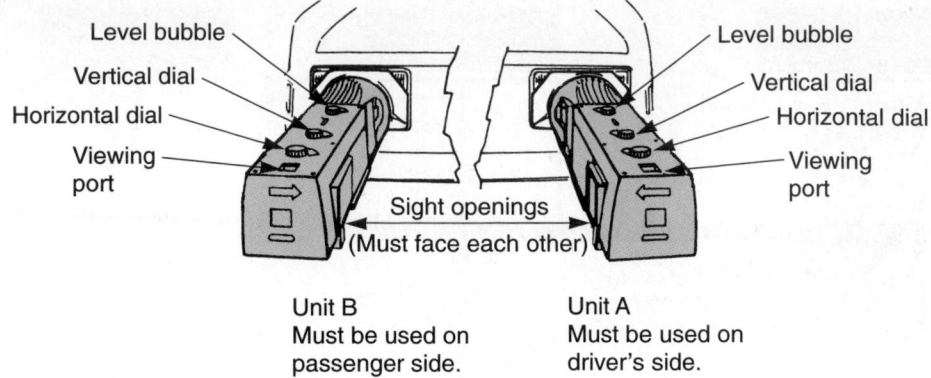

Figure 37-24. These headlight aimers mount on lamps with suction cups. Leveling bubbles show vertical adjustment. Reflection from mirrors shows a horizontal adjustment. With new cars again using round headlights, headlight aimers are gaining renewed popularity. (Chrysler)

A *headlight aiming screen* is a series of measured lines marked on a shop wall for aiming a car's headlamps. One is shown in **Figure 37-25.** Note that lines are drawn on the wall at specific horizontal and vertical locations depending upon vehicle dimensions.

The car is normally located 25′ (7.6 meters) from the screen on a level floor. When the headlights are turned on, the highest points of light intensity (brightness) should be as shown in **Figure 37-25.** Aiming the low beams also aims the high beams. Refer to a service manual for added details.

Headlight adjusting screws are provided to alter the direction of the headlamp beams. One screw provides vertical adjustment. Another provides horizontal adjustment. Turn the screws until the aimer or screen show correct beam alignment. See **Figure 37-26.**

Headlight leveling bubbles are built into many late-model headlight assemblies to simplify headlight aiming, **Figure 37-27A.** You simply adjust the housing screws until the leveling bubble is centered in its indicator, **Figure 37-27B.**

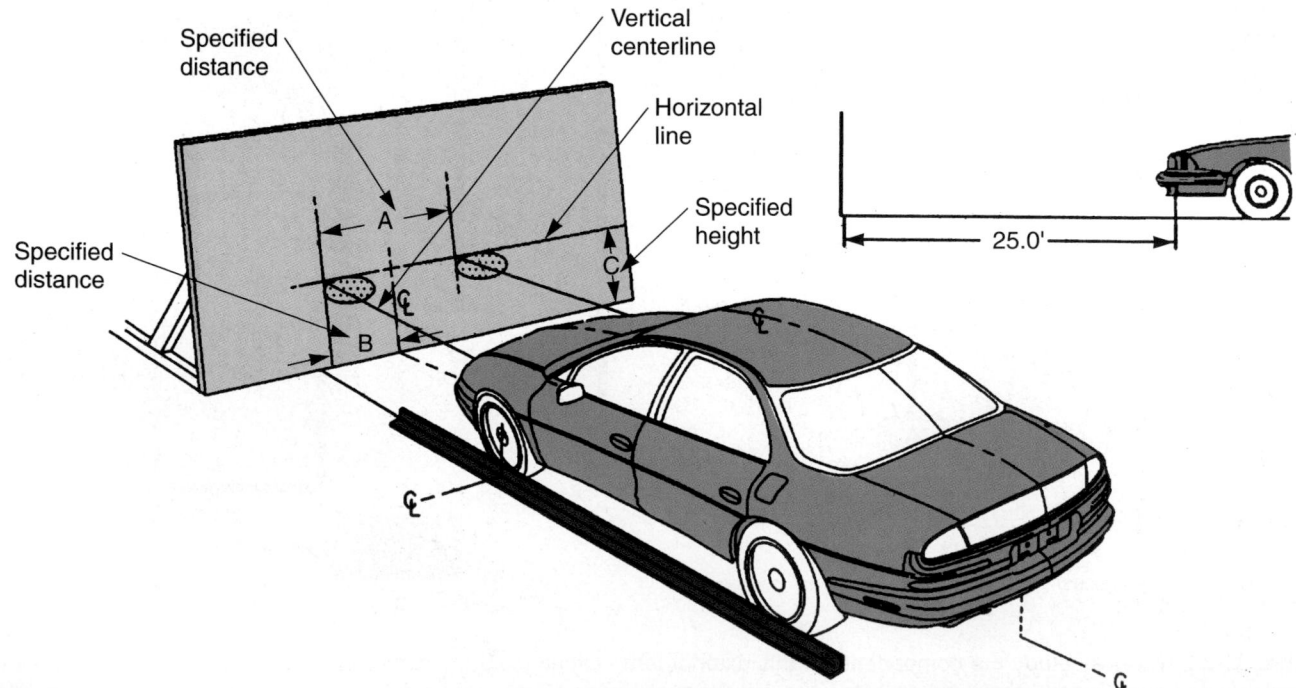

Figure 37-25. Lines can be marked on the shop wall according to vehicle specifications to aim headlights. The car must be located a prescribed distance from the wall. Adjust headlamp low beams until they shine in designated areas on the wall. (Oldsmobile)

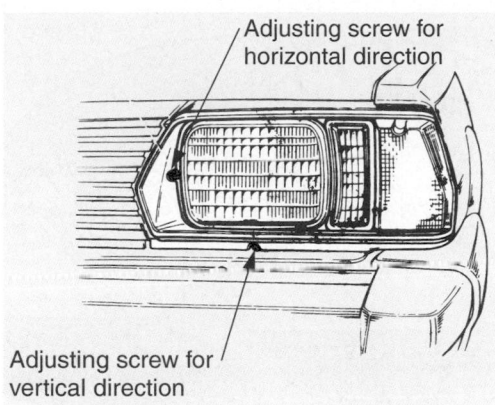

Figure 37-26. Note the location of headlight adjustment screws. They can often be adjusted without removing the trim. (Subaru)

A

Instrumentation

Instrumentation is used to inform the driver of various operating conditions—oil pressure, engine temperature, computer trouble codes, air bag system status, etc. Today's vehicles use a wide range of instruments, some simple and others complex. It is important that you understand the basics of instrumentation because they must be serviced in the field when problems develop.

Analog instruments use rotating needles or dials to indicate operating conditions. An analog speedometer, for example, uses a large needle that rotates around to show vehicle speed. An analog tachometer gives the same type of display for engine speed. See **Figure 37-28A.**

Digital instruments uses various types of lights and electronic displays to show operating conditions. They do not have the mechanical parts, as analog instruments do. Look at **Figure 37-28B.** Digital displays are more complex than analog displays because an electronic control unit is needed to act as an interface between the sensor and the digital readout panel. Digital instruments can use vacuum fluorescent or liquid crystal displays.

Vacuum fluorescent displays are small glass tubes filled with neon or argon gas. They glow when electrically energized, making them very visible and easy to read.

Liquid crystal displays (LCD) are semiconductor panels that will pass light when electrically energized and block light when not energized. An LCD display is often backlighted to make it easier to read.

Instrument Cluster

The *instrument panel* is the general area of the dash behind the steering wheel that holds most displays. It is usually a removable assembly that holds the gas gauge, oil pressure gauge, speedometer, tach, and other displays.

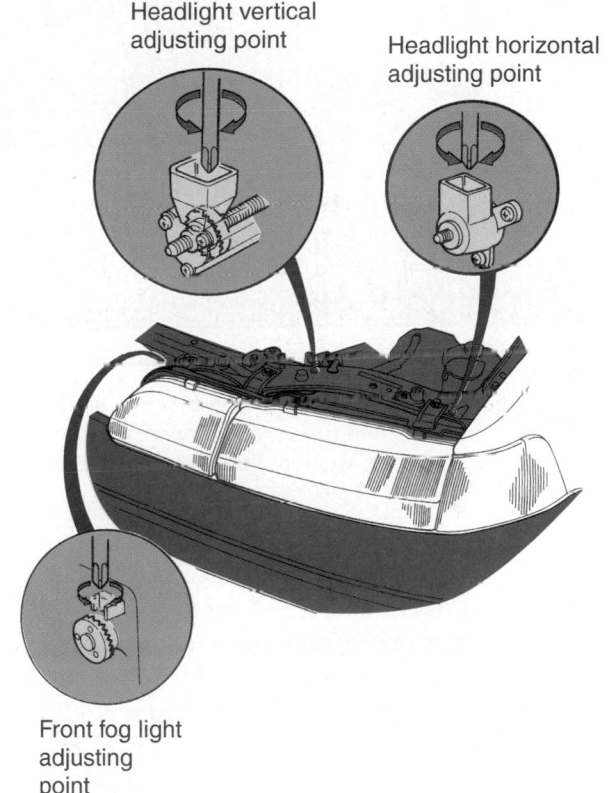

B

Figure 37-27. Many new vehicles with halogen insert bulbs have aiming screws on housing. A—This car also has a leveling bubble to aid in the adjustment of headlamps. B—Note adjustments for headlamps and fog lamps on this vehicle. (Honda)

The *instrument cluster* is the housing and clear plastic cover that holds the gauges, indicator lights, speedometer head, and bulbs.

To service an instrument cluster, remove the small screws around the outside of instrument panel. Often, you must remove a cover before accessing the screws that hold the instrument cluster in place. See **Figure 37-29A.**

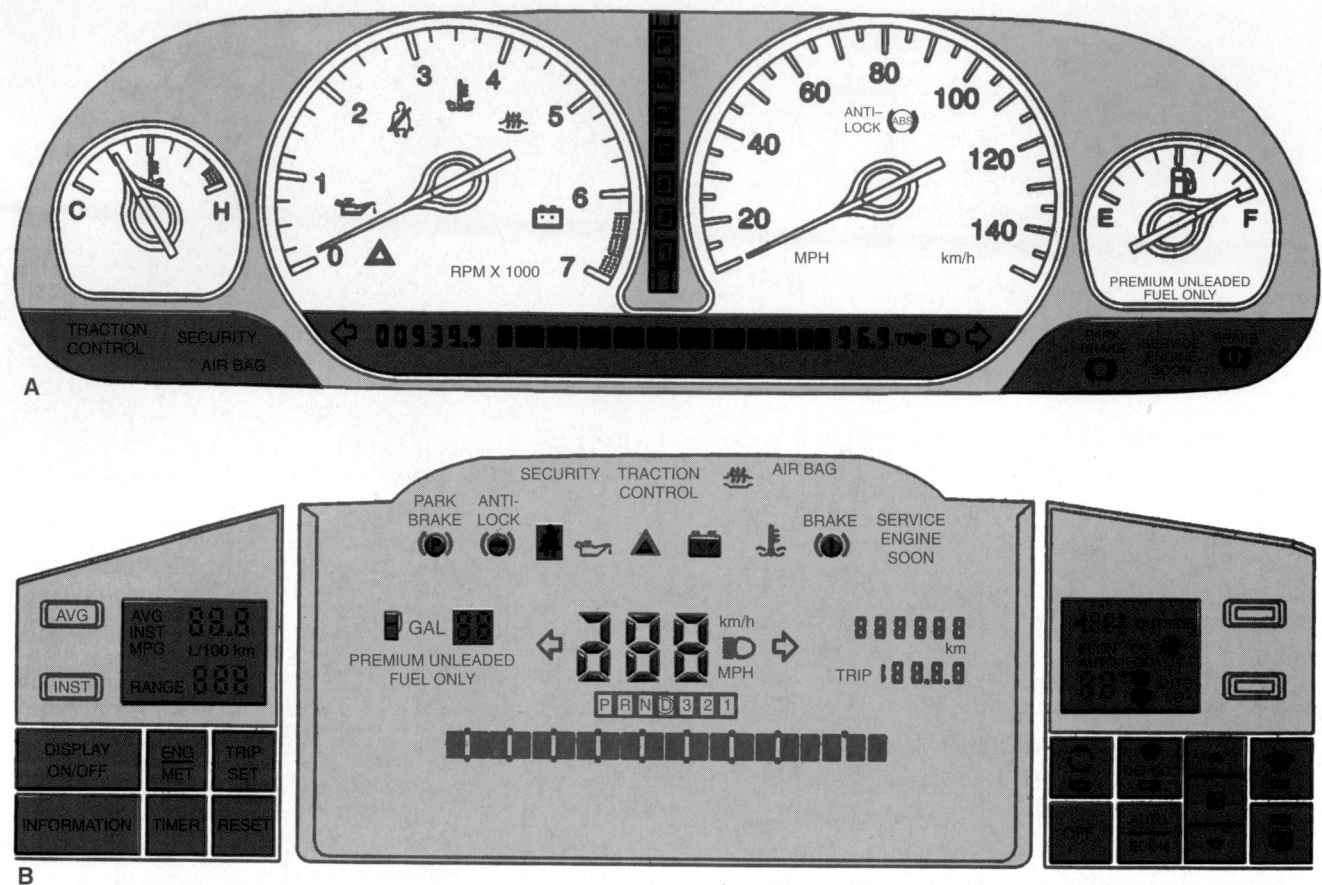

Figure 37-28. Compare analog and digital instrument clusters. Both can use conventional incandescent bulbs and light-emitting diodes. A—An analog instrument cluster uses needles that sweep around gauges. Speedometer can be mechanical or electronic. B—A digital cluster uses vacuum fluorescent or liquid crystal displays to show conditions. (Oldsmobile)

Before completely removing the instrument cluster, disconnect the wires and speedometer cable (if used). If you try to force the cluster out without disconnecting everything on the rear of the cluster, the wiring and cluster could be damaged. Look at **Figure 37-29B.**

With the cluster out, it is easy to replace burned-out bulbs and inoperative gauges. **Figure 37-30** shows the rear of one make of instrument cluster. Note how the service manual illustration identifies the bulbs.

To replace gauges, you must open up the instrument cluster. Small screws hold the front glass or plastic lens to the housing. Remove these screws and remove the lens. The gauges may also be held in place by small screws through the rear of the cluster, **Figure 37-31.**

Tech Tip

When handling the gauges and lens, keep fingerprints off everything. Dust will collect on the skin oil of fingerprints and cause customer complaints at a later date. Wipe everything clean before reassembly or wear clean gloves while handling instrument cluster components.

Dash Gauges

There are basically two types of gauges—balancing coil gauges and bimetal gauges. Both are commonly used and illustrated in **Figure 37-32.**

A *balancing coil gauge* uses two electric coils to cause needle deflection. The sending unit changes the current flow through the coils to deflect the gauge needle right or left using the magnetic field generated by each coil.

A *bimetal gauge* uses two dissimilar metals bonded together to cause gauge needle deflection. Generally, as current flow from the sending unit increases, the bimetal strip heats and defects more. Less current flow causes less bimetal arm bending and less needle deflection.

Sending units are used to control current flow to gauges. Many sending units are simply variable resistors. For example, with an oil pressure gauge, the sending unit is mounted on the engine to sense oil pressure. If oil pressure is low, the sending units resistance may be high to limit current flow to the oil pressure gauge. The gauge does not deflect to the right and the driver knows there is

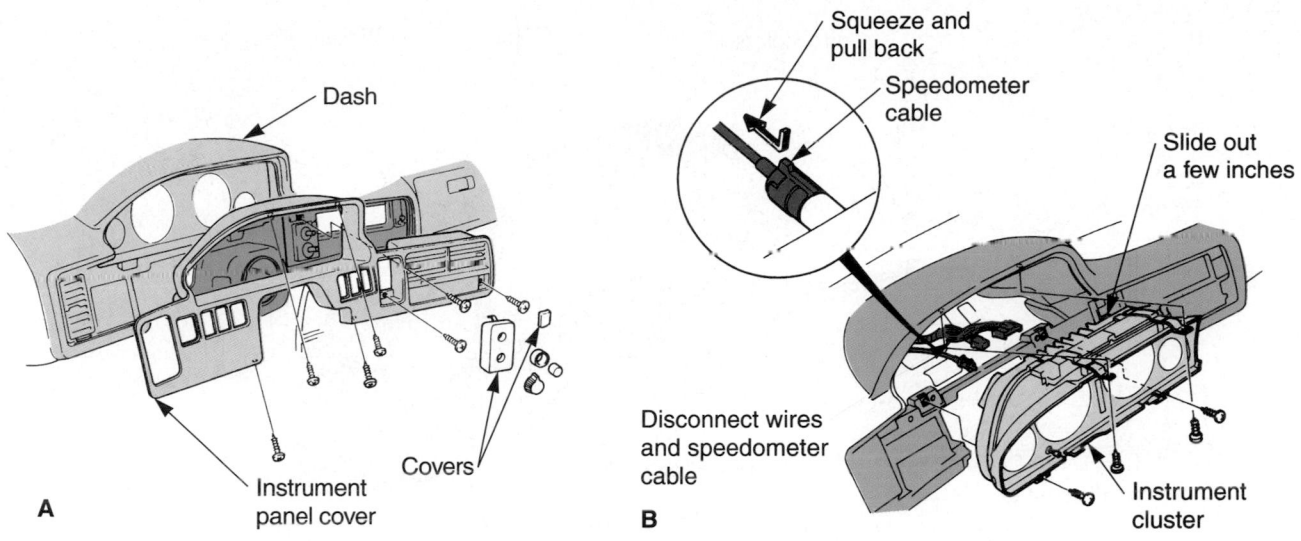

Figure 37-29. Study the basic steps for removal of the instrument cluster. A—First, remove the small screws that hold the covers over the cluster. B—Remove the screws that go through the edges of cluster and pull it out a little. You can then disconnect wires and the speedometer cable.

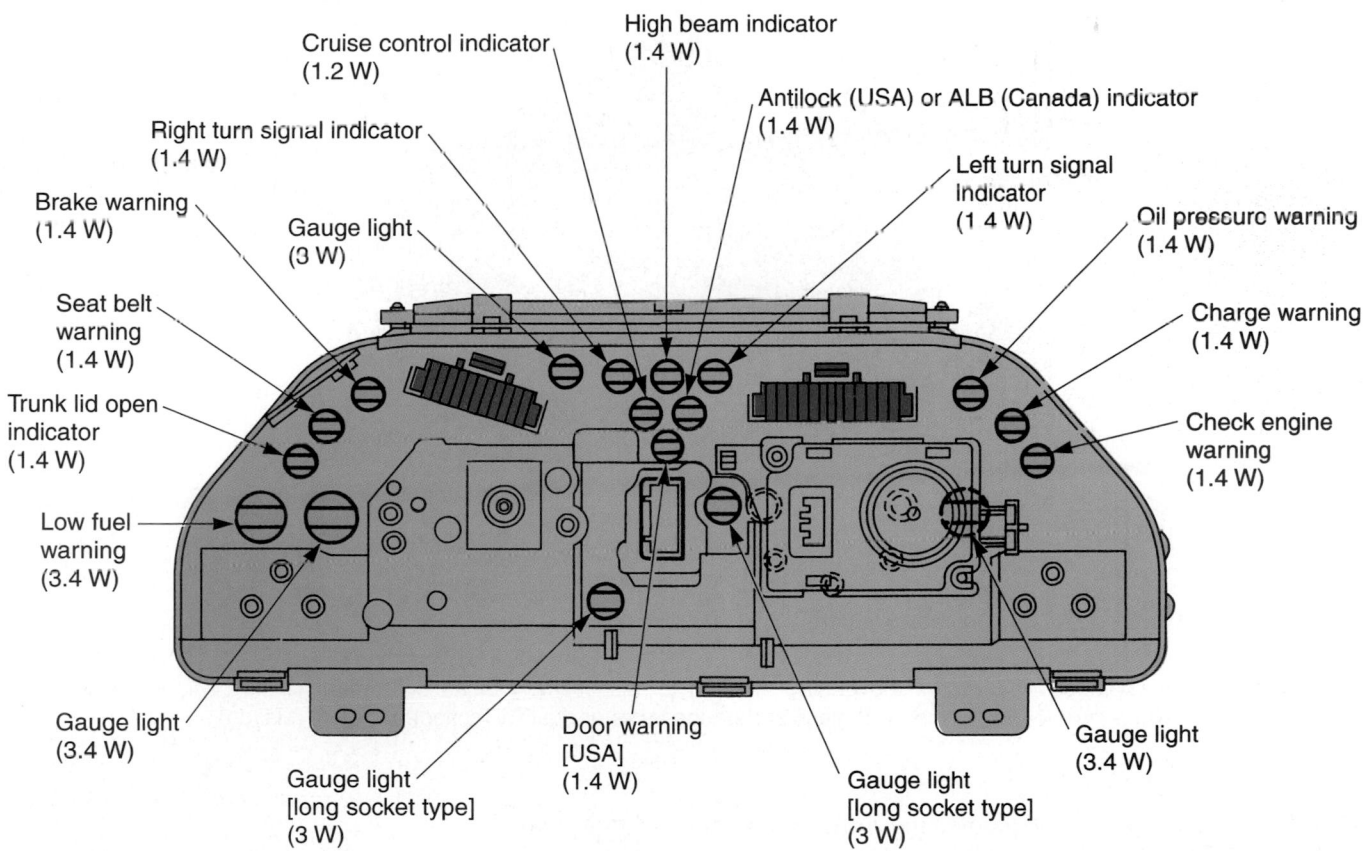

Figure 37-30. Here is rear view of one instrument cluster. Note how lamps are labeled for easier replacement. (Honda)

low oil pressure. This same principle also applies to other gauges.

If a gauge is not working, check the sending unit first. The sending unit may not be working and may be preventing gauge operation. Special testers are available for checking the sending unit current going to the gauge. For example, the sending unit wire may need to be grounded. This should make the gauge deflect completely

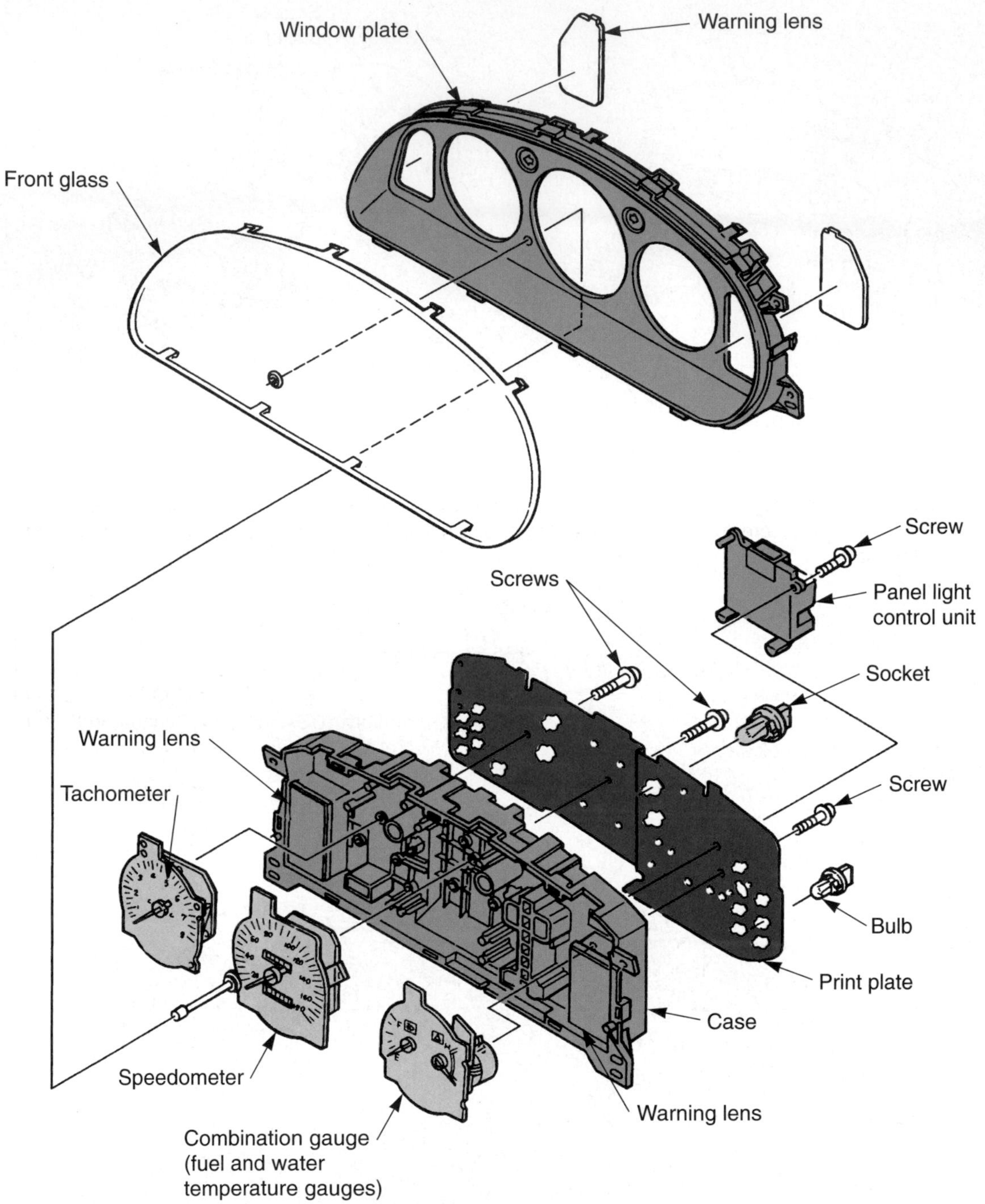

Warning lens

Window plate

Front glass

Screw

Panel light control unit

Screws

Socket

Warning lens

Tachometer

Screw

Bulb

Print plate

Case

Speedometer

Warning lens

Combination gauge (fuel and water temperature gauges)

Figure 37-31. This exploded view shows the general method for disassembly of an instrument cluster. This is required when replacing gauges. Keep fingerprints off the front glass and gauge faces or dust will collect on them. (Mazda)

to the right or left. If not, something other than the sending unit is at fault (wiring, fuse, gauge, connector).

Caution
Do not ground a sending unit wire unless told to do so in a service manual. Some digital display circuits can be damaged by grounding the sending unit wire.

Note
Gauges are also discussed in other applicable chapters. For example, the oil pressure gauge circuit is explained in the chapters on lubrication systems. The tachometer is summarized in the chapters on ignition systems. The engine temperature gauge is detailed in the chapters on cooling systems. Refer to the index for more information as needed.

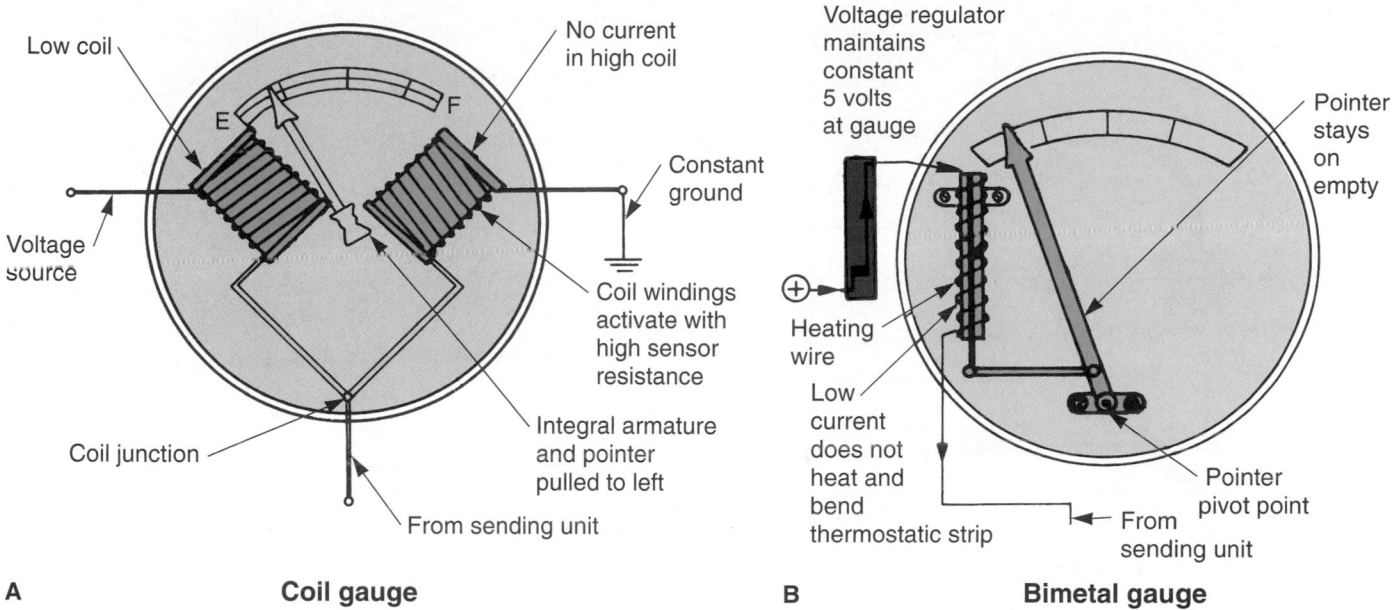

A   **Coil gauge**   **B**   **Bimetal gauge**

Figure 37-32. Compare the two types of gauges. Note the gauge voltage regulator in the circuit on the right. It maintains constant supply voltage to the gauge so it reads accurately. A—Coil gauge uses two coil windings to magnetically attract the indicating needle. With high current flow through the sending unit (low sending unit resistance), the left coil produces a stronger magnetic field, which moves the needle to the left. With low current through the sending unit (high sending unit resistance), more current flows through the coil on the right and the needle moves to the right. B—A bimetal gauge uses a thermostatic strip to move the indicator needle. With low current flow through the sending unit, the strip stays cool and does not bend, so the needle stays to the left. With higher current flow through the sending unit (low sending unit resistance), high current heats the bimetal strip and bends It. This bending action acts on the indicator needle to move it to the right.

Speedometers

There are two basic types of speedometers—mechanical and electronic. The electronic speedometer is becoming more common on modern vehicles. **Figure 37-33** compares simplified mechanical and electronic speedometers.

Mechanical Speedometers

A **mechanical speedometer** uses plastic transmission/transaxle drive gears, a metal speedometer cable, a cable housing, and a mechanical speedometer head. When the vehicle is moving, a drive gear on the transmission output shaft turns a speedometer cable gear. This gears spins the metal speedometer cable in its housing. This rotating motion is transferred up onto the dash to the speedometer head. Depending upon rotating speed, the needle on the speedometer head is deflected to the right to indicate vehicle speed.

A dry speedometer cable will stick in its housing and cause the indicator needle to jump back and forth. The speedometer cable can also break, preventing speedometer operation. These are common problems with high-mileage vehicles.

To service a mechanical speedometer cable, disconnect the cable housing at the instrument cluster. You must normally depress a plastic latch that holds the cable housing to the cluster. Then, you can slide the speedometer cable out of the rear of the cluster.

To remove the speedometer cable, grasp and slide it out of its housing using a pair of pliers. If it is dry, wipe grease on the cable and slide it back into its housing. If it is broken, disconnect the cable housing at the transmission. You may need to use needle nose pliers to remove the small broken end from the transmission. Install a new speedometer cable of the original length.

If the speedometer needle does not return to zero when stopped, the speedometer head is usually bad. It should be replaced.

Electronic Speedometers

An **electronic speedometer** uses the vehicle speed sensor, a electronic control module, and an electronic or electric speedometer head or display. The vehicle speed sensor on the transmission or transaxle sends a signal representing road speed to the control module. The module can then produce an output signal to drive the digital display or servo motor in the speedometer head.

Electronic speedometers are now more common than mechanical speedometers. They are more dependable and consume less energy. Instead of spinning a heavy steel

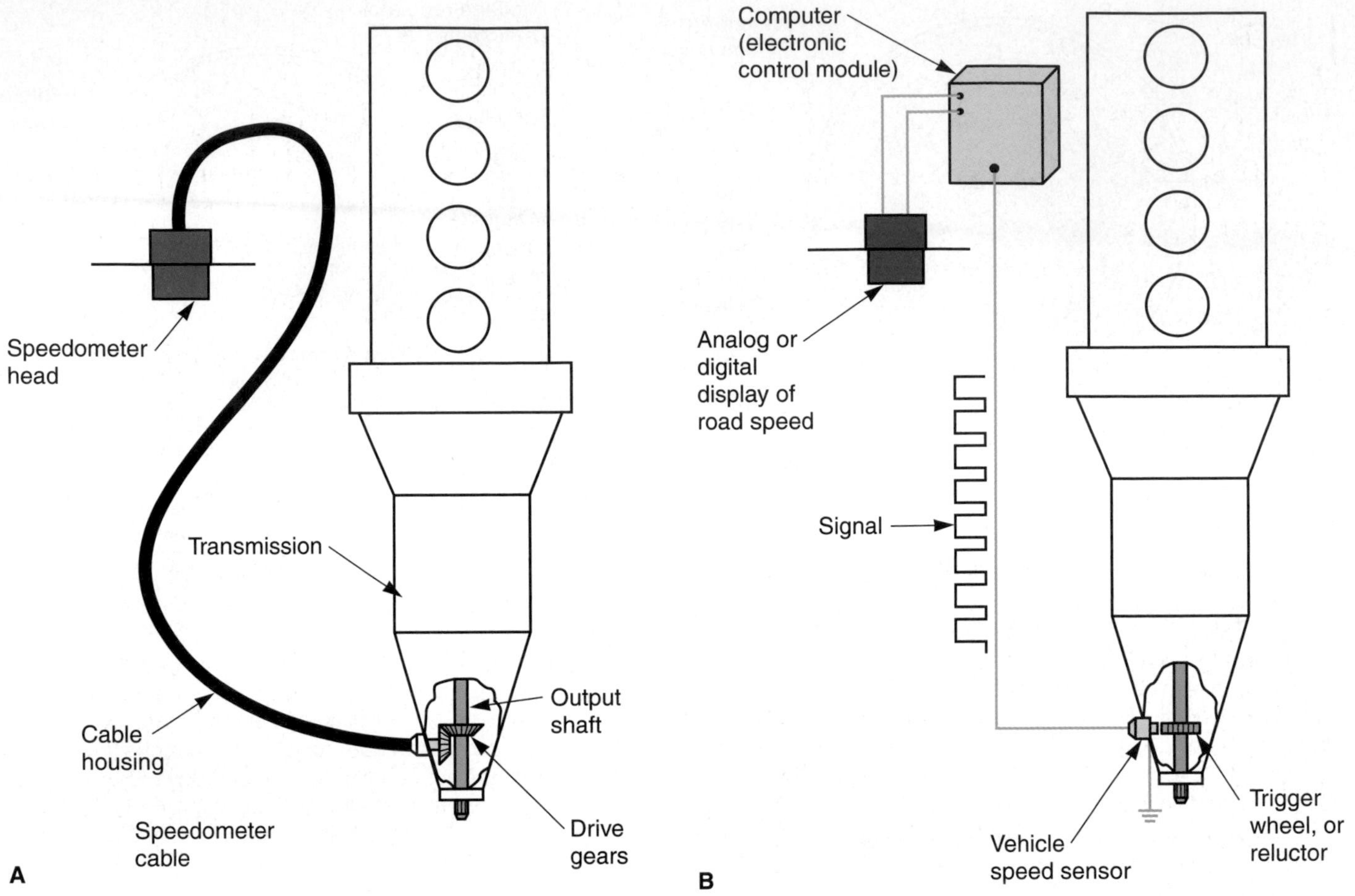

Speedometer head

Transmission

Cable housing

Speedometer cable

Output shaft

Drive gears

A

Computer (electronic control module)

Analog or digital display of road speed

Signal

Vehicle speed sensor

Trigger wheel, or reluctor

B

Figure 37-33. Compare mechanical and electronic speedometers. A—A mechanical speedometer is driven by small gears inside the transmission or transaxle. A gear on the output shaft spins another gear on the end of the speedometer cable. The cable spins inside the housing to transfer motion to the speedometer head to deflect the needle. B—An electronic speedometer commonly uses a vehicle speed sensor on the side of the transmission or transaxle. Trigger wheel or reluctor on the transmission output shaft sends a frequency signal into a sensor. The sensor signal is then used by the electronic control module to power the speedometer display.

cable, the speed sensor electrical signal is used to drive the electronic speedometer more efficiently.

Many electronic speedometers will only read up to a certain speed (85 mph or 145 km/h). Even though the vehicle can be driven faster, the ECM will not display a higher speed. This is normal on most vehicles.

Tech Tip

When servicing an electronic speedometer, never change the *odometer* (mileage) reading. You must often reuse the old *odometer chip* (miniaturized circuit that stores mileage of the vehicle). The chip must be removed from the old cluster and installed in the new one. For more information on chip or PROM replacement, refer to Chapter 19, *Computer System Service.*

Figure 37-34 shows an electronic speedometer that uses a coil and pulse motor. **Figure 37-35** shows how a digital display operates for several types of vehicle conditions.

If an electronic speedometer fails to work properly, first check for stored trouble codes. Then, check for a good signal from the vehicle speed sensor. If the signal is reaching the ECM, check for a control signal going to the digital display. If only part of the display is working, you probably have a bad *segment* (part of display number or letter) and the display should be replaced.

Driver Information Center

The *driver information center* is a dash mounted keyboard-display for inputting and reading data about the vehicle. It allows you to send data to the instrumentation computer to find miles-to-empty, average fuel consumption, estimated time of arrival, and other information. The information center uses signals from many sensors to calculate this information.

Driver information centers are very complex and vary with make and model. Always refer to the service manual for testing and repair directions.

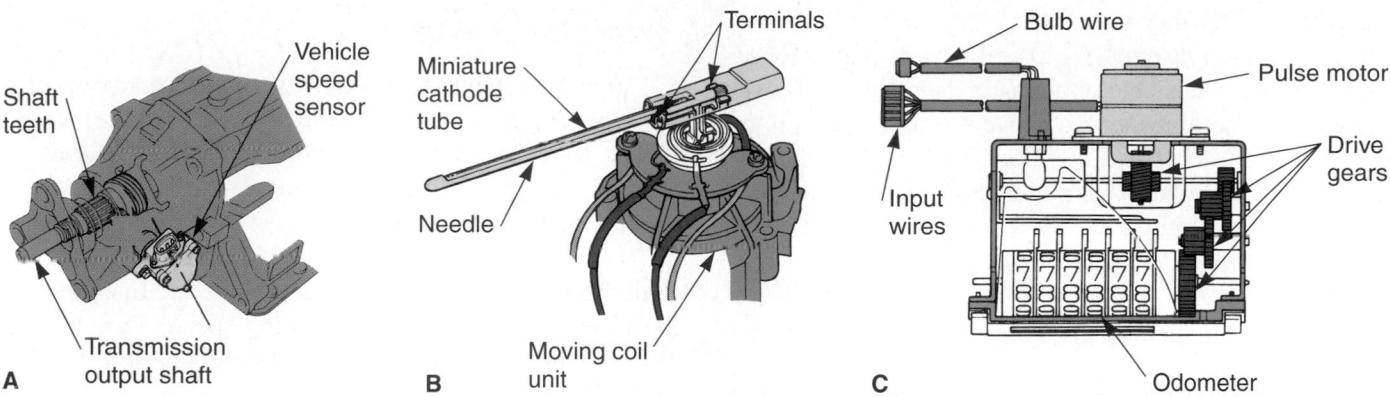

Figure 37-34. Study the parts of an electronic speedometer-odometer system. A—The speed sensor is mounted on the side of the automatic transmission. It feeds signals to the ECM. Note the teeth on the output shaft. B—The speedometer unit uses a small electric motor to move an analog needle. This indicator needle is a cathode tube that glows in the dark. The ECM powers the speedometer unit. C—A pulse motor is used to drive a mechanical odometer, which shows vehicle mileage. The pulse motor is also driven by the ECM. (Lexus)

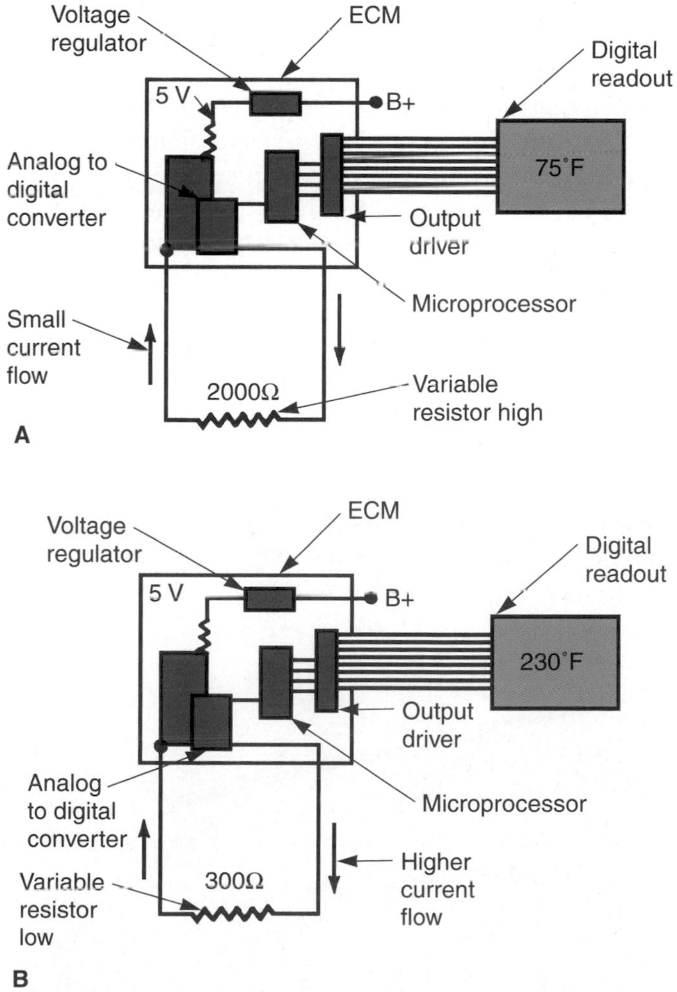

Sensor	Condition	Value	Readout
Engine temperature	Cold Normal	2000 Ω 200 Ω	75°F 230°F
Oil pressure	High Low	100 Ω 5 Ω	65 psi 15 psi
Fuel level	Full Half Empty	100 Ω 50 Ω 10 Ω	Full 1/2 E
Engine speed	Idle Cruising	5000 cycles/sec. 50,000 cycles/sec.	1000 rpm 5000 rpm
Oil level	Normal Low	Infinite Ω Zero Ω	Bulb out Bulb on
Vehicle speed	High Low	High frequency Low frequency	10 mph 55 mph

Figure 37-35. This simplified illustration shows how different types of sensors can be used to indicate different conditions. The ECM converts sensor signals into output signals for displays. A—The engine is cold and sending unit internal resistance is high. The ECM shows a cold temperature readout. B—The engine is warmer and the sending unit resistance drops. The ECM uses a different amount of current flow to show the warmer engine temperature.

Heads-Up Display

A *heads-up display* (*HUD*) reflects display information onto the windshield or a plastic dash panel for easier viewing. It compliments the conventional displays in the instrument cluster. The driver does not have to take his or her eye off the road to view the HUD. One is shown in **Figure 37-36**. **Figure 37-37** shows how one type of reflective display operates.

Figure 37-36. This heads-up display reflects images onto the windshield so readings are easy to see while driving. (GM Hughes Electronics)

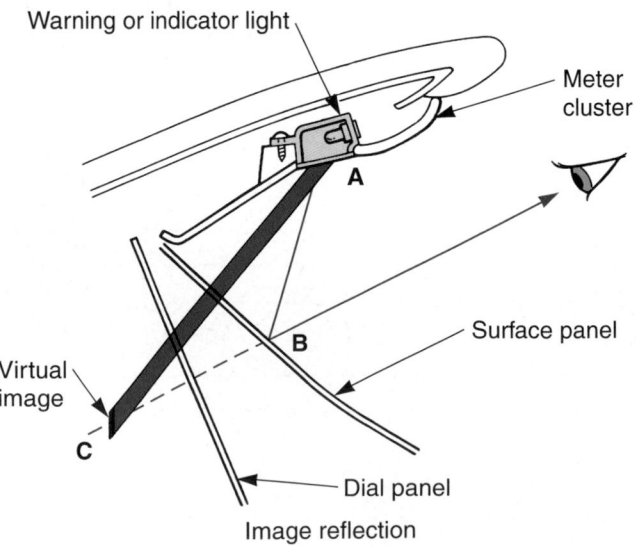

Figure 37-37. Drawing shows how an image is reflected onto a plastic surface on the dash panel. The LED at A is reflected to B before reaching the driver's eye. The driver actually views a virtual image at point C, a few inches behind the panel. (Lexus)

Windshield Wipers

A typical *windshield wiper system* is made up of a switch, wiper motor assembly, wiper linkage, wiper arms, wiper blades, and usually a windshield washer system. Either a fuse or circuit breaker protects the system. See **Figure 37-38**.

The *windshield wiper switch* is a multiposition switch that sometimes contains a rheostat. Each switch position provides a different wiping speed. The rheostat operates the delay mode for slow wiping action. A relay is frequently used to complete the circuit between the battery and the wiper motor. A typical windshield wiper-washer circuit diagram is shown in **Figure 37-39**.

The *wiper motor assembly* consists of a permanent-magnet motor and a transmission. The *wiper motor transmission* changes rotary motion into a back-and-forth wiping motion. The transmission is normally a set of plastic gears, an end housing, and a crank.

Figure 37-40 shows the parts of a typical wiper motor assembly. The drive crank on the transmission connects to the wiper linkage.

The *wiper linkage* is a set of arms that transfers motion from the wiper motor transmission to the wiper arms. The rubber wiper blades fit on the wiper arms. Refer back to **Figure 37-38**.

Windshield Washer

A *windshield washer* consists of a solvent reservoir, pump, rubber hoses, connections, and washer nozzles. As shown in **Figure 37-41**, the solvent reservoir, located in the engine compartment, holds a supply of water and solvent. When the washer switch or button is activated, the wiper motor and the washer pump turn on. Solvent is forced out of the reservoir and onto the windshield.

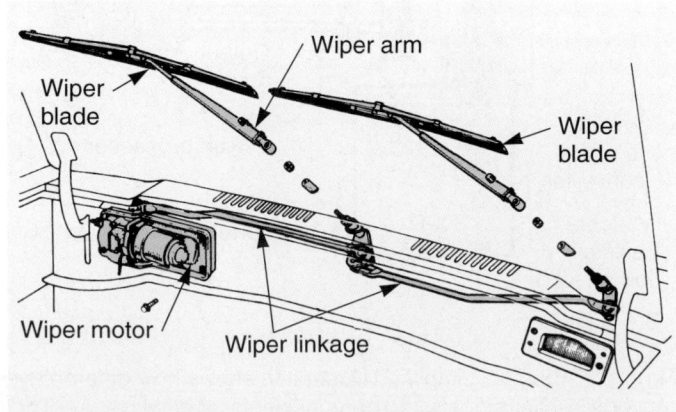

Figure 37-38. Study the basic parts of a typical windshield wiper system. (Toyota)

There are two common types of pumps used with windshield washer systems: a rotary pump and a bellows (diaphragm) type pump. Most new cars use a rotary pump mounted in the solvent reservoir, **Figure 37-41.** A tiny electric motor spins an impeller, which forces the washer solution onto the windshield. A bellows or diaphragm type pump is normally mounted on and powered by the wiper motor.

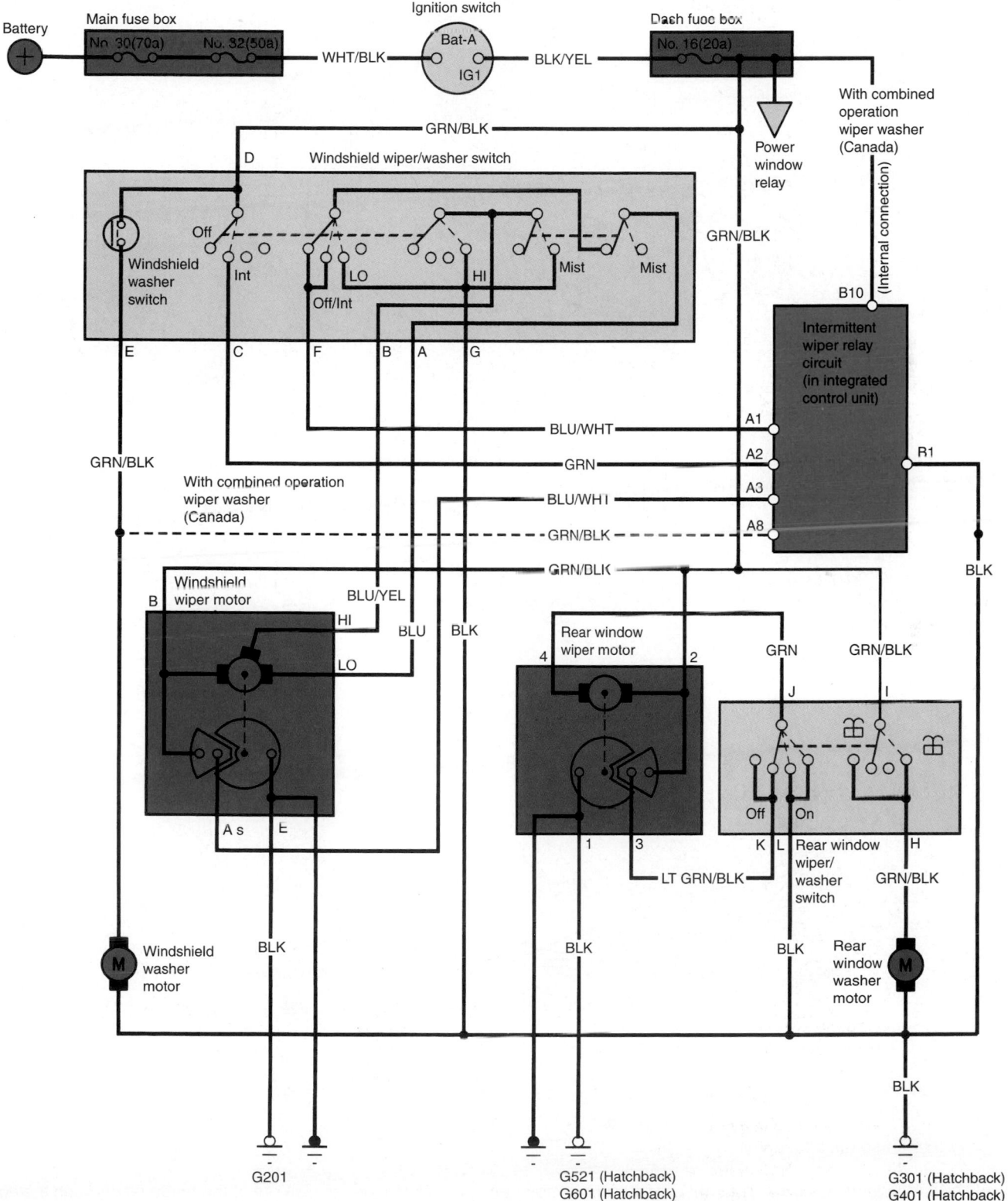

Figure 37-39. Study the wiring diagram of wiper-washer system. Trace through and find all major parts. (Honda)

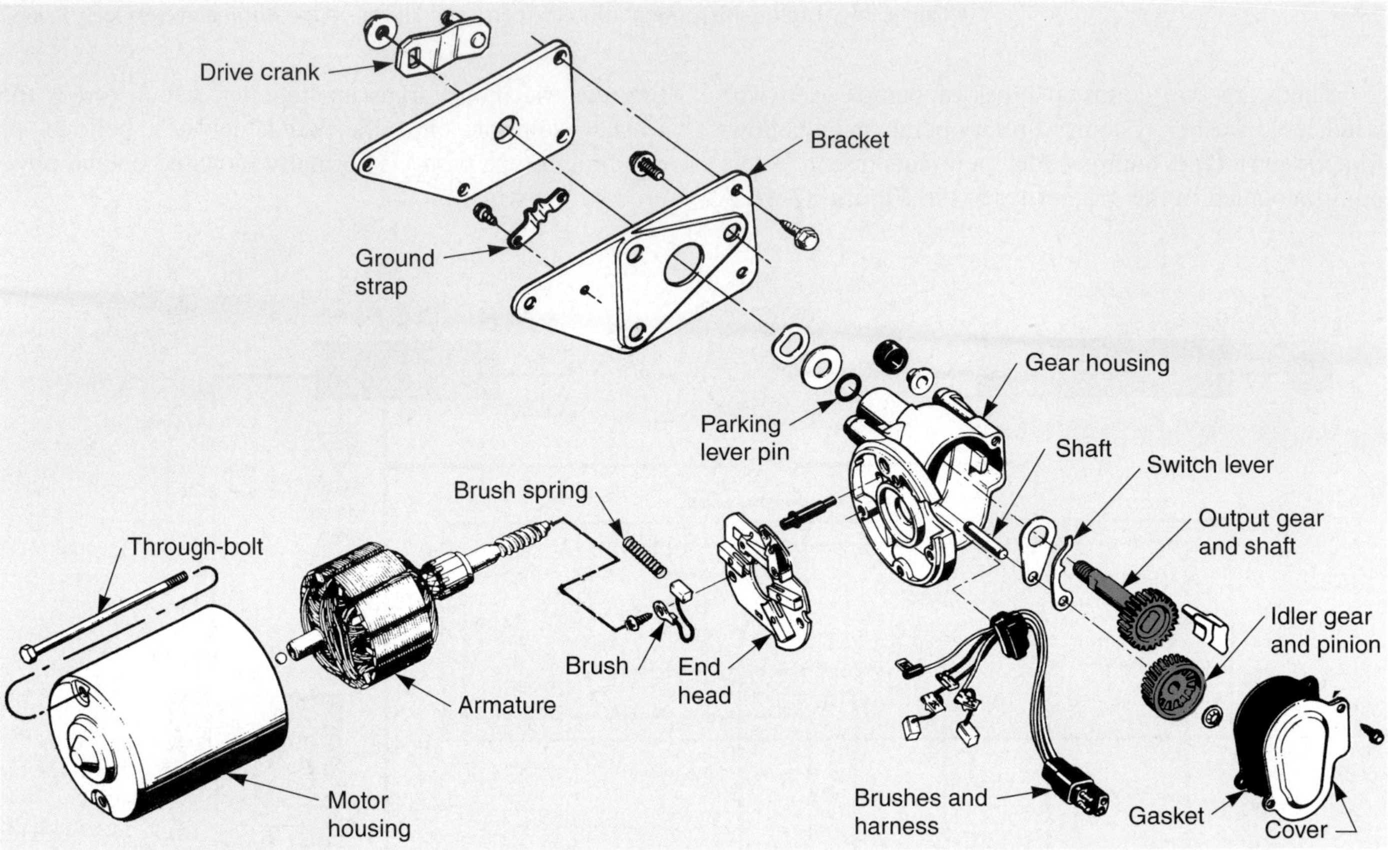

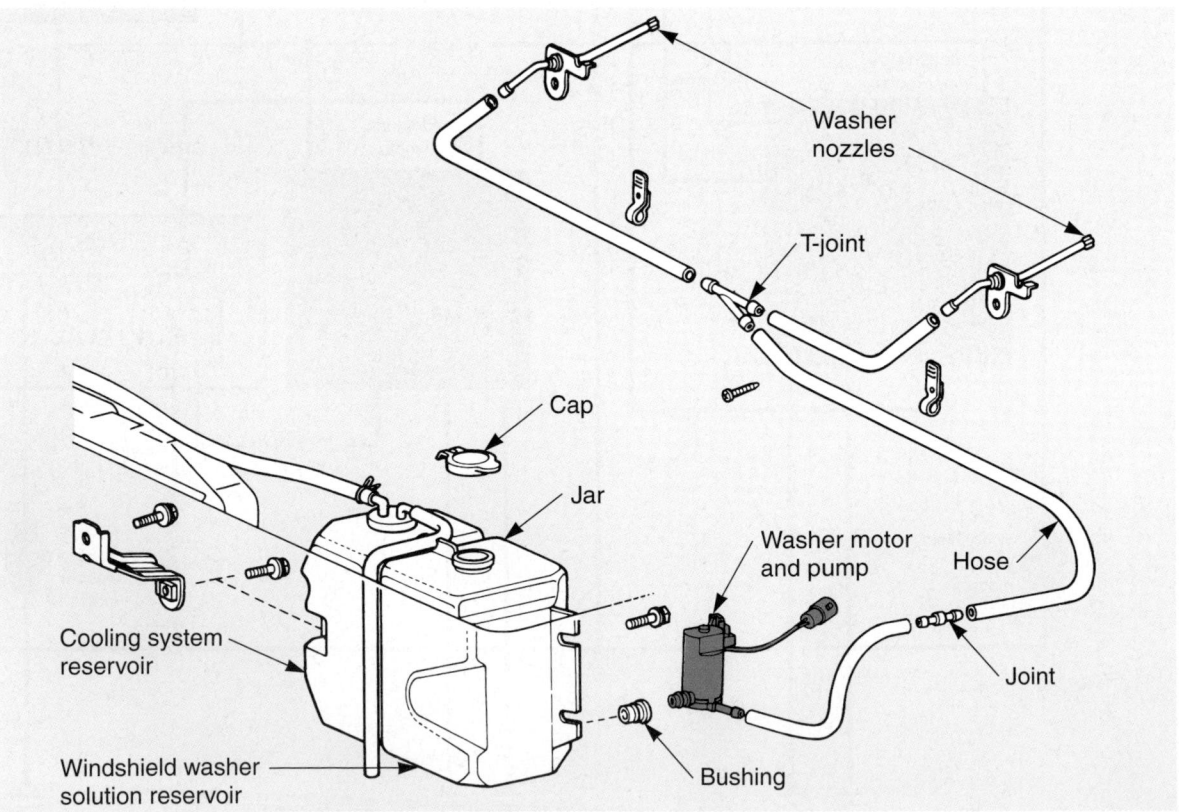

Figure 37-40. An exploded view of a wiper motor assembly. The two plastic gears are the most common problem. They are replaced when stripped or broken.

Figure 37-41. Windshield washer system. A small electric motor and pump force the solution out of the reservoir, through the hoses, and out of the nozzles. Check for debris in reservoir strainer, hoses, and nozzles when the system is not working. Then, check pump operation.

Rain-Sensing Wiper System

A **rain-sensing wiper system** automatically turns the windshield wipers on when water is detected on the windshield. The system uses light-emitting diodes and a light sensor or an infrared system to detect water. When water is on the windshield, less light is refracted back to the windshield or water sensor. This allows an electronic control module to turn the wipers on to remove rain or snow. In heavy rain, the wipers wipe across the windshield more quickly. Manual override is also possible.

Figure 37-42 shows the basic operation of a rain-sensing wiper system.

Windshield Wiper Service

Windshield wiper blades should be inspected periodically. If they are hardened, cut, or split, replace them. **Figure 37-43** shows common wiper blade service methods.

With electrical problems in a wiper system, use a service manual and its wiring diagram of the circuit. First, check the fuses and electrical connections. If they are good, use a test light to check for power to the wiper motor.

If power is being fed to the wiper motor, either the motor or transmission may be at fault. Before replacing

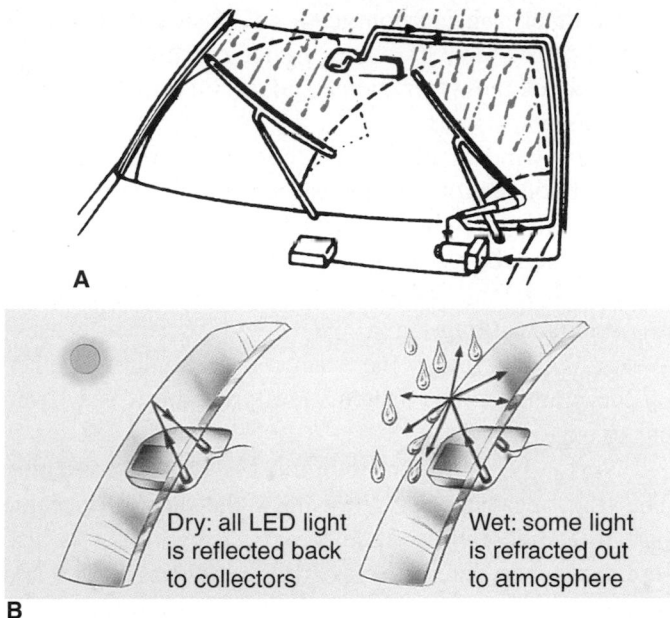

Figure 37-42. Basic operation of a rain-sensing wiper system. A—A sensor on the back of the windshield and the ECM operate the wiper motor. B—When the windshield is clear, LED light is reflected back to a photo sensor and the wipers stay off. When the windshield has water or snow on it, some light is reflected away from the sensor. The sensor signal tells the ECM to turn on the wipers.

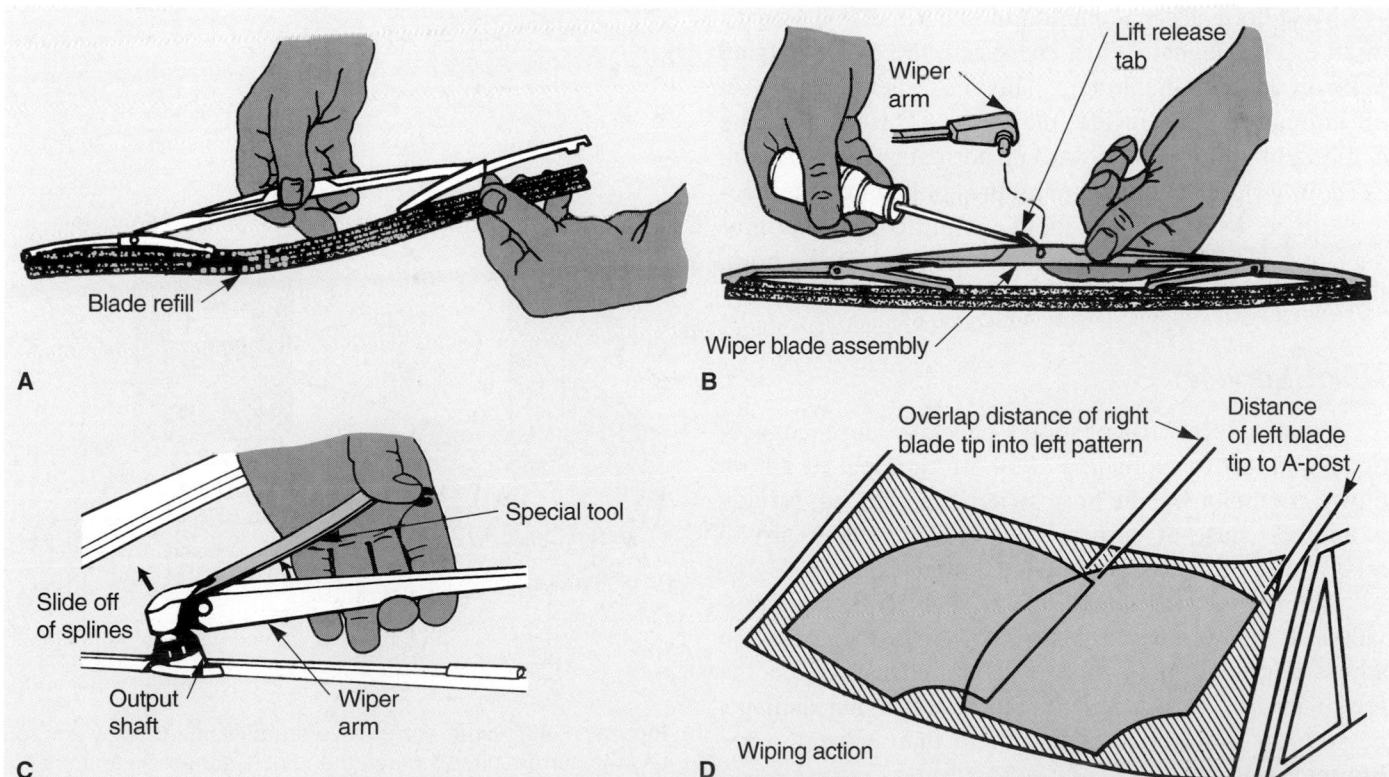

Figure 37-43. Study basic wiper service. A—A new rubber wiper refill is slid into place. B—To replace the complete wiper blade assembly, free it from the wiper arm. C—The arm can sometimes be removed by prying up as shown. D—Typical pattern for positioning arms on output shafts. Refer to a service manual for exact procedures. (Chrysler)

the motor or transmission gears, make sure the motor is properly grounded. If power is not reaching the wiper motor, check the wiper switch and circuit connections for openings.

If the windshield washer does not work, check the fuse and connections. If the washer pump is in the reservoir, use a test light to check for power going to the motor. When the test light does not glow (washer on), the wiper switch may be bad.

When working on a windshield wiper or washer system, always follow the exact recommendations given by the manufacturer. Systems and procedures vary from car to car.

Normally, the wiper motor must be replaced as a unit. The transmission gears are usually the only serviceable part in the assembly. An electric washer motor is also replaced when defective. It cannot be disassembled and repaired.

Horns

Today's *horn systems* typically include a fuse, horn button switch, relay, horn assembly, and related wiring. When the driver presses the horn button, it closes the horn switch and activates the horn relay. This completes the circuit. Current then flows through the relay circuit, and to the horn.

Most horns have a diaphragm that vibrates by means of an electromagnet. When energized, the electromagnet pulls on the horn diaphragm. This movement opens a set of contact points inside the horn. This allows the diaphragm to flex back toward its normal position. Again, the points close and the diaphragm is pulled into the electromagnet. As a result, a rapid vibrating action is produced. A honking sound is transmitted out of the horn, **Figure 37-44.**

Horn Service

When a horn will not sound, check the fuse, connections, and test for voltage at the horn terminal. If a horn blows continuously, the horn switch may be bad. A relay is another cause of horn problems. The contacts in the relay could be burned, or may stick together.

A *horn current adjusting screw* is sometimes provided on the horn to set the amp draw through horn. To adjust horn current, connect an ammeter between the feed wire and horn terminal. To prevent meter damage, be sure the ammeter can read more than 30 amps. See **Figure 37-45.**

Have someone sound the horn while you read the meter. If the current is not within specifications (typically 4–5 amps), turn the amps screw on the horn until the

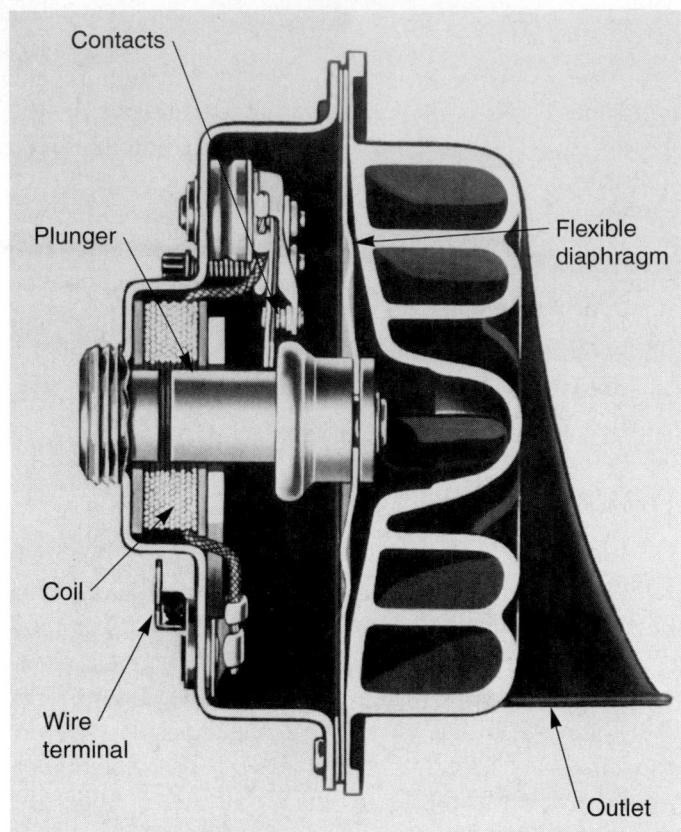

Figure 37-44. Horn contains a coil, points, and a flexible diaphragm. Coil and point action makes the plunger slide in and out of the coil. The plunger moves the diaphragm to produce horn sound. (Deere & Co.)

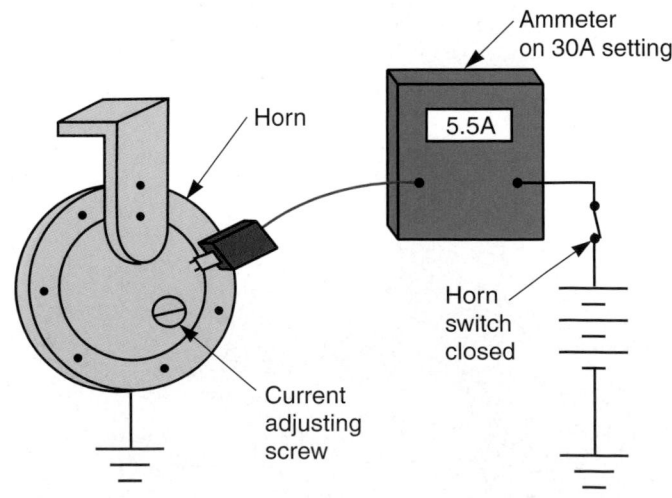

Figure 37-45. Some horns have an adjustment screw for setting the current draw through the horn. Connect an ammeter to measure the current draw. Press the horn button and adjust the screw for the specified current draw. If you cannot correctly adjust the current, check the supply circuit before replacing the horn.

meter reads properly. Also make sure you are getting adequate supply current/voltage and there is not a high resistance in the horn circuit. If you cannot get the horn to read within current specifications, replace it or isolate the circuit problem.

Refer to a shop manual when diagnosing and repairing a horn system. Although many horns are adjustable, they are not serviceable. Use your knowledge of the system and basic testing methods to locate troubles.

Finding Common Electrical Problems

There are several classifications of electrical problems. These are briefly covered in Chapter 8, *Basic Electricity and Electronics*. Problems include short circuits, open circuits, and high resistance in circuits. It is important for you to be able to quickly diagnose and locate these troubles when working on light, instrumentation, wiper, and horn circuits.

Short Circuit

A *short circuit* will show up as a blown fuse, a burned fusible link, or an open circuit breaker. A short circuit results when a *hot wire* (current-carrying wire) is touching ground. A wire's insulation may be cut, allowing the wire to touch body sheet metal, or a wire may be pinched between two parts, rupturing the insulation.

One way to locate a short circuit is to connect a test light across the blown fuse or open circuit breaker connections. This is shown in **Figure 37-46.**

The test light will glow as long as there is a short in the circuit. Disconnect components or sections of the circuit while watching the test light. The test light will stop glowing when you remove the short from the circuit. Keep tracing the circuit (checking each wire connection and part) until the source of the problem is located.

After you have found and corrected the short, you can disconnect the test light and install a new fuse. Always install a replacement fuse with the same current rating as the blown fuse, **Figure 37-47.**

 Caution
Never install a fuse with a higher current rating to try to keep it from blowing. This can cause excess circuit current that could start an electrical fire.

Open Circuit

An *open circuit* will keep the electrical components in that section of the circuit from working. An open circuit may be caused by a broken wire, disconnected

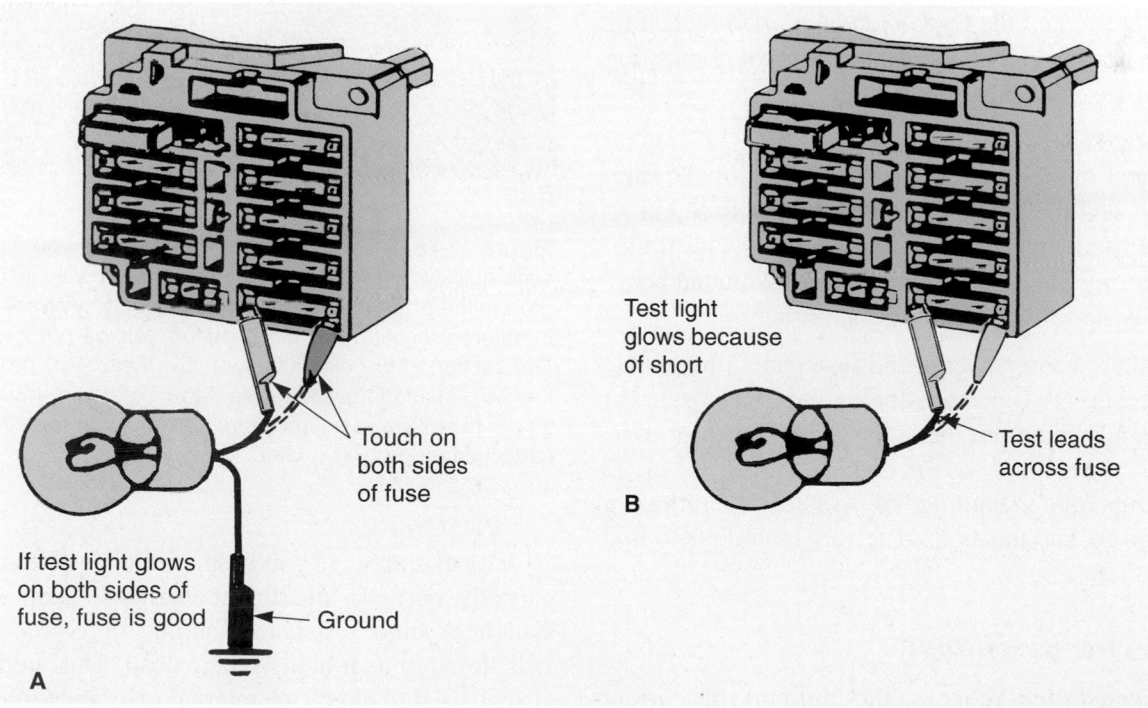

Figure 37-46. A—To check a fuse, touch a test light on both sides of the fuse. It should glow on both sides if the fuse is not blown. B—To help find the short circuit that blows the fuse, connect a test light across the fuse as shown. The test light will glow if a short exists. Unplug wires and components while watching the test light. When the light goes out, you have disconnected the circuit section with the short. (Peerless)

Figure 37-47. Always install a fuse with same current rating as the blown fuse. If the fuse still blows, you have short in the circuit.

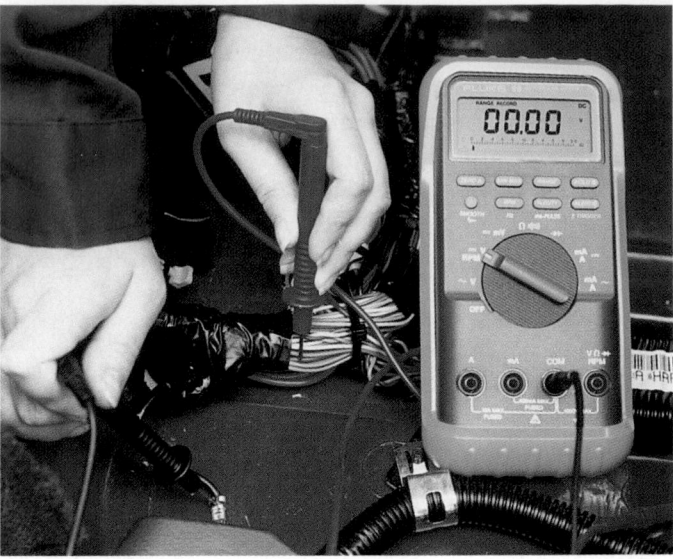

A

B

Figure 37-48. A voltmeter or test light will let you find open circuits quickly. A—Push pointed tip through wire insulation and ground the other lead. This will let you quickly check for voltage in different sections of the circuit. If you do not have voltage, that section of the circuit is open. B—When you probe through the wire insulation, always seal the hole with liquid electrical tape. This will keep out moisture that could corrode the wire and cause circuit problems later.

electrical connection, bad switch, or another problem that prevents current flow.

An open circuit is easily located using a test light or a voltmeter, **Figure 37-48.** Start at each end of the circuit. Check for power at the supply (fuse panel). Then, go to the other end of the circuit and check for power at the load (a light socket, for example). If you have power at the fuse but no power at the load (socket), the open is between the fuse and the load. Trace the open by testing for power at the switch and at each electrical connection between the fuse panel and load.

Tech Tip

If you probe through wire insulation, make sure you seal the hole in the insulation with liquid electrical tape. This will prevent moisture from entering the puncture in the insulation and possibly upsetting circuit operation later.

When there is no power at the fuse panel, the fusible link may be burned. If power is being fed to a switch, but power does not come out the switch, the switch or electrical connector is bad.

Use your understanding of system operation, a wiring diagram, and this type of testing technique to find an open circuit.

High Circuit Resistance

High resistance reduces the amount of current flowing through a circuit. For example, high resistance in a light circuit can cause the lamps to be dimmer than normal. To find a high resistance, measure the voltage drop across possible problem components.

For example, if you suspect that a switch might be partially burned, measure the voltage drop across the switch. A high resistance (burned or corroded switch) will show up as a high voltage drop. This technique can also be used to check the internal resistance of other electrical components (electrical plug connectors, wires, and relays).

Typically, the voltage drop across an electrical conductor should not exceed 0.5 volts. Remember that this applies to wires, switches, and connectors, but not to electric motors, lightbulbs, or other loads.

Using Wiring Diagrams

Wiring diagrams are drawings that show the relationship of the electrical components and wires in a circuit. They are useful when an electrical problem is difficult to locate and correct.

Wiring diagrams normally give the following information:

- Wire color coding—special color markings on wire insulation for tracing wires through the vehicle.
- Component location—number-letter sequence around the border of the diagram for quickly finding parts in circuit.
- Wire splices and connections—where and how wires are joined together.
- Electrical symbols—simple drawings showing minor components in the circuit.
- Component drawings—simplified illustrations of major circuit components for easy location on diagram.

To use a wiring diagram, locate the parts being tested. Then, follow the lines that show where the wires go to the next component in the circuit.

If, for example, you are having trouble with a horn circuit, find the horn circuit diagram in a service manual. Trace the horn wires. You would then find a relay in the circuit between the horn and horn switch. It may be at fault.

Relay Problems

Relay problems usually result from worn, burned, or sticking contact points that prevent normal circuit operation. The small coil windings in a relay can also break, preventing the points from closing.

Many circuits are controlled by relays. Always look for relays in wiring diagrams and check them if power is not reaching a circuit on the output side of a relay.

Relays can be located almost anywhere on a vehicle—in the engine compartment, under the dash, on a fuse box, or under the rear seat cushion. The service manual will give relay locations for the vehicle being repaired.

To test a relay, first make sure a voltage signal is being fed to the relay windings. You can check this with a test light or voltmeter. If power is reaching the relay, check that the relay points are closing. Check for voltage leaving the closed relay points. If not, replace the relay, **Figure 37-49.**

Figure 37-49. Relays are a common source of trouble in all types of circuits. A service manual will give their locations and terminal connections for testing and replacement. This relay was found under the rear seat cushion.

Summary

- The lighting system consists of the components (fuses, wires, switches, and relays) that operate the interior and exterior lights on a vehicle.
- The headlamp system generally includes the battery, headlamp-related wiring, fuse panel, light switch, dimmer switch, headlamps, taillights, marker lights, and instrument lights.
- A multifunction switch is capable of controlling several circuits—headlights (bright and dim circuits) and turn lights, for example—simultaneously.
- Most late-model cars use halogen headlamp bulb inserts, which are small bulbs that fit into a larger plastic housing.
- A dimmer switch controls the high and low headlamp beam function.
- An automatic headlight dimmer system uses a light sensor, amplifier, and relay to control the high and low beams.
- The turn light system consists of a fuse, turn light switch, flasher unit, turn lightbulbs, indicator bulbs, and related wiring.
- Light system service involves changing burned-out bulbs, testing bulb sockets, checking fuses, finding shorted and open circuits, aiming headlights, and other similar types of tasks.
- Aiming headlights involves adjusting the beams so they shine in the proper direction.
- Instrumentation is used to inform the driver of various operating conditions—oil pressure,

engine temperature, computer trouble codes, air bag system armed, etc.

- Analog instruments use rotating needles on dials to indicate operating conditions.

- Digital instruments use various types of lights or electronic displays to show operating conditions.

- The instrument cluster is the housing and clear plastic cover that holds the gauges, indicator lights, speedometer head, and bulbs.

- A balancing coil gauge uses two electric coils to cause needle deflection.

- A bimetal gauge uses two dissimilar metals bonded together to cause gauge needle deflection.

- A sending unit or sensor is used to control current flow to the gauge.

- A mechanical speedometer uses plastic transmission/transaxle drive gears, a metal speedometer cable, a cable housing, and a mechanical speedometer head.

- An electronic speedometer uses the vehicle speed sensor, an electronic control module, and an electronic or electric speedometer head or display.

- A typical windshield wiper system comprises a switch, wiper motor assembly, wiper linkage, wiper arms, wiper blades, and usually a windshield washer system.

- Horn systems typically include a fuse, horn button switch, relay, horn assembly, and related wiring.

Important Terms

Lighting system	Headlight lens
Exterior lights	Dimmer switch
Interior lights	Automatic headlight
Headlamp system	dimmer system
Headlamp switch	Auto-off headlamp
Rheostat	system
Multifunction switch	Delayed exit system
Flash-to-pass feature	Daylight running lights
Headlamp bulbs	Turn signal system
Low beams	Turn signal flasher
High beams	Emergency light system
Bulb element	Hazard warning system
Filament	Emergency flasher
Sealed-beam headlight	Backup light system
bulbs	Backup lamp switch
Number 1 lamp	Illuminated entry system
Number 2 lamp	Light system service
Halogen headlamp	Burned-out bulb
bulb inserts	No-light problem

Flickering lights	Odometer chip
Light switch problems	Segment
Aiming headlights	Driver information
Headlight aimers	center
Headlight aiming	Heads-up display
screen	(HUD)
Headlight adjusting	Windshield wiper
screws	system
Headlight leveling	Windshield wiper
bubbles	switch
Instrumentation	Wiper motor assembly
Analog instruments	Wiper motor
Digital instruments	transmission
Vacuum fluorescent	Wiper linkage
displays	Windshield washer
Liquid crystal displays	Rain-sensing wiper
Instrument panel	system
Instrument cluster	Horn systems
Balancing coil gauge	Horn current adjusting
Bimetal gauge	screw
Sending unit	Short circuit
Mechanical	Hot wire
speedometer	Open circuit
Electronic	High resistance
speedometer	Wiring diagrams
Odometer	Relay problems

Review Questions—Chapter 37

Please do not write in this text. Place your answers on a separate sheet of paper.

1. What is the function of a rheostat?

2. How does a halogen headlamp differ from a conventional headlamp?

3. A dimmer switch can be located on the car's floor or on the steering column. True or False?

4. An automatic headlight dimmer system consists of a _____ sensor, _____, and _____ to control the high and low beams.

5. Explain the construction and operation of a flasher unit.

6. A vehicle has a problem causing the left headlight to flicker on and off. The right headlight works properly. The light seems to flicker when the car strikes bumps in the road. Technician A says that the light system relay may be bad. Vibration could be opening the points in the relay. The relay should be tested and replaced if needed. Technician B says that the problem involves just the left headlight section of the circuit. This technician suggests checking the

headlamp socket and electrical connections to that bulb. Who is right?
(A) A only.
(B) B only.
(C) Both A and B.
(D) Neither A nor B.

7. One burned-out bulb cannot keep the turn signals from flashing on and off. True or False?

8. List four methods of aiming headlights.

9. Describe the operation of both mechanical and electronic speedometers.

10. List and explain the major parts of a windshield wiper system.

11. How does a rain-sensing wiper system work?

12. Explain the operation of a horn circuit.

13. Voltage drop across an electrical conductor should not exceed:
(A) 0.1 volts.
(B) 0.5 volts.
(C) 0.10 volts.
(D) 0.12 volts.

14. Wiring diagrams show:
(A) wire color coding.
(B) electrical symbols.
(C) splices and connections.
(D) All of the above.

15. The 12-volt circuit between a particular automotive turn signal switch and the turn signal lamps has a current flow of 2.3 amps. What is the approximate resistance in this circuit?
(A) 5.22 ohms.
(B) 7.35 ohms.
(C) 2.67 ohms.
(D) None of the above.

● ASE-Type Questions

1. Which of the following is often used by automatic headlight dimmers to control beams?
(A) *Relay.*
(B) *Amplifier.*
(C) *Light sensor.*
(D) *All of the above.*

2. The right turn signals work but the left ones do not. Technician A says to check the flasher unit. Technician B says the turn signal switch could be bad. Who is right?
(A) *A only.*
(B) *B only.*
(C) *Both A and B.*
(D) *Neither A nor B.*

3. Each of these conditions should exist during a headlight aiming test *except:*
(A) *panel lights dimmed.*
(B) *tires inflated correctly.*
(C) *car has half a tank of gas.*
(D) *spare tire and jack in trunk.*

4. A headlight aiming screen is a:
(A) *veil covering headlights.*
(B) *series of lines on shop wall.*
(C) *transparent paper partition.*
(D) *None of the above.*

5. Which of the following is *not* a type of pump used with windshield washer systems?
(A) *Rotary.*
(B) *Bellows.*
(C) *Rheostat.*
(D) *Diaphragm.*

6. Generally, the *only* serviceable part of a windshield wiper motor assembly is the:
(A) *wiper motor.*
(B) *wiper blades.*
(C) *wiper switch.*
(D) *transmission gears.*

7. When correcting a horn system problem, Technician A says horns are not serviceable and can only be adjusted. Technician B says horns are serviceable but cannot be adjusted. Who is right?
(A) *A only.*
(B) *B only.*
(C) *Both A and B.*
(D) *Neither A nor B.*

8. A short circuit will show up as a:
(A) *blown fuse.*
(B) *burned fusible link.*
(C) *open circuit breaker.*
(D) *All of the above.*

9. There is a short in a light circuit that keeps blowing the circuit fuse. Technician A says to install a larger fuse to keep it from blowing. Technician B says to connect a test light across the fuse holder. By disconnecting parts and circuit wires, you can find the short when the test light goes out. Who is right?
(A) *A only.*
(B) *B only.*
(C) *Both A and B.*
(D) *Neither A nor B.*

10. The fog lights on a new car will not work. No power is reaching the bulb sockets. Technician A says to check the operation of the fog light

relay shown on the wiring diagram. Technician B says the problem could also be the fog light switch. Who is right?

(A) A only.
(B) B only.
(C) Both A and B.
(D) Neither A nor B.

Activities—Chapter 37

1. Select a specific vehicle make and model that is at least two years old. Go to the automotive department of a discount or department store. Find the store's replacement lamp identifier system (it may be a book or large card or even a computer screen). Use it to identify the proper replacements for headlights, taillights, turn signals, brake lights, backup lights, and side marker lights (if used). List all the part numbers.

2. Demonstrate the proper procedure for checking headlight aiming, using marks on a wall. Show how to correct the aim of the lights, if necessary.

3. Find a wiring diagram for a car or truck, and make a photocopy. Use a brightly colored pencil or marker to trace the wiring that makes up the windshield wiper and washer circuit.

Headlamp Diagnosis

Condition	Possible causes	Correction
Headlamps are dim with the engine off or idling.	1. Loose or corroded battery cables. 2. Loose or worn alternator drive belt. 3. Charging system output too low. 4. Battery has insufficient charge. 5. Battery is sulfated or shorted. 6. Faulty lighting circuit. 7. Both headlamp bulbs defective.	1. Clean and tighten battery cable clamps and posts. 2. Adjust or replace alternator drive belt. 3. Test and repair charging system. 4. Test battery state-of-charge. Recharge or replace battery. 5. Perform load test. Recharge or replace battery. 6. Test and repair circuit. 7. Replace both bulbs.
Headlamp bulbs burn out frequently.	1. Charging system output too high. 2. Loose or corroded terminals or splices in circuit.	1. Test and repair charging system. 2. Inspect and repair all connectors and splices.
Headlamps are dim with engine running above idle.	1. Charging system output too low. 2. Faulty headlamp circuit. 3. High resistance in headlamp circuit. 4. Both headlamp bulbs defective.	1. Test and repair charging system. 2. Test and repair circuit as necessary. 3. Test amperage draw of headlamp circuit. 4. Replace both bulbs.
Headlamps flash randomly.	1. Poor headlamp circuit ground. 2. High resistance in headlamp circuit. 3. Faulty headlamp switch. 4. Loose or corroded terminals or splices in circuit.	1. Repair circuit ground. 2. Test amperage draw of headlamp circuit. 3. Replace headlamp switch. 4. Repair terminals or splices.
Headlamps do not illuminate.	1. No voltage to headlamps. 2. No ground at headlamps. 3. Faulty headlamp switch. 4. Faulty headlamp dimmer switch. 5. Faulty headlamp circuit.	1. Replace fuse. 2. Repair circuit ground. 3. Replace headlamp switch. 4. Replace headlamp dimmer switch. 5. Repair circuit.

Turn Signal Diagnosis

Condition	Possible causes	Correction
Turn signal flashes at twice the normal rate.	1. Faulty external lamp. 2. Poor ground at lamp. 3. Open circuit in wiring to external lamp. 4. Faulty contact in turn signal switch.	1. Replace lamp. 2. Check and repair wiring. 3. Repair wiring harness. Check connectors. 4. Replace switch.
Dash indicator lamp illuminated brightly; external lamp glows dimly and flashes at a rapid rate.	1. Loose or corroded external lamp connection. 2. Poor ground circuit at external lamp.	1. Tighten or replace connection. 2. Check and repair wiring.
Hazard warning system does not flash.	1. Faulty fuse. 2. Faulty flasher. 3. Open circuit in wiring to turn signal switch. 4. Faulty contact in turn signal switch. 5. Open or grounded circuit in wiring to external lamps.	1. Replace fuse. 2. Replace flasher. 3. Repair wiring. 4. Replace switch. 5. Repair wiring.

(Continued)

Turn Signal Diagnosis		
Condition	**Possible causes**	**Correction**
Indicator lamp illuminates brightly, but external lamp does not light.	1. Open circuit in wire to external lamp. 2. Burned out lamp.	1. Repair wiring. 2. Replace lamp.
System does not flash on either side.	1. Faulty fuse. 2. Faulty flasher unit. 3. Loose bulkhead connector. 4. Loose or faulty rear wiring harness or terminals. 5. Open circuit to flasher unit. 6. Open circuit in feed wire to turn signal switch. 7. Faulty switch connection in turn signal switch. 8. Open or grounded circuit in wiring to external lamps.	1. Replace fuse. 2. Replace flasher. 3. Tighten connector. 4. Repair wiring harness. 5. Check repair wiring harness. 6. Check and repair wiring harness. 7. Test and replace switch. 8. Check and repair wiring harness.
System does not cancel after completion of the turn.	1. Broken canceling finger on turn signal switch. 2. Broken or missing canceling cam on clockspring.	1. Replace switch. 2. Replace clockspring.
External lamps operate properly; no indicator lamp operation.	1. Faulty indicator lamp in instrument cluster.	1. Replace lamp.

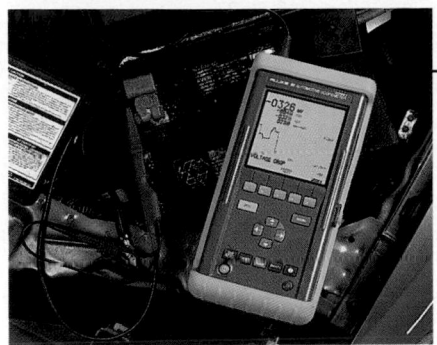

Chapter 38

Sound Systems and Power Accessories

After studying this chapter, you will be able to:

■ Describe the operating principles of a radio.

■ Explain the basic difference between AM and FM radios.

■ Diagnose basic sound system problems.

■ Explain the operation and service of power windows.

■ Sketch a rear window defogger circuit.

■ Describe and repair a power lock system.

■ Summarize the operation and testing of a speed control system.

■ Describe safety practices that must be followed when working with electrical accessory circuits.

■ Correctly answer ASE certification test questions on the service of sound systems and power accessories.

Many new cars are ordered from the factory with a wide array of power options. Options include premium sound systems, power seats, power windows, power door locks, and other convenience devices. These power options are very common; understanding their operating principles and service methods is important, even to the general technician.

After studying this chapter, you will be well-prepared to diagnose and repair optional power accessories. Remember that power accessory designs and circuits vary. Therefore, you must refer to a service manual to get specific diagrams and specifications for the exact system being repaired. This chapter will lay the groundwork so you understand manual instructions more clearly.

Note
Air bag systems, automatic seat belts, and many other power systems are explained in other chapters. Refer to the index for more information as needed.

Sound Systems

A basic *radio system* consists of the antenna, radio (receiver-amplifier), power supply circuit, and speakers.

A basic operation of a radio is illustrated in **Figure 38-1.** The *radio station* sends out an electromagnetic signal from a large broadcasting tower. When this signal moves past the vehicle's antenna, tiny electrical modulations (fluctuations) are induced in the antenna.

The radio amplifies and alters these "weak" radio signals into stronger current pulses that operate the speakers. Each speaker diaphragm moves back and forth, producing air pressure waves. We hear these air pressure waves as sounds (voices or music).

A *sound system* is more elaborate than the basic radio system just described and often includes an AM/FM stereo radio, a tape player, and sometimes an amplifier, a sub-woofer speaker, a CD player, and a power antenna. These parts are shown in **Figure 38-2.**

Radio

A modern *radio* contains complex electronic circuits that receive and amplify the radio signal to operate the speakers. It also contains a *tuner,* which allows the driver to select different *radio frequencies* (stations).

There are two types of radio signals: AM (amplitude modulating) and FM (frequency modulating).

An *AM radio* is designed to pick up radio signals that vary in amplitude (strength). AM signals, which operate on a frequency of 530-1,610 kilohertz (kHz), are reflected off the ionosphere (upper atmosphere). This gives them a longer broadcasting range than FM signals.

An *FM radio* is designed to receive a radio signal that varies in frequency (fluctuating speed). The FM band is from 88-108 megahertz (MHz). Since the FM radio wave is not reflected off the ionosphere, it has a short broadcasting range (approximately 35 miles or 56 kilometers). FM radio is capable of producing stereo

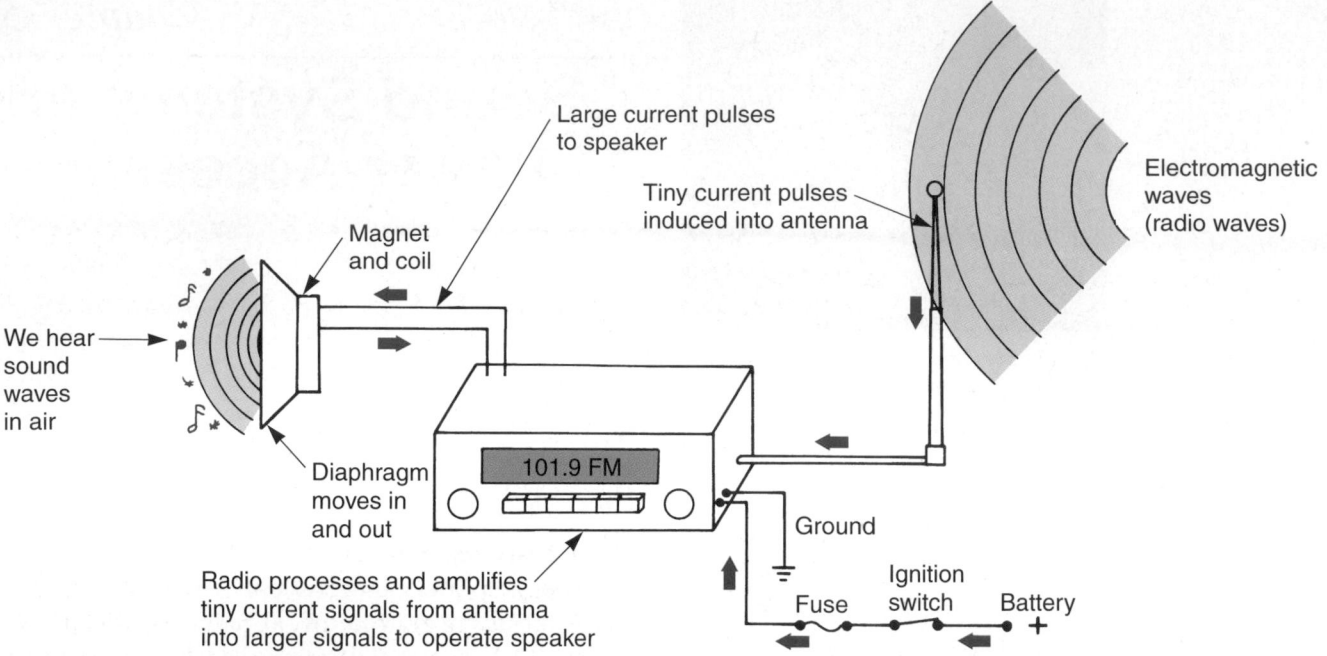

Large current pulses
to speaker

Tiny current pulses
induced into antenna

Electromagnetic
waves
(radio waves)

Magnet
and coil

We hear
sound
waves
in air

101.9 FM

Diaphragm
moves in
and out

Ground

Ignition
switch

Fuse

Battery
+

Radio processes and amplifies
tiny current signals from antenna
into larger signals to operate speaker

Figure 38-1. Study the basic principle of a car radio. Radio waves are tiny electromagnetic pulses picked up by the antenna. The radio alters and amplifies these waves into strong current signals for speakers. The physical movement of each speaker diaphragm causes air pressure pulses, which we hear as sound.

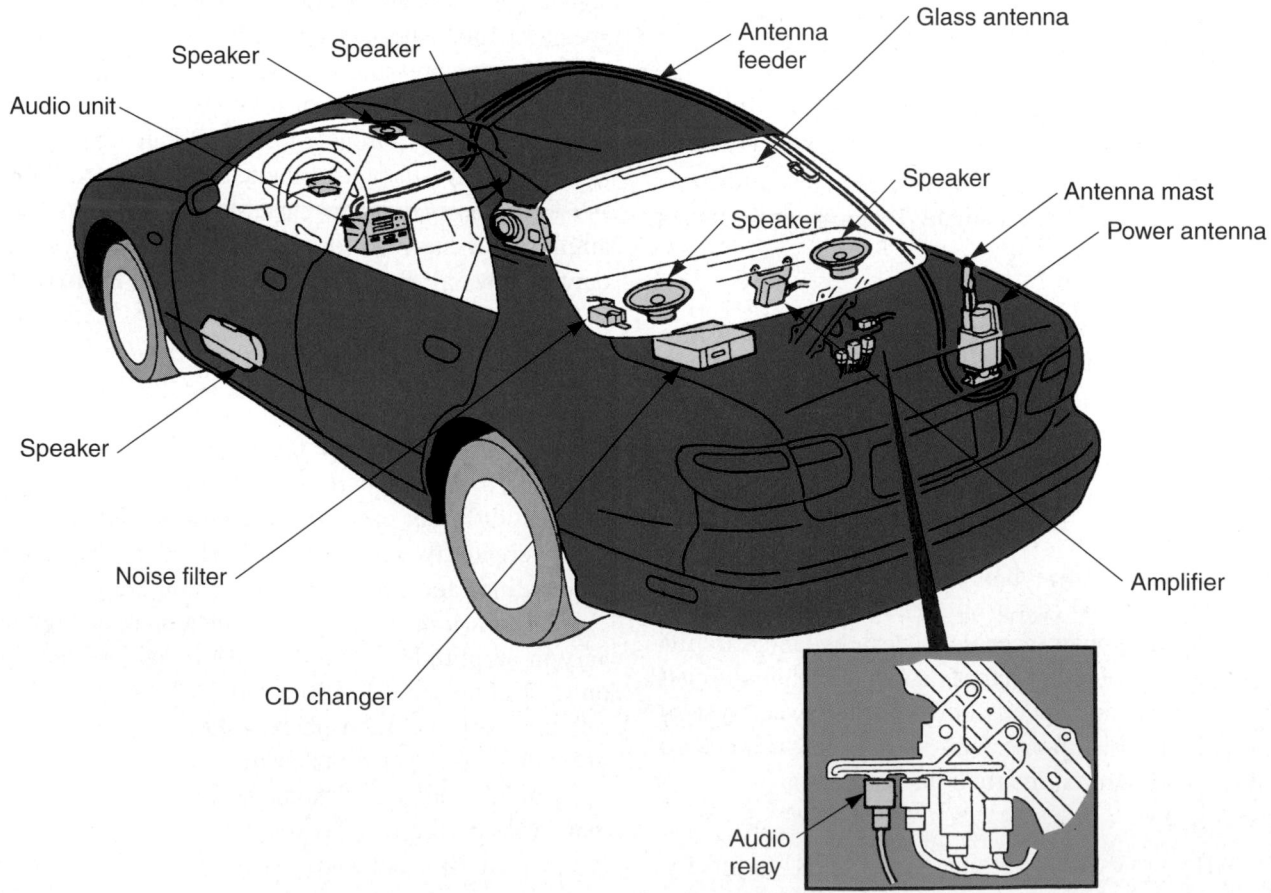

Speaker

Speaker

Audio unit

Antenna
feeder

Glass antenna

Speaker

Speaker

Antenna mast

Power antenna

Speaker

Noise filter

Amplifier

CD changer

Audio
relay

Figure 38-2. Late-model vehicles normally come equipped with elaborate sound systems. Multiple speakers, amplifier or power booster, CD player, and power antenna add to service complexity. (Mazda)

(stereophonic) sound. A stereo uses at least two speakers and has different sounds coming from each speaker. **Figure 38-3** illustrates the differences between AM and FM radio signals.

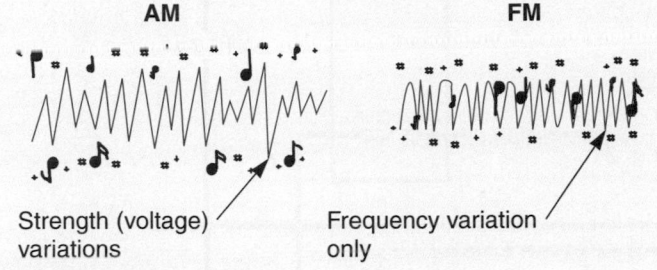

Figure 38-3. Note the difference between AM and FM signals. Radios must have circuitry that can "read" these different radio waves. (Chrysler)

Radio Service

If a radio fails to work, check its fuse. You may have external problems: a blown fuse, an open antenna or power supply lead, or bad speakers. Do not condemn the radio until all other problem sources have been eliminated. System designs vary; use the information in a service manual to diagnose radio problems.

Tech Tip

Just because the lights in a radio are working does not mean the radio circuitry is getting power. Quite often, two power leads feed the radio. One lead is for the circuitry and the other is for the face lights. Make sure both leads are providing electrical power to the radio.

Figure 38-4 illustrates a wiring diagram for a sound system circuit. Trace the wiring from the battery through each component. A wiring diagram can be used to find difficult problems.

When the radio is found to be faulty, it should be removed and sent to an authorized repair technician. Various methods are used to secure radios. Some radios can be removed by simply removing the screws on the faceplate and pulling the radio out far enough to disconnect the wires and antenna lead. Others require partial dash disassembly for removal. In some cases, you may need to lie on the floor to remove a rear support bracket and the wiring.

If provided, an **antenna trimmer screw** should be adjusted when the radio has been removed for repairs or after antenna replacement. After reconnecting the radio, set the tuner to a weak station. Then, as shown in **Figure 38-5,** turn the trimmer screw until the weak station comes in as loud and clear as possible.

Antennas

An **antenna** picks up the broadcast signal and feeds it through the antenna lead to the radio. A very fine piece of wire mounted in the windshield glass can serve as the antenna. Other antennas are a metal **mast** (rod) mounted on the body.

A **power antenna** is a telescoping antenna that is extended and retracted by an electric motor. The electric motor turns a gear. The gear then operates a cable or slide mechanism on the antenna mast. Most power antennas go up and down automatically when the stereo is turned on or off. In some cases, however, a dash-mounted switch actuates the electric motor to move the antenna up or down. Look at **Figure 38-6.**

Antenna Service

If you suspect a bad antenna (no signal coming from antenna lead), connect a known good antenna to the radio. If the new antenna causes the stereo to work, replace the old antenna.

If a power antenna does not go up or down, check for power going to the motor. Use a test light or voltmeter. If power is reaching the antenna motor, the motor is probably bad. If you are not getting power, trace back through the circuit to find the problem.

Speakers

A **speaker** uses a permanent magnet and a coil of wire mounted on a flexible diaphragm to convert electricity into motion and sound. When current passes through the coil of wire, the resulting magnetic field pulls the coil and diaphragm toward the permanent magnet. Rapid movement of the speaker diaphragm causes pressure waves in the air. We hear these pressure waves as sound. Look at **Figure 38-7.**

The speakers may be mounted in the doors, dash, or behind the rear seat. AM radios normally have only one speaker in the top of the dash. FM stereo radios have two or four speakers.

Speaker Service

A faulty speaker will usually distort the sound of the radio. The speaker may rattle, especially when the volume is adjusted to a higher output level. A broken coil winding or terminal-to-coil wire can render the speaker inoperable.

Speakers are not usually repairable and should be replaced when defective. However, terminal-to-coil wires can sometimes be soldered when broken. The speaker mounting screws can also vibrate loose, creating a sound similar to a blown speaker. Tighten the speaker mounting screws and recheck audio quality.

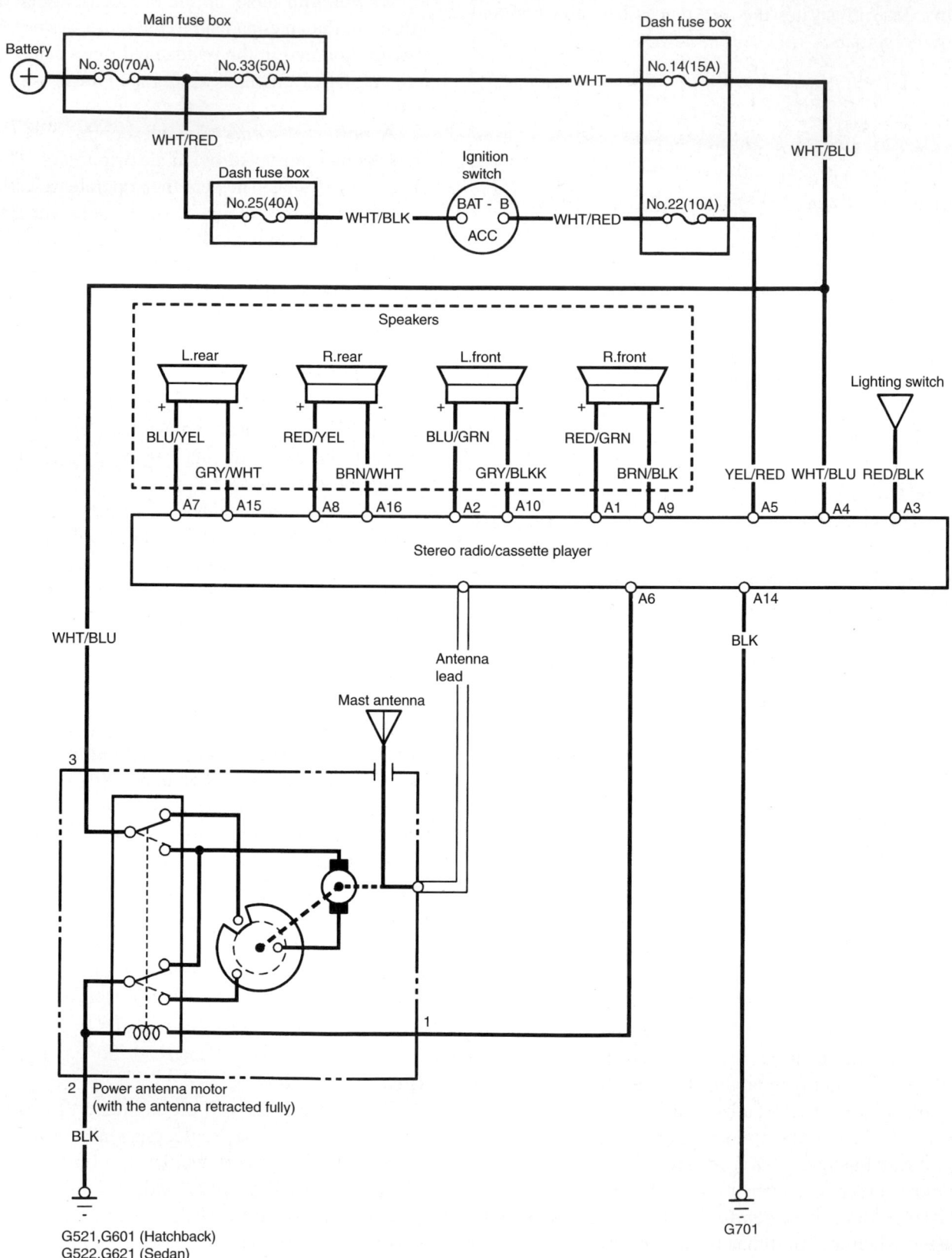

Figure 38-4. Study connections and components in this stereo system wiring diagram. (Honda)

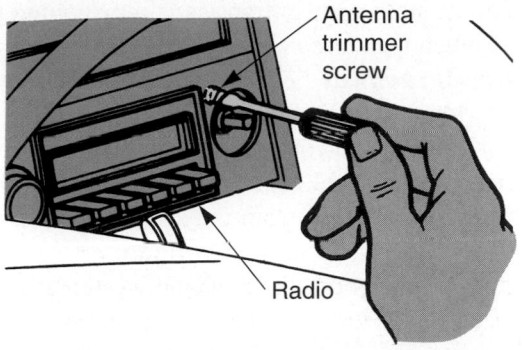

Figure 38-5. If provided, an antenna trim screw provides one of the few adjustments that can be done on a radio. The radio is set on weak station, then the screw is turned for best reception. (Buick)

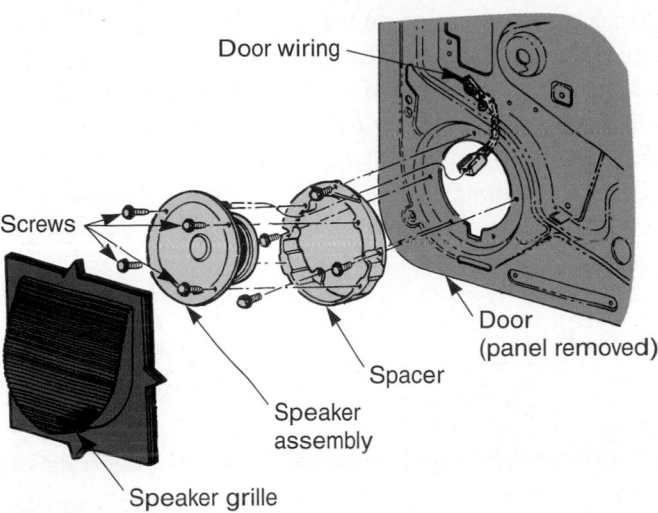

Figure 38-7. Speakers can be mounted in doors, in the upper dash, in interior side panels, or behind the rear seat. Mounting is similar to this door-mounted speaker. The rattle of loose mounting screws will sound like a blown speaker diaphragm or separated coil windings, which require speaker replacement. (Ford)

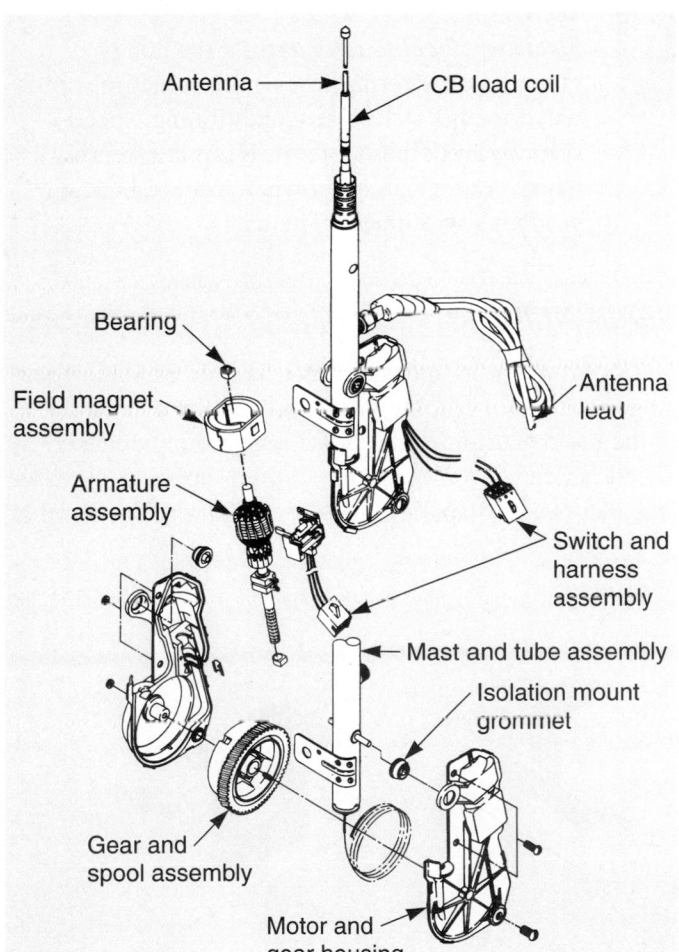

Figure 38-6. Most power antennas use these basic parts. If adequate voltage is reaching motor windings and the unit does not work, the unit is usually replaced with a new one. Internal repairs are seldom done. (Oldsmobile)

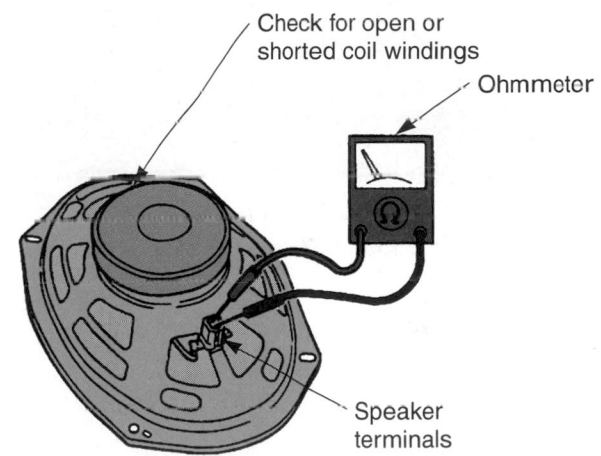

Figure 38-8. To verify an internal problem with a speaker coil, check it with an ohmmeter. If shorted, resistance will be almost zero. If the windings are open, the ohmmeter reading will be very high. (Mazda)

Tape Player

A cassette *tape player* is often incorporated into the radio, **Figure 38-9.** As with radios, internal problems with a tape player require a specialized electronic technician. However, an automotive technician can perform external repairs: removing broken recording tape and cleaning the tape head.

CD Players

CD players, also termed *compact disc players*, use a plastic disc to store and play music digitally. They are

Figure 38-8 shows how to check for shorted or open speaker coil windings. Most auto speakers should have a resistance of either 4 or 8 ohms.

also a common option on today's vehicles. The CD player can be made as part of the stereo in the dash, or a remote CD player can be located in the trunk, **Figure 38-10.**

Sometimes, a *stereo amplifier,* or *"power booster,"* is added to the sound system to increase volume without sound distortion, **Figure 38-11.** It can be located in the passenger compartment or trunk. It simply increases the electrical power available to drive the speakers. Without an amplifier, the stereo power transistors may go into thermal runaway when the stereo is played at high volumes.

Thermal runaway occurs when a transistor is driven too hard (too much output current is controlled by the transistor). The output of the transistor will fluctuate as the transistor's thermal capacity is exceeded, and severe distortion will be heard from the speakers. With enough thermal runaway, the transistor will fail.

When troubleshooting a sound system, determine which sections of the system are working and which are not. For example, if the radio and tape player work but the CD player does not, check the part of the circuit related only to the CD player. If tests find something wrong with the CD player, send it to a specialized shop for repairs.

Tech Tip

Steering wheel touch controls (switches mounted on steering wheel) are sometimes provided for the stereo, air conditioning, speed control, and other systems. Keep the steering wheel controls in mind when troubleshooting problems with these systems.

Radio Noise

Radio noise is undesired interference or static (popping, clicking, or crackling) obstructing the normal sound of the radio station. Radio noise is commonly caused by a bad antenna, an open or shorted noise suppressor (capacitor), a bad spark plug wire, a faulty radio, or other

Figure 38-9. When found to be defective, a receiver is often sent out to a specialty electronics shop for repairs.

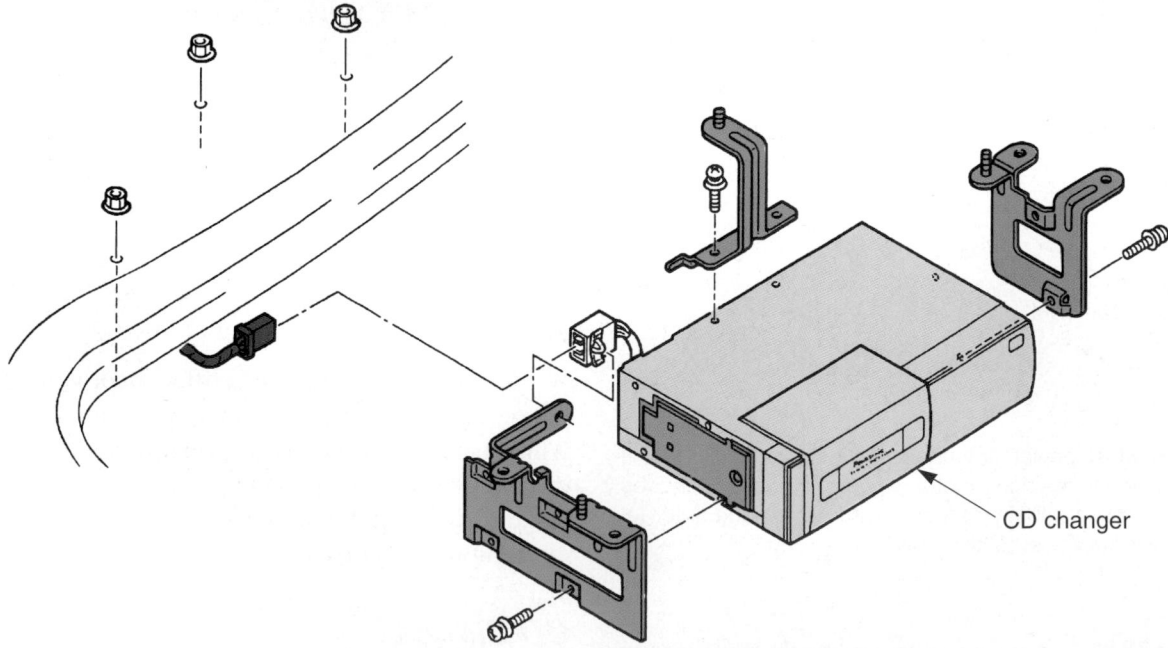

CD changer

Figure 38-10. A CD changer normally mounts in the trunk. Check for stuck CDs, dirty heads, and power feed problems caused by bad wiring before sending the unit out for repairs. (Mazda)

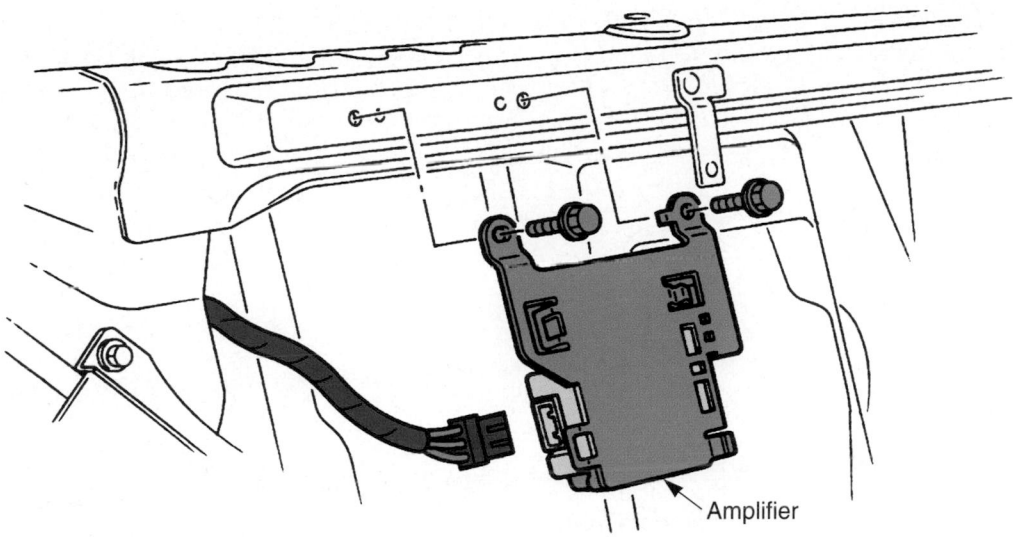

Figure 38-11. A stereo amplifier may be located in the trunk or under a seat. (Mazda)

problems. If the stations are too far away, noise will also interfere with the signal.

Study the sound of the radio noise to determine its source. For example, a low-pitched clicking that changes with engine speed may be caused by the ignition system (open spark plug wire). A high-pitched whirring sound that also changes with engine speed could be from the electrical system (bad capacitor or alternator diode), **Figure 38-12.**

One of the first parts to check when radio noise is present is the antenna. Plug a known good antenna into the radio. Ground the antenna base and note any changes in radio output. If the noise is eliminated, the old antenna is faulty. If the noise remains the same, check the noise suppressors.

Noise suppressors are capacitors that absorb voltage fluctuations in the car's electrical system. They result in smoother dc current entering the radio, which reduces radio noise.

Noise suppressors can be located at the alternator, voltage regulator, ignition coil, distributor, and heater blower motor. All these components can produce voltage fluctuations and noise. Refer to a service manual for exact suppressor locations.

A clip-on capacitor can be used to test noise suppressors, **Figure 38-13.** Connect the test capacitor across or in place of the suppressor. If the radio static is reduced, install a new suppressing capacitor.

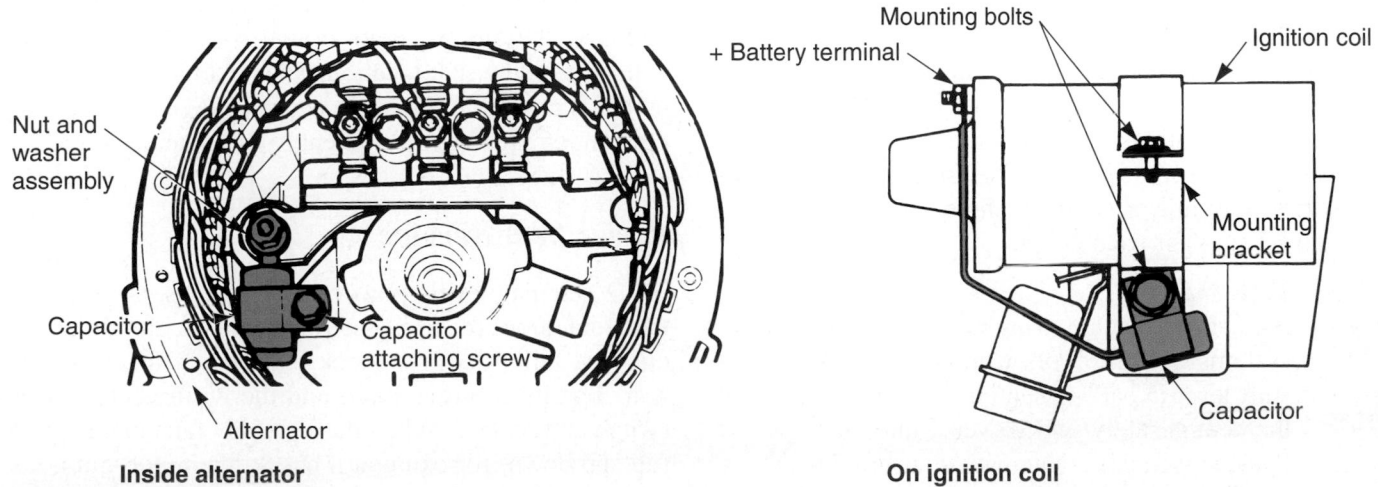

Inside alternator   **On ignition coil**

Figure 38-12. Two common locations for radio noise suppressors. (Chrysler)

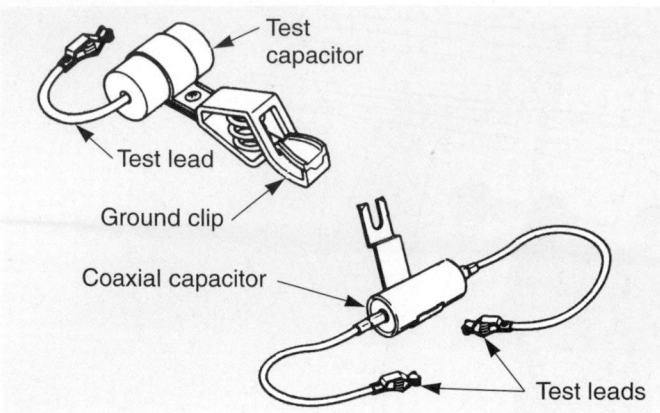

Figure 38-13. Test capacitors can be used to check existing noise suppressors. If radio static quiets when the test unit is connected to the circuit, the old capacitor is bad or there may be other electrical problems. (Pontiac)

Power Seats

Power seats use several switches, electric motors, and drive assemblies to change the seat positions. The power seat switches can be on the inner door panels or on the sides of the seats. The power seat motors and drive assemblies can be under or in the seats, **Figure 38-14.**

When activated by a switch, the reversible dc motors operate a gear mechanism. The gear mechanism changes the rotating motion of the motor armature into a linear motion that positions the seat. Most power seats have more than one seat motor. Power is sometimes transmitted from the power seat motors to the tracks or adjustment mechanisms by steel cables inside metal housings. **Figure 38-15** shows how a simple power seat circuit can create several seat positions.

Figure 38-16 is a typical wiring diagram for a power seat system. Trace the wiring from the seat position sensors, power seat module, switches, power seat motor, and other components.

Memory seats use a small computer, or module, to "remember" seat positions for several people. Some computer memory systems are networked and will remember not only seat positions but also steering wheel tilt positions, rearview mirror settings, and even electronic suspension settings for each driver.

 Tech Tip
As a safety feature, many power seat memory systems will only work with the transmission shift lever in park or neutral. They are wired to the neutral safety switch. Make sure the transmission is in park (automatic transmission or transaxle) when checking memory seat operation.

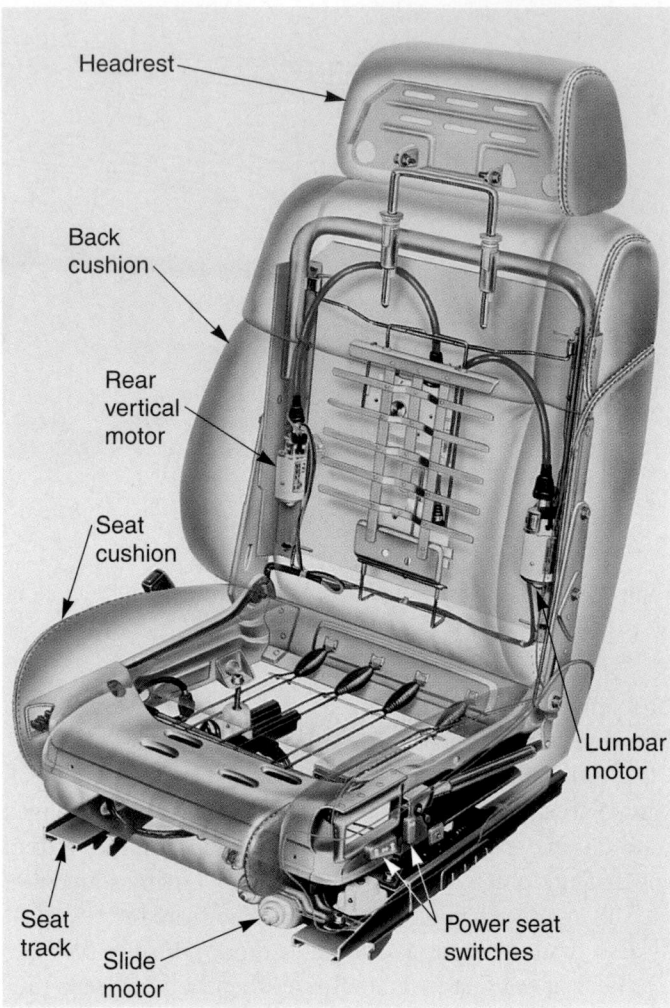

Figure 38-14. Cutaway view of a power seat shows the seat motors, transmission cables, tracks, and other parts. Seat removal may be needed to service seat motors. (General Motors)

A *power seat module* can be programmed to return the seat to a desired position for different drivers. The seat is placed in the desired position and then one of two memory buttons is held down so the module can "remember" the seat location. The next time the memory button is pushed, the module can use the seat sensors and power motors to return the seat to the same location.

Power Seat Service

When diagnosing power seat problems, try to isolate the problem source to a specific area of the circuit. If only one seat fails to work, check parts that affect only that seat. Test its control switch and the wiring between the switch and motors. When the seat only fails in one mode (up and down, for example), check the motor and transmission (gear-cable mechanism) providing that action.

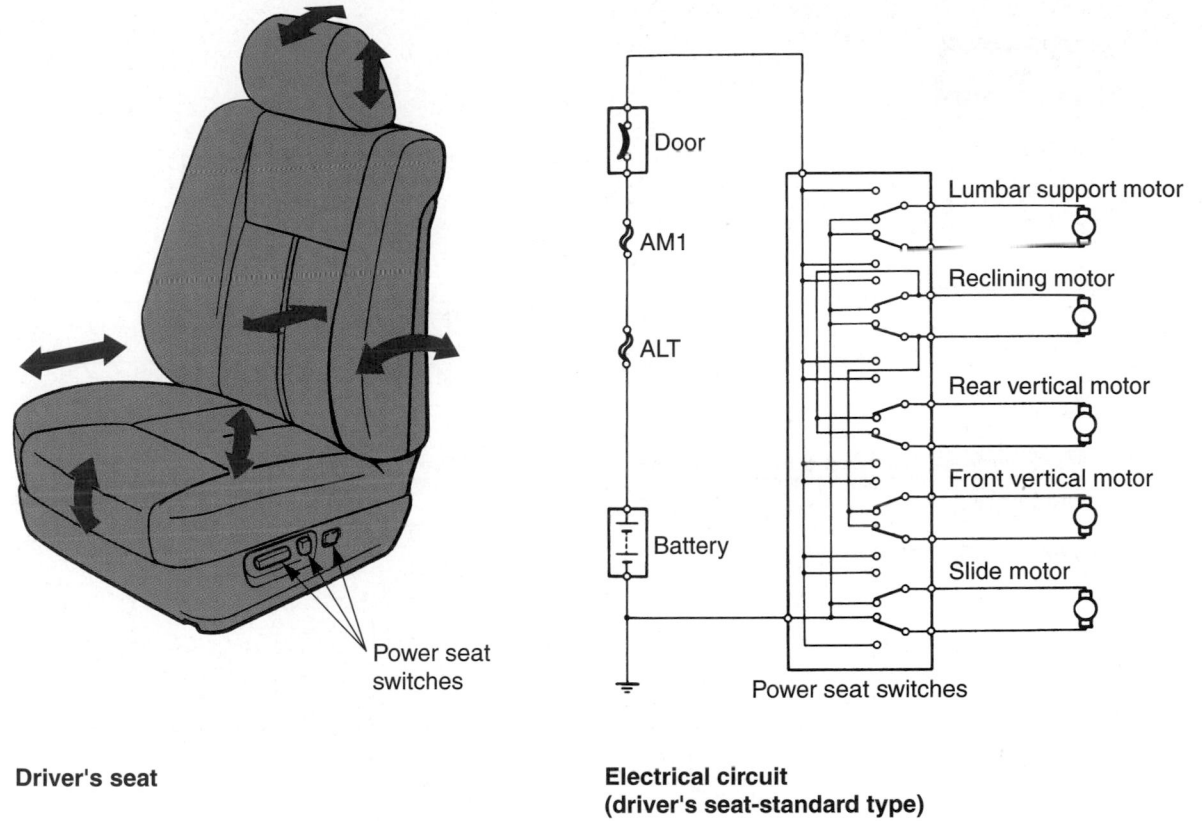

Driver's seat

**Electrical circuit
(driver's seat-standard type)**

Figure 38-15. Note how the wiring diagram shows how switches and each motor operates different seat adjustment. (Lexus)

If both front power seats fail to function, check the common section of the circuit. Inspect the fuse, circuit breakers, wire connections, and any other components affecting both seats.

The most common reasons for power seat problems are the switches and drive motors. Remove and replace any switch or motor that does not work properly. See **Figure 38-17.** If you have difficulty repairing a power seat, read the information in a service manual. It will give directions for servicing the particular unit.

Heated Seats

Heated seats use a heating element (or multiple heating elements), switch, relay, and related wiring to warm the seat cushions in cold weather. Most systems have two large-resistance wires or heating elements routed through the seat cushions. See **Figure 38-18.**

When you press the seat warmer button or switch, it energizes the warmer relay. The relay is a holding relay that stays on once energized. The relay then sends high current through the heating elements, which makes them get warm. The seat warmers stay on until you press an off button or until you shut the ignition switch off.

To service heated seats, check the fuse first. Next, test the warmer switch and then the relay. If you are getting power to the heating element but it does not get hot, the element is probably open and must be replaced.

Power Windows

A **power window** uses a control switch, reversible electric motor, circuit breaker, fuse, and related wiring to operate the door windows, **Figure 38-19.**

A small electric **power window motor** is located inside each door to operate the **window regulator** (up-down mechanism for the glass). The motors have a **gearbox,** or **transmission** (usually worm and ring gear), that changes the rotating motion of the motor armature into a partial rotation of a larger gear. This action pushes the window open or closed. See **Figure 38-20.**

A circuit breaker protects the window motor from overheating damage. The breaker can open if the switch is held in one position too long. The circuit breaker can be located inside the motor or elsewhere in the circuit. A basic power window circuit is shown in **Figure 38-21.**

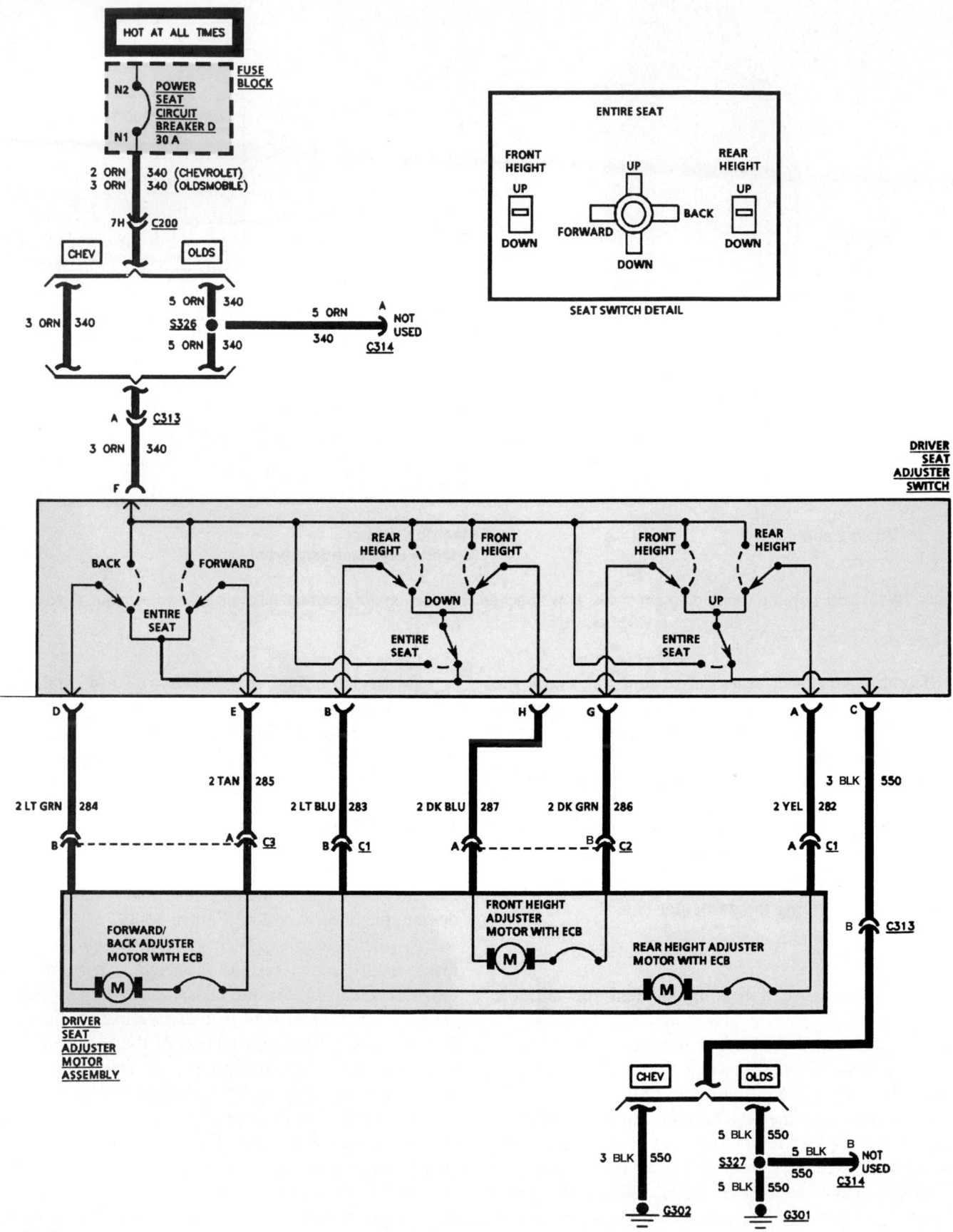

Figure 38-16. Study the wiring diagram for a typical power seat system. Note the various switches and motors. (GM)

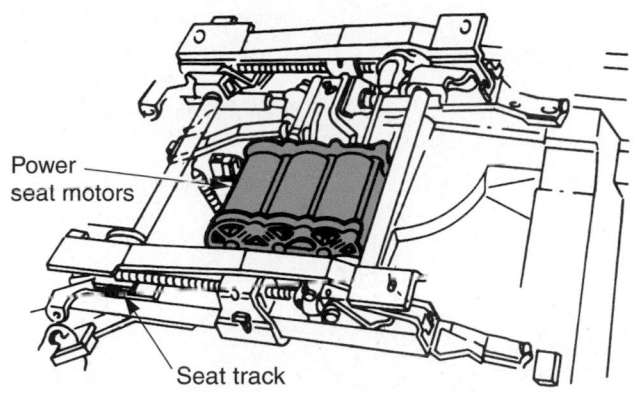

Figure 38-17. This seat has been removed to service a power seat motor. If the seat works in all but one position, only service the motor that operates that seat adjustment. (Ford)

Power Window Service

When none of the power windows work, first check the fuse or circuit breaker for the whole system. If only one of the windows is inoperative, use a test light to check for power to its switches and motor.

If you hear a humming sound when a window switch is pressed, the motor gearbox may have stripped gear teeth. The plastic gears in the window motor gearbox can strip after prolonged service. The motor will spin, but movement will not be transferred to the window. If the motor or the switches are found to be bad, they should be replaced.

With hard-to-find problems, refer to the service manual wiring diagram for the power windows. A typical

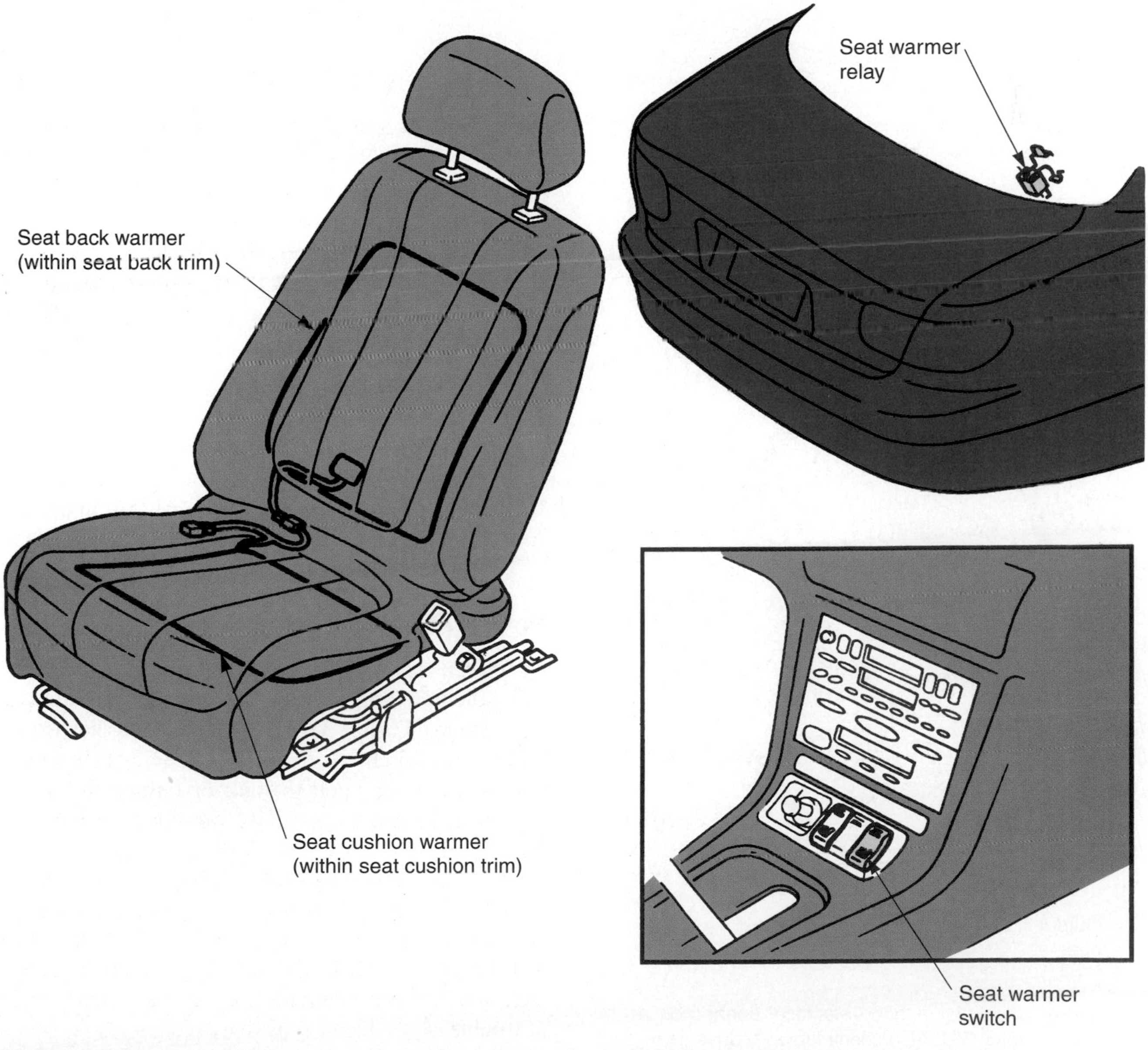

Figure 38-18. A heated seat simply uses two wires threaded inside seat cushions. First check for power to the resistance wires from the switch and relay. Then verify problem with resistance wire by testing it with ohmmeter. (Mazda)

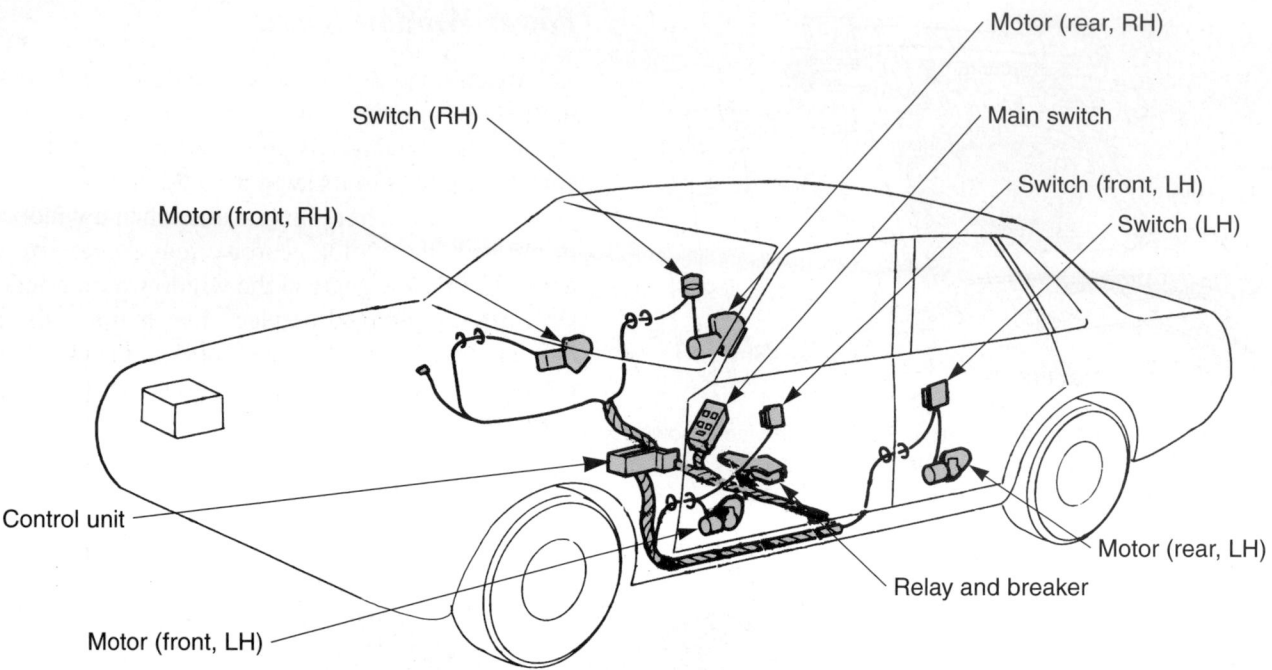

Figure 38-19. Study the basic layout of a complete power window system. Switches operate electric motors inside each door. With inoperative power windows, check fuses and switches first. Then, remove door panels to test and service window motors. (Subaru)

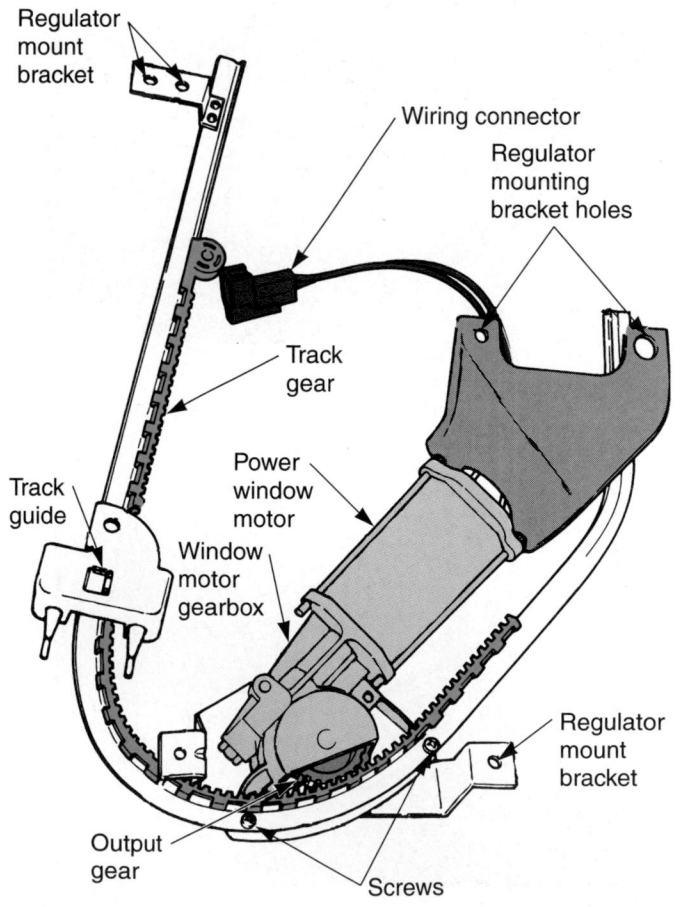

Figure 38-20. Check for a good electrical connection at the power window motor first. Also check for worn drive gear, track damage, and bad motor. Loose mounting bolts will cause a clunking sound when window changes direction. (Chrysler)

example is shown in **Figure 38-21.** The circuit will show all the components that could affect power window operation and help with troubleshooting.

 **Warning**
Be careful because the sheet metal edges on and in doors can cause serious cuts or lacerations. Also, never wear a wristwatch or rings when servicing inside a door. The metal could short across wires or get caught, causing injury or part damage.

Power Door Locks

Power door locks typically use electric switches and solenoids or motors to operate the door lock mechanisms, **Figure 38-22.** When the door key is turned, it closes a switch. A motor or solenoid then moves an arm on the door latch to lock or unlock the door. An additional switch is normally provided on the driver's door panel.

If all the power door locks fail to function, check the main fuse, electrical connections, and other components common to the whole circuit. If only one door lock is bad, check its switch and solenoid or motor. The door lock switches and solenoids are the most common causes of trouble.

Figure 38-23 is a typical power door lock circuit. Study its connections and components.

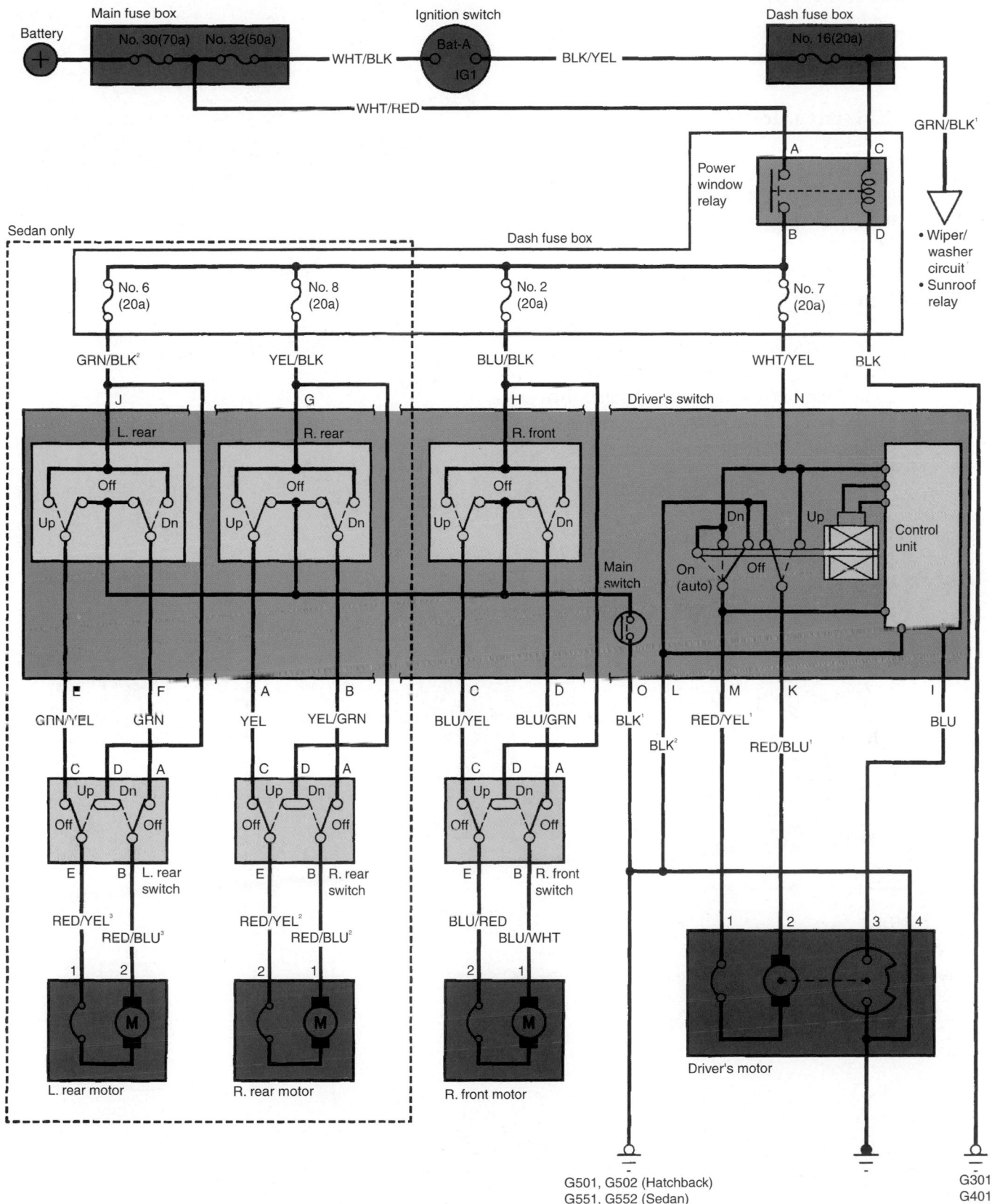

Figure 38-21. Study this wiring diagram for a typical power window circuit. Note how the electronic control unit operates the driver's window. It causes the window to go all the way down when the button is briefly pushed. The driver can release the switch right away to again grasp the steering wheel. This feature is handy at tollbooths. (Honda)

Door Panel Removal

Door panel removal may be needed to service power windows, seat control switches, door latches, or radio speakers.

The basic procedure for removing a door panel is as follows:

1. Remove all screws that hold the door panel to the door frame. Some are visible; others may be covered with small plastic buttons or covers.
2. Unscrew and remove the door lock button.
3. Remove the inner door handle and window crank, if used.
4. Once all major parts have been removed, you often have to pop out spring clips around the outside of the door panel. Be careful not to bend or tear the panel. Refer to a manual if you have difficulty. See **Figure 38-24.**

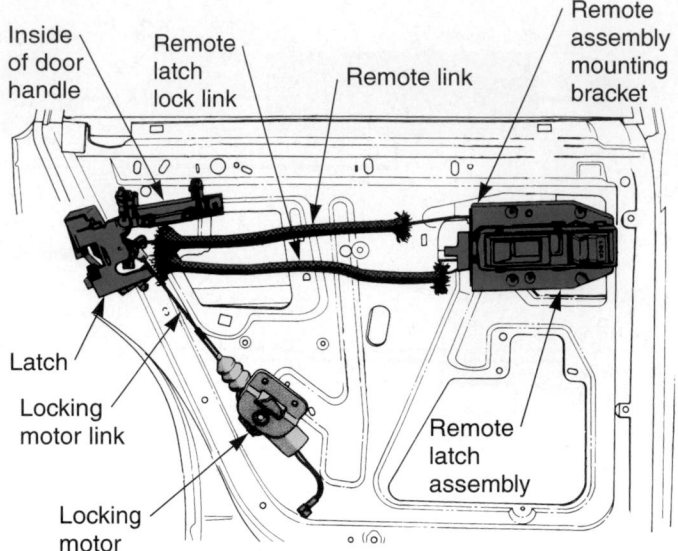

Figure 38-22. Most power door lock systems use a large solenoid to activate the latch. Study construction of this system. (Chrysler)

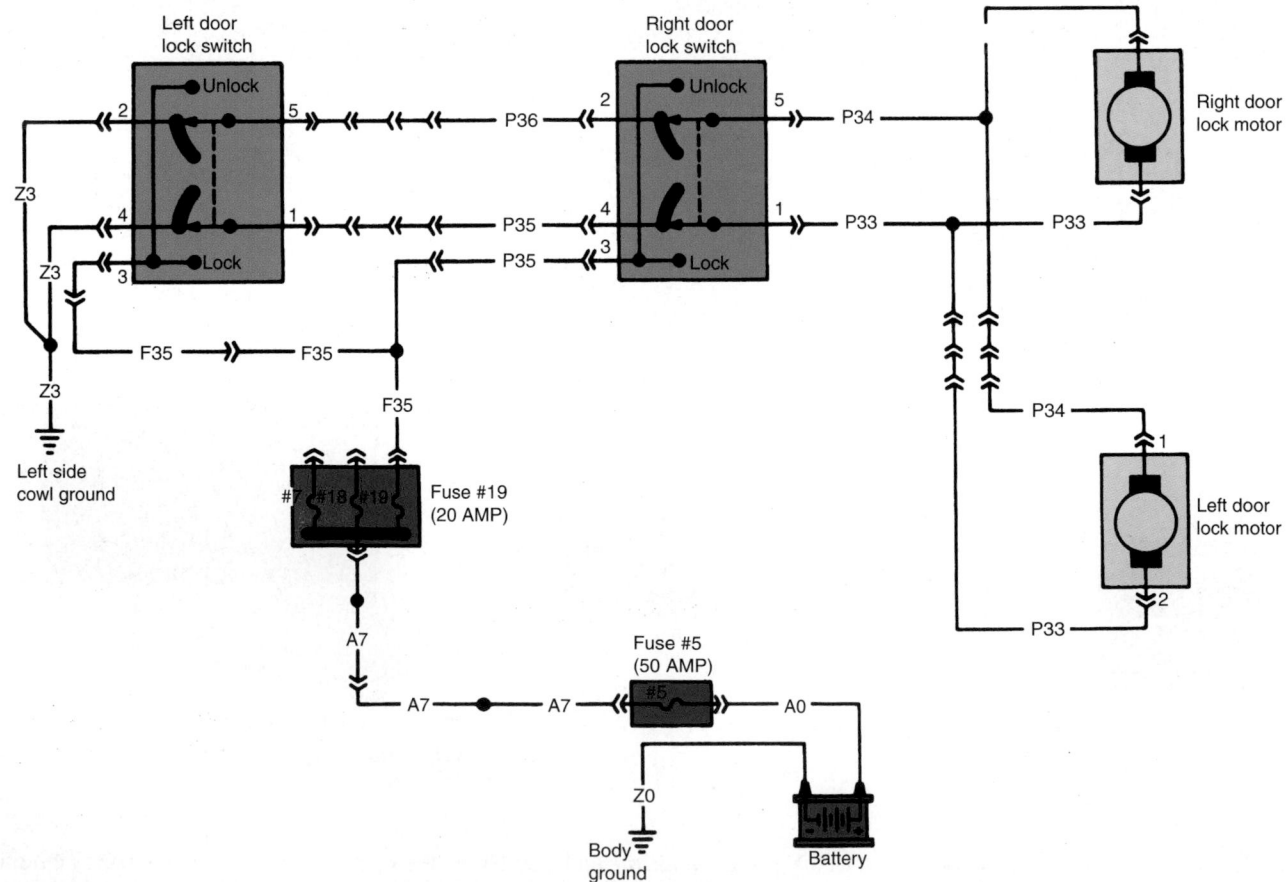

Figure 38-23. Basic door lock circuit connections and components. (Chrysler)

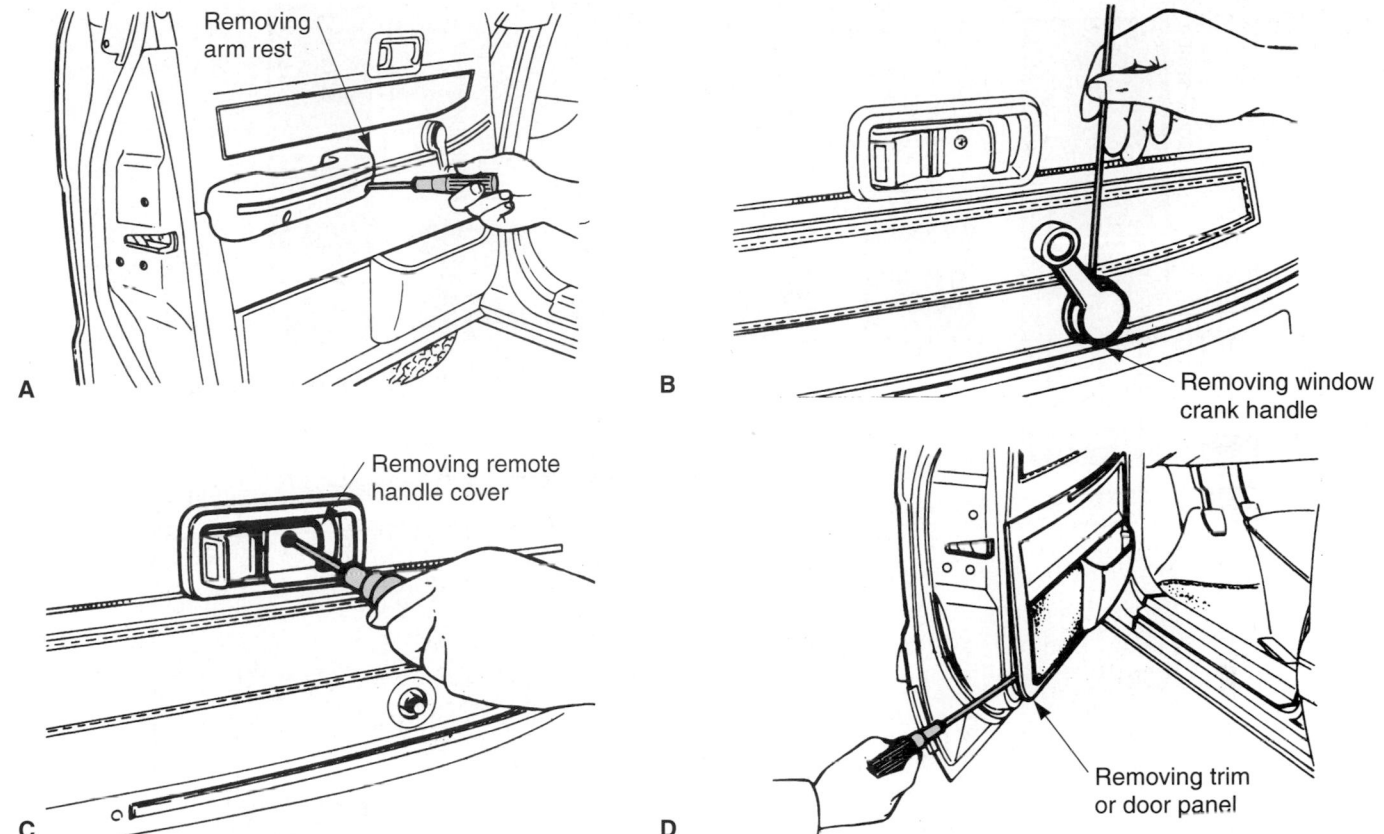

Figure 38-24. Basic procedure for removing a door panel. A—Remove screws from panel and arm rest, if used. Screws may be under pop-off cover or plugs. B—Remove window crank handles. They may be held with screws or you may need to release a special clip from behind the handle, as shown. C—Remove other hardware that is screwed into the body of the door: lock mechanism, lock knob, speaker cover, etc. D—Use a standard screwdriver or a special tool to pop clips out of the door to free the panel. Be careful to pry directly under each clip, or the door panel will tear. (Subaru)

Tech Tip
The wiring going through the body and into the door can break internally after the door has been opened and closed thousands of times. Keep this in mind when troubleshooting any power device inside or on the door.

Power Trunk Release

A *power trunk release* is comprised of a solenoid mounted on the trunk latch, a fuse, a switch, and related wiring. When you close the trunk release switch, it sends power to the solenoid. The solenoid plunger then acts on the trunk latch and the trunk pops open.

Service of this circuit is fairly simple. Most problems are in the switch or solenoid. Use basic testing methods to isolate the problem.

Power Steering Wheel

A *power steering wheel* uses a computer and various switches, sensors, and motors to automatically tilt and telescope (extend or retract) the steering wheel. Like memory seats, most systems can be programmed to position themselves automatically. Some are tied to the memory seat circuit and use the same computer.

Figure 38-25 shows a power steering wheel circuit. Study its parts and electrical connections.

Rear Window Defogger

A *rear window defogger,* also called a *rear window defroster,* commonly uses a switch, a relay, an indicating light, and a window heating grid. See **Figure 38-26.**

When the switch is turned on, it allows current to flow to the indicator light and to the heating grid. The heating grid is resistance wire, usually mounted on or in the window glass. Current flowing through the grid causes the wire to heat up and defog or deice the vehicle's window.

Some rear windows have *zone clearing defoggers,* which heat and clear one area before another. Most have higher-resistance wire in the middle to heat and clear the center area of the glass first. They provide for more rapid

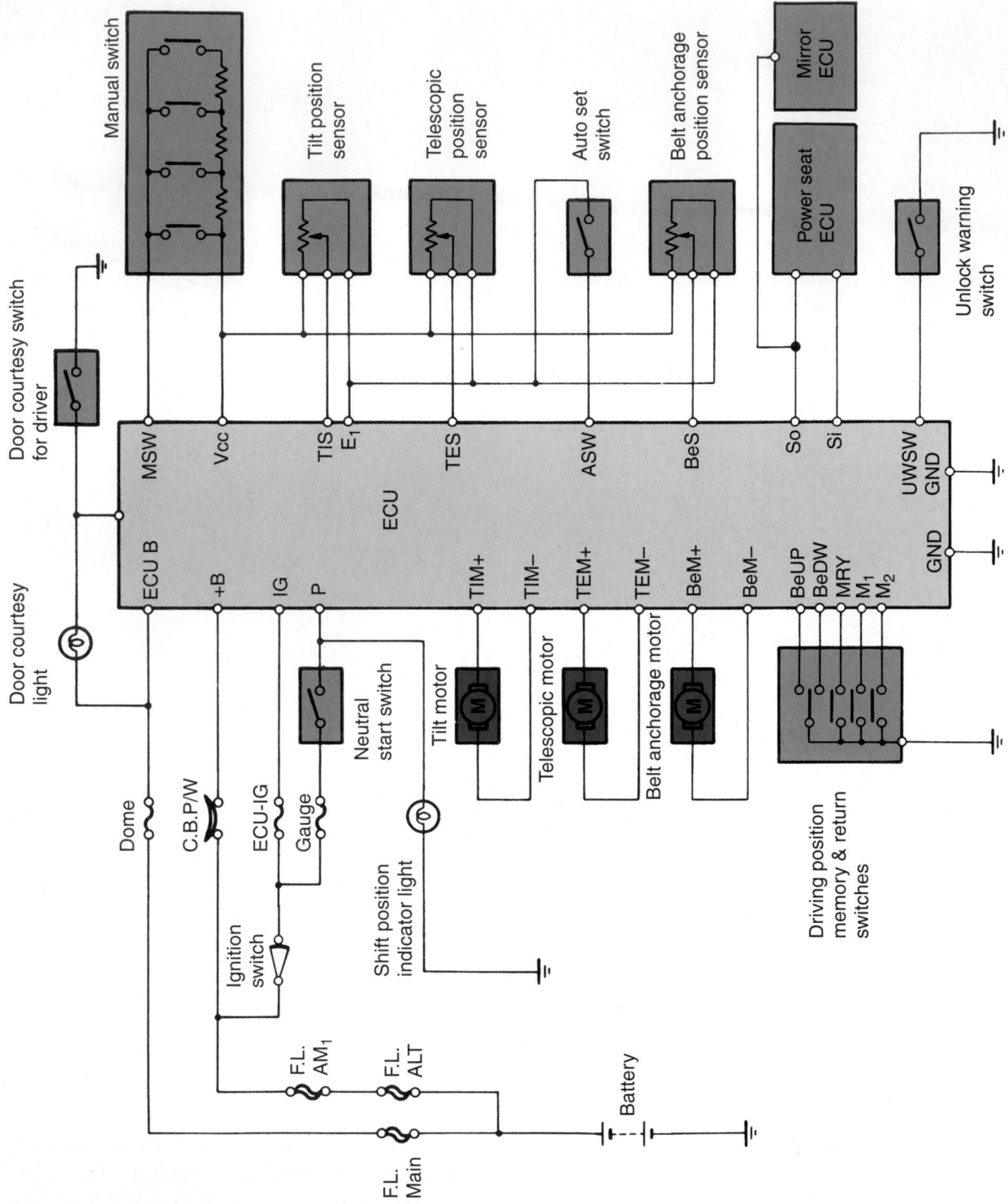

Figure 38-25. This wiring diagram shows how the ECU can control steering wheel tilt, telescoping action, and the seat belt anchoring motor. Note sensors and switches that feed data to the ECU. Also note the power seat ECU is also networked to this ECU. (Lexus)

rearview mirror use in less time. Then, the rest of the window heats and clears.

A few vehicles use a more conventional blower or fan to defog and deice the rear window. Its operation is similar to a heater blower.

Window Defogger Service

When the grid type rear window defogger does not work, check the fuse first. Then, check for adequate voltage going to the grid with the power switch closed.

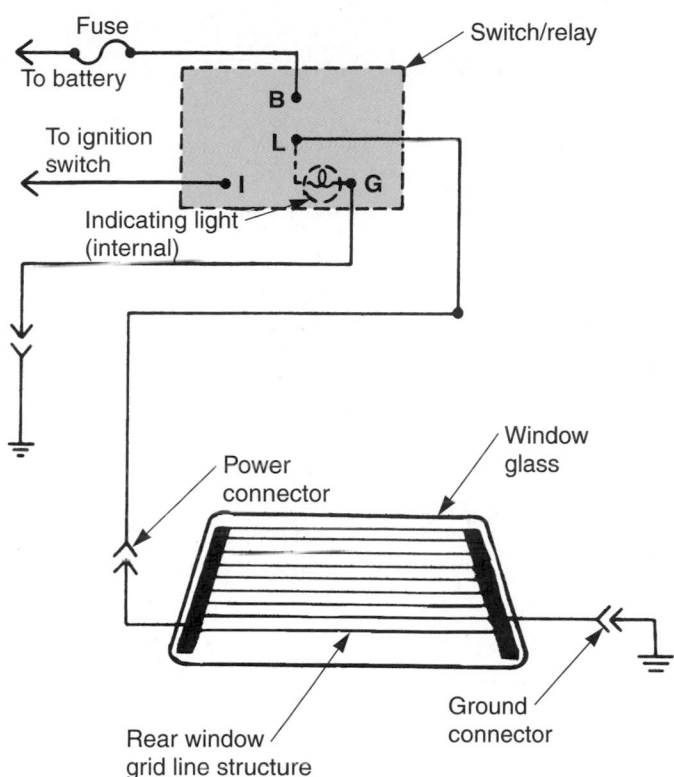

Figure 38-26. This is a circuit for a rear window defogger. When fed current from through a switch and relay, the resistance wire on the glass heats up. (Chrysler)

If voltage is low at the grid input, test the circuit for opens. If power is being fed to the grid, the grid may be bad. Test the grid as described by the manufacturer. Some allow the repair of a broken heating grid using a special *grid repair agent,* which conducts electricity.

Heated Windshield

A *heated windshield system* uses a special conductive film sandwiched inside the windshield glass. The invisible film in the windshield glass is usually a zinc and tin oxide material. When high voltage (typically 70–90 volts) is passed through this clear conductive film, the glass heats up and melts ice or snow.

Most systems use a control module to full-field the alternator and bypasses the rest of the electrical system so that enough energy is available to rapidly warm the windshield glass. The other electrical systems operate off of battery voltage when the windshield deicing system is turned on. If the system senses a low battery charge (below 11 volts), the system automatically shuts off to prevent excessive battery discharge.

Warning
Use caution when working on heated windshield systems. When turned on, there is enough voltage and current to cause a serious electrical shock. Electrocution could result from being shocked by this system's high power output!

Reminder System

A *reminder system* (warning system, chime system) makes an audible signal (buzz, chime, voice) if:

- Seat belts are not fastened when the engine is started.
- Keys are left in the ignition with the engine off.
- The headlamps are left on with the engine off and a door is open.

An illustration of a three chime reminder or warning system is shown in **Figure 38-27.** Note how the tone generator can be activated by the seat belts, ignition key, and headlamp switch.

Refer to a service manual when troubleshooting and repairing a reminder system. Numerous design variations are used by different vehicle makers.

Cruise Control Systems

An electronic *cruise control system* uses a computer, sensors, and a throttle actuator to maintain vehicle speed for highway driving. The major parts of a modern cruise control system include:

- *Power switches* feed current to computer to activate and ready system for operation.
- *Control switch* signals computer to maintain present vehicle speed.
- *Vehicle speed sensor* feeds pulsing signal into the computer that represents velocity of the car.
- *Cruise computer* uses input signals to control outputs to throttle actuator.
- *Throttle actuator* physically moves the engine throttle lever to control engine power and resulting vehicle speed.
- *Brake light switch* signals computer to shut off cruise control when brakes are applied.
- *Clutch switch* signals computer to deactivate cruise control when clutch pedal is depressed.
- *Neutral safety switch* signals computer to shut off cruise when shift lever is moved out of drive.

When set on cruise using the power and control switches, the vehicle speed sensor feeds an ac signal into

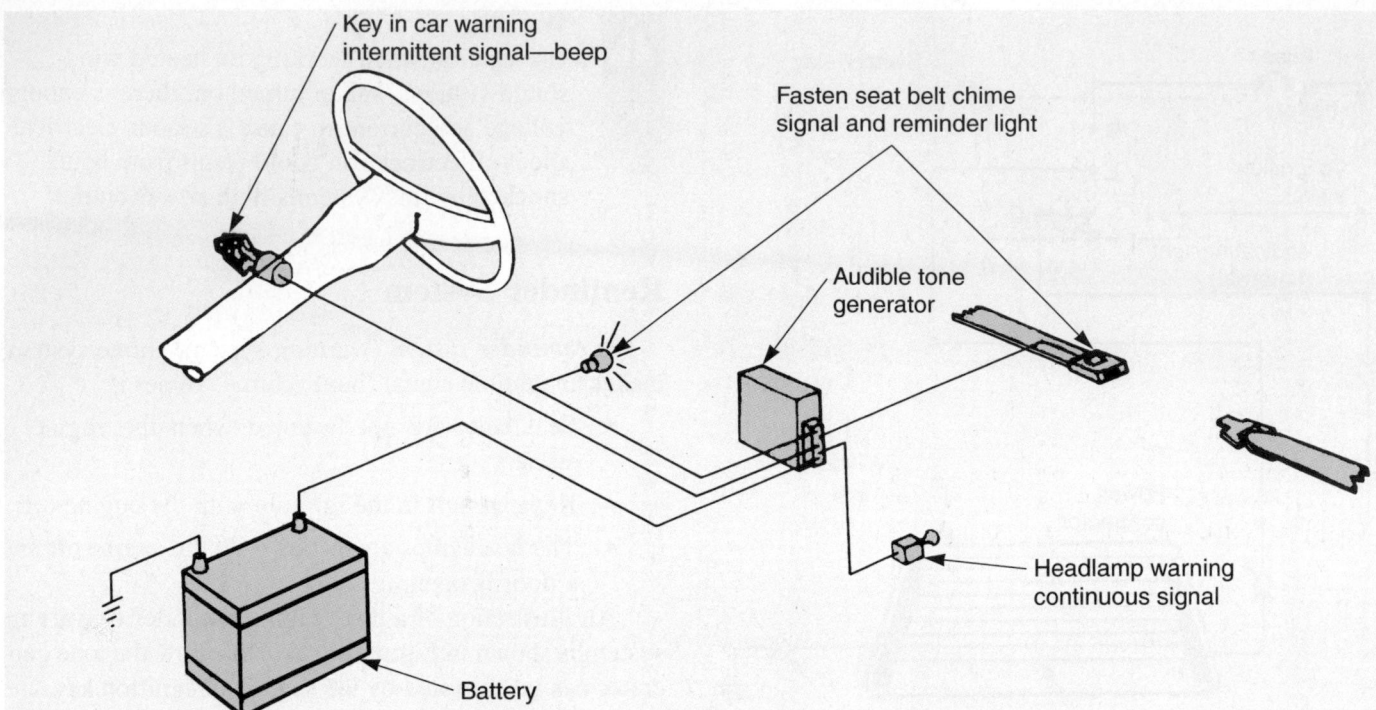

Figure 38-27. A warning or reminder system makes a sound when the key is left in the ignition, when seat belts are not fastened, and when lights are left on. Circuit in tone generator is the "heart" and "brain" of the system. (Plymouth)

the computer. The computer circuits use this signal to move the throttle actuator back and forth to maintain the same speed or vehicle sensor frequency.

The cruise control senses engine speed and controls the throttle valve opening on the engine. In this way, the driver can set the driving speed of the vehicle and the cruise control system will maintain that road speed.

For example, if the car starts to go up a hill, the vehicle speed will start to drop. The computer will detect a slower frequency signal from the speed sensor. It can then move the throttle actuator for more engine power to keep the car traveling at the preset speed. The opposite occurs if the car starts down a hill.

Although systems vary, a typical cruise control system consists of a speed control amplifier, speed control servo, speed control switch, speed sensor, and brake switch. See **Figure 38-28.**

When the driver activates the speed control, power is fed to the *speed control amplifier.* The amplifier then activates the *speed control servo*. The servo pulls on the throttle linkage to maintain engine power and vehicle speed. The *speed sensor* sends electrical pulses (speed information) to the amplifier.

If the vehicle starts to slow down when climbing a hill, the slower speed pulses cause the amplifier to make the servo open the throttle wider. This keeps the vehicle cruising at the correct speed. The opposite is true if speed increases (moving down a hill).

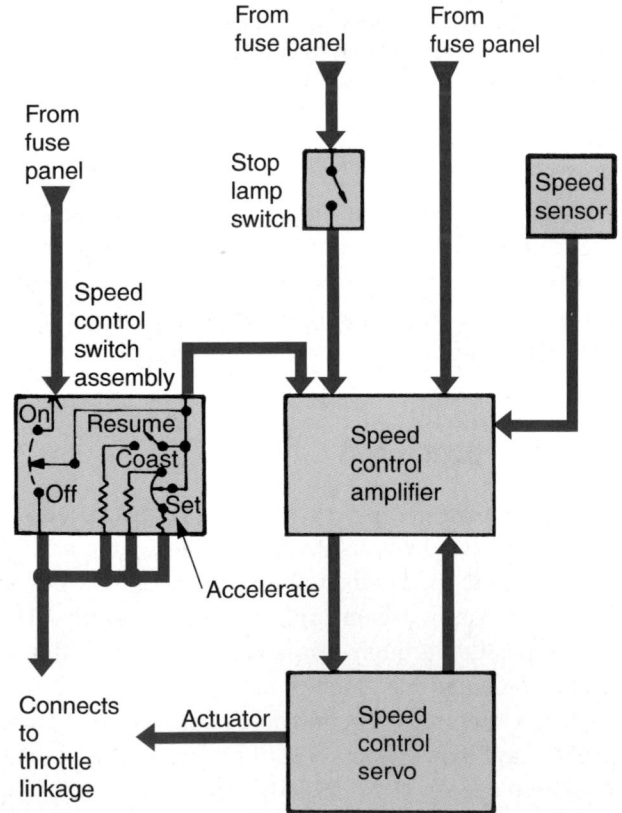

Figure 38-28. Layout of an electronic speed control system. Speed sensors send electrical information to control unit. Control unit can then operate servo, which connects to the throttle linkage on the engine. Also note the circuit in the speed control switch. (Ford)

When the driver presses on the brakes, the *brake switch* deactivates the speed control amplifier and the system. The *resume switch* allows the driver to reset the same cruising speed when desired.

Cruise Control System Service

Since speed or cruise control system designs vary so much, only general service methods can be given. Refer to a service manual for details.

If the speed control will not work at all, check the fuse and power feed wires to the system. Do not forget the simple things. For example, if a brake light switch is shorted on all the time, the speed control will not engage. The on and resume switches can also wear and fail. Test all simple things before suspecting the ECU and actuator.

Figure 38-29 is a typical electronic speed-control circuit diagram. When troubleshooting, you should study the one for the specific vehicle being repaired.

If the system has a vacuum actuator, check all vacuum lines for leaks. Use a piece of vacuum hose to listen for the hissing sound of a vacuum leak. You can apply a vacuum with a hand pump to the actuator diaphragm assembly. The diagram may rupture and leak after prolonged use. Any solenoid valves that control vacuum can also be tested using basic methods. See **Figure 38-30**.

Figure 38-31 shows a cutaway view of a motor-type speed-control actuator. It uses a small dc motor and gearbox to move the control arm and throttle cable. Check the motor for normal operation by connecting power to its windings. On older vehicles, the gears may be worn or stripped.

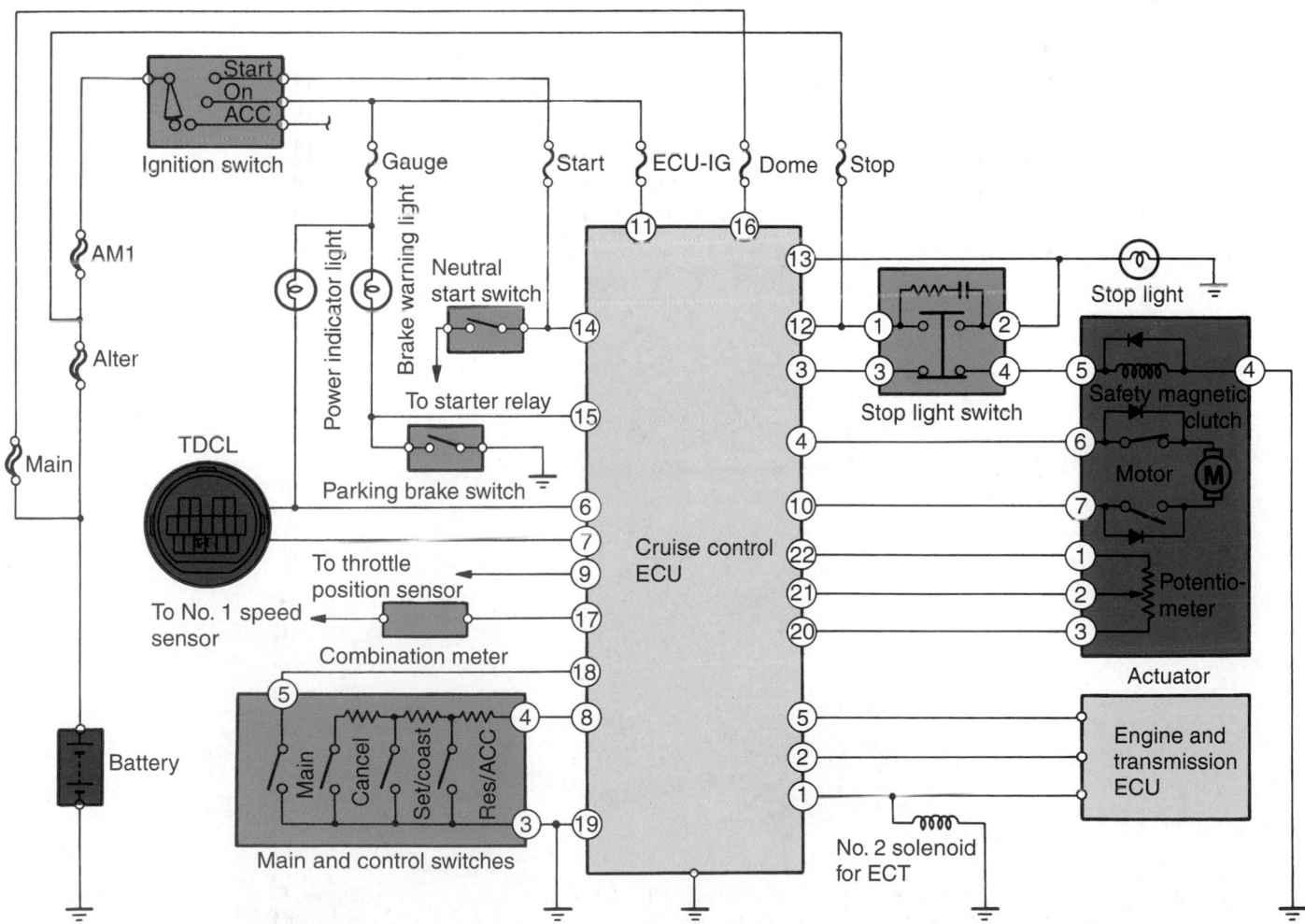

Figure 38-29. Study the operation of a typical electric cruise control system. Note the various inputs (on left) to the ECU and how the ECU can then operate the actuator (on right), which acts on the engine throttle to control vehicle speed. The vehicle speed sensor signal allows the ECU to maintain required road speed. (Lexus)

Warning
Use only recommended high-impedance testing instruments and factory-recommended procedures when servicing a cruise control system. If you do something wrong and the speed control will not disengage or not hold the correct speed, it could cause an accident. People could be killed by your mistake!

Power Mirrors

Power mirrors commonly use tiny reversible electric motors to tilt the side view mirror glass into different positions. A multiposition "joy switch" is normally used to energize each mirror motor. The mirror motors often operate a screw mechanism that can push or pull on the backside of the mirror glass, **Figure 38-32.**

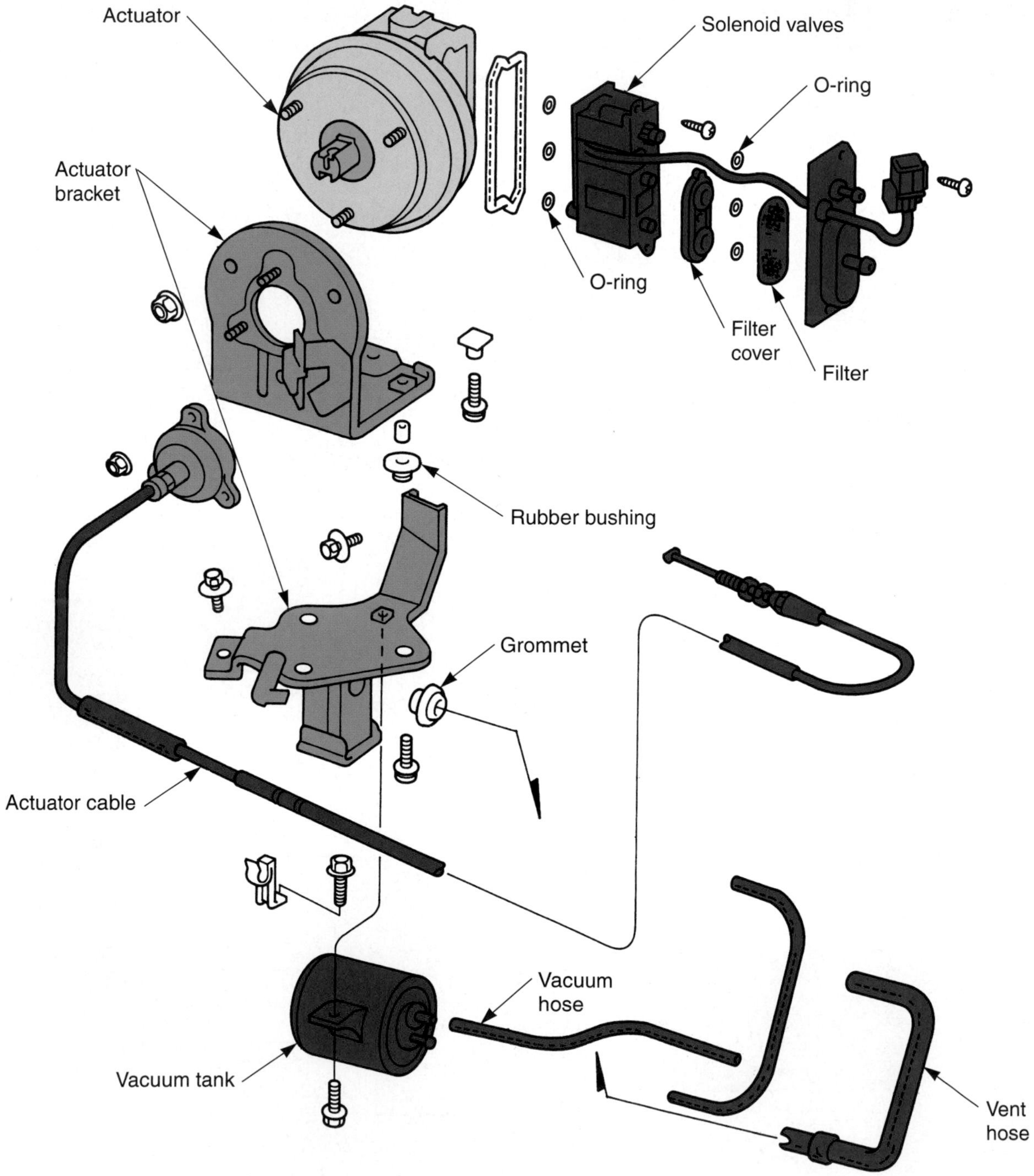

Actuator

Solenoid valves

O-ring

Actuator bracket

O-ring

Filter cover

Filter

Rubber bushing

Grommet

Actuator cable

Vacuum hose

Vacuum tank

Vent hose

Figure 38-30. This is an electro-pneumatic cruise control actuator. It uses small electric solenoids to control vacuum application to the actuator diaphragm. The diaphragm can then pull on the engine throttle lever to alter engine power output and vehicle road speed. Test the solenoids and test for a ruptured diaphragm if the system is inoperable. (Honda)

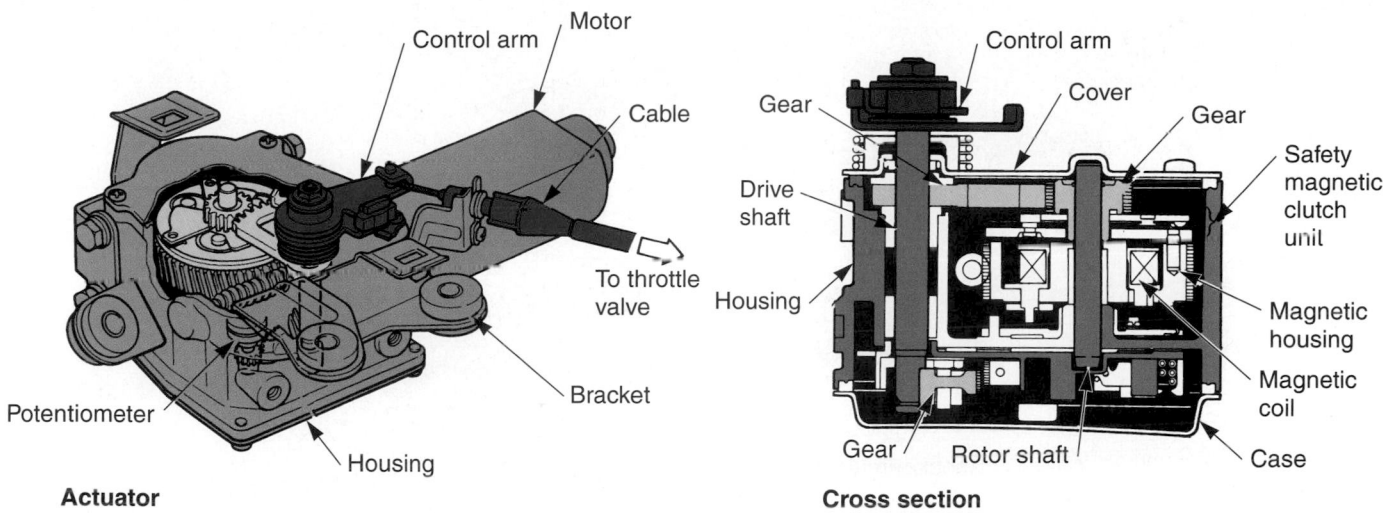

Actuator   **Cross section**

Figure 38-31. Cutaway view shows an electric motor-operated cruise control actuator. A reversible dc motor turns the worm gear. The larger driven gear has a lever that can pull or release a cable going to the engine throttle valve. The ECU can then spin the motor in the needed direction to maintain the desired road speed. (Lexus)

To service power mirrors, first check the fuse and power supply circuit if neither mirror works. If only one mirror fails to adjust, check its switch and motors. Usually, you will have to remove the door panel to test the switch, wiring, and motors.

A heat gun is often used to warm and remove the rearview mirror glass if cracked or if the motors can be replaced. Many manufacturers recommend replacement of the whole power mirror assembly if anything is wrong with it.

Cellular Mobile Telephone

A *cellular mobile telephone* uses a transceiver to send and pick up radio waves to communicate with other people using phone station towers. This is a common factory-installed accessory on many high-end vehicles. Some mobile phones have a small microphone so the phone can be used hands-free. A speaker can also be added so the phone handset does not have to be held to your ear.

Figure 38-33 shows the major parts of a mobile phone system. Refer to the service manual for details of working on mobile phones. Many are repaired by a specialized phone technician.

Driver Information Center (Voice Alert System)

A *driver information center* uses computer, a small speaker or digital display, and numerous sensors and switches to inform the driver of various conditions. The computer is programmed to output a display or spoken words when a specific sensor detects a bad condition. This might include displaying—or even saying—"Your washer fluid is low," "Check your brake fluid," "Your door is ajar," and similar messages.

Figure 36-34 shows the major parts of a driver information center. Note the numerous sensors that monitor vehicle conditions so the computer can display information in the monitor.

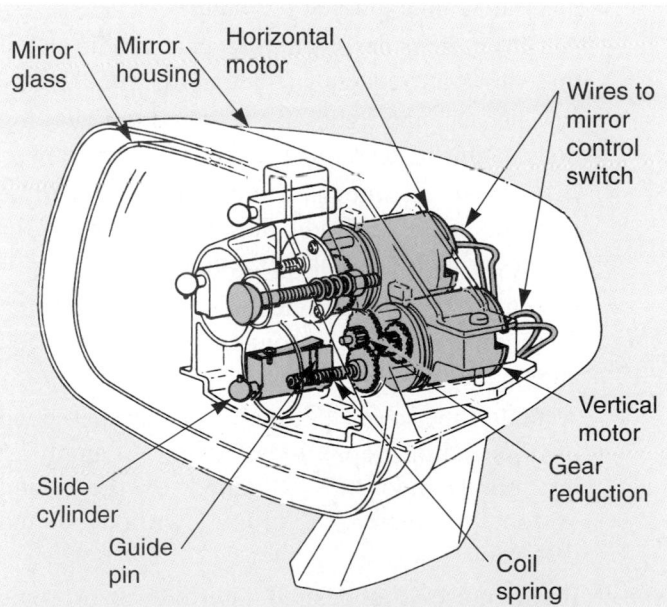

Figure 38-32. This power mirror uses tiny dc motors that drive screw mechanisms to adjust the rearview mirror angle. It is often serviced as a complete assembly and internal repairs are not done. (Chrysler)

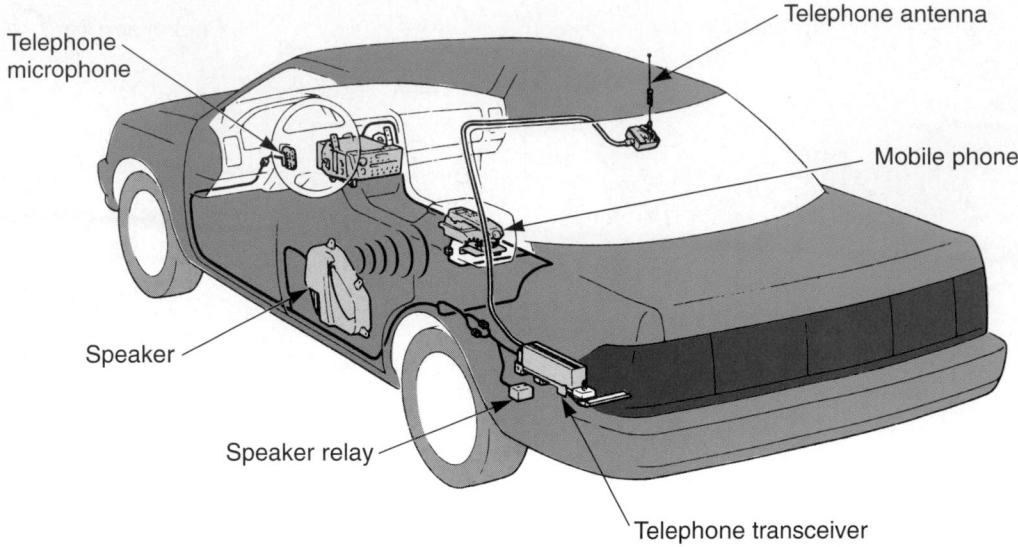

Figure 38-33. Note the major parts of a cellular mobile telephone system. The transceiver is located in the trunk. Speaker and microphone allow for hands-free use of telephone. (Lexus)

Figure 38-35 gives a basic wiring diagram for a vehicle monitoring system. It is also capable of relaying audio messages stored in computer memory.

Summary

- A basic radio system consists of an antenna, radio (receiver-amplifier), speaker(s), and power supply circuit.
- An AM radio is designed to pick up a radio signal that varies in amplitude (strength).
- An FM radio is designed to receive a radio signal that varies in frequency (fluctuating speed).
- Just because the lights in the stereo are working does not mean the radio circuitry is getting power.
- If provided, an antenna trimmer screw should be adjusted when the radio has been removed for repairs or after antenna replacement.
- If you suspect a bad antenna (no signal coming from antenna lead), connect a known good antenna to the radio.
- A faulty speaker will usually distort the sound of the radio. Sometimes, a stereo amplifier or "power booster" is added to the sound system to increase volume without sound distortion.
- Radio noise is undesired interference or static (popping, clicking, or crackling) obstructing the normal sound of the radio station.
- Power seats use several switches, electric motors and drive assemblies to change the front seat positions.

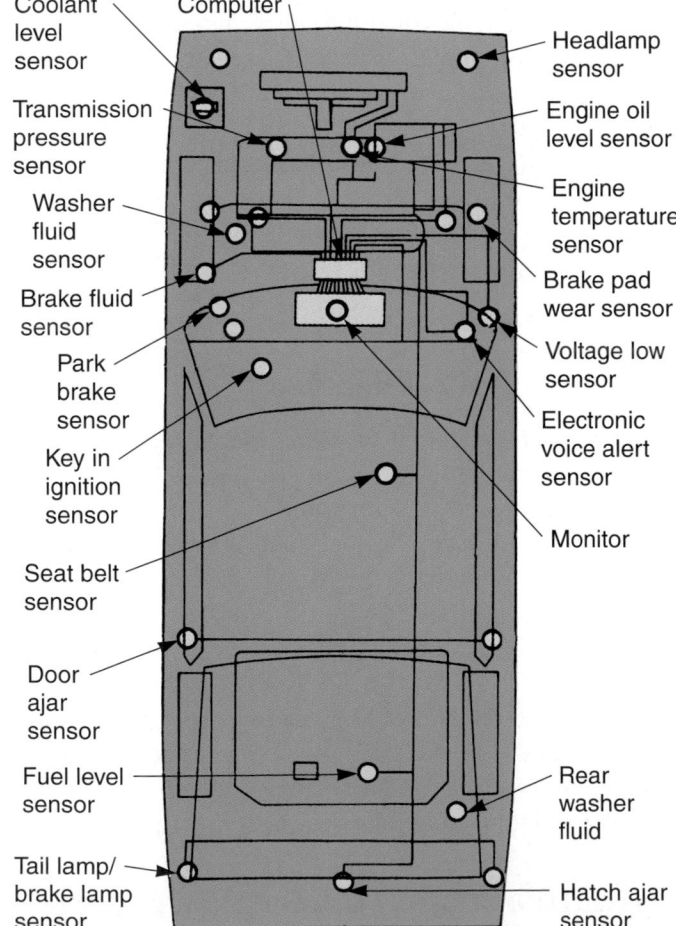

Figure 38-34. Top view of car shows general locations of components in this driver information center. A keyboard is often provided so driver can request information. An on-board computer can then use sensor data to calculate answer to be displayed on the monitor or display window.

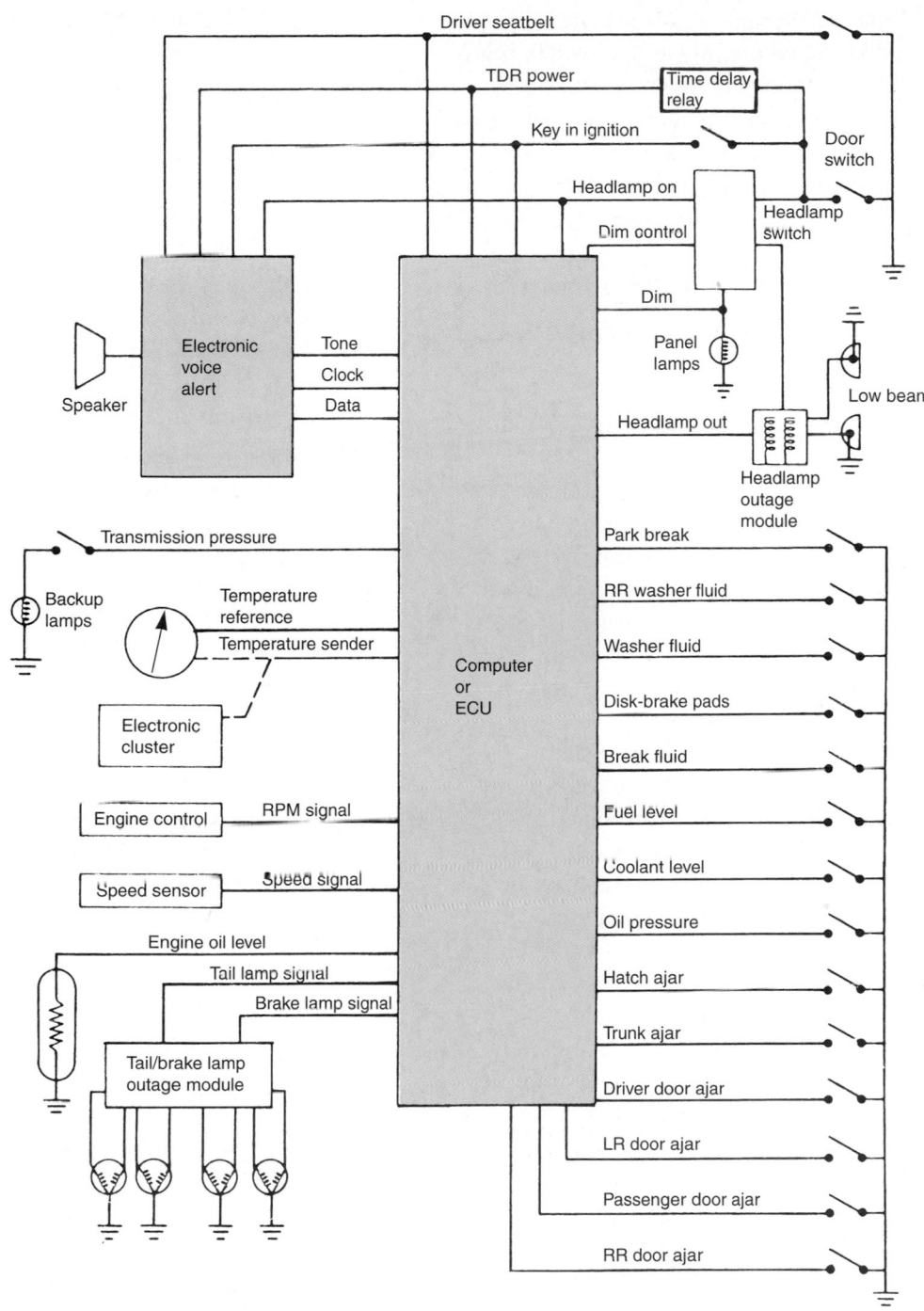

Figure 38-35. This simplified wiring diagram shows how the ECU processes inputs and produces outputs for voice alert system. Note the speaker at upper left. With this system, your car will actually talk to you to inform you of many operating conditions or possible problems, such as low fluid levels, fluid change needed, parking brake on, door ajar, and low oil pressure. (Chrysler)

- Memory seats use a small computer or ECU which can be programmed to remember seat positions for several people.
- A power window basically uses a control switch, reversible electric motor, circuit breaker, fuse, and related wiring to operate the door windows.
- Power door locks typically use an electric switch and a solenoid or motor to operate the door lock mechanisms.
- A power steering wheel uses an ECU, various switches, sensors, and motors to automatically tilt and telescope (extend or retract) the steering wheel.

- A rear window defogger, also called a rear window defroster, commonly uses a switch, relay, indicating light, and a window heating grid.
- A modern electronic cruise control system uses a computer, sensors, and a throttle actuator to maintain vehicle speed when highway driving.
- A driver information center uses numerous sensors, switches, a computer, and a small speaker or digital display to inform the driver of various conditions.

Important Terms

Radio system	Transmission
Radio station	Power door locks
Sound system	Power trunk release
Radio	Power steering wheel
Tuner	Rear window defogger
Radio frequencies	Rear window defroster
AM radio	Zone clearing defoggers
FM radio	Grid repair agent
Antenna trimmer	Heated windshield
screw	system
Antenna	Reminder system
Mast	Cruise control system
Power antenna	Power switches
Speaker	Control switch
Tape player	Vehicle speed sensor
Compact disc (CD)	Cruise computer
players	Throttle actuator
Stereo amplifier	Brake light switch
Power booster	Clutch switch
Steering wheel touch	Neutral safety switch
controls	Speed control amplifier.
Radio noise	Speed control servo
Noise suppressor	Speed sensor
Power seats	Brake switch
Memory seats	Resume switch
Power seat module	Power mirrors
Heated seats	Cellular mobile
Power windows	telephone
Power window motor	Driver information
Window regulator	center
Gearbox	

Review Questions—Chapter 38

Please do not write in this text. Place your answers on a separate sheet of paper.

1. What are the basic parts of a basic radio system?

2. Explain the difference between AM and FM signals.

3. A customer complains of poor radio reception (static in speakers). The radio has just been replaced. All electrical leads, including the antenna, are properly connected. Technician A says that the antenna trimmer screw may need adjustment. Technician B says that the new radio must be defective. Who is right?
   (A) A only.
   (B) B only.
   (C) Both A and B.
   (D) Neither A nor B.

4. A(n) _____ uses a permanent magnet and coil of wire mounted on a flexible diaphragm.

5. What are some of the symptoms of a faulty speaker?

6. Why are noise suppressors used?

7. If one power window fails to go up and down, check the circuit fuse first. True or False?

8. When a power window switch is pressed, a whirring sound can be heard from inside the door. Technician A says that the motor windings are open and the motor is bad. Technician B states that the power window gearbox might have stripped plastic gear teeth. Who is right?
   (A) A only.
   (B) B only.
   (C) Both A and B.
   (D) Neither A nor B.

9. How does a heated windshield system work?

10. Explain the operation of a modern cruise control system.

● ASE-Type Questions

1. Technician A says an AM radio is designed to pick up a radio signal that varies in amplitude. Technician B says an AM radio is designed to pick up a radio signal that varies in frequency. Who is right?
   (A) A only.
   (B) B only.
   (C) Both A and B.
   (D) Neither A nor B

2. Technician A says an FM radio has a longer broadcasting range than an AM radio. Technician B says an AM radio has a longer broadcasting range than an FM radio. Who is right?
   (A) A only.
   (B) B only.
   (C) Both A and B.
   (D) Neither A nor B.

3. Technician A says an AM radio is capable of producing stereophonic sound. Technician B says an FM radio is capable of producing stereophonic sound. Who is right?
   (A) A only.
   (B) B only.
   (C) Both A and B.
   (D) Neither A nor B.

4. Technician A says that the circuits feeding power should be considered when troubleshooting an automotive radio system. Technician B says a receiver-amplifier is a basic circuit inside a radio. Who is right?
   (A) A only.
   (B) B only.
   (C) Both A and B.
   (D) Neither A nor B.

5. A car's radio is being reinstalled after servicing. Technician A adjusts the radio's antenna trimmer screw after installation. Technician B says the antenna trimmer screw should be left in its original position before it was removed from the car. Who is right?
   (A) A only.
   (B) B only.
   (C) Both A and B.
   (D) Neither A nor B.

6. Which of the following cannot cause radio noise?
   (A) Bad alternator diodes.
   (B) Open spark plug wire.
   (C) Bad noise suppressor.
   (D) All of the above.

7. A car's power antenna is inoperative. Technician A checks for vacuum leaks which may affect the antenna's operation. Technician B checks the antenna's motor and electrical circuit. Who is right?
   (A) A only.
   (B) B only.
   (C) Both A and B.
   (D) Neither A nor B.

8. Technician A says an AM car radio normally has two speakers mounted in the top of the dash. Technician B says an AM car radio normally has one speaker mounted in the top of the dash. Who is right?
   (A) A only.
   (B) B only.
   (C) Both A and B.
   (D) Neither A nor B.

9. An automobile's radio is producing "radio noise." Technician A checks the radio's antenna. Technician B looks for a bad spark plug wire. Who is right?
   (A) A only.
   (B) B only.
   (C) Both A and B.
   (D) Neither A nor B.

10. Technician A says an automotive "noise suppressor" can be mounted inside the car's alternator. Technician B says an automotive "noise suppressor" can be mounted on the car's ignition coil. Who is right?
    (A) A only.
    (B) B only.
    (C) Both A and B.
    (D) Neither A nor B.

11. One of the power seats in an automobile is not operating. Technician A tests the seat's electrical switch. Technician B checks the operation of the seat's electrical motor. Who is right?
    (A) A only.
    (B) B only.
    (C) Both A and B.
    (D) Neither A nor B.

12. One of an automobile's power windows is not working. Technician A checks the operation of the window's electric motor. Technician B checks for a blown fuse at the car's fuse box. Who is right?
    (A) A only.
    (B) B only.
    (C) Both A and B.
    (D) Neither A nor B

13. All of an automobile's power door locks are inoperative. Technician A looks for a blown fuse at the fuse box. Technician B checks for a faulty solenoid in the left-rear door panel. Who is right?
    (A) A only.
    (B) B only.
    (C) Both A and B.
    (D) Neither A nor B.

14. A car's rear window defogger is malfunctioning. Technician A tests the operation of the defogger's "heating grid." Technician B checks the defogger's electrical switch. Who is right?
    (A) A only.
    (B) B only.
    (C) Both A and B.
    (D) Neither A nor B.

15. One of a car's door panels must be removed. Technician A removes the window crank handle. Technician B removes the arm rest screws. Who is right?
    (A) A only.
    (B) B only.
    (C) Both A and B.
    (D) Neither A nor B.

Activities—Chapter 38

1. Use a car radio to "hunt" for radio stations broadcasting from other communities, on both the AM and FM bands. Note the call letters and towns, then mark them on a map. Check the distance the farthest signal traveled. Was it AM or FM?

2. Take a survey among students at your school to determine which vehicle music system options are most popular. Determine how many student cars have just radios, radios with cassette players, radios with CD changer/players, etc. Make a bar graph to show your results.

Sound System Diagnosis		
Condition	**Possible causes**	**Correction**
Nothing works.	1. Blown fuse. 2. Bad switch. 3. Broken feed wire. 4. Broken ground wire. 5. Bad radio.	1. Replace fuse. 2. Replace switch. 3. Repair wire. 4. Repair wire. 5. Send out for service.
Radio powers up, static or no sound.	1. Poor antenna connection. 2. Misadjusted trim screw. 3. Disconnected speakers. 4. Bad radio.	1. Repair antenna. 2. Adjust antenna trim screw. 3. Fix wiring. 4. Send out for service.
Distorted sound from speakers.	1. Blown speakers. 2. Loose speaker mounting. 3. Speaker polarity reversed. 4. Incorrect speaker ohms. 5. Bad radio.	1. Replace speakers. 2. Tighten fasteners. 3. Reverse wires to speaker. 4. Match speaker resistance to radio. 5. Send out for service.
Poor reception.	1. Poor antenna connection. 2. Misadjusted trim screw. 3. Bad radio.	1. Repair antenna. 2. Adjust trim screw. 3. Send out for service.
Popping or clicking noise in speakers.	1. Radio interference. 2. Bad antenna. 3. Bad spark plug wires. 4. Missing ground straps.	1. Replace noise suppressors. 2. Replace antenna. 3. Replace spark plug wires. 4. Install ground straps.

Power Accessory Diagnosis		
Condition	**Possible causes**	**Correction**
Nothing works.	1. Blown fuse. 2. Open switch. 3. No power to circuit.	1. Replace fuse. 2. Replace switch. 3. Repair feed wire,
Intermittent problem.	1. Loose wires. 2. Bad switch.	1. Repair wiring. 2. Replace switch.

Cooling and Lubrication Systems

The cooling and lubrication systems are designed to prevent engine damage and wear. The cooling system removes excess combustion heat and maintains a constant engine operating temperature. The lubrication system reduces friction and wear between internal engine parts. If these systems are not operating properly, an engine will destroy itself in a matter of minutes. Aluminum pistons can literally melt and weld themselves to the cylinder walls, bearings can seize, and major parts can crack or warp.

This section details the operation, construction, diagnosis, and repair of modern cooling and lubrication systems. Study it carefully! This information will help you pass ASE Test A1, *Engine Repair,* and Test A8, *Engine Performance.*

Chapter 39

Cooling System Fundamentals

After studying this chapter, you will be able to:

- Summarize the functions of a cooling system.
- Explain the operation and construction of major cooling system components.
- Compare cooling system design variations.
- Explain the importance of antifreeze.
- Discuss safety procedures to follow when working with cooling systems.
- Correctly answer ASE certification test questions on cooling system construction and operation.

This chapter explains the design, construction, and operation of cooling systems. You must fully understand how different cooling systems work before learning to service and repair them.

A *cooling system* must control the operating temperature of the engine. During startup, it must help the engine warm to operating temperature quickly to lower emissions. After the engine has warmed up, the cooling system must maintain a constant engine operating temperature for maximum efficiency.

Without a properly operating cooling system, an engine can "self-destruct" in a matter of minutes. Internal engine parts can overheat and partially melt, causing the engine to "lock up," or seize (reciprocating assembly no longer moves freely). For this and other reasons, it is important that you fully understand the information in this chapter.

Cooling System Functions

A cooling system has several functions. It must remove excess heat from the engine, maintain a constant engine operating temperature, increase the temperature of a cold engine quickly, and provide a means for warming the passenger compartment.

Removing Engine Heat

The burning air-fuel mixture produces a tremendous amount of heat. Combustion flame temperatures can reach 4500°F (2500°C). This is enough heat to melt metal parts.

Some combustion heat is used to produce expansion and pressure for piston movement. Most combustion heat flows out with the exhaust gas or flows into the metal parts of the engine. Without removal of this excess heat, the engine would be seriously damaged in a matter of minutes.

Maintaining Operating Temperature

Engine operating temperature is the temperature the engine coolant (water and antifreeze mixture) reaches under normal running conditions. Typically, an engine's operating temperature is between 180°F and 210°F (80°C and 100°C).

When an engine warms to operating temperature, its parts expand. This ensures that all part clearances are correct. It also ensures proper combustion, emission output levels, and engine performance.

Reaching Operating Temperature Quickly

An engine must warm up rapidly to prevent poor combustion, part wear, oil contamination, reduced fuel economy, increased emissions, and other problems. A cold engine suffers from several problems.

For instance, the aluminum pistons in a cold engine will not be expanded by heat. This can cause too much clearance between the pistons and the cylinder walls. The oil in a cold engine will also be very thick. This can reduce lubrication and increase engine wear. The air-fuel mixture will not vaporize and burn as efficiently in a cold engine.

Heater Operation

A cooling system commonly circulates coolant to the vehicle's heater. Since the engine coolant is warm, its heat can be used to warm the passenger compartment. See **Figure 39-1.**

Note

Refer to Chapter 75, *Heating and Air Conditioning Fundamentals,* and Chapter 76, *Heating and Air Conditioning Service,* for more information on heaters.

Cooling System Operation

When the engine is running, the *water pump* forces coolant to circulate through the engine *water jackets* (internal passages in the engine). A drive belt often powers the water pump. The water pump can also be gear-driven off the crankshaft.

While the engine is cold, the thermostat remains closed, so coolant circulates inside the engine. This helps warm the engine quickly.

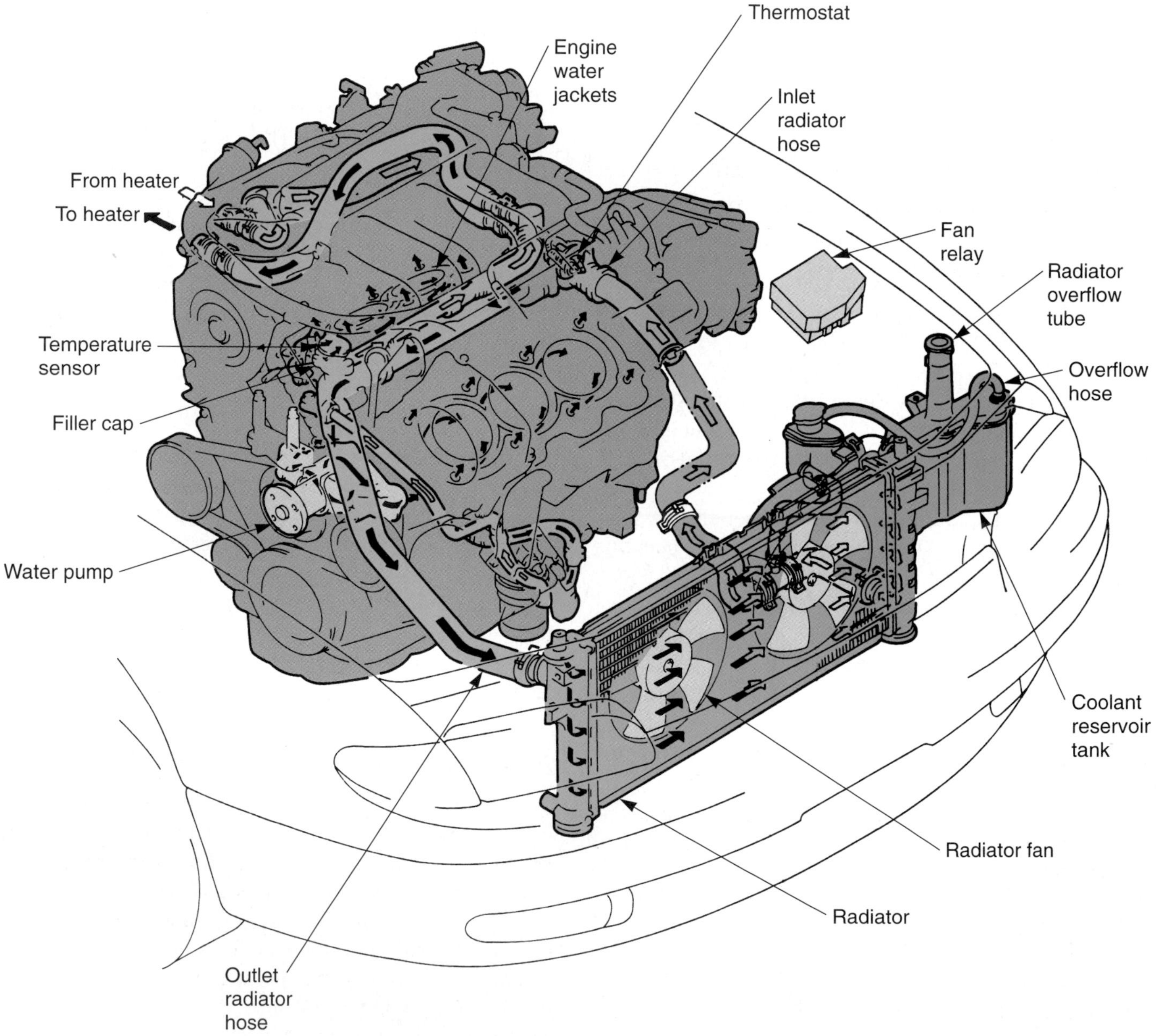

Figure 39-1. Study the basic names and locations for parts of a cooling system. This will help you as each part is explained in detail. (Mazda)

When the engine reaches operating temperature, the thermostat opens. Heated coolant then flows through the radiator. Excess heat is transferred from the coolant to the air flowing through the radiator.

Cooling System Types

There are two major types of automotive cooling systems: air cooling systems and liquid cooling systems.

Air Cooling Systems

An *air cooling system* uses large cylinder cooling fins and outside air to remove excess heat from the engine. The *cooling fins* increase the surface area of the metal around the cylinder. This allows enough heat to transfer from the cylinder to the outside air. Look at **Figure 39-2.**

An air cooling system commonly uses plastic or sheet metal ducts and *shrouds* (enclosures) to route air over the cylinder fins. Thermostatically controlled flaps regulate airflow and engine operating temperature.

Tech Tip
Air-cooled automotive engines are rare. Most late-model vehicles have liquid-cooled engines.

Liquid Cooling Systems

A *liquid cooling system* circulates coolant (a solution of water and antifreeze) through the water jackets. The coolant then collects excess heat and carries it out of the engine.

Figure 39-2 compares liquid and air cooling. **Figure 39-3** shows how combustion heat is transferred into the coolant.

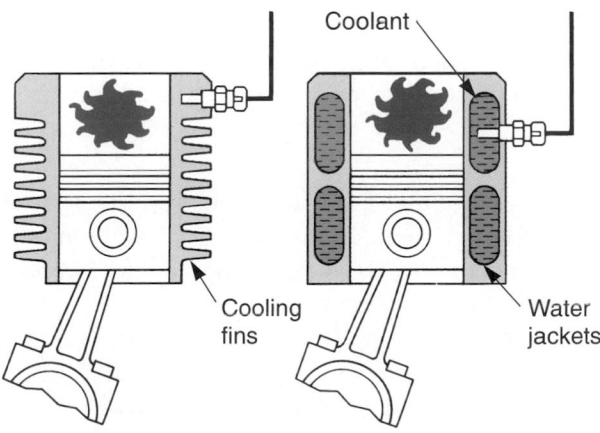

Figure 39-2. An air-cooled engine has large fins on the cylinder to dissipate heat into surrounding air. A water-cooled engine has water jackets around each cylinder to collect heat. (Robert Bosch)

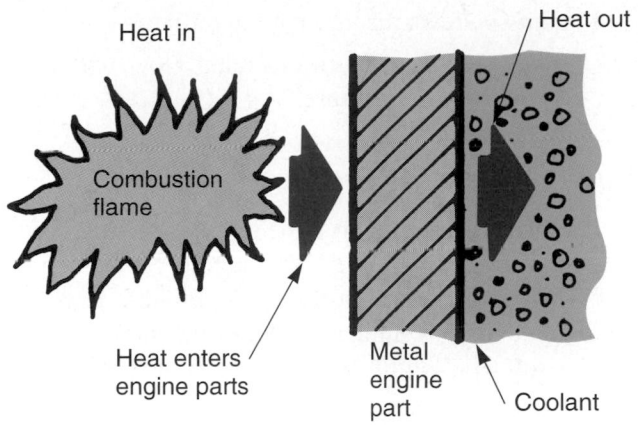

Figure 39-3. Combustion heat transfers into the cylinder wall and then into the coolant. Coolant carries heat away from the engine.

A liquid cooling system has several advantages over an air cooling system, including:

- More precise control of engine operating temperature.
- Less temperature variation inside engine.
- Reduced exhaust emissions because of better temperature control.
- Improved heater operation to warm passengers.

Conventional and Reverse Flow Cooling

With *conventional coolant flow,* hot coolant flows from the cylinder head to the radiator. After being cooled in the radiator, the coolant flows back into the engine block. This is the most common coolant flow direction.

Reverse flow cooling follows the opposite course: cool coolant enters the head and hot coolant exits the block to return to the radiator. Reversing the flow of the coolant helps keep a more uniform temperature throughout the engine, especially around the hot exhaust valves. Reverse flow cooling can be found on high-performance engines.

Basic Cooling System

The basic parts of a cooling system are shown in **Figure 39-1.** Refer to this illustration as each part is introduced.

- *Water pump*—Forces coolant through the engine and other system parts.
- *Radiator hoses*—Connect the engine to the radiator.
- *Radiator*—Transfers engine coolant heat to outside air.

- *Fan*—Draws air through the radiator.
- *Thermostat*—Controls coolant flow and engine operating temperature.

Water Pump

The *water pump* is an impeller or centrifugal pump that forces coolant through the engine block, cylinder head, intake manifold, hoses, and radiator. It is often driven by a belt running off the crankshaft pulley. In some cases, the pump is gear-driven directly off the crankshaft.

Water pump impellers can be made of steel or plastic. The impeller blades can be curved or straight. Straight blades, like paddle wheels, are sometimes used to reduce engine power consumption. Look at **Figure 39-4.**

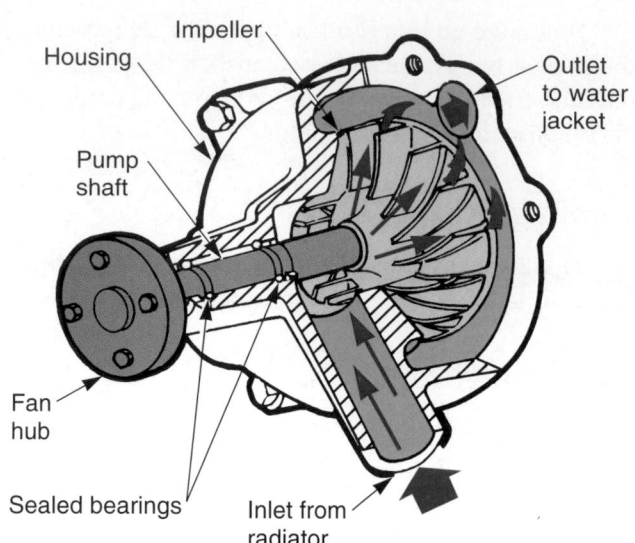

Figure 39-5. Cutaway of a simplified water pump. Note how the spinning impeller throws coolant outward to produce pressure and flow. (Mopar)

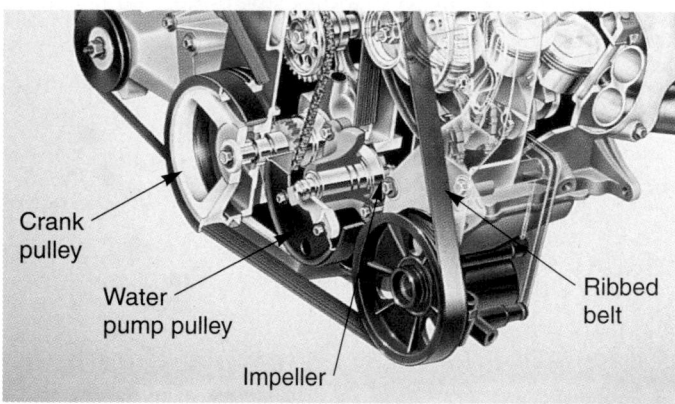

Figure 39-4. The fan belt turns the water pump pulley to operate the pump. This is a modern ribbed belt that powers all accessory units. (Ford)

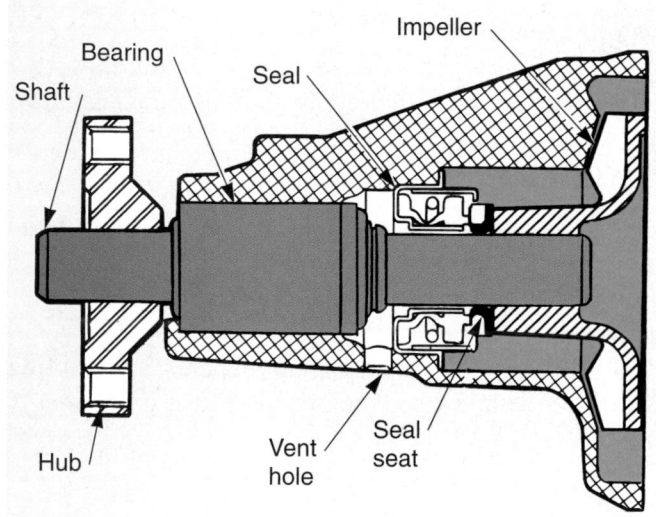

Figure 39-6. Side cutaway view of water pump shows how the seal keeps coolant from leaking out of the vent hole. (Chevrolet)

The major parts of a typical water pump include the:

- *Water pump impeller*—Disk with fan-like blades, the impeller spins and produces pressure and flow, **Figure 39-5.**
- *Water pump shaft*—Steel shaft that transfers turning force from the hub to the impeller.
- *Water pump seal*—Prevents coolant leakage between pump shaft and pump housing, **Figure 39-6.**
- *Water pump bearings*—Plain or ball bearings that allow the pump shaft to spin freely in housing.
- *Water pump hub*—Provides mounting place for belt pulley and fan.
- *Water pump housing*—Iron or aluminum casting that forms the main body of pump.

The water pump normally mounts on the front of the engine. With some transverse (sideways-mounted) engines, it bolts to the side of the engine and extends toward the front.

A *water pump gasket* fits between the engine and pump housing to prevent coolant leakage. RTV sealer or a rubber seal may be used instead of a gasket.

Water Pump Operation

Figure 39-7 illustrates water pump action and coolant flow through an engine. The spinning engine

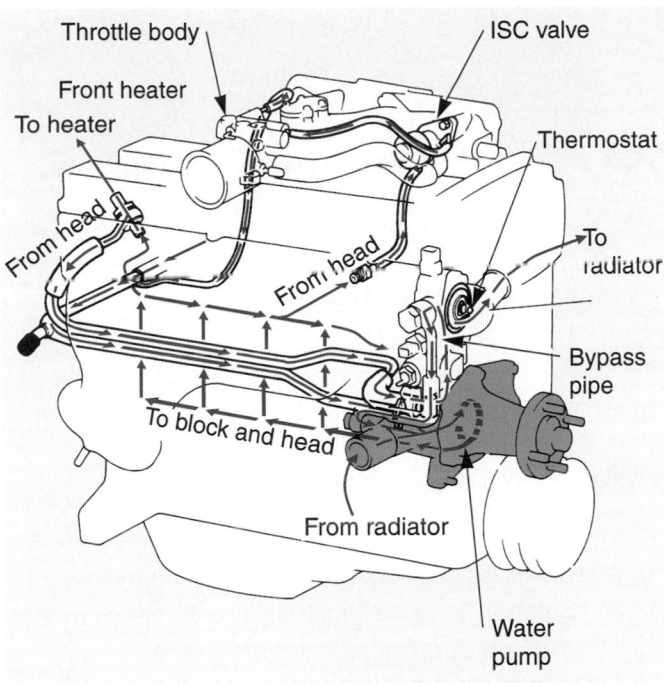

Figure 39-7. The water pump pulls coolant out of the bottom of the radiator and through the engine block, heads, and intake manifold. Hot coolant then reenters the radiator for cooling. (Ford)

crankshaft pulley causes the drive belt to turn the water pump pulley, pump shaft, and impeller. The coolant trapped between the impeller blades is thrown outward by centrifugal force. This produces suction (vacuum) in the central area of the pump housing. It also produces pressure in the outer area of the housing.

Since the pump inlet opening is near the center, coolant is pulled out of the radiator, through the lower hose, and into the engine. After being thrown outward and pressurized, the coolant flows into the engine. It circulates through the block, around the cylinders, up through the cylinder head(s), through the thermostat, and back into the radiator.

Radiator and Heater Hoses

Radiator hoses carry coolant between the engine water jackets and the radiator. Being flexible, hoses can withstand the vibrating and rocking of the engine on its motor mounts without breakage. Look at **Figure 39-8.**

The upper radiator hose normally connects to the thermostat housing on the intake manifold or cylinder head. Its other end fits on the radiator. The lower hose often connects the water pump inlet and the radiator.

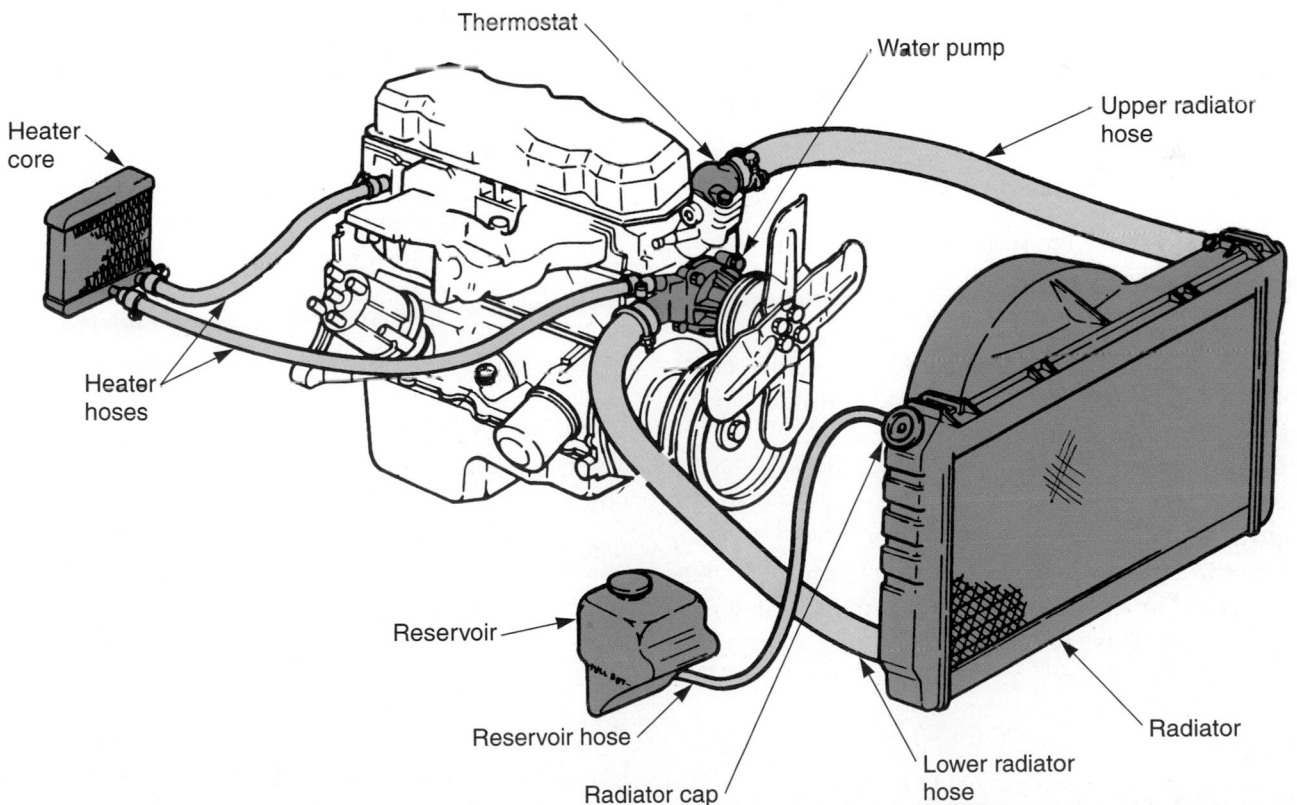

Figure 39-8. Radiator hoses carry coolant between the engine and the radiator. Heater hoses carry hot coolant to heater core in the passenger compartment, under the dash. (Peugeot)

A *molded hose* is manufactured in a special shape, with bends to clear the cooling fan and other parts. It must be purchased to fit the exact year and make of car. See **Figure 39-9.**

A *flexible hose* has an accordion shape and can be bent to different angles. The pleated construction allows the hose to bend without collapsing and blocking flow. The flexible hose is also called a universal-type radiator hose.

A *hose spring* is frequently used in the lower radiator hose to prevent its collapse. The lower hose is exposed to suction from the water pump. The spring ensures that the inner lining of the hose does not tear away, close up, and stop circulation.

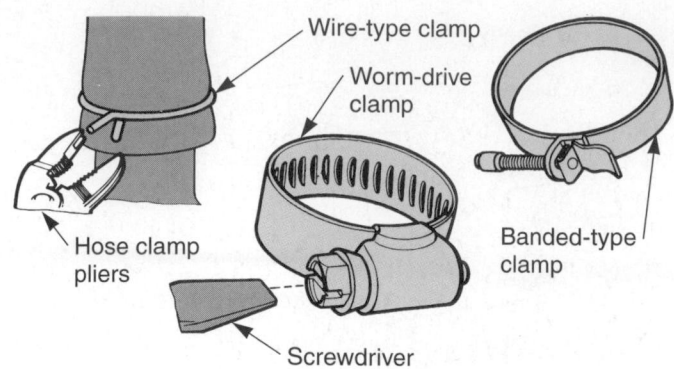

Figure 39-10. Three basic types of hose clamps. Worm drive clamp is the most common. Spring-type clamp requires hose clamp pliers with a groove cut in the jaws. (Mopar)

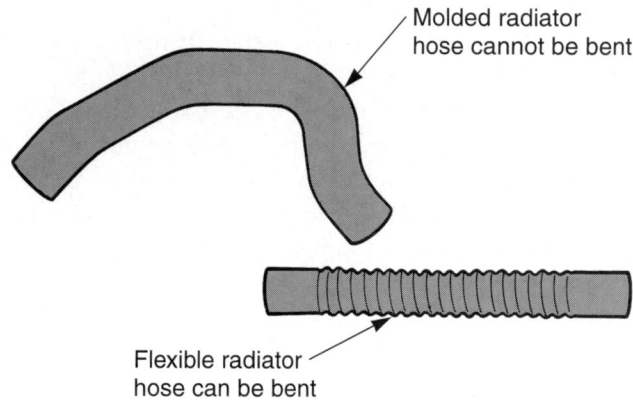

Figure 39-9. Two basic types of radiator hoses. (Chrysler)

Caution
Never remove the spring from the inside of a radiator hose. If you do, the hose can collapse and cause engine overheating damage.

Heater hoses are small-diameter hoses that carry coolant to the *heater core* (small radiator-like device under the dash). Refer to **Figure 39-8.**

Hose clamps hold the radiator hoses and heater hoses on their fittings. Three types of hose clamps are pictured in **Figure 39-10.**

A *worm-drive hose clamp* uses a worm gear that engages slots in the clamp strap to allow tightening around the hose. It is the most common type of replacement hose clamp.

Radiator

The *radiator* transfers coolant heat to the outside air. The radiator is normally mounted in the front of the engine. Cool outside air can then flow freely through it. See **Figure 39-11.**

A radiator typically consists of five components:

- *Radiator core*—Center section of the radiator. Made up of tubes and cooling fins.
- *Radiator tanks*—Metal or plastic ends that fit over the core tube ends to provide storage for coolant and fittings for hoses.
- *Radiator filler neck*—Opening for adding coolant. Also holds the radiator cap and overflow tube.
- *Transmission oil cooler*—Inner tank for cooling automatic transmission or transaxle fluid.
- *Radiator petcock*—Fitting on the bottom of the tank for draining coolant.

Radiator Action

Under normal operating conditions, hot engine coolant circulates through the radiator tanks and core tubes. Heat transfers into the core's tubes and fins. Cooler air flows over and through the radiator fins, so heat is removed from the radiator. This reduces the temperature of the coolant before it flows back into the engine.

Radiator Types

The two types of radiators are the crossflow and the downflow. Both are shown in **Figure 39-12.**

The tanks on a *downflow radiator* are on the top and bottom of the core, and the core tubes run vertically between the tanks. Hot coolant from the engine enters the top tank. The coolant flows downward through the core tubes. After cooling, the coolant flows out of the bottom tank and back into the engine.

A *crossflow radiator* is a more modern design that has its tanks on the sides of the core. The core tubes are arranged for horizontal coolant flow. The tank with the radiator cap is normally the outlet tank. A crossflow radiator can be shorter than a downflow radiator, allowing for a lower hood line. Look at **Figure 39-12.**

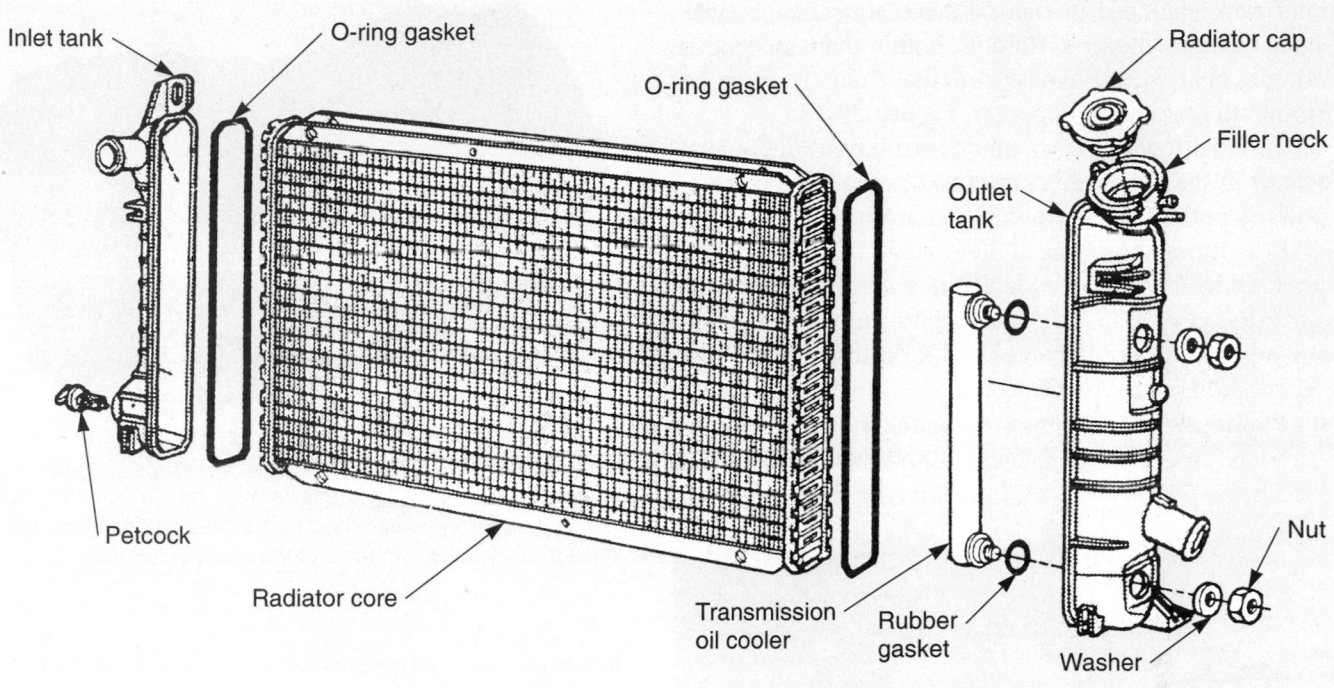

Figure 39-11. Exploded view of the major parts of a cooling system. (General Motors)

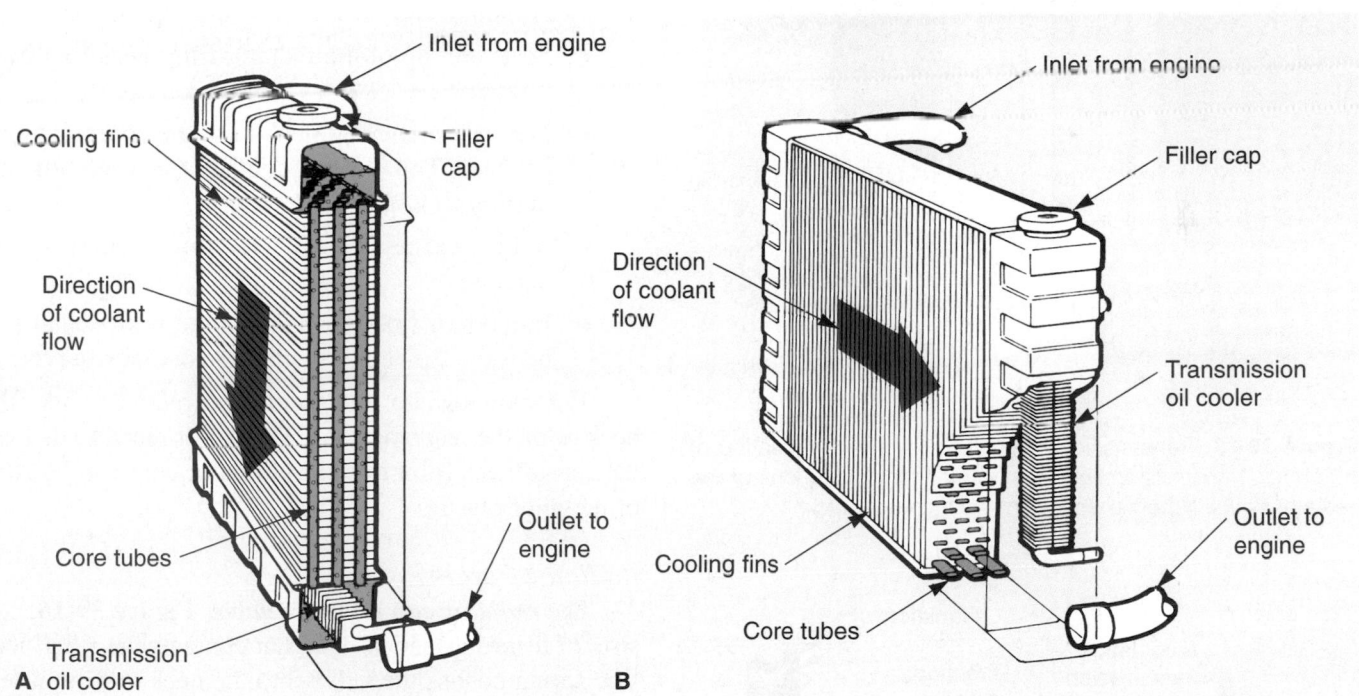

Figure 39-12. Two types of radiators. A—A downflow radiator has core tubes running up and down. B—A crossflow radiator has cooling tubes running horizontally. The crossflow radiator is more common on late-model cars. (Chrysler)

Radiator tanks can be made of metal or plastic. With metal radiator tanks, the core and tanks are soldered together. With plastic radiator tanks, rubber seals fit between the tanks and core to prevent leakage.

Transmission Oil Cooler

A *transmission oil cooler* is often placed in the radiator on cars with automatic transmissions or transaxles to prevent the transmission fluid from overheating. It is a small tank enclosed in one of the main radiator tanks. Since the transmission fluid is hotter than the engine coolant, heat is removed from the fluid as it passes through the radiator and cooler, **Figure 39-13.**

In downflow radiators, the transmission oil cooler is located in the lower tank. In crossflow radiators, the oil cooler is in the tank having the radiator cap.

Line fittings from the cooler extend through the radiator tank to the outside. Metal lines from the automatic transmission or transaxle connect to these fittings. The transmission oil pump forces the fluid through the lines and cooler, **Figure 39-14.**

Figure 39-15 shows how the radiator can be mounted next to the air-conditioning condenser. With other vehicles,

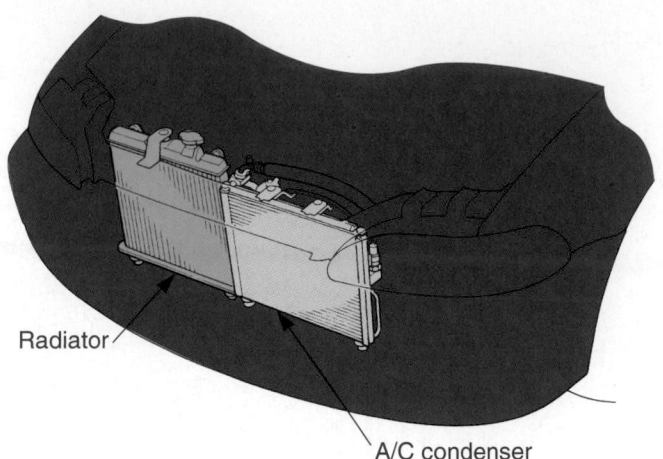

Figure 39-15. Note how this vehicle has a radiator and air-conditioning condenser mounted side-by-side. Many vehicles have the condenser in front of the radiator. (Honda)

however, the condenser is usually mounted in front of the radiator. With side-by-side mounting, cooler air flows through the radiator.

Radiator Cap

The *radiator cap* performs several functions:

- Seals the top of the radiator filler neck to prevent leakage.
- Pressurizes the system to raise the boiling point of coolant. This keeps coolant from boiling and turning to steam.
- Relieves excess pressure to protect against system damage.
- In modern closed systems, it allows coolant flow between the radiator and the coolant reservoir.

The radiator cap locks onto the radiator tank filler neck or on the reservoir tank. Rubber or metal seals make the cap-to-neck joint airtight. Radiator caps can be made of metal or plastic.

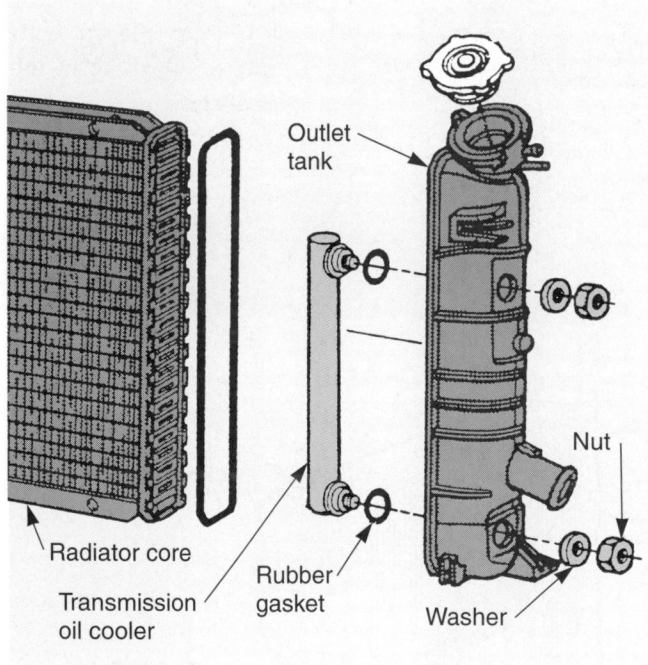

Figure 39-13. Transmission oil cooler prevents overheating of automatic transmission fluid. It is a small tank inside one of the radiator tanks. Note transmission line fittings. (Buick)

Radiator Cap Pressure Valve

The *radiator cap pressure valve,* **Figure 39-16,** consists of a spring-loaded disk that contacts the filler neck. The spring pushes the valve into the neck to form a seal.

Under pressure, water's boiling point increases. Normally, water boils at 212°F (100°C). However, for every pound of pressure increase, water's boiling point goes up about 3°F (-16°C). The radiator cap works on this principle.

Typical *radiator cap pressure* is 12–16 psi (83–110 kPa). This raises the boiling point of the engine coolant to 250–260°F (121–127°C). Many surfaces inside the engine's water jackets can be above 212°F (100°C).

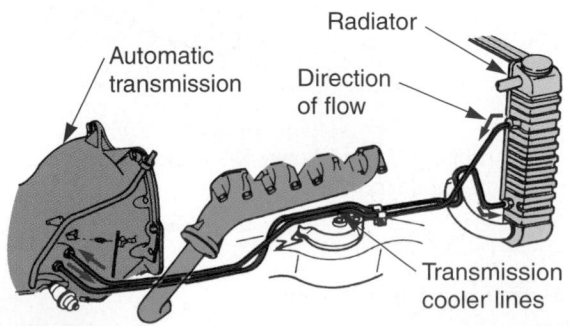

Figure 39-14. Automatic transmission lines run from the transmission to the transmission oil cooler fittings. (Cadillac)

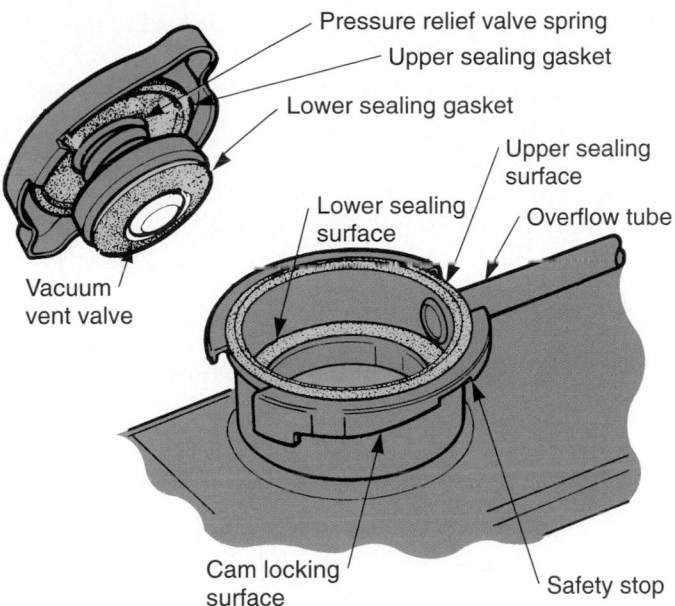

Figure 39-16. A radiator pressure cap screws onto the radiator filler neck or reservoir tank. Rubber or metal seals prevent leakage. (Mopar)

If the engine overheats and pressure exceeds the cap rating, the pressure valve opens. Excess pressure and steam force the coolant out of the overflow tube and into the reservoir or onto the ground. This prevents high pressure from rupturing the radiator, gaskets, seals, or hoses.

Radiator Cap Vacuum Valve

The **radiator cap vacuum valve** opens to allow flow back into the radiator when the coolant temperature drops after engine operation. It is often a small valve located in the center of the bottom of the cap. Look at **Figure 39-17.**

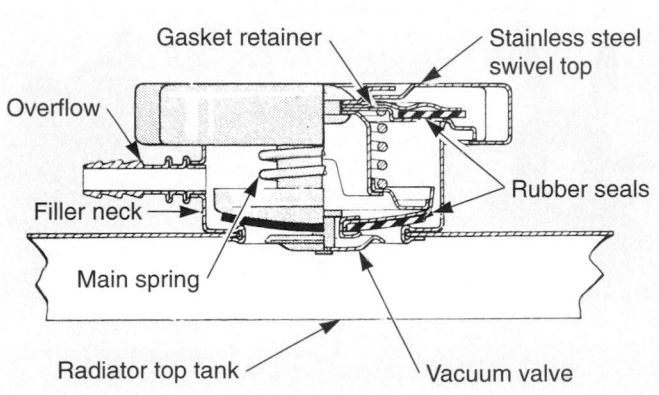

Figure 39-17. This cutaway view shows how the pressure cap installs and seals on the radiator filler neck. (Chrysler)

The cooling and contraction of the coolant and air in the system decrease the coolant volume and pressure. Without a cap vacuum valve, the radiator hoses and radiator tanks could collapse from outside pressure.

Closed and Open Cooling Systems

A *closed cooling system* uses an expansion tank, or reservoir, and a radiator cap with pressure and vacuum valves. The overflow tube is routed into the bottom of the reservoir tank. Pressure and vacuum valve action pull coolant in and out of the reservoir tank as needed. This keeps the cooling system correctly filled at all times.

Figure 39-18 shows the operation of a closed cooling system. When the engine heats up, the coolant expands and opens the cap pressure valve. Instead of leaking onto the ground, the coolant flows into the reservoir.

After the engine has been shut off, the coolant temperature drops and its volume decreases. This causes the vacuum valve to open. Atmospheric pressure (system suction) then forces coolant back into the radiator. This

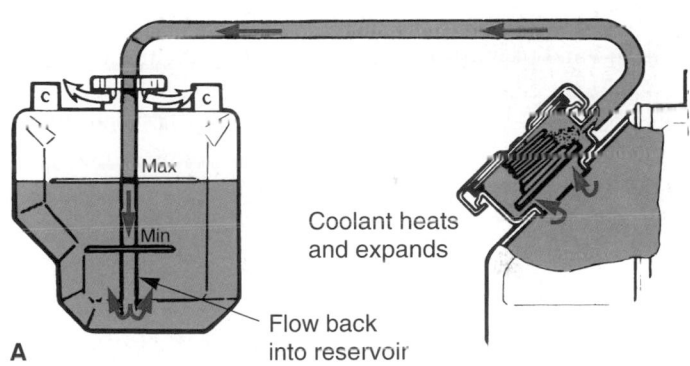

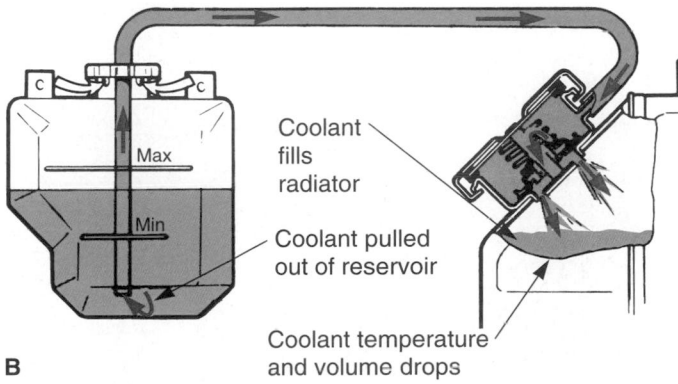

Figure 39-18. Study pressure cap operation. A—When the engine heats up, coolant expands. Excess fluid opens the cap pressure valve and coolant enters reservoir for reuse. B—When the engine is shut off, coolant temperature drops. This causes coolant to reduce in volume. Cap vent valve opens to let coolant flow back into the radiator. (Ford)

compensates for any small system leak, keeping the system properly filled.

An *open cooling system* does not use a coolant reservoir. The overflow tube allows excess coolant to leak onto the ground. Also, it does not provide a means of adding fluid automatically.

The open cooling system is no longer used on automobiles. It has been replaced by the closed system, which requires less maintenance.

Cooling System Fans

A *cooling system fan* pulls air through the core of the radiator and over the engine to help remove heat. It increases the volume of air flowing through the radiator, especially when the car is standing still. The fan is driven by a fan belt or an electric motor.

Engine-Powered Fans

An *engine-powered fan* bolts to the water pump hub and pulley. Sometimes, a *fan spacer* fits between the fan and pulley to move the fan closer to the radiator.

A *flex fan* has thin, flexible blades that alter airflow with engine speed. At low speeds, the fan blades remain curved and pull air through the radiator. At higher engine speeds, the blades flex until they are almost straight. This reduces fan action and saves engine power, **Figure 39-19.**

A *fluid coupling fan clutch* is designed to slip at higher engine speeds. It performs the same function as a flexible fan. The clutch is filled with silicone-based oil.

At a specific fan speed, there is enough load to make the clutch slip, **Figure 39-20.**

A *thermostatic fan clutch* has a temperature-sensitive, bimetal spring that controls fan action. The spring controls oil flow in the fan clutch. When cold, the spring causes the clutch to slip, speeding engine warm-up. After reaching operating temperature, it locks the clutch, providing forced-air circulation. See **Figure 39-21.**

Electric Cooling Fans

An *electric cooling fan* uses an electric motor and a thermostatic switch (coolant temperature sensor) to

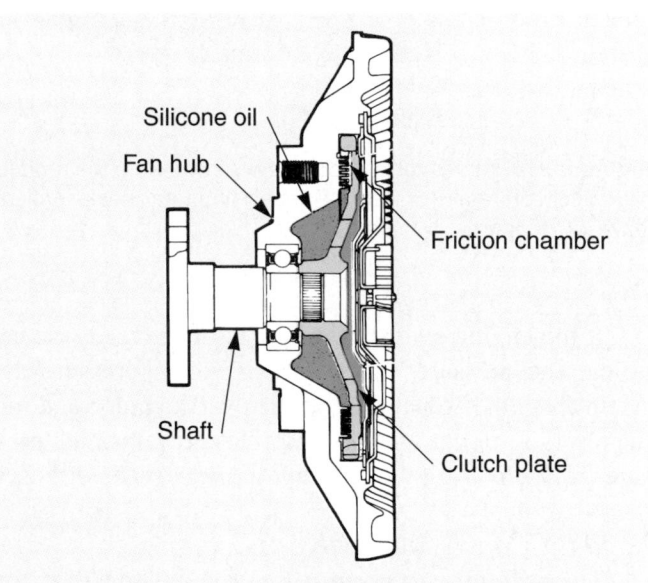

Figure 39-20. This is a fluid coupling fan clutch. A clutch plate operating in silicone-based oil causes enough friction at low speeds to turn fan. A high-speed load overcomes the friction, and the fan slips to save energy. (Chrysler)

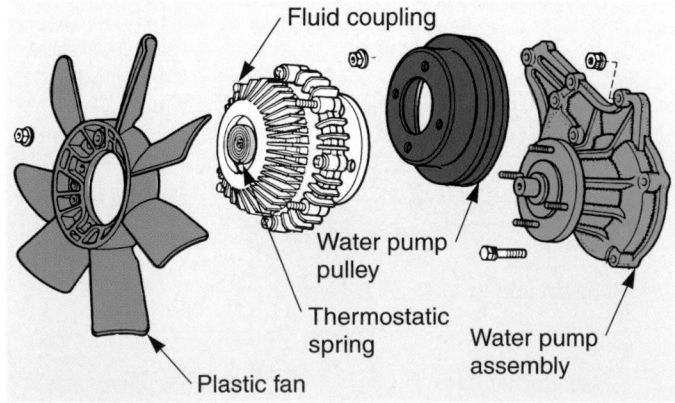

Figure 39-21. A thermostatic fan clutch is similar to a fluid coupling fan clutch. A bimetal spring is used to control clutching action. The fan only operates when the engine is hot and when the spring activates the clutch mechanism. (Toyota)

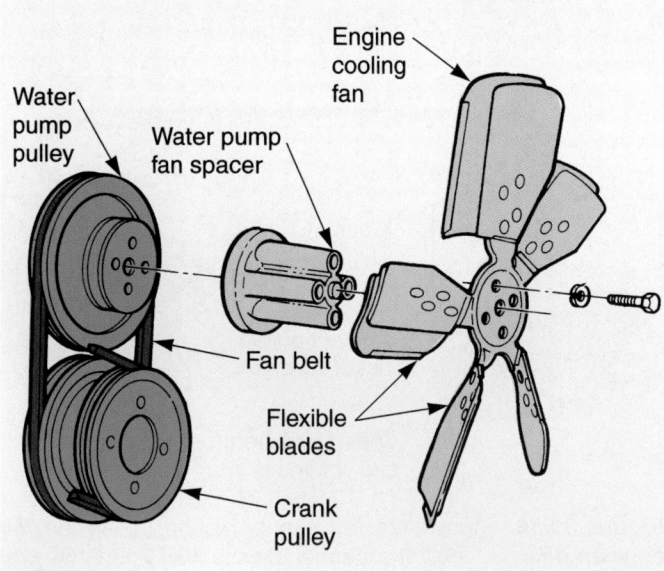

Figure 39-19. Note the construction of a flex-type radiator fan. High rpm causes fan blades to flex (bend), reducing blowing action. Note how a spacer is used to move fan closer to radiator. (Ford)

provide cooling action. An electric fan is needed on front-wheel-drive cars having transverse-mounted engines. In these vehicles, the water pump is normally located away from the radiator. Nevertheless, electric engine fans can be used on any engine/transmission layout. Look at **Figure 39-22.**

The *fan motor* is a small dc (direct current) motor. It mounts on a bracket secured to the radiator. A metal or plastic fan blade mounts on the end of the motor shaft to cause airflow.

An electric fan saves energy and increases cooling system efficiency. Because it only functions when needed, it helps speed engine warm-up. This reduces emissions and fuel consumption. In cold weather, the electric fan may shut off at highway speeds. There may be enough cool air rushing through the grille to provide adequate cooling.

Electric Engine Fan Circuits

The *fan switch* (*thermo switch*) is a temperature-sensitive switch that controls fan motor operation. With

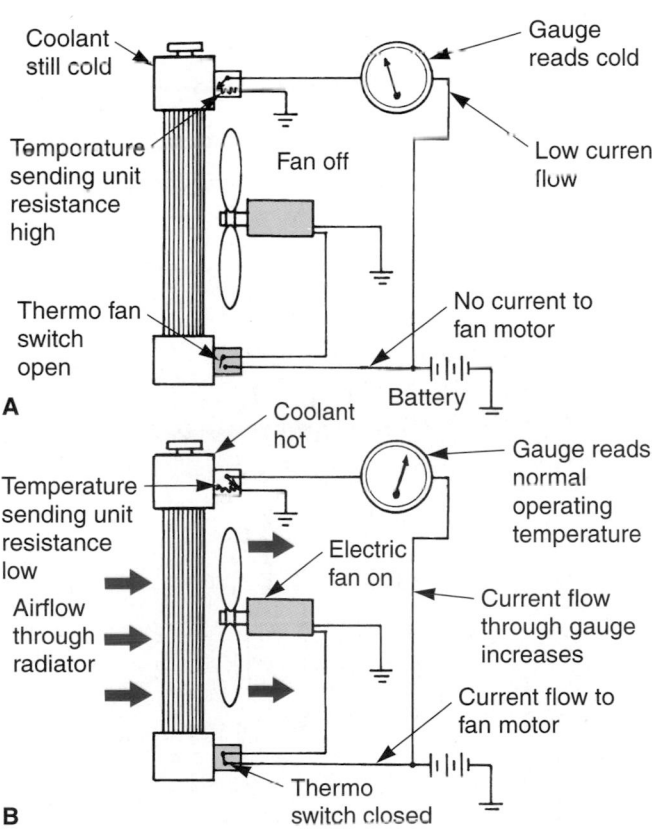

Figure 39-22. Study electric cooling fan operation. A—When the engine is cold, the thermo switch is open to prevent electric fan operation. This speeds engine warm-up. B—When the engine is at full operating temperature, the thermo switch closes. Current then flows to the fan motor to remove heat from the radiator.

late-model vehicles, the *coolant temperature sensor* (switch), relays, and engine control module (or power train control module) operate the engine cooling fan motors, **Figure 39-23.**

When the engine is cold, the coolant sensor signals the ECM that the engine is cold. The ECM does not energize the fan relays. This keeps the cooling fan from spinning and speeds engine warm-up.

After warm-up, the sensor resistance changes enough to signal the ECM of the need for cooling fan operation. The ECU sends current to the fan relay coils. This closes the relay contact, and high current flows to the cooling fans to prevent engine overheating. This is illustrated in **Figure 39-23.**

Radiator Shroud

The *radiator shroud* helps ensure that the fan pulls air through the radiator. It fastens to the rear of the radiator and surrounds the area around the fan. See **Figure 39-24.**

When the fan is spinning, the plastic shroud keeps air from circulating between the back of the radiator and the front of the fan. As a result, a huge volume of air flows through the radiator core. Without a fan shroud, the engine could overheat.

Thermostat

The *thermostat* senses engine temperature and controls coolant flow through the radiator. It reduces coolant flow when the engine is cold and increases coolant flow when the engine is hot.

The thermostat normally fits under a thermostat housing between the engine and the end of the upper radiator hose. Thermostats can be located at either the coolant inlet or outlet on the engine.

The thermostat has a wax-filled pellet, **Figure 39-25.** The pellet is contained in a cylinder-and-piston assembly. A spring holds the piston and valve in a normally closed position.

When the thermostat is heated, the pellet expands and pushes the valve open. As the pellet and thermostat cool, spring tension overcomes pellet expansion and the valve closes. **Figure 39-26** shows the basic action of an engine thermostat.

A *thermostat rating* is stamped on the thermostat to indicate the operating (opening) temperature of the thermostat. Normal ratings are between 180°F and 195°F (82°C and 91°C). High thermostat heat ranges are used in modern automobiles because they reduce exhaust emissions and increase combustion efficiency.

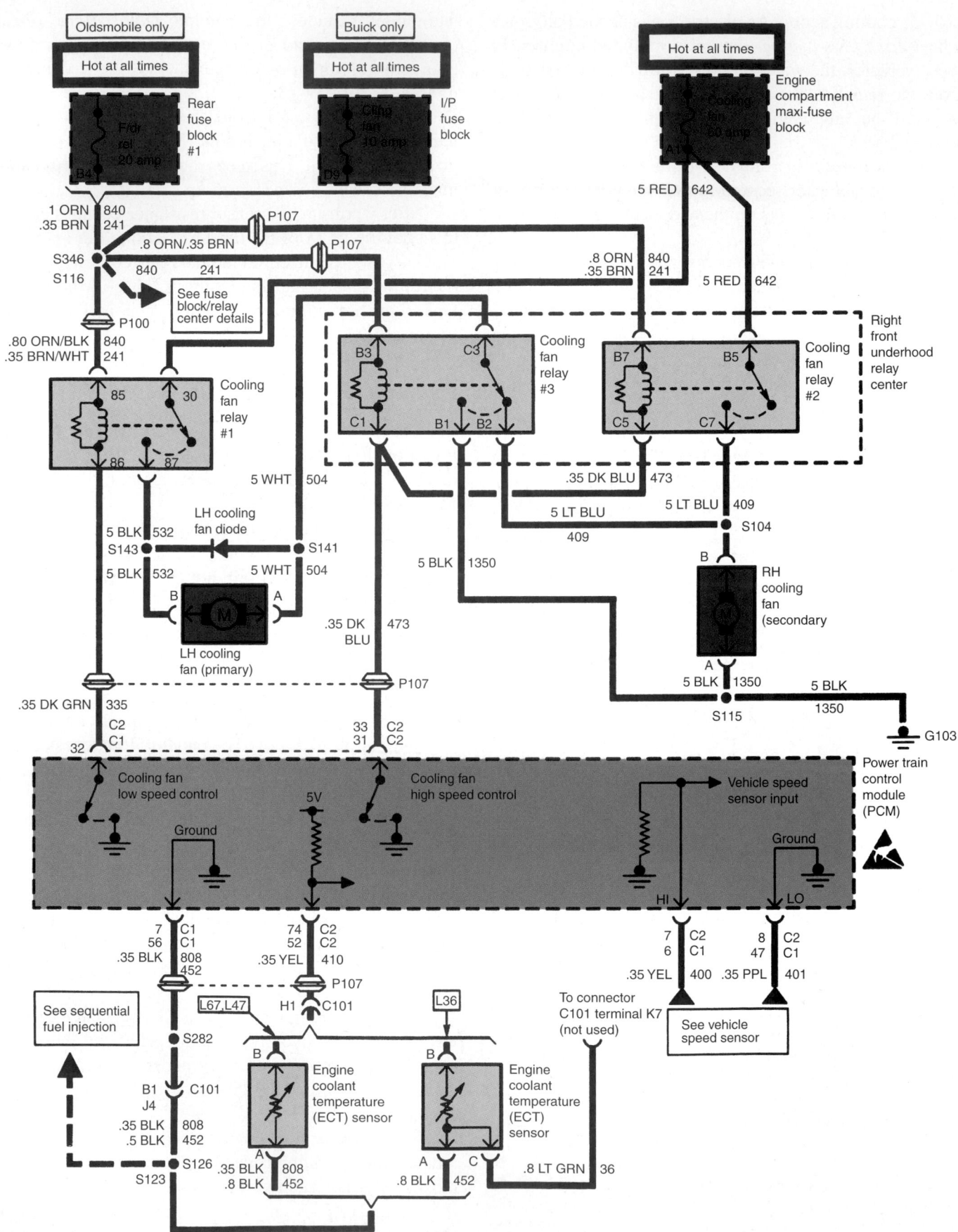

Figure 39-23. This wiring diagram shows modern electric engine fan circuit controlled by an electronic control unit. Note how the engine coolant temperature sensor (bottom) signals the power train control module whether to turn fans on or off. To turn fans on, the ECU sends a low-current signal to relays. The relay contact points then close to send higher current to the fan motors.

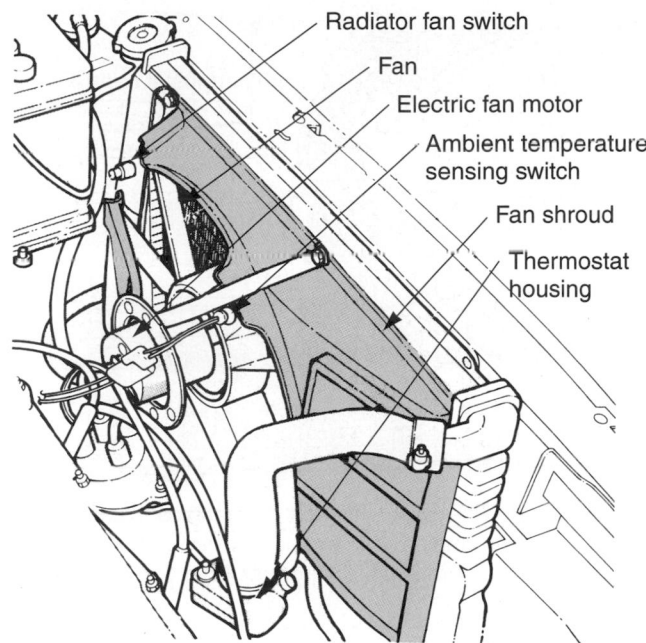

Figure 39-24. A fan shroud ensures that the fan pulls air through the radiator core. Without the shroud, air could circulate between the fan and the back of the radiator. Engine overheating could result. (Chrysler)

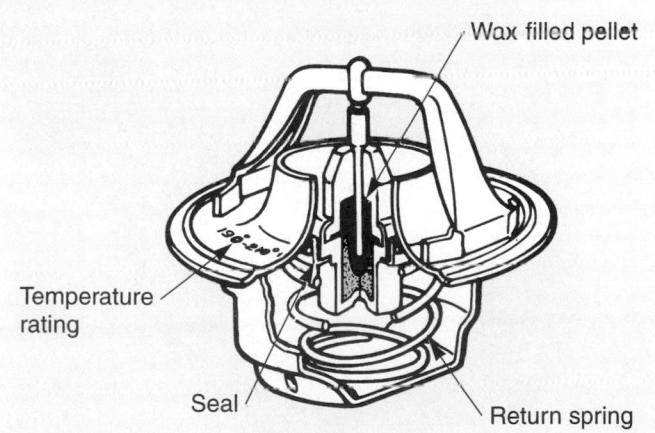

Figure 39-25. The thermostat is a temperature-sensitive valve. Note the pellet of wax enclosed in a cylinder-piston chamber. When heated, the pellet expands and pushes against spring tension to open the valve. (Mopar)

Thermostat Operation

When the engine is cold, the thermostat will be closed and coolant cannot circulate through the radiator. Instead, the coolant circulates around inside the engine block, cylinder head, and intake manifold until the engine is warm, **Figure 39-27A.**

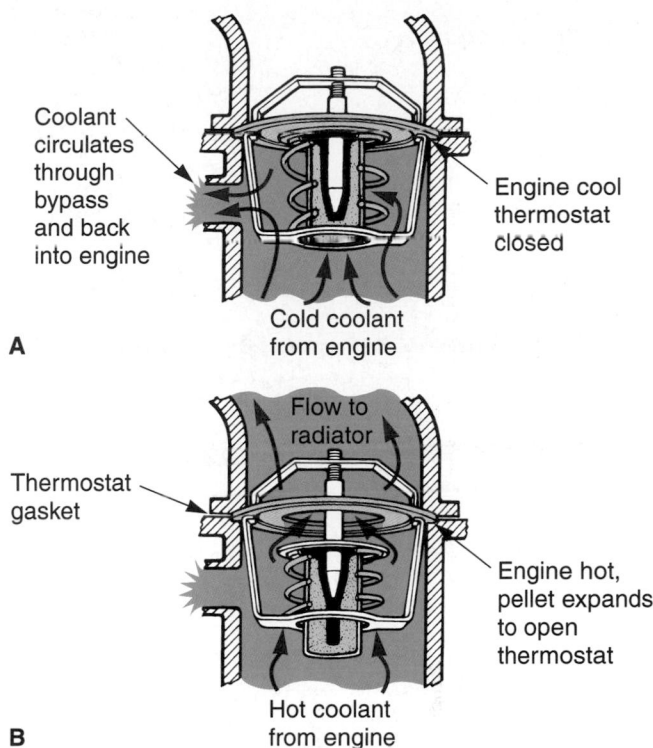

Figure 39-26. Study thermostat action. A—When coolant is cold, the thermostat remains closed due to spring tension. The water pump forces coolant to circulate in engine, but not through the radiator. B—When coolant is hot, the thermostat opens. The pump can then push coolant through the engine and the radiator. (Chrysler)

As the heat range of the thermostat is reached, the hot engine coolant causes the pellet inside the thermostat to expand. The thermostat gradually opens and allows coolant to flow through the system, **Figure 39-27B.**

Since the amount of thermostat opening is dependent on engine temperature, the exact operating temperature of the engine can be precisely controlled.

A *bypass valve,* **Figure 39-28,** and a bypass hose or passage permit coolant circulation through the engine when the thermostat is closed. If the coolant cannot circulate, hot spots could develop inside the engine.

A *bypass thermostat* has a second valve for routing *all* the hot coolant through the radiator, not just most of the hot coolant. The main thermostat valve regulates flow through the engine and radiator like a conventional thermostat. The added valve blocks off the bypass once the engine has reached operating temperature. See **Figure 39-29.**

A *thermostat jiggle valve* is a small valve fit into a hole formed in the thermostat. It helps prevent air pockets from forming in the housing.

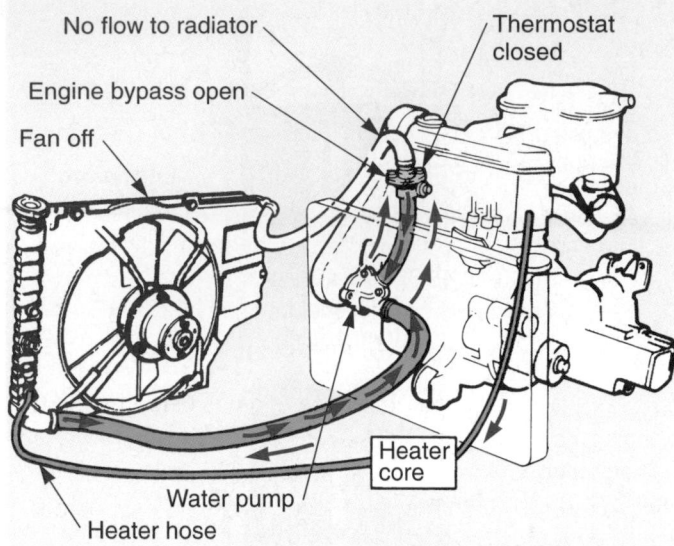

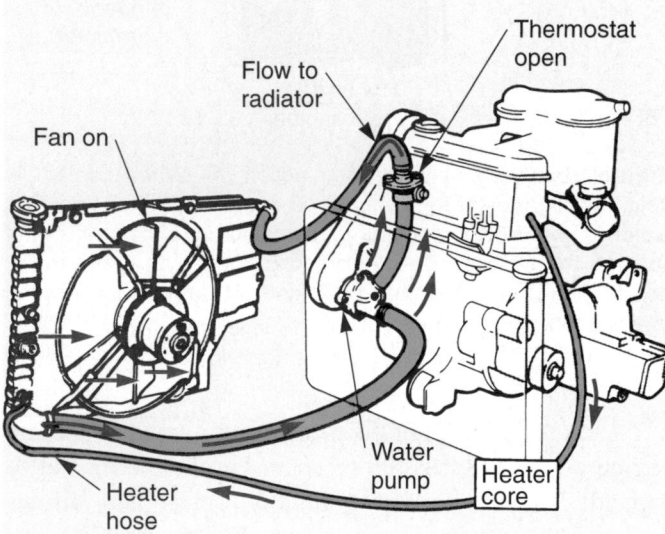

Figure 39-27. Thermostat operation. A—The thermostat does not allow coolant to enter the radiator when the engine is below operating temperature. B—When the engine is at operating temperature, the thermostat opens and allows flow into radiator. The thermostat moves open and closed different amounts to maintain correct engine operating temperature. (Dodge)

Cooling System Instrumentation

Most vehicles are equipped with a temperature warning light. Some vehicles also have an engine temperature gauge. It is important that you understand the operation of both.

Temperature Warning Light

A *temperature warning light* informs the driver when the engine is overheating. When the coolant

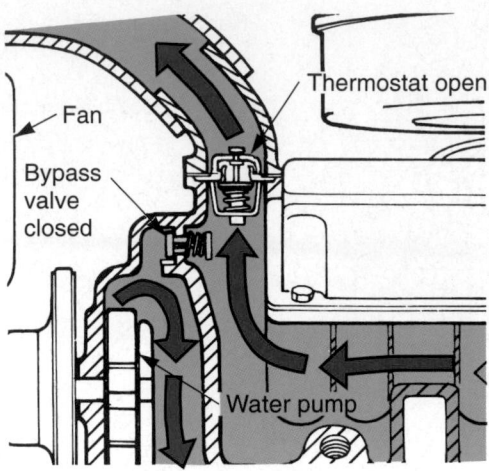

Figure 39-28. A bypass valve is sometimes used to allow circulation in the engine. It only opens when the thermostat is closed and when pressure is stronger than the bypass valve spring.

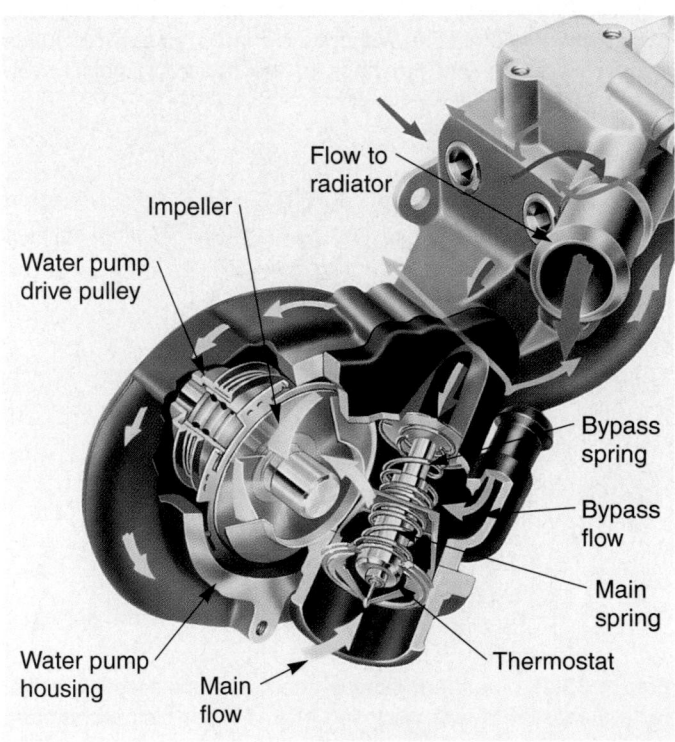

Figure 39-29. This thermostat has a conventional valve and bypass valve on the body of the thermostat. Note that it has two springs and valves built into one assembly. (Cadillac)

becomes too hot, a temperature sending unit (switch) in the engine block closes. This completes the circuit, and the indicator light on the dash glows, **Figure 39-30.**

When the engine is cold or at normal operating temperature, the sending unit circuit is open and the light remains off.

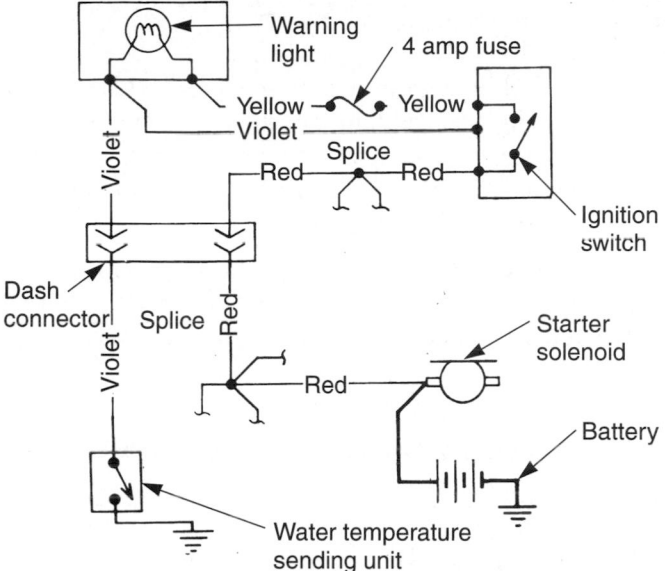

Figure 39-30. The circuit diagram for a simple engine temperature warning light. The sending unit screws into the engine water jacket. It closes when the engine overheats to light indicator bulb. Ignition switch lights the bulb when the engine is started. This lets driver know the bulb is not burned out. (Chrysler)

With many late-model vehicles, the engine temperature warning light is energized by the engine control module. If the sensor detects an overheating engine, the ECU sends current to the warning light.

Engine Temperature Gauge

An *engine temperature gauge* shows the exact operating temperature of the engine coolant. A variable resistance sending unit and a gauge are used in the circuit.

When the engine is cold, the gauge sending unit has high resistance and current does not flow to the gauge. The temperature gauge reads cold.

As engine temperature increases, the resistance in the sending unit drops. Current increases in the gauge circuit. Current causes the gauge needle to deflect to the right, showing engine temperature.

Again, the engine control module often acts as an interface between the sending unit and the gauge. This was explained in the previous chapter.

Antifreeze

Antifreeze, usually ethylene glycol, is mixed with water to produce engine coolant. Antifreeze has several functions. It prevents winter freeze up, prevents rust and corrosion, lubricates the water pump, and cools the engine.

Prevents Winter Freeze Up

Antifreeze keeps the coolant from freezing in very cold weather (outside temperature below 32°F or 0°C). Coolant freezing can cause serious cooling system and engine damage. As ice forms, it expands. This expansion can produce great force. The water pump housing, cylinder head, engine block, radiator, or other parts could be cracked and ruined by this force.

Prevents Rust and Corrosion

Antifreeze also prevents rust and corrosion inside the cooling system. It provides a protective film on part surfaces, **Figure 39-31.** Even in hot climates, antifreeze should be used to protect internal parts from corrosion.

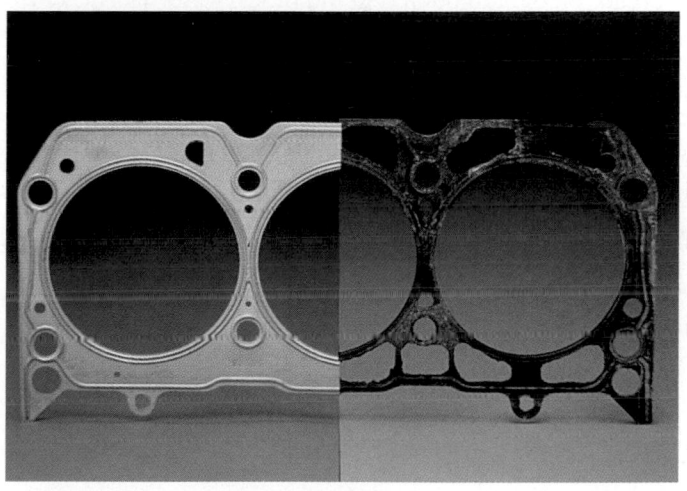

Figure 39-31. Antifreeze protects internal parts from rust and corrosion. One example, when steel head gasket is coated with antifreeze (left), it is protected. With only water (right), rust occurs very quickly on the steel surface. (Fel-Pro)

Lubricates the Water Pump

Antifreeze acts as a lubricant for the water pump and thermostat. It increases the service life of the water pump bearings and seals. It also prevents thermostat wear and corrosion.

Cools the Engine

Antifreeze conducts heat better than water and, therefore, cools the engine better. It is normally recommended in hot weather. For example, using the air conditioning system increases the temperature of the air flowing through the radiator. Antifreeze can help prevent overheating in very hot weather when the air conditioning is on.

Antifreeze/Water Mixture

For ideal cooling and protection from freeze up, a 50/50 mixture of water and antifreeze is usually recommended. It will provide protection from ice formation to about −34°F (−37°C). Higher ratios of antifreeze may produce even lower freezing temperatures, but this much protection is not normally needed.

 Caution
Plain water should never be used in a cooling system or the four antifreeze functions just discussed will not be provided.

Block Heater

A *block heater* may be used on an engine to aid engine starting in cold weather. It is simply a 120-volt heating element mounted in the block water jacket. Look at **Figure 39-32.**

The heater power cord is plugged into a wall outlet. This keeps the engine warm when the vehicle is not being used. Then, when the owner cranks the engine, it will start more easily.

A block heater is most commonly used on diesel engines. They are harder to start in cold weather than gasoline engines because of their compression ignition.

Figure 39-33 shows a cutaway of a diesel engine. Note the names of the cooling system parts.

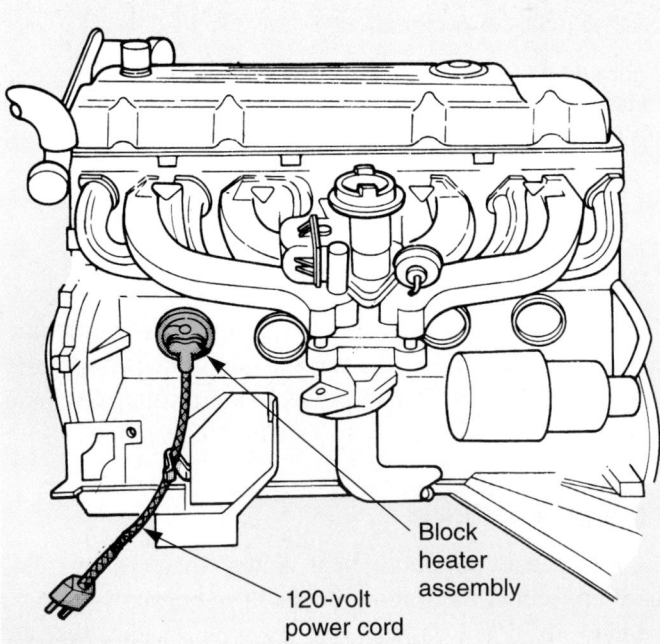

Figure 39-32. Block heaters plug into a home wall outlet. It heats coolant to aid starting in cold weather. Block heaters are common on diesel engines. (Chrysler)

Summary

- A cooling system must remove excess heat from the engine, maintain a constant engine operating temperature, increase the temperature of a cold engine quickly, and provide a means for warming the passenger compartment.

- Engine operating temperature is the temperature the engine coolant (water and antifreeze solution) reaches under normal running conditions. Typically, an engine's operating temperature is between 180°F and 210°F (82°C and 99°C).

- A liquid cooling system circulates a solution of water and antifreeze through the water jackets.

- In reverse flow cooling, cool coolant enters the head and hot coolant exits the block to return to the radiator.

- The water pump is an impeller or centrifugal pump that forces coolant through the engine block, cylinder head, intake manifold, hoses, and radiator.

- A water pump gasket fits between the engine and pump housing to prevent coolant leakage.

- Radiator hoses carry coolant between the engine water jackets and the radiator.

- Heater hoses are small diameter hoses that carry coolant to the heater core (small radiator-like device under the car dash).

- The radiator transfers coolant heat to the outside air. The radiator is normally mounted in front of the engine.

- A transmission oil cooler is often placed in the radiator on cars with automatic transmissions to prevent transmission fluid overheating.

- A closed cooling system uses an expansion tank or reservoir and a radiator cap with pressure and vacuum valves.

- A cooling system fan pulls air through the core of the radiator and over the engine to help remove heat.

- The thermostat senses engine temperature and controls coolant flow through the radiator.

- A thermostat rating is stamped on the thermostat to indicate the operating (opening) temperature of the thermostat. Normal ratings are between 180°F and 195°F (82°C and 91°C).

- Antifreeze, usually ethylene glycol, is mixed with water to produce the engine coolant.

- A block heater may be used on a diesel engine to aid engine starting in cold weather.

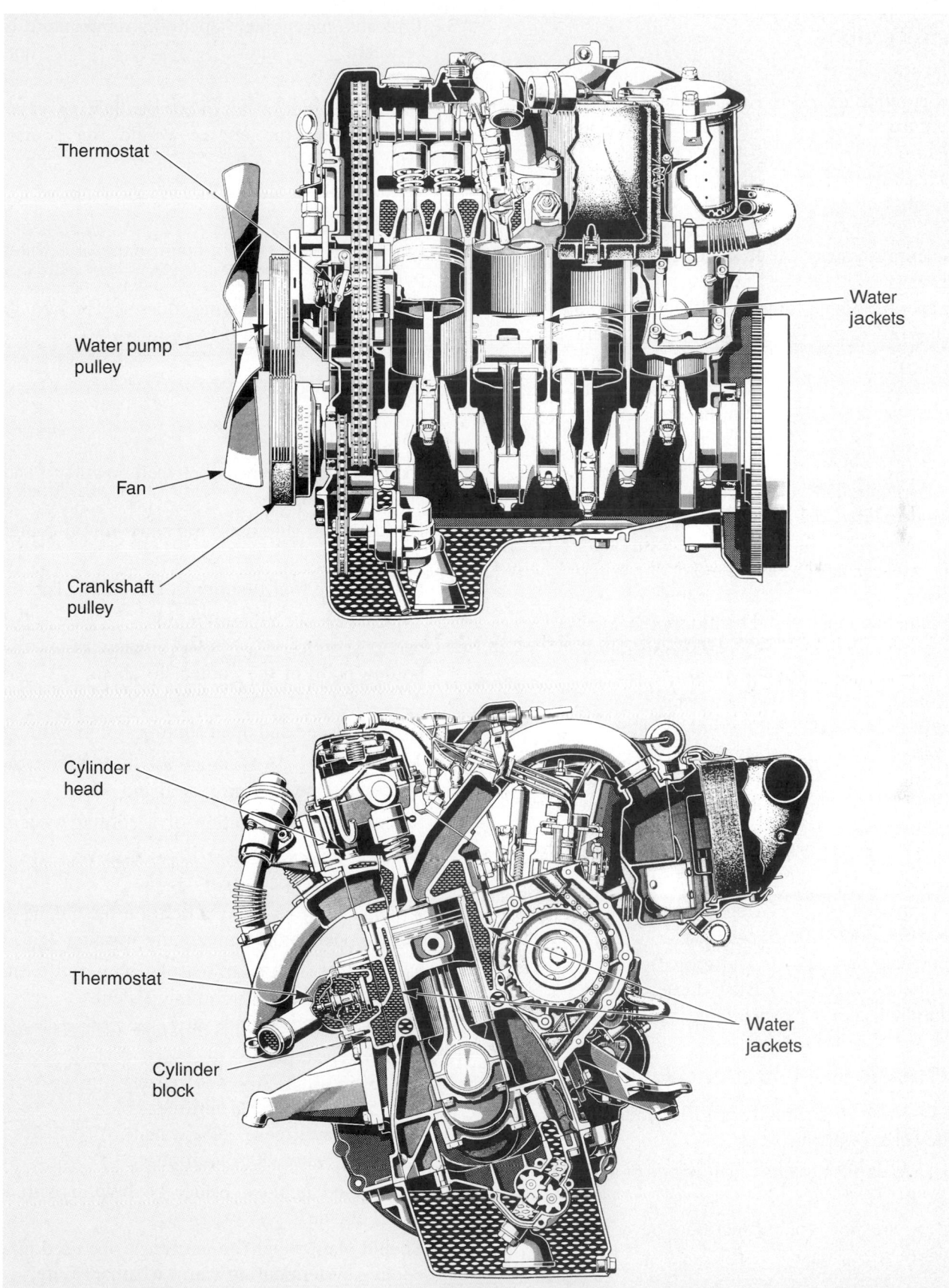

Figure 39-33. Study the side and front views of this modern, four-cylinder, diesel engine. It uses an overhead camshaft to operate the valves. Also, note cooling system water jackets in the cylinder head and cylinder block. The thermostat is located at the front, center of engine. (Mercedes Benz)

Important Terms

Cooling system
Engine operating
  temperature
Water pump
Air cooling system
Cooling fins
Shrouds
Liquid cooling system
Water jackets
Conventional coolant
  flow
Reverse flow cooling
Radiator hoses
Radiator
Fan
Thermostat
Water pump impeller
Water pump shaft
Water pump seal
Water pump bearings
Water pump hub
Water pump housing
Water pump gasket
Molded hose
Flexible hose
Hose spring
Heater hoses
Heater core
Hose clamps
Worm-drive hose
  clamp
Radiator core
Radiator tanks
Radiator filler neck
Radiator oil cooler
Radiator petcock
Downflow radiator

Crossflow radiator
Radiator tanks
Transmission oil cooler
Radiator cap
Radiator cap pressure
  valve
Radiator cap pressure
Radiator cap vacuum
  valve
Closed cooling system
Open cooling system
Cooling system fan
Engine-powered fan
Fan spacer
Flex fan
Fluid coupling fan
  clutch
Thermostatic fan clutch
Electric cooling fan
Fan motor
Fan switch
Thermo switch
Coolant temperature
  sensor
Thermostat
Thermostat rating
Bypass valve
Bypass thermostat
Thermostat jiggle valve
Radiator shroud
Temperature warning
  light
Engine temperature
  gauge
Antifreeze
Block heater

Review Questions—Chapter 39

Please do not write in this text. Place your answers on a separate sheet of paper.

1. List and explain the five major parts of a cooling system.

2. What are the four functions of a cooling system?

3. Typically, an engine's operating temperature is between _____ and _____ °F (_____ and _____ °C).

4. Not using a thermostat in hot weather is acceptable because the engine would run cooler. True or False?

5. Why has the liquid cooling system replaced the air types?

6. List and explain the six major parts of a water pump.

7. Which of the following does *not* relate to radiator construction?
   (A) Core.
   (B) Filler neck.
   (C) Tanks.
   (D) Impeller.

8. Explain the differences between downflow and crossflow radiators.

9. How does an automatic transmission oil cooler work?

10. Describe the four functions of a radiator cap.

11. Typical radiator cap pressure is _____ to _____ psi (_____ to _____ kPa), which raises the boiling point of the coolant to about _____ to _____ °F (_____ to _____ °C).

12. How do closed and open cooling systems differ?

13. A(n) _____ _____ is commonly used to turn an electric engine cooling fan on and off.

14. Summarize the operation of a cooling system thermostat.

15. Why is a radiator shroud used?

16. A temperature _____ (switch) on the engine is used to operate the temperature warning light.

17. List and explain four reasons why antifreeze should be used in the cooling system.

18. For ideal cooling, this mixture of water and antifreeze is typical.
    (A) 30% water, 70% antifreeze.
    (B) 50% water, 50% antifreeze.
    (C) 80% antifreeze, 20% water.
    (D) 70% water, 30% antifreeze.

19. Why could a block heater be helpful with a diesel engine?

20. Should plain water (no antifreeze) be used in a cooling system during warm weather? Why?

⬤ ASE-Type Questions

1. Which of the following system parts controls coolant flow?
   - (A) Fan.
   - (B) Radiator.
   - (C) Thermostat.
   - (D) Temperature sensor.

2. An engine's operating temperature is usually between:
   - (A) 82°F and 99°F.
   - (B) 100°F and 120°F.
   - (C) 125°F and 150°F.
   - (D) 180°F and 210°F.

3. A water pump normally mounts:
   - (A) under the engine.
   - (B) on the back of the engine.
   - (C) on the front of the engine.
   - (D) Any of the above.

4. Which of the following may be used to prevent coolant leakage between the water pump housing and engine?
   - (A) Gasket.
   - (B) RTV sealer.
   - (C) O-ring seal.
   - (D) None of the above.

5. Each of the following is a radiator component except:
   - (A) core
   - (B) petcock.
   - (C) oil cooler.
   - (D) bypass valve.

6. Which of the following is *not* a radiator cap function?
   - (A) Absorb heat.
   - (B) Seal radiator top.
   - (C) Pressurize system.
   - (D) Relieve excess pressure.

7. Which fan type is used on front-wheel-drive cars with transverse engines?
   - (A) Flexible fan.
   - (B) Electric engine fan.
   - (C) Engine powered fan.
   - (D) None of the above.

8. Which of the following controls coolant flow through a radiator?
   - (A) Shroud.
   - (B) Fan switch.
   - (C) Thermostat.
   - (D) Temperature sensor.

9. When a thermostat is closed, which of these permits coolant circulation through the engine?
   - (A) Bypass hose.
   - (B) Bypass valve.
   - (C) Both of the above.
   - (D) None of the above.

10. Antifreeze serves each of these functions *except:*
    - (A) lubricate water pump.
    - (B) prevent winter freeze up.
    - (C) prevent rust and corrosion.
    - (D) control engine temperature.

Activities—Chapter 39

1. Using the information contained in the chart on most antifreeze containers, construct a line graph to show the relationship between water/ antifreeze mixture and the low-temperature protection provided. (For example, the usual 50/50 mixture protects against freezing to temperatures as low as –34°F.)

2. Develop sketches to show how the expansion tank (coolant reservoir) functions in a closed cooling system. Make overhead transparencies from your sketches and use them to describe the expansion tank function to the class.

Chapter 40

Cooling System Testing, Maintenance, and Repair

After studying this chapter, you will be able to:

- List common cooling system problems and their symptoms.
- Describe the most common causes of system leakage, overheating, and overcooling.
- Perform a combustion leak test and a system pressure test.
- Check the major parts of a cooling system for proper operation.
- Replace faulty cooling system components.
- Drain, flush, and refill a cooling system.
- Describe safe working practices to use when testing, maintaining, or repairing a cooling system.
- Correctly answer ASE certification test questions on cooling system troubleshooting and repair.

A cooling system is extremely important to the performance and service life of an engine. Major engine damage can occur in minutes without the proper removal of excess combustion heat.

Combustion heat could collect in the metal engine parts. The heat can melt pistons, crack or warp the cylinder head or block, cause valves to burn, or "blow" the head gasket. To prevent these costly problems, the cooling system must be kept in good condition.

As an auto service technician, you must be able to locate and correct cooling system problems quickly and accurately. It is equally important that you know how to maintain a cooling system. This chapter will help you develop these skills.

Note

For more information on cooling-system-related problems and service, refer to the index. This subject is covered in the chapters on engine mechanical problems, performance problems, and engine rebuilding.

Cooling System Problem Diagnosis

The first step toward diagnosing cooling system problems involves gathering information. Talk to the vehicle owner or the service writer to find out as much as possible about the symptoms of the problem.

For example, you might ask the following questions:

- Can you describe the cooling system problem (temperature light on, overheating, or coolant loss)?
- When does the problem seem to occur (all the time, at highway speeds, or when idling only)?
- How long have you had the problem?
- When was the last time the coolant was replaced?
- Have any other repairs been performed (new thermostat, hoses, or engine repairs)?
- Have you noticed any coolant leaks (puddles on the ground, wetness around the engine)?
- Are there any unusual noises that might be related to the cooling system (grinding at the front of the engine, hissing)?

The answers to these kinds of questions will be very useful. It will help you eliminate the least likely sources so that you can concentrate on the most probable causes of the malfunction.

After gathering information, verify the complaint. Test drive the car. Inspect the engine compartment. Listen to engine noises. Do what is needed to make sure the symptoms have been properly described.

Cooling System On-Board Diagnostics

Many on-board diagnostic systems will trip a trouble code when certain cooling-related circuits are operating out of range. OBD II systems monitor coolant temperature, coolant level, engine oil temperature, belt tension, and other related functions. Use your scan tool to analyze the system when the malfunction indicator light is on.

For example, with some systems, if you fail to bleed air pockets out of the water jackets, a diagnostic trouble code can be tripped. The engine temperature sensor can be fooled by the air pocket and falsely indicate a circuit problem. The real problem is that the cooling system is not full of coolant and has an air pocket. The system must be bled to correct this false code.

Diagnosis Charts

A *cooling system diagnosis chart* should be used when problems are difficult to locate. A service manual will give a chart for the particular type of engine and cooling system. It will be very accurate and will help you decide what tests and repairs are needed.

Inspecting Cooling System

A visual inspection will frequently reveal the source of the cooling system problem. As shown in **Figure 40-1,** look for obvious troubles:

- Coolant leaks.
- Loose or missing fan belts.

- Low coolant level.
- Abnormal water pump noises.
- Leaves and debris covering the outside of the radiator.
- Coolant in the oil (oil looks milky).
- Combustion leakage into the coolant (air bubbles in coolant).

Warning
Keep your hands and tools away from a spinning engine fan. Wear eye protection and stand behind–not over–the spinning fan blade. Then, if tools are dropped into the fan or a fan blade breaks off, you are not likely to be hit and injured by flying parts.

Cooling System Problems

Cooling system problems can be grouped into three general categories:

- *Coolant leaks*—Crack or rupture, allowing pressure cap action to push coolant out of the system.
- *Overheating*—Engine operating temperature is too high, warning light is on, temperature gauge

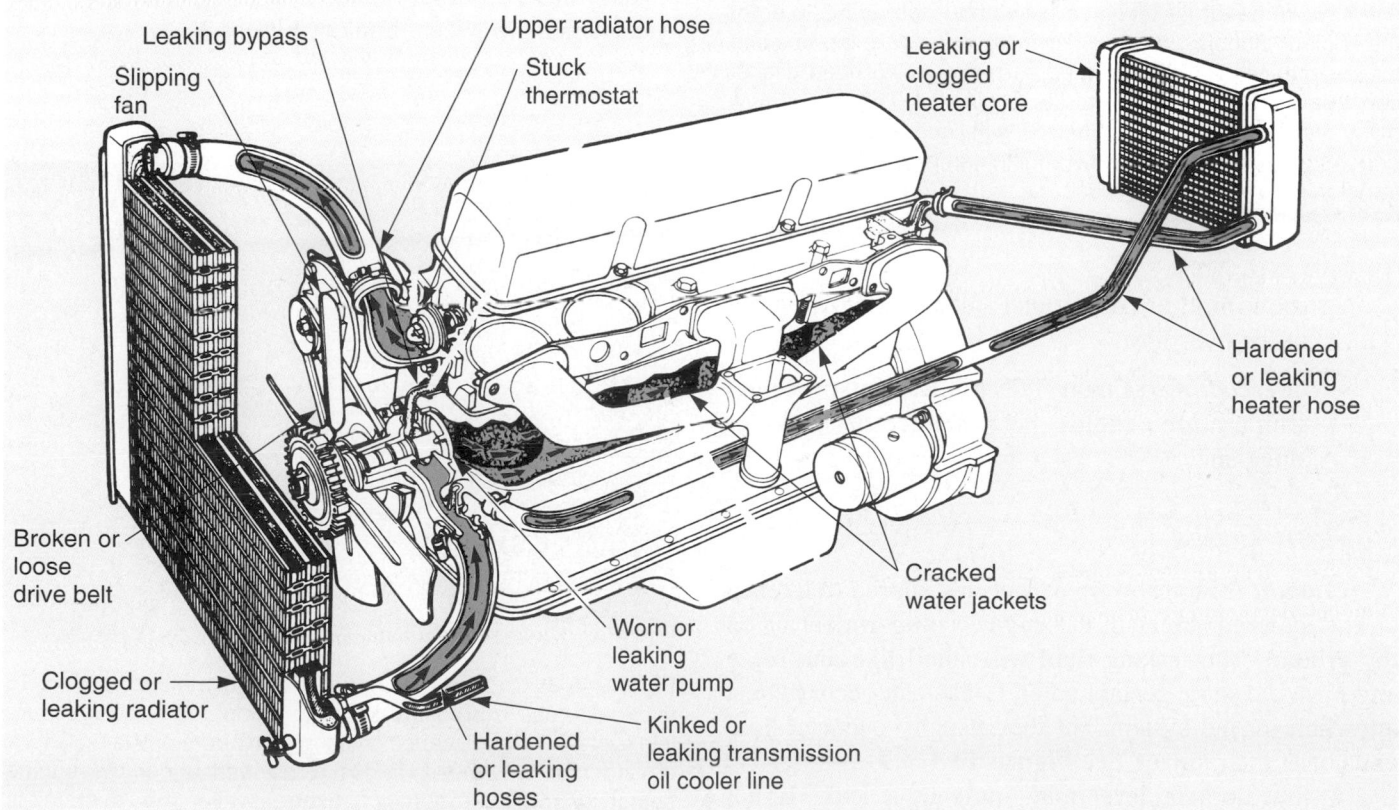

Figure 40-1. These are common problem areas in a cooling system. Note leakage points. (Ford Motor Co.)

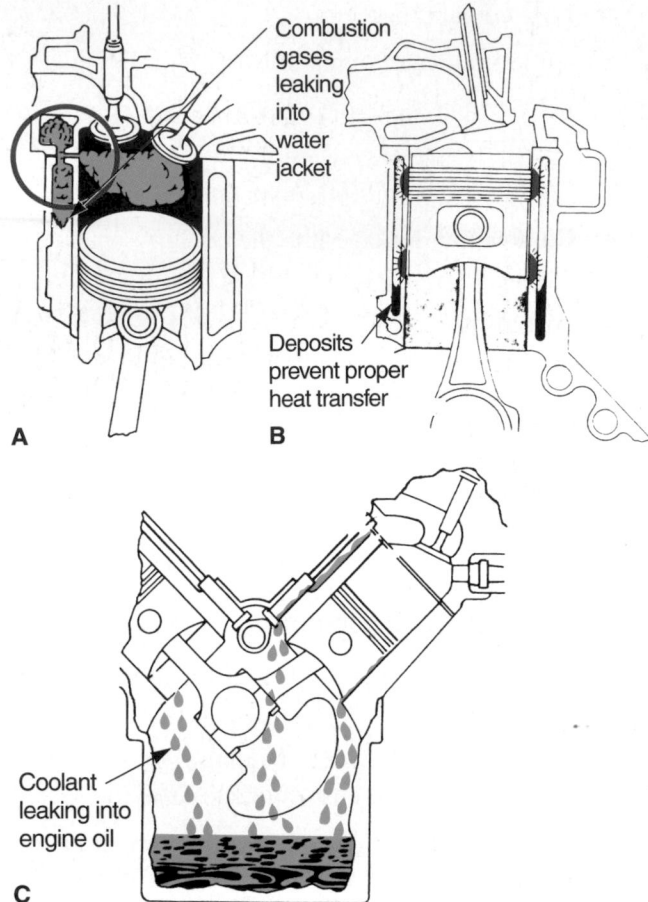

Figure 40-2. Engine problems can affect the cooling system. A—A blown head gasket can allow combustion gases to enter the coolant. Bubbles may be seen in the radiator or blowing into the reservoir tank. B—A cracked part or a blown head gasket can allow coolant to leak into the engine oil. The oil will turn milky white when contaminated with coolant. C—Mineral deposits in water jackets of the engine can prevent proper heat transfer. Overheating can result. (Deere & Co.)

shows hot, or coolant and steam are blowing out of the overflow.

- *Overcooling*—Engine fails to reach full operating temperature, engine performance is poor or sluggish.

Coolant Leaks

Coolant leaks show up as wet, discolored (darkened or rust colored) areas in the engine compartment or on the ground. The leaking fluid will smell like antifreeze and have the same general color. Leaks can occur almost anywhere in the system, but they usually occur at hose ends or at the radiator. See **Figure 40-1.**

A low coolant level may indicate a leak. If not visible, the leak may be an *internal engine leak* (cracked

engine block, cracked cylinder head, or blown head gasket). This is illustrated in **Figure 40-2.**

Remember to check the coolant level in late-model systems at the overflow tank, or reservoir. Do not remove the radiator cap. Only on open systems (no reservoir) must you remove the pressure cap to check coolant level.

⚠️ **Warning**
Never remove a radiator cap when the engine is hot. The pressure release can make the coolant begin to boil and expand. Boiling coolant could spurt out of the filler neck or reservoir, causing severe burns!

💡 **Tech Tip**
With today's sloped, rounded hood lines, the trend is to put the radiator filler neck on or near the engine or on the overflow tank instead of on the radiator. The higher filler location is needed to help purge air from the cooling system.

Cooling System Pressure Test

A *cooling system pressure test* is used to quickly locate leaks. Low air pressure is forced into the system. This will cause coolant to pour or drip from any leak in the cooling system.

A *pressure tester* is a hand-operated air pump used to pressurize the cooling system for leak detection. It is one of the most commonly used and important cooling system testing devices. Look at **Figure 40-3.**

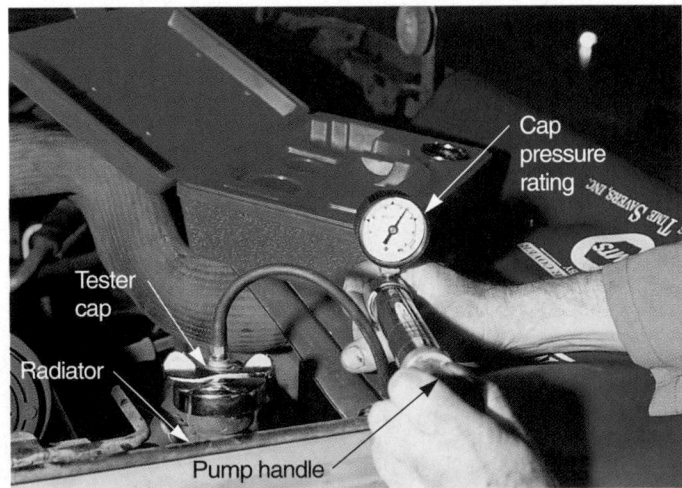

Figure 40-3. Coolant leakage is a very common problem. To find leaks, use a pressure tester to pump cap-rated pressure into system. This will cause coolant to drip from any leak. Check around radiator, hoses, freeze plugs, and under the engine compartment for coolant leakage while the system is pressurized. (Snap-On Tools)

Install the pressure tester on the radiator filler neck or the reservoir. Then pump the tester until the pressure gauge reads radiator cap pressure or maximum allowable pressure (around 14 psi, or 96 kPa).

Caution

Do not pump too much pressure into the cooling system, or part (radiator, hose, or gasket) damage may result. Never exceed radiator cap or system operating pressure when testing.

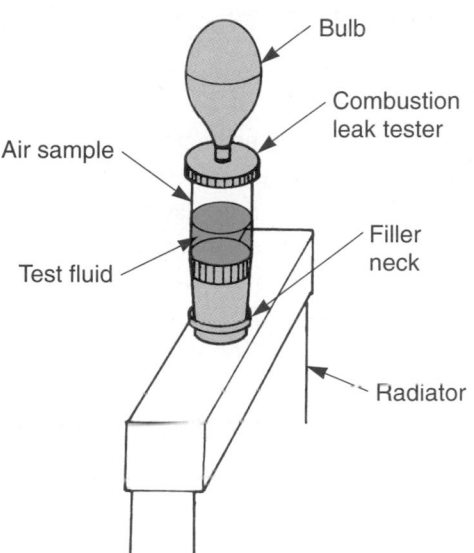

Figure 40-4. Use bulb to draw a sample of radiator air into the combustion leak tester. If the test fluid turns yellow, engine problems are allowing combustion gas into the cooling system. Combustion leakage can make the engine overheat.

With pressure in the system, inspect all parts for coolant leakage. Check all hose fittings, gaskets, and engine freeze (core) plugs. Look under the water pump and around the radiator. If a leak is found, tighten, repair, or replace parts as needed.

Combustion Leak Test

A *combustion leak test* checks for the presence of combustion gases in the coolant. It should be performed when signs point to a blown head gasket, cracked block, or cracked cylinder head (overheating, bubbles in the coolant, or a rise in the coolant level upon starting). Refer back to **Figure 40-2.**

A *block tester,* sometimes called a *combustion leak tester,* is placed in the radiator filler neck or the reservoir. The engine is started and the tester bulb is squeezed and then released. This will pull air from the cooling system through the test fluid, **Figure 40-4.**

The fluid in the block tester is normally blue. The chemicals in exhaust gases cause a reaction in the test fluid, changing its color. A combustion leak will turn the fluid yellow. If the fluid remains blue, there is no combustion leakage.

If combustion leakage is indicated, short out spark plugs one at a time. Test the cooling system with each plug shorted. When the fluid does not change color, the cylinder being checked (shorted) has a combustion leak.

Combustion leakage into the cooling system is very damaging. Exhaust gases mix with the coolant and form very corrosive acids. The acids can eat holes in the radiator and corrode other components.

Figure 40-5 shows how an exhaust gas analyzer will also check for combustion gases in the cooling system. Exhaust gas analyzer use is detailed in later chapters.

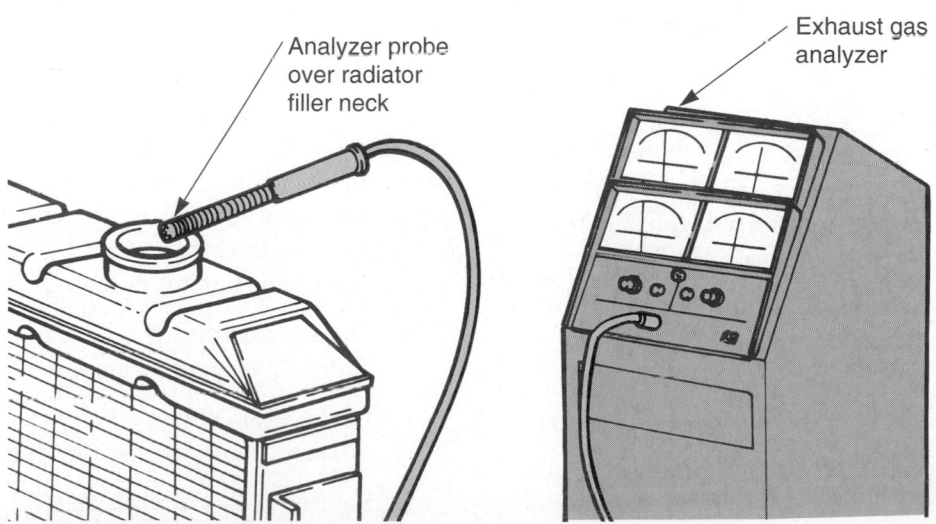

Figure 40-5. An exhaust gas analyzer will detect combustion leakage into coolant. Place probe over the filler neck and accelerate the engine. Hydrocarbon (HC) reading indicates internal combustion leakage. (Chrysler)

Coolant in Oil

When water, antifreeze, and oil mix, the solution turns *milky white* in color. If a milky white solution is found in the engine oil or in valve covers, it is an indication of a coolant leak. The cause may be one of the same problems that create combustion leakage (blown head gasket, cracked head or block, or leaking intake manifold gasket on V-type engines). Refer back to **Figure 40-2.**

It is possible to have both combustion leakage into the coolant and coolant leakage into the engine oil. When mixed with engine oil, antifreeze can cause engine damage. The antifreeze can collect and burn on the cylinder walls, causing piston and cylinder gumming or scoring. Always correct an engine problem causing internal leakage.

Overheating

Engine *overheating* is a serious problem that can cause major engine damage. The driver may notice the engine temperature light glowing, the temperature gauge reading high, or the coolant boiling. Boiling coolant will expand and blow out through the overflow as steam.

Tech Tip
Modern scan tools and OBD II systems will give temperature sensor readings, or operating values. This might be helpful when trying to solve engine overheating and other related problems.

There are many common causes of engine overheating:

- Low coolant level—Leak or lack of maintenance has allowed the coolant level in the engine and radiator to drop too low.

- Rust or scale—Mineral accumulations in the system have clogged the radiator core or built up in the water jackets, **Figure 40-2.**

- Stuck thermostat—Thermostat fails to open normally, restricting coolant flow.

- Retarded ignition timing—Late ignition timing allows combustion flame to blow out through an open exhaust valve, transferring too much heat into exhaust valves, ports, and manifold.

- Loose fan belt—Water pump drive belt slips under load and reduces coolant circulation.

- Bad water pump—Broken pump shaft or damaged impeller blades prevent normal pumping action.

- Collapsed lower hose—Suction from water pump may flatten the hose if the spring is missing or the hose is badly deteriorated.

- Missing fan shroud—Air circulates between the fan and the back of the radiator, reducing airflow through radiator.

- Ice in coolant—Coolant frozen due to lack of antifreeze can block circulation and cause overheating.

- Engine fan problems—Fan clutch or electric fan troubles can prevent adequate airflow through the radiator.

Any of these problems, or others, can make the engine overheat. You must use your knowledge of system operation and basic testing methods to find the problem's source. Methods for locating specific troubles will be covered later in this chapter.

Tech Tip
Some engines are protected from overheating damage caused by a loss of coolant. The engine control module uses data from the engine temperature sensor to detect overheating. The computer then cuts off spark to one cylinder at a time in a controlled sequence. It also retards ignition timing to limit top speed. The outside air pulled into the "dead cylinders" cools the engine and prevents overheating damage.

Overcooling

Overcooling may be indicated by slow engine warm-up, insufficient warmth from the heater, low fuel economy, sluggish engine performance, or a low reading from the coolant temperature sensor.

Overcooling can cause increased part wear. Because parts are not at full operating temperature, their clearances will be too great. The parts will not expand enough to produce the correct fit.

Overcooling also reduces fuel economy because more combustion heat transfers into the metal parts of the engine. Less heat remains to produce expansion of gases and pressure on the pistons.

The following conditions can cause overcooling:

- Stuck thermostat—Thermostat stuck open, allowing too much circulation.

- Locked fan clutch—Fan operates all the time, causing excess airflow through radiator.

- Shorted fan switch—Electric fan runs all the time, increasing warm-up time.

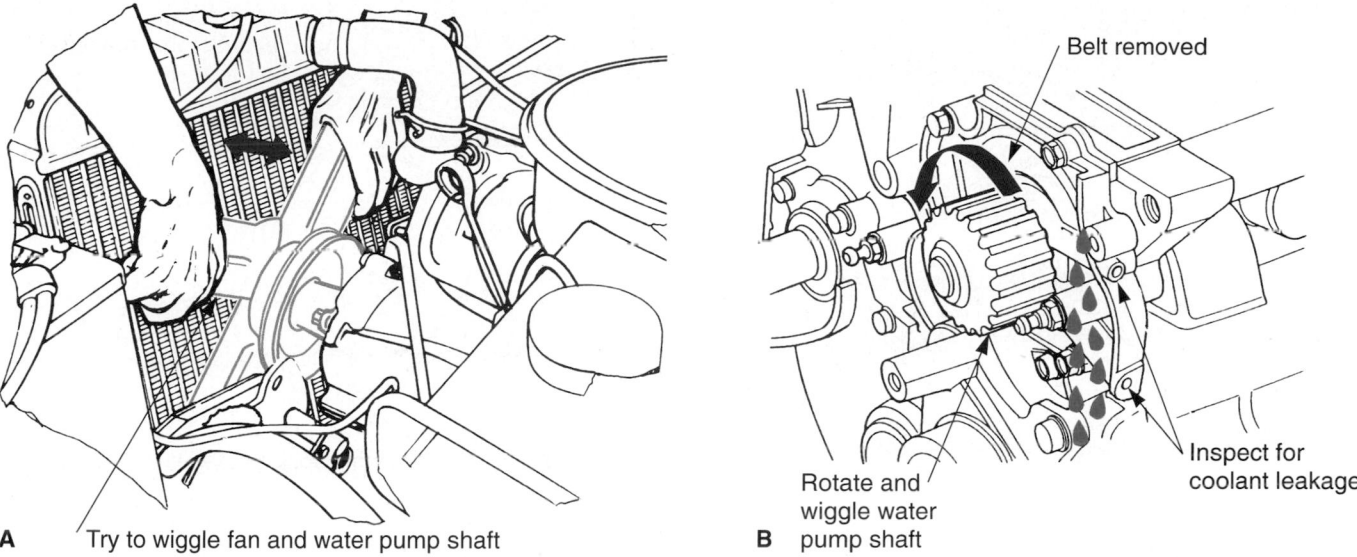

A Try to wiggle fan and water pump shaft

Belt removed

Rotate and wiggle water pump shaft

Inspect for coolant leakage

B

Figure 40-6. Visual inspection can find engine water pump problems. A—Wiggle the engine fan to check for water pump bearing wear. Pump shaft should not wiggle and coolant should not leak. B—With late-model engines using large drive belts and electric fans, you may need to remove the belt to check water pump condition. Turn pump pulley by hand to check for roughness or looseness. Note location of the bleed holes on this water pump. (Chrysler and Honda)

Water Pump Service

A bad water pump may leak coolant (worn seal), fail to circulate coolant (broken shaft or damaged impeller), or produce a grinding sound (faulty pump bearings).

Rust in the cooling system and lack of coolant are common reasons for pump failure. These conditions could speed seal, shaft, and bearing wear. An overtightened drive belt is another common cause for premature water pump failure. Excess belt tension will overheat the water pump bearings and make them fail prematurely.

Checking a Water Pump

To check for a bad water pump seal, pressure test the system and watch for leakage at the pump. Coolant will leak out of the small drain hole at the bottom of the pump or at the end of the pump shaft. Replace or rebuild a leaking pump.

To check for worn water pump bearings, try to wiggle the fan or pump pulley up and down. On vehicles equipped with a serpentine belt, you may need to remove the drive belt and turn the water pump pulley by hand to check for problems. Look at **Figure 40-6.**

If the pump shaft is loose in its housing, the pump bearings are badly worn and pump replacement is usually necessary. A stethoscope can also be used to listen for worn, noisy water pump bearings.

To check water pump action, warm the engine. Squeeze the top radiator hose while someone starts the engine. You should feel a pressure surge (hose swelling) if the pump is working. If not, pump shaft or impeller problems are indicated. You can also watch for coolant circulation in the radiator with the engine at operating temperature.

Removing a Water Pump

To remove the water pump, unbolt all brackets and other components (air-conditioning compressor, power steering pump, alternator, etc.) preventing pump removal. Then, unscrew the bolts holding the pump to the engine. Keep all bolts organized to aid reassembly.

Never use excessive force when trying to remove an old water pump. It is easy to overlook hidden bolts that secure the pump. Make sure all bolts are out before lightly tapping the pump housing with a mallet to free the pump.

Scrape off old gasket or sealer material. The engine-to-pump mating surfaces must be perfectly clean to prevent coolant leakage. On soft aluminum parts, be careful not to gouge or scratch the sealing surfaces.

Refer to the service manual if you are not sure how to remove the water pump. It will tell you exactly what parts must be removed to access the water pump bolts.

Water Pump Rebuild

A water pump rebuild involves pump disassembly, cleaning, part inspection, worn part replacement, and

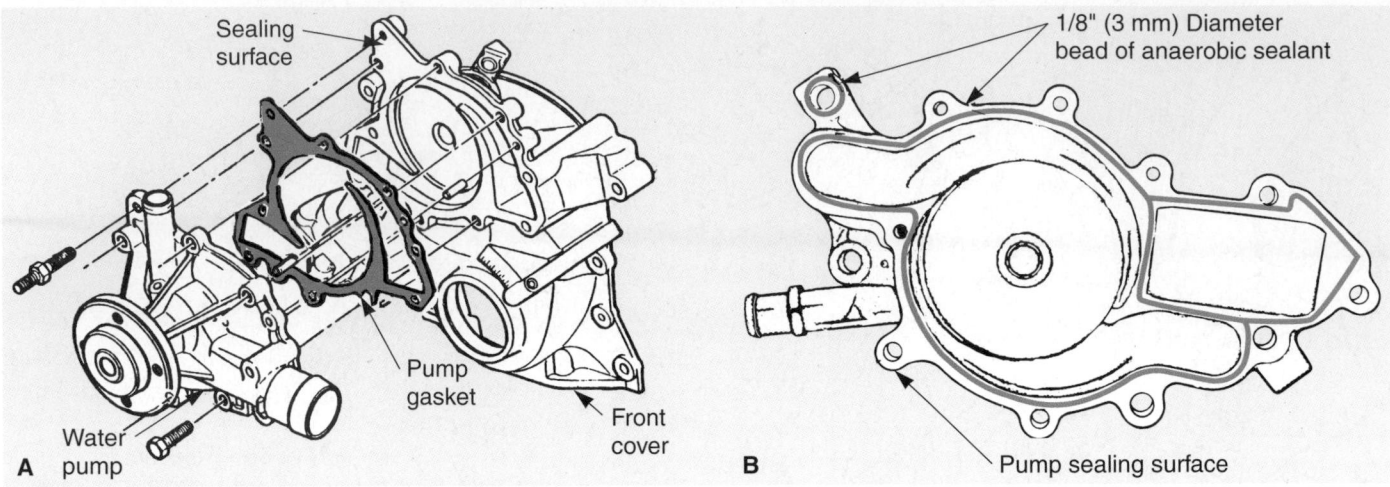

Figure 40-7. Two methods of sealing water pump-to-engine surfaces. A—A gasket is held in place with adhesive-sealer during assembly. B—Sealer can be used instead of a water pump gasket. Form a continuous bead and do not break bead when installing pump. Both part surfaces must be clean for sealer to work properly. (Buick)

reassembly. Few technicians rebuild water pumps. Most purchase new or factory-rebuilt pumps. Rebuilding takes too much time and would not be cost effective. **Figure 40-10** shows an exploded view of a water pump.

Installing a Water Pump

To install a water pump gasket, use an approved sealer to adhere the new gasket to the pump. This will keep the gasket in alignment over the bolt holes during pump installation. Look at **Figure 40-7A.**

To use a chemical gasket (sealer used in place of fiber gasket), squeeze out a bead of approved sealer (usually anaerobic or RTV) around the pump sealing surface. Form a continuous bead of consistent width (about 1/8″ or 3 mm). This is illustrated in **Figure 40-7B.**

If the water pump uses an O-ring seal, work the new seal down into the bottom of its groove in the pump or engine. Leakage will result if the seal doesn't reach the bottom of the groove. See **Figure 40-8.**

 After installing the gasket or sealer, the water pump should be installed as follows:
1. Fit the pump onto the engine. Move it straight into place. Do not shift the gasket or break the sealant bead.
2. Start all the bolts by hand. Screw them in two turns. Check that all bolt lengths are correct. Each bolt should be sticking out the same amount.
3. Torque all the fasteners a little at a time in a crisscross pattern. Go over the bolts several times to ensure correct tightening.
4. Install the other components and tighten the belt properly, **Figure 40-9.**

If needed, refer to a shop manual. It will give detailed directions on pump service for the exact make and model of car.

Thermostat Service

A stuck thermostat can cause engine overheating or engine overcooling. If the thermostat is stuck shut, coolant will not circulate through the radiator. As a result, overheating could make the coolant boil.

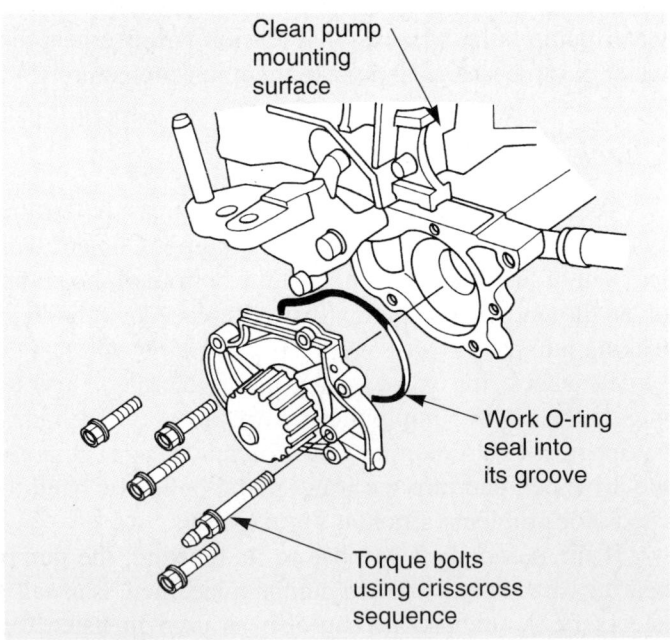

Figure 40-8. This water pump uses O-ring seal instead of gasket or sealer. Fit new seal down into its groove in part. Make sure seal is in place to prevent leakage after assembly.

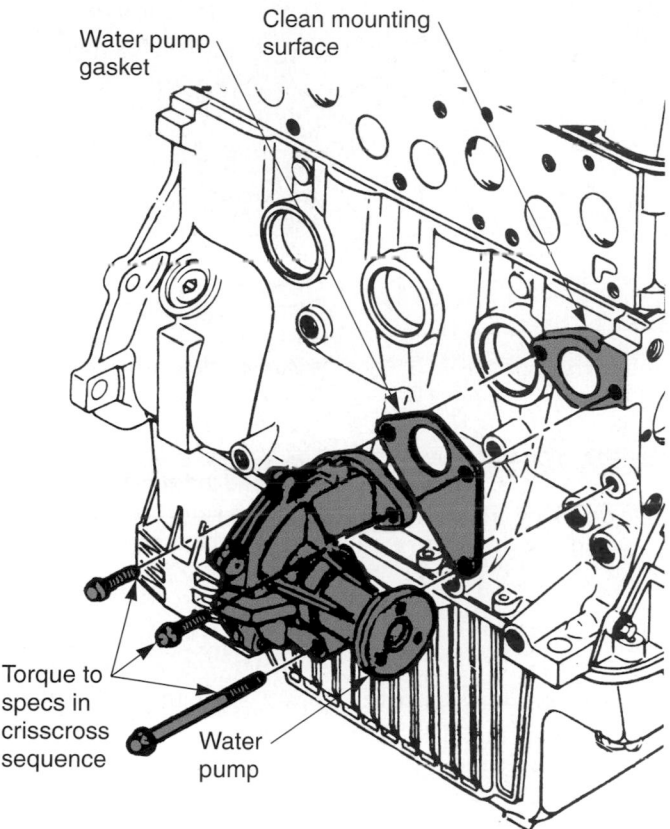

Figure 40-9. When tightening water pump bolts, use a crisscross pattern to compress the gasket evenly. Use a torque wrench if there is enough room to use one. (Honda Motor Co.)

If a thermostat is stuck open, too much coolant may circulate through the radiator. The engine may not reach proper operating temperature, and it may run poorly for extended periods in cold weather. Consequently, engine efficiency (power, gas mileage, and driveability) will be reduced.

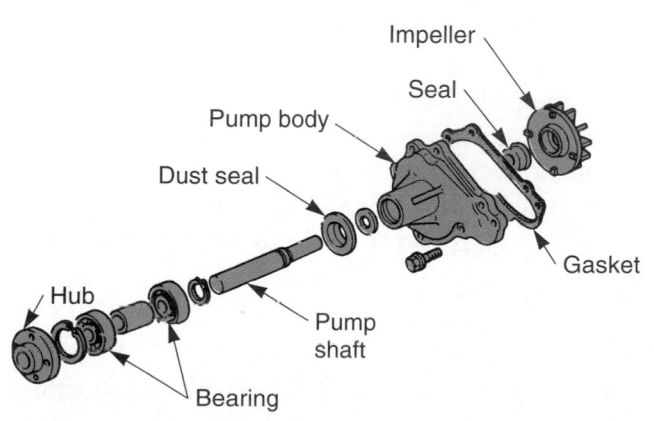

Figure 40-10. This exploded view shows the major parts of a water pump. A pump rebuild typically involves replacing pump bearings, seals, shaft, and impeller (sometimes). Most technicians install new or factory-rebuilt water pumps to save time and money. (Mazda)

Thermostat Testing

To check a thermostat, watch the coolant through the opening in the radiator neck (if provided). When the engine is cold, coolant should not flow through the radiator. When the engine warms, the thermostat should open and the coolant should begin to circulate through the radiator. If this action does not occur, the thermostat may be defective.

In some instances, the thermostat may have to be removed from the engine for testing in a container of water on a hot plate. The thermostat should open when heated to its operating temperature.

You can also use a digital thermometer to check part temperatures. For example, touch the thermometer probe on the engine near the thermostat and on the thermostat outlet hose. If the engine is reaching operating temperature but the outlet hose stays cool, the thermostat is not opening. If the thermostat does not open at the correct temperature, it is defective and should be replaced, **Figure 40-11.**

Thermostat Replacement

The thermostat is normally located on top of the engine, under the ***thermostat housing*** (fitting for the upper radiator hose). With newer vehicles, it can also be located on the side of the engine. The service manual will give the exact location.

To remove the thermostat, unscrew the bolts holding the thermostat housing to the engine. Tap the housing free with a rubber or plastic hammer. Lift off the housing and thermostat, **Figure 40-12A.**

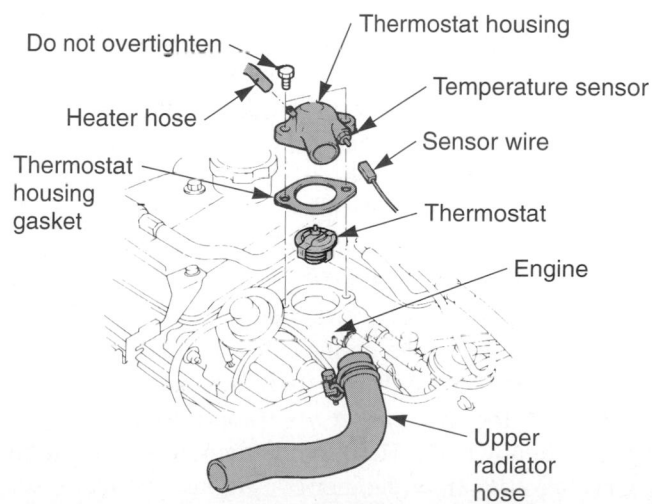

Figure 40-11. The thermostat is normally in housing at the engine end of the upper radiator hose. Remove the housing bolts and pop out the old thermostat. Be careful not to damage the thermostat housing. (Toyota)

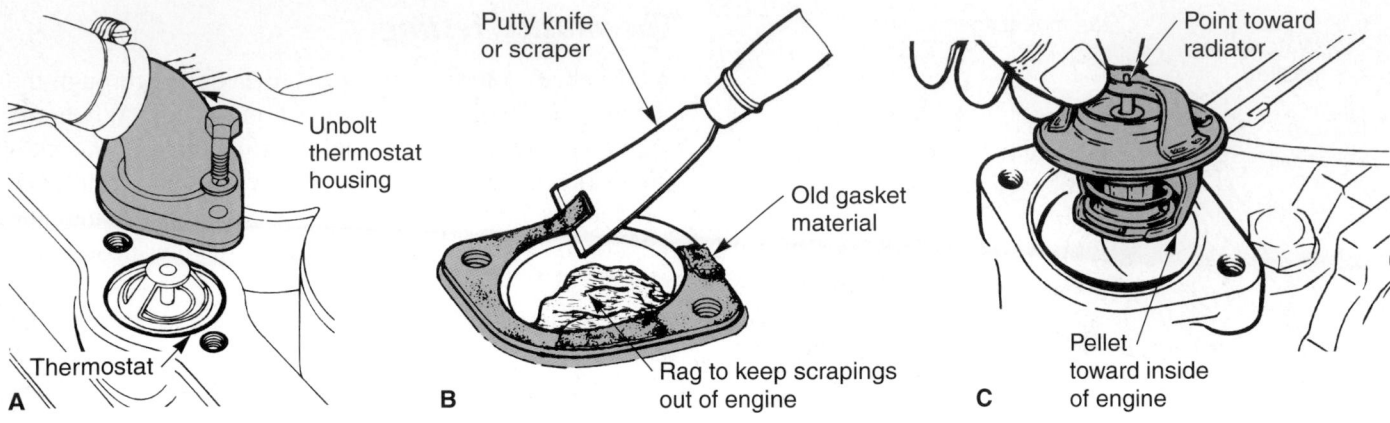

Figure 40-12. Study the basic steps for thermostat replacement. A—The thermostat is removed by unbolting the thermostat housing. Light taps or prying will free the housing. B—Scrape off all the old gasket or sealer from the engine and thermostat housing. Do not gouge the thermostat housing. C—Install the new thermostat with pellet toward inside of engine. Make sure it is centered in the housing. Torque thermostat housing bolts to specifications. If you overtighten the bolts, the housing will crack or warp easily. (Mopar and Ford)

Caution

Be careful not to damage the thermostat housing. It is often made of aluminum or "pot metal" and will break easily. Use light mallet taps and prying to free the gasket.

Scrape all the old gasket material off the thermostat housing and the sealing surface on the engine. Do not gouge or nick the sealing surface or leakage may result. See **Figure 40-12B.**

Make sure the thermostat housing is not warped. Place it on a flat surface and check for gaps between the housing and surface. If the housing is warped, file or sand the surface flat. This will prevent coolant leakage.

Make sure the temperature rating of the new thermostat is correct. Install the thermostat in the engine, **Figure 40-12C.** Normally, the rod (pointed end) on the thermostat should face the radiator hose. The pellet chamber should face the inside of the engine.

Make sure you purchase an exact replacement. Just because the thermostat will fit in its housing does not mean it will function as designed. The wrong thermostat can cause slow engine warm-up. It may also cause engine overheating in warm weather.

Some thermostats have a check ball or bleed pin to help purge air from the coolant. Make sure the replacement thermostat has either of these devices.

Position the new gasket with approved sealer. Start the fasteners by hand. Then torque them to specifications in a crisscross pattern. Do not overtighten the thermostat housing bolts, or warpage may result.

A rubber **thermostat housing seal** is used instead of a gasket with some engine designs. Place the new seal over or in its groove. Do not shift the seal out of place when installing the thermostat housing. See **Figure 40-13.**

Caution

With some engines, the thermostat housing, cooling system filler neck, radiator hose nipple, and overflow tube are all combined into one housing made of plastic. Do not damage it while working.

Bleeding the Cooling System

A **cooling system bleed screw,** or bleed valve, is sometimes provided to help remove trapped air when refilling the cooling system. Many late-model cars with low hood lines require a bleed screw to empty air pockets formed in areas of the system. See **Figure 40-14.** Some systems have more than one bleed screw.

Air trapped in the cooling system can cause engine overheating or damage (cracking or warpage) to the parts near the air pocket and hot spot. A **hot spot** is an area in the engine suffering from a buildup of combustion heat. This is often due to an air pocket in the water jacket.

Bleed the cooling system with the following procedure:
1. Fill the system with coolant.
2. Start and warm the engine to full operating temperature.
3. Crack open the bleed screw until all air is purged from the system and coolant leaks from the valve.

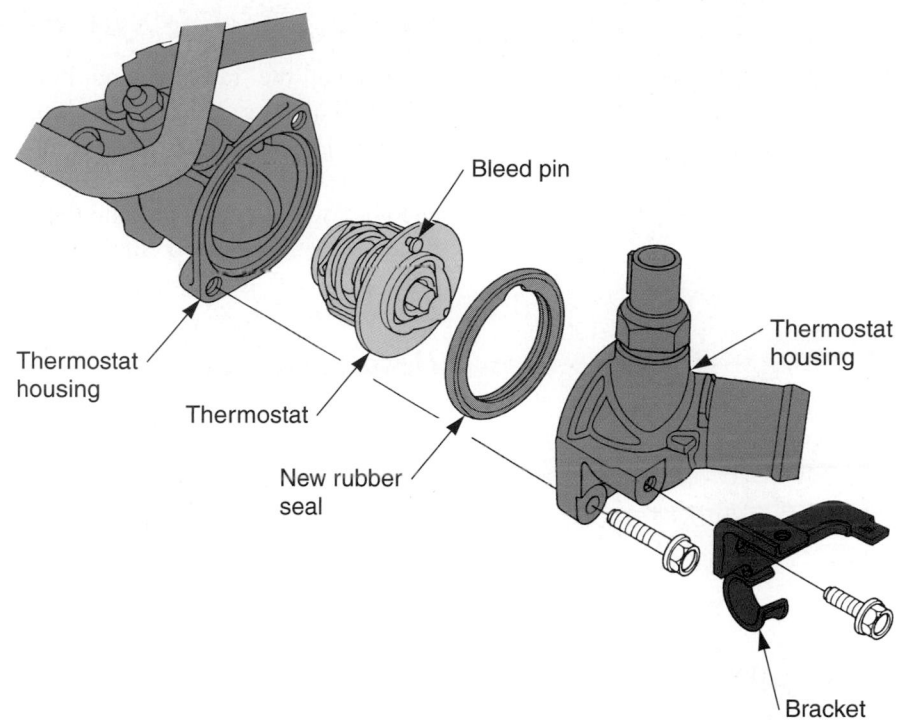

Figure 40-13. This modern thermostat uses a special O-ring seal. A new seal should be installed anytime the thermostat is removed. (Honda)

Warning
Never fully remove a cooling system bleed screw or any other cooling system component (hose, fitting, plug, sensor) with the engine at full operating temperature. Steaming hot coolant could spray out. Wear safety glasses and gloves when working with a hot cooling system.

Figure 40-14. If provided, use the bleed screw to remove trapped air from the cooling system. Fill the system with coolant. Install the radiator cap and start the engine. Crack open the bleed screw until all air is purged. (Snap-On Tools)

Cooling System Hose Service

Old radiator hoses and heater hoses are frequent causes of cooling system problems. After a few years of use, hoses deteriorate. They may become soft and mushy or hard and brittle. Cooling system pressure can rupture the hoses, resulting in coolant loss.

Hardened hoses become very brittle and crack from engine vibration. To check for hardened hoses, squeeze the hoses with your hand. If you cannot squeeze the hose with your fingers, replace it.

Hardened hoses often crack and leak where they connect to other parts. Engine heat is the most common cause of hose hardening. The heat "cooks" the rubber and removes its elasticity.

Softened hoses may have been contaminated with oil or other fluids that break down the rubber in the hose. A softened hose will lose much of its strength and can rupture and leak. The hose will usually swell if it has been softened.

A softened lower radiator hose can collapse from the suction of the water pump. The collapsed hose will restrict coolant circulation and cause overheating. The spring inside the lower radiator hose normally prevents hose collapse. It should never be removed.

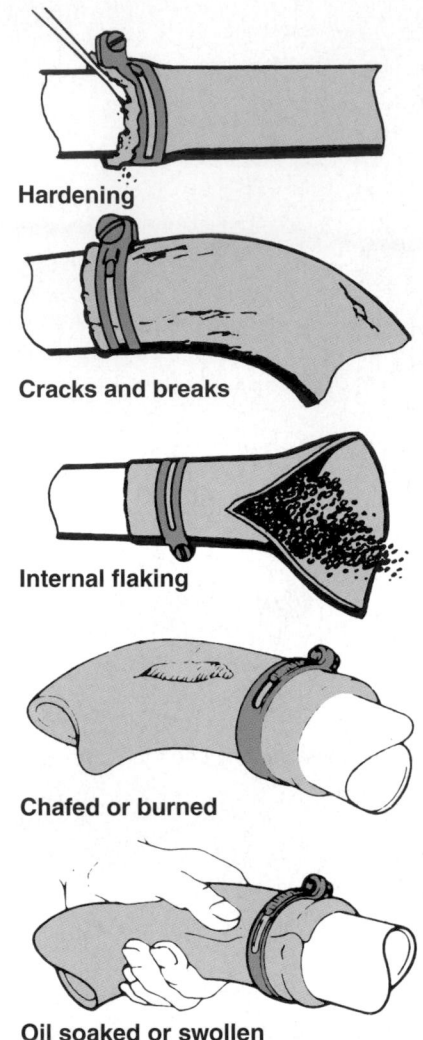

Hardening

Cracks and breaks

Internal flaking

Chafed or burned

Oil soaked or swollen

Figure 40-15. Check cooling system hoses for these kinds of problems. (Gates Rubber Co. and Ford)

Checking Cooling System Hoses

Inspect the radiator and heater hoses for cracks, bulges, cuts, or any other signs of deterioration or damage. Look at **Figure 40-15.**

Squeeze the hoses to check whether they are hardened or softened. Flex or bend the heater hoses and watch for surface cracks. If any problem is detected, the affected hoses should be replaced.

Hose Replacement

To remove a hose, loosen the hose clamps. Twist the hose while pulling it from the fittings, **Figure 40-16A.** If a new hose is to be installed, you can cut a slit in the end of the old hose to aid removal.

Clean the metal hose fittings. If the fittings are badly corroded and pitted, coat them with a nonhardening sealer, **Figure 40-16B.** Fit the hose clamps over the hose and install the new hose on the fittings.

Position the hose clamps so they are over the metal hose fitting, **Figure 40-16C.** Then, tighten the clamps. Install coolant and pressure test the system. Check all fittings for leaks.

Radiator and Pressure Cap Service

If overheating problems occur and a pressure test shows that the system is not leaking, check the radiator and the pressure cap. They are common sources of overheating. The pressure cap could have bad seals, allowing pressure loss. The radiator may be clogged and not permitting adequate air or coolant flow.

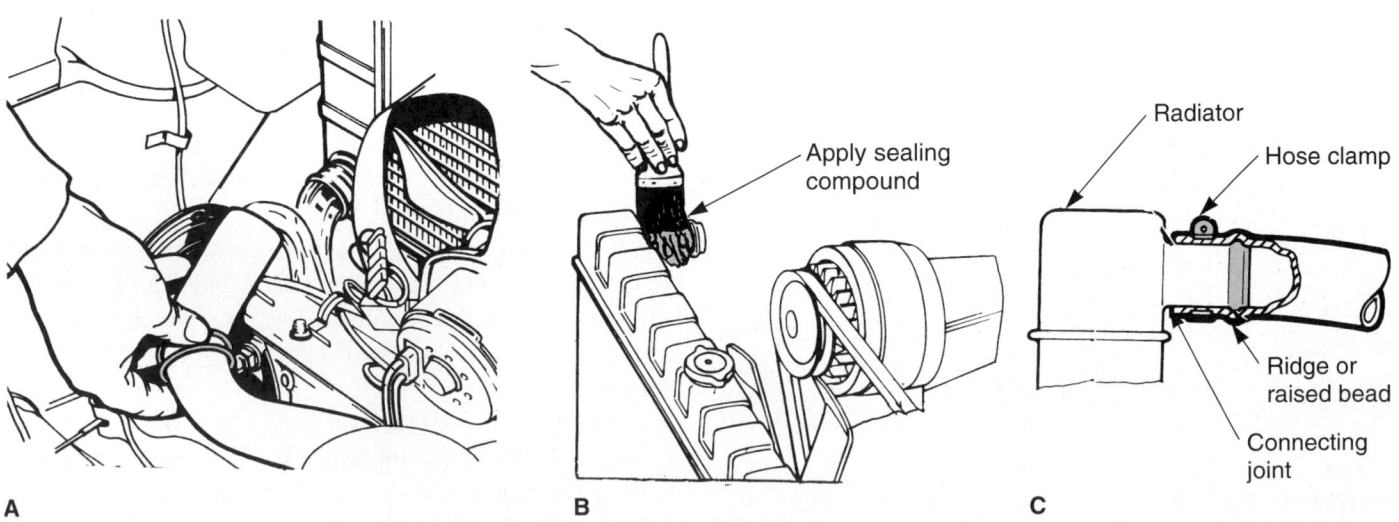

A   **B**   **C**

Figure 40-16. Study the basic steps for hose replacement. A—Loosen hose clamp. Twist and pull hose off the fitting. Cut off the old hose, if needed. B—Clean the fitting and coat it with nonhardening sealer (if the fitting is pitted). C—Slide on a new hose and clamp. Make sure the clamp is positioned inside the bead on the fitting. Tighten the clamp and check for leaks. (Ford and Chrysler)

When tests reveal that the radiator is leaking, it must normally be removed from the vehicle for service. Most radiator repair work is performed in specialized radiator repair shops.

Inspecting the Radiator and Pressure Cap

Inspect the outside of the radiator for debris, such as leaves and road dirt. Also, make sure the radiator shroud is in place and unbroken. These problems can limit air circulation through the core.

If needed, use a water hose to wash debris out of the core. Spray water from the back of the radiator to push debris out of the front. You may also use compressed air if the pressure is low enough not to damage the core (below or equal to rated cap pressure).

Inspect the radiator cap and filler neck. Check for cracks or tears in the cap seal. Check the filler neck sealing surfaces for nicks or dents. Replace the cap or have the neck repaired as needed.

Pressure Testing Radiator Cap

A *radiator cap pressure test* measures the cap opening pressure and checks the condition of the sealing washer. To perform this test, install the cap on a cooling system pressure tester.

Pump the tester to pressurize the cap, **Figure 40-17.** The cap should release air when its rated pressure (pressure stamped on cap) is exceeded. It should also hold the rated pressure for at least one minute. If not, install a new cap.

Radiator Removal

As previously mentioned, a leaking radiator must generally be removed and rebuilt or replaced. Place a catch pan under the cool radiator and open the radiator's petcock. After all the coolant has drained from the radiator, disconnect the hoses and oil cooler lines. You may also have to disconnect wires going to sensors and any electric cooling fans. Sometimes the fans must be removed from the radiator, or they may lift out while still attached to the radiator. Brackets over the top of the radiator or small bolts on the sides of the radiator secure it to its mounting.

Figure 40-18 shows an exploded view of electric fans, hoses, and related parts that must be disconnected for radiator service.

Radiator Repair

Once out of the vehicle, the radiator can be sent to a radiator shop for rebuilding. A *radiator shop* specializes in radiator repair. It has the facilities to properly

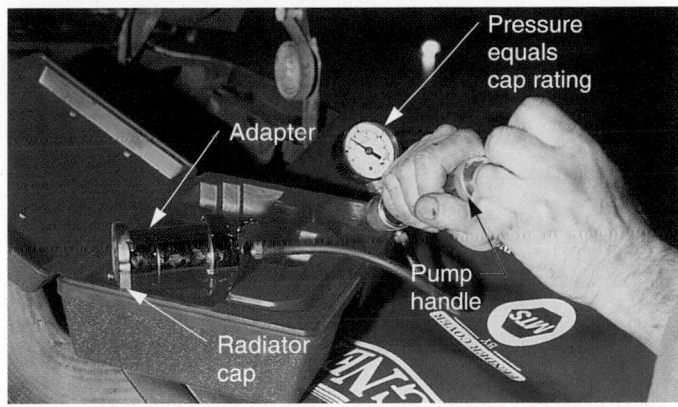

Figure 40-17. A pressure tester can be used to check the setting of the radiator cap. It should open at a temperature within specifications and hold pressure without leaking. If the pressure gauge drops or does not reach rated cap pressure, replace the radiator cap.

disassemble, clean (rod out), repair, reassemble, and pressure test a radiator. Few technicians try to fix a radiator in the general repair shop.

Radiator shops can solder pinhole leaks. They have special cleaning tanks for loosening and removing scale built up inside the radiator. A radiator shop can also remove the tanks and solder in a new core, if needed.

> **Tech Tip**
> Most shops do not repair plastic radiators. It is generally cheaper to replace a plastic radiator than to rebuild it. Also, avoid installing used radiators. With a used unit, you must be concerned with radiator condition. The used radiator could be rusted and ready to leak.

When installing a radiator, make sure the rubber mounts are in place in their brackets. Carefully lower the radiator into place without hitting and damaging it on the engine or body structure.

Fan Belt Service

A loose fan belt will slip and squeal, and may rotate the water pump and fan too slowly. As a result, the engine may overheat. Always inspect the condition and tension (tightness) of fan belts when servicing a cooling system. If a fan belt is cracked, frayed, glazed (hard, shiny surface), or oil soaked, it should be replaced.

Most manufacturers recommend the use of a *belt tension gauge* to measure belt tightness. Use of the gauge ensures that the belt is not too tight or too loose. An overtightened belt will fail quickly and may damage other parts. A loose belt will slip and squeal, or it could fly off its pulleys.

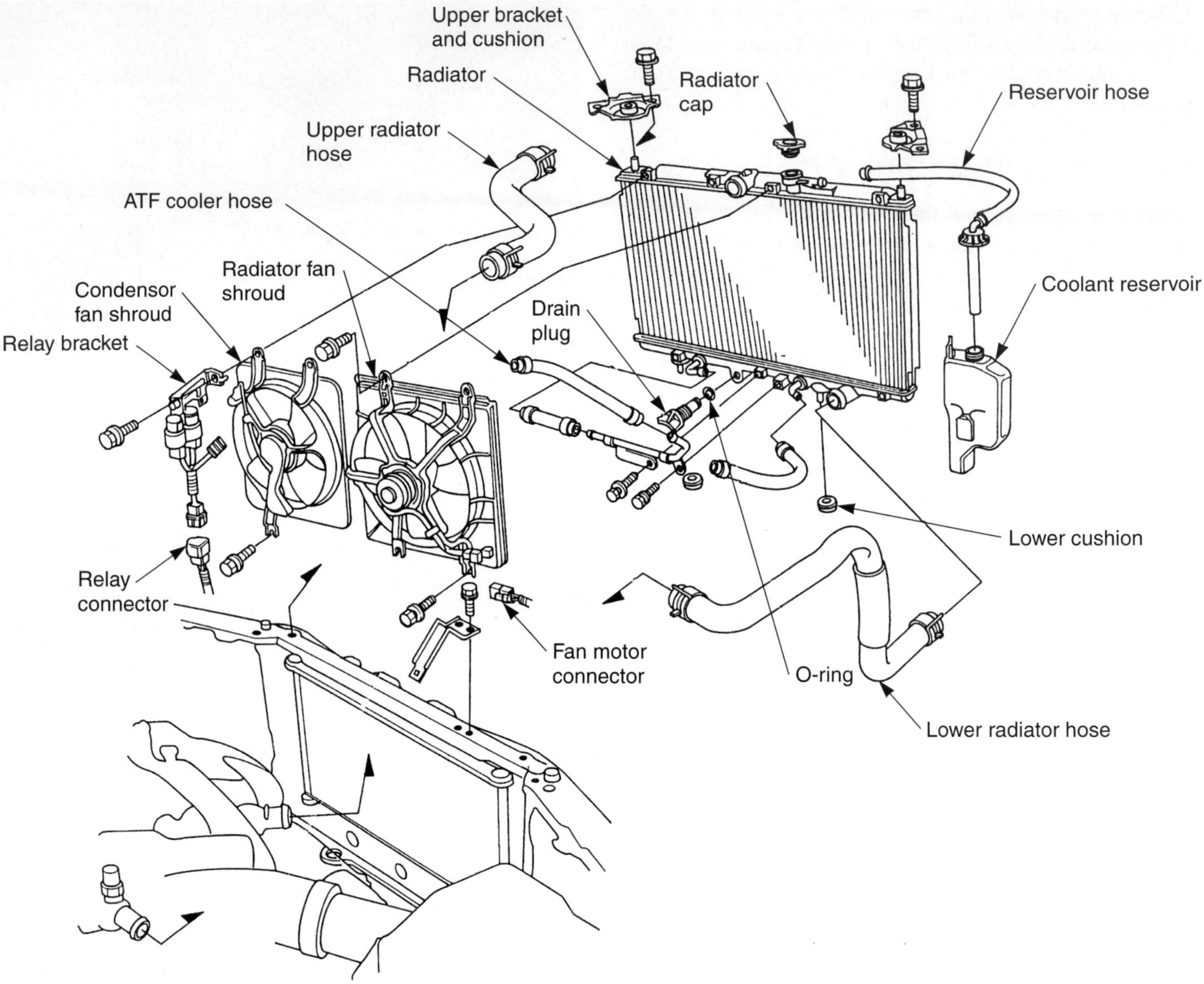

Upper bracket
and cushion

Radiator

Upper radiator
hose

ATF cooler hose

Radiator fan
shroud

Condensor
fan shroud

Relay bracket

Relay
connector

Radiator
cap

Drain
plug

Reservoir hose

Coolant reservoir

Lower cushion

Fan motor
connector

O-ring

Lower radiator hose

Figure 40-18. This exploded view shows how electric fans, radiator, and hoses fit together. (Honda)

Warning
Keep your hands away from a moving engine belt. The belt can pull your fingers into the pulleys, causing severe hand injuries.

Engine Fan Service

A faulty engine fan can cause overheating, over-cooling, vibration, and water pump damage. Always check the fan for bent blades, cracks, and other problems. Flexible fans are especially prone to these problems. If any troubles are found, replace the fan.

Warning
A fan with cracked or bent blades is extremely dangerous. Broken blades can be thrown out with great force, causing severe lacerations!

Testing a Fan Clutch

To test a thermostatic fan clutch, start the engine. The fan should slip when cold. When the engine warms, the clutch should engage. Air should begin to flow through the radiator and over the engine. You will be able to hear and feel the rush of air when the fan clutch locks up.

If the fan clutch is locked all the time (cold and hot), it is defective and must be replaced. Excessive play or oil leakage also indicates fan clutch failure.

Electric Cooling Fan Service

Most electric cooling fans are controlled by a heat-sensitive switch or sensor located somewhere in the cooling system (radiator, engine block, thermostat housing). When the engine is cold, the switch keeps the

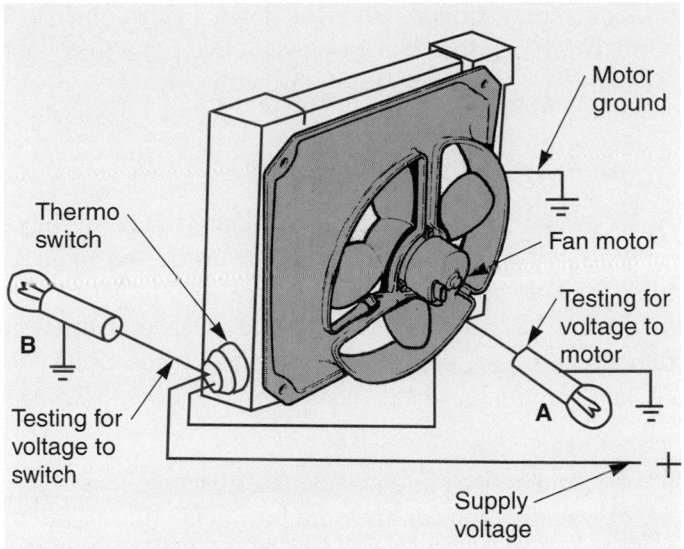

Figure 40-19. Testing a basic electric fan circuit. A—Check for power to the fan with the engine warm. The light should glow. B—With no power to the fan, check the action of the fan switch. The switch should be open when cold and closed when hot. If the relay and ECM are in the fan circuit, test them individually. (Honda)

electric fan motor off to speed engine warm-up. Then, when a predetermined temperature is reached, the switch closes and the fan begins to cool the engine.

Testing an Electric Cooling Fan

To test an electric cooling fan, observe whether the fan turns on when the engine is warm. Make sure the fan motor is spinning at a normal speed and is forcing enough air through the radiator.

If the fan does not function, check the fuse, electrical connections, relay, and supply voltage to the motor. Refer to **Figure 40-19.**

If the fan motor fails to operate with voltage applied, the motor should be replaced. If the engine is warm and no voltage is supplied to the fan motor, check the action of the fan switch or sensor, fan relay, and ECM. Use either a voltmeter or high-impedance test light. The switch or sensor should have factory-specified resistances at specific temperatures.

If these tests do not locate the trouble, refer to a factory service manual for instructions. There may be a defective relay, connection, ECM, or other problem.

Freeze Plug Service

A leaking engine *freeze plug* (core plug) is a frequent cause of coolant loss and overheating. Since the engine's freeze plugs are thinner than the metal in the engine

block or head, they will rust through before the other parts of the engine.

 To replace a freeze plug, use the following procedure:
1. Drive a drift or large full-shank screwdriver through the plug, **Figure 40-20A.**
2. Pry sideways, without scraping the engine block or cylinder head. The plug should pop out.
3. Sand the core plug hole in the engine and wipe it clean.
4. Coat the plug hole and plug with non-hardening sealer.
5. Drive a new freeze plug squarely into position, **Figure 40-20B.**

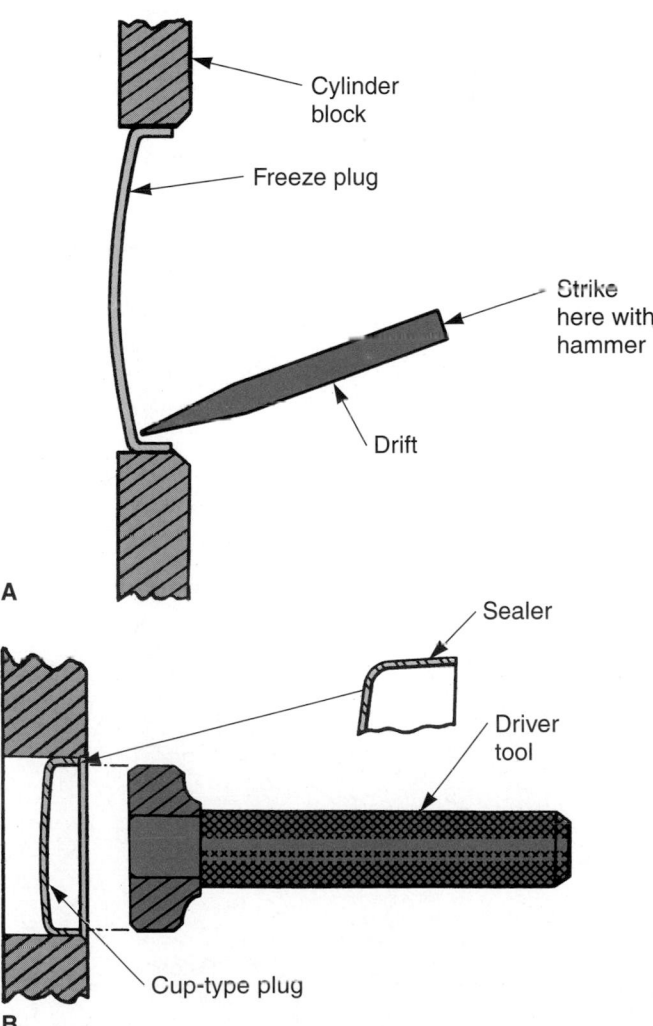

Figure 40-20. Freeze plug replacement. A—Drive a drift or full-shank screwdriver through old plug and pry it out. B—After cleaning and coating the hole with sealer, drive the new freeze plug into place. Drive the plug in squarely and to the proper depth. (Ford Motor Co.)

Figure 40-21. To check for coolant contamination, wipe a finger inside the filler neck or reservoir tank. Badly rusted coolant requires replacement of coolant and possibly system flushing. (Chrysler)

Expansion freeze plugs are available for tight quarters. They are installed by tightening a nut, which causes the plug to expand and lock into the hole. This allows the plug to be installed without hammering.

Coolant Service

The coolant, or antifreeze solution, should be checked and changed at regular intervals. After

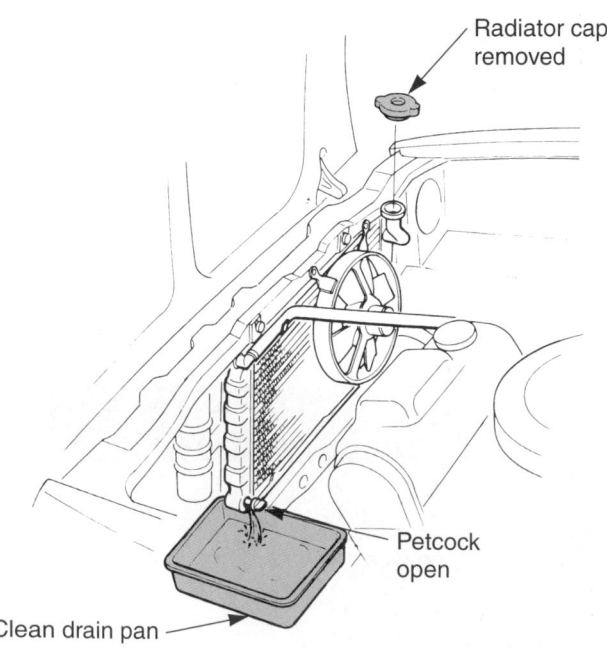

Figure 40-22. To drain coolant, remove radiator or reservoir cap. Place a pan under the drain fitting. Then, turn the petcock. Many have left-hand threads and must be turned clockwise to open. (Honda)

prolonged use, coolant will break down and become very corrosive. It can lose its rust preventative properties, and the cooling system can rapidly fill with rust.

Inspecting Coolant

A visual inspection of the coolant will help determine its condition. Rub your finger inside the radiator filler neck or reservoir tank, **Figure 40-21.** Check for rust, oil (internal engine leak), scale, or transmission fluid (leaking oil cooler). Also, find out how long the coolant solution has been in service.

 Tech Tip
Old antifreeze and water will become very acidic. You can measure how acidic the solution is with a voltmeter. Ground one voltmeter lead and submerse the other lead in the antifreeze solution. If the solution generates a voltage above 0.5 volts, the mixture is contaminated or deteriorated and should be drained and replaced.

If the coolant is contaminated or too old, it should be replaced. If badly rusted, you may also need to flush (clean) the system.

Changing Coolant

Coolant should be changed when contaminated or when two years old. Check a service manual for exact change schedules.

With the system cool and the pressure cap removed, loosen the petcock on the bottom of the radiator. Allow the old coolant to drain into a pan, **Figure 40-22.**

If the coolant is not filled with rust, you may refill the system. Tighten the petcock. Pour in the needed amount (about two gallons, or 7.6 liters) of antifreeze and the same amount of water.

Start and warm the engine. The coolant level may drop when the thermostat opens. Add more coolant, if needed. Then install the radiator or reservoir cap.

Figure 40-23 shows how to tell when the cooling system is full. Note the difference between checking closed and open cooling systems.

If the system has a bleed screw, crack the screw open until all air is purged. As soon as coolant leaks out, tighten the fitting or screw.

Testing Coolant Strength

Coolant strength is a measurement of the concentration of antifreeze compared to water. It determines the freeze-up protection of the solution.

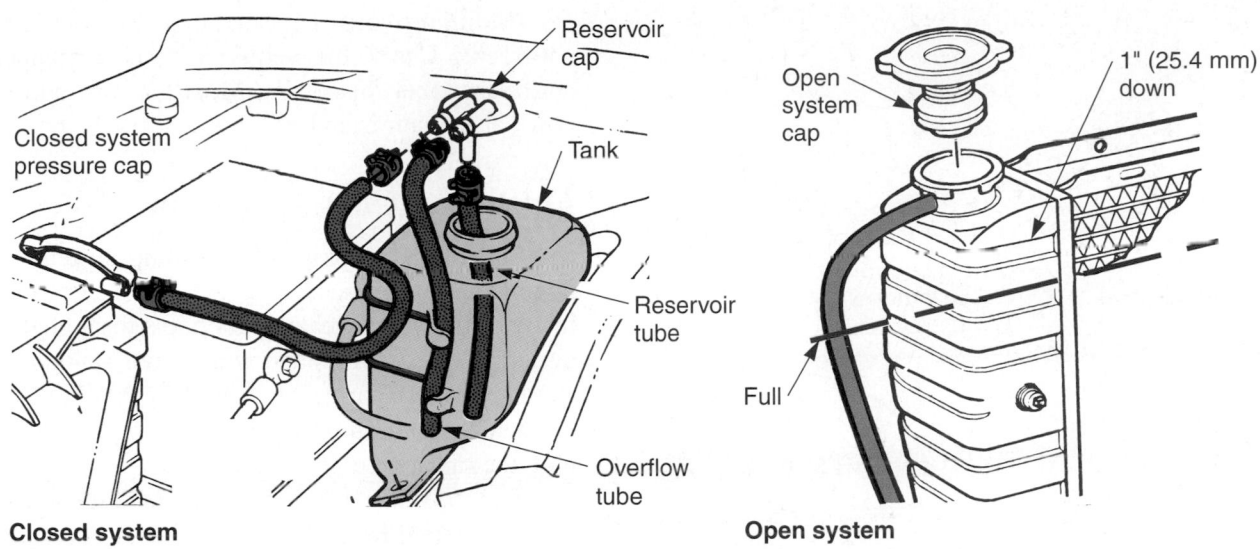

Figure 40-23. Checking the coolant level. With a closed system, coolant should be even with correct marking on reservoir with coolant at operating temperature. With an open system, coolant should be about 1″ (25 mm) below the top of the tank. (GMC and Ford)

A *cooling system hydrometer* is commonly used to measure the freezing point of the cooling system antifreeze solution. One type is shown in **Figure 40-24.** Submerse the tester inlet tube in the coolant. Squeeze and release the tester bulb. This will fill the tester with coolant. Note the indicator reading on the tester. It will give freeze-up protection in degrees. Add more antifreeze to the solution if freeze-up protection is too low.

 Caution
The most common reason for cracked blocks and cylinder heads is improper coolant protection. If the coolant in the engine freezes in cold weather, the ice will expand and break major engine parts.

A *refractometer* is another type of coolant strength measuring device. Draw coolant into the tester, as in **Figure 40-25.** Place a few drops of coolant on the measuring window (surface). Aim the tester at light and view through the tester. The scale in the refractometer will show freeze protection.

Minimum Coolant Strength

Minimum coolant strength should be several degrees lower than the lowest normal temperature for the climate of the area. For example, if the lowest normal temperature for the area is 10°F (–23°C), the coolant should test to –20°F (–29°C).

A *50/50 mix* of antifreeze and water is commonly used to provide protection for most weather conditions.

For example, two gallons (7.6 L) of antifreeze is mixed with two gallons of water in a four-gallon cooling system.

Corrosion of Aluminum

Many late-model vehicles use aluminum cooling system and engine parts. Radiators, water pumps,

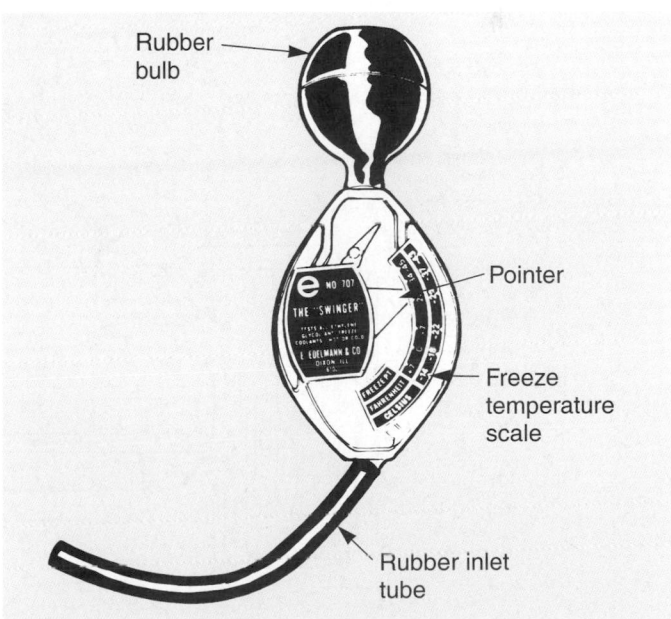

Figure 40-24. A cooling system hydrometer will check the freeze-up protection temperature. Squeeze and release the bulb to draw coolant into the tester. The needle will float and show freeze protection. Some testers must be corrected for coolant temperature differences. (Edelmann)

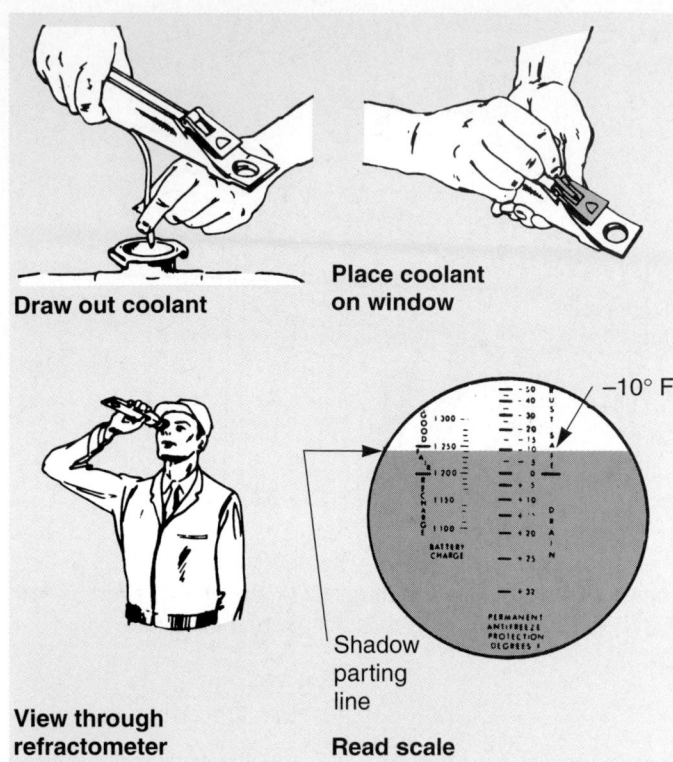

Figure 40-25. A refractometer can be used to measure coolant protection. Parting line of shadow on scale equals reading. (Oldsmobile)

cylinder heads, blocks, and intake manifolds can be made of aluminum. Antifreeze designed for aluminum components must be used in these systems.

Aluminum can be corroded by some types of antifreeze. Check the vehicle's service manual or the antifreeze label for details. Most types of antifreeze are now "aluminum friendly."

Flushing a Cooling System

Flushing (cleaning) of a cooling system should be done when rust or scale is found in the system. Flushing involves running a cleaning chemical through the cooling system. This dissolves and washes out contaminants.

Rust is very harmful to a cooling system. It can cause premature water pump wear. Rust can also collect in and clog the radiator or heater core tubes.

Fast flushing is a common method of cleaning a cooling system because the thermostat does not have to be removed from the engine. Look at **Figure 40-26.**

A water hose is connected to a heater hose fitting. The radiator cap is removed and the drain cock is opened. When the water hose is on and water flows into the system, rust and loose scale are removed.

Reverse flushing of a radiator requires a special adapter that is connected to the radiator outlet tank by a piece of hose. Another hose is attached to the inlet tank. Compressed air, under low pressure, is used to force a cleaning solution through the core backwards. This can be done on the engine block as well. See **Figure 40-27.**

Chemical flushing is needed when scale buildup in the system is causing engine overheating. A chemical cleaner is added to the coolant. The engine is operated for

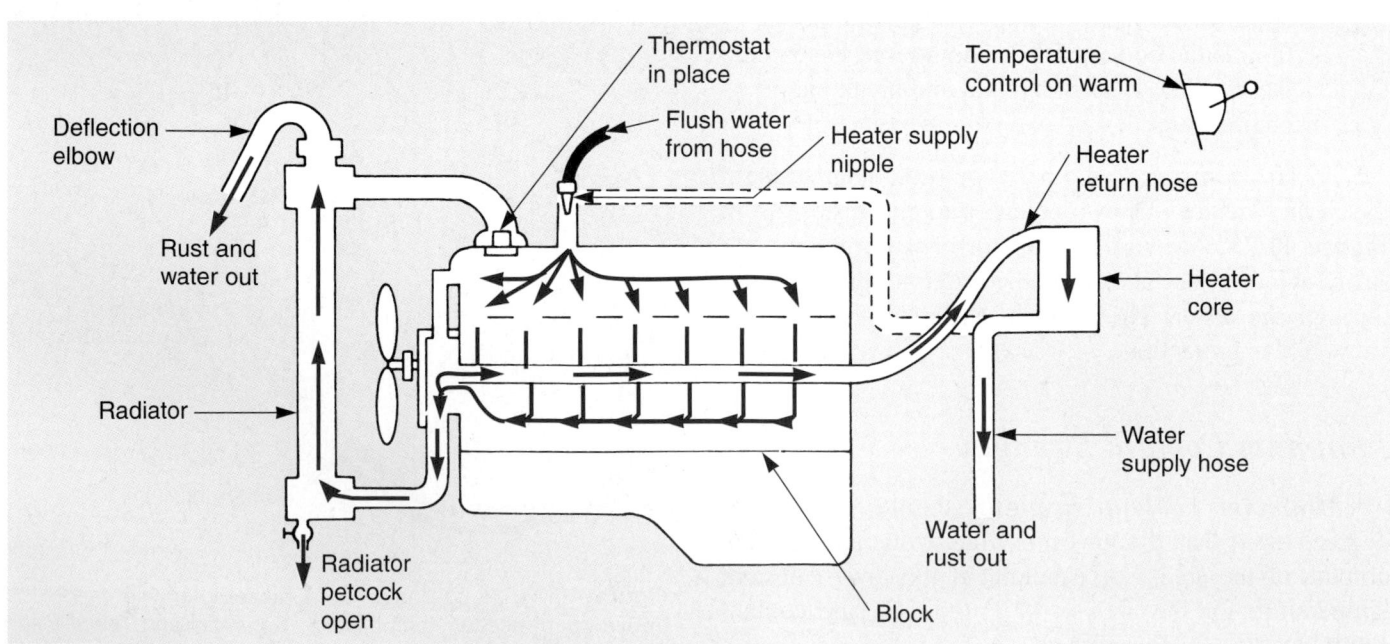

Figure 40-26. Study fast flushing a cooling system. A water hose is connected to the heater hose fitting. This will force water and rust out of the heater hose and the top of the radiator. (Union Carbide Corp.)

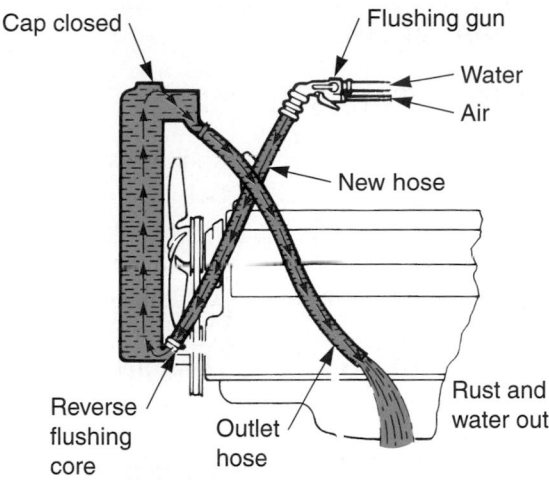

Reverse flushing a radiator

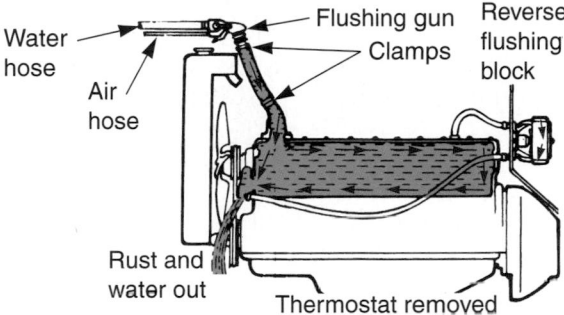

Reverse flushing a block

Figure 40-27. Study reverse flushing of a radiator and engine block. (Chrysler)

a specific amount of time to allow the chemical to act on the scale. Then the system is flushed with water.

 **Warning**
Always follow manufacturer's instructions when using a cooling system cleaning agent. The chemical may cause eye and skin burns. Wear rubber gloves and full face protection.

After flushing, always add the recommended type and amount of antifreeze. Antifreeze has rust inhibitors and lubricants for the water pump. Never leave plain water in the system.

Temperature Gauge Service

A defective temperature gauge may read hot or cold when the engine is actually at its proper operating temperature. The customer may complain about the gauge always reading cold or hot, or the complaint may be erratic movement of gauge pointer.

To quickly test a temperature gauge, disconnect the wire going to the temperature gauge sending unit. Shown

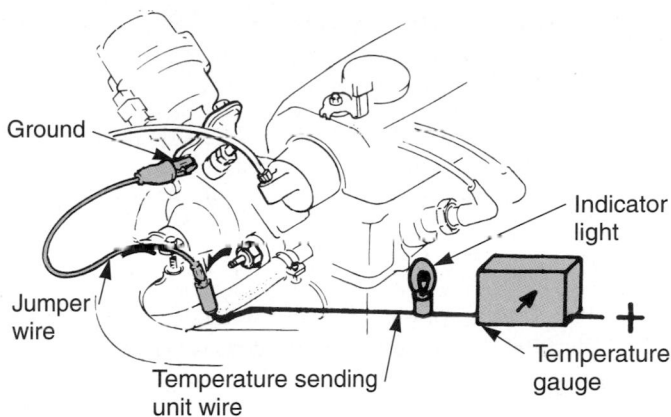

Figure 40-28. To check the action of a temperature gauge or indicator light, you can sometimes ground the wire to the temperature sending unit. This should cause the gauge to read hot or cause the light to glow. If not, the circuit before the sending unit is faulty. If the gauge or light functions, the sending unit may be bad. (Honda)

in **Figure 40-28,** the sending unit is normally located on the engine.

Using a jumper wire, ground the gauge wire to the engine block. Then, turn the ignition key switch on and watch the temperature gauge. It will normally swing to hot when the wire is grounded. It should return to cold when the wire is ungrounded.

If the gauge begins to function when grounded, the sending unit is defective and should be replaced. If the gauge does not function when grounded, either the gauge circuit or the gauge is faulty.

A **gauge tester** can also be used to check gauge and sending unit operation. It is a special testing device with a variable resistor. Set the tester to a specified resistance and the temperature gauge should read as specified.

To test a temperature indicating light, perform the same basic operation. The light should glow when the sending unit wire is grounded. It should go out when the wire is ungrounded.

 Caution
If available, use a gauge tester to check gauge and sending unit operation. Some temperature gauges could be damaged by grounding the sending unit wire.

Summary

- A cooling system is extremely important to the performance and service life of an engine.
- Many on-board diagnostic systems will trip a trouble code when certain cooling related circuits are operating or seem to be operating out of range.

- Coolant leaks show up as wet, discolored (darkened or rust colored) areas in the engine compartment or on the ground.

- Overcooling may be indicated by slow engine warm-up, insufficient warmth from the heater, low fuel economy, sluggish engine performance, or a low reading from the coolant temperature sensor.

- A cooling system pressure test is used to quickly locate leaks. Low air pressure is forced into the system.

- A combustion leak test checks for the presence of combustion gases in the engine coolant.

- When water, antifreeze, and oil mix, the solution turns milky white in color.

- A bad water pump may leak coolant, fail to circulate coolant, or produce a grinding sound.

- A stuck thermostat can either cause engine overheating or engine overcooling.

- A cooling system bleed screw is sometimes provided to help remove trapped air when refilling the system.

- Hardened hoses become very brittle and crack from engine vibration.

- Softened hoses feel very mushy and have been contaminated with oil or other system fluid that breaks down the rubber in the hose.

- A radiator cap pressure test measures cap opening pressure and checks the condition of the sealing washer.

- A loose fan belt will slip and squeal, and may rotate the water pump and fan too slowly.

- A faulty engine fan can cause overheating, overcooling, vibration, and water pump damage.

- A leaking engine freeze plug is a frequent cause of coolant loss and overheating.

- Coolant strength is a measurement of the concentration of antifreeze compared to water.

- A cooling system hydrometer is commonly used to measure the freezing point of the cooling system antifreeze solution.

- A 50/50 mix of antifreeze and water is commonly used to provide protection for most weather conditions.

- Flushing (cleaning) of a cooling system should be done when rust or scale is found in the system.

Important Terms

Cooling system
  diagnosis chart
Coolant leaks
Overheating
Overcooling
Coolant leaks
Internal engine leak
Cooling system
  pressure test
Pressure tester
Combustion leak test
Block tester,
Milky white
Overheating
Overcooling
Thermostat housing
Thermostat housing
  seal
Cooling system bleed
  screw

Radiator cap pressure
  test
Radiator shop
Belt tension gauge
Freeze plug
Expansion freeze plugs
Coolant strength
Cooling system
  hydrometer
Refractometer
Minimum coolant
  strength
50/50 mix
Flushing
Fast flushing
Reverse flushing
Chemical flushing
Gauge tester

Review Questions—Chapter 40

Please do not write in this text. Place your answers on a separate sheet of paper.

1. An engine can still operate for an extended period without a cooling system. True or False?

2. List seven checks that should be done when inspecting a cooling system.

3. Why should you stand to one side of a spinning engine fan?

4. What can happen if you remove a radiator cap with the coolant at operating temperature?

5. List and explain ten common causes of engine overheating.

6. A cooling system _____ _____ is used to quickly find leaks in the system.

7. A(n) _____ _____ test checks for the presence of combustion gases in the engine coolant, indicating an engine problem.

8. Which of the following is *not* a typical cause of engine overcooling?
   (A) Stuck thermostat.
   (B) Locked fan clutch.
   (C) Ice in cooling system.
   (D) Shorted electric fan switch.

9. When water, antifreeze, and oil mix, the solution turns _____ _____ in color.

10. A customer complains of sluggish engine performance and a lack of adequate warmth from the heater. Technician A says that this could not be caused by the cooling system. There may be separate problems with the engine and heating system. Technician B says that a missing or stuck open thermostat might cause these symptoms. The thermostat should be checked first before checking other possible components. Who is right?
    (A) A only.
    (B) B only.
    (C) Both A and B.
    (D) Neither A nor B.

11. How do you replace an engine freeze plug?

12. A(n) _____ is used to measure the freezing point of cooling system coolant.
    (A) voltmeter
    (B) refractometer
    (C) analyzer probe
    (D) temperature gauge
    (E) None of the above.

13. Which of the following cooling system cleaning methods involves the use of water or special chemicals?
    (A) Fast flushing.
    (B) Reverse flushing.
    (C) Chemical flushing.
    (D) All of the above.

14. How can you quickly determine if a dash temperature gauge is functioning?

15. While testing the strength of an automobile's coolant, it was found that the coolant tested at 33°F. What is this equivalent in Celsius?
    (A) 0.56°C.
    (B) 8.32°C.
    (C) 0.63°C.
    (D) 11.63°C.

✹ ASE-Type Questions

1. Which of the following is *not* a common indicator that an engine is overheating?
   (A) *Boiling coolant.*
   (B) *Slow engine warm-up.*
   (C) *Glowing temperature light.*
   (D) *High temperature gauge reading.*

2. A cooling system pressure tester is a(n):
   (A) *electric tester bulb.*
   (B) *leak detecting fluid.*
   (C) *hand-operated air pump.*
   (D) *None of the above.*

3. Overcooling may be caused by each of these *except:*
   (A) *ice in coolant.*
   (B) *stuck thermostat.*
   (C) *locked fan clutch.*
   (D) *shorted fan switch.*

4. When water, antifreeze, and oil mix, it turns:
   (A) *pale yellow.*
   (B) *milky white.*
   (C) *transparent.*
   (D) *rusty brown.*

5. A bad water pump may:
   (A) *leak coolant.*
   (B) *fail to circulate coolant.*
   (C) *produce a grinding sound.*
   (D) *All of the above.*

6. While discussing thermostat service, Technician A says a stuck thermostat can cause engine overheating. Technician B says a stuck thermostat can cause engine overcooling. Who is right?
   (A) *A only.*
   (B) *B only.*
   (C) *Both A and B.*
   (D) *Neither A nor B.*

7. Each of these is a hose problem that may effect cooling system performance *except:*
   (A) *shrinking.*
   (B) *swelling.*
   (C) *cracking.*
   (D) *hardening.*

8. A radiator shop will properly:
   (A) *solder a radiator.*
   (B) *rod out a radiator.*
   (C) *disassemble a radiator.*
   (D) *All of the above.*

9. A faulty engine fan may cause each of these engine problems *except:*
   (A) *revving.*
   (B) *vibrating.*
   (C) *overcooling.*
   (D) *overheating.*

10. Coolant should be changed when:
    (A) *two years old.*
    (B) *contaminated.*
    (C) *Both A and B.*
    (D) *Neither A nor B.*

Activities—Chapter 40

1. Obtain a cooling system thermostat and test it for proper opening at its rated temperature using a pan of water and stove or hot plate.

2. Demonstrate the use of a hydrometer to test coolant strength. If a refractometer is available, demonstrate its use as well.

Cooling System Diagnosis		
Condition	**Possible cause**	**Correction**
Temperature gauge reads low.	1. Thermostat is stuck open. 2. Temperature gauge is not connected to the coolant sensor. 3. Faulty temperature gauge 4. Coolant level low during cold ambient temperature.	1. Replace thermostat if necessary. 2. Check the connector at the engine coolant sensor. Repair as necessary. 3. Check gauge operation. Repair as necessary. 4. Check coolant level in the overflow tank and at the radiator. Inspect the system for leaks. Repair as necessary.
Temperature gauge reads high or engine coolant warning lamp illuminates. Coolant may or may not be lost from system.	1. Trailer being towed, steep hill being climbed, vehicle being operated in slow moving traffic, or engine idling during high ambient temperatures with air conditioning on. High altitudes can aggravate these conditions. 2. Faulty temperature gauge. 3. Temperature warning lamp illuminating unnecessarily. 4. Low coolant in overflow/reserve tank and radiator. 5. Pressure cap not installed tightly. 6. Poor seals at radiator cap. 7. Coolant level low in radiator but not in coolant overflow/reserve tank. This means the radiator is not drawing coolant from the coolant overflow/reserve tank as the engine cools. 8. Freeze point of coolant not correct. Mixture may be to rich. 9. Coolant not flowing through system. 10. Radiator or A/C condenser fins are dirty. 11. Radiator core is plugged or corroded. 12. Fuel or ignition system problems. 13. Dragging brakes. 14. Bug screen is being used, causing reduced airflow. 15. Thermostat partially or completely shut. 16. Electric cooling fan not operating properly. 17. Cylinder head gasket leaking. 18. Heater core leaking.	1. This may be a temporary condition and repair is not necessary. Turn off the air conditioning and drive the vehicle without any of the previous conditions. The gauge should return to the normal range. If the gauge does not return to the normal range, determine the cause of overheating and repair. 2. Check gauge. Repair as necessary. 3. Check warning lamp operation. Repair as necessary. 4. Check for cooling leaks and repair as necessary. 5. Tighten cap. 6. (a) Check condition of cap and cap seals. Replace cap, if necessary. (b) Check condition of filler neck. If neck is bent or damaged, replace neck. 7. (a) Check condition of radiator cap and cap seals. Replace cap, if necessary. (b) Check condition of filler neck. If neck is damaged, replace filler neck. (c) Check condition of hose from filler neck to coolant tank. It should be tight at both ends without any kinks or tears. Replace hose, if necessary. (d) Check coolant overflow/reserve tank and tank hoses for blockage. Repair as necessary. 8. Check coolant. Adjust mixture as required. 9. Check for coolant flow. If flow is not observed, determine reason for lack of flow and repair as necessary. 10. Clean insects or debris from fins. 11. Replace or re-core radiator. 12. Check systems and repair as necessary. 13. Inspect brake system and repair as necessary. 14. Remove bug screen. 15. Check thermostat operation and replace as necessary. Refer to thermostats in this group. 16. Check electric fan operation and repair as necessary. 17. Check cylinder head gasket for leaks. Repair as necessary. 18. Check heater core for leaks. Repair as necessary.

(Continued)

Cooling System Diagnosis		
Condition	**Possible cause**	**Correction**
Temperature gauge reading is inconsistent (fluctuates, cycles or is erratic).	1. Temperature gauge or engine-mounted gauge sensor defective or shorted. Also, corroded or loose circuit wiring. 2. Coolant level low in radiator (air will build up in the cooling system causing the thermostat to open late). 3. Cylinder head gasket leaking, allowing exhaust gas to enter cooling system and causing thermostat to open late. 4. Water pump impeller loose on shaft. 5. Loose accessory drive belt (water pump slipping). 6. Air leak on the suction side of water pump allows air to build up in cooling system. This causes thermostat to open late.	1. Check operation of gauge and repair as necessary. 2. Check and correct coolant leaks. 3. (a) Check for cylinder head gasket leaks with a commercially available Block Leak Tester. Repair as necessary. (b) Check for coolant in the engine oil. Inspect for white steam emitting from exhaust system. Repair as necessary. 4. Check water pump and replace as necessary. 5. Check belts and correct as necessary. 6. Locate leak and repair as necessary.
Pressure cap is blowing off steam and/or coolant to coolant tank. Coolant level may be high in coolant overflow tank.	1. Pressure relief valve in radiator cap is defective.	1. Check condition of radiator cap and cap seals. Replace cap as necessary.
Coolant loss to the ground without pressure cap blowoff. Gauge is reading high.	1. Coolant leaks in radiator, cooling system hoses, water pump, or engine.	1. Pressure test and repair as necessary.
Detonation or pre-ignition (not caused by ignition system). Gauge may or may not be reading high.	1. Engine overheating. 2. Freeze point of coolant not correct. Fuel mixture is too rich or too lean.	1. Check reason for overheating and repair as necessary. 2. Check the freeze point of the coolant. Adjust the glycol to water ratio as required.
Hose or hoses collapse when engine is cooling.	1. Vacuum created in cooling system on engine cool-down is not being relieved through coolant reserve/overflow system.	1. (a) Radiator cap relief valve stuck. Replace cap, if necessary. (b) Hose between coolant reserve/overflow tank and radiator is kinked. Repair as necessary. (c) Vent at coolant reserve/overflow tank is plugged. Clean vent and repair as necessary. (d) Overflow tank is internally blocked or plugged. Check for blockage and repair as necessary.
Electric radiator fan runs all the time.	1. Defective fan relay, control module, or engine coolant temperature sensor. 2. Low coolant level.	1. Repair as necessary. 2. Repair as necessary.
Electric radiator fan will not run. Gauge reading high or hot.	1. Fan motor defective. 2. Fan relay, powertrain control module (PCM), or engine coolant temperature sensor defective. 3. Blown radiator fan fuse.	1. Test motor and repair as necessary. 2. Test components and repair as necessary. 3. Determine reason for blown fuse and repair as necessary.

Chapter 41

Lubrication System Fundamentals

After studying this chapter, you will be able to:

- List the basic parts of a lubrication system.
- Summarize the operation of a lubrication system.
- Describe the construction of lubrication system parts.
- Compare different lubrication system designs.
- Explain the characteristics and ratings of engine oil.
- Discuss safety procedures that should be followed when working with the lubrication system.
- Correctly answer ASE certification test questions that require a knowledge of lubrication system construction and operation.

The *lubrication system* forces oil to high friction points in the engine to protect moving parts from friction, wear, and damage. It is one of the most important engine systems affecting engine service life.

Without a lubrication system, friction between rapidly moving, heavily loaded parts would destroy an engine in a matter of minutes. Many engine parts would quickly overheat and score or partially melt from this friction. Engine bearings, piston rings, cylinder walls, and other parts could be ruined. Without circulating engine oil, engine parts will lock up or weld together and the crankshaft will no longer rotate freely. This makes it critical that you understand the design and construction of modern engine lubrication systems.

Note

If needed, review the material in Chapters 1 and 10 that introduces lubrication system operation and maintenance.

Lubrication System Functions

A lubrication system has several important functions. The system:

- Reduces friction and wear between moving parts.
- Helps transfer heat away from engine parts.
- Cleans the inside of the engine by removing contaminants (metal, dirt, plastic, rubber, and other particles).
- Cuts power loss and increases fuel economy.
- Absorbs shocks between moving parts to quiet engine operation and increase engine life.

The design of modern engines and the properties of engine oil allow the lubrication system to accomplish these functions.

Lubrication System Operation

Shown in **Figure 41-1,** a lubrication system consists of the following:

- *Engine oil*—lubricant for moving parts in engine.
- *Oil pan*—reservoir or storage area for engine oil.
- *Oil pump*—forces oil throughout the inside of the engine.
- *Pressure relief valve*—limits maximum oil pump pressure.
- *Oil filter*—strains out impurities in the oil.
- *Oil galleries*—oil passages through the engine.

With the engine running, the oil pump pulls engine oil out of the oil pan. Before oil enters the pump, a screen on the pickup tube removes large particles. The pump then pushes the oil through the oil filter and oil galleries.

The oil filter cleans the oil, removing very small particles. The filtered oil then flows to the camshaft, crankshaft, balancer shafts, lifters, rocker arms, and other moving parts.

When oil leaks out of the engine bearings, it sprays on the outside of internal engine parts. For example, when oil leaks out of the connecting rod bearings, it sprays on the cylinder walls. This lubricates the piston

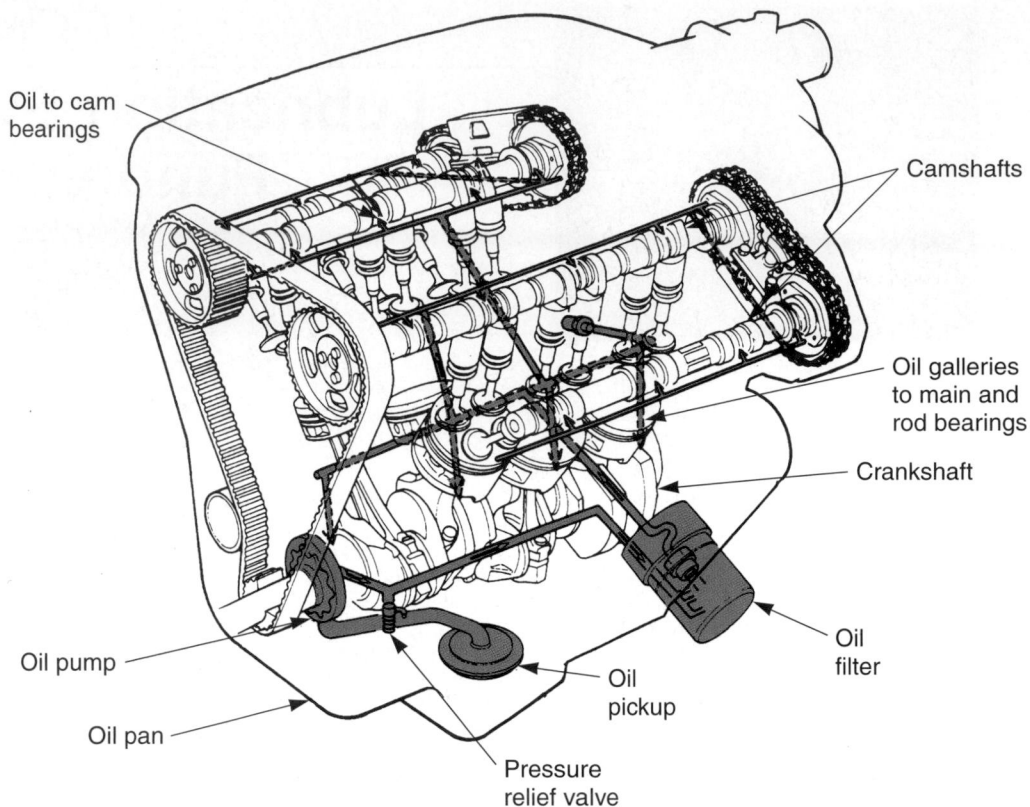

Figure 41-1. Study the basic parts of a typical lubrication system. Also, trace the flow of oil from the pan through the engine. (Ford)

rings, pistons, wrist pins, and cylinders. Oil finally drains back into the oil pan for recirculation.

Figure 41-2 is an exploded view of the major parts of a lubrication system.

Engine Oil

Engine oil, also called *motor oil,* is needed to keep moving parts in an engine from making direct contact. Its main purpose is to reduce friction. Engine oil is commonly refined from *petroleum* (crude oil), which is extracted from deep within the earth.

Synthetic oils are manufactured oils made from substances other than crude oil. They can be made from vegetable oils, for example.

An *oil film* (thin layer of oil) separates engine parts to prevent metal-on-metal contact. Without the oil film, the parts would rub together and wear rapidly. This is shown in **Figure 41-3.**

Oil Clearance

Oil clearance, or *bearing clearance,* is the small space between moving engine parts for the lubricating oil film. The clearance allows oil to enter the bearing to prevent part contact. See **Figure 41-4.**

For example, a connecting rod bearing typically has an oil clearance of about 0.002″ (0.05 mm). This clearance is large enough to allow oil entry. However, it is also small enough to keep the parts from "hammering together" and knocking during engine operation (reciprocating action).

Bearing Types

There are two basic types of engine bearings: friction bearings and antifriction bearings. Each can be used in many automotive assemblies, not only engines.

A *friction bearing,* or *plain bearing,* has two smooth surfaces sliding on each other. It is the most common type of bearing used in an engine. Look at **Figure 41-5.**

Crankshaft main bearings, connecting rod bearings, and cam bearings are normally friction bearings. They require a constant supply of oil under pressure for proper service life.

An *antifriction bearing* uses balls or rollers to avoid a sliding action between the bearing surfaces. They are only used in a few places in an engine. See **Figure 41-5.**

A good example of an engine antifriction bearing is a roller lifter in a diesel engine. The roller cuts high friction and wear between the camshaft lobe and the bottom of the lifter.

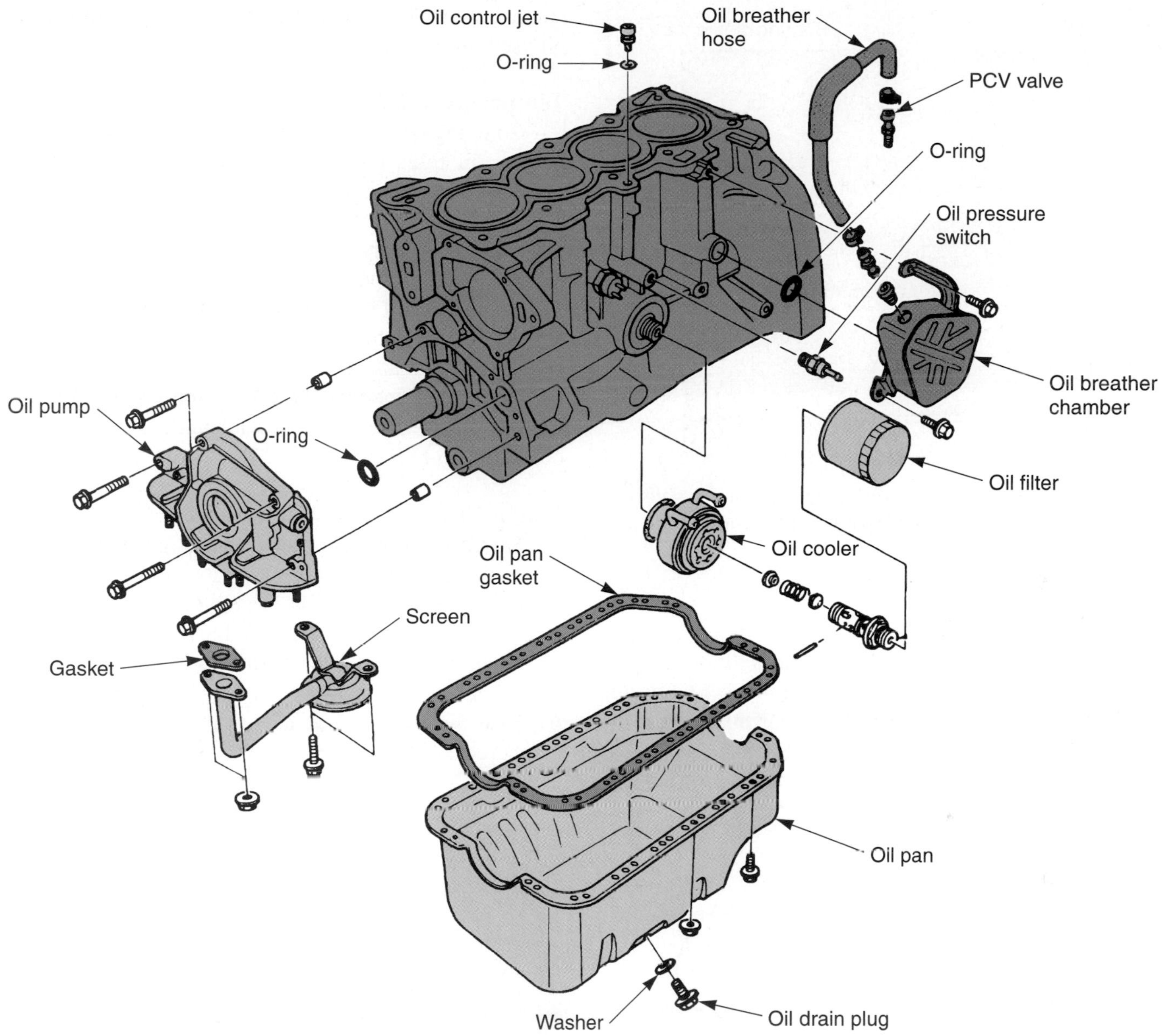

Oil control jet

Oil breather hose

O-ring

PCV valve

O-ring

Oil pressure switch

Oil pump

O-ring

Oil breather chamber

Oil filter

Oil cooler

Oil pan gasket

Screen

Gasket

Oil pan

Washer

Oil drain plug

Figure 41-2. This exploded view shows the major parts of a lubrication system. It also shows the crankcase ventilation system parts and oil cooler. (Honda)

Antifriction bearings do not require as much lubrication as friction bearings. Usually, splash oiling is sufficient.

Oil Viscosity (Weight)

Oil viscosity, also called *oil weight,* is the thickness or fluidity (flow ability) of the engine oil. A high viscosity oil is very thick and resists flow. A low viscosity oil is thin and flows easily.

A *viscosity numbering system* is used to rate the thickness of engine oil. A high viscosity number would indicate relatively thick oil. A lower viscosity number would denote a thinner oil. Look at **Figure 41-6.**

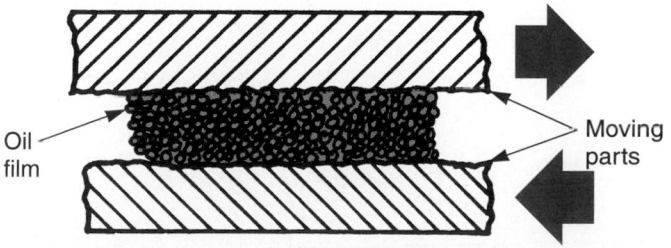

Oil film

Moving parts

Figure 41-3. Close-up view of the clearance between moving parts shows how an oil film keeps parts from touching and rubbing together.

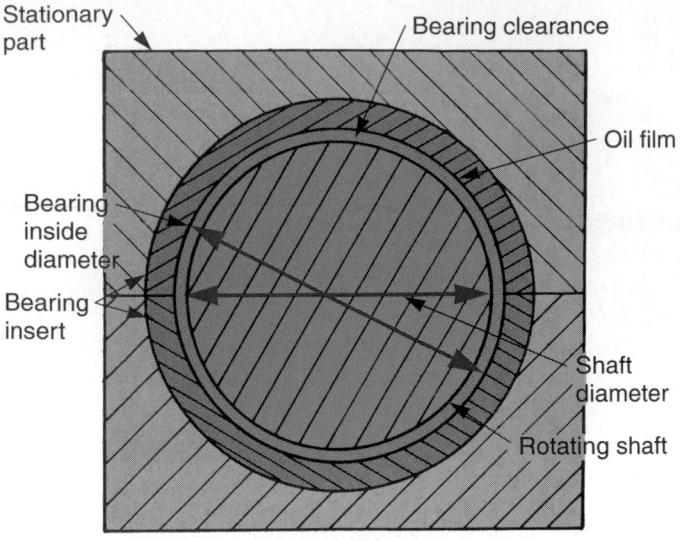

Figure 41-4. Oil clearance, also called bearing clearance, allows the oil film to hold the spinning shaft away from the stationary part.

The oil's *viscosity number* is printed on the oil container. The SAE (Society of Automotive Engineers) standardized this numbering system. For this reason, oil viscosity is written SAE 10, SAE 20, SAE 30, etc.

Engine oil viscosity commonly ranges from a thin SAE 10 weight to a thick SAE 50 weight. Auto manufacturers recommend a specific SAE viscosity for oil used in their engines.

Temperature Effects on Oil

Cold engine oil is very thick and resists flow. When heated, oil thins and becomes runny. This can pose a problem. The oil in a cold engine may be so thick that engine starting is difficult. The oil will not pump through the engine properly. This may increase starting motor drag and result in poor lubrication.

Hot engine oil tends to become thinner and less resistant to flow. If it becomes too hot and thin, the oil film can break down and part contact can result.

It is important that the oil be thin enough for easy starting but thick enough to maintain lubrication when hot. Refer to **Figure 41-7.**

Single-Viscosity and Multiviscosity Oils

Single-viscosity oil is rated and designed for a limited range of operating temperatures. Its viscosity is not as stable as multiviscosity oils. Examples of single-viscosity oils include SAE 20, SAE 30, and SAE 40 weight.

Multiviscosity oil, or *multiweight oil,* will exhibit operating characteristics of a thin, light oil when cold and a thicker, heavy oil when hot. A multiweight oil can be numbered SAE 10W-30, 10W-40, or 20W-50.

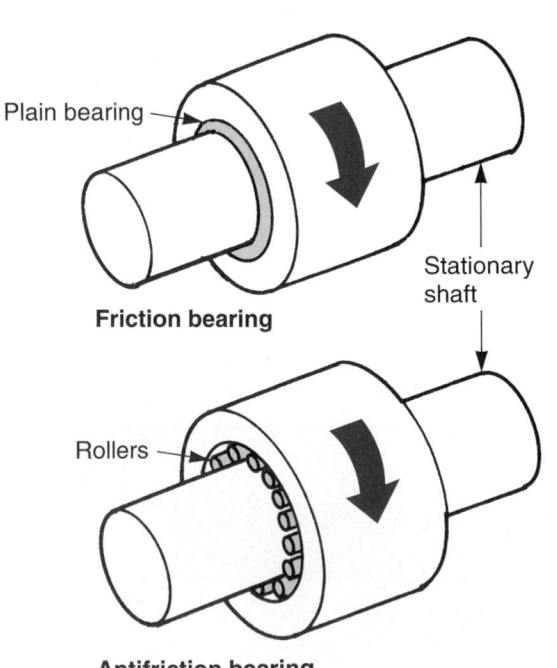

Figure 41-5. Compare basic bearing types. A friction bearing has two smooth surfaces sliding together. An antifriction bearing uses balls or rollers to prevent sliding (rubbing) action.

Figure 41-6. The oil container will give viscosity and service ratings. This is multiweight oil that has passed strict service rating tests. Oil used in late-model cars should have the statement "meets or exceeds car manufacturer's warranty requirements."

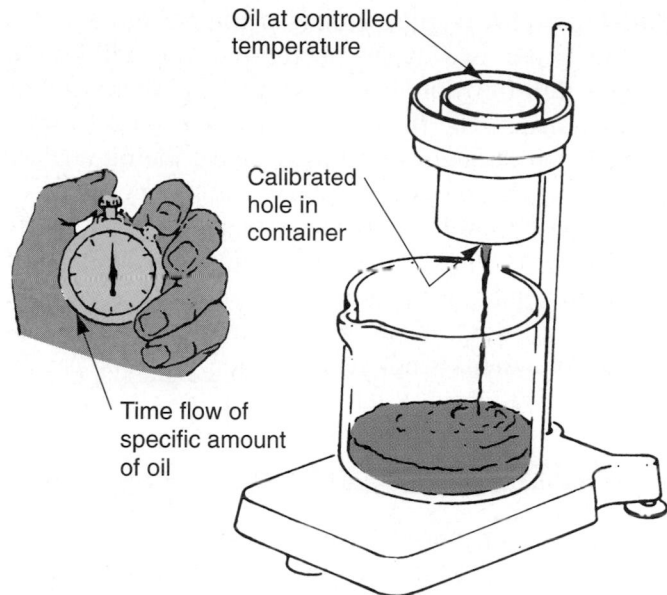

Figure 41-7. Engine oil viscosity rating is determined by measuring how long oil takes to flow through specific opening at a specific temperature. The longer the oil takes to flow into container, the higher the viscosity rating. (Binks)

For example, a 10W-30 weight oil will flow easily (like a 10W oil) when starting a cold engine. It will then act as a thicker oil (30 weight) when the engine warms to operating temperature. This will help the engine start more easily in cold weather. It will also provide adequate film strength (thickness) when the engine is at full operating temperature.

Selecting Oil Viscosity

Normally, you should use the oil viscosity recommended by the automaker. However, in a worn, high-mileage engine, higher viscosity oil may be beneficial. Thicker oil will tend to seal the rings and provide better bearing protection. It may also help reduce oil consumption and smoking.

Figure 41-8 is one automaker's chart showing recommended SAE viscosity numbers for different temperatures.

Oil Service Rating

An *oil service rating* is a set of letters printed on the oil bottle to denote how well the oil will perform under operating conditions. This is a performance standard set by the *American Petroleum Institute* (API), an association of oil-related companies that sets industry standards. The API service rating categories are:

- *SA oil rating*—lowest quality oil, should not be used in automotive engines.
- *SB oil rating*—minimum quality oil for automotive gasoline engines under mild service conditions. Not normally recommended.

- *SC oil rating*—meets oil warranty requirements for 1964–1967 automotive gasoline engines.
- *SD oil rating*—meets oil warranty requirements for 1968–1970 automotive gasoline engines.
- *SE oil rating*—meets oil warranty requirements for 1972–1979 automotive gasoline engines.
- *SF oil rating*—meets oil warranty requirements for 1980–1990 passenger car engines.
- *SG oil rating*—often recommended for 1991–1993 gasoline engines; contains more additives than SF oils; also can be used as CC or diesel-type oils.
- *SH oil rating*—recommend for 1993–1996 gasoline engines; corresponds to SG rating but has more stringent process requirements for evaluation.
- *SJ oil rating*—describes oils available since 1996; for use in current and earlier passenger cars, sport utility vehicles, vans, and light trucks operated under the manufacturer's recommended maintenance procedures.
- *CA* through *CG oil ratings*—recommended for diesel engines.

The service manual will give the service rating recommended for a specific vehicle. You can use a higher

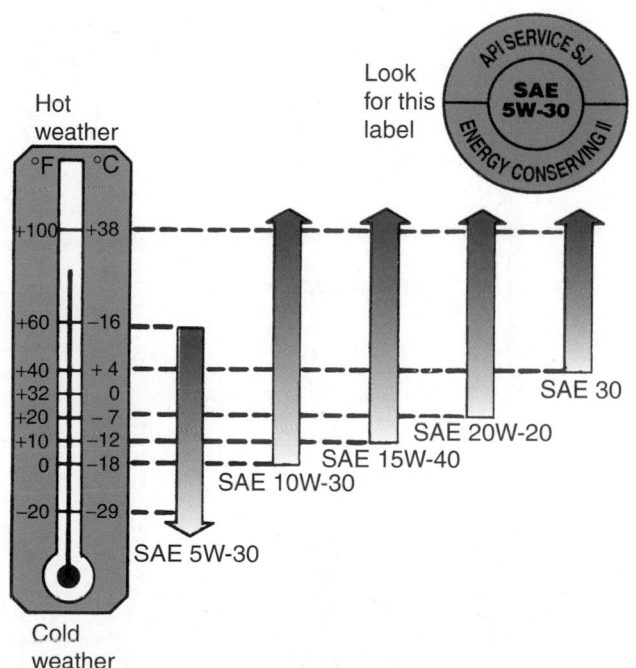

Figure 41-8. Typical recommended SAE viscosity oil rating is affected by climate or average outside temperatures. Note how thicker oil is specified for higher outside temperatures. Many automakers also warn against mixing different brands of oil. Different oil brands often use different additives and they may not be compatible. (General Motors)

service rating than recommended, but never a lower rating.

An oil with a high service rating (SJ, for example) can withstand higher temperatures and loads than an oil with a low rating. It will have more *oil additives* (extra chemicals) to prevent oil oxidation (gumming), engine deposits (sludging), breakdown (oil changes chemically), foaming (air bubbles form in oil), and other problems.

Engine Oiling Methods

There are two methods for lubricating engine components: pressure-fed oiling and splash oiling. See **Figure 41-9** for an oil flow diagram.

Pressure-fed oiling is provided by the oil pump to the crankshaft bearings, camshaft bearings, lifters, and rocker arm assemblies. This type of oiling is needed where load and friction are very high.

Splash oiling occurs when oil sprays out and on moving parts to provide lubrication. This type of oiling is used between parts with moderate load. For instance, splash oiling is used on the piston rings, cylinders, camshaft lobes, timing chains, and many other parts.

Full-Flow and Bypass Lubrication Systems

There are two types of full-pressure lubrication systems: full-flow lubrication systems and bypass lubrication systems. The *full-flow lubrication* system forces all of the oil through the oil filter before the oil reaches the parts of the engine. It is the most common type of lubrication system for automotive engines.

The *bypass lubrication system* does not filter all of the oil that enters the engine bearings. It filters some of the extra oil not needed by the bearings. The bypass lubrication system is not very common. It is not as efficient as the full-flow system.

Oil Pan and Sump

The *oil pan,* normally made of thin sheet metal or aluminum, bolts to the bottom of the engine block. It holds an extra supply of oil for the lubrication system. Refer to **Figure 41-10.**

The *oil pan drain plug* allows removal of the old oil during oil changes. The *oil pan gasket* prevents oil leakage between the engine block and pan. Holes in the

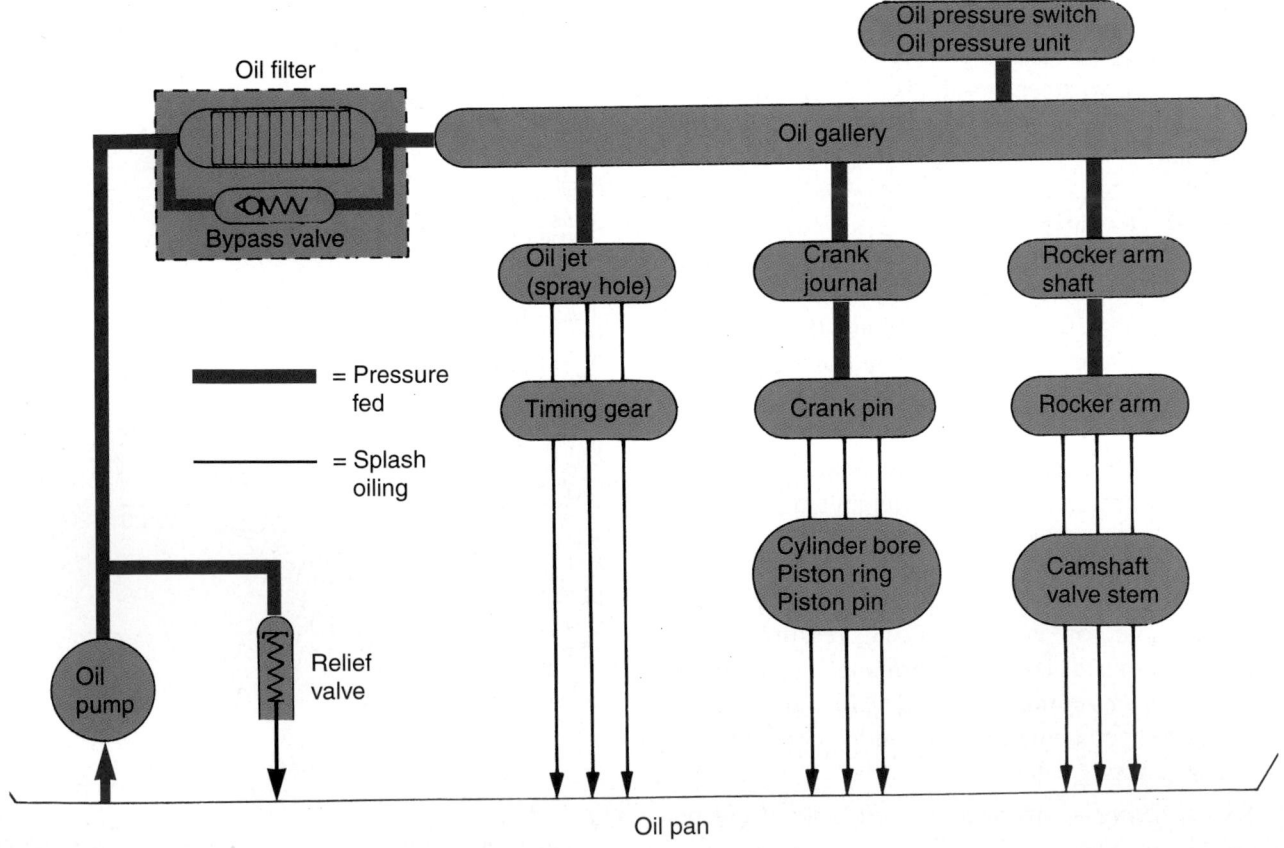

Figure 41-9. Trace through the engine oil flow diagram. This full-flow system requires all oil to pass through the filter before entering the gallery. Thicker lines represent oil under pressure. Thinner lines stand for oil draining and splashing on parts. Note the filter bypass valve and pressure valve. (GMC)

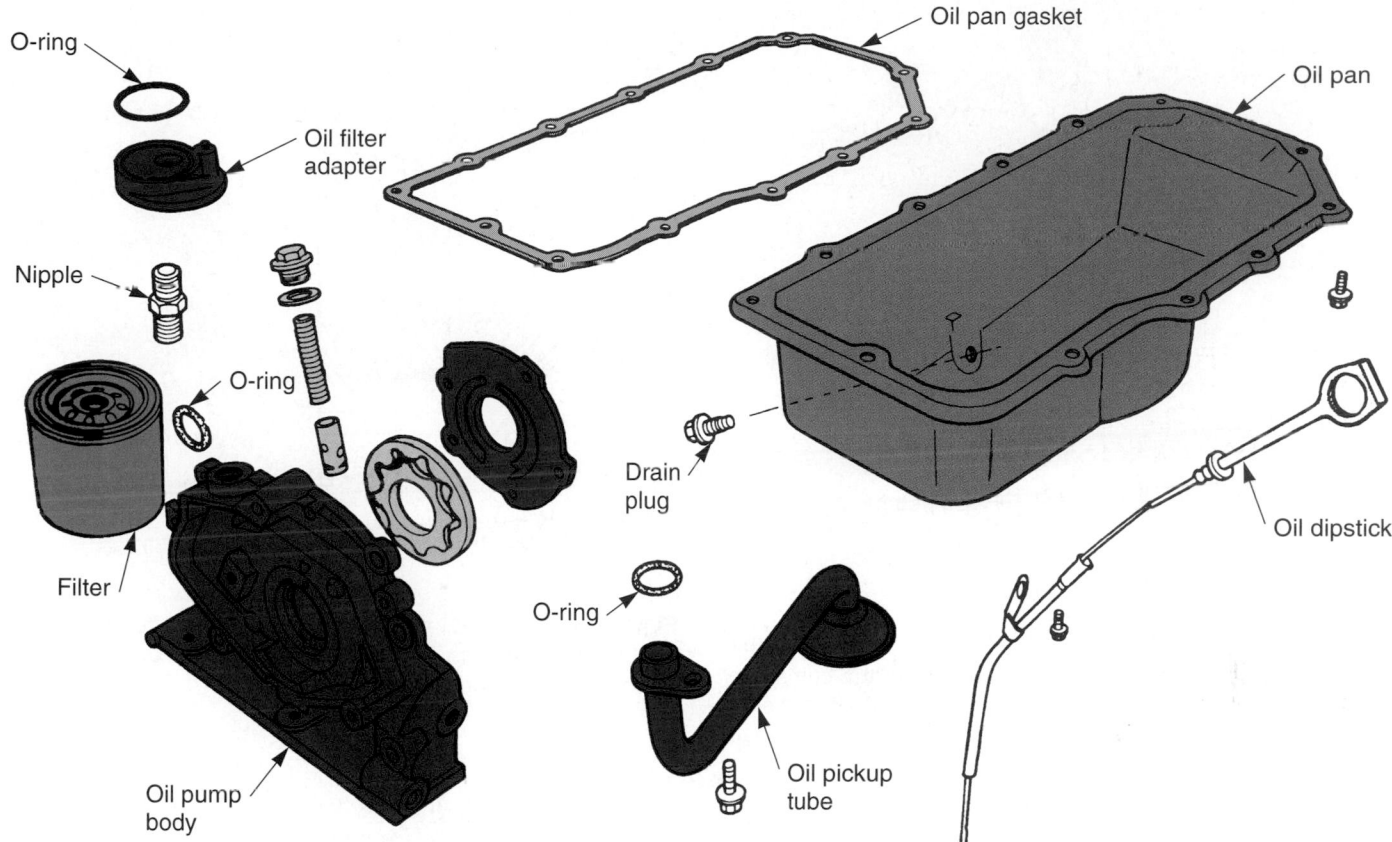

Figure 41-10. Study lubrication system parts. Note how modern oil pumps are often mounted in front cover. (Chrysler)

gasket are provided for the *oil pan bolts,* which secure the pan to the engine block.

The *sump* is the lowest area in the oil pan, **Figure 41-11.** As oil drains out of the internal parts of the engine, it fills the sump. Then the oil pump can pull oil out of the pan for recirculation.

A sheet metal *oil pan baffle* may be used to keep the oil from splashing up on the spinning crankshaft. This keeps the crank from stirring air into the oil (oil foaming) and reduces a fractional horsepower loss. The baffle may be an integral part of the pan or a separate part installed above the pan.

A *structural oil pan* is designed to add strength to the engine bottom end and cylinder block. Most of these oil pans have thick ribs and bosses to help support the crankshaft and the main bearings. Structural oil pans can be made of thick aluminum or composite materials that are light yet strong. In some designs, the main cap bolts go through the oil pan.

Oil Pickup and Screen

The *oil pickup* is a tube extending from the oil pump to the bottom of the oil pan. One end of the pickup bolts

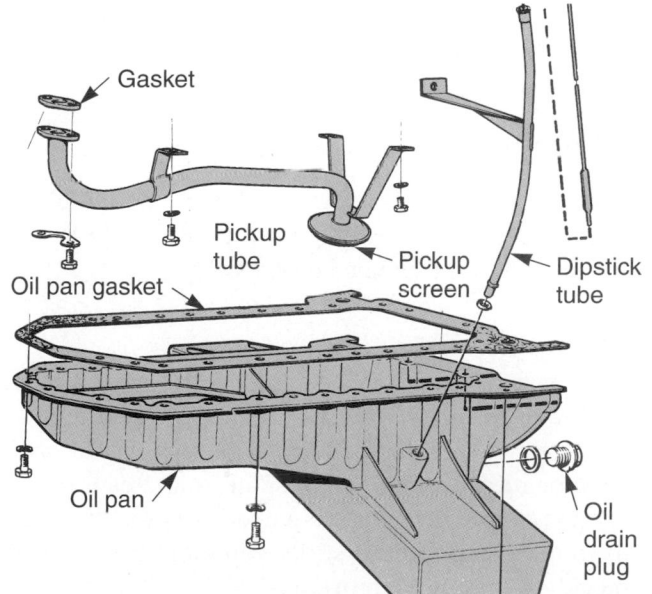

Figure 41-11. Oil pan forms lower crankcase and sump. A gasket seals the mating surface between the block and the pan. Also note the drain plug, pickup tube and screen, and dipstick assembly. (Volvo)

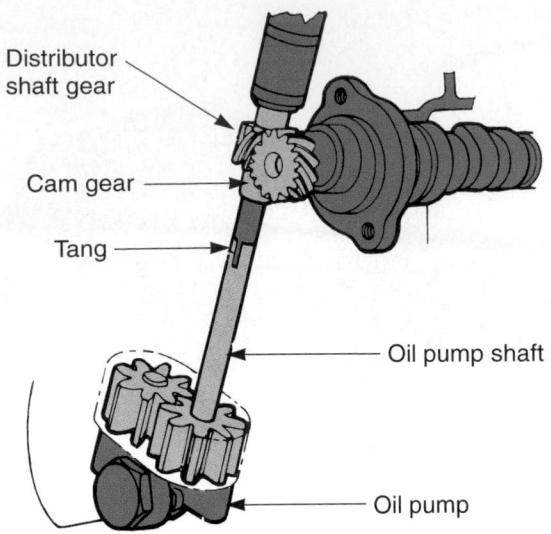

Shaft-driven oil pump

Figure 41-12. This oil pump is shaft driven. A gear on the bottom of the distributor meshes with a gear on the camshaft. The oil pump shaft extends from the distributor shaft to the oil pump. With the engine running, the shaft turns at one-half engine speed. (Chrysler)

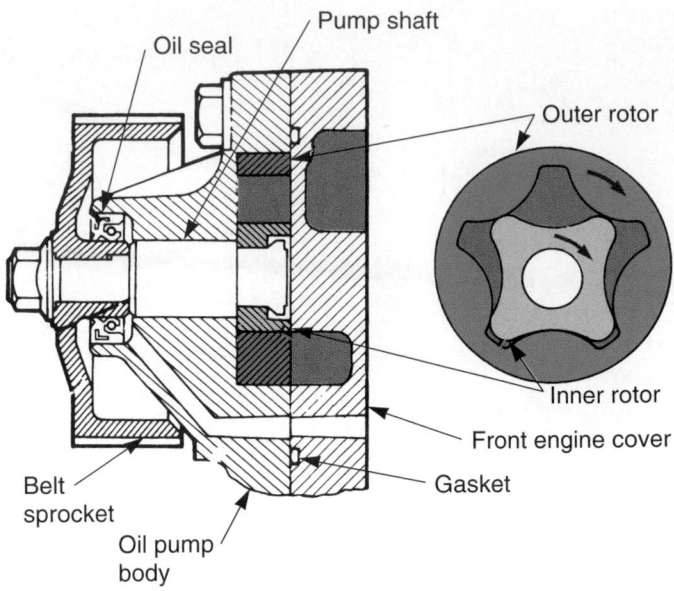

Figure 41-13. This oil pump is belt driven. The pump bolts to the front of the engine. This type is common with overhead cam engines using timing belt.

or screws into the oil pump or to the engine block. The other end holds the pickup screen.

The ***pickup screen*** prevents large particles from entering the pickup tube and oil pump. The screen is usually part of the pickup tube. Without the screen, the oil pump could be damaged by bits of valve stem seals and other debris flushed out of the engine. Refer back to **Figures 41-10** and **41-11**.

Oil Pumps

The ***oil pump*** is the heart of the engine lubrication system; it forces oil out of the pan, through the filter and galleries, and to the engine bearings.

Oil Pump Drives

There are several methods used to drive an oil pump. A ***shaft-driven oil pump*** uses a camshaft gear and distributor gear to spin the shaft and oil pump. A gear on the distributor meshes with and is spun by the cam gear. A metal oil pump shaft transfers power from the distributor shaft to the pump. This is still a common drive on cam-in-block engines. See **Figure 41-12.**

A ***belt-driven oil pump*** uses a cogged belt running off the crankshaft pulley for power. This is commonly used in overhead cam engines using a timing belt. See **Figure 41-13.**

A ***gear-driven oil pump*** uses the engine timing gears for power. Some oil pumps are gear-driven off oil pan–mounted balancer shafts. This design can vary considerably from engine to engine. Gear-driven oil pumps are very dependable but costly to manufacture, making them rare.

A ***crankshaft-driven oil pump*** is directly engaged to and powered by the crankshaft snout. The inner pumping element fits over and is keyed or locked to the crank snout. This type of oil pump drive is becoming more common because of its simplicity and dependability. See **Figure 41-14.**

Oil Pump Types

There are two basic types of engine oil pumps: rotary and gear. These are illustrated in **Figure 41-15.**

Rotary Oil Pumps

A ***rotary oil pump*** uses a set of star-shaped rotors in a housing to pressurize the engine oil. Look at **Figure 41-16.**

As the oil pump shaft turns, the inner rotor causes the outer rotor to spin. The eccentric action of the two rotors forms pockets that change in size. A large pocket is formed on the inlet side of the pump. As the rotors turn, the oil filled pocket becomes smaller as it nears the outlet of the pump. This squeezes the oil and makes it spurt out under pressure. As the pump spins, this action is repeated over and over to produce a relatively smooth flow of oil.

Figure 41-17 shows two common variations of rotary oil pumps. Compare them.

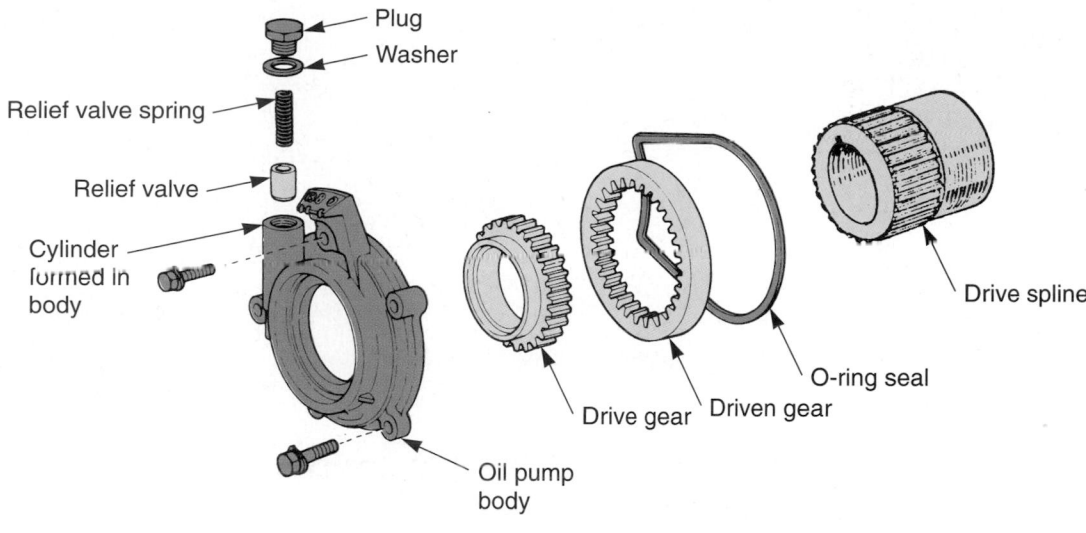

Crankshaft-driven oil pump

Figure 41-14. This oil pump is driven directly off the crankshaft snout. The drive spline on the crank turns the inner gear. The outer gear walks around to pump oil into block. (Toyota)

Gear Oil Pumps

A **gear oil pump** uses a set of gears to produce lubrication system pressure. See **Figure 41-18.**

Oil drawn into the inlet side of the pump is caught in the gear teeth and carried around the outer wall inside the pump housing. When the oil reaches the outlet side of the pump, the gear teeth mesh and seal. Oil caught in each gear tooth is forced into the pocket at the pump outlet and pressure is formed. Oil squirts out of the pump and to the engine bearings.

Figure 41-19 shows two gear oil pump variations. Study them.

Pressure Relief Valve

A **pressure relief valve** limits maximum oil pressure. It is a spring-loaded bypass valve in the oil pump, engine block, or oil filter housing. Refer to **Figure 41-20.**

The pressure relief valve consists of a small piston, spring, and cylinder. Under normal pressure conditions,

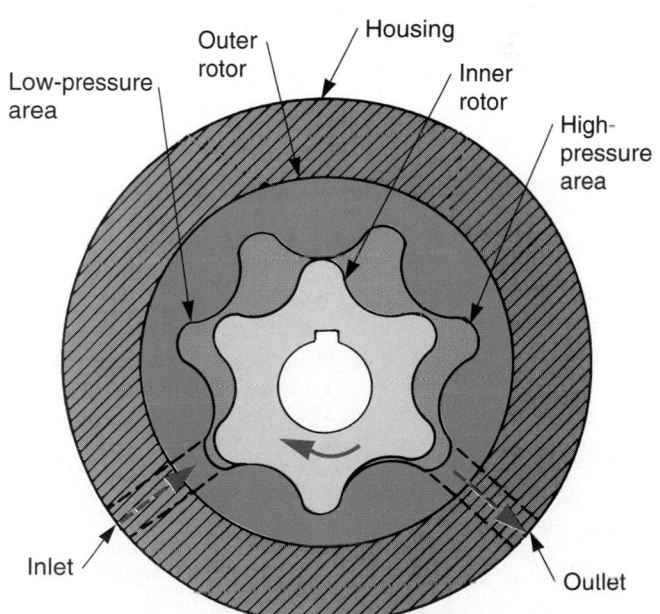

Figure 41-16. Study rotary oil pump operation. The inner rotor is driven by the pump shaft and turns the outer rotor. This causes the outer rotor to walk around the inner rotor. The space on one side of the rotor enlarges and pulls oil into pump. The space on the other side of the rotor is reduced to compress and force oil out. (Deere & Co.)

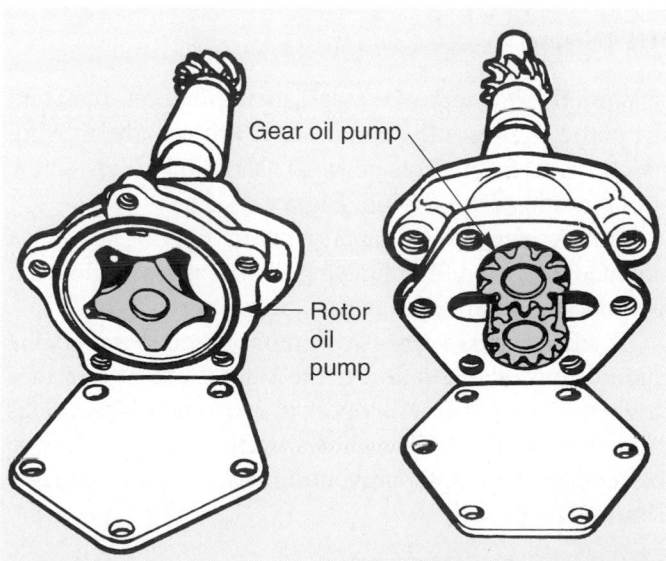

Figure 41-15. Compare the two basic types of oil pumps: rotary pump and gear pump. (Mopar)

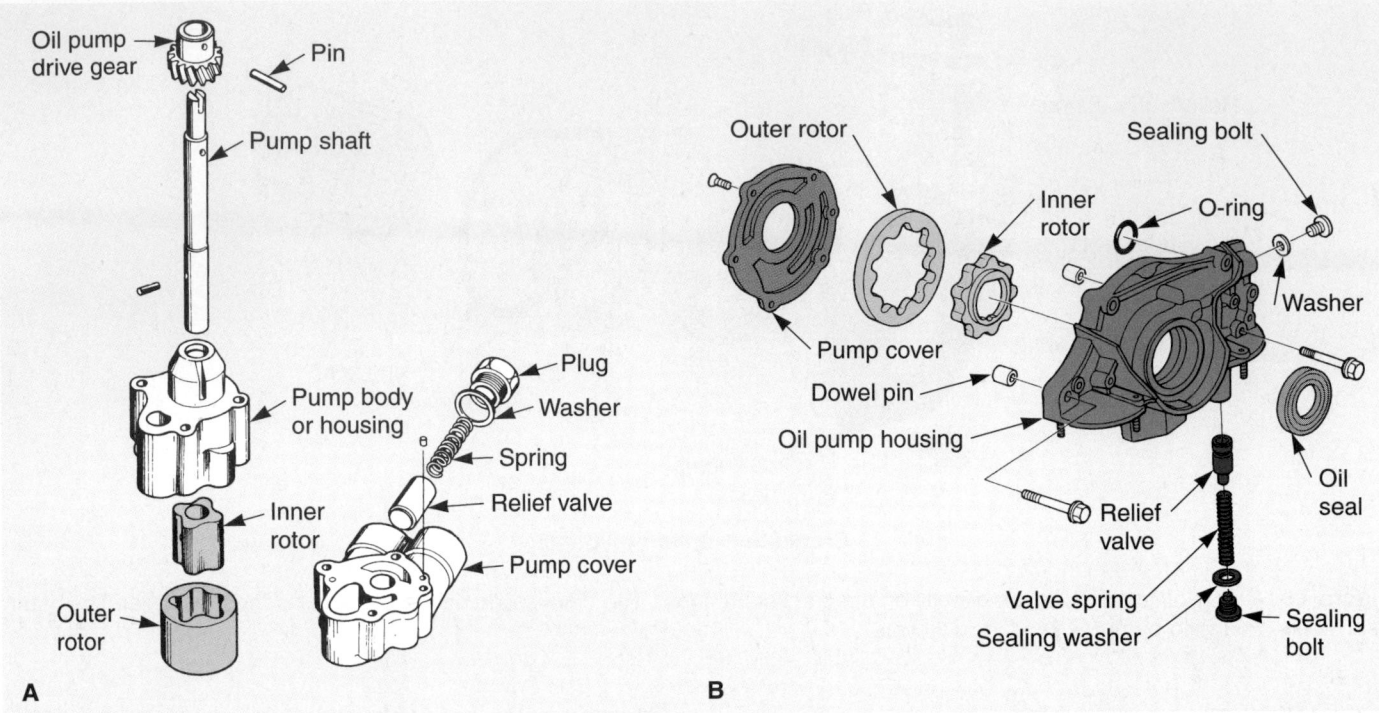

Figure 41-17 Compare the construction of two variations of rotary oil pumps. A—This pump bolts to the bottom of the cylinder block. The pressure relief valve is inside the pump body. This design is still used on late-model, cam-in-block engines. B—This pump is driven by the crankshaft snout. The assembly fits over the front of engine. It is becoming more common on new engines. (GMC and Honda)

the spring holds the relief valve closed. All the oil from the pump flows into the oil galleries and to the bearings.

However, under abnormally high oil pressure conditions (cold, thick oil for example), the pressure relief valve opens. Oil pressure pushes the small piston back in its cylinder by overcoming spring tension. This allows some oil to bypass the main oil galleries and pour back into the oil pan. Most of the oil still flows to the bearings, and a preset pressure is maintained.

Some pressure relief valves are adjustable. By turning a bolt or screw, or by changing spring shim thickness, the engine oil pressure setting can be altered.

Oil Filters

An *oil filter* removes small metal, carbon, rust, and dirt particles from the engine oil. It protects the moving engine parts from abrasive wear. Most oil filters screw onto the side of the engine, **Figure 41-21.**

A *filter element* is a paper or cotton filtering substance mounted inside the filter housing. It will allow oil flow but will block and trap small debris. See **Figure 41-22.**

A *filter bypass valve* is commonly used to protect the engine from oil starvation if the filter element becomes clogged. The valve will open if too much pressure is formed in the filter. This allows unfiltered oil to flow to the engine bearings, preventing major part damage, **Figure 41-23.**

Oil Filter Types

There are two classifications of engine oil filters: spin-on filters and cartridge filters.

Figure 41-18. Study gear oil pump operation. The oil pump shaft turns one gear, which drives the other gear. Oil is trapped in the teeth of the gears and carried around housing wall. On the outlet side, oil is trapped and pressurized. (Deere & Co.)

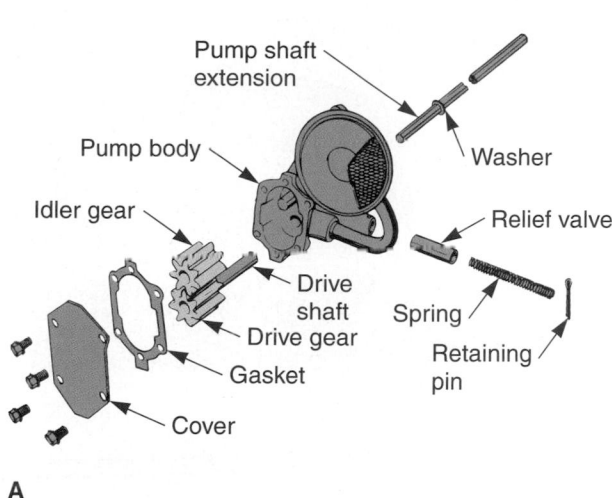

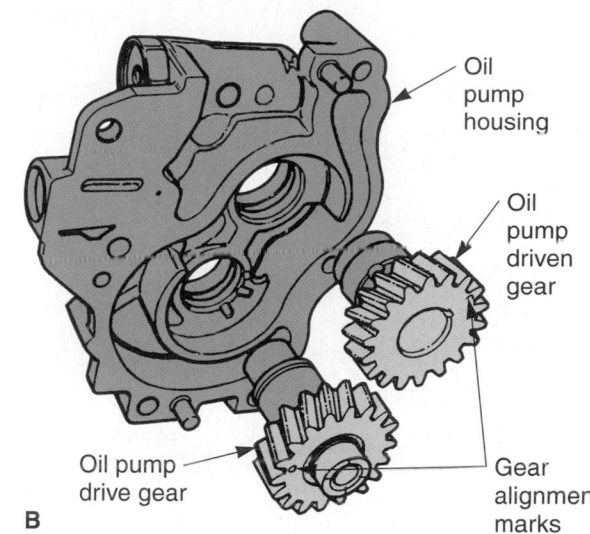

Figure 41-19. Compare construction of two common gear oil pumps. A—This gear oil pump has a pickup tube press-fit into the housing. A hex shaft drives the pump. Although still used on cam-in-block engines, this pump design is not as common as it was in the past. B—This gear oil pump has a remote-mounted pickup tube on the block. Note alignment marks on the pump gears. Gears were lapped together at the factory for more precision. (GMC and Chrysler)

The *spin-on oil filter* is a sealed unit. The filter element is permanently enclosed in the filter body. When it must be serviced, a new filter is simply screwed into place. This is the most common type of oil filter.

The *cartridge oil filter* has a separate element and housing. To service this type of filter, the housing is removed. Then, a new element is installed inside the existing housing. A cartridge oil filter is sometimes used for heavy-duty or diesel applications.

 Tech Tip

The number one reason for premature engine wear is failure to change the engine oil and filter. Decomposed oil results from excess heat and mechanical stress. The oil actually changes from a liquid to a solid, "chocolate pudding-like" substance. If you ever find extremely dirty or badly decomposed oil, warn the owner of the consequences of not changing the oil and filter at regular intervals.

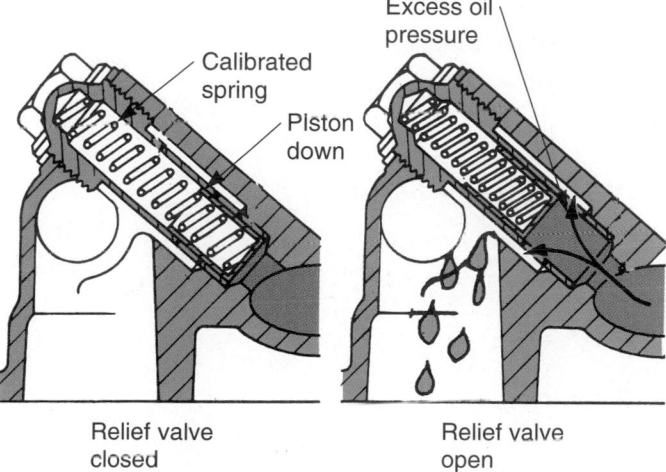

Figure 41-20. Cutaway view shows construction of a pressure relief valve. A small piston or valve fits in a small cylinder. A spring holds the valve in a normally closed position. When oil pressure is too high, the spring compresses to open the valve. Excess oil bypasses the system and drains into the oil pan.

Figure 41-21. The oil filter normally mounts on one side of the engine block or on the filter housing. Oil filter condition is critical to the service life of the engine. (Oldsmobile)

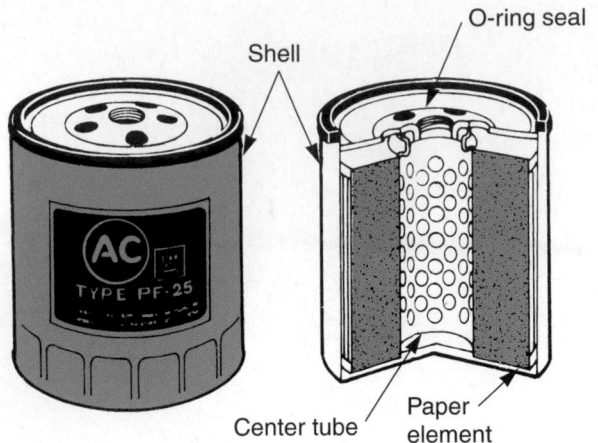

Figure 41-22. Spin-on oil filter is the most common type. The O-ring seal prevents leakage when screwed onto its housing. The element traps debris and keeps it out of the engine. (AC-Delco)

Oil Filter Housing

The *oil filter housing* is a metal part that bolts to the engine and provides a mounting place for the oil filter. The housing may also have a fitting for the oil pressure sending unit. Refer to **Figure 41-24.**

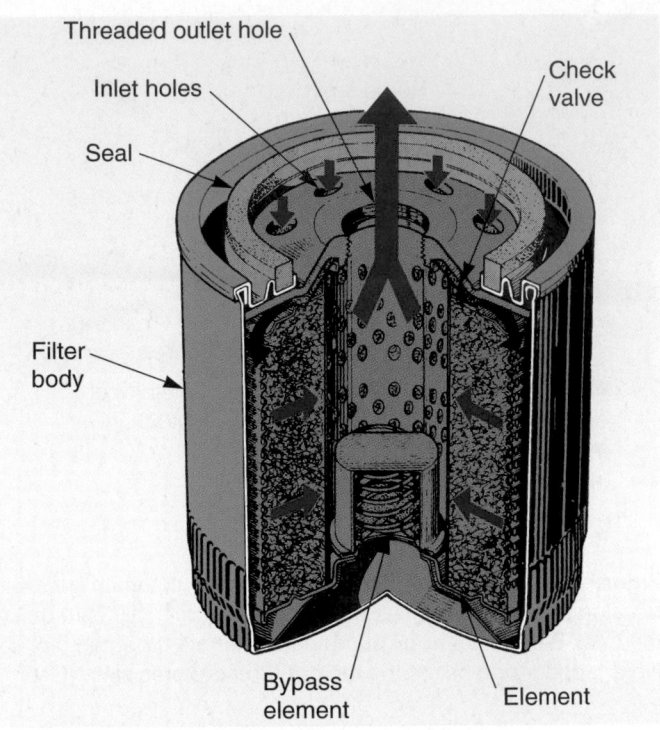

Figure 41-23. Cutaway of a spin-on oil filter shows oil flow. Oil enters small holes, passes through the element, and then flows out of the center hole to the engine. (Saab)

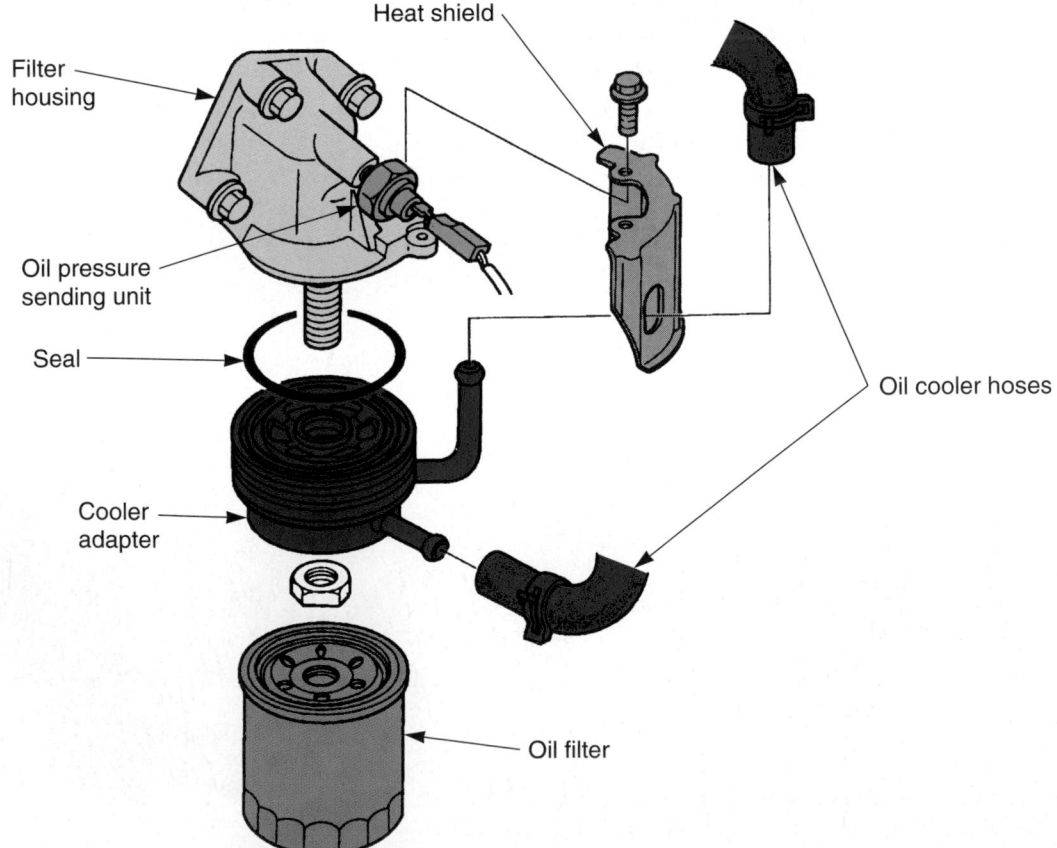

Figure 41-24. The oil filter housing often bolts to side of engine block. Note the sending unit for the oil pressure gauge. A small oil cooler adapter can mount between the filter and the housing. Hoses go to heat exchanger for removal of heat from oil. (Mazda)

A gasket normally fits between the engine and the oil filter housing to prevent leakage. Sometimes, the pressure relief valve, filter bypass valve, or oil pump is inside this housing.

Oil Cooler

An *oil cooler* may be used to help lower and control the operating temperature of the engine oil. It is a *heat exchanger* (radiator-like device) connected to the lubrication system. Oil is pumped through the cooler before flowing back into the engine.

Figure 41-24 shows a small *oil cooler adapter* that fits between the oil filter housing and oil filter. It provides hose connections for the lines leading to heat exchanger.

Figure 41-25 shows an oil cooler that consists of oil cooler lines running to a small heat exchanger. Airflow through the cooler removes heat and lowers the temperature of the oil.

Oil coolers are frequently used on turbocharged engines or heavy-duty applications (trailer towing package, for instance).

Oil Galleries

Oil galleries are passages through the cylinder block and head for lubricating oil. They are cast or machined passages that allow oil to flow to the engine bearings and other moving parts. Refer to **Figure 41-26.**

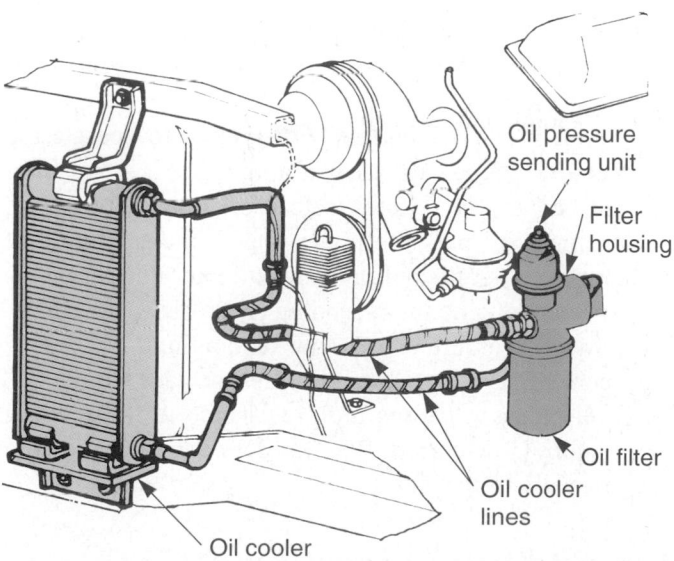

Figure 41-25. Oil coolers allow transfer of heat out of the oil and into the surrounding air, like a cooling system radiator. They are used on high performance or heavy-duty applications. High-pressure lines and hoses carry oil to and from the cooler. (Mopar)

The *main oil galleries* are large passages through the center of the block. They feed oil to the crankshaft bearings, camshaft bearings, and lifters. The main galleries also feed oil to smaller passages running up to the cylinder heads.

Oil Spray Nozzles

Oil spray nozzles are used to direct a stream of engine oil onto the bottoms of the engine pistons, timing gears or chain, and other moving engine parts. The extra oil spray nozzle provides added lubrication and also helps to carry heat away from the parts. See **Figure 41-27.**

Positive Crankcase Ventilation System

The *positive crankcase ventilation system,* abbreviated PCV system, draws fumes out of the engine crankcase and burns them inside the engine. The PVC system primarily prevents toxic vapors from entering and polluting the atmosphere. This system also helps prevent *engine sludging* (chocolate pudding–like oil formation), which could restrict oil circulation.

 Note
For details of a PCV system, refer to Chapter 43, *Emission Control Systems.*

Oil Pressure Indicator

An *oil pressure indicator* warns the driver of a low oil pressure problem. The circuit activates a warning light in the vehicle's dash. A basic oil pressure light circuit is shown in **Figure 41-28.**

An *oil pressure switch* is opened and closed by oil pressure so that it can operate the dash indicator light. It screws into the engine and is exposed to one of the oil galleries.

When oil pressure is low (engine is off or has a mechanical problem), the spring in the sending unit holds a pair of contacts closed. This completes the circuit and the indicator light glows.

When oil pressure is normal, oil pressure acts on a diaphragm in the sending unit. Diaphragm deflection opens the contact points to break the circuit. This causes the warning light to go out, informing the driver of good oil pressure.

 Tech Tip
With many low oil pressure warning lights, engine oil pressure must be very low (under about 10 psi at idle) before the light will go on. Light engine bearing knock or lifter noise will often result before the warning light is energized.

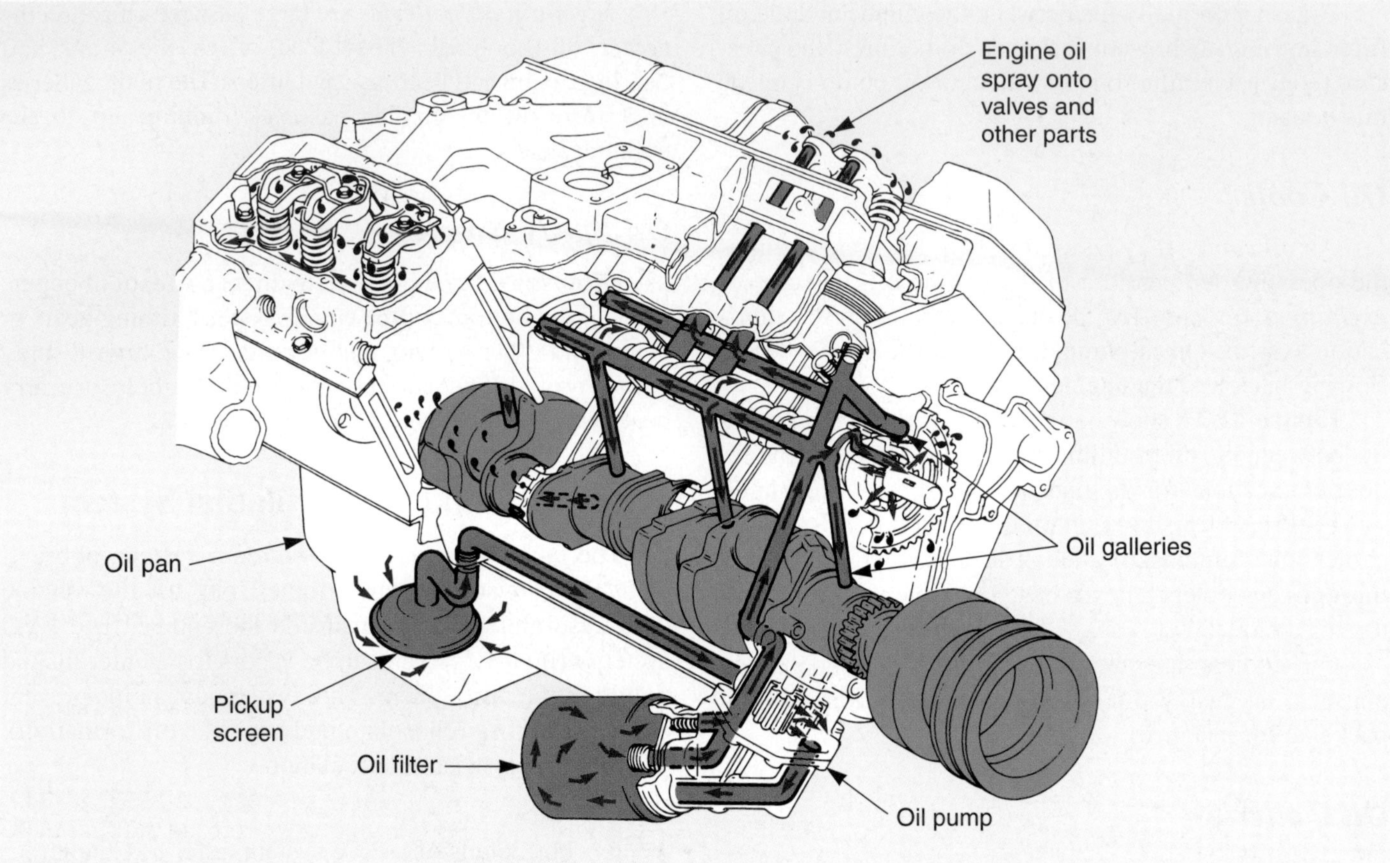

Engine oil spray onto valves and other parts

Oil galleries

Oil pan

Pickup screen

Oil filter

Oil pump

Figure 41-26. Oil galleries allow oil to pass through the engine. The main galleries are larger passages in the block. Note how oil flows through hollow push rods to rockers. (Chrysler)

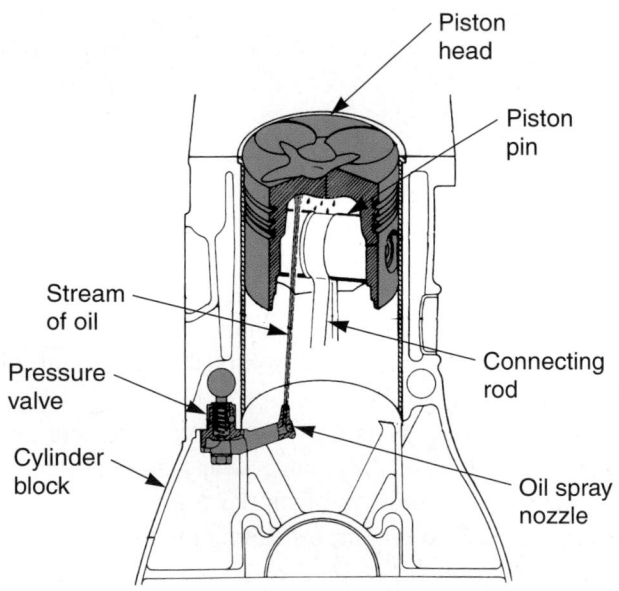

Piston head

Piston pin

Stream of oil

Connecting rod

Pressure valve

Cylinder block

Oil spray nozzle

Figure 41-27. Oil spray nozzle squirts engine oil onto the bottom of the piston to cool the piston head and add lubrication for the wrist pin. A pressure valve in this nozzle only opens when a specific oil pressure is reached. (Mercedes Benz)

Oil Pressure Gauge Circuit

Some cars are equipped with an *oil pressure gauge* that registers the actual oil pressure in the engine. It is similar in operation to the oil pressure indicating light. See **Figure 41-29.**

An *oil pressure sending unit* is used to operate the oil pressure gauge and the warning light. The sending unit uses a variable resistor instead of contact points.

As more oil pressure is developed, the sending unit diaphragm is deflected a proportional amount. This causes an equal amount of sending unit resistance change.

Low oil pressure causes low sending unit resistance, high current flow, and low oil pressure gauge readings. High oil pressure causes high resistance in the sending unit to allow low current flow for deflecting the pressure gauge needle to the right.

Low Oil Pressure Safety Circuit

A *low oil pressure safety circuit* can be used to shut off the engine if oil pressure drops too low. This will protect the engine from major damage in case of a lubrication system failure or loss of oil pressure. The circuit

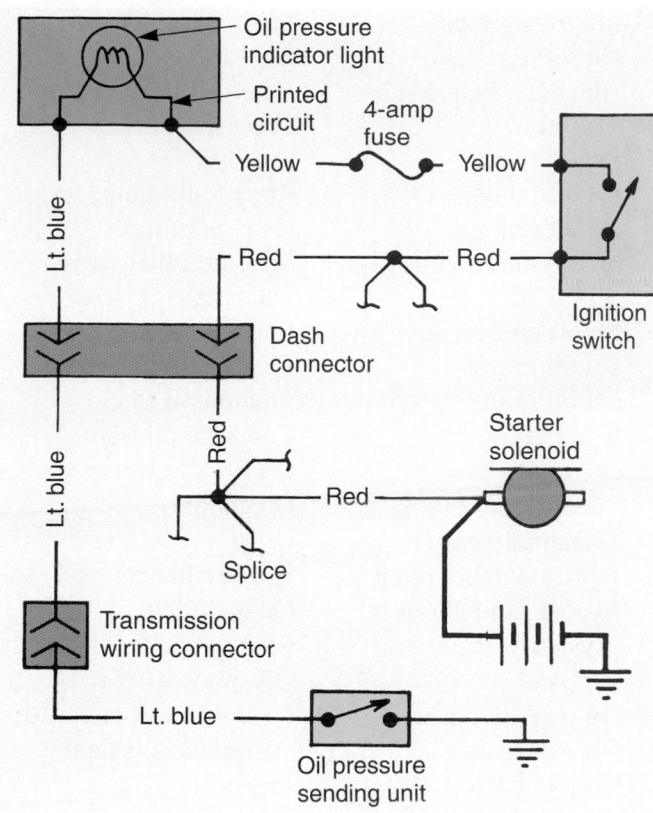

Figure 41-28. A wiring diagram for an oil pressure warning light. The sending unit controls current flow through the dash indicator light. Note wire color coding for tracing wires.

usually disables the ignition system to keep the engine from starting or running.

The oil pressure switch or sending unit is wired to the engine control module. If the ECM detects a dangerously

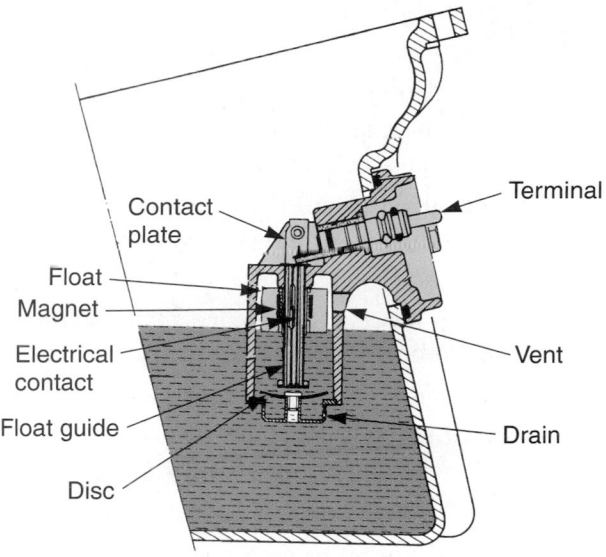

Figure 41-30. This low oil level sending unit uses a float and switch to operate a dash indicator. If the oil level becomes low from leakage or consumption, the float drops to close the switch and turn the dash light on. The driver then knows to add oil before the level is low enough to cause damage. (Mercedes Benz)

low oil pressure, it disables the engine ignition system. This prevents the "uninformed driver" from continuing to drive the vehicle without oil pressure, preventing destruction of the engine.

Oil Level Sensor

An oil level sensor is used to warn the driver to add oil to the engine. The sensor is usually mounted in the oil pan. The oil level sensor can be a float-variable resistance type or an optical type, **Figure 41-30.**

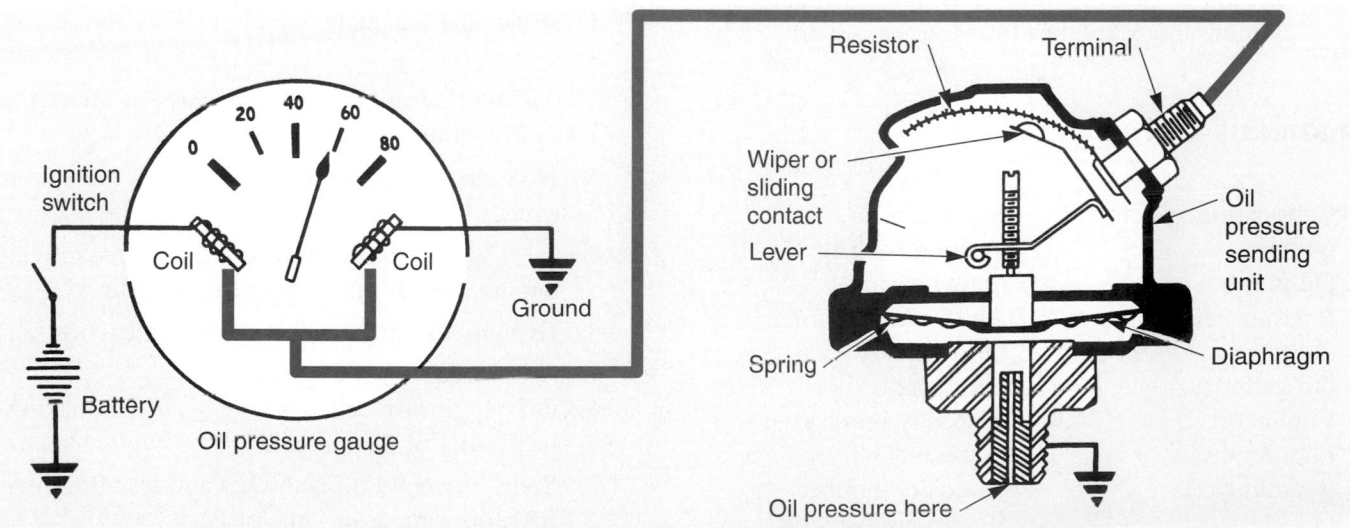

Figure 41-29. An oil pressure gauge circuit uses a variable resistance-type sending unit. Changes in oil pressure cause different amounts of diaphragm deflection. This moves the sliding contact on the resistor. Changes in resistance and current make the gauge show engine oil pressure. (Deere & Co.)

Summary

- The lubrication system forces oil to high friction points in the engine to protect moving parts from friction, wear, and damage.

- Engine oil, also called motor oil, is needed to keep moving parts in an engine from making direct contact.

- A friction bearing, also called plain bearing, has two smooth surfaces sliding on each other.

- An antifriction bearing uses balls or rollers to avoid a sliding action between the bearing surfaces.

- Oil viscosity, also called oil weight, is the thickness or fluidity (flow ability) of the engine oil.

- Multiviscosity oil or multiweight oil will exhibit operating characteristics of a thin, light oil when cold and a thicker, heavy oil when hot.

- An oil service rating is a set of letters printed on the oil bottle or can to denote how well the oil will perform under operating conditions.

- The oil pickup is a tube extending from the oil pump to the bottom of the oil pan.

- The oil pump is the "heart" of the engine lubrication system: it forces oil out of the pan, through the engine filter, galleries, and to the engine bearings.

- A pressure relief valve limits maximum oil pressure. It is a spring-loaded bypass valve in the oil pump, engine block, or oil filter housing.

- An oil filter removes small metal, carbon, rust, and dirt particles from the engine oil. It protects the moving engine parts from abrasive wear.

- Oil galleries are small passages through the cylinder block and head for lubricating oil.

Important Terms

Lubrication system	Bearing clearance
Oil pan	Friction bearing
Oil pump	Plain bearing
Pressure relief valve	Antifriction bearing
Oil filter	Oil viscosity
Oil galleries	Oil weight
Engine oil	Viscosity numbering
Petroleum	system
Synthetic oils	Viscosity number
Oil film	Single-viscosity oil
Oil clearance	Multiviscosity oil

Multiweight oil	Shaft-driven oil pump
Oil service rating	Belt-driven oil pump
American Petroleum	Gear-driven oil pump
Institute	Crankshaft-driven oil
SA oil rating	pump
SB oil rating	Rotary oil pump
SC oil rating	Gear oil pump
SD oil rating	Pressure relief valve
SE oil rating	Oil filter
SF oil rating	Filter element
SG oil rating	Filter bypass valve
SH oil rating	Spin-on oil filter
SJ oil rating	Cartridge oil filter
Oil additives	Oil filter housing
Pressure-fed oiling	Oil cooler
Splash oiling	Oil cooler adapter
Full-flow lubrication	Heat exchanger
Bypass lubrication	Oil galleries
system	Main oil galleries
Oil pan	Oil spray nozzles
Oil pan drain plug	Positive crankcase
Oil pan gasket	ventilation system
Oil pan bolts	Engine sludging
Sump	Oil pressure indicator
Oil pan baffle	Oil pressure switch
Structural oil pan	Oil pressure gauge
Oil pickup	Oil pressure sending unit
Pickup screen	Low oil pressure safety
Oil pump	circuit

Review Questions—Chapter 41

Please do not write in this text. Place your answers on a separate sheet of paper.

1. What are the main functions of a lubrication system?

2. List and describe the major parts of an engine lubrication system.

3. How does engine motor oil protect parts from excess wear?

4. _____ _____ is the small space between moving engine parts for the lubricating oil film.

5. Explain the difference between a friction and antifriction bearing.

6. Oil _____, also called oil _____, is the thickness or fluidity of the motor oil.

7. SAE 30 weight oil is thicker and less fluid than SAE 40 weight oil. True or False?

8. What is a multiweight oil?

9. Pressure-fed oiling is *not* commonly used with:
   (A) Crankshaft bearings.
   (B) Camshaft bearings.
   (C) Hydraulic lifters.
   (D) Cylinder walls.

10. There are two basic types of engine oil pumps: _____ and _____ types.

11. How does a pressure relief valve work?

12. Why is an oil filter very important to engine service life?

13. Why is an oil cooler sometimes used?

14. _____ _____ are small passages through the cylinder head and block for lubricating oil.

15. Summarize the operation of an oil pressure gauge circuit.

ASE-Type Questions

1. Each of the following is an engine lubrication system function *except:*
   (A) *clean inside of engine.*
   (B) *help cool engine parts.*
   (C) *prevent heat transfer.*
   (D) *increase fuel economy.*

2. Which of the following separates engine parts to prevent metal-on-metal contact?
   (A) *Oil film.*
   (B) *Bearings.*
   (C) *Galleries.*
   (D) *Additives.*

3. Which type of bearings use balls or rollers?
   (A) *Cam.*
   (B) *Plain.*
   (C) *Friction.*
   (D) *Antifriction.*

4. Oil "viscosity" refers to a motor oil's:
   (A) *fluidity.*
   (B) *thickness.*
   (C) *flow ability.*
   (D) *All of the above.*

5. The highest API oil service rating is:
   (A) *CA.*
   (B) *SE.*
   (C) *SJ.*
   (D) *CD.*

6. Oil additives are used to prevent:
   (A) *deposits.*
   (B) *foaming.*
   (C) *oxidation.*
   (D) *All of the above.*

7. Pressure-fed oiling is provided by the oil pump to each of the following *except:*
   (A) *lifters.*
   (B) *timing chain.*
   (C) *camshaft bearings.*
   (D) *crankshaft bearings.*

8. Which of the following holds an extra supply of oil for a lubrication system?
   (A) *Oil pan.*
   (B) *Oil pump.*
   (C) *Oil pickup.*
   (D) *Oil galleries.*

9. While discussing oil pump drive methods, Technician A says the oil pump can be driven by a gear on the camshaft. Technician B says the oil pump can be driven by a cogged belt. Who is right?
   (A) *A only.*
   (B) *B only.*
   (C) *Both A and B.*
   (D) *Neither A nor B.*

10. A pressure relief valve can be located in the:
   (A) *oil pump.*
   (B) *engine block.*
   (C) *oil filter housing.*
   (D) *Any of the above.*

Activities—Chapter 41

1. Visit an auto supply store. List the API service ratings (SA, SE, and so on) for several brands of 10W-30 oil. Do the ratings differ, or do all brands have the same ratings? List the price per quart for each brand that has an SH or SJ rating. Which brand represents the best buy?

2. Using either a CAD program on a computer or conventional drafting equipment, draw a cross sectional view of a typical spin-on oil filter. Label the parts and use arrows to show oil movement through the filter.

3. Interview the manager or owner of a business that performs "while you wait" oil changes. Has he or she noticed any changes in customer habits (waiting longer between oil changes or changing more frequently, changing brands of oil, or purchasing additional services such as an air-conditioning recharge or transmission fluid change)? Does he or she expect the business to be larger, smaller, or about the same in five years? Would she or he suggest the oil change business as an opportunity for a young entrepreneur? Report to the class on your interview.

Chapter 42

Lubrication System Testing, Service, and Repair

After studying this chapter, you will be able to:

- List common lubrication system problems and symptoms.
- Diagnose lubrication system troubles.
- Measure engine oil pressure.
- Change engine oil and filter.
- Remove and install an oil pan.
- Service an oil pump.
- Test and repair an oil pressure indicating light and gauge.
- Describe safe working practices to use when testing, servicing, or repairing a lubrication system.
- Correctly answer ASE certification test questions on the testing, service, and repair of engine lubrication systems.

Without a properly functioning lubrication system, an engine will "self-destruct" in a matter of minutes. Major damage to the engine bearings, crankshaft, piston rings, and other parts can result. Loss of oil pressure can result from many causes, such as a low oil level, broken pump drive, clogged strainer, or stuck pressure relief valve.

This chapter summarizes the most common service and repair operations performed on an engine lubrication system. It prepares you for typical repair tasks. Study this chapter thoroughly.

 Note

For more information on lubrication system service, refer to the index. Additional information on this subject can be found in several other chapters.

Lubrication System Problem Diagnosis

To troubleshoot an engine lubrication system, begin by gathering information on the trouble. Ask the vehicle owner or service writer questions like these:

- Does the oil light flicker or stay on?
- Do you hear any abnormal engine noises?
- Has oil been dripping on your garage floor?
- Do you have to add oil frequently?
- When was the last time you changed the oil and filter?
- Does blue smoke come out of the exhaust pipe?

This type of questioning may give you essential information for problem diagnosis. Analyze the symptoms using your understanding of lubrication system and engine operation. You should then be able to arrive at some logical possible causes of the problem. Further inspection and testing will then pinpoint the problem.

Lubrication System Problems

The problems found in a lubrication system are limited in number. They include:

- High oil consumption—oil must be added to engine frequently.
- Low oil pressure—gauge reads low, indicator light glows, or abnormal engine noises are present.
- High oil pressure—gauge reads high, oil filter swelled.
- Defective indicator or gauge—inaccurate or no indicator readings.

Figure 42-1 shows several possible problem areas relating to an engine's lubrication system. Study them carefully.

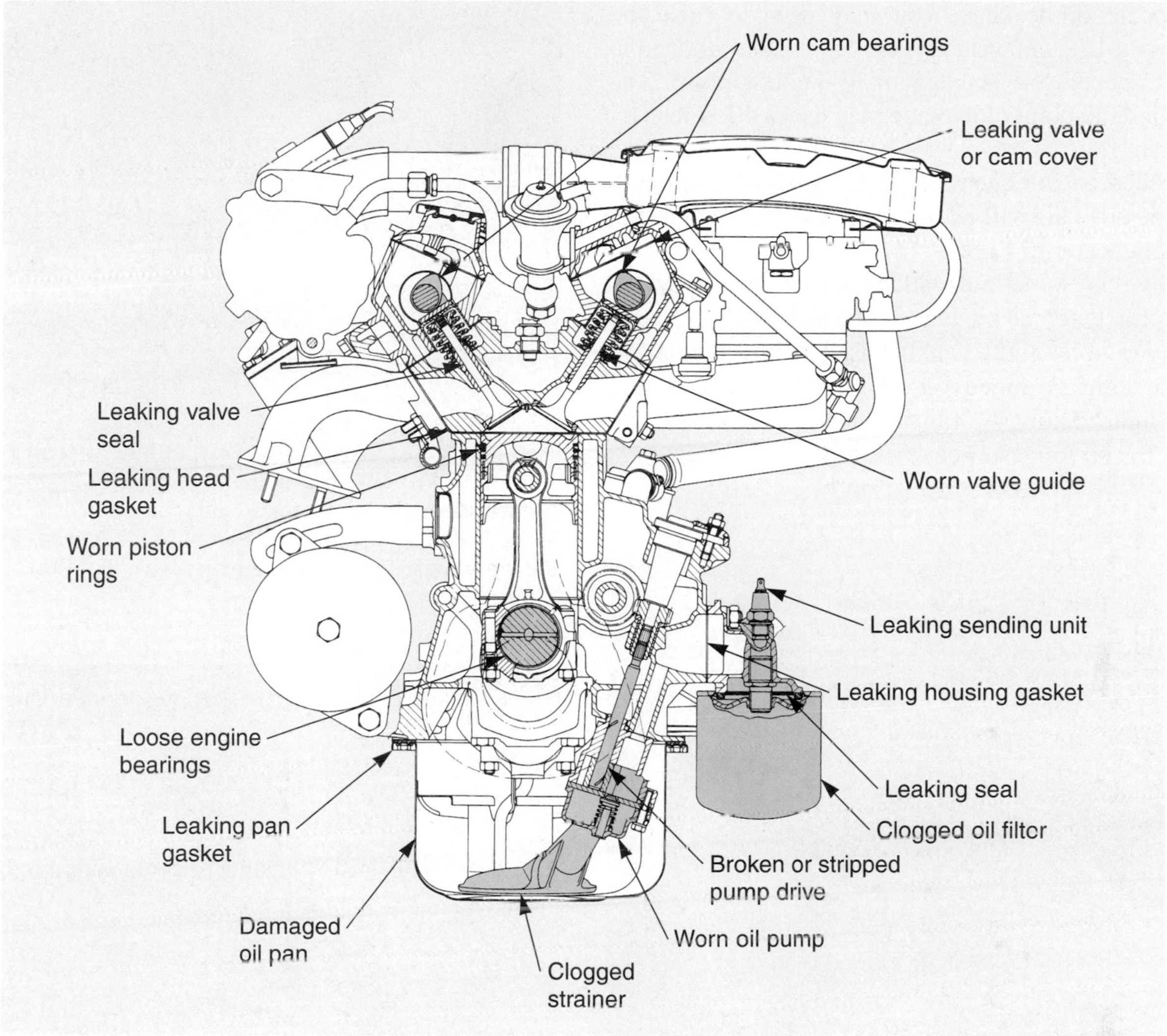
Worn cam bearings
Leaking valve or cam cover
Leaking valve seal
Leaking head gasket
Worn piston rings
Worn valve guide
Leaking sending unit
Leaking housing gasket
Loose engine bearings
Leaking seal
Clogged oil filter
Leaking pan gasket
Broken or stripped pump drive
Damaged oil pan
Worn oil pump
Clogged strainer

Figure 42-1. Note typical problem areas relating to the engine lubrication system. (Fiat)

When diagnosing the lubrication system troubles, visually inspect the engine for obvious problems. Check for oil leakage, a disconnected sending unit wire, a low oil level, a damaged oil pan, or other troubles that would relate to the symptoms. See **Figure 42-2.**

16VALVE
MITSUBISHI
DOHC 2000
OIL

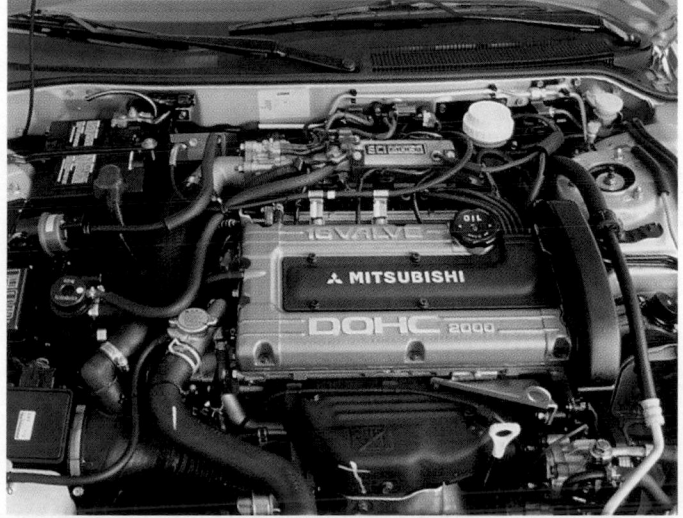

Figure 42-2. Inspect the engine compartment to locate oil leaks. The valve cover is a common leakage point. To check for a low oil level, find and remove the dipstick. (Mitsubishi)

High Oil Consumption

High oil consumption is caused by external oil leakage out of the engine or by internal oil leakage into the combustion chambers. If the vehicle owner must frequently add oil to the engine, this is a symptom of high oil consumption.

External Oil Leaks

External oil leakage is easily detected as darkened, oil-wet areas on or around the engine. Oil may also be found in small puddles under the vehicle. Leaking gaskets or seals are usually the source of external engine oil leakage.

To locate oil leakage, you may need to raise the vehicle on a lift and visually look for leaks under the engine. Trace the oil leakage to its highest point. The parts around the point of leakage may be washed clean by the constant flow of oil.

If needed, steam clean or power wash the engine to remove the oil. This will let you trace the source of a leak more easily. Refer to **Figure 42-3.**

Oil dye is sometimes added to engine oil when checking for external oil leaks. The dye is fluorescent, and an ultraviolet light can be used to make it glow. When the light is aimed on an oil leak, the dye will appear a light yellowish-orange color. The leak can then be easily traced to its source.

Figure 42-4. Oil seals are a common source of external oil leaks. Note how this engine rear oil seal has worn a groove in the crankshaft. This caused a major leak. (Fel-Pro)

Figure 42-3. If the engine is covered with oil, you may want to steam clean or pressure wash it to help find the source of oil leakage. (The Eastwood Co.)

Figure 42-5. Internal oil leaks allow oil to be burned in the combustion chambers. This deteriorated, broken valve seal caused internal leakage and engine smoking, especially at startup. (Fel-Pro)

After locating the point of leakage, you must service the gasket, seal, or component at fault. Oil seals are a common source of leaks. Sometimes a seal can wear a groove into its mating part, **Figure 42-4.**

Internal Oil Leakage

Internal oil leakage shows up as blue smoke coming out of the exhaust. For example, if the engine piston rings and cylinders are badly worn, oil can enter the combustion chambers and be burned during combustion.

Figure 42-5 gives one example of a problem that can cause internal oil leakage and engine smoking.

Tech Tip
Do not confuse black smoke (excess fuel in cylinder) or white smoke (water leakage into gasoline engine cylinder) with the blue smoke caused by motor oil.

Low oil pressure

Low oil pressure is indicated when the oil indicator light glows, the oil gauge reads low, or the engine lifters or bearings rattle.

There are several causes of low oil pressure:

- Low oil level—oil not high enough in the pan to cover the oil pickup.
- Worn connecting rod or main bearings—pump cannot provide enough oil volume.

- Thin or diluted oil—low oil viscosity or gasoline in the oil.
- Weak or broken pressure relief valve spring—valve opening too easily.
- Cracked or loose oil pump pickup tube—air being pulled into the oil pump.
- Worn oil pump—excess clearance between the rotor or gears and the housing.
- Clogged oil pickup screen—reduced amount of oil entering the pump.

A low oil level is a common cause of low oil pressure. Always check the oil level first when troubleshooting a low pressure problem. **Figure 42-6.**

High Oil Pressure

High oil pressure causes the gauge to read higher than normal and can cause the oil filter to rupture. This is seldom a problem. There are four common causes of high oil pressure:

- Pressure relief valve stuck closed—valve not opening at the specified pressure.
- High relief valve spring tension—strong spring, or spring has been improperly shimmed.
- High oil viscosity—excessively thick oil or use of oil additive that increases viscosity.
- Restricted oil gallery—defective block casting or debris in oil passage.

Defective Indicator or Gauge

A *defective oil pressure indicator* or gauge may appear to be a low or high oil pressure problem. The sending unit, circuit wiring, or gauge may be at fault. This topic is covered later in the chapter.

Oil Pressure Test

An *oil pressure test* uses a test gauge installed on the engine to measure actual lubrication system pressure. The pressure gauge is screwed into the hole for the oil pressure sending unit. The gauge may also be installed in one of the lines to the oil cooler, if used. See **Figure 42-7.**

Run the engine at the speed recommended in the service manual while testing. Read the test pressure gauge and compare to specifications. If oil pressure is too low or high, you must make repairs as needed.

Depending upon the type of engine and number of miles of use, oil pressure should be at least 20–30 psi (140–200 kPa) at idle and 40–60 psi (280–410 kPa) at cruising speeds. Check service manual specifications when testing because values can vary.

Note
For more information on oil pressure problems and symptoms, refer to Chapter 48, *Engine Mechanical Problems*.

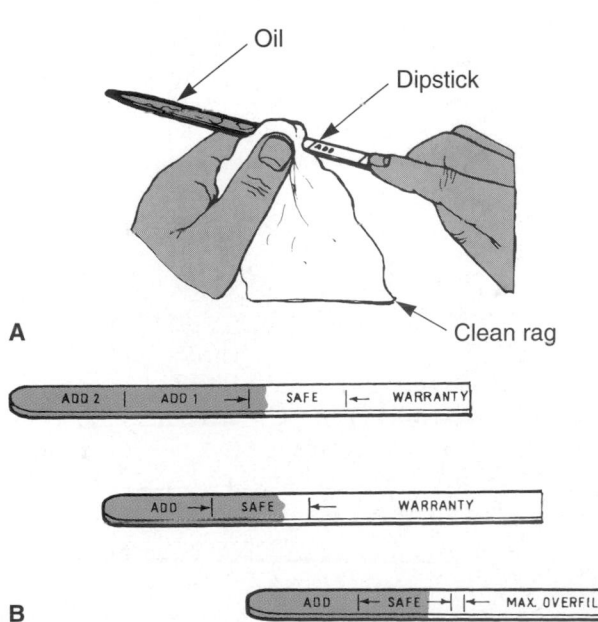

Figure 42-6. Review steps for checking engine oil level A—Pull out and wipe off dipstick. Then, reinsert fully and pull stick back out. B—Note height of oil on stick. Compare oil to markings on dipstick. (Chrysler)

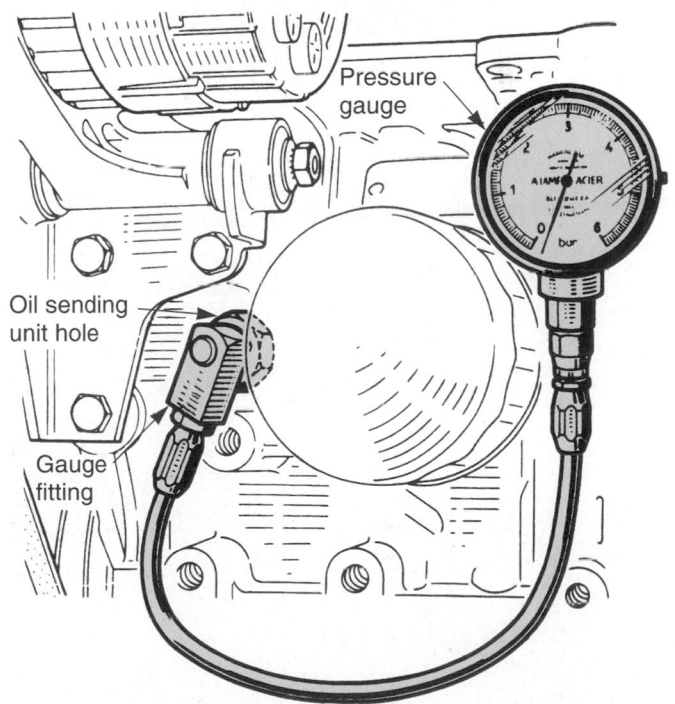

Figure 42-7. A test gauge can be used to measure actual engine oil pressure. The gauge is normally screwed into the hole for the oil sending unit. Measure the oil pressure with engine warm and at the specified engine speed.

Warning
Wear safety glasses when checking engine oil pressure. If hot oil squirts into your eye, a painful injury could result.

Engine Oil and Filter Service

It is extremely critical that the engine's oil and oil filter are serviced regularly. Lack of oil and filter maintenance can greatly shorten engine service life.

Used oil will be contaminated with dirt, metal particles, carbon, gasoline, ash, acids, and other harmful substances. Some of the smallest particles and corrosive chemicals are not trapped in the oil filter. They will circulate through the engine, increasing part wear and corrosion.

Oil and Filter Change Intervals

Automakers give *oil change intervals,* the maximum number of miles (or kilometers) a vehicle can be driven between oil changes. If the oil is not changed at this interval, the vehicle's warranty will become void.

New vehicles can go 3000–7500 miles (4800–12,000 km) between oil changes. Older vehicles should have oil changed more often. Refer to a manual for exact values.

An older, worn engine will contaminate the oil more quickly than a new engine. More combustion byproducts will blow past the rings and enter the oil. Also, engine bearing clearance will be larger, requiring more of the oil and lubrication system.

Also, if a vehicle is only driven for short periods and then parked, its oil should be changed more often. Since the engine may not be reaching full operating temperature, the oil can be contaminated with fuel, moisture, and other substances more quickly.

 Tech Tip
Diesel engines and turbocharged engines usually require more frequent oil and filter service than naturally aspirated (non-supercharged) engines.

Changing Engine Oil and Filter

To change the engine oil, warm the engine to full operating temperature. This will help suspend debris in the oil and make the oil drain more thoroughly. Then, follow the basic steps given in **Figure 42-8.**

There are a few rules to remember when changing engine oil and the oil filter:

- Keep the vehicle level so all oil drains from the pan.

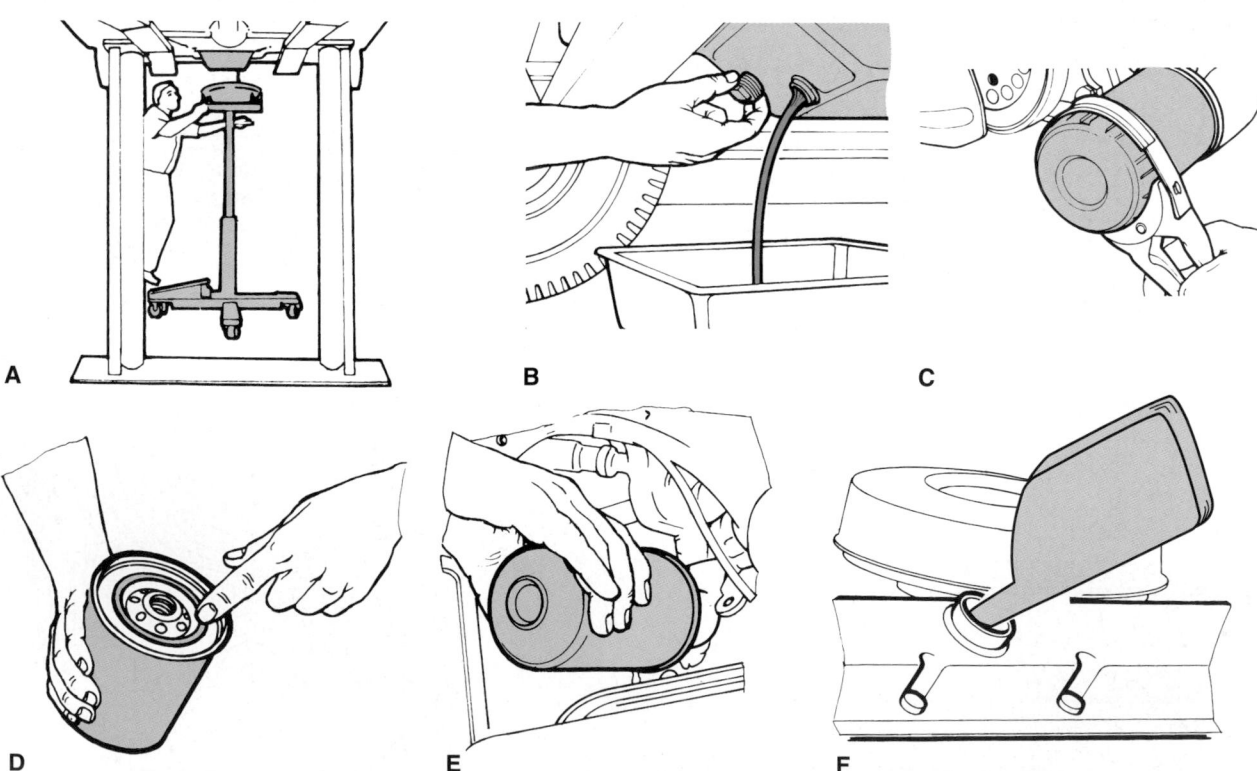

Figure 42-8. Review the steps for changing engine oil and filter. A—Use a lift or floor jack and jack stands to raise the vehicle in a level position. Place a catch pan under the drain plug. B—Unscrew the drain plug and allow oil to pour into the pan. Be careful of hot oil. It can cause painful burns. C—Use a filter wrench to remove the old filter. Turn counterclockwise. D—Wipe clean oil on the new filter O-ring. This will ensure proper tightening. E—Install and tighten the new oil filter by hand. Hands and filter should be clean and dry. Use a rag if needed. F—Install the correct type and quantity of oil. Pour oil into filler or breather opening in valve cover or intake manifold. (Mopar)

- Do not let hot oil pour out on your hand or arm!

- Check the condition of the drain plug threads and the O-ring washer. Replace them if needed.

- Do not overtighten the oil pan drain plug. It will strip very easily, possibly causing pan damage.

- Wipe clean oil on the filter O-ring before installation.

- Hand tighten the oil filter. Do not use the filter wrench because the filter canister could distort and leak.

- Fill the engine with the correct amount and type of oil. Remember that diesel engines usually hold more oil than gasoline engines. Check manufacturer recommendations.

- Check for oil leaks with the engine running before releasing the vehicle to the customer.

Oil Pan Service

An engine oil pan may need to be removed for various reasons: to service engine bearings or an oil pump, repair a damaged pan or to fix a stripped drain plug, or during an engine overhaul.

Some engine oil pans can be removed with the engine in the vehicle, as shown in **Figure 42-9.** After removing any bolt-on cross members and other obstructions, you must raise the engine off its motor mounts. This is normally needed to give enough clearance to unbolt and slide off the oil pan.

Other vehicle designs do not allow in-car oil pan removal. The engine must be lifted from the frame before the pan will come off. Check the service manual for details.

 To remove an oil pan:
1. Drain out the motor oil.
2. Reinstall the drain plug.
3. Unscrew the bolts around the outside of the pan flange. A swivel socket and long extension usually works well. Look at **Figure 42-10.**
4. Tap on the pan lightly with a rubber hammer to free it from the cylinder block. If stuck tight, carefully pry between the pan flange and the block.
5. Using a gasket scraper, remove all old gasket or silicone material from the pan and engine block. With an aluminum pan, be careful not to nick or dig into the sealing surface.

Special *gasket cutters* are available for freeing stuck gaskets. The tool blade is wedged into the gasket material and tapped with a plastic mallet. This will cut through the

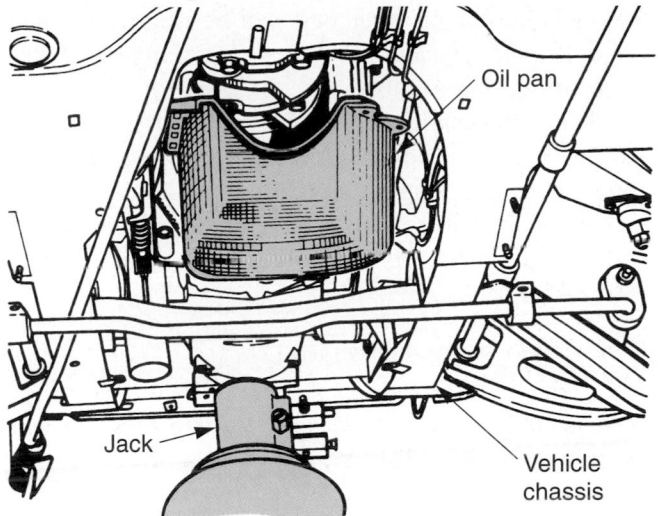

Figure 42-9. With some vehicles, the oil pan can be removed with the engine in vehicle. Use a jack to raise the engine off motor mount. Then, slide the pan down and out.

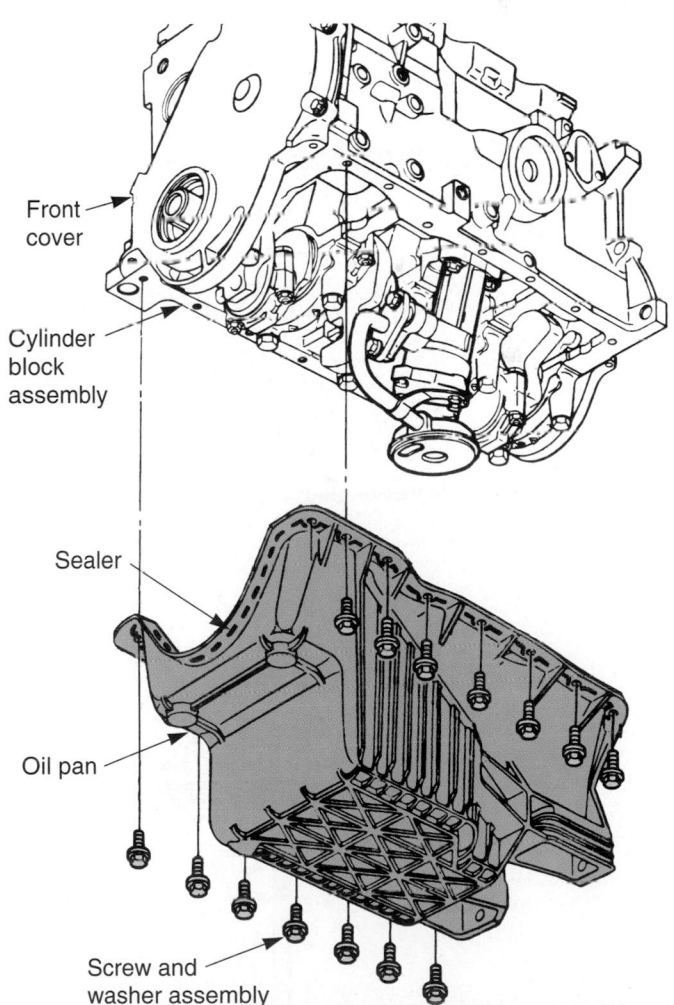

Figure 42-10. The oil pan is secured by a number of bolts around the pan flange. Be careful not to bend the flange during removal. (Ford)

gasket so the pan or other part can be removed without flange damage. Do not bend the oil pan flange or leakage may result upon assembly.

Check the inside of the pan for debris. Metal bits (bearing particles), plastic pieces (timing gear teeth), rubber particles (valve stem seals) indicate engine mechanical problems. Also inspect the oil pump and pickup screen for debris that will help problem diagnosis, **Figure 42-11.**

Prelubricator

A *prelubricator* can be used to help pinpoint worn engine bearings or other worn parts that lower oil pressure. Look at **Figure 42-12.**

A prelubricator is a metal pressure tank with special gauges and fittings. It can be used to force oil through the lubrication system without running the engine.

To check for worn bearings, remove the engine oil pan. Partially fill the prelubricator tank with oil. Then, connect the tank line to the engine, following equipment

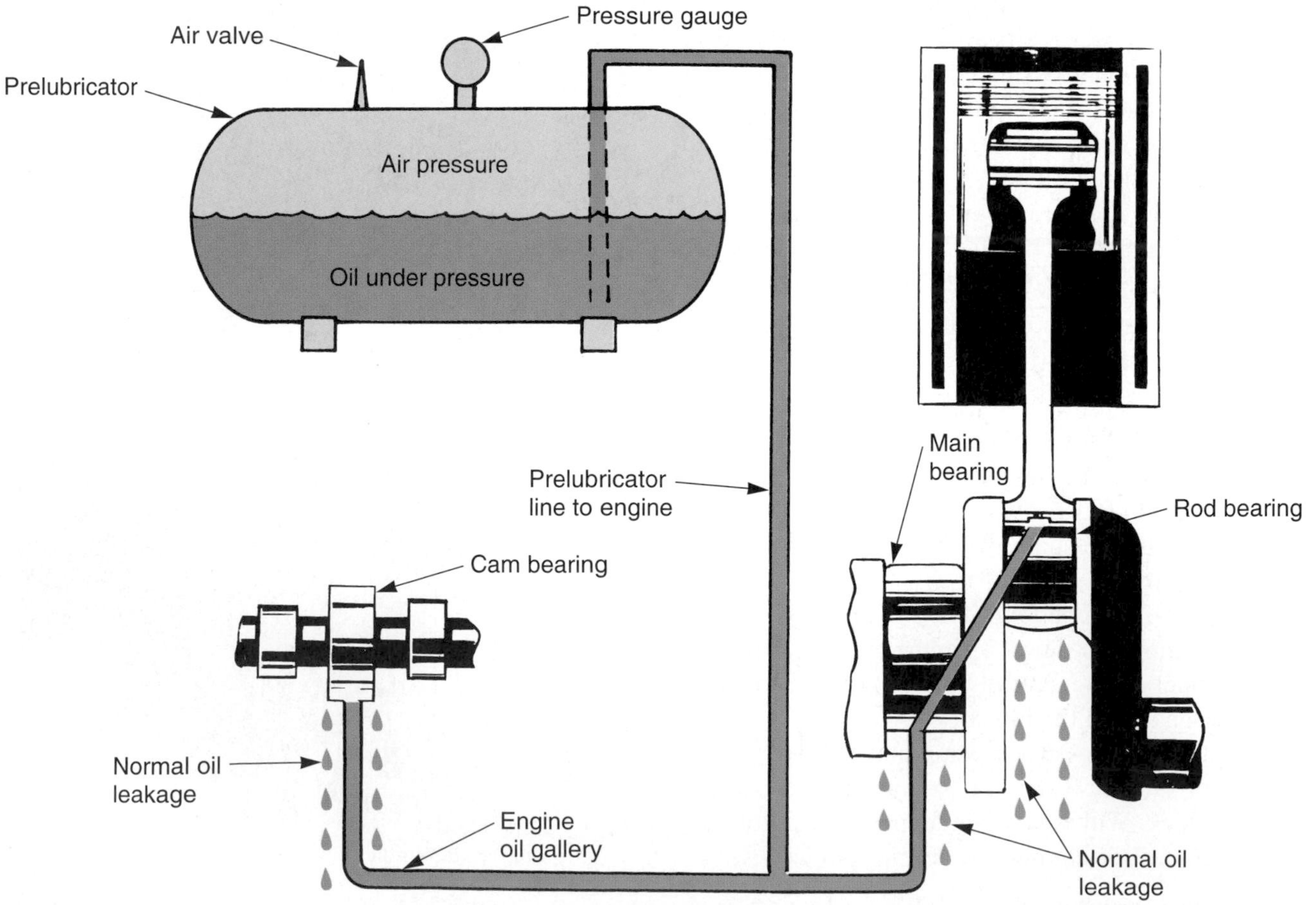

Figure 42-11. After removing the oil pan, inspect it for debris that could indicate problems. Note how this oil pickup screen has picked up pieces of debris. Debris can tell you whether you have spun bearings, hardened or broken valve seals, or broken piston skirts. The type of material will give hints about trouble sources. (Fel-Pro)

Figure 42-12. A prelubricator can be used to check for worn bearings. Air pressure pushes oil through the system. With the oil pan off, you can watch for too much oil (wear) at the engine bearings. (Dana Corp.)

instructions. Charge the tank with the recommended amount of air pressure. Place a large catch pan under the engine.

Open the valve on the prelubricator while watching the engine bearings. If an excess amount of oil pours out of any bearing, that bearing is bad. If oil flow out of all bearings is normal, remove and check the condition of the oil pump.

A prelubricator can also be used to prime the oil galleries with oil after an overhaul. It will ensure instant oil pressure when first starting a rebuilt engine.

Installing an Engine Oil Pan

To install an engine oil pan, wash the pan thoroughly in cold soak cleaner. Check the drain plug hole for stripped threads. Also, lay the pan upside down on a flat workbench. Make sure that the pan flange is not bent. Straighten the flange, if needed, with light hammer blows. See **Figure 42-13.**

Either a gasket or chemical sealer may be recommended for the oil pan. Use an approved gasket adhesive to position a new gasket. If rubber seals are used on each end, press them into their grooves, **Figure 42-14.**

Place silicone sealer where the gaskets meet the rubber seals. If you fail to do this, oil leakage will result, **Figure 42-15.**

To use a chemical or silicone gasket, clean the pan and mating block surfaces with a suitable solvent. Then, run a uniform bead (about 1/8″ or 3 mm) of sealer around the

pan or block flange as recommended. Form a continuous bead and place extra sealer at part or gasket-seal joints.

Fit the pan carefully into place on the block. Start all of the pan bolts by hand. Check that the gasket or sealer has not been shifted or smeared. Then, tighten each bolt a little at a time in a crisscross pattern. Torque the bolts to specifications.

> ⚠ **Caution**
> A common mistake by the novice is overtightening of the oil pan bolts. This can crush and split the pan gasket, causing oil leakage. Only moderate tightening is needed to compress the gasket and make it seal properly.

Double-check drain plug torque. Fill the engine with oil. Inspect for oil leaks with the engine running.

Oil Pump Service

A bad oil pump will cause low or no oil pressure and possibly severe engine damage. When inner parts wear, the pump may leak internally and have reduced output. The pump drive shaft can also strip in the pump or distributor, preventing pump operation. With belt drive, the belt might slip over a locked oil pump.

If pressure and prelubricator tests point to a faulty oil pump, remove and replace or rebuild the pump. However, make sure worn, loose engine bearings are not causing the problem. An open pressure relief valve will also prevent normal oil pressure.

Oil Pump Removal

Some oil pumps are located inside the engine oil pan. Others are on the front of the engine under a front cover or on the side of the engine. Removal procedures vary; refer to a shop manual for direction.

Figure 42-16 shows how a shaft driven oil pump is secured by two case hardened bolts. **Figure 42-17** shows the parts that must be removed when the oil pump mounts on the front of the engine.

Oil Pump Rebuild

Most mechanics install a new or factory-rebuilt pump when needed. It is usually too costly to completely rebuild an oil pump in-shop. However, you should have a general understanding of how to overhaul oil pumps. Large dealerships sometimes rebuild oil pumps when the vehicle is still under warranty.

Figures 42-18 and **42-19** give the basic steps for servicing a rotary oil pump. **Figure 42-20** summarizes the rebuilding procedures for a gear oil pump. A manual will

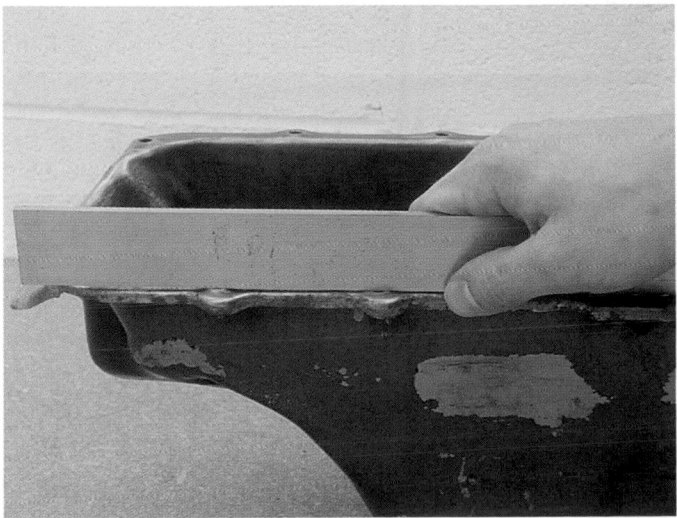

Figure 42-13. If the oil pan is leaking or if you had trouble getting it off engine, use a straightedge to check for flange warpage. If the bolts were overtightened, dents can be formed around bolt holes. Light hammer blows will straighten the flange. (Fel-Pro)

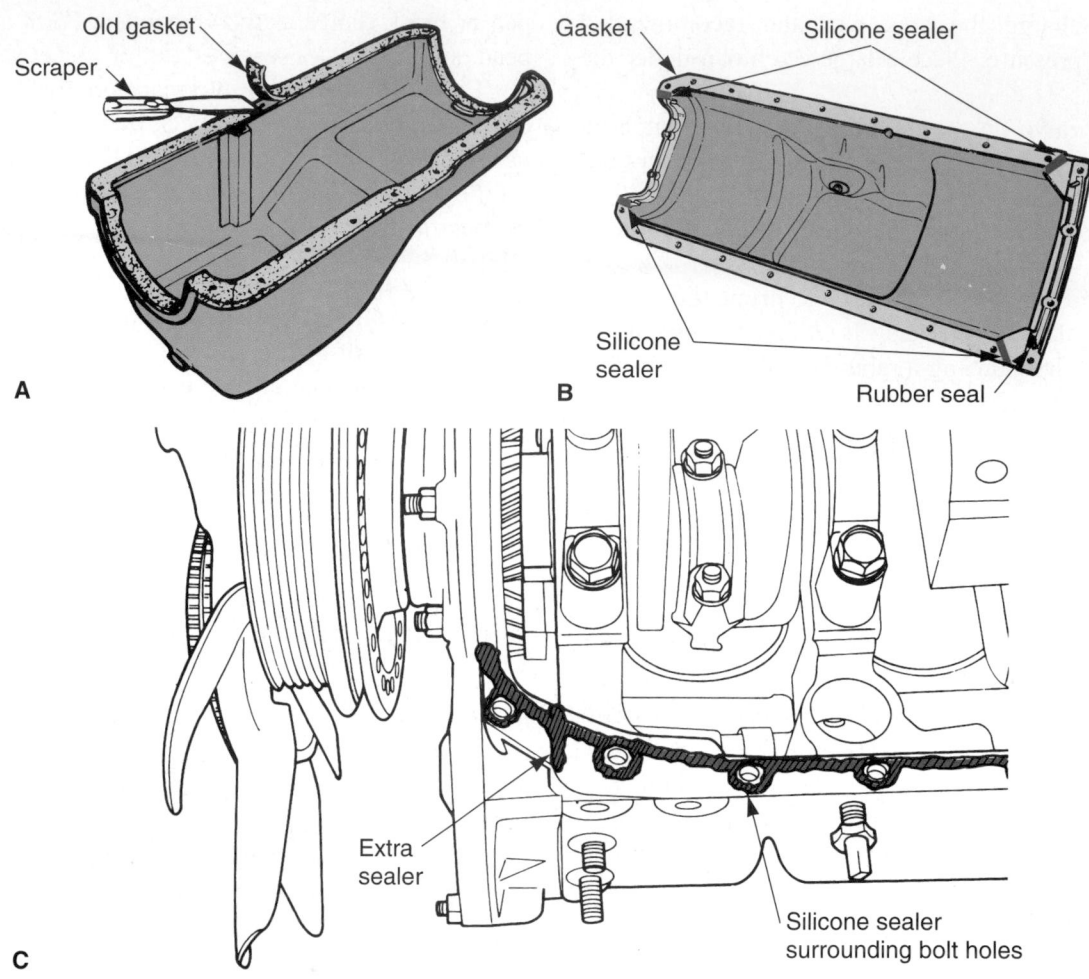

Figure 42-14. Study the basic steps for cleaning and installing gasket, seals, and sealer before oil pan installation. A—Scrape all old gasket or silicone sealer off the pan and block. Do not nick the surface of aluminum pans. B—If a gasket is used, adhere it to the flange and align bolt holes. Fit any seals into place and squirt sealer where the gasket meets the seal. C—To use silicone sealer only, clean the flange with recommended solvent. Then, carefully run a bead of sealer around the flange and all bolt holes. Use extra sealer where parts or gasket and seals butt together. (Chrysler and Ford)

Figure 42-15. A technician is installing RTV sealer where the gasket contacts the end of the rubber seal. This will prevent leakage. (Fel-Pro)

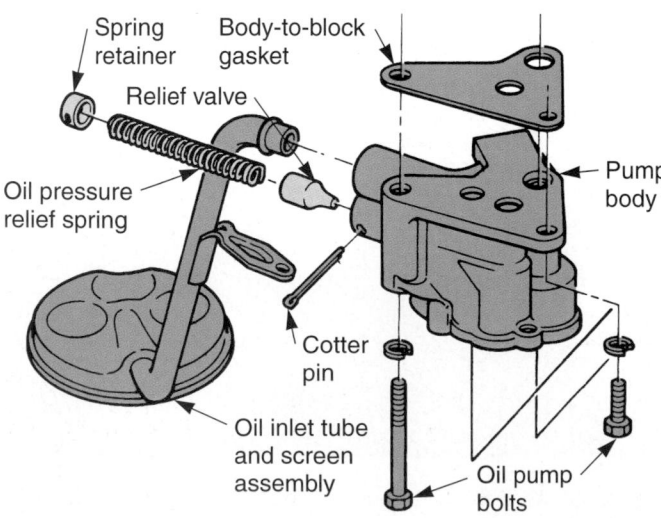

Figure 42-16. When servicing an oil pump, also service the in-housing pressure relief valve. Torque pump mounting to specifications. Make sure body-to-block gasket holes are aligned.

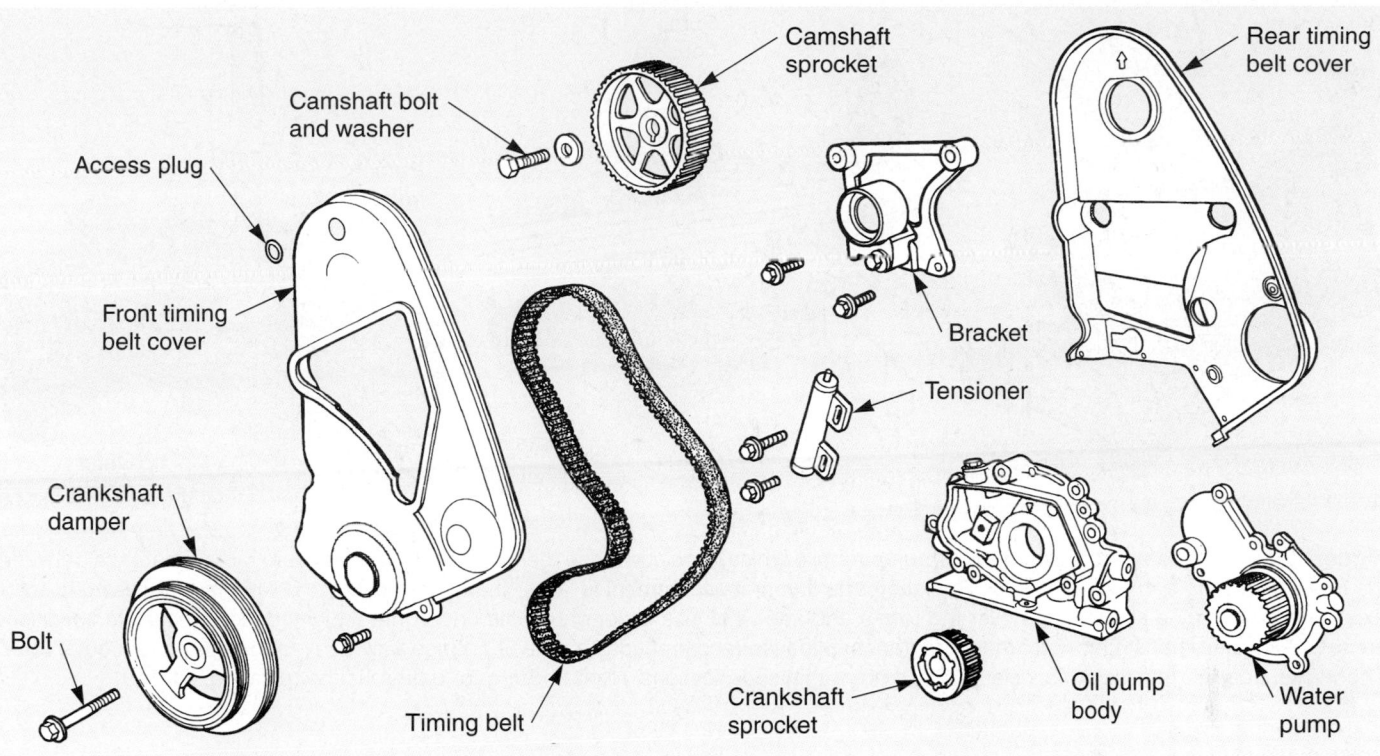

Figure 42-17. When servicing a belt-driven or crankshaft-driven oil pump, you will need to remove other parts to gain access to the pump—crank damper, timing belt cover, timing belt, belt sprocket, and other units. (Chrysler)

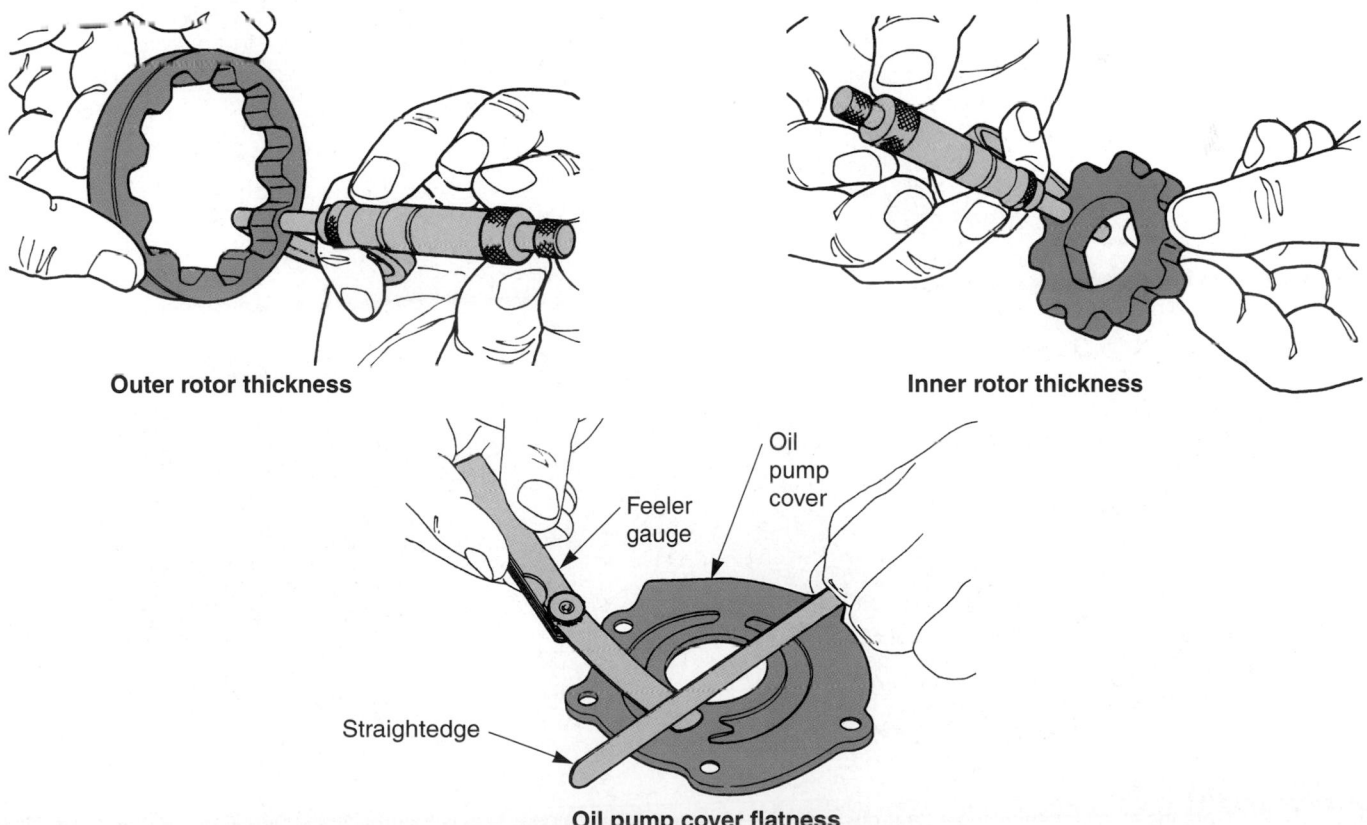

Outer rotor thickness

Inner rotor thickness

Oil pump cover flatness

Figure 42-18. To begin rebuild of a rotary oil pump, measure outer and inner rotor thickness. Check oil pump cover wear with a straightedge and feeler gauge. The thickest gauge that slides under the straightedge equals wear. Compare to wear limits in manual.

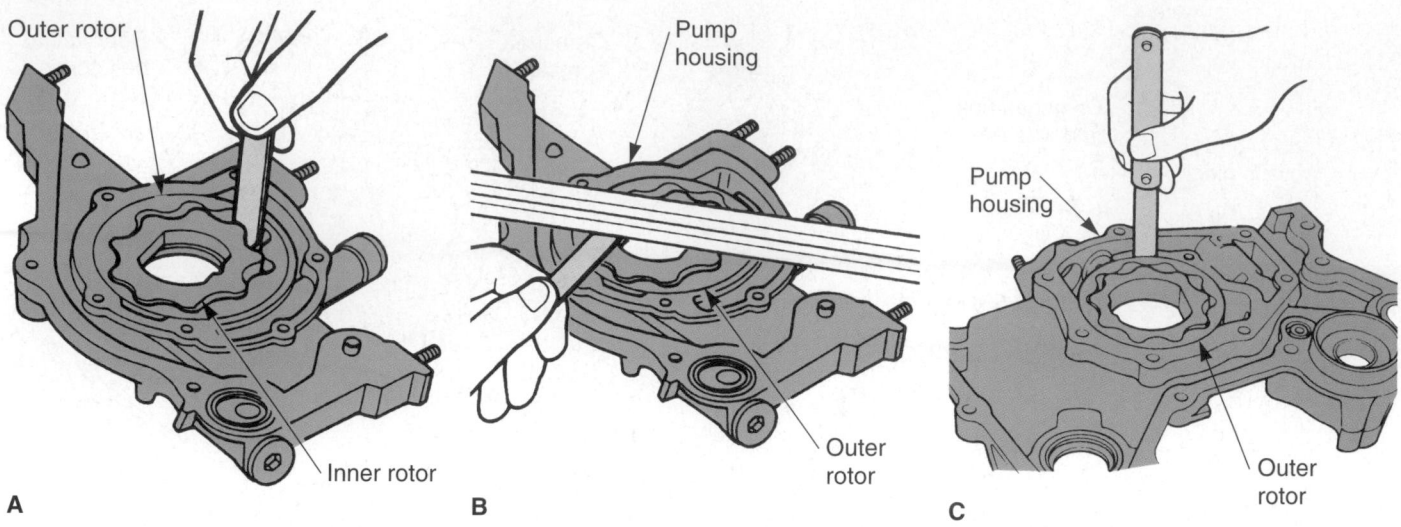

Figure 42-19. Other steps to finish rotary oil pump rebuild typically include these measurements. A—Measure radial clearance between the inner rotor lobe and the outer rotor. Replace parts if your measurement is larger than specifications. B—Measure housing-to-rotor axial clearance. Lay a straightedge over the rotors and see what size feeler gauge fits under the straightedge. Compare to specifications. C—Measure housing-to-rotor radial clearance. Slide the recommended size feeler gauge between pump housing and outer rotor. The thickest gauge that fits equals clearance. If not within specifications, rotor, housing, or both must be replaced. (Honda)

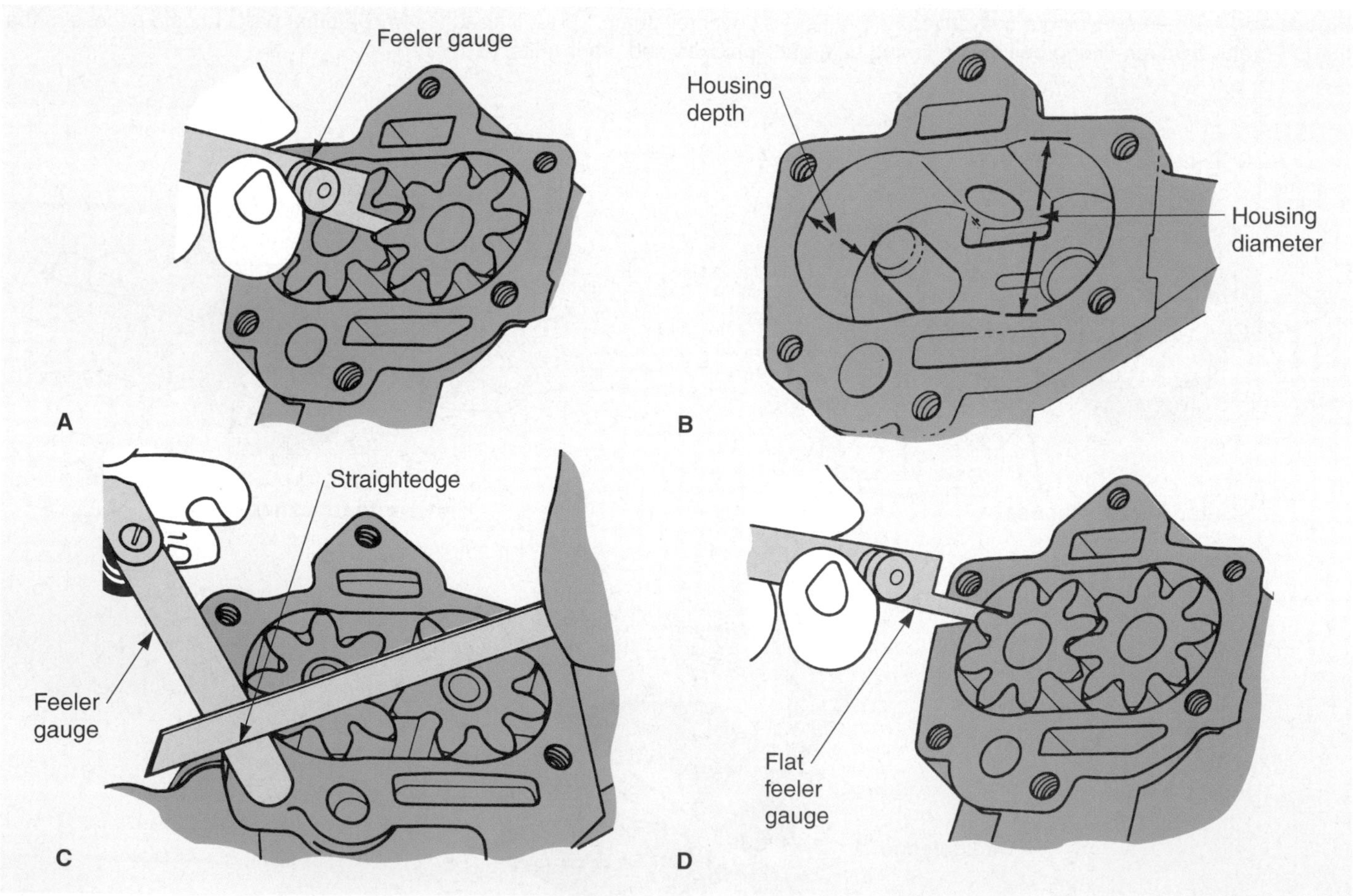

Figure 42-20. Note the steps for servicing a gear oil pump. A—Measure oil pump gear lash with small feeler gauge. B—Measure housing bore diameter and depth with telescoping gauge, depth gauge, and outside mike. Also, measure gear length or thickness and teeth diameter. C—Use a straightedge and feeler gauge to check end clearance. D—Measure gear side clearance with feeler gauge. If any measurement is not within specifications, replace parts as needed. (Buick)

give full instructions and specifications for the exact type of oil pump.

Oil Pump Installation

Before installation, prime the pump by filling it with motor oil. Pour oil into the pump inlet while turning its rotors or gears. This will ensure proper initial operation and prevent wear upon engine starting.

Install the pump in reverse order of removal. Double-check gasket position and hole alignment. Torque all bolts to specifications. If an oil pickup is used, make sure it is installed and secured properly.

Figure 42-21 shows how to install a shaft-driven oil pump. It bolts to the bottom of the block. When installing the drive shaft, make sure the small clip on the shaft is positioned correctly. The clip keeps the shaft from falling

out when removing or installing the distributor or cam sensor.

When installing a front-mounted oil pump, replace all oil seals to prevent leakage. Use a seal driver to force the new seal into place without damaging it. See **Figure 42-22.**

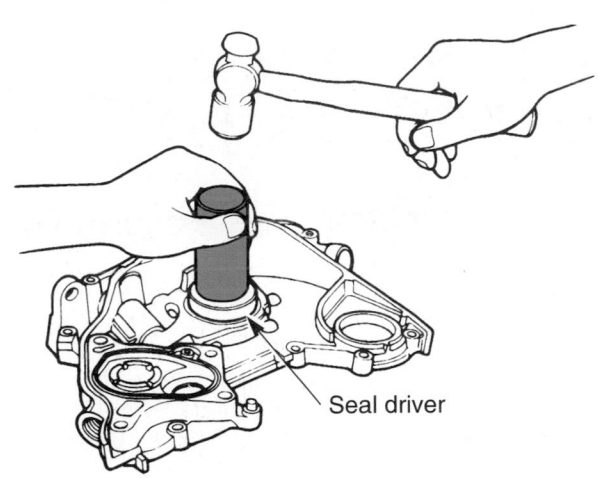

Figure 42-22. With front-mounted oil pumps, always install new seals when the pump is removed. Use a seal driver to force the seal into the housing. (Honda)

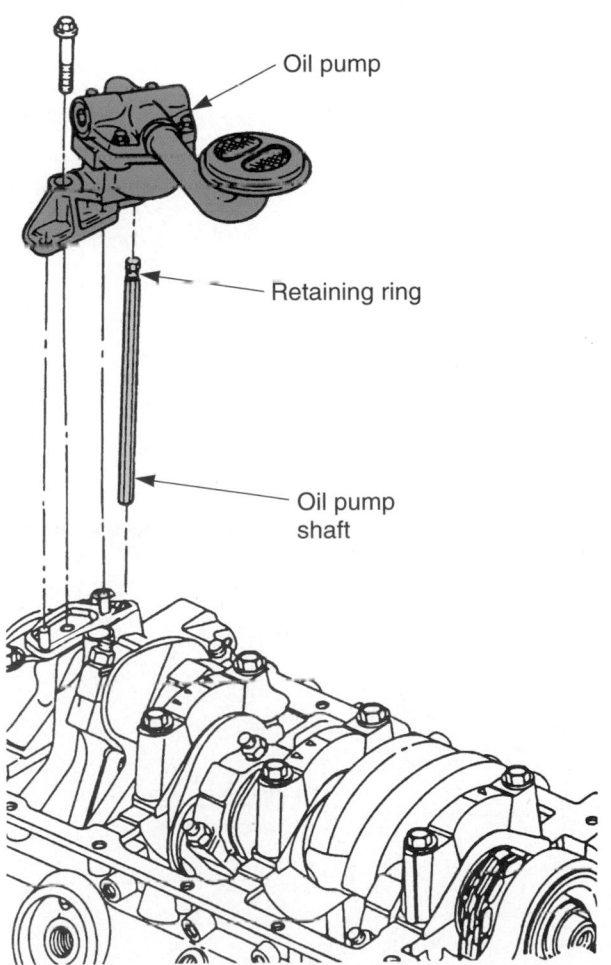

Figure 42-21. When installing a shaft-driven oil pump, be certain the shaft is installed into the block correctly and check any spring clips on the shaft. Note that some oil pumps require a gasket and others do not. Use the correct oil pump bolts and use a torque wrench to tighten to specifications. (Ford)

Pressure Relief Valve Service

A faulty pressure relief valve can produce oil pressure problems. The valve may be located in the oil pump, filter housing, or engine block.

If symptoms point to the pressure relief valve, it should be disassembled and serviced. It is also serviced during an engine overhaul. Relief valves are pictured in **Figures 42-16** and **42-23.**

Remove the cup or cap holding the pressure relief valve. Then, slide the spring and piston out of their bore. If needed, hold a rag over the valve and direct air pressure into the pump to remove the valve.

Measure *spring free length* (length of extended spring) and compare to specifications. If the spring is too short or too long, install a new spring. Spring tension can also be checked on a spring tester. Shims can be used to increase spring tension. An adjustment screw may be provided for changing opening pressure.

Use a micrometer and small hole gauge to check valve and valve bore wear. Also, check the sides of the valve for scratches or scoring. Replace parts if any problems are found.

Assemble the pressure relief valve. Make sure the valve is facing correctly in its bore. Slide the spring into place. Install any shims and the cover plug or cap. Refer to a manual for details.

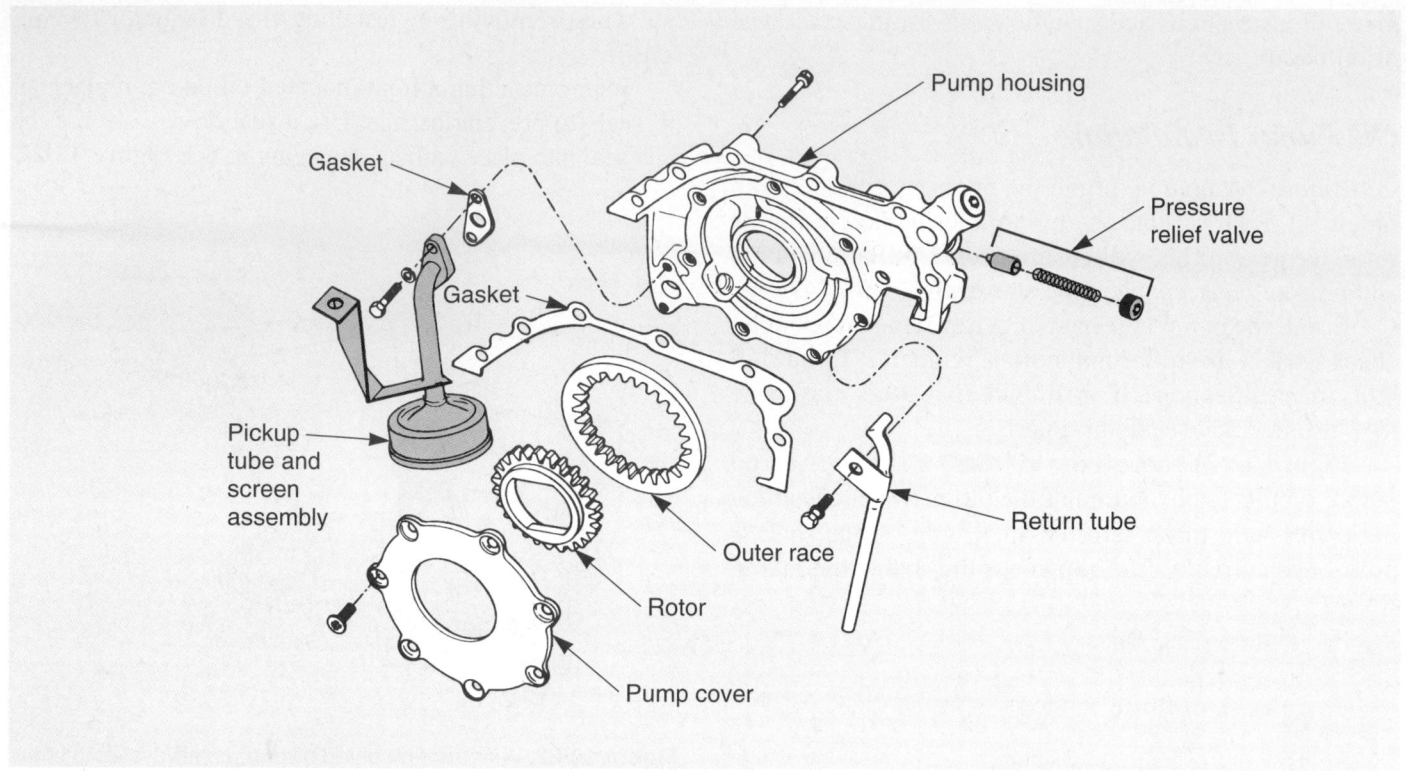

Figure 42-23. Exploded view shows parts of an oil pump that mounts on the front of the engine. Note pressure relief valve, oil pickup, and other parts. Refer to a shop manual for instructions on servicing. (Ford)

Oil Pressure Indicator and Gauge Service

A bad oil pressure indicator circuit can scare the customer into thinking there are major engine problems. The gauge can read low or high, which may or may not indicate a lubrication system problem.

A faulty oil sending unit can make the indicator light glow or gauge read low when oil pressure is normal. A flickering indicator light is often caused by a bad oil pressure sending unit. A sending unit can also leak oil between its body and plastic housing.

Inspect the indicator or gauge circuit for problems. The wire going to the sending unit may have fallen off (no light, flickering light, improper gauge readings). The sending unit wire may also be shorted to ground (light stays on or gauge always reads high).

To check the action of the indicator or gauge, remove the wire from the sending unit. Touch it on a metal part of the engine. This should make the indicator light glow or the oil pressure gauge read maximum. If it does, the sending unit may be bad. If it does not, then the circuit, indicator, or gauge may be faulty. Look at **Figure 42-24.**

Always check the service manual before testing an indicator or gauge circuit. Some automakers recommend a special gauge tester. This is especially important with some computer-controlled systems. The tester will place

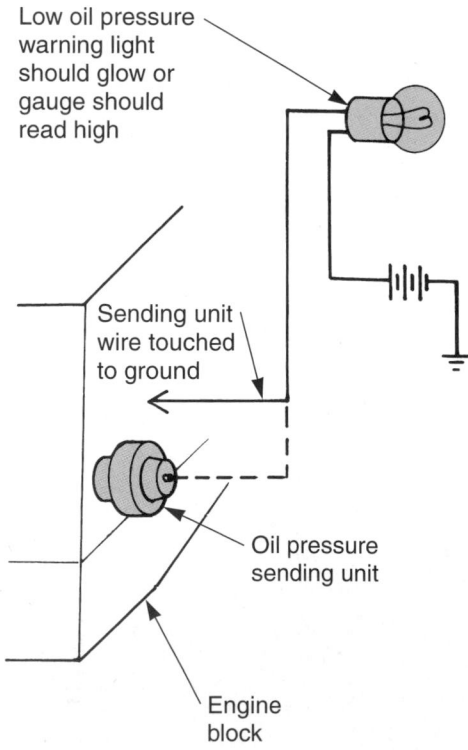

Figure 42-24. Checking the action of the oil pressure indicator circuit. When the sending unit wire is grounded, the indicator light should glow or the gauge should read maximum. When wire is disconnected, light should go out or gauge should read zero.

a specific resistance in the circuit to avoid circuit damage. Refer to a manual for instructions.

Normally, a low oil pressure light will glow when oil pressure is below a certain value. Typically, minimum oil pressure at idle is 5 psi–10 psi (35-70 kPa). This can vary with the engine design.

PCV Valve Service

A PCV (positive crankcase ventilation) valve should be cleaned or replaced at regular intervals. The valve can become clogged or restricted with dirt and sludge. This can cause engine sludging and wear.

The PCV valve is usually located in the valve cover. It may also be in a breather opening in the intake manifold. Many new vehicles do not use a PCV valve. They simply route a vacuum hose from the valve cover or crankcase to an air inlet duct on the engine.

Note
For details of PCV service, look in Chapter 44, *Emission Control System Testing, Service, and Repair.*

Summary

- Without a properly functioning lubrication system, an engine will "self-destruct" in a matter of minutes.
- High oil consumption is caused by external oil leakage out of the engine or by internal oil leakage into the combustion chambers.
- External oil leakage is easily detected as darkened, oil-wet areas on or around the engine.
- Internal oil leakage shows up as blue smoke coming out of the exhaust.
- Low oil pressure is indicated when the oil indicator light glows, oil gauge reads low, or when the engine lifters or bearings rattle.
- An oil pressure test uses a test gauge installed on the engine to measure actual lubrication system pressure.
- Used oil will be contaminated with dirt, metal particles, carbon, gasoline, ash, acids, and other harmful substances.
- Automakers specify oil change intervals.
- A prelubricator can be used to help pinpoint worn engine bearings or other worn parts that lower oil pressure.
- A common mistake by the novice is overtightening the oil pan bolts.
- A bad oil pump will cause low or no oil pressure and possibly severe engine damage.

- Before installation, prime the pump by filling it with motor oil.
- A faulty pressure relief valve can produce oil pressure problems.

Important Terms

High oil consumption
External oil leakage
Oil dye
Internal oil leakage
Low oil pressure
High oil pressure
Defective oil pressure indicator
Oil pressure test
Oil change intervals
Gasket cutters
Prelubricator
Spring free length

Review Questions—Chapter 42

Please do not write in this text. Place your answers on a separate sheet of paper.

1. List four common lubrication system problems.
2. High engine oil consumption *cannot* be caused by:
   (A) oil leakage into combustion chambers.
   (B) leaking valve cover gaskets.
   (C) leaking exhaust manifold gaskets.
   (D) worn piston rings or cylinders.
3. List seven common causes of low engine oil pressure.
4. How do you measure actual engine oil pressure?
5. How often should a new vehicle's oil and filter be changed?
6. Explain eight rules to follow when changing an engine's oil and oil filter.
7. Some engine oil pans should have an oil pan gasket while others may use a sealant. True or False?
8. A(n) _____ can be used to help find worn engine bearings or other parts that lower engine oil pressure.
9. Most technicians rebuild worn oil pumps in-shop. True or False?
10. A driver complains that her oil pressure warning light flickers on and off at stoplights. Technician A says that the oil level in the engine should be checked. Technician B says that the indicator light circuit could also be at fault. Who is correct?
    (A) A only.
    (B) B only.
    (C) Both A and B.
    (D) Neither A nor B.

● ASE-Type Questions

1. Each of these problems may be found in a lubrication system *except:*
   (A) low oil pressure.
   (B) high oil pressure.
   (C) low oil consumption.
   (D) high oil consumption.

2. External oil leakage may be detected by/as:
   (A) high oil consumption.
   (B) oil puddles under vehicle.
   (C) oil-wet areas around engine.
   (D) Any of the above.

3. Oil dye is sometimes added to motor oil to:
   (A) detect leaks.
   (B) detect contaminants.
   (C) destroy corrosive chemicals.
   (D) determine frequency of oil changes.

4. Internal oil leakage will appear as what color smoke?
   (A) Blue.
   (B) Black.
   (C) White.
   (D) None of the above.

5. Which of the following is *not* a common cause of low oil pressure?
   (A) Worn oil pump.
   (B) Thin or diluted oil.
   (C) Leaking valve stem seal.
   (D) Clogged oil pickup screen.

6. Upon encountering an uncommon high oil pressure problem, Technician A says its cause is low oil viscosity. Technician B says the cause is low relief valve spring tension. Who is right?
   (A) A only.
   (B) B only.
   (C) Both A and B.
   (D) Neither A nor B.

7. At idle, oil pressure should be at least:
   (A) 0–10 psi.
   (B) 20–30 psi.
   (C) 40–60 psi.
   (D) Over 100 psi.

8. Each of these should be followed when changing engine oil and the oil filter *except:*
   (A) keep the vehicle level.
   (B) do not overtighten the pan drain plug.
   (C) use the correct amount and type of oil.
   (D) tighten the oil filter with a filter wrench.

9. An engine oil pan may need to be removed to:
   (A) service an oil pump.
   (B) repair a damaged pan.
   (C) perform an engine overhaul.
   (D) All of the above.

10. A prelubricator is used to perform each of these tasks *except*:
    (A) locate worn engine bearings.
    (B) keep contaminants out of oil.
    (C) prime oil galleries after overhaul.
    (D) force oil through system without engine on.

Activities—Chapter 42

1. Use the school's video equipment to develop a short demonstration on how to properly check a vehicle's oil level.

2. Observe an oil change being done at a dealership service department or other facility. Take notes of the steps involved, from moving the vehicle into the bay to completion of the job. Use your notes to write a set of step-by-step instructions for a person doing an oil change for the first time.

Lubrication System Diagnosis		
Condition	**Possible causes**	**Correction**
Low oil pressure.	1. Oil level low. 2. Oil too thin. 3. Defective oil gauge. 4. Bad sending unit. 5. Excessive bearing clearance. 6. Worn oil pump. 7. Loose or disconnected oil pickup. 8. Open pressure relief valve. 9. Clogged pickup screen.	1. Add oil to engine. 2. Add heavier oil. 3. Replace gauge. 4. Replace sending unit. 5. Replace bearings. 6. Replace oil pump. 7. Repair oil pickup. 8. Repair valve in pump or block. 9. Clean or replace screen.
High oil pressure.	1. Stuck pressure relief valve. 2. Incorrect relief valve spring. 3. High oil viscosity. 4. Oil gallery restriction. 5. Inaccurate indicator.	1. Repair valve. 2. Replace spring. 3. Drain oil. Replace with recommended viscosity oil. 4. Remove blockage. 5. Repair circuit. Replace gauge or sending unit, if necessary.
Swelled oil filter.	1. High oil pressure.	1. See *High oil pressure.*

Emission Control Systems

43. Emission Control Systems

44. Emission Control System Testing, Service, Repair

Emission control systems protect our environment from chemical damage. They are designed to reduce the amount of toxic substances emitted by an automobile. Some systems prevent fuel vapors from entering the atmosphere; others remove unburned or partially burned fuel from the engine exhaust.

This section provides the information needed to repair emission control systems so that a vehicle operates as cleanly as possible. Not only will this information make you more employable as a technician, but it will help make our earth a pleasant, healthy place to live and breathe.

The information in this section will help you pass ASE Test A6, *Electrical/Electronic Systems,* and Test A8, *Engine Performance.*

Chapter 43

Emission Control Systems

After studying this chapter, you should be able to:

- Define the fundamental terms relating to automotive emission control systems.
- Explain the sources of air pollution.
- Describe the operating principles of emission control systems.
- Compare design differences in emission control systems.
- Explain how a computer or engine control module can be used to operate emission control systems.
- Summarize how OBD II systems use multiple oxygen sensors to check air-fuel mixture and catalytic converter efficiency.
- Correctly answer ASE certification test questions that require a knowledge of emission control system operation and construction.

Emission control systems are used on cars and light trucks to reduce the amount of harmful chemicals released into the atmosphere. These systems help to keep the air clean.

This chapter introduces the terminology, parts, and systems that control automobile emissions. This is an important chapter that prepares you for Chapter 44, *Emission Control System Diagnosis, Service, and Repair.*

Air Pollution

Air pollution is caused by an excess amount of harmful chemicals in the atmosphere. It is caused by a number of factors, some by nature and some man-made.

Natural air pollution is produced by erupting volcanoes, forest fires, wind-blown dust, and decay of vegetation. *Man-made air pollution* is produced by factories, home furnaces, fireplaces, and internal combustion engines. All of these sources contribute to air pollution, **Figure 43-1.**

The federal government passed strict laws to reduce air pollution. These laws, which are enforced by the *Environmental Protection Agency* (*EPA*) and local authorities, limit the amount of emissions that can be emitted. Auto manufacturers have implemented a number of design changes and added systems that reduced emissions to meet or exceed EPA standards.

Smog

Smog is a nickname given to a visible cloud of airborne pollutants. It is a word derived from the words "smoke" and "fog." Smog is common in large cities and industrialized areas. If dense enough, smog can be harmful to humans, animals, and vegetation. It can even damage paint, rubber, and other materials. See **Figure 43-2.** Smog is formed when airborne pollutants combine with oxygen, nitrogen, and other elements in the presence of sunlight. The resulting yellow-gray cloud of pollution is referred to as *photochemical smog.*

Exhaust Emissions

Exhaust emissions are pollutants produced by cars, trucks, buses, and motorcycles. There are four basic types

Figure 43-1. Air pollution comes from many sources, some natural and some man-made. (American Petroleum Institute)

Figure 43-2. Smog and other visible forms of air pollution, is common in large cities and industrial areas where there is a concentration of factories, homes, and vehicles. (American Petroleum Institute)

of vehicle exhaust emissions: hydrocarbons, carbon monoxide, oxides of nitrogen, and particulates.

Hydrocarbons (HC)

Hydrocarbons, abbreviated *HC*, is a form of emission resulting from the release of unburned fuel into the atmosphere. As you learned in earlier chapters, all petroleum (crude oil) products are made of hydrocarbons (hydrogen and carbon compounds). This includes gasoline, diesel fuel, LP-gas, and motor oil.

Hydrocarbons are produced by incomplete combustion or by fuel evaporation. For example, hydrocarbons are produced when unburned fuel escapes from the exhaust system of a poorly running engine. They can also be produced by fuel vapors escaping from a vehicle's fuel system. Some of the additives used in gasoline are extremely reactive, which allow the unburned fuel vapors to easily combine photochemically with other elements in the air to form smog.

Hydrocarbon emissions are a hazardous form of air pollution. They can contribute a variety of illnesses, including eye, throat, and lung irritation, and, possibly cancer.

Carbon Monoxide (CO)

Carbon monoxide (CO) is an extremely toxic emission resulting from the release of partially burned fuel. It

is created as a result of the incomplete combustion of a petroleum-based fuel.

Carbon monoxide is a colorless, odorless, and deadly gas. CO prevents human blood cells from carrying oxygen to body tissues. It can cause death if inhaled in large quantities. Symptoms of carbon monoxide poisoning include headaches, nausea, blurred vision, and fatigue.

Any factor that reduces the amount of oxygen present during combustion increases carbon monoxide emissions. For example, a rich air-fuel mixture would increase CO. As the mixture is leaned, CO emissions are reduced. See **Figure 43-3.**

Oxides of Nitrogen (NO_x)

Oxides of nitrogen, abbreviated NO_x, are emissions produced by extremely high temperatures during combustion. Air consists of about 79% nitrogen and 21% oxygen. With enough heat (above approximately 2500°F or 1370°C), nitrogen and oxygen in the air-fuel mixture combine to form NO_x emissions.

Oxides of nitrogen contribute to the dirty brown color in smog. They also produce ozone in smog, which causes an unpleasant smell and is an eye and respiratory irritant. Oxides of nitrogen is also harmful to many types of plants and rubber products.

An engine with a high compression ratio, lean air-fuel mixture, and high-temperature thermostat will produce high combustion heat, resulting in the formation of NO_x. This poses a problem, as these same factors tend to improve gas mileage and reduce HC and CO emissions. As a result, emission control systems must interact to lower each form of pollution.

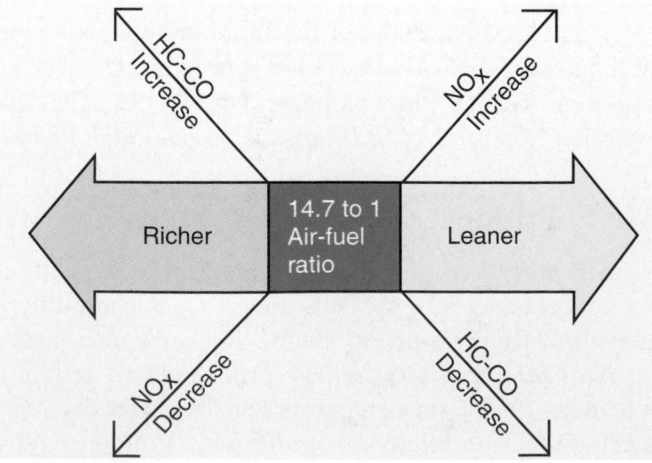

Figure 43-3. Study the relationship between air-fuel ratio and emissions. Close control of the engine's air-fuel mixture helps to keep exhaust emissions at a minimum.

Particulates

Particulates are solid particles of carbon soot and fuel additives that blow out a vehicle's tailpipe. Carbon particles make up the largest percentage of these emissions. The rest of the particulates consist of other additives sometimes used to make gasoline and diesel fuel.

While particulate emissions is rarely a problem with gasoline engines, it is a serious problem with diesel engines. You have probably seen a diesel truck or bus emitting black smoke out of its exhaust. Diesel particulates are normally caused by an extremely rich air-fuel mixture or a mechanical problem in the injection system.

About 30% of all particulate emissions are heavy enough to settle out of the air in a relatively short period of time. The other 70%, however, can float in the air for extended periods.

Sources of Vehicle Emissions

Shown in **Figure 43-4,** the majority of vehicle emissions come from three basic sources:

- Engine crankcase blowby fumes—chemicals that form in the engine bottom end from heating of oil and burning of fuel that blows past piston rings and into the crankcase.

- Fuel vapors— various chemicals that enter the air as fuel evaporates.

- Engine exhaust gases— harmful chemicals produced and blown out the tailpipe when an engine burns a hydrocarbon-based fuel (or most other fuels).

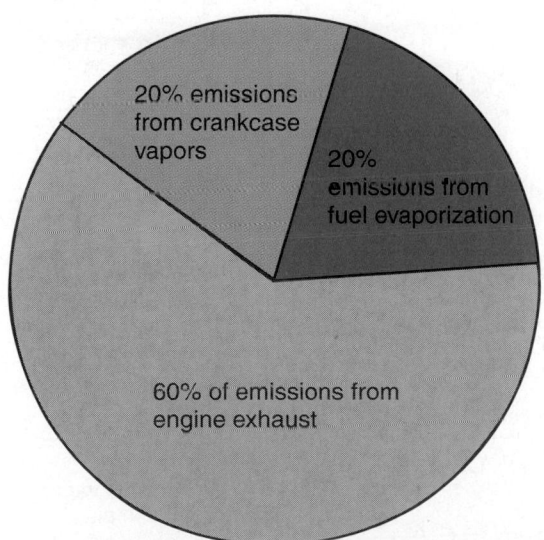

Figure 43-4. These are the three general sources and percentages of auto air pollution.

Various engine modifications and emission control systems are used to reduce air pollution from these sources.

Engine Modifications Related to Emission Control

Auto manufacturers agree that the best way to reduce exhaust emissions is to burn all of the fuel inside the engine. For this reason, several engine modifications have been introduced to improve combustion efficiency. Modern engines have the following modifications to lower emissions:

- Lower compression ratios allow the use of unleaded gasoline. Unleaded fuel permits the use of catalytic converters and burns completely to lower HC emissions. Eliminating the lead additives also reduces particulate emissions. Lower compression ratios also reduce combustion temperatures and NO_x emissions. Although improvements in fuel control have allowed new engines to utilize slightly higher compression ratios (over 10:1 in some cases), they are still lower than the compression ratios used in the "muscle car" engines of the 1960s (over 11:1).

- Smaller combustion chamber surface volumes are used to reduce HC emissions. A smaller surface area in the combustion chamber increases combustion efficiency by lowering the amount of heat dissipated out of the fuel mixture. Less combustion heat enters the cylinder head and more heat is left to burn the fuel. Also, modern combustion chambers have simple shapes to reduce the chance of fuel condensation. A hemispherical combustion chamber typically has the smallest surface volume.

- Reduced quench areas in the combustion chambers are used to lower HC and CO emissions. A quench area is produced when the engine pistons move too close to the cylinder head. When the distance between these metal parts is too close, it tends to quench (put out) combustion and increase emissions due to unburned fuel. Modern engines have redesigned cylinder heads and pistons which prevent high quench areas.

- Decreased valve overlap is used to reduce exhaust emissions and to increase engine smoothness. A camshaft with more overlap will provide greater power at higher speeds. However, the increased overlap dilutes the incoming air-fuel mixture with burned exhaust gases, requiring a richer air-fuel mixture at idle and low speeds, which increased HC and CO formation.

- Hardened valves and seats are used to prevent excessive valve wear from the use of unleaded fuel. Lead additives in fuel, besides increasing octane, also acted as a high-temperature lubricant. They reduced wear at the valve faces and seats.

- Higher operating temperatures are used to reduce HC and CO emissions. Today's engines have thermostats with higher temperature ratings. If the metal parts in an engine are hotter, less combustion heat will transfer out of the burning fuel. More heat will remain in the burning mixture to produce gas expansion, piston movement, and more complete combustion.

- Leaner air-fuel mixtures are used to lower HC and CO emissions. Leaner mixtures have more air to help all the fuel burn. More air also helps to cool the cylinder, reducing the formation of NO_x.

- Wider spark plug gaps are used to properly ignite the lean air-fuel mixtures. Wider gaps produce a hotter spark, which can ignite hard to burn, lean air-fuel mixtures. A wide gap also reduces spark plug fouling.

- Alcohols and other clean-burning substances are added to gasoline during the refinement process. In areas where high concentrations of air pollution are present, a specially formulated, or "oxygenated," gasoline is sold. These additives provide additional oxygen, or a "chemical enleanment" of the air-fuel mixture, which reduces CO.

Other external devices and methods are used to further reduce engine emissions. Many of these are covered later in this chapter.

Vehicle Emission Control Systems

Several different emission control systems are used to reduce the amount of air pollution produced by the automobile. The major ones found on late-model vehicles include:

- Positive crankcase ventilation system—recirculates engine crankcase fumes back into the combustion chamber.

- Evaporative emissions control system—closed vent system that stores fuel vapors and prevents them from entering the atmosphere.

- Exhaust gas recirculation system—injects burned exhaust gases into the engine to lower combustion temperatures and prevent the formation of NO_x.

- Air injection system—forces outside air into the exhaust system to help burn unburned fuel.

- Thermostatic air cleaner system—maintains a constant temperature of the air entering the engine for improved combustion and performance in cold weather.

- Catalytic converter—chemically changes exhaust byproducts into harmless substances.

- Computer control system—Electronic controls are used to monitor and interface with various systems to increase overall engine efficiency and reduce emissions.

These systems are used to make the modern vehicle very efficient.

Positive Crankcase Ventilation (PCV)

A *positive crankcase ventilation (PCV) system* uses engine vacuum to draw blowby gases into the intake manifold for reburning in the combustion chambers. Look at **Figure 43-5.** A number of years ago, crankcase fumes were simply vented into the atmosphere. A *road draft tube* vented crankcase fumes and blowby gases out the back of the engine. This contributed to air pollution.

Engine *blowby* is caused by pressure leakage past the piston rings on the power strokes. A small percentage of combustion gases can flow through the ring end gaps of the piston ring grooves and into the crankcase. If not reburned in the engine, these fumes would contribute to

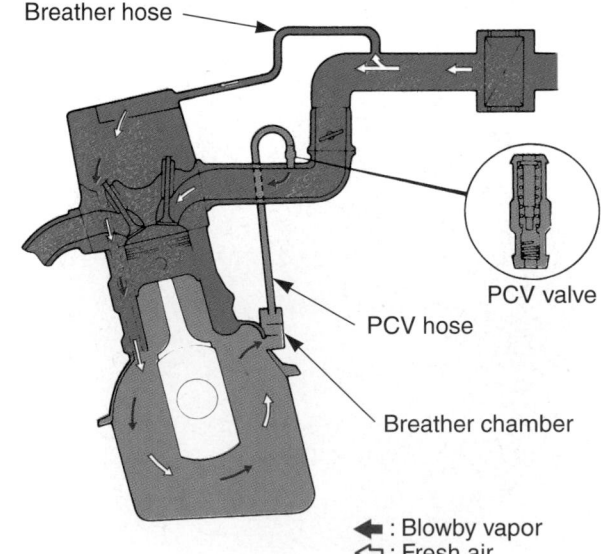

Figure 43-5. The PCV system draws vapors out of the crankcase and routes them into the engine to be burned. (Honda)

air pollution if vented to the atmosphere. If not vented from the crankcase, the gases would build to a point where engine damage would occur.

Engine blowby gases contain unburned fuel (HC); partially burned fuel (CO); particulates; and small amounts of water, sulfur, and acid. For this reason, blowby gases must be removed from the engine crankcase. Blowby gases can cause:

- Air pollution (if released into the atmosphere).
- Corrosion of engine parts.
- Engine oil dilution.
- Sludge formation.

A PCV system keeps the inside of the engine clean and reduces air pollution. Older engines used an open PCV system. This system was not sealed and gases could leak out when the engine was shut off. These systems have been totally replaced by the closed PCV system.

A *closed PCV system* uses a sealed oil filler cap, a sealed oil dipstick, ventilation hoses, and either a PCV valve or a flow restrictor. The gases are drawn into the engine and burned. The system stores the gases when the engine is not running.

PCV System Operation

Although designs vary and can use either vacuum or electronic control, the operation of all PCV systems is basically the same. Look at **Figure 43-5.**

A hose usually connects the intake manifold to the PCV valve. With the engine running, vacuum acts on the engine's crankcase. Air is drawn in through the engine's air cleaner, through a vent hose into a valve cover, and down into the crankcase.

After the fresh air mixes with the crankcase gases, the mixture is pulled by vacuum past the PCV valve, through the hose, and into the engine intake manifold. The crankcase gases are then drawn into the combustion chambers for burning.

An electronically controlled PCV system often uses a solenoid valve in the vacuum line leading to the valve. The computer system can energize or de-energize the solenoid to block or pass vacuum. This allows the computer to help control PCV operation.

PCV Valve

A *PCV valve* is used to control the flow of air through the PCV system. It may be located in a rubber grommet in a valve cover, in a breather opening in the intake manifold or plenum, or on the side of the engine block. The PCV valve varies the flow of air for idle, cruise, acceleration, wide open throttle, and engine-off conditions.

Figure 43-6 shows the action of a PCV valve under various operating conditions. At idle, the PCV valve is pulled toward the intake manifold by high vacuum. This restricts the flow of air and prevents a lean air-fuel mixture. When cruising, lower intake manifold vacuum allows the spring to open the PCV valve. However, enough vacuum is present to keep the PCV valve from

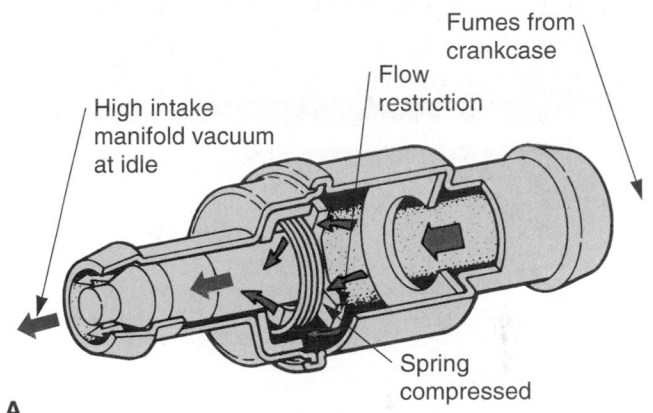

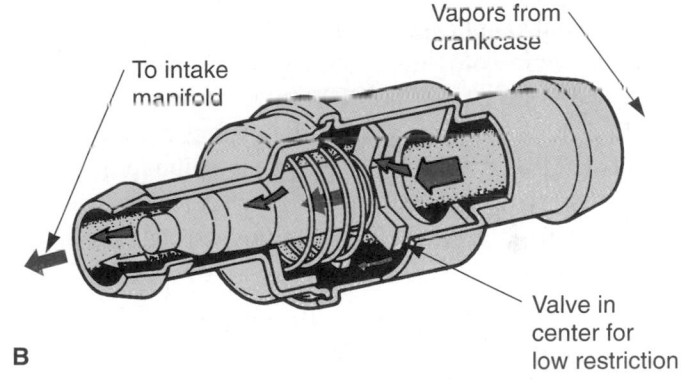

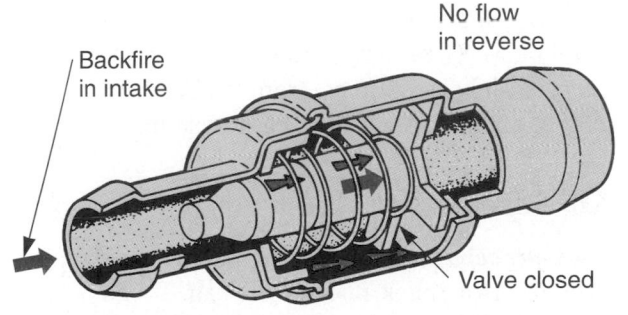

Figure 43-6. Study PCV operation under various operating conditions. A—At idle, high manifold vacuum pulls the plunger for minimum vapor flow. Only a small amount of vapor is drawn through the valve. B—During acceleration, intake manifold vacuum decreases. This allows the PCV valve to move to a center position for maximum flow. C—With engine off, a spring pushes the valve against its seat. This closes the valve. A backfire will also force the valve in this position. (AC Spark Plug)

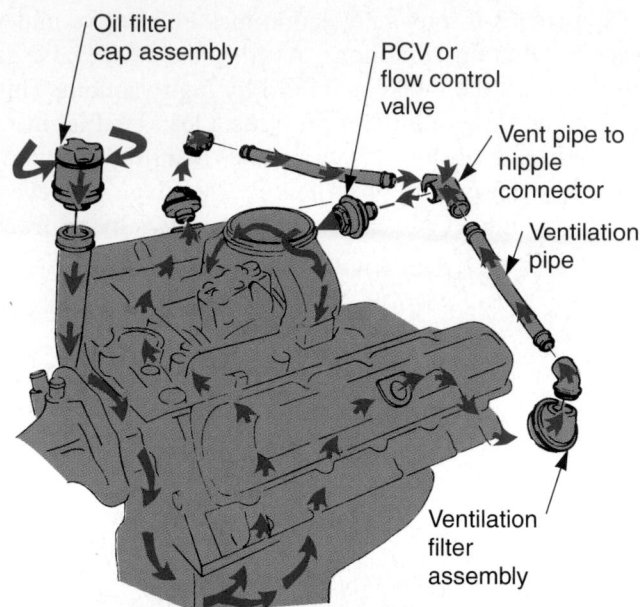

Figure 43-7. Trace PCV system vapor flow in this diesel engine. (Cadillac)

completely closing. More air can flow through the system to clean out crankcase fumes. At wide open throttle or with the engine off (low or no intake manifold vacuum), spring pressure closes the PCV valve completely.

In case of an engine backfire (air-fuel mixture in the intake manifold ignites), the PCV valve plunger is seated against the body of the valve. This keeps the backfire (burning) from entering and igniting the fumes in the engine crankcase.

Diesel Engine PCV System Variations

A PCV system for a diesel engine is illustrated in **Figure 43-7.** Note how outside air is drawn in through the oil filter cap. The breather cap contains a check valve that keeps crankcase fumes from leaking out of the engine. This maintains a closed PCV system.

Oil-Air Separators

An *oil-air separator* is a device that makes oil vapors condense and flow back into the oil pan. It can be used instead of a PCV system to reduce emissions and prevent oil sludging. The separator simply allows oil mists and vapors to settle out into a liquid so that they do not continue to circulate through the engine.

Evaporative Emissions Control Systems

The *evaporative emissions control (EVAP) system* prevents toxic fuel system vapors from entering the

atmosphere. As mentioned earlier, gasoline and many of its additives evaporate easily, especially if exposed to the atmosphere.

Pre-emission-control vehicles used vented gas tank caps. Carburetor bowls were also vented to the atmosphere, which caused a considerable amount of hydrocarbon emissions from unburned fuel. Modern vehicles commonly use an evaporative emissions control system to prevent this source of air pollution.

Evaporative Emissions System Components

The components of a typical evaporative emissions system are shown in **Figure 43-8.** The *non-vented fuel tank cap* prevents fuel vapors from entering the atmosphere through the tank's filler neck. It may contain pressure and vacuum valves that open in extreme cases of pressure or vacuum to prevent tank bulging or collapsing. When the fuel expands (from warming), tank pressure forces fuel vapors out of a vent line or lines at the top of the fuel tank, not out of the tank cap. The *fuel tank vent line* carries fuel vapors to a charcoal canister.

An *air dome* is a hump formed in the top of the fuel tank to allow for fuel expansion and tank filling without spillage. The dome normally provides about 10% air space to allow for fuel heating and the resulting volume increase.

A *liquid-vapor separator* is sometimes used to keep liquid fuel from entering the evaporative emission system. It is simply a small valve located above the main fuel tank. Liquid fuel condenses on the walls of the liquid-vapor separator because of the difference in temperature between the separator and the fuel. The liquid fuel then flows back into the fuel tank.

A *rollover valve* is used in the vent line from the fuel tank. It keeps liquid fuel from entering the vent line in case the vehicle rolls over during an auto accident. The valve contains a metal ball or a plunger valve that blocks the vent line when the valve is turned over.

The *charcoal canister* stores fuel vapors when the engine is *not* running. The metal or plastic canister is filled with activated charcoal granules. The charcoal is capable of absorbing fuel vapors. See **Figure 43-9.**

The top of the canister has fittings for the fuel tank vent line and the purge (cleaning) line. The bottom of the canister may have an inlet filter that cleans the outside air entering the canister.

On older vehicles, a carburetor vent line connects the carburetor fuel bowl with the charcoal canister. Bowl vapors flow through this line and into the canister.

A *purge line* is used for removing the stored vapors from the charcoal canister. It connects the canister to the

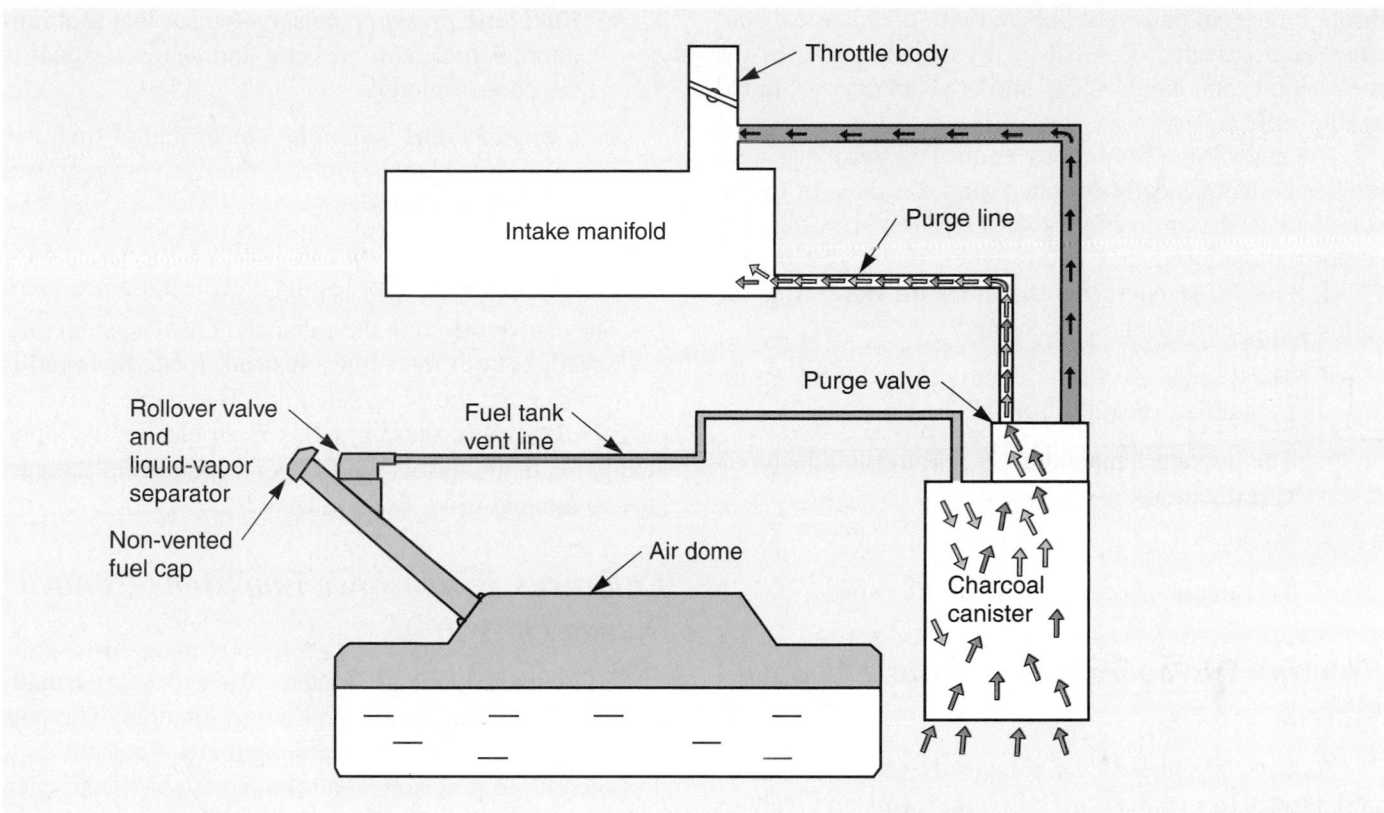

Figure 43-8. Note the parts of an evaporative emissions control system. This system draws excess fuel vapors into the engine for burning.

engine intake manifold. When the engine is running, engine vacuum draws the vapors out of the canister and through the purge line.

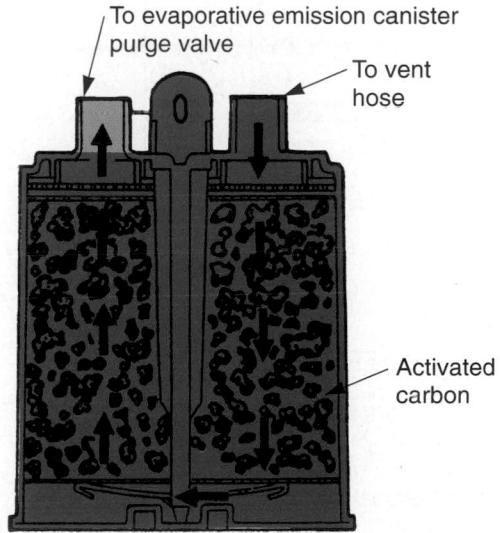

Figure 43-9. Cutaway view of a charcoal canister. The carbon granules store gasoline vapors when the engine is off. Then, when the engine starts, vacuum draws air through the bottom of the canister. This pulls vapors into the engine. (Ford)

A ***purge valve*** controls the flow of vapors from the canister to the intake manifold. This vacuum- or electrically operated valve is located on the top of the canister or in the purge line. Purge valves generally allow flow when the engine reaches operating temperature and is operating above idle speed. This helps minimize emissions when the engine is cold and prevents rough idle.

Evaporative Emissions Control System Operation

Figure 43-10 illustrates evaporative emissions system operation. Note that this system contains a vacuum-operated purge valve. When the engine is operating above idle speed, intake manifold vacuum causes the purge valve to open. This allows gases to flow through the purge line and causes fresh air to be drawn through the filter in the bottom of the canister. The incoming fresh air picks up the stored fuel vapors and carries them through the purge line. The vapors enter the intake manifold and are pulled into the combustion chambers for burning.

When the engine is shut off, gasoline slowly evaporates, producing unwanted vapors. These vapors flow through the fuel tank vent line and into the charcoal canister. The activated charcoal in the canister absorbs the fuel vapors and holds them until the engine is started again.

An evaporative emissions control system that contains an electronically-operated purge valve, or purge solenoid, is shown in **Figure 43-11**. The purge solenoid is normally closed and opens when energized by the ECM. The ECM energizes the solenoid only after the following conditions have been met:

- The vehicle has been operating in closed loop for a specified period of time.

- The coolant temperature is within manufacturers specifications.

- Vehicle speed is above approximately 15 mph.

- The engine is operating above idle speed.

Enhanced Evaporative Emissions Control System

As its name implies, an enhanced evaporative emission control system has several components and features not found on conventional EVAP systems. The enhanced system not only provides better control of fuel vapors, but it also monitors the condition of the fuel system. In addition to the components found in a conventional evaporative emission system, the enhanced EVAP system contains the following:

- *Fuel tank pressure sensor*—sensor that monitors internal fuel tank pressure and sends a signal to the control module.

- *Canister vent solenoid*—electrically operated vacuum valve that replaces the fresh air vent used on older canisters.

- *Service port*—fitting that allows the connection of service tools for testing and cleaning purposes.

The canister used in the enhanced EVAP system does not have a bottom inlet filter. Instead, fresh air is fed to the canister by the vent solenoid. The purge valve, or purge solenoid, in these systems is an electrically operated valve that controls the flow of vapors from the canister to the manifold. See **Figure 43-12.**

Enhanced Evaporative Emissions Control System Operation

The enhanced EVAP system often uses a normally closed, pulse-width modulated purge solenoid. The control module can send different length electrical pulses to the solenoid to precisely control vapor flow. When energized, the purge solenoid opens to allow vapors to be pulled into the engine.

The canister vent solenoid is normally open, allowing fresh air into the canister during purge modes. The vent can be closed during vacuum diagnostic modes to make sure the system is not leaking.

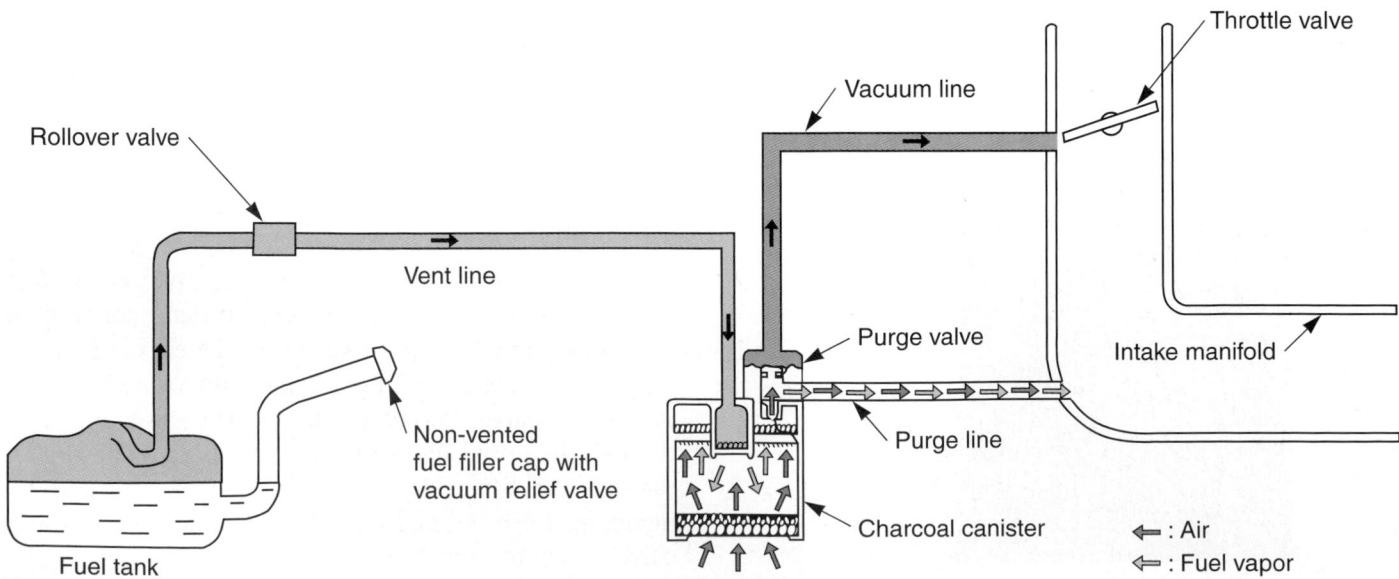

Figure 43-10. Study the operation of this evaporative emissions control system and its related components. Fuel tank vapors flow to the canister with the engine off. When the engine starts, vacuum pulls vapors out of the charcoal canister into the cylinders. The rollover valve will close during an auto accident where the vehicle rolls. This prevents fuel leakage and a possible fire.

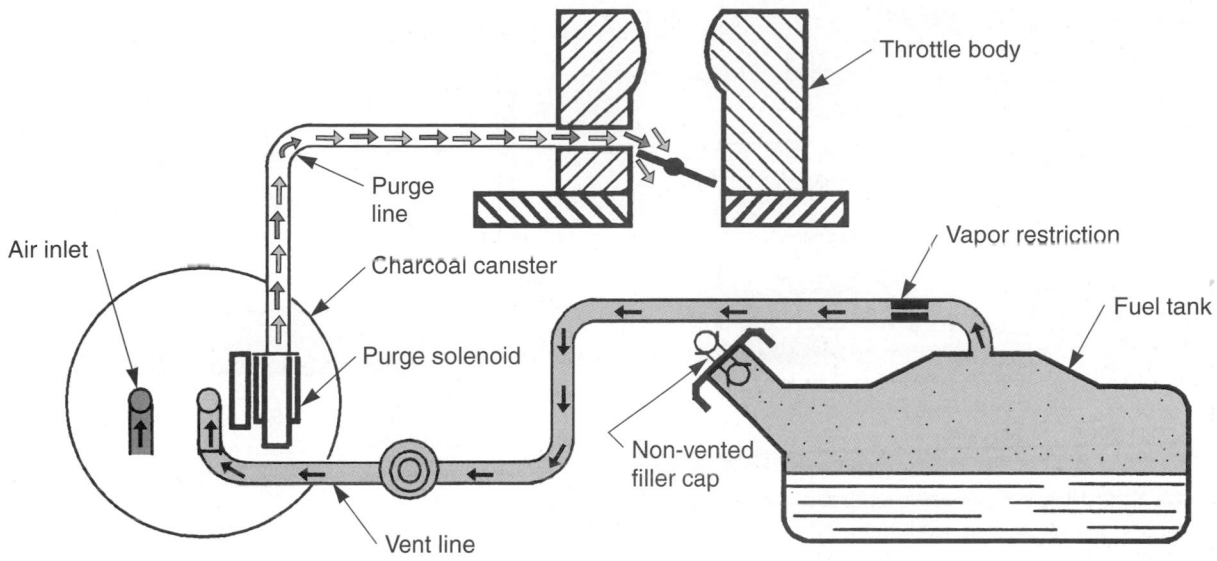

Figure 43-11. This evaporative emissions control system has an electronically controlled purge solenoid. The ECM opens the solenoid to allow vapors to flow from the canister to the throttle body. Compare this system to the one shown in Figure 43-10. (General Motors)

Air Cleaner Carbon Element

An *air cleaner carbon element* may be used to absorb fuel vapors when the engine is shut off. After a hot engine is turned off, hot soak fumes (engine heat causes excess formation of fuel vapors) can collect in the throttle body or carburetor air horn. The carbon element attracts and stores these fumes. When the engine is started, airflow through the element pulls these fumes into the engine for burning.

Exhaust Gas Recirculation (EGR)

The *exhaust gas recirculation system*, or *EGR system*, allows burned exhaust gases to enter the engine intake manifold to help reduce NO_x emissions. When exhaust gases are added to the air-fuel mixture, they decrease peak combustion temperatures (maximum temperature produced when the air-fuel mixture burns). For this reason, an exhaust gas recirculation system lowers the amount of NO_x in the engine exhaust. EGR systems

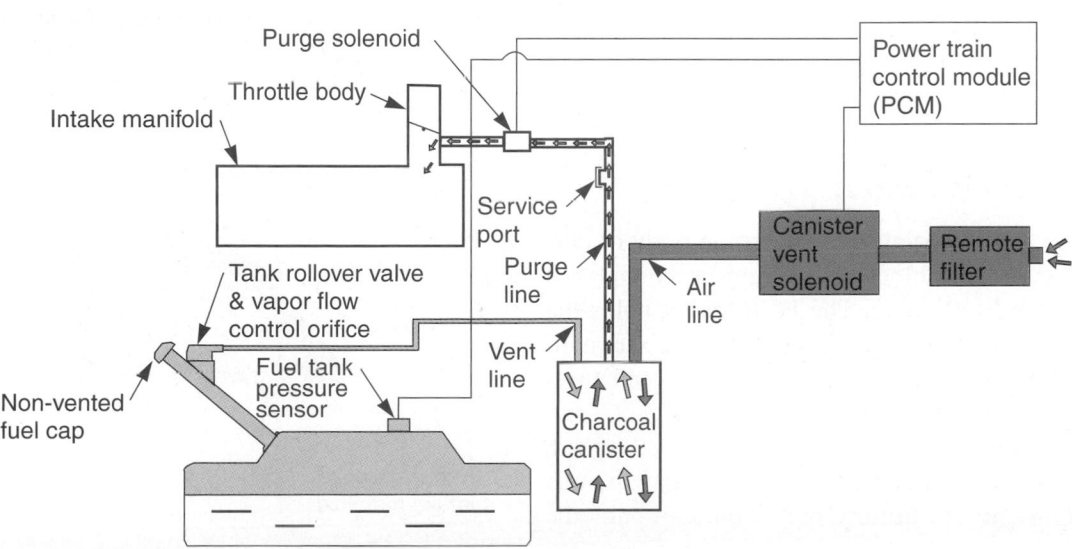

Figure 43-12. Typical enhanced evaporative emission system. Note the location of the purge solenoid, the canister vent solenoid, and the service port. This type of system is found on many late-model vehicles.

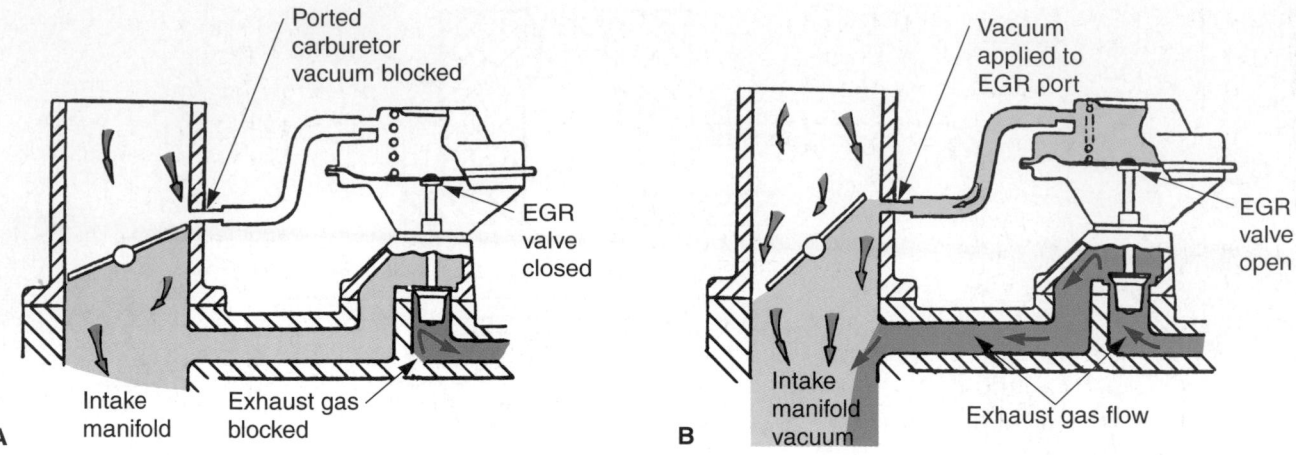

Figure 43-13. Study basic EGR valve operation. This is an older vacuum-operated EGR but it still shows valid principles. A—With the throttle closed at idle speed, vacuum to the EGR valve is blocked and the valve remains closed to prevent rough idling. B—When the throttle opens for more engine speed, the vacuum port to the EGR valve is exposed to vacuum. The EGR valve diaphragm is pulled up and exhaust gases enter the engine intake manifold without adversely affecting engine operation.

can be controlled by engine vacuum or by the engine control module.

Vacuum-Controlled EGR

A *vacuum-controlled EGR system* uses engine vacuum to operate the EGR valve. This system can be found on millions of older vehicles still on the road.

A basic vacuum EGR system is simple. It consists of a vacuum-operated EGR valve and a vacuum line from the throttle body or carburetor. The EGR valve usually bolts to the engine intake manifold or a carburetor plate. Exhaust gases are routed through the cylinder head and intake manifold to the EGR valve, **Figure 43-13.**

The EGR valve consists of a vacuum diaphragm, spring, plunger, exhaust gas valve, and a diaphragm housing. It is designed to control exhaust flow into the intake manifold. See **Figure 43-14.**

Vacuum EGR Operation

At idle, the throttle plate in the throttle body or carburetor is closed. This blocks off engine vacuum so it cannot act on the EGR valve. The EGR spring holds the valve shut and exhaust gases do *not* enter the intake manifold. If the EGR valve were to open at idle, it could upset the air-fuel mixture and the engine could stall.

When the throttle plate opens to increase speed, engine vacuum is applied to the EGR hose. Vacuum pulls the EGR diaphragm up. In turn, the diaphragm pulls the valve open.

Engine exhaust can then enter the intake manifold and combustion chambers. At higher engine speeds, there

is enough air flowing into the engine that the air-fuel mixture is not upset by the open EGR valve.

Electronic-Vacuum EGR Valves

An electronic-vacuum EGR valve uses both engine vacuum and electronic control for better exhaust gas metering. An EGR position sensor is located in the top of the valve and sends data back to the ECM. This allows

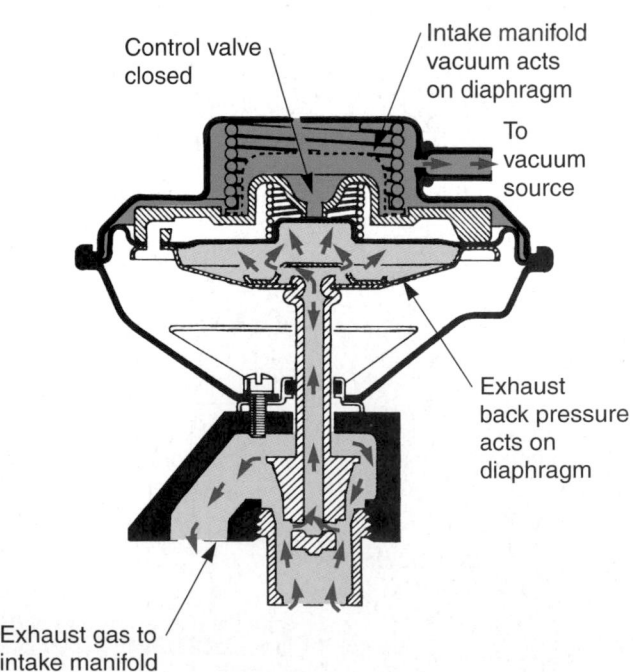

Figure 43-14. A back pressure EGR valve uses engine vacuum and pressure in the exhaust system to control valve operation. (Buick)

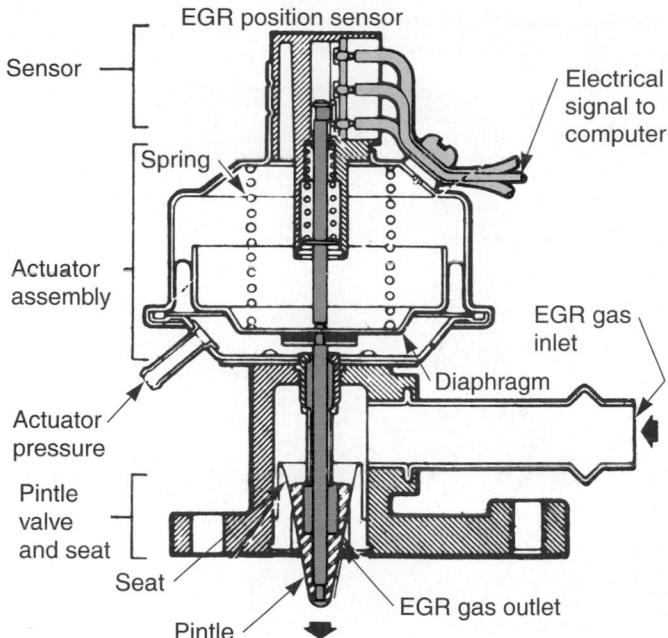

Figure 43-15. This EGR valve has a sensor connected to the vehicle's on-board computer or ECM. EGR pintle movement changes current flow through the sensor, allowing the ECM to alter engine operation. (Ford)

the computer to determine how much the EGR valve is opened. Refer to **Figure 43-15.**

With some systems, *EGR vacuum valves* are used to more closely control EGR opening. The vacuum valves use electric solenoids to block or pass airflow to the EGR valve. They are located in one or more of the vacuum lines going to the EGR valve. The ECM can then energize the solenoids to alter when and how fast the EGR valve opens or closes to improve efficiency. See **Figure 43-16.**

EGR System Variations and Components

There are several EGR system variations that you might encounter, including:

- A back pressure EGR valve that uses both engine vacuum and exhaust back pressure to control valve action. This provides accurate control of EGR valve operation.

- An engine coolant temperature switch may be used to prevent exhaust gas recirculation when the engine is cold. A cold engine does not have extremely high combustion temperatures, so production of NO_x is minimal. By blocking vacuum to the EGR valve when the engine operating temperature is below 100°F (38°C), the driveability and performance of the cold engine is improved.

- The vacuum line to the EGR valve is sometimes connected into a wide open throttle valve (WOT valve). It opens under full acceleration to provide venturi vacuum to the EGR valve. At wide open throttle, intake manifold vacuum is very low, but venturi vacuum is high.

- Small *EGR jets* have been used in the bottom of a few intake manifolds to replace the EGR valve. The jets meter a small amount of exhaust gases into the airflow before it enters the engine cylinder head ports. The jets are small enough that they do not upset the idle air-fuel mixture.

Electronic EGR System

An *electronic EGR system* uses vehicle sensors, the ECM, and a solenoid-operated exhaust gas recirculation valve to reduce NO_x emissions. This is the most common type of EGR system used on late-model engines.

The ECM uses input data from the EGR position sensor, engine coolant temperature sensor, mass airflow sensor, throttle position sensor, crankshaft position sensor, and other sensors. The sensor signals allow the ECM to determine how much duty cycle should be sent to open and close each valve for maximum efficiency and minimum exhaust emissions, **Figure 43-17.**

The *EGR duty cycle* is a measurement of control current on and off time sent from the ECM. The ECM can precisely control duty cycle to meter just the right amount of exhaust gases needed to reduce NO_x emissions, **Figure 43-18.**

Electronic EGR Valves

An *electronic EGR valve*, sometimes termed a *digital EGR valve,* uses one or more electric solenoids to open and close its exhaust passages. It works *without* engine vacuum.

A *single-stage EGR valve* only uses one solenoid and valve. It is a simple, dependable EGR design. One is shown in **Figure 43-19A.** To open one of the exhaust passages in the EGR valve, the ECM energizes its solenoid. When control current is sent to the solenoid windings, it pulls up on the metal armature connected to the valve. This lifts up the valve to open an exhaust recirculation passage. Exhaust gases flow through orifices (metered openings) to limit engine combustion temperatures and prevent NO_x pollution. When the ECM stops current flow to the EGR solenoid, spring tension closes the valve to prevent exhaust gas flow into the engine.

A *multi-stage EGR valve* uses more than one (usually three) solenoid valves to more closely match exhaust gas

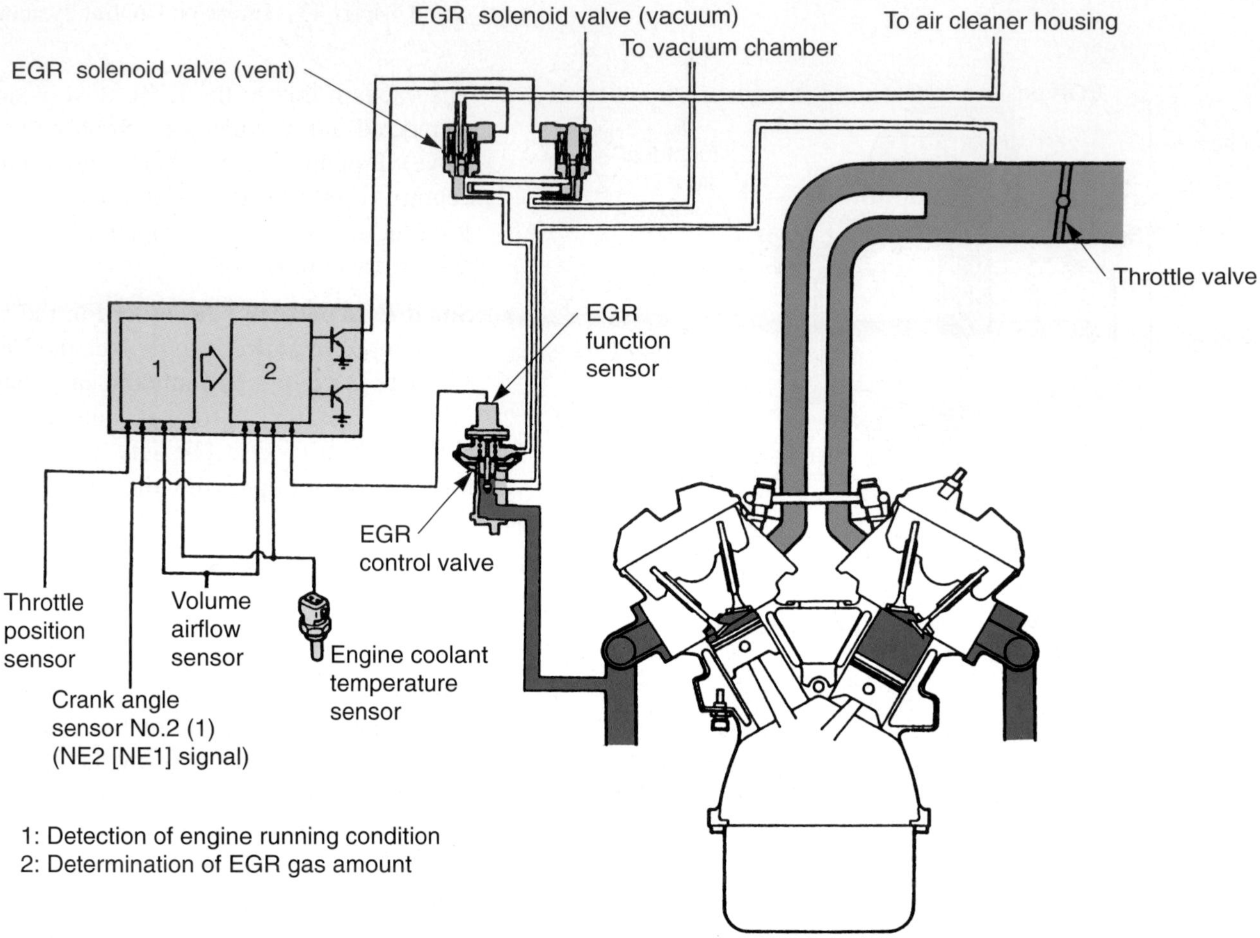

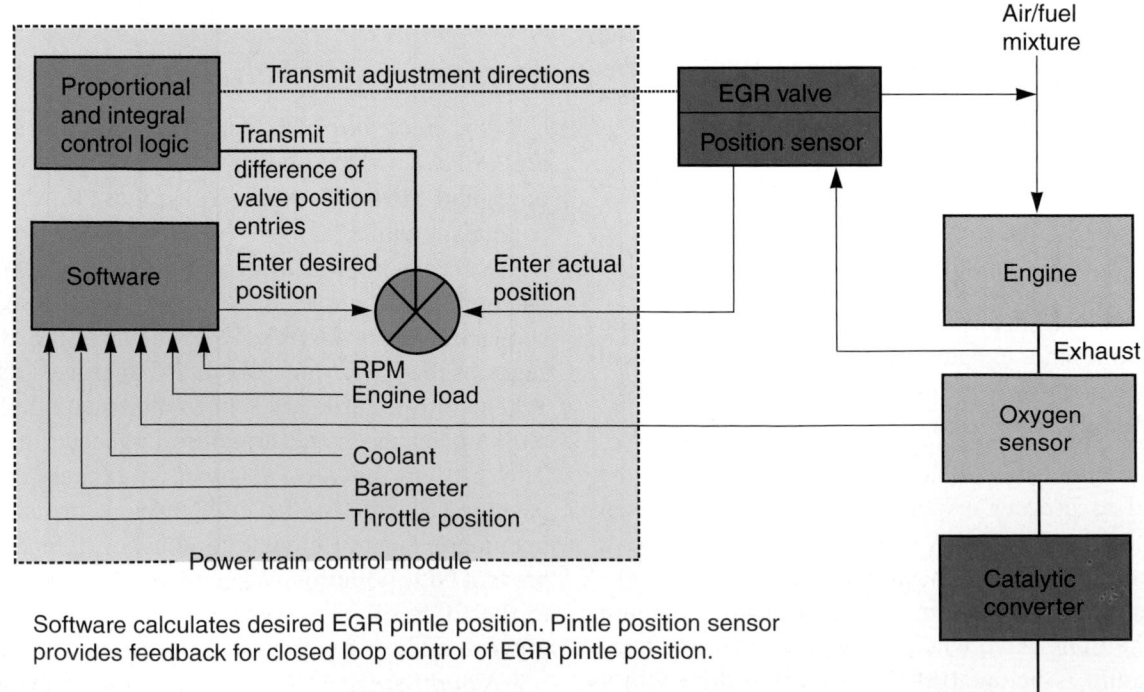

EGR solenoid valve (vacuum)

EGR solenoid valve (vent)

To vacuum chamber

To air cleaner housing

Throttle valve

EGR function sensor

EGR control valve

Throttle position sensor

Volume airflow sensor

Crank angle sensor No.2 (1) (NE2 [NE1] signal)

Engine coolant temperature sensor

1: Detection of engine running condition
2: Determination of EGR gas amount

Figure 43-16. This diagram shows how the control module can be used to monitor and control a vacuum-operated EGR valve. The electric solenoids can block or allow flow in the vacuum line going to the EGR valve, providing computer control for this system. The engine coolant temperature sensor allows the control to keep the EGR valve closed when the engine is cold and NO$_x$ emissions are not a problem. (Mazda)

Proportional and integral control logic

Transmit adjustment directions

EGR valve

Position sensor

Air/fuel mixture

Transmit difference of valve position entries

Software

Enter desired position

Enter actual position

Engine

RPM

Engine load

Coolant

Barometer

Throttle position

Exhaust

Oxygen sensor

Catalytic converter

Power train control module

Software calculates desired EGR pintle position. Pintle position sensor provides feedback for closed loop control of EGR pintle position.

Figure 43-17. Block diagram represents the relationship between a power train control module and the EGR system. Programmed data in the PCM allow it to change the EGR opening cycle with various conditions for maximum efficiency. (Mazda)

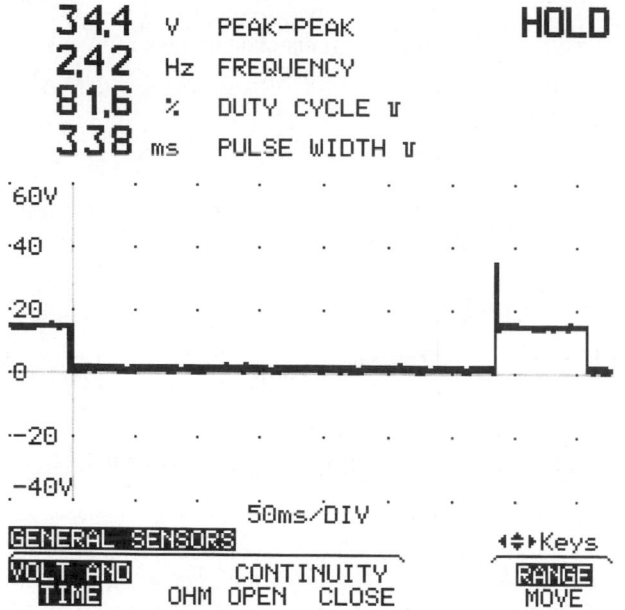

```
34.4  V   PEAK-PEAK          HOLD
2.42  Hz  FREQUENCY
81.6  %   DUTY CYCLE ↧
338   ms  PULSE WIDTH ↧

 60V
 ·40
 ·20
 ·0
 -20
 -40V
          50ms/DIV
GENERAL SENSORS              ◄♦►Keys
VOLT AND    CONTINUITY       RANGE
TIME     OHM OPEN  CLOSE     MOVE
```

Figure 43-18. The pulse width time of an electronic EGR valve can be monitored with an oscilloscope or a meter with scope feature. The duty cycle, pulse width, and waveform of the valve's operation are shown. (Fluke)

flow to engine needs. The solenoids mount on a base plate. When the valves are closed, they contact and seal against the base plate seats. See **Figure 43-19B.**

If only a small amount of exhaust recirculation gas is needed (combustion temperatures only slightly too hot), the ECM will only energize one of the EGR solenoids.

If combustion temperatures become hotter (engine conditions like speed, load, outside air temperature increase), the ECM will energize the other EGR solenoids as needed to increase exhaust gas recirculation flow. The added exhaust gases will decrease combustion temperatures to reduce NO_x.

Air Injection System

An *air injection system* forces fresh air into the exhaust ports or catalytic converter to reduce HC and CO emissions. The exhaust gases leaving an engine can contain unburned and partially burned fuel. Oxygen from the air injection system causes this fuel to continue to burn in the exhaust manifold or the catalytic converter. See **Figure 43-20.**

The *air injection pump* is belt-driven and forces air at low pressure into the system. The spinning vanes, or blades, pull air in one side of the pump. The air is trapped and compressed as the vanes rotate. As rotation continues, air is forced out of a second opening in the pump. See **Figure 43-21.**

Electric air injection pumps are driven by a small dc motor, instead of being engine driven. This reduces emissions at start-up and with high engine temperatures

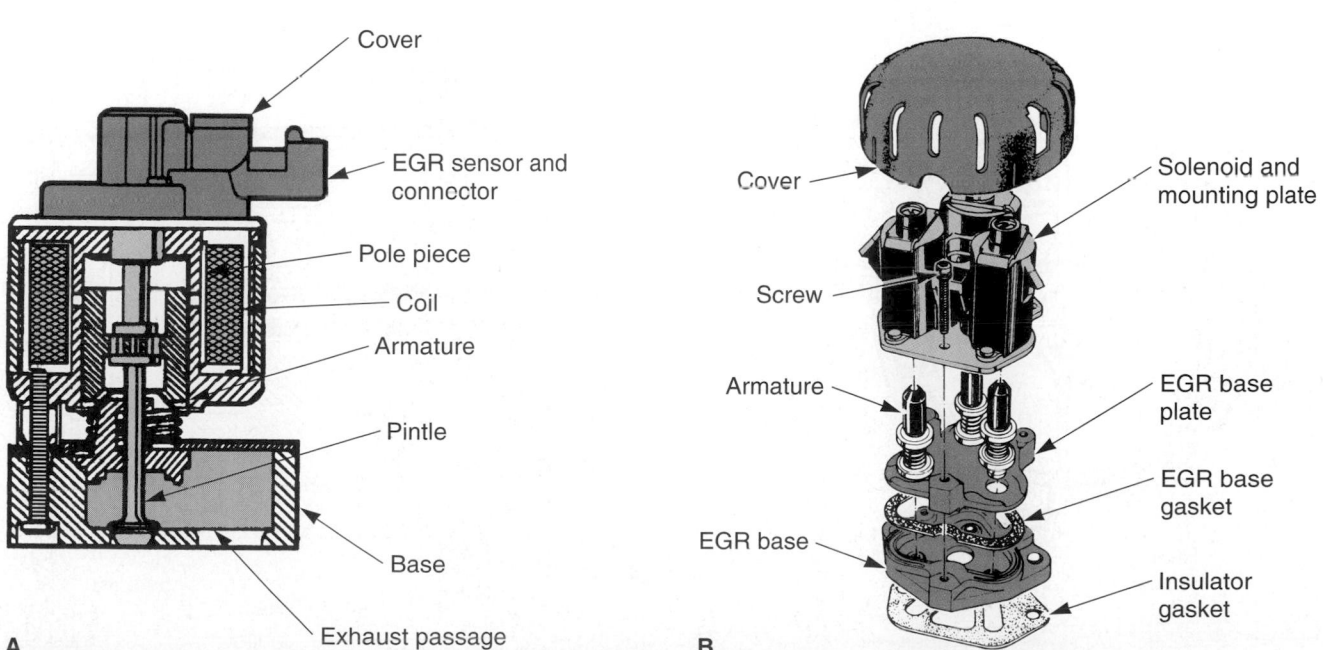

Figure 43-19. A—This cutaway view shows the components of an electronic or digital EGR valve. When the ECM sends current to the EGR solenoid, the windings produce a magnetic field that pulls the pintle up to open the EGR valve. B—This is a multistage EGR valve that uses three separate solenoids and valves. (Oldsmobile)

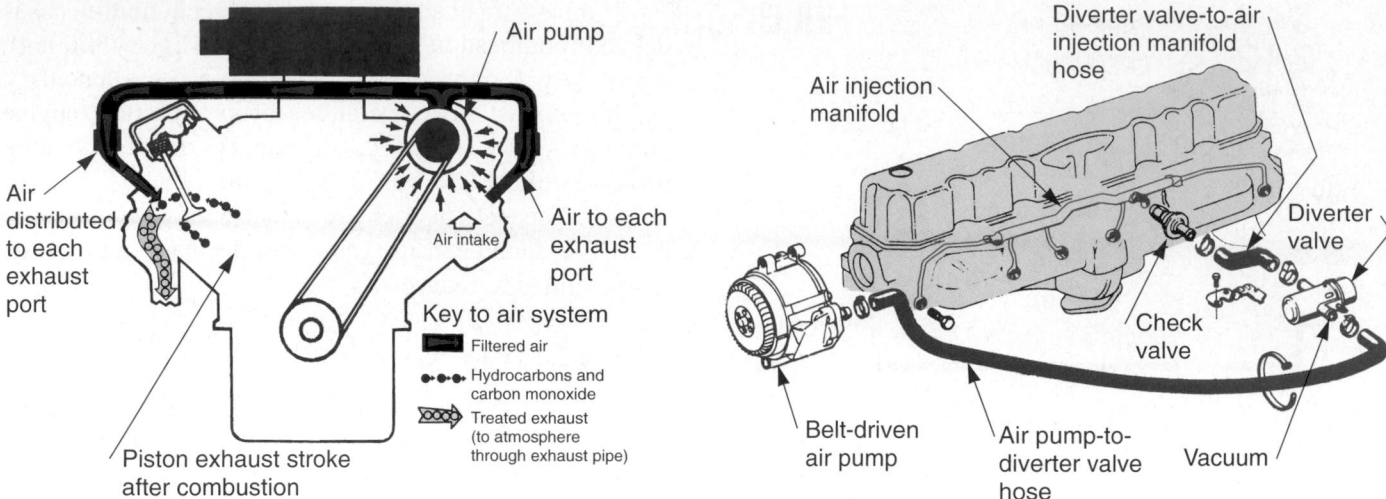

Figure 43-20. An air injection system is sometimes used to help burn excess fuel vapors that enter the exhaust system by forcing low pressure air into the exhaust stream. The air pump forces outside air into the exhaust manifolds. The air helps hot exhaust gases burn as they blow out of the open exhaust valves. (GMC)

Figure 43-22. The basic parts of air injection system. The air pump normally bolts to the front of the engine. A hose carries air to the diverter valve. Air then enters the exhaust manifold through a check valve and distribution manifold. The diverter valve prevents backfire during engine deceleration. (Chrysler)

because a more constant flow of air is produced by the electric motor. Air pump speed does not change with engine speed.

Air Injection System Components

Figure 43-22 shows the major parts of an air injection system. The *diverter valve* keeps air from entering the exhaust system during deceleration. This prevents backfiring (explosive burning of the air-fuel mixture outside the combustion chamber) in the exhaust system. The diverter valve also limits maximum system air pressure. It releases excessive pressure through a silencer or muffler. A plastic or rubber hose connects the pump output to a diverter valve.

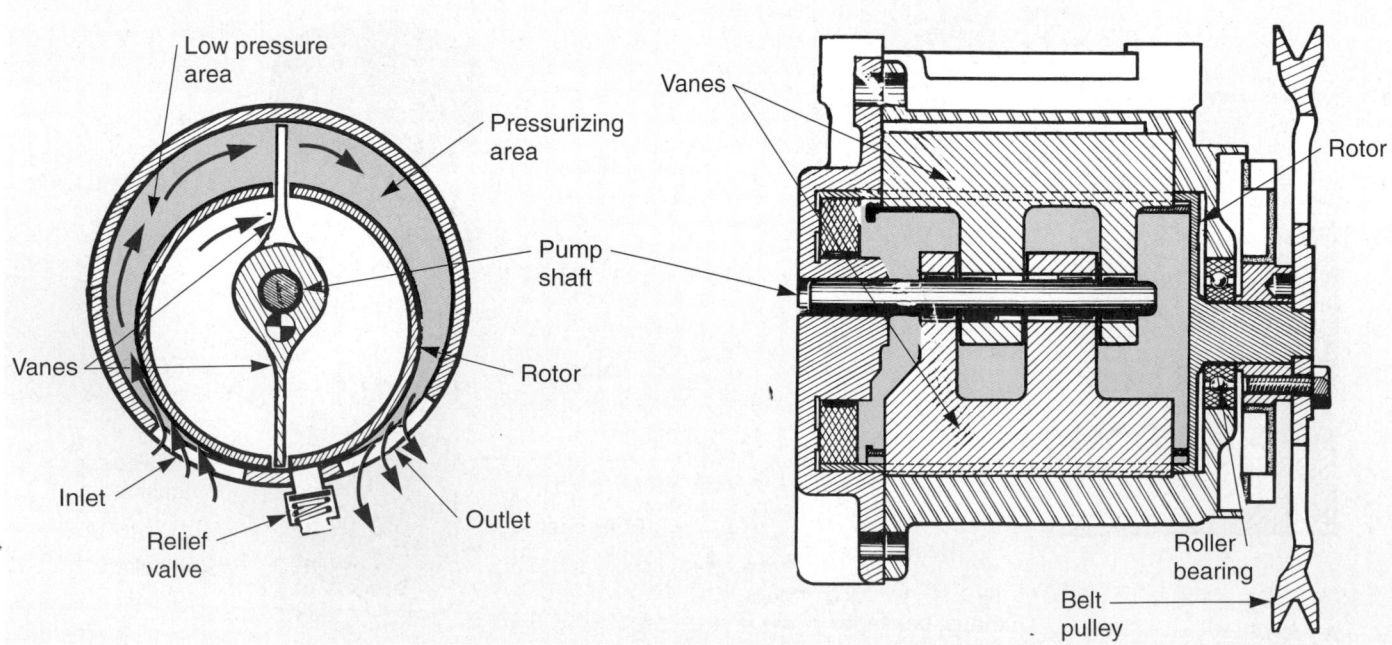

Figure 43-21. Cutaway shows the action inside an air pump used in the air injection system. A belt turns the pump pulley, shaft, and rotor. The vanes trap and pressurize air.

An *air distribution manifold* is used in air injection systems to direct a stream of air toward each engine exhaust valve. Fittings on the air distribution manifold screw into threaded holes in the exhaust manifold or cylinder head. **Figure 43-22** shows a typical air distribution manifold.

An *air check valve* is usually located in the line between the diverter valve and the air distribution manifold. It keeps exhaust gases from entering the air injection system.

Air Injection System Operation

When the engine is running, the spinning vanes in the air pump force air into the diverter valve. If not decelerating, the air is forced through the diverter valve, check valve, air injection manifold, and into the engine exhaust ports. The fresh air blows on the engine exhaust valves to keep any fuel burning as it leaves the engine. Look at **Figure 43-23.**

During periods of deceleration, the diverter valve blocks airflow into the engine exhaust manifold. This prevents a possible backfire, which could damage the

vehicle's exhaust system. When needed, the diverter valve's relief valve releases excess pressure.

Late-model vehicles can also use the air injection system to force air into the catalytic converter. This is done to help the converter burn, or oxidize, the partially burned fuel more completely. A metal line runs from the air pump to the catalytic converter. Air from this line is forced into the converter.

Thermostatic Air Cleaner System

The *thermostatic air cleaner system* speeds engine warm-up and keeps the air entering the engine warm. By maintaining a more constant inlet air temperature, a carburetor can be calibrated leaner at startup to reduce emissions.

Thermostatic air cleaners are not needed with modern fuel injection systems. An electronic fuel injection system can alter its operation with cold air entering the engine more efficiently than a carburetor system.

A *thermal vacuum valve* is normally located in the air cleaner to control the vacuum motor and heat control door. A vacuum supply is connected to the thermal

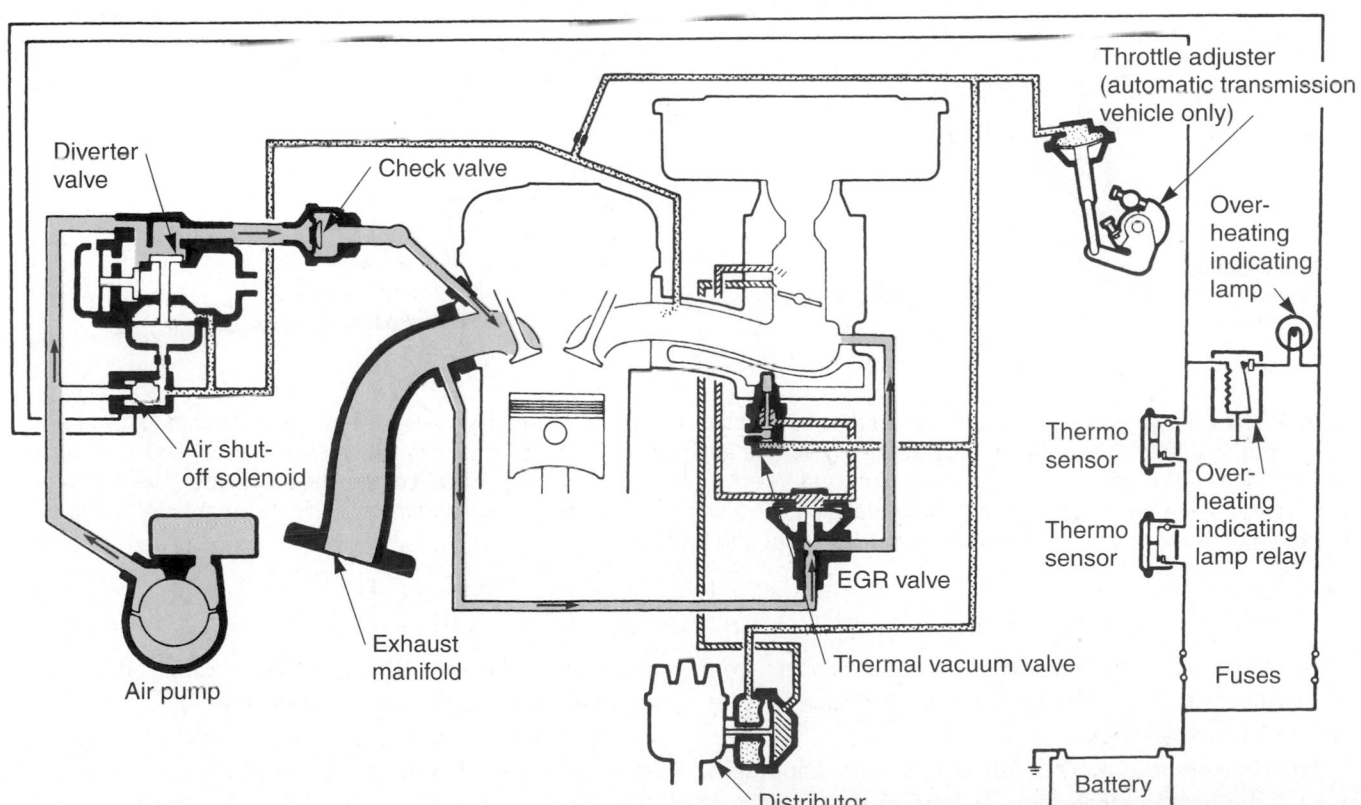

Figure 43-23. Note the relationship of components in the air injection and EGR systems. Also note the thermal vacuum valve that alters the ignition vacuum advance and EGR valve. (Chrysler)

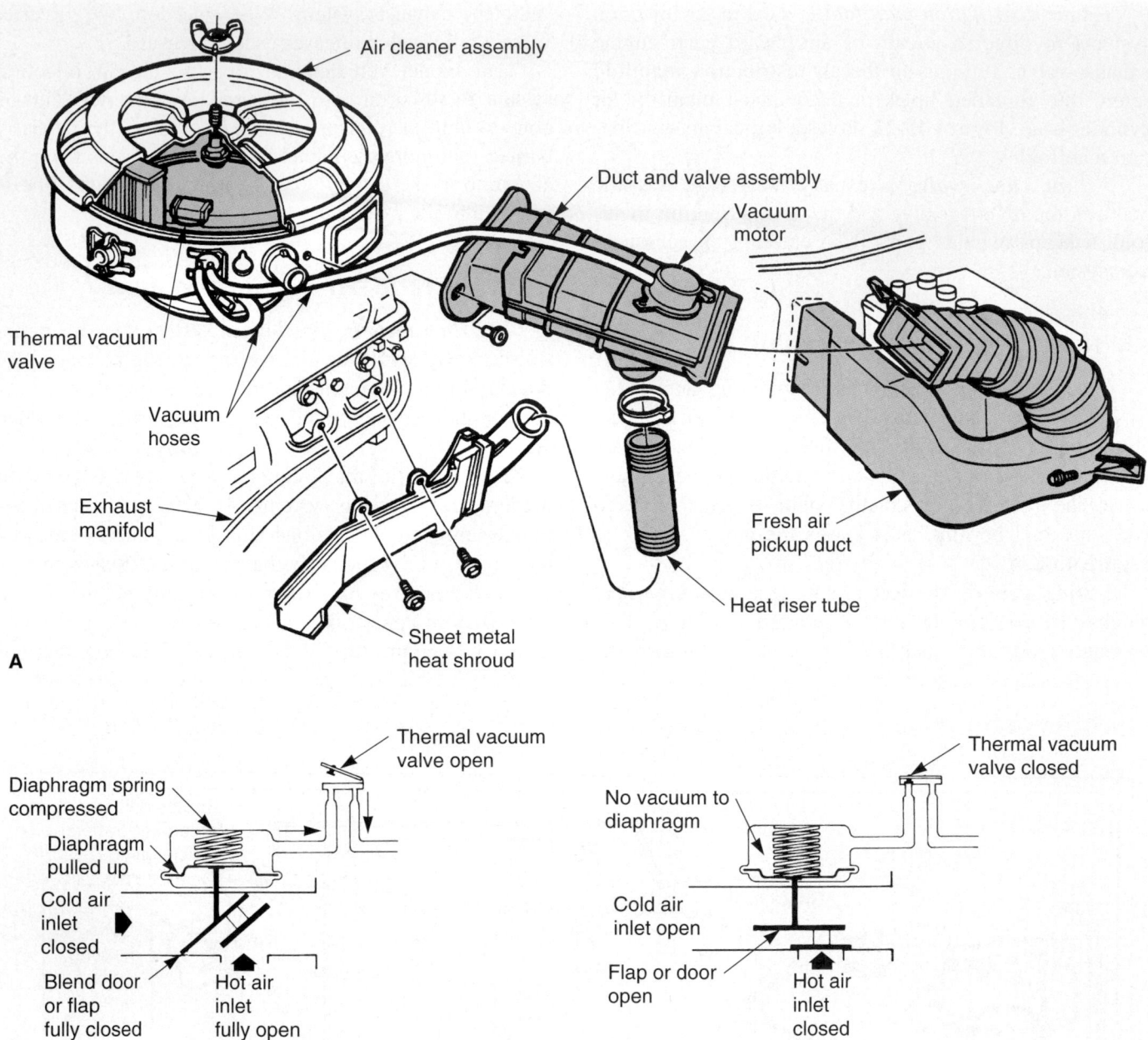

Figure 43-24. Study the basic parts and action of a thermostatic air cleaner assembly. A—Note how the heat shroud fits around the exhaust manifold for warming the air charge. B—When the engine is cold, the thermal vacuum valve is open. Vacuum deflects the diaphragm in the vacuum motor to pull the flap closed. Warm air from the exhaust manifold enters the air cleaner. C—As the engine warms to operating temperature, the thermal vacuum valve closes, which removes vacuum from the diaphragm. A spring pushes the diaphragm down to block the entry of warm air. (Ford and Buick)

vacuum valve from the engine. Another hose runs from the thermal valve to the vacuum motor (diaphragm). Refer to **Figure 43-24A.**

The *vacuum motor,* also called a vacuum diaphragm, operates the heat control door, or flap, in the air cleaner inlet. The vacuum motor consists of a flexible diaphragm, spring, rod, and diaphragm chamber. When vacuum is

applied to the unit, the diaphragm and rod are pulled upward, moving the heat control door.

The *heat control door* can be opened or closed to route either cool or heated air into the air cleaner. When the door is closed, hot air from the exhaust manifold shroud enters the engine. When the door is open, cooler outside air enters the engine. See **Figure 43-24B** and **C.**

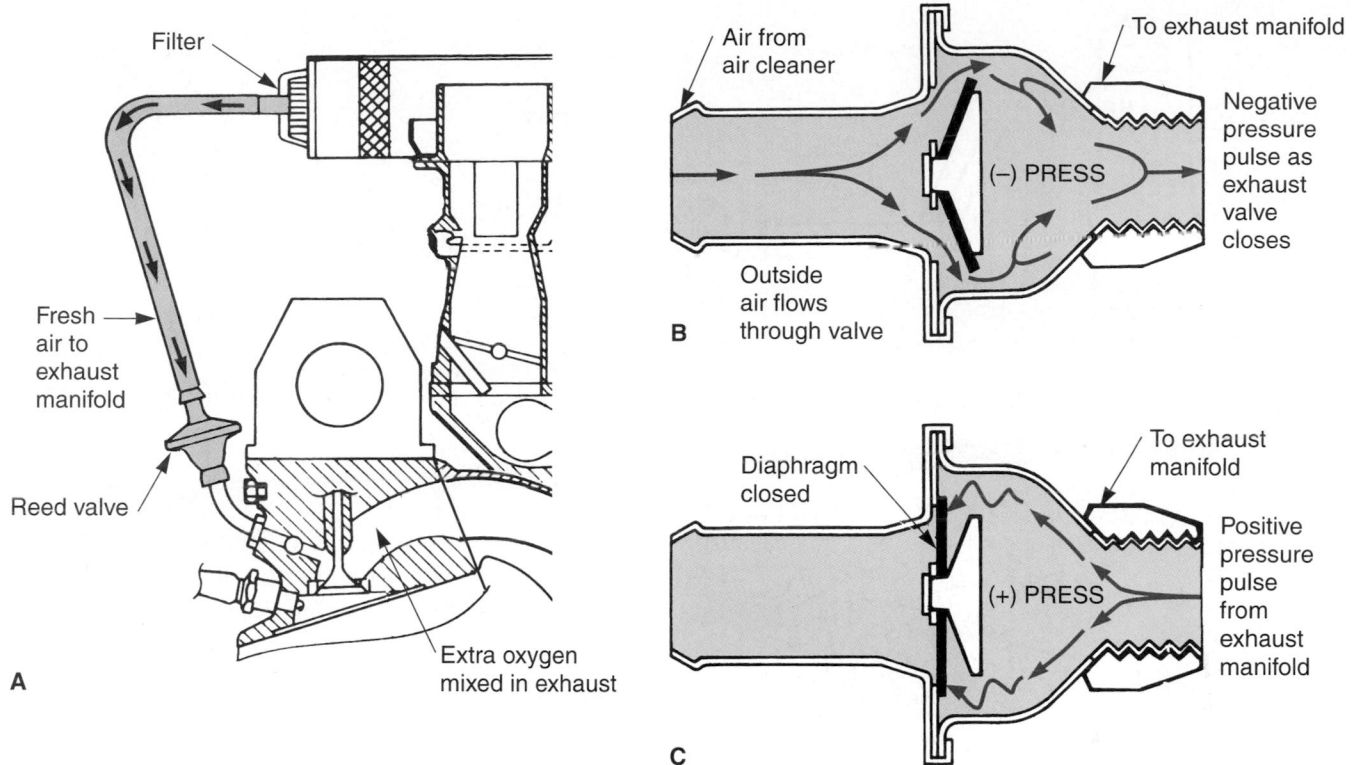

Figure 43-25. A—Pulse air injection systems do not use an air pump. Instead, aspirator valves act as check valves. B— When the exhaust valves close, they produce a vacuum pulse that allows air to enter the exhaust manifold. C—Aspirators block airflow when the exhaust valves open and produce a pressure pulse.

Thermal Vacuum Valves

A thermal vacuum valve is a temperature-sensitive valve that is installed in a vacuum line. It can block or pass airflow with changes in temperature. These devices can be found in various vacuum circuits.

Pulse Air System

A *pulse air system* performs the same function as an air injection system. However, instead of an air pump, it uses natural pressure pulses in the exhaust system to operate aspirator valves. **Figure 43-25** shows one type of pulse air system. Note how the pulse air system lines and aspirator valves are positioned on the engine.

The *aspirator valves,* also called *check valves*, *gulp valves*, or *reed valves*, block airflow in one direction and allow airflow in the other direction. This is illustrated in **Figure 43-25.** Pulse air systems can be found on older vehicles.

Pulse Air System Operation

Pressure in the exhaust manifold fluctuates as the engine valves open and close. The aspirator valves allow fresh air to enter the exhaust manifold on the low-pressure pulses (when exhaust valve first closes). However, they block backflow on the high-pressure pulses (as exhaust valve opens). This causes fresh air to flow through the system and into the engine exhaust manifold.

Catalytic Converter

A *catalytic converter* oxidizes (burns) the remaining HC and CO emissions that pass into the exhaust system. Extreme heat (temperatures of approximately 1400°F, or 760°C), ignite these emissions and change them into harmless carbon dioxide (CO_2) and water (H_2O). Look at **Figure 43-26.**

A *catalyst* is any substance that speeds a chemical reaction without itself being changed. The converter's catalyst agent is coated on either a ceramic honeycomb-shaped block or on small ceramic beads. The catalyst is encased in a stainless steel housing that is designed to resist heat. See **Figure 43-27.**

A catalytic converter contains a catalyst substance, usually the elements platinum, palladium, rhodium, or a mixture of these materials. Platinum and palladium treat the HC and CO emissions. Rhodium acts on the NO_x emissions. Some newer converters also use cerium to

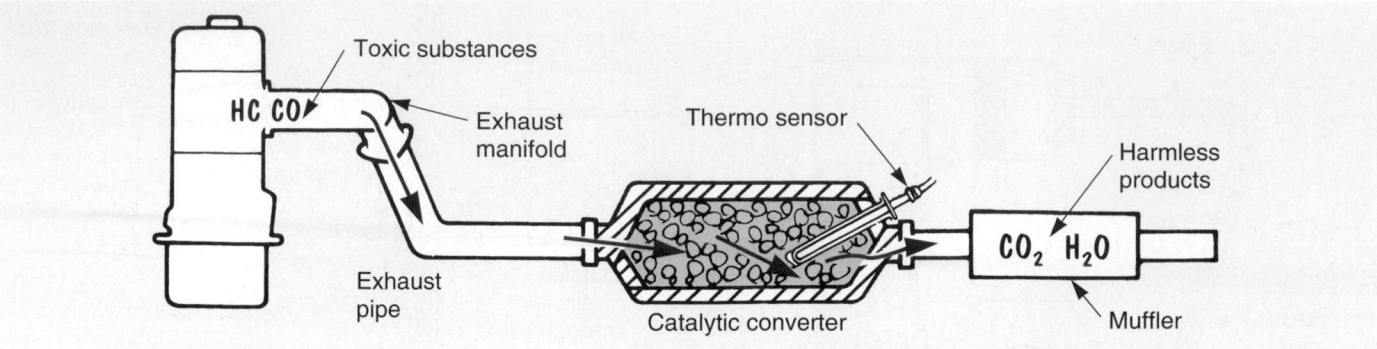

Figure 43-26. A catalytic converter burns and treats exhaust emissions and changes them into carbon dioxide and water. (Toyota)

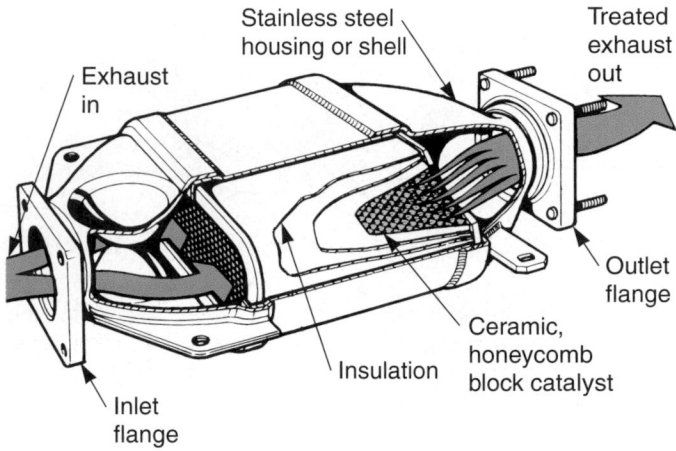

Figure 43-27. Monolithic catalytic converters use a honeycomb shaped block of ceramic material covered with catalytic elements to treat exhaust gases. The catalyst is enclosed in a stainless steel housing.

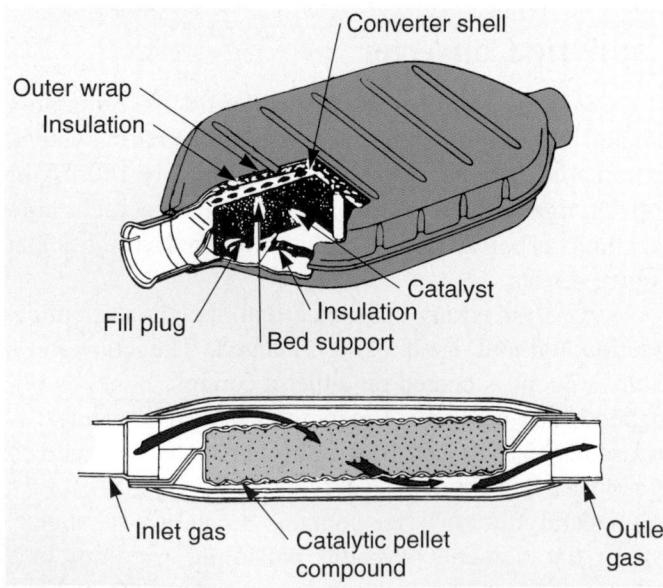

Figure 43-28. Older pellet-type catalytic converters use beads coated with catalyst elements. Study the construction and compare it to the monolithic converter in **Figure 43-27.** (GMC)

attract and release additional oxygen into the exhaust stream. The catalyst operating temperature is attained when the catalysts agents are hot enough (above 300°F, or 149°C) to start treating emissions.

Types of Catalytic Converters

A catalytic converter using a ceramic honeycomb catalyst is often termed a ***monolithic converter***. Look at **Figure 43-27.** When small ceramic beads are used, it is called a ***pellet catalytic converter***. Refer to **Figure 43-28.**

A *mini catalytic converter* is a very small converter placed close to the engine exhaust manifold. It heats up quickly to reduce emissions during engine warm-up. A mini catalytic converter is used in conjunction with a larger, main converter. See **Figure 43-29.**

A *two-way catalytic converter*, sometimes called an *oxidation converter*, can only reduce two types of exhaust emissions (HC and CO). A two-way converter is normally coated with platinum.

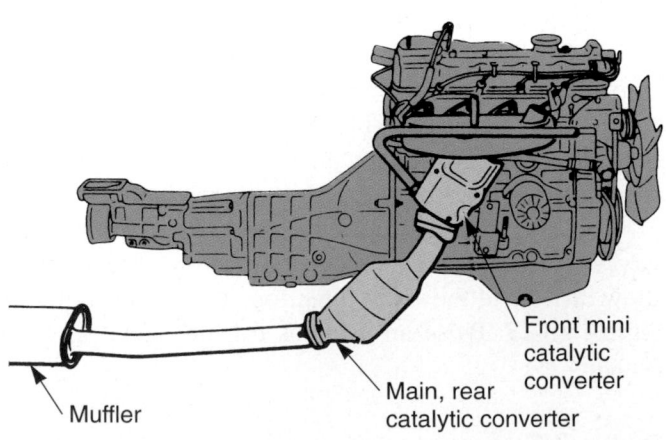

Figure 43-29. This engine uses a mini catalytic converter and a main converter. The mini converter functions right after engine startup. It heats up quickly to reduce emissions when the engine and main converter are cold. (Chrysler)

A *three-way catalytic converter*, also termed a *reduction type converter*, is capable of reducing all three types of exhaust emissions (HC, CO, and NO_x). A three-way converter is usually coated with rhodium and platinum.

A *dual-bed catalytic converter* contains two separate catalyst units enclosed in a single housing. A dual-bed converter normally has both a three-way (reduction) catalyst and a two-way (oxidation) catalyst. A mixing chamber is provided between the two. Air is forced into the mixing chamber to help burn the HC and CO emissions. Look at **Figure 43-30.**

Dual-Bed Catalytic Converter Operation

When the engine is cold (below approximately 128°F, or 52°C), the air injection system routes air into the exhaust manifold. Exhaust heat and the injected air are used to burn exhaust emissions. When the engine warms, the system forces air into the catalytic converter, **Figure 43-31.**

First, the exhaust gases pass through the front three-way catalyst that removes HC, CO, and NO_x. Then, the exhaust gas flows into the area between the two catalysts. The oxygen in the air flowing into the chamber causes the gases to continue to burn. The exhaust flows into the rear two-way catalyst which removes even more HC and CO.

Tech Tip
Some catalysts are coated with a material that absorbs and temporarily stores NO_x emissions. When a saturation level is reached, the on-board computer temporarily enriches the fuel mixture. This causes the converter's internal honeycomb block to glow red hot, breaking up the stored NO_x emissions into harmless by-products.

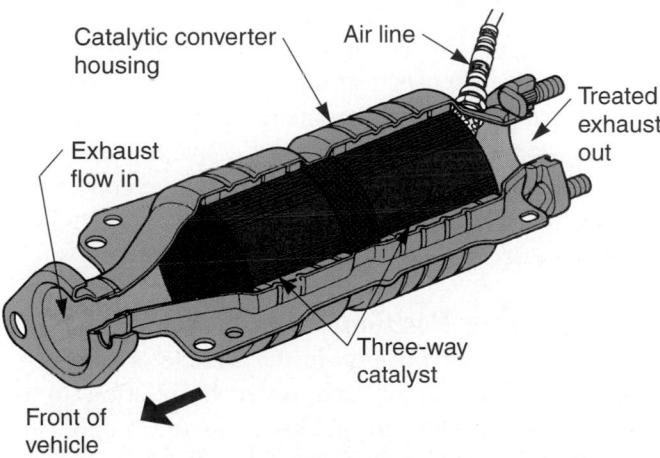

Figure 43-30. Dual-bed catalytic converter has two ceramic elements. One is a three-way catalyst. Air from air pump is forced into center of converter to aid burning and reaction. (Honda)

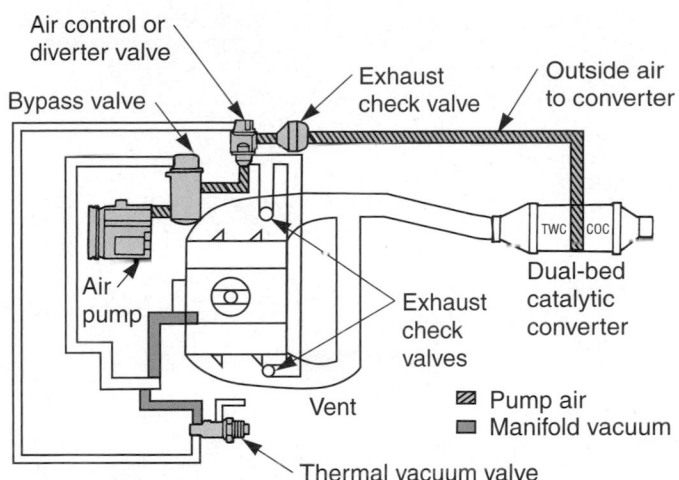

Figure 43-31. Diagram shows how air pump forces oxygen into dual-bed catalytic converter. Thermal-vacuum switch operates air control valve. It only lets air flow to converter when engine coolant is above set temperature.

Computerized Emission Control Systems

A *computerized emission control system* uses various engine sensors, a three-way catalytic converter, an ECM, electronic fuel injection, and other computer-controlled components to reduce pollution levels from the vehicle.

The ECM analyzes data from the many vehicle systems (engine, ignition, emission, fuel system) to closely monitor and control any function that can affect emissions, **Figure 43-32.** Note how almost all of the emission related components can be connected to computer control.

Note
For more information on closed and open loop computer system operation, refer to Chapter 22, *Gasoline Injection Fundamentals*. This subject is also discussed in several other locations. Refer to the index as necessary.

Oxygen Sensors

The *oxygen sensor* monitors the exhaust gases for oxygen content. The amount of oxygen in the exhaust gases is a good indicator of the engine's operational state. The oxygen sensor's voltage output varies with any changes in the exhaust's oxygen content. For example, an increase in oxygen, which would indicate a lean mixture, will make the sensor output voltage decrease. A decrease in oxygen, which occurs during rich mixture conditions,

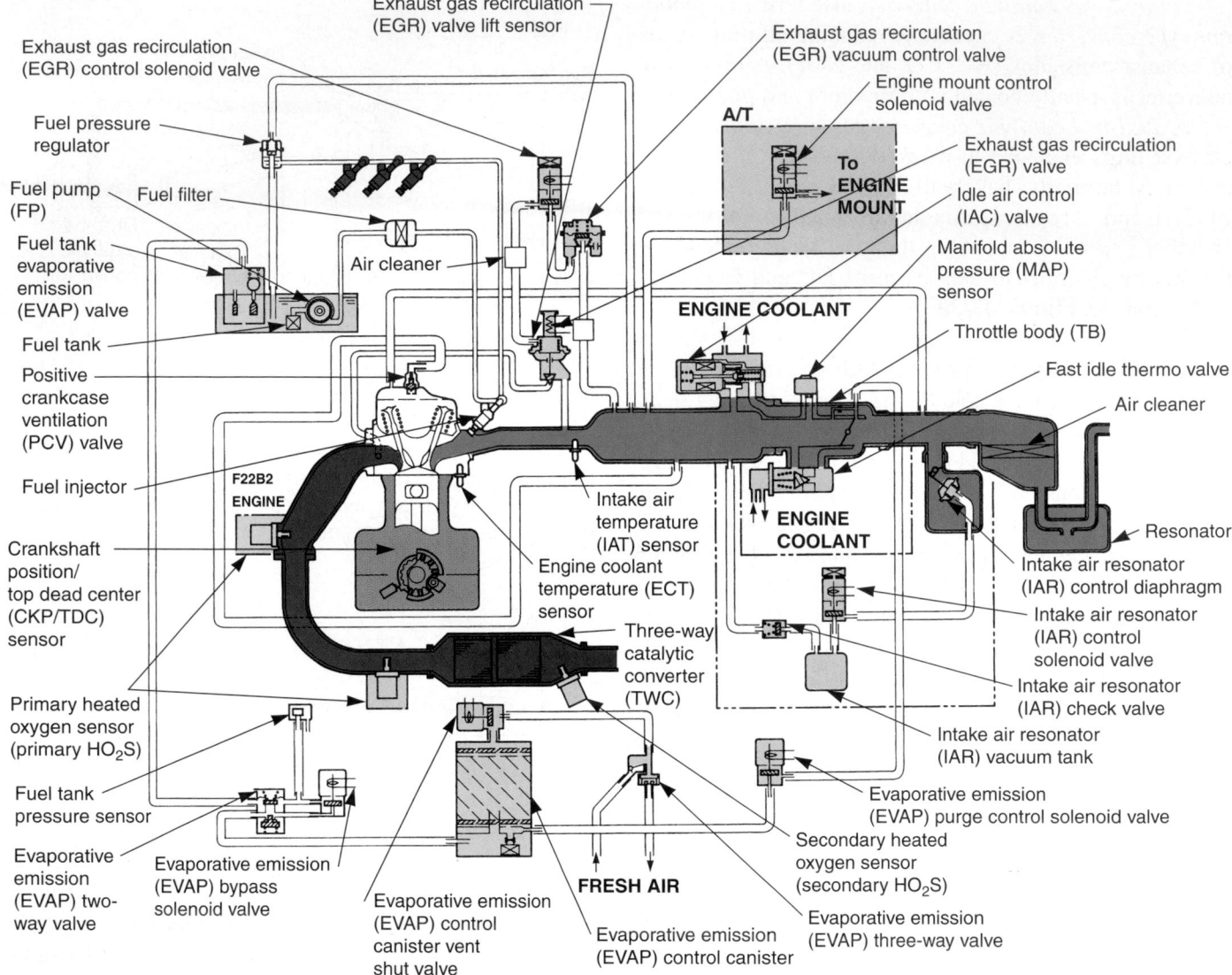

Figure 43-32. Note various sensors used in modern computer controlled emission systems. (Honda)

causes the sensor output voltage to increase. This is shown in **Figure 43-33.**

In this way, the oxygen sensor supplies data (different current levels) to the computer. The computer can then alter the opening and closing of the injectors to maintain a correct air-fuel ratio for maximum efficiency.

Primary and Secondary Oxygen Sensors

Vehicles may be equipped with up to five or more oxygen sensors, depending on the engine and application. The *primary oxygen sensor,* also termed *front O₂ sensor,* is used to monitor the oxygen in the exhaust gases as it leaves the engine. This indicates whether the engine's air-fuel mixture is too lean or rich. It is located before, or in front of, the catalytic converter, usually as close to the engine as possible. The *secondary oxygen sensor,* also termed the *rear O₂ sensor,* is mounted downstream in the exhaust system. Depending on its location downstream, the rear oxygen sensor can either be used to check the exhaust gas for oxygen content before it enters the catalytic converter or monitor the converter for proper operation. Any O_2 sensor mounted after the converter is referred to as a catalyst monitor. Its operation will be explained later in this chapter.

Oxygen Sensor Position

Oxygen sensor position in the vehicle is assigned a number by its location and order in relation to the engine's banks. The sensor closest to the number one cylinder is denoted as *Oxygen sensor, Bank 1, Sensor 1.* If the engine is equipped with only one oxygen sensor, which is the case for OBD I vehicles, it is referred to as *Oxygen sensor, Bank 1, Sensor 1,* no matter where it is located in the exhaust system. If the engine is a V-type,

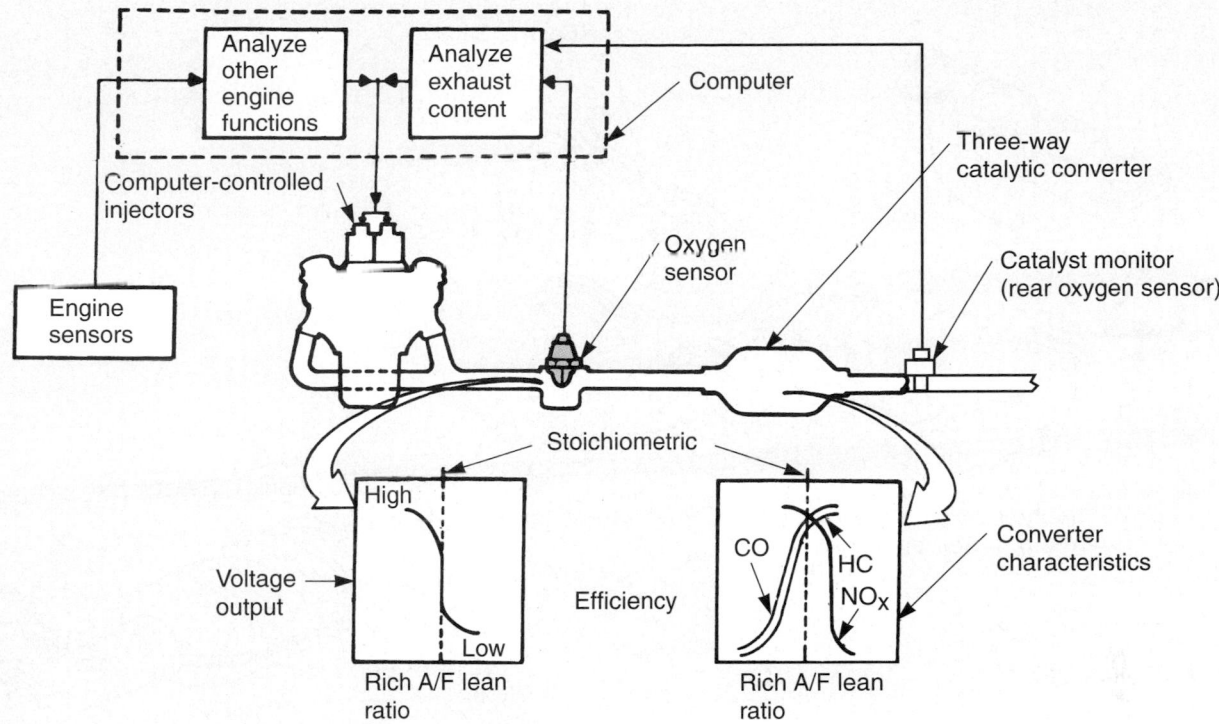

Figure 43-33. Oxygen sensors, also called exhaust gas sensors, monitor the amount of oxygen in the exhaust gases. A high oxygen content indicates a lean air-fuel mixture. If the oxygen content is too low, the computer can lean the engine's rich air-fuel mixture to reduce exhaust emissions. (Oldsmobile)

sensors located in the other bank are considered to be located in *Bank 2*. Sensors further down the exhaust stream from the engine are consecutively numbered as *Sensor 2, Sensor 3,* and so on, **Figure 43-34.** In almost all cases, the sensor with the highest number, such as *Sensor 3,* is the catalyst monitor.

Heated Oxygen Sensors

A *heated oxygen sensor,* abbreviated *HO₂S,* uses an electric heating element to quickly warm the sensor material to operating temperature. The heating element also stabilizes the temperature and operation of the sensor. The heating element allows the computer system to use the input sensor signals sooner.

Zirconia Oxygen Sensors

Most heated O₂ sensors are also called *zirconia oxygen sensors* because of their active materials. Zirconia and platinum are commonly used to produce the voltage output that represents oxygen in the exhaust gases. The platinum coating on the sensor surface causes any unburned fuel to ignite, which helps the sensor to maintain a high operating temperature. At an operating temperature of about 600°F (315°C), the oxygen sensor's element becomes a semiconductor and generates a small voltage. See **Figure 43-35.**

The zirconia oxygen sensor has an inner cavity that is exposed to the atmosphere. Since the earth's atmosphere is comprised of approximately 21% oxygen, this percentage serves as a reference for the amount of oxygen in the exhaust gases. The outer surface of the oxygen sensor is exposed to the exhaust gases. The outer surface serves as the positive connection of the sensor circuit. The inner cavity of the sensor serves as the negative connection, or ground.

The difference between the oxygen content in the inner cavity and the oxygen content of the exhaust gases flowing over the sensor's outer surface causes the sensor to generate a voltage. The ECM compares the voltage produced by the sensor to a reference voltage of approximately 450 millivolts (.45 volts).

For example, if the engine's air-fuel mixture is too rich, there will be almost no oxygen in the exhaust gases. This creates a large difference in oxygen content between the sensor's surfaces and causes the sensor to generate a voltage of about 600 millivolts (0.6 volts). This would inform the ECM to lean the mixture to reduce emissions. Refer to **Figure 43-36A.**

With a lean air-fuel mixture going to the engine, there will be a smaller difference in oxygen content between the sensor's inner and outer surfaces. The sensor will generate a weaker voltage signal of about 300 millivolts (0.3 volts),

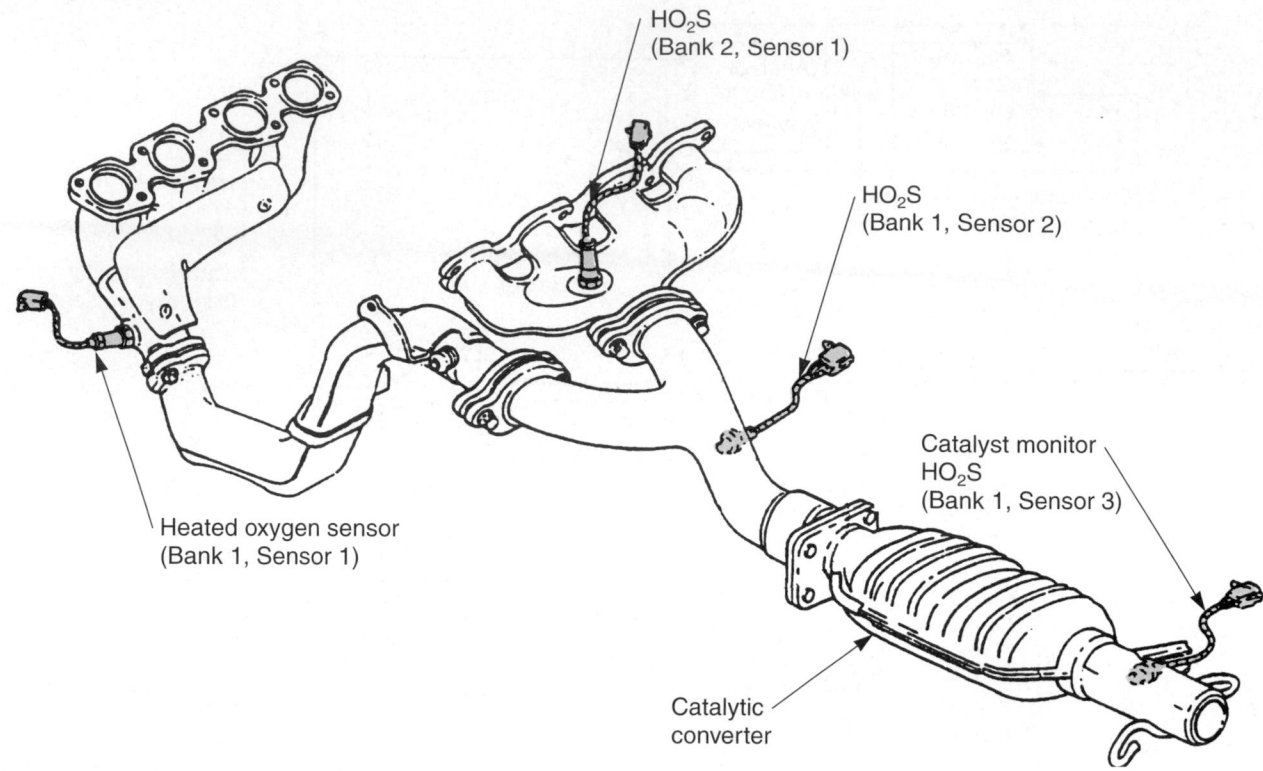

HO₂S
(Bank 2, Sensor 1)

HO₂S
(Bank 1, Sensor 2)

Catalyst monitor
HO₂S
(Bank 1, Sensor 3)

Heated oxygen sensor
(Bank 1, Sensor 1)

Catalytic
converter

Figure 43-34. Several oxygen sensors are used to monitor the exhaust gases as they travel through the exhaust system. Each sensor is identified by its position in relation to the engine banks and order in the exhaust system. (General Motors)

Figure 43-36B. The ECM will then enrich the fuel mixture and try to maintain a *stoichiometric* (chemically correct) air-fuel mixture. See **Figure 43-37.**

Note

See Chapter 20, *Automotive Fuels, Gasoline and Diesel Combustion,* for more information on stoichiometric air-fuel mixtures.

Titania Oxygen Sensors

A few late-model vehicles are equipped with titania oxygen sensors. The main difference between titania sensors and zirconia sensors is the way they produce their output signals. As previously discussed, zirconia sensors generate their own voltage signals. *Titania oxygen sensors,* on the other hand, vary their internal resistance to modify a reference voltage.

Titania sensors offer several advantages over zirconia sensors. They provide an oxygen content signal almost immediately after cold startup, eliminating the need for a heating element. Titania sensors are smaller than zirconia sensors and are manufactured as sealed units, making them less susceptible to outside contamination.

During operation, a constant reference voltage is sent from the ECM to the titania sensor's positive terminal, **Figure 43-38.** As the oxygen content of the exhaust changes, the resistance of the sensor also changes. The amount of resistance formed in the sensor determines the sensor's voltage drop. The ECM compares the sensor's voltage drop to a predetermined value. If the sensor's voltage drop is greater than this value, the ECM knows that the air-fuel mixture is too rich. If the sensor's output is below the predetermined value, the ECM knows the mixture is too lean. In either case, the control module can adjust fuel injection pulse width accordingly.

OBD II Emission System Monitoring

OBD II computers have the ability to monitor many functions that affect emissions. Most OBD II systems will monitor catalytic converter efficiency, engine misfire, O_2 sensor output, EGR valve action, fuel injection system performance, air injection system operation, and evaporative emissions system operation.

If any problems are detected, the ECM will turn on the malfunction indicator light (MIL) to warn the driver and technician of a problem. If the problem could damage the catalytic converter, the ECM flashes the MIL once per second while the problem is occurring. **Figure 43-39** shows how emission systems are monitored by OBD II diagnostics.

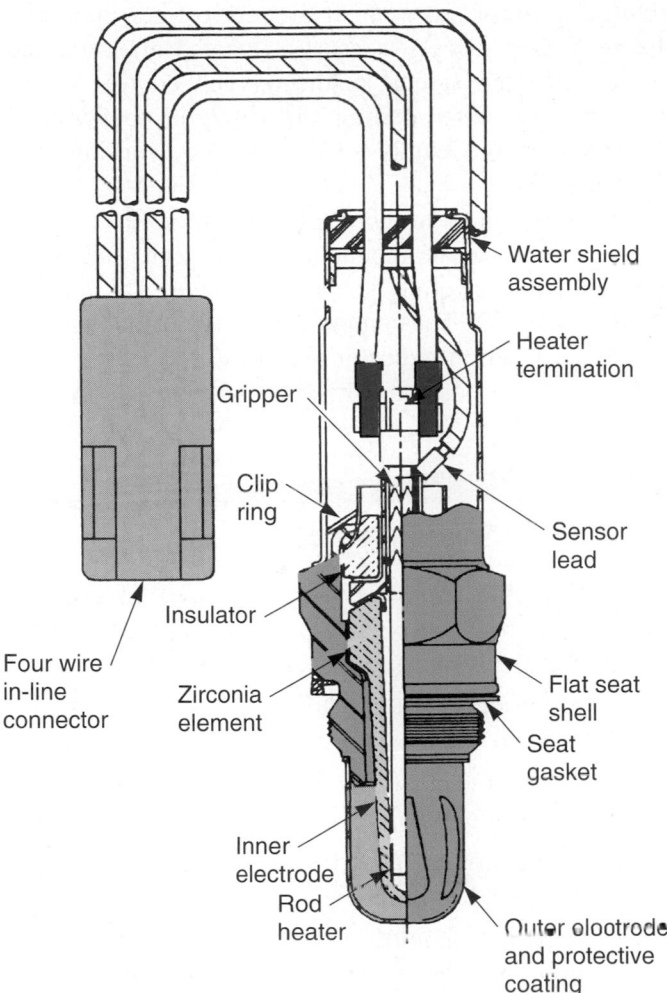

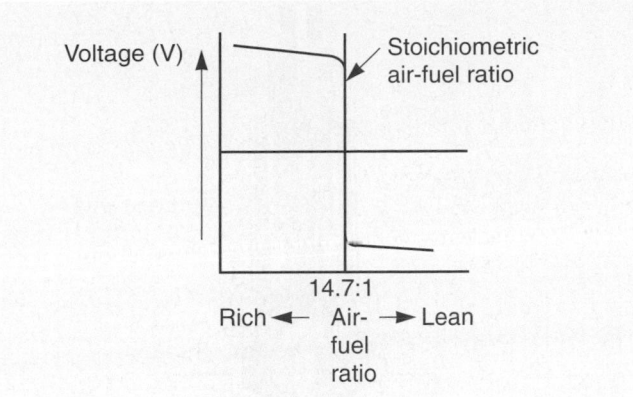

Figure 43-37. A stoichiometric air-fuel ratio is chemically correct. The oxygen sensor provides input to the ECM, which tries to maintain this ideal condition. This graph shows the relationship between oxygen sensor voltage and air-fuel ratio. (Honda)

Figure 43-35. Cutaway view shows the internal parts of a heated zirconia oxygen sensor. Study its construction.

OBD II Evaporative Emissions System Monitoring

Evaporative emissions system monitoring checks components for leakage and restrictions that could increase emissions. The computer energizes the solenoid valves to seal the system. This allows the computer to detect leaks or blockages in hoses and components. If the system does not pressurize and depressurize normally, it sets a trouble code to warn of a evaporative emissions system problem.

OBD II EGR Monitoring

EGR monitoring is done when the computer turns the EGR off while checking O_2 sensor readings. Changes in EGR valve opening and closing affect the air-fuel mixture and resulting O_2 sensor readings. If changes in the EGR valve do not affect O_2 sensor readings normally, a trouble code is produced. This is explained in more detail later.

Air Injection System Monitoring

Air injection system monitoring uses data from the rear O_2 sensor to determine if the right amount of air (oxygen) is being injected into the engine's exhaust stream. A low amount of air (oxygen) would trip a trouble code.

Catalyst Monitor

Oxygen sensors are used to monitor the content of the waste gases in the exhaust system to better control engine operation, which reduces emissions.

OBD II systems use at least two oxygen sensors—one before the catalytic converter and one after it. The

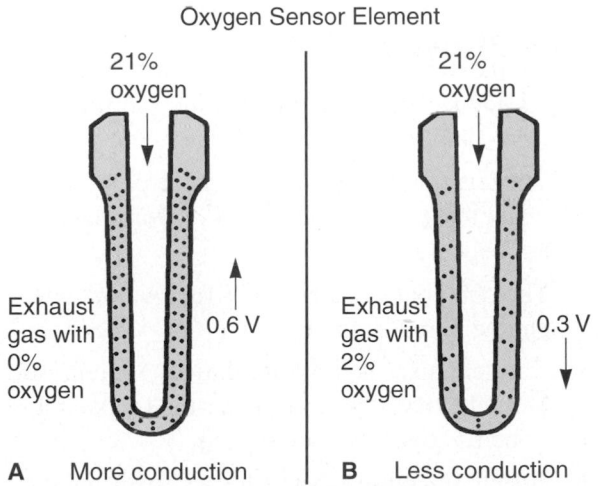

Figure 43-36. Study the operation of a heated oxygen sensor. A—There is greater voltage generation when the exhaust gas has a low oxygen content, indicating a rich mixture. B—There is less voltage generated when the exhaust has more oxygen, indicating a lean mixture. (Chevrolet)

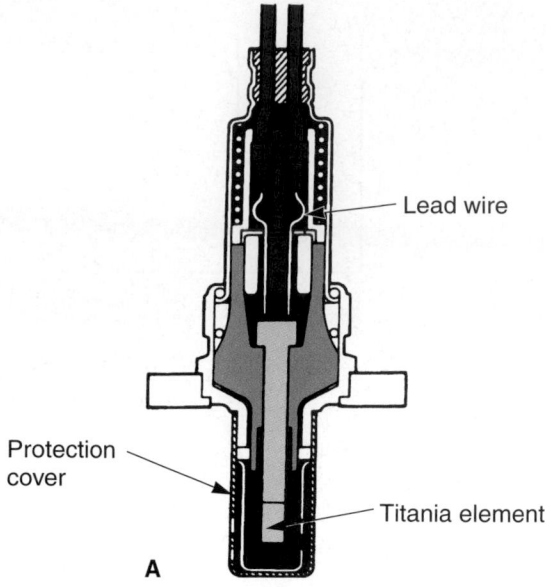

Lead wire

Protection cover

Titania element

A

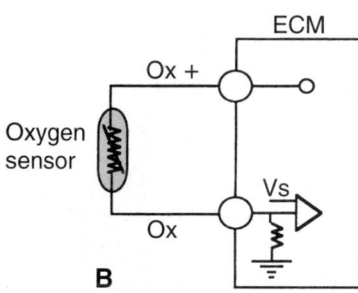

ECM

Ox +

Oxygen sensor

Vs

Ox

B

Figure 43-38. A—A titania oxygen sensor heats up quickly and does not require an electric heating element. Its internal resistance changes with the engine exhaust's oxygen content. B—Basic circuit from the titania oxygen sensor to the ECM. (Snap-On Tool Corp.)

OBD II MONITORS	ECM INPUT	DATA FROM
Air-injection system	Oxygen content leaving the catalyst	Rear oxygen sensor
Fuel-injection system	Change in the air-fuel-ratio control limits	Front oxygen sensor
Catalyst efficiency	Catalytic converter's oxygen storage capacity	Front and rear oxygen sensors
Engine misfires	Crankshaft speed and road conditions	Crank-position sensors, vehicle accelerometers, ABS sensors
Oxygen sensors	Sensor switching frequency	Front and rear oxygen sensors
Exhaust-gas-recirculation system	Change in the air-fuel-ratio control limits	Front oxygen sensor
Fuel vapor leaks	Pressure capacity of the fuel tank and lines	Leak-detection pump

Figure 43-39. This chart shows how modern on-board diagnostic systems check the operation of major emission control systems.

catalyst monitor, also called the rear oxygen sensor or the secondary oxygen sensor, is located after the catalytic converter. It checks the oxygen content of the exhaust gases after exiting the catalytic converter to determine if the catalyst elements are working. See **Figure 43-34.**

If the signal from the catalyst monitor becomes too similar to the engine-mounted oxygen sensor signal(s), the catalytic converter is not cleaning up the exhaust gases as it should. The computer or ECM would then turn on the malfunction indicator light to warn the driver and technician of a possible catalytic converter failure. See **Figure 43-40.**

Summary

- Emission control systems are used on cars and light trucks to reduce the amount of harmful chemicals released into the atmosphere.

- Air pollution is caused by an excess amount of harmful chemicals in our atmosphere.

- Smog is a nickname given to a visible cloud of airborne pollutants.

- Hydrocarbons result from the release of unburned fuel in the atmosphere.

- Carbon monoxide is a toxic emission that results from the release of partially burned fuel.

- Oxides of nitrogen are produced by extremely high combustion temperatures.

- Particulates are the solid particles of carbon soot and fuel additives that blow out of a vehicle's tailpipe.

- A positive crankcase ventilation system uses engine vacuum to draw toxic blowby gases into the intake manifold for reburning in the combustion chambers.

- The evaporative emissions control system prevents toxic fuel system vapors from entering the atmosphere.

- The charcoal canister stores fuel vapors when the engine is not running.

- The exhaust gas recirculation system allows burned exhaust gases to enter the intake manifold to help reduce NO_x emissions.

- An air injection system forces fresh air into the exhaust ports or the catalytic converter to reduce HC and CO emissions.

- A catalytic converter oxidizes the HC and CO emissions that pass into the exhaust system.

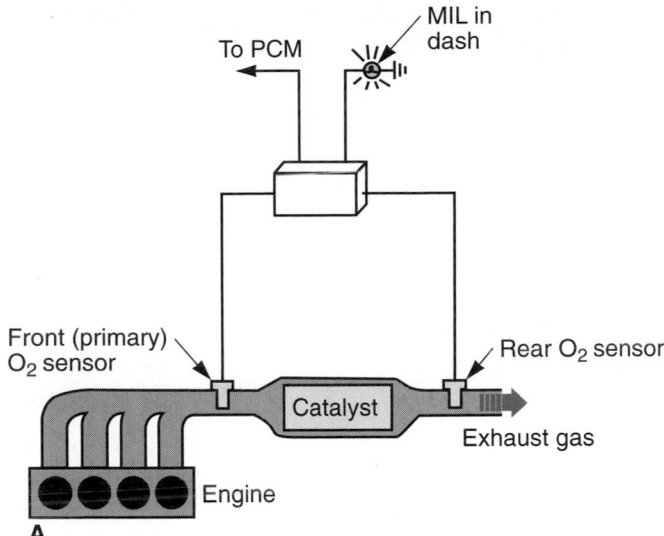

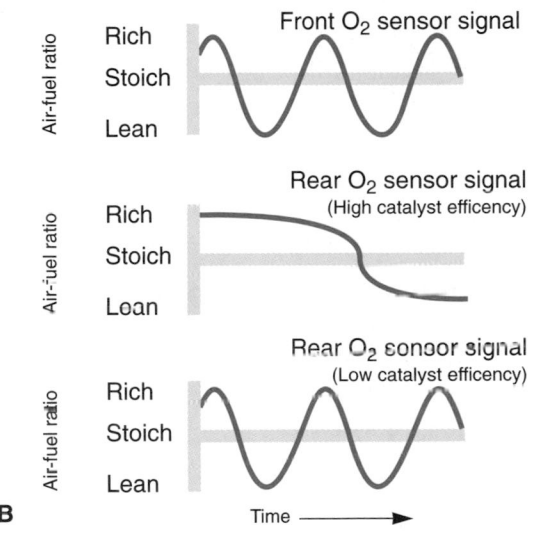

Figure 43-40. A—With late-model on-board diagnostic systems, two oxygen sensors are used with each cylinder bank. The primary, or front, oxygen sensor is used to determine whether the air-fuel mixture that is being fed to the engine is correct. The secondary, or rear, sensor monitors the efficiency of the catalytic converter. If the converter is not working normally, it will turn on the malfunction indicator light in the dash. B—If the rear sensor output is the same as the front sensor output, the computer knows the catalytic converter is not working properly.

- Oxygen sensors monitor the oxygen content of the exhaust gases to better control engine operation and reduce emissions.

- The primary oxygen sensor is used to monitor the oxygen content of the exhaust gases to determine whether the air-fuel mixture is too lean or too rich.

- The secondary oxygen sensor checks the oxygen content after the exhaust gases have been treated to determine if the catalytic converter is working normally.

Important Terms

Emission control
  systems
Air pollution
Environmental
  Protection Agency
Smog
Exhaust emissions
Hydrocarbons
Carbon monoxide
Oxides of nitrogen
Particulates
Positive crankcase
  ventilation (PCV)
  system
Blowby
PCV valve
Oil-air separator
Evaporative emissions
  control system
Non-vented fuel tank
  cap
Fuel tank vent line
Air dome
Liquid-vapor separator
Rollover valve
Charcoal canister
Purge line
Purge valve
Fuel tank pressure
  sensor
Canister vent solenoid

Service port
Air cleaner carbon
  element
Exhaust gas
  recirculation system
EGR jets
Electronic EGR system
EGR duty cycle
Air injection system
Air injection pump
Diverter valve
Air distribution
  manifold
Air check valve
Thermostatic air cleaner
  system
Thermal vacuum valve
Vacuum motor
Heat control door
Pulse air system
Aspirator valves
Catalytic converter
Catalyst
Computerized emission
  control system
Oxygen sensor
Heated oxygen sensor
Zirconia oxygen sensors
Titania oxygen sensor
Catalyst monitor

Review Questions—Chapter 43

Please do not write in this text. Place your answers on a separate sheet of paper.

1. What are some causes of air pollution?

2. Which of the following agencies enforces air pollution standards?
   (A) EGR.
   (B) SAE.
   (C) ASE.
   (D) EPA.

3. Define the term "smog."

4. List and explain the four kinds of vehicle emissions.

5. CO emissions are caused by partially burned fuel. True or False?

6. Increasing peak combustion temperature tends to reduce NO_x emissions. True or False?

7. _____ are the solid particles of carbon soot that exit a vehicle's tailpipe. They are more of a problem in diesel engines.

8. What are the three basic sources of vehicle emissions?

9. Which of the following is *not* a typical engine modification for reducing emissions?
   (A) Lower compression ratios.
   (B) Leaner air-fuel ratios.
   (C) Increased valve overlap.
   (D) Wider spark plug gaps.

10. Explain six major emission control systems.

11. A(n) _____ _____ _____ system uses engine vacuum to draw toxic blowby gases into the intake manifold for burning in the combustion chambers.

12. The _____ _____ _____ system speeds engine warm-up and keeps the temperature of the air entering the engine warm.

13. Explain the operation of the charcoal canister in an evaporative emissions control system.

14. How does an EGR system work?

15. An air injection system forces fresh air into the _____ _____ of the engine to reduce HC and CO emissions.

16. The diverter valve keeps air from entering the exhaust system during engine deceleration, preventing backfiring. True or False?

17. What is a catalytic converter?

18. Which of the following does *not* relate to catalytic converters?
    (A) Monolith.
    (B) Pellet.
    (C) Stores unburned fuel.
    (D) Oxidizes or burns emissions.

19. Why is a mini catalytic converter sometimes used?

20. Summarize the operation of a computer-controlled emission control system.

★ ASE-Type Questions

1. Each of the following is a natural cause of pollution *except:*
   (A) wind-blown dust.
   (B) volcanoes.
   (C) engine exhaust.
   (D) forest fires.

2. Strict laws which reduce air pollution are enforced by the:
   (A) EPA.
   (B) ASE.
   (C) SAE.
   (D) All of the above.

3. A nickname given to a visible cloud of airborne pollutants is:
   (A) fog.
   (B) smog.
   (C) smoke.
   (D) vapors.

4. Which of these is *not* a type of vehicle emissions?
   (A) Particulates.
   (B) Hydrocarbons.
   (C) Photochemicals.
   (D) Carbon monoxide.

5. Which of the following provides the highest percentage of vehicle emissions?
   (A) Fuel vapors.
   (B) Engine exhaust gases.
   (C) Lean air-fuel mixtures.
   (D) Engine crankcase blowby fumes.

6. Decreased valve overlap is used to reduce:
   (A) hydrocarbons.
   (B) carbon monoxide.
   (C) oxides of nitrogen.
   (D) All of the above.

7. Each of these is used to reduce the amount of air pollution produced by automobiles *except:*
   (A) catalytic converter.
   (B) air injection system.
   (C) heated air inlet system.
   (D) H_2O system.

8. While discussing the operation of the evaporative emissions control system, Technician A says a charcoal canister stores fuel vapors when the engine is not running. Technician B says the canister absorbs fuel vapors only when the engine is running. Who is right?
   (A) A only.
   (B) B only.
   (C) Both A and B.
   (D) Neither A nor B.

9. Which of the following is *not* an EGR valve component?
   (A) Spring.
   (B) Plunger.
   (C) Purge line.
   (D) Vacuum diaphragm.

10. In a pulse air system, which of the following blocks airflow in one direction and allows airflow in the other direction?
    (A) Gulp valve.
    (B) Reed valve.
    (C) Check valve.
    (D) All of the above.

Activities—Chapter 43

1. On a late-model vehicle designated by your instructor, try to locate various parts of the emission control system and explain their purpose.

2. Research the chemical makeup of smog and prepare a report on it.

3. Write a report comparing the emissions produced by late-model automobile engines to the higher emissions produced by two-stroke engines used in weed eaters, snow throwers, snowmobiles, watercraft, etc. Discuss what can be done to reduce the emissions produced by two-stroke engines. Send the report to your congressman or to the EPA. Include a letter asking that more stringent regulations be placed on two-stroke engines to help minimize their negative impact on our environment.

Chapter 44

Emission Control System Testing, Service, and Repair

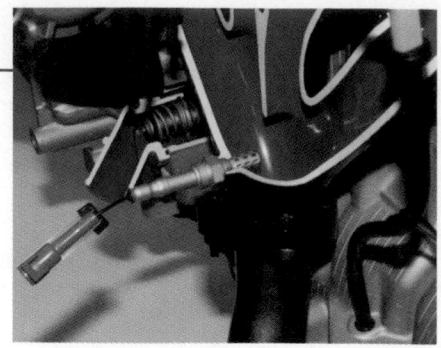

After studying this chapter, you will be able to:

- Explain the use of exhaust gas analyzers.
- Inspect and troubleshoot emission control systems.
- Perform periodic service operations on emission control systems.
- Test individual emission control components.
- Replace or repair major emission control components.
- Demonstrate and practice safe work procedures.
- Correctly answer ASE certification test questions on emission control system testing and service.

Emission control systems, like any automotive system, can malfunction. When this happens, excessive exhaust emissions and fuel vapors escape into the atmosphere. Engine performance may also suffer when emission system problems develop. However, in many cases, an emissions problem is not apparent to the vehicle's driver. It is very important that you, as an automotive technician, be able to correctly diagnose and repair emission control systems.

This chapter will summarize the most common tests and repairs done on today's emission control systems. It will explain how to diagnose problems using a scan tool and an exhaust gas analyzer. The chapter will then cover pinpoint tests and replacement procedures on individual parts.

 **Tech Tip**
Studies have shown the emissions from passenger vehicles have dropped over 90% in the last 15 years.

Computer-Controlled Emission System Service

Computer-controlled engine and emission systems can cause a wide range of problems. The computer controls the fuel injection system, EGR valve, evaporative emissions control, and other systems. Any of these systems can affect the vehicle's operation and can increase emissions.

It is important to remember that most emissions problems are caused by a malfunction in a system unrelated to the emission control system. The fuel, ignition, exhaust, computer control, and electrical systems should be checked before suspecting emissions control devices.

OBD II Engine Misfire Monitoring

Engine misfire monitoring uses the engine crankshaft position sensor to detect changes in crankshaft speed, which may indicate the engine is missing and not firing its air-fuel mixture properly. If crankshaft speed fluctuations are detected, the ECM trips a trouble code. The misfire monitor is covered in more detail in Chapter 45.

Sensor Monitoring

Sensor monitoring uses the ECM software to compare known normal sensor signal variations to actual operating values. If the sensor signal goes out of range or normal parameter, a trouble code will be produced and set in the ECM's memory. With OBD II system, almost every sensor can be monitored for normal operation. This helps the technician keep the vehicle running at maximum efficiency with minimum emissions.

Actuator Monitoring

Actuator monitoring uses the ECM software data to determine if an actuator is drawing too little or too much current. In this way, the computer system can detect actuators and actuator circuits that have problems. Fuel injectors, electronic EGR valves, evaporative solenoids, and similar actuators can be monitored in this way on OBD II systems.

For example, *fuel injector monitoring* has the ECM check the coil windings of each injector to check for opens, shorts, or an abnormal voltage feedback signal. For example, if an injector is shorted, the current pulse sent to open the injector will not generate voltage across the injector coil winding. The ECM can sense this lower than normal voltage and trip a code pinpointing the bad injector.

Scanning Emission Systems

Emission system scanning involves using a scan tool to check the condition of monitored parts of the various emission control systems.

Modern vehicles have elaborate on-board diagnostic capabilities that will find troubles in almost all emission-related components.

Explained in detail in Chapter 18, a scan tool will help you troubleshoot late-model vehicles quickly and easily. Connect the scan tool to the vehicle's data link connector. If needed, install the correct scan tool cartridge for the specific make vehicle and connect power to the tool, **Figure 44-1.** First check for stored trouble codes and retrieve any scan tool information that might help fix the problem. An example of scan tool readouts are in **Figure 44-2.**

The scan tool readouts can tell you in which circuit the fault is located. For example, if the scan tool says there is a constant low voltage signal from the catalyst monitor, you would know where to start your tests, but you would not know exactly what is wrong. You could have a defective oxygen sensor, a poor connection, or a malfunctioning catalytic converter. A trouble code from the catalyst monitor could be due to the sensor itself, its

A
```
13
O2 CKT OPEN
Code 01 of 30
Press ENTER for menu
```

B
```
Oxygen in the exhaust
reacts with the O2 sensor
to produce a voltage. The
ECM monitors the voltage
```
to determine the fuel mix. Some O_2 sensors are equipped with a heating element.

C
```
Faulty connections
(O2 Circuit), O2 sensor,
ECM.
```

Figure 44-2. These are sample displays from a typical scan tool. A—This screen indicates that a trouble code is present. It can list all the possible codes that can be produced. B—A description of the trouble code is shown to help analyze the problem. C—Some scan tools can also list possible causes of each trouble code. (OTC)

wiring, or a defective catalytic converter. Keep this in mind when scanning a vehicle's computer system. Think how the system is supposed to operate and if other components could cause the abnormally high or low sensor signal.

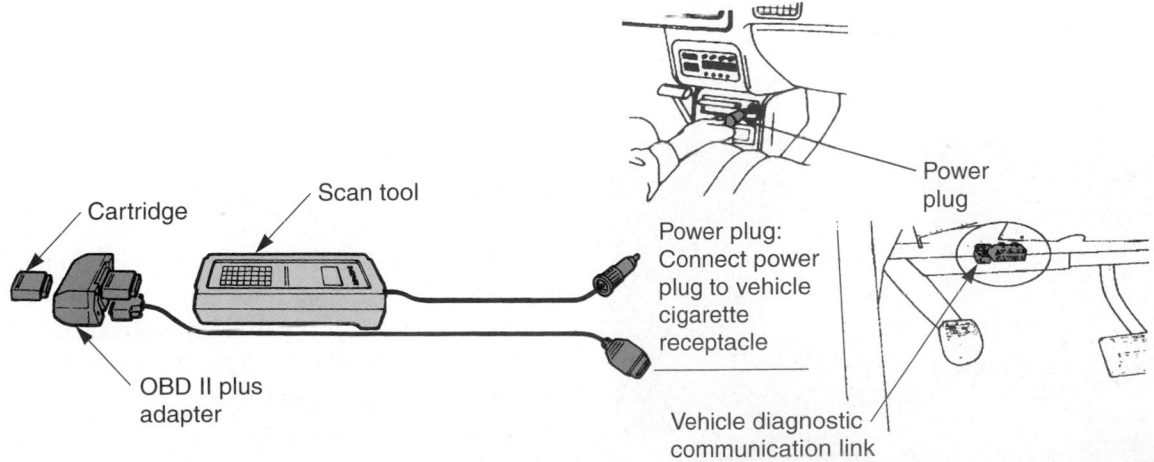

Cartridge

Scan tool

OBD II plus adapter

Power plug: Connect power plug to vehicle cigarette receptacle

Vehicle diagnostic communication link

Power plug

Figure 44-1. A scan tool is an essential diagnostic tool when troubleshooting emission control systems. It will communicate with the vehicle's computer(s) to retrieve and explain trouble codes. (OTC)

Emission Maintenance Reminder

The *emission maintenance reminder* is a circuit that automatically turns on a dash light to indicate the need for emission control system service. Some vehicles use a mechanical flag that is located inside the speedometer cluster.

After the prescribed adjustments and repairs are done, you must turn off the emission maintenance light. There are numerous methods to turn this light off. You might have to remove the speedometer cluster in order to move a small lever hidden in the dash, jump across a specific connector, etc. Since there are so many variations, refer to the service manual for specific instructions.

Inspecting Emission Control Systems

Using information from the scan tool, you must find the source of the problem. Start out by inspecting all engine vacuum hoses and wires. A leaking vacuum hose or disconnected wire could trip a trouble code and upset the operation of the engine and emission control systems, **Figure 44-3.**

A section of vacuum hose can be used as a listening device. Place one end of the hose next to your ear. Move the other end of the hose around the engine compartment, along vacuum hoses and connections. When the hose nears a vacuum leak, you will be able to hear a loud hissing sound. You can also remove the metal probe from your stethoscope instead of using a piece of vacuum hose. See **Figure 44-4.**

Look for evidence of disconnected wires, oil leaks onto the oxygen sensor, extremely dirty engine oil, and blowby, all of which could trip a trouble code. Also, inspect the air cleaner for clogging. Check that the air pump belt is properly adjusted. Try to locate any obvious problems. If nothing is found during your inspection, each system should be checked and tested.

Figure 44-3. Begin inspecting the emission control system by looking for disconnected vacuum hoses and wires.

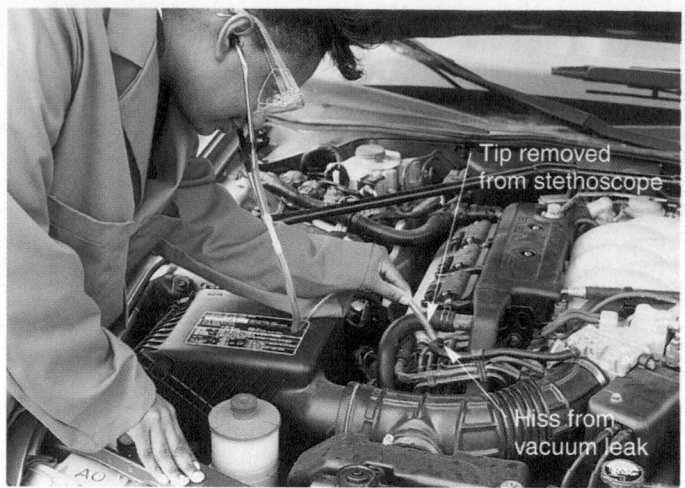

Figure 44-4. Use an automotive stethoscope with the metal tip removed to listen for vacuum leaks. (Snap-on Tool Corp.)

Engine Exhaust Gases

As discussed in the previous chapter, engine exhaust gases contain chemical substances that change with combustion efficiency. *Combustion efficiency* is the ratio of heat actually developed in the combustion process to the heat that would be released if the combustion were perfect.

Engine exhaust gases and the various substances in them are the result of combustion. Some of these substances, such as hydrocarbons (HC), carbon monoxide (CO), and oxides of nitrogen (NO_x), are harmful. Other byproducts of combustion, such as carbon dioxide (CO_2), oxygen (O_2), and water (H_2O), are not harmful. By measuring HC, CO, O_2, CO_2, and NO_x, we can find out how efficiently the engine and emission systems are working.

Exhaust Gas Analyzer

An *exhaust gas analyzer* is a test instrument that measures the chemical content of the engine's exhaust gases. See **Figure 44-5.**

With the engine running, the exhaust analyzer will sample, analyze, and indicate the amount of pollutants and other gases in the exhaust. The technician can use this information to determine the condition of the engine and other systems affecting emissions. An exhaust gas analyzer is an excellent diagnostic tool that will indicate excessive emissions caused by:

- Fuel metering problems.
- Engine mechanical problems.
- Vacuum leaks.
- Ignition system problems.

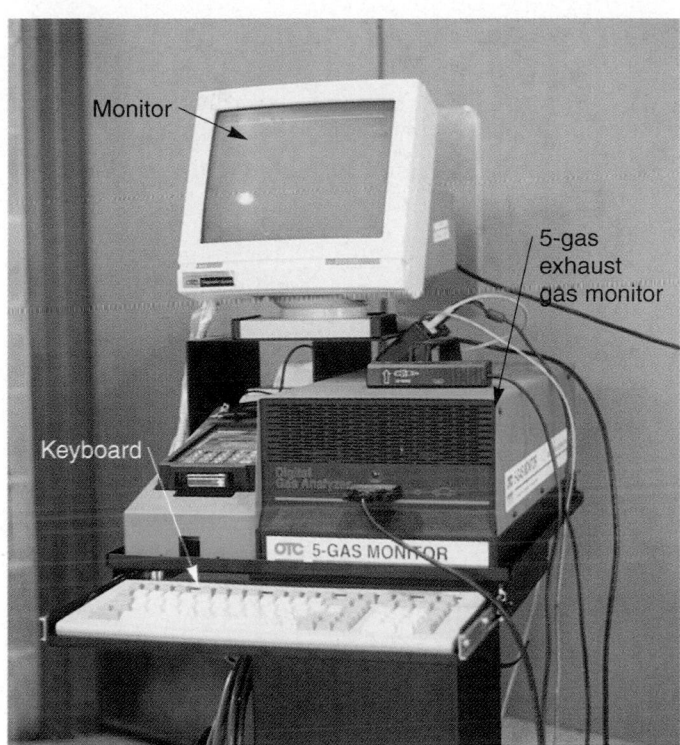

Figure 44-5. Five-gas analyzers are replacing older two- and four-gas analyzers. They are more sensitive than older analyzers and can provide added information for finding emission system problems. (OTC)

- PCV troubles.
- Clogged air filter.
- Faulty air injection system.
- Evaporative emissions control system problems.
- Computer control system troubles.
- Catalytic converter condition.

Types of Exhaust Gas Analyzers

There are three different kinds of exhaust gas analyzers: two-gas analyzers, four-gas analyzers, and five-gas analyzers.

The **two-gas exhaust analyzer** can measure the amount of hydrocarbons (HC) and carbon monoxide (CO) in a vehicle's exhaust gases. This common analyzer has been used for a number of years. However, the two-gas analyzer cannot accurately analyze the exhaust gases from newer engines; therefore, it is being replaced by four- or five-gas analyzers.

The **four-gas exhaust analyzer** measures the quantity of hydrocarbons (HC), carbon monoxide (CO), carbon dioxide (CO_2), and oxygen (O_2), in an engine's exhaust. Most state air quality agencies use the four-gas analyzer.

Although carbon dioxide and oxygen are not toxic emissions, they provide useful data about the engine's operating efficiency. Late-model engines are so clean burning, a four-gas exhaust analyzer is needed to accurately evaluate the makeup of the exhaust gases. It provides extra information for diagnosing problems and making adjustments.

The **five-gas exhaust analyzer** will measure hydrocarbons, carbon monoxide, carbon dioxide, oxygen, and oxides of nitrogen. It is the most modern and informative type of exhaust gas analyzer. Oxides of nitrogen is a toxic pollutant that should be measured, if possible, as a means of diagnosis.

Using an Exhaust Gas Analyzer

To use an exhaust gas analyzer, plug the machine in and allow it to warm up as described by the manufacturer. After warm-up, zero and calibrate the analyzer. Exhaust gas analyzer *calibration* involves zeroing the meter scales while sampling clean air (no exhaust gases present in room) with the analyzer. Newer analyzers sample a *calibration gas* (mixture of several gases) to adjust the meter readings for accuracy. The gas is automatically metered while the meter scales are calibrated, **Figure 44-6.** In most cases, any calibration adjustment is done by the analyzer itself automatically or when commanded by the technician. Older analyzers may have to be manually adjusted.

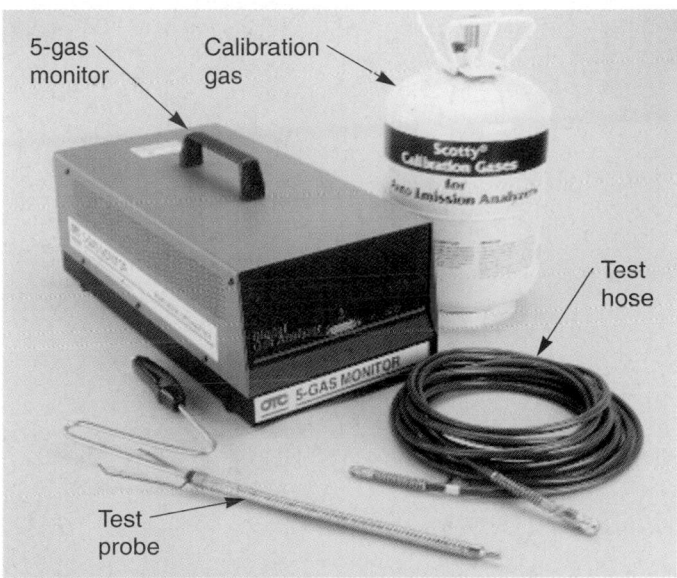

Figure 44-6. Calibration gas is used to set or zero meters on four- and five-gas analyzers. One or more tanks of calibration gas is connected to the analyzer. The gas is drawn through a pickup tube while the analyzer adjusts the meters to specifications for the gas. (OTC)

Warning

Never inhale exhaust gas analyzer calibration gas. The gas contains emission gases that can be harmful to your health.

Before testing the vehicle, take it on a thorough road test. This allows the vehicle to reach the proper operating temperature. Never test a vehicle with a cold engine, as inaccurate readings will result. To begin the test, install the probe in the vehicle's tailpipe. If working in an enclosed shop, slide the probe through a hole in the shop's vent hose, **Figure 44-7.** Since exact procedures vary, always follow the operating instructions for the particular exhaust analyzer. This will ensure accurate measurements.

Warning

When using an exhaust analyzer, do *not* let engine exhaust fumes escape into an enclosed shop area. Engine exhaust can kill. Use the shop exhaust ventilation system to trap and remove the toxic fumes.

Most analyzers measure hydrocarbons and carbon monoxide at idle and approximately 2500 rpm. If you have a five-gas analyzer, it will also measure oxygen, oxides of nitrogen, and carbon dioxide. Compare the analyzer readings with specifications.

When testing some electronic fuel injection systems without a load, only idle readings on the exhaust analyzer will be accurate. A dynamometer must be used to load the engine to simulate actual driving conditions.

Exhaust Gas Analyzer Readings

Exhaust gas analyzer readings are used to determine the chemical content of the engine's exhaust gases. See **Figure 44-8.** Your test readings must be within state-regulated specifications for the geographic area and the year of the vehicle. Generally, new vehicles have stricter specifications and require lower analyzer readings than older vehicles.

Although most states provide emission specifications in parts per million and percentage, some states give emission specifications in grams per mile. When this is the case, the vehicle must be operated on a chassis dyno so that the amount of emissions produced per mile driven can be calculated.

Hydrocarbon Readings

An exhaust gas analyzer measures hydrocarbons (HC) in parts per million (PPM) by volume. For example, an analyzer reading of 10 PPM means there are 10 parts of hydrocarbons for every million parts of exhaust gas.

Figure 44-7. Note how the exhaust analyzer probe is installed in a vehicle's tailpipe. The adapter on the vent hose prevents toxic vapors from entering the shop. Warm-up the engine and calibrate the analyzer before placing the test probe in the tailpipe. Compare the readings to specifications for the vehicle's model year. (Saab)

EXHAUST GAS DATA		
Engine Speed	RPM	750
Engine Temp.	°F	190
Hydrocarbons	HC	212 PPM
Carbon Monoxide	CO	0.93%
Oxygen	O_2	1.2%
Carbon Dioxide	CO_2	14.4%
Oxides of Nitrogen	NO_X	19 PPM

Figure 44-8. Exhaust gas analyzer readings. Note that the readings for hydrocarbons and oxides of nitrogen are given in parts per million. Carbon monoxide, oxygen, and carbon dioxide readings are given as percentages.

A vehicle that is 10–15 years old, for example, will have a relatively high hydrocarbon specification, such as 900 PPM. A newer vehicle, having stricter emission requirements, could have a 220 PPM hydrocarbon specification. If a vehicle's hydrocarbon reading is higher than the standard permits, the vehicle's hydrocarbon emissions (unburned fuel) are excessive.

Tech Tip
Always refer to the emission control sticker in the engine compartment or a service manual for emission level specifications. Values vary year by year.

A rotten egg smell from the exhaust is often an indication of the presence of unburned fuel and, therefore, excessive hydrocarbons. Higher-than-normal HC readings can be caused by one or more of the following conditions:

- Fuel system malfunction—leaking fuel injector, faulty pressure regulator, improper fuel pressure, or carburetor problems.
- Improper ignition timing—distributor, computer, or adjustment problem.
- Engine problems—blowby, worn rings, blown head gasket.
- Faulty emission control system—bad PCV, catalytic converter, EGR valve, evaporative control system.
- Ignition system troubles—fouled spark plug, cracked distributor cap, open spark plug wire.
- Computer control system problems—defective input sensor, output actuator, or ECM.

CO Readings

An exhaust analyzer measures carbon monoxide (CO) in percentage by volume. For instance, a 1.2% analyzer reading would mean that 1.2% of the engine exhaust is made up of carbon monoxide. The other 98.8% consists of other substances. High carbon monoxide is caused by an incomplete burning of fuel or a lack of air (oxygen) during the combustion process.

If the exhaust analyzer reading is higher than specifications, the engine is producing too much carbon monoxide. You would need to locate and correct the cause of the problem.

The exhaust analyzer's carbon monoxide reading is related to the air-fuel ratio. A *high* carbon monoxide reading would indicate an over-rich mixture (too much fuel compared to air). If a high carbon monoxide reading is accompanied by a high HC reading, the problem is related to something that will make the engine run rich.

A *low* or *no* carbon monoxide reading would indicate a lean air-fuel mixture (too much air compared to fuel). Typical causes of high carbon monoxide readings are:

- Fuel system problems—sticking or leaking injector, leaking fuel pressure regulator, high float setting, clogged carburetor air bleed, restricted air cleaner, choke out of adjustment, defective input sensor, computer control problem.
- Emission control system troubles—almost any emission control system problem can upset the carbon monoxide readings.
- Incorrect ignition timing—timing too far advanced or improper vacuum going to the vacuum advance unit.

Oxides of Nitrogen Readings

A five-gas analyzer can measure oxides of nitrogen, while a four-gas analyzer cannot. Since oxides of nitrogen are toxic, some state air quality agencies have made exhaust emission measurements with a five-gas analyzer mandatory.

Typical causes of high NO_x emissions include:

- High combustion chamber temperatures—excessively high engine compression ratio, carbon deposits in the combustion chambers, low cooling system, blocked water jackets, etc.
- EGR system problems—burned gases are not being injected into the intake manifold and combustion flame temperature is too high.

Carbon Dioxide Readings

Four- and five-gas exhaust analyzers measure carbon dioxide (CO_2) in percent by volume. Typically, CO_2 readings should be above 8%. CO_2 readings provide more data for checking and adjusting the air-fuel ratio.

Carbon dioxide is a byproduct of combustion. It is produced when one carbon molecule combines with two oxygen molecules in the combustion chamber. Carbon dioxide is not toxic at low levels. When you breathe, for example, you exhale carbon dioxide.

Normally, oxygen and carbon dioxide levels are compared when evaluating the content of the engine exhaust. For example, if the percent of carbon dioxide exceeds the percent of oxygen, the air-fuel ratio is on the rich side of a stoichiometric (chemically correct) mixture. It is also a good indicator of possible dilution of the exhaust gas sample through an exhaust leak.

Oxygen Readings

Four- and five-gas exhaust analyzers measure oxygen (O_2) in percentage by volume. Typically, oxygen readings should be between 1% and 7%. Oxygen is needed for the catalytic converter to burn HC and CO emissions. Without oxygen in the engine exhaust, exhaust emissions can pass through the converter and out the vehicle's tailpipe.

As detailed in Chapter 43, there are two systems that add oxygen to the engine exhaust: the air injection and pulse air systems. As air is added to the exhaust, CO and HC emissions decrease. As a result, oxygen readings can be used to check the operation of the fuel injection system, air injection system, catalytic converter, and computer.

The oxygen level in the engine exhaust sample is an accurate indicator of a vehicle's air-fuel mixture. It is also a good indicator of a possible exhaust leak, which can dilute the exhaust gas sample. When an engine is running lean, oxygen increases proportionately with the air-fuel ratio. As the air-fuel mixture becomes lean enough to cause a *lean misfire* (engine miss), oxygen readings rise dramatically. This provides a very accurate method of measuring lean and efficient air-fuel ratios.

If you find any exhaust gas analyzer reading to be abnormally high or low, use your knowledge of system operation to pinpoint the trouble. By knowing which emissions are affected by which engine problem or emission system trouble, you can narrow down the source of the problem to specific components. You would then need to test each component or circuit to verify your conclusions.

Note

Several other textbook chapters discuss how exhaust analyzer readings can be used. Refer to index for more information.

Tech Tip

An engine with a defective thermostat can fail an emission test. If the engine operating temperature is too high, it can affect engine combustion efficiency and the operation of the computer control systems, which will try to compensate for the overheating engine. If the engine cannot reach the proper operating temperature, the computer will not be able to go into closed loop mode. Keep these basic system malfunctions in mind when diagnosing problems.

Emissions Testing Programs

Many states have some form of *emissions testing program.* These programs, often referred to as *inspection and maintenance programs,* generally involve taking exhaust gas readings as a vehicle's engine operates at idle and at a set rpm. They may also involve checking the vehicle for the presence of a catalytic converter and a fuel inlet restrictor.

Exact test procedures vary from state to state. However, the federal government has passed laws that require stricter emission testing of vehicles in areas with air pollution problems. These tests are referred to as enhanced emissions tests.

IM 240

IM 240 is an enhanced emissions test that requires the vehicle to be operated on a dynamometer at speeds of up to 55 mph (89 kmh) for 240 seconds while exhaust gas emissions are measured. Two additional tests—the evaporative emissions system purge test and the evaporative emission system pressure test—may be required in some areas.

Evaporative Emissions System Purge Test

The *evaporative emissions system purge test* measures the flow of fuel vapors into the engine while performing the IM 240 test. A flow meter transducer is installed into the system purge line between the charcoal canister and the engine intake manifold fitting. A personal computer connects to the flow transducer to analyze data. The computer can then detect if there is adequate purge flow to remove fumes from the canister to burn them in the engine.

Evaporative Emissions System Pressure Test

An *evaporative emissions system pressure test* checks the system for leaks into the atmosphere. It is performed during the IM 240 test. Pressure test equipment is connected to the evaporative emission system's vapor vent line. A computer then meters low pressure nitrogen into the system (evaporative emission system, fuel lines, fuel tank, filler neck, etc.).

When about 0.5 psi (3.4 kPa) pressure is reached, the computer closes off the system and checks for a pressure drop for two minutes. If pressure remains above recommendations, the evaporative emission system passes the pressure test. If pressure drops too much, repairs would be needed to fix the leakage.

The EPA estimates that approximately 25% of the vehicles tested will fail their emission tests. Repair costs to customers to pass the emission tests is capped at a fixed amount in most areas, usually between $100–$500. If the capped repair amount is spent and the vehicle still does not pass, the owner will receive a waiver on further testing for a specified period of time. The emission testing facilities provide a printout of the emission fail records to help technicians repair the problem(s).

Vacuum Solenoid Service

Various vacuum solenoids are used to interface emission system electronics with the devices that operate off of engine vacuum. They can be used in almost all emission control systems.

When trying to find problems, you should refer to a *vacuum hose diagram,* which shows the routing of all

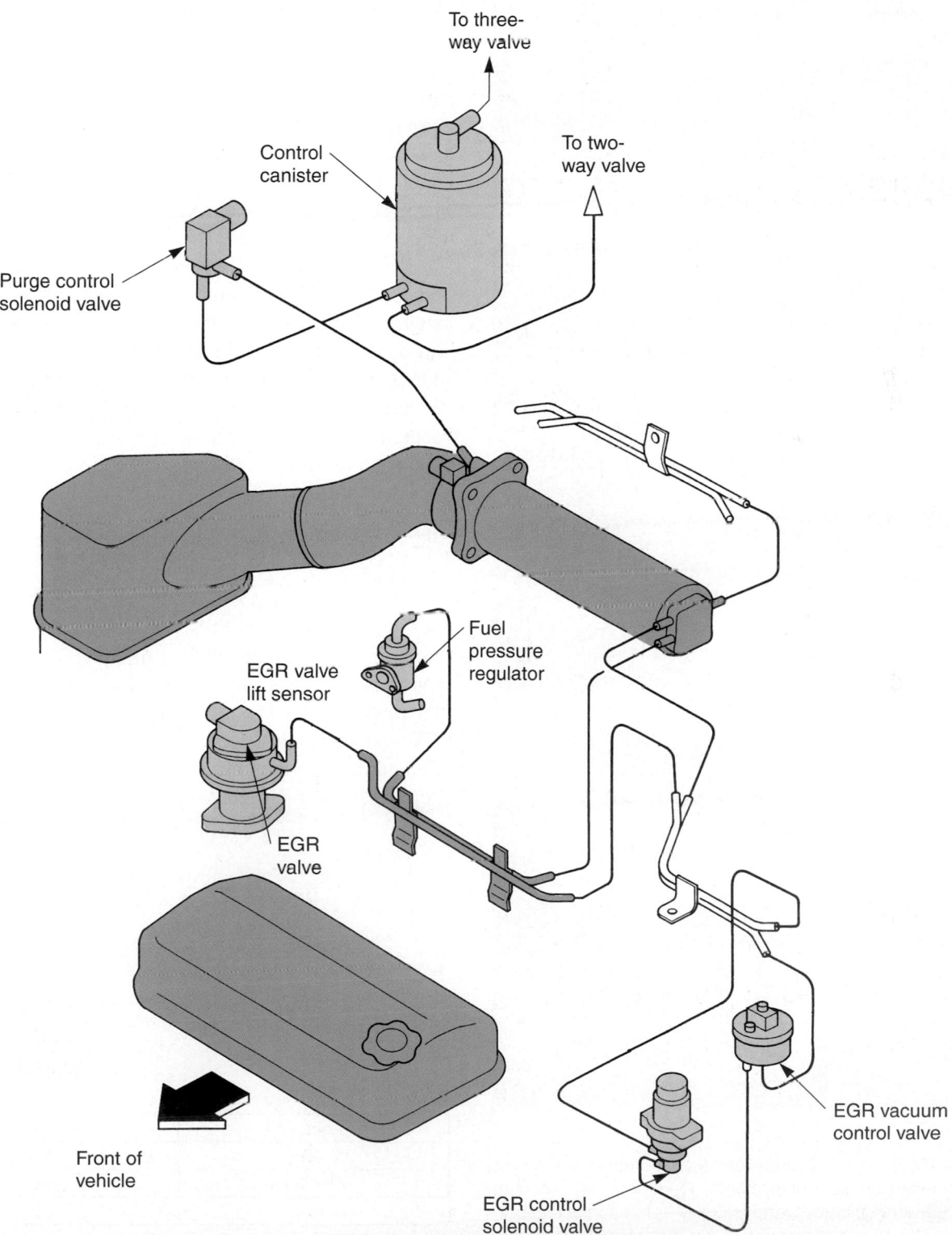

Figure 44-9. Service manual vacuum hose routing diagrams like this one will help you if vacuum lines have been disconnected or come off. They will also identify the purpose of each vacuum solenoid. (Honda)

A

B

C

Figure 44-10. Solenoid vacuum valves are common sources of trouble in emission control systems. A—Check the vacuum hoses for hardening, cracks, and leaks. B—Use a multimeter to check for voltage to the solenoid. If you are not getting control voltage, trace the wiring back to the ECM. C—A vacuum pump should be used to check that the solenoid is turning the vacuum source on and off as designed. (Fluke)

vacuum hoses. Just as a wiring diagram helps you trace circuit problems, a vacuum hose diagram will give useful information on finding incorrectly routed hoses, leaking or restricted hoses, and bad vacuum components. **Figure 44-9** is a sample vacuum diagram from a service manual. Note how the emission devices are connected. The service manual will explain the function and testing of each device.

When troubleshooting vacuum solenoids, check for hard, brittle hoses that can leak and prevent normal operation of parts, **Figure 44-10A.** If the vacuum solenoid is electrically powered, check it for voltage. Connect a voltmeter to the solenoid terminals and start the engine. Make sure you are getting voltage to the unit when needed, **Figure 44-10B.**

You may also need to check that the solenoid valve opens and blocks vacuum as designed. Connect a vacuum gauge or hand pump to the vacuum connections on the unit. When the solenoid is energized and de-energized, it should switch vacuum on and off. See **Figure 44-10C.**

You can also connect a remote source of voltage to a vacuum solenoid to check its operation. When voltage is connected to the solenoid, it would switch vacuum on or off. Look at **Figure 44-11.**

PCV System Service

An inoperative PCV system can increase exhaust emissions. It can also cause engine sludging and wear, rough engine idle, and other problems. A leaking PCV system can cause a vacuum leak and produce a lean air-fuel mixture, causing a rough engine idle. A restricted PCV system can enrich the fuel mixture, affecting engine idle and causing the engine to surge (idle speed goes up and down) and emit black smoke.

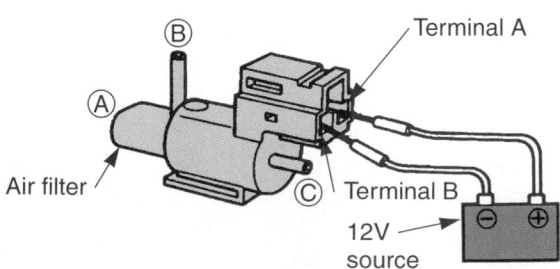

Specification

Port	Airflow
A—B	Yes
A—C	No
B—C	No

Figure 44-11. This illustration shows how you can connect battery voltage to a vacuum solenoid to check its operation. (Mazda)

PCV System Maintenance

Most auto manufacturers recommend periodic maintenance of the PCV system. Inspect the condition of the PCV hoses, grommets, fittings, and breather hoses. Replace any hose that shows signs of deterioration. Clean or replace the breather filter if needed. Also, check or replace the PCV valve. Since replacement intervals vary, always refer to the vehicle's service manual.

PCV System Testing

To quickly test a PCV valve, pull the valve out of the engine and shake it. If the PCV valve does not rattle when shaken, replace the valve. With the engine idling, place your finger over the end of the valve. With airflow stopped, you should feel suction on your finger and the engine idle speed should drop about 40–80 rpm, **Figure 44-12.**

If you cannot feel vacuum, the PCV valve or hose might be plugged with sludge. If engine rpm drops more than 40–80 rpm and the engine begins to idle smoothly, the PCV valve could be stuck open.

A *PCV valve tester* will measure the exact amount of airflow through the system. To use a tester, make sure engine intake manifold vacuum is correct. Then, connect the tester to the engine as described in the operating instructions. Start and idle the engine. Observe the airflow rate on the tester. Replace the PCV valve if airflow is not within specified limits.

Some auto manufacturers suggest placing a piece of paper over the PCV breather opening to test the PCV system. After sealing the dipstick tube with tape, start and idle the engine. After a few minutes of operation, the piece of paper should be pulled down against the breather opening by crankcase vacuum. If suction does not develop, there is a leak in the system (ruptured gasket, cracked hose, etc.) or the system may be plugged.

A four- or five-gas exhaust analyzer can also be used to check the general condition of a PCV system. Measure and note the analyzer readings with the engine idling. Then, pull the PCV valve out of the engine, but not off the hose. Compare the readings after the PCV valve is removed.

A plugged PCV system will show up on the exhaust analyzer when oxygen and carbon monoxide do *not change*. *Crankcase dilution* (excessive blowby or fuel in the oil) will usually show up as an excessive (1% or more) increase in oxygen or a 1% or more decrease in carbon monoxide. This is because the excess crankcase fumes will be pulled into and burned in the engine, affecting your readings.

Evaporative Emissions Control System Service

A faulty evaporative emissions control system can cause fuel odors, fuel leakage, fuel tank collapse (vacuum buildup), excess pressure in the fuel tank, or a rough engine idle. These problems usually stem from a defective fuel tank pressure-vacuum cap, leaking charcoal canister valves, deteriorated hoses, or incorrect hose routing.

Evaporative Emissions Control System Maintenance and Repair

Maintenance on an evaporative emissions control system typically involves cleaning or replacing the filter in the charcoal canister. Service intervals for the canister filter vary. However, if the vehicle is operated on dusty roads, clean or replace the filter more often. Look at **Figure 44-13.**

Also inspect the condition of the fuel tank filler cap. Make sure the cap is installed properly and the seals are in good condition. Special testers are available for checking the opening of the pressure and vacuum valves in the cap. The cap should be tested when excessive pressure or vacuum problems are noticed.

All hoses in the evaporative emissions system should also be inspected for signs of deterioration (hardening, softening, cracking). When replacing a hose, make sure you use special fuel-resistant hose. Vacuum hose can be quickly ruined by fuel vapors.

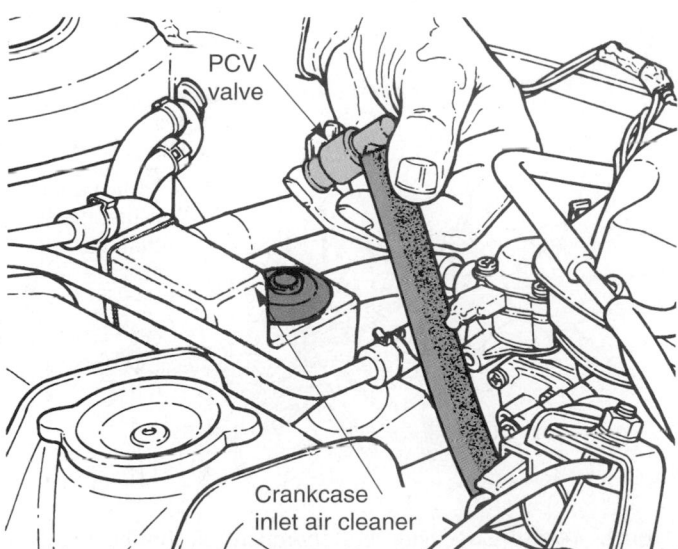

Figure 44-12. With the engine running, place your finger over the PCV valve. There should be suction present; if not, the hose may be plugged. (Chrysler)

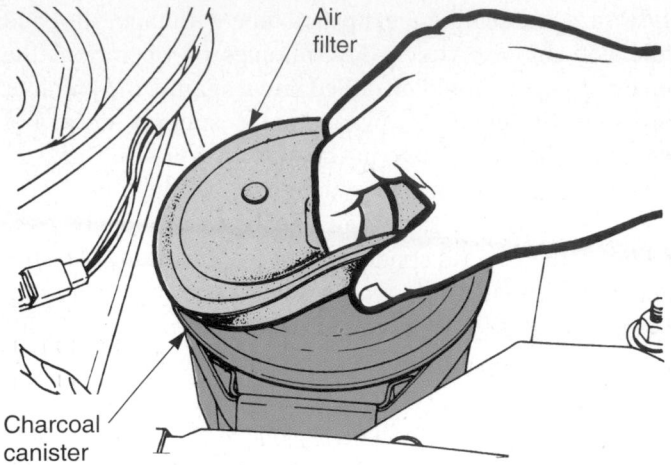

Figure 44-13. Some charcoal canisters have air filters. The filter should be removed and cleaned or replaced at periodic intervals. (Plymouth)

Use a hand vacuum pump to test the charcoal canister vacuum purge solenoids for diaphragm leakage. If a diaphragm will not hold a vacuum, it is ruptured and must be replaced. You can also use the vacuum gauge to check for a vacuum supply to any canister vacuum solenoid. See **Figure 44-14.**

Thermostatic Air Cleaner System Service

An inoperative thermostatic air cleaner system (heated air inlet) can cause several engine performance problems. If the air cleaner flap remains in the *open position* (cold air position), the engine could miss, stumble, stall, and warm up slowly. If the air cleaner flap stays in the *closed position* (hot air position), the engine could perform poorly when at full operating temperature.

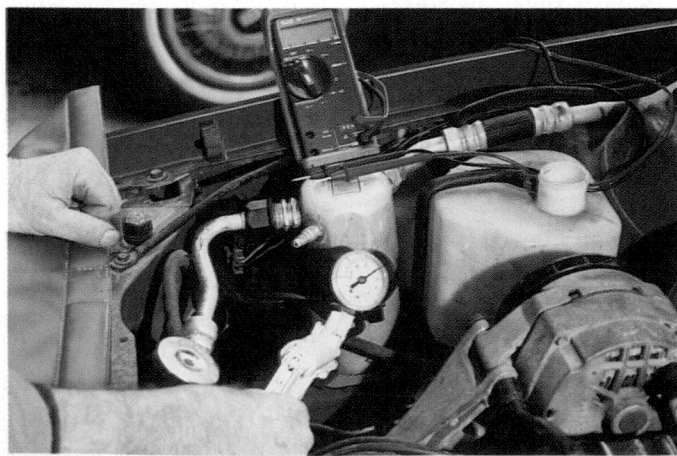

Figure 44-14. A hand vacuum pump can be used to check the operation of canister purge valves. Follow service manual directions as procedures will vary. (Snap-on Tool Corp.)

Thermostatic Air Cleaner System Maintenance

The thermostatic air cleaner system requires very little maintenance. You should inspect the condition of the vacuum hoses and hot air tube from the exhaust manifold heat shroud. The hot air tube is frequently made of heat resistant paper and metal foil. It will tear very easily. If torn or damaged, replace the hot air tube.

Testing Thermostatic Air Cleaner System

For a quick test of the thermostatic air cleaner system, watch the action of the heat control door in the air cleaner snorkel.

Start and idle the engine. When the air cleaner temperature sensor is cold, the door should be open. Place an ice cube on the sensor, if needed. Then, when the engine

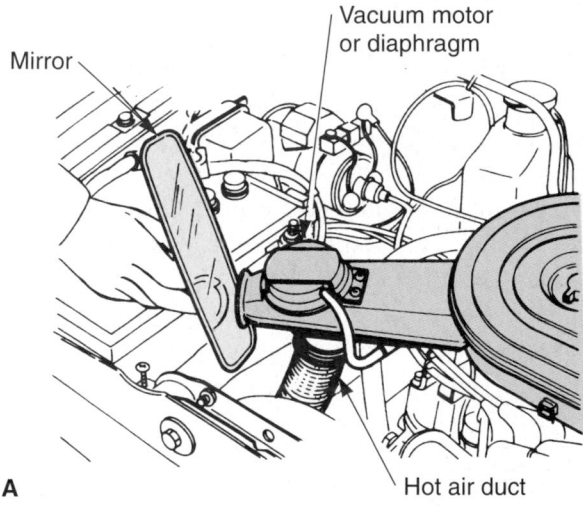

A

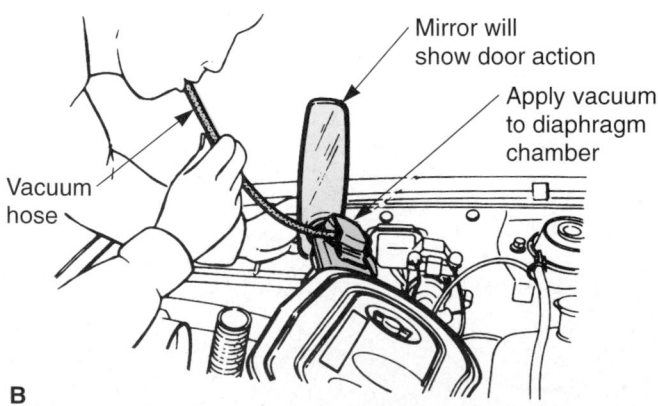

B

Figure 44-15. Checking the operation of thermostatic air cleaner. A—A mirror can be used to watch the action of the door in the air cleaner snorkel. The door should allow heated air to enter during engine warm-up. It should close as the engine warms. B—When you apply vacuum to the vacuum motor, the door should function. (Subaru)

and sensor warm to operating temperature, the door should swing closed. If the air cleaner flap does not function, test the vacuum motor and the temperature sensor. See **Figure 44-15A.**

To test the vacuum motor, apply vacuum to the motor diaphragm with a hand vacuum pump. With the prescribed amount of vacuum, the motor should pull the heat control door open. If the motor leaks or does not open the door, it should be replaced. After replacing the motor, recheck the thermostatic air cleaner system operation to make sure the air temperature sensor is working properly, **Figure 44-15B.**

To test the thermal vacuum valve in the air cleaner, place a thermometer next to the unit. With the valve cooled below its closing temperature, apply vacuum to the thermal vacuum valve. It should pass vacuum to the vacuum motor and the heat control door should open.

Then, warm the thermal vacuum valve to its closing temperature. A heat gun (hair dryer) can be used to heat the unit. When warm, the valve should block vacuum and the heat control door should close. Replace the thermal vacuum valve if the door fails to open and close properly.

EGR System Service

EGR system malfunctions can cause engine stalling at idle, rough idle, detonation, and poor fuel economy. If the EGR valve sticks open, it will cause a lean air-fuel mixture. The engine will run rough at idle or stall. If the EGR fails to open or the exhaust passage is clogged, higher combustion temperatures can cause abnormal combustion (detonation) and knocking.

 Tech Tip
It is fairly common for small rocks to fly into the engine compartments of four-wheel-drive vehicles during off-road driving. If the right size, these small rocks may become lodged in the EGR valve and cause it to stick open. The engine will have a low idle, run rough, or stall. This may or may not trip a trouble code.

EGR System Maintenance

Maintenance intervals for the EGR system vary with vehicle manufacturer. Refer to a service manual for exact mileage intervals. Some vehicles have a reminder light in the dash. The light will glow when EGR maintenance is needed. Also, check that the vacuum hoses in the EGR system are in good condition. They can become hardened, which can cause leakage. Also check for proper wire routing and for good electrical connections on digital EGR valves.

EGR System Testing (Vacuum Type)

To test a vacuum EGR system, allow the engine to warm to operating temperature. Operating the accelerator linkage by hand, increase engine speed to 2000–3000 rpm very quickly. If visible, observe the movement of the EGR valve stem. The stem should move as the engine is accelerated. If it does not move, the EGR system is not functioning.

Sometimes the EGR valve stem is not visible. You will then need to test each EGR system component separately. Follow the procedures described in a service manual.

To test the EGR valve, idle the engine. Connect a hand vacuum pump to the EGR valve. Plug the supply vacuum line to the EGR valve. When vacuum is applied to the EGR valve with the pump, the engine should begin to miss or stall. This lets you know that the EGR valve is opening and that exhaust gases are entering the intake manifold, **Figure 44-16.**

If the EGR valve operation does not affect the engine idle, remove the valve. The valve or the exhaust manifold passage could be clogged with carbon. If needed, clean the EGR valve and exhaust passage. When the EGR valve does not open and close properly, replace the valve.

EGR System Testing (Electronic Type)

Most problems with late-model electronic, or digital, EGR valves will trip a trouble code. Your scan tool with then let you isolate most problems quickly and easily. EGR valves that provide electrical data to a computer control system require special testing procedures. Refer

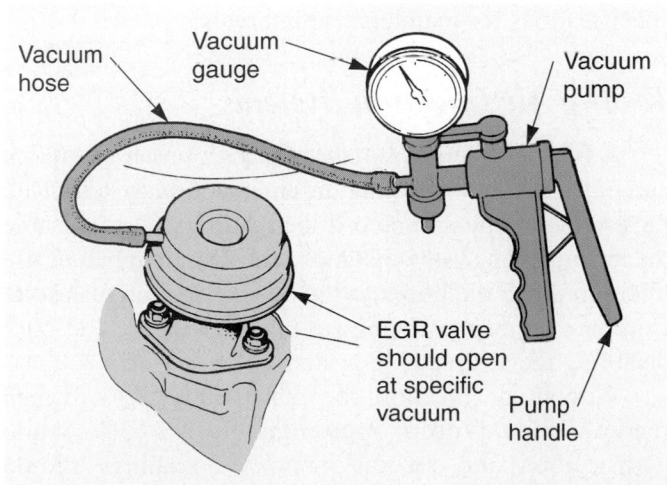

Figure 44-16. With the engine idling, apply vacuum to the EGR valve. If the EGR valve is working, the engine should miss or stall. (Honda)

to a shop manual covering the specific system. Component damage could result from using an incorrect testing method.

The problem symptoms described earlier also apply to a digital EGR. If not working normally, it can cause a rough engine idle, high oxides of nitrogen, and other problems.

To pinpoint test a digital EGR valve, connect a hand-held scope to the wires going to the valve. The scope's waveform will measure the voltage applied to the EGR from the ECM and it will also check the condition of the EGR windings. See **Figure 44-17.** If you do not have voltage to the EGR, check for a bad electrical connection. You could also have an ECM problem in the control circuit to the EGR valve.

Air Injection System Service

Air injection system problems can cause engine backfiring (loud popping sound), other noises, and increased HC and CO emissions. Remember, air injection is used to help burn any fuel that enters the exhaust manifolds and exhaust system. Without this system, the fuel could ignite all at once (backfire) with a loud bang. Insufficient air from the air injection system could also prevent the catalytic converter from functioning properly.

Air Injection System Maintenance

Air injection system maintenance typically includes replacing the pump inlet filter (if used), adjusting pump belt tension, and inspecting the condition of the hoses and lines.

If the pump belt or any hoses show signs of deterioration, they should be replaced. Refer to shop manual specifications for maintenance intervals.

Testing Air Injection Systems

A four- or five-gas exhaust analyzer provides a quick and easy method of testing an air injection system. Run the engine at idle and record the readings. Then, disable the air injection system. Remove the air pump belt or use pliers to pinch the hoses to the air distribution manifold. Compare the exhaust analyzer readings before and after disabling the air injection system.

Without air injection, the exhaust analyzer's oxygen reading should drop approximately 2%–5%, while hydrocarbon and carbon monoxide readings should increase. This would show that the air injection system is forcing air (oxygen) into the exhaust system. If the analyzer readings do *not* change, the air injection system is

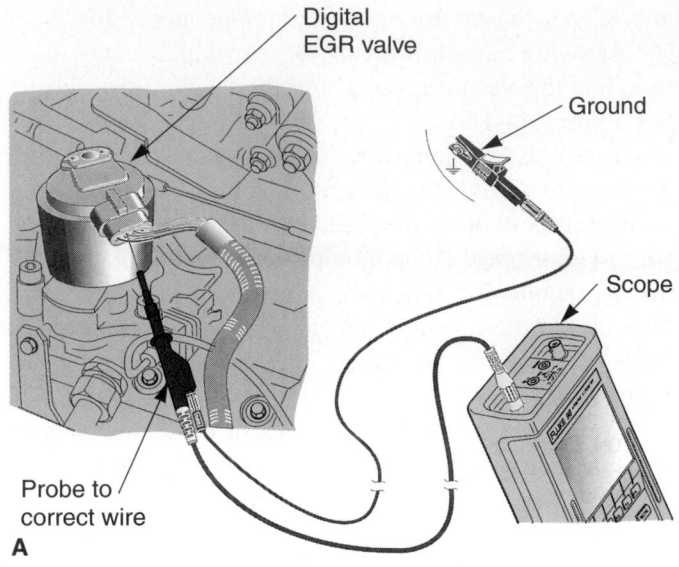

A

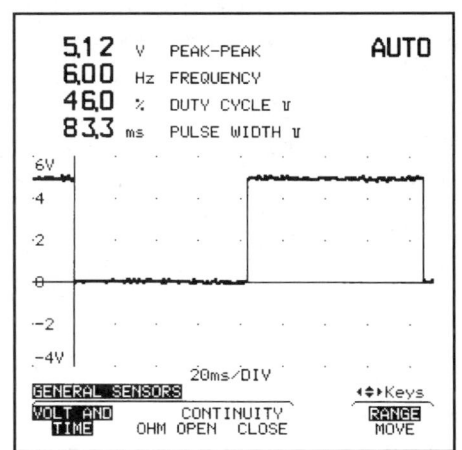

B

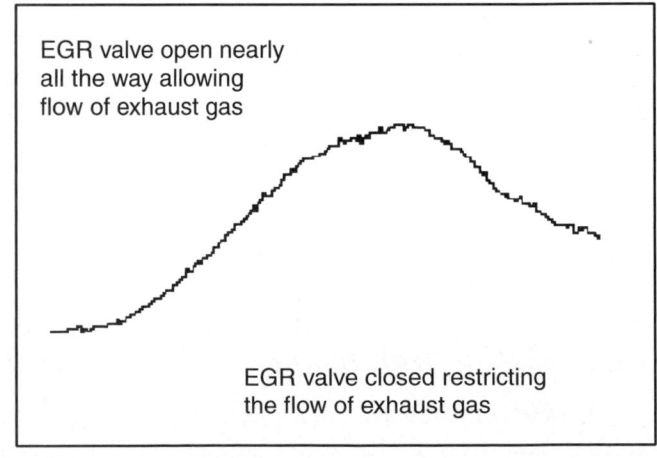

C

Figure 44-17. An oscilloscope can be used to check digital EGR valves and their ECM control circuits. A—Connect the scope to ground and probe through the EGR valve connector. The service manual wiring diagram will tell you which wires to probe. B—The scope should show a normal signal going to the EGR valve. C—A scope can also be connected to read the return signal from the EGR position sensor. (Fluke)

not functioning. Test each component until the source of the problem is found.

To test the air pump, remove the output line from the pump. Use a low-pressure gauge to measure the amount of pressure developed by the pump at idle. Typically, an air pump should produce about 2-3 psi (14-21 kPa) of pressure. See **Figure 44-18.**

If a low-pressure gauge is not available, place your finger over the line and check for pressure. Replace the pump if faulty. When testing the diverter valve or other air injection system valves, refer to a service manual. It will explain testing procedures for the specific components.

Pulse Air System Service

Many of the maintenance and testing methods discussed for an air injection system apply to the pulse air system. Inspect all hoses and lines and measure the exhaust's oxygen content with a four- or five-gas analyzer. Exhaust analyzer oxygen readings should drop when the pulse air system is disabled. If readings do not drop, check the action of the aspirator (reed) valves.

With the engine running, you should be able to feel vacuum pulses on your fingers. However, you should *not* feel exhaust pressure pulses trying to blow back through the valves. Replace the valves if they do not function as designed.

Catalytic Converter Service

Catalytic converter problems are commonly caused by contamination, overheating, and extended service. A *clogged catalytic converter*, resulting from deposits or overheating, can increase exhaust system back pressure. High back pressure decreases engine performance because gases cannot flow freely through the converter, **Figure 44-19.**

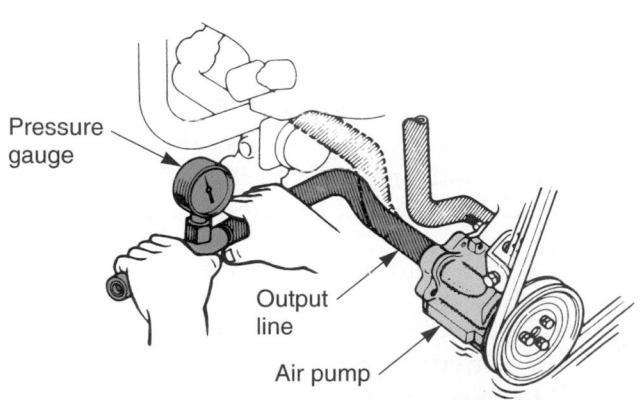

Figure 44-18. A pressure gauge can be used to check air pump output. If the pump's pressure is not within specifications, replace the pump. (Toyota)

Pressure gauge

Output line

Air pump

Figure 44-19. This badly damaged catalytic converter would block the flow of exhaust gas through the system and prevent normal engine operation. (Champion Spark Plug Co.)

 Tech Tip
A clogged catalytic converter is a fairly common problem. The increased back pressure will reduce engine power considerably. You may notice a rotten egg smell at the tailpipe. A clogged converter can also overheat, possibly causing a fire.

An *exhaust back pressure test* will check for a clogged catalytic converter and other system parts. Remove the front oxygen sensor and install a pressure gauge into the threaded hole, **Figure 44-20.**

Start the engine. Read the pressure gauge at idle and at higher speeds. If the pressure gauge reads too high, the converter, muffler, or an exhaust pipe is restricted. To isolate the exhaust restriction, disconnect parts one at a time. When the back pressure drops, you have found the source of the restriction.

After extended service, the catalyst in the converter can become coated with deposits. These deposits can keep the catalyst from acting on the hydrocarbon, carbon monoxide, and oxides of nitrogen. The inner baffles and shell can also deteriorate. With a pellet-type catalytic converter, this can allow BB-size particles to blow out the tailpipe.

Pellet catalytic converters normally have a plug that allows replacement of the catalyst agent. The old pellets can be removed and new ones installed. If the converter housing is damaged or corroded, replace the

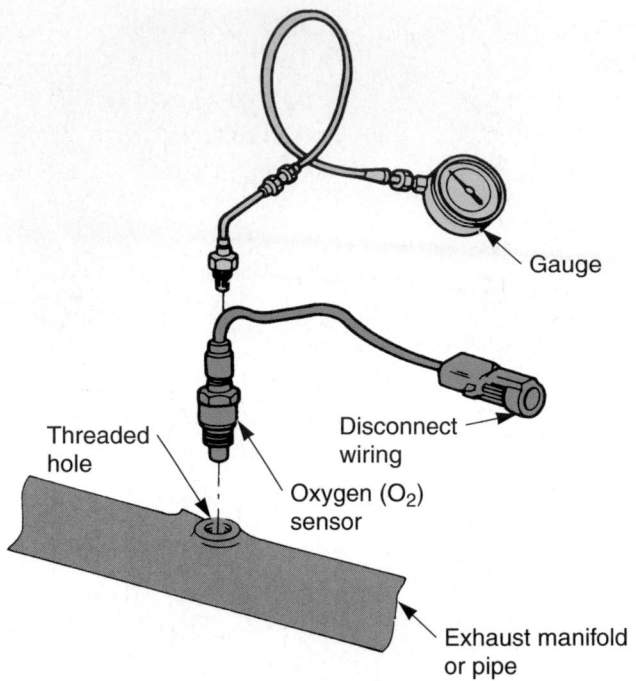

Figure 44-20. If you suspect a clogged catalytic converter or other exhaust system obstruction, test the exhaust system's back pressure. Remove an oxygen sensor before the converter and install a pressure gauge into the sensor hole. Start the engine and compare the pressure readings to specifications. (Snap-on Tool Corp.)

converter. Monolithic (honeycomb) catalytic converters must be replaced when the catalyst becomes damaged or contaminated.

Tech Tip

Before condemning a catalytic converter, refer to a factory service manual. It will give added information on checking other systems before converter replacement.

Testing Catalytic Converter Efficiency

An exhaust gas analyzer can be used to check the general condition of the catalytic converter. Follow the specific directions provided with the analyzer. Warm and idle the engine. With some systems, you may need to disable the air injection or pulse air system before performing this test. Measure the oxygen and carbon monoxide at the tailpipe.

If oxygen readings are above approximately 5%, you know there is enough oxygen for the catalyst to burn the emissions. However, if the carbon monoxide readings are still above about 0.5% (other systems operating properly), then the catalytic converter is not oxidizing the

emissions from the engine and the converter or catalyst requires replacement.

Tech Tip

Your scan tool can be used to diagnose catalytic converter problems on OBD II vehicles. The catalytic converter's condition is monitored by measuring its oxygen storage capacity using two oxygen sensors—a pre-converter sensor and a post-converter sensor. Under normal conditions, the pre-converter oxygen sensor switches frequently and the post-converter sensor seldom switches. The pre-converter sensor switches more frequently because it "smells emissions," while the post-converter sensor "sniffs" cleaner gases.

Catalyst Replacement

To install new pellets in a catalytic converter, follow service manual instructions. You must use a special vibrating tool to shake the old pellets out of a hole in the converter. Then, new pellets are installed and the service plug is replaced in the converter housing. This procedure is not used frequently, since it is faster and easier to simply replace the catalytic converter.

Warning

Remember that the operating temperature of a catalytic converter can be over 1400°F (760°C). This is enough heat to cause serious burns. Do *not* touch a catalytic converter until you are sure it has cooled.

Catalytic Converter Replacement

On many vehicles, the converter can be unbolted from the exhaust system. Remove the clamps that secure the converter to the exhaust pipes. Then use a muffler cutter or a chisel to cut and loosen the old converter from the exhaust pipes. Hammer blows to the converter should then free it from the vehicle.

Sometimes the catalytic converter is an integral part of the header pipe. With this design, the converter and pipe may have to be replaced together. When installing the new converter, use new gaskets and reinstall all heat shields, as in **Figure 44-21.**

Tech Tip

After replacing a catalytic converter, turn in the old converter to be recycled. The platinum and other precious elements are very expensive and are becoming increasingly difficult to find.

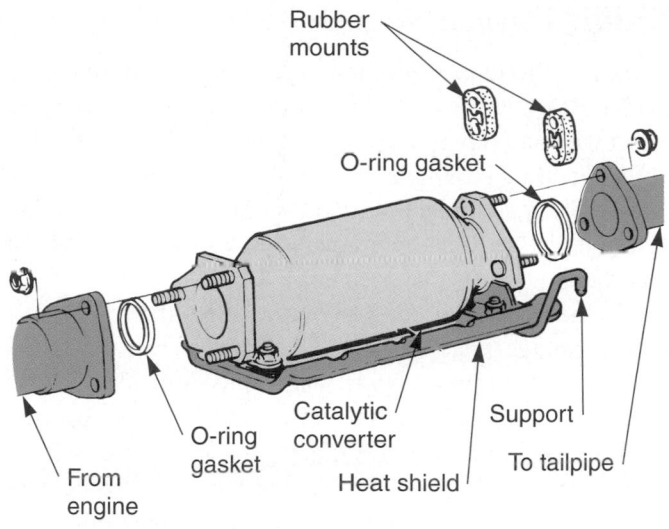

Figure 44-21. When installing a new catalytic converter, replace all gaskets, rubber mounts, mounting hardware, and make sure heat shield is in place. (Honda)

Oxygen Sensor Service

After prolonged service, oxygen sensors become coated or fouled with exhaust byproducts. As this happens, fuel economy and emissions may be adversely affected. If gas mileage is 10% to 15% lower than normal, suspect a lazy oxygen sensor. A *lazy O_2 sensor* will not alter its output signal fast enough to maintain an efficient air-fuel ratio. The sensor will be slow to change its voltage or resistance with changes in exhaust content.

A bad oxygen sensor will affect engine performance and emissions. If it does not work properly, fuel metering will be too lean or rich. A bad rear oxygen sensor (catalyst monitor) may not detect an inoperative catalytic converter.

A *dead O_2 sensor* has little or no resistance or voltage output change. Even when the exhaust content changes, the signal remains almost constant.

Depending on the driving conditions, the O_2 voltage will rise and fall, but it usually averages around 0.450V dc.

1. Shut off the engine and insert test lead in the input terminals shown.
2. Set the rotary switch to volts dc.
3. Manually select the 4V range by depressing the range button three times.
4. Connect the test leads as shown
5. Start the engine. If the O_2 sensor is unheated, fast-idle the car for a few minutes. Then press MIN MAX to select MIN MAX recording.
6. Press MIN MAX button to display maximum (MAX) O_2 voltage; press again to display minimum (MIN) voltage; press again to display average (AVG) voltage; press and hold down MIN MAX for 2 seconds to exit.

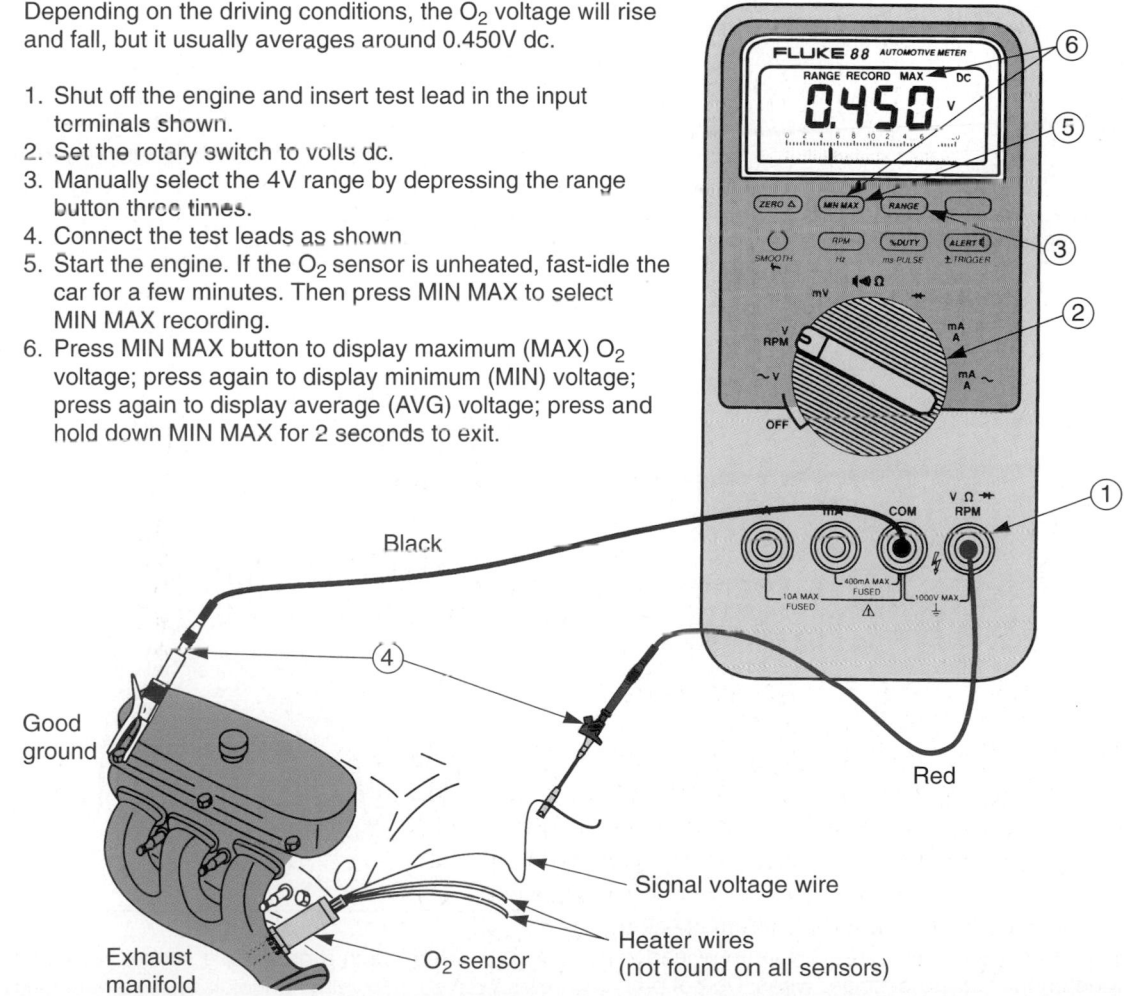

Figure 44-22. If the scan tool indicates a possible oxygen sensor problem, you should use a multimeter to check the sensor's actual output. Refer to the service manual for detailed procedures. (Fluke)

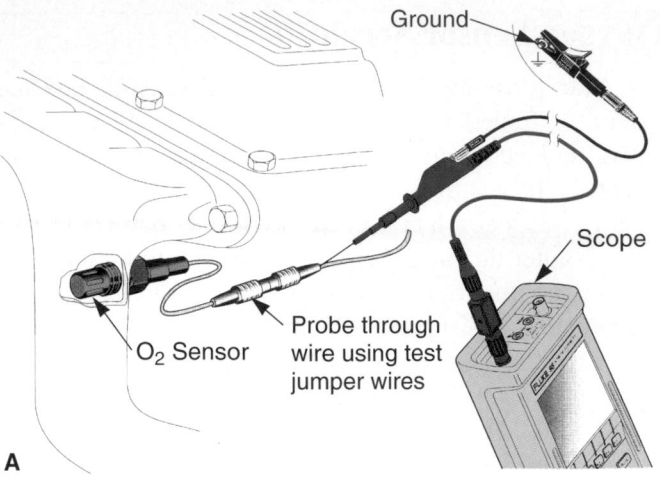

A

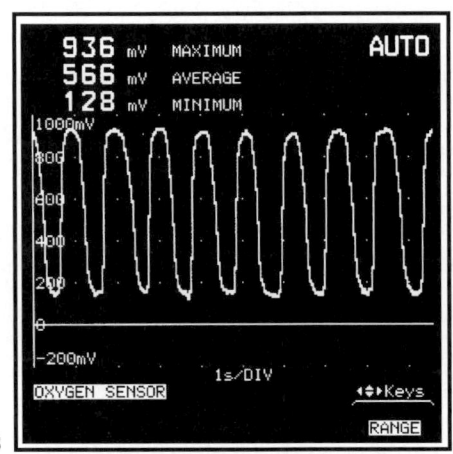

B

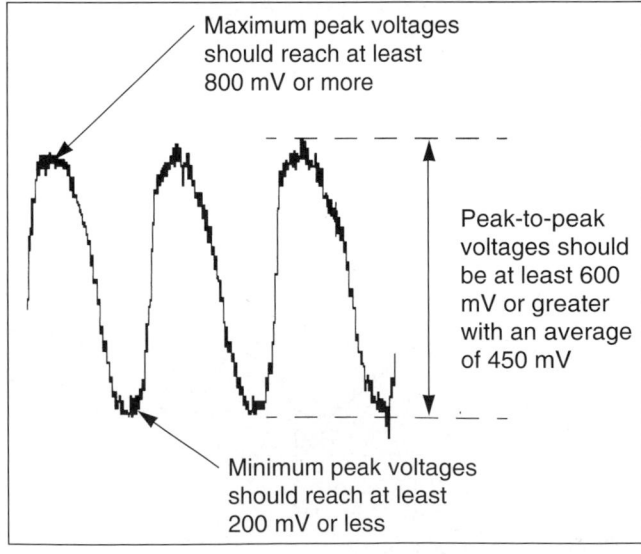

Maximum peak voltages should reach at least 800 mV or more

Peak-to-peak voltages should be at least 600 mV or greater with an average of 450 mV

Minimum peak voltages should reach at least 200 mV or less

C

Figure 44-23. An oscilloscope can also be used to check the output signal from an oxygen sensor. A—Ground one lead and probe through the connector using jumper wires. Do not pierce the oxygen sensor wiring. B—The oxygen sensor's signal output should vary as shown. Also note the millivolt readout on the scope face. C—This is a typical waveform for a zirconia O_2 sensor. (Fluke)

Testing Oxygen Sensors

Most O_2 sensor problems will trip a trouble code with OBD I and OBD II systems. However, there are times when an oxygen sensor will be close to but not out of its operating parameter and will not trip a trouble code. Even if the scan tool shows a problem with the O_2 sensor, pinpoint tests will also be needed to verify the source of the trouble.

If the scan tool readout shows that the O_2 sensor output voltage is abnormal, you might want to measure the sensor's output voltage with a multimeter. This is shown in **Figure 44-22.** An oscilloscope can also be used to check the signal leaving the O_2 sensor, **Figure 44-23.** When testing a titania oxygen sensor, it may also be necessary to check the reference voltage supplied by the ECM, as well as the sensor's resistance or voltage drop. Refer to the service manual for details.

By comparing actual voltage (zirconia-type sensor) or resistance levels (titania-type sensor) to scan tool readout values and manufacturer's specifications, you can determine whether the sensor, wiring, or ECM is at fault.

If you are using a dual trace scope (scope has two test leads and can display two separate waveforms at once), you can compare the signals from the front and rear oxygen sensors on OBD II vehicles. If the voltage levels from the sensors are too similar, you may have a faulty or inert catalytic converter, **Figure 44-24.**

If you have trouble isolating an oxygen sensor-related problem, refer to the factory service manual. It will give specific information to help find the source of the problem, **Figure 44-25.**

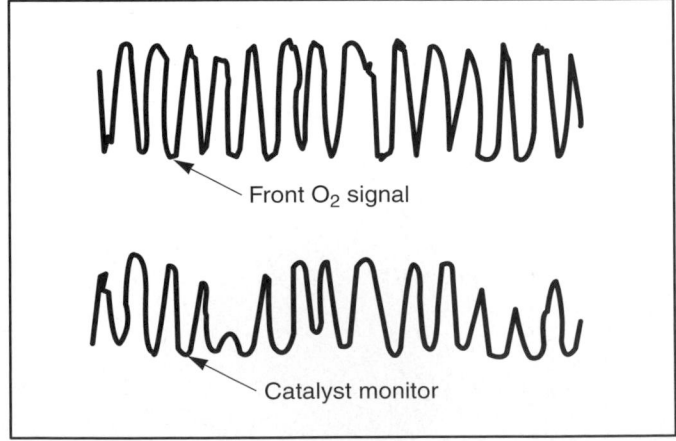

Front O_2 signal

Catalyst monitor

Figure 44-24. With a dual trace scope, you can compare signals from each oxygen sensor in systems with multiple sensors. If the waveform from the catalyst monitor is too similar to the signal produced by the front sensor(s), the catalytic converter is not functioning and should be replaced. You should also investigate common causes of converter damage. (Oldsmobile)

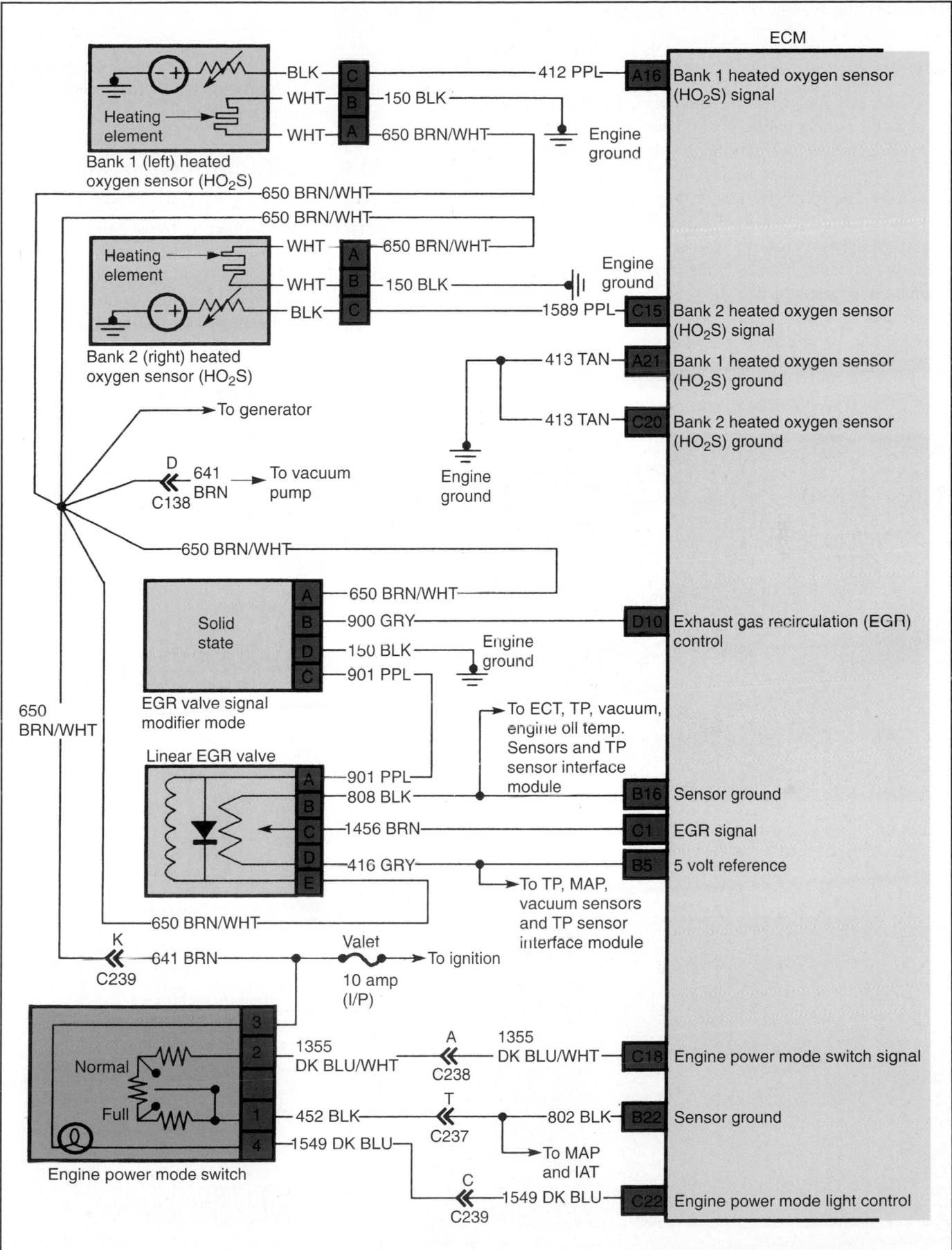

Figure 44-25. When emission system problems are hard to find, refer to the service manual wiring diagrams. They will allow you to see the components in the system as well as provide pinpoint tests of wiring and individual components. (Chevrolet)

Oxygen Sensor Replacement

If the oxygen sensor is defective, first disconnect the negative battery cable. Then, separate the sensor from the wiring harness by unplugging the connector. Never pull on the wires themselves, as damage may result, **Figure 44-26.** Spray the sensor threads with a generous coat of penetrating oil. Use a special sensor socket to remove the sensor from the exhaust system. Use care to avoid thread damage. Inspect the sensor for signs of contamination, **Figure 44-27.**

Obtain and install the correct replacement oxygen sensor. Coat the threads of the new sensor with antiseize compound and start the sensor by hand. Then tighten the sensor with a wrench or socket. Do not overtighten and damage the sensor during installation. Reconnect the wire connector and check system operation.

Emission Control Information Sticker

The *emission control information sticker,* or label, gives important engine data, evaporative emissions information, schematics, and other specifications for complying with EPA regulations. Study the information given on the label or sticker. The emission control sticker is normally located on the underside of the hood, on the radiator support, or on a valve cover. One is shown in **Figure 44-28.**

OBD II Drive Cycle

The *OBD II drive cycle* is normally performed whenever the battery or ECM has been disconnected, or after diagnostic trouble codes have been erased. Additionally, some states require the drive cycle to be performed before an emissions test.

Figure 44-26. Care must be taken when disconnecting the oxygen sensor connector from the wiring harness. (Fel-Pro)

Figure 44-27. This oxygen sensor became lazy because of a coating of carbon soot caused by a fuel injector that was leaking.

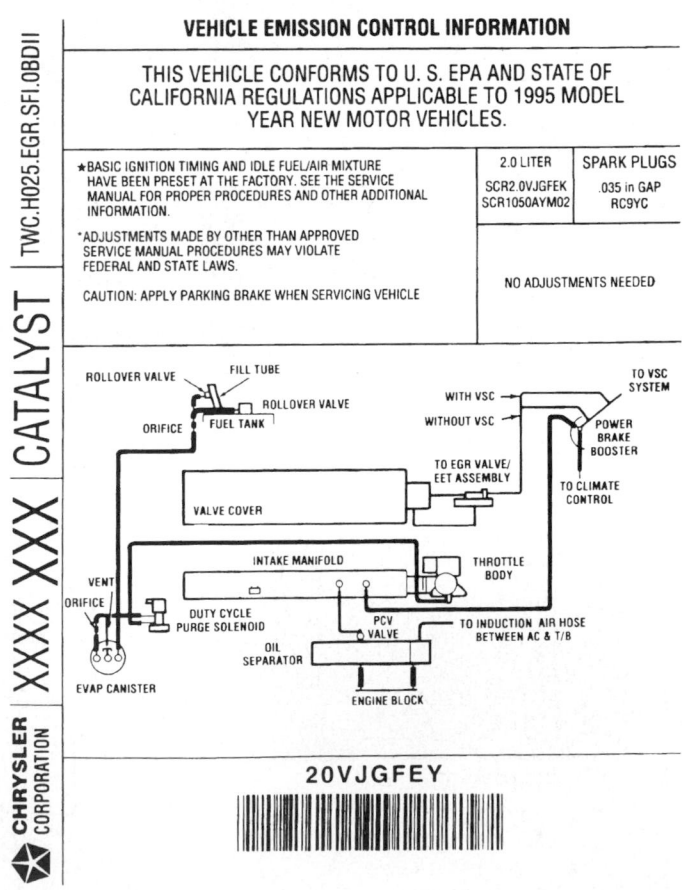

Figure 44-28. Typical emission control information sticker found in the engine compartment. (Chrysler)

The drive cycle is designed to tell the technician whether the OBD II system is operating properly. It involves attaching a scan tool to the vehicle and driving the vehicle for a specified period of time. The drive cycle will include periods of acceleration, cruising, and deceleration.

Performing the OBD II Drive Cycle

 Tech Tip
Carefully study the drive cycle procedure and scan tool operation before starting the drive cycle. You should be reasonably sure that you can complete the drive cycle from beginning to end. If the drive cycle has to be aborted for any reason, the engine must be allowed to cool, which can cause a considerable delay.

The drive cycle is different for each vehicle, so check the service manual for specific instructions. To begin the drive cycle, check that the coolant temperature is low enough to allow the ECM to start in the open loop mode. On most engines, coolant temperature should be below 120°F (49°C). Be sure to allow for variations in coolant temperature sensor calibration. In warm climates, the cooldown period can take as long as six hours. If possible, allow the vehicle to sit overnight before starting the drive cycle.

Attach the scan tool, and set the tool to record the ECM status as the engine operates. Some scan tools have a dedicated drive cycle option. A typical drive cycle will cover engine warm-up, idling, accelerating, decelerating,

and cruising, in a specific order, **Figure 44-29.** This order must be followed as outlined in the service manual. Some scan tools will prompt the technician throughout the drive cycle. After ensuring that the engine temperature is low enough to start the drive cycle, start the vehicle and complete the drive cycle sequence as outlined in the service manual.

A typical drive cycle will take from 8-15 minutes to complete, depending on the manufacturer. Some drive cycles require the technician to turn the air conditioning on and off at certain times, to decelerate without braking, and to decelerate with the manual transmission clutch engaged or released, depending on the portion of the cycle being performed. Some state air quality programs eliminate the warm-up portion of the drive cycle, as it is impractical to allow the vehicle to cool off before testing.

Performing the drive cycle with the vehicle on a chassis dynamometer will allow the scan tool to gather readings in the shortest possible time. If the vehicle is driven on the road, it may be impossible to complete the drive cycle exactly as designed. Therefore, some scan tools can be paused when the drive cycle must be delayed. However, if the engine is shut off for any reason, the drive cycle must be restarted from the beginning.

The scan tool will indicate when the drive cycle is complete, *not* whether the vehicle passed or failed. Any malfunctions will be stored as trouble codes in the scan tool. Check for stored trouble codes and make further diagnostic checks and repairs as needed. After repairs are complete, repeat the drive cycle to ensure that the vehicle is repaired.

TYPICAL OBD II DRIVE CYCLE

Diagnostic time schedule for I/M readiness	
Vehicle drive status	**What is monitored?**
Cold start, coolant temperature less than 122°F (50°C)	—
Idle 2.5 minutes in drive (auto) neutral (man), A/C and rear defogger ON	HO$_2$S heater, misfire, secondary air, fuel trim, EVAP purge
A/C off, accelerate to 90 km/h (55mph), 1/2 throttle	Misfire, fuel trim, purge
3 minutes of steady state–cruise at 90 km/h (55 mph)	Misfire, EGR, secondary air, fuel trim, HO$_2$S, EVAP purge
Clutch engaged (man), no braking, decelerate to 32 km/h (20 mph)	EGR, fuel trim, EVAP purge
Accelerate to 90–97 km/h (55–60 mph), 3/4 throttle	Misfire, fuel trim, EVAP purge
5 minutes of steady state cruise at 90–97 km/h (55–60 mph)	Catalyst monitor, misfire, EGR, fuel trim, HO$_2$S, EVAP purge
Decelerate, no braking. End of drive cycle	EGR, EVAP purge
Total time of OBD II drive cycle 12 minutes	—

Figure 44-29. This chart shows the OBD II drive cycle for a particular vehicle. Each vehicle and manufacturer has its own drive cycle procedure. Be sure to read and understand each part of the procedure before starting the drive cycle. (General Motors)

Summary

- The vehicle's computer or ECM will monitor many functions that affect emissions.

- Evaporative emission system monitoring checks components for leakage and restrictions that could increase emissions.

- Air injection system monitoring uses data from the rear O_2 sensor to determine if the right amount of air (oxygen) is being injected into the engine's exhaust stream.

- Fuel injector monitoring has the ECM check the coil windings of each injector to check for opens, shorts, or an abnormal voltage feedback signal.

- Emission system scanning involves using a scan tool to check the condition of monitored parts of the various systems.

- The emission maintenance reminder is a circuit that automatically turns on a dash light to indicate the need for emission control system service.

- An exhaust analyzer is a testing instrument that measures the chemical content of engine exhaust gases.

- The five-gas exhaust analyzer will measure HC, CO, CO_2, O_2, and NO_x.

- Exhaust gas analyzer calibration involves adjusting the needles or digital readouts to measure correctly.

- When using an exhaust analyzer, do not let engine exhaust fumes escape into an enclosed shop area.

- The evaporative system purge test measures the flow of fuel vapors into the engine while performing the IM 240 test.

- EGR system malfunctions can cause engine stalling at idle, rough engine idling, detonation (knock), and poor fuel economy.

- Air injection system problems can cause engine backfiring (loud popping sound), other noises, and increased hydrocarbon and carbon monoxide emissions.

- A clogged catalytic converter can increase exhaust system back pressure.

- With OBD II equipped vehicles, the catalytic converter's condition is monitored by measuring its oxygen storage capacity using two oxygen sensors.

Important Terms

Engine misfire monitoring
Sensor monitoring
Actuator monitoring
Emission system scanning
Emission maintenance reminder
Combustion efficiency
Exhaust gas analyzer
Calibration
Calibration gas
Emissions testing program
Inspection and maintenance programs
IM 240
Evaporative emissions system purge test
Evaporative emissions system pressure test
Vacuum hose diagram
PCV valve tester
Crankcase dilution
Exhaust back pressure test
Lazy O_2 sensor
Dead O_2 sensor
Emission control information sticker
OBD II drive cycle

Review Questions—Chapter 44

Please do not write in this text. Place your answers on a separate sheet of paper.

1. A(n) _____ _____ is a testing instrument that measures the chemical content of the engine exhaust gases.

2. What is the difference between two-, four- and five-gas exhaust analyzers?

3. Name five reasons that HC readings can be higher than normal.

4. High CO readings can be caused by fuel system problems. True or False?

5. With a five-gas exhaust analyzer, oxygen in the engine exhaust is an accurate indicator of air-fuel mixture. True or False?

6. Typically, if the levels of CO exceeds the O_2 levels, the air-fuel mixture is:
    (A) lean.
    (B) rich.
    (C) stoichiometric.
    (D) None of the above.

7. How do you check the general operation of a PCV valve with your finger and with an exhaust gas analyzer?

8. If a vacuum pump is used to activate an EGR valve at idle, the engine should *not* be affected. True or False?

9. An air pump in an air injection system should produce about _____ to _____ psi or _____ to _____ kPa.

10. Describe the results of a clogged catalytic converter.

11. The operating temperature of a catalytic converter may be over _____ or _____.

12. A driver complains that a malfunction indicator light glows in the dash. Technician A says that an ohmmeter should be used to measure the resistance of the computer system oxygen sensor. Technician B says that a scan tool should be used to check for stored trouble codes. Who is right?
    (A)   A only.
    (B)   B only.
    (C)   Both A and B.
    (D)   Neither A nor B.

13. What is the purpose of the OBD II drive cycle?

14. How do modern OBD II systems check the operation of the EGR valve?

15. The emission control _____ _____ gives important instructions, diagrams, and specifications for complying with EPA regulations.

✹ ASE-Type Questions

1. An exhaust gas analyzer will indicate each of the following *except:*
   (A)   *ignition system problems.*
   (B)   *faulty air injection system.*
   (C)   *stored trouble code.*
   (D)   *engine mechanical problems.*

2. Which of the following is *not* measurable by a four-gas exhaust analyzer?
   (A)   *Oxygen.*
   (B)   *Hydrocarbons.*
   (C)   *Carbon dioxide.*
   (D)   *Oxides of nitrogen.*

3. An exhaust gas analyzer measures hydrocarbons by volume in :
   (A)   *percentages*
   (B)   *parts per million*
   (C)   *grams per mile*
   (D)   *All of the above.*

4. The term "stoichiometric" indicates that a fuel mixture is:
   (A)   *chemically correct.*
   (B)   *too rich.*
   (C)   *too lean.*
   (D)   *contaminated.*

5. While using an exhaust gas analyzer, Technician A says that hydrocarbons and carbon monoxide must be measured at idle and at about 2500 rpm. Technician B says oxygen and carbon dioxide must be measured at these specifications. Who is right?
   (A)   *A only.*
   (B)   *B only.*
   (C)   *Both A and B.*
   (D)   *Neither A nor B.*

6. A faulty evaporative emissions control system can cause:
   (A)   *fuel odors.*
   (B)   *fuel leakage.*
   (C)   *rough engine idle.*
   (D)   *All of the above.*

7. This is *not* something the engine may do if t he air cleaner flap of a thermostatic air cleaner system remains in the open position.
   (A)   *Miss.*
   (B)   *Stall.*
   (C)   *Stumble.*
   (D)   *Not start.*

8. Air injection system problems can cause each of these *except:*
   (A)   *engine backfiring.*
   (B)   *increased hydrocarbon emissions.*
   (C)   *increased carbon dioxide emissions.*
   (D)   *increased carbon monoxide emissions.*

9. Technician A says that catalytic converter problems are commonly caused by contamination. Technician B says that catalytic converter problems are often caused by extended service. Who is right?
   (A)   *A only.*
   (B)   *B only.*
   (C)   *Both A and B.*
   (D)   *Neither A nor B.*

10. An oxygen sensor will not alter its output signal fast enough to maintain an efficient air-fuel ratio. Technician A says the sensor is shorted. Technician B says the sensor is lazy. Who is right?
    (A) A only.
    (B) B only.
    (C) Both A and B.
    (D) Neither A nor B.

Activities—Chapter 44

1. Set up an exhaust gas analyzer correctly, following the manufacturer's instructions. Demonstrate to your instructor and to the class the proper procedure for checking emission levels. If there are high emission levels, write your recommendations for service of any problems.

2. If a flat rate manual and parts catalog are available, prepare a bill for the service and repair in Activity 1. Use a labor rate suggested by your instructor.

Emission Control System Diagnosis		
Condition	**Possible Causes**	**Correction**
Excessive hydrocarbon reading.	1. Poor cylinder compression. 2. Leaking head gasket. 3. Ignition misfire. 4. Poor ignition timing. 5. Defective input sensor. 6. Defective output sensor. 7. Defective ECU. 8. Open EGR valve. 9. Sticking or leaking injector. 10. Improper fuel pressure. 11. Leaking fuel pressure regulator. 12. Oxygen sensor contaminated or responding to artificial lean or rich condition. 13. Improperly installed fuel filler cap.	Test components. Service or replace as necessary.
Excessive carbon monoxide reading.	1. Plugged air filter. 2. Engine carbon loaded. 3. Defective sensor. 4. Defective ECM. 5. Sticking or leaking injector. 6. Higher than normal fuel pressure. 7. Leaking fuel pressure regulator. 8. Oxygen sensor contaminated or responding to artificial lean or rich condition.	Test components. Service or replace as necessary.
Excessive hydrocarbon and carbon monoxide readings.	1. Plugged PVC valve or hose. 2. Fuel-contaminated oil. 3. Heat riser stuck open. 4. Air pump disconnected or defective. 5. Evaporative emissions canister saturated. 6. Evaporative emissions purge valve stuck open. 7. Defective throttle position sensor.	Test components. Service or replace as necessary.
Excessive oxides of nitrogen.	1. Vacuum leak. 2. Leaking head gasket. 3. Engine carbon loading. 4. EGR valve not opening. 5. Injector not opening. 6. Low fuel pressure. 7. Low coolant level. 8. Defective cooling fan or fan circuit. 9. Oxygen sensor grounded or responding to an artificial rich condition. 10. Fuel contaminated with excess water.	Test components. Service or replace as necessary.
Excessively low carbon dioxide reading.	1. Exhaust system leak. 2. Defective input sensor. 3. Defective ECU. 4. Sticking or leaking injector. 5. Higher-than-normal fuel pressure. 6. Leaking fuel pressure regulator.	Test components. Service or replace as necessary.

(Continued)

Emission Control System Diagnosis		
Condition	**Possible Causes**	**Correction**
Low oxygen reading.	1. Plugged air filter. 2. Engine carbon loaded. 3. Defective input sensor. 4. Defective ECU. 5. Sticking or leaking injector. 6. Higher-than-normal fuel pressure. 7. Leaking fuel pressure regulator. 8. Oxygen sensor contaminated or responding to artificial lean condition. 9. Defective evaporative emission system valve.	Test components. Service or replace as necessary.
High oxygen reading.	1. Vacuum leak. 2. Low fuel pressure. 3. Defective input sensor. 4. Exhaust system leak.	Test components. Service or replace as necessary.

Many vehicles use multiple catalytic converters. Note the two converters by the transfer case. (Mercedes-Benz)

Section 8

Engine Performance

45. Engine Performance Problems
46. Analyzers, Scope Testing
47. Engine Tune-Up, Advanced Diagnosis

Peak engine performance not only keeps the customer happy by providing maximum engine power, but it is also essential to reducing engine-related emissions.

Section 8 will describe common engine performance problems, detail their causes, and explain how each can be corrected. It explains how to use advanced diagnostic tools to find problems in the engine and related systems. The final chapter in this section explains how to tune an engine and troubleshoot complex electric-electronic problems.

The information in this section will give you the "state-of-the-art" skills needed to succeed as an automotive technician. It will also help you pass ASE Test A6, *Electrical/Electronic Systems,* and Test A8, *Engine Performance.*

Chapter 45

Engine Performance and Driveability

After studying this chapter, you will be able to:

- List the most common engine performance problems.
- Describe the symptoms for common engine performance problems.
- Explain typical causes of engine performance problems.
- Use a systematic approach when diagnosing engine performance problems.
- Correctly answer ASE certification test questions on problems affecting engine performance.

An *engine performance problem* is any trouble that affects the power, fuel economy, emission output levels, and dependability of the vehicle. Any problem that affects the performance of the vehicle is often referred to as a *driveability problem.* Driveability problems can be caused by the computer system, ignition system, fuel system, emission control systems, or the engine itself. This can make diagnosis very challenging. You must be able to narrow down the causes until the faulty system and part is found, **Figure 45-1.**

This chapter reviews typical performance problems that are fully explained in other textbook chapters. This summary is needed to prepare you for the next chapters on advanced diagnostics and engine tune-up.

Locating Engine Performance Problems

Use a systematic approach when trying to locate performance problems. A systematic approach, also termed **strategy-based diagnostics,** involves using your knowledge of automotive systems and a logical process of elimination to find the source of a problem.

Think of all of the possible systems and components that could upset engine operation. Then, one by one, mentally throw out the parts that could *not* produce the

Figure 45-1. Engine performance problems reduce engine power, lower fuel economy, engine smoothness, and increase emissions. (Saab)

symptoms. See **Figure 45-2.** The *problem symptoms* are the noticeable changes in performance to the system or part: abnormal noise, loss of power, rough idle, hesitation, etc. To troubleshoot driveability problems properly, you need to train yourself to always ask the following questions:

- What are the symptoms (noise, miss, smoke, etc.)?
- When do the symptoms occur (at idle, when accelerating, when engine is cold or hot, etc.)?
- What system could be producing the symptoms (ignition, fuel, engine, etc.)?
- Where is the most logical place to start testing?
- Which part is most likely to cause the symptoms?

In every case, you must look for the actual cause of the problem, also referred to as the *root cause of failure.* For example, suppose a fuel-injected engine misses and emits heavy black smoke shortly after cold starting. The heavy black smoke would tell you that too much fuel is

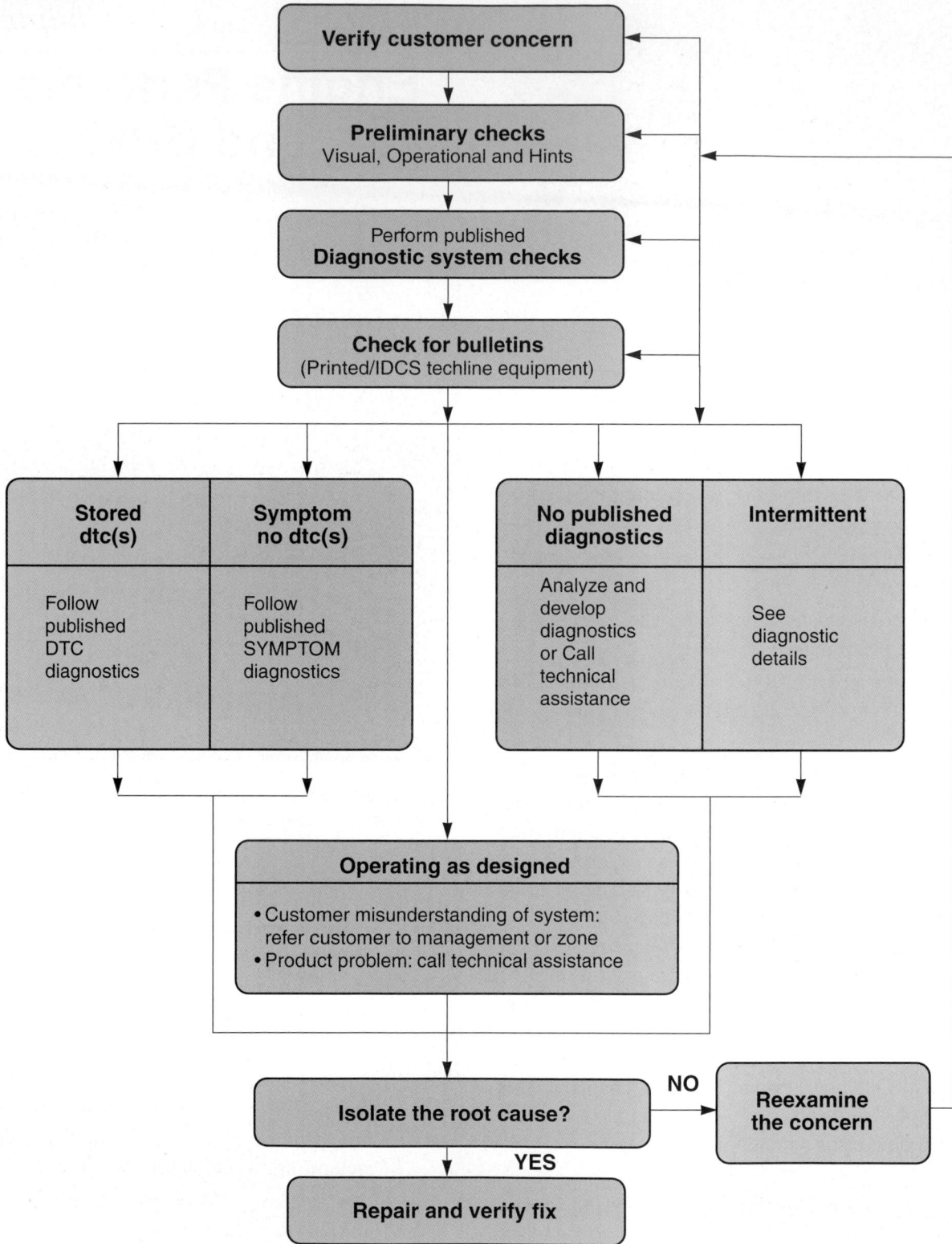

Figure 45-2. This chart shows the basic steps for strategy-based diagnostics as used by one manufacturer. Read through the chart's steps carefully. (Oldsmobile)

entering the engine. Since the engine only runs poorly when cold, the problem is temperature related.

Through simple deduction, you would probably think of the engine coolant temperature sensor. Based on read-

ings from the coolant temperature sensor, the ECM controls the fuel mixture and lengthens injector pulse width when the engine is cold. Possibly, the sensor has failed and is sending a false signal to the computer or ECM.

As you can see, logical thought will help you find the most likely problem. In this example, the rich fuel mixture was causing the miss and black smoke, but was not the actual cause of the problem. The defective coolant temperature sensor caused the computer to pulse the injectors for a longer period than necessary and is, therefore, the root cause of failure.

If your first idea is incorrect, rethink the problem and check the next most likely trouble source, **Figure 45-3.** If you approach automotive problems in a haphazard manner and simply install parts without taking time to properly diagnose the problem, you will be in for a long, frustrating day.

Performance Problem Troubleshooting Charts

If you have trouble locating an engine performance problem, refer to a service manual diagnosis chart, or ***troubleshooting chart.*** It will list problem causes and corrections. A service manual chart is written for a particular make and model vehicle, making it very accurate. A service manual troubleshooting chart is given in **Figure 45-4.**

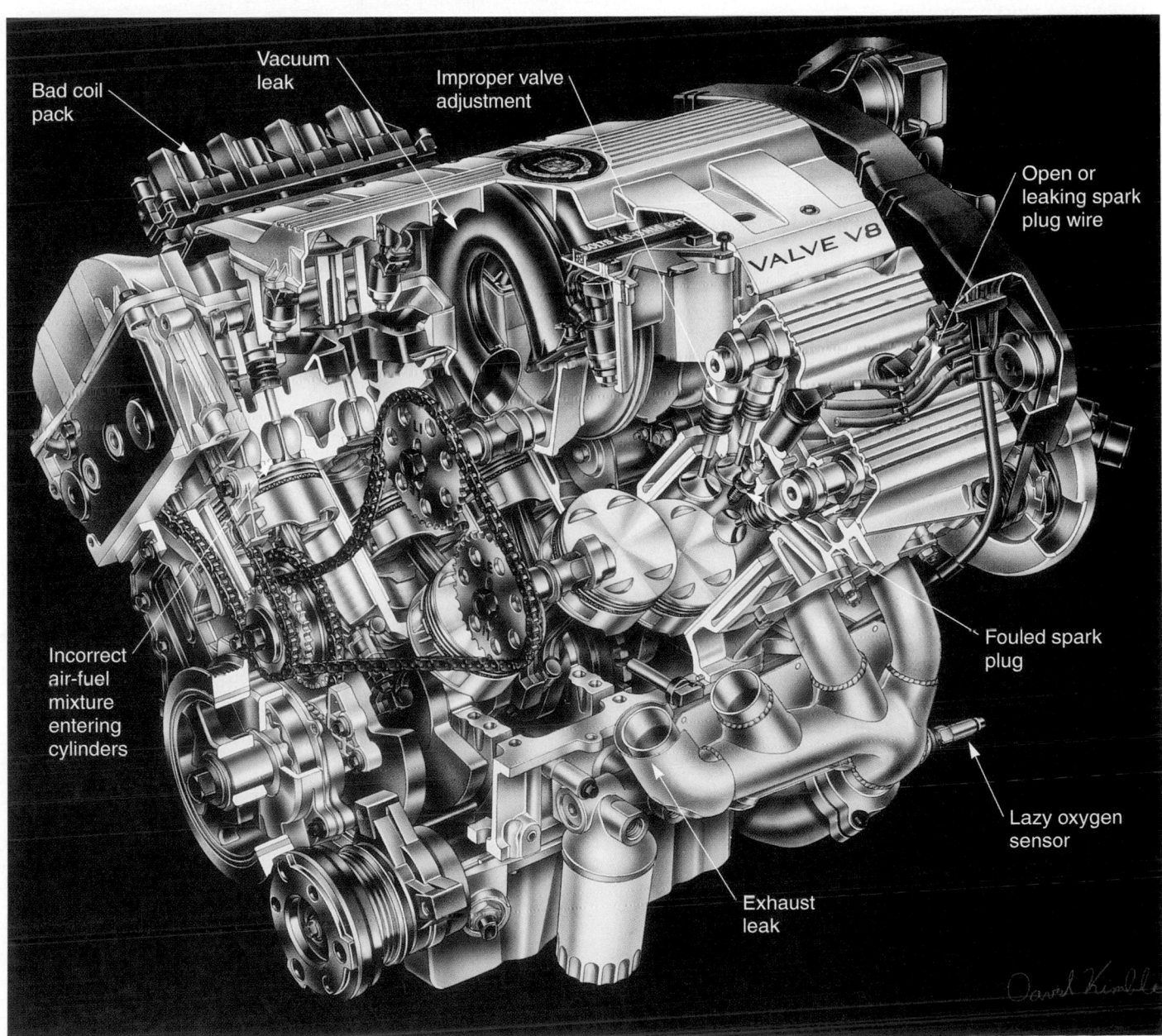

Figure 45-3. Note a few of the causes of engine performance problems. (Cadillac)

Fuel Supply System

System Troubleshooting Guide

Note: Across each row in the chart, the systems that could be sources of a symptom are ranked in the order they should be inspected starting with ①. Find the symptom in the left column, read across to the most likely source, then refer to the page listed at the top of that column. If inspection shows the system is OK, try the next most likely system ②, etc.

Page / Symptom	Sub system	Fuel injector	Injector resistor	Pressue Regulator (Pressure Regulator Cut-off Solenoid Valve)	Fuel Filter	Fuel Pump	Main Relay	Contaminated Fuel
		78	83	84	88	89	91	*
Engine won't start			③		③	①	②	
Difficult to start engine when cold or hot		③		③ (When hot)	②	①		
Rough idle		①		②				③
Frequent stalling	While warming up	①			②			
	After warming up	①			③	②		
Poor performance	Misfire or rough running	①		②				③
	Fails emission test	①		②				
	Loss of power				①	③		②

* Fuel with dirt, water or a high percentage of alcohol is considered contaminated.

Figure 45-4. This is a troubleshooting chart for the fuel supply system of one make of vehicle. (Honda)

Technical Service Bulletins

Technical service bulletins (TSBs) explain problems that frequently occur in one make or model vehicle. They are published by the auto manufacturers and explain the symptoms of, as well as the tests and corrections for, common problems. Auto manufacturers keep records of specific problems that have been fixed in the past. If one or more problems happen regularly, they describe and publish them in service bulletins sent out to their dealerships.

Always obtain and read technical service bulletins. They give helpful information on fixing common problems with specific makes and models of cars and light trucks.

Other Sources of Service Information

There are several other sources of service information that can be used when a problem is difficult to locate and correct. Trade magazines often discuss common and difficult-to-locate failures. Always read these types of magazines.

If you can access the Internet, you can search for service information and review problems encountered by other technicians. You can type in information about the vehicle or problem and quickly search for additional information. Some vehicle manufacturers and part manufacturers provide service data on the Internet.

Typical Performance Problems

It is important that you understand the most common engine performance problems. The following sections will help you use test equipment, troubleshooting charts, and service manuals during diagnosis.

No-Crank Problem

A *no-crank problem* occurs when the starter fails to turn the crankshaft. This condition is generally caused by starting system malfunction or an engine mechanical problem. Check the starter and solenoid, as well as all related wiring. If the starter spins but the engine does not turn over, the starter drive pinion may not be engaging with the flywheel or the flywheel teeth may be damaged.

 If a mechanical problem is suspected, try to turn the engine over with a flywheel turner. If the engine will not turn, or if it turns with difficulty, the engine has mechanical problems. If the engine turns over easily, the starting system is at fault.

No-Start Problem

A *no-start problem* occurs when the engine cranks but fails to start. This is the most obvious and severe performance problem. There is a defective part in either the fuel system, ignition system, a major engine part, or another related system.

To check for *spark*, remove the spark plug wire or coil assembly (direct ignition) from one of the plugs. Install a spark tester on the plug wire or coil assembly and ground the tester to the engine, **Figure 45-5.** A bright spark should jump the gap in the spark tester when the engine is cranked. If not, the problem is in the ignition system.

If you have spark, check for *fuel* by making sure the fuel system is feeding gasoline to the engine. With throttle body fuel injection, watch the injector outlet for a spray of gasoline during engine cranking. With multiport injection, check for fuel pressure at the test fitting on the fuel rail. You can also pull an injector to check for fuel spray.

Check the wiring harness going to the injectors for pulse from the ECM. Disconnect a wiring harness from one of the injectors and install a *noid light,* **Figure 45-6.** If the light flashes, the ECM is sending a signal to the injectors. If the noid light does not flash, there is a problem in the computer control system, its inputs, or the wiring.

With a carburetor fuel system, operate the throttle lever. Watch for fuel squirting out the accelerator pump discharge. If you do not have fuel, then something is wrong with the fuel system.

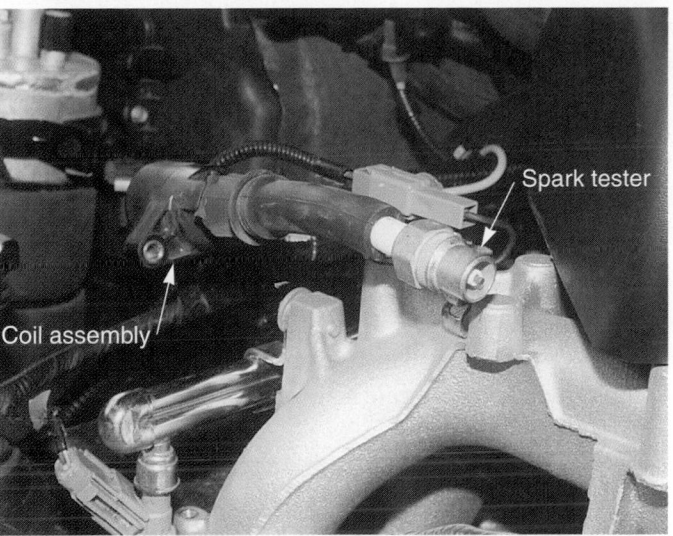

Figure 45-5. A spark tester should be used to check the ignition system for the presence of a spark. Do not use a screwdriver to ground the ignition wire as this will expose you to high voltages.

Figure 45-6. An injector noid light can quickly determine the presence of injector pulse from the ECM.

By checking for fuel and spark, you have narrowed down the cause of the problem to components that are part of the computer, fuel, or ignition systems. If you have both fuel and spark, then check engine compression. Excessively low compression will prevent the engine from starting. A jumped timing chain or belt could be keeping the engine from starting. With a diesel, slow cranking speed can prevent starting.

Hard Starting

Hard starting is due to the partial failure of a system. The starting motor could be dragging and not spinning the engine quickly enough. There could be a vacuum leak, an engine mechanical problem, a fuel or ignition system malfunction, or a sensor problem. When hard starting is encountered, you should check all the components that have some effect upon initial engine starting.

Stalling

During a *stall,* the engine stops running, or dies. This may occur at idle, after cold starting, or after warm-up. There are many causes of stalling: low idle speed, injection system or carburetor problem, ignition system trouble, severe vacuum leaks, etc.

Misfiring

Engine *misfiring* is a performance problem resulting from one or more cylinders failing to fire (produce normal combustion) normally or not at all. The engine may miss at idle, under acceleration, or at cruising speeds.

If an engine only misses at idle, for example, check the components that affect idle. With multiport fuel injection, an injector may not be opening. A fouled spark plug, an open plug wire, a cracked distributor cap, corroded terminals, and vacuum leaks are a few other possible causes for a miss. If the engine has a carburetor, the idle circuit may be clogged with debris.

OBD II Cylinder Misfire Monitoring

An engine misfire occurs when the fuel mixture fails to ignite and burn properly. The unburned fuel is pushed out of the engine and into the exhaust system. It can then damage the catalytic converter and pollute the environment.

Discussed briefly in earlier chapters, OBD II engine misfire is monitored using the engine crankshaft sensor. When a cylinder fails to fire, the rate of crankshaft acceleration or rotation will slow down slightly with the misfire. The crankshaft sensor and computer interpret this fluctuation in crankshaft speed as an indication of a poorly running engine. A misfire rate of less than 2% is acceptable because the catalytic converter can easily handle this amount of pollutants. See **Figure 45-7.**

A few OBD II systems detect misfire by electronically measuring the ionization at the spark plug electrodes. An OBD II scan tool can produce the following *misfire data*:

- *Misfire data values*—scan tool readouts that indicate something is causing an engine cylinder not to fire its fuel mixture properly. The misfire data can be recorded by the vehicle computer and stored in memory. The scan tool will retrieve this data and help you find problem sources.

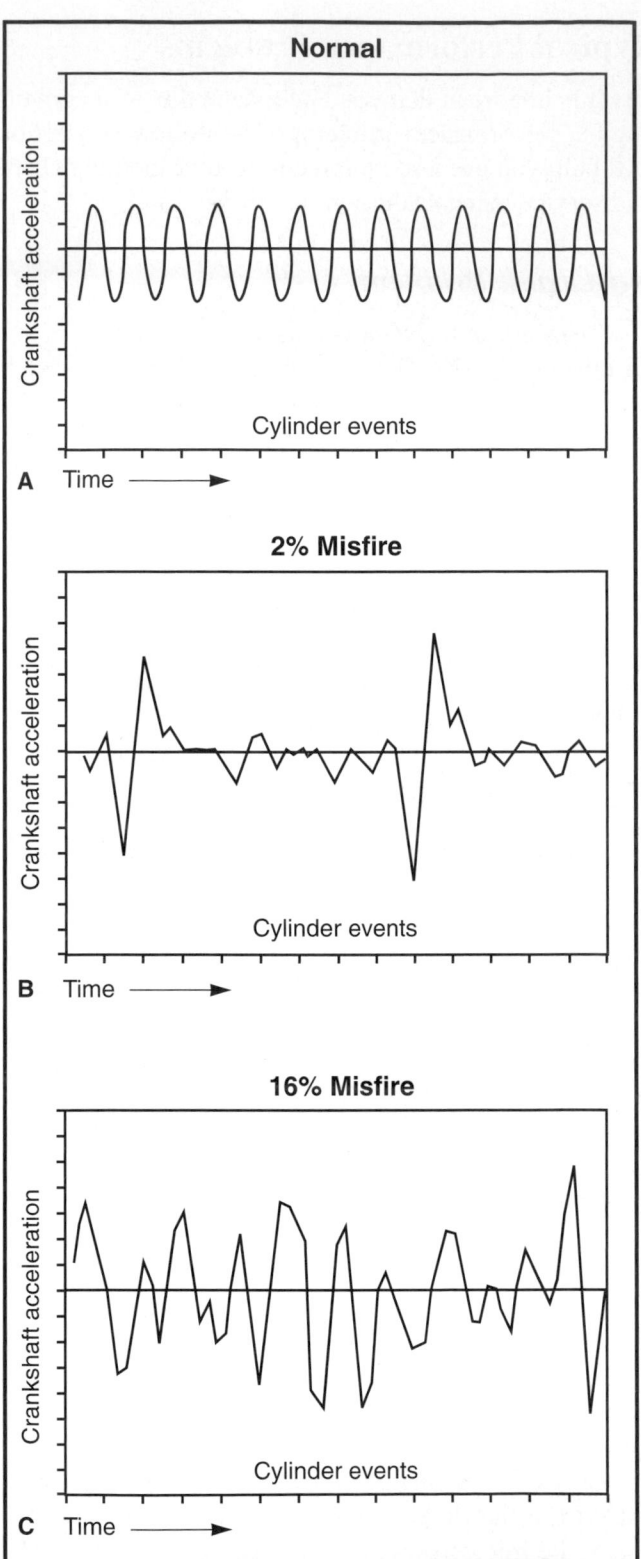

Figure 45-7. Misfire monitoring senses changes in crankshaft acceleration. When a cylinder misfires, the rate of acceleration slows momentarily. A—Normal crankshaft acceleration. B—Crankshaft acceleration with a 2% misfire rate. C—Crankshaft acceleration with a 16% misfire rate.

- *Misfire history*—indicates which cylinder was misfiring and how badly it has been misfiring.
- *Misfire passes*—shows how many times the cylinder has not misfired.
- *Misfire failures*—indicates how many misfire tests have been recorded.
- *Misfire rev. status*—shows accepted misfires (real misfires) and rejected misfires (false data caused by rough road or other cause).
- *Total misfires*—a readout averaging the number of misfires recorded during the last 200 crankshaft revolutions.
- *Misfiring cylinder*—shows the primary or worst missing cylinder and the secondary or second worst missing cylinder by cylinder number.
- *RPM at misfire*—shows the engine rpm when the computer detected a cylinder misfire. This is handy for further scope diagnosis since you know the engine speed when the problem occurs.
- *Load at misfire*—gives a load in percent when the engine miss happened. This is usually information gathered from the manifold pressure sensor which measures engine load.

Other misfire data can also be produced depending upon vehicle make, model, and year. Refer to the service manual for more information.

Rough Idle

A *rough idle* occurs when the engine seems to vibrate on its mounts. You can also hear a popping noise at the tailpipe from some of the cylinders not firing and burning their fuel normally.

A *vacuum leak* a common cause of rough idling. If a vacuum hose hardens and cracks, it will allow outside air to enter the engine intake manifold, bypassing the throttle body or carburetor. This can cause a lean air-fuel mixture, preventing normal combustion.

Usually, a vacuum leak will produce a *hissing sound*. The engine roughness will smooth out when rpm is increased. A section of vacuum hose can be used to locate vacuum leaks. Place one end of the hose next to your ear. Move the other end around the engine. When the hiss becomes very loud, you have found the leak.

Other causes for a rough idle include bad spark plugs, faulty plug wires, a lean mixture, and similar conditions.

Hesitation

Hesitation is a condition in which the engine does not accelerate normally when the gas pedal is pressed.

The engine may almost stall and has a flat spot before developing power, **Figure 45-8**. The condition is similar to stumbling.

A hesitation is usually caused by a temporary lean air-fuel mixture. With fuel injection, the throttle position sensor may be bad. With a carburetor, the accelerator pump may not be functioning. Check the parts that aid engine acceleration.

Stumbling

Stumbling is similar to a hesitation, but the engine misses instead of temporarily losing power and causes jerking upon acceleration. When accelerating, a vibration can be felt as the engine misfires. See **Figure 45-8**. When traveling down the highway, the engine may temporarily miss or stumble.

Stumbling can be caused by a bad oxygen sensor, throttle position sensor, engine temperature sensor, or any part that leans or richens the air-fuel mixture too much. Stumbling can also be caused by intermittent electrical problems that cut current flow to the ignition and fuel systems.

Surging

Surging is a condition in which engine power fluctuates up and down with a continuous, soft jerking motion. When driving at a steady speed, the engine seems to speed up and slow down, without movement of the accelerator, **Figure 45-8**.

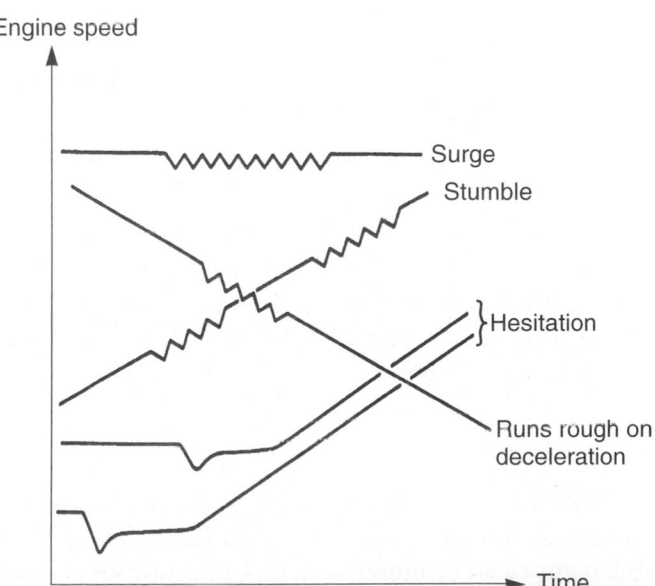

Figure 45-8. Graph represents several engine performance problems. It compares engine speed over time when each problem occurs. (Mazda)

Surging is sometimes caused by an extremely lean fuel injection or carburetor setting. Surging can also be caused by ignition or computer control system problems, an electronically controlled automatic transmission that is changing gears abruptly, or intermittent electrical troubles that affect engine and drivetrain systems.

Backfiring

Backfiring is caused by the air-fuel mixture igniting in the intake manifold or the exhaust system. A loud *bang* or *pop* sound can be heard when the mixture ignites and burns.

Backfiring can be caused by incorrect ignition timing, crossed spark plug wires, a cracked distributor cap, a defective ignition coil, a bad throttle position sensor or carburetor accelerator pump, exhaust system leakage, a faulty air injection system, a computer system malfunction, or other system faults.

A mild backfire in the engine intake manifold is sometimes referred to as a *cough.* Instead of a loud bang or pop, the mild backfire will sound like a rush of air or a cough.

Dieseling (After-Running, Run-On)

Dieseling, also called **after-running** or **run-on,** occurs when the engine fails to shut off. The engine keeps firing, coughing, and producing power. The air-fuel mixture ignites spontaneously, forcing the pistons down without the spark plugs firing. Dieseling is usually caused by high idle speed, carbon buildup in the combustion chambers, low octane fuel, or an overheated engine.

Pinging (Spark Knock)

Pinging, or **spark knock,** is a metallic tapping or light knocking sound, that usually occurs when the engine accelerates under load. Pinging is caused by abnormal combustion (*preignition* or *detonation*). Pinging is normally caused by low-octane fuel, advanced ignition timing, carbon buildup in combustion chambers, or engine overheating.

Vapor Lock

Vapor lock occurs when the fuel is overheated, forming air bubbles that upset the air-fuel mixture. Vapor lock can cause engine stalling, lack of power, hard starting, and no starting. Vapor lock is usually caused by too much engine heat transferring into the fuel. A fuel line may be touching a hot engine part or the fuel return

system may be plugged. You would need to locate any condition that could overheat the fuel.

Fuel Line Freeze

Fuel line freeze results when moisture in the fuel turns to ice. The ice will block fuel filters and prevent engine operation. This problem does not occur as frequently as in the past, due to the use of alcohol and other fuel additives. With diesel fuel, the overcooled fuel can form wax, which blocks the filters.

To correct fuel line freeze, replace clogged fuel filters. You may need to place the car in a warm garage until the water in the fuel thaws.

Poor Fuel Economy

Poor fuel economy is a condition causing the car to use too much fuel for the miles driven. Fuel economy can be measured by comparing the miles that can be driven on one gallon (3.79 L) of fuel to the manufacturers estimated MPG rating. Poor fuel economy can be caused by a wide range of problems. Some of these include rich air-fuel mixture, engine miss, incorrect ignition timing, or leakage in the fuel system.

Lack of Engine Power

Lack of engine power, also termed a sluggish engine, causes the vehicle to accelerate slowly. When the gas pedal is pressed, the car does not gain speed properly.

As with poor fuel economy, there are many troubles that can reduce engine power: fuel system problems, ignition system problems, emission control system problems, engine mechanical problems, etc.

Other Performance Problems

There are many other more specific engine performance problems. Many of these are covered in other chapters of this book. Use the index to locate more information if needed.

Summary

- An engine performance problem is any trouble that affects the power, fuel economy, emission output levels, and dependability of the engine.

- A systematic approach, also termed strategy based diagnostics, involves using your knowledge of automotive service and a logical process of elimination to find the source of a problem.

- Problem symptoms are the noticeable changes in performance to the system or part: abnormal noise, loss of power, rough idle, hesitation, etc.

- Engine misfiring is a performance problem resulting from one or more cylinders failing to fire.

- A vacuum leak is a common cause of rough idling. If a vacuum hose hardens and cracks, it will allow outside air to enter the intake manifold, bypassing the throttle body or carburetor.

- A hesitation is a condition in which the engine does not accelerate normally when the gas pedal is pressed.

- Stumbling is similar to a hesitation, but the engine misses instead of temporarily losing power and jerking upon acceleration.

- Surging is a condition in which engine power fluctuates up and down with a continuous soft jerking motion.

Important Terms

Engine performance problem	Stall
	Misfiring
Driveability problem	Rough idle
Strategy- based diagnostics	Hesitation
	Stumbling
Problem symptoms	Surging
Root cause of failure	Backfiring
Troubleshooting chart	Dieseling
Technical service bulletins	Pinging
	Vapor lock
No-crank problem	Fuel gas line freeze
No-start problem	Poor fuel economy
Noid light	Lack of engine power
Hard starting	

Review Questions—Chapter 45

Please do not write in this text. Place your answers on a separate sheet of paper.

1. How do you use a systematic approach to troubleshooting?

2. If you have difficulty locating an engine performance problem, refer to a service manual _____ _____.

3. A car is towed into a shop with a no-start problem. The engine cranks but will not fire and run on its own power. The car has a throttle body, fuel injected gasoline engine. Technician A says to check for spark and fuel first. Technician B says that the spark plugs should be removed first. Who is right?
   (A) A only.
   (B) B only.
   (C) Both A and B.
   (D) Neither A nor B.

4. Define the term "misfiring."

5. A vacuum leak is a common cause of rough idling. True or False?

6. A(n) _____, also called a(n) _____, is a condition where the engine does not accelerate normally when the gas pedal is pressed.

7. What causes "backfiring"?

8. Define the term "dieseling."

9. A(n) _____ or _____ _____ is a metallic tapping or light knocking sound, usually when accelerating under load.

10. Vapor lock occurs when the gasoline is cooled and forms a gel, preventing fuel flow and engine operation. True or False?

● ASE-Type Questions

1. Technician A says when attempting to locate an engine performance problem, you should observe the engine's exhaust smoke. Technician B says when attempting to locate an engine performance problem, you should listen for any abnormal engine noises. Who is right?
   (A) A only.
   (B) B only.
   (C) Both A and B.
   (D) Neither A nor B.

2. A fuel injected engine misses and emits heavy black smoke. Technician A says this problem normally indicates an emission control system malfunction. Technician B says this problem may indicate a faulty engine temperature sensor. Who is right?
   (A) A only.
   (B) B only.
   (C) Both A and B.
   (D) Neither A nor B.

3. An automotive engine has a no-start problem. Technician A says this condition may indicate an ignition system malfunction. Technician B says this problem may indicate a fuel system problem. Who is right?
   (A) A only.
   (B) B only.
   (C) Both A and B.
   (D) Neither A nor B.

4. An engine has poor fuel economy. Technician A says this problem may be the result of a leaky fuel line. Technician B says this problem may be caused by a rich air-fuel mixture. Who is right?
   (A) A only.
   (B) B only.
   (C) Both A and B.
   (D) Neither A nor B.

5. All of the following are common engine performance problems *except:*
   (A) hard starting.
   (B) misfiring.
   (C) oil leak.
   (D) dieseling.

6. An automotive engine won't start. Technician A says you should first look in the engine compartment for obvious signs of trouble. Technician B says an engine analyzer should be used first to diagnose the engine problem. Who is right?
   (A) A only.
   (B) B only.
   (C) Both A and B.
   (D) Neither A nor B.

7. A car has a hard starting problem. Technician A says the engine may have a faulty throttle position sensor. Technician B says this engine malfunction may be caused by a faulty coolant temperature sensor. Who is right?
   (A) A only.
   (B) B only.
   (C) Both A and B.
   (D) Neither A nor B.

8. Technician A says stalling may occur while the engine is idling. Technician B says stalling may occur after cold starting. Who is right?
   (A) A only.
   (B) B only.
   (C) Both A and B.
   (D) Neither A nor B.

9. An automotive engine has a stalling problem. Technician A says this problem may be due to a low idle speed setting. Technician B says this problem may be due to a severe engine vacuum leak. Who is right?
   (A) A only.
   (B) B only.
   (C) Both A and B.
   (D) Neither A nor B.

10. All of the following may cause an engine to surge *except:*
    (A) lean air-fuel mixture.
    (B) ignition system problem.
    (C) computer system malfunction.
    (D) starter solenoid problem.

11. An automobile engine has a high speed misfiring problem. Technician A says this problem may be caused by a fouled spark plug. Technician B says engine misfiring may be the result of a vacuum leak. Who is right?
    (A) A only.
    (B) B only.
    (C) Both A and B.
    (D) Neither A nor B.

12. A vacuum leak is detected in an automotive engine. Technician A says the leak should be most noticeable at highway speeds. Technician B says the vacuum leak should be most noticeable at idle. Who is right?
    (A) A only.
    (B) B only.
    (C) Both A and B.
    (D) Neither A nor B.

13. Technician A says engine hesitation is normally caused by a temporary rich air-fuel mixture. Technician B says engine hesitation is usually caused by a temporary lean air-fuel mixture. Who is right?
    (A) A only.
    (B) B only.
    (C) Both A and B.
    (D) Neither A nor B.

14. An engine is brought into the shop with a backfiring problem. Technician A says this problem can be the result of a cracked distributor cap. Technician B says this problem can be the result of a faulty carburetor accelerator pump. Who is right?
    (A) A only.
    (B) B only.
    (C) Both A and B.
    (D) Neither A nor B.

15. An engine is brought into the shop with a dieseling problem. Technician A says this may be caused by a low idle speed setting. Technician B says this may be the result of an overheated engine. Who is right?
    (A) A only.
    (B) B only.
    (C) Both A and B.
    (D) Neither A nor B.

Activities—Chapter 45

1. Check shop manuals for a no-start diagnosis chart and demonstrate on a vehicle engine how you would troubleshoot a no-start problem.

2. Invite an experienced driveability technician to speak to your class about troubleshooting procedures. Prepare a list of questions that you might ask the technician. Discuss the questions with your instructor beforehand.

Engine Performance Diagnosis		
Condition	**Possible Causes**	**Correction**
Engine will not start.	1. Weak battery. 2. Corroded or loose battery connections. 3. Defective starter. 4. Moisture on ignition wires. 5. Faulty ignition cables. 6. Faulty coil or ignition control module. 7. Incorrect spark plug gap. 8. Dirt or water in fuel system. 9. Faulty fuel pump.	1. Charge or replace battery as necessary. 2. Clean and tighten battery connections. 3. Test starter. Rebuild or replace as necessary. 4. Wipe wires and dry. 5. Replace any damaged or worn cables. 6. Test and replace, if necessary. 7. Check gap and reset as necessary. 8. Clean system and replace fuel filter. 9. Install new fuel pump.
Engine stalls or rough idle.	1. Low idle speed. 2. Idle mixture too lean or too rich. 3. Vacuum leak. 4. Incorrect ignition wiring. 5. Faulty coil.	1. Check and adjust idle speed. 2. Test and repair as necessary. 3. Inspect intake manifold gasket and vacuum hoses. Replace, if necessary. 4. Check and correct wiring. 5. Test and, if necessary, replace coil.
Loss of power.	1. Dirty or incorrectly gapped spark plugs. 2. Dirt or water in fuel system. 3. Faulty fuel pump. 4. Incorrect valve timing. 5. Blown cylinder head gasket. 6. Low compression. 7. Burned, warped or pitted valves. 8. Restricted exhaust system. 9. Faulty ignition cables. 10. Faulty coil.	1. Clean plugs and set gap. 2. Clean system and replace fuel filter. 3. Install new fuel pump. 4. Adjust valve timing. 5. Replace head gasket. 6. Repair mechanical problem. 7. Install new valves. 8. Install new parts, as necessary. 9. Replace cables. 10. Replace coil.
Engine misses on acceleration.	1. Dirty spark plug or improper gap. 2. Dirt in fuel system. 3. Burned, warped or pitted valves. 4. Faulty coil.	1. Clean spark plugs and set gap. 2. Clean fuel system and replace fuel filter. 3. Install new valves. 4. Test and, if necessary, replace coil.
Engine misses at high speed.	1. Dirty spark plug or improper gap. 2. Faulty coil. 3. Dirty fuel injectors. 4. Dirt or water in fuel system.	1. Clean spark plugs and set gap. 2. Test and, if necessary, replace coil. 3. Clean injectors. 4. Clean system and replace fuel filter.

Chapter 46

Advanced Diagnostics

After studying this chapter, you will be able to:

- Use advanced diagnostic techniques to trouble-shoot difficult problems.
- Use a scan tool to find problems not tripping trouble codes by using snapshot and datastream values.
- Use a breakout box to measure circuit values.
- Explain the principles of an oscilloscope.
- Summarize how to use waveforms to analyze the operation of sensors, actuators, ECU outputs, and other electrical-electronic devices.
- Evaluate ignition system waveforms.
- Summarize how to use an engine analyzer.

Earlier chapters explained how to diagnose problems using a scan tool, a multimeter, and various specialized testing devices. If these basic troubleshooting tools fail to pinpoint the source of the problem, you may need to use more complex diagnostic methods to find and fix the trouble.

This chapter summarizes advanced methods used to troubleshoot difficult-to-locate problems. It also intro-duces the operating principles of vehicle analyzers, with emphasis on using the oscilloscope.

Advanced Diagnostics

At one time or another, every technician encounters a problem that seems impossible to fix. He or she might replace a part that seems bad, only to find the same annoying symptoms when the repair is complete. The new part must be removed and further diagnosis com-pleted until the real "culprit" is found. This is when advanced diagnostic techniques come in handy.

By learning a few advanced "tricks of the trade" (using datastream values, a breakout box, an oscillo-scope, and an analyzer), even the most difficult problems can be located and corrected with minimal frustration.

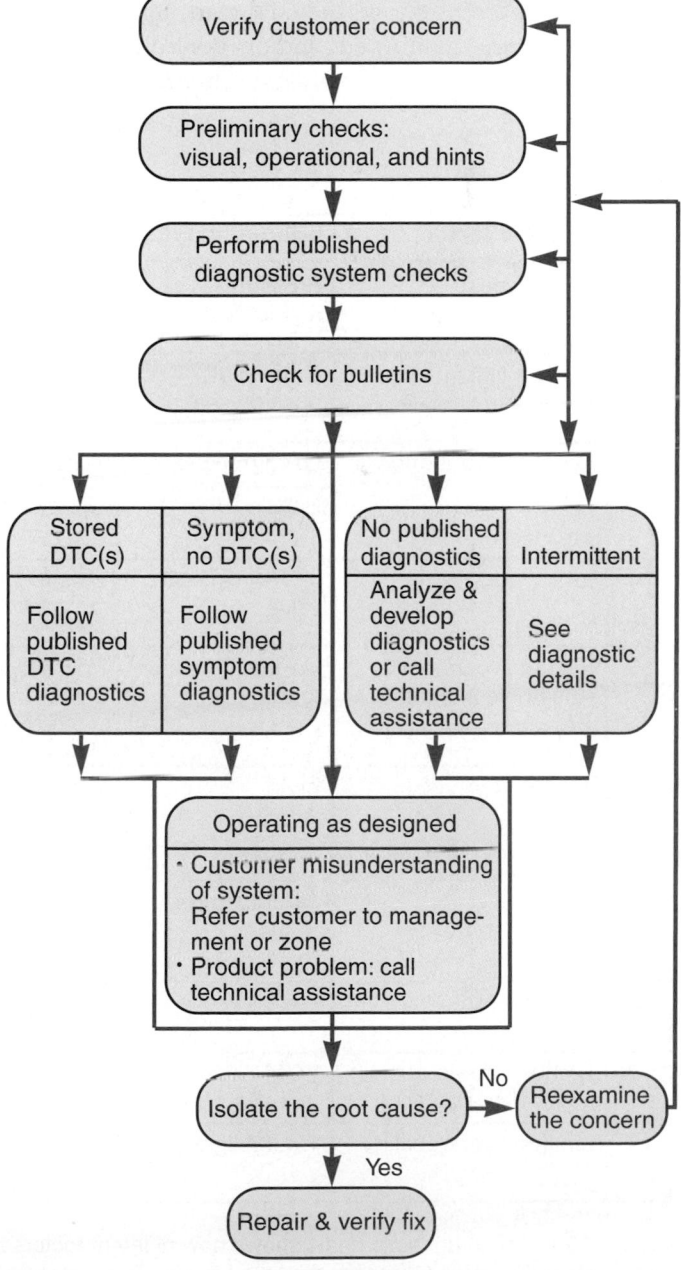

Figure 46-1. This chart shows basic steps of strategy-based diagnostics. Study the chart carefully. (Oldsmobile)

Strategy-based diagnostics involves using a consistent, logical procedure to narrow down the possible problem sources. Basically, procedure involves verifying the complaint, making preliminary checks, reading service bulletins, performing service manual–recommended checks, repairing the problem, and rechecking system operation. **Figure 46-1** gives a chart showing the basic flow as you use strategy-based diagnosis.

When diagnosing problems, use your knowledge of system operation to determine which part could be malfunctioning and causing the symptoms.

For example, if an engine misses only when cold, think of which parts affect cold engine operation. You should think of the coolant temperature sensor, the intake air temperature sensor, and the cold start injector. These components are monitored and/or controlled by the ECM, which enriches the air-fuel mixture when the engine is cold.

Service manuals contain information to help facilitate troubleshooting. The chart shown in **Figure 46-2** summarizes the basic service procedures that can be performed to help find intermittent problems in one particular vehicle.

Vacuum and Pressure Gauge Tests

A *vacuum gauge* measures negative air pressure (pressure lower than atmospheric pressure) produced by the engine, fuel pump, vacuum pump, and other components. It is a valuable tool for determining engine condition and testing vacuum-actuated devices.

A *pressure gauge* measures positive pressures (pressures higher than atmospheric pressure) produced by the engine, turbocharger, fuel pump, or other device. It can be used to check for high exhaust system back pressure (restricted converter or system), abnormal fuel pressure,

	Variable factor	Influential part	Target condition	Service procedure
1	Mixture ratio	Pressure regulator	Made lean	Remove vacuum hose and apply vacuum.
			Made rich	Remove vacuum hose and apply pressure.
2	Ignition timing	Crankshaft position sensor	Advanced	Rotate distributor clockwise.
			Retarded	Rotate distributor counterclockwise.
3	Mixture ratio feedback control	Oxygen sensor	Suspended	Disconnect oxygen sensor harness connector.
		ECM	Operation check	Perform on-board diagnostic system (On-board Diagnostic Test Mode II) at 2000 rpm.
4	Idle speed	IAC valve-AAC valve	Raised	Turn idle adjusting screw counterclockwise.
			Lowered	Turn idle adjusting screw clockwise.
5	Electrical connection (Electric continuity)	Harness connectors and wires	Poor electrical connection or improper wiring	Tap or wiggle.
				Race engine rapidly. See if the torque reaction of the engine unit causes electric breaks.
6	Temperature	ECM	Cooled	Cool with an icing spray or similar device.
			Warmed	Heat with a hair drier. **[WARNING: Do not overheat the unit.]**
7	Moisture	Electric parts	Damp	Wet. **[WARNING: Do not directly pour water on components. Use a mist sprayer.]**
8	Electric loads	Load switches	Loaded	Turn on headlamps, air conditioning, rear defogger, etc.
9	Closed throttle position switch condition	ECM	ON-OFF switching	Rotate throttle position sensor body.
10	Ignition spark position	Timing light	Spark power check	Try to flash timing light for each cylinder using ignition coil adapter (SST).

Figure 46-2. This diagnostic chart shows how different factors and parts can cause abnormal operating conditions. The technician can perform the service procedures listed to simulate intermittent problems. (Nissan)

incorrect turbocharger boost pressure, and other problems. You must use your knowledge of system operation, the problem symptoms, and basic pressure testing methods to isolate hard-to-find problems.

A *vacuum-pressure gauge* reads both negative and positive pressures. It can be used to check all the previously mentioned systems.

To use a vacuum-pressure gauge (or a vacuum gauge) to check the engine, connect the gauge to a vacuum fitting on the intake plenum or manifold. Start the engine and note the reading on the gauge. Compare the gauge readings to normal readings. **Figure 46-3** shows typical vacuum gauge readings and explains what they mean.

To help pinpoint problems that cannot be duplicated in the shop, technicians will often mount the vacuum-pressure gauge in the passenger compartment and run a long hose to the vacuum fitting on the engine. This allows the vehicle to be driven while checking for intermittent vacuum or pressure problems. It also allows the technician to monitor engine and other system values under load. See **Figure 46-4.**

Figure 46-4. A pressure gauge with a long section of vacuum hose is needed for road testing. The vacuum gauge will read turbocharger or supercharger boost pressures while vehicle is driven.

Normal engine reading
Vacuum gauge should have reading of 18-22 inches of vacuum. The needle should remain steady.

Burned or leaky valves
Burned valve will cause pointer to drop every time burned valve opens.

Weak valve springs
Vacuum will be normal at idle but pointer will fluctuate excessively at higher speeds.

Worn valve guides
If pointer fluctuates excessively at idle but steadies at higher speeds, valves may be worn allowing air to upset fuel mixture.

Choked muffler
Vacuum will slowly drop to zero when engine speed is high.

Intake manifold air leak
If pointer is down 3 to 9 inches from normal at idle, throttle valve is not closing or intake gaskets are leaking.

Carburetor or fuel injection problem
A poor air-fuel mixture at idle can cause needle to slowly drift back and forth.

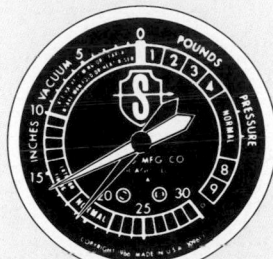

Sticking valves
A sticking valve will cause pointer to drop intermittently.

Figure 46-3. Typical vacuum gauge readings and possible causes. (Sonco)

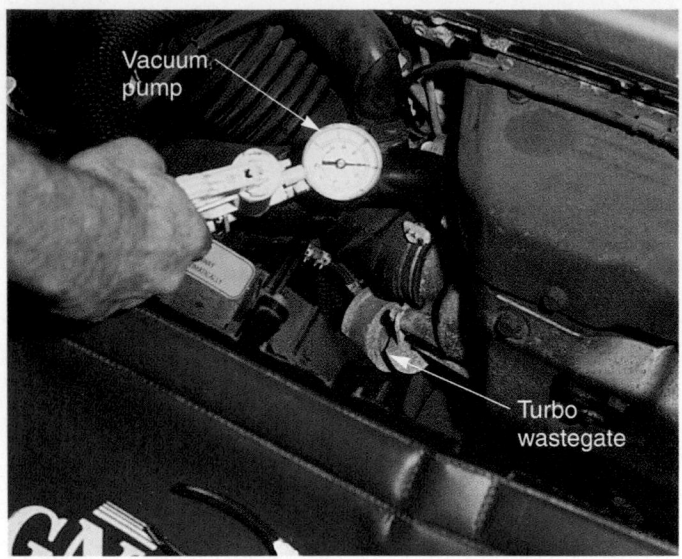

Figure 46-5. Always keep vacuum diaphragms in mind as a source of trouble while troubleshooting. They can rupture, leak air and not operate properly. A vacuum pump can be used to check the condition of vacuum diaphragm.

A *hand vacuum pump* is commonly used to check vacuum-actuated devices and vacuum diaphragms. Since vacuum diaphragms are made of rubber, which can rupture or leak, they are a common source of performance problems.

To check a vacuum-actuated device, connect the vacuum pump to the fitting. Pump the handle and see if the device will hold a vacuum. If it leaks, the diaphragm or the device should be replaced, **Figure 46-5.**

Diesel Engine Testers

A *diesel injection tester* is a set of pressure gauges and valves used to measure injection system pressure. This tester will check fuel pump pressure and volume, injector operation (out of engine), lubrication system pressure, and other functions. Refer to **Figure 46-6.** If diesel injection pressures are not within specifications, repairs or adjustments are needed.

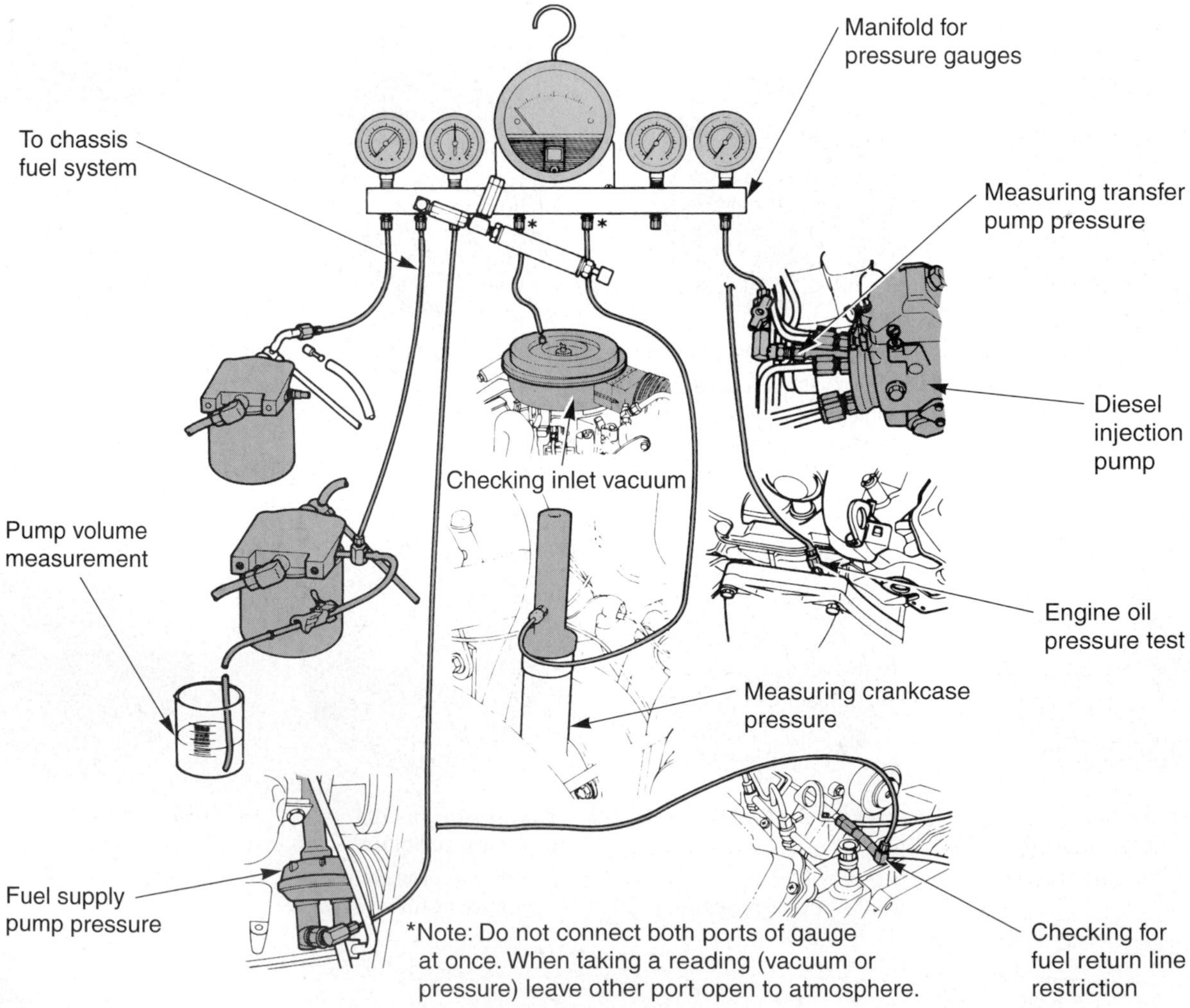

*Note: Do not connect both ports of gauge at once. When taking a reading (vacuum or pressure) leave other port open to atmosphere.

Figure 46-6. Diesel injection system testers can perform several tests. Note the various test connections. (Ford)

Warning
The operating pressures in a diesel injection system are high enough to cause serious injury. Even a small fitting leak could allow high-pressure fuel to spray out and puncture your skin or eyes.

A *glow plug test harness* can be used to find the cause of a cold start or rough idle problem in a diesel engine. The harness is connected to each glow plug, one at a time. Then, the ohmmeter is used to check the resistance of each glow plug. After a period of engine operation, the resistance of each glow plug is checked again. Combustion will increase glow plug temperature, affecting its resistance. An unequal change in glow plug resistance (temperature) indicates a cylinder is not firing.

Advanced Scan Tool Tests

In addition to retrieving trouble codes, modern scan tools can be used for advanced diagnostic procedures. For example, the scan tool can take a "picture" of operating parameters at the moment a problem occurs, display "live" operating values as an engine is running, and check actuators for proper operation.

Scan Tool Snapshot

A scan tool *snapshot* is an instantaneous reading of operating parameters present when a problem occurs. This feature is often used when a problem is hard to find or when intermittent troubles are present. Most scan tools can be programmed to automatically take a snapshot of operating parameters whenever a diagnostic trouble code is set. If desired, the snapshot feature can be triggered manually, allowing the use of the snapshot even when the vehicle does not generate a trouble code. The manual capture feature requires the technician to monitor operating conditions and to press a button on the scan tool when the problem occurs. For example, if a car only acts up when driving at a specific highway speed, drive the vehicle at the trouble-causing speed and scan under these conditions. When the symptom occurs (engine misses), press the appropriate button to capture the operating values while the problem is happening.

After returning to the shop, look for any operating parameter that is almost out-of-specs. The operating parameter may not be tripping a trouble code, but it may be affecting vehicle operation. Sometimes, two or more electrical values can be almost out-of-specs.

By using the information provided by the snapshot feature, you can often make further conclusions about what is causing the problem. You might have two sensors ready to fail, a poor electrical connection in combination with a mechanical failure, etc.

Scan Tool Datastream Values

Scan tool datastream values are "live" electrical values measured with the vehicle running or driving. They almost eliminate the need for a breakout box or pinpoint measurements of electrical values. You can read the scan tool screen to see weak values or values that are almost out of specs. In most cases you can choose which datastream values you want the scan tool to display. For example, you may want to look only at the inputs to the ECM.

If you have a performance problem but no trouble codes have been set, study the datastream values. Values that are almost out of specifications may signal a problem area. Datastream values give added information for finding troublesome problems. **Figure 46-7** gives a few datastream values that can be read by a scan tool.

Scan Tool Actuator Tests

Most scan tools can switch computer-controlled actuators on and off. This allows the technician to verify the operation of these components. For example, a scan tool can be used to fire an ignition coil, control the idle speed motor, or disable a fuel injector.

The scan tool can also be used to perform a power balance test. This test involves disabling a fuel injector or spark plug in a specific cylinder while monitoring the corresponding rpm drop. Power balance tests are detailed later in this chapter.

Checking Computer Terminal Values

Computer terminal values can be tested at the metal pins of the ECM. A digital VOM can be used to read terminal voltage and resistance values. These readings can then be compared to known good values. Often, the readings are taken with one or more wiring harnesses connected to the ECM. This eliminates the need to unplug connectors when making electrical measurements.

Figure 46-8 shows ECM terminal voltages for a specific vehicle. Note how the pin numbers correspond to certain circuits and electrical values.

Caution
Never connect a low-impedance (resistance) analog meter or test light to a computer system unless instructed to do so by the service manual. A low-impedance meter or tester could draw enough current to damage delicate electronic devices.

Using a Breakout Box

A *breakout box* allows you to test electrical values at specific pins on an ECM or in the system the ECM

DATASTREAM VALUES

For Cold Key On, Cold Idle and Hot Idle: Vehicle in PARK, A/C turned OFF, no power steering load, all ACC's OFF, Brake Pedal Released. For 55 MPH Cruise: Vehicle in Drive 4, A/C turned ON and no power steering load, compare data after driving for approximately 1 mile.

Scan Tool Parameter	Display Units	Data List	Cold Key ON	Cold Idle	Hot Idle	55 MPH Cruise
Engine Speed	RPM	ENG 1	0	Within 80 RPM of Desired Idle	Within 80 RPM of Desired Idle	1730
Desired Idle	RPM	ENG 1	0	700 to 1200	550 to 675	720
MAF	gms-sec	ENG 1	0.0	9.8 to 11.0	5.0 to 6.0	20 to 28
TP Sensor	V/°	ENG 1	0.63/1.7	.60 mV/ 0.8°	.60 mV/ 0.8°	1.06/11.0
ECT	°C	ENG 1	80° C	−20° C to 50° C	90° C to 110° C	90° C to 110° C
IAT	°C	ENG 1	80° C	−20° C to 50° C	0° C to 90° C	0° C to 90° C
MAP	kPa/V	ENG 1	97/4.63	30 to 50 kPa 1.50 V @ 38 kPa	30 to 50 kPa 1.50 V @ 38 kPa	64/2.88
BARO	kPa/V	ENG 1	97/4.65	85 to 103 kPa	85 to 103 kPa	98/4.69
TP Angle	%/°	ENG 1	0%/0.0°	0%/0.0°	0%/0.0°	11%/8.6°
Engine Load	%	ENG 1	0%	1 to 5%	1 to 5%	13%
Engine Speed	RPM	ENG 1	0	Within 80 RPM of Desired Idle	Within 80 RPM of Desired Idle	1730
IAC Position	counts	ENG 1	160	Varies	30 to 80	100
Inj. PWM Bank 1	ms	ENG 1	0.0	3.75 to 4.50	3.20 to 3.75	5.1 ms
Inj. PWM Bank 2	ms	ENG 1	0.0	3.75 to 4.50	3.20 to 3.75	5.2 ms
HO₂S Bn 1 Sen. 1	mV	ENG 1	67	Varies	Varies	Varies
HO₂S Bn 2 Sen. 1	mV	ENG 1	111	Varies	Varies	Varies
Rich to Lean Status Bn 1 Sen. 1	Lean/Rich	ENG 1	Lean	Varies	Varies	Varies
Rich to Lean Status Bn 2 Sen. 1	Lean/Rich	ENG 1	Lean	Varies	Varies	Varies
HO₂S Bn 1 Sen. 2	mV	ENG 1	45	Varies	Varies	Varies
HO₂S Bn 1 Sen. 3	mV	ENG 1	156	Varies	600 mV or more	600 mV or more
Rich to Lean Status Bn 1 Sen. 2	Lean/Rich	ENG 1	Lean	Varies	Varies	Varies
Rich to Lean Status Bn 1 Sen. 3	Lean/Rich	ENG 1	Lean	Varies	Varies	Varies
Fuel Trim Cell	Number	ENG 1	0	16	16	5
Fuel Trim Learn	Disabled/ Enabled	ENG 1	Disabled	Disabled	Enabled	Enabled
Shrt Term FT Bn 1	%	ENG 1	0%/128	−2.0 to 2.0	−3.0 to 3.0	−6.0 to 6.0

Figure 46-7. Scan tool datastream values can be helpful when you have performance problems but no trouble codes. Datastream values are electrical values detected by the ECM. If values from a pinpoint test do not match datastream values, suspect wiring or ECM problems.

controls. It is one of the last tools used in diagnostics, as it is time consuming, **Figure 46-9.** It is often used to locate problems in systems that do not have self-diagnostic capabilities.

The breakout box is connected in place of the ECM in the wiring harness. Then, a multimeter is used to touch specific terminals on the breakout box. The measured circuit values are compared to the manufacturer's

ECM TERMINAL VOLTAGE
5.0L (V.I.N. H)

This ECM voltage chart is for use with a digital voltmeter to further aid in diagnosis. These voltages were derived from a known good car. The voltages you get may vary due to low battery charge or other reasons, but they should be very close.

The following conditions must be met before testing:
• Engine at operating temperature • Closed loop • Engine idling (for "engine run" column)
• Test terminal not grounded • Scanner not installed

Voltage Key "On"	Voltage Engine Run	Voltage Circuit Open	Description	Term	Term	Description	Voltage Key "On"	Voltage Engine Run	Voltage Circuit Open
0	0	0	Sensor return	22	1	Baro sensor signal decreases with altitude	4.75	4.75	*.5
5	5	5	5V reference	21	2	TPS sensor signal	*1.0 †5.0	*1.0	5.0
.5–.65	3–5	*.5	Vacuum sensor output	20	3	Coolant temp. sensor signal	*2.5	*2.5	5.0
12	4–7 (var.)	*.5	PWM EGR solenoid	19	4	Air control solenoid	12.5	*1.0	*.5
12	5–10 (var.)	*.5	M/C solenoid	18	5	Diagnostic test term	5	5	5
*.5	*.5	12	3rd Gear switch	17	6	A/C W.O.T. cutout (5.7L)	12	14	*.5
**10	**11	10	VSS signal	16	7	Coolant temp. sensor return	0	0	0
			Not used	15	8	Not used			
*.5	*.5	1.7	Oxygen sensor–LO	14	9	Oxygen sensor–HI	.3–.45	.1–.9 (var.)	.3–.45
*.5	*.5	*1.0	Dist. ref. pulse–LO	13	10	Dist. ref. pulse–HI	*.5	1–2 (var.)	*.5
*.5	1–2 (var.)	*.5	ECT	12	11	IGN module bypass	*.5	3.7	*.5

Voltage Key "On"	Voltage Engine Run	Voltage Circuit Open	Description	Term	Term	Description	Voltage Key "On"	Voltage Engine Run	Voltage Circuit Open
			Not used	J	K	Not used			
*.5 P/N 12 D/R	*.5 P/N 14 D/R	12	Park/neutral switch	H	L	ESC (5.0L)	7–10	7–10	*.5
10	*.5	*.5	"Check engine" lamp	G	M	Not used			
			Not used	F	N	4th gear switch if used	*.5	*.5	12
12	*1.0	*.5	Throttle kicker	E	P	Trans converter clutch solenoid	12	14	*.5
			Not used	D	R	Trouble code memory power	12	14	*.5
12	14	*.5	IGN 1 power	C	S	Not used			
12	14	*.5	Air switching solenoid	B	T	Not used			
0	0	0	Ground (to engine)	A	U	Ground (to engine)	0	0	0

* = Value shown or less than that value
† = Wide open throttle
(var.) = variable

P/N = Park or Neutral
D/R = Drive or Reverse
** = If less than 1V rotate drive wheel to verify

Figure 46-8. The service manual will usually specify the electrical values that should be present at each terminal of the computer connector. (GM)

Figure 46-9. A breakout box is usually the last tool used to find performance problems. It is connected to the wiring harness in computer system. Then a multimeter can be used to check terminals on the breakout box for actual operating voltages, resistance, and current values. They can be compared to known good values or to datastream values to find the cause of the problem. (OTC Div. of SPX Corp.)

specifications. If the measured values are not within specifications, you usually have a defective component or a wiring problem. If the values are within specifications, the problem is in the ECM itself.

Isolating Electromagnetic Interference

Electromagnetic interference (EMI), or radiation interference, occurs when an induced voltage enters another system's wiring. Sources of EMI can include loose, misrouted, or unshielded spark plug wires; police and CB radios; and aftermarket accessories.

In the past, electromagnetic interference was limited to noise in the radio speakers. In late-model vehicles, EMI can cause a computer-controlled system to malfunction.

For example, induced voltage from a loose spark plug wire could enter a sensor wire. This unwanted voltage then enters the computer as false data. Numerous computer malfunctions or false outputs can result. See **Figure 46-10.**

To isolate the source of electromagnetic interference, try turning off or disabling circuits or devices. If, for example, removing the drive belt from the alternator

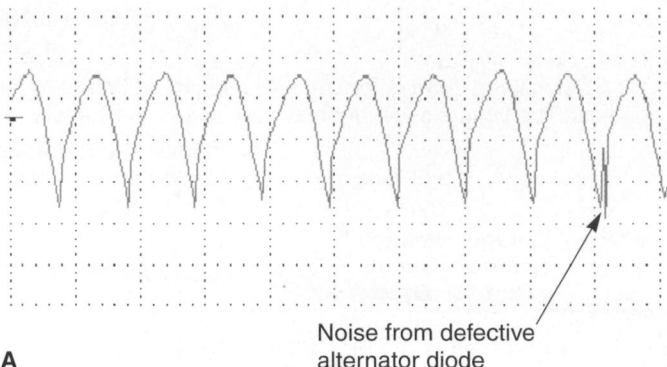

A

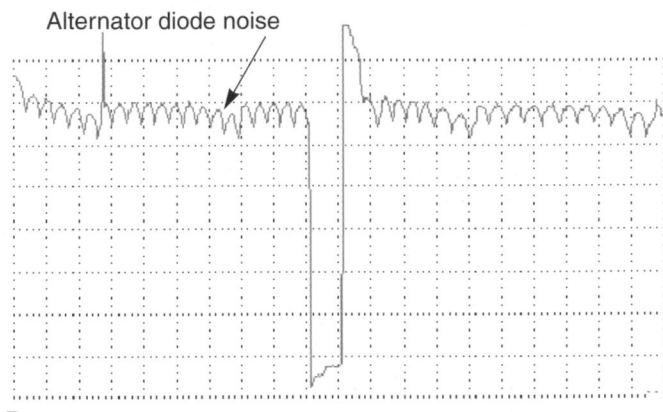

B

Figure 46-10. A—Electromagnetic interference can come from a variety of sources and can cause major problems. This waveform is caused by a defective alternator diode. B—This is the injector pulse waveform from the same vehicle. Note the hump pattern similar to the alternator pattern. (IATN)

corrects the problem, suspect problems inside the alternator. If the problem only occurs with the heated windshield turned on, check components within the heated windshield circuit.

A small transistor radio can be used to find induced voltage sources. Turn the radio on and set it on the AM band, but do not tune it to a station. If the shop is equipped with fluorescent lights, turn them off or test the vehicle outside, away from power lines and any other sources of EMI.

Move the radio around the engine compartment and under the dash with the engine running. If EMI noise (static) is present, a popping or cracking noise will be produced by the transistor radio. You can also use a car antenna cable and the car radio as a "noise sniffer." See **Figure 46-11.**

To correct an EMI problem, you must stop the source of the interference (replace leaking spark plug wire, use suppressing condenser, etc.) or shield the affected system's wiring from the interference (reroute the sensor wire, or wrap the wire with foil-type tape, for example).

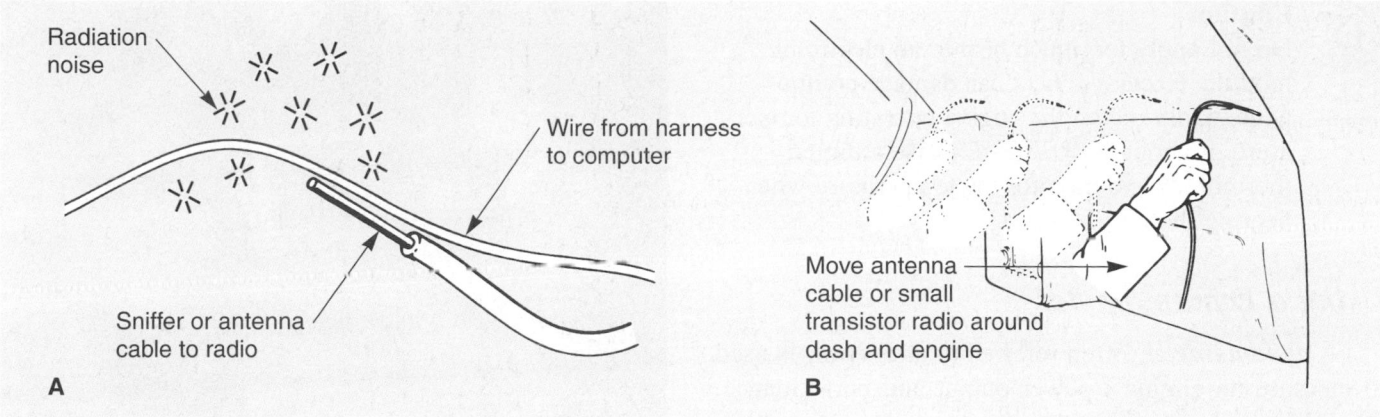

Figure 46-11. A—Electromagnetic interference can be caused by ignition secondary voltage, leaking diode in alternator, and other sources of voltage spikes or magnetic field. B—A cheap transistor radio or an extra antenna cable connected to vehicle's radio will "listen" or "sniff" for source of interference. Radiation can upset operation of computer sensor signals and car radio. (General Motors)

Using a Digital Pyrometer

A *digital pyrometer* is an electronic device used to measure temperature. Measuring temperature can help you verify scan tool readouts and find hard-to-locate problems. It is handy for advanced diagnosis of various systems and components. A digital pyrometer can be used to check:

- Engine operating temperature.
- Exhaust temperature.
- Coolant temperature.
- Sensor temperature.
- Ambient temperature.
- Air conditioning outlet temperature.

You can test a temperature sensor while it is still in the engine by checking pyrometer readings against sensor resistance when the engine is cold and again after it warms. Touch the pyrometer's probe to the sensor to get a reading of its operating temperature. This will let you compare sensor temperature and resistance readings with manual specifications.

Finding Temperature-Related Performance Problems

When an engine performance problem only occurs at a specific temperature, suspect electronic parts. Electronic circuits, especially ignition control modules, can be affected by temperature extremes.

To check a component for problems affected by temperature, use a heat gun to warm the component or a can of freeze spray to cool the unit. If the problem occurs with the temperature change, the unit is at fault and should be replaced. See **Figure 46-12.**

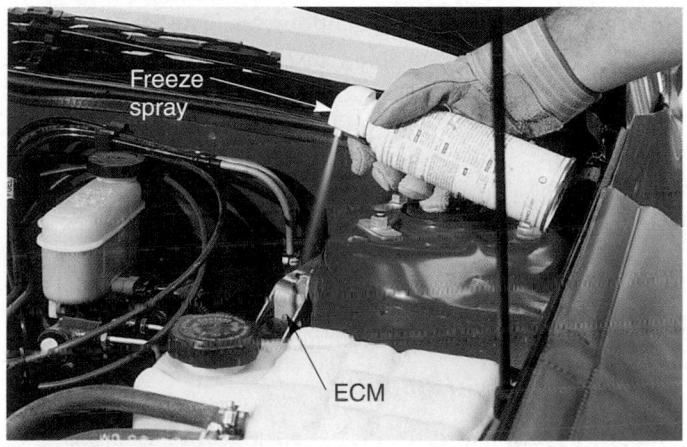

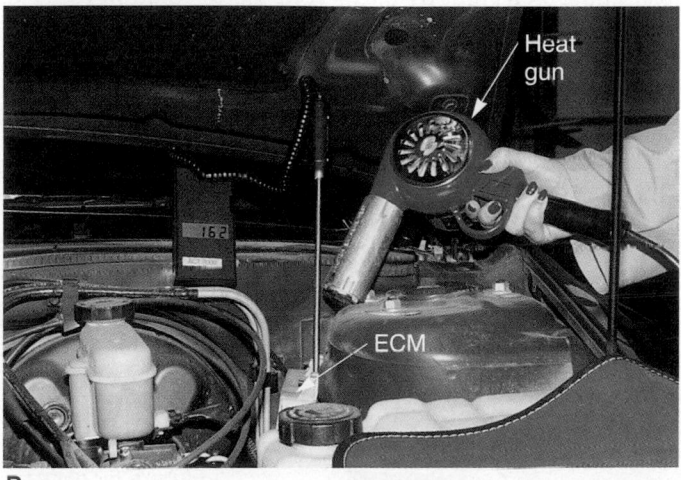

Figure 46-12. When intermittent engine problems appear to be caused by heat, cold, or period of engine operation, suspect electronic control circuits in ECMs. A—Freeze spray directed onto the ECM may cause or solve engine performance problem. If it does, replace the ECM. B—If the engine stops running when hot, direct air from heat gun onto ECM. If engine stops running, you have found problem sources. Do not overheat ECM, however, or you could damage it.

Caution
Do not apply too much heat to an electronic module. Excessive heat can damage components. Only match the engine operating temperature of about 200°F. (93°C). Use a digital thermometer to monitor the temperature when heating the unit.

Using a Dynamometer

A **dynamometer,** often referred to as a **dyno,** is used to measure an engine's power output and performance. By loading the engine, the dynamometer can check engine acceleration, maximum power output, and on-the-road performance characteristics. **Figure 46-13** shows a chassis dynamometer.

If you are having trouble finding a driveability problem, you might perform diagnostic tests while operating the vehicle on a dynamometer. This will let you simulate any condition that causes the problem.

For example, you could connect a five-gas exhaust analyzer to the tailpipe and operate the vehicle under load, **Figure 46-14.** You can also shift the vehicle through each gear, accelerate to the speed at which the problem occurs, walk around the vehicle to listen for abnormal noises, or connect listening devices—all while simulating driving conditions.

Using an Oscilloscope

An **oscilloscope,** often called a **scope,** is a piece of test equipment that displays voltages in relation to time. When connected to circuit voltage, the scope produces a line on a cathode ray tube or a liquid crystal screen. The line illustrates the various voltages present in the circuit over short periods of time.

By comparing the scope pattern (line shape) to a known good pattern, the technician can determine whether something is wrong in the circuit. An oscilloscope is usually a major component of an analyzer.

Figure 46-14. A five-gas analyzer is often used to check the content of engine exhaust gases. This will give added information for finding source of performance problem. (OTC Div. of SPX Corp.)

However, it may be mounted by itself on a small, roll-around cart, or it may be part of a hand-held scan tool or multimeter. See Figure **46-15.**

Reading the Scope Screen

The **scope screen** can give instructions, display voltages as a trace, or give other values as digital displays. The oscilloscope's ability to draw a **trace,** or pattern of circuit voltages, for very short time spans makes it very useful for testing ignition and computer system performance. You should learn to recognize good scope patterns. Then, you can easily detect scope patterns that indicate problems.

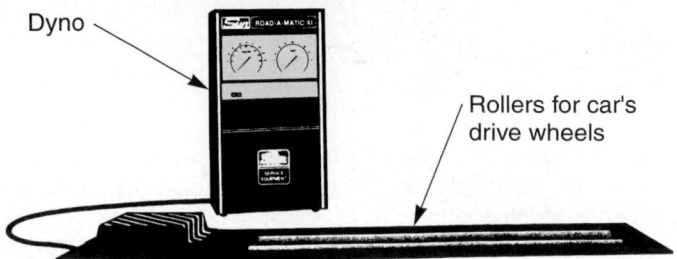

Figure 46-13. A chassis dynamometer will measure engine power output under road conditions. It will also load the engine while other tests are performed. (Sun Electric Corp.)

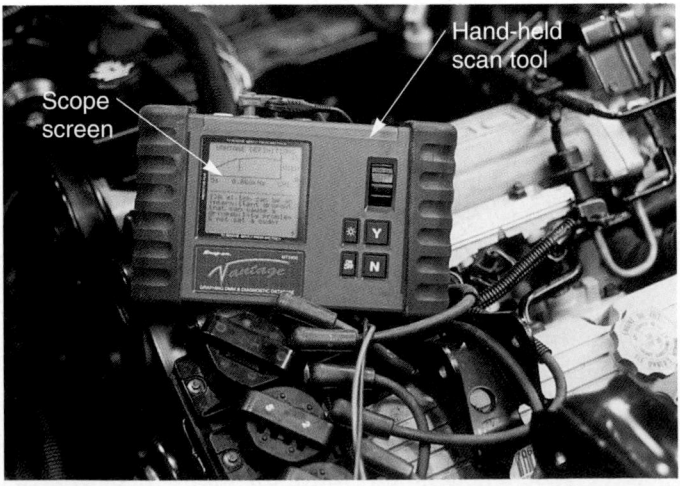

Figure 46-15. This hand-held scan tool also functions as an oscilloscope. It can display voltages in relation to time. Note the scope pattern on the screen.

Voltage is shown on the scope screen along the vertical (up and down) axis, or scale. Voltage values are given on the right and left borders of the screen. See **Figure 46-16.**

With the controls set on kV, the numbers on the screen represent kilovolts. One kV equals 1000 volts; 5 kV equals 5000 volts; etc. If a line on the scope screen extends from zero to 7 kV, the scope is reading 7000 volts.

If the scope is set to read 0-10 volts for checking the ECM and its sensors, a line five divisions tall would indicate 5 volts. Similarly, a waveform five divisions tall would be a reading of 5 volts peak-to-peak (from the top of the positive trace to the bottom of the negative trace).

Voltage is the most commonly used value on a scope screen. As voltage increases, the trace line on the scope moves up. As voltage drops, the trace line moves down a proportionate amount.

Scope time is given on the horizontal scale of the scope screen in degrees, milliseconds, or duty cycle.

Different scales may be given on the bottom of the screen for four-, six-, or eight-cylinder engines. These scales are calibrated in degrees of distributor rotation. Degrees may also be given as a percentage, for quick reference to any number of cylinders.

The scope screen may also have a milliseconds scale for measuring actual time. This makes it possible to measure how long each spark plug fires in milliseconds.

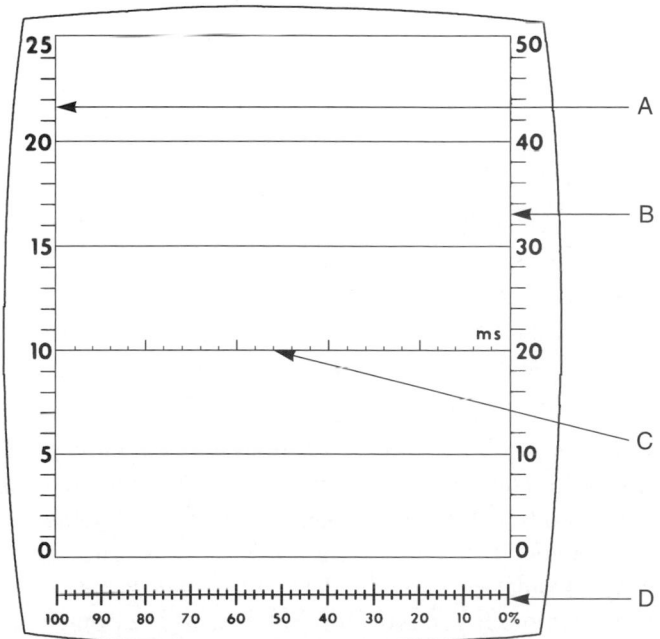

Figure 46-16. Scales on an oscilloscope screen allow you to measure voltage and time accurately. A—0–25,000 volt scale. B—0–50,000 volt scale. C—Scale for measuring time in milliseconds. D—Scale for measuring in degrees. (Sun Electric Corp.)

A certain amount of time is needed to properly ignite and burn the air-fuel mixture.

Scope Sweep Rate

Scope sweep rate is the frequency or time division shown on the screen during each test. The sweep rate adjustment affects the horizontal, or time, measurement. The scope sweep rate must be set to match the waveform frequency to be analyzed. Sweep rate is commonly given in milliseconds (ms).

A *low scope sweep rate* will compress the waveform, and too much information will be shown at once. A *high sweep rate* will expand the waveform, and only a small section of the complete duty cycle will be displayed.

Trial and error adjustment of sweep rate is commonly used. The sweep rate knob, or sweep knob, (time/division) on the scope is turned until the desired waveform is displayed on the screen. Compare the waveform pattern on the scope to a known good pattern.

Ignition System Patterns

A vehicle's ignition system is designed to produce wide fluctuations in voltage. When an ignition system is functioning properly, these voltages are within specifications.

A component with higher-than-normal resistance (open spark plug wire, for example) would be indicated on the scope as higher-than-normal voltage trace. The high resistance would produce a high voltage drop. A shorted component (fouled spark plug) would have low resistance and would produce lower-than-normal voltage trace.

An oscilloscope's controls allow it to display either the primary (low voltage) pattern or the secondary (high voltage) pattern of the ignition system. The scope patterns are similar, but important differences should be understood.

To introduce the basic sections of a scope pattern, the primary and secondary patterns for *one cylinder* will be explained. More complex patterns for specialized tests will be covered later in this chapter.

Primary Scope Pattern

The **primary scope pattern** shows the low voltage, or primary voltage, changes in an ignition system. A primary scope pattern is shown in **Figure 46-17.**

The primary ignition pattern has three sections: firing, intermediate, and dwell. Note how the voltages change in each section of the pattern.

The ignition secondary circuit cannot work properly unless the primary circuit is in good condition. A problem in the primary circuit will usually affect the secondary circuit. For this reason, the secondary circuit pattern is *checked more often* than the primary pattern.

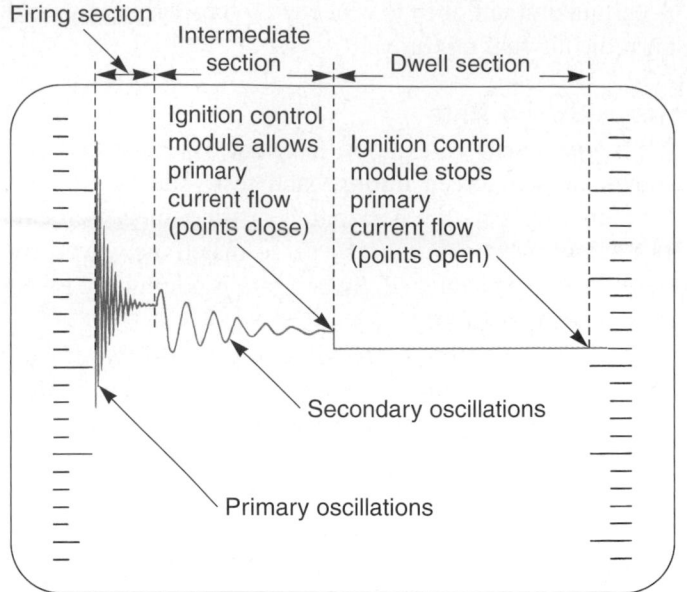

Figure 46-17. Typical primary waveform for an ignition system. Study the various sections of the trace.

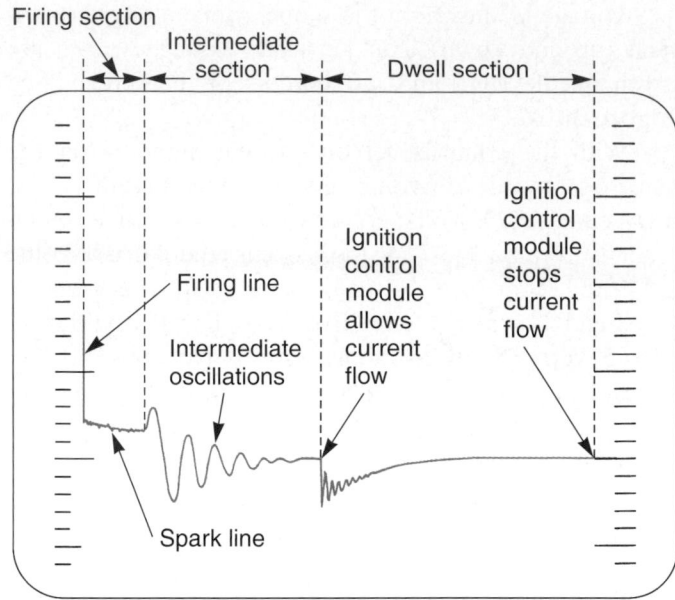

Figure 46-18. A secondary waveform for one cylinder. The firing line is voltage needed to fire the spark plug. The spark line is voltage needed to maintain the spark across the plug gap. Intermediate oscillations show the coil and condenser action. Dwell is the amount of time primary current flows through the ignition coil.

Secondary Scope Pattern

The *secondary scope patterns* show the high voltages needed to fire the spark plugs. **Figure 46-18** illustrates the secondary pattern for one cylinder.

Secondary Firing Section. The secondary pattern starts on the left with the firing section. The *firing section* will pinpoint problems with the spark plugs, the plug wires, the distributor rotor, and the distributor cap, **Figure 46-18.**

The *firing line* is the tall spike or line representing the amount of voltage needed to cause the electric arc to jump across the spark plug gap. It is normally the peak voltage in the ignition system, **Figure 46-18.**

The *spark line* shows the voltage used to maintain the arc across the spark plug electrodes, **Figure 46-18.** Once the spark is started, less voltage is needed to maintain the arc. The spark line should be almost straight, clean, and about one-fourth as high as the firing line.

Secondary Intermediate Section. The secondary pattern's *intermediate section,* or *coil oscillations section,* shows voltage fluctuations after the spark plug stops firing. Typically, the voltage should swing up and down four times (four waves) at low engine speeds. This section of the pattern will indicate problems with the ignition coil or coil pack. See **Figure 46-18.**

The voltage oscillations will disappear at the end of the intermediate section as the ignition amplifier begins to conduct or the breaker points close.

Secondary Dwell Section. The secondary pattern's *dwell section* starts when the ignition module conducts primary

current through the ignition coil. In a contact point system, it is the time when the points are closed. The ignition coil is building up a magnetic field during the dwell section.

The dwell section will indicate problems such as a faulty ignition module, burned contact points, or a leaking condenser. Contact point dwell (related to point gap) can be read by measuring the length of the dwell section along the bottom scale of the scope screen.

 **Tech Tip**
An electronic ignition can have different dwell periods from cylinder to cylinder. However, if the dwell varies in a contact-point type ignition, it indicates distributor wear or damage.

The scope pattern for an electronic ignition will vary from the pattern of a contact point–type ignition. The firing and intermediate sections are similar, but the dwell sections differ. Instead of mechanical contact points, an ignition module operates the ignition coil. The circuit design inside the module determines the shape of the dwell section. If you are not familiar with electronic ignition waveforms, they can be easily misinterpreted.

Scope Test Patterns

There are five scope test patterns commonly used by the technician when checking ignition system operation:

primary superimposed, secondary superimposed, parade (display), raster (stacked), and expanded display (cylinder select).

As you will learn, each of these patterns is capable of showing certain types of problems.

Primary Superimposed Pattern

The **primary superimposed pattern** shows the low voltages in the primary system—the ignition module or the condenser, coil primary windings, and points.

Superimposed means that the patterns for all the cylinders are placed on top of one another. This makes the trace line thicker than the single cylinder pattern discussed earlier.

Sometimes, an experienced technician will inspect the primary superimposed pattern before going to the more informative secondary pattern. Look at **Figure 46-19A.**

Secondary Superimposed Pattern

The **secondary superimposed pattern** places all the cylinder waveforms on top of each other, but it also shows the high voltages produced by the ignition coil. It is one of the most commonly used scope patterns. The superimposed secondary waveform allows you to quickly check the operating condition of all cylinders.

For example, if one spark plug is not firing properly, the waveform for that cylinder (spark plug) will not align with the others. The abnormal trace will stand out because the firing voltage is higher or lower than normal.

The secondary superimposed pattern is used to check for general problems in the ignition system. If one of the waveforms is out of place, the other scope patterns may be used to find exactly which component is causing the problem.

Parade Pattern

The **parade pattern**, also called the **display pattern,** lines up the waveform for each cylinder side by side across the screen. The number one cylinder is on the left. The other cylinders are displayed in firing order going to the right, **Figure 46-19B.**

The parade pattern takes the superimposed waveforms and separates each along a horizontal axis. This makes the parade pattern useful for comparing firing voltages of each spark plug. If one or more firing lines are too tall or short, a problem is present in those cylinders.

During normal operating condition, secondary voltages will vary from 5-12 kV for contact point–type ignitions and from 7-25 kV for electronic ignitions. The electronic ignition normally produces higher voltages because of the wider spark plug gaps needed to ignite lean fuel mixtures.

A *tall firing line* on the parade pattern indicates high resistance in the ignition secondary caused by an open spark plug wire, a wide spark plug gap, a burned distributor cap side terminal, or a burned secondary connection in a distributorless ignition. High resistance requires higher voltage output from the ignition coil.

A *short firing line* indicates low resistance in the ignition secondary, which may be an indication of

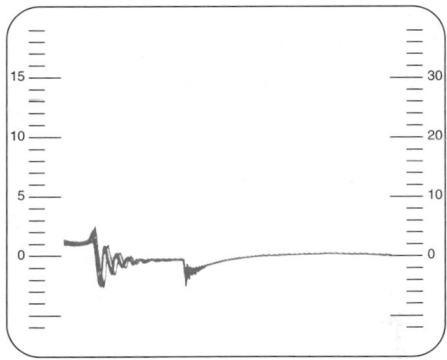

A—Superimposed display has all patterns on top of each other. It checks that all patterns are uniform.

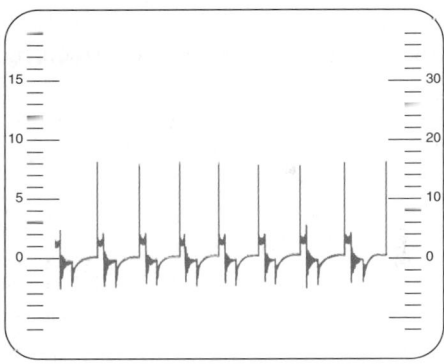

B—Parade display has cylinder patterns side by side in firing order. It is useful for comparing firing voltages. Number one cylinder is on left, with its firing line on the right.

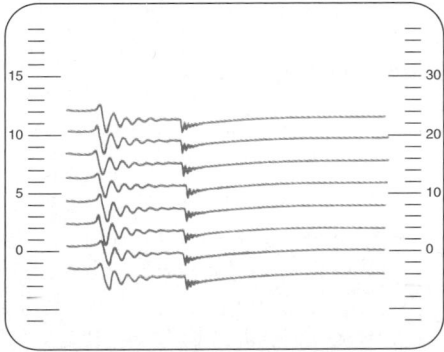

C—Stacked or raster has all cylinders one above the other. It is useful for comparing duration of events. Number one is on bottom. Others are in firing order.

Figure 46-19. Three common scope test patterns. A—Superimposed. B—Parade pattern. C—Stacked or raster pattern. (FMC)

leaking spark plug wire insulation, oil-fouled spark plugs, carbon tracking on the distributor cap or coil pack, or similar problems. Not as much voltage would be needed.

Raster Pattern

In a **raster pattern,** or **stacked pattern,** the voltage waveforms are placed one above the other as shown in **Figure 46-19C.** The bottom waveform is the number one cylinder. The other cylinders are arranged in firing order from the bottom up.

The raster pattern is normally used to check timing or dwell variations between cylinders.

Expanded Display

Some oscilloscopes have a control that allows one cylinder waveform to be displayed above the parade pattern. This arrangement is called an ***expanded display,*** or ***cylinder select.*** If a problem is located in one trace, that trace can be expanded (enlarged and moved up on screen) for closer inspection.

Reading Oscilloscope Patterns

To read a scope pattern, inspect the waveform for abnormal shapes (high or low voltages, incorrect dwell or time periods).

Since there are so many variations of electronic ignition waveforms, refer to the scope operating manual or another reference. Locate an illustration of a good scope pattern for the particular ignition system and compare it to the test pattern.

Figure 46-20 shows electronic ignition waveforms from several manufacturers. They can be used as a guide when troubleshooting. **Figure 46-21** shows the true spark and the waste spark that occur when one ignition coil in a coil pack fires two spark plugs at once. **Figure 46-22** gives several faulty scope patterns.

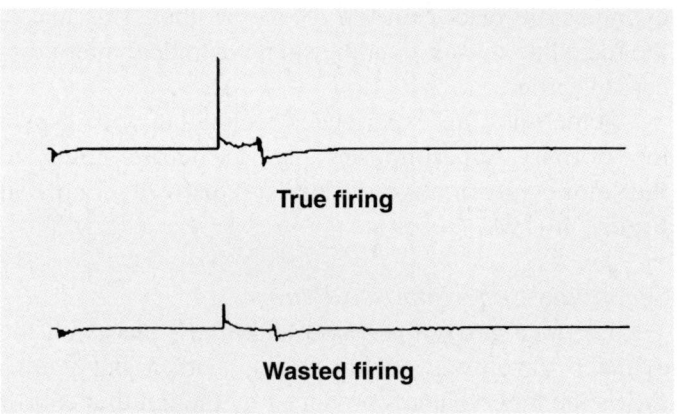

Figure 46-21. Note the differences between true firing (actual spark) and wasted firing (waste spark) when one coil fires two spark plugs at the same time. True firing starts the air-fuel mixture burning while the waste firing does nothing since the cylinder is on its exhaust stroke. (Snap-on Tool Corp.)

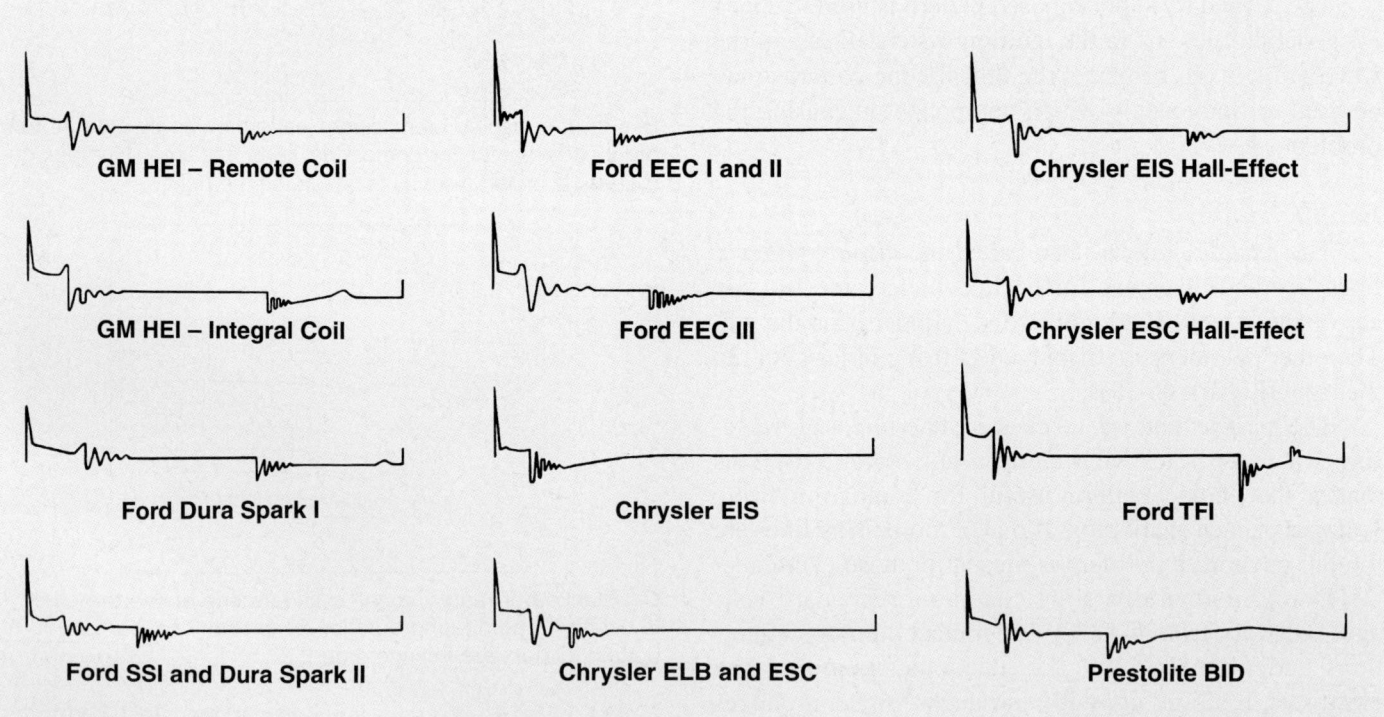

Figure 46-20. Study the differences in secondary waveforms from various auto manufacturers. (Snap-on Tool Corp.)

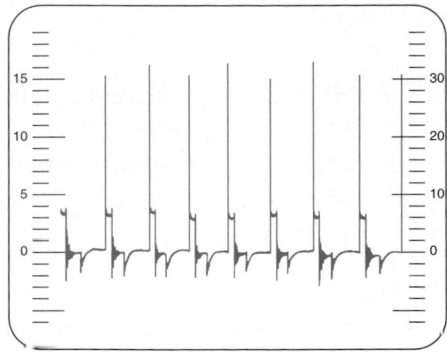

All firing lines fairly even but too high. Look for problems common to all cylinders: worn spark plug electrodes, excessive rotor gap, coil high-tension wire broken or not seated fully, late timing, excessively lean air/fuel mixture, or air leaks in intake manifold.

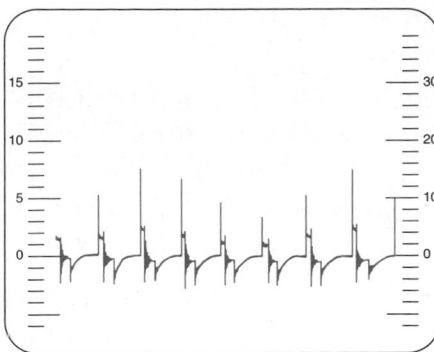

Uneven firing lines. Can be caused by worn electrodes, a cocked or worn distributor cap, fuel mixture variations, vacuum leaks, or uneven compression.

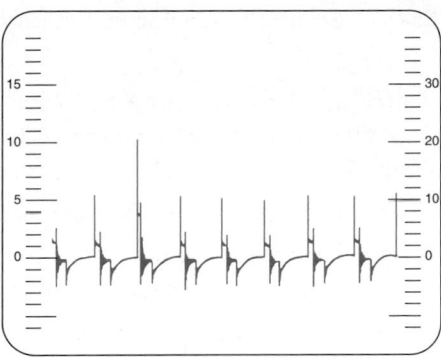

Consistently high firing line in one or more cylinders. Caused by a broken spark plug wire, a wide spark plug gap, or a vacuum leak.

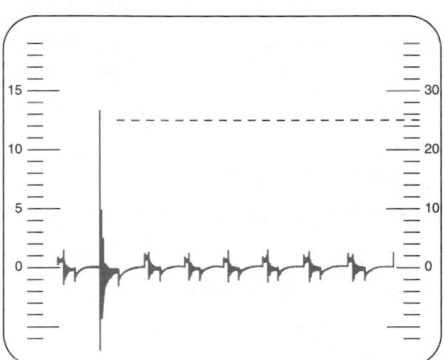

Maximum available voltage during coil test should be within the manufacturer's specifications. Disconnect plug wire to check maximum coil output.

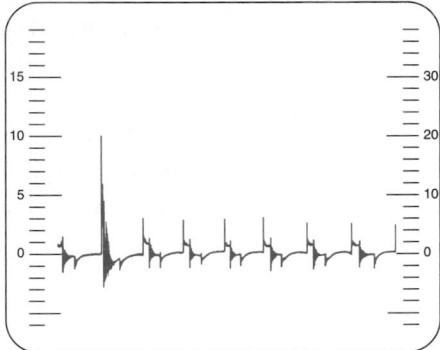

With plug wire removed for coil output test, a short, intermittent, or missing lower spike indicates faulty insulation. This is usually caused by a defective spark plug wire, distributor cap, rotor, coil wire, or coil tower.

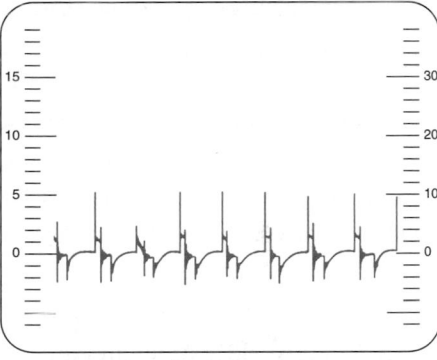

Consistently low firing line in one or more cylinders. Caused by fouled plug, shorted wire, low compression (valve not closing), or rich mixture.

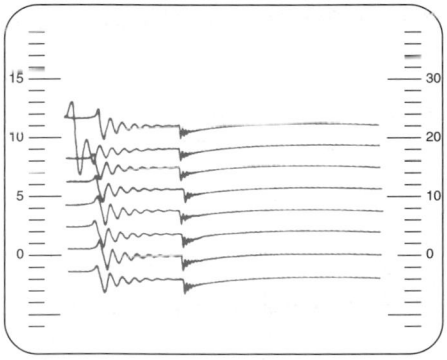

No spark line. Caused by complete open in cable or connector.

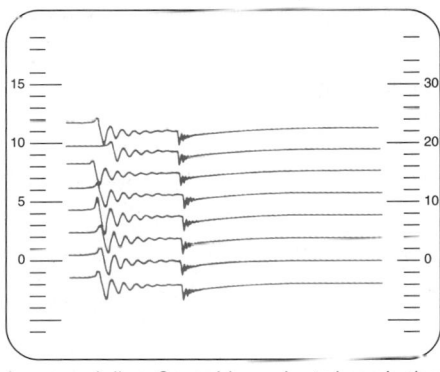

Long spark line. Caused by a shorted spark plug or partially grounded plug wire.

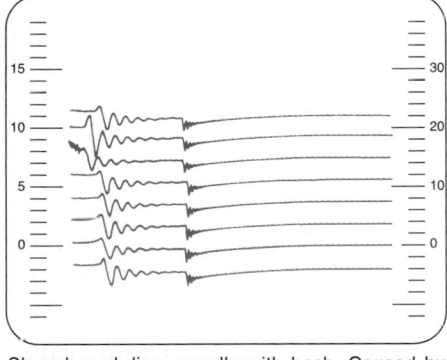

Sloped spark line, usually with hash. Caused by fouled spark plug.

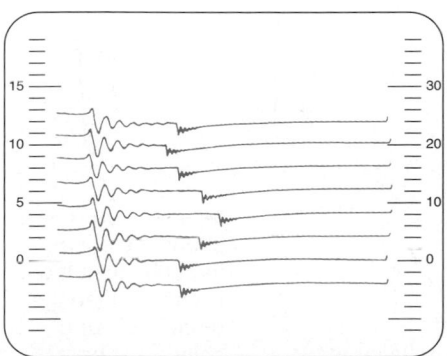

Poor vertical alignment of point-open spikes. Caused by worn or defective distributor shaft, bushings, cam lobes, or breaker plate.

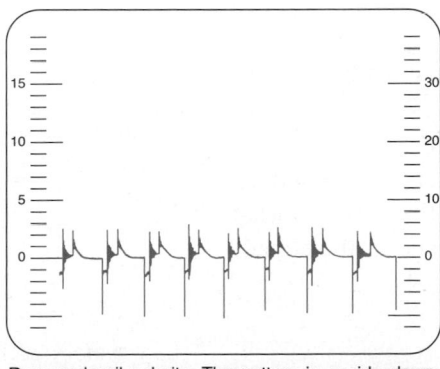

Reversed coil polarity. The pattern is upside down. This problem is usually caused by someone accidentally connecting the primary leads to the coil backwards. The ignition will still work, but not as well.

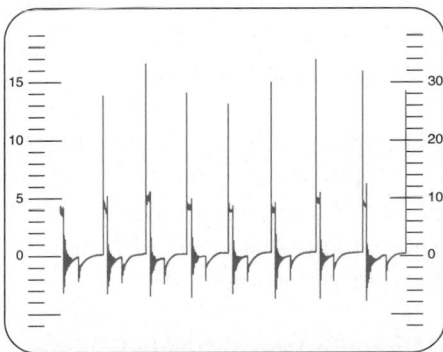

Run the engine at about 1000 rpm. While watching firing lines on scope, snap throttle fully open, then quickly release it. Highest firing line peaks should not be more than 75% of coil output.

Figure 46-22. Examples of bad scope patterns. Study the shape of each trace and the problems that cause each waveform. (FMC)

Analyzing Square and Sine Wave Signals

When analyzing a square wave, there are several things you should check. They include:

- The *base line* is the reference line, or zero volts.
- The *rising edge,* or *leading edge,* is where the square wave goes from zero to high voltage.
- The *on-time,* or *high-time,* is the part where the square wave stays at maximum voltage.
- The *trailing edge,* or *falling edge,* is the drop in voltage back to zero.
- The *off-time,* or *low-time,* is where the square wave stays on the baseline.
- The *amplitude,* or *peak-to-peak voltage,* of a square wave is determined by the horizontal distance from the baseline to the high-time.

You can inspect these sections of the waveform to determine if there is a problem. Some common problems that can affect a square wave include:

- Low or high resistance in the circuit or its components.
- Faulty electronic circuit.
- Circuit contaminated by moisture.

When analyzing sine waves, check the following:

- Analog peak-to-peak voltage—Is the waveform voltage strong from top to bottom?
- Analog wave shape—Is the trace normal for a known good component?
- Analog wave frequency—Is the distance between waves normal?
- Analog wave smoothness—Is there unwanted hair or static on sine wave?

Computer System Scope Tests

An oscilloscope can be used to help you find computer system problems. When the scan tool does not find anything and you still have performance problems, you may need to check sensor and ECM signals with a scope.

Distributor Pickup Coil Scope Testing

An oscilloscope can also be used to check the output signal of a distributor pickup coil. It will not only measure voltage, but it will also show the shape of the signal leaving the pickup coil.

Magnetic Sensor Testing

A *magnetic sensor test* is done by measuring the output voltage from the sensor with the engine cranking. With a magnetic sensor, connect the scope primary leads to the pickup coil. Set the selector to primary and the

primary height control to 40V. Adjust the pattern length to minimum.

With the engine cranking, an ac (alternating current) signal about 1.5V peak-to-peak should be generated, **Figure 46-23.**

Hall-Effect Sensor Testing

A *Hall-effect sensor test* is best done by checking the sensor's output waveform with an oscilloscope. Without disconnecting the circuit reference voltage, probe the output wire at the sensor connector. The service manual will give pin numbers for probing. See **Figure 46-24A.**

A Hall-effect sensor waveform should switch rapidly, have vertical sides, and have the specified voltage output (typically about 4-5 volts peak-to-peak). The top of the square wave should reach reference voltage and the bottom should almost reach ground, or zero. Signal frequency should change with engine cranking speed or engine rpm. Hall-effect pickups can be found in distributors and some crankshaft position sensors. Since specifications vary for Hall-effect sensors, refer to the service manual for that vehicle. See **Figure 46-24B.**

Optical Sensor Testing

An optical sensor can also be tested with an oscilloscope. You can probe the output wires from the sensor and compare the waveform to specifications.

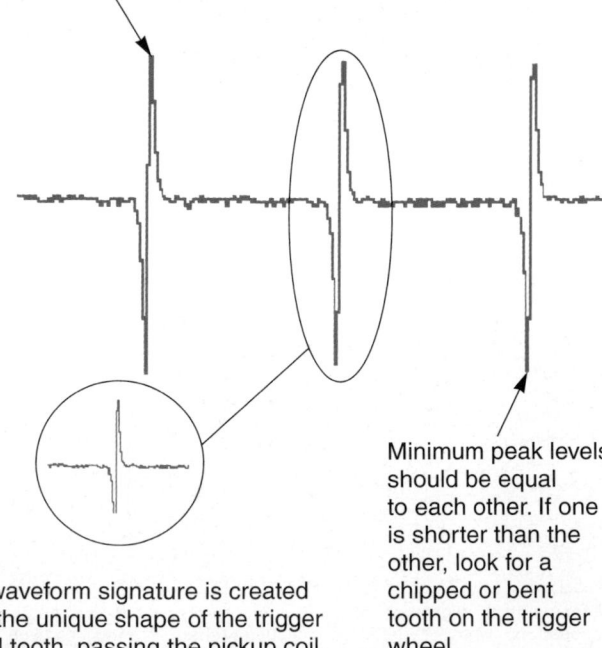

Maximum peak levels should be equal to each other. If one is shorter than the other, look for a chipped or bent tooth on the trigger wheel.

Minimum peak levels should be equal to each other. If one is shorter than the other, look for a chipped or bent tooth on the trigger wheel.

The waveform signature is created from the unique shape of the trigger wheel tooth, passing the pickup coil.

Figure 46-23. Typical waveform from a magnetic distributor pickup. (Fluke)

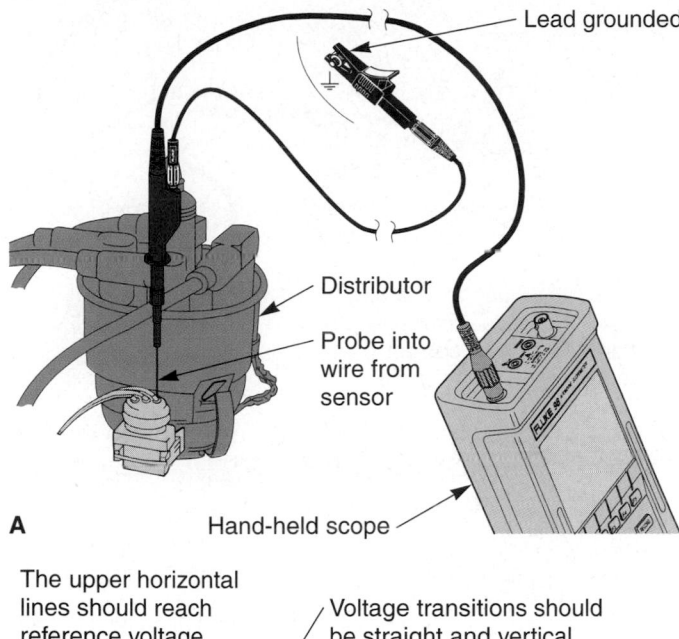

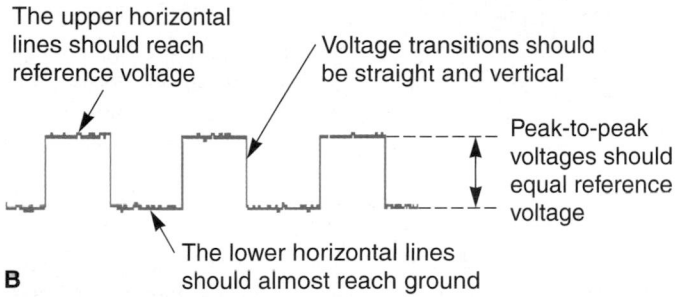

Figure 46-24. A—This scope is being used to check the signal from a Hall-effect sensor. B—Hall-effect sensor signal. The frequency of the signal should increase as engine speed increases. (Fluke)

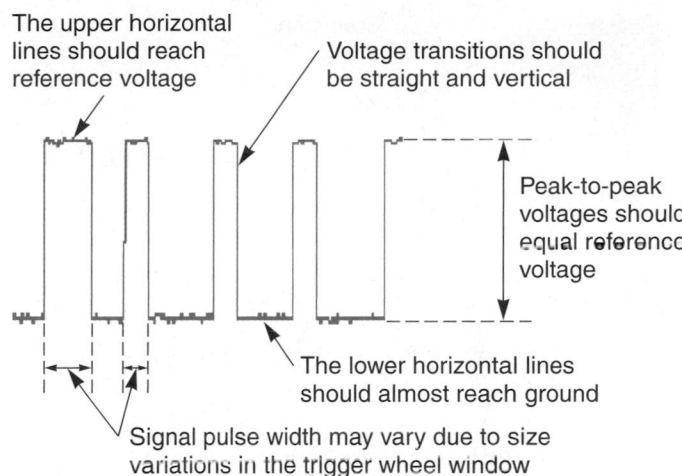

Figure 46-25. Typical waveform generated by an optical sensor. If the shutter blade widths vary, the pulse width will also vary. (Fluke)

A *optical pickup test* measures the output generated by the photo diodes as they are energized by the LEDs. It is also easily done with a hand-held scope probing into the sensor's electrical connector. Again, refer to the service manual to find the connector pin numbers for the optical pickup's output wire. Optical sensors are used in a few distributor designs and are never used in crankshaft sensors.

An optical sensor's waveform should have straight sides and adequate voltage output. The upper horizontal line on the waveform should almost reach reference voltage. The bottom horizontal line should almost reach ground, or zero. See **Figure 46-25.**

Remember that optical sensors are susceptible to dirt. An oil mist or a film of dirt can prevent light transfer from the LEDs to the photo diodes. Again, refer to the manufacturer's service information for specific information.

Crankshaft Position Sensor Testing

Figure 46-26A shows how to use a hand-held scope to test a crankshaft position sensor. You can use the needle probe on the scope lead to check for an output

signal without disconnecting wires. This scope will show both ac output and a trace for voltage signal variations. Note that this testing method would also work on engine block-mounted magnetic crankshaft position sensors.

Tech Tip
Some electrical connectors are sealed and do not allow easy probing. You may need to install a test connector or jumper wires between the two halves of the connector to probe sensor voltages.

When reading the sensor waveform, make sure the peak voltage levels are equal to each other. If one is short or missing, inspect the trigger wheel for a broken tooth. Peak-to-peak voltage levels should be within specifications. See **Figure 46-26B.**

As with any sensor, reference voltages, wiring, and other criteria will vary. If in doubt, always refer to the service manual for the vehicle being tested to get accurate electrical values.

Throttle Position Sensor Testing

To scope test a throttle position sensor (TPS), connect the test leads to the sensor output wire and to ground. Voltage should still be fed to the sensor from the ECM. Move the throttle open and closed. The TPS waveform should show a smooth curve, without any spikes. See **Figure 46-27.**

Manifold Absolute Pressure Sensor Testing

A scope can also be used to test the operation of both analog and digital manifold absolute pressure sensors. Accelerate the engine and note the changes in airflow signals going to the ECM. Compare the amplitude and

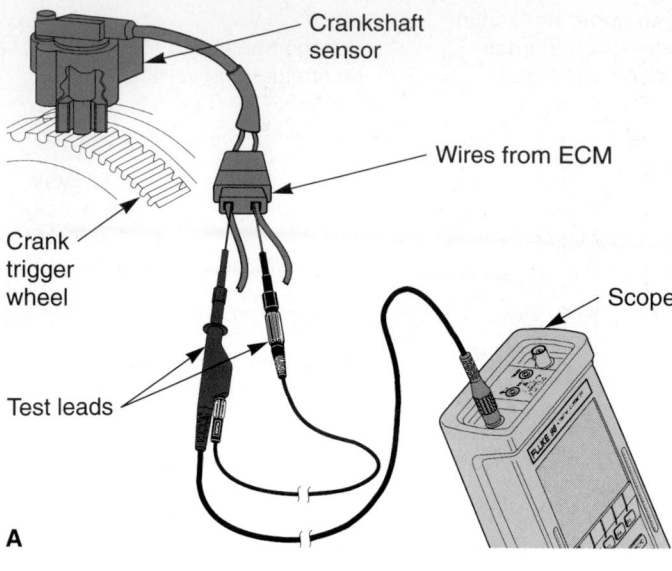

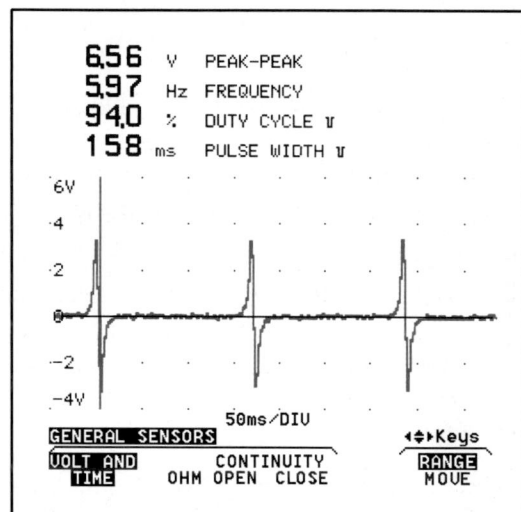

Figure 46-26. Scope testing crankshaft position sensors is similar to testing magnetic distributor sensors. A—Since crankshaft sensors generate their own voltage signal, connect scope to terminals as specified. B—Note the resulting display. Compare the waveform to the service manual description.

shape of the waveform to known good patterns. This is shown in **Figure 46-28.**

Mass Airflow Sensor Testing

To test analog or digital mass airflow sensors using a scope, probe the connector as recommended in the service manual. Compare your scope readings to factory specifications and known good readings. See **Figure 46-29.**

Knock Sensor Testing

To test a knock sensor with a scope, connect the scope test leads to the sensor. Then tap on the engine next to the sensor with a small hammer or a wrench.

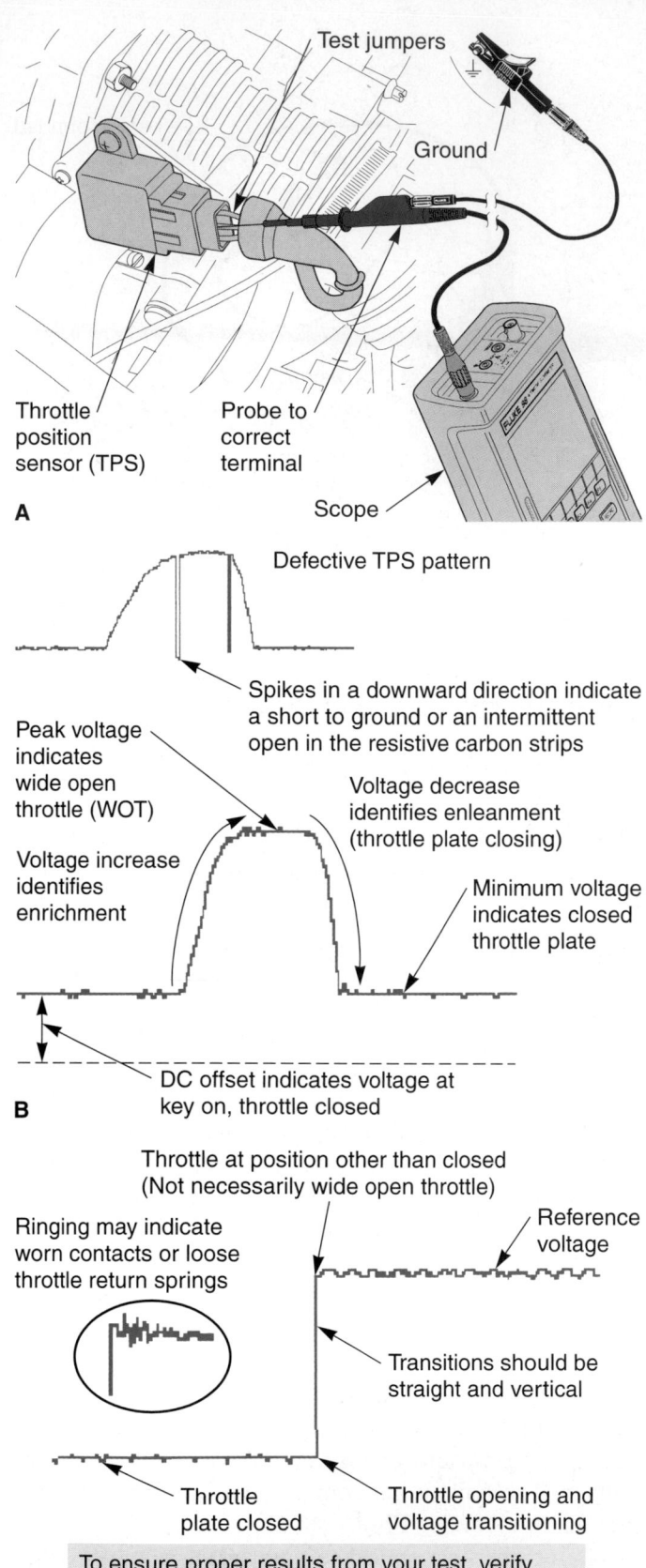

Figure 46-27. Throttle position sensor can also be checked with a scope. A—Probe through wires or use jumpers so power can be connected to sensor. B—Potentiometer, or variable-resistor, TPS should produce smooth curve as throttle is moved open and closed. Spikes indicate sensor problem. C—Switching-type TPS should produce good square wave without ringing. (Fluke)

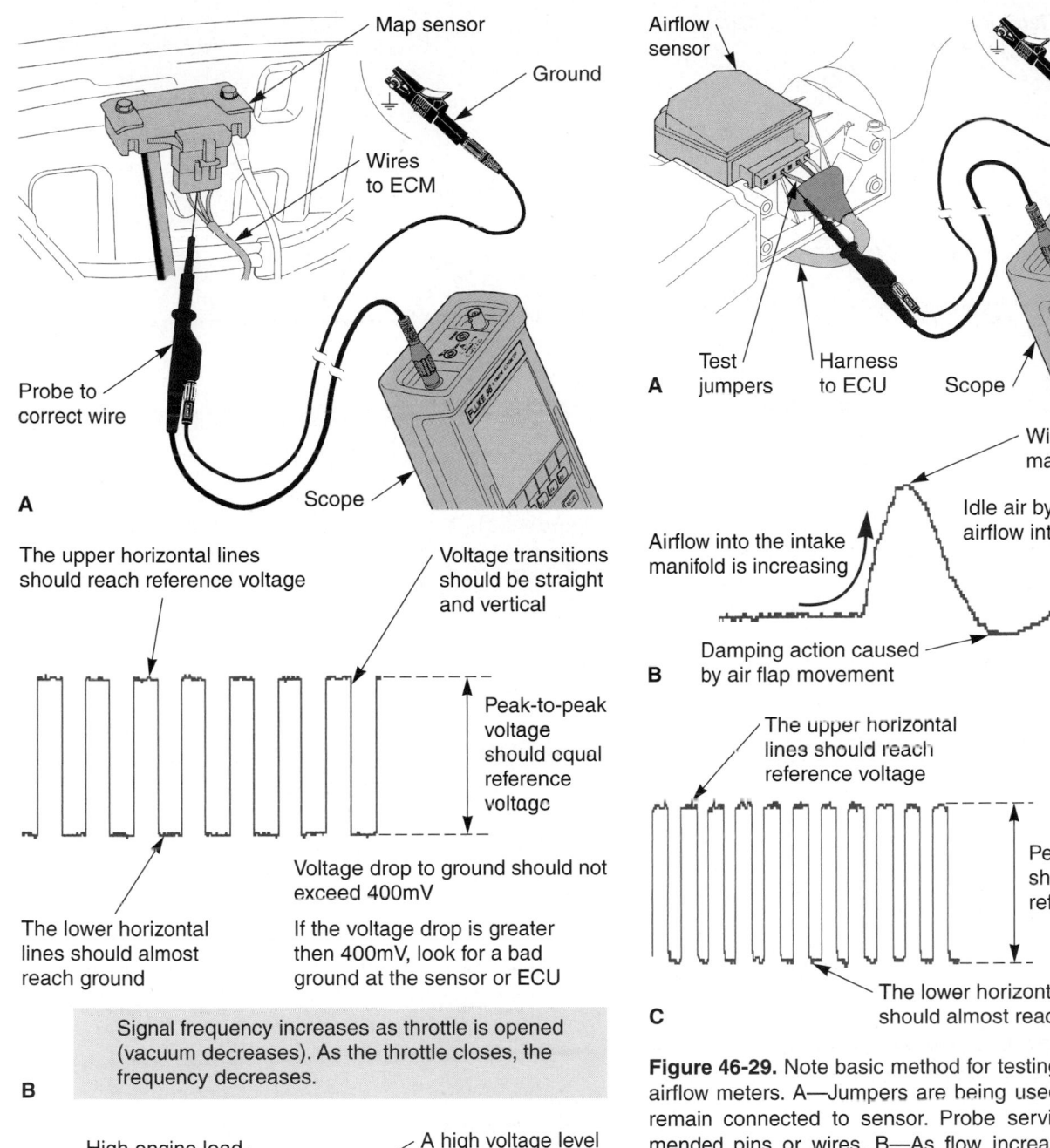

A

The upper horizontal lines should reach reference voltage

Voltage transitions should be straight and vertical

Peak-to-peak voltage should equal reference voltage

The lower horizontal lines should almost reach ground

Voltage drop to ground should not exceed 400mV

If the voltage drop is greater then 400mV, look for a bad ground at the sensor or ECU

Signal frequency increases as throttle is opened (vacuum decreases). As the throttle closes, the frequency decreases.

B

High engine load

A high voltage level indicates high intake manifold pressure (low vacuum)

Low engine load

As the throttle plate opens, manifold pressure rises (manifold vacuum lowers)

A low voltage level indicates low intake manifold pressure (high vacuum)

C

Figure 46-28. Manifold absolute pressure sensor can also be checked with scope. A—Here scope is probing through connector to test MAP sensor. Other lead is grounded. B—Signal frequency should increase with engine speed with digital MAP. C—Amplitude should increase with engine speed with analog MAP. (Fluke)

Airflow sensor

Ground

A Test jumpers   Harness to ECU   Scope

Wide open throttle, maximum acceleration

Idle air bypass compensating airflow into intake manifold

Airflow into the intake manifold is increasing

Damping action caused by air flap movement

B

The upper horizontal lines should reach reference voltage

Peak-to-peak voltages should equal reference voltage

The lower horizontal lines should almost reach ground

C

Figure 46-29. Note basic method for testing analog and digital airflow meters. A—Jumpers are being used to allow power to remain connected to sensor. Probe service manual recommended pins or wires. B—As flow increases, analog airflow meter should produce more voltage. C—With digital airflow meter, signal frequency usually increases with engine speed and airflow. (Fluke)

See **Figure 46-30A.** This should make the sensor produce a signal that is similar to the one shown in **Figure 46-30B.**

Another way to check a knock sensor and the ECM is to measure ignition timing while tapping on the engine next to the sensor. The ECM should retard ignition timing when you tap on the engine.

Alternator Diode Testing

Most analyzers are capable of checking alternator diode condition. The scope will display the alternator's

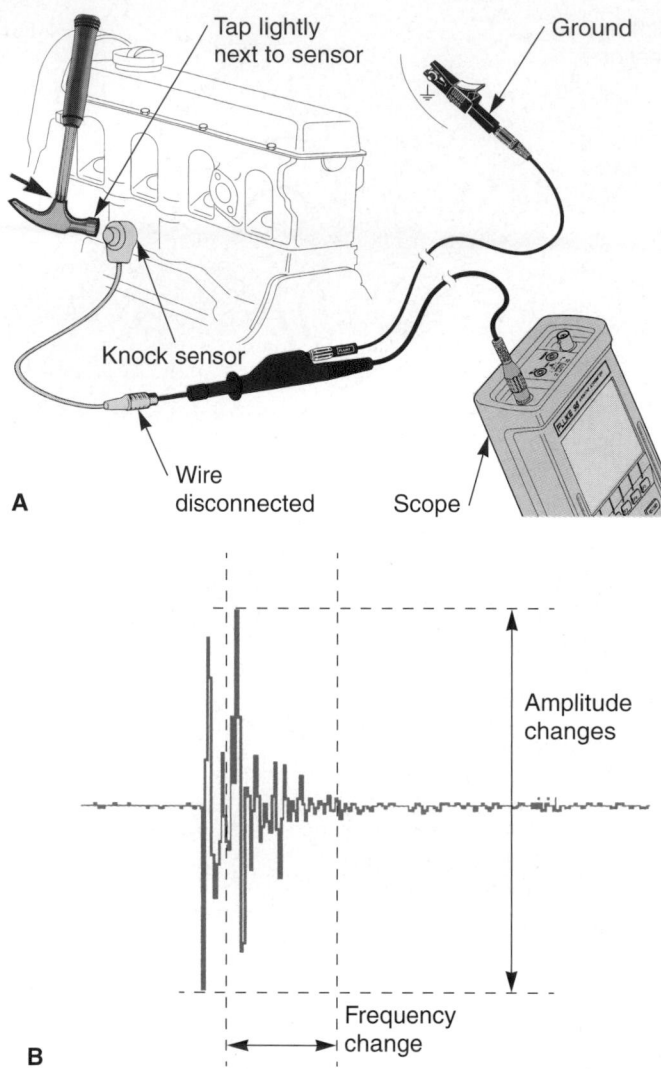

Figure 46-30. Knock sensor signal can also be analyzed with scope. A—Connect scope lead to knock sensor and other lead to ground. Tap on engine with small hammer or wrench to produce output signal. B—Knock sensor should produce normal frequency and amplitude signal when engine is tapped on. (Fluke)

voltage output. If the alternator diodes are good, the pattern should be wavy but almost even. This was detailed in Chapter 34, *Charging System Diagnosis, Testing, and Repair.*

Electronic Fuel Injector Testing

Oscilloscopes can also be used to check injector operation in an electronic fuel injection system. Refer to equipment and service manual instructions for details. **Figure 46-31** shows typical waveforms for good and defective fuel injectors.

Oxygen Sensor Testing

An oscilloscope can also be used to check the signal produced by an oxygen sensor. Oxygen sensor testing

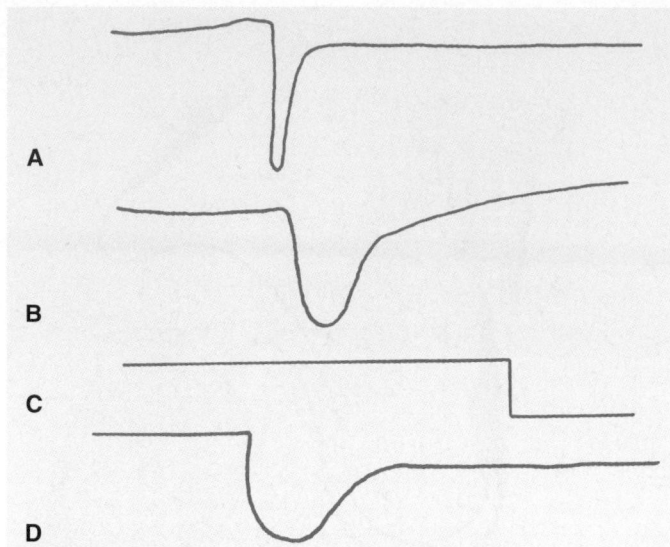

Figure 46-31. These are waveforms for electronic injectors. A—Normal injector pattern. B—Stuck injector. C—Open injector. D—Partially shorted injector. (Snap-on Tool Corp.)

and service is covered in detail in Chapter 44, *Emission Control System Testing, Service, and Repair.*

ECM Scope Testing

A scope can be used to check the output pulses leaving an electronic control module. You can measure and observe the pulses going to fuel injectors, solenoids, and servo motors. You can also check the reference voltage being sent to sensors, **Figure 46-32.**

Since ECM testing varies and is complex, always refer to the service manual for detailed instructions. You must compare your test waveform or voltage to known correct values. If the ECM fails to produce a good pulse or reference voltage, it should be replaced.

 Note
For more information on ECM service, refer to the index. This subject is covered in several other chapters.

Flight Record Test

A *flight record test* stores the sensor waveform in the scope's memory when a problem occurs. For example, when trying to check an intermittent problem, connect the hand-held scope to the sensor and test drive the vehicle. When the problem occurs, press the storage button on the scope. The scope will then store a picture of the sensor output for analysis, **Figure 46-33.**

Engine Analyzer (Computer Analyzer)

An *engine analyzer,* also called a *vehicle analyzer* or computer analyzer, consists of a group of test instruments

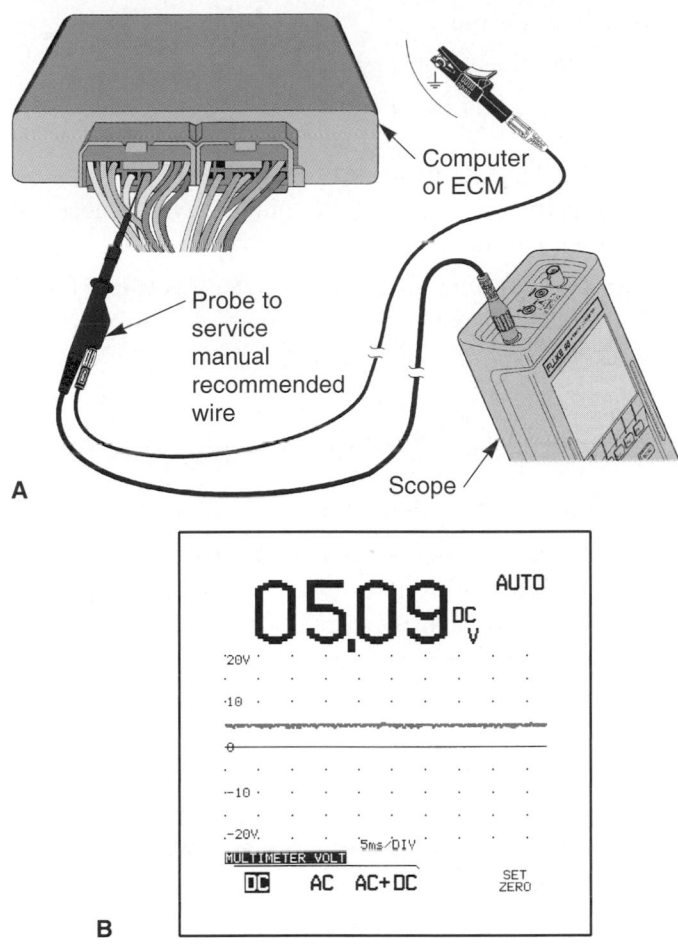

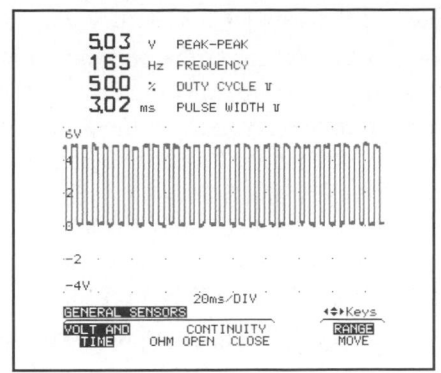

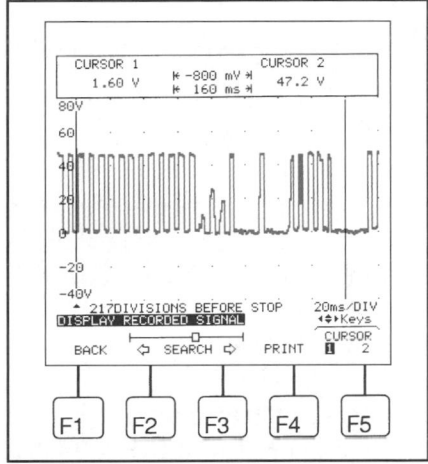

Figure 46-32. A—A scope will also check reference voltage going to sensors and the control pulses from the ECM to the actuators. Compare readings and waveforms to service manual specifications. B—Reference voltages should meet specifications and the waveforms should reflect smooth dc voltage. (Fluke)

Figure 46-33. The flight record is a feature on some small hand-held scopes. Connect to a suspect sensor using long test leads and place the scope on seat during test drive. A—Normal, consistent signals from magnetic sensor. B—When the problem occurs, the scope will store a picture of the sensor signals. Note how each signal varies, possibly from intermittent open sensor coil windings or loose mounting. (Fluke)

that includes a scope, a tach-dwell meter, a VOM, an exhaust gas analyzer, and, sometimes, a scan tool. These tools are mounted in a large, roll-around cabinet. The operation of each instrument is often controlled by a computer that interfaces all the testing devices. See **Figure 46-34.**

When connected to the vehicle, the analyzer will help you check the condition of the engine and its support systems. An engine analyzer will help find problems when a scan tool does not show a trouble code or an out-of-parameter operating value.

For example, if an engine misfire is being diagnosed, the analyzer will help find which parts are defective. It will pinpoint fouled spark plugs, open plug wires, rich or lean fuel mixtures, inoperative fuel injectors, and other problems, even before removing and inspecting parts.

Figure 46-34. This technician has test driven a vehicle with a hand-held scan tool connected to a flight recorder tester. At the shop, the data collected can be fed into a computerized analyzer for further evaluation. This is useful on difficult-to-find problems. (OTC Div. of SPX Corp.)

Modems

Some analyzers can transmit data over telephone lines for comparison to information stored in a larger mainframe computer by using a modem. A *modem* is an electronic device that allows computer data to be sent and received over telephone lines.

Data can be sent back and forth between modems. This allows the technician to access information that can be used to troubleshoot difficult problems. Most dealerships have modem-equipped computer analyzers. The analyzer is plugged into the vehicle's data link connector and the information is sent by modem to the mainframe computer.

A *mainframe computer* is a very large computer that can store tremendous amounts of data. It can also do multiple tasks or transfer information to several computer analyzers at the same time.

The auto or equipment manufacturer's mainframe computer may contain information about common problems. Steps for finding problems, specific voltages, and other electrical values for each model may also be stored in the mainframe's memory.

Engine Analyzer Differences

There are a number of different makes of analyzers on the market. The controls and meter faces may be organized differently, but the basic test equipment and operation of each are almost the same. See **Figure 46-35.** Most analyzers will check:

- Battery, charging, and starting systems.
- Ignition system.
- Engine condition.

Figure 46-35. An analyzer can perform different tests and measurements. It is like having a multimeter, tach-dwell, exhaust analyzer, timing light, and other testers connected at once for problem evaluation.

- Fuel system.
- Emission control systems.
- Sensor and ECM signals.

Analyzer Test Equipment

Typically, an analyzer will contain several pieces of test equipment, including:

- *Oscilloscope*—high-speed meter that uses a liquid crystal display or a television picture tube.
- *Voltmeter, ammeter, and ohmmeter*—meters used to measure electrical values.
- *Tachometer*—meter used to measure engine speed in rpms. It is commonly used when adjusting fuel injection, ignition timing, and other.
- *Dwell meter*—instrument that measures ignition module or contact point conduction time in degrees of distributor rotation. It will detect point misadjustment and other problems.
- *Timing light*—strobe light for ignition timing adjustment. Most analyzer timing lights have a degree meter for measuring distributor advance.
- *Vacuum gauge*—gauge used to measure vacuum when checking the operation of the engine operation and various vacuum-operated devices.
- *Vacuum pump*—pump capable of producing a supply vacuum for operating and testing vacuum devices.
- *Cylinder power balance tester*—unit for electrically shorting out one or more fuel injectors or spark plugs. It will determine if a cylinder is firing properly and producing power.
- *Exhaust gas analyzer*—measures the chemical content and amount of pollution in the vehicle's exhaust.
- *Scan tool*—often incorporated into analyzers for retrieving trouble codes and circuit operating values.
- *Digital display*—displays operating values for various components in alpha-numeric form. Modern analyzers display readings of various test values—engine rpm, charging system voltage, exhaust gas content, etc. See **Figure 46-36.**
- *Printer*—prints information about ignition timing, dwell, engine speed, emission levels, and other values on paper. If repairs are needed, the technician can show the customer the improper readings on the printout. If the vehicle is in good condition, the printout can serve as a record if later repairs are needed. See **Figure 46-37.**

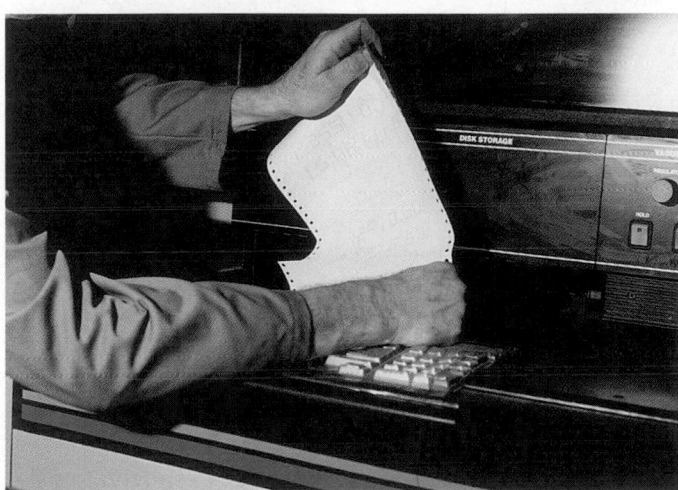

```
          CRANKING/PINPOINT TESTS        HOLD
    . . . . . ! . . . ! . . . . ! . . . . . ! . . . . ! 210
    0    100   200   300   400   500 RPM
         ENGINE      210 RPM
         CURRENT     170 AMPS
         BATTERY     10.2 VOLTS
         DIST RES    0.35 VOLTS
         DWELL       31.5 DEG. (70.0%)
         TIMING      0.0 DEG      TEST
         HC          610 PPM      DATA
         VACUUM                   SAVE D
    . . . . . ! . . . . . ! . . . ! . . . . ! . . . . . !   4.5
    0    5    10    15    20    25   "HG
    RESISTANCE   250 K OHMS

                        ENGINE KILL
A      8 CYL    TEMP  19°C   TDC        - 9.5°
```

```
              RUNNING TESTS
    . ! . . . . ! . . . . . ! . . . . ! . . . . ! . . . . -|. ! 2385
    0    5    10    15    20   2500 RPM
    RPM                2385   1550    520
    DWELL   DEG        31.8   31.8   31.5
    DWELL   %          70.7   70.7   70.0
    TIMING  DEG        19.5   17.5    5.5
    HC      PPM        640    825    900
    CO      %          3.85   4.25    4.0
    AMPS    A          22.3   32.4   35.0
    VOLTAGE V          14.0   14.6   14.2
    VACUUM  "HG        15.7   14.6    4.5
    . ! . . . . ! . . . . ! . . . ! |. . . . . ! . . . . ! .   15.7
    0    5    10    15    20    25   "HG
B      8 CYL    TEMP  19°C       TDC      00.0°
```

Figure 46-36. Many modern analyzers are equipped with a digital display or extra screen. A—Digital display for cranking tests. B—Digital display for running tests. (Snap-on Tool Corp.)

Figure 46-37. A printer will type or print analyzer test results.

Figure 46-38. Most late model analyzers will give detailed instructions for connecting the various leads to the vehicle and for doing each test. This simplifies analyzer operation considerably. (Snap-on Tool Corp.)

Analyzer Connections

Analyzer connections differ with each type and model. Nevertheless, most have the same general test connections. Modern analyzers will give you directions for connecting the test leads to the vehicle, **Figure 46-38.** If not, test leads should be connected as described in the user's manual. Special leads and hoses may be provided for measuring starting current, charging voltage, engine vacuum, fuel pump pressure, sensor signals, and exhaust gas content. These leads are generally in the same manner as those covered in other chapters.

Figure 46-39 shows how to connect an analyzer to conventional and unitized ignition coils. An adapter may be needed to connect the analyzer to a distributorless ignition system, **Figure 46-40A.** You must install secondary jumper wires on some direct ignition systems so the inductive test leads can be clamped around them to read voltages, **Figure 46-40B.**

Using an Analyzer

To use an analyzer, plug the electrical cord into a wall outlet. Set the controls and connect the test leads to the vehicle. If needed, read the operating manual for the analyzer.

Caution

Before starting the engine, make sure all leads are away from hot or moving parts. The analyzer leads are very expensive and can be easily damaged by contact with a hot exhaust manifold or a spinning fan blade or pulley. See **Figure 46-41.**

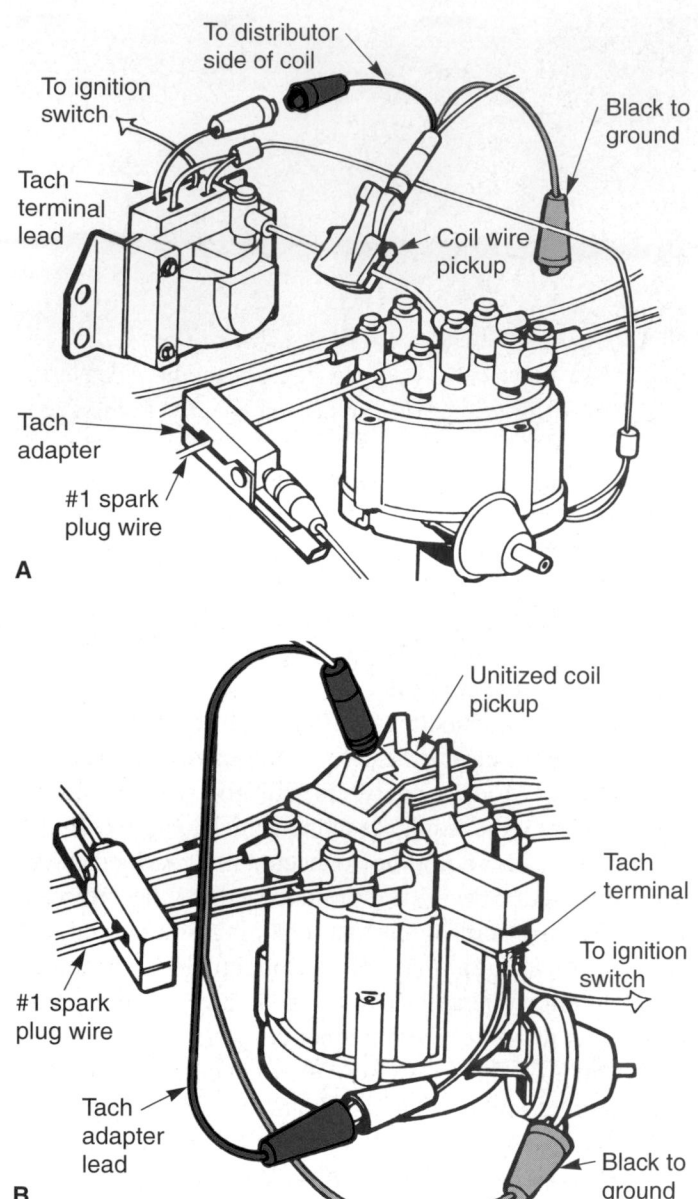

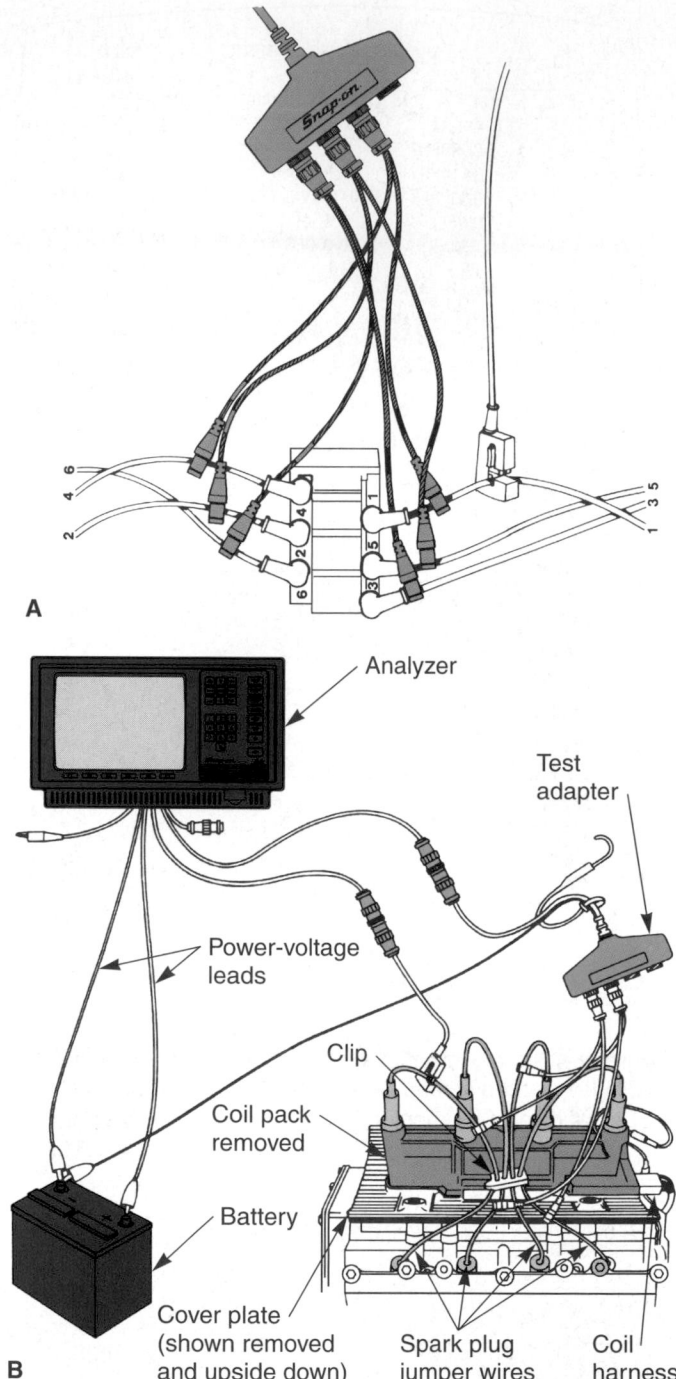

Figure 46-39. Analyzer connections to distributor ignition systems. A—Connection to an ignition system in which the coil is separate from the distributor. B—Connection to a distributor with unitized coil.

Figure 46-40. This is operating manual illustration for connecting an analyzer to distributorless and integrated direct ignition systems. A—Separate inductive leads are needed for each wire on this distributorless ignition system. B—Note jumper spark plug wires between coil pack and spark plugs. (Snap-on Tool Corp.)

Set the parking brake and start the engine. Many analyzer manufacturers recommend increasing engine idle speed to around 1500 rpm during scope tests.

Ignition Coil Output Test

A scope *ignition coil output test* measures the maximum available voltage produced by the ignition coil. A spark plug only requires about 5-20 kV for operation. However, the ignition coil should have a higher reserve voltage. Without this extra voltage, the spark plugs could misfire under load or at high engine speeds when voltage requirements are greater.

To perform the coil output test, set the analyzer controls and display to the highest kV range. Run the engine at 1000-1500 rpm. Using insulated pliers, disconnect a spark plug wire. Hold the end of the wire away from ground while watching the scope screen.

Figure 46-41. When connecting an analyzer to a vehicle, keep all cables away from hot or moving parts. Test cables are very expensive to replace.

Tech Tip

With a coil pack, you must test each coil's output voltage separately. Just because one coil passed its tests does not mean the others will. By using the coil pack firing order and the secondary pattern, you can tell which coil should be tested.

With the spark plug wire removed, a tall firing line should stand out from the others. Look over to the scope scale on the side of the screen. Read the voltage even with the top of the spike. This value should equal the capacity of the ignition coil.

Caution

A few electronic ignitions may be damaged by disconnecting spark plug wires while the engine is running. Be sure to check manufacturer's directions.

With older electronic ignitions, coil output voltage should range between 30,000-45,000 volts. However, some electronic ignition coils are able to produce up to 100,000 volts.

Warning

Even though ignition coil or coil pack current is too low to normally cause electrocution, the high voltage could injure you or cause a potentially deadly heart attack.

If the ignition coil voltage is below specifications, do not condemn the coil until completing further tests. Low coil output could be due to low primary supply voltage, leaking secondary wires, or similar problems. Eliminate these as sources of the problem before replacing the ignition coil.

Load Test

A *load test,* or *acceleration test,* measures the spark plug firing voltages when engine speed is rapidly increased. When an engine is accelerated, higher voltage is needed to fire the spark plugs. While a defective component may not produce an abnormal scope pattern at idle, it may not operate properly under load.

To perform a load test, set the scope on parade and idle the engine between 1000-1200 rpm. While watching the firing lines on the scope, quickly open the injection throttle valve (or carburetor throttle plate) and release it. The firing voltage should increase, but it must not exceed certain limits.

The highest firing line should not be more than 75% of actual coil output. Typically, voltage should *not* exceed 15 kV in a contact point ignition or 20 kV in an electronic ignition. The upward movement of the firing lines during the load test should be the same. If any of the firing lines are high or low, a defect is present.

Cylinder Balance Test

A *cylinder balance test,* also called a *power balance test,* measures the power output from each of the engine's cylinders. As each cylinder is shorted, the tachometer should indicate an rpm drop. During a cylinder balance test, all cylinders should have the same percentage of rpm drop (within 5%). If a shorted cylinder does not produce an adequate amount of rpm drop, the cylinder is not firing properly. Study **Figure 46-42.**

Caution

Never short a cylinder in a vehicle with a catalytic converter for more than 15 seconds; converter damage could result.

If the rpm drop in one or more cylinders is below normal, a problem common to those cylinders is indicated. The cylinders could have low compression (burned valve, blown head gasket, worn piston rings), a lean mixture (vacuum leak, faulty fuel injector, computer malfunction), or other problems.

Cranking Balance Test

A *cranking balance test* is done to check the engine's mechanical condition. It can be used to isolate a cylinder with low compression due to a burned valve, worn piston rings, or other problems. The analyzer will show how much current is drawn by the starter motor as each cylinder goes through its compression stroke. High current draw means high compression stroke pressure. Low current draw (low display line) means that cylinder has low compression. Look at **Figure 46-43.**

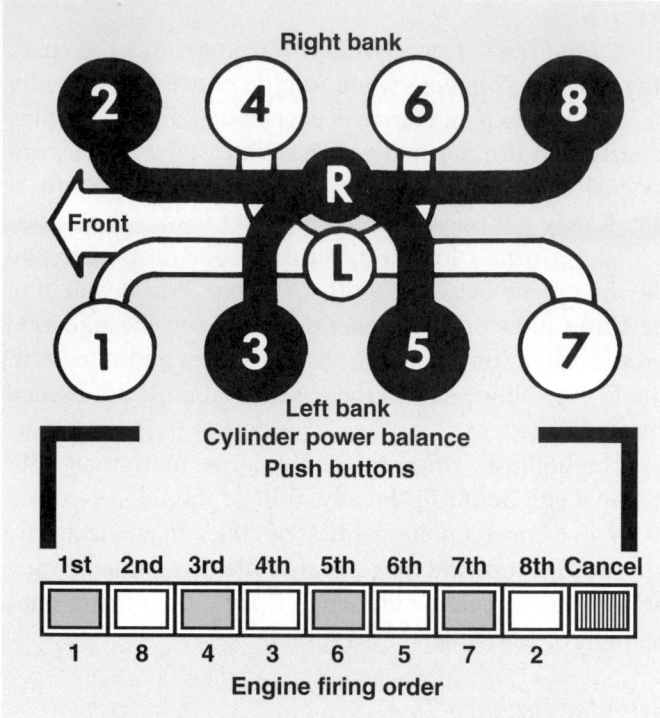

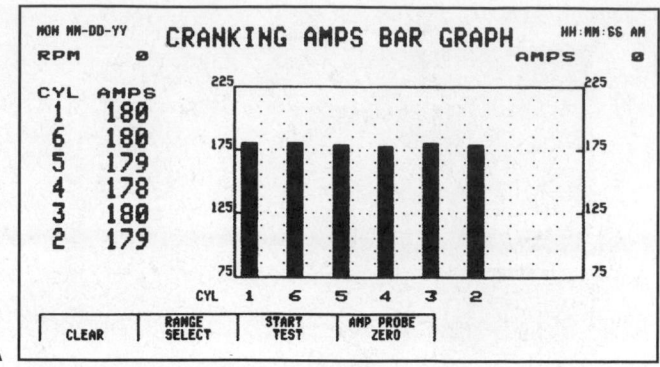

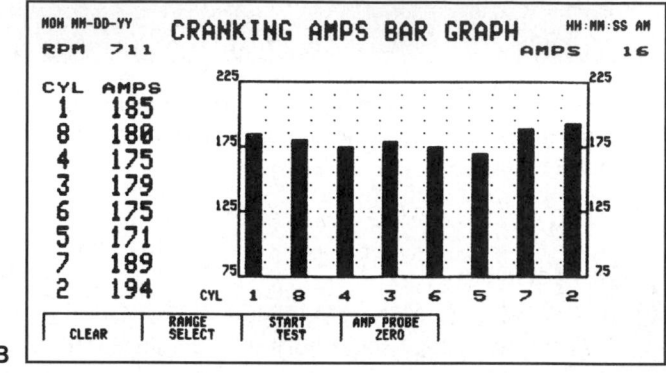

Figure 46-42. A cylinder balance test is done by pressing buttons on analyzer control panel. Each button will short and disable one cylinder. If the engine rpm does not drop sufficiently, that cylinder is not producing enough power. Note how each button corresponds to the firing order. A cylinder balance test will find any engine cylinder that is not producing power. Some analyzers will short each cylinder automatically and keep a record of the results.

Figure 46-43. Cranking balance tests are used to check general engine compression. If any cylinder does not load the starter motor as much as the others, it has low compression pressure and a possible leak. A—All bar graphs are at the same height, so all cylinders have same compression. B—The number 5 cylinder has a low bar graph indicating less compression pressure. (Snap-on Tool Corp.)

Other Analyzer Tests

An analyzer is usually capable of performing other tests besides those discussed in this chapter. These include starter cranking amps, charging voltage, and exhaust gas analysis. Such tests are almost identical to those done with individual instruments explained in other chapters.

Summary

- Strategy-based diagnostics involves using a consistent, logical procedure to narrow down possible problem sources.

- A vacuum gauge measures negative air pressure produced by the engine, fuel pump, vacuum pump, and other components.

- A pressure gauge measures positive pressure produced by the engine, turbocharger, fuel pump, or other device.

- A diesel injection tester is a set of pressure gauges and valves for measuring system pressure.

- A glow plug test harness can be used for checking a diesel engine rough idle problem.

- Scan tool datastream values are live electrical values measured with the vehicle running or driving.

- A breakout box allows you to pinpoint test electrical values at specific pins on the ECM or in the computer system.

- Electromagnetic interference results from induced voltage into wires and can cause a computer to malfunction.

- Scope voltage is shown on the scope screen along the vertical (up and down) axis or scale.

- Scope time may be given on the scope screen on the horizontal scale in degrees, milliseconds, or duty cycle.

- A primary pattern shows the low voltage or primary voltage changes in an ignition system.

- The secondary scope pattern shows the actual high voltages needed to fire the spark plugs.

- The term superimposed means that all of the cylinders are placed one on top of the other.
- The parade pattern, also called the display pattern, lines up the waveform for each cylinder side by side across the screen.
- A magnetic sensor scope test is done by measuring the output voltage from the sensor with the engine cranking.
- Most electrical connectors are sealed and do not allow easy probing. You may need to install a test connector or jumper wires between the two halves of the connector to probe sensor voltages.
- An oscilloscope can be used to check the output pulses leaving an ECM or ignition module.
- A flight record test stores the sensor waveform in the scope memory when a problem occurs.
- An engine analyzer, also called a vehicle analyzer, consists of a group of test instruments including an oscilloscope, tach-dwell, VOM, exhaust gas analyzer, and sometimes a scan tool.
- A modem allows a shop-owned analyzer to communicate over telephone lines with a larger mainframe computer.
- An analyzer displays operating values for various components in number form.
- An ignition coil output test measures the maximum available voltage produced by the ignition coil.
- A load or acceleration test measures the firing voltage of the spark plugs when engine speed is rapidly increased.

Important Terms

Strategy-based diagnostics
Vacuum gauge
Pressure gauge
Vacuum-pressure gauge
Hand vacuum pump
Diesel injection tester
Glow plug test harness
Snap-shot
Scan tool datastream values
Breakout box
Electromagnetic interference (EMI)
Digital pyrometer
Dynamometer
Oscilloscope
Scope screen
Trace
Scope time
Scope sweep rate
Primary scope pattern
Secondary scope patterns
Firing section
Firing line
Spark line
Intermediate section
Coil oscillations section
Dwell section
Primary superimposed pattern

Secondary superimposed pattern
Parade pattern
Display pattern
Raster pattern
Stacked pattern
Expanded display
Cylinder select
Base line
Rising edge
Leading edge
On-time
High-time
Trailing edge
Falling edge
Off-time
Low-time
Amplitude
Peak-to-peak voltage
Flight record test
Engine analyzer
Vehicle analyzer
Modem
Mainframe computer
Ignition coil output test
Load test
Acceleration test
Cylinder balance test
Power balance test
Cranking balance test

Review Questions Chapter 46

Please do not write in this text. Place your answers on a separate sheet of paper.

1. Define strategy-based diagnostics.
2. A vacuum gauge measures _____ air pressure.
3. A scan tool _____ _____ is a record of the operating parameters present at the moment a problem occurs.
4. Electromagnetic interference can be caused by _____.
   (A) loose wires
   (B) unshielded secondary wires
   (C) aftermarket accessories
   (D) All of the above.
5. A(n) _____ is one of the last tools used when diagnosing computer system problems.
6. On the scope screen, _____ is given on the vertical scale and _____ is given on the horizontal scale.
7. One kV equals _____.
   (A) 00 volts
   (B) 1000 volts
   (C) 10,000 volts
   (D) None of the above.
8. If a scope waveform is higher or taller than normal, this indicates a higher-than-normal _____.
9. The _____ scope pattern shows the actual voltages needed to fire the spark plugs.
10. Sketch and explain the three major parts of a scope secondary waveform.
11. How do you read a scope pattern?

12. Electronic ignition system waveforms will vary depending upon the make and model of vehicle. True or False?

13. When analyzing a square wave, what six things should be checked?

14. Summarize how you use a scope to test computer system sensors.

15. With the engine cranking, a magnetic sensor should commonly produce:
    (A) 5 volts peak-to-peak.
    (B) 0.5 volts peak-to-peak.
    (C) 1.5 volts peak-to-peak.
    (D) 15 volts peak-to-peak.

16. How do you scope test a Hall-effect sensor?

17. Which of the following is commonly used as part of an analyzer?
    (A) Tach-dwell.
    (B) Oscilloscope.
    (C) Multimeter.
    (D) All of the above.

18. What is a scope digital display?

19. Most analyzers recommend that engine idle speed be increased to about _____ rpm during scope tests.

20. Which of the following tests measures the power output from each of the engine's cylinders?
    (A) Load test.
    (B) Cylinder balance test.
    (C) Ignition coil output test.
    (D) EFI injector test.

● ASE-Type Questions

1. Technician A says that strategy-based diagnostics involves using a logical procedure to narrow down possible problem sources. Technician B says that advanced diagnostic techniques are used when conventional tests fail to pinpoint a problem. Who is right?
    (A) A only.
    (B) B only.
    (C) Both A and B.
    (D) Neither A nor B.

2. Technician A says computer terminal values can be measured with a low-impedance meter. Technician B says computer terminal values should be measured with a digital VOM. Who is right?
    (A) A only.
    (B) B only.
    (C) Both A and B.
    (D) Neither A nor B.

3. A scan tool has the capability to perform each of the following tasks except:
    (A) display datastream values.
    (B) capture a snap shot of operating parameters.
    (C) measure exhaust emissions.
    (D) switch actuators on and off.

4. Technician A says an oscilloscope's primary pattern represents the high voltage changes in an engine's ignition system. Technician B says an oscilloscope's primary pattern represents the low voltage changes in an engine's ignition system. Who is right?
    (A) A only.
    (B) B only.
    (C) Both A and B.
    (D) Neither A nor B.

5. An oscilloscope is connected to an automobile engine to check ignition system operation. The scope's primary pattern indicates a malfunction in this section of the ignition system. Technician A says this malfunction should show up in the scope's secondary pattern. Technician B says this malfunction will not show up in the scope's secondary pattern. Who is right?
    (A) A only.
    (B) B only.
    (C) Both A and B.
    (D) Neither A nor B.

6. The spark line on an oscilloscope's secondary pattern is almost straight and about one-fourth as high as the firing line. Technician A says this indicates normal ignition system operation. Technician B says this indicates an ignition system malfunction. Who is right?
    (A) A only.
    (B) B only.
    (C) Both A and B.
    (D) Neither A nor B.

7. All of the following are oscilloscope test patterns used by auto technicians *except:*
   (A) *raster.*
   (B) *parade.*
   (C) *secondary imposed.*
   (D) *cylinder select.*

8. Technician A says low circuit resistance can affect a square waveform. Technician B says high circuit resistance can affect a square waveform. Who is right?
   (A) *A only.*
   (B) *B only.*
   (C) *Both A and B.*
   (D) *Neither A nor B.*

9. Technician A says an oscilloscope is one type of test equipment normally used in an engine analyzer. Technician B says a timing light is one type of test equipment normally used in an engine analyzer. Who is right?
   (A) *A only.*
   (B) *B only.*
   (C) *Both A and B.*
   (D) *Neither A nor B.*

10. A vehicle is brought into the shop with fuel injector problems. Technician A says an engine analyzer can be used to detect certain fuel injector malfunctions. Technician B says an engine analyzer is not capable of testing fuel injector operation. Who is right?
   (A) *A only.*
   (B) *B only.*
   (C) *Both A and B.*
   (D) *Neither A nor B.*

11. An engine analyzer normally contains all the following instruments and test equipment *except:*
   (A) *multimeter.*
   (B) *dwell meter.*
   (C) *test light.*
   (D) *vacuum gauge.*

12. An oscilloscope is being used to test an automobile's ignition system. Technician A increases engine idle speed to about 950 rpm during this test. Technician B increases engine idle speed to about 1500 rpm during this test. Who is right?
   (A) *A only.*
   (B) *B only.*
   (C) *Both A and B.*
   (D) *Neither A nor B.*

13. An ignition coil output test is being performed on an automobile. Technician A sets the function controls to the lowest kV range and to raster. Technician B sets the function controls to the highest kV range and to display. Who is right?
   (A) *A only.*
   (B) *B only.*
   (C) *Both A and B.*
   (D) *Neither A nor B.*

14. Technician A says during a scope load test, a defective ignition system component will always produce an abnormal scope pattern at idle speeds. Technician B says during a scope load test, a defective ignition system component may not produce an abnormal pattern at idle speeds. Who is right?
   (A) *A only.*
   (B) *B only.*
   (C) *Both A and B.*
   (D) *Neither A nor B.*

15. A cylinder balance test is being performed on an automotive engine equipped with a catalytic converter. Technician A says during this test, each cylinder should be shorted for at least 45 seconds. Technician B says during this test each cylinder should not be shorted for more than 15 seconds. Who is right?
   (A) *A only.*
   (B) *B only.*
   (C) *Both A and B.*
   (D) *Neither A nor B.*

Activities—Chapter 46

1. Study the instruction manual for an analyzer and demonstrate how to hook it up for a test designated by your instructor.

2. With the analyzer hooked up and working, point out the three sections of the trace.

3. Interpret the trace patterns of an analyzer scope set up to test the ignition system.

4. Scope test several sensors. Make a sketch of the waveform produced by each with a written explanation of your results.

Chapter 47

Engine Tune-Up

After studying this chapter, you will be able to:

- Describe the typical difference between a minor tune-up and a major tune-up.
- List the basic steps for an engine tune-up.
- Explain service operations commonly performed during a tune-up.
- List the safety precautions that should be remembered during a tune-up.
- Correctly answer ASE certification test questions on engine tune-up and engine problem diagnosis.

An engine tune-up was, at one time, performed to restore an engine to peak performance. It involved replacing worn parts, performing basic maintenance tasks, making adjustments, and, sometimes, making minor repairs.

Tune-up procedures have changes in recent years and are done as part of routine maintenance, rather than to correct a driveability problem. New engines can run for longer periods between tune-ups. Most new engines do not require spark plug replacement for up to 100,000 miles (160 000 km). However, filters, belts, hoses, and other parts do deteriorate and require service. This chapter covers tune-up procedures. It also reviews and expands upon test and service information given in other textbook chapters.

Engine Tune-Up

An *engine tune-up* ensures that the engine, ignition system, fuel system, and emission control systems are within factory specifications. Auto manufacturers normally recommend a maintenance tune-up after a specific engine operating period, based on mileage and vehicle use.

The exact procedures for a tune-up will vary from vehicle to vehicle and from shop to shop. In one shop, a tune-up may only consist of spark plug replacement. In another shop, it may involve a long list of tests and repairs.

Refer to the vehicle manufacturer's recommendations for parts service during a tune-up. They will give service intervals for all parts that wear or deteriorate and require adjustment or replacement. See **Figure 47-1.**

Minor Tune-Up

A *minor tune-up* primarily involves parts inspection, fluid replacement, filter and spark plug replacement, and minor adjustments on a low mileage engine. For example, a vehicle that has only been driven 30,000–40,000 miles (48,279–64,372 km) may only require a minor tune-up. Most of the engine systems would be in satisfactory condition, with little or no wear. A minor tune-up typically involves these tasks:

- Visual inspection for obvious problems affecting engine performance, such as loose belts, hoses, and wires; vacuum leaks; coolant leaks; etc.
- Checking for trouble codes on computer-controlled vehicles.
- Replacing spark plugs.
- Ignition service including checking the secondary components and adjusting the ignition timing, if applicable.
- Replacement of air and fuel filters.
- Oil change and chassis lube.
- Emission control system tests and service, reset emission maintenance reminder circuit.
- Fuel system adjustments, if any are available.
- Clean throttle body and idle air bypass on fuel injected engines.
- Road test before and after tune-up to check engine performance.

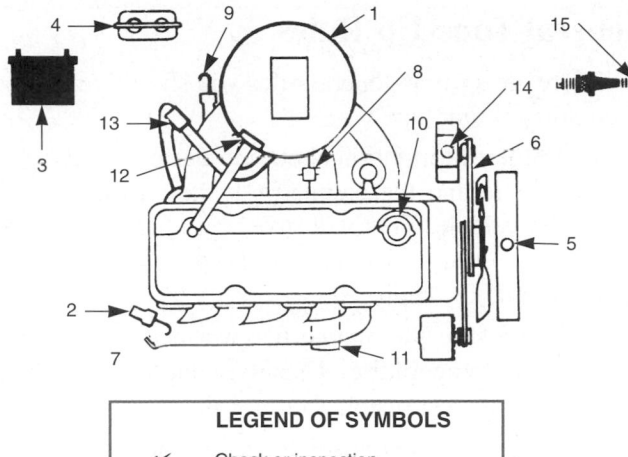

Footnotes

a. For normal use, no automatic transmission maintenance is required except regular fluid checks. For heavy-duty operation, change fluid and filter and adjust bands at the interval shown.

b. Change coolant initially at 25 months or 25,000 miles (40,000 km), whichever occurs first, then at the start of each winter season.

c. Perform service initially at 5000 miles (8000 km), then at 15,000 miles (24,000 km) and every 15,000 miles (24,000 km) thereafter.

d. Four-cylinder engines are not equipped with exhaust heat valves.

e. Clean filter with kerosene. See Positive Crankcase Ventilation in the Owner's Manual for details.

f. California cars equipped with automatic transmissions only.

g. Eight-cylinder cars only. (Eight-cylinder equipped cars in California do not require this service.)

LEGEND OF SYMBOLS

- ✔ Check or inspection
- Drain and replace fluid
- Engine tune-up
- Heavy-duty operation
- Lubrication
- Service component replacement

COMPONENT	SERVICE	INTERVAL	
		Four-cylinder	**Six- and eight-cylinder**
1. Air cleaner		30,000 mi (48,000 km)	30,000 mi (48,000 km)
		15 mo/15,000 mi (24,000 km)	15 mo/15,000 mi (24,000 km)
2. Automatic transmission	✔ a.	5 mo/5000 mi (8000 km)	7 mo/7,500 mi (12,000 km)
		15 mo/15,000 mi (24,000 km)	15 mo/15,000 mi (24,000 km)
3. Battery	✔	15 mo/15,000 mi (24,000 km) and start of each winter season	15 mo/15,000 mi (24,000 km) and start of each winter season
4. Brake master cylinder	✔	5 mo/5000 mi (8000 km)	7 mo/7500 mi (12,000 km)
5. Coolant (radiator)	✔ b.	5 mo/5000 mi (8000 km)	7 mo/7500 mi (12,000 km)
		25 mo/25,000 mi (40,000 km)	25 mo/25,000 mi (40,000 km)
6. Drive belts	✔	5000 mi (8000 km) to 50,000 c.	5000 mi (8000 km) to 50,000
7. Exhaust heat valve		– d.	30,000 mi (48,000 km)
8. Fuel filter		15,000 mi (24,000 km)	15,000 mi (24,000 km)
9. Oil dipstick (oil level)	✔	At each fuel fill	At each fuel fill
10. Oil (filler cap)		5 mo/5000 mi (8000 km)	7 mo/7500 mi (12,000 km)
		2.5 mo/2500 mi (4000 km)	3.5 mo/3700 mi (6000 km)
11. Oil filter		5 mo/5000 mi (8000 km)	7 mo/7500 mi (12,000 km)
12. PCV filter	✔ e.	30,000 mi (48,000 km)	30,000 mi (48,000 km)
13. PCV valve		30,000 mi (48,000 km) f.	30,000 mi (48,000 km)
14. Power steering pump	✔	5 mo/5000 mi (8000 km)	7 mo/7500 mi (12,000 km)

COMPONENT	SERVICE	INTERVAL	
		Four-cylinder	**Six- and eight-cylinder**
15. Tune-up	✔ c.	5000 mi (8000 km) Retorque cylinder head bolts, adjust engine valves, check and adjust curb and high idle speeds.	5000 mi (8000 km) Check and adjust curb and high idle speeds.
	✔	15,000 mi (24,000 km) Check and reset ignition timing as required. Replace ignition points and condenser.	15,000 mi (24,000 km) g. Check the following and correct as required. Choke system, idle mixture, ignition timing and vacuum fittings hoses and connections.
		30,000 mi (48,000 km) to 100,000 mi. Complete engine tune-up.	30,000 mi (48,000 km) to 100,000 mi. Complete engine tune-up.
		Engine Mechanical Systems Inspect: Air guard system hoses, vacuum lines, and fittings. Exhaust gas recirculation lines, hoses, and connections. Also: Retorque cylinder head bolts. Adjust engine valves. Ignition System Inspect: Coil and spark plug wires, distributor–cap and rotor, vacuum and centrifugal advance mechanisms, distributor shaft and cam lobes, transmission-controlled spark system (TCS), if equipped. Replace ignition points condenser, and spark plugs. Fuel System Inspect: Fuel tank, cap, lines, and connections. Air cleaner, thermostatic control system (TAC). choke linkage for free movement. PCV system hoses. Clean PCV filter in air cleaner. Replace PCV valve. f. Replace charcoal canister air inlet filter. Final Adjustment. Ignition timing. Idle mixture. Curb and high idle speeds.	Engine Mechanical Systems Inspect: Air guard system hoses, vacuum lines, and fittings Exhaust gas recirculation lines, hoses, and connections. Ignition System Inspect: Coil and spark plug wires, distributor–cap and rotor, vacuum and centrifugal advance mechanisms, transmission controlled spark system (TCS), if equipped. Replace spark plugs. Fuel System Inspect: Fuel tank, cap, lines, and connections. Air cleaner thermostatic control systems (TAC) choke linkage for free movement. PCV system hoses. Clean PCV filter (6-cyl in air cleaner, 8-cylinder in oil filler cap). Replace PCV valve and charcoal canister air inlet filter. Final Adjustment Ignition timing. Idle mixture. Curb and high idle speeds.

Figure 47-1. Specific tune-up intervals and special recommendations will be in the service manual. Read through this example and note each recommendation, symbol, and footnote. (Chrysler)

Major Tune-Up

A *major tune-up* is more thorough than a minor tune-up and is done when parts in the ignition, fuel, and emissions control systems are worn. After prolonged use, these systems can deteriorate, requiring a greater amount of service.

Besides the steps listed for a minor tune-up, a major tune-up typically includes more diagnostic tests using a scan tool, an exhaust gas analyzer, a compression gauge, a multimeter, an oscilloscope, a vacuum gauge, and other tools to determine engine and system condition. It may also include fuel injector service or carburetor system repairs, a distributor rebuild, and other more time-consuming tasks.

A major tune-up might involve the tasks listed for a minor tune-up, plus:

- Scope analysis of engine systems.
- Spark plug wire testing and replacement.
- Test and replace fuel pump.
- Replacement of drive belts.
- Replace brake, transmission, differential, and other fluids.
- Throttle body replacement or rebuild.
- Distributor rebuild or replacement.
- Valve train adjustment (some engines).
- Replacement of weak sensors: oxygen sensor, cam position sensor, crankshaft position sensor, etc.

Importance of a Tune-Up

A tune-up is very important to the operation of an engine. It can affect:

- Engine power and acceleration.
- Fuel consumption.
- Exhaust pollution.
- Smoothness of engine operation.
- Ease of starting.
- Engine service life.

You must make sure you return every engine-related system to peak operating condition during a tune-up. If you overlook just one problem, the engine will not perform properly.

Tech Tip
If a computer-controlled vehicle has a performance problem, such as rough idle, stalling, surging, etc., a tune-up will not correct the condition in most cases. Most performance problems in newer vehicles are caused by the failure of an engine component, a sensor, an actuator, or the ECM.

General Tune-Up Rules

There are several general rules you should remember when doing a tune-up:

- Gather information about the performance of the vehicle and the engine. Ask the customer about the vehicle. This will give you clues about what systems and components should be investigated, tested, and replaced.
- Make sure the engine has warmed to full operating temperature. Usually, you cannot obtain accurate test results from a cold engine.
- Make sure the tools and test equipment are accurate and will give precise readings.
- Refer to the service manual or emission control information sticker for specifications and procedures. Modern vehicles are so complex that the slightest mistake could cause major damage.
- Use quality parts. Quality parts will ensure that your tune-up lasts a long time.
- Keep accurate service records. You should document all the operations performed on the vehicle. This will give you and the customer a record for future reference. If a problem develops, you can check your records to help correct the trouble.
- Make sure you complete all basic maintenance service. Lubricate door, hood, and trunk latches and hinges during a tune-up. Also, check all fluid levels, belts, and hoses. Make sure that the tire pressure is correct. This will build good customer relations and ensure vehicle safety and dependability.

Tune-Up Safety Rules

Safety rules for the shop were discussed in Chapter 5. The following list highlights several safety rules that you should know and practice while performing a tune-up:

- Wear eye protection at all times.
- If the engine is to be running during your tests, set the parking brake and block the wheels. Place an automatic transmission in park or manual transmission in neutral.
- Place an exhaust hose over the tailpipe when running the engine in an enclosed shop.
- Keep your clothing, hands, tools, and equipment away from the engine fan.
- Disconnect the battery when recommended in the service manual. This will help prevent accidental engine cranking or an electrical fire.

- Be careful not to touch the hot exhaust manifold when removing spark plugs. Keep test equipment leads away from the exhaust manifolds.

- Keep a fire extinguisher handy when performing fuel system tests and repairs.

- Relieve residual fuel pressure from an injection system before opening the system.

- Disable the injection pump when removing a diesel fuel injector line. Diesel fuel under pressure can puncture your skin and eyes.

- Read the operating instructions for all test equipment. Failure to follow directions could cause bodily injury and severe damage to the part or instrument.

- Never look into a carburetor or throttle body when cranking or running the engine. Do not cover the air inlet with your hand. If the engine were to backfire, it could cause severe burns.

Typical Tune-Up Procedures

The following is a summary of the most common procedures for an engine tune-up. They are typical and apply to most makes and models of vehicles.

Preliminary Inspection

To begin a tune-up, most technicians inspect the engine compartment for problems. The inspection includes looking for:

- Battery problems—dirty case top, corroded terminals, and physical damage.

- Air cleaner problems—clogged or contaminated filter, inoperative air flap, or disconnected vacuum hoses, **Figure 47-2.**

- Belt problems—looseness, fraying, and slippage.

- Low fluid levels—engine oil, brake fluid, power steering fluid, etc.

- Deteriorated hoses—hardened or softened cooling system, fuel system, and vacuum hoses.

- Poor electrical connections—loose or corroded connections; frayed, pinched, or burned wiring and terminals.

If any problems are found, correct them before beginning the tune-up. Many of these problems can affect engine performance.

Tune-Up Parts Replacement

Depending on the age and condition of the vehicle, any number of parts may need replacement during a

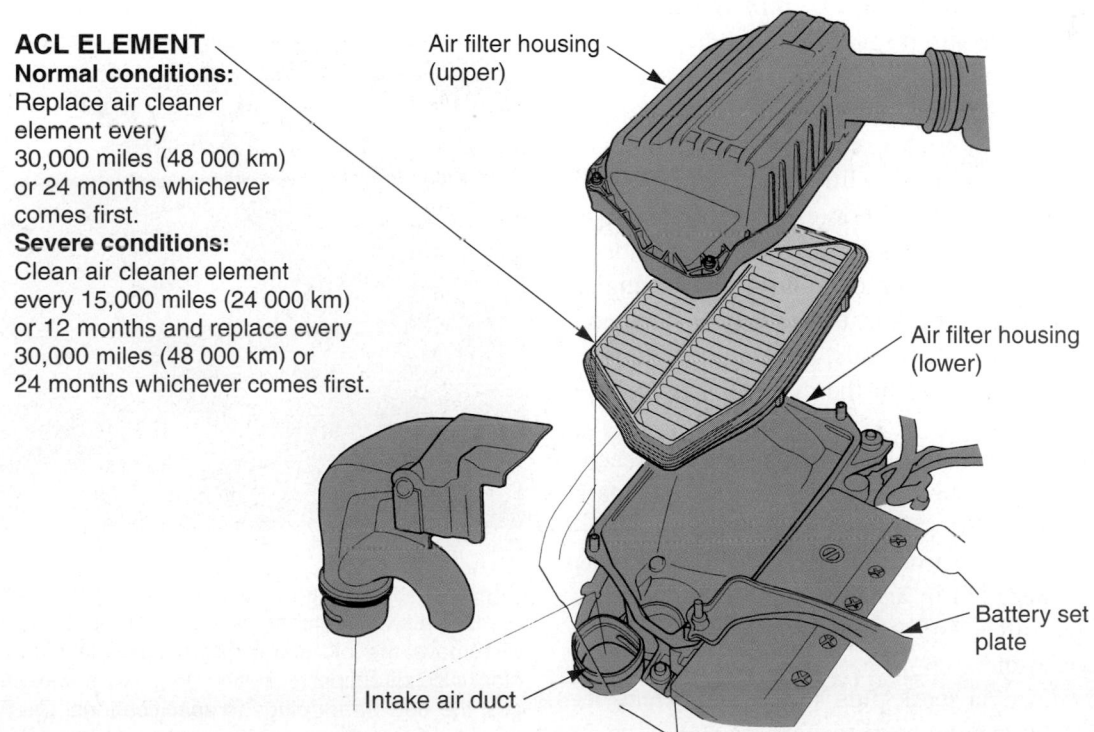

Figure 47-2. The air cleaner element is often replaced during a tune-up. A dirty element will reduce engine performance. Small clips or screws are used to secure the cover over the element. (Honda)

tune-up. With a late-model vehicle, you may only need to replace the spark plugs and filters. With an older, high mileage vehicle, you may have to replace the spark plugs, injectors, distributor components, spark plug wires, or other worn parts.

Spark Plug Replacement

To remove the spark plugs, first check that the spark plug wires are numbered or located correctly in their clips. Grasping the spark plug wire *boot,* pull the wire off the plug. Twist the boot back and forth if it sticks.

If the engine has the ignition coils mounted over the spark plugs, the coils must be removed before the plugs can be replaced. Normally, a few small bolts secure the coil or coils. After removing the bolts, you can lift off the ignition coils to gain access to the spark plugs.

 Caution
If the engine is equipped with an aluminum head, make sure the engine is cool before removing the spark plugs. Removing spark plugs from a warm aluminum head can damage the threads in the spark plug holes.

Blow debris away from the spark plug holes with compressed air. This will prevent particles from falling into the engine cylinders when the plugs are removed. Using a spark plug socket, a ratchet, and an extension, if needed, remove each spark plug, **Figure 47-3A.**

As you remove each plug, lay it in order on the fender cover or workbench. Do not mix up the plugs. After all the plugs are out, inspect them to diagnose the condition of the engine.

To **read the spark plugs**, closely inspect and analyze the tip and insulator of each used spark plug. This will give you information on the condition of the engine, fuel system, and ignition system. For example, a properly burning plug should have a *brown* to *grayish-tan* color. A *black* or *wet* plug indicates that the plug is NOT firing or that there is an engine problem (worn piston rings and cylinders, leaking valve stem seals, low engine compression, or a rich fuel mixture) in the cylinder from which the plug was removed.

Obtain the replacement plugs (heat range and reach) recommended by the manufacturer. Then, set the spark plug gap by spacing the side electrode the correct distance from the center electrode, **Figure 47-3B.** If the new plugs are to be installed in an aluminum cylinder head, coat the threads with antiseize compound or clean engine oil before installation.

Use your fingers, a spark plug socket and extension, or a short piece of vacuum hose to start the new plugs in their holes. See **Figure 47-3C.**

A

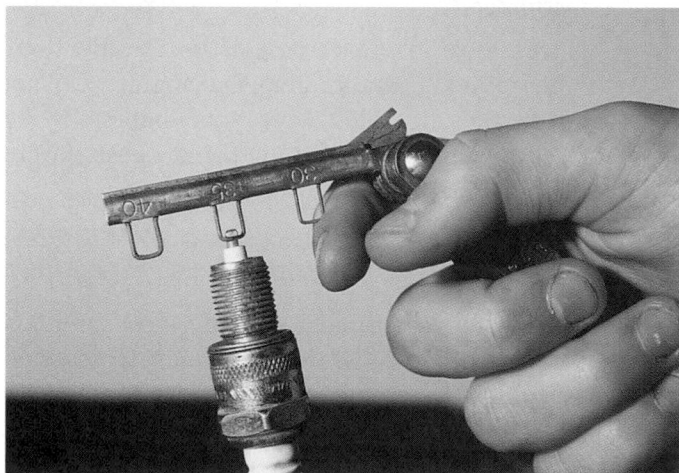
B

C

Figure 47-3. Note the basic steps for spark plug replacement. A—Use a swivel ratchet, an extension, and a spark plug socket to remove the old spark plugs. Turn the plugs in a counter-clockwise direction to loosen. B—Use a wire feeler gauge to gap the new spark plugs to specifications. C—Start the spark plug by hand. This will help prevent cross threading, which could damage the cylinder head. (Snap-on Tool Corp.)

Tech Tip

Spark plug sockets contain a rubber grommet, which holds the spark plug in the socket during installation. In many cases, the spark plug has a tendency to stick to the rubber grommet, pulling the socket off the extension. This can be a problem in newer engines equipped with spark plugs located in recessed holes in the cylinder head. Spark plug sockets with permanently fixed extensions are available to make replacing spark plugs on these engines easier.

With the spark plugs threaded into the head a few turns by hand, spin them in the rest of the way with your ratchet. Tighten the spark plugs to specifications. Some auto manufacturers give spark plug torque specifications. Others recommend bottoming the plugs on the seat and then turning an additional one-eighth to one-quarter turn. Refer to a service manual for exact procedures.

Note

For more information on replacing filters, refer to the text index under air filter service, fuel filter service, oil filter service, automatic transmission service, and charcoal canister service.

Caution

When servicing computer-controlled vehicles, never reroute secondary wires. Keep all wires in their original locations and reinstall them in all clips or harness positions. If you reroute spark plug wires or other high current carrying wires, current can be induced into low current sensor or computer feedback wires. This could upset computer system operation and cause a "troubleshooting nightmare!"

Oxygen Sensor Replacement

Oxygen sensors should be replaced at periodic intervals. After prolonged service, they become coated or fouled with exhaust byproducts. As this happens, fuel economy and emissions will be adversely affected. If gas mileage is 10–15 percent lower than normal, suspect the oxygen sensor of slow response. One- and two-wire sensors should be replaced at about 50,000–60,000 miles. Heated three-wire oxygen sensors should be replaced at about 100,000 miles.

Disconnect the negative battery cable. Then, separate the sensor from the wiring harness by unplugging the connector. Never pull on the wires themselves, as damage may result.

Use a ratchet and sensor socket wrench to unscrew the sensor. Inspect its condition. Some sensors may be difficult to remove at temperatures below 120°F. Use care to avoid thread damage.

Coat the threads of the new sensor with antiseize compound. Start the sensor by hand. Then screw in and tighten the sensor with a wrench. Do not overtighten or the sensor may be damaged. Check fuel injection system operation after sensor installation.

Engine Compression Testing

A *compression test* is frequently made during a tune-up to check the engine's mechanical condition. It is impossible to tune an engine that is not in good mechanical condition. If the engine fails the compression test, mechanical repairs must be made *before* the tune-up.

Note

Procedures for compression testing engines are covered in Chapter 48, *Engine Mechanical Problems.*

Valve Adjustment

When mechanical (solid) lifters are used, the valves require periodic adjustment. Proper *valve adjustment* is important to the performance of the engine.

Incorrect valve adjustment will upset the amount of air-fuel mixture pulled into the cylinder. It also affects valve lift and duration. This will affect combustion and reduce engine efficiency.

Note

Valve adjustment is covered in Chapter 51, *Engine Top End Service.*

Tune-Up Adjustments

The adjustments needed during a tune-up will vary with the make, model, and condition of the vehicle. A few of the most common tune-up adjustments include:

- *Pickup coil air gap adjustment*—Use a nonmagnetic feeler gauge to set the gap between the trigger or reluctor wheel and the pickup to specifications.

- *Idle speed adjustment*—After turning off all accessories, turn the idle speed adjusting screw on the throttle body or carburetor until the engine shows the specified rpm on the tachometer. Note that some engines do not have an idle speed adjustment because the computer monitors and controls idle speed. See **Figure 47-4.**

- *Ignition timing adjustment* (gasoline engine)—Ignition timing adjustment varies. Basically, you must set the engine to run on base timing (disconnect vacuum hose to distributor advance, trigger computer by disconnecting timing plug,

WARNING
The ignition system can create a potential shock hazard. Ensure that the engine is off before connecting or removing the pickup.

1. Turn engine off.
2. Connect output plug of inductive pickup in the input terminals shown. Make sure the black plug is in COM and the red is in RPM. If your pickup has a dual banana connector, the plug with the GND (ground) tab goes in COM.
3. Turn rotary switch to volts dc.
4. For 4-cycle engines that fire once every two revolutions, press RPM once to select RPM(2). For systems that fire every revolution (2-cycle engines), and for waste spark DIS systems, press RPM twice to select RPM(1).
5. Clamp the inductive pickup to a plug wire near the spark plug. (Make sure that the jaws are closed completely and the side labeled SPARK PLUG SIDE faces the spark plug.)
6. Turn engine on. Read RPM on the display. Turn engine off before removing pickup.

Notes: 1. If meter reading is too high or is unstable, move to the 40V range by pressing RANGE once.
2. On some systems with nonresistor plugs, the pickup may need to be moved away from plug.
3. On waste spark systems, the pickup may need to be reversed, depending on what side of the coil the plug is on.

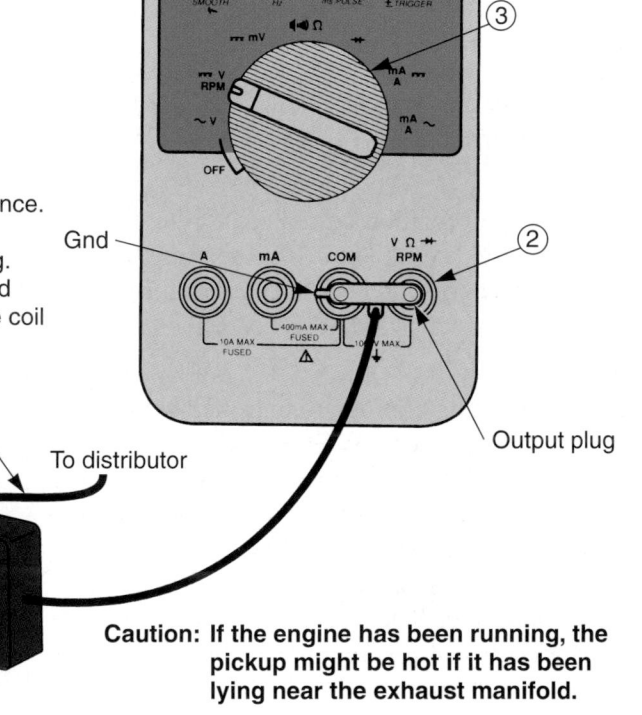

Caution: If the engine has been running, the pickup might be hot if it has been lying near the exhaust manifold.

Figure 47-4. A tachometer with an inductive pickup is commonly used to check engine idle speed. (Fluke)

etc.). Then, aim a timing light on the timing marks on the front damper or flywheel. The correct timing mark numbers must align. If not, rotate the distributor to change timing. Many late-model engines do not provide a method of changing ignition timing. Parts must be replaced (distributor, timing chain, etc.) to restore correct ignition timing. See **Figures 47-5** through **47-8.**

Other adjustments may also be needed. Follow the specific directions and specifications outlined in an appropriate service manual. This will ensure a thorough and long-lasting tune-up.

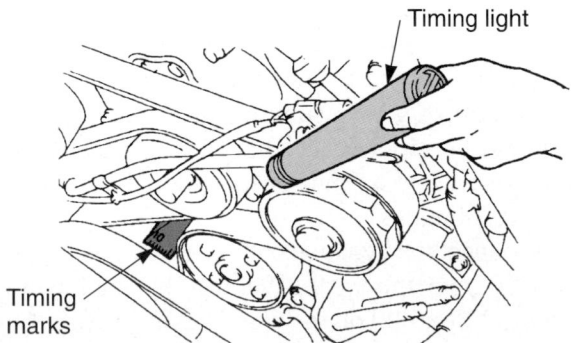

Figure 47-5. If possible, check the ignition timing with a timing light. (Mazda)

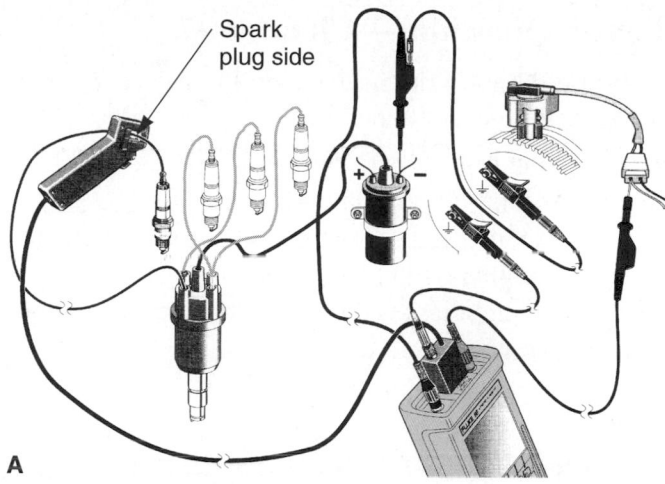

Spark plug side

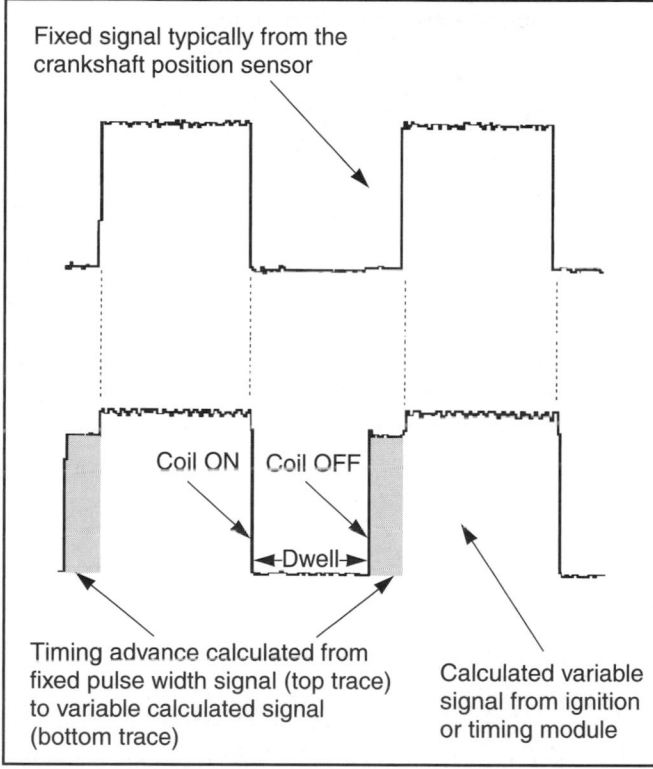

Fixed signal typically from the crankshaft position sensor

Coil ON   Coil OFF

←Dwell→

Timing advance calculated from fixed pulse width signal (top trace) to variable calculated signal (bottom trace)

Calculated variable signal from ignition or timing module

B

Figure 47-6. If you suspect a problem with the ignition timing and no provision for checking the timing is given, a hand-held scope can be used to check ignition timing advance. A—Connect the scope inductive lead around the spark plug wire. Connect probes to the crankshaft position sensor and the coil primary. B—The waveform allows you to watch and calculate timing advance changes with engine speed and other variables.

Diesel Engine Tune-Up (Maintenance)

Diesel engines do *not* require tune-ups like gasoline engines. A diesel does not have spark plugs to replace or an ignition system to fail. The diesel injection system is very dependable and only requires major service when problems develop.

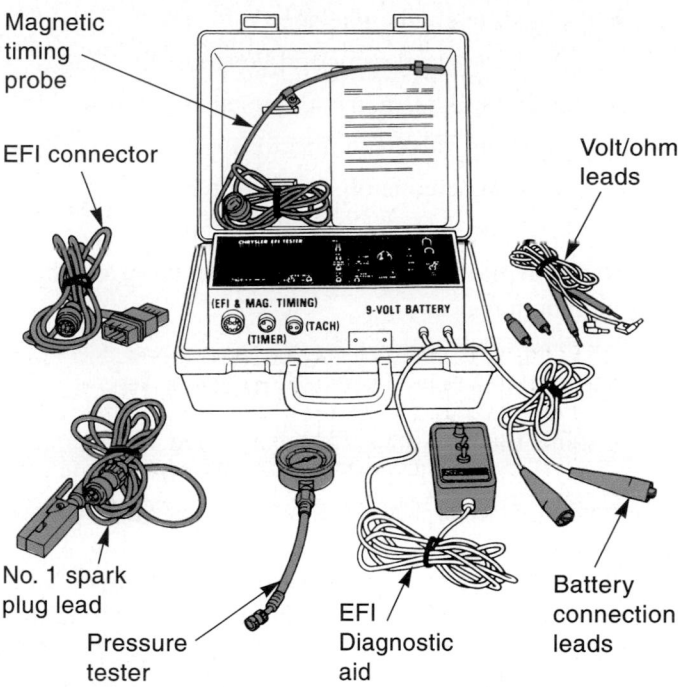

Magnetic timing probe

EFI connector

Volt/ohm leads

No. 1 spark plug lead

Pressure tester

EFI Diagnostic aid

Battery connection leads

Figure 47-7. This tester is capable of checking timing magnetically. It also has tools that can be used to perform several other tests. Readings are given by numbers on the face of the tester. (Chrysler)

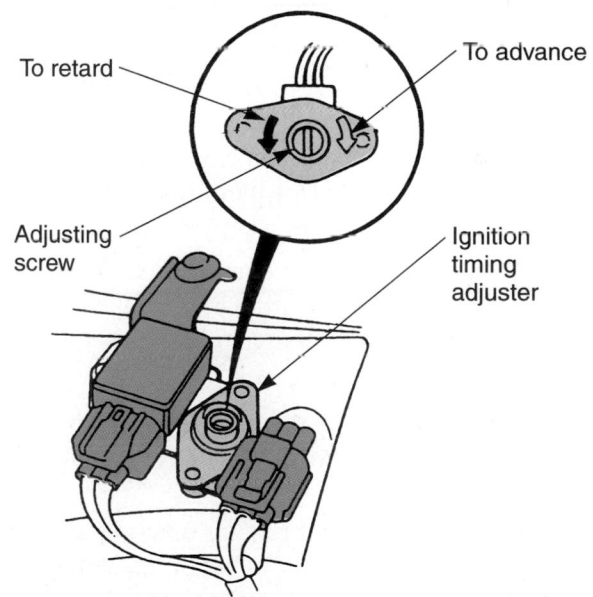

To retard

To advance

Adjusting screw

Ignition timing adjuster

Figure 47-8. Some vehicles with computer controlled ignition timing provide an adjusting screw to alter ignition timing. (Snap-on Tool Corp.)

A diesel engine tune-up, which is more accurately called *diesel engine maintenance,* typically involves:

- Replacing the air filter element.
- Cleaning, draining, or replacing the fuel filters.

- Adjusting engine idle speed.
- Adjusting accelerator and governor linkage.
- Inspecting the engine and related systems.
- Adjusting valve and injector timing.
- Changing engine oil and oil filter.
- Servicing emission control systems.

Refer to the index for more information on diesel engine maintenance.

Engine Tune-Up (Maintenance) Intervals

Engine tune-up intervals, also called *maintenance intervals,* are specific periods (in miles or months) for component service. They are given in the vehicle's service manual.

When doing an actual tune-up, it is very important to check in a manual for interval information. Recommended service intervals vary.

Summary

- An engine tune-up returns the engine to a condition of peak performance.
- A minor tune-up is a basic tune-up done when the engine is in good operating condition.
- A major tune-up is more thorough and is done when the engine systems (ignition, fuel, emission) are in worn condition.
- Depending upon the age and condition of the vehicle, any number of parts may need replacement during a tune-up.
- With modern computer-controlled vehicles, never reroute wires. Keep all wires in their original locations and reinstall them in all clips or harness positions.
- A compression test is frequently made during a tune-up to check the engine's mechanical condition.
- Diesel engines do not require tune-ups like gasoline engines. A diesel does *not* have spark plugs to replace or an ignition system to fail.

Important Terms

Engine tune-up	Idle speed adjustment
Minor tune-up	Ignition timing
Major tune-up	adjustment
Compression test	Diesel engine
Valve adjustment	maintenance
Pickup coil air gap	Maintenance intervals
adjustment	

Review Questions—Chapter 47

Please do not write in this text. Place your answers on a separate sheet of paper.

1. What is an engine tune-up?
2. A tune-up ensures that the _____, _____ system, _____ system, and _____ systems are operating within factory specifications.
3. List eight tasks commonly done during a minor tune-up.
4. What is a major tune-up?
5. List eight safety rules to remember during a tune-up.
6. A tune-up usually begins with:
   (A) inspection of engine compartment.
   (B) compression test.
   (C) spark plug replacement.
   (D) throttle body injector service.
7. Why is a compression test frequently done during a tune-up?
8. How can improper valve adjustment affect engine performance?
9. Name the eight steps typically done during diesel engine maintenance.
10. Engine tune-up or maintenance _____ are specific periods (in miles or months) for component service.

● ASE-Type Questions

1. A low-mileage automobile with an engine miss is brought into the shop. The owner of the vehicle wants a tune-up performed on the engine. Technician A says the car probably needs a minor tune-up. Technician B says a major tune-up should be performed on this engine. Who is right?
   (A) A only.
   (B) B only.
   (C) Both A and B.
   (D) Neither A nor B.
2. Technician A says an engine tune-up restores the ignition system to factory specifications. Technician B says an engine tune-up restores the fuel system to factory specifications. Who is right?
   (A) A only.
   (B) B only.
   (C) Both A and B.
   (D) Neither A nor B.

3. A multiport fuel injected engine with 25,000 miles (40,098 km) is brought into the shop for a tune-up. Technician A says this engine would normally require a major tune-up including injector replacement. Technician B says this engine would normally require a minor tune-up. Who is right?
   (A) A only.
   (B) B only.
   (C) Both A and B.
   (D) Neither A nor B.

4. Ignition timing adjustment and spark plug replacement are performed on an automobile engine. Technician A says these procedures would be classified as a major tune-up. Technician B says these procedures would be classified as a minor tune-up. Who is right?
   (A) A only.
   (B) B only.
   (C) Both A and B.
   (D) Neither A nor B.

5. Which of the following is *not* usually performed during a major tune-up?
   (A) Throttle body cleaning.
   (B) Air cleaner replacement.
   (C) Cylinder head reconditioning.
   (D) Distributor rotor cap replacement.

6. A vehicle's engine requires a major tune-up. Technician A says a compression test is normally performed during this procedure. Technician B says a compression test is not usually performed during this procedure. Who is right?
   (A) A only.
   (B) B only.
   (C) Both A and B.
   (D) Neither A nor B.

7. An automobile is brought into the shop for a tune-up. Technician A tells the owner of the vehicle that a tune-up will probably improve the engine's power and acceleration. Technician B tells the owner of the vehicle that a tune-up will normally improve the quality of the engine's exhaust emissions. Who is right?
   (A) A only.
   (B) B only.
   (C) Both A and B.
   (D) Neither A nor B.

8. Technician A says a tune-up will usually improve the fuel efficiency of an engine. Technician B says a tune-up normally has no effect on engine fuel consumption. Who is right?
   (A) A only.
   (B) B only.
   (C) Both A and B.
   (D) Neither A nor B.

9. A tune-up is going to be performed on an automobile engine in an enclosed shop. Technician A says the vehicle's emergency brake should be set before starting tune-up procedures. Technician B says an exhaust vent hose should be installed in the vehicle's tailpipe. Who is right?
   (A) A only.
   (B) B only.
   (C) Both A and B.
   (D) Neither A nor B.

10. Which of the following safety rules should be followed when performing an engine tune-up?
    (A) Wear eye protection at all times.
    (B) Keep a fire extinguisher handy when repairing fuel system.
    (C) Disconnect battery when recommended in service manual.
    (D) All of the above.

11. A tune-up is being performed on a diesel engine. Technician A disables the injection pump before removing an injection line. Technician B does not disable the injection pump before removing an injection line. Who is right?
    (A) A only.
    (B) B only.
    (C) Both A and B.
    (D) Neither A nor B.

12. Technician A says to begin a tune-up, you should inspect the battery for problems. Technician B says to begin a tune-up, you should inspect the condition of the engine belts. Who is right?
    (A) A only.
    (B) B only.
    (C) Both A and B.
    (D) Neither A nor B.

13. An automobile is brought into the shop for a tune-up. The engine has a cracked vacuum hose. Technician A says a tune-up should be performed on the engine before the vacuum hose is replaced. Technician B says the vacuum hose should be replaced before performing a tune-up. Who is right?
    (A) A only.
    (B) B only.
    (C) Both A and B.
    (D) Neither A nor B.

14. Technician A says if an automobile engine fails a compression test, a tune-up should be performed before finding the source of the compression loss. Technician B says the source of compression loss must be found before a proper tune-up can be performed. Who is right?
    (A) A only.
    (B) B only.
    (C) Both A and B.
    (D) Neither A nor B.

15. Which of the following procedures is *not* performed during a diesel engine tune-up?
    (A) *Air filter element replacement.*
    (B) *Idle speed adjustment.*
    (C) *Pickup coil air gap adjustment.*
    (D) *Check injection timing.*

Activities—Chapter 47

1. Visit the service manager of a local auto repair shop or auto dealer. Determine what the shop would include in a minor tune-up and a major tune-up.

2. Taking the information gathered in Activity 1, using a flat rate manual and a parts catalog, determine the cost of a minor tune-up and a major tune-up at a labor rate of $56 per hour.

Front view of a modern V-8 engine. This engine has a plastic intake plenum. (BMW)

Engine Repair

Engine mechanical problems generally result from abuse or prolonged service. As an engine runs, friction slowly wears pistons, piston rings, cylinder walls, bearings, journals, valves, and other components.

This section will give you the skills needed to properly troubleshoot and rebuild automotive engines. Chapter 48 describes engine problems and summarizes repair procedures. Chapter 49 explains how to properly tear down an engine while gathering information about its condition. Chapters 50 52 detail how to measure, inspect, and assemble the parts of an engine. The information in this section will help you pass ASE Test A1, *Engine Repair.*

Engine Mechanical Problems

After studying this chapter, you will be able to:

- Explain why proper diagnosis methods are important to engine repair.
- List common symptoms of engine mechanical problems.
- Discuss how to find abnormal engine noises.
- Summarize procedures for gasoline and diesel engine compression testing.
- Explain when and how to do a wet compression test.
- Summarize common causes of engine mechanical problems.
- Discuss safety practices to follow while performing engine inspections.
- Correctly answer ASE certification test questions on engine mechanical problems.

Engine mechanical problems generally result from physical part wear and breakage after prolonged use or abuse. As an engine runs, part friction slowly wears pistons, piston rings, cylinder walls, bearings, journals, valves, and other components. This constant wear increases clearances between precise parts. With enough clearance increase, engine parts can leak pressure or fluid, "hammer together," or even break.

In earlier chapters, you learned about engine part construction and operation. To further your understanding of automotive engines, this chapter will explain the most common types of engine mechanical problems. It will introduce symptoms and describe the basic inspections and tests needed to find the source of these problems. As a result, you should be well prepared for the next chapters on engine removal and engine overhaul.

Why Is Diagnosis Important?

If a technician does not know how to properly diagnose engine problems, a great deal of time, effort, and money will be wasted. In fact, an untrained technician may unknowingly rebuild an engine when a minor repair would have corrected the fault.

For example, a worn or stretched timing chain can cause the engine valves to open and close at the wrong times in relation to piston movement. This could cause low compression stroke pressure in all cylinders. The engine would have low power output and low compression readings.

The technician could incorrectly diagnose the trouble (worn timing chain) as more major problems (worn piston rings, scored cylinders, etc.). The engine could be overhauled when a new timing chain would have corrected the problem. Unfortunately, the customer would have to pay for the technician's lack of training.

Symptoms of Engine Mechanical Problems

A few common engine mechanical problems include leaking gaskets, worn piston rings, burned and leaking valves, loose or worn engine bearings, worn timing chains, and damaged (cracked, broken, scored) engine parts. See **Figure 48-1**.

These problems can occur after extended service (high mileage), engine overheating, lack of periodic maintenance, and other types of abuse. The symptoms (signs) that result from these types of problems include:

- Excessive oil consumption—engine oil must be added too often.
- Crankcase blowby—combustion pressure blows past piston rings into crankcase, and out breather.
- Abnormal engine noises—knocking, tapping, hissing, rumbling.
- Engine smoking—blue-gray, black, or white smoke blows out tailpipe.
- Poor engine performance—rough idle, vehicle accelerates slowly, engine vibrates.

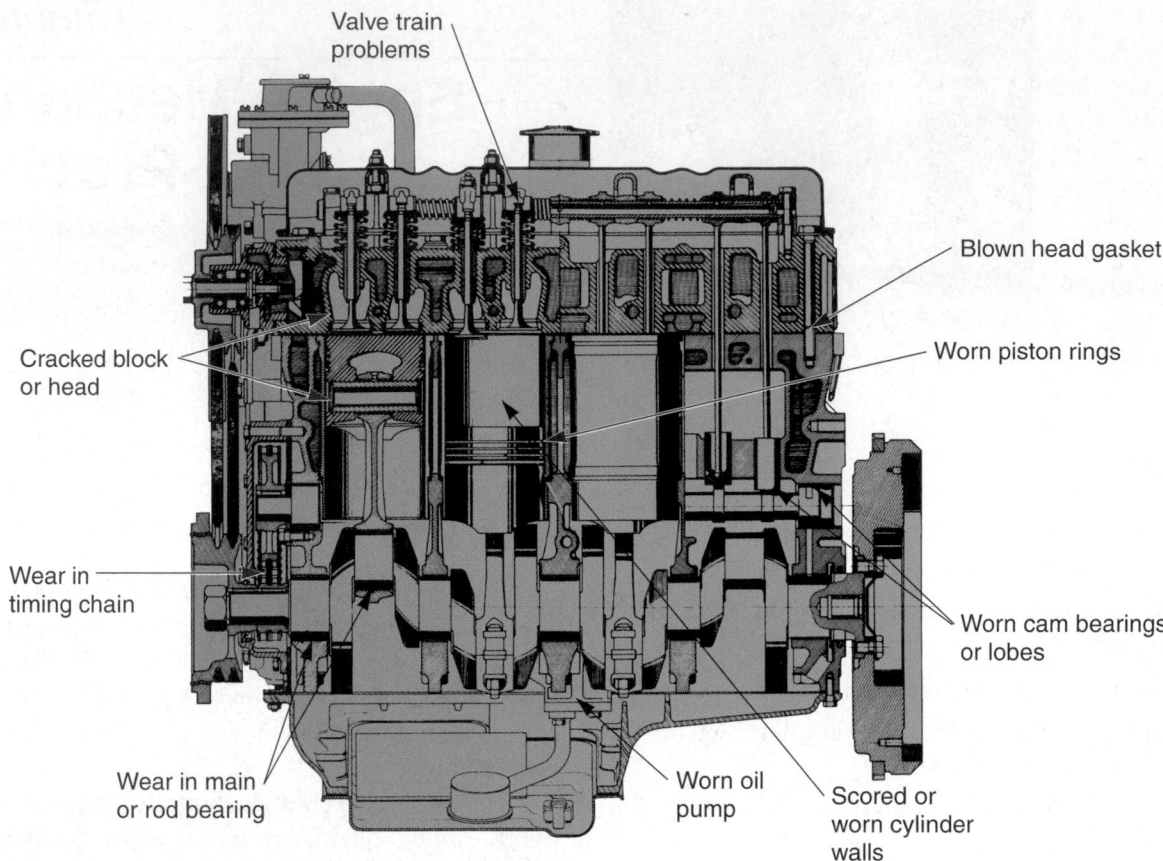

Valve train problems

Blown head gasket

Worn piston rings

Cracked block or head

Wear in timing chain

Worn cam bearings or lobes

Wear in main or rod bearing

Worn oil pump

Scored or worn cylinder walls

Figure 48-1. Many mechanical problems can occur in an engine.

- Coolant in engine oil—oil has white, milky appearance.
- Engine lockup—crankshaft will not rotate.

With any of these troubles, inspect and test the engine to determine the exact problem source. You must find out what repair is needed. Then, you can determine whether the engine can be repaired in the vehicle or if it must be removed for more major repairs.

Engine Pre-Teardown Inspection

After gathering information from the customer or service writer, start the vehicle and inspect the engine using your senses (sight, smell, hearing, touch). Look for external problems (oil leaks, vacuum leaks, part damage, contaminated oil).

If a leak is found, smell the fluid to determine if it is oil, coolant, or another type of fluid. Listen for unusual noises, which may indicate part wear or damage.

Increase engine speed while listening and watching for problems. The engine may run fine at idle but act up at higher speeds. Several engine problems can be identified through simple inspection.

Coolant in oil will show up as white or milky oil. It is caused by a mechanical problem that allows engine coolant to leak into the engine crankcase. There may be a cracked block or head, leaking head gasket, leaking intake manifold gasket (V-type engine), or similar troubles. Look at **Figure 48-2.**

Oil-fouled spark plugs point to internal oil leakage into the engine combustion chambers. They are an indication that the engine has worn rings, worn or scored cylinder walls, or bad valve seals. You will need to

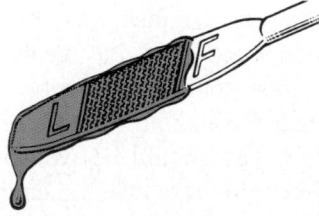

Figure 48-2. During engine inspection, check the condition of oil on the dipstick. Check dipstick for coolant, gasoline, or other contaminants in oil. The condition of the oil may reflect the condition of engine. (Toyota)

perform additional tests to find the source of the problem. See **Figure 48-3.**

Oil in the coolant is usually not an engine problem. It is commonly caused by a leak in the radiator (transmission) oil cooler. However, on rare occasions, a faulty head gasket, a cracked block, or cracked head can allow oil to seep into coolant passages, contaminating the coolant with oil.

Engine oil leaks occur when gaskets harden and crack, when seals wear, when fasteners work loose, or when there is part damage (warped surfaces, cracked parts, etc.). To find oil leaks, clean the affected area on the outside of the engine. Then, trace the leak upward to its source. Oil will usually flow down and to the rear of the engine because of cooling fan action.

External coolant leaks will show up as a puddle of coolant under the engine. Leaks can be caused by hose problems; rusted freeze (core) plugs; or warped, worn, or damaged parts. Use a pressure tester (see Chapter 40) to locate external coolant leaks.

Engine blowby occurs when combustion pressure blows past the piston rings into the lower block and oil pan area of the engine. Pressure then flows up to the valve covers and out through the breather. Excessive blowby will show up as an oily area around the breather. With the valve cover breather removed, oil vapors may blow out when the engine speed is increased. See **Figure 48-4.**

Engine vacuum leaks show up during inspection as a hissing sound, like air leaking out of a tire. Vacuum leaks are loudest at idle and temporarily quiet down as the engine is accelerated (manifold vacuum drops with engine acceleration). Very rough engine idling usually accompanies a vacuum leak.

Engine exhaust leaks show up as a clicking sound around the engine. Combustion pressure will blow out the leak and make a metallic noise. A ruptured exhaust manifold gasket, warped exhaust manifold, or loose header nuts are the most common reasons for exhaust

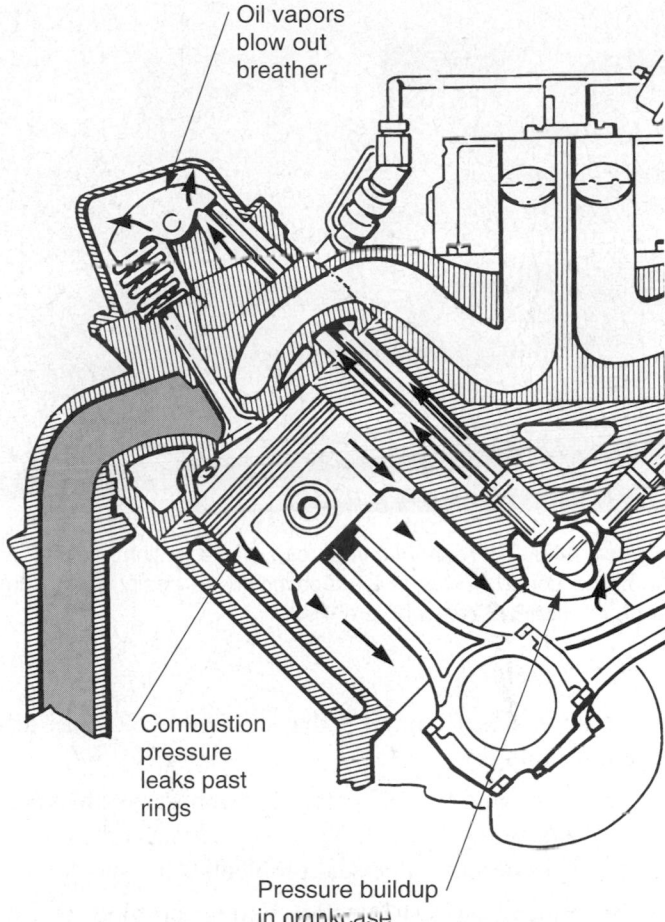

Figure 48-4. Blowby allows combustion pressure to enter the engine crankcase past the piston rings. Pressure buildup in lower end of the engine can cause oil mist to blow out breather openings. (Cadillac)

leakage. The clicking sound will tend to decrease upon engine deceleration, because combustion pressure is lessened as soon as the throttle is closed and fuel metering is decreased. Refer to **Figure 48-5.**

Engine smoking is normally noticed at the tailpipe when the engine is accelerated or decelerated. The color of the smoke can be used to help diagnose the source of the problem, **Figure 48-6.**

With a gasoline engine, the exhaust smoke may be three different colors:

- Blue-gray smoke indicates motor oil is entering the combustion chambers. It may be due to worn rings, worn cylinders, leaking valve stem seals, or other troubles.

- Black smoke is caused by an extremely rich air-fuel mixture (not an engine mechanical problem).

- White smoke, if not water condensation on a cool day, may be due to internal coolant leakage into cylinders.

Figure 48-3. Check spark plugs for oil fouling. This would point to bad rings, cylinder walls, or valve seals causing oil leakage into the combustion chambers. (Champion Spark Plugs)

Figure 48-5. An exhaust leak can make a light ticking or knocking sound. Exhaust leakage could be easily seen after exhaust manifold removal. (Fel-Pro)

With a diesel engine, exhaust smoke color has different meanings:

- Blue smoke—oil entering combustion chambers and being burned. This may be caused by ring, cylinder, or valve seal problems.
- Black smoke—injection system problem or low compression is keeping fuel from burning.
- White smoke—unburned fuel, cold engine, or coolant leaking into combustion chambers.

Note
Refer to Chapter 27, *Diesel Injection Diagnosis, Service, and Repair,* for more information on diesel exhaust smoke.

Abnormal Engine Noises

Abnormal engine noises (hisses, knocks, rattles, clunks, popping) may indicate part wear or damage. You must be able to quickly locate and interpret engine noises and decide what repairs are needed.

A *stethoscope* is a listening device used to find internal sounds in parts. Like a doctor's stethoscope for listening to your heart, it will amplify (increase) the loudness of noises.

To use a stethoscope, place the headset in your ears. Then, touch the probe on different parts around the noise. When the sound becomes the loudest, you have located the part producing the abnormal noise.

A long screwdriver can be used when a stethoscope is not available. Sounds will travel through the screwdriver, as they do with a stethoscope.

A section of hose can be used to locate vacuum leaks and air pressure leaks. Place one end of the hose next to your ear. Move the other end around the engine compartment. The hiss will become loudest when the hose is near the leaking part.

Warning
When using a listening device, keep it away from the spinning engine fan or belts. Severe injury could result if the stethoscope, screwdriver, or hose were hit or pulled into the fan or belts!

Compression Test

A *compression gauge* is used to measure the amount of pressure during the engine compression stroke. It provides a means of testing the mechanical condition of the engine. If the compression gauge readings are not within specifications, something is mechanically wrong in the engine.

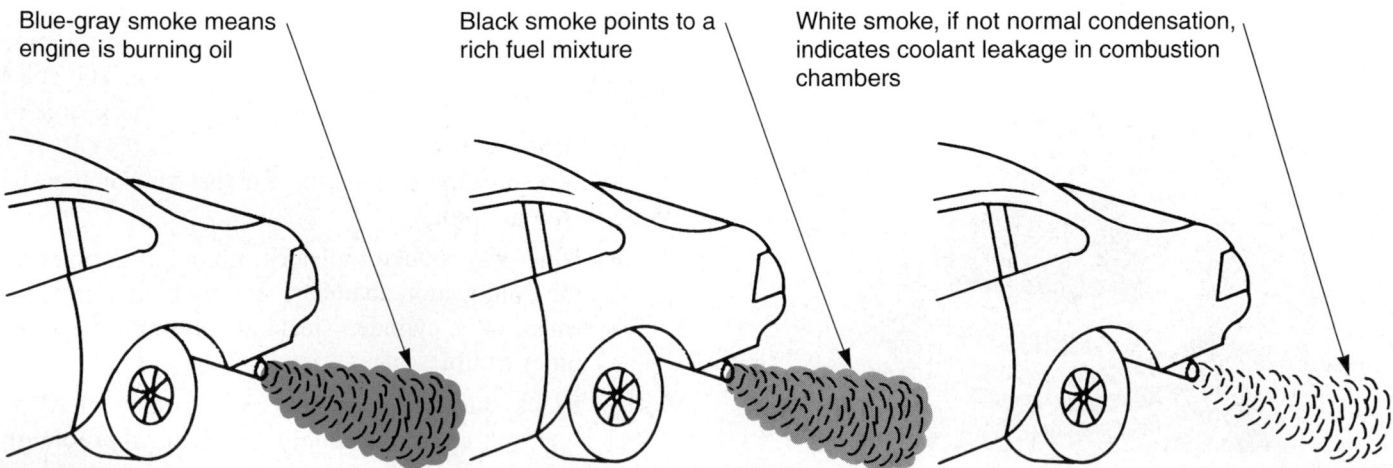

Blue-gray smoke means engine is burning oil

Black smoke points to a rich fuel mixture

White smoke, if not normal condensation, indicates coolant leakage in combustion chambers

Figure 48-6. Engine smoke may indicate major problems. Blue smoke points to oil entering combustion chambers. Black smoke suggests rich fuel mixture. White smoke, if not condensation, may indicate coolant leakage into cylinders.

A *compression test* is one of the most common methods of determining engine mechanical condition. It should be done anytime symptoms point to cylinder pressure leakage. An extremely rough idle, a popping noise from air inlet or exhaust, excessive blue smoke, and blowby are all reasons to consider a compression test.

Figure 48-7 shows several mechanical problems that could cause compression leakage. **Figure 48-8** shows a blown head gasket that would cause low compression on the two adjacent cylinders.

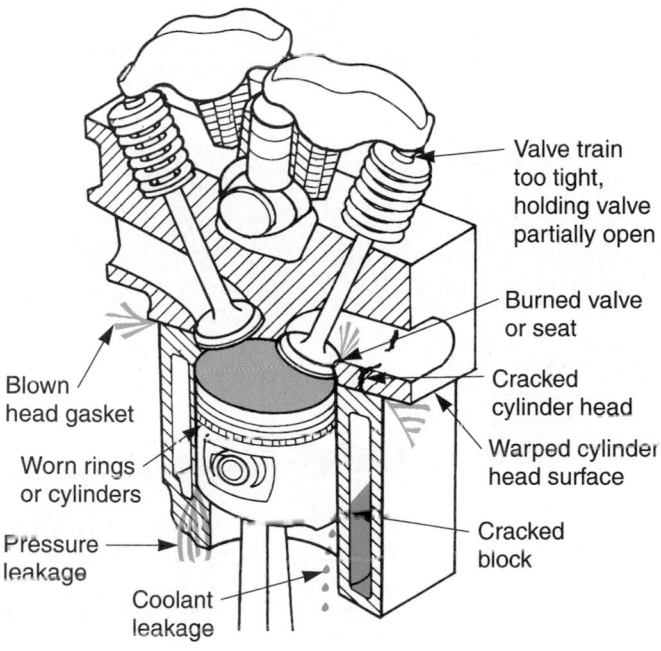

Figure 48-7. Typical reasons for combustion pressure leakage.

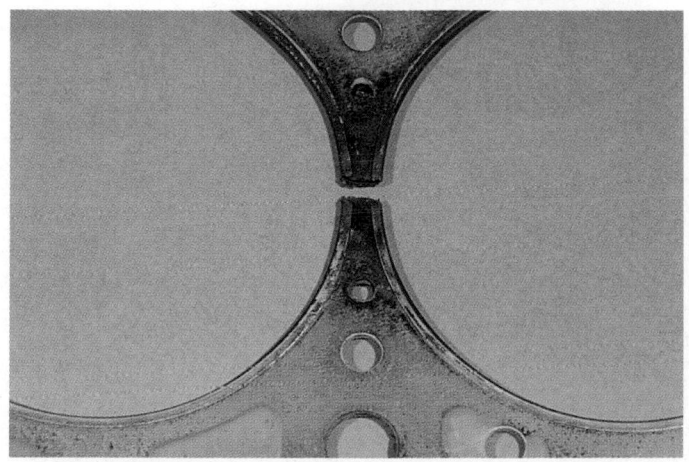

Figure 48-8. Blown head gasket can cause pressure, coolant, and oil leakage. This blown gasket allowed compression stroke pressure to flow into adjacent cylinder. The engine ran rough and made puffing sounds. (Fel-Pro)

Gasoline Engine Compression Testing

To perform a compression test on a gasoline engine, remove all the spark plugs so that the engine will rotate easily. Block open the throttle. This will prevent a restriction of airflow into the engine.

Disable the ignition system to prevent sparks from arcing out of the disconnected spark plug wires. Usually, the feed wire going to the ignition coil can be removed to disable the system.

If electronic fuel injection is used, it should also be disabled so that fuel will not spray into the engine. Check your service manual for specific directions.

Screw the compression gauge into one of the spark plug holes. Crank the engine, allowing it to rotate for about four to six compression strokes (compression gauge needle moves four to six times). Write down the gauge readings. Repeat these steps for each cylinder and compare the gauge readings to specifications. See **Figure 48-9.**

 Tech Tip

Discussed in Chapter 46, engine analyzers will allow you to check the relative compression of cylinders electronically. Then, you may only need to check actual compression pressure in the lowest cylinder, not in all of them. This can save time.

Diesel Engine Compression Test

A diesel engine compression test is similar to a gasoline engine compression test. However, a compression gauge intended on a gasoline engine must never be used on a diesel engine. The gauge can be damaged by the high compression stroke pressure. A diesel compression gauge must read up to approximately 600 psi (4000 kPa).

 To perform a diesel compression test, use the following procedure:

1. Remove either the injectors or the glow plugs. Refer to a service manual for instructions.
2. Install the compression gauge in the recommended hole.
3. Use a heat shield to seal the gauge when the gauge is installed in place of an injector.
4. Disconnect the fuel shutoff solenoid to disable the injection pump.
5. Crank the engine and note the highest reading on the gauge.

Wet Compression Testing

A *wet compression test* should be completed if cylinder pressure reads below specifications during a standard compression test. It will help you determine which engine parts are causing the problem.

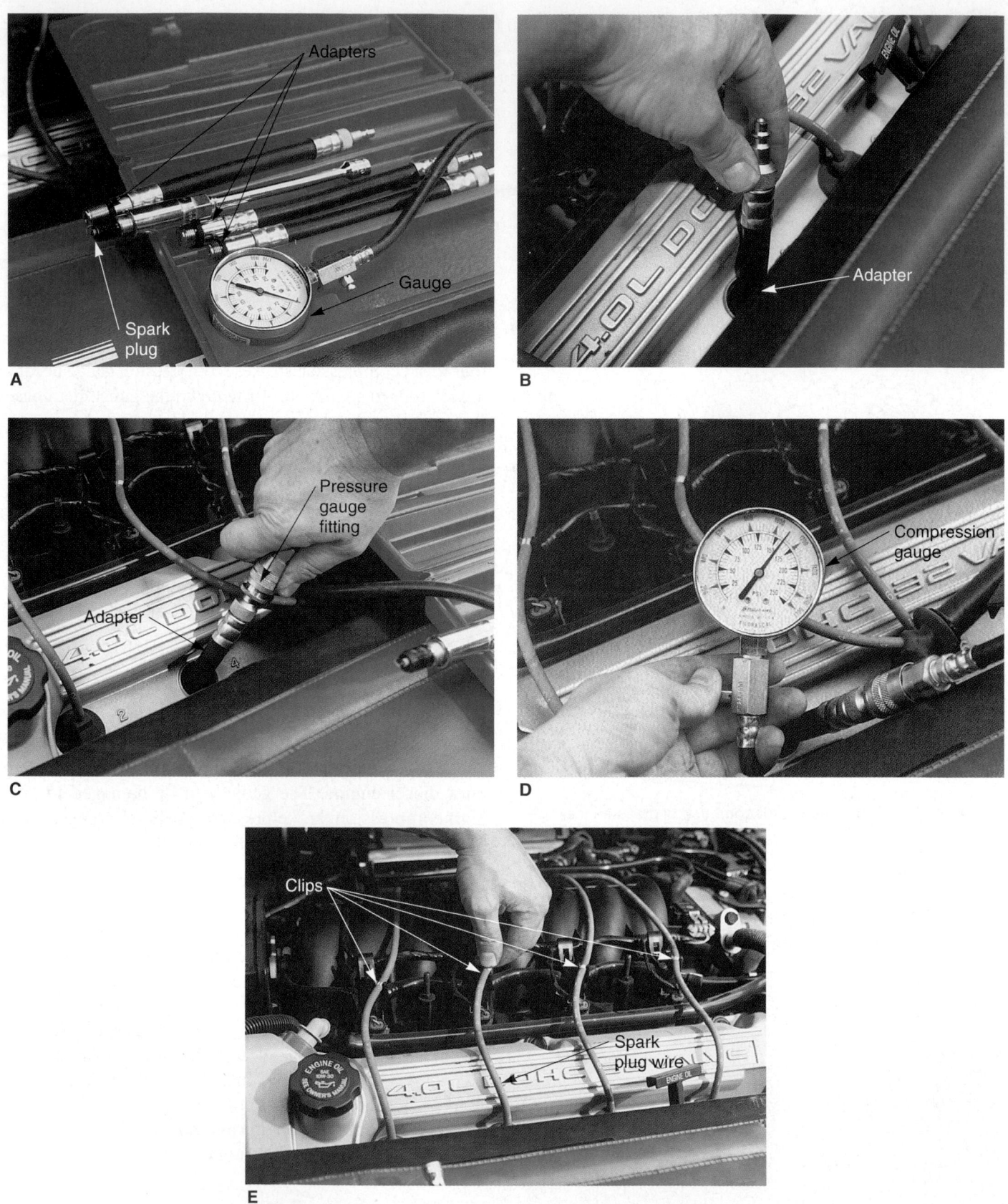

Figure 48-9. Study the major steps for doing a compression test to check the mechanical condition of the engine. A—Select the correct adapter to match spark plug thread size. Remove all spark plugs. B—Screw the compression test hose adapter into the spark plug hole. C—Install a pressure gauge onto the adapter by pulling back on the sleeve and pushing down. D—Crank the engine until the pressure gauge reading levels off at maximum. Repeat on other cylinders. E—After the compression test, replace the spark plugs and install spark plug wires in their clips to keep them off hot or moving parts. (Snap-on Tool Corp.)

Squirt a tablespoon of 30-weight motor oil into the cylinder with the low pressure reading. Install the compression gauge and recheck cylinder pressure, **Figure 48-10.**

Tech Tip

Do not squirt too much oil into the cylinder during a wet compression test, or a false reading will result. With excessive oil in the cylinder, compression readings will go up even if the piston rings and cylinders are in good condition. Oil, like any liquid, will not compress. It will take up space in the cylinder, raising the compression ratio and gauge readings.

If the compression gauge reading increases with oil in the cylinder, the piston rings or cylinders may be worn and leaking pressure. The oil will temporarily coat and seal bad compression rings to increase pressure.

If the pressure reading remains the same with oil in the cylinder, the engine valves or head gasket may be leaking.

The engine oil will seal the rings but will not seal a burned valve or blown head gasket. In this way, a wet compression test will help diagnose low-compression problems.

Caution

Some automakers warn against performing a wet compression test on a diesel engine. If too much oil is squirted into the cylinder, hydraulic lock and part damage could result because the oil will not compress in the small cylinder volume.

Compression Test Results

A normal compression reading will make the gauge increase evenly to specifications. The pressure in each cylinder should not vary more than 10%.

Look for cylinder pressure variation during an engine compression check. If some cylinders have normal pressure and one or two have low readings, engine performance will be reduced. The engine will idle roughly and lack power.

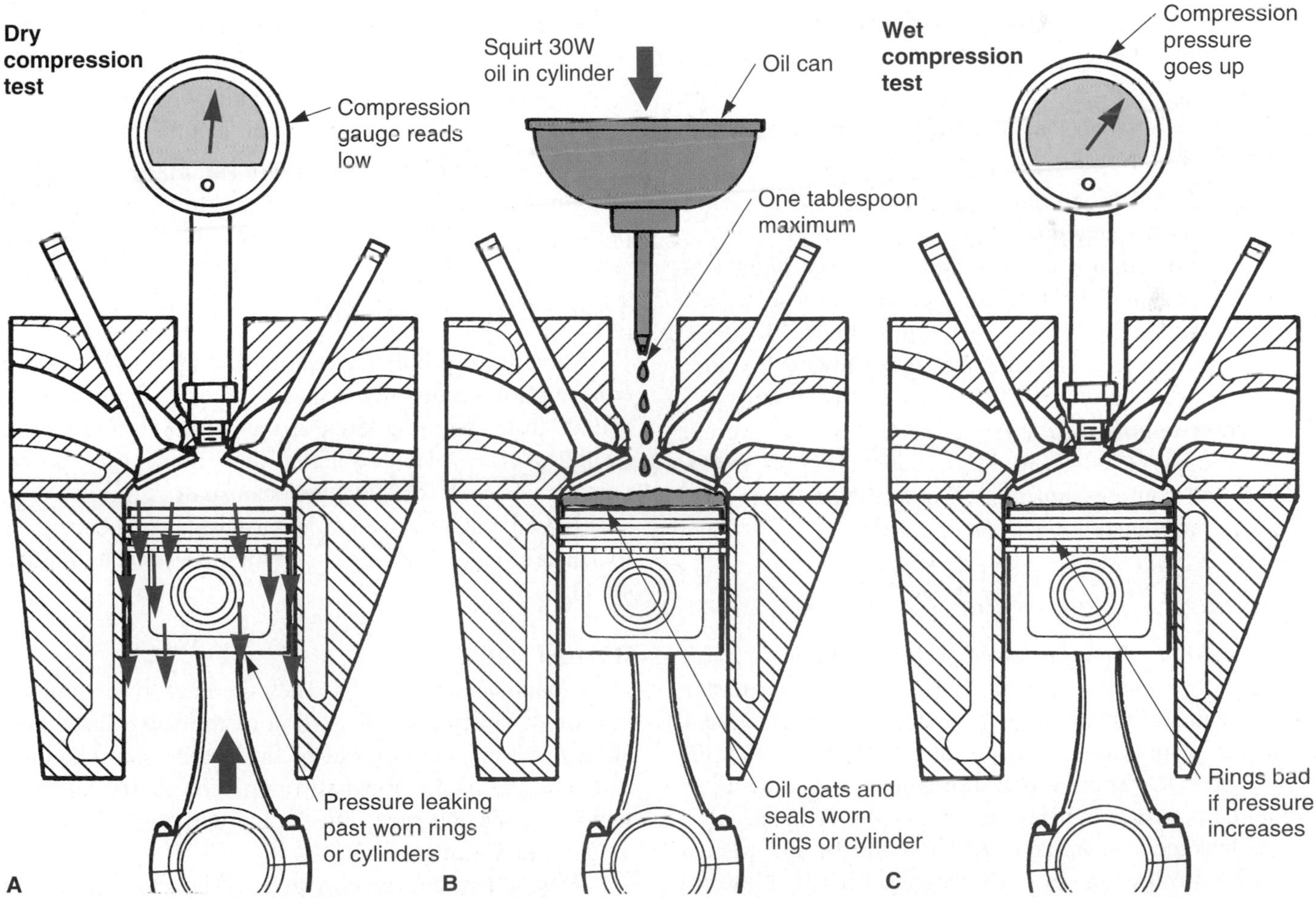

Dry compression test — Compression gauge reads low — Pressure leaking past worn rings or cylinders

Squirt 30W oil in cylinder — Oil can — One tablespoon maximum — Oil coats and seals worn rings or cylinder

Wet compression test — Compression pressure goes up — Rings bad if pressure increases

A   B   C

Figure 48-10. Basic wet compression test. A—Complete conventional, dry compression test with pressure gauge. Record all readings and compare to specifications. B—If compression is low, squirt a tablespoon of oil into the cylinder. This will temporarily seal rings. C—Measure compression pressure again. If pressure reading goes up, that cylinder may have bad rings or a worn cylinder. The same pressure reading would indicate a burned valve or blown head gasket.

If all the cylinders have low compression readings (worn timing chain for example), the engine may run smoothly but lack power and get poor gas mileage.

If two adjacent cylinders read low, it might point to a blown head gasket between the two cylinders. A blown head gasket will sometimes produce a louder than normal puffing noise from the spark plug, injector, or glow plug holes with the gauge removed.

Low engine compression can be caused by:

- Burned valve—valve face damaged by combustion heat.

- Burned valve seat—cylinder head seat damaged by combustion.

- Physical engine damage—hole in piston, broken valve, etc.

- Blown head gasket—head gasket ruptured, **Figure 48-8.**

- Worn rings or cylinders—part wear prevents ring-to-cylinder seal.

- Valve train troubles—valves adjusted with insufficient clearance (keeps them from fully closing), broken valve springs, etc.

- Jumped timing chain or belt—loose or worn chain or belt has jumped over teeth, upsetting valve timing.

For other, less common sources of low compression, refer to a service manual troubleshooting chart.

Gasoline engine compression readings should be 125–175 psi (860–1200 kPa). Generally, the compression pressure should not vary more than 15–20 psi (100–140 kPa) from the highest to the lowest cylinder. Readings must be within 10% to 15% of each other.

Diesel engine compression readings will average 275–400 psi (1900–2750 kPa), depending on engine design and compression ratio. Compression levels must not vary more than about 10% to 15%.

Cylinder Leakage Tester

A *cylinder leakage tester* performs about the same function as a compression gauge: it measures the amount of air leakage out of the engine combustion chambers. External air pressure is forced into the cylinder with the piston at TDC. Then, a pressure gauge is used to determine the percent of air leakage out of the cylinder.

If leakage is severe enough, you will be able to hear and feel air blowing out of the engine. Air may blow out of the intake manifold (bad intake valve), exhaust system (burned exhaust valve), or breather (bad rings, piston, or cylinder). It may also blow into an adjacent cylinder (blown head gasket).

Decide What Type of Engine Repair Is Needed

After performing all the necessary inspections and tests, decide what part or parts must be repaired or replaced to correct the engine mechanical problem. Evaluate all data from your pre-teardown diagnosis. If you still cannot determine the exact problem, the engine may have to be partially disassembled for further inspection.

Evaluating Engine Mechanical Problems

Before you can properly repair engine problems, you must be able to:

- Explain the function of each engine part.

- Describe the construction of each engine part.

- Explain the cause of engine problems.

- Describe the symptoms of major engine problems.

- Select specific methods to pinpoint specific problems.

- Know which parts must be removed for certain repairs.

- Know whether the engine must be removed from the vehicle before the repair can be made.

This section of the chapter will summarize important information about major engine problems. This should prepare you for following chapters.

Valve Train Problems

Valve train problems can cause engine missing, oil consumption, blue-gray exhaust smoke, light tapping sounds from the upper area of the engine, rough idling, overall performance problems, and cylinder or piston damage. A worn or damaged camshaft; timing chain, gear or belt; valve; valve guide; push rod; cam bearing; or rocker arm can upset engine operation. Refer to **Figure 48-11.**

Burned Valve

A *burned valve* results when the heat from combustion blows away a small portion of the valve face. This allows pressure to leak out of the combustion chamber and enter the intake or exhaust port. The air-fuel mixture will not ignite and burn. The engine will miss, especially at idle. See **Figure 48-12.**

With a burned valve, you may be able to hear a puffing sound as pressure blows past the valve. There may be a popping sound at the carburetor or throttle body (bad intake valve) or at the exhaust system tailpipe (burned exhaust valve).

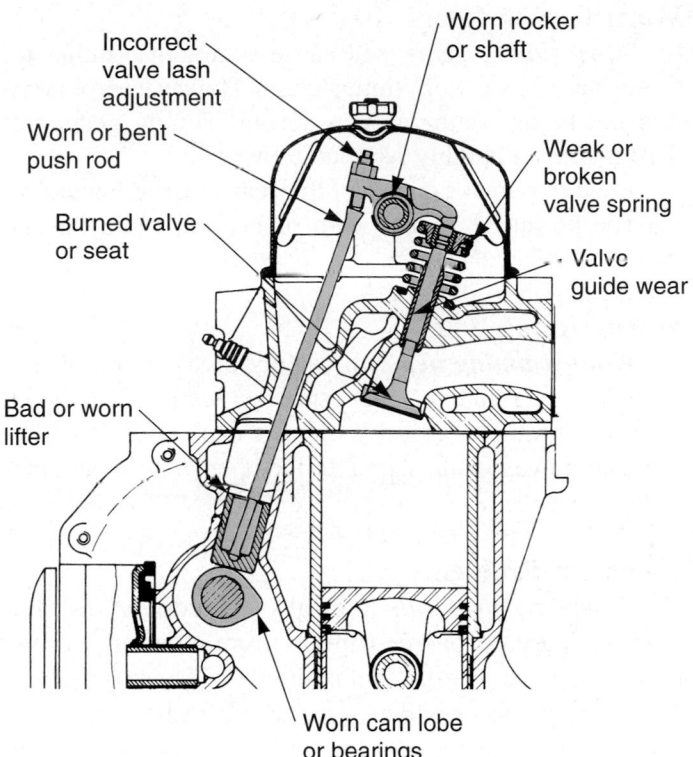

Figure 48-11. Study these typical valve train problems.

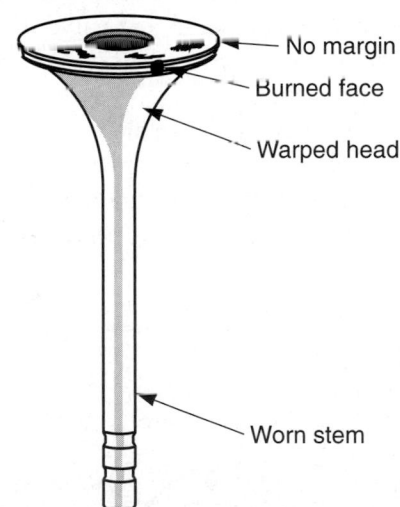

Figure 48-12. Note common problems that can develop with an engine valve. (Sioux Tools)

To fix a burned valve, you must remove the cylinder head from the engine. Then, as described in Chapter 51, all the valves and seats should be ground (machined). New valves are normally needed to replace any that are burned.

Worn Valve Guides and Stems

A *worn valve guide* or stem will allow the valve to rock or tip sideways in the cylinder head. Excess clearance

between the stem and guide will result. This can cause a tapping noise, oil consumption, spark plug fouling, or valve stem breakage, as shown in **Figure 48-13.**

To check for worn valve guides, remove the valve cover. Use a large screwdriver to pry sideways on the valve stem. If the valve wiggles in its guide, remove the cylinder head for guide repairs.

Leaking Valve Stem Seals

Leaking valve stem seals will let oil drain through the clearance between the valve stem and guide. Oil will be pulled into the intake or exhaust port and burned. The engine will emit blue smoke, especially after initial starting or upon deceleration.

Valve seals can usually be replaced without removing the cylinder head. Air pressure is used to hold the valve against its seat in the head, while a special tool is used to compress the valve spring. The keepers, springs, and seals can then be removed for service.

Valve Breakage

Valve breakage may be caused by valve stem fatigue or by a broken or weak valve spring that allow the piston to hit the valve head. When the head of a valve breaks off, it usually causes severe damage to the piston, cylinder wall, and combustion chamber. Major engine repairs are normally needed.

Stuck Valve

A *stuck valve* results when the valve stem rusts or corrodes and locks in the valve guide. This can happen when the engine sits in storage for an extended period.

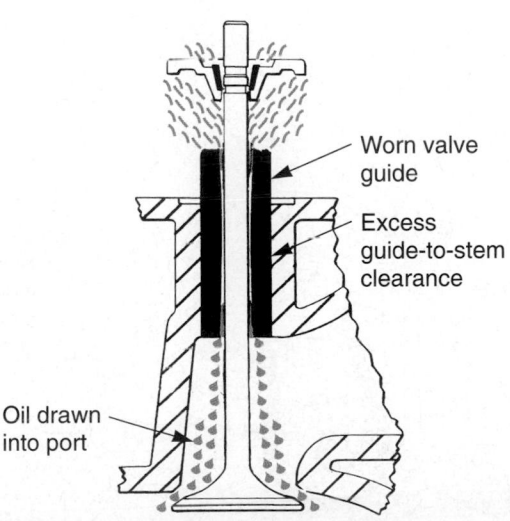

Figure 48-13. A worn valve guide will let the valve wiggle sideways in the head, causing oil to be drawn into the port. Bad valve seals can also cause oil consumption. (Dana Corp.)

A stuck or frozen valve can sometimes be fixed by squirting rust penetrant around the top of the valve guide. If this fails to correct the problem, remove the cylinder head for service.

Valve Float

Valve float is a condition in which excess engine speed, weakened valve springs, or hydraulic lifter problems cause the valves to remain partially open. This problem usually occurs at higher engine speeds. The engine may begin to miss, pop, or backfire as the valves float.

Weak valve springs are the result of prolonged use. The springs lose some of their tension. The springs may become too weak to close the valves properly.

A *broken valve spring* will frequently let the valve hang partially open. Excess valve-to-rocker clearance may cause valve train clatter (light tapping noise). Popping or backfiring can also result.

Valve springs can be replaced without cylinder head removal. As with valve seal replacement, air pressure and a special tool will permit spring replacement.

Worn Timing Chain

A *worn timing chain* will upset valve timing, reducing compression stroke pressure and engine power. Wear can cause slack between the crankshaft sprocket and camshaft sprocket. The camshaft and valves will no longer be kept in time with the pistons. Look at **Figure 48-14.**

To check for a worn timing chain, rotate the crankshaft back and forth while watching the distributor rotor or the rocker arms. If you can turn the crankshaft several degrees without rotor or valve movement, the timing chain is worn and should be replaced.

Worn Timing Gears

Worn timing gears will cause symptoms similar to those caused by a worn timing chain. However, problems will not be as common or as severe. Timing gears are dependable and highly resistant to wear.

If a tooth breaks on one of the gears, a growling noise may be produced. When worn or broken, timing gears must be replaced as a set.

Worn Timing Belt

A *worn timing belt* will usually break, jump off its sprockets, or skip over a few sprocket teeth. Severe performance problems or valve damage can result. The pistons can move up and slam into the open valves, bending or breaking them.

Camshaft Problems

Camshaft problems typically include worn cam lobes or journals, broken cams, a worn distributor drive gear, or a loose or worn fuel pump drive eccentric.

Cam lobe wear will reduce valve lift (distance valve slides open). This reduces engine power and can cause a rough idle. See **Figure 48-15.**

Camshaft breakage, though not common, will keep some of the valves from operating. The lobes on one end of the broken cam will not rotate and open their valves. Severe performance problems or valve damage can result. Removal of a valve cover will let you check valve train and camshaft action.

Worn cam bearings or journals will reduce engine oil pressure. Generally, this only happens after prolonged engine service. Normally, other engine parts will fail before journal or cam bearing wear becomes critical.

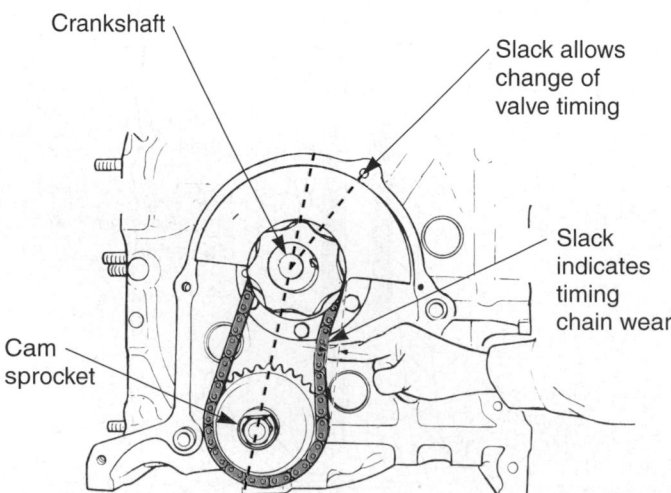

Figure 48-14. Timing chain wear can let the camshaft rotate out of time with the crankshaft. Valves do not open when they should, reducing engine power and efficiency. (Mazda)

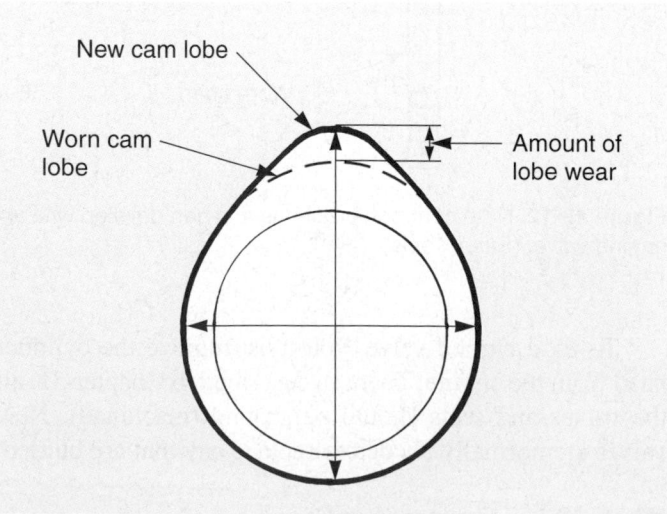

Figure 48-15. Cam lobe wear will reduce valve lift. Note the difference in lift between new and worn lobe. (Ford)

Rocker Arm and Push Rod Problems

Worn rocker arms can cause valve clatter (light tapping noise) by upsetting valve clearance. Worn rocker arms have little effect on engine performance. Rocker arm wear usually results from lubrication system problems (dirty oil, clogged oil passages, etc.) that increase friction, **Figure 48-16.**

A rocker arm will wear at the points of contact with the valve, push rod, and rocker shaft or ball socket. After removal, the rocker arms should be inspected closely for indentations or roughness, which indicate wear.

Worn or bent push rods can also cause valve clatter. To check for wear, measure the push rod length and compare your measurements to specifications.

To check for bent push rods, roll them on a flat work surface. A bent push rod will wobble up and down when rolled, and you will able to see under it. When a push rod is bent, check the opening action of that push rod's valve. The valve may be stuck and could bend a new push rod if not repaired. Replace bent push rods. Do not try to straighten them.

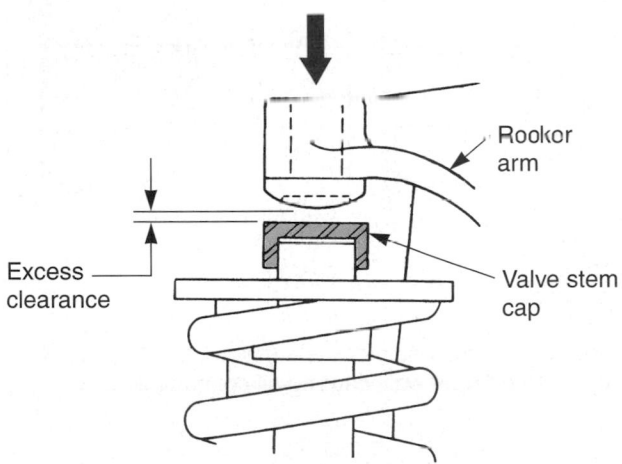

Figure 48-16. Excess clearance in the valve train can produce a clattering sound from under the valve cover. The rocker will clack as it strikes the valve stem tip or cap. (Deere & Co.)

Hydraulic Lifter Problems

A worn or defective hydraulic lifter may produce valve clatter identical to that produced by a maladjusted, worn, or loose rocker arm; a worn valve guide; or a bent push rod. It will sound like a small ball peen hammer tapping on the cylinder head.

After removing the valve cover to check the valve noise, try adjusting the valves. If adjustment will not quiet the valve noise, check for valve train wear. If other valve train parts are good, the valve lifter is probably bad.

Low engine oil pressure can cause hydraulic lifter clatter. Check the oil level and oil pressure before condemning the lifters. Contaminated or dirty oil can also cause lifter noise.

Engine Gasket Problems

A *blown head gasket* can cause a wide range of problems: overheating, missing, coolant or oil leakage, engine smoking, even head or block damage (burned mating surfaces). Quite often, a blown head gasket will show up during a compression test. Two adjacent cylinders, usually the two center cylinders, will have low pressure. Refer back to **Figure 48-8.**

A *leaking intake manifold gasket* can cause a vacuum leak, with resulting rough idle. To check for an intake gasket leak, squirt oil along the edge of the gasket. The oil may temporarily seal the leak, showing an intake gasket rupture. A low vacuum gauge reading can also indicate an intake manifold gasket leak.

A *leaking exhaust manifold gasket* will show up as a clicking-type sound. As combustion gases blow into the manifold and out the bad gasket, a metallic-like rap is produced.

Part warpage is a common cause of gasket failure. Always check for warpage when servicing a bad gasket.

Piston and Cylinder Problems

Piston and cylinder problems are major and usually require engine removal. It is important for the technician to be able to detect and diagnose the source of piston and cylinder-related troubles.

Piston Knock (Slap)

Piston knock, or *slap,* is a loud metallic knock produced when the piston flops back and forth inside its cylinder. It is caused by excess piston skirt or cylinder wear, excess clearance, and possibly damage. Refer to **Figure 48-17.**

Piston slap is normally louder when the engine is cold and tends to quiet down as the engine reaches operating temperature. Heat expansion of the aluminum piston takes up some of the clearance.

Piston Pin Knock

Piston pin knock occurs when too much clearance exists between the piston pin and the piston pin bore or connecting rod bushing. Excessive clearance allows the pin to hammer against the rod or piston as the piston changes direction in the cylinder.

Piston pin knock will usually make a double knock (two rapid knocks and then a short pause). It does not change much with engine load.

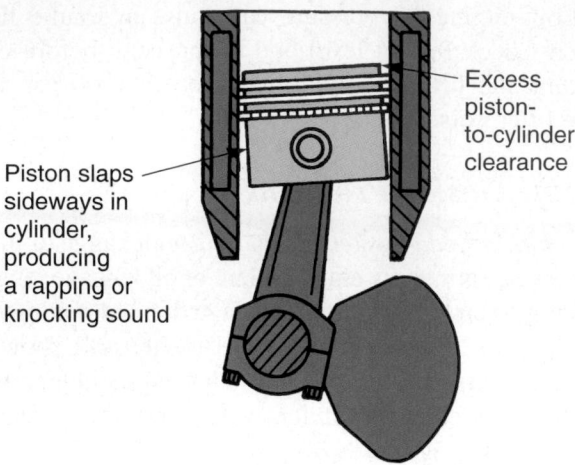

Figure 48-17. A worn cylinder wall or piston skirt can let the piston flop back and forth. This can produce a knocking sound and oil consumption. Piston knock, or slap, is loudest when the engine is cold. It tends to quiet down as the engine warms and the piston expands.

Worn Piston Rings and Cylinder

Worn piston rings or cylinders result in blowby, blue-gray engine smoke, low engine power, spark plug fouling, and other problems caused by poor ring sealing. See **Figure 48-18.**

To check for ring and cylinder problems, increase engine speed while watching the tailpipe and valve cover breather opening. If blue-gray smoke pours out of the vehicle's exhaust under load, the oil rings and cylinders may need service. If excessive oil vapors and air blow out the valve cover breather, blowby is entering the crankcase. Compression ring or cylinder problems are indicated. Worn ring grooves can also cause oil consumption, **Figure 48-19.**

If, after disassembly, the engine cylinders are found to be worn, the engine must be removed from the vehicle. The block must be sent to a machine shop for *boring* (cylinders machined oversize) or *sleeving* (liners installed to restore cylinders).

If only the rings are worn (and not the cylinders), the engine may be rebuilt while still installed in the vehicle. Check a manual for details.

Burned Piston

A *burned piston* is often a result of preignition or detonation damage. Abnormal combustion, excessive pressure, and excessive heat melt and blow a hole in the piston crown or the area around the ring lands. The engine may smoke, knock, and have excessive blowby or other symptoms, **Figure 48-20.**

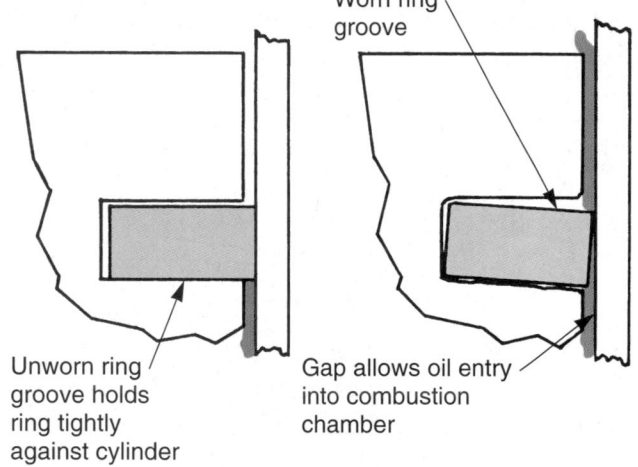

Figure 48-19. If the ring grooves are worn, they can allow rings to tip on the cylinder wall, causing oil consumption.

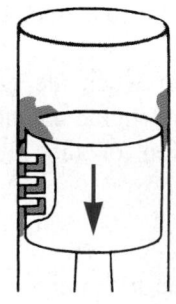

Intake stroke

Downstroke forces oil into ring grooves.

Compression stroke

Upstroke forces rings to bottom of ring grooves, trapping oil.

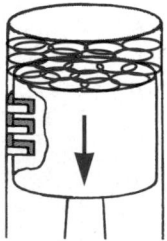

Power stroke

Downstroke transfers oil to cylinder wall. Piston ring is held down against ring land by force of expansion.

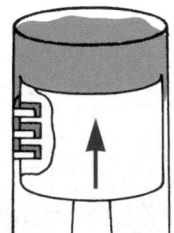

Exhaust stroke

Upstroke forces oil into combustion chamber, where it is burned and causes smoking.

Figure 48-18. Study how oil is trapped in worn rings and then burned on the power stroke. (Deere & Co.)

Figure 48-20. A burned piston usually results from prolonged preignition or detonation. Symptoms would be low compression, blowby, smoking, and rough idle. (Fel-Pro)

A compression test or cylinder leakage test (discussed earlier) may indicate a burned piston, but engine disassembly is usually needed to verify the problem. If the cylinder wall is not damaged, the repair can be an in-vehicle operation.

Crankshaft Problems

Crankshaft problems include journal wear, main bearing wear, rod bearing wear, and resulting low oil pressure.

Rod Bearing Knock

Connecting rod bearing knock is caused by wear and excessive rod bearing-to-crankshaft clearance. It is a light, regular, rapping noise with the engine floating (point at which throttle is held constant and engine is not accelerating). It is loudest after engine warm-up. In a cold engine, thickened oil tends to cushion and quiet rod knock.

To locate a bad rod bearing, short out or disconnect each spark plug wire one at a time. The loose, knocking rod bearing may quiet down or change pitch when its spark plug is disabled.

Main Bearing Knock

Main bearing knock is similar to rod bearing knock but is slightly deeper or duller in pitch. It is usually more pronounced when the engine is pulling or lugging under a load. Worn bearings and journals are letting the crankshaft move up and down in the cylinder block.

Main bearing wear will usually reduce oil pressure significantly. To verify main bearing noise, remove the oil pan and pressure test the lubrication system. Excessive oil flow out of one or more of the main bearings implies too much bearing clearance. If a pressure tester is not available, remove and inspect each of the main bearings. If the crankshaft is not worn, a bearing insert replacement should correct the problem.

Excess Crankshaft End Play

Excess crankshaft end play is caused by a worn main thrust bearing. In vehicles with a manual transmission, thrust bearing wear can produce a deep knock, usually when applying and releasing the clutch. With an automatic transmission, the end play problem may only show up as a single thud or knock during acceleration or deceleration.

 **Tech Tip**
A knock occurring with clutch or torque converter action could also be caused by loose flywheel bolts or other drivetrain problems. Check out all possible causes before beginning repairs.

Broken Engine Mounts

Broken engine mounts allow the engine to move in the vehicle chassis and can cause other part damage. The rubber part of the mount can rip apart, allowing the engine to move excessively with engine torque. With enough movement, the engine fan can hit the radiator shroud, the alternator or air filter can hit the hood, or wires and hoses can bind, **Figure 48-21.**

To check for broken engine mounts, open the hood. Engage the parking brake and place the transmission in drive or in gear. While holding down on the brake pedal, slowly increase engine speed or release the clutch pedal. This will twist the engine and cause it to move excessively if the mounts are broken.

Figure 48-21. A broken or deteriorated engine mount will allow the engine to move excessively in the engine compartment. The broken mount may not be noticed during visual inspection. Load the engine and drivetrain while checking for engine movement on the mounts. (Fel-Pro)

Service Manual Troubleshooting Charts

Service manual troubleshooting charts give lists of possible problems and needed repairs. Refer to these charts when you have difficulty locating or correcting an engine mechanical problem. A troubleshooting chart will be written for the specific make and model of engine, making it very accurate.

Summary

- If a technician does not know how to properly diagnose (locate) engine problems, a great deal of time, effort, and money will be wasted.

- After gathering information from the customer or service writer, inspect the engine using your senses (sight, smell, hearing, touch).

- Coolant in oil will show up as white or milky oil.

- Oil-fouled spark plugs indicate internal oil leakage into the engine combustion chambers.

- Engine blowby occurs when combustion pressure blows past the piston rings into the lower block and oil pan area of the engine.

- Engine smoking is normally noticed at the tailpipe when the engine is accelerated or decelerated.

- Abnormal engine noises (hisses, knocks, rattles, clunks, popping) may indicate part wear or damage.

- A compression test is one of the most common methods of determining engine mechanical condition.

- A wet compression test should be completed if cylinder pressure reads below specifications. It will help you determine what engine parts are causing the problem.

- A normal compression reading will make the gauge increase evenly to specifications. The pressure in each cylinder should not vary more than about 10%.

- Valve train problems can cause engine missing, oil consumption, blue-gray exhaust smoke, light tapping sounds from the upper area of the engine, rough idling, overall performance problems, and even cylinder and piston damage.

- A worn timing chain will upset valve timing, reducing compression stroke pressure and engine power.

- Worn rocker arms can cause valve clatter (light tapping noise) by upsetting valve clearance.

- A blown head gasket can cause a wide range of problems: overheating, missing, coolant or oil leakage, engine smoking, even head or block damage (burned mating surfaces).

- Piston knock or slap is a loud metallic knocking sound produced when the piston flops back and forth inside its cylinder.

- Worn piston rings or cylinders result in blowby, blue-gray engine smoke, low engine power, spark plug fouling, and other problems caused by poor ring sealing.

- Connecting rod bearing knock is caused by wear and excessive rod bearing-to-crankshaft clearance.

- Main bearing knock is similar to rod bearing knock but is slightly deeper or duller in pitch.

Important Terms

Coolant in oil	Weak valve springs
Oil-fouled spark plugs	Broken valve spring
Oil in the coolant	Worn timing chain
Engine oil leaks	Worn timing gears
External coolant leaks	Worn timing belt
Engine blowby	Camshaft problems
Engine vacuum leaks	Cam lobe wear
Engine exhaust leaks	Camshaft breakage
Engine smoking	Worn cam bearings
Stethoscope	Worn rocker arms
Compression gauge	Blown head gasket
Compression test	Leaking intake manifold gasket
Wet compression test	
Low engine compression	Leaking exhaust manifold gasket
Gasoline engine compression readings	Part warpage
	Piston knock
Diesel engine compression readings	Slap
	Piston pin knock
Cylinder leakage tester	Boring
Burned valve	Sleeving
Worn valve guide	Burned piston
Leaking valve stem seals	Connecting rod bearing knock
Valve breakage	Main bearing knock
Stuck valve	Excess crankshaft end play
Valve float	

Review Questions—Chapter 48

Please do not write in this text. Place your answers on a separate sheet of paper.

1. List seven symptoms of engine mechanical problems.

2. Oil-fouled spark plugs are an indication of worn _____, worn or scored _____ _____ or bad _____ _____.

3. With a gasoline engine, blue-gray smoke indicates:
   (A) Rich fuel mixture.
   (B) Coolant leakage into combustion chambers.
   (C) Oil leakage into combustion chambers.
   (D) Lean fuel mixture.

4. A section of vacuum hose can be used to find engine vacuum leaks. True or False?

5. Low cylinder compression *cannot* be caused by:
   (A) Worn camshaft bearings.
   (B) Blown head gasket.
   (C) Burned valve.
   (D) Worn piston rings.

6. Why is a wet compression test helpful?

7. What are typical gasoline engine and diesel engine compression readings?

8. Valve breakage can be caused by valve _____ _____ or by a broken or weak _____ _____.

9. Define the term "valve float."

10. How can you tell if an engine has a worn timing chain?

11. What are the problems resulting from camshaft lobe wear?

12. A leaking intake manifold gasket can cause a _____ leak, with resulting _____ idle.

13. Explain the symptoms of piston pin knock.

14. Why is piston slap loudest with the engine cold?

15. What are the symptoms of worn piston rings and cylinders?

❋ ASE-Type Questions

1. An automobile is brought into the shop. A compression test on the engine indicates low compression stroke pressure in all the engine's cylinders. Technician A says a complete engine overhaul is required to repair this problem. Technician B says before overhauling the engine, you should look for other mechanical problems that would affect engine compression. Who is right?
   (A) A only.
   (B) B only.
   (C) Both A and B.
   (D) Neither A nor B.

2. Technician A says when inspecting an engine for mechanical troubles, you should look for engine oil leaks. Technician B says when inspecting an engine for mechanical troubles, you should look for vacuum leaks. Who is right?
   (A) A only.
   (B) B only.
   (C) Both A and B.
   (D) Neither A nor B.

3. Technician A says a leaking gasket is a common engine mechanical problem. Technician B says a burned valve is a common engine mechanical problem. Who is right?
   (A) A only.
   (B) B only.
   (C) Both A and B.
   (D) Neither A nor B.

4. Which of the following is a common cause of engine mechanical problems?
   (A) Engine overheating.
   (B) Lack of periodic maintenance.
   (C) High engine mileage.
   (D) All of the above.

5. The oil in an automotive engine has a white, milky appearance. Technician A says this indicates engine coolant is probably leaking into the engine. Technician B says this indicates transmission fluid is probably leaking into the engine. Who is right?
   (A) A only.
   (B) B only.
   (C) Both A and B.
   (D) Neither A nor B.

6. An engine is brought into the shop with oil-fouled spark plugs. Technician A says this problem is probably caused by an ignition system malfunction. Technician B says this problem is probably caused by internal oil leakage into the engine's combustion chambers. Who is right?
   (A) A only.
   (B) B only.
   (C) Both A and B.
   (D) Neither A nor B.

7. Oil is present in an automobile engine's coolant. Technician A says this problem is normally the result of worn valve seals. Technician B says this problem is usually caused by a leak in the radiator transmission oil cooler. Who is right?
   (A) A only.
   (B) B only.
   (C) Both A and B.
   (D) Neither A nor B.

8. An automobile's engine is producing "crankcase blowby." Technician A says the solution to this problem is simple valve adjustment. Technician B says the solution to this problem normally requires extensive engine repair. Who is right?
   (A) A only.
   (B) B only.
   (C) Both A and B.
   (D) Neither A nor B.

9. Technician A says engine vacuum leaks are normally loudest at highway speeds. Technician B says engine vacuum leaks are normally loudest during engine cranking. Who is right?
   (A) A only.
   (B) B only.
   (C) Both A and B.
   (D) Neither A nor B.

10. Blue-gray smoke is coming out of an automobile's exhaust. Technician A says this condition normally indicates water in the engine's oil supply. Technician B says this condition normally indicates internal engine mechanical problems. Who is right?
   (A) A only.
   (B) B only.
   (C) Both A and B.
   (D) Neither A nor B.

11. Black smoke is coming out of a gasoline engine's tailpipe. Technician A says this normally indicates an engine mechanical problem. Technician B says this condition normally indicates a fuel system problem. Who is right?
   (A) A only.
   (B) B only.
   (C) Both A and B.
   (D) Neither A nor B.

12. White exhaust smoke is coming out of a gasoline engine's tailpipe. Technician A says white exhaust smoke may indicate an internal coolant leak into the engine's cylinders. Technician B says white exhaust smoke may be due to water condensation on a cool day. Who is right?
   (A) A only.
   (B) B only.
   (C) Both A and B.
   (D) Neither A nor B.

13. An automobile equipped with a diesel engine is brought into the shop. The customer says there is blue smoke coming out of the engine's exhaust. Technician A says this problem may be due to worn engine cylinders. Technician B says

this problem may be due to worn valve seals. Who is right?
   (A) A only.
   (B) B only.
   (C) Both A and B.
   (D) Neither A nor B.

14. Technician A says a gasoline engine compression gauge can be used to perform a compression test on a diesel engine. Technician B says a gasoline engine compression gauge should not be used to perform a diesel engine compression test. Who is right?
   (A) A only.
   (B) B only.
   (C) Both A and B.
   (D) Neither A nor B.

15. Diesel engine compression readings should average approximately:
   (A) 125–175 psi.
   (B) 180–200 psi.
   (C) 150–175 psi.
   (D) 275–400 psi.

16. Valve train problems can cause all of the following *except:*
   (A) oil consumption.
   (B) rough idle.
   (C) crankshaft damage.
   (D) engine missing.

17. Technician A says that leaking valve stem seals will cause the engine to emit blue smoke after initial starting. Technician B says that the cylinder head must generally be removed to replace the valve stem seals. Who is right?
   (A) A only.
   (B) B only.
   (C) Both A and B.
   (D) Neither A nor B.

18. Technician A says that valve float can be caused by hydraulic lifter problems. Technician B says that valve float can be the result of weak valve springs. Who is right?
   (A) A only.
   (B) B only.
   (C) Both A and B.
   (D) Neither A nor B.

19. Technician A says that worn rocker arms can upset valve clearance. Technician B says that worn rocker arms have little effect on engine performance. Who is right?
   (A) A only.
   (B) B only.

(C) *Both A and B.*
(D) *Neither A nor B.*

20. All of the following are caused by a blown head gasket *except:*
    (A) *overheating.*
    (B) *head damage.*
    (C) *excess valve clearance.*
    (D) *coolant leaks.*

21. Technician A says that a leaking intake manifold gasket can cause a vacuum leak. Technician B says that a high vacuum gauge reading indicates a blown head gasket. Who is right?
    (A) *A only.*
    (B) *B only.*
    (C) *Both A and B.*
    (D) *Neither A nor B.*

22. Technician A says that piston slap is generally loudest when the engine is warm. Technician B says that piston slap is caused by excessive piston skirt wear. Who is right?
    (A) *A only.*
    (B) *B only.*
    (C) *Both A and B.*
    (D) *Neither A nor B.*

23. Technician A says that rod knock is a light, regular rapping noise that occurs when the engine is accelerating. Technician B says that rod knock is loudest when the engine is cold. Who is right?
    (A) *A only.*
    (B) *B only.*
    (C) *Both A and B.*
    (D) *Neither A nor B.*

24. Technician A says that main bearing knock is duller in pitch than rod bearing knock. Technician B says that main bearing wear will significantly reduce oil pressure. Who is right?
    (A) *A only.*
    (B) *B only.*
    (C) *Both A and B.*
    (D) *Neither A nor B.*

25. Excess crankshaft end play is caused by:
    (A) *worn rod journals.*
    (B) *worn crank throws.*
    (C) *worn thrust bearings.*
    (D) *worn connecting rods.*

Activities—Chapter 48

1. Demonstrate the steps for making a dry compression test on a gasoline engine. Compare your readings to specifications.

2. Make a poster listing the probable causes of blue-gray, black, and white exhaust emissions.

Compression Test Results	
Condition	**Possible Causes**
All cylinders at normal pressure (no more than 10–15% difference between cylinders).	None. Engine is in good condition.
All cylinders low (more than 20% below specifications).	1. Burned valve or valve seat. 2. Blown head gasket. 3. Worn rings or cylinder. 4. Maladjusted valves. 5. Jumped timing chain or belt. 6. Physical engine damage.
One or more cylinders low (more than 20% below specifications).	1. Burned valve or valve seat. 2. Damage or wear on affected cylinder(s).
Two adjacent cylinders low (more than 20% below specifications).	1. Blown head gasket. 2. Cracked block or head.
One or more cylinders high (more than 20% above specifications).	1. Carbon buildup in cylinder.

Cylinder Leakage Test Results	
Condition	**Possible Causes**
No air escapes.	Normal condition. No leakage.
Air escapes from carburetor or throttle body.	1. Intake valve damaged or not properly seated. 2. Valve train mistimed. Possible jumped timing belt or chain. 3. Broken or damaged valve train component.
Air escapes from tailpipe.	1. Exhaust valve damaged or not seated properly. 2. Valve train mistimed. 3. Broken or damaged valve train part.
Air escapes from dipstick tube or oil fill opening.	1. Worn piston rings. 2. Worn cylinder walls. 3. Damaged piston. 4. Blown head gasket.
Air escapes from adjacent cylinder.	1. Blown head gasket. 2. Cracked head or block.
Air bubbles in radiator coolant.	1. Blown head gasket. 2. Cracked head or block.
Air heard around outside of cylinder.	1. Cracked or warped head or block. 2. Blown head gasket.

Chapter 49

Engine Removal, Disassembly, and Parts Cleaning

After studying this chapter, you will be able to:

- Determine if engine removal is needed to make specific engine repairs.
- List the preparations for engine removal.
- Describe the general safety rules pertaining to engine removal, disassembly, and parts cleaning.
- Explain the use of an engine lifting fixture or chain, and an engine crane.
- Summarize how to properly disassemble an engine.
- Describe typical inspections that should be made during engine disassembly and cleaning.
- List various methods for cleaning engine parts.
- Describe safety practices to follow when cleaning parts.
- Correctly answer ASE certification test questions on engine removal, disassembly, and cleaning procedures.

Engine removal and disassembly procedures vary from vehicle to vehicle. However, there are many general rules and methods that apply to all cars and small trucks. This chapter will outline the most important steps for engine removal, teardown, and cleaning. This should make an *engine R&R* (engine remove and repair) job much easier.

Many engine repairs can be made with the engine block still mounted in the chassis. Repairs to the cylinder head, valve train, and other external parts are normally in-vehicle operations.

Engines are removed when the cylinder block or crankshaft is badly damaged. Depending on the year, make, and model of the vehicle, engine removal may be required for other repairs.

When in doubt, always refer to a manufacturer's service manual. It will give directions for the exact vehicle and engine.

Engine Removal

 To prepare for engine removal, use the following general steps and a shop manual:

1. Park the vehicle so there is plenty of work space on both sides and in front of the engine compartment.
2. Use fender covers to protect the paint.
3. Scribe the hood along the hinges to aid realignment, **Figure 49-1.** Then have someone help you remove the hood. Store the hood in a safe place where it cannot be damaged.
4. Disconnect the battery to prevent an electrical short and fire. Remove the battery if it is in the way.
5. Drain the engine oil and coolant.
6. Unplug all electrical wires between the engine and chassis. If needed, use masking tape to label or identify the wires. This will simplify reconnection. See **Figure 49-2.**
7. Remove all coolant and vacuum hoses that prevent engine removal. Label vacuum hoses if necessary for proper reconnection.
8. Disconnect the throttle cable at the throttle body, **Figure 49-2.**
9. When disconnecting fuel lines, be careful not to let fuel spray out. Wrap a shop rag around the hose or fitting during disconnection.
10. Keep fasteners organized in several different containers. For instance, keep all of the bolts and nuts from the front of the engine in one container. Keep engine top end and bottom end fasteners separate in two more containers. This will speed reassembly. Look at **Figure 49-3.**

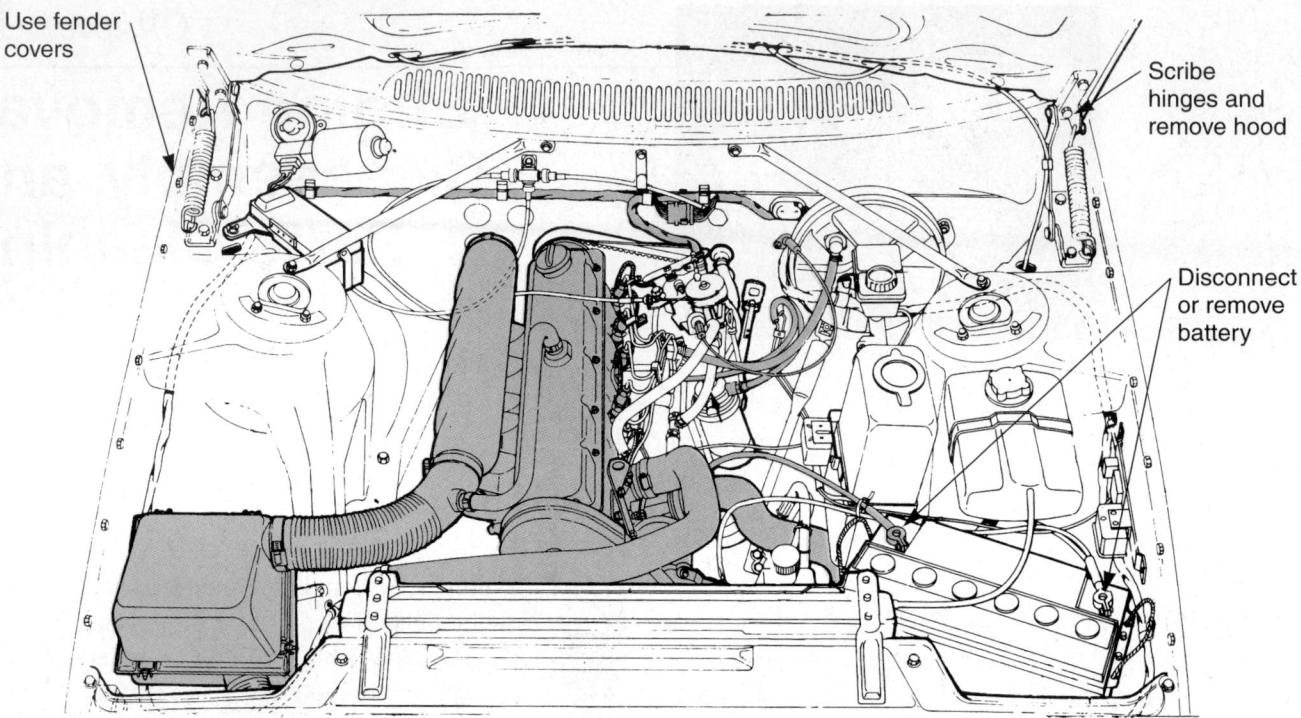

Use fender covers

Scribe hinges and remove hood

Disconnect or remove battery

Figure 49-1. Before removing the engine, the hood must often be removed. To ensure proper realignment, scribe the hood along the top and outside edges of each hinge before loosening the mounting bolts. (Volvo)

11. Do not disconnect any power steering or air conditioning lines or hoses unless absolutely necessary. Usually, the power steering pump or air conditioning compressor can be unbolted and placed on one side of the engine compartment.

12. Remove the radiator, fan, and other accessory units in front of the engine. Be careful not to hit or drop the radiator. Only remove parts that hinder engine removal.

13. Remove any other part that prevents engine removal: exhaust header pipe, automatic transmission flywheel fasteners, and bell housing bolts. Refer to **Figure 49-4.**

Before engine removal, double-check that everything is disconnected or removed. For example, check:

1. Behind and under the engine for hidden wires or ground straps.

2. That all bell housing bolts are out.

3. To see that you have removed the torque converter bolts (automatic transmission to stay in vehicle).

4. That all fuel lines are disconnected and plugged.

5. That a floor jack is supporting the transmission.

6. That motor mounts are unbolted, **Figure 49-5.**

 Tech Tip

Some engine motor mounts are computer-controlled hydraulic units that reduce engine noise. The computer can alter the stiffness of the mounts by using solenoid valves that control hydraulic flow in the mounts. These are expensive. Do not damage them during engine removal or installation.

Transmission Removal

It is sometimes necessary to remove the engine and transmission together. Some front-wheel-drive vehicles that use a transaxle require that the two be removed as a unit. Check the service manual for details.

You may want to drain the fluid from the transmission or transaxle if it is to be removed. With rear-wheel drive, the drive shaft, transmission and clutch linkage, speedometer cable, rear motor mount, and other parts must also be removed. With a transaxle, the axle shafts must be disconnected.

Installing a Lifting Fixture

Connect the lifting fixture or chain to the engine. Position the fixture at recommended lifting points. Sometimes brackets are provided on the engine.

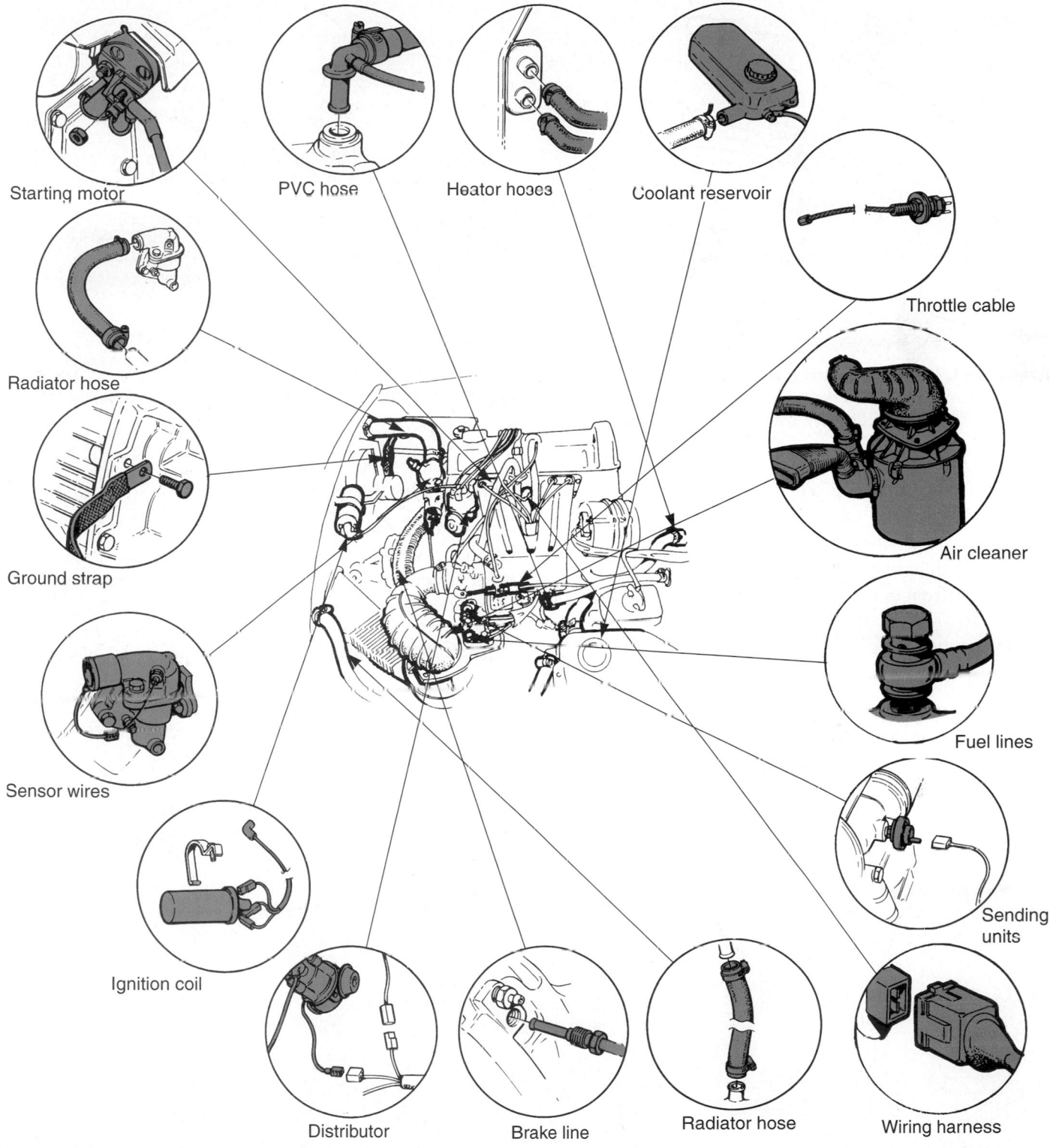

Starting motor

PVC hose

Heater hoses

Coolant reservoir

Throttle cable

Radiator hose

Air cleaner

Ground strap

Fuel lines

Sensor wires

Sending units

Ignition coil

Distributor

Brake line

Radiator hose

Wiring harness

Figure 49-2. Close inspection is needed to find all the parts that must be disconnected before engine removal. Study this illustration closely. (Saab)

If a lifting chain is to be used, fasten it to the engine. Install a bolt, nut, and washers on the chain to keep it from slipping and dropping the engine, as shown in **Figure 49-6.**

If bolts are used to secure the chain to the engine, make sure they are large enough in diameter and that they are fully installed. The bolts must not be too long (stick out from chain) or too short (they must thread into the

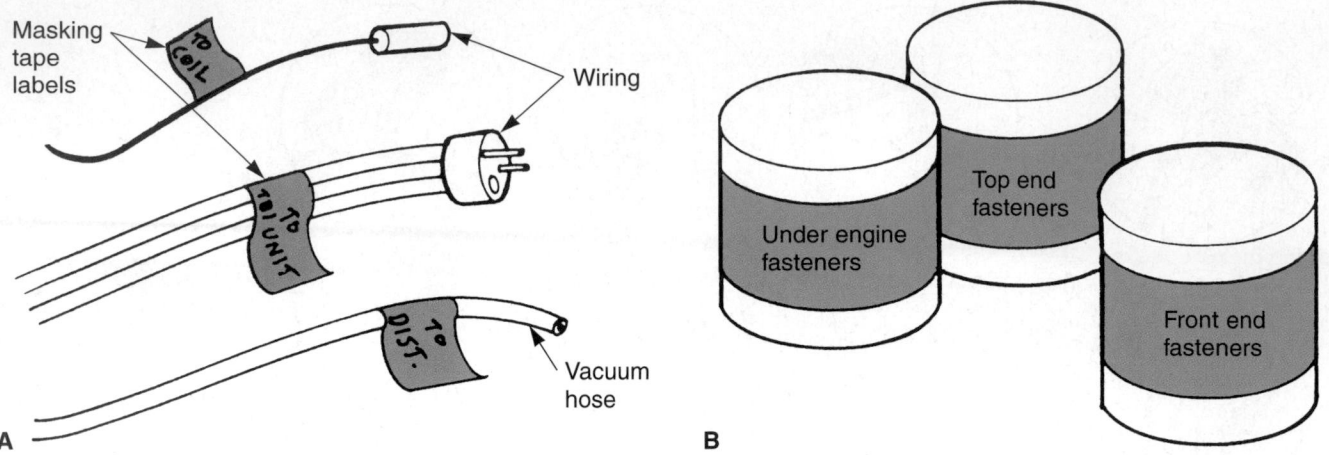

Figure 49-3. To save time and prevent confusion, stay as organized as possible during engine removal and teardown. A—Label wires and hoses to speed reconnection. B—Use several different cans to hold bolts, nuts, and small parts. Each container should hold components from different sections of the engine.

hole a distance that is equal to one and one-half times thread diameter).

Generally, position the fixture or chain so that it will raise the engine in a level manner. If one lifting point is at the right-front of the cylinder head, the other should be on the left-rear of the head. Use common sense and follow manufacturer's instructions.

Lifting the Engine

Attach the lifting device (crane or hoist) to the fixture or chain on the engine. Make sure the crane boom or hoist is centered directly over the engine. Place a floor jack under the transmission.

Slowly raise the engine about an inch or two. Then check that everything is out of the way and disconnected.

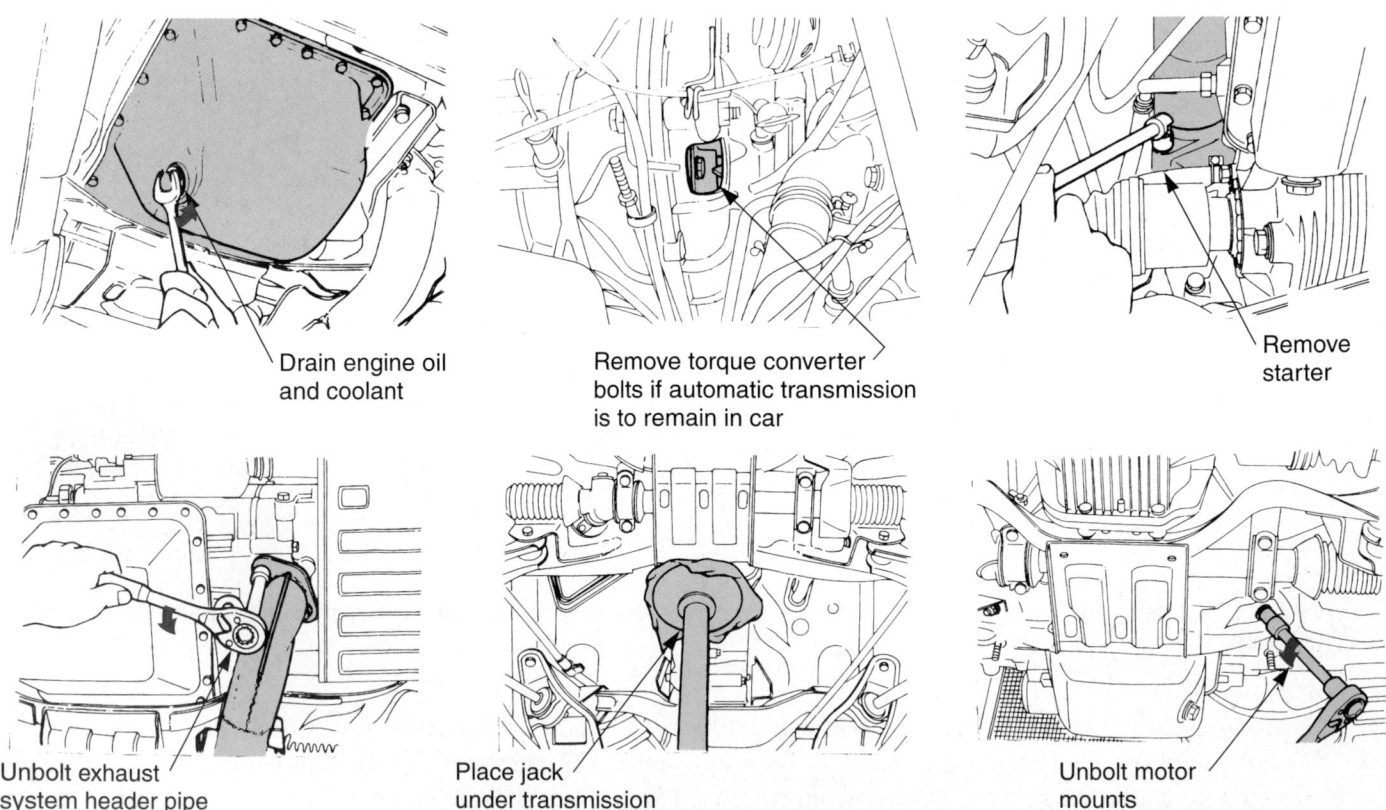

Drain engine oil and coolant

Remove torque converter bolts if automatic transmission is to remain in car

Remove starter

Unbolt exhaust system header pipe

Place jack under transmission

Unbolt motor mounts

Figure 49-4. When under the vehicle, these are some of the parts that must be disconnected before engine removal. Check a service manual for details. (Subaru)

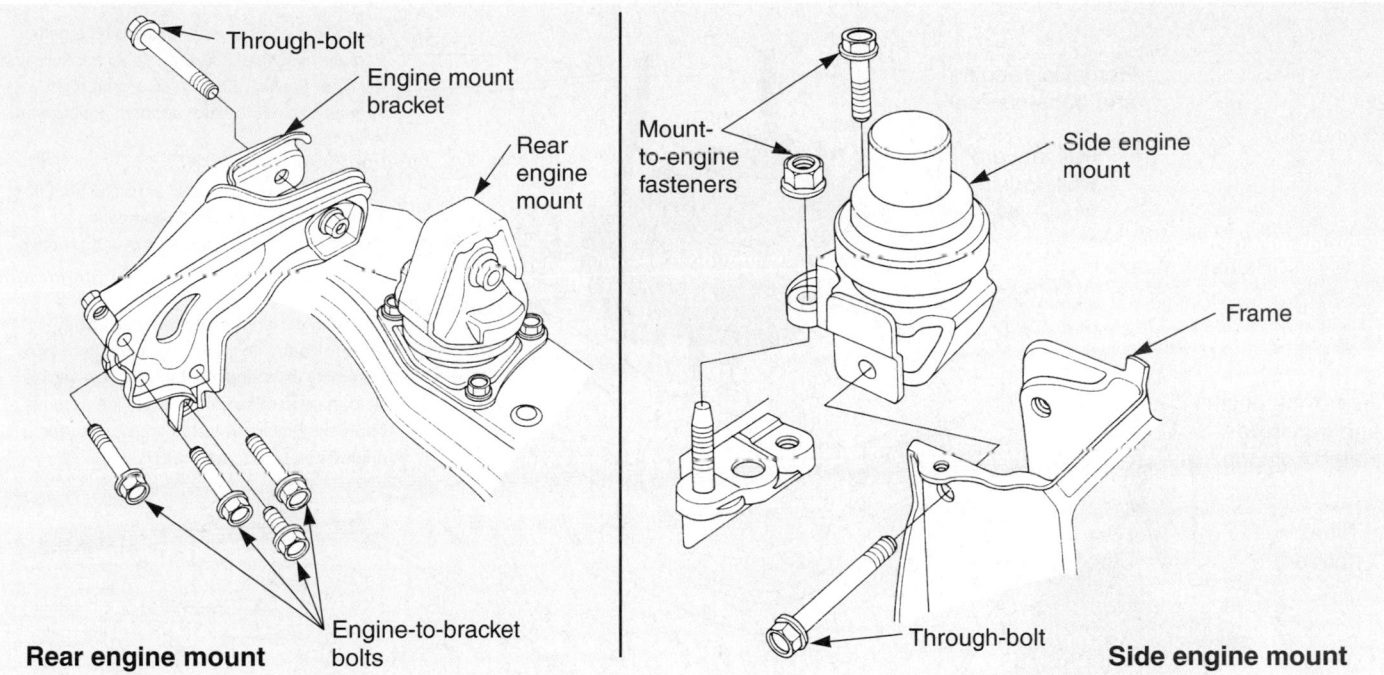

Figure 49-5. Various engine or motor mount designs are used. Sometimes, only one through-bolt has to be removed from each mount to free the engine from the vehicle. (Honda)

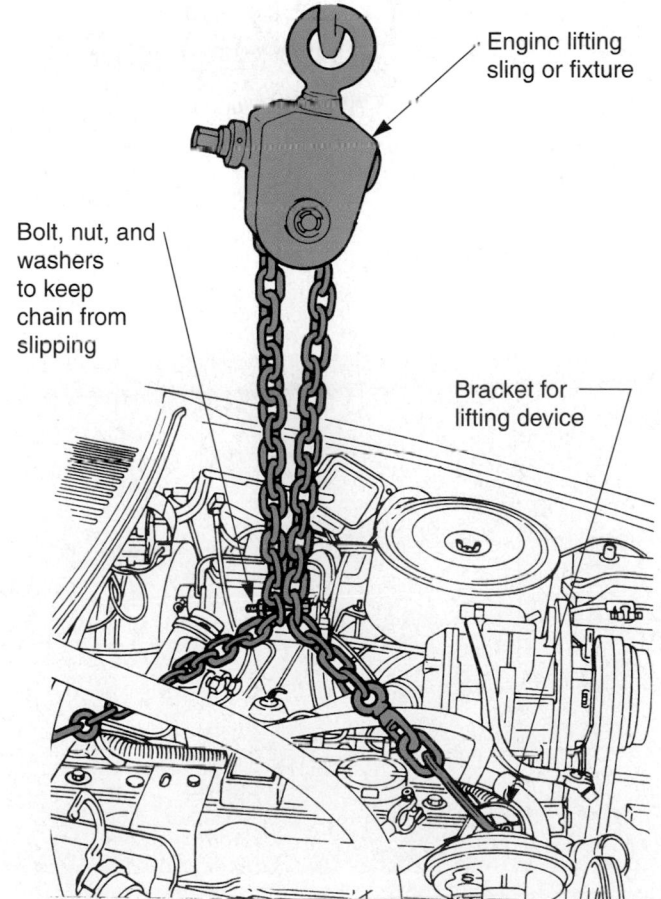

Figure 49-6. Make sure the lifting chain or fixture is attached properly. Note how this engine has brackets for engine removal. If the chain is attached with bolts, make sure the bolts are strong enough and fully threaded into holes. (Ford)

Warning

Never place any part of your body under an engine held in the air. A heavy engine can chop off fingers, cripple, or even kill you if dropped!

Continue raising the engine while pulling forward. This will separate the engine from the transmission or slide the transmission out from under the firewall. Do not let the engine bind or damage parts.

When the engine is high enough to clear the radiator support, roll the crane and engine straight out and away from the vehicle. With a stationary hoist, roll the vehicle out from under the engine.

With some rear-engine and front-wheel-drive vehicles, the engine and transaxle are removed from below the vehicle. With some vans, the engine must be removed through the large door in the side of the body. Again, check a service manual for details, **Figure 49-7.**

Do not let a transmission hang unsupported after engine removal. This could damage the rear rubber mount or the drive shaft (rear-wheel drive vehicle). See **Figure 49-8** for one method of supporting a transmission.

As soon as you can, lower the engine to the ground or mount it on an engine stand. See **Figure 49-9.**

Warning

Never work on an engine that is held by a crane or hoist. The engine could shift and fall, damaging the engine or causing serious injury!

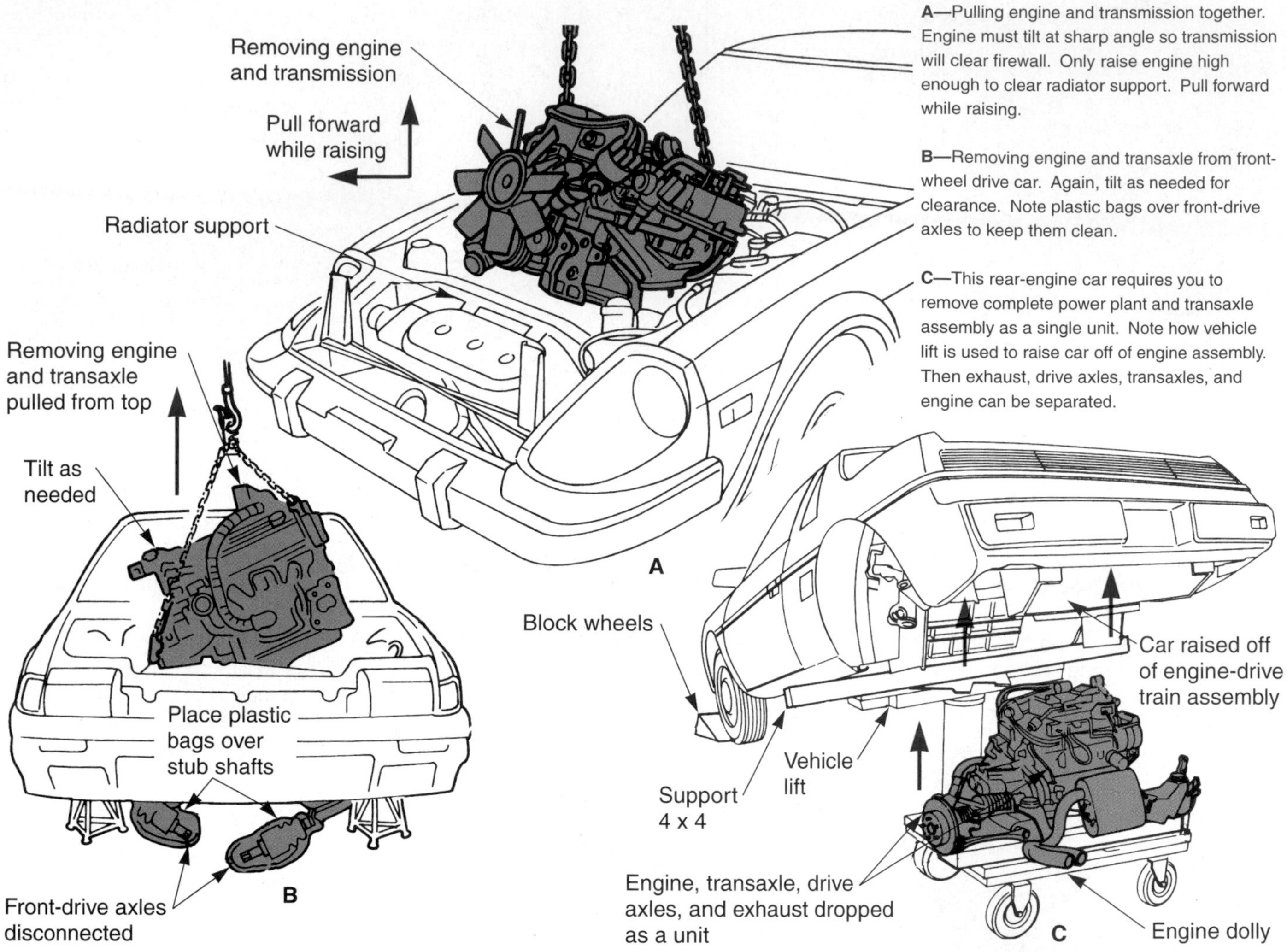

Removing engine and transmission

Pull forward while raising

Radiator support

A—Pulling engine and transmission together. Engine must tilt at sharp angle so transmission will clear firewall. Only raise engine high enough to clear radiator support. Pull forward while raising.

B—Removing engine and transaxle from front-wheel drive car. Again, tilt as needed for clearance. Note plastic bags over front-drive axles to keep them clean.

C—This rear-engine car requires you to remove complete power plant and transaxle assembly as a single unit. Note how vehicle lift is used to raise car off of engine assembly. Then exhaust, drive axles, transaxles, and engine can be separated.

Removing engine and transaxle pulled from top

Tilt as needed

Place plastic bags over stub shafts

Front-drive axles disconnected

B

A

Block wheels

Car raised off of engine-drive train assembly

Vehicle lift

Support 4 x 4

Engine, transaxle, drive axles, and exhaust dropped as a unit

Engine dolly

C

Figure 49-7. Note different methods of removing engines from vehicles. (Nissan, Honda, Pontiac)

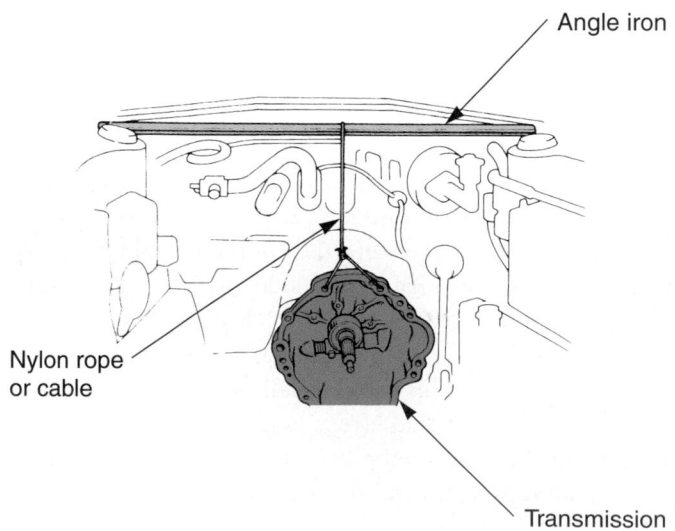

Angle iron

Nylon rope or cable

Transmission

Figure 49-8. If the transmission is not removed, it must be supported. Note the use of an angle iron and rope to hold up on transmission, protecting the rear mount and driveline. (Toyota)

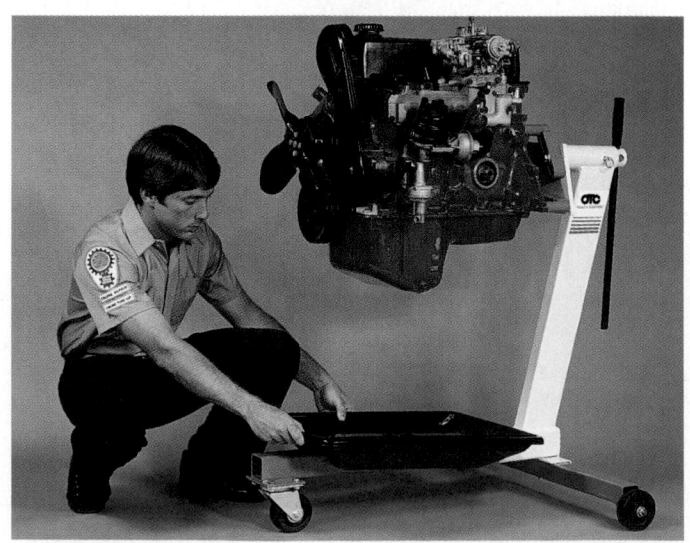

Figure 49-9. An engine stand makes engine repair much easier. The engine can be rotated into different positions. Also note the catch pan for dripping oil and coolant. (OTC Div. of SPX Corp.)

Tech Tip
Coolant and engine oil will usually drip onto the shop floor during engine removal. To prevent an accident, wipe up spills as soon as they occur. There is nothing professional about trying to work in a "grease pit."

Engine Disassembly

With the engine bolted to an engine stand or sitting on blocks, you are ready to begin teardown. During engine teardown, go slowly and inspect each part for

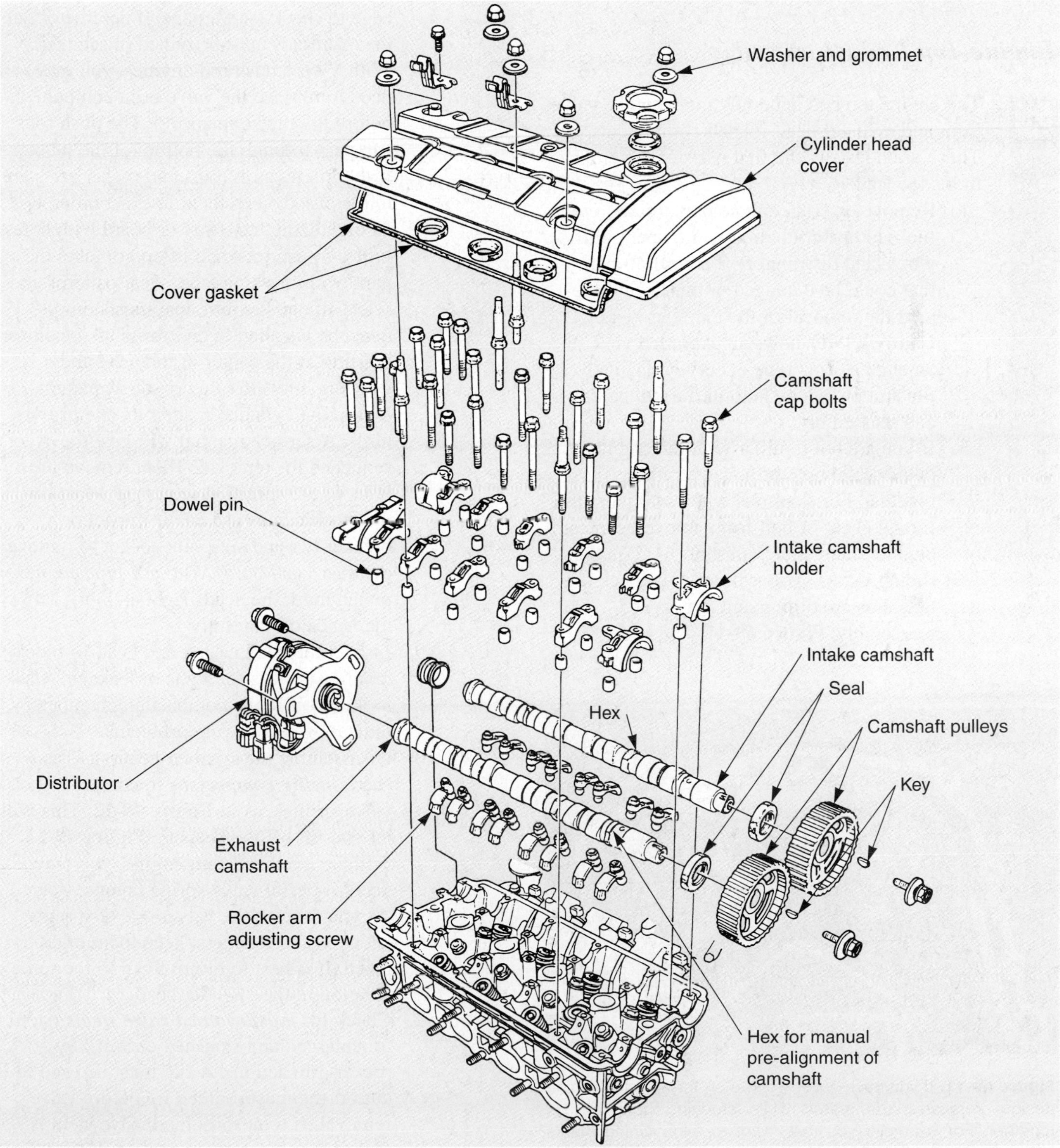

Washer and grommet

Cylinder head cover

Cover gasket

Camshaft cap bolts

Dowel pin

Intake camshaft holder

Intake camshaft

Seal

Camshaft pulleys

Distributor

Hex

Key

Exhaust camshaft

Rocker arm adjusting screw

Hex for manual pre-alignment of camshaft

Figure 49-10. An exploded view of the engine may be helpful during disassembly and reassembly. Find one for your engine in service manual. (Honda)

signs of trouble. Look for wear, cracks, damage, seal leakage, and gasket leakage.

Remember, if you overlook one problem, your engine repair may fail in service. All of your work could be for nothing.

Teardown methods vary somewhat from engine to engine. However, general procedures are similar and apply to all engines. The following will serve as a guide.

Engine Top End Disassembly

The engine top end generally includes the valve train and cylinder head-related components. These are normally the first parts of the engine to be serviced.

1. Remove external engine parts (carburetor, fuel rail or throttle body unit, spark plug wires, and distributor). Take off all parts that could be damaged or that would prevent the removal of the cylinder head.

2. **Figure 49-10** shows an exploded view of an engine. This type of service manual illustration can be helpful during teardown and reassembly.

3. If you are not familiar with the engine, take special note of how everything fits together. For example, when removing the timing chain or belt from an overhead cam engine, rotate the crankshaft to align the timing marks. This will let you check how the cams are timing and may speed reassembly, **Figure 49-11.**

Figure 49-11. If you have never worked on the specific engine design, remove parts slowly while studying how they fit together. For example, you may want to align timing marks before removing the timing belt or chain. This will let you become familiar with how the cams are timing with the crank.

4. Unbolt the valve cover(s), exhaust manifold(s), and intake manifold. If light prying is needed, be careful not to damage the mating surfaces.

5. Keep groups of fasteners organized in different containers. Note odd bolt lengths. Inspect the gaskets and mating surfaces for signs of leakage. If leaking, use a straight-edge to check for warpage. If not true or flat, the manifolds must be milled (machined).

6. With V-type push rod engines, you may need to remove the valve train components before the intake manifold. The push rods can pass through the bottom of the intake.

7. If the lifters, push rods, and rocker arms are to be reused, keep them in exact order. Use an organizing tray (tray or board with holes in it for push rods and lifters) or label these parts with masking tape. Wear patterns and select-fit parts require that most components be installed in their original locations.

8. Starting at the center of the head and working outward in a crisscross pattern, loosen the cylinder head bolts one or two turns. A service manual will give the exact sequence for removal. Then, remove the bolts in the specified sequence. Most service technicians use an air impact or a breaker bar and six-point socket to remove cylinder head bolts. With a V-type engine, punch mark the heads right and left. Lift the heads off carefully.

9. Inspect the head gasket and head-to-block mating surfaces for signs of leakage. Also look for oil in the combustion chambers, indicating seal or ring problems.

10. Disassemble the cylinder head(s). Use a **valve spring compressor** to compress the valve springs, as in **Figure 49-12.** This will let you lift off the keepers, **Figure 49-13.** With an overhead cam engine, you may need a special valve spring compressor.

11. As you remove the valves, valve springs, keepers, and retainers, keep them organized. It is best to return them to the same location in the cylinder head.

12. Check for **mushroomed valve stems** (stem tip enlarged and smashed outward by rocker arm action). A file must be used to cut off the mushroomed tip before valve removal. If a mushroomed valve stem is forced out, it can score and crack the valve guide and head.

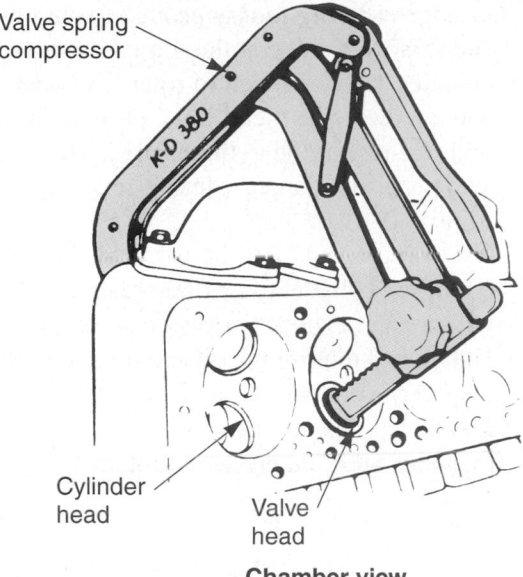

Chamber view

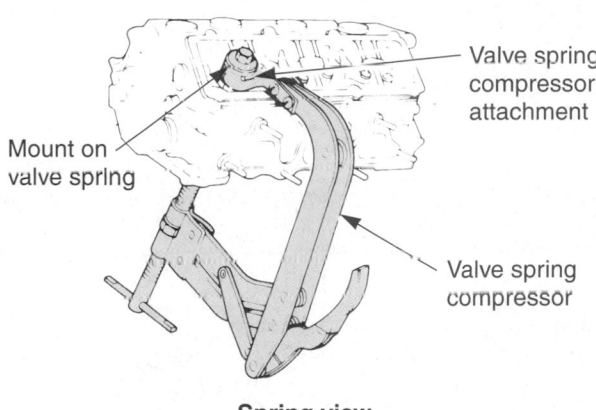

Spring view

Figure 49-12. To disassemble a cylinder head, use a valve spring compressor. One end of the tool fits on the valve head. The other end fits over the valve spring retainer. (K-D Tools and Honda)

Figure 49-13. With the spring compressed, the keepers can be lifted from their grooves. Release compressor and valve assemblies can be removed from the head. (G.M.)

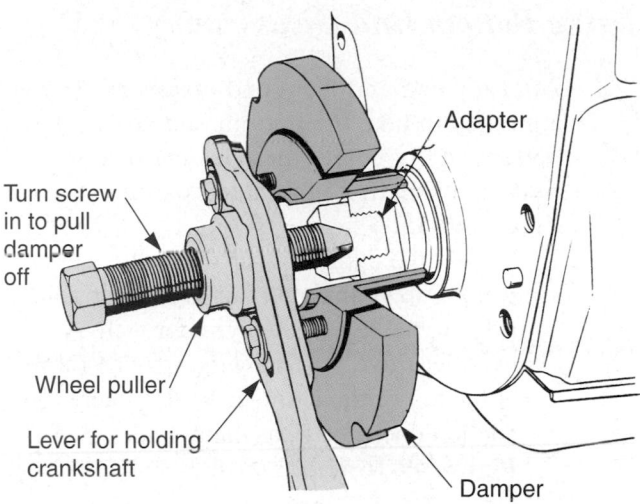

Figure 49-14. A wheel puller is normally needed to force the damper off of the crankshaft snout. Use a wrench to turn the screw inward. This will force the damper outward. (Buick)

Engine Front End Disassembly

 Engine front end disassembly is simple if a few basic rules and service manual instructions are followed.

1. Remove the water pump and any other parts bolted in front of the engine timing cover. If a timing belt is used, remove the belt cover. Loosen the tensioner and slip off the belt (this would have to be done before cylinder head removal).

2. Do not attempt to rotate the crankshaft of an overhead cam engine with the timing belt off (cylinder head still in place). The pistons could slide up and bend the valves.

3. A *wheel puller* is normally needed to remove the harmonic balancer or damper. The balancer is commonly press-fit onto the crankshaft. **Figure 49-14** shows how to use a wheel puller.

4. Unbolt and remove the timing chain or gear cover. If prying is necessary, do it lightly while tapping with a rubber hammer. Do not bend or scar mating surfaces.

5. Remove the oil slinger and timing mechanism. Usually, the timing gears or sprockets will slide off after light taps with a brass hammer. If not, use a wheel puller.

6. If the oil pump or other components are mounted in the front cover, refer to a service manual for directions.

Engine Bottom End Disassembly

After top end and front end disassembly, you are ready to take the bottom end apart. The bottom end typically includes the pistons, rods, crankshaft, and related bearings. See **Figure 49-15.**

1. Inspect the cylinders for signs of excess wear. Use your fingernail to feel for a lip or ridge at the top of the cylinder wall. A cylinder ridge or *ring ridge* may be formed at the top of the cylinder walls, where ring friction does not wear the cylinder. See **Figure 49-16A.**

2. A *ridge reaming tool* is needed to cut out and remove a ridge at the top of a worn cylinder. Use a wrench to rotate the reamer and cut away the metal lip. Cut until flush with the rest of the cylinder wall. This will prevent piston damage during removal, **Figure 49-16B.**

3. Use compressed air to blow metal shavings out of the cylinder after ridge reaming. This will prevent cylinder or piston scoring.

4. Unbolt and remove the oil pan and oil pump. Inspect the bottom of the pan for debris. Metal chips and plastic bits may help you diagnose and find engine problems.

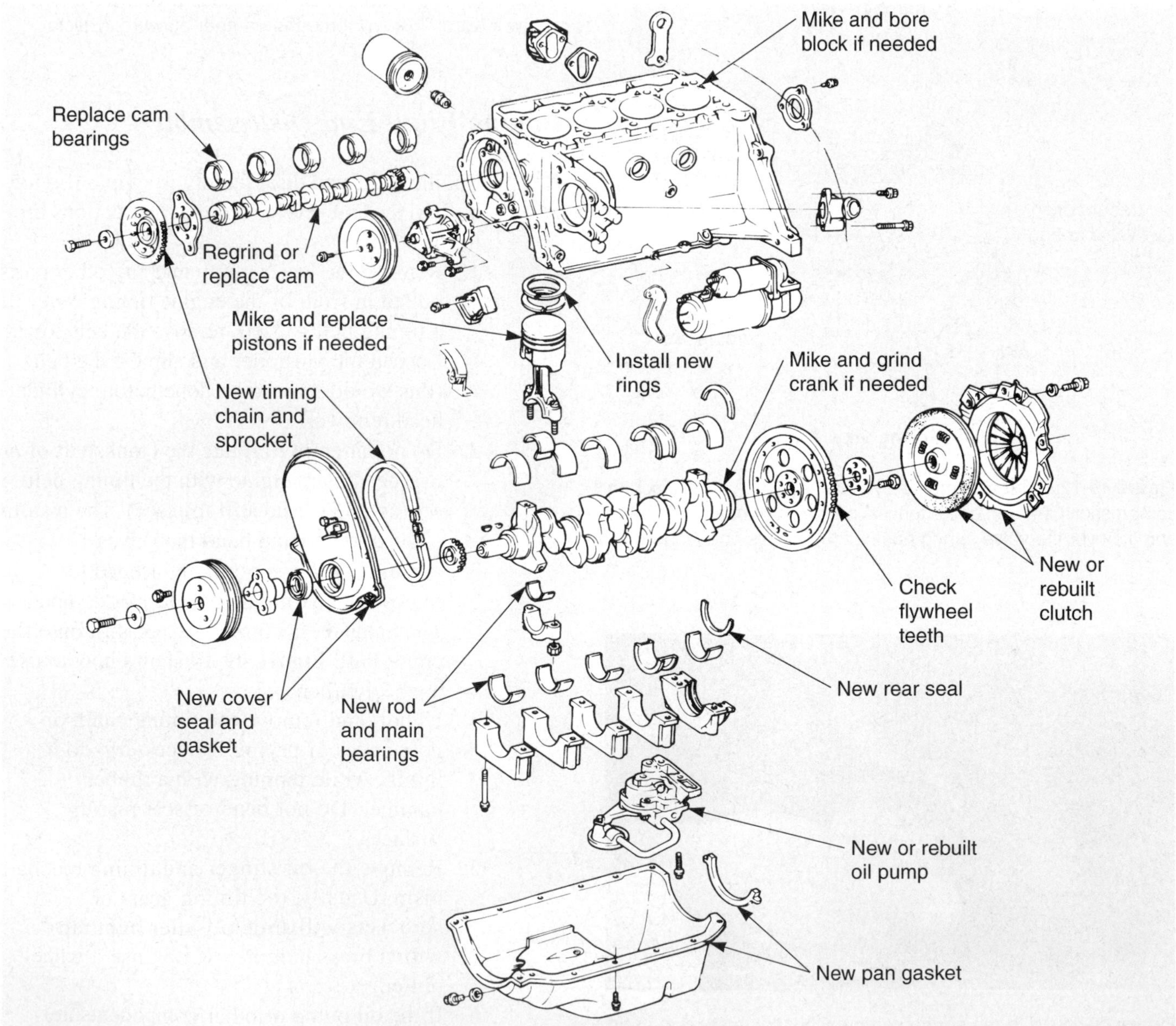

Figure 49-15. Major parts that are removed and cleaned during engine bottom end service. (Buick)

5. Unbolt one of the connecting rod caps. Then, use a wooden hammer handle to tap the piston and rod out of the cylinder. Refer to **Figure 49-17.**

6. As soon as the piston is out, replace the rod cap. Also, check the piston head and connecting rod for identification markings. The piston will usually have an arrow pointing to the front of the engine. The connecting rod and rod cap should have numbers matching the cylinder number. This is shown in **Figure 49-18.**

7. If needed, mark the piston heads with arrows or numbers. Also, if needed, number the connecting rods. If you mix up the pistons or rod caps, severe problems can develop when trying to reassemble the engine.

8. Remove the other piston and rod assemblies one at a time. Reinstall each cap on its rod. Mark them if needed.

9. Remove all of the old rings from their pistons. Spiral the rings off with your fingers or use a ring expander, **Figure 49-19.**

10. If the vehicle has a manual transmission, check for flywheel warpage. Use a dial indicator setup, as illustrated in **Figure 49-20.** Turn the crankshaft while noting the indicator reading. The crank will turn easily with all of the pistons out of the block. If runout is beyond specifications, send the flywheel to a machine shop for resurfacing.

11. Before removing the main bearing caps, check that they are numbered. Normally, numbers and arrows are cast on each cap. The number one cap is at the front of the engine. See **Figure 49-21.**

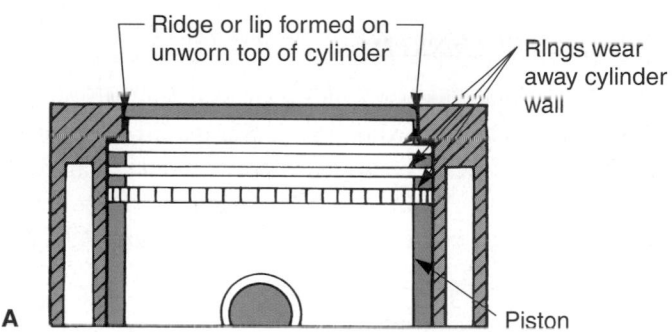

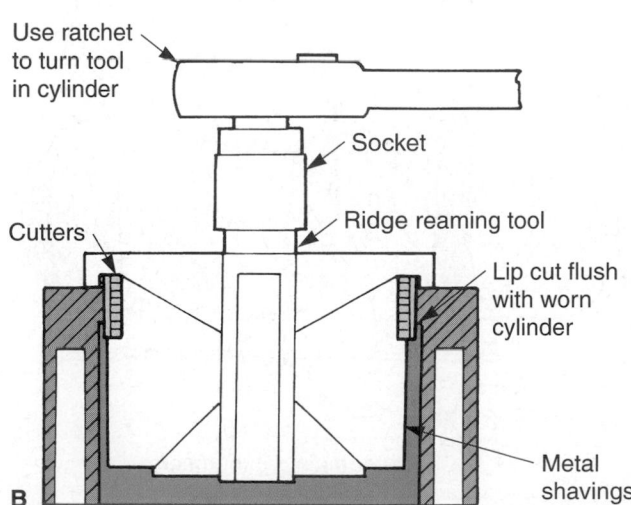

Figure 49-16. Cylinder problems. A—If the cylinder is badly worn, a ridge forms at the top of cylinder wall. B—A ridge reamer cuts off this lip so rings will not catch and damage the piston during removal.

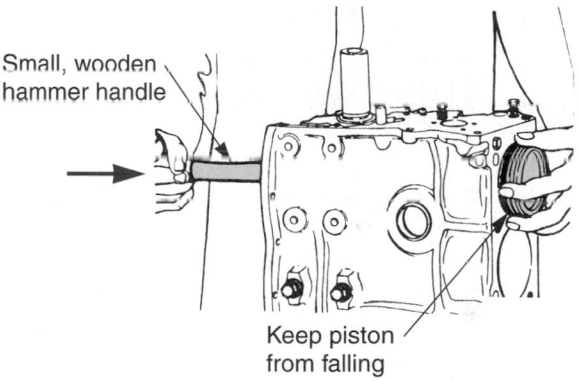

Figure 49-17. To remove pistons, unbolt rod cap. Use a wooden hammer handle to carefully tap assembly out of the block. (Nissan)

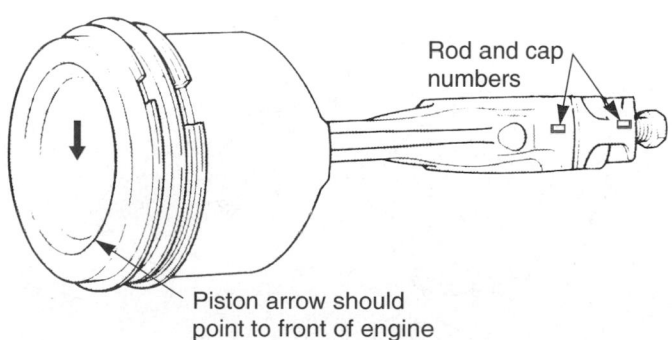

Figure 49-18. As soon as piston and rod assembly is removed, check it for markings. Numbers on cap and rod are identical to make sure caps are not mixed up. The arrow on the piston shows how the piston goes back on the rod and in the block. Number or mark parts if needed. (Chrysler)

12. If needed, use a **number set** (punch set for indenting numbers in metal parts) or a center punch to mark the main caps. If the caps are mixed up, the crank bore may be misaligned and the crank can lock in the bore and main bearings during reassembly.

13. Unbolt the main caps. To remove the caps, wiggle them back and forth while pulling. Then, lift the crankshaft carefully out of the block. Do not hit and nick the journals.

14. If the engine is old, pry out the block and head core or freeze plugs. They rust out and will leak after prolonged service. This also must be done if the block is going to be **boiled** (cleaned in strong chemicals at machine shop to remove mineral deposits in water jackets).

15. If the cylinders have deep ridges, the block must be sent to a machine shop for boring and hot tank cleaning. Make sure all external hardware (motor mount brackets, oil and coolant temperature sending units) are removed.

Cleaning Engine Parts

After you have removed all of the parts from the engine block and cylinder head, everything should be cleaned. Different cleaning techniques are needed, depending upon part construction and type of material.

Closer part inspection can be done during and after part cleaning. Problems can be hard to see when a part is covered with oil, grease, or carbon deposits.

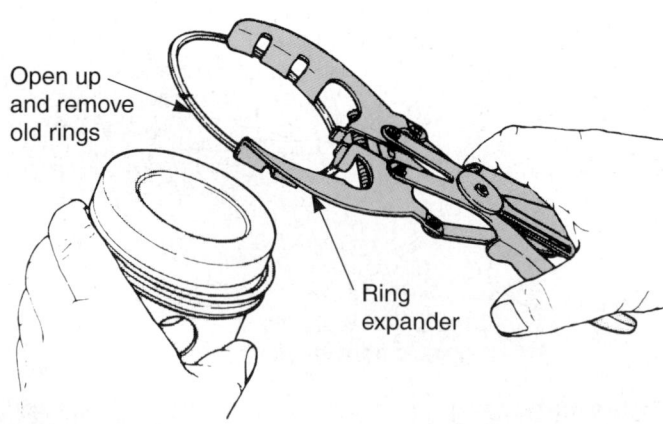

Figure 49-19. You can remove old rings by hand or with ring expander. Inspect for groove wear and damage as you work. (Chrysler)

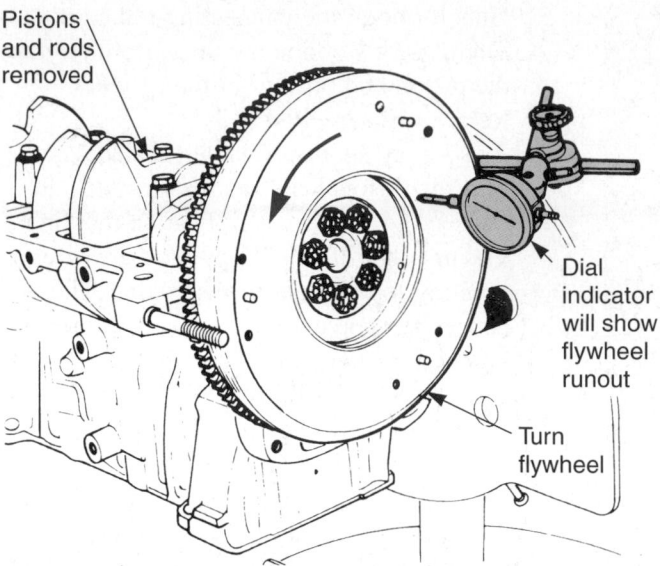

Figure 49-20. With pistons out of the block, check for flywheel warpage. Mount a dial indicator on the block. Position the pointer parallel with the surface of the flywheel. Rotate the flywheel and the indicator will read runout or warpage. If beyond specifications, send the flywheel to a machine shop. (Renault)

Scrape Old Gaskets

Begin engine parts cleaning by scraping off all old gasket material and hard deposits. Scrape off the gaskets for the valve covers, head, front cover, intake manifold, oil pan, and other components. Also scrape off as much hardened oil and carbon as you can. See **Figure 49-22.**

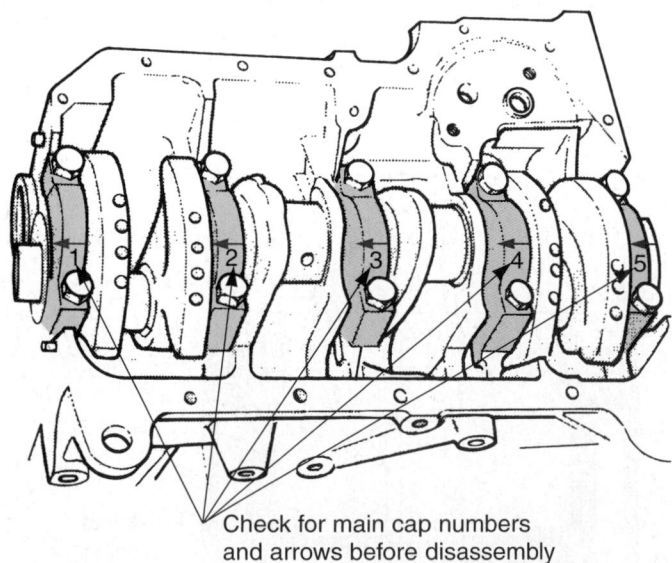

Figure 49-21. Before removing block main caps, make sure they are identified with arrows and numbers. Number them if needed. Main caps must be reinstalled in same location or the block bore will be misaligned. (Chrysler)

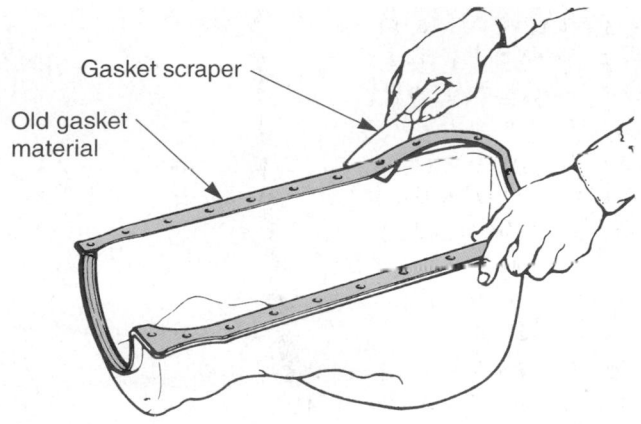

Figure 49-22. Use a scraper to remove old, thick gaskets from mating surfaces. (Ford)

Figure 49-24. Gasket remover can speed part cleanup if gasket material has become very hard and difficult to remove. After the chemical has softened the gasket material, scrape off without damaging part. (Fel-Pro)

Use a dull scraper and work carefully when cleaning soft aluminum parts. The slightest nick could cause leakage when returned to service, **Figure 49-23.**

 **Warning**
When using a gasket scraper, push the scraper *away* from your body, not toward your body. A scraper can inflict serious cuts.

Gasket remover is a chemical that dissolves old gasket material and sealer to aid in part cleaning. If needed, spray the remover over the material after scraping. Wait a few minutes and then scrape the dissolved residue off of the part. See **Figure 49-24.**

Power Cleaning Tools

There are several power cleaning tools used by the professional mechanic. If used properly, they can speed up and ease engine repairs.

A *power brush* is driven by an air or electric drill to remove hard carbon. It is especially handy inside hard-to-reach areas—combustion chambers, for example.

 **Warning**
Always wear eye protection when cleaning parts with power tools. Metal bristles, bits of carbon, or metal chunks from tool or part breakage can fly into your face.

A *wire wheel* on a grinder is another common method of cleaning engine parts. Frequently, it is used to clean carbon off valves. Keep the safety shield and tool rest in place.

Scuff pads are hard plastic cleaning wheels that mount in a drill for cleaning parts. They are better to use on aluminum and plastic engine parts because they will not erode metal as easy as a steel cleaning brush will. See **Figure 49-25.**

Cleaning Solvent

After scraping off the gaskets, use cleaning solvent to remove hard-to-reach deposits.

A *cold soak tank* is a cleaning machine for removing oil and grease from parts. It will not remove hard carbon or mineral deposits. Most auto shops have a cold cleaning tank or machine, **Figure 49-26.**

Cold soak tanks have a pump and filter that circulates clean solvent out of a spout. To wash off parts, direct the stream of solvent on the part while rubbing with a soft bristle brush.

Figure 49-23. A dull scraper should be used on soft aluminum parts to avoid nicks, which could cause leakage after reassembly. (Fel-Pro)

Figure 49-25. Scuff pads are a fast way to remove old gasket material. They will not damage soft aluminum. (Fel-Pro)

Warning
Never use gasoline to clean parts. The slightest spark or flame could ignite the fumes, causing a deadly fire!

A **hot tank** is a large cleaning machine filled with strong, corrosive chemicals. It will remove mineral deposits in the water jackets, hard carbon deposits, oil, grease, and even paint. Automotive machine shops normally have a hot tank, **Figure 49-27.**

Caution
Aluminum components can be corroded or etched by soaking in a hot tank. Clean only cast iron and steel parts in a hot tank.

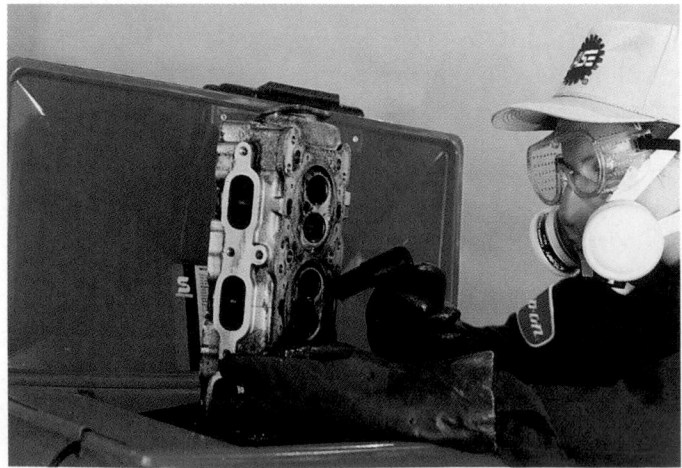

Figure 49-26. A cold soak cleaning machine is common in most auto shops. It will remove oil and grease, but not carbon deposits. Wear eye protection, rubber or plastic gloves, and a respirator. (Snap-on Tools)

Figure 49-27. Large, hot tank cleaners are found in automotive machine shops. A complete block can be lowered and cleaned, removing hard carbon and mineral deposits in water jackets. (Miller Engineering)

Special Cleaning Tools

Other special cleaning tools may also be needed for engine components.

A **ring groove cleaner** is used to scrape carbon from inside piston ring grooves. It is a special cleaning tool commonly used during piston service. The groove cleaner is rotated around the piston. A scraper bit, the same size as the groove, removes the carbon, **Figure 49-28.**

A **valve guide cleaner** is another special tool for use on engine cylinder heads. One is pictured in **Figure 49-29.** It is inserted into each valve guide. An electric or air drill spins the tool to remove deposits.

Air Blow Gun

An **air blow gun** is normally the last method of cleaning parts. It uses pressure from the shop's air compressor to blow off small bits of dirt, solvent, water, and other debris.

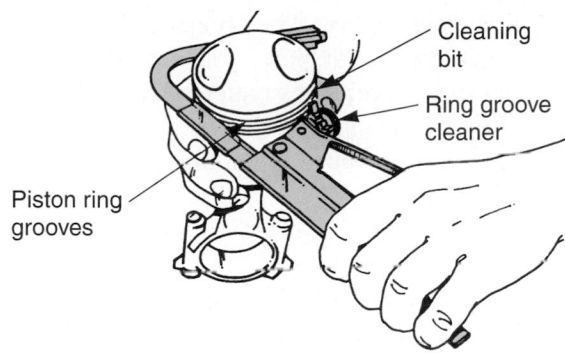

Figure 49-28. Always use a ring groove cleaner to scrape carbon from piston grooves. If grooves are not cleaned, new rings can be forced out against the cylinder wall, scoring rings and cylinder walls as soon as the engine starts. (Lisle Tools)

When using an air blow gun, direct the blast of air into all pockets and holes in parts. This will prepare the parts for reassembly.

 **Warning**
Use extreme care when using an air gun. Wear goggles and avoid aiming the gun at your body. If air bubbles enter your bloodstream, the result could be death!

Summary

- Engine removal and disassembly procedures vary from vehicle to vehicle. However, there are many general rules and methods that apply to all cars and small trucks.

- Many engine repairs can be made with the engine block mounted in the chassis. Repairs to the cylinder head, valve train, and other external parts are normally in-vehicle operations.

- Before engine removal, make sure all necessary parts are disconnected or removed.

- It is sometimes necessary to remove the engine and transmission together. Some front-wheel-drive vehicles that use a transaxle require that the two be removed as a unit.

- Position a lifting fixture or chain so that it will raise the engine in a level manner.

- With some rear-engine and front-wheel-drive vehicles, the engine and transaxle are removed from below the vehicle.

- Never let a transmission hang unsupported after engine removal. This could damage the rear rubber mount or the drive shaft (rear-wheel-drive vehicle).

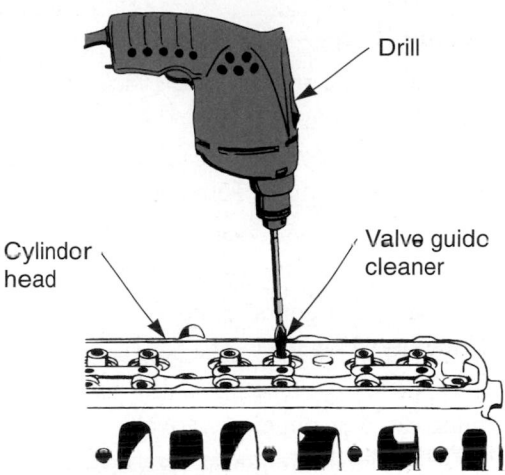

Figure 49-29. A valve guide cleaner is needed to remove carbon. Use an electric drill to spin the tool while pulling up and down. Do not let the tool come completely out of the valve guide. (Oldsmobile)

- As soon as you can, lower the engine to the ground or mount it on an engine stand.

- Disassemble an engine slowly and inspect each part for signs of trouble.

- Teardown methods vary somewhat from engine to engine. However, general procedures are similar and apply to all engines.

- A ridge reaming tool is needed to remove the ridge at the top of a worn cylinder.

- After you have removed all the parts from the engine block and cylinder head, everything should be cleaned. Different cleaning techniques are needed, depending on part construction and material type.

Important Terms

Engine R&R	Gasket remover
Valve spring	Power brush
compressor	Wire wheel
Mushroomed valve	Scuff pads
stems	Cold soak tank
Wheel puller	Hot tank
Ring ridge	Ring groove cleaner
Ridge reaming tool	Valve guide cleaner
Number set	Air blow gun
Boiled	

Review Questions—Chapter 49

Please do not write in this text. Place your answers on a separate sheet of paper.

1. Repairs to the cylinder head and valve train normally require engine removal. True or False?

2. List six things to double-check right before lifting an engine out of its chassis.

3. How do you keep a chain from slipping in its lifting hook?

4. Why should you support the transmission or transaxle after engine removal?

5. Which of the following statements does *not* apply to proper engine disassembly methods?
   (A) Note bolt lengths and organize bolts in containers.
   (B) Inspect gasket mating surfaces.
   (C) Keep critical parts (lifter, valves, etc.) organized so they can be installed in same location.
   (D) All of the above apply to proper disassembly methods.

6. A _____ _____ tool is used when the cylinders are badly worn. It will remove a lip at the top of the cylinders.

7. When servicing a vehicle with a manual transmission, the _____ should be checked for warpage.
   (A) flywheel
   (B) torque converter
   (C) clutch apply piston
   (D) None of the above.

8. The number one main bearing cap is located at the _____ of the engine.

9. A wire wheel is commonly used to remove carbon from valves. True or False?

10. A cold soak tank will not remove hard _____ or _____ deposits.

ASE-Type Questions

1. An automobile engine's oil pan must be removed. Technician A says when removing the oil pan on any type of automobile, the engine must be removed from the vehicle. Technician B says on certain automobiles, the oil pan can be removed while the engine is still in the vehicle. Who is right?
   (A) A only.
   (B) B only.
   (C) Both A and B.
   (D) Neither A nor B.

2. Technician A says when preparing to remove an engine from a vehicle, you should always disconnect the vehicle's battery. Technician B says when preparing to remove an engine from a vehicle, you must drain the engine's coolant. Who is right?
   (A) A only.
   (B) B only.
   (C) Both A and B.
   (D) Neither A nor B.

3. An engine must be removed from an automobile for major servicing. Technician A says before removing this engine, all air conditioning hoses should be disconnected. Technician B says before removing this engine, all power steering hoses must be disconnected. Who is right?
   (A) A only.
   (B) B only.
   (C) Both A and B.
   (D) Neither A nor B.

4. The piston rings need to be replaced on a particular automotive engine. Technician A says on certain automobiles, the piston rings can be replaced while the engine is still in the vehicle. Technician B says when replacing automotive piston rings, the vehicle's engine must always be removed first. Who is right?
   (A) A only.
   (B) B only.
   (C) Both A and B.
   (D) Neither A nor B.

5. Which of the following automotive components should normally be removed during engine removal?
   (A) Engine fan.
   (B) Radiator.
   (C) Air conditioning compressor.
   (D) All of the above.

6. Technician A says when engine removal is necessary, the vehicle's transmission must also be removed. Technician B says when engine removal is necessary, transmission removal is not always required. Who is right?
   (A) A only.
   (B) B only.
   (C) Both A and B.
   (D) Neither A nor B.

7. An engine and a manual transmission must be removed as a unit from an automobile. Technician A says during this procedure, the clutch linkage must be disconnected. Technician B says during this procedure, the clutch linkage should remain in place. Who is right?
   (A) A only.
   (B) B only.
   (C) Both A and B.
   (D) Neither A nor B.

8. Technicians are going to remove an engine from a vehicle. Technician A says all fuel lines must be disconnected and plugged. Technician B says all engine ground straps must be disconnected. Who is right?
   (A) A only.
   (B) B only.
   (C) Both A and B.
   (D) Neither A nor B.

9. A lifting chain is being used to remove an engine from an automobile. Technician A says a bolt should be used on the chain to keep it from slipping. Technician B says a bolt, nut, and washers should be used on the chain to keep it from slipping. Who is right?
   (A) A only.
   (B) B only.
   (C) Both A and B.
   (D) Neither A nor B.

10. The transmission is left in a vehicle during engine removal. Technician A says supports shouldn't be used on the transmission or drive shaft damage can occur. Technician B says supports should be used on the transmission to avoid drive shaft damage. Who is right?
    (A) A only.
    (B) B only.
    (C) Both A and B.
    (D) Neither A nor B.

11. An engine and transaxle must be removed from a vehicle for major servicing. Technician A says with some rear-engine vehicles, the engine and transaxle must be removed from below the vehicle. Technician B says on all front-wheel drive vehicles, the engine and transaxle are raised and removed from the bottom of the engine compartment. Who is right?
    (A) A only.
    (B) B only.
    (C) Both A and B.
    (D) Neither A nor B.

12. A "mushroomed" valve stem is detected during engine top end disassembly. Technician A says this valve must be removed with a hammer and punch. Technician B says the "mushroomed" valve stem tip must be filed before removing the valve. Who is right?
    (A) A only.
    (B) B only.
    (C) Both A and B.
    (D) Neither A nor B.

13. Technician A says a pry bar should be used to remove an engine's harmonic balancer. Technician B says a wheel puller should be used to remove an engine's harmonic balancer. Who is right?
    (A) A only.
    (B) B only.
    (C) Both A and B.
    (D) Neither A nor B.

14. A "ring ridge" is detected at the top of an engine's cylinder. Technician A says the piston in this particular cylinder should be removed before attempting to cut out the "ring ridge." Technician B says the "ring ridge" must be cut out of the cylinder before attempting to remove the piston. Who is right?
    (A) A only.
    (B) B only.
    (C) Both A and B.
    (D) Neither A nor B.

15. An engine's aluminum cylinder heads must be cleaned before they are reconditioned. Technician A says these aluminum heads should be cleaned in a "hot tank." Technician B says aluminum heads can be damaged if they are soaked in a "hot tank." Who is right?
    (A) A only.
    (B) B only.
    (C) Both A and B.
    (D) Neither A nor B.

Activities—Chapter 49

1. Review the safety tips in this chapter. Use the information to create a small brochure or flyer that could be distributed to automotive technology students.

2. Show classmates how to remove piston rings with a ring expander, or by spiraling them off manually.

Chapter 50

Engine Bottom End Service

After studying this chapter, you will be able to:

- Explain how to measure cylinder bore wear.
- Hone cylinder walls.
- Check block main bore straightness.
- Measure block, head, and manifold warpage.
- Measure piston wear and piston-to-cylinder clearance.
- Explain how to assemble a rod and piston.
- Describe how to install piston rings
- Check piston ring end gap and piston ring side clearance.
- Measure crankshaft journal wear and crankshaft straightness.
- Install a rear main oil seal.
- Use Plastigage to measure rod and main bearing clearance.
- Measure rod and crank side clearance.
- Properly assemble an engine bottom end.
- Describe safety practices to be followed when performing engine bottom end service.
- Correctly answer ASE certification test questions on engine bottom end repair.

In previous chapters, you learned how to diagnose engine problems; remove an engine from a vehicle; and disassemble, clean, and inspect engine parts. This chapter will continue your study of engine service by detailing bottom end overhaul. This chapter, along with Chapters 51 and 52, will complete your study of engines. After studying these chapters, you should be able to rebuild, or overhaul, a complete engine.

Engine bottom end service, or short block service, is needed after extended engine operation. Ring and cylinder wear can cause engine smoking and oil consumption.

Bearing wear can cause low oil pressure, bearing knock, or complete part failure. See **Figure 50-1.**

An *engine overhaul* involves the service of the engine's bottom end, top end, and front end. All the internal parts are serviced.

Do You Need Review?

You may want to refer to Chapter 49, which covers engine disassembly and the specialized tools (cylinder ridge reamer, piston ring groove cleaner, etc.) needed to properly service the engine's bottom end parts.

Cylinder Block Service

Cylinder block service commonly includes checking the block for cracks and distortion, inspecting the cylinder walls for damage, measuring the cylinders for wear, honing or deglazing the cylinder walls, cleaning the cylinders after honing, and installing core plugs. All these operations are essential to satisfactory engine operation.

Some cylinder block repairs, such as cylinder boring (machining) and crack repair, may require special equipment found in a machine shop. These repairs will be detailed later.

Block Pressure Testing

Make sure the block is not cracked before continuing engine block service. **Block pressure testing** involves blocking all passages and forcing air into the passages while the block is submerged in water. Any cracks or pores will show up where air bubbles leak out of the block. Other ways to check the block for cracks are dye penetrant testing and magnetic testing, which was covered in the chapter on cylinder head service. If cracks are found, discard the cylinder block.

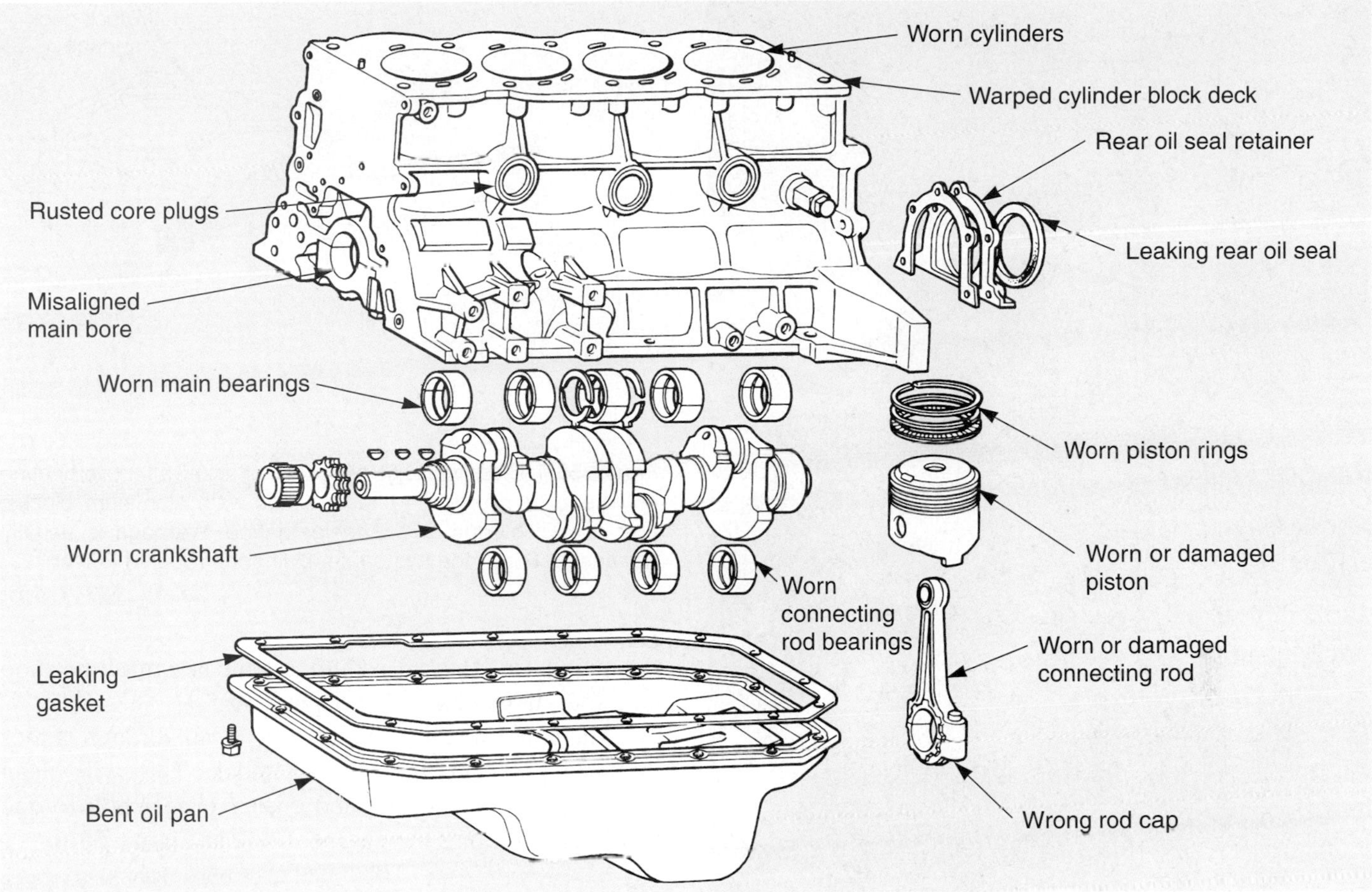

Worn cylinders

Warped cylinder block deck

Rear oil seal retainer

Leaking rear oil seal

Rusted core plugs

Misaligned main bore

Worn main bearings

Worn piston rings

Worn crankshaft

Worn or damaged piston

Worn connecting rod bearings

Worn or damaged connecting rod

Leaking gasket

Bent oil pan

Wrong rod cap

Figuro 50 1. These are the major engine bottom end parts that require service. Do you romombor the function and construction of each? If not, review earlier chaptors. Note the common problems. (Toyota)

Checking Main Bores

After repeated heating and cooling or overheating, the main bearing bores in the cylinder block can warp or twist. This will affect main bearing insert alignment and crankshaft fit in the block. In severe cases of main bearing bore misalignment, the crankshaft can lock up when the main caps are torqued.

Main bearing bore alignment can be checked with a straightedge and a feeler gauge. Lay the straightedge on the bores; then determine the thickest feeler gauge blade that will slide between the straightedge and any of the bores. The size of this blade indicates the amount of main bearing bore misalignment. *Always* check bore misalignment after main bearing failure. See **Figure 50-2.**

Block line boring is used to straighten, or true, misaligned main bearing bores. A machine shop will have a boring bar for machining the bearing bores back into alignment. Line boring is becoming more common with today's thin-walled, lightly constructed blocks, **Figure 50-3.**

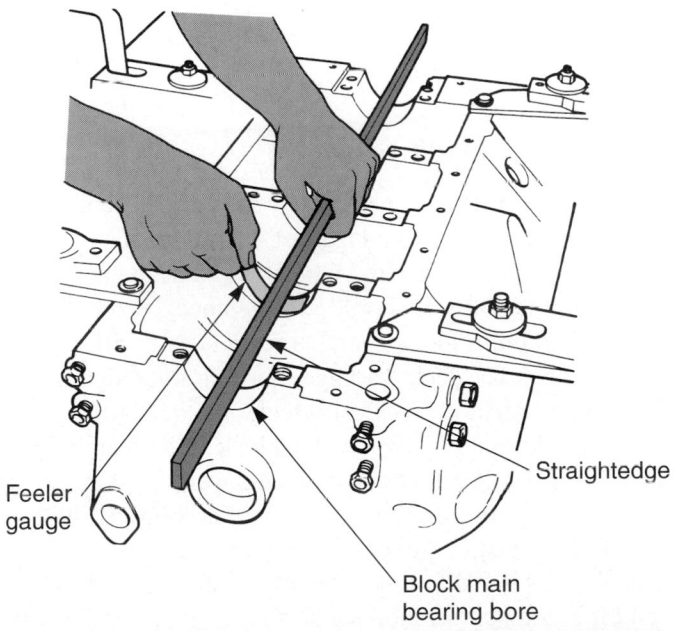

Feeler gauge

Straightedge

Block main bearing bore

Figure 50-2. After severe block overheating or physical damage, check main bearing bore alignment. The size of the feeler gauge that fits under the straightedge equals misalignment. (Federal Mogul)

Figure 50-3. This machinist is using a large boring bar to true up the block main bore. If this is not done when needed, the crankshaft will lock up in the new main bearings. (Sunnen)

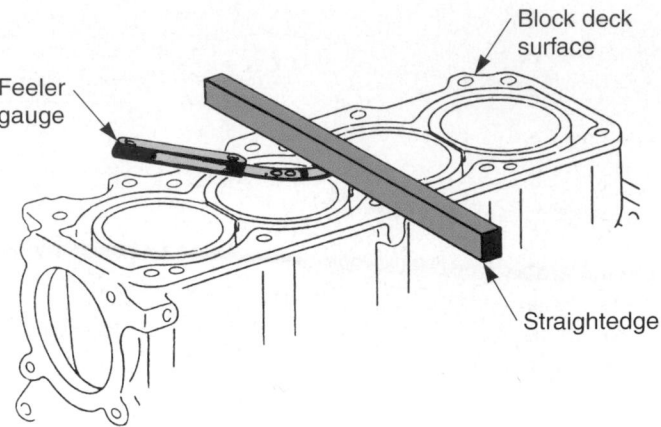

Figure 50-4. Block warpage should be checked after overheating or whenever overhauling today's thin-wall, aluminum blocks. Use a straightedge and a feeler gauge. Warpage is usually greatest between the two center cylinders. (Sealed Power)

Measuring Deck Warpage

Deck warpage is measured with a straightedge and feeler gauge on the head gasket sealing surface of the block. It should be checked when the old head gasket was blown and leaking. It should also be checked on all late-model aluminum blocks, since they distort easily.

To check for deck warpage, lay a straightedge on the clean block surface. Try to slip feeler gauge blades of different thicknesses between the block and straightedge. The thickest blade that fits indicates warpage. Check for deck warpage in different locations. See **Figure 50-4.**

If warpage exceeds specifications (about 0.003″–0.005″ [0.08 mm–0.13 mm]), replace the block or send it to a machine shop for surface milling, or decking.

Decking a block, or squaring the block, involves machining the cylinder head mounting surfaces until they are parallel to the main bore. On V-type engines, both mounting surfaces must also be the same distance from the main bore. This procedure is done at a machine shop on a milling machine. See **Figure 50-5.**

Many technicians like to run a tap through major threaded holes in the cylinder block. This will clean debris out of the threads and ensure proper bolt torque when reassembling the engine. Look at **Figure 50-6.**

Inspecting Cylinder Walls

Using a droplight, closely inspect the surface of each cylinder wall. Look for vertical scratches, scoring, or

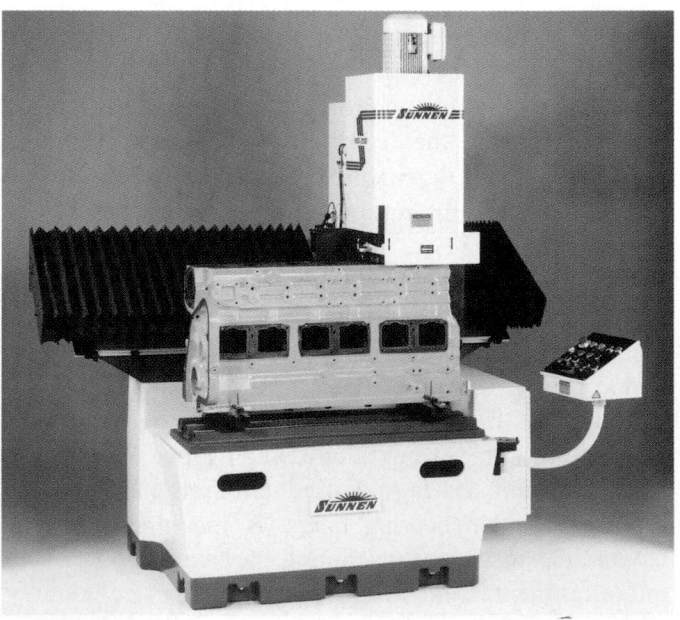

Figure 50-5. This specialized milling machine is designed to resurface cylinder block decks and cylinder heads. If the deck is not flat, head gasket leakage can result. (Sunnen)

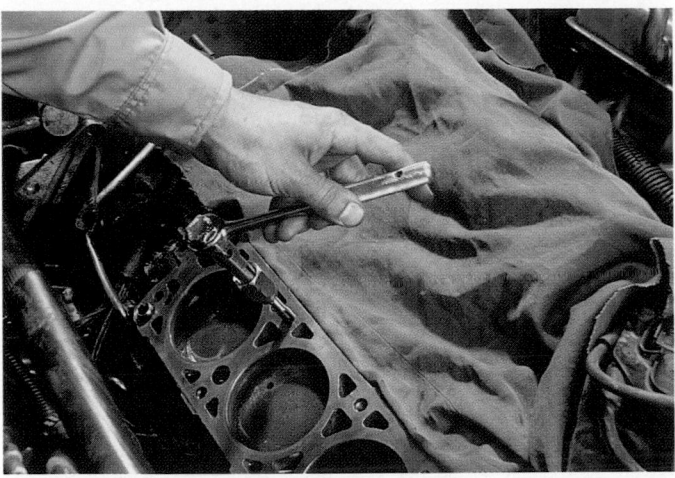

Figure 50-6. It is wise to run a tap down through the threaded holes in the cylinder block. This will remove debris and help locate damaged threads. (Fel-Pro)

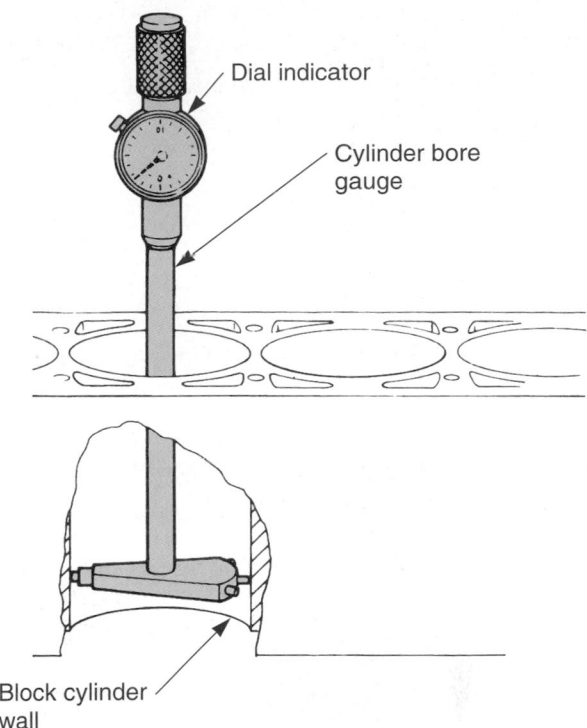

Figure 50-7. Using a cylinder bore gauge is a quick and easy way to check for wear in a cylinder. Slide the gauge up and down. Indicator movement indicates differences in diameter. (Chrysler)

cracks. Rub your fingernail around the cylinder wall. This will help you locate problems.

If you had to remove a large ring ridge or you find deep scratches, the cylinder is badly worn. Cylinder boring and oversize pistons may be required.

Measuring Cylinder Wear

If the cylinder is not badly scratched or scored, measure cylinder wear to ensure that the new rings will seal properly. Obviously, new rings cannot seal in a worn, tapered, or out-of-round cylinder. Also, the cylinder measurement will let you determine piston-to-cylinder clearance.

Cylinder taper is a difference in the diameter measured at the top of the cylinder and bottom of the cylinder. It is caused by less lubricating oil at the top of the cylinder. More oil splashes on the lower area of the cylinder. As a result, the top of the cylinder wears faster (larger) than the bottom, producing taper.

Cylinder out-of-roundness is a difference in cylinder diameter when measured front-to-rear and side-to-side in the block. Piston thrust action normally makes the cylinders wear more at right angles to the centerline of the crankshaft.

Cylinder taper should not exceed approximately 0.005″ (0.13 mm). Some manufacturers do not permit even this much wear. Refer to a service manual for exact specifications. If taper is beyond limits, replace the block or bore the block and install oversize pistons.

A *dial bore gauge* is a tool used to quickly and accurately measure cylinder taper. Slide the bore gauge up and down in the cylinder, **Figure 50-7.**

Check both parallel and perpendicular to the crank bore centerline. **Figure 50-8** shows how to measure cylinder bore wear.

An *inside micrometer* can also be used to measure cylinder wear. This is much more time-consuming, however. If needed, refer to Chapter 6, *Automotive Measurement and Math,* for a review of measuring tools.

 Tech Tip
Always make sure the block is in acceptable condition. Are the cylinders round, smooth, and untapered? Is the crank bore aligned and in good condition? Are deck surfaces flat and true? Remember that the block is the foundation of an engine rebuild. If the block has problems, you are "building your house on quicksand."

Cylinder Honing

Cylinder honing is used to true worn cylinders and to break the glaze (polished surface) on used cylinders before installing new piston rings. It must also be used to smooth rough cylinders after boring. Most ring manufacturers recommend deglazing; some do not. Check the instructions provided with the new piston rings for

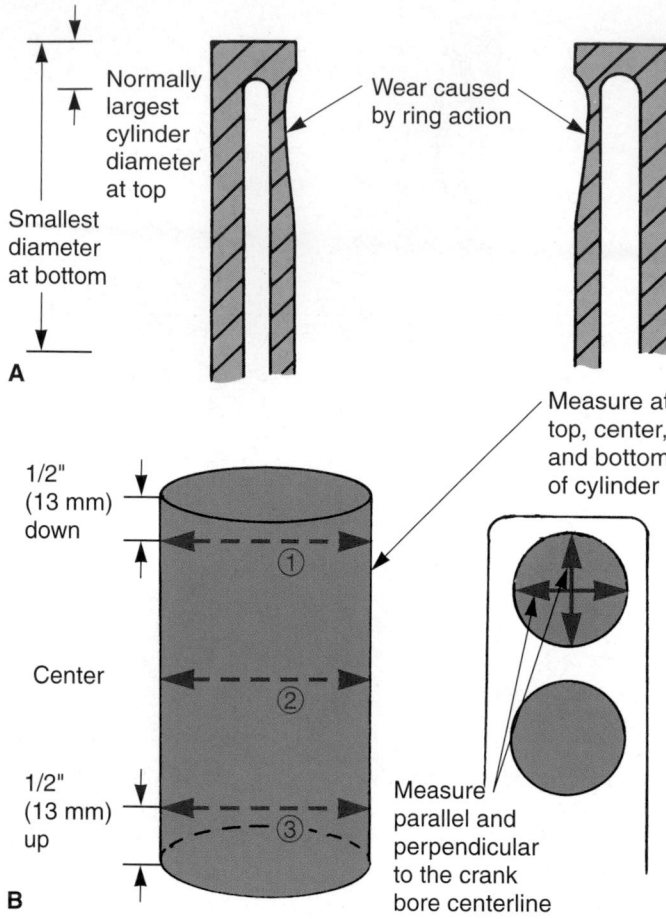

Figure 50-8. A—The cylinder will wear more at the top than at the bottom. There is more oil splashing on the bottom portion of the cylinder. B—Measure sideways and forward at the top, center, and bottom of the cylinder. This will let you detect taper and out-of-round.

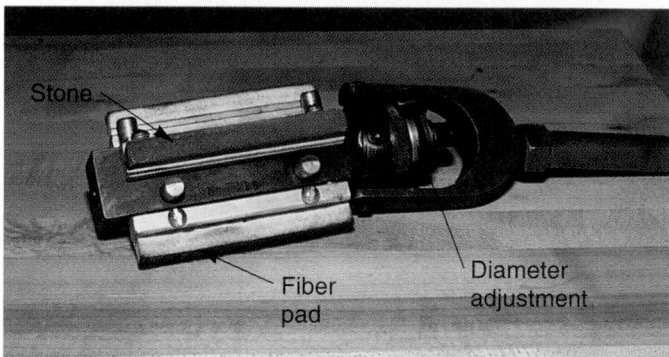

Figure 50-9. When honing or deglazing a cylinder, pull the drill up and down, but do not let the hone come out of the cylinder. Check the manual and equipment instructions to determine how long to hone each cylinder. Hone more at bottom than top to help remove taper. A—A brush hone leaves the proper texture on cylinder walls to aid ring break-in and sealing. It removes little material and should be lubricated during use. B—Rigid hone will remove metal to help eliminate minor scratches. It should be used dry. (Goodson Auto Machine Shop Supplies)

details. The term *deglazing* is generally used when referring to very light honing that simply scuffs the cylinder wall to aid ring break-in.

A *cylinder hone* produces a precisely textured, cross-hatched pattern on the cylinder wall to aid ring seating and sealing. Tiny scratches from the hone cause initial ring and cylinder wall break-in wear. This makes the ring fit in the cylinder perfectly after only a few minutes of engine operation.

There are several types of engine cylinder hones:

- *Brush hone*—has small balls of abrasive material formed on the ends of round metal brush bristles. It is desirable when the cylinder is in good condition and requires very little honing, **Figure 50-9A.**

- *Flex hone*—has hard, flat, abrasive stones attached to spring-loaded, movable arms. It is used when the cylinder wear is slight and a moderate amount of honing is needed.

- *Rigid hone (Sizing hone)*—has adjustable stones that lock into a preset position. It will remove a small amount of cylinder taper or out-of-roundness, **Figure 50-9B.** A rigid hone will accurately remove more cylinder material than a flex hone. This type hone should be used in badly worn cylinders that are still within spec wear limits.

- *Honing machine*—a large piece of equipment used to rigid hone the cylinders. This type of machine is often found in a machine shop and is used after boring, **Figure 50-10.**

Figure 50-10. A machine shop will have the heavy equipment for power honing and boring a cylinder block. This is a honing machine. (Sunnen)

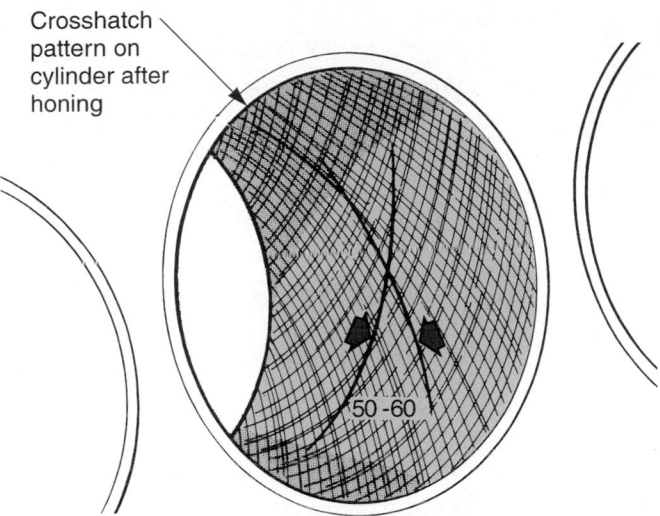

Figure 50-11. When honing a cylinder, try to produce a 50°–60° crosshatch pattern. Moving the drill up and down faster or slower alters angle. (Chrysler)

A ridged hone or a honing machine can be used like a boring bar to true a cylinder when wear does not exceed acceptable limits. Do *not* hone more than tool manufacturer recommendations, however.

To hone a cylinder, follow the equipment manufacturer's instructions. Install the hone in a large, low-speed electric drill. Compress the stones (squeeze inward) and slide them into the cylinder. Be careful not to scratch the cylinder. Turn on the electric drill and move the spinning hone up and down in the cylinder.

> **Warning**
> Make sure you do *not* pull the hone completely out of the cylinder while honing. The hone could break and bits of stone fly out, causing serious injury.

Move the hone up and down in the cylinder fast enough to produce a 50°–60° crosshatch pattern. Moving the hone up and down faster or slower will change the pattern, **Figure 50-11.**

When honing a moderately worn or freshly bored cylinder, it is recommended that you work in stages. For example, start honing with coarse-cut stones (#133, for example). Then use rough-cut stones (#525). Finally, finish the cylinders with fine-cut stones (#625). If you plan to use molybdenum or chrome rings, you may want to use superfine stones (#820) to ensure rapid ring seating and sealing.

Cleaning Cylinder Walls

After honing, it is very important to remove all **honing grit** (bits of stone and metal) from inside the engine. If this grit is not removed, it will act like grinding compound on bearings, rings, and other vital engine parts.

First, wash out the cylinders with a warm solution of *water* and *soap* (detergent). A soft bristle brush, *not* a wire brush, will quickly loosen particles inside the honing marks on the cylinder walls. After washing, rinse the block thoroughly with clean, hot water and blow the cylinders dry with compressed air.

Next, place engine oil on a clean shop rag. Wipe the cylinder down thoroughly with the *oil-soaked rag*. The heavy oil will pick up any remaining grit embedded in the cylinder's honing marks. Wipe the cylinders down until the rag comes out perfectly clean.

After cleaning and oiling, recheck the cylinder for scoring or scratches. If honing did *not* clean up all the vertical scratches in the cylinder, cylinder boring or sleeving may be needed.

> **Tech Tip**
> New rings will *not* seal deep, vertical marks in the cylinder. Engine oil consumption and compression leakage may result.

Cylinder Block Boring

Cylinder boring is needed to remove deep scratches, scoring, or excess wear from the cylinder walls. It involves machining the cylinders to a larger diameter. After boring, oversize pistons must be installed in the engine.

The block is generally sent to a machine shop for boring. The machine shop will use a large boring bar (machine tool) to cut a thin layer of metal off the cylinder walls. Normally, a cylinder block is bored in increments of 0.010″ (0.25 mm).

Tech Tip

Always use a reputable machine shop. Your work is only as good as the machine work done on your parts. If the machine shop bores the engine block improperly, the overhauled engine will quickly fail. You will pay the price by having to remove and rebuild the engine a second time. You are responsible for checking the work of the machine shop before reassembling the engine.

The *overbore limit* (typically 0.030″–0.060″) is the largest possible diameter increase to which a cylinder can be bored. It is specified by the engine manufacturer and can vary with block design. If the overbore limit is exceeded, the cylinder wall can become too thin. The wall can distort or crack in service from combustion heat and pressure.

Oversize pistons and rings are required to fit a cylinder block that has had its cylinders bored out. The pistons must be purchased to match the oversize of the cylinders.

Boring cylinders and installing oversize pistons will help restore the engine to like-new condition. New pistons and rings will operate on freshly machined cylinder surfaces, providing excellent ring sealing and service life.

Cylinder sleeving involves machining one or more of the cylinders oversize and pressing in a cylinder liner. Sleeving is needed when the damage to the cylinder wall is too severe to clean up with boring.

Sleeving also allows the bad cylinder to be restored to its original diameter. The same size pistons can be reused. If only one piston is damaged, for instance, all of the other pistons and cylinders may be good and usable. This would save the customer money on the repair.

Core Plug Service

To replace a freeze plug:
1. Drive a drift or a large screwdriver (full shank type) through the plug.
2. Pry sideways on the drift or screwdriver. Avoid scraping the engine block or cylinder head. The plug should pop out.
3. Sand the core plug hole in the engine and wipe it clean.
4. Coat the plug hole and plug with a nonhardening sealer.
5. Drive the new core plug squarely into position.

Balancer Shaft Service

Many late-model four- and six-cylinder engines use balancer shafts to help smooth engine operation. They are usually mounted in and serviced with the cylinder block.

Servicing a balancer shaft involves the same basic tasks as servicing a camshaft or crankshaft. You must measure shaft runout, journal wear, and bearing clearance. Replace components that are worn beyond specs. See **Figures 50-12 and 50-13.**

Piston Service

Pistons are made of aluminum, which is prone to wear and damage. It is critical that each piston be checked thoroughly. Look for cracked skirts, worn ring grooves, cracked ring lands, worn pin bores, and other problems. You must find any trouble that could affect piston performance and engine service life.

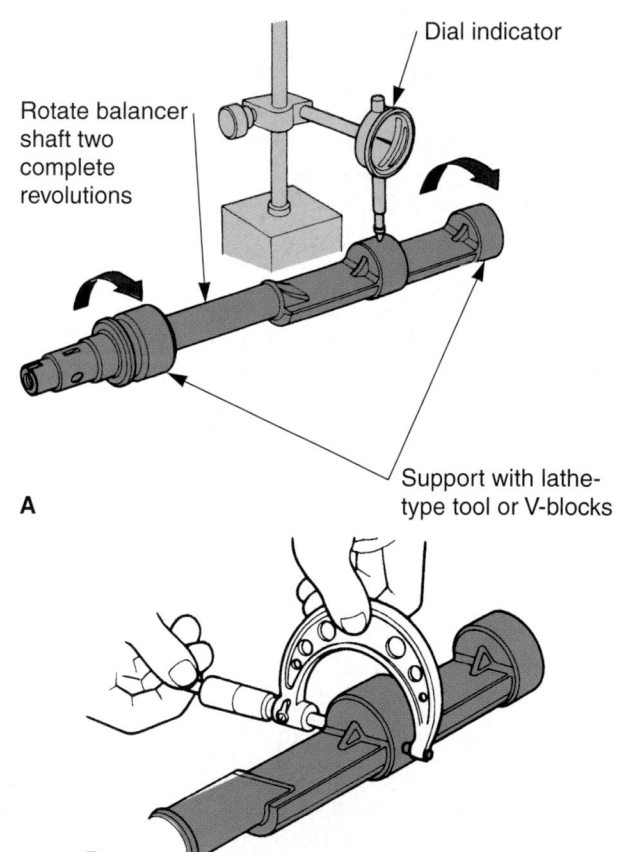

Figure 50-12. Note the basic steps for servicing balancer shafts. A—Use a dial indicator to check for a bent balancer shaft. Mount the shaft on V-block. Place the indicator tip on the journals. Rotate the shaft and read the indicator. Replace the balancer shaft if runout exceeds specifications. B—Use a micrometer to measure balancer shaft journal wear. Compare readings to specs.

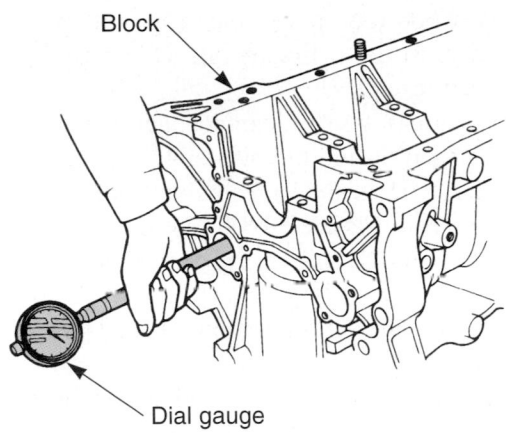

Figure 50-13. A hole or dial gauge is often used to measure balancer shaft bearing diameters.

Cleaning piston ring grooves (using a ring groove cleaner) was covered in Chapter 49, *Engine Removal, Disassembly, and Parts Cleaning.* Review this material if needed.

Measuring Piston Wear

A large outside micrometer is used to measure piston wear. Micrometer readings are compared to specs to determine the amount of wear.

Piston size is measured on the skirt, just below the piston pin hole, **Figure 50-14.** Adjust the micrometer for a slight drag as it is pulled over the piston. If piston wear exceeds specifications, replace or knurl (denting operation that raises surface) the piston(s).

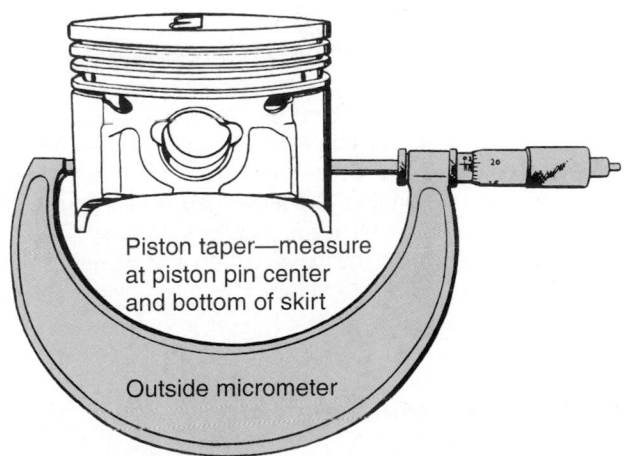

Figure 50-14. An outside micrometer is used to measure piston diameter and wear. To measure piston size, measure about 3/4″ (19 mm) down from pin hole on skirt. To measure taper, measure piston diameter and the top and bottom of the skirt. (Pontiac)

Piston taper is measured by comparing piston diameter at the top (even with pin hole) and bottom of the skirt. The difference between the two measurements equals piston taper. If taper is not within the service manual limits, replace or knurl the piston.

Knurling a Piston

Piston knurling can be used to increase the diameter of the skirt a few thousandths of an inch (hundredths of a mm). Knurling makes dents in the skirts, pushing up the metal next to the dents. This increases piston diameter. A machine shop can usually knurl pistons using the operation shown in **Figure 50-15.**

Measuring Piston Clearance

To find *piston clearance,* subtract piston diameter from cylinder diameter. The difference between the cylinder diameter measurement and the piston diameter measurement will equal piston clearance.

Average piston-to-cylinder clearance is about 0.001″ (0.025 mm). Since specifications vary, always refer to the service manual.

Figure 50-16 shows another way of measuring piston clearance. A long, flat feeler gauge is placed on the piston skirt. Then the gauge and piston are pushed into the cylinder. A spring scale is used to pull the feeler

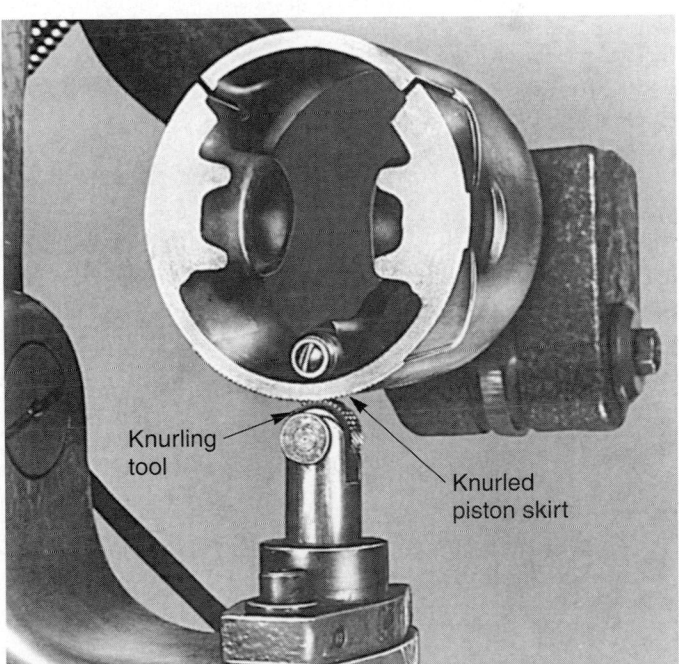

Figure 50-15. Piston knurling increases piston skirt diameter. It can be used to correct small amounts of wear and restore proper piston dimensions. (Deere & Co.)

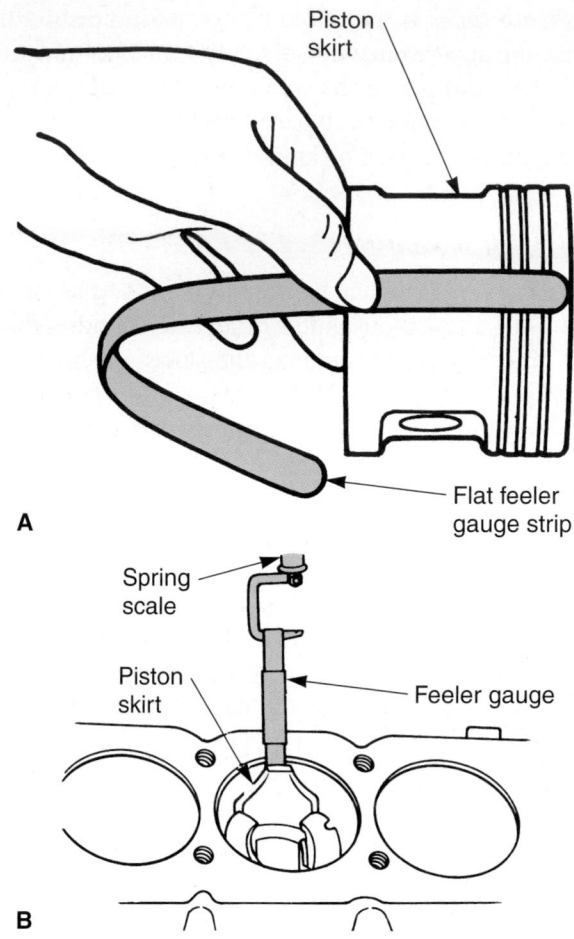

Figure 50-16. Measuring piston clearance can be done with feeler gauge and a spring scale. A—Place a long strip feeler gauge on the piston skirt. Insert the piston and the gauge into cylinder. B—Use a spring scale to pull the piston out of the cylinder. When the spring scale reads within specifications, the feeler gauge size equals piston clearance. Refer to the service manual for details. (Chevrolet)

gauge out of the cylinder. When the spring scale reading equals specifications, the size of the feeler gauge equals piston clearance.

When piston-to-cylinder clearance is excessive, you must either:

- Knurl the pistons.
- Install new standard-size pistons (providing cylinders are not worn beyond specs).
- Bore the cylinders and install oversize pistons.
- Sleeve the cylinders.

Measuring Piston Ring Side Clearance

Piston ring side clearance is the space between the side of a compression ring and the inside of the piston groove. Ring groove wear increases this clearance. If the groove is worn too much, the ring will not be held squarely against the cylinder wall. Oil consumption and smoking can result.

To measure ring side clearance, obtain the new piston rings to be used during the overhaul. Insert the new ring into its groove. Then slide a feeler gauge between the ring and groove, as in **Figure 50-17.** The largest feeler gauge that fits between the ring and groove indicates ring side clearance. The top ring groove is usually checked because it suffers from more combustion heat and wear than the second groove.

If ring side clearance is beyond specs, either replace the piston or have a machine shop fit ring spacers in the grooves. *Ring spacers* are thin steel rings that fit next to the compression rings. The piston groove is machined wider to accept the spacer. This will restore ring side clearance to desired limits.

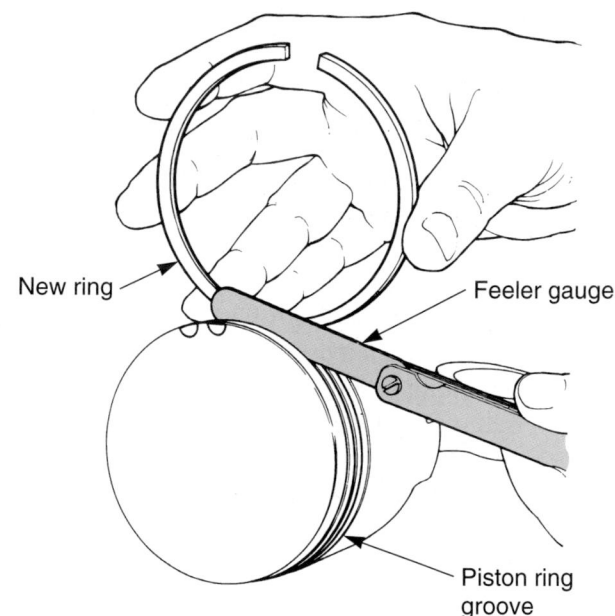

Figure 50-17. To measure ring side clearance, fit the ring into its groove. Then, find the feeler gauge that will fit snugly into groove next to the piston ring. The feeler gauge size equals ring side clearance. If side clearance is greater than specs, the ring groove may be worn or you have wrong piston ring set for engine. (Chrysler)

Measuring Piston Ring Gap

Discussed in earlier chapters, *piston ring gap* (clearance between ends of ring when installed in cylinder) is very important. If the gap is too small, the ring could lock up or score the cylinder as it heats up and expands. If the ring gap is too large, ring tension against the cylinder wall may be too low, causing blowby.

To check ring gap:
1. Compress a compression ring and place it in its cylinder.
2. Push the ring to the bottom of normal ring travel with the head of a piston. This will square the ring in the cylinder and locate it at the smallest cylinder diameter. See **Figure 50-18A.**
3. Measure the ring gap with a flat feeler gauge.
4. Compare the measurements to specifications.

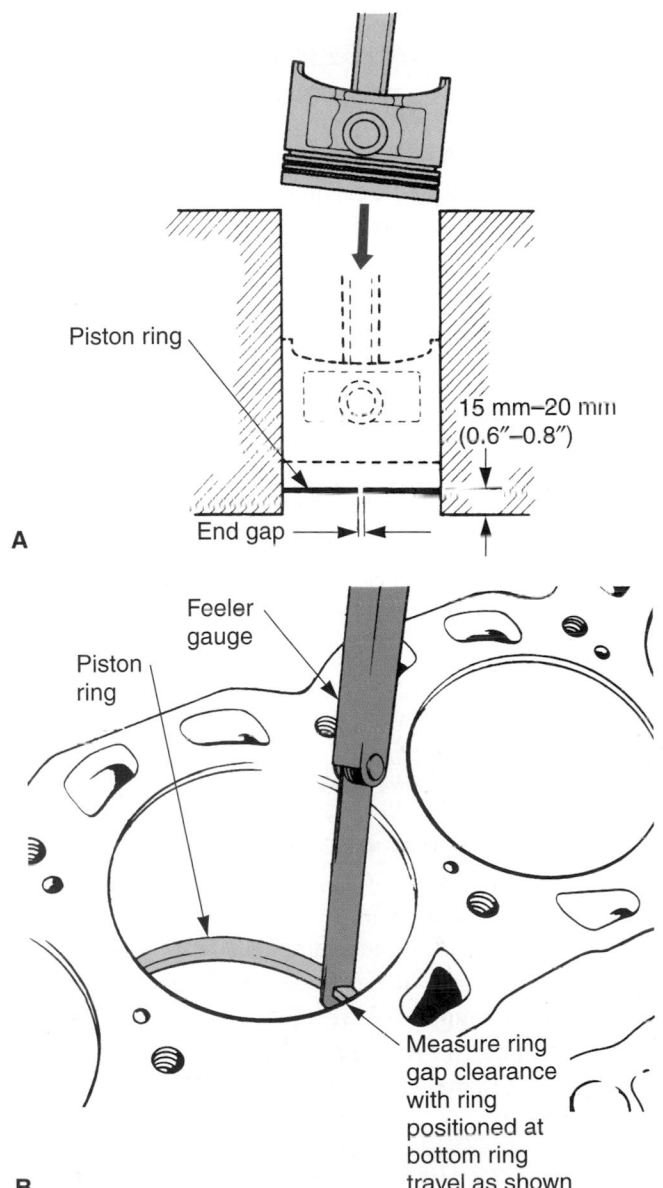

Figure 50-18. A—To measure piston ring end gap, install the ring and push it to the bottom of ring travel with the head of a piston. B—Use a feeler gauge to measure the gap. If ring end gap is too small, check the ring size or file the ends. If the gap is too large, double-check cylinder dimensions and the size of the rings. (Cadillac and Honda)

If the ring gap is not correct, you may have the wrong piston ring set or the cylinder dimensions may be off. Look at **Figure 50-18B.**

Some manufacturers allow *ring filing* (using a special grinding wheel to remove metal from the rings) to increase piston ring gap.

Piston Pin Service

Depending on the type and make of engine, the piston pin may either be *free-floating* (pin will turn in both rod and piston) or *press-fit* (pin force-fit in rod but turns in piston). Other setups have been manufactured but are not common.

During piston and rod service, check the pin clearance on both free-floating and press-fit pins. Check pin-to-connecting-rod fit on free-floating piston pins. With press-fit pins, the piston pin should be locked tightly in the connecting rod.

To check for excessive piston pin clearance, clamp the connecting rod I-beam lightly in a vise. Holding the piston straight up, rock the piston against normal pin movement. If play can be detected, the pin, rod bushing, or piston bore is worn. A small telescoping gauge and an outside micrometer should be used to determine exact clearance after pin removal.

Free-Floating Pin Service

To remove a free-floating pin from the piston, use snap ring pliers to compress and lift out the snap rings on each end of the pin. Then, push the pin out of the piston with your thumb. In some cases, a brass drift and light hammer blows may be needed to drive the pin from the piston. Refer to **Figure 50-19.**

When the pin is worn, it should be replaced. If the pin bore in the piston measures larger than specs, the piston must generally be replaced. In some cases, the pin bores can be reamed larger and oversize piston pins can be used. Pin bore reaming is usually done by a machine shop.

Pressed-In Pin Service

To remove a pressed-in piston pin, you will need to use a press and a driver setup similar to the one shown in **Figure 50-20.**

⚠️ **Warning**
When pressing out piston pins, wear eye protection and make sure the piston is mounted properly.

Measure pin and pin bore wear. Compare your measurements to specs and replace or repair parts as needed.

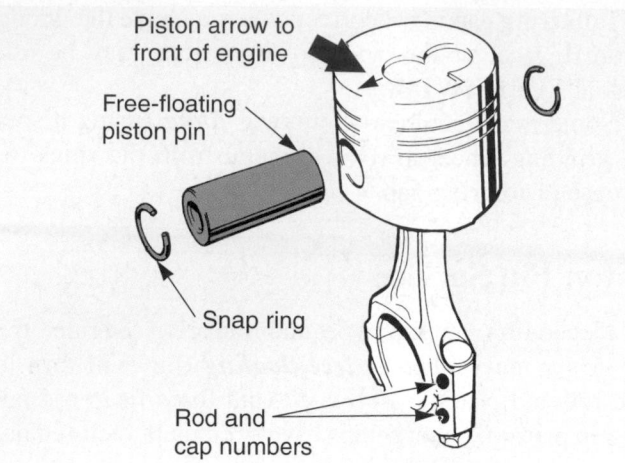

Figure 50-19. A free-floating piston pin is held in place by snap rings. The pin should slide in and out of piston and rod with finger pressure or with light taps with brass drift and a hammer. Make sure the piston notch or arrow and the rod numbers are facing the proper direction. Most pistons can only be installed on the rod in one direction. Piston pin offset or valve reliefs must be positioned correctly. (Volvo)

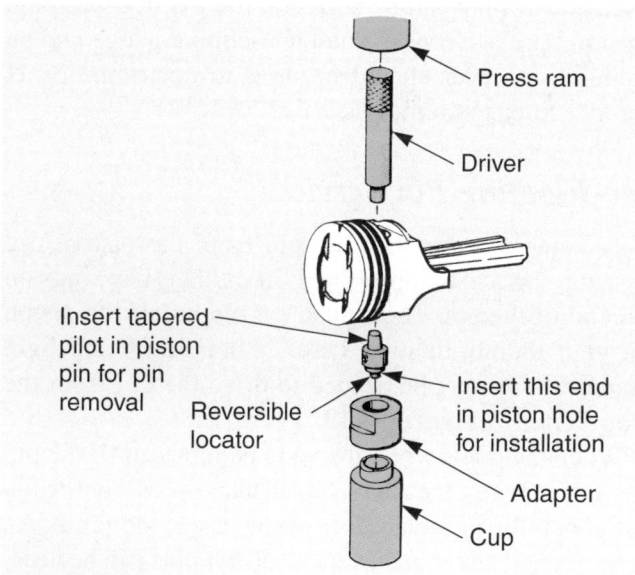

Figure 50-20. A press-fit piston pin must be forced out with a press. Use suitable adapters to prevent piston damage. (Ford)

If needed, send new pistons and pins to a machine shop for fitting (honing piston pin bores to correct clearance).

Piston Pin Installation

Before installing a piston pin, make sure the piston is facing in the right direction in relation to the connecting rod. Normally, a piston will have a marking on its head that should point toward the front of the engine.

One edge of the connecting rod's big end bore may be chamfered. The chamfered edge of the rod big end must face the outside of the crank journal during piston installation. The rod may also have an oil spray hole or rod numbers that must face in a specified direction. Check the vehicle's shop manual for directions.

To start a pressed-in piston pin, tap it into the piston bore with a brass hammer. Then, use a press to force the pin into the piston. The connecting rod small end must be centered on the pin.

After pushing a free-floating piston pin into the piston, install the snap rings to secure the pin. Make sure the snap rings are fully seated in their grooves.

Connecting Rod Service

Connecting rods are subjected to tons of force during engine operation. As a result, they can wear, bend, or even break. If piston or bearing wear abnormalities are found, there may be a problem with the corresponding connecting rod.

For example, if one side of the bearing is worn, the connecting rod may be bent. If the back of the bearing insert has marks on it, the big end of the rod may be distorted, allowing the insert to shift inside the bore.

Rod Small End Service

Measure the rod's small end bore with a telescoping gauge and a micrometer. If worn beyond specs, have a machine shop replace the rod bushing. The pin will have to be *fitted* (bushings reamed for proper clearance) in the rod.

Rod Big End Service

To check the connecting rod big end bore for problems, remove the bearing insert. Reinstall the rod cap and torque the cap to specs. Then, measure the rod bore diameter on both edges and in both directions.

Any difference in edge diameter equals rod big end taper. Any difference in the cross diameters equals rod big end out-of-roundness. If taper or out-of-round is greater than specs, have a machine shop rebuild the rod or purchase a new rod. Refer to **Figure 50-21.**

Checking Rod Straightness

To determine if a rod is bent, a special *rod alignment fixture* is needed. It will check whether the rod's small end and its big end are perfectly parallel. Follow the operating instructions provided with the particular fixture.

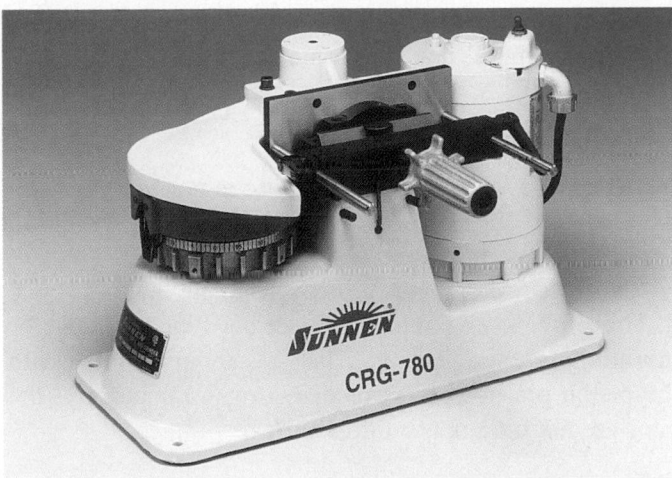

Figure 50-21. If not badly damaged, connecting rods can be rebuilt using this machine. It is often found in a machine shop. (Sunnen)

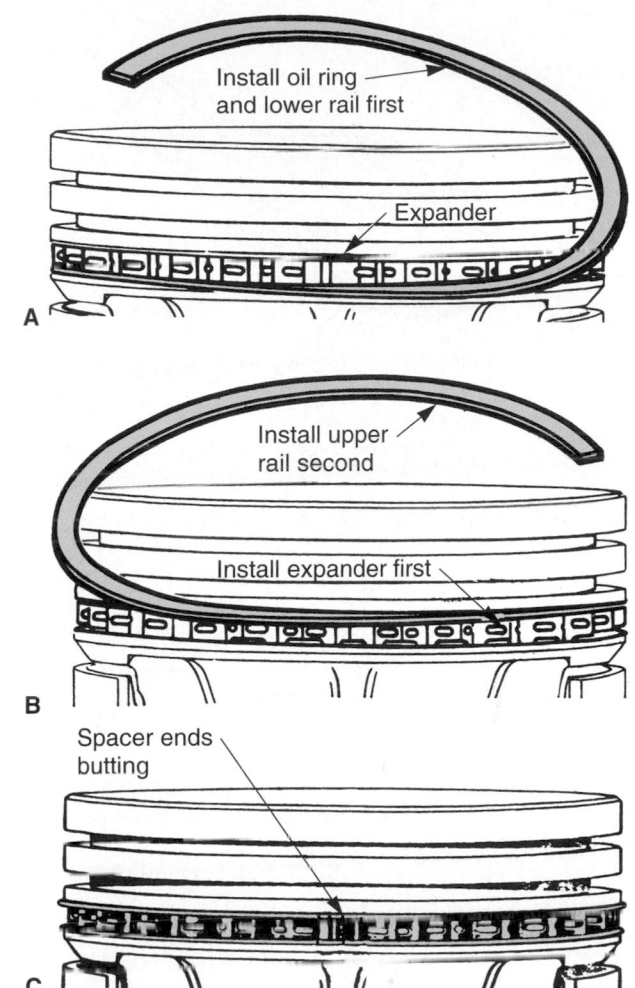

Figure 50-22. A—Install the oil ring first. Fit the expander-spacer into its groove. Then spiral the bottom oil rail around bottom of expander-spacer. B—Spiral the rail top in above the expander-spacer. C—Double-check that the expander-spacer ends are butted. They must not overlap.

Piston Ring Service

After checking the cylinders, pistons, and connecting rods, you are ready to install the new piston rings on the pistons.

Installing Oil Ring

Before installing the compression rings, install the oil ring into the piston's bottom groove. Wrap the expander spacer around the groove and *butt* the ends of the spacer together. Spiral the bottom rail into the bottom groove below the expander-spacer. Look at **Figure 50-22.** Then, spiral in the top oil ring rail above the expander-spacer. Make sure the ends of the expander-spacer do *not* overlap—they must butt together.

Make sure the oil ring assembly will rotate on the piston. There will be a moderate drag as the oil ring is turned. One ring gap should be almost aligned with the end of the piston pin. The other gap should be at the opposite end of the pin.

Installing Compression Rings

Read the instructions provided with the new piston ring set. There will usually be hints on proper ring installation. Usually, compression rings have a top and a bottom. If installed upside down, ring failure or leakage may result.

Ring markings are usually provided to show how compression rings should be installed. The markings identify the top of each ring and indicate which ring goes into the top or second piston groove. See **Figure 50-23.**

Figure 50-23. Compression ring markings and instructions provided with the ring set will tell you how and in which groove each ring should be installed.

 Caution
Compression rings are made of cast iron, which is very brittle. They will *break easily* if expanded or twisted too much.

Using a *piston ring expander* (special tool for spreading and installing rings), slip the compression rings into their grooves. See **Figure 50-24.** If a ring

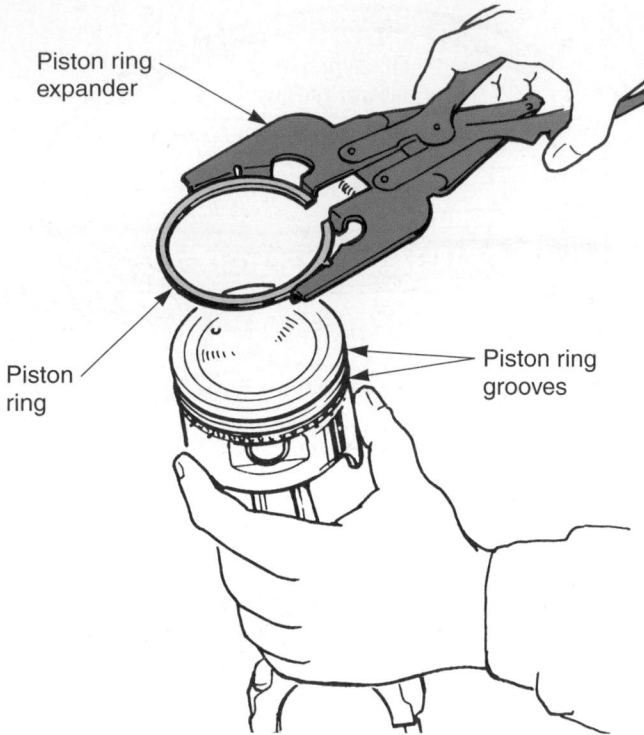

Figure 50-24. Be careful not to open the compression rings too much when installing them on the piston. They are brittle and can break easily. (Honda)

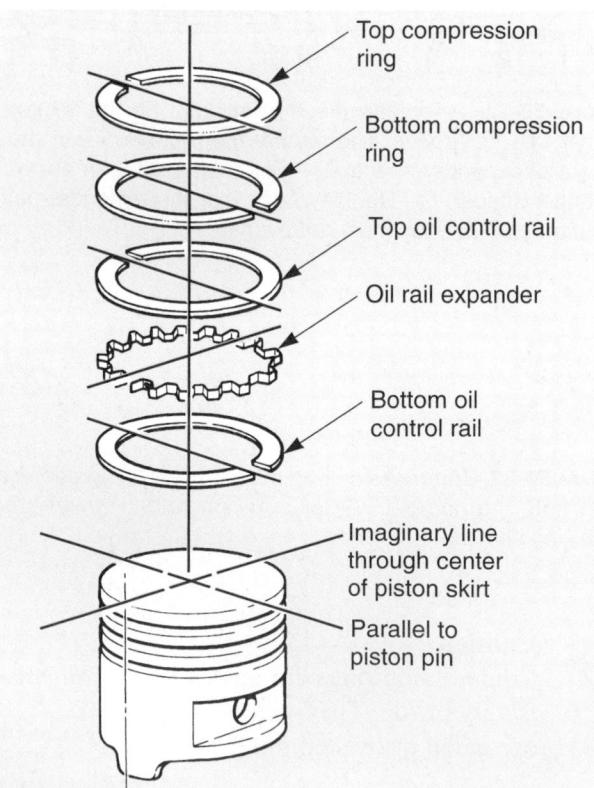

Figure 50-25. Align the piston ring end gaps as shown. They should be staggered and face the ends of the piston pin. (Chrysler)

expander is not available, use your fingers to carefully spiral the compression rings onto the piston.

Piston Ring Gap Spacing

Specific *piston ring gap spacing* is normally recommended to reduce blowby and ring wear. **Figure 50-25** shows a typical method.

Note that each gap is directly opposite the one above it. This provides maximum distance between each gap for minimum pressure leakage. The gaps are also in line with the piston pin hole. This reduces ring wear because the gaps are not on a major thrust surface.

Crankshaft Service

Before final inspection, make sure the crankshaft is perfectly clean. Use compressed air to blow out all the oil passages. Study each connecting rod journal and main journal closely. Look for scratching, scoring, and any other signs of wear. The slightest nick or groove can be serious. Very fine crocus cloth may be used to clean up minor burrs or marks.

Checking Crankshaft Straightness

A *bent crankshaft* can ruin new main bearings or cause the engine to lock up when the main caps are tightened. To measure crankshaft straightness, place the crankshaft in the block main bearings or on a V-block and mount a dial indicator against the center main bearing journal. See **Figure 50-26.**

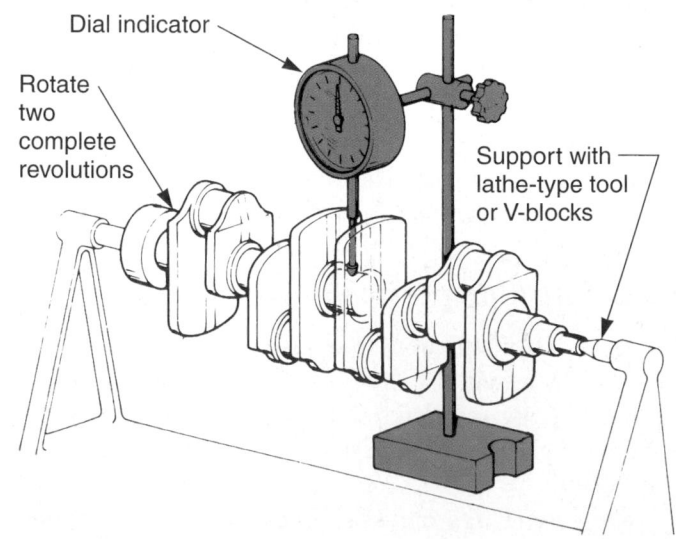

Figure 50-26. Crankshaft straightness can be measured as shown. If a special holding tool or V-blocks are not available, lay the crank in the block main bearings, leaving the center main cap off. (Honda)

Slowly turn the crankshaft while watching the indicator. Indicator movement equals crankshaft bend. If not within limits, replace the crank or have it straightened and turned by a machine shop.

Measuring Journal Taper

If one side of a crankshaft journal is worn more than the other, the journal is *tapered*, **Figure 50-27.** To measure journal taper, use an outside micrometer. Measure both ends of each journal, **Figure 50-28.** Any difference indicates taper. Taper beyond recommended limits requires crankshaft turning.

Measuring Journal Out-of-Round

When you measure for journal taper, also measure *journal out-of-roundness* (journal worn more on top than bottom), as was shown in **Figure 50-27.**

Measure across the journal from side to side and then from top to bottom. If not within spec limits, send the crankshaft to a machine shop for turning.

Turning or Grinding a Crankshaft

Turning a crankshaft involves grinding the rod and main journals to a smaller diameter to fix journal wear or damage. Crankshaft turning is done by a machine shop.

Undersize bearings are needed after the crankshaft has been turned. Since the crank journals are smaller in diameter, the new bearings must be thicker to provide the correct bearing-to-journal fit.

Note
For more details on standard and undersize bearings, refer to Chapter 14, *Engine Bottom End Construction.*

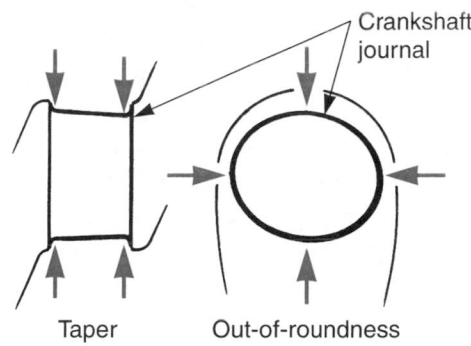

Figure 50-27. When measuring crankshaft journal wear, check as shown to detect taper and out-of-roundness. (Volvo)

Undersize Bearing Markings

Always look at the old crankshaft bearings to determine whether they are undersize. Undersize bearings may be used in rebuilt engines and even in a few new engines.

Inspect the back of the old bearings for an *undersize number.* The number stamped on the bearing back denotes bearing undersize (.010″ for example). Look at **Figure 50-29.**

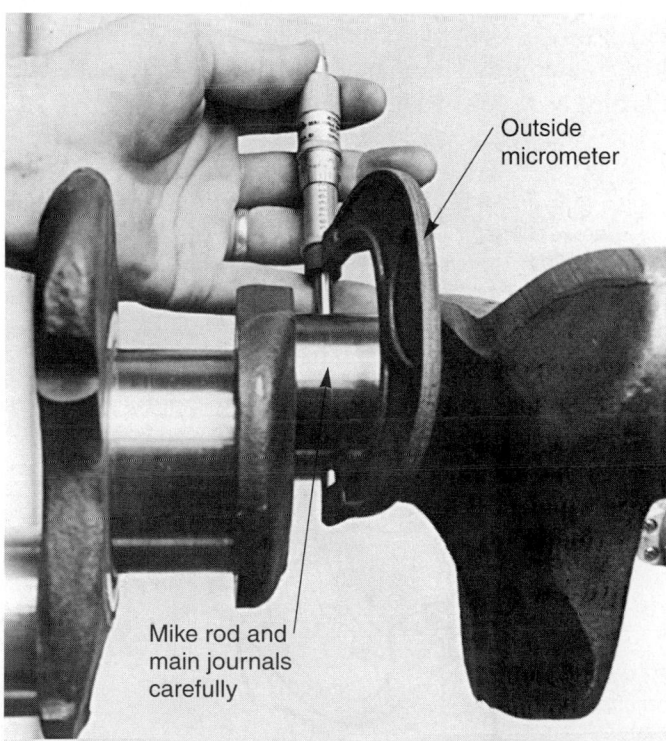

Figure 50-28. An outside micrometer is used to measure crankshaft journal wear. If wear is excessive, the crankshaft must be replaced or turned to accept undersize bearings. (Goodson)

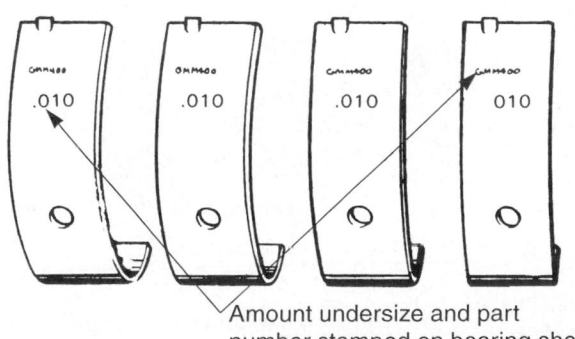

Figure 50-29. Inspect old crankshaft bearings for undersize markings. A 0.010″ on the back of a bearing shell would indicate that the crankshaft journals have been machined 0.010″ smaller in diameter. If the crankshaft is in good condition, purchase bearings that are the same size as the old ones. (Oldsmobile)

If, during engine repair, you find that the old crank-shaft is in good condition, install new bearings that are the same size bearings as the old ones. However, if the crankshaft is badly worn or damaged, replace or turn the shaft. Then, install the appropriate size bearings.

Tech Tip
Sometimes the engine manufacturer will stamp bearing sizes on the cylinder block in the form of a code. You must refer to the service manual to find out what size bearings the code represents.

Installing Crankshaft Bearings

Purchase the correct (standard or undersize) main bearings. With the main bearing bores and the back of each bearing *clean* and *dry,* fit the bearing inserts into place. One bearing insert goes into the block, and the matching insert goes into the corresponding main cap. Also install the main thrust bearing into the correct posi-tion in the block and cap. Make sure the oil holes in the bearings align with the oil holes in the block!

Figure 50-30 shows how a crankshaft can be replaced without cylinder head, piston, or connecting rod removal.

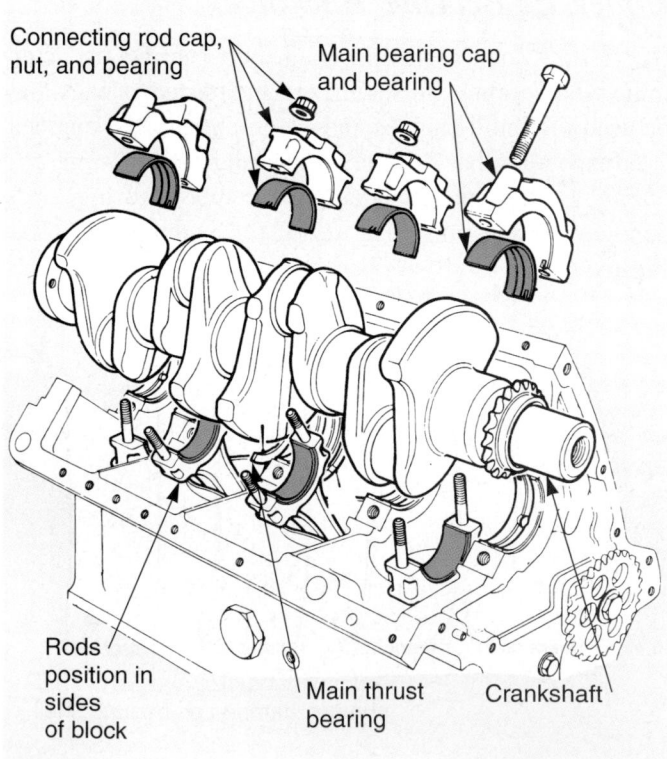

Figure 50-30. Crankshaft and engine bearings can be replaced without removing the heads, pistons, and rods. Unbolt the rod caps and the main caps. Carefully lift the crankshaft out of block.

Installing the Rear Main Oil Seal

There are three basic types of rear main oil seals: two-piece synthetic rubber seals, two-piece rope (wick) seals, and one-piece synthetic rubber seals. Each requires a different installation technique.

The *two-piece synthetic rubber seal* is very easy to install. Simply press it into place in the block and the rear main cap. The sealing lip on the rear main seal must point toward the *inside* of the engine. If installed backwards, oil will pour out of the back of the seal when the engine is started. Lubricate the sealing lip with motor oil prior to installation, **Figure 50-31.**

A *two-piece rope seal,* or *wick seal,* must be worked down into the rear main cap and the block carefully. Use a special seal installation tool and light hammer blows. If necessary, hand pressure and a smooth steel bar can be used to install the seal. Use a single-edge razor to cut the rope seal flush with the cap and block parting line.

Frequently, *silicone sealer* is recommended on the rear main cap to prevent oil leakage. The sealer keeps oil from seeping between the main cap and block mating surfaces, **Figure 50-31.**

If additional side seals are provided for the cap, follow the instructions with the gasket set. Sealer is com-monly recommended on main cap side seals.

A *one-piece synthetic rubber seal* is installed after the rear main cap has been bolted to the block. It is driven into position from the rear of the engine using a seal driver. After installation, make sure the seal is square and undamaged. Look at **Figure 50-32.**

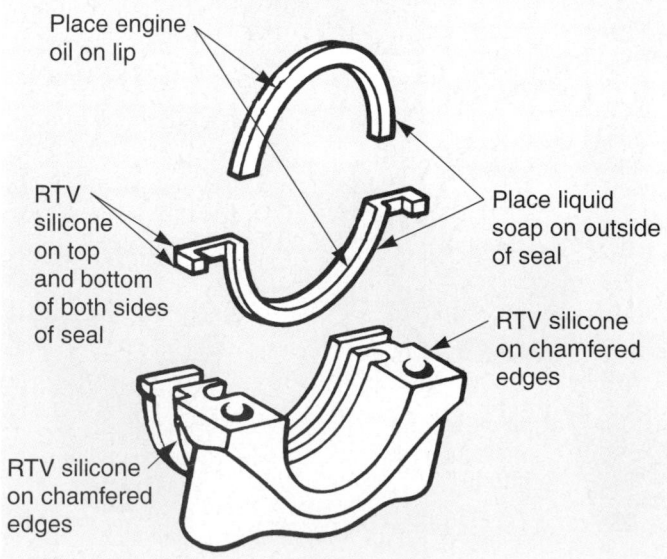

Figure 50-31. Common synthetic rubber rear main bearing oil seal. Note the manufacturer's installation recommendations. The oil seal lip must face the inside of the engine.

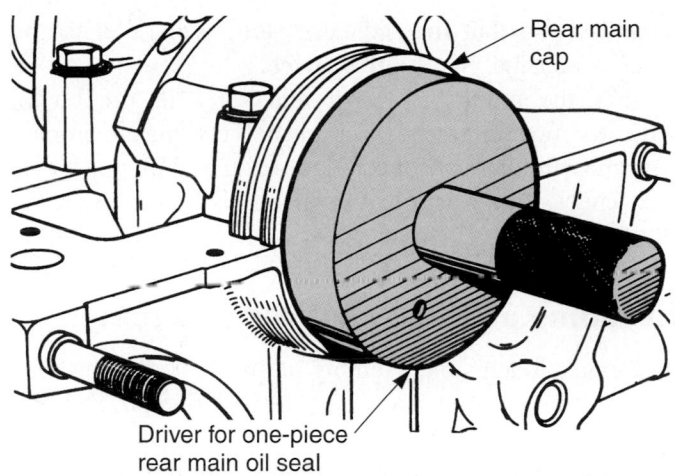

Figure 50-32. This one-piece, synthetic rubber rear seal being installed with special driver. Make sure the seal is perfectly square and fully seated. This type of seal is usually installed after the crankshaft is in place. (Renault)

Installing the Crankshaft

After main bearing installation, coat the bearings in the block with clean motor oil. Carefully lay the crankshaft into place, **Figure 50-33.**

Caution
Do not rotate the crankshaft without the main caps bolted to the block or the inserts may spin and be damaged.

Checking Main Bearing Clearance

To check the oil clearance between the crank journal and main bearing:

1. Place a small bead of *Plastigage* (clearance-measuring material) on the unoiled crankshaft journal.
2. Install the proper rod cap and torque the cap bolts to specifications, **Figure 50-34.**
3. Remove the cap.
4. Compare the smashed Plastigage to the paper scale. The width of the smashed Plastigage will let you determine bearing clearance. See **Figure 50-35.** If clearance is not correct, check bearing sizes and crankshaft journal measurements.

An average main bearing clearance is about 0.002″ (0.05 mm).

Torquing Main Bearing Caps

After checking clearance, oil the crank journals and bearing faces. Place each main bearing cap into place in

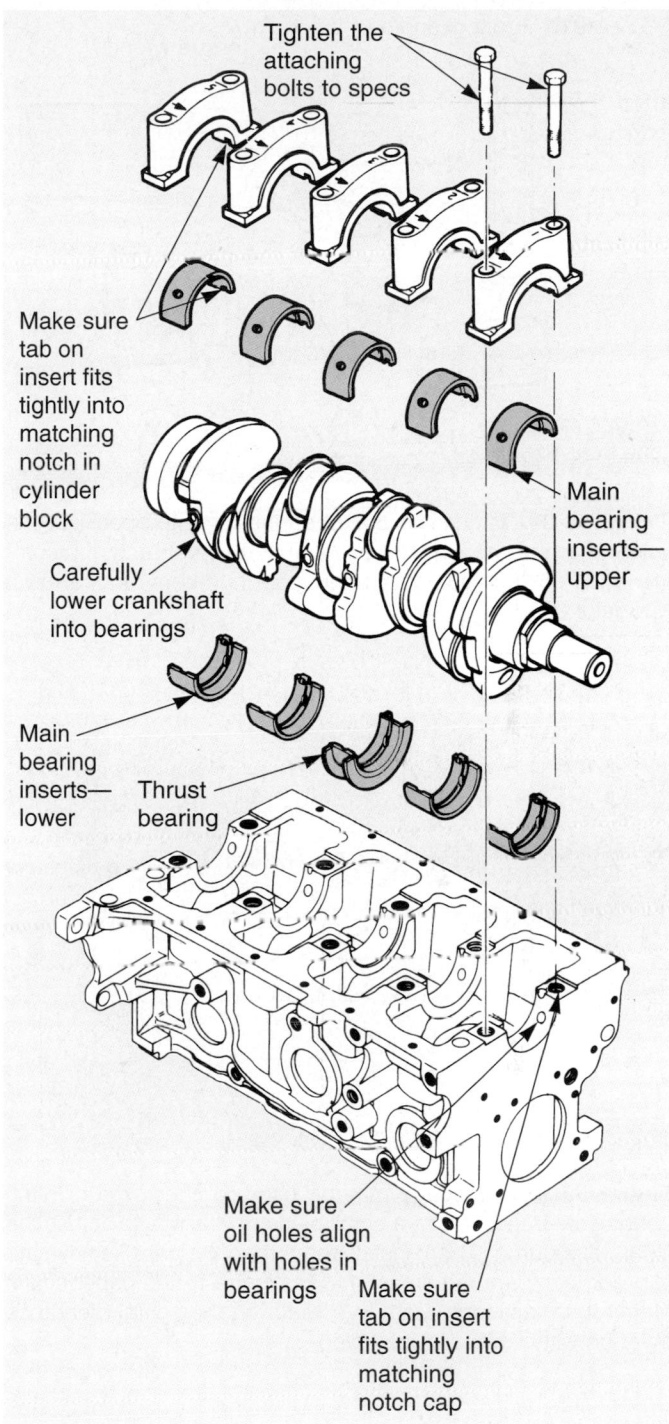

Figure 50-33. Fit each main bearing insert into its bore. Make sure the bearing tab fits fully into the notch in the block or cap. The back of each bearing insert should be clean and dry. Oil the front, or face, of the bearing with motor oil. Carefully lower the crank into the block without bumping the bearings or journals. Make sure the rear seal is installed correctly. (Ford)

the block. Double-check bearing installation. Make sure that the arrows and numbers on each cap are correct. They should usually read 1, 2, 3, etc., going from the front to the rear of the block.

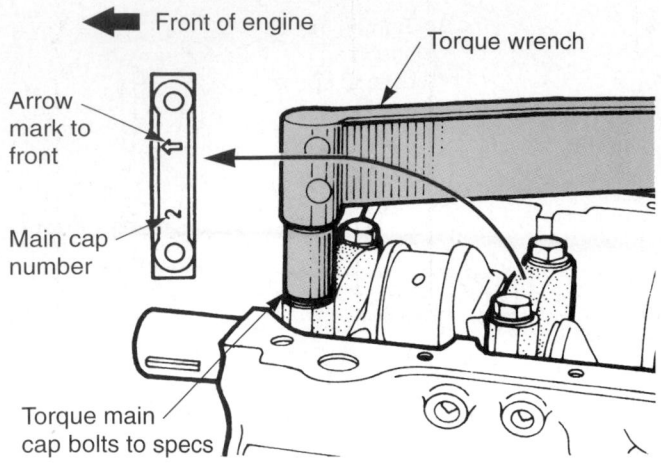

Front of engine

Torque wrench

Arrow mark to front

Main cap number

Torque main cap bolts to specs

Figure 50-34. Torque the main bearing caps to specifications using a torque wrench. The bolts must be clean and, if recommended, oiled. Also, check the cap numbers and arrows. (Chrysler)

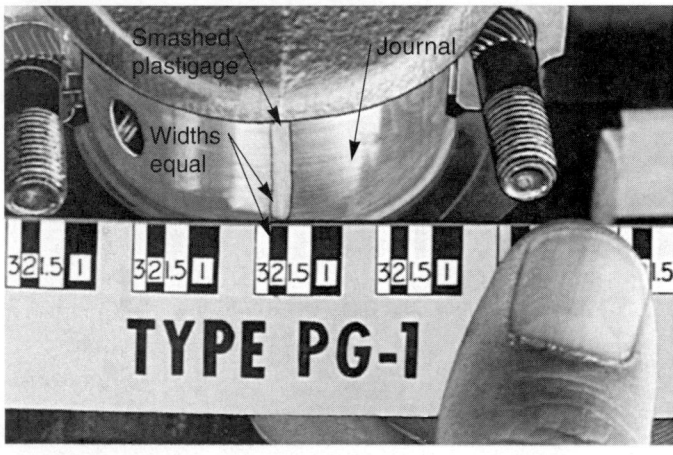

Smashed plastigage

Journal

Widths equal

TYPE PG-1

Figure 50-35. Plastigage may be used to check rod or main bearing clearance. Place a small piece of Plastigage across the clean, dry journal. Then, install and torque the cap. Remove the cap and compare the smashed Plastigage to the paper scale. Match the Plastigage width to a width on the scale. Clearance must be within specs. (Chevrolet)

Tighten each cap bolt a little at a time. This will pull the cap squarely down onto the block. Then, use a torque wrench to tighten the main cap bolts to factory specs.

Checking Crankshaft End Play

Crankshaft end play is the amount of front-to-rear movement of the crankshaft in the block. It is controlled by the clearance between the main thrust bearing and the crankshaft thrust surface or journal.

To measure crankshaft end play, mount a dial indicator on the block. Position the dial indicator against the crankshaft so that the indicator stem is parallel to the crank centerline. See **Figure 50-36.**

Pry the crankshaft back and forth in the block. Indicator movement equals crankshaft end play. Compare your measurements to specifications. If end play is incorrect, check the thrust bearing insert size and the crankshaft thrust journal width.

Installing a Piston and Rod Assembly

To install a piston assembly in the block, dip the head of the piston and new rings into clean engine oil. Double-check that the ring gaps are still spaced properly.

Fit the unoiled rod bearing insert into the connecting rod and the matching insert into the correct rod cap. Then, wipe a generous layer of oil on the face of the bearings. Refer to **Figure 50-37.**

Clamp a ring compressor around the piston and rings. The small indentations around the edge of the compressor should face the bottom of the piston, **Figure 50-38.** While tightening the compressor, hold the tool squarely on the piston.

Protect the Crankshaft

Slide *rod bolt covers* (plastic or rubber hoses) over the connecting rod bolts. This will prevent the rod bolts from scratching the crankshaft journal, **Figure 50-39.**

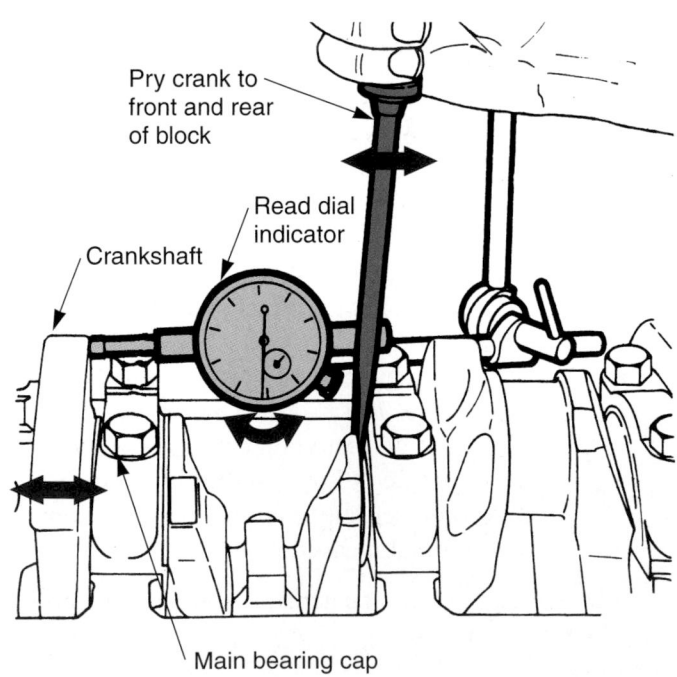

Pry crank to front and rear of block

Read dial indicator

Crankshaft

Main bearing cap

Figure 50-36. Use dial indicator to measure crankshaft end play. Pry back and forth on the crankshaft while reading the indicator. If end play not within specs, check thrust bearing and thrust surfaces on crankshaft. (Chrysler)

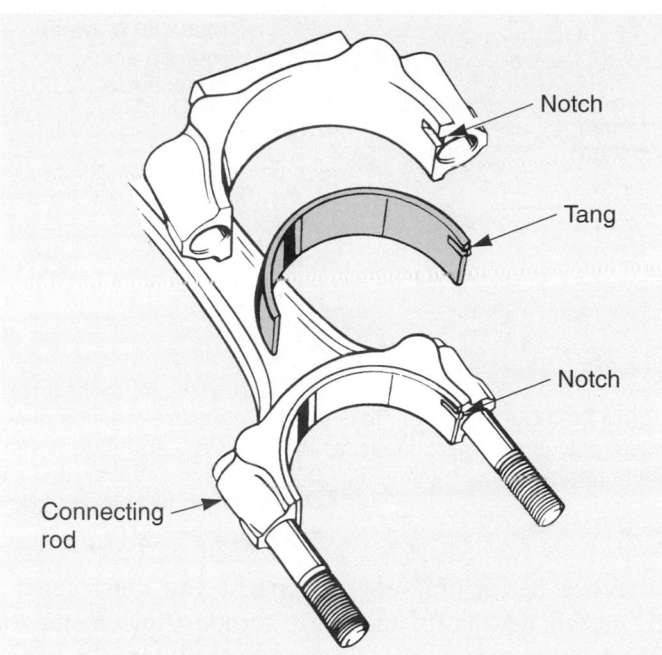

Figure 50-37. Install clean, dry bearings in the connecting rod and the cap. Keep your fingers off the bearing faces. Fit the bearing tangs into the rod notches. After installing, coat bearing faces with motor oil. (Oldsmobile)

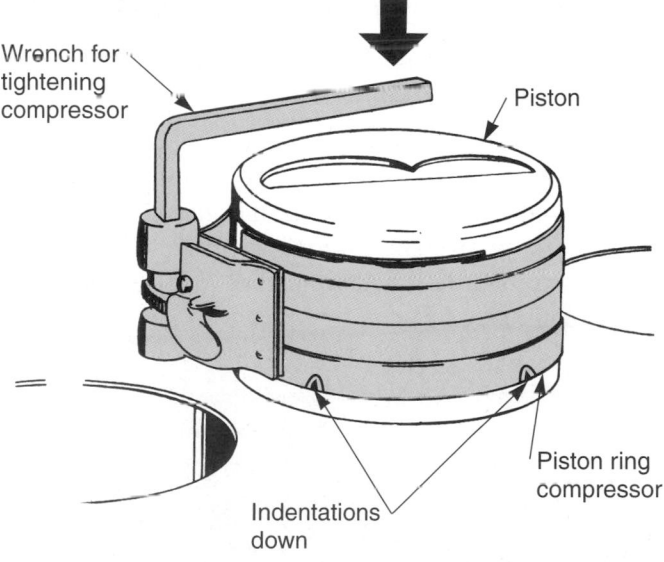

Figure 50-38. After checking ring gap spacing and placing oil on the rings and the piston, install the ring compressor. Tighten the compressor handle firmly. (Cadillac)

Double-check the markings on the rod and the piston. Make sure the rod is facing the right direction and that the cap number matches the rod number. Also check that the piston notch or arrow is facing the front of the engine.

For example, if the rod is marked with the number one, it would normally go in the very front cylinder with

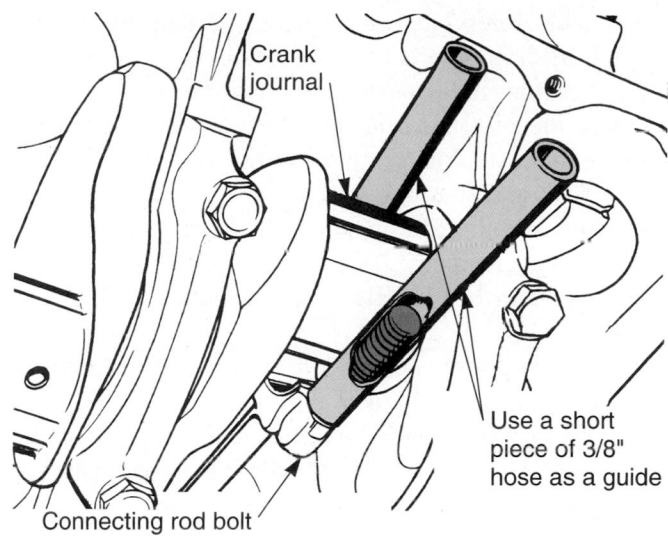

Figure 50-39. Be careful not to let the rod bolts nick the crankshaft journals during installation. Sections of rubber hose or soft plastic tubing will protect the crankshaft. (Buick)

the piston marking to the front. Check the service manual if in doubt.

Installing Piston and Rod in Block

To install the piston and rod assembly in its cylinder, turn the crankshaft until the corresponding crank journal is at BDC. Place the piston and rod assembly into its cylinder. Look at **Figure 50-40.**

While guiding the rod bolts over the crankshaft journal with one hand, tap the piston into the cylinder with a *wooden hammer handle.* A soft wooden or plastic hammer handle will not mar or dent the head of the piston. Hold the ring compressor squarely against the

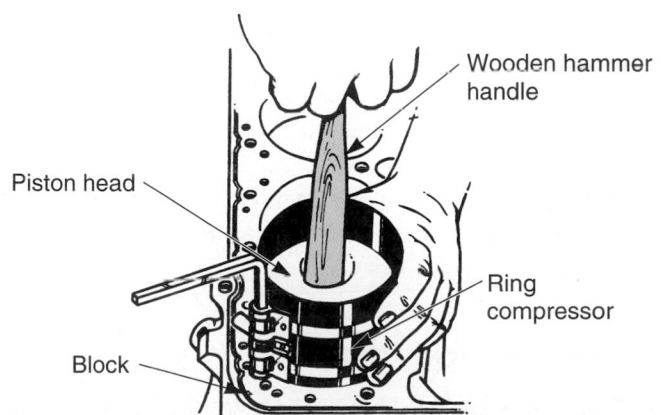

Figure 50-40. Use a wooden hammer handle to tap the piston and rod assembly into the cylinder block. The rod journal for the piston being installed should be at BDC. Carefully guide rod over crankshaft as you tap piston into cylinder. (Lisle Tools)

block deck. Keep tapping until the rod bearing bottoms around the crank journal.

If a piston ring pops out of the compressor, do *not* try to force the piston down into the cylinder. This would damage the piston rings or piston. Instead, loosen the ring compressor and start over.

After installing the piston, use Plastigage to check rod bearing clearance. This is done the same way as checking main bearing clearance, which was explained earlier.

When installing a connecting rod cap, make sure the numbers or marking on the rod and the cap *are the same.* If the rod is marked with the number five, the cap should also have a five stamped on it. Mixing up rod caps will damage the bearings or crankshaft.

Torquing the Connecting Rods

It is very important for you to properly torque each rod nut or bolt to specifications.

If a rod is overtightened, the rod bolt could break during engine operation. Severe block, crank, piston, and cylinder head damage could result as parts fly around inside the engine.

If a rod is under-tightened, the bolts could stretch under load, allowing the bearing to spin or hammer against the crankshaft. Again, serious engine damage could result.

Using a *torque wrench,* tighten each rod fastener a little at a time. This will pull the rod cap down squarely against the rod. Gradually increase torque until full specifications are reached. Double-check each rod bolt or nut.

Checking Rod Side Clearance

Connecting rod side clearance is the distance between the side of the connecting rod and the side of the crankshaft journal or another rod. To measure rod side clearance, insert different size feeler gauge blades into the gap between the rod and crank. See **Figure 50-41.** The largest feeler blade that slides between the rod indicates side clearance.

Compare your measurements to specifications. If side clearance is not within specs, crankshaft journal or connecting rod width is incorrect.

Torque-to-Yield Bolts

Torque-to-yield bolts are tightened to a preset yield, or stretch, point. This preloads the fastener for better clamping under varying conditions. Some cylinder head bolts and main bearing cap bolts are of the torque-to-yield variety.

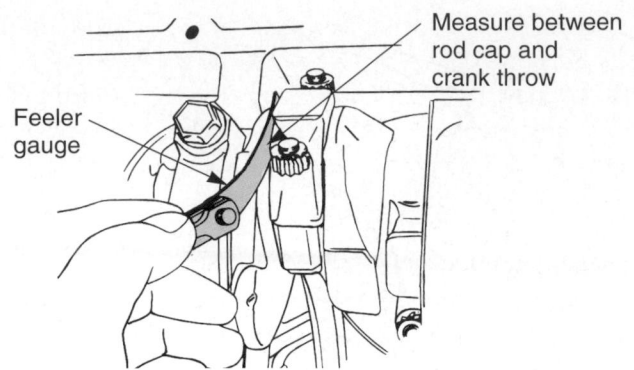

Figure 50-41. A feeler gauge is used to measure rod side play. If side play is incorrect, measure the rod width and the journal width. (Buick)

When performing engine repairs, you must generally install new bolts and use a torque angle meter to tighten them properly. A *torque angle meter* is a numbered wheel used to measure fastener rotation in degrees. It can be mounted on a ratchet or a torque wrench. See **Figure 50-42.**

To tighten a bolt using a torque angle meter, torque the bolt to specifications. Then, install the meter on the torque wrench and zero the pointer. Finally, turn the bolt until the meter reads as specified by the manufacturer.

Engine Balancing

Engine balancing may be needed to prevent engine vibration when the weight of the pistons, connecting rods, or crankshaft is altered during an engine service. For example, if oversize pistons are installed and they weigh more than the standard pistons, engine balancing will be required.

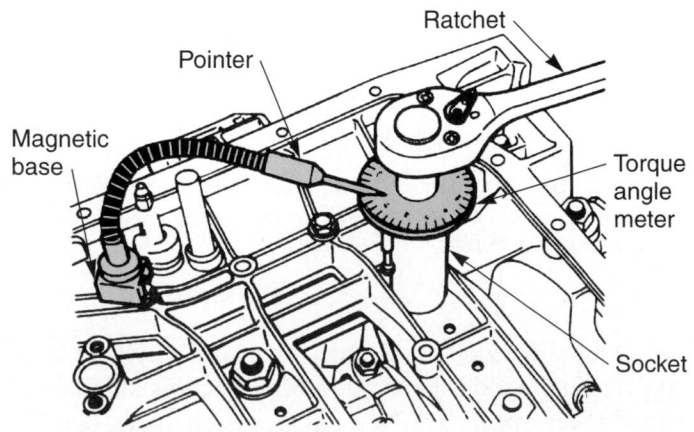

Figure 50-42. To use a torque angle meter, first tighten the bolt to specs. Then, zero the pointer and tighten the bolt to the meter reading specified by the manufacturer. (Chrysler)

Most large automotive machine shops have engine balancing equipment. Basically, the pistons, rings, piston pins, connecting rods, and rod bearings are weighed on an accurate scale. Material is machined or ground off the pistons and rods until all pistons weigh the same and all rods weigh the same. All rod big ends must weigh the same, and all small ends should also be equal in weight.

Then, bob weights comparable to the weight of each piston and rod assembly are bolted to the crankshaft rod journals. The crankshaft, front damper, and flywheel are bolted together and rotated on the engine balancing machine. The balancing machine will show where weight should be added (metal welded on) or removed (metal drilled out) from the crankshaft counterweights, damper, and flywheel.

Proper engine balance is very critical with today's small, high-rpm engines. Keep engine balancing in mind when major engine modifications are made.

Final Assembly of Engine

With all the pistons and rods installed and torqued, you can now install all the other parts on the block: oil pump, oil pan, cylinder heads, camshaft drive, manifolds, etc. The procedures for installing these engine parts are described elsewhere in this text. Use the index to find this information if needed.

Summary

- An engine overhaul involves the service of the engine bottom end, top end, and front end.
- Cylinder block service commonly includes measuring the cylinders for wear, inspecting the cylinder walls for damage, honing or deglazing the cylinder walls, cleaning the cylinders after honing, and installing core plugs.
- Block line boring is used to straighten misaligned main bearing bores.
- Decking a block involves machining the cylinder head mounting surfaces until they are parallel and equal distance from the main bore.
- Cylinder taper is a difference in the diameter at the top and bottom of the cylinder.
- Cylinder out-of-roundness is a difference in cylinder diameter when measured front-to-rear and side-to-side in the block.
- Cylinder honing is used to break the glaze on a used cylinder.
- After honing, it is very important to clean and remove all honing grit from inside the engine.

- Cylinder boring is done by machining the cylinders larger in diameter to make the cylinder walls perfectly smooth and straight.
- Piston size is measured on the skirts, just below the piston pin hole.
- To find piston clearance, subtract piston diameter from cylinder diameter.
- Piston ring side clearance is the space between the side of a compression ring and the inside of the piston groove.
- Piston ring gap is the clearance between the ends of ring when it is installed in the cylinder. If the gap is too small, the ring could lock up or score the cylinder upon heating and expanding. If the gap is too large, ring tension against the cylinder wall may be low, causing blowby.
- Ring markings are usually provided to show how compression rings should be installed.
- Undersize bearings are needed after the crankshaft has been turned.
- To check the oil clearance between the crank journal and main bearing, place a small bead of Plastigage on the unoiled crankshaft journal.
- Crankshaft end play is the amount of front-to-rear movement of the crankshaft in the block.
- Connecting rod side clearance is the distance between the side of the connecting rod and the side of the crankshaft journal or another rod.

Important Terms

Engine overhaul	Piston knurling
Cylinder block service	Piston clearance
Block pressure testing	Piston ring side
Block line boring	clearance
Deck warpage	Ring spacers
Decking a block	Piston ring gap
Cylinder taper	Ring filing
Cylinder out-of-	Free-floating
roundness	Press-fit
Dial bore gauge	Rod alignment fixture
Cylinder honing	Ring markings
Deglazing	Piston ring expander
Cylinder hone	Piston ring gap spacing
Honing grit	Tapered journal
Cylinder boring	Journal out-of-
Overbore limit	roundness
Oversize pistons	Turning a crankshaft
Cylinder sleeving	Undersize bearings
Piston size	Two-piece synthetic
Piston taper	rubber rear seal

Two-piece rope seal
Silicone sealer
One-piece synthetic
    rubber seal
Plastigage
Crankshaft end play

Rod bolt covers
Connecting rod side
    clearance
Torque-to-yield bolts
Torque angle meter
Engine balancing

Review Questions—Chapter 50

Please do not write in this text. Place your answers on a separate sheet of paper.

1. What is an "engine overhaul?"

2. How do you measure main bearing bore alignment?

3. Define the term "cylinder taper."

4. A dial bore gauge is a quick and accurate tool for measuring cylinder taper. True or False?

5. Explain the purpose of honing a cylinder.

6. Describe three types of cylinder hones.

7. _____ _____ is done by machining the cylinders larger in diameter to make the cylinder walls perfectly smooth and straight.

8. Normally, a cylinder is bored in increments of:
   (A) 0.050″ (1.27 mm).
   (B) 0.010″ (0.254 mm).
   (C) 0.005″ (0.127 mm).
   (D) 0.001″ (0.025 mm).

9. _____ pistons and rings are needed in a cylinder block that has had its cylinders bored out.

10. Why is cylinder sleeving used?

11. Piston size should be measured on the piston _____, just below the _____ _____.

12. How do you find piston clearance?

13. What can happen if the piston ring gap is too large or too small?

14. To measure piston ring end gap, you would use a:
    (A) feeler gauge.
    (B) micrometer.
    (C) dial bore gauge.
    (D) telescoping gauge.

15. A specific piston ring gap spacing is recommended to reduce _____ and _____ _____.

16. How do you check crankshaft straightness?

17. What are undersize main and rod bearings?

18. When installing a rear main seal, the sealing lip should point toward the inside of the engine. True or False?

19. Explain how Plastigage is used to measure rod and main bearing clearance.

20. Which of the following is *not* a recommended practice when installing a piston and rod assembly?
    (A) Cover the rod bolts with rubber hose or plastic tubing.
    (B) Use a ring compressor to squeeze the rings into the grooves.
    (C) Drive the pistons into block with blows from a hammer head.
    (D) Check that the rod cap numbers align and are correct.

⬤ ASE-Type Questions

1. A vehicle is brought into the shop with low oil pressure. Technician A says this problem can be the result of worn crankshaft bearings. Technician B says this problem can be caused by worn piston rings. Who is right?
   (A) A only.
   (B) B only.
   (C) Both A and B.
   (D) Neither A nor B.

2. Technician A says main bearing bore alignment should be measured with an outside micrometer and a feeler gauge. Technician B says main bearing bore alignment should be measured with a straightedge and feeler gauge. Who is right?
   (A) A only.
   (B) B only.
   (C) Both A and B.
   (D) Neither A nor B.

3. An engine's block deck appears to be warped beyond specifications. Technician A says the block should be replaced. Technician B says the block should be sent to a machine shop for surface milling. Who is right?
   (A) A only.
   (B) B only.
   (C) Both A and B.
   (D) Neither A nor B.

4. The cylinder taper in a particular engine cylinder is 0.020″. Technician A says this measurement is within specs. Technician B says this measurement exceeds typical factory specifications. Who is right?
   (A) A only.
   (B) B only.
   (C) Both A and B.
   (D) Neither A nor B.

5. Cylinder taper in an engine cylinder is being measured. Technician A says a dial bore gauge should be used to measure cylinder taper. Technician B says an outside micrometer should be used to measure cylinder taper. Who is right?
   (A) A only.
   (B) B only.
   (C) Both A and B.
   (D) Neither A nor B.

6. Cylinder taper in a particular engine block exceeds specifications. Technician A says a brush hone will repair the cylinders. Technician B says the cylinders must be bored to repair the block. Who is right?
   (A) A only.
   (B) B only.
   (C) Both A and B.
   (D) Neither A nor B.

7. Technician A says a "brush" hone should be used to true an engine cylinder when wear is within specs. Technician B says a "brush" hone should be used when the cylinder is in good condition and requires very little honing. Who is right?
   (A) A only.
   (B) B only.
   (C) Both A and B.
   (D) Neither A nor B.

8. Technician A says when honing engine block cylinders, you should use a high-speed electric drill. Technician B says when honing engine block cylinders, you should use a low-speed electric drill. Who is right?
   (A) A only.
   (B) B only.
   (C) Both A and B.
   (D) Neither A nor B.

9. Vertical scratches remain in an engine's cylinders after honing. Technician A says cylinder sleeving may be needed to repair the block's cylinders. Technician B says cylinder boring may be needed to repair the cylinders. Who is right?
   (A) A only.
   (B) B only.
   (C) Both A and B.
   (D) Neither A nor B.

10. Technician A says a cylinder block is normally bored in increments of 0.010″ until it is perfectly smooth and straight. Technician B says a cylinder block is normally bored in increments of 0.040″ until the desired diameter is achieved. Who is right?
    (A) A only.
    (B) B only.
    (C) Both A and B.
    (D) Neither A nor B.

11. Technician A says the "overbore limit" for an engine's cylinder block is the same for all automotive engines. Technician B says the "overbore limit" for a cylinder block will vary with engine design. Who is right?
    (A) A only.
    (B) B only.
    (C) Both A and B.
    (D) Neither A nor B.

12. A customer wants to know the advantages of having worn engine cylinders sleeved. Technician A says "cylinder sleeving" can restore a bad cylinder to its original diameter. Technician B says "cylinder sleeving" allows the engine's original size piston to be placed back into the sleeved cylinder. Who is right?
    (A) A only.
    (B) B only.
    (C) Both A and B.
    (D) Neither A nor B.

13. Piston-to-cylinder clearance in a particular engine cylinder is 0.030″. Technician A says this measurement is within specs. Technician B says this measurement exceeds factory specs. Who is right?
    (A) A only.
    (B) B only.
    (C) Both A and B.
    (D) Neither A nor B.

14. A free-floating piston pin must be removed. Technician A says a press is normally used to remove this type of piston pin. Technician B says snap ring pliers are normally used to remove this type of piston pin. Who is right?
    (A) A only.
    (B) B only.
    (C) Both A and B.
    (D) Neither A nor B.

15. Technician A says crankshaft end play should be measured with Plastigage. Technician B says crankshaft end play should be measured with a dial indicator. Who is right?
    (A) A only.
    (B) B only.
    (C) Both A and B.
    (D) Neither A nor B.

Activities—Chapter 50

1. Use a cylinder bore gauge to measure wear in a cylinder, making and recording at least five measurements. Then, use a computer-aided drafting program (or manual drafting) to draw a precise, full-size cross section of the cylinder. Note the bore dimensions on your drawing.

2. If a video camera is available, arrange to visit a machine shop and videotape the process of boring cylinders. Play the tape for the class and explain what they see happening.

Chapter 51

Engine Top End Service

After studying this chapter, you will be able to:

- Check for cylinder head damage, valve guide wear, and other engine top end problems.
- Describe how to correct worn valve guides, warped cylinder heads, damaged valve seats, and other troubles.
- Grind valve seats and valves.
- Remove and install diesel engine precombustion chambers.
- Test and shim valve springs.
- Assemble a cylinder head.
- Inspect, test, and service valve lifters, push rods, and rocker assemblies.
- Reassemble the top end of an engine.
- Adjust engine valves.
- Describe safety practices that must be followed while performing engine top end service.
- Correctly answer ASE certification questions on engine top end service.

Engine top end service, commonly referred to as a *valve job,* typically involves servicing the cylinder head and valve train. A specialized engine technician must be capable of quickly and accurately servicing any of these parts. This chapter discusses engine top end service and repair. **Figure 51-1** shows the basic components involved and lists typical part failures.

Cylinder Head Service

Cylinder head service is very critical to engine performance and service life. The cylinder head, valves, and head gasket work together to contain the tremendous heat and pressure of combustion. If the technician makes the slightest mistake when working on the head or the valve train, the repair can fail in a very short period of time.

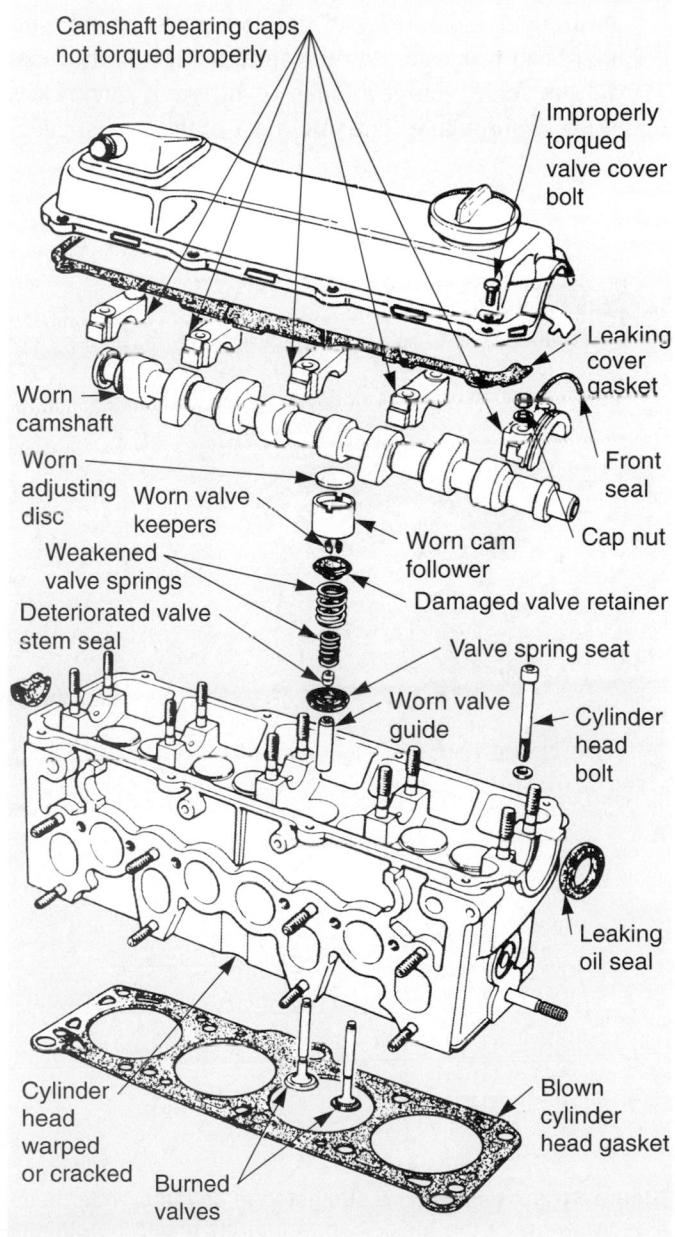

Figure 51-1. These are typical parts of a modern engine's top end assembly. (Chrysler)

Inspecting a Cylinder Head

A *cylinder head stand* is used to hold the head off the workbench surface. After cleaning the head, mount it on a stand. Then, inspect the head closely for problems. Look for cracks, burning, and erosion between combustion chambers. Also, check the valve guides and seats for wear or damage.

Measuring Cylinder Head Warpage

A *warped cylinder head* has a bent or curved deck surface. Head warpage is usually caused by engine overheating. This is a common and serious problem with today's aluminum heads.

A straightedge and feeler gauge are used to measure cylinder head warpage. Lay the straightedge on the head. Try to slip feeler gauge blades of different thicknesses under the straightedge. The thickness of the largest gauge that fits between the straightedge and the head equals head warpage, **Figure 51-2A.**

Check warpage in different positions across the head surface, **Figure 51-2B.** The most common place warpage shows up is between the two center combustion chambers.

A straightedge and a feeler gauge can also be used to check the cam bore in an OHC engine for misalignment.

Milling a Cylinder Head

Cylinder head milling is a machining operation in which a thin layer of metal is removed from the deck surface of the cylinder head. It is done to correct head warpage. See **Figure 51-3.**

Check the manufacturer's specifications to see how much metal can be milled from a head. Milling increases

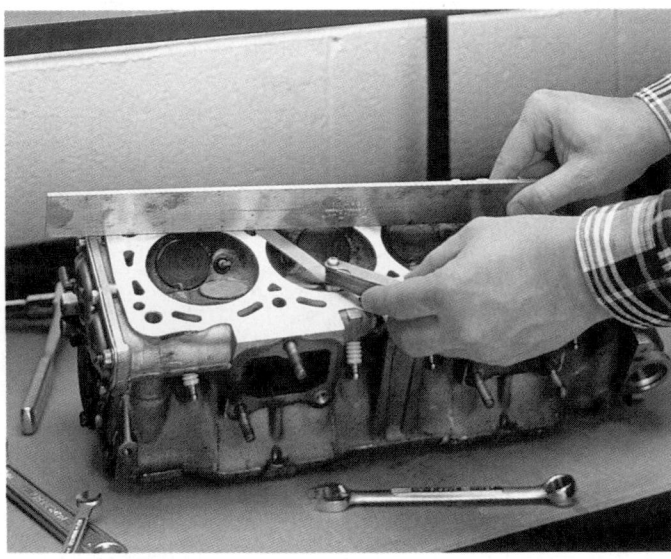

A

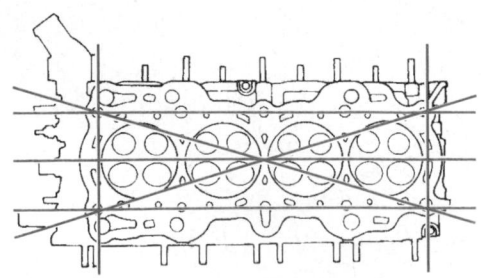

B

Figure 51-2. A—Position a straightedge on the cylinder head. The thickness of the feeler gauge that fits under the straightedge equals head warpage. B—Check across the head at these angles. If warpage exceeds specs, mill or replace the head. (Fel-Pro and Honda)

A

B

Figure 51-3. A—If you find head warpage, a machine shop can mill the deck surface flat using a large machine tool. B—This will correct warpage and allow for good head gasket sealing. (Fel-Pro)

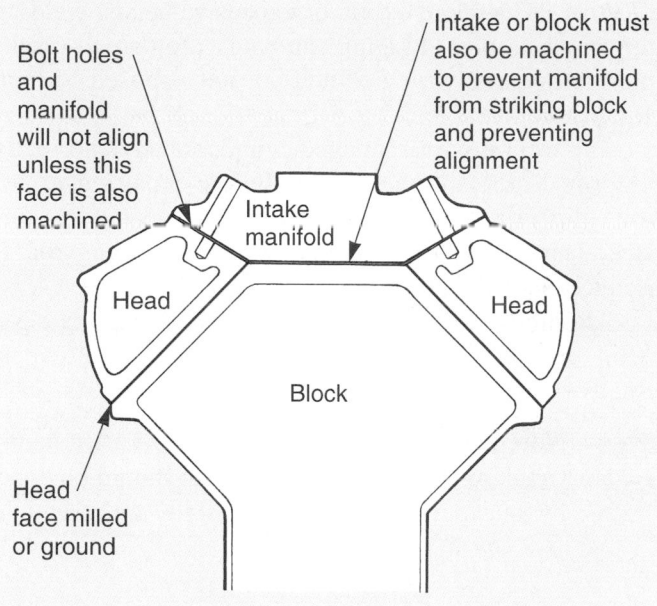

Figure 51-4. If excessive head milling is required on a V-type engine, you may also need to mill intake manifold or the block. Check the service manual for details. (Sealed Power)

the engine's compression ratio and can also affect valve train geometry (relationship or angles between parts).

On V-type engines, the intake manifold will require milling if an excess amount of metal is milled from the head. This is illustrated in **Figure 51-4.**

With a diesel engine, cylinder head milling is very critical. A diesel engine compression ratio is so high that any milling can reduce clearance volumes and increase compression pressures beyond acceptable limits.

Some diesel cylinder heads are case hardened and cannot be milled. They must be replaced when warped. Check with the manufacturer when in doubt.

Crack Detection

Always check the cylinder head for cracks after severe engine overheating or physical damage. *Magnafluxing* is commonly used to find cracks in cast iron parts (cylinder heads, blocks, manifolds). It is a simple process that involves the use of a large magnet and a metal powder to highlight cracks. Some automotive shops have magnafluxing equipment. In many cases, however, the part is sent to a machine shop for crack detection.

To magnaflux a cast iron head, position the large magnet near the suspected crack. Then, sprinkle the metal powder over the area. If there is a crack, the magnet's magnetic field will make the powder collect in the crack, making it visible. See **Figure 51-5.**

Dye penetrant is normally used to find cracks in aluminum components (aluminum heads, blocks, and manifolds). It can also be used on cast iron when magnafluxing is not possible (crack area inside hole or pocket).

The special dye penetrant is sprayed on the part. Then, a chemical developer is sprayed over the penetrant. The powder-like developer will turn the penetrant inside a crack *red*. This makes the crack show up.

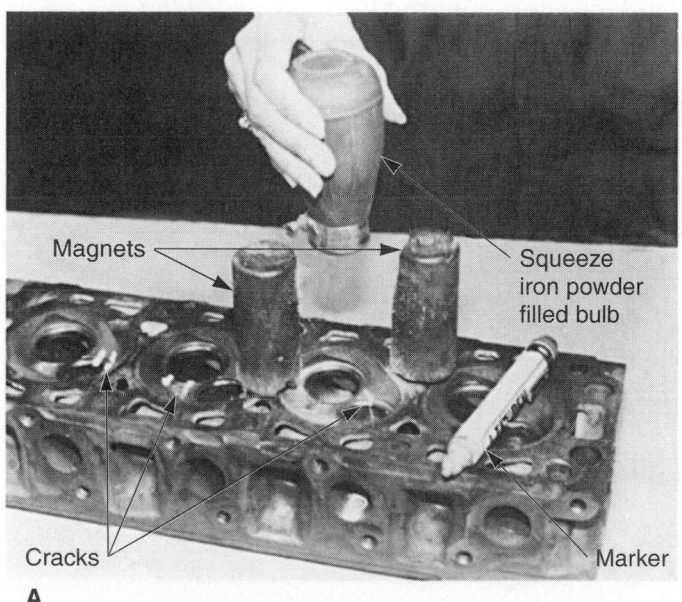

A

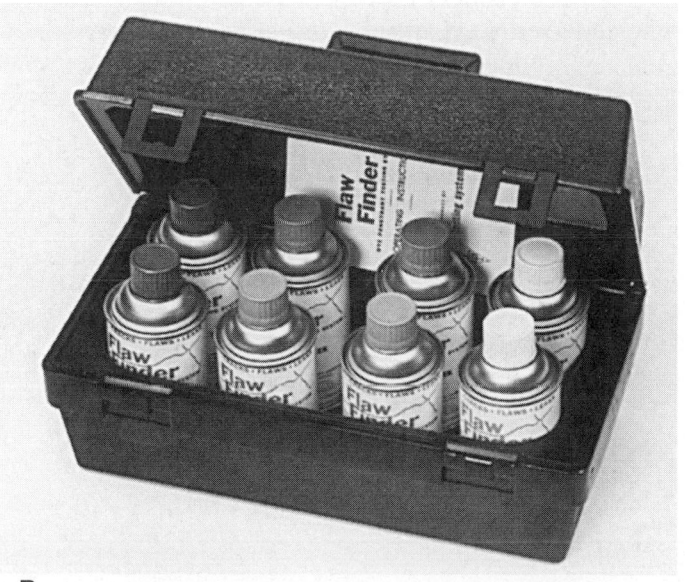

B

Figure 51-5. A—Magnafluxing will make metal powder collect in cracks, making crack detection easy. B—Dye penetrant is used to find cracks in aluminum parts. (Magnaflux Corp.)

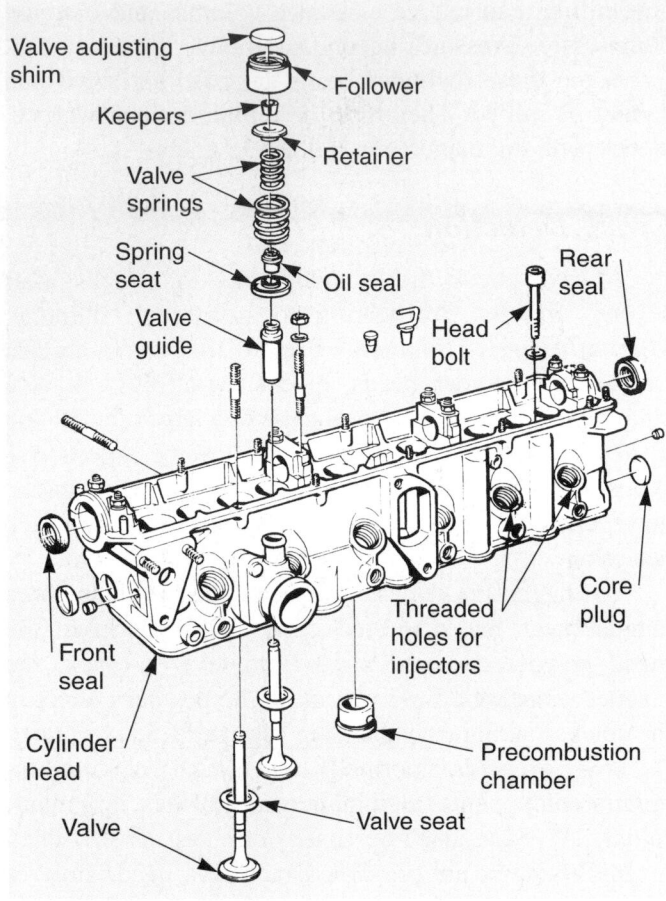

Figure 51-6. Exploded view of a cylinder head assembly. Note relationship of parts. (Volvo)

Repairing a Cracked Cylinder Head

When a cylinder head is cracked, it must be welded, plugged (series of metal plugs used to fix crack), or replaced. With most heads, replacement is more cost effective. With more exotic or expensive heads, welding may be desirable. Plugging can be used on small cracks that are not easily welded and are not exposed to high combustion heat.

The welding of a damaged cylinder head is normally performed at a specialty shop. Before repairing a cast iron cylinder head, the head should be preheated in a furnace. Then, a NI (high nickel content) welding rod is used to repair the crack.

Aluminum cylinder heads can be welded using a gas metal arc welding (GMAW) process. Preheating is usually *not* needed when welding aluminum heads.

Note
Crack repair on heads and blocks should be done by an expert. Special welding skills are essential.

Diesel Precombustion Chamber Service

Automotive diesel engines use precombustion chambers. They are small chambers pressed into the cylinder head. The tips of the diesel injectors and glow plugs extend into the precombustion chambers. Look at **Figure 51-6.**

After prolonged use, the precombustion chambers may require removal for cleaning or replacement because of damage.

Precombustion Chamber Removal and Installation

A brass drift and a hammer are commonly used to remove a precombustion chamber from the cylinder head. The drift, **Figure 51-7A,** is inserted through a hole in the head. Light blows with the hammer will drive the chamber out of the head.

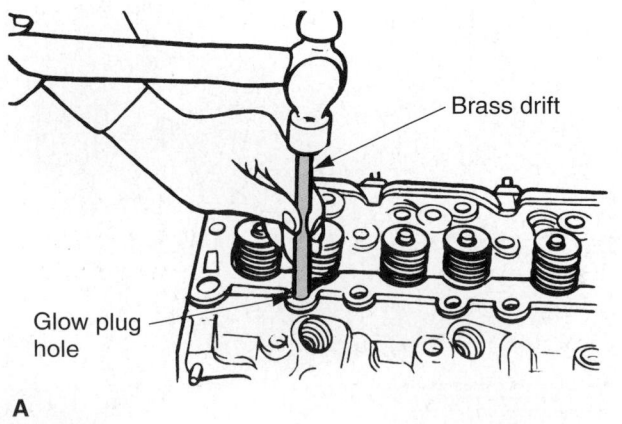

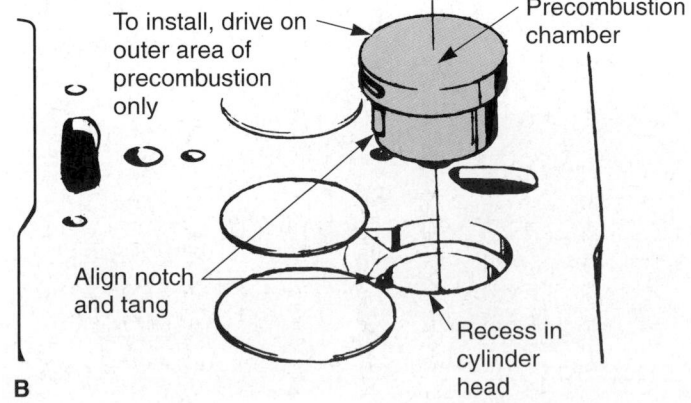

Figure 51-7. A—If precombustion chambers are damaged or filled with carbon, you may need to drive them out with a brass drift. Do not use a steel drift. B—When installing a diesel precombustion chamber, use a special driver, if available. Do not hammer on the inner portion of the chamber. (Ford and Oldsmobile)

When installing a precombustion chamber, be careful not to damage the chamber. Use a special driver or brass hammer to tap the unit back into the head, **Figure 51-7B.** Hammer only on the outer edge of the precombustion chamber. Careless hammering may dent the center area of the chamber.

Make sure the precombustion chamber is perfectly *flush* with the deck surface of the cylinder head. If it is not, head gasket leakage can result. See **Figure 51-8.**

Valve Guide Service

Valve guide wear is a common problem; it allows the valve to move sideways in its guide during operation. This can cause oil consumption (oil leaks past valve seal and through guide), burned valves (poor seat-to-valve face seal), or valve breakage. Refer to **Figure 51-9.**

Measuring Valve Guide and Stem Wear

To check for valve guide wear, slide the valve into its guide. Pull it open about 1/2″ (12.7 mm). Then, try to wiggle the valve sideways in its guide. If the valve moves sideways in any direction, the guide or the stem is worn.

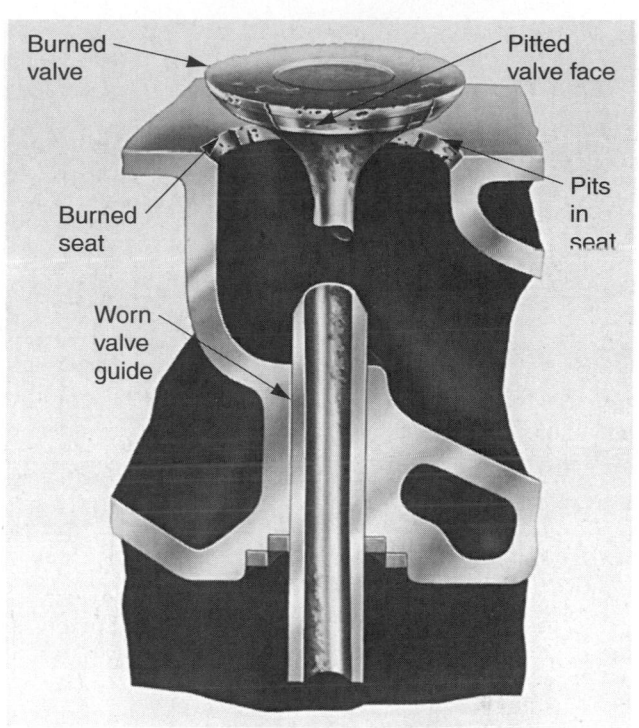

Figure 51-9. A valve job is mainly needed to recondition the valve faces and valve seats. However, other repairs may also be needed, such as valve guide replacement. (Sioux Tools)

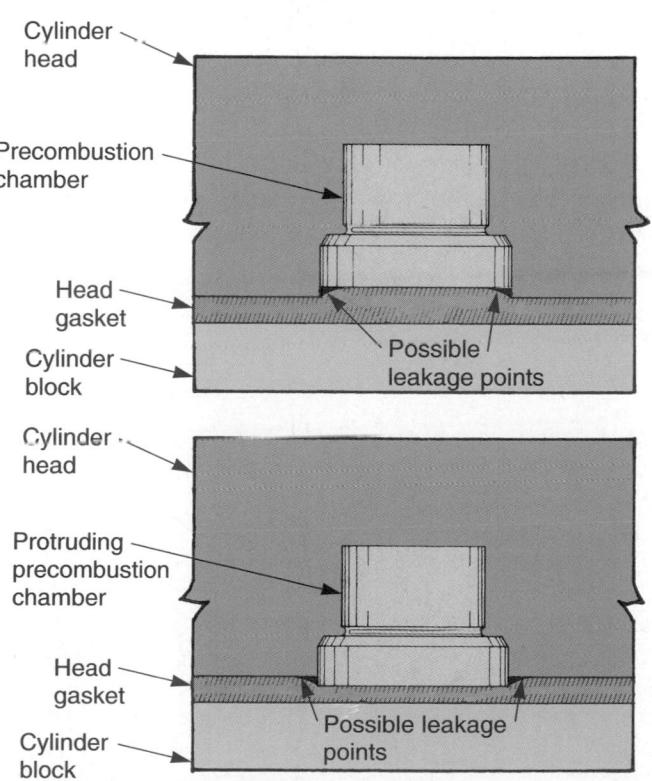

Figure 51-8. After installation, the precombustion chamber must be perfectly flush with the cylinder head gasket surface. If not, the head gasket could leak. (Fel-Pro)

Figure 51-10 shows how a small hole gauge and an outside micrometer are used to measure valve guide and valve stem wear.

Figure 51-11 pictures how a dial indicator is used to measure valve stem clearance. If clearance is not within specs, part replacement or repair is needed.

Repairing Valve Guide Wear

There are three common methods used to repair worn valve guides. These include:

- Knurling the valve guide—a machine shop tool is used to press indentations in the guide to reduce its inside diameter, **Figure 51-12.**

- Reaming the valve guide—the guide is reamed to a larger diameter and new valves with oversize stems are installed, **Figure 51-13.**

- Installing a valve guide insert—the old guide is pressed or machined out and a new guide is pressed into the head, **Figure 51-14.**

Valve Grinding

Valve grinding is done by machining a fresh, smooth surface on the valve faces and the valve stem tips. Valve faces can burn, pit, and wear as the valves open and close

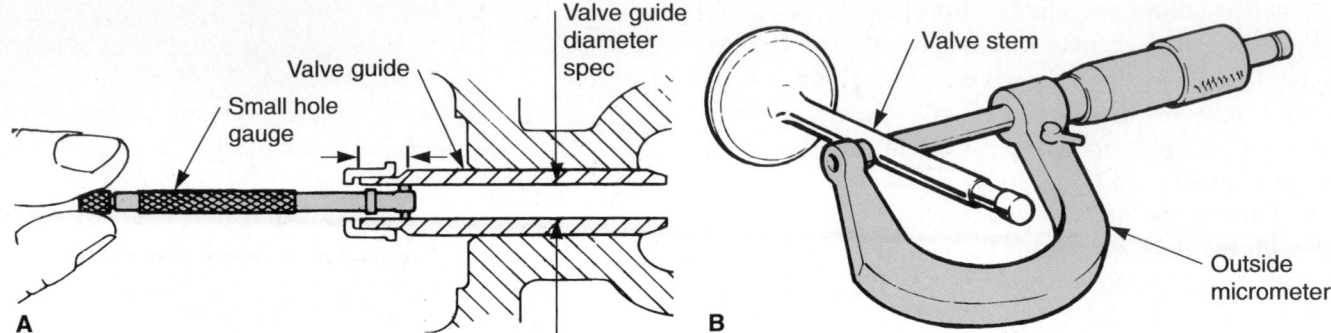

Figure 51-10. A—Hole gauge can be used to measure the inside diameter of a valve guide. B—Outside mike being used to measure the diameter of a valve stem. If not within specs, repairs to guide or replacement of guide and valves are required. (Chrysler and Honda)

during engine operation. Valve stem tips wear because of friction from the rocker arms or followers.

Before grinding, inspect each valve face for burning and each stem tip for wear. Replace valves that are badly

Figure 51-11. A dial indicator can also be used to measure valve guide wear. Mount the indicator stem against the side of the valve head. Wiggle the valve sideways and read the indicator. Check in different positions and compare readings to specs. (Honda)

burned or worn. Grind a new valve along with the old, used valves.

Warning
Wear a face shield when grinding valves. The grinding stone could shatter, throwing debris into your face.

Valve Grind Machine

A *valve grind machine* is used to resurface valve faces and stems. Although there are some variations in design, most valve grind machines are similar. They use a grinding stone to remove a thin layer of metal from the valve face and the valve stem tip. See **Figure 51-15.**

Before grinding a valve, *dress* the stone by using a diamond-tipped cutting attachment to true the grinding surface. The cutting tool is generally provided with the valve grind machine. Follow equipment manufacturer's instructions.

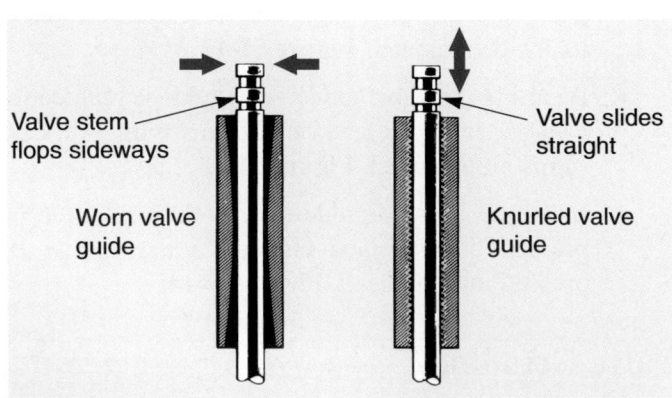

Figure 51-12. Knurling can be used to decrease the inside diameter of a valve guide, restoring proper stem-to-guide clearance. (TRW)

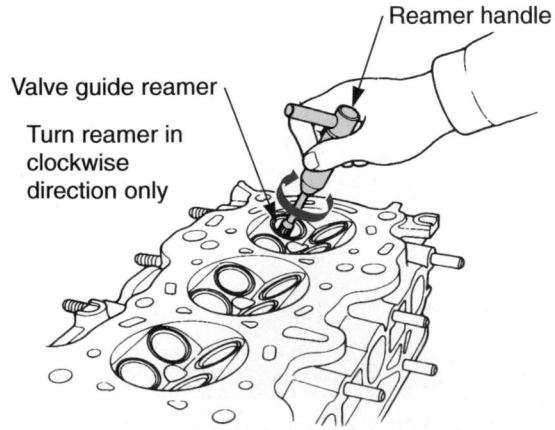

Figure 51-13. A reamer can be used to enlarge valve guide diameters. Then, new valves with oversize stems can be installed. (Honda)

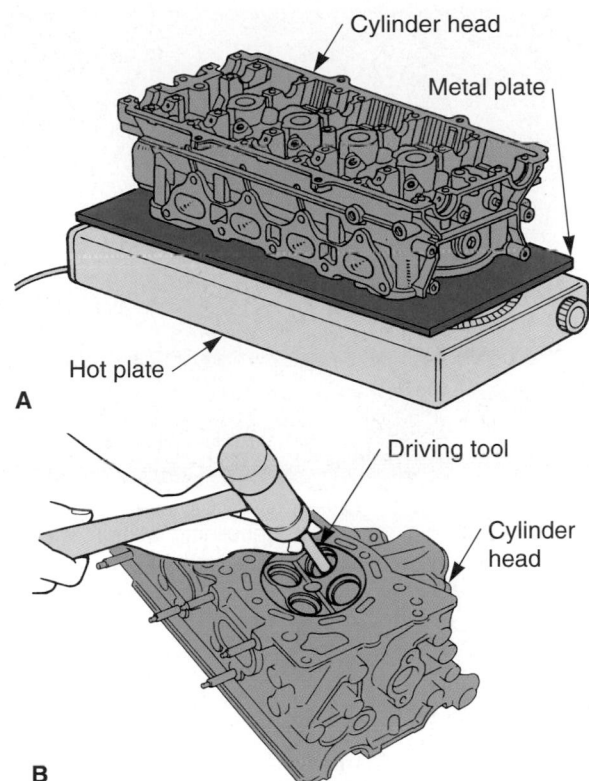

Figure 51-14. Note the major steps for installing a valve guide insert. A—Heating the head on a hot plate will expand the head and aid guide removal. The metal plate between the hot plate and the head helps to distribute heat evenly. B—Use a special driving tool and a hammer to drive the guide out in the right direction, usually through the port. Drive the new guide in the other way. (Honda)

> **Warning**
> Be very careful when using a diamond cutting tool to dress a stone. Wear eye protection and feed the tool into the stone *slowly*. If fed in too fast, tool or stone breakage may result.

The ***chuck angle*** is generally set by loosening a locknut and rotating the grinding machine's chuck assembly to the desired cutting angle. A degree scale is provided so that the angle can be precisely adjusted.

The chuck angle is generally set to produce an ***interference angle*** (a 1° difference between valve face angle and valve seat angle). If, for example, the valve seat angle is 45°, the chuck is set to grind the valve face to 44°. The interference angle provides a thin line of contact between the valve face and the valve seat, reducing the valve's break-in time, **Figure 51-16.**

Chuck the valve in the valve grind machine by inserting the valve stem into the chuck. Make sure the stem is inserted so that the chuck grasps the machined surface nearest the valve head. The chuck must *not* clamp onto an unmachined surface or runout will result. Look at **Figure 51-17.**

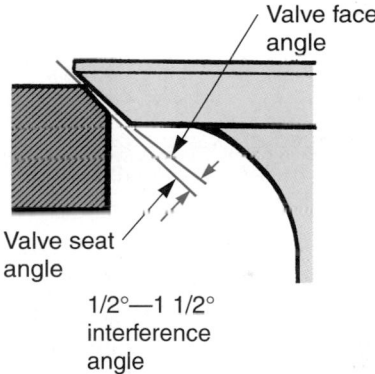

Figure 51-16. Valve grind chuck is commonly set for a one degree interference angle. The valve is ground one degree less than the valve seat. This aids valve seating and sealing. (Sioux Tools)

Figure 51-15. Study the various parts of this valve grind machine. (Sioux Tools)

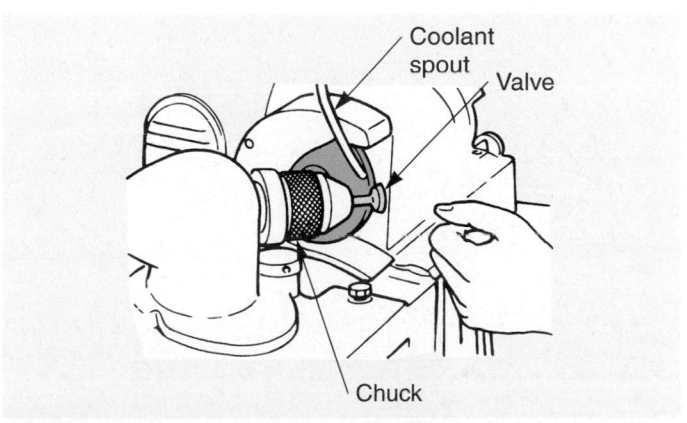

Figure 51-17. Direct coolant on the valve head while grinding. Move the valve into the stone slowly to avoid valve or stone damage. (Subaru)

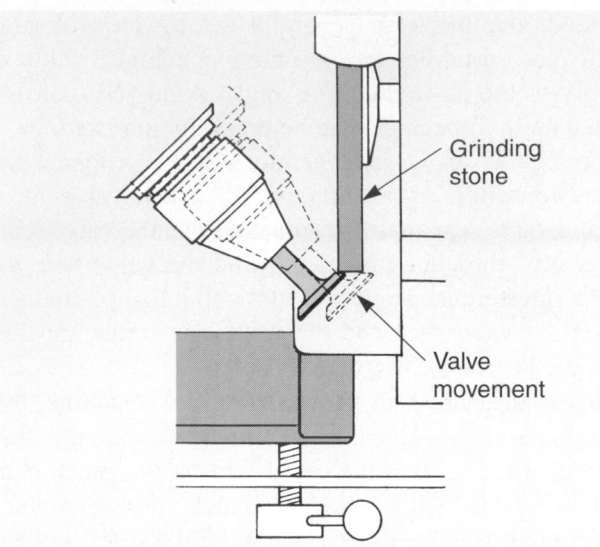

Figure 51-18. Use the lever to move the valve back and forth on the stone, but do not let the valve lose contact with the stone. (TRW)

Grinding the Valve Face

Turn on the valve grind machine and the cooling fluid. *Gradually* feed the valve face into the stone. At the same time, slowly move the valve back and forth over the stone. Use the full face of the stone, but do *not* let the valve face move out of contact with the stone while grinding. See **Figures 51-17** and **51-18.**

Grind the valve only long enough to "clean up" its face. When the full face looks shiny, with no darkened pits, shut off the machine and inspect the face. Look carefully for pits or grooves.

Figure 51-19. Only cut or grind a valve face enough to remove dark pits. If the valve face is shiny all the way around, stop cutting. (Neway Manufacturing Co.)

Figure 51-20. A cylinder head rebuilding station will speed work. (Sunnen)

 Tech Tip
Some valve refacing machines are equipped with a carbide tip instead of a grinding wheel. See **Figure 51-19.**

Grinding, by removing metal from the valve face, will increase valve stem height (distance the valve stem extends above the surface of the head). This affects spring tension and valve train geometry. Grind the face of each valve as little as possible.

If the valve head wobbles as it turns on the valve grind machine, the valve is either bent or chucked improperly. Shut off the machine and find the cause.

Figure 51-20 shows a large cylinder head rebuild station. This type machine is found in machine shops and a few larger garages.

A *sharp valve margin* indicates excess valve face removal and requires valve replacement. Manufacturers give a specification for minimum valve margin thickness. Refer to **Figure 51-21.** If the margin is too thin, it will not dissipate heat fast enough. The head of the valve can actually begin to melt, burn, and blow out the exhaust port.

If not noticed during initial inspection, a *burned valve* will show up during grinding operations. Excess grinding will be needed to remove the deep pits or grooves found on a burned valve. Replace the valve if it is burned.

Repeat the grinding and inspecting operation on the other valves. Return each ground valve to its place in an organizing tray. Used valves should be returned to the same valve guide in the cylinder head. The stems may have been select-fit at the factory.

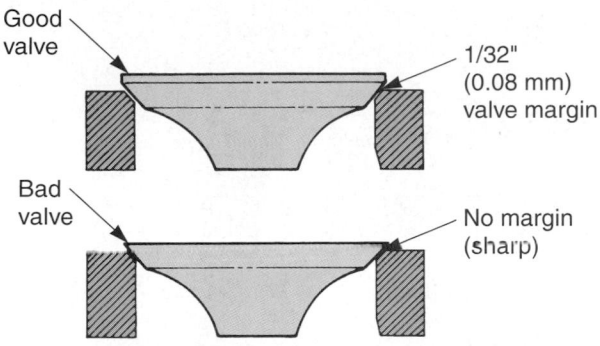

Figure 51-21. The valve must have a margin to be used in an engine. Too much grinding can remove the margin and sharpen the valve head. Without a margin, the valve could overheat and burn.

Grinding the Valve Stem Tip

A second stone on the valve grind machine is normally provided for truing the valve stem tips. This is pictured in **Figure 51-22.** Note how the valve is chucked in the machine.

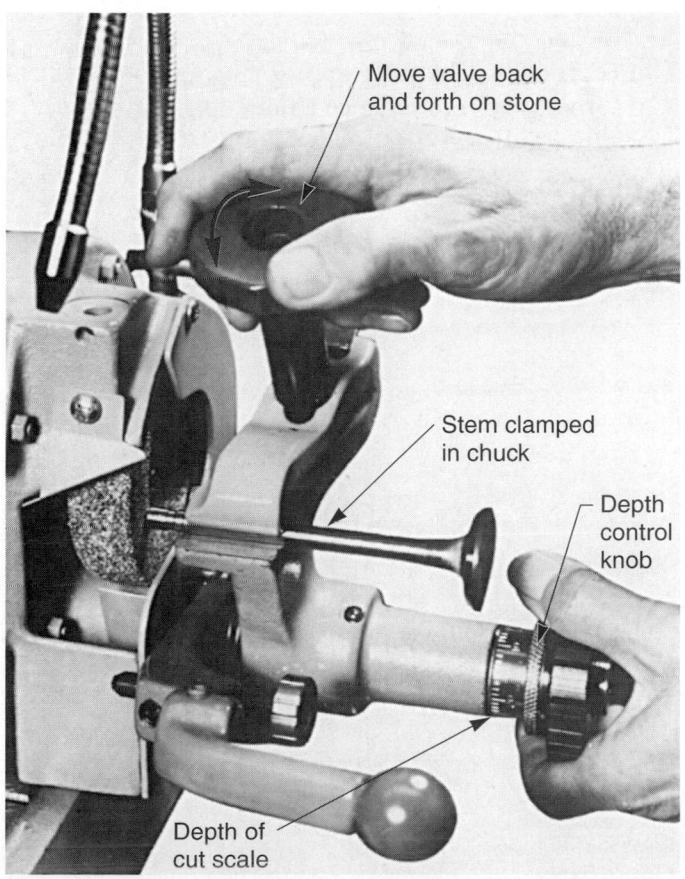

Figure 51-22. Grind the valve stem as little as possible. Wear eye protection. (Sioux Tools)

Grind as little off the stem tip as possible. Many stems are hardened. Too much grinding will cut through the hardened layer and result in rapid wear when the valve is returned to service.

An indicator is provided on the valve grind machine to show the depth of cut for both the valve face and valve stem tip. Generally, the same amount of metal should be removed from the face and the tip. This will help keep valve train geometry correct.

Valve Seat Reconditioning

Valve seat reconditioning involves grinding (using a stone) or cutting (using carbide cutter) the cylinder head valve seats to remove damage or wear. Like the valves, the seats are exposed to tremendous heat, pressure, and wear.

Replacing Valve Seats

Valve seat replacement is needed when a valve seat is cracked, burned, pitted, or recessed (sunk) in the cylinder head. Replacement is only needed when wear or damage is severe. Normally, valve seats can be machined and returned to service.

Technicians generally send the cylinder head to an automotive machine shop for valve seat replacement. Most repair shops do *not* have the specialized tools required for seat removal and installation.

To remove a pressed-in valve seat, split the old seat with a sharp chisel. Then, pry the seat out of the cylinder head. To remove an integral seat, use a seat-cutting tool to machine the seat from the cylinder head. Extreme care must be taken not to damage the head.

To install a valve seat, some machinists shrink the seat by chilling it in dry ice. The seat will expand when returned to room temperature. This helps lock the seat in the cylinder head. Use a driving tool to force the seat into the recess in the head. Seat installation tools vary. Follow the manufacturer's directions.

Staking the valve seat involves placing small dents in the cylinder head next to the seat. The stakes force the head metal over the seat and keep the seat from falling out. *Top cutting* may be needed to machine the top of the seat flush with the surface of the combustion chamber.

Grinding Valve Seats

After installing new valve seats, or when the old seats are in serviceable condition, grind or cut the faces of the seats. The equipment needed to grind valve seats is shown in **Figure 51-23.**

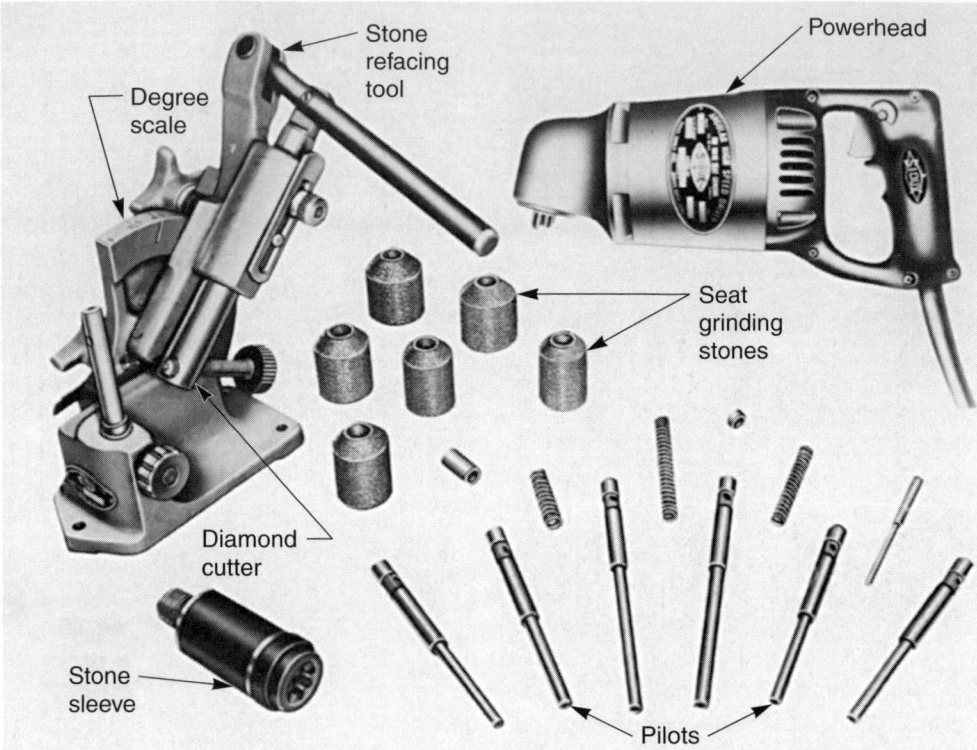

Figure 51-23. One type of valve seat grinding equipment. Note the different tool names. (Sioux Tools)

To grind a valve seat, install the correct size *pilot* (metal shaft that fits into the valve guide and supports the cutting stone or the carbide cutter). The pilot should fit snugly in the valve guide and should not wiggle, **Figure 51-24.**

Select the correct stone for the valve seat. It must be slightly larger in diameter than the seat and must also have the correct face angle.

Dress the stone using the diamond cutter provided with the grinding equipment. Set the cutter to the correct angle (usually 45° or 30°). Slowly feed the diamond cutter into the stone while spinning the stone with the power head (electric drive motor). Cut only enough to clean up and true the stone. **Figure 51-25** shows the basic steps for grinding a valve seat.

A hand-operated valve seat cutter set is shown in **Figure 51-26.** To use hand cutters, follow the same general procedures explained for grinding stones. Fit the correct size pilot securely into the guide, **Figure 51-27A.** Select the correct diameter and angle cutter. Fit the cutter down over the pilot. Then, while applying a very light downward force, turn the cutter on the seat. Make sure you turn the cutter in the direction indicated by the arrow on the tool. Only remove enough material to clean up the seat and make it totally shiny. See

Figure 51-27B. The procedure for using a hand-operated carbide cutter to produce a three-angle valve seat is shown in **Figure 51-28.**

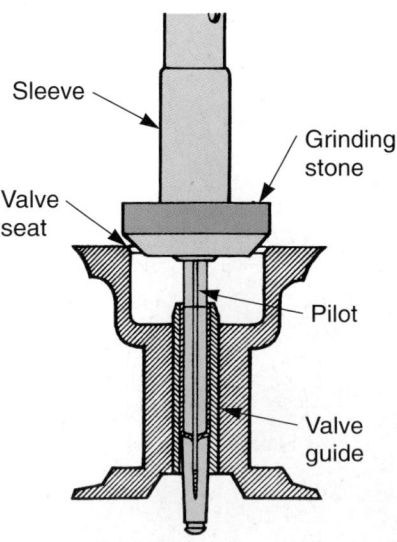

Figure 51-24. This cutaway view shows how a pilot, stone, and sleeve fit on the cylinder head for valve seat grinding.

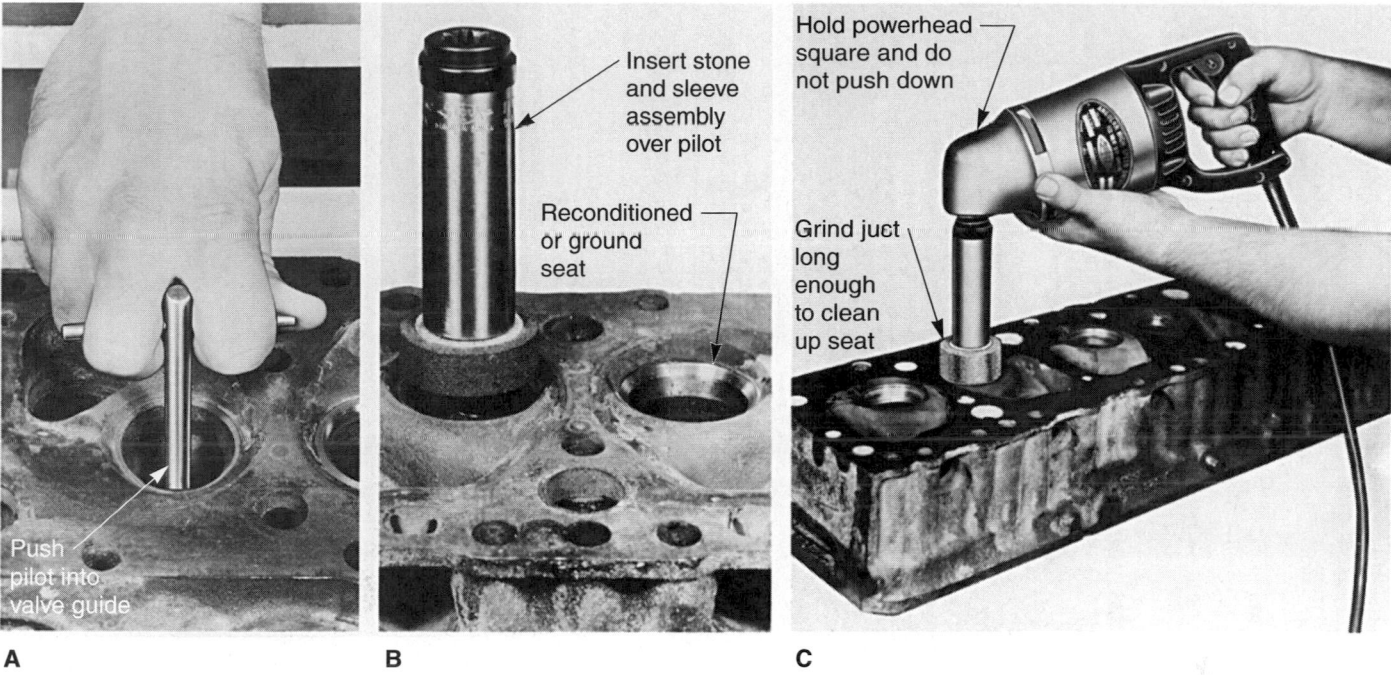

A   **B**   **C**

Figure 51-25. Grinding a valve seat. A—Push the pilot securely into the valve guide. The pilot must be tight in the guide. B—Slip the stone and sleeve assembly over the pilot. Make sure the stone has the right angle and is slightly larger than the seat. C—Use the power head to spin the stone. Support the weight of the power head. Grind only long enough to clean up pits in the seat.

Narrowing Valve Seats

Narrowing a valve seat, or *positioning a valve seat,* is needed to center or locate the valve-to-seat contact area. If the seat does not touch near the center of the valve face, or if the width of the contact area is incorrect, valve service life can be reduced. **Figure 51-29** illustrates seat contact patterns.

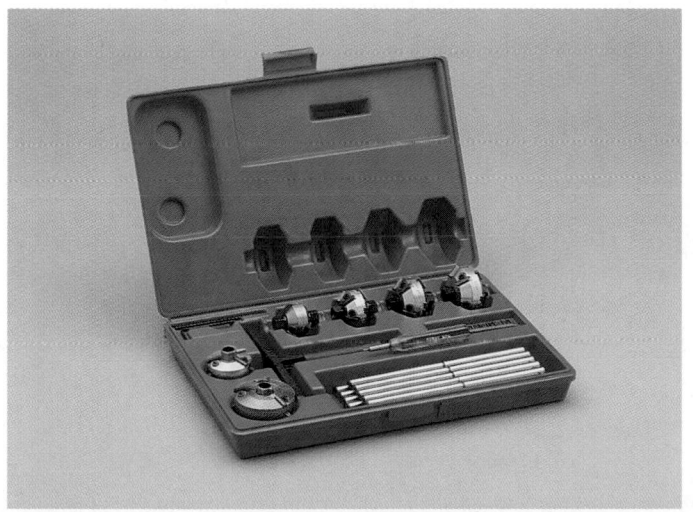

Figure 51-26. This is a hand-operated valve seat cutter set. It uses carbide cutters instead of stones. Stone dressing is not necessary. (Neway Manufacturing Co.)

Typically, an intake valve should have a valve-to-seat contact width of about 1/16″ (1.6 mm). An exhaust valve should have a contact width of approximately 3/32″ (2.4 mm). Check manual specifications for exact values.

When the valve seat does *not* touch the valve face properly (wrong width or location on valve), regrind the seat using different stone or cutter angles, usually 15° and 60°.

To narrow the valve seat contact area and move it inward (closer to the valve stem), grind the seat with a 15° stone or carbide cutter. This will remove metal from around the top of the seat. See **Figure 51-30.**

To narrow the valve seat contact area and move it outward (toward the outer edge of the valve), machine the valve seat with a 60° stone or cutter. This will cut metal away from the inner edge of the seat. See **Figure 51-30.**

If you narrow a seat too much, simply hit it with the original stone or cutter to widen the seat back to specs.

Checking Seat Runout

Valve seat runout occurs when the seat is not centered around the valve guide. Some automakers suggest checking runout after seat grinding. A special dial indicator setup can be used to measure valve seat runout, **Figure 51-31.** If runout exceeds specifications, regrind the seat or check guide installation.

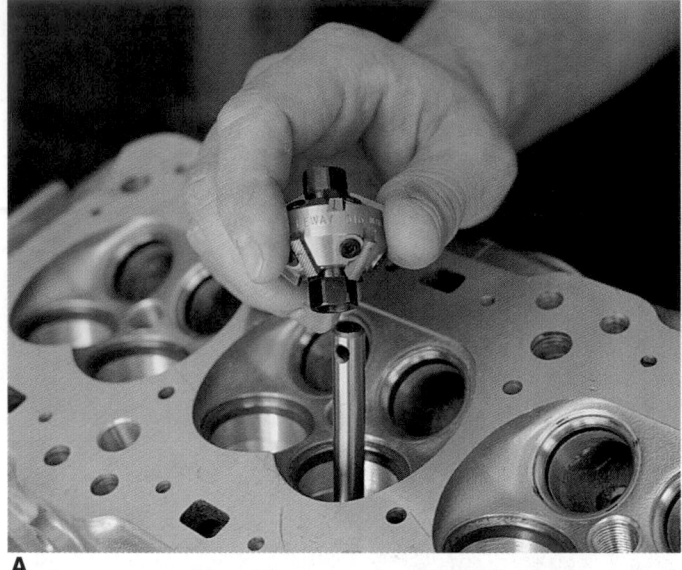
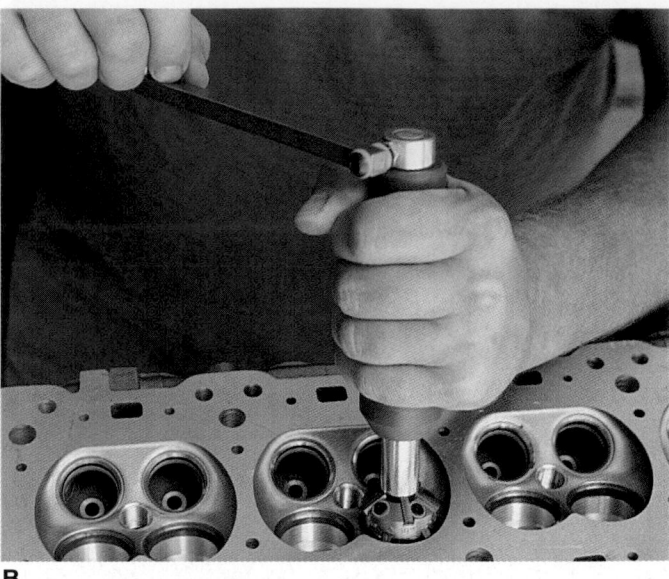

A **B**

Figure 51-27. A—Install the correct size pilot into the valve guide. B—Rotate the cutter until the seat is cleaned up. (Neway Manufacturing Co.)

Lapping Valves

Lapping valves is done to check the location of the valve-to-seat contact area and to smooth the mating surfaces.

Grinding compound (abrasive paste) is dabbed on the valve face. The valve is then installed in the cylinder head and rotated with a *lapping stick* (wooden stick with a rubber plunger that holds valve head).

Rub your hands back and forth on the lapping stick to spin the valve on its seat. This will rub the grinding compound between the valve face and seat.

Remove the valve and check the contact area. A *dull gray stripe* around the seat and the face indicates the valve-to-seat contact area. This will help determine whether the contact area should be narrowed or moved.

Some manufacturers do *not* recommend valve lapping. Refer to a service manual for details.

 Caution
Make sure you clean all the valve grinding compound off the valve and the cylinder head. The compound can cause rapid part wear after ignition startup.

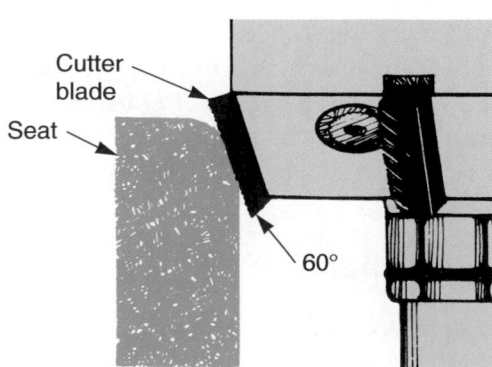

First cut cleans and reconditions area below seat.

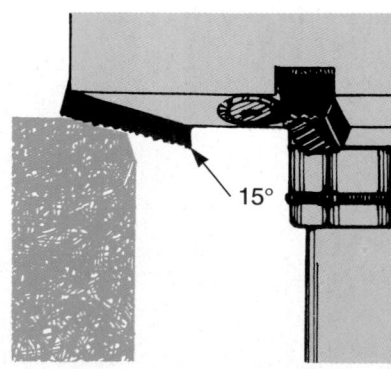

Second cut cleans and reconditions area above seat.

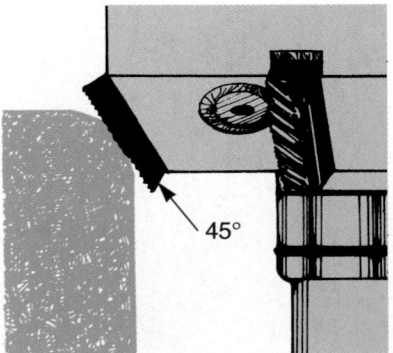

Three or four revolutions of cutter produce a precision seat.

Figure 51-28. The basic steps for using a carbide cutter to produce a three-angle valve job are shown. This procedure will produce a very accurate valve seat. (Neway Manufacturing Co.)

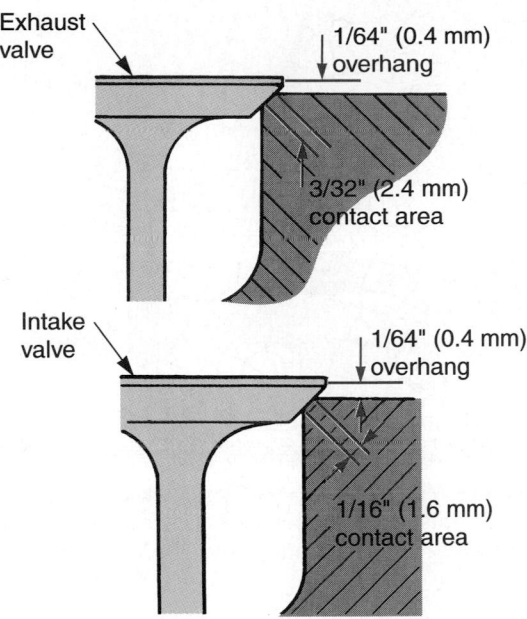

Figure 51-29. The general rules for positioning valve-to-seat contact on both intake and exhaust valves are shown. The exhaust seat must be slightly wider than the intake seat to help dissipate heat. Note typical contact width and overhang values. Refer to a service manual for exact specifications. (TRW)

Using Prussian Blue

Prussian blue is a metal stain that is often used to check the contact point between the valve faces and valve seats. Apply a small amount of Prussian blue on the valve seat or face. Tap the valve down on the seat to mark the Prussian blue and make the contact point visible. See **Figure 51-32.**

Testing Valve Springs

After prolonged use, valves springs tend to weaken, lose tension, or even break. During engine service, always test each valve spring to make sure it is usable.

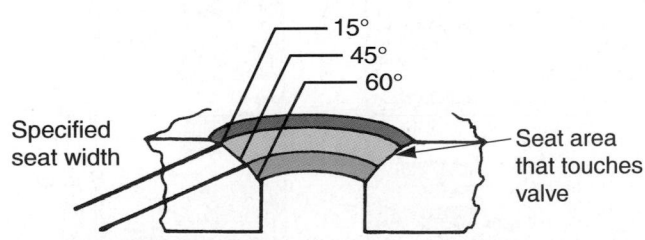

Figure 51-30. Different stone or cutter angles can be used to move or narrow a seat. A 60° cut would narrow the seat and move it up on the valve face. A 15° cut would narrow the seat and move it down on the valve face.

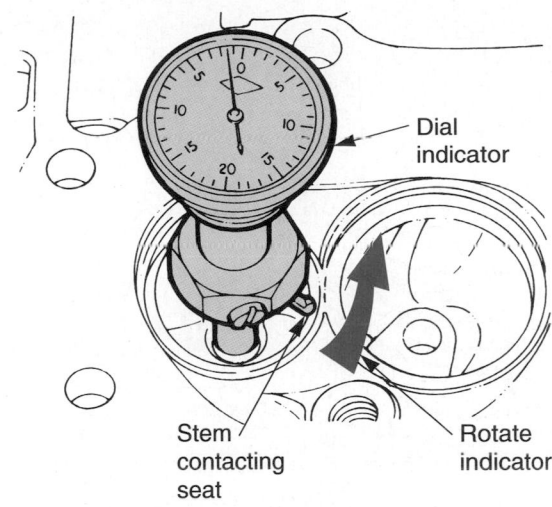

Figure 51-31. Some automakers suggest a check of valve seat runout. The indicator reading equals runout.

Valve spring squareness can be checked with a combination square. Place the spring next to the square on a flat work surface. Rotate the spring while checking for a gap between the side of the spring and the square. Replace the spring if it is not square. Look at **Figure 51-33A.**

Valve spring free height can be measured with the combination square or with a valve spring tester. Simply measure the length of each spring in a normal, uncompressed condition. If too long or too short, replace the spring.

Valve spring tension, or pressure, is measured with a valve spring tester. Compress the spring to the specified height and read the scale. Spring pressure must be within specifications. If spring pressure is too low, the spring has weakened and needs replacement or shimming. See **Figure 51-33B.**

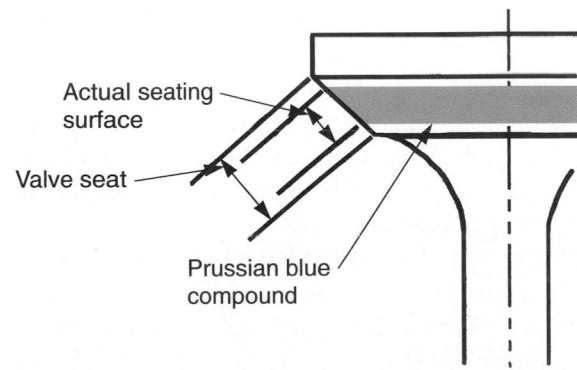

Figure 51-32. Other automakers recommend that you use Prussian blue on the face of the valve to check its contact on the seat. The valve is tapped or rotated on seat. Marks on the Prussian blue will show the contact area. (Oldsmobile)

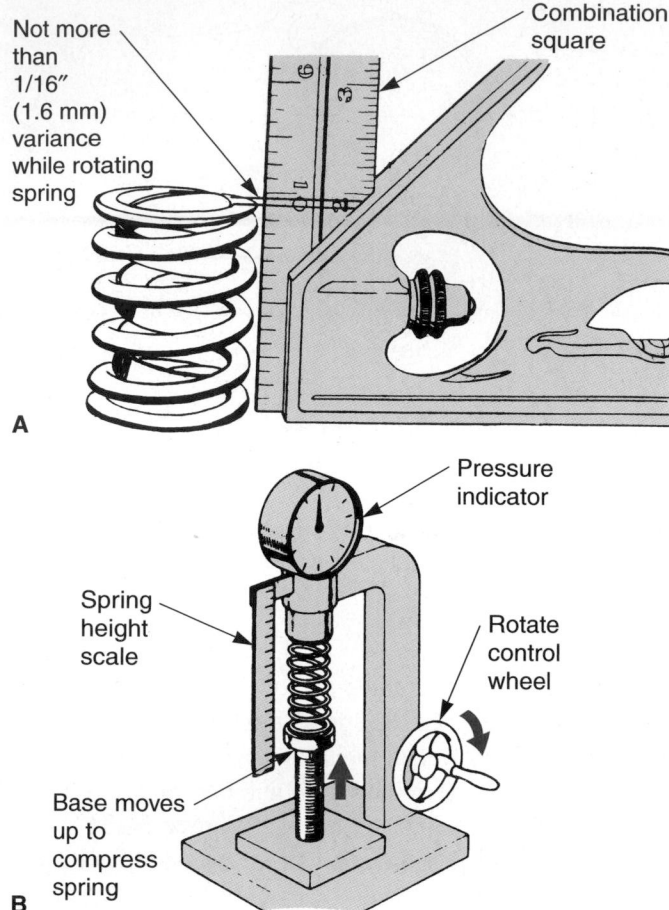

Not more than 1/16″ (1.6 mm) variance while rotating spring

Combination square

A

Pressure indicator

Spring height scale

Rotate control wheel

Base moves up to compress spring

B

Figure 51-33. A—A combination square can be used to check valve spring squareness. Replace springs that are not square. B—A valve spring tester will measure spring pressure, or tension, at a specific spring height or length. Shim or replace weak springs. (Cadillac and Toyota)

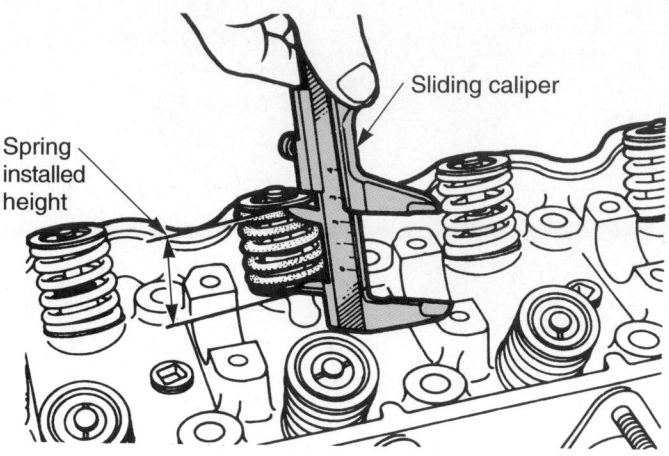

Sliding caliper

Spring installed height

Figure 51-34. Measuring valve spring installed height. Grinding valves and seats tends to increase installed height. Shims are needed to restore correct spring pressure. (Chrysler)

Assembling Cylinder Head

After cleaning, inspecting, measuring, and reconditioning, the cylinder head is ready to be assembled. Place a drop or two of oil on each valve stem and slide the valves into their valve guides.

Installing Valve Seals and Spring Assemblies

To install umbrella-type valve seals, simply slide the seals over the valve stems. Look at **Figures 51-36**

Valve Spring Shimming

Valve spring shimming is used to keep correct tension when the springs are installed on the cylinder heads. When valves and seats are ground, the valve stem height is increased. This increases the valve spring installed height and reduces spring pressure.

Valve spring installed height is the distance from the top of the valve spring to the bottom with the spring installed on the cylinder head. It can be measured with a sliding caliper, **Figure 51-34.**

If valve spring installed height is greater than specs, add shims to reduce installed height and return spring pressure to normal. Place the shim(s) under the valve spring, **Figure 51-35.**

Generally, you should never shim a valve spring more than 0.060″ (1.5 mm). Thicker shimming could cause *spring bind* (spring fully compresses and locks valve train, damaging components).

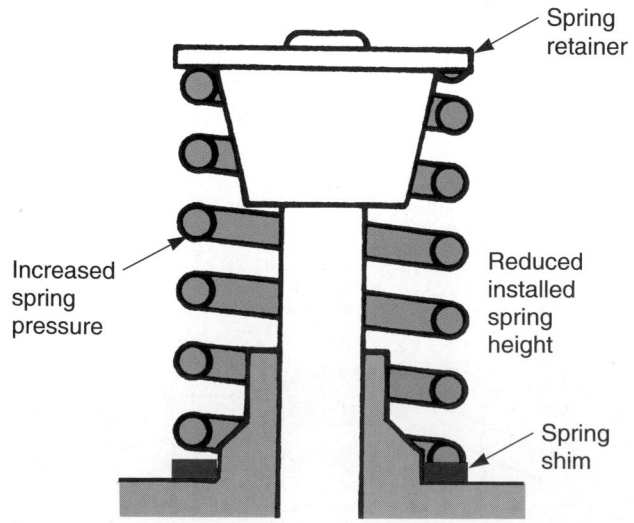

Spring retainer

Increased spring pressure

Reduced installed spring height

Spring shim

Figure 51-35. A valve spring shim is installed under the valve spring. This restores spring pressure, preventing valve float. (TRW)

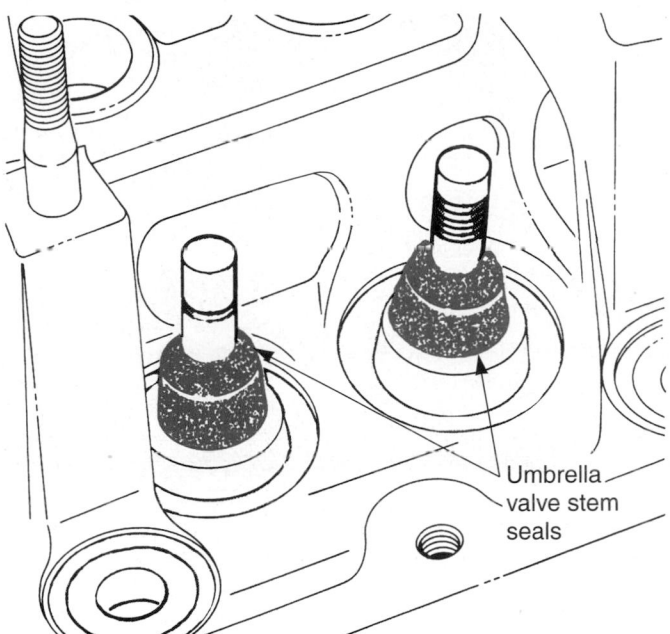

Figure 51-36. To install umbrella valve stem seals, oil the valve stems and slide the seals into place. Then, compress and install the springs. (Chrysler)

Figure 51-38. Use a plastic shield over the valve stem to protect a hard plastic umbrella seal from damage during installation. (Fel-Pro)

and **51-37.** With some hard plastic umbrella seals, you should use a plastic sleeve to protect the seal from damage during installation. See **Figure 51-38.**

Some locking-type seals require a special installation tool to force the seal around the upper end of the valve guide.

 Tech Tip

When installing O-ring type valve seals, compress the valve spring *before* fitting the seal on the valve stem. If you install the seal first, it will be cut, split, or pushed out of its groove when the spring is compressed. Engine oil consumption and smoking will result. See **Figures 51-39** and **51-40.**

After installing the valve seal, place the valve spring over the valve stem. Compress the spring using an appropriate spring compressor. **Figure 51-41** shows how

Figure 51-37. This technician is sliding an umbrella-type valve seal over the valve stem. (Fel-Pro)

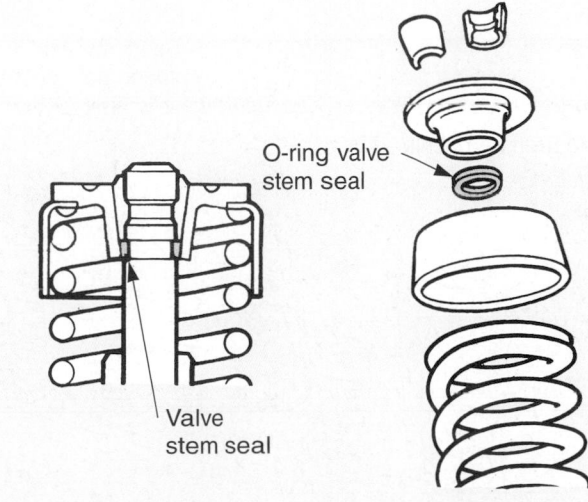

O-ring valve stem seal

Valve stem seal

Figure 51-39. Note! When installing an O-ring type valve seal, compress the spring and retainer first. Then, fit the valve seal into its groove, install the keepers, and release the spring. If seal is installed first, it can be ruined when the spring is compressed. Engine smoking or oil consumption can result. (Buick)

Figure 51-40. Use your fingers to fit a small O-ring seal over the stem and down into its groove. Make sure the seal is in the correct groove, not in the groove for the keepers. (Fel-Pro)

a special valve spring compressor is used to install the springs on an OHC engine. **Figure 51-42** shows an air-powered valve spring compressor. It saves time and effort, especially on heads with four or five valves per cylinder.

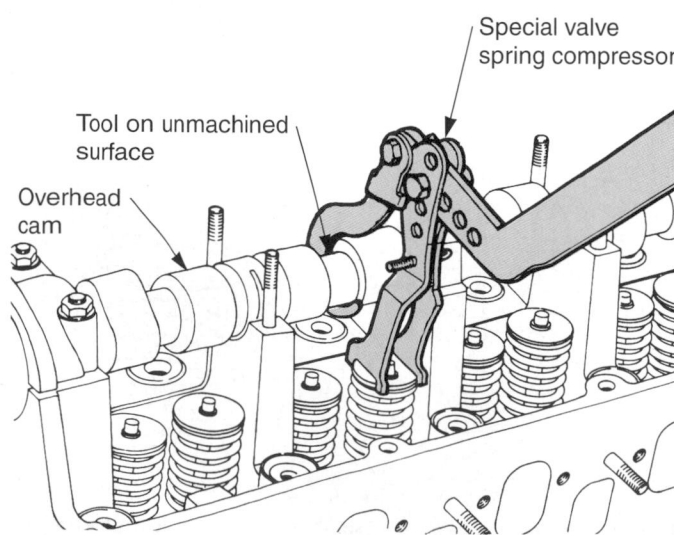

Figure 51-41. A special valve spring compressor may be needed on some OHC engines. Tap the valve stems lightly with a brass hammer to seat the keepers. (Chrysler)

Figure 51-42. This air-powered valve spring compressor is handy for high-production work. (Fel-Pro)

With the spring compressed, fit the retainer and keepers into place on the valve, **Figure 51-43.** Release the compressor and lightly tap on the valve stem with a brass hammer to seat the keepers in their groove.

Checking for Valve Leakage

To check for valve leakage after head reconditioning and assembly, lay the cylinder head on its side. Pour clean cold-soak solvent or water into the intake and exhaust ports. With the fluid in the ports, watch for leakage around the valve heads. Solvent or water drippings from around a valve indicates leakage. Remove any leaking valves from the head and check for problems.

Figure 51-43. With the seal in place, install the keepers into their grooves. Then release the compressor and check that the keepers are fully seated in their grooves. (Fel-Pro)

In-Car Valve Seal Service

In-car valve seal service involves replacing the valve seals with the heads and engine still installed in the vehicle. The procedure for in-vehicle valve seal replacement is as follows:

1. Remove the spark plugs, valve covers, and rocker arms or followers.
2. Install a compression gauge–type air hose into the spark plug hole in one cylinder.
3. Connect shop air pressure to the hose. This will hold the valves in the cylinder up against their seats, **Figure 51-44A.**
4. Use a spring compressor to remove the keepers, retainers, and springs from each valve in the cylinder, **Figure 51-44B.**
5. Replace the valve seal on each valve.
6. Reinstall the springs, retainers, and keepers.
7. Remove the air hose from the cylinder.
8. Repeat this process on the other cylinders.

A

B

Figure 51-44. Valve seals can be replaced with the cylinder heads and engine still in the vehicle. A—Use a hose screwed into the spark plug hole to inject air pressure into the cylinder. This will keep the valves from falling down into the engine. B—Use a small screw-type spring compressor to remove the keepers, retainers, and seal. (Fel-Pro)

Camshaft Service

Camshaft service involves measuring cam lobe and journal wear. It also includes distributor-oil pump gear inspection and cam bearing measurement or replacement.

Measuring Camshaft Wear

Cam lobe wear can be measured with a dial indicator when the camshaft is installed in the engine. Refer to **Figure 51-45.** When the camshaft is out of the engine, an outside micrometer is used, **Figure 51-46A.** If lobe lift or micrometer readings are not within specs, the cam is worn and should be reground or replaced.

Cam journal wear is measured with an outside micrometer, **Figure 51-46B.** A worn cam is usually replaced or reground. Journal wear lowers engine oil pressure.

Camshaft straightness is checked with V-blocks and a dial indicator. This is illustrated in **Figure 51-46C.** If the dial indicator reading exceeds specs, the camshaft is bent and must be replaced.

Camshaft end play is also measured with a dial indicator. Position the indicator to measure front-to-rear shaft movement, **Figure 51-46D.** Pry the camshaft back and forth. If cam end play is not within specs, check any part affecting end play (camshaft, thrust washer, or end plate).

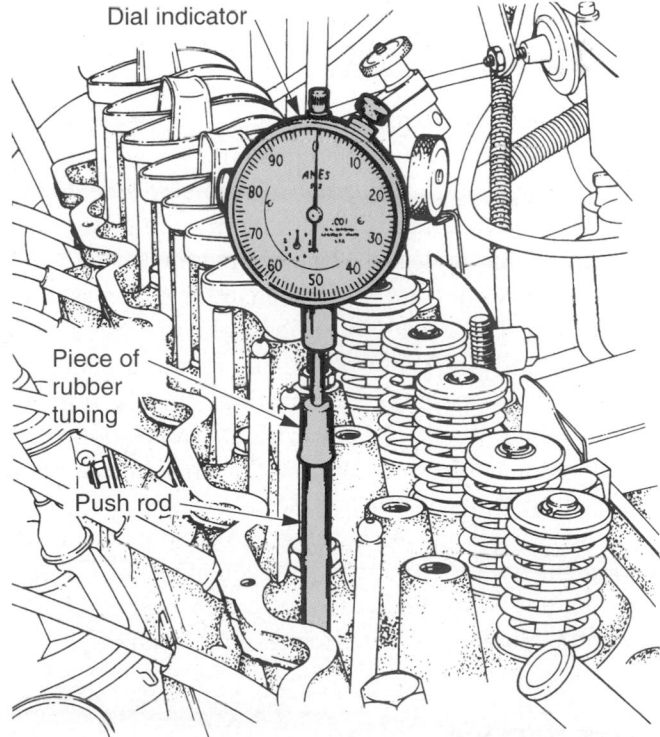

Figure 51-45. Cam lobe lift and wear can be measured before engine tear down. Rotate the engine while reading a dial indicator. If lift is less than specifications, the camshaft is worn.

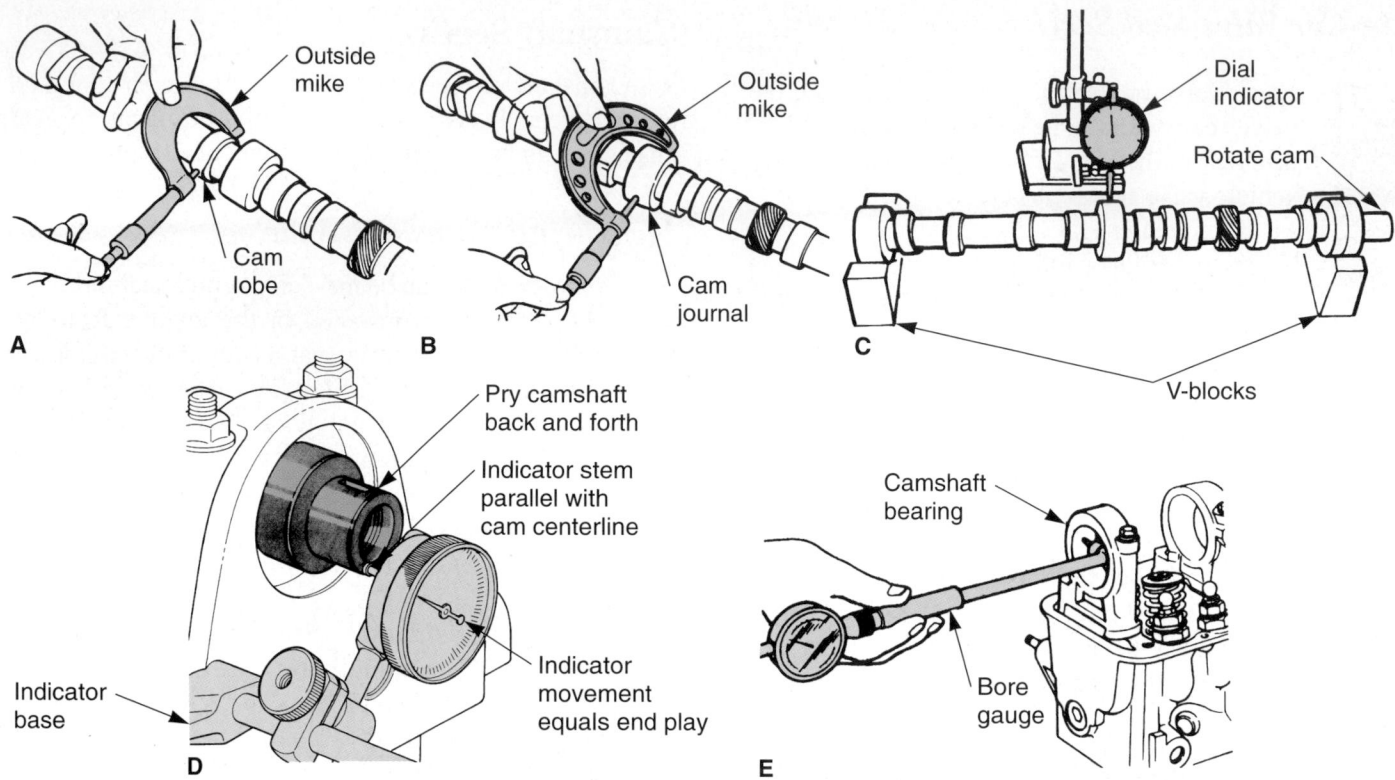

Figure 51-46. Measuring camshaft wear. A—Mike the cam lobe and compare your reading to specs. B—Measure the cam journal diameter and compare measurements to specs. C—Mount the camshaft in V-blocks. Position the dial indicator on the center journal. Turn the cam and read the dial indicator. If the camshaft is not straight, replace it. D—To check camshaft end play, install the cam in the engine. Position the dial indicator as shown. Pry back and forth on the camshaft while reading the indicator. E—A dial bore gauge can be used to check cam bearing or cam bore diameter. (Ford, Chrysler, and Nissan)

Cam Bearing Service

Cam bearing diameter indicates cam bearing wear. Measure bearing diameter with a bore gauge or a telescoping gauge and an outside micrometer. See **Figure 51-46E.** If the bearings are worn, they must be replaced. Most technicians replace the cam bearings during an engine overhaul, since they are critical to engine oil pressure.

The two-piece cam bearings used on many OHC engines simply snap into place, like rod and main bearings. However, one-piece cam bearings must be forced in and out of the block or head with a special tool. Many technicians send the head or block to a machine shop for cam bearing replacement.

When installing cam bearings, do not dent or mar the bearing surfaces. Also, make sure you align the oil holes in the engine with the holes in the cam bearings. Since exact procedures vary, refer to a shop manual for details.

Lifter Service

The contact surface between a lifter and a cam lobe is one of the highest friction and wear points in an engine.

Hydraulic lifters can also wear internally, causing valve train clatter (tapping noise).

Inspecting Lifters

Inspect the bottom of each lifter (surface that contacts cam lobe) for wear. An unworn lifter will have a slight hump, or convex shape, on the bottom. A worn lifter will be flat or concave on the bottom. Replace the lifters if the bottoms are worn. See **Figure 51-47.**

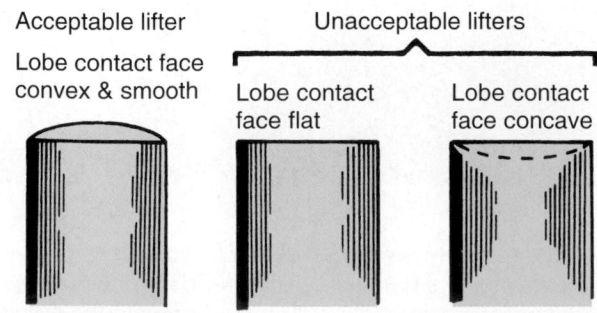

Figure 51-47. The bottoms of the lifters should be checked for wear as shown. (Ford)

Never install used lifters on a new camshaft. Used lifters will cause rapid cam lobe wear and additional lifter wear. Install new lifters whenever a camshaft is replaced.

Testing Lifter Leak-Down Rate

Lifter leak-down rate is measured by timing how long it takes to push the lifter plunger to the bottom of its stroke under controlled conditions. A lifter tester is pictured in **Figure 51-48.**

Generally, fill the tester with a special test fluid. Place the lifter in the tester. Then, follow specific instructions to determine lifter leak-down rate. If leak-down is too fast or too slow, replace or rebuild the lifter.

Rebuilding Hydraulic Lifters

Rebuilding hydraulic lifters typically involves disassembling, cleaning, measuring, and reassembling lifter components. All worn or scored parts must be replaced following manufacturer instructions.

Many shops do *not* rebuild hydraulic lifters. If the lifters are defective, new ones are installed. This can save time and money under most circumstances.

Push Rod Service

Check each push rod tip for wear. The center of a worn push rod tip may be pointed where it fits into the oil hole in the rocker arm. Also, make sure that none of the push rods is bent. Check the push rods for straightness by rolling them on a flat surface. If the push rods are hollow, look through each one to make sure it is clean (not clogged).

Rocker Arm Assembly Service

Inspect the rocker arms for wear, clogged oil holes, and other problems. If wear is indicated inside the rocker bore, measure the bore with a bore gauge or a telescoping gauge and a micrometer. Replace any rocker arm showing wear. See **Figure 51-49A.**

Also inspect the rocker arm shaft for wear. A worn rocker arm shaft will have indentations where the rocker arms swivel on the shaft. Wear will usually be greatest on the bottom of the shaft. Mike the shaft to determine if wear is within specifications, **Figure 51-49B.**

When reassembling a rocker arm shaft, make sure the oil holes are facing in the proper direction. The oil holes normally face down to lubricate the loaded side of the shaft.

With a ball- or stand-type valve train, inspect the ball or stand for wear. If the rocker stand is made of aluminum, check it closely for grooves, which indicate wear. Replace any part worn beyond specifications.

Engine Top End Reassembly

Proper engine assembly procedures are critical to good engine operation. The slightest mistake can result in major engine problems. The next section will outline the most important facts to remember when reassembling the top end of an engine.

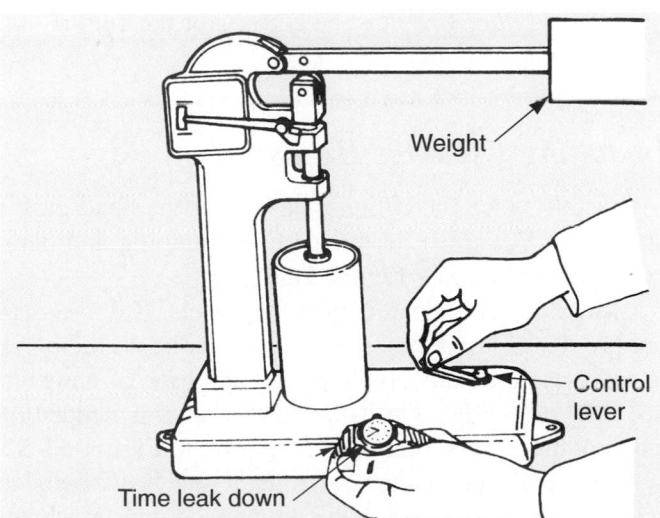

Figure 51-48. Special equipment is used to measure hydraulic lifter leak-down rate. Follow the directions provided with the tester. Replace or rebuild the lifter if it fails leak-down test. (Buick)

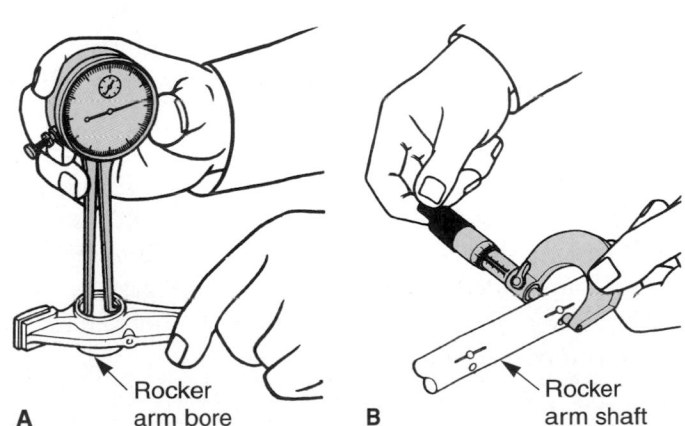

Figure 51-49. A—Using a bore gauge to measure the diameter of the rocker arm bore. B—Checking rocker arm shaft diameter with a micrometer. If measurements are not within specifications, replace the worn components. (Toyota)

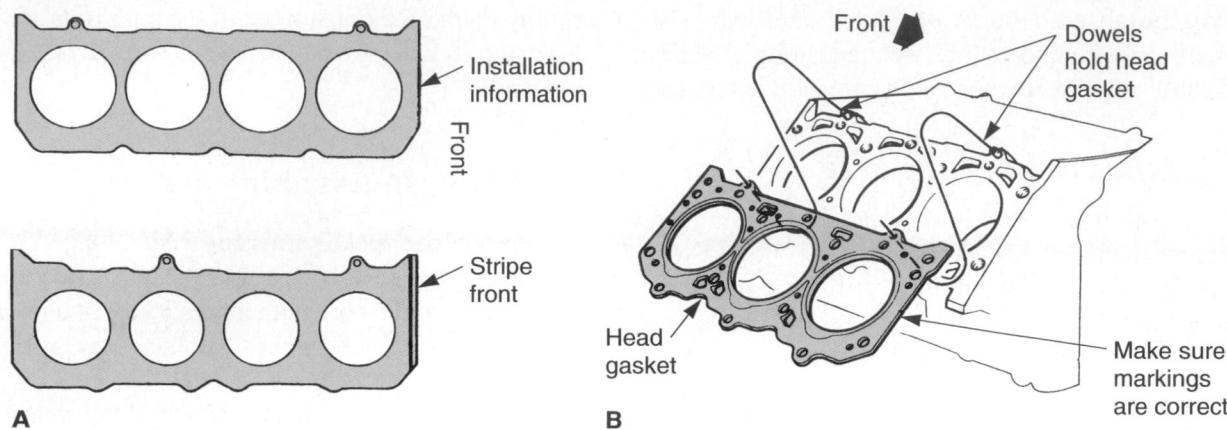

Figure 51-50. A—Cylinder head gaskets are normally marked in some manner to indicate the installation direction. This helps ensure proper coolant flow through block and heads. B—Fit the new head gasket over the dowels. Sealer is not required with many modern head gasket designs. Double check that gasket is installed properly. (Oldsmobile and Buick)

Valve Grind Gasket Set

A *valve grind gasket set* includes all the gaskets and seals required to reassemble an engine's top end. It will typically include the head gasket(s), intake and exhaust manifold gaskets, valve seals, valve cover gasket(s), and other gaskets, depending on engine design.

It is normally cheaper to purchase a full valve grind gasket set than to request individual gaskets. A gasket set commonly has instructions summarizing specific gasket and seal installation methods. Always follow them closely!

Installing Cylinder Head Gaskets

Usually, a head gasket can be installed only one way. If it is installed backwards, coolant and oil passages may be blocked, causing serious problems.

Cylinder head gasket markings are normally provided to show the front or top of the head gasket. The gasket may be marked with the word "*top*" or "*front*," or it may have a line to show installation direction. Metal dowels are often provided to align the head gasket on the block. See **Figure 51-50.**

Most modern, Teflon®-coated, permanent-torque (retorquing is not needed after engine operation) cylinder head gaskets should be installed clean and dry. Sealer is *not* recommended. However, some head gaskets may require retorquing and sealer. When in doubt, refer to manufacturer's instructions.

Diesel Engine Head Gaskets

With diesel engines, head gasket thickness and bore size are very critical. Head gaskets are provided in different thicknesses to allow for cylinder head milling or varying block deck heights. Gasket thickness may be denoted with a color code, a series of notches or holes, or another marking method.

A dial indicator can be used to check TDC piston height in the block. The difference between block deck height and piston head height is measured. This measurement will let you use service manual information to determine the required head gasket thickness.

When a diesel engine block is bored oversize, it also requires a special gasket. You must request an over-bore gasket from the parts supplier. A standard-bore gasket will usually stick out into the cylinder, causing problems.

 **Note**
When buying a diesel engine head gasket, make sure you have the right one. Gasket *thickness* and *bore size* must be correct for the engine being repaired.

Installing Cylinder Heads

Gently place the cylinder head over the head gasket and block. You must do this without bumping and damaging the gasket. See **Figure 51-51.**

Make sure the head is over its dowels. If dowels are *not* provided, stud bolts should be screwed into the block to serve as guide dowels for gasket and head alignment.

Start all the head bolts by hand. Sealer is needed on any bolt that enters a water jacket. Refer to **Figure 51-52.**

Use a torque wrench to tighten the head bolts to specs. Tighten in a factory-recommended crisscross pattern. A service manual will give the exact sequence. One example is given in **Figure 51-53.**

Generally, tighten the bolts starting in the middle of the head. Then, work your way outward.

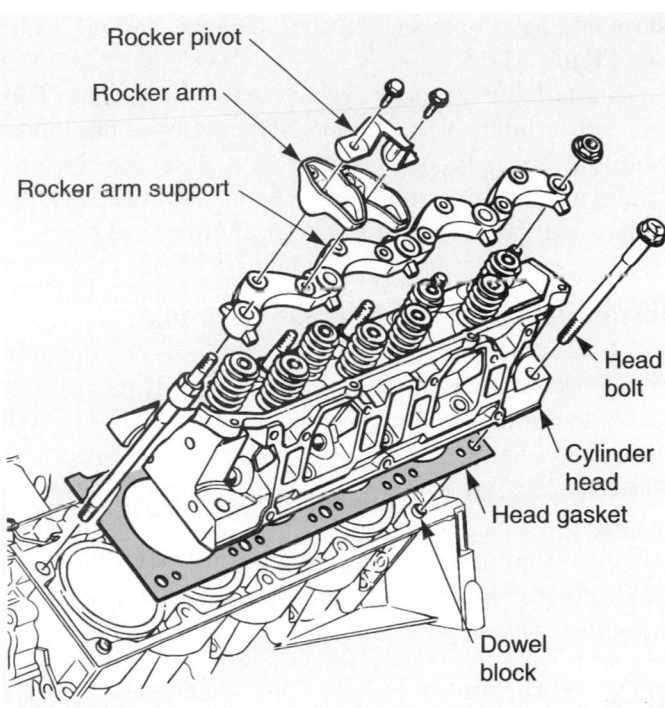

Figure 51-51. When installing a cylinder head, do not bump the head gasket or the gasket could be damaged. Slowly lower the head onto the dowels. (Cadillac)

Tighten each bolt a little at a time (1/2 torque, 3/4 torque, and then full torque). For example, if the head bolt torque is 100 ft lb, (130 N·m), tighten all the bolts to 50 ft lb (68 N·m), then to 75 ft lb (101 N·m), and finally to 100 ft lb (130 N·m). To ensure accuracy, double check that each bolt is tightened to full torque.

Figure 51-52. The threads of any head bolt that protrudes into a water jacket should be coated with a nonhardening sealer before installation. Use a service manual to find out which bolts should be sealed. If the bolts are not coated with sealer, coolant could leak out around the threads and the bolt head. (Fel-Pro)

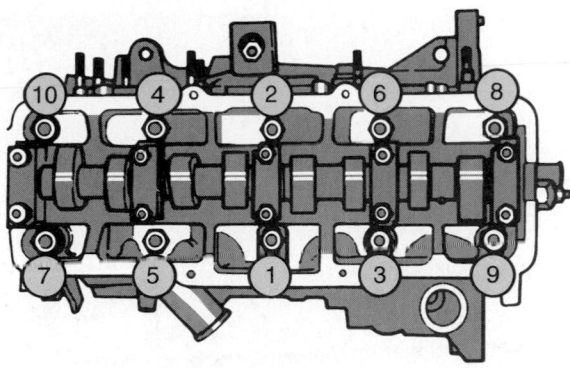

Figure 51-53. Torque a cylinder head in the exact torque pattern recommended by manufacturer. Generally, use a criss-cross pattern. Start in the middle and work your way to the outer bolts. Tighten the fasteners in steps. Start 1/2 torque, 3/4 torque, and full torque. (Chrysler)

Installing the Cam Housing

Some overhead cam engines are equipped with a cam housing that bolts to top of the cylinder head. O-ring seals often fit between the housing and head. Work the seal into its groove carefully. Then, lower the cam cover onto the head. Dowel pins usually align the housing on the head. See **Figure 51-54.**

With the cover torque in place, install the cam followers into their bores. Lubricate the followers with engine lube or heavy engine oil before installation. Next, install the valve or cam cover. It may use an O-ring seal or conventional gasket.

 Tech Tip
Reed valves are sometimes used in the intake runners of 4-stroke-cycle engines. These valves help prevent gas flow back into the intake manifold during valve overlap. They snap shut and only allow flow into the engine combustion chambers. Always inspect reed valves and, if needed, replace them.

Installing Intake Manifolds

Position the intake manifold gasket on the engine. Dowels may be provided to hold the gasket in place during manifold installation.

The use of silicone sealer may be recommended, especially where the front and rear seals meet the side gaskets on V-type engines. Look at **Figure 51-55.** With some inline engines, sealer is not needed on the intake gasket.

Installing an intake manifold on V-type engines can be difficult. It is critical that you use sealer properly on

A

B

C

Figure 51-54. On engines equipped with a cam housing, the housing and related parts must be installed after torquing cylinder head. A—Carefully place the gasket on top of the cylinder head. B—With the gasket in place, lower the cam housing onto the head. C—Lubricate the cam followers and place them in their bores.

these engines to prevent vacuum, coolant, and oil leaks. See **Figure 51-56.**

Start *all* the fasteners by hand before tightening. This will aid in bolt hole alignment. Then, torque the intake manifold fasteners to specifications. Use the specific torque pattern given in the service manual. Basically, the pattern will be a crisscross pattern, **Figure 51-57.**

Installing a Two-Piece Intake Manifold

To install a two-piece intake manifold, use the general procedures just discussed. First, install the gaskets and seals for the lower half of the assembly. Then, install the lower section and torque its fasteners to specifications. Next, fit the gaskets for the top half of the intake manifold, or plenum, into place. Without disturbing the gaskets, install the plenum on the lower half of the manifold. Finally, torque the plenum fasteners to specs in a crisscross pattern.

Tech Tip

Modern two-piece intake manifolds can be made of thin aluminum or soft plastic. Overtightening the fasteners can cause manifold warpage or other damage. Make sure you tighten late-model intake manifolds to factory specifications.

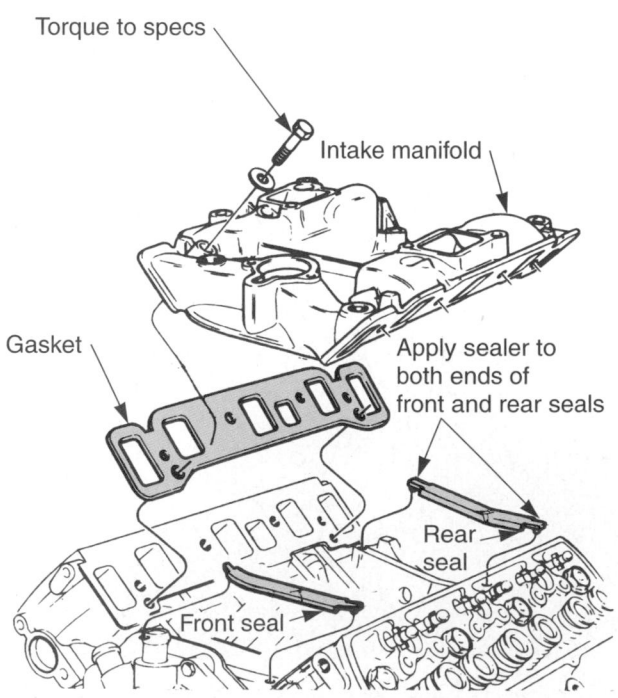

Figure 51-55. When rubber end seals butt against a fiber gasket in the corners of V-type engines, use silicone sealer. Place a dab of sealer where two join. (Oldsmobile)

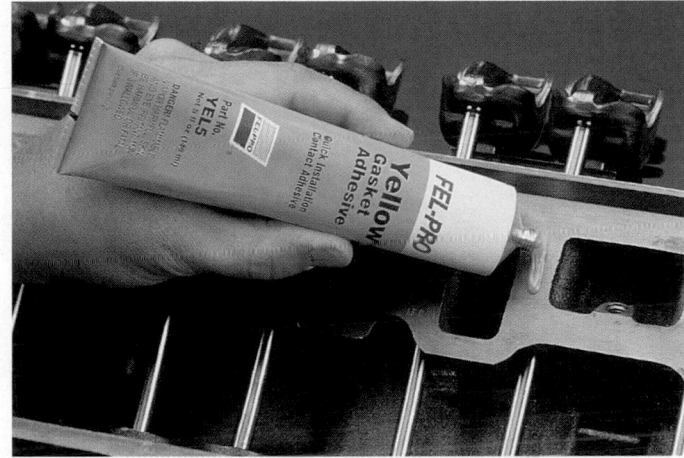

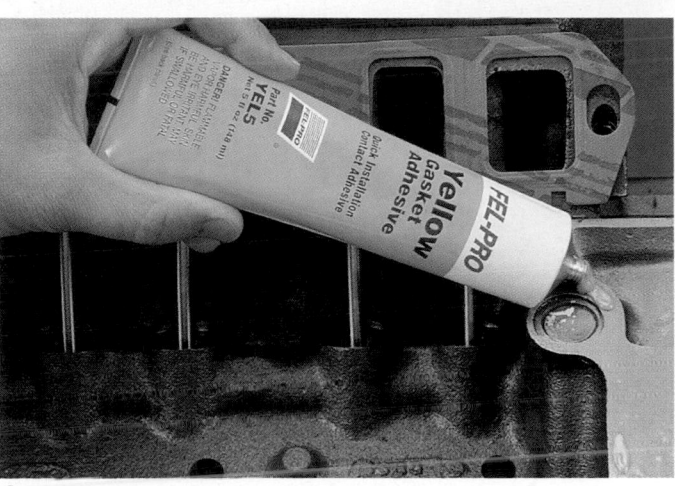
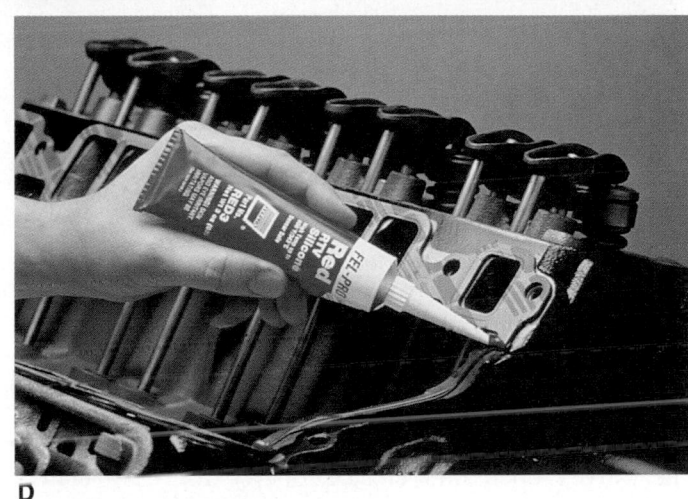

Figure 51-56. Note the basic steps for sealing a V-type intake manifold gasket. A—Read specific instructions that come with the gasket set. B—If recommended, use gasket sealer around the ports and water jackets. C—Gasket adhesive may be recommended to hold the front and rear rubber seals in place. D—Silicone sealer is usually needed where gaskets and seals fit over each other. This will prevent oil seepage. (Fel-Pro)

Installing Exhaust Manifolds

Most used exhaust manifolds require a gasket. Sealer is *not* commonly recommended because of the extreme heat at the exhaust manifold. Hold the gasket in place by hand as you start the bolts, **Figure 51-58.** Torque the bolts to specs in a crisscross sequence.

Installing the Rocker Assembly

If a shaft and rocker assembly is used, fit assembly on the cylinder head. Start the bolts by hand; then torque them to specs. Use the tightening sequence given for the particular engine. **Figure 51-59** shows a rocker shaft tightening sequence for a modern OHC engine.

If nonadjustable individual rocker stands or studs are used, torque the bolts or nuts to specs. With adjustable rockers, start the nuts but do *not* tighten them. You will have to adjust the valves. This is also true with solid lifters.

Valve Adjustment

Valve adjustment, also called ***tappet clearance adjustment*** or ***rocker adjustment,*** is critical to the performance and service life of the engine.

If the valve train is too tight (inadequate clearance), a valve may be held open. This can allow combustion heat to blow over and *burn* the valve.

If the valve train is too loose (too much clearance), it can cause ***valve train noise*** (tapping or clattering noise from the rockers striking the valve stems). This can increase part wear and cause part breakage.

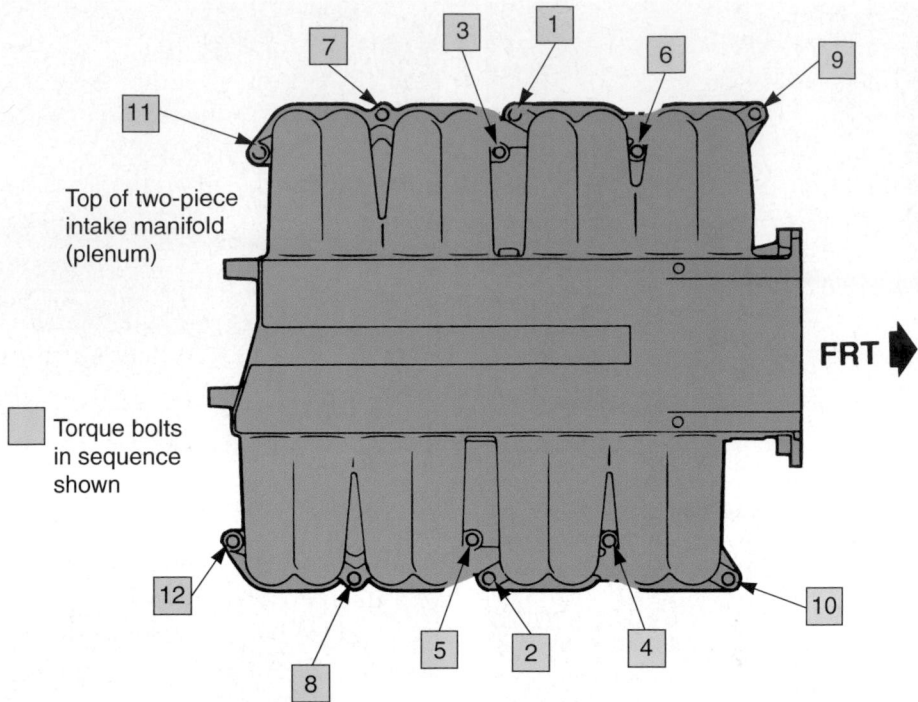

Figure 51-57. Torque the intake manifold or plenum fasteners in a crisscross pattern. The sequence for one particular engine is shown. Lubricate the bolt threads with a few drops of motor oil. (Chevrolet)

Valve Adjustment—Push Rod Engines

Valve adjustment on push-rod engines ensures that the valves open and close properly for easy engine starting after the valve job or top end service. If the valves are *too tight* (adjusted down too much), they may not close fully. The cylinder will lack compression and will not fire normally. If *too loose* (adjusted too loose), the valve may not open far enough for good performance. In some cases, the valve may not open at all.

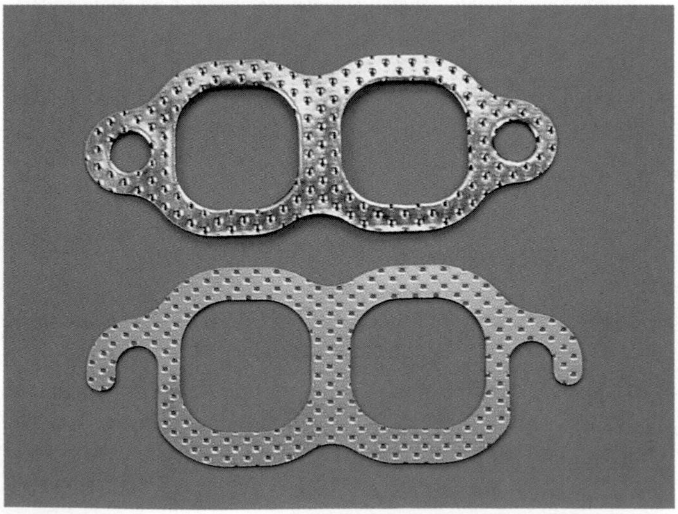

Figure 51-58. Often, you must hold the exhaust gasket on the head while starting the bolts. To simplify installation, some manufacturers provide exhaust manifold gaskets with slots, as shown at bottom. This allows you to start the bolts before slipping the gasket into place. (Fel-Pro)

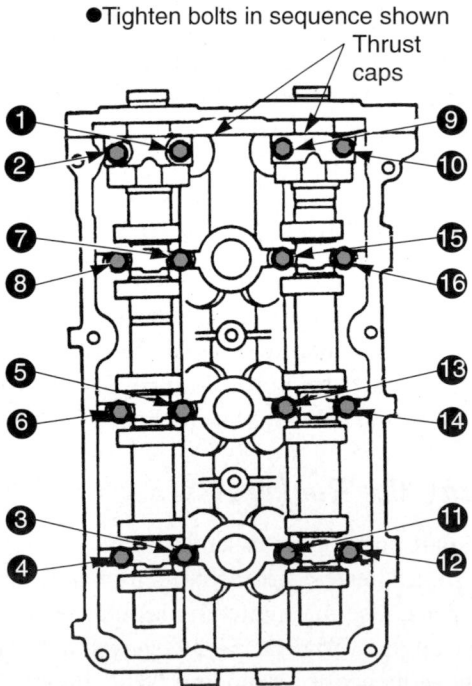

Figure 51-59. Tighten the rocker arm shaft assembly or the caps as described in the service manual. Tighten each bolt a little at a time to prevent bending of the shaft(s). (Ford)

Nonadjustable Rocker Arms

Nonadjustable rocker arms are used on many push rod engines with hydraulic (self-adjusting) lifters. Hydraulic lifters automatically compensate for changes in valve train clearance, maintaining *zero valve lash* (no clearance in valve train for quiet operation). They adjust valve train clearance as parts wear, temperature changes (part contraction or expansion), or oil thickness changes.

If adjustment is needed because of valve grinding, head milling, or other conditions, shorter or longer push rod can be installed with nonadjustable rocker arms. Refer to the service manual for details.

Adjusting Hydraulic Lifters

Hydraulic lifter adjustment is done to center the lifter plunger in its bore. This will let the lifter automatically take up or allow more valve train clearance. Some manuals recommend adjustment with the engine off. However, many technicians adjust hydraulic lifters with the engine running.

To adjust hydraulic lifters with the engine off, turn the crankshaft until the lifter is on the camshaft base circle (not on the lobe). The valve must be fully closed, as illustrated in **Figure 51-60.**

Loosen the rocker adjusting nut until you can wiggle the push rod up and down. Then, slowly tighten the adjusting nut until all play is out of the valve train (cannot wiggle push rod).

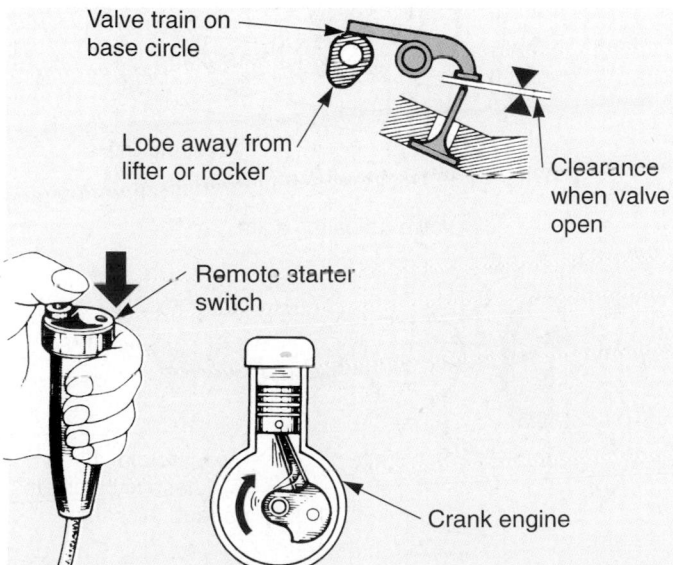

Figure 51-60. When adjusting valves, crank the engine until the lifter or rocker for the valve being adjusted is on its base circle. This will ensure that the valve is fully closed and that the proper clearance adjustment will be made. (Renault)

To center the lifter plunger, tighten the adjusting nut about *one more turn*. Refer to a service manual for exact details. The adjustment procedure can vary with engine design. Repeat the adjustment procedure on the other rockers.

To adjust hydraulic lifters with the engine running, install special oil shrouds, clothespins, or other devices to catch oil spray off the rockers. Start the engine and allow it to reach operating temperature.

Tighten all the rockers until they are quiet. One at a time, loosen a rocker until it *clatters*. Then, tighten the rocker slowly until it *quiets down*. This will be zero valve lash.

To set the lifter plunger halfway down in its bore, tighten the rocker about *one-half to one more turn*. Tighten the rocker slowly to give the lifter time to leak down and prevent engine missing or stalling. Repeat the adjustment on the other rockers.

Other adjustment methods may also be recommended. Check the manual for more detailed information.

Adjusting Mechanical Lifters

Mechanical lifters, also called solid lifters, are adjusted to ensure proper valve train clearance. Since mechanical lifters cannot automatically compensate for changes in valve train clearance, they must be adjusted periodically. Check the vehicle's service manual for adjustment intervals and clearance specifications. Typical clearance is approximately 0.014″ (0.35 ram) for the intake valves and 0.016″ (0.40 ram) for the exhaust valves.

Unlike hydraulic lifters, mechanical lifters make a clattering or pecking sound during engine operation. This is normal. Mechanical lifters are used on heavy-duty engines (taxi cabs, pickup trucks, diesel engines) and high-performance engines (early model, high-horsepower engines, for example).

To adjust a mechanical lifter, position the lifter on its base circle (valve fully closed). This can be done by cranking the engine until the piston in the corresponding cylinder is at TDC on its *compression stroke* (you can feel air blow out of spark plug hole). With the piston at TDC on the compression stroke, all valves in the cylinder can be adjusted.

Slide a flat feeler gauge of the correct thickness between the rocker arm and the valve stem. When valve clearance is properly adjusted, the feeler gauge will slide between the valve and the rocker with a slight drag, as shown in **Figure 51-61.**

If needed, adjust the rocker to obtain the specified valve clearance. You will normally have to loosen a lock nut and turn an adjusting screw. Then, tighten the lock nut and recheck clearance. Repeat this procedure on the other lifters.

Tech Tip

In some cases, valve train clearance is adjusted when the *engine is cold.* In other cases, the engine must be *hot* (at operating temperature) when clearance is adjusted. Check a manual for specific information. A change in engine temperature will cause part expansion or contraction. This, in turn, will cause a change in valve train clearance.

Valve Adjustment—Overhead Cam Engines

There are several different methods of adjusting the valves on an overhead cam engine. In many overhead cam designs, the valves are adjusted like the mechanical lifters in a push rod engine. A rocker arm adjustment screw is turned until the correct size feeler gauge fits between the cam lobe and the follower, valve shim, or valve stem.

Valve adjusting shims may also be used on modern OHC engines to allow valve clearance adjustment. Measure valve clearance with a feeler gauge. Then, if needed, remove and change shim thickness. Look at **Figure 51-62.**

Other OHC engines have an adjusting screw in each cam follower. Turning the screw changes valve clearance. Always refer to a shop manual for detailed directions.

Installing Valve Covers

If not installed properly, valve covers, or rocker covers, can leak oil. It is important for you to realize how easily a valve cover will leak. This may help prevent an incorrect installation technique and a "comeback" (customer returns to shop after failure of repair).

Some valve covers use a cork or synthetic rubber gasket. A few late-model valve covers are factory sealed using silicone sealer. When reinstalling the valve cover, either sealer or gaskets may be used, depending on the cover design.

Checking the Valve Cover Sealing Surface

Before installing a valve cover, make sure the cover is *not* warped or bent. Lay the cover on a flat workbench. Sight between the gasket surface and the workbench to detect gaps (dents, bends, or warpage).

Thin sheet metal covers can be straightened with taps from a small ball peen hammer. Cast aluminum covers can sometimes be sent to a machine shop for resurfacing. Warped plastic valve cover must be replaced.

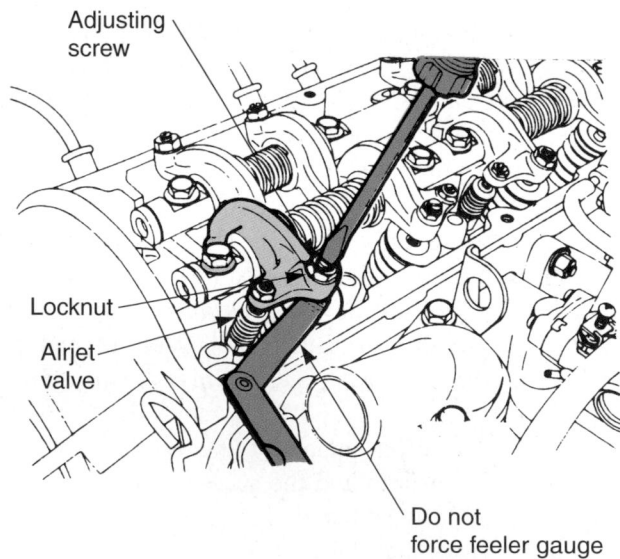

Figure 51-61. To adjust mechanical lifters, bump the engine until the cam lobe is away from the lifter or rocker. Insert the specified thickness feeler gauge between the rocker and the valve stem. If needed, loosen the lock nut and turn the adjusting screw to obtain the correct clearance. The feeler gauge should have a light drag when pulled back and forth. (Chrysler)

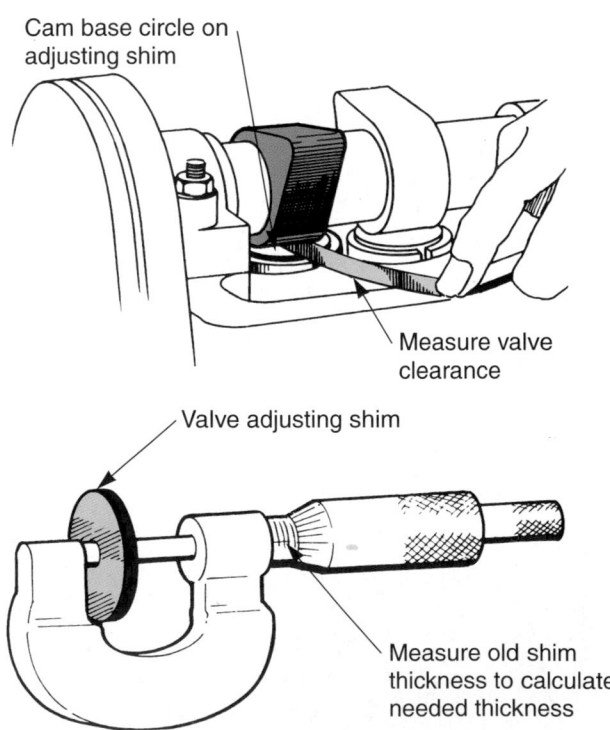

Figure 51-62. When valve adjusting shims are used, measure the clearance with feeler gauge. If clearance is incorrect, remove the old shim and measure its thickness. Calculate the required thickness of the new shim. Install the new shim and recheck clearance. (Volvo)

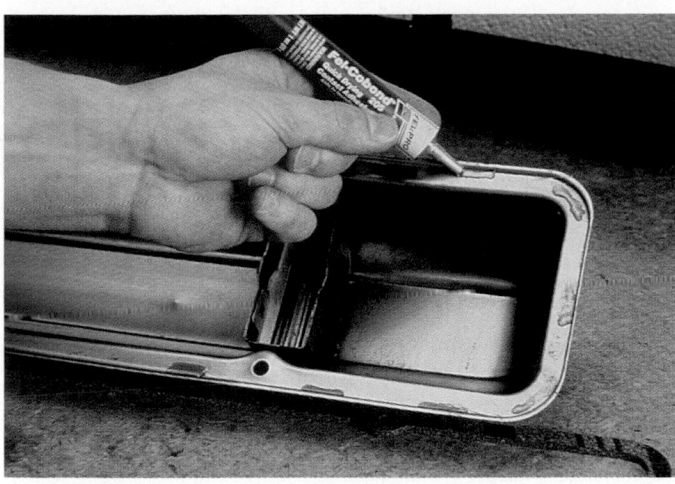

Figure 51-63. Apply a bead of quick-drying adhesive to the valve cover. This will hold the gasket in place during assembly. (Fel-Pro)

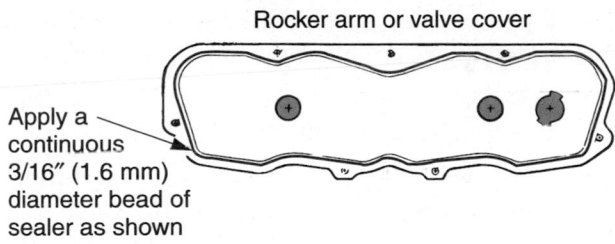

Rocker arm or valve cover

Apply a continuous 3/16″ (1.6 mm) diameter bead of sealer as shown

Figure 51-64. To use silicone sealer on a valve cover, make sure the sealing surface is perfectly clean. Run a uniform bead of specified thickness all the way around the cover. Install the cover without bumping or breaking the bead. (Chevrolet)

Installing the Valve Cover Gasket

To install a valve cover gasket, place a very light coat of approved sealer or adhesive around the edge of the valve cover. This is mainly needed to hold the gasket in place during assembly. Fit the gasket on the cover and align the bolt holes. Refer to **Figure 51-63.**

After letting the sealer cure slightly, place the cover and gasket on the cylinder head and start *all* the bolts by hand. Tighten the valve cover bolts to specifications using a crisscross pattern.

Tech Tip

A very common mistake is to *overtighten* valve cover bolts. Overtightening can smash and split the gasket. It can also bend the valve cover, causing oil leakage. Torque the bolts to specs, generally just enough to lightly compress the gasket.

Installing a Valve Cover With Silicone Sealer

To use silicone sealer, or RTV sealer, on the valve cover, double-check that the cover and cylinder head surfaces are *perfectly clean.* Sealer will *not* bond and seal on a dirty, oily surface.

Apply a continuous bead of sealer around the valve cover sealing surface. Typically, the bead should be about 3/16″ (1.6 mm) wide. Carefully install the valve cover on the cylinder head and start the valve cover bolts by hand. Finally, torque the bolts to specifications. Look at **Figure 51-64.**

Summary

- A valve job typically involves servicing the cylinder head and the valve train.
- A warped cylinder head has a bent or curved deck surface, usually from engine overheating.
- Cylinder head milling is a machine shop operation in which a thin layer of metal is removed from the gasket surface (deck) of the cylinder head.
- Magnafluxing is commonly used to find cracks in *cast iron parts* (cylinder heads, blocks, manifolds).
- Dye penetrant is used to find cracks in aluminum components (aluminum heads, blocks, and manifolds).
- Valve guide wear is a common problem; it allows the valve to move sideways in its guide during engine operation.
- Valve grinding is done by machining a fresh, smooth surface on the valve faces and stem tips.
- An interference angle (normally one degree difference in valve face angle and valve seat angle) is set on the valve grind machine.
- Valve seat reconditioning involves grinding (using a stone) or cutting (using carbide cutter) to resurface the cylinder head valve seats.
- Narrowing a valve seat centers the valve-to-seat contact point.
- Cam lobe wear can be measured with a dial indicator with the camshaft installed in the engine.
- A worn rocker shaft will have indentations where the rocker arms swivel on the shaft.
- A valve grind gasket set includes all the gaskets and seals required to reassemble an engine top end.

- Cylinder head gasket markings are normally provided to show the front or top of the gasket.

- Some overhead cam engines use a cam housing that bolts on top of the cylinder head.

- Valve adjustment is critical to the performance and service life of an engine. If the valve train is too tight, the valves may be held open. If the valve train is too loose, valve train noise will occur.

- Valve adjusting shims may be used on modern OHC engines to allow the adjustment of cam-to-valve clearance.

Important Terms

Engine top end service	Prussian blue
Valve job	Valve spring squareness
Cylinder head stand	Valve spring free height
Warped cylinder head	Valve spring tension
Cylinder head milling	Valve spring shimming
Magnafluxing	Valve spring installed
Dye penetrant	height
Valve grinding	Spring bind
Valve grind machine	Lifter leak-down rate
Interference angle	Valve grind gasket set
Valve seat	Cylinder head gasket
reconditioning	markings
Staking	Valve adjustment
Top cutting	Tappet clearance
Pilot	adjustment
Valve seat runout	Rocker adjustment
Lapping valves	Valve train noise
Grinding compound	Zero valve lash
Lapping stick	Valve adjusting shims

Review Questions—Chapter 51

Please do not write in this text. Place your answers on a separate sheet of paper.

1. How can you tell if a cylinder head is warped?

2. _____ is commonly used to find cracks in cast iron engine parts.

3. _____ _____ is normally used to find cracks in aluminum engine components.

4. Typically, how do you remove and install a pre-combustion chamber in an automotive diesel engine?

5. Which of the following is *not* a common method of repairing worn valve guides?
   (A) Shimming the existing guide.
   (B) Reaming the old guide for an oversize stem.
   (C) Knurling the guide to reduce the inside diameter.
   (D) Installing a new guide insert.

6. Valve grinding is done by machining a fresh, smooth surface on the valve faces and the valve stem tips. True or False?

7. How do you "dress the stone" on a valve grind machine?

8. What is an interference angle?

9. Explain what can happen if a valve margin is too thin.

10. When grinding a valve, the same amount of metal should be removed from the _____ and the _____.

11. Typically, an intake valve should have a valve-to-seat contact width of _____. The valve-to-seat contact width of an exhaust valve should be about _____.

12. How do you move the valve-to-seat contact point in and out on the valve face?

13. Which of the following is *not* a normal check done on valve springs?
    (A) Squareness.
    (B) Free height.
    (C) Horizontal dimension.
    (D) Tension.

14. Valve spring _____ is used to maintain the correct tension when the springs are installed on the engine.

15. An O-ring type valve seal should be installed before compressing the valve spring over the stem. True or False?

16. Valve springs and valve seals can be replaced without cylinder head removal. True or False?

17. Lifter _____ _____ _____ is measured by timing how long it takes to push the plunger to the bottom of its stroke under controlled conditions.

18. What is included in a valve grind gasket set?

19. What is the purpose of cylinder head gasket markings?

20. Explain why head gasket selection on a diesel engine is extremely critical.

21. Generally, tighten cylinder head bolts starting on each end and working your way to the middle. True or False?

22. If the valve train is too tight, the valves can _____.

23. How can you adjust valve clearance on an engine with hydraulic lifters while the engine is running?

24. How do you adjust valve clearance on an engine with mechanical lifters?

25. When using RTV sealer on a valve cover gasket, typically run a bead of sealer about _____ wide all the way around the cover. All sealing surfaces must be perfectly _____ and dry.

ASE-Type Questions

1. An automobile's cylinder head is believed to be warped. Technician A says a feeler gauge and a dial indicator are needed to measure the head for warpage. Technician B says a straightedge and feeler gauge are needed to measure the head for warpage. Who is right?
   (A) A only.
   (B) B only.
   (C) Both A and B.
   (D) Neither A nor B.

2. Technician A says a dial indicator and inside micrometer are used to measure cam bore alignment in an OHC engine. Technician B says a straightedge and feeler gauge are used to measure cam bore alignment in an OHC engine. Who is right?
   (A) A only.
   (B) B only.
   (C) Both A and B.
   (D) Neither A nor B.

3. An engine's cylinder head is brought to the machine shop for milling. Technician A says cylinder head milling normally decreases an engine's compression ratio. Technician B says cylinder head milling normally increases an engine's compression ratio. Who is right?
   (A) A only.
   (B) B only.
   (C) Both A and B.
   (D) Neither A nor B.

4. An aluminum engine block may have a crack in it. Technician A says "magnafluxing" can be used to determine whether or not the block is cracked. Technician B says dye penetrant can be used to determine whether or not the block is cracked. Who is right?
   (A) A only.
   (B) B only.
   (C) Both A and B.
   (D) Neither A nor B.

5. Technician A says a brass drift and a hammer are normally used to remove a diesel engine precombustion chamber. Technician B says a hydraulic press is normally used to remove a diesel engine precombustion chamber. Who is right?
   (A) A only.
   (B) B only.
   (C) Both A and B.
   (D) Neither A nor B.

6. An engine has an oil consumption problem. Technician A says this condition may be caused by worn valve guides. Technician B says this condition may be caused by worn valve seals. Who is right?
   (A) A only.
   (B) B only.
   (C) Both A and B.
   (D) Neither A nor B.

7. A cylinder head's valve guides are worn. Technician A says "knurling" will repair the damaged valve guides. Technician B says valve guide inserts can be installed to repair the cylinder head. Who is right?
   (A) A only.
   (B) B only.
   (C) Both A and B.
   (D) Neither A nor B.

8. A valve head wobbles as it turns on the valve grind machine. Technician A says the valve may have been improperly "chucked." Technician B says the valve may be bent. Who is right?
   (A) A only.
   (B) B only.
   (C) Both A and B.
   (D) Neither A nor B.

9. An engine's cylinder head has a severely burned valve seat. Technician A says machining will repair the seat. Technician B says the seat must be replaced. Who is right?
   (A) A only.
   (B) B only.
   (C) Both A and B.
   (D) Neither A nor B.

10. Technician A says an intake valve should have a valve-to-seat contact width of approximately 1/16″. Technician B says an intake valve should have a valve-to-seat contact width of approximately 1/32″. Who is right?
    (A) A only.
    (B) B only.
    (C) Both A and B.
    (D) Neither A nor B.

11. Technician A says "lapping" valves is done to check the location of the valve-to-seat contact point. Technician B says "lapping" valves is done to smooth the valve-to-seat mating surfaces. Who is right?
    (A) A only.
    (B) B only.
    (C) Both A and B.
    (D) Neither A nor B.

12. Camshaft service includes all of the following except:
    (A) inspecting the distributor-oil pump gear.
    (B) checking cam lobe wear.
    (C) measuring camshaft length.
    (D) checking cam bearing diameter.

13. Technician A says that camshaft journal wear lowers engine oil pressure. Technician B says that a worn camshaft can sometimes be reground. Who is right?
    (A) A only.
    (B) B only.
    (C) Both A and B.
    (D) Neither A nor B.

14. A head gasket is being installed on an engine. Technician A says sealer may or may not be required, depending on the type of gasket used. Technician B says sealer is never used on a head gasket. Who is right?
    (A) A only.
    (B) B only.
    (C) Both A and B.
    (D) Neither A nor B.

15. An engine's valve train clearance is too small. Technician A says this condition will produce valve train noise during engine operation. Technician B says this condition may cause some valves to burn. Who is right?
    (A) A only.
    (B) B only.
    (C) Both A and B.
    (D) Neither A nor B.

Activities—Chapter 51

1. Use a dial indicator to measure valve guide wear on all the valves of an engine. Make a chart showing your readings, and compare them to specs. If one or more guides is out of spec, determine which technique should be used to correct the problem.

2. Demonstrate the steps required to set up and operate a valve grind machine. Grind the face of an actual valve, if one is available.

3. Micrometers are used to make many measurements while servicing engine top ends. For practice in reading a standard outside micrometer, use it to check a variety of known thicknesses, such as the blades of a feeler gauge set. Further practice can be done by checking such components as valve stems, cam journals, rocker shafts, or gaskets.

Chapter 52

Engine Front End Service and Engine Installation

After studying this chapter, you will be able to:

- Inspect a timing chain and sprockets for wear.
- Service a chain tensioner and a timing chain assembly.
- Properly align timing marks.
- Check timing gears for wear or damage.
- Remove and install timing gears.
- Measure timing gear runout and backlash.
- Install a timing chain or timing gear cover.
- Remove and install a front cover oil seal.
- Service a timing belt.
- Summarize engine installation procedures.
- Describe safety practices to be followed while servicing engine front ends and installing engines.
- Correctly answer ASE certification test questions about servicing engine front end assemblies.

Modern cars and trucks use several types of engine front end designs. Many late-model engines use a timing belt to drive an overhead camshaft. Others use a timing chain to transfer turning force to the camshaft. Some heavy-duty gasoline engines and diesel engines use timing gears to operate the camshaft and valve train.

It is very important that you know how to service and repair all types of camshaft drive mechanisms. This chapter summarizes the most essential information on this subject.

Earlier textbook chapters covered subjects relating to engine front end service. Review these chapters if needed. For example, Chapter 49 explained engine front end disassembly (crankshaft damper, front cover, and camshaft drive mechanism removal) and part cleaning methods.

Timing Chain Service

Timing chains can be used on both push rod engines and overhead cam engines. An exploded view of a timing chain and front end assembly for a push rod engine is given in **Figure 52-1. Figure 52-2** pictures a timing chain setup for an overhead cam engine. Study the parts and the part relationships.

Inspecting the Timing Chain, Chain Guides, and Sprockets

You should inspect the timing chain for looseness during engine disassembly. Excess *slack,* or *play,* in the chain normally requires replacement of the chain and the sprockets.

Also check the crankshaft key or the camshaft dowel for wear or damage. If used, check the chain tensioner and the chain guides for wear, **Figure 52-2.** Replace any components showing wear or damage.

Installing the Timing Chain Assembly

Timing marks—lines, circles, dots, or other shapes—will be indented or cast into the timing sprockets (also used on timing gears and belt sprockets). These markings must be pointing in the correct direction to time the camshaft (valves) with the crankshaft (pistons).

To align the timing marks, fit the sprockets on the camshaft and the crankshaft. Do not install the timing chain. Rotate the cam and the crankshaft by hand until the *timing marks align.* Refer to a service manual if needed.

Without turning the camshaft or the crankshaft, remove the sprockets. Then, fit the chain over the sprockets so that the timing marks are aligned correctly. Install the chain and sprockets on the engine as a unit. Double check the alignment of the timing marks. See **Figure 52-3.**

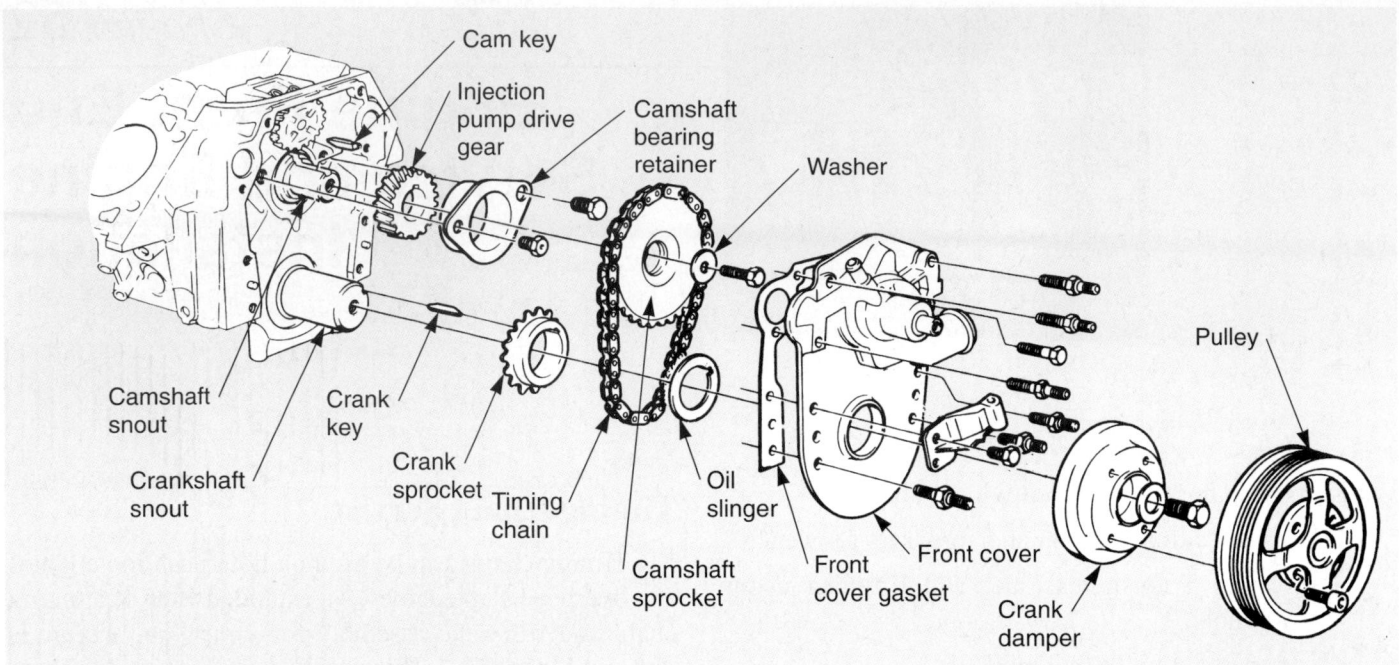

Figure 52-1. Review the parts of an engine front end assembly. This is a V-type, diesel engine with a timing chain. Note the extra gear, which drives the fuel injection pump. (Buick)

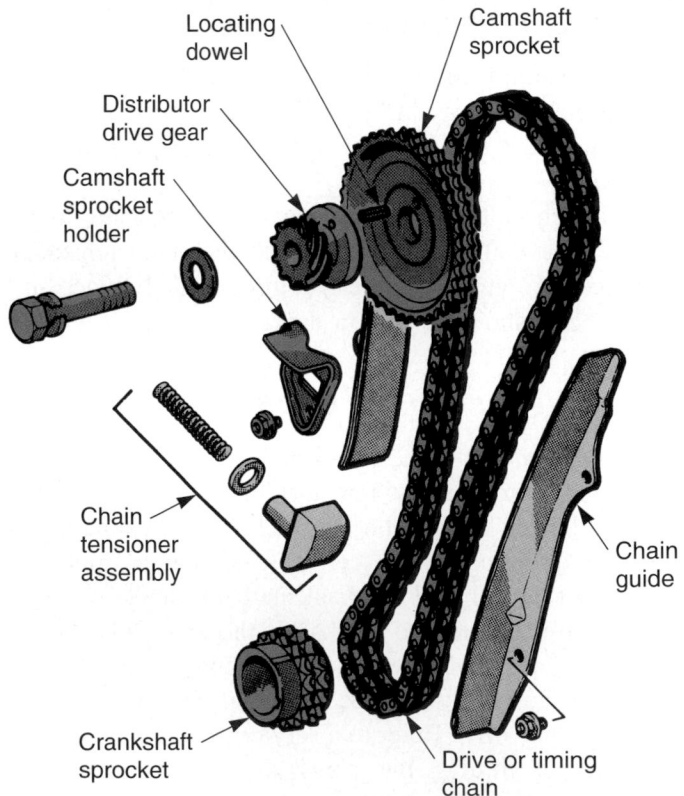

Figure 52-2. Overhead cam timing chains commonly use a tensioner to compensate for chain wear and stretch. Chain guides prevent chain vibration, or slap. The tensioner and guides frequently have a fiber or plastic facing that can wear. (Dodge)

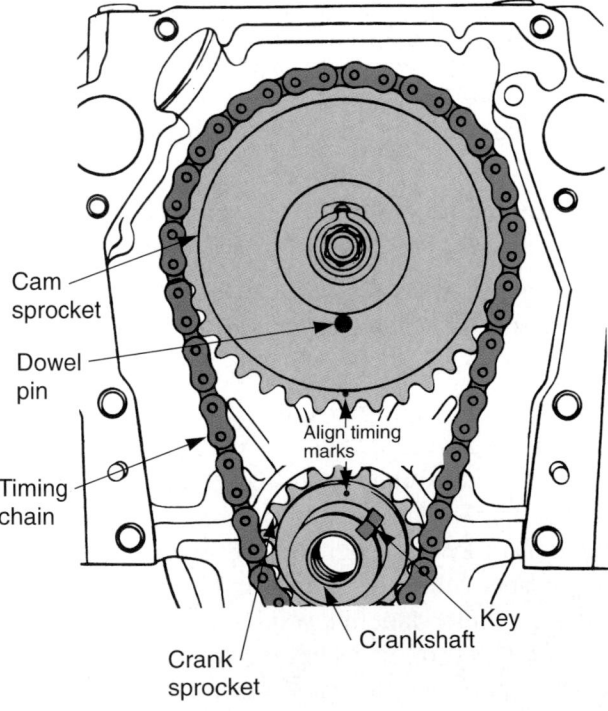

Figure 52-3. A key aligns the crankshaft sprocket. A dowel aligns the camshaft sprocket. You must align marks on the sprockets to time the crankshaft with the camshaft. (General Motors)

If used, install the chain tensioner, the chain guides, and any other parts. Torque all fasteners to exact specs. Bend any locking tabs against the bolt heads to prevent loosening. Install the oil slinger on the crankshaft snout. See **Figure 52-4.**

Tech Tip
Make sure you install the oil slinger in the right direction. If you install an offset slinger backwards, it will rub on the timing cover and make a loud noise. You would have to completely disassemble the front end to fix the mistake.

When installing complex OHC timing chains, double check everything. See **Figure 52-5.** Make sure all timing marks align. Check that the guides and tensioners are correctly in place. Also, make sure any hidden rear cover bolts are in place.

Timing Gear Service

Timing gear service is similar to timing chain service. Timing gears are normally more dependable than timing chains. They provide thousands of miles of trouble-free operation. However, after prolonged use, timing gear teeth can wear or become chipped and damaged.

Inspect the used timing gears carefully. Look for any signs of wear or other problems. Replace the gears as a set, if needed.

Measuring Timing Gear Backlash

Timing gear backlash is the amount of clearance between the timing gear teeth. It can be measured to

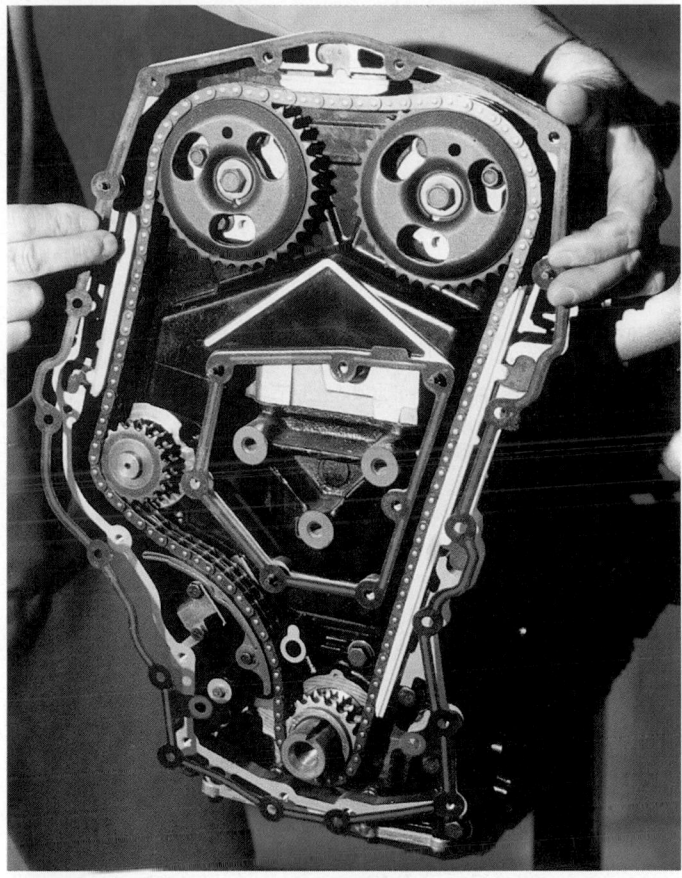

Figure 52-5. Take your time when assembling the timing chain assembly on an OHC engine. Double check all timing marks, chain tension, and chain guides. Note that this engine uses a two-piece timing cover. Inspect the back cover closely to make sure you have not missed any hidden bolts.

determine timing gear wear. A dial indicator is an accurate tool for measuring backlash.

To measure timing gear backlash:
1. Set the dial indicator stand on the engine.
2. Locate the indicator stem on one of the cam gear teeth. The stem must be parallel with gear tooth travel. Look at **Figure 52-6.**
3. Rotate the cam gear one way and then the other, without turning the crankshaft.
4. Read the indicator at the end of tooth travel in each direction to determine backlash.

If timing gear backlash is greater than specs, the gears are worn. They should be replaced. Refer to the service manual for backlash specifications.

Replacing Timing Gears

Timing gears are usually press-fit on the crankshaft and camshaft. A puller is normally needed to remove the crankshaft gear. **Figure 52-7** shows how a press is used

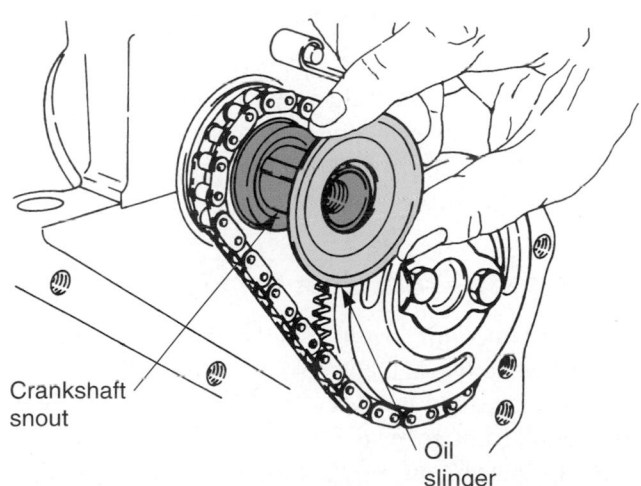

Crankshaft snout

Oil slinger

Figure 52-4. The oil slinger fits next to the crankshaft sprocket. It helps keep oil from leaking out the front seal and sprays oil on the chain and sprockets to aid lubrication. (Ford)

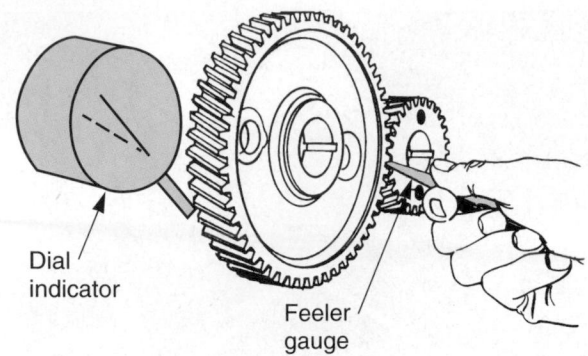

Figure 52-6. Either a dial indicator or a feeler gauge can be used to measure timing gear backlash. Position the indicator as shown. Rotate the cam gear back and forth while reading the indicator to find clearance. The largest feeler gauge that fits properly in the gear teeth also indicates clearance. If backlash is too great, gears are worn. (Deere & Co.)

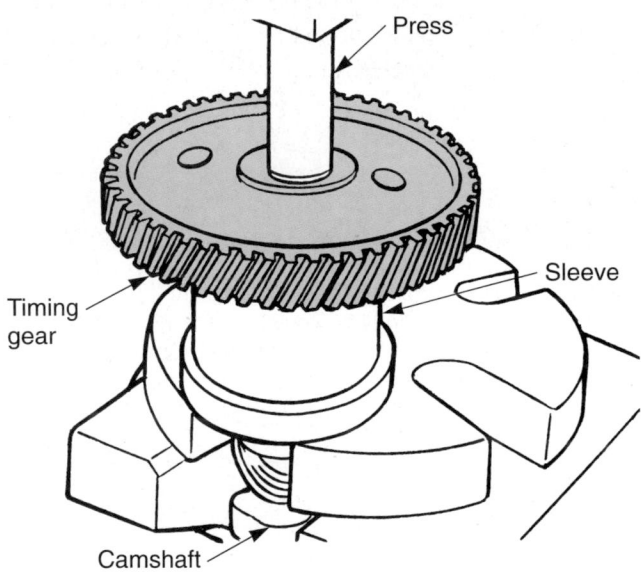

Figure 52-7. Some camshaft timing gears must be pressed on and off. The camshaft must be removed from engine. Other engine designs allow the gear to be driven off the installed camshaft with light taps from a brass hammer. (Buick)

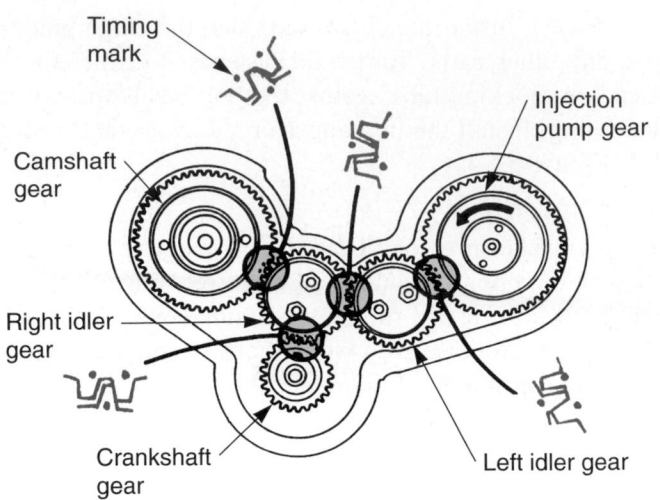

Figure 52-8. Study the timing marks on the timing gears in this diesel engine. (Ford)

to remove the cam gear. With this design, the camshaft must be removed from the engine for timing gear replacement.

Timing gears can often be installed with light blows from a brass hammer. Make sure the keyseats and keyways are aligned. Tap in a circular motion to force the gears squarely into position. A press may also be needed to install the cam gear on the camshaft.

As with a timing chain and sprockets, double check the alignment of the timing marks. The timing marks must be positioned properly to time the camshaft with the crankshaft. **Figure 52-8** shows the manufacturer's

recommended timing gear alignment on a small truck diesel engine.

Measuring Timing Gear Runout

Timing gear runout, or *wobble,* is unwanted gear wobble, or side movement, during rotation and is measured with a dial indicator.

To measure runout:
1. Position the indicator stand on the engine block.
2. Place the indicator stem on the outer edge of the camshaft timing gear. The stem should be parallel with the camshaft centerline, **Figure 52-9.**
3. Turn the crankshaft while noting indicator needle movement. The indicator reading denotes runout.

If runout is greater than specs, remove the timing gear and check for problems. The gear may not be fully seated or it may be machined improperly. Also check camshaft straightness.

Crankshaft Front Seal Service

The *crankshaft front seal* keeps engine oil from leaking out between the crankshaft snout and the front cover. Replace the front seal whenever it is leaking or anytime the front cover is removed.

Seal replacement requires only partial engine disassembly and can usually be performed without removing the engine from the vehicle. Typically, you must remove the radiator and other accessory units on the front of the

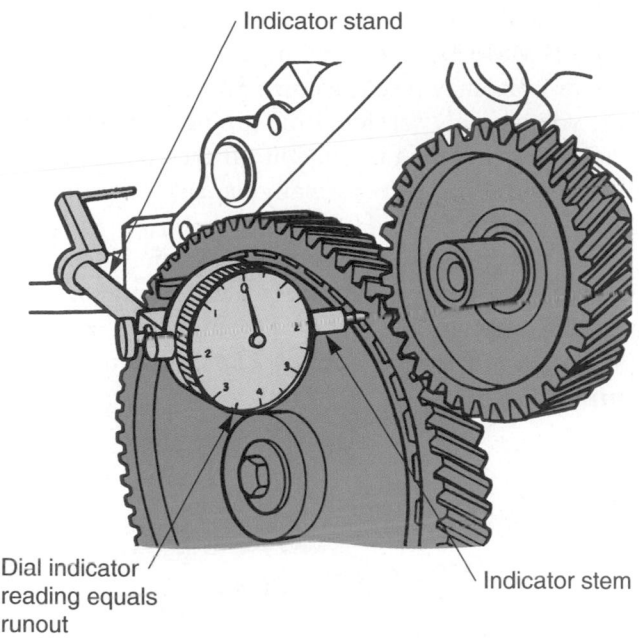

Figure 52-9. After installing the timing gears, check gear runout, or wobble, as shown. Turn the crankshaft and read the indicator. Correct the problem if runout is beyond specs. (Ford)

engine. A puller can then be used to remove the crankshaft damper, **Figure 52-10.**

Sometimes, the front crankshaft seal can be replaced without front cover removal. A special seal puller may be used to remove the old seal, as shown in **Figure 52-11A.**

Figure 52-10. To service a front cover seal, the crankshaft damper or its hub must be removed with a wheel puller. Threaded holes are normally provided in the damper or hub for puller bolts. Do not hammer or pry on the outer edge, or damage will result. (OTC)

Before installing the new front seal, coat the outside diameter of the seal with nonhardening sealer. This will prevent oil seepage between the seal body and the front cover. Also, coat the rubber sealing lip with engine oil. This will lubricate the seal during initial engine startup. To install the new seal, use a seal driver to squarely seat the seal in its bore, **Figure 52-11B.**

Tech Tip
In some cases, the crankshaft front seal can only be serviced from the rear of the front cover. Cover removal is necessary when replacing the seal. When in doubt, check the service manual for detailed instructions.

Engine Front Cover Service

The *engine front cover,* also called the *timing cover,* encloses the timing chain or the timing gears. It is usually made of thin sheet metal or cast aluminum. Make sure a sheet metal cover is not warped.

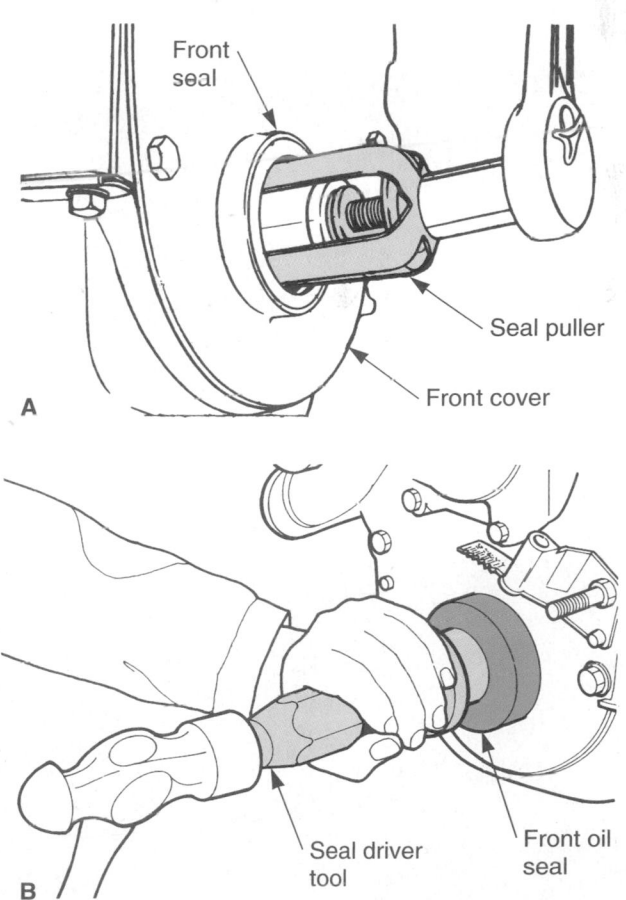

Figure 52-11. When the seal is pressed in from the front, it can be replaced without front cover removal. A—Special seal puller will pull out the old seal. B—Use a seal driver to force the new seal squarely into the bore in the front cover. Coat the outside diameter of the seal with a nonhardening sealer. (GMC)

Installing the Front Cover Gasket or Silicone Sealer

A *front cover gasket set,* or *front end gasket set,* includes the parts needed (gaskets, seals, sealers) to service the front cover. If you are only servicing the timing mechanism, purchase a front cover gasket set. One is shown in **Figure 52-12.**

To install the front cover gasket, coat the flange on the cover or the gasket with an approved adhesive-sealer, **Figure 52-13.** This will hold the gasket in place while you fit the cover on the engine. Allow the adhesive to become *tacky* (partially dried, but still sticky and moist). Then carefully place the gasket on the cover flange. Make sure that all the bolt holes in the gasket and the cover are aligned properly.

To use silicone sealer instead of a gasket, apply an even bead of sealer. The bead should be of specified thickness and should surround all bolt holes and coolant passages. Make sure the sealing surfaces are perfectly clean, or leaks may result, **Figure 52-14.**

If you are only servicing the timing mechanism and are not removing the oil pan, cut off the ends of the old oil pan gasket. They are normally damaged during cover removal, **Figure 52-15.**

The front cover gasket set often contains small pieces of gasket material for the protruding end of the oil pan. After cleaning the surfaces, apply sealer to the oil pan flange and fit the new gasket pieces into place, **Figure 52-16.**

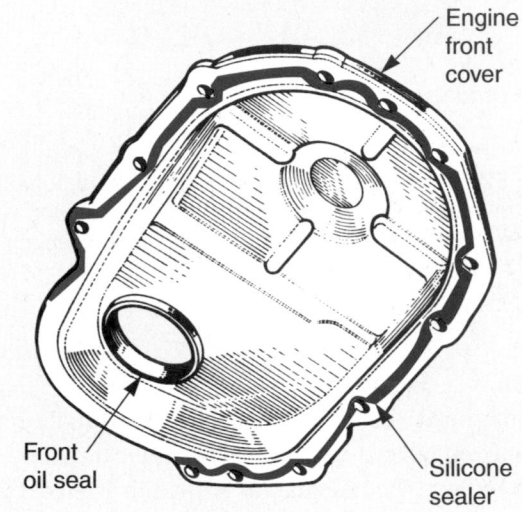

Figure 52-14. To use silicone sealer on a front cover, make sure the cover is perfectly clean and dry. Coat the flange with approved sealer. Use the correct diameter bead and surround all bolt holes and water passages. (Renault)

Figure 52-12. Note the parts often included in a high-quality engine front end gasket set. (Fel-Pro)

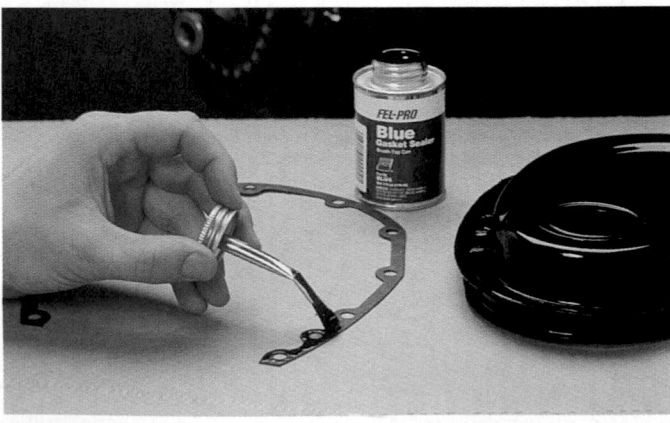

Figure 52-13. Adhesive-sealer is commonly used to hold a gasket in place during assembly and to help prevent leakage. Make sure you use an approved product. (Fel-Pro)

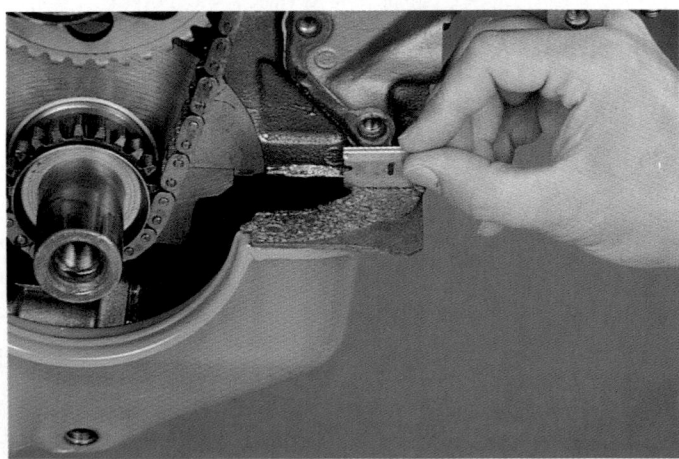

Figure 52-15. When you are servicing a timing cover only, the oil pan gasket may be damaged. Cut off the old gasket with razor blade. Clean the surfaces on the pan and the cover thoroughly. (Fel-Pro)

Figure 52-16. Extra silicone sealer is normally recommended where the front cover meets the oil pan. Without sealer, a gap could form, allowing oil leakage. A small piece of pan gasket can then be placed on the protruding flange of the oil pan. (Fel-Pro)

Installing the Engine Front Cover

Carefully fit the front cover into place without disturbing the gasket or silicone sealer. With some engine designs, you must loosen the oil pan bolts and partially drop (lower) the pan. Also, you may need to center the seal around the crankshaft using a *seal alignment tool,* **Figure 52-17.**

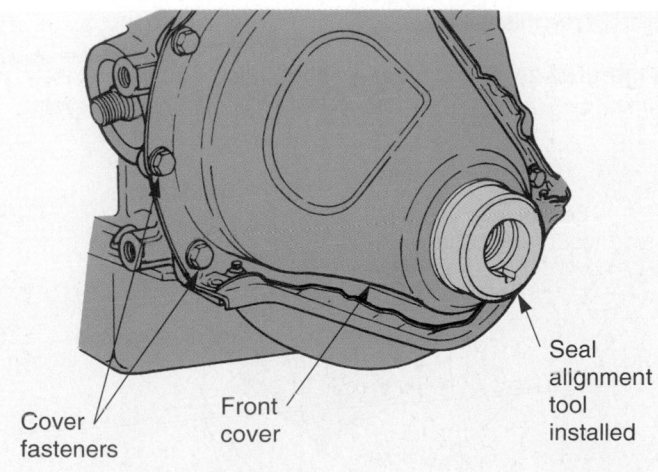

Figure 52-17. An alignment sleeve is sometimes needed to center the seal around the crankshaft. Fit the sleeve over the crankshaft snout during front cover installation. Remove the sleeve after tightening the cover fasteners. (Buick)

Start all the front cover fasteners by hand. Do *not* tighten any fastener until all are started. Double check that the gasket is properly aligned and that bolt lengths are correct. Start the oil pan bolts into the bottom of the cover, if needed.

Torque all fasteners to specs in a crisscross pattern. Remove seal alignment tool (if used) and reinstall all the components on the front of the engine.

Figure 52-18 shows a front cover for an engine using a timing belt. The cover simply houses the crankshaft oil seal. Note how sealer is needed on one of the bolts that extends into a water jacket. A service manual is needed to obtain this kind of detailed information.

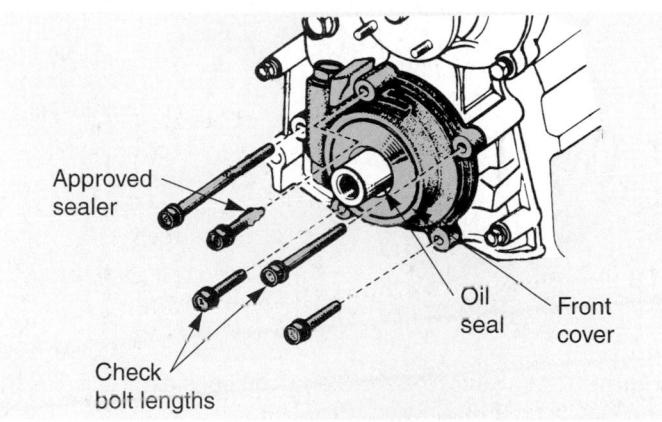

Figure 52-18. This overhead cam engine has a small front cover. Since an external timing belt is used, the cover does not have to enclose a timing chain or gears. Note that the lengths of the cover bolts differ and that one bolt requires sealer because it enters the water jacket. (Toyota)

Timing Belt Service

Many late-model OHC engines use a synthetic rubber belt to operate the camshaft. The cogged (toothed) belt provides an accurate, quiet, light, and dependable means of turning the camshaft, **Figure 52-19.**

Tech Tip

If the timing belt jumps a tooth or two, some on-board diagnostic systems can detect probable belt failure. By using data from the crankshaft and camshaft position sensors, it can warn the driver of the timing belt problem, illuminating the Check Engine Light or the Malfunction Indicator Light before major engine damage occurs.

Timing belt service is very important. If the timing belt were to break or be timed improperly, engine valves,

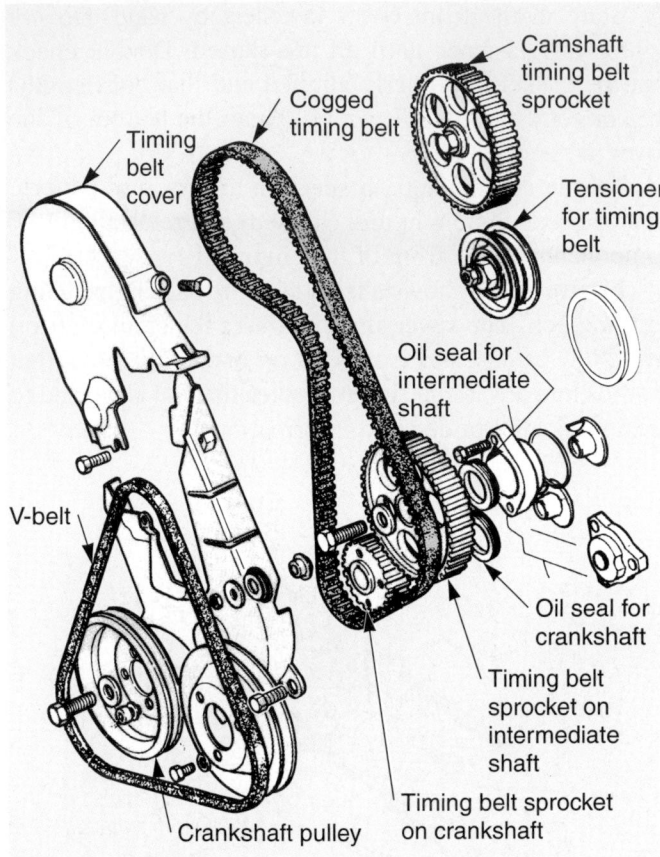

Figure 52-19. Study the front end components of this timing belt–equipped OHC engine. (Plymouth)

pistons, and other parts could be damaged. If the camshaft is out of time with the crankshaft, the pistons could slide up and hit the open valves.

 Caution
Never use the starting motor to crank an engine when the timing belt, chain, or gears are removed. Valves could be bent or broken.

Inspecting the Timing Belt

Inspect the engine timing belt for signs of deterioration (cracks, hardening, softening, and fraying). Look for problems illustrated in **Figure 52-20.**

Most manufacturers recommend timing belt replacement at intervals of about 50,000-100,000 miles (80,000–160,000 km). Refer to the owner's manual or the service manual for exact replacement intervals.

When severe timing belt damage is found, inspect for mechanical problems. Check sprocket condition and installation. Make sure that coolant or engine oil is not leaking onto the belt. Turn the tensioner wheel to make sure its bearing spins freely. Replace the tension wheel if it feels dry and rough when rotated, **Figure 52-21.**

A–Do not bend, twist, or turn belt excessively. Oil and water will deteriorate belt. Fix all engine leaks.

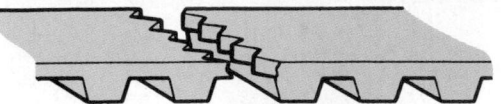

B–Belt breakage may be caused by sprocket or tensioner problem. Check these parts before installing new belt.

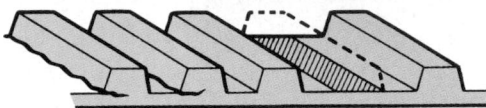

C–If timing belt teeth are missing, check for locked component. Oil pump, injection pump, etc., could be frozen.

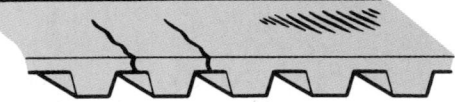

D–If there is wear or cracks on smooth side of belt, check idler or tensioner pulley.

E–With damage or wear on one side of belt, check belt guide and pulley alignment.

F–Wear on timing belt teeth may be caused by timing sprocket problem. Inspect sprockets carefully. Oil contamination will also cause this trouble.

Figure 52-20. Always inspect the timing belt very closely. A timing belt failure can cause major engine damage. (Toyota)

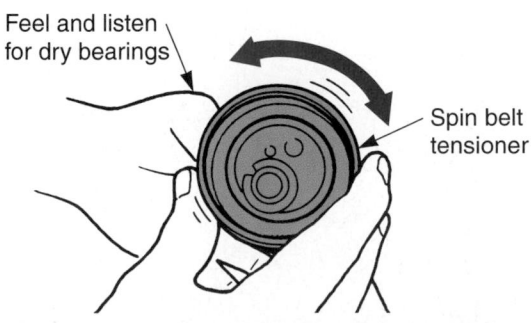

Figure 52-21. Always check the timing belt tensioner bearings. Spin the bearing by hand. It should turn smoothly. Replace the unit if it seems to have dry or rough bearings. (Mazda)

Caution
Never submerse a tensioner wheel in solvent or steam clean it. This could wash the grease out of the bearings and cause rapid failure when the tensioner is returned to service.

Installing the Timing Belt Sprocket

The camshaft timing belt sprocket is normally secured to the camshaft with a hex bolt. Hold the camshaft stationary by fitting a wrench or a special holding tool on a flat formed on the camshaft or on another unmachined surface. Be careful not to damage the cam lobes. Use another wrench to loosen the bolt holding the sprocket. This is shown in **Figure 52-22.**

When installing a timing belt sprocket, align any key or dowel that positions the sprocket. Make sure you have the right washer and bolt for the sprocket. Install and torque the bolt to specs.

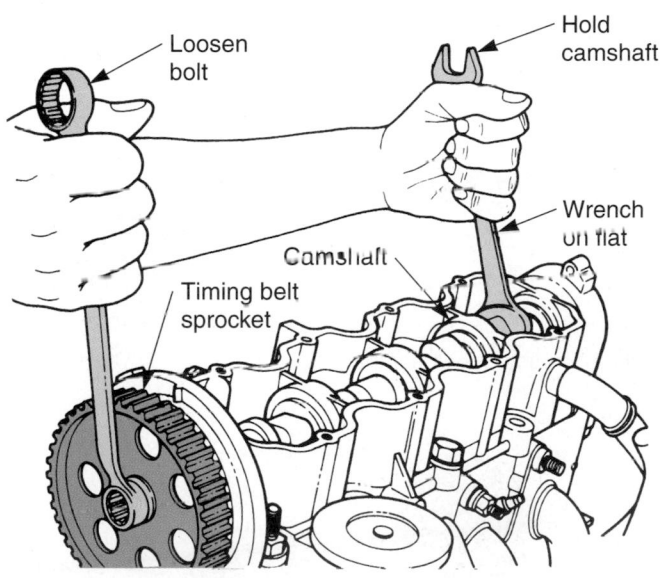

Figure 52-22. To remove a belt sprocket, hold the camshaft stationary while loosening the sprocket bolt. A special holding tool may also be needed. (Buick)

Installing the Timing Belt

To install a timing belt, line up the timing marks on the camshaft and crankshaft sprockets. Refer to the service manual for specific details, **Figure 52-23.** When the distributor is driven by an accessory sprocket, it will also have timing marks that must be aligned properly.

With the sprocket marks positioned correctly, slip the belt over the sprockets. Move the tensioner into the belt to hold the belt on its sprockets. See **Figure 52-24.**

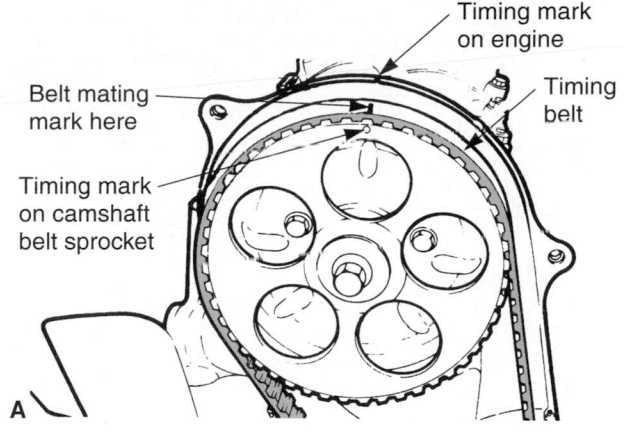

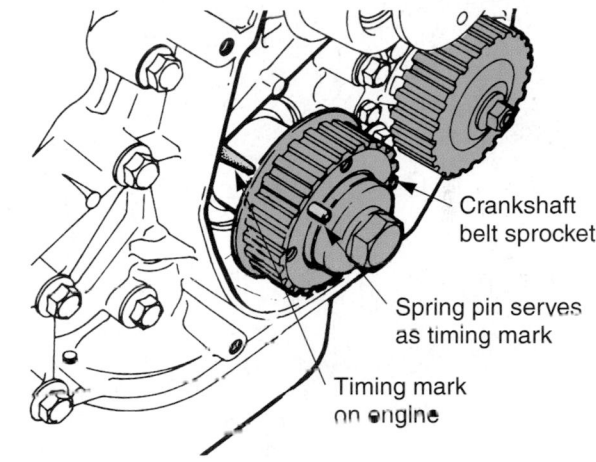

Figure 52-23. When installing a new timing belt, align the timing marks properly. Sometimes, the belt is also marked. Do not crank the engine with belt *off,* or the valves may be bent. Turn the sprockets by hand. A—Camshaft belt sprocket mark is aligned with the belt marking and the timing mark on the engine. B—This crankshaft sprocket uses a dowel as a timing reference. Align the dowel with the timing pointer on the engine. The types of markings vary. Check the service manual. (Chrysler)

Adjusting Timing Belt Tension

Proper *timing belt tension* (tightness) is very important to belt service life. If the belt is too tight, it can wear out quickly or break. If it is too loose, the belt can flap, vibrate, or fly off the sprockets.

When automatic adjustment is not provided, a pry bar is used to adjust the belt tensioner. Sometimes, a special tool is needed to measure tension. See **Figure 52-25.**

Figure 52-26 shows a simple way to check timing belt tension for many OHC engines. Adjust the timing belt until moderate finger and thumb pressure is needed to twist the belt about one-quarter turn.

After you are sure belt tension is adjusted to specs, install the timing belt cover. The *timing belt cover* is a

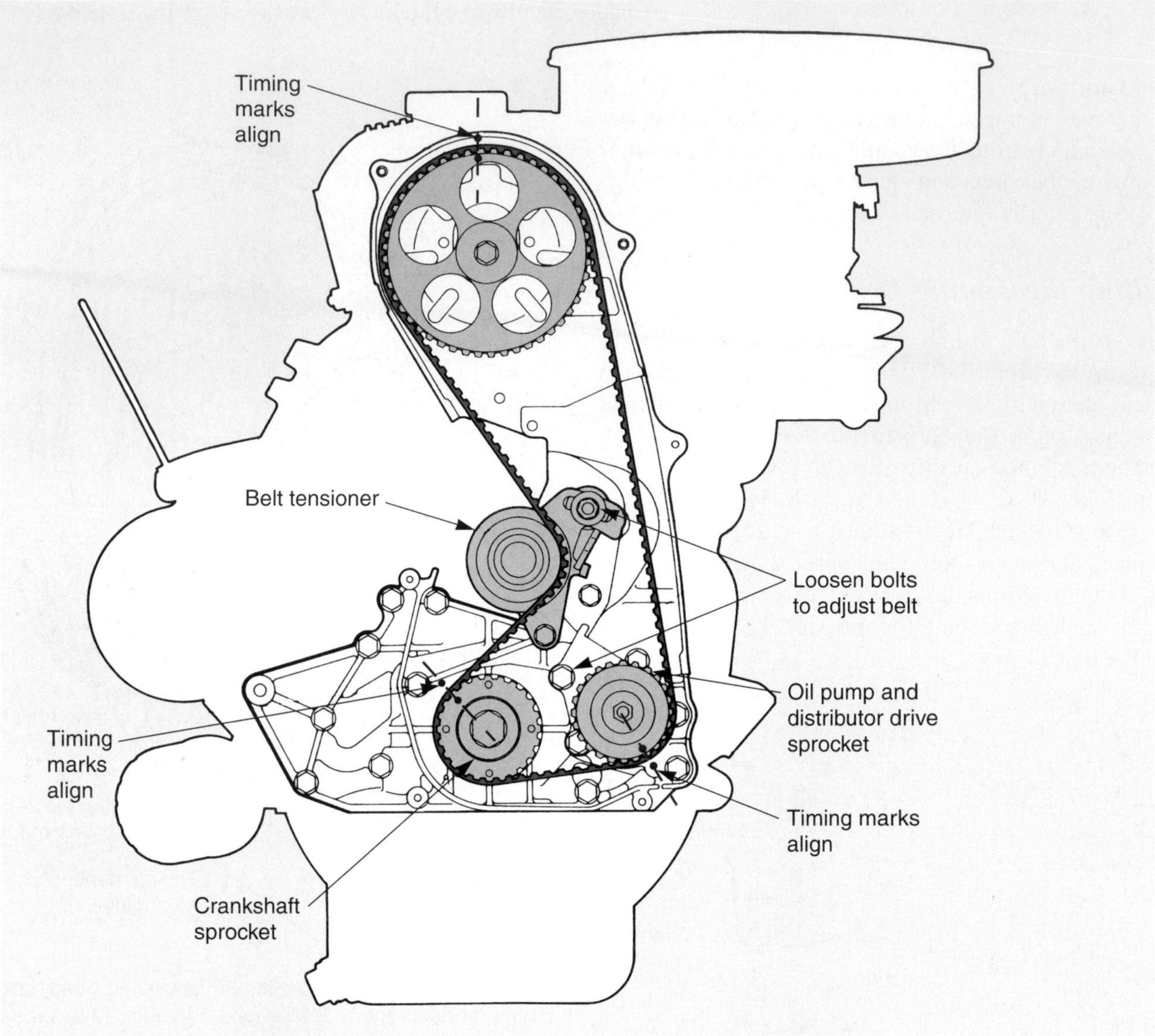

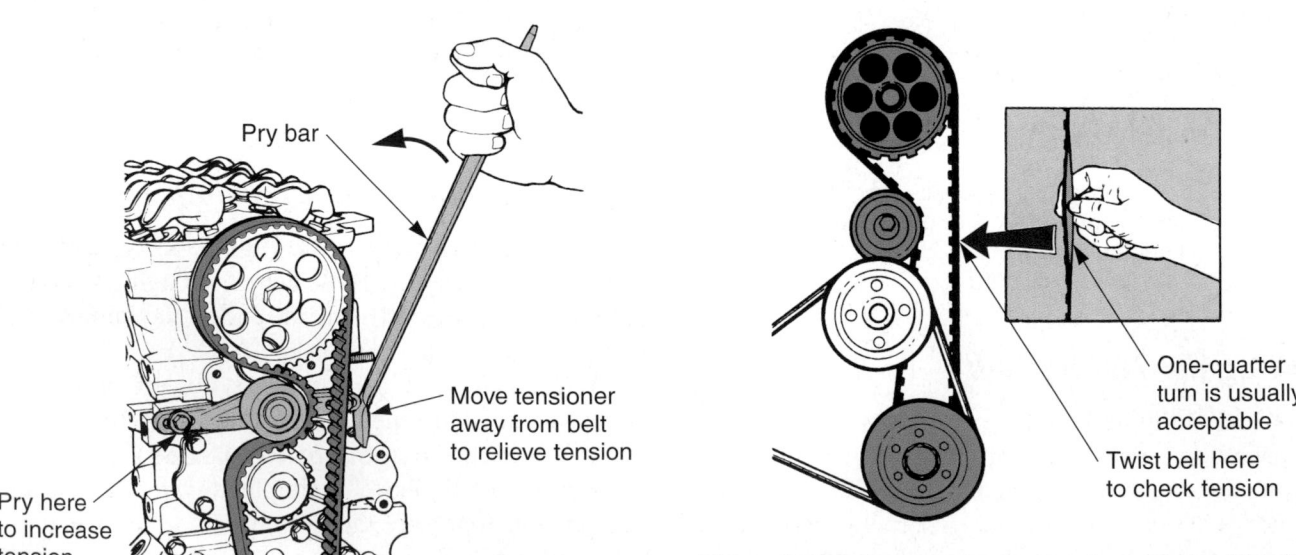

Figure 52-24. Study the timing marks and how the tensioner can be moved to adjust belt tightness. (Dodge)

Timing marks align

Belt tensioner

Loosen bolts to adjust belt

Timing marks align

Oil pump and distributor drive sprocket

Timing marks align

Crankshaft sprocket

Pry bar

Move tensioner away from belt to relieve tension

Pry here to increase tension

Figure 52-25. A pry bar is commonly used to shift the tensioner and adjust the timing belt when it is not adjusted automatically. (Ford)

One-quarter turn is usually acceptable

Twist belt here to check tension

Figure 52-26. As a general rule, the belt should twist one-quarter turn with moderate finger pressure when belt tension is correct. However, refer to the service manual for the exact testing method. (Chrysler)

sheet metal or plastic cover that surrounds the belt and the sprockets. It does *not* contain an oil seal, nor does it require a gasket. A typical timing belt cover is pictured in **Figure 52-27.**

Complete Engine Assembly

Information summarizing complete engine assembly has been given in the previous chapters. Refer to the index for more information. Look at **Figure 52-28.**

Generally, to complete your engine rebuild, install all remaining parts that do not prevent engine installation. Check that all sensors, sending units, brackets, motor mounts, heat shrouds, and similar external parts are in place. It is much easier to install them with the engine out of the vehicle. In fact, some parts are almost impossible to install after the engine is back in the engine bay!

Installing the Engine

Installing an engine in a vehicle is about the opposite of removing the engine. Engine removal was covered in Chapter 49. It explains the use of an engine hoist, a lifting chain or fixture, organizing part containers, and special equipment. Refer to this chapter as needed.

Figure 52-28. Before installing an engine in a vehicle, install all the parts that will not obstruct engine installation. Spark plug wires, some hoses, sending units, and lifting brackets are a few examples. (Subaru)

A few rules to remember when installing an engine include:

- Keep your hands and feet out from under the engine.
- The hoist should be used for engine removal and installation *only*. Never work on an engine raised on a hoist. Never move the hoist when the engine is raised high in the air. The hoist could flip over, damaging the engine or causing serious injury.
- With many automatic transmission–equipped cars, the engine's flywheel must be turned so that the large hole in the flywheel faces straight down. Rotate the transmission torque converter until its drain plug is also facing down. This will make it easier to align the engine and transmission.
- On vehicles with manual transmissions, check the condition of the pilot bearing (bushing in the end of the crankshaft). Replace the bearing if it is worn (See Chapter 54, *Clutch Diagnosis and Repair*).
- Position the lifting chain or fixture so that the engine will be raised in a *level position*. If the engine is *not* level, sliding it onto the transmission will be difficult.
- Slowly lower engine into vehicle while watching for clearance all around the engine compartment. Position the engine so that its crankshaft center-line is aligned with the transmission input shaft centerline.

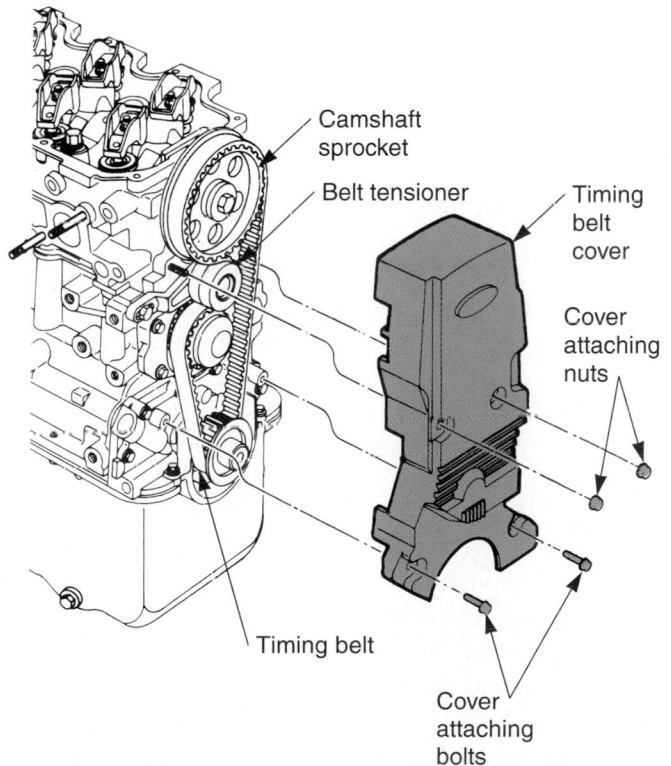

Figure 52-27. Double check the timing marks and the belt tension. Then install the belt cover. Tighten the fasteners to specs. (Ford)

- Push the engine back and align the engine dowel pins with the holes in the transmission. Use a large bar to shift the engine, if necessary.

- As soon as the dowel pins slide *fully* into their holes, install an engine-to-transmission bolt, but do *not* tighten it. Start another bolt on the other side of the engine.

- Check that the torque converter is properly lined up with the holes in the flywheel. Turn the converter as needed. Tighten the bell housing bolts (engine-to-transmission bolts).

- Finish installing the other components on the engine: motor mounts, oil filter, fuel lines, coolant hoses, throttle linkage, battery ground cable, and fan belts. Refer to **Figure 52-29.**

- Fill the engine with oil and the radiator with coolant.

- Start the engine and run it at a fast idle until warm while watching for adequate oil pressure.

- Shut off the engine and recheck fuel injection or carburetor adjustments, ignition timing, and other related adjustments.

- Look for fluid leaks or any other signs of trouble.

Engine Break-In

Engine break-in is done mainly to seat and seal new piston rings. It also aids initial wearing in of other components under controlled conditions.

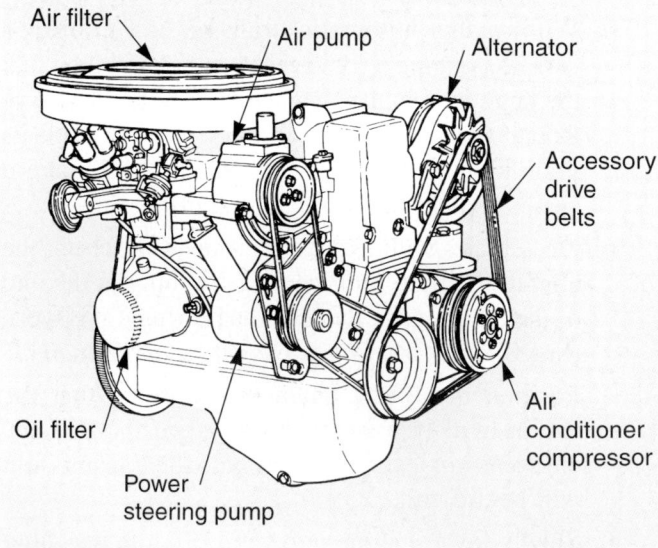

Figure 52-29. After engine installation, install all other accessory units, wires, and hoses. Fill the engine with oil and the radiator with coolant. Double check everything before starting the engine.

After allowing the engine to warm-up at a fast idle, most technicians road test the vehicle. At the same time, they use moderate acceleration and deceleration to break in the engine.

Generally, accelerate the car to about 40 mph (65 km/h). Then, release the accelerator pedal fully and let the car coast down to about 20 mph (32 km/h). Do this several times while carefully watching engine temperature and oil pressure.

Do *not* allow the engine to overheat during break in; ring and cylinder scoring may result.

 **Warning**
When breaking in an engine, drive the car on a road that is free of traffic. Do *not* exceed posted speed limits and conform to safe driving standards.

Inform the vehicle's owner of the following rules concerning the operation of a freshly overhauled engine.

- Avoid prolonged highway driving during the first 100-200 miles (161-322 km). This will prevent ring friction from overheating the rings and cylinders, possibly causing damage.

- Do not worry about oil consumption until after the first 2000 miles (3220 km) of engine operation. It will take this long for full ring seating.

- Check the engine oil and other fluid levels frequently.

- Change the engine oil and filter after approximately 2000 miles (3220 km) of driving. This will help remove any particles in the oil.

- Inform the customer of any problems not corrected by the engine overhaul. For example, if the radiator is in poor condition (has been previously repaired or was filled with rust), tell the customer about the consequences of *not* correcting the problem. The problem may upset engine performance or reduce engine service life. Having the customer sign a release form will protect you if the faulty part fails.

Summary

- You should inspect the timing chain for looseness during engine disassembly. Excess slack, or play, in the chain normally requires replacement of the chain and the sprockets.

- Timing marks will either be indented or cast into the timing sprockets.

- Timing gear backlash is the amount of clearance between the timing gear teeth. Backlash can be measured to determine timing gear wear.

- Timing gear runout, or wobble, is measured with a dial indicator.
- A front cover gasket set includes the parts needed (gaskets, seals, sealers) to service the time cover.
- Never use the starting motor to crank an engine when the timing belt, chain, or gears are removed. Valves could be bent or broken.
- Proper timing belt tension (tightness) is very important to belt service life.
- Installing an engine in a vehicle is about the opposite as removing the engine.
- Keep your hands and feet out from under the engine. Slowly lower engine into vehicle while watching for clearance all around the engine compartment.
- Engine break-in is done to seat and seal new piston rings. It also aids initial wearing in of other components under controlled conditions.

Important Terms

Timing marks	Front cover gasket set
Timing gear backlash	Front end gasket set
Timing gear runout	Seal alignment tool
Crankshaft front seal	Timing belt tension
Engine front cover	Timing belt cover

Review Questions—Chapter 52

Please do not write in this text. Place your answers on a separate sheet of paper.

1. Timing chains can be used on both push rod engines and overhead cam engines. True or False?
2. Excess slack in a timing chain requires chain adjustment. True or False?
3. Timing marks on the timing chain sprockets are given as:
   (A) lines.
   (B) circles.
   (C) dots.
   (D) All of the above.
4. Explain how to align timing chain marks.
5. Timing gear _____ is the amount of clearance between the timing gear teeth.
6. Summarize the basic procedure for replacing a front crankshaft seal.
7. What is a seal alignment tool?

8. Most automakers recommend timing belt replacement every _____ miles (_____ km).
9. List fourteen rules to remember when installing an engine in a car.
10. Engine _____ is done mainly to seat and seal the new piston rings.

ASE-Type Questions

1. A push rod engine is brought into the shop for timing mechanism replacement. Technician A says most push rod engines are equipped with a timing chain. Technician B says most push rod type engines are equipped with a timing belt. Who is right?
   (A) A only.
   (B) B only.
   (C) Both A and B.
   (D) Neither A nor B.
2. A timing chain is being installed on an engine. Technician A says the timing marks on the crank and cam sprockets must face away from each other when installing the timing chain. Technician B says the timing marks on the crank and cam sprockets must face each other when installing the timing chain. Who is right?
   (A) A only.
   (B) B only.
   (C) Both A and B.
   (D) Neither A nor B.
3. An oil slinger on an engine's front end is damaged. Technician A says the oil slinger must be replaced in order to help keep oil from leaking out the front seal. Technician B says the oil slinger must be replaced in order to lubricate the front end components. Who is right?
   (A) A only.
   (B) B only.
   (C) Both A and B.
   (D) Neither A nor B.
4. Technician A says timing chains are normally more dependable than timing gears. Technician B says timing gears are usually more dependable than timing chains. Who is right?
   (A) A only.
   (B) B only.
   (C) Both A and B.
   (D) Neither A nor B.

5. Which of the following instruments should be used to measure timing gear backlash?
   (A) Outside caliper.
   (B) Inside micrometer.
   (C) Dial indicator.
   (D) Straightedge.

6. A particular engine's timing gear backlash is greater than specs. Technician A says the engine's timing gears must be adjusted. Technician B says the engine's timing gears must be replaced. Who is right?
   (A) A only.
   (B) B only.
   (C) Both A and B.
   (D) Neither A nor B.

7. An automobile engine's timing gears are being replaced. Technician A says the timing gears can usually be removed by lightly tapping them with a brass hammer. Technician B says a wheel puller and press may be required to remove the timing gears. Who is right?
   (A) A only.
   (B) B only.
   (C) Both A and B.
   (D) Neither A nor B.

8. Timing gears are being installed on a taxi cab engine. Technician A says a hammer and punch should be used to install these gears. Technician B says timing gears on this type of engine can be installed with a hydraulic press. Who is right?
   (A) A only.
   (B) B only.
   (C) Both A and B.
   (D) Neither A nor B.

9. An engine's timing gear runout is being checked. Technician A says an inside micrometer is normally used to check timing gear runout. Technician B says an outside caliper should be used to check timing gear runout. Who is right?
   (A) A only.
   (B) B only.
   (C) Both A and B.
   (D) Neither A nor B.

10. An engine's timing gear runout is beyond specs. Technician A says the problem may be caused by a defective timing gear. Technician B says the problem may be due to an improperly seated timing gear. Who is right?
    (A) A only.
    (B) B only.
    (C) Both A and B.
    (D) Neither A nor B.

11. An engine's crankshaft front seal must be replaced. Technician A says you must always remove the engine's front cover when replacing this seal. Technician B says the front cover does not have to be removed when replacing this seal on some engines. Who is right?
    (A) A only.
    (B) B only.
    (C) Both A and B.
    (D) Neither A nor B.

12. A crankshaft front seal is being installed. Technician A says non-hardening sealer should be used on the outside diameter of this seal before installation. Technician B says sealer should never be used when installing a crankshaft front seal. Who is right?
    (A) A only.
    (B) B only.
    (C) Both A and B.
    (D) Neither A nor B.

13. Technician A says an engine's front cover encloses the timing belt and its tensioner. Technician B says an engine's front cover encloses the timing chain or timing gears. Who is right?
    (A) A only.
    (B) B only.
    (C) Both A and B.
    (D) Neither A nor B.

14. A vehicle is brought into the shop with a damaged timing belt. Technician A tells the owner that most timing belts should be replaced every 25,000 miles. Technician B says most timing belts should be replaced every 50,000 miles. Who is right?
    (A) A only.
    (B) B only.
    (C) Both A and B.
    (D) Neither A nor B.

15. A rebuilt engine has been reinstalled in a vehicle. Technician A tells the car's owner to avoid prolonged highway driving for the first 100–200 miles. Technician B tells the owner not to worry about oil consumption until after the first 2000 miles of engine operation. Who is right?
    (A) A only.
    (B) B only.
    (C) Both A and B.
    (D) Neither A nor B.

Activities—Chapter 52

1. Examine service manuals for as many different makes of cars or light trucks as you can find. Determine whether the engine uses timing gears, a timing chain, or a timing belt. Are certain sizes or types of engines more likely to use a particular timing device?

2. Demonstrate the proper use of a wheel puller in removing a pulley or a crankshaft damper (harmonic balancer).

Section 10

Drive Trains and Axles

The drive train transfers turning force from the engine crankshaft to the drive wheels. Drive train configurations vary, depending on vehicle design. The drive train parts commonly found on a front-engine, rear-wheel-drive vehicle include the clutch, transmission, drive shaft, and rear axle assembly. The drive train parts used on most front-engine, front-wheel-drive vehicles include the clutch, transaxle, and drive axles.

Section 10 explains the operation, construction, and repair of all types of automotive drive trains. You will learn about front-wheel drive, rear-wheel drive, and all-wheel drive. This will give you the knowledge needed to work on any make or model car, truck, or sport utility vehicle.

This section will help you pass two ASE certification tests: Test A2, *Automatic Transmission/Transaxle,* and Test A3, *Manual Drive Train and Axles.*

Chapter 53

Clutch Fundamentals

After studying this chapter, you will be able to:

- List the basic parts of an automotive clutch.
- Explain the operation of a clutch.
- Describe the construction of major clutch components.
- Compare clutch design differences.
- Explain the different types of clutch release mechanisms.
- Correctly answer ASE certification test questions that require a knowledge of clutch designs and operation.

An *automotive clutch* is used to connect and disconnect the engine and manual (hand-shifted) transmission or transaxle. The clutch is located between the engine and the transmission.

Figure 53-1 shows the basic parts of drive trains for both rear-wheel-drive and front-wheel-drive vehicles. Review the relationship of the parts.

This chapter begins your study of a vehicle's drive train. The clutch is the first drive train component powered by the engine crankshaft. The clutch allows the driver to control power flow between the engine and the transmission or transaxle (transmission-differential assembly).

The next chapters cover clutch service, transmissions, drive shafts, transfer cases, differentials, transaxles, and other drive train units. To fully understand these later chapters, you must first understand clutch operation and design. Study carefully!

Clutch Principles

Only vehicles with manual transmissions require a clutch. Vehicles with automatic transmissions do not need one. They have a fluid coupling, or torque converter, which automatically disengages the engine and transmission at low engine speeds.

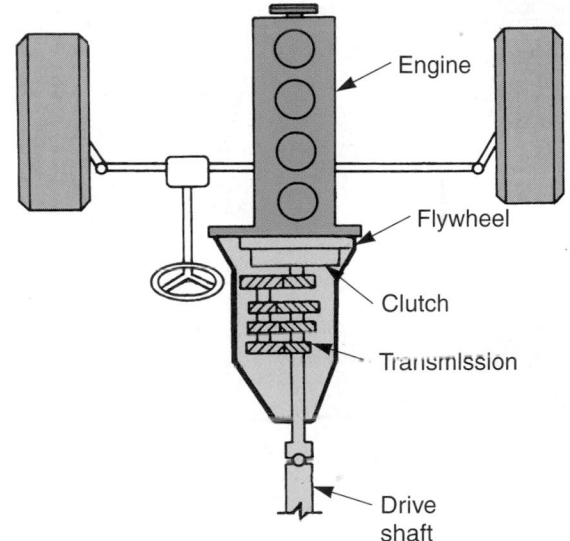

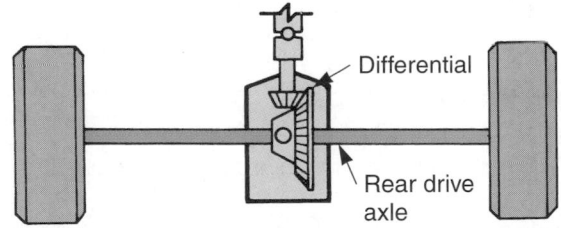

Rear-wheel drive

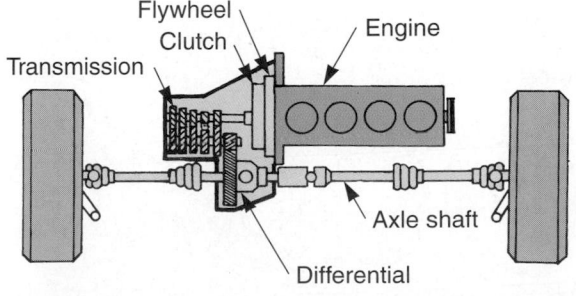

Front-wheel drive

Figure 53-1. Compare drive trains for front-wheel and rear-wheel drives. The clutch mounts on the engine flywheel.

Power flow from one unit to another can be controlled with a drive disc and a driven disc. Relating to an automotive clutch, one disc is fastened to the rear of the engine crankshaft. The other disc is attached to the input shaft of the transmission. Look at **Figure 53-2.**

When the discs do not touch, the crankshaft can rotate while the transmission input shaft remains stationary. However, when the transmission disc is forced into the spinning disc on the crankshaft, both spin together. Power flow is then transferred out of the engine and into the transmission.

Figure 52-3 shows a simplified, exploded view of a clutch. Note how the simplified parts fit together.

Basic Clutch Parts

Figure 53-4 shows the major parts of an automotive clutch. Study the names, locations, and relationship of the components.

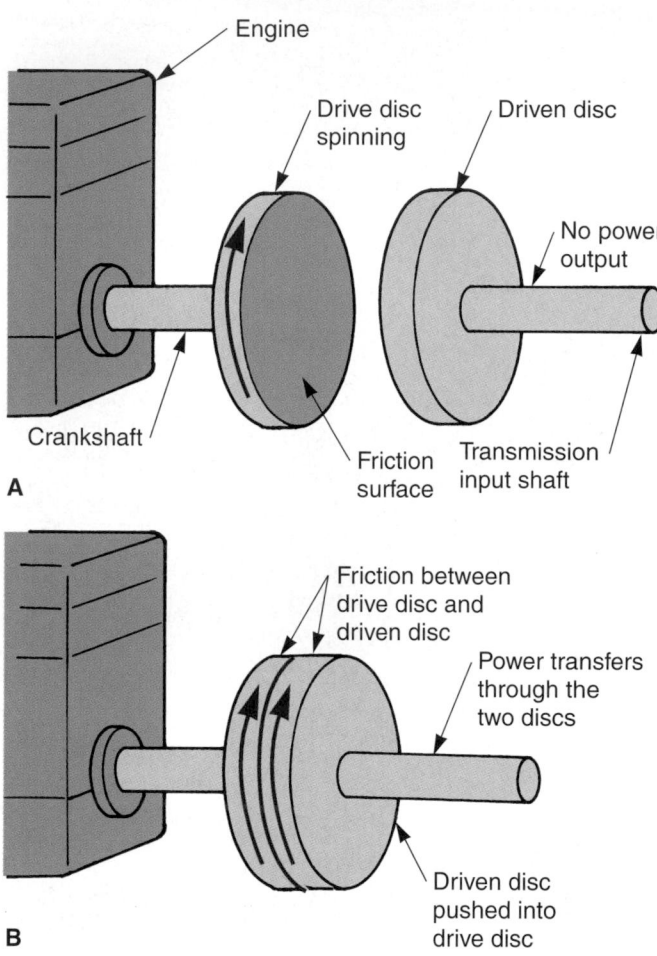

Figure 53-2. Rotating discs demonstrate the action of a clutch. A—The crankshaft spins the drive disc. The driven disc is not in contact with the drive disc. No power transfers. B—The two discs are pushed together. Friction causes the crankshaft disc to turn the other disc connected to the transmission input shaft. Power is transferred through the clutch.

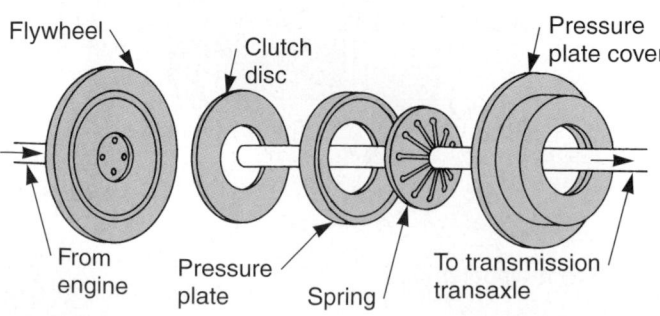

Figure 53-3. Study basic clutch parts. (Deere & Co.)

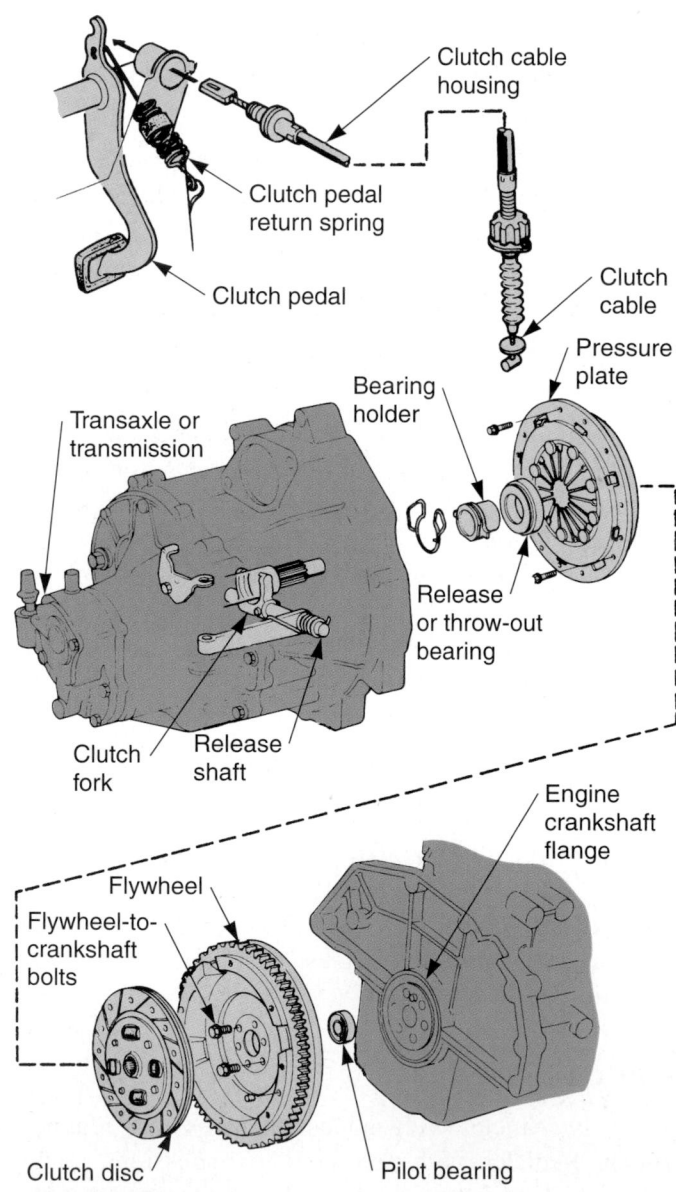

Figure 53-4. Note the fundamental components of a clutch and how parts are located in relation to each other. (Honda)

- *Clutch release mechanism*—this cable, linkage, or hydraulic system allows the driver to disengage the clutch with a foot pedal.
- *Clutch fork*—lever that forces release bearing into the pressure plate.
- *Release bearing*—bearing that reduces friction between clutch fork and pressure plate.
- *Pressure plate*—spring-loaded device that presses the clutch disc against the flywheel.
- *Clutch disc*—friction disc splined (fastened) to the transmission input shaft and pressed against the face of flywheel.
- *Flywheel*—provides mounting place for clutch and friction surface for clutch disc.
- *Pilot bearing*—bushing or bearing that supports forward end of transmission input shaft.

Clutch Action

When the driver presses the clutch pedal, the clutch release mechanism pulls or pushes on the clutch fork. See **Figure 53-5A.** The fork moves the release bearing into the center of the pressure plate. This causes the pressure plate face to pull away from the clutch disc, releasing the disc from the flywheel. The engine crankshaft can then turn without turning the clutch disc and the transmission input shaft.

When the clutch pedal is released by the driver, spring pressure inside the pressure plate pushes forward on the clutch disc. It locks the flywheel, disc, pressure plate, and transmission input together. The engine again rotates the transmission input shaft, transmission gears, drive train, and wheels. Refer to **Figure 53-5B.**

Clutch Construction

Now that you understand the basic action of a clutch, we will discuss, in more detail, how each part is made. This information will be useful when learning to diagnose and repair a clutch. Refer to **Figure 53-6** throughout the discussion.

Pilot Bearing

A *pilot bearing,* or *pilot bushing,* is pressed into the end of the crankshaft to support the end of the transmission input shaft. Usually, the pilot is a solid bronze bushing. It may also be a roller or ball bearing.

The end of the transmission input shaft has a small journal machined on its end. This journal slides inside the pilot bearing. The pilot prevents the transmission shaft and clutch disc from wobbling up and down when the clutch is released. It helps the input shaft center the disc on the flywheel.

Flywheel

The *flywheel* is the mounting place for the clutch. The pressure plate bolts to the flywheel face. The clutch disc is pinched and held against the flywheel by the spring action of the pressure plate. Look at **Figures 53-5 and 53-6.**

The face of the flywheel is precision machined to a smooth surface. Normally, the face of the flywheel that touches the clutch disc is made of iron. Even if the flywheel is aluminum, the face is iron because it wears well and dissipates heat well.

Clutch Disc

The *clutch disc,* also called *friction disc,* consists of a splined hub and a round metal plate covered with friction material (lining). One is pictured in **Figure 53-7.**

The splines (grooves) in the center of the clutch disc mesh with splines on the transmission input shaft. This makes the input shaft and disc turn together. However, the disc is free to slide back and forth on the shaft.

Clutch Disc Torsion Springs

Clutch disc *torsion springs,* also termed *damping springs,* help absorb some of the vibration and shock produced by clutch engagement. They are small coil springs located between the clutch disc splined hub and the friction disc assembly.

When the clutch is engaged, the pressure plate jams the stationary disc against the spinning flywheel. The torsion springs compress and soften the shock as the disc first begins to turn with the flywheel, **Figure 53-7.**

Clutch Disc Facing Springs

Clutch disc *facing springs,* also called *cushioning springs,* are flat, metal springs located under the disc's friction material. These springs have a slight wave (curve). They allow the friction material to flex inward slightly during initial clutch engagement. This also smoothes engagement.

Clutch Disc Friction Material

The clutch disc *friction material,* also called *disc lining* or *facing,* is made of heat-resistant substances. Refer to **Figure 53-7.**

Grooves are cut in the friction material to aid cooling and release of the clutch disc. Rivets are used to bond the friction material to both sides of the metal body of the disc.

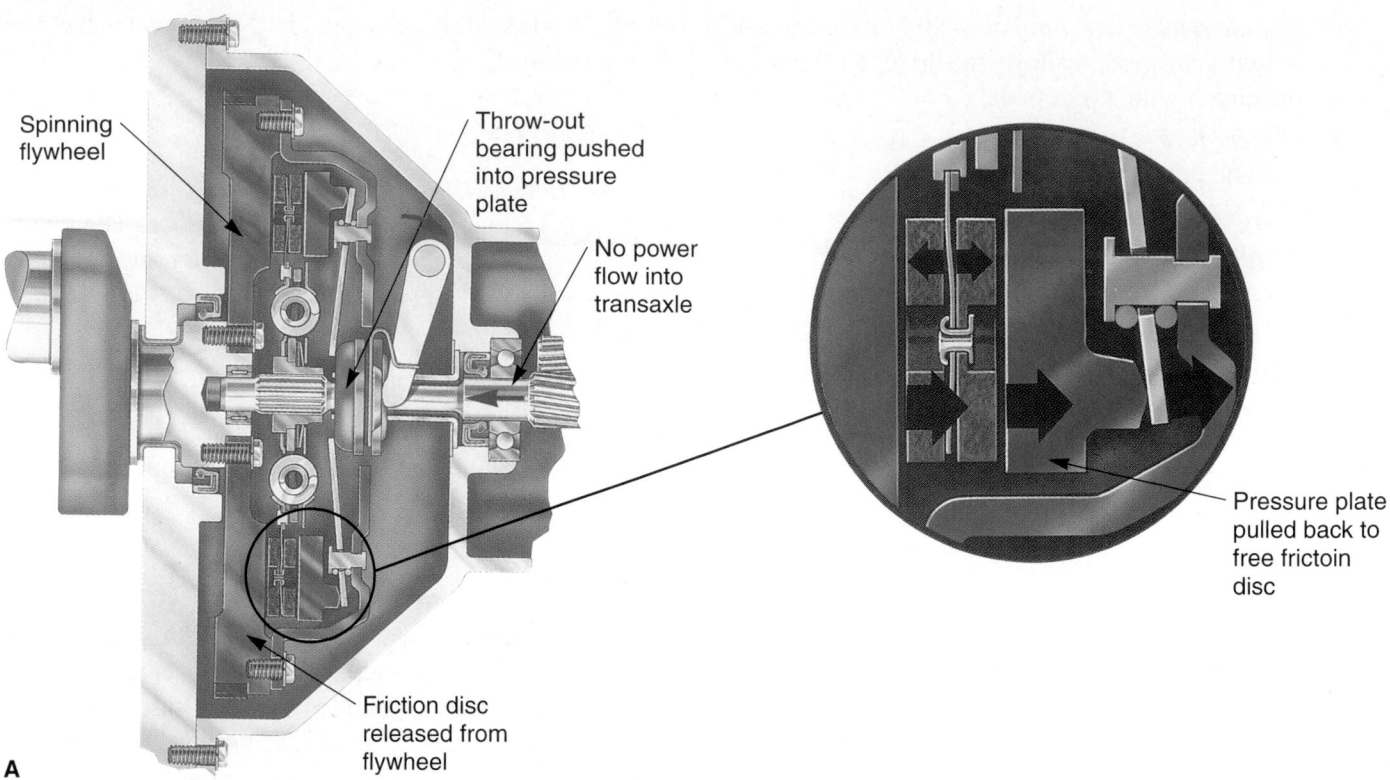

Spinning flywheel

Throw-out bearing pushed into pressure plate

No power flow into transaxle

Pressure plate pulled back to free frictoin disc

Friction disc released from flywheel

A

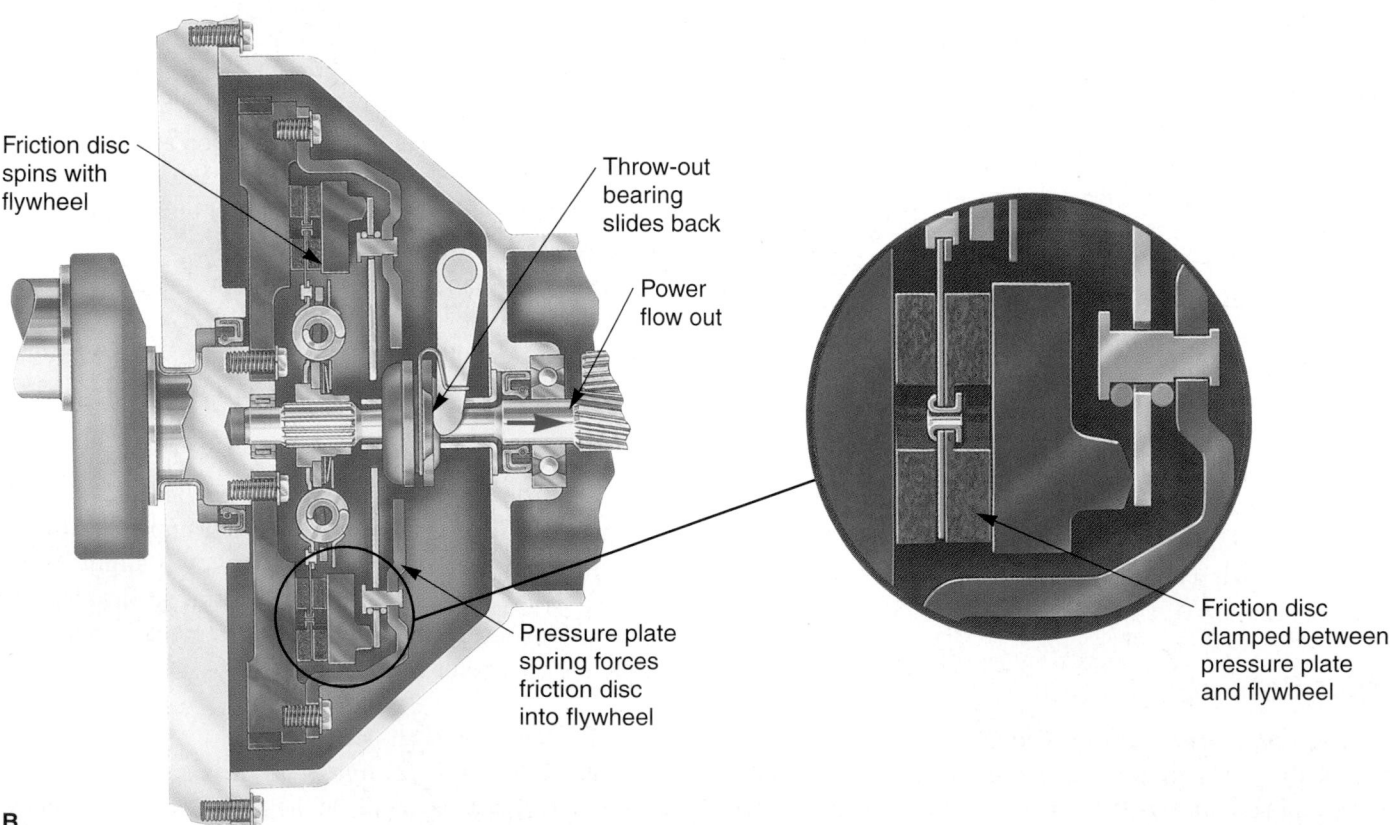

Friction disc spins with flywheel

Throw-out bearing slides back

Power flow out

Pressure plate spring forces friction disc into flywheel

Friction disc clamped between pressure plate and flywheel

B

Figure 53-5. Study what happens when you press and release the clutch pedal. A—The clutch pedal is pressed. The release mechanism forces the throw-out bearing into the pressure plate. This pushes in on the center spring of the pressure plate to pull it away from friction disc. The disc is then free of the spinning flywheel and no engine power transfers into transmission or transaxle. B—When the driver releases the clutch pedal, the throw-out bearing moves away from pressure plate. The pressure plate spring presses the friction disc up against the flywheel. Power is transferred into the transmission or transaxle to propel the vehicle. (LUK)

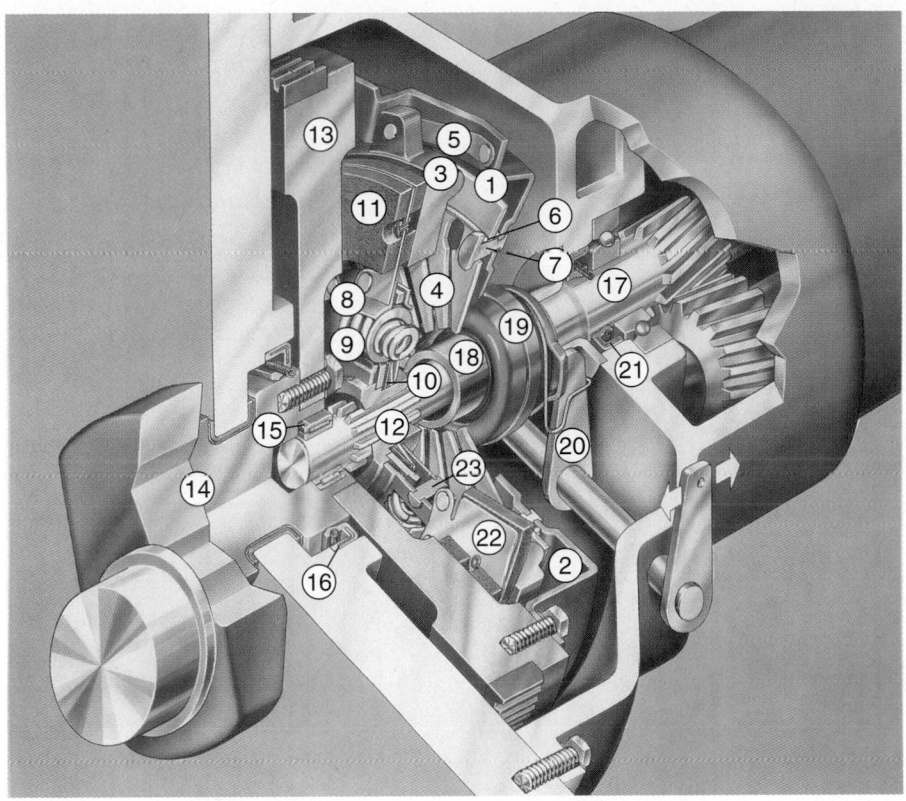

Figure 53-6. Cutaway view shows an assembled clutch. Note how the transmission input shaft extends through the clutch and into the pilot bearing in the crankshaft. 1. Clutch pressure plate. 2. Clutch cover. 3. Pressure plate. 4. Diaphragm spring. 5. Leaf springs/straps. 6. Pivot ring. 7. Diaphragm rivet. 8. Disc plate. 9. Torsion damper. 10. Friction device. 11. Clutch facing. 12. Hub. 13. Flywheel 14. Crankshaft. 15. Pilot bearing. 16. Main seal (crank). 17. Transmission shaft. 18. Quill. 19. Throw-out bearing. 20. Clutch fork. 21. Shaft seal 22. Cushion segment. 23. Stop pin. (LUK)

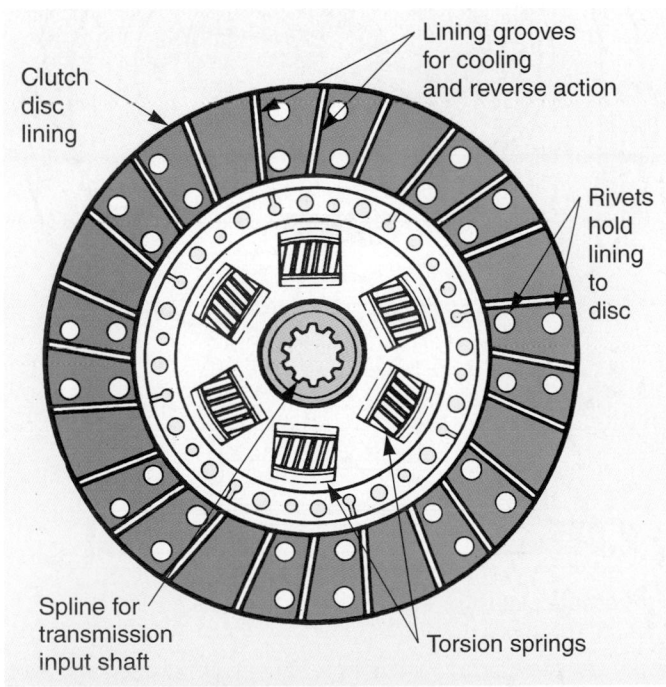

Figure 53-7. Clutch disc construction. Friction material is held on metal disc by rivets. Splines in center of the disc fit over splines on the transmission input shaft. (Chrysler Corp.)

Pressure Plate

The **pressure plate** is a spring-loaded device that can either lock or unlock the clutch disc and the flywheel. It bolts to the flywheel. The clutch disc fits between the flywheel and pressure plate. Refer to **Figure 53-6.**

There are two basic types of pressure plates: coil spring and diaphragm spring.

Coil Spring Pressure Plate

A **coil spring pressure plate** uses small coil springs, similar to valve springs. Study **Figure 53-8.**

The **pressure plate face** is a large ring that contacts the friction disc during clutch engagement. It is normally made of iron. The backside of the pressure plate face has pockets for the coil springs and brackets for hinging the release levers. During clutch action, the pressure plate face moves back and forth inside the clutch cover.

Pressure plate release levers are hinged inside the pressure plate to move the pressure plate face away from the disc and flywheel. Small clip springs fit around the release levers to keep the levers in a fully retracted (released) position. The springs also keep the levers from rattling.

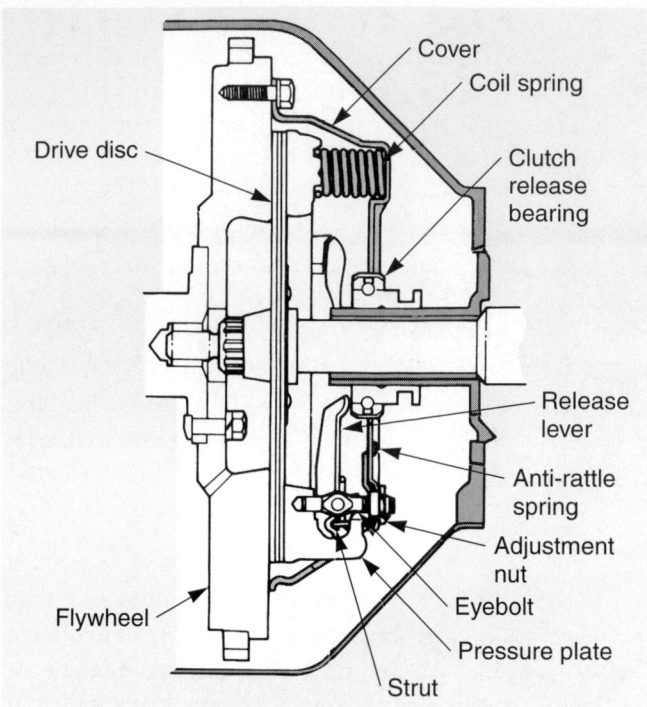

Figure 53-8. Cutaway view of coil spring pressure plate. Study the parts. (GMC)

The **pressure plate cover** fits over the springs, release levers, and pressure plate face. Its main purpose is to hold the parts of the pressure plate assembly together. Holes around the outer edge of the cover are for bolting the pressure plate to the flywheel.

Coil Spring Pressure Plate Action

When the clutch is disengaged, the release levers are pushed forward, toward the flywheel. This pries the pressure plate face away from the flywheel, compressing the coil springs. The clutch disc slides back and power is not transferred into the transmission.

When the clutch is engaged, the release bearing moves away from the release levers. Then, the pressure plate springs force the face and disc forward, into the rotating flywheel. The disc and transmission input shaft begin to spin and transmit power.

Centrifugal Action of Pressure Plate

A **semi-centrifugal pressure plate** uses weighted release levers or rollers, and the resulting centrifugal force, to increase clamping pressure on the clutch disc. The weights on the release levers or the rollers are positioned so that, as engine speed increases, their outward thrust acts on the release levers. The extra force helps keep the clutch from slipping. It also permits the use of slightly weaker coil springs to reduce the amount of foot pedal pressure required to disengage the clutch.

Diaphragm Spring Pressure Plate

A **diaphragm spring pressure plate** uses a single diaphragm spring instead of several coil springs. This type of pressure plate functions similarly to a coil spring pressure plate. See **Figure 53-9.**

The **diaphragm spring** is a large, round disc of spring steel. The spring is bent (dished) and has pie-shaped segments running from the outer edge to the center opening. The diaphragm spring is mounted in the pressure plate with the outer edge touching the back of the pressure plate face.

A **pivot ring** mounts behind the diaphragm spring. It is located part way in from the outer edge of the diaphragm spring.

Diaphragm Spring Pressure Plate Action

When the center of the diaphragm spring is pushed toward the engine, its outer edge bends back away from the engine. This lets the pressure plate face and clutch disc slide away from the spinning flywheel.

When the center of the diaphragm spring is released, the spring tries to return to its normal dished shape. As a result, the outer edge of the spring pushes the pressure plate face into the clutch disc.

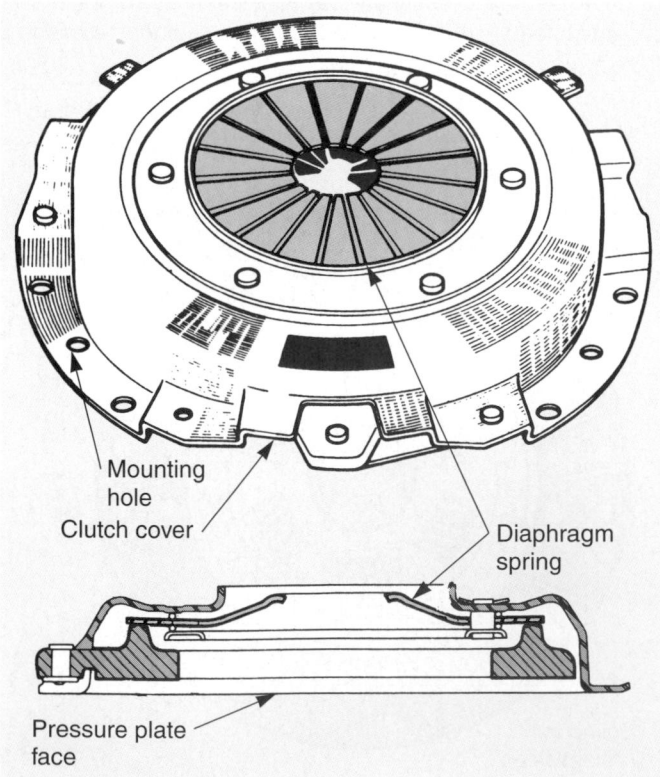

Figure 53-9. A diaphragm spring pressure plate uses a single spring instead of several small coil springs. Pushing in on the center of the spring bends the outer edge away from the drive disc. This releases the clutch disc. (Renault and Chrysler)

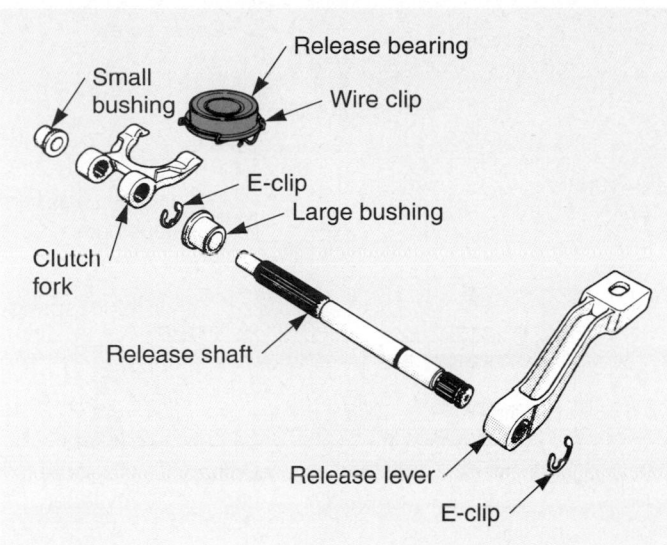

Figure 53-10. The throw-out bearing acts on the center of the pressure plate. It is an antifriction bearing that cuts down rubbing contact between the clutch fork and the pressure plate. Note other parts that operate this throw-out bearing. (Chrysler)

Release Bearing

The *release bearing,* also called a *throw-out bearing,* is usually a ball bearing and collar assembly that reduces friction between the pressure plate levers and the clutch fork. The release bearing is a sealed unit packed with grease. It slides on a hub or sleeve extending out from the front of the manual transmission or transaxle. Refer to **Figure 53-10.**

A few vehicles, especially foreign vehicles, use a graphite throw-out bearing. The ring-shaped block of friction resistant graphite presses on a smooth, flat plate on the clutch release levers.

The throw-out bearing usually snaps over the end of the clutch fork. Small spring clips hold the bearing on the fork (throw-out lever). Then, fork movement in either direction slides the throw-out bearing along the transmission hub sleeve.

Clutch Housing

The *clutch housing,* sometimes called a *bell housing,* bolts to the rear of the engine, enclosing the clutch assembly. It can be made of aluminum, magnesium, or cast iron. The manual transmission bolts to the back of the clutch housing. Look at **Figure 53-11.**

A hole is provided in the side of the clutch housing for the clutch fork. The fork or fork shaft sticks through the housing. A bracket or ball is needed to hold an arm-type fork.

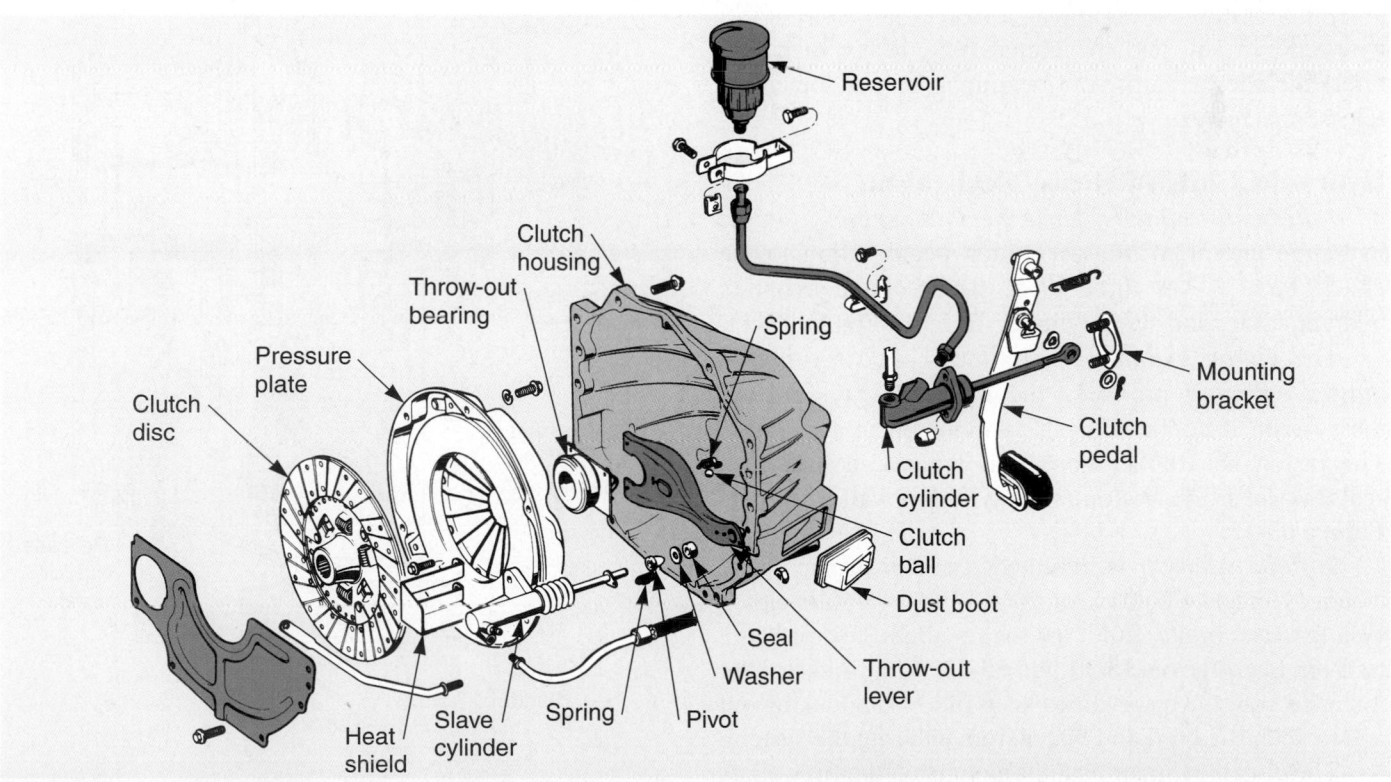

Figure 53-11. This diaphragm clutch is operated by a hydraulic release mechanism (slave cylinder). When the driver presses the clutch pedal, the clutch master cylinder develops pressure in the system. Pressure actuates the slave cylinder piston and operates the clutch fork to release the clutch. (Chrysler)

The lower front of the clutch housing usually has a thin, sheet metal cover. It can be removed for flywheel ring gear inspection or when the engine must be separated from the clutch assembly.

Clutch Fork

The **clutch fork,** also called a **clutch arm, throw-out lever,** or **release arm,** transfers motion from the clutch release mechanism to the throw-out bearing and pressure plate. There are several basic types of clutch forks.

A lever-type clutch fork sticks through a square hole in the bell housing and mounts on a pivot. When moved by the release mechanism, the clutch fork pries on the throw-out bearing to disengage the clutch. See **Figure 53-11.**

A rubber dust boot fits over the pivot-type clutch fork. The boot keeps road dirt, rocks, oil, water, and other debris from entering the clutch housing.

The second type of clutch fork has a round shaft. When the lever on the outer end of the assembly is moved, the shaft rotates. This swings the fork to push on the throw-out bearing, releasing the pressure plate.

Clutch Release Mechanisms

A **clutch release mechanism** allows the driver to operate the clutch. Generally, it consists of clutch pedal assembly, either mechanical linkage, a cable, or a hydraulic circuit, and the clutch fork. Many manufacturers include the throw-out bearing as part of the clutch release mechanism.

Hydraulic Clutch Release Mechanism

A **hydraulic clutch release mechanism** uses a simple hydraulic circuit to transfer clutch pedal action to the clutch fork. It has three basic parts: clutch cylinder, hydraulic line, and slave cylinder. Refer to **Figure 53-11.**

The **clutch cylinder,** sometimes called the **clutch master cylinder,** produces the hydraulic pressure for the system. It contains a piston mounted in a cylinder. The piston has rubber cups that produce a leakproof seal between the piston and cylinder wall. Look at **Figure 53-12.**

A **fluid reservoir** is mounted above or on top of the clutch cylinder to hold extra fluid. Most hydraulic clutch systems use brake fluid as the medium for pressure transfer. See **Figures 53-11** and **53-12.** A cap and seal are threaded onto the reservoir to keep fluid from leaking out and to keep road dirt and water from entering the system.

The clutch cylinder usually mounts on the firewall. A push rod links the clutch pedal and the cylinder piston. When the clutch pedal is pressed, the push rod moves the piston to produce pressure in the cylinder.

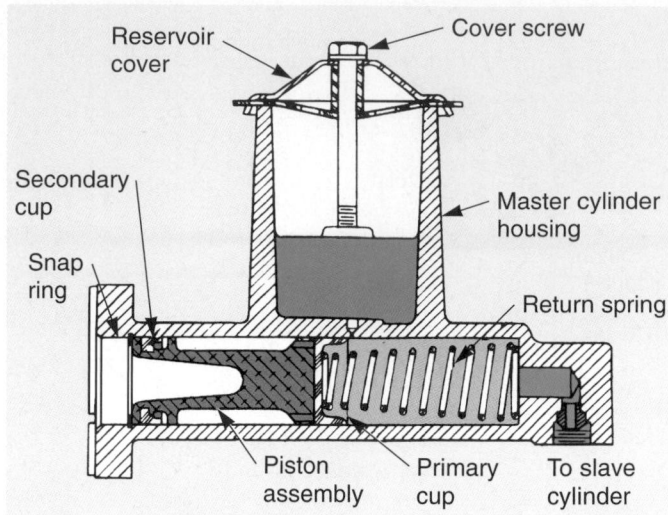

Figure 53-12. Cutaway view shows the inside of a clutch master cylinder. The clutch pedal and linkage push the piston and cup into the cylinder, producing hydraulic pressure.

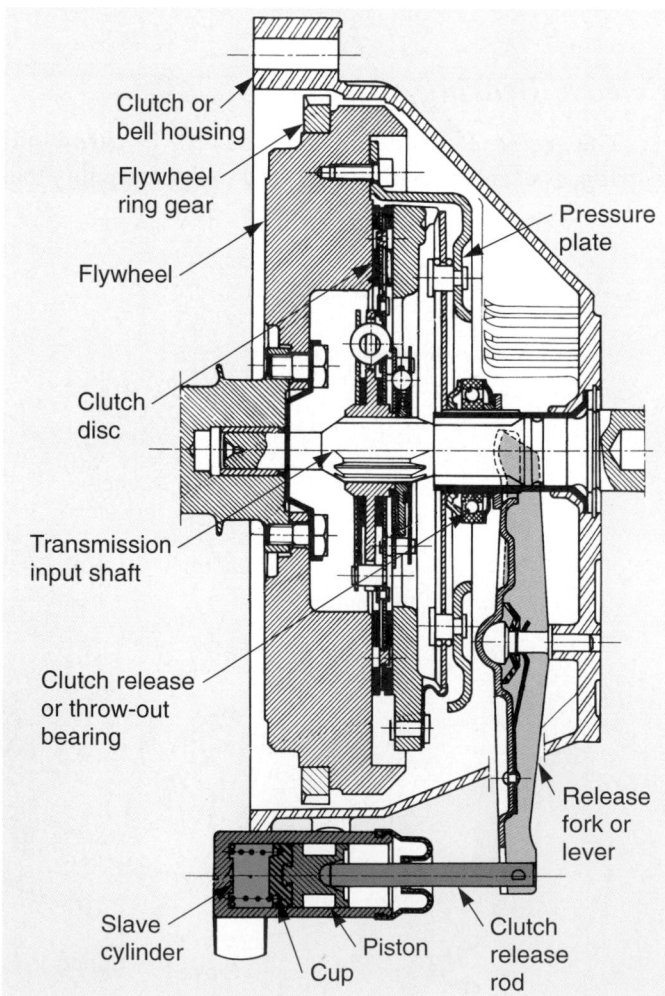

Figure 53-13. The slave cylinder releases this clutch. Pressure from the master cylinder enters the slave cylinder, moving the small piston toward the clutch fork. Study the other components. (Peugeot)

The *hydraulic line* is an assembly comprising rubber hose and metal tubing that moves high-pressure fluid from the clutch cylinder to the slave cylinder. When pressure is produced in the clutch cylinder, fluid flows through the hydraulic line, **Figure 53-11.**

The *slave cylinder* uses the system's hydraulic pressure to cause clutch fork movement. It contains a piston assembly inside a cylinder. When the master cylinder forces fluid into the slave cylinder, pressure pushes the piston outward, as in **Figure 53-13.**

Hydraulic Clutch Action

When the clutch pedal is depressed, linkage pushes on the piston in the clutch cylinder. A check valve in the clutch cylinder keeps fluid from entering the reservoir. As a result, fluid flows into the hydraulic line and slave cylinder. Pressure forms in the system and the slave cylinder piston is slid outward. The slave cylinder piston and push rod then act on the clutch fork to disengage the clutch.

When the clutch pedal is released, a spring on the clutch pedal pulls it back. Other springs inside the two cylinders push the pistons back into their retracted positions. Brake fluid flows back through the line and into the reservoir through the check valve.

Clutch Linkage Mechanism

A *clutch linkage mechanism* uses levers and rods to transfer motion from the clutch pedal to the clutch fork. When the pedal is pressed, a push rod shoves on the bellcrank. One arrangement is shown in **Figure 53-14.**

The *bellcrank* reverses the forward movement of the clutch pedal. The other end of the bellcrank is connected to a release rod. The *release rod* transfers bellcrank movement to the fork and usually provides a method of adjustment.

Clutch Cable Mechanism

A *clutch cable mechanism* uses a steel cable inside a flexible housing to transfer pedal movement to the

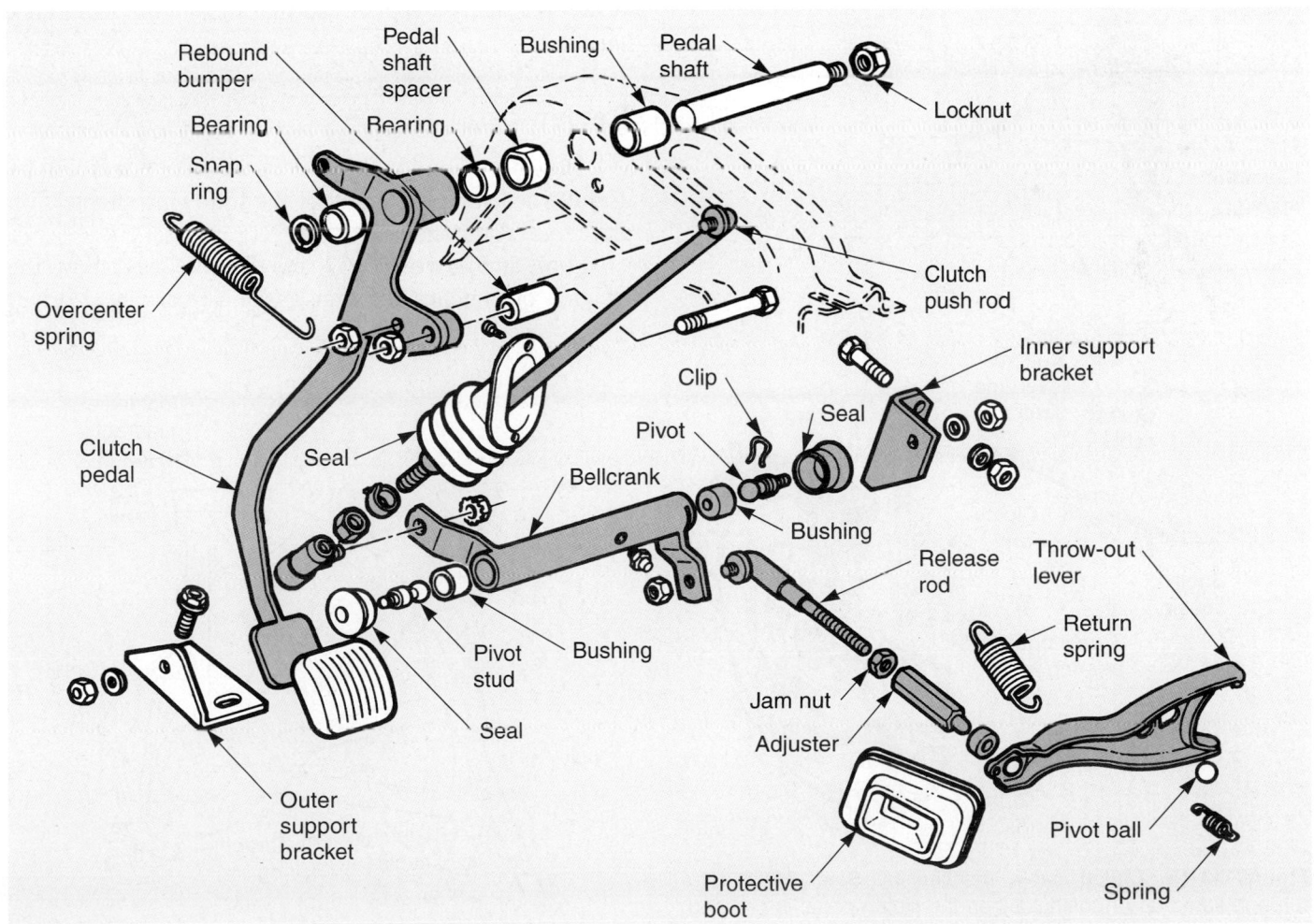

Figure 53-14. Clutch linkage release mechanism. Arms and rods transfer clutch pedal action to clutch fork. (Chrysler)

clutch fork. This is a simple mechanism, as shown in **Figure 53-15.**

The cable is usually fastened to the upper end of the clutch pedal. The other end of the cable connects to the clutch fork. The cable housing is mounted in a stationary position. This causes the cable to slide inside the housing whenever the clutch pedal is moved.

When the clutch pedal is depressed, the cable pulls on the clutch fork to disengage the clutch. When the clutch pedal is released, a strong spring pulls back on the pedal, cable, and fork to engage the clutch. One end of the clutch cable housing usually has a threaded sleeve for clutch adjustment.

Automatic Clutch Adjuster

An **automatic clutch adjuster** removes play from the clutch cable as components wear. The clutch pedal has a quadrant and pawl device. A spring inside the quadrant

applies light tension on the cable to take up extra slack. Refer to **Figures 53-15** and **53-16.**

When the clutch is applied, the pawl locks into one of the quadrant teeth and the clutch cable is activated. Then, if there is too much play in the cable, the pawl will ratchet over the quadrant teeth when the clutch pedal is released. This positions the pawl in another quadrant tooth and takes up any slack in the clutch cable, **Figure 53-16.**

Clutch Start Switch

The **clutch start switch** prevents the engine from cranking (starting motor operation) unless the clutch pedal is depressed. It serves as a safety device that keeps the engine from starting while in gear. The clutch start switch is usually mounted on the clutch pedal assembly.

Wires from the ignition switch feed starter solenoid current through the clutch start switch. Unless the switch is closed (clutch pedal depressed), the switch prevents current from reaching the starter solenoid.

When the transmission is in neutral, the clutch pedal switch is usually bypassed so that the engine will crank and start.

Review

The information you just covered on clutches will be useful when studying many of the following chapters on drive train components.

Look at **Figures 53-17** and **53-18.** They show the location of clutches for transmission (rear-wheel drive)

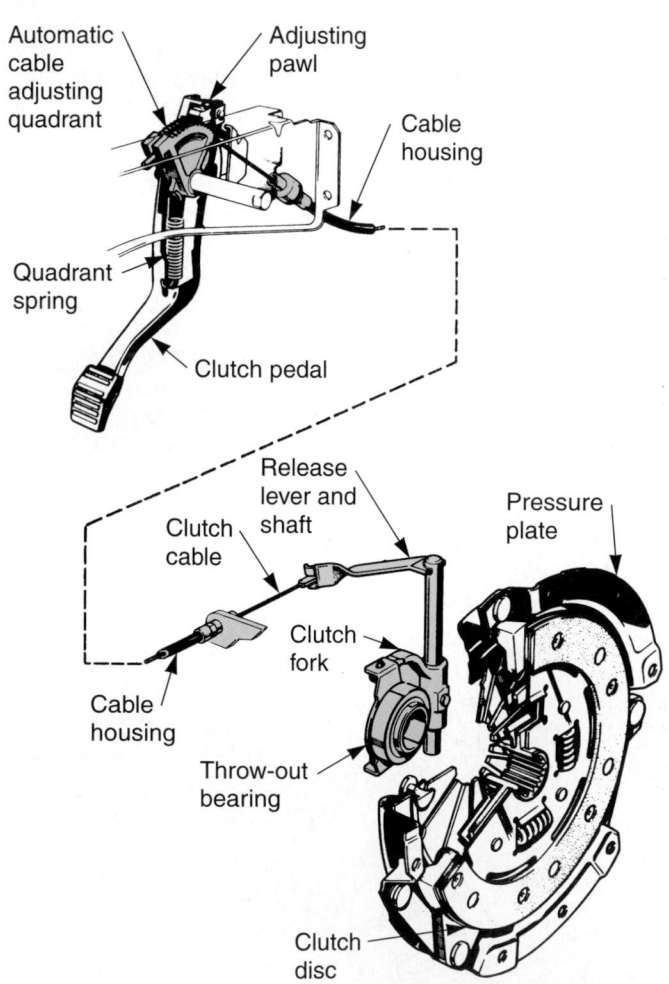

Figure 53-15. Clutch cable mechanism. Steel cable runs through stationary housing. When the clutch pedal is pressed, the cable slides in the housing to operate the release lever and throw-out bearing. Also note automatic cable adjuster on foot pedal assembly. (Ford)

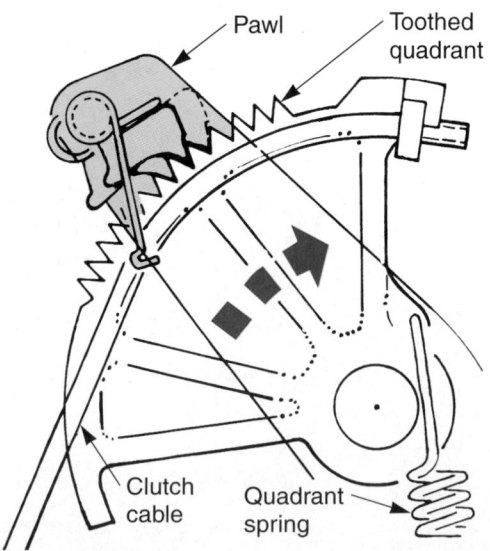

Figure 53-16. Automatic clutch cable adjusting device. With excess slack in the cable, the pawl ratchets over teeth on the quadrant. This takes up play in the cable as parts wear. (Ford)

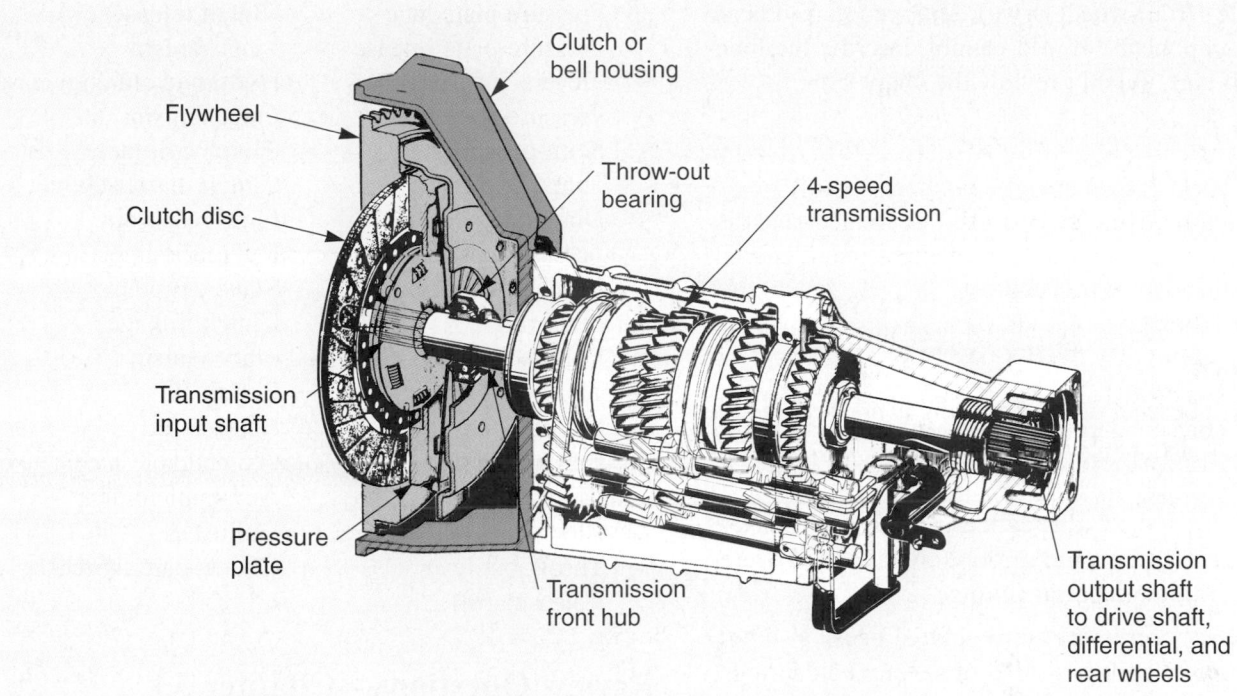

Figure 53-17. Note how the clutch installs in the bell housing. The transmission bolts to the rear of the bell housing. Manual transmissions are covered in Chapters 55 and 56. (Peugeot)

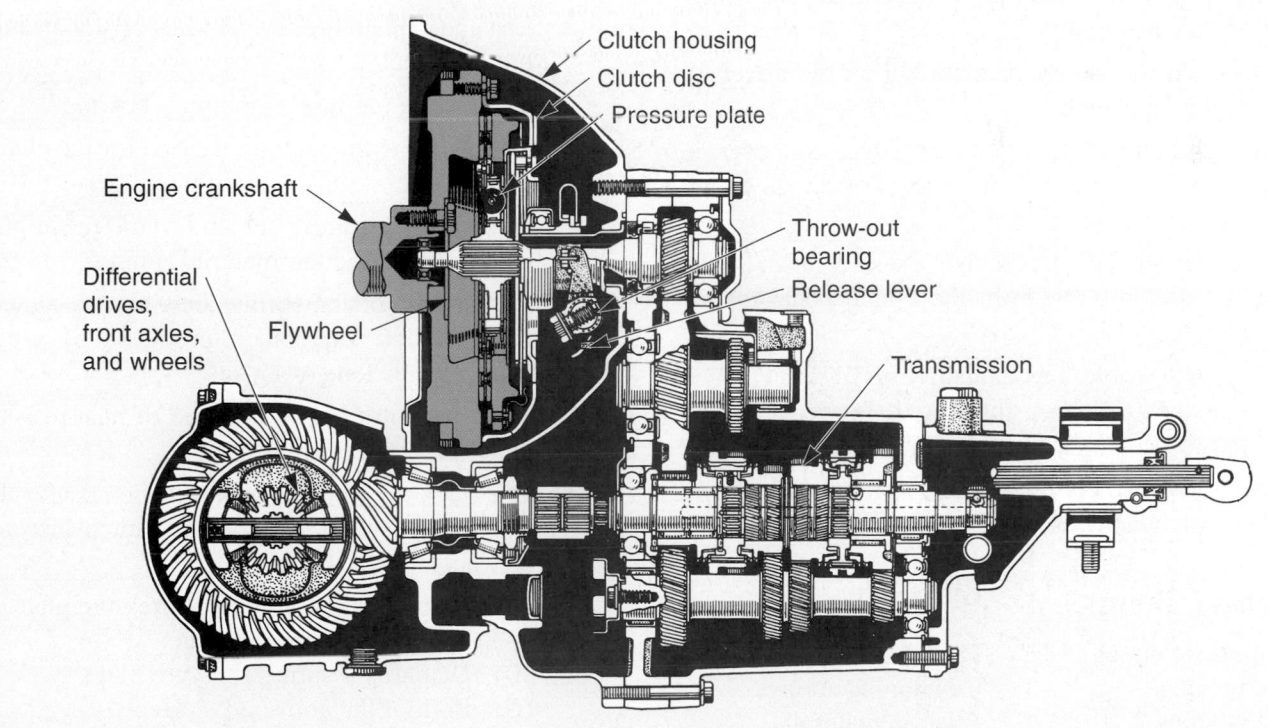

Figure 53-18. See how the clutch is located in relation to the manual transaxle for this front-wheel drive vehicle. Transaxles are covered in Chapters 63 and 64. (Toyota)

and transaxle (front-wheel drive) equipped cars. Locate the clutch components. If you cannot describe the functions of each part, quickly review the chapter.

Summary

- An automotive clutch is used to connect and disconnect the engine and manual (hand-shifted) transmission or transaxle.

- Only vehicles with manual transmissions require a clutch.

- A pilot bearing or pilot bushing is pressed into the end of the crankshaft to support the end of the transmission input shaft.

- The clutch disc, also called friction disc, consists of a splined hub and a round metal plate covered with friction material (lining).

- A diaphragm pressure plate uses a single diaphragm spring instead of several coil springs.

- The release bearing, also called a throw-out bearing, is usually a ball bearing and collar assembly that reduces friction between the pressure plate levers and the clutch fork.

- The clutch housing, sometimes called a bell housing, bolts to the rear of the engine, enclosing the clutch assembly.

- A clutch release mechanism allows the driver to operate the clutch.

- A hydraulic clutch release mechanism uses a simple hydraulic circuit to transfer clutch pedal action to the clutch fork.

- A clutch linkage mechanism uses levers and rods to transfer motion from the clutch pedal to the clutch fork.

- A clutch cable mechanism uses a steel cable inside a flexible housing to transfer pedal movement to the clutch fork.

- An automatic clutch adjuster removes play from the clutch cable as components wear.

Important Terms

Automotive clutch	Facing springs
Pilot bearing	Cushioning springs
Pilot bushing	Friction material
Flywheel	Disc lining
Clutch disc	Facing
Friction disc	Pressure plate
Torsion springs	Coil spring pressure
Damping springs	plate

Pressure plate face	Clutch release
Pressure plate release	mechanism
levers	Hydraulic clutch release
Pressure plate cover	mechanism
Semi-centrifugal	Clutch cylinder
pressure plate	Clutch master cylinder
Diaphragm spring	Fluid reservoir
pressure plate	Hydraulic line
Diaphragm spring	Slave cylinder
Pivot ring	Clutch linkage
Release bearing	mechanism
Throw-out bearing	Bellcrank
Clutch housing	Release rod
Bell housing	Clutch cable mechanism
Clutch fork	Automatic clutch
Clutch arm	adjuster
Throw-out lever	Clutch start switch
Release arm	

Review Questions—Chapter 53

Please do not write in this text. Place your answers on a separate sheet of paper.

1. An automotive _____ connects and disconnects the engine and manual transmission or transaxle.

2. List and explain the seven basic parts of an automotive clutch.

3. What is the purpose of the pilot bearing?

4. The _____ is the mounting place for the clutch.

5. The clutch _____, also called _____ _____, consists of a splined hub and round metal plate covered with friction material (lining).

6. Clutch disc torsion springs help absorb some of the vibration and shock produced by clutch engagement. True or False?

7. What part of a clutch is made of heat-resistant material?

8. The _____ _____ is a spring-loaded device that can either lock or unlock the clutch disc and flywheel.

9. This is *not* a common type of pressure plate.
   (A) Coil spring pressure plate.
   (B) Diaphragm spring pressure plate.
   (C) Both of the above are correct.
   (D) None of the above are correct.

10. How does the release or throw-out bearing work?

11. What is the clutch or bell housing?

12. Explain clutch fork action.

13. A _____ clutch release mechanism uses a clutch cylinder to operate a slave cylinder.

14. A linkage or cable release mechanism can be used to transfer pedal movement to the clutch fork. True or False?

15. What is a clutch start circuit?

● ASE-Type Questions

1. Technician A says automobiles equipped with an automatic transmission require the use of a clutch. Technician B says automobiles equipped with an automatic transmission do not require the use of a clutch. Who is right?
   (A) A only.
   (B) B only.
   (C) Both A and B.
   (D) Neither A nor B.

2. Technician A says an automotive clutch connects and disconnects the engine and drive shaft. Technician B says an automotive clutch connects and disconnects the engine and torque converter. Who is right?
   (A) A only.
   (B) B only.
   (C) Both A and B.
   (D) Neither A nor B.

3. Technician A says a clutch fork is a linkage that allows the driver to disengage the clutch by pressing down on the foot pedal. Technician B says a clutch fork is a lever that forces the release bearing into the pressure plate. Who is right?
   (A) A only.
   (B) B only.
   (C) Both A and B.
   (D) Neither A nor B.

4. Technician A says a pilot bearing is a bushing that supports the forward end of the transmission input shaft. Technician B says a pilot bearing is a bushing that reduces friction between the clutch fork and pressure plate. Who is right?
   (A) A only.
   (B) B only.
   (C) Both A and B.
   (D) Neither A nor B.

5. Technician A says the engine's flywheel is not utilized during the operation of an automotive clutch assembly. Technician B says the engine's flywheel is a basic component utilized during the operation of an automotive clutch assembly. Who is right?
   (A) A only.
   (B) B only.
   (C) Both A and B.
   (D) Neither A nor B.

6. While discussing the functions of a pressure plate, Technician A says the purpose of a pressure plate is to press the pilot bearing against the engine's flywheel. Technician B says the pressure plate is used to press the clutch disc against the engine's flywheel. Who is right?
   (A) A only.
   (B) B only.
   (C) Both A and B.
   (D) Neither A nor B.

7. Technician A says the clutch disc is bolted to the transmission input shaft. Technician B says the clutch disc is splined to the transmission input shaft. Who is right?
   (A) A only.
   (B) B only.
   (C) Both A and B.
   (D) Neither A nor B.

8. Technician A says a pilot bearing is normally an aluminum bushing. Technician B says a pilot bearing is usually a solid bronze bushing. Who is right?
   (A) A only.
   (B) B only.
   (C) Both A and B.
   (D) Neither A nor B.

9. Technician A says the face of the flywheel that touches the clutch disc is normally made of iron. Technician B says the face of the flywheel that touches the clutch disc is normally made of aluminum. Who is right?
   (A) A only.
   (B) B only.
   (C) Both A and B.
   (D) Neither A nor B.

10. Which of the following is another name for an automotive clutch disc?
    (A) Throw-out plate.
    (B) Friction disc.
    (C) Pressure plate.
    (D) None of the above.

11. Technician A says clutch disc torsion springs are used to fasten the clutch disc to the transmission input shaft. Technician B says clutch disc torsion springs help absorb some of the vibration produced by clutch engagement. Who is right?
    (A) A only.
    (B) B only.
    (C) Both A and B.
    (D) Neither A nor B.

12. Technician A says snap rings are normally used to fasten the clutch disc friction material to the body of the clutch disc. Technician B says rivets are normally used to fasten the clutch disc friction material to the body of the clutch disc. Who is right?
    (A) A only.
    (B) B only.
    (C) Both A and B.
    (D) Neither A nor B.

13. Technician A says the clutch disc is mounted behind the flywheel and pressure plate. Technician B says the clutch disc fits between the flywheel and pressure plate. Who is right?
    (A) A only.
    (B) B only.
    (C) Both A and B.
    (D) Neither A nor B.

14. Which of the following is another name for a "throw-out bearing?"
    (A) Pilot bushing.
    (B) Fork bushing.
    (C) Release bearing.
    (D) Pressure plate hub.

15. Technician A says an automobile's clutch housing is sometimes made of aluminum. Technician B says an automobile's clutch housing is sometimes made of cast iron. Who is right?
    (A) A only.
    (B) B only.
    (C) Both A and B.
    (D) Neither A nor B.

Activities—Chapter 53

1. Identify principles of fluids that underlie the operation of a hydraulic clutch; explain its operation to the class.

2. Sketch out a simple clutch, label its parts, and explain its operation.

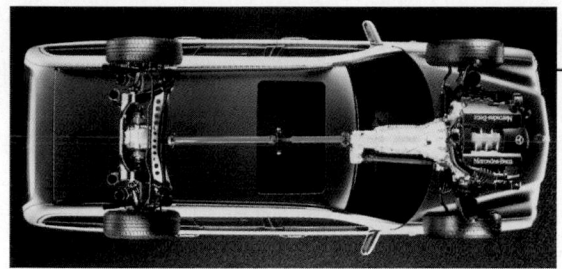

Chapter 54

Clutch Diagnosis and Repair

After studying this chapter, you will be able to:

- Troubleshoot common clutch problems.
- Describe symptoms of typical clutch troubles.
- Adjust a clutch.
- Remove, repair, and install a clutch.
- Inspect clutch parts for wear and damage.
- Cite safety rules and demonstrate safe work procedures.
- Correctly answer ASE certification test questions on clutch diagnosis and repair.

An automotive technician must be able to quickly and accurately diagnose and repair clutch problems. Simply replacing parts is not sufficient. A technician must know why the old clutch failed. If she or he installs new parts without properly diagnosing the cause of the failure, the new parts could also fail in a short period of time.

This chapter will help develop the skills needed to service automotive clutches. It will also provide you with the background needed to use a service manual properly.

Warning

Clutch disc lining or friction material sometimes contains asbestos. This is especially true of older clutch discs. Asbestos is a known cancer-causing substance. Do not breathe asbestos dust. Avoid using an air hose to blow asbestos dust off clutch parts. Wear a respirator or use an enclosed vacuum system.

Diagnosing Clutch Problems

An automobile clutch normally provides dependable service for thousands of miles. However, the service life of clutch components vary greatly from vehicle to vehicle. One vehicle's clutch might last 100,000 miles, while another's could fail in only 50,000 miles.

Stop-and-go city traffic will wear out a clutch quicker than highway driving. Every time the clutch is engaged, the clutch disc and other components are subjected to considerable friction, heat, and wear.

Driver abuse commonly causes premature clutch trouble. For instance, "riding the clutch" (overslipping the clutch and resting a foot on the clutch pedal while driving) can cause early clutch failure.

Verify the Clutch Problem

After talking to the service writer or the customer about the problem, verify the complaint. Test drive the vehicle and make your own decisions about the clutch troubles. Gather as much information as you can. Check the action of the clutch pedal. Listen for unusual noises. Feel for clutch pedal vibrations.

Use this information, your knowledge of clutch operating principles, and a service manual troubleshooting chart to decide which components are at fault. You must determine whether the clutch failure was due to normal wear, improper driving techniques, incorrect clutch adjustment, or other problems.

Common Clutch Part Failures

Damaged or worn clutch parts can cause a variety of clutch problems: slipping, grabbing, dragging, abnormal noises, and vibration. It is important to know the symptoms produced by these problems and the parts that might be the cause. See **Figure 54-1.**

A worn clutch disc will cause clutch slippage and, sometimes, damage to the flywheel and pressure plate. See **Figure 54-2.**

An overheated flywheel can have surface cracks and hardened or warped areas that upset clutch operation. Cracks in the surface of a flywheel can cause rapid clutch disc wear. If the flywheel is warped, the clutch may grab or vibrate upon acceleration.

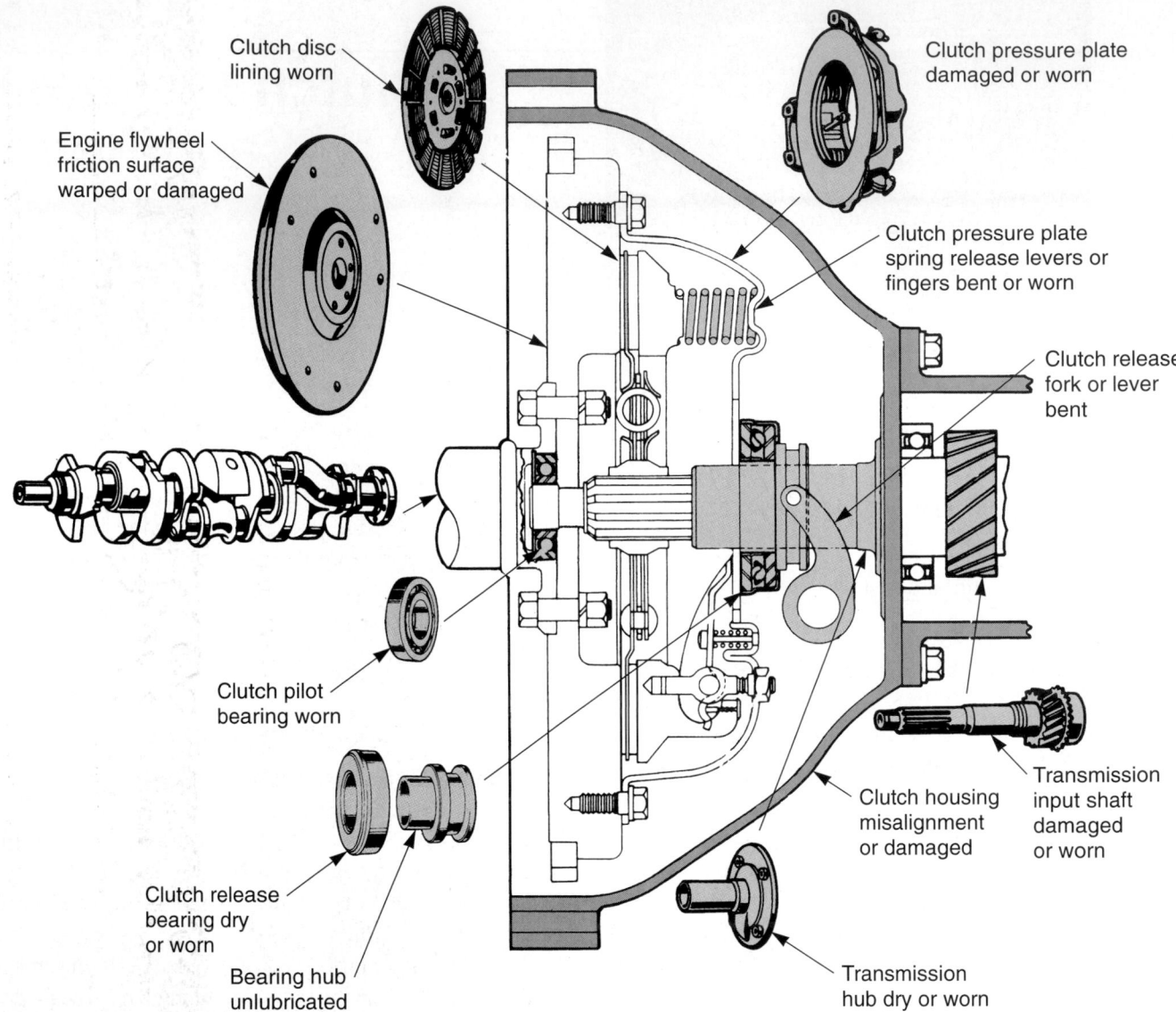

Clutch disc
lining worn

Engine flywheel
friction surface
warped or damaged

Clutch pressure plate
damaged or worn

Clutch pressure plate
spring release levers or
fingers bent or worn

Clutch release
fork or lever
bent

Clutch pilot
bearing worn

Clutch release
bearing dry
or worn

Bearing hub
unlubricated

Clutch housing
misalignment
or damaged

Transmission
input shaft
damaged
or worn

Transmission
hub dry or worn

Figure 54-1. Study typical problems that can develop in a clutch. Keep these problems in mind during clutch diagnosis and repair.

A bad throw-out bearing will produce a grinding noise whenever the clutch pedal is pushed down. The roller bearings may be dry (out of grease). See **Figure 54-3.**

A bad pressure plate can cause clutch slippage, as well as clutch release problems (stiff clutch pedal, abnormal noises, clutch grabbing, or clutch dragging). The springs inside the clutch could lose tension or break. The release levers could be bent or out of adjustment. The face of the pressure plate could also be scored. Look at **Figures 54-4** and **54-5.**

A worn pilot bearing will allow the transmission input shaft and clutch disc to wobble up and down. This can cause clutch vibration, abnormal noises, and damage to the transmission. A bent or worn clutch fork can prevent the clutch from releasing properly.

Free travel (free play) is the distance the clutch pedal or clutch fork moves before the throw-out bearing acts on the pressure plate. Free travel is needed to ensure complete clutch engagement. See **Figure 54-6.**

Excess clutch pedal free travel is often due to part wear that reduces clutch release action. This can cause the clutch drag even when the pedal is fully pushed to the floor. *Insufficient clutch pedal free travel* can be due to bent parts and similar mechanical problems. Insufficient free travel in the clutch release mechanism can cause the clutch to stay partially disengaged and result in clutch slippage.

Clutch Slippage

Typically, *clutch slippage* is noticed when the engine *races* (engine speed increases quickly) without an

A

Separated
lining

B

C

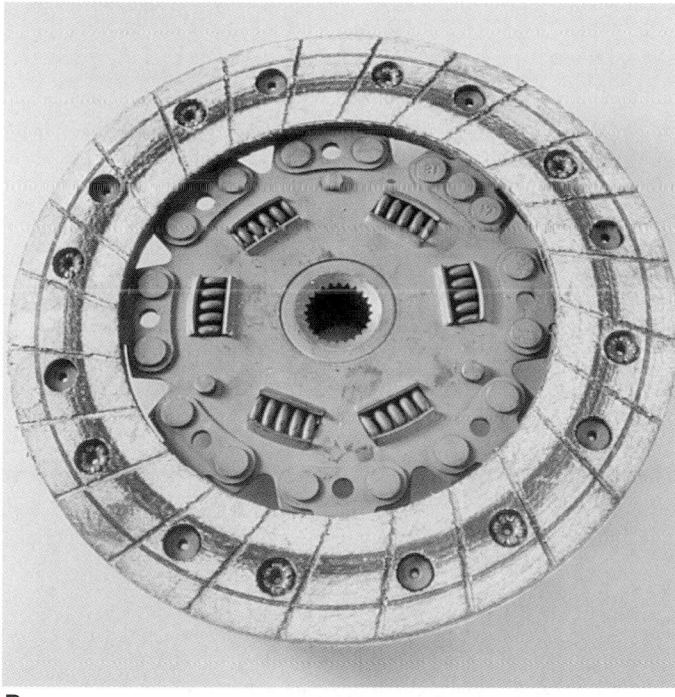

D

Figure 54-2. Study clutch disc problems and their causes. A—This clutch disc has worn friction material from extended use, a slip-ping clutch, or weak pressure plate springs. B—This clutch disc exploded from high-speed shifts. Note how friction material has fallen off the disc. This clutch slipped so badly that the vehicle would not move. C—Clutch friction material has been overheated and burned because of oil contamination, not enough release mechanism clearance, or from the driver riding the clutch pedal. D—This clutch disc has scored friction material from an unmachined flywheel or the reuse of a badly scored pressure plate. (LUK)   *(Continued)*

E

F

Figure 54-2. *Continued.* E—Friction disc hub splines were badly damaged when the transmission was drawn into place with bolts. The transmission input shaft was forced through the splines. F—This clutch disc hub wear was caused by installing the disc backwards. The metal hub rubbed on the flywheel and held friction material away from the flywheel. (LUK)

A

B

Figure 54-3. Common throw-out bearing related failures. A—This dry throw-out bearing made a loud grinding noise any time the clutch pedal was pressed. The bearings had run dry of grease and started rapid wear and high friction. B—This transmission front bearing retainer hub was badly worn and damaged by a dry, locked, spinning throw-out bearing. (LUK)

increase in the vehicle's road speed. This is caused by the clutch friction disc sliding between the flywheel and the pressure plate. Clutch slippage usually occurs as the vehicle is accelerated from a standstill, when shifting, or when under a heavy load (climbing a hill or pulling a trailer).

To test the clutch for slippage, set the emergency brake and start the engine. Place the transmission or transaxle in high gear. Then, try to drive the vehicle forward by slowly releasing the clutch pedal.

A clutch in good condition should lock up, killing (stalling or stopping) the engine immediately. A badly

A

B

Figure 54-4. Common pressure plate friction surface failures. A—This pressure plate surface was badly scored. Causes could have been a worn friction disc, the driver slipping the clutch, or binding linkage. B—Pressure plate shows heat-checking damage. This required massive slippage that generated high friction and heat. Causes could be oil or grease contamination, poor clutch adjustment (too tight), binding linkage, or engaging the clutch at high engine speed. (LUK)

A

B

Figure 54-5. Pressure plate spring and release finger problems. A—This pressure plate has broken diaphragm springs. This could have happened because of installation error, poor adjustment (too tight, forcing the throw-out bearing too far into pressure plate), or clutch spring problems. B—The release levers on this pressure plate are broken. This could be due to no free play in the release mechanism, bad release bearing, or improper part alignment.

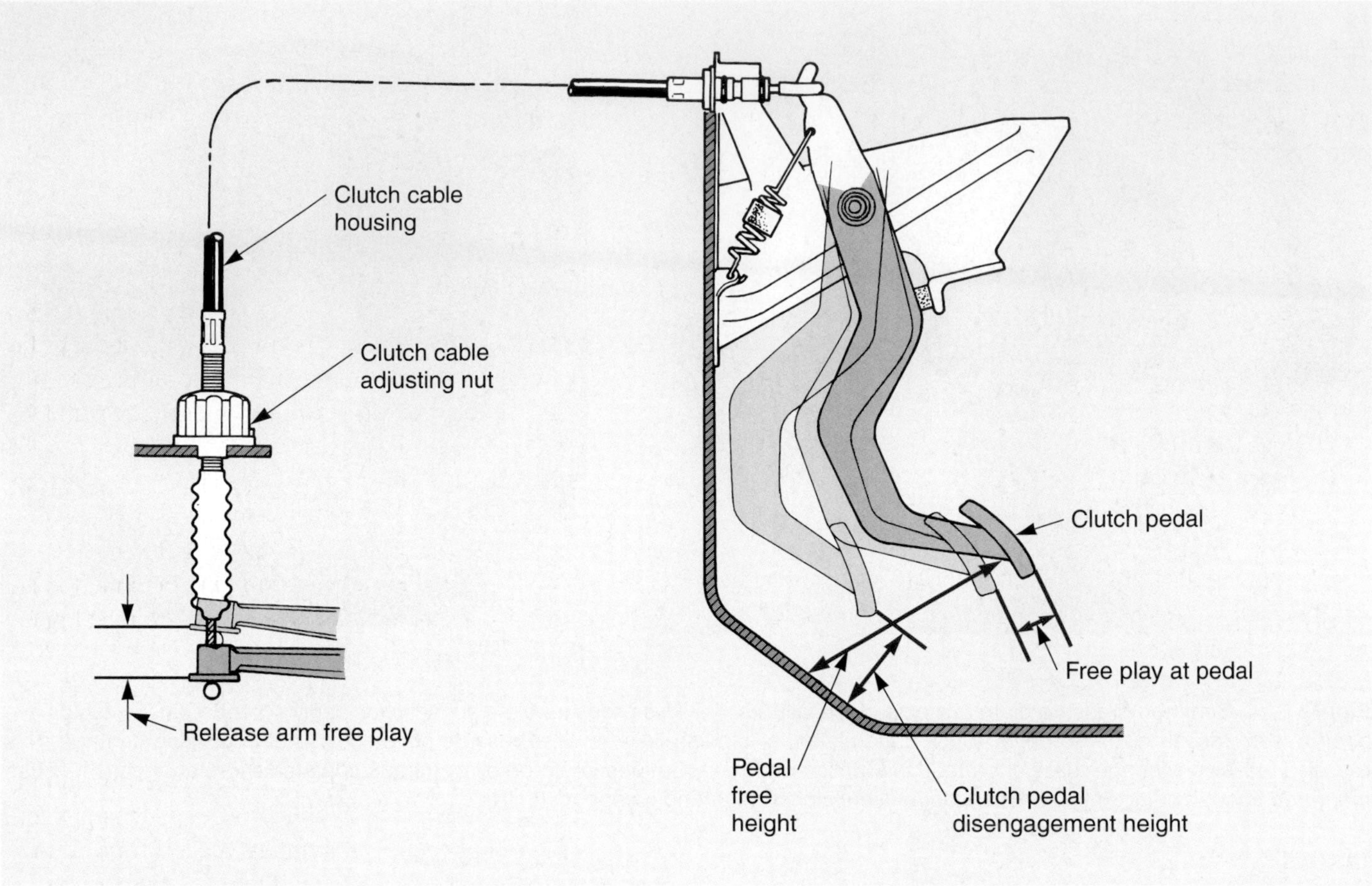

Figure 54-6. Clutch pedal free play is the distance the pedal moves until the throw-out bearing touches the pressure plate. Release arm or fork free play is the distance the end of the arm moves back and forth with the clutch released. Note adjuster nut for changing cable free play. (Honda)

slipping clutch may allow the engine to run, even with the clutch pedal fully released. Partial clutch slippage could let the engine run momentarily before stalling.

 Caution
Never let a clutch slip for more than a second or two. The extreme heat generated by slippage can damage the flywheel or pressure plate faces.

Some common causes of clutch slippage include a misadjusted clutch, binding clutch linkage or cable, worn clutch disc, broken motor mount, and oil or grease on the clutch disc (leaking oil seal).

Improper clutch adjustment can cause slippage by keeping the throw-out bearing in contact with the pressure plate in the released position. Even with your foot off the pedal, the release mechanism will act on the clutch fork and throw-out bearing. This can make the clutch slip under load.

A binding clutch release mechanism can also cause clutch slippage. Parts can become rusted, bent, misaligned, or damaged.

A broken motor mount (engine mount) can cause clutch slippage by allowing engine movement to bind the clutch linkage. Under load, the engine can lift up in the engine compartment. This can shift the clutch linkage and push on the clutch fork.

If clutch slippage is not caused by a problem with the clutch release mechanism, then the trouble is normally inside the clutch housing. You would need to remove the transmission and clutch components for further inspection.

Grabbing Clutch

A *grabbing,* or *chattering, clutch* will produce a very severe vibration or jerking motion when the vehicle is accelerated from a standstill. Even though the driver is slowly releasing the clutch pedal, it will feel as if the clutch pedal is being rapidly pumped up and down. A loud banging or chattering sound may be heard as the vehicle's body vibrates.

Normally, clutch grabbing is caused by problems with components inside the clutch housing (friction disc, flywheel, or pressure plate). The clutch will usually require disassembly to correct these problems.

However, a broken motor mount can also cause erratic clutch linkage operation. Check the engine mounts before removing the clutch from the vehicle.

Dragging Clutch

A *dragging clutch* will normally make the transmission or transaxle grind when trying to engage and shift gears. Something is causing the friction disc to stay engaged to the flywheel. This keeps the transmission input shaft spinning, even when the clutch is disengaged. Severe clutch drag will make the vehicle move forward whenever the engine is running and the transmission is in gear.

One of the most common causes of a dragging clutch is too much pedal free travel. With excessive free travel, the pressure plate will not be fully released when the pedal is pushed to the floor. Always check clutch adjustment when symptoms point to a dragging clutch.

A dragging clutch can also be caused by a warped or bent friction disc, oil or grease on the friction surfaces, rusted or damaged transmission input shaft splines, or other problems inside the clutch housing.

Abnormal Clutch Noises

Various noises can be made by faulty clutch parts. To diagnose noises, note when the clutch noise is produced. Does the sound occur when the pedal is moved, when in neutral, when in gear, or when the pedal is held to the floor? This will help determine which parts are producing the abnormal noises.

A worn, unlubricated, or dry clutch release mechanism will produce odd sounds (squeaks, clunks, scrapes) whenever the clutch pedal is moved up or down. With the engine shut off, pump the clutch pedal while listening for the sound.

If needed, have a helper work the pedal while you locate the source of the noise. Use a stethoscope or a section of vacuum hose as a listening device. Clean, lubricate, and replace parts as required.

Sounds produced when the clutch is initially engaged are normally due to friction disc problems. The lining could be worn, causing an abrasive, metal-on-metal grinding sound. If the friction disc damper springs are weak or broken, a knocking or rattling sound may be produced.

Abnormal sounds from the clutch that occur when the clutch is disengaged may be from a bad throw-out bearing. It may be dry and badly worn.

A worn pilot bearing in the crankshaft may also produce noises during clutch disengagement. The worn pilot can let the transmission input shaft and clutch disc vibrate up and down.

Abnormal sounds that are heard only in neutral and disappear when the clutch pedal is pressed are usually caused by problems inside the transmission. The manual transmission input shaft is still spinning whenever the clutch is engaged. However, the input shaft stops turning when the clutch is disengaged. The front input shaft bearing could be worn, for example.

Pulsating Clutch Pedal

A *pulsating clutch pedal* is normally caused by the runout (wobble or vibration) of one of the rotating components of the clutch assembly. Slight up and down movements of the clutch pedal can be felt with light foot pressure. The flywheel may be warped. The clutch housing might not be properly aligned with the engine. The pressure plate release levers could be bent or maladjusted.

To correct a pedal pulsation problem, the clutch must be removed and inspected. Then, the faulty or misaligned parts can be replaced or repaired.

Stiff Clutch Pedal

A *stiff clutch pedal* results from a problem with one of the parts relating to the clutch release mechanism: linkage, cable, hydraulic components, clutch fork, throw-out bearing, or pressure plate. One of these parts is resisting normal movement and is increasing the amount of pedal pressure needed to release the clutch. A part may be worn or lacking lubrication.

If the clutch fork has fallen off its pivot ball or bracket, the clutch pedal can be very hard to push down. Normally, bell housing removal is needed to reinstall the clutch fork or repair the pivot.

Servicing a Clutch

Clutch service is fairly common. After prolonged service or abuse, the clutch parts wear and fail in service. The friction disc can wear out, causing clutch slippage and, sometimes, damage to the pressure plate and flywheel. The throw-out bearing can run dry and start making noise whenever the clutch pedal is pressed. This next section of this chapter will explain how to repair these and other clutch problems.

Adjusting the Clutch

Clutch adjustment involves setting the correct amount of free play in the release mechanism. Too much free play could cause the clutch to drag during clutch disengagement. Too little free play could cause clutch slippage.

It is important for you to know how to adjust the three basic types of clutch release mechanisms.

Clutch Linkage Adjustment

Mechanical clutch linkage is usually adjusted at the push rod attached to the clutch fork. One end of the fork push rod is threaded. The effective length of the rod can be increased to raise the clutch pedal (decrease free travel). Rod length can also be shortened to lower the clutch pedal (increase free travel). See **Figure 54-7.**

To change the clutch adjustment, loosen the push rod nuts. You may need to grasp the unthreaded portion of the push rod with vise grips to free them. Turn the nuts on the threaded push rod until you have the correct pedal free travel.

Note

Specific adjustment procedures and free travel specifications vary. Always refer to a service manual for exact details.

Clutch Cable Adjustment

As with the linkage release mechanism, a clutch cable may require periodic adjustment to maintain the correct pedal height and free travel. Typically, the clutch cable housing will have an adjusting nut. When the nut is turned, the length of the cable housing increases or decreases. Look at **Figure 54-8.**

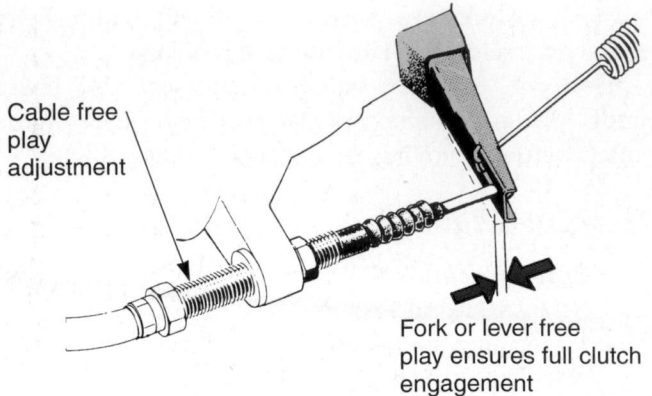

Cable free play adjustment

Fork or lever free play ensures full clutch engagement

Figure 54-8. Without free play in the clutch release mechanism, the fork could push the throw-out bearing into pressure plate, even with clutch pedal fully released. Clutch slippage could result. (Volvo)

In most cases, to increase clutch pedal free travel, turn the cable housing nut to shorten the housing. To decrease clutch pedal free travel, lengthen the clutch cable housing.

Some cable release mechanisms have an automatic adjusting mechanism. If the clutch requires adjustment, the automatic clutch adjuster may be faulty or the clutch may be badly worn.

Hydraulic Clutch Release Mechanism Adjustment

A hydraulic clutch release mechanism may need adjustment after prolonged clutch operation. Normal wear of the friction disc, throw-out bearing, and other parts can cause the pedal free travel to increase. The adjusting nut on a hydraulic clutch can be located on the fork push rod or the clutch master cylinder push rod. See **Figure 54-9.**

To adjust a hydraulic clutch, simply turn the nut on the push rod as needed. Generally, lengthening the rod decreases pedal free travel. Shortening the rod increases free travel. Check a service manual for specifications and procedures.

Removing the Clutch

Clutch removal procedures vary from one vehicle to another. However, general procedures and safety warnings should be understood.

Warning

Always disconnect the battery ground when removing a clutch assembly. This will prevent accidental cranking of the engine and possible injury. It can also prevent electric shorts that could damage the vehicle's wiring.

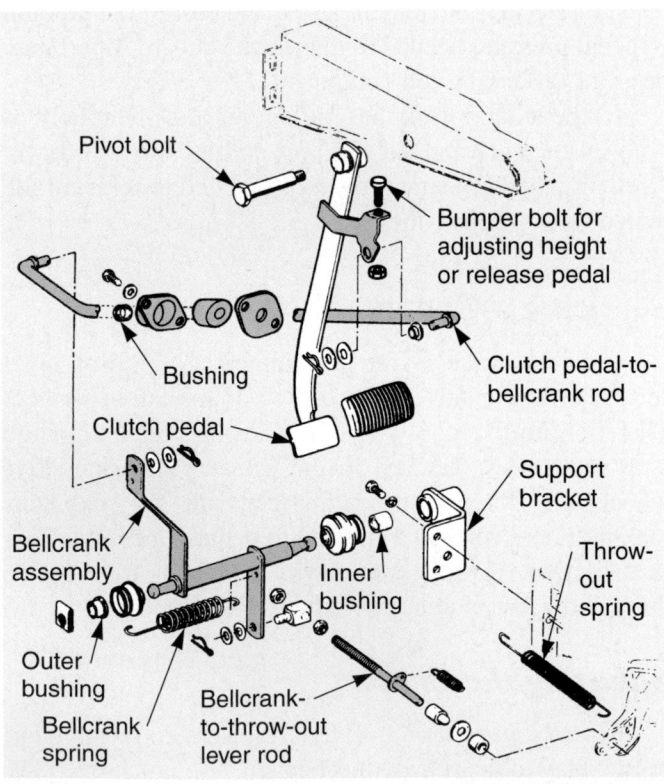

Pivot bolt

Bumper bolt for adjusting height or release pedal

Bushing

Clutch pedal-to-bellcrank rod

Clutch pedal

Support bracket

Bellcrank assembly

Inner bushing

Throw-out spring

Outer bushing

Bellcrank-to-throw-out lever rod

Bellcrank spring

Figure 54-7. Always check the clutch linkage when clutch pedal action is faulty. Look for bent rods, worn bushings, missing springs, unlubricated bearings, damaged bell crank, and other troubles. (Chrysler)

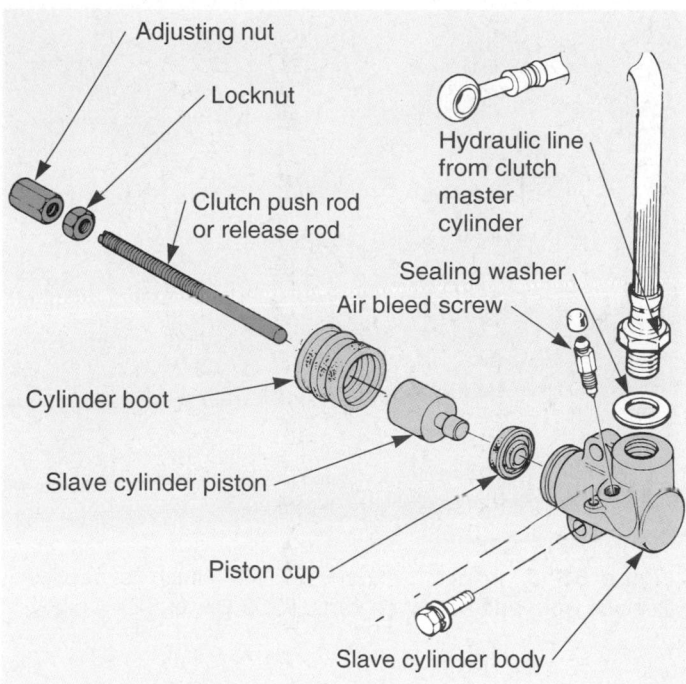

Figure 54-9. Exploded view of a clutch slave cylinder. To rebuild unit, hone cylinder and replace cup and boot. Also note adjustable push rod for setting free play. (Honda)

Transmission or transaxle removal is needed to service a clutch. Always follow the detailed directions in the service manual.

On a rear wheel drive vehicle, remove the drive shaft, clutch fork release rod or cable, and the transmission. Use a transmission jack when lifting the heavy transmission or transaxle out of the vehicle, **Figure 54-10.** Refer to Chapter 56 for general instructions for removing a transmission. Refer to Chapter 60 for more information on drive shaft removal. These chapters also contain information relating to clutch service.

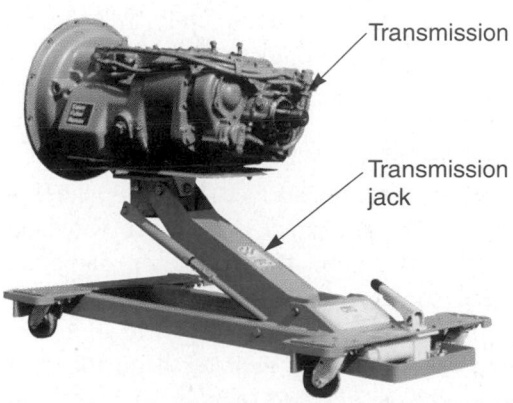

Figure 54-10. Transmissions and transaxles are very heavy. Use a transmission jack to remove the assembly during clutch repairs. (OTC Div. of SPX Corp.)

With a front-wheel-drive vehicle, the axle shafts (drive axles), transaxle, and sometimes the engine must be removed for clutch repairs. Use the specific instructions in a shop manual. Chapter 64 discusses general procedures for removing a transaxle assembly.

Caution

When removing a transmission or transaxle, support the weight of the engine. Never let the engine, transmission, or transaxle hang unsupported. The transmission input shaft, clutch fork, motor mounts, and other parts could be damaged.

After removing the transmission or transaxle, unbolt the bell housing from the rear of the engine. Support the housing as the last bolt is removed. Be careful not to drop the bell housing as you pull it off its dowel pins. See **Figure 54-11.**

Use a hammer and center punch to mark the pressure plate and flywheel. These marks may be needed if the same pressure plate is to be reinstalled. Lining up the marks ensures correct balancing of the clutch.

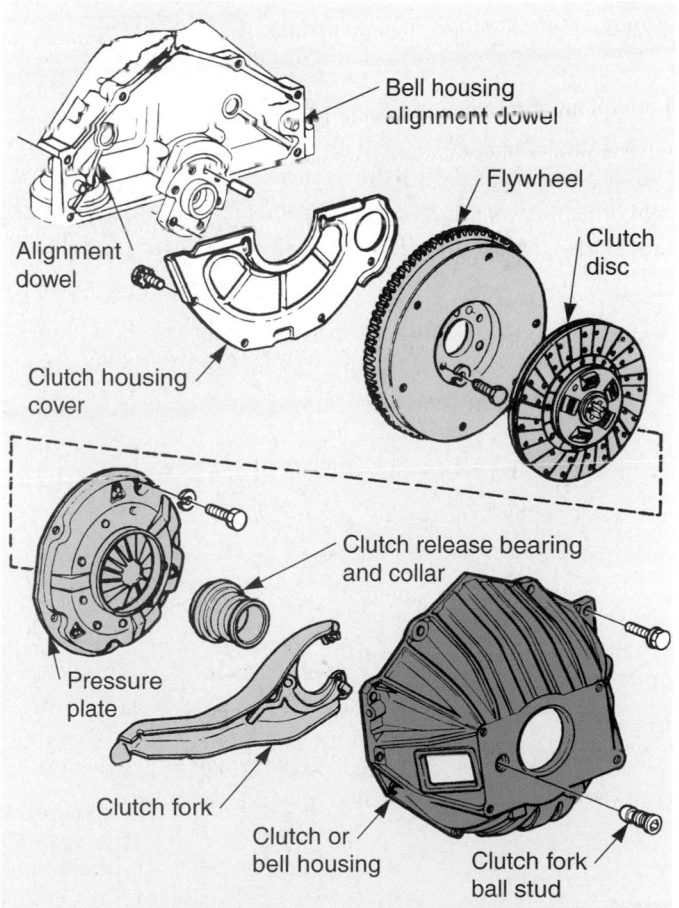

Figure 54-11. Inspect each clutch component as it is removed from the rear of the engine. If you overlook any trouble, your clutch repair may fail. (GMC)

Slide a *pilot shaft* (clutch alignment tool or old transmission input shaft) into the clutch. This will keep the clutch disc from falling as you unbolt the pressure plate. Loosen each pressure plate bolt a little at a time to avoid placing too much stress on any one bolt. Hold the pressure plate against the flywheel as the last bolt is removed. Look at **Figure 54-12.**

Lift the pilot shaft, pressure plate, and clutch disc off the back of the engine without dropping them. They are fairly heavy. Be prepared to support their weight.

With the clutch removed, check the rear of the engine and the front of the transmission for oil leaks. If oil were to leak on the new clutch, it would be ruined. Check the crankshaft rear main seal and the transmission front seal.

Inspecting and Cleaning Clutch Parts

With the clutch removed, each component must be cleaned and carefully inspected for wear and damage, **Figure 54-13.**

Warning

Be careful how you clean the parts of a clutch. Avoid using compressed air to blow clutch dust off the parts. A clutch disc often contains asbestos, a powerful cancer-causing substance.

Wipe the clutch parts down with a clean rag. Use sandpaper to deglaze and polish the surface of the flywheel and the face of the pressure plate. Keep cleaning solvent, which might contain traces of oil, off the friction surfaces (clutch disc, flywheel, and pressure plate faces).

Caution

Do *not* wash the throw-out bearing in solvent. This could wash the grease out of the bearing and ruin it.

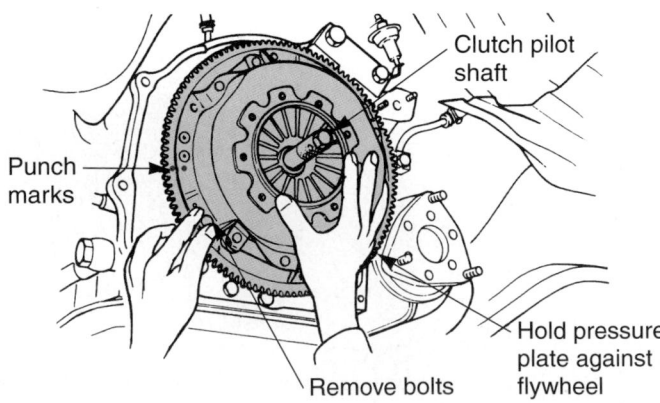

Figure 54-12. Punch mark the pressure plate and flywheel before disassembly. You will need to align these marks to reuse the old pressure plate. A pilot shaft will keep the disc from falling as pressure plate bolts are removed. (Mazda)

Clutch pilot shaft

Punch marks

Remove bolts

Hold pressure plate against flywheel

Figure 54-13. Inspect all parts as the clutch is serviced. Replace worn and damaged parts. (OTC Div. of SPX Corp.)

Pilot Bearing Service

Closely inspect the pilot bearing or bushing. Using a telescoping gauge and micrometer, measure the amount of wear in the bushing. If a roller bearing is used, turn the bearing with your finger. Feel for roughness and wear. If needed, replace the bearing.

The pilot bearing can be removed from the crankshaft with a slide hammer puller. A few light blows will drive the bearing out of the crank.

If a puller is not available, fill the inside of the bearing with heavy grease. Then, insert the metal pilot shaft in the bearing. Tap on the shaft with a mallet and the grease will force the pilot bearing out of the crankshaft.

Check the fit of the new pilot bearing by sliding it over the input shaft of the transmission. Then, drive the new pilot bearing into the end of the engine crankshaft, **Figure 54-14.**

If needed, place a small amount of grease in the pilot bearing cavity. Do not completely fill the cavity, or grease can squirt out onto the friction disc when you install the transmission or transaxle.

Flywheel Service

Closely inspect the surface of the flywheel. Look for discolored areas and cracks. Measure flywheel runout with a dial indicator. If warped or damaged, either replace the flywheel or have it resurfaced at a machine shop.

Also, check the ring gear teeth on the flywheel. If they are worn or chipped, a new ring gear should be installed. Heat the old ring gear with an acetylene torch. This will expand the gear, allowing it to be easily knocked off the flywheel with a hammer and punch.

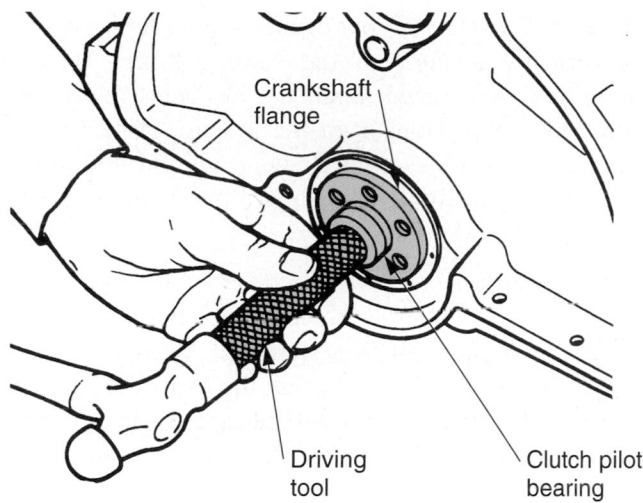

Figure 54-14. Always check the pilot bearing for wear during clutch service. If worn, remove the old bearing. Install a new pilot bearing as shown. (Ford)

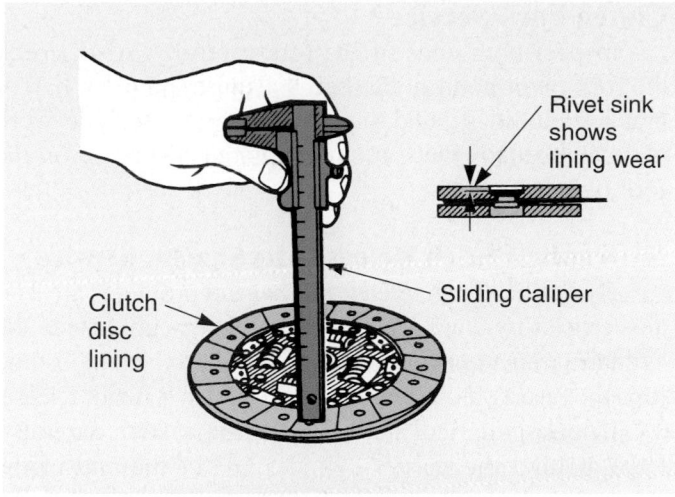

Figure 54-15. Measure clutch disc lining wear at the rivets. If the distance from the surface of the lining to the top of the rivet head is too small, install a new clutch disc. Refer to a service manual for specifications. Only when a disc is relatively new should it be reused. Keep grease and oil off friction surfaces. (Chrysler)

To install the new ring gear, heat the ring gear with your torch. Then, carefully position the ring gear and drive it on with light hammer blows.

Clutch Disc Service

To check disc wear, inspect the depth of the rivet holes. The closer the rivets are to the surface of the friction material, the more worn the disc. Look at **Figure 54-15**.

Normally, the friction disc is replaced anytime the clutch is torn down for repairs. The disc is reasonably inexpensive and highly prone to wear.

Pressure Plate Service

Inspect the pressure plate closely using the information in a service manual. The manual will describe various measurements to determine its condition.

Modern practice is to replace worn or defective pressure plates. Most technicians no longer rebuild or repair them. Considering the cost of labor, it is normally cheaper to purchase and install a new or rebuilt unit.

Throw-Out Bearing Service

To check the action of the throw-out bearing, insert your fingers into the bearing. Then, turn the bearing while pushing in on it. Try to detect any roughness. The throw-out bearing should rotate smoothly.

Also, inspect the spring clips on the throw-out bearing or fork. These clips hold the bearing on the end of the clutch fork. If bent, worn, or fatigued, the bearing collar or fork must be replaced.

To replace the throw-out bearing, drive it off its collar. Use a vise and hammer or a hydraulic press.

Follow the detailed procedures in a service manual to avoid damaging the collar and new bearing.

The throw-out bearing is subject to considerable wear and is a frequent cause of clutch problems. Therefore, most technicians replace the throw-out bearing anytime the clutch is disassembled for repairs. Before installation, lubricate the throw-out bearing collar as shown in **Figure 54-16**.

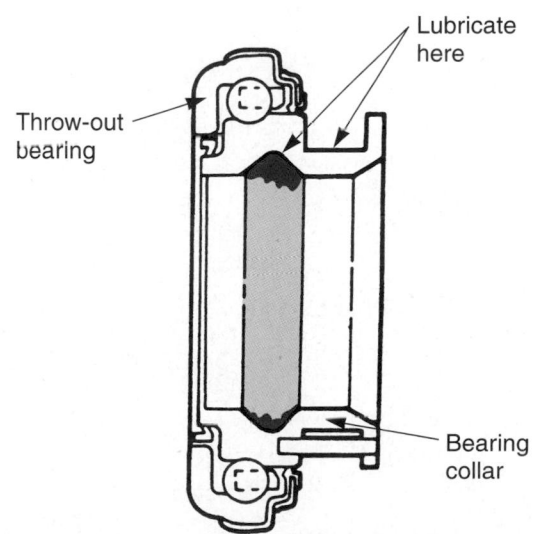

Figure 54-16. When installing a new throw-out bearing, place hot grease on the areas shown. This will allow the bearing collar to freely slide on the transmission hub. Do not use too much grease; it could ruin the clutch disc. (Oldsmobile)

Clutch Fork Service

Inspect both ends of the fork carefully. Also, check the fork pivot point in the bell housing. The pivot ball or bracket should be undamaged and tight. Replace worn parts as needed. Place a small amount of grease on the fork pivot point.

Hydraulic Clutch Release Mechanism Service

Hydraulic clutch release mechanism problems are usually caused by fluid leakage. The rubber cups inside the cylinders can wear and begin to leak. After enough fluid loss, the reservoir can empty and the clutch will not release.

If leakage is indicated, inspect the system carefully. Look behind the master cylinder and at the end of the slave cylinder. If leaks are found, replace or repair the components as needed. Refer to **Figure 54-17.**

After reassembly, the hydraulic clutch will require **bleeding** (removal of air from inside the system). Air is compressible and will cause the clutch pedal to be very soft and spongy.

To bleed a hydraulic clutch release, use the same procedures outlined in the section on bleeding brakes. Use a pressure bleeder or manually pump the clutch pedal to force fluid and any air out of the system. A bleeder screw is normally located on the slave cylinder. When the system is pressurized, open the bleeder screw to allow fluid and air to flow from the system. Repeat the step until all air bubbles stop flowing out and you only get clear hydraulic fluid. This should purge all air from the lines and cylinders.

 Caution

Install only the recommended type of fluid in a hydraulic clutch system. Oil, kerosene, and grease must *never* enter the hydraulic system. These substances will cause the rubber cups to swell and deteriorate. *Keep your hands clean!*

Servicing a hydraulic clutch (bleeding, honing cylinders, replacing cups, etc.) is very similar to servicing a hydraulic brake system. For more information on how to work on hydraulic components, refer to Chapter 72, *Brake System Diagnosis and Repair.*

Installing the Clutch

Install the clutch in the reverse order of removal. Mount the clutch disc and pressure plate on the flywheel. Use a clutch alignment tool (pilot shaft) to center the disc. Align any punch marks.

Make sure the friction disc is facing the proper direction. Usually, the disc's offset center section (hub and torsion springs) fits into the pressure plate. Start all the pressure plate bolts by hand.

 Caution

Never let oil or grease contact with the clutch friction surfaces. The slightest amount of oil or grease could cause clutch slippage or grabbing. *Keep your hands and tools clean!*

Tighten each pressure plate bolt a little at a time in a crisscross pattern, **Figure 54-18.** This will apply equal stress on each bolt as the pressure plate springs are compressed. When all the bolts are snug, torque them to specifications.

Never replace clutch pressure plate bolts with weaker bolts. Always install the special case-hardened bolts recommended by the manufacturer. Weaker bolts could break, causing severe part damage.

With the pressure plate bolts properly torqued, slide the pilot out of the clutch. The pilot ensures that the clutch friction disc is centered on the flywheel. If a pilot is not used, the transmission input shaft will not slide into the crankshaft pilot bearing. It would be impossible to install the transmission or transaxle.

Next, install the clutch fork and throw-out bearing in the bell housing. Fit the bell housing over the rear of the

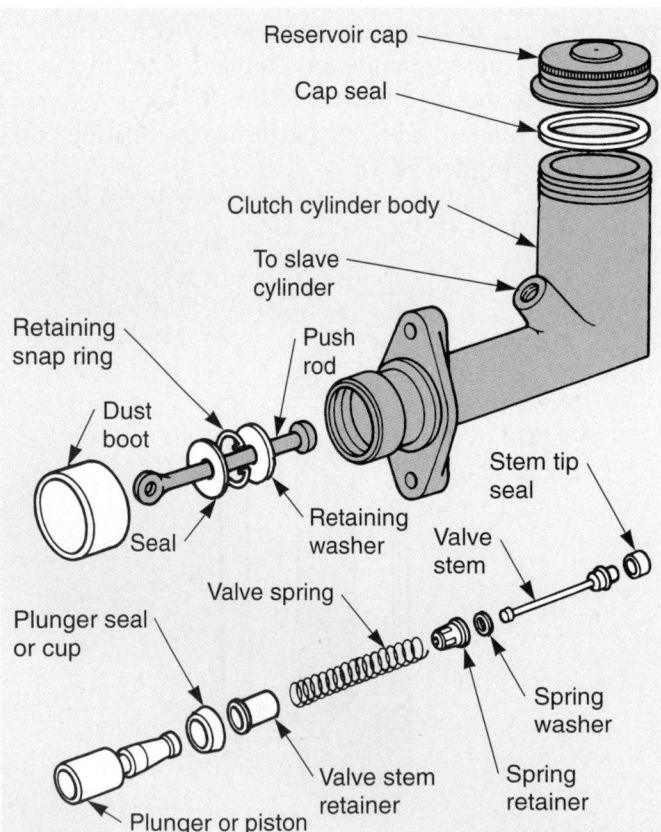

Figure 54-17. A clutch master cylinder is serviced much like a brake system master cylinder. When leaking, either rebuild or replace the unit. Study the part names. (Chrysler)

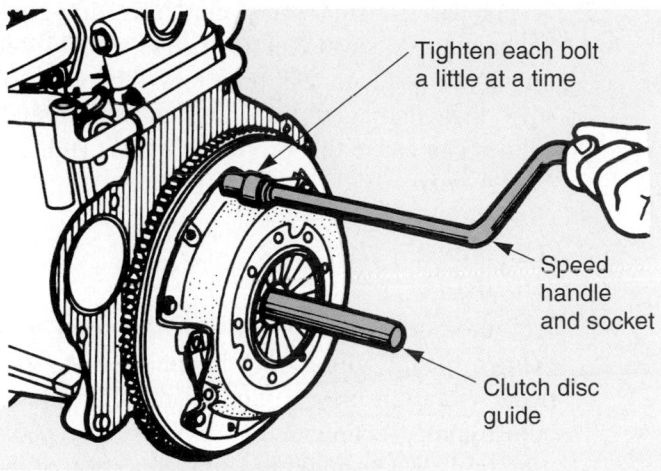

Figure 54-18. Tighten each pressure plate bolt a little at a time. Then, torque them to specifications. Remove the pilot after the bolts are fully torqued. Slide the shaft in and out of pilot bearing to double-check disc alignment. (Dodge)

engine. Large dowels are provided to align the housing on the engine. Install and tighten the bell housing bolts in a crisscross pattern.

Finally, install the transmission and drive shaft or the transaxle assembly and axle shafts. Reconnect the linkage, cables, wires, battery, and other parts. Adjust clutch pedal free travel as described earlier. Test drive the vehicle.

Summary

- Stop-and-go city traffic will wear out a clutch quicker than highway driving.
- A worn clutch disc will cause clutch slippage and, sometimes, damage to the flywheel and pressure plate.
- A bad throw-out bearing will produce a grinding noise whenever the clutch pedal is pushed down.
- A bad pressure plate can also cause clutch slippage and clutch release problems (stiff clutch pedal, clutch grabbing, abnormal noises, clutch dragging).
- A worn pilot bearing will allow the transmission input shaft and clutch disc to wobble up and down. This can cause clutch vibration, abnormal noises, and damage to the transmission.
- Clutch slippage is noticed when the engine races (engine speed increases quickly) without an increase in the vehicle's road speed.

- Free travel (free play) is the distance the clutch pedal or clutch fork moves before the throw-out bearing acts on the pressure plate. Free travel is needed to ensure complete clutch engagement.
- A grabbing or chattering clutch will produce a very severe vibration or jerking motion when the vehicle is accelerated from a standstill.
- A dragging clutch will normally make the transmission or transaxle grind when trying to engage or shift gears.
- A pulsating clutch pedal is normally caused by the runout (wobble or vibration) of one of the rotating components of the clutch assembly.
- A stiff clutch pedal results from a problem with one of the parts relating to the clutch release mechanism: linkage, cable, hydraulic components, over-center spring, clutch fork, throw-out bearing, or pressure plate.
- Clutch adjustment involves setting the correct amount of free play in the release mechanism.
- After reassembly, the hydraulic clutch will require bleeding (removal of air from inside hydraulic system).
- Slide a pilot shaft (clutch alignment tool or old transmission input shaft) into the clutch. This will keep the clutch disc from falling as you unbolt the pressure plate.
- Never let oil or grease come in contact with the friction surfaces of a clutch. The slightest amount of oil or grease could cause clutch slippage or grabbing.

Important Terms

Clutch slippage	Pulsating clutch pedal
Races	Stiff clutch pedal
Free travel	Clutch adjustment
Grabbing	Pilot shaft
Chattering clutch	Bleeding
Dragging clutch	

Review Questions—Chapter 54

Please do not write in this text. Place your answers on a separate sheet of paper.

1. Clutch disc lining is commonly made of asbestos which is a known _____ causing substance.
2. What are the symptoms of clutch slippage?
3. How do you check for clutch slippage?

4. Which of the following could cause clutch slippage?
   (A) Improper clutch adjustment.
   (B) Binding release mechanism.
   (C) Broken motor mount.
   (D) All of the above.

5. Describe some causes of a dragging clutch.

6. A dragging clutch makes the transmission or transaxle gears grind. True or False?

7. A driver complains of a grinding sound when the clutch pedal is pressed to release the clutch. Otherwise, clutch operation is normal. Technician A says that the clutch lining is worn, resulting in metal-on-metal contact. The transmission and clutch will require removal for disc replacement. Technician B says that the problem could be a worn, dry throw-out bearing. The bearing is grinding when the clutch fork pushes the bearing into the spinning pressure plate. Who is correct?
   (A) A only.
   (B) B only.
   (C) Both A and B.
   (D) Neither A nor B.

8. What commonly causes a pulsating clutch pedal?

9. Define the term clutch pedal "free play."

10. A hydraulic clutch release mechanism requires bleeding after major repairs. True or False?

11. Why should you disconnect the car battery when servicing a clutch?

12. What precautions should you take when cleaning clutch parts prior to reinstalling them?

13. How do you remove and install a pilot bearing?

14. A worn _____ _____ will cause clutch slippage and, sometimes, damage to the flywheel and pressure plate.

15. Why do you need an alignment or pilot shaft when installing a clutch?

⬟ ASE-Type Questions

1. A car is brought into the shop with a worn clutch. Technician A tells the owner of the car that highway driving will wear out a clutch faster than city driving. Technician B tells the owner that city driving wears out a clutch faster than highway driving. Who is right?
   (A) A only.
   (B) B only.
   (C) Both A and B.
   (D) Neither A nor B.

2. A manual transmission's clutch is dragging. Technician A says that excessive clutch free travel can produce this problem. Technician B says that damaged transmission input shaft splines can cause this problem. Who is right?
   (A) A only.
   (B) B only.
   (C) Both A and B.
   (D) Neither A nor B.

3. An automobile is brought into the shop with "clutch pedal vibration." Technician A says pedal vibration may be due to improper clutch adjustment. Technician B says "clutch pedal vibration" is a normal operating condition of the clutch assembly. Who is right?
   (A) A only.
   (B) B only.
   (C) Both A and B.
   (D) Neither A nor B.

4. A manual transmission has a "clutch slippage" problem. Technician A says this condition may be caused by a damaged clutch linkage. Technician B says this problem may be caused by a worn pressure plate. Who is right?
   (A) A only.
   (B) B only.
   (C) Both A and B.
   (D) Neither A nor B.

5. A clutch is being tested for slippage. Technician A says when performing this test, the emergency brake should be engaged. Technician B says when testing a clutch for slippage, the car's transmission should be placed in high gear. Who is right?
   (A) A only.
   (B) B only.
   (C) Both A and B.
   (D) Neither A nor B.

6. A clutch is being tested for slippage. Technician A says you shouldn't let the clutch slip for more than one or two seconds during this test. Technician B says you shouldn't let the clutch slip for more than ten to fifteen seconds during this test. Who is right?
   (A) A only.
   (B) B only.
   (C) Both A and B.
   (D) Neither A nor B.

7. The shop writer wants a clutch slippage problem diagnosed. Technician A checks for a broken motor mount. Technician B checks for a binding clutch linkage. Who is right?
   (A) A only.
   (B) B only.
   (C) Both A and B.
   (D) Neither A nor B.

8. An automobile with a clutch "free play" problem is brought into the shop. Technician A checks clutch "free play" at the clutch fork. Technician B checks clutch "free play" at the clutch pedal. Who is right?
   (A) A only.
   (B) B only.
   (C) Both A and B.
   (D) Neither A nor B.

9. An automobile has a "stiff" clutch pedal. Technician A says this problem can be caused by a faulty throw-out bearing. Technician B says this problem can be caused by a pressure plate malfunction. Who is right?
   (A) A only.
   (B) B only.
   (C) Both A and B.
   (D) Neither A nor B.

10. An automobile's clutch has too much "free play." Technician A says this problem can cause clutch slippage. Technician B says this problem can cause clutch dragging. Who is right?
    (A) A only.
    (B) B only.
    (C) Both A and B.
    (D) Neither A nor B.

11. A manual transmission's input shaft and clutch disc wobble up and down. Technician A says this problem is probably the result of a damaged clutch housing cover. Technician B says this problem is probably the result of a worn pilot bearing. Who is right?
    (A) A only.
    (B) B only.
    (C) Both A and B.
    (D) Neither A nor B.

12. An automotive clutch assembly needs to be removed from an automobile. Technician A disconnects the car's battery before performing this task. Technician B disconnects the clutch linkage before performing this task. Who is right?
    (A) A only.
    (B) B only.
    (C) Both A and B.
    (D) Neither A nor B.

13. The parts of a clutch assembly are being inspected and cleaned. Technician A says during this procedure you should clean the throw-out bearing in cleaning solvent. Technician B says during this procedure the throw-out bearing should be cleaned only with a shop rag. Who is right?
    (A) A only.
    (B) B only.
    (C) Both A and B.
    (D) Neither A nor B.

14. Technician A is using compressed air to remove clutch dust from a clutch disc. Technician B is using a shop rag to remove clutch dust from a clutch disc. Who is right?
    (A) A only.
    (B) B only.
    (C) Both A and B.
    (D) Neither A nor B.

15. A clutch grabs when the car is accelerated. Technician A checks the condition of the engine's flywheel. Technician B inspects the clutch pedal return spring. Who is right?
    (A) A only.
    (B) B only.
    (C) Both A and B.
    (D) Neither A nor B.

Activities—Chapter 54

1. Study damaged and worn clutch parts and try to determine why they failed.

2. Study a shop manual for manual clutch service; demonstrate to your instructor the proper procedure for removal and disassembly.

Clutch Diagnosis		
Condition	**Possible Causes**	**Correction**
Clutch slips.	1. Insufficient pedal free travel. 2. Disc facing soaked with oil or grease. 3. Broken or weak pressure plate spring or springs. 4. Worn clutch disc facing. 5. Sticking hydraulic or mechanical linkage.	1. Adjust free travel. 2. Correct source of oil contamination. Clean clutch and pressure plate. Replace disc. 3. Rebuild or replace pressure plate. 4. Replace clutch disc. 5. Clean, align, and lubricate linkage.
Clutch chatters and/or grabs.	1. Oil- or grease-soaked clutch disc facing. 2. Burned clutch disc facing. 3. Warped or worn clutch disc. 4. Warped pressure plate. 5. Scored pressure plate or flywheel surface. 6. Binding pressure plate fingers. 7. Clutch housing-to-transmission surface out of alignment with crankshaft centerline. 8. Sticking linkage. 9. Worn pilot bearing. 10. Improperly adjusted pressure plate release fingers. 11. Loose or worn engine mounts. 12. Loose transmission. 13. Loose rear spring shackles or axle housing control arms. 14. Worn splines on transmission input shaft. 15. Faulty throw-out bearing.	1. Replace clutch disc. Correct source of leak. 2. Replace clutch disc. 3. Replace clutch disc. 4. Grind or replace pressure plate. 5. Grind or replace pressure plate or flywheel. 6. Free fingers. 7. Align or replace housing. 8. Free linkage. 9. Install new pilot bearing. 10. Adjust fingers. 11. Tighten or replace mounts. 12. Tighten transmission fasteners. 13. Tighten shackles or replace control arm insulators and tighten. 14. Replace input shaft. 15. Replace throw-out bearing.
Clutch will not release properly.	1. Excessive pedal free travel. 2. Warped clutch disc. 3. Clutch facing torn loose and folded over. 4. Warped pressure plate. 5. Misaligned clutch housing. 6. Clutch disc hub binding on transmission input shaft. 7. Worn pilot bearing. 8. Faulty throw-out bearing. 9. Throw-out fork off pivot. 10. Clutch disc is frozen (corroded) to flywheel and pressure plate. 11. Excessive idle speed.	1. Adjust pedal travel. 2. Replace clutch disc. 3. Replace clutch disc. 4. Grind or replace. 5. Align housing. 6. Free hub. 7. Replace pilot bearing. 8. Replace throw-out bearing. 9. Install fork properly. 10. Replace disc and clean flywheel and pressure plate. 11. Adjust idle speed.
Clutch is noisy when pedal is depressed—engine running.	1. Dry or worn throw-out bearing. 2. Worn pilot bearing. 3. Excessive total pedal travel. 4. Throw-out fork off pivot. 5. Misaligned clutch housing. 6. Excessive crankshaft end play.	1. Replace bearing. 2. Replace pilot. 3. Adjust pedal travel. 4. Install fork correctly. 5. Align housing. 6. Correct end play.
Clutch is noisy when pedal is depressed—engine not running.	1. Dry, sticking linkage. 2. Dry or scored throw-out bearing sleeve. 3. Pressure plate drive lugs rubbing clutch cover.	1. Lubricate and align linkage. 2. Lubricate or replace. 3. Lubricate with high temperature grease.

Clutch Diagnosis		
Condition	**Possible Causes**	**Correction**
Clutch noisy when pedal is fully released—engine running.	1. Insufficient pedal free travel. 2. Worn clutch disc. 3. Broken clutch disc springs. 4. Misaligned clutch housing. 5. Worn clutch disc hub splines. 6. Worn input shaft splines. 7. Sprung input shaft. 8. Worn input shaft transmission bearing.	1. Adjust free travel. 2. Replace clutch disc. 3. Replace clutch disc. 4. Align housing. 5. Replace clutch disc. 6. Replace input shaft. 7. Replace input shaft. 8. Replace transmission bearing.
Excessive pedal pressure.	1. Dry, sticking linkage. 2. Binding pressure plate release fingers. 3. Linkage misaligned. 4. Throw-out bearing sleeve binding on transmission bearing retainer. 5. Sticking linkage in master or slave cylinder.	1. Lubricate linkage. 2. Free and lubricate. 3. Align linkage. 4. Free and lubricate retainer. 5. Clean or replace as needed.
Rapid clutch disc wear.	1. Insufficient pedal free travel. 2. Scored flywheel or pressure plate. 3. Driver "rides" the clutch. 4. Driver races engine and slips clutch excessively during starting. 5. Driver holds vehicle on hill by slipping clutch. 6. Weak pressure plate springs.	1. Adjust free travel. 2. Resurface or replace. 3. Advise driver. 4. Advise driver. 5. Advise driver. 6. Rebuild or replace pressure plate assembly.

Chapter 55

Manual Transmission Fundamentals

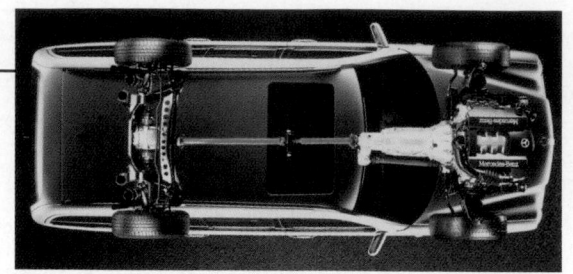

After studying this chapter, you will be able to:

- Describe gear operating principles.
- Identify and define all the major parts of a manual transmission.
- Explain the fundamental operation of a manual transmission.
- Trace the power flow through transmission gears.
- Compare the construction of different types of manual transmissions.
- Explain the purpose and operation of a transmission overdrive ratio.
- Correctly answer ASE certification test questions requiring a knowledge of manual transmission operating principles.

A *manual transmission* must be shifted by hand to change the amount of torque going to various parts of the drivetrain. It is normally bolted to the clutch housing at the rear of the engine.

The clutch disc rotates the transmission input shaft. Gears inside the transmission transfer engine power to the drive shaft and the rear wheels. A shift lever allows the driver to select which set of transmission gears to engage.

A manual transmission should not be confused with an automatic transmission or automatic transaxle. A foot-operated friction clutch is used to disengage the engine.

Automatic transmissions (covered in Chapters 57 and 58) use hydraulic pressure and sensing devices to shift gears. They detect engine speed and load to determine shift points. Automatic transmissions also use a fluid coupling instead of a dry-friction clutch.

A *transaxle* combines both the transmission and the differential into a single housing. It is commonly used in front-wheel-drive vehicles. A transaxle can contain either a manual or an automatic transmission. Transaxles are covered in Chapters 63 and 64.

Basic Transmission Parts

To understand later sections of the chapter, study the parts of the manual transmission in **Figure 55-1.** Learn to identify and locate the fundamental components. This knowledge will prepare you for more specific details of transmission construction and operation. The basic parts of a manual transmission include:

- *Transmission input shaft*—a shaft, operated by the clutch, that turns the gears inside the transmission.
- *Transmission gears*—provide a means of changing output torque and speed.
- *Synchronizers*—devices for meshing (locking) gears into engagement.
- *Shift forks*—pronged units for moving gears or synchronizers on their shafts for gear engagement.
- *Shift linkage*—arms or rods that connect the shift lever to the shift forks.
- *Gear shift lever*—lever allowing the driver to change transmission gears.
- *Output shaft*—shaft that transfers rotating power out of the transmission to drive shaft.
- *Transmission case*—housing that encloses transmission shafts, gears, and lubricating oil.

Purpose of a Manual Transmission

A manual transmission is designed to change the vehicle's drive wheel speed and torque in relation to engine speed and torque. Without a transmission, the engine would not develop enough power to accelerate the vehicle from a standstill. The engine would stall as soon as the clutch was engaged.

With a transmission in low (first) gear, the engine crankshaft must turn several times to make the drive shaft

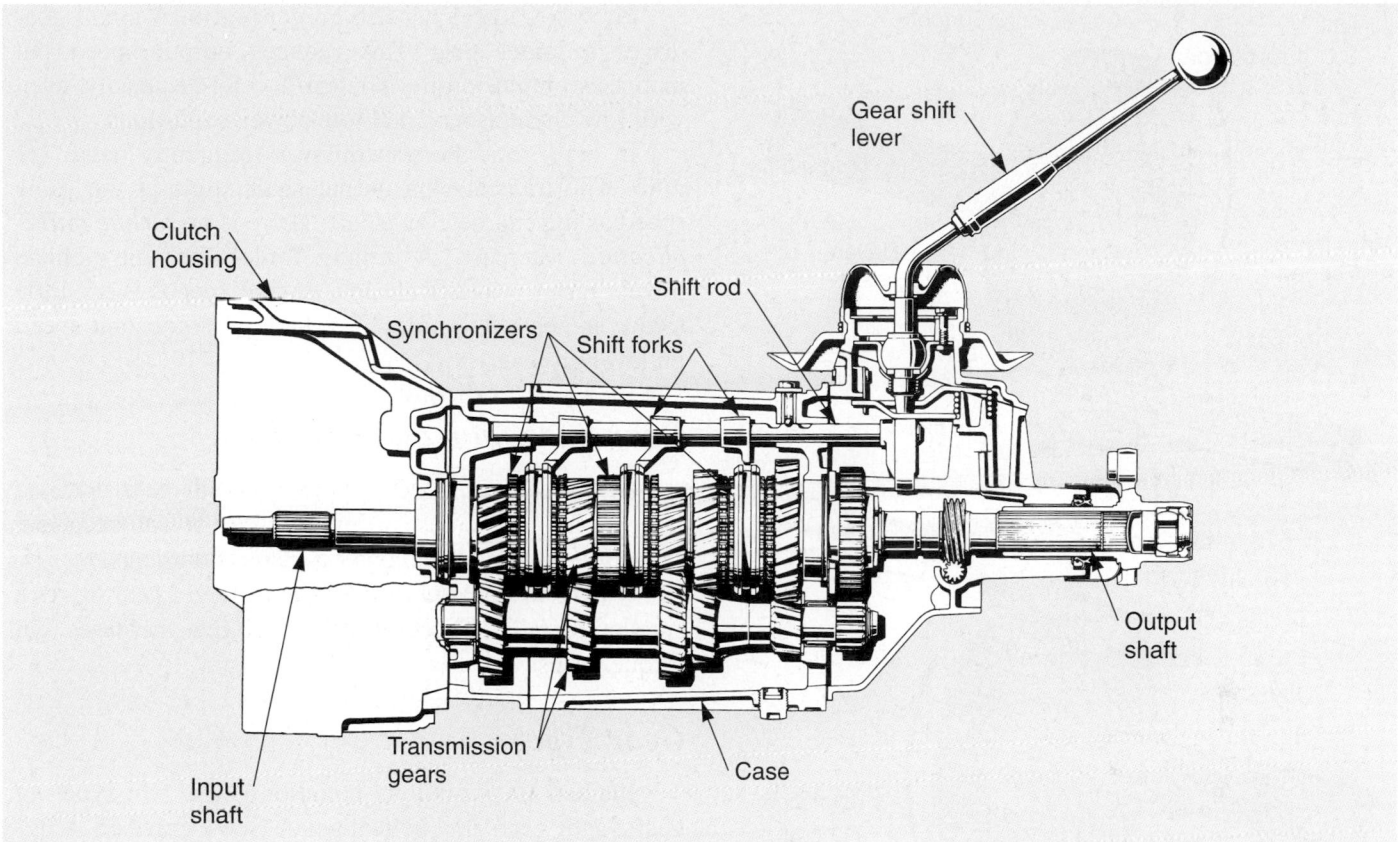

Figure 55-1. Study the basic names and locations of manual transmission parts. This will help you as you learn about each part in more detail. (Fiat)

and wheels turn once. This increases the torque going to the wheels, but reduces vehicle speed.

Then, as the transmission is shifted through the gears and into high, the engine and drive shaft begin to turn at approximately the same speed. Wheel and vehicle speed increases, while engine speed drops.

A manual transmission in proper operating condition should:

- Be able to increase torque going to the drive wheels for quick acceleration.
- Supply different gear ratios to match different engine load conditions.
- Have a reverse gear for moving backwards.
- Provide the driver with an easy means of shifting transmission gears.
- Operate quietly, with minimum power loss.

Gear Fundamentals

Gears are round wheels with teeth machined on their perimeters (rims). They are used to transmit turning effort from one shaft to another. Basically, one gear is used to turn another gear. When the gears are different sizes, the output speed and torque (turning power) change. This is illustrated in **Figure 55-2.**

Gear Ratios

A *gear ratio* is the number of revolutions a drive gear must turn before the driven gear completes one revolution. Gear ratio is calculated by dividing the number of teeth on the driven gear by the number of teeth on the drive gear. See **Figure 55-3.**

If the drive gear has 12 teeth and the driven gear has 24 teeth, the gear ratio is *two-to-one* (24 divided by 12), written 2:1.

In this example, the drive gear would have to revolve two times to turn the driven gear once. As a result, the speed of the larger, driven gear would be half that of the drive gear. However, the torque on the shaft of the larger gear would be twice that of the input shaft.

Various sizes of drive and driven gears can be used to produce any number of gear ratios. As the number of teeth on the driven gear increase in relation to the number of teeth on the drive gear, the gear ratio increases. For example, a gear ratio of 10:1 is larger than a ratio of 5:1, for example.

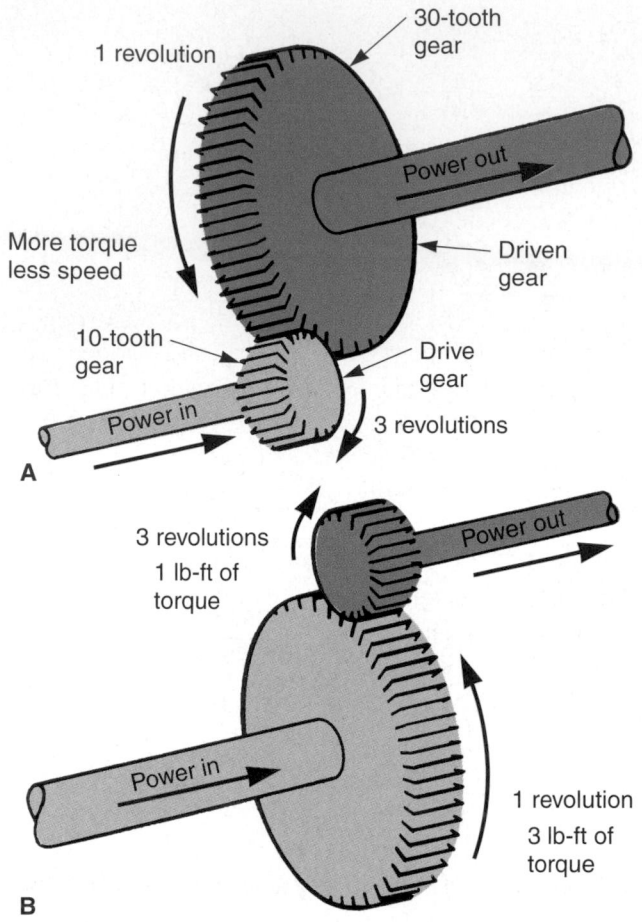

Figure 55-2. A—When a small gear drives a larger gear, it increases torque output but reduces rotating speed of output. B—When a large gear drives a smaller gear, torque is reduced but rotating speed is increased.

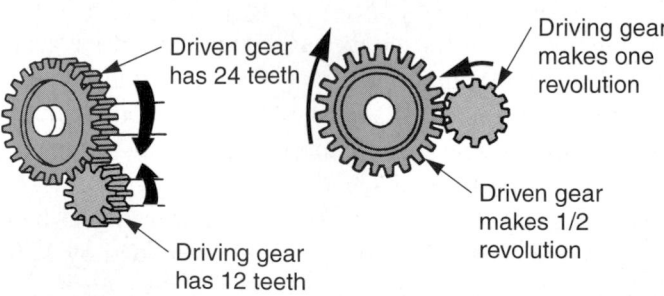

Figure 55-3. Gear ratio is determined by the number of teeth on the gears. If the drive gear has half as many teeth as the driven gear, a two-to-one ratio is produced. (Deere & Co.)

Transmission Gear Ratios

Transmission gear ratios vary from one manufacturer to another. However, approximate gear ratios are 3:1 for first gear; 2:1 for second gear; 1:1 for third, or high, gear; and 3:1 for reverse gear.

First (low) gear has a high gear ratio. A small gear drives a larger gear This reduces output speed but increases output torque. The car accelerates easily, even with low engine speed and low power conditions.

In high gear, the transmission frequently has a 1:1 ratio. The transmission output shaft spins at the same speed as the engine crankshaft. There is no *torque multiplication* (increase in turning force), but the vehicle travels faster at relatively low engine speeds. Very little torque is needed to propel a vehicle at a constant speed on level ground.

Gear Reduction and Overdrive

Gear reduction occurs when a small gear drives a larger gear to increase turning force, or torque. Gear reduction is used in the lower transmission gears.

An *overdrive ratio* results when a larger gear drives a smaller gear. The speed of the output gear increases, but torque drops.

Gear Types

Manual transmissions commonly use two types of gears: spur gears and helical gears. See **Figure 55-4.**

Spur gears have their teeth cut parallel to the centerline of the gear shaft. They are sometimes called straight-cut gears. Spur gears are somewhat noisy and are no longer used as the main drive gears in a transmission. They may be used for the sliding (moves sideways) reverse gear, however.

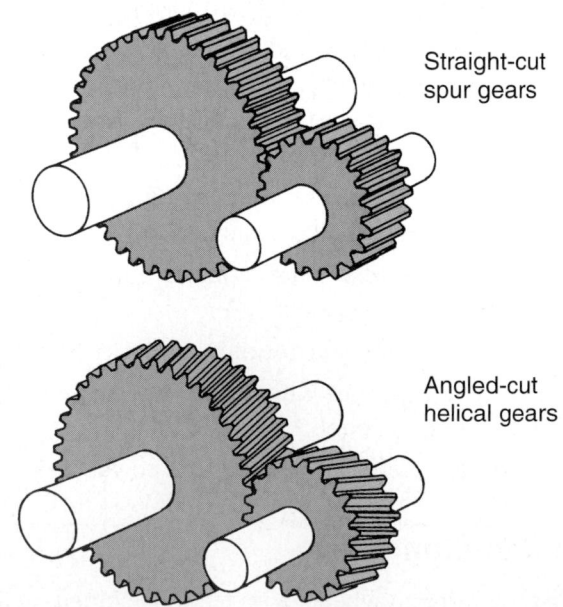

Figure 55-4. The two basic types of gears used in manual transmissions are straight-cut spur gears and angled-cut helical gears. (Deere & Co.)

Helical gears have their teeth machined at an angle to the centerline of the gear shaft. Modern transmissions commonly use helical gears as the main drive gears. Helical gears are quieter and stronger than spur gears.

Gear Backlash

Gear backlash is the small clearance between the meshing gear teeth. Backlash allows lubricating oil to enter the high-friction area between the gear teeth. This reduces friction and wear. Backlash also allows the gears to expand during operation without binding or damage.

Manual Transmission Lubrication

The bearings, shafts, gears, and other moving parts in a transmission are lubricated by oil throw-off, or *splash lubrication.* As the gears rotate, they sling oil around inside the transmission case.

Typically, 80W or 90W *gear oil* is recommended for use in a manual transmission. However, follow the manufacturer's recommendations.

Transmission Bearings

Bearings are used to reduce the friction between the surfaces of rotating parts in the transmission. Manual transmissions normally use three basic types of bearings: ball bearings, roller bearings, and needle bearings. These three types are shown in **Figure 55-5.**

Bearings are lubricated by oil spray from the spinning transmission gears. Typically, *antifriction bearings* (bearing using a rolling action) fit between the transmission shafts and the housing or between some of the gears and the shafts. These are high-friction points that must be capable of withstanding the engine's power.

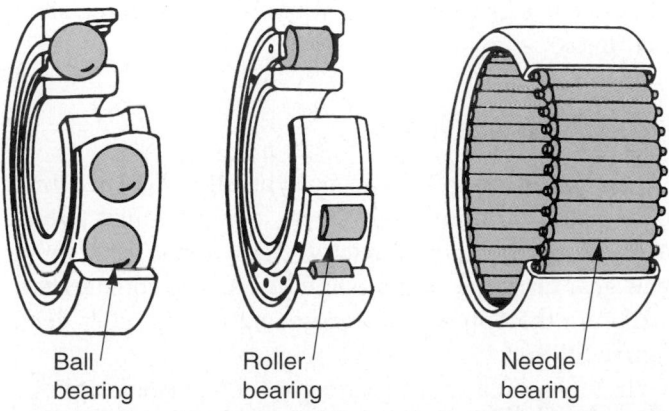

Ball bearing   Roller bearing   Needle bearing

Figure 55-5. Three types of antifriction bearings are found in transmissions: ball, roller, and needle. (Deere & Co.)

Manual Transmission Construction

Now that you have a general grasp of gear and transmission principles, we will assemble each part of a working transmission. We will start out with the case and then install the shafts, gears, bearings, and other parts.

Transmission Case

The *transmission case* must support the transmission bearings and shafts and provide an enclosure for gear oil. A manual transmission case is usually made of cast iron or aluminum. Aluminum is becoming more common because of its light weight. Refer to **Figure 55-6.**

A drain plug and a fill plug are usually provided in the transmission case. The drain plug is on the bottom of the case. The fill plug is on the side of the case.

The fill plug also serves as a means of checking the oil level in the transmission. Typically, the oil level should be even with the bottom of the fill plug hole when the transmission is at operating temperature.

Extension Housing and Front Bearing Hub

The *extension housing,* also called the *tail shaft housing,* bolts to the rear of the transmission case. It encloses the transmission output shaft and holds the rear oil seal. See **Figure 55-6.**

A flange on the bottom of the extension housing provides a base for the rubber transmission mount, or rear motor mount. A gasket usually seals the mating surfaces between the transmission case and the extension housing.

A *front bearing hub,* sometimes called *front bearing cap,* covers the front transmission bearing and acts as a sleeve for the clutch throw-out bearing. It bolts to the transmission case. A gasket fits between the front hub and the case to prevent oil leakage.

Transmission Shafts

A manual transmission normally has four steel shafts mounted inside its case: an input shaft, a countershaft, a reverse idler shaft, and an output shaft. **Figure 55-7** shows the general location and shape of these shafts.

The *input shaft,* often termed *clutch shaft,* transfers rotation from the clutch disk to the countershaft gears in the transmission. The outer end of the shaft is splined. The inner end of the shaft has a gear machined on it. See **Figure 55-8.**

A bearing in the transmission case supports the input shaft in the case. Anytime the clutch disc turns, the input shaft gear turns.

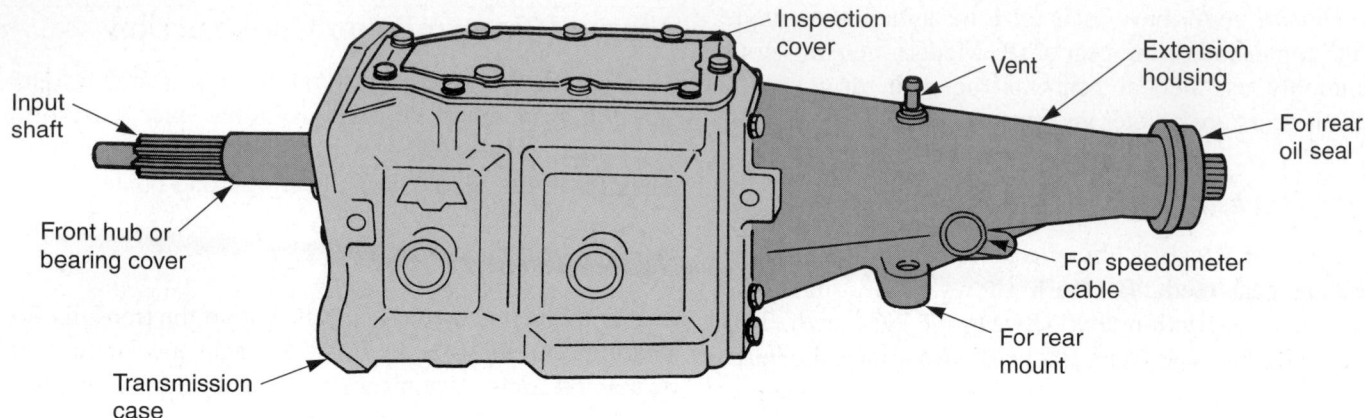

Figure 55-6. The case is the center section of a transmission. The extension housing bolts to the rear of the case. The front bearing cover bolts to the front of the case, enclosing the front output shaft bearing and supporting the clutch throw-out bearing. This transmission also has a sheet metal inspection cover bolted to the top of the case. (GMC)

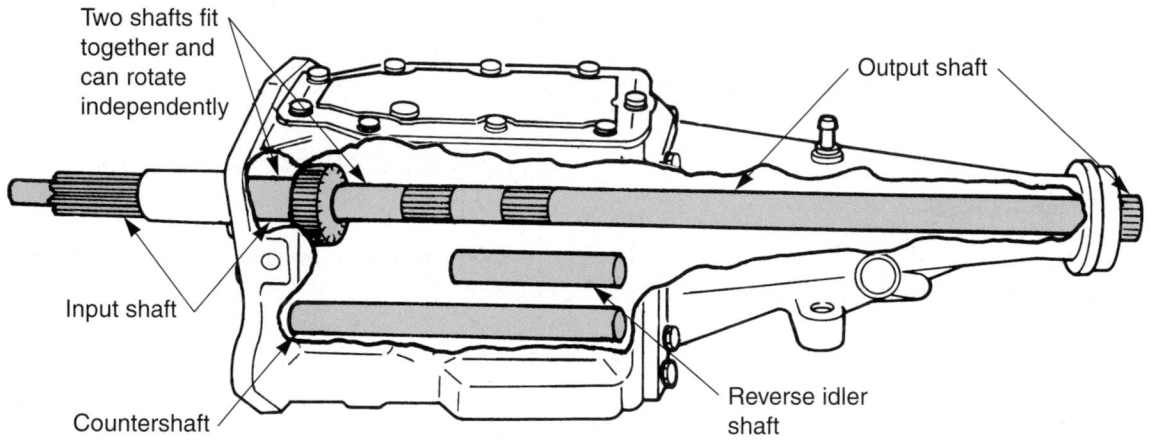

Figure 55-7. Note how transmission shafts are located in the transmission case. The input shaft is driven by the clutch. The output shaft is on same centerline as input shaft. The countershaft and reverse idler shaft mount below and to one side in case.

The ***countershaft,*** also called ***cluster gear shaft,*** holds the countershaft gear into mesh with the input gear and other gears in the transmission. It is located slightly below and to one side of the input shaft, **Figure 55-9.**

Normally, the countershaft does not turn. It is locked in the case by either a steel pin, a force fit, or lock nuts. Refer to **Figure 55-8.**

The ***reverse idler shaft*** is a short shaft that supports the reverse idler gear. It normally mounts in the case, midway between the countershaft and the output shaft. Then, the reverse idler gear can mesh with gears on both the countershaft and output shaft. Refer to **Figures 55-7** and **55-8.**

The transmission ***output shaft,*** or ***main shaft,*** holds the output gears and synchronizers. The rear of this shaft extends to the back of the extension housing. It connects to the drive shaft to turn the wheels of the vehicle.

The output shaft is splined in the center. In modern transmissions, the gears are free to revolve on the output shaft, but the synchronizers are locked on the shaft by splines. The synchronizers will only turn when the shaft itself turns.

Transmission Gears

Transmission gears can be typically classified into four groups: input shaft gear, countershaft gear, reverse idler gear, and output shaft gears. The input shaft gear turns the countershaft gears. The countershaft gears turn the output shaft gears and reverse idler gear, **Figure 55-9.**

In low gear, a small gear on the countershaft drives a larger gear on the output shaft. This provides a high gear ratio for accelerating, **Figure 55-10A.**

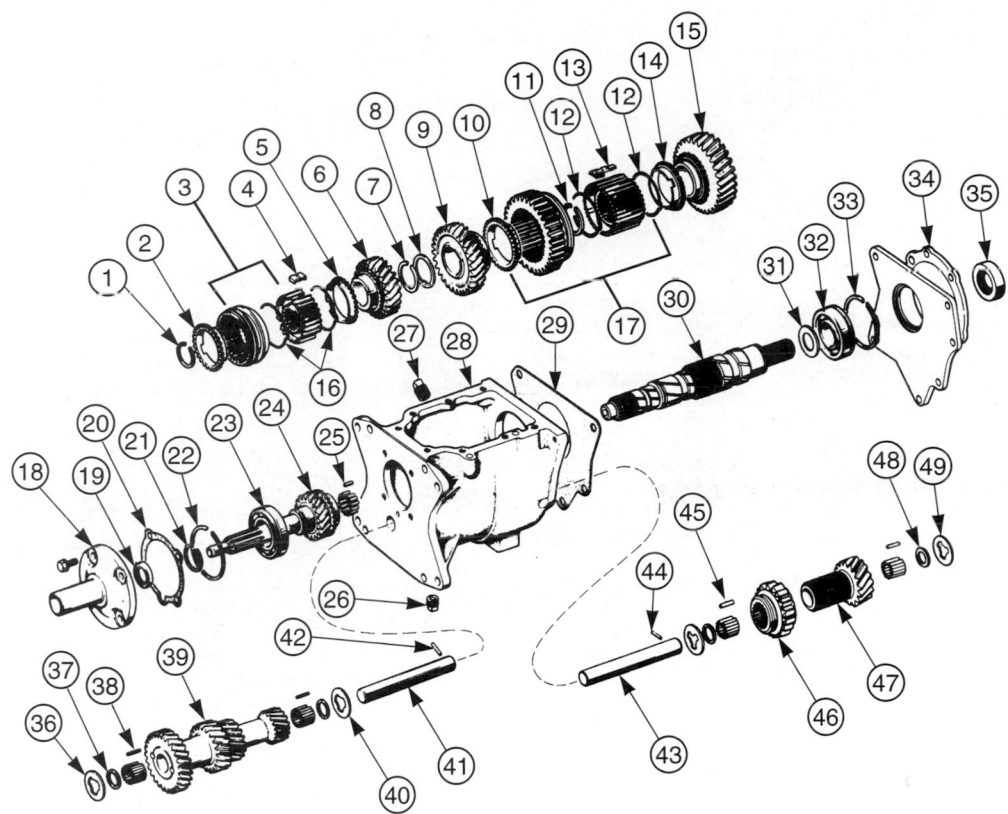

1. 3rd–4th gear snap ring	18. Front bearing cap	35. Adapter seal
2. 4th gear synchronizer ring	19. Oil seal	36. Front countershaft gear thrust washer
3. 3rd–4th gear clutch assembly	20. Gasket	37. Roller washer
4. 3rd–4th gear plate	21. Snap ring	38. Rear roller bearing
5. 3rd gear synchronizer ring	22. Lock ring	39. Countershaft gear
6. 3rd speed gear	23. Front ball bearing	40. Rear countershaft thrust washer
7. 2nd gear snap ring	24. Input shaft	41. Countershaft
8. 2nd gear thrust washer	25. Roller bearing	42. Pin
9. 2nd speed gear	26. Drain plug	43. Reverse idler shaft
10. 2nd gear synchronizer ring	27. Fill plug	44. Pin
11. Main shaft snap ring	28. Case	45. Idler gear roller bearing
12. 1st-2nd synchronizer spring	29. Gasket	46. Reverse idler sliding gear
13. Low–2nd plate	30. Output shaft	47. Reverse idler gear
14. 1st gear synchronizer ring	31. 1st gear thrust washer	48. Idler gear washer
15. 1st gear	32. Rear ball bearing	49. Idler gear thrust washer
16. 3rd–4th synchronizer spring	33. Snap ring	
17. 1st–2nd gear clutch assembly	34. Adapter plate	

Figure 55-8. Exploded view shows the major parts of a typical transmission. Note the four shafts and components. (Chrysler)

In high gear, a larger countershaft gear drives an equal-size or smaller output shaft gear. This reduces the gear ratio and the vehicle moves faster. See **Figure 55-10B.**

When in reverse, power flows from the countershaft gear to the reverse idler gear. Power is then transferred from the reverse idler gear to the engaged gear on the output shaft. This reverses output shaft rotation, as shown in **Figure 55-10C.**

Input Gear

The transmission *input gear* is a machined part of the steel input shaft. The input gear drives the forward gear on the countershaft gear. A small spur gear is usually located next to the main helical drive gear. This small gear is used for synchronizer engagement.

Figure 55-11 shows an input gear with its related parts. Study the shape and relationship of each component carefully.

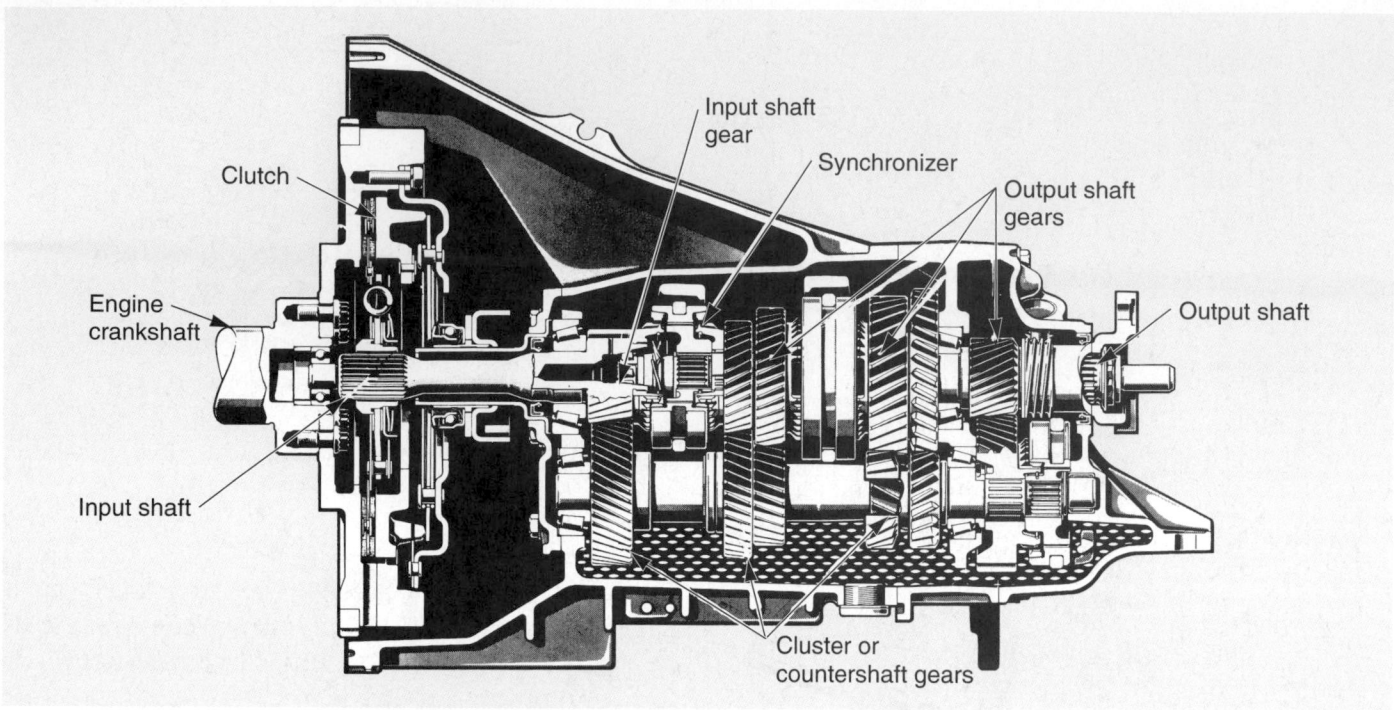

Figure 55-9. Cutaway view of a transmission shows gears assembled on their shafts. The gear on the input shaft drives the countershaft gears. Countershaft gears turn the gears on the output shaft. (Mercedes Benz)

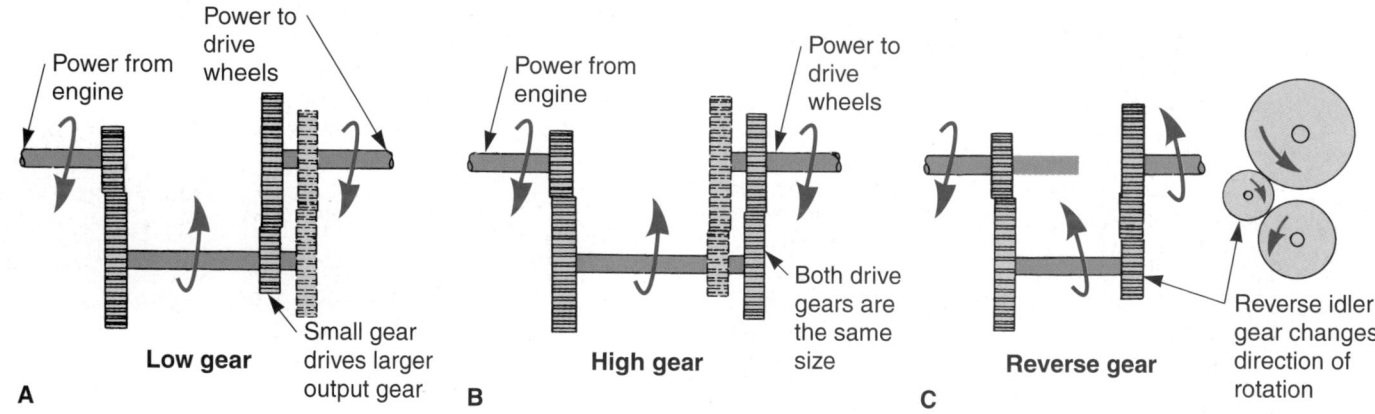

Figure 55-10. Simplified transmission action. A—Low gear. The input shaft gear turns the countershaft gears. A small countershaft gear drives a larger output shaft gear, producing gear reduction. B—High gear. Engaged gears are the same size. Less torque is needed. C—Reverse. The reverse idler gear is used between the countershaft gear and the output shaft gear. This reverses the direction of rotation of the output shaft. (Deere & Co.)

Countershaft Gear

The *countershaft gear,* or *counter gear,* turns the gears on the output shaft. This gear is actually several gears machined out of a single piece of steel. It is often called the *cluster gear.*

The countershaft gear rides on roller bearings. Thrust washers fit on each end of the gear to set end play or case-to-gear clearance. See **Figure 55-12.**

When the input gear drives the engaged countershaft gear, all the countershaft gears turn as a unit. However,

since each forward gear is a different size, the countershaft gear unit is capable of providing several gear ratios.

Reverse Idler Gear Assembly

A *reverse idler gear assembly* changes the direction of gear rotation so the vehicle can be moved in reverse. Note that the reverse idler gear assembly is constructed like the other transmission shaft-gear assemblies just discussed. See **Figure 55-13.**

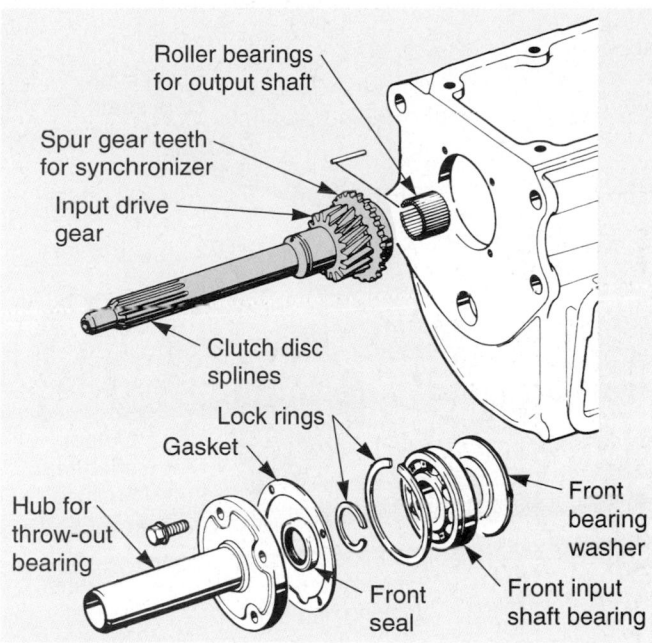

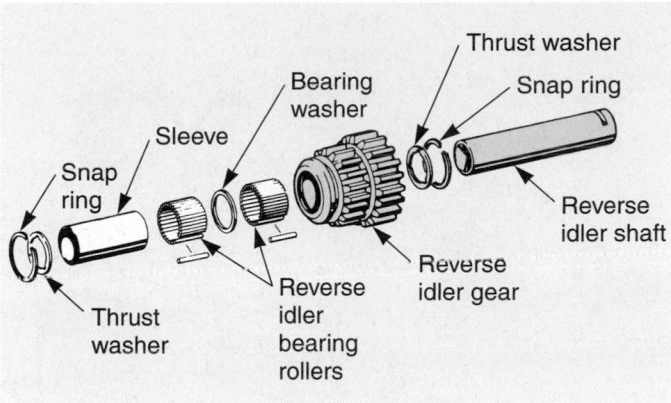

Figure 55-13. Reverse idler gear and shaft assembly. (Chrysler)

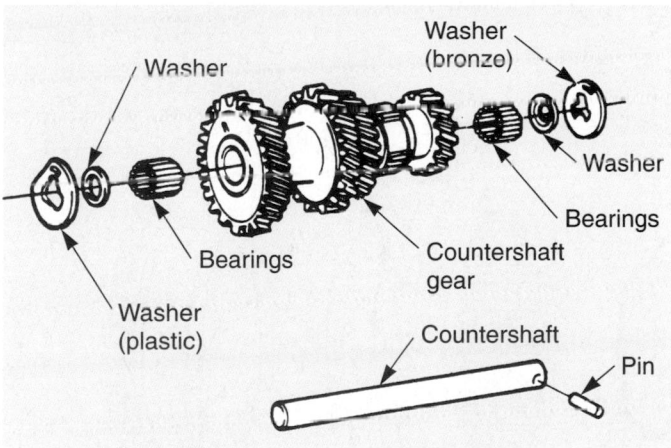

Figure 55-11. An exploded view of the input shaft and gear. The gear is normally a machined part of the shaft. A large bearing supports the shaft in front of the transmission case. Individual roller bearings support the rear of the shaft. Snap rings secure the assembly in the case. (Chrysler)

Figure 55-12. Countershaft assembly. A countershaft gear has several gears formed as a single unit. They mount on roller bearings and the countershaft. Washers control end play of the unit. (Ford)

Output Shaft Gears

The *output shaft gears,* or *main shaft gears,* transfer rotation from the countershaft gears to the output shaft. Only one of the output shaft gears is normally engaged and locked to the shaft at a time.

A set of output shaft gears is shown in **Figure 55-14.** Notice how each consists of a main drive gear (helical gear) and a smaller synchronizer gear (spur gear).

The inside bore of each output shaft gear is smooth so that it can spin freely on its shaft when not engaged. Normally, one output shaft gear will be provided for each transmission speed, including reverse.

Transmission Synchronizers

A *transmission synchronizer* has two functions:

- Prevents the gears from clashing (grinding) during engagement.
- Locks the output gear to the output shaft.

When the synchronizer is away from an output gear, the output gear freewheels (spins) on the output shaft. When the synchronizer slides against the output gear, the gear is locked to the synchronizer and to the output shaft. Power is then sent out of the transmission and to the rear wheels.

Synchronizer Construction

The most popular synchronizer consists of an inner splined hub, inserts, insert springs, an outer sleeve, and blocking rings. See **Figure 55-15.**

The synchronizer hub is splined on the output shaft. It is held in a stationary position between the transmission gears. Inserts fit between the hub and sleeve. The springs push the inserts into the sleeve. This helps hold and center the sleeve on its hub. The blocking rings fit on the outer ends of the hub and sleeve.

Synchronizer Operation

When the driver shifts gears, the synchronizer sleeve slides on its splined hub toward the output gear.

First, the *blocking ring* cone rubs on the side of the drive gear cone, causing friction between the two. This causes the output gear, the synchronizer, and the output shaft to begin to spin at the same speed, **Figure 55-16.**

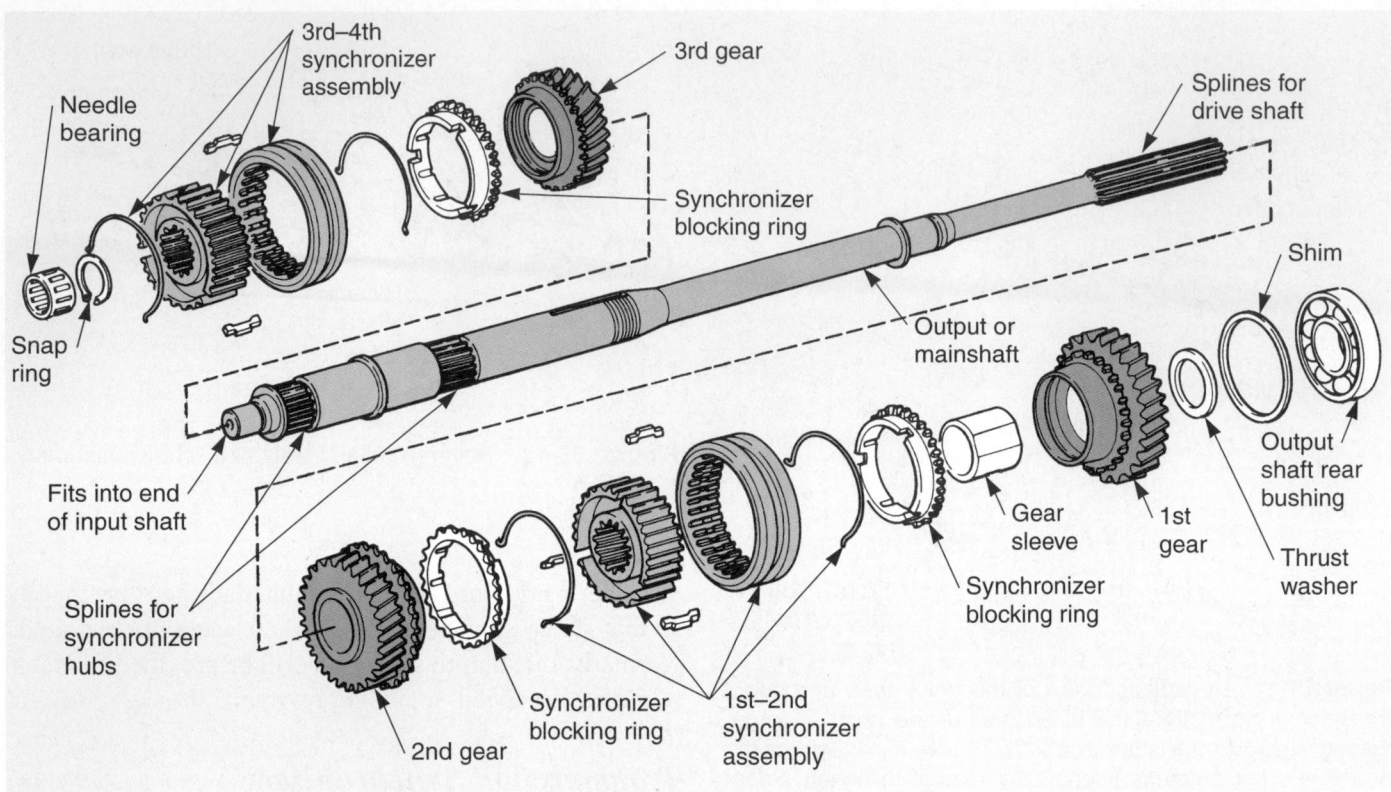

Figure 55-14. The output shaft is a long shaft extending through the transmission extension housing. The drive shaft is splined to the rear of this shaft. (Mazda)

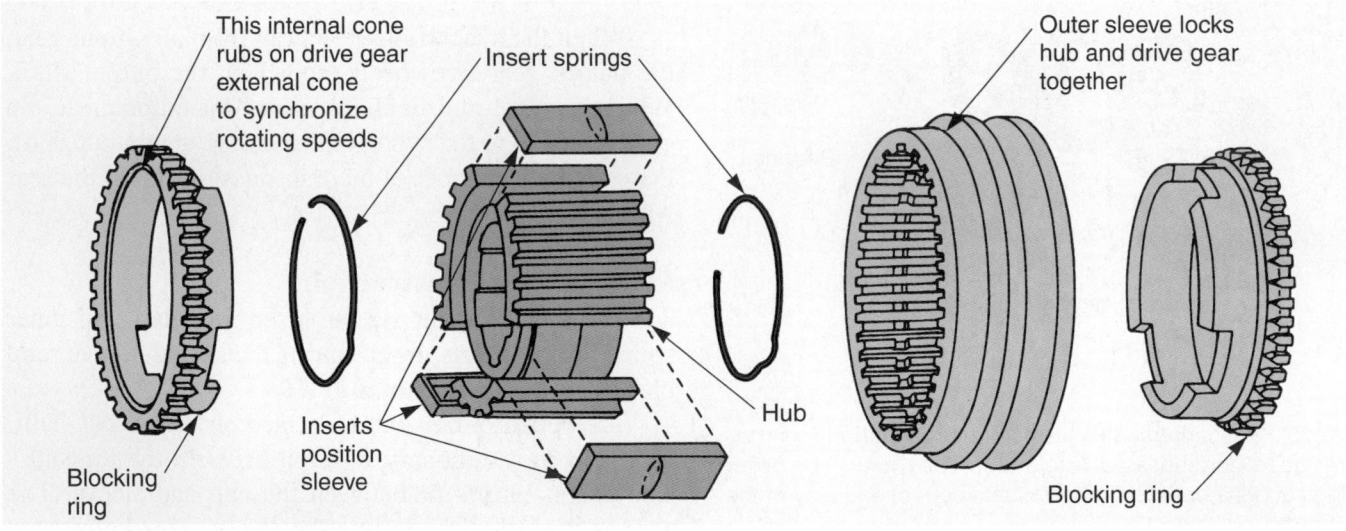

Figure 55-15. Basic synchronizer components. The hub is splined to the output shaft. It will slide but not turn on the shaft. A sleeve fits over the hub. Shifter plates position the sleeve. Blocking rings allow the sleeve to slide into and mesh with the output gear without clashing or grinding. (Deere & Co.)

As soon as the speed is equalized (synchronized), the sleeve can slide completely over the blocking ring and over the small spur gear teeth on the output gear. This locks the output gear to the synchronizer hub and to the shaft. Power then flows through that gear to the rear wheels.

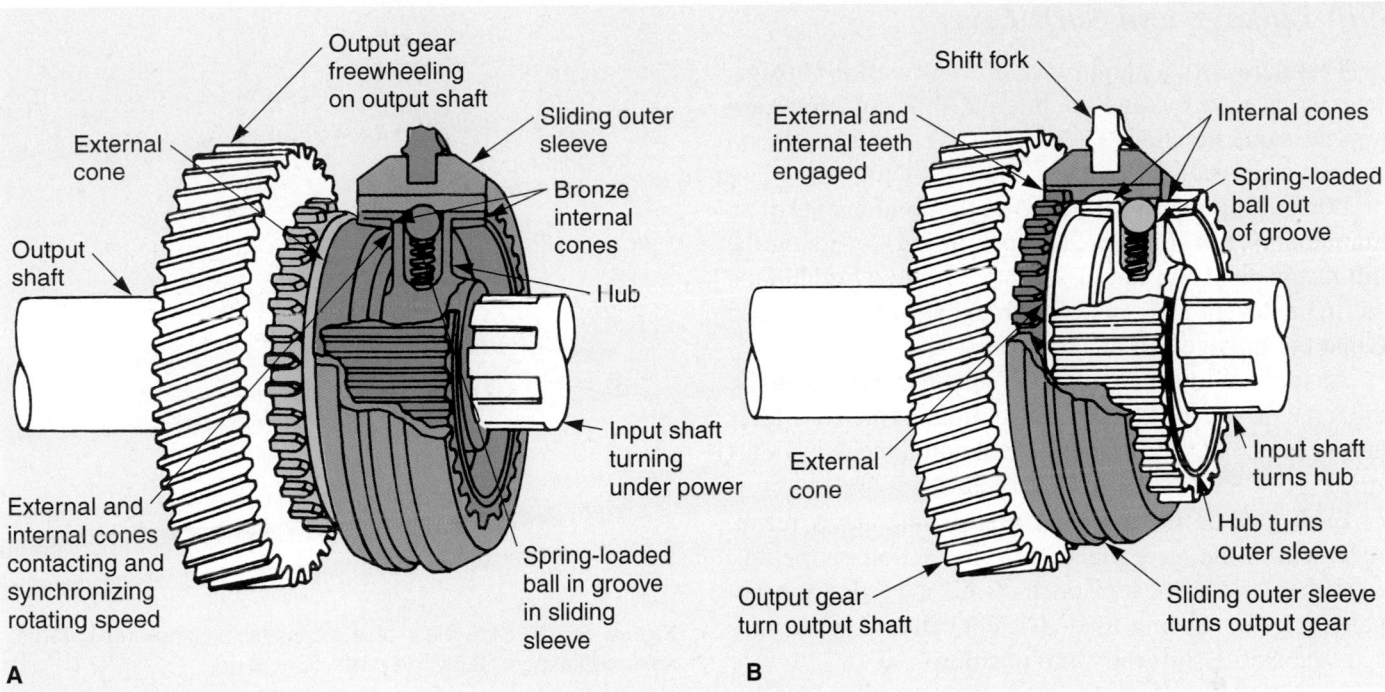

Figure 55-16. Synchronizer operation. A—As the synchronizer sleeve moves into the output gear, the cone on the sleeve rubs against the cone on the gear. Friction makes the gear and sleeve begin to turn at the same speed. B—When at same speed, the sleeve can slide over and mesh with the small spur teeth on the side of the output gear. This locks the output gear, sleeve, hub, and input shaft together. The output gear is then engaged to the output shaft.

Fully Synchronized Transmission

In a *fully synchronized transmission,* all the forward output gears use a synchronizer. This allows the driver to downshift into any lower gear (except reverse) while the car is moving. Most modern manual transmissions are fully synchronized.

Many older three-speed transmissions did not have a synchronized first gear. In this case, the driver must wait until the vehicle comes to a complete stop before downshifting into first. Trying to shift into first while the vehicle is in motion will cause first gear to grind.

> **Tech Tip**
> A *clutchless manual transmission* uses computer control and transmission sensors to allow shifting gears without clutch disengagement. Computer control synchronizes the speed of the engine with the transmission so that no load is sent through the transmission while shifting.

Shift Forks

Shift forks fit around the synchronizer sleeves to transfer movement from the gear shift linkage to the sleeves. This is illustrated in **Figure 55-17.**

The shift fork fits into a groove cut into the synchronizer sleeve. A shift rail, or linkage rod, connects the fork to the driver's shift lever. When the shift lever moves, the rail moves the shift fork and the synchronizer sleeve to engage the correct transmission gear.

Figure 55-18 pictures a typical shift fork assembly. Study the parts and how they fit together.

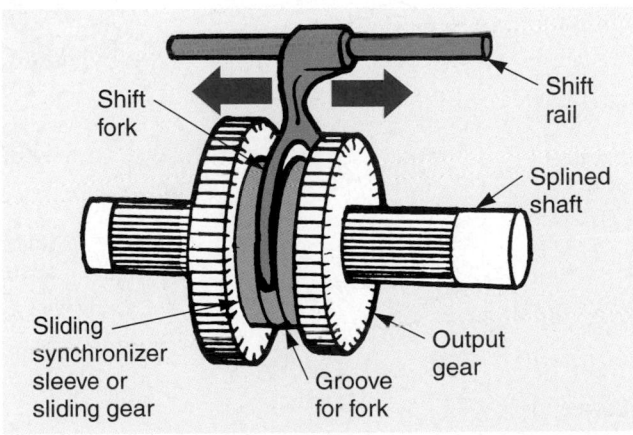

Figure 55-17. A shift fork is used to move the synchronizer sleeve. Motion is transferred to the shift fork through the shift rail.

Shift Linkage and Shift Lever

There are two general types of transmission linkage: the external shift rod and the internal shift rail. Both perform the same function. They connect the shift lever with the shift fork mechanism.

Look at **Figure 55-19.** It shows the components of an external shift rod linkage. The rods fit into levers on the shift mechanism and fork assembly. Spring clips hold the rods in the levers. One end of each linkage rod is threaded so that the linkage can be adjusted.

An internal shift rail linkage is shown in **Figure 55-20.** When the driver shifts gears, the bottom of the shift lever catches in one of the *gates* (notched unit attached to shift rail), **Figure 55-21.**

Each gate is mounted on a shift rail. As a result, movement of the lever places a prying action on the rail. Since the fork is located on the rail, it is also moved, changing gears. Spring-loaded detent balls are sometimes used to lock the shift rails into position.

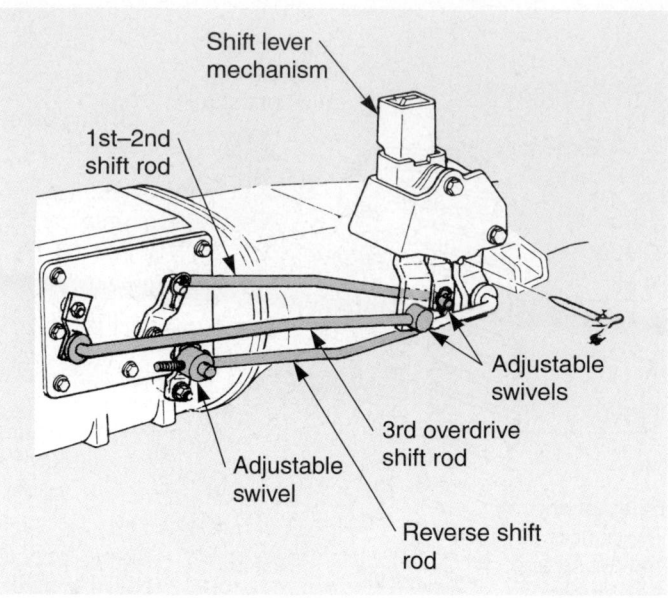

Figure 55-19. Side view of a transmission showing external shift rod linkage. Study the parts. (Chrysler)

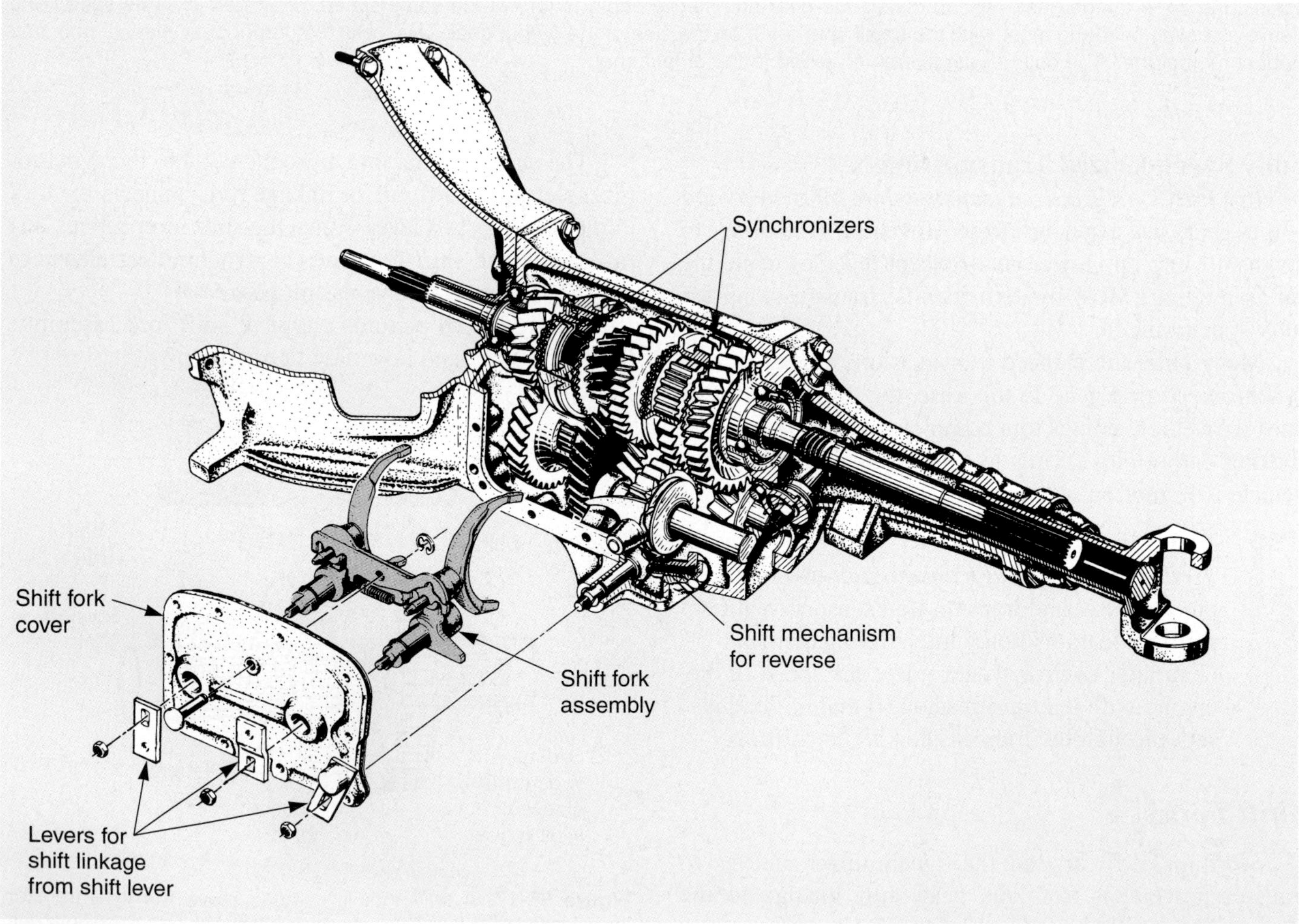

Figure 55-18. The shift fork fits over a groove in the center of the synchronizer sleeve. Movement of shift linkage moves the shift fork to engage different output gears. (Plymouth)

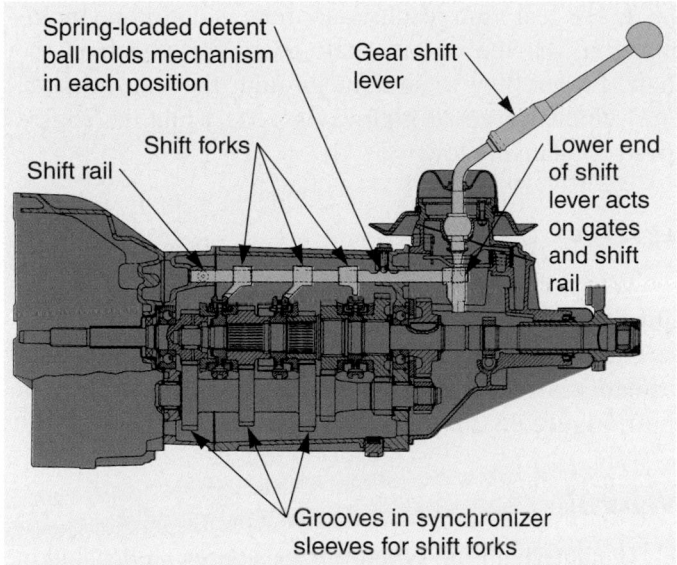

Figure 55-20. This transmission uses an internal shift rail mechanism instead of external shift rods. The shift lever acts on the rail. The rail then operates the shift forks and synchronizer sleeves. (Fiat)

Variations in shift rail linkages are also available. However, the basic construction and operation are similar for all systems. The transmission shift lever assembly is moved to cause movement of the shift linkage, shift forks, and synchronizers.

Figure 55-22 pictures a steering column–mounted shift lever. Study the parts and how they function.

Transmission Types

There are several types of manual transmissions: three-speed, four-speed, five-speed, etc. Some transmissions have an overdrive in high gear. Transmissions with more forward speeds provide a better selection of gear ratios.

Older vehicles were commonly equipped with three-speed transmissions. Modern vehicles, however, frequently have four- or five-speed transmissions. Extra gear ratios are needed for the smaller, low-horsepower, high-efficiency engines of today.

Transmission Power Flow

Now that you understand the basic parts and construction, we will cover the flow of power through manual transmissions.

First Gear

To get the vehicle moving from a standstill, the driver moves the gear shift lever into first gear. The clutch pedal

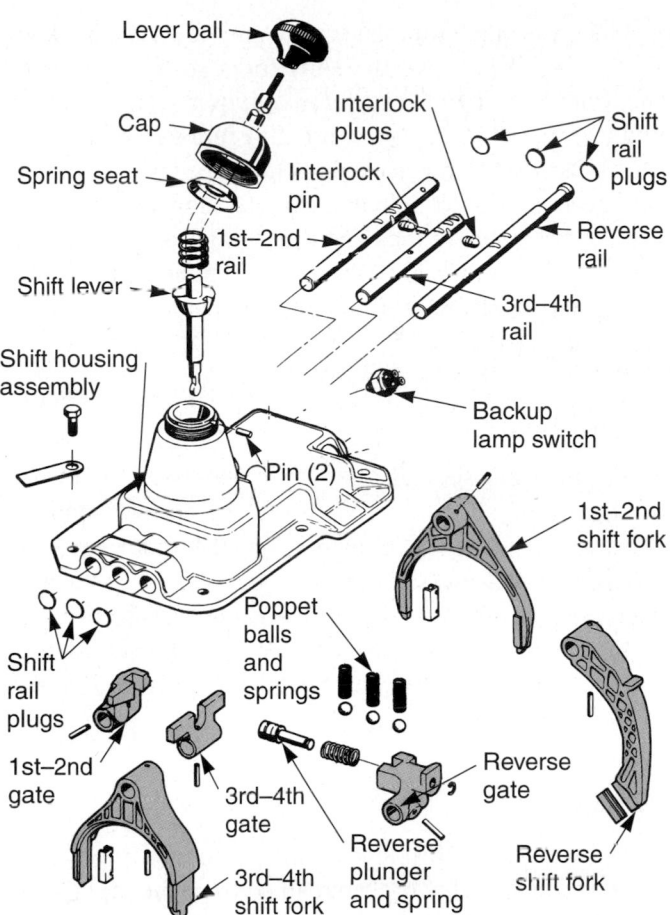

Figure 55-21. An exploded view of the shift mechanism for a late-model transmission. The shift lever acts on shift gates. Shift gates are attached to shift rails. The rails move shift forks for gear changes. Spring-loaded balls and plunger help position shift forks during each gear change. Note back-up lamp switch. (Chrysler)

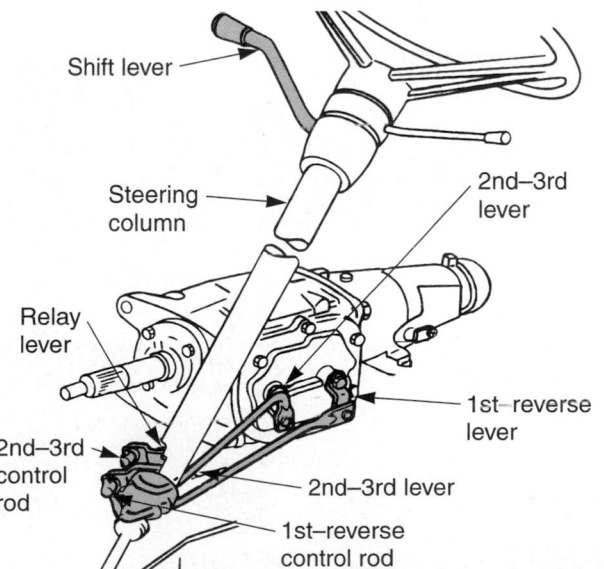

Figure 55-22. Column shift mechanism. The gear shift lever operates levers on the bottom of the column. Rods then transfer movement to the transmission. (GMC)

must be pressed to stop power flow into the transmission. The linkage rods move the shift forks so that first gear synchronizer is engaged to the first output gear. The other output gears are in neutral. Look at **Figure 55-23.**

As the driver releases the clutch pedal, the clutch shaft gear begins to spin the countershaft gears. Since only the first gear is locked to the output shaft, a small gear on the countershaft drives the larger gear on the output shaft. The gear ratio is approximately 3:1 and the vehicle accelerates easily.

Second Gear

To shift into second, the driver depresses the clutch and moves the shift lever. With the engine momentarily disconnected from the transmission, the first gear synchronizer is slid away from first gear. The second-third synchronizer is then engaged. See **Figure 55-24.**

Now power flow is through second gear on the output shaft. A gear ratio of about 2:1 is produced to give the vehicle a little more speed.

Third Gear

When the gear shift lever is moved into third gear, the synchronizer is slid over the small teeth on the input shaft gear. The synchronizer sleeve locks the input shaft directly to the output shaft. Refer to **Figure 55-25.**

A 1:1 gear ratio results, and there is no torque multiplication. All the output shaft gears freewheel on the shaft. Power flow is straight through the transmission. The vehicle travels at highway speeds while the engine speed is relatively low.

Reverse

When shifted into reverse, a synchronizer is moved into the reverse gear on the output shaft. This locks the gear to the output shaft. Power flows through the countershaft, reverse idler gear, reverse gear, and to the output shaft, **Figure 55-26.**

Neutral

In neutral, all the synchronizer sleeves are located in the center of their hubs. This allows all the output shaft gears to freewheel on the output shaft. No power is transmitted to the output shaft, **Figure 55-27.**

Overdrive Gear

In many modern transmissions, high gear is an overdrive. Either fourth (four-speed) or fifth (five-speed) has a ratio of less than 1:1 (0.87:1 for example) to increase fuel economy.

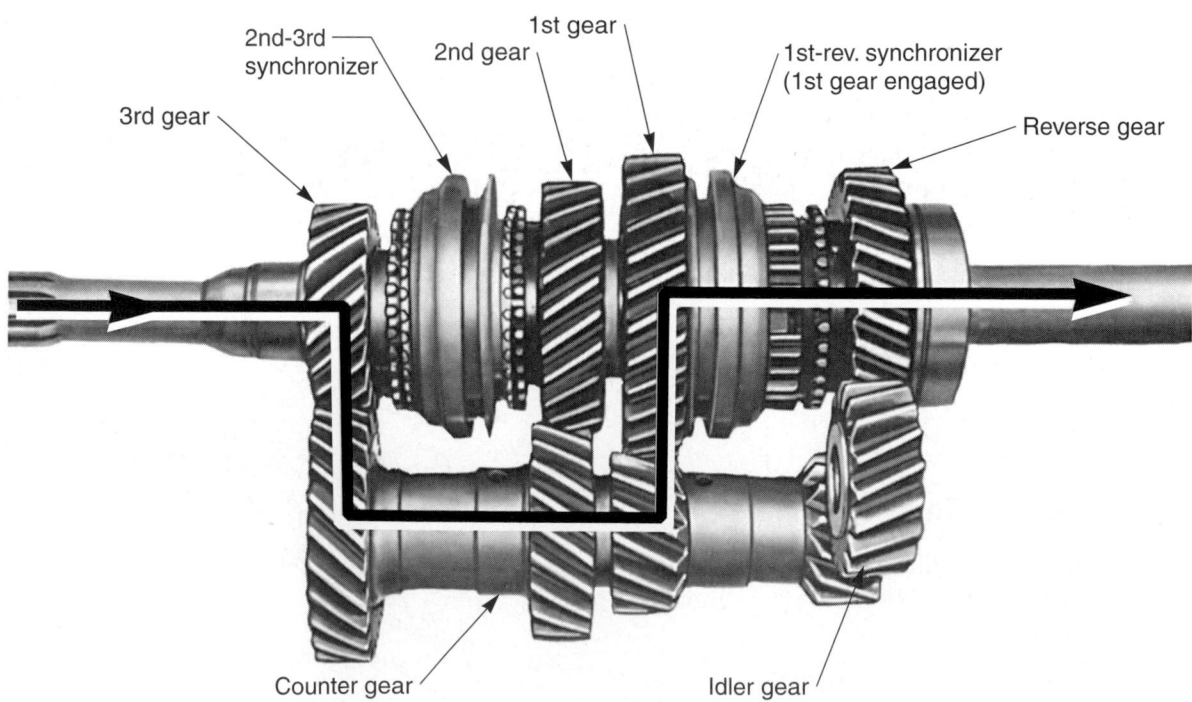

Figure 55-23. A transmission in first gear. The first-reverse synchronizer is engaged with the first output gear. The other synchronizer is in neutral position. The first output gear is locked to the output shaft and transfers high torque to drive shaft. (Chevrolet)

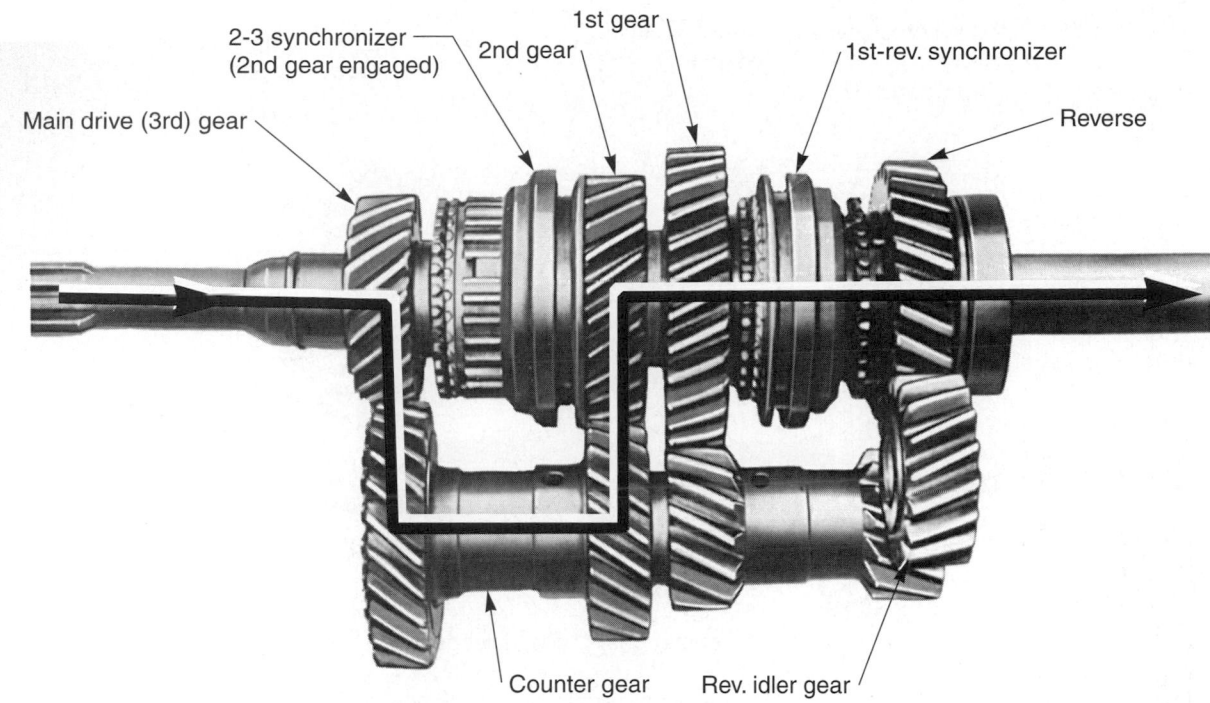

Figure 52-24. A transmission in second gear. The first-reverse synchronizer is moved into neutral. The second-third synchronizer is engaged with the second output gear, locking it to its shaft. (Chevrolet)

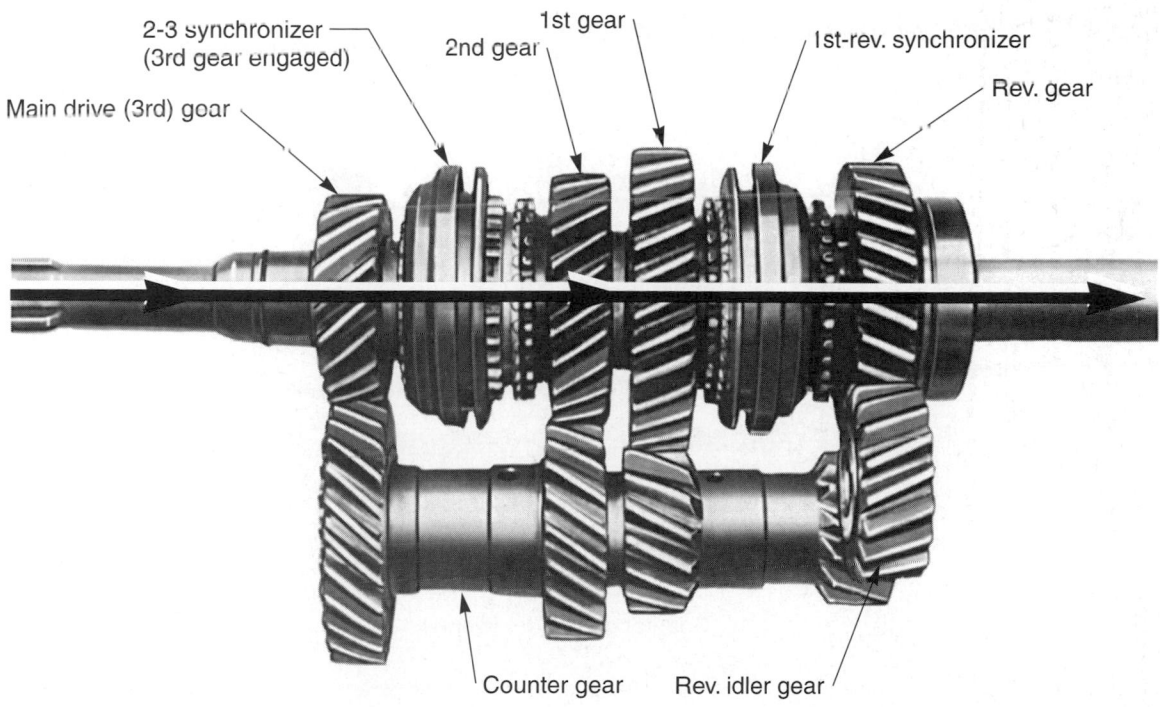

Figure 55-25. A transmission in third gear. The second-third gear synchronizer is slid to engage the input shaft gear. This locks the input shaft directly to the output shaft. Both shafts turn at the same speed for no gear reduction. (Chevrolet)

Figure 55-28 shows the power flow through a late-model, five-speed, overdrive transmission. It is designed for a low-horsepower, diesel engine. Trace flow through the transmission in each gear. The first four forward speeds allow the vehicle to accelerate quickly. The overdrive gear keeps engine speed down at highway speeds to increase fuel economy and engine service life.

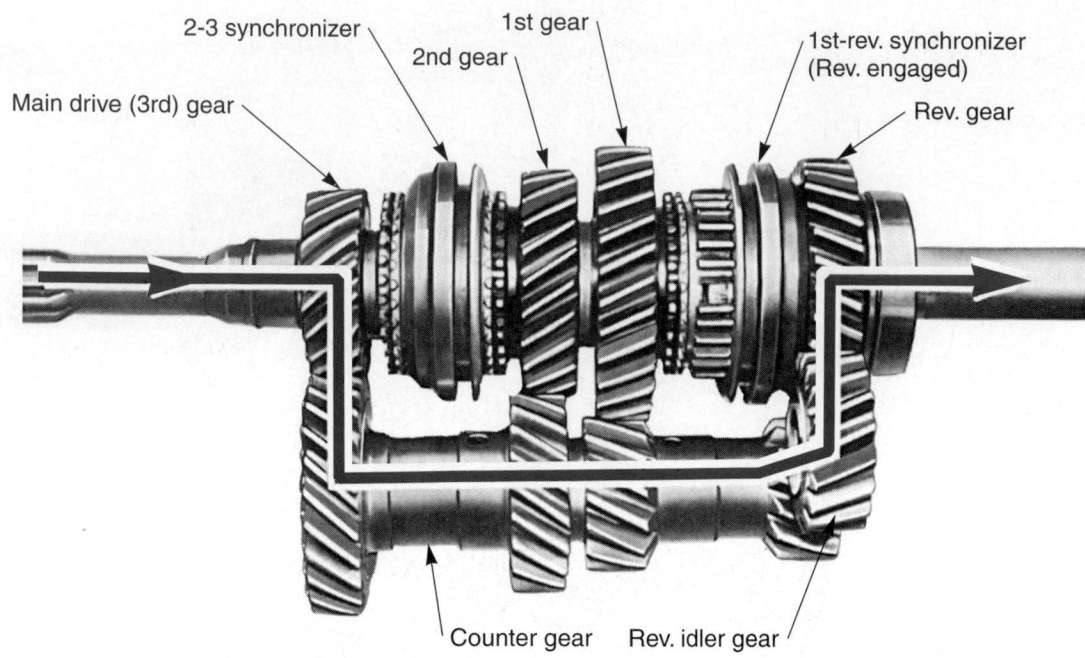

Figure 55-26. A transmission in reverse. The second-third synchronizer is moved into neutral. The first-reverse synchronizer meshes with the reverse output gear. The countershaft gear drives the reverse idler gear. The idler gear drives the output shaft backwards. (Chevrolet)

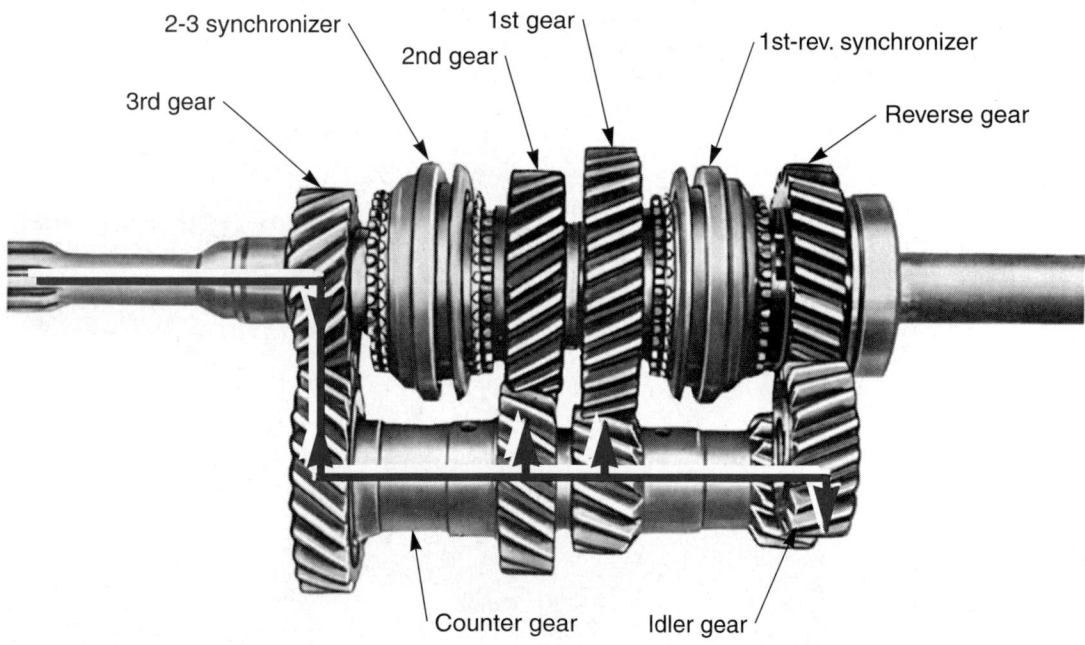

Figure 55-27. In neutral, both synchronizers are in center positions. No output gears are locked to output shaft. The output gears freewheel and do not transfer power to drive shaft. (Chevrolet)

Other Transmission Designs

Many transmission design variations are used by the numerous auto manufacturers. However, all transmissions use the basic operation and construction principles just explained.

Review the transmission parts in **Figures 55-29** and **55-30.** Can you explain the basic function of each part?

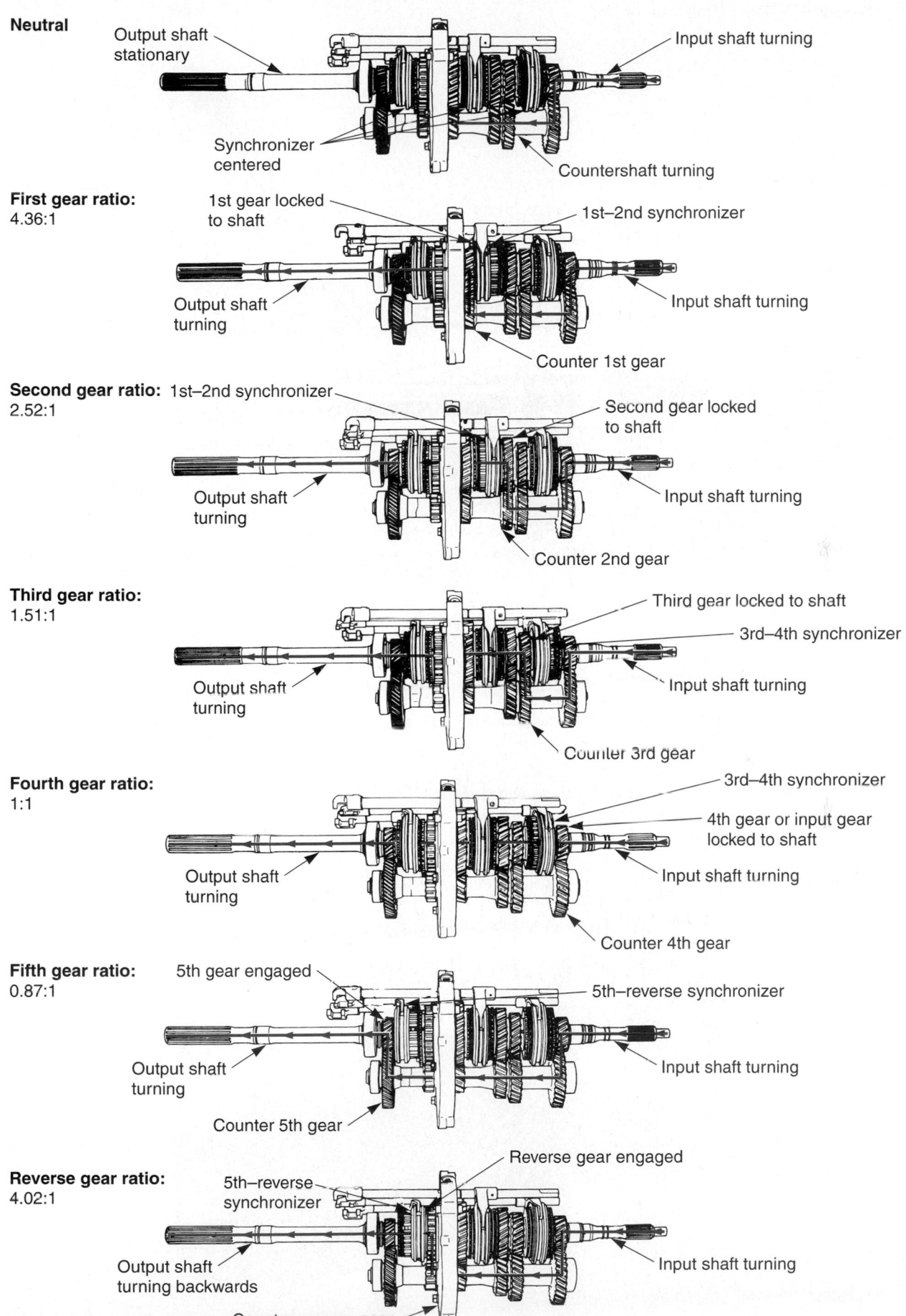

Neutral
Output shaft stationary — Input shaft turning
Synchronizer centered — Countershaft turning

First gear ratio: 4.36:1
1st gear locked to shaft — 1st–2nd synchronizer
Output shaft turning — Input shaft turning
Counter 1st gear

Second gear ratio: 2.52:1
1st–2nd synchronizer — Second gear locked to shaft
Output shaft turning — Input shaft turning
Counter 2nd gear

Third gear ratio: 1.51:1
Third gear locked to shaft
3rd–4th synchronizer
Output shaft turning — Input shaft turning
Counter 3rd gear

Fourth gear ratio: 1:1
3rd–4th synchronizer
4th gear or input gear locked to shaft
Output shaft turning — Input shaft turning
Counter 4th gear

Fifth gear ratio: 0.87:1
5th gear engaged — 5th–reverse synchronizer
Output shaft turning — Input shaft turning
Counter 5th gear

Reverse gear ratio: 4.02:1
Reverse gear engaged
5th–reverse synchronizer
Output shaft turning backwards — Input shaft turning
Counter reverse gear

Figure 55-28. Power flow through a five-speed transmission with overdrive in high gear. Study each illustration carefully.

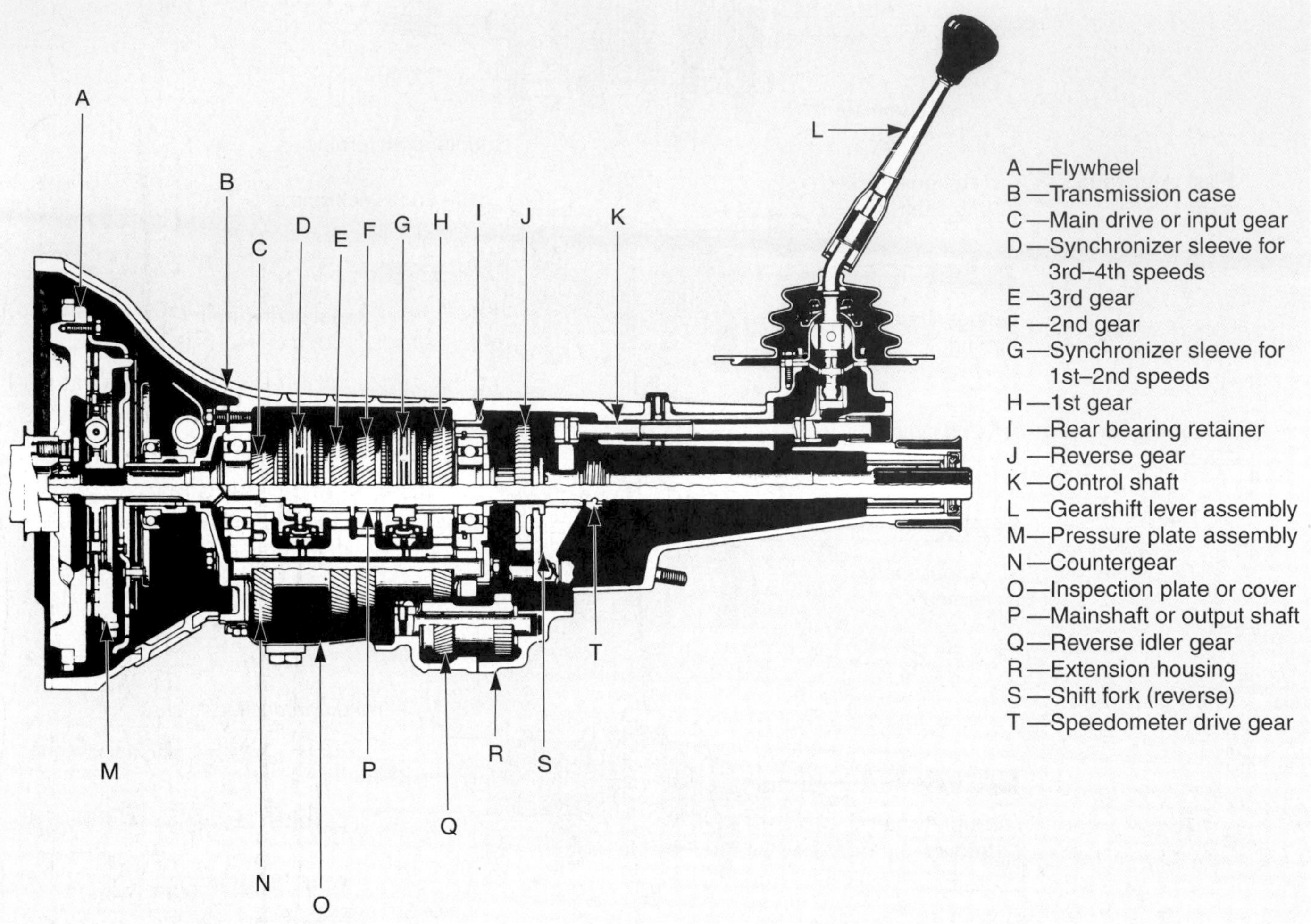

A — Flywheel
B — Transmission case
C — Main drive or input gear
D — Synchronizer sleeve for
     3rd–4th speeds
E — 3rd gear
F — 2nd gear
G — Synchronizer sleeve for
     1st–2nd speeds
H — 1st gear
I — Rear bearing retainer
J — Reverse gear
K — Control shaft
L — Gearshift lever assembly
M — Pressure plate assembly
N — Countergear
O — Inspection plate or cover
P — Mainshaft or output shaft
Q — Reverse idler gear
R — Extension housing
S — Shift fork (reverse)
T — Speedometer drive gear

Figure 55-29. Cutaway view of a late-model four-speed transmission. Note part names and locations. (Chrysler)

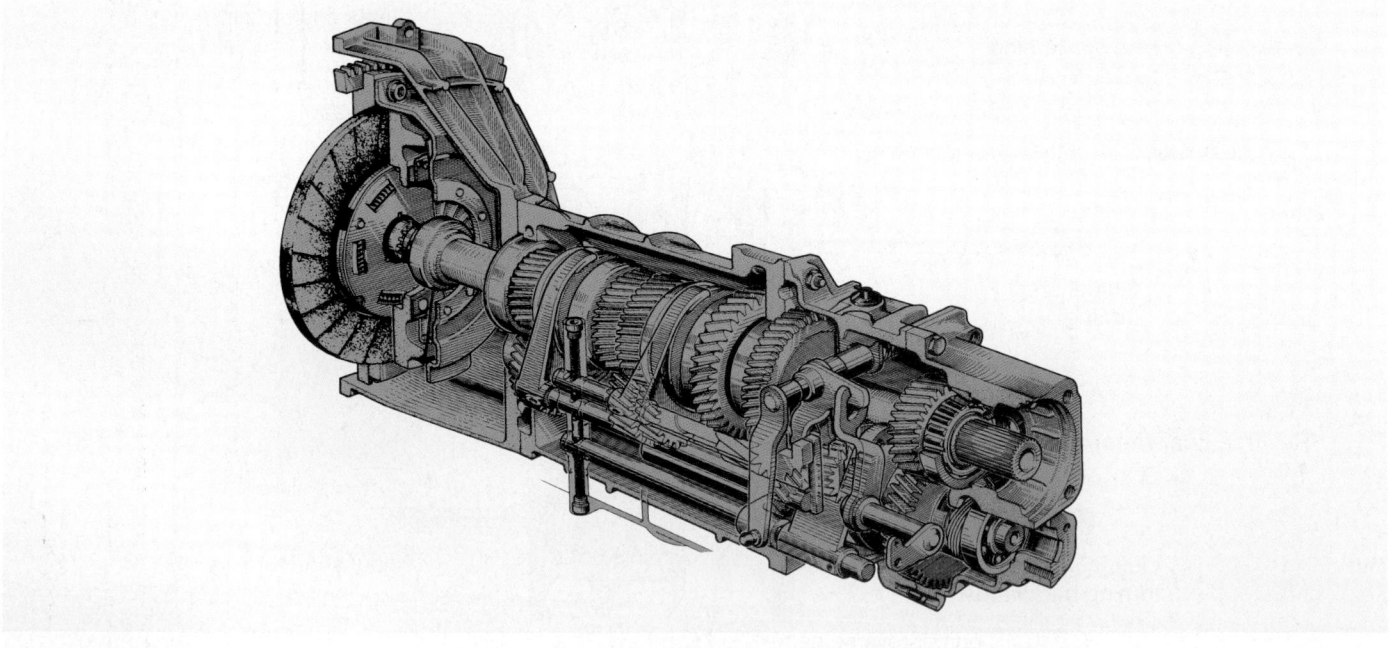

Figure 55-30. A five-speed, manual transmission with clutch installed on the input shaft. Can you describe function of major parts?

Speedometer Drive

Normally, a manual transmission has a worm gear on the output shaft that drives the speedometer gear and cable. See **Figure 55-29.**

The gear on the output shaft turns a plastic gear on the end of the speedometer cable. The cable runs through a housing up to the speedometer head (speed indicator assembly) in the dash. A retainer and bolt hold the cable assembly in the transmission extension housing.

Whenever the output shaft turns, the speedometer cable turns. This makes the speedometer head register the road speed of the vehicle.

Manual Transmission Switches

Two types of electric switches are sometimes mounted on the manual transmission: the back-up light switch and the ignition spark switch.

The **back-up light switch** is an electric switch closed by the action of the reverse gear shift linkage. When shifted into reverse, the linkage closes the switch to connect the back-up lamps to the battery. Refer back to **Figure 55-21.**

A few manual transmissions have an **ignition spark switch,** which allows distributor vacuum advance only in high gear. The switch usually mounts in the side of the transmission. It is normally closed until activated in high gear. This retards the ignition timing in lower gears to reduce exhaust pollution.

Figure 55-31 shows the driveline for a front-engine, four-wheel drive vehicle. The transmission is located directly behind the engine.

Summary

- A manual transmission must be shifted by hand to change the amount of torque going to the other parts of the drive train.
- An automatic transmission uses hydraulic pressure and sensing devices to shift gears.
- A manual transmission is designed to change the vehicle's drive wheel speed and torque in relation to engine speed and torque.
- A gear ratio is the number of turns a driving gear must turn before the driven gear turns one complete revolution.
- Gear reduction occurs when a small gear drives a larger gear to increase turning force.
- An overdrive ratio results when a larger gear drives a smaller gear.
- Gear backlash is the small clearance between meshing gear teeth.
- Typically, 80W or 90W gear oil is recommended for use in a manual transmission.
- The transmission case must support the transmission bearings and shafts and provide an enclosure for gear oil.

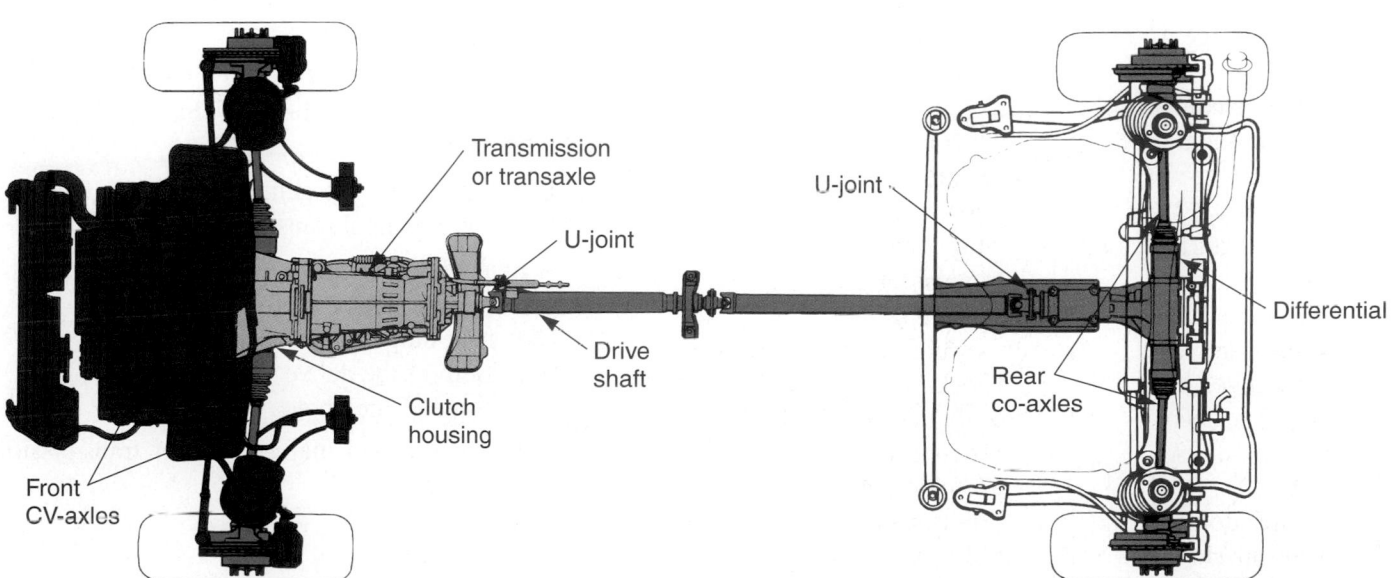

Figure 55-31. Note the parts of this four-wheel drive system. The engine and transaxle are in front; a drive shaft transmits power to the rear differential assembly. (Subaru)

- The extension housing, also called the tail shaft housing, bolts to the rear of the transmission case.

- A front bearing hub, sometimes called front bearing cap, covers the front transmission bearing and acts as a sleeve for the clutch throw-out bearing.

- The input shaft, often termed clutch shaft, transfers rotation from the clutch disc to the countershaft gears in the transmission.

- The countershaft, also called cluster gear shaft, holds the countershaft gear into mesh with the input gear and other gears in the transmission.

- The transmission output shaft, also called main shaft, holds the output gears and synchronizers.

- A transmission synchronizer has two functions: it prevents the gears from grinding or clashing during engagement and locks the output gear to the output shaft.

- The shift forks fit around the synchronizer sleeves to transfer movement to the sleeves from the gear shift linkage.

Important Terms

Manual transmission
Transaxle
Transmission input
  shaft
Transmission gears
Synchronizers
Shift forks
Shift linkage
Gear shift lever
Output shaft
Transmission case
Gears
Gear ratio
Torque multiplication
Gear reduction
Overdrive ratio
Spur gears
Helical gears
Gear backlash
Splash
Gear oil
Antifriction bearings
Transmission case
Extension housing
Tail shaft housing
Front bearing hub
Front bearing cap

Input shaft
Clutch shaft
Countershaft
Cluster gear shaft
Reverse idler shaft
Output shaft
Main shaft
Transmission input gear
Countershaft gear
Counter gear
Cluster gear
Reverse idler gear
  assembly
Output shaft gears
Main shaft gears
Transmission
  synchronizer
Blocking ring
Fully synchronized
  transmission
Clutchless manual
  transmission
Shift forks
Gates
Back-up light switch
Ignition spark switch

Review Questions—Chapter 55

Please do not write in this text. Place your answers on a separate sheet of paper.

1. List and explain the eight major parts of a manual transmission.

2. Define the term "gear ratio."

3. How do you find the gear ratio of two gears?

4. A certain manual transmission's drive gear has 12 teeth and the driven gear has 72 teeth. What is the gear ratio in this situation?
   (A) 3:1.
   (B) 6:1.
   (C) 2:1.
   (D) 4:1.

5. Approximate manual transmission gear ratios are _____ for first, _____ for second, _____ for high, and _____ for reverse.

6. A gear reduction results when a small gear drives a larger gear to increase turning force. True or False?

7. This would be an overdrive ratio.
   (A) 1:1.
   (B) 0.87:1.
   (C) 1:0.87.
   (D) 3:1.

8. _____ _____ is the small clearance between the meshing gear teeth for lubrication and heat expansion.

9. Typically, _____ or _____ gear oil is used in a manual transmission.

10. What is the transmission extension housing?

11. Name and describe the four shafts in a manual transmission.

12. List and explain the general gear classifications found in a manual transmission.

13. A manual transmission synchronizer is used to:
    (A) prevent gear clashing or grinding.
    (B) lock output gear to output shaft.
    (C) Both of the above.
    (D) None of the above.

14. Describe the two major types of transmission shift linkages.

15. Why is an overdrive ratio used?

ASE-Type Questions

1. Technician A says a manual transmission normally bolts to the engine's torque converter. Technician B says a manual transmission normally bolts to the automobile's clutch housing. Who is right?
   (A) A only.
   (B) B only.
   (C) Both A and B.
   (D) Neither A nor B.

2. Technician A says a transaxle is an automatic transmission and a differential combined in a single housing. Technician B says a transaxle is a manual transmission and a differential combined in a single housing. Who is right?
   (A) A only.
   (B) B only.
   (C) Both A and B.
   (D) Neither A nor B.

3. Which of the following shafts are normally mounted inside a manual transmission's case?
   (A) Input shaft.
   (B) Countershaft.
   (C) Reverse idler shaft.
   (D) All of the above.

4. While discussing the functions of the basic parts of a manual transmission, Technician A says the input shaft transfers rotating power out of the transmission to the drive shaft. Technician B says the input shaft is used to turn the gears inside the transmission. Who is right?
   (A) A only.
   (B) B only.
   (C) Both A and B.
   (D) Neither A nor B.

5. Technician A says a manual transmission's synchronizers are devices used to disengage the transmission gears. Technician B says a manual transmission's synchronizer's are devices used to engage the transmission's output shaft. Who is right?
   (A) A only.
   (B) B only.
   (C) Both A and B.
   (D) Neither A nor B.

6. Which of the following is a basic function of an automotive manual transmission?
   (A) *Increase drive wheel torque for quick acceleration.*
   (B) *Supply different gear ratios to match engine load.*
   (C) *Operate efficiently, with minimum power loss.*
   (D) *All of the above.*

7. A particular transmission's drive gear has 10 teeth and the gear being driven has 30 teeth. Technician A says the gear ratio in this situation would be 3:1. Technician B says the gear ratio in this situation would be 4:1. Who is right?
   (A) A only.
   (B) B only.
   (C) Both A and B.
   (D) Neither A nor B.

8. Technician A says a gear ratio of 10:1 is smaller than a gear ratio of 5:1. Technician B says a gear ratio of 10:1 is larger than a 5:1 gear ratio. Who is right?
   (A) A only.
   (B) B only.
   (C) Both A and B.
   (D) Neither A nor B.

9. Technician A says a 1:1 gear ratio is generally used for a manual transmission's first gear. Technician B says a 3:1 gear ratio is generally used for a manual transmission's first gear. Who is right?
   (A) A only.
   (B) B only.
   (C) Both A and B.
   (D) Neither A nor B.

10. Technician A says "gear reduction" is commonly utilized in a manual transmission's lower gears. Technician B says "gear reduction" is commonly utilized in a manual transmission's high gear. Who is right?
    (A) A only.
    (B) B only.
    (C) Both A and B.
    (D) Neither A nor B.

11. Technician A says "spur" gears are commonly used in a manual transmission. Technician B says "helical" gears are commonly used in a manual transmission. Who is right?
    (A) A only.
    (B) B only.
    (C) Both A and B.
    (D) Neither A nor B.

12. Technician A says a manual transmission's "gear back lash" helps lubricate the rotating gears in the transmission. Technician B says a manual transmission's "gear backlash" helps prevent gear damage during transmission operation. Who is right?
    (A) A only.
    (B) B only.
    (C) Both A and B.
    (D) Neither A nor B.

13. Technician A says normally, 90W gear oil is recommended for use in a manual transmission. Technician B says 50W gear oil is normally recommended for use in a manual transmission. Who is right?
    (A) A only.
    (B) B only.
    (C) Both A and B.
    (D) Neither A nor B.

14. Technician A says a transmission's synchronizer prevents the gears from grinding during engagement. Technician B says a transmission's synchronizer locks the output gear to the output shaft. Who is right?
    (A) A only.
    (B) B only.
    (C) Both A and B.
    (D) Neither A nor B.

15. Technician A says when a manual transmission is in neutral, all of the synchronizer sleeves are located at the end of the countershaft. Technician B says when a manual transmission is in neutral, all of the synchronizer sleeves are located in the center of their hubs. Who is right?
    (A) A only.
    (B) B only.
    (C) Both A and B.
    (D) Neither A nor B.

Activities—Chapter 55

1. Locate a drive gear and a driven gear on a manual transmission and determine the gear ratio when the two gears are meshed. Show your calculations.

2. Locate the transmission section in a shop manual and study the names of the gears. Locate these same gears on an actual transmission.

3. Demonstrate the adjustment procedures for a manual transmission linkage.

Chapter 56

Manual Transmission Diagnosis and Repair

After studying this chapter, you will be able to:

- Diagnose common manual transmission problems.
- Remove a standard transmission from a vehicle.
- Disassemble and inspect a manual transmission.
- Assemble a manual transmission.
- Install a manual transmission.
- Adjust manual transmission linkage.
- Cite and observe safety rules for transmission service.
- Correctly answer ASE certification test questions on manual transmission diagnosis and repair.

Most makes of manual transmissions use similar construction methods. Therefore, if you learn the common methods for disassembling, inspecting, and rebuilding a typical unit, you will have most of the knowledge needed to repair any manual transmission.

This chapter discusses basic methods and rules that apply to the service of any type of manual transmission. It will give you the information needed to understand and use a service manual when working on any manual transmission. The service manual gives specific instructions and specifications for the exact make and model transmission.

Manual Transmission Problem Diagnosis

Normally, a manual transmission will provide thousands of miles of trouble-free service. Quite often, it will last the life of the vehicle without major repairs. However, driver abuse and normal wear after prolonged service can cause transmission failure.

A technician's first step toward fixing a transmission is to determine why the problem developed. Was it because of driver abuse (speed shifting, drag racing, lack of maintenance), normal wear (extremely high mileage), or another cause?

After proper diagnosis, the service technician can decide whether the transmission must be removed for major repairs. Sometimes, a simple linkage or clutch adjustment may correct the problem.

To begin diagnosis, gather information on the transmission trouble. Then, test drive the vehicle to verify the complaint.

Find out which gears in the transmission act up: first, second, high, all forward gears. Does it happen at specific speeds? This information will help determine which parts are at fault.

Figure 56-1 shows typical transmission troubles. Study them carefully.

Manual Transmission Problems

Many transmission problems are a result of driver abuse. Speed shifting, down shifting at high speeds, and improper clutch use can damage internal manual transmission parts. With high mileage, parts can also wear and fail. The next section of this chapter will summarize common manual transmission problems.

Gears Grind When Shifting

A grinding sound (gear clashing) when shifting is frequently caused by incorrect transmission linkage adjustment. If the transmission linkage is badly worn, the gears inside the transmission may not engage properly. If the clutch is dragging, the synchronizer teeth can grind while trying to equalize gear and output shaft speed, especially when shifting out of neutral.

Problems inside the transmission can also cause gear grinding during shifts. Worn or damaged synchronizers, shift forks or rails, and excessive wear in bearings, cones, and shafts may all prevent the gears from engaging smoothly.

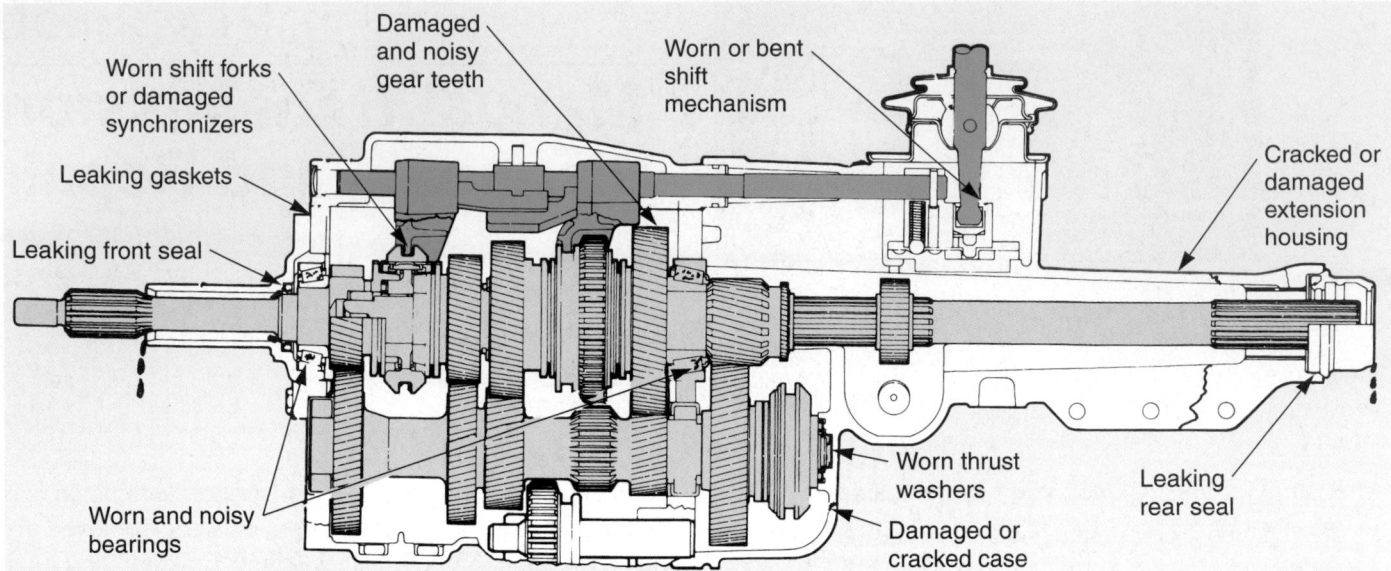

Figure 56-1. Common problems found in manual transmissions.

Transmission Noise

When a manual transmission is noisy (roaring, humming, or whirring sound), first check the transmission lubricant. It may be low or contaminated with metal particles.

Transmission noises will usually tell the experienced technician where the problem is. For example, if the transmission is noisy in all gears, something common to all of the gears is at fault. Transmission bearings or shaft end play spacers may be worn, or a shaft may be damaged.

On the other hand, if there is only a noise in one gear (first, second, third), the problem is due to components related to that gear.

Transmission Difficult to Shift

When the hand shifter is difficult to move through the gears, check the linkage or shift mechanism. Make sure the linkage is lubricated and moving freely. A bent or misaligned shift rod will cause hard shifting. Also, inspect the operation of the clutch linkage. If the clutch is not releasing completely, the transmission can be hard to shift.

Transmission Jumps Out of Gear

When a transmission jumps out of gear, the driver's shift lever "pops" or moves into neutral while driving.

First, check the transmission linkage and shift lever arms. If the shifter assembly is badly worn, it should be rebuilt or replaced.

A worn clutch pilot bearing may also cause the transmission to jump out of gear. Severe vibration, caused by the wobbling transmission input shaft, can wiggle and move the shift forks and synchronizers.

Other problems inside the transmission can cause jumping out of gear. They include worn synchronizer inserts and springs, worn shift fork assembly or shift rails, and wear and excessive play in the countershaft and output shaft assemblies.

Transmission Leaks Lubricant

Manual transmission lubricant leaks are caused by ruptured gaskets, worn seals, loose fasteners, or damage to the case, housings, or covers. When this problem occurs, check the lubricant level in the transmission. If the lubricant level is too high, lubricant may leak from the transmission vent or from otherwise good seals. Also check that all housing or cover bolts are tight.

 Note
If needed, refer to Chapter 10, *Vehicle Maintenance, Fluid Service, and Recycling,* for a review of manual transmission oil service.

When a seal leaks, always check the shaft bearing or bushing. A worn bearing or bushing and the wobbling action of the shaft can make a new seal leak.

Some of the gaskets and seals in a transmission can be replaced without removing the transmission from the vehicle. For example, the rear housing seal and gasket can normally be installed with the transmission in the vehicle.

Transmission Locked in Gear

When the shifter is locked in one gear, check the transmission shifter assembly and linkage. Look for bent shift rods, worn linkage, bushings, and shifter arms. Also check linkage adjustment. With a shift rail mechanism, worn or damaged rails, detents, or forks could be the cause.

A transmission can also become locked in gear when drive gear teeth are broken. The teeth can jam together and be locked by bits of metal from chipped gear teeth.

Diagnosis Chart

Refer to a transmission diagnosis chart in a service manual when a problem is difficult to locate. It will be written for the exact type of transmission.

Transmission Identification

When repairing a manual transmission, you must be able to identify the exact type of transmission. Usually, there will be an ID tag (identification label) or stamped set of numbers on the transmission. These numbers can be given to the parts counter person when you are ordering new parts.

Manual Transmission Service

Many problems that seem to be caused by the transmission are caused by clutch, linkage, or driveline

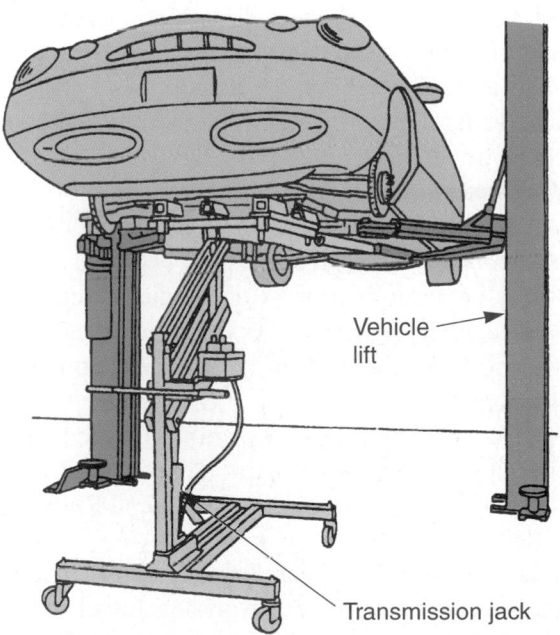

Figure 56-2. A transmission jack should be used to avoid back injuries when transmissions are removed or installed by hand. (Ford Motor Co.)

Vehicle lift

Transmission jack

problems. Keep this in mind before removing and disassembling a transmission.

Manual Transmission Removal

When removing a transmission, use a transmission jack, if one is available. A *transmission jack* has a special saddle and chains for securing the transmission to keep it from falling during removal and installation. See **Figure 56-2.**

Warning

A manual transmission is heavy and clumsy. If you are not using a transmission jack, ask another technician to help you lift the transmission out of the vehicle. A transmission jack is needed when working alone to prevent injury.

Use the following procedure to remove a manual transmission:

1. Secure the vehicle on a hoist or set of jack stands. A hoist is better because it allows you to stand while working.
2. Remove the transmission drain plug and drain the oil into a catch pan, **Figure 56-3A.**
3. Remove the drive shaft, **Figure 56-3B.**
4. Install a plastic cap over the end of the transmission shaft, **Figure 56-3C.** This will help keep oil from dripping out.
5. Disconnect the transmission linkage at the transmission.
6. Unbolt and pull the speedometer cable out of the extension housing.
7. Remove all electrical wires going to switches on the transmission.
8. Often, a cross member (transmission support bolted to frame) must be removed. Support the transmission with a jack and use another jack under the rear of the engine. Operate the jack on the engine to take the weight off the transmission. Be careful not to crush the oil pan. Never let the engine hang suspended by only the front motor mounts.
9. Depending upon what is recommended in the service manual, remove either the transmission-to-clutch cover bolts or the bolts going into the engine from the clutch cover.
10. Slide the transmission straight back, holding it in alignment with the engine. You may have to wiggle the transmission slightly to free it from the engine.
11. Clean the outside of the transmission and take it to a workbench.

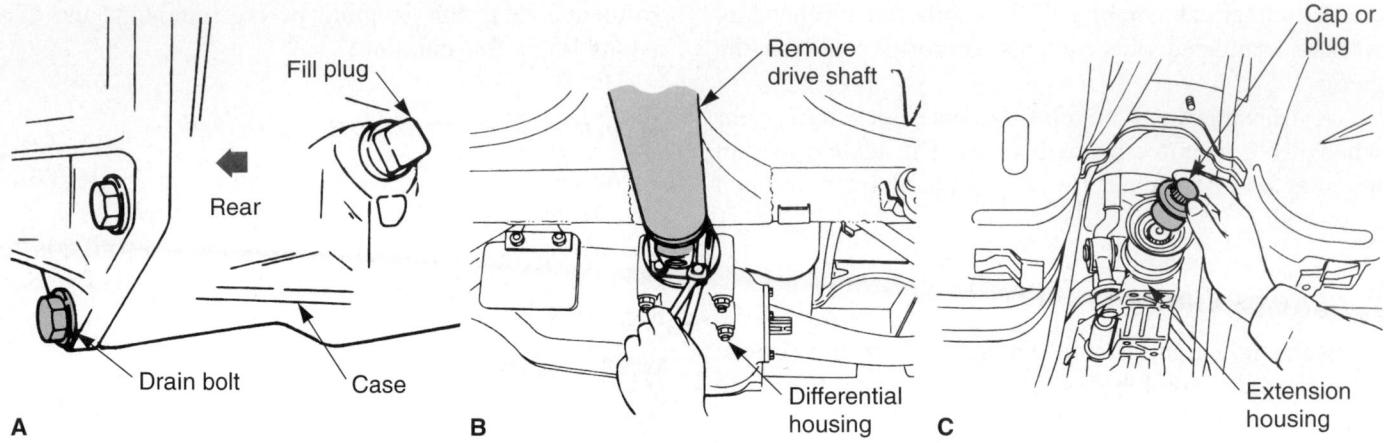

Figure 56-3. Preparing for transmission removal. A—Drain gear oil. B—Remove the drive shaft. C—If you do not drain the transmission, install a plastic cap to prevent oil leakage. (GMC and Subaru)

Manual Transmission Disassembly

Teardown procedures will vary from one transmission to another. Always consult a service manual. Improper disassembly methods could cause major part damage.

Basically, remove the shift fork assembly and cover. With a shift rail type, remove the shift lever assembly.

If the transmission has an inspection cover, observe transmission action with the cover removed. Shift the transmission into each gear by moving the small levers on the shift forks. At the same time, rotate the input shaft while inspecting the condition of the gears and synchronizers.

Unbolt the rear extension housing. Tap the extension housing off with a brass hammer, as shown in **Figure 56-4A.**

Going to the front of the transmission, remove the front bearing cover and any snap rings. Carefully, pry the input shaft and gear forward far enough to free the main output shaft.

Next, use a *dummy shaft* or *arbor shaft* (shaft tool designed for driving) to drive out the countershaft and reverse idler shaft. See **Figure 56-4B.**

Now you can remove the input shaft and the output shaft assemblies. Slide the output shaft and gears out of the back or top of the transmission as a unit. Be careful not to nick the gears on the case, **Figure 56-4C.**

Cleaning and Inspecting Parts

With all of the parts removed from the case, inspect everything closely. First check the inside of the case for metal shavings. If brass-colored particles are found, one or more of the synchronizers or thrust washers are damaged. These are normally the only parts in the transmission made of this material.

If iron chips are found, the output drive gears are probably damaged. After checking the case, clean the inside with solvent. Then, blow it dry with compressed air while wearing eye protection. Also, clean the transmission bearings.

 **Warning**
When blowing bearings dry with compressed air, do not allow the bearing to spin. Air pressure can make the bearing whirl at tremendously high speed. The bearing can explode and fly apart with lethal force.

Next, inspect all of the output gears. Look for wear patterns or chips on the gear teeth. The gears are usually case-hardened. If wear is more than a few thousandths of an inch, the hard outer layer will be worn through and the gear must be replaced.

Transmission shaft runout is the amount of wobble produced when a bent or worn shaft is rotated. If gear tooth wear is uneven, check the shaft bearings and shafts. They may be worn or bent. A dial indicator can be used to check the transmission shafts for straightness. Refer to specifications for the amount of allowable runout. Look at **Figure 56-5.**

Inspect the synchronizer assemblies, especially if the transmission had gear shifting problems. Check the teeth, splines, and grooves on the synchronizers. Replace parts as needed, **Figure 56-6.**

When removing the gears from the output shaft, keep everything organized on your workbench. All snap rings, spacers, and other parts should be installed exactly as removed. If synchronizers are to be reused, scribe alignment marks on the sleeve and hub. This will let you realign the same splines during reassembly.

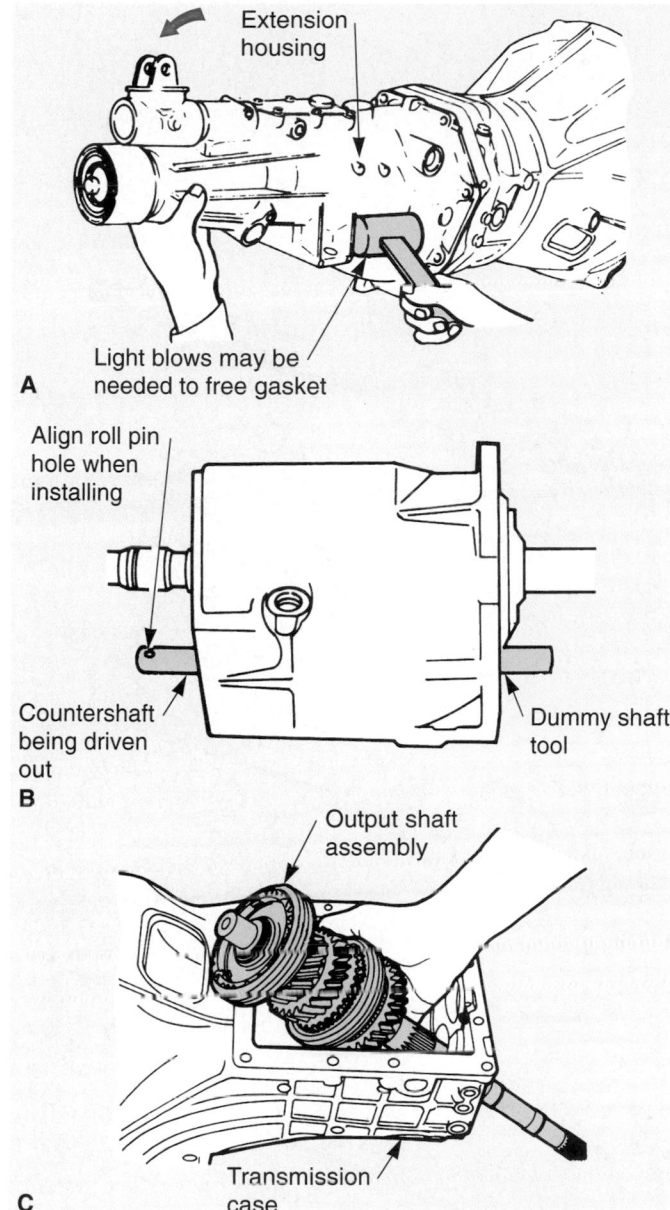

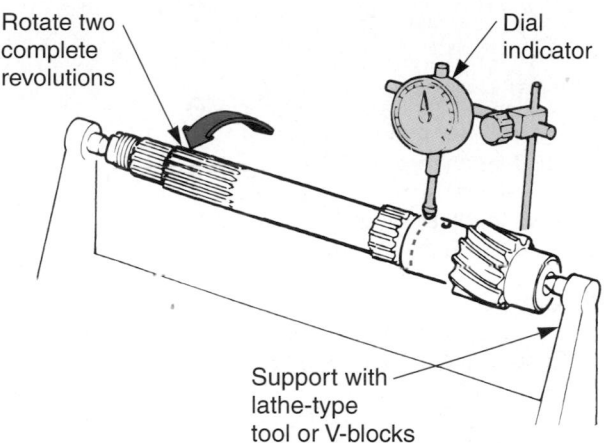

Figure 56-5. If gear wear is irregular, check shafts for runout. Use a lathe-type support or V-blocks to hold the shaft. Use a dial indicator to measure runout. (Honda)

Figure 56-4. Follow manual directions during transmission disassembly. A—A soft mallet may be needed to free the extension housing from the case. B—A dummy shaft is used to drive out the countershaft and reverse idler shaft. C—After removing snap rings and front bearing cover, lift the output shaft and other components out of the case. (Nissan, GMC, Chrysler)

Figure 56-7 shows an exploded view of a transmission. Since it is typical, study how all of the parts are positioned and held on their shafts.

Replace Worn or Damaged Parts

Any worn or damaged part in the transmission must be replaced. This is why your inspection is very important. If any trouble is not corrected, the transmission

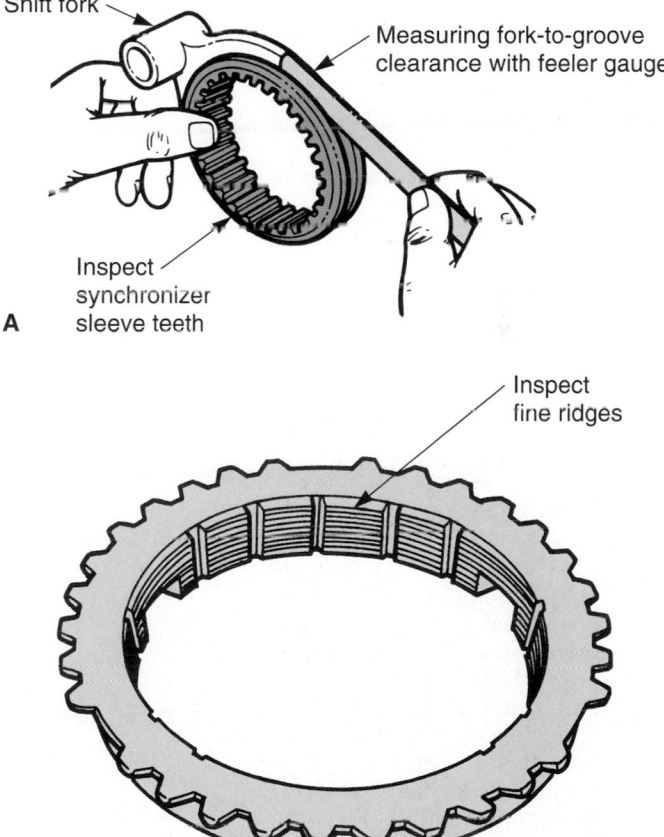

Figure 56-6. A—Check shift forks and synchronizers for wear. Use a feeler gauge between the fork and the groove. If more than specifications, replace parts as needed. B—Ridges on the inside of blocking rings must be sharp. If rounded off and worn, replace blocker rings to prevent gear clashing when shifting gears.

1. Mainshaft pilot bearing roller spacer
2. Third-fourth blocking ring
3. Third-fourth retaining ring
4. Third-fourth synchronizer snap ring
5. Third-fourth shifting plate (C)
6. Third-fourth clutch hub
7. Third-fourth clutch sleeve
8. Third gear
9. Mainshaft snap ring
10. Second gear thrust washer
11. Second gear
12. Second gear blocking ring
13. Output shaft
14. First-second clutch hub
15. First-second shifting plate (C)
16. Poppet ball
17. Poppet spring
18. First-second insert ring
19. First-second clutch sleeve
20. Countershaft gear thrust washer
    (steel, rear)
21. Counter shaft gear thrust washer
    (steel backed bronze, rear)
22. Countershaft gear bearing washer
23. Countershaft gear bearing rollers (88)
24. Countershaft gear bearing spacer
25. Countershaft gear bearing spacer
26. Countershaft gear thrust washer (front)
27. Rear bearing
28. Rear bearing locating snap ring
29. Rear bearing spacer ring
30. Rear bearing snap ring
31. Adapter plate seal
32. Adapter plate-to-transmission gasket
33. Adapter to transmission
34. Countershaft-reverse idler shaft lockplate
35. Reverse idler gear shaft
36. Reverse idler gear snap ring
37. Reverse idler gear thrust washer
38. Reverse idler gear
39. Reverse idler gear bearing rollers (74)
40. Reverse idler gear bearing washer
41. Reverse idler shaft sleeve
42. Countershaft
43. Front bearing retainer washer
44. Front bearing
45. Front bearing locating snap ring
46. Front bearing lock ring
47. Front bearing cap gasket
48. Front bearing cup seal
49. Front bearing cap
50. Mainshaft pilot bearing rollers (22)
51. Clutch shaft
52. Drain plug
53. Filler plug
54. Transmission case

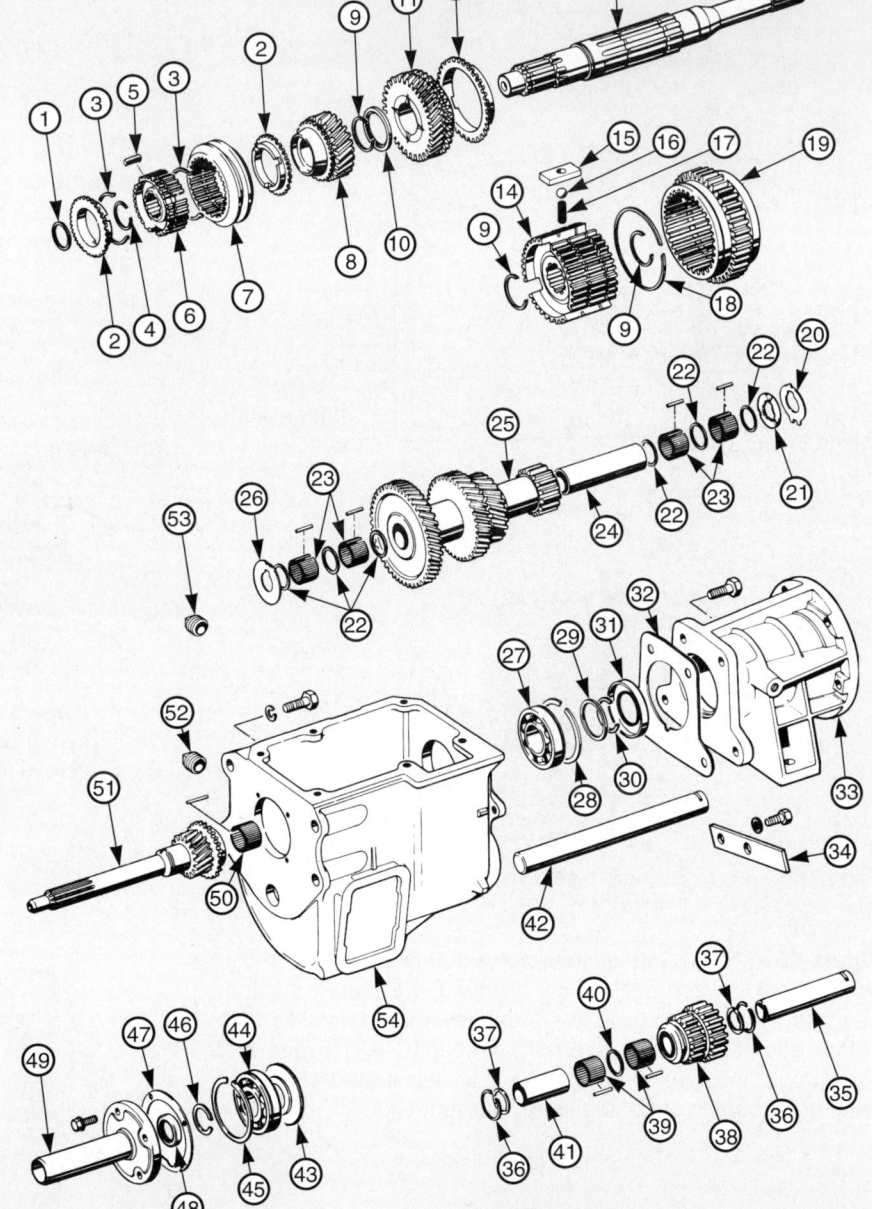

Figure 56-7. Check every part closely for wear or damage. A service manual will normally provide an illustration like this one for the exact transmission being repaired. This can be helpful during teardown and reassembly. (Chrysler Corp.)

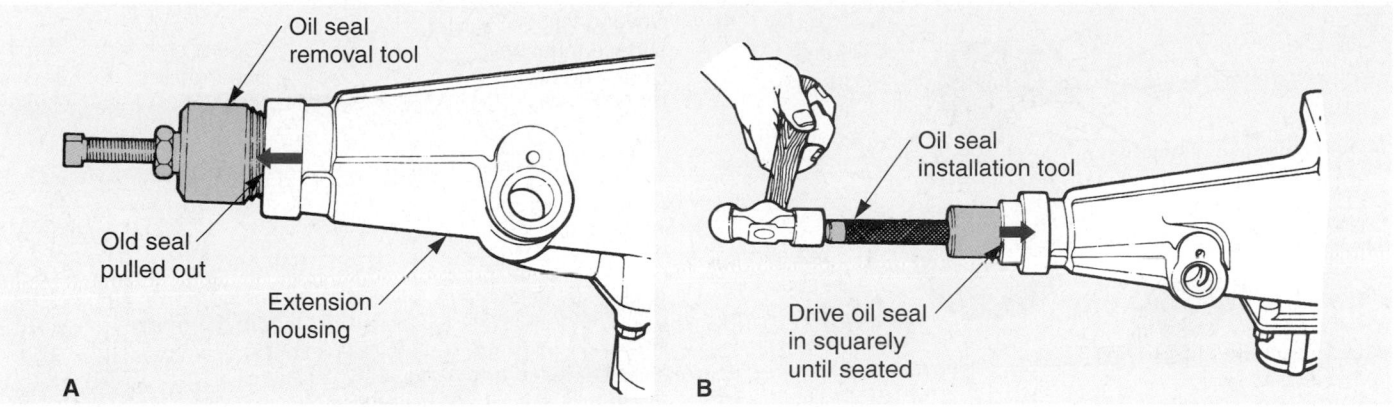

Figure 56-8. All seals should be replaced during a transmission rebuild. The rear seal can be removed and installed with the transmission still in the vehicle. A—Removing the oil seal. B—Driving in new seal. Coat the outside of the new seal with nonhardening sealer before installing. (Ford)

rebuild may fail. You would have to complete the job a second time—probably free of charge.

Always replace all gaskets and seals in the transmission. Even though a seal or gasket might not leak before teardown, it could start to leak after assembly. **Figure 56-8** shows a common way of replacing a rear seal.

When replacing a gear on the output shaft, you should also replace the matching gear on the countershaft. If a new gear is meshed with an old worn gear, gear noise can result.

Frequently, you will need to replace input shaft bearings, output shaft bearings, and sometimes countershaft bearings. These bearings are prone to wear because they support a great amount of load, **Figure 56-9.**

Some transmissions use metric fasteners. If a new bolt or nut is needed, make sure it is the correct thread type and length. Mixing threads will cause part damage.

Transmission Reassembly

After obtaining new parts to replace the old worn ones, you are ready for transmission assembly. Generally, the transmission is assembled in reverse order of disassembly. Again, refer to a service manual for exact directions.

The service manual will usually have an exploded view of the transmission. It will show how each part is located in relation to the others. Step-by-step instructions will accompany the illustrations.

To hold the needle bearings in countershaft gears, coat the bearings with heavy grease. Then, fit each bearing into position. The grease will hold the bearings as you slide the countershaft into the gear. This is illustrated in **Figure 56-10.**

Also, following the manufacturer's instructions, measure the end play or clearance of the gears and synchronizers as needed. Look at **Figure 56-11.**

The end play between the countershaft gear and case should be checked. If excessive, thicker thrust washers are required.

After the transmission shafts and gears are in place, pour the recommended quantity of oil into the case.

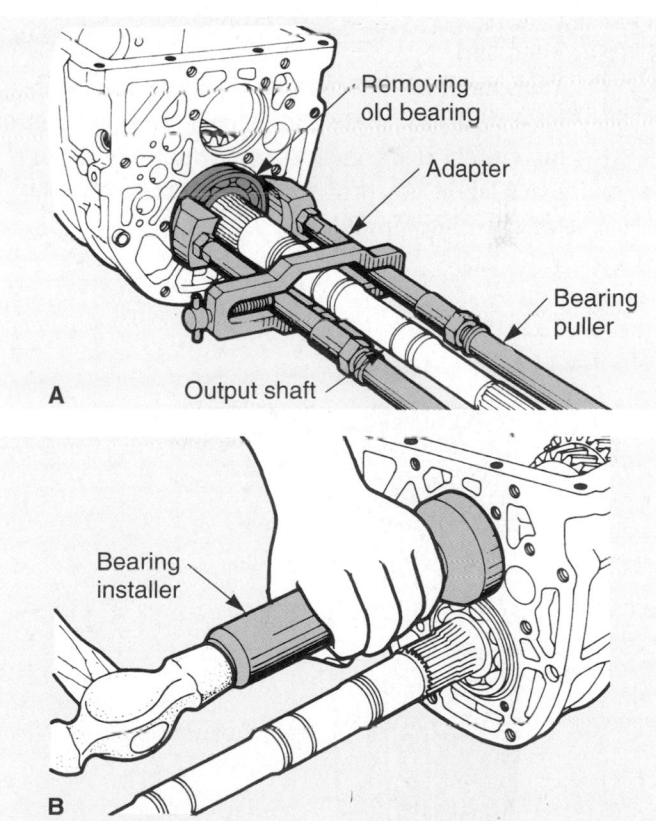

Figure 56-9. If bearings show signs of wear or feel rough when turned by hand, replace them. A—A special puller may be needed on some bearings. B—Use a driver to install new bearings. Do not hammer on the inner portion of the bearing or damage will result. (Chrysler)

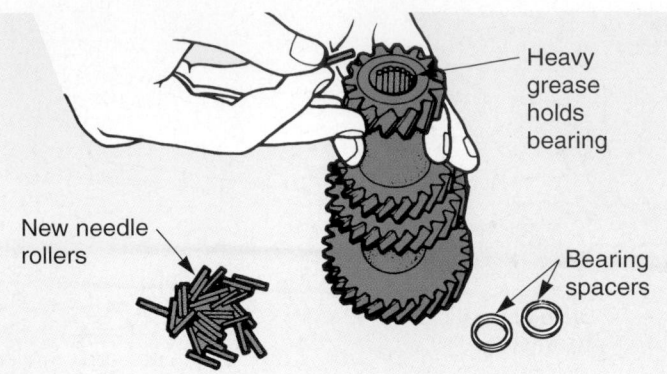

Figure 56-10. Heavy grease is commonly used to hold small needle bearings during reassembly. The grease will hold the bearings in the countershaft gear as the countershaft is slid into place. (Dodge)

Assemble the shift fork mechanism. Then, with the synchronizers and shift forks in neutral, fit the shift fork assembly on or in the case. Check the action of the shift forks. Look at **Figure 56-12.**

Make sure the transmission shifts properly before installing it. This could save you from having to remove the transmission if there are still problems.

Tech Tip

With the transmission out of the vehicle, it is wise to disassemble and inspect the condition of the clutch. If the customer will not pay for the extra labor, you will no longer be responsible for any clutch problems.

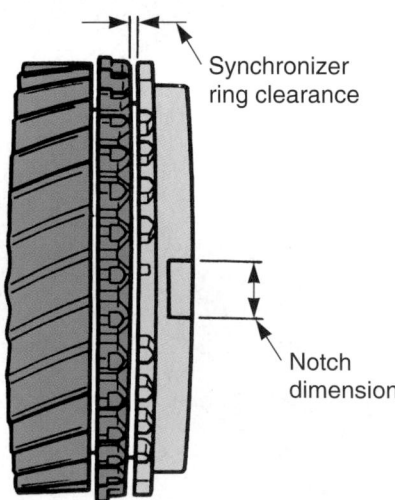

Figure 56-11. The service manual will give specifications for important transmission measurements. Synchronizer ring clearance is an important measurement. Replace parts if not within factory specifications. (General Motors)

Figure 56-12. Make sure nothing in the shift mechanism is worn or damaged. Check shift rail, fork-to-synchronizer contact points, shift gates, and other components. Use an approved method of sealing the shift cover on the case. Refer to a service manual for details. (Chrysler)

Transmission Installation

Before transmission installation, place a small amount of grease in the pilot bearing and on the inner surface of the throw-out bearing. Do not place lubricant on the end of the clutch shaft, input shaft splines, or pressure plate release levers. Grease in these locations can spray onto the clutch friction disk, causing clutch slippage and failure.

Shift the transmission into high gear. This will help position the input shaft into the clutch disk during transmission installation.

Place the transmission on the transmission jack. Position it behind the engine. Double-check that the throw-out bearing is in place on the clutch fork. Carefully align the transmission with the engine.

The input and output shaft must line up perfectly with the centerline of the engine crankshaft. If the transmission is tilted, even slightly, it will not fit into place.

With the transmission in high gear to hamper input shaft rotation, slowly push the transmission into the clutch housing. You may need to raise or lower the transmission slightly to keep it in alignment. When the transmission is almost in place, wiggle the extension housing in a circular pattern while pushing toward the engine. This should help start the input shaft in the crankshaft pilot bearing. If the clutch and pilot bearing are installed correctly, the transmission should slide fully into place by hand.

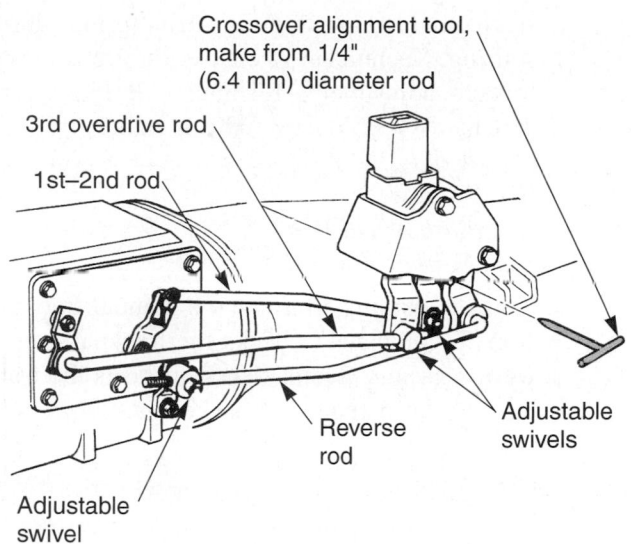

Crossover alignment tool, make from 1/4" (6.4 mm) diameter rod

3rd overdrive rod

1st–2nd rod

Adjustable swivels

Reverse rod

Adjustable swivel

Figure 56-13. External linkage is sometimes adjusted by installing a pin through holes in the shifter arms. Transmission arms and shifter must be in neutral. Adjust rod length until tool easily passes through holes in each shift lever arm. (Ford)

 Caution
Do not use the transmission bolts to draw the transmission into the clutch housing. The transmission input shaft could be smashed into the crankshaft pilot bearing. Serious part damage may result.

With the transmission bolted to the clutch cover, install the rear cross member and motor mount. Reinstall the clutch linkage, transmission linkage, and other parts. Adjust the clutch linkage.

Adjusting Transmission Linkage

To adjust many types of transmission linkage, place the gear shift lever and transmission levers in neutral. Then, insert an *alignment pin* (special diameter tool or rod) through the linkage arms. The pin must fit through the holes in the shifter levers, as in **Figure 56-13.**

If the pin will not fit through the hole, lengthen or shorten the linkage rods. Adjust the rods so that the alignment pin fits easily through the hole in the shifter assembly.

This basic procedure will vary with different types of gear shift mechanisms. When in doubt, refer to the specific directions in a manual.

After adjustment, lower the vehicle to the ground. Road test the vehicle and check for leaks.

Summary

- Normally, a manual transmission will provide thousands of miles of trouble-free service.
- A technician's first step toward fixing a transmission is to determine why the problem developed.
- When a manual transmission is noisy (roaring, humming, or whirring sound), first check the transmission lubricant.
- Manual transmission lubricant leaks are caused by ruptured gaskets, worn seals, loose fasteners, or damage to the case, housings, or covers.
- When the shifter is locked in one gear, check the transmission shifter assembly and linkage.
- When blowing bearings dry with compressed air, do not allow the bearing to spin.
- If gear tooth wear is uneven, check the shaft bearings and shafts.
- Inspect the synchronizer assemblies, especially if the transmission had shifting problems.
- Any worn or damaged part in the transmission must be replaced.

Important Terms

Transmission jack	Transmission shaft
Dummy shaft	runout
Arbor shaft	Alignment pin

Review Questions—Chapter 56

Please do not write in this text. Place your answers on a separate sheet of paper.

1. A grinding or gear clashing noise when shifting out of neutral could be caused by:
   (A) incorrect linkage adjustment.
   (B) worn synchronizers.
   (C) dragging clutch.
   (D) All of the above.
   (E) None of the above.

2. If a transmission is only noisy in second gear, what does that tell you about the problem?

3. How do you identify a manual transmission?

4. A(n) _____ _____ is commonly used to remove a manual transmission because it is heavy and clumsy.

5. A(n) _____ is commonly used to drive the reverse idler shaft out of the case during transmission disassembly.

6. If you find bronze- or brass-like metal shavings in the case, synchronizers or thrust washers may be damaged. True or False?

7. What can happen if you spin a ball or roller bearing with compressed air?

8. When replacing a gear on the output shaft, you must also replace the output shaft. True or False?

9. Heavy _____ will hold needle bearings in place during transmission assembly.

10. Which of the following is *not* a normal procedure during transmission linkage adjustment?
    (A) Measure linkage-to-transmission clearance.
    (B) Place all shift rods in neutral.
    (C) Insert an alignment pin through linkage arms.
    (D) Refer to service manual directions.

11. A dial indicator shows 0.001″ runout on a transmission shaft. Specs allow for 0.005″ runout. How much is the shaft in or out of specifications?
    (A) 0.006″.
    (B) 0.004″.
    (C) 0.040″.
    (D) 0.060″.

⚙ ASE-Type Questions

1. A manual transmission is not operating properly. Technician A's first step is to remove the transmission from the vehicle and begin teardown. Technician B's first step is to determine the cause of the problem. Who is right?
   (A) *A only.*
   (B) *B only.*
   (C) *Both A and B.*
   (D) *Neither A nor B.*

2. All of the following are common manual transmission problems *except:*
   (A) *Bent shift mechanism.*
   (B) *Worn thrust washers.*
   (C) *Clogged modulator valve.*
   (D) *Damaged extension housing.*

3. A manual transmission's gears grind during shifting. Technician A checks the transmission linkage adjustment. Technician B looks for a damaged synchronizer. Who is right?
   (A) *A only.*
   (B) *B only.*
   (C) *Both A and B.*
   (D) *Neither A nor B.*

4. Noise is coming from a car's manual transmission. Technician A checks the transmission for worn bearings. Technician B inspects the transmission for damaged or broken gear teeth. Who is right?
   (A) *A only.*
   (B) *B only.*
   (C) *Both A and B.*
   (D) *Neither A nor B.*

5. An automobile's manual transmission is noisy in all gears. Technician A looks for a damaged transmission shaft. Technician B checks the condition of the first output gear. Who is right?
   (A) *A only.*
   (B) *B only.*
   (C) *Both A and B.*
   (D) *Neither A nor B.*

6. A manual transmission is hard to shift into gear. Technician A looks for a bent shift rod. Technician B checks the condition of the clutch. Who is right?
   (A) *A only.*
   (B) *B only.*
   (C) *Both A and B.*
   (D) *Neither A nor B.*

7. An automobile's manual transmission keeps "jumping out of gear." Technician A thinks the transmission's shift lever arms are not operating properly. Technician B looks for a worn clutch pilot bearing. Who is right?
   (A) *A only.*
   (B) *B only.*
   (C) *Both A and B.*
   (D) *Neither A nor B.*

8. Lubricant is leaking from a manual transmission. Technician A looks for cracks in the transmission case. Technician B checks the condition of the front seal. Who is right?
   (A) *A only.*
   (B) *B only.*
   (C) *Both A and B.*
   (D) *Neither A nor B.*

9. Lubricant is leaking out of a manual transmission. Technician A says this problem may be caused by an excessive amount of oil in the transmission case. Technician B checks the condition of the converter gasket. Who is right?
   (A) *A only.*
   (B) *B only.*
   (C) *Both A and B.*
   (D) *Neither A nor B.*

10. All of the following problems can cause a manual transmission to "lock in gear" *except:*
    (A) *Bent shift rod.*
    (B) *Faulty shift mechanism adjustment.*
    (C) *Valve body malfunction.*
    (D) *Broken drive gear teeth.*

11. Certain parts in a manual transmission need to be replaced. Technician A says that to identify the particular transmission being serviced, you should look for an ID tag located on the transmission. Technician B says that to identify the type of transmission being serviced, you should look for a stamped set of numbers located on the transmission. Who is right?
    (A) *A only.*
    (B) *B only.*
    (C) *Both A and B.*
    (D) *Neither A nor B.*

12. Technician A mounts a car on jack stands to make it easier to work on during transmission removal. Technician B mounts a car on a hoist to make it easier to work on during transmission removal. Who is using the most efficient method?
    (A) *A only.*
    (B) *B only.*
    (C) *Both A and B.*
    (D) *Neither A nor B.*

13. Brass-colored particles are discovered in a manual transmission's case during disassembly. Technician A checks the condition of the thrust washers. Technician B believes one or more synchronizers could be damaged. Who is right?
    (A) *A only.*
    (B) *B only.*
    (C) *Both A and B.*
    (D) *Neither A nor B.*

14. A manual transmission's shafts are being checked for runout. Technician A is going to use a dial indicator to perform this task. Technician B is going to use an inside micrometer to perform this task. Who is right?
    (A) *A only.*
    (B) *B only.*
    (C) *Both A and B.*
    (D) *Neither A nor B.*

15. A manual transmission is being installed back into a vehicle. Technician A places a small amount of lubricant on the input shaft splines before installing the transmission. Technician B places a small amount of lubricant on the pressure plate release levers before installing the transmission. Who is right?
    (A) *A only.*
    (B) *B only.*
    (C) *Both A and B.*
    (D) *Neither A nor B.*

Activities—Chapter 56

1. Disassemble a defective transmission assigned by your instructor. Clean and examine the parts, listing those that are defective.

2. Remove a transmission from a vehicle.

3. Reassemble a transmission and reinstall it in the vehicle.

4. Operate a transmission selected by your instructor; identify noise and possible causes.

Manual Transmission/Transaxle Diagnosis		
Condition	**Possible Causes**	**Correction**
Shifts hard—all gears.	1. Excessive clutch pedal free travel. 2. Worn or defective clutch. 3. Failure to fully depress clutch pedal when shifting. 4. Loose shift cover. 5. Worn or loose shift fork, shafts, levers, or detents. 6. Improper shift linkage adjustment. 7. Linkage needs lubrication. 8. Binding, bent, or loose linkage. 9. Wrong transmission lubricant. 10. Insufficient lubricant. 11. Excess lubricant. 12. Misaligned transmission. 13. Loose or cracked input shaft bearing retainer. 14. Worn, damaged, or improperly assembled synchronizer.	1. Adjust free travel. 2. Replace worn parts. 3. Advise driver. 4. Tighten cover. 5. Tighten or replace. 6. Adjust linkage. 7. Lubricate linkage. 8. Free, straighten, or tighten as needed. 9. Drain and fill with recommended lubricant. 10. Add lubricant to filler plug level. 11. Drain excess lubricant. 12. Correct transmission alignment. 13. Tighten or replace retainer. 14. Replace or reassemble synchronizer.
Gear clash during downshifting.	1. Worn, damaged, or improperly assembled synchronizer. 2. Shifting too fast (ramming into lower gear). 3. Shifting to a lower gear when vehicle speed is too high. 4. Clutch not releasing properly. 5. Excessive output shaft end play.	1. Replace or reassemble synchronizer. 2. Force into gear with a smooth, slower shift. 3. Slow down to appropriate speed before shifting. 4. Adjust or repair as needed. 5. Adjust end play.
Jumps out of gear.	1. Loose or misaligned transmission. 2. Loose or misaligned clutch housing. 3. Improperly adjusted shift linkage. 4. Worn shift rail detents or weak detent springs. 5. Worn synchronizer clutch sleeve teeth. 6. Loose shifter cover. 7. Worn shift fork, shaft, or levers. 8. Worn clutch teeth on input shaft or other gears. 9. Worn gear teeth. 10. Worn counter gear bearings and/or thrust washers. 11. Worn reverse idler gear bushing or bearings. 12. Worn output shaft pilot bearing. 13. Loose input shaft bearing retainer. 14. Other parts striking shift linkage. 15. Worn input or output shaft bearings. 16. Worn input shaft bushing in flywheel. 17. Bent output shaft.	1. Tighten or align transmission. 2. Tighten or align clutch housing. 3. Adjust linkage. 4. Replace rail detents and/or springs. 5. Replace synchronizer. 6. Tighten shifter cover. 7. Replace worn parts. 8. Replace input shaft or gears. 9. Replace gears. 10. Replace counter gear shaft, bearings, and washers. 11. Replace gear, bearings, and shaft. 12. Replace rollers. Replace shafts, if necessary. 13. Tighten bearing retainer. 14. Make adjustments to provide clearance. 15. Replace bearings. 16. Replace input shaft bushing or bearing. 17. Replace output shaft.
Noise in all gears.	1. Insufficient lubrication. 2. Worn or damaged bearings. 3. Worn or damaged gears. 4. Wrong lubricant. 5. Excessive synchronizer wear. 6. Defective speedometer drive gears.	1. Fill to filler plug. 2. Replace bearings. 3. Replace gears. 4. Drain and fill with recommended lubricant. 5. Replace synchronizer. 6. Replace speedometer drive gears.

(Continued)

Manual Transmission/Transaxle Diagnosis		
Condition	**Possible Causes**	**Correction**
Noise in all gears. *(Continued)*	7. Misaligned transmission. 8. Excessive input or output shaft and/or counter gear end play. 9. Contaminated lubricant.	7. Correct alignment. 8. Adjust end play. 9. Disassemble, clean, and repair transmission.
Noise in neutral with engine running.	1. Worn or damaged input shaft bearing. 2. Worn or damaged gears. 3. Lack of lubrication. 4. Worn or damaged countershaft bearings. 5. Worn or damaged output shaft pilot bearing. 6. Worn or damaged counter gear anti-lash plate. 7. Lubricant contaminated with broken metal.	1. Replace input shaft bearing. 2. Replace gears. 3. Fill transmission to the proper level. 4. Replace bearings, counter gear, and shaft. 5. Replace all rollers. 6. Replace plate or counter gear as required. 7. Disassemble, clean, and repair transmission.
Noise in direct-drive gear.	1. Defective input shaft bearing. 2. Defective output shaft bearing. 3. Defective synchronizer. 4. Defective speedometer drive gears.	1. Replace input shaft bearing. 2. Replace output shaft bearing. 3. Replace synchronizer. 4. Replace speedometer drive gears.
Noise in reduction or overdrive gear.	1. Counter gear rear bearings worn or damaged. 2. Defective synchronizer. 3. Constant mesh gear loose on shaft. 4. Constant mesh gear teeth worn or chipped.	1. Replace counter gear bearings. Replace shaft and counter gear, if needed. 2. Replace synchronizer. 3. Replace gear and/or shaft. 4. Replace gear.
Noise in reverse.	1. Worn reverse idler bushings. 2. Worn or damaged reverse idler gear. 3. Worn or damaged counter gear reverse gear. 4. Defective reverse sliding gear (synchromesh low gear).	1. Replace idler gear or bushings. 2. Replace reverse idler gear. 3. Replace counter gear. 4. Replace reverse sliding gear.
Sticks in gear.	1. Insufficient lubricant. 2. Burred synchronizer clutch sleeve teeth. 3. Sticking shift rails. 4. Synchronizer blocking ring stuck to mating gear. 5. Defective shift linkage. 6. Insufficient clutch pedal free travel. 7. Misaligned transmission.	1. Fill to filler plug. 2. Replace synchronizer. 3. Free and lubricate shift rails. 4. Free, lubricate, or replace blocking ring or mating gear. 5. Repair or replace linkage. 6. Adjust clutch pedal free travel. 7. Correct alignment.
Gear clash when shifting from neutral to low or reverse.	1. Insufficient clutch pedal free travel. 2. Wrong lubricant. 3. Engine rpm too high. 4. Insufficient time between depressing clutch and shifting. 5. Sticking input shaft clutch pilot bearing.	1. Adjust free travel. 2. Drain and fill with correct lubricant. 3. Set to correct idle rpm. 4. Advise driver. 5. Replace pilot bearing.

Chapter 57

Automatic Transmission Fundamentals

After studying this chapter, you will be able to:

- ■ Identify the basic components of an automatic transmission.
- ■ Describe the function and operation of the major parts of an automatic transmission.
- ■ Trace the flow of power through an automatic transmission.
- ■ Explain how an automatic transmission shifts gears.
- ■ Compare the different types of automatic transmissions.
- ■ Correctly answer ASE certification test questions requiring a knowledge of automatic transmission operation and construction.

An *automatic transmission* performs the same functions as a manual transmission, but it "shifts gears" and "releases the clutch" automatically. Most modern vehicles use an automatic transmission (or transaxle) because it saves the driver from having to move a shift lever and depress a clutch pedal.

As you will learn, an automatic transmission normally senses engine speed and engine load (engine vacuum or throttle position) to determine shift points. It then uses internal oil pressure to shift gears. Computers can also be used to sense or control automatic transmission shift points.

Basic Automatic Transmission

Before detailing the construction and operation of individual parts, it is important for you to have a general idea of how an automatic transmission works. Then, you will be able to relate the details of each part to the complete transmission assembly.

Refer to **Figure 57-1** as the basic parts of an automatic transmission are introduced.

- *Torque converter*—fluid coupling that connects and disconnects engine and transmission.
- *Input shaft*—transfers power from the torque converter to the internal drive members and gearsets.
- *Transmission oil pump*—produces pressure to operate the hydraulic components in the transmission.
- *Valve body*—operated by shift lever and sensors; controls oil flow to the pistons and servos.
- *Pistons and servos*—actuate bands and clutches.
- *Planetary gearsets*—provide different gear ratios and reverse gear.
- *Bands and clutches*—apply clamping pressure on different parts of the planetary gearsets to operate them.
- *Output shaft*—transfers engine torque from the gearsets to the drive shaft.

Transmitting Power

As you will see, an automatic transmission uses three methods to transmit power: fluids, friction, and gears. This is illustrated in **Figure 57-2.**

The torque converter uses *fluid* to transfer power. The bands and clutches use *friction*. The transmission *gears* not only transmit power, but they can increase or decrease speed and torque.

Transmission Housings and Case

An automatic transmission is normally constructed with four main components: bell housing, case, pan, and extension housing. These parts support and enclose all the other components in the transmission. Refer to **Figure 57-3.**

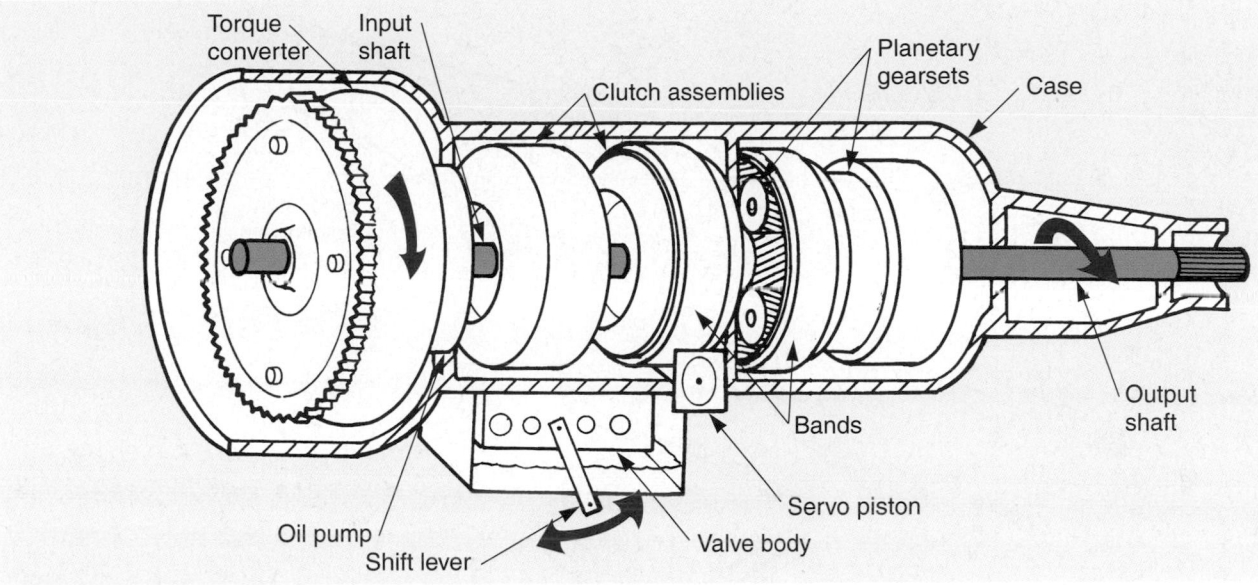

Figure 57-1. Study the basic parts of a simplified automatic transmission. Note the general shape and location of components. This will help prepare you to learn details of each part.

The ***bell housing*** surrounds the torque converter and holds the transmission against the engine. It is usually made of aluminum.

Bolts fit through holes in the bell housing and attach to the engine block. The bell housing also keeps road dirt, rocks, and other debris off the spinning torque converter and flywheel.

The ***transmission case*** encloses the clutches, bands, gearsets, and inner ends of the transmission shafts. It may be made of aluminum or cast iron. The bell housing bolts to the front of the case. The extension housing bolts to the rear of the case. The valve body and pan bolt to the bottom of the case.

The ***oil pan,*** also called the ***transmission pan,*** collects and stores a supply of transmission oil. It is usually made of thin, stamped steel or cast aluminum. The pan fits over the valve body. A gasket or sealant prevents leakage between the case and the oil pan.

The ***extension housing*** slides over and supports the output shaft. The housing uses a gasket on the front and a seal on the rear to prevent oil leakage. It is made of aluminum or cast iron.

Torque Converter

The ***torque converter*** is a fluid clutch that performs the same basic function as a manual transmission's dry-friction clutch. It provides a means of uncoupling the engine from the transmission. It also provides a means of coupling the engine for acceleration.

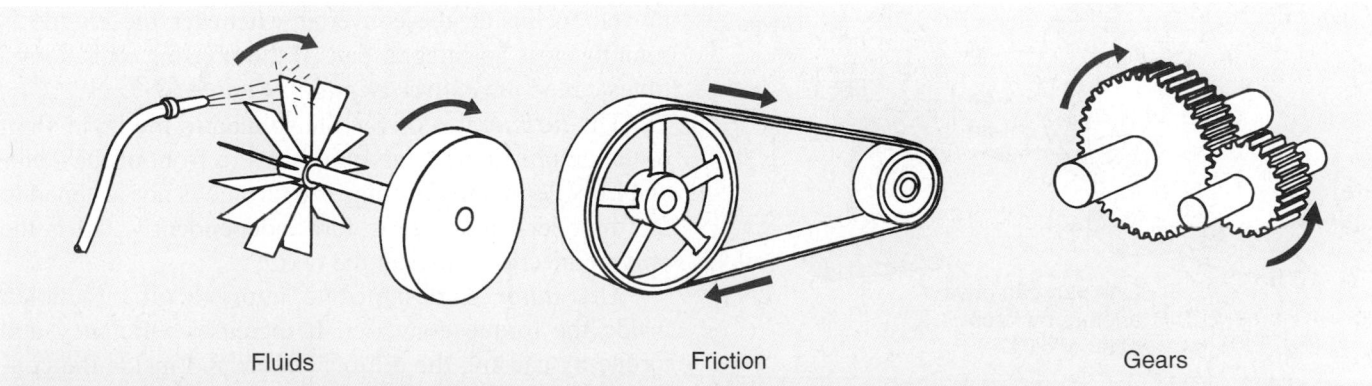

Fluids   Friction   Gears

Figure 57-2. An automatic transmission uses these methods of transmitting power. (Deere & Co.)

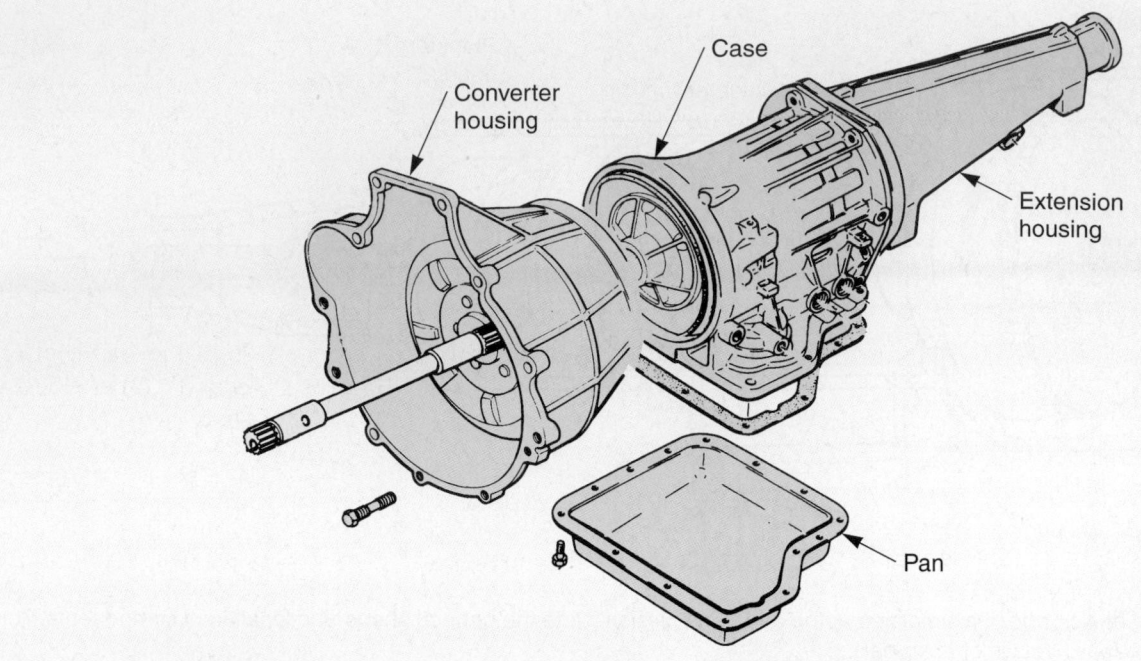

Figure 57-3. Most automatic transmissions are constructed with a front bell housing, central case, rear extension housing, and lower pan. Most parts are located inside the case.

Torque Converter Principles

Two house fans can be used to demonstrate the basic action inside a torque converter. Look at **Figure 57-4.** One fan is plugged in and is spinning. The other fan is not plugged into electrical power. Since the spinning fan is facing the other, it can be used to spin the unplugged fan, transferring power through a fluid (air). This same principle applies inside a torque converter, but oil is used instead of air.

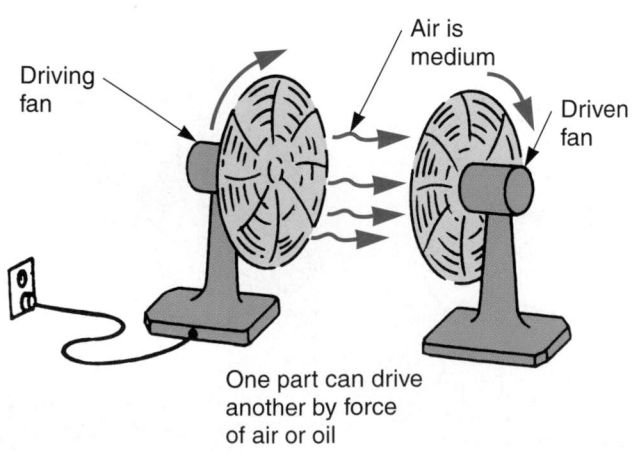

Figure 57-4. Two fans demonstrate the principle of torque converter operation. (Deere & Co.)

Torque Converter Construction

A torque converter consists of four basic parts: the outer housing, an impeller or pump, a turbine, and a stator. These parts are shown in **Figure 57-5A.**

The impeller, stator, and turbine have curved fan blades. They work like our simple example of one fan driving another. The impeller drives the turbine, as shown in **Figure 57-5B.**

Converter Housing

The impeller, stator, and turbine are contained inside a doughnut-shaped housing. This housing is normally made of two pieces of steel welded together. The housing is filled with transmission oil (fluid). See **Figure 57-6.**

The *impeller* is the driving fan that produces oil movement inside the converter whenever the engine is running. It is an integral part of the housing. It is sometimes called the *converter pump,* **Figure 57-7.**

The *turbine* is a driven fan splined to the input shaft of the automatic transmission. It fits in front of the stator and impeller in the housing. The turbine is not fastened to the impeller—it is free to turn independently. Oil is the only connection between the two.

The *stator* is designed to improve oil circulation inside the torque converter. It increases efficiency and torque by causing the oil to swirl around inside the converter housing. This helps make use of nearly all the force produced by the moving oil.

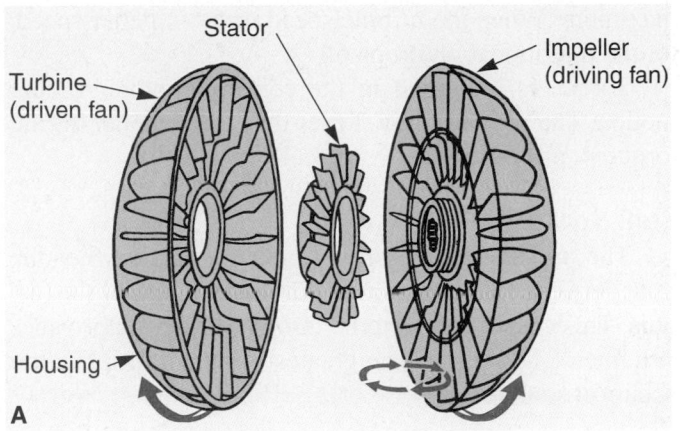

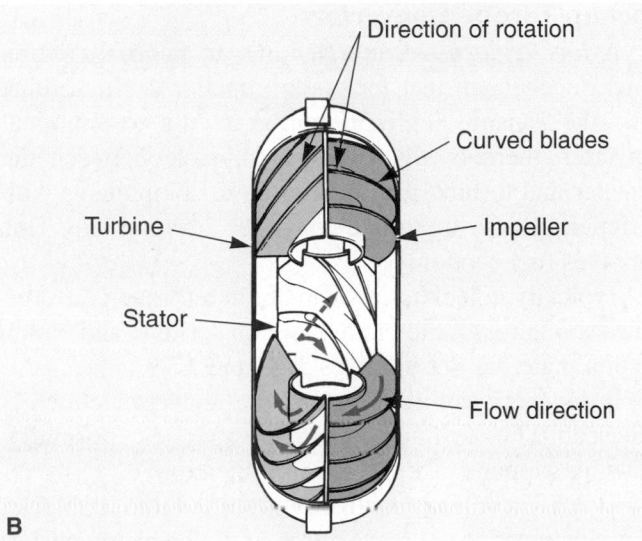

Figure 57-5. A—The four major parts of a torque converter: housing, impeller, stator, and turbine. B—Blades on the impeller and stator direct oil circulation onto blades of the turbine. The impeller is driven by the engine. The turbine is driven by the impeller. (Texaco and Subaru)

Figure 57-6. The torque converter fits inside a housing at the rear of the engine.

Flywheel Action

The torque converter is very large and heavy. This allows it to serve as a flywheel to smooth out engine power pulses. Its inertia reduces vibration entering the transmission and driveline.

An automatic transmission uses a very thin and light flywheel. It is simply a stamped disc with a ring gear for the starting motor. If the ring gear is on the torque converter, a flex plate, without a ring gear, can be used to connect the crankshaft to the torque converter.

The crankshaft bolts to one side of the flywheel or flex plate. The torque converter bolts to the other side.

Torque Converter Operation

With the engine idling, the impeller spins slowly. Only a small amount of oil is thrown into the stator and turbine. Not enough force is developed inside the torque converter to spin the turbine. The car remains stationary with the transmission in gear.

During engine acceleration, the engine crankshaft, converter housing, and impeller begin to spin faster. More oil is thrown out of the impeller by centrifugal force. This makes the turbine begin to turn. As a result,

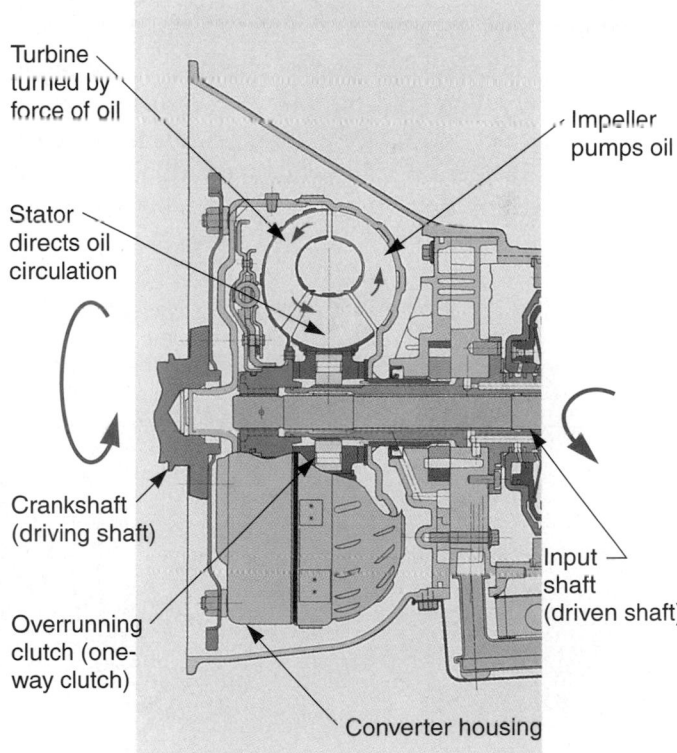

Figure 57-7. The crankshaft is fastened to the converter housing and impeller. The stator is mounted on a one-way clutch. When the engine crankshaft spins fast enough, oil movement rotates the turbine and transmission input shaft. (Ford)

the transmission input shaft and vehicle start to move, but with some slippage.

At cruising speeds, the impeller and turbine spin at almost the same speed, with very little slippage.

Converter One-Way Clutch

A *one-way clutch* (overrunning clutch) allows the stator to turn in only one direction. The stator mounts on the one-way clutch mechanism. Stator action is only needed when the impeller and turbine are turning at different speeds. See **Figure 57-8.**

The one-way clutch locks the stator when the impeller is turning faster than the turbine. This causes the stator to route oil flow over the impeller vanes properly. Then, when turbine speed almost equals impeller speed, the stator can freewheel on its shaft, so as not to obstruct oil flow.

Torque Multiplication

Torque multiplication refers to the ability of a torque converter to increase the amount of engine torque applied to the transmission input shaft. Just as a small gear driving a large gear increases torque, the torque converter can act as several different gear ratios to alter torque output. Torque can be doubled by the converter under certain conditions.

Torque multiplication occurs when the impeller is spinning faster than the turbine. For instance, if the engine is accelerated quickly, the engine and impeller speed might increase rapidly while the turbine is almost stationary. At this time, torque multiplication would be maximum. When the turbine speed nears impeller speed, torque multiplication drops off.

Torque is increased in the converter by sacrificing motion. The turbine spins slower than the impeller during torque multiplication.

Stall Speed

The *stall speed* of a torque converter occurs when the impeller is at maximum speed without rotation of the turbine. This causes the oil to be thrown off the stator vanes at tremendous speeds. The greatest torque multiplication occurs at stall speed.

Lockup Torque Converters

A *lockup torque converter* has an internal friction clutch mechanism that locks the impeller to the turbine when the transmission is in high gear. In a conventional converter, there is always some slippage between the impeller and turbine. By locking these components with a friction clutch, the torque converter does not slip. This improves fuel economy.

Typically, a lockup mechanism in a torque converter consists of a hydraulic piston, torsion springs, and clutch friction material. See **Figures 57-8** and **57-9.**

In lower transmission gears, the converter clutch is released. The torque converter operates normally, allowing slippage and torque multiplication.

Then, when the transmission is shifted into high gear, oil is channeled to the converter piston. The piston pushes the friction discs together to lock the converter. The torsion springs help dampen engine power pulses entering the drive train.

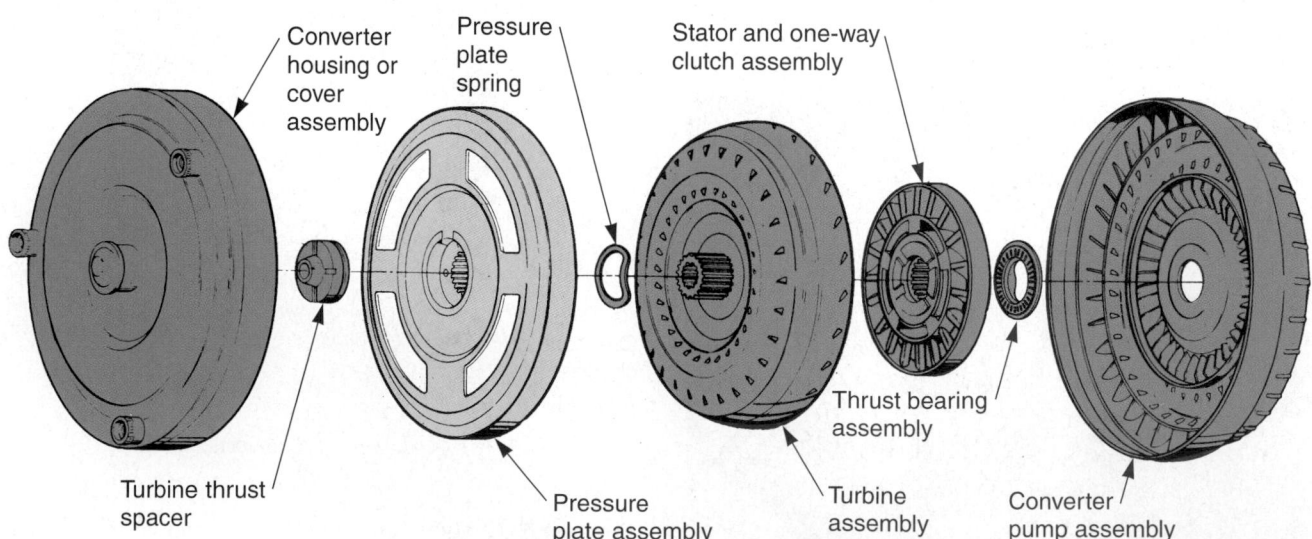

Figure 57-8. This lockup torque converter is a conventional converter with a friction pressure plate added. The pressure plate can be used to lock the turbine to the converter housing, eliminating slippage and increasing fuel economy. The converter also has a one-way clutch attached to the stator. (Oldsmobile)

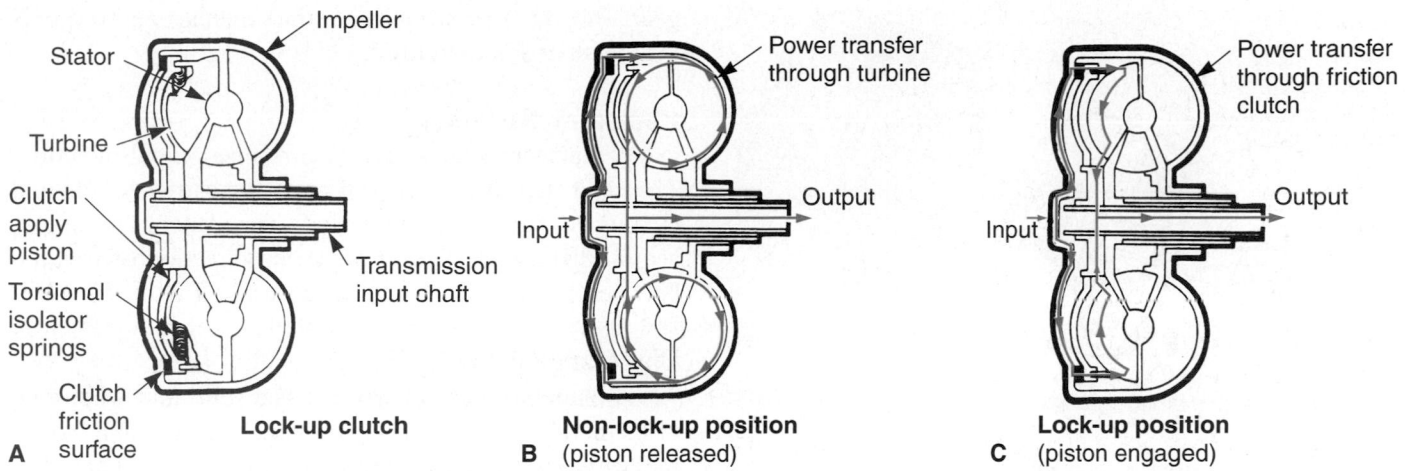

Figure 57-9. Lockup torque converter operation. A—Parts of a lockup converter. B—In lower gears, no oil pressure acts on clutch apply piston. The torque converter operates like a conventional unit, and the impeller drives the turbine. C—In high gear, oil is transferred into the piston chamber. The clutch apply piston forces friction surfaces together. The turbine is mechanically locked to the converter housing and impeller. The crankshaft drives the transmission input shaft directly, without slippage.

Automatic Transmission Shafts

Typically, an automatic transmission has two main shafts: the input shaft and the output shaft.

An automatic transmission *input shaft,* or *turbine shaft,* connects the torque converter with the driving components in the transmission. Look at **Figure 57-10.**

Each end of the input shaft has male (external) splines. These splines fit into splines in the torque converter turbine and a driving unit in the transmission. The

input shaft rides on bushings. Transmission oil lubricates the shaft and bushings.

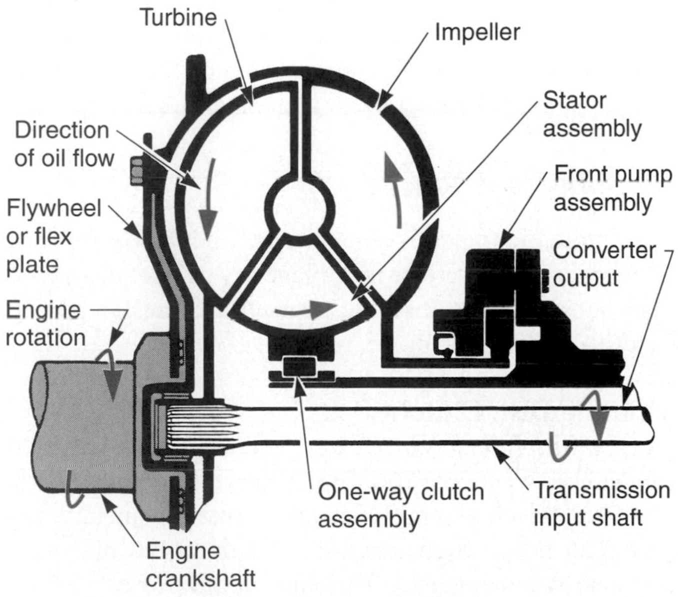

The *output shaft* connects the driving components in the transmission with the drive shaft. This shaft runs in the same centerline as the input shaft. Its front end almost touches the input shaft.

Stator Support

The *stator support,* also called *stator shaft,* is usually a stationary shaft splined to the stator assembly. It is a tube that extends forward from the front of the transmission and surrounds the input shaft.

Planetary Gears

A *planetary gearset* consists of a sun gear, several planet gears, a planet gear carrier, and a ring gear. A simple planetary gearset is shown in **Figure 57-11.**

The name planetary gearset refers to our solar system. Just as our planets circle the sun, the planet gears revolve around the sun gear.

As you can see, a planetary gearset is always in mesh. It is very strong and compact. An automatic transmission will commonly use two or more planetary gearsets.

By holding or releasing the components of a planetary gearset, it is possible to:

- Reduce output speed and increase torque (gear reduction).

- Increase output speed while lowering torque (overdrive).

- Reverse output direction (reverse gear).

Figure 57-10. The transmission input shaft extends through the stator support. The shaft is splined to the turbine. Also note how the stator mounts on a one-way clutch. (Ford)

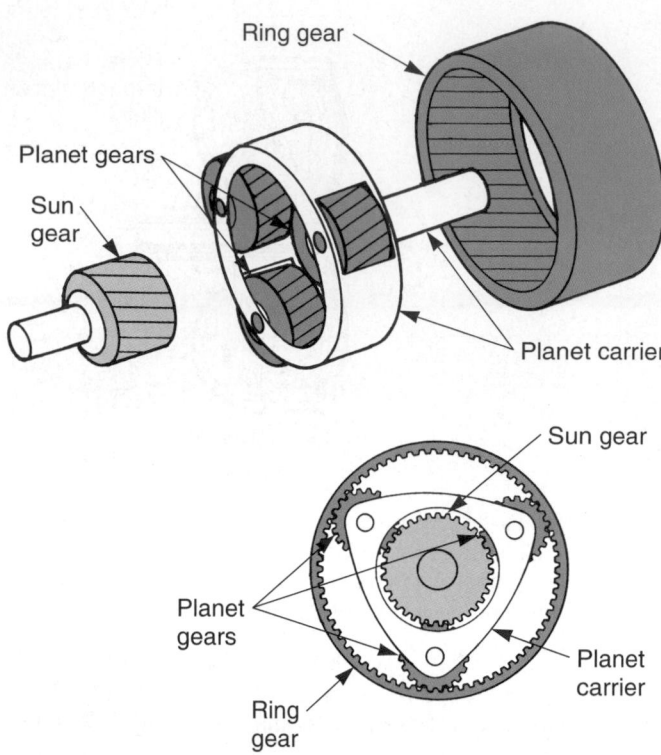

Figure 57-11. Simplified planetary gearset. Planet gears fit between a ring gear and a sun gear. The planet gears are mounted on planet carrier. Gears are always in mesh, making a compact, strong, and dependable assembly.

- Serve as a solid unit to transfer power (one-to-one ratio).
- Freewheel to stop power flow (park or neutral).

Planetary Reduction

One method of obtaining a gear reduction and torque increase is to hold the sun gear (stop it from turning) while driving the ring gear. This makes the planet carrier the output member. Refer to **Figure 57-12A.**

When power turns the ring gear, the planet gears "walk" (rotate) around the locked sun gear. The planet gears move in the same direction as the ring gear, but not as fast. As a result, more torque is applied to the output member (planet carrier) and output shaft.

Gear reduction can also be produced in the planetary gearset by turning the sun gear and holding the ring gear.

Planetary Overdrive

Driving the carrier while holding the ring gear achieves an overdrive ratio in a planetary gearset. Driving the carrier while holding the sun gear will also produce an overdrive ratio. Look at **Figure 57-12B.**

The input shaft powers the planet carrier. The sun gear is the output member driving the output shaft. The planet gears "walk" in the ring gear and power the sun

gear. The sun gear spins faster than the carrier. Torque is lost but speed is increased.

Planetary Reverse

A planetary gearset can also reverse output direction. The input shaft drives the sun gear. The carrier is held and the ring gear turns the output shaft. The planet gears simply act as idler gears. They reverse the direction of rotation between the sun gear and ring gear. See **Figure 57-12C.**

Planetary Direct Drive

A planetary gearset will act as a solid unit when two of its members are held. This causes the input and output members to turn at the same speed, **Figure 57-12D.**

Planetary Neutral

When none of the planetary members are held, the unit will not transfer power. This freewheeling condition is used when an automatic transmission is placed in neutral or park.

Compound Planetary Gearset

A *compound planetary gearset* combines two planetary units in one housing or ring gear, **Figure 57-13.** A compound planetary gearset is used because it can provide more forward gear ratios than a simple planetary gearset.

Some compound planetary gearsets contain two sun gears. In this design, short planet gears engage the forward sun gear. Long planet gears mesh with the rear sun gear. The ring gear engages both sets of planet gears.

The Simpson compound gearset is a compound planetary gearset that uses a single sun gear to operate two sets of planet gears on the same ring gear. The Simpson compound gearset is the most common type of compound planetary gearset used in automatic transmissions.

Clutches and Bands

Automatic transmission *clutches* and *bands* are friction devices that drive or lock planetary gearset members. They are used to cause the gearsets to transfer power. Refer to **Figure 57-14.**

Multiple Disc Clutches

A *multiple disc clutch* has several clutch discs that can be used to couple (hold) planetary gearset members. The front clutch assembly usually drives a planetary sun gear. The next clutch transmits power to the planetary ring gear when engaged. This can vary, however.

A clutch assembly generally consists of a drum, a hub, an apply piston, spring(s), driving discs, driven discs, a pressure plate, and snap rings.

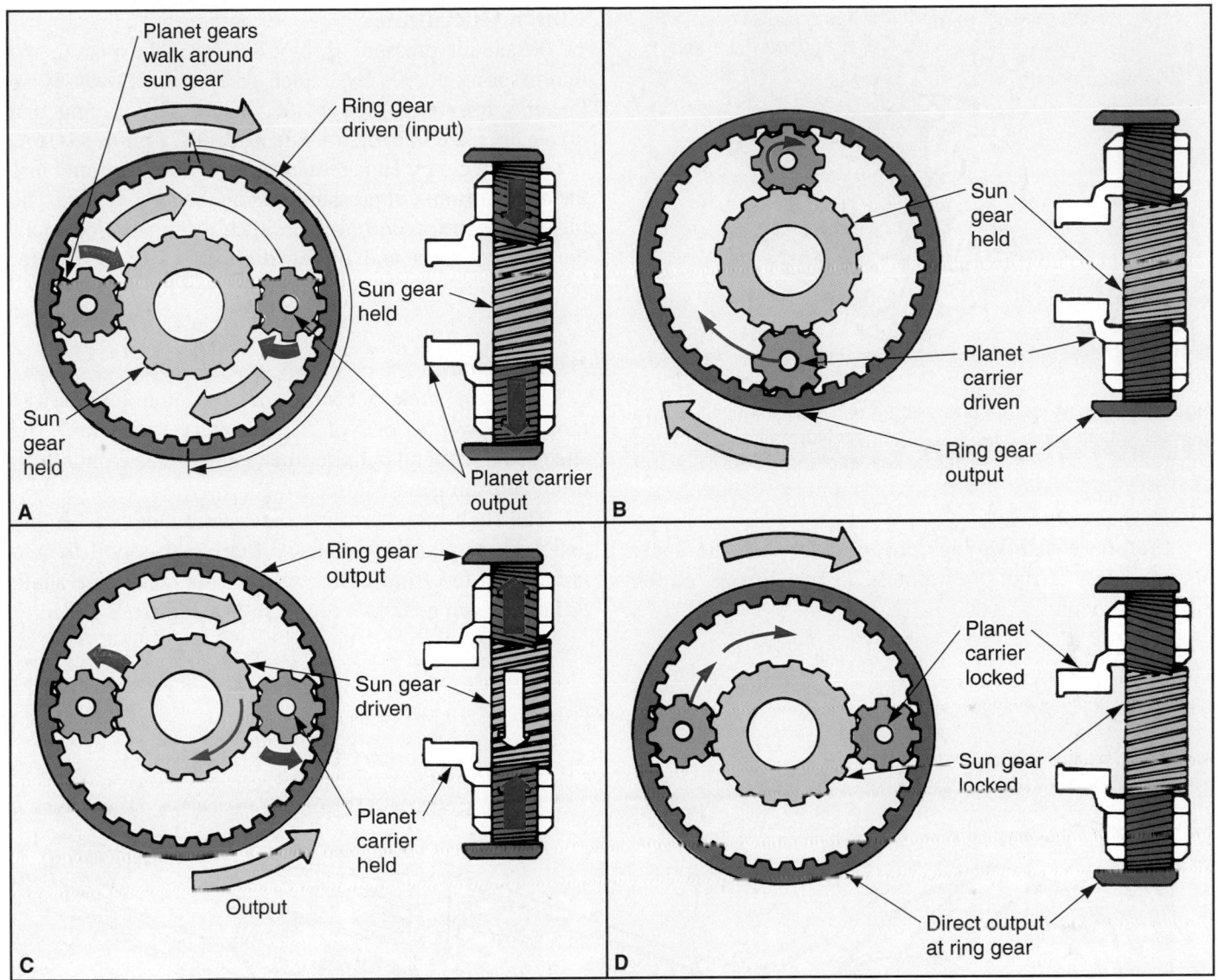

Figure 57-12. Study how different planetary gearset members can be held to provide different gear ratios and reverse. A—Simple gear reduction. The sun gear is stationary, the ring gear is the driving gear. The planet carrier is the output. Input torque increases and speed decreases. B—Overdrive. The sun gear is held stationary. The planet carrier is the drive gear. The ring gear, which turns faster than the input, is the output. C—Reverse gear. The planet carrier is held and the sun gear is the drive gear. The ring gear turns backwards as output. D—Direct drive results when any two members of the planetary gearset are held, or by driving any two members from same input. (Ford)

Clutch Construction

The ***clutch drum,*** also called the ***clutch cylinder,*** encloses the apply piston, pressure plate, discs, seals, and other parts of the clutch assembly, **Figure 57-15.**

The ***clutch hub*** fits inside the clutch discs and clutch drum. It has teeth on its outer surface that engage the teeth on the driving discs. The front clutch hub is also splined to the transmission input shaft.

The ***driving discs*** are usually covered with friction lining. They have teeth on their inside diameter that engage the clutch hub.

The ***driven discs*** are steel plates that have outer tabs that lock into the clutch drum. Driven discs alternate with the driving discs inside the clutch drum. This enables the hub and driving discs to turn the driven discs and drum when the clutch is activated.

The ***clutch apply piston*** slides back and forth inside the clutch drum to clamp the driving discs and the driven discs together. Seals fit on the piston to prevent fluid leakage during clutch application.

The ***pressure plate*** serves as a stop for clutch discs when the piston is applied. The piston pushes the discs against the pressure plate. Look at **Figure 57-15.**

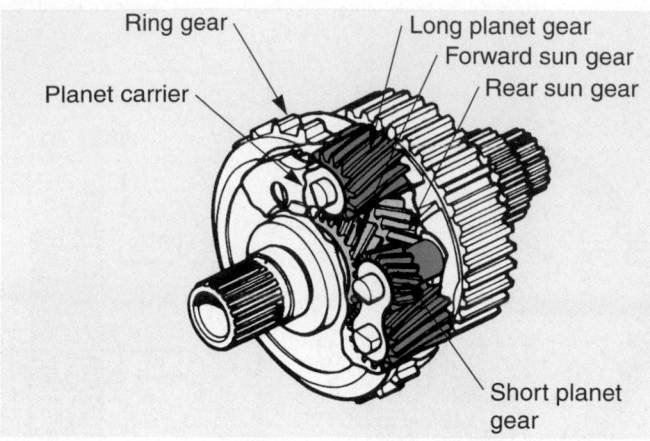

Figure 57-13. A compound planetary gearset acts like two gearset assemblies mounted together. Normally, a common ring gear is used for two separate sets of planet gears. (Subaru)

Clutch release springs are used to push the apply piston away from the clutch discs during clutch disengagement.

Clutch Operation

When oil pressure is blocked from the piston, the return spring pushes the clutch discs apart. Power is no longer transferred through the clutch. The driving and driven discs are free to turn independently, **Figure 57-16A.**

During clutch engagement, oil pressure is routed into the clutch drum. Oil pressure acts on the large piston. The piston is then forced into the clutch discs. Friction locks the driving discs and driven discs together to transfer power through the clutch assembly. See **Figure 57-16B.**

Driving Shell

A *driving shell,* or *clutch shell,* is commonly used to transfer power to one of the planetary sun gears. The shell is a thin metal cylinder that connects the front clutch drum and sun gear, **Figure 57-17.**

The shell may surround the second clutch assembly and forward planetary gearset. Tabs on the shell fit into notches on the front clutch drum. This makes the shell, drum, and sun gear turn together. See **Figure 57-18.**

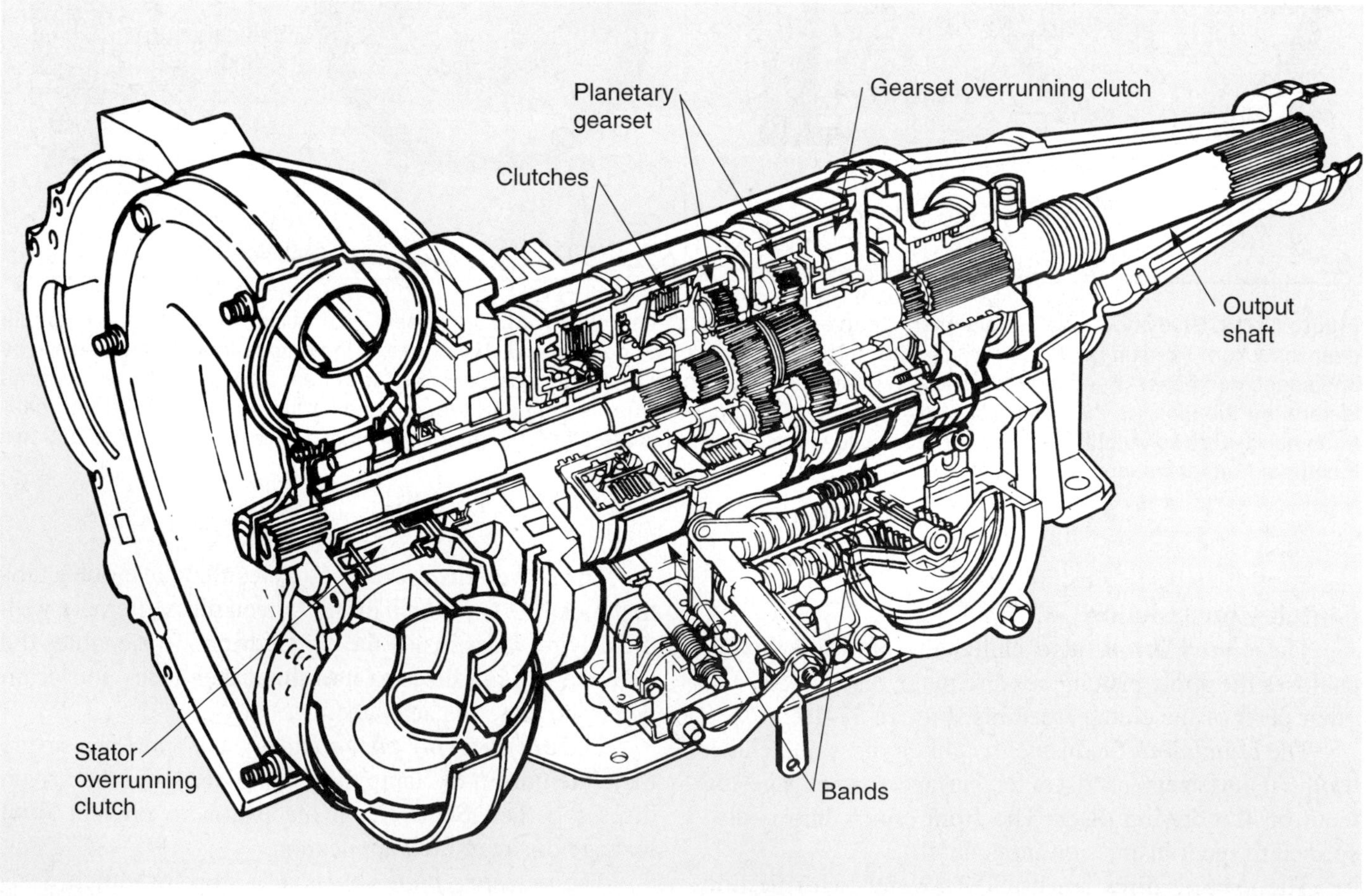

Figure 57-14. Study the location of gearsets and holding devices. Clutches, bands, and the rear one-way clutch operate the gearsets. Another one-way clutch operates the torque converter stator. (Ford)

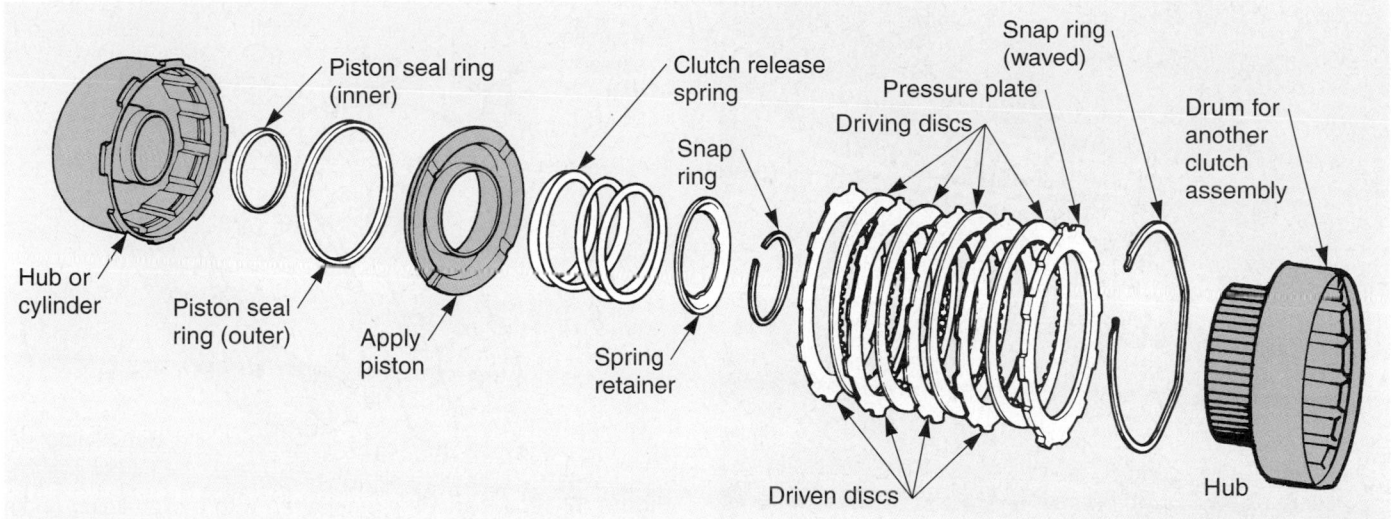

Figure 57-15. Study the construction of a clutch assembly for an automatic transmission. Clutch parts fit inside the clutch drum. Also note difference in the clutch discs. Driving discs are splined to hub. Driven discs are locked in the drum by tabs. (Chrysler)

Bands and Servos

Automatic transmission **bands** are friction devices for holding members of the planetary gearsets. Two or three bands are commonly used in modern transmissions. See **Figure 57-19.** *Servos* are apply pistons that operate the bands.

Band and Servo Construction

A band is a steel strap with a lining of friction material on its inner surface. The band's lining can be clamped around the clutch drum to stop drum rotation.

The friction material on the inside of the band is designed to operate in automatic transmission oil. It resists the lubricating qualities of the oil.

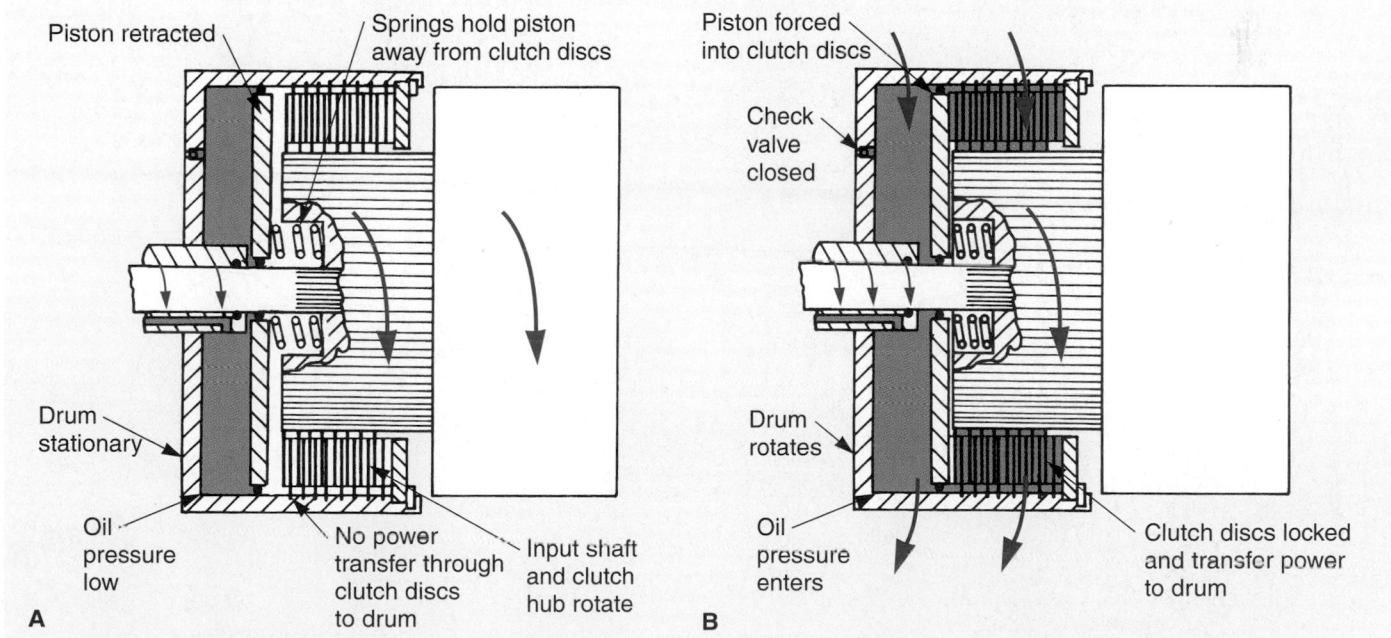

Figure 57-16. Basic clutch operation. A—No oil pressure enters drum. Springs hold the piston away from the clutch discs. The input shaft turns the clutch hub and the driving discs, but the driven discs and drum remain stationary. B—Oil is routed into the clutch drum. Oil pressure pushes the piston into the clutch discs, forcing the discs into the pressure plate. This locks the discs, hub, and drum together. Power is then transferred from the input shaft to the drum.

Figure 57-17. The drive shell is a thin metal cylinder that transfers power.

The *servo piston* is a metal plunger that operates in a cylinder machined in the transmission case. Rubber seals fit around the outside of the piston to prevent oil leakage. See **Figure 57-20.**

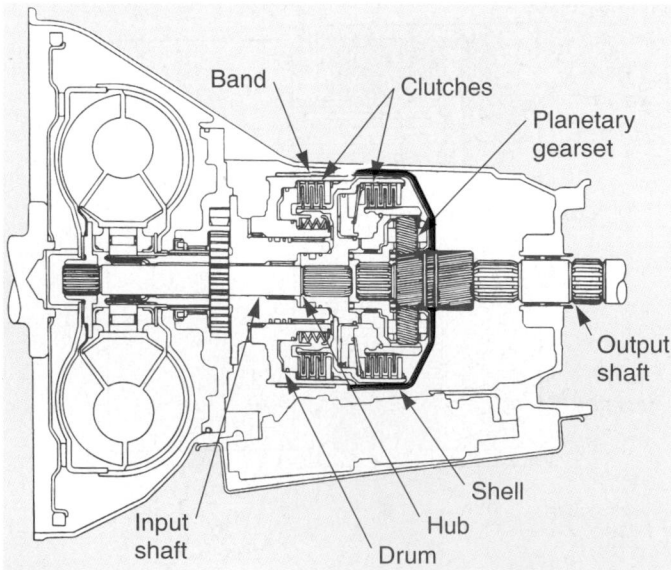

Figure 57-18. This drive shell connects the front drum to the sun gear. Note how the shell surrounds the second clutch assembly and front planetary gearset. When the front clutch is locked, the shell turns the sun gear. Also note the band used to hold the front drum and sun gear stationary.

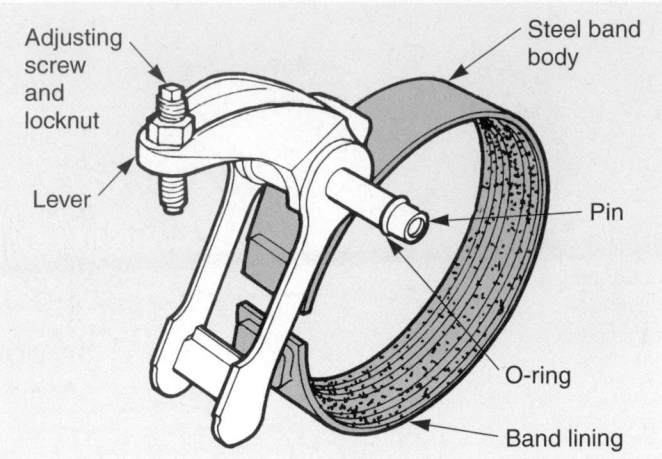

Figure 57-19. A band is a steel strap with friction lining on its inner surface. One end of the band is anchored in the case. (Dodge)

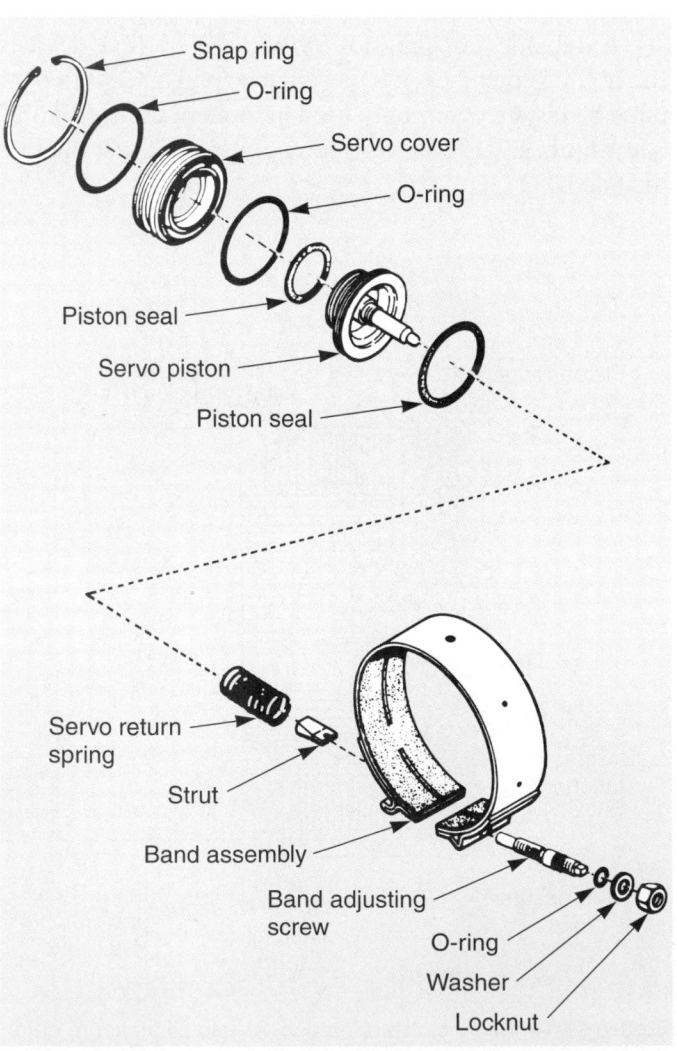

Figure 57-20. Exploded view of a band and servo assembly. Note the relationship between the parts. (Subaru)

A rod on the servo piston attaches to one end of the band. The other end of the band is anchored to the transmission case.

A *band adjustment screw* provides a means of adjusting the band-to-drum clearance. It moves the band closer to the drum as the friction material wears.

Servo seals prevent oil leakage around the servo piston and cylinder. Snap rings hold the piston in its cylinder.

Transmission Band Operation

To activate a band, oil pressure is sent into the servo cylinder. Pressure acts on the servo piston. The piston then slides in the cylinder and pushes on one end of the band, as in **Figure 57-21.**

Since the other end of the band is anchored, the band tightens around the drum. The friction material rubs on the drum and stops it from turning. This keeps one of the planetary components from revolving.

When the oil flow to the piston servo is blocked, the servo spring pushes on the piston. This slides the piston rod away from the band. The band then releases the drum and the planetary gearset member.

Accumulator

An *accumulator* is used in the apply circuit of a band or clutch to cushion the initial application. It temporarily absorbs some of the oil pressure to cause slower movement of the apply piston.

Overrunning Clutches

Besides the bands and clutches, an *overrunning clutch* can be used to hold a planetary gearset member. It is a one-way roller clutch that locks in one direction and freewheels in the other.

An overrunning clutch for the planetary gears is similar to the ones used in a torque converter stator or an electric starting motor drive gear. The typical locations of these clutches (stator clutch and gearset clutch) are illustrated in **Figure 57-14.**

A planetary gearset overrunning clutch consists of an inner race, rollers, a set of springs, and an outer race, **Figure 57-22.**

Hydraulic Valve Action

The basic action of a hydraulic valve is illustrated in **Figure 57-23.** Valves like this are used to operate the band servos and clutch pistons.

Transmission oil pump pressure causes oil to flow through the spool valve pressure lines to the piston cylinder. This pushes the piston. When the spool valve is

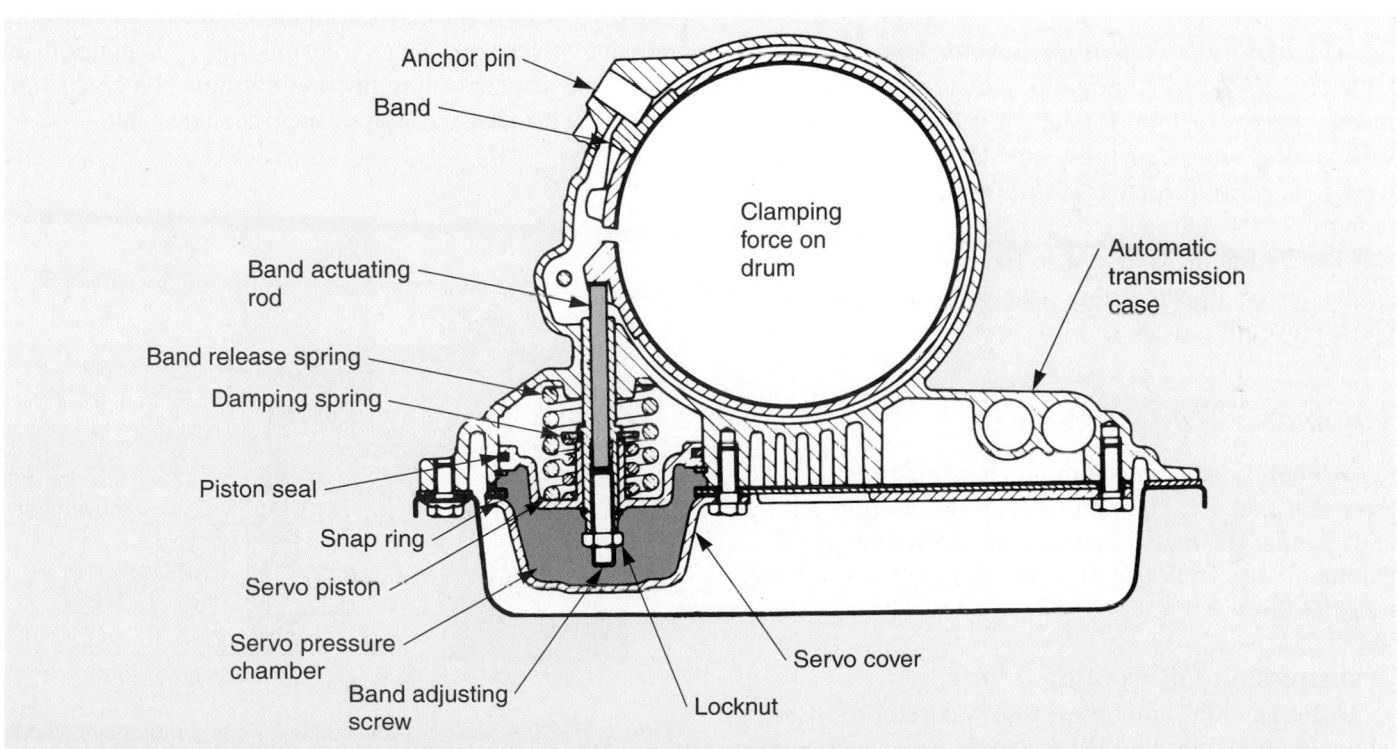

Figure 57-21. Servo piston and band action. When oil pressure enters the servo pressure chamber, the servo piston slides up in the cylinder. An actuating rod then pushes to squeeze the band inward on the drum.

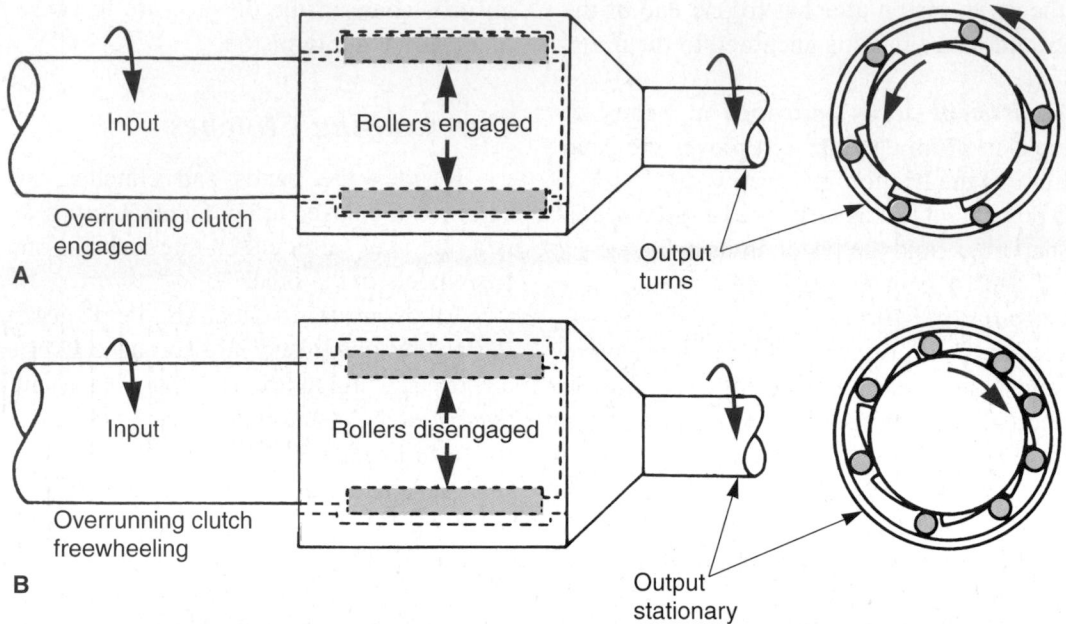

Figure 57-22. Overrunning clutch action A—When driven in one direction, rollers lock between the ramps on the inner race and the outer race. Both races turn together. This action can be used to stop movement of planetary member. B—When turned in the other direction, the two races are free to turn independently. (Deere & Co.)

moved the other way, pump pressure is not sent to the piston. The piston is then forced back into its cylinder.

Hydraulic System

The **hydraulic system** for an automatic transmission typically consists of the pump, pressure regulator valve, manual valve, vacuum modulator valve, governor valve, shift valves, servos, pistons, and valve body. These parts work together to form the "brain" (sensing) and "muscles" (control) of the automatic transmission. See **Figure 57-24.**

The hydraulic system also forces oil to high-friction points in the transmission. This lubricates the moving parts, preventing wear and overheating.

Automatic Transmission Oil

Automatic transmission oil (fluid) has several additives that make it compatible with the friction clutches and bands in the transmission. Different types of automatic transmission oils are required for different transmissions.

Transmission Oil Cooling

A tremendous amount of heat is developed inside an automatic transmission. When the torque converter slips, friction heats the oil. This heat must be removed, or transmission failure could result.

Many transmissions have an oil cooling system that includes external oil lines and a cooling tank inside the engine radiator. Look at **Figure 57-25.**

When the engine is running, the transmission oil pump forces oil through the cooling lines and into the radiator cooler tank. Since transmission oil is hotter than the engine coolant, oil temperature drops. The cooled oil returns to the transmission through the other line.

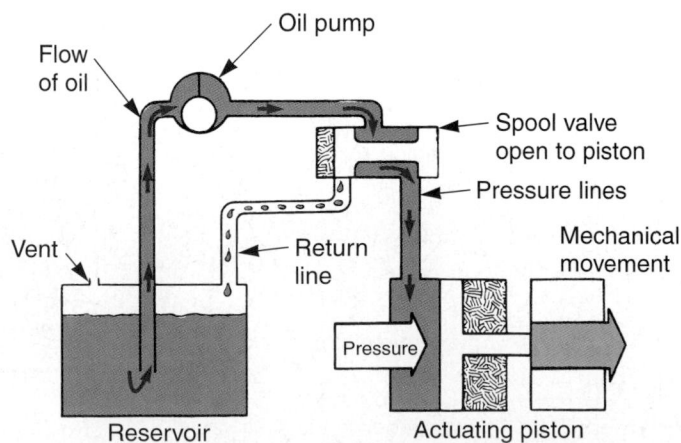

Figure 57-23. A basic hydraulic circuit. The pump draws oil out of the reservoir and forces it through a spool valve. In this position, the spool valve routes oil to the piston. The piston uses oil pressure to produce clamping pressure. (Chrysler)

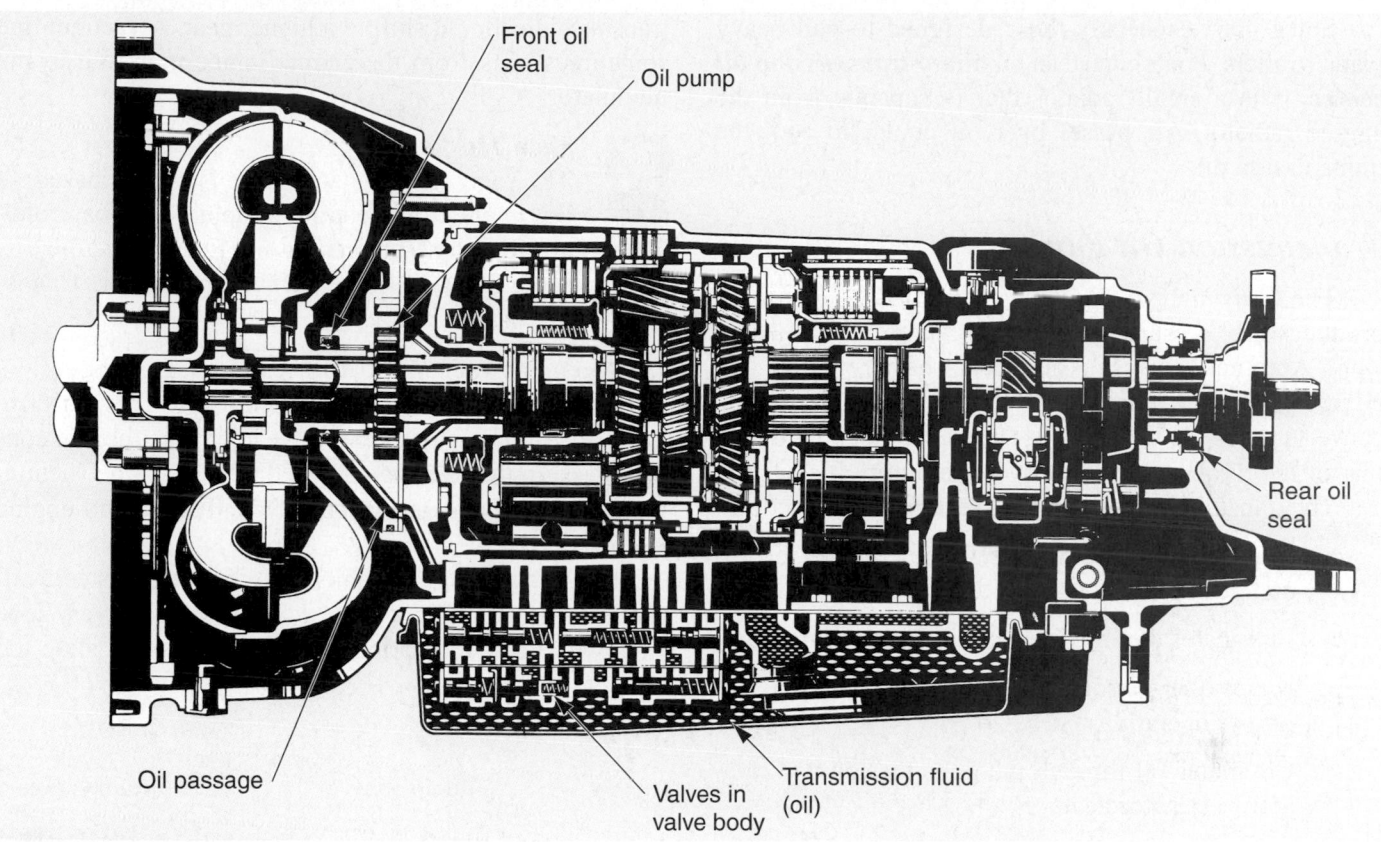

Figure 57-24. The transmission oil pump is normally located in the front of the case. Oil is drawn out of the pan and circulated through passages to hydraulic components. Also note the location of oil seals. (Mercedes Benz)

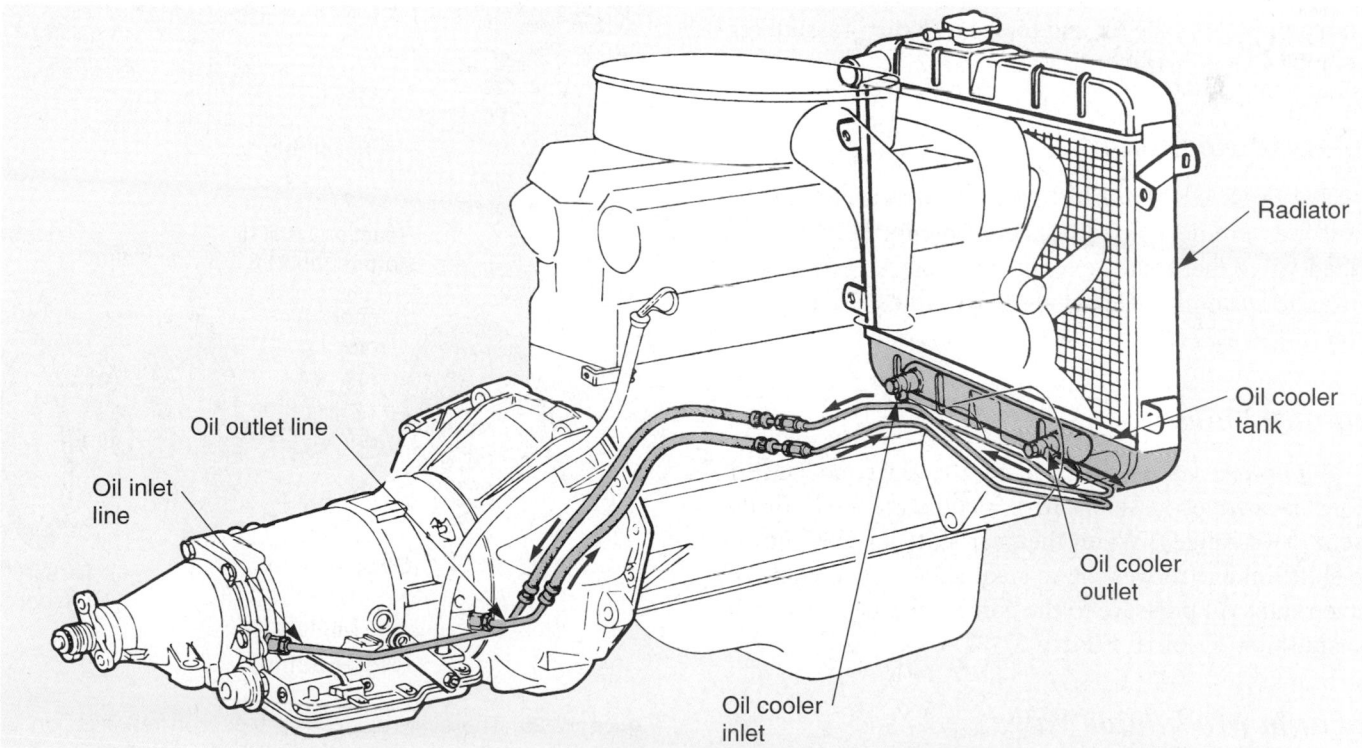

Figure 57-25. An oil cooler tank is commonly used in transmissions. The oil pump pushes oil through lines to the cooler tank to maintain acceptable oil temperature.

Some cars, especially those designed to pull heavy loads (trailers, boats), have an auxiliary *transmission oil cooler.* It is a small radiator that is separate from the engine radiator. Air passes over oil cooler to cool the transmission oil.

Transmission Oil Pump

The *transmission oil pump,* also called the *oil pump,* produces the pressure to operate an automatic transmission. Automatic transmissions can have one or two pumps. The pumps are often located behind the torque converter or in the valve body. The sleeve or collar on the rear of the torque converter drives the pump.

The transmission oil pump has several basic functions:

- Produces pressure to operate the clutches, bands, and gearsets.

- Lubricates the moving parts in the transmission.

- Keeps the torque converter filled with oil for proper operation.

- Circulates oil through the transmission and radiator to transfer heat.

- Operates hydraulic valves in the transmission.

There are two commonly used oil pumps: the gear type and the rotor type, **Figure 57-26.**

When the torque converter spins the oil pump, transmission oil is drawn into the pump from the pan. The pump compresses the oil and forces it to the pressure regulator. This is illustrated in **Figure 57-27.**

Pressure Regulator

The *pressure regulator* limits the maximum amount of oil pressure developed by the oil pump. It is a spring-loaded valve that routes excess pump pressure out of the hydraulic system. This ensures proper transmission operation, **Figure 57-27.**

Manual Valve

A *manual valve,* operated by the shift mechanism, allows the driver to select park, neutral, reverse, or different drive ranges. When the gear shift lever is moved, the shift linkage moves the manual valve. As a result, the valve routes oil pressure to the correct components in the transmission. Look at **Figure 57-27.**

Vacuum Modulator Valve

The *vacuum modulator valve,* or *throttle valve,* senses engine load (vacuum) and determines when the

transmission should shift to a higher gear. A vacuum line sometimes runs from the engine intake manifold to the modulator.

Tech Tip
The vacuum modulator valve is being phased out in electronic control systems. The manifold absolute pressure sensor and ECM serve the same function as the vacuum modulator. This is discussed shortly.

As engine vacuum (load) rises and falls, it moves the diaphragm inside the vacuum modulator. This, in turn, moves the rod and hydraulic valve to change throttle control pressure in the transmission. In this way, the vacuum modulator can match transmission shift points to engine loading.

For example, if a vehicle is climbing a steep hill (under a heavy pull), engine vacuum will be very low. This will allow the spring in the modulator to slide the

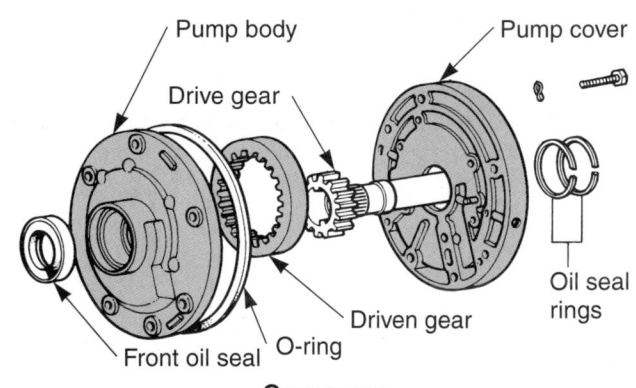

Gear pump

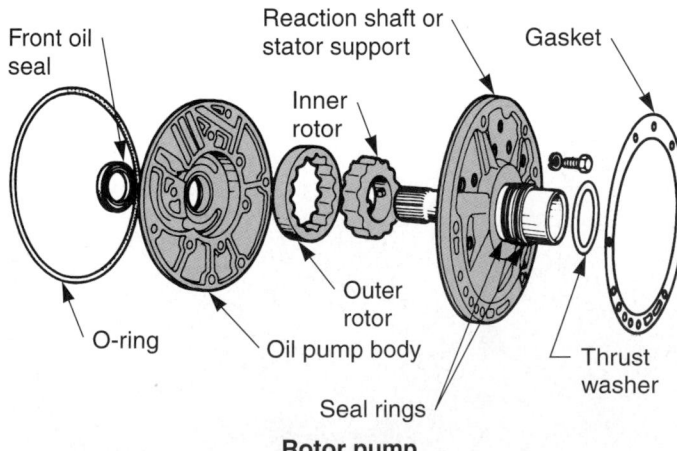

Rotor pump

Figure 57-26. There are two basic types of transmission oil pumps: rotor and gear. Study the similarities and differences. The torque converter normally drives the pump. (Chrysler and Toyota)

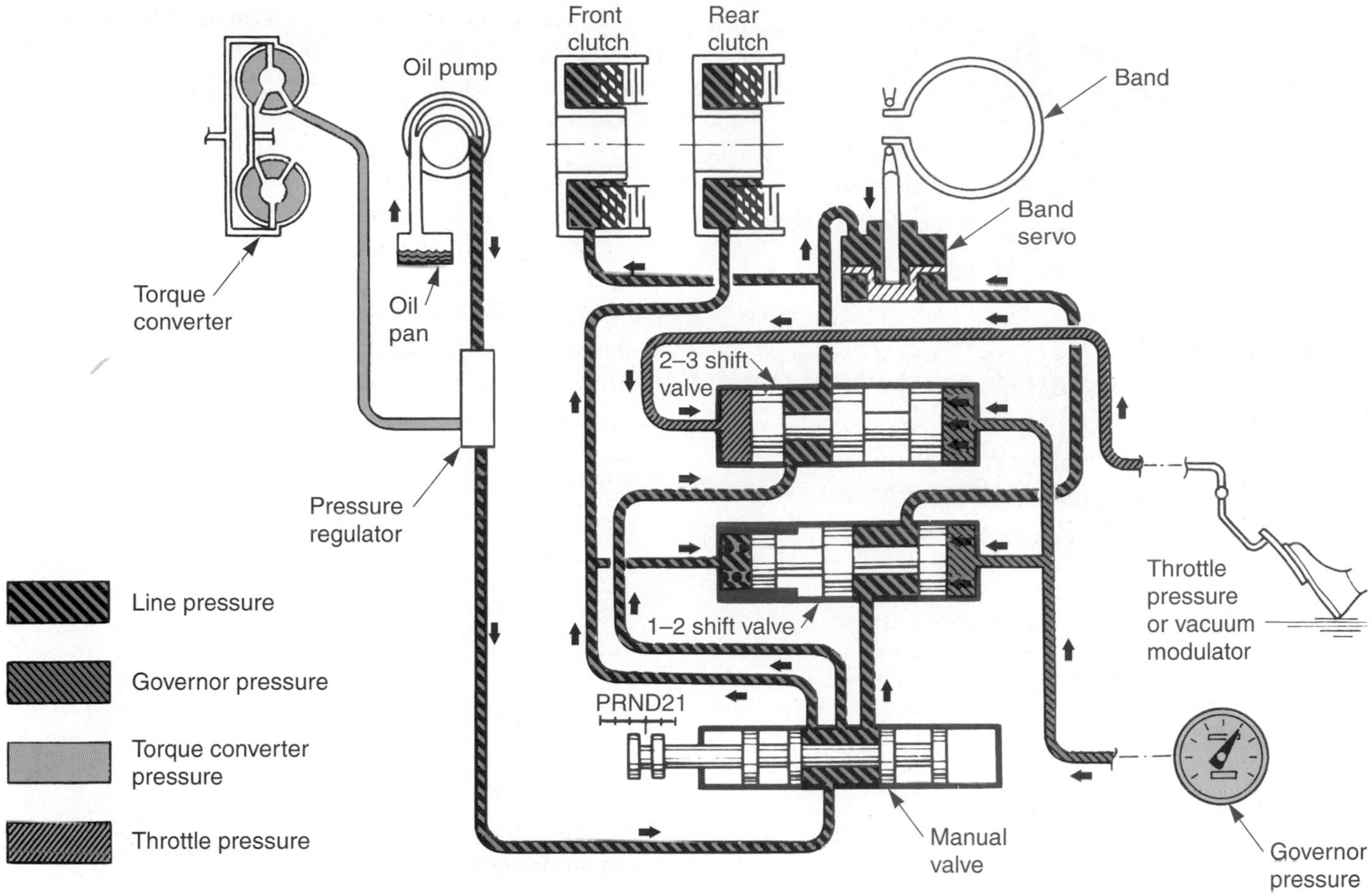

Front clutch Rear clutch Band

Oil pump

Torque converter

Oil pan

Band servo

2–3 shift valve

Pressure regulator

1–2 shift valve

Throttle pressure or vacuum modulator

PRND21

Manual valve

Governor pressure

Line pressure

Governor pressure

Torque converter pressure

Throttle pressure

Figure 57-27. Simplified circuit showing hydraulic action in an automatic transmission. Manual valve pressure, throttle valve pressure, and governor valve pressure operate shift valves. Shift valves then direct oil pressure to correct clutch or band pistons. Study this diagram carefully. (Nissan)

modulator valve further into the transmission. The valve then directs oil pressure to delay the upshift. The transmission stays in a lower gear longer to allow the car to accelerate up the hill. Look at **Figure 57-28.**

Governor Valve

The **governor valve** senses vehicle speed to help control gear shifting. The vacuum modulator and governor work together to determine shift points. See **Figure 57-29.**

A basic mechanical governor valve consists of a drive gear, centrifugal weights, springs, a hydraulic valve, and a shaft. The governor gear is usually meshed with a gear on the transmission output shaft. Whenever the car and output shaft are moving, the centrifugal weights rotate.

When the output shaft and weights are spinning slowly, the weights are held in by the governor springs. This causes a low pressure output and the transmission remains in a low gear.

As engine and shaft speed increase, the weights are thrown out further and governor pressure increases. This

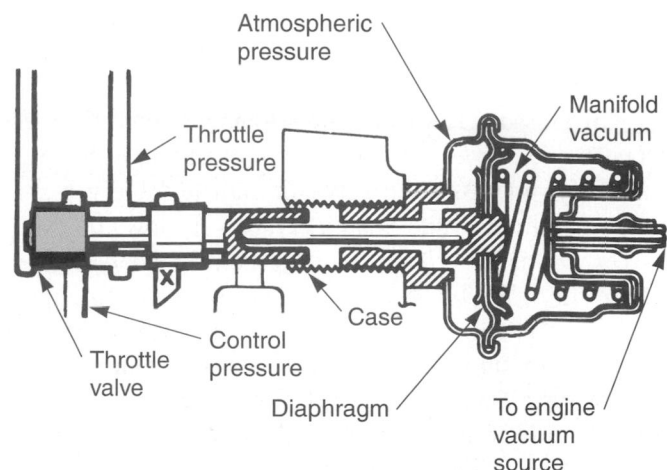

Atmospheric pressure

Throttle pressure

Manifold vacuum

Throttle valve

Control pressure

Case

Diaphragm

To engine vacuum source

Figure 57-28. The vacuum modulator operates the throttle valve. Engine vacuum allows the modulator to sense engine load. For example, with engine acceleration and high load, vacuum drops. The vacuum modulator spring overcomes vacuum pull on the diaphragm. The spring pushes the valve to the left. This alters throttle oil pressure, keeping the transmission in a lower gear. (Ford)

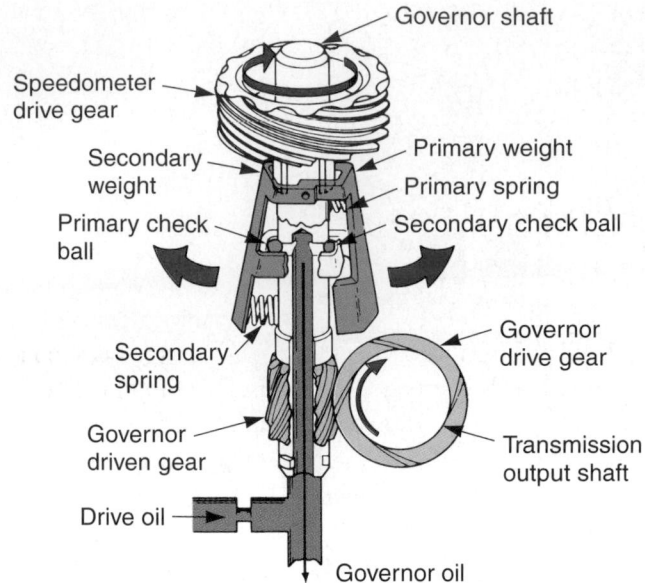

Figure 57-29. The governor senses vehicle speed. A gear on the transmission output shaft spins the governor. As speed increases, centrifugal weights are thrown outward. This opens the governor valve enough to change governor pressure and cause an upshift. (Cadillac)

moves the shift valve and causes the transmission to shift to a higher gear.

Other types of governor valves are also used. They do the same job.

Shift Valves (Balanced Valves)

Shift valves, also called **balanced valves,** use control pressure (oil pressure from regulator, governor valve, throttle valve, and manual valve) to operate the bands, servos, and gearsets.

Oil pressure from the other transmission valves act on each end of the shift valves. For example, if the pressure from the governor valve is high and the pressure from the throttle and manual valves is low, the shift valves will be moved sideways in their cylinders.

In this way, the shift valves are sensitive to engine load (throttle valve oil pressure), engine speed (governor valve oil pressure), and gear shift position (manual valve oil pressure). The shift valves move according to these forces and keep the transmission shifted into the correct gear for the driving conditions.

Kickdown Valve

A **kickdown valve** causes the transmission to shift into a lower gear during fast acceleration. A rod or cable links the throttle body or carburetor to a lever on the transmission.

When the driver presses down on the accelerator, the lever moves the kickdown valve. This causes hydraulic pressure to override normal shift control pressure, and the transmission downshifts.

Valve Body

The **valve body** contains many of the hydraulic valves (pressure-regulating valve, shift valves, manual valve, etc.) of an automatic transmission. The valve body bolts to the bottom of the transmission case. It is housed in the transmission pan. A filter or screen is usually attached to the bottom of the valve body. See **Figure 57-30.**

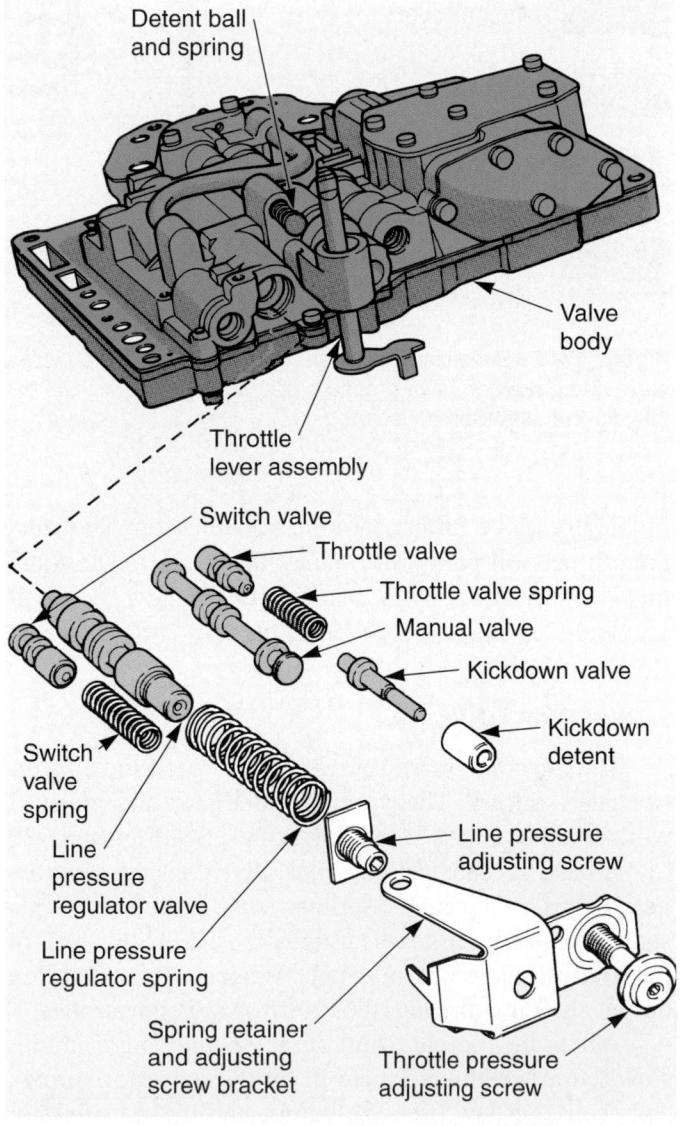

Figure 57-30. The valve body bolts to the bottom of the transmission case. It houses the manual valve, pressure regulator valve, kickdown valve, and other valves. (Plymouth)

Passages in the valve body route oil from the pump to the valves and then into the transmission case. Passages in the case carry oil to the other hydraulic components.

Parking Pawl

A *parking pawl* is used to lock the transmission output shaft and keep the car from rolling when not in use. Its basic action is shown in **Figure 57-31.**

Automatic Transmission Power Flow

The flow of power through an automatic transmission depends on its specific design. However, you should have a general understanding of how power is transmitted through the major parts of transmissions.

Figure 57-32 shows how torque moves from the input shaft to the output shaft. This is a typical three-speed transmission. Study each illustration carefully, noting which clutches, bands, and gearset members are activated.

Overdrive Power Flow

Figure 57-33 shows the power flow through a late-model, four-speed, overdrive automatic transmission in high gear. This design uses two input shafts (turbine shaft and direct input shaft). Trace the power flow and compare it to the other, more conventional transmissions covered earlier.

Electronic Transmission Control

Electronic transmission control involves using sensors, actuators, and a computer (transmission control module or power train control module) to control shift points, torque converter lockup, downshifts, and other functions for more efficient operation.

A *transmission control module* (TCM) monitors and controls the functions of the transmission. A *power train control module* (PCM) monitors and controls the engine, transmission, fuel injection system, emission control system, and other units. It is not as specialized as a transmission control module.

Basically, solenoids on the transmission are used to move hydraulic valves. This allows the transmission control module (TCM) to control internal hydraulic circuit action and automatic transmission operation.

A diagram of a modern electronic system for operating an automatic transmission is shown in **Figure 57-34.** Note how various vehicle sensors feed data to the TCM. The control module can then use preprogrammed information to activate the shift solenoids and torque converter lockup solenoid. The solenoids open and close fluid pressure passages to operate the transmission or transaxle.

A *shift schedule* stored in the TCM memory determines when an electronically controlled transmission should change gears. The TCM uses input signals from various sensors and the shift schedule to know when to upshift or downshift.

For example, if the driver presses down quickly on the gas pedal for rapid acceleration, the throttle position sensor would send a signal to the TCM. At the same time, the engine manifold pressure sensor would report a low-vacuum, high-load condition. The vehicle speed sensor would signal low road speed, so the TCM would know to keep the transmission from upshifting into a higher gear. This type of logic is used for precise control of automatic transmissions.

Transmission control sensors have replaced some of the mechanical components in electronically controlled automatic transmissions. For example, the speed sensor replaces the governor, the MAP sensor replaces the vacuum modulator, and the throttle position sensor replaces the kickdown rod.

Transmission solenoids are actuators found on electronically controlled automatic transmissions. When the TCM sends control current through one of these solenoids, the solenoid's magnetic field acts on a control valve. In this way, the electronic system can change hydraulic pressure in the transmission to control its operation. See **Figure 57-35.**

Automatic transmission solenoids are used to control torque converter lockup, shift points, reverse, and other functions. This will vary with the make and model of the transmission. The operation of an automatic transmission solenoid is illustrated in **Figure 57-36.**

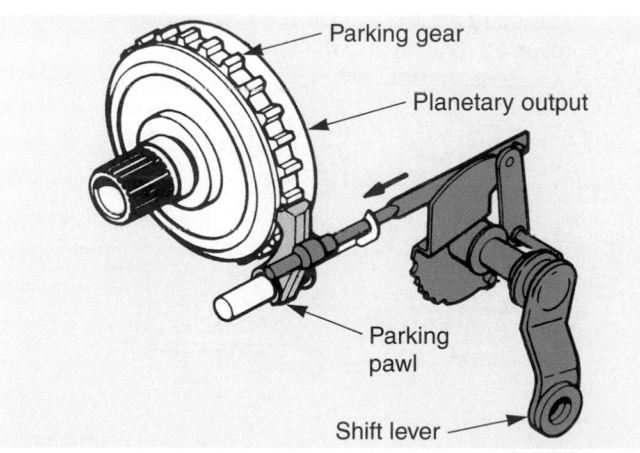

Figure 57-31. A parking pawl is simply a latch that locks into large teeth on the parking gear. Since the pawl is mounted on the transmission case, this locks the parking gear and output shaft. (Subaru)

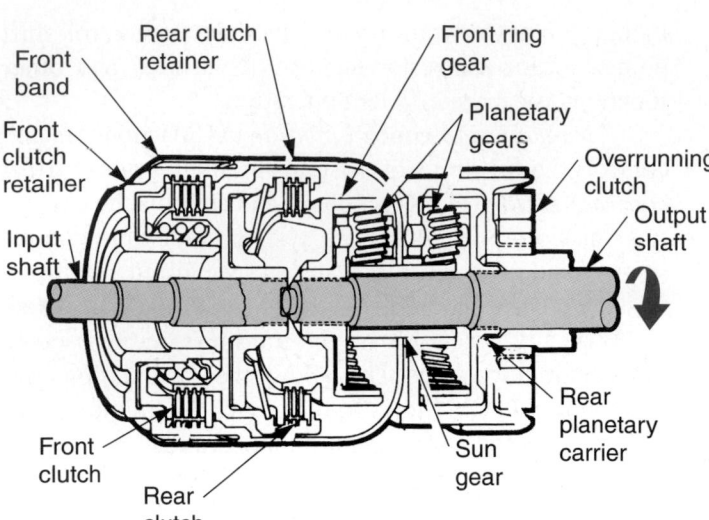

Rear clutch retainer

Front band

Front clutch retainer

Input shaft

Front ring gear

Planetary gears

Overrunning clutch

Output shaft

Front clutch

Rear clutch

Sun gear

Rear planetary carrier

A—Study parts relating to power flow.

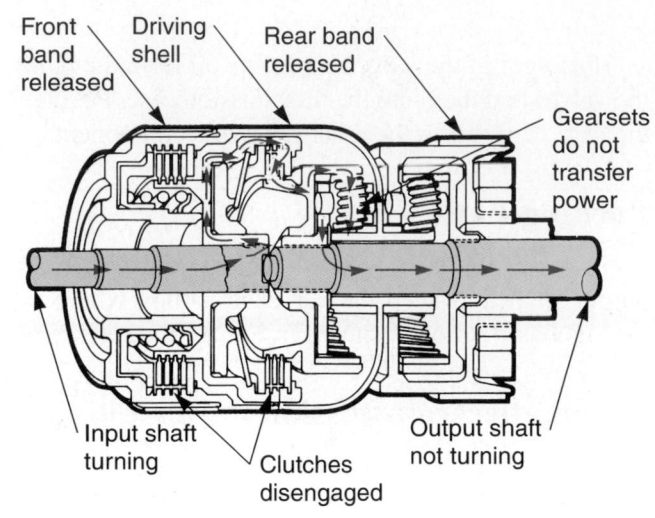

Front band released

Driving shell

Rear band released

Gearsets do not transfer power

Input shaft turning

Clutches disengaged

Output shaft not turning

B—Neutral. Clutches and bands disengaged. Input shaft and hub turn, but power does not flow through clutches or drum. Output stationary.

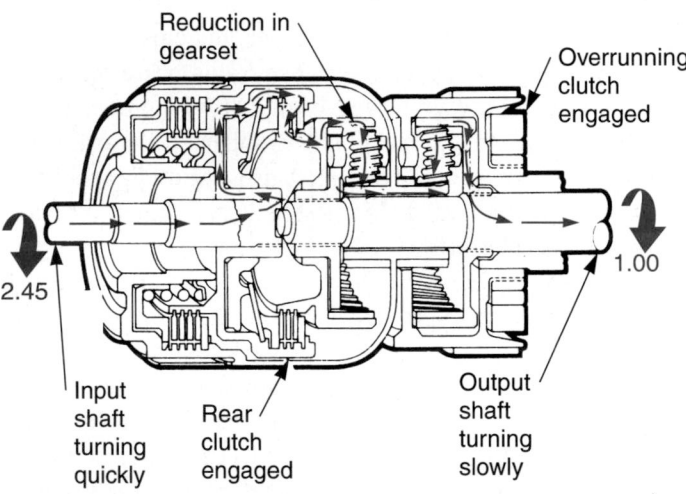

Reduction in gearset

Overrunning clutch engaged

2.45

Input shaft turning quickly

Rear clutch engaged

Output shaft turning slowly

1.00

C—First gear. Rear clutch and overrunning clutch engaged. Gear reduction through planetary gearsets results in high torque output to driveline.

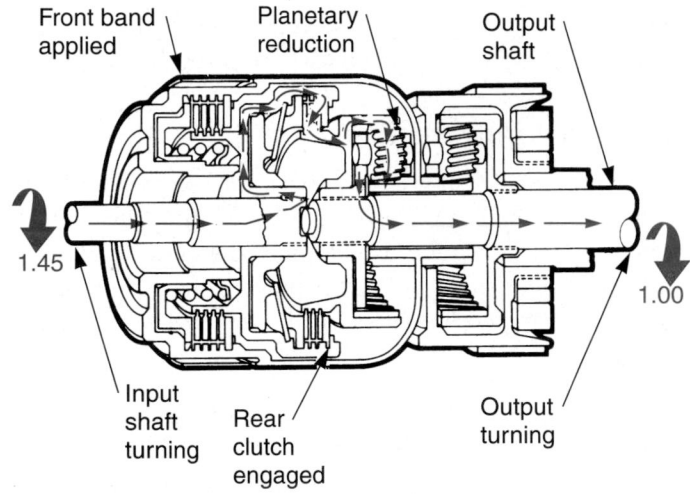

Front band applied

Planetary reduction

Output shaft

1.45

Input shaft turning

Rear clutch engaged

Output turning

1.00

D—Second gear. Front band applied. Rear clutch engaged. Power flows through input, hub, clutch, drum, and front gearset to output. Less reduction results.

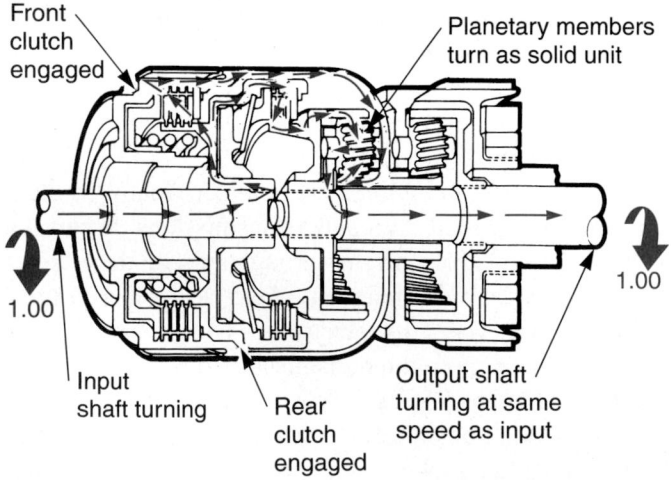

Front clutch engaged

Planetary members turn as solid unit

1.00

Input shaft turning

Rear clutch engaged

Output shaft turning at same speed as input

1.00

E—Third gear. Front and rear clutches engaged. Planetary members locked for direct drive. One-to-one ration for higher vehicle speeds results.

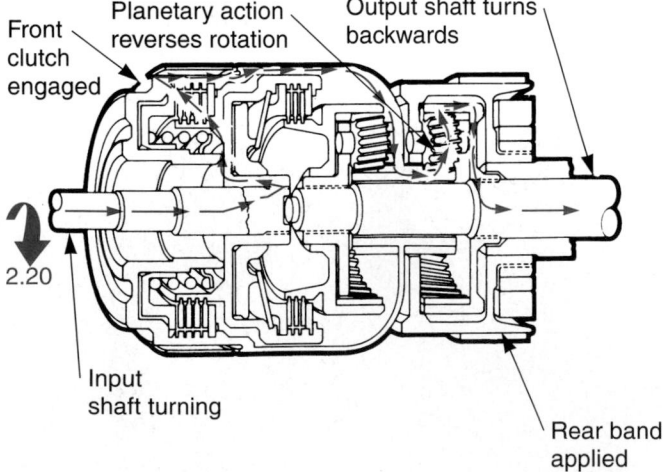

Front clutch engaged

Planetary action reverses rotation

Output shaft turns backwards

2.20

Input shaft turning

Rear band applied

F—Reverse. Front clutch and rear band applied. Power flows through clutch, shell, and sun gear to rear planetary gearset, which reverses rotation.

Figure 57-32. Study power flow through a typical automatic transmission. (Chrysler)

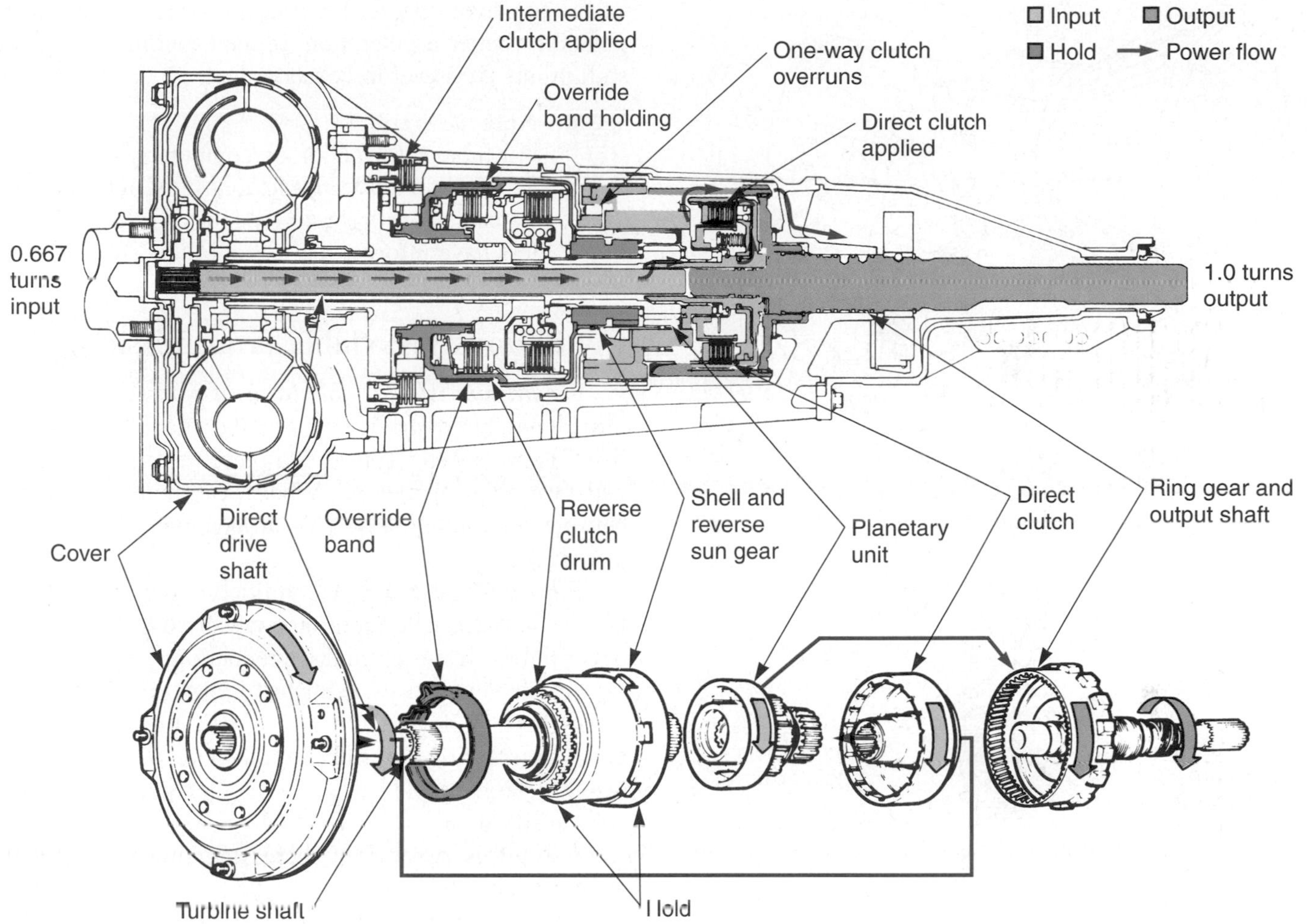

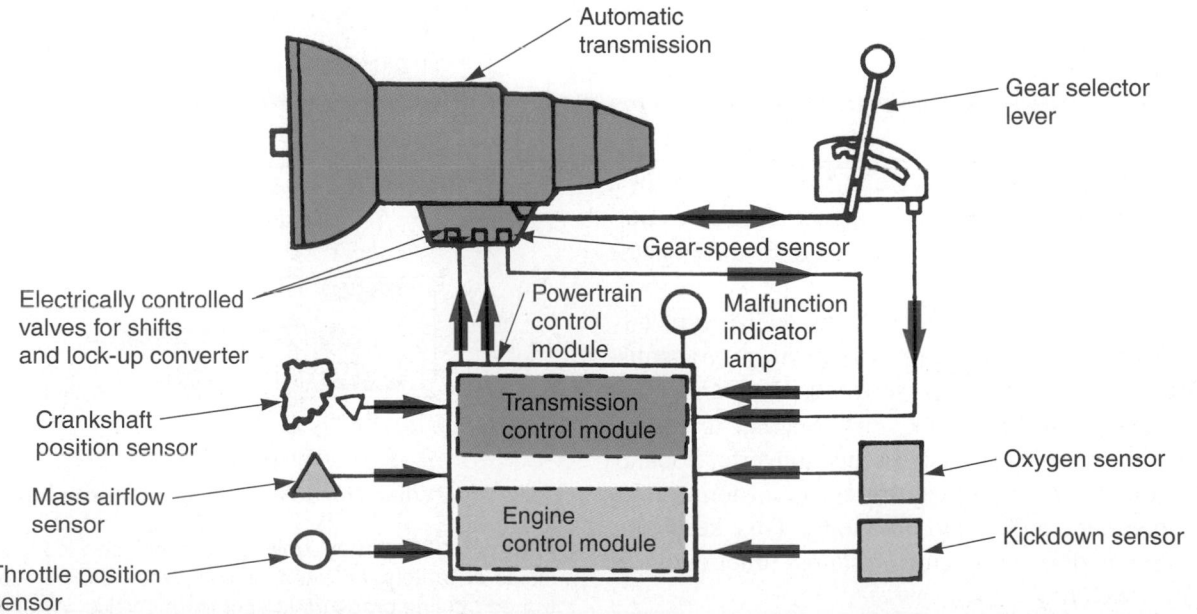

Intermediate clutch applied
Override band holding
One-way clutch overruns
Direct clutch applied

□ Input   □ Output
□ Hold   → Power flow

0.667 turns input
1.0 turns output

Cover
Direct drive shaft
Override band
Reverse clutch drum
Shell and reverse sun gear
Planetary unit
Direct clutch
Ring gear and output shaft

Turbine shaft
Hold

Figure 57-33. Power flow in high gear of a four-speed automatic with overdrive. Study the differences with transmissions already covered in the chapter. (Ford)

Automatic transmission
Gear selector lever
Gear-speed sensor

Electrically controlled valves for shifts and lock-up converter
Powertrain control module
Malfunction indicator lamp

Crankshaft position sensor
Transmission control module
Oxygen sensor

Mass airflow sensor

Throttle position sensor
Engine control module
Kickdown sensor

Figure 57-34. Late-model automatic transmissions use a computer to help control shift points. Note the flow of data to and from the computer and transmission.

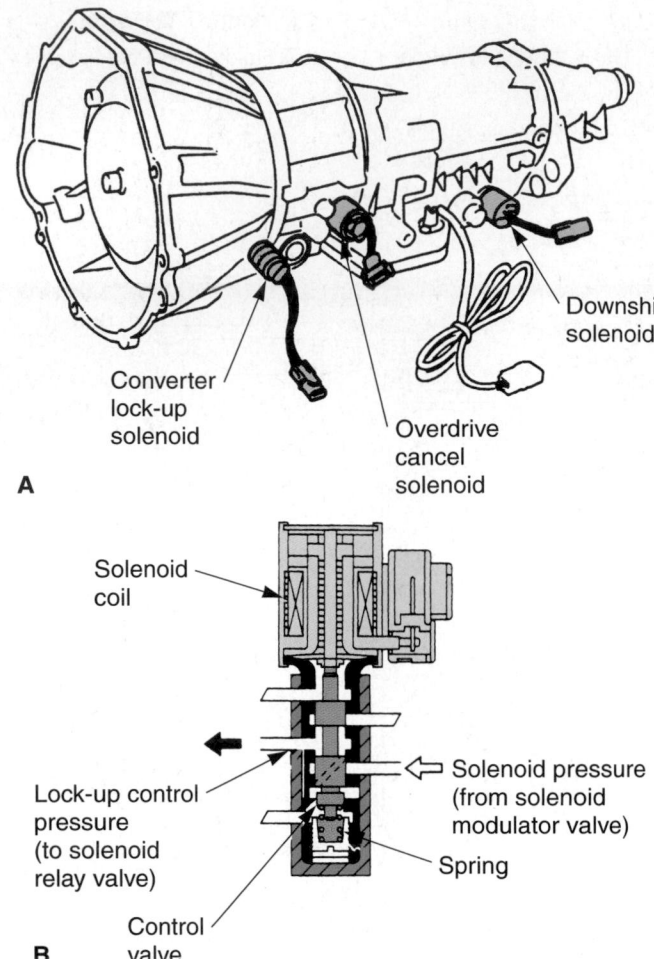

A

B

Figure 57-35. Transmission solenoids. A—Solenoids are commonly used on late-model automatic transmissions for computer control. B—When the ECM sends current through solenoid windings, the solenoid moves a spool valve to alter hydraulic pressure and flow in the transmission circuit. In this way, the ECM and solenoids can replace older mechanical and vacuum-operated devices, improving efficiency. (Nissan and Lexus)

Figure 57-37 shows a service manual diagram for a complete computer control system, including the automatic transmission. Study the flow of data through the system.

The TCM, in conjunction with the ECM, typically monitors engine speed, load, throttle position, transmission output shaft speed, gear shift position, and other variables. The control modules work together to control for transmission shift points, torque converter lockup, ignition timing, fuel injection timing, emission control system operation, and other functions. This keeps the transmission and other engine systems functioning at maximum efficiency.

Some transmission control systems allow for economy and sport settings. By moving a switch in the dash, the driver can set the transmission to higher shift points for better acceleration. In the economy mode, the shift points are lower to conserve fuel.

Note

For more information on computers, sensors, actuators, and electronic control of automotive transmissions, refer to the index. Several chapters have information useful to this subject.

Continuously Variable Transmission

A *continuously variable transmission* (CVT) has an infinite number of driving ratios rather than three, four, or five forward speeds. It uses two-piece centrifugal pulleys with variable diameters. V-belts (normally two) run between the pulley sets. This arrangement replaces the planetary gearsets.

Figure 57-38 shows a simplified drawing of a CVT. During initial acceleration, a small drive pulley turns a larger pulley, resulting in drive reduction.

As speed increases, centrifugal force pushes the two-piece drive pulley together. The belt rides out in the pulley, increasing the pulley's effective diameter. As a result, a larger pulley drives a smaller pulley for more vehicle speed.

A CVT is capable of increasing fuel economy by approximately 25% because it keeps the engine at its most efficient operating speed. Engine speed can be kept

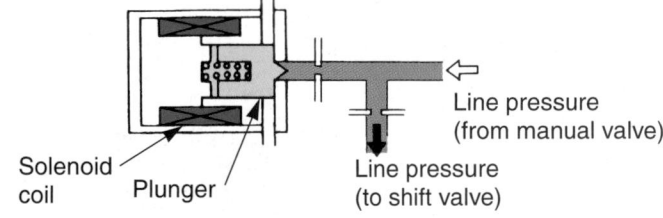

When the solenoid valve is turned off

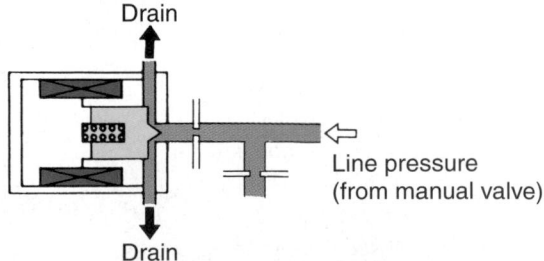

When the solenoid valve is turned on

Figure 57-36. Study how a solenoid can control pressure inside an automatic transmission. When the solenoid valve is turned off, line pressure is sent to shift valve. When solenoid is energized, the valve is pulled back to drain line pressure and keep pressure from acting on the shift valve. (Lexus)

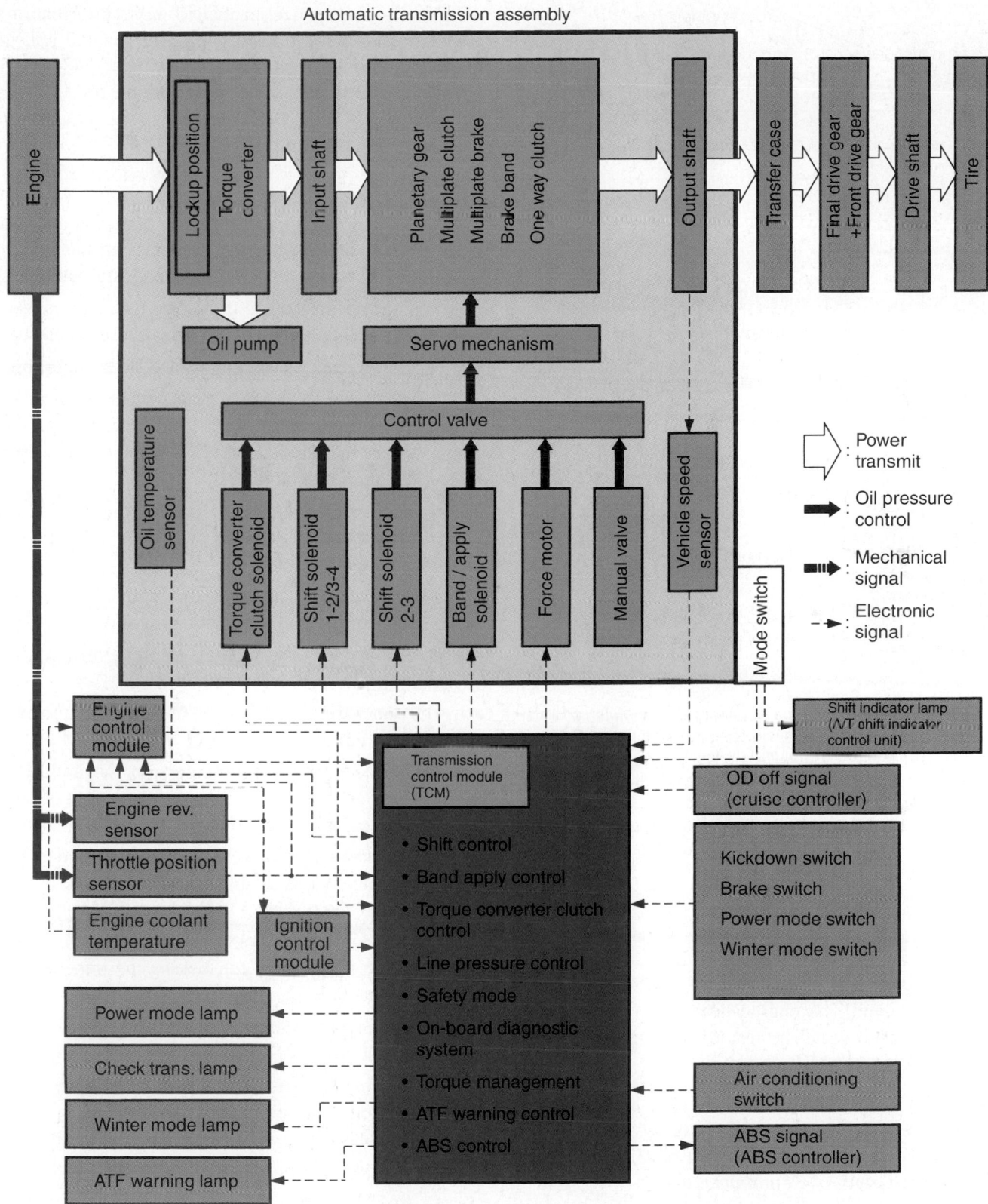

ABS: Anti-lock brake system

Figure 57-37. Study the controls in a complete computer system. Note how the transmission, ABS, and engine controls all work together. (General Motors)

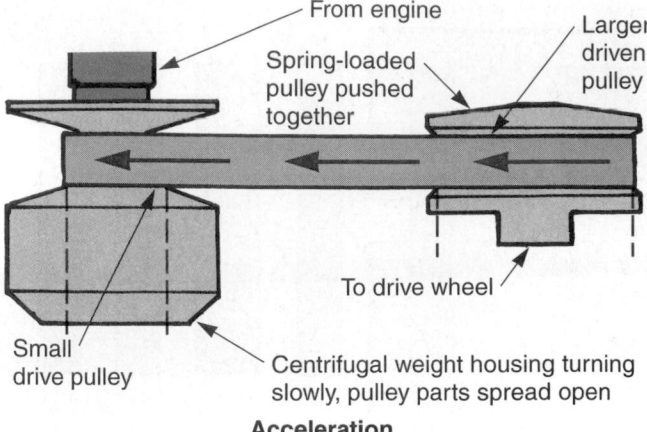

Acceleration

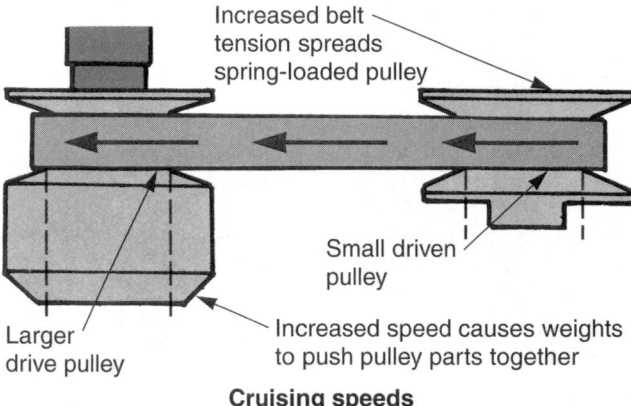

Cruising speeds

Figure 57-38. Basic action of a continuously variable transmission. Centrifugal weights in the housing cause pulley diameters to change with vehicle speed. Two drive belt mechanisms are commonly used. A—Upon initial acceleration, the drive pulley has a small diameter and the driven pulley has a larger diameter. This provides gear reduction for rapid acceleration. B—As vehicle and pulley speed increase, centrifugal weights push the drive pulley together, increasing its diameter. This increases belt tension, pulling the other pulley apart. As a result, the ratio decreases with increase in speed.

relatively constant. The engine does not have to accelerate through each gear. The result is an almost perfectly smooth increase in vehicle speed.

 Note
More information on continuously variable transaxles is given in Chapter 63, *Transaxle and Front Drive Axle Fundamentals.*

Complete Transmission Assemblies

Figures 57-39 through **57-41** show different types of automatic transmissions. Study each closely. As you look at each part, try to remember its function.

Refer to the service manual for more information on a specific automatic transmission. The manual will give hydraulic circuit diagrams, illustrations, and detailed operating and construction descriptions for the major components.

Summary

- An automatic transmission performs the same functions as a manual transmission but it "shifts gears" and "releases the clutch" automatically.
- The transmission case encloses the clutches, bands, gearsets, and inner ends of the transmission shafts.
- The extension housing slides over and supports the output shaft.
- The torque converter is a fluid clutch that performs the same basic function as a manual transmission's dry friction clutch.
- The impeller is the driving fan that produces oil movement inside the converter whenever the engine is running.
- The turbine is a driven fan splined to the input shaft of the automatic transmission.
- The stator is designed to improve oil circulation inside the torque converter.
- A one-way clutch allows the stator to turn in one direction but not the other.
- Torque multiplication refers to the ability of a torque converter to increase the amount of engine torque applied to the transmission input shaft.
- The stall speed of a torque converter occurs when the impeller is at maximum speed without rotation of the turbine.
- A lockup torque converter has an internal friction clutch mechanism for locking the impeller to the turbine in high gear.
- A planetary gearset consists of a sun gear, several planet gears, a planet gear carrier, and a ring gear.
- Automatic transmission clutches and bands are friction devices that drive or lock planetary gearset members.
- Servos are apply pistons that operate the bands.
- An accumulator is used in the apply circuit of a band or clutch to cushion initial application.
- The hydraulic system for an automatic transmission typically consists of a pump, pressure regulator valve, manual valve, vacuum modulator valve, governor valve, shift valves, servos, pistons, and valve body.

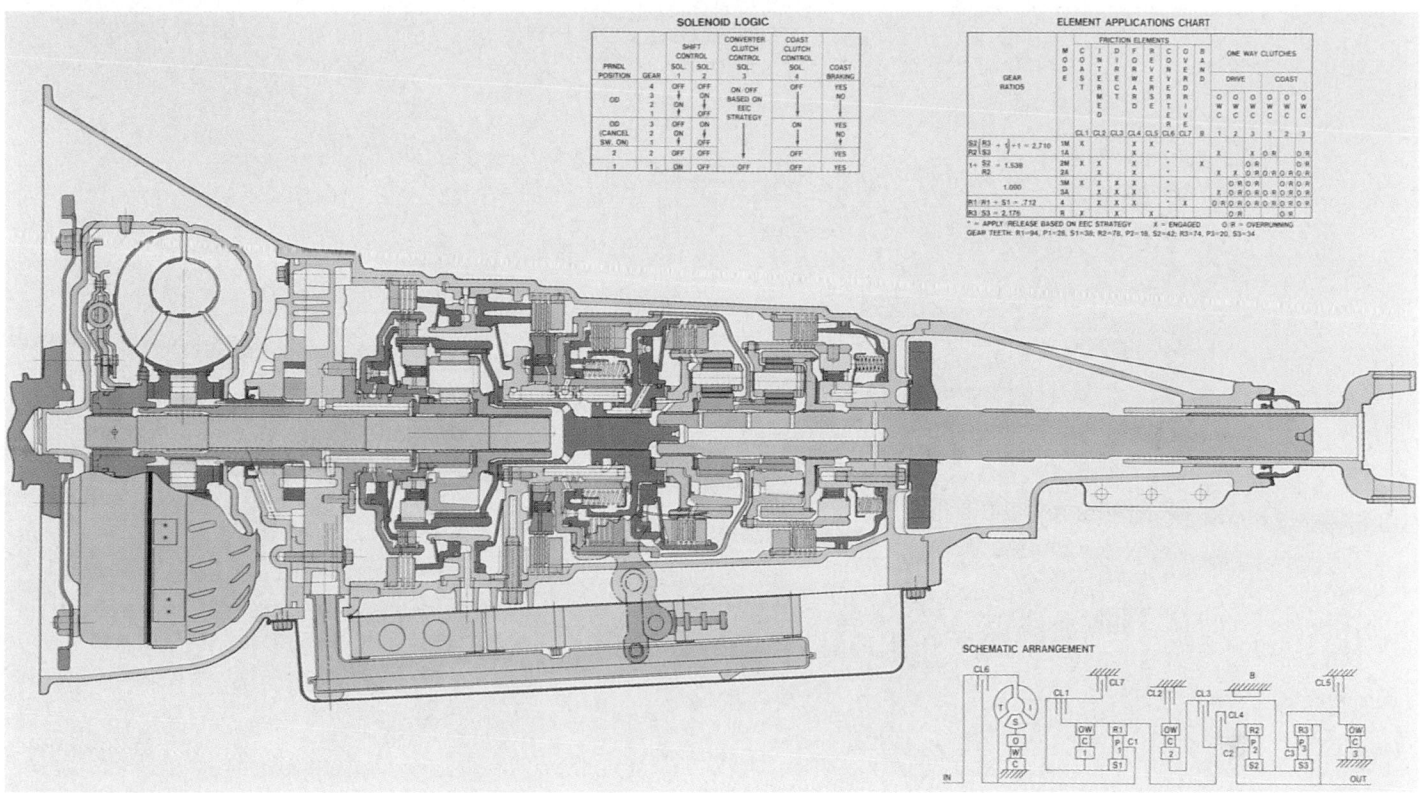

Figure 57-39. This side view shows the major parts of a modern automatic transmission. Note the charts that explain its operation. (Ford)

- Shift valves use pressure to operate the bands, servos, and gearsets.

- The valve body contains many of the hydraulic valves in an automatic transmission.

- Automatic transmission oil is a special type oil having several additives that make it compatible with the friction clutches and bands in the transmission.

- A parking pawl is used to lock the transmission output shaft and keep the car from rolling when not in use.

- Electronic transmission control uses sensors, actuators, and a computer to control shift points, torque converter lockup, downshifts, and other functions.

- A transmission solenoid is an actuator commonly found on modern electronically controlled automatic transmissions.

- A continuously variable transmission has an infinite number of driving ratios.

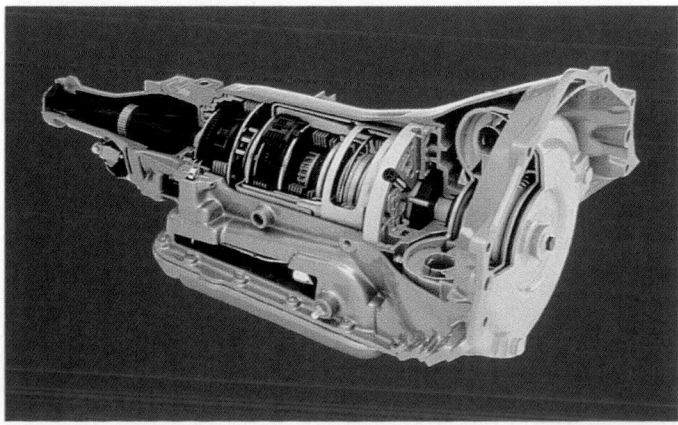

Figure 57-40. This is an electronically controlled automatic transmission. (Oldsmobile)

Important Terms

Automatic transmission	Torque converter
Input shaft	Impeller
Output shaft	Converter pump
Bell housing	Turbine
Transmission case	Stator
Oil pan	One-way clutch
Transmission pan	Torque multiplication
Extension housing	Stall speed

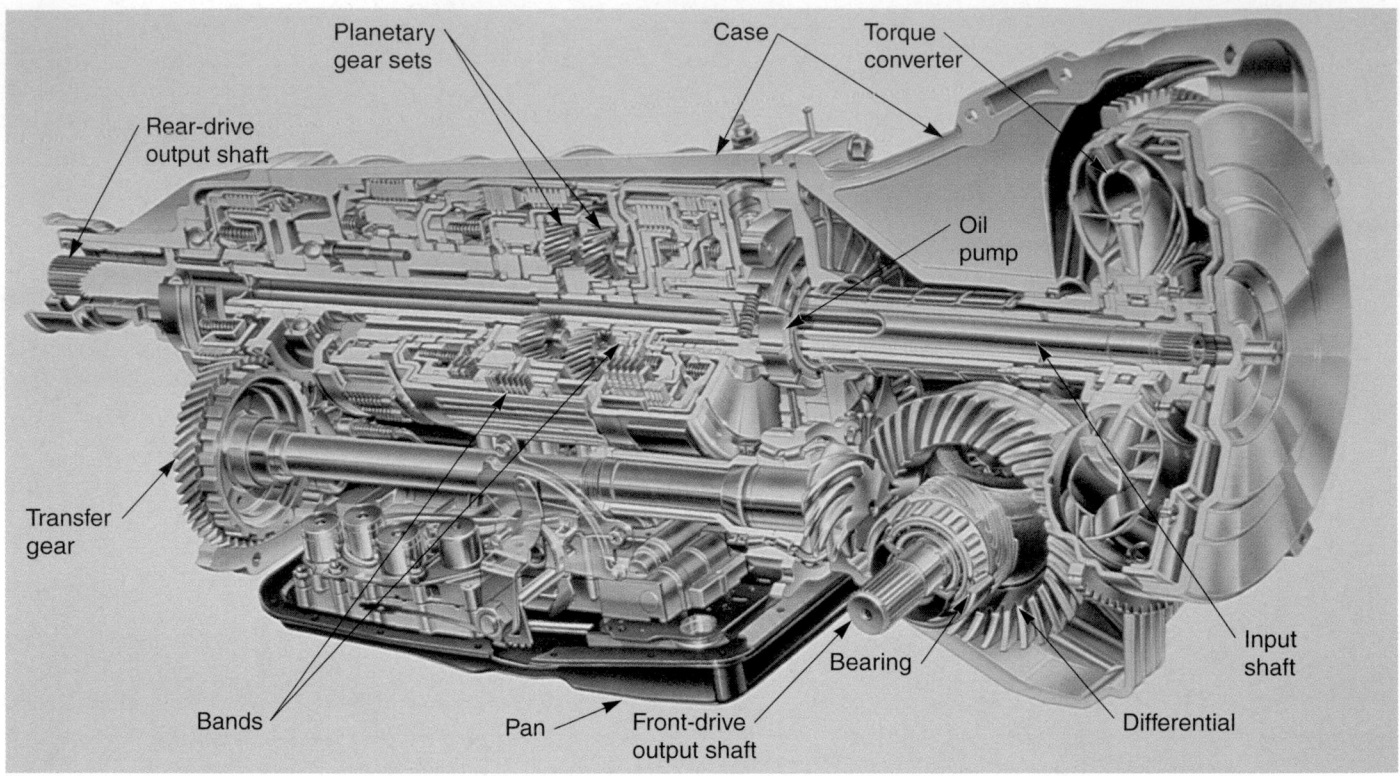

Figure 57-41. Cutaway view shows internal parts of an all-wheel drive type automatic transmission. Note that it has output shafts for front and rear wheels. A transfer gear and shaft send rear output shaft torque to front differential. (Subaru)

Lockup torque
  converter
Stator support
Stator shaft
Turbine shaft
Output shaft
Planetary gearset
Compound planetary
  gearset
Clutches
Bands
Multiple disc clutch
Clutch drum
Clutch cylinder
Clutch hub
Driving discs
Driven discs
Clutch apply piston
Pressure plate
Clutch release springs
Driving shell
Clutch shell
Servo piston
Band adjustment
  screw
Servo seals

Accumulator
Overrunning clutch
Hydraulic system
Automatic transmission
  oil (fluid)
Transmission oil cooler
Transmission oil pump
Pressure regulator
Manual valve
Vacuum modulator
  valve
Throttle valve
Governor valve
Shift valves
Balanced valves
Kickdown valve
Valve body
Parking pawl
Electronic transmission
  control
Shift schedule
Transmission control
  sensors
Transmission solenoids
Continuously variable
  transmission (CVT)

Review Questions—Chapter 57

Please do not write in this text. Place your answers on a separate sheet of paper.

1. List and explain the eight major parts of an automatic transmission.

2. An automatic transmission uses the following methods of transferring power.
   (A) Friction.
   (B) Fluids.
   (C) Gears.
   (D) All of the above.

3. Describe the four major housings or components of an automatic transmission.

4. A(n) _____ _____ is a fluid clutch that provides a means of coupling and uncoupling the engine and transmission.

5. Which of the following is *not* part of a torque converter?
   (A) Band.
   (B) Stator.
   (C) Impeller.
   (D) Turbine.

6. _____ _____ refers to the ability of a torque converter to increase the amount of engine torque applied to the transmission's input shaft.

7. Define the term "stall speed."

8. Why do many late-model vehicles use a lockup torque converter?

9. A planetary gearset consists of a _____ gear, several _____ gears, _____ gear _____, and a _____ gear.

10. List five functions of a planetary gearset.

11. Automatic transmission _____ and _____ are friction devices that drive and lock planetary gearset members.

12. Explain the operation of a clutch apply piston.

13. What is a servo?

14. A(n) _____ is used in the apply circuit of a band or clutch to cushion initial application.

15. An overrunning clutch locks in one direction and freewheels in the other. True or False?

16. List and describe the major parts of the hydraulic system in an automatic transmission.

17. List five functions of the oil pump in an automatic transmission.

18. This valve senses engine speed (transmission output shaft rpm) to help control gear shifting.
    (A) Vacuum modulator valve.
    (B) Governor valve.
    (C) Regulator valve.
    (D) Manual valve.

19. Engine oil is compatible with the friction material in an automatic transmission. True or False?

20. How do the shift valves work?

ASE-Type Questions

1. Technician A says an automatic transmission's valve body connects and disconnects the engine and transmission. Technician B says an automatic transmission's torque converter connects and disconnects the engine and transmission. Who is right?
   (A) A only.
   (B) B only.
   (C) Both A and B.
   (D) Neither A nor B.

2. Technician A says an automatic transmission's input shaft transfers power from the valve body to the gearsets. Technician B says an automatic transmission's input shaft transfers power from the gearsets to the drive shaft. Who is right?
   (A) A only.
   (B) B only.
   (C) Both A and B.
   (D) Neither A nor B.

3. While discussing the operation of a modern automatic transmission's valve body, Technician A says a shift lever is used to control valve body action. Technician B says valve body operation is controlled by a shift lever and sensors. Who is right?
   (A) A only.
   (B) B only.
   (C) Both A and B.
   (D) Neither A nor B.

4. Which of the following is *not* a basic component of a modern automatic transmission?
   (A) Pistons and servos.
   (B) Planetary gearsets.
   (C) Pinion seal.
   (D) Transmission bands.

5. Technician A says the oil pump in an automatic transmission is used to produce pressure to operate the transmission's hydraulic components. Technician B says an automatic transmission's oil pump is used to lubricate the transmission counter gears. Who is right?
   (A) A only.
   (B) B only.
   (C) Both A and B.
   (D) Neither A nor B.

6. Technician A says the torque converter housing bolts to the rear of the transmission case. Technician B says the converter housing bolts to the front of the automatic transmission's extension housing. Who is right?
   (A) A only.
   (B) B only.
   (C) Both A and B.
   (D) Neither A nor B.

7. Technician A says an automatic transmission's oil pan is located at the bottom of the extension housing. Technician B says an automatic transmission's oil pan fits over the valve body. Who is right?
    (A) A only.
    (B) B only.
    (C) Both A and B.
    (D) Neither A nor B.

8. Technician A says an automatic transmission's extension housing supports the transmission input shaft. Technician B says an automatic transmission's extension housing supports the transmission output shaft. Who is right?
    (A) A only.
    (B) B only.
    (C) Both A and B.
    (D) Neither A nor B.

9. Which of the following is another name for a torque converter impeller?
    (A) Stator.
    (B) Pump.
    (C) Turbine.
    (D) None of the above.

10. Technician A says the term "torque multiplication" refers to the ability of a torque converter to increase the amount of engine torque applied to the transmission output shaft. Technician B says the term "torque multiplication" refers to the ability of a torque converter to increase the amount of engine torque applied to the transmission input shaft. Who is right?
    (A) A only.
    (B) B only.
    (C) Both A and B.
    (D) Neither A nor B.

11. While discussing the construction of a modern automatic transmission, Technician A says the input and output shafts are the main shafts in this type of transmission. Technician B says the counter shaft and output shaft are the main shafts in an automatic transmission. Who is right?
    (A) A only.
    (B) B only.
    (C) Both A and B.
    (D) Neither A nor B.

12. Technician A says only one planetary gearset is commonly used in an automatic transmission. Technician B says an automatic transmission commonly uses two or more planetary gearsets. Who is right?
    (A) A only.
    (B) B only.
    (C) Both A and B.
    (D) Neither A nor B.

13. Technician A says a planetary gearset will act as a solid unit when two of its members are held. Technician B says a planetary gearset will act as a solid unit when one of its members is held. Who is right?
    (A) A only.
    (B) B only.
    (C) Both A and B.
    (D) Neither A nor B.

14. Technician A says "bands" are used in an automatic transmission to cause the gearsets to transfer power. Technician B says "clutches" are used in an automatic transmission to cause the gearsets to transfer power. Who is right?
    (A) A only.
    (B) B only.
    (C) Both A and B.
    (D) Neither A nor B.

15. Which of the following automatic transmission components senses engine load and determines when the transmission should shift to a higher gear?
    (A) Oil pump.
    (B) Vacuum modulator valve.
    (C) Pressure regulator.
    (D) None of the above.

Activities—Chapter 57

1. Disassemble an automatic transmission and identify the parts.

2. Demonstrate to the class how an automatic transmission works.

3. Design and set up a demonstration that will help explain the operating principle of an automatic transmission.

Chapter 58

Automatic Transmission Service

After studying this chapter, you will be able to:

- Troubleshoot an automatic transmission.
- Explain the types of problems common to an automatic transmission.
- Describe the tests needed to locate automatic transmission problems.
- Change automatic transmission oil and filter.
- Make basic external adjustments on an automatic transmission.
- Locate and repair automatic transmission leaks.
- Cite and observe safety rules while working on transmissions.
- Troubleshoot electronically controlled automatic transmissions.
- Remove and replace an automatic transmission.
- Correctly answer ASE Certification test questions about automatic transmission service.

Although automatic transmission service is usually done by specialists, it is very important that every technician have some knowledge of basic service methods. Problems with an automatic transmission can affect, or appear to affect, the operation of other vehicle systems. For example, a torque converter malfunction (inoperative stator or frozen lockup clutch) can appear to be an engine performance problem. The faulty converter could consume a tremendous amount of engine power or not allow engine speed to increase normally.

A technician without some training in automatic transmissions might think the engine lacked sufficient power. The untrained technician could be led to think the vehicle had a clogged fuel filter, a worn engine timing chain, or a similar engine performance problem.

This chapter will introduce the most important tests and service tasks performed on automatic transmissions. It will give you the essential information needed to understand more specific service manual instructions.

Automatic Transmission Identification

The exact type of transmission can usually be found in the VIN (vehicle identification number) or on a number on the transmission case. You can refer to the service manual to find out what the number code says about the transmission type.

Automatic Transmission Diagnosis

Several problems are common to an automatic transmission. A few of these are illustrated in **Figure 58-1.**

Automatic transmission slippage is often caused by a low oil level, misadjusted linkage, worn clutches or bands, or valve body problems. With partial slippage, the engine may briefly race as the transmission shifts to a higher gear. With severe slippage, the vehicle might not move.

Incorrect shift points are sometimes caused by a low oil level; a faulty vacuum modulator circuit; an engine performance problem; a damaged governor; a bad vehicle speed sensor circuit; or trouble with hydraulic valves, servos, or pistons. The transmission might shift too soon (engine lugs in higher gear) or too late (engine races in lower gear). In some cases, it may not shift at all (locked in one gear or fails to upshift).

Mushy shifts or *harsh shifts* are normally caused by the same types of problems that cause slippage and incorrect shift points. A mushy shift is noticed when the transmission takes too much time changing gears. A harsh shift is just the opposite. It occurs when the transmission changes gears to quickly, causing the vehicle to jerk during shifts.

A noisy transmission (whining, whirring, grinding) might result from an improper oil level, planetary gear troubles, damaged bearings, a faulty torque converter, loose components, or other problems.

If the abnormal sound occurs in every gear, the problem might be in the torque converter, the transmission oil

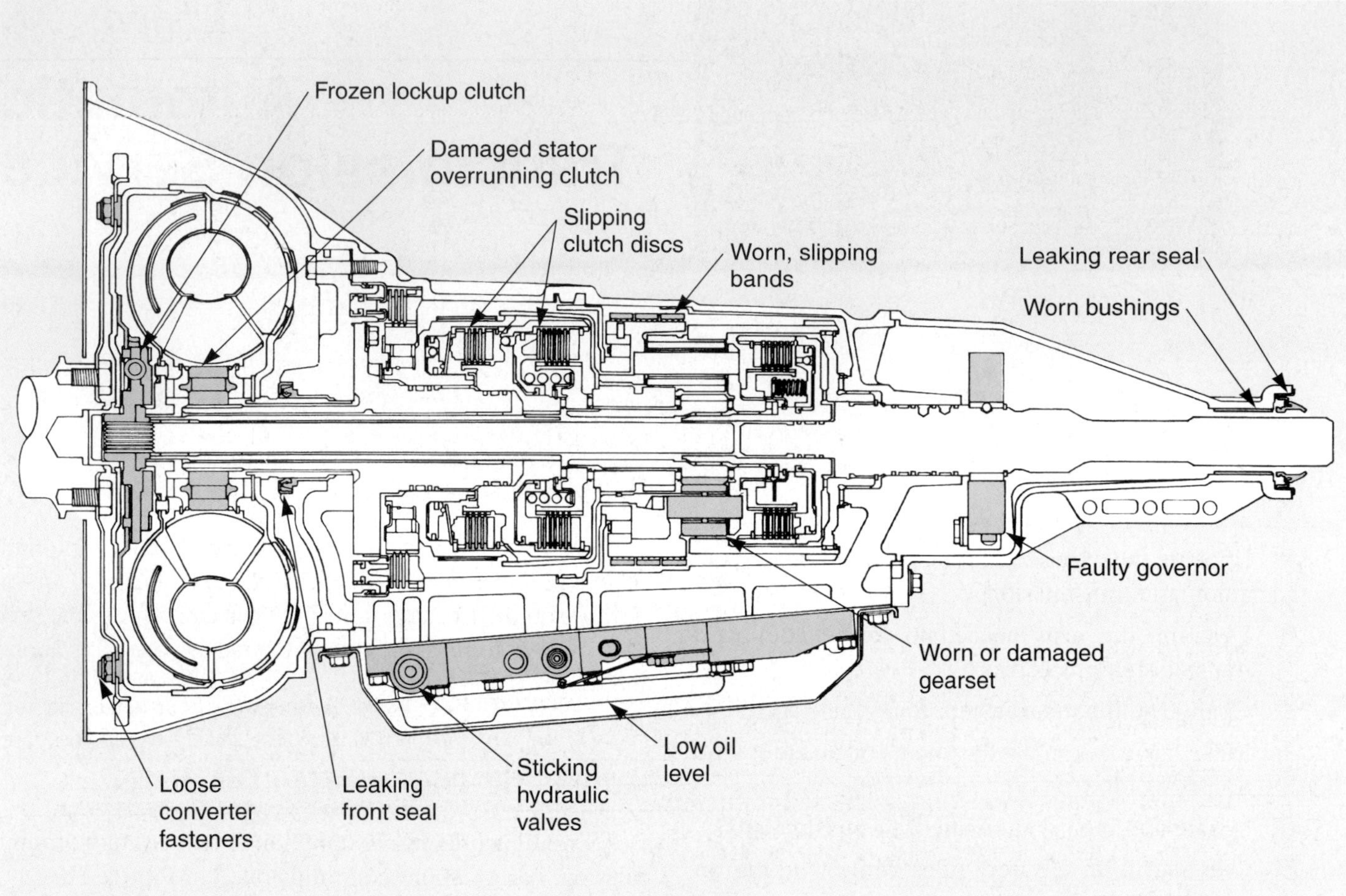

Figure 58-1. Visualize the kinds of symptoms these problems could cause in an automatic transmission.

pump, or another part common to all gears. If the transmission makes the noise in only one gear, parts operating in only that gear are at fault.

Tech Tip
When trying to find the source of transmission noises, make sure the problem is inside the transmission before attempting transmission repairs. Worn wheel bearings, dry universal joints, and engine problems can all produce noises that seem like they are coming from the transmission.

Follow an orderly procedure when diagnosing transmission troubles. Suspect the most common problems and causes. Begin with basic checks of the engine and transmission; then perform the recommended shop tests. See **Figure 58-2.**

Refer to a service manual troubleshooting chart when more specific problems cannot be located. The service manual will have symptoms applicable to the specific transmission model.

Warning
Servicing a transmission often requires the vehicle to be raised and supported. Make sure to use proper jacks and supporting tools. Stay clear of drivelines when testing a transmission. Remove rings and jewelry when working. To prevent shorts, disconnect the battery when it is not needed.

Preliminary Checks

Before road testing the car, there are several checks you should make: oil level, oil condition, linkage operation, and engine condition. You might find a problem and quickly fix the transmission.

Check Oil Level

Low oil level is normally caused by a leak and may cause the oil pump to pump air. Air is compressible and can prevent the oil from acting like a solid when pressurized in the hydraulic system. Pressure may build up too slowly, causing various engagement and shifting problems.

General diagnosis sequence

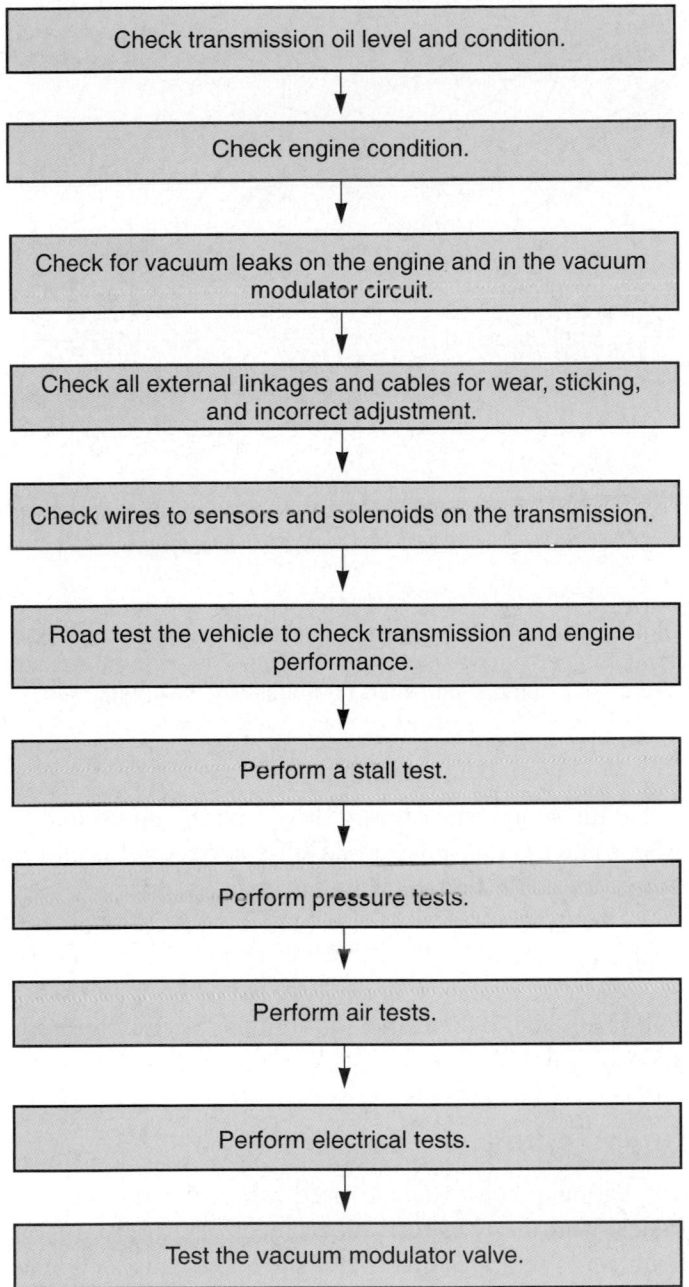

Check transmission oil level and condition.

↓

Check engine condition.

↓

Check for vacuum leaks on the engine and in the vacuum modulator circuit.

↓

Check all external linkages and cables for wear, sticking, and incorrect adjustment.

↓

Check wires to sensors and solenoids on the transmission.

↓

Road test the vehicle to check transmission and engine performance.

↓

Perform a stall test.

↓

Perform pressure tests.

↓

Perform air tests.

↓

Perform electrical tests.

↓

Test the vacuum modulator valve.

Figure 58-2. Procedure for diagnosing automatic transmission troubles.

A **high oil level** can produce symptoms similar to those produced by a low oil level. The transmission can churn the oil into a foam. The air bubbles in the foam will make the oil compressible, upsetting normal operation.

Check Oil Condition

Transmission oil condition can tell you a great deal about the condition of the transmission. Whenever you pull out the transmission dipstick, wipe the oil off on your finger or a white paper towel. Inspect the oil closely for signs of foreign matter or an unusual smell. See **Figure 58-3.**

Burned transmission oil will be black or dark brown and will have a burned odor. The darkness is normally caused by band and clutch friction material failure. The friction material has been slipping and overheating.

Usually, when you find burned transmission oil accompanied by slipping or shifting problems, serious damage has occurred. Typically, the transmission will need major repairs.

Milky transmission oil (white or nontransparent pink appearance) is normally caused by engine coolant mixing with the transmission oil. The oil cooler in the engine radiator is leaking and allowing antifreeze to enter the transmission lines. Coolant in the transmission oil can sometimes cause oil seals and friction material (clutches and bands) to deteriorate. The seals can swell and leak. The clutch and band material can soften and wear quickly.

Transmission oil varnish is evident when a light brown coating is found on the dipstick. The transmission oil has broken down, coating the internal parts of the transmission with a sticky, glue-like substance.

The oil varnish can cause a wide range of transmission problems. It can cause hydraulic valves, servos, and

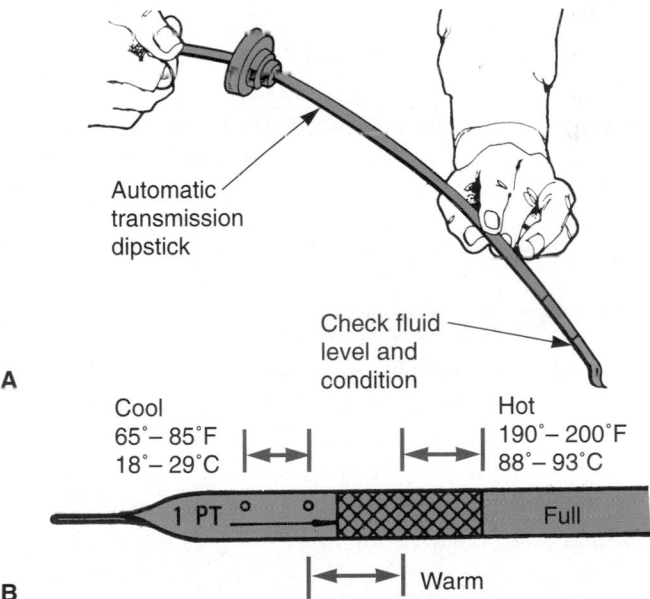

Note: Do not overfill. It takes only one pint to raise level from add to full with a hot transmission.

Figure 58-3. Always check oil level and condition when automatic transmission problems occur. A—Feel, inspect, and smell fluid. If burned or contaminated, problems are indicated. B—Make sure fluid is to proper level on stick. Follow instructions for each transmission. (AC-Delco and GMC).

pistons to stick. With an extreme case of varnish buildup, the transmission filter can clog.

With a light varnish buildup, changing the transmission fluid may prevent further problems. With a heavy varnish buildup, the transmission must often be rebuilt.

Warning
Let the vehicle cool before draining transmission oil. Hot oils can cause serious burns. Wear safety glasses or a face shield to protect your eyes.

Check Engine Condition

During your preliminary checks, always inspect the operating condition of the engine. The engine should start properly, idle smoothly, and perform normally under acceleration.

Check for Vacuum Leaks

When shift problems occur, check for vacuum leaks on the engine and in the vacuum modulator circuit. If a hose, fitting, or gasket is leaking, it can upset vacuum modulator operation.

The vacuum modulator senses engine vacuum (load) and helps to determine shift points. If the modulator is not receiving the correct amount of vacuum, the transmission will not shift gears properly.

Check Shift Linkage and Cables

Inspect the operation of the shift linkage. Move the shift lever through the gears while feeling the transmission click into each gear. Make sure the shift linkage is not worn.

Also, check the kickdown rod or cable, if used. Make sure it is free to move. Move the throttle lever while watching the kickdown rod. If the kickdown rod is locked in the full-throttle position, the transmission will be slow to upshift. See **Figure 58-4.**

Check Electrical Connections

Inspect any electrical connections on the transmission. Many late-model automatic transmissions have sensors and actuators on them. A poor connection can upset the operation of the transmission and related systems. Look for disconnected or frayed wires, corroded connectors, and other basic problems.

Road Testing

If you do not find the source of the transmission problem through your preliminary checks, road test the vehicle. Drive the vehicle while checking the general operation of the transmission: shift points, noises, etc. The test route should be a smooth road with little traffic. This will help reduce outside noise and distractions.

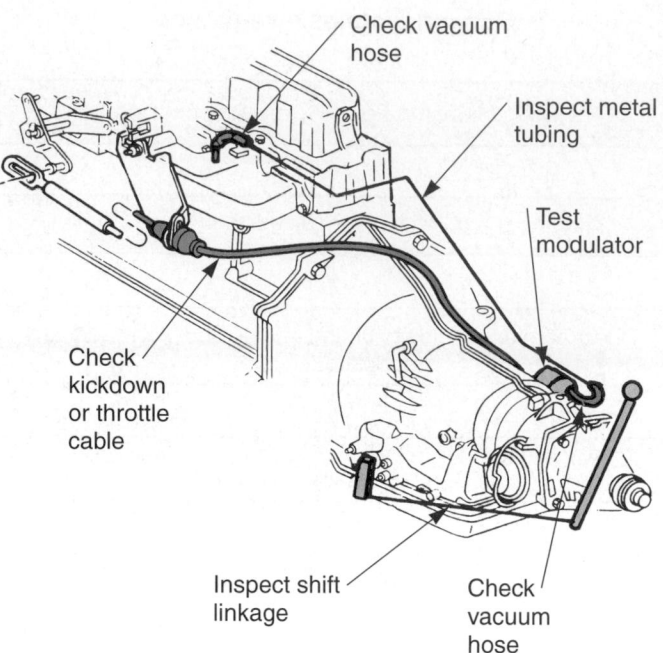

Figure 58-4. Check kickdown cable or rod, shift linkage, and vacuum line to modulator for problems. They are a frequent source of automatic transmission troubles. (Chevrolet)

With the transmission in drive, accelerate normally. Make sure the transmission upshifts correctly. Listen for noises in each gear and try to detect any slippage.

Then, manually shift the transmission through the gears. For example, if you think you heard a noise in second gear, manually shift into second (Low 2). This will give you more time to evaluate performance in second gear.

Shop Testing

Various shop tests are used when the preliminary checks and the road test fail to locate the transmission problem. There are three main shop tests: the stall test, the pressure test, and the air test.

Stall Test

A *stall test* can be used to detect transmission slippage or a malfunctioning torque converter. To perform a stall test, connect a tachometer to measure engine speed. Apply the emergency brake and press down firmly on the brake pedal. Start the engine and place the transmission in low. Refer to **Figure 58-5.**

When the accelerator is slowly pressed to the floor, the engine should "stall" (stop increasing) at a specified rpm. If the tachometer doesn't register the specified stall speed, a problem exists. The transmission should be tested in every gear position.

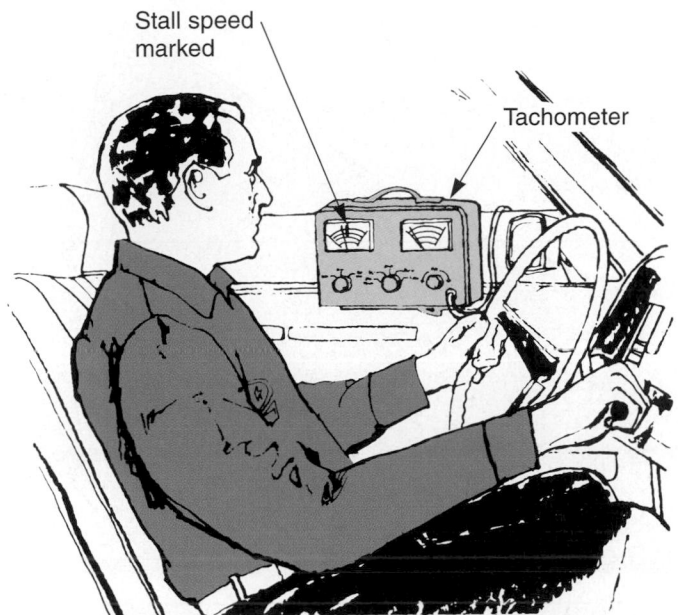

Stall speed marked

Tachometer

Figure 58-5. To do stall test, connect a tachometer to the engine. Apply the emergency brake and the foot brake. Place the transmission in an appropriate gear and accelerate. The maximum speed (stall speed) should meet specifications. It not, a problem is indicated. (Ford)

Clutch and band application chart

Low(D) (Breakaway) 2.74 RWD 2.69 FWD	Low(1) (Manual) 2.74 RWD 2.69 FWD	Second 1.54 RWD 1.55 FWD	Direct 1.0:1	Reverse 2.22 RWD 2.10 FWD
Rear clutch Drives front ring gear	**Rear clutch** Drives front ring gear	**Rear clutch** Drives front ring gear	**Rear clutch** Drives front ring gear	**Front clutch** Drives sun gear
Overrunning clutch Holds rear planet carrier	**Low and reverse band** Holds rear planet carrier	**Kickdown band** Holds sun gear	**Front clutch** Drives sun planet	**Low and reverse band** Holds rear carrier

Figure 58-6. A clutch and band application chart shows which components are slipping in each gear during a stall test. It is also useful during road and pressure tests. A service manual will give a chart for the specific transmission.

If engine speed is high, the transmission is slipping and could quickly overheat, burning friction material. Release the accelerator immediately to prevent further transmission damage. The problem may be low hydraulic control pressure, reducing the clamping force of the pistons and servos. Friction material could be worn.

If the stall speed is too low, the engine might have a performance problem or the torque converter stator could be inoperative.

Caution
Some automakers do not recommend a stall test. It could damage motor mounts, clutches, and bands.

The service manual will give more details on performing a stall test. You might need to disconnect the kickdown lever or other components to keep the transmission in each gear.

Figure 58-6 shows a transmission clutch and band application chart. It can be used to determine which parts are faulty when a transmission slips.

Pressure Tests

Pressure tests are used to determine whether oil pressure in the various transmission circuits is normal. Plugs are provided on the outside of the transmission case for pressure tests, **Figure 58-7.**

Follow this procedure to perform a pressure test:
1. Connect a 300 psi (2000 kPa) gauge to the line pressure port on the transmission. See **Figure 58-8.**
2. Run the engine until it is at operating temperature. Typically, the engine should be at curb idle.
3. While holding your foot on the brake, shift through all the gears while noting the reading on the pressure gauge.

Compare your pressure readings to those in a service manual The service manual will tell you which components might be leaking or frozen. The pressure gauge can then be installed in the other plug holes to check the pressure in other areas.

Air Test

An *air test* is used to further isolate problems in automatic transmission circuits. After removing the oil pan and valve body, a rubber-tipped nozzle is used to blow air pressure into the transmission passages. The air pressure used is normally 25–35 psi (170-240 kPa). Look at **Figure 58-9.**

The air pressure should activate the pistons, servos and other components. This will let you detect leaks, stuck components, bad check valves, and blocked passages.

To perform an air test, look up the exact procedure in the service manual. The manual will show which

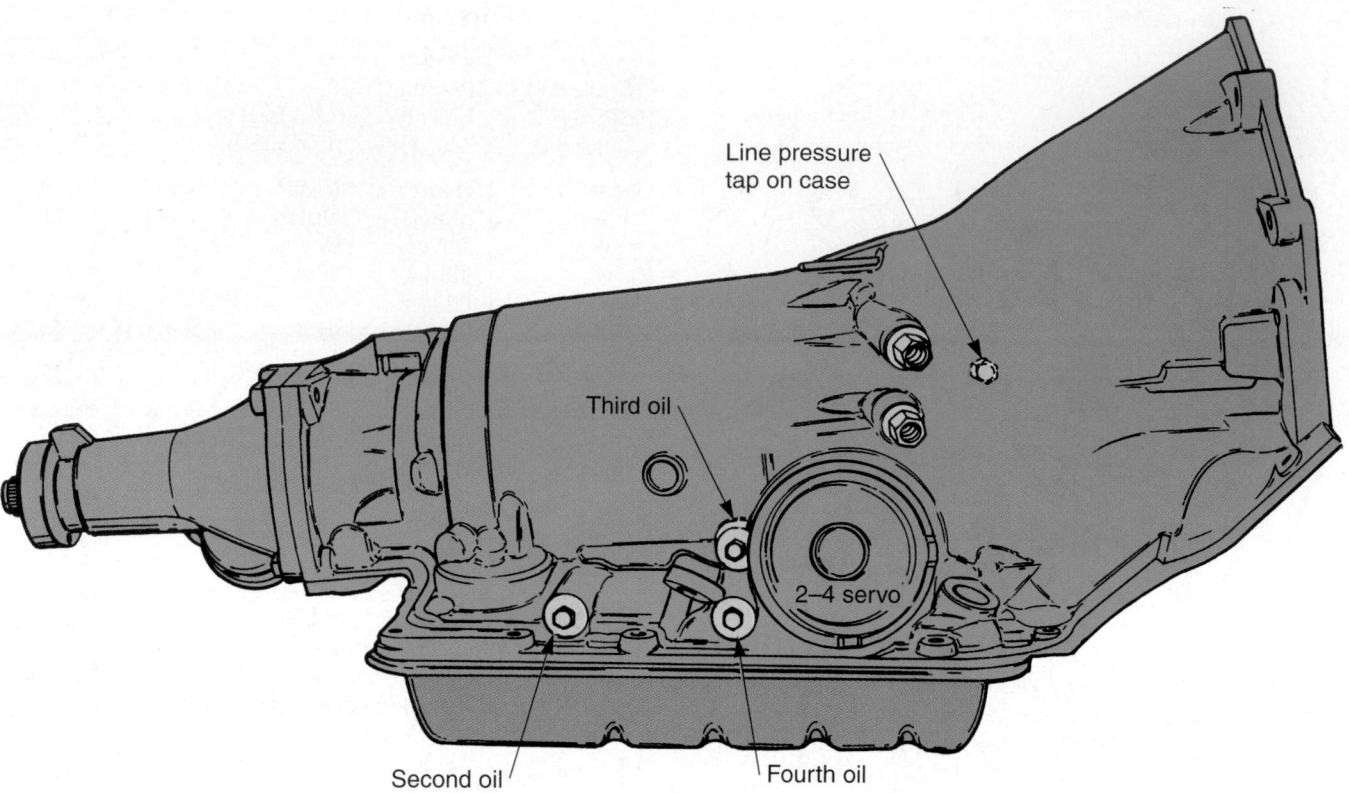

Figure 58-7. Plugs on the side of the transmission case cover ports to hydraulic components. To pressure test, install a pressure gauge in a port. Compare the pressure to specifications to determine the condition of various components. (General Motors)

transmission case passages lead to particular components. An example is shown in **Figure 58-10.**

When air pressure is blown into a piston or servo passage, a dull thud should be heard. This sound is made as the piston slides in its cylinder and bottoms. A hissing sound indicates a leak. No sound indicates a blockage or

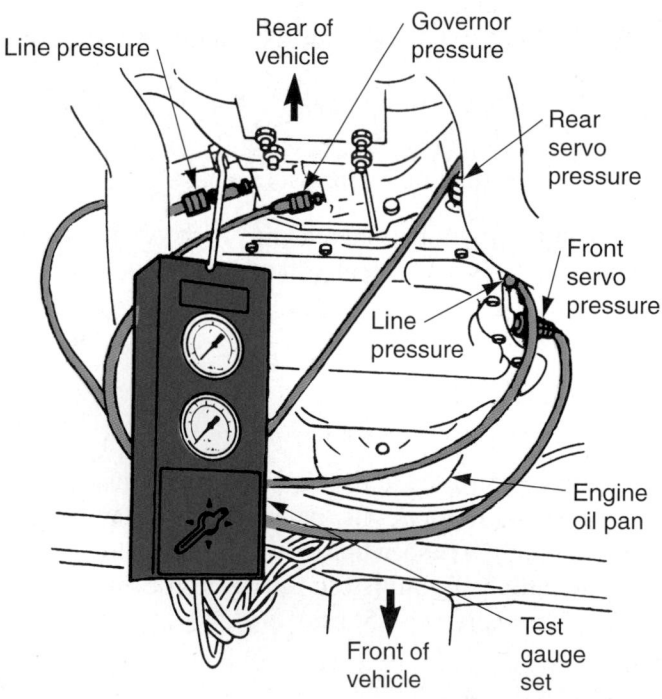

Figure 58-8. A special gauge set like this one or pressure and vacuum gauges are connected to the transmission. Gauges can also be placed in the passenger compartment for watching pressures during a road test. (Chrysler)

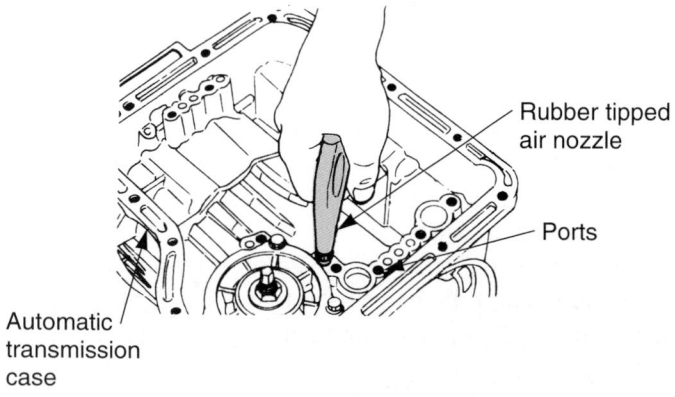

Figure 58-9. An air pressure test involves injecting air pressure into transmission passages. Air pressure should activate pistons and servos, making a clunking or thudding sound. Other sounds can be made when air is injected into ports to other parts. Refer to a service manual for details. (Nissan)

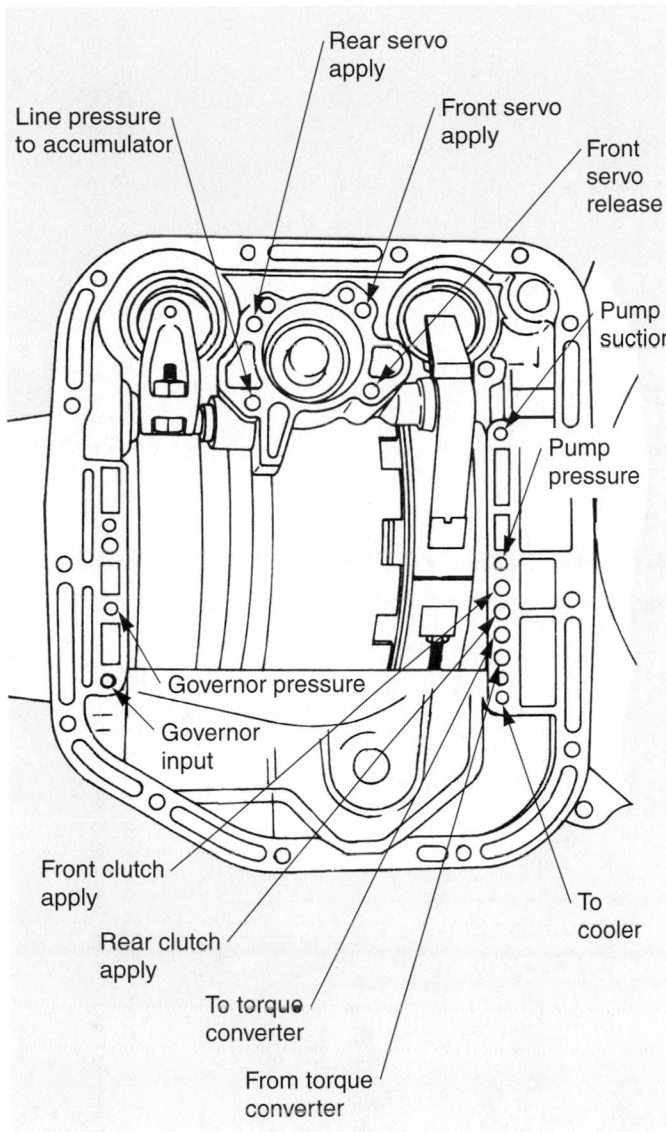

Line pressure
to accumulator

Rear servo
apply

Front servo
apply

Front
servo
release

Pump
suction

Pump
pressure

Governor pressure

Governor
input

Front clutch
apply

Rear clutch
apply

To torque
converter

From torque
converter

To
cooler

Figure 58-10. This service manual illustration shows the function of each port in the bottom of a specific transmission case. Look up this type of illustration in the service manual when performing an air test. (Chrysler)

a frozen piston. Some governors may produce a whistling sound under pressure.

Hydraulic Circuit Diagrams

Hydraulic circuit diagrams show how the oil passages inside an automatic transmission are connected to each component. A hydraulic circuit diagram is frequently used when tracing hard-to-find problems or when doing pressure or air tests.

One hydraulic diagram is often given for each gear position. This lets you know how oil, air, and electrical energy should flow through the transmission in each gear. Look at **Figure 58-11.**

Electrical Tests

Electrical tests on an automatic transmission involve checking sensors, actuators, and wiring for basic troubles. For example, if an automatic transmission has a solenoid, you can use an ohmmeter to check its windings. A vehicle speed sensor can be tested with a multimeter. Compare your test values to service manual specifications.

Note

For more information on working with electronic components, refer to the index.

Vacuum Test

A **vacuum test** is used to check the operation of the vacuum modulator valve. It measures the amount of supply vacuum reaching the valve. If the valve is inoperative or is not receiving a correct vacuum signal (broken, leaking, or kinked vacuum line), the transmission cannot shift properly.

To perform the test, connect a vacuum gauge to the modulator valve line with a T-fitting. Then, start the engine. The vacuum gauge should read within specifications (normally the full-engine vacuum).

If the modulator supply vacuum is low, there may be a vacuum leak, a blockage in the supply line, or a hole in the modulator diaphragm. If the supply vacuum is normal, you may need to adjust or replace the valve.

Automatic Transmission Maintenance

Maintenance is very important to the life of an automatic transmission. Automatic transmission oil, just like engine oil, can become filled with foreign matter after prolonged operation. Bits of metal, friction material, water, dust, and other substances can circulate through the hydraulic system, causing premature wear.

Note

If needed, refer to Chapter 10, *Vehicle Maintenance, Fluid Service, and Recycling,* for a review of automatic transmission oil service.

When checking transmission oil, apply the parking brake and block the wheels. Make sure the oil is between add and full on the dipstick with oil with the engine running at full operating temperature.

If the dipstick indicates low level, add more oil. Use the transmission oil type (Type-F, Dexron, etc.) specified in the service manual.

To add oil, insert a long funnel in the dipstick tube. Make sure the funnel is clean. Do not overfill the transmission; add a partial quart if needed. Add a little and recheck the dipstick.

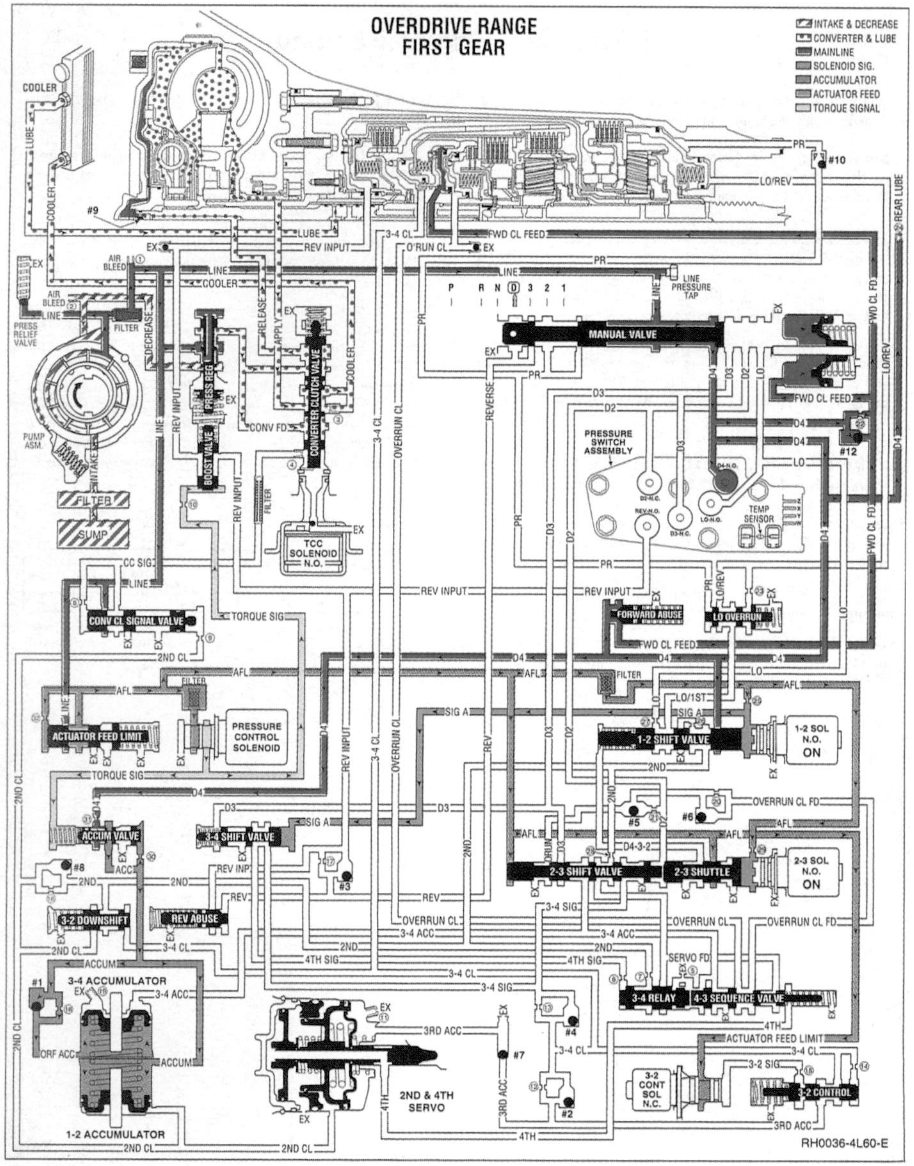

OVERDRIVE RANGE – FIRST GEAR

When the gear selector lever is moved to the Overdrive (D) position, from the Neutral position, the following changes occur to the transmissions hydraulic and electrical systems:

Manual Valve: Line pressure flows through the manual valve and fills the D4 fluid circuit. All other fluid circuits remain empty with the manual valve in the Overdrive position.

FORWARD CLUTCH APPLIES

Forward Clutch Accumulator Checkball (#12): D4 fluid pressure seats the ball and is orificed (#22) into the forward clutch feed fluid circuit. This orifice helps control the forward clutch apply rate.

Forward Clutch Piston: Forward clutch feed fluid pressure moves the piston to apply the forward clutch plates and obtain First gear.

Forward Clutch Accumulator Piston: Forward clutch feed fluid pressure moves the piston against spring force. This action absorbs some of the initial increase of forward clutch feed fluid pressure to cushion the forward clutch apply.

Forward Clutch Abuse Valve: D4 fluid pressure acts on the valve opposite of spring force. At engine speeds greater than idle, D4 fluid pressure increases and moves the valve against spring force (as shown). D4 fluid can then quickly fill the forward clutch feed fluid circuit, thereby bypassing the control of orifice #22 and providing a faster apply of the forward clutch. Otherwise, with increased throttle opening and engine torque, the clutch may slip during apply.

Pressure Switch Assembly (PSA): D4 fluid pressure is routed to the PSA and closes the normall open D4 fluid pressure switch. This signals the PCM that the transmission is operating in Overdrive range.

1-2 Shift Solenoid: Energized (ON) as in Neutral, the normally open solenoid is closed and blocks signal "A" fluid from exhausting through the solenoid. This maintains pressure in the signal "A" fluid circuit.

2-3 Shift Solenoid: Energized (ON) as in Neutral, the normally open solenoid is closed and blocks actuator feed limit (AFL) fluid from exhausting through the solenoid. This maintains AFL fluid pressure at the solenoid end of the 2-3 shift valve.

2-3 Shift Valve Train: AFL fluid pressure at the solenoid end of the 2-3 shift valve holds the valve train in the downshifted position against AFL fluid pressure acting on the 2-3 shuttle valve. In this position, the 2-3 shift valve blocks AFL fluid from entering the D432 fluid circuit. The D432 fluid circuit to open to an exhaust port past the valve.

1-2 Shift Valve: Signal "A" fluid pressure holds the valve in the downshifted position against spring force. In the First gear position the valve blocks D4 fluid from entering the 2nd fluid circuit.

Accumulator Valve: Biased by torque signal fluid pressure, spring force and orificed accumulator fluid pressure at the end of the valve, the accumulator valve regulates D4 fluid into accumulator fluid pressure. Accumulator fluid is routed to both the 1-2 and 3-4 accumulator assemblies in preparation for the 1-2 and 3-4 upshifts respectively.

Rear Lube: D4 fluid is routed through an orifice cup plug (#24) in the rear of the transmission case to feed the rear lube fluid circuit.

Pressure Control Solenoid: Remember that the pressure control solenoid continually varies torque signal fluid pressure in relation to throttle position and vehicle operating conditions. This provided a precise control of line pressure.

3-2 Control Solenoid: The PCM keeps the solenoid OFF in First gear and the normally closed solenoid blocks AFL fluid from entering the 3-2 signal fluid circuit.

SUMMARY

SHIFT SOLENOID	2-4 BAND	REVERSE INPUT CLUTCH	OVERRUN CLUTCH	FORWARD CLUTCH ASSEMBLY* APPLIED	FORWARD SPRAG CL. ASSEMBLY* HOLDING	3-4 CLUTCH	LO/REVERSE CLUTCH ASSEMBLY* HOLDING	LO/REV. CLUTCH
1-2 2-3								
ON ON								

B

OVERDRIVE RANGE FIRST GEAR

RH0036-4L60-E

A

Figure 58-11. A—A service manual will have hydraulic circuit diagrams similar to this one. The diagram shows how each part and passage interconnect in each gear position. This is very useful during diagnosis. B—This description applies to diagram on the opposite page. Read it as you trace the flow and action of parts in the diagram. (Chevrolet)

Changing Transmission Oil

Refer to service manual to determine when automatic transmission oil should be changed. A typical oil change interval is 15,000–20,000 miles (24,000–32,000 km).

Use the following procedure to change transmission oil and filter:

1. Drain the oil from the transmission pan. Be careful, the oil may be hot!
2. On some transmissions, the torque converter must also be drained before changing the filter. Refer to the service manual for information on draining a torque converter.
3. Remove the pan. Keep it level and lower it slowly to prevent spilling.
4. Inspect the pan for debris. Bits of metal or friction material may indicate transmission problems.
5. Clean the pan thoroughly.
6. Change the filter, **Figure 58-12.**
7. If sealer is used on the pan, make sure you do not use too much. If sealer is squeezed into the pan during pan installation, it may block oil passages and upset transmission operation.
8. Install the pan. Torque the pan bolts to specifications in a crisscross pattern.
9. Fill the transmission with the correct amount and type of oil. Capacity will vary depending on whether the torque converter was drained.
10. Start the engine. Shift the transmission through its gears. Check for leaks.

Warning

When draining automatic transmission oil, be careful not to be burned by hot oil.

Fluid Oil Leaks

Automatic transmission oil leaks commonly occur at the rear seal, front seal, oil pan gasket, extension housing gasket, and shift lever shaft seal. Whenever the oil level is excessively low, always inspect the transmission for leaks. Raise the car on a hoist and check for automatic transmission oil, **Figure 58-13.**

Normally, the leaking transmission oil will be cleanest near the source of the leak. The clean oil is normally red. The leaking oil tends to wash road dirt off the outside of the transmission case, causing it to darken as it flows along the case.

As mentioned in other chapters, you can add a leak detection dye to the transmission fluid. This will help you find a leak because of its bright color. In some cases, the dye agent will cause the transmission fluid to glow when illuminated with a black light.

If the leak is found at the rear seal, pan gasket, or extension housing, you can normally repair the leak without removing the transmission.

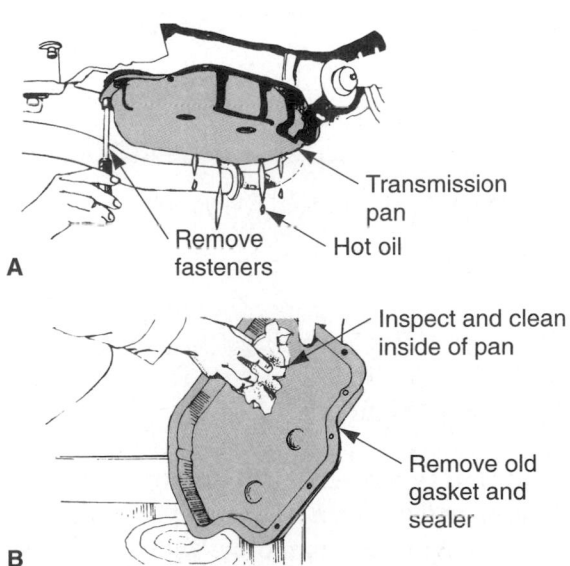

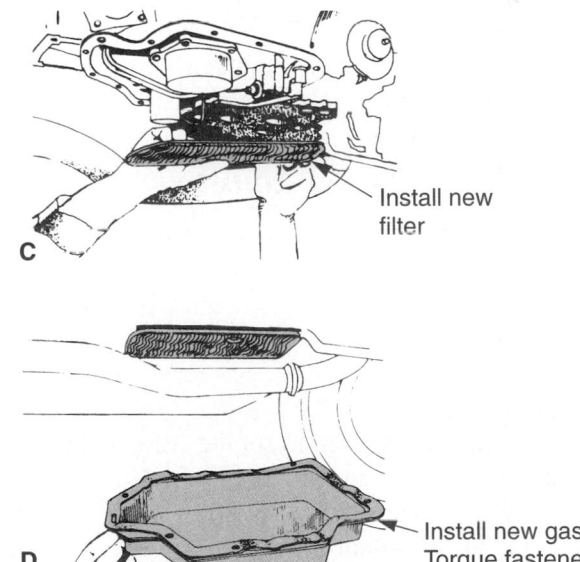

Figure 58-12. Replacing an automatic transmission filter. A—Be careful not to be burned by hot oil when removing the pan. B—Inspect the pan for debris before cleaning. C—Install a new filter. D—Install a new gasket or sealer and torque pan fasteners. (AC-Delco)

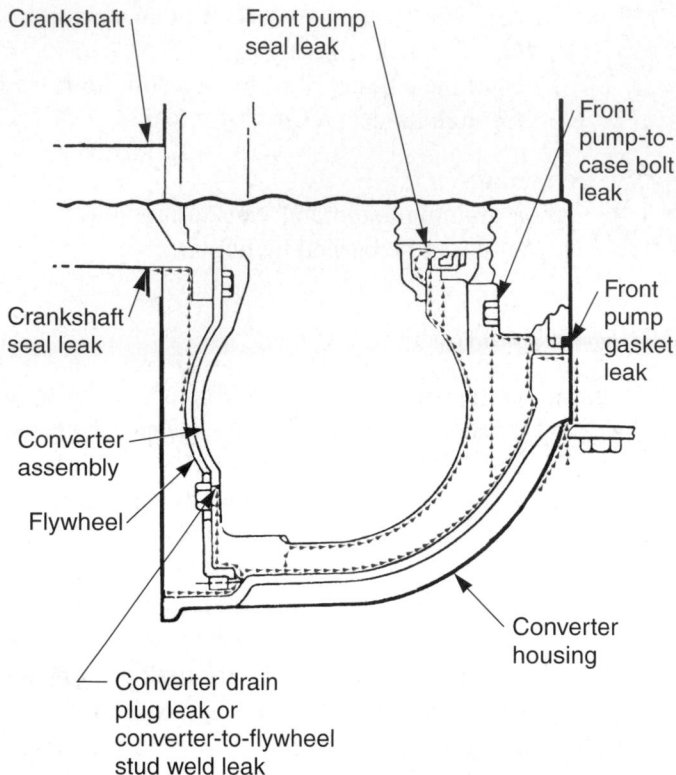

Figure 58-13. Diagnose automatic transmission leaks properly. Note how the engine rear main seal leak and transmission front seal leak will drip out of same location. (Ford)

Seal Replacement

To remove a rear transmission seal, use a seal removing tool. This is covered in Chapter 56.

To install the new seal, force it into place with a seal driving tool. Drive the seal into the housing squarely until seated. The front transmission seal is usually replaced in a similar manner. However, the transmission must be removed from the vehicle to replace the front seal.

Automatic Transmission Adjustments

There are several adjustments that can be made with the transmission installed in the vehicle. The most common of these are band adjustment, shift linkage adjustment, and neutral safety switch adjustment.

Band Adjustment

Band adjustment is needed to set the correct amount of clearance between the band and the drum. If the clearance is too large, the band could slip. If it is too tight, the band could drag and burn up.

To adjust a transmission band, loosen the locknut and turn the adjustment screw on the side of the transmission case. Turning the screw in (clockwise) normally tightens the band.

Typically, a service manual will require that you tighten the adjustment screw to a specific torque value. After torquing, the screw is backed off (one turn, typically) to provide band-to-drum clearance. The locknut is then tightened to lock the setting.

Many modern automatic transmissions do not need band adjustment. They have improved friction material that is extremely resistant to wear. If a band slips in a late-model transmission, major repairs are normally needed.

Shift Linkage Adjustment

Exact procedures for adjusting the shift linkage on an automatic transmission vary. Generally, make sure the lever going into the valve body is synchronized with the shift selector in the driver's compartment. If the selector is set to drive, the lever on the transmission must also be centered in the drive mode.

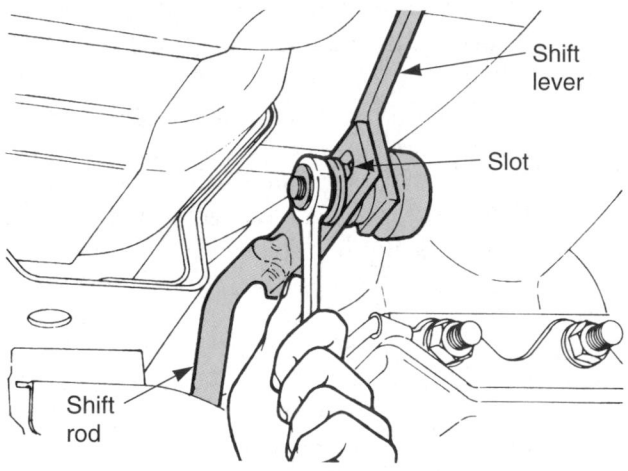

Figure 58-14. To adjust automatic transmission linkage, loosen the nut on the linkage rod. Position the driver's shift lever and the lever on the transmission in the same gear, usually park. Then tighten the nut and check. (Chrysler)

To adjust most shift linkages, a locknut is loosened on the shift rod. Then, the rod can be shortened or lengthened as needed, **Figure 58-14.**

Neutral Safety Switch Adjustment

Neutral safety switch adjustment is needed when the engine does not crank (turn over) with the shift selector in park. You might have to wiggle the shifter or hold it forward before the engine cranks. Either the linkage has worn, upsetting the neutral switch setting, or the switch itself is faulty.

The procedure for adjusting a neutral safety switch is fairly simple:

1. Loosen the bolts that hold the switch in place.
2. Position the shift selector in park.
3. While holding the ignition key to start, slide the switch toward the park position.
4. As soon as the engine begins to crank, release the key and tighten the switch bolts. Be careful not to move the switch while tightening.
5. Double-check your adjustment by starting the engine with the selector in park and neutral.

Testing the Neutral Safety Switch

To check for a bad (open or shorted) neutral safety switch, connect an ohmmeter across the switch, as shown in **Figure 58-15.**

The ohmmeter should read zero ohms (closed) with the switch in park and neutral. It should read infinite resistance (open) in all other gear positions.

A test light can also be used to check and adjust a neutral safety switch. Connect the test light to the output wire from the switch. The test light should glow when shift lever is in park or neutral but not in any gear position. If needed, loosen the switch mounting screws and slide the switch until the test light just turns on with the shifter in park. If the neutral safety switch still fails to turn the test light on when the shift lever is in park and neutral, replace the switch.

Backup Light Switch

Sometimes, the neutral safety switch also operates the backup lights. This side of the switch can also be checked with an ohmmeter or a test light. The backup light circuit should only have continuity (zero ohms or glowing test light) when the shift lever is in reverse.

Electronic Control System Service

Electronically controlled automatic transmissions can suffer conventional problems, as well as electronic problems. You must keep mechanical, hydraulic, and electrical malfunctions in mind when troubleshooting these transmissions.

To begin electronic diagnosis, mentally compare each component to its function. Try to think of which parts might be causing the trouble. For example, if the transmission does not shift from 1st to 2nd normally, consider the shift solenoid or the vehicle speed sensor. Both control the shift points of an electronically controlled automatic transmission.

A chart listing the function of major parts of an electronically controlled automatic transmission is given in **Figure 58-16.**

Most electronically controlled transmissions have a *limp-in mode,* which allows the transmission to function if there is an electronic control system failure. If the TCM detects something wrong in the electronic control circuit (bad sensor, solenoids, wiring), it switches into a preprogrammed limp-in mode. In the limp-in mode, the transmission or transaxle may have only second gear, reverse, neutral, and park. However, it will work well enough to allow the vehicle to be driven to a repair facility.

A scan tool can be used to analyze late-model electronic transmissions. If the malfunction indicator light is glowing, connect the scan tool to the computer's data link connector. Check for stored trouble codes or abnormal electrical values. If the scan tool finds a circuit problem, you will need to perform specific tests to verify the exact problem source.

 Note
For more information on using a scan tool, refer to Chapter 18, *On-Board Diagnostics and Scan Tools,* and Chapter 46, *Advanced Diagnostics.*

Automatic transmission solenoids and sensors can be located on the side of the transmission case or on the valve body. If on the case, you can test and replace them

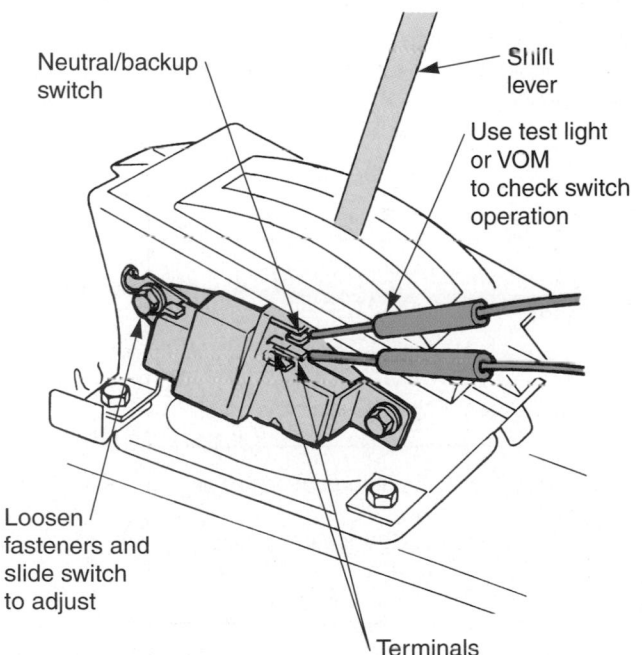

Figure 58-15. To adjust the neutral safety switch, loosen the fasteners. Turn key with transmission in park while sliding the switch forward. As soon as the engine cranks, lock down the switch. A test light or an ohmmeter can be used to check the switch while moving the shifter. (Honda)

Electronically Controlled Automatic Transmission

Components	Functions
Kickdown switch	Detects the accelerator pedal position depressed beyond full throttle valve opening.
Engine speed sensor	Detects the engine speed.
Neutral safety switch	Detects the shift lever position.
OD direct clutch speed sensor	Detects OD input shaft speeds from 1st through 3rd gears.
No. 1 and No. 2 solenoid valves	Control the hydraulic pressure applied to each shift valve, and control gear shift position and timing.
Cruise control ECM	Prohibits vehicle running in OD gear and lockup control when vehicle speed drops lower than a predetermined level of the auto drive set speed.
No. 3 solenoid valve	Controls the hydraulic pressure applied to the lockup clutch and controls lockup timing.
No. 4 solenoid valve	Controls hydraulic pressure acting on accumulator back chamber during gear shifting to smoothly engage clutches and brakes.
Stoplight switch	Detects the brake pedal depression.
No. 1 and No. 2 speed sensors	Detect the vehicle speed. Ordinarily, ECT control uses signals from the No. 2 speed sensor, and the No. 1 speed sensor is used as a backup.
Pattern select switch	Selects the shift and lockup timings by the Power mode or the Normal mode.
OD OFF indicator light	Blinks and warns the driver, while the OD main switch is pushed in, when the electronic control circuit is malfunctioning.
OD switch	Prevents up shift to the OD gear if the OD switch is off.
Engine and transmission ECM	Controls the engine and transmission actuators based on signals from each sensor.
Throttle position sensor	Detects the throttle valve opening angle.
Water temperature sensor	Detects the engine coolant temperature.

Figure 58-16. This chart shows components and their functions. If you understand what components are supposed to do, you will be able to analyze them when they fail to work. (Lexus)

without disassembly, **Figure 58-17.** If solenoids are located on the valve body, you must remove the pan to gain access to them, **Figure 58-18.**

For example, some modern automatic transmissions use an oil temperature sensor. The *oil temperature sensor* is used to compensate for changes in oil viscosity to maintain normal shifting. Cold transmission oil is thicker and affects transmission operation. The oil temperature sensor and TCM or PCM work together to allow for good cold transmission operation.

If the transmission acts up when cold or hot only, suspect the oil temperature sensor. If tests indicate a bad sensor, replace it, **Figure 58-19.** This same type of logic applies to other sensors and solenoids as well.

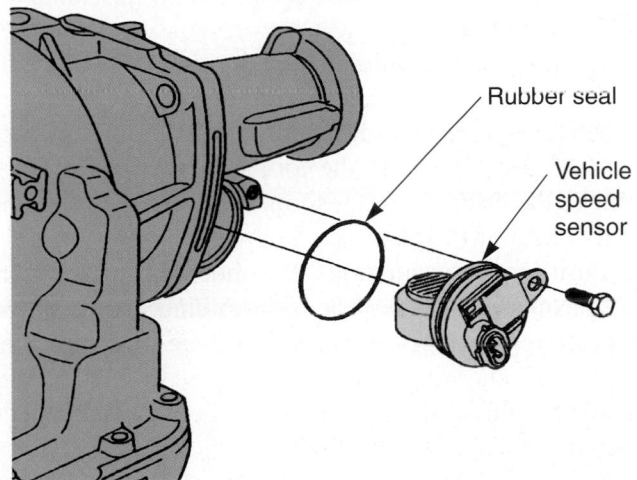

Figure 58-17. Some electronic parts, like this vehicle speed sensor, can be serviced without transmission removal and disassembly. (Chevrolet)

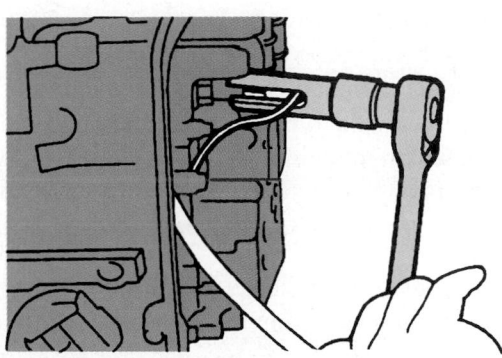

Figure 58-19. A special sensor socket is needed to remove some transmission sensors. Wires can fit through the side of this socket. (Mazda)

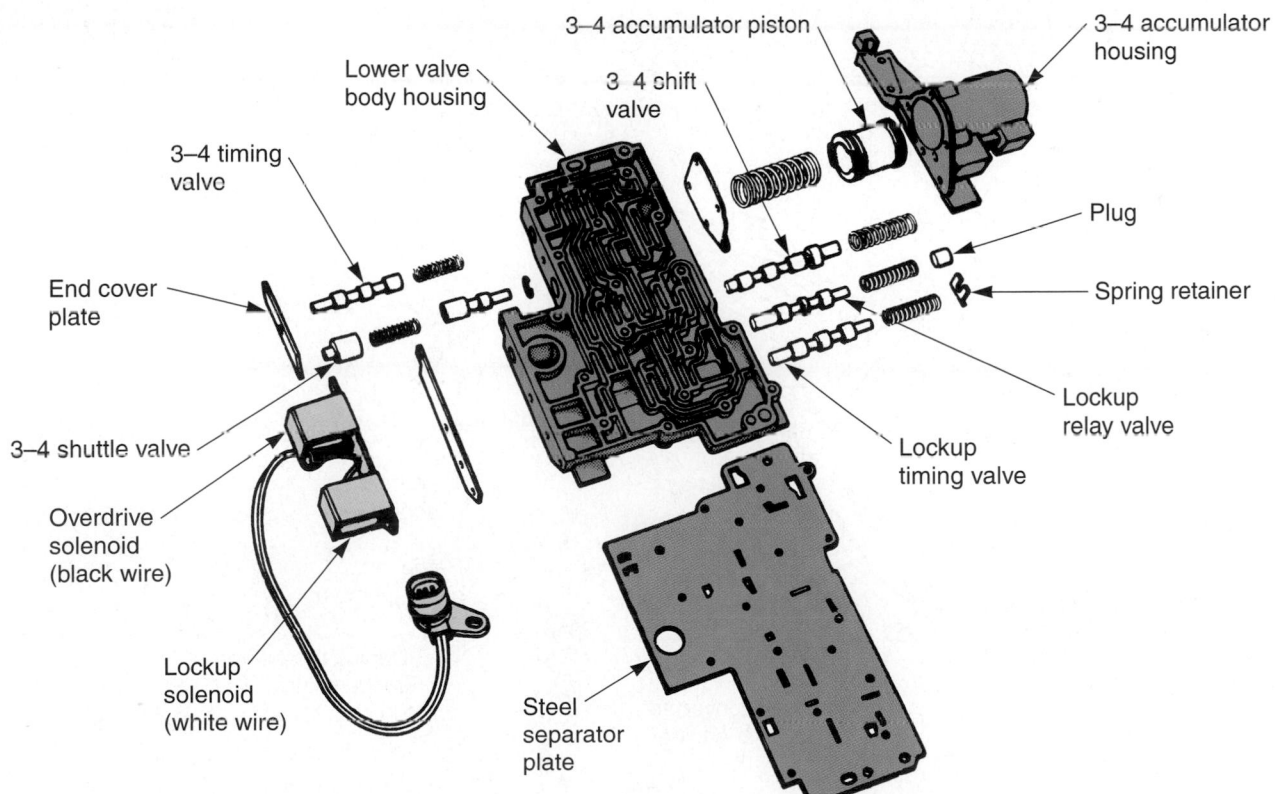

Figure 58-18. This transmission has overdrive and lockup solenoids mounted on the valve body. You would have to remove the pan to replace them. (Chrysler)

Major Transmission Service

If tests point to major transmission problems, the transmission may need to be removed, repaired or replaced, and reinstalled. Transmission designs vary greatly, so transmission repair is normally performed by specialized technicians. However, a general technician may be required to remove and replace automatic transmissions.

 Note
Automatic transmissions are very complex assemblies. Their construction and service methods vary widely. For this reason, it is not within the scope of this textbook to fully explain automatic transmission rebuilding. We will cover the types of tasks that a general technician might complete in a service facility.

Automatic Transmission Removal

When your tests indicate major internal problems, the automatic transmission must be removed from the car. Automatic transmission removal is similar to manual transmission removal. The service manual will contain removal instructions for the specific transmission.

To remove an automatic transmission, remove the drive shaft. Then remove all wires, lines, and cables between the transmission and the vehicle. Disconnect the rear transmission mount and remove any parts that obstruct removal (front of exhaust system, for example). See **Figure 58-20.**

Remove the transmission bell housing bolts that hold the transmission to the engine. The bolts along the top of the bell housing may be difficult to find. Remove the rock shield to gain access to the torque converter fasteners. Remove the fasteners so the torque converter is free from the flywheel or flexplate.

Position a transmission jack under the pan. Secure the transmission to the jack with holding chains or use the clamping device on the jack. Raise the jack to take the weight off the rear mount. Remove the rear cross member. Slide the transmission straight back and lower it from the vehicle, **Figure 58-21.**

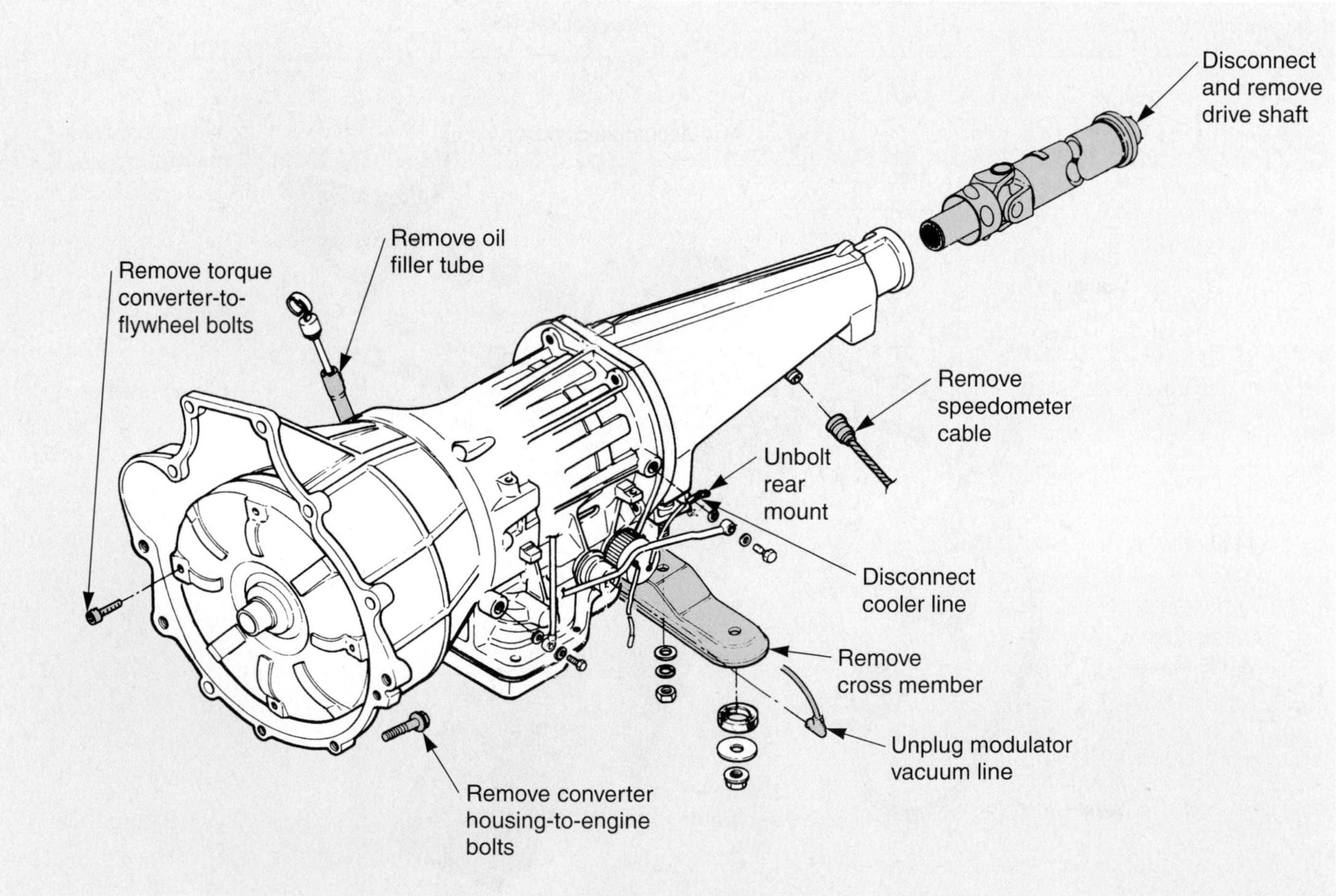

Figure 58-20. Disconnect parts before removing an automatic transmission. The procedure is similar to the removal of a manual transmission. (Mazda)

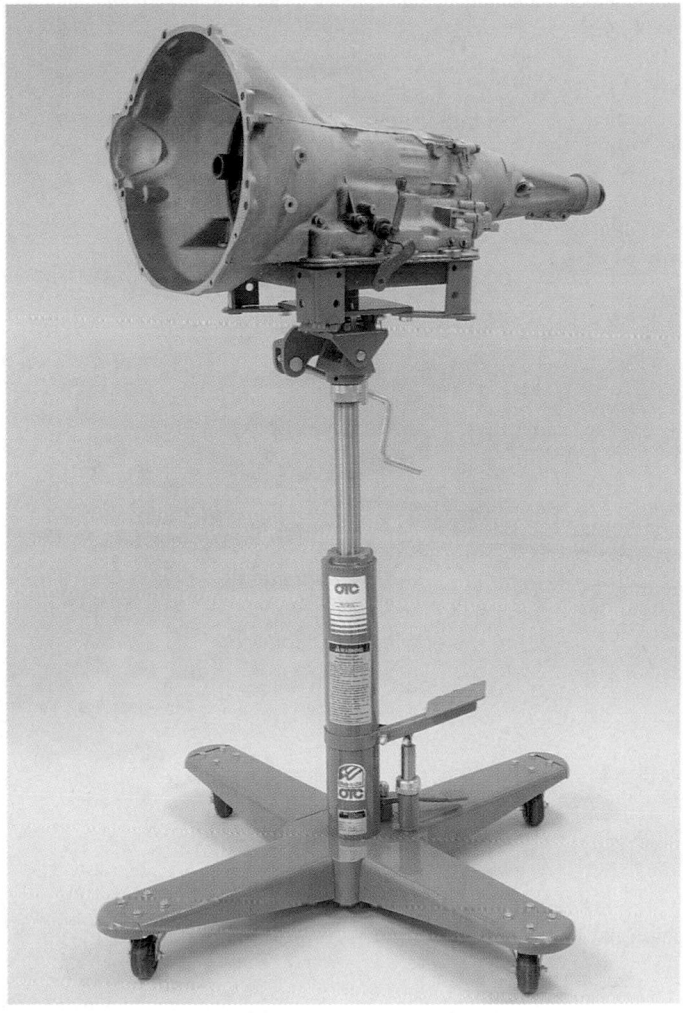

Figure 58-21. A transmission jack will safely hold and move a heavy automatic transmission. (OTC)

Figure 58-22. A specialized technician rebuilds a transmission.

Automatic Transmission Rebuild

A general technician can remove and install a transmission, but the damaged transmission must be sent to a specialized transmission shop for repair. Normally, the transmission is rebuilt. Specialized technicians perform the work, **Figure 58-22.**

When rebuilding an automatic transmission, the technician must refer to the service manual. It will give the needed details and specifications. See **Figure 58-23.**

After a transmission is rebuilt, it is tested using a transmission dynamometer, **Figure 58-24.** After testing, the transmission can be reinstalled.

Automatic Transmission Installation

Before installing an automatic transmission, make sure the torque converter is fully in place. Quite often, the converter will catch and can only be partially installed into the front pump. If you force the transmission against

the engine with the mounting bolts, severe transmission damage will result, **Figure 58-25.**

Raise or lower the transmission jack until the transmission centerline and crankshaft centerline are perfectly aligned. Then, push the transmission into the engine while wiggling the tailshaft. Make sure the torque converter studs fit through the flywheel holes.

When the transmission bell housing is touching the engine block, you can safely install the bell housing bolts and other parts. See **Figure 58-26.**

Summary

- Although automatic transmission service is usually done by specialists, it is very important that every technician has some knowledge of basic service methods.

- Automatic transmission slippage is often caused by low transmission oil level, misadjusted

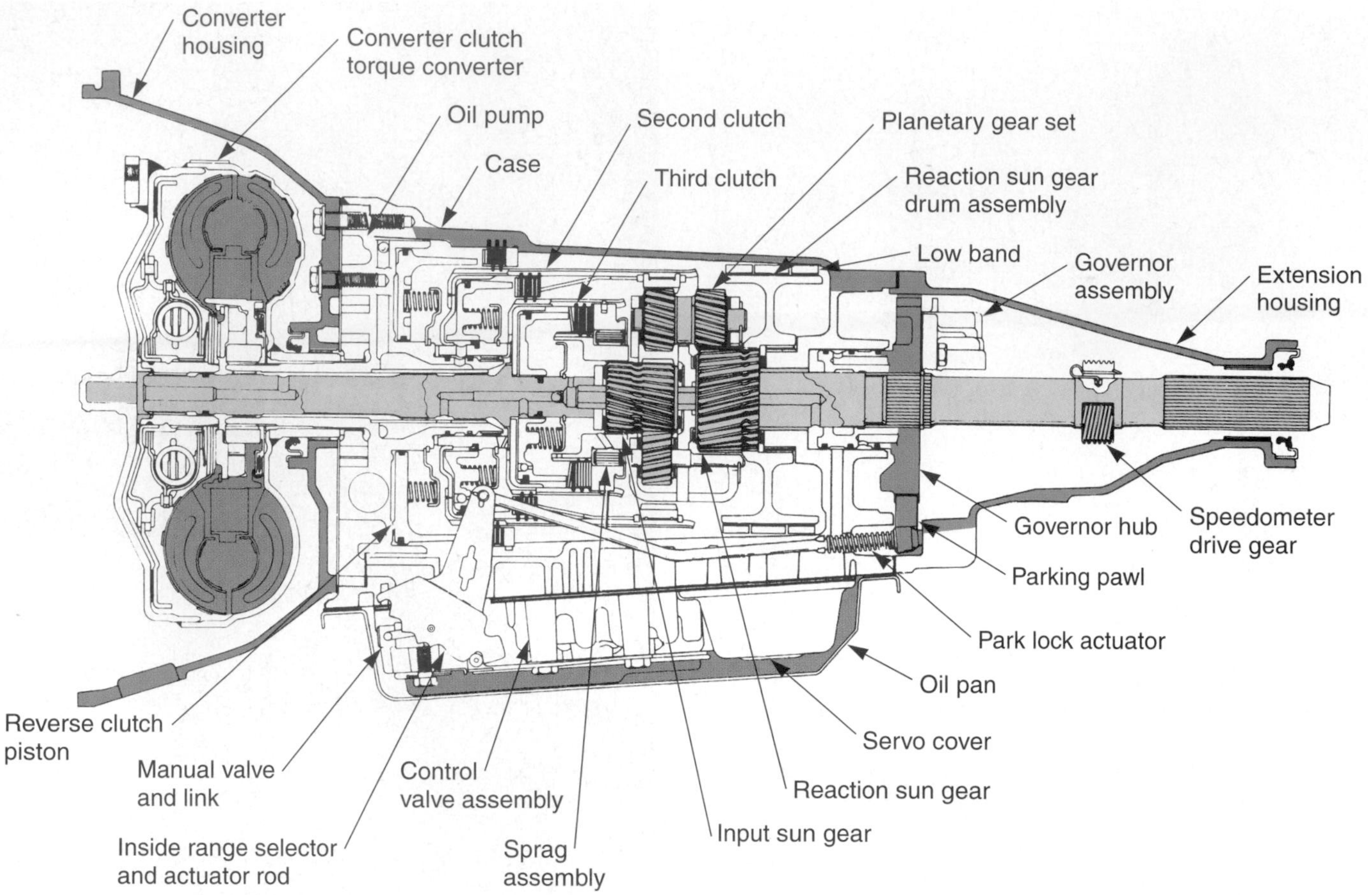

Converter housing

Converter clutch torque converter

Oil pump

Second clutch

Case

Third clutch

Planetary gear set

Reaction sun gear drum assembly

Low band

Governor assembly

Extension housing

Governor hub

Speedometer drive gear

Parking pawl

Park lock actuator

Oil pan

Servo cover

Reaction sun gear

Input sun gear

Sprag assembly

Control valve assembly

Inside range selector and actuator rod

Manual valve and link

Reverse clutch piston

Figure 58-23. Service manual illustrations aid in automatic transmission service. (General Motors)

Figure 58-24. Specialty transmission shops often have a dynamometer for testing a transmission before it is installed in a vehicle.

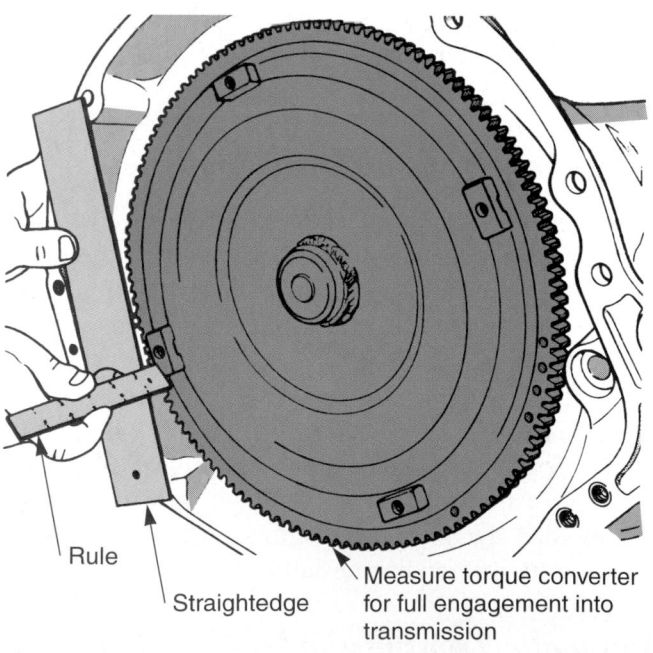

Rule

Straightedge

Measure torque converter for full engagement into transmission

Figure 58-25. A common, damaging mistake is to not have torque converter fully installed. Usually, the converter must slide over internal parts twice to be fully inserted. If not all the way in, the front pump and the converter can be damaged when the bell housing bolts are tightened. (Chrysler)

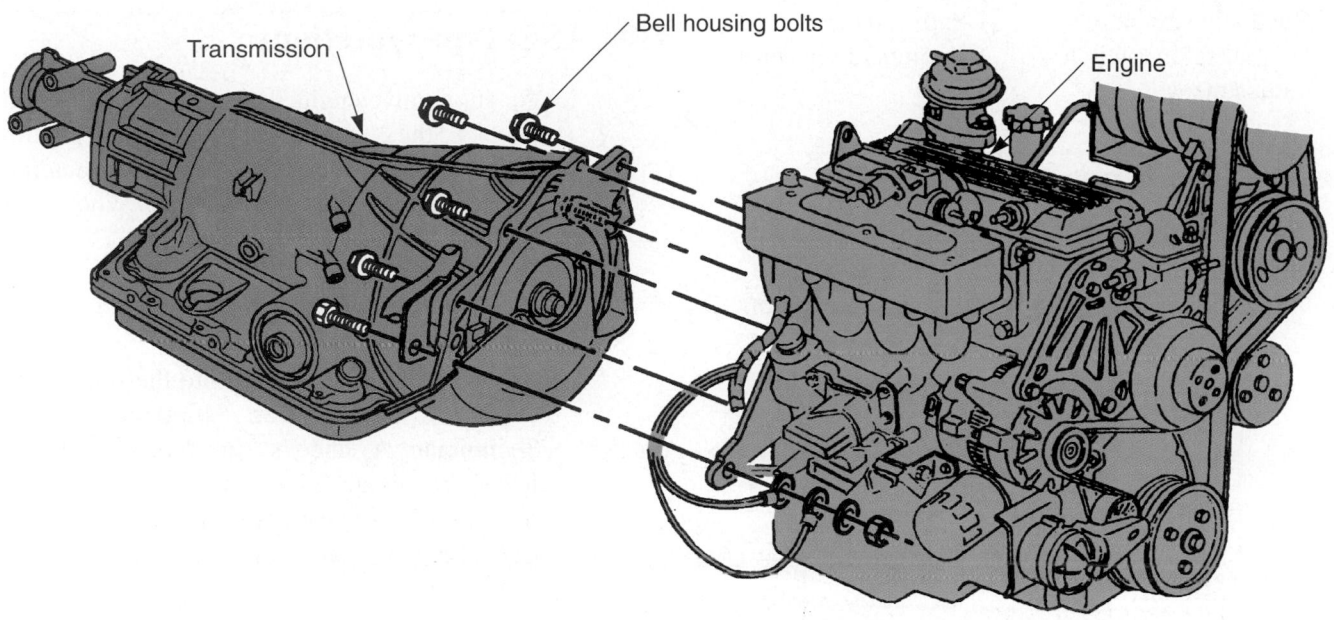

Figure 58-26. Part alignment is very important when installing a transmission. The transmission must be on same plane and engine. Also, torque converter bolts must align with flywheel holes. Make sure the transmission is fully against engine by hand before tightening the bell housing bolts. (General Motors)

linkage, worn clutches or bands, or valve body problems.

- Incorrect shift points are sometimes caused by a low oil level, faulty vacuum modulator circuit, engine performance problem, damaged governor, bad vehicle speed sensor circuit, or trouble with hydraulic valves, servos, or pistons.

- A noisy transmission (whining, whirring, grinding) may result from an improper oil level, planetary gear troubles, damaged bearings, faulty torque converter, loose components, or other troubles.

- Before road testing the car, there are several checks you should make: oil level, oil condition, linkage operation, and engine condition.

- Burned transmission oil will be dark or black, and have a burned odor.

- Milky transmission oil is normally caused by engine coolant mixing with the transmission oil.

- Transmission oil varnish is evident when a light brown coating is found on the dipstick.

- If you do not find the source of the transmission problem through your preliminary checks, road test the vehicle.

- Various shop tests are used when the preliminary checks and the road test fail to locate the transmission problem. These tests can be divided into three classifications: stall test, pressure test, and air test.

- Hydraulic circuit diagrams show how the oil passages inside an automatic transmission are connected to each component.

- Automatic transmission oil leaks commonly occur at the rear seal, front seal, oil pan gasket, extension housing gasket, and shift lever shaft seal.

- Transmission band adjustment is needed to set the correct amount of clearance between the band friction material and the drum.

- Electronically controlled automatic transmissions can suffer both conventional and electronic problems.

- A scan tool can be used to analyze late-model transmissions.

- The manufacturer's service manual will give accurate instructions on how to remove, disassemble, inspect, rebuild, and install the transmission.

Important Terms

Automatic transmission slippage	Milky transmission oil
Incorrect shift points	Transmission oil varnish
Mushy shifts	Stall test
Harsh shifts	Air test
Low oil level	Hydraulic circuit diagrams
High oil level	Electrical tests
Burned transmission oil	Vacuum test

Band adjustment
Neutral safety switch
  adjustment

Limp-in mode
Oil temperature sensor

Review Questions—Chapter 58

Please do not write in this text. Place your answers on a separate sheet of paper.

1. Which of the following can cause automatic transmission slippage?
   (A) Low oil level.
   (B) Misadjusted linkage.
   (C) Worn clutches or bands.
   (D) Valve body problems.
   (E) All of the above.

2. What is burned transmission oil and what does it tell you?

3. Engine vacuum leaks can affect automatic transmission operation. True or False?

4. What is a stall test?

5. _____ tests are used to determine whether oil pressure in the various circuits is normal.

6. _____ _____ _____ show how the oil passages inside the automatic transmission are connected to each component.

7. List thirteen steps you should carefully follow when servicing the oil and filter for an automatic transmission.

8. Where do automatic transmission oil leaks commonly occur?

9. Summarize the general adjustment of an automatic transmission band.

10. A car fails to crank (starting motor operation) when in park. It will only crank when the shift lever is in neutral. Technician A says that the neutral safety switch must be bad and should be replaced. Technician B says that the neutral safety switch could require adjustment. Who is correct?
    (A) A only.
    (B) B only.
    (C) Both A and B.
    (D) Neither A nor B.

ASE-Type Questions

1. An automotive engine lacks power. Technician A checks the engine for any malfunctions. Technician B checks the engine and automatic transmission for any malfunctions. Who is right?
   (A) A only.
   (B) B only.
   (C) Both A and B.
   (D) Neither A nor B.

2. An automobile is brought into the shop with an automatic transmission "slippage" problem. Technician A checks the transmission's oil level. Technician B inspects the condition of the transmission's valve body first. Who is right?
   (A) A only.
   (B) B only.
   (C) Both A and B.
   (D) Neither A nor B.

3. A car's automatic transmission shifts too soon. Technician A looks for a faulty servo. Technician B inspects the transmission's vacuum modulator circuit. Who is right?
   (A) A only.
   (B) B only.
   (C) Both A and B.
   (D) Neither A nor B.

4. An automatic transmission is producing an abnormal sound in all gears. Technician A checks the transmission's oil pump for problems. Technician B inspects the transmission's reverse shift fork. Who is right?
   (A) A only.
   (B) B only.
   (C) Both A and B.
   (D) Neither A nor B.

5. An automatic transmission has shifting problems. Technician A says low oil level can produce this problem. Technician B says high oil level can produce this problem. Who is right?
   (A) A only.
   (B) B only.
   (C) Both A and B.
   (D) Neither A nor B.

6. An automobile is brought into the shop with an automatic transmission shift problem. Technician A inspects the transmission's extension housing. Technician B looks for a faulty shift fork. Who is right?
   (A) A only.
   (B) B only.
   (C) Both A and B.
   (D) Neither A nor B.

7. An automatic transmission has burned oil and a slippage problem. Technician A believes serious transmission damage is indicated. Technician B thinks this condition requires simple band adjustment. Who is right?
   (A) A only.
   (B) B only.
   (C) Both A and B.
   (D) Neither A nor B.

8. An automatic transmission's oil has a "milky" color. Technician A checks to see if engine oil is leaking into the transmission. Technician B checks to see if engine coolant is leaking into one of the transmission lines. Who is right?
   (A) A only.
   (B) B only.
   (C) Both A and B.
   (D) Neither A nor B.

9. All of the following transmission troubleshooting procedures should be utilized before a diagnostic road test is performed except:
   (A) check transmission oil level.
   (B) check transmission linkage operation.
   (C) check engine operating conditions.
   (D) check synchronizer operation.

10. A stall test is being performed on an automatic transmission. Technician A places the transmission in high gear during this test. Technician B places the transmission in neutral during this test. Who is right?
    (A) A only.
    (B) B only.
    (C) Both A and B.
    (D) Neither A nor B.

11. Technician A says all automakers recommend performing a stall test on an automatic transmission to determine torque converter operating conditions. Technician B says some automakers do not recommend performing a stall test because possible transmission damage can result. Who is right?
    (A) A only.
    (B) B only.
    (C) Both A and B.
    (D) Neither A nor B.

12. Pressure tests are being performed on an automatic transmission. Technician A inserts the pressure gauge in the ports located on the outside of the torque converter to perform these tests. Technician B inserts the pressure gauge in the ports located on the outside of the transmission case to perform these tests. Who is right?
    (A) A only.
    (B) B only.
    (C) Both A and B.
    (D) Neither A nor B.

13. The check valves in an automatic transmission are possibly malfunctioning. Technician A uses an ohmmeter to check the operation of these valves. Technician B performs an air test to check the operation of these valves. Who is right?
    (A) A only.
    (B) B only.
    (C) Both A and B.
    (D) Neither A nor B.

14. An automatic transmission's oil level needs to be checked. Technician A checks the transmission's oil level at operating temperature with the engine running in neutral. Technician B checks the transmission's oil level when the oil is cool with the engine off and in neutral. Who is right?
    (A) A only.
    (B) B only.
    (C) Both A and B.
    (D) Neither A nor B.

15. Technician A normally turns an automatic transmission band adjustment screw clockwise to tighten the band. Technician B normally turns the adjustment screw counterclockwise to tighten the band. Who is right?
    (A) A only.
    (B) B only.
    (C) Both A and B.
    (D) Neither A nor B.

Activities—Chapter 58

1. Demonstrate the proper method of checking transmission oil level.

2. Check and diagnose the condition of the transmission oil for a vehicle in the shop for service.

3. Demonstrate the procedure for band adjustment.

Automatic Transmission and Transaxle Diagnosis

Condition	Possible Causes	Correction
Fluid leaks.	1. Defective gaskets or seals. 2. Loose bolts. 3. Porous or cracked case. 4. Leaking vacuum modulator diaphragm. 5. Overfilled transmission.	1. Replace defective parts. 2. Tighten bolts. 3. Repair or replace case. 4. Replace vacuum modulator. 5. Reduce fluid level.
Slipping in gear.	1. Low fluid level. 2. Clogged filter. 3. Stuck valve. 4. Burned holding members. 5. Misadjusted bands (when used). 6. Internal leaks.	1. Add fluid and check for leaks. 2. Replace filter. 3. Remove valve body and free sticky valves. 4. Disassemble transmission, replace burned holding members. 5. Adjust bands, recheck operation. 6. Disassemble transmission and correct leaks.
No upshifts or downshifts.	1. Misadjusted linkage. 2. Stuck governor. 3. Stuck valves. 4. Defective or disconnected vacuum modulator (when used). 5. Internal leaks.	1. Readjust linkage. 2. Remove and free sticky governor or replace. 3. Remove valve body and free sticky valves. 4. Replace modulator, check vacuum lines. 5. Disassemble transmission and correct leaks.
Noises.	1. Clogged filter. 2. Pump or torque converter defective. 3. Defective gears.	1. Replace filter. 2. Replace pump or torque converter. 3. Replace gears.

Chapter 59

Drive Shafts and Transfer Cases

After studying this chapter, you will be able to:

- Identify and describe the parts of a modern drive shaft assembly.
- Explain the functions of a drive shaft.
- Describe the different types of universal joints.
- List the different types of drivelines.
- Identify the major parts of a four-wheel-drive driveline.
- Explain the basic operation of a transfer case.
- Correctly answer ASE certification test questions that require a knowledge of drive shafts and transfer cases.

The term *driveline* generally refers to the parts that transfer power from a vehicle's transmission to its drive wheels. Front-engine, rear-wheel-drive vehicles, such as large luxury cars, pickup trucks, and full-size vans, have a driveline that contains a drive shaft. The *drive shaft* is a long shaft that transfers power from the transmission back to the rear axle assembly. The drive shaft has universal joints at its ends.

Front-wheel-drive and rear-engine or mid-engine vehicles do not use a transmission and a drive shaft. Instead, they use a *transaxle* (transmission-differential assembly) and two drive axle assemblies (swing axles). The short axle assemblies extend directly out of the transaxle to power the drive wheels.

Drive Shaft Assembly

A *drive shaft assembly* is illustrated in **Figure 59-1**. The drive shaft assembly typically consists of the:

- *Slip yoke*—Connects transmission output shaft to front universal joint.

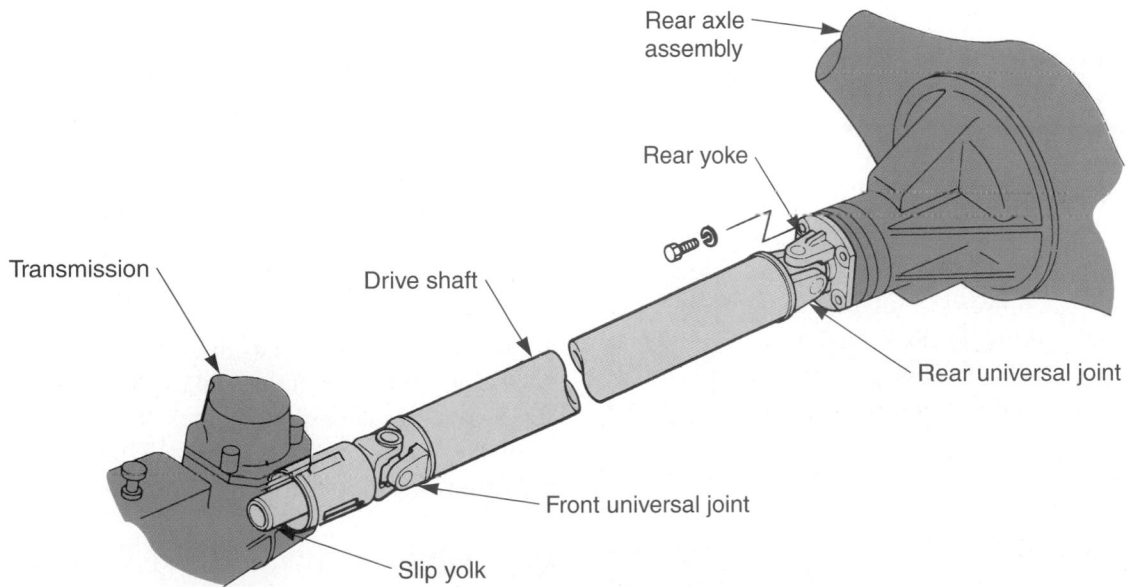

Figure 59-1. This drive shaft assembly connects the transmission output shaft with the rear axle assembly. Note the major parts. (Mazda)

- *Front universal joint*—Swivel connection that fastens slip yoke to drive shaft.
- *Drive shaft*—Hollow metal tube that transfers turning power from the front universal joint to the rear universal joint.
- *Rear universal joint*—Another flex joint connecting the drive shaft to rear yoke.
- *Rear yoke*—Holds rear universal joint and transfers torque to gears in the rear axle assembly.

This drive shaft design is the most common type used on front-engine, rear-wheel-drive vehicles. A few variations are sometimes used to satisfy special applications or to improve the smoothness of power transfer. These are discussed later in the chapter.

Functions of Drive Shaft Assembly

The drive shaft assembly has several important functions:

- Sends turning power from the transmission to the rear axle assembly.
- Flexes and allows vertical (up and down) movement of the rear axle assembly.
- Provides a sliding action to adjust for changes in driveline length.
- Provides smooth power transfer.

Operation of a Drive Shaft Assembly

When the vehicle is moving, the transmission output shaft turns the slip yoke. The slip yoke then turns the front universal joint, drive shaft, rear universal joint, and rear yoke on the differential. The differential contains gears that transfer power to the rear drive axles. The axles rotate the wheels.

Driveline Flex

When the tires strike a bump in the road, the rear suspension moves upward and the springs are compressed. When nonindependent rear suspension is used, this pushes the rear axle upward in relation to the body.

The *universal joints* let the driveline flex as the rear axle moves up and down. This protects the drive shaft from damage that could be caused by the bending. Refer to **Figure 59-2.**

Changes in Driveline Length

The movement of the rear axle assembly also causes the distance between the rear axle and the transmission to change. The slip yoke allows for this change of driveline length.

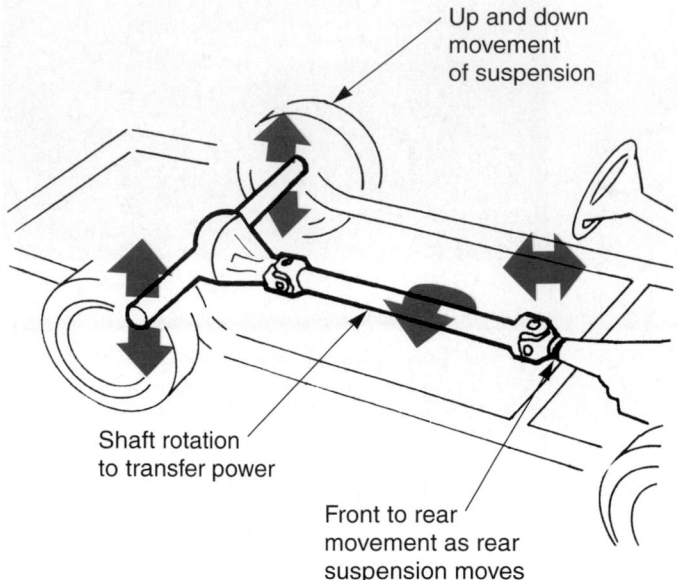

Figure 59-2. The drive shaft universal joints allow the driveline to bend and flex as the rear axle moves up and down over bumps in road. Most types also allow for length changes to allow suspension action.

Slip Yoke

The *slip yoke,* or *slip joint,* is splined to the transmission output shaft. It allows for changes in driveline length by sliding in and out of the transmission.

Cutaway views of a slip yoke are shown in **Figure 59-3.** Note how the inside of the slip joint has splines that fit over the transmission output shaft splines. This causes the two to rotate together. It also permits the yoke to slide on the splines.

The outer diameter of the yoke is machined smooth. This smooth surface provides a bearing surface for the bushing and oil seal in the transmission. The *extension housing bushing* supports the slip yoke as it spins in the transmission.

The *transmission rear seal* rides on the slip yoke and prevents fluid leakage from the rear of the transmission. The seal also keeps road dirt out of the transmission and off the slip yoke.

Normally, the outside of the slip joint is lubricated by the transmission oil. Transmission oil prevents bushing, yoke, and seal wear. Some yokes, however, require special heavy grease on their splines. The splines are sealed from the transmission lubricant. This keeps the oil or fluid from washing the grease from the splines.

Rear Yoke

The *rear yoke* is the yoke bolted to the outer end of the pinion (drive) gear on the rear axle assembly. It

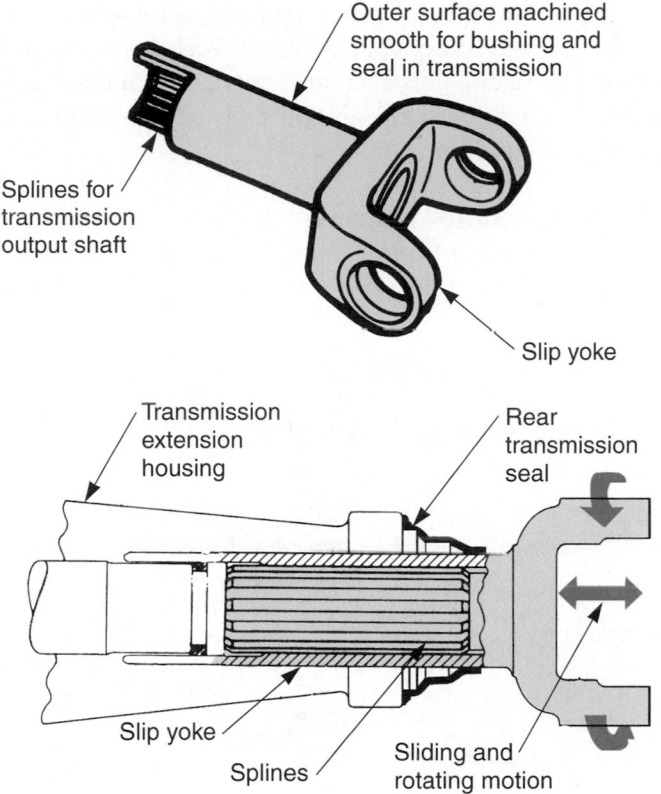

Outer surface machined
smooth for bushing and
seal in transmission

Splines for
transmission
output shaft

Slip yoke

Transmission
extension
housing

Rear
transmission
seal

Slip yoke

Splines

Sliding and
rotating motion

Figure 59-3. The slip yoke is splined to the transmission output shaft and fits inside the transmission extension housing. The rear transmission seal contacts the slip yoke. The yoke rides on a bushing in the extension housing. It rotates with the output shaft but is free to slide in and out of the transmission. (Ford)

transfers torque to the gears in the rear axle assembly. The rear universal joint is held by this yoke.

Drive Shaft

The *drive shaft,* or *propeller shaft,* is commonly a hollow steel tube with permanent yokes welded on each end. See **Figure 59-4.**

The tubular design makes the drive shaft very strong and light. Since the drive shaft spins much faster than the wheels and tires, it must be straight and perfectly balanced.

As mentioned, most front-engine, rear-wheel-drive vehicles use a single, one-piece drive shaft. However, a few large passenger cars and some pickup trucks have a two-piece drive shaft. This cuts down the length of each shaft, reducing driveline vibration.

 Tech Tip
Some drive shafts are designed for increased performance. Lightweight drive shafts are made of thin-wall aluminum with longitudinally aligned graphite fibers for added strength.

Drive Shaft Balance

In high gears, the drive shaft spins at the same speed as the engine. Therefore, the drive shaft must be perfectly balanced, with its weight evenly distributed around its centerline. If the drive shaft is not balanced, the shaft could vibrate violently.

Drive shaft balancing weights are frequently welded to the shaft to reduce vibration. The drive shaft is spun on a balancing machine at the factory. If needed, small metal weights are attached to the light side of the shaft. This counteracts the heavy side to smooth operation. See **Figure 59-5.**

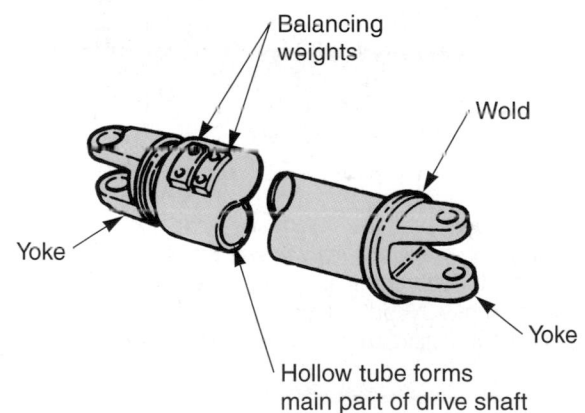

Balancing
weights

Weld

Yoke

Yoke

Hollow tube forms
main part of drive shaft

Figure 59-5. A drive shaft is normally a hollow tube with yokes welded to each end. Balancing weights are welded to the shaft to prevent vibration. (Ford)

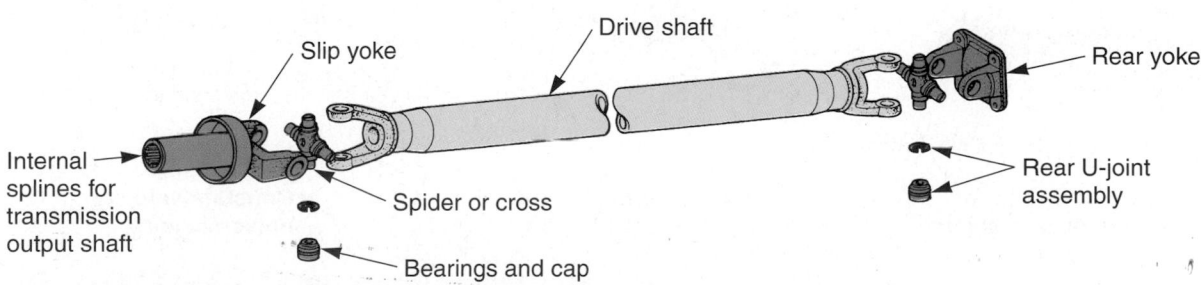

Slip yoke

Drive shaft

Rear yoke

Internal
splines for
transmission
output shaft

Spider or cross

Bearings and cap

Rear U-joint
assembly

Figure 59-4. A typical drive shaft assembly. Note the basic components. (Toyota)

Sometimes, the drive shaft is equipped with a ***drive shaft vibration damper.*** This is a large, ring-shaped weight mounted on rubber. The vibration damper helps keep the shaft spinning smoothly by absorbing torsional (twisting) vibration. See **Figure 59-13.**

Universal Joints

A ***universal joint,*** or ***U-joint,*** is a swivel connection capable of transferring a turning force between shafts at an angle to one another. A simple universal joint is made of two Y-shaped yokes (knuckles) that are connected by a cross (spider). Bearings on each end of the cross allow the yokes to swing into various angles while turning. See **Figure 59-6.**

Today, most drive shafts use two universal joints. However, extra universal joints are sometimes needed in very long drive shafts.

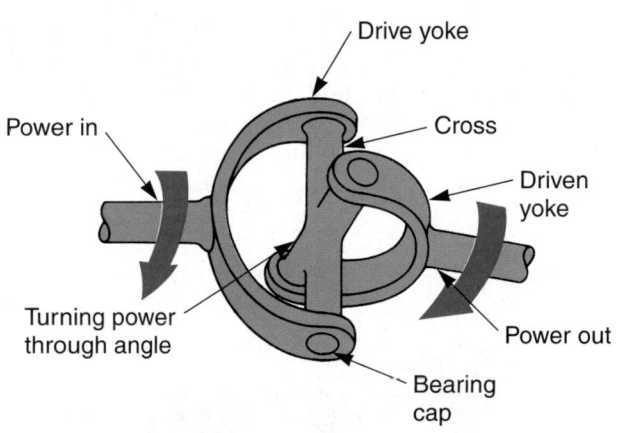

Figure 59-6. A universal joint will swivel to allow for changes in transmission-to-differential alignment. Two yokes are connected to a central cross. Needle bearings fit between the yokes and the cross. Each yoke can be swiveled in relation to the other.

There are three common types of automotive drive shaft universal joints used on rear-wheel-drive vehicles: cross-and-roller joints (cardan joints), constant velocity joints (double-cardan joints), and ball-and-trunnion joints. Look at **Figure 59-7.**

Cross-and-Roller Universal Joint

The ***cross-and-roller joint,*** also called a ***cardan universal joint,*** is the most common type of drive shaft

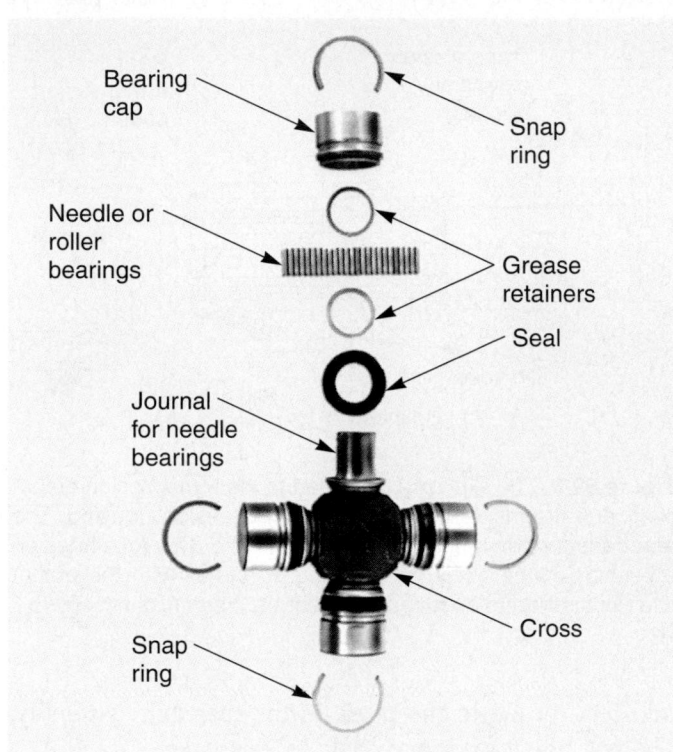

Figure 59-8. A partially disassembled view of a cross-and-roller universal joint. The needle bearings are packed with grease. The snap rings hold the bearing caps in their yokes. A rubber cup (boot) keeps grease inside the joint. (Oldsmobile)

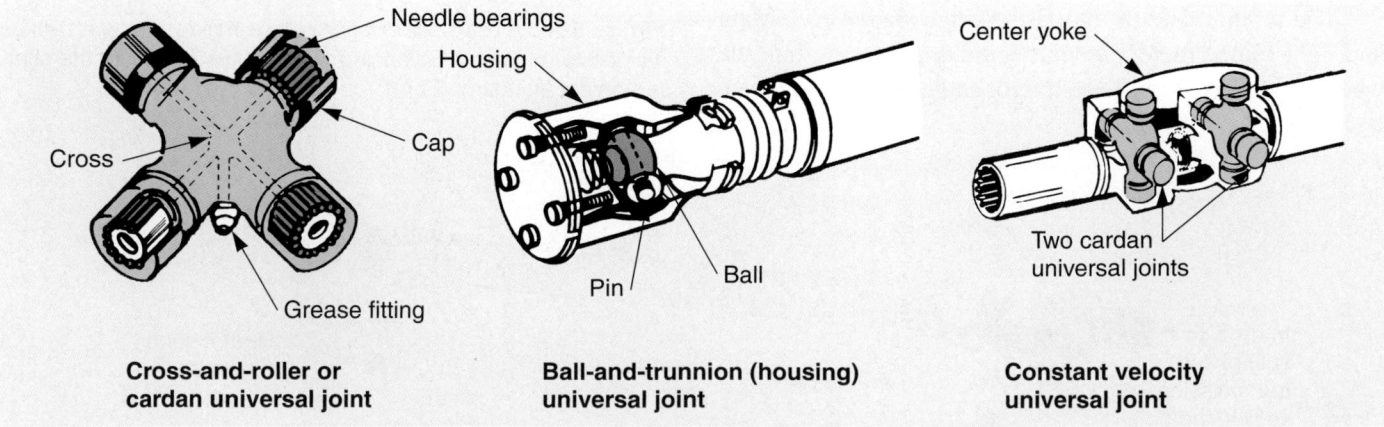

Cross-and-roller or cardan universal joint

Ball-and-trunnion (housing) universal joint

Constant velocity universal joint

Figure 59-7. Three basic types of universal joints are used in drive shafts going to rear-axle assemblies. Other types are used in front drive axles.

universal joint. It consists of four bearing caps, four needle roller bearings, a cross, grease seals, grease retainers, and snap rings. See **Figure 59-8.**

The bearing caps are held stationary in the drive shaft yokes. Roller bearings fit between the caps and cross to reduce friction. The cross is free to rotate inside the caps and yokes.

Snap rings usually fit into grooves cut in the caps or the yoke bores. There are several other methods of securing the bearing caps in the yokes. Sometimes, bearing covers, U-bolts, or injected plastic rings keep the caps and rollers from flying out of the spinning drive shaft assembly. See **Figure 59-9.**

Figure 59-10 shows an exploded view of a cross-and-roller drive shaft assembly. Note how the universal joint components fit together.

Constant Velocity Universal Joint

When a cross-and-roller universal joint is driven at a sharp angle, its output speed tends to accelerate and decelerate during each revolution. This can set up tiny torque fluctuations and torsional vibrations. One-piece drive shaft vibration problems can be reduced by using constant velocity universal joints.

A *constant velocity universal joint,* or CV-joint, normally has two cross-and-roller joints connected by a centering socket and center yoke. Another name for this type of joint is the *double-cardan joint.* See **Figure 59-11.**

With two universal joints operating together on one end of the drive shaft, the output shaft speed fluctuations are counteracted. The action of the second joint cancels the shaft speed changes produced by the first joint. Some

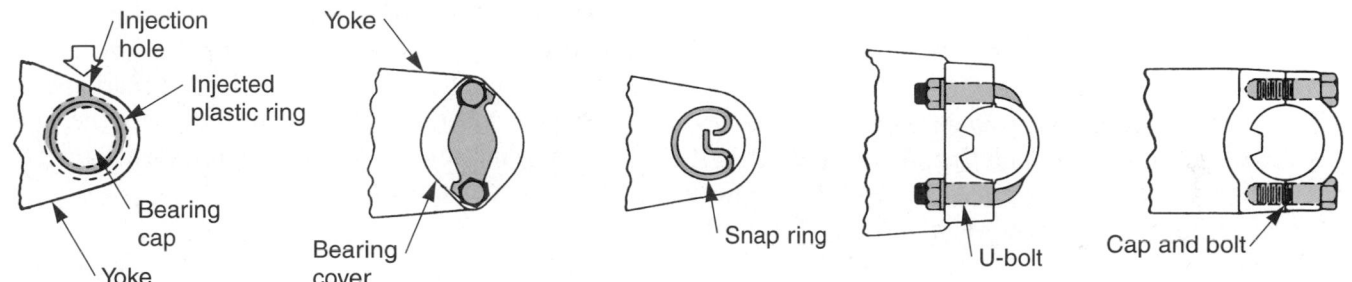

Figure 59-9. Several methods can be used to hold universal joint caps in the yoke. (Dana Corp.)

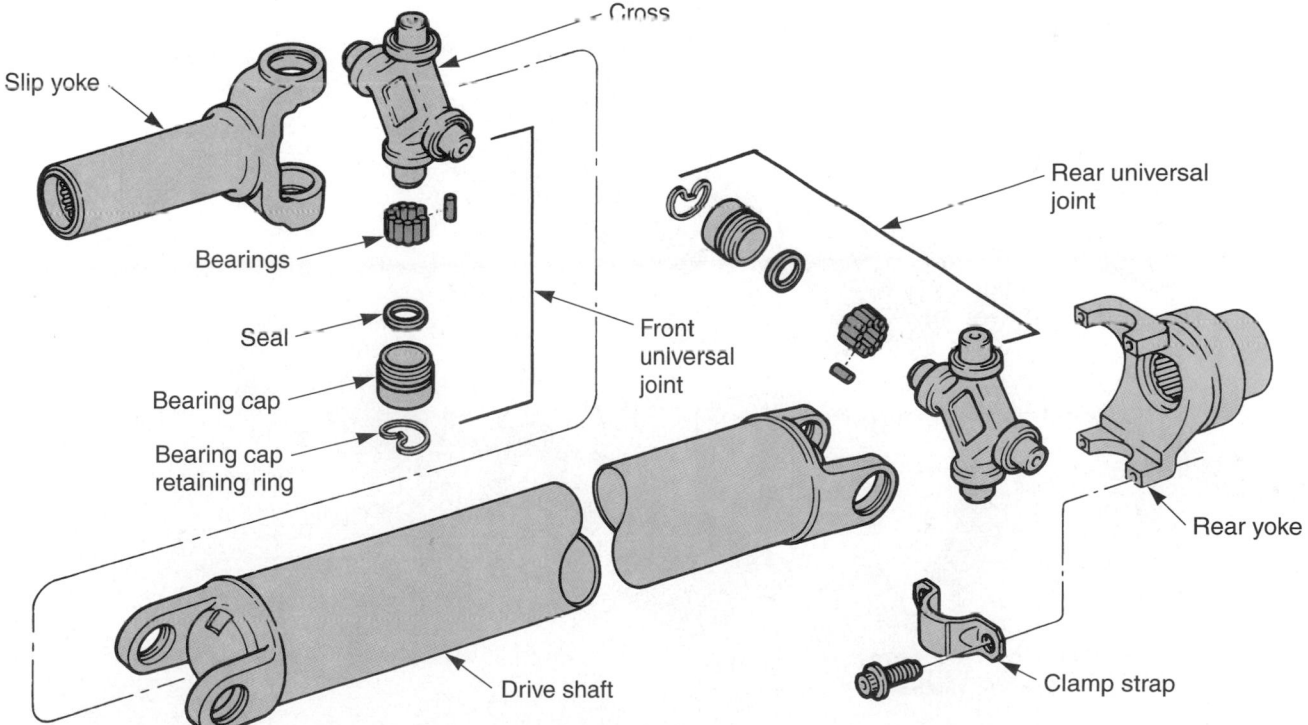

Figure 59-10. An exploded view of a drive shaft with cross-and-roller universal joints. Note how the parts fit together. (Chrysler)

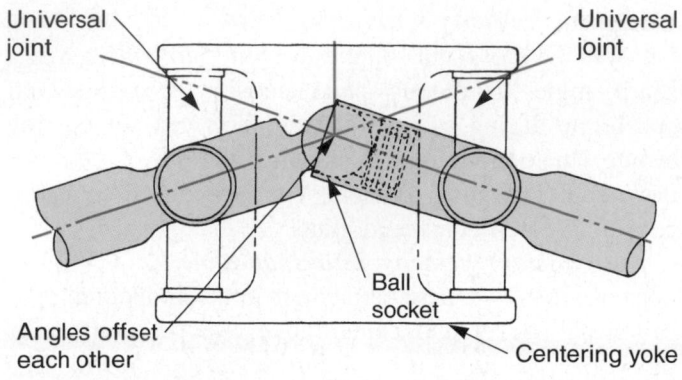

Figure 59-11. Simplified illustration of a constant velocity universal joint. With two cross-and-roller joints connected to same center yoke, speed variations are counteracted. Speed changes at the output of the first joint are offset by speed changes at the other joint. (GMC)

drive shafts use only one constant velocity universal joint. Others use more than one, **Figure 59-12.**

Note
Refer to Chapter 63, *Transaxle and Front Drive Axle Fundamentals,* and Chapter 64, *Transaxle and Front Drive Axle Diagnosis and Repair,* for more information on constant velocity universal joints.

Ball-and-Trunnion Universal Joint

The *ball-and-trunnion joint* is another joint designed for constant velocity. It not only eliminates shaft speed fluctuations, but can also allow for slight length changes in the driveline. It is seldom used, however. Refer back to **Figure 59-7.**

Center Support Bearing

A *center support bearing* is needed to hold the middle of a two-piece drive shaft. The center bearing bolts to the vehicle's frame or underbody. It supports the center of the drive shaft where the two shafts come together. See **Figure 59-13.**

Pickup trucks commonly use a center support bearing. A two-piece drive shaft is required in these vehicles because of the large distance between the transmission and rear axle.

A cutaway view of a center support bearing is shown in **Figure 59-14.** A sealed ball bearing allows the drive shaft to spin freely. The outside of the ball bearing is held by a thick, rubber, doughnut-shaped mount. The rubber mount prevents noise and vibration from transferring into the driver's compartment.

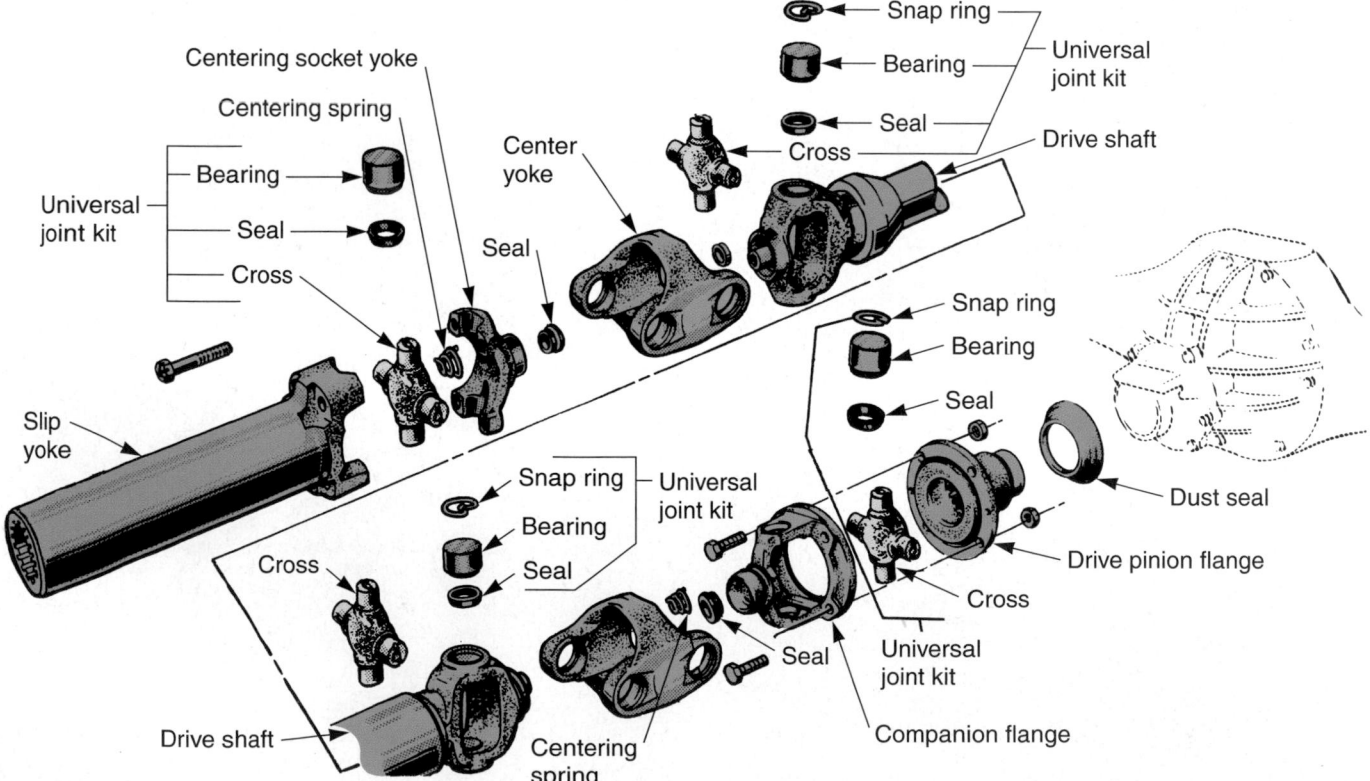

Figure 59-12. Note the construction of a drive shaft using two constant velocity universal joints. Two center yokes and four cross-and-roller joints are needed. (Ford)

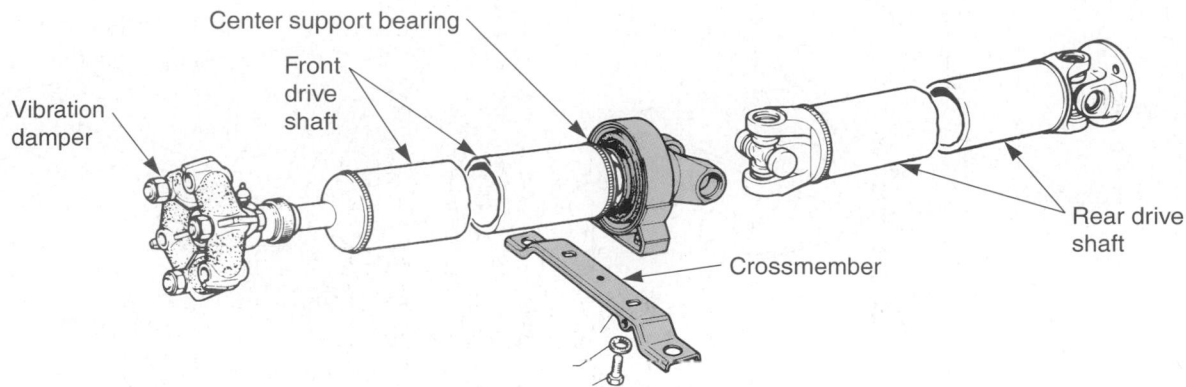

Figure 59-13. The center support bearing holds the center of a two-piece drive shaft. It is a roller bearing mounted in rubber. Also note the rubber torsion damping ring. (Fiat)

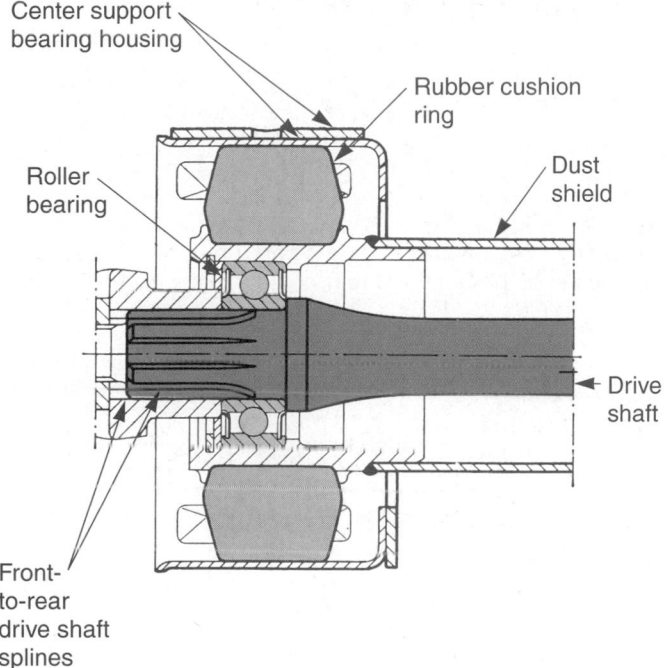

Figure 59-14. Cutaway view of a center support bearing. Study the parts. (Fiat)

Drivelines

There are several variations of rear-wheel drivelines in current use. However, most can be broken down into one of two categories: hotch kiss drivelines and torque tube drivelines.

Hotch Kiss Driveline

A *hotch kiss driveline* has an exposed drive shaft that operates a rear axle assembly mounted on springs. This is the most common rear-wheel driveline. It usually has cross-and-roller universal joints.

A hotch kiss driveline is pictured in **Figure 59-10.** It has almost totally replaced the torque tube setup.

Torque Tube Driveline

The *torque tube driveline* uses a solid steel drive shaft enclosed in a large hollow tube. Only one swivel joint is used at the front. The rear of the torque tube is formed as a rigid part of the rear axle housing.

Transfer Cases

A *transfer case* sends power to both the front and rear axle assemblies in a four-wheel-drive vehicle. The transfer case usually mounts behind and is driven by the transmission, **Figure 59-15.** Two drive shafts normally run from the transfer case, one to each drive axle. Look at **Figure 59-16.**

Transfer Case Construction

A transfer case is constructed something like a transmission. It uses shift forks, splines, gears, shims, bearings, and other components found in manual and automatic transmissions.

A transfer case has an outer case made of cast iron or aluminum. It is filled with lubricant (oil) that cuts friction on all moving parts. Seals hold the lubricant in the case and prevent leakage around shafts and yokes. Shims provide the proper clearances between the internal components and the case. **Figure 59-17** shows the major parts of a transfer case assembly.

Note
If needed, review Chapters 55 through 58. They cover transmission principles, which are related to transfer cases. Also, refer to a service manual for details of the particular unit.

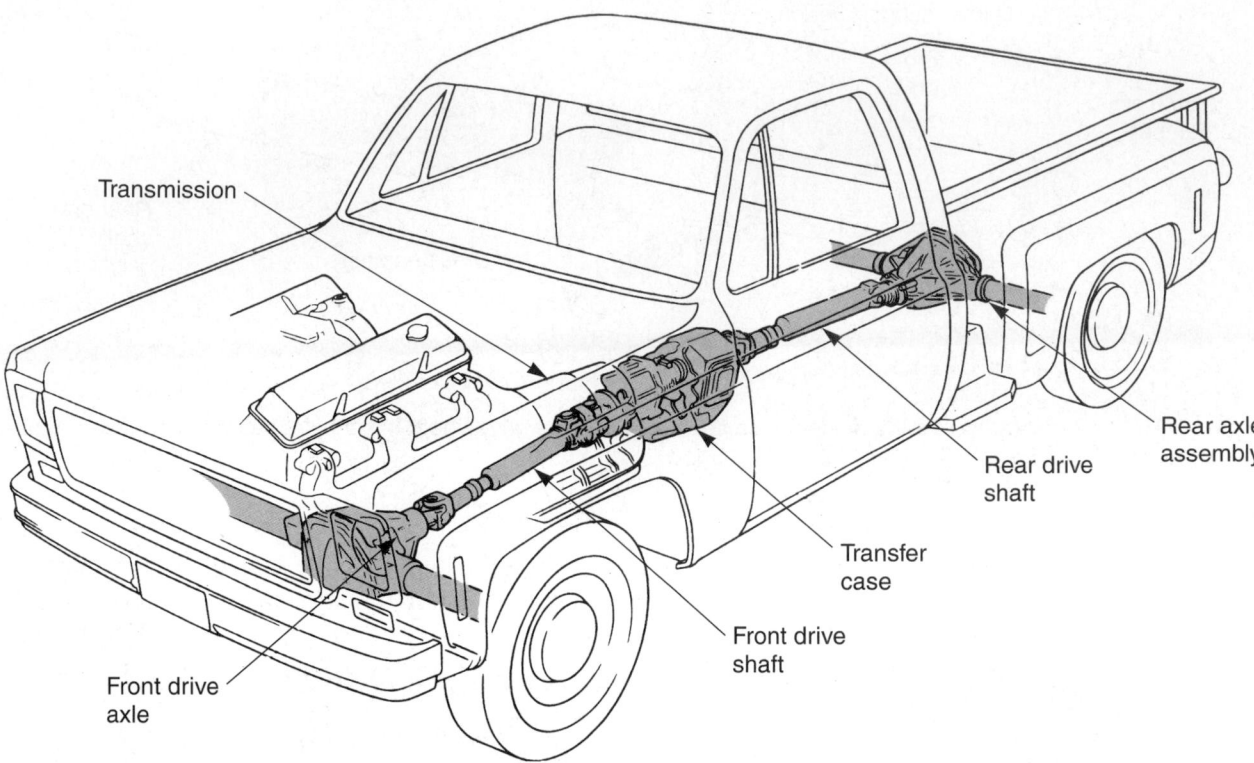

Figure 59-15. Small trucks and passenger cars commonly use four-wheel drive. The transfer case is a power takeoff unit that sends power to both the front and rear drive axle assemblies. Drive shafts extend out of the front and rear of the transfer case. (GMC)

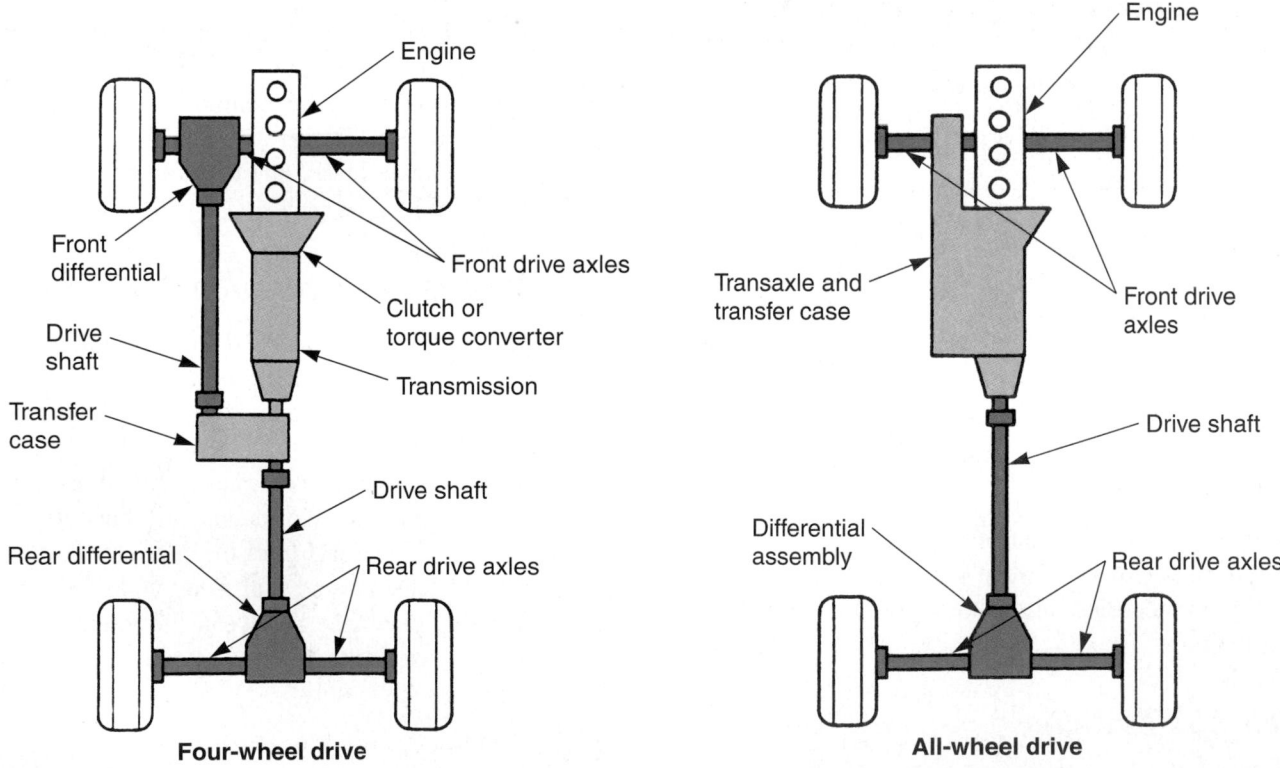

Four-wheel drive

All-wheel drive

Figure 59-16. A four-wheel-drive system has a transfer case separate from the transmission. Various drive ranges, such as 2H, 4H, and 4L, are normally provided. An all-wheel-drive system has the transfer case included as part of the transaxle.

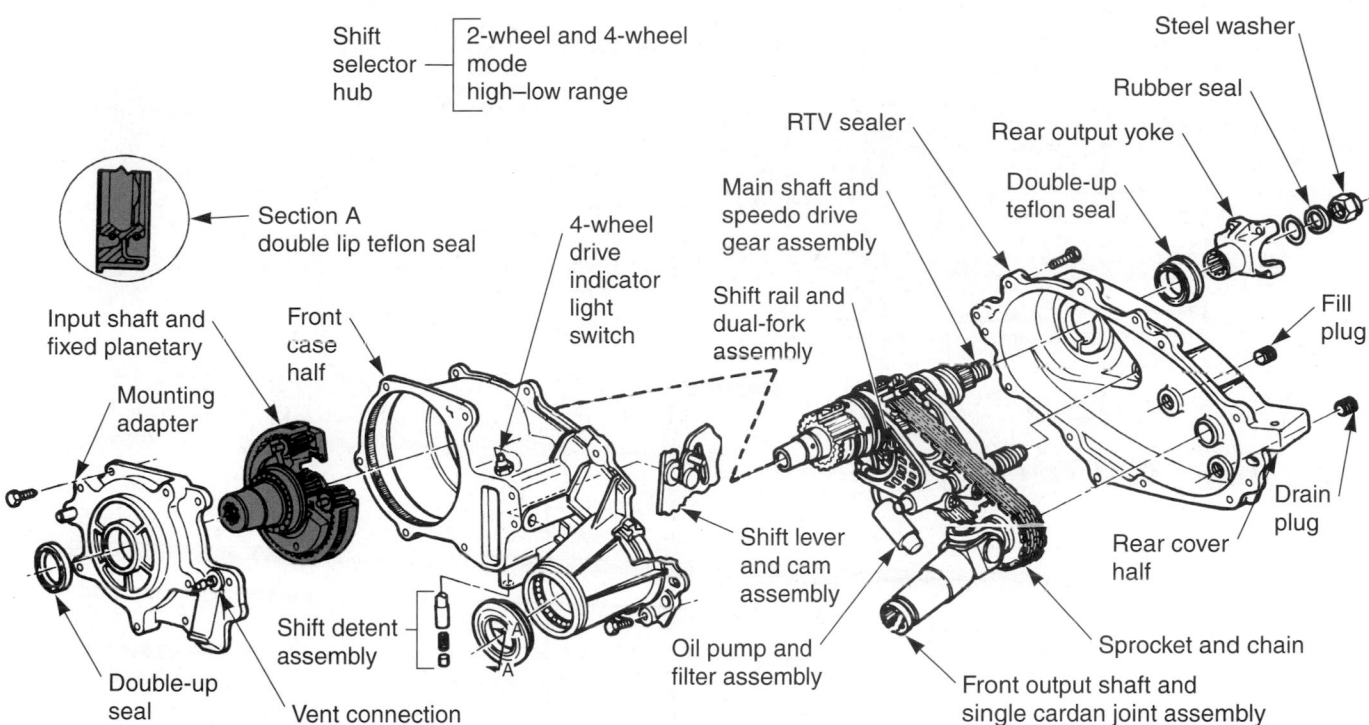

Figure 59-17. Major parts of modern transfer case. The planetary gearset provides high and low ranges. The large chain sends power to the front output shaft. A shift rail and fork assembly is activated to control two-wheel-drive or four-wheel-drive mode. (Ford)

Transfer Case Ranges

Most modern transfer cases provide a two-wheel drive, high range (2H), a four-wheel-drive, high range (4H), and a four-wheel-drive, low range (4L). High range normally has a gear ratio of 1:1. Low range typically has a gear ratio of 2:1 for climbing steep hills or pulling heavy loads. **Figure 59-18** shows how power flows through a transfer case for the different ranges.

Two-Wheel Drive, High Range (2H)

The 2H range is provided for normal driving when four-wheel-drive traction is not needed. In 2H range, torque flows from the input gear to the locked planet gears and ring gear, which rotate as a single unit. Torque is transferred to the main shaft through the planet carrier splined to the main shaft. Power finally flows out the rear yoke, through the rear drive shaft, and to the rear differential. Refer to **Figure 59-18.**

In 2H, the sliding clutch remains in the neutral position. As a result, torque is not transferred to the front axle assembly.

Four-Wheel Drive, High Range (4H)

In 4H, torque flows through the input gear, the planet gears, and ring gear in the same fashion as in 2H. However, the sliding clutch is shifted into the main shaft

clutch gear. Torque then flows through the drive chain, the front output yoke, and the front drive axle assembly. Both the front and rear axles drive the vehicle.

Four-Wheel Drive, Low Range (4L)

In 4L, torque transfer is almost the same as in 4H. However, the ring gear is shifted forward into the lock plate. This holds the ring gear stationary. As a result, the planet gears walk inside the ring gear, producing a gear reduction.

All-Wheel Drive

All-wheel drive refers to a four-wheel-drive system that does not use a conventional transfer case. It is a system designed for a front-wheel-drive transaxle or a transmission. See **Figure 59-19.**

Basically, the transmission or transaxle in an all-wheel-drive system is modified to allow power flow to the front and rear drive axles. The drive shaft, drive axles, and differentials are the same as the units previously discussed.

A fluid coupling, or viscous coupling, controls the power split to the front and rear axle assemblies. Usually, more drive torque is sent to the rear wheels than to the front.

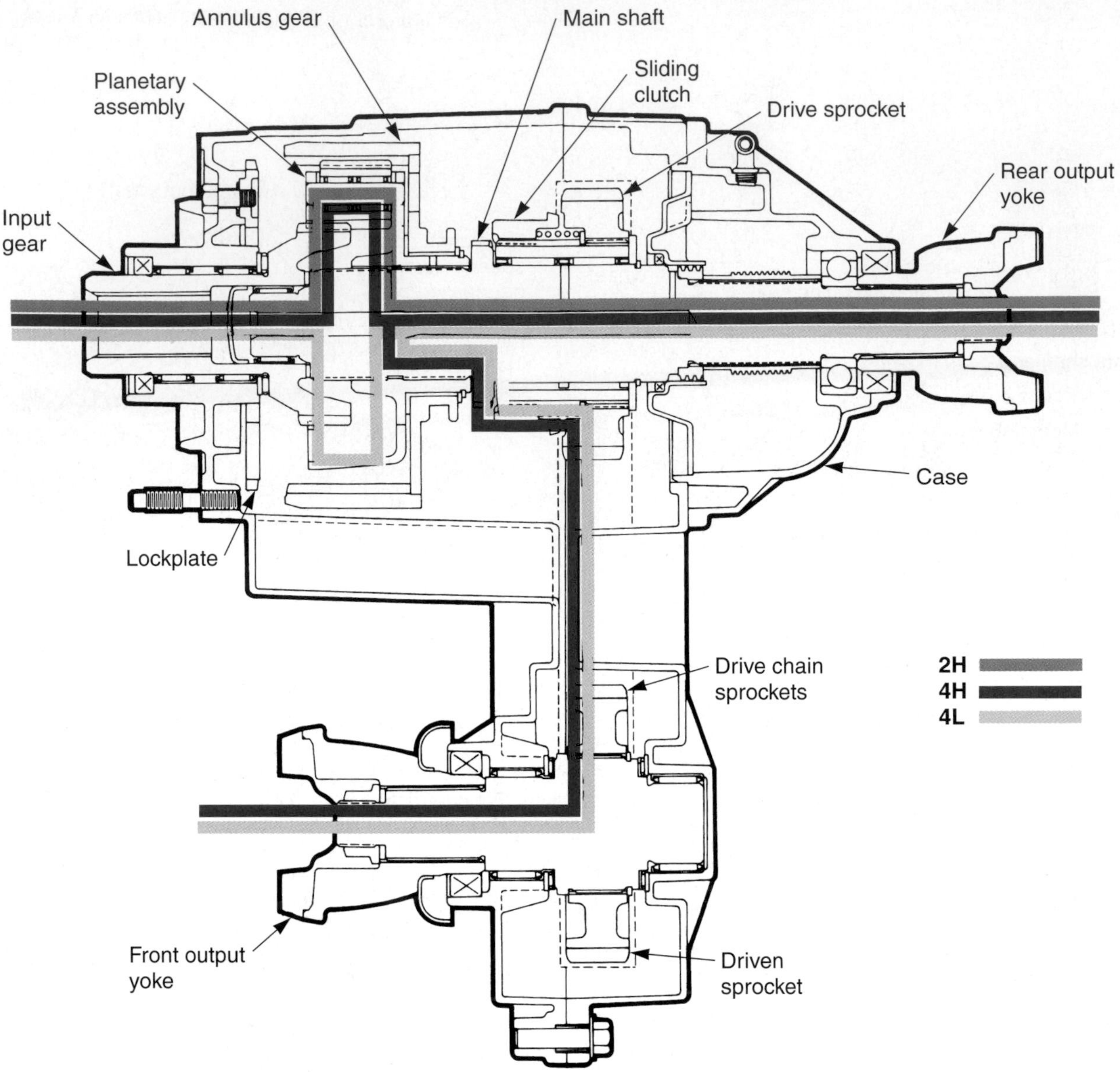

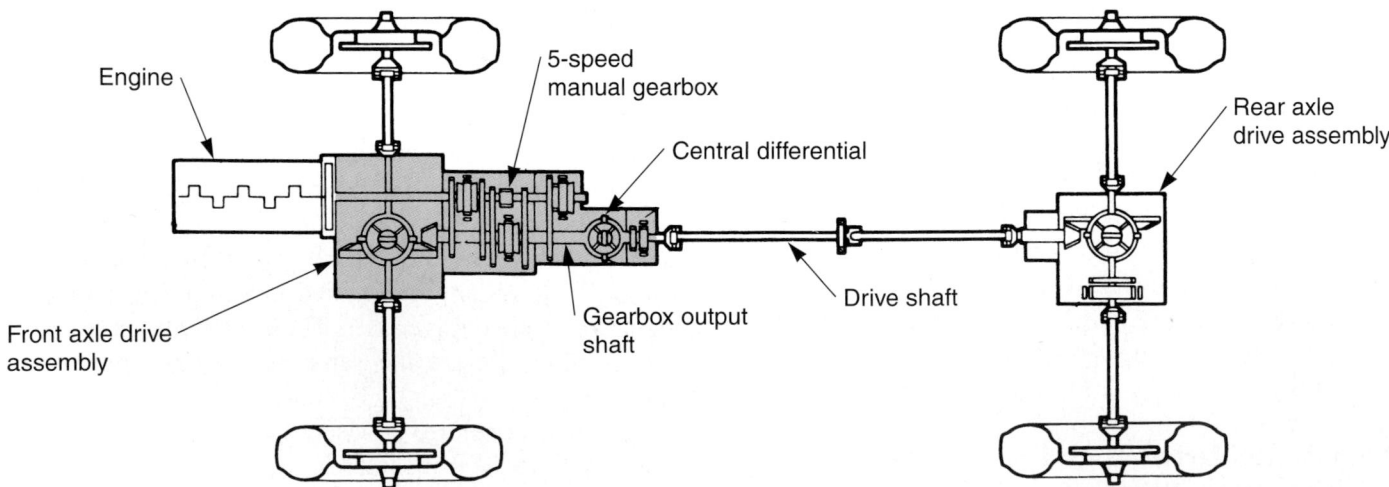

Figure 59-18. Trace power flow through a transfer case in 2H, 4H, and 4L range. Notice that this unit uses a planetary gearset to produce the two gear ratios. A hand-shifted sliding clutch regulates power transfer for two-wheel and four-wheel drive. (Chrysler)

Figure 59-19. All-wheel drive uses a variation of a standard transmission and transaxle. The main gearbox shaft drives the front differential directly. The rear of the gearbox shaft turns the drive shaft for the rear axle assembly. A conventional transfer case is not needed. (Porsche-Audi)

Summary

- The term driveline generally refers to the parts that transfer power from the transmission to the drive wheels.

- A drive shaft generally refers to a long shaft with universal joints that transfer power from the transmission back to the rear axle assembly.

- A universal joint allows the driveline to flex as the rear axle moves up and down.

- The slip yoke, splined to the transmission output shaft, allows for any changes in driveline length by sliding in and out of the transmission.

- The extension housing bushing supports the slip yoke as it spins in the transmission.

- The transmission rear seal rides on the slip yoke and prevents fluid leakage.

- The drive shaft is commonly a hollow steel tube with permanent yokes welded on each end.

- Drive shaft balancing weights are frequently welded to the shaft to avoid vibration.

- The cross-and-roller (cardan) universal joint is the most common type of drive shaft universal joint.

- A cardan constant velocity joint normally has two cross-and-roller joints connected by a centering socket and center yoke.

- The ball-and-trunnion joint is designed for constant velocity.

- A center support bearing is needed to hold the middle of a two-piece drive shaft.

- A transfer case sends power to both the front and rear axle assemblies in a four-wheel-drive vehicle.

- All-wheel drive refers to a four-wheel-drive system that does not use a conventional transfer case.

Important Terms

Drive line	Extension housing
Drive shaft	bushing
Transaxle	Transmission rear seal
Drive shaft assembly	Rear yoke
Slip yoke	Drive shaft
Front universal joint	Propeller shaft
Drive shaft	Drive shaft
Rear universal joint	balancing weights
Rear yoke	Drive shaft vibration
Universal joints	damper
Slip yoke	Universal joint
Slip joint	U-joint

Cross-and-roller
Cardan universal joint
Constant velocity
universal joint
Double-cardan joint
Ball-and-trunnion
joint

Center support bearing
Hotch kiss driveline
Torque tube driveline
Transfer case
All-wheel drive

Review Questions—Chapter 59

Please do not write in this text. Place your answers on a separate sheet of paper.

1. The term _____ generally refers to the parts that transfer power from the transmission to the drive wheels.

2. List and explain the five major parts of a drive shaft.

3. What are four functions of a drive shaft?

4. The movement of the rear axle assembly also causes the distance between the rear axle and transmission to change. True or False?

5. The _____ _____ or _____ _____ is splined to the transmission output shaft.

6. Describe the construction of a typical drive shaft, not including the universal joints or other parts.

7. Which of the following does *not* attach to an assembled drive shaft?
   (A) Balance weights.
   (B) Yokes.
   (C) Universal joints.
   (D) All of the above.
   (E) None of the above.

8. How does a constant velocity universal joint (double-cardan joint) work?

9. When is a center support bearing needed and why?

10. Explain the basic operation of a transfer case.

ASE-Type Questions

1. Technician A says the term *driveline* normally refers to the components that transfer power from the engine crankshaft to the drive wheels. Technician B says the term *driveline* normally refers to the parts that transfer power from the transmission to the drive wheels. Who is right?
   (A) A only.
   (B) B only.
   (C) Both A and B.
   (D) Neither A nor B.

2. Technician A says front-wheel-drive cars use a driveline with a single, long drive shaft. Technician B says front-wheel-drive cars are usually equipped with two drive axle shafts. Who is right?
   (A) A only.
   (B) B only.
   (C) Both A and B.
   (D) Neither A nor B.

3. While discussing the functions of a drive shaft assembly's components, Technician A says a slip yoke connects the transmission output shaft to the front universal joint. Technician B says a slip yoke connects the transmission input shaft to the rear axle assembly. Who is right?
   (A) A only.
   (B) B only.
   (C) Both A and B.
   (D) Neither A nor B.

4. Technician A says a drive shaft assembly's rear yoke holds the rear universal joint. Technician B says a rear yoke transfers torque to gears in the rear axle assembly. Who is right?
   (A) A only.
   (B) B only.
   (C) Both A and B.
   (D) Neither A nor B.

5. Technician A says a drive shaft assembly normally has one universal joint. Technician B says a drive shaft assembly is normally equipped with four universal joints. Who is right?
   (A) A only.
   (B) B only.
   (C) Both A and B.
   (D) Neither A nor B.

6. Which of the following is *not* a drive shaft assembly component?
   (A) Universal joint.
   (B) Slip yoke.
   (C) Transmission input shaft.
   (D) Rear yoke.

7. Technician A says a drive shaft assembly sends turning power from the transmission to the rear axle assembly. Technician B says a drive shaft assembly must flex and allow up and down movement of the rear axle assembly. Who is right?
   (A) A only.
   (B) B only.
   (C) Both A and B.
   (D) Neither A nor B.

8. Technician A says during drive shaft assembly operation, the transmission input shaft turns the rear yoke. Technician B says during drive shaft assembly operation, the transmission output shaft turns the slip yoke. Who is right?
   (A) A only.
   (B) B only.
   (C) Both A and B.
   (D) Neither A nor B.

9. Technician A says a slip yoke allows the distance between the rear axle and transmission to change. Technician B says a slip yoke allows the driveline to flex without damaging the drive shaft. Who is right?
   (A) A only.
   (B) B only.
   (C) Both A and B.
   (D) Neither A nor B.

10. Technician A says a drive shaft assembly's slip yoke is normally splined to the transmission output shaft. Technician B says the drive shaft assembly's slip yoke is usually bolted to the transmission output shaft. Who is right?
    (A) A only.
    (B) B only.
    (C) Both A and B.
    (D) Neither A nor B.

11. Technician A says a rear yoke is bolted to the outer end of the rear axle assembly's ring gear. Technician B says a rear yoke is bolted to the outer end of the rear axle assembly's pinion gear. Who is right?
    (A) A only.
    (B) B only.
    (C) Both A and B.
    (D) Neither A nor B.

12. Technician A says drive shaft balancing weights are usually bolted to the drive shaft. Technician B says drive shaft balancing weights are normally welded to the drive shaft. Who is right?
    (A) A only.
    (B) B only.
    (C) Both A and B.
    (D) Neither A nor B.

13. Technician A says a cardan universal joint is the most common type of drive shaft universal joint used on rear-wheel-drive vehicles. Technician B says the ball-and-trunnion universal joint is the most common type of universal joint used on rear-wheel-drive vehicles. Who is right?
    (A) A only.
    (B) B only.
    (C) Both A and B.
    (D) Neither A nor B.

14. Technician A says a transfer case sends power to the front axles on front-wheel-drive cars. Technician B says a transfer case sends power to the rear axle assembly on rear-wheel-drive automobiles. Who is right?
    (A) A only.
    (B) B only.
    (C) Both A and B.
    (D) Neither A nor B.

15. Which of the following is a basic component of a modern transfer case?
    (A) Input shaft and fixed planetary.
    (B) Oil pump and filter assembly.
    (C) Shift rail and dual-fork assembly.
    (D) All of the above.

Activities—Chapter 59

1. Identify the components of a drive train on a vehicle in the automotive shop for service.

2. Sketch and label the parts of a universal joint.

3. Prepare an overhead transparency from the sketches made in Activity 2; use it to explain the parts of a universal joint.

Chapter 60

Drive Shaft and Transfer Case Diagnosis, Service, and Repair

After studying this chapter, you will be able to:

- Troubleshoot common drive shaft problems.
- Check universal joint wear.
- Measure drive shaft runout.
- Remove and replace a drive shaft assembly.
- Replace universal joints.
- Perform basic service operations on a transfer case.
- Cite and practice good safety procedures.
- Correctly answer ASE certification test questions on the service and repair of drive shafts and transfer cases.

A drive shaft is subjected to very high loads and rotating speeds. When a vehicle is cruising down the highway, the drive shaft and universal joints may be spinning at full engine speed. They are sending high torque to the rear axle assembly.

To function properly, the drive shaft must be almost perfectly straight. The universals must be unworn. If any component allows the drive shaft to wobble, severe vibration, abnormal noises, and major damage can result. See **Figure 60-1.**

Drive Shaft Problems

When driving the vehicle to verify a drive shaft–related complaint, keep in mind that other components could be at fault. A worn wheel bearing, squeaking spring, defective tire, transmission problems, or differential troubles could be at fault. You must use your knowledge of each system to detect which component is causing the trouble.

Drive shaft problems can normally be divided into two categories: drive shaft noise and drive shaft vibration.

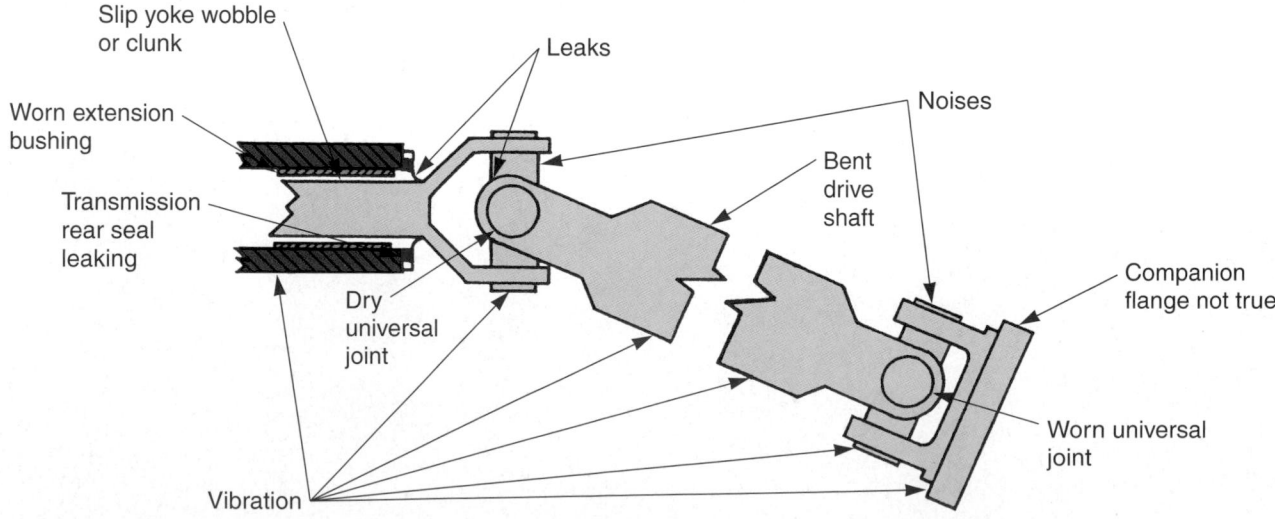

Figure 60-1. A drive shaft inspection will frequently detect troubles. Wiggle joints to check for wear and looseness.

Drive Shaft Noise

Drive shaft noise is usually caused by worn universal joints, a worn slip joint, or a faulty center support bearing. Refer to **Figure 60-1.**

Grinding or squeaking from the drive shaft is frequently caused by worn universal joints. The joints may become dry, causing the rollers to wear. The damaged rollers then produce a grinding or squeaking sound as they operate on the scored cap and cross surfaces.

Tech Tip

Use special care in diagnosing squeaking noises that seem to come from the universal joints. Wheel covers can make similar noises as tires flex against them. Remove covers if uncertain of the noise.

A clunking sound during acceleration or deceleration may be caused by slip yoke problems or extension housing bushing wear. The slip yoke splines may be worn. This will allow the yoke to flop up and down as driveline torque changes. An excessively worn universal joint or differential problems can also cause similar symptoms.

A whining noise from the drive shaft is sometimes caused by a dry, worn center support bearing. Since this bearing makes complete revolutions, it will make a different sound than a bad universal joint. A faulty center support bearing will usually produce a high-pitched, constant whine.

Other abnormal sounds should be traced using your knowledge of mechanical principles, a stethoscope, and a service manual troubleshooting chart.

Drive Shaft Vibration

Drive shaft vibration can be caused by any problem that affects drive shaft balance, runout (straightness), and angle. The vibration is usually more rapid than vibration caused by wheels and tires. Drive shaft vibration is similar to the vibration produced by an unbalanced clutch, flywheel, or engine crankshaft.

When test driving, drive in high gear at the engine speed that causes the most vibration. Then shift into different gears or neutral while maintaining that speed. If there is not a change, the vibration may be in the drive shaft. A vibration change indicates the engine, clutch, torque converter, or transmission is at fault. When vehicle speed remains the same, drive shaft speed remains the same.

Drive Shaft Inspection

To inspect the drive shaft, raise the vehicle on a hoist. Look for undercoating or mud on the drive shaft that could cause vibration. Check for a sharp driveline angle, missing balance weights, cracked welds, and other drive shaft problems.

To check for worn universal joints, wiggle and rotate each universal joint back and forth while watching for side movement. Try to detect any play between the cross and yoke. If the cross moves inside the yoke, the universal joint is worn and should be replaced.

Also, wiggle the slip yoke up and down. If it moves excessively in the transmission bushing, either the yoke or the bushing is worn. Inspect the rear yoke bolts for tightness. Make sure the rear motor mount is not broken. Look for any condition that could upset the operation of the drive shaft.

If you fail to find a problem, you may need to measure drive shaft runout (wobble) and check balance.

Measuring Drive Shaft Runout

Drive shaft runout is caused by a bent drive shaft, damaged yokes, or worn universal joints. A dial indicator is normally used to measure drive shaft runout.

First, sand and clean around the front, center, and rear of the drive shaft. This will give the dial indicator a smooth surface for accurate measurements. See **Figure 60-2.** Mount the dial indicator perpendicular to the shaft. The indicator base must be placed on a rigid surface (differential, floor pan, transmission, or special post stand). The drive shaft must not be at a sharp angle during runout measurement.

With the transmission in neutral, turn the drive shaft. Measure runout at the front, center, and rear of the shaft. Compare your measurements to specifications. Generally, drive shaft runout should not exceed 0.010″–0.030″ (0.25–0.75 mm).

If drive shaft runout is beyond specifications, try removing and rotating the shaft 180° in the rear yoke. This may shift select-fit parts back into alignment and correct the runout problem. Also, make sure the universal joints are in good condition and that the yokes are not damaged.

If runout is still excessive, replace the drive shaft or send it to a machine shop for repairs. Some specialized shops can replace the center tube of drive shaft.

Balancing a Drive Shaft

If the drive shaft is within recommended runout limits but still vibrates, the drive shaft may need to be balanced. Place the vehicle on a twin post lift so that the rear axle housing is supported. The rear axle must be held up in its normal position with the wheels free to rotate. If needed, use long hoist jack stands, **Figure 60-3.**

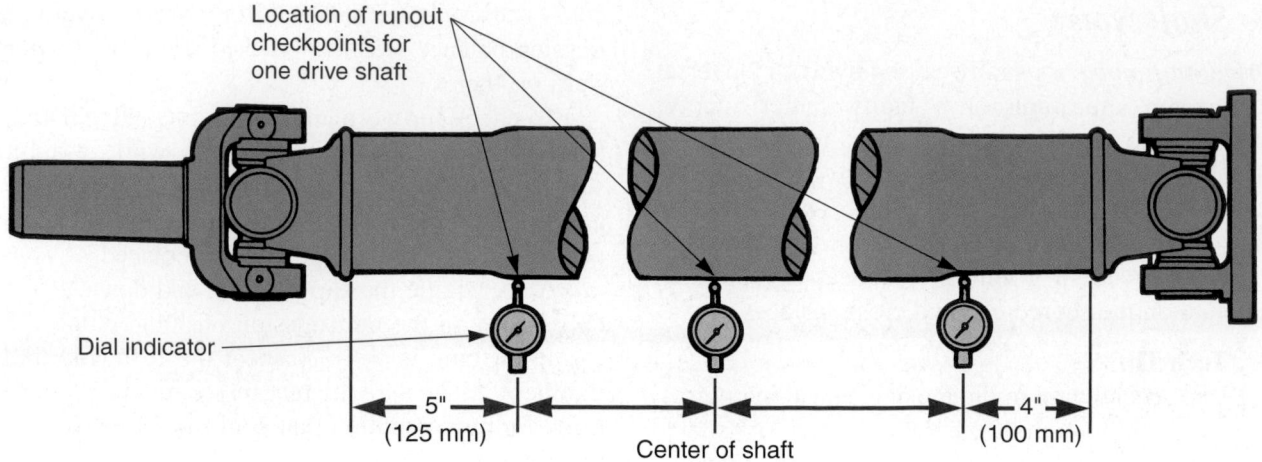

Figure 60-2. A dial indicator can be used to measure drive shaft runout. Check in the middle and on each end of the shaft. If runout is more than specifications, the shaft is bent or yokes are damaged. (Cadillac)

Ask another worker to start the engine and engage the transmission in high gear. With the drive shaft rotating at a speedometer reading of 40–50 mph (64–81 km/h), carefully bring a crayon or pencil up to the drive shaft a few inches away from the rear universal joint.

As soon as the crayon touches the spinning drive shaft, pull it back from the shaft. A mark on only one side of the drive shaft indicates the heavy side of the shaft. See **Figure 60-4A.**

Warning

Be extremely careful not to let any part of your body contact the spinning drive shaft or rear wheels. If your hair, for example, touches a spinning universal joint, severe injuries can result.

If needed, install two screw or worm-type hose clamps on the drive shaft as in **Figure 60-4B.** The screw heads on the clamps should be located opposite the marks made by the crayons. The weight of the screws will offset the heavy side of the drive shaft. Tighten the clamps securely.

After installing the clamps, start the engine and run the speedometer back up to the vibration speed. If the vibration is gone, lower the vehicle and take it for a test drive. If an imbalance still exists, rotate the clamps 45° away from each other and recheck vibration. Continue rotating the clamps apart or together in smaller increments until the vibration is gone, **Figure 60-4C.**

If required, repeat this balancing operation on the front of the drive shaft. If you cannot balance the shaft, send it to a machine shop that has specialized shaft balancing equipment.

Measuring Driveline Angle

When your runout and balance checks fail to find a problem, you may need to check the *driveline angle.* If the angle is too sharp, the universal joints can cause speed fluctuations and vibration.

There are various methods used to check the driveline angle. Some automakers recommend a bubble gauge that reads in degrees. One is shown in **Figure 60-5.**

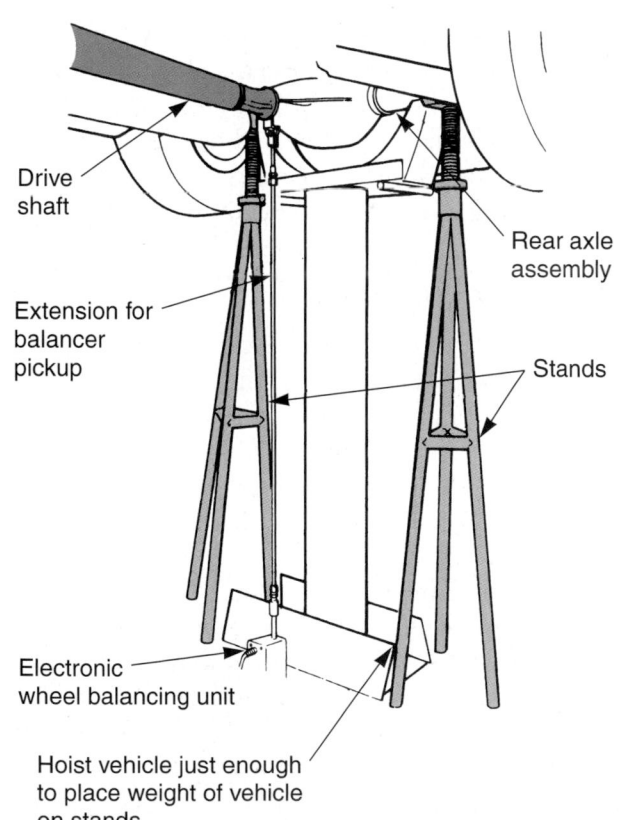

Figure 60-3. When checking the drive shaft, support the rear axle assembly. This will keep the transmission and rear axle in normal alignment. Also note the use of an electronic wheel balancing unit for checking shaft vibration. (Buick)

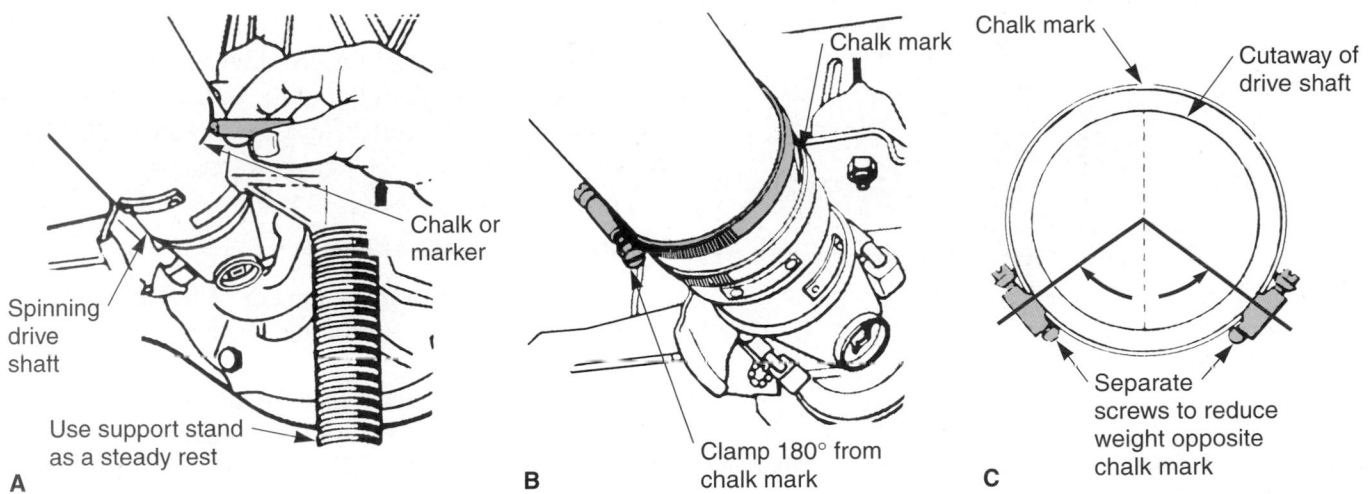

Figure 60-4. Checking for drive shaft imbalance. A—Idle engine in gear while bringing marker up to drive shaft slowly. Move marker away as soon as it touches shaft. B—Place hose clamps around the drive shaft. Screw heads should be opposite the mark. C—Move the screw heads away from each other as needed to obtain correct balance. (Ford)

Simply place the gauge on the drive shaft (with the car in a level position and the rear axle supporting the vehicle weight) and read the driveline angle. If the angle is incorrect, you can adjust the driveline angle by placing shims under the rear axle housing or transmission mount. Refer to a service manual for exact procedures.

Drive Shaft Maintenance

Normally, very little maintenance is required on modern drive shafts. The universal joints are usually sealed units that are filled with grease at the factory. However, some universal joints, especially aftermarket types, have grease fittings, allowing lubrication. A *grease gun* can be used to lubricate universal joints that have fittings, **Figure 60-6.**

Drive Shaft Service

Drive shaft service requires that the drive shaft be removed for repairs. The universal joints may be worn and require replacement, or the drive shaft itself may be damaged. Although some design variations are used, general service procedures are the same for most rear-wheel-drive vehicles.

Drive Shaft Removal

To remove the drive shaft, raise the vehicle on a hoist. Scribe marks on the differential yokes and the universal

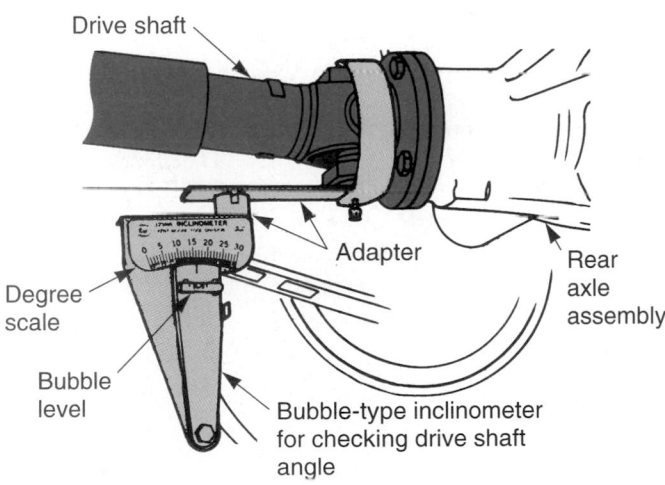

Figure 60-5. An angle gauge or inclinometer can be used to check driveline alignment. Shim the rear axle or transmission, or replace springs and other parts to correct misalignment problem. Misalignment can cause vibration and universal joint wear. (Cadillac)

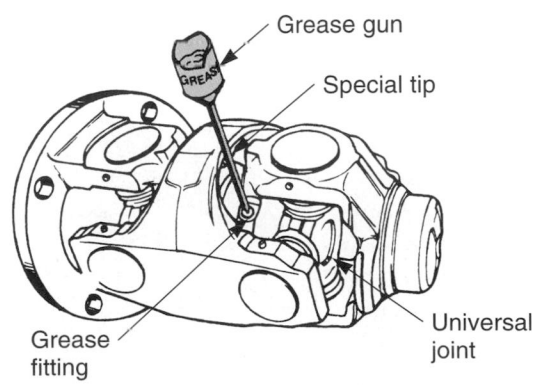

Figure 60-6. Some universal joints have grease fittings. Use a grease gun and a special long stem, if needed, to keep universal joint bearings lubricated. (GMC)

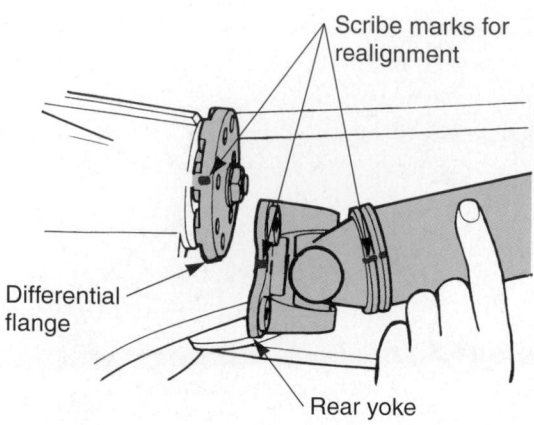

Figure 60-7. Scribe mark the drive shaft, rear yoke, and universal joints before disassembly. This will help maintain shaft balance when reassembling. (Ford)

joints. This will help ensure proper drive shaft balance upon reassembly. See **Figure 60-7.**

Unbolt the rear joint from the differential. Pry the shaft forward and lower the shaft slightly. Do not allow the full weight of the drive shaft to hang on the slip yoke. Support the drive shaft to prevent damage to the extension housing, rear bushing, and front universal joint.

Wrap tape around the two caps on the rear universal joint if needed. This will keep the caps from falling off and spilling the small roller bearings. Also, unbolt the center support bearing (if used), **Figure 60-8.**

Slide the drive shaft out of the transmission. If transmission lubricant begins to leak out, install a plastic plug or an old slip yoke in the extension housing.

Universal Joint Service

A **worn universal joint,** the most common drive shaft problem, can cause squeaking, grinding, clunking, or clicking sounds. The grease inside the joint can dry out. The roller bearings can then wear small indentations (dents) in the cross. When the bearings try to roll over these indentations, a loud metal-on-metal grinding or chirping sound can result.

Quite often, a worn universal joint will be noticed when the vehicle is in reverse gear. When the car is backed up, the roller bearings are forced over the wear indentations against normal rotation. The rollers will catch on the sharp edges in the worn joint, causing an even louder sound.

Universal Joint Disassembly

Before disassembling the universal joint, scribe mark each component. The marks will show you how to reassemble the joint. This is especially important when working with constant velocity universal joints.

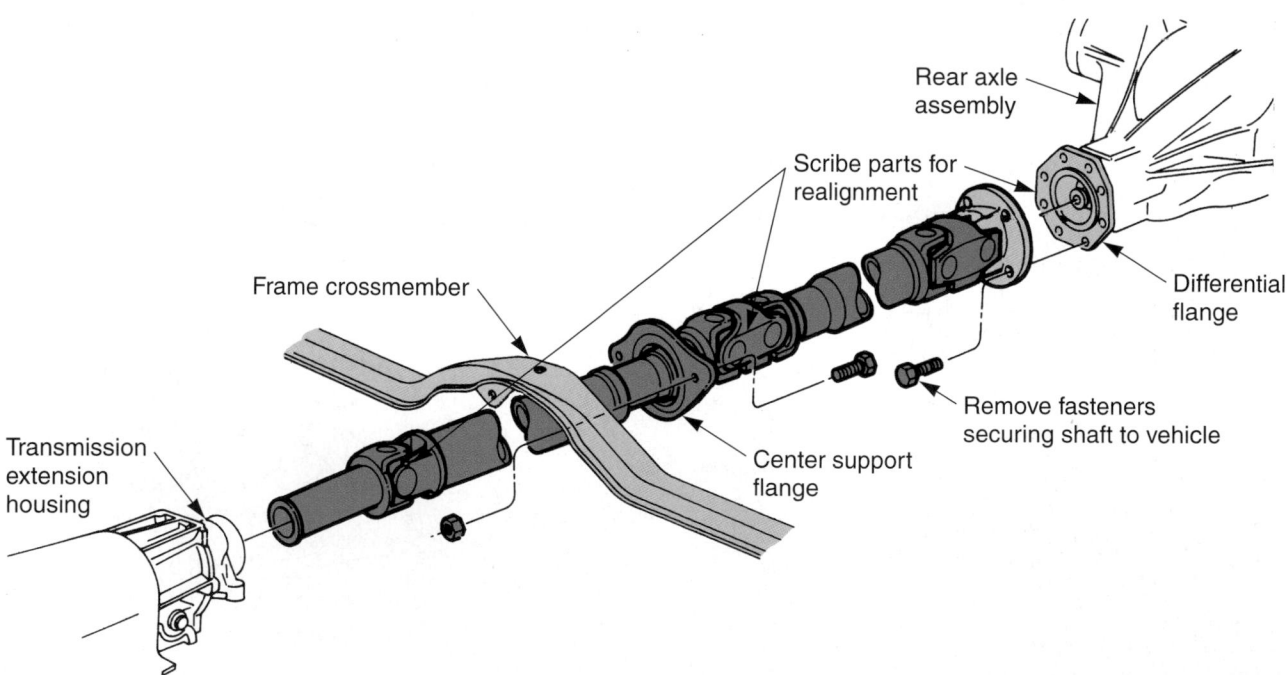

Figure 60-8. To remove a drive shaft, unbolt the rear yoke from the differential. You may also need to remove the center support bearing or flange. Slide the drive shaft out of transmission. If applicable, wrap tape around universal joints to keep caps from falling off. (Cadillac)

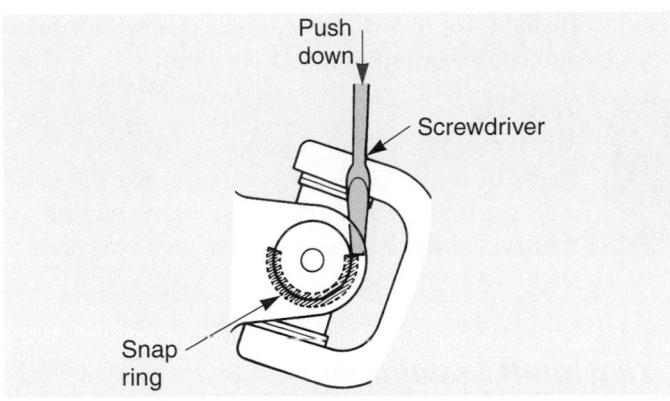

Figure 60-9. To disassemble a cross and roller universal joint, remove the snap rings using a screwdriver or snap ring pliers. (Toyota)

Clamp the drive shaft yoke in a vise. Do not clamp the weaker center section of the drive shaft or it may be bent.

Tech Tip
Avoid excessive force when holding a drive shaft in a vise. If the shaft or yokes are bent, the drive shaft may vibrate when returned to service.

If used, remove the snap rings from the universal joint caps or yokes. Use a screwdriver, snap-ring pliers, or needle nose pliers to remove the rings. Look at **Figure 60-9.**

Warning
Wear safety glasses to protect your eyes in case the snap rings fly out of the joint during removal.

Figure 60-10. With the snap rings removed, press the cross and bearing caps out of the yokes. Use a small socket as the driver. Use a large socket to accept the bearing cap on the other side. (Chrysler)

To remove the universal joint from the yokes, you can use a vise and two sockets, **Figure 60-10.**

Place a socket *smaller* than the bearing cap on one end of the universal joint. Place a socket with an inside diameter *larger* than the bearing cap on the opposite side of the joint.

Fit both sockets and the universal joint inside a vise. Slowly tighten the vise to force the bearing caps out of the yoke. Repeat this operation on the other yoke if needed.

Special C-clamp *universal joint tools* are also available for pressing universal joints apart and together. These tool are used in the same way as the vise.

Universal Joint Replacement

Normally, a universal joint is replaced whenever it is disassembled. However, if the joint is relatively new, you can inspect, lubricate, and reassemble it.

During inspection, clean the roller bearings and other parts in solvent. Then, check the cross and rollers for signs of wear. If the slightest sign of roughness or wear is found on any part, replace the universal joint.

Universal Joint Assembly

To assemble a universal joint, use the following procedure:
1. Pack the roller bearings with high-temperature grease.
2. Position the cross inside one of the yokes. Align your punch marks.
3. Fit the bearing caps into each end of the yoke. Center the cross partially into each cap to keep the roller bearings from falling out.
4. Place the assembly in a vise.
5. Tighten the vise so that the bearing caps are forced into the yoke.
6. Place a small socket on one bearing cap and tighten the vise until the cap is pushed in far enough for snap ring installation. Install the snap ring. See **Figure 60-11.**
7. With one snap ring in place, use the socket to force the other cap into place. Install its snap ring.

Repeat this procedure on the other universal joint and yoke, if needed. Look at **Figure 60-12.**

Caution
If the bearing cap fails to move into place with normal pressure, disassemble the joint and check the roller bearings. It is easy for a roller bearing to fall down and block cap installation. If you try to force the cap with excess pressure, the universal joint and drive shaft could be ruined.

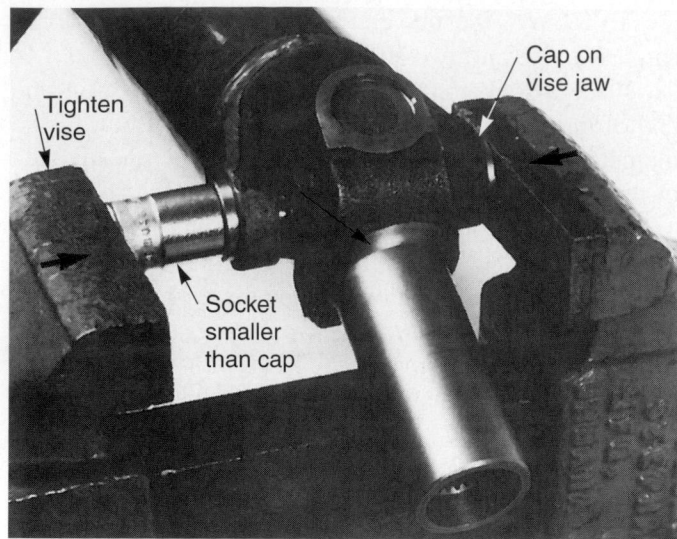

Figure 60-11. To install a universal joint, use one small socket and vise to press the caps into the yoke. Make sure the small needle bearings are in place or part damage can result. Press in until you can install one snap ring. Then, press the other cap in and install the other snap ring. (Chrysler)

After assembly, check the action of the universal joint. Swing it back and forth into various positions. The joint should move freely, without binding. Double-check that all the snap rings have been installed properly. Refer to **Figure 60-13.**

Some caps are secured by a special plastic resin that is injected into holes in the universal joint. When dry, the plastic forms a ring to hold the joint together. Refer to a service manual for exact procedures. Allow adequate drying time before installing the drive shaft.

Tech Tip

To service the universal joints on the drive shafts of four-wheel-drive vehicles, use the general instructions given in the previous section. A four-wheel-drive system simply uses two drive shafts instead of one.

Drive Shaft Installation

To install the drive shaft, wipe the slip yoke clean and place a small amount of grease on its internal splines (if recommended). Align your marks and slide the yoke into the rear of the transmission.

Push the shaft all the way into the extension housing and position the rear joint in the differential. Pull the shaft back and center the rear universal joint properly. Check your rear alignment marks.

Install the U-bolts, bearing caps, or yoke bolts to secure the rear universal joint to the differential. Lower the car to the ground.

Transfer Case Service

Transfer case service is required when the unit makes abnormal noises (grinding, whining), fails to engage properly, or is locked in a range.

When diagnosing transfer case problems, check the condition and level of the oil in the unit first. See **Figure 60-14.**

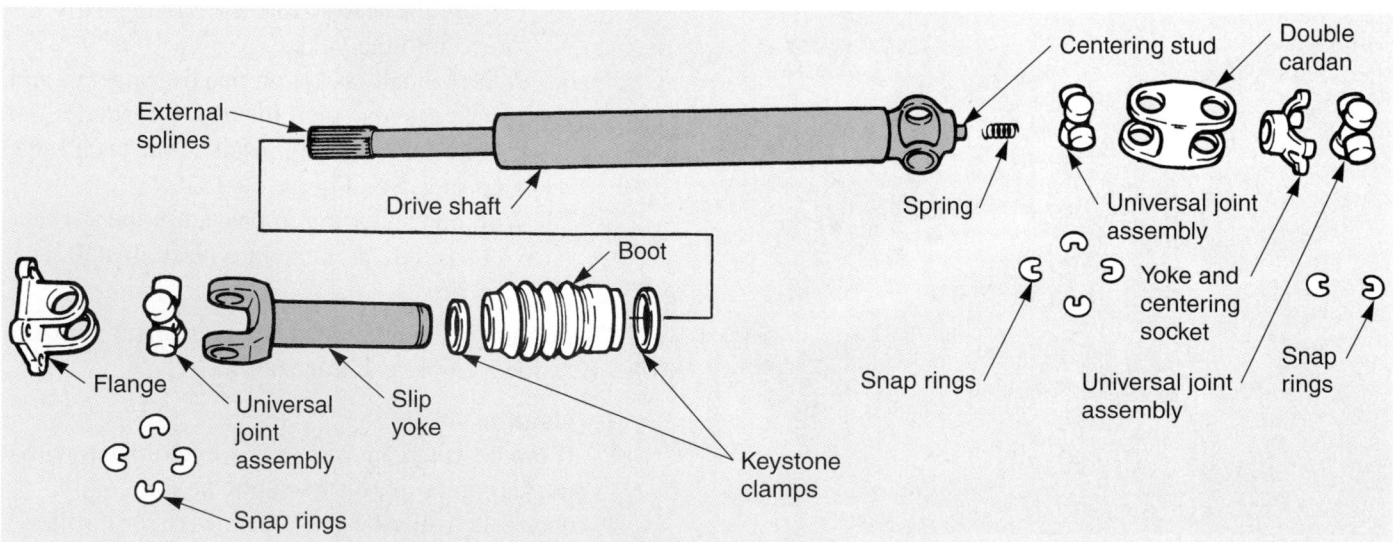

Figure 60-12. Inspect all drive shaft assembly components carefully. In particular, check snap ring grooves, yokes, and universal joints. Replace any part in poor condition. Also, make sure you mark position of all parts. They must be installed in the same position to prevent possible shaft vibrations. (Ford)

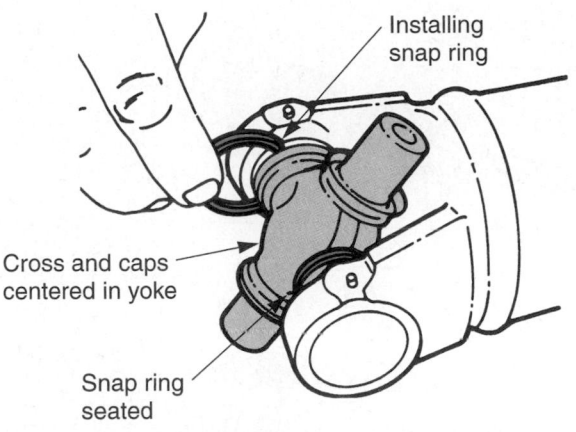

Figure 60-13. Make sure snap rings are fully installed in their grooves. Start them by hand. Then use small hammer and drift to seat them, if needed. (Cadillac)

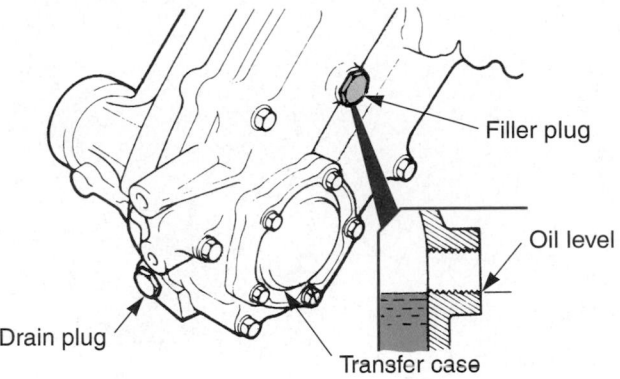

Figure 60-14. Remove the filler plug to check the amount of oil in a transfer case. The fluid level should be about even with the filler hole. (Chrysler)

If the oil is dirty or contaminated with water, drain and replace it. Make sure you install the fluid type and viscosity recommended by the manufacturer.

When removing a transfer case, use a transmission jack. A transfer case is heavy and can cause injury or part damage if it is dropped.

Figure 60-15 shows the parts of a modern transfer case. To repair a transfer case, follow the procedures outlined in a service manual. It will give directions for repairing the particular make and model. There are so many design variations, it is not in the scope of this book to explain the rebuilding procedures for all types of units

Summary

- To function properly, the drive shaft must be straight and its universal joints must be unworn.
- Drive shaft noises are usually caused by worn universal joints, slip joint wear, or a faulty center support bearing.
- Grinding and squeaking from the drive shaft is frequently caused by worn universal joints.
- A clunking sound during acceleration or deceleration may be caused by slip yoke problems or extension housing bushing wear.
- A whining noise from the drive shaft is sometimes caused by a dry, worn center support bearing.
- Drive shaft vibration can be caused by any problem that affects drive shaft balance, runout, and angle.
- To check for worn universal joints, wiggle and rotate each universal joint back and forth while watching for side movement.

- If the drive shaft is within recommended runout limits and still vibrates, the drive shaft may need to be balanced.
- Be extremely careful not to let any part of your body come in contact with the spinning drive shaft or rear wheels.
- A grease gun can be used to lubricate universal joints that have fittings
- A worn universal joint, the most common drive shaft problem, can cause squeaking, grinding, clunking, or clicking sounds.
- Wear safety glasses to protect your eyes in case the snap rings fly out of the joint during removal.
- If the bearing cap fails to press into place with normal pressure, disassemble the joint and check the roller bearings.
- When diagnosing transfer case problems, check the condition and level of oil.

Important Terms

Drive shaft noise	Grease gun
Drive shaft vibration	Worn universal joint
Drive shaft runout	Universal joint tools
Driveline angle	

Review Questions—Chapter 60

Please do not write in this text. Place your answers on a separate sheet of paper.

1. Drive shaft problems can normally be classified into two categories: drive shaft _____ and drive shaft _____.

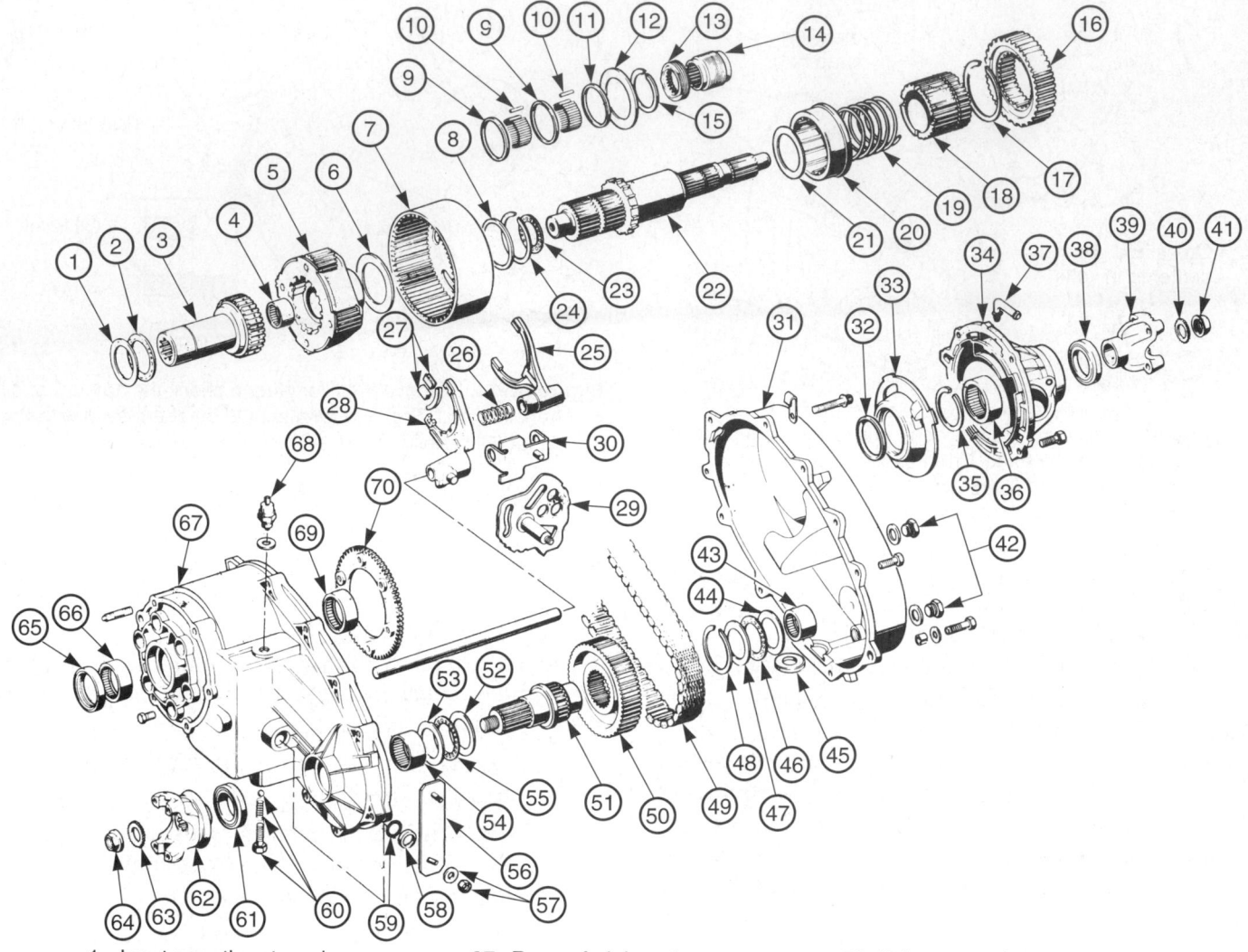

1. Input gear thrust washer
2. Input gear thrust bearing
3. Input gear
4. Main shaft pilot bearing
5. Planetary assembly
6. Planetary thrust washer
7. Annulus gear
8. Annulus gear thrust washer
9. Needle bearing spacers
10. Main shaft needle bearings (120)
11. Needle bearing spacer
12. Thrust washer
13. Oil pump
14. Speedometer gear
15. Drive sprocket retaining ring
16. Drive sprocket
17. Sprocket carrier stop ring
18. Sprocket carrier
19. Clutch spring
20. Sliding clutch
21. Thrust washer
22. Main shaft
23. Main shaft thrust bearing
24. Annulus gear retaining ring
25. Mode fork
26. Mode fork spring

27. Range fork inserts
28. Range fork
29. Range sector
30. Mode fork bracket
31. Rear case
32. Seal
33. Pump housing
34. Rear retainer
35. Rear output bearing
36. Bearing snap ring
37. Vent tube
38. Rear seal
39. Rear yoke
40. Yoke seal washer
41. Yoke nut
42. Drain and fill plugs
43. Front output shaft rear bearing
44. Front output shaft rear thrust bearing race (thick)
45. Case magnet
46. Front output shaft rear thrust bearing
47. Front output shaft rear thrust bearing race (thin)
48. Driven sprocket retaining ring
49. Drive chain

50. Driven sprocket
51. Front output shaft
52. Front output shaft front thrust bearing race (thin)
53. Front output shaft front thrust bearing race (thick)
54. Front output shaft front thrust bearing
55. Front output shaft front bearing
56. Operating lever
57. Washer and locknut
58. Range sector shaft seal retainer
59. Range sector shaft seal
60. Detent ball, spring, and retainer bolt
61. Front seal
62. Front yoke
63. Yoke seal washer
64. Yoke nut
65. Input gear oil seal
66. Input gear front bearing
67. Front case
68. Lock mode indicator switch and washer
69. Input gear rear bearing
70. Lockplate

Figure 60-15. During transfer case repairs, refer to instructions and illustrations in the service manual. It will give details of the particular unit being serviced. Study the construction of this transfer case. (Chrysler)

2. What are some common causes of drive shaft noise?

3. Grinding or squeaking from the drive shaft is frequently caused by:
   (A) worn slip joint.
   (B) bent drive shaft.
   (C) worn, dry universals.
   (D) None of the above.

4. List three common causes of drive shaft vibration.

5. How do you check for worn universal joints?

6. How do you check drive shaft runout?

7. How do you balance a drive shaft in-shop?

8. Summarize the procedure for disassembly of a cross-and-roller universal joint.

9. If a universal joint fails to press together with normal force, it is possible that one of the needle bearings has fallen out of place. True or False?

10. A transmission jack is commonly used when removing a transfer case. True or False?

● ASE-Type Questions

1. A grinding sound is coming from an automobile's drive shaft assembly. Technician A inspects the condition of the slip yoke. Technician B checks for a bad universal joint. Who is right?
   (A) A only.
   (B) B only.
   (C) Both A and B.
   (D) Neither A nor B.

2. A squeaking noise is believed to be coming from underneath an automobile. Technician A says this noise may indicate a bad universal joint. Technician B says the car's wheel covers may be producing this noise. Who is right?
   (A) A only.
   (B) B only.
   (C) Both A and B.
   (D) Neither A nor B.

3. A whining noise is coming from an automobile's drive shaft assembly. Technician A says this problem is the result of a bad transmission input shaft bushing. Technician B looks for a worn center support bearing. Who is right?
   (A) A only.
   (B) B only.
   (C) Both A and B.
   (D) Neither A nor B.

4. A customer complains that a vibration-type noise is coming from the bottom of his/her car. Technician A checks the drive shaft for runout. Technician B inspects the condition of the universal joint. Who is right?
   (A) A only.
   (B) B only.
   (C) Both A and B.
   (D) Neither A nor B.

5. Vibration is coming from underneath an automobile. Technician A checks the condition of the drive shaft. Technician B checks for an unbalanced engine flywheel. Who is right?
   (A) A only.
   (B) B only.
   (C) Both A and B.
   (D) Neither A nor B.

6. All of the following are common drive shaft problems *except*:
   (A) *drive shaft warpage.*
   (B) *worn universal joint.*
   (C) *damaged center support bearing.*
   (D) *carrier warpage.*

7. A car's drive shaft needs to be inspected for problems. Technician A inspects the balance weights. Technician B inspects the condition of the rear yoke bolts. Who is right?
   (A) A only.
   (B) B only.
   (C) Both A and B.
   (D) Neither A nor B.

8. A car's drive shaft needs to be checked for runout. Technician A is going to use an inside micrometer to perform this task. Technician B is going to use a dial indicator to perform this task. Who is right?
   (A) A only.
   (B) B only.
   (C) Both A and B.
   (D) Neither A nor B.

9. A car is brought into the shop to have the drive shaft inspected. Technician A places the transmission in neutral when measuring for shaft runout. Technician B places the transmission in park when measuring for shaft runout. Who is right?
   (A) A only.
   (B) B only.
   (C) Both A and B.
   (D) Neither A nor B.

10. An automobile's drive shaft has excessive runout. Technician A believes a bent drive shaft may have caused this problem. Technician B thinks a damaged slip yoke could have caused this problem. Who is right?
    (A) A only.
    (B) B only.
    (C) Both A and B.
    (D) Neither A nor B.

11. Drive shaft runout on a particular automobile exceeds 0.030″. Technician A says this measurement indicates that runout is still within specifications. Technician B says this measurement indicates that runout is beyond specifications. Who is right?
    (A) A only.
    (B) B only.
    (C) Both A and B.
    (D) Neither A nor B.

12. An automobile is brought into the shop with an incorrect drive shaft angle. Technician A adjusts the drive shaft angle by placing shims under the rear axle housing. Technician B adjusts the drive shaft angle by placing shims under the transmission mount. Who is right?
    (A) A only.
    (B) B only.
    (C) Both A and B.
    (D) Neither A nor B.

13. A customer wants his/her automobile's universal joints greased. Technician A tells the customer that certain modern universal joints are normally sealed units that are filled with grease at the factory. Technician B tells the customer that some universal joints have grease fittings that allow for lubrication. Who is right?
    (A) A only.
    (B) B only.
    (C) Both A and B.
    (D) Neither A nor B.

14. A drive shaft needs to be removed from a vehicle. To begin this procedure, Technician A removes the drive shaft's slip yoke mounting bolts. To begin this procedure, Technician B removes the drive shaft's rear yoke mounting bolts. Who is right?
    (A) A only.
    (B) B only.
    (C) Both A and B.
    (D) Neither A nor B.

15. A transfer case's lubricant level is being checked. Technician A checks the lubricant level by removing and reading the dipstick. Technician B checks the lubricant level by removing the fill plug. Who is right?
    (A) A only.
    (B) B only.
    (C) Both A and B.
    (D) Neither A nor B.

Activities—Chapter 60

1. Inspect a universal joint on a vehicle in the shop. State its condition and indicate if replacement is required.

2. Explain and demonstrate a procedure for checking runout on a drive shaft.

3. Replace a universal joint using the appropriate procedure.

4. Demonstrate the proper technique for removing and replacing a drive shaft.

Drive Line Problem Diagnosis

Condition	Possible Causes	Correction
Noisy operation.	1. U-joint fasteners (U-bolts, cap screws) loose. 2. Lack of lubricant in U-joints. 3. Worn U-joint. 4. Worn center support bearing. 5. Loose center support. 6. Joint or shaft striking some part of vehicle underbody. 7. Worn CV-joint (front-wheel drive).	1. Tighten fasteners. 2. Lubricate U-joints. 3. Replace U-joint. 4. Replace support bearing. 5. Tighten support fasteners. 6. Shim, tighten, or replace center mount. Check for debris in frame tunnel and for worn mounts. 7. Replace CV-joint.
Propeller shaft vibration or shudder.	1. U-joint fasteners loose. 2. Worn U-joint. 3. Shaft sprung or dented. 4. Undercoating on shaft. 5. Joint flange surface nicked or burred. 6. Worn slip joint splines. 7. Dry slip joint. 8. Shaft yokes out of phase (not aligned). 9. Shaft yoke and slip yoke assembled wrong. 10. Shaft yoke and pinion flange yoke assembled wrong. 11. Cross not centered in yoke. 12. U-joints tight. 13. Roller U-bolts over tightened. 14. Drive angle wrong. 15. Loose center support. 16. Center support rubber insulator deteriorated. 17. Worn center support bearing. 18. Loose rear spring U-bolts. 19. Loose rear axle housing control arm bolts. 20. Loose pinion companion flange retaining nut. 21. Weak springs. 22. Rear spring center bolt sheared, axle housing shifted.	1. Tighten U-joint fasteners. 2. Replace U-joint. 3. Replace drive shaft. 4. Remove undercoating. 5. Disassemble. File off burrs. 6. Replace slip yoke and/or stub shaft. 7. Clean and lubricate. 8. Align yokes as required. 9. Disconnect slip yoke. Rotate 180° and reconnect. 10. Disconnect flange yoke. Rotate 180° and reconnect. 11. Strike yoke to move rollers out against snap rings. 12. Strike yoke lugs to free. Replace joint if needed. 13. Loosen and torque properly. 14. Check and adjust as required. 15. Tighten center support. 16. Replace center support bearing. 17. Replace support bearing. 18. Torque bolts. 19. Tighten bolts. Replace bushings if worn. 20. Tighten nut. 21. Replace springs. 22. Realign axle housing (if needed) and replace center bolt.
CV-axle shaft vibration or shudder.	1. Worn CV-joint. 2. Bent CV-axle shaft. 3. Worn front wheel bearings.	1. Replace CV-joint. 2. Replace shaft. 3. Replace bearings.

Transfer Case Diagnosis
Part-Time Drive

Condition	Possible Causes	Correction
Jumps out of gear in two-wheel drive.	1. Shift lever detent spring weak or broken. 2. Sliding clutch spline engaging surface worn or tapered.	1. Replace spring. 2. Replace worn parts.
Noise. Note: Transfer cases using a gear drive produce considerable gear whine, which is normal.	1. Worn bearings, splines, chipped gears, or worn shafts. 2. Low lubrication level. 3. Loose or broken mounts.	1. Rebuild unit. 2. Fill to proper level. 3. Tighten or replace mounts.

(Continued)

Transfer Case Diagnosis		
Part-Time Drive (Continued)		
Condition	**Possible Causes**	**Correction**
Jumps out of gear in four-wheel drive.	1. Shift lever interference with floor pan. 2. Excessive transfer case movement. 3. Sliding clutch engaging surfaces tapered or worn. 4. Bent shift fork. 5. Shift rod detent spring weak or broken. 6. Shift lever torsion spring (where used) not holding. 7. Worn bearings, gear teeth, or shafts.	1. Provide proper clearance. 2. Check and replace transfer case mounts. 3. Replace worn parts. 4. Replace shift fork. 5. Replace detent spring. 6. Replace torsion spring. 7. Overhaul unit.

Transfer Case Diagnosis		
Full-Time Drive		
Condition	**Possible Causes**	**Correction**
Noisy operation.	1. Low lubrication level. 2. Operating in "lockout" on hard, dry surface roads. 3. Improper lubricant. 4. "Slip-stick" condition ("Quadra-Trac" type). Makes a grunting, pulsating, rasping sound. 5. Excessive wear on gears, chains, or differential unit. 6. Loose or deteriorated mounts.	1. Fill to correct level. 2. Shift out of "lockout." Advise driver. 3. Drain and fill with recommended lubricant. 4. Normal if vehicle has not been driven for a week or two. Should stop after some usage. If it persists, drain fluid and refill. Use special additive, if required. Make certain tire sizes are the same and pressures are equal. 5. Rebuild as needed. 6. Tighten or replace.
Jumps out of low range and/or is hard to shift into or out of low range.	1. Shift linkage improperly adjusted, bent, or broken. 2. Shift rails dry or scored. 3. Improper driver operation. 4. Reduction unit parts worn or damaged.	1. Adjust correctly. Straighten or replace. 2. Clean, polish, or lubricate or replace as needed. 3. Follow shift procedure recommended by manufacturer. 4. Repair as needed.
Lockout will not engage.	1. Lockout parts damaged. 2. Defective vacuum control. Loose or damaged vacuum lines. 3. Defective shift linkage.	1. Repair as needed. 2. Replace control. Replace or connect vacuum hoses ("Quadra-Trac"). 3. Repair or replace.
Will not engage in two-wheel drive.	1. No vacuum. Loose or broken hoses. 2. Defective shift motor (axle). 3. Defective shift motor (transfer case).	1. Replace hoses. Secure all loose connections. 2. Replace shift motor. 3. Replace shift motor.
Will not engage in four-wheel drive.	1. No vacuum. Loose or broken hoses. 2. Defective axle shift motor. 3. Binding or broken transfer case shift linkage. 4. Defective axle shift linkage. 5. Damaged transfer case.	1. Replace hoses. Secure all loose connections. 2. Replace shift motor. 3. Repair or replace shift linkage. 4. Repair or replace shift linkage. 5. Repair or replace transfer case.
Vehicle wanders when driving straight ahead.	1. Improperly matched tire size. 2. Uneven tire pressure.	1. Use a matched set of tires. 2. Adjust air pressure to recommended levels.

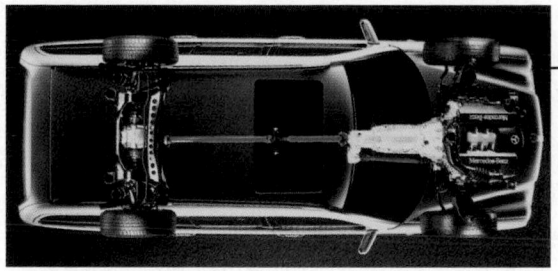

Chapter 61

Differential and Rear Drive Axle Fundamentals

After studying this chapter, you will be able to:

- Identify the major parts of a rear drive axle assembly.
- List the functions of a rear axle assembly.
- Describe the operation of a differential.
- Explain differential design variations.
- Compare different types of axles.
- Describe the principles of a limited-slip differential.
- Relate rear axle ratios to vehicle performance.
- Correctly answer ASE certification test questions requiring a knowledge of differential and rear drive axle fundamentals.

Many new cars and most trucks use a front-engine, rear-wheel-drive setup. After engine power flows through the transmission and drive shaft, it enters the rear axle assembly. The rear axle assembly's differential and axle shafts then transfer torque to the rear drive wheels.

As you will learn in later chapters, front-engine, front-wheel-drive vehicles also use a differential, but it is located inside the transaxle. This chapter will prepare you to understand transaxles and front-wheel-drive systems.

Basic Rear Drive Axle Assembly

A simple *rear drive axle assembly* is shown in **Figure 61-1.** It consists of the following parts:

- *Pinion drive gear*—transfers power from the drive shaft to the ring gear.
- *Ring gear*—transfers turning power to the differential case assembly.
- *Differential case assembly*—holds the ring gear and other components that drive the rear axles.

- *Rear drive axles*—steel shafts that transfer torque from the differential assembly to the drive wheels.
- *Rear axle bearings*—ball or roller bearings that fit between the axles and the inside of the axle housing.
- *Axle housing*—metal body that encloses and supports parts of the rear axle assembly.

Rear Axle Power Flow

Power enters the rear axle assembly from the drive shaft. The drive shaft spins the pinion gear. The pinion gear turns the larger ring gear to produce a gear reduction, **Figure 61-1.**

The ring gear is bolted to the differential case, so the case rotates with the ring gear. Small gears inside the differential case send torque to each axle.

The axles extend beyond the axle housing. The axles normally hold and turn the rear wheels to propel the vehicle.

Functions of a Rear Drive Axle Assembly

A *rear drive axle assembly* has several functions. These include:

- Sending power from the drive shaft to the rear wheels.
- Providing a final gear reduction.
- Transferring torque through a 90° angle.
- Splitting the amount of torque going to each wheel.
- Allowing the wheels to rotate at different speeds in turns.
- Supporting the rear axles, brake assemblies, suspension components, and chassis.

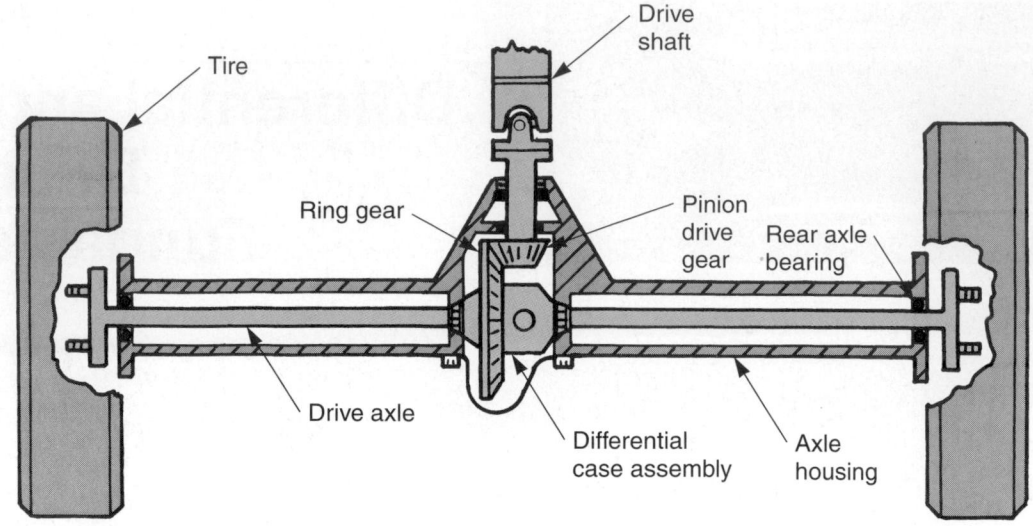

Figure 61-1. Study the fundamental components of a rear drive axle assembly. The drive shaft turns the pinion gear. The pinion gear turns the ring gear and differential case assembly. The differential transfers power to the drive axles and wheels.

Differential Construction

A *differential assembly* uses drive shaft rotation to transfer power to the axle shafts. The differential must be capable of providing torque to both axles, even when they are turning at different speeds (when the vehicle is turning a corner, for example).

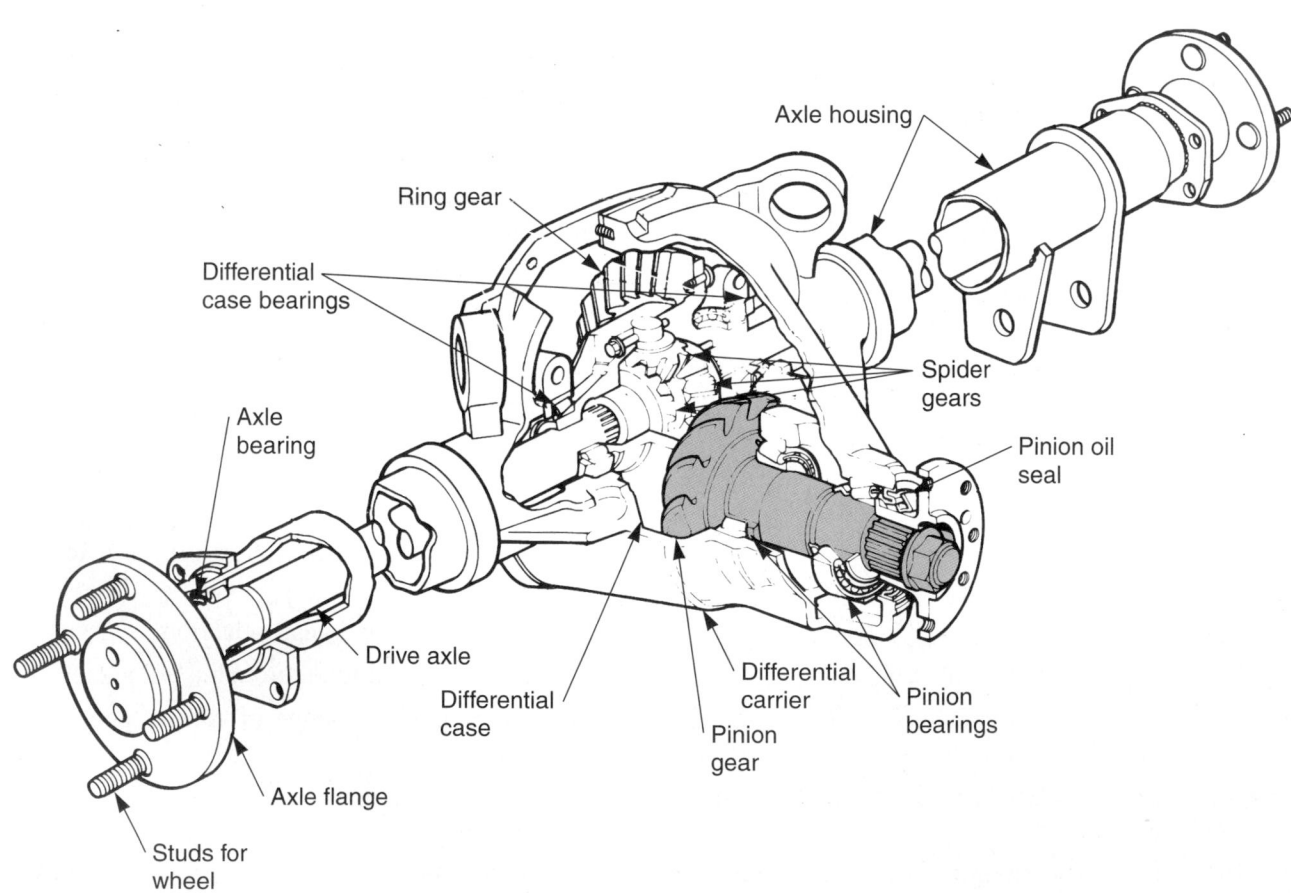

Figure 61-2. This cutaway shows a detailed view inside the axle housing. Note the relationship of parts. The axle housing fastens to suspension components. (Ford)

Pinion Gear and Ring Gear

The *pinion gear* turns the ring gear when the drive shaft is rotating. One is shown in **Figure 61-2.**

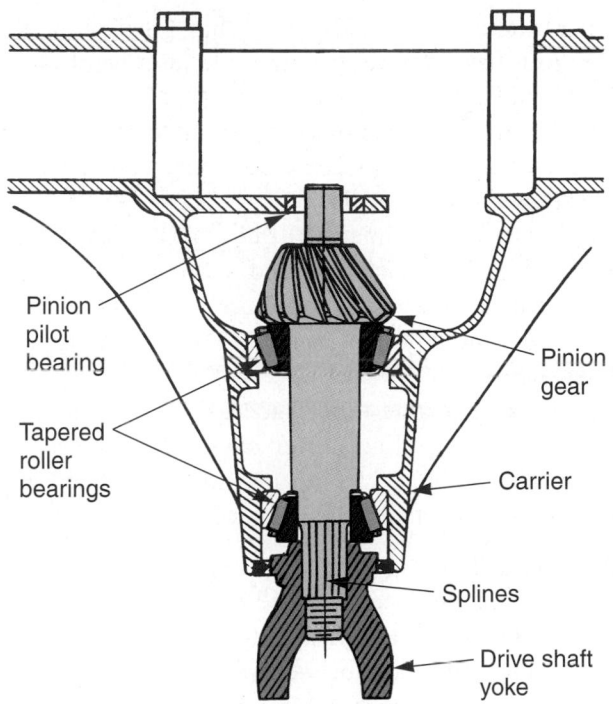

Figure 61-3. The pinion gear is mounted on tapered roller bearings and sometimes a pinion pilot bearing. The drive shaft yoke is splined to the pinion gear. (Ford)

The outer end of the pinion drive gear is splined to the rear universal joint companion flange or yoke. The inner end of the pinion gear meshes with the teeth on the ring gear. The pinion gear is normally mounted on tapered roller bearings. They allow the pinion gear to revolve freely in the carrier. Either a crushable sleeve or shims are used to preload the pinion gear bearings. *Gear preload* is a small amount of pressure applied to the bearings to remove play and excess clearance.

With some differentials, the extreme inner end of the pinion gear is supported by a *pinion pilot bearing.* This is a straight roller bearing. The pinion pilot bearing helps the two tapered roller bearings support the pinion gear during periods of heavy load. See **Figure 61-3.**

The *ring gear* is driven by the pinion gear and transfers rotating power through an angle change of 90°. The ring gear has more teeth than the pinion gear. Bolts hold the ring gear securely to the differential case. Refer to **Figure 61-4.**

The ring gear and pinion gear are commonly a matched set. They are *lapped* (meshed and spun together with abrasive compound on their teeth) at the factory. Then, one tooth on each gear is marked to show the correct tooth engagement. Lapping produces quieter operation and ensures longer gear life.

Hunting and Nonhunting Gears

A *hunting gearset* does not mesh the same gear teeth during each revolution of the ring gear. The number of

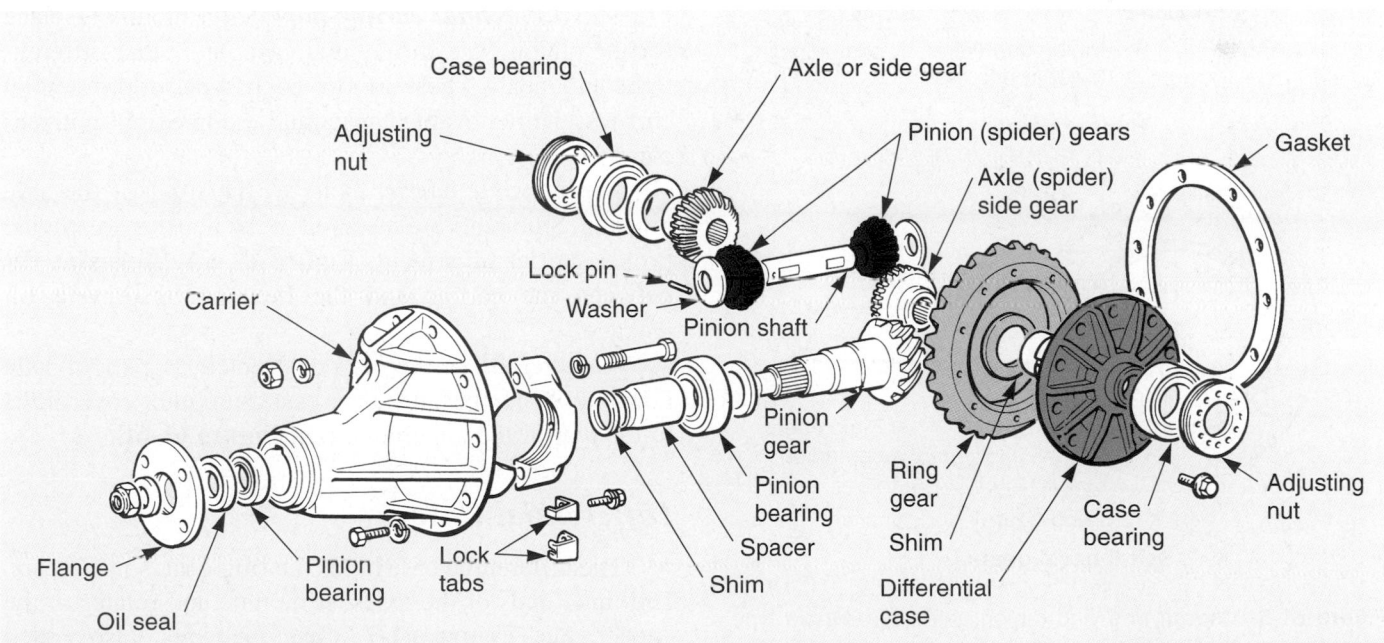

Figure 61-4. This exploded view shows all the major parts of a differential assembly. The ring gear bolts to the differential case. The case mounts in bearings, which are adjusted using large nuts. Study the part names and how they fit together. (Chrysler)

teeth on the pinion compared to the number or teeth on the ring gear causes different teeth to mesh.

A *nonhunting gearset* meshes the same gear teeth over and over during gearset operation. Most ring and pinion gears are nonhunting types. They will have markings that must be aligned during assembly.

Hypoid and Spiral Bevel Gears

Hypoid gears have the driving pinion centerline offset, or lowered, from the centerline of the ring gear. Modern differential ring and pinion gears are hypoid type. Refer to **Figure 61-5.**

Spiral bevel gears have curved gear teeth with the pinion and ring gears on the same centerline. This type gear setup is no longer used.

Hypoid gears have replaced spiral bevel gears because they allow for a lower hump in the vehicle floor and improve gear meshing action. Hyphoid gears have a larger gear tooth contact area than spiral bevel gears. This helps increase gear life and reduce gear noise.

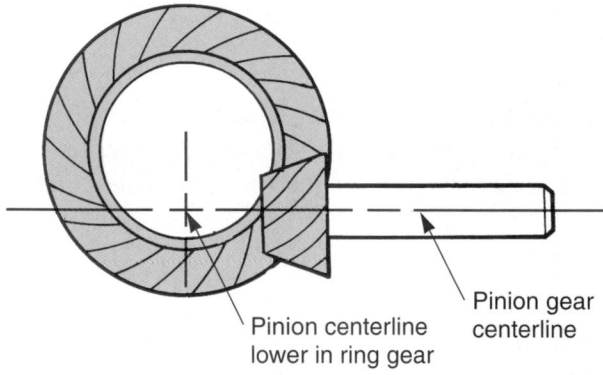

Hypoid gears

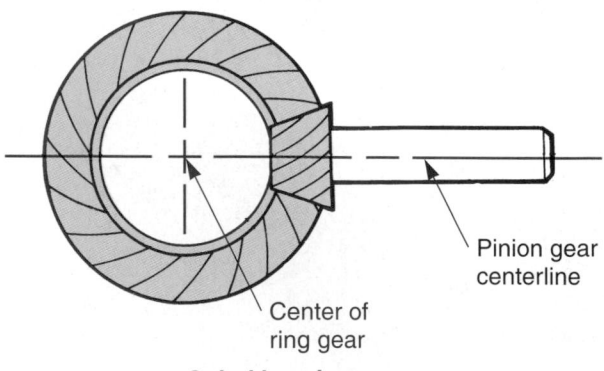

Spiral bevel gears

Figure 61-5. Modern ring and pinion gears are hypoid type. Note how hypoid lowers the centerline of the pinion gear. This improves gear tooth contact and lowers the drive shaft hump in the floor of the vehicle.

Rear Axle Ratio

Rear axle ratio, or *differential ratio,* is the ratio of ring gear teeth to pinion gear teeth. To calculate rear axle ratio, count the number of teeth on each gear. Then divide the number of ring gear teeth by the number of pinion teeth.

For example, if the ring gear has 30 teeth and the pinion gear has 10 teeth, the rear axle ratio would be 3:1 (30 divided by 10).

Generally, automakers use a rear axle ratio that provides a compromise between performance and economy. An average ratio is 3.50:1.

A higher axle (numerical) ratio, 4.11:1 for instance, would increase acceleration and pulling power, but it would also decrease fuel economy. The engine would have to run at a higher speed to maintain an equal cruising speed.

A lower rear axle ratio, such as 3:1, would reduce acceleration and pulling power but would increase fuel economy. The engine would run at a lower speed while maintaining a constant vehicle speed.

Tech Tip

Do not become confused about the terms "rear axle ratio" and "final drive ratio." Rear axle ratio is the ratio between the differential pinion and ring gears. Final drive ratio, or overall drive ratio, includes the axle ratio and the transmission/transaxle high-gear ratio.

Differential Carrier

The *differential carrier* provides a mounting place for the pinion gear, differential case, and other differential components. There are two basic types of differential carriers: the removable carrier and the integral (unitized) carrier.

A *removable carrier* bolts to the front of the axle housing. Stud bolts are installed in the housing to provide proper carrier alignment, **Figure 61-6A.** A gasket fits between the carrier and the housing to prevent oil leakage.

An *integral carrier* is constructed as part of axle housing. A stamped metal or cast aluminum cover bolts to the rear of the integral carrier, **Figure 61-6B.**

Differential Case

The differential case holds the ring gear, spider gears, and inner ends of the axles. It mounts and rotates in the carrier. See **Figure 61-7.** *Case bearings,* also called *carrier bearings,* fit between the outer ends of the differential case and the carrier.

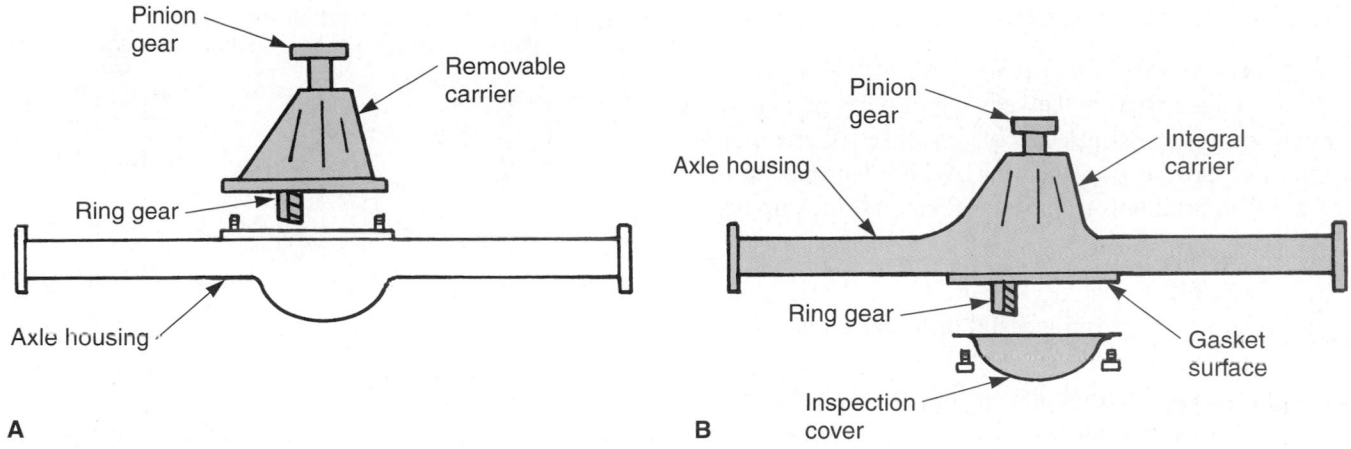

Figure 61-6. Two basic types of rear axle carriers. A—A removable carrier is handy because it can be serviced at a workbench. B—An integral or unitized carrier is formed as part of the axle housing.

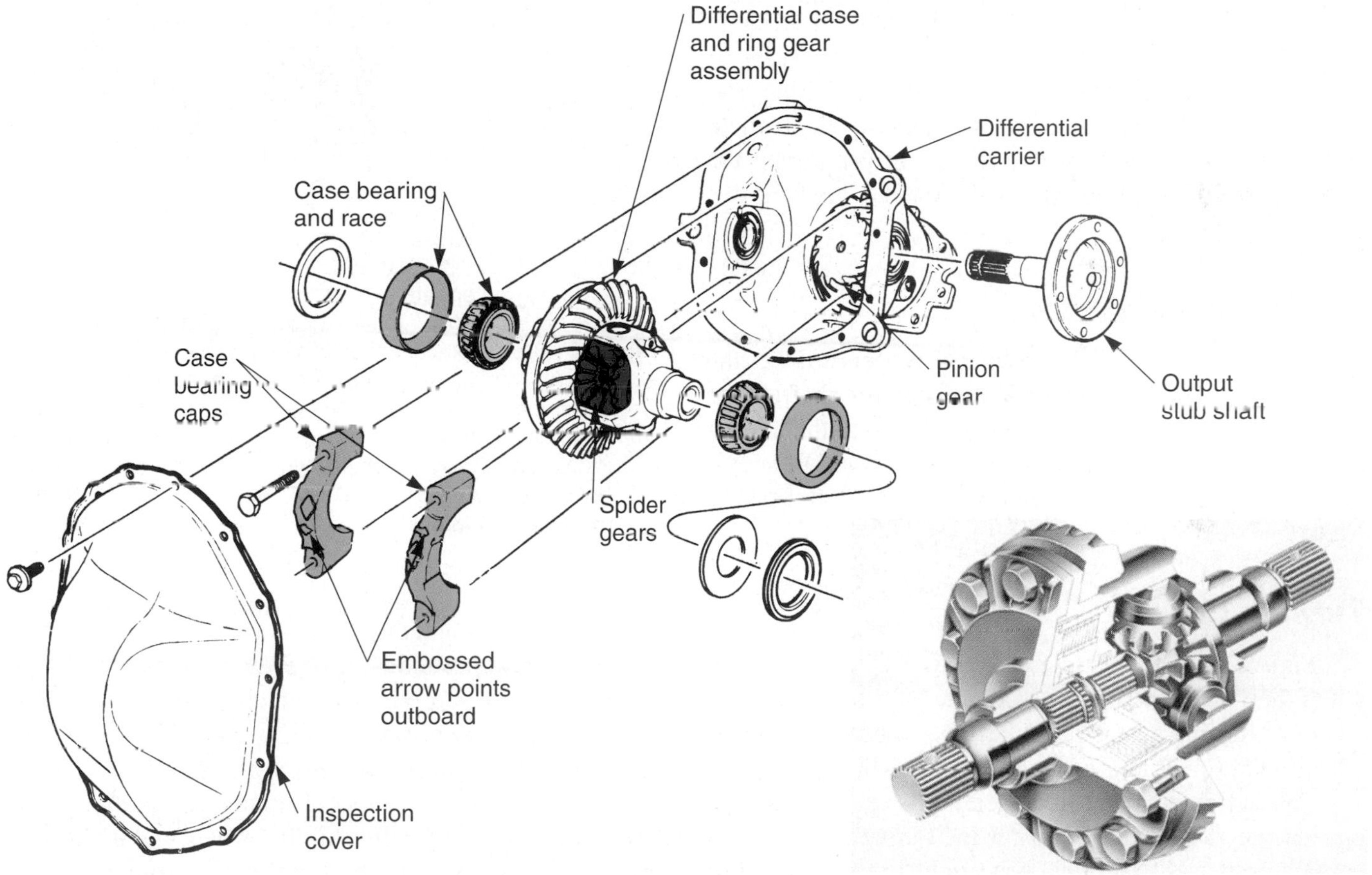

Figure 61-7. The differential case mounts in the carrier on tapered roller bearings. Races fit between the bearings and the carrier. Large caps secure bearing assemblies. Inset—Cutaway of rear differential. (Cadillac, Subaru)

Spider Gears

The **spider gears,** or **differential gears,** include two **axle gears** (differential side gears) and two **pinion gears** (differential idler gears). The spider gears mount inside

the differential case. They are small bevel gears. Refer again to **Figure 61-4.**

A **pinion shaft** passes through the two pinion gears and the case. The two axle gears are splined to the inner ends of the axles.

Differential Lubricant

Differential lubricant, usually SAE 80W-90 gear oil, is used to reduce friction between the moving parts in the rear axle assembly. Ring gear rotation splashes the oil on all moving parts to prevent wear. Without proper lubricant, a differential will usually overheat and lock up very quickly.

Tech Tip
A clutch-type limited-slip differential usually requires a special gear lubricant for the clutch pack. The friction discs will not function properly with regular gear oil.

Differential Action

The rear wheels of a car do not always turn at the same speed. When the car is turning or when tire diameters differ slightly, the rear wheels must rotate at different speeds.

If there were a solid connection between each rear axle and the differential case, the tires would tend to slide, squeal, and wear whenever the driver turned the steering wheel. A differential is designed to prevent this problem.

Driving Straight Ahead

When driving straight, both rear wheels turn at the same speed. The spinning case and pinion shaft rotate the differential pinion gears. The teeth on the pinion gears apply torque to the axle gears and axles. Balanced forces make the differential appear to be locked. Look at **Figure 61-8A.**

Turning Corners

Figure 61-8B illustrates the action of a differential when the vehicle is turning a corner. Note how the outer wheel is turning faster than the inner wheel. The outer wheel must travel farther than the inner wheel.

The action of the spider gears allows each axle to change speed while still transferring torque to propel the car. Without a differential, you could break an axle or wear out tires because of the different turning speeds of the rear wheels.

Limited-Slip Differentials

With a conventional differential, there may not be adequate traction on slippery pavement, in mud, or during rapid acceleration.

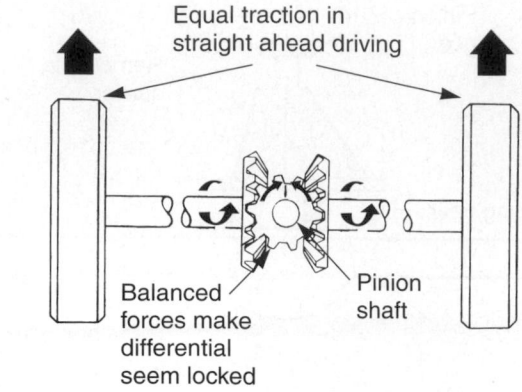

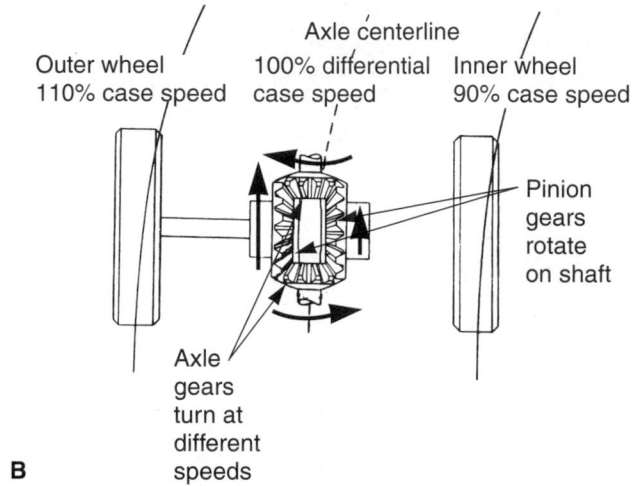

Figure 61-8. Differential action allows wheels to turn at different speeds. A—Vehicle traveling straight. The differential spider gears are inactive. Spider gears seem to be locked. B—With the car turning a corner, the outer wheel must travel farther and turn faster than the inner wheel. Pinion spider gears can rotate on their shaft, allowing axle side gears to turn at different speeds. (General Motors)

When one wheel of a conventional rear axle assembly lacks traction (on ice for example), the other wheel will not propel the vehicle. Torque will flow through the spider gears and to the axle that turns easiest.

A *limited-slip differential* provides driving force to both rear wheels at all times. It transfers a portion of the driving torque to both the slipping wheel and the driving wheel. This helps prevent the vehicle from becoming stuck in mud or snow.

Other names for a limited-slip differential are positraction, sure-grip, equal-lock, and no-spin.

Clutch Pack Differential

The most popular type of limited-slip differential uses a *clutch pack* (set of friction discs and steel plates).

The friction discs are sandwiched between the steel plates inside the differential case. See **Figure 61-9.**

The friction discs are usually splined to the differential side gears. The steel plates have tabs that lock into notches in the differential case. The friction discs turn with the axle side gears. The plates turn with the case, **Figure 61-10.**

Springs (belleville springs, coil springs, or a leaf spring) force the friction discs and steel plates together. As a result, both rear axles try to turn with the differential case.

The thrust action of the spider gears normally helps the clutch spring(s) apply the clutch pack. Under high-torque conditions, the rotation of the differential pinion gears pushes out on the axle side gears. The axle side gears then push on the clutch discs. This action helps lock the discs and keep both rear wheels turning.

However, when driving normally, the vehicle can turn a corner without both wheels rotating at the same speed. The clutch pack will slip in turns, **Figure 61-11.**

Cone Clutch Differential

A *cone clutch differential* uses the friction produced by cone-shaped axle gears to provide improved traction. See **Figure 61-12.** Springs are used to force the cones against the ends of the differential case. With the axles splined to the cone gears, the axles tend to rotate with the case.

Under rapid acceleration, the differential pinion gears push outward on the cone gears. This increases friction between the cones and differential case, and the drive wheels are turned with even greater torque.

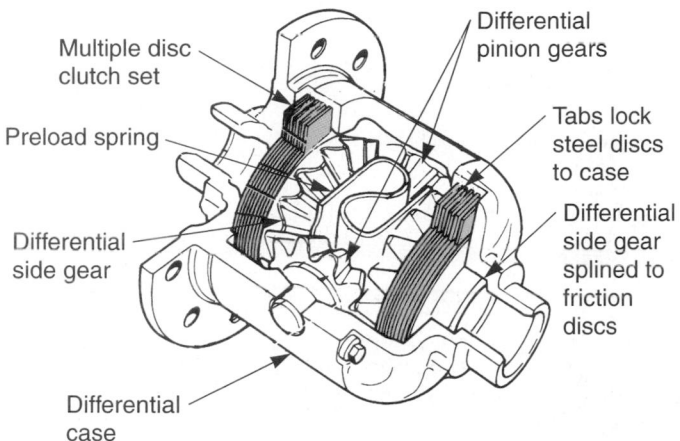

Figure 61-9. This section view of a limited-slip differential shows how clutch discs and spider gears fit in the case. A spring pushes the side gears and clutch discs together. This produces friction, causing both axles to drive the vehicle. (Ford)

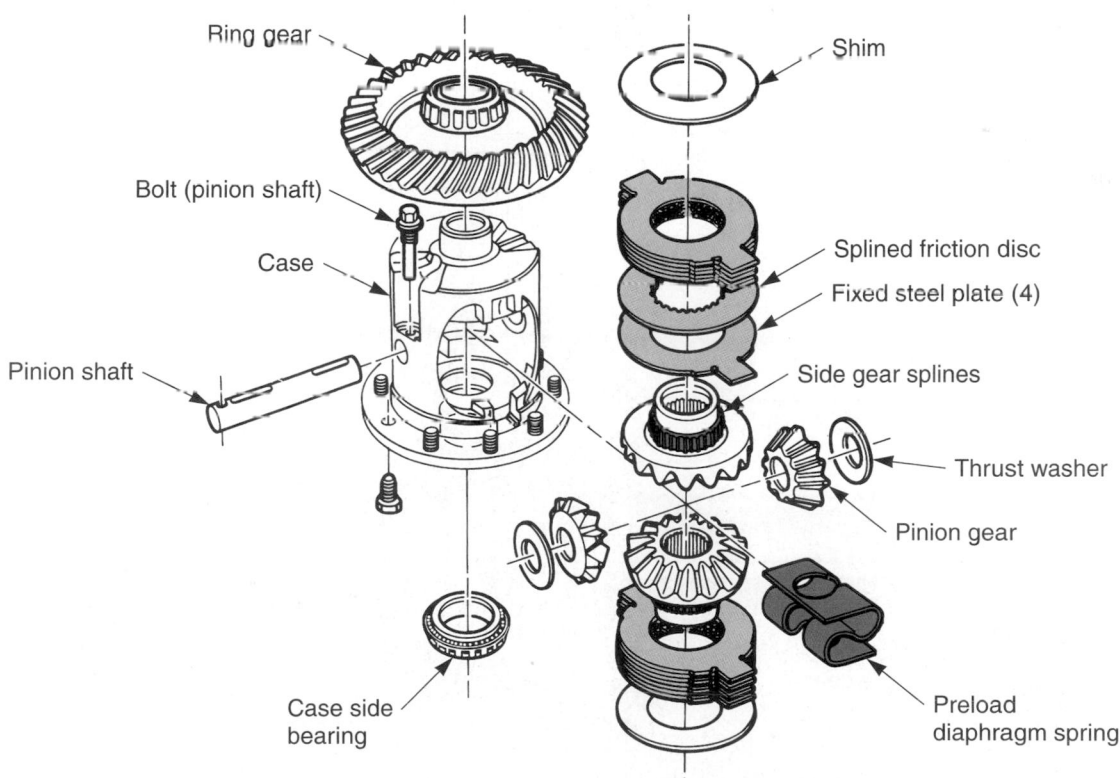

Figure 61-10. This limited-slip differential uses belleville or diaphragm springs to preload the clutch discs. The friction discs are splined to the axle side gears. Steel discs are locked to the case by large tabs. (Chrysler)

Ratchet Differential

A *ratchet differential,* also called a *Detroit locker,* uses a series of cams and ramps to direct torque to the drive axle with the most traction. Its operation is derived from relative wheel speed, rather than traction. The ratchet differential sends power through sets of teeth (*dog clutch*) that can engage and disengage to keep torque applied to the slowest turning axle. The operation of a ratchet differential is illustrated in **Figure 61-13.**

Torsen Differential

A Torsen® differential is a locking differential that uses complex worm gearsets. The gearsets include worms (drive gears) and worm wheels (driven gears).

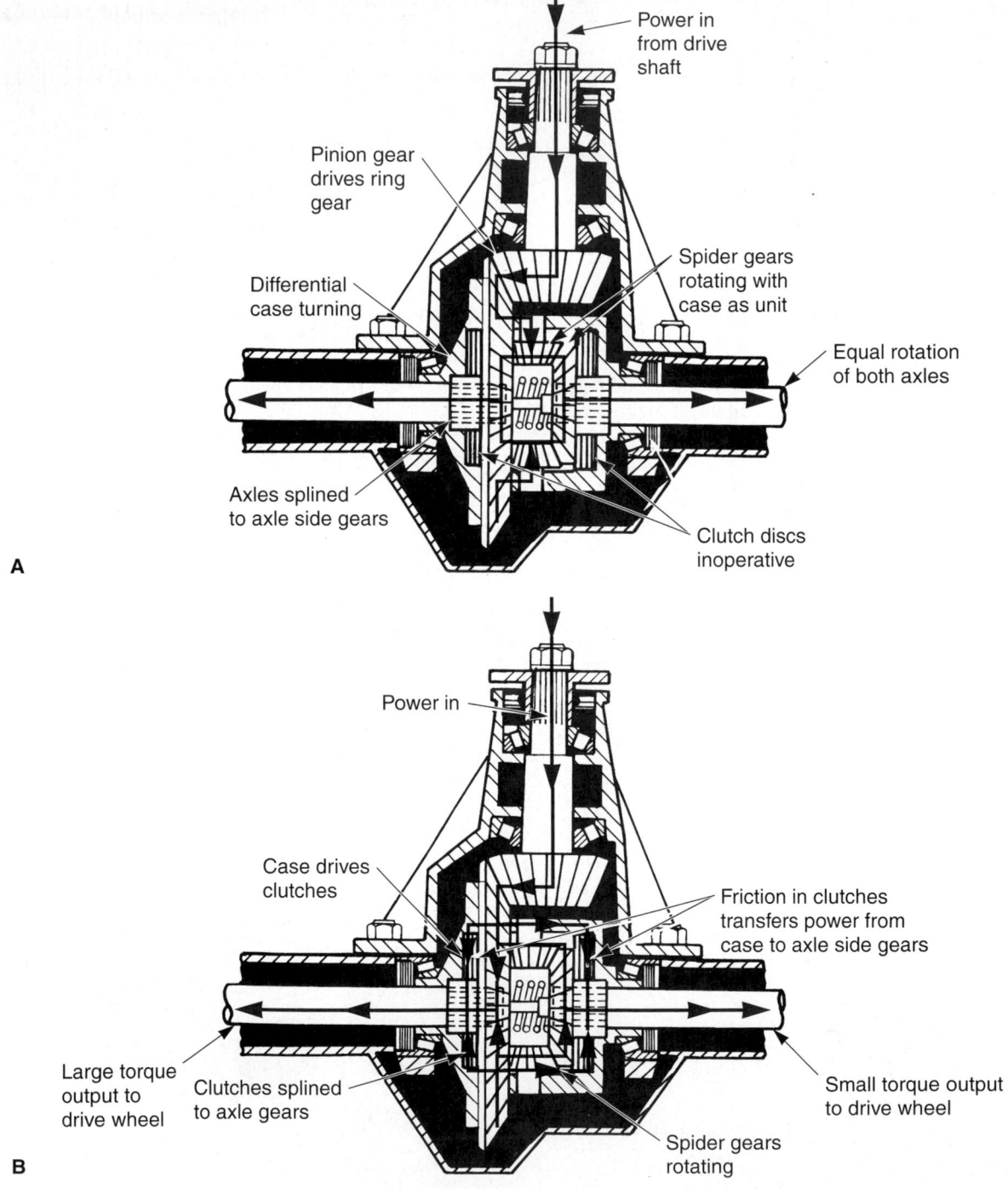

Figure 61-11. Power flow through a limited-slip or posi-traction differential. A—Car is traveling straight ahead and on dry pavement. Spider gears transfer power normally. Clutch discs turn together and do not slip. B—One wheel spins on slippery pavement. Friction between clutches still transfers torque to the other axle. This gives the vehicle more traction than with a conventional differential.

The basic principle behind this differential design is that worm gears can drive other gears but cannot be driven. The construction of a Torsen differential is shown in **Figure 63-14.**

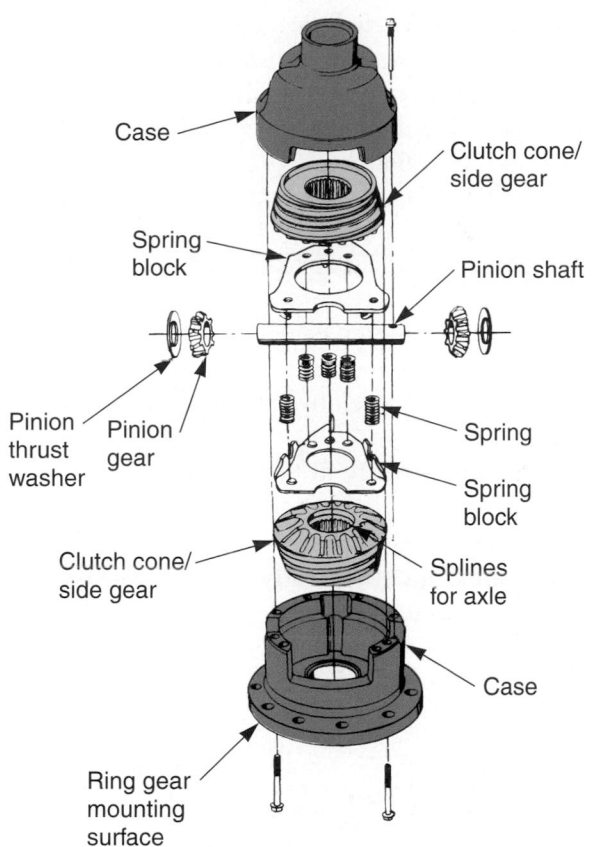

Figure 61-12. A cone clutch limited-slip differential. The cone surface on the axle side gears serves as friction surface to drive both axles. Increased engine torque pushes the side gears and cones outward to lock the axles. In turns, the side thrust on axles helps release one axle. (Oldsmobile)

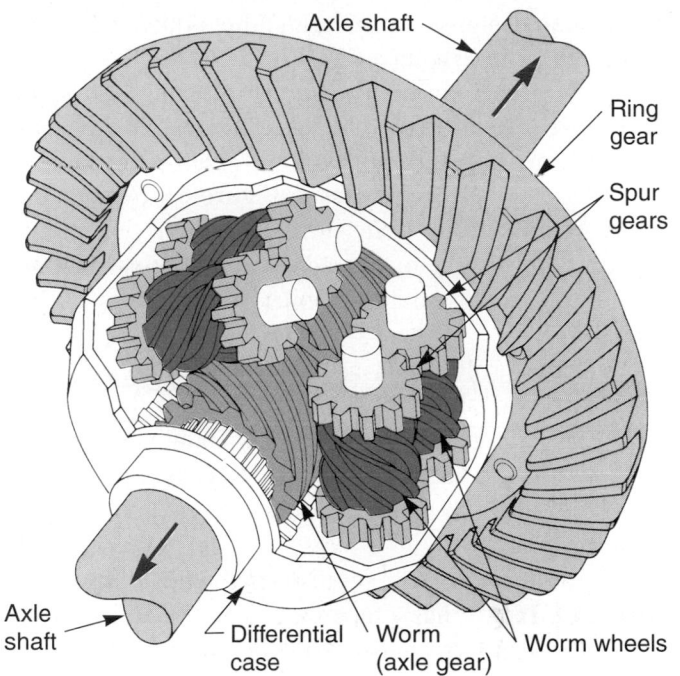

Figure 61-14. A Torsen differential uses a complex series of gears to send the required torque to both drive axles. This differential has been available for years but is just now being used in some exotic, high-priced production vehicles. (Torsen)

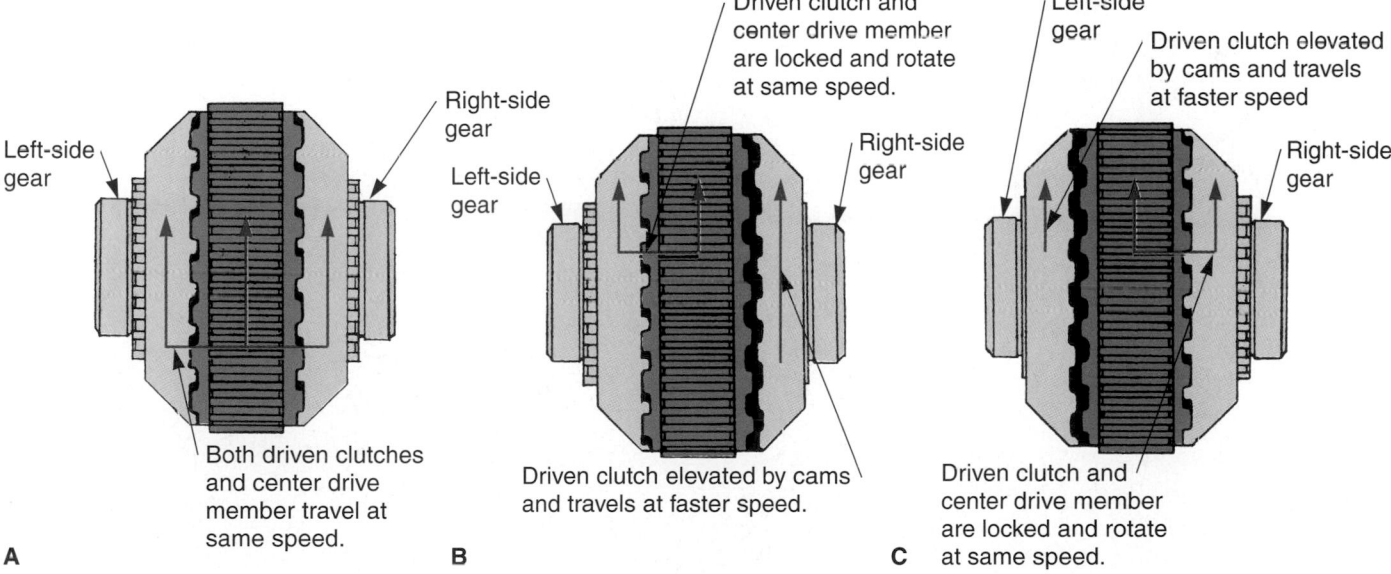

Figure 61-13. A ratchet differential uses matching sets of teeth on each side of the differential case. Teeth are engaged and disengaged to transfer power. A—The differential during straight ahead operation. Teeth are engaged on both sides of the case, and power is transferred equally to each wheel. B—When the vehicle makes a left turn, the greater speed of the right wheel causes the internal cam on the right side of the case to disengage the right-side teeth. All power goes through the left axle and wheel. C—When making a right turn, the left-side cam disengages the left-side teeth. (Ford)

Rear Drive Axles

The *rear drive axles* connect the differential side gears to the drive wheels. They normally support the weight of the vehicle. Rear axles are usually induction-hardened for increased strength. There are several types of rear axle designs: semifloating, three-quarter floating, full-floating, and swing axles, **Figure 61-15.**

Modern cars normally use semifloating and swing axles. Four-wheel-drive vehicles and some pickup trucks use three-quarter and full-floating axles. Compare the construction of the axle types in **Figure 61-16.**

A *semifloating axle* turns the drive wheel and supports the weight of the car. It is the most common type of axle found on automobiles. See **Figures 61-16A** and **61-16B.**

A ball or roller bearing fits between the axle shaft and the axle housing. Splines on the inner end of the axle fit into matching splines in the differential side gears. A flange is usually machined on the outer end of the axle shaft. A collar may be used to hold the axle bearing on the axle.

A variation of the semifloating axle is shown in **Figure 61-16B.** It has a tapered end that accepts a wheel hub. A key and large nut lock the hub to the axle.

The *rear wheel bearings* reduce friction between the axle and axle housing. They allow the axle to turn freely. The inner bearing race fits against the axle. The outer bearing race fits into the machined end of the axle housing. Ball or roller bearings can be used.

The *rear axle seals* usually press into the axle housing, as shown in **Figure 61-16A.** The seal lips contact the axle or axle collar to prevent lubricant leakage from the housing.

Axle retainer plates can bolt to the outside of the axle housing to keep the axles from sliding out. They are normally used with a removable carrier differential. See **Figure 61-17.**

Special bolts usually fit through holes in the housing flange and retainer. Nuts screw on these bolts to secure the axle into the housing.

Axle shims are frequently used between the axle retainer plate and the housing to limit axle end play. The thinner the shim, the greater the end play. Refer to **Figure 61-18.**

Swing Axles

Swing axles are used when the differential is rigidly mounted on the car's frame. Universal joints in the axles

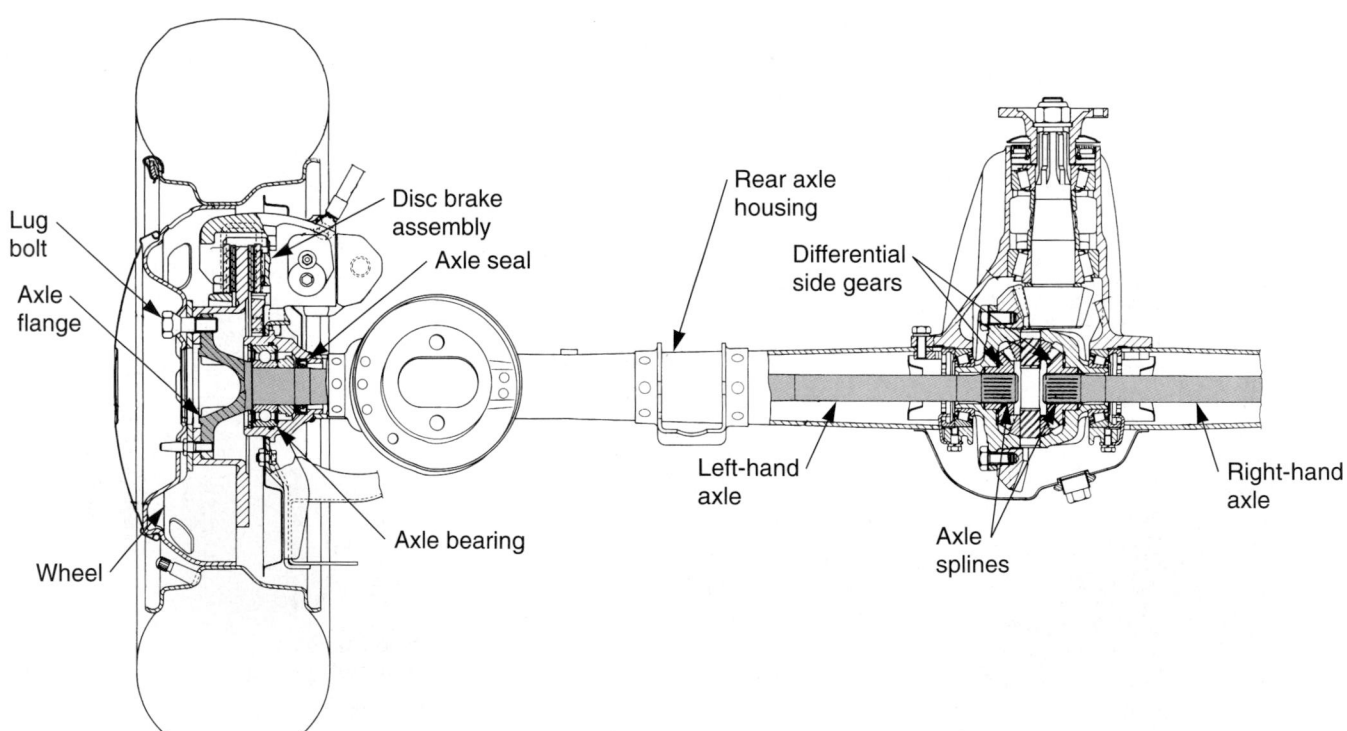

Figure 61-15. A solid steel, case-hardened rear drive axle extends beyond the differential to the outside of the axle housing. The case and side gear support the inner end of the axle. The rear wheel bearing supports the outer end of the axle. Also note the flange on the axle for the wheel.

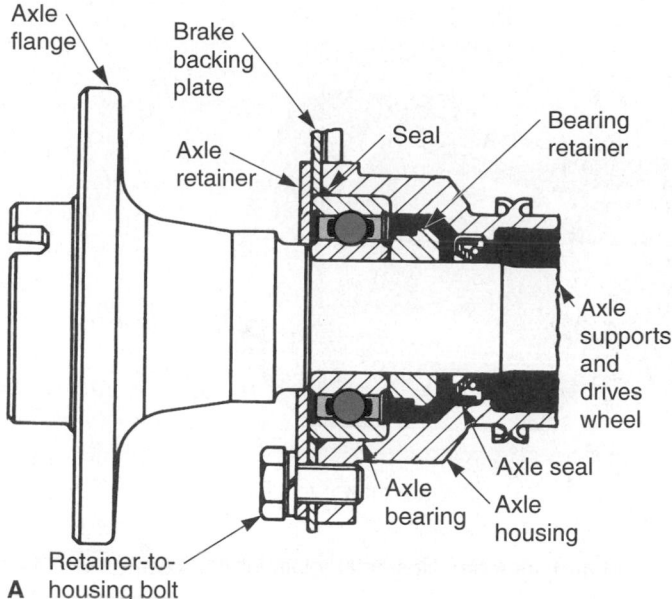

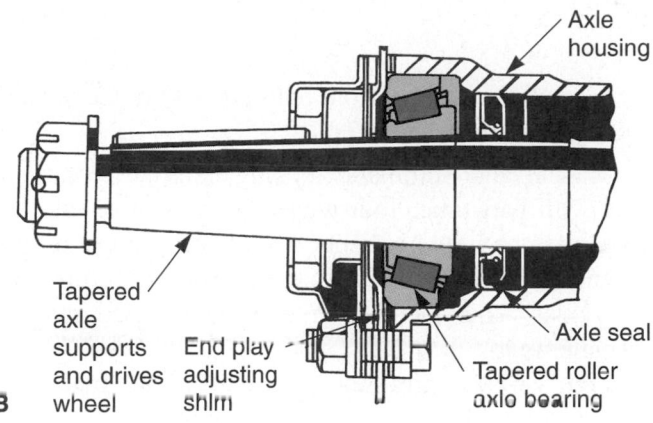

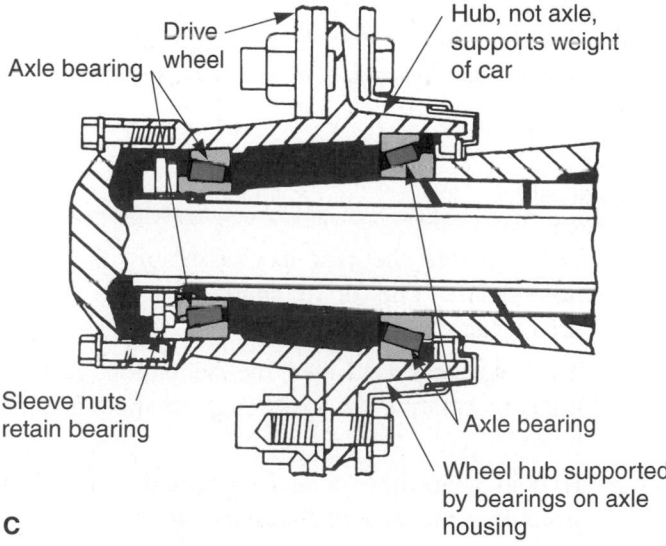

Figure 61-16. Three common rear wheel bearing variations: A—Semifloating, ball bearing type. B—Semifloating, roller bearing type. C—Full-floating axle, used on heavy-duty, pickup truck, and large truck applications. (Fiat and Deere & Co.)

are needed to allow for up and down suspension action. **Figure 61-19** illustrates a rear drive axle assembly using swing axles.

The differential works like a conventional unit. However, the drive axles are not solid, steel shafts. They are flexible. Each has two universal joints, one of them mounted on each end.

Note

For more information on swing axles, refer to Chapter 63, *Transaxle and Front Drive Axle Fundamentals.*

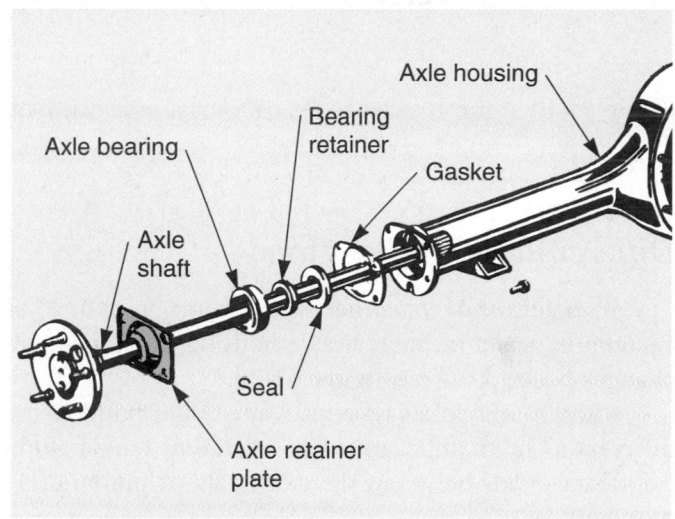

Figure 61-17. This rear axle assembly has a removable carrier, semifloating axles, and conventional differential. (Ford)

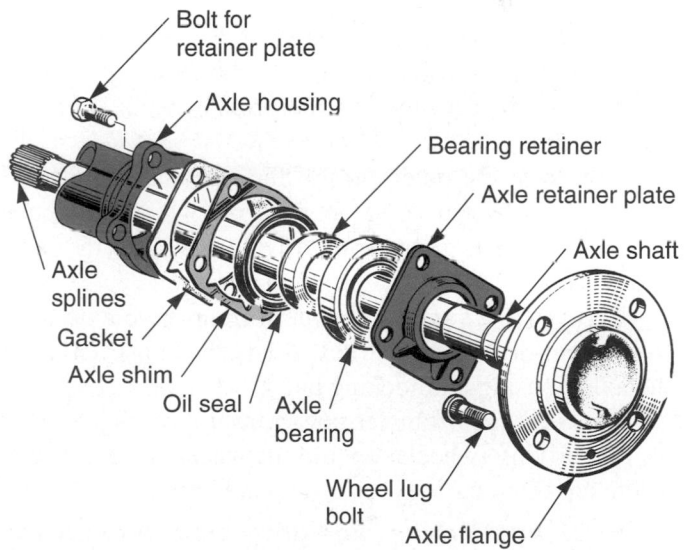

Figure 61-18. Most rear axles with removable carriers have an axle retainer plate that bolts to the axle housing. Shims are used to adjust axle end play. A gasket and oil seal prevent leakage of differential oil. (Toyota)

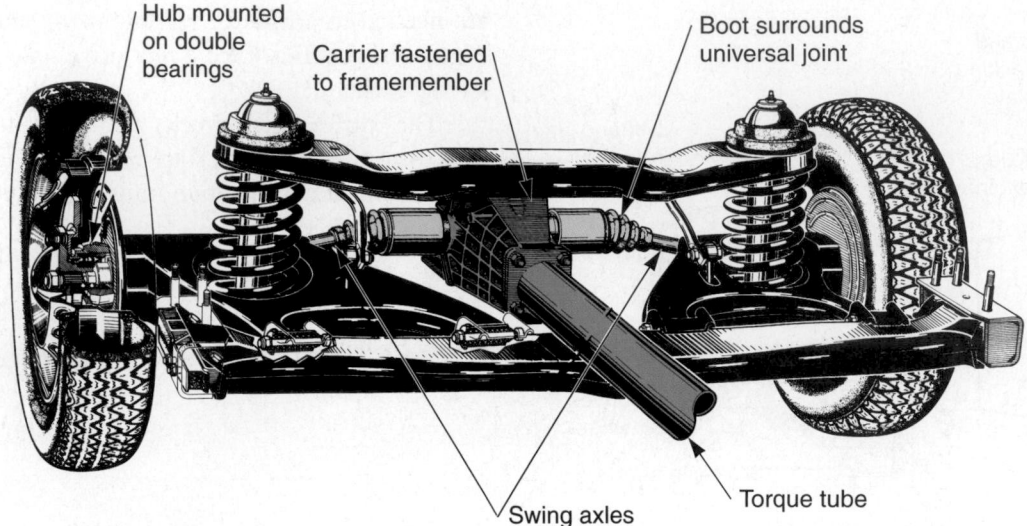

Hub mounted on double bearings

Carrier fastened to framemember

Boot surrounds universal joint

Swing axles

Torque tube

Figure 61-19. A swing axle has the differential assembly mounted on a frame member. Universal joints let the axles and wheels move up and down with suspension action.

Differential Breather Tube

A *differential breather tube* vents pressure or vacuum in or out of the rear axle housing as temperature changes occur. Look at **Figure 61-20.**

Without a breather tube, pressure could build as the differential lubricant warmed to operating temperature. Lubricant could blow out the axle seals or pinion drive gear seal.

Front Drive Axle—Four-Wheel-Drive Vehicles

The *front drive axle* used in four-wheel-drive systems is similar to a rear drive axle; however, provisions must be made for steering the front wheels. Look at **Figure 61-21.**

Note how the outer ends of the axles have universal joints. The universal joints let the front wheels and hubs swivel while still transferring driving power to the hubs and wheels.

Figure 61-22 shows a modern front-drive axle for a four-wheel-drive pickup truck. Study the construction of the axle housing and locking hubs.

Locking hubs transfer power from the driving axles to the driving wheels on a four-wheel-drive vehicle. There are three basic types of locking hubs:

- *Manual locking hub*—driver must turn a latch on the hub to lock the hub.
- *Automatic locking hub*—hub locks the front wheels to axles when the driver shifts into four-wheel drive.

- *Full time hub*—front hubs are always locked and drive front wheels.

Manual and automatic locking hubs are common. Used with part-time, four-wheel drive, they enable the driveline to be in two-wheel drive for vehicle use on dry pavement. The front wheels can turn without turning the front axles. This increases fuel economy and reduces driveline wear.

Figure 61-22 shows the basic parts of automatic and manual locking hubs.

Note
For more information on four-wheel drive operation and service, refer to Chapters 59 and 60.

Summary

- Power enters the rear axle assembly from the drive shaft. The drive shaft spins the pinion drive gear.
- The ring gear is driven by the pinion gear; it transfers rotating power through an angle change of 90°.
- Hypoid gears have a driving pinion gear offset from the centerline of the ring gear.
- Rear axle ratio is determined by comparing the number of teeth on the pinion drive gear and on the ring gear.
- The differential carrier provides a mounting place for the drive pinion gear, differential case, and other differential components.

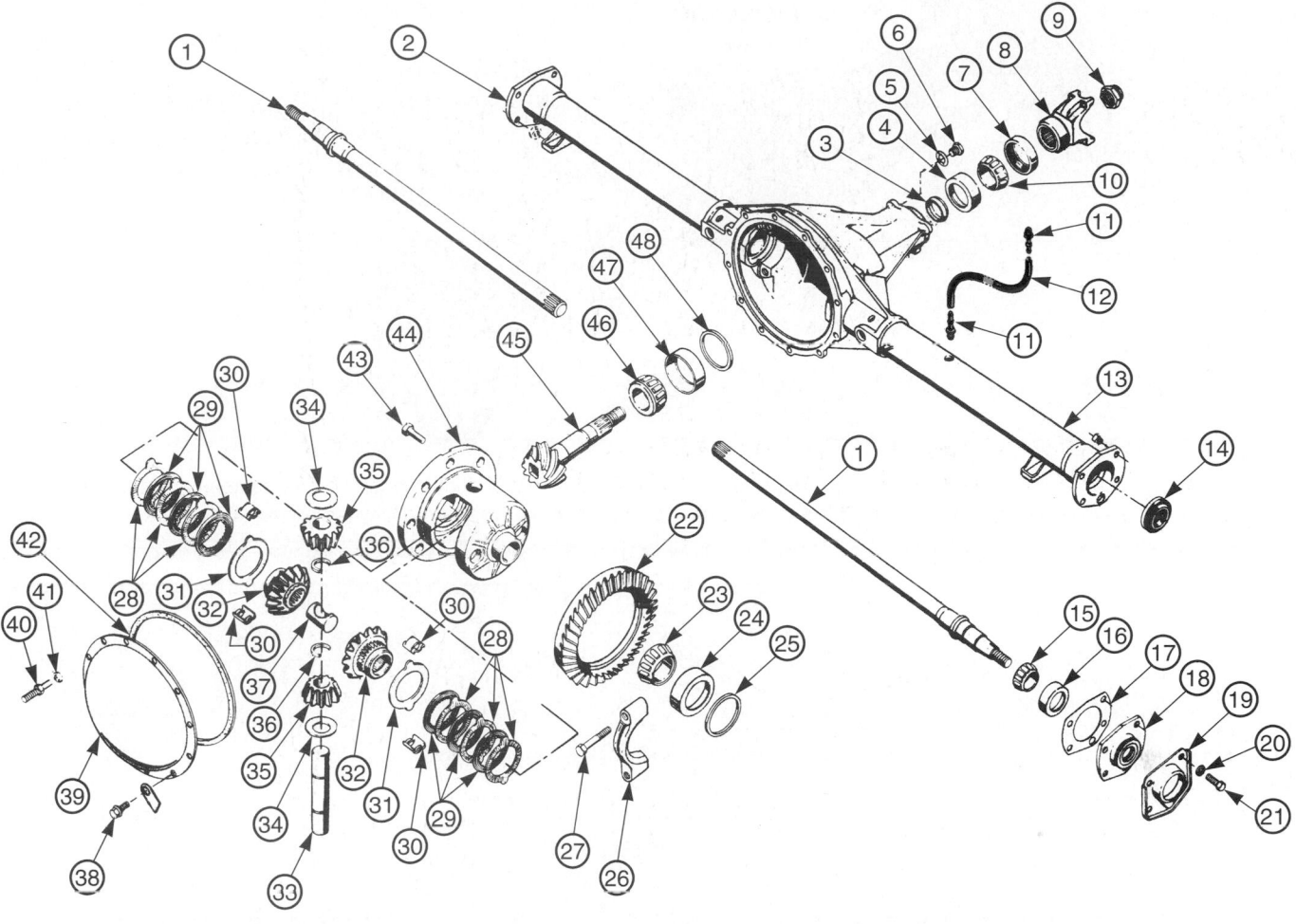

1. Axle Shaft	18. Axle Shaft Oil Seal	34. Differential Pinion Thrust Washer
2. Rear Axle Housing	19. Axle Shaft Oil Seal Retainer	35. Differential Pinion
3. Pinion Front Bearing Cup	20. Washer	36. Differential Pinion Shaft Snap Ring
4. Collapsible Spacer	21. Bolt	37. Differential Pinion Shaft Thrust
5. Filler Plug Gasket	22. Ring Gear	Block
6. Filler Plug	23. Differential Bearing	38. Bolt
7. Pinion Oil Seal	24. Differential Bearing Cup	39. Housing Cover
8. Universal Joint Yoke	25. Differential Bearing Shim	40. Stud
9. Pinion Nut	26. Differential Bearing Cap	41. Washer
10. Front Pinion Bearing	27. Bolt	42. Housing Cover Gasket
11. Breather (2)	28. Clutch Plates	43. Bolt
12. Breather Hose	29. Clutch Discs	44. Differential Case
13. Nut	30. Clutch Retainer Clip	45. Pinion Gear
14. Axle Shaft Inner Oil Seal	31. Clutch Belleville	46. Rear Pinion Bearing
15. Axle Shaft Bearing	Spring	47. Rear Pinion Bearing Cup
16. Axle Shaft Bearing Cup	32. Differential Side Gear	48. Pinion Depth Adjusting Shim
17. Axle Shaft Bearing Shim	33. Differential Pinion Shaft	

Figure 61-20. Disassembled view of a complete rear axle assembly. Note the breather hose assembly. (Chrysler)

- A removable carrier bolts to the front of the axle housing. An integral carrier is constructed as part of the axle housing.

- The spider gears include two axle gears and two pinion gears.

- A limited-slip differential provides driving force to both rear wheels at all times.

- The most popular type of limited-slip differential uses a clutch pack (set of friction discs and steel plates).

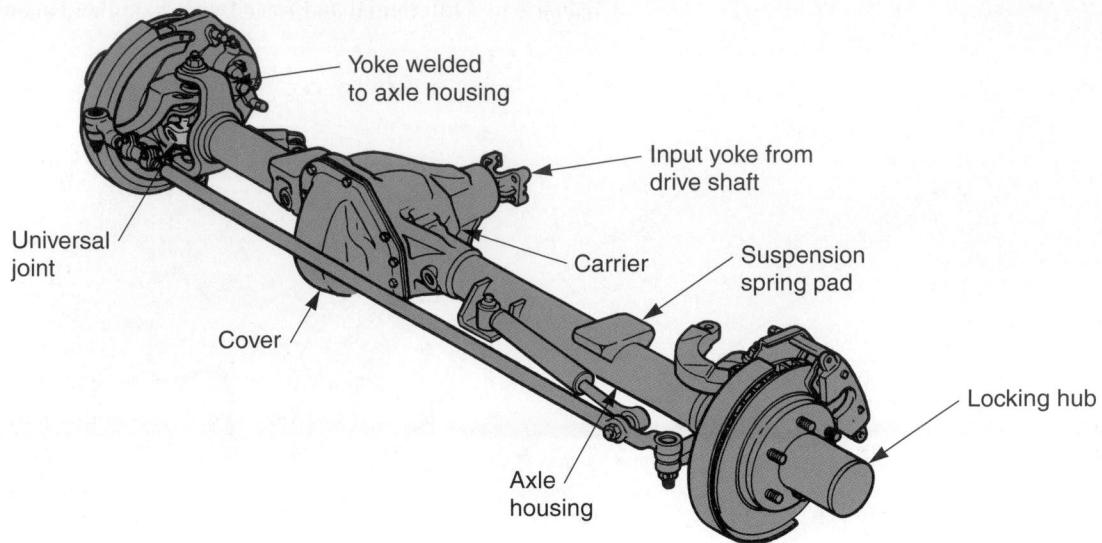

Figure 61-21. This front drive axle for a four-wheel-drive vehicle has a conventional differential. Universal joints on the outer end of the axles allow for steering. Special hubs lock the drive axle to hub and wheel when in four-wheel drive. (GM Trucks)

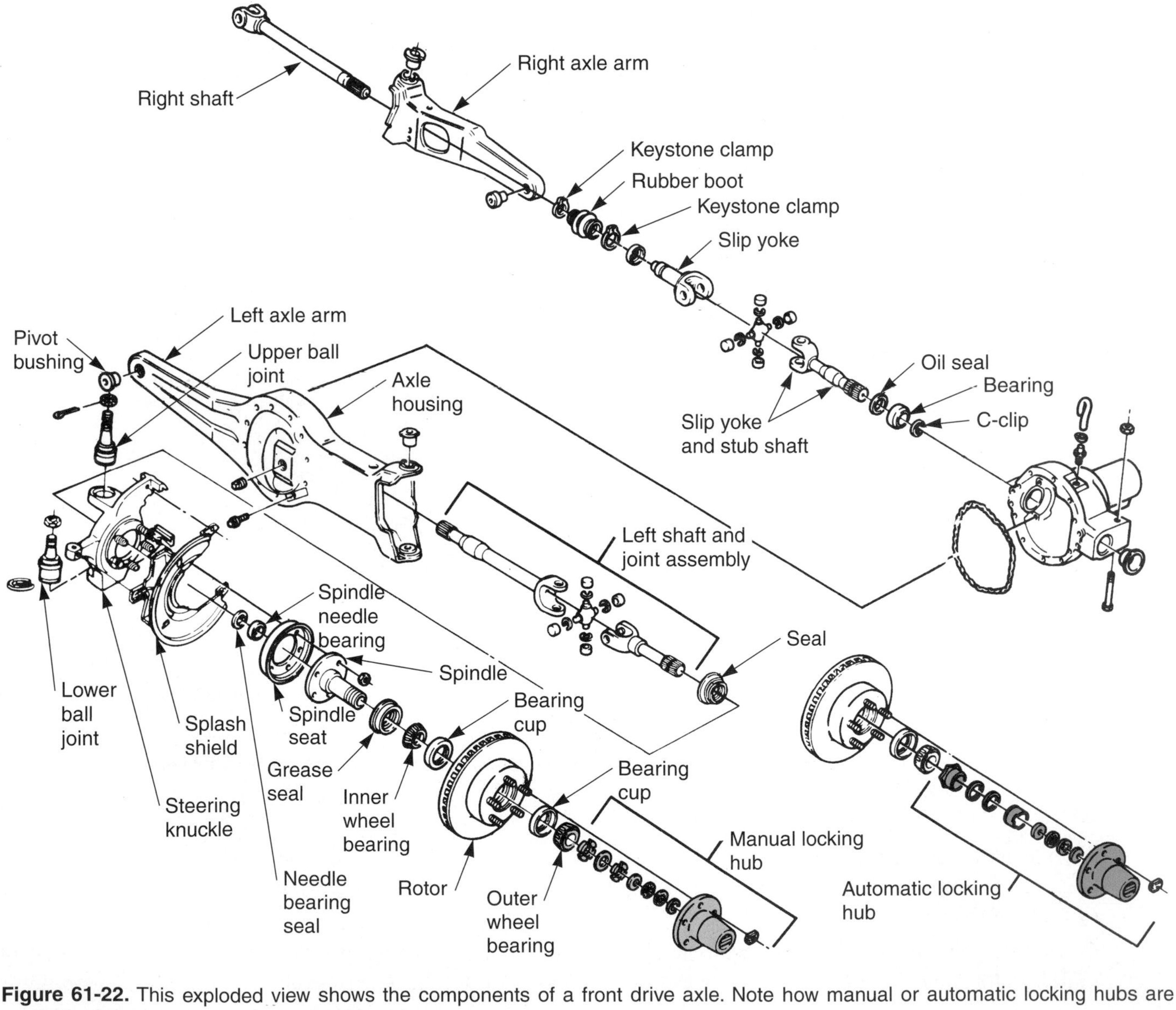

Figure 61-22. This exploded view shows the components of a front drive axle. Note how manual or automatic locking hubs are available. Axles are exposed on rear of housing.

- A cone clutch limited-slip differential uses the friction produced by cone shaped axle gears to provide improved traction.

Important Terms

Rear drive axle assembly	Axle gears
Differential assembly	Pinion gears
Ring gear	Pinion shaft
Pinion drive gear	Differential lubricant
Pinion pilot bearing	Limited-slip differential
Lapped	Clutch pack
Hunting gearset	Cone clutch limited-slip differential
Nonhunting gearset	
Hypoid gears	Ratchet differential
Spiral bevel gears	Detroit locker
Rear axle ratio	Dog clutch
Differential ratio	Semifloating axle
Differential carrier	Rear wheel bearings
Removable carrier	Rear axle seals
Integral carrier	Axle retainer plates
Case bearings	Axle shims
Carrier bearings	Swing axles
Spider gears	Differential breather tube
Differential gears	Locking hubs

Review Questions—Chapter 61

Please do not write in this text. Place your answers on a separate sheet of paper.

1. Which of the following is *not* a basic part of the rear axle assembly?
   - (A) Differential case assembly.
   - (B) Axle housing.
   - (C) Pinion drive gear.
   - (D) None of the above.

2. The _____ must be capable of providing torque to both axles when turning corners.

3. The purpose of the pinion gear is to transfer power from the ring gear to the axle. True or False?

4. Explain the difference between a hunting gearset and a nonhunting gearset.

5. Rear axle ratio is determined by comparing the number of teeth on the _____ gear to the number of teeth on the _____ gear.

6. An integral carrier is constructed as part of the axle housing. True or False?

7. What major problem is a differential designed to prevent?

8. A limited-slip differential usually uses 80W-90 gear oil as a lubricant. True or False?

9. Which of the following statements best describes a clutch pack for a limited-slip differential?
   - (A) Set of friction discs and steel plates usually splined to the differential side gears.
   - (B) Friction-producing cone shaped axle gears that are splined to the axles.

10. Swing axles are found on a differential that is rigidly mounted on the auto frame. True or False?

ASE-Type Questions

1. Technician A says an automobile's rear axles transfer torque to the differential pinion gear. Technician B says an automobile's rear axles transfer torque to the rear wheels. Who is right?
   - (A) *A only.*
   - (B) *B only.*
   - (C) *Both A and B.*
   - (D) *Neither A nor B.*

2. Technician A says a rear drive axle assembly's pinion drive gear transfers power from the transmission input shaft to the ring gear. Technician B says the pinion drive gear transfers power from the drive shaft to the ring gear. Who is right?
   - (A) *A only.*
   - (B) *B only.*
   - (C) *Both A and B.*
   - (D) *Neither A nor B.*

3. Technician A says an automotive differential's ring gear transfers turning power to the differential case assembly. Technician B says the ring gear transfers turning power to the drive shaft's rear yoke. Who is right?
   - (A) *A only.*
   - (B) *B only.*
   - (C) *Both A and B.*
   - (D) *Neither A nor B.*

4. Which of the following is not considered an automotive rear drive axle component?
   - (A) *Spider gears.*
   - (B) *Pilot shaft.*
   - (C) *Drive axle.*
   - (D) *Rear axle bearing.*

5. Technician A says a rear drive axle assembly's ring gear is bolted to the rear axle flange.

Technician B says the ring gear is bolted to the differential case. Who is right?

(A) *A only.*
(B) *B only.*
(C) *Both A and B.*
(D) *Neither A nor B.*

6. Technician A says a rear drive axle assembly splits the amount of torque going to each wheel. Technician B says a rear drive axle assembly transfers torque through a 90° angle. Who is right?

(A) *A only.*
(B) *B only.*
(C) *Both A and B.*
(D) *Neither A nor B.*

7. Technician A says a differential's pinion gear is normally mounted on tapered roller bearings. Technician B says a differential's pinion gear is normally mounted on the differential case. Who is right?

(A) *A only.*
(B) *B only.*
(C) *Both A and B.*
(D) *Neither A nor B.*

8. Which of the following is a major part of a differential assembly?

(A) *Inboard universal joint.*
(B) *Outer race.*
(C) *Carrier.*
(D) *None of the above.*

9. Technician A says a differential's ring and pinion drive gears are never manufactured as a matched set. Technician B says a differential's ring and pinion gear are commonly manufactured as a matched set. Who is right?

(A) *A only.*
(B) *B only.*
(C) *Both A and B.*
(D) *Neither A nor B.*

10. Technician A says a hunting gearset meshes the same gear teeth over and over during operation. Technician B says a hunting gearset does not mesh the same gear teeth during operation. Who is right?

(A) *A only.*
(B) *B only.*
(C) *Both A and B.*
(D) *Neither A nor B.*

11. Technician A says most ring and pinion gears are classified as a hunting gearset. Technician B says most ring and pinion gearsets are non-hunting types. Who is right?

(A) *A only.*
(B) *B only.*
(C) *Both A and B.*
(D) *Neither A nor B.*

12. Technician A says modern differential ring and pinion gears are hypoid type. Technician B says modern differential ring and pinion gears are classified as a spiral bevel gearset. Who is right?

(A) *A only.*
(B) *B only.*
(C) *Both A and B.*
(D) *Neither A nor B.*

13. Technician A says rear axle ratio is determined by comparing the number of teeth on the spider gears and the ring gear. Technician B says rear axle ratio is determined by comparing the number of teeth on the pinion drive gear and the ring gear. Who is right?

(A) *A only.*
(B) *B only.*
(C) *Both A and B.*
(D) *Neither A nor B.*

14. Which of the following is another name for a limited-slip differential?

(A) *Posi-traction.*
(B) *Sure-grip.*
(C) *Equal-lock.*
(D) *All of the above.*

15. Technician A says several modern automobiles use the semifloating rear axle design. Technician B says several modern automobiles use the swing type rear axle design. Who is right?

(A) *A only.*
(B) *B only.*
(C) *Both A and B.*
(D) *Neither A nor B.*

Activities—Chapter 61

1. Using a mock-up of a differential provided by your instructor, identify its various parts.

2. Count the number of teeth on a ring gear and on its mating pinion gear. Write down these numbers and then determine the gear ratio.

3. Demonstrate to the rest of the class how power is transmitted from the drive pinion through the rest of the differential to the drive wheels.

Chapter 62

Differential and Rear Drive Axle Diagnosis and Repair

After studying this chapter, you will be able to:

- Diagnose common differential and rear drive axle problems.
- Explain the basic service and repair of a differential assembly.
- Remove and replace axles, axle bearings, and seals.
- Describe limited-slip differential testing and service.
- Adjust ring and pinion gears.
- Check and replace rear axle lubricant.
- Cite safety rules and practice safe work habits.
- Correctly answer ASE certification test questions on differential and rear drive axle repairs.

Differentials and rear drive axles are normally very dependable. However, if operated dry (without lubricant), abused, or used for prolonged service, parts can wear and fail. This chapter will introduce the most important steps for diagnosing and repairing a differential and rear axle assembly.

Note
Much of the information in this chapter will help you when studying about the service of a transaxle, which also contains a differential.

Differential and Rear Axle Problem Diagnosis

Rear end (rear axle assembly) problems usually show up as abnormal noises. It is critical that correct procedures be used when trying to find the source of these noises. Other problems (worn wheel bearings, universal joints, or transmission gears) can produce symptoms similar to those caused by faulty rear drive axle components.

To begin diagnosis, gather information. Determine when the noise or condition occurs—when accelerating, coasting at certain speed, or rounding corners. Use common sense and your understanding of operating principles to narrow down the possible sources.

Road test the vehicle on a smooth road surface. Listen for changes in the noise under different driving conditions. **Figure 62-1** lists some typical problems you might find.

Eliminate Other Problem Sources

After your road test to verify the complaint, use your knowledge of noise diagnosis to determine which components should be checked further. If the noise could possibly be tire related, increase inflation pressure in the tires as described in a service manual. If the abnormal noise changes, the tires may be at fault.

If the sound originates from the middle or front of the car, listen for noises in the transmission or transaxle. Also, check for bad front wheel bearings.

Problem Isolation

A *stethoscope* (listening device) can often be used to isolate the source of a differential or rear axle bearing problem. Raise the vehicle on a lift. Ask another technician to start the engine, place the transmission in gear, and accelerate to 30 mph (48 km/h).

Touch your stethoscope on the ends of the axle housing and on the housing near the carrier bearings. The area producing the loudest noise contains the faulty parts.

Warning
When using a stethoscope to listen for rear axle noise, stay away from the spinning tires and driveline. Serious injury could result if you touch these parts.

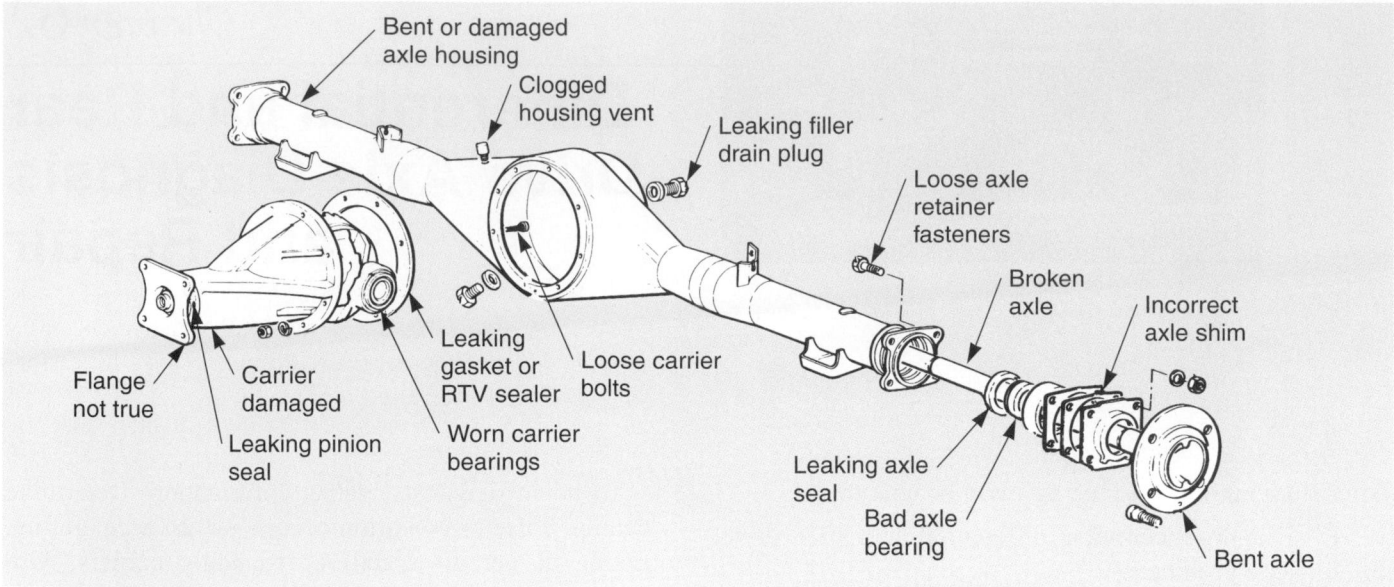

Figure 62-1. These are typical problems that can develop in a rear axle assembly.

Common Rear Axle Assembly Problems

Rear axles assemblies are normally very dependable units. However, with abuse or prolonged use, gear teeth can wear, bearings can fail, and seals can leak.

Ring and Pinion Problems

Ring and pinion problems usually show up as a howling or whining noise that changes when going from acceleration to deceleration. Since the ring gear and pinion gear transfer engine power through a 90° angle, they are very sensitive to load. The sound of bad ring and pinion gears will usually change pitch (noise frequency or tone) as you press and release the accelerator while driving.

The ring and pinion gears can become worn, scored, out of adjustment, or damaged. These problems can result from prolonged service, fatigue, and lack of lubricant. You need to inspect the differential to determine whether adjustment or part replacement is required. See **Figure 62-2.**

If the backlash (clearance) between the ring and pinion is too great, a clunking sound can be produced by the gears. For example, when an automatic transmission is shifted into drive, the abrupt rotation of the drive shaft could bring the gears together with a loud thump.

Rear Axle Leaks

Rear axle lubricant leaks can occur at the pinion gear seal, the carrier or inspection cover gaskets, and the two axle seals. A leak will show up as an oily or dirty area below the pinion gear, under the carrier, or on the inside of the wheel and brake assembly.

Always make sure that a possible axle seal leak is not a brake fluid leak. Touch and smell the wet area to determine the type of leak. Refer to **Figure 62-1.**

Axle and Carrier Bearing Problems

Worn or damaged bearings in the carrier or on the axles often produce a constant whirring or humming sound. When bad, these bearings make about the same sound whether accelerating, decelerating, or coasting.

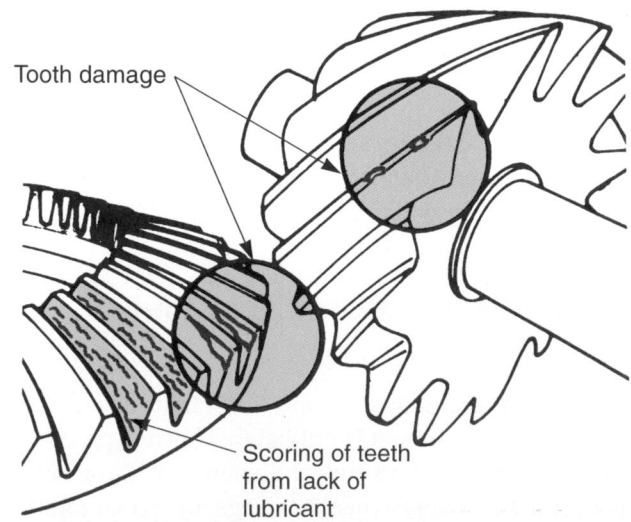

Figure 62-2. Damaged pinion and ring gear teeth cause abnormal noise that can be affected by load. Check each tooth closely during inspection. (Ford)

Figure 62-3 shows some typical bearing failures. These apply to both differential and axle bearings. When diagnosing and repairing bearing failures, do the following:

1. Check the general condition of all parts during disassembly, not just the most badly worn or damaged parts.

2. Compare the failure to any information in the service manual and your knowledge of component operation.

3. Determine the cause of the part failure. This helps ensure that the problem does not reoccur.

4. Make all repairs following manufacturer's recommendations and specifications.

Differential Case Problems

Differential case problems frequently show up when rounding a corner. With the rear wheels turning at different speeds, any problem (damaged spider gears or a grabbing limited-slip clutch pack, for example) will usually show up as an abnormal sound (clunking or clattering) from the rear of the vehicle.

A limited-slip differential (clutch-type) can sometimes make a chattering sound when turning a corner. The clutches are sticking to each other and then releasing. Many automakers recommend that the differential fluid be drained and replaced when this occurs.

Differential Maintenance

Many automobile manufacturers recommend that the differential fluid be checked or replaced at specific intervals. See **Figure 62-4.** Refer to Chapter 10 for more information on checking and changing differential lubricant.

Tech Tip
Always install the correct type of differential fluid. Limited-slip differentials often require a special type of fluid intended for the friction clutches.

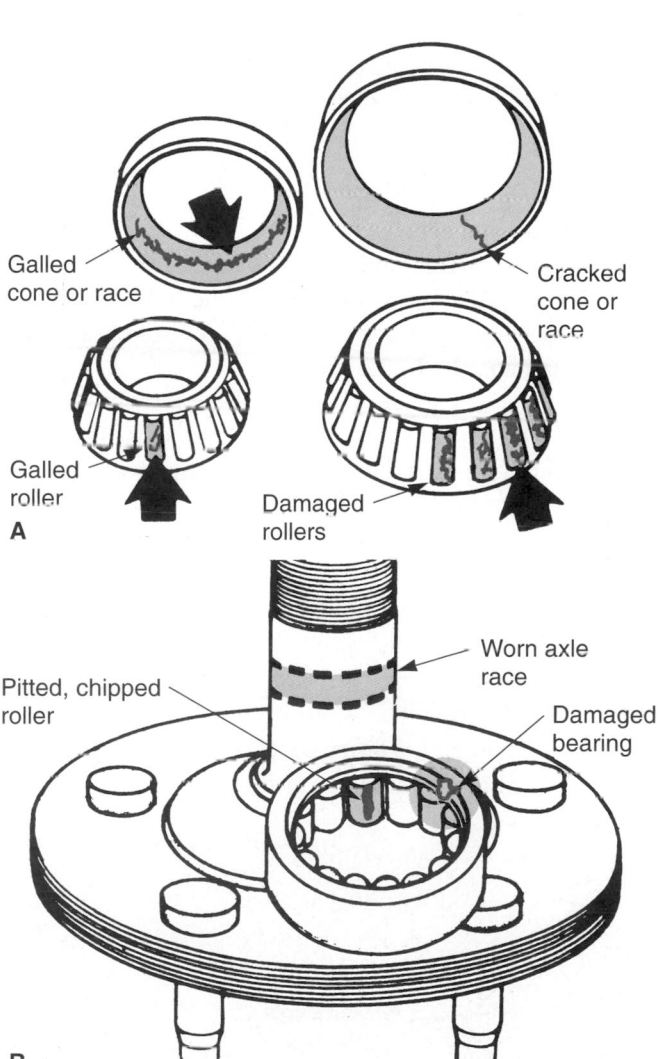

Figure 62-3. Bearing problems usually produce noise that does not change very much with load. A—Pinion gear and case bearings can fail from prolonged service or lack of lubricant. B—Axle bearings are a common source of trouble. They are not as well lubricated as bearings in the carrier. (Ford)

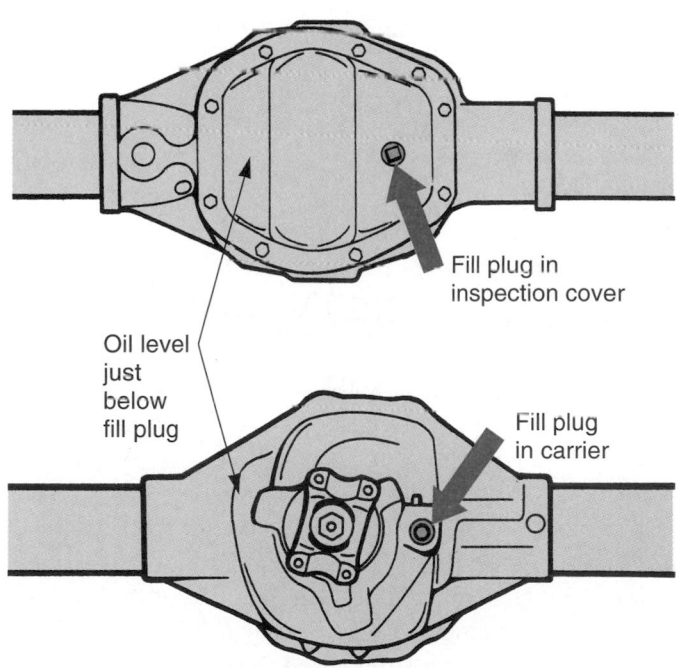

Figure 62-4. Check the oil level when you detect problems in the rear drive axle assembly. Lubricant should be almost even with the fill hole when the oil is warm. If a drain plug is not provided, you must remove the cover or carrier or use a suction gun to remove the old oil. (Chrysler)

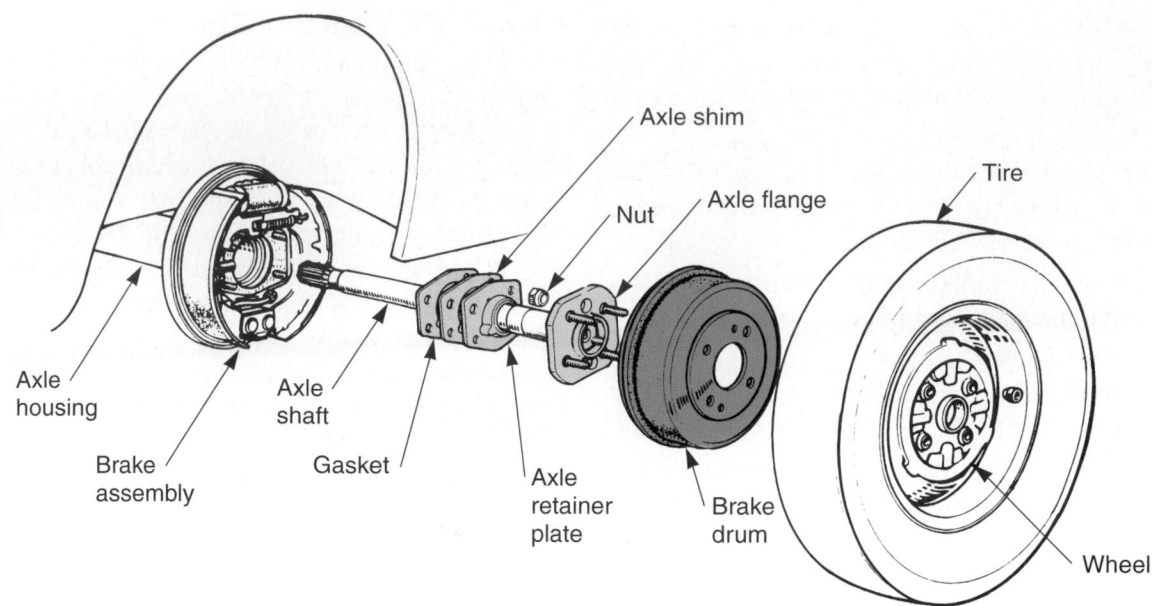

Figure 62-5. To remove an axle with a retainer plate, remove the four nuts on the plate and end of housing. Reach through the axle flange with a socket and extension to loosen the fasteners. Pull the axle out without moving the brake backing plate assembly. (Toyota)

Rear Axle Service

Rear axle service is needed when an axle bearing is noisy; when an axle is broken, bent, or damaged; or when an axle seal is leaking. As you will learn later, the rear axles must also be removed to allow removal and repair of the differential assembly.

Axle Removal

Generally, two methods are used to hold axles in their housing: a retainer plate on the outside of the housing or C-clips on the inner ends of the axles. A *retainer plate type axle* is commonly used with a removable carrier. A *C-clip type axle* is frequently used with an integral carrier.

 To remove an axle that is secured with a retainer plate, refer to **Figure 62-5** and the following procedure:
1. Support the car on jack stands.
2. Unbolt the rear wheel and slide off the brake drum.
3. Unscrew the nuts on the ends of the axle housing. A socket, extension, and ratchet can normally be used to reach through a hole in the axle flange. You may need to install a slide hammer puller on the axle studs, **Figure 62-6.**
4. Tap the axle out of the housing. Repeat on the other axle if needed.

 Caution
While pulling an axle, be careful not to pull the brake backing plate off with the axle. If the backing plate is pulled off, you could bend and damage the brake line.

 To remove an axle that is held in place with a C-clip, a different procedure is needed:
1. Drain the axle lubricant.
2. Unbolt the cover on the rear of the axle housing.
3. Remove the pinion shaft bolt and pinion shaft, if necessary.
4. Slide the axle inward.
5. Remove the C-clip from the groove in the axle end, **Figure 62-7.**
6. Slide the axle out of the housing.

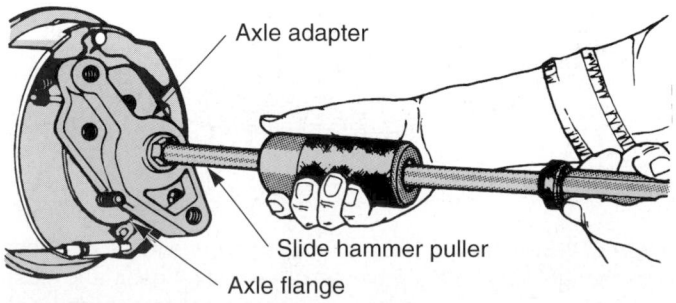

Figure 62-6. If the axle is stuck in the housing, use a slide hammer puller as shown. (Plymouth)

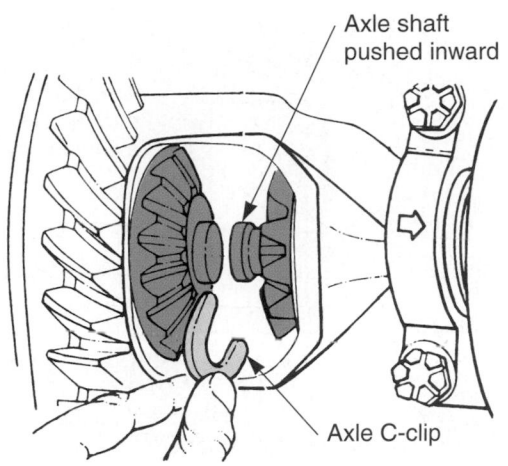

Axle shaft
pushed inward

Axle C-clip

Figure 62-7. With many integral carriers, you must remove the pinion shaft bolt and pinion shaft. Then slide axle inward and remove the C-clip. This will let you pull the axle out of the housing. (Ford)

With the axle removed, inspect the bearings and splines for damage. Rotate the bearings by hand while feeling for roughness. Some vehicles use axles with tapered ends. This type of axle may require a puller for removal of the hub. Refer to a service manual for details.

Axle Bearing Service

When an axle bearing is faulty, it must be removed from the axle or housing carefully and a new bearing must be installed. Most axle bearings are press fit on the axle. A collar may also be pressed on the axle to help secure the bearing. To remove this type of axle bearing, carefully cut the collar off with a grinder and sharp chisel. See **Figure 62-8.**

Caution
Do *not* use a cutting torch to remove an old collar and axle bearing. The heat can weaken and ruin the axle.

With the collar cut off, place the axle in a hydraulic press, as shown in **Figure 62-9.** The driving tool should be positioned so that it contacts the *inner* bearing race. Never press a bearing off using the outer race surface: bearing damage will result.

Warning
Wear eye and face protection when pressing a bearing on or off an axle. The tremendous pressure could cause the bearing to shatter and fly into your face with deadly force.

To install the new bearing, slide the retainer and bearing onto the axle. Make sure the bearing is facing in

the proper direction. Sometimes, a chamfer on the inner bearing race must face the axle flange. Refer again to **Figure 62-9.**

Applying force on the inner bearing race, press the bearing fully into place. Then, if needed, press the bearing collar or retaining ring onto the axle.

Caution
Do not attempt to press the bearing and collar on at the same time. Bearing or collar damage could result.

Rear Axle Seal Service

Anytime the axle is removed for service, it is wise to install a new axle seal. The axle seal is normally press-fit in the end of the axle housing.

To remove a housing-mounted seal, use a slide hammer puller equipped with a hook nose. Place the hook on the metal part of the seal. Jerk outward on the puller slide to pop out the seal. A large screwdriver will also work. Be careful not to scratch the bearing bore in the axle housing, **Figure 62-10.**

Make sure that you have the correct new seal. Its outside and inside diameters must be the same as the old seal. A part number is normally stamped on the side of a seal. This number and the name of the seal manufacturer may help when ordering the replacement.

Before installing the new seal, coat its outer diameter with nonhardening sealer. Coat the inside of the seal with

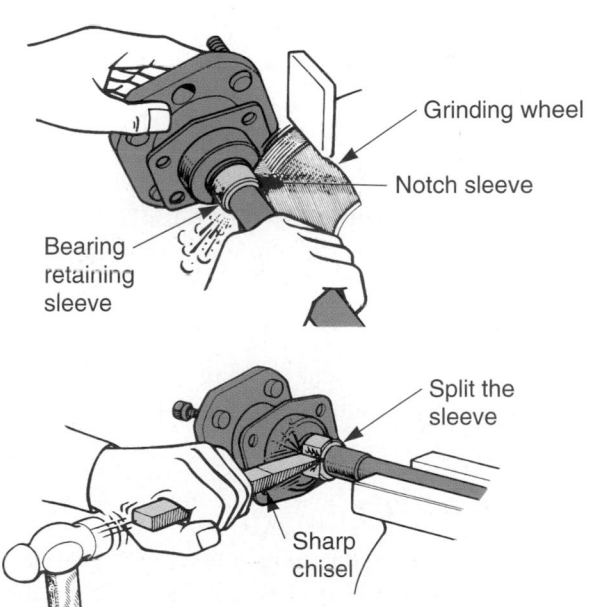

Grinding wheel

Notch sleeve

Bearing retaining sleeve

Split the sleeve

Sharp chisel

Figure 62-8. To prepare for axle bearing removal, grind a notch in the bearing retainer. Then, split the retainer with a sharp chisel. Wear safety glasses. (Toyota)

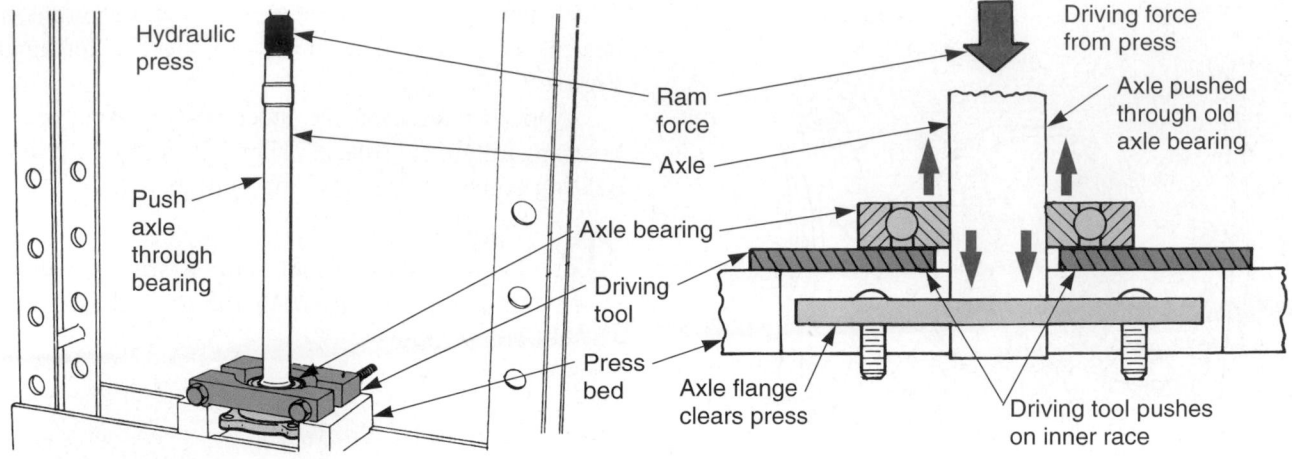

Axle bearing removal

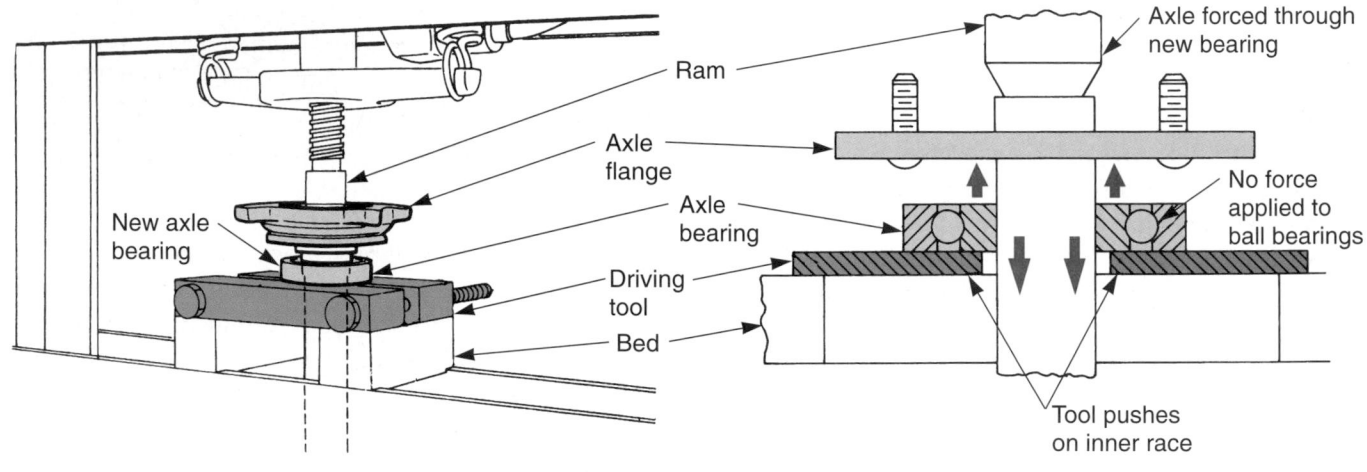

Axle bearing installation

Figure 62-9. Axle bearing removal and installation. When removing a bearing, position the driving tool so it contacts the inner bearing race. Push the axle through the bearing. When installing a bearing, again position the tool to contact the inner race. Press the axle back through the bearing. Wear a face shield and do not exceed recommended ram pressure. (Dodge)

lubricant. With the seal facing in the proper direction (sealing lip toward the inside of the housing), drive the seal squarely into place. Use a seal driving tool, **Figure 62-10.**

 Caution
When installing an axle seal, be careful not to bend the metal seal housing or a leak could result. Make sure the seal is fully seated.

Axle Stud Service

Before installing the axle, make sure that the lug studs on the flange are in good condition. If the threads are damaged or the lugs are loose, install new lugs. Drive or press out the damaged studs. Then, force new ones into the axle flange.

Rear Axle Installation

Before installation, wipe the axle clean. Then, carefully slide the axle into the housing.

To prevent seal damage, support the weight of the axle as it slides over the seal. Do not allow the axle to rub on the new seal, or the seal could be ruined.

Wiggle the axle up and down and move it around until its splines fit into the splines in the differential side gear. Align the axle bearing and push the axle fully into place.

If the axle has a retainer, tighten the retainer nuts. Install the brake drum and wheel. If the axle has C-clips, fit the clips over the inner axle ends. Then, install the pinion shaft and differential cover. Fill the differential with lubricant.

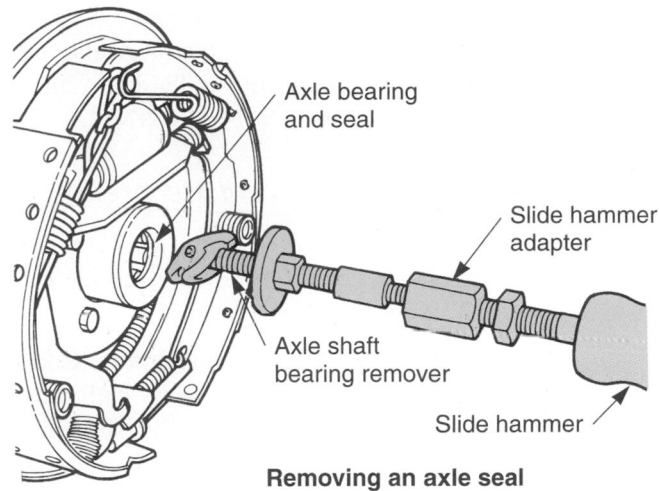

Removing an axle seal

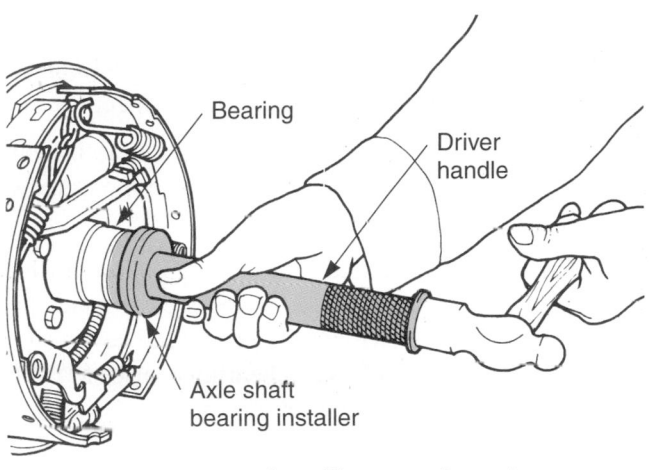

Installing an axle seal

Figure 62-10. Servicing an axle bearing that is press-fit into the axle housing. Use hook-nose adapter on a slide hammer to remove the old seal and bearing. Use a special driver to force a new axle bearing and seal into the housing. Be careful not to damage the housing bore, bearing, or seal. (Cadillac)

Measuring Axle End Play

Excessive *axle end play* can cause a clunking sound as the vehicle rounds a corner. The axle can slide one way and then the other, producing an audible clunk or knock. Insufficient end play can result in axle bearing, retainer, or differential side gear failure.

To measure axle end play, mount the indicator so that the indicator plunger is parallel with the axle centerline. Set the indicator needle on zero. Pull the axle in and out while watching the indicator. Compare the readings to factory specifications. A special adapter is sometimes installed on the axle when taking end play measurements. See **Figure 62-11.**

If there is too much end play, you may need to add shims next to the axle retainer plate. If the end play is too

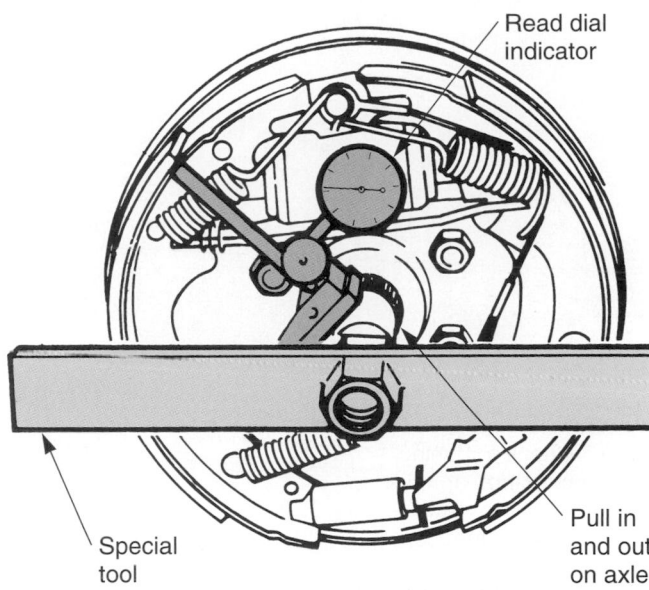

Figure 62-11. Many automakers recommend measurement of axle end play. Mount the adapter over the axle. Then use a dial indicator to measure in and out movement of the axle. Add or remove shims or replace parts as needed to correct end play. (Chrysler)

small, install thinner shims. With a C-clip type axle, you need to replace parts or install shims behind the side gears in the differential to adjust axle end play. Check a service manual for details.

Differential Service

When symptoms point to differential troubles, remove the differential carrier or rear inspection cover. Inspect the ring gear, pinion drive gear, bearings, and spider gears.

A *differential ID number* (identification number) is provided to show the exact type of differential for ordering parts and looking up specifications. The number may be on a tag under one of the carrier or inspection cover fasteners. It may also be stamped on the axle housing or carrier. Use the ID number to find the axle type, axle ratio, make of unit, and other information.

Differential Removal

 To remove a separate carrier differential, use the following procedure:
1. Remove the drive shaft.
2. Unbolt the nuts around the outside of the carrier.
3. Place a drain pan under the differential.
4. Force the differential away from the housing.
5. Drain the lubricant into the pan.

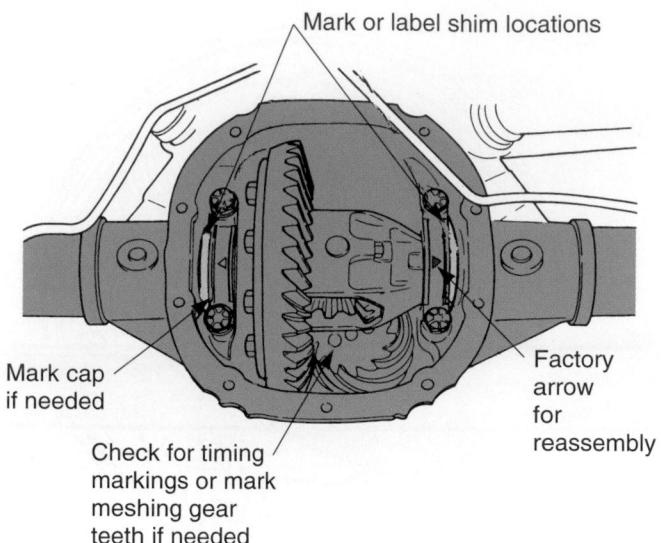

Mark or label shim locations

Mark cap
if needed

Check for timing
markings or mark
meshing gear
teeth if needed

Factory
arrow
for
reassembly

Figure 62-12. As you disassemble the differential, make sure you check markings and mark critical components. Most parts must be installed in the same location during reassembly. Note the markings on this unit. (Ford)

Warning
A differential carrier can be surprisingly heavy. Grasp it securely during removal. A carrier can cause painful injuries if dropped.

To remove an integral differential, remove the cover on the rear of the axle housing. Drain the lubricant. With the cover off, inspect and mark the individual components as they are removed. Look at **Figure 62-12.**

Differential Disassembly and Reassembly

Procedures for repairing a differential will vary with the particular unit. However, the following service rules relate to almost all types of differentials:

- Check for markings before disassembly. Carrier caps, adjustment nuts, shims, ring and pinion gears, spider gears, and the pinion yoke and flange should be reinstalled exactly as they are removed. If needed, punch mark, label, or scribe these components so they can be reassembled properly.

- Clean all parts carefully. Then, inspect them closely for wear or damage. **Figure 62-13** shows the most important parts needing inspection.

- Use a holding fixture for the differential, if available. It will make your work easier. One is shown in **Figure 62-14.**

- Rotate the pinion and case bearings by hand while checking for roughness. Inspect each roller and race. Replace the bearing and race as a set if faulty. To install new bearings, use a press, **Figure 62-15,** or a puller, if required.

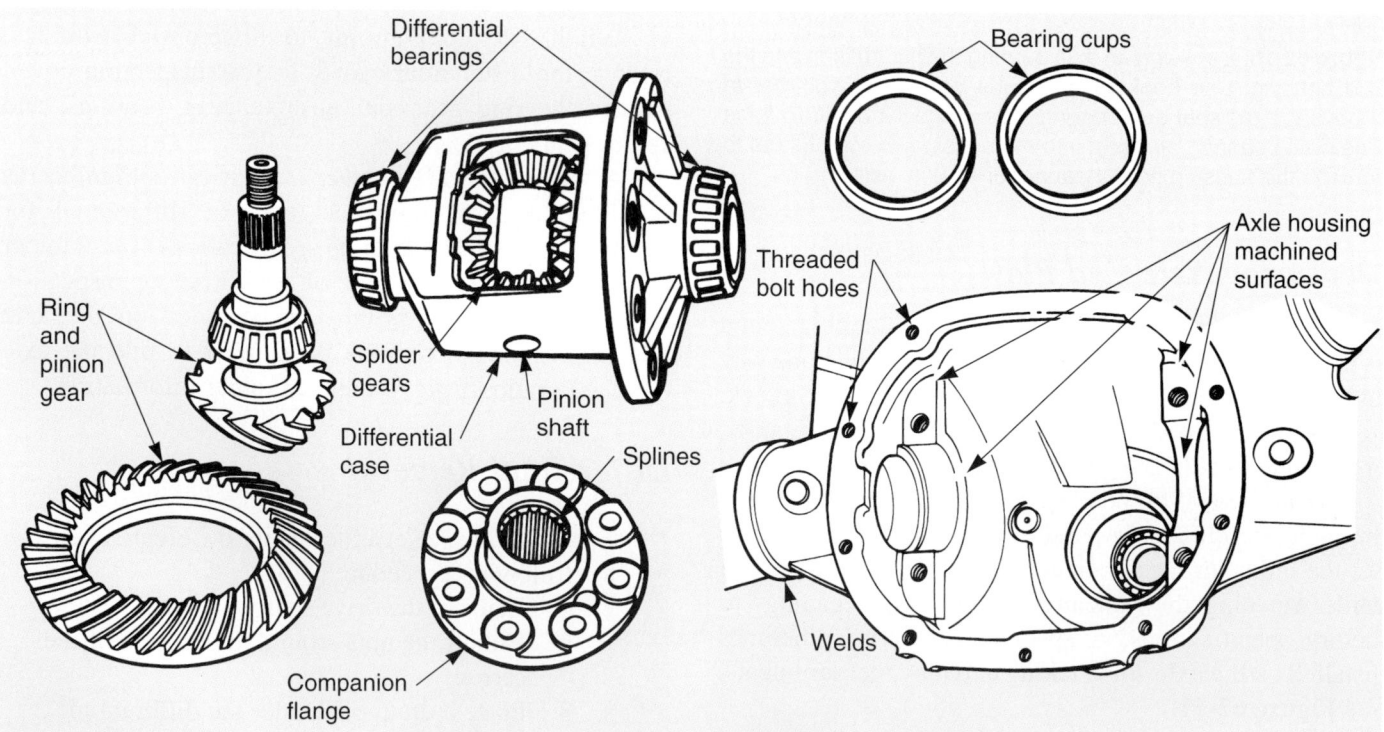

Differential bearings

Bearing cups

Threaded
bolt holes

Axle housing
machined
surfaces

Ring
and
pinion
gear

Spider
gears

Differential
case

Pinion
shaft

Splines

Companion
flange

Welds

Figure 62-13. Inspect all differential components carefully. If you overlook even one bad part, your repair could fail. (Ford)

- If the pinion gear has a collapsible spacer (device for preloading the pinion bearings), always replace it. See **Figure 62-16.**

- Coat the outside of all seals with nonhardening sealer. Lubricate the seal's inside diameter. Make sure the sealing lip faces the inside of the differential. To avoid seal damage, use a seal driver to install the seal, **Figure 62-17.**

- When tightening the pinion yoke nut, clamp the yoke in a vise or use a special holding bar, **Figure 62-18.**

- Replace the ring and pinion gears as a set. Mesh or align the gear timing markings (painted lines or other marks) on the ring and pinion gears (if used). This will properly match the teeth, which have been lapped together at the factory, **Figure 62-19.**

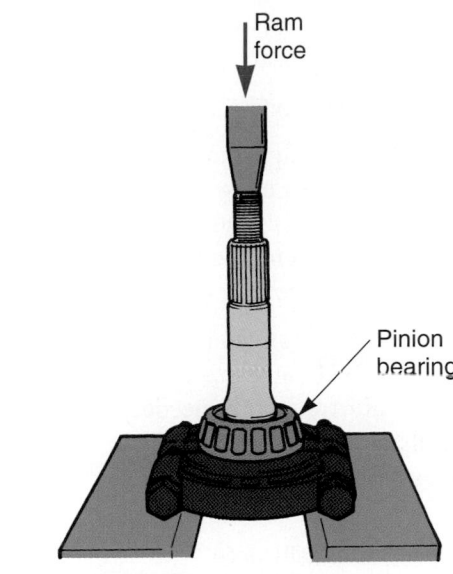

Figure 62-15. Use a press to remove pinion bearing. Stand back and use the recommended driving tools. (General Motors)

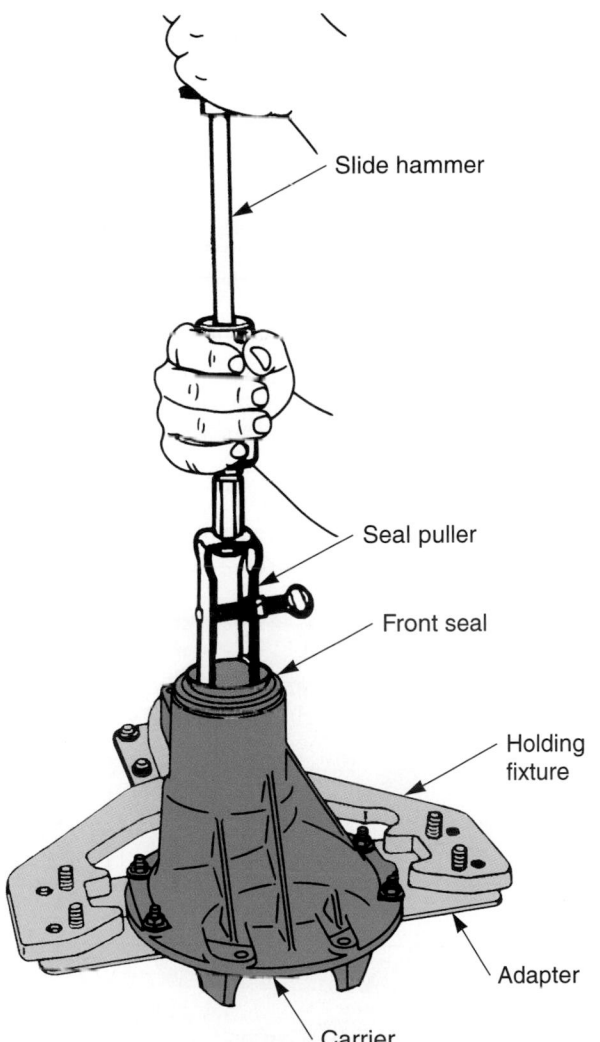

Figure 62-14. A differential holding fixture allows you to swivel the heavy differential carrier into different positions while working. Here, a pinion seal is being removed. (General Motors)

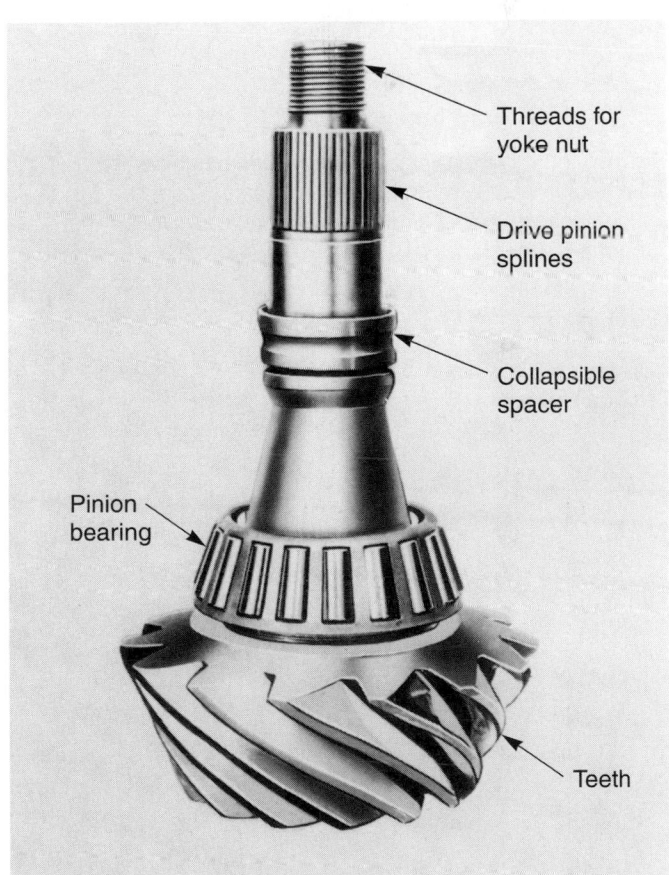

Figure 62-16. If the pinion gear uses a collapsible spacer, install a new one anytime the bearings are disassembled. If an old spacer is used, the bearing preload cannot be accurately attained. (Oldsmobile)

- Torque all fasteners to specifications. Refer to the service manual for torque values.

- Use new gaskets and approved sealer.

- Align all markings during reassembly. If you install the carrier caps backwards, for example, the caps could crush and damage the bearings and races. The differential could fail as soon as it is returned to service, **Figure 62-20.**

- Differential designs and repair procedures vary, **Figures 62-21** and **62-22.** Special tools and methods are frequently needed. Refer to a shop manual for detailed instructions.

Limited-Slip Differential Service

Limited-slip differential repair may be required after prolonged service or after damage from abuse or lack of maintenance. The clutch discs can wear, losing much of their frictional quality. This can make the differential act like a conventional unit.

A limited-slip differential (clutch and cone types) should be tested as follows:

1. Bolt a special tool to the wheel. The tool mounts over the lug studs so that a torque wrench can be used to rotate the axle.

2. Raise the tire off the shop floor.

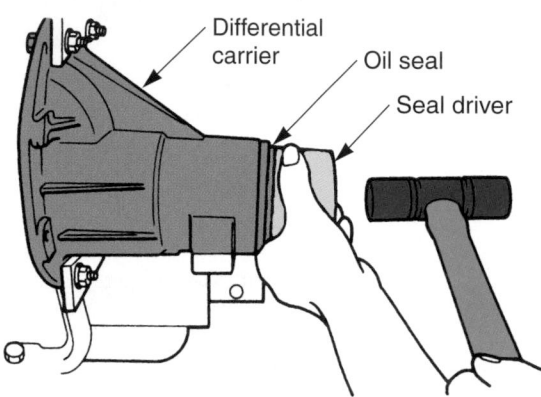

Figure 62-17. When installing pinion or other seals, coat the outside diameter with nonhardening sealer. Coat the inside lip with lubricant. Use a seal driver to squarely install the seal. (General Motors)

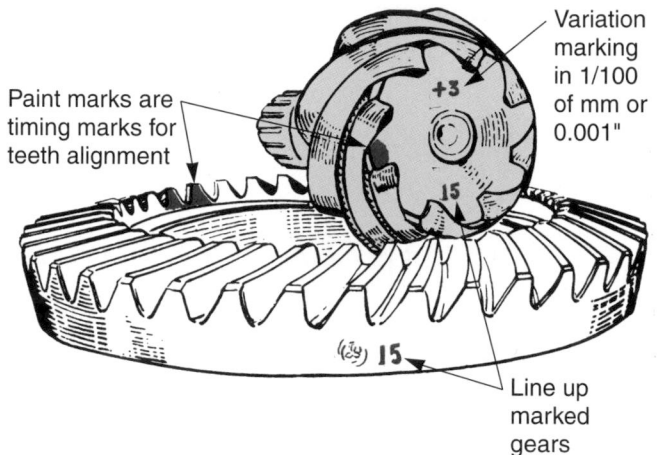

Figure 62-19. Replace ring and pinion gears as a set. Note the markings. Some give information for adjustment. Others show which teeth must be meshed together during installation. (Chrysler)

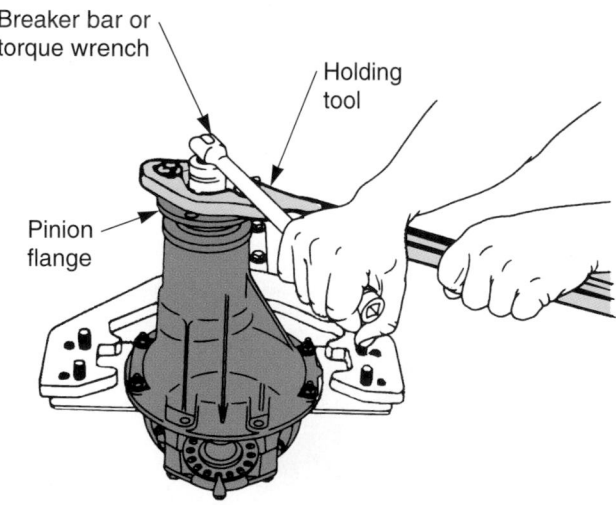

Figure 62-18. Using a large holding bar is the best way to keep the yoke from turning when tightening the drive pinion nut. If a collapsible spacer is used, tighten the nut in small increments and measure the preload. Without collapsible spacer, you must torque nut to specifications. (General Motors)

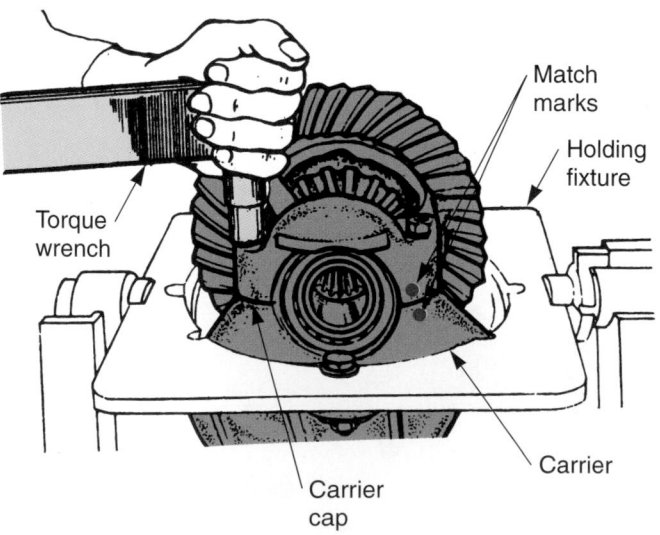

Figure 62-20. Make sure you align markings during assembly and torque all fasteners to specifications. (Plymouth)

3. Place the transmission in neutral.

4. Turn the wheel and axle with a torque wrench. Note the torque reading when the wheel begins to turn. This is the clutch pack or cone break-away torque, **Figure 62-23.**

Break-away torque is the amount of torque needed to make one axle or differential side gear rotate the limited-slip differential clutches. If the break-away torque is too low (worn clutches or cones, weakened clutch or cone springs, or part damage) or too high (shimmed improperly, part damage), repairs are needed.

If a limited-slip differential needs to be serviced, replace the clutch discs and springs. See **Figure 62-24.**

Differential Measurements and Adjustments

There are several measurements and adjustments that must be made when assembling a differential. When "setting up" (measuring and adjusting) a differential, correct bearing preloads and gear clearances are extremely critical.

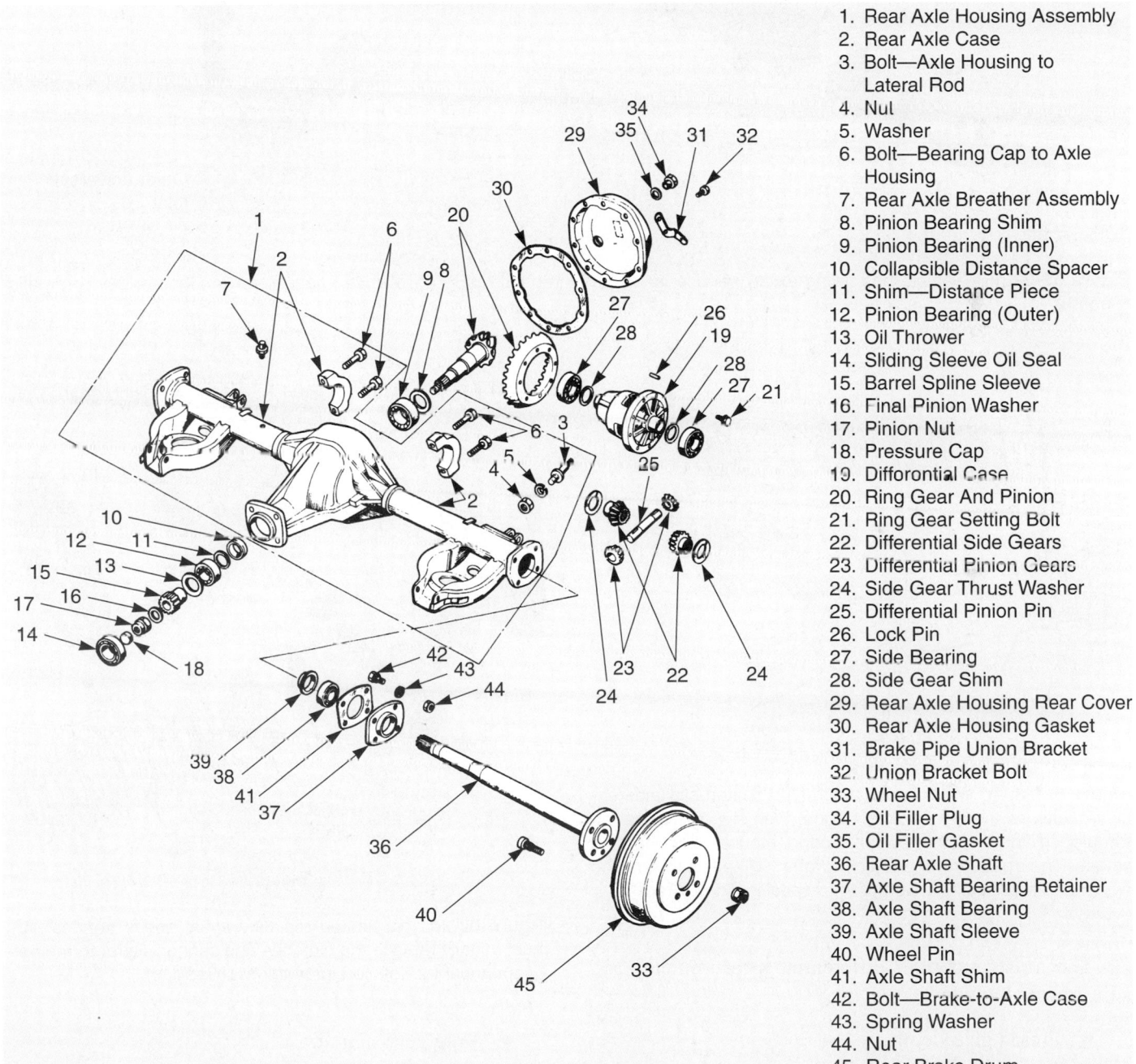

1. Rear Axle Housing Assembly
2. Rear Axle Case
3. Bolt—Axle Housing to Lateral Rod
4. Nut
5. Washer
6. Bolt—Bearing Cap to Axle Housing
7. Rear Axle Breather Assembly
8. Pinion Bearing Shim
9. Pinion Bearing (Inner)
10. Collapsible Distance Spacer
11. Shim—Distance Piece
12. Pinion Bearing (Outer)
13. Oil Thrower
14. Sliding Sleeve Oil Seal
15. Barrel Spline Sleeve
16. Final Pinion Washer
17. Pinion Nut
18. Pressure Cap
19. Differential Case
20. Ring Gear And Pinion
21. Ring Gear Setting Bolt
22. Differential Side Gears
23. Differential Pinion Gears
24. Side Gear Thrust Washer
25. Differential Pinion Pin
26. Lock Pin
27. Side Bearing
28. Side Gear Shim
29. Rear Axle Housing Rear Cover
30. Rear Axle Housing Gasket
31. Brake Pipe Union Bracket
32. Union Bracket Bolt
33. Wheel Nut
34. Oil Filler Plug
35. Oil Filler Gasket
36. Rear Axle Shaft
37. Axle Shaft Bearing Retainer
38. Axle Shaft Bearing
39. Axle Shaft Sleeve
40. Wheel Pin
41. Axle Shaft Shim
42. Bolt—Brake-to-Axle Case
43. Spring Washer
44. Nut
45. Rear Brake Drum

Figure 62-21. Use a service manual illustration like this one to help you with differential reassembly. (GMC)

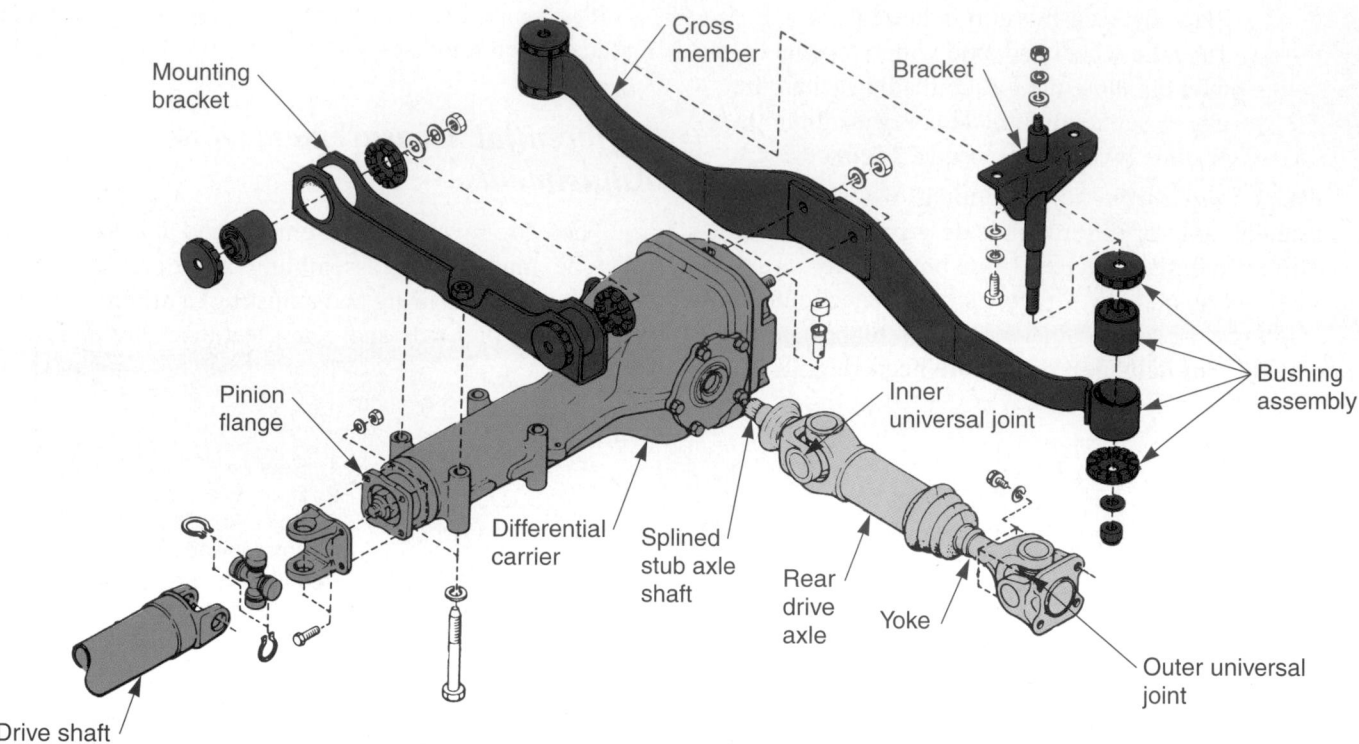

Figure 62-22. If a vehicle has swing axles, service is similar to repairs made on a drive shaft. Refer to Chapter 60 for details. (Subaru)

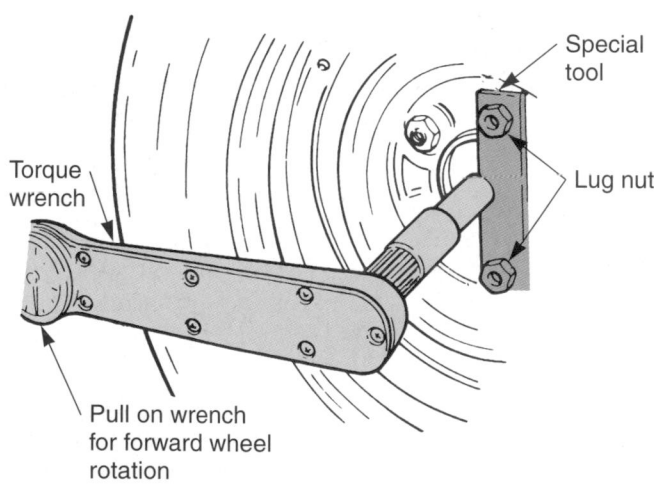

Figure 62-23. To check most clutch-type limited-slip differentials, measure the break-away torque. With the test wheel off the ground and the other on the ground, mount the tool on the wheel. Then measure how much torque is needed to turn the wheel and slip clutches. Compare to specifications. (Ford)

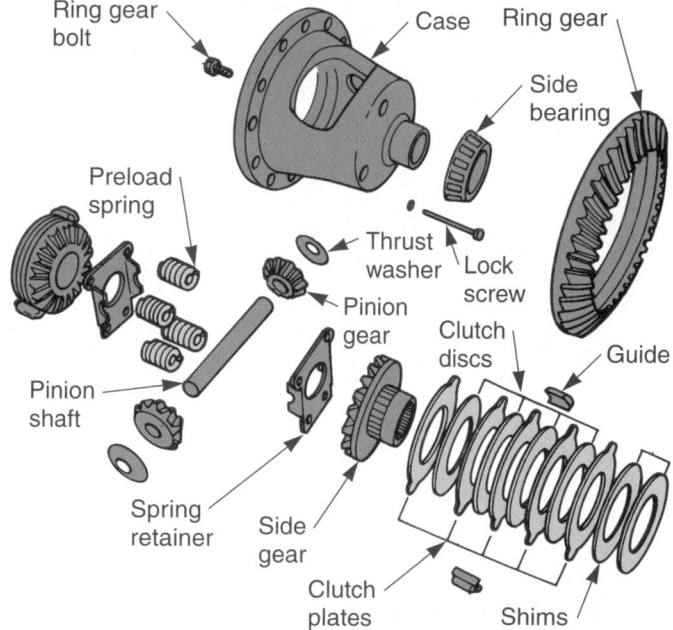

Figure 62-24. If a limited-slip differential needs service, you must usually replace the clutches and springs. Refer to the service manual for detailed instructions. (Cadillac)

The most important differential measurements and adjustments include:

- Pinion gear depth.
- Pinion bearing preload.
- Case bearing preload.
- Ring gear runout.
- Ring and pinion backlash.
- Ring and pinion contact pattern.

Pinion gear depth refers to the distance the pinion gear extends into the carrier. Pinion gear depth affects where the pinion gear teeth mesh with the ring gear teeth. Pinion gear depth is commonly adjusted by varying shim thickness on the pinion gear and bearing assembly. **Figure 62-25** illustrates how shims affect pinion gear depth in both removable and integral carriers.

Pinion bearing preload is frequently adjusted by torquing the pinion nut to compress a collapsible spacer. The more the pinion nut is torqued, the more the spacer will compress to increase the preload (tightness) of the bearings.

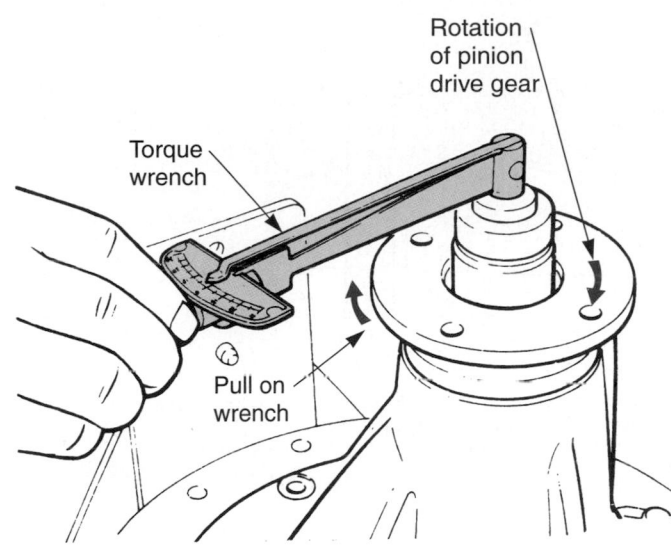

Figure 62-26. Measure pinion bearing preload with a small torque wrench. Compare the measurement to specifications and adjust using service manual directions. (Dodge)

When a solid spacer is used, *pinion gear shims* commonly control pinion bearing preload. The pinion nut is usually torqued to a specific value.

To set pinion gear or bearing preload, a holding tool is used to keep the pinion gear stationary. Then, a breaker bar or torque wrench can be used to tighten the pinion nut.

With a collapsible spacer, only tighten the nut in small increments. Then measure pinion preload by turning the pinion nut with a torque wrench. See **Figure 62-26.**

Case bearing preload is the amount of force pushing the differential case bearings together. As with pinion bearing preload, it is critical.

If case bearing preload is too low (bearings too loose), differential case movement and ring and pinion gear noise can result. If preload is too high (bearings too tight), bearing overheating and failure can result.

When adjusting nuts are used to set case bearing preload, the nuts are typically tightened until all the play is out of the bearings. Then, each nut is tightened a specific portion of a turn to preload the bearings. This is done when adjusting backlash (covered shortly).

When shims are used to adjust preload, you may need to use a feeler gauge to check side clearance between the case bearing and carrier. This will let you calculate the correct shim thickness to preload the case bearings. Refer to a shop manual for specific adjustment procedures.

Ring gear runout is the amount of wobble (side-to-side movement) produced when the ring gear is rotated. Ring gear runout must be within specifications.

To measure ring gear runout, mount a dial indicator against the back of the ring gear. The indicator stem

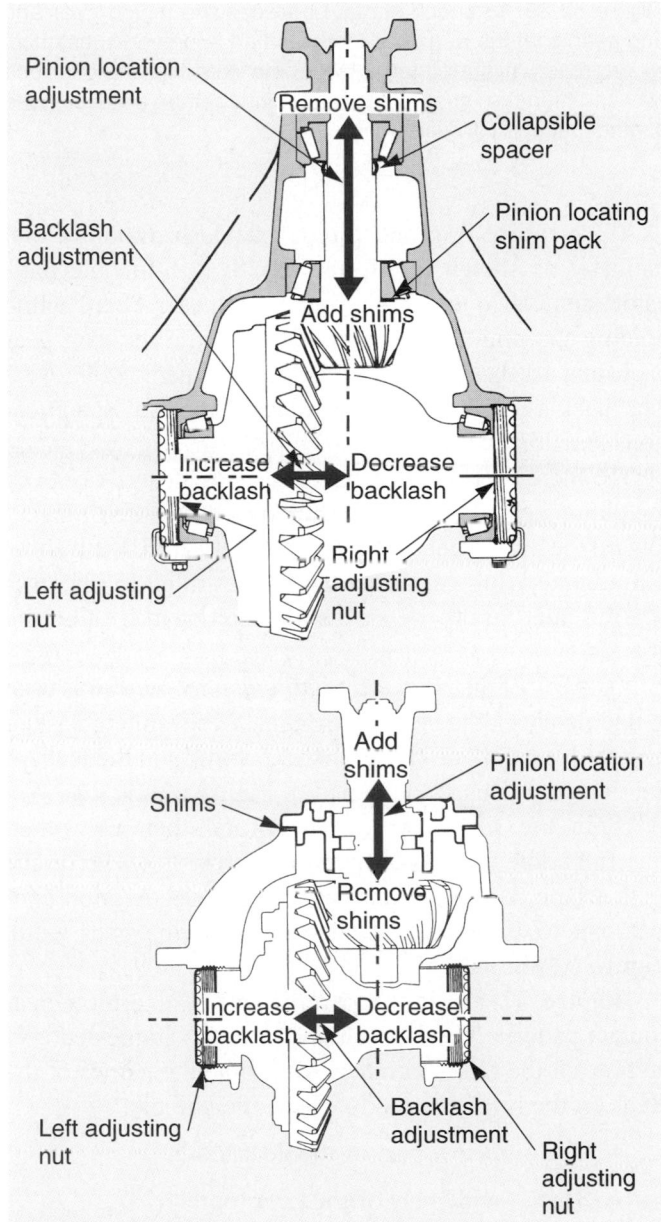

Figure 62-25. Study how the movement of the pinion and ring gears adjusts differentials. Shim thickness change affects each differently. (Ford)

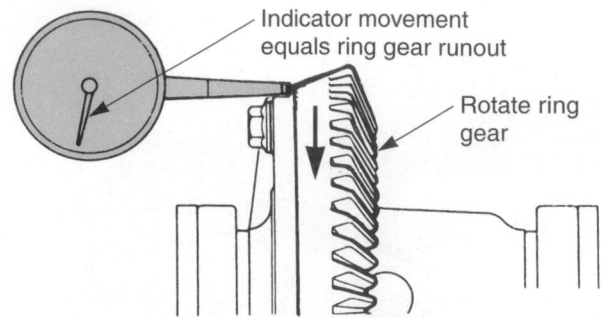

Figure 62-27. If gears seem to tighten up and loosen when turned, measure ring gear runout. Indicator needle movement equals runout. (Plymouth)

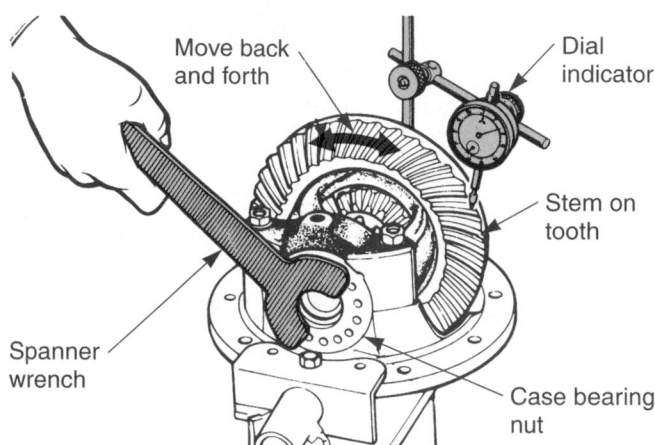

Figure 62-28. After setting pinion bearing preload and depth, adjust backlash and case bearing preload. This unit is adjusted by turning nuts. The indicator is mounted with the direction of gear rotation, touching the outer end of the tooth. Hold the pinion stationary and move the ring gear back and forth while watching the indicator.

should be perpendicular to the ring gear surface. Then, turn the ring gear and note the indicator reading. See **Figure 62-27.**

If ring gear runout is excessive, check the ring gear mounting and the differential case runout. If the ring gear mounting is not a problem, replace either the ring gear (and pinion) or the case as needed.

Ring and pinion backlash refers to the amount of space between the meshing teeth of the gears. Backlash is needed to allow for heat expansion and lubrication.

As the gears operate, they produce friction and heat. This makes the gears expand, reducing the clearance between meshing teeth. Without sufficient backlash, the ring and pinion teeth could jam into each other as they reach operating temperature. The gears could fail in a short time. Too much ring and pinion backlash could cause gear noise (whirring, roaring, or clunking).

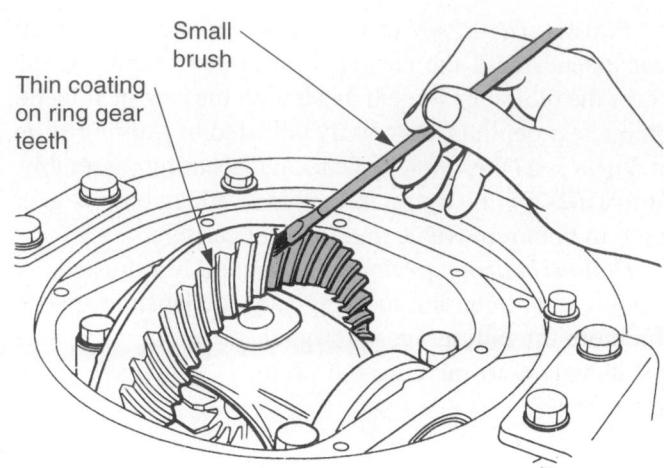

Figure 62-29. To check contact between the pinion gear and ring gear, coat the ring gear teeth with an approved substance: white grease, hydrated ferric oxide, etc. Turn the ring gear one way and then the other to rub the teeth together, producing the contact pattern. (Chrysler)

To measure ring and pinion backlash, mount a dial indicator as shown in **Figure 62-28.** Position the indicator stem on one of the ring gear teeth. Then, while holding the pinion gear stationary, wiggle the ring gear back and forth. Indicator needle movement will reveal gear backlash. Compare your measurement to specifications and adjust backlash as needed.

To increase backlash, move the ring gear away from the pinion gear. To decrease backlash, move the ring gear toward the pinion gear. Refer to **Figure 62-25.** With some differentials, ring gear position is controlled by the case bearing nuts. In others, shims are used to move the ring gear.

A check of the ***ring and pinion contact pattern*** is used to double-check the gear adjustment. Wipe white grease or hydrated ferric oxide (yellow oxide of iron) on the teeth of the ring gear. This is illustrated in **Figure 62-29.**

Spin the ring gear one way and then the other. Carefully note the contact pattern, which shows up on the teeth where the grease has been wiped off. A good contact pattern is located in the center of the gear teeth, **Figure 62-30.**

Figure 62-31 shows several ring and pinion gear contact patterns. Study each and note the suggested correction for the faulty contact. Also, note the names of the areas on the ring gear teeth. These include the:

- Toe—narrow part of the gear tooth.

- Heel—wide part of the gear tooth.

- Pitch line—imaginary line along the center of the tooth.

- Face—area on the tooth above the pitch line.

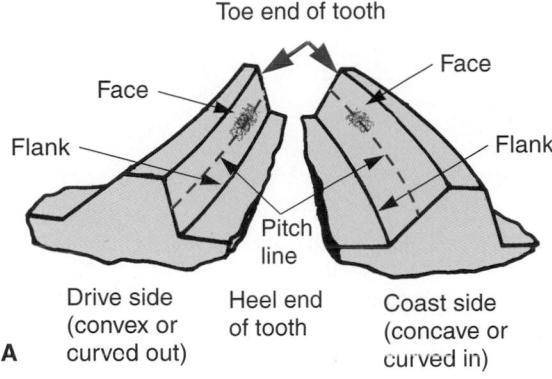

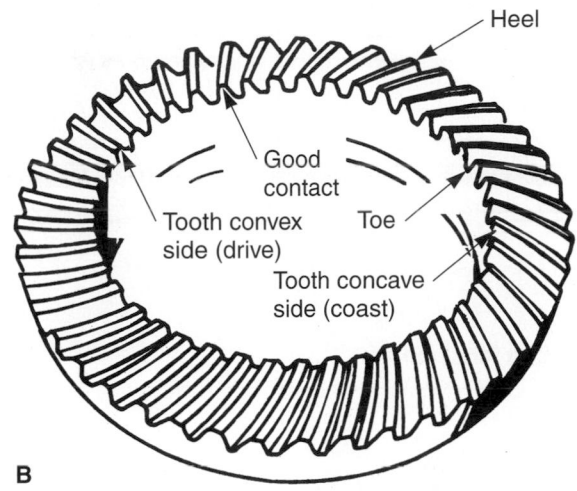

Figure 62-30. Ring gear teeth nomenclature. Study the names. They are needed when reading or interpreting gear contact patterns. (Oldsmobile)

- Flank—area on the tooth below the pitch line.
- Drive side—convex side of the tooth.
- Coast side—concave side of the tooth.

Summary

- Differentials and rear drive axles are normally very dependable. However, if operated dry (without lubricant), abused, or used for prolonged service, parts can wear and fail.
- Ring and pinion problems usually show up as a howl or whining noise that changes when going from acceleration to deceleration.
- Rear axle lubricant leaks can occur at the pinion gear seal, carrier or inspection cover gaskets, and at the two axle seals.
- Worn or damaged bearings in the carrier or on the axles often produce a constant whirring or humming sound.

- Differential case troubles frequently show up when rounding a corner. With the rear wheels turning at different speeds, any problem will usually show up as an abnormal sound from the rear of the vehicle.
- A stethoscope can often be used to isolate the source of a differential or rear axle bearing problem.
- Many automobile manufacturers recommend that the differential fluid be checked or replaced at specific intervals.
- A retainer plate type axle is commonly used with a removable carrier. A C-clip type axle is frequently used with an integral carrier.
- Wear eye and face protection when pressing a bearing on or off an axle.
- Excessive axle end play can cause a clunking sound as the car rounds a corner.
- Insufficient axle end play can result in axle bearing, retainer, or differential side gear failure.
- Break-away torque is the amount of torque needed to make one axle rotate the limited-slip differential clutches.
- Pinion gear depth refers to the distance the pinion gear extends into the carrier.
- Pinion gear preload is frequently adjusted by torquing the pinion nut to compress a collapsible spacer. When a solid spacer and pinion nut torque is used, pinion gear shims commonly control pinion bearing preload.
- Case bearing preload is the amount of force pushing the differential case bearings together.
- Ring gear runout is the amount of wobble produced when the ring gear is rotated.
- Ring and pinion backlash refers to the amount of space between the meshing teeth of the gears.
- A check of the gear tooth contact pattern in a differential is used to double-check ring and pinion adjustment.

Important Terms

Rear end	Pinion bearing preload
Stethoscope	Pinion gear shims
Retainer plate type axle	Case bearing preload
C-clip type axle	Ring gear runout
Axle end play	Ring and pinion backlash
Break-away torque	
Pinion gear depth	Ring and pinion contact pattern

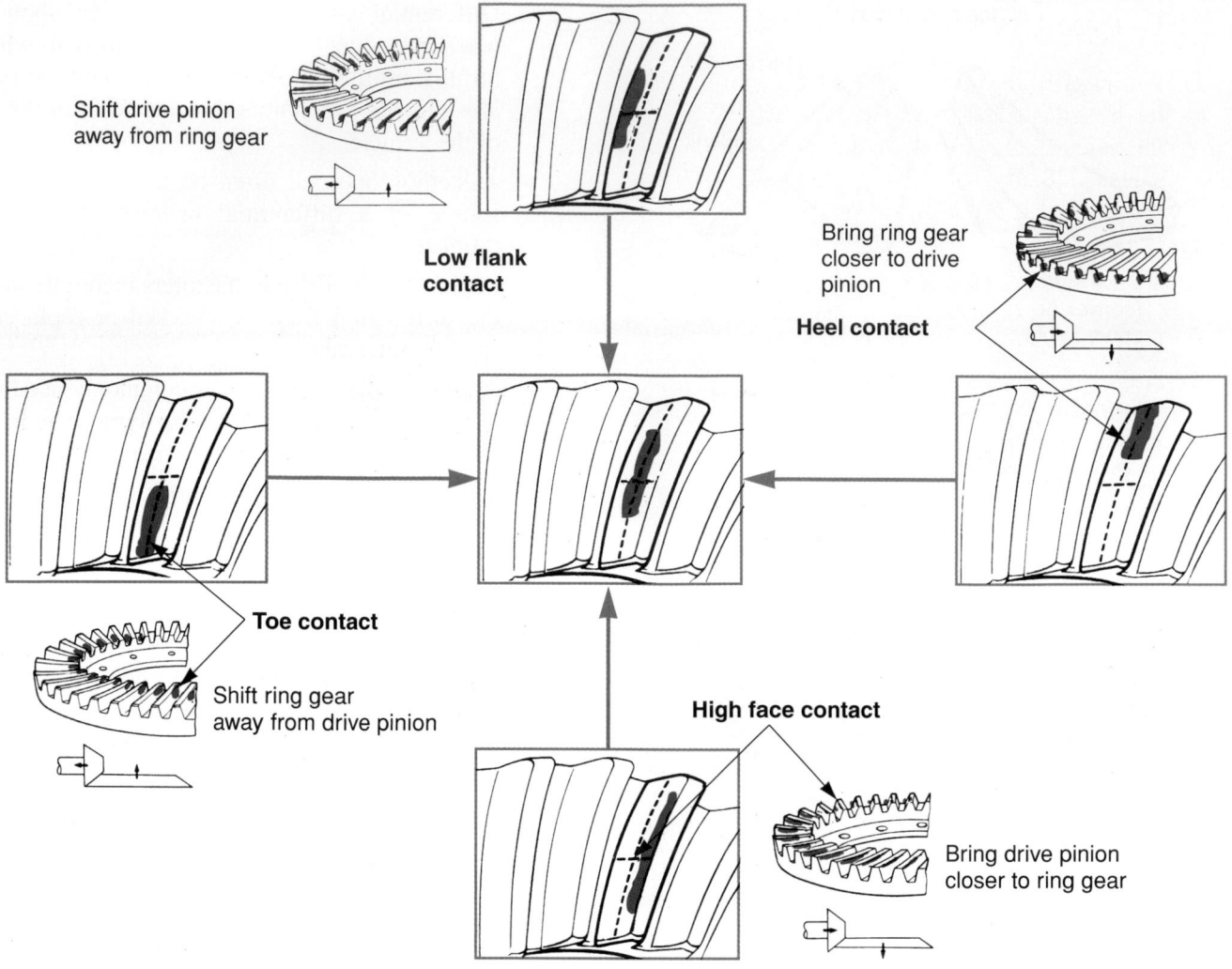

Figure 62-31. The area where grease is rubbed off the ring gear indicates contact point. It must be centrally located on the teeth. Note typical methods of correcting patterns. (General Motors and Toyota)

Review Questions—Chapter 62

Please do not write in this text. Place your answers on a separate sheet of paper.

1. A(n) _____ is a listening device that can be used to isolate differential and axle noises.

2. Describe ring and pinion noise.

3. Excess ring and pinion backlash can cause a "clunk" sound when the transmission is placed in drive. True or False?

4. Worn or damaged bearings in the carrier or on the axles produce a humming sound that changes with load. True or False?

5. List four points to consider when diagnosing bearing failures.

6. _____ _____ often require a lubricant that is compatible with friction clutches.

7. Explain the two methods commonly used to hold a rear drive axle in its housing.

8. How do you measure axle end play?

9. List twelve service rules for differentials.

10. Which of the following is *not* a common differential adjustment?
    (A) Pinion gear depth.
    (B) Pinion gear preload.
    (C) Case bearing preload.
    (D) Ring and pinion preload.

⬤ ASE-Type Questions

1. An automobile rear drive axle assembly needs to be road tested for diagnostic purposes. Technician A road tests the car on a smooth road surface. Technician B road tests the vehicle rounding corners. Who is right?
   - (A) A only.
   - (B) B only.
   - (C) Both A and B.
   - (D) Neither A nor B.

2. Noise is coming from a car's rear drive axle assembly. Technician A checks for broken pinion gear teeth. Technician B inspects the condition of the ring gear. Who is right?
   - (A) A only.
   - (B) B only.
   - (C) Both A and B.
   - (D) Neither A nor B.

3. A whining noise is coming from a car's differential case when driving straight. Technician A inspects the ring and pinion gear for problems. Technician B replaces the axle shims. Who is right?
   - (A) A only.
   - (B) B only.
   - (C) Both A and B.
   - (D) Neither A nor B.

4. A customer wants to know what caused his/her car's ring and pinion gears to malfunction. Technician A says prolonged service may have caused this problem. Technician B says low differential lubricant level may have caused this problem. Who is right?
   - (A) A only.
   - (B) B only.
   - (C) Both A and B.
   - (D) Neither A nor B.

5. A customer complains of a lubricant leak near the rear axle assembly. Technician A checks the housing's axle seals. Technician B checks the housing's filler drain plug. Who is right?
   - (A) A only.
   - (B) B only.
   - (C) Both A and B.
   - (D) Neither A nor B.

6. A rear axle is leaking lubricant. Technician A checks the pinion gear seal. Technician B checks the drive shaft slip yoke seal. Who is right?
   - (A) A only.
   - (B) B only.
   - (C) Both A and B.
   - (D) Neither A nor B.

7. A humming sound is coming from a car's rear axle's carrier when driving straight ahead. Technician A wants to replace the carrier bearings. Technician B wants to replace the spider gears. Who is right?
   - (A) A only.
   - (B) B only.
   - (C) Both A and B.
   - (D) Neither A nor B.

8. The bearings in a rear axle assembly need replacing. Before performing this task, Technician A checks the condition of all axle assembly parts. Before performing this task, Technician B wants to determine the cause of the original bearing failure. Who is right?
   - (A) A only.
   - (B) B only.
   - (C) Both A and B.
   - (D) Neither A nor B.

9. Noise is coming from a car's differential case in turns. Technician A looks for damaged spider gears. Technician B inspects the condition of the axle mounting plates. Who is right?
   - (A) A only.
   - (B) B only.
   - (C) Both A and B.
   - (D) Neither A nor B.

10. The lubricant in a limited-slip differential needs changing. Technician A tells the owner of the vehicle that all differential lubricant is the same. Technician B tells the owner of the car that certain limited-slip differentials require a special type of lubricant. Who is right?
    - (A) A only.
    - (B) B only.
    - (C) Both A and B.
    - (D) Neither A nor B.

11. A rear axle must be removed from its housing. Technician A says a retainer plate may be used to hold this axle in its housing. Technician B says a C-clip may be used to hold this particular axle in its housing. Who is right?
    (A) A only.
    (B) B only.
    (C) Both A and B.
    (D) Neither A nor B.

12. An automobile's retainer plate type axle needs replacing. Technician A begins this repair procedure by disassembling the differential carrier. Technician B begins this procedure by removing the spider gears. Who is right?
    (A) A only.
    (B) B only.
    (C) Both A and B.
    (D) Neither A nor B.

13. A press-fit rear axle bearing must be removed and replaced. Technician A wants to use a cutting torch to remove the bearing. Technician B wants to use a grinder, chisel, and hydraulic press to remove this bearing. Who is right?
    (A) A only.
    (B) B only.
    (C) Both A and B.
    (D) Neither A nor B.

14. A rear axle assembly differential is being serviced. Technician A checks pinion gear depth during this service procedure. Technician B checks ring gear runout during this service procedure. Who is right?
    (A) A only.
    (B) B only.
    (C) Both A and B.
    (D) Neither A nor B.

Activities—Chapter 62

1. Check out a rear axle assembly on a shop vehicle and report on its condition.

2. Remove an axle of the retainer plate type.

3. Remove an axle of the C-clip type.

4. Clean a differential and replace the fluid.

Rear Axle Diagnosis		
Condition	**Possible Causes**	**Correction**
Noise during straight ahead driving.	1. Insufficient lubricant.	1. Fill housing to correct level.
	2. Worn improper lubricant.	2. Drain. Flush and fill with correct lubricant.
	3. Differential case bearings.	3. Replace bearings.
	4. Worn drive pinion shaft bearings.	4. Replace pinion bearings.
	5. Worn ring and pinion.	5. Replace ring and pinion.
	6. Excessive backlash.	6. Adjust backlash.
	7. Insufficient backlash.	7. Adjust backlash.
	8. Excessive ring and pinion backlash.	8. Adjust backlash.
	9. Insufficient ring and pinion backlash.	9. Adjust backlash.
	10. Pinion shaft or differential case bearings improperly preloaded.	10. Preload as specified by the manufacturer.
	11. Excessive ring gear runout.	11. Remove ring, clean, and check flange runout. Reinstall and check runout. Replace ring or case as needed.
	12. Loose ring gear fasteners.	12. Torque fasteners.
	13. Unmatched ring and pinion.	13. Install a matched ring and pinion set.
	14. Loose differential case bearing cap fasteners.	14. Torque fasteners.
	15. Warped housing.	15. Replace housing.
	16. Loose pinion shaft companion flange retaining nut.	16. Torque flange nut.
	17. Incorrect tooth (ring and pinion) contact pattern.	17. Adjust as needed.
	18. Loose wheel.	18. Tighten wheel lugs.
	19. Wheel hub loose on tapered axle.	19. Inspect, if not damaged, torque retaining nut.
	20. Wheel hub key (on tapered axle) sheared.	20. Install new key.
	21. Worn wheel (axle) bearing.	21. Replace axle bearing and seal.
	22. Bent axle.	22. Replace axle.
	23. Worn wheel hub or axle keyway.	23. Replace axle or hub as needed.
	24. Dry pinion shaft seal.	24. Replace pinion shaft seal.
	25. Loose universal joint retainers.	25. Tighten universal joint retainers.
	26. Damaged universal joint.	26. Replace universal joint.
	27. Worn or broken front-wheel drive front suspension parts.	27. Repair or replace as necessary.
	28. Worn or broken transaxle unit.	28. Repair or replace as needed.
Noise when rounding a curve.	1. Worn or broken differential pinion gears.	1. Replace gears.
	2. Worn differential pinion shaft.	2. Replace pinion shaft.
	3. Worn or broken axle side gears.	3. Replace side gears.
	4. Excessive axle side gear or pinion gear end play.	4. Install new thrust washers or replace case and/or gears.
	5. Excessive axle end play.	5. Adjust end play.
	6. Improper type of lubricant.	6. Drain. Flush and fill with correct lubricant.
	7. Loose or broken suspension parts (front-wheel drive).	7. Repair or replace parts as needed.
	8. Loose or broken universal joints.	8. Tighten or replace universal joints.

Rear Axle Diagnosis		
Condition	**Possible Causes**	**Correction**
Clunking sound when engaging clutch, accelerating, or decelerating.	1. Excessive ring and pinion backlash. 2. Excessive end play in pinion shaft. 3. Worn axle side gears and pinions. 4. Worn differential bearings. 5. Worn side gear thrust washers. 6. Differential pinion shaft loose in case or pinions. 7. Worn axle shaft splines. 8. Worn wheel hub or axle keyway. 9. Loose wheel or hub. 10. Loose or broken universal joints.	1. Adjust backlash. 2. Preload bearings. 3. Replace worn gears. 4. Replace bearings. 5. Replace thrust washers. 6. Replace pinion shaft, gears, or differential case. 7. Replace axle. 8. Replace hub or axle. 9. Tighten fasteners. 10. Tighten or replace universal joints.
Axle leaking lubricant.	1. Clogged breather. 2. Worn seals. 3. Loose carrier-to-housing or inspection cover. 4. Carrier or inspection cover gasket damaged. 5. Lubricant level too high. 6. Wrong type of lubricant. 7. Porous housing (standard and transaxle). 8. Stripped fill plug threads. 9. Cracked housing (standard and transaxle).	1. Open breather. 2. Install new seals. 3. Tighten fasteners. 4. Install new gasket or sealer. 5. Drain lubricant to proper level. 6. Drain. Flush and install correct lubricant. 7. Repair or replace housing. 8. Repair or replace as needed. 9. Repair or replace housing.
Noises that may be confused with drive axle assembly.	1. Low air pressure in tires. 2. Road surface. 3. Transmission. 4. Bent propeller shaft. 5. Loose U-joints. 6. Engine. 7. Front wheel bearings. 8. Tire tread. 9. Dragging brakes. 10. Excessive front wheel end play.	1. Inflate tires to proper pressure. 2. Test on several different road surfaces. 3. Check transmission. 4. Replace shaft. 5. Tighten or replace U-joints. 6. Check engine. 7. Replace bearings. 8. Inflate temporarily to 50 psi (34.5 kPa) for road test only. 9. Adjust brakes. 10. Adjust wheel bearings.
Rear axle overheating.	1. Wrong type of lubricant. 2. Insufficient lubricant. 3. Overloading (pulling heavy trailer). 4. Worn gears. 5. Excessive bearing preload. 6. Insufficient backlash between ring and pinion.	1. Drain, flush, and fill with correct lubricant. 2. Fill lubricant to proper level. 3. Reduce vehicle load. Advise driver. 4. Replace gears. 5. Adjust preload. 6. Adjust backlash.

Chapter 63

Transaxle and Front Drive Axle Fundamentals

After studying this chapter, you should be able to:

- Identify the major parts of a transaxle assembly.
- Explain the operation of a manual transaxle.
- Explain the operation of an automatic transaxle.
- Trace the flow of power through manual and automatic transaxles.
- Describe design differences in transaxles.
- Identify the parts of constant velocity drive axles.
- Compare design differences in CV-joints.
- Correctly answer ASE certification test questions requiring a knowledge of transaxle and front drive axle designs.

This chapter describes the construction and operating principles of both manual and automatic transaxles and front drive axle assemblies. It relies and builds on the information given in previous chapters on clutches, manual transmissions, automatic transmissions, and differentials.

Note

In this text, the term *front drive axle assembly* is used when referring to an axle assembly comprised of multiple drive shafts connected by universal joints. Most of these axle assemblies are found in front-wheel-drive vehicles, but rear-engine vehicles use similar axle assemblies.

Transaxle

A *transaxle* is a transmission and a differential combined in a single assembly. See **Figure 63-1.** Short drive axle assemblies can be used to connect the transaxle output to the hubs and drive wheels. Look at **Figure 63-2.**

A transaxle is commonly used in late-model front-wheel drive cars. However, a few rear-engine or mid-engine sports cars and some older rear-engine economy cars also use a transaxle.

A front-engine vehicle having a transaxle and front-wheel drive has several advantages over a front-engine vehicle with rear-wheel drive:

- Reduced drive train weight and improved efficiency.
- Improved traction on slippery pavement because of more weight on the drive wheels.
- Increased passenger compartment space (no hump in the floor for the drive shaft).
- Smoother ride because of less unsprung weight (weight that must move with suspension action).
- Quieter operation because engine and drive train noise is centrally located in the engine compartment (no transmission, drive shaft, and rear axle under passenger compartment).

Figure 63-1. A transaxle is used in modern front-wheel-drive vehicles. It combines a transmission and a differential for driving the front axles. (Audi)

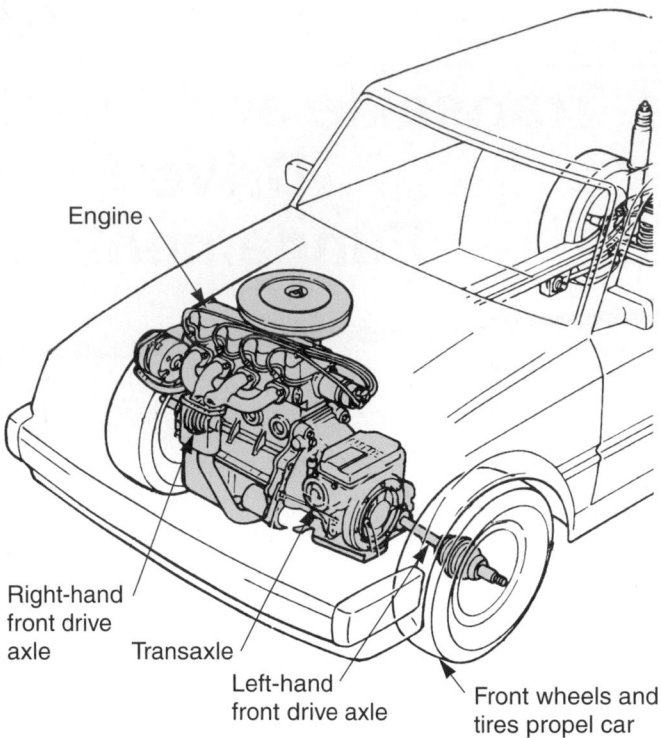

Figure 63-2. Note how the transaxle bolts to the side of the engine and CV-axles extend out to front-drive wheels. (Ford)

- Improved safety because of increased mass in front of passengers.

However, a front-wheel-drive vehicle also has the following disadvantages:

- *Torque steer* causes the front wheels to pull to one side when engine power acts on the steering

system through the front drive axles. Upon rapid acceleration, the vehicle tends to steer to one side of the road.

- Loss of directional control under hard acceleration. If the front wheels spin from high engine power output, the vehicle can skid out of control, especially when turning while accelerating.

- Less steering wheel feel of road because steering system must overcome torque steer. *Steering feel* is the feedback of road imperfections that is sent into the steering wheel.

- Loss of traction upon acceleration on dry pavement. This is especially true for vehicles with high-performance engines. Upon acceleration, vehicle weight transfer is to the rear wheels, which lifts weight off the front drive wheels and tires. Drag race cars are rear-wheel-drive for this reason.

Both manual and automatic transaxles are available. A manual transaxle uses a friction clutch and a standard transmission gearbox. An automatic transaxle commonly uses a torque converter and a hydraulic system to control gear engagement. Compare the manual and automatic transaxles shown in **Figure 63-3**.

Transverse Engine Transaxle

Most transaxles are designed so that the engine can be *transverse* (sideways) in the engine compartment, **Figure 63-4A**. The engine crankshaft centerline points at

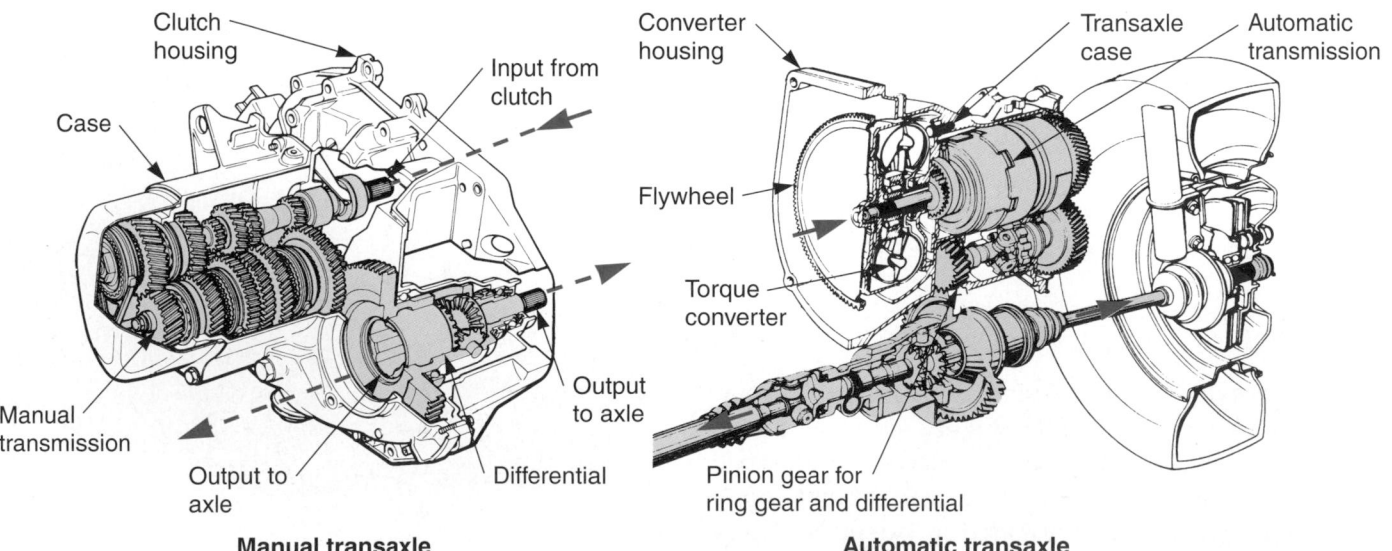

Manual transaxle   **Automatic transaxle**

Figure 63-3. A manual transaxle is a manual transmission and a differential in single assembly. An automatic transaxle is an automatic transmission and differential combined. (Renault and Chrysler)

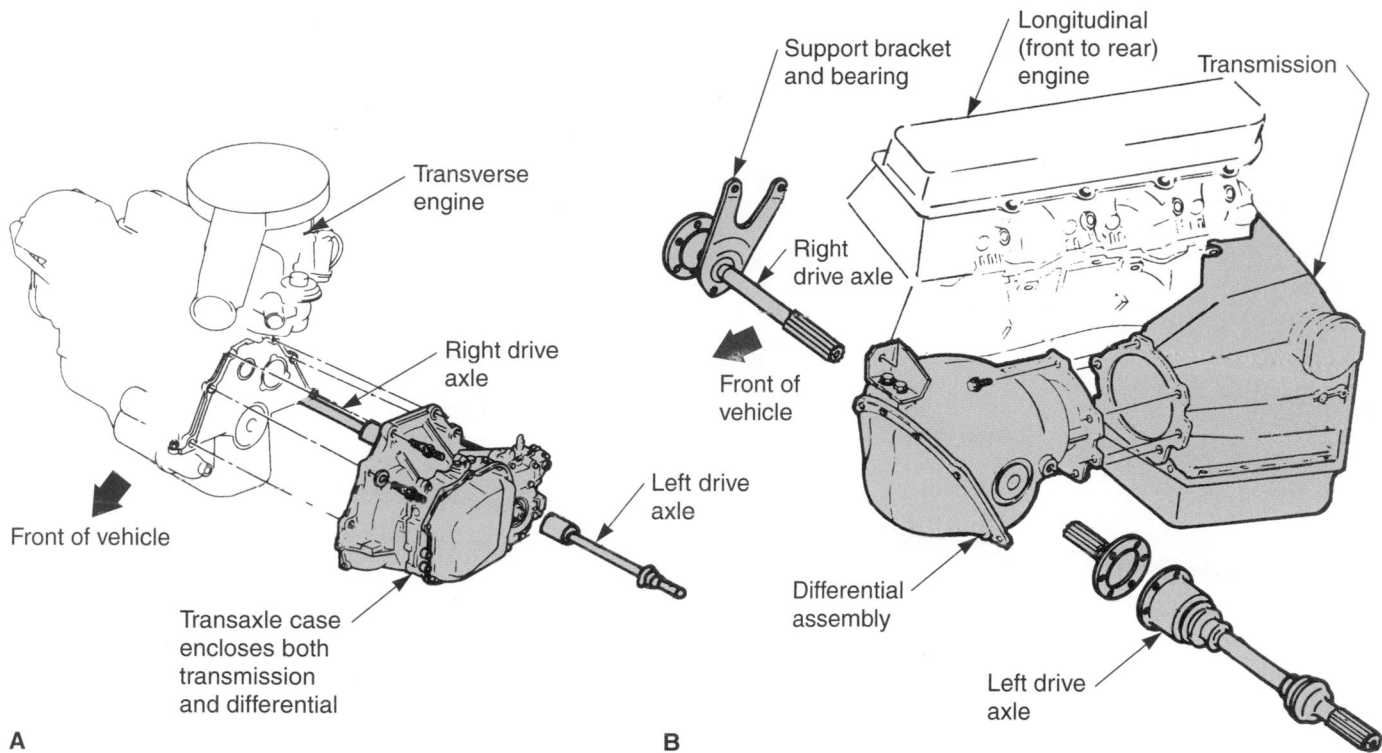

Figure 63-4. There are two basic transaxle differential design variations. A—With a transverse engine, the engine crankshaft centerline and axle centerline are on same plane. B—With a longitudinal engine, the differential must change power flow 90°, as with rear-wheel drive. (Ford and Cadillac)

both drive wheels. The transaxle bolts to the rear of the engine. This produces a very compact unit.

Engine torque enters the clutch and transmission. The transmission transfers power into the differential. Then, the differential turns the drive axles, which rotate the front wheels, **Figure 63-5A.**

Longitudinal Engine Transaxle

A few transaxles are made so that the engine is mounted *longitudinally* (lengthwise). The crankshaft centerline points toward the front and rear of the vehicle. Look at **Figure 63-4B.**

A transaxle for a longitudinally mounted engine frequently uses a more conventional differential with helical ring and pinion gears. See **Figure 63-5B.**

The flow of engine torque must be changed 90° in order to turn the drive axles. This type of transaxle uses the same principles as a conventional transmission and rear drive axle assembly. However, the parts are arranged differently.

Manual Transaxle

A *manual transaxle* uses a manual clutch and transmission. A foot-operated clutch engages and disengages the engine and transaxle. A hand-operated shift lever allows the driver to change gear ratios.

As the basic parts relating to a manual transaxle are introduced, refer to **Figure 63-6.**

- *Input shaft*—main shaft splined to the clutch disc; it turns gears in the transaxle.

- *Input gears*—either freewheeling or fixed gears on the input shaft; they mesh with output gears.

- *Output gears*—either free wheeling or fixed gears driven by input gears.

- *Output shaft*—a pinion shaft that transfers torque to the differential ring and pinion gears.

- *Synchronizers*—splined hub assemblies that can be used to lock freewheeling gears to their shafts for engagement.

- *Differential*—transfers gear box torque to driving axles and allows axles to turn at different speeds.

- *Transaxle case*—aluminum housing that encloses and supports the parts of the transaxle.

Manual Transaxle Clutch

A *manual transaxle clutch* is almost identical to the clutch used with a manual transmission for a rear-wheel-

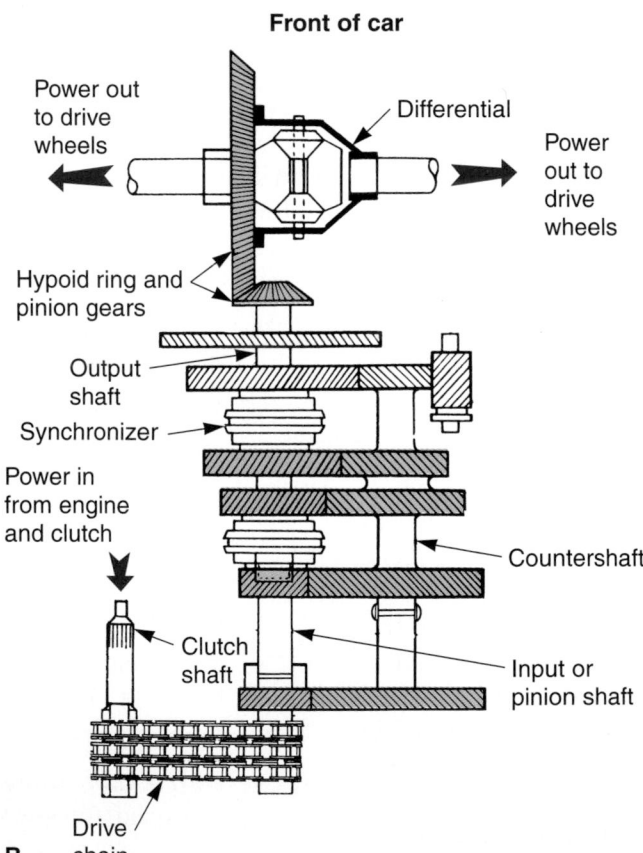

Front of car

Power in from engine and clutch

Input or main shaft

Pinion gear

Helical gears

Power out to wheels

Synchronizers

Output or pinion shaft

Power out to wheels

Differential

Ring gear

A

Front of car

Power out to drive wheels

Differential

Power out to drive wheels

Hypoid ring and pinion gears

Output shaft

Synchronizer

Power in from engine and clutch

Countershaft

Clutch shaft

Input or pinion shaft

Drive chain

B

Figure 63-5. Study the flow of power through the transaxles. A—In a transverse engine, pinion and ring gears are helical. They are positioned in same direction as transmission gears. B—In a longitudinal engine, the differential uses hypoid gears to change the direction of output. Also note the drive chain, which transfers power from the crankshaft and clutch to the input shaft. (Saab)

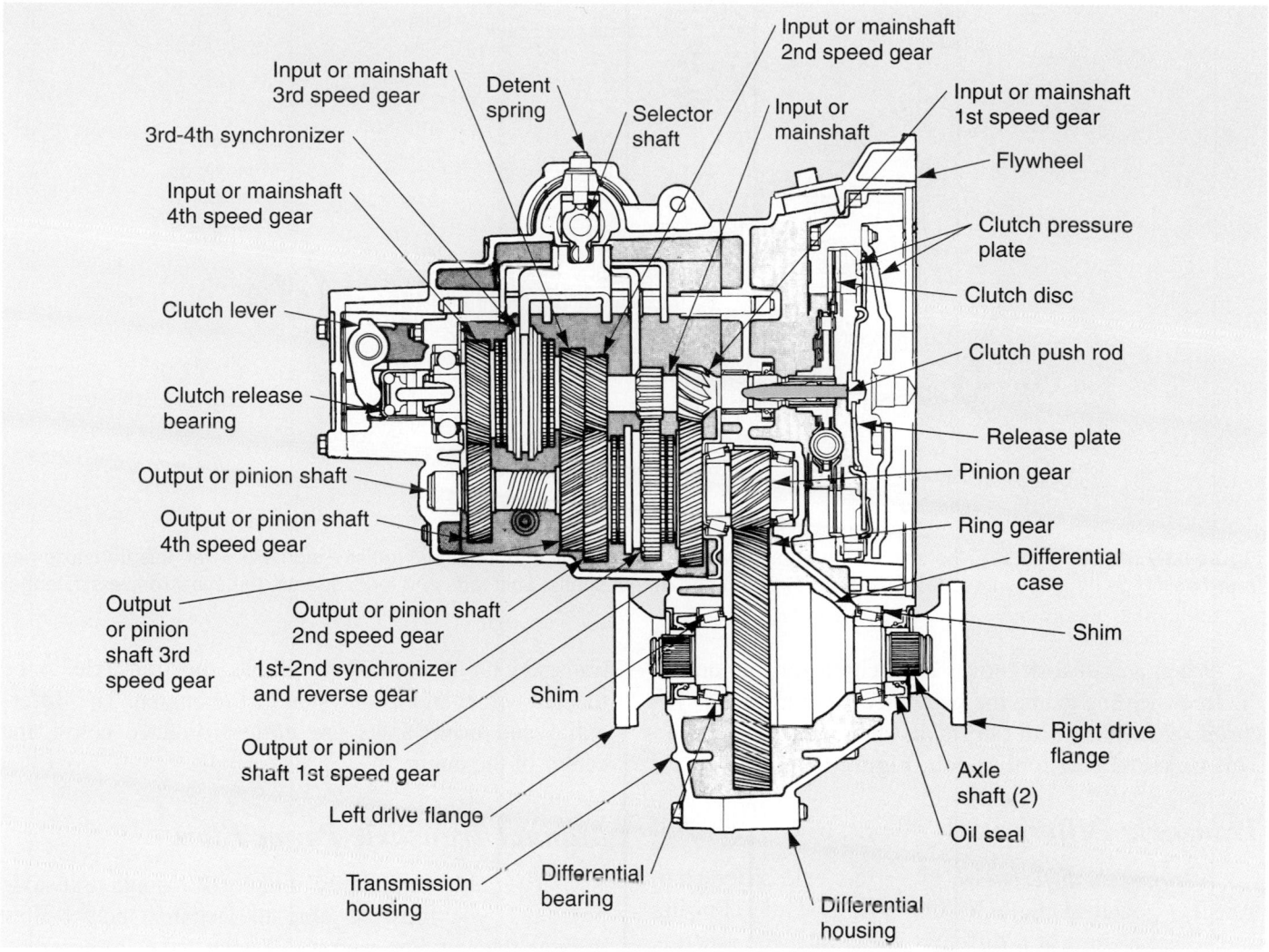

Figure 63-6. This detailed view shows all major parts of a manual transaxle. Study the names and locations carefully. In this transaxle, a long push rod extends through the center of the input shaft to actuate clutch. (Plymouth)

drive vehicle. It uses a friction disc and a spring-loaded pressure plate bolted to a heavy flywheel, **Figure 63-6.**

Some transaxles use a conventional clutch release mechanism (throw out bearing and fork). Others use a long push rod that passes through the input shaft.

Manual Transaxle Transmission

A *manual transaxle transmission* provides several (usually four or five) forward gear ratios and reverse. Sometimes, high gear can provide an overdrive ratio for increased fuel economy.

As you read service manuals, you will find that the names of the shafts, gears, and other transaxle parts will vary. This will depend on the location and function of the components. For example, look at **Figures 63-6** and **63-7.**

The input shaft is sometimes called the *mainshaft.* The output shaft is often called the *pinion shaft* because it drives the ring and pinion gears.

Sometimes, the input or output shaft gears are called the *cluster gear assembly* or the *countershaft gear assembly.* Like a manual transmission countershaft gear, several gears in the transaxle can be machined together as a unit.

The *transaxle shafts* are normally mounted in either tapered roller bearings or ball bearings. The shaft bearings fit into the transaxle case. The output shaft usually has a gear or sprocket for driving the differential ring gear, as in **Figure 63-7.**

The *transaxle synchronizers* are almost identical to those used in many manual transmissions. The inside hub of the synchronizer is splined to a transaxle shaft. The outer sleeve is free to slide on the hub. See **Figure 63-8.**

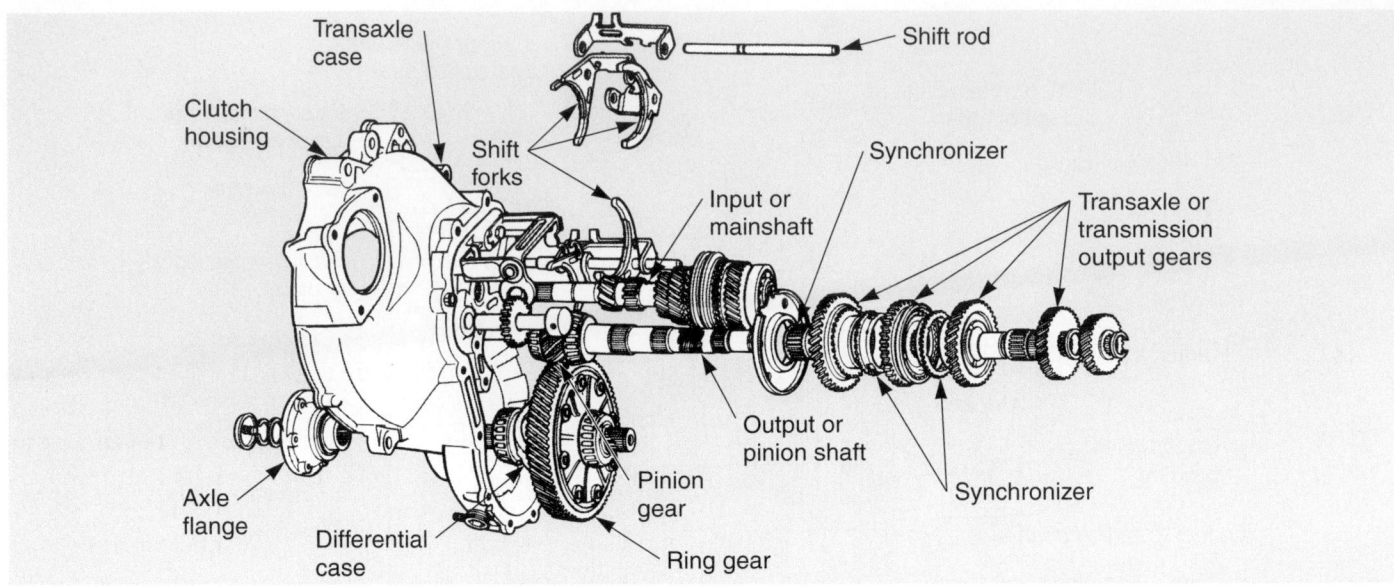

Figure 63-7. As you can see, the inside of manual transaxle is very similar to a manual transmission. However, this transaxle has freewheeling gears and synchronizers on both the input and output shafts. Shift rods and forks operate the synchronizers. (Dodge)

When a shift fork moves a synchronizer into one of the freewheeling gears, the outer sleeve (ring) of the synchronizer meshes with the small, outer teeth on the gear. This locks the gear to the shaft, **Figure 63-8.**

Transaxle Differential

A **transaxle differential,** like a rear axle differential, transfers power to the axles and wheels while allowing one wheel to turn at a different speed than the other. A small pinion gear on the output shaft turns the differential ring gear. Look at **Figures 63-9** and **63-10.**

The differential ring gear is fastened to the differential case. The case holds the spider gears (pinion gears and axle side gears) and a pinion shaft. The axle shafts are splined to the differential side gears.

When the gearbox spins the differential ring gear and case, the pinion shaft and pinion gears also revolve. The pinion gear teeth transfer power to the side gears. This causes the axle shafts to rotate. See **Figure 63-10.**

The transaxle case is normally made of cast aluminum. The shaft and differential bearing races are press-fit into the case. The case also has provisions for mounting the shift forks and other components.

Drive Chain

A **drive chain,** or **drive link,** is sometimes used to transfer crankshaft power to the transaxle gearbox, or transmission. It is used with longitudinal engines and some transverse engines.

One sprocket is mounted on a shaft connected to the engine crankshaft. The other sprocket drives the

transaxle input shaft. This allows the transaxle to be located below and to one side of the engine. The differential and drive axles are almost directly below the center of the engine.

Manual Transaxle Power Flow

To be able to quickly diagnosis manual transaxle problems, you must be able to visualize power flow through the unit. For example, if a noise develops in only one gear position, you can picture which parts might be causing the trouble.

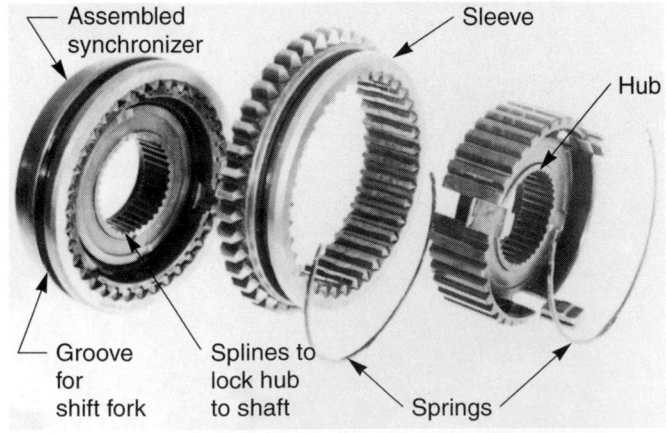

Figure 63-8. This manual transaxle synchronizer is similar to those used in manual transmissions. The inner hub is splined to the shaft. The outer sleeve can slide on the hub to engage the small teeth on the side of a freewheeling gear. This locks the gear to the hub and shaft for engagement. (Chevrolet)

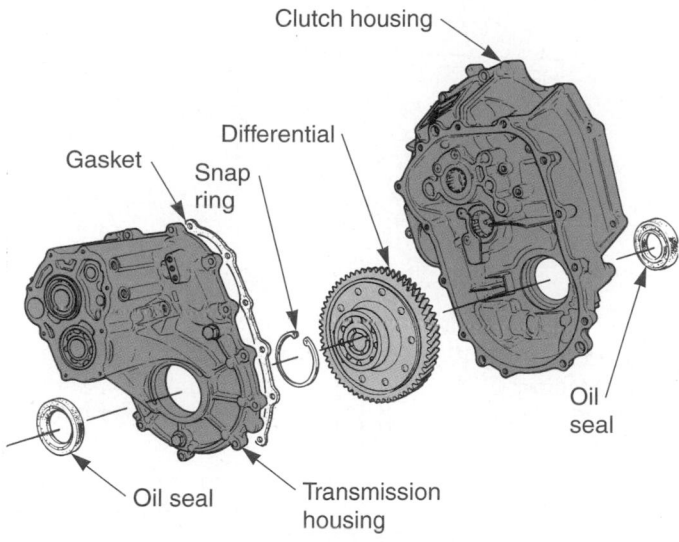

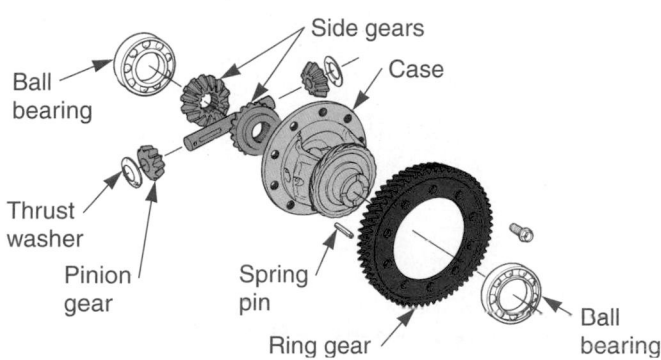

Figure 63-9. The differential in a transaxle uses spider gears, a pinion shaft, and the case to provide turning power to the drive axles. Note that this unit for a transverse engine uses helical ring and pinion gears, rather than hypoid gears. (Plymouth)

Transaxle in Neutral

With the transaxle in neutral, the engine spins the input shaft. However, since the synchronizers do not lock any gears to the output shaft, power is not transferred to the differential. See **Figure 63-11A.**

Transaxle in First Gear

When the driver shifts the transaxle into first gear, a shift fork slides the first-second synchronizer into mesh with first gear. This locks the synchronizer teeth with the small teeth on the side of first gear. The first gear is now locked to its shaft.

Power flows through the input shaft first gear, output shaft first gear, pinion gear, ring gear, and differential spider gears to the drive axles, **Figure 63-11B.**

Since the input shaft first gear is much smaller than the output shaft first gear, a gear reduction is produced. Torque is increased because the engine must rotate several times to produce one axle shaft rotation.

Transaxle in Second Gear

When the transaxle is shifted into second gear, the first-second synchronizer is moved into mesh with second gear on the input shaft. The second gear can no longer freewheel on its shaft. Power then flows through the second gears and into the differential. From the differential, power flows to the axle shafts and drive wheels. Refer to **Figure 63-11C.**

Transaxle in Third Gear

With the transaxle in third gear, the first-second synchronizer is shifted into neutral so that the first and

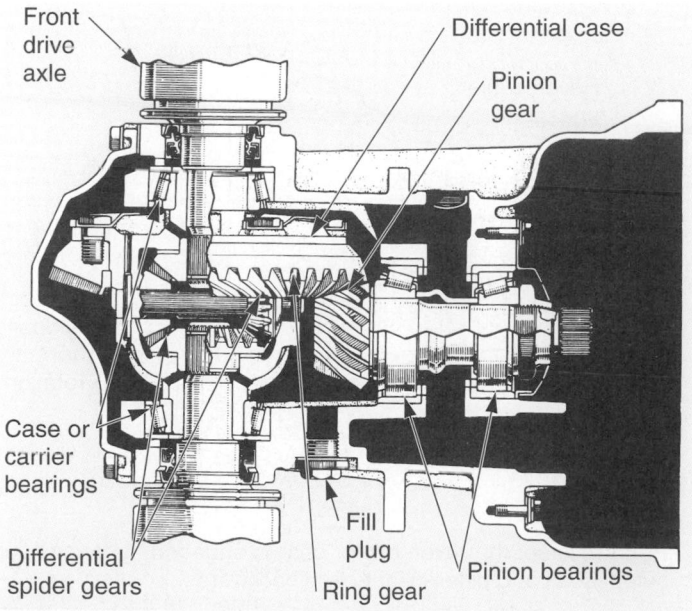

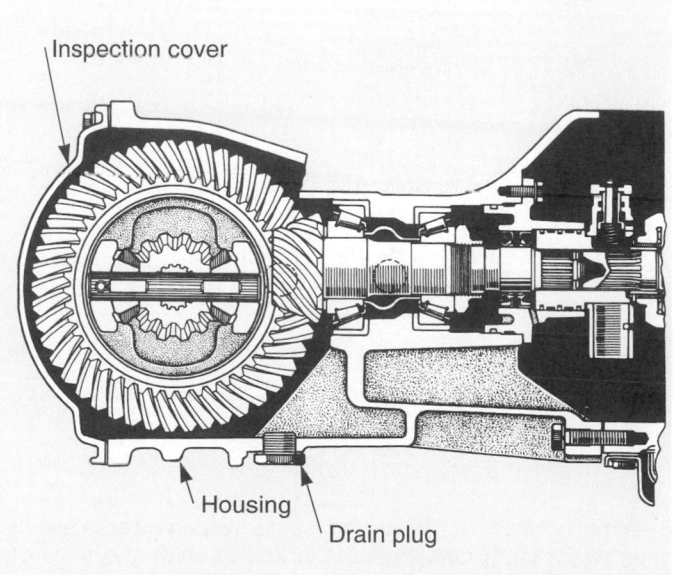

Figure 63-10. This transaxle for a longitudinal engine is almost identical to a differential in a rear-wheel drive axle. Hypoid gears transfer driving power. (Toyota)

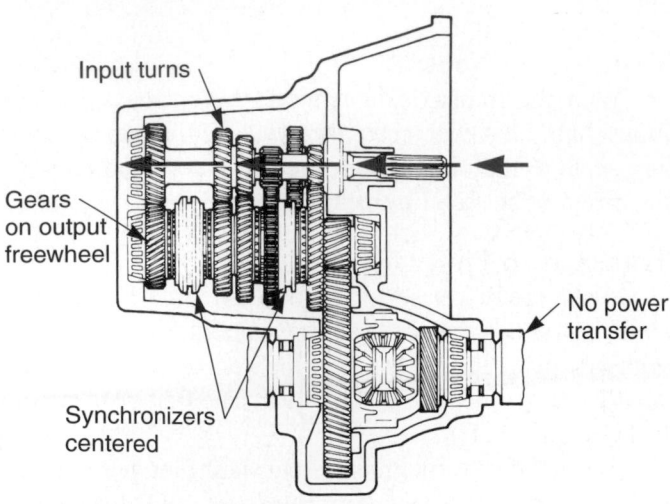

Input turns

Gears on output freewheel

Synchronizers centered

No power transfer

A—In neutral, no synchronizers are engaged with gears. Input shaft spins but gears freewheel and do not transfer power to output.

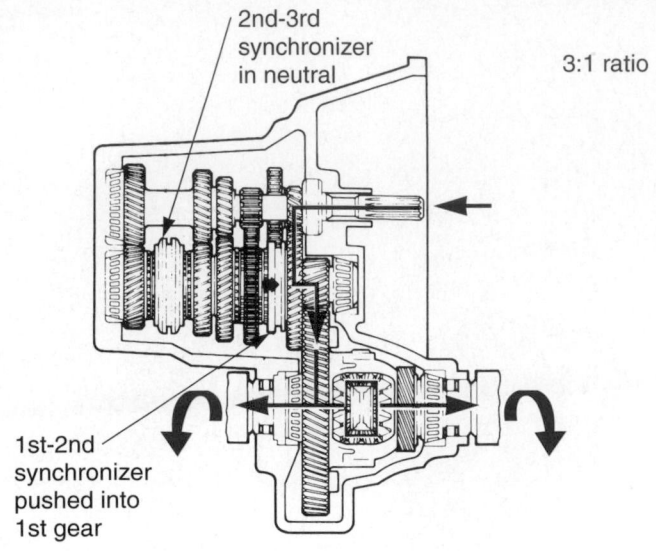

2nd-3rd synchronizer in neutral

3:1 ratio

1st-2nd synchronizer pushed into 1st gear

B—In first gear, first-second synchronizer slides to right, engaging first output gear. This locks gear to shaft and power flows to output shaft and differential.

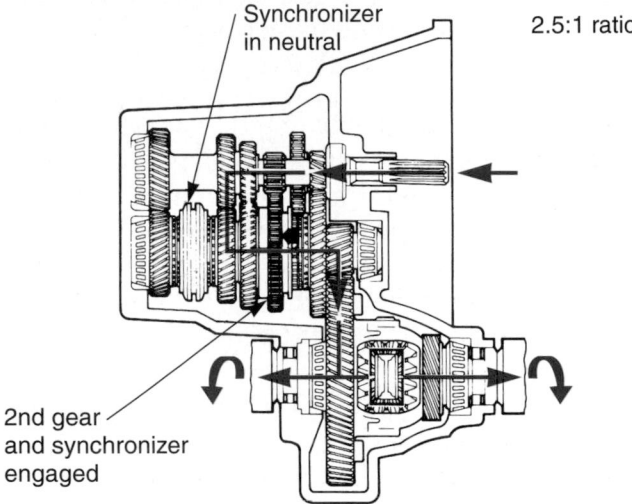

Synchronizer in neutral

2.5:1 ratio

2nd gear and synchronizer engaged

C—In second gear, same sychronizer slides to left. This locks output second gear to shaft. Power flows through second gears and to differential.

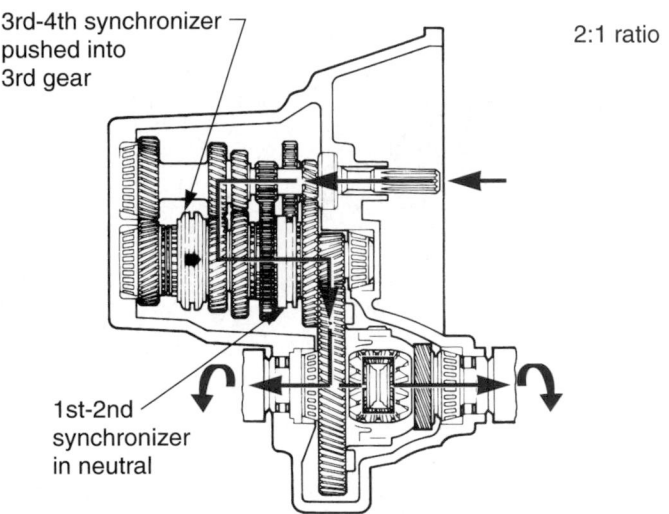

3rd-4th synchronizer pushed into 3rd gear

2:1 ratio

1st-2nd synchronizer in neutral

D—In third gear, first-second synchronizer is centered. Third-fourth synchronizer moves to right, engaging third output gear to its shaft.

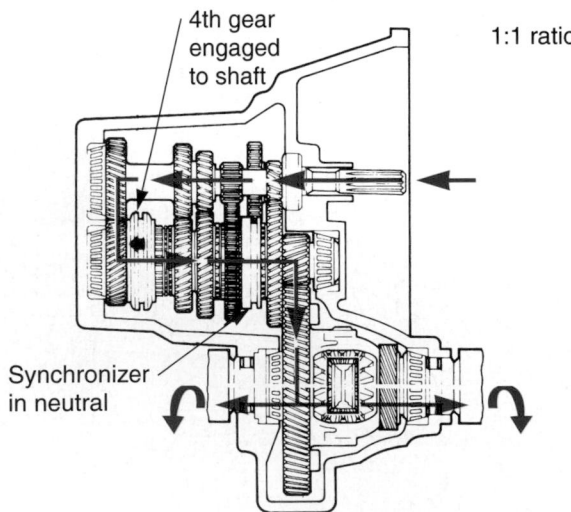

4th gear engaged to shaft

1:1 ratio

Synchronizer in neutral

E—When shifted into fourth, the same synchronizer is slid the other way. Fourth output gear is locked to shaft and transmits power.

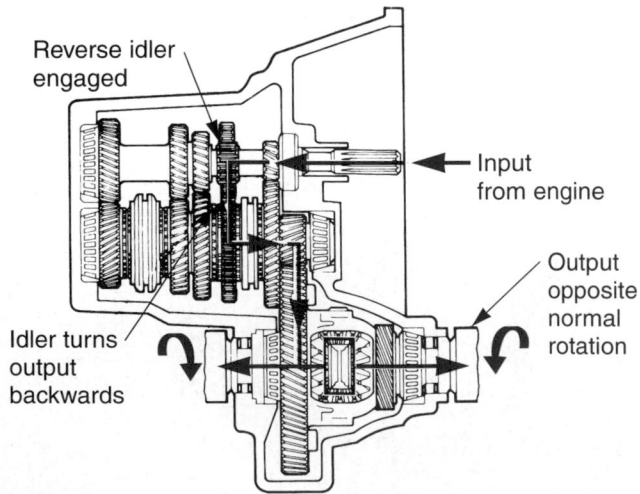

Reverse idler engaged

Input from engine

Output opposite normal rotation

Idler turns output backwards

F—In reverse, the reverse idler gear is engaged. It causes the output shaft and differential to turn backwards.

Figure 63-11. Trace the power flow through a typical manual transaxle. (Ford)

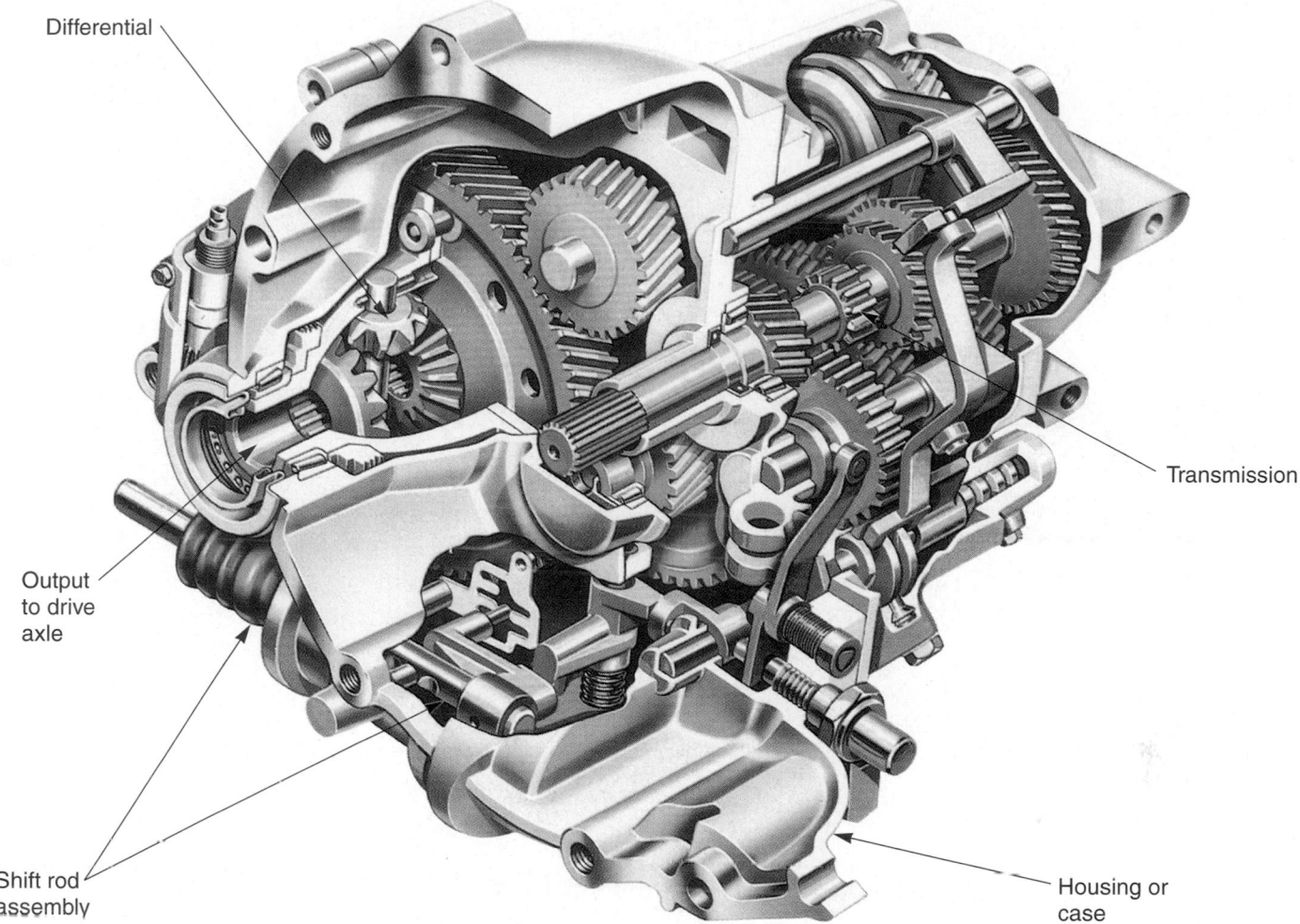

Differential

Transmission

Output
to drive
axle

Shift rod
assembly

Housing or
case

Figure 63-12. Cutaway view shows inside of five-speed manual transaxle with overdrive in high gear. Study how the shift levers and rods connect to shift forks. (Ford)

second gears freewheel. At the same time, the third-fourth synchronizer is slid into mesh with third gear. Third gear is then locked to its shaft. Power flows through the transaxle as shown in **Figure 63-11D.**

Transaxle in Fourth Gear

When the transaxle is in fourth gear, the three-four synchronizer is moved into contact with fourth gear. Power flows through the fourth gears, into the differential, and to the front wheels. **Figure 63-11E.**

Since the fourth gear on the input shaft and the fourth gear on the output shaft are the same size, the gear ratio is reduced. Compared to its rotation in the other gears, the engine turns slowly while the differential case and axles spin at a relatively high speed. This allows the vehicle to cruise at highway speeds with the engine running at low speeds.

Transaxle in Reverse

When the transaxle is shifted into reverse, the reverse idler gear is moved into mesh with the reverse gears on

the input shaft and output shaft. The idler gear reverses the direction of rotation. As a result, the differential and axle shafts are turned backwards and the car moves in reverse. Look at **Figure 63-11F.**

Figures 63-12 and **63-13** show two more manual transaxles. Compare these to the ones shown earlier.

Automatic Transaxle

An *automatic transaxle* is a combination of an automatic transmission and a differential. See **Figure 63-14.** It consists of seven main components:

- *Torque converter*—fluid-type clutch that slips at low speeds but locks up and transfers full engine power at a predetermined speed; it couples the engine crankshaft to the transaxle input shaft and gear train.

- *Oil pump*—produces hydraulic pressure to operate, lubricate, and cool the automatic transaxle; its pressure activates the pistons and servos.

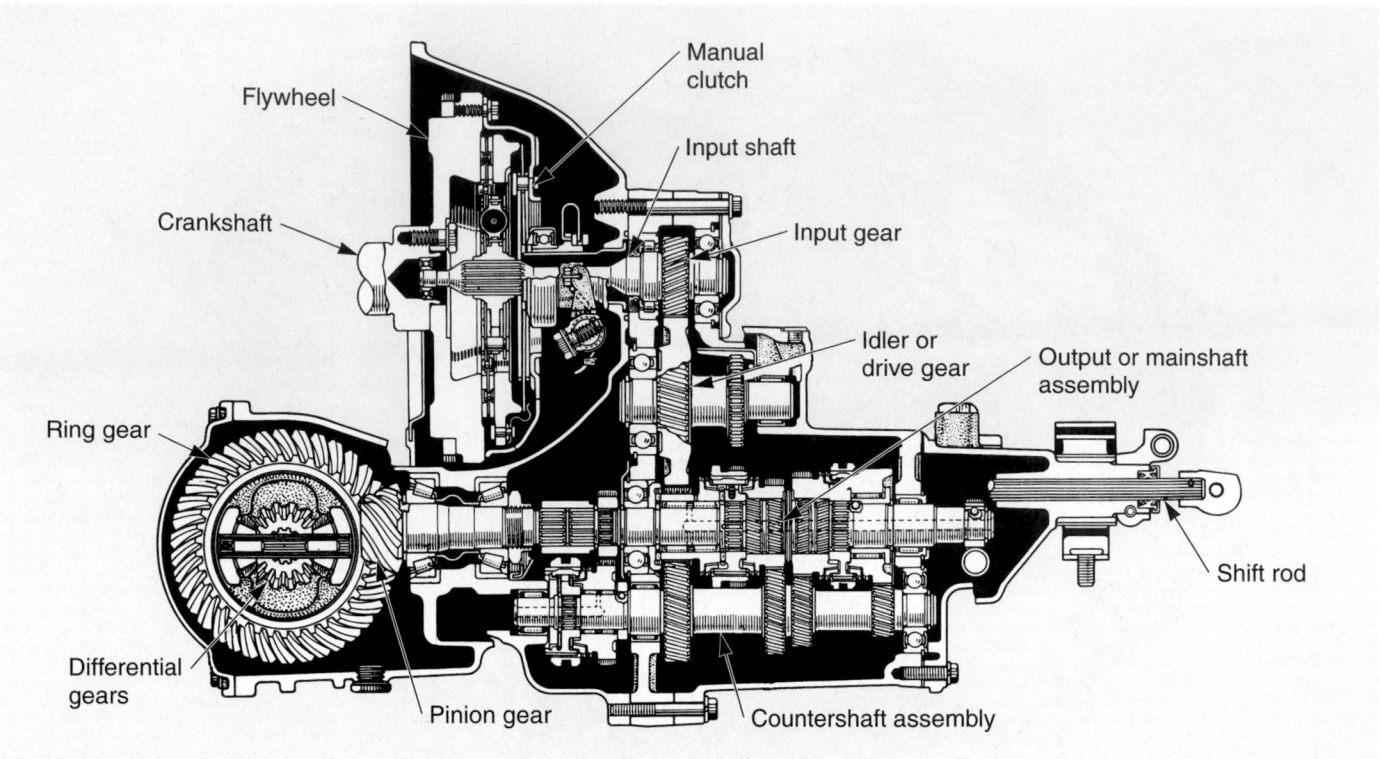

Figure 63-13. A late-model manual transaxle for a longitudinally mounted engine. Note how gears, rather than a drive chain, transfer power to the transaxle gearbox.

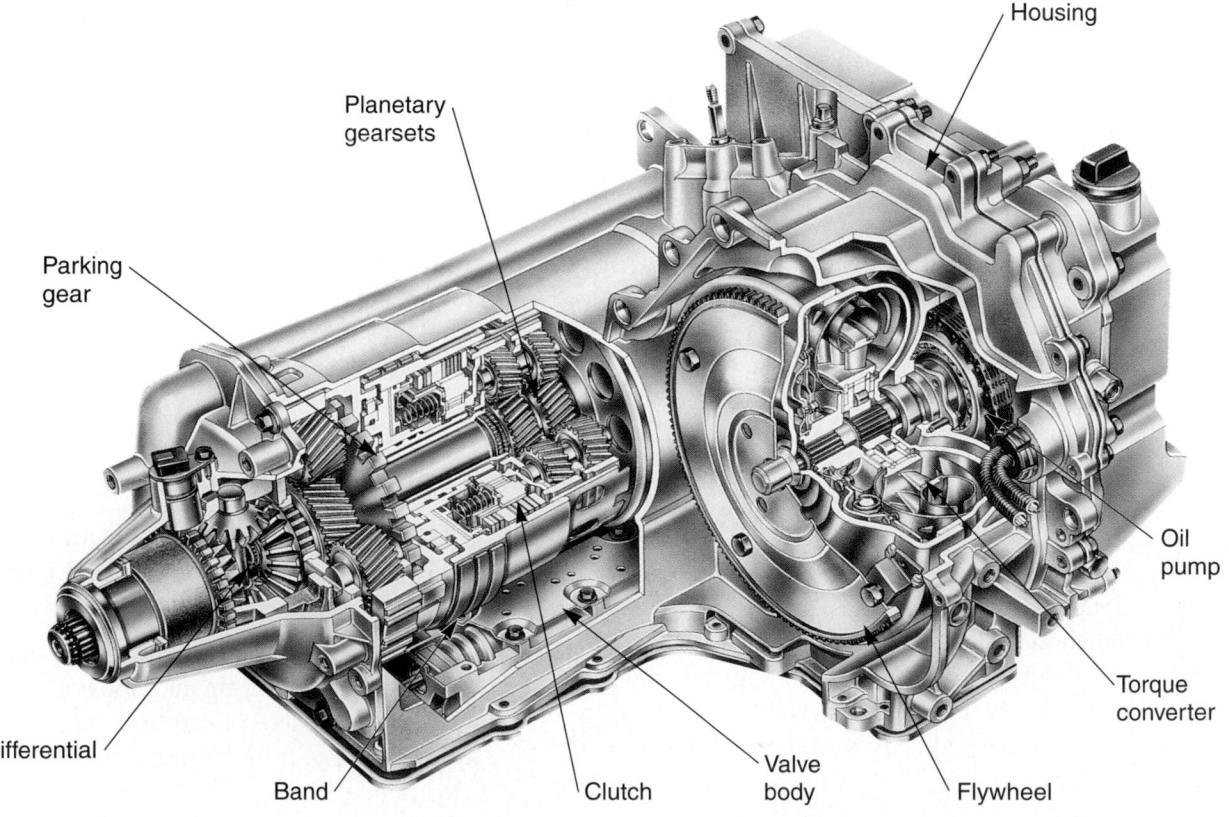

Figure 63-14. Study the basic parts of an automatic transaxle. It uses the same parts as an automatic transmission. (General Motors)

- *Valve body*—controls oil flow to pistons and servos in the transaxle; it contains hydraulic valves that are operated by shift linkage, engine speed, and load-sensing devices.

- *Pistons and servos*—operate clutches and bands; they are activated by oil pressure from the valve body.

- *Bands and clutches*—apply planetary gears in transaxle; different bands and clutches can be activated to operate different units in gearsets.

- *Planetary gearsets*—provide different gear ratios and reverse gear.

- *Differential*—transfers power from the transmission to the axle shafts.

As you can see, an automatic transaxle uses many of the same parts found in an automatic transmission. For a review of the operating principles of these components, turn back to Chapter 57, *Automatic Transmission Fundamentals.*

Figure 63-15 shows some of the parts of a transaxle hydraulic system. Note that the valve body on this particular unit bolts to the top of the transaxle case. The oil filter and sump are located on the bottom of the case. Gaskets seal the valve body cover and oil pan to the case.

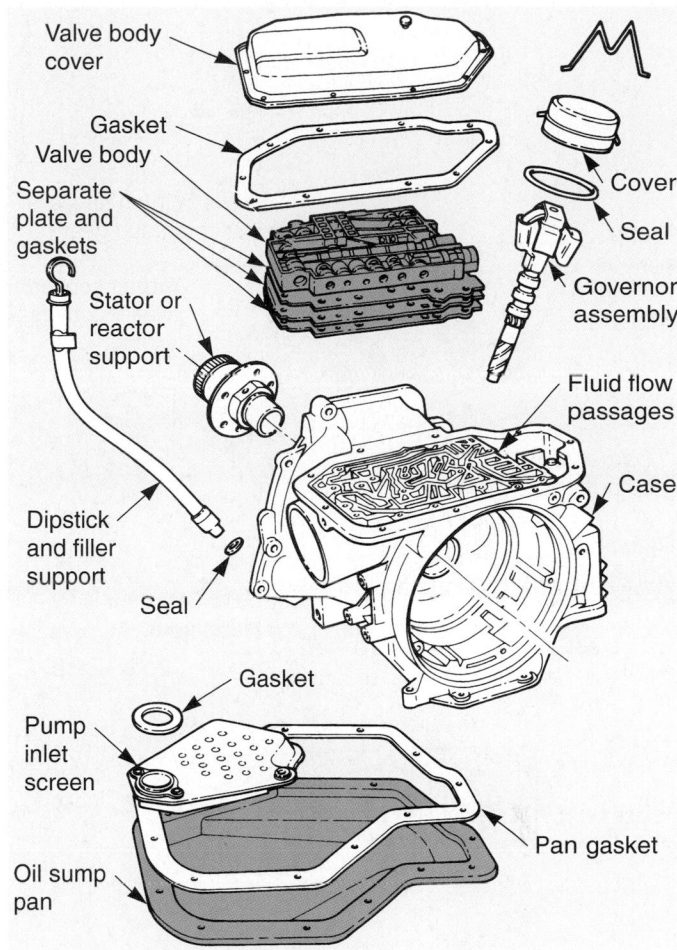

Figure 63-15. General arrangement of major hydraulic components in an automatic transaxle. With this particular unit, the valve body is on top and the sump and pan are on the bottom. Case construction is similar to an automatic transmission. (Ford)

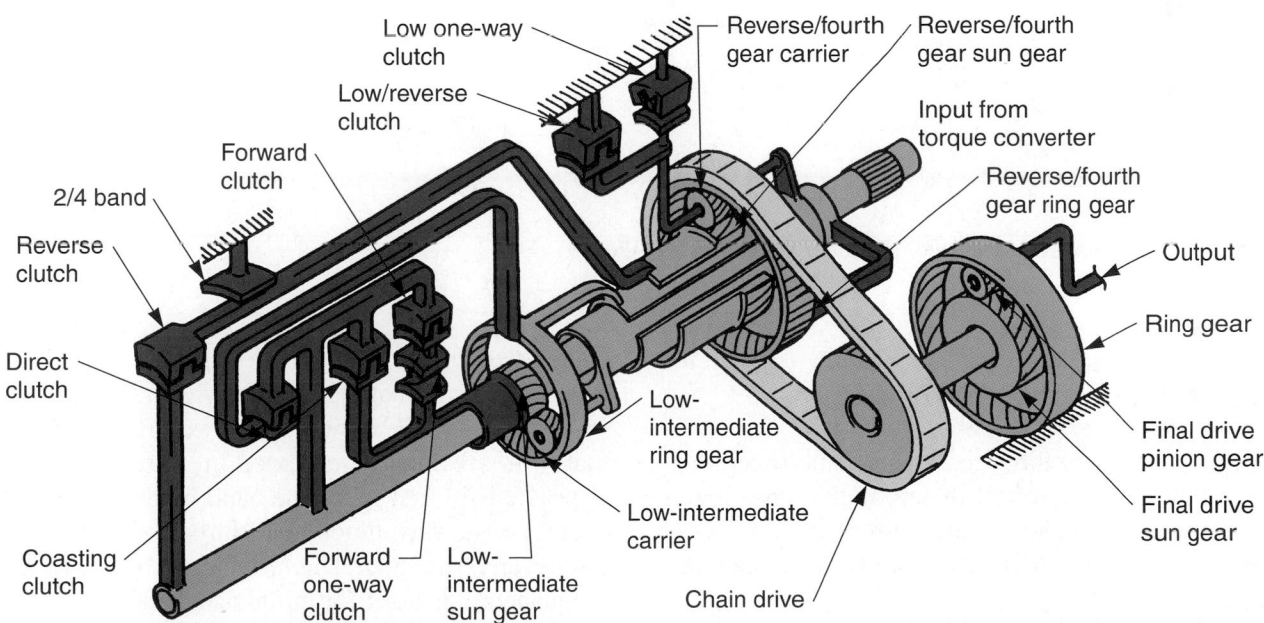

Figure 63-16. This drawing shows how major parts transfer power through one automatic transaxle design. Study power flow and part names. Note the drive chain, which transfers power from the planetary gearset to the final drive gears. (Mazda)

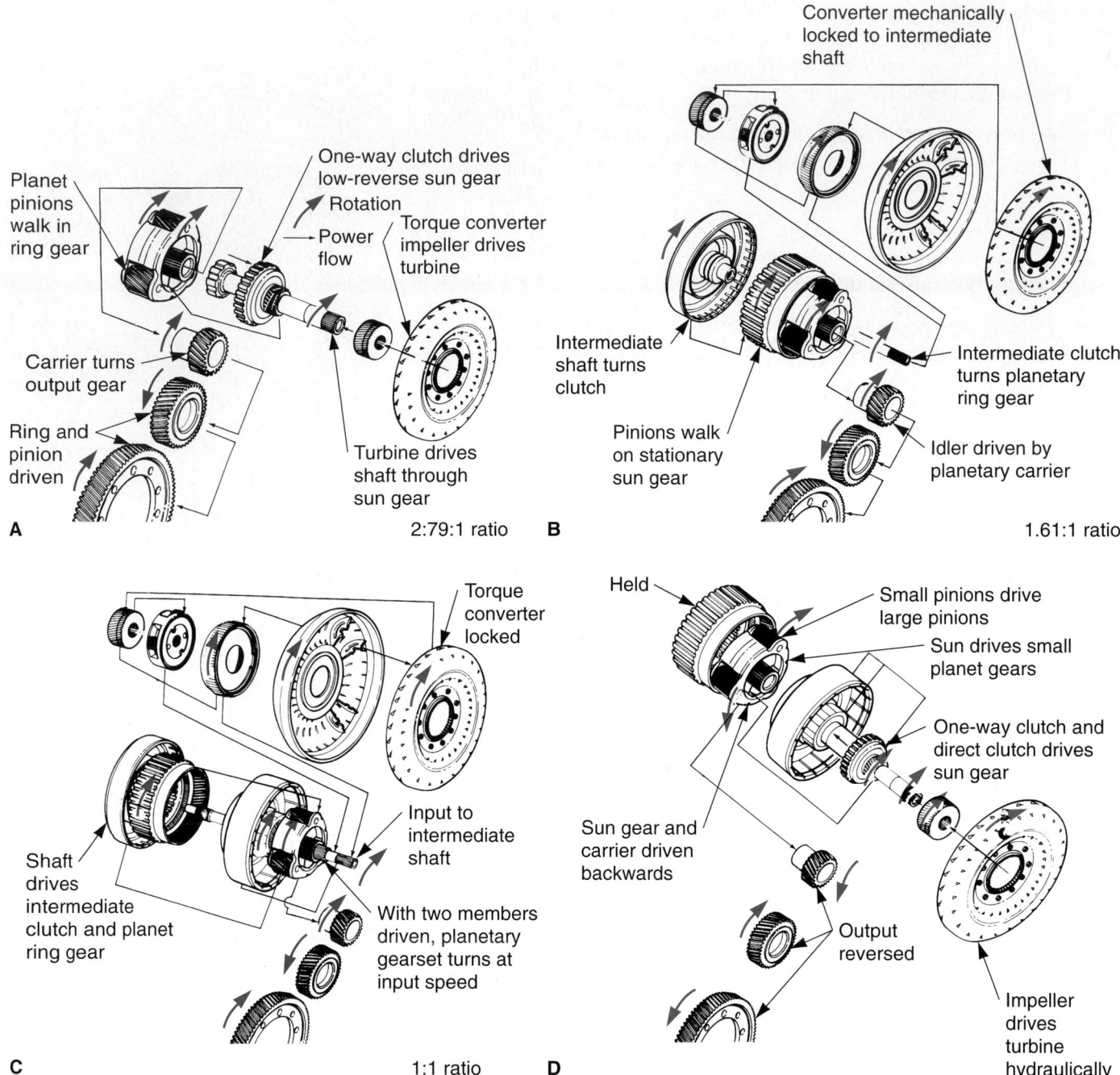

Figure 63-17. Basic power flow through an automatic transaxle. (Ford)

Automatic Transaxle Power Flow

The flow of power through an automatic transaxle is similar to power flow through an automatic transmission. Engine torque enters the torque converter. The torque converter then turns the input shaft and planetary gearsets.

Depending upon which bands and clutches hold the gearset members, power flows through the planetary gearsets to the ring and pinion gears. The differential powers the axle shafts and front wheels, **Figure 63-16.**

Figure 63-17 shows the power flow through one make of automatic transaxle. In **Figure 63-17A**, the transaxle is in first gear. The band holds the forward sun gear. The one-way clutch sends turbine shaft torque to the low-reverse sun gear. This produces a gear reduction in the planetary gearset for initial acceleration.

In **Figure 63-17B**, the transaxle is in second gear. The band remains applied and holds the forward sun gear.

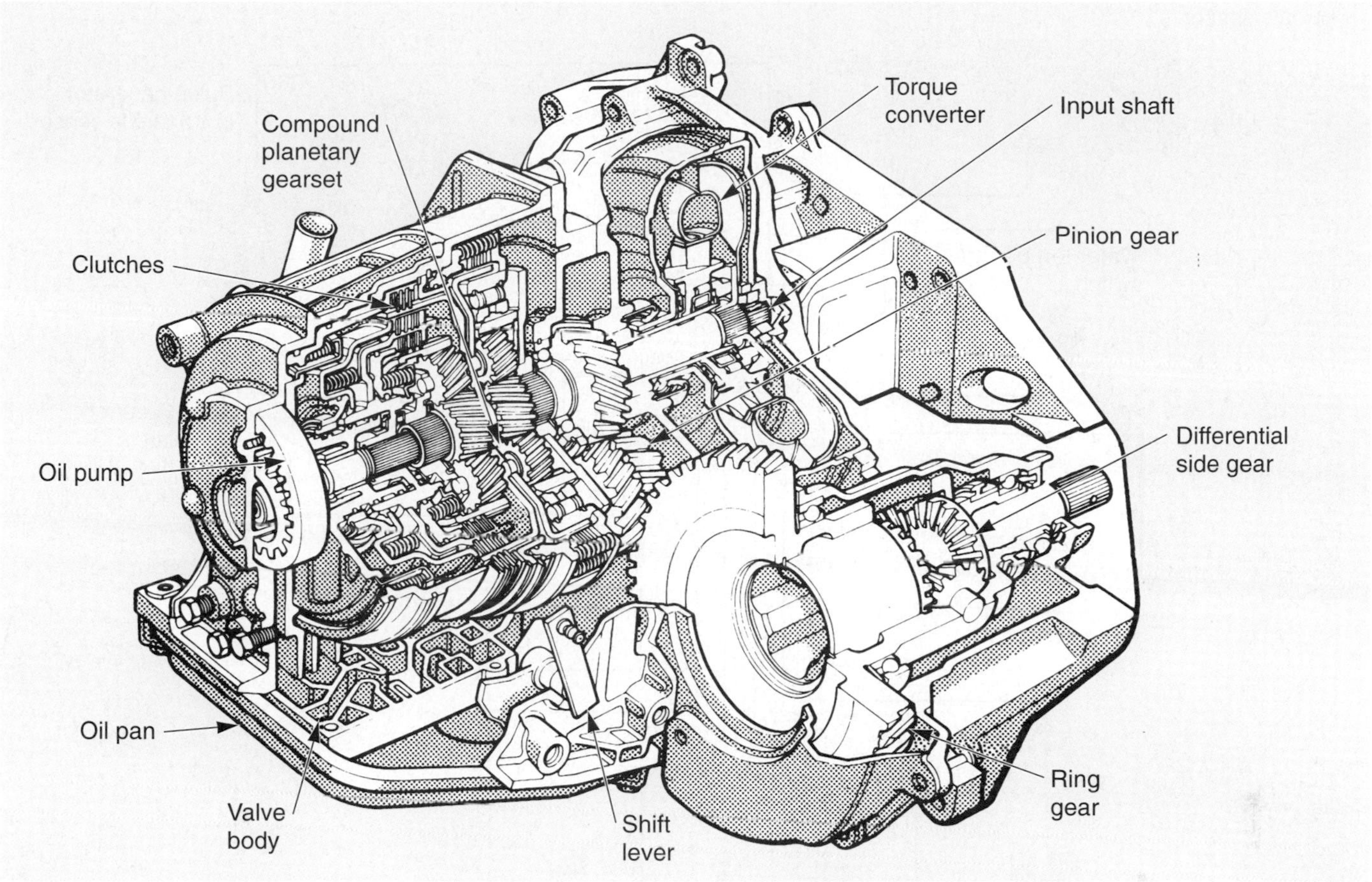

Torque
converter

Input shaft

Compound
planetary
gearset

Clutches

Pinion gear

Oil pump

Differential
side gear

Oil pan

Ring
gear

Valve
body

Shift
lever

Figure 63-18. Another automatic transaxle. This is a typical unit that uses a compound planetary gearset, bands, and clutches. (Chrysler)

The intermediate clutch is applied, locking the intermediate shaft to the ring gear. This causes slightly less gear reduction than first gear.

In **Figure 63-17C**, the transaxle is in third gear. Both clutches are applied. This locks both members of the planetary gearset and the unit turns as a single member for direct drive to the differential.

In **Figure 63-17D**, the transaxle is in reverse. The reverse clutch holds the planetary ring gear stationary. The direct clutch locks the turbine shaft to the low-reverse sun gear. The one-way clutch allows the turbine shaft to turn the low-reverse sun gear clockwise. Then, the output from the planetary gearset is reversed.

Figures 63-18 and **63-19** illustrate other automatic transaxle design variations. Compare these transaxles. Make sure you can identify all of the major components.

Transaxle Electronic Controls

Modern transaxles use computer control to increase efficiency. The PCM can monitor various vehicle functions

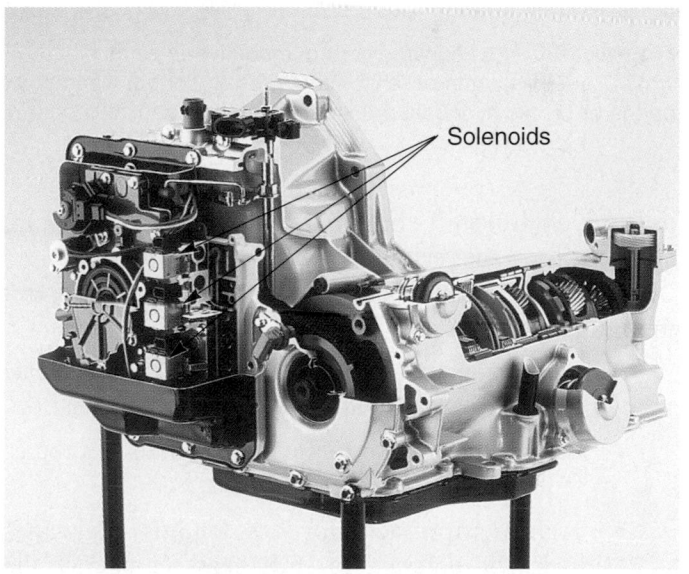

Solenoids

Figure 63-19. This photo shows modern electronically controlled automatic transaxle. Solenoids on the transaxle provide an interface between the computer system electronics and transmission hydraulics. (Ford)

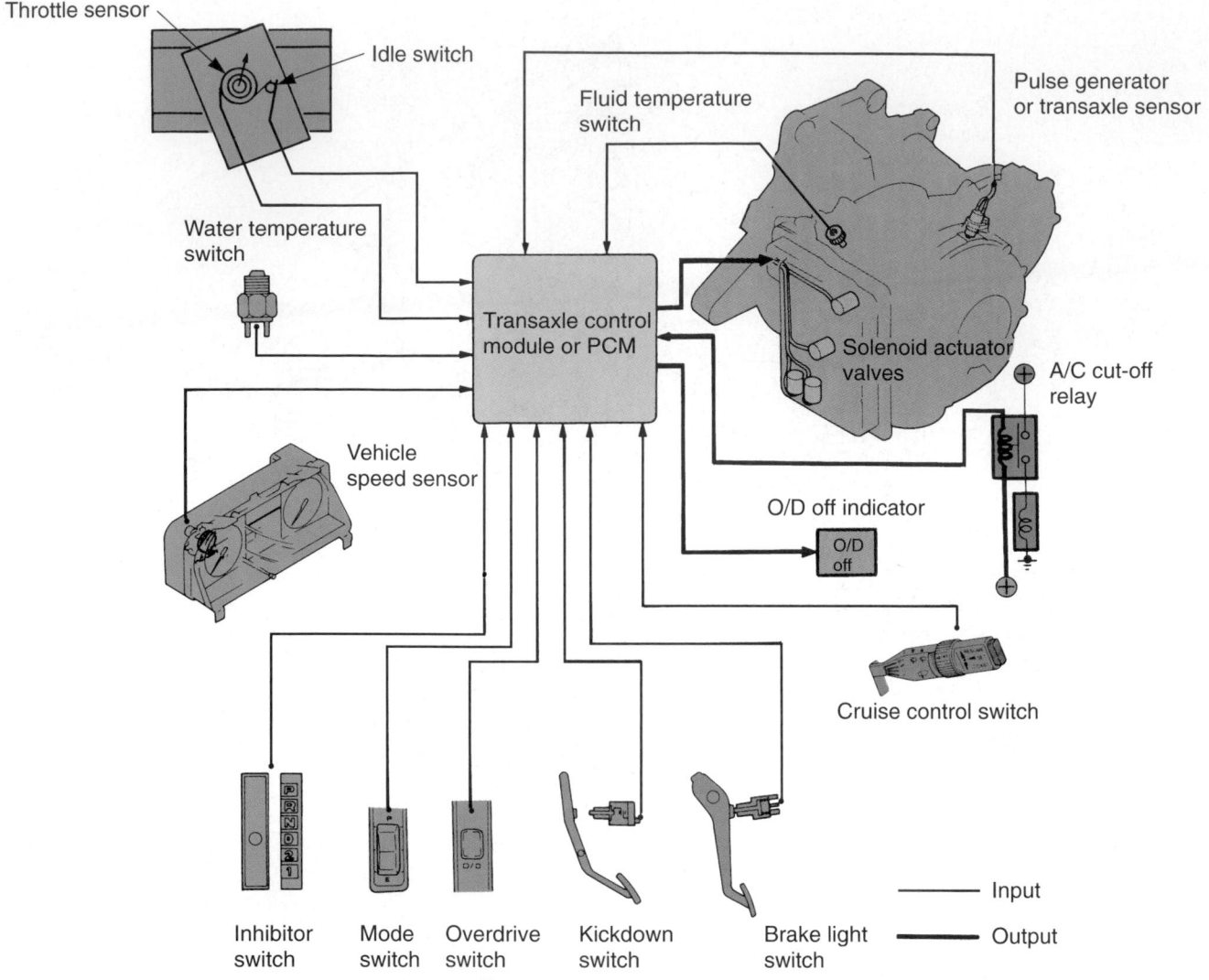

Figure 63-20. Note how the control module serves as the "brain" to determine shift points. Various sensors feed data to the module, which uses programmed shift schedules to control when power is sent to each shift solenoid. When the solenoid is energized, it opens or blocks hydraulic pressure to change transaxle gear ratios.

to better control shift points and torque converter lockup. This increases fuel economy and saves energy.

A basic transaxle electronic control system includes the following elements:

- Control module—computer that uses sensor signals to adjust output voltages to operate the transaxle.

- Vehicle sensors—provide input signals that represent various conditions.

- Transmission actuators—solenoids that alter hydraulic pressure to shift gears or operate the lockup torque converter.

Some vehicles use a separate transaxle control module to analyze sensor inputs and determine the outputs needed to operate the automatic transaxle. Others use a power train control module, which controls the functions of both the engine and transaxle. See **Figure 63-20.**

An electronically controlled transaxle uses solenoids to alter hydraulic pressure to each circuit. Sensors on the transaxle provide feedback concerning fluid temperature, transaxle speed, vehicle speed, and other variables. Look at **Figure 63-21.**

Electronic controls can also be used on transaxles using manual-type gearsets. The solenoids move the synchronizers to shift gears. See **Figure 63-22.**

An exploded view of an electronically controlled transaxle is shown in **Figure 63-23.**

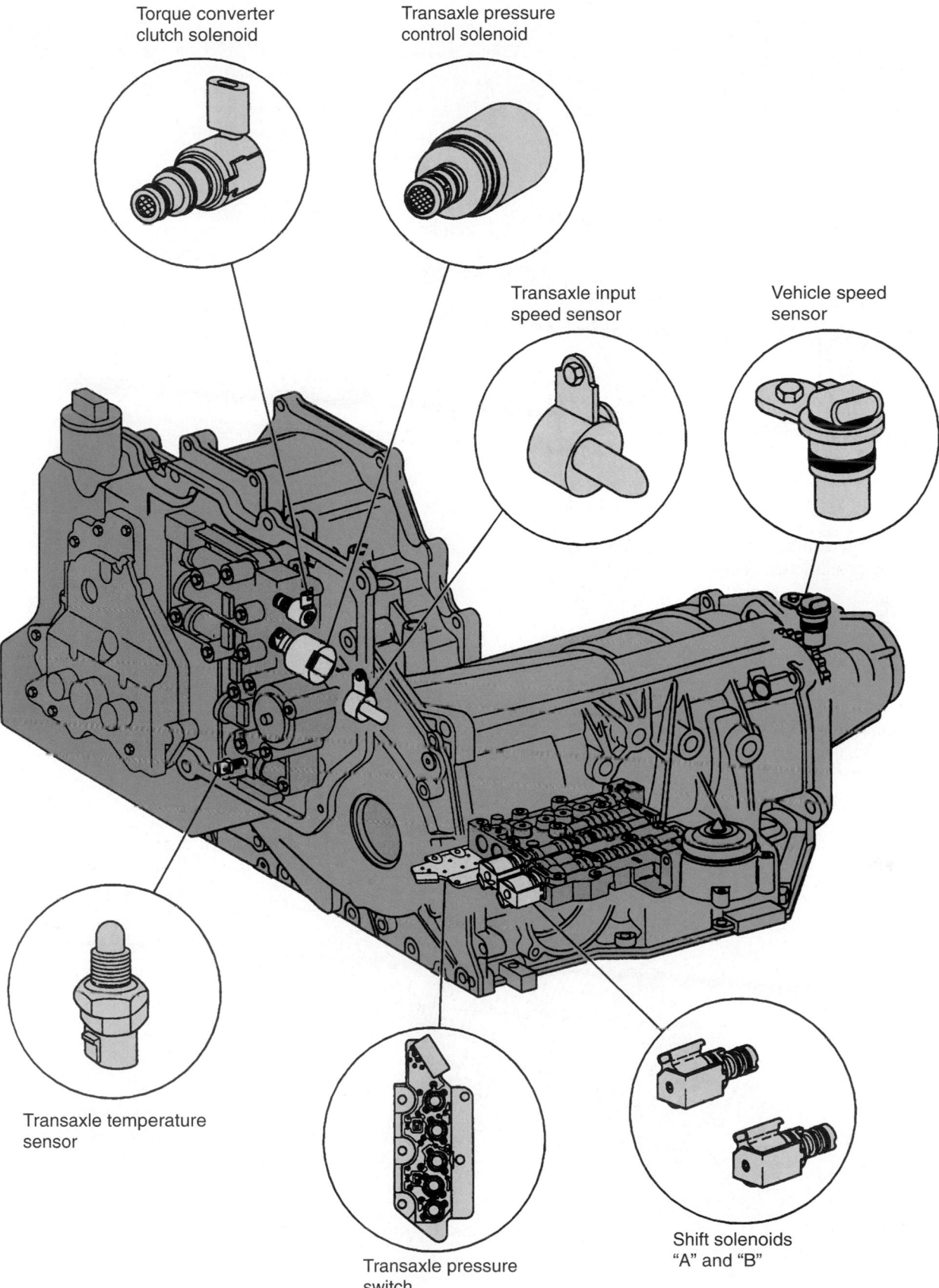

Torque converter
clutch solenoid

Transaxle pressure
control solenoid

Transaxle input
speed sensor

Vehicle speed
sensor

Transaxle temperature
sensor

Transaxle pressure
switch

Shift solenoids
"A" and "B"

Figure 63-21. Note sensors, switches, and solenoids used in this electronic-controlled transaxle. (General Motors)

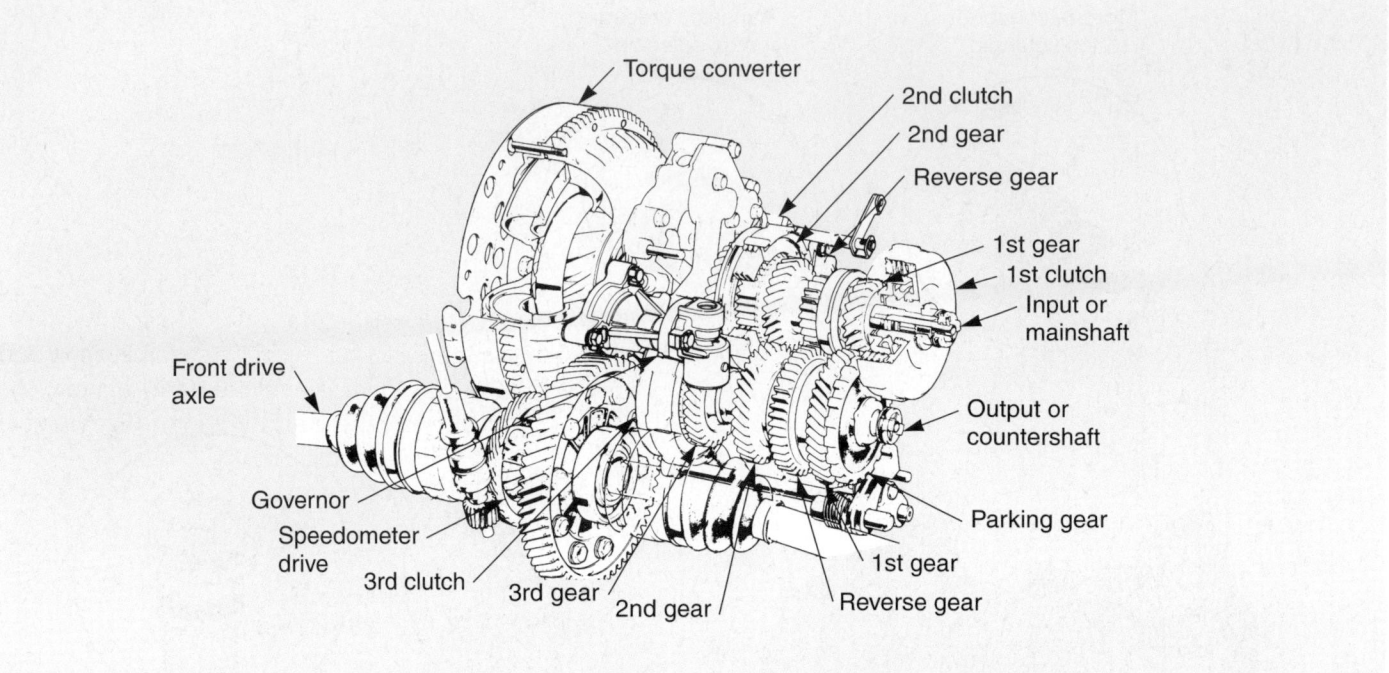

Figure 63-22. This automatic transaxle does not use a planetary gearset. It uses hydraulically operated clutches to activate helical gears. Study its construction. (Honda)

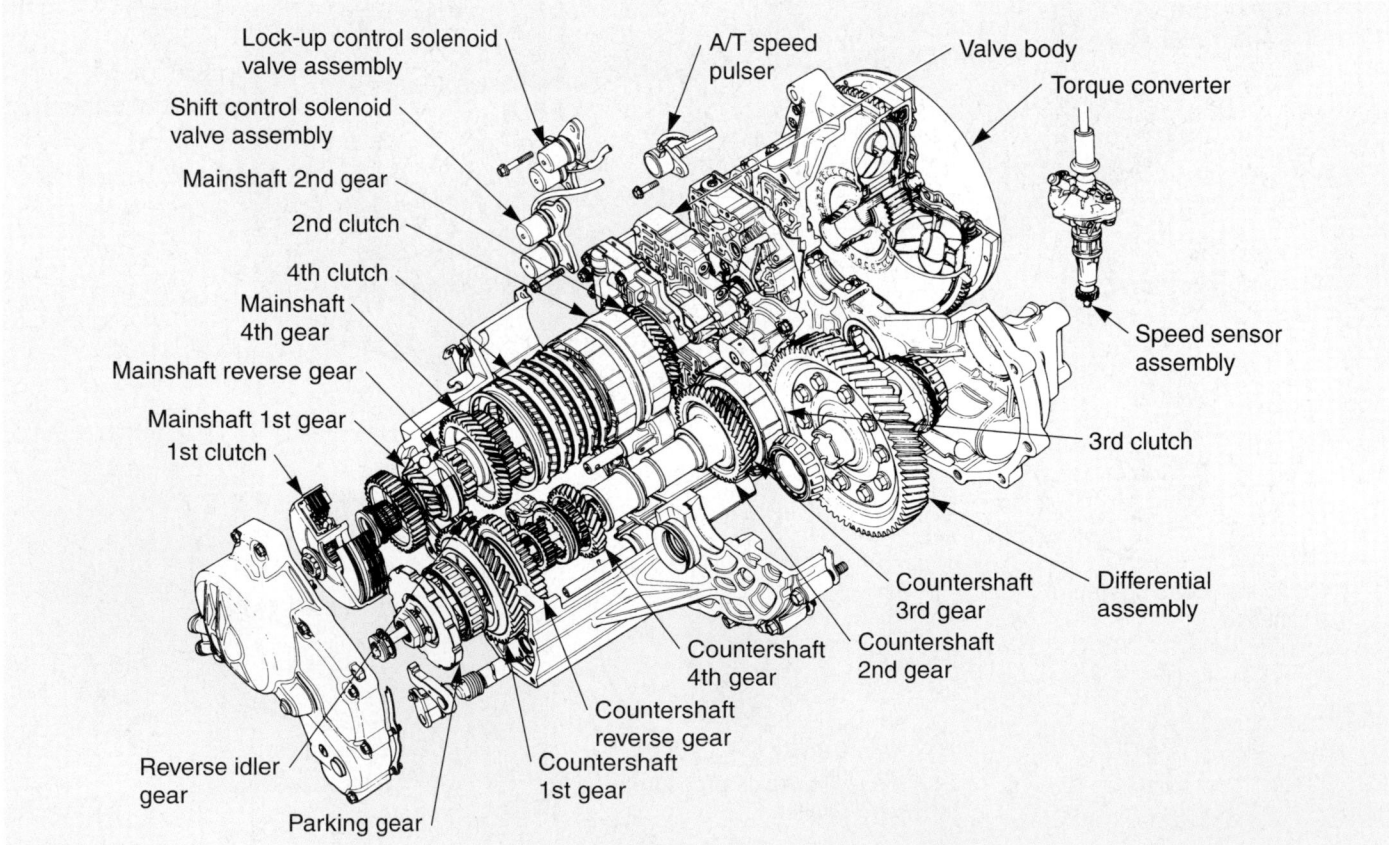

Figure 63-23. This exploded view shows the unique construction of this automatic transaxle. It uses solenoids to shift helical gears and lock torque converter clutch. Conventional automatic-type clutches are also used in this design. (Honda)

Figure 63-24. This cutaway view shows the internal parts of a constantly variable transaxle. Using moveable pulleys and a steel drive belt, it can provide a smooth transition to various gear ratios without shifting gears. (Honda)

Continuously Variable Transaxle

Discussed briefly in Chapter 57, a *continuously variable transaxle* (or transmission), abbreviated CVT, has an infinite number of driving ratios. It uses movable two-piece pulleys that have variable diameters. V-belts run between the pulley sets. This arrangement takes the place of the planetary gearsets or helical manual transmission gears.

Constantly variable transaxles are now being used by several automakers. For this reason, you should have a basic understanding of their construction and operation. See **Figure 63-24.**

Basically, the parts of the pulleys can be moved closer together to make the effective pulley diameter larger or moved apart to make the effective pulley diameter smaller. In this way, the ratio produced by the two pulleys can be changed.

The main advantage of a continuously variable transaxle is that the engine can remain at its most efficient operating speed at all times. The CVT changes gear ratios to maintain a relatively constant engine speed, even when the vehicle is accelerating. This can increase fuel economy. Also, you do not feel the jerk as the transmission changes gears.

A CVT can be computer controlled for even greater efficiency, **Figure 63-25.** Note how the control module receives input signals from the shift selector, accelerator, brake switch, engine sensors, vehicle speed sensor, and other sensors. It can then use solenoids to operate the hydraulic control valve systems to move the variable pulley in or out, changing gear ratios.

Another illustration of a CVT is shown in **Figure 63-26.** Study its construction.

Front Drive Axle Assembly

The *front drive axle assembly* transfers power from the differential to the hubs and wheels of the vehicle, **Figure 63-27.**

Most modern front drive axle assemblies consist of two or three separate shafts and two universal joints. This enables the assembly to transfer power smoothly as the front wheels move up and down over bumps and from side-to-side while turning.

Front drive axles turn much slower than a drive shaft for a rear-wheel drive vehicle. They turn about one-third the speed of a rear drive shaft. They are connected directly to the drive wheels and do not have to act through the reduction of the rear axle ring and pinion gears.

Axle Shafts

Three *axle shafts* are typically used in the front drive axle assembly, **Figure 63-28**:

- *Inner stub shaft*—a short shaft splined to the side differential gear and connected to the inboard (inner) universal joint, **Figure 63-28.**

- *Outer stub shaft*—a short shaft connected to the outboard (outer) universal joint and front wheel hub.

- *Interconnecting shaft*—the center shaft that fits between the two universal joints.

The outer ends of each shaft are machined. They may have splines for meshing with the splines on mating parts. A portion of a universal joint may be machined as an integral part of the shaft. Grooves are also cut in the shafts for snap rings, boots, and other components.

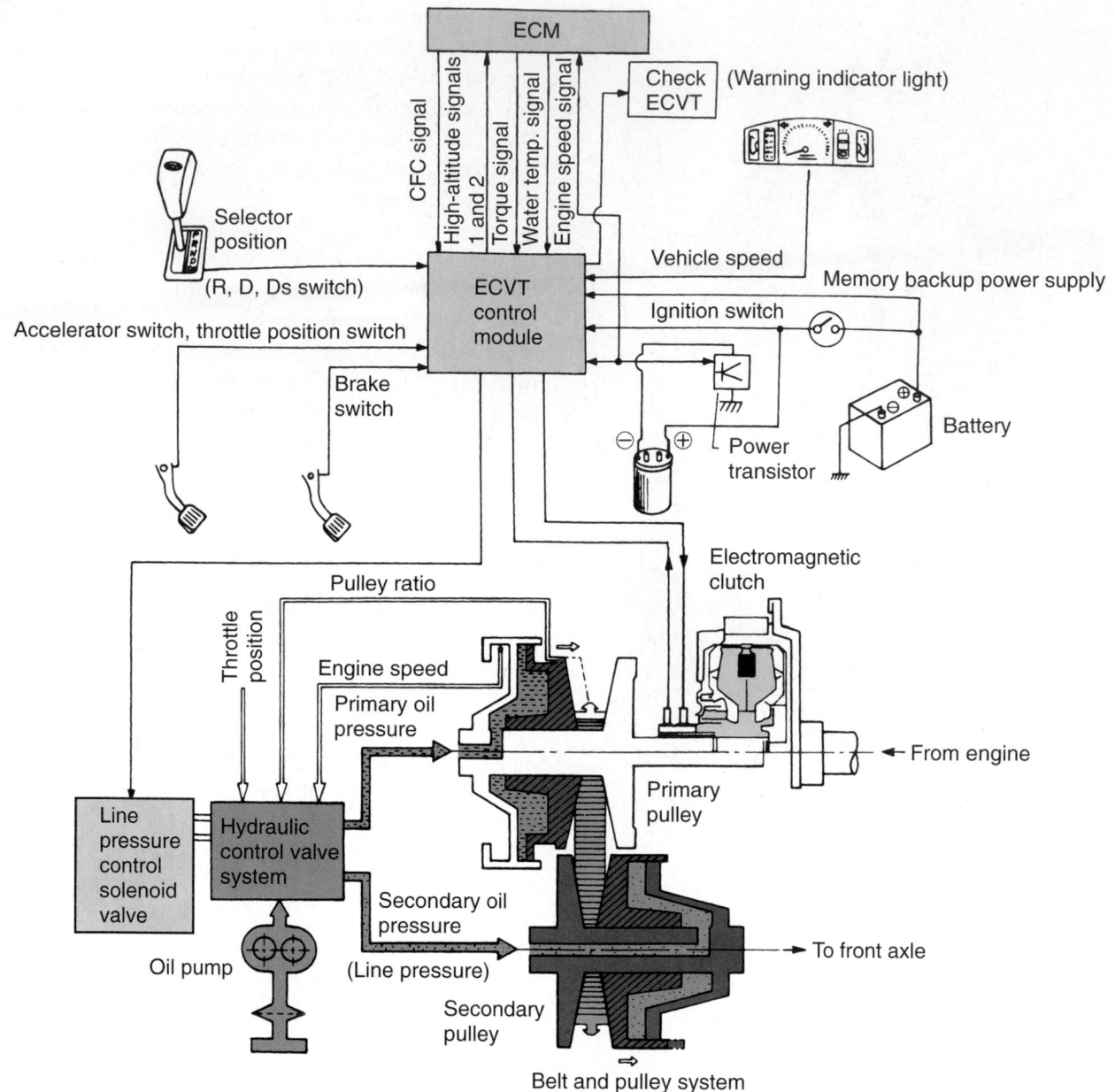

Figure 63-25. This diagram shows the basic operation of computer-controlled CVT. Trace the flow of data to and from the control modules. Also note how the line pressure control solenoid provides an interface between the control module and the hydraulic control valve system. An electromagnetic clutch replaces the conventional torque converter. (Subaru)

Universal Joints

Universal joints in the front drive axle assemblies allow the shafts to operate through an angle without damage. They are normally the *constant velocity joints (CV joints)*. Normally, either Rzeppa (ball-and-cage) or tripod (ball-and-housing) CV-joints are used in front drive axles. However, a double-cardan (cross-and-roller) joint may sometimes be used.

The *outboard CV-joint* is normally a fixed (non-sliding) Rzeppa joint, **Figure 63-29.** Sometimes, it is a fixed tripod type. The outboard CV-joint transfers rotating power from the axle shaft to the hub assembly, **Figure 63-28.**

The *inboard CV-joint* is commonly a *plunging* (sliding) tripod joint. It acts like a slip joint in a drive shaft for a rear-wheel drive vehicle. As the front wheels move up and down over bumps in the road, the length of the drive axle assembly must change. The plunging action of the inboard CV-joint allows for a change in distance between the transaxle and wheel hub. Look at **Figures 63-28** and **63-29.**

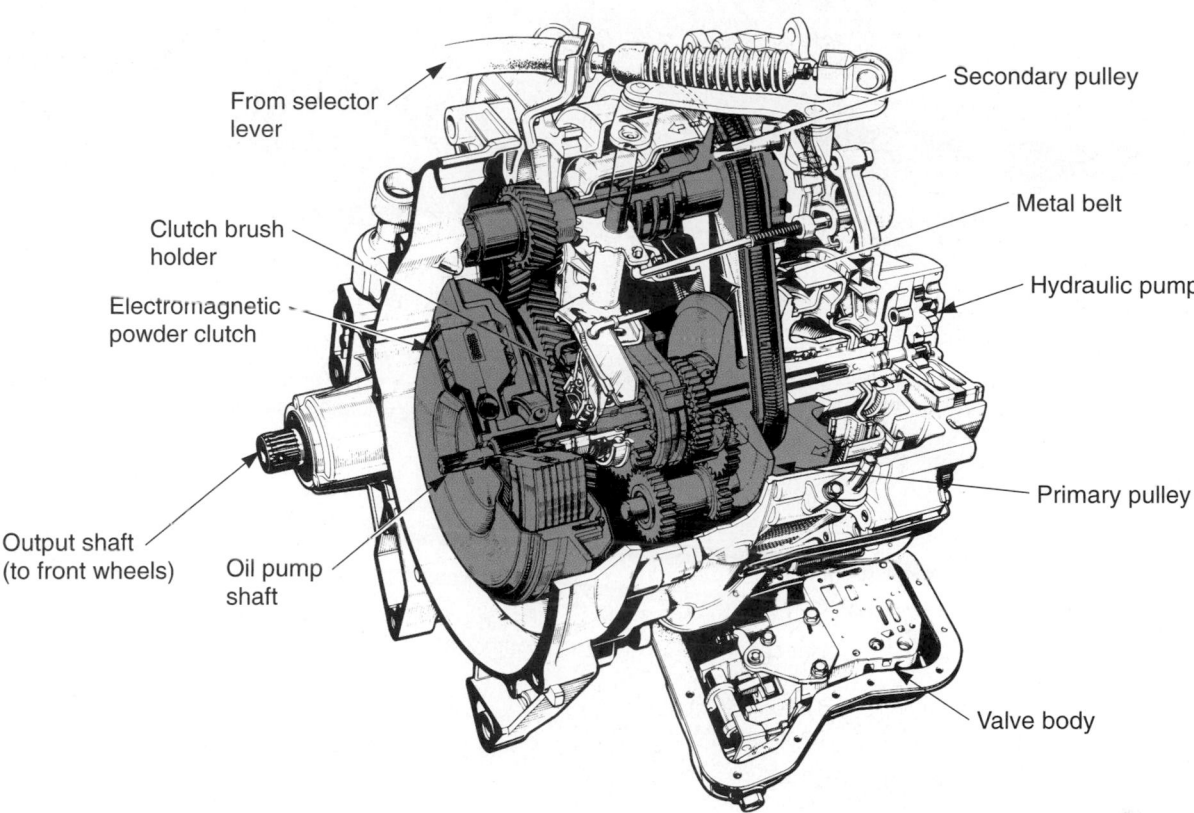

From selector lever

Secondary pulley

Clutch brush holder

Metal belt

Electromagnetic powder clutch

Hydraulic pump

Output shaft (to front wheels)

Oil pump shaft

Primary pulley

Valve body

Figure 63-26. Study the construction of this CVT. An electromagnetic clutch engages and disengages the engine and transmission like an air-conditioning compressor. (Subaru)

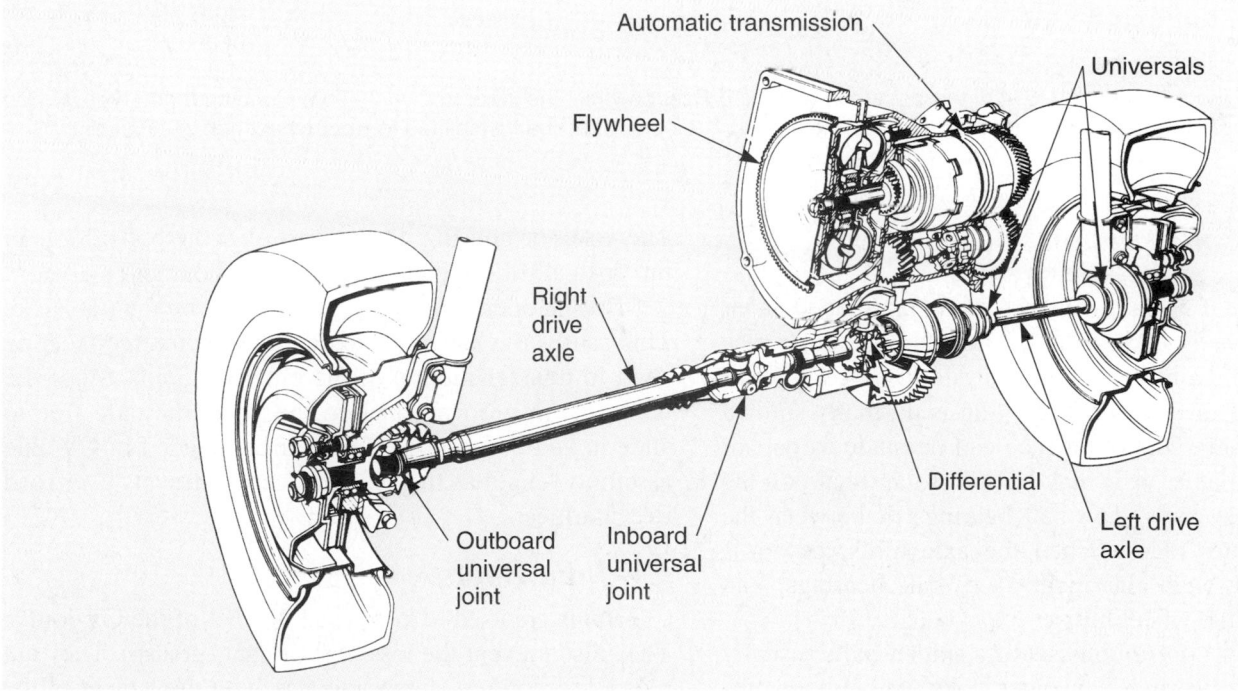

Automatic transmission

Universals

Flywheel

Right drive axle

Differential

Left drive axle

Outboard universal joint

Inboard universal joint

Figure 63-27. Front drive axles connect the differential side gears to wheel hubs. When differential side gears turn, the axles rotate the hubs and front wheels to propel the vehicle. (Dodge)

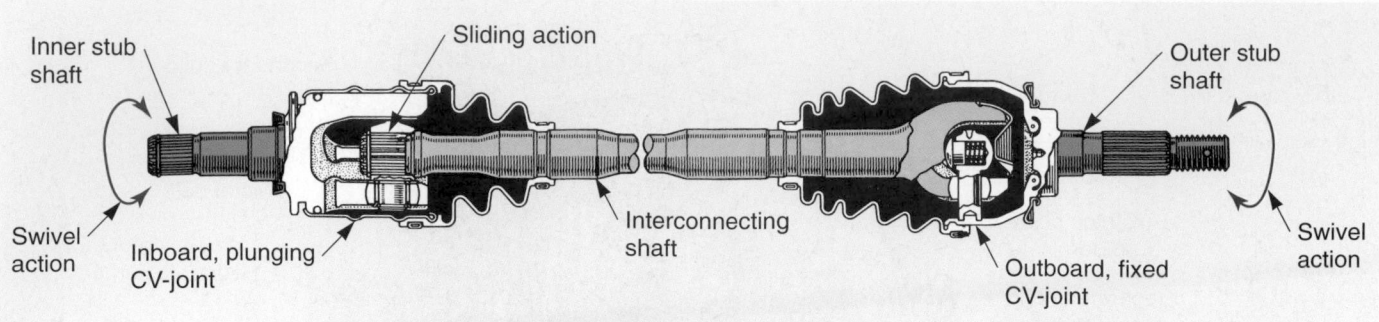

Figure 63-28. This is a three-piece drive axle: inner stub shaft, intermediate shaft, and outer stub shaft. The universal joints allow the drive axle to swivel into various angles. The inboard universal joint is normally a sliding joint to allow for the length changes caused by suspension and steering action. (Toyota)

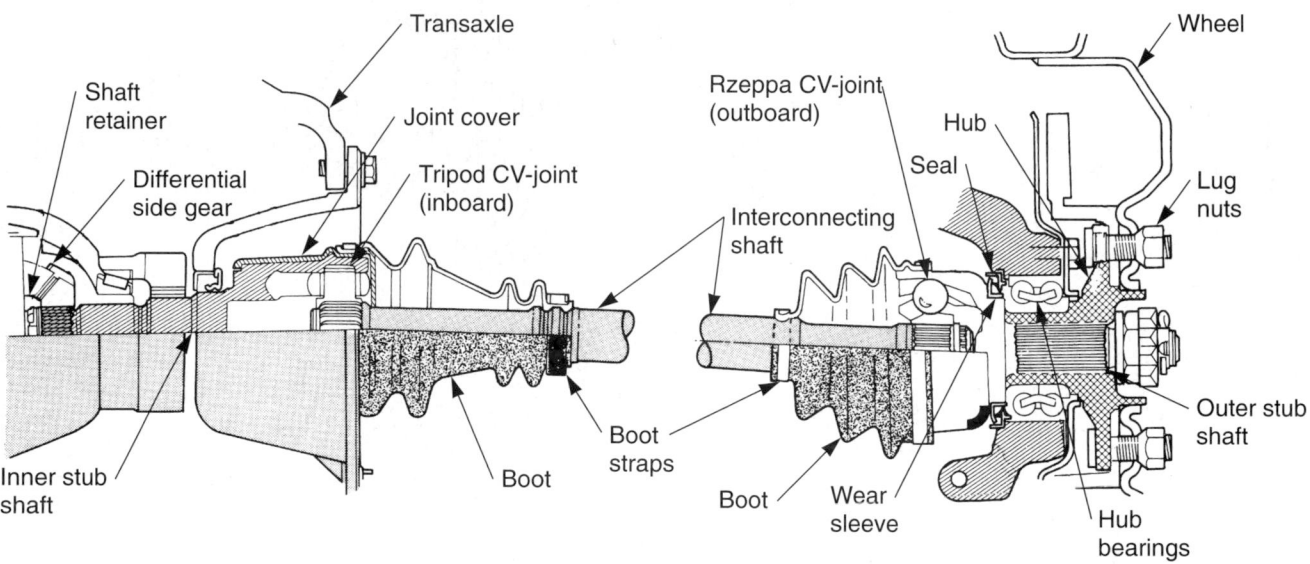

Figure 63-29. The outboard CV-joint of this assembly is a fixed Rzeppa joint. The inboard CV-joint is a plunging tripod type. Notice that the inner stub shaft is splined to the side differential gear. The outer stub shaft is splined to front wheel hub. (Chrysler)

CV-Joint Construction

A **Rzeppa CV-joint** consists of a star-shaped inner race, several ball bearings, a bearing cage, an outer race or housing, and a rubber boot. Refer to **Figure 63-30.**

The inner race of this type joint is normally splined to the axle shaft. The outer race can be made as part of the outer stub axle, or it may be splined and held on the axle with snap rings. The ball bearings fit between the inner and outer races. When the axle turns, power is transferred through the inner race, ball bearings, and outer race to the wheel hub.

A **tripod CV-joint** consists of a spider, balls (usually three), needle bearings, an outer yoke, and a boot. See **Figure 63-31.**

The inner spider is normally splined to the axle shaft. The needle bearings and three balls fit around the spider.

The yoke, or housing, then slides over the balls. Slots in the yoke allow the balls to slide in and out and swivel.

During operation, the yoke is driven by the transaxle. This causes the balls, the spider, and the interconnecting shaft to transfer motion to the outer CV-joint. Since the balls ride in grooves cut into the yoke, they are free to slide in and out to allow for changes needed in CV-axle assembly length as the vehicle steers or travels over road irregularities.

CV-Joint Boots

Boots are used to keep road dirt out of the CV-joints. They also prevent the loss of lubricant (grease). They are pleated (accordion-shaped) to flex with movement of the joint, **Figure 63-31.**

Retaining collars, or **straps,** secure the boots to the drive axle and the housing. They are usually plastic straps

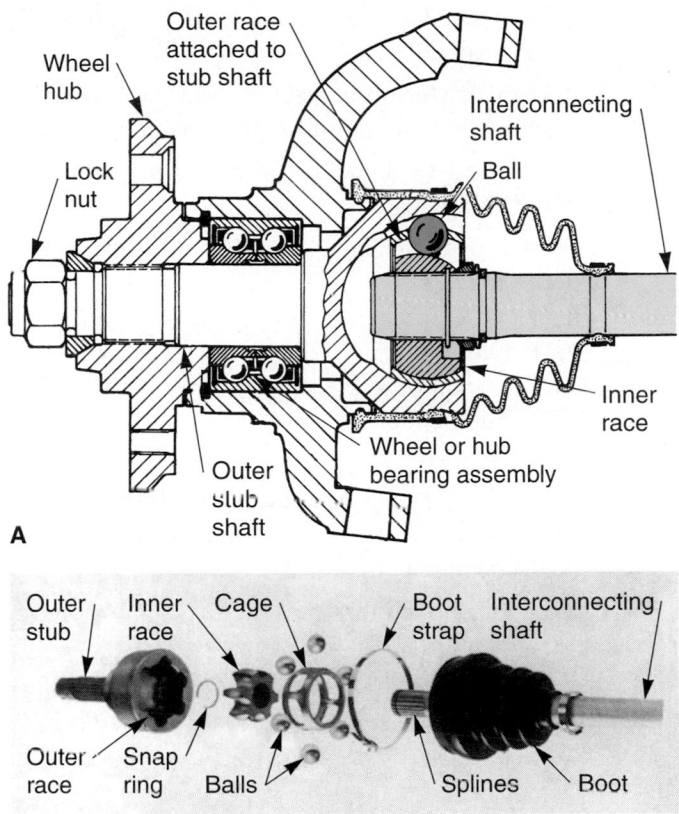

Figure 63-30. Study the construction of a Rzeppa CV-joint. A—The drive axle turns the inner race. The inner race turns the balls. The balls transfer turning force to the outer race and hub, rotating the wheels. B—Exploded view of a Rzeppa CV-joint. Study how the parts fit together. A plunging joint is used inboard; a fixed joint is used outboard. (Saab and Dana Corp.)

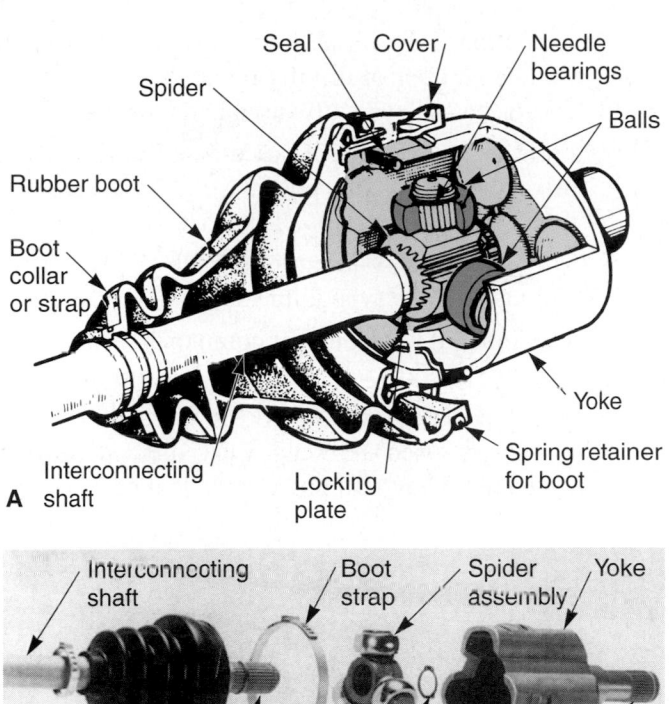

Figure 63-31. Tripod CV-joint construction. A—Cutaway view shows how a tripod joint fits together. The interconnecting shaft is splined to the spider. The spider rotates the balls and housing. Balls turn on needle bearings to allow swiveling action. B—Exploded view of tripod joint. Inboard joints are sliding tripods. Outboard joints are fixed tripods. Tripod joints are commonly used as inboard plunging joints. (Dana Corp.)

or metal spring clamps that hold the ends of the boot, providing a tight seal.

Summary

- A transaxle is a transmission and a differential combined in a single assembly.
- A torque steer problem causes the front wheels to pull to one side when the engine power acts on the steering system through the front-drive axles.
- A manual transaxle uses a friction clutch and standard transmission type gearbox.
- An automatic transaxle commonly uses a torque converter and a hydraulic system to control gear engagement.
- A few transaxles are made so that the engine is mounted longitudinally (lengthwise).
- Most transaxles are designed so that the engine can be transverse (sideways) in the engine compartment.

- Transaxle shafts are normally mounted in either tapered roller or ball bearings. The shaft bearings fit into the transaxle case.
- The transaxle synchronizers are almost identical to those used in manual transmissions.
- A transaxle differential, like a rear axle differential, transfers power to the axles and wheels while allowing one wheel to turn at a different speed than the other.
- The flow of power through an automatic transaxle is similar to power flow through an automatic transmission.
- Modern transaxles use computer control to increase efficiency. The computer can monitor various vehicle functions to better control shift points and converter lockup. This increases fuel economy and saves energy.
- An electronic-controlled transaxle uses solenoids to alter hydraulic pressure to each circuit.

- A continuously variable transaxle (CVT) has an infinite number of driving ratios. It uses movable two-piece pulleys with variable diameters.

- Front drive axles, also called axle shafts or front drive shafts, transfer power from the differential to the hubs and wheels of the car.

- The outboard CV-joint is normally a fixed (non-sliding) Rzeppa type joint.

- The inboard CV-joint is commonly a plunging (sliding) tripod joint. It acts like a slip joint in a drive shaft for a rear-wheel-drive vehicle.

- Boots are used to keep road dirt out of the CV-joints. They also prevent the loss of lubricant.

Important Terms

Transaxle	Drive link
Torque steer	Automatic transaxle
Steering feel	Torque converter
Transverse	Oil pump
Longitudinally	Valve body
Manual transaxle	Pistons and servos
Input shaft	Bands and clutches
Input gears	Planetary gearsets
Output gears	Differential
Output shaft	Continuously variable
Synchronizers	transaxle
Differential	Front drive axle
Transaxle case	assembly
Manual transaxle	Axle shafts
clutch	Inner stub shaft
Manual transaxle	Outer stub shaft
transmission	Interconnecting shaft
Mainshaft	Universal joint
Pinion shaft	CV-joint
Cluster gear assembly	Outboard CV-joint
Countershaft gear	Inboard CV-joint
assembly	Plunging
Transaxle shafts	Rzeppa CV-joint
Transaxle	Tripod CV-joint
synchronizers	Boots
Transaxle differential	Retaining collars
Drive chain	Straps

Review Questions—Chapter 63

Please do not write in this text. Place your answers on a separate sheet of paper.

1. What is a transaxle?

2. List six possible advantages of front-wheel drive.

3. Both manual and automatic transaxles are available. True or False?

4. Summarize the differences between transaxles for transverse and longitudinally mounted engines.

5. Name and explain the seven major parts of a manual transaxle.

6. Which of these parts is *not* found in a manual transaxle?
   (A) Synchronizers.
   (B) Differential.
   (C) Transmission.
   (D) Fluid coupling.
   (E) All of the above.

7. A drive _____ is sometimes used to send crankshaft power to the transaxle with a longitudinally mounted engine.

8. List and explain the seven major components of an automatic transaxle.

9. Describe the two common types of CV-joints used on front drive axles.

10. Which of the following is *not* part of a front drive axle assembly?
    (A) Boots.
    (B) CV-joints.
    (C) Stub shafts.
    (D) Interconnecting shaft.
    (E) Pivot shaft.

● ASE-Type Questions

1. Technician A says transaxles are used only on front-wheel-drive cars. Technician B says transaxles are used on front-wheel-drive cars and certain types of rear-engine cars. Who is right?
   (A) A only.
   (B) B only.
   (C) Both A and B.
   (D) Neither A nor B.

2. Technician A says one possible advantage of a front-wheel-drive car equipped with a transaxle is that it improves traction on a slippery pavement. Technician B says one possible advantage of a front-wheel-drive vehicle equipped with a transaxle is that it improves the safety of the passengers because there is increased mass in the front of the car. Who is right?
   (A) A only.
   (B) B only.
   (C) Both A and B.
   (D) Neither A nor B.

3. Technician A says cars equipped with a transaxle can use either a manual transmission or automatic transmission. Technician B says cars equipped with a transaxle can only use manual transmissions. Who is right?
   (A) A only.
   (B) B only.
   (C) Both A and B.
   (D) Neither A nor B.

4. Technician A says most transaxles are designed so that the engine can be mounted sideways in the engine compartment. Technician B says most transaxles are designed so that the engine can be transverse in the engine compartment. Who is right?
   (A) A only.
   (B) B only.
   (C) Both A and B.
   (D) Neither A nor B.

5. Which of the following are basic components of an automatic transaxle?
   (A) Torque converter.
   (B) Pinion gear.
   (C) Transaxle case.
   (D) All of the above.

6. Technician A says during transaxle operation, the differential transfers power into the transmission. Technician B says during transaxle operation, the transmission transfers power into the differential. Who is right?
   (A) A only.
   (B) B only.
   (C) Both A and B.
   (D) Neither A nor B.

7. Technician A says if an engine is mounted in an automobile longitudinally, the crankshaft centerline points to the front and rear of the vehicle. Technician B says if an engine is mounted longitudinally in an automobile, the crankshaft centerline points to both front drive wheels. Who is right?
   (A) A only.
   (B) B only.
   (C) Both A and B.
   (D) Neither A nor B.

8. Technician A says the input shaft on a manual transaxle turns the gears in the transaxle. Technician B says the input shaft on a manual transaxle powers the torque converter. Who is right?
   (A) A only.
   (B) B only.
   (C) Both A and B.
   (D) Neither A nor B.

9. Which of the following are basic components of a manual transaxle?
   (A) Ring gear.
   (B) Synchronizers.
   (C) Shift forks.
   (D) All of the above.

10. Technician A says all manual transaxles use a throw out bearing and fork to release the clutch. Technician B says all manual transaxles use a push rod mechanism to release the clutch. Who is right?
    (A) A only.
    (B) B only.
    (C) Both A and B.
    (D) Neither A nor B.

11. Technician A says a manual transaxle transmission's input shaft is sometimes called the pinion shaft. Technician B says a manual transaxle transmission's input shaft is sometimes called the mainshaft. Who is right?
    (A) A only
    (B) B only.
    (C) Both A and B.
    (D) Neither A nor B.

12. Technician A says manual transaxle transmission's shafts are normally mounted in tapered roller bearings. Technician B says manual transaxle transmission's shafts are normally mounted in ball bearings. Who is right?
    (A) A only.
    (B) B only.
    (C) Both A and B.
    (D) Neither A nor B.

13. Technician A says a manual transaxle transmission is always driven by a drive shaft powered by the engine's crankshaft. Technician B says a manual transaxle transmission is sometimes driven by a chain link powered by the engine's crankshaft. Who is right?
   (A) A only.
   (B) B only.
   (C) Both A and B.
   (D) Neither A nor B.

14. Which of the following is *not* a component of an automatic transaxle?
   (A) Valve body.
   (B) Oil pump.
   (C) Synchronizers.
   (D) Planetary gearsets.

15. Technician A says that screws are normally used to secure front drive axle CV-joint boots. Technician B says that straps are normally used to secure front drive axle CV-joint boots. Who is right?
   (A) A only.
   (B) B only.
   (C) Both A and B.
   (D) Neither A nor B.

Activities—Chapter 63

1. Identify the major parts of a transaxle chosen by your instructor.

2. Identify the various parts of front drive axle assembly.

3. Compare two different designs of front drive axle CV-joints.

Chapter 64

Transaxle and Front Drive Axle Diagnosis and Repair

After studying this chapter, you will be able to:

- Diagnose common transaxle and drive axle problems.
- Adjust transaxle shift linkage.
- Complete maintenance operations on a transaxle.
- Remove and install a transaxle assembly.
- Remove and install a front drive axle.
- Replace CV-joints on front drive axles.
- Correctly answer ASE certification test questions on the diagnosis and repair of manual and automatic transaxles.

Transaxles suffer from the same kinds of problems as transmissions and differentials. The gears, shafts, bearings, seals, and other parts can wear and fail after prolonged use. Because the drive axles follow the movement of the steering and suspension systems, they too can cause problems after extended service.

Since new vehicles commonly use front-wheel drive, it is very important for you to understand the basic service and repair of transaxles and front drive axles.

This chapter will explain the service procedures that are unique to transaxle-equipped vehicles. You may want to review earlier chapters on clutch, transmission, drive shaft, and differential repair.

 Note
It is beyond the scope of this textbook to explain the complete diagnosis, service, and repair of every transaxle ever produced. For this reason, we will explain the most common procedures so that you can successfully use a service manual to work on any transaxle.

Transaxle Problem Diagnosis

Correct diagnosis of transaxle and front drive axle problems can require a great deal of care. With the engine, clutch, transmission, differential, and drive axles close together in the engine compartment, noises and other symptoms can be difficult to pinpoint. However, if you use your knowledge of component operation and basic troubleshooting techniques, transaxle problems can be isolated without too much trouble.

Test drive the vehicle to verify the customer's complaint. Shift into each gear while accelerating, decelerating, coasting, and cruising at a constant speed. Check for abnormal noises and vibration under each driving condition.

Use your understanding of CV-joints, gears, bearings, shafts, synchronizers, and shift mechanisms to decide what might be the trouble. Approach diagnosis of a transaxle as you would problems in a transmission or a differential. See **Figure 64-1.**

On-Board Diagnostics

Late-model vehicles often use on-board diagnostic systems to monitor and control transaxle operation. With some OBD I systems and all OBD II systems, certain transaxle problems will trip a diagnostic code and light the dash malfunction indicator light.

You can connect a scan tool to the vehicle to check for diagnostic trouble codes and to read operating values in the computer control system. See **Figure 64-2A.** If you find trouble codes or abnormal operating parameters, you should use pinpoint tests to find the actual source of the trouble. See **Figure 64-2B.**

 Note
For more information on on-board diagnostics and scan tool use, refer to the index. Several chapters discuss this topic.

Manual Transaxle Problems

Manual transaxle problems include abnormal bearing and gear noises (whirring, howling, whining), shifting

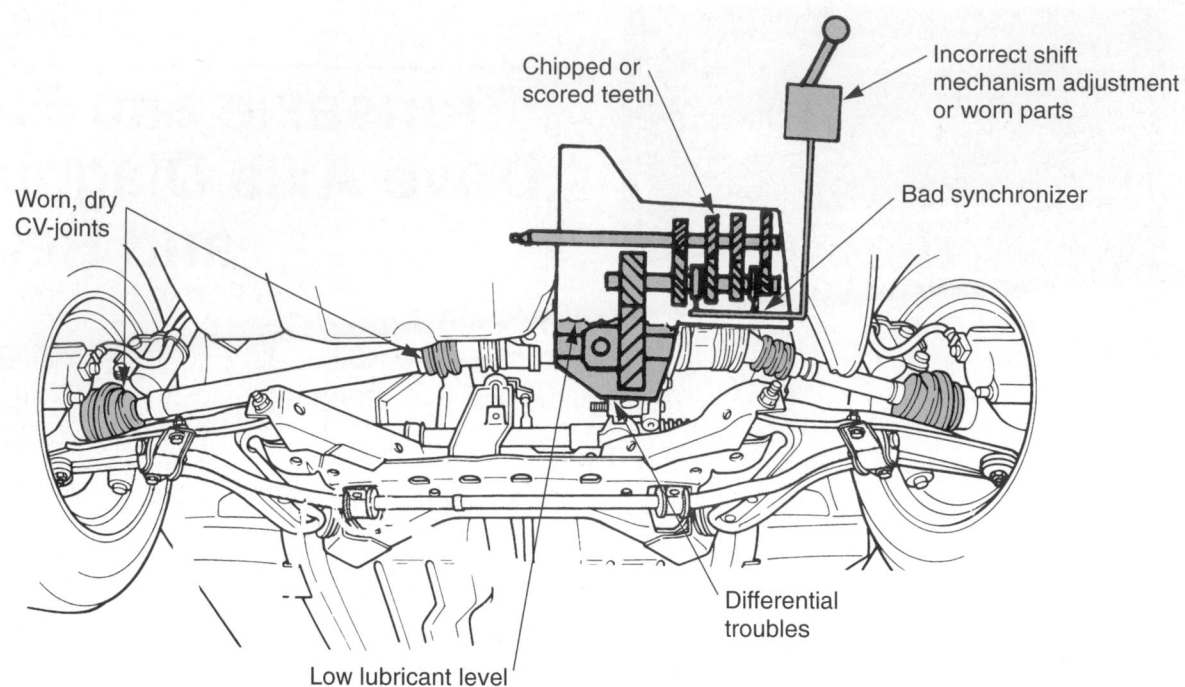

Figure 64-1. When diagnosing transaxle and drive axle problems, visualize the operation of all components. Compare this information to the symptoms. (Chrysler)

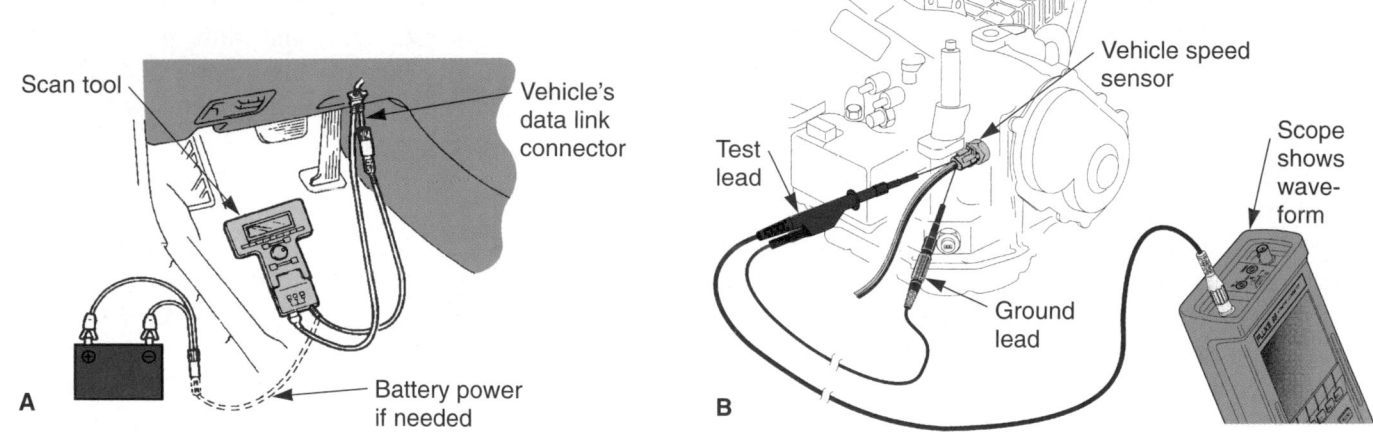

Figure 64-2. With OBD II vehicles, the ECM monitors and controls transaxle functions. A—If the MIL light is on, use a scan tool to retrieve stored trouble codes. The scan tool will also give other operating values, which can aid diagnosis. B—If the scan tool indicates a bad sensor or circuit, you must perform a pinpoint test to verify the source of trouble. Here, a hand-held scope is checking the signal from the vehicle speed sensor on the transaxle. (Mazda and Fluke)

problems (hard to shift into gear, grinds when shifted, will not shift, pops out of gear), and fluid leaks (ruptured gasket or RTV sealer, damaged seal).

If the transaxle is noisy, find out when the abnormal sounds occur. For example, if the transaxle is noisy in only second gear, the input second gear or the output second gear might be at fault.

If a transaxle noise occurs in all gears and under all driving conditions, a common problem exists (bad shaft bearings or scored ring and pinion gears, for example). Use this elimination technique to narrow down the possible sources of trouble.

A manufacturer's diagnosis chart may also be helpful in determining what parts are causing the symptoms. Use a chart that is written for the exact make and model of transaxle being serviced.

Automatic Transaxle Problems

Automatic transaxle problems, like automatic transmission problems, show up as slipping in gear (worn bands or clutches, low system pressure), abnormal noises (damaged planetary gears, bad bearings), and fluid leaks.

If the test drive or the preliminary inspection does not pinpoint the trouble, you may need to perform a stall test, pressure tests, or air tests. Look at **Figure 64-3.**

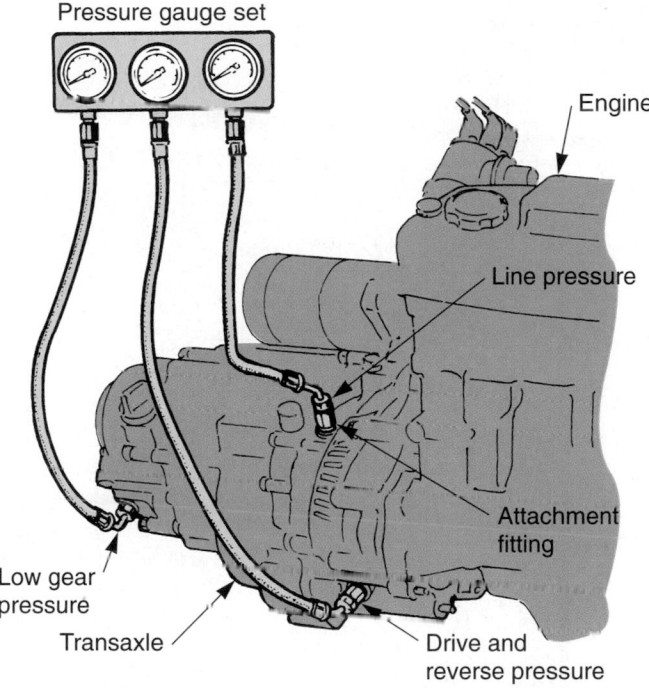

Figure 64-3. As with an automatic transmission, pressure gauges are used for measuring oil pressure to determine the condition of automatic transaxle components. (Honda)

 **Note**
Refer to Chapter 58, *Automatic Transmission Service,* for instructions on completing these tests.

A shop manual will provide detailed procedures and specifications for testing an automatic transaxle. It will give pressure values, passage locations, engine speed values for stall testing, and hydraulic circuit diagrams.

Figure 64-4 shows a hydraulic circuit diagram for a specific automatic transaxle. It is handy when you need to trace the passages in the transaxle to find out how each component is connected to fluid pressure. The service manual will contain one diagram for each gear position. Then, if you have trouble in just one gear, you can trace fluid pressure and see what each component should be doing in that gear.

Front Drive Axle Problems

Front drive axle problems are usually noticed as abnormal noises (clicking, grinding, clunking, humming), vibration, or excessive play in the universal joints. Drive axles are under stress from engine torque being transmitted at various angles. This can cause the CV-joints to fail after prolonged service. If the rubber boot around a CV-joint is ruptured, road dirt and water can enter the joint. A damaged boot will also allow lubricant to escape from the joint. Contamination and lubricant loss will cause joint failure in a very short time.

 Tech Tip
To check for bad front drive axles, drive the vehicle in reverse with the steering wheel turned sharply to the right and then to the left. This can be done in an empty parking lot. This test will make the noise from a worn CV-joint much louder and easier to detect.

When symptoms point to possible drive axle troubles, raise the vehicle on a lift. Inspect the axle shafts. Make sure the rubber boots are not torn or cracked. Wiggle and rotate the joints to check for excessive play and wear, **Figure 64-1.**

 Warning
Do *not* operate the engine and transaxle with the suspension system hanging unsupported on a lift. The CV-joints may be bent at a sharp angle. When spun, the joints could bind and fly apart with lethal force.

Transaxle Maintenance

Proper maintenance is very important to the service life of a transaxle. If the transaxle lubricant is not changed at recommended intervals, abnormal part wear and failure can result. The clutch, shift linkage, and bands must be adjusted periodically if the transaxle is to provide dependable operation.

Checking Transaxle Lubricant

You should check the lubricant level in both a manual and automatic transaxle at regular intervals. See **Figure 64-5.**

An *automatic transaxle* will normally have a dipstick for checking lubricant level. Normally, the transaxle should be at operating temperature with the engine running and the shift selector in park. If the lubricant level is low, add the recommended type of fluid.

A *manual transaxle* will have a fill hole or a dipstick for checking lubricant level. Some manufacturers

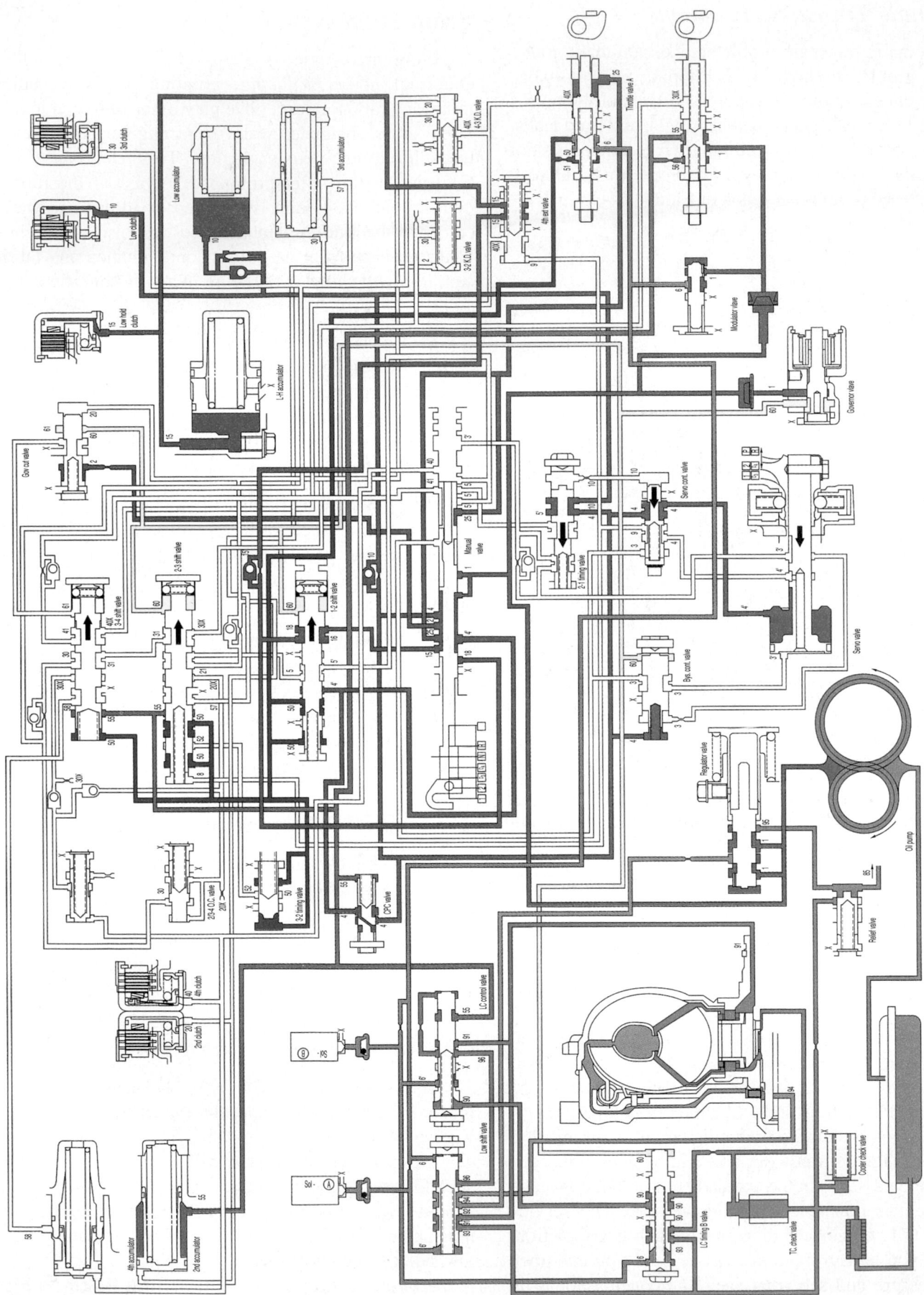

Figure 64-4. A hydraulic circuit diagram in the service manual will help trace passages to each component for diagnosis. The service manual will usually have a diagram for each shift selector position. Dark red shows high pressure going to components. (Oldsmobile)

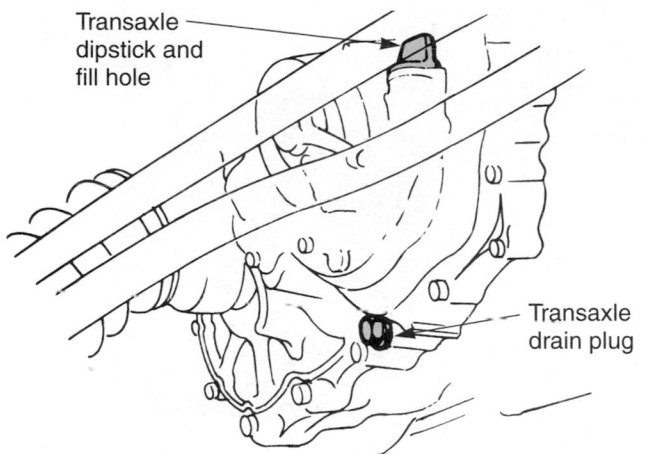

Figure 64-5. Check transaxle oil level using the dipstick (if provided) or at the fill hole. Use the manufacturer's recommended checking and changing procedures and the correct type of lubricant. (Honda)

recommend checking manual transaxle lubricant with the unit at room temperature. Others require that the lubricant be warmed to operating temperature. The lubricant should be almost even with the fill hole or between the prescribed lines on the dipstick.

Caution
Some manual transaxles use automatic transmission fluid, while others use manual transmission gear oil. Always refer to a service manual when in doubt. The incorrect type of lubricant could cause severe transaxle damage.

Changing Transaxle Fluid

To change the transaxle fluid, raise and secure the vehicle in a level position. The fluid should be warmed to operating temperature. Locate and remove the transaxle drain plug, **Figure 64-5.**

If a drain plug is not provided, you must unbolt and remove the sump pan (in an automatic transaxle). Pan removal is also needed if you are going to replace the automatic transaxle filter.

Warning
When at operating temperature, transaxle fluid can cause painful burns. Do not allow hot fluid to run down your arm when draining.

If you removed the sump pan, clean off the sealing surfaces on the pan and the transaxle case. Install a new filter, if needed. Install a new gasket or RTV sealer on the pan and bolt the pan to the transaxle. Tighten the pan bolts to specifications in a crisscross pattern.

Using a long funnel, fill the transaxle with the correct type and amount of fluid. A shop manual will indicate how much fluid the transaxle will hold. Lower the vehicle and start the engine. Shift the transaxle through each gear. Then, check under the vehicle for lubricant leaks.

External Transaxle Adjustments

Some automakers recommend clutch and shift linkage adjustments at regular intervals. Transaxle shift linkage or cables are adjusted using procedures similar to those for transmissions. See **Figure 64-6.** Check that the

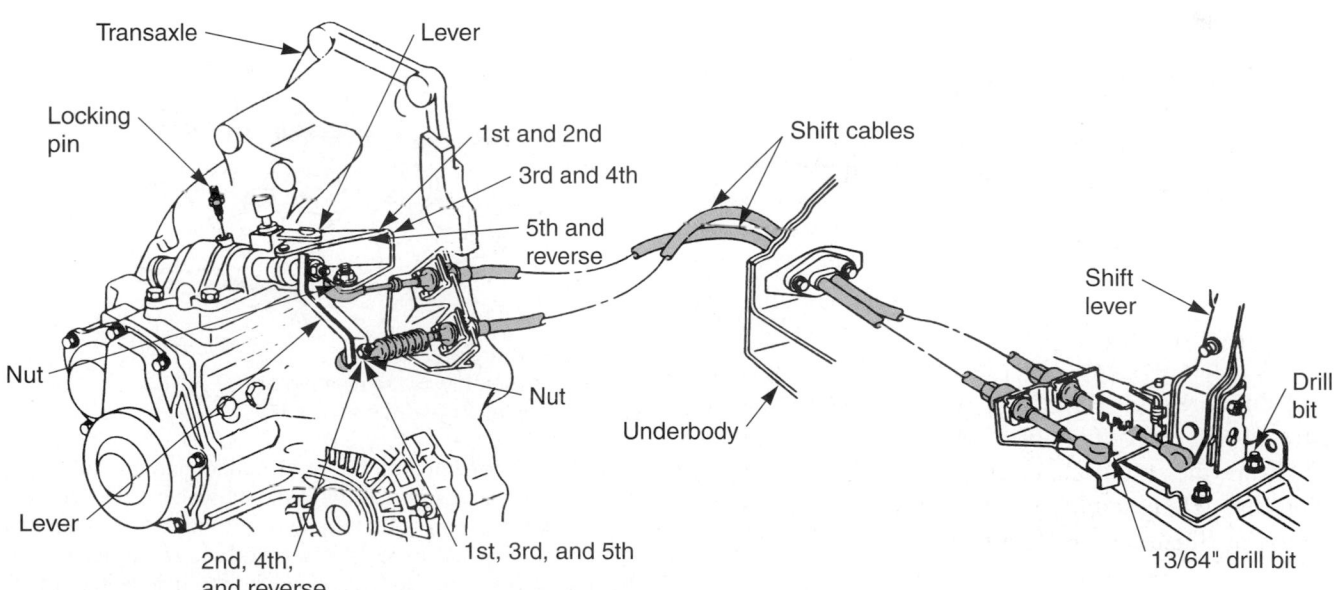

Figure 64-6. Transaxle shift cable adjustment is similar to transmission shift cable adjustment. With this particular unit, a lock pin is used to hold the transaxle in third gear. Two drill bits are then installed in the shifter to align the levers. Then, cable lengths are adjusted to position the levers on the transaxle properly. (Buick)

shift arms on the transaxle are in the proper position in relation to the driver's shift lever. Some transaxles require a special linkage holding fixture for adjustment.

Adjusting a manual transaxle clutch is similar to adjusting the clutch linkage on a manual transmission. Make sure the clutch pedal has the correct free play and applied height.

A few automatic transaxle bands have an adjusting screw inside the pan. By loosening the locknut, the screw can be turned to set the band-to-drum clearance. Normally, the screw is turned in until it bottoms at a specific torque. Then, the screw is turned out a specified number of turns and the locknut is tightened.

Front Drive Axle Service

Normally, when you find problems with a front drive axle assembly, it must be removed for service. The most common front drive axle problem is worn CV-joints.

Front Drive Axle Removal

To remove a front drive axle assembly, you must remove the hub nut. Either leave the vehicle on the ground or hold the brake disc with a large screwdriver. This will allow you to turn the nut without turning the hub and disc, **Figure 64-7A.**

On some cars, a puller must be used to free the outer stub shaft from the hub. The puller grasps the hub's lug studs and pushes in on the axle shaft. With other vehicles, the stub shaft may slide easily out of the hub. See **Figure 64-7B.**

Frequently, you must remove the lower ball joint before the drive axle will slide out of the transaxle. The stub axle shaft will not slide out far enough unless the steering knuckle is free to swing away from the transaxle, **Figure 64-7C.**

With some transaxles, the inner stub shaft is held in place by snap rings or circlips. To access these rings or clips, the transaxle differential cover must be removed. The snap rings or circlips can then be removed from the grooves in the stub shafts, and the axles can be pulled out of the transaxle, **Figure 64-8A.**

With other designs, the inner axle flange must be unbolted before removing the stub shaft. This is shown in **Figure 64-8B.** If required, use a slide hammer to pull the shaft from the transaxle. See **Figure 64-8C.** While holding the axle assembly together, pull it out or off the transaxle as illustrated in **Figure 64-9.**

Transaxle CV-Joint Replacement

In most cases, the entire drive axle assembly is replaced when a faulty CV-joint is found. However, the

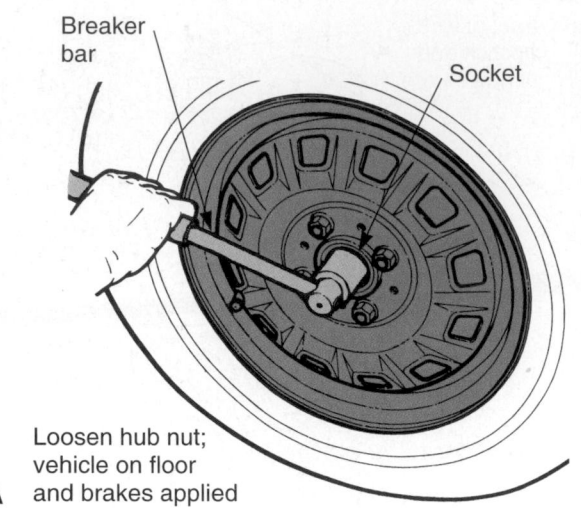

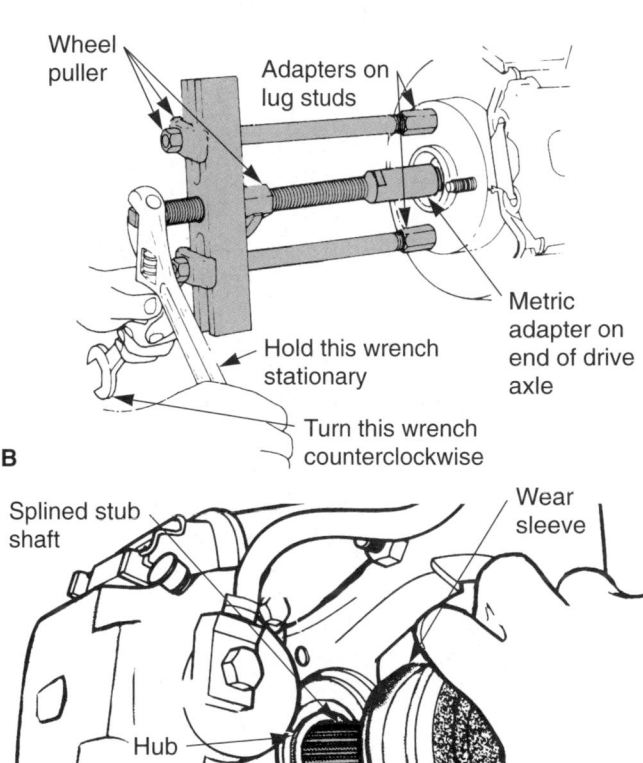

Figure 64-7. Typical method for disconnecting the outer end of a front drive axle assembly. A—With the vehicle on the ground, loosen hub nut with a breaker bar and socket. B—A puller is sometimes needed to force the outer stub shaft splines from the hub. Make sure you have removed the ball joint, lower control arm fastener, sway bar, and any other obstructing parts. C—With the lower ball joint removed, swing the steering knuckle outward so axle splines will slide out. Support the axle. (Ford and Dodge)

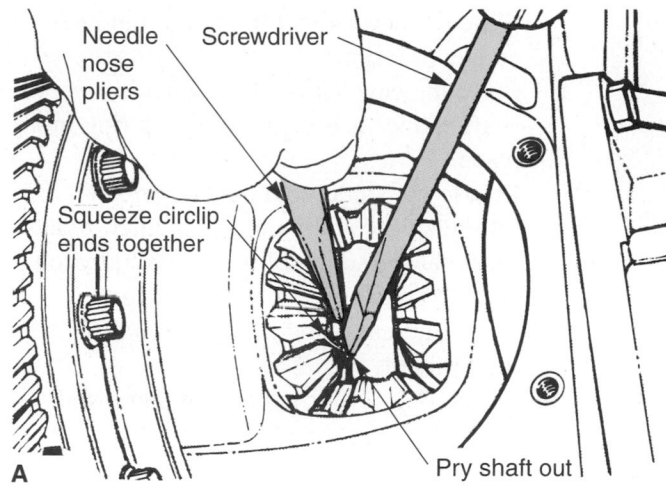

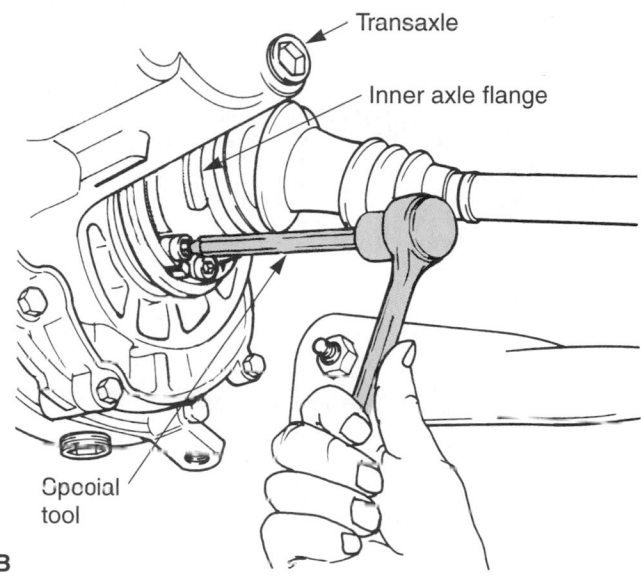

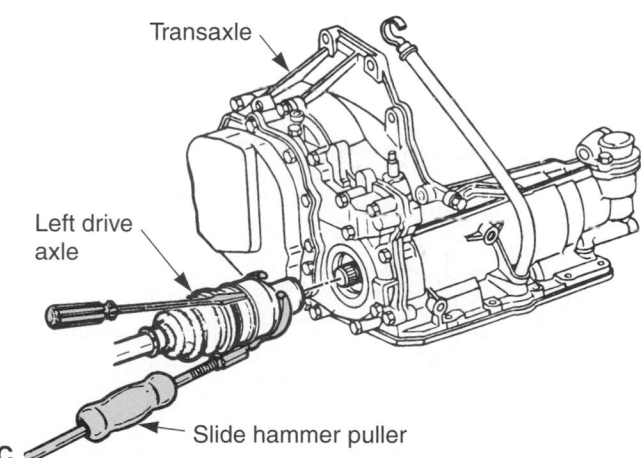

Figure 64-8. Inner stub shafts are commonly connected by snap rings or bolts. A—The snap ring must be removed from its groove before the shaft can be removed. B—When the inner axle flange is bolted to the drive flange, remove the fasteners and the shaft can be removed. C—A prying or slide hammer may be needed to free snap ring and the shaft from the differential. Check the service manual. (General Motors and Chrysler)

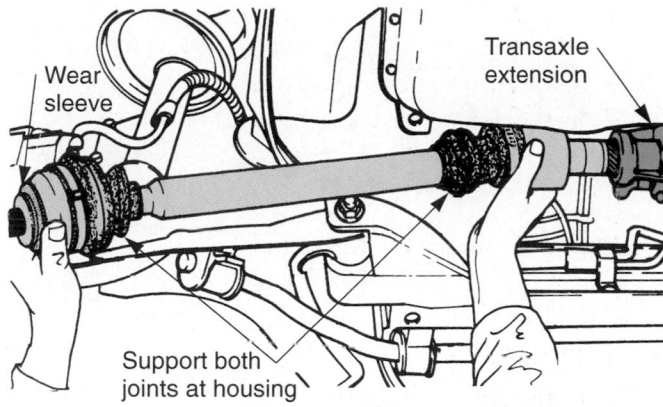

Figure 64-9. When removing the front drive axle assembly, support both ends and hold the plunging joint together. Never let the axle hang unsupported; joint or boot damage can result. (Dodge)

CV-joint can be replaced, if desired. A ***CV-joint kit*** usually includes new joint components, grease, a boot, and boot straps. To replace CV-joints, mount the drive axle assembly in a vise, **Figure 64-10.** If you have a tubular, or hollow, axle shaft (interconnecting shaft), be careful not to bend or dent the shaft. The slightest dent could cause vibration.

 To service a *tripod CV-joint:*
1. Cut (plastic type) or pry off (reusable spring type) the old boot straps and slide the boot off the joint. See **Figure 64-11A.**
2. Separate the tripod from the housing, **Figure 64-11B.**
3. Scribe mark the housing and interconnecting shaft to ensure proper alignment during reassembly, **Figure 64-11C.**
4. To remove the spider from the interconnecting shaft, remove the snap ring in the end of the shaft. Then, drive off the old spider. See **Figure 64-11D**

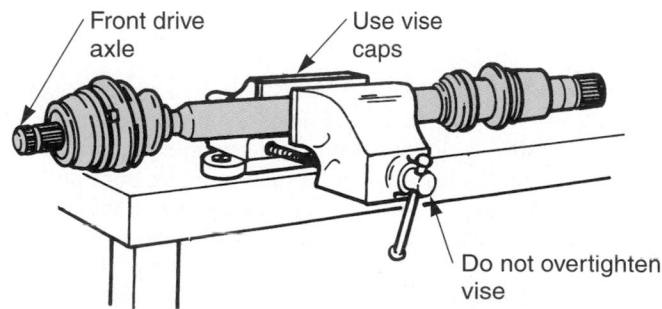

Figure 64-10. To service CV-joints, mount the axle in a vise. With hollow drive axles, be extremely careful not to bend or dent shaft. This could cause shaft vibration. (Dana Corp.)

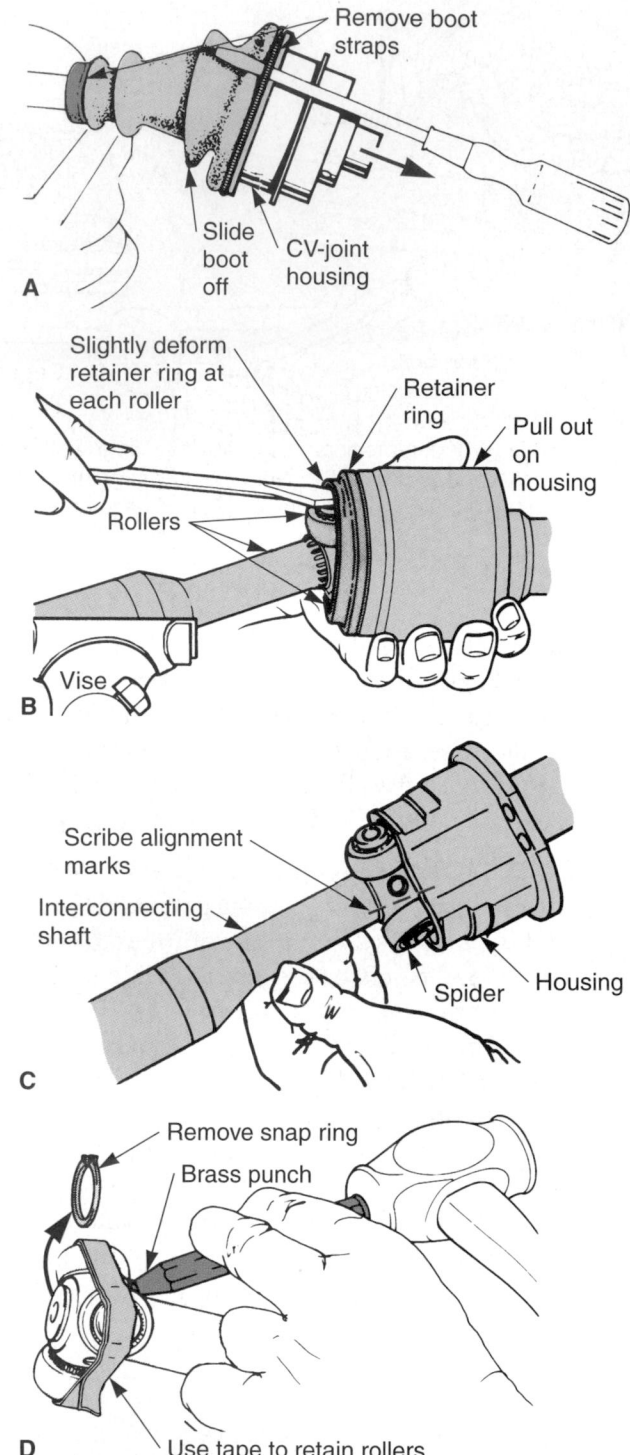

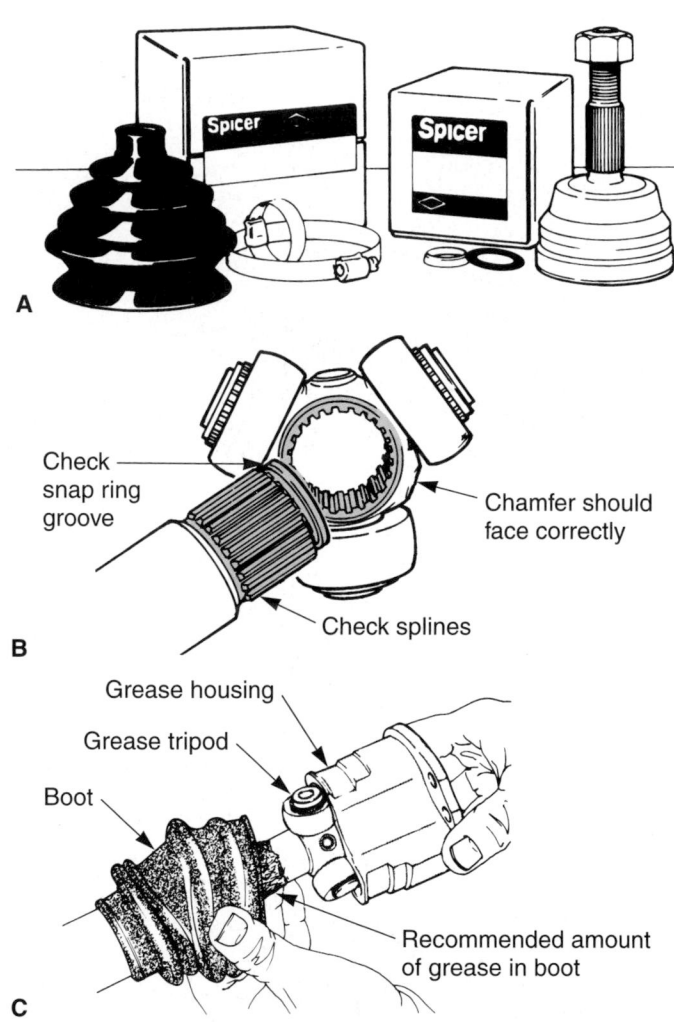

5. Slide the new rubber boot onto the interconnecting shaft.
6. Place the new spider on its interconnecting shaft splines and install the snap ring in place.
7. Grease the spider and the housing.
8. Fit the spider and housing together. Make sure that the scribed lines on the interconnecting shaft and the housing align.
9. Place the remaining kit grease in the rubber boot.
10. Fit the boot over the housing and install the boot straps. See **Figure 64-12.**

Figure 64-11. Basic steps for disassembling a tripod CV-joint. A—Remove straps or clamps holding the boot around the joint. If plastic straps, pry or cut them off. If spring straps, pry them off carefully because they can be reused. Slide boot off joint. B—Slide spider out of the housing. If needed, flex the retainer ring out of the way. C—If any joint components are to be reused, mark their alignment during disassembly. This will let you reassemble parts in same position. D—Remove the snap ring holding the spider on the axle splines. Then use a brass drift or punch to push the spider off the axle. Tape is used to keep rollers and needles from falling off. (Renault, Chrysler, Dana Corp.)

Figure 64-12. Fundamental steps for rebuilding a tripod CV-joint. A—A CV-joint kit normally contains a new joint spider, boot, rollers, needle bearings, snap rings, and boot straps. This will ensure quality repair. B—Inspect the axle shaft splines and snap ring grooves. Make sure spider is fitted on splines correctly. Chamfer may have to face certain direction. Install a new snap ring to hold the spider on the axle. C—Grease the spider and the housing. Place recommended amount of grease in boot. Then fit parts together. Install boot and boot straps. (Dana Corp. and Chrysler)

Caution

Always use the recommended type of grease on a CV-joint. Quality CV-joint kits will provide the correct type and amount of grease. The wrong type of grease can cause boot deterioration and joint failure.

Servicing a *Rzeppa CV-joint* is illustrated in **Figures 64-13** and **64-14.** Use the following procedure:

1. Remove the boot straps.
2. Slide the boot back and remove the retaining ring, **Figure 64-13A.**
3. Remove the joint from its shaft.

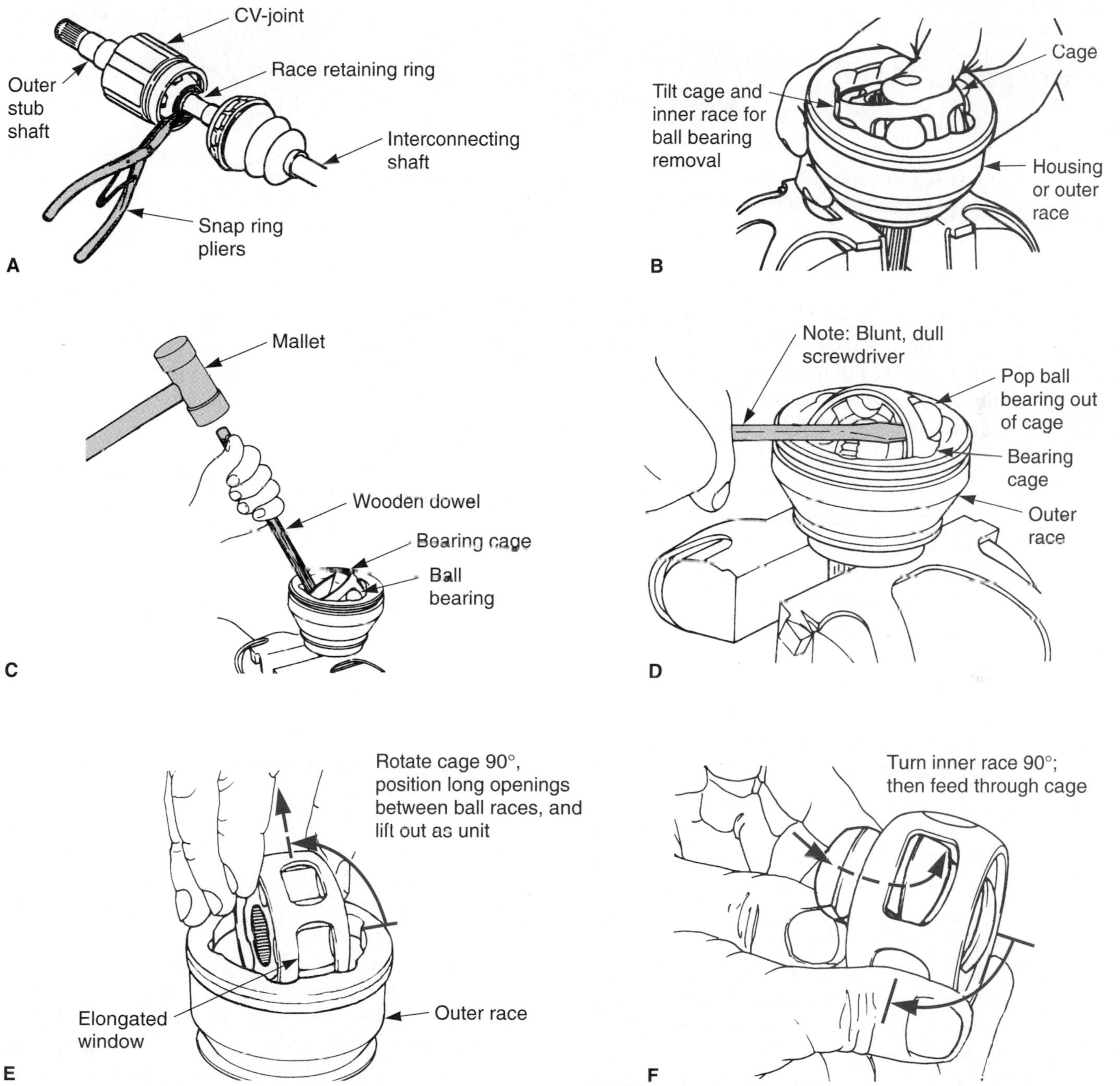

Figure 64-13. General procedures for disassembling a Rzeppa CV-joint. A—Remove the boot straps and pull back the boot. Use snap ring pliers to remove the retaining ring and then pull the axle shafts apart. B—Use thumb pressure to tilt the cage so the balls can be removed. C—You may need to use hammer and wooden dowel. D—Remove the balls with a dull-tip screwdriver. E—Swivel the cage and inner race 90° and remove from the housing. F—Rotate the inner race to remove it from the cage. (GMC, Ford, and Chrysler)

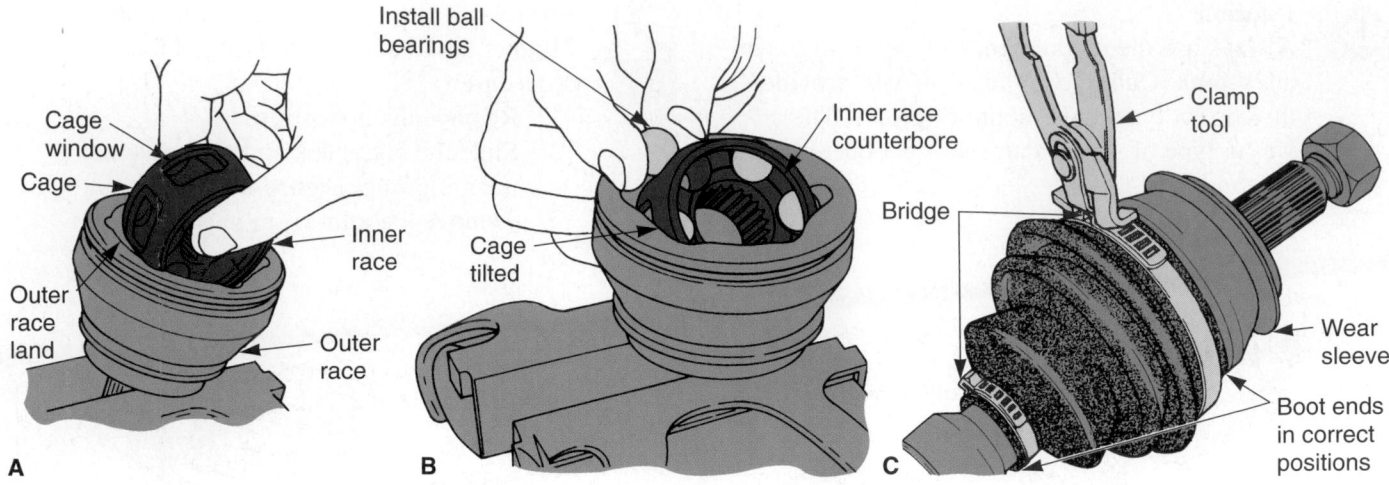

Figure 64-14. Reassembly steps for a Rzeppa or ball and cage CV-joint. A—Install the cage and inner race inside the housing. B—Tilt cage and install balls one at a time. With a sliding Rzeppa joint, balls, cage, and inner race will slide into the housing as a unit. C—Install the boot and boot straps. Note use of special pliers designed for boot installation. (Ford and Chrysler)

4. Tilt the cage and inner race using thumb pressure, **Figure 64-13B.** In some cases, light taps with a hammer and wooden dowel may be required to tilt the cage, **Figure 64-13C.** With the cage tilted, remove the balls from the cage. Tilt the cage in different directions to free each ball, **Figure 64-13D.**

5. Swivel the cage and inner race and remove them from the housing, **Figure 64-13E.**

6. Rotate the inner race and remove it from the cage, **Figure 64-13F.**

7. Clean and inspect the parts to be reused.

8. Obtain the correct CV-joint repair kit.

9. Grease the inner race and fit it into the cage.

10. Install the cage and inner race into the joint housing, **Figure 64-14A.**

11. Tilt the cage and position each ball, **Figure 64-14B.**

12. Assemble the joint on the axle shaft. Make sure all snap rings are in good condition. Sometimes, a press is needed to force the joint over its splines.

13. After reassembly, thoroughly grease all moving parts of the joint. If recommended, place extra grease in the boot for added lubrication.

14. Fit the boot over the joint, **Figure 64-14C.** Make sure the boot ends fit into their grooves.

15. Install the boot straps. Do not overtighten the straps. You may cut the boot or break the strap.

 Caution
If a hammer and dowel are used to tilt the cage when removing the balls, do not use a metal dowel. The housing could be damaged.

Figure 64-15 shows two typical front drive axle assemblies. Note how both have sliding inboard CV-joints. One is a Rzeppa joint and the other is a tripod joint. Both axles have fixed outer Rzeppa joints.

 Note
Refer to a service manual when servicing either type of CV-joint. The manual will give special, detailed directions.

Installing a Front Drive Axle

Basically, to install a front drive axle, reverse the directions used in axle removal. Fit or slide the inner end of the axle on or into the transaxle. The shaft may slide easily into the transaxle, or you may need to use light taps with a hammer and driving tool. If used, make sure the snap ring or spring clip locks in the differential by pulling on the joint, **Figure 64-16A.**

Reinstall any bolts or other parts removed during disassembly. Replace and lubricate the hub seal, or grease seal. Also lubricate the *wear sleeve* (bushing between axle and seal), if used. See **Figure 64-16B.** Hub seal replacement is normally recommended during front drive axle service.

Slide the outer end of the axle into the hub. You may need to use a puller to force the axle through the hub splines. If needed, reinstall the lower ball joint. Screw on the spindle nut and torque it to specifications.

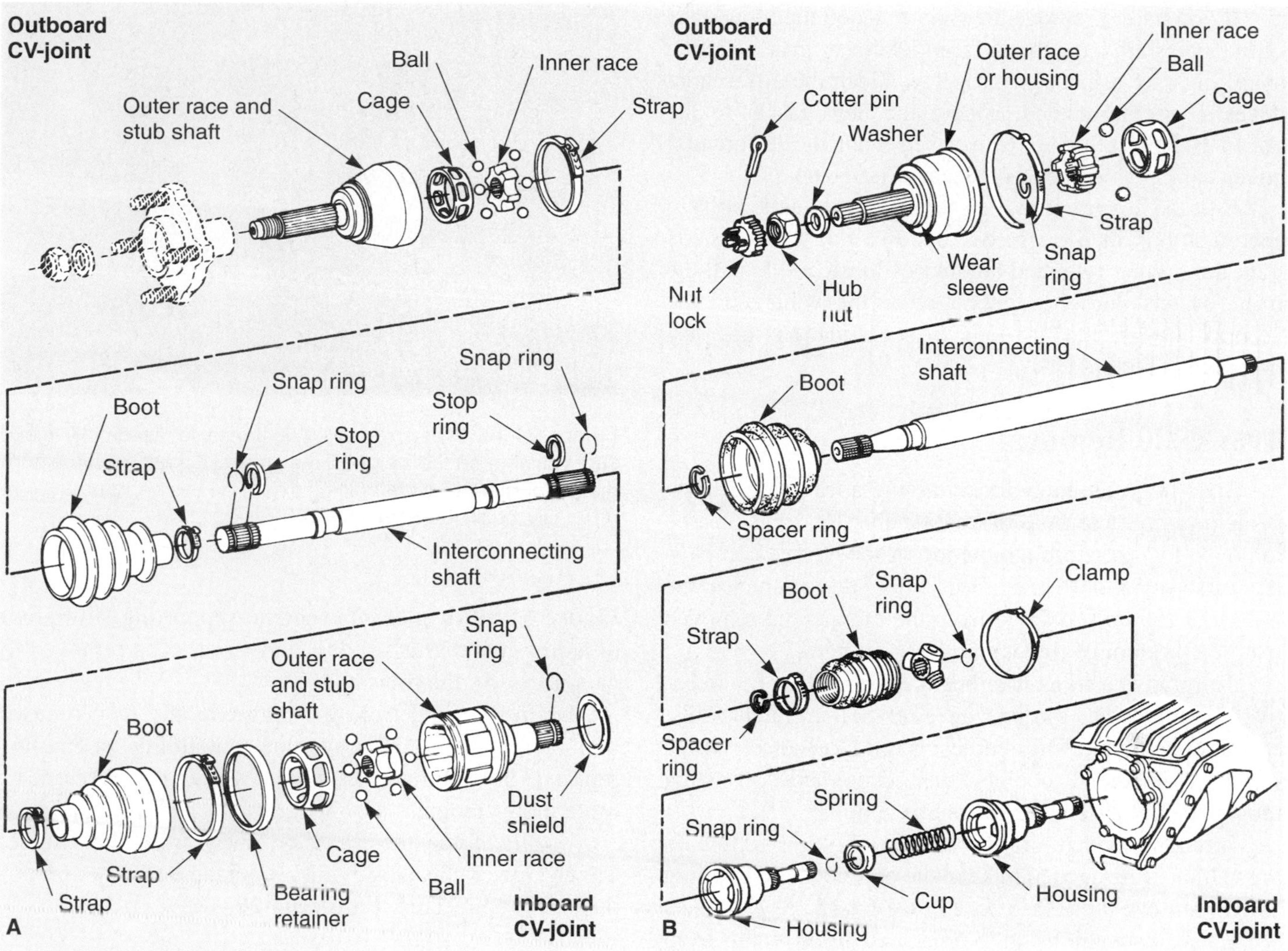

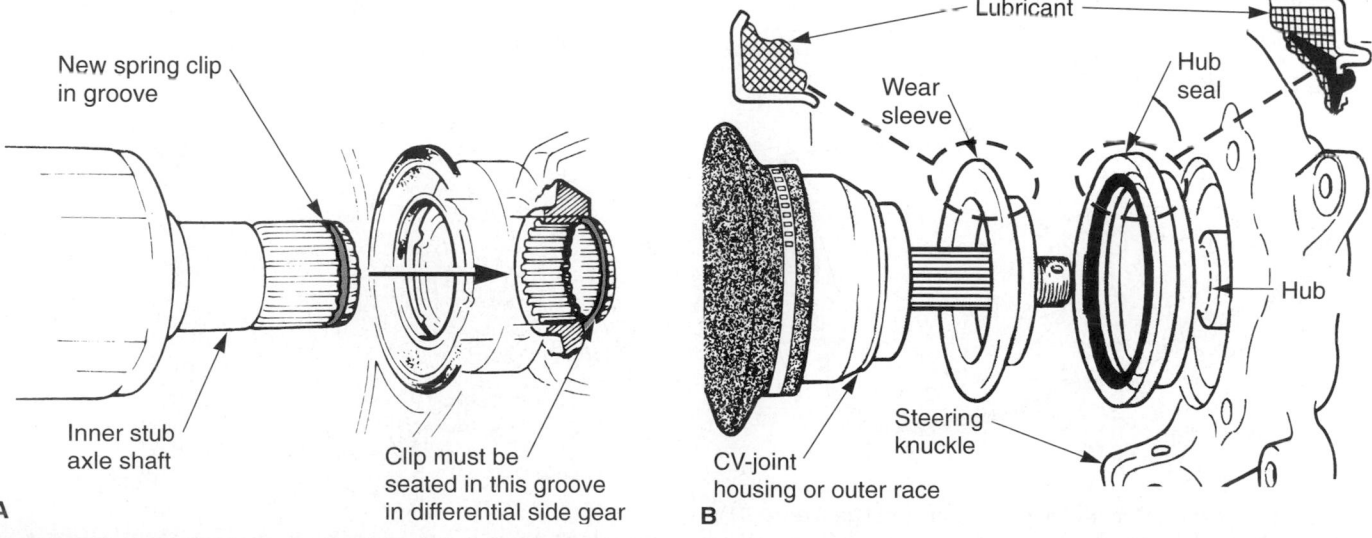

Outboard CV-joint

Outer race and stub shaft — Cage — Ball — Inner race — Strap

Outboard CV-joint

Cotter pin — Washer — Outer race or housing — Inner race — Ball — Cage — Strap — Snap ring — Wear sleeve — Hub nut — Nut lock

Boot — Strap — Snap ring — Stop ring — Stop ring — Snap ring — Interconnecting shaft

Boot — Spacer ring — Interconnecting shaft

Boot — Strap — Snap ring — Outer race and stub shaft — Inner race — Ball — Cage — Bearing retainer — Dust shield — Strap — Strap

Inboard CV-joint

Boot — Strap — Spacer ring — Snap ring — Clamp

Snap ring — Spring — Housing — Cup — Housing

Inboard CV-joint

Figure 64-15. Exploded view of two typical CV-axles. A—This axle has two Rzeppa joints. Outboard CV-joint is fixed. Inboard Rzeppa joint is sliding. B—This axle has a fixed Rzeppa outboard and a sliding tripod joint inboard. Study construction and location of parts.

New spring clip in groove — Inner stub axle shaft — Clip must be seated in this groove in differential side gear

A

Lubricant — Wear sleeve — Hub seal — Hub — CV-joint housing or outer race — Steering knuckle

B

Figure 64-16. Front drive axle installation. A—If a snap ring or spring clip holds the inner stub shaft in the differential, make sure it snaps into place. B—Lubricate parts as recommended. Note how this axle design requires grease on wear sleeve and seal. (Chrysler)

If you had to remove the snap rings on the inner ends of the axle shafts and the differential cover, make sure the snap rings are reinstalled properly. Clean the differential cover. If recommended, adhere the new gasket to the cover. If RTV sealer is to be used, coat the differential cover with an even bead of sealer, **Figure 64-17.**

Bolt the cover to the transaxle and torque the bolts to specifications in a crisscross pattern. Fill the transaxle with the correct type and amount of lubricant. Install the front wheels and axle nuts. Lower the vehicle to the ground. Torque the lug nuts to specifications and test drive the vehicle.

Transaxle Removal

Specific procedures for removing a transaxle from a vehicle vary. Sometimes the engine and transaxle must be removed together through the top of the engine compartment. However, with most front-wheel-drive vehicles, the transaxle can be separated from the engine and removed from the bottom of the engine compartment.

To remove a transaxle, begin by removing the negative battery cable, torque converter bolts, clutch cable, speedometer cable, shift cables or rods, sensor or switch wiring connectors, cooler lines, drive axles, starting motor, and upper engine-to-transaxle bolts.

If recommended, mount an engine holding fixture on the vehicle. It is needed to keep the engine from dropping as you remove the transaxle, **Figure 64-18.**

Raise the vehicle on a hoist and remove the front wheels. Place a transmission jack under the transaxle,

Figure 64-18. When removing a transaxle assembly, make sure you support the weight of the engine. It must not hang from the motor mounts. Note holding fixture on this vehicle. (OTC Div. of SPX Corp.)

Figure 64-19. If a *cradle* (engine-supporting subframe) or a crossmember is used, it may have to be removed to gain access to the transaxle.

Drain the fluid from the transaxle and then remove both drive axles. Remove the remaining engine-to-transaxle bolts. It may also be necessary to remove one or more motor mounts.

Finally, force the transaxle away from the engine. Slowly lower the jack while watching for any components still connected, **Figure 64-20.**

Transaxle Service

When repairing or rebuilding a transaxle, follow the detailed procedures given in a service manual.

A few general rules to remember include:

- Do not dent or nick the gasket sealing surfaces during disassembly or cleaning.

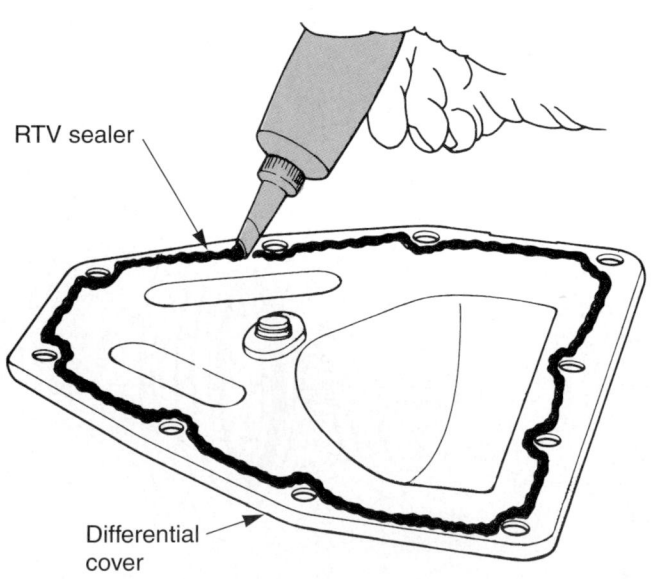

Figure 64-17. If it was removed, clean and apply gasket or RTV sealer to the differential cover. Install the cover and torque fasteners to specifications. (Chrysler)

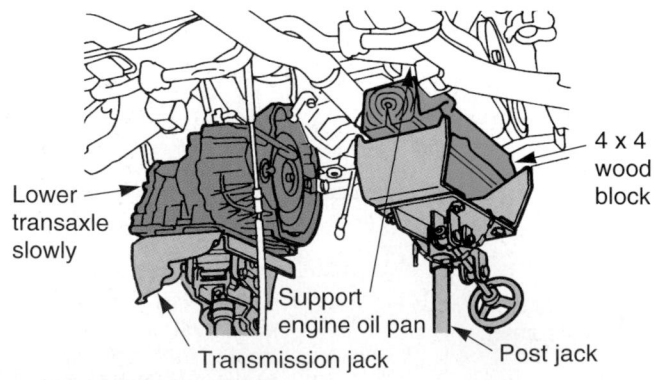

Figure 64-19. Use a transmission jack to remove a transaxle. A post jack supports the engine. (Nissan)

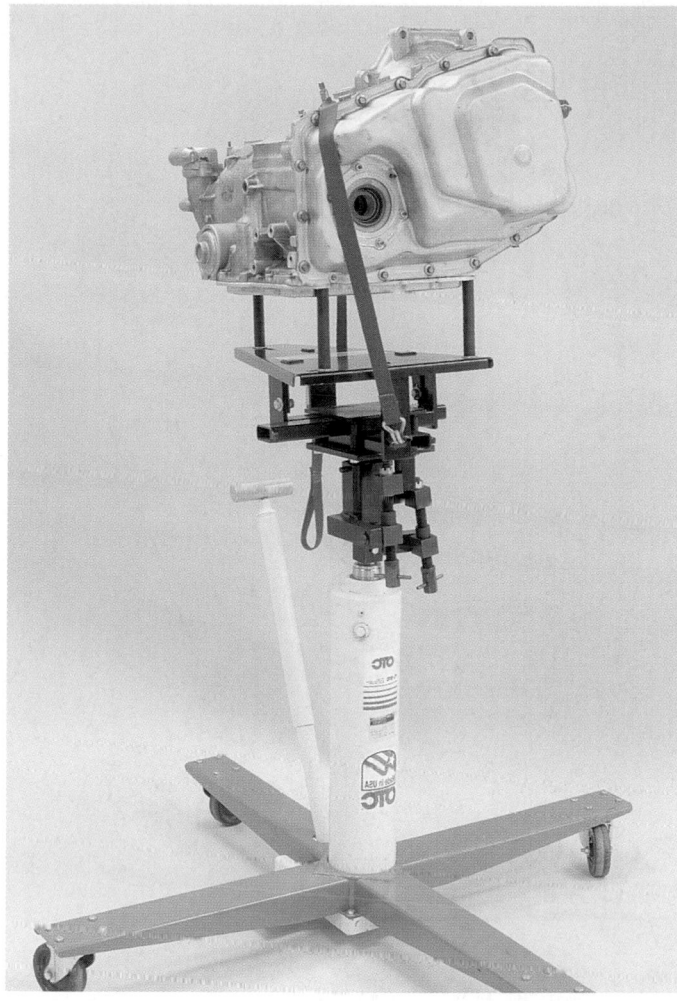

Figure 64-20. Note the large saddle on the transmission jack. It will safely secure and hold the transaxle during removal. (OTC)

- Keep all parts organized so that used parts can be returned to their original positions, **Figure 64-21.**
- Clean all parts and closely inspect them for damage. Replace any parts showing wear.
- Replace all gaskets and seals. See **Figure 64-22.**
- Use special tools when needed.
- During reassembly, measure all gears, shafts, bearings, and other critical components as described in a service manual, **Figure 64-23.**
- Measure bearing preload as described in the service manual, **Figure 64-24.**
- Make sure all snap rings, shims, and other parts are installed correctly. See **Figures 64-25** and **64-26.**
- Torque all bolts to specifications. Most transaxles have aluminum cases that are very sensitive to bolt torque.

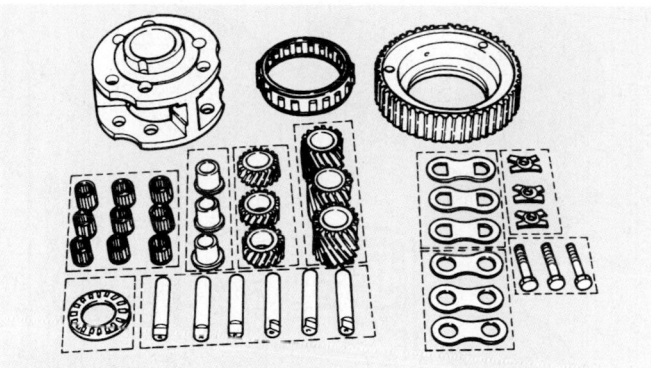

Figure 64-21. Clean and inspect all components very carefully. If you overlook just one problem, your repair may fail. Keep parts organized to simplify reassembly. (Subaru)

Transaxle Installation

Install a transaxle in the reverse order of removal. Use a jack to position the transaxle behind the engine. Align the engine crankshaft centerline with the transaxle input shaft centerline.

If you have a manual transaxle, make sure the throwout bearing is new or in good condition. Make sure it is installed correctly.

Slowly push the transaxle against the engine. Make sure the studs on the engine fit into the holes in the transaxle bell housing. With an automatic transaxle, you may also have to align the torque converter bolts with the flywheel while fitting the transaxle.

 Caution
As with a transmission, double-check that the torque converter is correctly installed. If the converter is not fully into position, part damage could result when you tighten the bolts holding the transaxle to the engine.

Start and torque the engine-to-transaxle bolts. Then, install the other components (cradle, starter, mounts, etc.). Fill the transaxle with lubricant and test drive the car.

Summary

- Transaxles suffer from the same kinds of problems as transmissions and differentials. The gears, shafts, bearings, seals, and other parts can wear and fail.
- The drive axles, as they follow the movement of the steering and suspension systems, can also cause problems after prolonged service.
- Late-model vehicles often use electronics to control and monitor transaxle operation. With

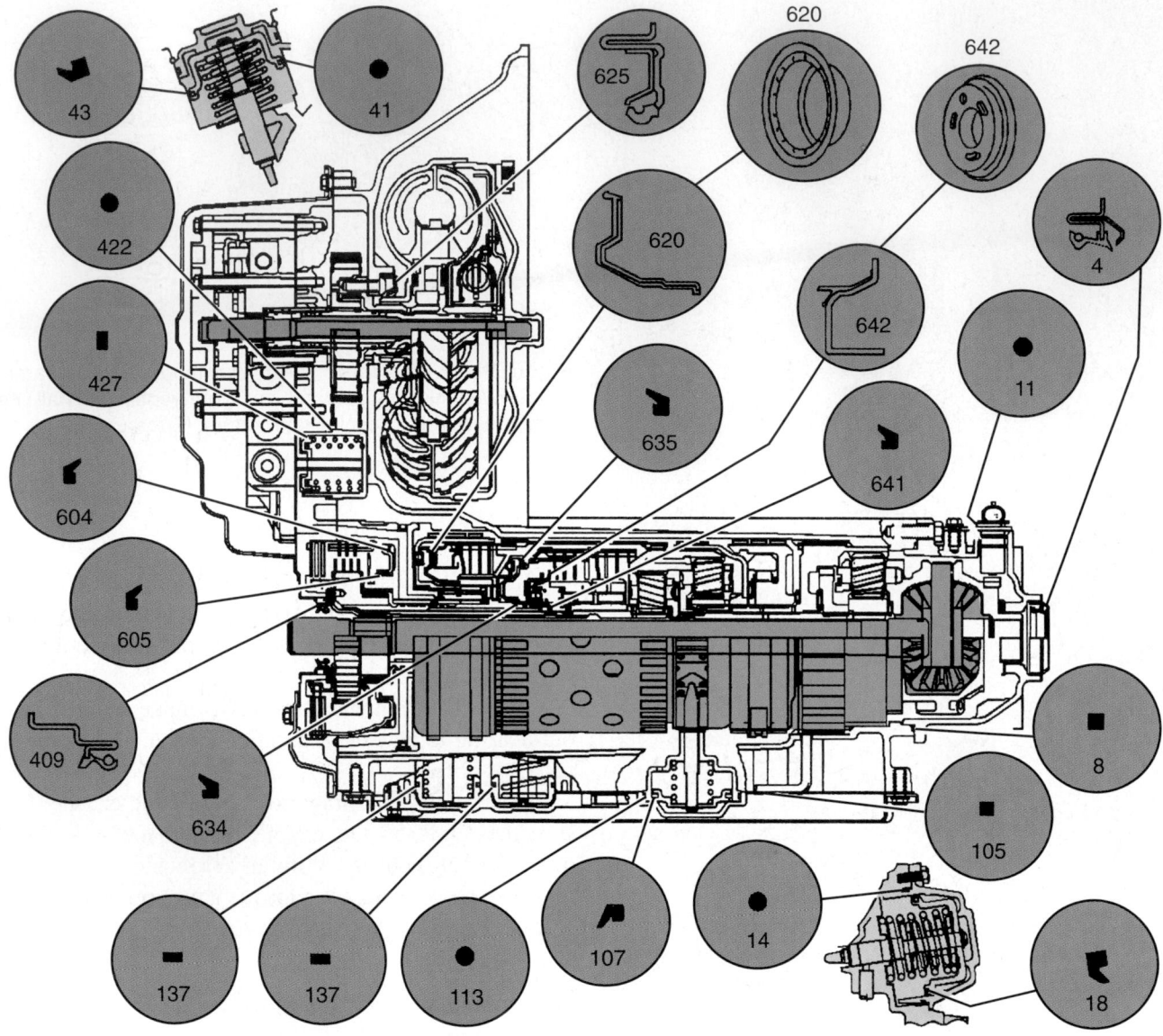

Legend

(4) Axle oil seal assembly	(409) Axle oil seal assembly (left side)
(8) Extension to case seal	(422) O-ring seal
(11) O-ring speed sensor seal	(427) Oil seal ring (3-4 accum. piston)
(14) O-ring seal (forward servo cover)	(604) 4th clutch piston seal (outer)
(18) Oil seal ring (forward servo piston)	(605) 4th clutch piston seal (inner)
(41) O-ring seal (reverse servo cover)	(620) 2nd clutch w/molded seal piston
(43) Oil seal ring (reverse servo piston)	(634) Input clutch piston seal (inner)
(105) Square cut seal (2/1 servo)	(635) Input clutch piston seal (outer)
(107) Lip seal (manual 2/1 servo piston)	(641) 3rd clutch piston seal (inner)
(113) O-ring seal (manual 2/1 servo cover)	(642) Seal and ball capsule ASM, 3rd clutch piston
(137) Oil seal accumulator piston ring	

Figure 64-22. A service manual will give details needed to do major transaxle repairs. Note how this manual illustration shows the location and cross section of all seals that must be replaced during a rebuild. (Oldsmobile)

OBD II, several transaxle problems will trip a diagnostic code and light the dash malfunction indicator light.

- If you find trouble codes or an abnormal operating parameter, you would need to use pinpoint tests to find the actual source of the trouble.

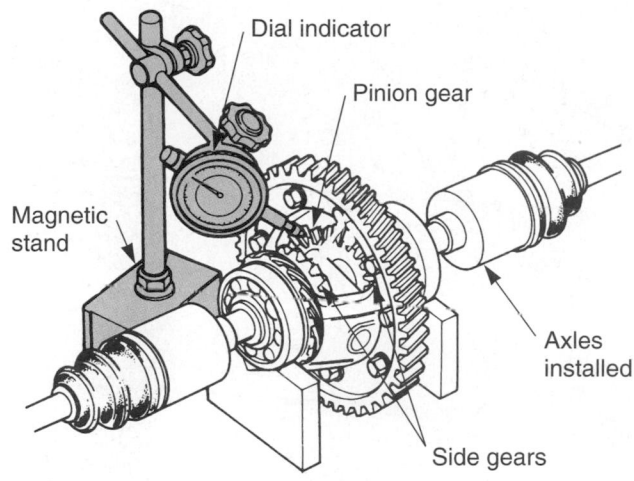

Figure 64-23. Measure all clearances as described in the service manual. This shows how you should measure differential pinion gear backlash. Many other measurements are also essential. (Honda)

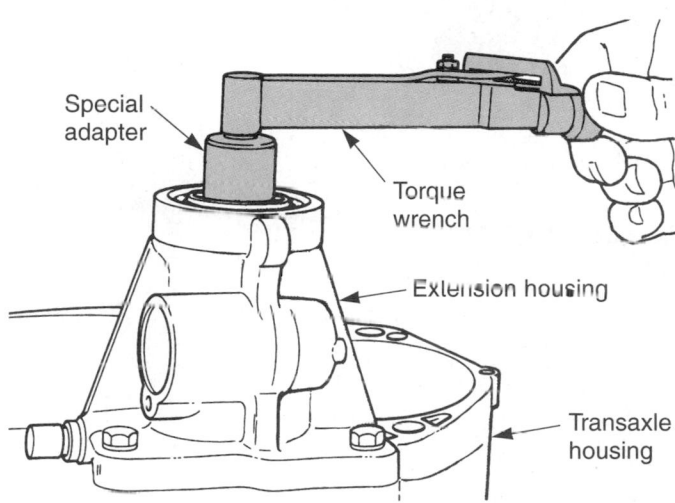

Figure 64-24. As with a rear axle differential, a torque wrench can be used to measure transaxle differential case bearing preload. Compare reading to specifications. (Dodge)

- A manufacturer's diagnosis chart may also be helpful in determining what parts might be causing the symptoms.
- Automatic transaxle problems, like automatic transmission problems, show up as slipping in gear (worn bands or clutches, low system pressure), abnormal noises (damaged planetary gears, bad bearings), and fluid leaks.
- Front drive axle assembly problems are usually noticed as abnormal noises (clicking, grinding, clunking, humming), vibration, or excessive play in the universals.
- To check for bad front drive axles, drive the vehicle in reverse with the steering wheel turned sharply right and left.
- When you find problems with a front drive axle shaft, the shaft must be removed for service.
- Refer to a service manual when servicing CV-joints.

Important Terms

Automatic transaxle	Rzeppa CV-joint
Manual transaxle	Wear sleeve
Tripod CV-joint	Cradle
CV-joint repair kit	

Review Questions—Chapter 64

Please do not write in this text. Place your answers on a separate sheet of paper.

1. What are three general manual transaxle problems?
2. What are three general automatic transaxle problems?
3. Explain three symptoms or problems noticed with front drive axle troubles.
4. Name the two types of lubricants used in manual transaxles.
5. How do you check the oil or lubricant level in automatic and manual transaxles?
6. This is *not* an adjustment relating to an automatic transaxle.
   (A) Clutch fork.
   (B) Bands.
   (C) Shift cable.
   (D) All of the above.
   (E) None of the above.
7. The most common problem with a front drive axle is worn CV-joints. True or False?
8. What components are normally included in a CV-joint repair kit?
9. Reuse the old hub seal when replacing a front drive axle. True or False?
10. Besides following the directions in a service manual, name nine general rules for servicing a transaxle.

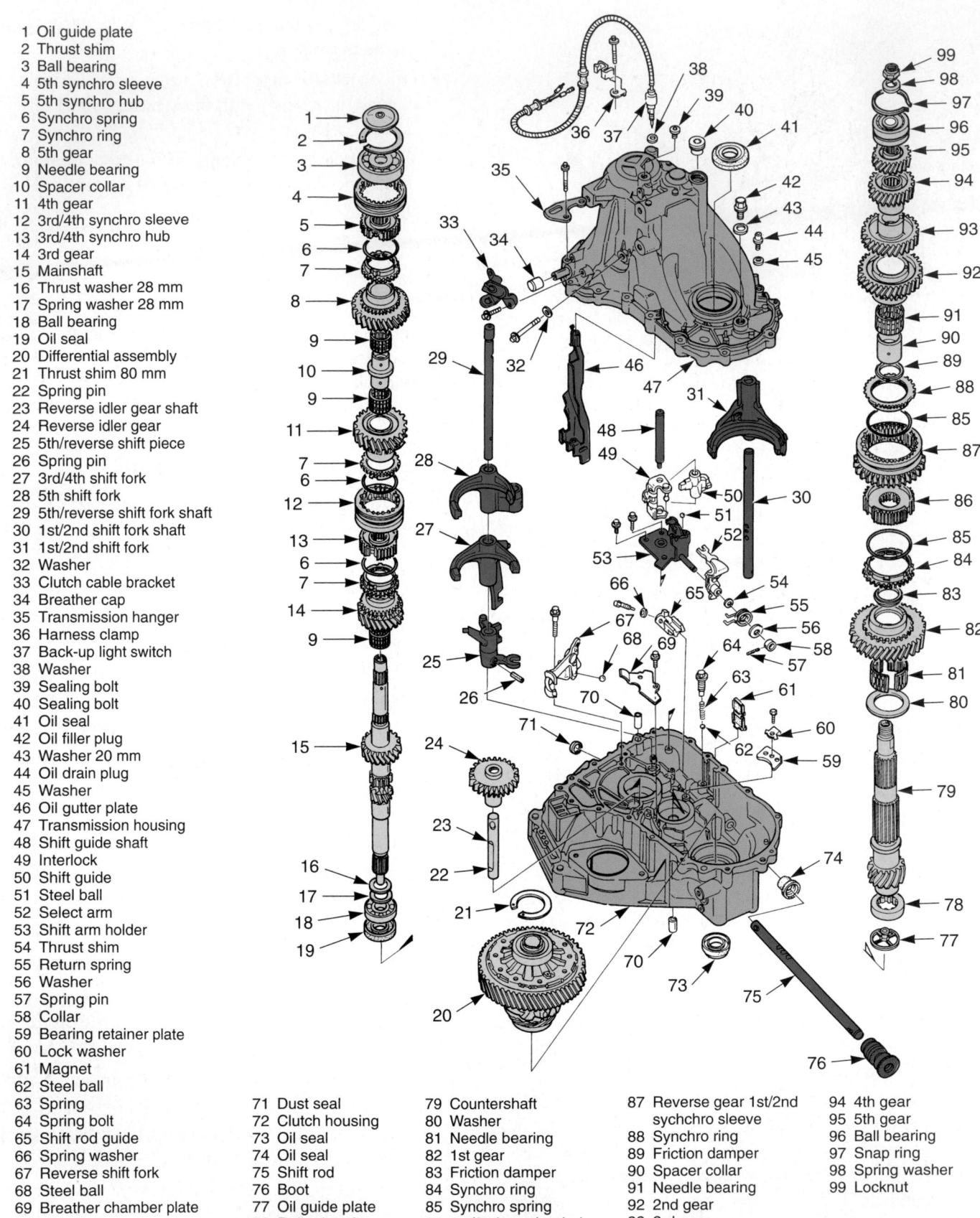

1 Oil guide plate
2 Thrust shim
3 Ball bearing
4 5th synchro sleeve
5 5th synchro hub
6 Synchro spring
7 Synchro ring
8 5th gear
9 Needle bearing
10 Spacer collar
11 4th gear
12 3rd/4th synchro sleeve
13 3rd/4th synchro hub
14 3rd gear
15 Mainshaft
16 Thrust washer 28 mm
17 Spring washer 28 mm
18 Ball bearing
19 Oil seal
20 Differential assembly
21 Thrust shim 80 mm
22 Spring pin
23 Reverse idler gear shaft
24 Reverse idler gear
25 5th/reverse shift piece
26 Spring pin
27 3rd/4th shift fork
28 5th shift fork
29 5th/reverse shift fork shaft
30 1st/2nd shift fork shaft
31 1st/2nd shift fork
32 Washer
33 Clutch cable bracket
34 Breather cap
35 Transmission hanger
36 Harness clamp
37 Back-up light switch
38 Washer
39 Sealing bolt
40 Sealing bolt
41 Oil seal
42 Oil filler plug
43 Washer 20 mm
44 Oil drain plug
45 Washer
46 Oil gutter plate
47 Transmission housing
48 Shift guide shaft
49 Interlock
50 Shift guide
51 Steel ball
52 Select arm
53 Shift arm holder
54 Thrust shim
55 Return spring
56 Washer
57 Spring pin
58 Collar
59 Bearing retainer plate
60 Lock washer
61 Magnet
62 Steel ball
63 Spring
64 Spring bolt
65 Shift rod guide
66 Spring washer
67 Reverse shift fork
68 Steel ball
69 Breather chamber plate
70 Dowel pin

71 Dust seal
72 Clutch housing
73 Oil seal
74 Oil seal
75 Shift rod
76 Boot
77 Oil guide plate
78 Roller bearing

79 Countershaft
80 Washer
81 Needle bearing
82 1st gear
83 Friction damper
84 Synchro ring
85 Synchro spring
86 1st/2nd synchro hub

87 Reverse gear 1st/2nd
   sychchro sleeve
88 Synchro ring
89 Friction damper
90 Spacer collar
91 Needle bearing
92 2nd gear
93 3rd gear

94 4th gear
95 5th gear
96 Ball bearing
97 Snap ring
98 Spring washer
99 Locknut

Figure 64-25. Exploded view of a manual transaxle. Study the location of the parts and visualize their reassembly. The service manual will contain an illustration for the exact transaxle being rebuilt. (Honda)

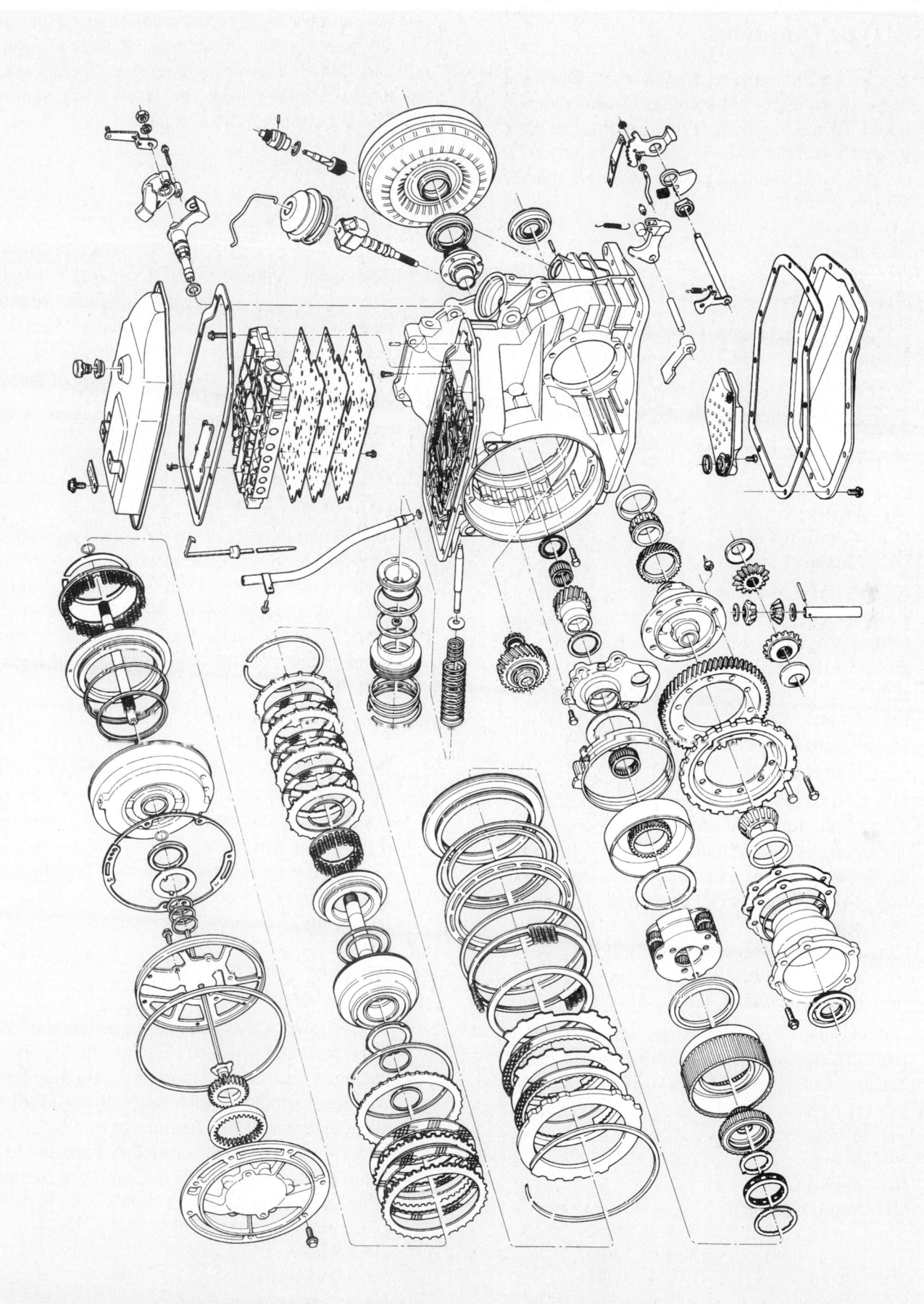

Figure 64-26. Exploded view of an automatic transaxle. The service manual will normally provide a similar illustration of the unit being repaired. Can you identify the major parts? (Ford)

⬥ ASE-Type Questions

1. A customer complains that his/her automatic transaxle equipped automobile is making an odd noise. Technician A asks the customer in what gear or when the noise is present. Technician B test drives the car in all gears at different speeds. Who is right?
   (A) A only.
   (B) B only.
   (C) Both A and B.
   (D) Neither A nor B.

2. A car is brought into the shop with possible front drive axle problems. Technician A believes this condition may be caused by worn CV-joint boots. Technician B says to drive the car in reverse with the wheels cut hard right and left. Who is right?
   (A) A only.
   (B) B only.
   (C) Both A and B.
   (D) Neither A nor B.

3. A car's manual transaxle produces a noise in every gear. Technician A inspects the condition of the valve body. Technician B checks the condition of the transaxle's shaft bearings. Who is right?
   (A) A only.
   (B) B only.
   (C) Both A and B.
   (D) Neither A nor B.

4. A manual transaxle is noisy in second gear. Technician A checks the condition of the input second gear. Technician B inspects the condition of the output second gear. Who is right?
   (A) A only.
   (B) B only.
   (C) Both A and B.
   (D) Neither A nor B.

5. An automatic transaxle slips when placed in gear. Technician A checks for worn bands. Technician B checks the transaxle's fluid level. Who is right?
   (A) A only.
   (B) B only.
   (C) Both A and B.
   (D) Neither A nor B.

6. A customer wants to know what caused the car's CV-joints to fail. Technician A says CV-joints can fail after prolonged service. Technician B says CV-joint failure can result from ruptured CV-joint boots. Who is right?
   (A) A only.
   (B) B only.
   (C) Both A and B.
   (D) Neither A nor B.

7. A car's drive axle has possible problems. Technician A inspects the drive axle while the car is raised on a lift with the engine running and the transaxle in gear. Technician B checks the drive axle while the car is raised on a lift with the engine off and the transaxle in park or neutral. Who is right?
   (A) A only.
   (B) B only.
   (C) Both A and B.
   (D) Neither A nor B.

8. Technician A checks an automatic transaxle's lubricant level while the lubricant is at operating temperature with the engine running. Technician B checks an automatic transaxle's lubricant level while the lubricant is at room temperature with the engine off. Who is right?
   (A) A only.
   (B) B only.
   (C) Both A and B.
   (D) Neither A nor B.

9. A manual transaxle's lubricant level needs to be checked. Technician A looks for the transaxle's dipstick or fill hole. Technician B says all manual transaxles use a dipstick for checking lubricant level. Who is right?
   (A) A only.
   (B) B only.
   (C) Both A and B.
   (D) Neither A nor B.

10. A customer wants the car's manual transaxle lubricant level checked. Technician A tells the customer that certain manual transaxle's lubricant level should be checked with the lubricant at room temperature. Technician B tells the customer that all manual transaxle's lubricant level should be checked with the lubricant at operating temperature. Who is right?
    (A) A only.
    (B) B only.
    (C) Both A and B.
    (D) Neither A nor B.

11. A manual transaxle's lubricant level is low. Technician A adds manual transmission gear oil to the transaxle. Technician B checks the manual to see if this transaxle uses automatic transmission fluid or manual transmission gear oil for lubrication. Who is right?
    - (A) *A only.*
    - (B) *B only.*
    - (C) *Both A and B.*
    - (D) *Neither A nor B.*

12. A car is brought into the shop to have its transaxle fluid changed. Technician A warms the transaxle's fluid to operating temperature before draining it. Technician B says the fluid should remain cool while draining it. Who is right?
    - (A) *A only.*
    - (B) *B only.*
    - (C) *Both A and B.*
    - (D) *Neither A nor B.*

13. A tripod type CV-joint needs to be replaced. Before installing the new CV-joint, Technician A inspects the axle shaft splines. Before installing the new CV-joint, Technician B checks the condition of the axle shaft snap ring groove. Who is right?
    - (A) *A only.*
    - (B) *B only.*
    - (C) *Both A and B.*
    - (D) *Neither A nor B.*

14. An automobile's CV-joints and boots need replacing. Technician A tells the owner of the car that improper type of grease may have damaged these components. Technician B tells the owner of the car that any type of axle grease can be used on automotive CV-joints without causing part damage. Who is right?
    - (A) *A only.*
    - (B) *B only.*
    - (C) *Both A and B.*
    - (D) *Neither A nor B.*

15. A front-wheel drive car's transaxle must be removed for major repairs. Technician A tells the owner of the vehicle that on most front-wheel drive cars, the transaxle can be disconnected from the engine and removed from the bottom of the engine compartment. Technician B tells the owner of the vehicle that on any front-wheel drive car, the transaxle and engine must be removed together out the top of the engine compartment. Who is right?
    - (A) *A only.*
    - (B) *B only.*
    - (C) *Both A and B.*
    - (D) *Neither A nor B.*

Activities—Chapter 64

1. Given a transaxle by your instructor, be prepared to show proper diagnosis procedure for that transaxle.

2. Remove, disassemble, and replace a CV-joint.

3. Change the transaxle lubricant in a transaxle assigned by your instructor.

4. Given a flat rate manual, a list of parts to be replaced, and a parts catalog, prepare a repair estimate for replacing CV-joints on a transaxle. Use an hourly rate suggested by your instructor.

	Automatic Transmission and Transaxle Diagnosis	
Condition	**Possible Causes**	**Correction**
Delayed engagement from neutral to drive.	1. Hydraulic pressure too low. 2. Valve body malfunction. 3. Low fluid level. 4. Incorrect gearshift linkage adjustment. 5. Oil filter clogged. 6. Faulty oil pump. 7. Worn input shaft seal rings. 8. Aerated fluid. 9. Engine idle speed too low. 10. Worn or faulty rear clutch.	1. Check fluid pressure at ports. 2. Inspect valve body and repair. 3. Fill transaxle to proper level. 4. Adjust gearshift linkage. 5. Replace oil filter. 6. Replace oil pump. 7. Replace input shaft seal rings. 8. Replace transaxle fluid. 9. Check engine. 10. Replace discs and seals at rear clutch.
Delayed engagement from neutral to reverse.	1. Low reverse band misadjusted. 2. Hydraulic pressures too low. 3. Low reverse band worn out. 4. Valve body malfunction. 5. Low reverse band, servo or linkage malfunction. 6. Low fluid level. 7. Incorrect gearshift linkage adjustment. 8. Oil filter clogged. 9. Faulty oil pump. 10. Worn input shaft seal rings. 11. Aerated fluid. 12. Engine idle speed too low. 13. Worn reaction shaft support seal rings. 14. Worn or faulty front clutch.	1. Adjust bands to specifications. 2. Check fluid pressure at ports. 3. Replace low reverse band. 4. Inspect valve body and repair. 5. Repair low reverse servo. Adjust reverse band and linkage. 6. Fill transmission to level. 7. Adjust gearshift linkage. 8. Replace oil filter. 9. Replace oil pump. 10. Replace input shaft seal rings. 11. Replace transmission fluid. 12. Set up engine to specifications. 13. Inspect and replace reaction shaft support seal rings. 14. Replace discs and seals at front clutch.
Harsh engagement from neutral to drive.	1. Engine idle speed too high. 2. Valve body malfunction. 3. Hydraulic pressure too high. 4. Worn or faulty rear clutch. 5. Rear clutch spring load high. 6. Engine performance.	1. Check engine curb idle speed. Correct as necessary. 2. Inspect valve body and repair. 3. Check fluid pressure at ports. 4. Replace discs and seals at rear clutch. 5. Replace rear clutch spring. 6. Check engine.
Harsh engagement from neutral to reverse.	1. Low/Reverse band misadjusted. 2. Engine idle speed too high. 3. Low/Reverse band worn out. 4. Low/Reverse band, servo or linkage malfunction. 5. Hydraulic pressure too high. 6. Worn or faulty rear clutch. 7. Engine performance.	1. Adjust band to specifications. 2. Set up engine to specifications. 3. Replace Low/Reverse band. 4. Repair Low/Reverse servo. Adjust band and linkage. 5. Check fluid pressure at ports. 6. Replace discs and seals at rear clutch. 7. Check engine.
Runaway upshift.	1. Hydraulic pressures too low. 2. Valve body malfunction. 3. Low fluid level. 4. Oil filter clogged. 5. Aerated fluid. 6. Incorrect throttle linkage. 7. Worn reaction shaft support seal rings. 8. Governor malfunction. 9. Kickdown band, servo or linkage malfunction. 10. Worn front clutch.	1. Check fluid pressure at ports. 2. Inspect valve body and repair. 3. Fill transmission to level. 4. Replace oil filter. 5. Replace transmission fluid. 6. Adjust throttle linkage. 7. Replace reaction shaft support seal rings. 8. Inspect and repair governor. 9. Inspect and repair kickdown band, servo, or linkage. 10. Replace discs and seals at front clutch.

(Continued)

Automatic Transmission and Transaxle Diagnosis		
Condition	**Possible Causes**	**Correction**
No upshift.	1. Hydraulic pressure too low. 2. Valve body malfunction. 3. Low fluid level. 4. Incorrect gearshift linkage adjustment. 5. Incorrect throttle linkage adjustment. 6. Governor support seal rings worn. 7. Worn reaction shaft support seal rings. 8. Governor malfunction. 9. Kickdown band, servo or linkage malfunction. 10. Worn front clutch. 11. Engine performance.	1. Check fluid pressure at ports. 2. Inspect valve body and repair. 3. Fill transmission to proper level. 4. Adjust gearshift linkage. 5. Adjust throttle linkage. 6. Replace governor support seal rings. 7. Replace reaction shaft support seal rings. 8. Inspect and repair governor. 9. Inspect and repair kickdown band, servo, or linkage. 10. Replace discs and seals at front clutch. 11. Set up engine to specifications.
Kickdown runaway.	1. Hydraulic pressure too low. 2. Valve body malfunction. 3. Low fluid level. 4. Aerated fluid. 5. Incorrect throttle linkage adjustment. 6. Kickdown band out of adjustment. 7. Governor support seal rings worn. 8. Kickdown band, servo or linkage malfunction. 9. Worn front clutch.	1. Check fluid pressure at ports. 2. Inspect valve body and repair. 3. Fill transmission to proper level. 4. Replace transmission fluid. 5. Adjust throttle linkage. 6. Adjust kickdown band. 7. Replace governor support seal rings. 8. Inspect and repair kickdown band, servo or linkage. 9. Replace discs and seals at front clutch.
No kickdown or normal downshift.	1. Valve body malfunction. 2. Incorrect throttle linkage adjustment. 3. Governor malfunction. 4. Kickdown band, servo or linkage malfunction.	1. Inspect valve body and repair. 2. Adjust throttle linkage. 3. Inspect and repair governor. 4. Inspect and repair kickdown band, servo, or linkage.
Erratic shifts.	1. Hydraulic pressure too low. 2. Valve body malfunction. 3. Low fluid level. 4. Incorrect gearshift linkage adjustment. 5. Oil filter clogged. 6. Faulty oil pump. 7. Aerated fluid. 8. Incorrect throttle linkage adjustment. 9. Governor support seal rings worn. 10. Worn reaction shaft support seal rings. 11. Governor malfunction. 12. Kickdown band, servo or linkage malfunction. 13. Worn front clutch. 14. Engine performance.	1. Check fluid pressure at ports. 2. Inspect valve body and repair. 3. Fill transmission to level. 4. Adjust gearshift linkage. 5. Replace oil filter. 6. Replace oil pump. 7. Replace transmission fluid. 8. Adjust throttle linkage. 9. Replace governor support seal rings. 10. Replace reaction shaft support seal rings. 11. Inspect and repair governor. 12. Inspect and repair kickdown band, servo, or linkage. 13. Replace discs and seals at front clutch. 14. Set up engine to specifications.
Slips in forward drive positions.	1. Hydraulic pressure too low. 2. Valve body malfunction. 3. Low fluid level. 4. Incorrect gearshift linkage adjustment. 5. Oil filter clogged. 6. Faulty oil pump. 7. Worn input shaft seal rings. 8. Aerated fluid. 9. Incorrect throttle linkage adjustment. 10. Overrunning clutch not holding.	1. Check fluid pressure at ports. 2. Inspect valve body and repair. 3. Fill transmission to proper level. 4. Adjust gearshift linkage. 5. Replace oil filter. 6. Replace oil pump. 7. Replace input shaft seal rings. 8. Replace transmission fluid. 9. Adjust throttle linkage. 10. Inspect and repair overrunning clutch.

(Continued)

Automatic Transmission and Transaxle Diagnosis		
Condition	**Possible Causes**	**Correction**
Slips in forward drive positions. *(Continued)*	11. Worn rear clutch. 12. Overrunning clutch worn, broken, or seized.	11. Replace discs and seals at rear clutch. 12. Replace overrunning clutch assembly.
Slips in reverse only.	1. Low reverse band misadjusted. 2. Hydraulic pressure too low. 3. Low reverse band worn out. 4. Valve body malfunction. 5. Low reverse band, servo or linkage malfunction. 6. Low fluid level. 7. Incorrect gearshift linkage adjustment. 8. Faulty oil pump. 9. Aerated fluid. 10. Worn reaction shaft support seal rings. 11. Worn front clutch.	1. Adjust low reverse band. 2. Check fluid pressure at ports. 3. Replace low reverse band. 4. Inspect valve body and repair. 5. Repair low reverse servo. Adjust reverse band and linkage. 6. Fill transmission to level. 7. Adjust gearshift linkage. 8. Replace oil pump. 9. Replace transmission fluid. 10. Replace reaction shaft support seal rings. 11. Replace discs and seals at front clutch.
Slips in all positions.	1. Hydraulic pressure too low. 2. Valve body malfunction. 3. Low fluid level. 4. Oil filter clogged. 5. Faulty oil pump. 6. Worn input shaft seal rings. 7. Aerated fluid.	1. Check fluid pressure at ports. 2. Inspect valve body and repair. 3. Fill transmission to level. 4. Replace oil filter. 5. Replace oil pump. 6. Replace input shaft seal rings. 7. Replace transmission fluid.
No drive in any position.	1. Hydraulic pressure too low. 2. Valve body malfunction. 3. Low fluid level. 4. Oil filter clogged. 5. Faulty oil pump. 6. Planetary gearsets broken or seized.	1. Check fluid pressure at ports. 2. Inspect valve body and repair. 3. Fill transmission to level. 4. Replace oil filter. 5. Replace oil pump. 6. Replace planetary gearsets.
No drive in forward drive positions.	1. Hydraulic pressure too low. 2. Valve body malfunction. 3. Low fluid level. 4. Worn input shaft seal rings. 5. Overrunning clutch not holding. 6. Worn rear clutch. 7. Planetary gearsets broken or seized. 8. Overrunning clutch worn, broken, or seized.	1. Check fluid pressure at ports. 2. Inspect valve body and repair. 3. Fill transmission to level. 4. Replace input shaft seal rings. 5. Inspect and repair overrunning clutch. 6. Replace discs and seals at rear clutch. 7. Replace planetary gearsets. 8. Replace overrunning clutch assembly.
No drive in reverse.	1. Hydraulic pressure too low. 2. Low reverse band worn out. 3. Valve body malfunction. 4. Low reverse band, servo or linkage malfunction. 5. Incorrect gearshift linkage adjustment. 6. Worn reaction shaft support seal rings. 7. Worn front clutch. 8. Worn rear clutch. 9. Planetary gearsets broken or seized.	1. Check fluid pressure at ports. 2. Replace low reverse band. 3. Inspect valve body and repair. 4. Repair low reverse servo. Adjust reverse band and linkage. 5. Adjust gearshift linkage. 6. Replace reaction shaft support seal rings. 7. Replace discs and seals at front clutch. 8. Replace discs and seals at rear clutch. 9. Replace planetary gearsets.
Drives in neutral.	1. Valve body malfunction. 2. Incorrect gearshift linkage adjustment. 3. Insufficient clutch plate clearance. 4. Worn rear clutch. 5. Rear clutch dragging.	1. Inspect valve body and repair. 2. Adjust gearshift linkage. 3. Check and adjust clutch plate clearance. 4. Replace discs and seals at rear clutch. 5. Inspect and repair rear clutch.

(Continued)

Automatic Transmission and Transaxle Diagnosis		
Condition	**Possible Causes**	**Correction**
Drags or locks.	1. Low reverse band worn out. 2. Kickdown band adjustment too tight. 3. Planetary gearsets broken or seized. 4. Overrunning clutch worn, broken or seized.	1. Replace low reverse band. 2. Adjust kickdown band. 3. Replace planetary gearsets. 4. Replace overrunning clutch assembly.
Grating, scraping, or growling noise.	1. Low reverse band worn out. 2. Kickdown band out of adjustment. 3. Replace drive shaft bushing. 4. Planetary gearsets broken or seized. 5. Overrunning clutch worn, broken, or seized.	1. Replace low reverse band. 2. Adjust kickdown band. 3. Drive shaft bushing damaged. 4. Replace planetary gearsets. 5. Replace overrunning clutch assembly.
Buzzing noise.	1. Valve body malfunction. 2. Low fluid level. 3. Aerated fluid. 4. Overrunning clutch inner race damaged.	1. Inspect valve body and repair. 2. Fill fluid to level. 3. Replace transmission fluid. 4. Replace overrunning clutch assembly.
Hard to fill, oil blows out filler hole.	1. Oil filter clogged. 2. Aerated fluid. 3. High fluid level.	1. Replace oil filter. 2. Replace transmission fluid. 3. Adjust fluid level to specifications.
Transaxle overheats.	1. Stuck cooler flow switch valve. 2. Engine idle speed too high. 3. Hydraulic pressures too low. 4. Low fluid level. 5. Incorrect gearshift linkage adjustment. 6. Faulty oil pump. 7. Kickdown band adjustment too tight. 8. Faulty cooling system. 9. Insufficient clutch plate clearance.	1. Replace switch valve behind oil pump housing. 2. Adjust engine idle to specifications. 3. Check fluid pressure at ports. 4. Fill transmission to level. 5. Adjust gearshift linkage. 6. Replace oil pump. 7. Adjust kickdown band. 8. Check cooling system temperature and repair as needed. 9. Check and adjust clutch plate clearance.
Harsh upshift.	1. Hydraulic pressures too low. 2. Incorrect throttle linkage adjustment. 3. Kickdown band out of adjustment. 4. Hydraulic pressure too high. 5. Engine performance.	1. Check fluid pressure at ports. 2. Adjust throttle linkage. 3. Adjust kickdown band. 4. Check fluid pressure at ports. 5. Set up engine to specifications.
Delayed upshift.	1. Incorrect throttle linkage adjustment. 2. Kickdown band out of adjustment. 3. Governor support seal rings worn. 4. Worn reaction shaft support seal rings. 5. Governor malfunction. 6. Kickdown band, servo or linkage malfunction. 7. Worn front clutch. 8. Engine performance.	1. Adjust throttle linkage. 2. Adjust kickdown band. 3. Replace governor support seal rings. 4. Replace reaction shaft support seal rings. 5. Inspect and repair governor. 6. Inspect and repair kickdown band, servo, or linkage. 7. Replace discs and seals at front clutch. 8. Set up engine to specifications.
No torque converter clutch application.	1. Stuck cooler flow switch valve. 2. Hydraulic pressures too low. 3. Low fluid level. 4. Faulty oil pump. 5. Worn input shaft seal rings. 6. Aerated fluid.	1. Replace switch valve behind oil pump housing. 2. Check fluid pressure at ports. 3. Fill transmission to level. 4. Replace oil pump. 5. Replace input shaft seal rings. 6. Replace transmission fluid.

Suspension, Steering, and Brakes

The suspension, steering, and brake systems literally "take a pounding" during the life of a motor vehicle. The suspension will go over thousands of bumps and holes in the road surface. The steering system is also subjected to this pounding and constant motion while reacting to driver control. The brake system must be capable of bringing thousands of pounds of steel and plastic to a quick, controlled halt.

Section 11 details the operation, construction, service, and repair of everything from tires to anti-lock brake systems. It contains information that will make you a better chassis technician. This section will also help you pass the two ASE certification tests: Test A4, *Suspension and Steering,* and Test A5, *Brakes.*

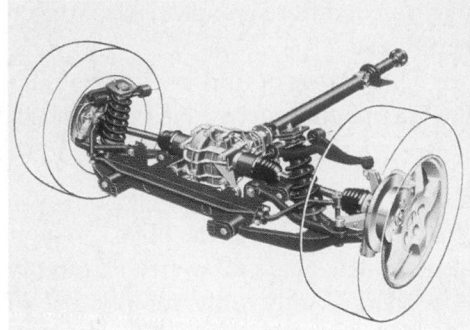

Chapter 65

Tire, Wheel, and Wheel Bearing Fundamentals

After studying this chapter, you will be able to:

- Identify the parts of a tire and wheel.
- Describe different methods of tire construction.
- Explain tire and wheel sizes.
- Describe tire ratings.
- Identify the parts of driving and nondriving hub and wheel bearing assemblies.
- Correctly answer ASE certification test questions requiring a knowledge of tires, wheels, hubs, and wheel bearings.

This chapter introduces the various tire, wheel, and hub designs used on modern vehicles. It explains how tires and wheels are constructed to give safe and dependable service. The chapter also covers hub and wheel bearing construction for both rear- and front-wheel-drive vehicles. As a result, you will be better prepared to study later chapters on brakes, suspension systems, and wheel alignment.

Tires

Automobile *tires* perform two basic functions: they act as a soft *cushion* between the road and the metal wheel, and they provide adequate *traction* (friction) with the road surface.

Tires must transmit driving, braking, and cornering forces to the road in both good and bad weather. At the same time, they should resist punctures and wear.

The tires used on early vehicles were solid rubber. Today's automotive tires are *pneumatic,* which means they are filled with air. Internal air pressure pushes out on the inside of the tire to support the weight of the vehicle.

Today's vehicles use *tubeless tires,* which do *not* have a separate inner tube. The tire and wheel form an airtight unit. Refer to **Figure 65-1.** Older vehicles used *inner tubes* (soft, thin, leakproof rubber liner) that fit inside the tire and wheel assemblies.

Parts of a Tire

Although there are several tire designs, all tires have the same basic parts, **Figure 65-2:**

- **Beads**—Two rings that are made of steel wire and encased in rubber. They hold tire sidewalls snugly against wheel rim.
- **Body plies**—Rubberized fabric and cords wrapped around beads. They form the carcass, or body, of the tire.
- **Tread**—Outer surface of the tire that contacts the road.
- **Sidewall**—Outer part of the tire that extends from the bead to the tread. Markings on the sidewall provide information about the tire.
- **Belts**—Sometimes used to strengthen the body plies and stiffen the tread. They lie between the tread and plies.
- **Liner**—Thin layer of rubber that is bonded to the inside of the plies. It provides a leakproof membrane for the modern tubeless tire.

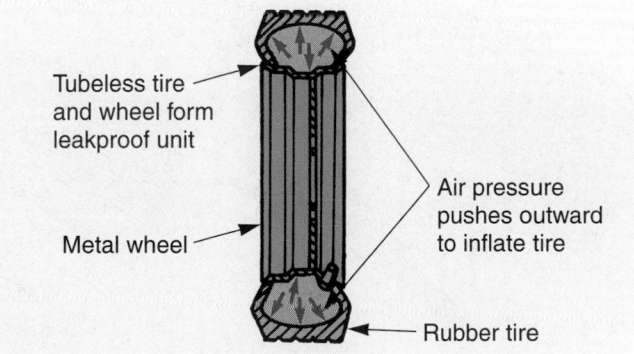

Figure 65-1. With tubeless tires, the tire and wheel form a leakproof unit. Air pressure pushes outward on the inside of the tire.

Figure 65-2. Study the basic parts of a tire. (Goodyear)

Tire Rolling Resistance

Tire rolling resistance is a measurement of the amount of friction produced as the tire operates on the road surface. A high rolling resistance would increase fuel consumption and wear. Typically, rolling resistance is reduced by higher inflation pressure, proper tire design, and a lighter vehicle.

Tire Construction

There are many construction and design variations in tires. A different number of plies may be used. The plies may run at different angles. Also, different materials may be utilized.

Three types of tires are found on late-model vehicles: the bias ply tire, belted bias tire, and radial tire.

Bias Ply Tire

In a *bias ply tire,* the plies run at an angle from bead to bead. The angle is reversed from ply to ply. The tread is bonded directly to the top ply. See **Figure 65-3A.**

A bias ply tire is one of the oldest designs, and it does *not* use belts. The position of the cords in a bias ply tire allows the body of the tire to flex easily. This tends to improve cushioning action. A bias ply tire provides a very smooth ride on rough roads.

One disadvantage of the bias ply tire is the weakness of the plies and tread, which reduces traction at high speeds and increase rolling resistance.

Belted Bias Tire

A *belted bias tire* is a bias ply tire with belts added to increase tread stiffness. The plies and belts normally run at different angles. The belts do *not* run around to the sidewalls—they lie under the tread area only. Usually, two stabilizer belts and two or more plies are used to improve tire performance. Look at **Figure 65-3B.**

A belted bias tire provides a smooth ride and good traction. It also offers some reduction in rolling resistance over a bias ply tire.

Radial Ply Tire

A *radial ply tire* has plies running straight across from bead to bead, with stabilizer belts directly beneath the tread. The belts can be made of steel, flexten, fiberglass, or other materials. A typical radial tire is illustrated in **Figure 65-3C.**

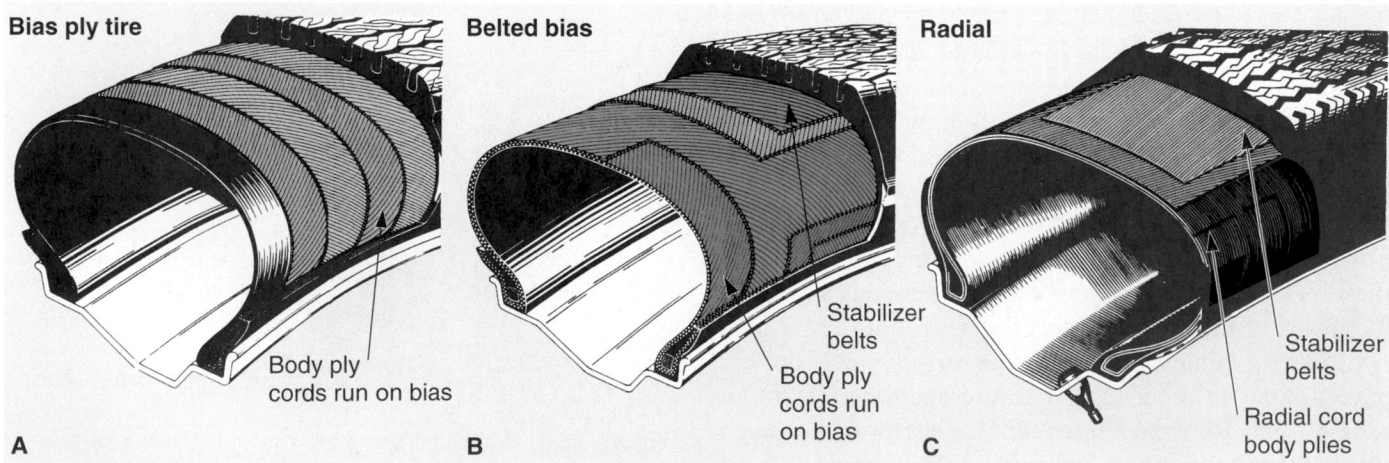

Figure 65-3. There are three basic tire types. A—Bias ply tire B—Belted bias tire. C—Radial ply tire. (Firestone)

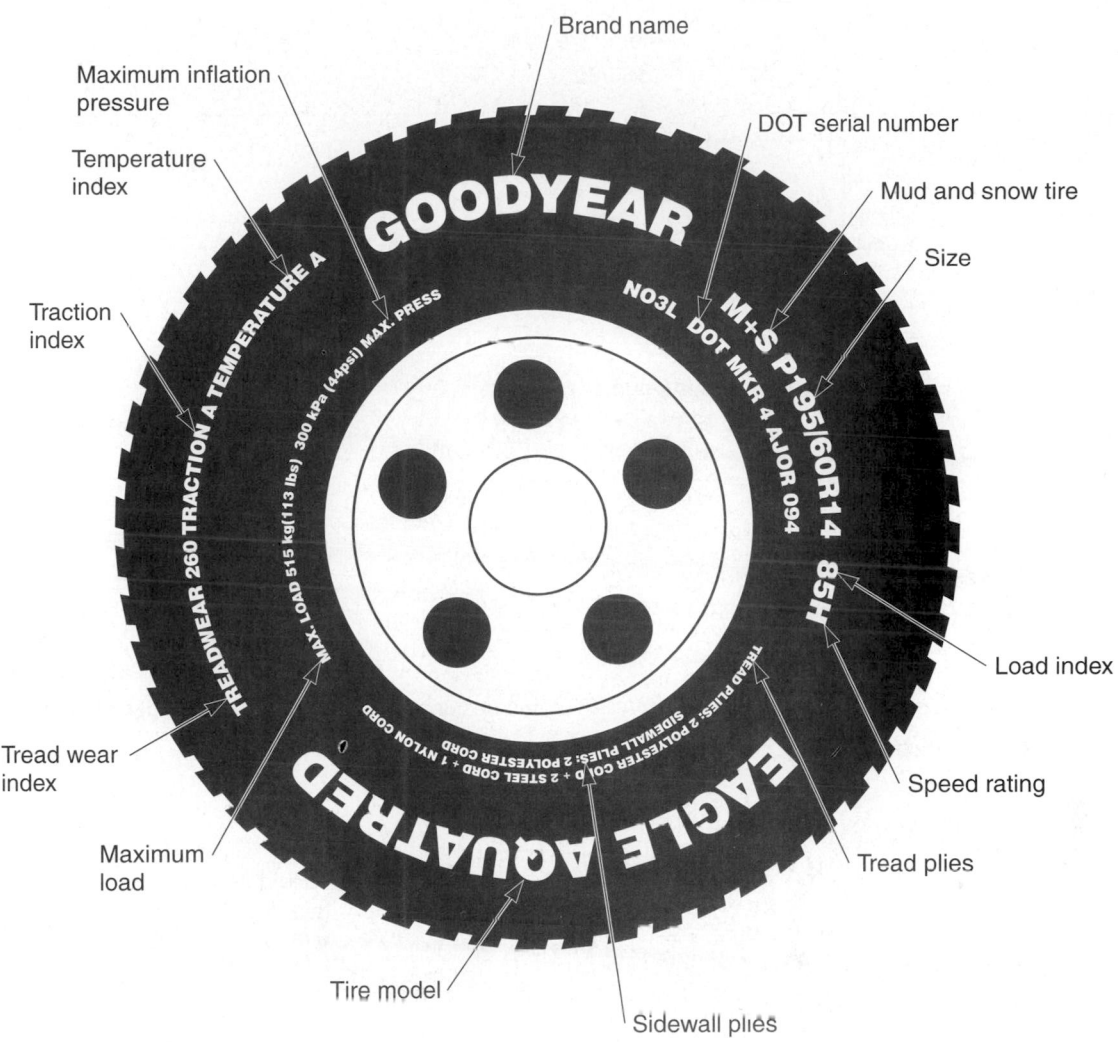

Figure 65-4. Sidewall markings contain information about a tire. Study the sidewall markings on this tire.

A radial tire has a very flexible sidewall *and* a stiff tread, giving it a very stable *footprint* (shape and amount of tread touching road surface). This improves safety, cornering, braking, and wear.

One possible disadvantage of a radial tire is that it may produce a harder, or harsher, ride at low speeds. The stiff tread area does *not* give or flex as much on rough roads.

Tire Markings

Tire markings on the sidewall of a tire give information about tire size, tire model, load-carrying ability, inflation pressure, number of plies, speed rating, manufacturer, etc. It is important that you understand these tire markings. Refer to **Figure 65-4.**

Tire Size

Tire size is given on the sidewall as a letter-number sequence. There are two common size designations:

alpha-numeric (conventional measuring system) and P-metric (metric measuring system).

The *P-metric* size designation is the newest tire identification system. It uses metric values and international standards. Look at **Figure 65-5A.**

The letter "P" indicates that the tire is designed for use on a passenger car. The first number, 155, gives section width in millimeters. The second number, 80, is the aspect ratio. The R indicates radial (B means belted bias and D means bias ply). The last number, 13, shows the rim diameter in inches (not metric values).

The *alpha-numeric* (alphabetical-numerical) tire size designation uses letters and numbers to denote tire size in inches and its load-carrying capacity in pounds. An example is given in **Figure 65-5B.**

The first letter, G, indicates the load and size relationship. The higher the letter, the larger the size and load-carrying ability of the tire. G is smaller than H, for example. The second letter, R, means that the tire is a radial. Bias and belted bias tires do not have a second

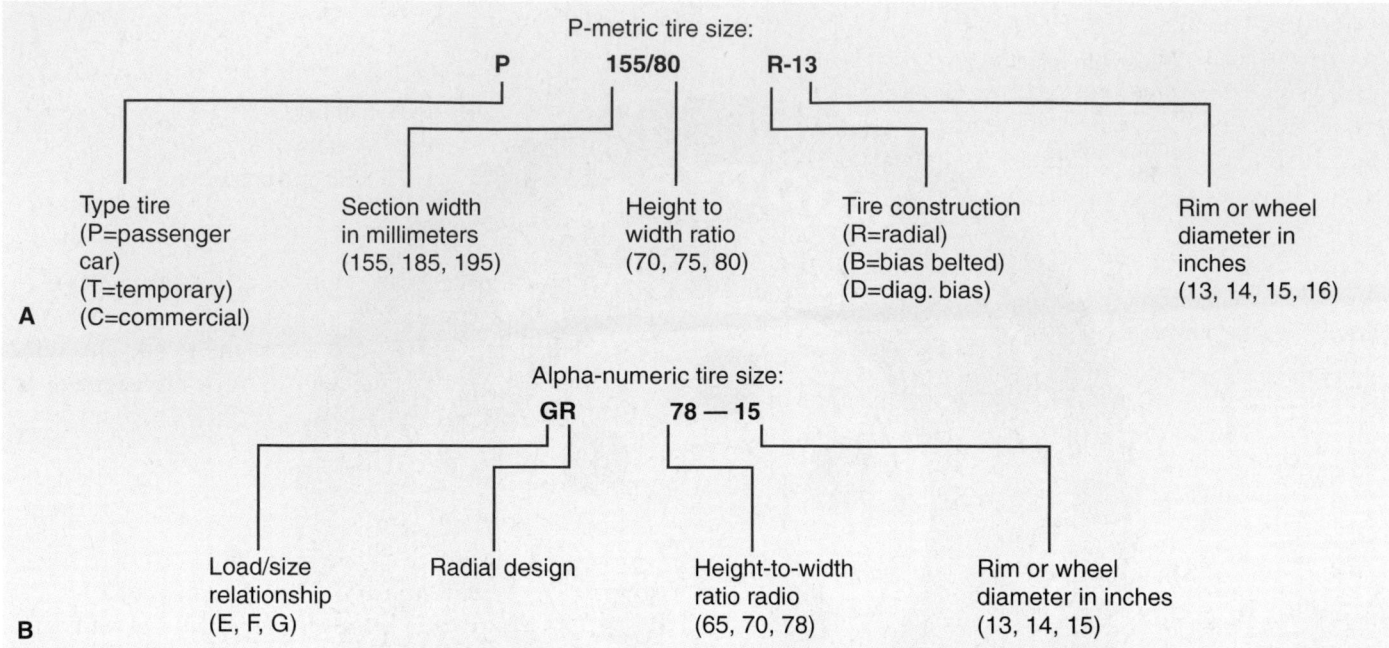

A

B

Figure 65-5. Note the two modern tire size designations. A—P-metric designation. B—Alpha-numeric designation.

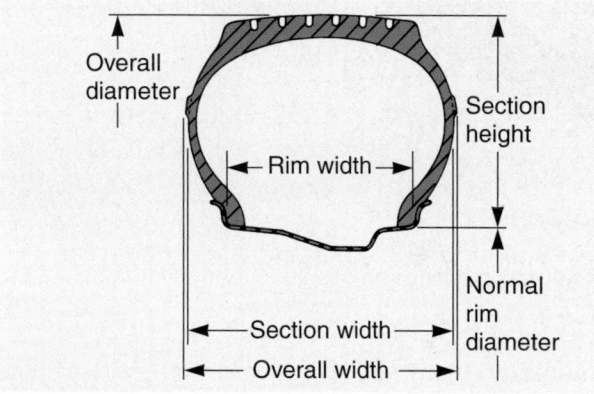

Figure 65-6. Points of measurement on a tire. These dimensions are important when ordering new tires or wheels. (B.F. Goodrich)

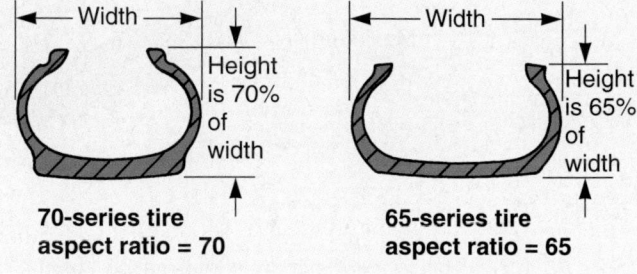

Figure 65-7. The aspect ratio is the height-to-width ratio of tire. Note that a 65 series tire is wider and shorter than a 70 series tire.

letter. The first number, 78, is the aspect, or height-to-width, ratio. The last number, 15, is the rim diameter in inches. The points of measure for a tire are given in **Figure 65-6.**

Aspect Ratio

The *aspect ratio,* or height-to-width ratio, in the tire size designation is the most difficult value to understand. See **Figure 65-7.**

Note that, as the number becomes smaller, the tire becomes more *squat* (wider and shorter). The aspect ratio is the comparison of the tire's height (bead-to-tread distance) and width (sidewall-to-sidewall distance).

A 70-series tire, for example, has an aspect ratio of 70; the height of the tire is 70 percent of the width. A 60 series tire would be "short" and "fat." A 78 tire would be "narrower" and "taller."

Maximum Load Rating

The *maximum load rating* of a tire indicates the amount of *weight* the tire can carry at the recommended inflation pressure. The maximum load value—1500 pounds, for example—is printed on the sidewall.

With P-metric tire designations, the load rating is simply given in kilograms and pounds. The alpha-numeric load rating is indicated by a letter. Most alpha-numeric size tires are load range B. They are restricted to the load specified at 32 psi (220 kPa). Where a greater load carrying ability is needed, load range C or D tires are used.

Maximum Inflation Pressure

The *maximum inflation pressure* is the highest air pressure that should be induced into the tire. Many tires have a maximum recommended inflation pressure of 32 psi (220 kPa). However, tires with higher load ranges can hold higher pressures and carry more weight.

Tread Plies

The tire sidewall also includes information on the number of plies and ply rating. For example, the tire may be a 2-ply tire, a 2-ply with a 4-ply rating (plies made stronger than normal), or 4-ply tire. A greater number of plies or higher ply rating generally indicates a greater load-carrying ability.

Sidewall Plies

The number of sidewall plies is also shown on the tire sidewall.

DOT Serial Number

DOT stands for *Department Of Transportation.* When you see "DOT" on the tire sidewall, the tire has passed prescribed safety tests.

Following the letters DOT is the *DOT serial number,* which identifies the particular tire manufacturer, plant location, construction, and date of manufacture. The DOT number is stamped into the tire sidewall.

Tire Grades

Tread wear, traction, and temperature grades are normally shown on the tire sidewall according to the Uniform Tire Quality Grading System.

Tread wear is given as a number. Tread wear ratings range from 100 to 500. The higher the number, the more resistant the tire is to wear.

Tire traction is given an A, B, or C rating. A tire with an A rating would provide the most traction, while a tire with a C rating would provide the least traction.

Tire temperature resistance is also given an A, B, or C rating. A tire with an A rating resists a temperature buildup better than a tire with a B or C rating.

Speed Rating

A *tire speed rating* is the maximum allowable sustained road speed a tire can safely withstand without failure. Speed ratings range from B (31 mph or 50 km/h) to Z (149 mph or 238 km/h).

It is important that the speed rating of a tire be higher than the speed at which the vehicle will be driven. If too low a speed rating is used, the tire could fail and cause a fatal accident. This is extremely important if the vehicle is to be driven in areas where there is no speed limit (some states and foreign countries).

Special Tires and Tire Features

You should be familiar with several types of special tires and tire features. The automotive technician will encounter compact spare tires, self-sealing tires, retreads, and run-flat tires.

Wear Bars

Wear bars are used to indicate a critical amount of tread wear. When too much tread has worn from the tires, solid rubber bars will show up across the tread. This tells the customer and the technician that tire replacement is needed. Look at **Figure 65-8.**

Spare Tires

A *spare tire,* or spare, is an extra wheel and tire assembly that can be installed if the vehicle has a flat tire. The spare can be a full-size tire or a compact tire.

A *full-size spare* is a tire of the same size and type used on the other four wheels of the vehicle. It can be mounted and used like any other tire. It is being replaced by compact, lighter spare tires.

Most new cars use a *compact spare tire,* or *space saver spare,* which is much smaller than the normal tire used on the vehicle. It saves space in the trunk or storage area. See **Figure 65-9.**

A *high-pressure spare* is a compact spare tire that holds higher-than-normal air pressure (typically about 60 psi). It is inflated when in the trunk and is ready for use.

Some spare tires are *not* inflated when in storage. A small bottle of compressed air may be used to inflate the tire when needed.

A *lightweight spare tire* has very thin sidewall and tread construction. Being light, it can be more easily picked up and mounted on the vehicle. However, the

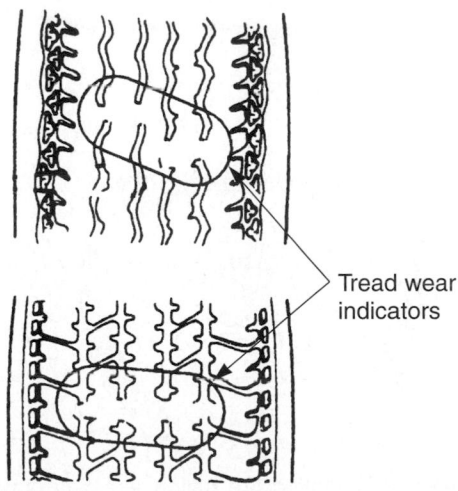

Tread wear indicators

Figure 65-8. When wear bars are visible, a tire is worn enough to be unsafe and should be replaced. (General Motors)

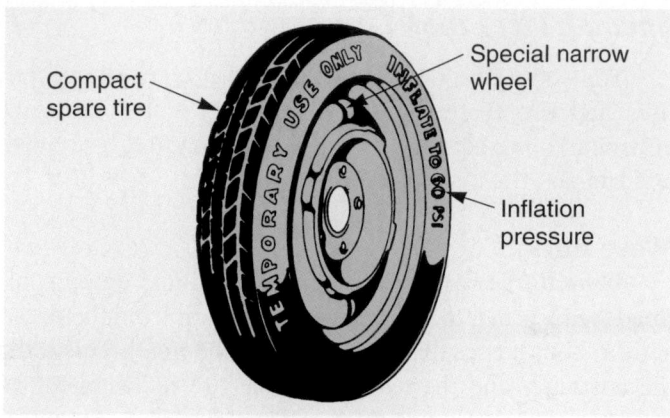

Figure 65-9. A compact spare tire is for temporary use only. This one should be inflated to 60 psi (415 kPa). It requires a special narrow wheel. (Oldsmobile)

driving range and maximum speed rating of a lightweight spare are generally lower than those for a compact spare.

Tech Tip

Most modern spare tires are designed for *temporary use only*. Refer to the manufacturer's specifications on inflation pressure, maximum driving speed, and the number of miles that can be driven. Most compact spares should not be used at speeds over 50 mph and will only last a few thousand miles.

Self-Sealing Tires

Some tires are *self-sealing* (seal small punctures). These tires have a coating of sealing compound applied to their liners. If a nail punctures a self-sealing tire, air pressure will push the soft compound into the hole to stop air leakage. Refer to **Figure 65-10.**

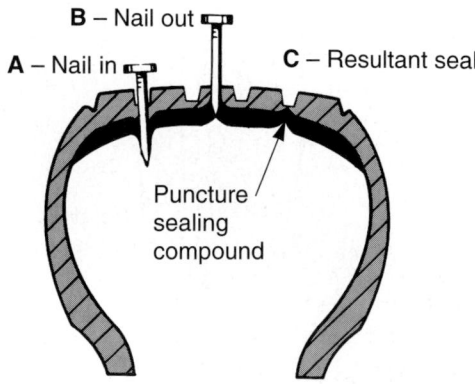

Figure 65-10. Self-sealing tire action. A—Nail punctures the tire. B—Nail is pulled out. C—Sealing compound flows into the hole to prevent air from leaking out. (GMC)

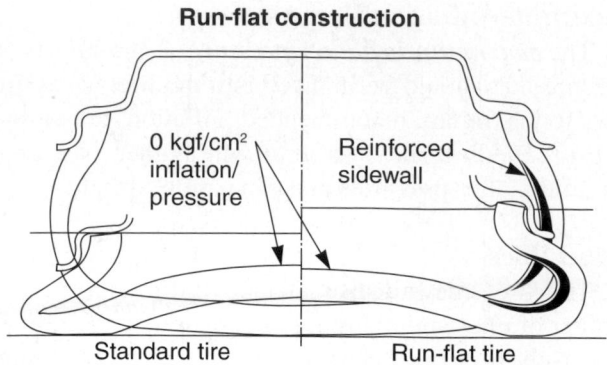

Figure 65-11. Compare the construction of a run-flat tire to that of a conventional tire. (Bridgestone)

Retreads

Retreads are used tires that have had a new tread vulcanized (applied using heat and pressure) to the old carcass, or body. Retreads, also called recaps, are seldom used on passenger cars. However, large truck tires are frequently recapped because of the high cost of new truck tires.

Run-Flat Tires

Run-flat tires have an extremely stiff sidewall construction so that they are still usable with a loss of air pressure. If the tire leaks, you can drive the vehicle to a repair shop without tire and wheel damage. See **Figure 65-11.**

The tire will retain most of its shape because the sidewall is strong enough to support vehicle weight. The tire uses a special rubber compound and a thick rubber sidewall support insert that helps support the weight of the vehicle.

Tire Inflation Monitoring System

Tire inflation monitoring systems can detect both high and low tire pressure and are often used with run-flat tires. In these systems, wheel sensors are mounted on each wheel, **Figure 65-12.** If tire pressure is not correct, the wheel sensor produces a radio signal. The signal is picked up by a small receiver that turns on a dash warning light to inform the driver of the inflation problem, **Figure 65-13.**

Wheels

Wheels must be designed to support the tire while withstanding loads from acceleration, braking, and cornering. Most wheels are made of steel. A few optional types are cast aluminum or magnesium. Refer to **Figure 65-14.**

Mag wheels, or *mags,* is a nickname for aluminum or magnesium wheels. They do not need wheel covers. See **Figure 65-15.**

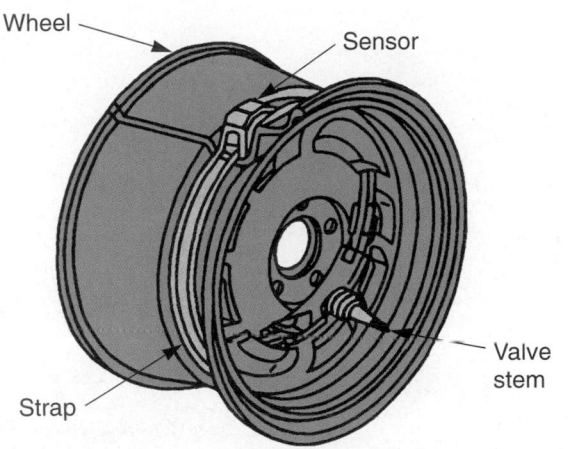

Figure 65-12. A sensor is strapped to the inside of the wheel to monitor inflation pressure. (Chevrolet)

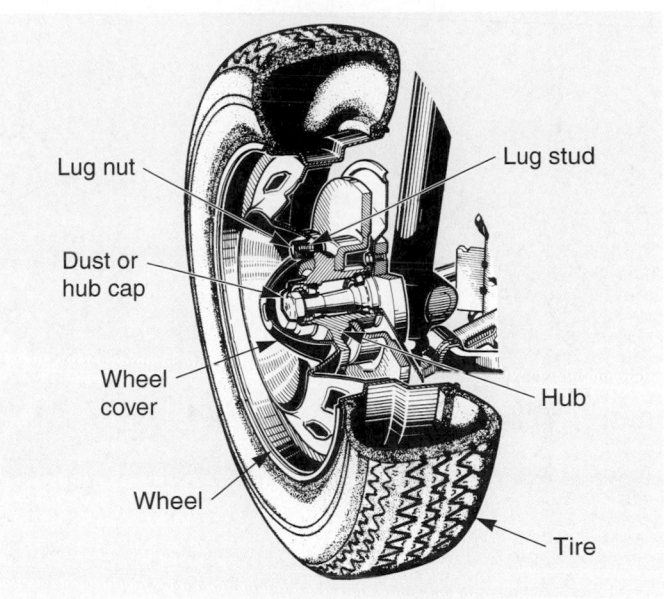

Figure 65-14. Cutaway shows many of the components relating to a wheel. (Peugeot)

A *drop-center wheel* is commonly used on passenger vehicles because it allows for easier installation and removal of the tire. Since the center of the wheel is smaller in diameter (dropped) than the rim, the tire bead can fall into the recess. Then, the other side of the tire bead can be forced over the rim for removal. See **Figure 65-16.**

A standard wheel consists of the *rim* (outer lip that contacts tire bead) and the *spider* (center section that bolts to vehicle hub). Normally, the spider is welded to the rim.

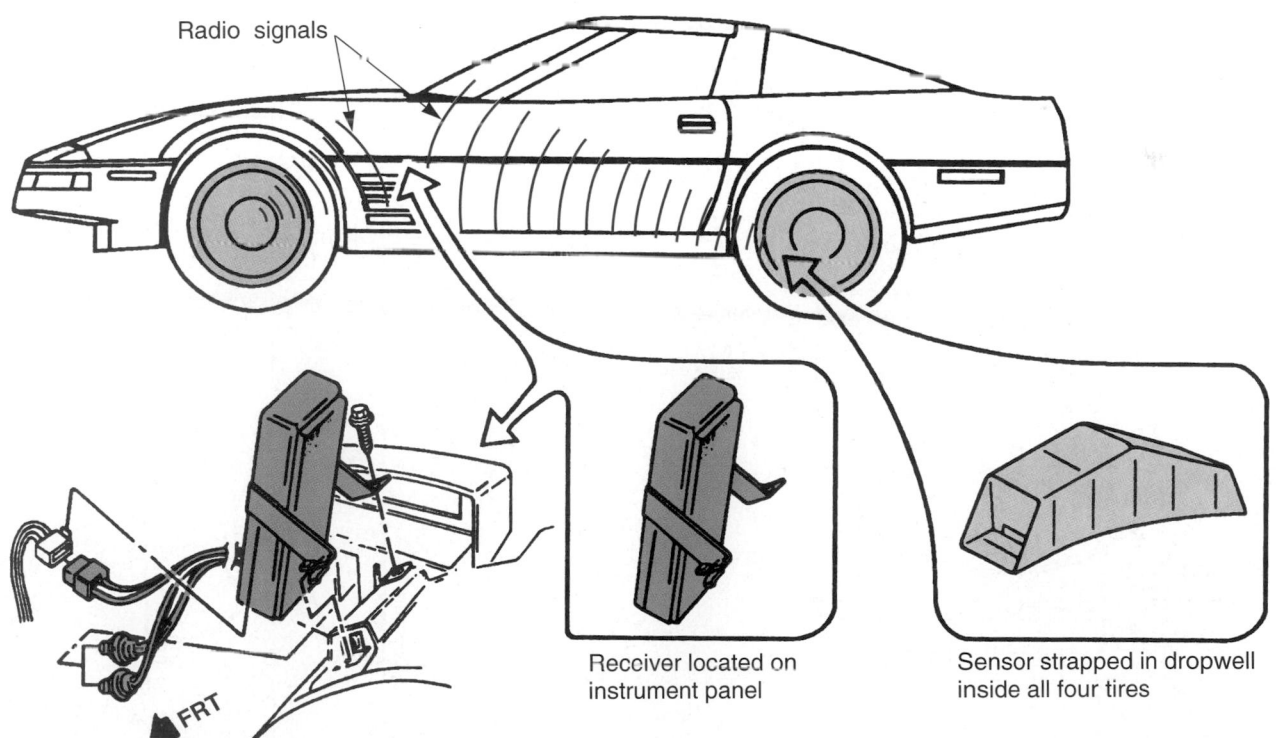

Figure 65-13. If the wheel sensor detects low tire pressure, it will produce a radio signal. A receiver in the passenger compartment is tuned to detect the signal and trigger a warning light in the dash. (Chevrolet)

Figure 65-15. Aluminum or magnesium wheels are often called "mags." They do not need wheel covers. (Dodge)

Figure 65-17 illustrates the various dimensions of a wheel. Compare this illustration with **Figure 65-16.**

Safety Rims

A *safety rim* has small ridges that hold the tire beads on the wheel during a tire *blowout* (instant rupture and air loss) or a *flat* (slow leak reduces inflation pressure).

Small raised lips around the rim keep the tire beads from sliding into the drop-center section. This improves safety by keeping the tire from coming off the wheel. See **Figure 65-18.**

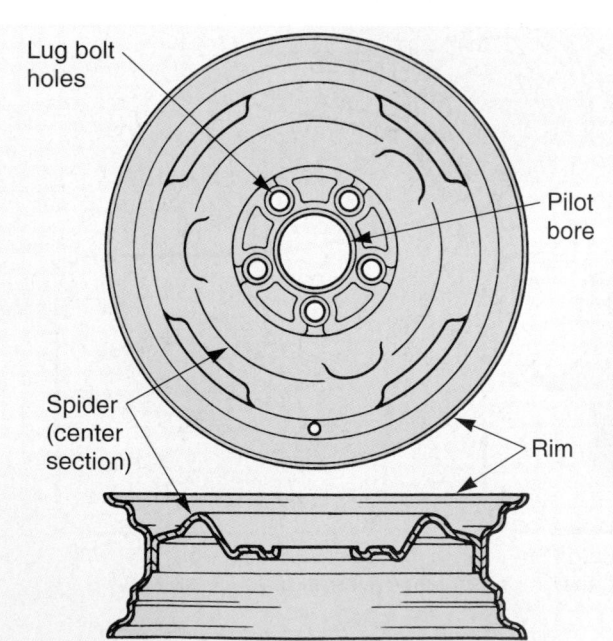

Figure 65-16. Study the parts of this conventional drop-center wheel.

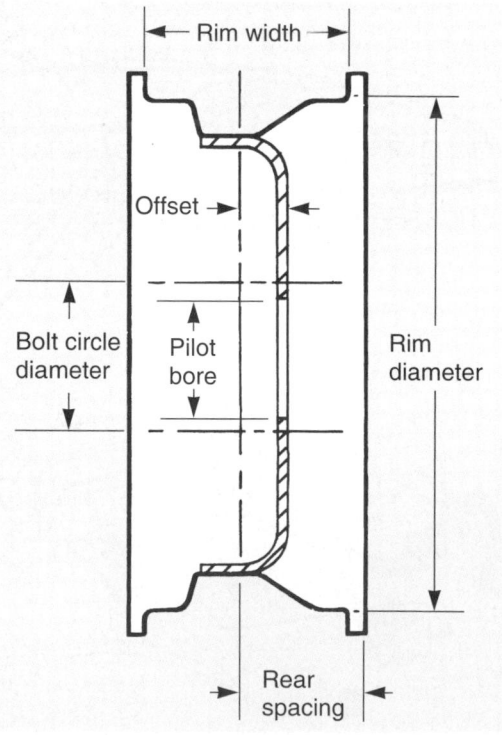

Figure 65-17. Study the basic dimensions of a wheel. (Goodyear)

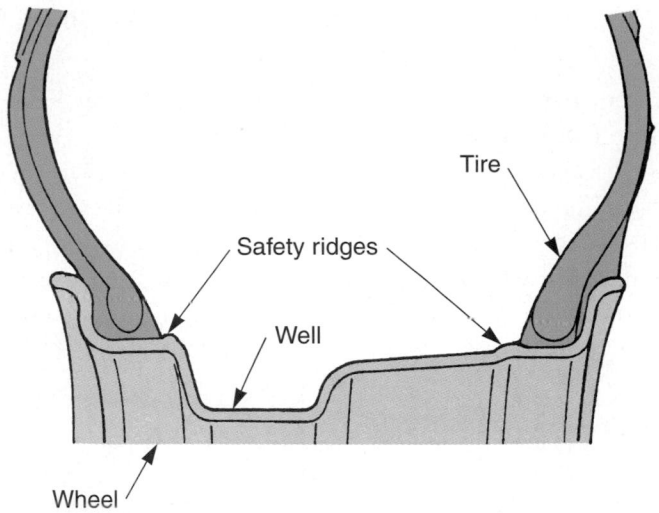

Figure 65-18. Safety ridges prevent the tire from dropping into the center well in the event of a flat or a blowout. (Chrysler)

Valve Stems and Cores

A *valve stem* is pressed into a hole in the wheel of a tubeless tire to allow inflation and deflation. The stem is made of rubber. A threaded metal tube is formed in the end of the stem. **See Figure 65-19.**

The *valve core* is a spring-loaded air valve that is threaded into the valve stem, **Figure 65-19.** The valve core allows air to be added to the tire when an *air chuck* (tool for filling tire with air) is placed over the valve stem. When the air chuck is removed from the valve stem, the spring pushes the valve closed to prevent air leakage.

To remove air from the tire, push the center of the valve core inward. The valve will open, allowing air to escape. A *valve stem cap* screws over the threaded valve stem to

protect the air valve and stem threads from dirt, moisture, and damage. It also helps prevent accidental depression of the valve and the resulting loss of air pressure.

Lug Nuts, Studs, and Bolts

Lug nuts hold the wheel and tire assembly on the vehicle. They fasten to special studs. The inner face of the lug nut is tapered to help center the wheel on the hub. Refer to **Figure 65-20A.**

Lug studs are the special studs that accept the lug nuts. The studs are pressed through the back of the hub or axle flange. See **Figure 65-20B.**

Normally, the lug nuts and studs have right-hand threads (turn clockwise to tighten). When left-hand threads are used, the nut or stud will be marked with an "L." Metric threads will be identified with an "M" or the word "Metric."

A few cars use *lug bolts* instead of lug nuts. The bolts screw into threaded holes in the hub or axle flange.

Wheel Weights

Wheel weights are small lead weights that are attached to the wheel rim to balance the wheel-and-tire assembly, preventing vibration. The weights are used to offset a heavy area of the wheel and tire.

Hub and Wheel Bearing Assemblies

Wheel bearings allow the wheel to turn freely around the spindle, in the steering knuckle, or in the bearing support. Most wheel bearings are either tapered roller bearings or ball bearing, **Figure 65-21.**

The wheel bearings are lubricated with heavy, high-temperature grease. This lets the elements (rollers or balls) operate with very little friction and wear.

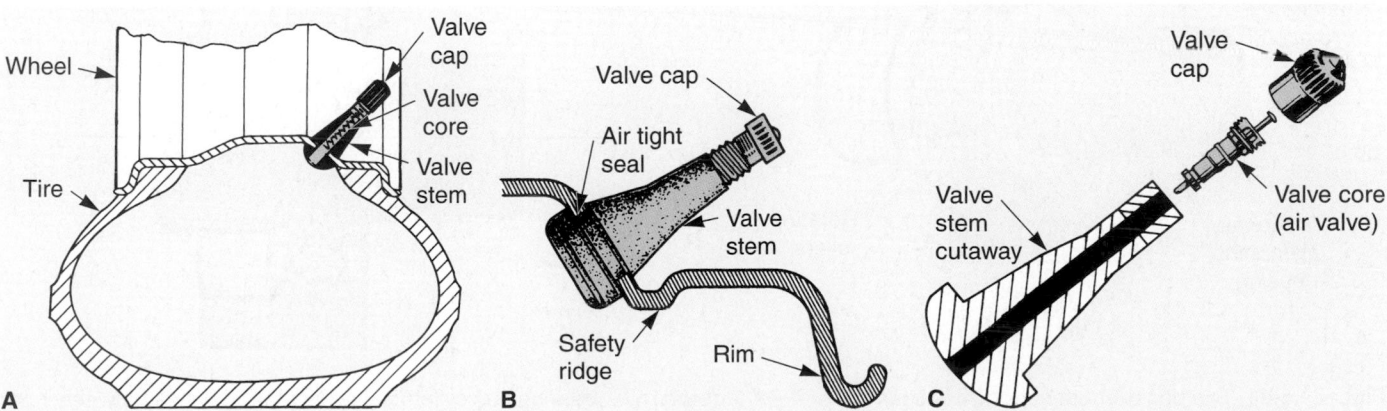

Figure 65-19. A—Valve stem snaps into a hole in the wheel. B—The press fit between the stem and the wheel forms an airtight seal. C—The valve core is an air valve that screws into the valve stem body. The valve cap screws over the end of the stem. (Toyota)

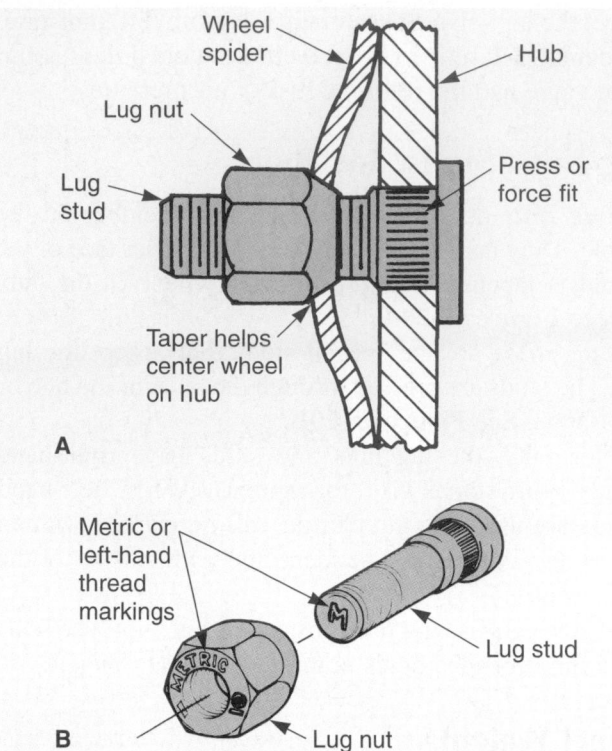

Figure 65-20. A—The lug nut screws onto a lug stud to secure the wheel to the hub. The tapered end of the nut must contact the wheel. The stud is pressed into the hub. B—If metric or left-hand threads are used, markings will normally be given on the nut or stud. (Cadillac)

The basic parts of a wheel bearing are:

- *Outer race*—cup or cone pressed into hub, steering knuckle, or bearing support.

- *Balls or rollers*—antifriction elements that fit between inner and outer races.

- *Inner race*—cup or cone that rests on spindle or drive axle shaft.

There are two basic hub and wheel bearing designs: those for the nondriving wheels and those for driving wheels. The front wheels on a rear-wheel-drive car are nondriving wheels. The front wheels on a front-wheel-drive car are driving wheels (hubs transfer power to the wheels and tires).

Hub and Wheel Bearing Assembly— Nondriving Wheels

The basic parts found in a *hub and wheel bearing assembly* for a vehicle's *nondriving wheels* are illustrated in **Figure 65-22:**

- *Spindle*—stationary shaft extending outward from the steering knuckle or the suspension system.

- *Wheel bearings*—usually tapered roller bearings mounted on the spindle and in the wheel hub.

- *Hub*—outer housing that holds the brake disc or drum, front wheel, grease, and wheel bearings.

- *Grease seal*—seal that prevents the loss of lubricant from the inner end of the spindle and hub.

- *Safety washer*—flat washer that keeps the outer wheel bearing from rubbing on and turning the adjusting nut.

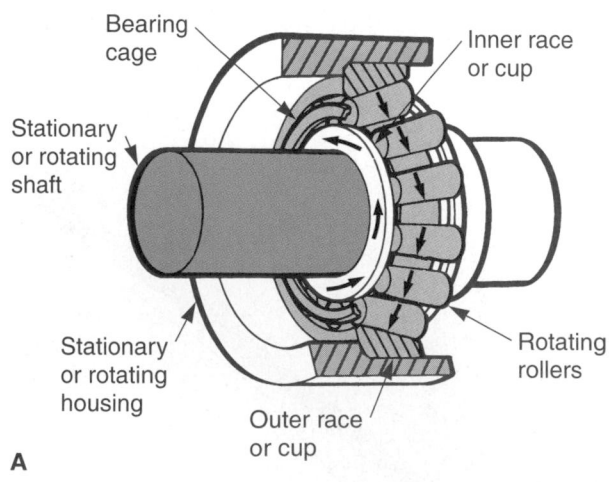

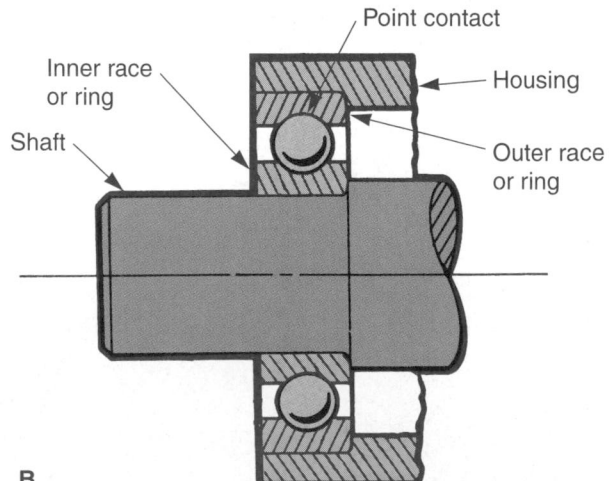

Figure 65-21. Two basic wheel bearing configurations. A—A tapered roller bearing has cylindrical rollers that operate between inner and outer races. If the shaft is stationary, the bearing will allow the outer housing or hub to turn. If outer bearing mount is stationary, the shaft or axle can turn in the bearing. B—Ball bearings are also used as wheel bearings, especially on a vehicle's driving wheels. The balls allow parts to rotate with a minimum amount of friction and wear. (Federal Mogul)

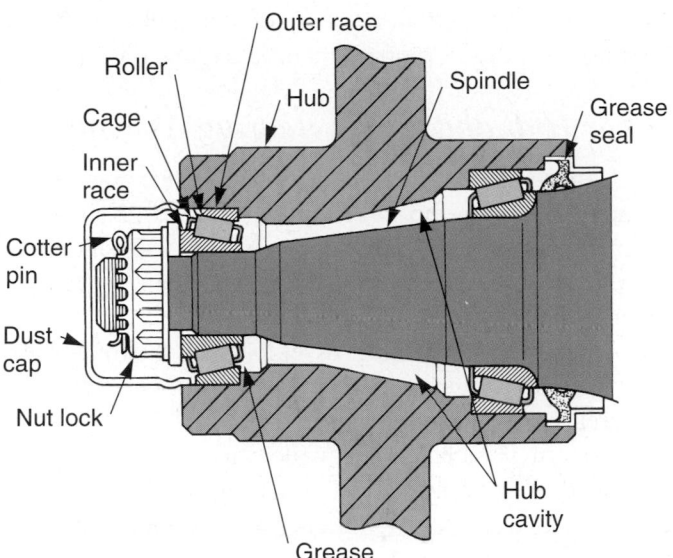

Figure 65-22. Typical nondriving, or freewheeling, hub and wheel bearing assembly for the front or rear of a car. Two tapered roller bearings allow the hub and wheel to revolve around the stationary spindle. Grease partially fills the hub to lubricate the bearings. An inner seal prevents the loss of grease. A nut on the end of the spindle allows adjustment of the bearing preload. (Chrysler)

- *Spindle adjusting nut*—nut that is threaded on end of the spindle for adjusting the wheel bearing.
- *Nut lock*—thin, slotted nut that fits over the main spindle nut.
- *Cotter pin*—soft metal pin that fits through a hole in the spindle, adjusting nut, and nut lock to keep the adjusting nut from turning in service.
- *Dust cap*—metal cap that fits over the outer end of the hub to keep grease in and road dirt out of bearings.

Since this hub and wheel bearing assembly does *not* transfer driving power, the spindle is stationary. It simply extends outward and provides a mounting place for the wheel bearings, hub, and wheel. When the vehicle is moving, the wheel and hub spin on the bearings and spindle. The hub simply freewheels. The hub is partially filled with grease to lubricate the bearings.

Figure 65-23 shows a disassembled view of a nondriving front bearing and hub assembly. Compare these parts to the ones in **Figure 65-22.**

Hub and Wheel Bearing Assembly— Driving Wheels

Figure 65-24 illustrates the basic parts of a typical hub and wheel bearing assembly for a vehicle's driving wheels:

- *Outer drive axle*—stub axle shaft that extends through the bearings and is splined to the hub.
- *Ball or roller bearings*—antifriction elements that allow the drive axle to turn in the steering knuckle or bearing support.
- *Steering knuckle* or *bearing support*—steering or suspension component that holds the wheel bearings, axle stub shaft, and hub.
- *Drive hub*—mounting place for the wheel; transfers driving power from the stub axle to the wheel.
- *Axle washer*—special washer that fits between the hub and locknut.
- *Hub* or *axle locknut*—nut that screws on the end of the drive axle stub shaft to secure the hub and other parts of assembly.

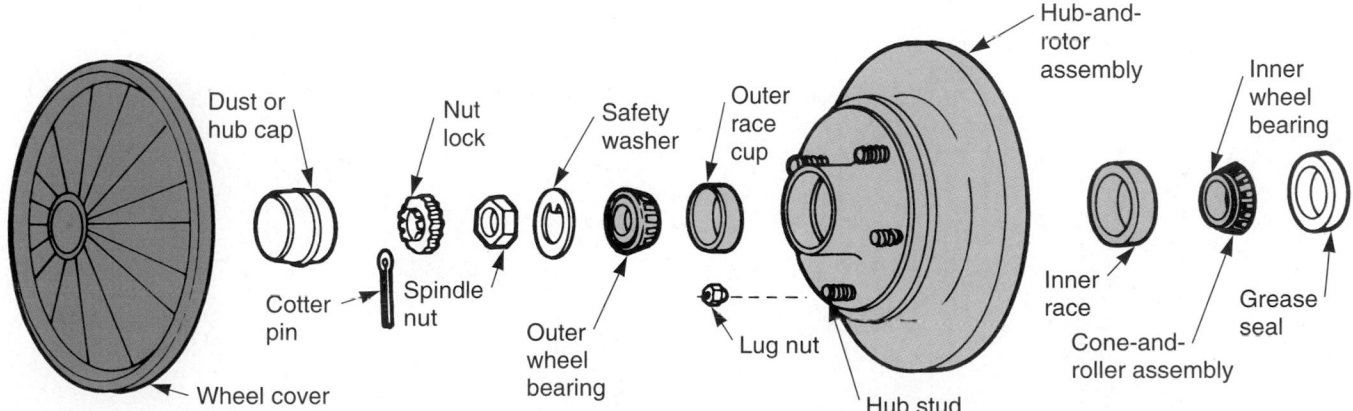

Figure 65-23. Disassembled view of a nondriving hub and wheel bearing assembly. Note the part names and relationships. This type of assembly can be used on the front of rear-wheel-drive cars or the rear of front-wheel-drive vehicles. (Florida Dept. of Voc. Ed.)

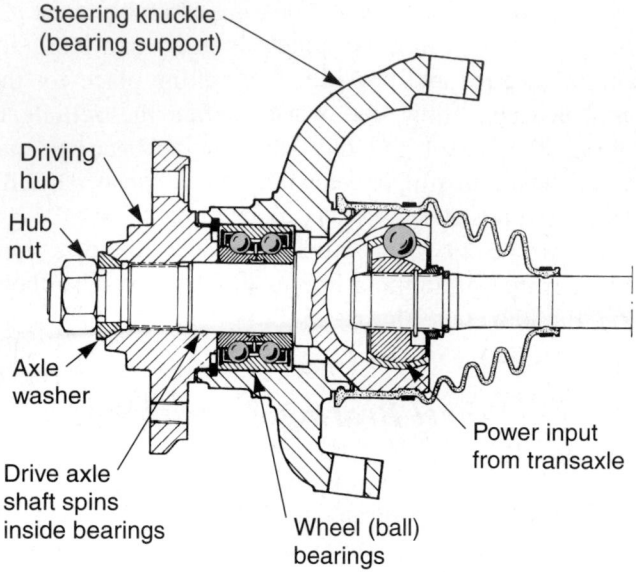

Figure 65-24. Driving hub and wheel bearing assembly has bearings mounted in a stationary steering knuckle or a bearing support. The drive axle shaft fits through the center of the bearings. The hub is splined to axle shaft. The ball bearings are lubricated by thick, high-temperature grease.

- *Grease seal*—prevents lubricant loss between the inside of the axle and the knuckle or bearing support.

As you can see in **Figure 65-24,** a hub and wheel bearing assembly for a driving wheel is very different from a nondriving unit. Instead of a stationary spindle, the axle shaft spins inside a stationary support. **Figure 65-25**

shows an exploded view of a driving hub and wheel bearing assembly. Compare them to **Figure 65-24.**

Other Hub and Wheel Bearing Assemblies

A four-wheel drive hub and wheel bearing assembly is shown in **Figure 65-26.** Note that it has a driving axle extending through a stationary spindle. A special freewheel, or locking, hub transfers power from the axle to the hub-disc assembly.

A rear wheel bearing assembly for a front-wheel drive vehicle is almost identical to a front wheel bearing for a rear-wheel drive vehicle.

Modern vehicles use a wide variation of hub and wheel bearing assemblies. This is due to the increased use of front-wheel drive. When you need more information on a specific vehicle, always refer to the factory service manual. It will explain and illustrate the hub and wheel bearing assembly clearly. Most designs, however, will be similar.

Summary

- Automobile tires perform two basic functions: they act as a soft *cushion* between the road and the metal wheel. Tires must also provide adequate *traction* (friction) with the road surface.

- Car tires are pneumatic which means they are filled with air.

- Today's vehicles use tubeless tires that do *not* have a separate inner tube. The tire and wheel form an airtight unit.

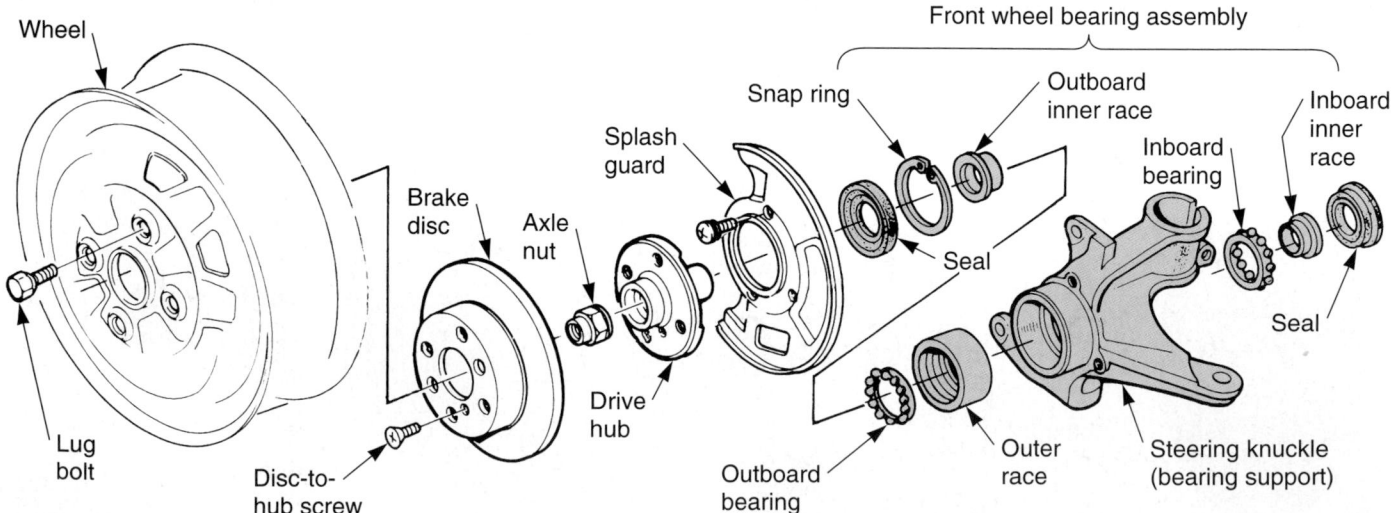

Figure 65-25. Disassembled view of a driving hub and wheel bearing assembly. Study the names of the parts. This type of assembly is commonly used on the front of front-wheel-drive vehicles. However, it can also be found on rear-engine, rear-wheel-drive sports cars. (Honda)

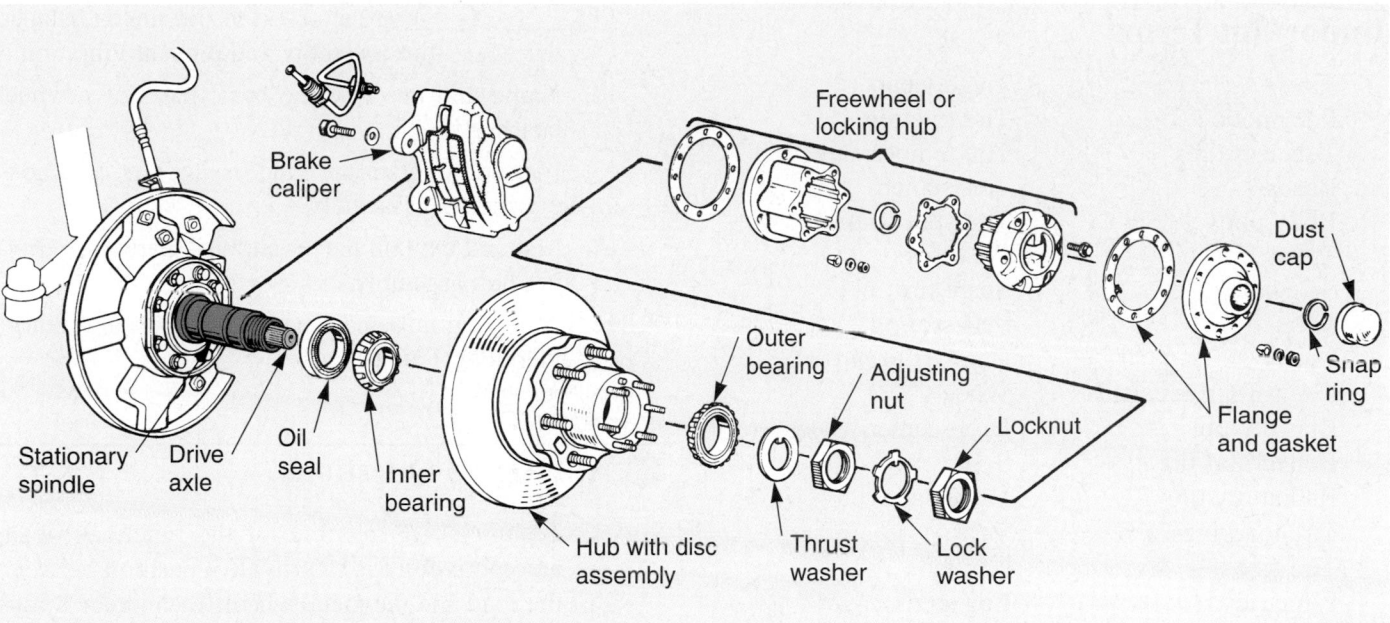

Figure 65-26. This front hub and wheel bearing assembly is used in a four-wheel-drive vehicle. Compare this unit to those shown earlier. Note how the drive axle sticks through a stationary spindle. An adjustable hub allows the drive axle to be connected and disconnected from the hub and wheel assembly for two- or four-wheel-drive operation. (Toyota)

- Tire rolling resistance is a measurement of the amount of friction produced as the tire operates on the road surface.

- A bias ply tire has plies running at an angle from bead to bead.

- A belted bias tire is a bias tire with belts added to increase tread stiffness.

- A radial ply tire has plies running straight across from bead to bead with stabilizer belts directly beneath the tread.

- Tire markings on the sidewall of a tire give information about tire size, load carrying ability, inflation pressure, number of plies, identification numbers, quality ratings, and manufacturer.

- Tire size is given on the sidewall as a letter-number sequence.

- The aspect ratio, or height-to-width ratio, is a comparison of the tire's height and width.

- The maximum load rating of a tire indicates the amount of *weight* the tire can carry at recommended inflation pressure.

- The maximum inflation pressure, printed on the tire sidewall, is the highest air pressure that should be induced into the tire.

- A tire speed rating is the maximum allowable sustained road speed the tire can safely withstand without failure.

- Some tires are self-sealing (seal small punctures) because of a coating of sealing compound applied to the tire liner.

- Retreads are old, used tires that have had a new tread vulcanized (applied using heat and pressure) to the old carcass or body.

- Run-flat tires have an extremely stiff sidewall construction so that they are still usable with a loss of air pressure.

- A tire inflation monitoring system can detect high or low tire pressure and is often used with run-flat tires.

- A drop-center wheel is commonly used on passenger vehicles because it allows for easier installation and removal of the tire.

- A valve stem snaps into a hole in the wheel of a tubeless tire to allow inflation and deflation.

- Lug nuts hold the wheel and tire assembly on the vehicle.

- Wheel weights are small lead weights attached to the wheel rim to balance the wheel-tire assembly and prevent vibration.

Important Terms

Tires
Pneumatic
Tubeless tires
Beads
Body plies
Tread
Sidewall
Belts
Liner
Tire rolling resistance
Bias ply tire
Belted bias tire
Radial ply tire
Tire markings
Tire size
P-metric
Alpha-numeric
Aspect ratio
Maximum load rating
Maximum inflation
  pressure
DOT serial number

Tread wear
Tire traction
Tire temperature
  resistance
Tire speed rating
Wear bars
Spare tire
Self-sealing
Run-flat tires
Wheels
Drop-center wheel
Safety rim
Valve stem
Valve core
Lug nuts
Lug studs
Lug bolts
Wheel weights
Wheel bearings
Hub and wheel bearing
  assembly

Review Questions—Chapter 65

Please do not write in this text. Place your answers on a separate sheet of paper.

1. What are the two basic functions of a tire?

2. Car tires are _____ which means that they are filled with air.

3. List and explain the six major parts of a tire.

4. Tire _____ _____ is a measurement of the amount of friction produced as the tire operates on the road surface.

5. Which of the following is *not* a type of tire commonly used on modern passenger cars?
   (A) Radial.
   (B) Bias ply.
   (C) Lateral ply.
   (D) Belted bias.

6. What information is commonly given on the tire sidewall?

7. A typical tire inflation pressure would be 22 psi (152 kPa). True or False?

8. How does a self-sealing tire work?

9. A(n) _____ _____ has small ridges that hold the tire on the wheel during a blowout.

10. Explain why a valve core is needed.

11. _____ _____ are attached to the rim to balance the wheel-tire assembly and prevent vibration.

12. Name and describe the basic parts of a wheel bearing.

13. List and explain the nine basic parts of a non-driving hub assembly.

14. List and explain the seven basic parts of a driving hub assembly.

15. A driving hub and a nondriving hub are almost identical. True or False?

✺ ASE-Type Questions

1. Technician A says one of the functions of an automotive tire is to provide a cushion between the road and the metal wheel. Technician B says one of the functions of an automotive tire is to provide adequate traction with the road surface. Who is right?
   (A) A only.
   (B) B only.
   (C) Both A and B.
   (D) Neither A nor B.

2. Technician A says automotive tires are pneumatic. Technician B says modern automotive tires are not pneumatic. Who is right?
   (A) A only.
   (B) B only.
   (C) Both A and B.
   (D) Neither A nor B.

3. Technician A says older vehicles use tires equipped with inner tubes. Technician B says older vehicles came from the factory equipped with tubeless tires. Who is right?
   (A) A only.
   (B) B only.
   (C) Both A and B.
   (D) Neither A nor B.

4. Technician A says an automotive tire's beads are used to stiffen the tire treads. Technician B says an automotive tire's beads are used to hold the sidewalls against the wheel rim. Who is right?
   (A) A only.
   (B) B only.
   (C) Both A and B.
   (D) Neither A nor B.

5. Technician A says an automotive tire's body plies provide a leakproof membrane for modern tubeless tires. Technician B says an automotive

tire's liner provides a leakproof membrane for modern tubeless tires. Who is right?
(A) A only.
(B) B only.
(C) Both A and B.
(D) Neither A nor B.

6. Technician A says an automotive tire's belts form the body of the tire. Technician B says an automotive tire's belts are sometimes used to strengthen the plies and stiffen the treads. Who is right?
(A) A only.
(B) B only.
(C) Both A and B.
(D) Neither A nor B.

7. Technician A says tire "rolling resistance" is normally reduced by lower inflation pressure. Technician B says tire "rolling resistance" is normally reduced by higher inflation pressure. Who is right?
(A) A only.
(B) B only.
(C) Both A and B.
(D) Neither A nor B.

8. Technician A says "belted bias" tires are used on late-model vehicles. Technician B says "radial" tires are used on late-model vehicles. Who is right?
(A) A only.
(B) B only.
(C) Both A and B.
(D) Neither A nor B.

9. Technician A says a "bias ply tire" is one of the newest tire designs and is equipped with belts. Technician B says a "bias ply tire" is one of the oldest tire designs and does not use belts. Who is right?
(A) A only.
(B) B only.
(C) Both A and B.
(D) Neither A nor B.

10. Technician A says a radial ply tire's belts are sometimes made of steel. Technician B says a radial ply tire's belts are sometimes made of fiberglass. Who is right?
(A) A only.
(B) B only.
(C) Both A and B.
(D) Neither A nor B.

11. Technician A says tire markings on the sidewall of a tire give information about tire load carrying ability. Technician B says tire markings on the sidewall of a tire provide the recommended inflation pressure. Who is right?
(A) A only.
(B) B only.
(C) Both A and B.
(D) Neither A nor B.

12. Technician A says most automotive wheels are made of steel. Technician B says most automotive wheels are made of aluminum. Who is right?
(A) A only.
(B) B only.
(C) Both A and B.
(D) Neither A nor B.

13. Technician A says wheel bearings in several car models are tapered roller bearing type. Technician B says wheel bearings in certain car models are ball bearing type. Who is right?
(A) A only.
(B) B only.
(C) Both A and B.
(D) Neither A nor B.

14. While discussing the basic parts of an automotive wheel bearing, Technician A says the bearing's outer race is sometimes pressed into the automobile's steering knuckle. Technician B says a wheel bearing's outer race is sometimes pressed into the hub. Who is right?
(A) A only.
(B) B only.
(C) Both A and B.
(D) Neither A nor B.

15. Technician A says a steering knuckle is a basic component in a driving hub and wheel bearing assembly. Technician B says an inner drive axle is a basic component in a driving hub and wheel bearing assembly. Who is right?
(A) A only.
(B) B only.
(C) Both A and B.
(D) Neither A nor B.

Activities—Chapter 65

1. Using a section of an old worn tire, prepare a cutaway teaching aid showing the different parts and layers of the tire.

2. Prepare an overhead transparency that explains tire and wheel sizes and tire ratings.

3. As a classroom demonstration, disassemble and identify the parts of a hub.

4. Examine the tires on a vehicle in the shop for service. Report their condition to your instructor.

Chapter 66

Tire, Wheel, and Wheel Bearing Service

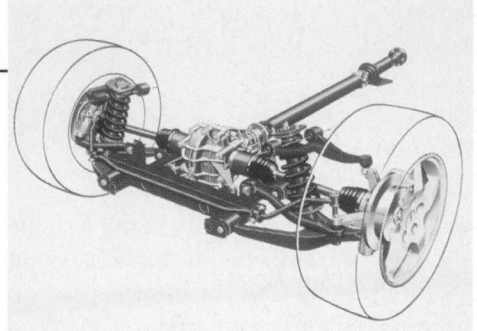

After studying this chapter, you will be able to:

- Diagnose common tire, wheel, and wheel bearing problems.
- Describe tire inflation and rotation procedures.
- Measure tire and wheel runout.
- Explain static and dynamic wheel balance.
- Summarize different methods of balancing wheels and tires.
- Explain service procedures for wheel bearings.
- Use safe practices while servicing tires and wheels.
- Correctly answer ASE certification test questions requiring a knowledge of the service and repair of tires, wheels, and wheel bearings.

Tires are a common source of trouble. Normal wear, damage from road debris, and improper inflation pressure commonly cause tire failure. Wheels can be damaged by impact with curbs and other objects, causing vibration or tire deflation. Wheel bearings can wear and fail when operated without grease.

This chapter will present the information needed to service and repair tires, wheels, and wheel bearings. You will learn how to use a tire changer, wheel balancer, wheel bearing packer, and other related equipment. This knowledge will help you when working in almost any auto repair facility.

Tire, Wheel, and Wheel Bearing Diagnosis

Tire problems usually show up as vibrations, abnormal tread wear patterns, steering wheel pull, abnormal noises, and other symptoms. In some cases, you may need to test drive the vehicle to verify the customer complaint. Make sure the symptoms are *not* being caused by steering, suspension, or front wheel alignment problems.

Closely inspect the tires. Check for bulges, splits, cracks, chunking, cupping of the tread, and other signs of damage or abnormal wear. Look closely at the outer sidewall, the tread area, and the inner sidewall. If problems are found, determine what caused the damage before replacement or repair, **Figure 66-1.**

Tire Impact Damage

Tire impact damage, also termed *road damage,* includes punctures, cuts, tears, and other physical tire injuries. Depending upon the severity of the damage, you must either repair or replace the tire. Some examples of tire failure are illustrated in **Figure 66-2.**

Tire Wear Patterns

A *tire wear pattern* (area of tread worn *off*) can usually be studied to determine the cause of the abnormal wear. Look at **Figure 66-3.**

Figure 66-1. Tire service normally begins with a thorough inspection. Look for damage and wear on all surfaces of the tire. Rotate and wiggle the tire to check for dry, rough, or loose wheel bearings.

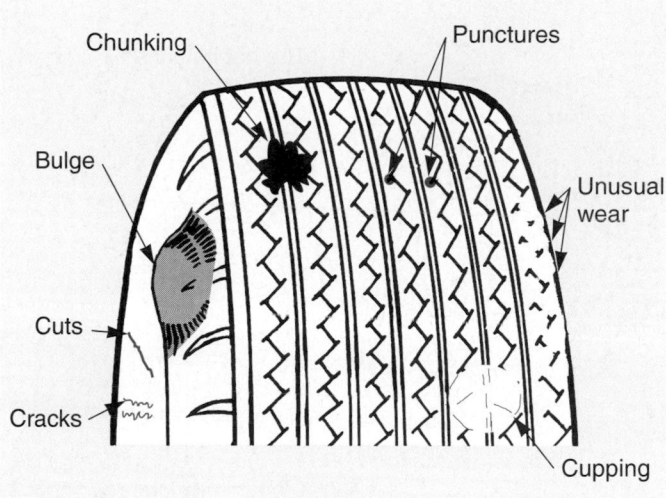

Figure 66-2. These are typical types of damage found on used tires. (Florida Dept. of Voc. Ed.)

Improper tire inflation, ply separation, incorrect wheel alignment, lack of periodic tire rotation, and an out-of-balance condition can cause excessive tire wear. By carefully inspecting the tire tread, a technician can determine what parts should be serviced or repaired.

Note
Chapter 74, *Wheel Alignment,* contains more information on tire wear patterns.

Tire Inflation Problems

Correct tire inflation pressure is important to the service life of a tire. Proper inflation is needed so the full tire tread contacts the road, **Figure 66-4.**

Underinflation (low air pressure) is a common and destructive problem that wears the outer corners of the tread area. The low pressure allows the tire sidewalls to flex, building up heat during operation. The center of the tread flexes upward and does *not* wear, **Figure 66-4A.**

Underinflation can cause rapid tread wear, loss of fuel economy, and, possibly, *ply separation* (plies tear away from each other). It will cause the tire sidewalls to bulge outward near the road surface.

Overinflation (too much air pressure) causes the center area of the tread to wear. The high pressure causes the body of the tire to stretch outward. This pushes the center of the tread against the road surface but lifts the outer edges of the tread off the road, **Figure 66-4B.**

An overinflated tire will produce a rough or hard ride. It will also be more prone to impact damage. With

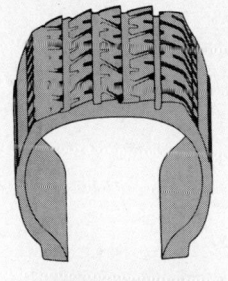

A–Feathering. This is caused by erratic scrubbing against road when tire is in need of toe-in or toe-out alignment correction.

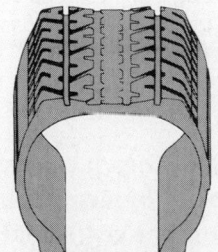

B–Overinflation. Over-inflation can cause fast centerline wear in bias and bias belted tires. In this case, center ribs get more contact with road than they should and wear much faster than outer ribs.

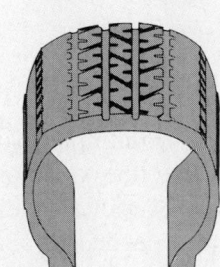

C–Underinflation. When a tire is underinflated, most of its contact with road is on outer tread rib, or shoulder, causing faster wear here than in middle. Be sure to check tire's air pressure.

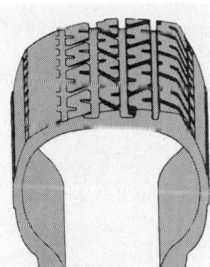

D–One-side wear. Here's another type of alignment problem–excessive camber, which means tire is leaning too much to inside or outside of tread, and placing all work on one side of tire.

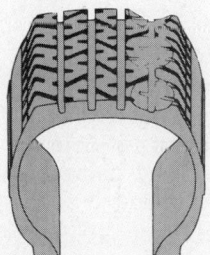

E–Cupping. This means the car may need wheels balanced, or possibly new shock absorbers or ball joints, or both.

Figure 66-3. Tire tread wear patterns will help you identify the cause of rapid or abnormal wear. Study the patterns and the causes. (Goodyear)

practice, you can detect overinflation by pressing in on the tire sidewall with your thumb. If the tire is overinflated, it will feel too hard.

Proper tire inflation makes the full tread area of the tire touch the road. The tire will wear evenly across the tread. This increases tire life and improves handling and safety, **Figure 66-4C.**

Steering wheel pull can be caused by uneven tire inflation. For example, if the left front tire (driver's side) is underinflated and the right front tire is properly inflated, the car may pull to the left. The tire with low air pressure will have more rolling resistance. It will tend to pull the steering wheel away from the normally inflated tire.

Tech Tip
Keep tires inflated properly. Low tire pressure increases tire rolling resistance, fuel consumption, and tire wear. Low pressure is not obvious with modern radial tires. Their sides bulge out when correctly inflated. Use a pressure gauge to check inflation pressure.

Tire Vibration Problems

Tire vibration is commonly caused by an out-of-balance condition, ply separation, excessive tire runout, a bent wheel, tire stiffness variation, or tire cupping. Refer to **Figure 66-5.**

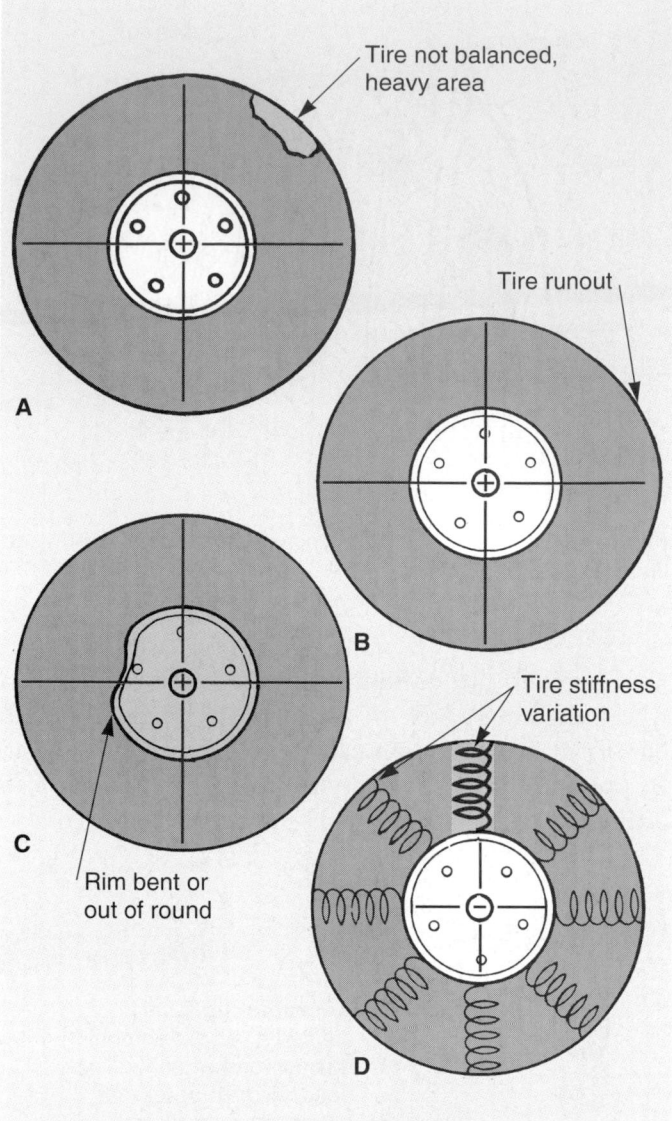

Figure 66-5. Four common causes of tire-related vibration. A—Wheel-tire assembly not balanced. B—Excessive tire runout. C—Bent rim or excessive rim runout. D—Stiffness variation in rubber. (Buick)

When one of the *front* tires is vibrating, it can usually be felt in the *steering wheel*. When one of the *rear* tires is vibrating, the vibration will be felt in the *center* and *rear* of the car.

Tire and Wheel Bearing Noise

Tire noise usually shows up as a thumping sound caused by ply separation or as a whine due to abnormal tread wear. Inspect the tire for an out-of-round condition or tread cupping. Tire replacement is needed to correct these problems.

Wheel bearing noise is normally produced by a dry, worn wheel bearing. The balls or rollers are damaged

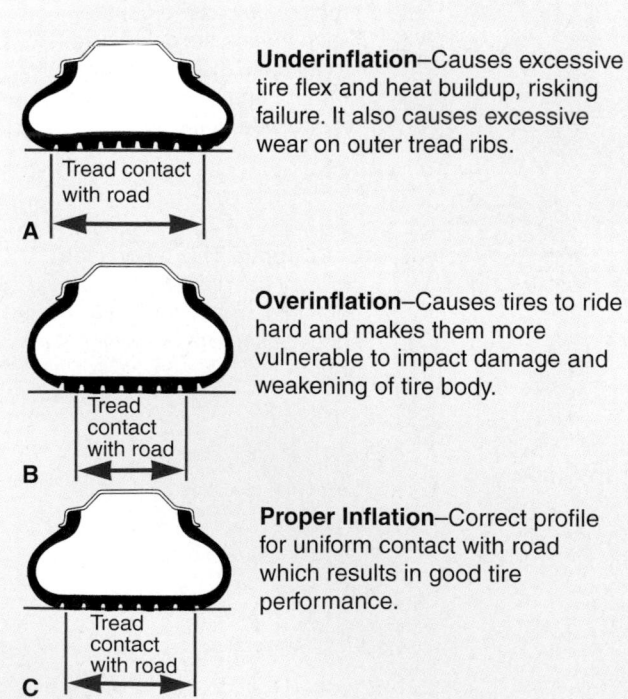

Underinflation—Causes excessive tire flex and heat buildup, risking failure. It also causes excessive wear on outer tread ribs.

Overinflation—Causes tires to ride hard and makes them more vulnerable to impact damage and weakening of tire body.

Proper Inflation—Correct profile for uniform contact with road which results in good tire performance.

Figure 66-4. Tire inflation pressure is critical to tire life, vehicle handling, and ride.

from lack of lubrication and are no longer smooth. The weight of the vehicle pressing down on the chipped, pitted bearings causes them to emit a noticeable humming or growling sound.

To check for a worn or loose wheel bearing, raise and secure the vehicle. Rotate the tire by hand. Feel and listen carefully for bearing roughness. Also, wiggle the tire back and forth to check for bearing looseness. You may have to disassemble the wheel bearing to verify the problem.

Wheel Cover Removal and Installation

A *wheel cover* is the large metal or plastic decorative cover for the wheel. It is press fit over the wheel rim to hide the lug nuts, dust cap, and other parts. To remove the cover, use the flat end of a lug wrench or a large screwdriver. Working carefully, pry between the wheel and cover at four alternating points as shown in **Figure 66-6A.**

To install a wheel cover, use a rubber hammer. Hold the wheel cover snugly in place with the valve stem

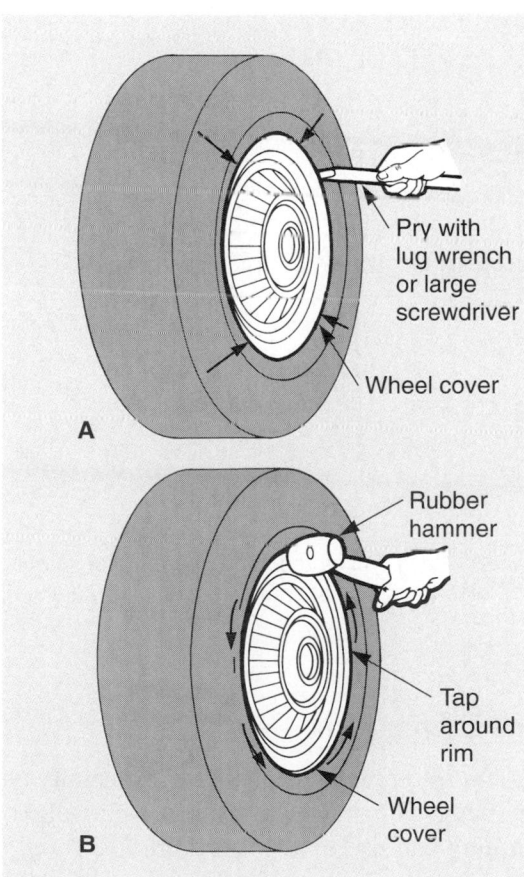

Figure 66-6. A—Pry off the wheel cover carefully, working from side to side as shown by colored arrows. B—Use a rubber hammer and an alternating or circular tapping motion to reinstall the cover. (Florida Dept. of Voc. Ed.)

aligned and sticking through the cover. Tap in a circular or alternating pattern, **Figure 66-6B.**

Be careful not to drop, bend, or dent wheel covers during removal or installation. They are very thin and can be easily damaged.

Tire Maintenance

Tire maintenance involves periodic inspection, checking of inflation pressure, and rotation. These preventive maintenance steps will help ensure vehicle safety and longer tire life.

Periodic Inspection

Periodic tire inspection involves visually checking the condition of the tires. Look for anything that could affect the dependability of the tires—tread wear, cracks, cuts, tears, etc.

Checking Tire Inflation Pressure

 A *tire pressure gauge* is used to measure tire inflation pressure, **Figure 66-7.** The procedure for checking tire inflation pressure is as follows:
1. Remove the valve stem cap.
2. Press the tire gauge squarely over the valve stem.
3. Read the air pressure indicated on the gauge.
4. Compare your reading to the recommended maximum tire pressure printed on the tire's sidewall.
5. If the tire pressure is low, add air. If the pressure is high, press in on the valve core stem to release air from the tire.
6. Recheck tire pressure and add or release air as necessary.
7. Reinstall the valve stem cap.

 Tech Tip
Remember that the tire inflation pressure on the sidewall is the maximum recommended pressure. The ideal tire pressure for the vehicle is given in the owner's manual. This pressure is dependent on several variables, such as curb weight, suspension stiffness, etc.

Most tire manufacturers recommend a *cold inflation pressure* that is 1 to 3 psi (21 kPa) below the maximum listed air pressure. This allows for tire heating, air expansion, and pressure increase without exceeding the maximum pressure limits.

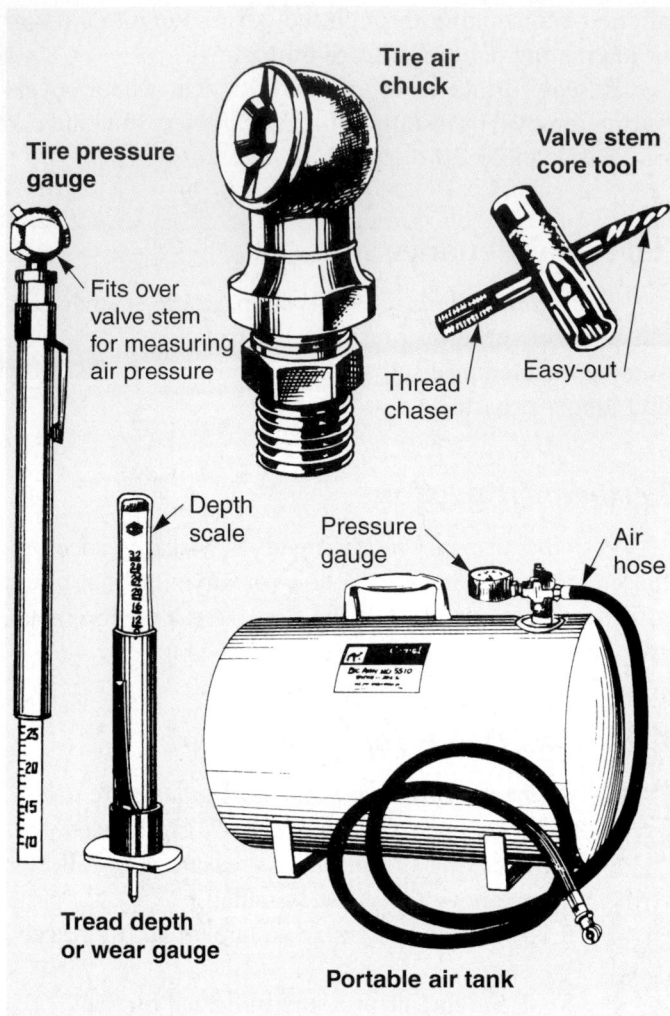

Figure 66-7. Common tire service tools: Pressure gauge to measure tire inflation. Depth gauge will accurately check tread depth and wear. Tire air chuck for filling tire with air. Portable air tank for filling car tire in remote areas. Core tool for removing and installing core in valve stem. (Camel and Snap-On)

Tech Tip

A *tire load index number* represents the weight (in pounds or kilograms) the tire can carry without failure. For example, a load index number of 91 means the tire can carry a maximum load of 1356 pounds (610 kilograms) at its highest rated speed. This index number, as well as the tire's load-carrying ability, is molded on the side of the tire for easy reference.

Rotating Tires

Tire rotation is needed to ensure maximum tire life. Normally, the front or rear tires wear differently. Rotation helps *even out* tire wear and prevents premature failure of any one tire. Generally, tires should be rotated at intervals

suggested by the tire manufacturer (typically every 3000 miles or 4827 km) or sooner if irregular wear develops.

Figure 66-8 shows the tire rotation patterns for bias and radial tires. Note that bias and radial tires have different patterns. Bias tires use an x-type (cross rotation) pattern. Radial tires often require a front-to-rear and rear-to-front rotation pattern.

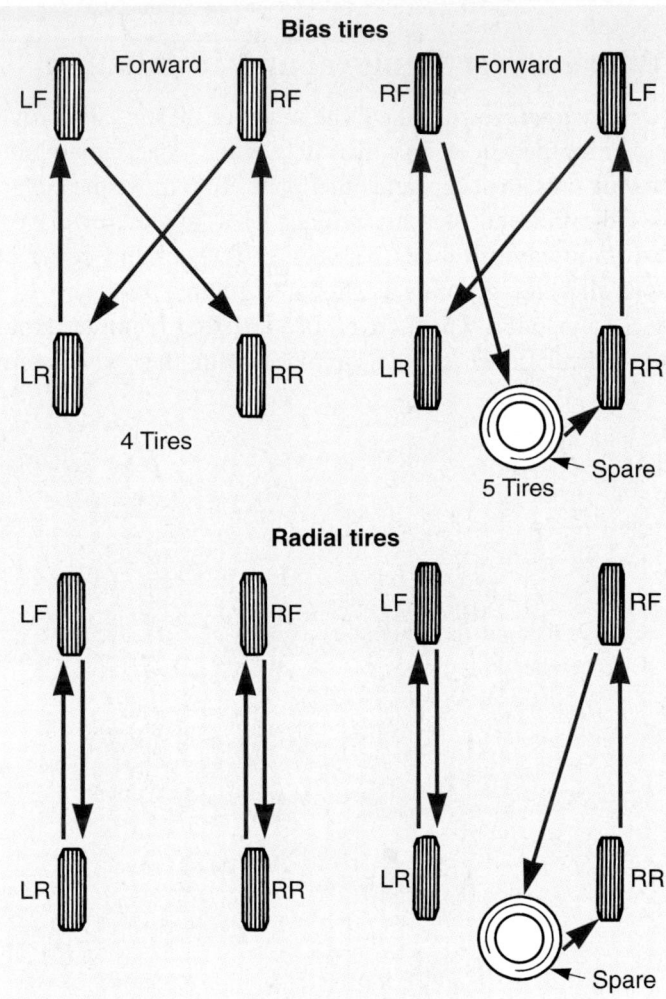

Figure 66-8. Study typical recommendations for rotating other radial and bias tires. Do not change sides with radials! (Dodge)

Torquing Lug Nuts

Lug nut torque is very important, especially on vehicles equipped with mag wheels and lightweight hubs. Overtorquing can cause wheel or hub distortion, excessive runout, and vibration. Undertorquing might allow the lug nuts to loosen and the wheel to fall off.

Automakers suggest that lug nuts be tightened to specs with a torque wrench. Tighten the nuts in a crisscross pattern as in **Figure 66-9.**

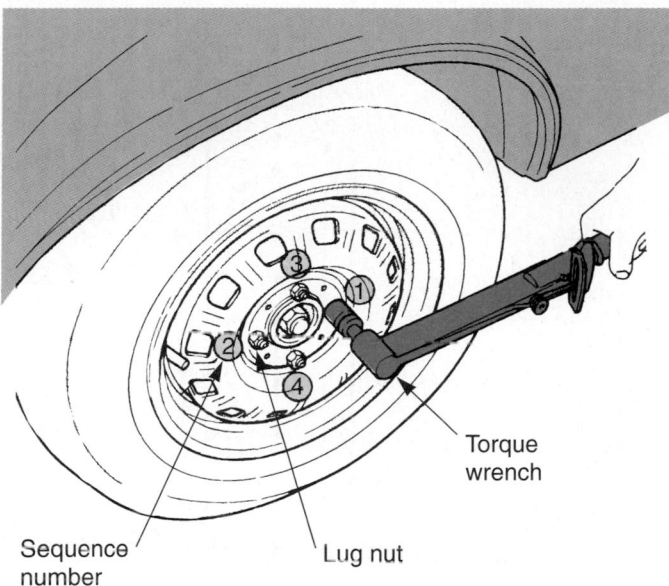

Figure 66-9. Use a torque wrench to tighten lug nuts to specifications in crisscross pattern. Modern lightweight wheels and hubs require exact lug nut torque. (Dodge)

Replacing Lug Studs

Lug studs can become stripped or damaged. To replace them, force out the old stud with a special pressing tool or a hydraulic press. Do not use a hammer to drive the studs out because the hammer blows could damage the wheel bearing.

To install the new stud, use flat washers and a lug nut. You can usually draw the new stud into place by tightening the nut on the washers. Double-check that the stud is fully seated before installing the wheel.

Measuring Tire and Wheel Runout

Tire runout or wobble is caused by a faulty tire (ply separation or manufacturing defect).

Wheel runout, or wobble, is caused by impact damage or incorrect welding of the spider and rim. When you suspect a minor runout problem, use a dial indicator to measure tire and wheel runout.

Lateral runout is side-to-side movement. It is measured by placing a dial indicator against the side of the rim or on the tire sidewall. A special indicator with a small wheel or roller is commonly used to measure lateral runout. The technician turns the tire by hand while noting the indicator reading. See **Figure 66-10A.**

Radial runout is caused by a difference in radius from the center axis of rotation. It is measured by placing the dial indicator on the tire tread and on the inner part of the rim. Again, the tire is turned by hand and the indicator reading is noted. This is illustrated in **Figure 66-10B.**

Compare your dial indicator readings to specifications. If either lateral or radial runout is beyond specs, replace the wheel or tire.

Sometimes, it is possible to reduce tire runout by rotating the tire 180° on the wheel. Typically, tire radial runout should *not* exceed 0.060″ (1.5 mm). Tire lateral runout should be under 0.090″ (2.0 mm). Wheel radial runout should *not* be over 0.035″ (0.9 mm). Wheel lateral runout should *not* exceed 0.045″ (1.0 mm). However, always refer to exact specifications before condemning a wheel or tire.

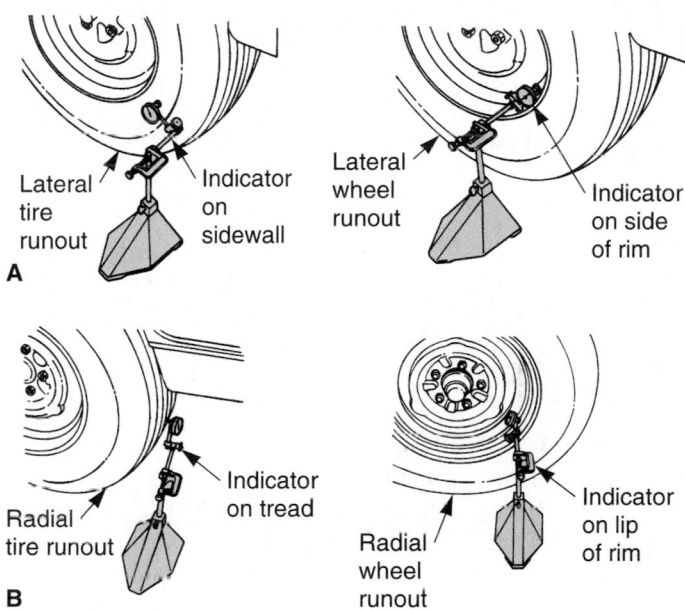

Figure 66-10. Using a dial indicator and a large base to measure tire and wheel runout (wobble). A—To measure radial runout, or out-of-round, mount the indicator on the tread and on the wheel inner lip as shown. Turn the tire by hand and read the indicator. B—To measure lateral runout, mount the indicator on the tire sidewall and the side of rim. Rotate the tire and read the indicator. (Chrysler)

Wheel Balance

Improper wheel balance is one of the most common causes of tire and steering wheel vibration. When one side of the tire is heavier than the other, centrifugal force tries to throw the heavy side outward when the tire is spinning. There are two types of tire imbalance: static imbalance and dynamic imbalance.

Static imbalance, called *wheel tramp* or *hop,* causes the tire to vibrate up and down. For a wheel and tire assembly to be in static balance, its weight must be evenly distributed around the axis of rotation. See **Figure 66-11A.**

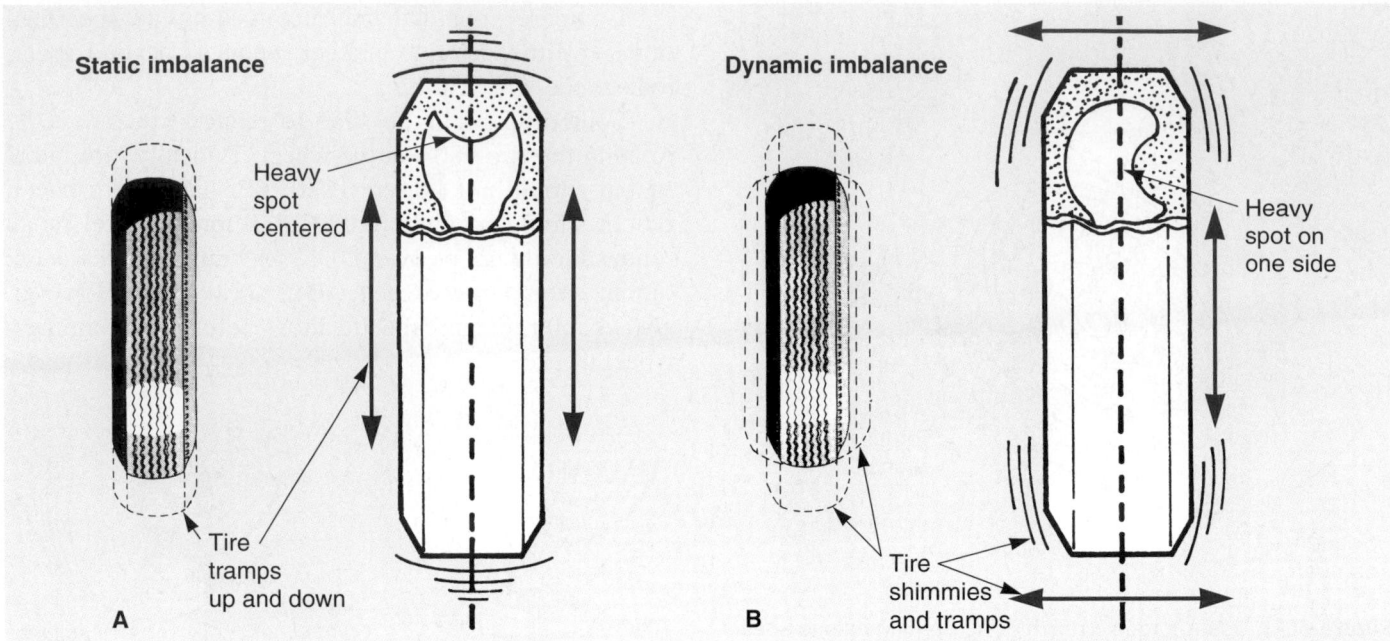

Figure 66-11. Static and dynamic imbalance are common causes of vibration. A—Static imbalance will cause tire to vibrate up and down. It is caused by a heavy spot located in the center of the tire tread. B—Dynamic imbalance will make tire vibrate from side to side and up and down. The heavy spot on tire is to one side of the tread or on sidewall.

Dynamic imbalance causes both *wheel hop* (up-and-down movement) and *wheel shimmy* (side-to-side movement). To be in dynamic balance, the top-to-bottom weight and side-to-side weight must all be equal, **Figure 66-11B.**

When *not* in dynamic balance (one side heavier than the other), centrifugal force tries to throw the heavy area outward. As a result, the tire and wheel assembly is pushed one way (thrown down and to center) and then the other way (thrown up and to center).

Balancing a Wheel Assembly

A wheel assembly is balanced by adding *wheel weights* (special lead weights that clip on rim) to the side opposite the heavy area. There are various types and sizes of wheel weights. Most press-fit onto the wheel. Weights for some mag wheels, however, stick onto the wheel with an adhesive backing. See **Figure 66-12.**

To static balance a wheel and tire assembly, add wheel weights opposite the heavy area of the wheel. If a large amount of weight is needed, add half to the inside of the wheel and the other half to the outside of the wheel. This helps keep dynamic balance correct. Look at **Figure 66-13A.**

To dynamically balance a wheel and tire, you must add weights exactly where needed, **Figure 66-13B.** The dynamic balancing machine will indicate where and how

much weight to add to bring the assembly into dynamic balance.

When balancing a wheel assembly, follow these safety rules:

- Wear eye protection when spinning the wheel and tire assembly.
- Remove rocks and other debris from tire tread before balancing the wheel assembly.

Figure 66-12. Wheel weights are used to counteract a heavy area on the wheel and tire assembly. Weights are often kept on the wheel balancing machine. (Hunter Engineering Co.)

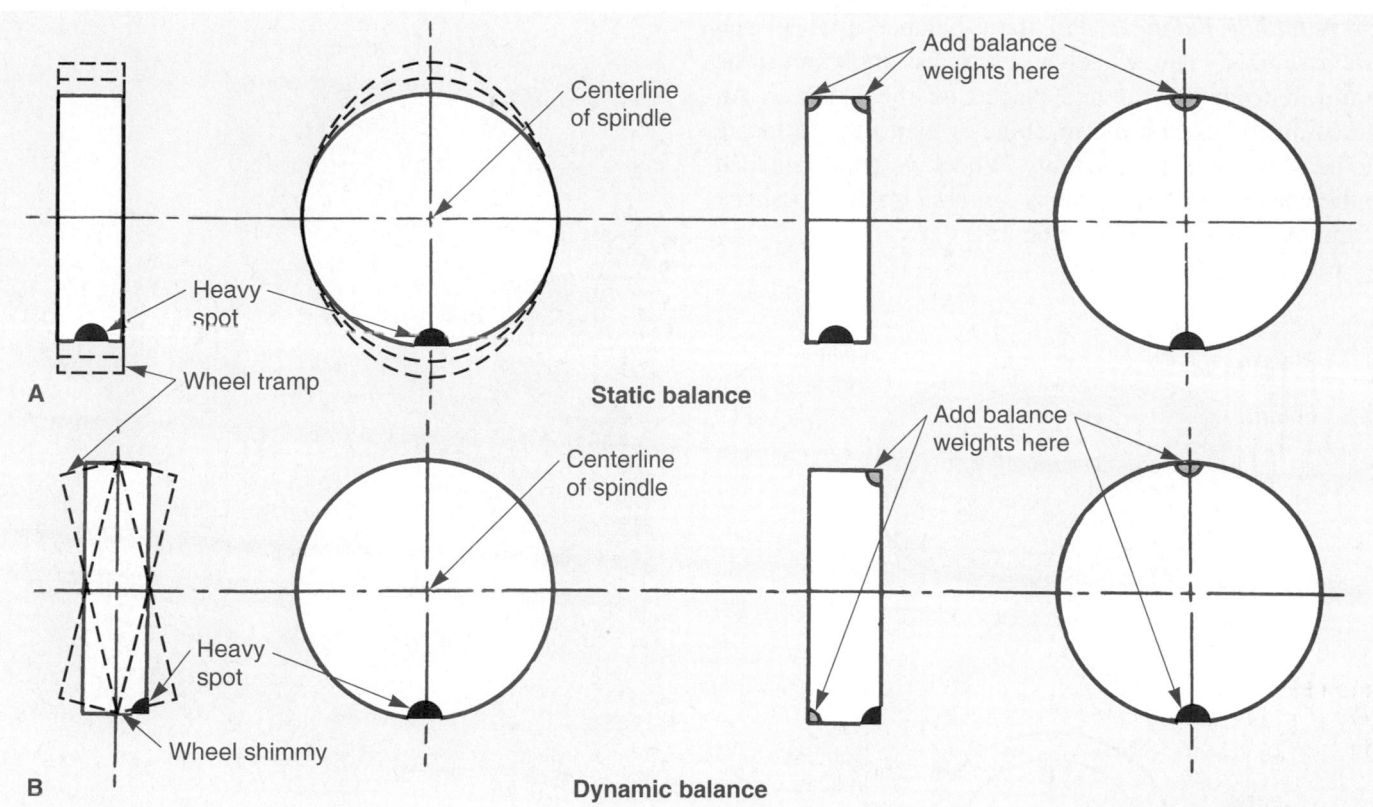

Figure 66-13. Note the basic procedure for correcting static and dynamic imbalance. A—Static imbalance will cause wheel tramp. Add weight opposite the heavy area on both sides of the wheel. Each weight should be half as heavy as the heavy spot on the tire. B—Dynamic imbalance will cause wheel shimmy. Add weight in two locations to counteract the heavy area. (Buick)

- Place a jack stand under the vehicle when using an on-car balancer.

- When using the engine to spin the rear wheels, do *not* exceed 35 mph (40 to 56 km/h) on the speedometer. When one wheel is on the ground, the free wheel will spin at *twice indicated speed*. If the speedometer reads 60 mph, for example, the tire will be spinning at a *very dangerous* 120 mph (193 km/h).

- With a limited-slip differential, raise both rear wheels off the ground when using an on-car balancer to balance a rear wheel assembly. Refer to **Figure 66-14.**

- Follow the operating instructions provided with the wheel balancing equipment. Machine designs vary and so do operating procedures.

Wheel Balancing Machines

A **wheel balancing machine** is used to determine which part of a wheel assembly is heavy. There are three basic types of wheel balancing machines: the bubble balancer, off-car balancer, and on-car balancer.

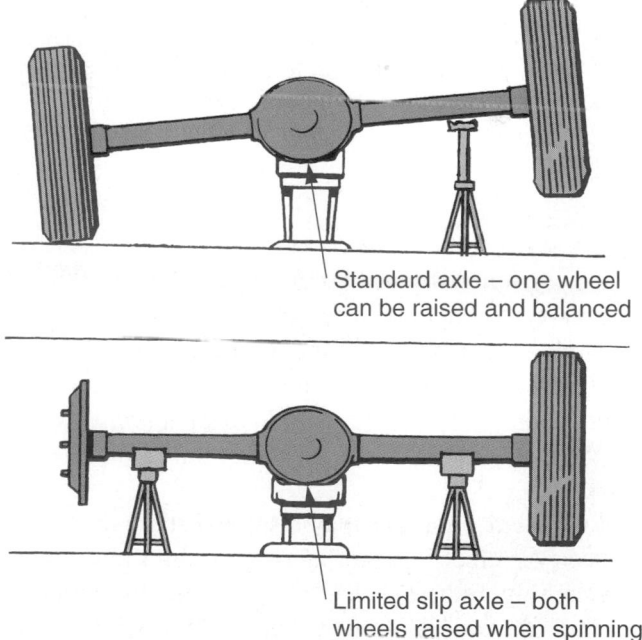

Figure 66-14. Be careful when spinning rear wheels during on-car wheel balancing. If the car has a limited-slip differential, do *not* leave one wheel on ground while spinning the other with the engine. The car could drive off the jack stands. Raise both wheels and remove one tire while balancing the other. (Pontiac)

A *bubble balancer* will static balance a wheel and tire assembly. The wheel and tire assembly must be removed from the car and placed on the balancer. An indicating bubble on the machine is then used to locate the heavy area of the assembly. Wheel weights are added to the wheel until the bubble is *centered* on the crosshairs of the machine. See **Figure 66-15.**

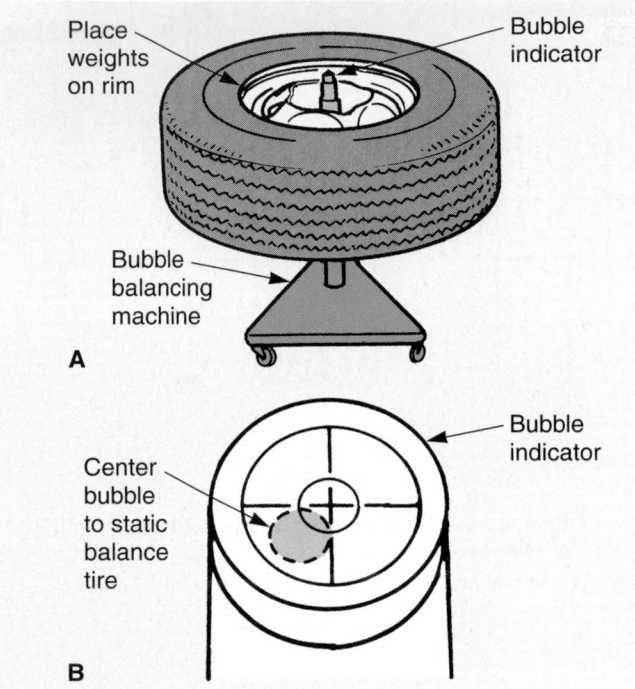

Figure 66-15. A bubble balancer will correct static imbalance only. A—The tire is mounted on a bubble balancer as shown. B—Use balancer instructions to place weights on the wheel until the bubble centers in the circle on the balancer head. (Ammco)

An *off-car balancer* can either be a static- or dynamic-type machine. The wheel and tire assembly is mounted on the balancer and spun. The machine will detect vibration of the assembly and indicate where wheel weights should be added. After adding weights, spin the tire again to check for vibration. Look at **Figure 66-16.**

An *on-car balancer* may also be a static or dynamic balancer. An electric motor is used to spin the wheel and tire assembly. Either an electronic pick-up unit or a hand-operated device is used to determine where wheel weights are needed.

An on-car balancer is sometimes desirable because it can balance the wheel cover, brake disc, and lug nuts along with the tire and wheel. Everything is rotated and balanced as a unit.

Figure 66-16. The off-car wheel balancer is very common. This unit will correct both static and dynamic imbalance. Always use a safety guard when spinning the tire. (Hunter Engineering Co.)

Mounting and Dismounting Tires

When mounting or dismounting a tire on its wheel, a *tire changing machine* is used to force the tire on and off the wheel. One type of tire changer is illustrated in **Figure 66-17.**

The bead breaker is used first to force the tire bead away from the wheel rim, **Figure 66-18.** Then, a special bar is attached to the drive head. The head turns the bar, which pries the bead over and off the rim. The other end of the bar is used when installing the tire on the wheel.

A few rules to follow when mounting or dismounting tires include:

- Wear eye protection and remove the valve core before breaking the bead away from the wheel.

- Keep your fingers out of the way when removing a tire from a wheel.

- Never mount a tire on a rim that is *not* smooth and clean.

- Always lubricate the tire bead and wheel flange with proper lubricant (vegetable oil-soap solution) before mounting a tire. *Never* use antifreeze, motor oil, or any other nondrying petroleum-

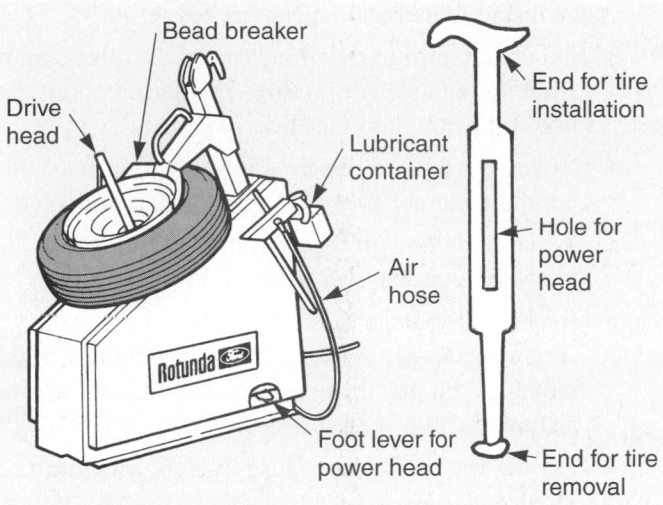

Figure 66-17. Modern tire changers use air pressure to force the tire on and off the wheel. Follow all safety rules provided with the equipment. The bar is mounted on the power head. The head turns the bar to force the beads over the rim. (Ford)

based substance as a lubricant. Tire deterioration or movement on the wheel may result.

- Do *not* inflate a tire when it is lying on the floor. When the beads seat, the tire could fly dangerously into the air.

- Stand away from the tire when adding air after mounting. Usually, a loud "pop" sound will be made when the beads seat.

- Do *not* exceed 40-50 psi (276-345 kPa) when initially inflating a tire.

- After initial inflation, install the core and reduce tire pressure to recommended limits.

Some modern, automated tire changers use a rubber wheel to force the bead back over the wheel. See **Figure 66-19.**

Tire Puncture Repair

To find a leak, fill the tire with air. Then, place the tire in a drum full of water or wet the tire with a hose. Look for air bubbles forming on the tire. Bubbles indicate a leak. Mark the leak with a crayon or chalk.

In past years, puncture repairs were attempted by inserting a rubber plug in the puncture without dismounting the tire. Because of serious safety concerns, this is *no longer recommended.*

Figure 66-18. A tire changer exerts tremendous force to stretch the tire bead on and off the wheel. Use care. (Hunter Engineering Co.)

Figure 66-19. Automated tire changers greatly simplify equipment operation. Make sure you read the operating instructions for the particular unit before use. (Hunter Engineering Co.)

According to the Rubber Manufacturer's Association (RMA), using a plug to attempt tire repair without dismounting is effective only 80% of the time. The remaining 20% of such repairs will result in tire failure, which may take the form of a dangerous sudden deflation.

 RMA lists two requirements for correct puncture repair: the injury to the tire must be filled and the inner liner must be soundly patched. The safe, correct procedure for tire repair is as follows:

1. Remove the tire from the rim.
2. After dismounting the tire, *inspect* the inside surface carefully to locate the puncture and to determine the nature and extent of the damage.
3. Fill the injury to the tire using a recommended plug or liquid sealant.
4. Select a patch that extends well beyond the damaged area, so that it will adhere properly to the inner liner and withstand the heat and mechanical stresses of tire use.
5. Scuff (roughen) the area that the patch will cover, so that it will adhere tightly.
6. Apply the proper cement (adhesive) to the inner liner, following the repair kit directions.
7. Remove the covering from the adhesive side of the patch and carefully place the patch on the inner liner.
8. Use a stitcher tool to tightly bond the patch to the inner liner. See **Figure 66-20.**

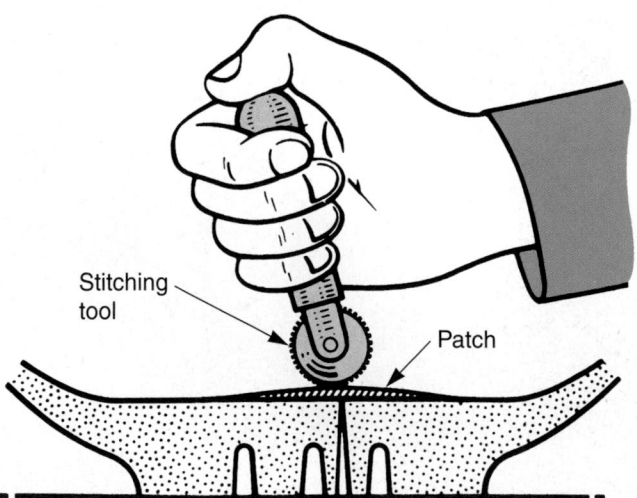

Figure 66-20. Patch on inside of tire is needed for all repairs. Basically, scuff the area around the injury. Apply a coat of approved cement. Remove the cover from the patch and apply the patch over the damage. Use a stitcher to adhere the patch to the tire liner.

A few basic rules for tubeless tire repair are:

- Do *not* attempt to repair a puncture by plugging it from the outside of the tire. *Always* dismount the tire and patch the inner liner.
- Never attempt to repair sidewalls or tires with punctures larger than 1/2″ (13 mm) in diameter.
- When removing an object from the tire, reduce air pressure to at least 15 psi (103 kPa).
- Broken strands in steel belted tires may indicate serious damage. Also, the broken strands could puncture the patch or plug, causing a serious tire failure.
- Follow the exact procedures given with the tire repair kit.

Wheel Bearing Service

Wheel bearings are normally filled with grease. If this grease dries out, the bearing will fail. It is wise to check wheel bearings for wear when performing wheel- or tire-related service. Some wheel bearings can be disassembled and packed (filled) with grease. Others are sealed units that require replacement when worn.

As you will learn, service methods for vehicles with rear-wheel drive are quite different than those for vehicles with front-wheel drive.

Servicing Wheel Bearings (Nondriving Wheels)

An exploded view of a typical nondriving wheel bearing assembly is shown in **Figure 66-21.**

To disassemble the bearings, partially loosen the lug nuts. Then, raise the vehicle and secure it on jack stands. Remove the wheel, grease cap, cotter pin, adjusting nut, and safety washer. Wiggle the hub and pull out the outer wheel bearing.

Screw the adjusting nut back onto the spindle. Unbolt the brake caliper and secure it to one side, if needed. Slide the hub outward on the spindle. When the inner bearing catches on the adjusting nut, the grease seal and the inner wheel bearing will pop out.

Wipe the bearings and races clean. Keep the bearings in *order,* because they must be reinstalled in the same races. Closely inspect the bearings and races for damage. If problems are found, you must replace both the bearing and race as a *set.* See **Figure 66-22.**

To replace a bearing race, drive out the old race with a large drift punch and hammer, **Figure 66-23.** Be careful not to damage the hub. To install the new race, use a driving tool, as shown in **Figure 66-24.**

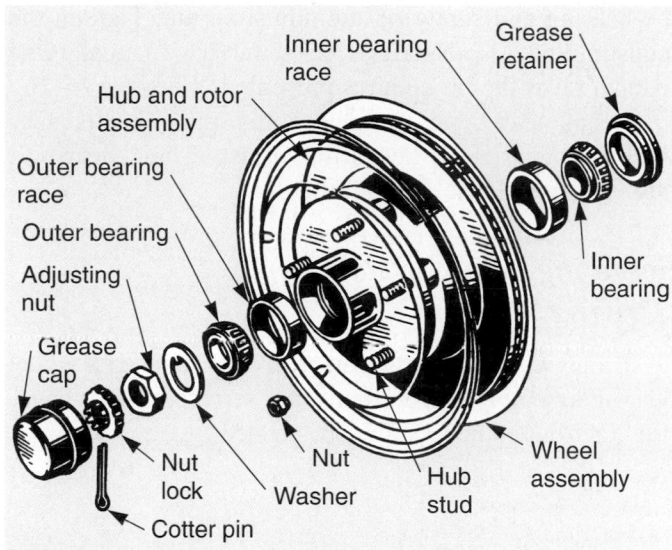

Figure 66-21. Service of a nondriving hub and wheel bearing assembly involves cleaning, inspection, lubrication, and adjustment. Study how parts fit together. (Ford)

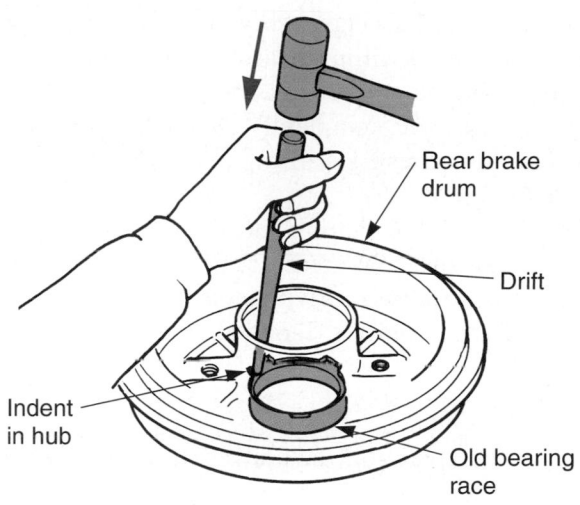

Figure 66-23. If a bearing is bad, the race must also be replaced. Use a flat-nose drift to force the old race out of the hub. (Honda)

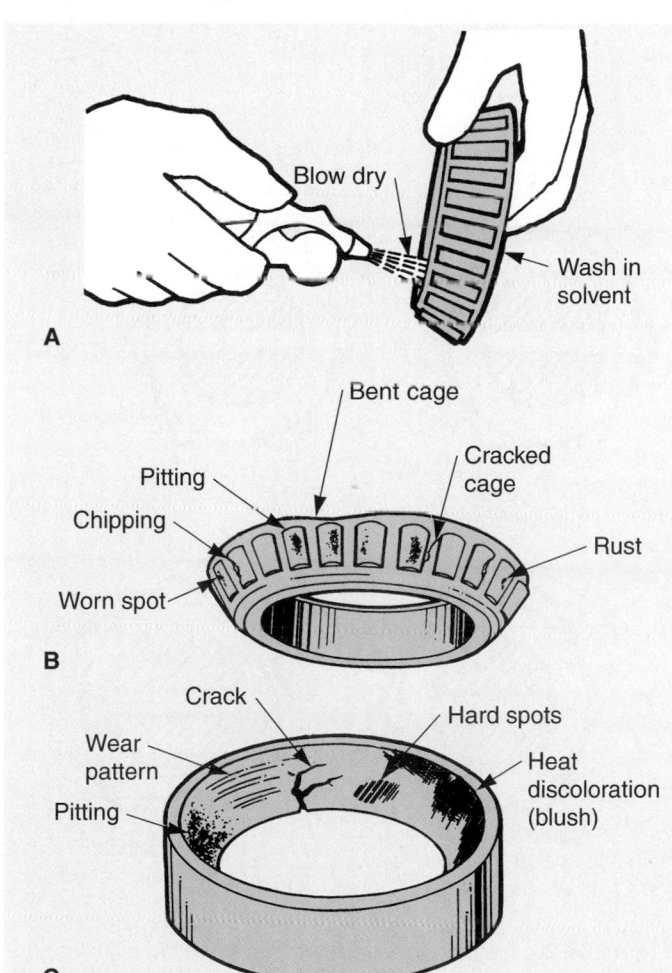

Figure 66-22. A—Wash and dry bearings. Do *not* let bearings spin while blowing them dry. B and C—Inspect the bearing and race for these kinds of troubles. (Florida Dept. of Voc. Ed.)

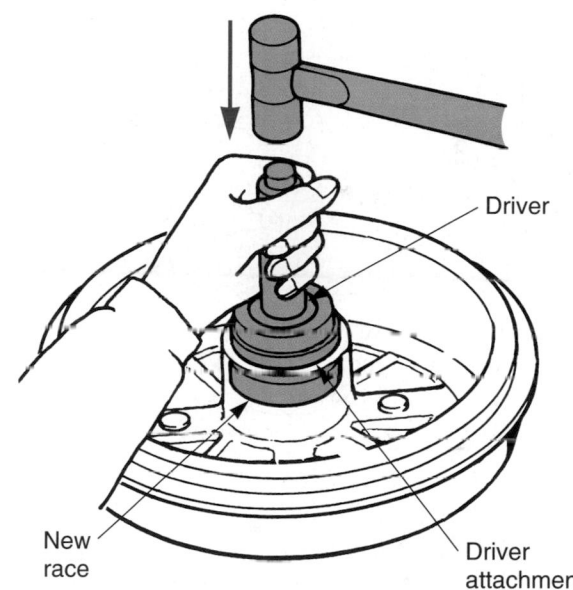

Figure 66-24. Use a driver to squarely seat the new bearing race. Be careful not to damage the bearing contact surface. (Honda)

Wipe out all the old grease from the inside of the hub. Partially fill the cavity with new wheel-bearing grease. Use the recommended type of grease. High-temperature wheel-bearing grease is recommended in most cases.

 Caution
Do *not* use all-purpose wheel-bearing grease in wheel bearings on cars with disc brakes. The heat generated by the brakes can liquefy the grease and cause leakage out of the grease seals.

To pack the bearings, use your hands or a *bearing packer* (device for filling bearing cage with grease) to properly lubricate the bearing assemblies. Make sure grease is worked completely through each bearing cage and around every ball or roller. This is shown in **Figure 66-25.**

Place the inner bearing into its race. Using a seal driver, install the new grease seal in the hub. If a seal driver is not available, drive the seal into the hub with light taps from a hammer. Tap in a circular pattern around the seal. Make sure you do not dent the seal.

Caution

Keep grease off the brake disc or drum when servicing the wheel bearings. The slightest amount of grease could ruin the brake pads or shoes or cause squeaking.

Wipe the spindle clean. Slide the hub into position and install the outer bearing. Fit the safety washer against the bearing and screw on the adjusting nut. Tighten the adjusting nut as described in a service manual. One manual gives the procedures presented in **Figure 66-26.**

Install and bend the *new* cotter pin. This is very important because it keeps the adjusting nut, bearings, hub, and front wheel from falling off.

Servicing Wheel Bearings (Driving Wheels)

Front wheel bearings on a driving hub and wheel bearing assembly are normally not serviced unless specified or when major repairs are needed.

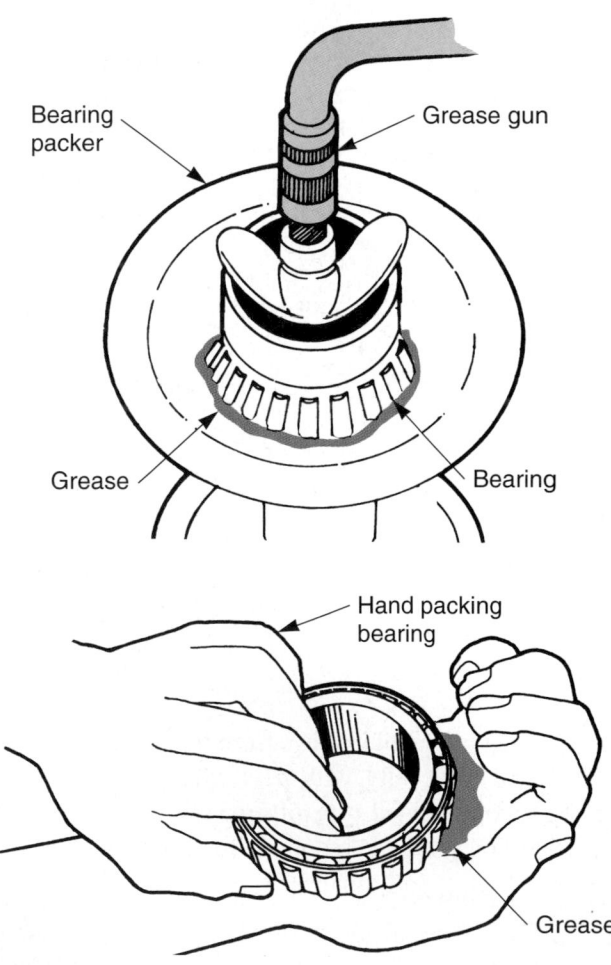

Figure 66-25. Pack bearings with quality, high-temperature wheel-bearing grease. Use either a bearing packer or the palm of your hand to work grease into cage. (Florida Dept. of Voc. Ed.)

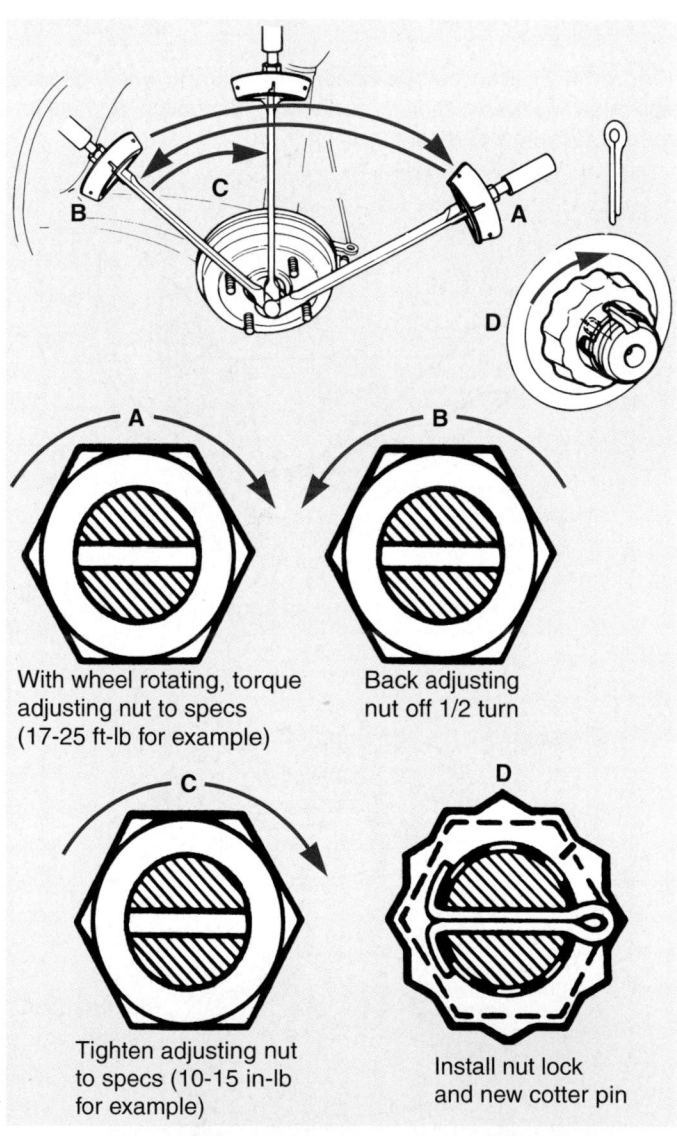

Figure 66-26. Typical procedure for tightening the adjusting nut on a nondriving wheel bearing assembly. A—Torque the nut to specifications to seat the bearings. Rotate the hub. B—Back off the adjusting nut one-half turn. C—Tighten the adjusting nut to specs (about one foot-pound or less). D—Install a new cotter pin. (Honda and Ford)

Figure 66-27 shows an exploded view of a typical driving hub and wheel bearing assembly.

Note

For more information on front-wheel-drive bearing service, refer to Chapter 64, *Transaxle and Front Drive Axle Diagnosis and Repair.* It explains front drive axle shaft removal.

Generally, to disassemble a front-drive wheel bearing, loosen the wheel lug nuts and spindle nut. Raise the vehicle and secure it on jack stands. Remove the lug nut or bolts, the wheel, and the axle nut.

Remove the caliper and hang it to one side. Unbolt the brake disc from the hub, if needed. Remove the steering knuckle and hub assembly from the vehicle.

Depending on the design of the hub, you may need to remove the hub either before or after removing the steering knuckle. Refer to a service manual for detailed procedures.

Figure 66-28A shows a puller being used to remove the hub from a steering knuckle that is mounted in a vise. After hub and bearing removal, new bearings and a new seal must be installed.

Before installing new bearings, pack them with grease if they are not sealed units. Using a press or driving tool,

force the new bearings into place. Follow service manual procedures. See **Figure 66-28B.**

Warning

When pressing new front-wheel drive bearings into position, do *not* exceed manufacturer's press load limits. If excessive pressure is used, parts can shatter. Wear eye protection!

After pressing or driving the bearings into the steering knuckle or the bearing support, install the new grease seal. Coat the inner lip of the seal with grease and the outer diameter with sealer. Use a seal driving tool and light hammer taps to force the seal into the hub. Avoid excessive driving pressure that could bend or distort the seal body and cause leakage. Look at **Figure 66-28C.**

Press the hub into the steering knuckle and wheel bearing assembly, if needed. See **Figure 66-28D.**

When pressing the wheel bearing in or out, apply force to the correct bearing race (one press fit into part). Bearing damage will result if the incorrect race is pressed and jammed into the bearings. See **Figure 66-29.**

After assembling the wheel bearings, install the steering knuckle assembly on the vehicle. Also, install the brake disc, caliper, and other components. Make sure the spindle nut is tightened to specifications. You will

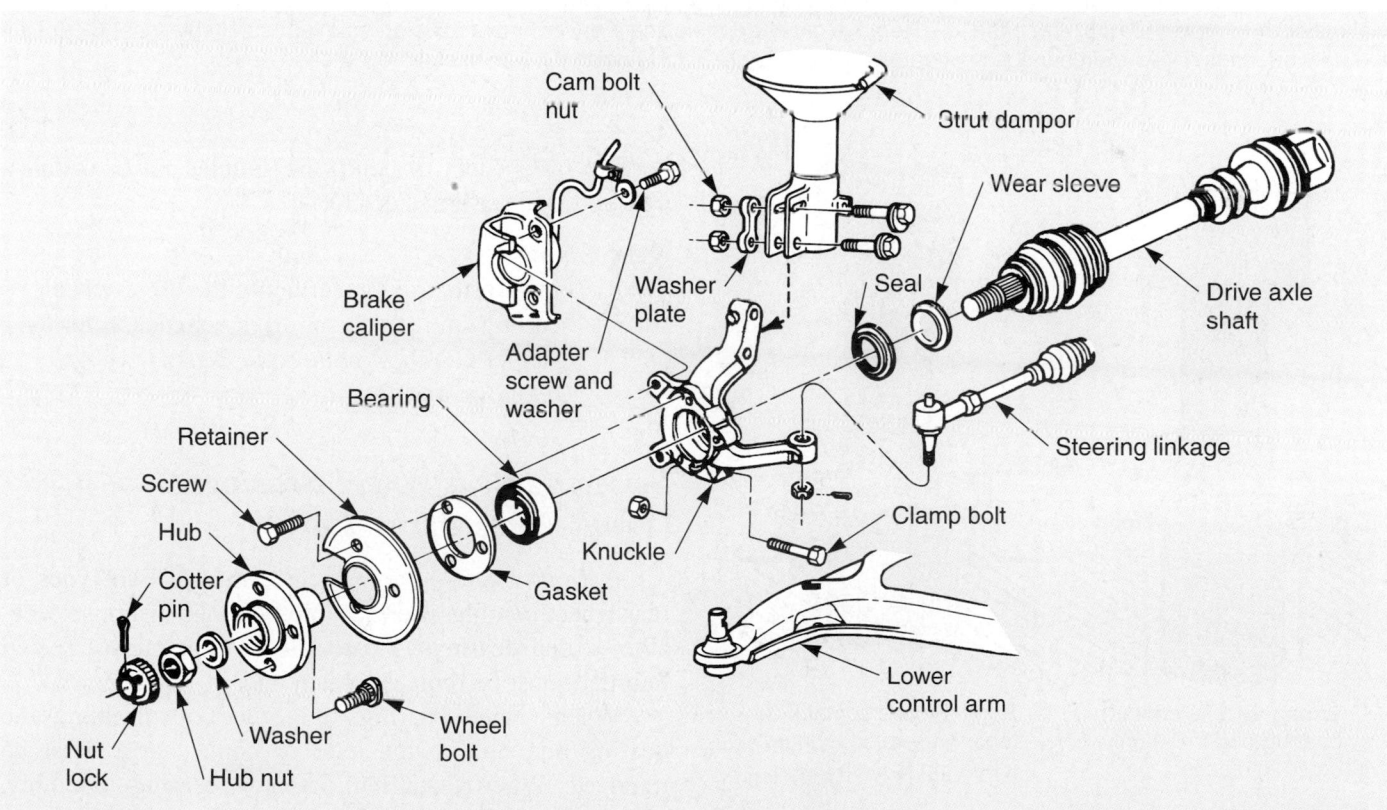

Figure 66-27. A driving hub assembly requires different service procedures than nondriving unit. The steering knuckle or bearing support must frequently be removed from car to service the bearings. (Plymouth)

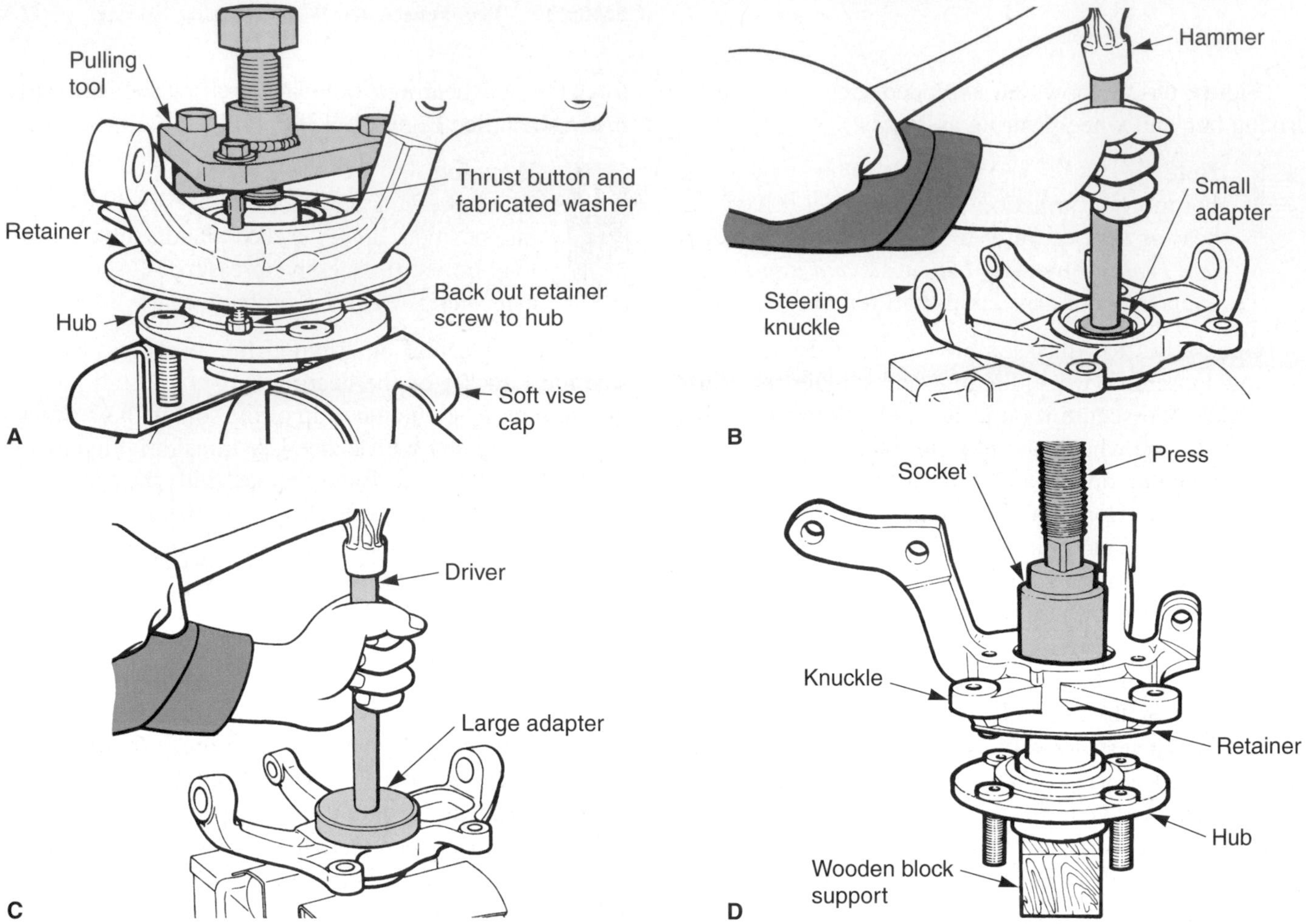

Pulling tool

Thrust button and fabricated washer

Retainer

Back out retainer screw to hub

Hub

Soft vise cap

A

Hammer

Small adapter

Steering knuckle

B

Driver

Large adapter

C

Socket

Press

Knuckle

Retainer

Hub

Wooden block support

D

Figure 66-28. Basic steps for replacing wheel bearings in driving hub assembly. A—After knuckle or support removal, use a special puller to remove the hub, if needed. B—Remove old bearings using a driver or hydraulic press. Wear safety glasses! C—Install the new bearings. Drive the bearings in squarely. D—Use a press to reinstall the hub. Install new grease seals.

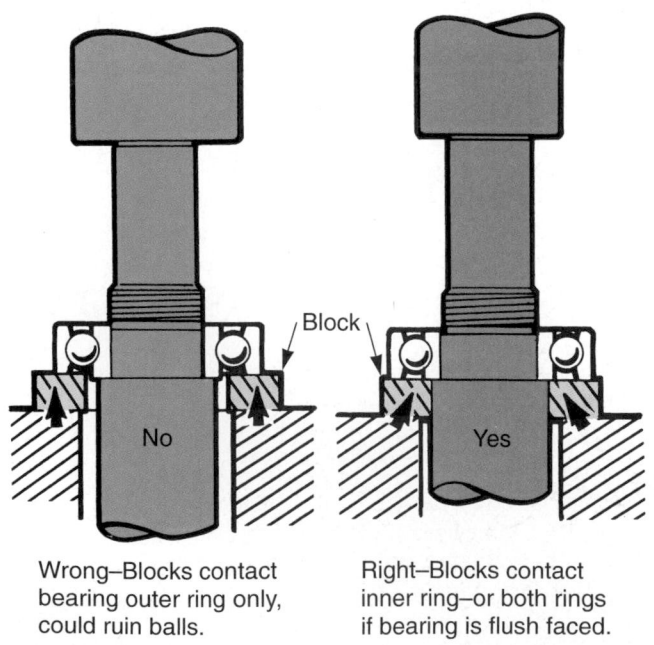

Block

No

Yes

Wrong–Blocks contact bearing outer ring only, could ruin balls.

Right–Blocks contact inner ring–or both rings if bearing is flush faced.

Figure 66-29. When pressing bearings in or out, apply driving force so that pressure is not applied to the balls or rollers. This will prevent bearing damage. (Federal Mogul)

need to *stake* (dent or bend) the spindle nut or install a new cotter pin, **Figure 66-30.**

Note

For information on servicing the drive-wheel bearings on rear-wheel-drive vehicles, refer to Chapter 62, *Differential and Rear Drive Axle Diagnosis and Repair.*

Servicing Rear Wheel Bearings (Front-Wheel-Drive Car)

A front-wheel-drive vehicle can have two types of rear wheel bearings: serviceable bearings or nonserviceable, sealed bearings. **Figure 66-31** illustrates a sealed unit that must be replaced when bad.

When sealed bearings fail, the bolts holding the bearing and hub to the axle or control arm must be removed. When installing the new bearing assembly, torque the bolts to specifications. Some manufacturers require the use of new fasteners when rear bearings and hubs are replaced.

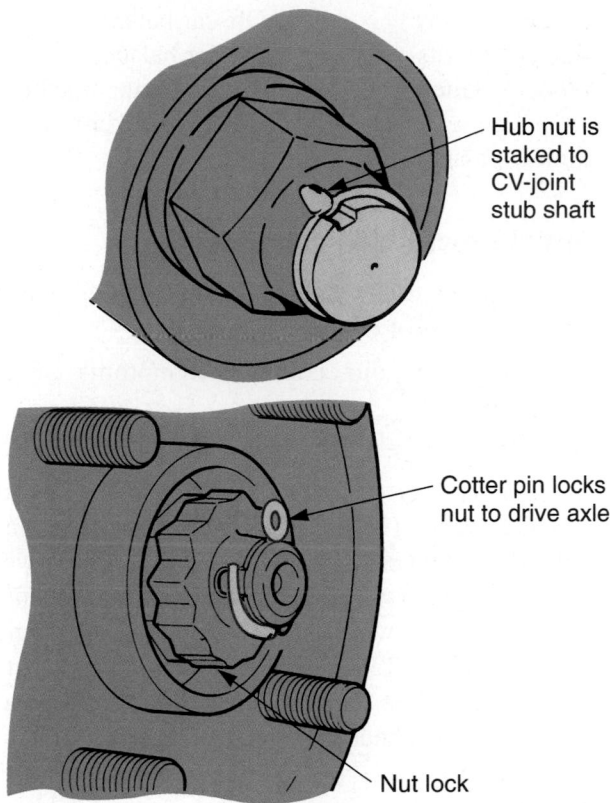

Figure 66-30. After reassembling the driving hub and bearings, make sure you torque the spindle nut to specifications. Stake the spindle nut or install a cotter pin as recommended. This keeps front wheel and hub from coming off, possibly causing a serious accident. (Ford and Dodge)

When a serviceable wheel bearing is used, the bearing can be disassembled and packed with grease. It is serviced like a front wheel bearing on a rear-wheel-drive car.

Remember! The procedures given in this chapter are general and they apply to most vehicles. When in doubt, always use service manual procedures, which are written for the specific year, make, and model car you are working on.

Summary

- Tires are a common source of trouble. Normal wear, damage from road debris, and improper inflation pressure commonly cause tire failures.

- Tire problems usually show up as vibration, abnormal tread wear patterns, steering wheel pull, abnormal noises, and other similar symptoms.

- A tire wear pattern (area of tread worn *off*) can usually be studied to determine the cause of the abnormal wear.

- Tire underinflation (low air pressure) is a very common and destructive problem that wears the outer corners of the tread.

- Tire overinflation (too much air pressure) causes the center area of the tread to wear.

- Tire vibration is commonly caused by an out-of-balance condition, ply separation, tire runout, a bent wheel, or tire cupping wear.

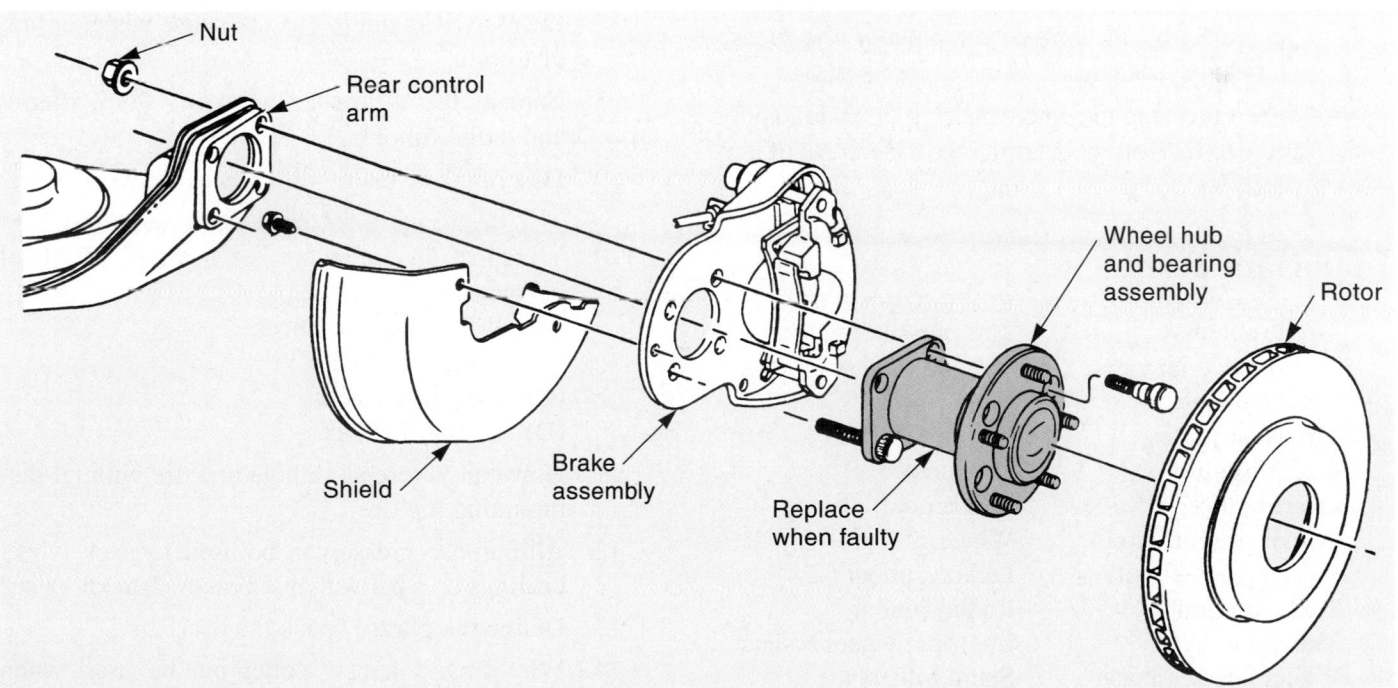

Figure 66-31. This rear wheel bearing and hub assembly is *not* serviceable. If bearings are dry and noisy when turned by hand, remove and replace the entire assembly. Torque all fasteners to specifications and follow shop manual instructions. (Cadillac)

- Tire noise usually shows up as a thumping sound caused by ply separation or as a whine caused by abnormal tread wear (cupping, for example).

- Wheel bearing noise is normally produced by a dry, worn wheel bearing. The bearing will make a steady humming sound.

- Tire rotation is also needed to ensure maximum tire life.

- Lug nut torque is very important, especially on vehicles using mag wheels and lightweight hubs.

- Lateral runout is side-to-side movement. It is measured by placing a dial indicator against the side of the rim or on the tire sidewall.

- Radial runout is caused by a difference in diameter from the center axis of rotation.

- Static imbalance causes the tire to vibrate up and down.

- Dynamic imbalance makes the tire vibrate up and down and from side to side.

- A wheel assembly is balanced by adding wheel weights to the side opposite the heavy area.

- When mounting or dismounting a tire, a tire changing machine is used to force the tire on and off the wheel.

- Wheel bearings are normally filled with grease. If the grease dries out, the bearing will fail.

- Never use all-purpose wheel-bearing grease in wheel bearings on cars with disc brakes. The heat generated by the brakes can liquefy the grease and cause leakage out of the grease seals.

- When pressing the wheel bearing in or out, apply force to the correct bearing race (one press fit into part).

Important Terms

Tire problems	Tire pressure gauge
Tire impact damage	Tire load index number
Road damage	Tire rotation
Tire wear pattern	Lug nut torque
Underinflation	Lug studs
Overinflation	Tire runout
Proper tire inflation	Wheel runout
Steering wheel pull	Lateral runout
Tire vibration	Radial runout
Tire noise	Improper wheel balance
Wheel bearing noise	Static imbalance
Wheel cover	Dynamic imbalance
Tire maintenance	Wheel hop

Wheel shimmy	Off-car balancer
Wheel weights	On-car balancer
Wheel balancing machine	Tire changing machine
Bubble balancer	Bearing packer
	Stake

Review Questions—Chapter 66

Please do not write in this text. Place your answers on a separate sheet of paper.

1. What are four common symptoms of tire problems?

2. Why is a tire wear pattern useful?

3. A customer complains of excessive right-front tire wear. The tread is worn along the outer edges. The center of the tread shows little wear. Technician A says that incorrect alignment is causing the wear and that the car needs a wheel alignment. Technician B says that underinflation could be the problem and that tire pressure should be checked. Who is correct?
   (A) A only.
   (B) B only.
   (C) Both A and B.
   (D) Neither A nor B.

4. What is ply separation?

5. A metal ball peen hammer should be used to install wheel covers. True or False?

6. Why is periodic tire rotation important?

7. Excessive lug nut _____ can cause wheel or hub distortion.

8. Explain the difference between lateral runout and radial runout.

9. Describe the major differences between static and dynamic tire imbalance.

10. Which of the following is *not* a type of wheel balancer?
    (A) Bob-weight balancer.
    (B) Bubble balancer.
    (C) On-car balancer.
    (D) Off-car balancer.

11. How can you repair a hole in a tire without dismounting the tire?

12. All-purpose grease can be used to pack wheel bearings on a car with disc brakes. True or False?

13. Define the phrase "pack the bearings."

14. Why should a new cotter pin be used when assembling a wheel bearing?

15. When pressing a front-wheel drive bearing in or out, what part of the bearing should contact the driving tool?

● ASE-Type Questions

1. Vibration is coming from an automobile's front end. Technician A checks the condition of the car's front tires. Technician B says automotive tire problems will not produce front-end vibration. Who is right?
   (A) A only.
   (B) B only.
   (C) Both A and B.
   (D) Neither A nor B.

2. A customer wants her car's tires checked. Technician A checks for cracks. Technician B inspects the inner sidewall condition. Who is right?
   (A) A only.
   (B) B only.
   (C) Both A and B.
   (D) Neither A nor B.

3. A car's left-front tire has a puncture hole. Technician A tells the owner of the car that this problem is called "tire impact damage." Technician B tells the owner of the car that this problem is called "tire road damage." Who is right?
   (A) A only
   (B) B only.
   (C) Both A and B.
   (D) Neither A nor B.

4. A car's front tires are worn excessively. Technician A checks the car's front-wheel alignment. Technician B checks the tire's inflation pressure. Who is right?
   (A) A only.
   (B) B only.
   (C) Both A and B.
   (D) Neither A nor B.

5. Bulges are discovered on an automobile's front tires. Technician A replaces the tires. Technician B is not going to replace the tires until after determining the cause of the tire damage. Who is right?
   (A) A only.
   (B) B only.
   (C) Both A and B.
   (D) Neither A nor B.

6. The outer corners of a tire's tread are worn. Technician A checks to see if the tire is underinflated. Technician B says an underinflated tire will not produce this type of problem. Who is right?
   (A) A only.
   (B) B only.
   (C) Both A and B.
   (D) Neither A nor B.

7. An automobile has a steering wheel pull problem. Technician A checks to see if the tires are inflated evenly. Technician B checks for loose shock absorbers. Who is right?
   (A) A only.
   (B) B only.
   (C) Both A and B.
   (D) Neither A nor B.

8. A car is brought into the shop with a tire vibration problem. Technician A looks for ply separation. Technician B checks to see if the wheel is damaged. Who is right?
   (A) A only.
   (B) B only.
   (C) Both A and B.
   (D) Neither A nor B.

9. Vibration is coming from a car's center and rear. Technician A checks the condition of the rear tires. Technician B checks the front ball joints. Who is right?
   (A) A only.
   (B) B only.
   (C) Both A and B.
   (D) Neither A nor B.

10. A car is brought in with a noise coming from the right front tire. Technician A checks the tire's tread. Technician B looks for a worn wheel bearing. Who is right?
    (A) A only.
    (B) B only.
    (C) Both A and B.
    (D) Neither A nor B.

11. An automobile is brought into the shop with damaged lug studs on the left wheel hub assembly. Technician A removes the damaged studs with a hammer and punch. Technician B uses a hydraulic press to remove the damaged studs. Who is right?
    (A) A only.
    (B) B only.
    (C) Both A and B.
    (D) Neither A nor B.

12. A customer wants his/her car's front wheel bearings greased. Technician A tells the customer that some automotive wheel bearings can be cleaned, packed with grease, and reinstalled. Technician B tells the customer that certain types of wheel bearings are sealed and must simply be removed and replaced. Who is right?
    (A) A only.
    (B) B only.
    (C) Both A and B.
    (D) Neither A nor B.

13. A car equipped with disc brakes is brought into the shop with worn front wheel bearings. Technician A replaces the bearings and packs the new bearings with all-purpose bearing grease. Technician B replaces the bearings and packs the new bearings with high-temperature wheel-bearing grease. Who is right?
    (A) A only.
    (B) B only.
    (C) Both A and B.
    (D) Neither A nor B.

14. Technician A says front wheel bearing on a driving hub and wheel bearing assembly are serviced during a brake job. Technician B says this type of wheel bearing assembly is not normally serviced. Who is right?
    (A) A only.
    (B) B only.
    (C) Both A and B.
    (D) Neither A nor B.

15. Technician A says some front-wheel-drive vehicles use servicable rear wheel bearings. Technician B says some front-wheel-drive vehicles are equipped with nonservicable rear wheel bearings. Who is right?
    (A) A only.
    (B) B only.
    (C) Both A and B.
    (D) Neither A nor B.

Activities—Chapter 66

1. On a vehicle assigned by your instructor, test a wheel bearing complaint and diagnose the problem.

2. Perform a tire rotation on a vehicle assigned by your instructor.

3. Demonstrate the procedure for balancing a wheel and tire. Explain each step of the procedure and why it is done.

4. Prepare a bill for a tire rotation using a flat hourly rate suggested by your instructor.

Tire, Wheel, and Bearing Diagnosis		
Condition	**Possible Causes**	**Correction**
Wheel tramp	1. Brake drum, rotor, wheel, or tire out of static balance. 2. Wheel or tire out-of-round (excessive radial runout). 3. Bulge on tire. 4. Loose or worn wheel bearings.	1. Balance assembly statically and dynamically. 2. Change tire position on wheel or discard tire or wheel as needed. 3. Replace tire. 4. Adjust or replace bearings.
Wheel shimmy	1. Wheel and tire assembly out of dynamic balance. 2. Uneven tire pressure. 3. Worn or loose front wheel bearings. 4. Excessive tire or wheel runout. 5. Worn tires. 6. Low tire pressure. 7. Loose wheel lugs. 8. Bent wheel. 9. Incorrect wheel alignment.	1. Balance assembly statically and dynamically. 2. Inflate both front tires to same pressure. 3. Adjust or replace bearings. 4. Correct by moving tire on rim or replace defective parts. 5. Move to rear if still serviceable. 6. Inflate tires to correct pressure. 7. Tighten lugs. 8. Replace wheel. 9. Correct alignment. *See* Chapter 74.
Poor recovery following turns and/or hard steering	1. Low tire pressure. 2. Lack of lubrication. 3. Incorrect wheel alignment.	1. Inflate to proper pressure. 2. Lubricate steering system. 3. Correct alignment. *See* Chapter 74.
Vehicle pulls to one side	1. Uneven tire pressure. 2. Tires not same size. 3. Worn or improperly adjusted wheel bearings. 4. Dragging brakes. 5. Incorrect wheel alignment.	1. Inflate both front tires to same pressure. 2. Install same size tires on both sides. 3. Adjust or replace bearings. 4. Adjust brakes. 5. Correct alignment. *See* Chapter 74.
Vehicle wanders from side to side	1. Low or uneven tire pressure. 2. Worn or improperly adjusted front wheel bearings. 3. Vehicle overloaded or loaded too much on one side. 4. Incorrect wheel alignment.	1. Inflate tires to recommended pressure. 2. Replace or adjust wheel bearings. 3. Advise owner regarding vehicle load limits. 4. Correct alignment. *See* Chapter 74.
Tire squeal on corners	1. Low tire pressure. 2. Excessive cornering speed. 3. Incorrect wheel alignment.	1. Inflate to recommended pressure. 2. Advise driver. 3. Correct alignment. *See* Chapter 74.
Loose, erratic steering	1. Loose front wheel bearings. 2. Loose wheel lugs. 3. Wheel out of balance.	1. Replace or adjust. 2. Tighten lugs. 3. Balance wheel assembly.
Hard riding	1. Excessive tire pressure. 2. Improper tire size. 3. Heavy-duty shock absorbers installed.	1. Reduce pressure to specifications. 2. Install correct size. 3. Advise driver and/or change shocks.
Noise from front or rear wheels	1. Loose wheel lugs. 2. Defective wheel bearings. 3. Loose wheel bearings. 4. Lack of lubrication. 5. Lump or bulge on tire tread. 6. Rock or debris stuck in tire tread. 7. Damaged wheel. 8. Wheel hub loose on axle taper (where used). 9. Worn or defective wheel bearing.	1. Tighten lugs. 2. Replace wheel bearings. 3. Adjust wheel bearings. 4. Lubricate bearings. 5. Replace tire. 6. Remove rock or debris. 7. Replace wheel. 8. Inspect and tighten. 9. Replace wheel bearing.

Tire, Wheel, and Bearing Diagnosis		
Condition	**Possible Causes**	**Correction**
Tires lose air	1. Puncture. 2. Bent, dirty, or rusty rim flanges. 3. Leaking valve core or stem. 4. Striking curbs with excessive force. 5. Flaw in tire casing. 6. Excessive cornering speed, especially with low tire pressure. 7. Porous wheel rim.	1. Repair puncture. 2. Clean or replace wheel. 3. Replace as needed. 4. Advise driver. 5. Repair or replace tire. 6. Advise driver. 7. Repair porous wheel rim.
Tire wears in center	1. Excessive pressure.	1. Reduce tire pressure to specifications.
Tire wears on one edge	1. Incorrect wheel alignment. 2. High speed cornering.	1. Correct alignment. *See* Chapter 74. 2. Advise driver.
Tire wears on both sides	1. Low tire pressure. 2. Overloading vehicle.	1. Inflate tires to specifications. 2. Advise driver.
Tire scuffing or feather edging	1. Excessive cornering speed. 2. Improper tire pressure. 3. Wheel shimmy. 4. Excessive runout. 5. Incorrect wheel alignment.	1. Advise driver. 2. Inflate tires to specifications. 3. Balance wheels statically and dynamically. 4. Correct or replace tire or wheel. 5. Correct alignment. *See* Chapter 74.
Tire cupping	1. Incorrect wheel alignment. 2. Improper tire pressure. 3. Excessive runout. 4. Wheel and tire assembly out of balance. 5. Worn or improperly adjusted wheel bearings. 6. Grabby brakes.	1. Correct alignment. *See* Chapter 74. 2. Inflate tires to specifications. 3. Correct or replace wheel or tire. 4. Balance both statically and dynamically. 5. Replace or adjust wheel bearings. 6. Repair brakes.
Heel and toe wear	1. Grabby brakes. 2. Heavy acceleration.	1. Repair brakes. 2. Advise driver.

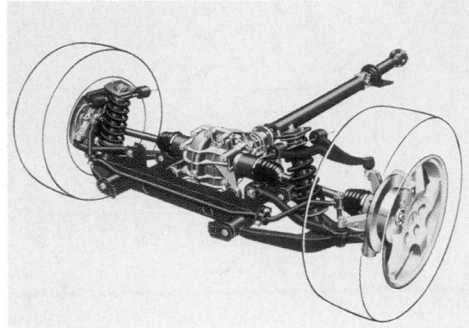

Chapter 67

Suspension System Fundamentals

After studying this chapter, you will be able to:

■ Identify the major parts of a suspension system.

■ Describe the basic function of each suspension system component.

■ Explain the operation of the four common types of springs.

■ Compare the various types of suspension systems.

■ Explain automatic suspension leveling systems.

■ Correctly answer ASE certification test questions requiring a knowledge of suspension system construction and design.

The *suspension system* allows a vehicle's tires and wheels to move up and down over bumps and holes in the road. It makes the vehicle ride more smoothly over rough roads. The suspension system consists of a series of arms, rods, ball joints (swivel joints), bushings, and other parts.

The suspension system works in unison with the tires, unibody or frame, wheels, wheel bearings, brake system, and steering system to provide a safe and comfortable means of transportation.

This chapter will summarize the most important parts of a basic suspension system and introduce the most common suspension system designs. It will also prepare you to better understand information on the diagnosis and repair of suspension systems presented in Chapter 68. Study carefully!

Functions of a Suspension System

A suspension system has several important functions:

• Supports the weight of the frame, body, engine, transmission, drive train, and passengers.

• Provides a smooth, comfortable ride by allowing the wheels and tires to move up and down with minimum movement of the vehicle body.

• Allows rapid cornering without extreme *body roll* (vehicle leans to one side).

• Keeps the tires in firm contact with the road, even after striking bumps or holes in the road.

• Prevents excessive *body squat* (body tilts down in rear) when accelerating or heavily loaded.

• Prevents excessive *body dive* (body tilts down in front) when braking.

• Allows the front wheels to turn from side-to-side for steering.

• Works with the steering system to help keep the wheels in correct alignment.

As you will learn, a suspension system uses springs, swivel joints, dampening devices, movable arms, and other components to accomplish these functions.

Tech Tip

Chassis stiffness is a primary factor affecting the quietness and smoothness of a vehicle's ride—the stiffer the chassis, the better. *Chassis hertz* is a measurement of the stiffness of a vehicle's structure. A high-hertz chassis (25 hertz, for example) is stiffer and stronger than a low-hertz chassis.

Basic Suspension System

Before discussing suspension system components in detail, you should be able to visualize each major part and understand how it functions in relation to the other parts. As each part is introduced, look at **Figure 67-1**.

• *Control arm*—movable lever that fastens the steering knuckle to the vehicle's body or frame.

• *Steering knuckle*—provides a spindle or bearing support for the wheel hub, bearings, and wheel assembly.

• *Ball joint*—swivel joint that allows the control arm and the steering knuckle to move up or down and from side to side.

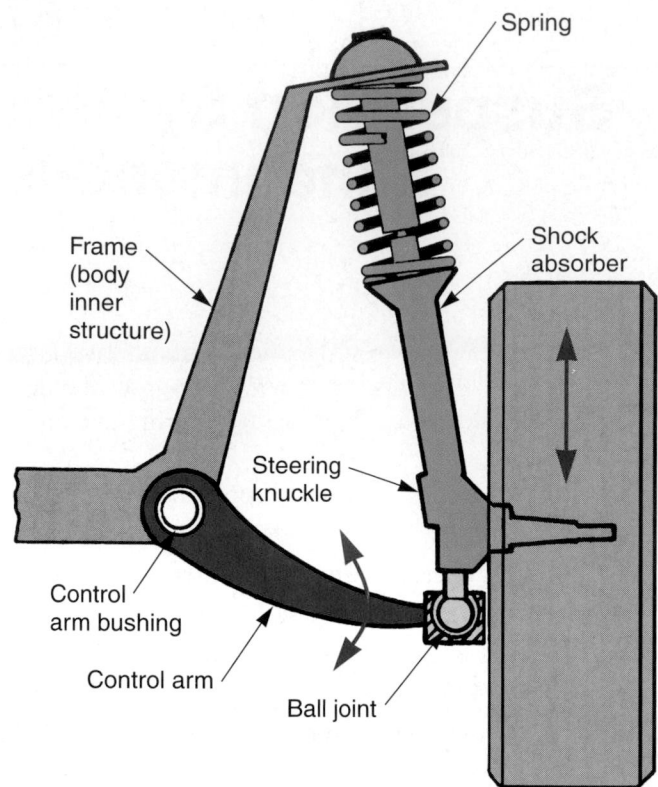

Figure 67-1. Elementary parts of a suspension system. Study the basic motion of the components.

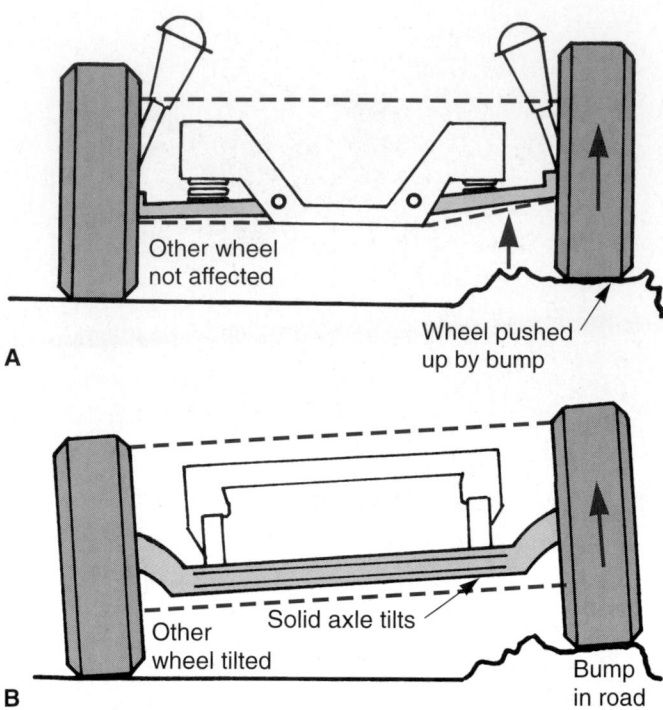

Figure 67-2. Comparison of independent and nonindependent suspension systems. A—Independent suspension allows one wheel to roll over a bump with minimal effect on the other wheel. B—Nonindependent suspension causes the action of one wheel to tilt and affect the other wheel.

- *Spring*—supports the weight of the vehicle; permits the control arm and wheel to move up and down.
- *Shock absorber* or *damper*—keeps the suspension from continuing to bounce after spring compression and extension.
- *Control arm bushing*—sleeve that allows the control arm to swing up and down on the frame.

Independent and Nonindependent Suspension Systems

Suspension systems may be grouped into two broad categories: independent and nonindependent. Both can be found on today's cars and trucks.

Independent Suspension

Independent suspension allows one wheel to move up and down with minimal effect on the other wheels. Look at **Figure 67-2A.** Since each wheel in an independent suspension system is attached to its own suspension unit, movement of one wheel does *not* cause direct movement of the wheel on the other side of the vehicle.

Detailed later in this chapter, there are many types of independent suspension systems. Independent suspension is widely used on modern vehicles, especially passenger cars.

Nonindependent Suspension

Nonindependent suspension has both the right and left wheels attached to the same solid axle. When one tire hits a bump in the road, its upward movement causes a slight upward *tilt* of the other wheel. Hence, neither wheel is independent of the other, **Figure 67-2B.**

Understeer and Oversteer

Understeer means that the vehicle is slow to respond to steering changes in a turn. The rear tires retain traction, but the front tires may slip on the road surface due to lack of downforce or other factors.

Oversteer basically means that the rear tires try to skid around sideways in a sharp or hard turn. The front tires retain traction, but the rear tires skid.

Suspension systems are designed to balance oversteer and understeer. The perfect suspension system will provide **neutral steering**—all four wheels have equal traction in turns.

Lateral Acceleration

Lateral acceleration is the amount of side force a vehicle can handle before its tires lose traction and skid in a sharp turn. It is measured in units of gravity, or "g-force" usually on a *skidpad* (round or circular driving course). The higher the "g's" the better. Passenger cars can attain a lateral acceleration of about 1.0 g, while race cars can produce more than 3.0 g's in turns.

Independent suspension systems generally obtain higher lateral acceleration than nonindependent designs. They are able to keep all four tires in full contact with the road surface better than older, heavier solid-axle designs.

Suspension System Springs

Suspension system springs must *jounce* (compress) and *rebound* (extend) as a vehicle travels over bumps and holes in the road surface. They must support the weight of the vehicle while still allowing suspension *travel* (movement). The most common types of springs are the coil spring, leaf spring, air spring, and torsion bar. See **Figure 67-3.**

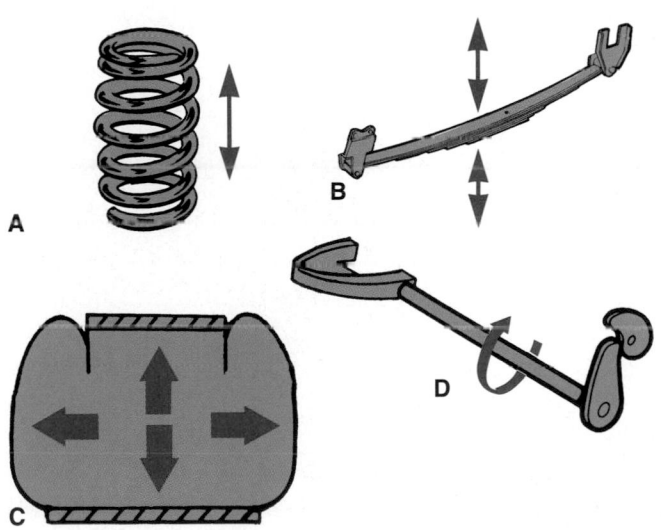

Figure 67-3. Suspension system springs. A—Coil spring. B—Leaf spring. C—Air spring. D—Torsion bar.

Coil Spring

A *coil spring* is a length of spring-steel rod wound into a spiral. It is the most common type of spring found in modern suspension systems. Coil springs may be used on both the front and rear of the vehicle. See **Figure 67-4.**

Leaf Spring

A *leaf spring* is commonly made of flat plates or strips of spring steel that are bolted together. A few are made of fiberglass. Although leaf springs were once used on front suspension systems, they are now limited to the rear of some cars. An exploded view of a leaf spring assembly is given in **Figure 67-5.**

A monoleaf spring is made of a single, thick leaf, which is usually made of reinforced fiberglass. Multiple-leaf springs have several thin steel leaves sandwiched together.

Each end of the leaf spring has an *eye* (cylinder shaped hole), which holds a bushing. The front spring eye normally bolts directly to the frame structure. Two large U-bolts secure the axle or axle housing to the leaf springs. A *shackle* fastens the rear leaf spring eye to the

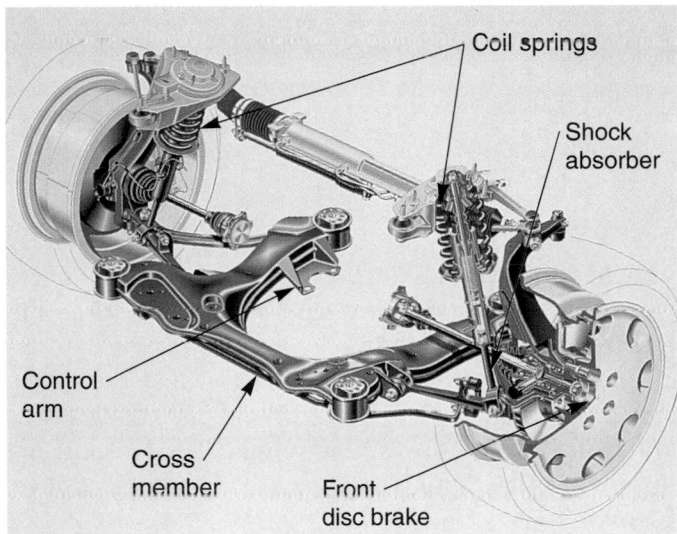

Front suspension

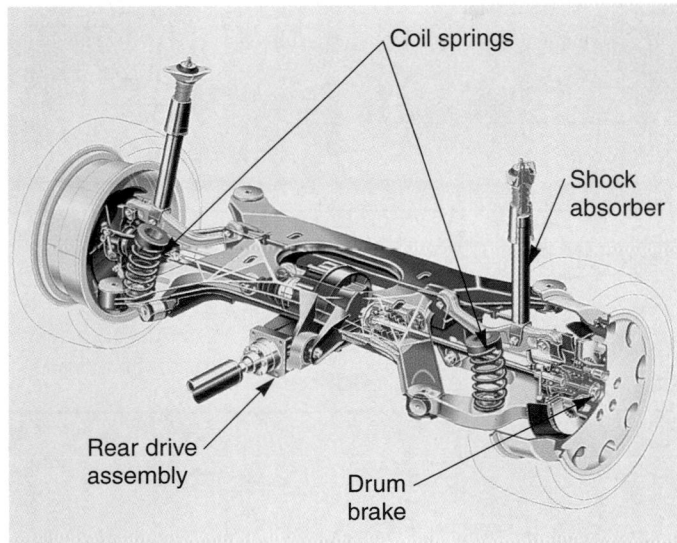

Rear suspension

Figure 67-4. Both the front and rear of the vehicle may use coil springs. They are the most common springs used on passenger cars and light trucks. (Audi)

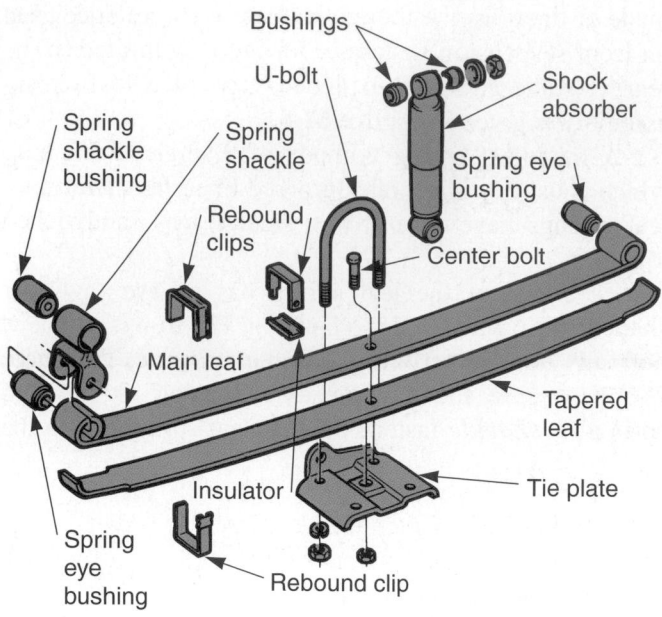

Figure 67-5. Exploded view of simple leaf spring assembly. (Chrysler)

Air Spring

An *air spring* is typically a two-ply rubber cylinder that is filled with air. End caps are formed on the air spring for mounting. Air pressure in the rubber cylinder gives the unit a spring action, similar to that of a coil spring. Refer to **Figure 67-7.**

An air spring is lighter than a coil spring. This gives it the potential to produce a smoother ride than a coil spring. Special synthetic rubber compounds must be used so the air spring can operate properly in cold weather. Low temperatures tend to stiffen rubber.

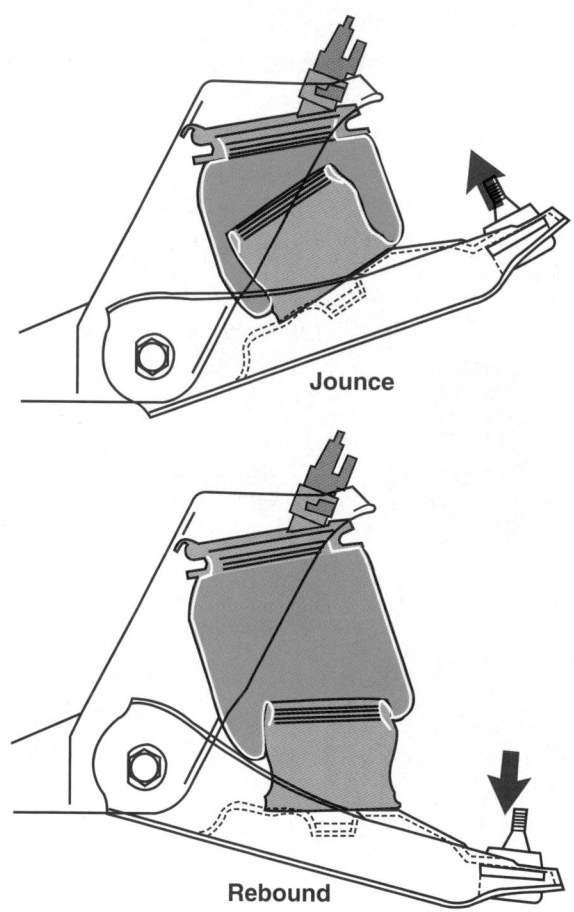

Figure 67-7. Air springs are used on some late-model cars. They are especially adaptable to automatic leveling systems. (Ford)

vehicle's frame and allows the spring to change length when bent. *Insulators* are placed between the springs to prevent squeaks and rattles.

Leaf spring windup is a condition that causes the rear leaf springs to flex when driving or braking forces are applied to the suspension system. The twisting and distortion of the spring can cause body squat and dive, **Figure 67-6.**

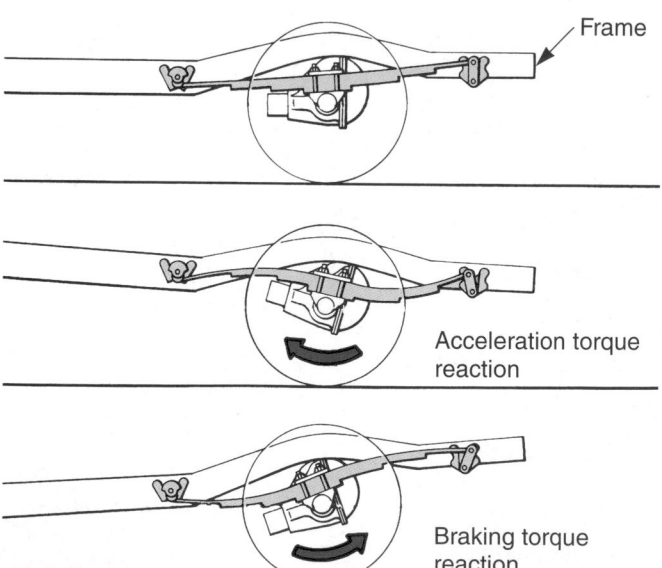

Figure 67-6. Leaf spring windup is a problem when leaves support driving axle. Torque tends to twist the springs. (Ford)

Torsion Bar (Spring)

A *torsion bar* is made of a large spring-steel rod. One end of the torsion bar is attached to the frame. The other end is fastened to the suspension system control arm. See **Figure 67-8.**

Up-and-down movement of the suspension system twists the torsion bar. It will then try to return to its original shape, moving the control arm back into place.

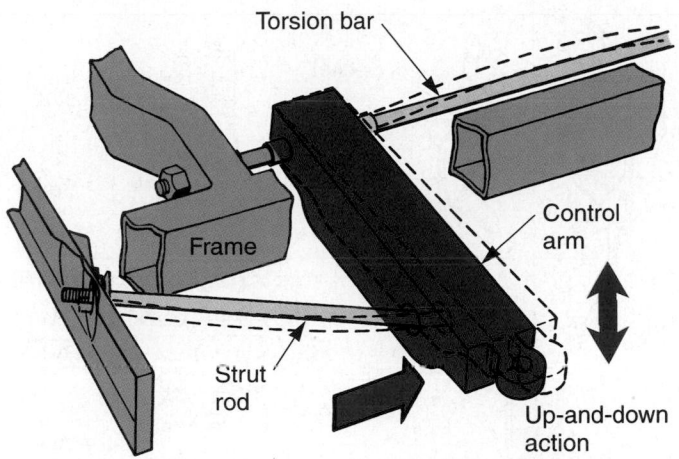

Figure 67-8. The torsion bar is twisted with control arm movement. The bar resists twisting action and acts like a conventional spring. (Moog)

Spring Terminology

There are several terms relating to springs that you should understand. A few of the most important ones are discussed in the following section.

Spring rate refers to the stiffness, or tension, of a spring. The rate of a spring is determined by the weight needed to bend it.

Sprung weight is the weight of the parts that are supported by the springs and suspension system. Sprung weight should be kept *high* in proportion to unsprung weight.

The *unsprung weight* is the weight of the parts that are *not* supported by the springs. The tires, wheels, wheel bearings, steering knuckles, and axle housing would be considered unsprung weight.

Unsprung weight should be kept *low* to improve ride smoothness. Movement of a high unsprung weight (heavy wheel and suspension components) would tend to transfer vibration to the passenger compartment.

Suspension System Construction

Now that you have been introduced to suspension system basics, we will cover the construction of each part in detail.

Control Arms

A *control arm* holds the steering knuckle, bearing support, or axle housing in position as the wheel moves up and down. Look at **Figure 67-9.**

The outer end of a control arm contains a ball joint; the inner end contains bushings. A rear suspension control arm may have bushings on both ends.

Control arm bushings act as bearings, allowing the arm to swing up and down on a shaft bolted to the frame or suspension unit. Bushings may either be pressed or screwed into holes in the control arm.

Strut Rod

A *strut rod* fastens to the outer end of the lower control arm and to the unibody or frame. It keeps the control arm from swinging toward the front or rear of the vehicle. See **Figure 67-8.**

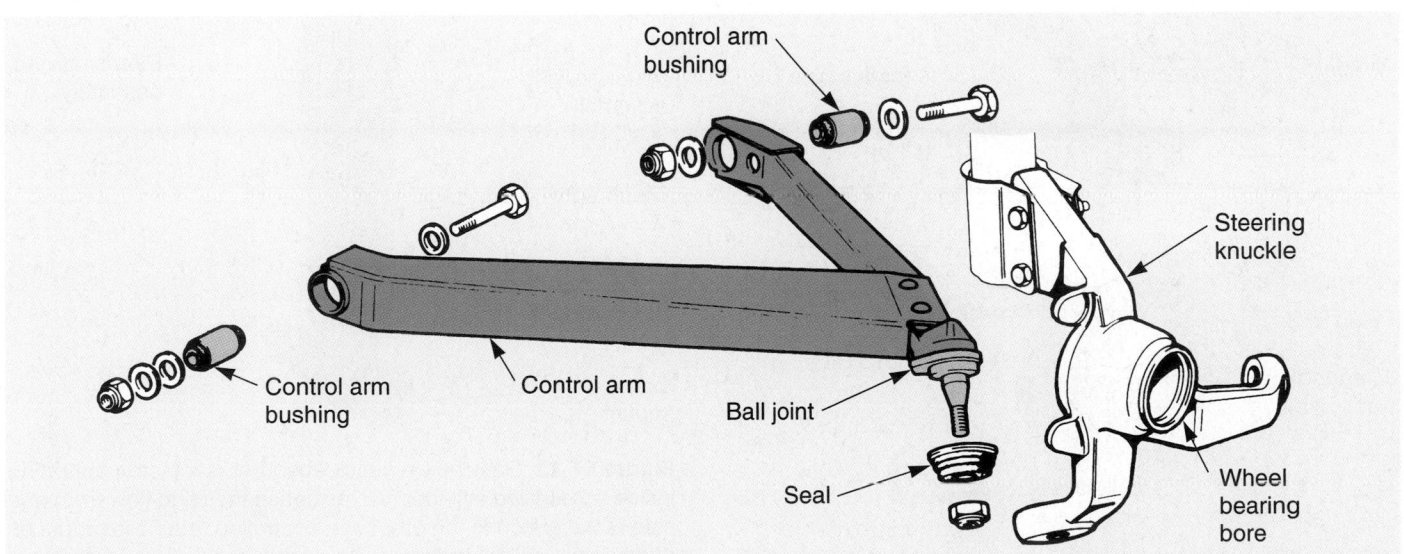

Figure 67-9. Study the basic parts of a control arm. Bushings fit into the inner ends of the arm. A ball joint fits into the outer end of the control arm. The ball joint is connected to the steering knuckle. (Fiat)

The front of the strut rod contains rubber bushings that soften the action of the rod. They permit a controlled amount of lower control arm front-to-rear flex while allowing full suspension travel.

Ball Joints

Ball joints (short for ball-and-socket joints) are connections that allow limited rotation in every direction. They connect the outer ends of control arms to the steering knuckle. See **Figures 67-9** and **67-10.**

Since the ball joint must be filled with grease, a grease fitting and a grease seal are normally placed on the joint, **Figure 67-11.** The end of the stud on the ball joint is threaded for a large nut. When the nut is tightened, it draws the tapered stud into the steering knuckle or bearing support, creating a force fit.

Shock Absorbers

Shock absorbers limit spring *oscillations* (compression-extension movements) to smooth the vehicle's ride. Without shock absorbers, the vehicle would continue to bounce up and down long after striking a dip or hump in the road. This would make the ride uncomfortable and unsafe.

Figure 67-12 shows the basic parts of a shock absorber. They include a piston rod, rod seal, piston, oil

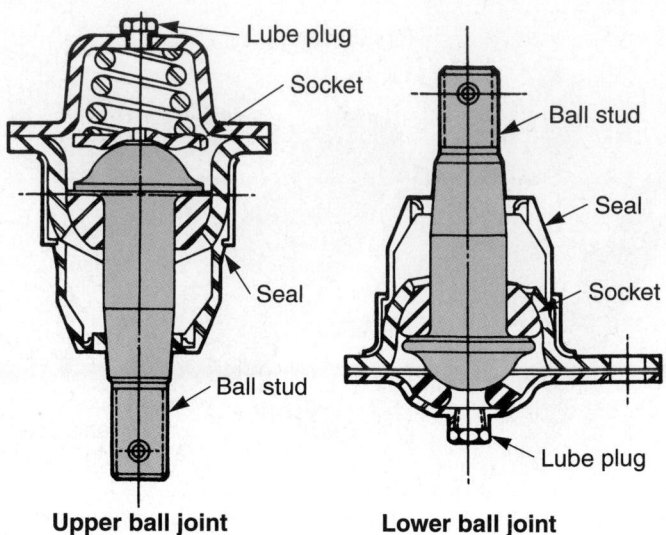

Upper ball joint   **Lower ball joint**

Figure 67-11. A ball joint is simply a ball in a socket. The ball stud is free to move in all directions. This allows the control arm and steering knuckle to move up and down freely. (Buick)

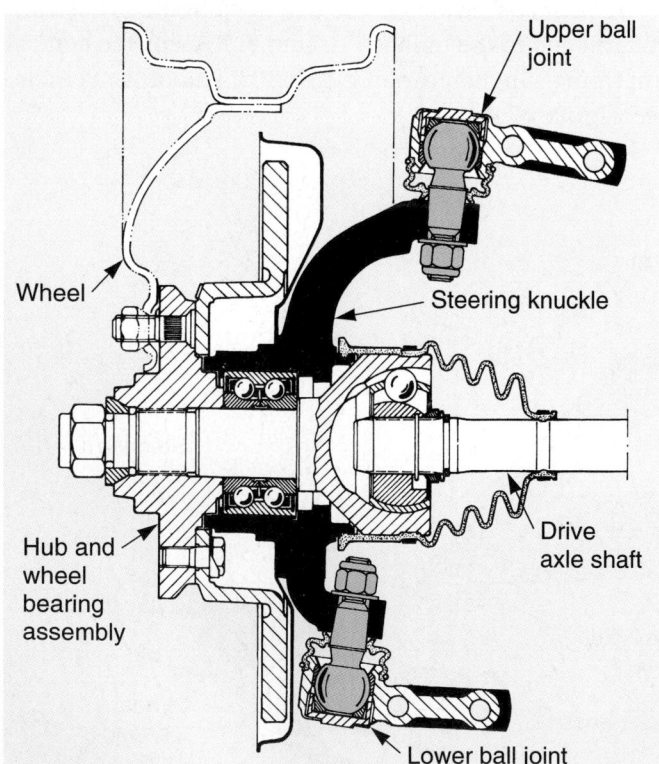

Figure 67-10. Cutaway view shows the ball joints, steering knuckle, and driving hub for a front-wheel-drive vehicle. Study the construction of the parts. (Chrysler)

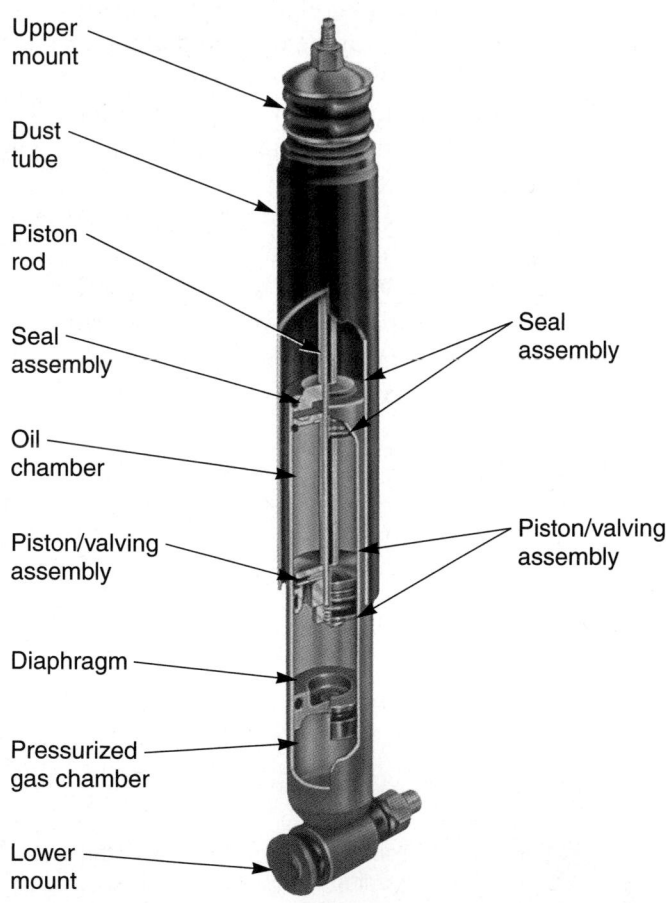

Figure 67-12. Basically, a shock absorber is a piston operating inside an oil-filled cylinder. Valves cause the oil to flow from one side of the piston to the other at a controlled rate. This produces dampening action that restricts spring oscillations. Gas-pressure dampened shocks operate like conventional oil-filled shocks. Gas is used to keep oil pressurized, which reduces oil foaming and increases efficiency on rough roads. (Pontiac)

chamber or reservoir, compression cylinder, extension cylinder, and flow control valves. Most shocks are filled with oil. Some are filled with air or gas and oil.

Whether the shock is compressed or extended, the oil causes resistance to movement. The piston rod tends to drag slowly in or out. This dampens spring and suspension system action.

One end of the shock absorber connects to a suspension component, such as a control arm. The other end of the shock fastens to the vehicle's body or frame. In this way, the shock rod is pulled in and out and resists suspension system movements.

Shock absorber operation is illustrated in **Figure 67-13.** *Shock absorber compression* occurs when the vehicle's tire is forced upward upon hitting a bump. *Shock absorber extension* is the outward movement of the piston and rod as the control arm moves down. This occurs right after a compression stroke or when the tire encounters a hole in the road.

Gas-Charged Shock Absorbers

Gas-charged shock absorbers use a low-pressure gas to help keep the oil in the shock from foaming. Refer back to **Figure 67-12.**

Usually, hydrogen gas is enclosed in a chamber separate from the main oil cylinder. The shock piston operates in the oil. The gas maintains constant pressure on the oil to stop air bubbles from forming.

Self-Leveling Shock Absorbers

A *self-leveling shock absorber* uses a special design that causes a hydraulic lock action to help maintain normal vehicle curb height.

A valve system in each shock retains hydraulic pressure when the shock is compressed near its minimum length setting. This hydraulic lock action helps keep the shock rod at the same length with changes in force or curb weight. This tends to help keep the vehicle level under varying conditions.

Adjustable Shocks

Adjustable shock absorbers provide a means of changing shock stiffness. Usually, by turning the shock outer body or an adjustment knob, you can set the shock soft for a smooth ride or stiff for better handling.

Some electronic (computer-controlled) suspension systems automatically change shock dampening stiffness. These are discussed on page 1258.

Strut Assembly

A *strut assembly* consists of a shock absorber, a coil spring (most types), and an upper damper unit. The strut often replaces the upper control arm. Only the lower

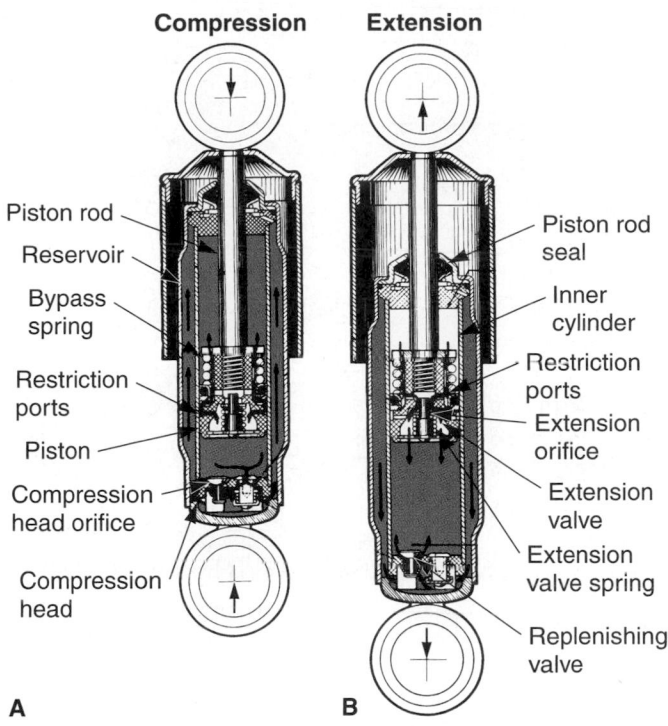

Figure 67-13. Cutaway view of shock absorber in action. A—Compression stroke forces the piston down in the cylinder. Oil is forced into the upper area of the shock. B—Extension stroke causes oil to be pulled back into the lower area. Note the valve action. (Gabriel)

control arm and the strut are needed to support the front wheel assembly. Look at **Figure 67-14.**

The basic parts of a typical strut assembly are shown in **Figure 67-15.** They include:

- *Strut shock absorber*—piston operating in an oil-filled (or oil and gas) cylinder to prevent coil spring oscillations.

- *Dust shield*—metal shroud or rubber boot that keeps road dirt off the shock absorber rod.

- *Lower spring seat*—lower mount formed around the shock body for the coil spring.

- *Coil spring*—supports the weight of the vehicle and allows suspension action.

- *Upper spring seat*—holds the upper end of the coil spring and contacts the strut bearing.

- *Strut bearing*—ball bearing that allows the shock-and-spring assembly to rotate for steering action; only used on the front of the vehicle.

- *Rubber bumpers*—jounce and rebound bumpers that prevent metal-to-metal contact during extreme suspension compression and extension.

- *Rubber isolators*—prevent noise from transmitting into the body structure of the vehicle.

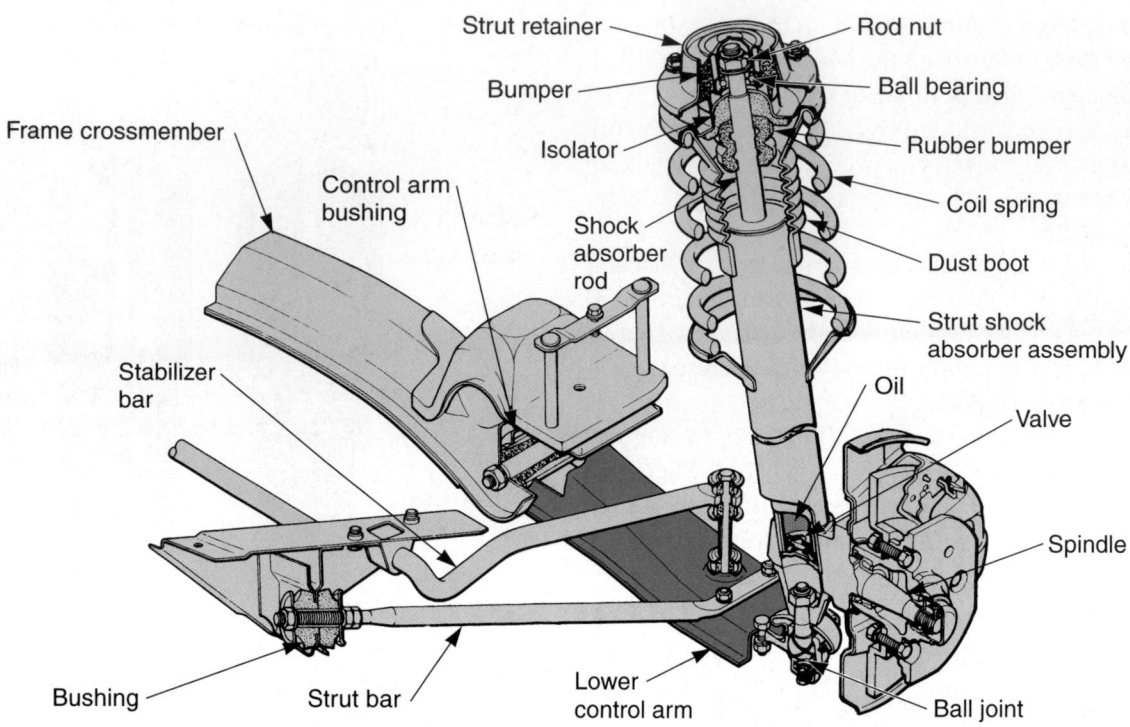

Figure 67-14. Study the parts of this strut assembly closely. This is one of the most modern suspension systems and is commonly used on today's vehicles. (Chrysler)

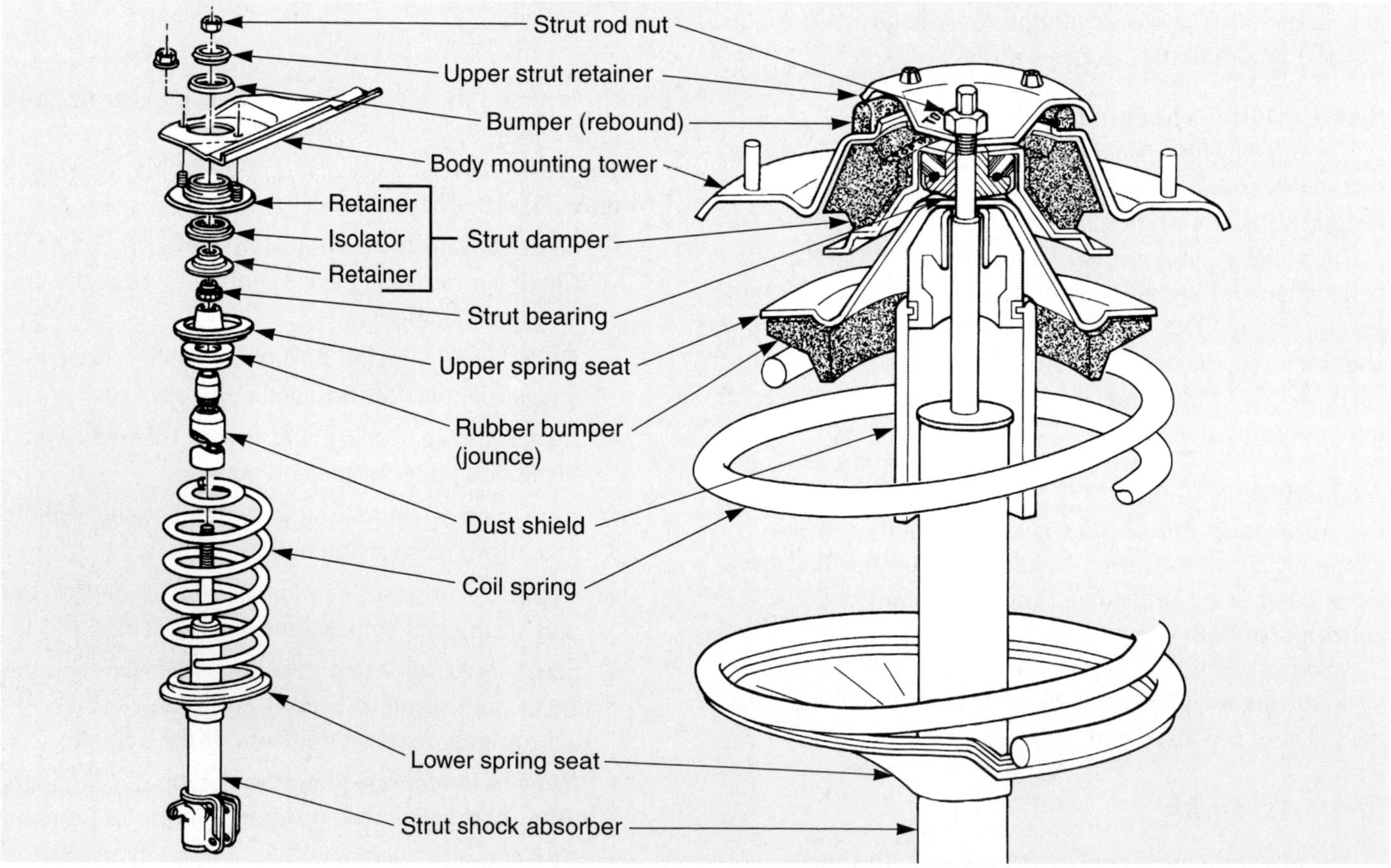

Strut retainer — Rod nut
Bumper — Ball bearing
Isolator — Rubber bumper
Shock absorber rod — Coil spring
Frame crossmember — Dust boot
Control arm bushing — Strut shock absorber assembly
Stabilizer bar — Oil — Valve
Bushing — Strut bar — Lower control arm — Spindle — Ball joint

Strut rod nut
Upper strut retainer
Bumper (rebound)
Body mounting tower
Retainer
Isolator — Strut damper
Retainer
Strut bearing
Upper spring seat
Rubber bumper (jounce)
Dust shield
Coil spring
Lower spring seat
Strut shock absorber

Figure 67-15. Exploded and cutaway views of a strut. Note the strut bearing, which allows the front wheel, steering knuckle, and strut to revolve for steering action. (Chrysler)

- *Upper strut retainer*—secures the upper end of the strut assembly to the frame or unibody.
- *Strut rod nut*—hex nut that holds the shock absorber rod in the upper strut retainer.
- *Damper unit*—shock that fits inside coil spring to prevent excessive jounce and rebound.

A strut shock absorber is similar to a conventional shock absorber. However, it is longer and has provisions (brackets or connections) for mounting and holding the steering knuckle (front of vehicle) or bearing support (rear of vehicle) and spring.

Sway Bar (Stabilizer Bar)

A *sway bar,* or *stabilizer bar,* is used to keep the body from leaning excessively in sharp turns. The sway bar is made of spring steel. It fastens to both lower control arms and to the frame. Rubber bushings fit between the bar and the control arms. Brackets and bushings are used to secure the bar to the frame. *Sway bar links* connect the sway bar to the control arms. See **Figure 67-16.**

When a vehicle rounds a corner, centrifugal force makes the outside of the body drop and the inside of the body rise. This twists the sway bar. The bar's resistance to this twisting motion limits body lean in corners.

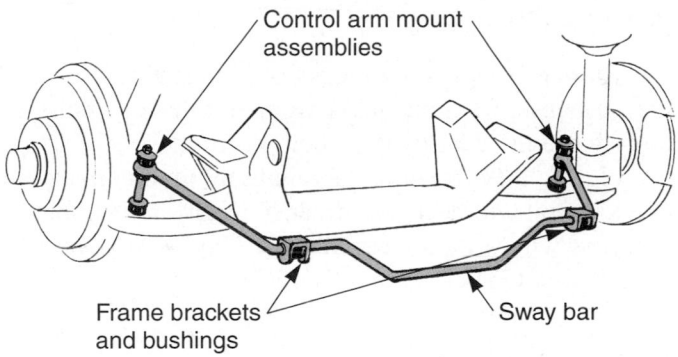

Figure 67-16. The sway bar attaches to both control arms. When the car rounds a corner, its body tends to lean to one side. This bends the bar. As a result, the bar lessens sway, or body lean, in turns. (Moog)

Track Rod (Lateral Control Rod)

A *track rod,* also known as a *lateral control rod,* is sometimes used on rear suspension systems to prevent side-to-side axle movement during cornering. The track rod is almost parallel to the rear axle. One end of the rod is fastened to the axle; the other end of the rod is fastened to the frame or body structure on the opposite side of the vehicle, **Figure 67-17.**

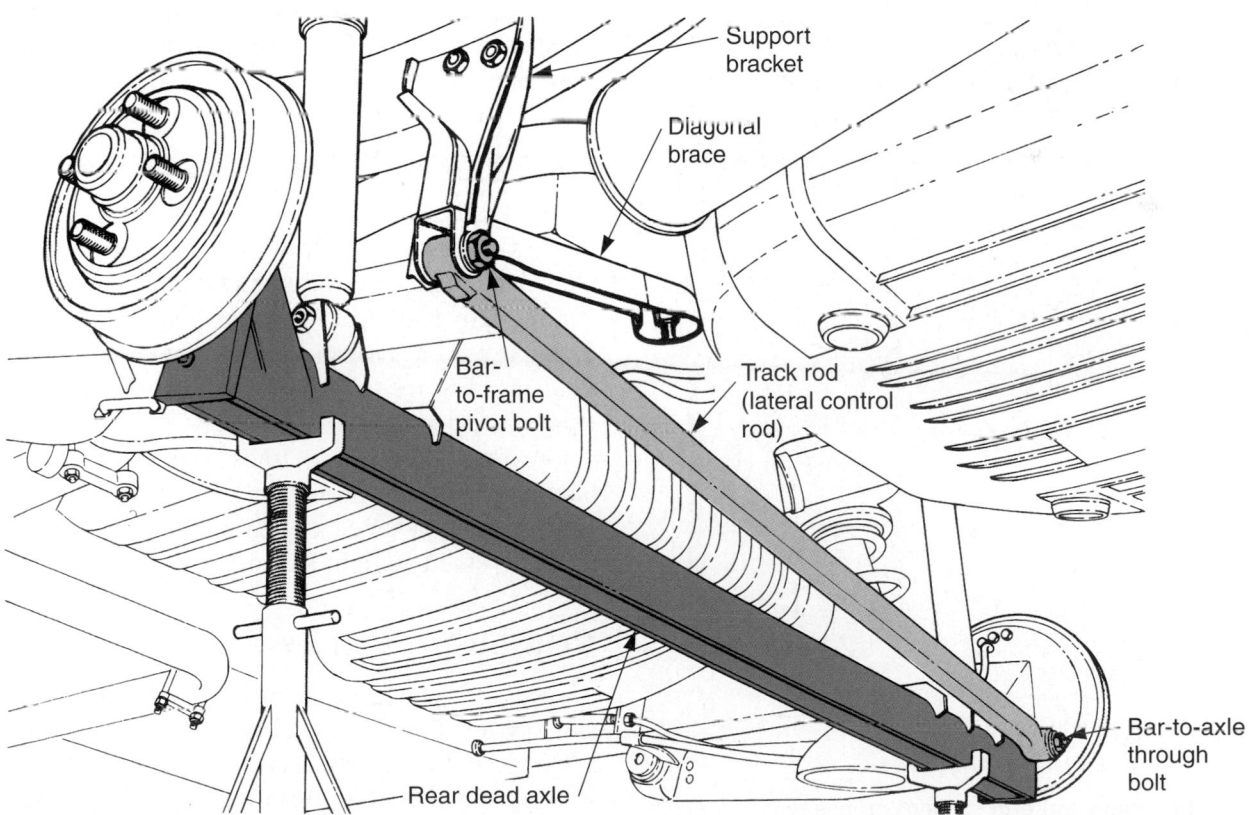

Figure 67-17. A track rod is commonly used on the rear axle to prevent side-to-side movement. Note how the rod connects to the frame and axle. (Chrysler)

Jounce Bumpers

Jounce bumpers are blocks of hard rubber that keep the suspension system parts from hitting the frame or body when the car hits large bumps or holes.

If you drive over a large hump in the road too quickly, you will hear the suspension arms hit the jounce bumpers with a loud bang or thud. This warns you that the tires are leaving the road surface.

Long-Short Arm Suspension

A *long-short arm suspension* uses control arms of different lengths to keep the tires from tilting with suspension action. The upper control arms are shorter than the lower control arms. See **Figure 67-18.**

If the control arms were the same length, the front tires would pivot outward at the top when the vehicle hits a bump. This would cause undue tire scuffing and wear.

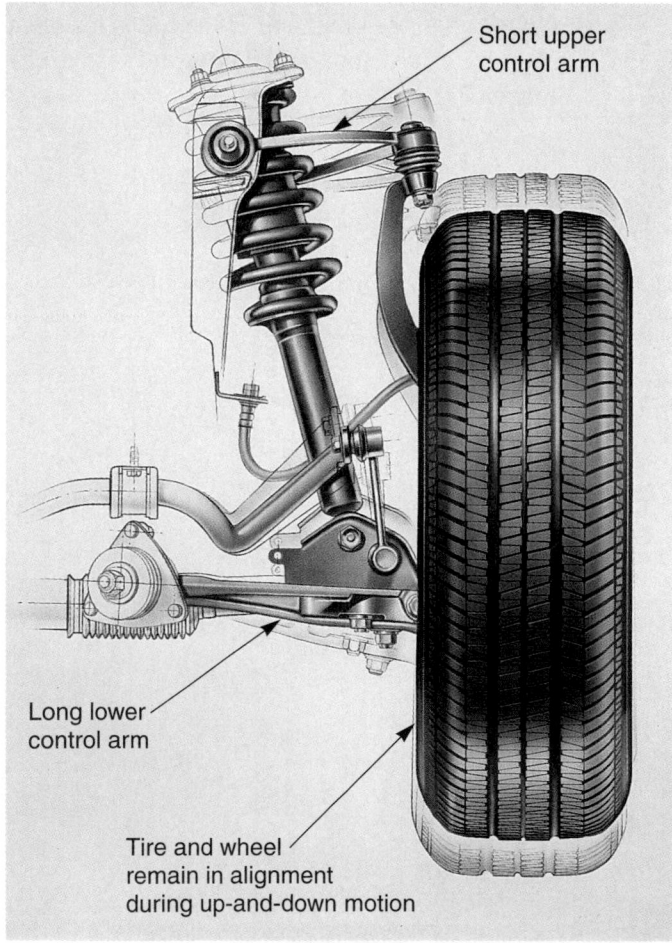

Figure 67-18. Long-short arm suspension has different length control arms. This helps keep the steering knuckle in alignment with suspension travel. (Lexus)

Torsion Bar Suspension

A *torsion bar suspension* is a suspension system that contains torsion bar springs instead of coil springs. Most torsion bar suspensions allow easy adjustment of *curb height* (distance from road up to specific point on car). By turning an adjustment bolt, you can increase or decrease the tension on the torsion bar. This will either raise or lower the frame and/or body of the vehicle. A torsion bar suspension was shown in **Figure 67-3.**

MacPherson Strut Suspension

A *MacPherson strut suspension* uses only *one* control arm and a *strut assembly* (spring, damper, and shock absorber unit) to support each wheel assembly, **Figure 67-19.** Note that the *modified strut suspension* has the coil spring mounted on top of the lower control arm, not around the strut.

A conventional lower control arm attaches to the frame and to the lower ball joint. The ball joint holds the control arm to the steering knuckle or bearing support. The top of the steering knuckle or bearing support is bolted to the strut assembly. The top of the strut is fastened to the reinforced body structure, **Figure 67-20.**

MacPherson strut suspension is the most common type of suspension found on late-model cars. It may be used on both the front or rear wheels. It reduces the number of parts in the suspension system, lowering the unsprung weight and smoothing the ride, **Figure 67-21.**

Pickup Truck Suspension Systems

Pickup trucks use numerous suspension system designs: long-short control arm; MacPherson strut; solid axle; and twin axle, or twin I-beam, suspension. The control arm and strut types are basically the same as those used on passenger cars, but are heavier and stronger. **Figure 67-22** shows a twin-axle, or *twin I-beam,* suspension.

A four-wheel drive truck can have a solid axle housing and differential in the front. The steering knuckles are mounted on the axle housing so that they will swivel left or right for cornering.

Rear Suspension Systems

Rear suspension systems are similar to front suspension systems, but they normally do *not* provide for steering. With a rear-wheel-drive vehicle, the rear axle housing may be solid, resulting in nonindependent suspension. However, rear swing axles and independent suspension can also be used.

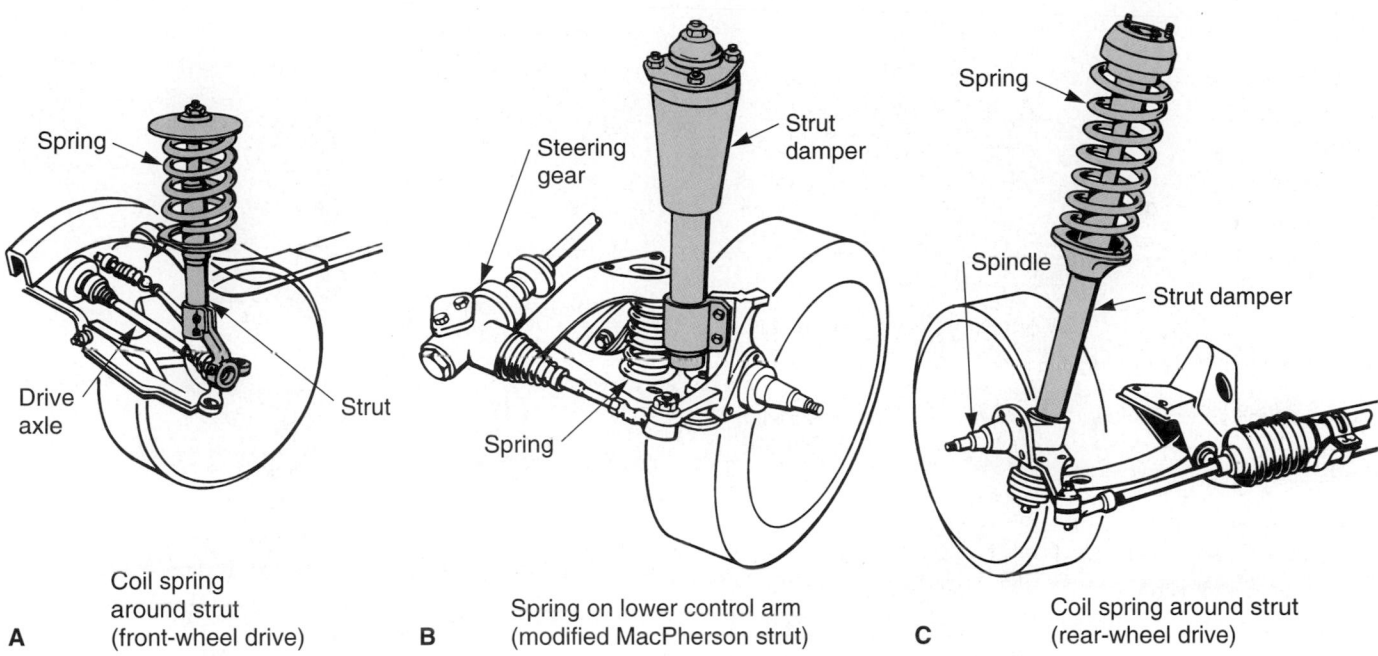

A Coil spring around strut (front-wheel drive)

B Spring on lower control arm (modified MacPherson strut)

C Coil spring around strut (rear-wheel drive)

Figure 67-19. MacPherson strut suspensions. A—Coil spring around strut, front-wheel drive. B—Modified strut has the coil spring mounted on control arms. C—Same as A, but without front-wheel drive. (Moog)

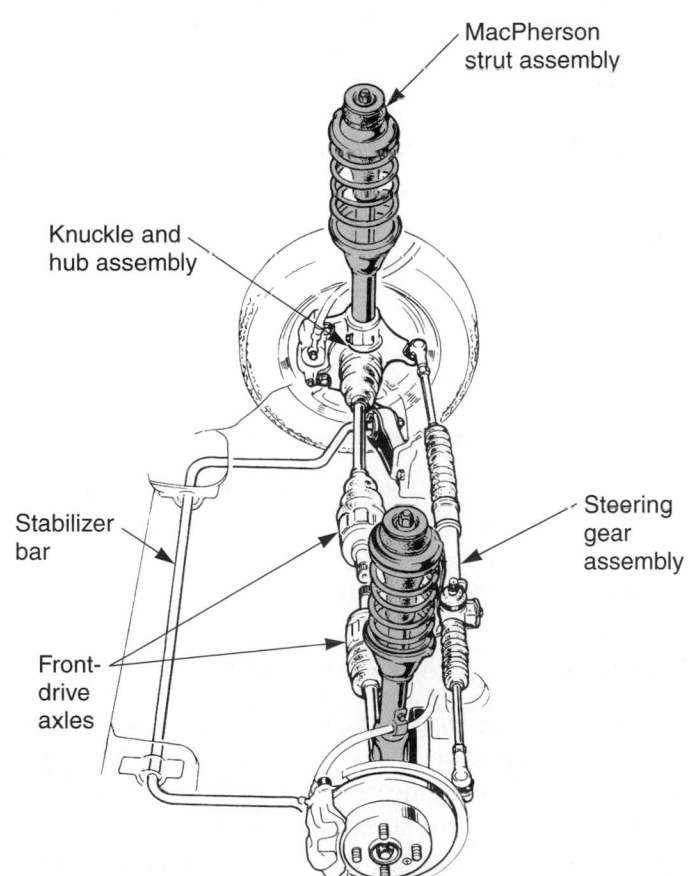

Figure 67-20. MacPherson strut suspension. (Honda)

Nonindependent Rear Suspension

Figure 67-23 shows a typical rear suspension setup for a rear-wheel-drive vehicle. It has a solid axle housing and vertically mounted shock absorbers. Note how the coil springs are mounted between the axle housing and frame of the vehicle.

Dead Axle

A *dead axle* is a term used to describe a solid rear axle on a front-wheel-drive vehicle. Since the front wheels transfer driving power to the road, the rear axle is simply a straight, or solid, axle, **Figure 67-24.**

Semi-Independent Suspension

Semi-independent suspension means that the right and left wheel are partially independent of each other. This type suspension uses a flexible axle, **Figure 67-24.**

When one tire hits a bump, its control arm moves up. Since the axle can flex or twist, the effect on the other tire is minimized.

Independent Rear Suspension

Many new cars use independent rear suspension. As with front suspension, independent suspension increases

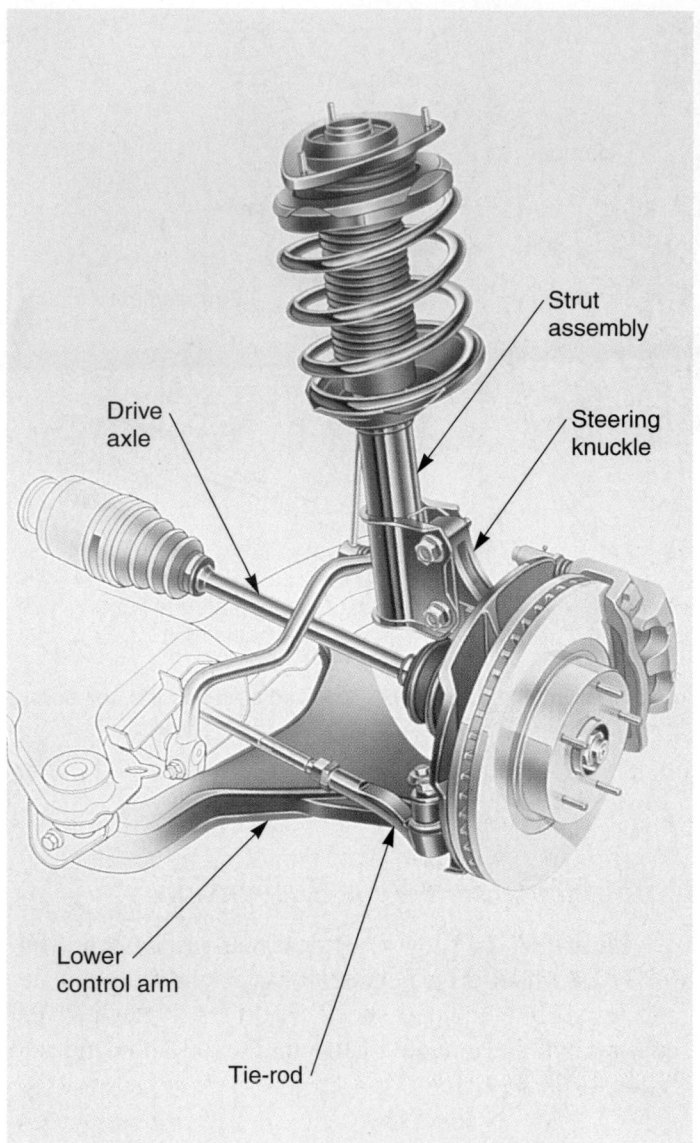

Front suspension

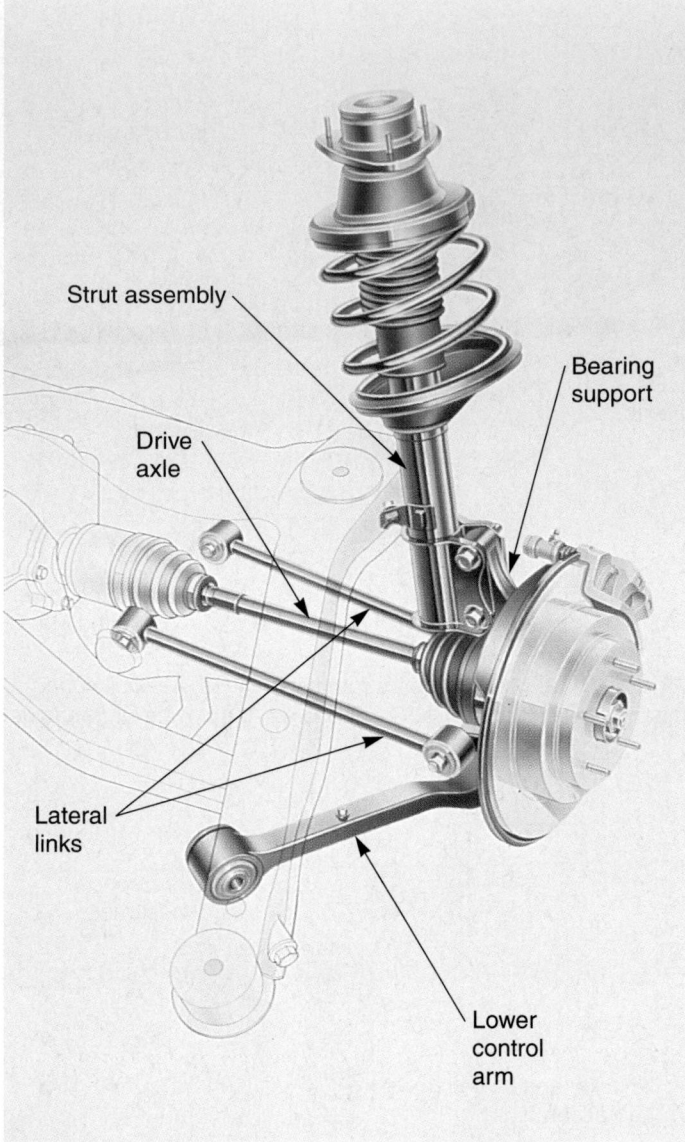

Rear suspension

Figure 67-21. Compare struts for the front and rear of a vehicle. (Subaru)

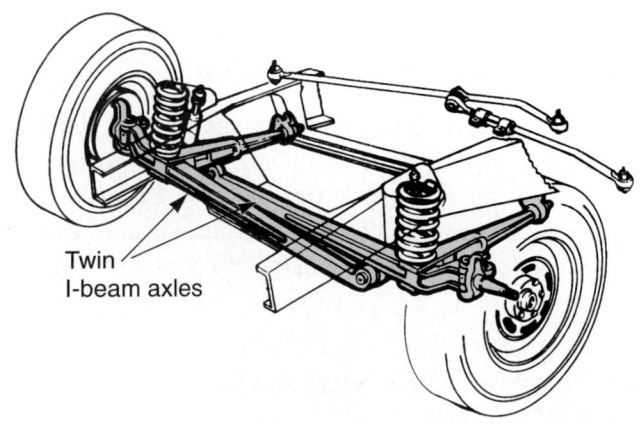

Figure 67-22. Twin I-beam suspension is used on a few pickup trucks. (Ford)

ride smoothness. This type of suspension can be used with either a front- or rear-wheel-drive vehicle.

Refer to **Figures 67-25** through **67-27.** They show typical suspension system designs used on modern vehicles.

Suspension Leveling Systems

A *suspension leveling system* is used to maintain the same vehicle *attitude* (body height) with changes in the amount of weight in the car. For example, if weight is added in the trunk, the suspension leveling system keeps the springs from compressing and lowering the body height.

There are two classifications of suspension leveling systems: manual and automatic.

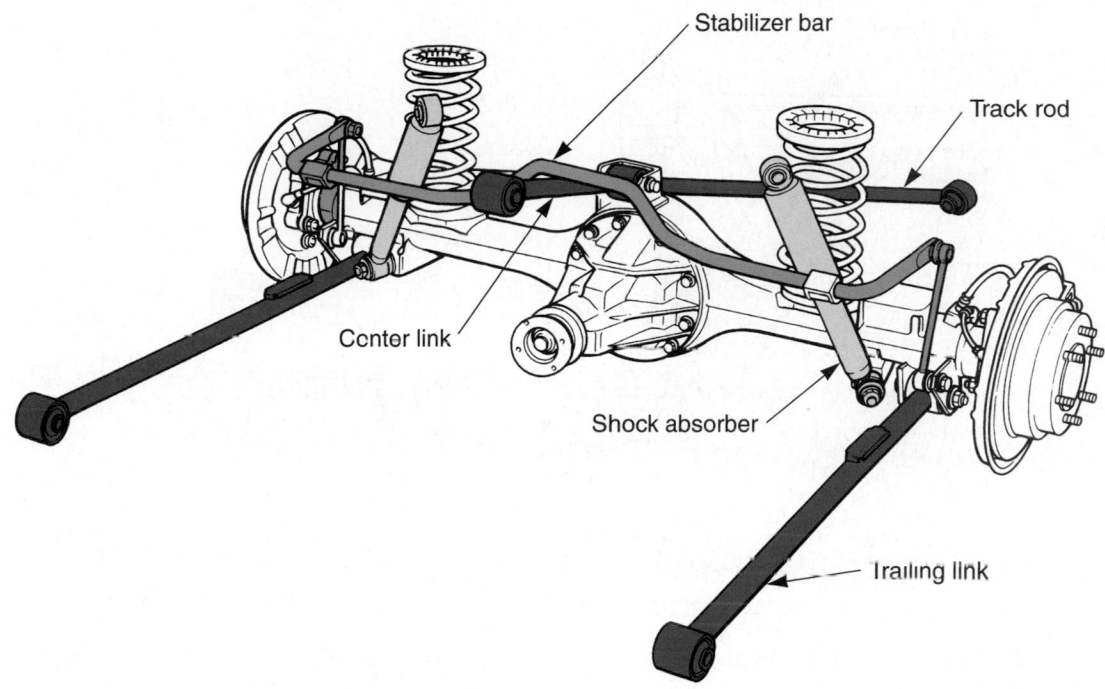

Figure 67-23. Solid axle housing rear suspension for rear-wheel-drive vehicle. Study the parts. (Isuzu, General Motors)

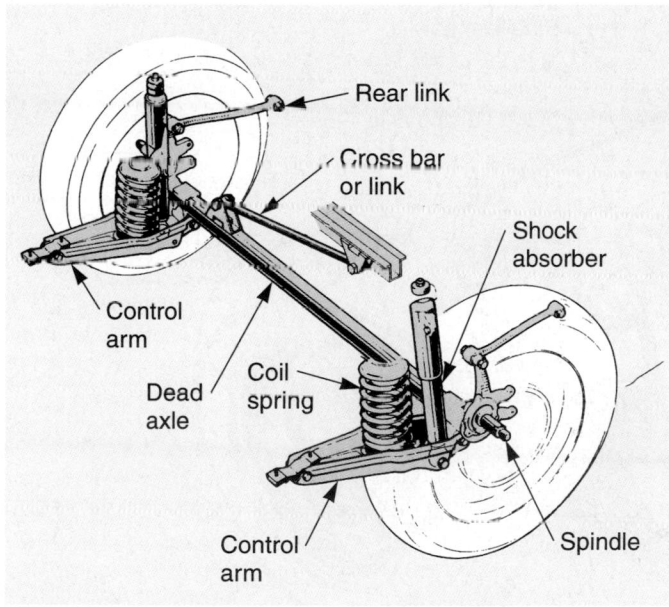

Figure 67-24. The term dead axle generally refers to a solid axle that does not drive wheels. (Saab)

Manual Suspension Leveling System

A *manual suspension leveling system* uses air shocks and an electric compressor to counteract changes in passenger and luggage weight. A manual switch is used to activate the compressor to alter air shock pressure and body height.

Automatic Suspension Leveling System

Automatic suspension leveling systems use air shocks or air springs, height sensors, and a compressor to maintain curb height. System designs vary.

Figure 67-28 shows an automatic suspension leveling system that uses air shock absorbers. A height sensor is connected to the frame and to the axle housing. If load changes, the sensor can turn the compressor on, increasing pressure in the air shocks to counteract the increased load. It can also bleed air out of the shocks to counteract decreased load.

Figure 67-29 shows an automatic air suspension system that uses *air springs* and height sensors on all four wheels. A microcomputer uses information (electric signals) from the sensors to operate the air compressor.

Electronic Height Control

An *electronic height control system* uses height sensors and an electronic control module to control the operation of a small electric air compressor, which maintains the correct ride height. This system is used on the rear of the car to compensate for loads placed in the trunk or for the weight of passengers in the back seat. The main parts of the electronic height control system include:

- *Height sensor*—lever-operated switch that reacts to changes in body height and suspension movement.

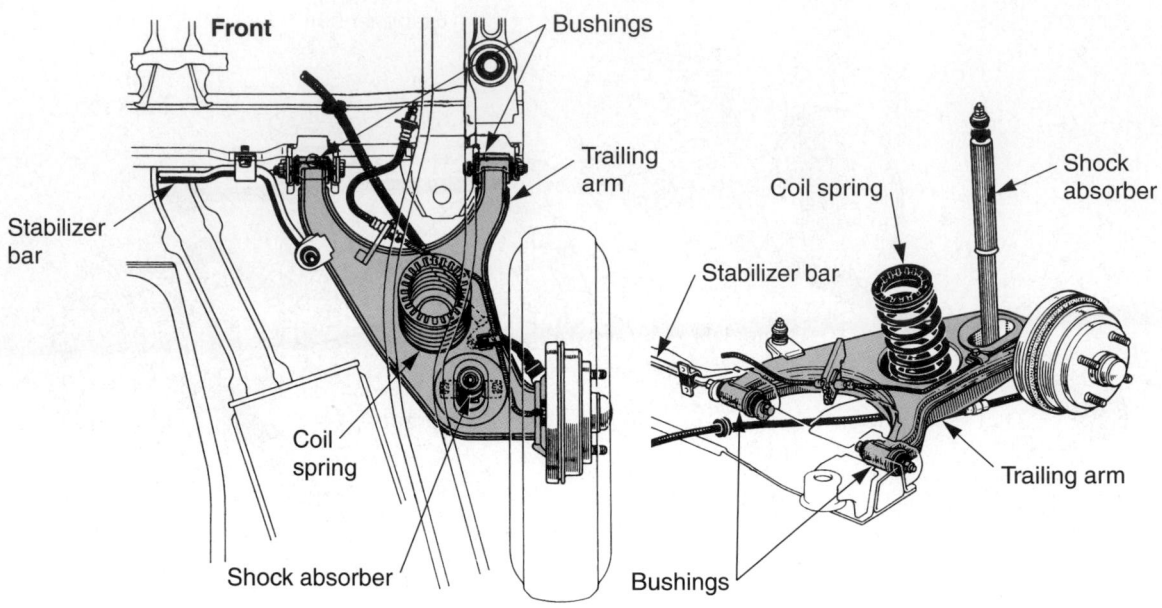

Figure 67-25. Top and side views of a trailing-arm independent rear suspension. Note the location of the bushings, the spring, and the shock absorber. (Toyota)

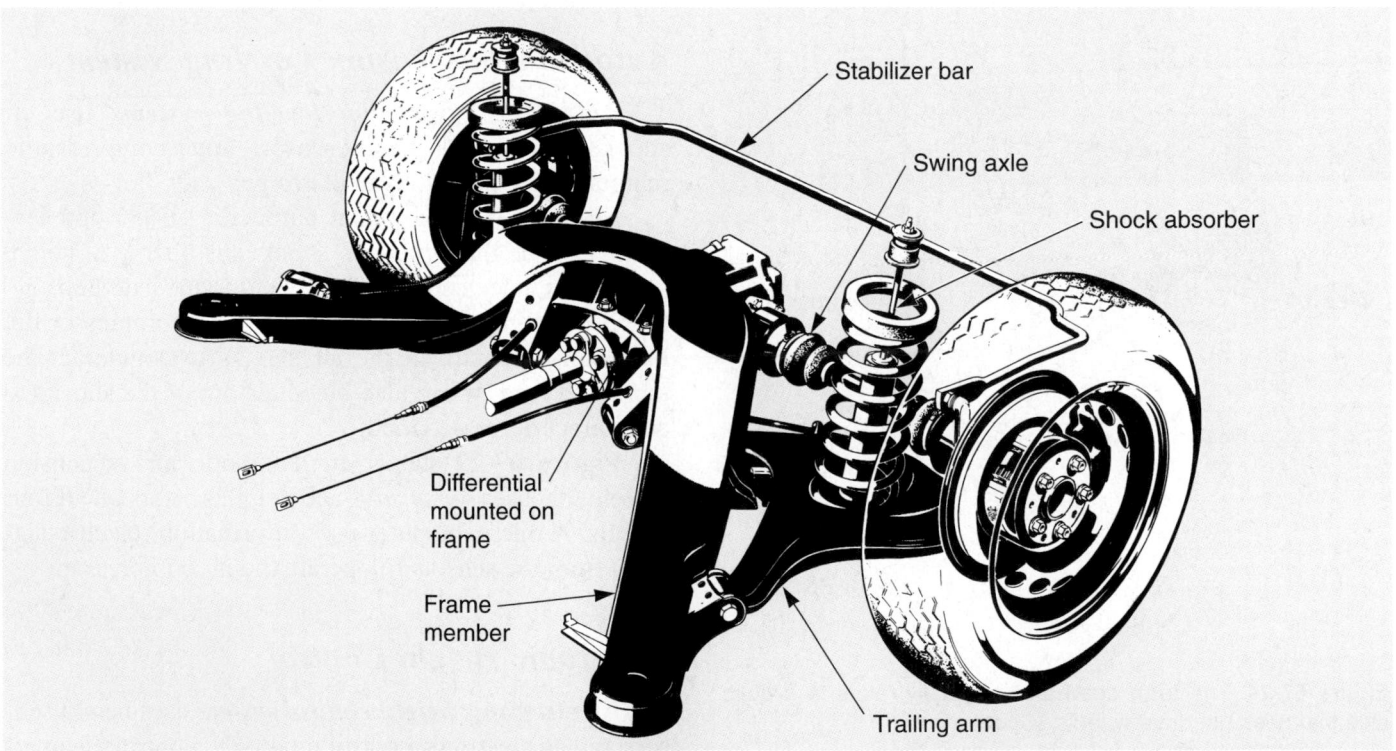

Figure 67-26. This rear drive axle uses a differential that is mounted on the frame. Swing axles extend out to the drive wheels. Note the trailing arms and the other components. (Mercedes Benz)

- *Compressor assembly*—motor-powered air pump that produces pressure for the system.
- *Pressure lines*—air hoses that connect the compressor to the air shock absorbers.

- *Air shocks*—air-filled shock absorbers that act on the suspension system to alter ride height.
- *Sensor link*—linkage rods that connect the height sensor to the suspension.

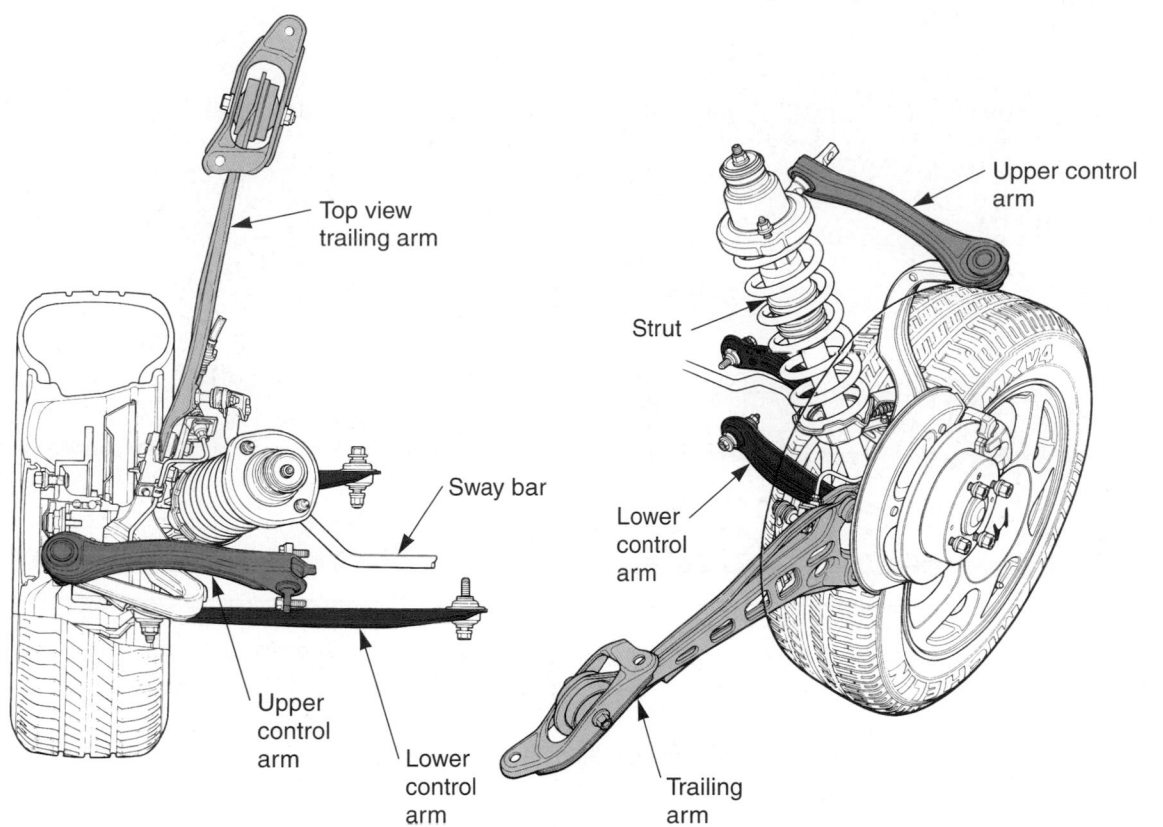

Figure 67-27. This double wishbone suspension system is a long-short arm design. Note how the lower trailing arms have been lightened by forming holes in them. This improves ride smoothness. (Honda)

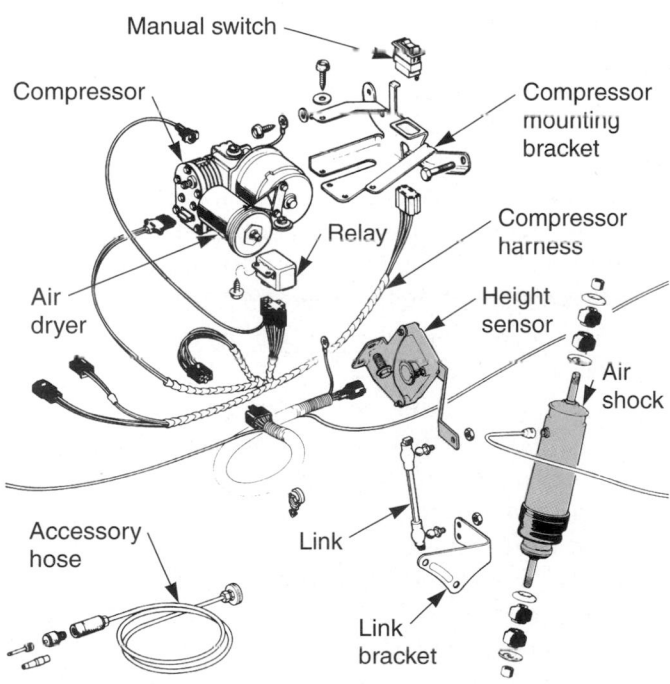

Figure 67-28. Major parts of suspension leveling system. Basically, the height sensor triggers the air pump. The air pump operates the air shocks to maintain the correct vehicle height. (Chrysler)

- **Solenoid valve**—solenoid-operated air valve that can release air pressure from the system.

- **Suspension control module**—small computer that operates the solenoid valve by responding to signals from the height sensor.

When the car body is at a normal riding height, the electronic height control system is off. Air pressure in the shocks is adequate to keep the car body the correct distance from the road surface. The height sensor does not feed current to the control module.

If the trunk is loaded with heavy luggage, the weight of the luggage will compress the rear air shocks. This will lower the ride height. When the car is started, the height sensor will be activated by the action of the sensor link and the sensor switch will close. The signal from the height sensor prompts the suspension control module to energize the compressor. The compressor will pump more air pressure into the rear shock absorbers, extending the shocks and raising the car body.

When the specific ride height is reached, the height sensor switch will open to turn off the compressor. This restores the vehicle to the correct ride height, even with extra weight in the trunk.

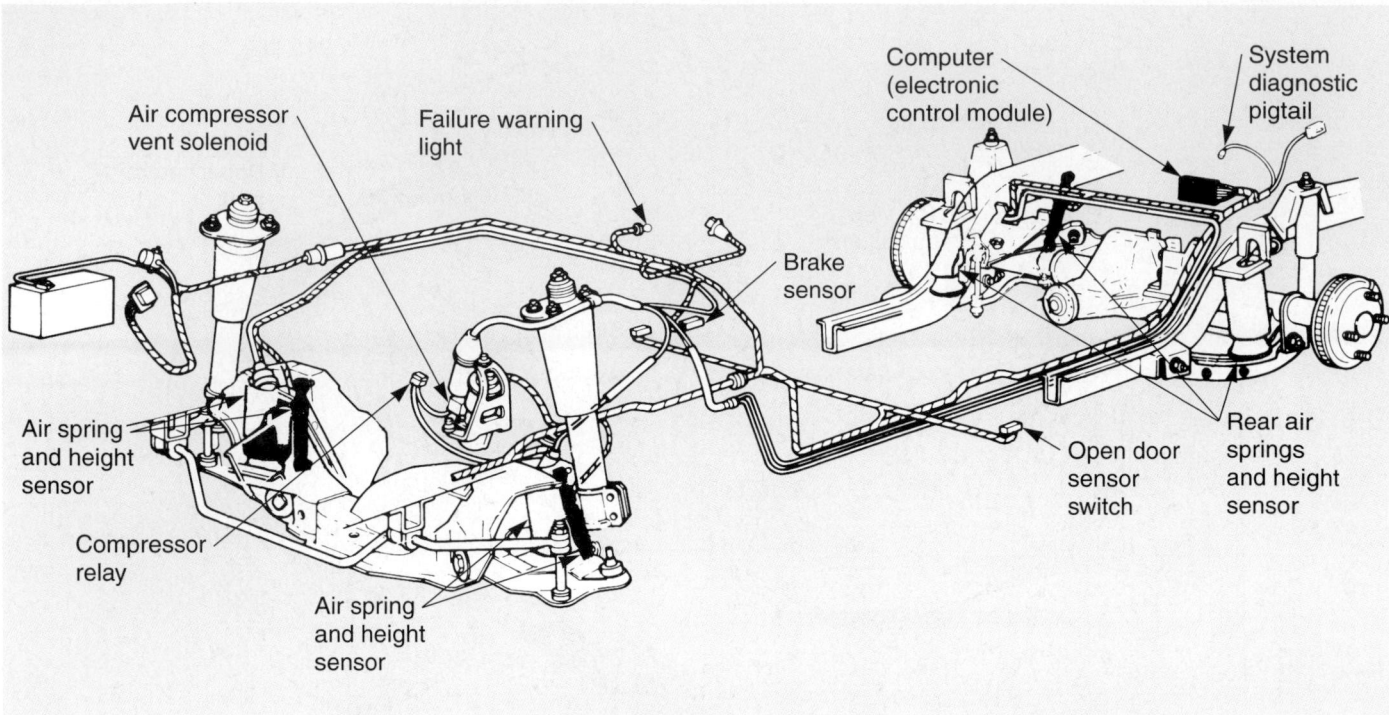

Figure 67-29. Suspension leveling system using air springs. Study the parts. (Ford)

When the weight is removed from the trunk, the car body will rise. The height sensor switch will then be moved in the other direction by the link. This closes another set of contacts in the switch, energizing the pressure release solenoid valve. Air pressure is then expelled from the rear shocks until the body drops down the correct ride height.

Electronic Suspension System

An *electronic suspension system* uses various sensors, a computer, and shock absorber actuators to control ride stiffness. It is designed to increase comfort and safety by matching suspension system action to driving conditions, **Figure 67-30.**

Although exact designs vary, the major components of a typical electronic shock absorber system include:

- *Steering sensor*—detects steering wheel rotational direction and speed and feeds data about vehicle direction to the computer.
- *Brake sensor*—usually, the brake light switch is used to report when brakes are applied.
- *Acceleration sensor*—usually, the throttle position sensor is used to detect when car is accelerating rapidly.
- *Mode switch*—dash-mounted switch that allows the driver to choose the desired shock action or ride stiffness.

- *Electronic control module*—small computer that uses sensor inputs to control the shock actuators.
- *Shock actuators*—solenoid-operated valves that control fluid flow inside the shock absorbers.

If a car is being driven on curving country roads, the driver might switch to a stiff setting with the mode switch. The electronic control module would then energize the shock actuators to close or restrict the shock absorber valves to increase dampening action. This would stiffen the ride and make the car corner better.

If driving on a rough highway, the mode switch might be moved to a soft setting. The electronic control module would then energize the shock actuators to open the valves more. This would soften the ride by allowing easier shock movement.

Under hard braking, the brake sensor would send a signal to the electronic control module. The control module could then stiffen the shocks to prevent the front of the car from diving.

With rapid turning or cornering, the steering sensor could also signal the electronic control module. The module could then stiffen the shocks to prevent excess body roll, or lean, in turns.

Figure 67-31 shows one type of shock actuator. Note how it uses a solenoid and a small DC motor to act upon the shock absorber control rod. The control rod can be moved up or down to control fluid flow resistance and shock stiffness or dampening.

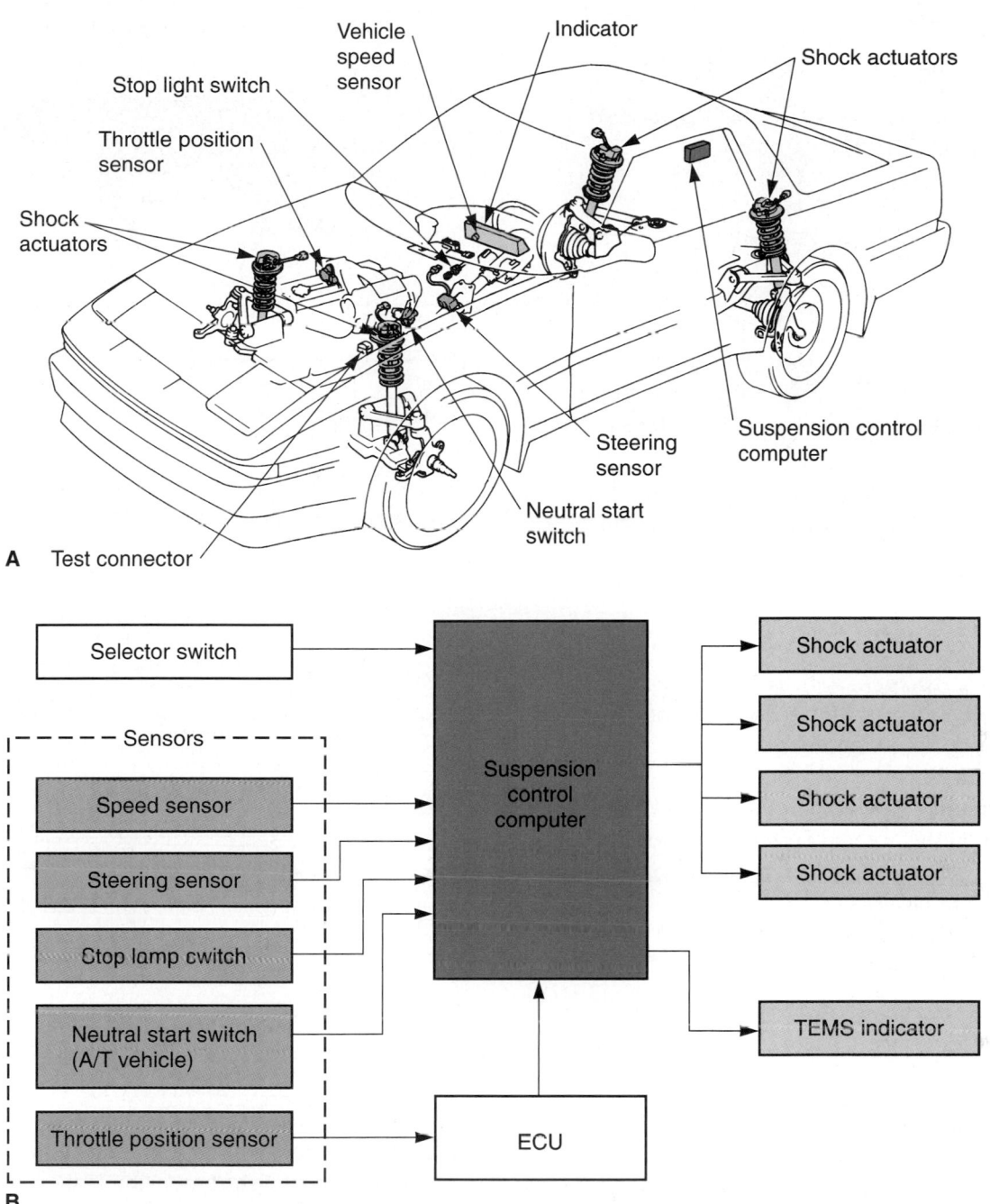

Figure 67-30. A—This electronic suspension system automatically adjusts shock stiffness to match driving and road conditions. For instance, it will produce a soft, smooth ride when traveling down a straight highway. However, it will stiffen shock dampening when cornering on a country road. B—This block diagram shows how various sensors feed electrical data to the suspension control computer. The computer can then energize the shock actuators to modify ride stiffness and shock action. (Toyota)

Some cars use air shocks or air bags instead of hydraulic shock absorbers. The operation of these systems is similar.

One type of electronic shock absorber system uses a sonar sensor to detect changes in road conditions. The sensor, which is mounted at the front of the vehicle, produces sound waves. These waves bounce off the road and back into the sensor. The sensor's output signal is determined by the amount of time needed for the waves to strike the road surface and bounce back. The computer uses the output signals from the sonar sensor to determine proper ride stiffness. If, for example, there is a dip in the road, the sound waves will take longer to return to the sensor. This will cause the sensor to modify its output signal. The computer analyzes the sensor's output signals and adjusts shock action as necessary.

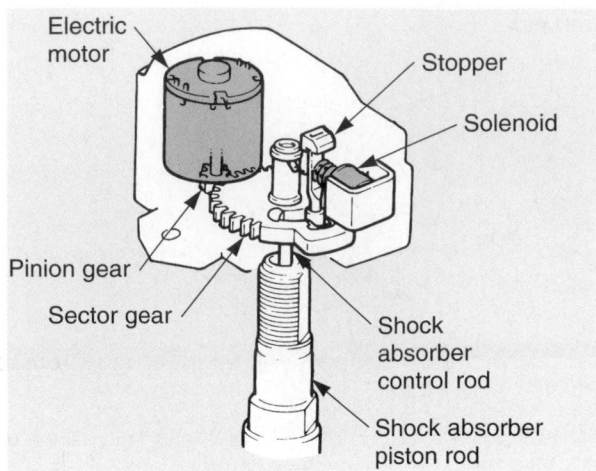

Figure 67-31. This shock actuator uses small electric motor and a solenoid to move the shock piston rod in and out. Piston rod movement alters shock dampening action.

Active Suspension System

An *active suspension system* uses computer-controlled hydraulic rams instead of conventional springs and shock absorber actuators to control ride characteristics, **Figure 67-32.**

The hydraulic rams support the weight of the car and react to the road surface and driving conditions. An active suspension system is similar to an electronic shock absorber system, but it is more complex.

Basically, pressure sensors on each hydraulic ram provide the main control for the system. They react to suspension system movement and send signals to the computer. The computer can then extend or retract each ram to match the road surface. A hydraulic pump provides pressure to operate the rams, **Figure 67-33.**

Pressure control valves are located on each ram. By opening and closing these valves, the computer can adjust the pressure of the rams and the resulting height of each corner of the vehicle.

For example, if one side of the vehicle travels over a bump in the road, the pressure sensors can instantly detect a rise in pressure inside the ram as the tire and wheel push up on the suspension and the hydraulic ram. Instead of the vehicle's body rising with spring action, the computer can release enough ram pressure to allow the suspension to move up over the bump without causing body movement. Then, as the tire travels back down over the bump, the sensor detects a pressure drop in the ram and the computer increases ram pressure so the tire follows the road surface.

The active suspension system can make a car feel as if it is floating on a cushion of air. It can theoretically eliminate most body movement as the car travels over

Active suspension compared to the cheetah

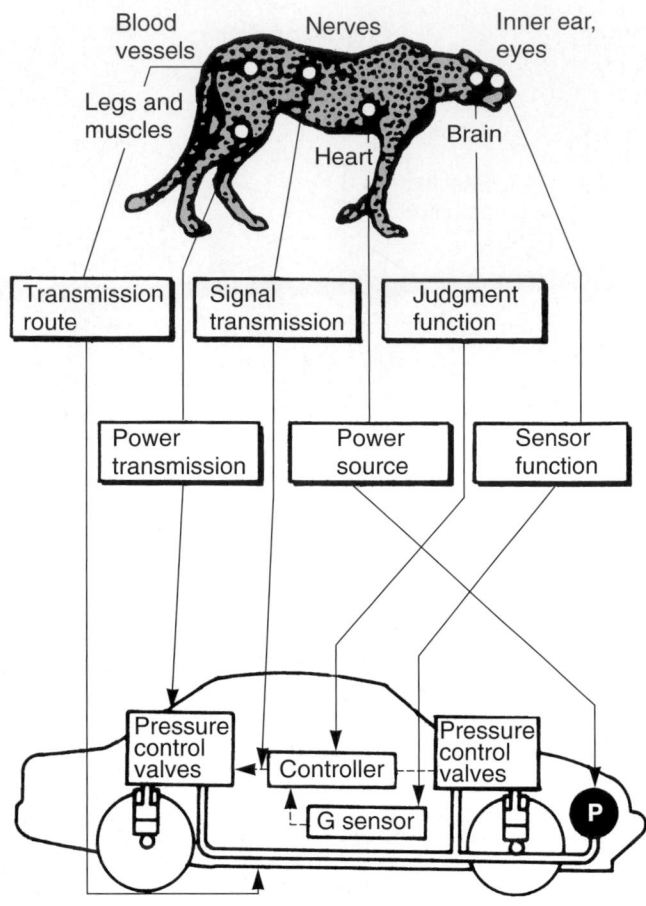

Figure 67-32. Fully active suspension is called "intelligent suspension" because a computer is used to control the hydraulic system. (Infiniti)

small dips and bumps in the road. It can also help keep the body level under various driving conditions. During hard braking, it can keep the front of the body from diving and the rear from rising, thus improving action.

In turns, the active suspension system's ability to prevent body roll (lean) can make the car stay level to increase cornering ability. It can tilt the vehicle's body against a turn to improve handling. It can also be used to lower the body for highway driving aerodynamics and to raise the body for added ground clearance during city driving.

Many auto manufacturers are experimenting with active suspension systems of this type. We may see them in the future, at first on more exotic sports cars, but someday on everyday passenger cars.

Summary

- The suspension system allows the tires and wheels to move up and down over bumps and holes in the road surface.

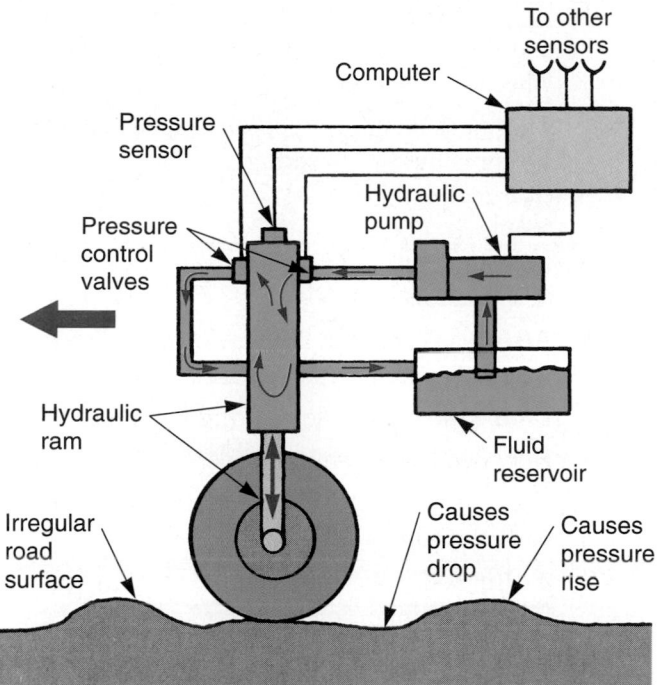

Figure 67-33. This simplified illustration shows the major components of an active suspension system. A pressure sensor on the hydraulic ram reacts to the up-and-down movement of ram and the resulting pressure changes. If pressure in the ram rises when the wheel hits a bump in the road, the sensor signals the computer. The computer can quickly react to release ram pressure so the suspension moves upward with the wheel. As the wheel travels down on the other side of the bump, the sensors make the computer increase ram pressure so that the suspension travels back down to the original road surface.

- Chassis stiffness is a primary factor affecting how quiet and smooth a vehicle drives—the stiffer the chassis, the better.
- Independent suspension allows one wheel to move up and down with little effect on the other wheels.
- Nonindependent suspension has both the right and left wheels attached to the same solid axle.
- Lateral acceleration is the amount of side force a vehicle can handle in a sharp turn or curve before its tires loose traction and skid.
- Suspension system springs must jounce (compress) and rebound (extend) with bumps and holes in the road surface.
- A control arm holds the steering knuckle, bearing support, or axle housing in position as the wheel moves up and down.
- A strut rod fastens to the outer end of the lower control arm and to the frame. It keeps the control arm from swinging toward the rear or front of the vehicle.

- Ball joints (short for ball-and-socket joints) are connections that allow limited rotation in every direction.
- Shock absorbers limit spring oscillations (compression-extension movements) to smooth the vehicle's ride.
- A strut assembly consists of a shock absorber, a coil spring (most types), and an upper damper unit.
- A sway bar, also called stabilizer bar, is used to keep the body from leaning excessively in sharp turns.
- Jounce bumpers are blocks of hard rubber that keep the suspension system parts from hitting the frame or body when the vehicle hits large bumps or holes.
- A long-short arm suspension uses control arms of different lengths to keep the tires from tilting with suspension action.
- A MacPherson strut suspension uses only *one* control arm and a strut assembly (spring, damper, and shock absorber unit) to support each wheel assembly.
- A dead axle is a term used to describe a solid rear axle on a front-wheel-drive vehicle.
- A suspension leveling system is used to maintain a constant vehicle attitude (body height) in the amount of weight in the vehicle changes.
- To maintain correct ride height, an electronic height control system uses height sensors and an ECM to control the operation of a small electric air compressor.
- An electronic shock absorber system uses various vehicle sensors, an electronic control module, and shock absorber actuators to control ride stiffness.
- An active suspension system uses computer-controlled hydraulic rams instead of conventional suspension system springs and shock absorbers.

Important Terms

Suspension system	Independent suspension
Body roll	Nonindependent
Body squat	suspension
Body dive	Understeer
Chassis stiffness	Oversteer
Chassis hertz	

Neutral steering
Lateral acceleration
Skidpad
Suspension system
  springs
Jounce
Rebound
Travel
Coil spring
Leaf spring
Insulators
Leaf spring windup
Air spring
Torsion bar
Spring rate
Sprung weight
Unsprung weight
Control arm
Control arm bushings
Strut rod
Ball joints
Shock absorbers
Oscillations
Shock absorber
  compression
Shock absorber
  extension
Gas-charged shock
  absorbers
Self-leveling shock
  absorber
Adjustable shock
  absorbers
Strut assembly
Strut shock absorber
Dust shield
Lower spring seat
Upper spring seat
Trut bearing
Rubber bumpers
Rubber isolators
Upper strut retainer
Strut rod nut
Damper unit
Sway bar

Stabilizer bar
Sway bar links
Track rod
Lateral control rod
Jounce bumpers
Long-short arm
  suspension
Torsion bar suspension
Curb height
Macpherson strut
  suspension
Modified strut
  suspension
Dead axle
Semi-independent
  suspension
Suspension leveling
  system
Attitude
Manual suspension
  leveling system
Automatic suspension
  leveling systems
Electronic height control
  system
Height sensor
Compressor assembly
Pressure lines
Air shocks
Sensor link
Solenoid valve
Suspension control
  module
Electronic suspension
  system
Steering sensor
Brake sensor
Acceleration sensor
Mode switch
Electronic control
  module
Shock actuators
Active suspension
  system

Review Questions—Chapter 67

Please do not write in this text. Place your answers on a separate sheet of paper.

1. List eight functions of a suspension system.
2. List and explain the six major parts of a suspension system.

3. _____ suspension allows one wheel to move up and down with a minimum effect on the other wheels.

4. The most common type of suspension system spring is the _____ spring.
   (A) leaf
   (B) coil
   (C) air
   (D) torsion

5. A(n) _____ fastens the rear of a leaf spring to the car frame.

6. Define the phrase "leaf spring windup."

7. How does a torsion bar work?

8. The _____ weight of a car is the weight of the parts *not* supported by the springs.

9. A strut rod is used to keep the steering knuckle from swiveling. True or False?

10. Why are ball joints needed?

11. Summarize the basic operation of a conventional shock absorber.

12. What is the advantage of gas-charged shocks?

13. List and explain the eleven major parts of a strut assembly.

14. The _____ is used to keep the car body from rolling or leaning excessively in turns or corners.
    (A) strut rod
    (B) jounce bumper
    (C) track rod
    (D) sway bar

15. Describe a MacPherson strut suspension.

⬤ ASE-Type Questions

1. Technician A says an automotive suspension system supports the weight of the engine. Technician B says an automotive suspension system supports the weight of the transmission. Who is right?
   (A) *A only.*
   (B) *B only.*
   (C) *Both A and B.*
   (D) *Neither A nor B.*

2. Which of the following is not a function of an automotive suspension system?
   (A) *Helps provide a smooth ride for passengers.*
   (B) *Keeps tires in firm contact with the road.*
   (C) *Allows the front wheels to turn from side to side for steering.*
   (D) *Allows the body to tilt when heavily loaded.*

3. Technician A says an automotive suspension system works with the steering system to help keep the wheels in correct alignment. Technician B says an automotive suspension system works independently of the steering system and has no affect on wheel alignment. Who is right?
   (A) A only.
   (B) B only.
   (C) Both A and B.
   (D) Neither A nor B.

4. Which of the following is not a basic component of an automotive suspension system?
   (A) Sway bar.
   (B) Strut.
   (C) Internal control shaft.
   (D) Stabilizer bar.

5. Technician A says a suspension system's control arm fastens the steering knuckle to the shock absorber. Technician B says the control arm fastens the steering knuckle to the frame. Who is right?
   (A) A only.
   (B) B only.
   (C) Both A and B.
   (D) Neither A nor B.

6. Technician A says a ball joint allows control arm movement. Technician B says a ball joint allows steering knuckle movement. Who is right?
   (A) A only.
   (B) B only.
   (C) Both A and B.
   (D) Neither A nor B.

7. Technician A says the control arm bushing allows the suspension system control arm to swing up and down on the vehicle frame. Technician B says the control arm bushing allows the control arm to move from side to side on the suspension system damper. Who is right?
   (A) A only.
   (B) B only.
   (C) Both A and B.
   (D) Neither A nor B.

8. Technician A says that in an independent suspension system, movement of one wheel does not cause direct movement of the wheel on the opposite side of the vehicle. Technician B says that in a nonindependent suspension system, movement of one wheel does not cause direct movement of the other wheel on the opposite side of the vehicle. Who is right?
   (A) A only.
   (B) B only.
   (C) Both A and B.
   (D) Neither A nor B.

9. Technician A says an independent suspension system has both the right and left wheels attached to the same solid axle. Technician B says a nonindependent suspension system has both the right and left wheels attached to the same solid axle. Who is right?
   (A) A only.
   (B) B only.
   (C) Both A and B.
   (D) Neither A nor B.

10. Technician A says a coil spring is the most common spring used on modern automotive suspension systems. Technician B says a leaf spring is the most common spring used on modern automotive suspension systems. Who is right?
    (A) A only.
    (B) B only.
    (C) Both A and B.
    (D) Neither A nor B.

11. Technician A says coil springs are used on either the front or the rear of an automobile. Technician B says coil springs are only used on the front of an automobile. Who is right?
    (A) A only.
    (B) B only.
    (C) Both A and B.
    (D) Neither A nor B.

12. Technician A says a torsion bar fastens to the vehicle's frame. Technician B says a torsion bar fastens to the suspension system control arm. Who is right?
    (A) A only.
    (B) B only.
    (C) Both A and B.
    (D) Neither A nor B.

13. Technician A says most automotive shock absorbers are filled with oil. Technician B says most automotive shock absorbers are filled with air. Who is right?
    (A) A only.
    (B) B only.
    (C) Both A and B.
    (D) Neither A nor B.

14. Technician A says an automotive strut assembly normally consists of a shock absorber, a coil spring, and an upper damper unit. Technician B says an automotive strut assembly normally consist of a shock absorber, a coil spring, and a control arm. Who is right?

    (A) A only.
    (B) B only.
    (C) Both A and B.
    (D) Neither A nor B.

15. Technician A says the term "dead axle" is used to describe a solid rear axle on a front-wheel-drive vehicle. Technician B says the term "dead axle" refers to a solid rear axle on an independent suspension system. Who is right?

    (A) A only.
    (B) B only.
    (C) Both A and B.
    (D) Neither A nor B.

Activities—Chapter 67

1. Prepare an overhead transparency (or transparencies) showing the basic parts of a suspension system. (You can trace and/or enlarge **Figure 67-1** onto the transparency material and label the parts.) Use the transparency to explain to the class the function of each part of the system.

2. Examine vehicles in the shop and identify the type of suspension in each.

3. Join a discussion on the merits of different suspension systems.

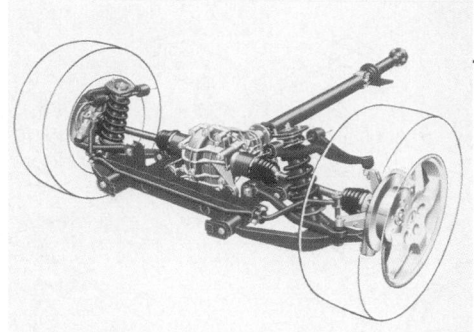

Suspension System Diagnosis and Repair

After studying this chapter, you will be able to:

- Diagnose problems relating to a suspension system.
- Replace shock absorbers and ball joints.
- Describe the removal and replacement of springs.
- Service a strut assembly.
- Replace control arm bushings.
- Use safe work procedures while repairing suspension systems.
- Diagnose and repair electronically-controlled suspension systems.
- Correctly answer ASE certification test questions on the diagnosis and repair of suspension systems.

A suspension system takes a tremendous "pounding" during normal vehicle operation. Bumps and potholes in the road cause constant movement, fatigue, and wear of shock absorbers, ball joints, bushings, springs, and other parts. After prolonged use, major repairs may be needed to restore the suspension system to its original condition.

This chapter covers the most common procedures for diagnosing and repairing suspension system problems. It will summarize how to inspect, remove, and replace parts that commonly wear and fail in service. Study this chapter carefully!

Suspension System Diagnosis

Suspension system problems usually show up as:

- *Abnormal noises*—pops, squeaks, and clunks from suspension parts.
- *Tire wear*—worn suspension system parts allow wheels to go out of alignment.

- *Steering wheel pull*—part wear and incorrect wheel alignment make vehicle wander or veer to one side of road.
- *Front end shimmy*—side-to-side vibration of the front wheels caused by part wear in suspension system.

Make sure that the trouble is in the suspension system, *not* in the steering system, the wheel bearings, the tires, or other related parts.

Suspension system wear can upset steering system operation and change wheel alignment angles. Worn ball joints may let the steering knuckles tilt sideways on their control arms. This, in turn, allows the wheels and tires to cock, or lean.

To begin diagnosis of a suspension system, talk to the customer or the service writer. Then, either inspect the parts that could cause the problem or test drive the vehicle.

Figure 68-1 shows typical problem areas. **Figure 68-2** is an exploded view of the parts of a common front-wheel suspension assembly. Study the relationship between the parts.

Shock Absorber Service

Worn shock absorbers will cause a vehicle to ride poorly on rough roads. When the tire strikes a bump, a bad shock will not dampen spring oscillations. The suspension system will continue to jounce and rebound. This movement is transferred to the frame, body, and passenger compartment.

A loose or damaged shock absorber may produce a loud clanking or banging sound. The rapid up-and-down suspension movement can hammer the loose shock absorber against the body, shock tower, or control arm.

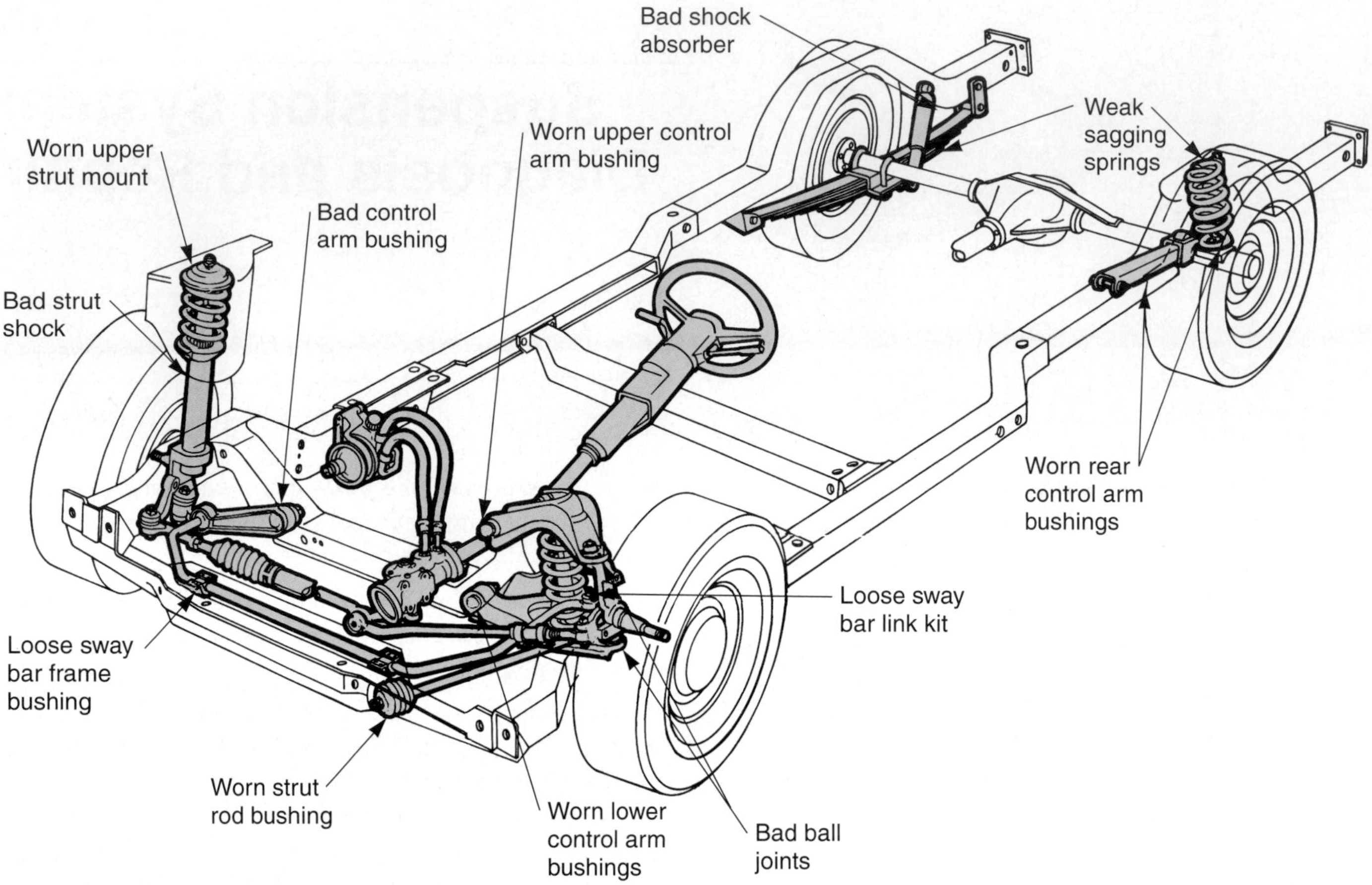

Figure 68-1. Study the types of problems that can develop in a suspension system. (Moog)

Checking Shock Absorber Condition

A shock absorber bounce test and visual inspection will normally locate shock absorber problems. To perform a *shock absorber bounce test,* simply push down on one corner of the vehicle's body. Then release the body and count the number of times the vehicle rebounds (moves up and down).

Generally, good shock absorbers should stop body movement in two or three rebounds. Bad shock absorbers will let the body bounce more than three times. The bounce test should be performed on each corner of the vehicle.

Also, inspect the shock absorbers for signs of leakage (oily wetness) and damage. If the shock is leaking oil, new shocks are needed.

Check the rubber bushings on each end of the shock. They should not be smashed or split. Make sure the shock absorber fasteners are tight.

Replacing Shock Absorbers

When shock absorbers are faulty, they must be replaced.

Warning

With many suspension systems, you must place jack stands or lifting devices under the control arms or the axle before removing the shock absorbers. This will keep the control arms or axle from flying downward when the shocks are unbolted.

The typical procedure for replacing shock absorbers is as follows:

1. Place the vehicle on jack stands or on a lift.
2. Remove the wheel-and-tire assemblies.
3. Place jack stands or other lifting devices under the control arms or the axle housing, if necessary. See **Figure 68-3A.**
4. Using an air impact gun, remove the old shock absorbers. If the fasteners are rusted, spray rust penetrant on the threads. When a threaded stud and a nut are used, you may need to hold the stud while turning the nut, **Figure 68-3B.**
5. Install the new shock in reverse order of removal.

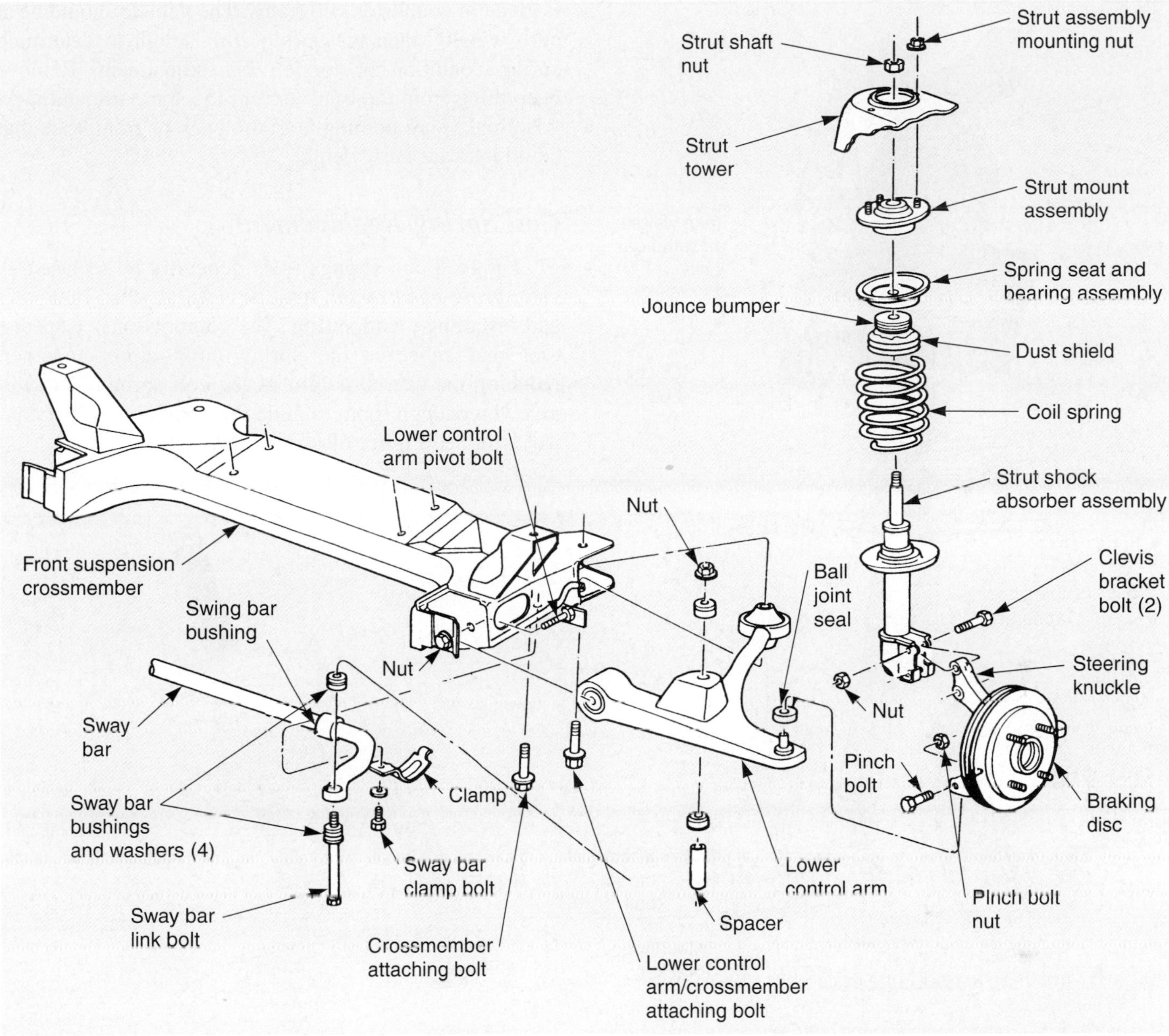

Strut shaft nut

Strut assembly mounting nut

Strut tower

Strut mount assembly

Jounce bumper

Spring seat and bearing assembly

Dust shield

Coil spring

Strut shock absorber assembly

Clevis bracket bolt (2)

Steering knuckle

Braking disc

Pinch bolt nut

Ball joint seal

Nut

Lower control arm

Spacer

Lower control arm/crossmember attaching bolt

Lower control arm pivot bolt

Nut

Front suspension crossmember

Swing bar bushing

Nut

Sway bar

Sway bar bushings and washers (4)

Sway bar clamp bolt

Clamp

Sway bar link bolt

Crossmember attaching bolt

Pinch bolt

Figure 68-2. Note the parts of this front suspension system.

Air- and Gas-Charged Shock Service

Air- and gas-charged shocks are replaced using the same general procedures described for conventional shocks. Gas-filled shocks must be replaced when faulty. Air shocks may be repairable.

The most common problem with air shocks is *air leakage.* The air lines, air valve, or shocks can develop pinhole leaks that allow air pressure to bleed off.

To find air leaks, wipe on a soap-and-water solution at possible leakage points, such as air line fittings, shock boots, and air valves. Bubbles are signs of leakage. If a line, fitting, or shock leaks, it should be tightened or replaced.

Caution

Never exceed the recommended maximum air pressure given for air shocks. If excess pressure is forced into the system, the shocks can be ruptured (blown apart) and ruined.

Suspension Spring Service

Spring fatigue (weakening) allows a vehicle's body to settle toward the axles, lowering vehicle height. This settling or sagging changes the position of the control arms, resulting in misaligned wheels. This condition also affects the ride and appearance of the vehicle. Fatigue can occur after prolonged service.

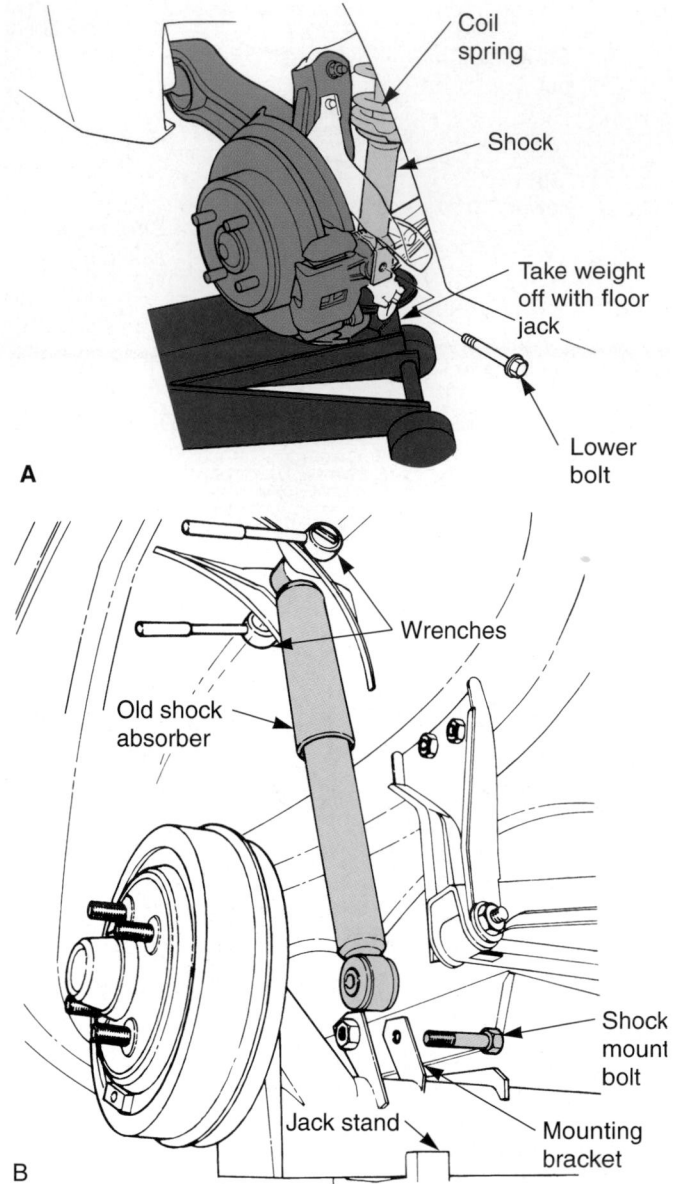

Figure 68-3. A—Before removing the bolts that secure the shock absorber, support the weight of the suspension system to keep the spring from forcing parts down violently. B—Conventional shock absorber removal simply involves unbolting the top and bottom of the shock. (Honda and Chrysler)

Curb Height and Curb Weight

To check spring condition or torsion bar adjustment, measure the vehicle's *curb height* (the distance from a point on the vehicle to the ground). Place the vehicle on a level surface. Then, measure from a specified point on the frame, body, or suspension down to the shop floor. Compare your measurements to specifications. If the curb height is too low (measurement too small), replace the fatigued springs or adjust torsion bar tension.

Curb weight is generally the weight of the vehicle with a full tank of fuel and no passengers or luggage. It is given in pounds or kilograms. The vehicle should be at curb weight when measuring curb height to determine spring condition or torsion bar adjustment. Remove everything from the trunk except the spare tire and jack. Also make sure nothing is in the back or front seats that could increase curb weight.

Coil Spring Replacement

Fatigued coil springs must generally be replaced. A *coil spring compressor* may be needed when removing and installing a coil spring. The compressor is a special tool that squeezes the spring coils closer together, reducing the overall height of the coil spring. This will give you enough room to slide the spring out of the control arm. See **Figure 68-4.**

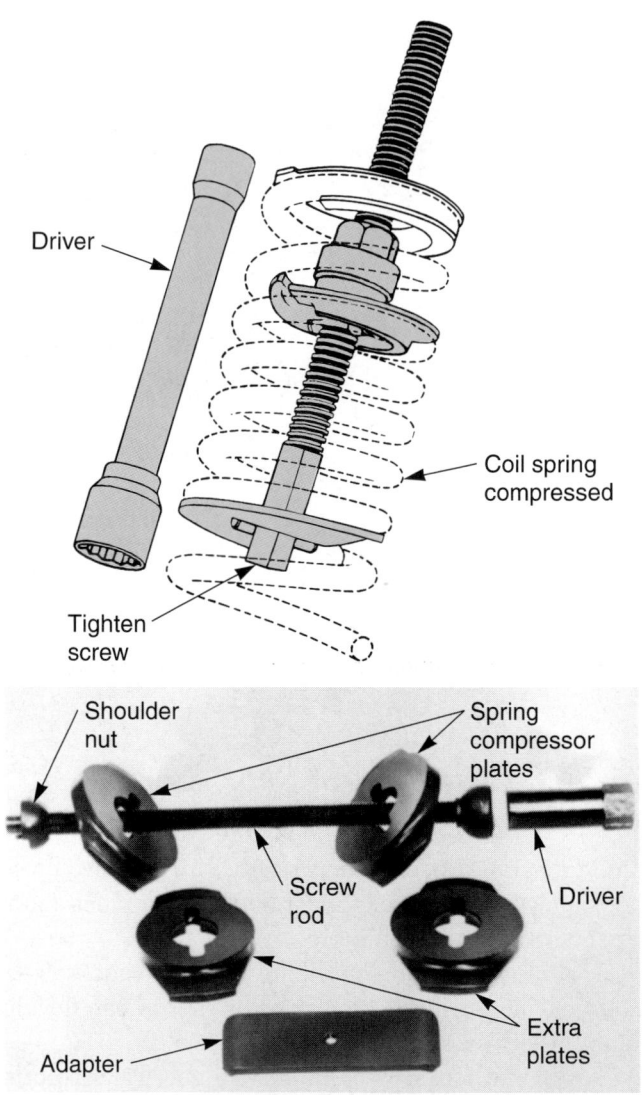

Figure 68-4. A coil spring compressor must be used to squeeze the coils together before unbolting the ball joints. This will keep the spring from flying out with deadly force. (OTC Div. of SPX Corp.)

Warning
A compressed coil spring has a tremendous amount of stored energy. N*ever* unbolt the ball joint without *first* compressing the coil spring. If the spring is *not* compressed, the lower control arm and the spring could shoot downward with *deadly force* when the ball joint is unbolted.

To remove a coil spring from most front suspensions:

1. Place the vehicle on jack stands and remove the shock absorber.
2. Install the spring compressor and compress the spring.
3. Depending on the suspension system design, you may have to separate the lower or upper ball joint. Refer to the service manual for details.
4. Use a *fork tool,* or *ball joint separator,* and hammer blows to remove the ball joint from the steering knuckle. Special pullers and drivers are also available for ball joint separation, **Figure 68-5.**
5. Remove any other components (brake line, strut rod, steering linkage) that could be damaged when the control arm is lowered.
6. Pull the spring and the compressor out as a unit.

To install a new coil spring in a front suspension:

1. Compress the new spring with the spring compressor.
2. Slip the spring into place and position the coil ends in the same location as the ends of the old spring.
3. Reassemble the ball joint and other components.

4. Unscrew the spring compressor while guiding the coil into its seats. Keep your fingers out from under the spring!

When replacing a rear coil spring, a spring compressor may *not* be needed. After shock removal, the axle should drop far enough to free the coil spring. Support the weight of the axle on jack stands or a floor jack, **Figure 68-6.** Always reinstall isolators or the new spring could squeak or rattle. Check a manual for exact procedures.

Leaf Spring Service

Leaf spring service usually involves spring or bushing replacement. Basically, for spring replacement, place jack stands under the frame. Then, use a floor jack to raise the weight of the rear axle off the leaf spring.

Use a box wrench and an impact to remove the nuts on each end of the leaf spring. Leave the through-bolts installed. Then remove the U-bolts that clamp around the middle of the spring and the axle. If the jack is adjusted properly, the through-bolts should now slide out with little resistance. You can then lift the old leaf spring off the vehicle.

Warning
Never drive out the leaf spring eye bolts or shackle bolts unless all force is released from the spring. The eye or shackle bolts should pull out or drive out easily. If the bolts are removed with weight on the spring, the spring could fly upward or downward with *lethal force.*

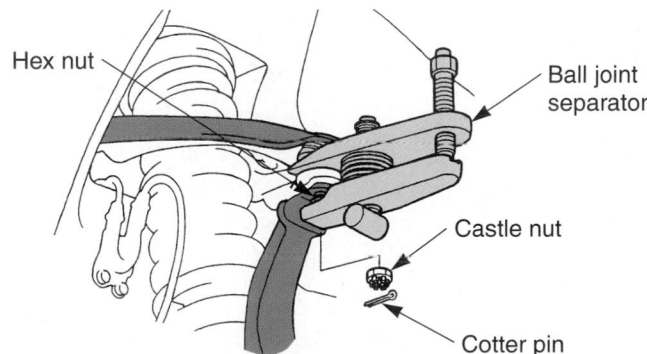

Figure 68-5. A fork or ball joint separator is used to force the ball joint stud out of the steering knuckle. (Honda)

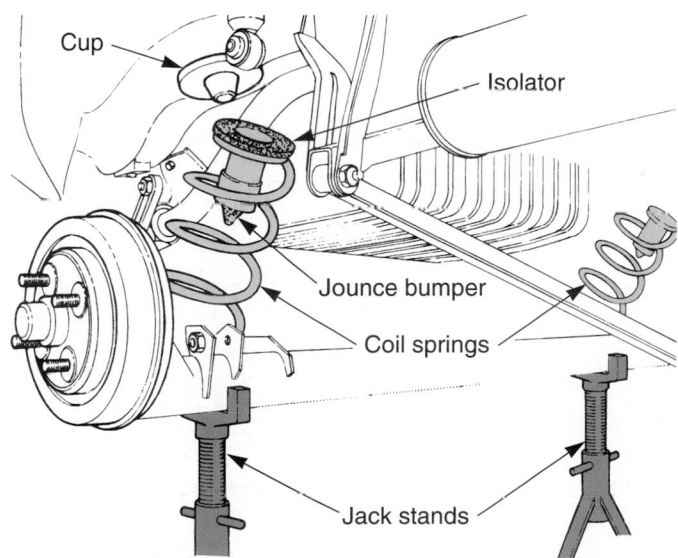

Figure 68-6. A rear coil spring can usually be removed by simply dropping the axle slowly after unbolting the shock. (Chrysler)

Install the new leaf spring in the opposite order. Make sure you position the axle assembly on the leaf spring correctly. A small guide pin may be provided to ensure proper rear axle alignment.

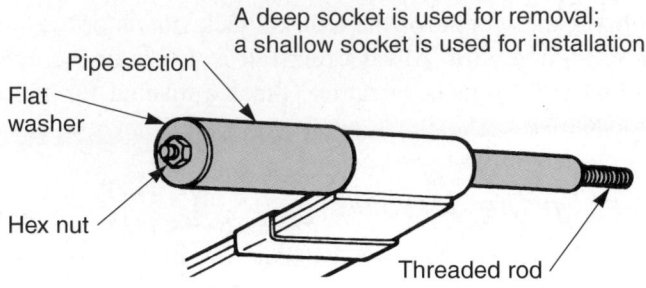

Figure 68-7. A driving tool makes worn bushing removal easy. When the hex nut is tightened, the tool pushes the bushing out of the spring. (Chrysler)

When just the leaf spring bushings are worn, they can be replaced without spring removal. With the through-bolt removed, select the correct diameter pulling-driving tool, **Figure 68-7.** One end of the tool should fit inside the spring eye, and the other end should contact the face of the eye. Tighten the tool nut to force out the old bushing. To install the new spring bushing, use tool to pull the bushing into the spring eye.

Torsion Bar Service

Most torsion bars are adjustable. Therefore, torsion bar replacement is *not* frequently needed. A torsion bar and its related components are illustrated in **Figure 68-8.**

To adjust a torsion bar, you must usually turn a bolt to increase or decrease the tension, or twist, on the bar. When curb height is too low, tension must be increased to raise the vehicle.

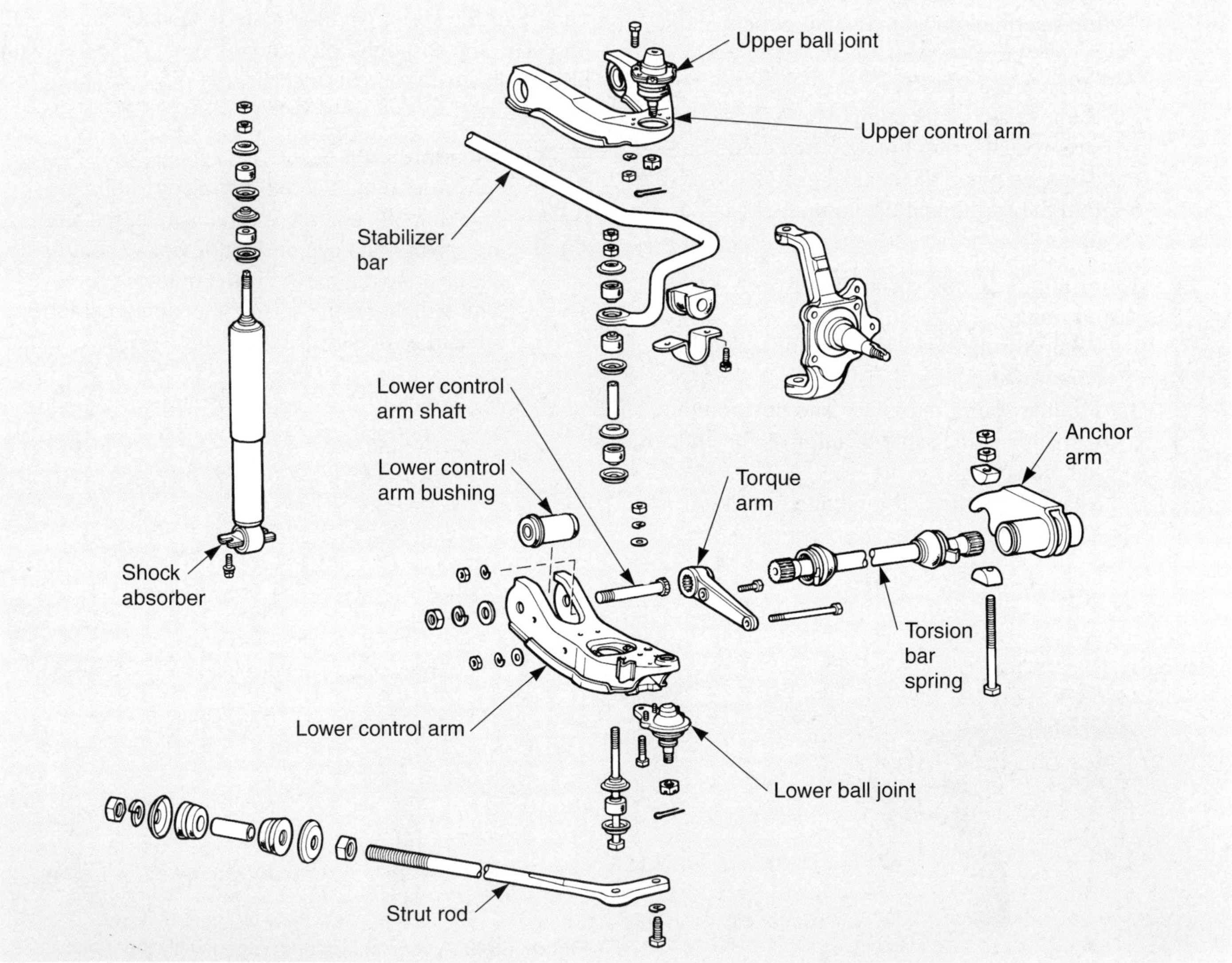

Figure 68-8. Exploded view of a torsion bar suspension. Study the part relationships. (Toyota)

Ball Joint Service

Worn ball joints cause the steering knuckle and wheel assembly to be loose on the control arms. A worn ball joint might make a clunking or popping sound when turning or when driving over a bump.

Ball Joint Lubrication

Ball joint wear is usually a result of improper lubrication or prolonged use. The load-carrying ball joints support the weight of the vehicle while swiveling into various angles. If dry, the joints can wear out quickly. Look at **Figure 68-9.**

Grease fittings or lube plugs are provided for ball joint lubrication. If plugs are provided, they must be replaced with grease fittings before the joint can be lubricated. A *grease gun* is used to inject chassis grease into the ball joint grease fittings. This is illustrated in **Figure 68-10A.**

> **Caution**
> When greasing a ball joint equipped with a balloon seal (airtight seal), be careful not to inject too much grease. Inject only enough grease to cause the seal to bulge slightly. Too much grease can rupture the rubber boot. See **Figure 68-10B.**

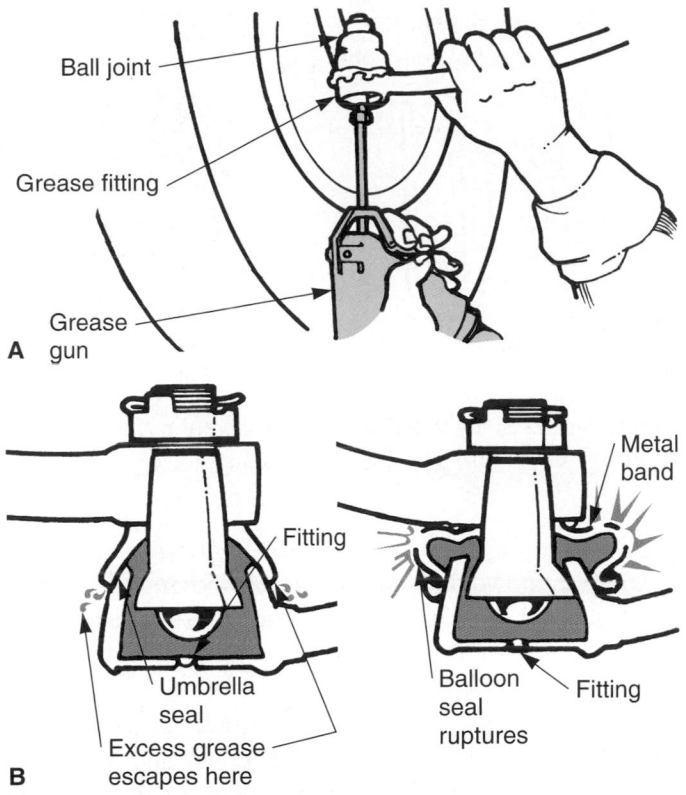

Figure 68-10. A—After cleaning the outside of the fitting, use a grease gun to force chassis grease into the ball joints or other fittings. B—Only install enough grease to fill the boot. Too much grease can rupture some types of boots. (Florida Dept. of Voc. Ed.)

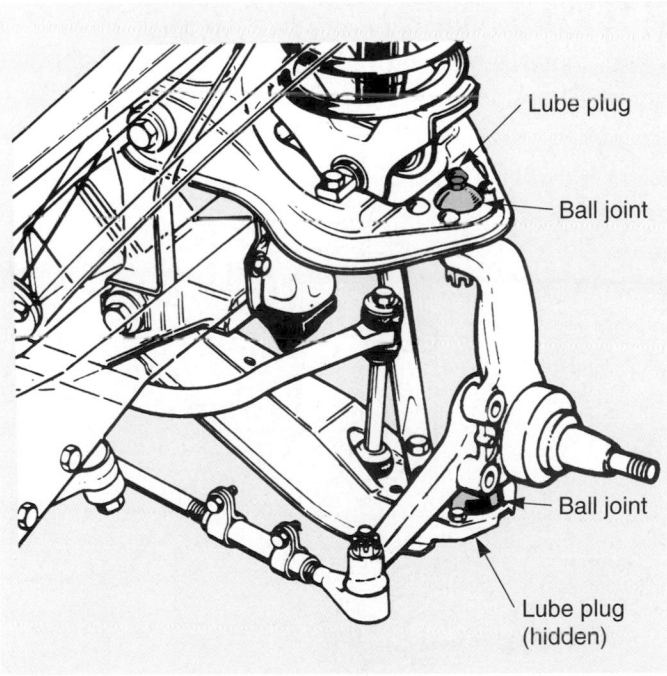

Figure 68-9. Grease fittings or lube plugs are normally provided at the upper and lower ball joints. Grease fittings can also be located on the upper shaft of the inner control arm and on the steering components. (Chrysler)

Checking Ball Joint Wear

To check for ball joint wear, inspect the ball joint wear indicator or measure the play in the joint. With a *ball joint wear indicator,* simply inspect the shoulder on the joint to determine ball joint's condition. If the shoulder on the joint is recessed, the joint should be replaced. This is shown in **Figure 68-11.**

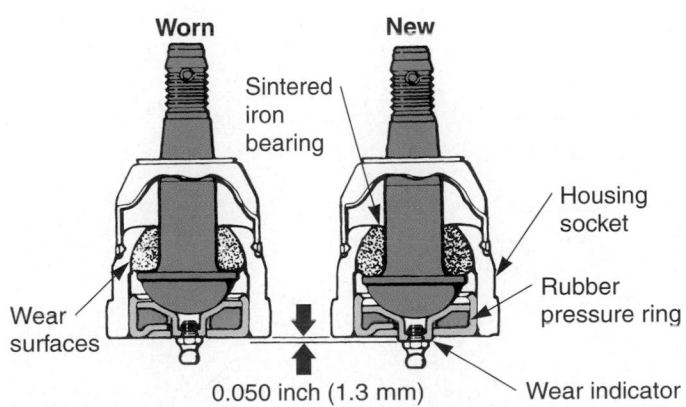

Figure 68-11. This ball joint has a wear indicator. When the shoulder recedes into the socket, a new joint is needed. (Moog)

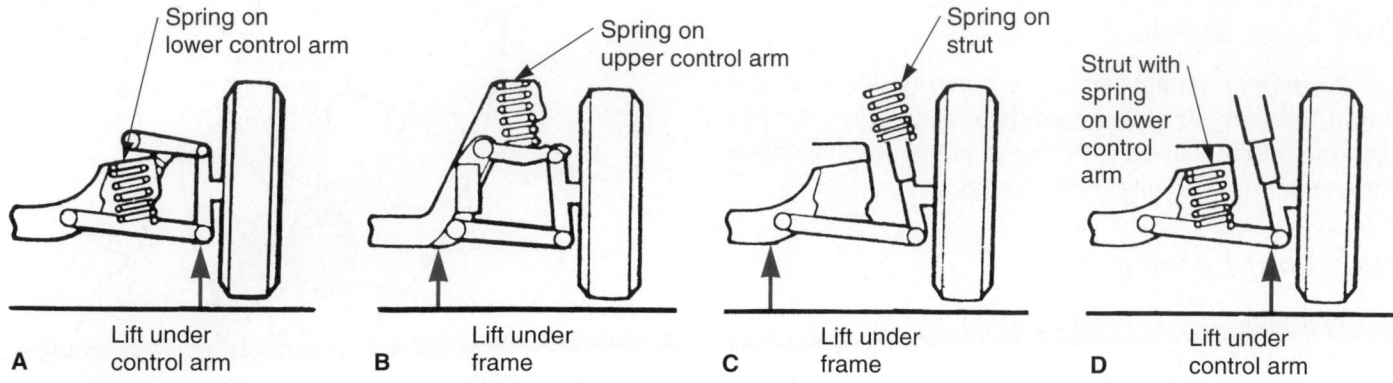

A   Lift under control arm

B   Lift under frame

C   Lift under frame

D   Lift under control arm

Spring on lower control arm

Spring on upper control arm

Spring on strut

Strut with spring on lower control arm

Figure 68-12. Study the lift points for different suspension systems. A specific lift point is needed when checking for ball joint and bushing wear. (Ford)

Another way to check ball joint wear involves jacking up the vehicle and physically moving the control arm and joint. Depending upon the type suspension, you may need to raise the vehicle by the frame or by the lower control arm, **Figure 68-12.**

With the suspension properly raised, use a long steel pry bar to wiggle the tire up and down and sideways. While wiggling, note the amount of movement in the ball joints. Refer to service manual specifications and replace any ball joint worn too much. See **Figure 68-13.**

Ball Joint Replacement

A ball joint can usually be replaced without removing the control arm from the vehicle. Place the vehicle on jack stands. Remove the shock absorber and install a spring compressor on the coil spring. Remove the nut securing the ball joint to the steering knuckle and separate the knuckle and the joint.

The ball joint may be pressed, riveted, bolted, or screwed into the control arm. A service manual will give the details for the specific make and model vehicle.

If pressed in, the ball joint can be hammered or driven out with a special tool. Be careful not to damage the control arm. **Figure 68-14** shows how to remove and install a pressed-in ball joint.

If riveted in place, drill out the old rivets with the correct diameter drill bit. Then, use hardened bolts that come with the kit to secure the new unit. **Figure 68-15** illustrates service of a riveted ball joint.

If screwed into place, use a large wrench to unscrew the old ball joint. If rusted, free the threads with rust penetrate or heat from a torch first. Clean the threads in the control arm before threading in the new ball joint. Torque the ball joint properly.

 Warning
Always install a *new cotter pin* when a ball joint nut is removed. An old cotter pin could break, causing a serious auto accident.

Move pry bar up and down

Watch for excess play in ball joints

Figure 68-13. With the vehicle jacked up properly, use a long pry bar to wiggle the tire up and down. This will help detect part wear.

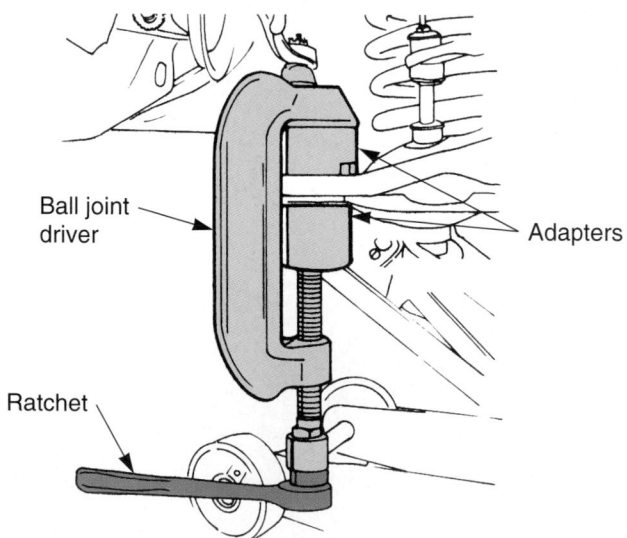

Ball joint driver

Adapters

Ratchet

Figure 68-14. A special ball joint driver is being used to force the ball joint out of the control arm. Wear eye protection when using this tool. (Buick)

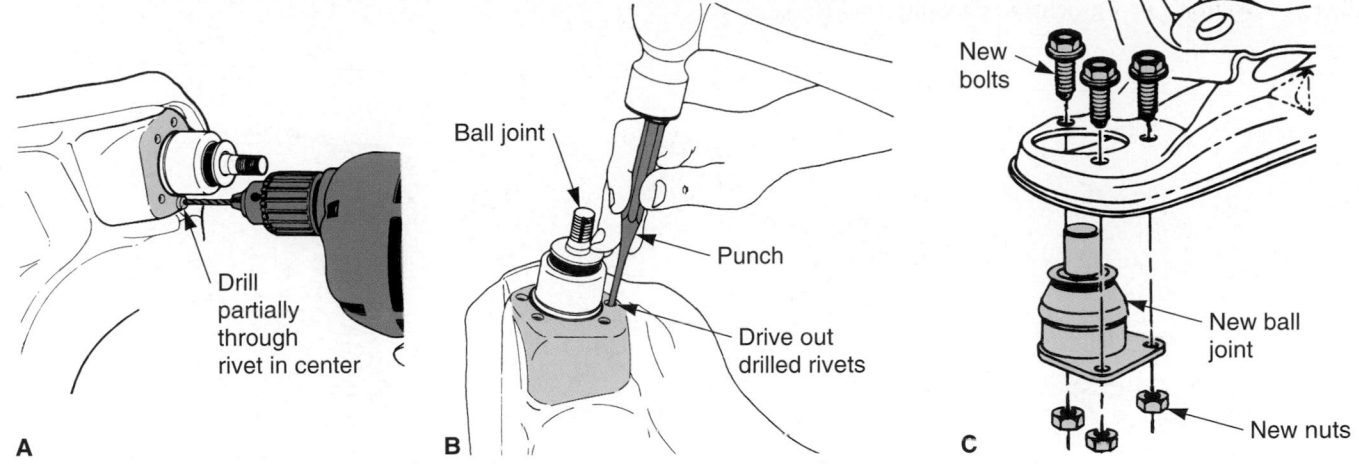

Figure 68-15. Replacement of a riveted ball joint. A—Drill out the rivet heads. B—Use a punch to drive out the rivets. C—Bolt on the new ball joint. (Oldsmobile)

Suspension Bushing Service

Rubber bushings are commonly used in the inner ends of front control arms, rear control arms, and other parts. These bushings are wear prone and should be inspected periodically.

Worn control arm bushings can let the control arms move sideways, causing tire wear and steering problems. **Figure 68-16** shows the various bushings used on a front suspension. **Figure 68-17** pictures the bushings used on a rear suspension system.

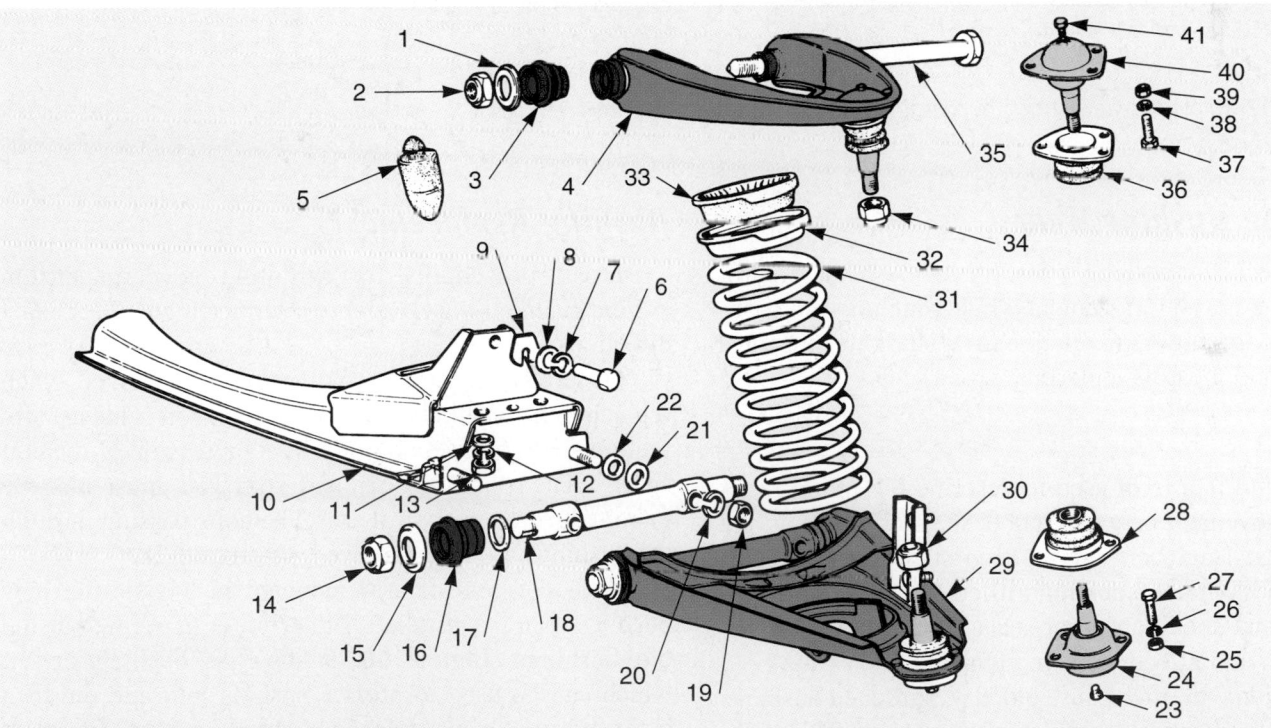

1. Cup. 2. Nut fixing upper control arm to body. 3. Resilient bushing. 4. Upper control arm. 5. Buffer. 6. Bolt. 7. Spring washer. 8. Flat washer. 9. Tab strip. 10. Crossmember. 11. Flat washer. 12. Spring washer. 13. Nut. 14. Nut fixing pivot bar 18 to lower control arm. 15. Cup. 16. Resilient bushing. 17. Flat washer. 18. Pivot bar. 19. Nut fixing lower control arm to crossmember 10. 20. Spring washer. 21. Flat washer. 22. Tab strip. 23. Plug. 24. Lower ball joint. 25. Nut. 26. Spring washer. 27. Bolt. 28. Seal. 29. Lower control arm. 30. Self-locking nut fixing steering knuckle to lower control arm. 31. Spring. 32. Spring seat. 33. Rubber pad. 34. Self-locking nut fixing steering knuckle to upper control arm. 35. Bolt. 36. Seal. 37. Bolt. 38. Spring washer. 39. Nut. 40. Upper ball joint. 41. Plug.

Figure 68-16. Study how ball joints fit into the upper and lower control arms. Bolts secure the joints. Note the location of the control arm bushings.

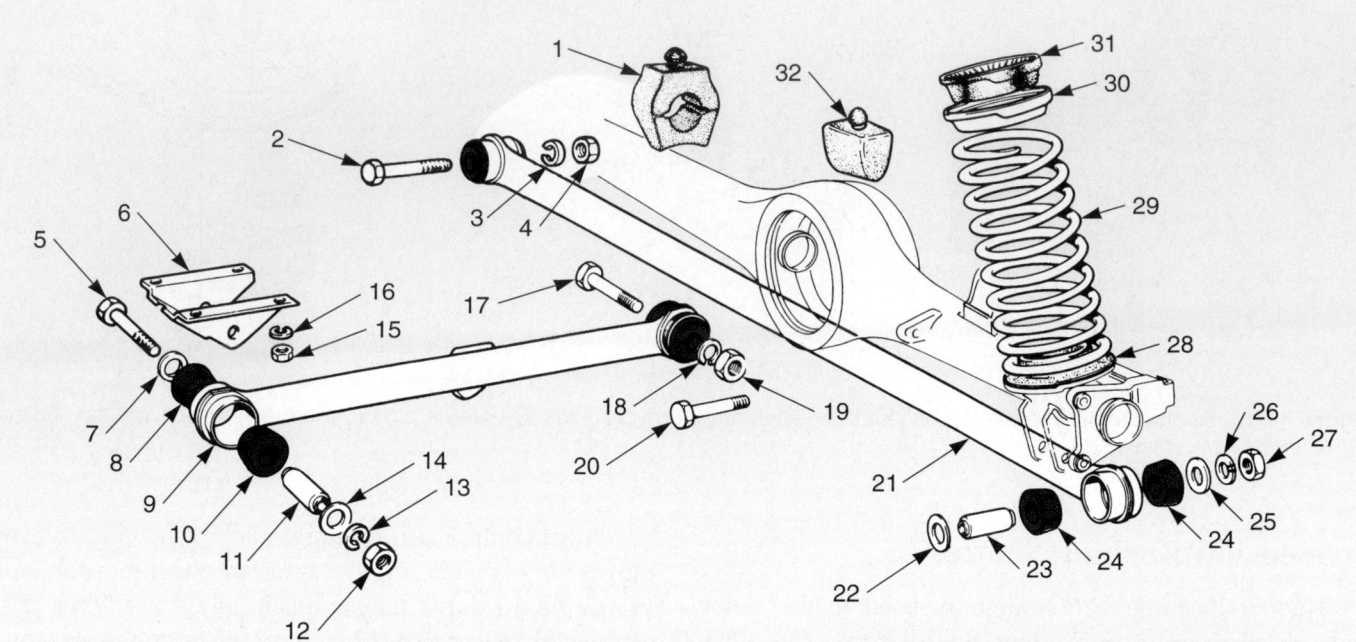

1. Rubber pad. 2. Bolt anchoring cross rod to body. 3. Lock washer. 4. Nut. 5. Bolt anchoring lower side rod to bracket. 6. Bracket. 7. Flat washer. 8. Rubber bushing. 9. Lower side rod. 10. Rubber bushing. 11. Spacer. 12. Nut. 13. Lock washer. 14. Flat washer. 15. Nut. 16. Lock washer. 17. Bolt anchoring lower side rod to axle housing. 18. Lock washer. 19. Nut. 20. Bolt anchoring cross rod to axle housing. 21. Cross rod. 22. Flat washer. 23. Spacer. 24. Rubber bushings. 25. Flat washer. 26. Lock washer. 27. Nut. 28. Lower ring-pad. 29. Coil spring. 30. Upper seating ring. 31. Upper rubber ring-pad. 32. Rubber buffer.

Figure 68-17. Exploded view shows bushings on one type of rear suspension. (Fiat)

Checking Bushing Wear

To check for bushing wear, try to move the control arm against normal movement while watching the bushings. If the arm moves in relation to its shaft, the bushings are worn and must be replaced.

Sticktion

Sticktion is a term used to describe the initial resistance to movement between rubber seals or bushings and metal parts. Engineers try to keep sticktion to a minimum. For example, sticktion afflicts suspension system bushings and shock absorber seals. Sticktion resists initial motion and affects ride smoothness.

Sticktion will often cause old, dry, hardened bushings to make a loud squeaking noise when moved. This noise is caused by the high frequency vibration set up in the bad bushing.

Tech Tip

With worn bushings, you might have heard sticktion. It makes a loud squealing sound when you bounce the body of the vehicle and slowly move the worn, sticking bushings.

Bushing Replacement

Exact procedures for installing new suspension system bushings vary. Refer to a service manual for exact directions.

Generally, the control arm must be removed when replacing the bushings in a front suspension. This usually requires ball joint separation and compression of the coil spring. The stabilizer bar and strut rod must also be unbolted from the control arm. The bolts passing through the bushings are then removed, **Figure 68-18.**

With the control arm mounted in a vise, the new bushings can be installed. Either press or screw out the old bushings. **Figure 68-19** shows a bushing driver, which can be used to press a bushing into and out of a control arm. Always refer to a service manual for exact directions and specifications, **Figure 68-20.** This will ensure a safe, quality repair.

Install the control arm in the reverse order of removal. Torque all bolts properly. Install the ball joint cotter pin and other components.

The procedure for preloading the control arm bushings will depend upon their design. Check the service manual for detailed instructions. Generally, you must

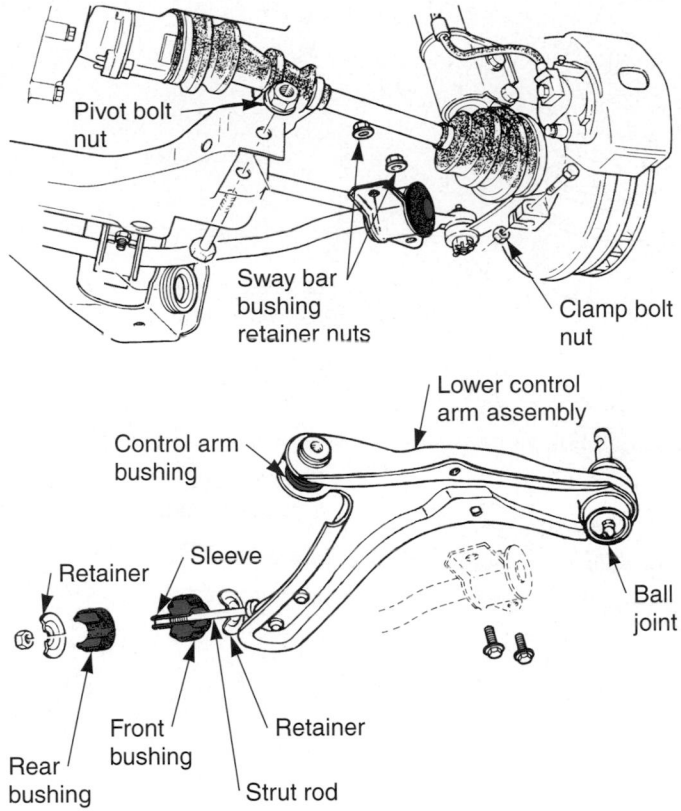

Figure 68-18. Control arm bushing replacement requires control arm removal. Note the parts that must be disconnected on a front-wheel-drive vehicle. (Chrysler)

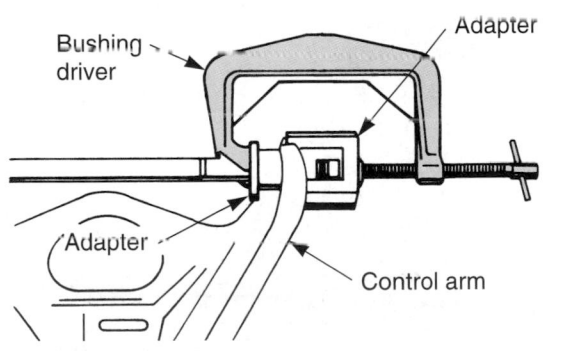

Figure 68-19. A bushing driver is helpful when removing or installing control arm bushings. (Buick)

install the bushing and lightly torque the nuts on the control arm shaft (if used). Then the weight of the vehicle is lowered onto the suspension system. This will place the control arms in their most common location or angle. Then, you must tighten and preload the control arm bushings fully.

Strut rod bushings and stabilizer (sway) bar bushings also require replacement when worn. See **Figure 68-18.**

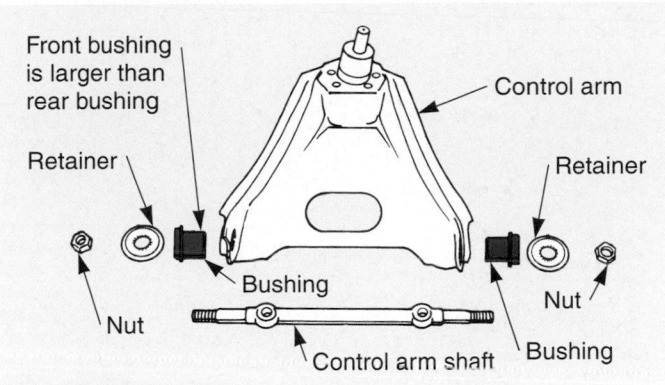

Figure 68-20. With this design, nuts are used to force new bushings into the control arm. Follow service manual directions, as a specific preload is normally needed. (Oldsmobile)

MacPherson Strut Service

The MacPherson strut suspension can wear and cause problems similar to those covered for other types of suspension systems. The major difference in service procedures relates to how the strut assembly is disassembled and reassembled.

The most common trouble with a MacPherson strut suspension is worn strut shock absorbers. Like conventional shocks, the pistons and cylinders inside the strut shocks can begin to leak. This reduces the dampening action and causes the vehicle to ride poorly.

Strut Removal

Basically, strut removal involves unbolting the steering knuckle (front) or bearing support (rear), the brake lines, and the upper strut assembly-to-body fasteners. Marking the cam bolt with a scratch awl or marker will help during reassembly. You can realign this mark to roughly adjust camber. See **Figure 68-21.**

Remove the strut assembly (coil spring and shock) as a single unit, **Figure 68-22.** Do not remove the nut on the end of the shock rod, or the unit could fly apart.

Strut Assembly Service

Strut assembly service generally involves replacing or rebuilding the strut shock absorber. Before removing the strut shock absorber, use a *strut spring compressor* to remove the coil spring from the strut. Look at **Figure 68-23.** After the coil spring has been squeezed together, remove the upper damper assembly. Then, release spring tension and lift the spring off the strut. Inspect all parts closely, **Figure 68-24.**

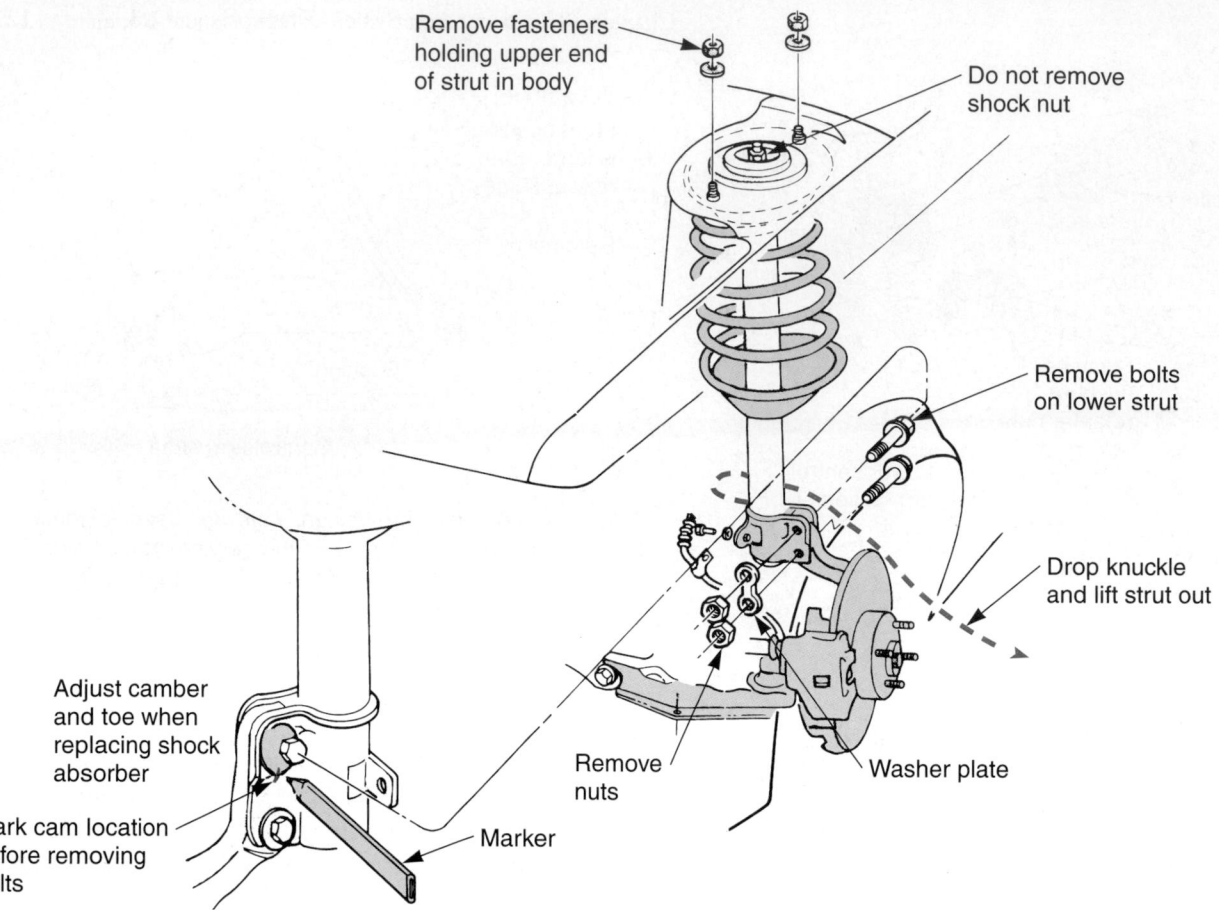

Figure 68-21. Before removing a strut assembly, mark the cam-type bolt, if used. Remove the steering knuckle bolts and the upper fasteners from strut. (Dodge)

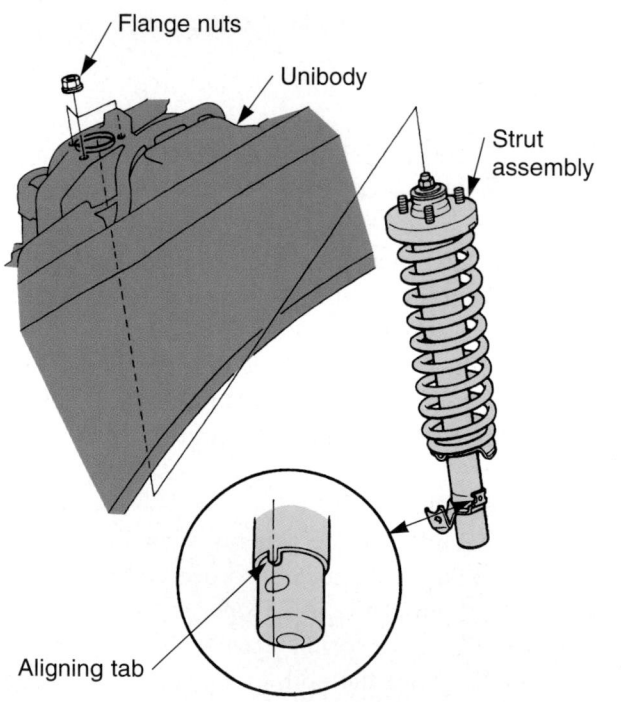

Figure 68-22. Carefully lift the strut out of the vehicle. Be ready to support its weight. Note how the unit aligns in the vehicle for easier reassembly. (Honda)

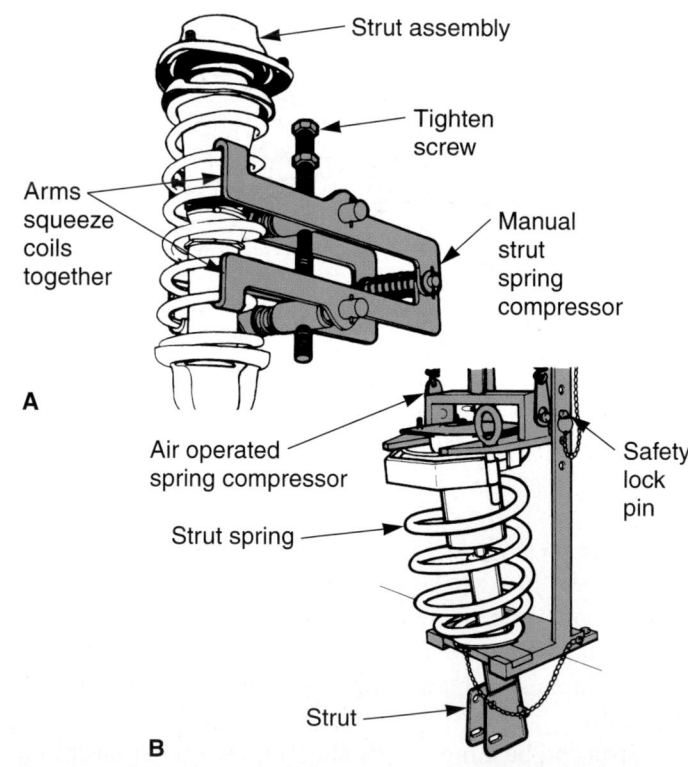

Figure 68-23. A spring compressor is needed to remove the spring from the strut assembly. A—Portable strut spring compressor. B—Bench-mounted strut spring compressor. (Moog)

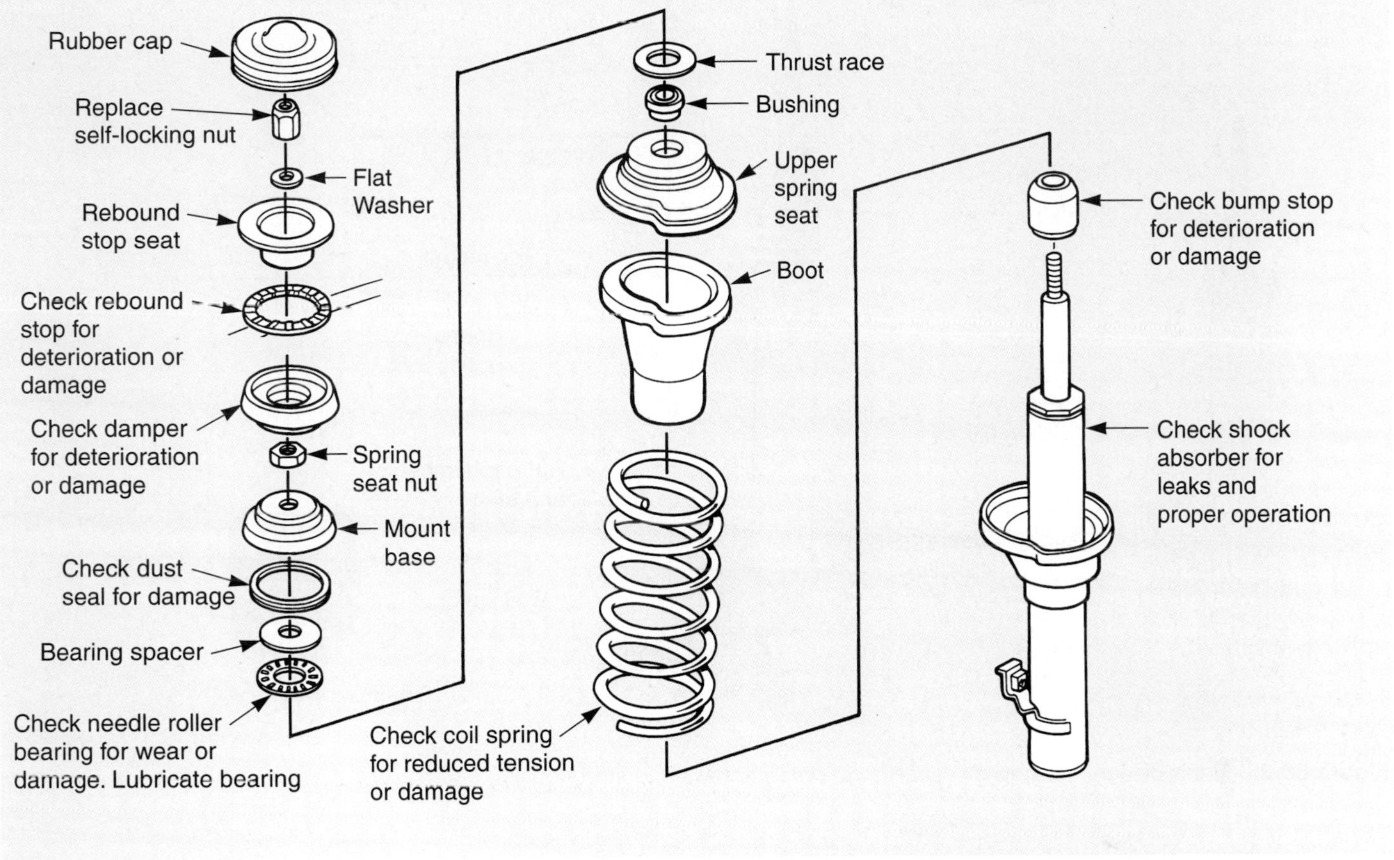

Rubber cap

Replace
self-locking nut

Flat
Washer

Rebound
stop seat

Check rebound
stop for
deterioration or
damage

Check damper
for deterioration
or damage

Spring
seat nut

Mount
base

Check dust
seal for damage

Bearing spacer

Check needle roller
bearing for wear or
damage. Lubricate bearing

Thrust race

Bushing

Upper
spring
seat

Boot

Check coil spring
for reduced tension
or damage

Check bump stop
for deterioration
or damage

Check shock
absorber for
leaks and
proper operation

Figure 68-24. Exploded view shows the parts of a strut assembly. Note the strut bearing, which should be in good condition. Inspect all parts carefully. (Honda)

Warning
When compressing any suspension system spring, be extremely careful to position the compressor properly. If the spring were to pop out of the compressor, serious injury or death could result.

Most automakers recommend replacing a faulty strut shock absorber (removable shock cartridge). The new shock simply screws into the strut's outer housing. A spanner wrench may be needed to turn the nut on the top of the cartridge. **Figure 68-25** shows a replacement shock absorber cartridge.

In some cases, the strut shock absorber must be rebuilt. A strut rebuild will be described in the service manual. Make sure you check the strut bearing.

After replacing or rebuilding the strut shock absorber, fit the strut into the compressor and compress the coil spring. Then, install the upper spring seat and related components. After releasing the spring compressor, the strut can be installed on the vehicle.

Tech Tip
To safely dispose of gas filled shocks, you must drill a small hole through the gas chamber. This will keep the shocks from possibly blowing up if burned or crushed. Wear eye protection when drilling to keep metal bits from flying in your face as the gas pressure is released from the chamber. Instructions for safely drilling the hole will be given in the service manual, **Figure 68-26.**

Strut Installation

Install the strut assembly in the reverse order of removal. See **Figure 68-27.** Lift the strut into position in the upper body mount. Attach the lower end of the strut to the steering knuckle or bearing support. Align your reference marks. Install the fasteners and torque them to specifications. Install any other parts and double-check your work.

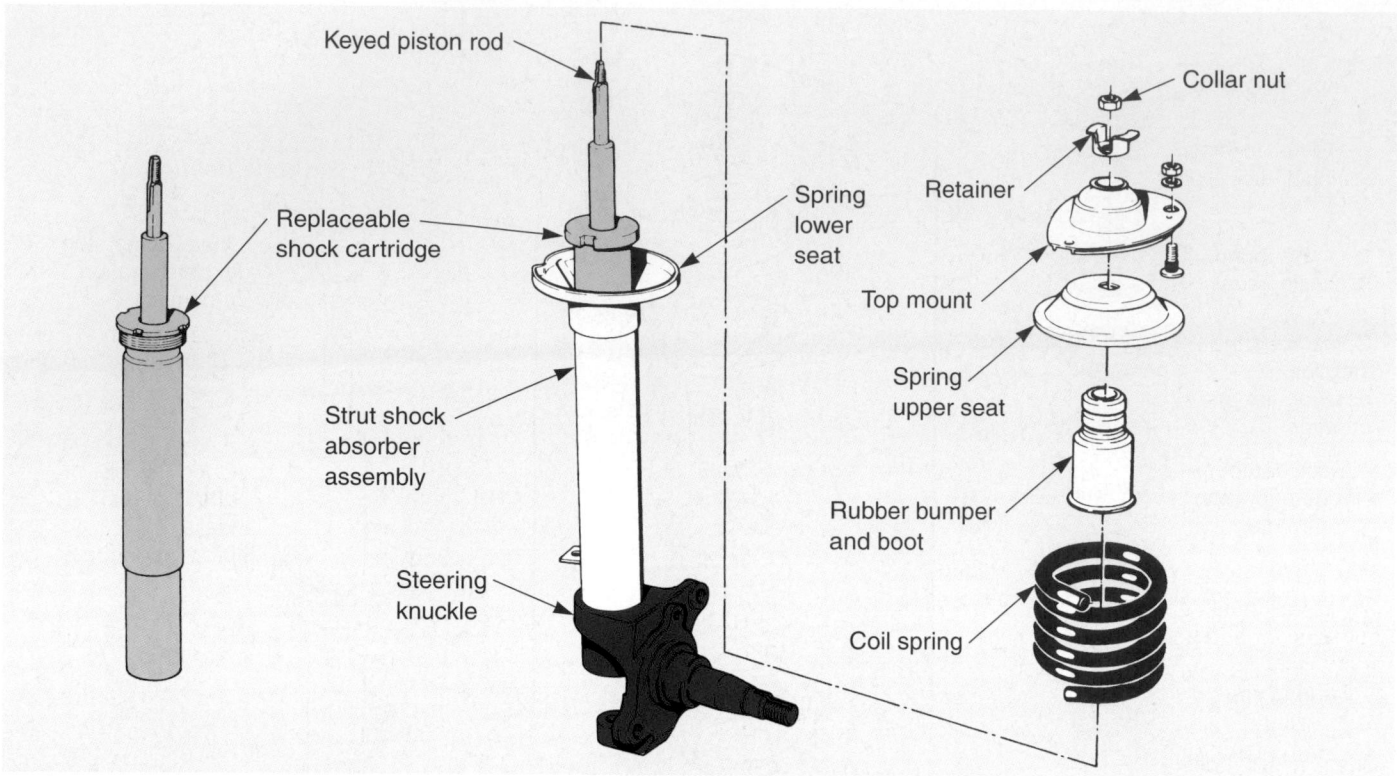

Figure 68-25. In this design, the new strut cartridge simply screws into strut housing. (Mercury)

Warning

Always torque all suspension systems bolts and nuts to factory specifications. This is basic to a competent, safe, dependable repair. If you fail to torque a suspension system fastener properly and the part disengages, a deadly auto accident could result. See **Figure 68-28.**

Wheel Alignment is Needed

After servicing ball joints, control arm bushings, strut rods, springs, strut assemblies, or other suspension

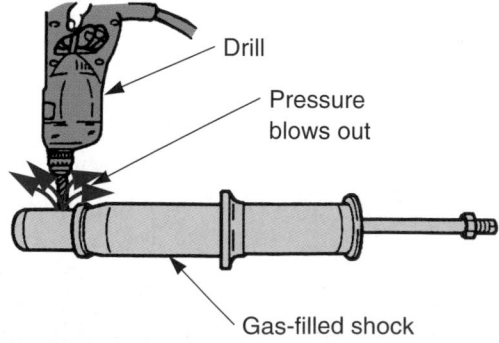

Figure 68-26. Before discarding gas-filled shocks, you should drill a small hole in the gas chamber to release the gas pressure. Wear face protection and follow service manual details. (Mazda)

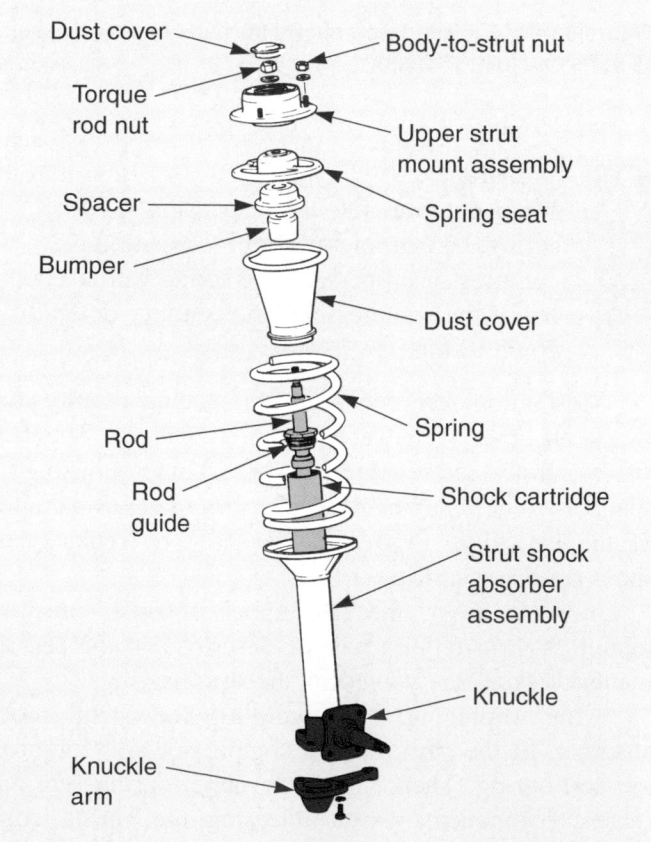

Figure 68-27. Exploded view shows how strut parts fit together. (Toyota)

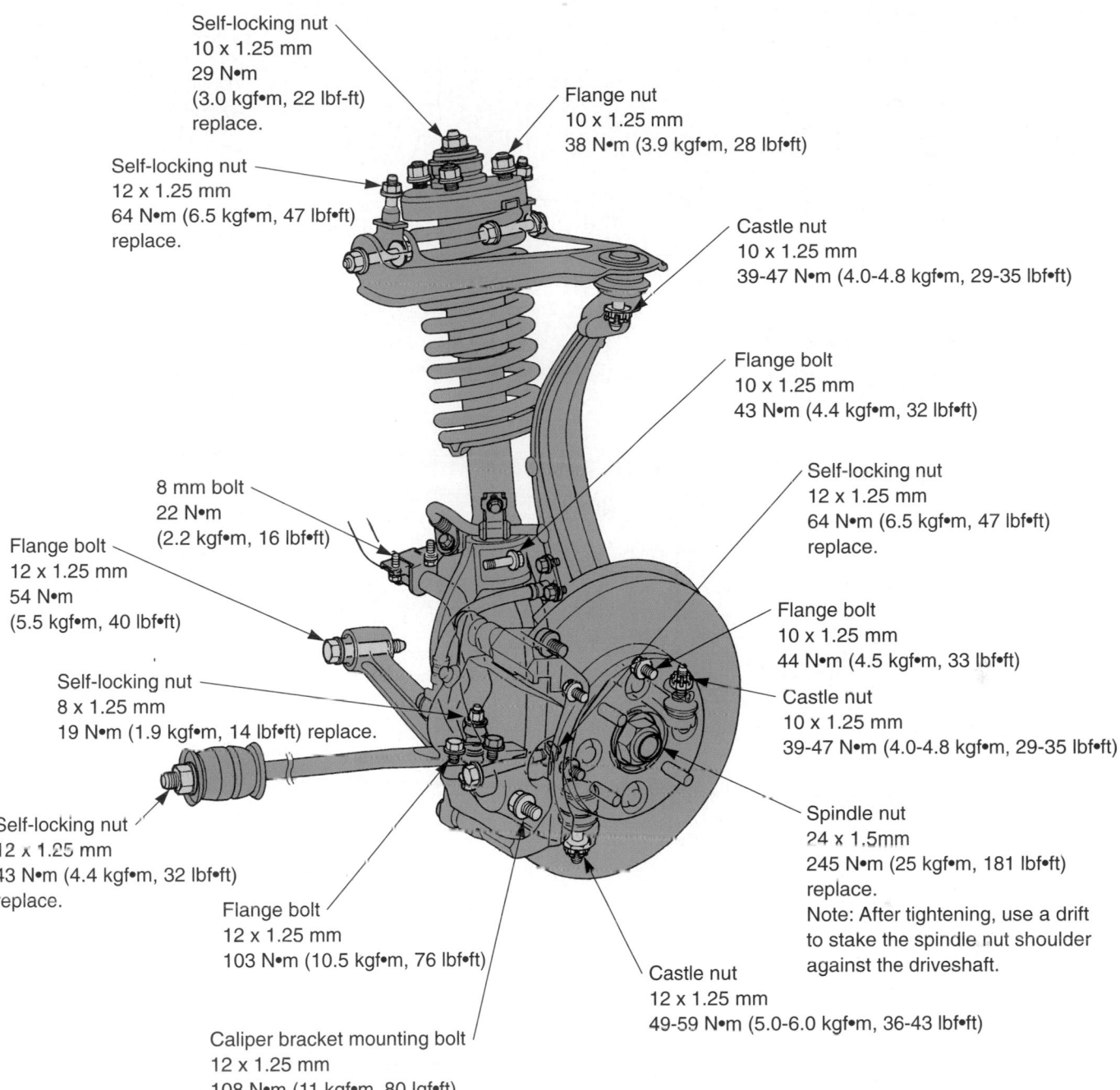

Self-locking nut
10 x 1.25 mm
29 N•m
(3.0 kgf•m, 22 lbf-ft)
replace.

Flange nut
10 x 1.25 mm
38 N•m (3.9 kgf•m, 28 lbf•ft)

Self-locking nut
12 x 1.25 mm
64 N•m (6.5 kgf•m, 47 lbf•ft)
replace.

Castle nut
10 x 1.25 mm
39-47 N•m (4.0-4.8 kgf•m, 29-35 lbf•ft)

Flange bolt
10 x 1.25 mm
43 N•m (4.4 kgf•m, 32 lbf•ft)

8 mm bolt
22 N•m
(2.2 kgf•m, 16 lbf•ft)

Self-locking nut
12 x 1.25 mm
64 N•m (6.5 kgf•m, 47 lbf•ft)
replace.

Flange bolt
12 x 1.25 mm
54 N•m
(5.5 kgf•m, 40 lbf•ft)

Flange bolt
10 x 1.25 mm
44 N•m (4.5 kgf•m, 33 lbf•ft)

Self-locking nut
8 x 1.25 mm
19 N•m (1.9 kgf•m, 14 lbf•ft) replace.

Castle nut
10 x 1.25 mm
39-47 N•m (4.0-4.8 kgf•m, 29-35 lbf•ft)

Self-locking nut
12 x 1.25 mm
43 N•m (4.4 kgf•m, 32 lbf•ft)
replace.

Spindle nut
24 x 1.5mm
245 N•m (25 kgf•m, 181 lbf•ft)
replace.
Note: After tightening, use a drift
to stake the spindle nut shoulder
against the driveshaft.

Flange bolt
12 x 1.25 mm
103 N•m (10.5 kgf•m, 76 lbf•ft)

Castle nut
12 x 1.25 mm
49-59 N•m (5.0-6.0 kgf•m, 36-43 lbf•ft)

Caliper bracket mounting bolt
12 x 1.25 mm
108 N•m (11 kgf•m, 80 lgf•ft)

Figure 68-28. Always torque suspension system fasteners to specs. This is a basic safety precaution that will help prevent accidents caused by loose or overtightened fasteners. The torque values shown are for this suspension only. Refer to manual to find the correct torque values for the vehicle at hand. (Honda)

parts, wheel alignment must be checked and adjusted. *Never* let a vehicle leave the shop without checking alignment. Rapid tire wear or handling problems could occur.

Note
Wheel alignment is covered in Chapter 74. You should, however, study steering systems before attempting wheel alignment procedures.

Computerized Suspension Diagnosis

Discussed in the previous chapter, many modern suspension systems use some means of computer control to maintain ride height, adjust suspension system stiffness, and improve handling. A wiring diagram for one computerized suspension system is given in **Figure 68-29.**

To start troubleshooting of a computer-controlled suspension, connect a scan tool to the vehicle's diagnostic

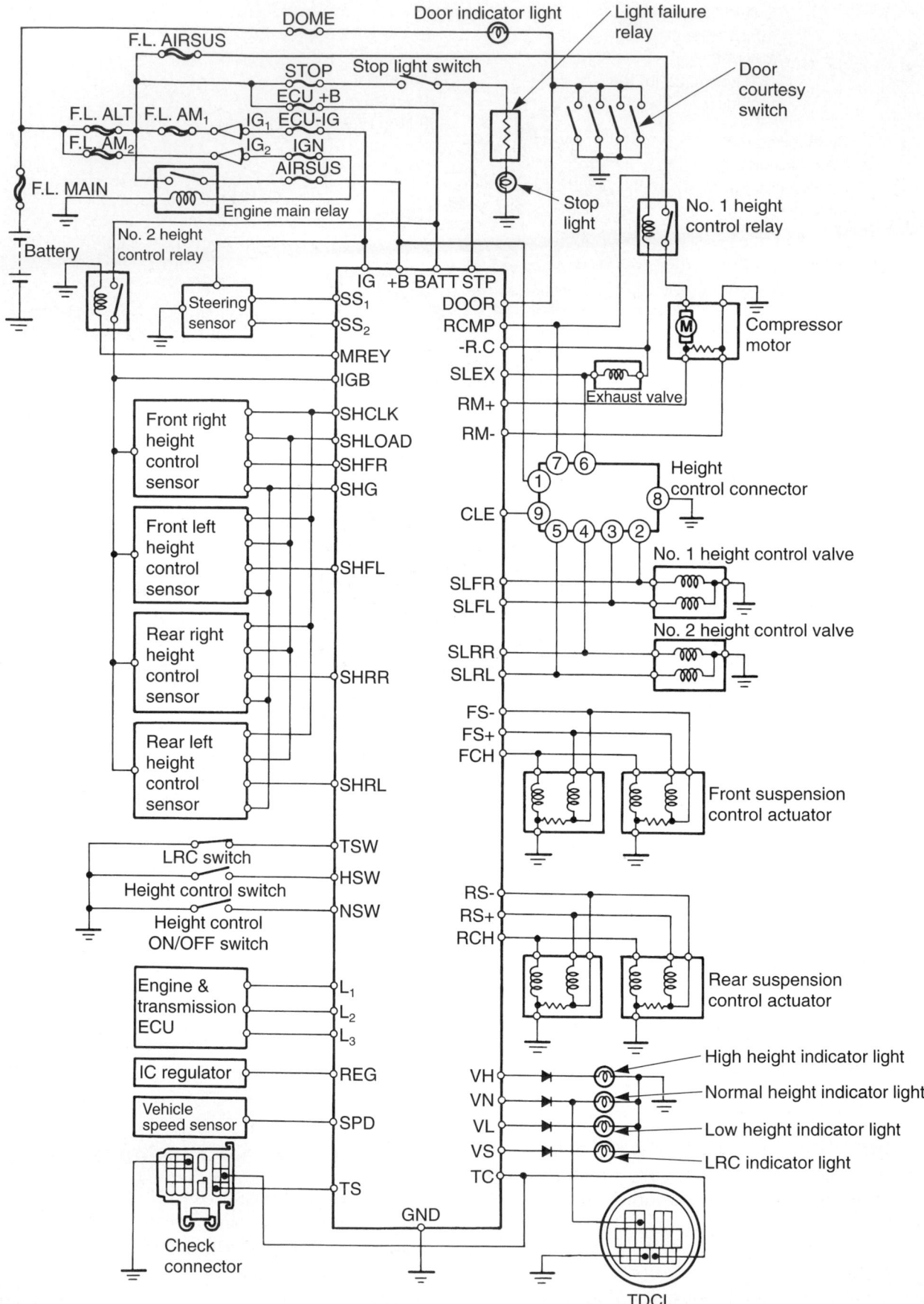

Figure 68-29. This wiring diagram shows the major connections in computer-controlled suspension system. It is helpful when problems are hard to diagnose. (Lexus)

connector. Read any stored diagnostic trouble codes and check operating values for all suspension-related components. Look at **Figure 68-30.**

Your scan tool might find problems with the following parts: solenoids, shock actuator motor, electrical relays, height sensors, compressor motor, and other computerized suspension components.

If the scan tool indicates a possible problem with a certain part or circuit, use pinpoint multimeter tests to verify the exact source of the trouble. For example, if the scan tool indicates a problem with the height sensor, use a VOM or a hand-held scope to test the unit's output, **Figure 68-31.**

Height Sensor Service

A faulty height sensor will fail to produce a normal output signal for each lever or height position. The sensor can fail mechanically due to worn parts, a bent arm, or broken parts. It can also fail electrically, ceasing to produce a normal signal. Replace the sensor if it does not work properly and cannot be adjusted, **Figure 68-32.**

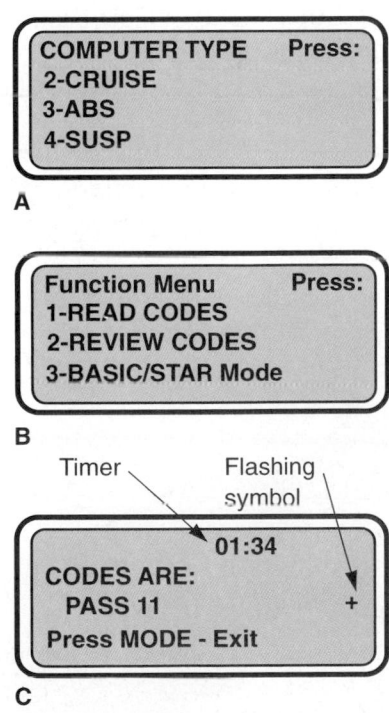

A

B

C

Figure 68-30. A scan tool will display trouble codes and operating values. Follow the directions on scan tool to help find problems with electrical parts in suspension system. A—Scanner will ask you to select the computer or control module to be analyzed. B—From the function menu, you can choose to read present codes, review old codes, or switch to a special factory test mode. C—Trouble codes for the suspension system are displayed. (OTC)

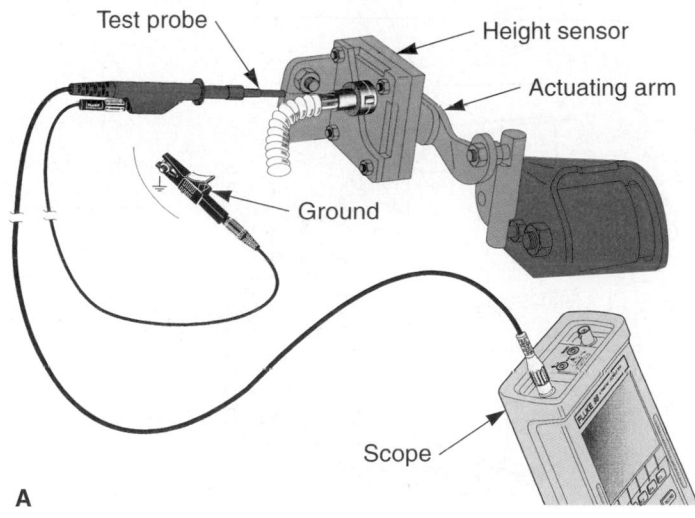

A

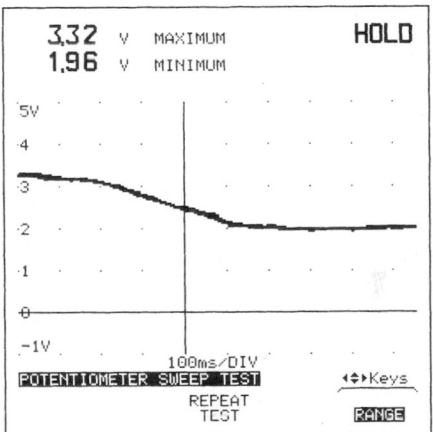

B

Figure 68-31. Pinpoint tests are needed to locate the exact source of trouble. (Fluke) A—A hand-held scope is being used to check the output signal from a height sensor. Voltage should steadily increase and decrease as the sensor arm is moved up and down. B—Voltage waveform for a typical height sensor. With many height sensors, the waveform voltage should go up and down smoothly as the arm is moved. If you have spikes or abnormal voltages, check the reference voltage before replacing the sensor.

Compressor Service

A bad compressor will not produce the air pressure needed to maintain the correct ride height. Check the compressor's electrical connections and its source of voltage first. Then, connect a pressure gauge to the output hose fitting to measure the compressor's pressure output. If the pressure reading is not within specs, replace the compressor, **Figure 68-33.**

Electronically-Controlled Shock Service

When replacing the shocks on electronic suspension systems, you may be able to transfer some of the electronic parts from the old units onto the new ones. If you

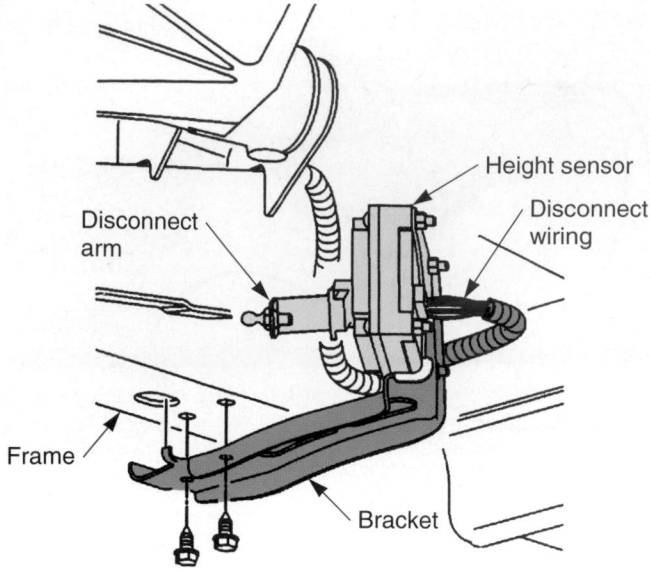

Figure 68-32. If wiring to a sensor is good but the sensor still does not produce a normal signal, replace the sensor. (Oldsmobile)

try to install conventional shocks to save the customer money, it could confuse the computer and cause constant trouble codes and other troubles. Do not try to modify electronic suspension systems without the manufacturer's guidance.

Summary

- A suspension system takes a tremendous "pounding" during normal vehicle operation. Bumps and potholes in the road cause constant movement, fatigue, and wear of shock absorbers, ball joints, bushings, springs, and other parts.

- Make sure that the suspected troubles are actually in the suspension system, *not* in the steering, wheel bearings, tires, or other related parts.

- Worn shock absorbers will cause a vehicle to ride poorly on rough roads.

- Generally, good shock absorbers should stop body movement in two or three rebounds.

- Spring fatigue (weakening) lowers the height of the vehicle, allowing the body to settle toward the axles.

- To check spring condition or torsion bar adjustment, measure curb height.

- A coil spring compressor may be needed when removing and installing a coil spring.

- Worn ball joints cause the steering knuckle to be loose on the control arms.

- A grease gun is used to inject chassis grease into the ball joint grease fittings.

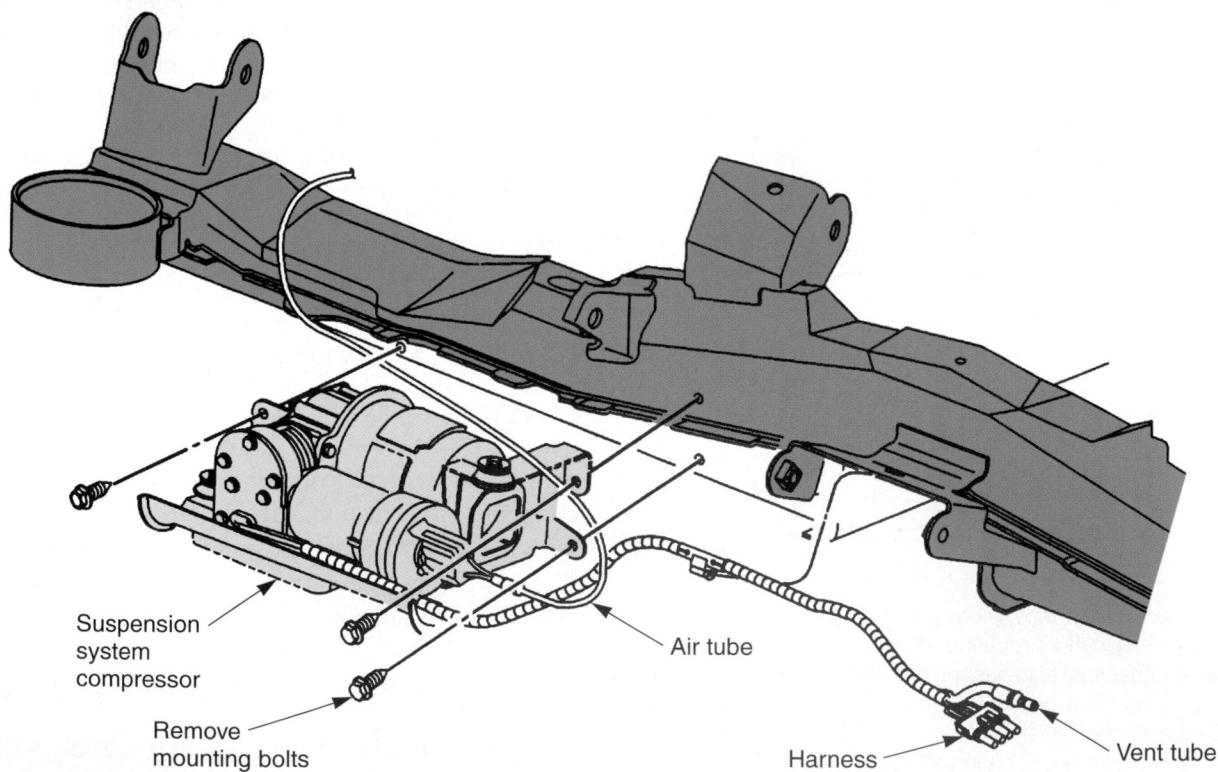

Figure 68-33. If the compressor does not produce normal pressure, the unit must be replaced. (Buick)

- Worn control arm bushings can allow the control arms to move sideways, causing tire wear and steering problems.
- Sticktion is a term used to describe the initial resistance to movement between rubber seals or bushings and metal parts.
- Most automakers recommend that faulty strut or shock cartridges be replaced.
- To safely dispose of gas filled shocks, you must drill a small hole through the gas chamber.
- Always torque all suspension system bolts and nuts to factory specifications.
- The first step in troubleshooting a computer-controlled suspension is to connect a scan tool to the vehicle's diagnostic connector.

Important Terms

Abnormal noises	Curb weight
Tire wear	Coil spring compressor
Steering wheel pull	Fork tool
Front end shimmy	Ball joint separator
Shock absorber bounce test	Grease gun
Air leakage	Ball joint wear indicator
Spring fatigue	Sticktion
Curb height	Strut spring compressor

Review Questions—Chapter 68

Please do not write in this text. Place your answers on a separate sheet of paper.

1. List several symptoms of common suspension system problems.
2. A customer complains that her vehicle's ride is very rough and the vehicle continues to rebound long after going over humps in the road. Technician A says that a bounce test is needed to check the shocks. Technician B says that a general inspection for worn suspension parts is also needed. Who is correct?
   (A) A only.
   (B) B only.
   (C) Both A and B.
   (D) Neither A nor B.
3. What happens with spring fatigue?
4. Define the terms "curb height" and "curb weight."
5. A(n) _____ _____ _____ is needed to remove a suspension system coil spring.
6. Why is suspension spring removal dangerous?
7. A(n) _____ tool or _____ _____ _____ is needed to force the ball joint stud from the steering knuckle.
8. How do you check for ball joint wear?
9. Generally, how do you remove a strut assembly?
10. Which of the following is *not* normally needed during strut service?
    (A) Shock or strut cartridge.
    (B) Floor jack.
    (C) Spring compressor.
    (D) Air chisel.

ASE-Type Questions

1. A car is brought into the shop with a "steering wheel pull" problem. Technician A checks the condition of the sway arm bushings. Technician B inspects the front control arm components. Who is right?
   (A) A only.
   (B) B only.
   (C) Both A and B.
   (D) Neither A nor B.
2. A car's wheel alignment angles are not within specifications. Technician A checks the condition of the ball joints. Technician B looks for worn control arms bushings. Who is right?
   (A) A only.
   (B) B only.
   (C) Both A and B.
   (D) Neither A nor B.
3. All the following are typical automotive suspension system problems *except:*
   (A) *bad control arm bushings.*
   (B) *damaged flex plate.*
   (C) *worn strut rod bushings.*
   (D) *bad ball joints.*
4. A customer brings his/her automobile into the shop with a rough ride complaint. Technician A looks for worn strut shock absorbers. Technician B checks the condition of the strut mounting nuts. Who is right?
   (A) *A only.*
   (B) *B only.*
   (C) *Both A and B.*
   (D) *Neither A nor B*

5. A vehicle's body bounces five times after a "shock bounce test" is performed. Technician A says this indicates that the shocks are in good condition. Technician B says the results of this test indicate bad shock absorbers. Who is right?
(A) A only.
(B) B only.
(C) Both A and B.
(D) Neither A nor B.

6. A small pickup truck is brought into the shop with a leaking shock absorber. Technician A says that the shock's oil seal must be replaced. Technician B says that the leaking shock must be replaced. Who is right?
(A) A only.
(B) B only.
(C) Both A and B.
(D) Neither A nor B.

7. A car's strut shock absorbers are being replaced. During the procedure, Technician A says jack stands must be placed under the vehicle's strut assemblies. Technician B says jack stands should be placed under the axles. Who is right?
(A) A only.
(B) B only.
(C) Both A and B.
(D) Neither A nor B.

8. One of an automobile's air shocks is leaking. Technician A says the shock must be replaced. Technician B says the shock's air line can be repaired if it is leaking. Who is right?
(A) A only.
(B) B only.
(C) Both A and B.
(D) Neither A nor B.

9. A customer wants to know why his car's wheels are out of alignment. Technician A says weak suspension system springs can affect an automobile's wheel alignment. Technician B says weak suspension system springs do not affect an automobile's wheel alignment. Who is right?
(A) A only.
(B) B only.
(C) Both A and B.
(D) Neither A nor B.

10. A car's "curb weight" is being measured. Technician A checks "curb weight" with the automobile's gas tank empty. Technician B checks "curb weight" with the automobile's gas tank full. Who is right?
(A) A only.
(B) B only.
(C) Both A and B.
(D) Neither A nor B.

11. Technician A always uses a spring compressor when replacing a rear coil spring. Technician B says a spring compressor is not always required when replacing a rear coil spring. Who is right?
(A) A only.
(B) B only.
(C) Both A and B.
(D) Neither A nor B.

12. An automobile's coil springs are being replaced. During this procedure, Technician A unbolts the ball joint before compressing the coil spring. Technician B compresses the coil spring before unbolting the ball joint. Who is right?
(A) A only.
(B) B only.
(C) Both A and B.
(D) Neither A nor B.

13. A car is brought into the shop with a possible ball joint problem. Technician A uses a ball joint wear indicator to check the condition of the automobile's ball joints. Technician B uses his hands to check the condition of the car's ball joints. Who is right?
(A) A only.
(B) B only.
(C) Both A and B.
(D) Neither A nor B.

14. A car's front tires are worn. Technician A tells the car's owner that worn control arm bushings may have caused this problem. Technician B tells the owner that worn control arm bushings do not affect tire wear. Who is right?
(A) A only.
(B) B only.
(C) Both A and B.
(D) Neither A nor B.

15. An automobile's ball joints have been replaced. After performing this procedure, Technician A says the vehicle should be road tested before further adjustments are made. Technician B says the vehicle's wheel alignment should be checked before the vehicle is road tested. Who is right?

    (A) A only.
    (B) B only.
    (C) Both A and B.
    (D) Neither A nor B.

Activities—Chapter 68

1. Measure a vehicle for proper curb height after checking curb weight.

2. Inspect ball joints for wear and report their condition to your instructor.

3. Inspect and test drive a vehicle with suspension problems. Diagnose the problem and suggest what service is needed.

4. Prepare a bill for a customer for replacement of front struts. Base your bill on an hourly charge rate established by your instructor. Find cost of parts from a parts catalog or by contacting an automotive parts store. Use a flat rate manual or a time suggested by your instructor.

Suspension System Diagnosis		
Condition	**Possible Causes**	**Correction**
Front end whines on turns.	1. Defective wheel bearing. 2. Incorrect wheel alignment. 3. Worn tires.	1. Replace wheel bearing. 2. Check and reset wheel alignment. 3. Replace tires.
Front end growl or grinding on turns.	1. Defective wheel bearing. 2. Engine mount grounding. 3. Worn or broken CV-joint. 4. Loose lug nuts. 5. Incorrect wheel alignment. 6. Worn tires.	1. Replace wheel bearing. 2. Check for motor mount hitting frame rail and reposition engine as required. 3. Replace CV-joint. 4. Verify wheel lug nut torque. 5. Check and reset wheel alignment. 6. Replace tires.
Front end clunk or snap on turns.	1. Loose lug nuts. 2. Worn or broken CV-joint. 3. Worn or loose tie-rod, ball joint, control arm bushing, sway bar, or upper strut attachment. 4. Loose crossmember bolts.	1. Verify wheel lug nut torque. 2. Replace CV-joint. 3. Tighten or replace tie-rod end or ball joint; replace control arm bushing; tighten sway bar or upper strut attachment to specified torque. 4. Tighten crossmember bolts to specified torque.
Front end whine with vehicle going straight at a constant speed.	1. Defective wheel bearing. 2. Incorrect wheel alignment. 3. Worn tires. 4. Worn or defective transaxle gears or bearings.	1. Replace wheel bearing. 2. Check and reset wheel alignment. 3. Replace tires. 4. Replace transaxle gears or bearings.
Front end growl or grinding with vehicle going straight at a constant speed.	1. Engine mount grounding. 2. Worn or broken CV-joint.	1. Reposition engine as required. 2. Replace CV-joint.
Road wander.	1. Incorrect tire pressure. 2. Incorrect wheel toe. 3. Worn wheel bearings. 4. Worn control arm bushings. 5. Excessive friction in steering gear. 6. Excessive friction in steering shaft coupling. 7. Excessive friction in strut upper bearing.	1. Inflate tires to recommended pressure. 2. Check and reset front wheel toe. 3. Replace wheel bearing. 4. Replace control arm bushing. 5. Replace steering gear. 6. Replace steering coupling. 7. Replace strut bearing.
Lateral pull.	1. Unequal tire pressure. 2. Radial tire lead. 3. Incorrect front wheel camber. 4. Power steering gear imbalance. 5. Wheel braking.	1. Inflate tire to recommended pressure. 2. Perform lead correction procedure. 3. Check and reset front wheel camber. 4. Replace power steering gear. 5. Correct braking condition causing lateral pull.
Excessive steering effort.	1. Low tire pressure. 2. Lack of lubricant in steering gear. 3. Low power steering fluid level. 4. Loose power steering pump belt. 5. Lack of lubricant in steering ball joints. 6. Steering gear malfunction. 7. Lack of lubricant in steering coupler.	1. Inflate all tires to recommended pressure. 2. Replace steering gear. 3. Fill power steering fluid reservoir to correct level. 4. Correctly adjust power steering pump belt. 5. Lubricate or replace steering ball joints. 6. Replace steering gear. 7. Replace steering coupler.

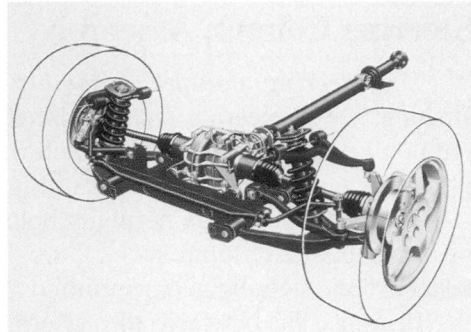

Chapter 69

Steering System Fundamentals

After studying this chapter, you will be able to:

▪ Identify the major parts of a steering system.

▪ Explain the operating principles of steering systems.

▪ Compare the differences between a linkage steering and a rack-and-pinion steering system.

▪ Describe the operation of hydraulic and electric-assist power steering systems.

▪ Explain the operation of four-wheel steering systems.

▪ Correctly answer ASE certification test questions requiring a knowledge of modern steering system designs.

The steering system is vital to the safety of an automobile. It allows the driver to safely maneuver the vehicle. To be a good diagnostic and repair technician, you must understand how various types of steering systems are constructed and operate.

Early steering systems were simple mechanical mechanisms. A series of shafts, rods, and gears transferred steering wheel motion to the wheels and tires. Today, steering systems are much more sophisticated. They use hydraulics and electronics to improve steering precision.

This chapter will build upon your knowledge of automotive technology by introducing late-model steering system variations. It will explain basic manual steering systems before progressing to more complex hydraulic-mechanical and electronic-mechanical designs.

Functions of a Steering System

The steering system must perform several important functions.

• Provide precise control of the front-wheel direction and, sometimes, the rear-wheel direction.

• Maintain the correct amount of effort needed to turn the wheels.

• Transmit **road feel** (slight steering wheel pull caused by road surface) to the driver's hands.

• Absorb most of the shock going to the steering wheel as the tires hit bumps and holes in the road.

• Allow for suspension action.

Basic Steering Systems

There are two basic kinds of steering systems in wide use today: linkage (worm gear) steering systems and rack-and-pinion steering systems. They may be operated manually or with power assist. See **Figure 69-1.**

Before studying the individual parts of these systems, you should have a basic understanding of both linkage and rack-and-pinion steering systems. This will allow you to develop a better "picture" of how each component operates.

Basic Linkage Steering

A linkage steering system consists of the following parts:

• **Steering wheel**—used by the driver to rotate a steering shaft that passes through the steering column, **Figure 69-1A.**

• **Steering shaft**—transfers turning motion from the steering wheel to the steering gearbox.

• **Steering column**—supports the steering wheel and the steering shaft.

• **Steering gearbox**—changes turning motion into a straight-line motion to the left or right.

• **Steering linkage**—connects the steering gearbox to the steering knuckles and the wheels.

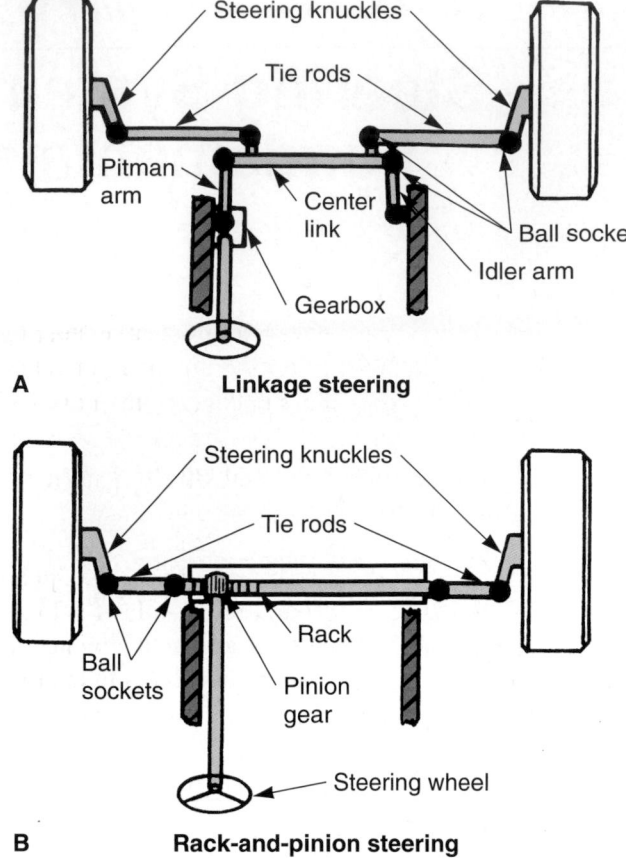

A   **Linkage steering**

B   **Rack-and-pinion steering**

Figure 69-1. There are two types of steering systems. Both are found on today's vehicles. Compare each type closely.

- **Ball sockets**—allow linkage arms to swivel up and down (for suspension action) and from left to right (for turning).

Basic Rack-and-Pinion Steering

A rack-and-pinion steering system also uses a steering wheel, steering column, and steering shaft. These components transfer the driver's turning effort to gears in the steering gear assembly. See **Figure 69-1B.**

Other than the parts just covered, the major components of a rack-and-pinion steering system are:

- **Pinion gear**—rotated by the steering wheel and steering shaft; its teeth mesh with the teeth on the rack.

- **Rack**—long steel bar with teeth along one section; slides sideways as the pinion gear turns.

- **Gear housing**—holds the pinion gear and the rack.

- **Tie-rods**—connect the rack with the steering knuckles.

Steering Column Assembly

The *steering column assembly* consists of the steering wheel, steering shaft, column (outer housing), ignition key mechanism, and, sometimes, a flexible coupling and a universal joint. Look at **Figure 69-2.**

The steering column normally bolts to the underside of the dash. The column sticks through the firewall and fastens to the steering gear assembly.

Bearings fit between the steering shaft and the column. They let the shaft rotate freely. The steering wheel is splined to the shaft. A large nut holds the steering wheel on the shaft splines.

Note

Air bags are enclosed in all late-model steering wheel assemblies. Refer to Chapter 77, *Restraint Systems,* and Chapter 78, *Restraint System Service,* for information on air bags.

Ignition Lock and Switch

The ignition lock and switch mechanism are mounted on the steering column in most late-model vehicles. The *ignition lock mechanism* is normally on the top, right-hand side of the column. The *ignition switch* is usually bolted inside the steering column.

Note

For more information on ignition switches, refer to the index.

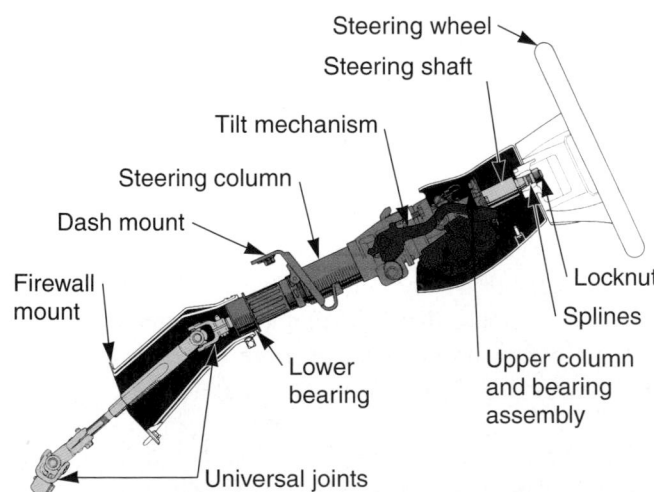

Figure 69-2. Note the steering column components. The steering wheel is splined to a shaft that extends through the column and down into the steering gearbox. This is a tilt column, which allows the steering wheel to be moved into different driving positions. (Lexus)

Locking Steering Wheel

To help prevent theft, late-model vehicles have a *locking steering wheel.* When the ignition key is off, the steering wheel cannot be turned.

Figure 69-3 shows a common method of locking the steering wheel. A rack and a sector are used to slide a steel pin into mesh with a slotted disc that is splined to the steering shaft. This locks the steering shaft to the column, preventing the steering wheel from being turned.

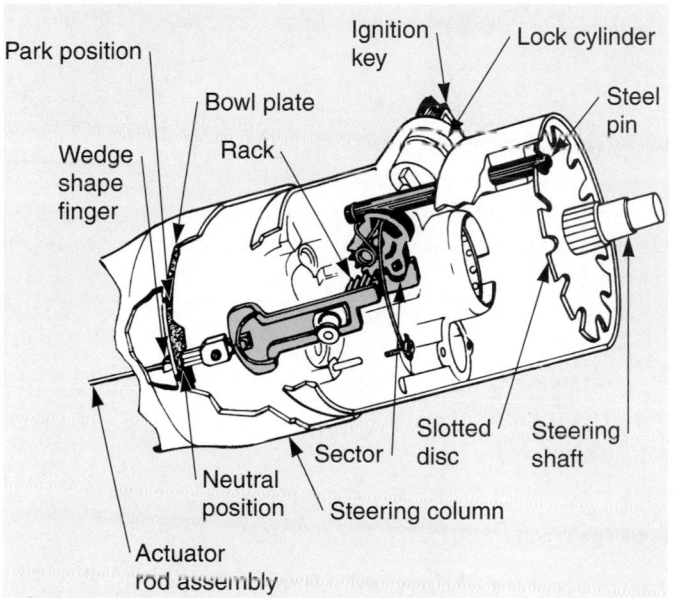

Figure 69-3. Cutaway view shows how a locking steering column functions. The steel pin slides into a slot in the disc to keep the wheel from turning after the key is removed. Also note the small rod that extends down to the ignition switch. (Buick)

Collapsible Steering Column

Most vehicles use a *collapsible steering column* to help prevent driver chest and face injury during an auto accident. The column is designed to crumple or slide together when forced forward during a collision. Look at **Figure 69-4.**

When a vehicle hits a stationary object, the engine and the front body structure can be pushed rearward into the steering column. At the same time, the driver can be thrown forward into the steering wheel. With a rigid steering column, the driver's chest could be injured.

There are several types of collapsible steering columns: steel mesh (crushing) columns, tube-and-ball (sliding) columns, and the shear capsule (break and slide) columns. In all types, the column is made up of two pieces.

Tilt Steering Columns

A *tilt steering column,* or *tilt steering wheel,* has a flex joint, or u-joint, that allows the top half of the column and the steering wheel to be positioned at different angles.

A *manual tilt column* uses a lever on the steering column to unlock the flex joint so the wheel can be moved up or down.

A *power tilt column* uses a small electric motor, a control switch, and a gear mechanism to change steering wheel angle or height. When the tilt switch is activated, current to the motor spins the small gears to move the steering column's upper tube. Moving the switch the other way reverses current flow through the motor and moves the wheel in the opposite direction.

Memory tilt wheels use an tilt control module (computer) to "remember" more than one steering wheel position. A sensor in the steering column provides feedback so the tilt control module knows where the wheel is located. When the driver presses the proper steering wheel position switch, the control module energizes the power tilt motor until the wheel has moved to the driver's preprogrammed position. Refer to **Figure 69-5.**

Many memory tilt wheels will move the steering wheel upward whenever the ignition switch is turned off or on. This gives the driver more room to enter and exit

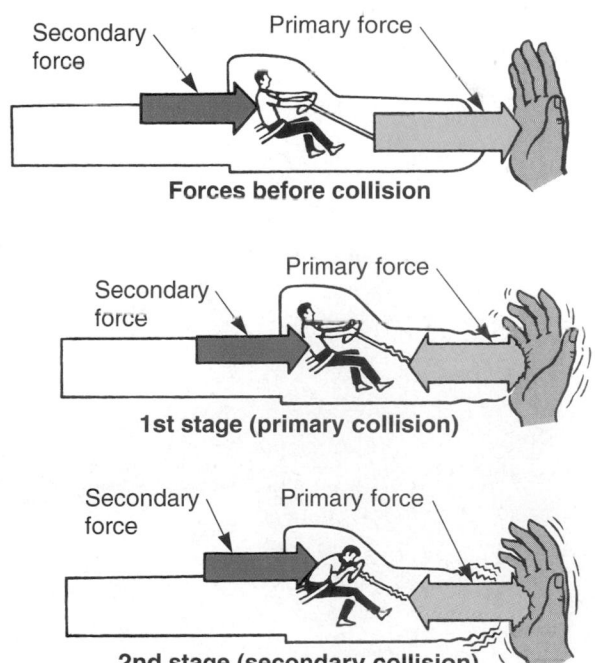

Figure 69-4. A collapsible steering column crushes, protecting the driver. (GMC)

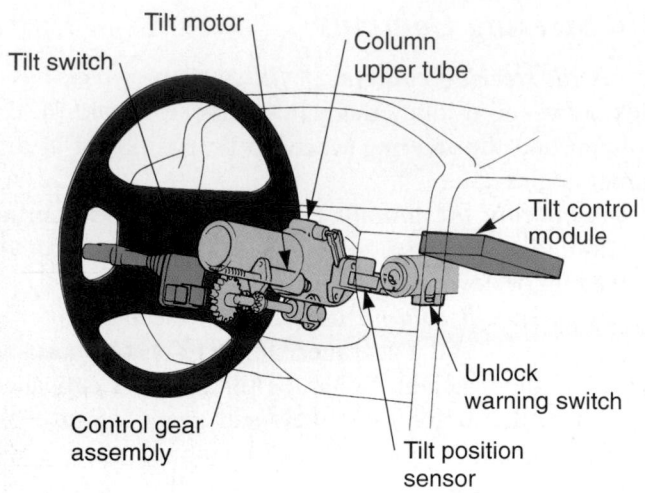

Figure 69-5. Note the parts of a power tilt steering wheel. This system can be programmed to remember the steering wheel position for more than one driver. The control module receives data on steering wheel position from the tilt position sensor. When the driver presses the appropriate tilt switch, the module operates the motor to move the steering wheel to the correct position. (Lexus)

the vehicle. One mechanism for a power tilt steering wheel is detailed in **Figure 69-6.**

Steering Gear Principles

Mentioned briefly, some steering systems use a worm-type steering gear assembly. Others use a pinion gear and a rack. These two gear principles are illustrated in **Figures 69-7.**

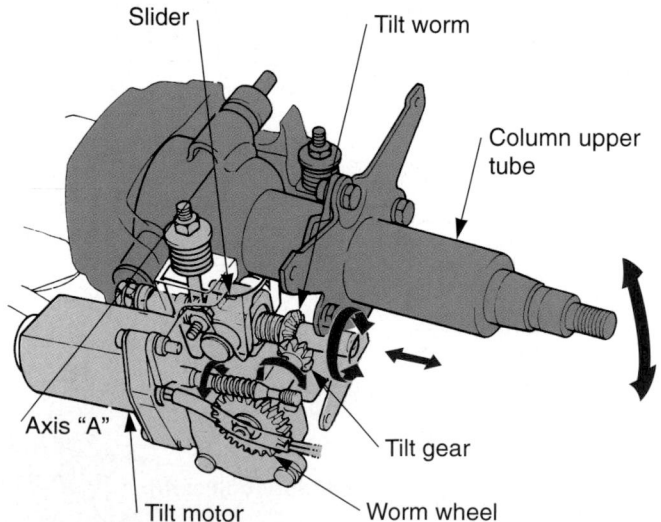

Figure 69-6. Note the gear mechanism in this power tilt steering wheel system. (Lexus)

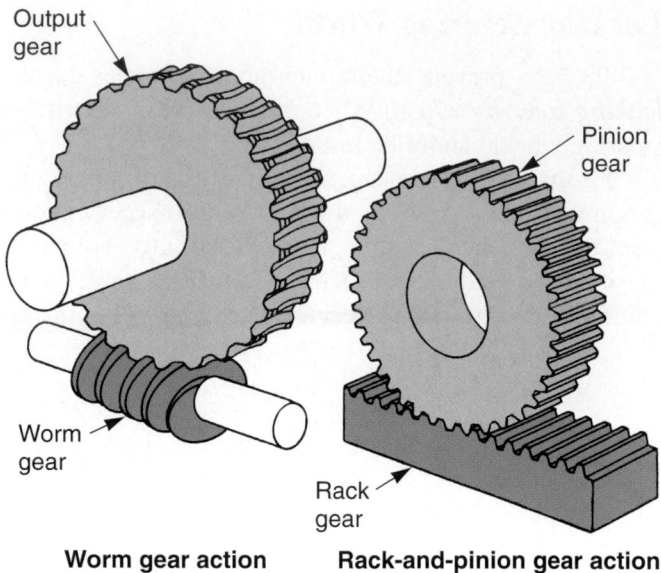

Worm gear action   **Rack-and-pinion gear action**

Figure 69-7. Two basic types of gear mechanisms are found in steering gearboxes: worm gears and rack-and-pinion gearset. (Deere & Co.)

Cutaway views of both rack-and-pinion and worm-type (recirculating-ball) steering mechanisms are given in **Figure 69-8.**

Recirculating-Ball Gearbox

The *recirculating-ball gearbox* is normally used with a linkage steering system. It has small steel balls that circulate between the gear members. See **Figure 69-9.**

A *worm shaft* is the input gear connected to the steering column shaft. The balls fit and ride in the grooves in the worm gear.

The *sector shaft* is the output gear from the steering gearbox. It transfers motion to the steering linkage. A sector gear is machined on the inner end of the sector shaft.

A *ball nut* rides on the ball bearings and the worm gear. Grooves are cut in the ball nut to match the shape of the worm gear. Since the ball nut cannot rotate, it slides up and down as the worm gear rotates. See **Figure 69-9.**

Ball guides route extra ball bearings in and out from between the worm and ball nut. These are pictured in **Figure 69-10.**

The worm shaft is mounted in either ball bearings or roller bearings. The sector shaft is also mounted on antifriction bearings.

A bearing *adjusting nut* is usually provided to set worm shaft bearing preload. An *adjusting screw* is used to set the sector shaft clearance.

The *gearbox housing* provides an enclosure for the other components. Seals press into the housing to prevent

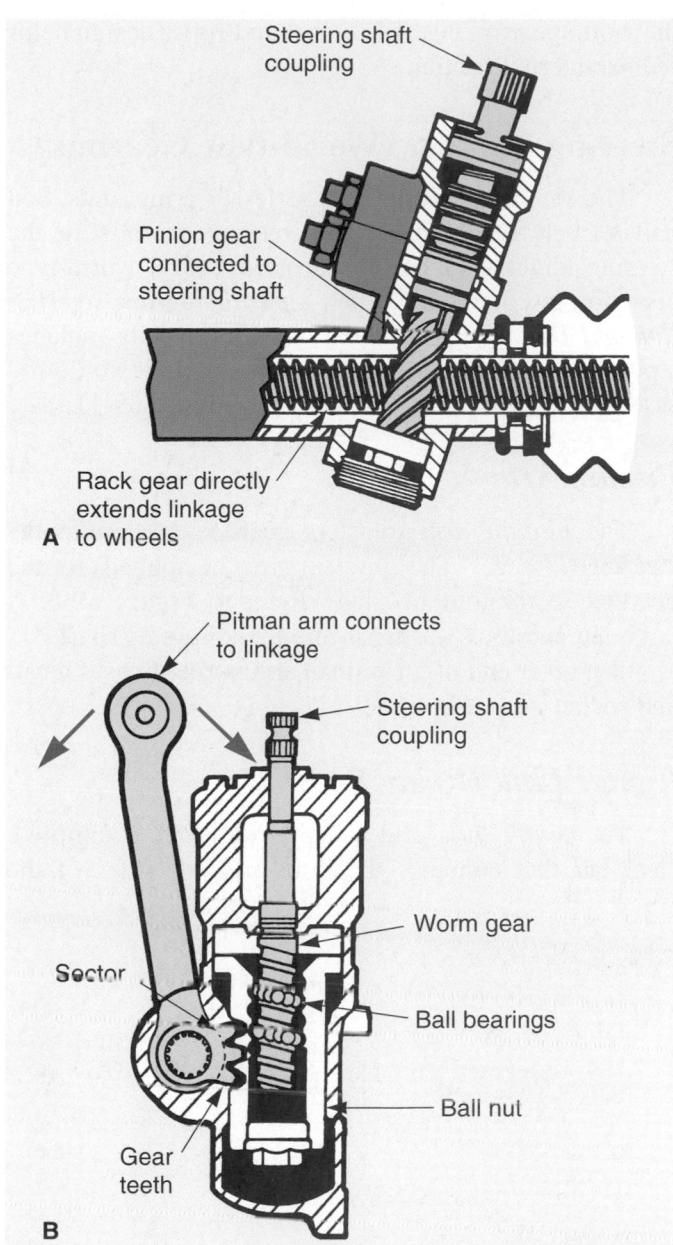

Figure 69-8. Another view of two types of steering gears. A—Rack-and-pinion gear. B—Worm steering gear. (Chrysler Corp.)

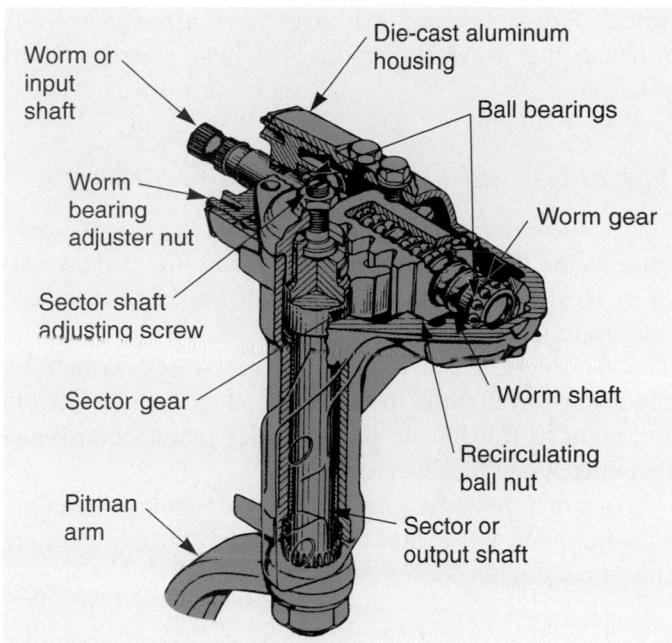

Figure 69-9. Cutaway view of a recirculating-ball steering gearbox. Study the part relationships. (Chrysler)

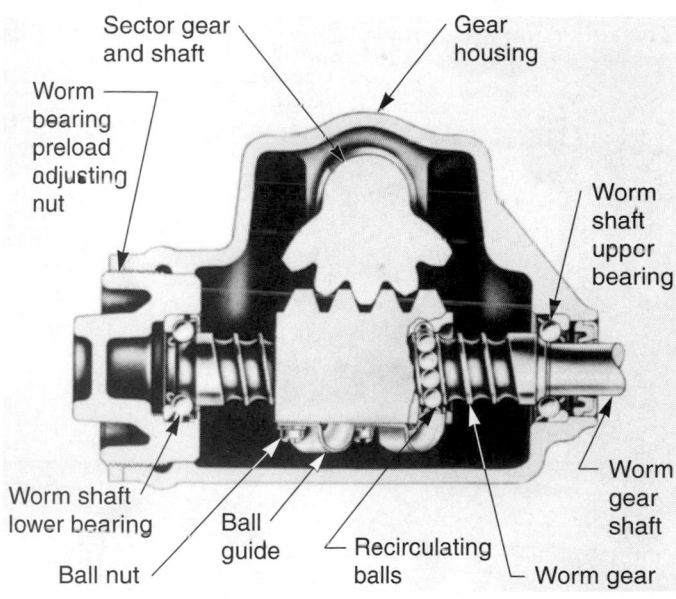

Figure 69-10. When the steering wheel is turned, the shaft and the worm gear rotate. This causes the balls and the ball nut to walk on the worm. As a result, the ball nut turns the sector gear and shaft. (Chrysler)

lubricant leakage at the worm and sector shafts. The shaft bearings also press into the gearbox housing.

The housing bolts to the vehicle frame or to a reinforced area on the unitized body. An *end cover* normally bolts on the housing to cover the end of the sector gear. It can be removed for gearbox service.

Gearbox Ratio (Steering Gear Reduction)

Gearbox ratio, also termed *steering ratio* or *steering gear reduction,* is basically a comparison between steering wheel rotation and sector shaft rotation.

Steering gearbox ratios range from 15:1 to 24:1. With a 15:1 ratio, the worm shaft turns 15 times to turn the sector shaft once.

A manual gearbox will have a *high gearbox ratio* to reduce the amount of effort needed to turn the steering wheel. Power steering gearboxes have a *lower gearbox ratio* so the wheel turn more with less steering wheel rotation.

Variable- and Constant-Ratio Gearboxes

A *variable-ratio gearbox* changes the internal gear ratio as the front wheels are turned from the center position. Most modern recirculating-ball gearboxes are variable-ratio designs.

Variable-ratio steering is faster when cornering, requiring fewer turns to move the steering wheel from full right to full left. It also provides better control and response when maneuvering.

A *constant-ratio gearbox* has the same gear reduction from full left to full right. The sector gear teeth are the same length.

Worm-and-Roller Steering Gearbox

A *worm-and-roller steering gearbox* contains a roller that is mounted on the pinion shaft and meshes with the worm gear. The roller replaces the ball bearings and the ball nut used in the recirculating-ball gearbox. The roller imparts a rotary force on the pitman shaft as

the worm gear rotates. The worm-and-roller design helps reduce internal friction.

Steering Linkage (Worm-type Gearbox)

The *steering linkage* is a series of arms, rods, and ball sockets that connect the steering gearbox to the steering knuckles. The linkage used with a worm-type gear box is commonly called a *parallelogram steering linkage.* This type of steering linkage typically includes a pitman arm, a center link, an idler arm, and two tie-rod assemblies. These parts are shown in **Figure 69-11.**

Pitman Arm

The *pitman arm* transfers gearbox motion to the steering linkage. The pitman arm is splined to the gearbox sector (output) shaft. Refer to **Figure 69-8.** A large nut and lock washer secure the arm to its shaft.

The outer end of the pitman arm normally uses a ball and socket joint, **Figure 69-11.**

Center Link (Relay Rod)

The *center link,* also called a *relay rod,* is simply a steel bar that connects the right and left sides of the

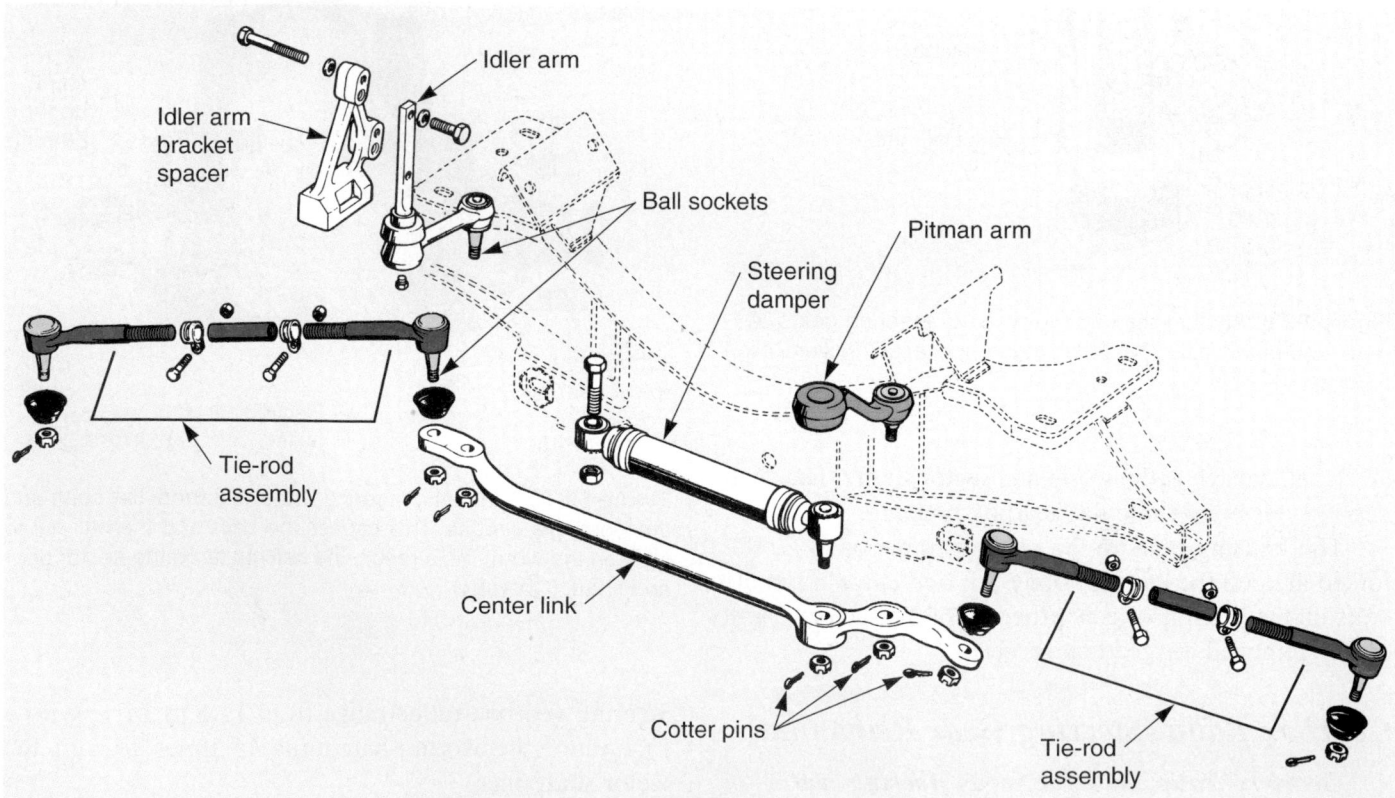

Figure 69-11. Study the arrangement of this linkage-type steering system. A pitman arm connects to the steering gearbox. The arm swings right or left and moves the other linkage components. (Chrysler)

steering linkage. It has holes that accept the pitman arm, tie-rod ends, and idler arm, as shown in **Figure 69-11.**

Idler Arm

The *idler arm* supports the end of the center link on the passenger side of the vehicle. The idler arm bolts to the vehicle's frame or subframe, **Figure 69-11.**

Ball Sockets

Ball sockets are like small ball joints; they provide for motion in all directions between two connected parts. Ball sockets are needed so the steering linkage is *not* bent and damaged when the wheels turn or move up and down over rough road surfaces. See **Figure 69-11.**

Cutaway views of various ball sockets are given in **Figure 69-12.** Note how a ball stud fits into a socket.

Ball sockets are filled with grease to reduce friction and wear. Some ball sockets are sealed. Others have a grease fitting that allows chassis grease to be inserted with a grease gun, **Figure 69-12.**

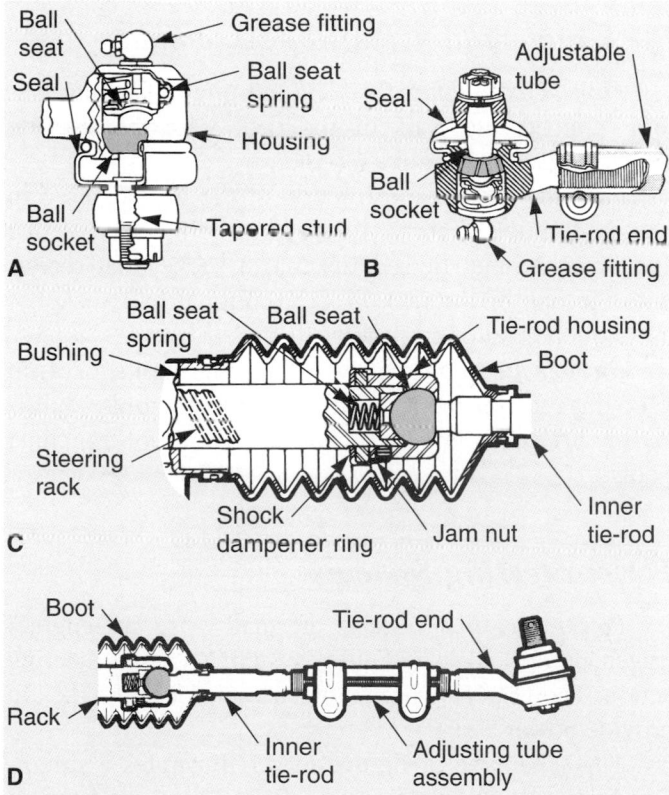

Figure 69-12. Ball sockets allow linkage components to swivel freely. They are commonly used on the end of the pitman arm, idler arm, and tie-rods. A—Ball socket for an idler arm. B—Ball socket for a tie-rod end. C—Ball socket for the inner end of a tie-rod on a rack-and-pinion setup. D—Ball sockets on both ends of a tie-rod for a rack-and-pinion steering system. (Ford and Chrysler)

Tie-Rod Assemblies

Two *tie-rod assemblies* are used to fasten the center link to the steering knuckles. Ball sockets are normally used on both ends of each tie-rod assembly. Look at **Figures 69-11** and **69-12.**

A tie-rod assembly typically consists of two tie rods and a toe adjustment sleeve. The tie rod that connects to the center link is called the inner tie rod. The tie rod that fastens to the steering knuckle is called the outer tie rod. The *toe adjustment sleeve* is provided for changing the length of the tie-rod assembly during wheel alignment.

Manual Rack-and-Pinion Steering

Rack-and-pinion steering is the most popular type of steering system on today's vehicles. For this reason, it is important that you understand its parts and operating principles.

Figure 69-13 pictures the external parts of a manual rack-and-pinion steering mechanism. Note that the steering gear is bolted to the frame crossmember.

Steering Shaft, Flexible Coupling, and Universal Joint

Many steering systems have a flexible coupling and/or a universal joint in the steering shaft. Look at **Figures 69-13** and **69-14.**

The *flexible coupling* helps keep road shock from being transmitted to the steering wheel. It also allows for slight misalignment of the steering shaft and steering gear input shaft (pinion shaft).

A *universal joint* allows for a change in the angle between the steering column and steering shaft. One is shown in **Figure 69-13.**

Rack-and-Pinion Steering Gear

A manual *rack-and-pinion steering gear* basically consists of a pinion shaft, a rack, a thrust spring, bearings, seals, and a gear housing. The assembly bolts to vehicle's frame or unibody structure. Large bolts with rubber bushings secure the unit and help absorb road shock. A cutaway view of a rack-and-pinion steering gear is shown in **Figure 69-14.**

When the steering shaft turns the pinion shaft, the pinion gear acts on the rack gear. The rack then slides sideways inside the gear housing, **Figure 69-15.**

The *thrust spring* preloads the rack-and-pinion gear teeth to prevent excessive gear backlash (play), **Figure 69-15.** Adjustment screws or shims may be used to set thrust spring tension.

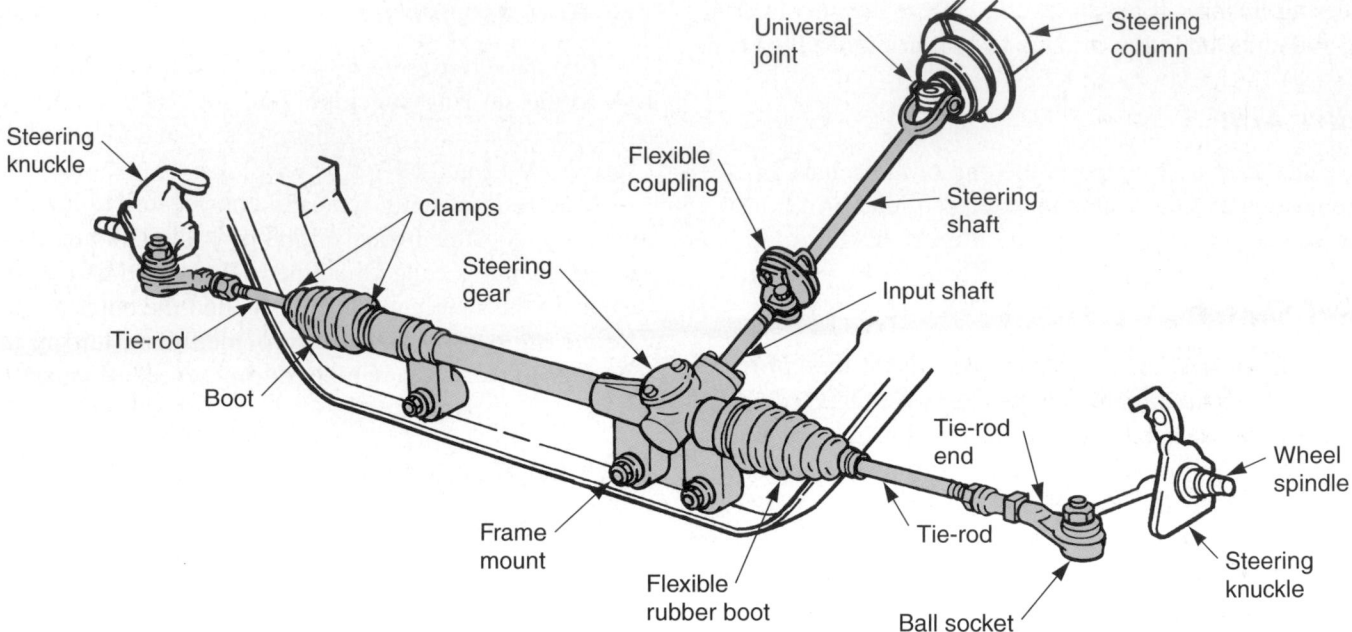

Figure 69-13. Study the components of this manual rack-and-pinion steering system. It uses fewer parts than a linkage system. (Ford)

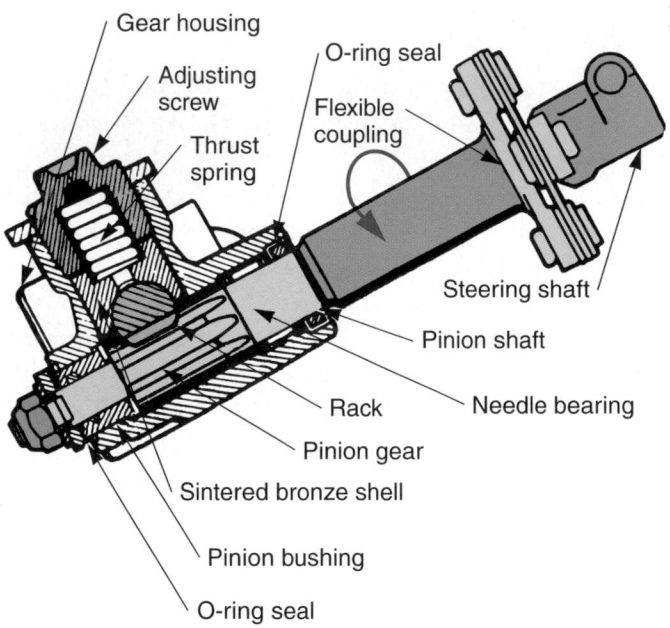

Figure 69-14. Cutaway shows how a pinion shaft gear rotates in its housing. The pinion gear teeth mesh with the rack gear teeth to slide the rack left or right for steering action. The thrust spring holds the rack-and-pinion gears in contact. (Ford)

Either bushings or roller bearings may be used on the pinion shaft and the rack. Frequently, the pinion shaft uses roller bearings and the rack uses plain bushings. This is pictured in **Figures 69-14, 69-15,** and **69-16.**

Rack-and-Pinion Tie-Rod Assemblies

Tie-rod assemblies for rack-and-pinion steering systems connect the ends of the rack with the steering knuckles. Look at **Figure 69-16.**

Note the unique construction of the tie-rods used in rack-and-pinion steering. A very large ball formed on one end of each inner tie-rod fits into a socket that screws onto the end of the rack. A conventional ball socket is used on one end of each outer tie-rod.

Rubber dust boots fit over the inner ball sockets to keep out road dirt and water, as well as to hold in lubricating grease. Clamps secure each end of the dust boots.

Power Steering Systems

Power steering systems normally use an engine-driven pump and a hydraulic system to assist steering action. They can also use an electric motor in the rack to provide power assist.

The schematic in **Figure 69-17** illustrates a simplified hydraulic power steering system. Pressure from an oil pump is used to operate a piston-and-cylinder assembly. When the control valve routes oil pressure into one end of the piston, the piston slides in its cylinder. Piston movement can then be used to help move the steering system components and the front wheels of the vehicle.

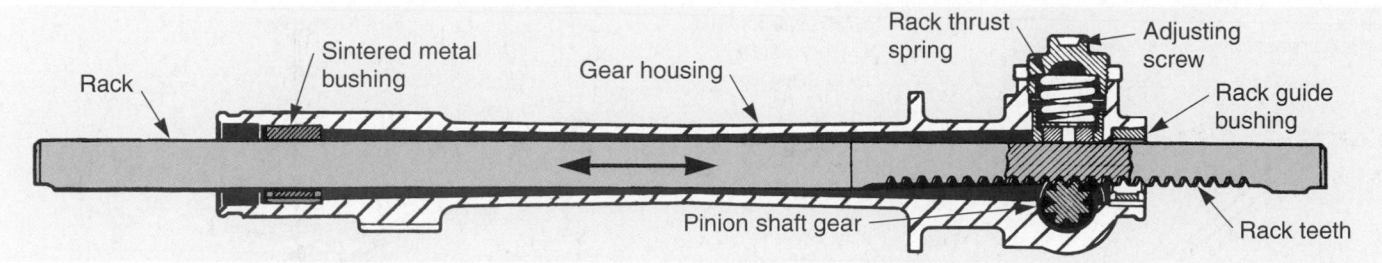

Figure 69-15. The rack slides sideways and pushes or pulls on the tie-rods. This rotates the steering knuckles and the front wheels. (Buick)

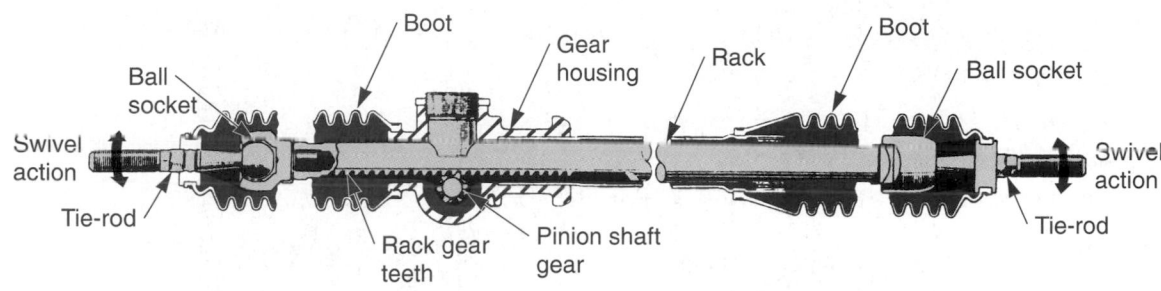

Figure 69-16. Note how tie-rods attach to the rack with ball-and-socket joints. (Toyota)

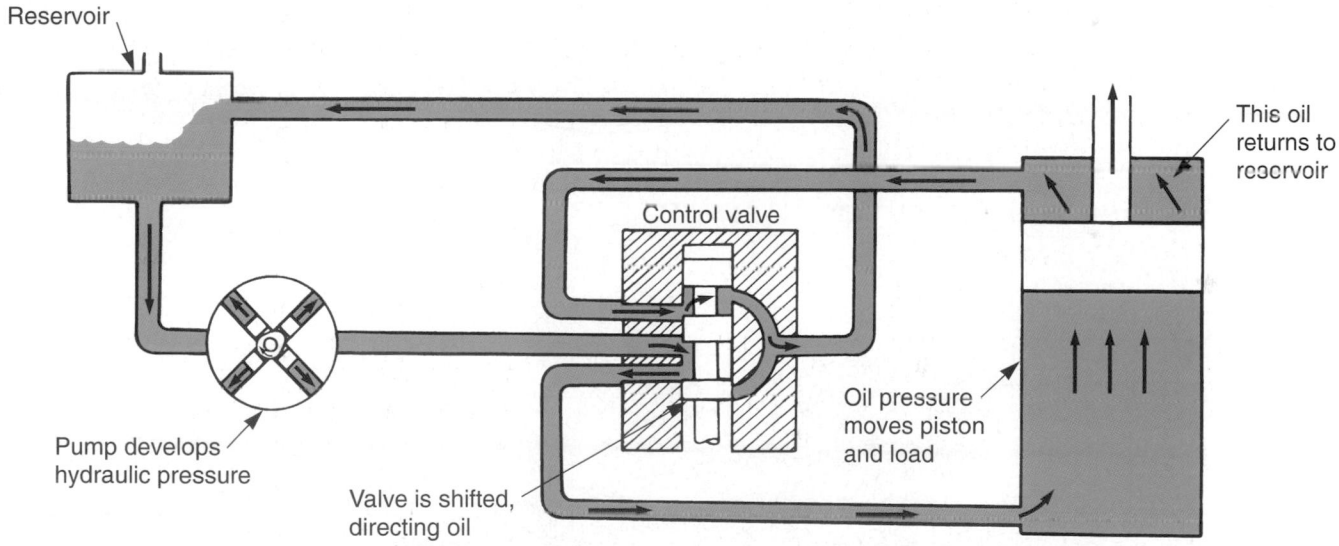

Figure 69-17. Basic components of a power steering system. A hydraulic (oil) pump pressurizes the system. The control valve routes oil to either side of the piston. Piston action can then be used to aid in moving the front wheels. (Deere & Co.)

There are three major types of power steering systems used on modern automobiles: integral-piston linkage systems, external cylinder power steering systems, and rack-and-pinion systems. The rack-and-pinion system is further divided into the integral and the external power piston systems. The integral rack-and-pinion power steering system is the most common.

Figure 69-18 shows the three main types of power steering systems. A cutaway view of a modern power rack-and-pinion mechanism is shown in **Figure 69-19.**

Power Steering Pumps

The **power steering pump** is engine driven and produces the hydraulic pressure for steering system

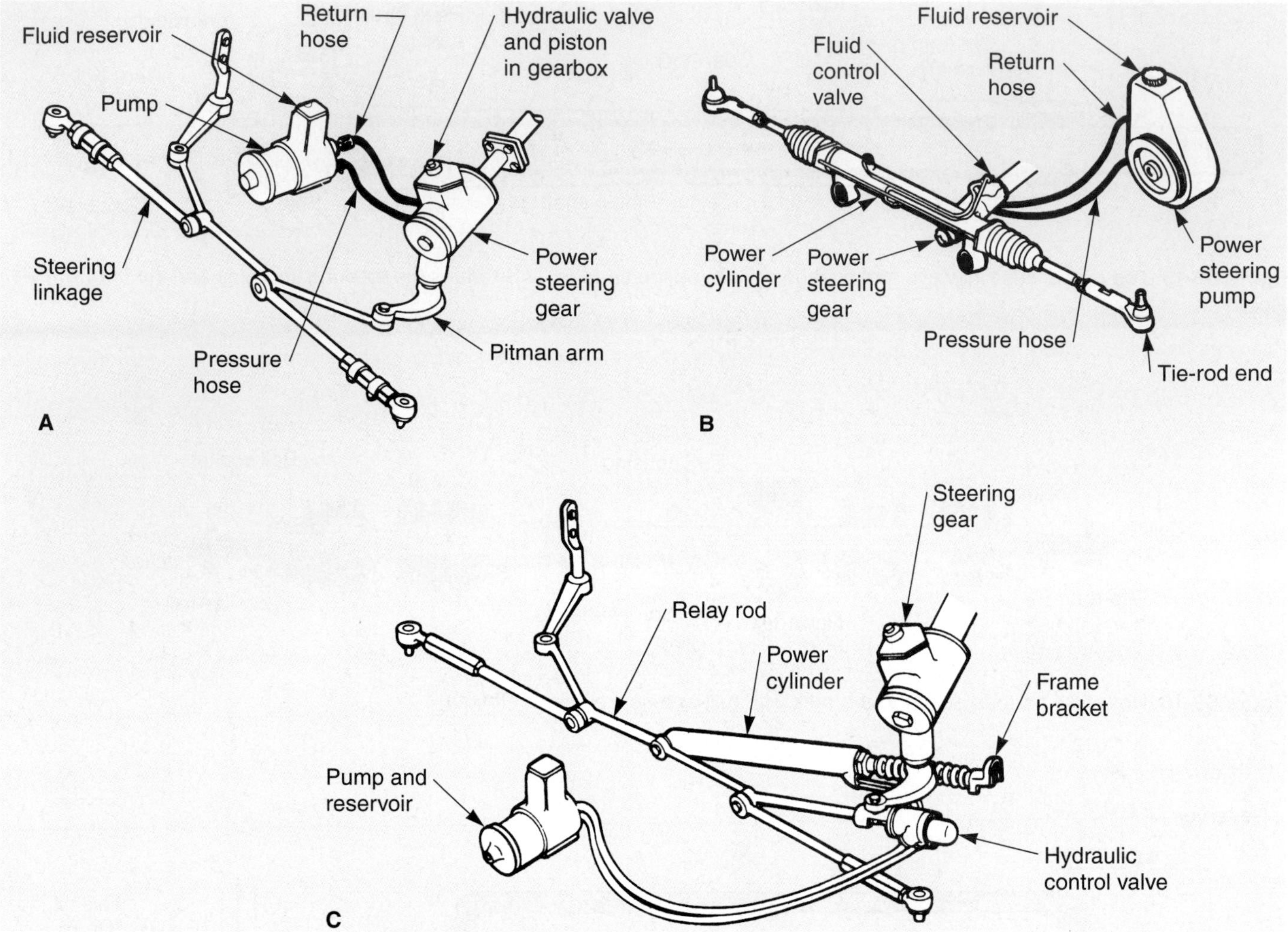

Fluid reservoir

Return hose

Hydraulic valve and piston in gearbox

Pump

Steering linkage

Pressure hose

Power steering gear

Pitman arm

A

Fluid reservoir

Fluid control valve

Return hose

Power cylinder

Power steering gear

Power steering pump

Pressure hose

Tie-rod end

B

Steering gear

Relay rod

Power cylinder

Frame bracket

Pump and reservoir

Hydraulic control valve

C

Figure 69-18. The three major power steering systems. A—Integral-piston linkage system. B—Rack-and-pinion system. C—External cylinder power steering. (Florida Dept. of Voc. Ed.)

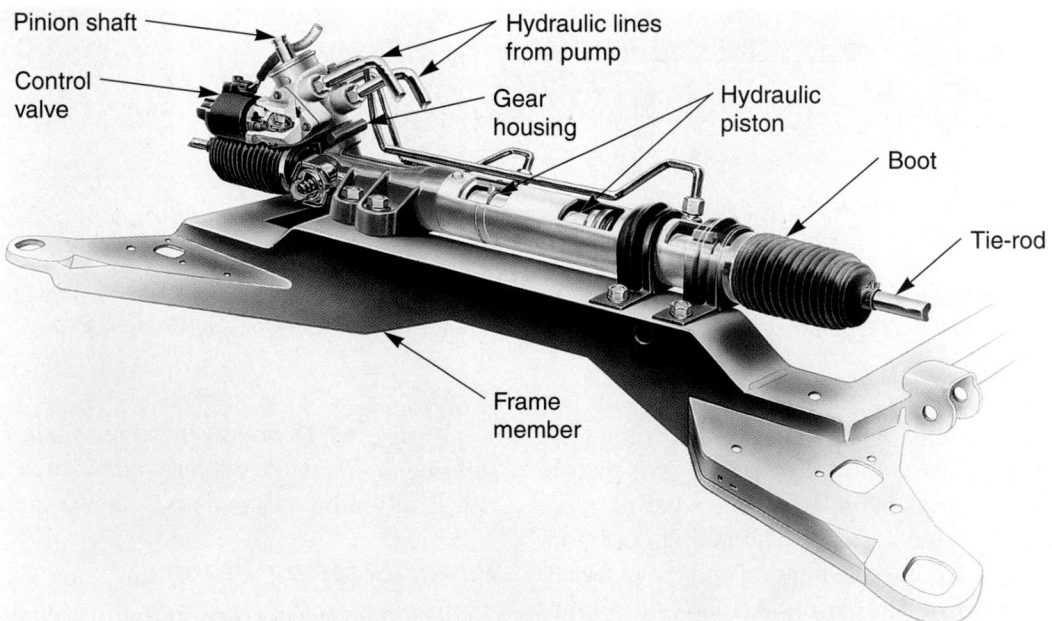

Pinion shaft

Control valve

Hydraulic lines from pump

Gear housing

Hydraulic piston

Boot

Tie-rod

Frame member

Figure 69-19. Cutaway of a power rack-and-pinion assembly. Study its parts carefully. (Ford)

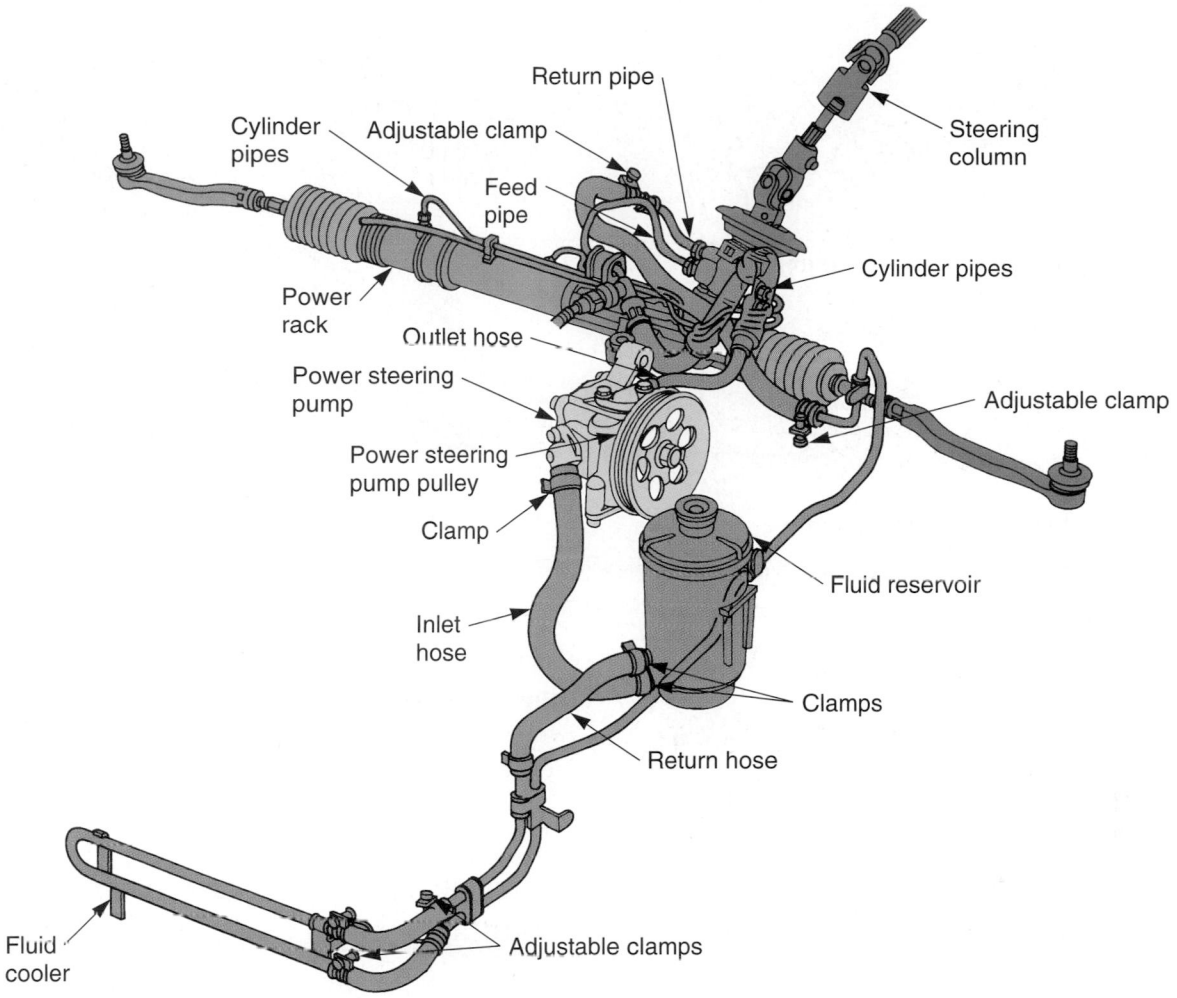

Figure 69-20. The power steering pump generally bolts to the front of the engine. The belt spins the pump. Power steering lines connect the parts of the system. Note the remote reservoir and the oil cooler. (Honda)

operation. In most cases, a belt running from the engine crankshaft pulley powers the pump.

A few power steering pumps are driven off the rear of the camshaft. This reduces the number of devices that must be mounted on the front of the engine. It also reduces the high bearing loads on the front pulleys of the engine.

The *power steering fluid reservoir* holds an extra supply of fluid. It can be formed as part of the pump body or it can be a separate container. Refer to **Figure 69-20.**

Figure 69-21 shows the four basic types of power steering pumps: roller, vane, slipper, and gear types. A cutaway view of a modern power steering pump is in **Figure 69-22.**

During pump operation, the drive belt turns the pump shaft and pumping elements. Oil is pulled into one side of the pump by vacuum. The oil is then trapped and squeezed into a smaller area inside the pump. This pressurizes the oil at the output, causing it to flow to the rest of the power steering system, **Figure 69-23.**

Pressure-Relief Valve

A *pressure-relief valve* is used in a power steering system to control maximum oil pressure. It prevents system damage by limiting pressure when needed.

Figure 69-24 shows the fundamental operation of a pressure-relief valve in a modern power steering pump.

Power Steering Hoses

Power steering hoses are high-pressure, hydraulic, rubber hoses that connect the power steering pump and the integral gearbox or the power cylinder. One line serves as the pressure feed line. The other acts as a return line to the reservoir. Metal lines are also used to carry fluid between parts where vibration is not a problem. Refer to **Figure 69-20.**

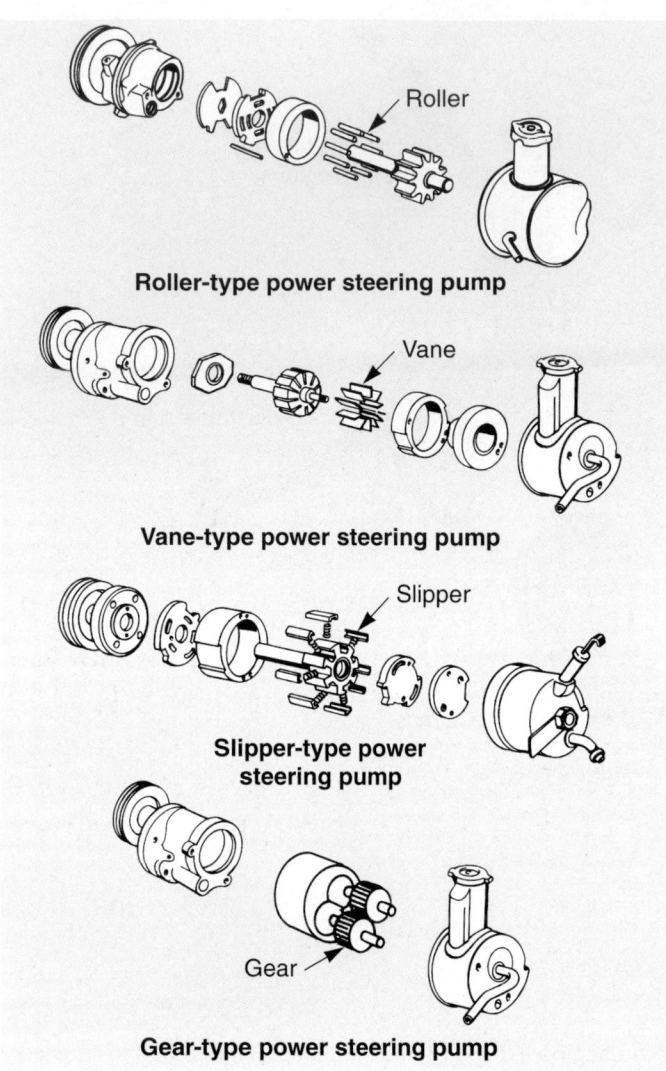

Figure 69-21. Study the four basic types of power steering pumps. (Moog)

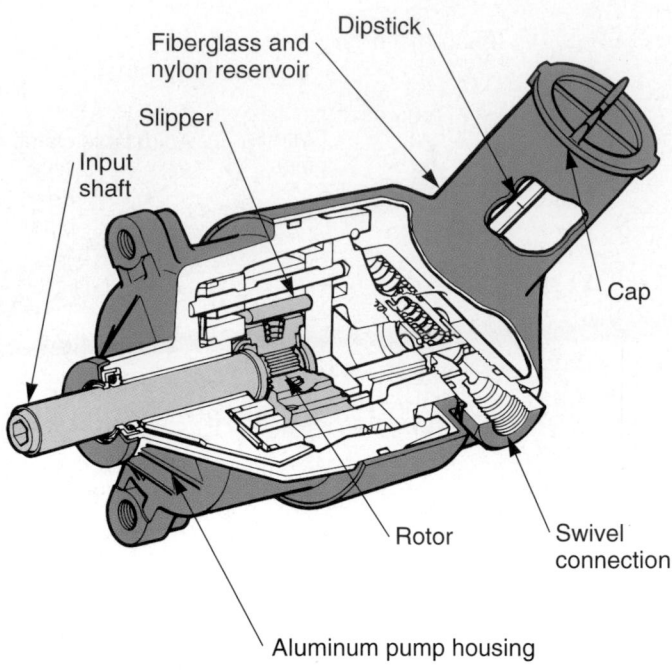

Figure 69-22. Rotation of the input shaft turns the rotor and slippers. Power steering fluid is forced out under pressure. (Ford)

Integral-Piston Power Steering System (Linkage Type)

An *integral-piston power steering system* has the hydraulic piston mounted inside the steering gearbox housing. It is common type of linkage power steering system. Basically, it consists of a power steering pump, hydraulic lines, and an integral power-assist gearbox.

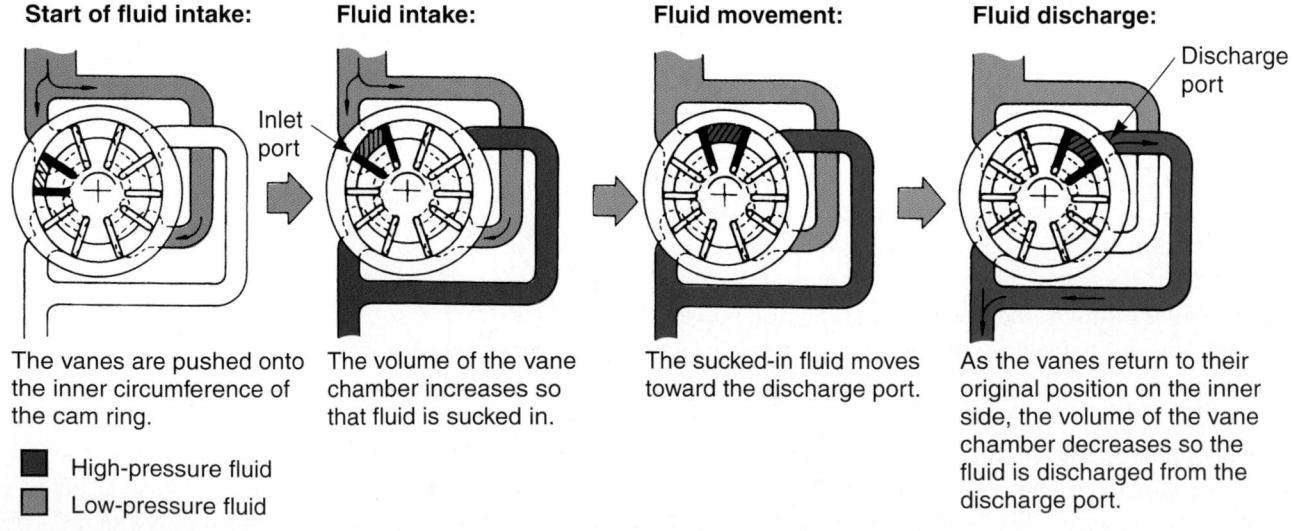

Start of fluid intake:

The vanes are pushed onto the inner circumference of the cam ring.

■ High-pressure fluid
■ Low-pressure fluid

Fluid intake:

The volume of the vane chamber increases so that fluid is sucked in.

Fluid movement:

The sucked-in fluid moves toward the discharge port.

Fluid discharge:

As the vanes return to their original position on the inner side, the volume of the vane chamber decreases so the fluid is discharged from the discharge port.

Figure 69-23. Note the pumping action inside this vane-type power steering pump. (Honda)

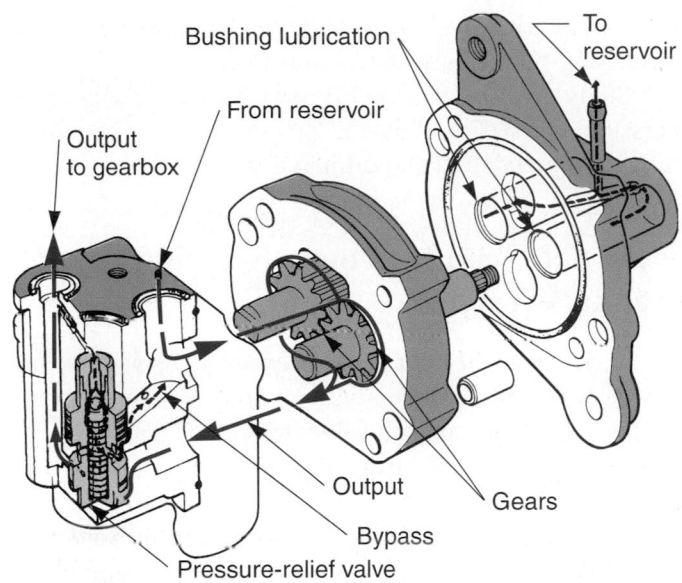

Bushing lubrication
To reservoir
From reservoir
Output to gearbox
Output
Gears
Bypass
Pressure-relief valve

Figure 69-24. Note the construction of this gear-type power steering pump. A pressure-relief valve is built into the body of the pump. The relief valve is designed to open when pressure becomes too high, such as when the steering wheel is turned to the full-right or full-left position.

The *integral power steering gearbox* contains a conventional worm-and-sector gear, a hydraulic piston, and a flow-direction valve. One type of integral power steering gearbox uses a spool valve. Another popular type is equipped with a rotary valve.

Figure 69-25 shows a spool valve–type power steering gearbox. Note that it uses a small *spool valve* to control the pressure entering the power chambers.

When the steering wheel is turned to the right, the pivot lever moves the spool valve so that pressure enters the right-turn chamber. This forces the power piston to the left and helps turn the sector shaft for a right turn. Pressure enters the opposite chamber when the steering wheel is turned to the left.

A rotary valve–type power steering gearbox has a small torsion bar to detect steering wheel turning direction and turning effort.

When the steering wheel is turned, the torsion bar twists and turns the rotary valve. The *rotary valve* then directs hydraulic pressure to the correct side of the power

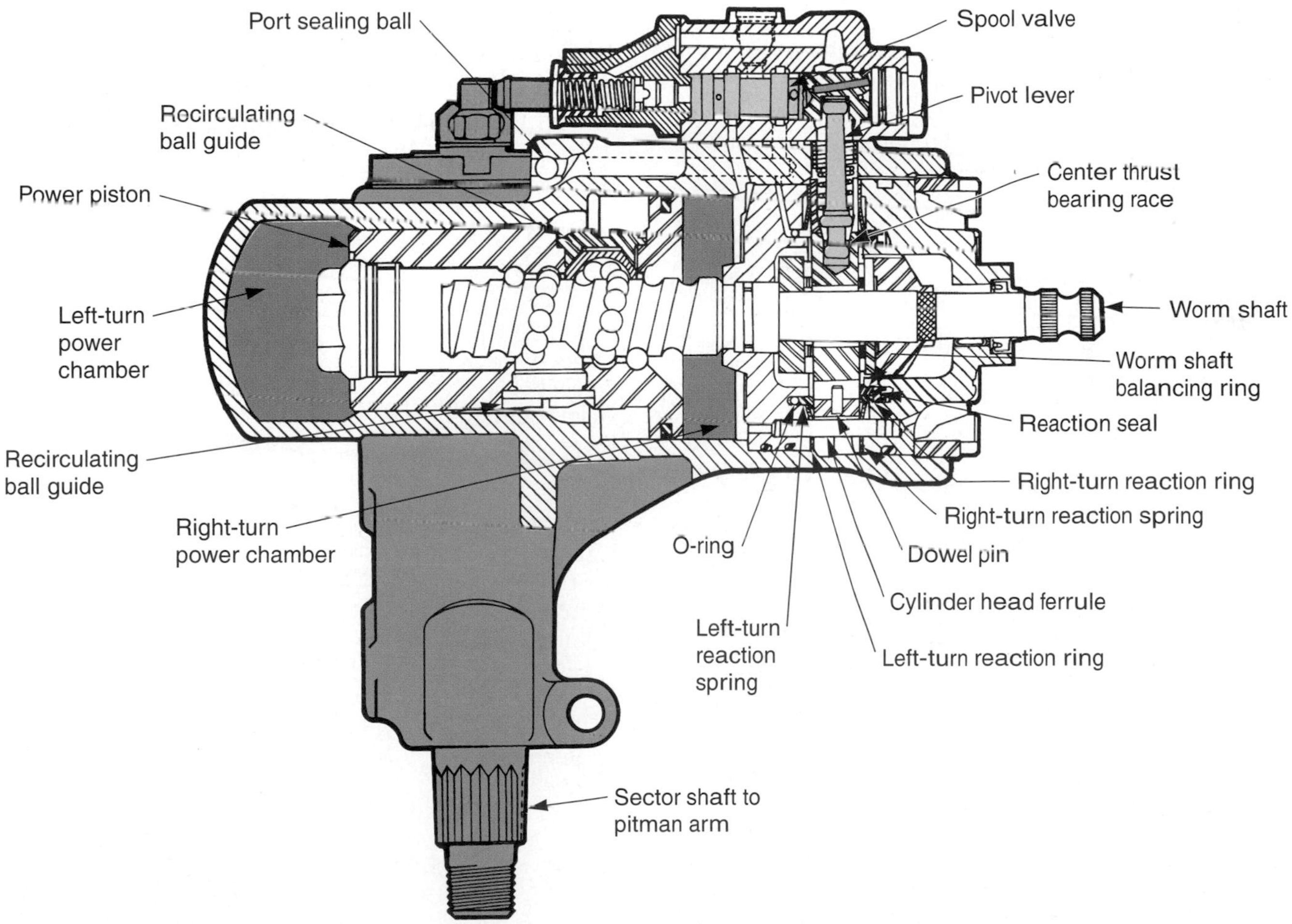

Port sealing ball
Spool valve
Recirculating ball guide
Pivot lever
Power piston
Center thrust bearing race
Worm shaft
Left-turn power chamber
Worm shaft balancing ring
Reaction seal
Recirculating ball guide
Right-turn reaction ring
Right-turn reaction spring
Right-turn power chamber
O-ring
Dowel pin
Left-turn reaction spring
Cylinder head ferrule
Left-turn reaction ring
Sector shaft to pitman arm

Figure 69-25. Cutaway of a power steering gearbox for a linkage steering system. The spool valve controls pressure on each side of the power piston. (Chrysler)

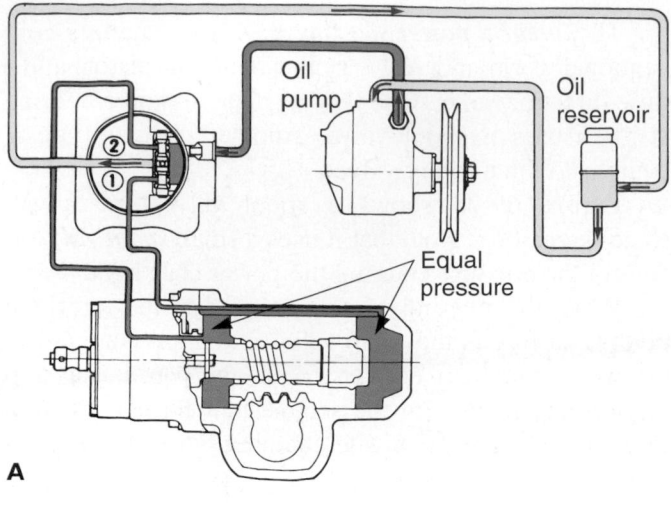

A

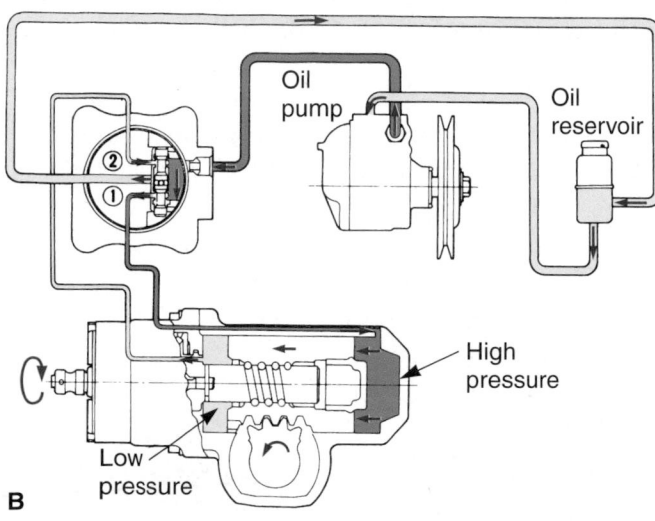

B

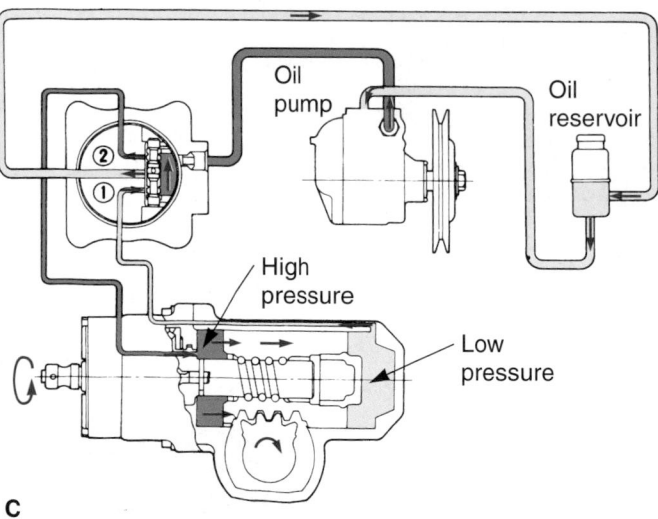

C

Figure 69-26. Integral power steering gear operation. A—Steering wheel held straight ahead, or neutral. The control valve balances pressure on both sides of the power piston. Oil returns to the pump reservoir from the valve. B—With a right turn, the control valve routes oil to one side of the power piston. The piston is pushed in the cylinder to aid pitman shaft rotation. C—With a left turn, the control valve routes oil to the other side of the power piston. Piston movement forces the oil on the nonpressure side of the piston back through the control valve and to pump. (Plymouth)

piston. Rotary valve action will be covered later, when discussing power rack-and-pinion steering.

Study the flow of oil in **Figure 69-26.** Notice how the pressure is used to slide the power piston. The piston turns the sector shaft and pitman arm.

External Cylinder Power Steering (Linkage Type)

In an *external cylinder power steering system,* the power cylinder is commonly bolted to the frame and the center link. The control valve may be located in the gearbox or on the steering linkage. This is shown in **Figure 69-27.**

External power steering cylinders can be found on older passenger cars and large industrial equipment. This design has been phased out for more compact arrangements.

Power Rack-and-Pinion Steering

Power rack-and-pinion steering uses hydraulic pump pressure to assist the driver in moving the rack and the front wheels. **Figure 69-28** shows this type of system.

The power steering pump is normally mounted on the front of the engine. A drive belt powers the pump. Power steering hoses and metal lines connect the pump with the rack-and-pinion gear.

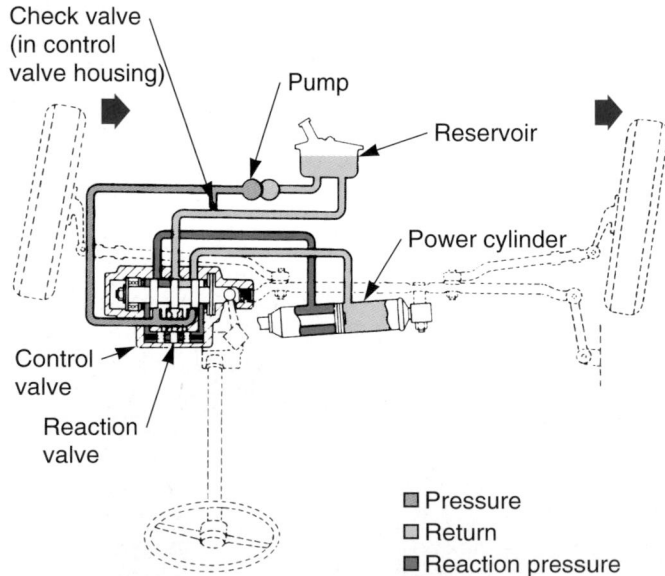

Figure 69-27. Power steering system using a power cylinder mounted on the steering linkage. Note the system pressures. (Ford)

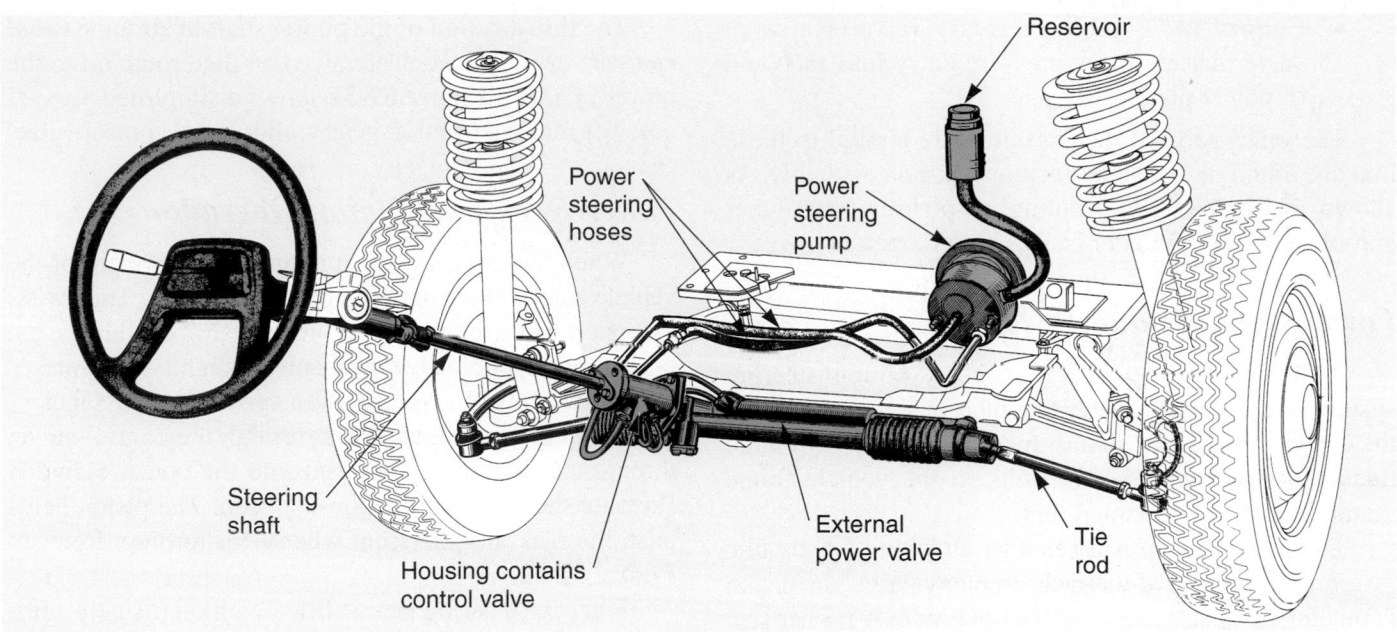

Figure 69-28. External view of a power rack-and-pinion setup. (Peugeot)

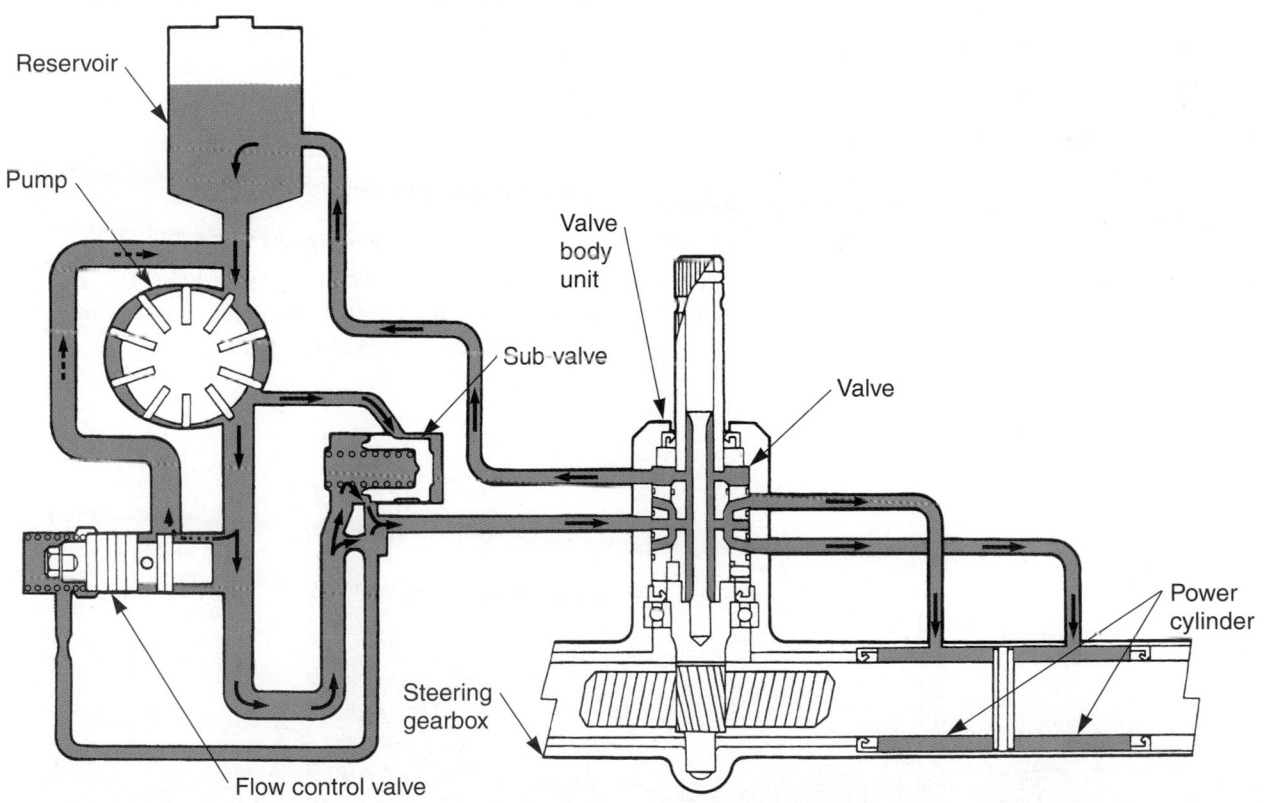

Figure 69-29. The power cylinder is formed around the rack. Pressure acts on the rack piston to help slide the rack in its housing. Note how the flow control valve can direct pressure to either side of the power piston. (Honda)

A power rack-and-pinion assembly basically consists the following:

- *Power cylinder*—hydraulic cylinder machined inside the rack or gear housing. See **Figure 69-29.**

- *Power piston*—hydraulic, double-acting piston formed on rack.

- *Hydraulic lines*—steel tubing connecting the control valve and the power cylinder.

- *Control valve*—either a rotary valve or a spool valve that regulates pressure entry into each end of power piston.

The other parts of the assembly are similar to those that are found on a manual rack-and-pinion assembly. As shown in **Figure 69-29,** routing oil pressure into either end of the power cylinder causes piston operation.

Power Cylinder and Piston

A power cylinder for a rack-and-pinion steering system is a precisely machined tube designed to accept the power piston. Provisions are made for the hydraulic lines. The cylinder housing bolts to the vehicle frame member, just like a manual unit.

The power piston is formed by attaching a hydraulic piston to the center of the rack. In many cases, the piston is machined as an integral part of the rack. A rubber seal fits around a groove in the piston. Seals are also used on each end of the piston to keep fluid from leaking out. This is shown in **Figure 69-30.**

Power Rack-and-Pinion Control Valves

There are two types of control valve mechanisms used on power rack-and-pinion gears: rotary control valves and spool control valve. The rotary is more common.

The *rotary control valve* is operated by a torsion shaft connected to the pinion gear. Study **Figure 69-30** closely. It illustrates the operation of a power rack-and-pinion gear using a rotary control valve. Today's vehicles frequently use this design.

The thrust action of the pinion shaft to shift the *spool control valve.* The control valve can then route oil to the power cylinder. **Figure 69-31** shows a simplified view of a power rack-and-pinion gear with a spool control valve.

Power Rack-and-Pinion Operation

When the steering wheel is turned, the weight of the vehicle causes the front tires to resist turning. This twists a torsion bar (rotary valve mechanism) or thrusts the pinion shaft (spool valve mechanism) slightly. This makes the control valve move and align specific oil passages.

Pump pressure than flows through the control valve, through the hydraulic line, and into the power cylinder. Pressure then acts on the power piston. The piston helps push the rack and the front wheels for turning. Refer to **Figure 69-32.**

Since the steering gear is filled with oil (usually automatic transmission fluid), the internal parts of the system are always lubricated. They slide or turn easily, with little friction and wear.

A *power steering oil cooler* is used to remove excess heat from the oil. It is usually a small metal line or radiator located in one of the power steering lines. Refer back to **Figure 69-20.**

Electronic Steering Assist

Electronic steering assist uses a small electric motor to help move the rack-and-pinion gearbox. The motor is mounted inside the rack housing. The motor acts upon the steering rack. See **Figure 69-33.**

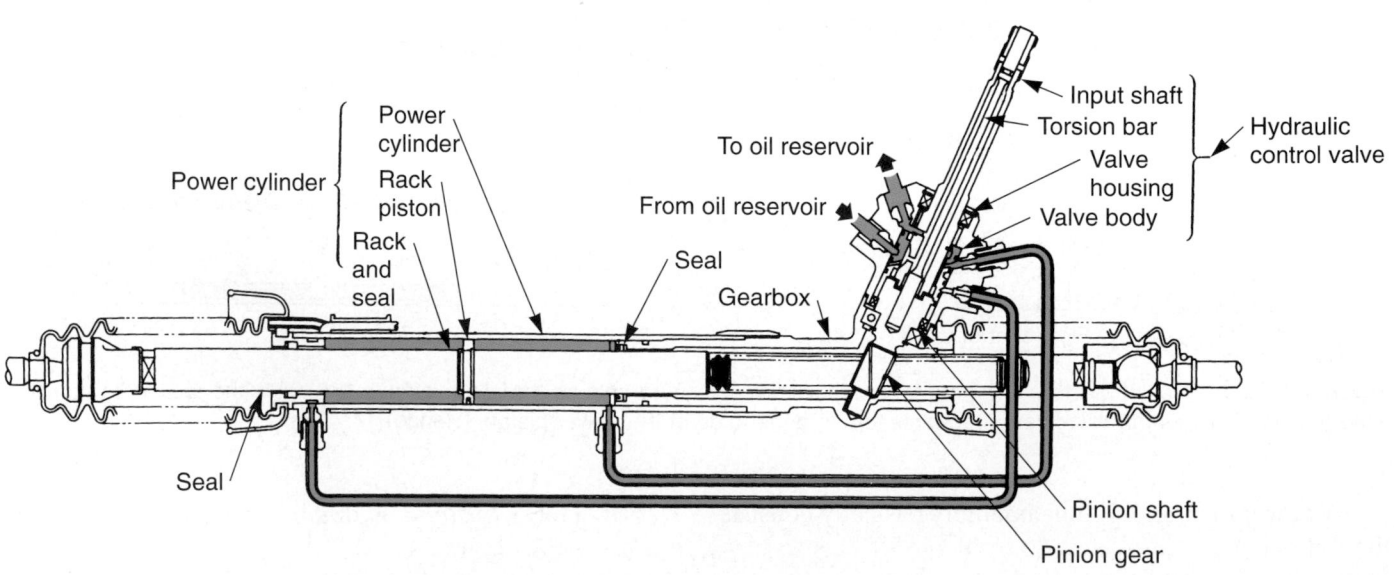

Figure 69-30. Study the parts of this modern power steering system. (Subaru)

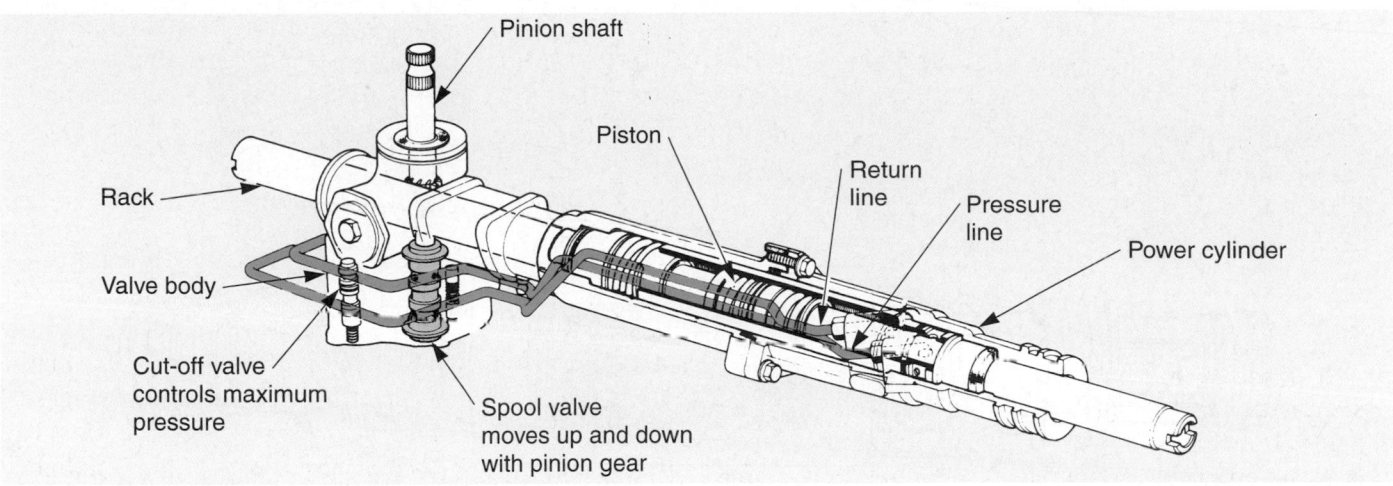

Figure 69-31. This power rack-and-pinion steering assembly uses a spool valve that detects thrust action of the helical pinion gear. It can then control oil pressure to the rack piston. (Honda)

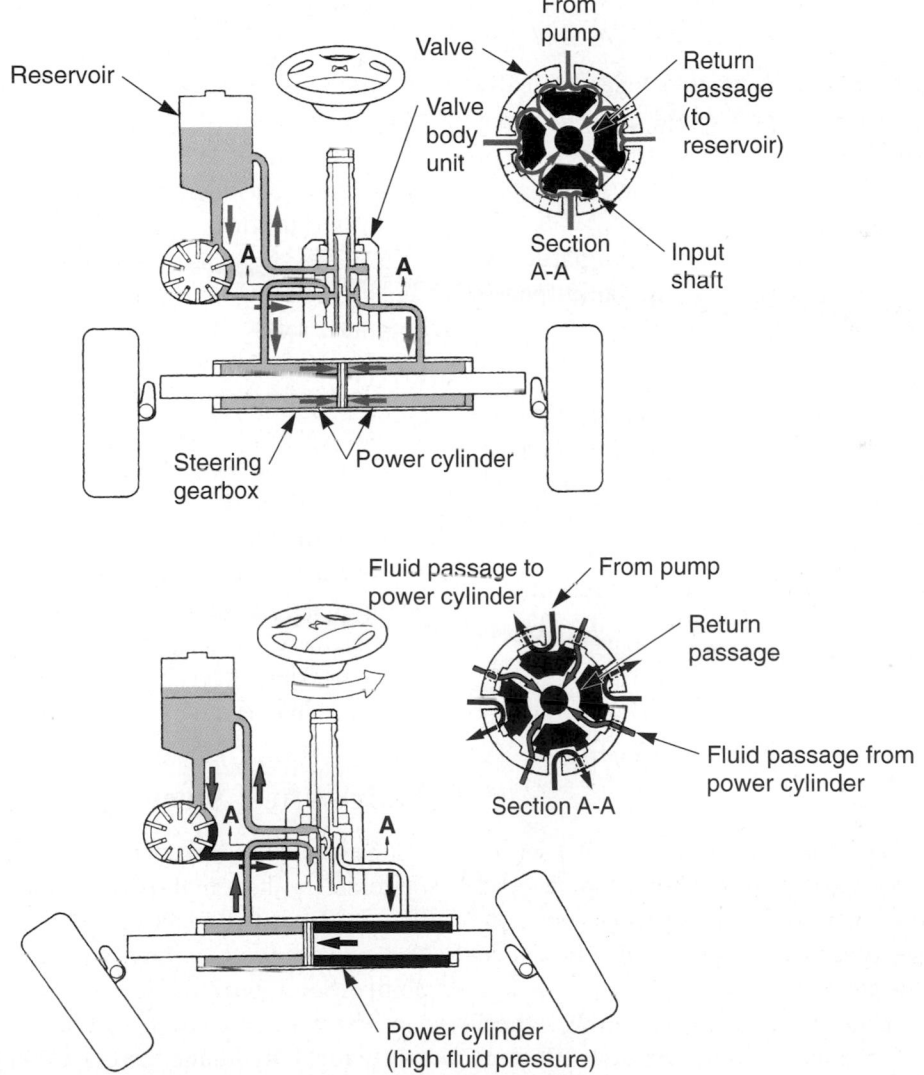

Figure 69-32. Study the operation of this speed-sensitive steering system. A—When driving at high road speeds, low power assist is desirable. Less force is needed to steer the wheels. B— When stopped or driving at low road speeds, more power assist is needed. Wheels resist turning more because of the high friction between the stationary tires and the road surface. (Honda)

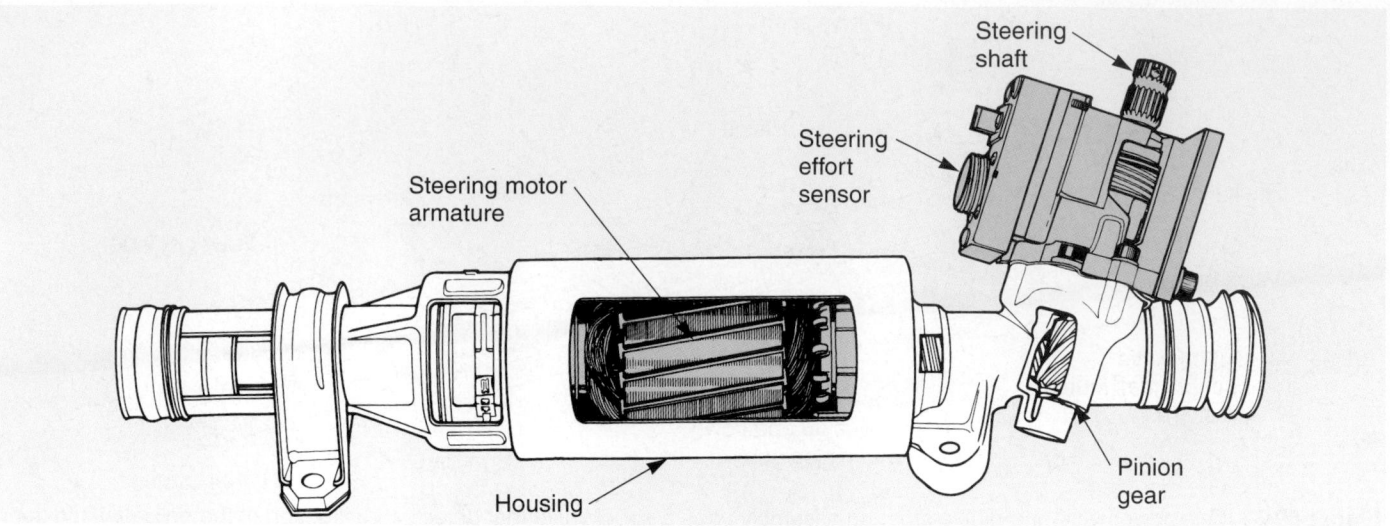

Figure 69-33. An electronic assist rack-and-pinion steering gear has a small electric motor inside the housing and a steering effort sensor on the steering shaft. The motor is used to supplement manual effort and provide power assist. This system is much lighter and more compact than a conventional hydraulic power steering system. (Moog)

A steering control module (computer) electronically reacts to steering pressure. It operates the electric motor in the rack assembly to help the driver steer the wheels of the vehicle. The control module can reverse motor rotation and alter motor speed as needed.

An electronic steering mechanism eliminates the need for hydraulic steering assist and the bulky power steering pump, hoses, and hydraulic cylinder.

Tech Tip

Idle speed can be affected by the signal from the power steering pressure sensor. Some systems signal the engine control module or the power train control module when the wheels are turned to full lock and higher-than-normal power steering system pressures are present. The control module can then operate the idle speed motor to prevent engine stalling.

Proportional Rack-and-Pinion Power Steering

Proportional rack-and-pinion power steering senses vehicle speed and steering load to ensure adequate road feel (small road imperfections felt in steering wheel for more feedback). This type of system can also be called *speed-sensitive power steering.*

Proportional steering systems alter steering wheel effort as road speed changes. They increase effort at higher speeds for more road feel. They also lower steering effort at low speeds to help the driver turn the steering wheel when parking.

In some systems a computer-controlled electromagnet is mounted around the steering gear stub shaft. The stub shaft acts on a small torsion bar. The stub shaft and torsion bar operate the power steering valve, which controls fluid flow to the rack for power assist. Look at **Figure 69-34.**

If the vehicle is not moving or is moving slowly, the computer sends current through the electromagnet in one direction to help attract and pull on the stub shaft and torsion bar, increasing valve output pressure to make the vehicle easier to steer. Then, when the vehicle reaches road speed, the computer reverses electrical flow through the electromagnet. This increases steering effort to improve road feel.

Other systems use a stepper motor or a solenoid to act upon the steering mechanism. The computer-operated stepper motor or solenoid helps shift the fluid valve to affect steering effort. The stepper motor is not as smooth or as variable as the electromagnetic-assist system.

Four-Wheel Steering Systems

Several auto makers now provide four-wheel steering systems on their high-performance vehicles. Instead of just the front two wheels, all four wheels change direction to improve handling, stability, feel, and maneuverability. See **Figure 69-35.**

There are three types of four-wheel steering systems: mechanical, hydraulic, and electronic.

The *mechanical four-wheel steering system* uses a special front rack-and-pinion gearbox with a transfer box. The transfer box operates a long shaft that extends

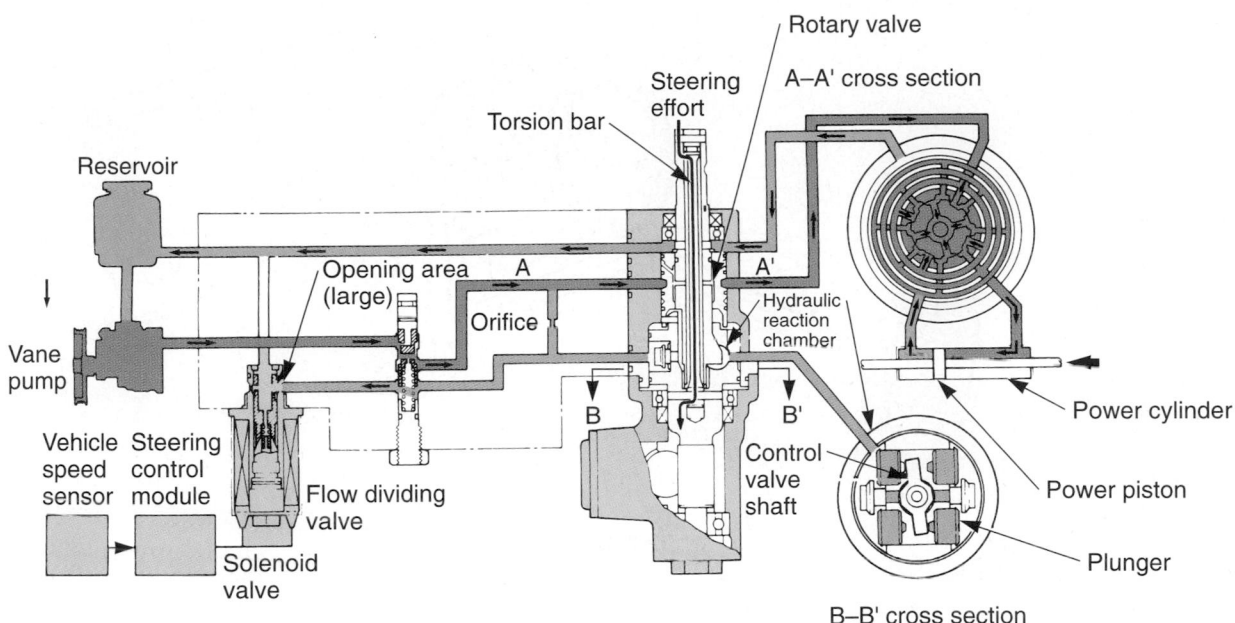

Figure 69-34. Diagram shows components of a proportional steering system. This design uses a steering control module (computer), a vehicle speed sensor, and a solenoid to control hydraulic pressure in the system. Study the parts and the flow. (Lexus)

back to the rear rack. When the front wheels are turned, the shaft rotates to turn the rear wheels. This was the first four-wheel steering system on the market.

A *hydraulic four-wheel steering system* uses a conventional power rack-and-pinion steering system up front. A vane pump forces fluid to the rack to provide power assist. Hydraulic lines extend back to a rear power steering pump, which is driven by the differential. Depending on vehicle speed, the rear pump forces fluid under pressure into a control valve.

When a specific road speed is reached, the control valve can then operate the rear steering system. With this design, the rear rack tie-rod ends are attached to the suspension system's trailing arm. When activated, the rack shifts the trailing arm to steer the rear wheels.

Modern *electronic four-wheel steering systems* have an electric-motor-driven power rack that acts upon the rear wheels via its own recirculating-ball drive and mechanical links. See **Figure 69-36.**

A computer controls rear-wheel steering angles. The computer analyzes signals from angle sensors in the front steering and from speed signals from wheel speed sensors in the anti-lock brake system.

The motor-driven power rack is then energized by the computer to move the rear wheels as needed. The computer energizes a dc motor inside the power rack. The motor powers a recirculating-ball drive to slide the rack right or left to respond to computer signals. Refer to **Figure 69-37.**

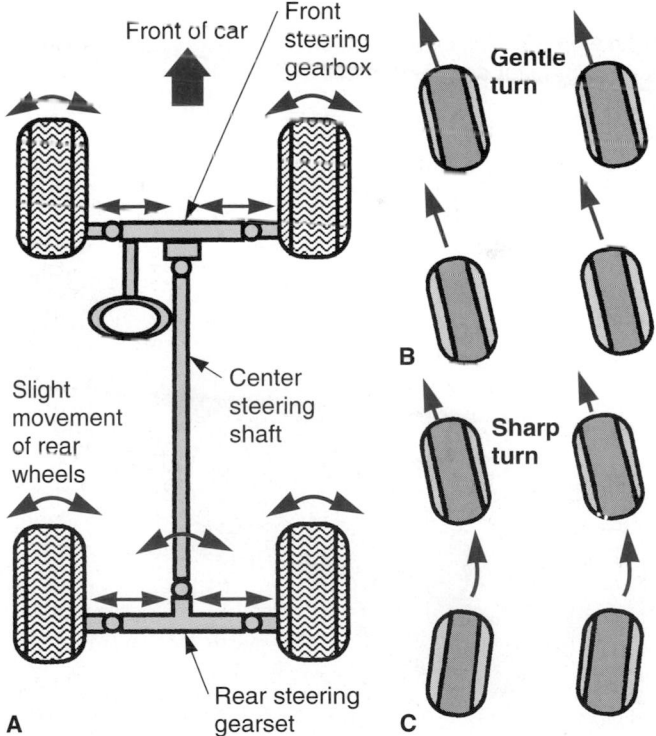

Figure 69-35. A—Note the basic parts of this four-wheel steering system. B—When the steering wheel is turned gently, the wheels pivot in the same direction. C—When the steering wheel is turned more sharply, the wheels pivot back and then turn in opposite directions.

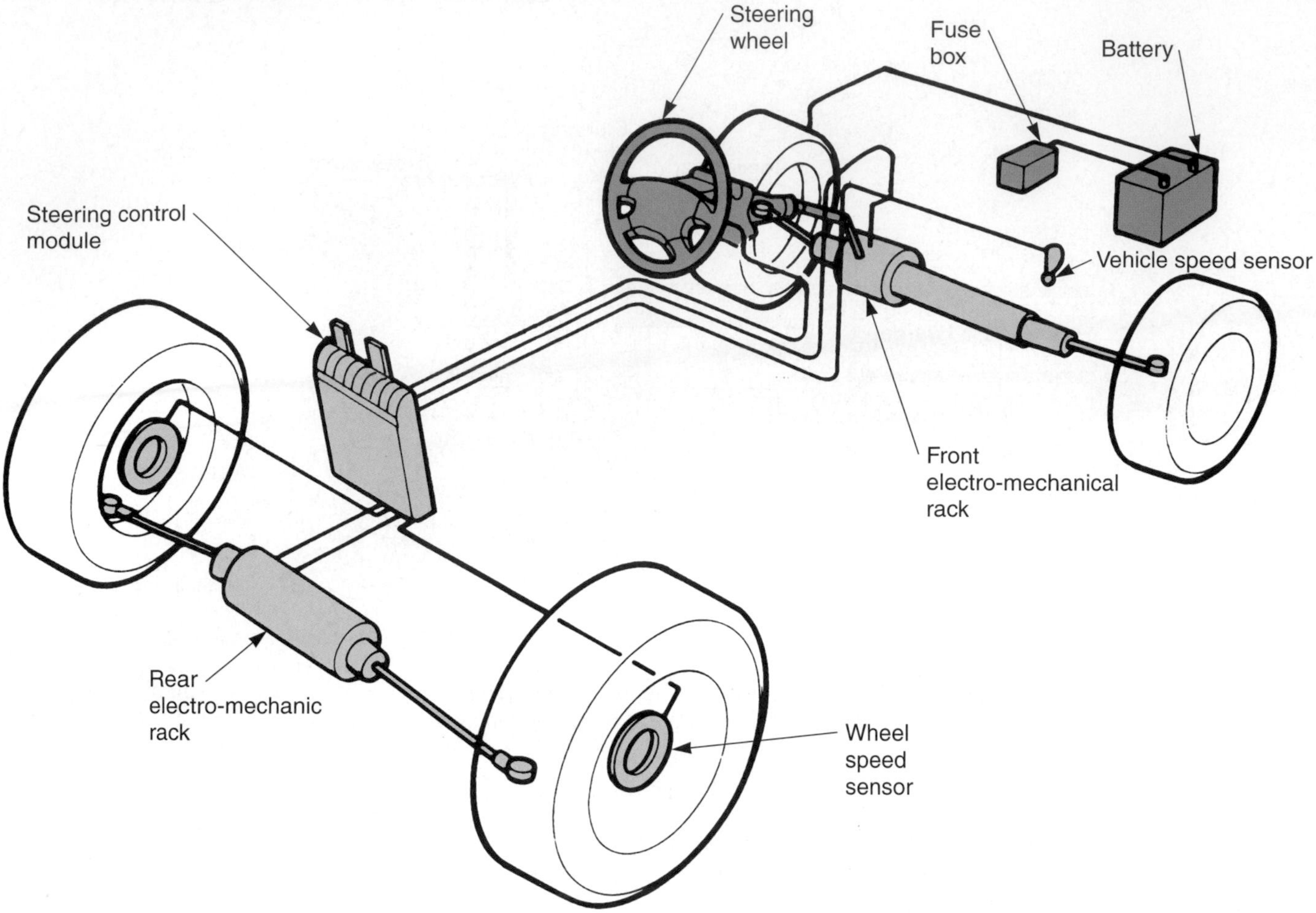

Steering wheel

Fuse box

Battery

Vehicle speed sensor

Steering control module

Front electro-mechanical rack

Rear electro-mechanic rack

Wheel speed sensor

Figure 69-36. This is an electronically-controlled four-wheel steering system. A dedicated computer, or steering control module, is used to monitor and control this system. (Honda)

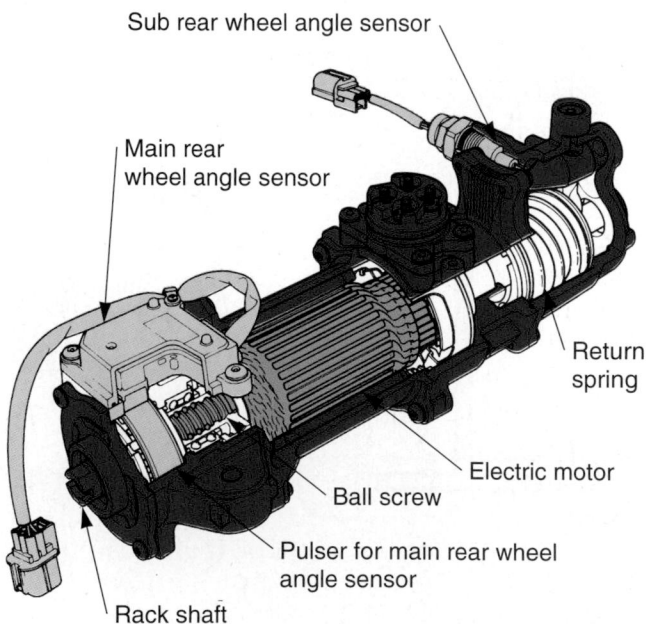

Sub rear wheel angle sensor

Main rear wheel angle sensor

Return spring

Electric motor

Ball screw

Pulser for main rear wheel angle sensor

Rack shaft

Figure 69-37. This is the rear actuator, or electric power steering assist unit, for a four-wheel steering system. Note how it has rear wheel angle sensor built into the assembly. The motor operates a ball screw, which transfers turning force to the rack shaft. (Honda)

With a slight turn of the steering wheel, the four-wheel steering system turns the rear wheels in the same direction as the front wheels. However, during a sharp turn, the special gear box straightens out the rear wheels and then turns them in the opposite direction of the front wheels. Electronic systems operate in a similar fashion, but they also react to changes in vehicle speed. Generally, when at high speeds, the wheels steer in the same direction to improve maneuverability. At low vehicle speeds, the wheels steer in opposite directions to reduce turning radius.

Summary

- The steering system is vital to the safety of an automobile. It allows the driver to safely maneuver the vehicle.

- There are two basic kinds of steering systems in use today: linkage (worm gear) steering systems and rack-and-pinion steering systems.

- The steering column assembly consists of the steering wheel, steering shaft, column (outer housing), ignition key mechanism, and, sometimes, a flexible coupling and universal joint.

- To help prevent theft, late-model vehicles also have a locking steering wheel.

- Some vehicles use a collapsible steering column to help prevent driver chest and face injury during an auto accident.

- A tilt steering column has a flex or U-joint that allows the top half of the column and the steering wheel to be positioned at different angles or heights.

- The recirculating-ball gearbox is normally used with a linkage steering system.

- Gearbox ratio, or steering gear reduction, is basically a comparison between steering wheel rotation and sector shaft rotation.

- The steering linkage is a series of arms, rods, and ball sockets that connects the steering gearbox to the steering knuckles.

- Ball sockets provide for motion in all directions between two connected parts.

- A manual rack-and-pinion steering gear basically consists of a pinion shaft, a rack, thrust spring, bearings, seals, and a gear housing.

- Tie-rod assemblies for rack-and-pinion steering connect the ends of the rack with the steering knuckles.

- Power steering systems normally use an engine-driven pump and a hydraulic system to assist steering action. They can also use an electric motor in the rack to provide power assist.

- A power cylinder for rack-and-pinion steering is precisely machined tube that accepts the power piston.

- Proportional rack-and-pinion power steering senses vehicle speed and steering load to ensure adequate road feel at the steering wheel.

- In a four-wheel steering system, all four wheels change direction to improve handling, stability, feel, and maneuverability.

Important Terms

Road feel	Relay rod
Steering wheel	Idler arm
Steering shaft	Ball sockets
Steering column	Tie-rod assemblies
Steering gearbox	Toe adjustment sleeve
Steering linkage	Flexible coupling
Ball sockets	Universal joint
Pinion gear	Rack-and-pinion
Rack	steering gear
Gear housing	Thrust spring
Tie-rods	Tie-rod assemblies
Steering column	Rubber dust boots
assembly	Power steering systems
Ignition lock	Power steering pump
mechanism	Power steering fluid
Ignition switch	reservoir
Locking steering	Pressure-relief valve
wheel	Power steering hoses
Collapsible steering	Integral-piston power
column	steering system
Tilt steering column	Integral power steering
Manual tilt column	gearbox
Power tilt column	Spool valve
Memory tilt wheels	Rotary valve
Recirculating-ball	External cylinder power
gearbox	steering system
Worm shaft	Power rack-and-pinion
Sector shaft	steering
Ball nut	Power cylinder
Ball guides	Power piston
Adjusting nut	Hydraulic lines
Adjusting screw	Control valve
Gearbox housing	Rotary control valve
End cover	Spool control valve
Gearbox ratio	Power steering oil
Steering ratio	cooler
Steering gear	Electronic steering assist
reduction	Proportional rack-and-
Variable-ratio gearbox	pinion power steering
Constant-ratio gearbox	Speed-sensitive power
Worm-and-roller	steering
steering gearbox	Mechanical four-wheel
Steering linkage	steering system
Parallelogram steering	Hydraulic four-wheel
linkage	steering system
Pitman arm	Electronic four-wheel
Center link	steering systems

Review Questions—Chapter 69

Please do not write in this text. Place your answers on a separate sheet of paper.

1. Name five functions of the steering system.

2. List and explain the six major parts of a linkage-type steering system.

3. List and explain the four major parts of a manual rack-and-pinion steering system.

4. Today's vehicles commonly use an _____ steering column to help prevent driver injury during an accident.

5. Which of the following is *not* a part in a recirculating-ball steering gearbox?
   (A) Roller.
   (B) Worm shaft.
   (C) Sector shaft.
   (D) Ball nut.

6. Define the term "gearbox ratio."

7. The idler arm supports the pitman arm on the passenger side of the vehicle. True or False?

8. _____-and-_____ is the most popular type steering system used on modern vehicles.

9. What is the purpose of tie-rods on a rack-and-pinion steering system?

10. Power steering systems normally use an engine-driven _____ and a(n) _____ system to assist steering action.

11. A pressure-relief valve is used in a power steering system to control maximum system pressure. True or False?

12. Describe an integral-linkage power steering system.

13. List and explain the four major parts of a power rack-and-pinion steering system.

14. Define the term "road feel."

15. How does a electric power rack work?

⬢ ASE-Type Questions

1. Technician A says "worm gear" steering systems are used on most modern automobiles. Technician B says rack-and-pinion steering systems are used on most late-model automobiles. Who is right?
   (A) A only.
   (B) B only.
   (C) Both A and B.
   (D) Neither A nor B.

2. Technician A says one of the functions of an automotive steering system is to transmit "road feel" to the driver's hands. Technician B says one of the functions of a steering system is to prevent excessive "body squat." Who is right?
   (A) A only.
   (B) B only.
   (C) Both A and B.
   (D) Neither A nor B.

3. Technician A says a steering shaft transfers turning motion from the steering wheel to the steering column. Technician B says a steering shaft transfers turning motion from the steering wheel to the steering gearbox. Who is right?
   (A) A only.
   (B) B only.
   (C) Both A and B.
   (D) Neither A nor B.

4. Technician A says an automotive steering linkage connects the steering gearbox to the steering wheel. Technician B says a steering linkage connects the steering gearbox to the steering knuckles and wheels. Who is right?
   (A) A only.
   (B) B only.
   (C) Both A and B.
   (D) Neither A nor B.

5. Which of the following is *not* a basic component of a linkage steering system?
   (A) Steering gearbox.
   (B) Pinion gear.
   (C) Gear housing.
   (D) Both B and C.

6. Technician A says tie rod assemblies are used in linkage steering systems. Technician B says tie rod assemblies are used in rack-and-pinion steering systems. Who is right?
   (A) A only.
   (B) B only.
   (C) Both A and B.
   (D) Neither A nor B.

7. Technician A says tie-rod assemblies for rack-and-pinion steering connect the ends of the pitman arms with the steering knuckles. Technician B says tie-rod assemblies for rack-and-pinion steering connect the ends of the rack with the steering knuckles. Who is right?
   (A) A only.
   (B) B only.
   (C) Both A and B.
   (D) Neither A nor B.

8. Technician A says a Woodruff key is used to lock the steering wheel to the steering shaft. Technician B says splines are used to lock the steering wheel to the steering shaft. Who is right?
   (A) A only.
   (B) B only.
   (C) Both A and B.
   (D) Neither A nor B.

9. Technician A says a modern automotive ignition switch is usually bolted to the steering column. Technician B says special clamps are normally used to secure an automotive ignition switch to the steering column. Who is right?
   (A) A only.
   (B) B only.
   (C) Both A and B.
   (D) Neither A nor B.

10. Technician A says a steel mesh collapsible steering column is a one-piece unit. Technician B says a steel mesh collapsible steering column is a two-piece unit. Who is right?
    (A) A only.
    (B) B only.
    (C) Both A and B.
    (D) Neither A nor B.

11. Technician A says a pitman arm is a basic component of a parallelogram linkage. Technician B says an idler arm is a basic component of a parallelogram linkage. Who is right?
    (A) A only.
    (B) B only.
    (C) Both A and B.
    (D) Neither A nor B.

12. Technician A says screws are used to secure a rack-and-pinion steering system's rubber dust boots. Technician B says clamps are used to secure a rack-and-pinion steering system's rubber dust boots. Who is right?
    (A) A only.
    (B) B only.
    (C) Both A and B.
    (D) Neither A nor B.

13. Technician A says a belt running from the engine's idler pulley normally powers an automobile's power steering pump. Technician B says a belt running from the engine's crankshaft normally powers an automobile's power steering pump. Who is right?
    (A) A only.
    (B) B only.
    (C) Both A and B.
    (D) Neither A nor B.

14. Technician A says an integral-piston linkage-type power steering system is used on certain vehicles. Technician B says an external-piston linkage-type power steering system is used on many types of automobiles. Who is right?
    (A) A only.
    (B) B only.
    (C) Both A and B.
    (D) Neither A nor B.

15. Technician A says a speed-sensitive steering system with an electric-motor-driven rack uses sensors. Technician B says a control module (computer) is also used in this type of system. Who is right?
    (A) A only.
    (B) B only.
    (C) Both A and B.
    (D) Neither A nor B.

Activities—Chapter 69

1. Examine a vehicle in the shop and determine what type of power steering system it has.

2. Identify the parts of a vehicle's power steering system.

3. On an overhead transparency, trace the flow of hydraulic fluid through a power steering system.

Chapter 70

Steering System Diagnosis and Repair

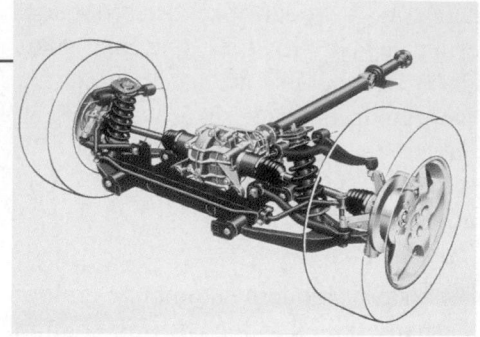

After studying this chapter, you will be able to:

- Describe common steering system problems.
- Properly inspect and determine the condition of a steering system.
- Explain basic steering column repair operations.
- Adjust both worm gears and rack-and-pinion gears.
- Describe service and repair procedures for a rack-and-pinion steering gear.
- Service power steering belts, hoses, and fluid.
- Explain how to complete basic power steering tests.
- Use safe work procedures.
- Correctly answer ASE certification test questions on the diagnosis and repair of today's steering systems.

After prolonged use, the steering system gearbox, ball sockets, belts, hoses, and other system parts can wear. This can cause play in the steering system. The vehicle can wander from side to side when the driver is holding the steering wheel straight. This can render the vehicle unsafe to drive. It is important that the steering system be kept in perfect working condition to prevent an accident.

As a technician, it is your job to find and correct steering system troubles. This chapter will help you develop the most important skills for troubleshooting and repairing all types of automotive steering systems.

Steering System Problem Diagnosis

The most common steering system problems are play in the steering wheel, hard steering, and abnormal noises when turning the steering wheel. These problems normally point to part wear, lack of lubrication, or an incorrect adjustment. You must inspect and test the steering system to find the source of the trouble. Refer to **Figure 70-1.**

Steering Wheel Play

The most frequent of all steering system problems is excessive play in the steering wheel. *Steering wheel play* is normally caused by worn ball sockets, a worn idler arm, or too much clearance in the steering gearbox (worm or rack-and-pinion types).

Typically, you should *not* be able to turn the steering wheel more than about 1 1/2″ (33 mm) without causing movement of the front wheels. If the steering wheel rotates excessively without moving the wheels, a serious steering problem exists.

The *dry park test* is an effective way to check play in the steering linkage or the rack-and-pinion mechanism. With the full weight of the vehicle on the wheels, ask someone to rock the steering wheel back and forth while you look for looseness in the steering system.

You can also place the vehicle on a lift to inspect the steering parts for wear. Have someone wiggle one of the tires right and left as you look for play between parts. See **Figure 70-2.**

Start your inspection at the steering column shaft and work your way out to the tie-rod ends. Make sure that movement of one part causes an equal amount of movement of the adjoining part.

In particular, watch for ball studs that wiggle in their sockets. With rack-and-pinion steering, you must squeeze the rubber boots and feel the inner tie-rod ends to detect wear. If the tie-rod moves sideways (parallel with rack), the socket is worn and should be replaced.

Hard Steering

Hard steering (steering wheel requires excessive turning effort) can be caused by problems with the steering gearbox, the rack-and-pinion steering gear, the

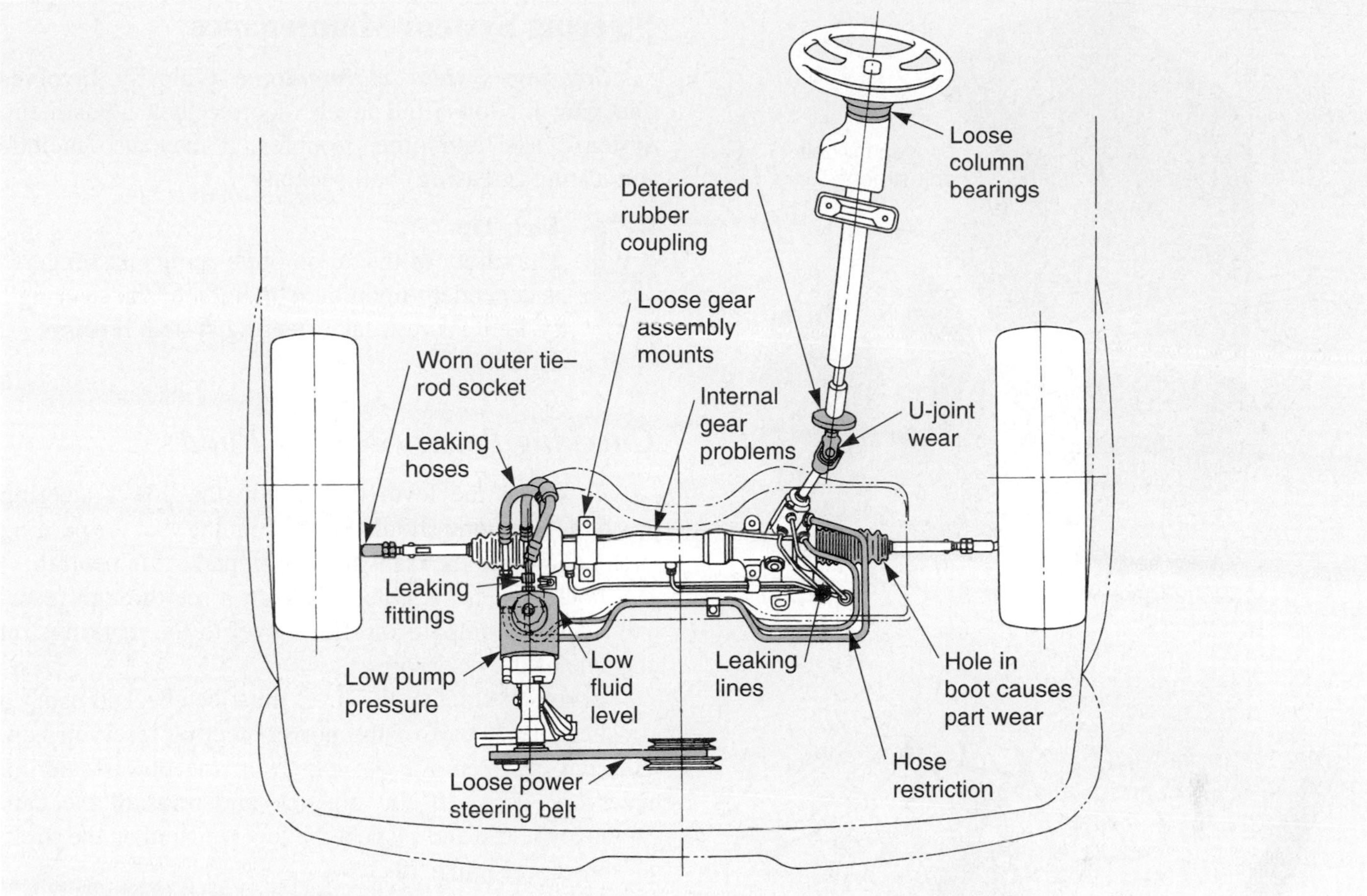

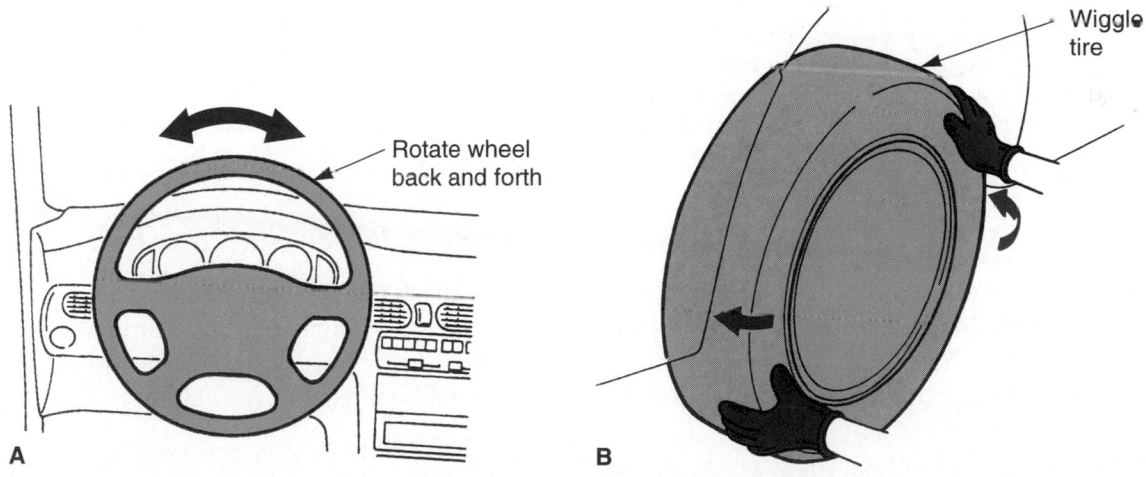

Figure 70-1. Study the types of problems that can develop in a modern steering system.

Figure 70-2. Check all steering parts for wear carefully. A—With the vehicle on the ground, rotate the steering wheel back and forth while someone watches for part wear. B—You could also raise the vehicle on a lift and wiggle the tires back and forth while watching for steering play. (Honda and Mazda)

power steering components, the ball sockets, and the suspension system.

Power steering systems commonly suffer from hard steering. When this occurs, check the fluid level in the power steering pump. If the fluid level is low, inspect the system for leaks. Also, check the power steering pump belt. If the belt is slipping, hard steering could result. Look at **Figure 70-3.**

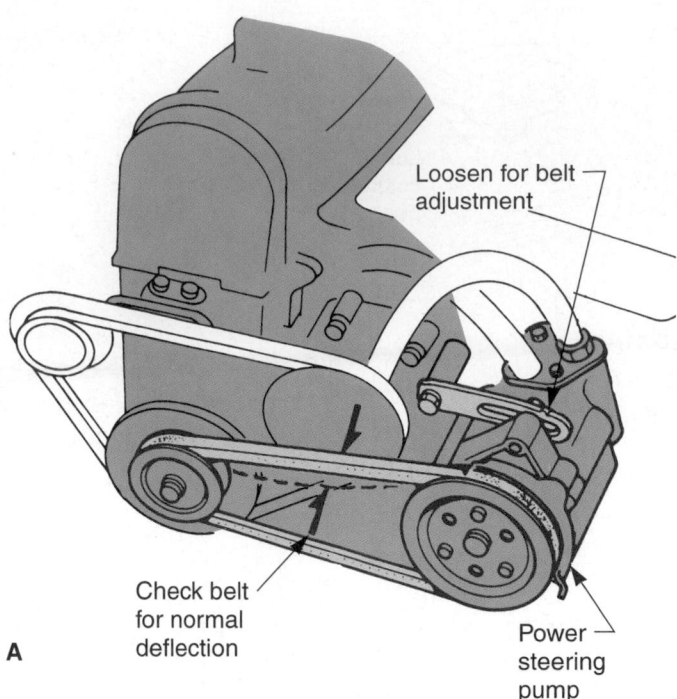

Loosen for belt adjustment

Check belt for normal deflection

A

Power steering pump

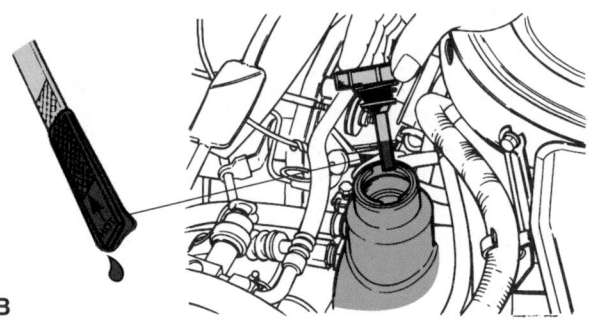

B

Figure 70-3. A—Always check power steering belt condition and tension. A slipping belt is a common problem. B—Also check the power steering fluid level. If low, check for leaks and add the correct type and amount of fluid. (Honda and Subaru)

Steering System Noise

Steering system problems can produce abnormal noises. *Steering system noise* can be a sign of worn parts, unlubricated bearings or ball sockets, loose parts, slipping belts, a low power steering fluid level, or other system troubles.

Belt squeal is a loud screeching sound produced by a slipping belt. A slipping power steering belt will usually show up when turning the steering wheel. Turning the steering wheel to the full right or full left position will increase system pressure and cause belt squeal.

Power steering pump noise is usually a loud whine that only occurs when you turn the steering wheel. It is often due to a low fluid level and air in the system. After adding fluid, the noise should go away after turning the wheels from full right to full left several times.

Steering System Maintenance

Steering system maintenance typically involves checking for low fluid level, incorrect belt adjustment, system leaks, and other troubles. It may also include lubricating (greasing) ball sockets.

 Tech Tip
The safety of the automobile and its passengers is dependent upon the condition of the steering system. Always take steering system repairs seriously.

Checking Power Steering Fluid

To check the level of fluid in the power steering system, the engine should *not* be running. Set the parking brake and place the transmission in park or in neutral.

If the vehicle is equipped with a see-through reservoir, simply compare the fluid level to the markings on the side of the reservoir.

In some vehicles, the level must be checked using a dipstick that is part of the power steering reservoir cap. Unscrew and remove the cap from the power steering reservoir. Wipe off the dipstick and reinstall the cap. Remove the cap and inspect the level of fluid on the stick, as shown in **Figure 70-3.**

Most power steering dipsticks have markings for checking the fluid when *hot* and *cold*. Make sure you read the correct marking on the dipstick. The fluid level will rise on the stick as the system warms.

If needed, add only enough fluid to reach the correct mark on the dipstick or the side of the reservoir. Do *not* overfill the system. Overfilling could cause fluid to spray out the top of the reservoir onto the engine and other components.

Servicing Power Steering Belts

A loose power steering belt can slip, squeal, and cause erratic or high steering effort. A worn or cracked belt may snap. This will cause a loss of power assist. Always inspect the belt very closely.

 Caution
When tightening a power steering belt, do *not* pry on the side of the pump. If the thin housing of the pump is dented, the pump can be ruined. Only pry on a reinforced flange or a recommended point.

To install a new power steering belt, loosen the bolts holding the pump to its brackets, **Figure 70-3.** Push inward on the pump to release tension; then remove the old belt.

Obtain the correct belt and install it in reverse order of removal. Pry on a recommended point when adjusting belt tension to specifications.

Steering Column Service

Steering column service is needed after a collision (crushing of collapsible steering column) or when internal parts of the column fail. Most steering column repairs can be done with the column mounted in the vehicle. However, some repairs require steering column removal. Refer to **Figure 70-4.**

Steering Wheel Removal and Replacement

A *wheel puller* is used to remove a steering wheel from its shaft. After removing the horn button and the steering shaft nut, scribe alignment marks on the steering wheel and the steering shaft. This will help you position the steering wheel correctly during reassembly.

Mount the wheel puller as shown in **Figure 70-5.** Screw the bolts into the threaded holes in the steering wheel. Make sure the bolts have the correct thread type. Using a wrench or ratchet, tighten the puller down against the steering shaft to remove the wheel.

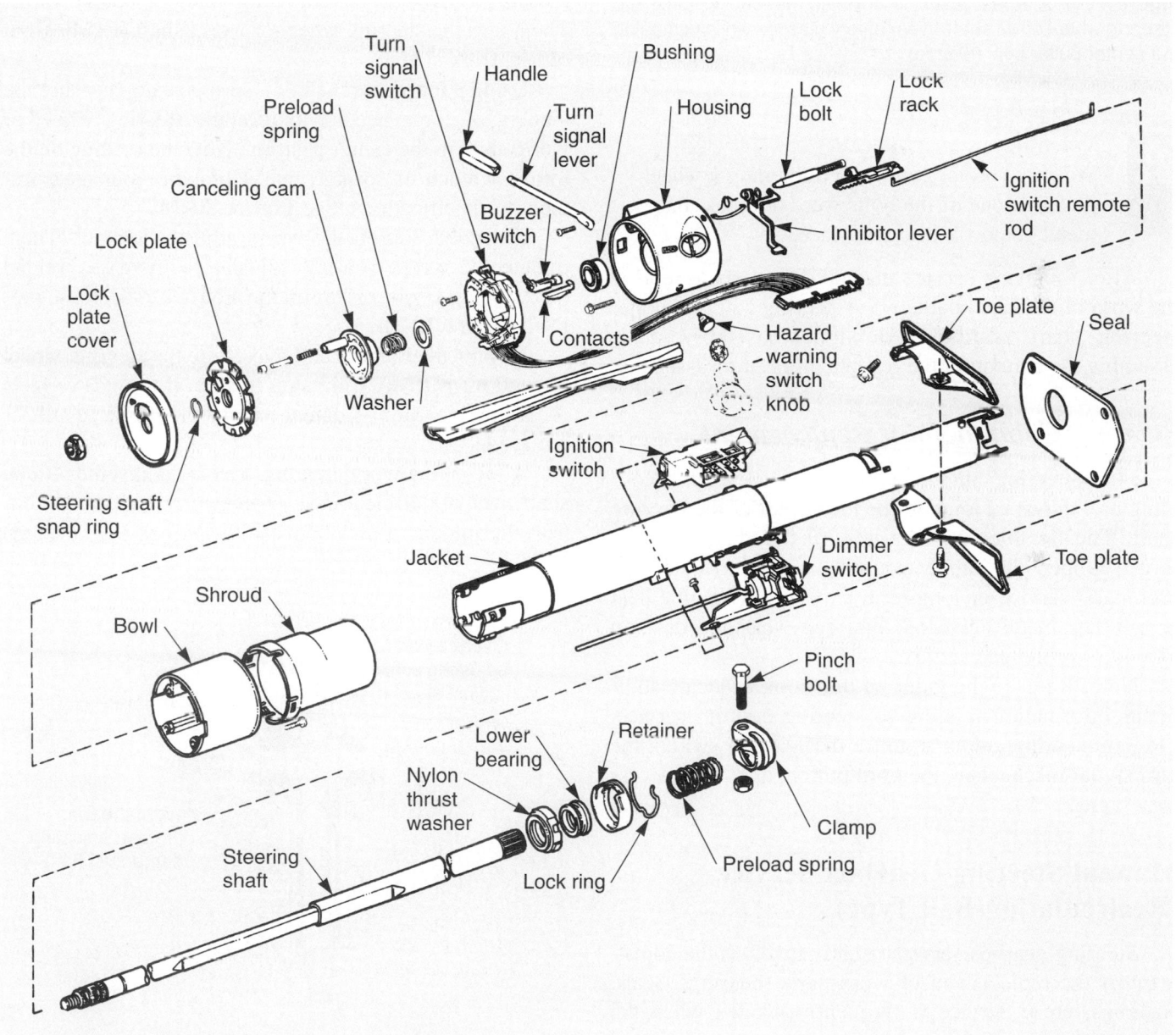

Figure 70-4. Exploded view shows the relationship between the parts a of steering column. Partial or complete disassembly may be needed, depending upon the problem. (Chrysler)

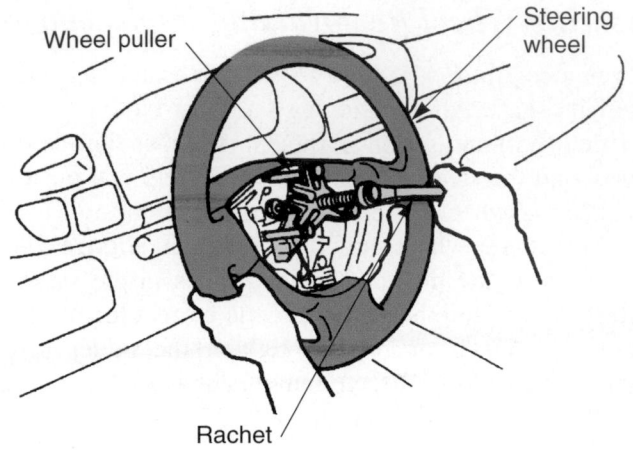

Figure 70-5. A wheel puller is normally needed to force the steering wheel off its shaft. Wear safety glasses while tightening the center puller bolt. (Mazda)

 **Warning**
Wear eye protection when tightening a wheel puller. If one of the bolts were to break, bits of metal could fly into your face.

After following service manual directions to replace the worn or damaged parts in the steering column (upper steering shaft bearing, turn signal assembly, etc.), assemble the column in the reverse order of disassembly.

Steering Column Joint Replacement

Many steering column assemblies use a flexible coupling or a universal joint on the lower part of the steering shaft. The flexible coupling is normally made of reinforced rubber. The rubber coupling can deteriorate after prolonged service or long-term exposure to engine heat or oil. The metal universal joint can wear and develop play after prolonged service.

Note! Refer to the index of this book to locate additional information relating to steering column service. Modern steering columns house the ignition switch, the turn signal mechanism, the horn button, and other accessory items.

Manual Steering Gearbox Service (Recirculating-Ball Type)

Steering gearbox service usually involves the adjustment or the replacement of worn parts (bearings, seals, bushings, etc.). Service is frequently needed when the worm shaft rotates back and forth *without* normal pitman shaft movement. This points to play (excess clearance) inside the gearbox. If adjustment does not correct this condition, the gearbox must be rebuilt or replaced.

Manual Steering Gearbox Adjustment

There are two basic adjustments on manual recirculating-ball gearboxes: worm bearing preload and over-center clearance. You should set the worm bearing preload first and the over-center clearance second. Refer to **Figure 70-6**.

Worm bearing preload ensures that the worm shaft is held snugly inside the gearbox housing. If the worm shaft bearings are too loose, the worm shaft could move sideways and up-and-down during operation.

To make a worm bearing preload adjustment, disconnect the pitman arm (if the gearbox is installed in the vehicle). Loosen the pitman shaft over-center locknut and loosen the over-center adjusting screw a couple turns. Then, turn the steering wheel or worm shaft (gearbox out of car) from side to side slowly.

Using a torque wrench or a spring scale, measure the amount of force needed to turn the steering wheel or worm shaft to the center position. Note the reading on the torque wrench or spring scale. Compare your measurement to specifications. See **Figure 70-7A**.

If needed, loosen the worm adjuster locknut. Then, tighten the worm bearing adjuster to increase preload (and turning effort) or loosen the adjuster to decrease preload, **Figure 70-7B**.

Tighten the locknut and make sure the steering wheel or shaft turns freely from stop to stop. If it binds or feels rough, the gearbox is damaged and should be rebuilt or replaced.

After setting worm bearing preload, adjust the pitman shaft over-center clearance. *Over-center clearance* controls the amount of play between the pitman shaft (sector)

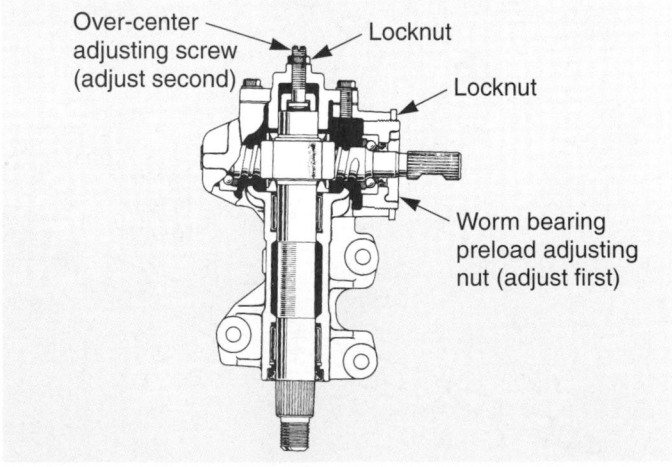

Figure 70-6. Worm bearing preload controls play in the worm shaft bearings. Over-center clearance controls play between pitman shaft sector gear and ball nut. Both adjustments are critical. (Toyota)

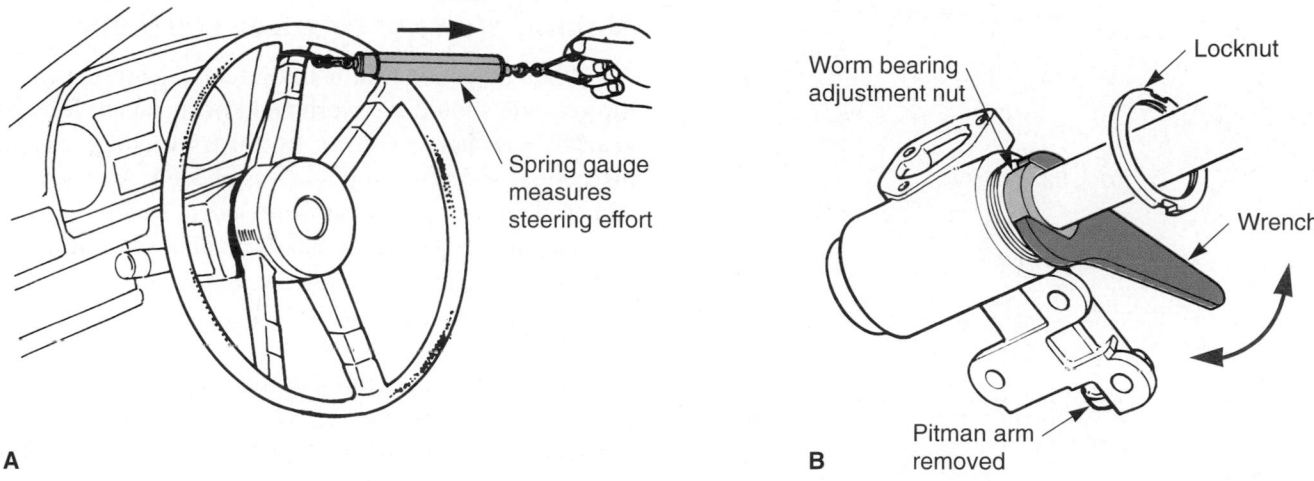

Figure 70-7. Worm bearing preload adjustment. A—Measure the pull required to turn the steering wheel with a spring scale. Compare your measurements to specs. B—If needed, tighten or loosen the large adjustment nut until the spring scale reading is correct. (Mazda)

gear and the teeth on the ball nut. It is the most critical adjustment affecting steering wheel play.

To make the over-center clearance adjustment, find the center position of the steering wheel or the worm shaft. Turn the wheel or shaft from full right to full left while counting the number of turns. Divide this number by two to find the number of turns required to move the wheel or shaft to the center position. Turn the steering wheel or worm shaft from full right or full left to the center position.

The steering wheel or shaft must be centered during over-center adjustment. Most gearboxes are designed to have more gear tooth backlash (clearance) when turned to the right or left. A slight preload is produced in the center position to avoid steering wheel play during straight-ahead driving.

After centering the steering wheel or worm shaft, loosen the adjustment screw locknut. Turn the over-center adjustment screw in until it bottoms lightly. This will remove the backlash, **Figure 70-8A.**

Using the specific instructions in a service manual, measure the amount of force needed to turn the steering wheel or the gearbox worm shaft, **Figure 70-8B.** Loosen or tighten the adjustment screw as needed to meet specs. Then, tighten the locknut and recheck gearbox action.

Power steering gearboxes sometimes require the adjustment of rack-piston preload. Since procedures differ, refer to a shop manual for specific instructions.

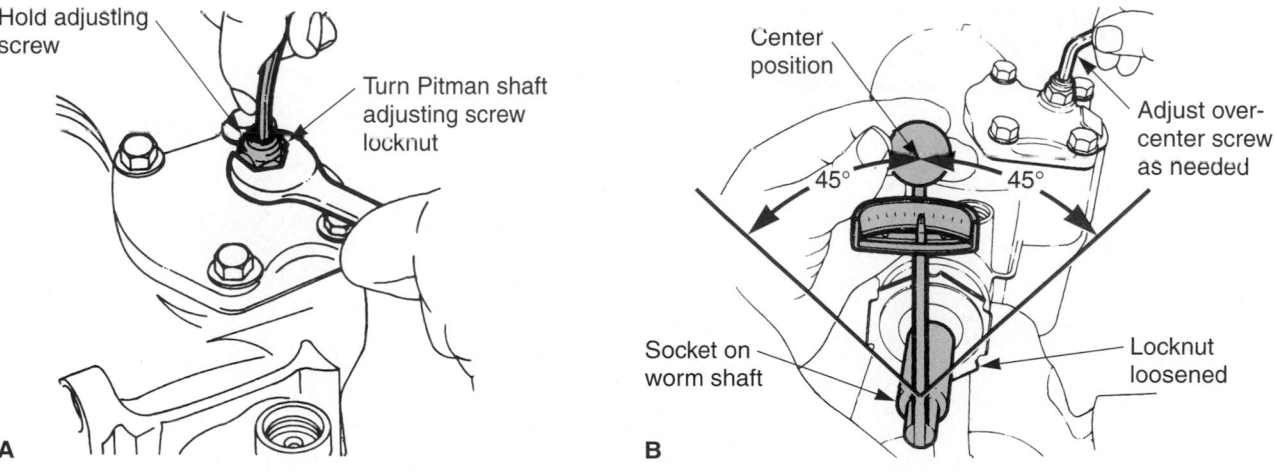

Figure 70-8. Pitman shaft over-center adjustment. A—Center the worm shaft or steering wheel. Loosen the locknut and bottom the adjustment screw lightly. B—Use a torque wrench or spring scale to measure amount of pull needed to move the shaft back and forth through center position. Tighten or loosen the adjustment screw until the amount of pull needed to move the shaft is within specs. Secure the locknut and recheck the adjustment. (Chrysler)

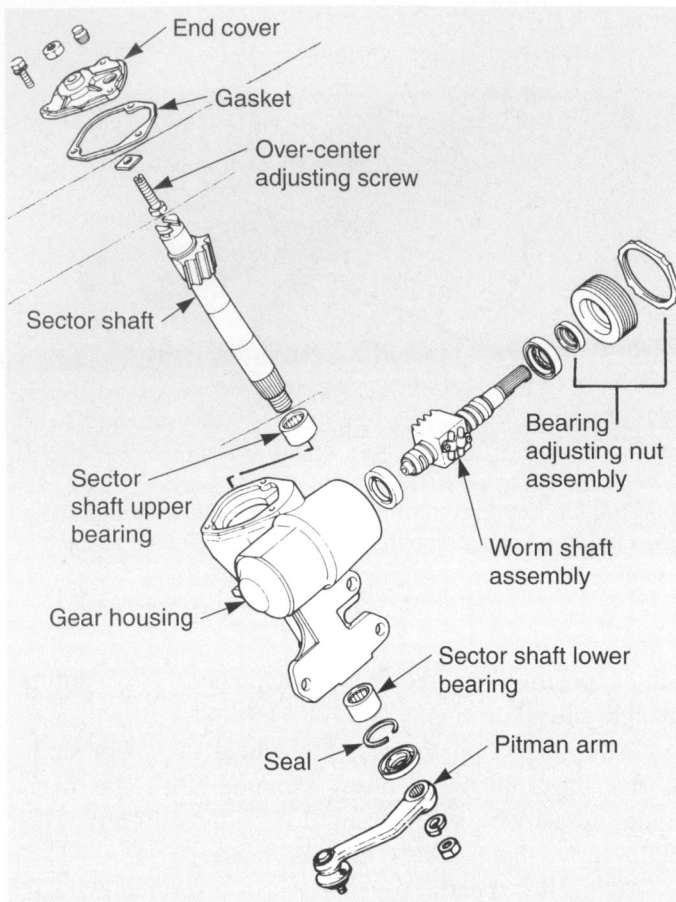

Figure 70-9. Service manual illustrations, such as this exploded view, are helpful during gearbox overhaul. (Toyota)

Manual Steering Gearbox Overhaul

When adjustment fails to correct a gearbox problem, the gearbox should be overhauled or replaced. Basically, **gearbox overhaul** is done by disassembling, cleaning, inspecting, and replacing parts as needed. All worn parts and used rubber seals must be replaced. Refer to a service manual for the particular gearbox since procedures and specs vary. **Figure 70-9** shows an exploded view of a typical manual steering gearbox.

After replacing worn parts, assemble and install the gearbox. Fill the housing with the correct lubricant. Most manual steering gearboxes use *SAE 90 gear oil*. Make sure you do *not* overfill the gearbox. A shop manual will give information on filling and correct oil type.

Steering Linkage Service

When your inspection finds worn steering linkage parts, new parts must be installed. **Figure 70-10** illustrates the parts of a typical linkage steering system. Study the types of problems you might find.

Idler Arm Service

A **worn idler arm** will cause play in the steering wheel. The front wheels, mainly the right wheel, can turn without causing movement of the steering wheel. An idler arm is a very common wear point in a linkage steering system. Check it carefully, **Figure 70-10.**

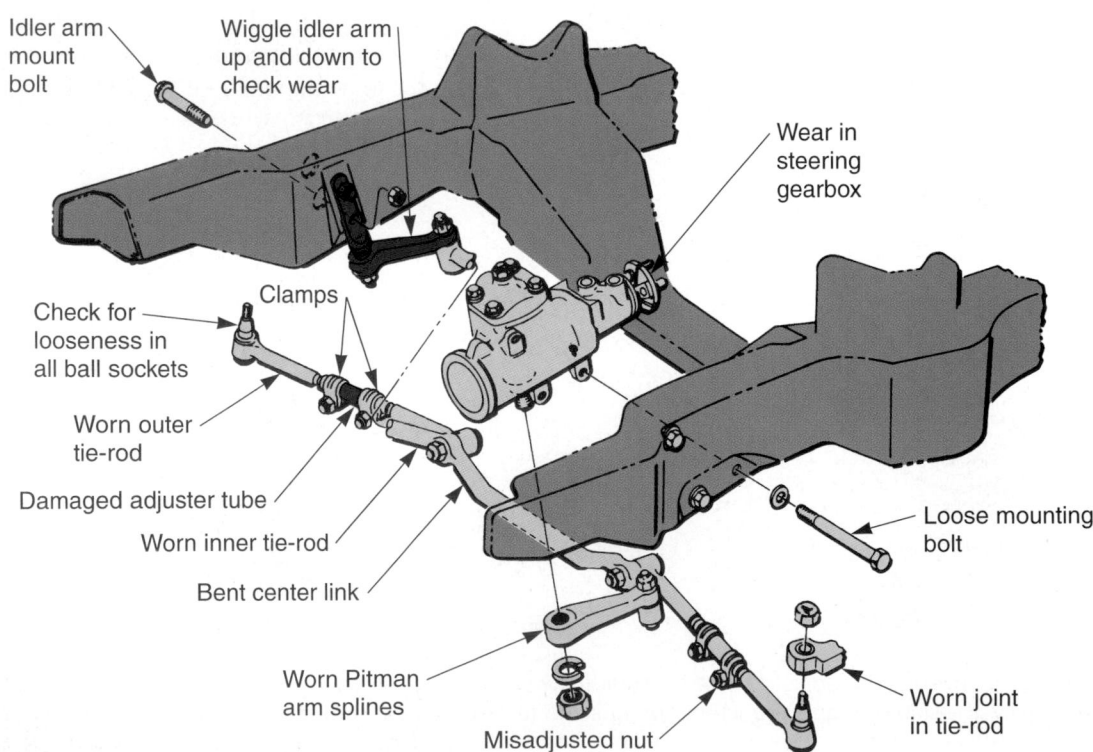

Figure 70-10. Every part of the linkage system must be checked carefully to find worn components. (Buick)

To check an idler arm for wear, grasp the outer end of the arm (end opposite frame). Force the idler arm up and down by hand. Note the amount of movement at the end of the arm. Compare this movement to specifications. Typically, an idler arm should *not* wiggle up and down more than about 1/4″ (6.5 mm).

A worn idler arm must be replaced. Remove the cotter pin and the castle nut from the outer end of the idler arm. Then, use a fork tool or a puller to force the arm's ball stud out of the hole in the center link. Finally, unbolt the idler arm from the frame, **Figure 70-10.**

Install the new idler arm in the reverse order of removal. Make sure you torque the idler arm fasteners properly. Install a new cotter pin and bend it properly.

Tie-Rod End Service

A *worn tie-rod end* will also cause steering play. When you detect movement between the ball stud and its socket, install a new tie-rod end.

To remove a tie-rod end, separate the tie-rod from the steering knuckle or center link. Use a fork or a puller as shown in **Figure 70-11.** Be careful *not* to damage any components.

Before loosening the adjustment sleeve, measure or mark tie-rod length. This will allow you to set the new tie-rod at about approximately the same length as the old one. The alignment of the front wheels is altered when the length of a tie-rod is changed.

💡 **Tech Tip**
Instead of measuring tie-rod length, you can count the number of turns required to remove the tie-rod from the adjustment sleeve. The new tie-rod end should be installed the same number of turns.

Loosen the tie-rod adjustment sleeve and unscrew the tie-rod end. Look at **Figure 70-12.**

Turn the new tie-rod end into the sleeve until it is the exact length of the old tie-rod. Install the tie-rod ball stud in the center link or the steering knuckle. Tighten the fasteners to specifications. Install new cotter pins and bend them correctly. Tighten the adjustment sleeve and check steering action. After tie-rod service, always check toe for proper adjustment.

Inner Tie-Rod End Service—Rack-and-Pinion Steering

The service of an inner tie-rod end on a rack-and-pinion assembly is similar to that for a conventional tie-rod end. However, the inner tie-rod end on a rack-and-pinion assembly is enclosed in a rubber boot. Various methods are used to secure the inner tie-rod end.

To remove an inner tie-rod end, remove the boot straps. You usually have to cut them off with side cut pliers, **Figure 70-13A.** Then, unscrew the inner end of the tie-rod from the rack. See **Figure 70-13B.**

When installing the new inner tie-rod end, install all flat and lock washers in their original locations. Torque the tie-rod nut to factory specs. Replace the boot and boot

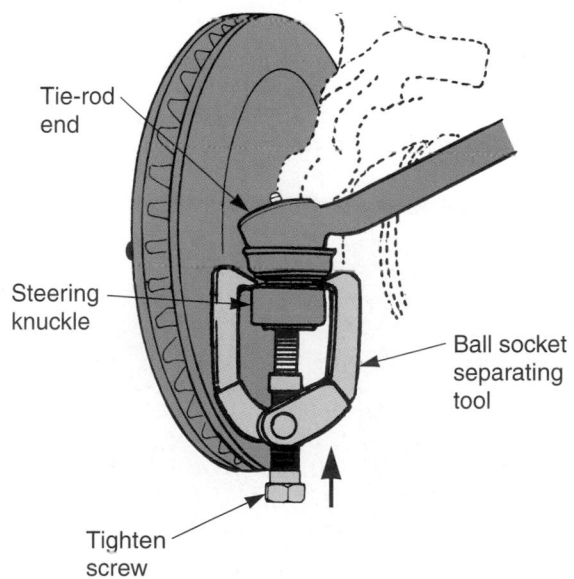

Figure 70-11. A special puller, such as the one shown above, or a fork tool can be used to separate the tie-rod end from the steering knuckle. This tool can also be used on other ball sockets. (Cadillac)

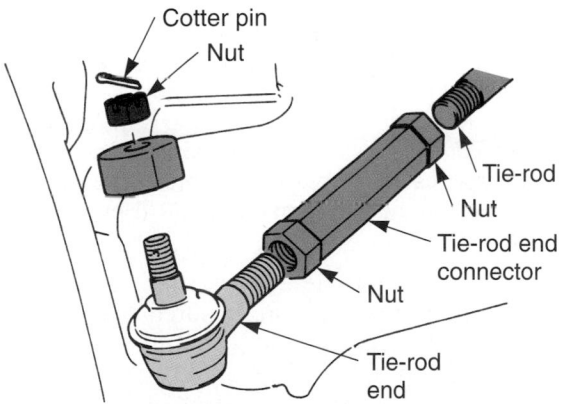

Figure 70-12. Once the outer end is separated from the knuckle, you can unscrew tie-rod end from the adjustment sleeve. Loosen the locking nut on the adjustment sleeve and rotate the tie-rod. Count the number of threads or the number of turns so the new unit will be installed in the same position as the old. This will help maintain correct wheel alignment or toe. Always install a new cotter pin. (General Motors)

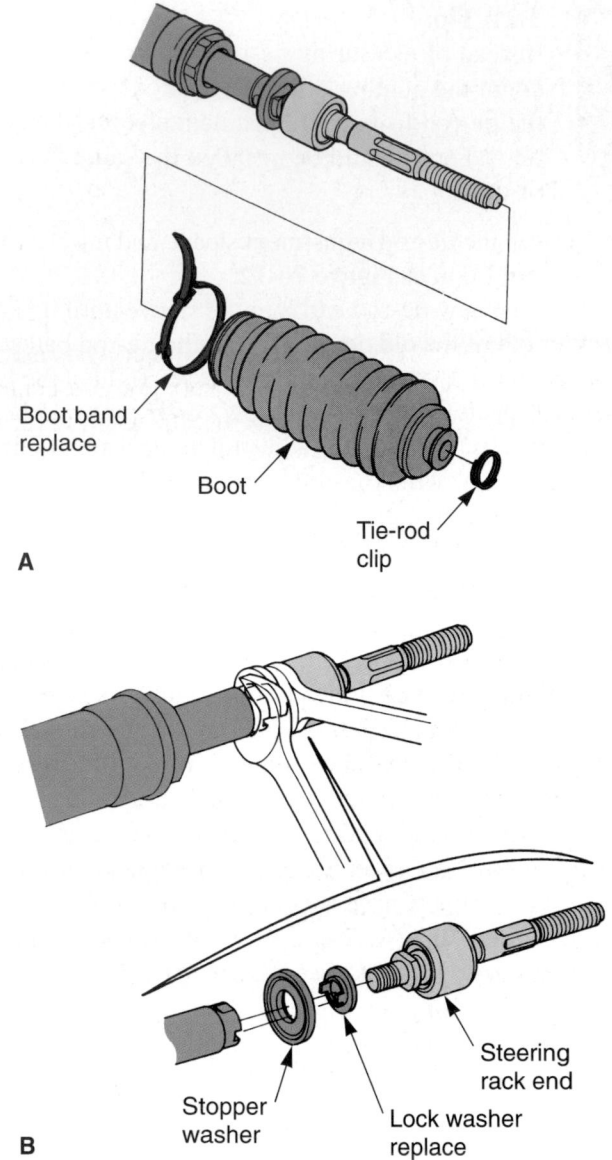

Figure 70-13. The inner end of the tie-rod on a rack-and-pinion is a common wear point. If wear is not excessive, you may have to remove boot to find play in the inner joint. A—Cut off the old boot straps and slide the boot out of the way. A new boot and straps are normally installed since the boot can rupture after prolonged service. B—Two wrenches are commonly needed to unscrew inner tie-rod end from rack shaft. (Honda)

straps, making sure they are secured into any grooves formed in the tie-rod end and rack.

Manual Rack-and-Pinion Service

A manual rack-and-pinion steering system can develop problems similar to those described for a linkage-type system. However, a rack-and-pinion system has few parts to fail. When problems do develop, they are frequently in the tie-rod ends. When *not* properly lubricated, the rack-and-pinion gear may also wear, causing problems.

Since so many new cars use rack-and-pinion steering, it is important that you learn common service and repair procedures for this type of system.

Manual Rack-and-Pinion Lubrication

Some manual rack-and-pinion steering gears require periodic lubrication. Others only need lubrication when the unit is torn down for repairs.

Figure 70-14 shows one type of rack-and-pinion steering gear that should be lubricated. Note the location of the grease fittings. Also, note that the technician has removed one of the rubber bellows for inspection.

A grease gun is used to place chassis grease into the fittings. Use only a small amount of grease, as described by the auto manufacturer.

Manual Rack-and-Pinion Gear Adjustment

Most manual rack-and-pinion steering gears have a rack guide adjustment screw. When there is play in the steering, try adjusting the steering gear, **Figure 70-15.**

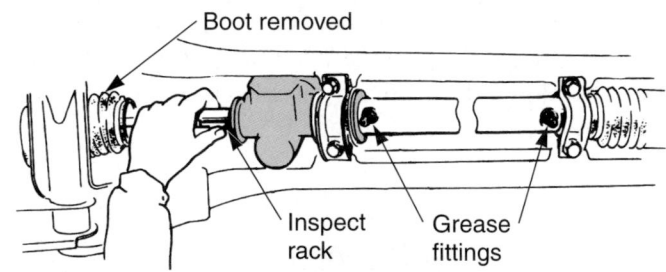

Figure 70-14. Many steering system components have grease fittings that require lubrication. This technician has removed the boot to inspect the inner tie-rod socket. (Honda)

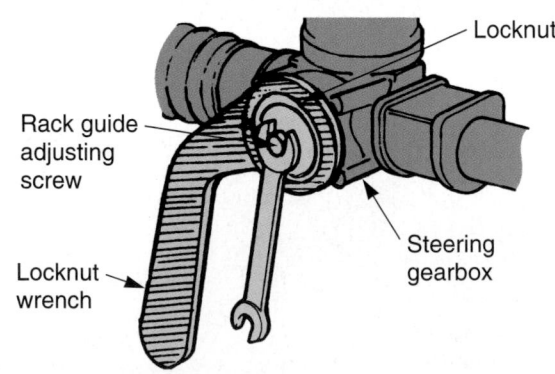

Figure 70-15. Many rack-and-pinion steering mechanisms have a rack guide adjustment screw. Follow the manufacturer's directions. Basically, hold the locknut while turning the screw. Bottom the adjustment screw and then back it off the prescribed amount. (Honda)

To adjust the rack guide screw, loosen the locknut on the screw. Then, turn the rack guide screw in until it bottoms lightly. Back off the screw the recommended amount (approximately 45%) or until a prescribed turning effort is obtained. Tighten the locknut. Recheck for tight or loose steering and measure steering effort. If not within specs, the steering gear may require major repairs.

Manual Rack-and-Pinion Gear Removal

To remove a manual rack-and-pinion steering gear, separate the outer tie-rod ends from the steering knuckles. Then, unbolt the steering gear mounting brackets from the frame or the crossmember. Also, disconnect the steering column coupler or universal joint. Rotate the steering gear and slide it out of the chassis, **Figure 70-16.**

Refer to a service manual for additional information. You may have to remove the front wheels or slide the gear out one particular side of the car.

Manual Rack-and-Pinion Gear Overhaul

Exact procedures for overhauling a manual rack-and-pinion steering gear will vary with vehicle make and model. Generally, however, you must disassemble the unit and check each part closely. Replace any part that shows signs of wear.

Figure 70-17 shows a disassembled manual rack-and-pinion steering gear. Study how the parts fit together.

Use a service manual to obtain the exact procedures for rebuilding a particular unit.

An example of a critical repair method is pictured in **Figure 70-18.** A special guide fixture is being used to drill a lock pin out of the gear assembly. The manual gives correct drill size and replacement pin number.

Power Steering System Service

Many of the parts of a power steering system are the same as those used on a manual steering system. However, a pump, hoses, a power piston, and a control valve are added. These parts can also fail, requiring repair or replacement.

Power Steering Leaks

Power steering fluid leakage is a common problem. With the extremely high pressures produced in power steering systems (over 1000 psi or 6895 kPa), leaks can easily develop around fittings, in hoses, at the gearbox seals, or at the rack-and-pinion assembly. **Figure 70-19** shows the common leakage points for two common power steering systems.

To check for leaks, wipe fluid-soaked areas clean. Then, have a helper start and idle the engine. Watch for leaks as your helper turns the steering wheel to the right and left. This will pressurize all parts of the system that might be leaking.

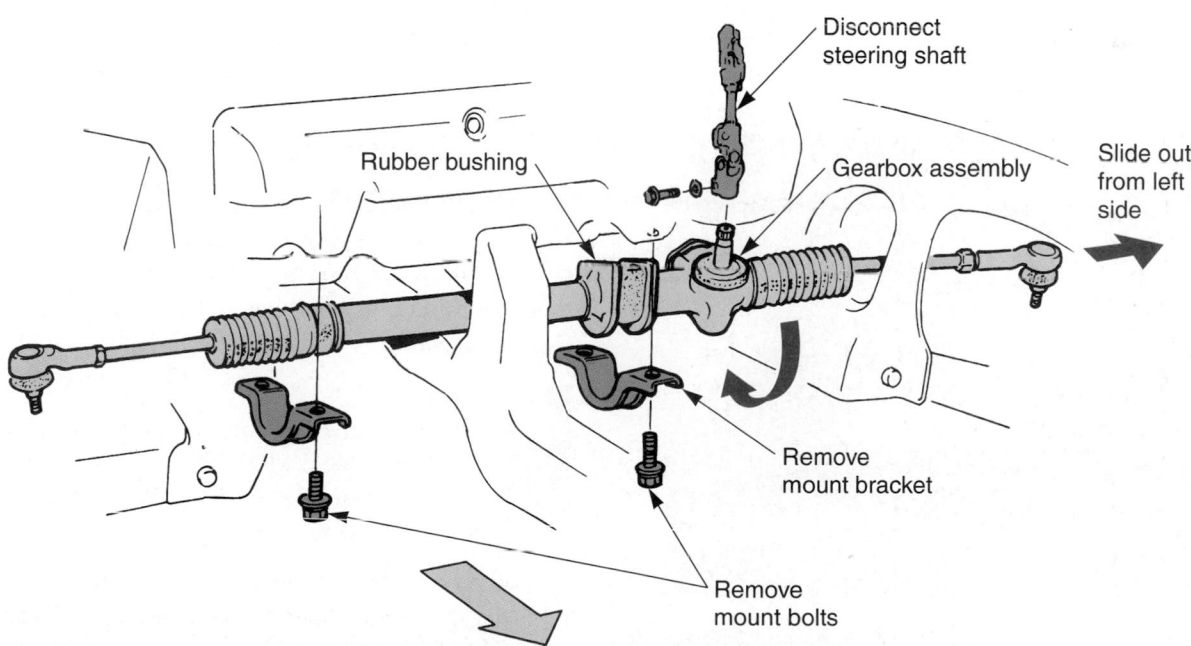

Figure 70-16. Removal of a rack-and-pinion gear mechanism usually involves disconnecting the parts shown. Slide the unit out one side of the vehicle. (Honda)

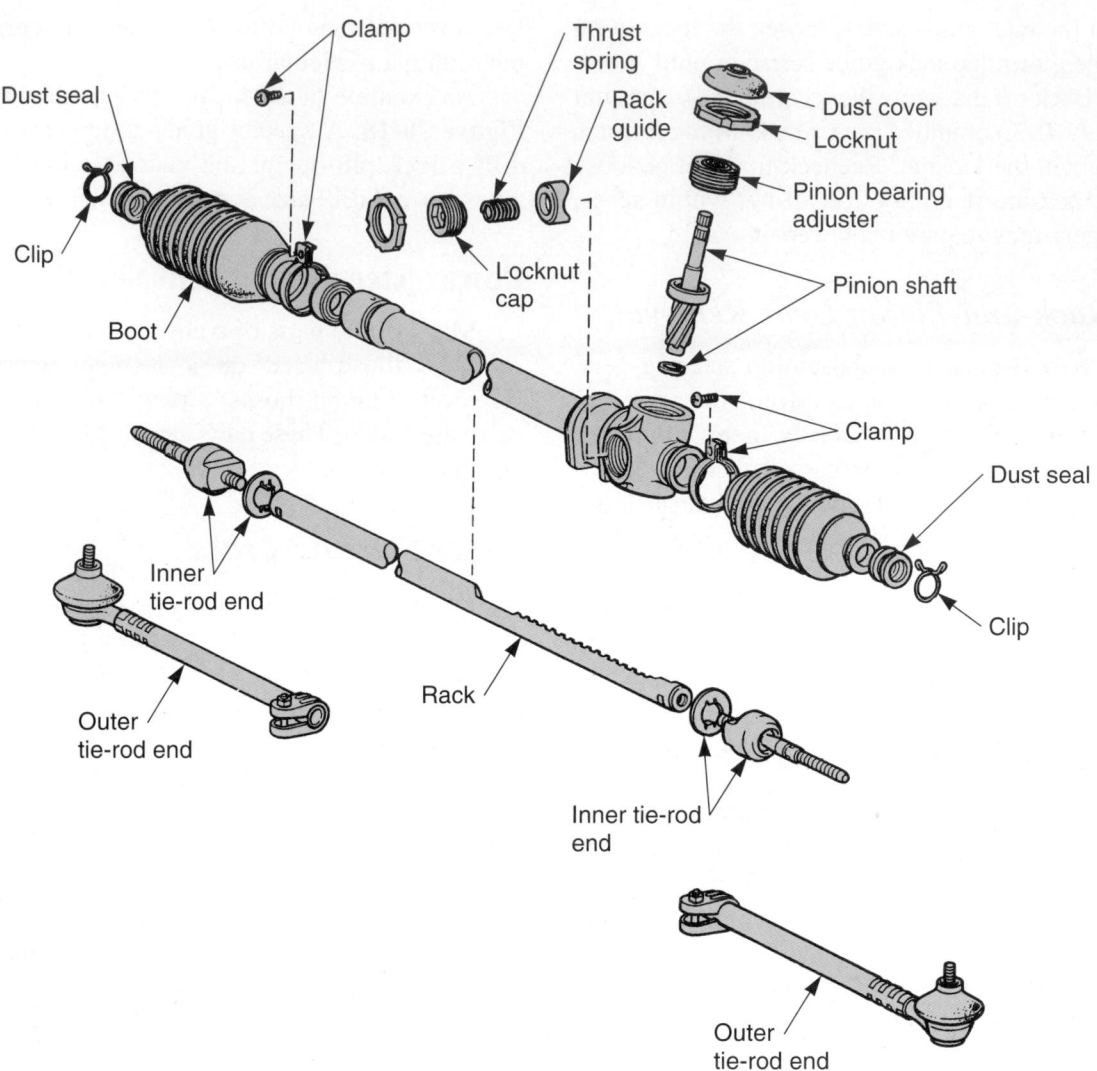

Figure 70-17. Exploded view of typical rack-and-pinion steering mechanism. During service, inspect each component and measure wear. Any part worn beyond specs must be replaced. All seals and other rubber or plastic parts must also be replaced. (Toyota)

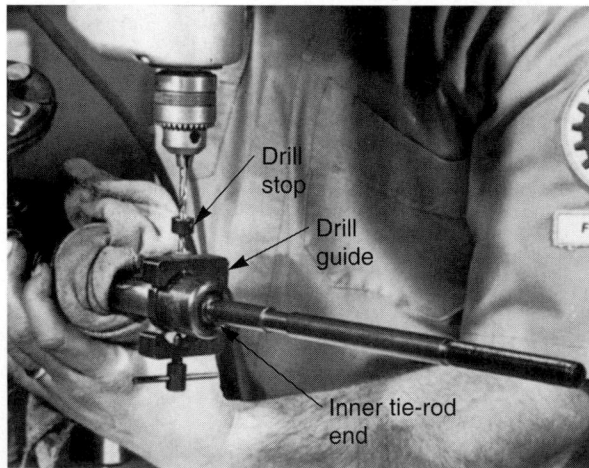

Figure 70-18. A service manual is very important when doing major steering system repairs. For example, it contains information about correct drill bit size for drilling out the pin in this tie-rod ball socket. (Moog)

Servicing Power Steering Hoses

Always check the condition of power steering hoses when checking a power steering system. The high-pressure hose can be exposed to tremendous pressures. If this hose ruptures, a sudden and dangerous loss of power assist can occur.

 **Warning**
Power steering pump pressure can exceed 1000 psi (6895 kPa). This is enough pressure to cause a serious eye injury. Wear eye protection when working on a power steering system.

When installing a new hose, start the new hose fittings by hand to avoid cross threading. Use a tubing wrench to tighten the hose fittings properly. Make sure the hose does *not* contact moving or hot parts. This could cause hose failure.

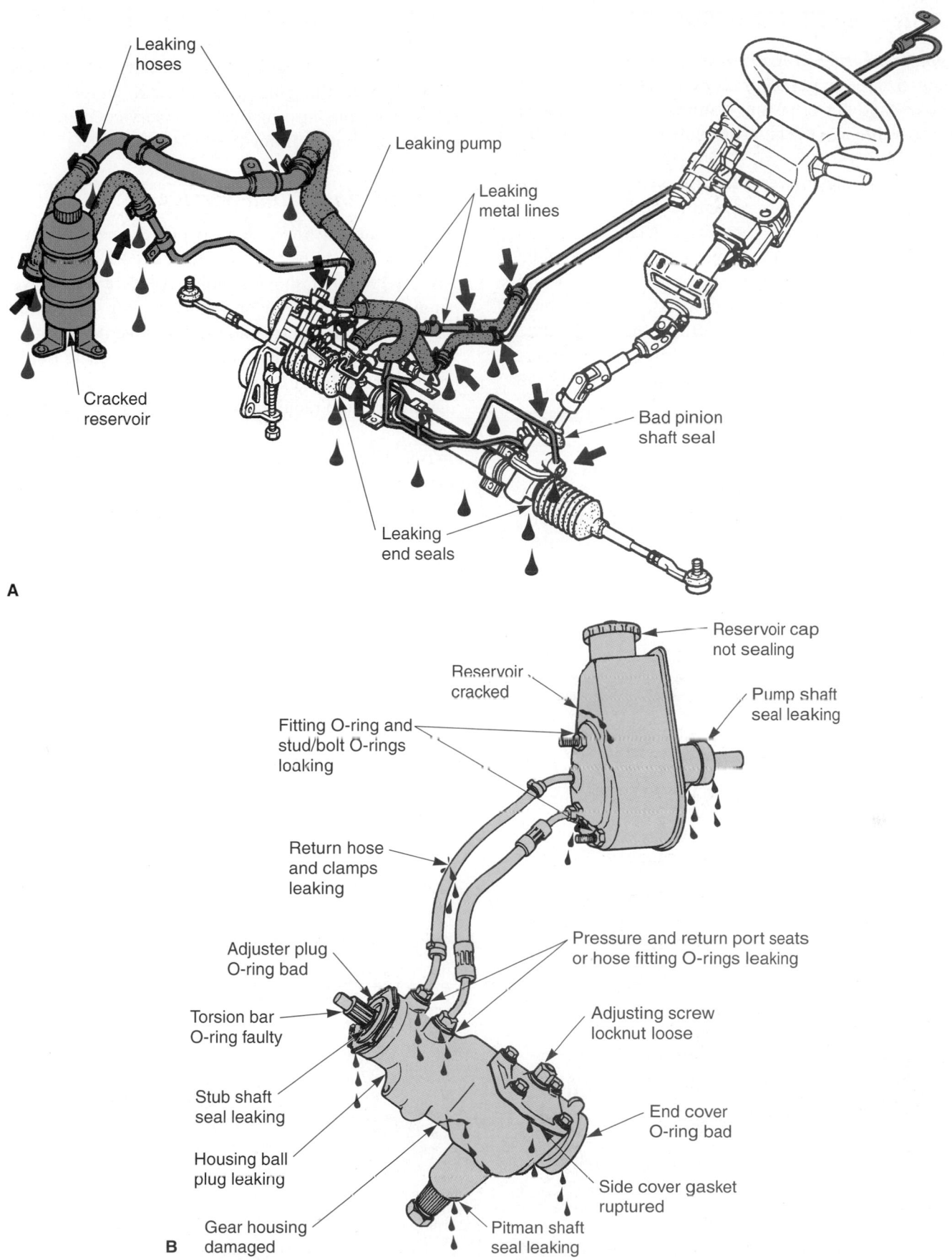

Figure 70-19. Note possible leakage points on power steering systems. A—Rack-and-pinion system. B—Linkage system. (Honda and Ford)

Power Steering Pressure Test

A *power steering pressure test* checks the operation of the power steering pump, pressure relief valve, control valve, hoses, and power piston. Connect a pressure gauge and shutoff valve into the high-pressure hose, as shown in **Figure 70-20.**

When checking power steering system pressure, follow the manufacturer's recommended procedures. Torque the hose fittings properly. Make sure the system is full of fluid. Start the engine and allow it to idle (test valve open) while turning the steering wheel back and forth. This will bring the fluid up to operating temperature.

To check system pressure, close the test valve. Compare your pressure reading with specifications. When the pressure is *not* within specs, check the condition of the pressure relief valve and the pump.

 **Caution**

Do *not* close the test valve for more than approximately *five seconds*. If the valve is closed longer, power steering pump overheating damage could result.

To check the action of the power piston, control valve, and hoses, measure system pressure as you turn the steering wheel to full lock (right and left) with the test valve open. Note the gauge readings and compare them to specs. Use the information in a service manual to determine the source of any trouble.

Power Steering Pump Service

When your tests indicate a bad power steering pump, the pump must be removed for repair or replacement. See **Figure 70-21.** Most shops simply replace a bad power steering pump with a *new* or factory *rebuilt unit*. A few shops disassemble and rebuild the defective pump.

Before attempting to rebuild a power steering pump, make sure a rebuild kit is available for the particular unit.

During a pump rebuild, clean and inspect all the parts. In particular, check the pump vanes, thrust plate, and pump ring. Replace all used O-rings, gaskets, and seals, as well as any other parts showing signs of wear or damage. Lubricate the parts with system fluid before assembly.

Power Steering Gear Service

The procedures for servicing power steering gears also vary with the make and model of the vehicle and the type of gear assembly. Follow service manual directions.

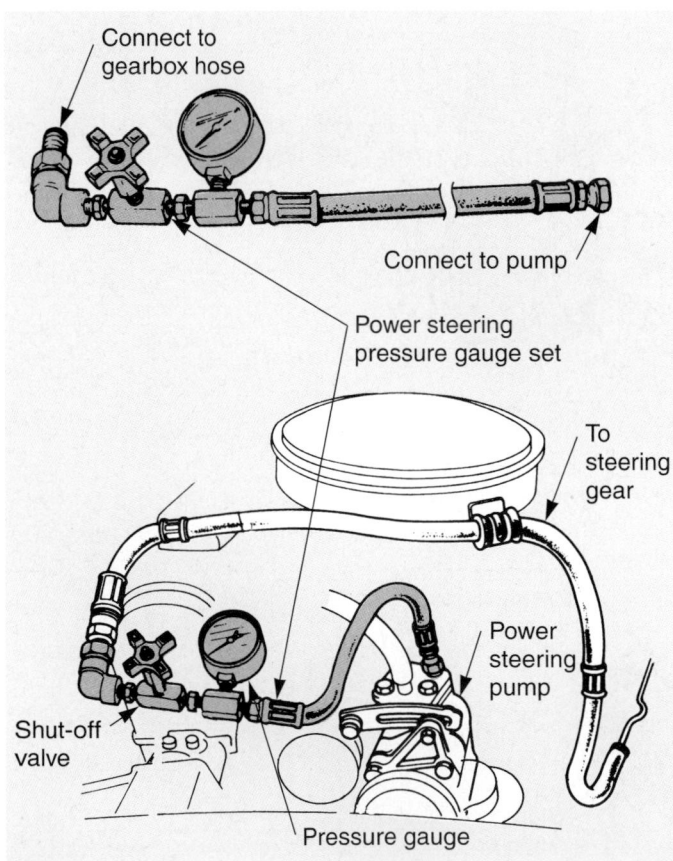

Figure 70-20. The pressure test gauge is connected to the high-pressure hose or line from the pump. Follow service manual instructions when testing a power steering system. (Honda)

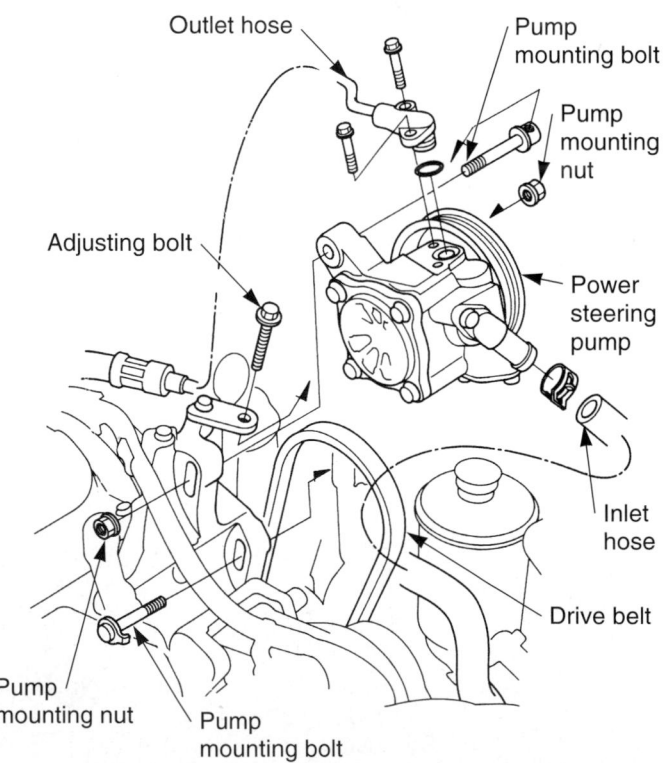

Figure 70-21. Note the bolts, hoses, and brackets that must be removed when replacing a power steering pump. (Honda)

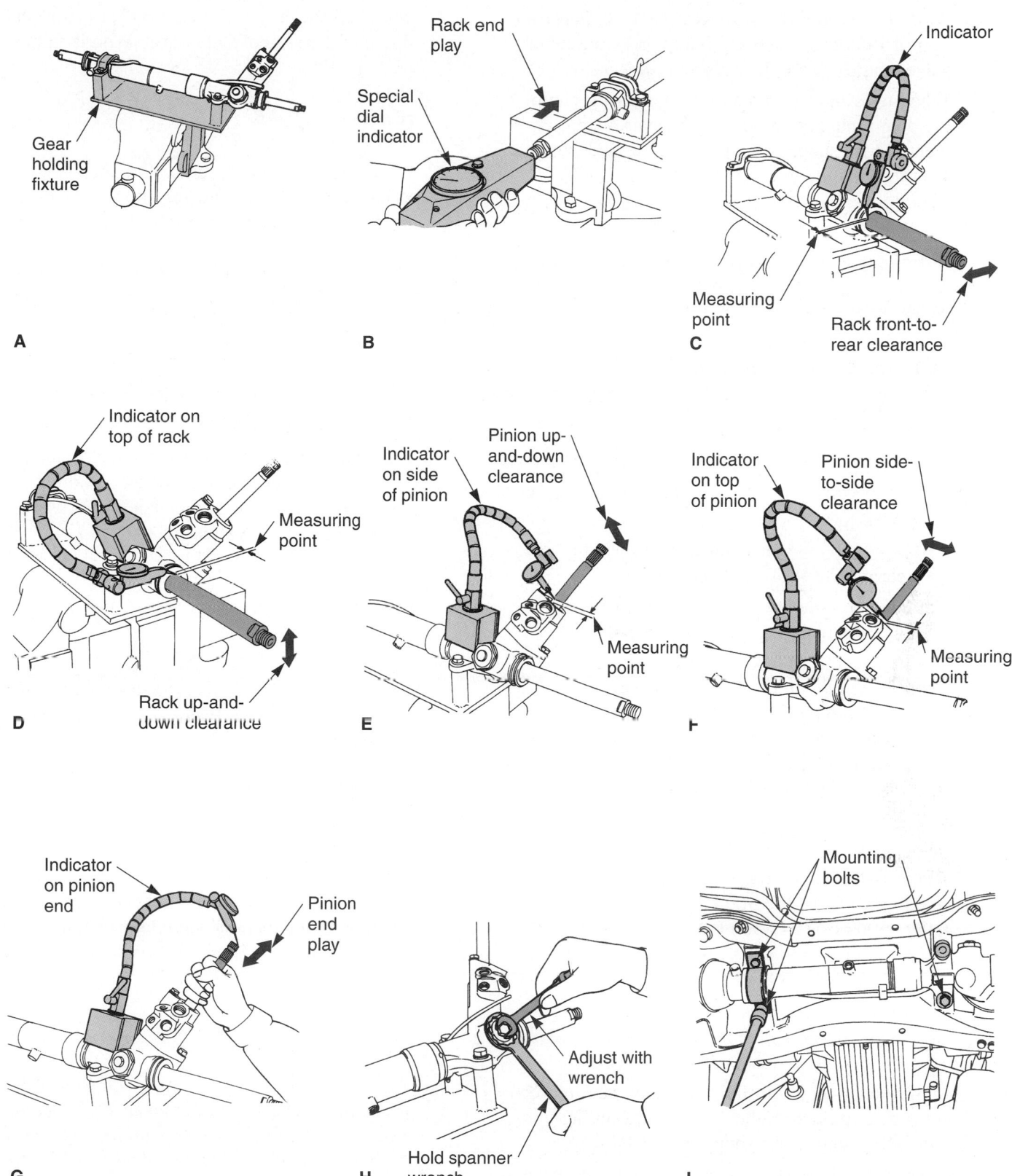

Figure 70-23. Fundamental measurements and service steps for a power rack-and-pinion steering gear. A—Mounting the unit in holding fixture. B—Using a special indicator to measure rack end play. C—Using a dial indicator to check front-to-rear clearance at the rack. D—Measuring rack up-and-down movement. E—Measuring pinion shaft up-and-down clearance. F—Measuring pinion shaft side-to-side clearance. G—Measuring pinion end play. H—Adjusting the rack guide screw. I—Installing the rack-and-pinion assembly in the vehicle. (Fiat)

- A power steering pressure test checks the operation of the power steering pump, pressure relief valve, control valve, hoses, and power piston.

- When tests indicate a bad power steering pump, the pump must be removed for repair or replacement. Most shops simply replace a bad power steering pump with a *new* or factory *rebuilt unit.*

- Bleeding a power steering system ensures that all the air is out of the lines, pump, and gearbox.

- If the vehicle is equipped with on-board diagnostics, steering system problems may trip trouble codes.

Review Questions—Chapter 70

Please do not write in this text. Place your answers on a separate sheet of paper.

1. What normally causes play in the steering wheel?

2. How do you do a "dry park test" of a steering system?

3. A(n) _____ _____ is commonly needed to force a steering wheel off its shaft.

4. Summarize the two major adjustments for a manual recirculating-ball steering gearbox.

5. Which of the following components can wear and cause play in the steering system?
   (A) Tie-rod end.
   (B) Idler arm.
   (C) Ball socket.
   (D) All of the above.

6. What adjustments are commonly done on a manual rack-and-pinion steering system?

7. Power steering pump pressure can exceed:
   (A) 10,000 psi (68,900 kPa).
   (B) 5000 psi (34,450 kPa).
   (C) 100 psi (689 kPa).
   (D) 1000 psi (6895 kPa).

8. A power steering pressure test checks the operation of the pump, relief valve, control valve, hoses, and power piston. True or False?

9. Most technicians install rebuilt or new power steering pumps rather than overhauling the defective unit in the shop. True or False?

10. After hydraulic steering components have been serviced, the system should be _____.

⬤ ASE-Type Questions

1. A car is brought into the shop with excessive steering wheel play. Technician A checks for worn ball sockets. Technician B checks the condition of the system's idler arm. Who is right?
   (A) A only.
   (B) B only.
   (C) Both A and B.
   (D) Neither A nor B.

2. A car's steering wheel turns more than two and a half inches before causing movement of the front wheels. Technician A tells the owner of the vehicle that this indicates normal play in the steering system. Technician B tells the owner of the vehicle that this indicates excessive steering wheel play. Who is right?
   (A) A only.
   (B) B only.
   (C) Both A and B.
   (D) Neither A nor B.

3. Lubricant is leaking from a car's power steering system. Technician A checks the condition of the pinion shaft seal. Technician B checks for a cracked reservoir. Who is right?
   (A) A only.
   (B) B only.
   (C) Both A and B.
   (D) Neither A nor B.

4. An automobile has a "hard steering" problem. Technician A checks the steering gearbox for problems. Technician B inspects the suspension system for problems. Who is right?
   (A) A only.
   (B) B only.
   (C) Both A and B.
   (D) Neither A nor B.

5. Noise is coming from a steering system's steering pump. Technician A checks to see if the steering pump belt is slipping. Technician B checks the pump's fluid level. Who is right?
   (A) A only.
   (B) B only.
   (C) Both A and B.
   (D) Neither A nor B.

6. An automotive steering system's fluid level needs to be checked. Technician A checks the steering system's fluid level with the engine off. Technician B checks the fluid level while the engine is running. Who is right?
   (A) A only.
   (B) B only.
   (C) Both A and B.
   (D) Neither A nor B.

7. A car's power steering system is low on fluid. Technician A adds automatic transmission fluid to this system. Technician B checks to see if a special power steering fluid is required in this system. Who is right?
   (A) A only.
   (B) B only.
   (C) Both A and B.
   (D) Neither A nor B.

8. A car's power steering belt is loose. Technician A pries on the power steering pump pulley when tightening the belt. Technician B pries on the side of the steering pump when tightening the belt. Who is right?
   (A) A only.
   (B) B only.
   (C) Both A and B.
   (D) Neither A nor B.

9. An automobile's steering column needs servicing. Technician A tells the owner of the car that most steering columns can be repaired with the column mounted in the car. Technician B tells the owner of the car that most steering columns must be removed from the vehicle when service is required. Who is right?
   (A) A only.
   (B) B only.
   (C) Both A and B.
   (D) Neither A nor B.

10. A car's steering wheel needs to be removed from its shaft. Technician A uses a hammer and brass punch to perform this service procedure. Technician B uses a wheel puller to perform this service procedure. Who is right?
    (A) A only.
    (B) B only.
    (C) Both A and B.
    (D) Neither A nor B.

11. An automobile is brought into the shop with a steering system problem. Technician A inspects the condition of the steering column's universal joint. Technician B checks the condition of the steering system's outer tie-rod sockets. Who is right?
    (A) A only.
    (B) B only.
    (C) Both A and B.
    (D) Neither A nor B.

12. An automobile's manual steering gearbox needs lubricant. Technician A adds SAE 40 gear oil to the gear-box. Technician B adds SAE 90 gear oil to the gearbox. Who is right?
    (A) A only.
    (B) B only.
    (C) Both A and B.
    (D) Neither A nor B.

13. A car equipped with manual rack-and-pinion steering is brought into the shop with excessive steering wheel play. Technician A adjusts the steering gear. Technician B adjusts the idler arm. Who is right?
    (A) A only.
    (B) B only.
    (C) Both A and B.
    (D) Neither A nor B.

14. An automobile's power steering pump is not operating properly. Technician A performs a power steering pressure test to check the operation of the power steering pump. Technician B says a power steering pressure test will not detect power steering pump malfunctions. Who is right?
    (A) A only.
    (B) B only.
    (C) Both A and B.
    (D) Neither A nor B.

15. A "buzzing" sound is coming from an automobile's power steering system. Technician A tells the owner of the vehicle that this problem may be the result of a damaged pitman arm. Technician B tells the owner of the vehicle that this problem may be caused by air in the power steering system. Who is right?
    (A) A only.
    (B) B only.
    (C) Both A and B.
    (D) Neither A nor B.

Activities—Chapter 70

1. Check the power steering lubricant level of a vehicle in the shop for service.

2. Check a shop manual for lube points and then lubricate a steering system.

3. Test a steering system and diagnose any problems detected.

4. Prepare a repair bill for a repair to the steering system. (Use a flat rate manual to determine time on the job or use an hourly rate set by your instructor. Check a parts catalog for cost of parts. Your instructor will suggest a flat rate charge for the labor.)

Manual and Power Steering Systems Diagnosis		
Condition	**Possible Causes**	**Correction**
Hard steering and poor recovery following turns.	1. Low tire pressure. 2. Defective power steering pump. 3. Power steering pump fluid level low. 4. Manual steering gear lubricant level low. 5. Incorrect front wheel alignment. 6. Dry ball joints. 7. Dry steering linkage sockets. 8. Binding linkage. 9. Damaged suspension arms. 10. Steering gear adjusted too tight. 11. Dry steering shaft bushing. 12. Binding steering shaft bushing or coupling. 13. Excessive caster. 14. Sagged front springs. 15. Bent spindle body. 16. Steering wheel rubbing steering column jacket. 17. Misaligned steering gear. 18. Sticky valve spool. 19. Loose steering pump belt. 20. Kinked or clogged power steering hose. 21. Different size front tires. 22. Malfunctioning steering gear pressure port poppet valve. 23. Maladjusted rack-and-pinion. 24. High internal leakage of rack-and-pinion assembly. 25. Rack-and-pinion mountings loose, causing binding. 26. Defective steering stabilizer.	1. Inflate to correct pressure. 2. Repair or replace pump. 3. Add fluid to reservoir. 4. Add lubricant. 5. Align front wheels. 6. Lubricate ball joints. 7. Lubricate linkage. 8. Relieve binding. 9. Replace arms. 10. Adjust gear correctly. 11. Lubricate bushings. 12. Align shaft or coupling. 13. Adjust caster. 14. Replace springs. 15. Replace spindle. 16. Adjust jacket; check for steering shaft damage. 17. Align gear. 18. Clean or replace spool valve. 19. Adjust belt tension. 20. Replace hose. 21. Install correct size tires on both sides. 22. Repair or replace. 23. Adjust to specifications. 24. Repair leaks or replace gear. 25. Tighten mountings to specifications. 26. Replace stabilizer.
Vehicle pulls to one side.	1. Uneven tire pressure. 2. Brakes dragging. 3. Improper front end alignment. 4. Improperly adjusted wheel bearings. 5. Damaged or worn steering valve assembly. 6. Tire sizes not uniform. 7. Broken or sagging spring. 8. Misaligned rear axle housing. 9. Bent spindle. 10. Frame sprung. 11. Radial tire problem.	1. Equalize pressure. 2. Adjust brakes. 3. Align front end. 4. Adjust bearings. 5. Replace steering valve assembly. 6. Install tires of the same size. 7. Replace spring. 8. Align rear housing. 9. Replace spindle. 10. Straighten frame. 11. Switch front tires. If vehicle now pulls in the other direction, tires are defective.
Vehicle wanders from side to side.	1. Weak shock absorber or strut. 2. Loose steering gear. 3. Loose rack-and-pinion mountings. 4. Ball joints and steering linkage need lubrication. 5. Steering gear not on high point. 6. Broken or missing stabilizer bar or link. 7. Improperly adjusted rack-and-pinion. 8. Loose tie-rod end.	1. Replace shocks or struts. 2. Torque mounting fasteners. 3. Tighten to specifications. 4. Lubricate suspension. 5. Adjust steering gear properly. 6. Replace stabilizer or link. 7. Adjust to specifications. 8. Tighten. Replace if worn.
Sudden increase in steering wheel resistance.	1. Slipping pump belt. 2. Internal leakage in gear. 3. Fluid level low in pump.	1. Adjust belt tension. 2. Overhaul or replace gear. 3. Add fluid.

(Continued)

Manual and Power Steering Systems Diagnosis		
Condition	**Possible Causes**	**Correction**
Sudden increase in steering wheel resistance. *(Continued)*	4. Low engine idle. 5. Air in system. 6. Low tire air pressure. 7. Insufficient pump pressure. 8. High internal leakage in rack-and-pinion. 9. Defective steering stabilizer.	4. Adjust idle. 5. Bleed system. 6. Inflate to recommended level. 7. Test and repair or replace as required. 8. Repair leaks or replace gear. 9. Replace stabilizer.
Steering wheel action jerky during parking.	1. Loose pump belt. 2. Oily pump belt. 3. Defective flow control valve. 4. Insufficient pump pressure.	1. Adjust belt tension. 2. Replace belt. Clean pulleys. Repair source of leak. 3. Replace flow control valve. 4. Test and repair.
No effort required to turn wheel.	1. Broken steering gear torsion bar. 2. Broken tilt column U-joint. 3. Missing steering wheel hub-to-shaft key. Stripped splines. Loose nut.	1. Replace spool valve and shaft assembly. 2. Replace U-joint. 3. Replace key, shaft, or wheel. Tighten nut to specifications.
Excessive wheel kickback and play.	1. Worn steering linkage. 2. Air in system. 3. Improperly adjusted front wheel bearings. 4. Loose gear over-center adjustment. 5. Worm gear not preloaded. 6. No worm to rack-piston preload. 7. Loose pitman arm. 8. Loose steering gear. 9. Steering arms loose on spindle body. 10. Excessive play in ball joints. 11. Defective rotary valve. 12. Worn steering shaft universal joint. 13. Extra large tires. 14. Defective steering stabilizer.	1. Replace worn linkage components. 2. Bleed and add fluid if needed. 3. Adjust front wheel bearings. 4. Make correct over-center adjustment. 5. Preload worm gear. 6. Install larger set of rack piston balls. 7. Torque pitman arm nut. 8. Tighten mounting fasteners. 9. Tighten arm fasteners. 10. Replace ball joints. 11. Replace rotary valve. 12. Replace joint and/or shaft. 13. Advise owner. Install a steering stabilizer or replace with a larger stabilizer. 14. Replace stabilizer.
No power assist in one direction.	1. Defective steering gear.	1. Overhaul or replace gear as needed.
Steering pump pressure low.	1. Loose pump belt. 2. Oily belt. 3. Worn pump parts. 4. Relief valve springs defective or stuck open. 5. Low fluid level in reservoir. 6. Air in system. 7. Defective hose. 8. Flow control valve stuck open. 9. Pressure plate not seated against cam ring. 10. Scored pressure plate, thrust plate, or rotor. 11. Vanes incorrectly installed. 12. Vanes sticking in rotor. 13. Worn or damaged O-rings.	1. Adjust belt. 2. Clean pulleys. Replace belt. Correct source of leak. 3. Overhaul pump. 4. Repair or replace as needed. 5. Add fluid. 6. Correct source of leak. Bleed system. 7. Replace hose. 8. Clean or replace valve. 9. Repair or replace cam ring and pressure plate. 10. Replace damaged parts and flush system. 11. Install vanes correctly. 12. Free vanes. Clean thoroughly. 13. Replace O-rings.
Steering pump noise.	1. Air in system. 2. Loose pump pulley. 3. Loose belt. 4. Glazed belt.	1. Correct leak and bleed system. 2. Tighten pulley. 3. Tension belt correctly. 4. Replace belt.

(Continued)

	Manual and Power Steering Systems Diagnosis	
Condition	**Possible Causes**	**Correction**
Steering pump noise. *(Continued)*	5. Hoses touching splash shield. 6. Low fluid level. 7. Clogged or kinked hose. 8. Scored pressure plate. 9. Scored rotor. 10. Vanes installed wrong. 11. Vanes sticking in rotor. 12. Defective flow control valve. 13. Loose pump. 14. Plugged reservoir vent. 15. Dirty fluid. 16. Worn pump bearing. 17. Chirp-type noise. 18. Whine or growl.	5. Reroute hose to prevent contact. 6. Add fluid, check for leaks. 7. Replace hose. 8. Polish. Replace if badly scored. 9. Polish. Replace if badly scored. 10. Install vanes correctly. 11. Free vanes and clean thoroughly. 12. Replace flow control valve. 13. Tighten pump mounting fasteners. 14. Clean vent. 15. Drain, flush, and refill. 16. Overhaul as needed. 17. Tighten loose belt. 18. Low fluid. Fill to proper level.
Steering gear dull rattle or chuckle.	1. Gear loose on frame. 2. Loose over-center adjustment. 3. No worm shaft preload. 4. Insufficient or improper lubricant (manual gear).	1. Tighten gear mounting fasteners. 2. Make correct over-center adjustment. 3. Adjust preload. 4. Fill with specified lubricant.
Hissing sound in gear.	1. Normal sound when turning wheel when vehicle is standing still or when holding wheel against stops. 2. Loose gear. 3. Noisy pressure control valve. 4. Intermediate shaft rubber plug missing.	1. Normal condition, advise driver. 2. Tighten mounting fasteners. 3. Replace valve. 4. Replace plug.
Tire squeal on turns.	1. Excessive speed. 2. Low air pressure. 3. Faulty wheel alignment. 4. Excessive load. 5. Loose rack-and-pinion mountings.	1. Advise driver. 2. Inflate to correct pressure. 3. Align wheels. 4. Advise driver. 5. Tighten mountings to specifications.
External fluid leaks.	1. Defective hose. 2. Loose hose connections. 3. Cracked hose connections. 4. Defective pitman shaft seal in gear. 5. Gear housing end cover O-ring seal leaking. 6. Leaking gear torsion bar seal. 7. Leaking adjuster plug seals. 8. Leaking side cover gasket. 9. Pump too full. 10. Pump shaft seal defective. 11. Scored shaft in pump. 12. Oil leaking out of reservoir from air contamination. 13. Pump assembly fasteners loose. 14. Leaking power cylinder (linkage type). 15. Cracked rack-and-pinion housing. 16. Extreme cam ring wear. 17. Scored pressure plate, rotor, or thrust plate. 18. Incorrectly installed vanes. 19. Cracked reservoir. 20. Leaking pump reservoir cap. 21. Defective rack-and-pinion stub shaft seal.	1. Replace hose. 2. Tighten to proper torque. 3. Replace hose. 4. Replace seal. Check bearing for excessive wear. 5. Replace seal. 6. Replace valve and shaft assembly. 7. Replace seals. 8. Replace gasket. 9. Reduce fluid level. 10. Replace shaft seal. 11. Replace shaft. 12. Correct source of air leak. 13. Torque fasteners. 14. Overhaul as needed. 15. Replace steering gear. 16. Replace parts. Flush system. 17. Replace parts. Flush system. 18. Install pump vanes properly. 19. Replace reservoir. 20. Repair or replace cap. 21. Replace stub shaft seal.

Chapter 71

Brake System Fundamentals

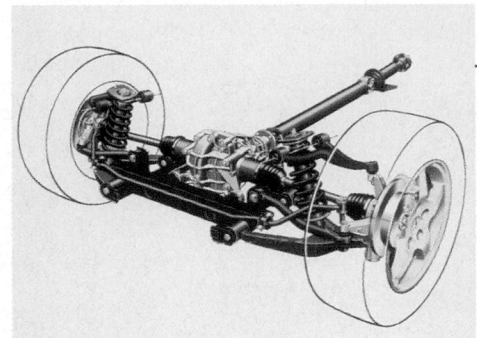

After studying this chapter, you will be able to:

- Explain the hydraulic and mechanical principles of a brake system.
- Identify the major parts of an automotive brake system.
- Define the basic functions of the major parts of a brake system.
- Compare drum and disc brakes.
- Describe the operation of parking brakes.
- Explain the operation of power brakes.
- Correctly answer ASE certification test questions requiring a knowledge of automotive brake systems.

Automotive brakes provide a means of using friction to either slow, stop, or hold the wheels of a vehicle. When a car is moving down the highway, it has a tremendous amount of stored energy in the form of ***inertia*** (tendency to keep moving). To stop the vehicle, the brakes convert ***kinetic*** (moving) energy into heat.

As you will learn, a modern automobile uses numerous devices to improve braking ability. For example, dual hydraulic brake systems, hydraulic valves, and electronic anti-lock braking systems are all found on today's vehicles.

Note
Anti-lock brakes are covered in Chapter 73, *Anti-Lock Brakes, Traction Control, and Stability Control.*

Basic Brake System

Before studying the construction and operation of individual brake system parts, you should have a general understanding of overall system design and operation. The basic parts of an automotive brake system include:

- ***Brake pedal assembly***—foot lever for operating the master cylinder and power booster. Look at **Figure 71-1.**
- ***Master cylinder***—hydraulic-piston pump that develops pressure for the brake system.
- ***Brake booster***—vacuum- or power steering–operated device that assists brake pedal application.
- ***Brake lines***—metal tubing and rubber hose that transmits pressure to the wheel brake assemblies.
- ***Wheel brake assemblies***—devices that use system pressure to produce friction for slowing or stopping wheel rotation.
- ***Emergency or parking brake***—mechanical system for applying rear brake assemblies.

When the driver pushes on the brake pedal, lever action pushes a rod into the brake booster and master cylinder. This produces hydraulic pressure in the master cylinder. Fluid flows through the brake lines to the wheel brake assemblies. The brake assemblies use this pressure to cause friction for braking.

An emergency brake system uses cables or rods to mechanically apply the rear brakes. This provides a system for holding the wheels when the vehicle is parked on a hill or stopping the vehicle during complete hydraulic brake system failure.

Drum and Disc Brakes

There are two common types of brake assemblies used on modern automobiles: disc brakes and drum brakes. Refer to **Figure 71-2.**

Disc brakes are frequently used on the two front wheels of a vehicle. Drum brakes are often used on the rear wheels. However, disc or drum brakes may be used on all four wheels.

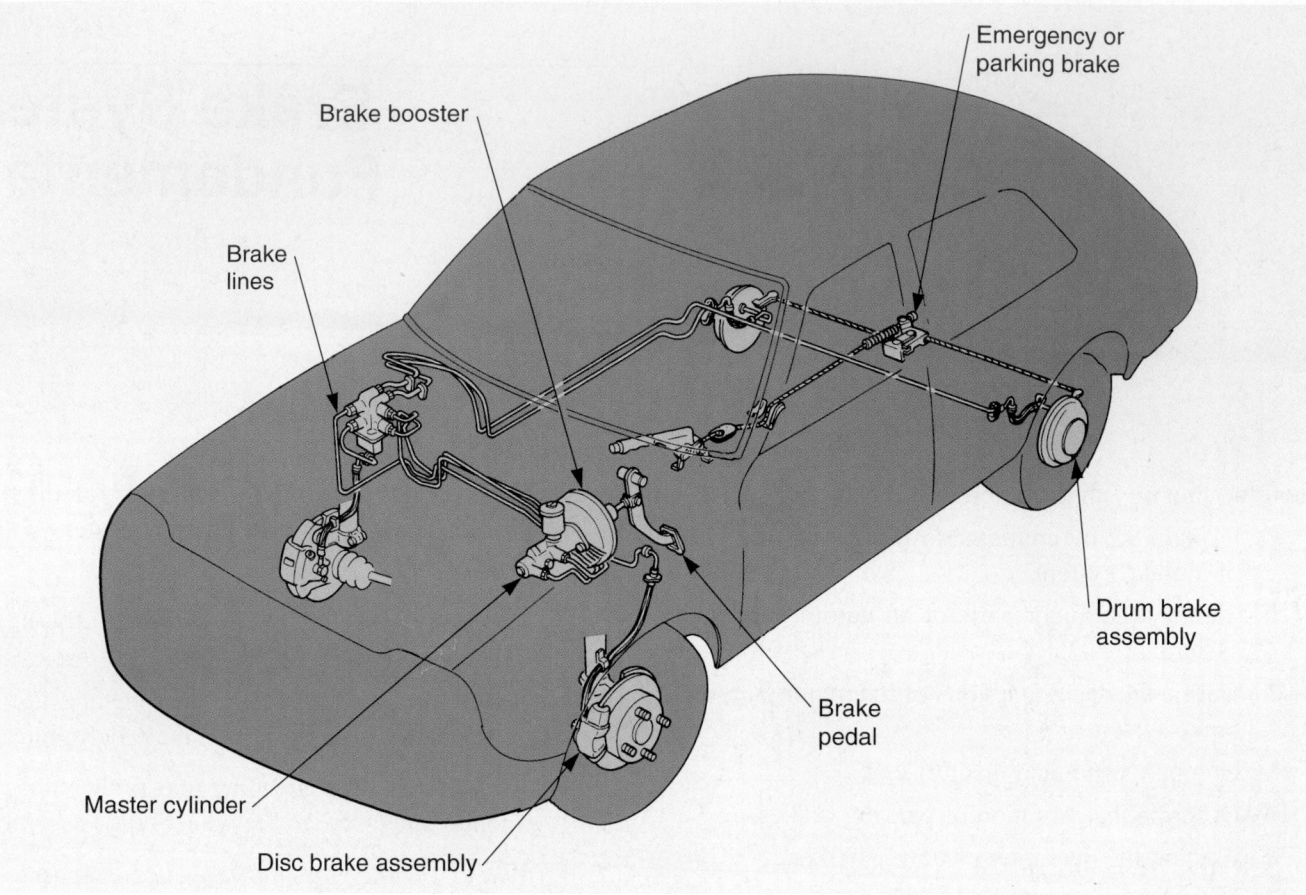

Figure 71-1. These are the basic parts of an automotive brake system. Study them. (Honda)

A *disc brake assembly* basically includes:

- *Caliper*—assembly that holds the wheel cylinder piston and brake pads.

- *Caliper cylinder*—machined hole in the caliper; the piston fits into this cylinder.

- *Brake pads*—friction members pushed against the rotor by the action of the master cylinder, caliper cylinder, and piston.

- *Rotor*—metal disc that uses friction from the brake pads to stop or slow wheel rotation.

A *drum brake assembly* basically includes:

- *Wheel cylinder assembly*—hydraulic piston forced outward by fluid pressure.

- *Brake shoes*—friction units that are pushed against the rotating brake drum by the action of the wheel cylinder assembly.

- *Brake drum*—rubs against the brake shoes to stop or slow wheel rotation.

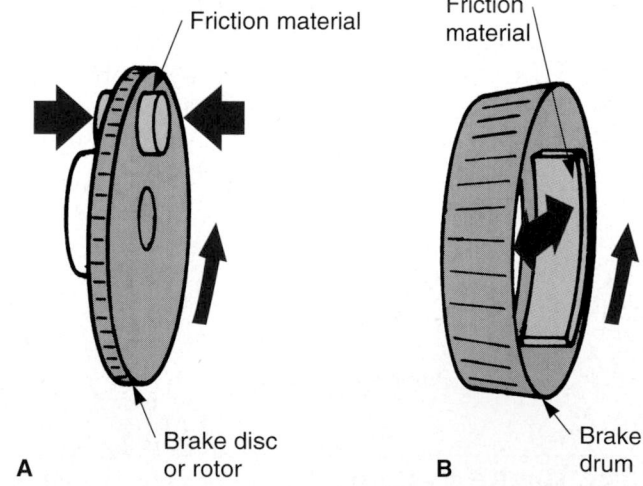

Figure 71-2. Compare disc brakes and drum brakes. A—Disc brakes are commonly used on the front of the vehicle and sometimes on the rear. B—Drum brakes are normally used on the rear of the vehicle.

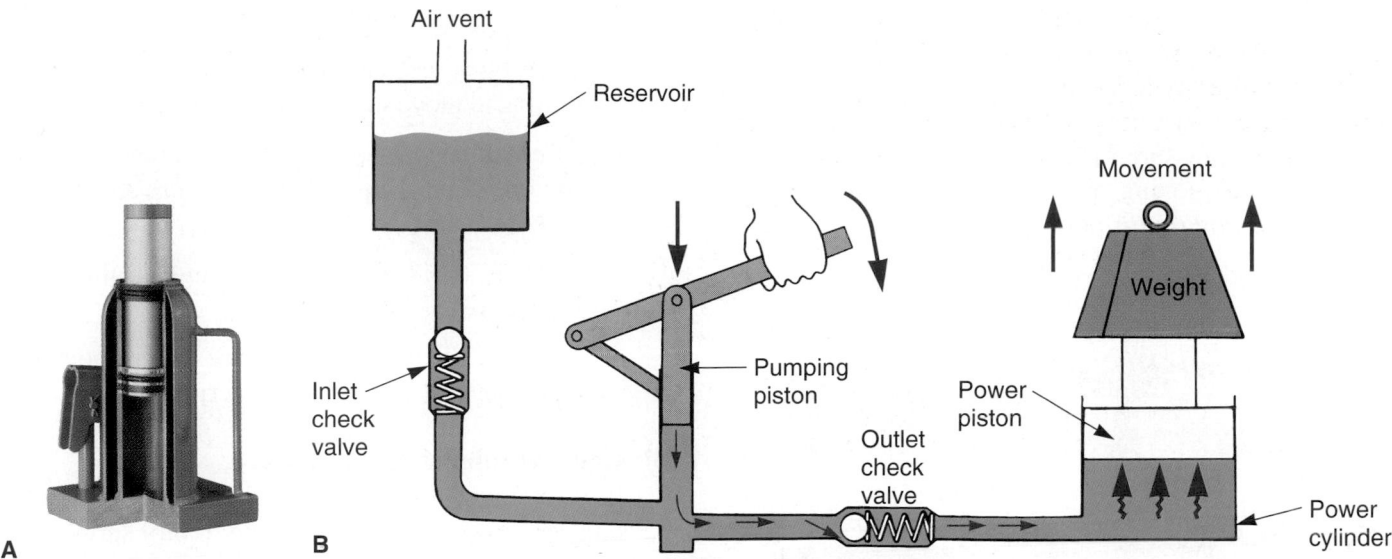

Figure 71-3. A hydraulic jack and a brake system use many of the same operating principles. A—Cutaway view shows how oil under pressure is used to raise the hydraulic jack when the handle is pumped. B—This drawing of a hydraulic jack demonstrates how a small piston acting on a large piston can increase force tremendously. Study how the check valves only allow oil flow in one direction. (OTC and Deere & Co.)

Braking Ratio

Braking ratio refers to the comparison of front wheel braking effort to rear wheel braking effort. When a vehicle stops, its weight tends to transfer onto the front wheels. The front tires are pressed against the road with greater force. The rear tires lose some of their grip on the road. As a result, the front wheels do more of the braking than the rear.

For this reason, many vehicles have disc brakes on the front and drum brakes on the rear. Disc brakes are capable of producing more stopping effort than drum brakes. If drum brakes are used on both the front and rear wheels, the front shoe linings and drums normally have larger surface areas than the rear linings and drums.

Typically, front wheel brakes handle 60% to 70% of the braking power. Rear wheel brakes handle 30% to 40% of the braking. Front-wheel-drive cars, having even more weight on the front wheels, can have even a higher braking ratio at the front wheels.

Brake System Hydraulics

A *hydraulic system* is basically a system that uses a liquid to transmit motion or pressure from one point to another. Modern brake systems are hydraulic. They use a confined brake fluid to transfer brake pedal motion and pressure to each of the wheel brake assemblies.

Several principles apply to the operation of a hydraulic system. These include:

- Liquids in a confined area will *not* compress. However, air in a confined area will compress.

- When pressure is applied to a closed system, pressure is exerted equally in all directions.

- A hydraulic system can be used to increase or decrease force or motion.

Hydraulic System Action

Suppose two cylinders of equal diameter are placed side by side with a tube connecting them. The system is filled with liquid. Each cylinder contains a piston. If you push down on one of the pistons, the other piston will move an equal distance and with equal force.

Since the liquid will not compress and the cylinders are the same diameter, the same amount of liquid, pressure, and motion is transferred from one cylinder to the other.

When pistons of different sizes are used, motion and force can be increased or decreased. Suppose a small piston acts on a larger piston. The larger piston will move with more force. However, it will move a shorter distance.

When a large diameter piston acts on a smaller piston, the opposite is true. The smaller piston slides farther in its cylinder than the large piston, but it moves with less force.

A simple hydraulic jack demonstrates the principles just discussed. Look at **Figure 71-3.**

Applying these hydraulic principles to a brake system, you can see how stopping force is transmitted from the master cylinder to each wheel brake assembly. The master cylinder acts as the pumping piston that supplies system pressure. The wheel cylinder acts as the power piston, moving the friction linings into contact with the rotating drums or discs, **Figure 71-4.**

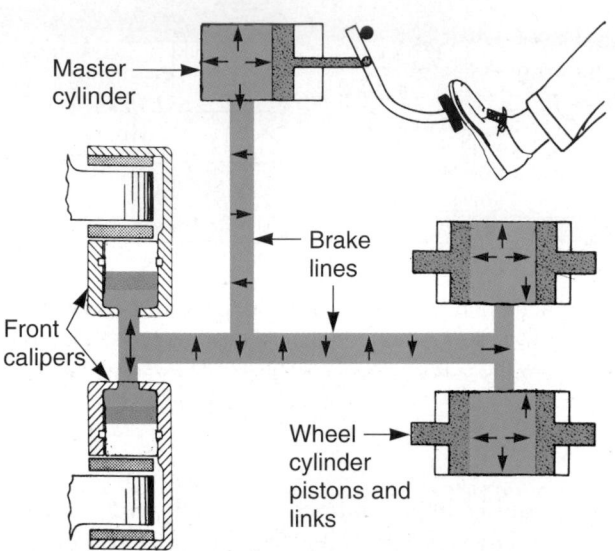

Figure 71-4. Like a basic hydraulic jack, the master cylinder acts as a pumping piston to move the pistons at the wheel brake assemblies. (GM)

Brake System Components

It is important for you to know the location and construction of major brake system components. This will better prepare you to troubleshoot and repair these important parts.

Brake Pedal Assembly

The **brake pedal assembly** acts as a lever arm to increase the force applied to the master cylinder piston. A manual master cylinder bolts directly to the engine's firewall. The brake pedal assembly bolts under the dash, **Figure 71-5.**

The pedal swings on a hinge in the pedal support bracket. A push rod connects the brake pedal to the master cylinder piston.

Master Cylinder

A **master cylinder** is a foot-operated pump that forces fluid into the brake lines and wheel cylinders. Refer to **Figure 71-6.**

A master cylinder can have four basic functions:

- It develops pressure, causing the wheel cylinder pistons to move toward the rotors or drums.

- After all the shoes or pads produce sufficient friction, it helps equalize the pressure required for braking.

- It keeps the system full of fluid as the brake linings wear.

- It can maintain a slight pressure to keep contaminants (air and water) from entering the system.

Master Cylinder Components

In its simplest form, a master cylinder consists of a housing, reservoir, piston, rubber cup, return spring, and rubber boot. Look at **Figure 71-7.**

A cylinder is precision machined in the housing of the master cylinder. The spring, cup, and metal piston slide in this cylinder. Two ports are drilled between the reservoir and cylinder.

The **cup** and **piston** in the master cylinder are used to pressurize the brake system. When they are pushed forward, they trap the fluid, building pressure.

The master cylinder **intake port,** or vent, allows fluid to enter the rear of the cylinder as the piston slides forward. Refer to **Figure 71-8A.** Fluid flows out of the reservoir, through the intake port, and into the area behind the piston and cup.

In **Figure 71-8B,** the cup has moved past the compensating port and is beginning to develop pressure to apply the brakes.

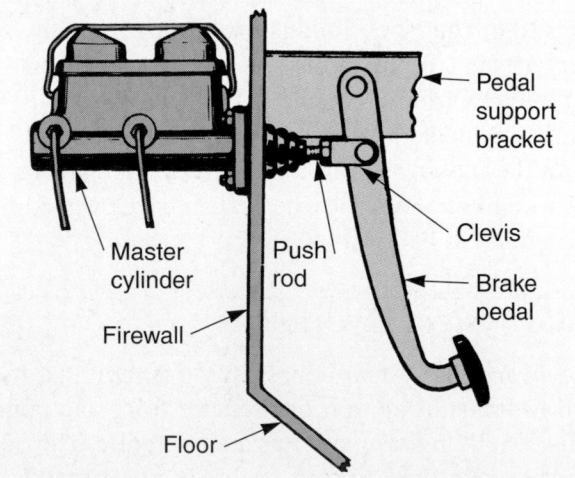

Figure 71-5. The brake pedal assembly bolts under the dash. A push rod transfers pedal movement into the master cylinder and operates the piston in the master cylinder. (Bendix)

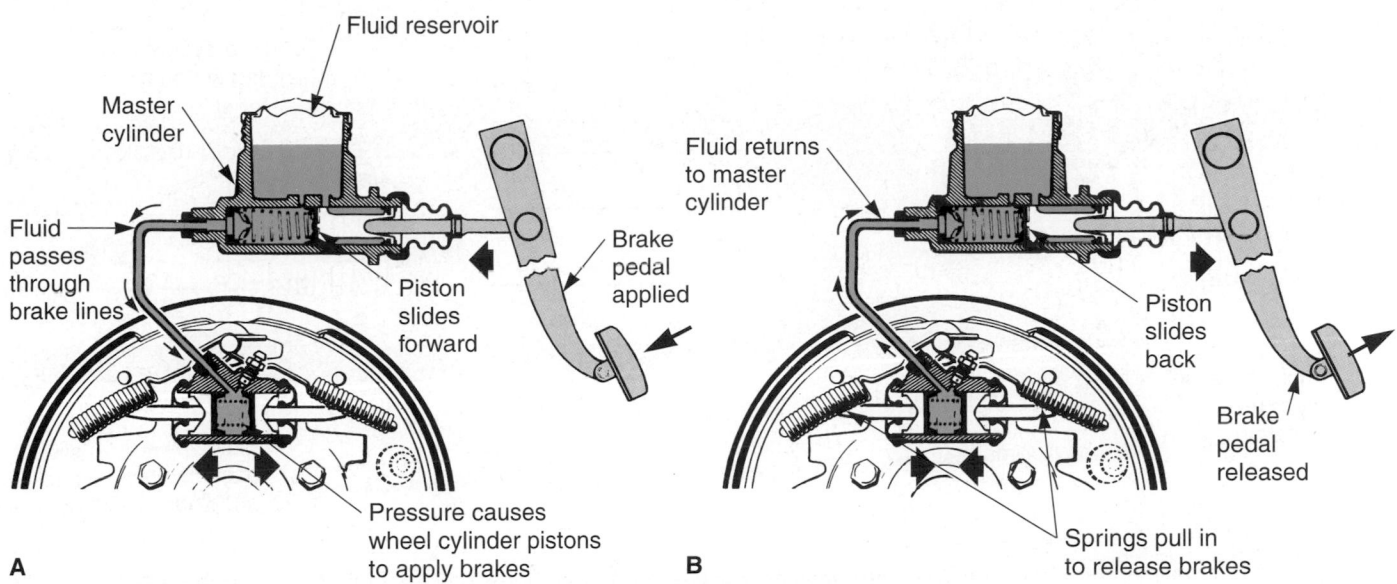

Figure 71-6. This single-piston master cylinder demonstrates the simplified action of a brake system. A—Application of pedal moves push rod into piston. Master cylinder piston pressurizes fluid in cylinder and line. Pressure pushes wheel cylinder pistons apart and brakes are applied. B—Release of brake pedal allows retracting springs to pull brake shoes away from brake drum. Fluid flows back through line and into master cylinder. (FMC)

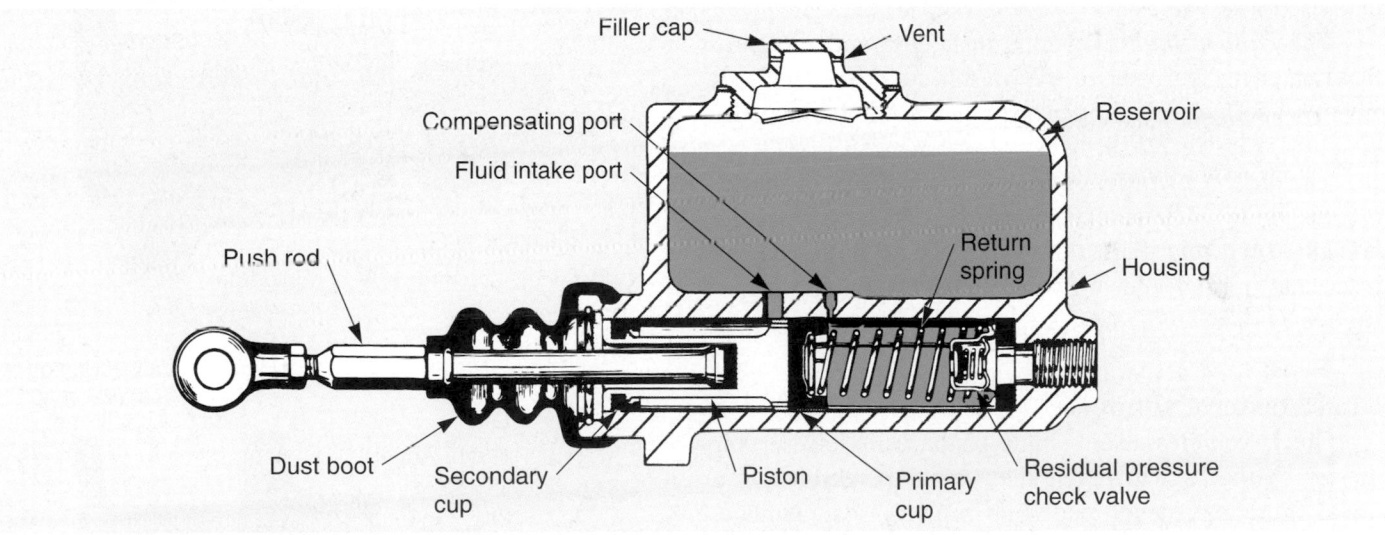

Figure 71-7. Study the basic parts of a master cylinder. (Bendix)

Then, when the brake pedal is released, the return spring forces the piston and cup back in the cylinder. If needed, the rubber cup flexes forward, allowing fluid to enter the area in front of the piston and cup. Usually, small holes are drilled in the edge of the piston so that fluid can flow past the cup. See **Figure 71-8C.**

The *compensating port* releases extra pressure when the piston returns to the released position. Fluid can flow back into the reservoir through the compensating port.

The action of the intake port and the compensating port keep the system full of fluid.

Residual pressure valves maintain residual fluid pressure of approximately 10 psi (69 kPa) to help keep contaminants out of the system. One valve is located in each outlet to the brake lines. A cutaway view of a typical residual pressure valve is given in **Figure 71-9.**

Note how the residual pressure valve allows fluid flow out of the master cylinder. However, it resists free

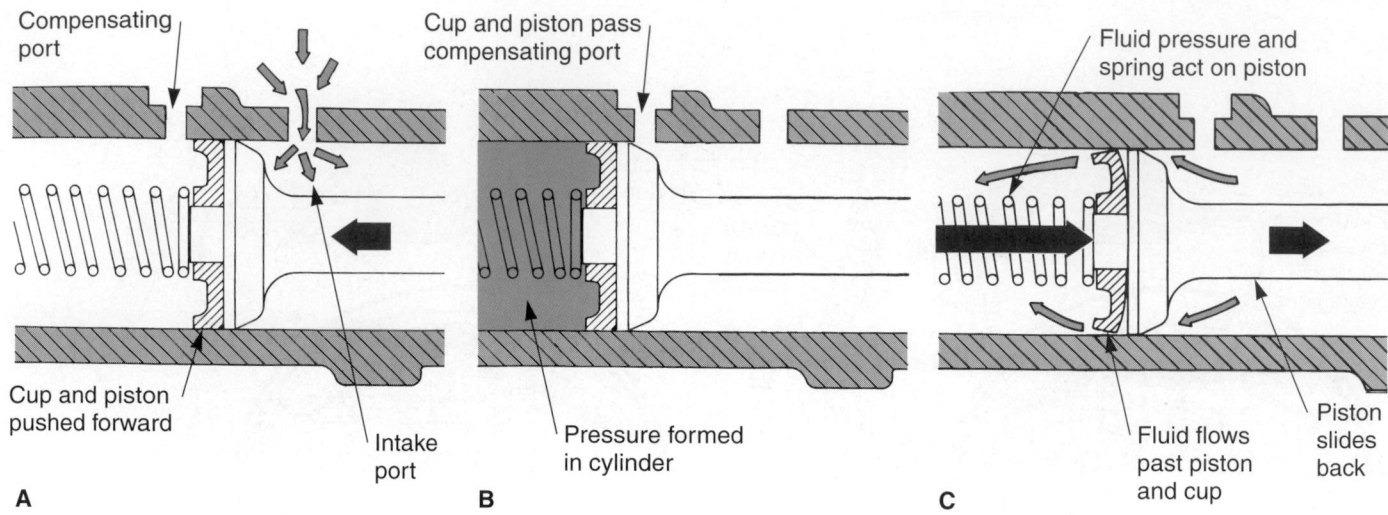

Figure 71-8. Note piston and cup action inside the master cylinder. A—Piston slides forward. Fluid flows into the area behind the piston. Excess fluid flows into the reservoir through the compensating port. B—Piston and cup move past the compensating port, and pressure forms in the front section of the cylinder to apply the brakes. C—When the brake pedal is released, the cup flexes forward so that fluid can flow to the front of the piston for release of the brakes. (EIS)

flow of fluid back into the cylinder. Many master cylinders use these valves.

The ***rubber boot*** prevents dust, dirt, and moisture from entering the back of the master cylinder. The boot fits over the master cylinder housing and the brake pedal push rod, **Figure 71-10.**

The ***master cylinder reservoir*** stores an extra supply of brake fluid. The reservoir may be cast as part of the housing, or it may be a removable plastic part. Today's reservoirs normally have two sections, or compartments, **Figure 71-10.**

Dual Master Cylinder

Older vehicles used single-piston, single-reservoir master cylinders. However, they were dangerous. If a brake fluid leak developed (line rupture, seal damage, crack in brake hose), a sudden loss of braking ability could occur. Modern vehicles use a dual master cylinder for added safety, **Figure 71-10.**

The ***dual master cylinder,*** also called the ***tandem master cylinder,*** has two separate hydraulic pistons and two fluid reservoirs. Each piston operates a hydraulic circuit that controls two wheel brake assemblies. Then, if there is a system leak in one of the hydraulic circuits, the other circuit can still provide braking action on two wheels. See **Figure 71-11.**

In the dual master cylinder, the rear piston assembly is called the ***primary piston.*** The front piston is termed the ***secondary piston.***

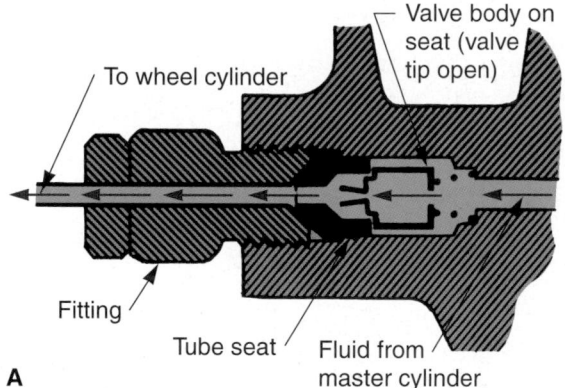

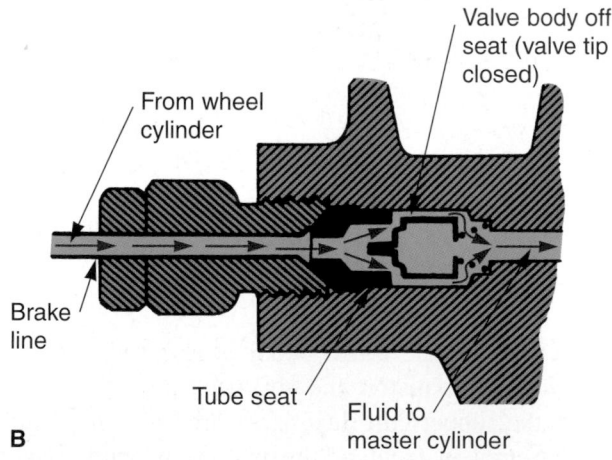

Figure 71-9. Residual pressure valves maintain some line pressure after the brakes are released. This helps keep air out of the system. A—When the brakes are applied, fluid flows freely through the valve. B—After the brakes are released, the valve closes to restrict the return of fluid to the master cylinder. (FMC)

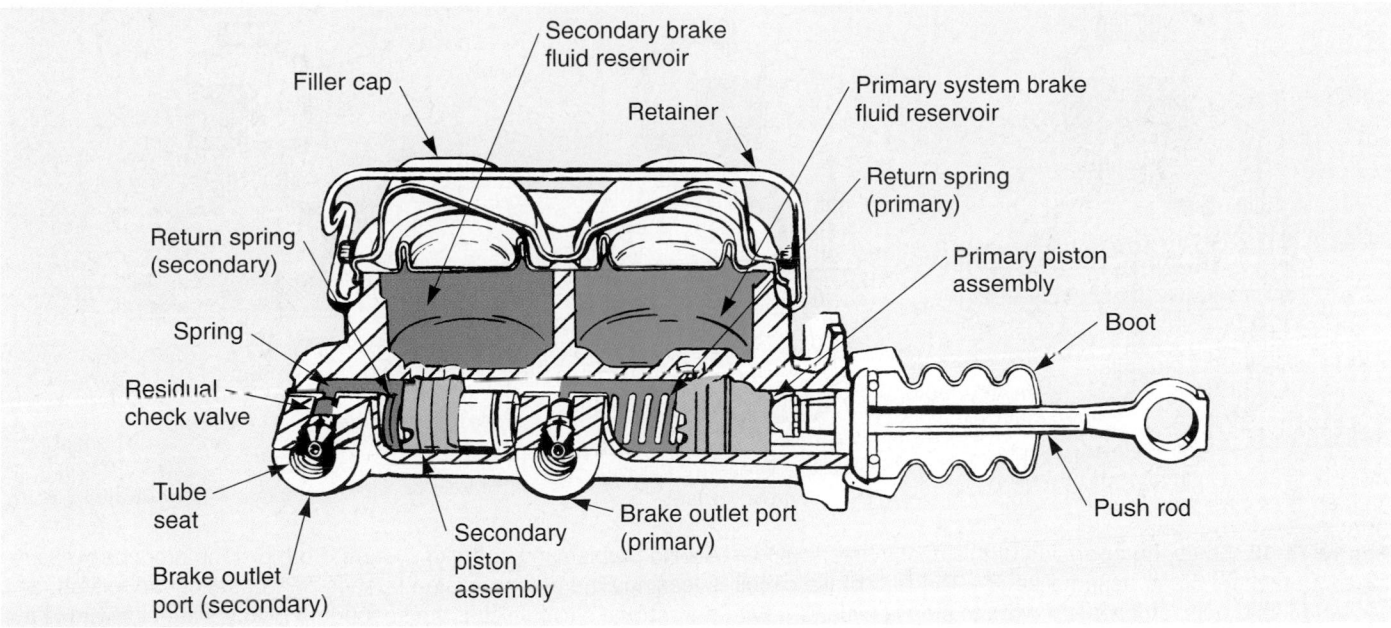

Figure 71-10. Dual master cylinders are now used because they will still provide braking action when a major hydraulic leak develops. With single-piston master cylinder, a leak could cause sudden and complete loss of brakes. (Delco)

Dual Master Cylinder Operation

The action of the pistons, cups, and ports in the dual master cylinder is similar to the action of those used in a single-piston unit. Look at **Figure 71-12A.**

When both systems are intact (no fluid leaks), each piston produces and supplies pressure to two wheel cylinders (all four wheel brake assemblies are applied)

If there is a pressure loss in the primary section of the brake system (rear section of master cylinder), the primary piston slides forward and pushes on the secondary piston. This forces the secondary piston forward mechanically, building pressure in two of the wheel brake assemblies, as shown in **Figure 71-12B.**

When a brake line, wheel cylinder, or other component leaks in the secondary circuit (parts fed by secondary piston), the secondary piston slides completely forward in the cylinder. Then, the primary piston provides hydraulic pressure for the other two brake assemblies. See **Figure 71-12C.**

It is very unlikely that both systems would fail at the same time. If they did, the mechanical parking brake would be needed to slow and stop the vehicle.

Power Brakes

Power brakes use a booster and engine vacuum or hydraulic pressure to assist brake pedal application. The booster (assisting mechanism) is located between the brake pedal linkage and the master cylinder. When the

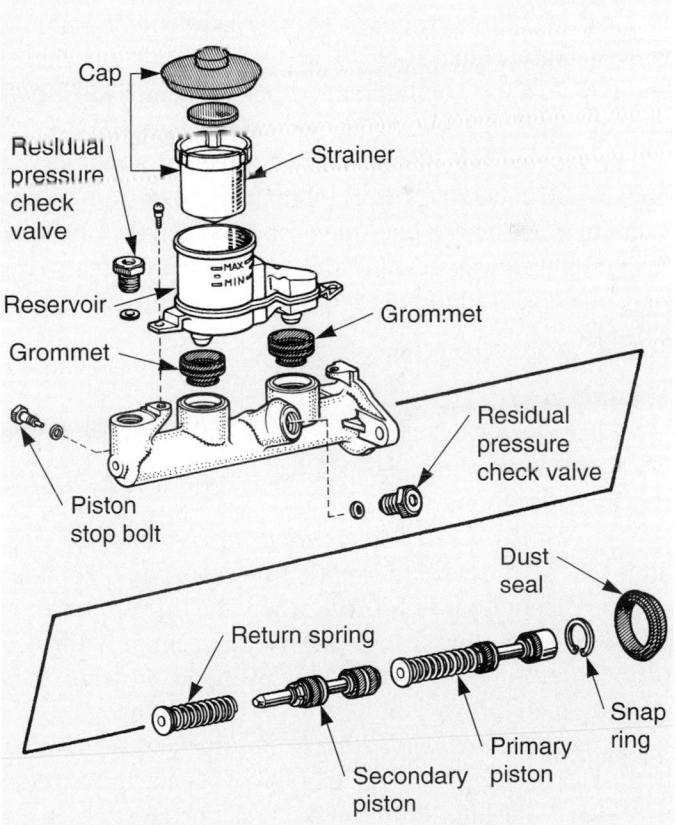

Figure 71-11. Note the major parts of this modern dual master cylinder. (Toyota)

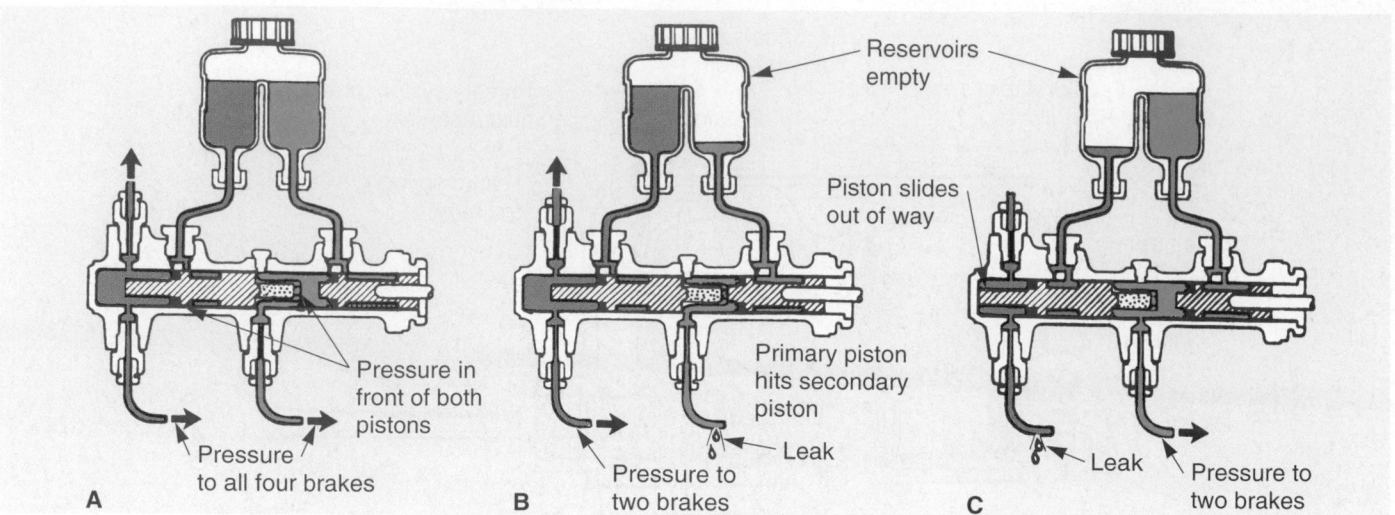

Figure 71-12. Study the operation of this dual master cylinder. A—No problem in the brake system. Both pistons produce pressure for all four wheel brake assemblies. B—The rear brake circuit is leaking. The primary piston pushes on the secondary piston, and two wheel brake assemblies still work to stop vehicle. C—With a front brake circuit leak, the secondary piston slides forward in the cylinder. The primary piston then operates normally to apply two wheel brake assemblies. (Delco)

driver presses the brake pedal, the brake booster helps push on the piston.

Power Brake Vacuum Boosters

A power brake *vacuum booster* uses engine vacuum (or vacuum created by a separate pump on diesel engines) to apply the hydraulic brake system. The principles of a vacuum booster are shown in **Figure 71-13.**

A vacuum booster basically consists of a cylindrical housing that encloses a diaphragm or a piston. When vacuum is applied to one side of the booster, the piston or diaphragm moves toward the low-pressure area. This movement is used to help force the piston into the master cylinder.

Vacuum Booster Types

There are two general types of vacuum brake boosters: atmospheric suspended boosters and vacuum suspended boosters.

An *atmospheric suspended brake booster* has atmospheric pressure (normal air pressure) on both sides of the diaphragm or piston when the brake pedal is released. As the brakes are applied, a vacuum is formed on one side of the booster. Atmospheric pressure then pushes on and moves the piston or diaphragm.

A *vacuum suspended brake booster* has vacuum on both sides of the piston or diaphragm when the brake pedal is released. Pushing down on the brake pedal releases the vacuum on one side of the booster. The difference in pressure pushes the piston or diaphragm for braking action.

Figures 71-14 and **71-15** show views of two vacuum brake boosters.

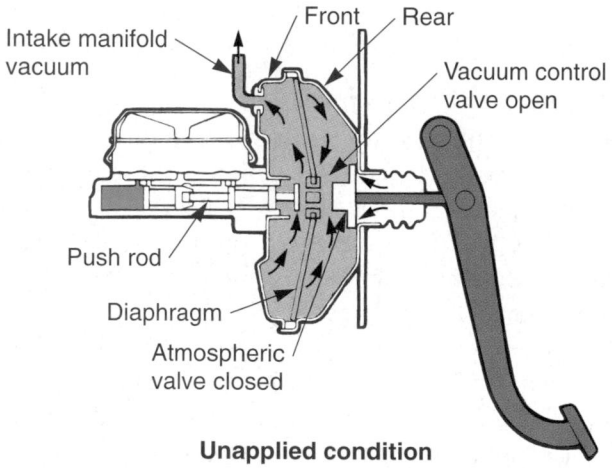

Unapplied condition

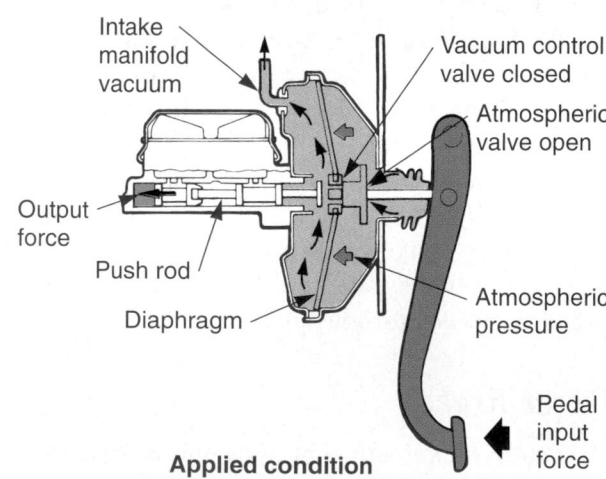

Applied condition

Figure 71-13. Vacuum booster uses engine vacuum or vacuum created by an auxiliary pump to help apply the brake pedal. The vacuum acts on a diaphragm, which pushes on a push rod. (Mopar)

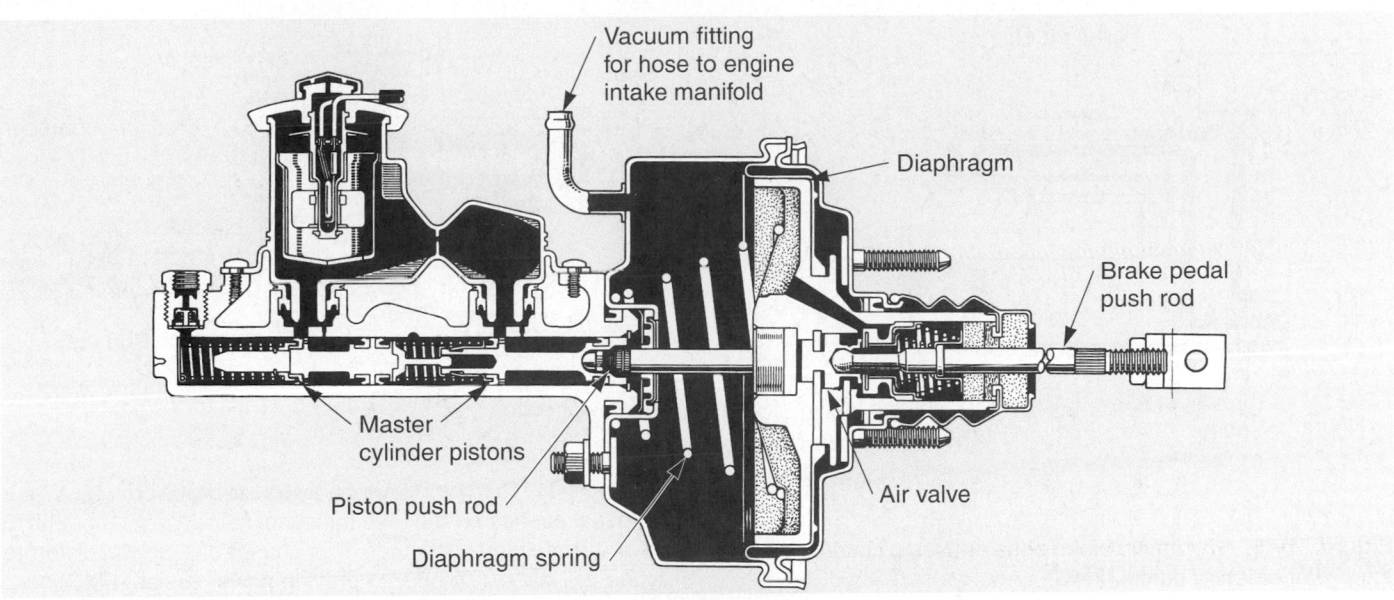

Figure 71-14. Study the internal parts of this brake booster and master cylinder. (Toyota)

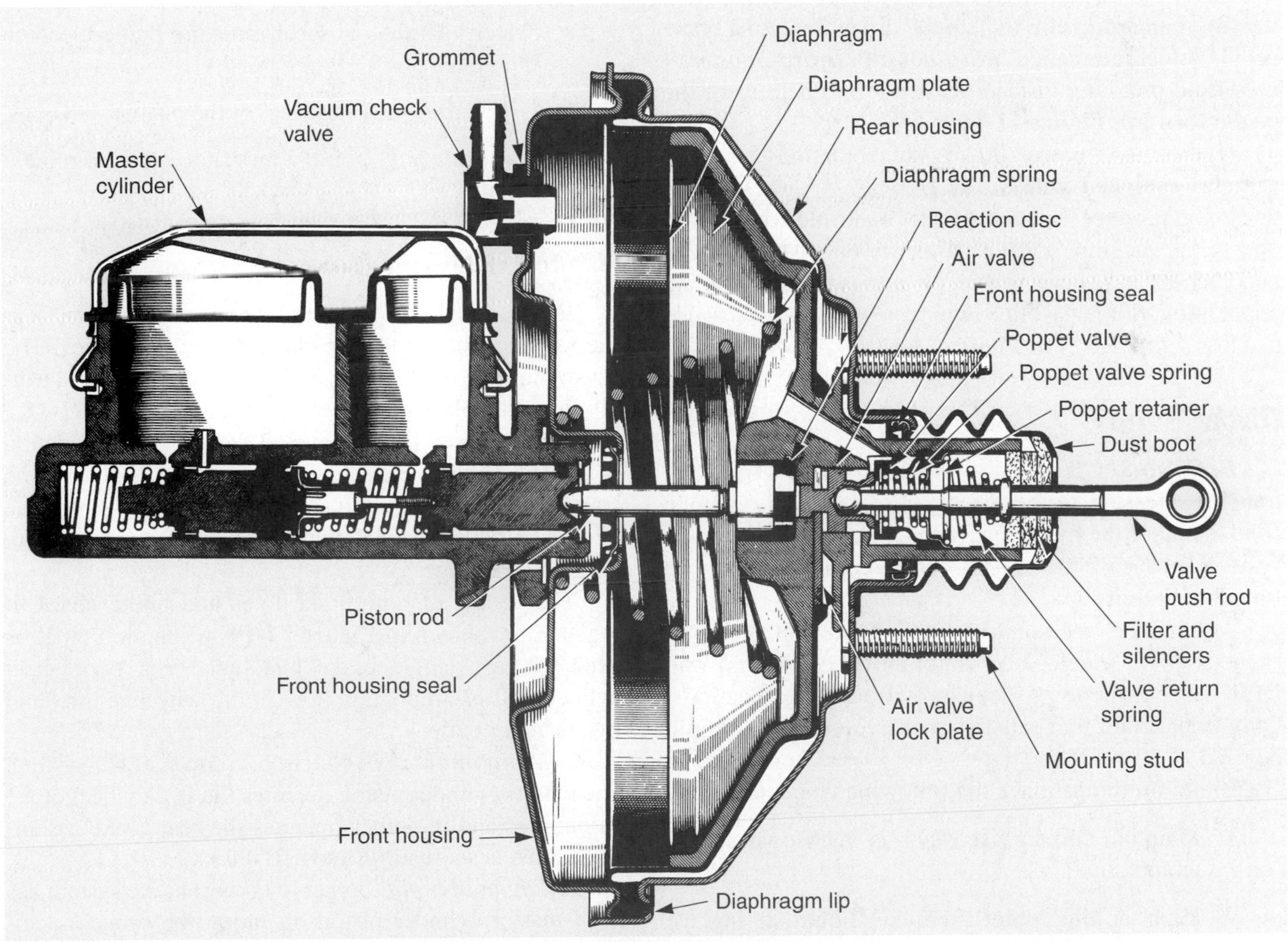

Vacuum fitting for hose to engine intake manifold

Diaphragm

Brake pedal push rod

Master cylinder pistons

Piston push rod

Diaphragm spring

Air valve

Grommet

Vacuum check valve

Master cylinder

Diaphragm

Diaphragm plate

Rear housing

Diaphragm spring

Reaction disc

Air valve

Front housing seal

Poppet valve

Poppet valve spring

Poppet retainer

Dust boot

Valve push rod

Filter and silencers

Valve return spring

Mounting stud

Air valve lock plate

Piston rod

Front housing seal

Front housing

Diaphragm lip

Figure 71-15. Another type of vacuum brake booster. (Bendix)

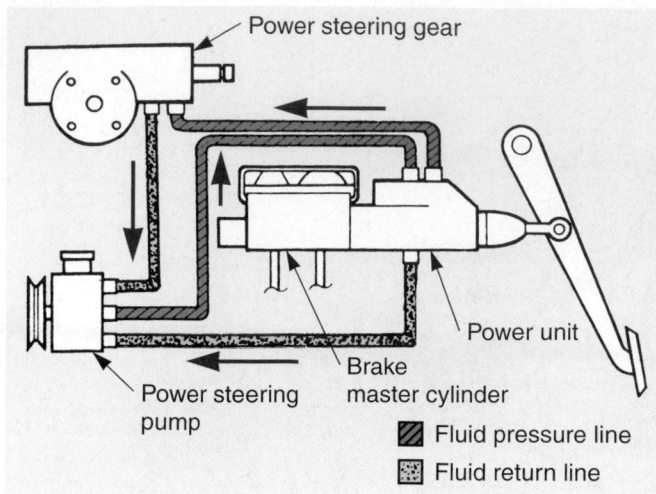

Figure 71-16. A hydraulic brake booster uses oil pressure from the power steering pump. (FMC)

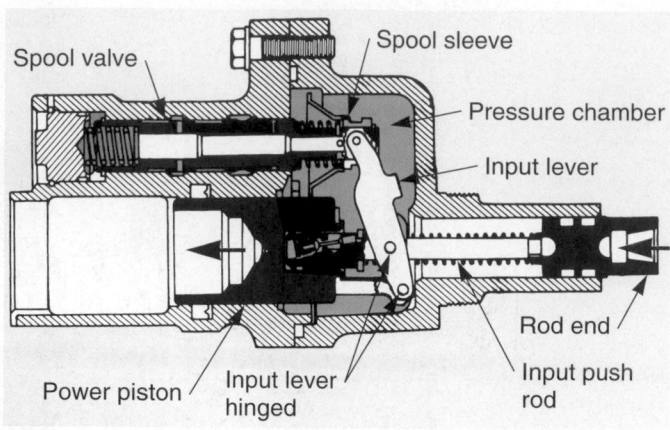

Figure 71-17. Cutaway view of a hydraulic brake booster. When the pedal pushes on the unit, the spool valve allows more oil to enter the pressure chamber. This causes the power piston to act on the master cylinder to provide power assist. (FMC)

Power Brake Hydraulic Boosters

A power brake *hydraulic booster* uses power steering pump pressure to help the driver apply the brake pedal. Sometimes called *hydro-boost* or *hydra-booster,* it uses fluid pressure, instead of vacuum, to help force the piston forward, **Figure 71-16.**

Hydro-boost power brakes are commonly used on vehicles equipped with *diesel engines.* A diesel engine does not produce enough intake manifold vacuum to operate a vacuum booster. The power steering pump becomes an excellent source of energy to assist brake application. Some gasoline engine powered vehicles also use this setup. Refer to **Figure 71-17.**

Brake Fluid

Brake fluid is a specially blended hydraulic fluid that transfers pressure to the wheel brake assemblies. Brake fluid is one of the most important components of a brake system because it ties all the other components into a functioning unit.

Automakers recommend brake fluid that meets or exceeds SAE (Society of Automotive Engineers) and DOT (Department of Transportation) specifications. Only brake fluid that satisfies their requirements should be used.

Brake fluid must have the following characteristics:

- Maintain correct viscosity—is free flowing at all temperatures.

- High boiling point—remains liquid at highest system operating temperature.

- Noncorrosive—does not attack metal or rubber brake system parts.

- Water tolerant—absorbs moisture that collects in the system.

- Lubricates—reduces wear of the pistons and cups.

- Low freezing point—does not freeze in cold weather.

Brake Lines and Hoses

Brake lines and *hoses* transfer fluid pressure from the master cylinder to the wheel brake assemblies. The brake lines are made of double-wall steel tubing and usually have double-lap flares on their ends, **Figure 71-18.**

Rubber brake hoses are used where a flexing action is needed. For instance, brake hose is used between the frame and the front caliper cylinders. This allows the wheels to move up and down or from side to side without brake line damage.

The details of how brake lines and brake hoses fit together are shown in **Figure 71-19.** A junction block is used where a single brake line must feed two wheel cylinders. It is simply a hollow fitting with one inlet and two or more outlets.

In a *longitudinally split* (front to rear) brake system, one master cylinder piston operates the front wheel brake assemblies and the other operates the rear brake assemblies. This is shown in **Figure 71-20A.**

A *diagonally split* (corner to corner) brake system has each master cylinder piston operating the brake assemblies on opposite corners of the vehicle, **Figure 71-20B.**

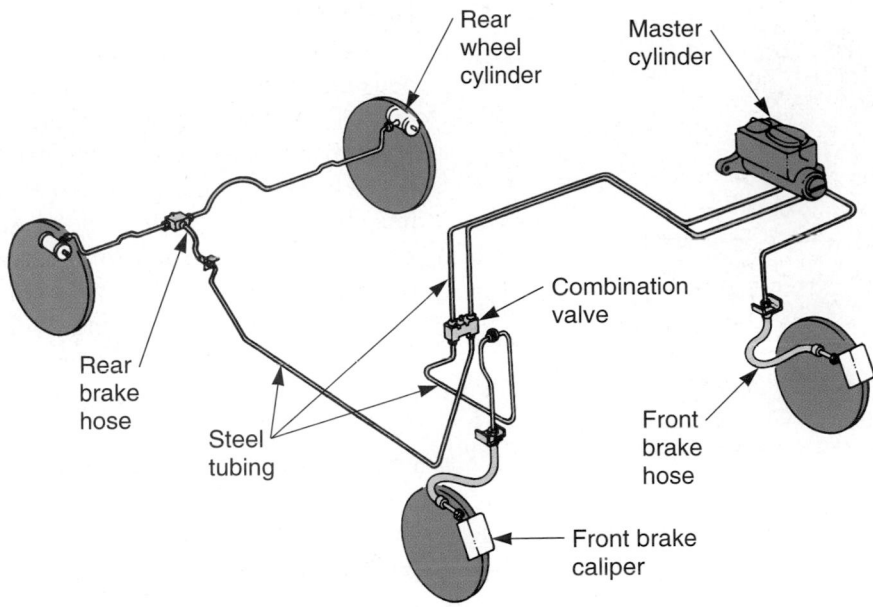

Figure 71-18. Note the routing and location of metal brake lines and flexible hoses. Hoses are needed because suspension movement would break steel tubing. (Chrysler)

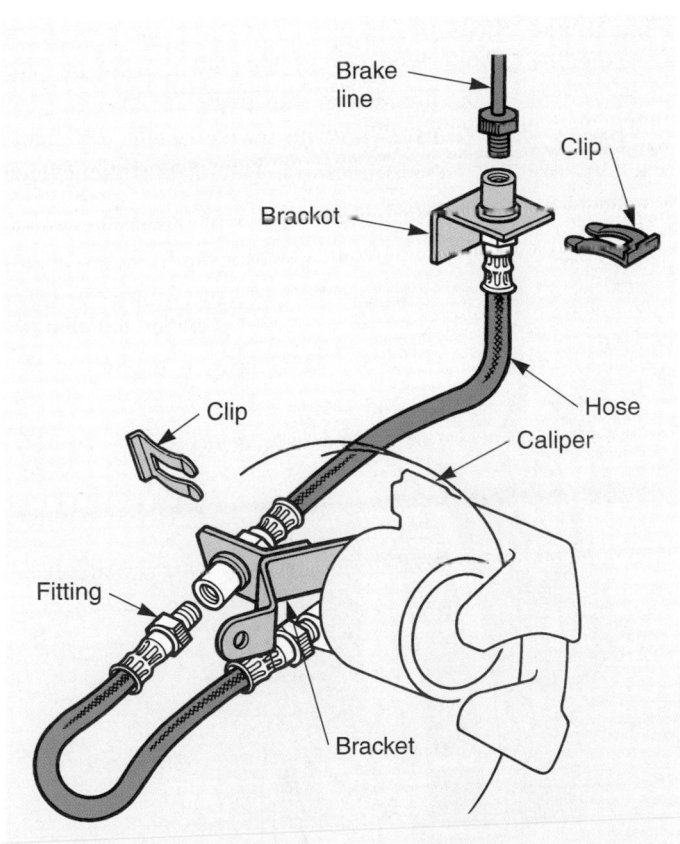

Figure 71-19. Brackets and clips are used to secure the brake hoses and lines to the frame or body. Lines must not be allowed to vibrate or they can fatigue and break. (Toyota)

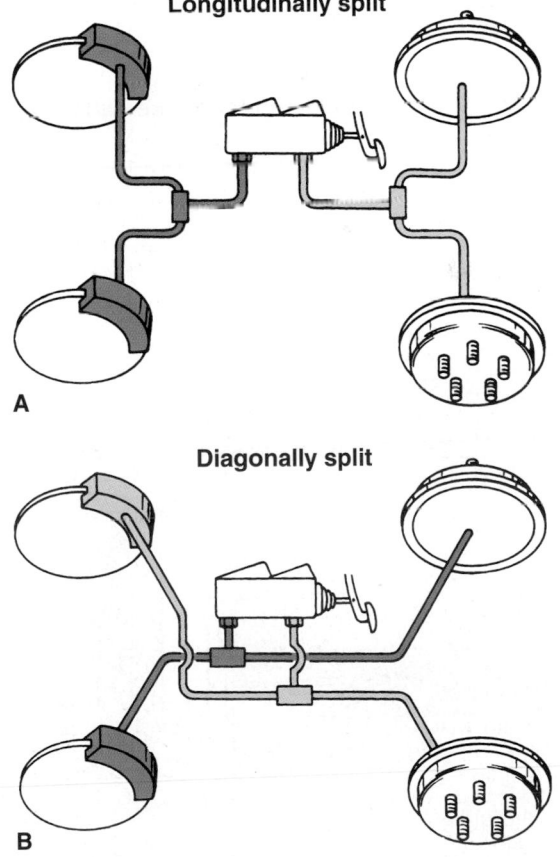

Figure 71-20. Study how each master cylinder piston operates different wheel brake assemblies. (EIS)

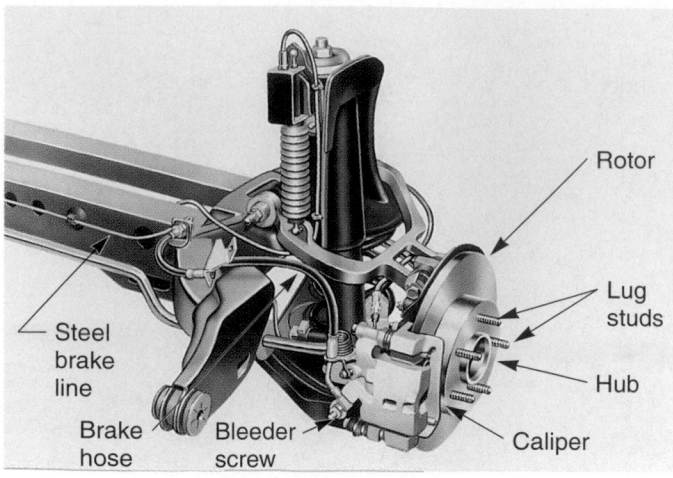

Figure 71-21. Note the parts of a disc brake unit. (Cadillac)

Disc Brake Assemblies

Disc brakes are basically like the brakes on a ten-speed bicycle. The friction elements are shaped like rectangular pucks and are squeezed inward to clamp against a rotating disc, or wheel.

A *disc brake assembly* consists of a caliper, brake pads, a rotor, and related hardware (bolts, clips, springs). See **Figure 71-21. Figure 71-22** shows sectioned views of typical disc brake assemblies. Note how the caliper pistons move inward to clamp the brake pads against the rotor.

Brake Caliper

The *brake caliper* assembly includes the caliper housing, piston, piston seal, dust boot, brake pads, special hardware (clips, springs), and a bleeder screw. These parts are pictured in **Figure 71-23.**

When the brake pedal is applied, brake fluid flows into the caliper cylinder. The piston is then pushed outward by fluid pressure, forcing the brake pads into the rotor. See **Figure 71-24.**

The *piston seal* in the caliper prevents pressure leakage between the piston and the cylinder. The piston seal also helps pull the piston back into the cylinder when the brakes are *not* applied. The elastic action of the seal acts as a spring to retract the piston.

The *piston boot* keeps road dirt and water off the caliper piston and the wall of the cylinder. The boot and seal usually fit into grooves cut in the caliper cylinder and piston, **Figure 71-24.**

A *bleeder screw* allows air to be removed from the hydraulic brake system. It is threaded into the side or top of the caliper housing. When the screw is loosened, hydraulic pressure can be used to force fluid and air out of the system, **Figure 71-23.**

Disc Brake Pads

Disc brake pads are steel plates to which linings are riveted. Brake pads are shown in **Figure 71-23.**

Brake pad linings are normally made of heat-resistant organic or semimetallic (metal-particle filled)

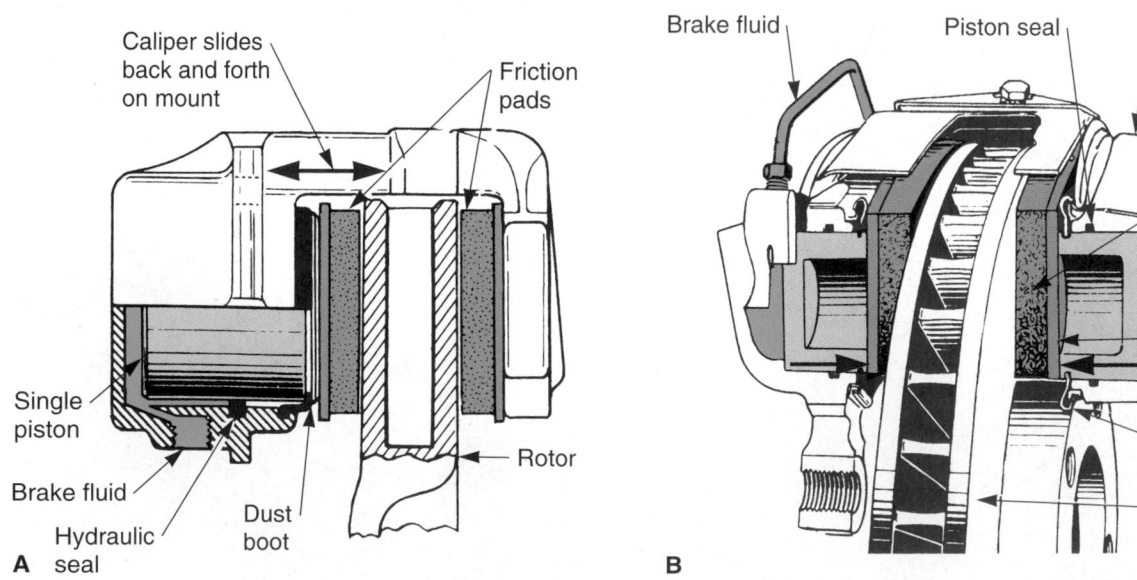

Figure 71-22. The caliper piston pushes the brake pad into a revolving disc to slow or stop the vehicle. A—The single-piston caliper is very common. The caliper floats, or slides, on its mount so both pads contact the disc. B—A fixed caliper uses pistons on both sides of disc. The caliper remains stationary on its mounting.

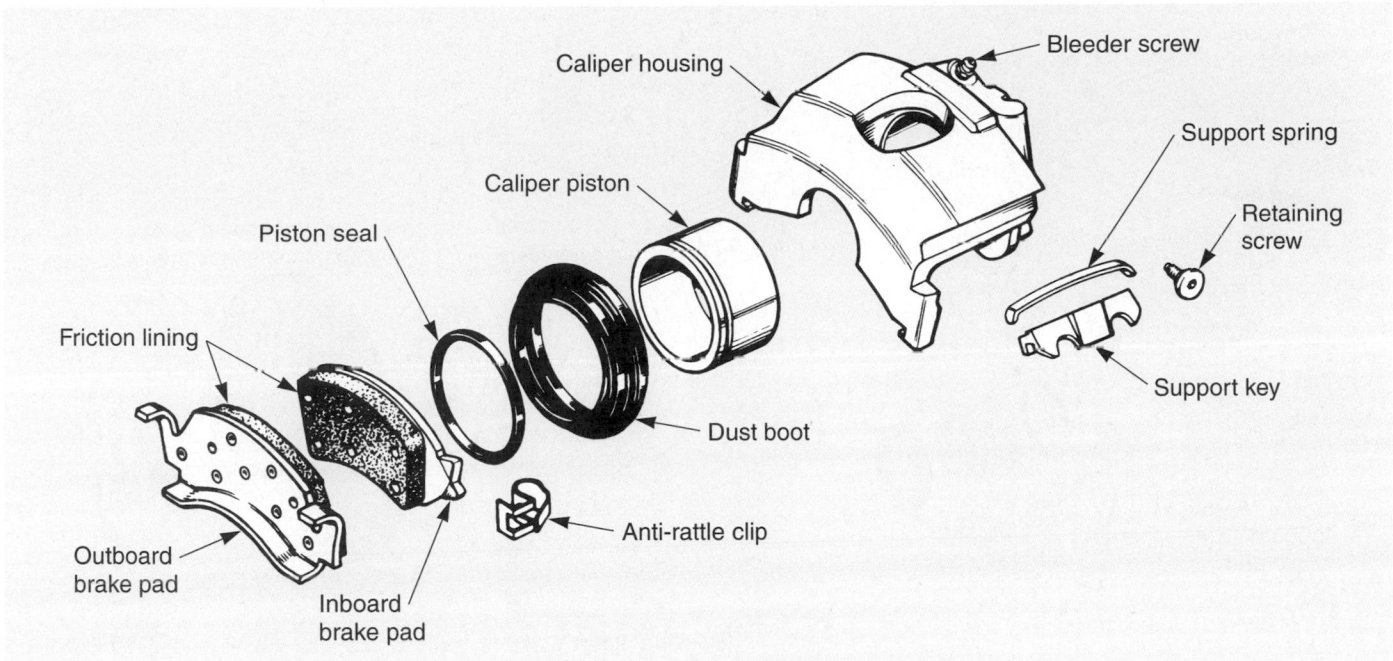

Figure 71-23. Disassembled view of a caliper. Note how the parts fit together. (Chrysler)

friction material. Older pad linings were made of asbestos, which has been phased out because of its cancer-causing properties. Many new vehicles, especially those with front-wheel drive, use semimetallic linings on the front. Semimetallic linings can withstand higher operating temperatures without losing their frictional properties.

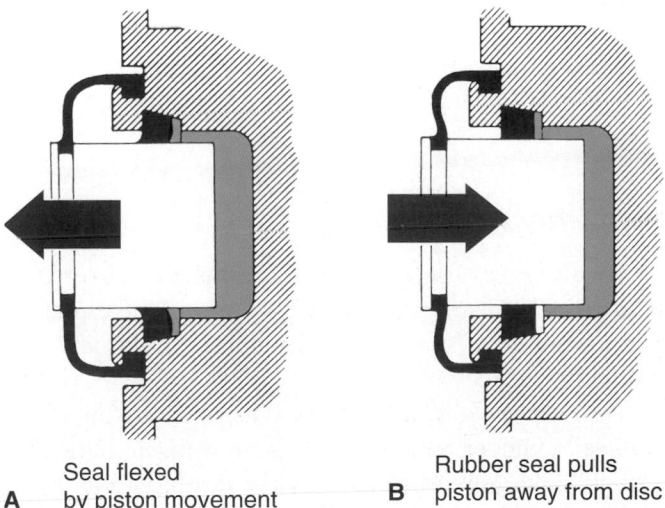

A   B
Seal flexed   Rubber seal pulls
by piston movement   piston away from disc

Figure 71-24. Note the operation of the caliper piston. A—The brakes are applied and the piston is pushed partially out of the cylinder. B—The stretched piston seal pulls the piston back after brake release. This keeps the pads from rubbing on the disc. (EIS)

Anti-rattle clips are frequently used to keep the brake pads from vibrating and rattling. The clip snaps onto the brake pad to produce a force fit in the caliper. Sometimes, an anti-rattle spring is used instead of a clip, **Figure 71-25A.**

A *pad wear sensor* is a metal tab on the brake pad that informs driver of worn brake pad linings. The wear sensor tab will emit a loud squeal or squeak when it scrapes against the brake disc. The sensor only touches the disc when the brake lining has worn too thin.

Brake Disc

The *brake disc,* also called *brake rotor,* uses friction from the brake pads to slow or stop wheel rotation. The brake disc is normally made of cast iron. It may be an integral part of the wheel hub. However, in many front-wheel-drive vehicles, the disc and the hub are separate units. Refer to **Figures 71-21** and **71-25A.**

The brake disc may be solid or ventilated. The ventilated rib disc is hollow, which allows cooling air to circulate inside the disc.

Disc Brake Types

Disc brakes can be classified as floating caliper brakes, sliding caliper brakes, and fixed caliper brakes. Floating and sliding calipers are common. Fixed calipers were used on a few older passenger cars.

The *floating caliper* disc brake is mounted on two bolts supported by rubber bushings. This one-piston caliper is free to shift, or float, in the rubber bushings.

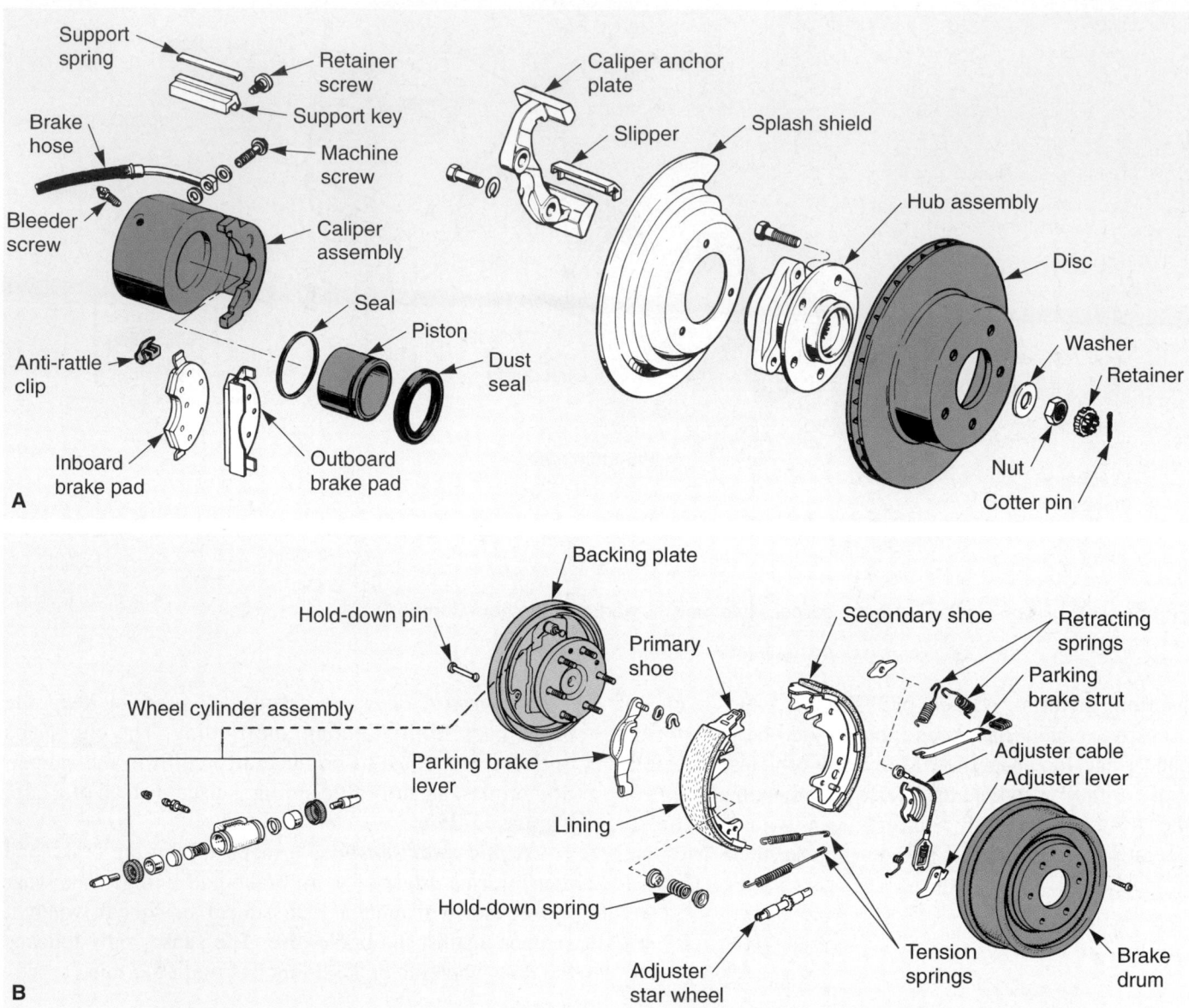

Figure 71-25. Compare complete disc and drum brake assemblies. A—Disc brake assembly. B—Drum brake assembly. (Chrysler and Toyota)

The *sliding caliper* disc brake is a one-piston caliper that is mounted in slots machined in the caliper adapter. The caliper is free to slide sideways in the slots or grooves as the linings wear. Look at **Figure 71-25A.**

The *fixed caliper* disc brake normally uses more than one piston and caliper cylinder. The caliper is bolted directly to the steering knuckle. It is *not* free to move in relation to the disc. Pistons on both sides of the disc push against the brake pads, **Figure 71-22B.**

Floating and sliding calipers are used to avoid vibration problems. With a fixed caliper, severe vibrations can occur with a slight *runout* (wobble) of the disc.

Drum Brake Assemblies

Drum brakes use many of the principles covered in the section on disc brakes. However, drum brakes have a large drum that surrounds the brake shoes and the hydraulic wheel cylinder.

A *drum brake assembly* consists of a backing plate, a wheel cylinder, brake shoes and linings, retracting springs, hold-down springs, a brake drum, and an automatic adjusting mechanism, **Figure 71-25B.**

Backing Plate

The *brake backing plate* holds the shoes, springs, wheel cylinder, and other parts inside the brake drum. It also helps keep road dirt and water off the brakes. The backing plate bolts to the axle housing or the spindle support, **Figure 71-25B.**

Wheel Cylinder Assembly

The *wheel cylinder assemblies* use master cylinder pressure to force the brake shoes out against the brake drums. The cylinders bolt to the top of the backing plates.

A *wheel cylinder* consists of a cylinder or housing, an expander spring, rubber cups, pistons, dust boots, and a bleeder screw. See **Figure 71-26.**

The *wheel cylinder housing* forms the enclosure for the other parts of the assembly. It has a precision hole, or cylinder, in it for the pistons, the cups, and the spring.

The *wheel cylinder boots* keep road dirt and water out of the cylinder. They snap into grooves on the outside of the housing, **Figure 71-26.**

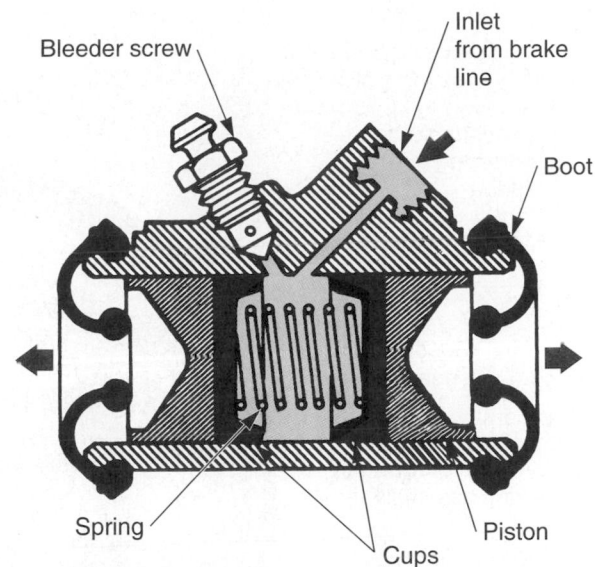

Figure 71-27. This cutaway view of a drum brake wheel cylinder shows the fluid passages to the cylinder and the bleeder screw. Pressure inside the cylinder pushes the cups and pistons outward to force the linings into the drum. (EIS)

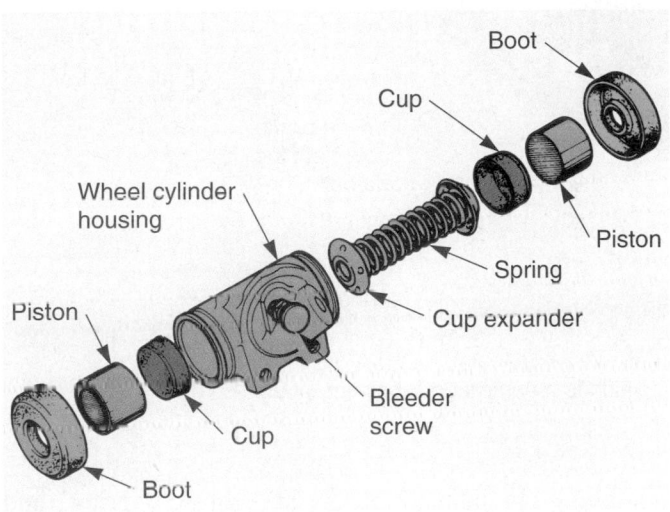

Figure 71-26. Disassembled view shows a wheel cylinder for a drum brake. The cups prevent fluid leakage out of the cylinder. The boots keep debris out of the cylinder. (Renault)

The *wheel cylinder pistons* are metal or plastic plungers that transfer force out of the wheel cylinder assembly. They act on push rods connected to the brake shoes or directly on the shoes.

The *wheel cylinder cups* are rubber seals that keep fluid from leaking past the pistons. They fit in the cylinder and against the pistons, as in **Figure 71-27.**

The *wheel cylinder spring* helps hold the rubber cups against the pistons when the wheel cylinder assembly is *not* pressurized. Sometimes, the ends of this spring have metal expanders. Called *cup expanders,* they help press the outer edges of the cups against the wall of the wheel cylinder, **Figure 71-26.**

The *bleeder screw* provides a means of removing air from the brake system. It threads into a hole in the back of the wheel cylinder. When the screw is loosened, hydraulic pressure can be used to force air and fluid out of the system. Refer to **Figure 71-27.**

Drum Brake Shoes

The *drum brake shoe assemblies,* or brake shoes, rub against the revolving brake drum to produce braking action. Drum brake shoe assemblies are made by fastening organic friction material onto a *metal shoe,* **Figure 71-25B.** The lining serves as a heat-resistant surface that contacts the brake drum.

The metal shoe supports and holds the soft lining material. The lining may be held on the shoe by either rivets or a bonding agent (glue).

The *primary brake shoe* is the front shoe, **Figure 71-28.** It normally has a slightly *shorter lining* than the secondary shoe.

The *secondary brake shoe* is the rear shoe, **Figure 71-28.** It has the *largest lining* surface area.

Retracting Springs and Hold-Down Springs

Retracting springs pull the brake shoes away from the brake drums when the brake pedal is released. This, in turn, pushes the wheel cylinder pistons inward. Usually, the retracting springs fit in holes in the shoes and around an anchor pin at the top of the backing plate. Refer to **Figure 71-28.**

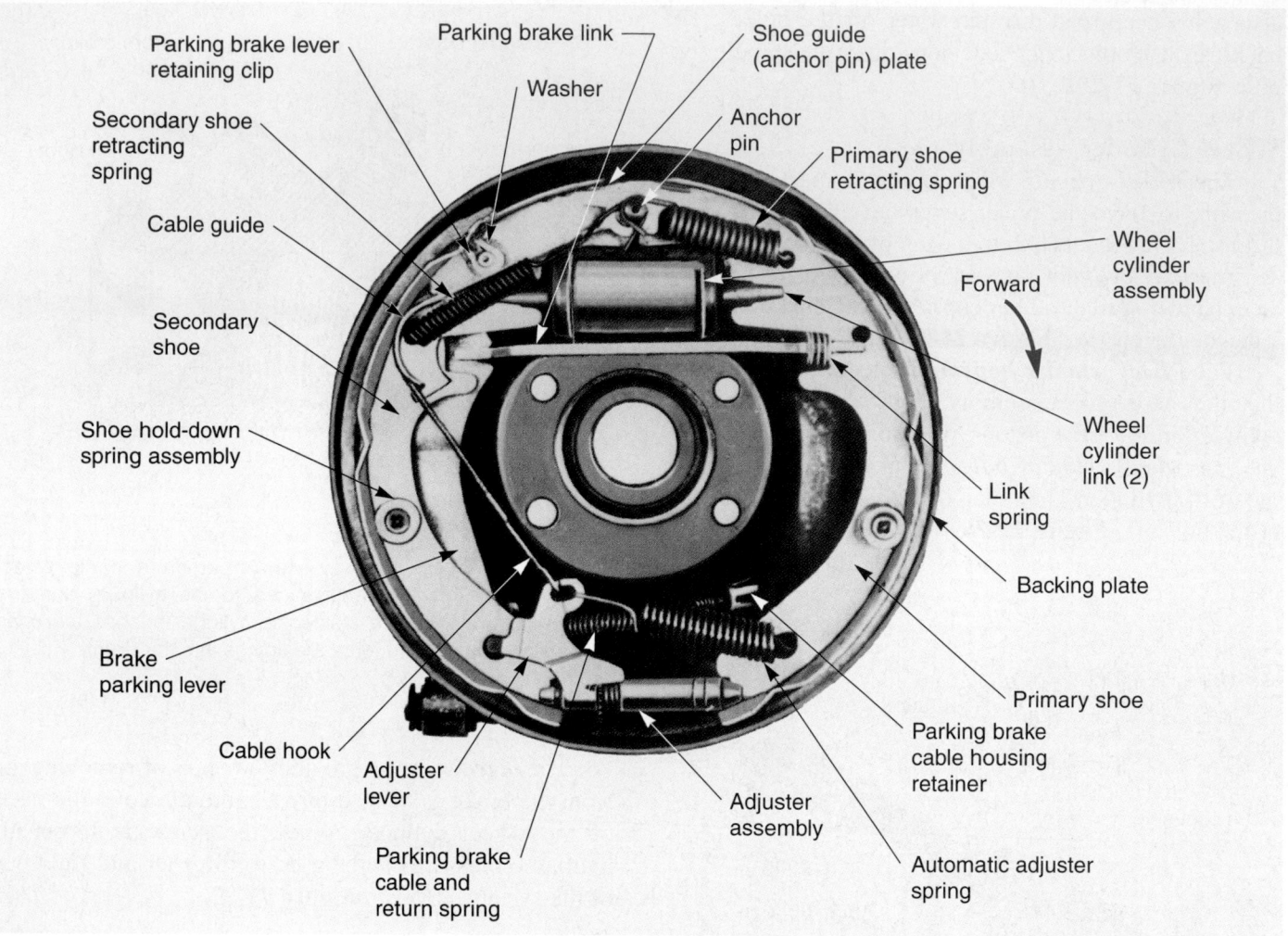

Figure 71-28. Study the parts of this common drum brake assembly, which is mounted on a backing plate. (Ford)

Hold-down springs hold the brake shoes against the backing plate when the brakes are in the released position. A hold-down pin fits through the back of the backing plate, **Figure 71-28.** A metal cup locks onto these pins to secure the hold-down springs to the shoes. Some manufacturers use spring clips instead of hold-down springs and locking cups. Others use a spring that is simply hooked to the shoe and the backing plate.

Springs are used on the automatic adjusting mechanism. Brake springs are made from high-quality steel that is capable of withstanding the high temperatures encountered inside the brake drum, **Figure 71-29.**

Brake Shoe Adjusters

Brake shoe adjusters maintain the correct drum-to-lining clearance as the brake linings wear. Look at **Figure 71-29.**

Many vehicles use a star wheel–type (screw-type) brake shoe adjusting mechanism. This type of adjusting mechanism includes a *star wheel* (adjusting screw

assembly), an adjuster lever, an adjuster spring, and either an adjuster cable, lever arm, or link (rods). See **Figure 71-29A** through **D.**

Automatic brake shoe adjusters normally function when the brakes are applied with the vehicle moving in reverse. If there is too much lining clearance, the brake shoes move outward and rotate with the drum enough to operate the adjusting lever. This lengthens the star wheel assembly. The linings are moved closer to the brake drum, maintaining the correct lining-to-drum clearance.

Figure 71-29 E and **71-29F** picture brake assemblies that use modern latch-type adjusters.

Brake Drums

Brake drums provide a rubbing surface for the brake shoe linings. The drum usually fits over the wheel lug studs. A large hole in the middle of the drum centers the drum on the front hub or the rear axle flange. The wheel and drum turn as a unit. See **Figure 71-30.**

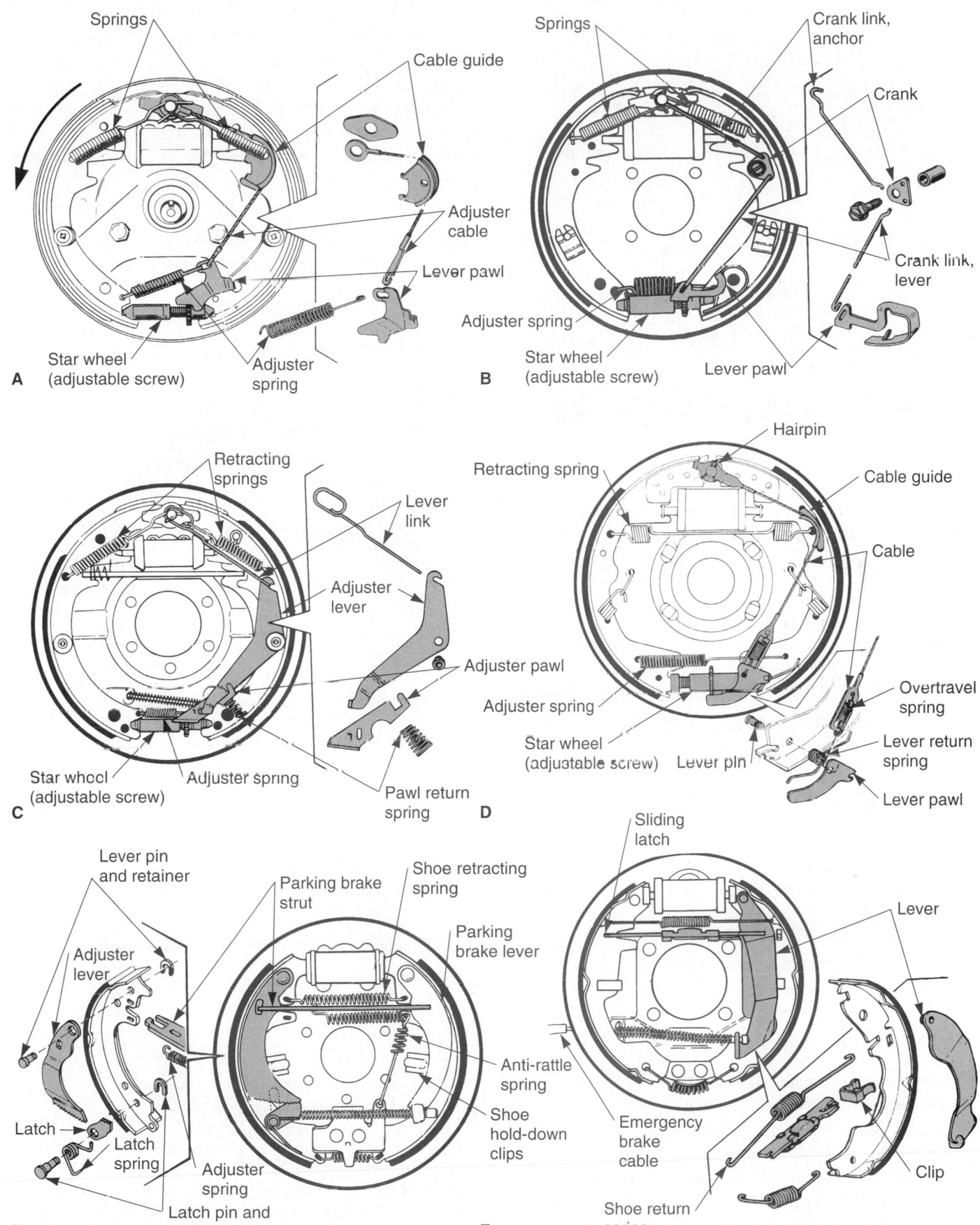

Figure 71-29. These are the most common brake shoe adjuster mechanisms. A—Cable-type star wheel adjuster. B—Link-type star wheel adjuster. C—Lever-type star wheel adjuster. D—Cable-type star wheel adjuster with an overtravel spring. E—Lever-latch adjuster. F—Sliding-latch adjuster. (FMC)

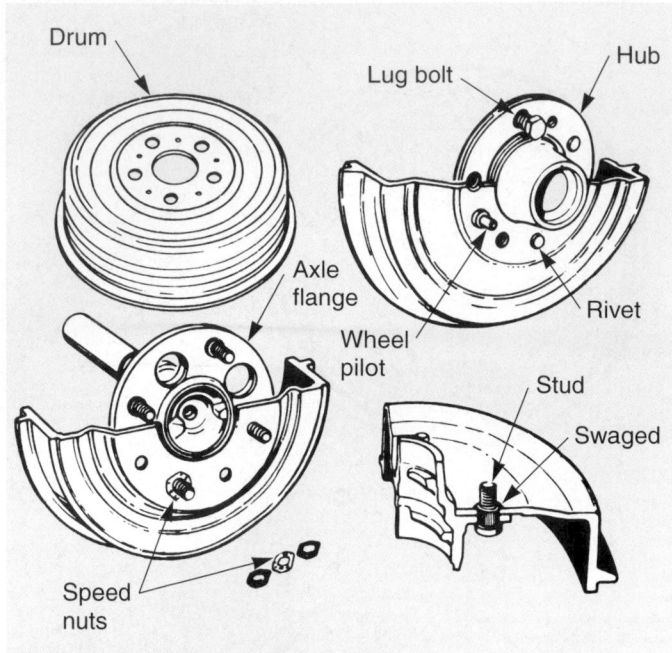

Figure 71-30. The brake drum provides a friction surface for the brake shoe linings. Study the construction of the drum. (Ford)

Brake Shoe Energization

When the brake shoes are forced against the rotating drum, they are pulled away from their pivot point by friction. This movement, called *self-energizing action,* draws the shoes tighter against the drum, **Figure 71-31A.**

With most drum brake designs, shoe energization is supplemented by servo action. *Servo action* results when the primary (front) shoe helps apply the secondary (rear) shoe. Look at **Figure 71-31B.**

The backing plate anchor pin holds the secondary shoe during brake application. However, the primary shoe is free to float out and push against the end of the secondary shoe through the star wheel assembly. This action presses the secondary shoe into the drum with extra force.

Less wheel cylinder hydraulic pressure is needed to apply the brakes because of servo action.

Brake System Switches

There are three types of switches commonly used in a brake system: the stoplight switch, brake warning light switch, and low-fluid warning light switch.

Stoplight Switch

The *stoplight switch* is a spring-loaded electrical switch that operates the rear brake lights of the vehicle. Most modern cars use a mechanical switch on the brake

pedal mechanism. The switch is normally open. When the brake pedal is pressed, it closes the switch and turns on the brake lights.

Hydraulically operated stoplight switches are used on some older vehicles. Brake system pressure pushed on a switch diaphragm and closed the switch that operated the brake lights.

 Note
Brake light circuits are covered in Chapter 37, *Lights, Instrumentation, Wipers, and Horns—Operation and Service.*

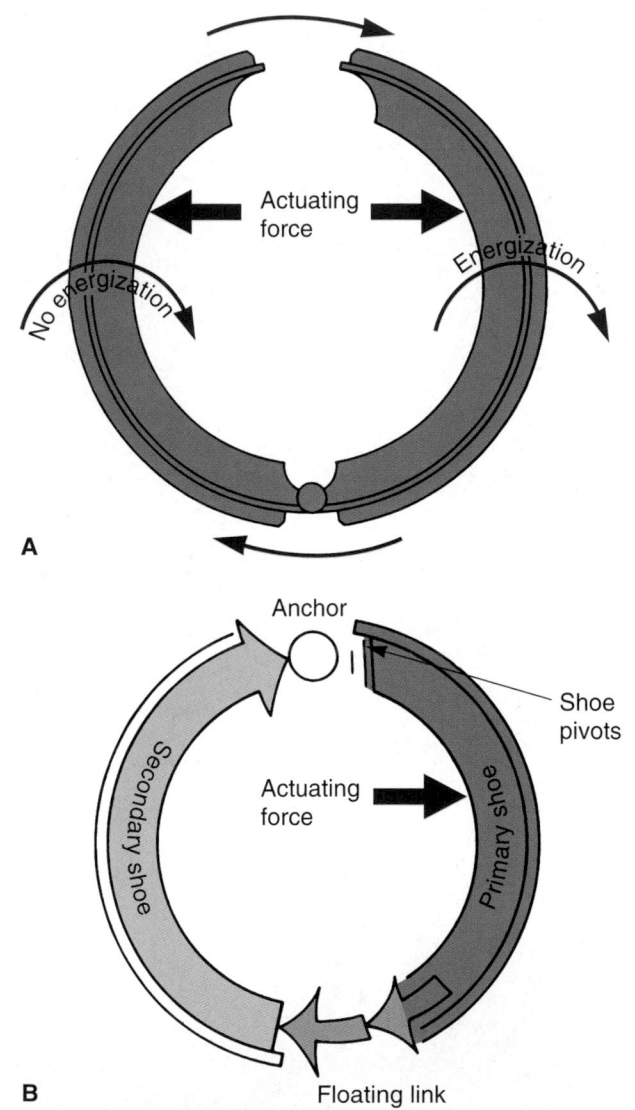

Figure 71-31. A—Self-energizing brakes use friction to force one brake shoe tighter against drum. B—Servo action results when both shoes are free to swing into the drum and the primary shoe helps apply the secondary shoe. (EIS)

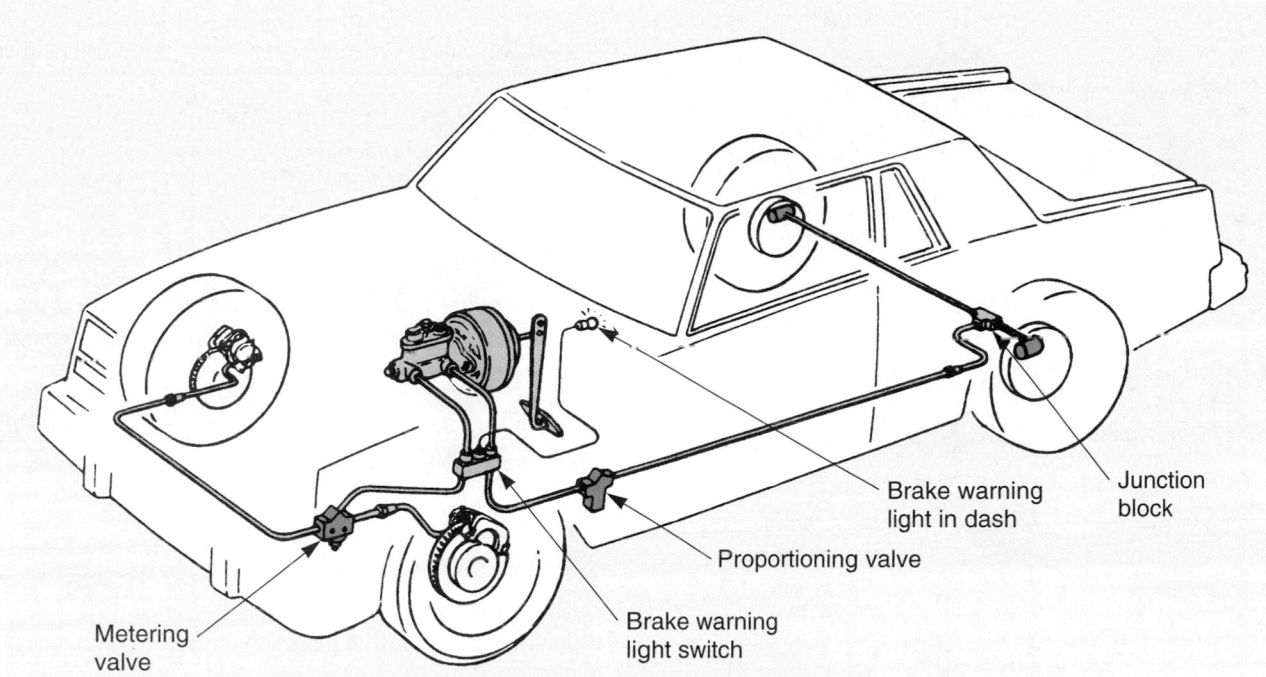

Figure 71-32. Note the location of the various valves used in a brake system. (EIS)

Brake Warning Light Switch (Pressure Differential Valve)

The *brake warning light switch,* also called the *pressure differential valve,* warns the driver of a pressure loss on one side of a dual brake system. Look at **Figure 71-32.**

If a leak develops in either the primary or secondary brake system, unequal pressure acts on the warning light switch piston. This pushes the piston to one side, grounding the indicator circuit, **Figure 71-33.**

Low-Fluid Warning Light Switch

The *low-fluid warning light switch* turns on a dash light if the brake fluid in the master cylinder becomes low. It is a small switch that often mounts in the master cylinder lid or cover.

A float sometimes operates the switch. If the level of brake fluid in the reservoir drops below a critical point, the float moves down and turns the switch on. This illuminates a warning light, alerting the driver to a brake system problem. If the fluid level becomes too low, air will be pumped into the system, possibly causing a loss of braking action.

Brake System Control Valves

Many brake systems use control valves to regulate the pressure going to each wheel cylinder. The three types of valves are the metering valve, the proportioning

valve, and the combination valve. **Figure 71-32** shows the general locations of these valves.

Metering Valve

A *metering valve* is designed to equalize braking action at each wheel during light brake applications. A metering valve is used on vehicles with front-wheel disc brakes and rear-wheel drum brakes. The metering valve is located in the line to the disc brakes, **Figure 71-32.**

The metering valve functions by preventing the front disc brakes from applying until approximately 75-135 psi (517-930 kPa) has built up in the system. This overcomes the rear drum brake return springs.

Proportioning Valve

A *proportioning valve* is also used to equalize braking action in systems with front disc brakes and rear drum brakes. It is commonly located in the brake line to the rear drum brakes. Look at **Figure 71-32.**

The function of the proportioning valve is to limit pressure at the rear drum brakes when high pressure is needed to apply the front disc brakes. Thus, the proportioning valve prevents rear-wheel lockup during heavy brake applications.

Combination Valve

A *combination valve* is a single unit that functions as a brake warning light switch, a metering valve, and/or a proportioning valve. Many late-model vehicles use a

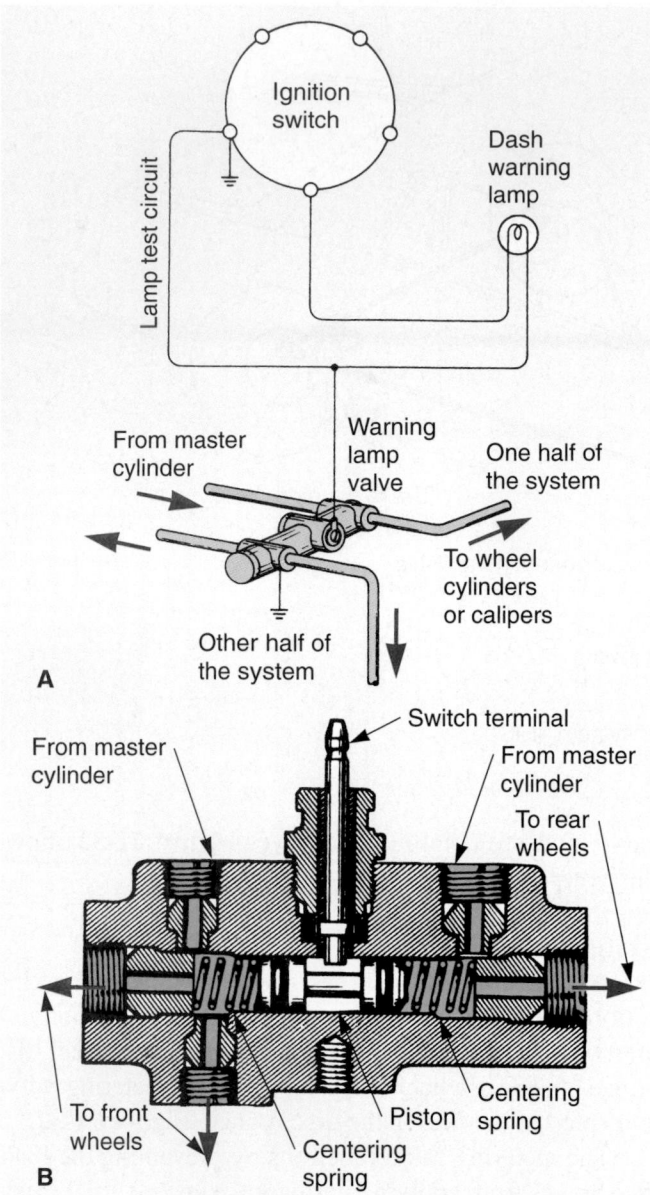

Figure 71-33. The brake warning light switch is activated by a difference in the pressures in the primary and secondary systems. A pressure difference pushes the small piston in the valve to close the warning lamp circuit. A—Brake warning lamp circuit. B—Cutaway view of a brake warning lamp switch. (Bendix)

combination valve. **Figure 71-34** is a cutaway view of a three-function combination valve. Study the three sections.

With some master cylinders, the proportioning and warning lamp valves are mounted inside the master cylinder housing. This design uses the same operating principles, **Figure 71-35.**

Parking Brakes

Parking brakes, also called *emergency brakes,* provide a mechanical means (cable and levers) of applying the brakes. One is pictured in **Figure 71-36.**

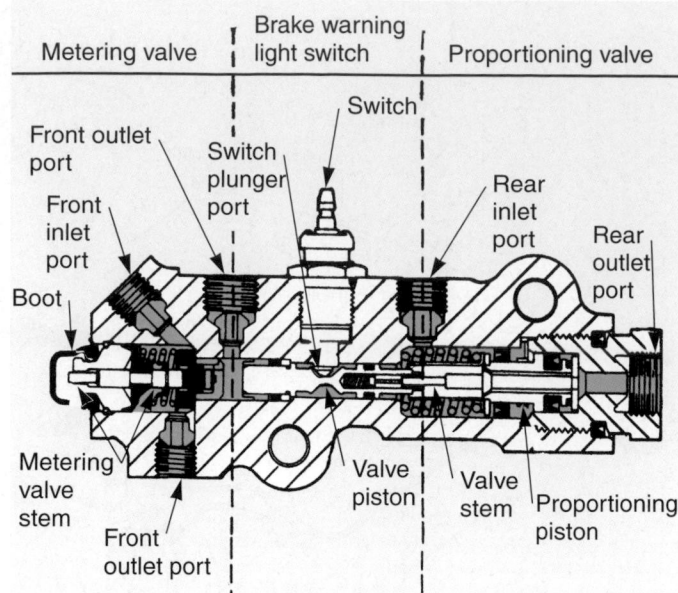

Figure 71-34. Study the sections of this combination valve. (Chrysler)

When the parking brake hand or foot lever is activated, it pulls a steel cable that runs through a housing. The movement of the cable pulls on a lever inside the drum or disc brake assembly. The lever action forces the brake linings against the rear drums or discs to resist vehicle movement.

Figure 71-37 shows a foot-operated parking brake unit.

When disc brakes are used on the rear of a vehicle, a thrust screw and a lever can be added to the brake caliper. Then, when the parking brake is applied, the cable pulls on the caliper lever. The caliper lever turns the large thrust screw, which pushes on the caliper piston and applies the brake pads to the disc. See **Figure 71-38.**

Summary

- Automotive brakes provide a means of using friction to either slow down, stop, or hold the wheels of the vehicle.
- Brake pedal assembly (foot lever for operating master cylinder and power booster).
- Master cylinder (hydraulic piston type pump that develops pressure for brake system).
- Brake booster (vacuum or power steering operated device for assisting brake pedal application).
- Brake lines (metal tubing and rubber hose for transmitting pressure to wheel brake assemblies).
- Wheel brake assemblies (devices that use system pressure to produce friction for slowing or stopping wheel rotation).

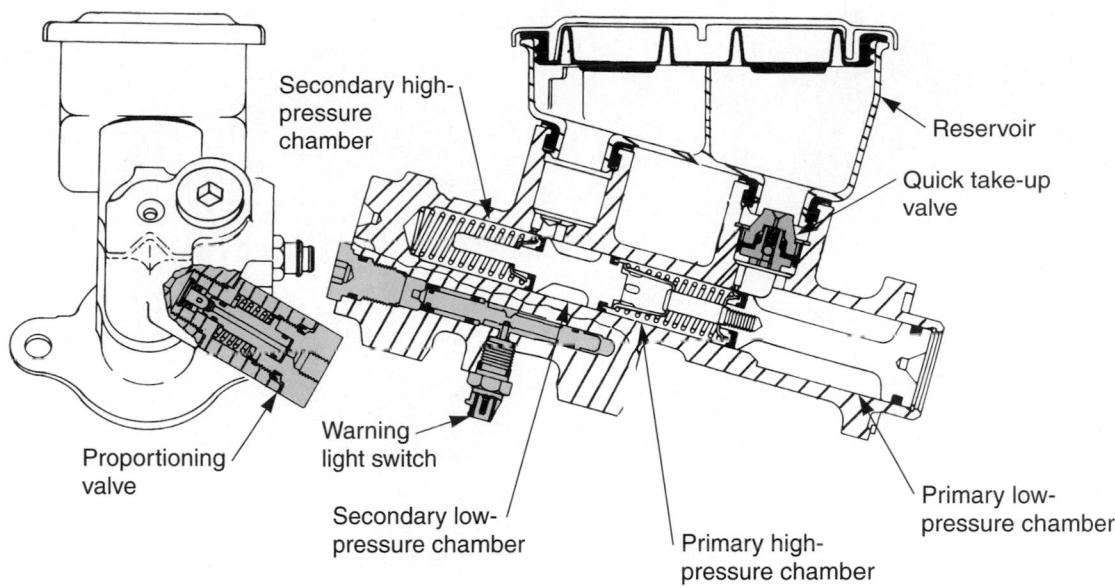

Figure 71-35. This master cylinder has a proportioning valve and a warning light switch mounted internally. (Delco)

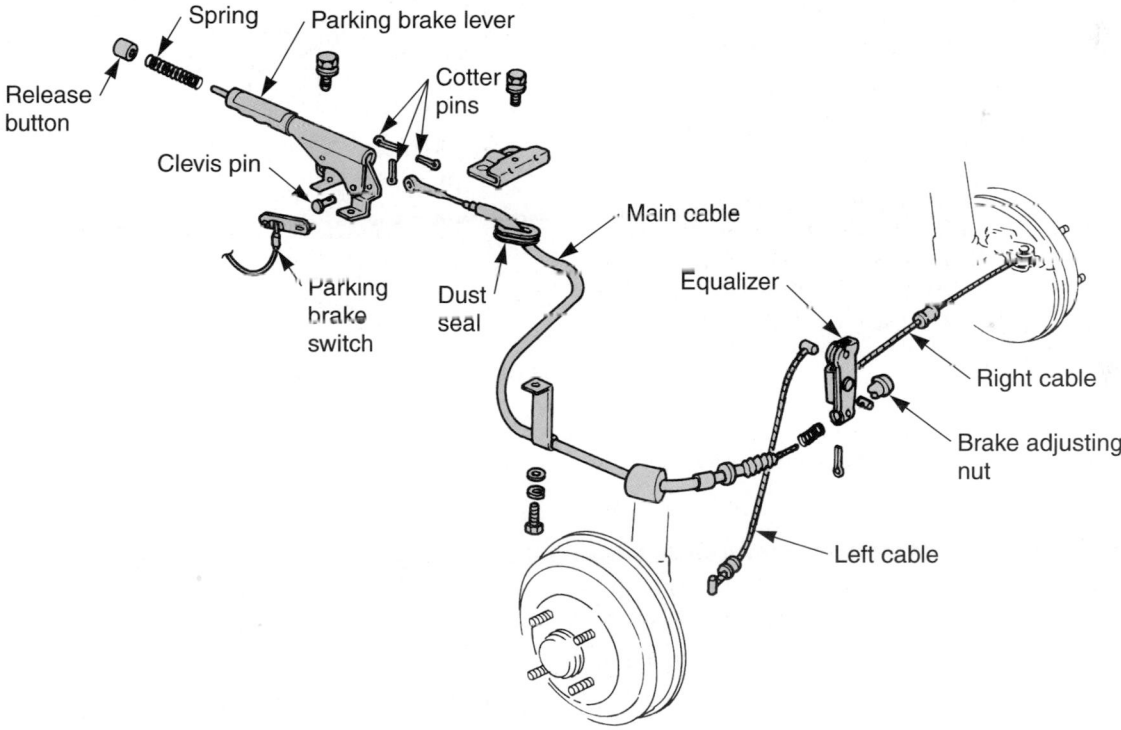

Figure 71-36. Study the parts of this typical parking brake mechanism. (Toyota)

- A hydraulic system is basically a system that uses a liquid to transmit motion or pressure from one point to another.
- The cup and piston in the master cylinder are used to pressurize the brake system.

- The dual master cylinder, also called a tandem master cylinder, has two separate hydraulic pistons and two fluid reservoirs.
- Power brakes use engine vacuum, a vacuum pump, or power steering pump pressure to assist brake pedal application.

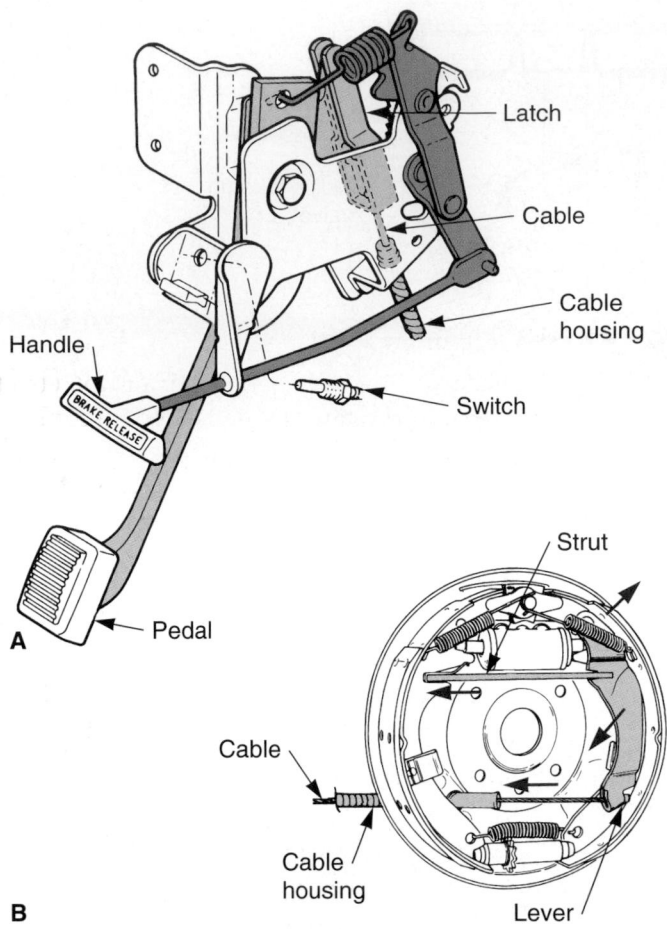

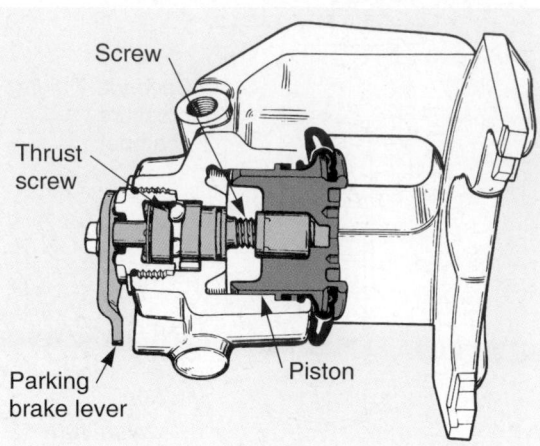

Figure 71-38. Cross-sectional view of a caliper that has a parking brake mechanism. A cable rotates the parking brake lever. The lever turns a screw that pushes the piston outward to apply the brake. (Bendix)

Figure 71-37. Parts of a parking brake system. A—Foot-operated parking brake pedal. Note the latch and the release handle. B—The cable activates the lever in the brake assembly that pushes the shoes out against the drum. (Chrysler and Ford)

- Braking ratio refers to the comparison of front wheel to rear wheel braking effort.
- A metering valve is designed to equalize braking action at each wheel during light brake applications.
- A proportioning valve is also used to equalize braking action with front disc and rear drum brakes.
- A combination valve serves as two or three valves in one.

- Brake fluid is a specially blended hydraulic fluid that transfers pressure to the wheel cylinders.
- Brake lines and hoses transfer fluid pressure from the master cylinder to the wheel cylinders.
- The piston seal in the caliper prevents pressure leakage between the piston and cylinder.
- The floating caliper disc brake is mounted on two bolts supported by rubber bushings.
- The bleeder screw provides a means of removing air from the brake system.
- The primary brake shoe is the front shoe. It normally has a slightly shorter lining than the secondary shoe.
- The secondary brake shoe is the rear shoe. It has the largest lining surface area.
- Emergency brakes, also called parking brakes, provide a mechanical means (cable and levers) of applying the brakes.

Important Terms

Automotive brakes	Secondary piston
Inertia	Power brakes
Kinetic	Vacuum booster
Braking ratio	Atmospheric suspended
Hydraulic system	brake booster
Brake pedal assembly	Vacuum suspended
Master cylinder	brake booster
Cup	Hydraulic booster
Piston	Brake fluid
Intake port	Brake lines
Compensating port	Hoses
Residual pressure	Longitudinally split
valves	Diagonally split
Rubber boot	Disc brake assembly
Master cylinder	Brake caliper
reservoir	Piston seal
Dual master cylinder	Piston boot
Primary piston	Bleeder screw

Disc brake pads
Brake pad linings
Anti-rattle clips
Pad wear sensor
Brake disc
Brake rotor
Floating caliper
Sliding caliper
Fixed caliper
Runout
Drum brake assembly
Brake backing plate
Wheel cylinder
assemblies
Wheel cylinder boots
Wheel cylinder pistons
Wheel cylinder cups
Wheel cylinder spring
Cup expanders
Bleeder screw
Drum brake shoe
assemblies

Primary brake shoe
Secondary brake shoe
Retracting springs
Hold-down springs
Brake shoe adjusters
Star wheel
Brake drums
Self-energizing action
Servo action
Stoplight switch
Brake warning light
switch
Pressure differential
valve
Low-fluid warning light
switch
Metering valve
Proportioning valve
Combination valve
Parking brakes
Emergency brakes

Review Questions—Chapter 71

Please do not write in this text. Place your answers on a separate sheet of paper.

1. List and explain the six major parts of a brake system.

2. Describe the four major parts of a disc brake assembly.

3. Which of the following is *not* part of a drum brake assembly?
   (A) Wheel cylinder.
   (B) Booster.
   (C) Shoes.
   (D) Drum.

4. A(n) _____ system uses a liquid to transmit motion or pressure from one part to another.

5. What are four functions of a master cylinder?

6. Why is a dual master cylinder used?

7. A power brake _____ uses pedal pressure and engine vacuum to assist brake application.

8. A hydro-boost power brake system uses pressure from the power steering pump. True or False?

9. In a diagonally split brake system, each master cylinder cup and piston assembly operates a brake assembly on opposite sides and corners of the vehicle. True or False?

10. Which of the following is *not* part of a brake caliper?
    (A) Piston seal.
    (B) Bleeder screw.
    (C) Piston boot.
    (D) All of the above.

11. What causes a brake caliper piston to retract away from the rotor after brake application?

12. Why are floating and sliding calipers more common than fixed calipers?

13. The _____ _____ provides a means of removing air from the wheel cylinder after repairs.

14. Explain the difference between a primary brake shoe and a secondary brake shoe.

15. The brake _____ _____ _____, also called the _____ _____ _____, warns the driver of a pressure loss on one side of a dual brake system.

16. Why is a metering valve used?

17. How does a proportioning valve equalize braking action?

18. Describe a combination valve.

19. How does the parking brake work?

20. A certain automotive brake system's residual pressure valve maintains residual fluid pressure of 9 psi. What is this approximate pressure in kilopascals?
    (A) 62 kPa.
    (B) 32 kPa.
    (C) 83 kPa.
    (D) 56 kPa.

⬤ ASE-Type Questions

1. Technician A says an automotive master cylinder is a device that uses brake system pressure to produce friction for stopping wheel rotation. Technician B says an automotive master cylinder holds the wheel cylinder piston and brake pads. Who is right?
   (A) A only.
   (B) B only.
   (C) Both A and B.
   (D) Neither A nor B.

2. Technician A says an automotive brake booster is a piston-type pump that develops pressure for the vehicle's brake system. Technician B says an automotive brake booster is a vacuum or power steering operated device that assists brake pedal application. Who is right?
   (A) A only.
   (B) B only.
   (C) Both A and B.
   (D) Neither A nor B.

3. Technician A says certain automobiles are equipped with disc brakes on all four wheels. Technician B says disc brakes are most commonly used on an automobile's rear wheels. Who is right?
   (A) A only.
   (B) B only.
   (C) Both A and B.
   (D) Neither A nor B.

4. Which of the following is not a basic component of an automotive disc brake assembly?
   (A) Rotor.
   (B) Brake drum.
   (C) Brake pads.
   (D) Caliper.

5. Technician A says a caliper cylinder holds the wheel cylinder piston on a drum brake assembly. Technician B says a caliper cylinder holds the wheel cylinder piston on a disc brake assembly. Who is right?
   (A) A only.
   (B) B only.
   (C) Both A and B.
   (D) Neither A nor B.

6. Technician A says a hydraulic system uses air to transmit motion or pressure from one point to another. Technician B says a hydraulic system uses a liquid to transmit motion or pressure from one point to another. Who is right?
   (A) A only.
   (B) B only.
   (C) Both A and B.
   (D) Neither A nor B.

7. Technician A says a manual master cylinder normally bolts to the side of the engine. Technician B says a manual master cylinder normally bolts to the engine firewall. Who is right?
   (A) A only.
   (B) B only.
   (C) Both A and B.
   (D) Neither A nor B.

8. Technician A says a push rod connects the brake pedal to the master cylinder piston. Technician B says a steel plate connects the brake pedal to the master cylinder piston. Who is right?
   (A) A only.
   (B) B only.
   (C) Both A and B.
   (D) Neither A nor B.

9. Technician A says an automotive master cylinder keeps the brake system full of fluid as the brake linings wear. Technician B says an automotive master cylinder maintains a slight amount of pressure to keep contaminants from entering the brake system. Who is right?
   (A) A only.
   (B) B only.
   (C) Both A and B.
   (D) Neither A nor B.

10. Technician A says a master cylinder's cup and piston are used to release brake system pressure. Technician B says a master cylinder's cup and piston are used to pressurize the brake system. Who is right?
    (A) A only.
    (B) B only.
    (C) Both A and B.
    (D) Neither A nor B.

11. Technician A says a dual master cylinder's rear piston assembly is called the primary piston. Technician B says a dual master cylinder's rear piston assembly is called the secondary piston. Who is right?
    (A) A only.
    (B) B only.
    (C) Both A and B.
    (D) Neither A nor B.

12. Technician A says certain automotive power brake systems use engine vacuum to assist brake pedal application. Technician B says certain automotive power brake systems use a vacuum pump to assist brake pedal application. Who is right?
    (A) A only.
    (B) B only.
    (C) Both A and B.
    (D) Neither A nor B.

13. Technician A says an atmospheric suspended vacuum brake booster is used on modern automobiles. Technician B says a vacuum suspended booster is used on modern automobiles. Who is right?
    (A) A only.
    (B) B only.
    (C) Both A and B.
    (D) Neither A nor B.

14. Technician A says a "longitudinally split" brake system has each master cylinder piston operating the brake assemblies on opposite corners of the vehicle. Technician B says a "longitudinally split" brake system has one master cylinder piston operating the front wheel brake assemblies and the other piston operating the rear wheel brake assemblies. Who is right?
    (A) A only.
    (B) B only.
    (C) Both A and B.
    (D) Neither A nor B.

15. Which of the following is not part of a combination valve?
    (A) Proportioning valve.
    (B) Warning light switch.
    (C) Fluid level sensor.
    (D) Metering valve.

Activities—Chapter 71

1. Identify the type of brake system—disc or drum—on a vehicle assigned by your instructor.

2. Examine the calipers of a disc brake system and identify the type: floating or fixed.

3. Prepare an overhead transparency from **Figure 71-1** and use it to explain to your class the function of the various parts of a brake system.

4. Using a small hydraulic jack, explain and demonstrate the principle of the hydraulic brake.

Chapter 72

Brake System Diagnosis and Repair

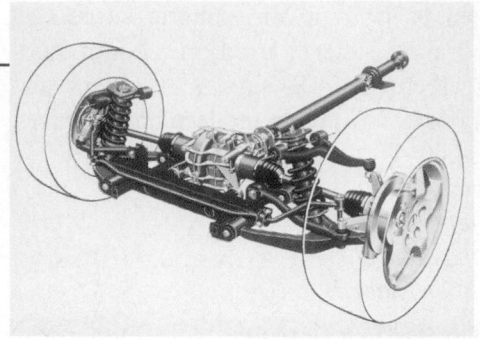

After studying this chapter, you will be able to:

- Diagnose common brake system problems.
- Inspect and maintain a brake system.
- Describe basic procedures for servicing a master cylinder and a brake booster.
- Explain how to service a disc brake assembly.
- Explain how to service a drum brake assembly.
- Describe the procedures for both manual and pressure bleeding of a brake system.
- Cite the safety rules that should be followed when servicing brake systems.
- Correctly answer ASE certification test questions on the diagnosis and repair of brake systems.

The brake system is the most important system on a vehicle from a safety standpoint. The customer trusts the technician to do every service and repair operation correctly because his or her life may depend upon it.

When working on a brake system, always keep in mind that a brake system failure could result in a deadly accident. It is up to you to make sure the vehicle's brake system is in perfect operating condition before releasing the vehicle to the customer.

This chapter will explain the most important procedures to follow when servicing automotive brakes. It will give you the fundamental information needed to use service manual directions and specifications for the exact year and make of vehicle.

 Note
The next chapter explains the operation, diagnosis, and repair of anti-lock brakes.

Brake System Problem Diagnosis

Use *symptoms* (noises, smells, abnormal brake pedal movements, improper braking action, etc.) when diagnosing

brake system problems. Listen to the customer's complaints or the service writer's explanation. Then, inspect the brake system or test drive the vehicle.

It is important that you be able to interpret the symptoms. You can then decide what adjustments or repairs are needed. Refer to **Figure 72-1.**

Almost all brake system problems can be sorted into the following basic categories: brake pedal vibration, grabbing brakes, excessive brake pedal effort, pulling brakes, spongy brake pedal, dropping brake pedal, low brake pedal, dragging brakes, no brake pedal, illuminated brake warning light, and braking noise.

On-Board Diagnostics

Many modern brake systems, primarily those with anti-lock brake systems, have self-diagnostic capabilities. If the computer detects an abnormal operating condition in the system, it will store one or more diagnostic trouble codes and illuminate a malfunction indicator light. You can then energize the self-diagnostic system or hook up a scan tool to retrieve the stored codes. These codes will indicate which circuit or component is causing the problem.

Refer to the text index for more information on on-board diagnostics. It was covered in an earlier chapter and will be discussed in the next chapter of this textbook.

Brake Vibration

Brake vibration shows up as a chatter, pulsation, or shake in the brake pedal or steering wheel. It happens only when the brake pedal is applied. The vibration may be felt mostly in the steering wheel (front brake assembly problems) or in the brake pedal itself (could be problems at any wheel). When problems are severe, the whole chassis and body of the vehicle may vibrate when braking.

Brake vibration is usually caused by a warped disc or an out-of-round brake drum. Hard spots on the disc or drum can also cause vibration.

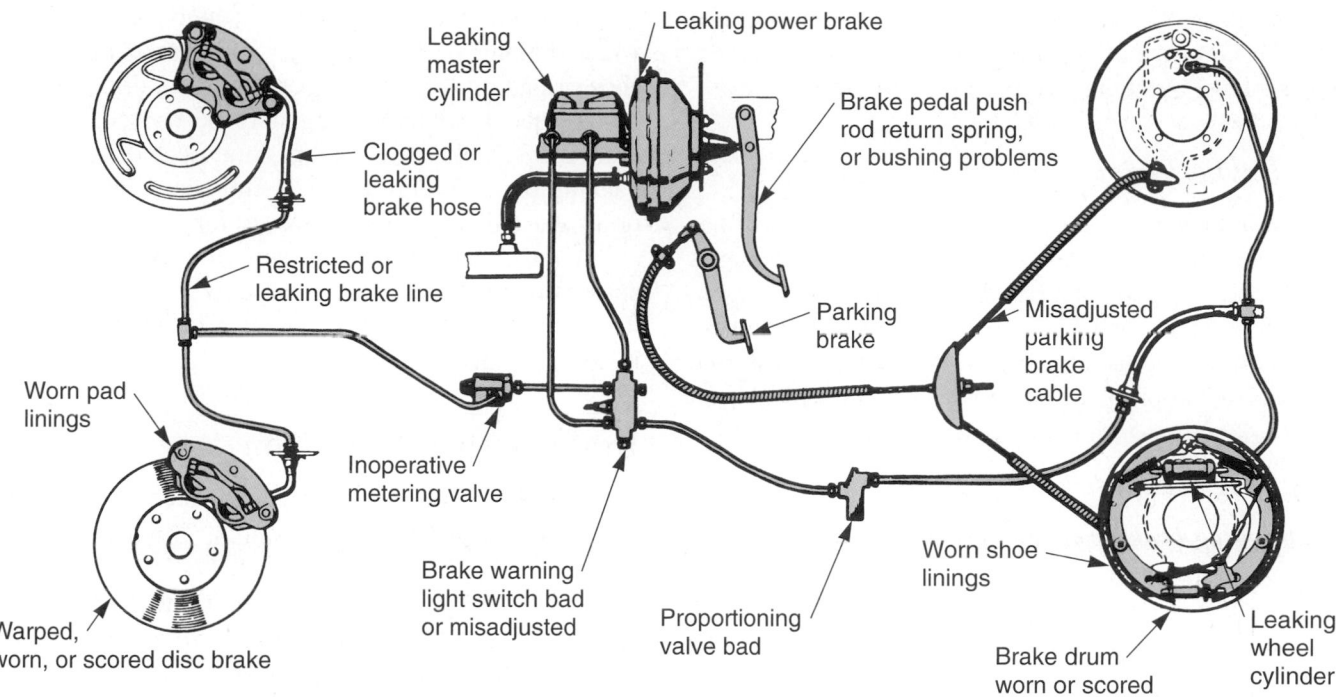

Figure 72-1. Note some of the common problems that can develop in a brake system. (Bendix)

Grabbing Brakes

Grabbing brakes apply too quickly, even with light brake pedal application. If the brakes are grabbing, it will be very difficult to stop slowly.

Grabbing is usually caused by a malfunctioning brake booster, brake fluid or grease on the linings, worn brake linings, a faulty metering valve, or a mechanical problem in the wheel brake assembly.

Excessive Brake Pedal Effort

Excessive brake pedal effort is a noticeable increase in the amount of foot pressure needed to apply the brakes. It can be caused by a variety of problems, such as a frozen wheel cylinder or caliper piston, a clogged brake hose or line, a faulty master cylinder, contaminated linings, a disconnected brake booster vacuum line, or a defective brake booster.

Pulling Brakes

Pulling brakes cause the vehicle to veer to the right or left when the brakes are applied. When the problem is in the front brakes, a strong pulling force will be noticed on the steering wheel. Rear brake pull is only noticeable during very hard braking—one of the rear brakes will lock up, causing tire skid and squeal.

Brake pull is usually caused by a frozen caliper or wheel cylinder piston, a grease- or fluid-coated lining, a leaking cylinder, a faulty automatic adjuster, or a buildup of brake lining dust. Incorrect front end alignment can also make the vehicle pull to one side during braking.

Spongy Brake Pedal

A *spongy brake pedal* feels like it is connected to a spring or a rubber band. The brakes will apply, but the pedal does not feel solid. It will travel farther toward the floor before full braking action occurs.

This condition is usually caused by *air* in the brake system. The air is compressing, so it takes up less space. A spongy pedal can also result from faulty residual pressure check valves in the master cylinder or maladjusted brake shoes.

Dropping Brake Pedal

A *dropping brake pedal* slowly moves all the way to the floor when steady pressure is applied to it. This condition usually shows up when the driver is stopped at a stoplight. The brake may apply normally at first. However, while holding the brakes, the pedal steadily creeps downward. Pumping usually restores pedal height momentarily.

A dropping brake pedal is usually caused by an internal leak in the master cylinder. Pressure is slowly leaking past the piston cups. Since fluid returns to the reservoir, the fluid level will not drop.

A fluid leak anywhere else in the system can also cause the same symptom. If fluid is leaking out, however, the fluid level in the master cylinder will drop.

Low Brake Pedal

A *low brake pedal* travels too far toward the floor before braking. The pedal is *not* spongy, and braking is normal once the pedal applies the brakes.

A low brake pedal can be caused by inoperative brake adjusters, a maladjusted master cylinder push rod, or a mechanical problem in the wheel brake assemblies.

Dragging Brakes

Dragging brakes remain partially applied when the brake pedal is released. The brakes will overheat if the vehicle is driven very far. To detect dragging brakes, feel each wheel assembly. The dragging brake or brakes will be abnormally *hot*.

Dragging brakes can be caused by frozen wheel cylinder pistons, an overadjusted parking brake, weak return springs, an overadjusted master cylinder push rod, brake fluid contamination, or a master cylinder problem.

No Brake Pedal

No brake pedal is a very dangerous condition in which the brake pedal moves to the floor with no braking action. Pumping the brakes does *not* help.

Lack of braking action is usually caused by a hydraulic problem. A system leak may have emptied the master cylinder reservoir.

With today's dual master cylinders, a complete loss of braking is unlikely. However, it can occur from driver neglect.

Brake Warning Light On

When the *brake warning light* is on, it indicates either an internal leak (master cylinder) or an external leak (brake line, hose, wheel cylinder) in the hydraulic system. Unequal pressure in the dual master cylinder system has caused the warning light switch (usually part of combination valve) to shift to one side. Inspect the system for leaks and check the action of the master cylinder.

Braking Noise

Braking noise can be grinding sounds, squeaks, rattles, and other abnormal noises. Use your understanding of system operation to determine the cause of braking noise.

A metal-on-metal grinding sound occurring only when braking may be due to worn brake linings. The shoe or pad may be rubbing on the metal drum or disc.

A squeak when braking may be caused by glazed (hardened) brake linings, an unlubricated brake drum backing plate, foreign material embedded in the linings, or a wear indicator rubbing on a rotor.

A rattle may be due to a missing anti-rattle clip or spring on the brake pads. Loose or disconnected parts in the drum brake assembly can also cause a rattle from inside the brake assembly.

Brake System Inspection

Most automobile manufacturers recommend a periodic inspection of the brake system. The inspection is a form of preventive maintenance that helps ensure safety. A typical brake system inspection involves checking brake pedal action, fluid level in the master cylinder, and the condition of the brake lines, hoses, and wheel brake assemblies, **Figure 72-1.**

Checking Brake Pedal Action

A fast and accurate way of checking many components of the brake system is the *brake pedal check.* This is done by applying the brake pedal and comparing its movement to specs. There are three brake pedal application specs (distances): pedal height, pedal free play, and pedal reserve distance.

Brake pedal height is the distance from the pedal to the floor with the pedal at rest. Refer to **Figure 72-2.** Incorrect pedal height is usually caused by problems in the pedal mechanism. There may be worn pedal bushings, a weak return spring, or a maladjusted master cylinder push rod.

Brake pedal free play is the amount of pedal movement before the beginning of brake application. It is the difference between the "at rest" position and the initially applied position, **Figure 72-2.**

Brake pedal free play is needed to prevent brake drag. If pedal free play is not correct, check the adjustment of the master cylinder push rod. A worn pedal bushing or a bad return spring can also increase pedal free play.

Brake pedal reserve distance is measured from the vehicle's floor to the brake pedal when the brakes are applied. Typically, brake pedal reserve distance should be 2″ (51 mm) for manual brakes and 1″ (25 mm) for power brakes. See **Figure 72-2.**

If brake pedal reserve distance is incorrect, check push rod adjustment. Also, check for air in the system or inoperative brake adjusters. Numerous other problems can cause incorrect pedal reserve distance.

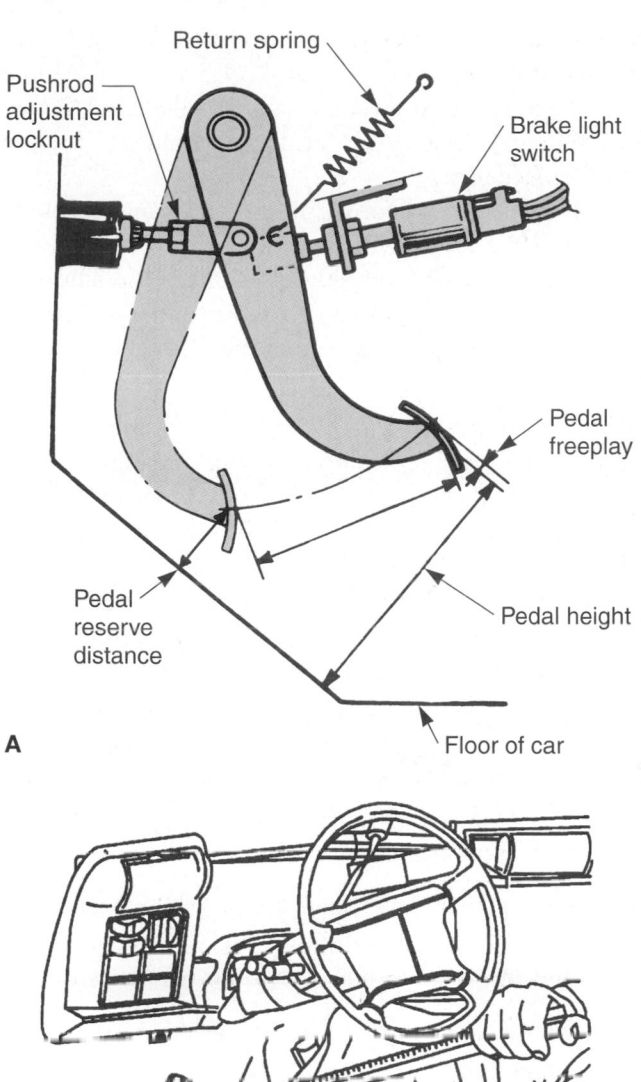

Figure 72-2. A—Brake pedal height, reserve distance, and free play are important. Some vehicles have an adjustable push rod. Also note the location of the brake light switch. B—One service manual method of measuring brake pedal height. A yardstick is used to measure from the steering wheel to the brake pedal. (Honda and General Motors)

When checking brake pedal action, apply and hold the brake pedal firmly for about 15 seconds. The engine should be running if the vehicle has power brakes. Try to detect any system leakage, which would cause the pedal to slowly move toward the floor. Also, make sure the pedal is firm and that it returns properly.

While checking the brake pedal, you should also make sure the brake lights are operating. If they do not work, check the bulbs, the fuses, and the switch.

Checking Brake Fluid

An important part of a brake system inspection involves checking the level and condition of the brake fluid. To check the fluid, remove the master cylinder cover. Pry off the spring clip or unbolt the cover.

Typically, the *brake fluid level* should be 1/4″ (6 mm) from the top of the reservoir. Add fluid as needed. See **Figure 72-3.**

> **Tech Tip**
> Only use the manufacturer's recommended type of brake fluid. Also, keep oil, grease, and other substances out of the brake fluid. Contamination can cause rapid deterioration of the master cylinder cups, resulting in a sudden loss of braking ability.

Checking For Brake System Leaks

When the fluid level in the master cylinder is low, you should inspect the brake system for leaks. Check all brake lines, hoses, and wheel cylinders.

Brake fluid leakage will show up as a darkened, damp area around one of the components. Verify brake system leakage by checking that the leaking fluid smells like brake fluid.

Checking the Parking Brake

Apply the parking brake. The pedal or lever should *not* move more than 2/3 of full travel. The parking brake

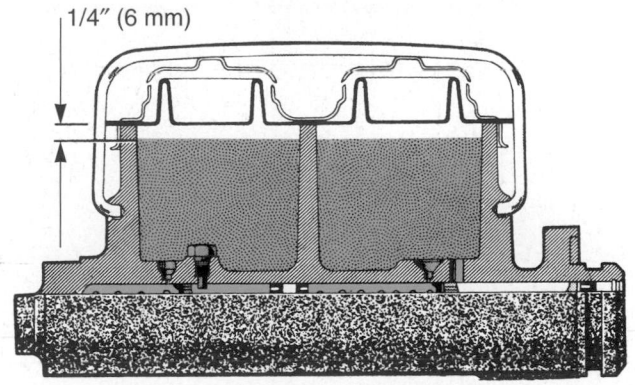

Figure 72-3. Brake fluid level should be checked periodically. The master cylinder reservoir must be kept full. Refer to the service manual for specifications on filling the reservoir. One-quarter inch down is average. (GM Trucks)

should keep the vehicle from moving with the engine idling and the transmission in drive.

The parking brake cables and linkage should also be inspected. The cables should *not* be frayed. The linkage should be tight, undamaged, and well lubricated.

Checking Wheel Brake Assemblies

When inspecting a brake system, remove one of the front and rear wheels. This will let you inspect the condition of the brake linings and other wheel brake assembly components.

When inspecting disc brakes, check the thickness of the brake pad linings. Pads should be replaced when the thinnest (most worn) part of the lining is no thicker than the metal shoe (approximately 1/8″ or 3 mm). **Figure 72-4** shows common disc brake problems. Study them closely.

Check the caliper piston for fluid leakage and inspect the disc for damage. The disc should not be scored, cracked, or **heat checked** (small hardened and cracked areas caused by overheating). The wheel bearings should be adjusted properly. To check for rattles, strike the caliper with a rubber mallet. Repair any problems according to service manual instructions.

When inspecting drum brakes, you must remove the brake drum. This will expose the brake shoes, the wheel cylinder, the braking surface of the drum, the adjuster mechanism, and other parts. See **Figure 72-5.**

The brake shoe linings must never be allowed to wear thinner than approximately 1/16″ (1.5 mm). The shoes should *not* be glazed or coated with brake fluid, grease, or differential fluid. Any of these problems requires lining replacement.

Pull back the wheel cylinder boots and check for leakage. If the boot is full of fluid, the wheel cylinder should be rebuilt or replaced. Also, check the automatic adjuster, the return springs, and the brake drum. The brake drum should *not* be scored, cracked, heat checked, or worn beyond specs.

⚠ Warning

Take every precaution to avoid breathing brake lining dust. If needed, wear an approved filter mask over your nose and mouth. *Never* use compressed air to blow brake dust off a wheel brake assembly. Use a special brake vacuum and a clean rag to remove the dust. Older brake linings were made of *asbestos,* a known cancer-causing substance. Newer linings are made from safer materials, but the brake dust created by these substances can be harmful to your health.

Vacuum Booster Service

When a vehicle has vacuum-type power brakes, you should inspect the brake booster and the vacuum hose. Make sure the vacuum hose from the engine is in good

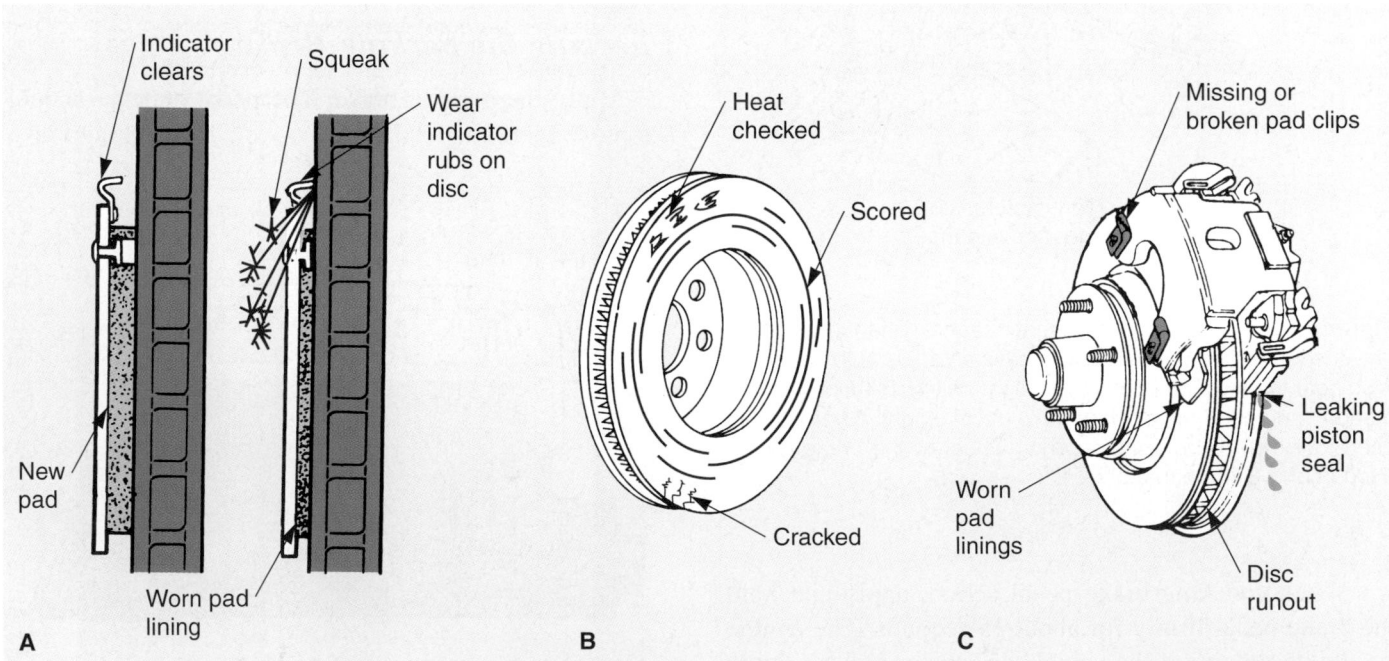

Figure 72-4. Study these disc brake problems. A—Wear indicator clip will produce a loud squeal when the lining wears enough to let the clip touch the rotor. B—Check the rotor for heat checking, cracks, and scoring. C— Also check for a leaking piston seal, worn pad linings, and missing clips. (Cadillac and FMC)

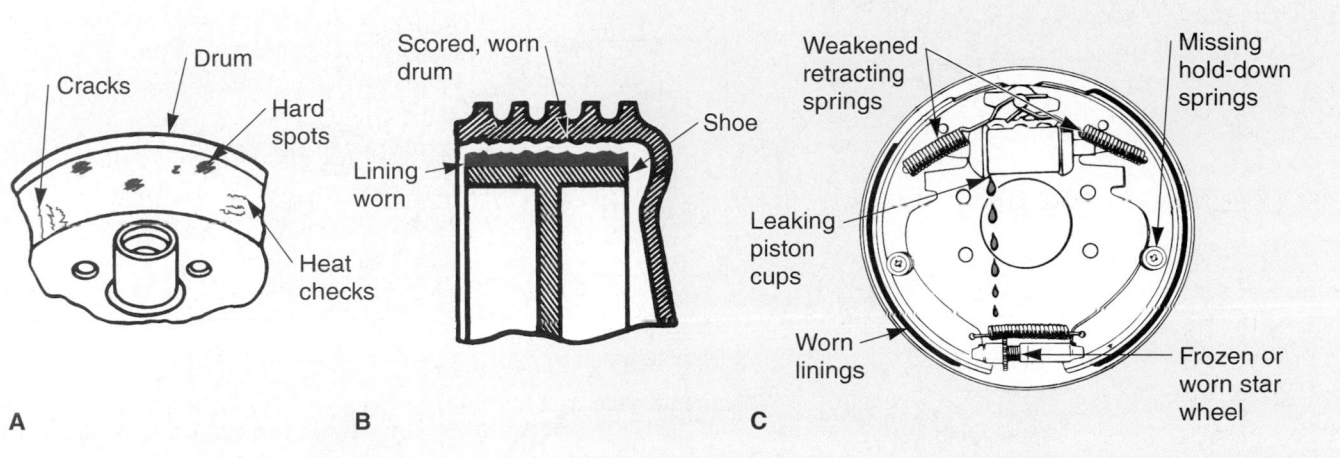

Figure 72-5. Study these common drum brake problems. A—Check the drum for cracks, heat checks, and hard spots. B—A badly scored drum must be machined. If the drum is worn too much, it must be replaced. C—Check for a leaking wheel cylinder, worn linings, and missing or damaged parts. (Bendix and FMC)

condition. It should *not* be hardened, cracked, or swollen. Also, check the hose fitting in the booster.

To test the vacuum booster, pump the brake pedal several times to remove any vacuum from the booster. Then, press down lightly on the brake pedal as you start the engine. If the vacuum booster is functioning, the brake pedal will *move downward slightly* as soon as the engine starts.

Many shops do *not* rebuild vacuum brake boosters. Instead, they install a new or factory rebuilt unit. Some boosters are sealed and cannot be disassembled.

Figure 72-6 shows the parts that must be removed when replacing a vacuum booster.

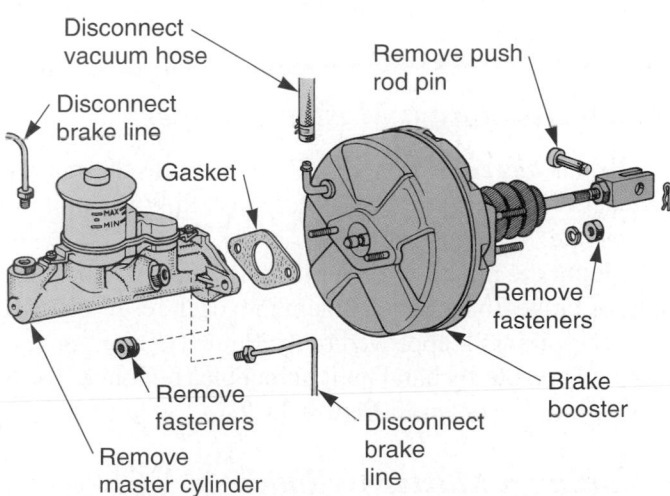

Figure 72-6. These components may need to be removed for service of the master cylinder and brake booster. (Toyota)

Figure 72-7 is an exploded view of a vacuum booster. This particular unit is *not* sealed and can be rebuilt. A rebuild normally involves replacing the diaphragm, the valves, and other plastic or rubber parts. Refer to a service manual for exact procedures on a particular booster.

Hydraulic Booster Service

A hydraulic brake booster should be checked when inspecting a brake system. Check all the hydraulic lines for signs of leakage. Tighten connections or replace any line that leaks.

If the booster is inoperative, check the fluid level in the power steering pump. A low fluid level can prevent hydro-boost operation. When the hydraulic booster is found to be faulty, it should be replaced. Most types are *not* repairable.

Refer to a shop manual when testing or servicing a hydraulic brake booster. System designs and repair procedures vary.

Master Cylinder Service

A faulty master cylinder usually leaks fluid past the rear piston or leaks internally. When fluid is leaking past the rear piston, you should find brake fluid in the rear boot or on the firewall. When the leak is internal, the brake pedal will slowly sink to the floor as pressure is applied. Inoperative valves in the master cylinder are also a reason for service.

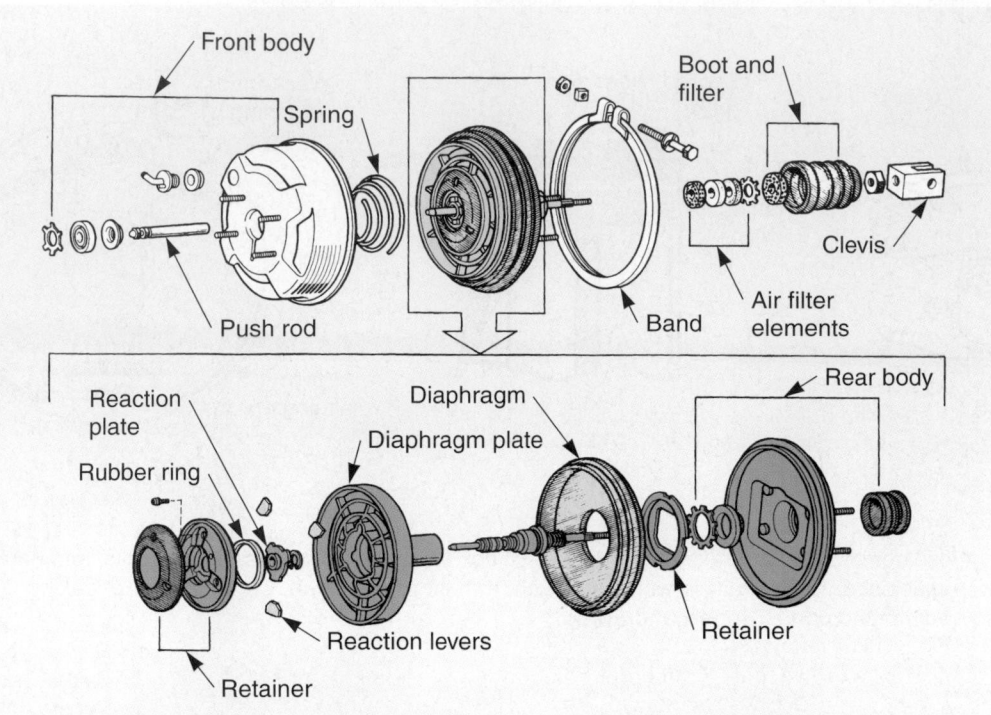

Figure 72-7. Exploded view shows a modern vacuum brake booster. Most technicians replace the unit when it is bad. (Toyota)

Master Cylinder Removal

To remove a master cylinder, disconnect the brake lines from the master cylinder using a tubing wrench. Then, unbolt the master cylinder from the brake booster or the firewall. Sometimes, the push rod must be disconnected from the brake pedal assembly.

Master Cylinder Rebuild

Many shops simply replace a faulty master cylinder with a new or factory rebuilt unit. A replacement cylinder may be cheaper than the cost of labor and parts for an in-shop rebuild. Further, the new or factory rebuilt master cylinder will have a remachined cylinder and a guarantee.

To rebuild a master cylinder, drain the fluid from the reservoir. Then, completely disassemble the unit, following the instructions in a service manual, **Figure 72-8.**

Basically, you must hone the cylinder and replace the piston cups and valves. Clean the parts in brake fluid or a recommended cleaner.

Warning
Do *not* clean the hydraulic parts of a brake system with conventional parts cleaners. These cleaners can destroy the special rubber cups in the brake system. Use only brake fluid or a manufacturer-suggested cleaner (denatured alcohol for example).

After cleaning, measure the piston-to-cylinder clearance. Use a telescoping gauge and an outside micrometer or a strip-type feeler gauge. The cylinder must *not* be tapered or worn beyond specifications. Also, make sure the cylinder is not corroded, pitted, or scored. Replace the master cylinder if the cylinder is not in perfect condition after honing.

Blow all parts dry with compressed air. Blow out the ports and check that they are unobstructed. Lubricate the parts with brake fluid and assemble the unit according to the manufacturer's instructions.

Figure 72-8 shows an exploded view of a typical dual master cylinder.

Bench Bleeding a Master Cylinder

Bench bleeding removes air from inside the master cylinder. This must be done before installing the unit on the vehicle.

Mount the master cylinder in a vise. Install short sections of brake line and bend them into each reservoir. Fill the reservoirs with approved brake fluid. Then, pump the piston in and out by hand until air bubbles no longer form in the fluid, as shown in **Figure 72-9.**

Installing a Master Cylinder

To install a master cylinder, replace the reservoir cover after bench bleeding. Bolt the master cylinder to

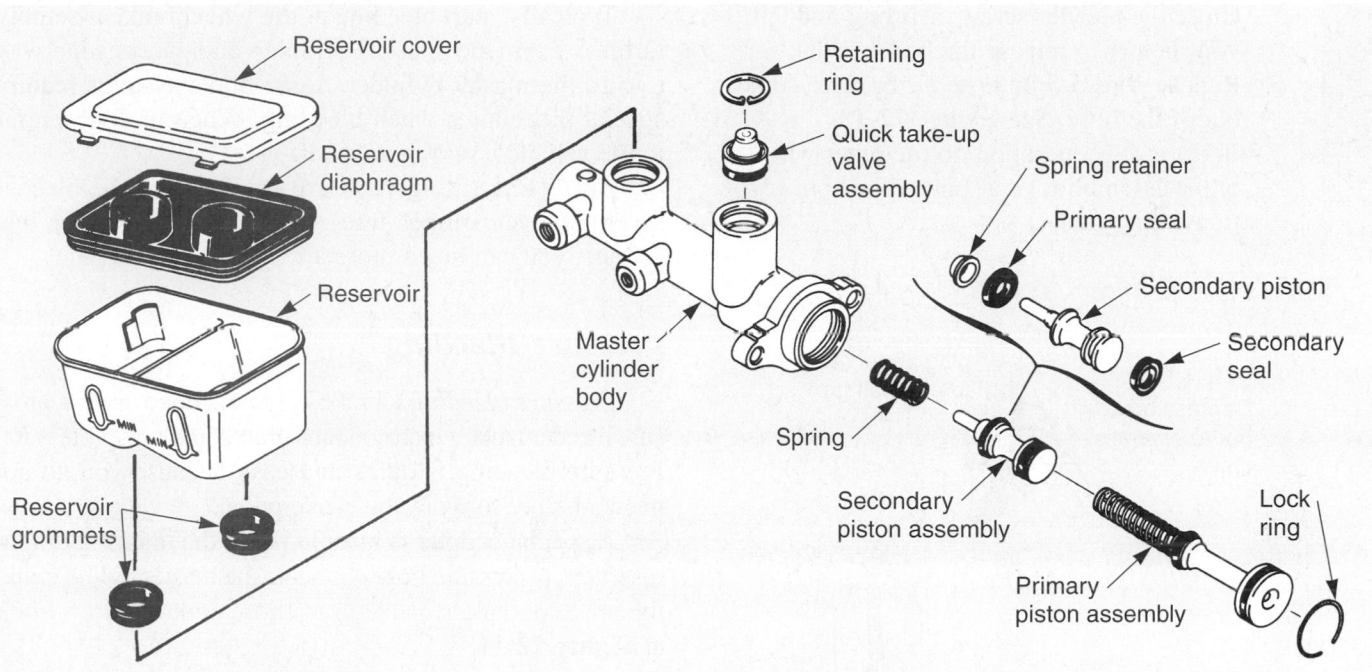

Figure 72-8. Exploded view shows a dual master cylinder with a removable reservoir. Note how the parts fit into the cylinder. (Oldsmobile)

the firewall or booster. Check the adjustment of the push rod if a means of adjustment is provided.

Without cross-threading the fittings, screw the brake lines into the master cylinder. Lightly snug the fittings. Then, bleed (remove air from) the system.

Tighten the brake line fittings. Fill the reservoir with fluid and check brake pedal feel. Test drive the vehicle.

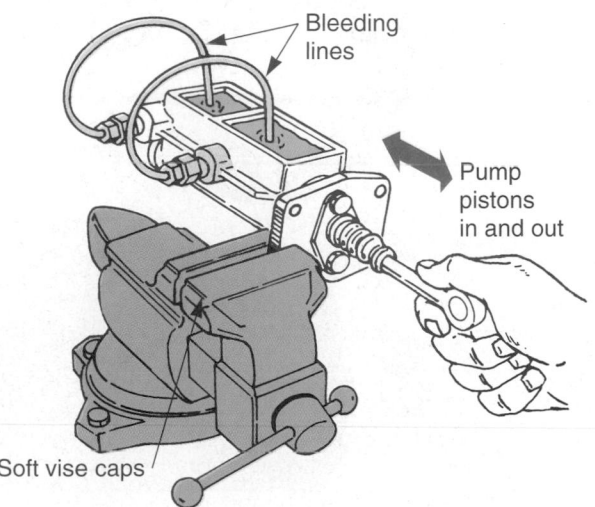

Figure 72-9. Before installing a new or rebuilt master cylinder, bench bleed the unit. Fill the reservoir with brake fluid and install bleeding lines. Then pump the piston in and out until no air bubbles are visible in the reservoir (EIS)

Brake System Bleeding

The brake system must be free of air to function properly. Air in the system will compress, causing a spongy brake pedal. Air can enter the system anytime a hydraulic component (brake line, hose, master cylinder, wheel cylinder, etc.) is disconnected or removed.

Brake system bleeding involves the use of fluid pressure to force air out of the brake line connections or wheel cylinder bleeder screws. There are two methods of bleeding brakes: manual bleeding and pressure bleeding.

Manual Bleeding

Manual bleeding is accomplished by using master cylinder pressure to force fluid and trapped air out of the system.

 To manually bleed a brake system:
1. Attach one end of a hose to the bleeder screw.
2. Place the other end of the hose in a jar partially filled with clean brake fluid. Make sure the end of the hose is submersed in the fluid.
3. Have another technician apply light foot pressure on the brake pedal.
4. Open the bleeder screw or fitting on the caliper or wheel cylinder while watching for air bubbles at the hose.

5. Close the bleeder screw or fitting and tell your helper to release the brake pedal.
6. Repeat steps 3-5 until no air bubbles come out of the hose. See **Figure 72-10.**
7. Perform this operation on the other wheel brake assemblies or at brake line connections, if needed.

Typically, start bleeding at the wheel brake assembly farthest from the master cylinder and work your way toward the master cylinder. Some brake systems require special procedures when bleeding. When in doubt, refer to the vehicle's service manual.

Brake line fittings that are higher than the bleeder screws can sometimes trap air. By loosening these line fittings, you can bleed more air out of the system.

Pressure Bleeding

Pressure bleeding a brake system is done using a pressure bleeder tank, which contains brake fluid under pressure. Pressure bleeding is quick and easy because you do not need a helper to work the brake pedal.

A special adapter is installed over the master cylinder reservoir. A pressure hose connects the master cylinder and the pressure tank. A valve in the hose controls flow. Look at **Figure 72-11.**

Note
Check the manufacturer's instructions. You may need to push or pull out the metering valve stem before bleeding the brake system.

Pour enough brake fluid in the bleeder tank to reach the prescribed level. Charge the tank with 10-15 psi (69-103 kPa) of air pressure. Fill the master cylinder with brake fluid. Install the adapter and hose on the master

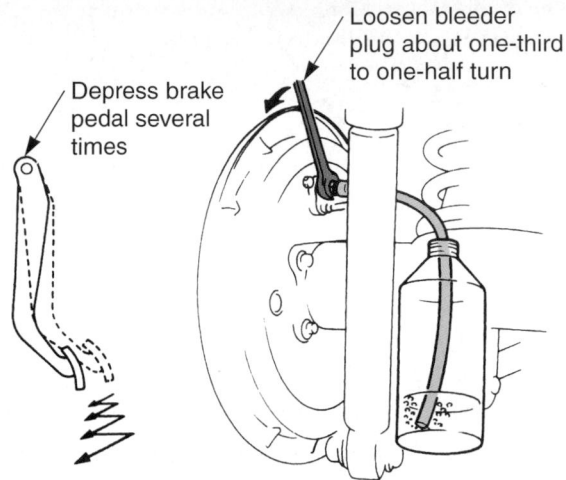

Figure 72-10. If you are not using a pressure bleeder, you can have a helper pump the brake pedal up and down lightly to build system pressure. Then, crack open the bleeder screw to purge air from the system. A container of brake fluid will keep air from being pulled back into the system. (General Motors)

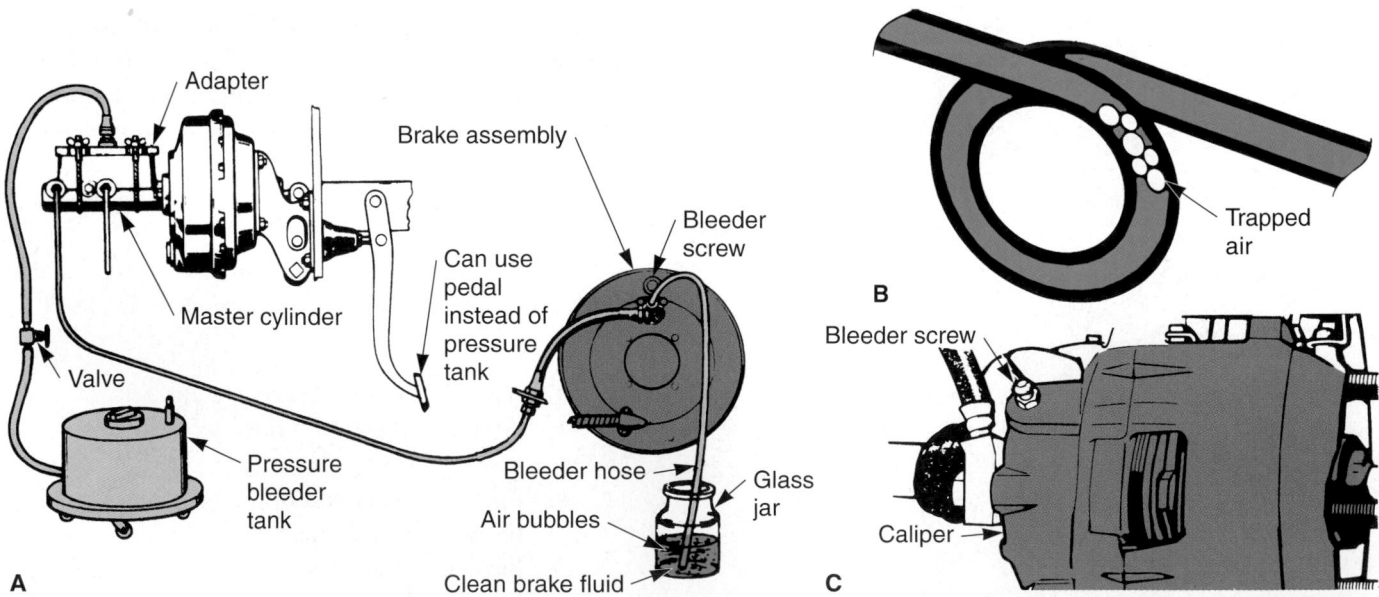

Figure 72-11. Bleeding a brake system involves forcing brake fluid through the lines and the wheel brake assemblies. This forces air out of the system. Note how a pressure bleeder is connected to the master cylinder. With pressure in the system, open each bleeder screw until the air is purged from the system. A—Complete setup. B—Trapped air. C—Bleed screw location. (Bendix and Chrysler)

cylinder. Open the valve in the hose. You are now ready to bleed the brakes.

Attach one end of a hose to a bleeder screw and place the other end of the hose in a jar containing clean brake fluid. Open the screw, **Figure 72-11.** As soon as the fluid entering the jar is clear, close the screw. Repeat the bleeding operation on the other wheel brake assemblies in the proper order.

Tech Tip

A special pressure bleeding adapter is needed on master cylinders using a *plastic reservoir.* Use an adapter that seals over the ports in the bottom of the master cylinder. This will prevent reservoir damage. A special vacuum bleeder that is mounted at each wheel bleeder screw will also work.

Flushing a Brake System

Brake system flushing is done by pressure bleeding *all* the old fluid out of the system. Flushing is needed when the brake fluid is **contaminated** (filled with dirt,

rust, corrosion, oil, or water). Bleed each wheel cylinder until clean fluid flows from the bleeder screw or fitting.

Brake Line and Hose Service

Brake lines and hoses can become damaged or deteriorated after prolonged service. When replacing a brake line, use approved double-wall steel tubing. Brake lines normally use double-lap flares.

Figure 72-12 shows common tools and procedures used when servicing brake lines and hoses.

Disc Brake Service

Complete **disc brake service** typically involves four major operations:

- Replacing worn brake pads.
- Rebuilding the caliper assembly.
- Turning (machining) the brake discs.
- Bleeding the system.

Depending on the condition of the disc brake assembly, the technician may need to do one or more of

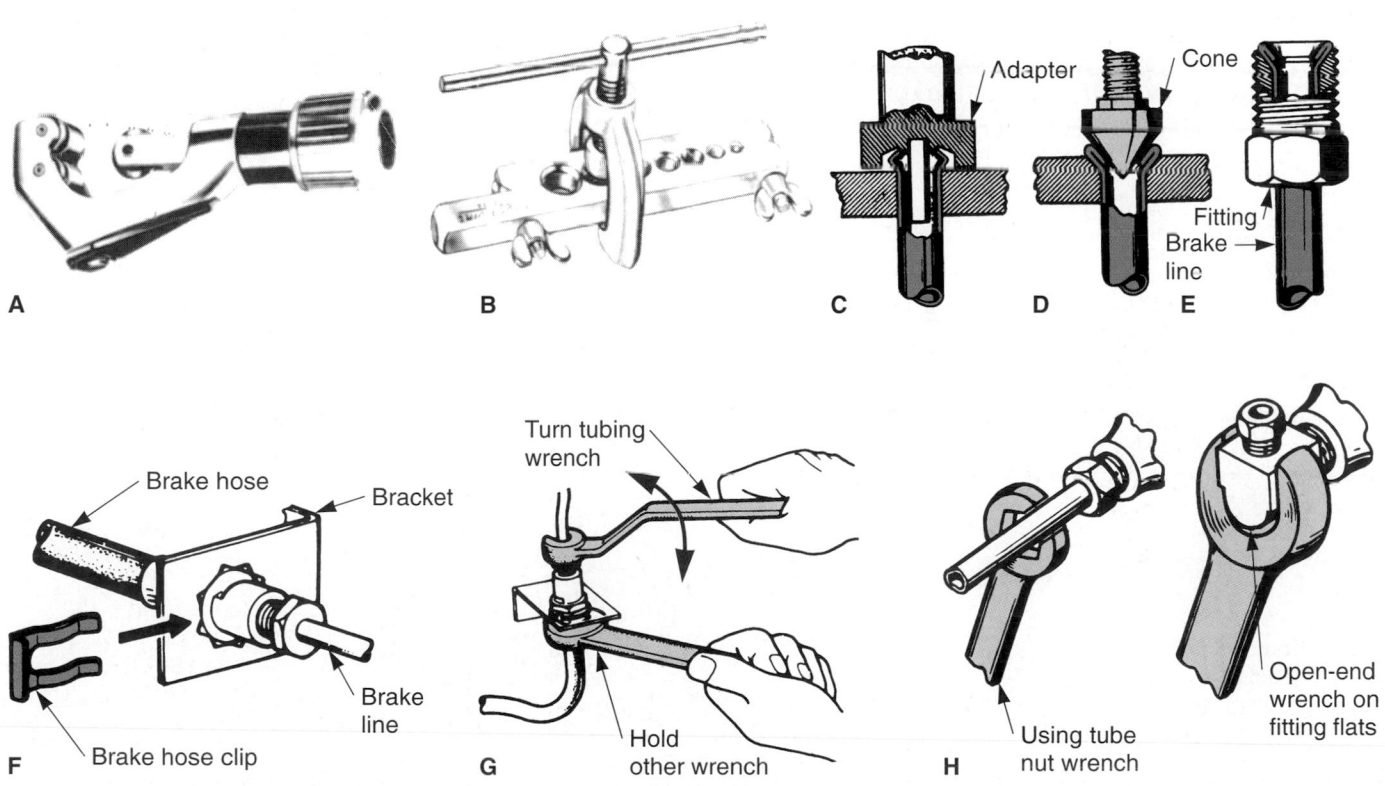

Figure 72-12. Proper service of brake lines and hoses is critical to vehicle safety. A—Tubing cutter. B—Flaring tool for forming ends on tubing. C—Using an adapter to fold the tubing inward. D—Using a cone on a flaring bar to form a double-lap. E—The fitting is installed over the double-lap flare. F—Clips are commonly used to secure the ends of a brake hose. G—Two wrenches are normally needed to loosen or tighten this type of tubing-to-hose connection. H—Use a tubing or a line wrench on fitting nuts. An open-end wrench will hold the flats of some fittings. (Bendix and Snap-on Tool Corp.)

these operations. Service manual procedures and shop policies vary. In any case, you must make sure the brake assembly is in sound operating condition.

Replacing Brake Pads

To replace worn brake pads on a floating caliper:

1. Loosen the wheel lug nuts.
2. Place the vehicle on jack stands and remove the tire-and-wheel assemblies, **Figure 72-13.**
3. Use a large C-clamp to push each piston back into its cylinder. Refer to **Figure 72-14A.**
4. Unbolt the calipers and slide them off the discs, **Figure 72-14B.**
5. Hang the calipers by a piece of wire (if the caliper is not to be removed). See **Figure 72-14C.**
6. Remove the old pads from the calipers.
7. Install anti-rattle clips on the new pads.
8. Fit the new pads into the calipers, **Figure 72-14D.**
9. Slide the caliper assemblies over the discs.
10. Assemble the caliper mounting hardware in reverse order of disassembly.
11. Make sure all bolts are torqued properly.
12. Install the wheel and tighten the lug nuts or bolts to specs.

Tech Tip

It is acceptable to service only the front or rear brakes. However, *never* service only the right or left brake assemblies. This could cause dangerous brake pull.

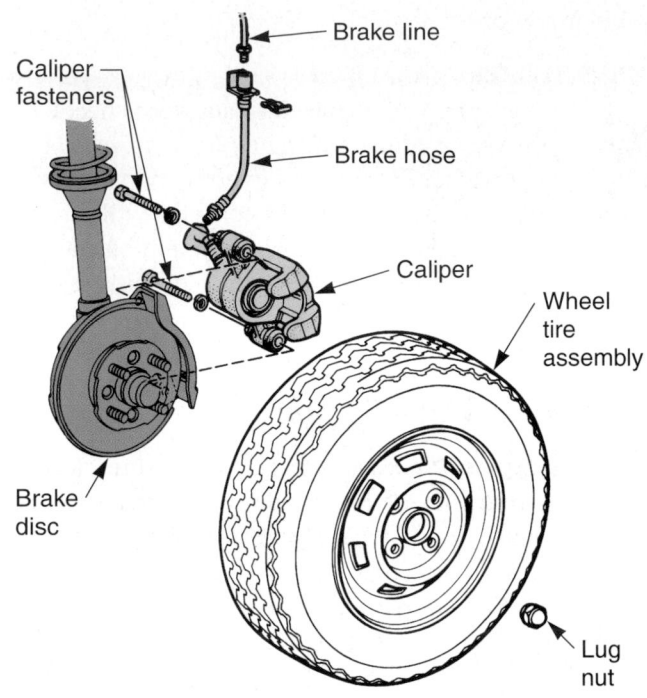

Figure 72-13. On modern vehicles, two bolts commonly hold the caliper on the steering knuckle. The brake hose connects the metal brake line to the caliper. (Toyota)

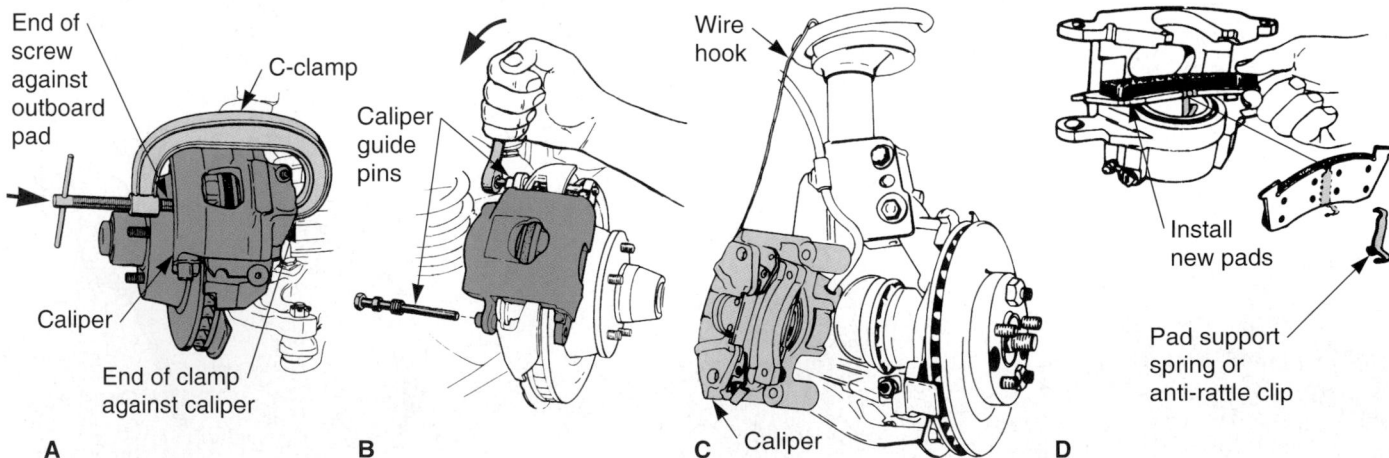

Figure 72-14. Study basic steps for floating caliper pad service. A—Use a C-clamp or a large screwdriver to force the piston back into the caliper. This will open the caliper wide enough for the new, thicker pad linings. B—Use a six-point socket or an Allen socket to unscrew the bolts holding the caliper. C—Lift the caliper off the knuckle and support it on a piece of wire. Do not let the unit hang by the brake hose, or the hose could be damaged. D—Remove the old pads. Note the position of the anti-rattle clips. The clips can be reused if they are in good condition. New pads can be installed without caliper service if the vehicle has low mileage and if the boot and seals are in good condition. (Bendix, Buick, EIS)

Rebuilding a Caliper Assembly

When a caliper piston is frozen, leaking, or has extremely high mileage, it must be rebuilt. Remove the caliper from the vehicle and take it to a clean work area.

To remove the piston from the caliper, apply just enough air pressure in the hose fitting hole to push the piston out of its cylinder, **Figure 72-15A**. Some automakers recommend using brake system hydraulic pressure to force the piston out of the caliper.

Warning

Keep your fingers out of the way when using compressed air to remove a caliper piston. A stuck piston could fly out with tremendous force. Serious hand injuries could result.

After piston removal, pry the old dust boot and the seal out of the caliper, **Figures 72-15B** and **72-15C.** Keep all parts organized on your workbench. Do not mix up right and left or front and rear parts. See **Figure 72-16.**

Check the caliper cylinder wall for wear, scoring, or pitting. Light surface imperfections can be cleaned up with a cylinder hone. Use brake fluid to lubricate the hone. If excessive honing is needed, replace the caliper. See **Figure 72-17A**.

Also, check the caliper piston for wear or damage. Install a new piston if you find any problems. The piston and the cylinder are critical and must be in perfect condition.

Clean all the caliper parts with an approved cleaner. Wipe the parts clean with a clean shop rag. Then, coat them with brake fluid.

Assemble the caliper in the reverse order of disassembly. Typically, work the new seal into the cylinder bore groove with your fingers, **Figure 72-17B.**

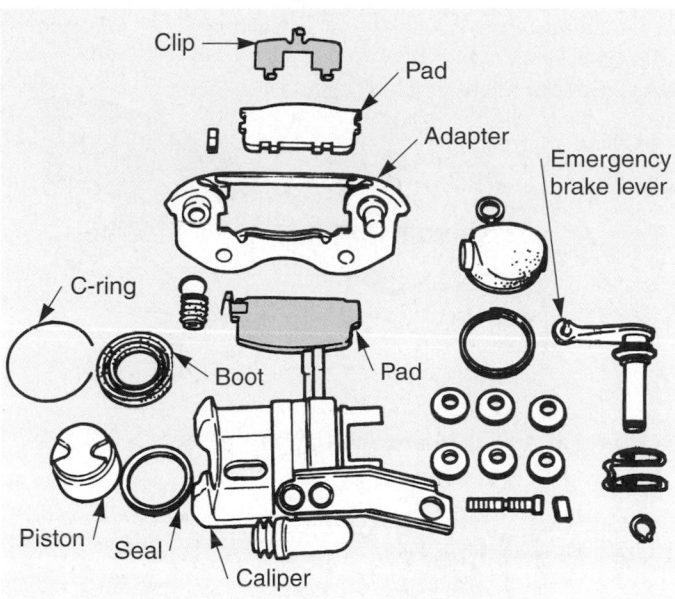

Figure 72-16. When servicing calipers, keep all parts organized on a workbench. Do not mix parts from the right and left sides of the caliper. Inspect all parts closely for signs of wear or damage. (Subaru)

Then, install the new boot in its groove. Coat the piston with more brake fluid. Spread the boot with your fingers and slide the piston into the cylinder, **Figure 72-17C.**

Brake Disc (Rotor) Service

When servicing a disc brake system, it is important to check the condition of the disc. Automakers provide specifications for maximum disc runout (wobble) and minimum disc thickness. Also, the disc must *not* be scored, cracked, or heat checked.

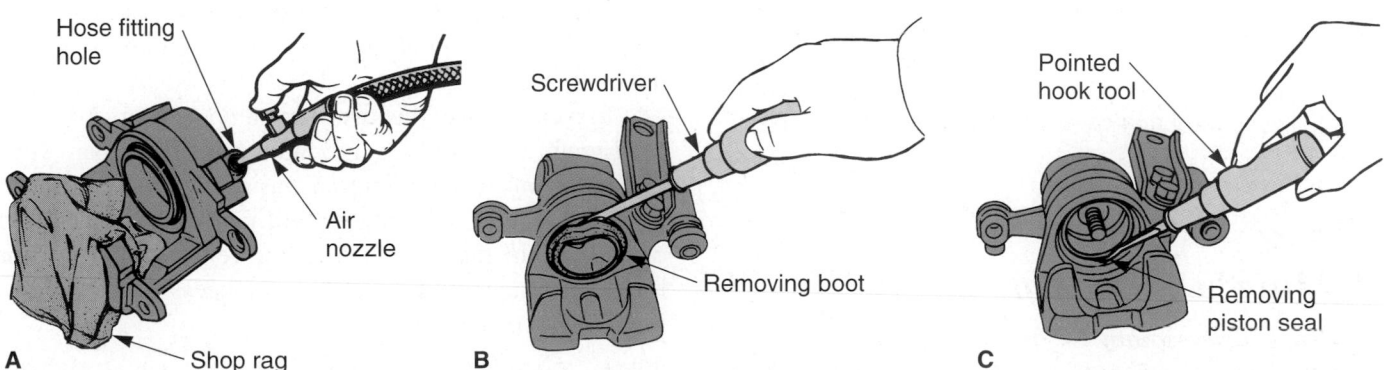

Figure 72-15. Note the fundamental steps for caliper disassembly. A— Use air pressure or brake system pressure to push the piston out of the caliper. Place a thick rag or a block of wood in the caliper and keep your hands out of the way. B—Use a screwdriver or a similar tool to pry the old boot out of the caliper. C—A pointed hook-type tool will make removal of the old seal easy. The seal is positioned inside the groove in cylinder. (Buick, Subaru)

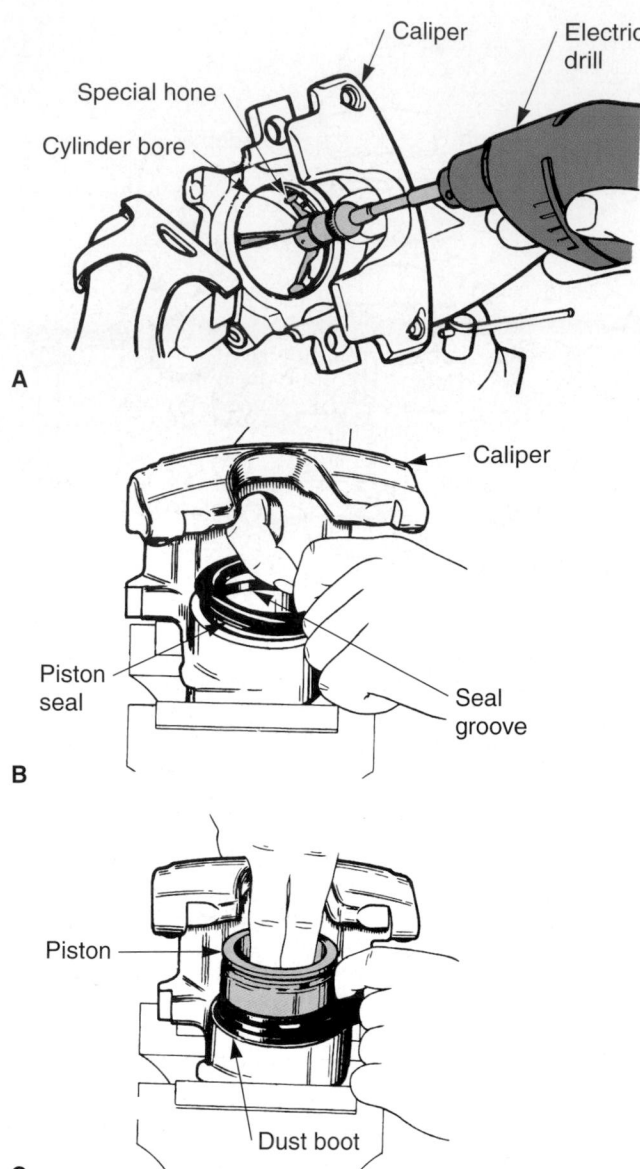

Figure 72-17. Proper honing and reassembling are critical to a caliper rebuild. A— Electric drill is used to spin a special hone inside the caliper cylinder. This will remove minor surface flaws. B—Fit the new seal and boot into position. The seal must be down in its groove. Sometimes, it is easier to install the boot after installing the piston. C—Slide the piston into the cylinder squarely. Make sure the seal does not come out of its groove. Fit the boot into the groove on the outside of the piston. (Chrysler and Bendix)

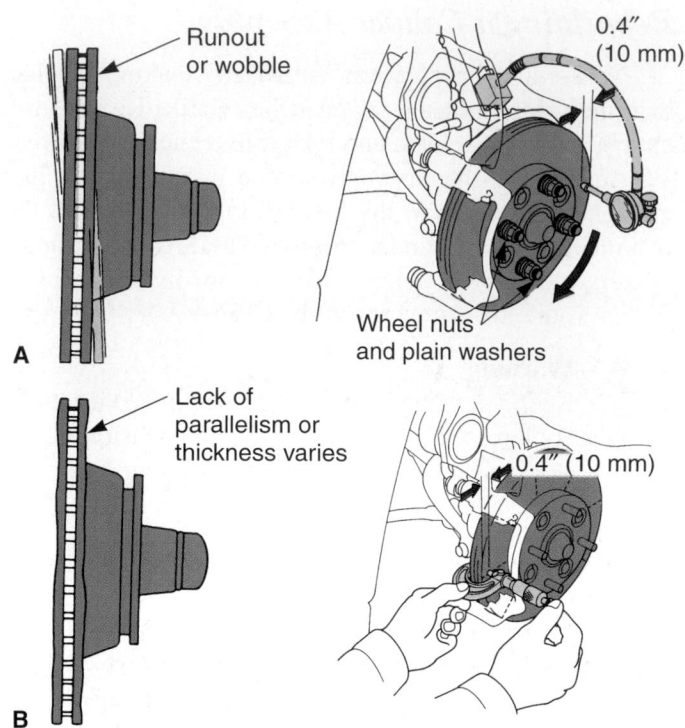

Figure 72-18. Checking for rotor runout and thickness variations is important. A—Mount a dial indicator as shown to measure disc runout, or wobble. Rotate the disc by hand and read the indicator. If runout is greater than specs, the rotor should be turned or replaced. B—An outside micrometer can be used to detect any thickness variations. Take thickness measurements at several places around the disc. If readings are not the same, or if they are not within specs, turn or replace the disc. (EIS and Honda)

Measuring Disc Runout

Brake disc runout is the amount of side-to-side movement measured near the outer edge of the disc's friction surface. A dial indicator is used to measure disc runout. See **Figure 72-18A.**

Compare your indicator reading to factory specifications. Typically, runout should not exceed 0.004"

(0.10 mm). If runout is beyond specs, turn (machine) the disc on a brake lathe to true its friction surfaces.

Measuring Disc Thickness

Brake disc thickness is measured across the two friction surfaces in several locations. Variation in disc thickness indicates wear.

To measure disc thickness, use an outside micrometer. Measure in several places around the disc. Compare your measurements to specifications, as in **Figure 72-18B.**

Minimum disc thickness will sometimes be printed on the side of the disc. If it is not, refer to a manual or brake specification chart. If disc thickness is under specifications, replace the disc.

Tech Tip

A thin disc cannot dissipate heat properly and may warp or fail in service. A warped disc can cause severe vibration upon braking. Complete failure of the disc may severely damage other parts.

Resurfacing a Brake Disc

Brake disc resurfacing involves machining a disc's friction surfaces on a brake lathe, **Figure 72-19**. Disc resurfacing is needed to correct runout, thickness variation, or scoring.

When a disc is in good condition, most manufacturers do *not* recommend resurfacing. Only machine a disc when it is absolutely necessary.

To resurface a brake disc, mount the disc on the brake lathe. Use the appropriate spacers and cones to position the disc on the machine's arbor. See **Figure 72-20**.

Follow the directions provided with the brake lathe. Wrap a *vibration damper* (spring or rubber damper) around the disc to prevent high-frequency vibration and squeal, **Figure 72-21**.

While wearing eye protection, adjust the cutting tools until they contact the friction surfaces on the disc. Then, with the machine feeds and controls set properly, machine smooth surfaces on the disc.

 **Warning**

Do not attempt to operate a brake lathe without first obtaining proper training. The machine could be damaged or you could be injured by incorrect operating procedures. Look at **Figure 72-22**.

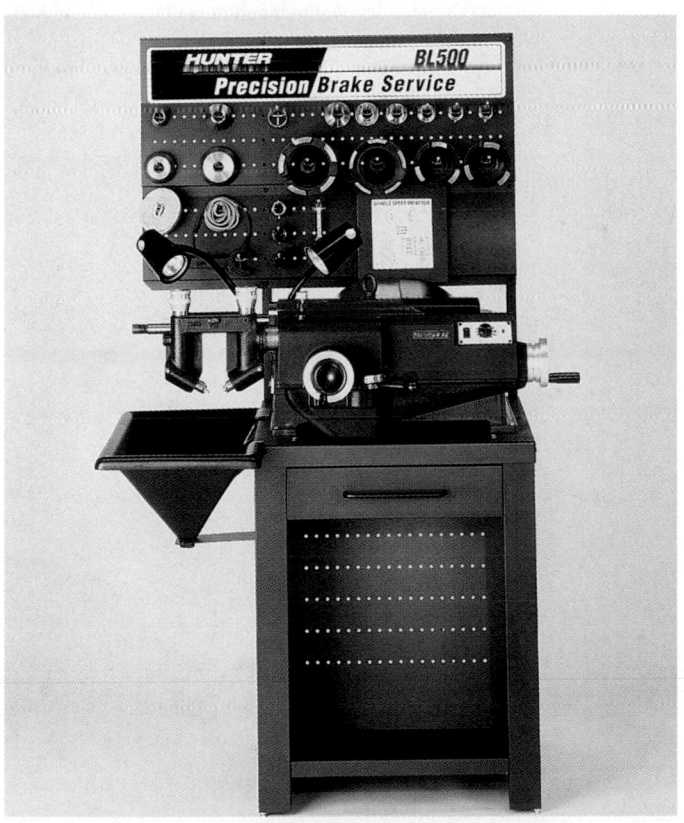

Figure 72-19. Always read the operating manual before trying to use a brake lathe. Note the controls. (Hunter)

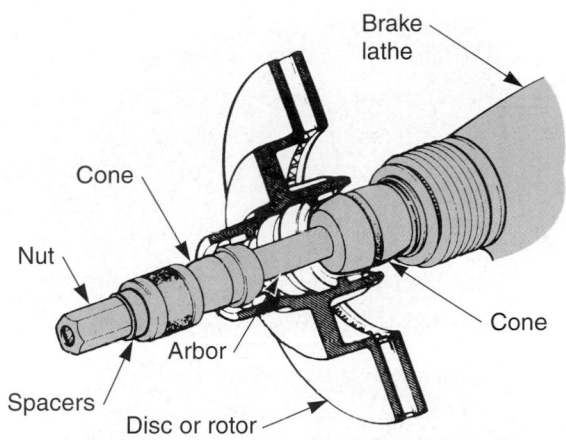

Figure 72-20. Special cones are used to mount the rotor on a brake lathe. The cones must contact the bearing races. Spacers and a nut then lock the assembly on the arbor. (EIS)

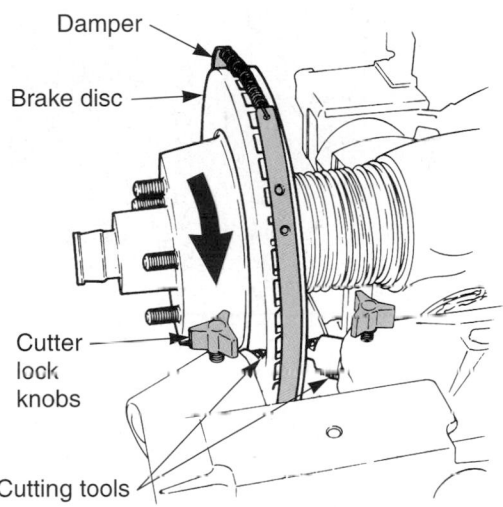

Figure 72-21. After installing a damper to prevent vibration, small cutters are fed into the rotor a prescribed depth. Then, the controls are set so the cutters advance over rotor's surface automatically. (Chrysler)

Only machine off enough metal to true the disc. Then, without touching the machined surfaces with your fingers, remove the disc from the lathe. Double-check disc thickness and install the disc on the vehicle.

On-Car Brake Lathes

Some European or Asian front-wheel drive vehicles have brake discs that are difficult to remove. With these vehicles, an on-car disc lathe (grinder) is a time saver because you do not have to remove the disc to machine it. The lathe mounts on the vehicle's steering knuckle. A large electric motor turns the disc during the machining process. See **Figure 72-23**.

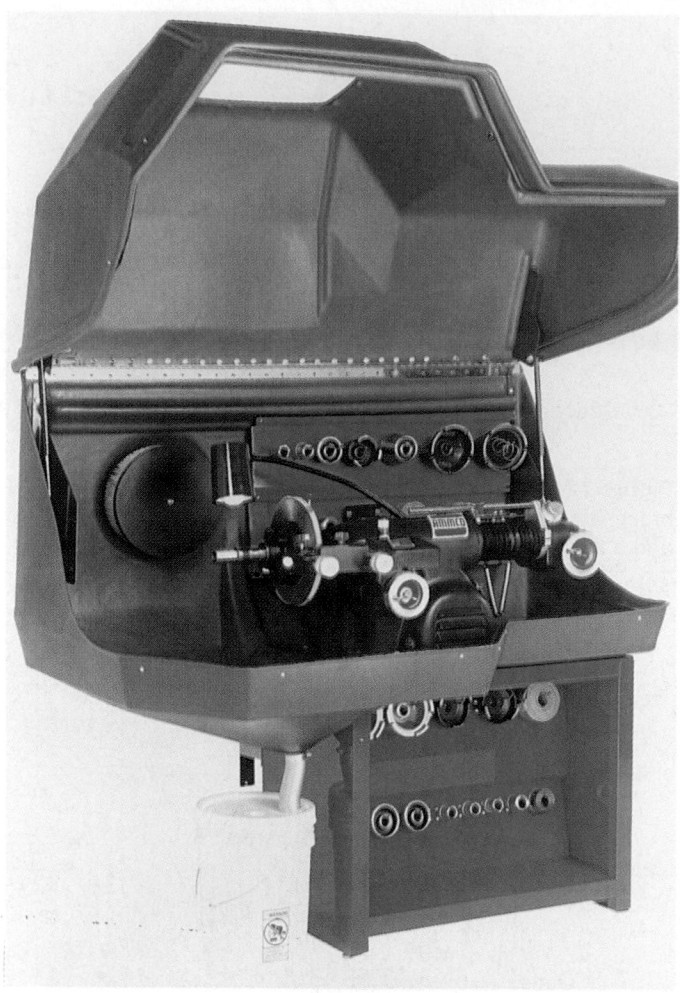

Figure 72-22. This brake lathe has a large guard, or cover, that should be closed during operation. It protects you and others in the shop from flying debris if something should go wrong. (Ammco)

A

B

Figure 72-23. An on-car brake lathe is handy when rotors are difficult to remove from a vehicle. A—The stand-mounted electric motor spins the disc. B—The lathe assembly has controls for feeding cutters into the disc surface. (RTI)

Disc Brake Reassembly

Reassemble the disc brake in the opposite order of disassembly. **Figure 72-24** is an exploded view of a complete disc brake assembly.

After installing the brake disc, fit the caliper assembly into place. Make sure the new pads are properly installed. In some cases, high-temperature silicone is used on the backs of the pads to help prevent brake rattle and squeal.

If needed, apply a small amount of grease to the caliper mounting bolt threads. This will ensure proper tightening and secure mounting of the caliper, **Figure 72-25.** Torque all fasteners to specs.

During assembly, keep your hands off the surface of the disc and the brake pad linings. Grease will burn and harden on the friction lining, causing brake squeak. Refer to **Figure 72-26.**

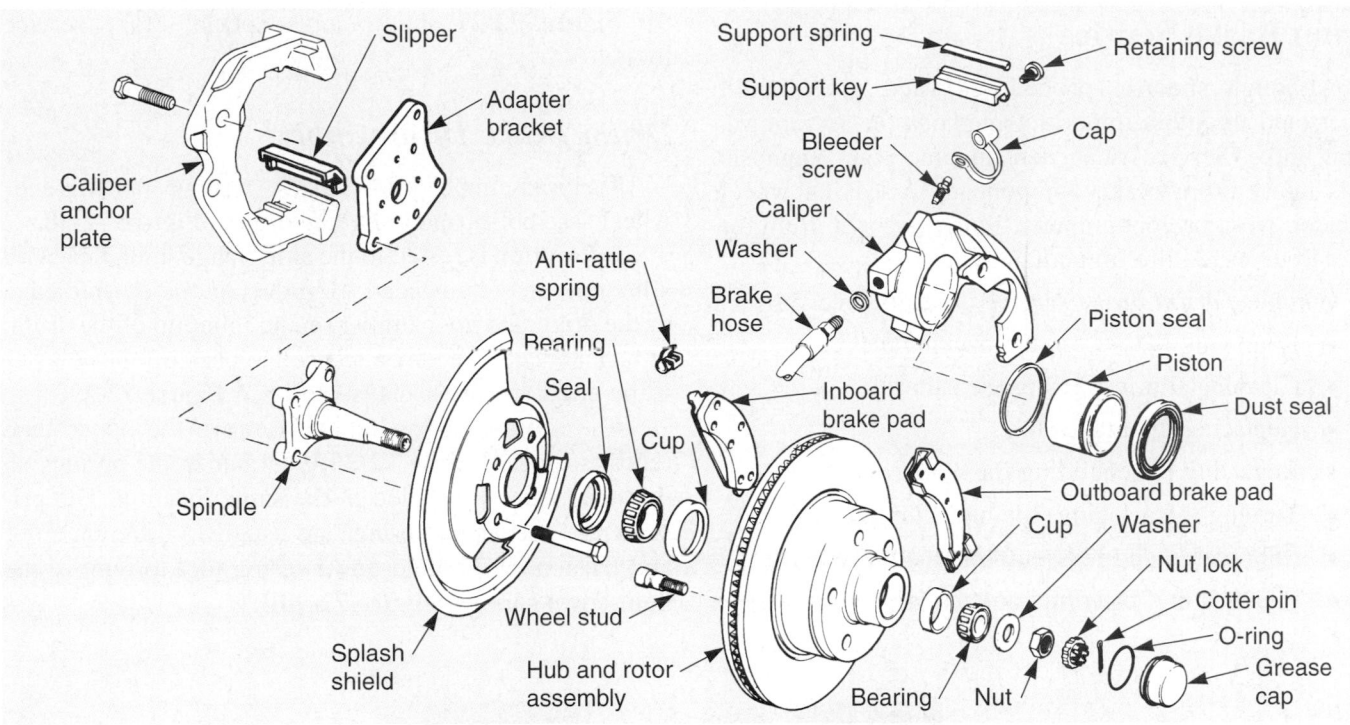

Figure 72-24. Disassembled view of a disc brake. Study how the parts fit together. (Chrysler)

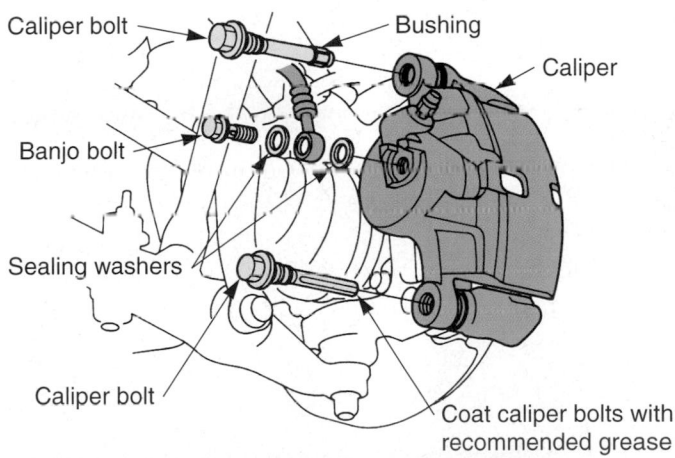

Figure 72-25. When reinstalling a caliper, lubricate the bolt threads. Make sure the special washers are in place on the brake hose, if used. (Honda)

 Tech Tip

Some disc brake rotors have their fins arranged to pull air through the inside of the rotor in only one direction. The spinning action of the rotor aids cooling. With these designs, the rotors are different on the right and left sides of the vehicle. If you accidentally reverse sides or purchase a rotor for the wrong side, rotor overheating and damage can result. Not enough air will be pulled through the rotor to prevent warpage under hard braking heat.

Figure 72-26. Rotor and lining surfaces should be perfectly clean after repairs are complete. If you accidentally get greasy fingerprints on the rotor, wipe off the surfaces with an approved solvent. (Volvo)

Drum Brake Service

Although specific procedures vary, you should understand the most important methods for servicing a drum brake. Service is needed anytime your diagnosis finds faulty drum brake components. A leaking wheel cylinder, worn or contaminated linings, scored drum, or other troubles require immediate repairs.

Complete *drum brake service* typically involves:

- Removing parts from the backing plate.
- Cleaning and inspecting the parts.
- Replacing the brake shoes.
- Replacing or rebuilding the wheel cylinders.
- Turning (resurfacing) the brake drums.
- Lubricating and reassembling the brake parts.
- Preadjusting, bleeding, and testing the brakes.

Figure 72-27 shows some special brake service tools.

Drum Brake Disassembly

To disassemble drum brakes, remove the tire-and-wheel assemblies and the brake drums, **Figure 72-28.**

If the drum is rusted to the axle flange, light taps with a hammer may be needed. Hammer on the outside edge of the drum. Do *not* hammer on the inner lip of the drum, or it can break. You may also need to back off the adjuster if the drum is worn. This is shown in **Figure 72-29.**

Use a brake spring tool to remove the upper shoe return springs, **Figure 72-30A.** Organize the springs so that they can be installed in the same location. The primary and secondary springs are usually a different color and tension. Use a hold-down spring tool to remove the hold-down springs, **Figure 72-30B.**

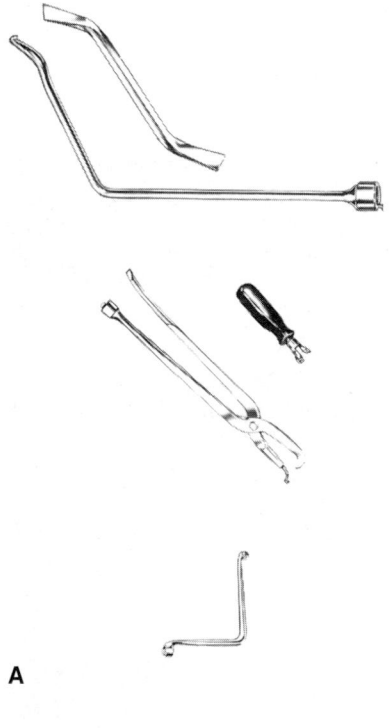

A

B

Figure 72-27. A—Special tools and equipment are available for servicing drum brakes. Small brake tools make the work fast and easy. B—A brake vacuum is used for cleaning off brake assemblies. It helps prevent you from breathing brake lining dust, which can contain asbestos—a cancer causing agent. (Snap-on Tool Corp. and Nilfisk)

Lift the brake shoes off the backing plate. Remove the automatic adjuster mechanism. See **Figure 72-31.**

Before disassembling the wheel cylinder, clean the backing plate. Wipe off the plate with a shop rag. If the backing plate is coated with brake fluid or axle lubricant, wash the plate with an approved cleaner.

Servicing Wheel Cylinders

Pull back the wheel cylinder boots. The wheel cylinder should be rebuilt or replaced if it shows signs of leakage or sticking. In fact, many shops service the wheel cylinders anytime the linings are replaced, **Figure 72-31.**

To disassembly a wheel cylinder, remove the boots, the pistons, the cups, and the spring. Usually, the wheel cylinder can be serviced while bolted to the backing plate. Sometimes, however, the wheel cylinder must be removed before disassembly.

A *wheel cylinder rebuild* normally involves honing the cylinder and replacing the rubber cups and boots.

It is very important that the inside of the cylinder be in good condition. Cylinders that are deeply scratched, scored, or pitted must be replaced.

To hone a wheel cylinder, mount the small hone in an electric drill. Insert the hone in the cylinder. Turn on the drill while moving the hone back and forth in the cylinder. Keep the hone lubricated with brake fluid.

Warning

When honing a wheel cylinder, do pull the hone out of the cylinder. The spinning hone could fly apart. Wear eye protection!

After honing, wash the wheel cylinder thoroughly with clean shop rags and brake fluid or an approved cleaning solvent. Make sure the cylinder is *clean* and in perfect condition before reassembly. The slightest bit of grit or roughness could cause cup leakage.

Make sure the new wheel cylinder cups are the same size as the old ones. Cup size is normally printed on the

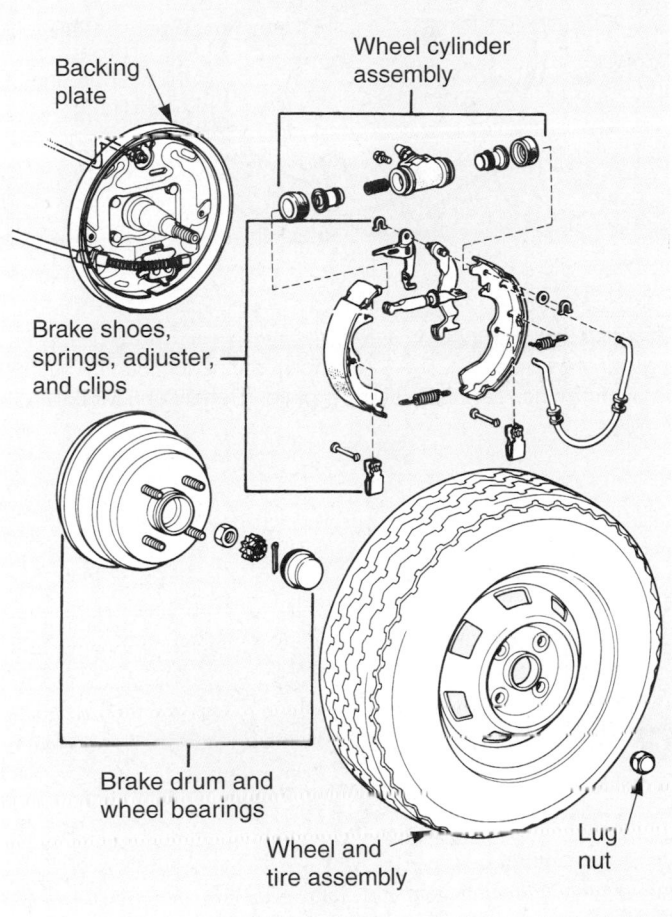

Figure 72-28. Disassembled view of drum brakes. Note the part relationships. (Toyota)

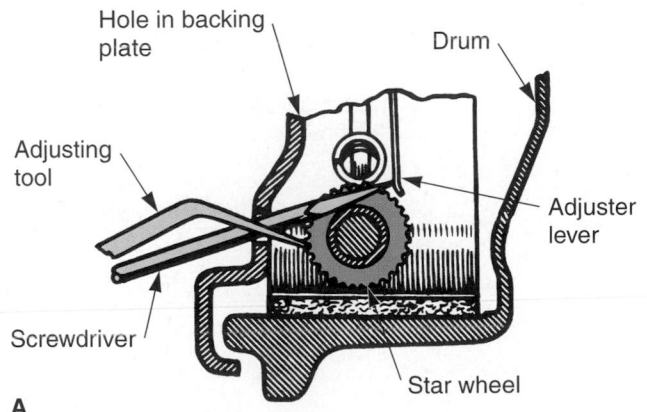

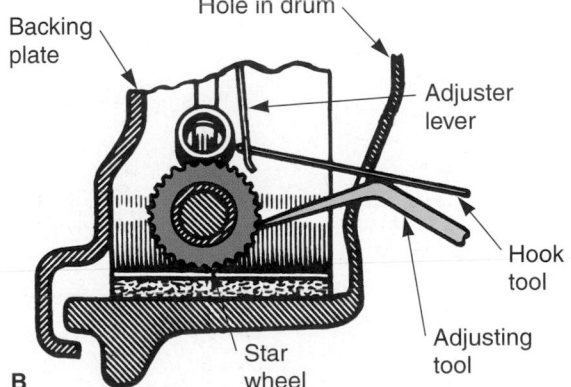

Figure 72-29. Backing off the star wheel will allow drum removal. A—When the access hole is in the backing plate, hold the lever and turn the star wheel as shown. B—Turning the star wheel when the access hole is in the drum. (Bendix)

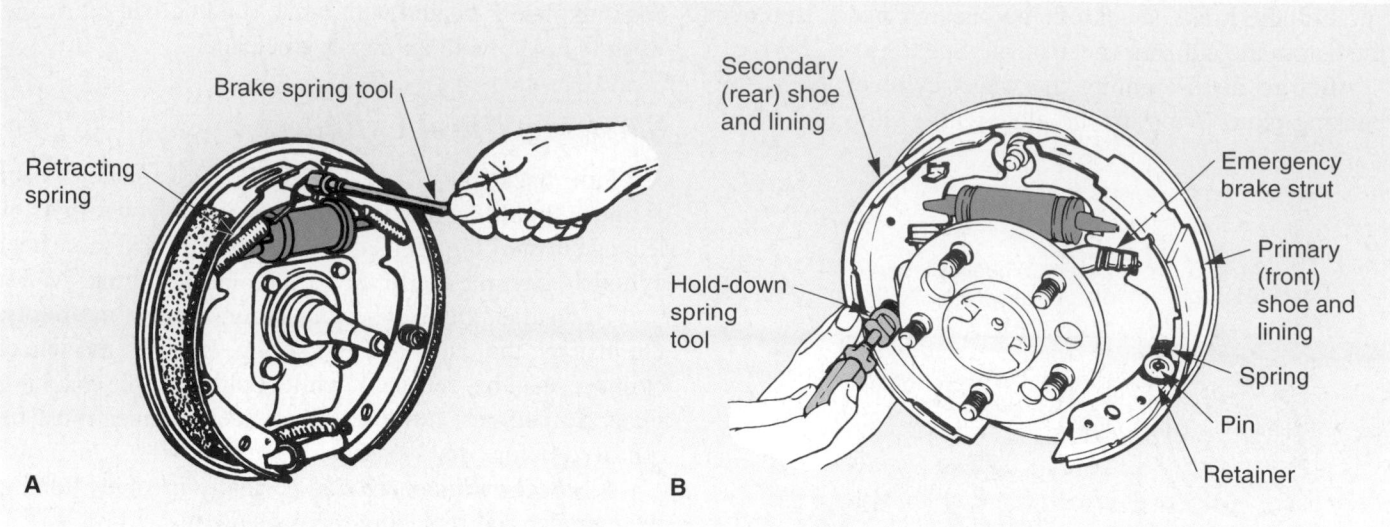

Figure 72-30. A—Brake retracting spring tool can be turned to pry springs off the anchor pin. B—A hold-down spring tool can be pushed in and turned to free the spring from the pin. Use your finger to hold the pin from the rear of the backing plate. (Bendix and Mopar)

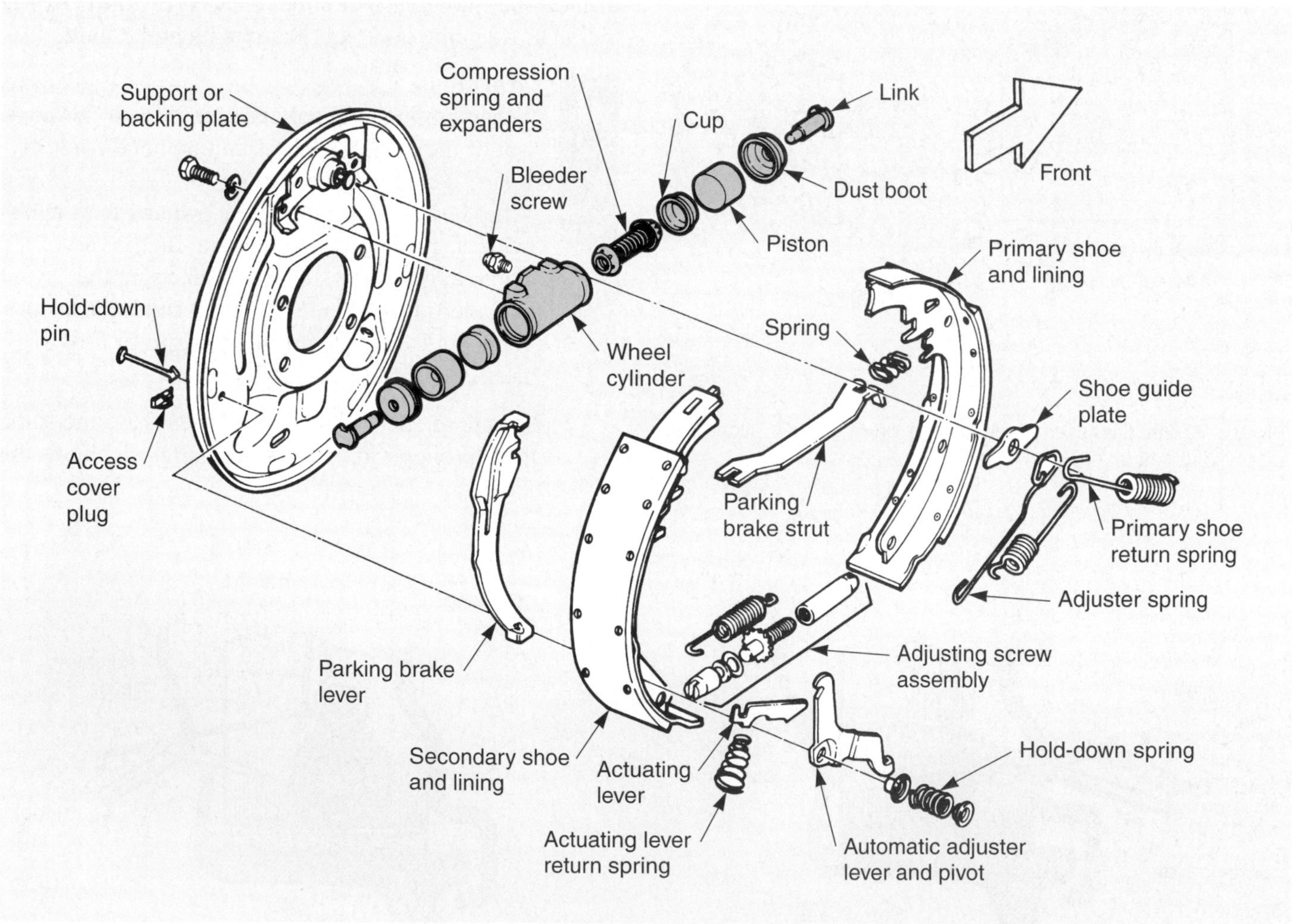

Figure 72-31. Disassembled view of typical drum brake unit. Study part names and locations. (Chrysler)

face of the cup. Coat the parts with brake fluid and fit them into the cylinder.

Brake Drum Resurfacing

Brake drum resurfacing, also called *turning,* involves machining the friction surface of the drum on a brake lathe. Look at **Figure 72-32.**

Resurfacing is needed when the drum is scored, out-of-round, or worn unevenly. Some shops machine a drum anytime the brake linings are replaced. Other shops only resurface the drum when needed. Refer to auto manufacturer suggestions when in doubt.

To resurface a brake drum, mount the drum on the lathe. Follow the operating instructions for the particular lathe. Wrap a silencing band (rubber or spring strap) around the outside of the drum. The band will prevent vibration that could affect drum surface smoothness, **Figure 72-33.**

Wearing eye protection, feed the cutting tool against the inner surface of the drum. Adjust the depth of the cut to lathe specs and activate the automatic feed.

When resurfacing a drum, machine off as little material as possible. Resurfacing thins the metal around the friction surface. As a result, the drum is more prone to overheating and warping.

Figure 72-32. When installing a brake drum on a lathe, you must use the correct collars and hubs. This will safely secure the drum to the lathe shaft.

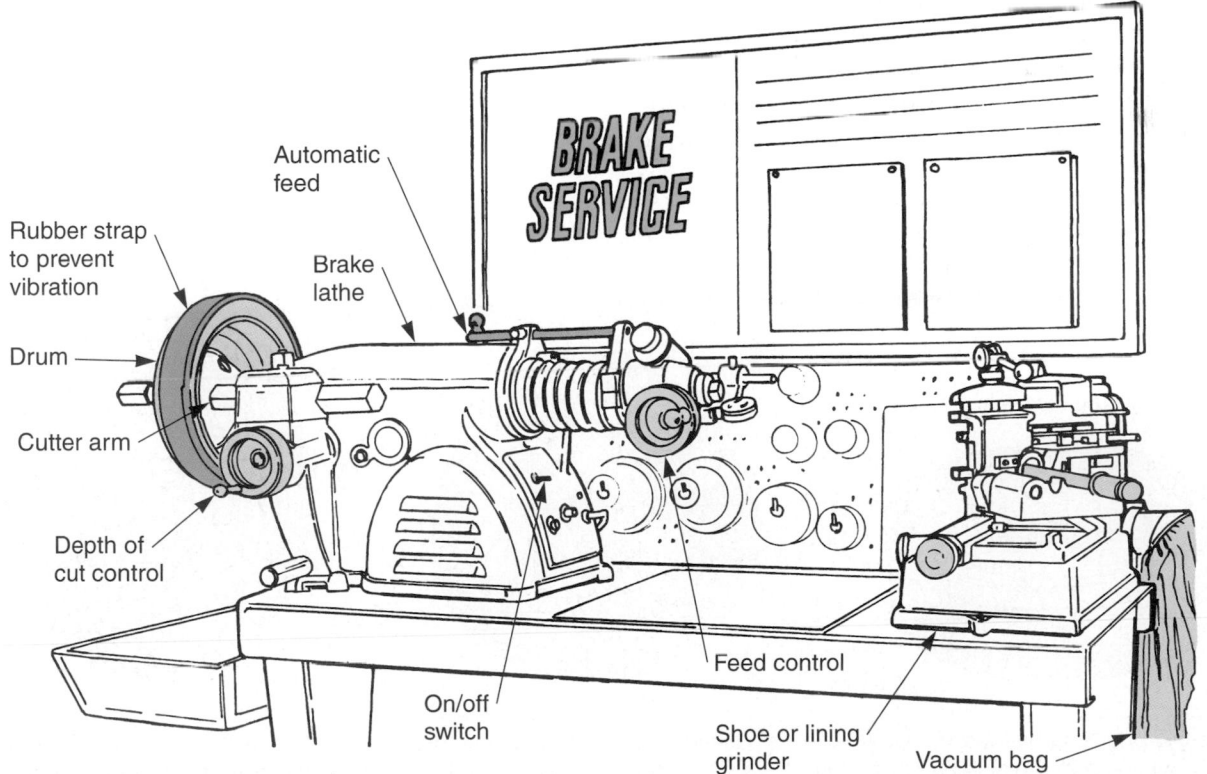

Figure 72-33. Note how this drum is mounted on the brake lathe. A shoe grinder is also shown. (EIS)

Machine the right- and left-hand drums to the *same diameter*. This will help ensure even straight-line braking.

A conventional cutting tool will not remove metal from hardened areas of the brake drum. A high spot will remain on the friction surface, and brake vibration will result. *Brake drum grinding* is sometimes needed to remove this hardened metal and true the friction surface accurately. The grinder is mounted on the lathe in place of the cutter arm.

Measuring Brake Drum Diameter

Typically, a brake drum should not be more than 0.060″ (1.5 mm) oversize. Look at **Figures 72-34A** and **72-34B**. For example, a drum that is 9″ (229 mm) in diameter when new must not be over 9.060″ (230 mm) in diameter after resurfacing. If larger in diameter, the drum is dangerously thin.

To measure brake drum diameter, use a special brake drum micrometer. It will measure drum diameter quickly and accurately, **Figure 72-34C.**

Replace the drum if it is worn beyond specs. Sometimes, maximum diameter is stamped on the side of the brake drum.

Drum Brake Reassembly

Before reassembling drum brakes, clean the wheel bearings and inspect them for wear or damage, **Figure 72-35.** Then, pack the bearings with grease and install new grease seals. (This was detailed in Chapter 66.)

Lubricate the small pads or bumps on the backing plate. This will keep the shoes from squeaking. Avoid using too much lubricant, or the linings can become contaminated and ruined, **Figure 72-36.** Also, lubricate the threads on the star wheel screw, if there is one.

Before installing the new shoes, check their fit inside the brake drum. There should be a small clearance between the ends of the lining and the drums. The shoes should rock slightly when moved in the drum. If the center of the linings are not touching the drum, the linings should be *arced* (ground).

When arcing brake shoe linings, follow the instructions provided by the equipment and vehicle manufacturer. Basically, you must grind the linings so that they are 0.35″ (0.89 mm) smaller in diameter than the drum. Without this clearance, the linings could chatter and vibrate when returned to service.

> **Warning**
> Make sure the vacuum system on the lining grinder is working properly. Remember! Asbestos brake lining dust can cause *cancer.*

Install the new shoes and the adjuster mechanism on the backing plate. Make sure all the parts are positioned correctly, **Figure 72-37.**

Ask yourself the following questions:

- Are the wheel cylinders in perfect condition and assembled properly?
- Did I lubricate the backing plate and the star wheel?
- Is the primary (smaller) lining facing the front of the vehicle and the secondary (larger) lining facing the rear of the vehicle?
- Are the shoes centered on the backing plate?

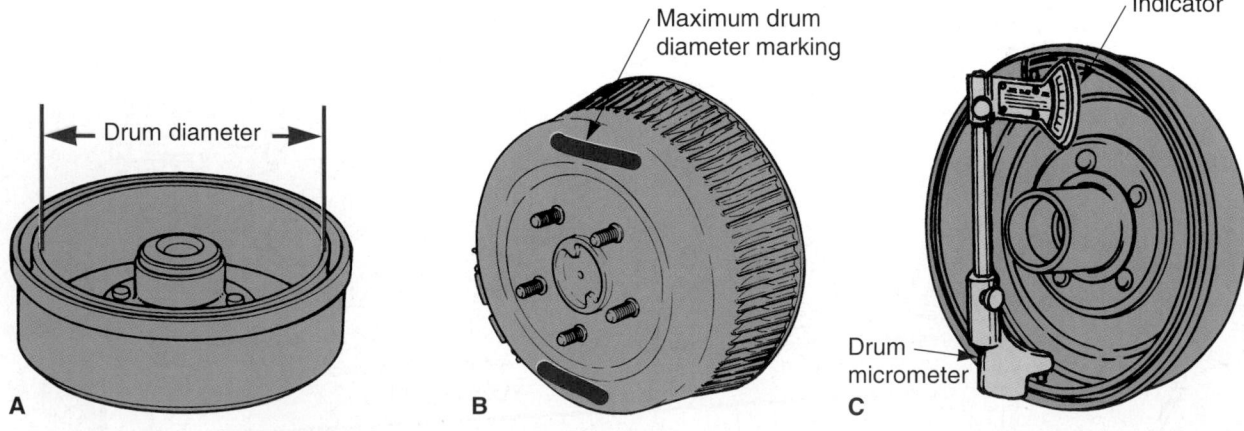

Figure 72-34. Checking brake drum wear. A—Drum diameter is the distance across the inside friction surface. B—Drum diameter is normally stamped on the outside of the drum. Oversize limits may also be given. C—A special drum micrometer will quickly and accurately measure drum diameter and wear. Generally, drums should not be more than 0.060″ (1.5 mm) oversize. (Toyota, Chrysler, FMC)

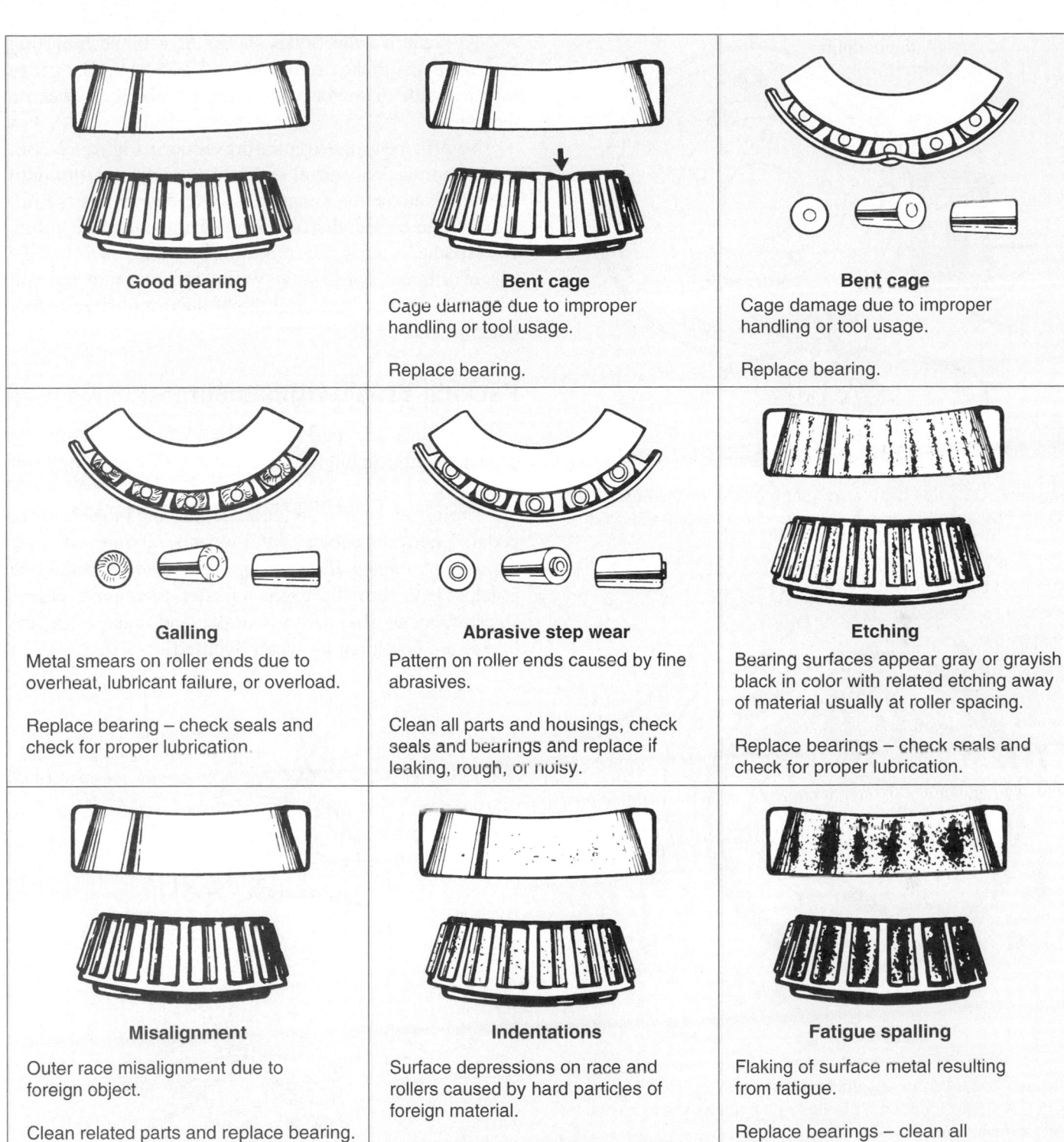

Figure 72-35. Common wheel bearing problems. You must usually service the wheel bearings when machining a drum or disc. Replace the bearings and their races if you find any of these problems. (Cadillac)

- Are the shoes contacting the anchor correctly?
- Are all the springs installed properly?
- Does the automatic adjuster work?

- Are the lining surfaces perfectly clean (sand if needed)?
- Do I need to bleed the brakes?

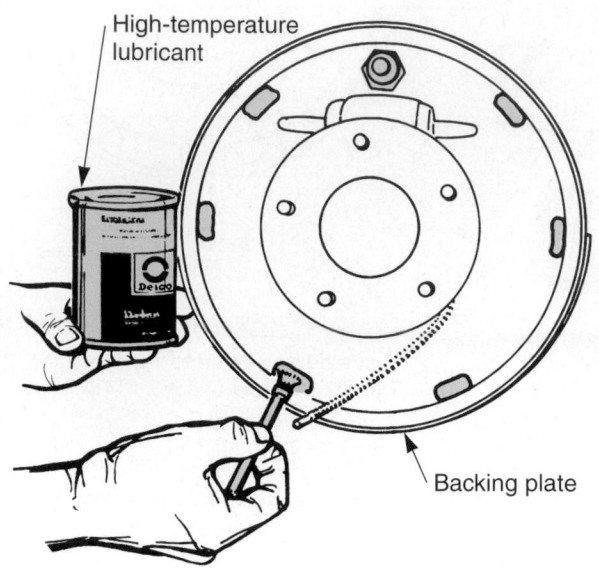

Figure 72-36. Wipe high-temperature grease on the raised pads on the backing plate and on the star wheel threads. (Bendix)

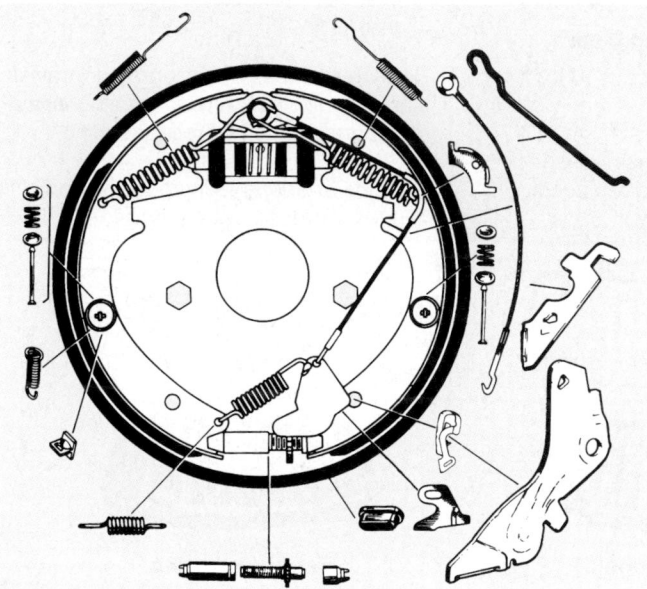

Figure 72-37. When assembling brakes, make sure every part is in the proper location. This illustration shows the locations for one specific unit.

Your service manual will have assembly illustrations for the particular brake design being serviced. Use them to help position the parts on the backing plate correctly.

Preadjusting Drum Brakes

Even though modern brakes have automatic adjusters, you should preadjust the brake shoes. This will ensure proper initial brake system operation.

To preadjust the brake shoes, fit a brake adjusting gauge into the brake drum, **Figure 72-38A.** Set the gauge for the inside diameter of the drum. Tighten the lock on the gauge.

Next, fit the gauge over the brake shoes, **Figure 72-38B.** Then, turn the star wheel or move the adjuster arm until the linings touch the gauge. This will ensure that the linings are the correct distance from the inside of the drum.

Another way to preadjust drum brakes involves the use of a brake spoon (star wheel tool) to turn the star wheel. Turn the wheel until the brake drums drag lightly when rotated by hand.

Parking Brake Adjustment

To adjust the parking brake, you must normally tighten an adjustment nut on the cable mechanism. Refer to **Figure 72-39.**

During adjustment, release the parking brake lever or pedal. Lubricate cables and linkages. To prevent over adjustment, engage the parking brake lever or pedal one notch. Then, turn the cable adjuster to remove excess slack. Operate the parking brake and make sure the brakes are not dragging when released.

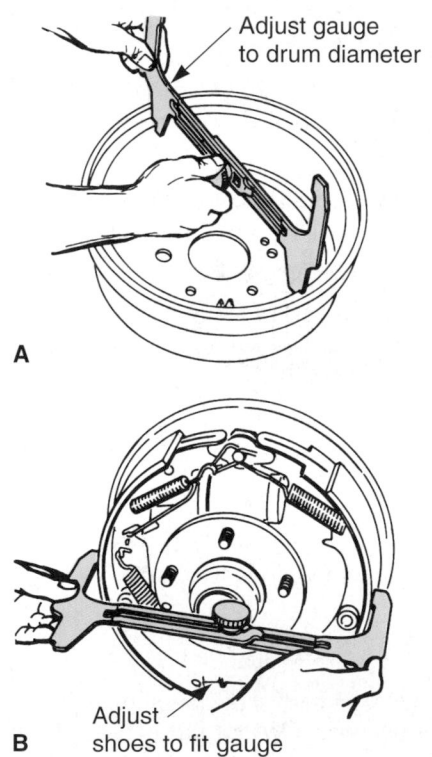

Figure 72-38. Study drum brake preadjustment. A—Fit the gauge into the brake drum and adjust it against the inside diameter of the drum. Lock the unit into position. B—Fit the gauge around the outside of the brake shoes. Adjust the shoes outward until they just touch the gauge. (Pontiac)

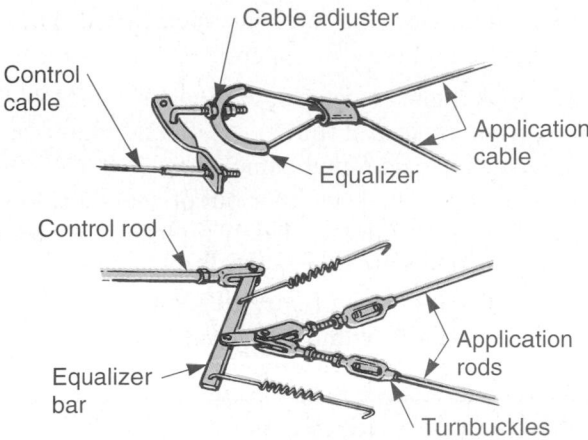

Figure 72-39. Parking brake adjustment. Simply turn the threaded fastener or the turnbuckle until excess slack is out of the cable. Do not overtighten the parking brake, or the brakes can drag and overheat. (FMC)

Summary

- Use symptoms (noises, smells, abnormal brake pedal movements, improper braking action) when diagnosing brake system problems.

- Many modern brake systems, especially those with ABS (anti-lock braking systems) provide on-board diagnostics.

- Brake vibration shows up as a chatter, pulsing, or shake. It happens only when the brake pedal is applied.

- Grabbing is a condition in which the brakes apply too quickly, with only light brake pedal application.

- Pulling brakes is a symptom in which the vehicle veers to the right or left when braking.

- A spongy brake pedal is a condition which causes the brake pedal to feel like it is connected to a spring or rubber band.

- A dropping brake pedal is a symptom where the brake pedal slowly moves all the way to the floor when steady pressure is applied to it.

- Dragging brakes are brakes that remain partially applied even though the brake pedal is released.

- Brake pedal height is the distance from the pedal to the floor with the pedal at rest.

- Brake pedal free play is the amount of pedal movement before the beginning of brake application.

- Typically, the brake fluid level should be 1/4″ (6 mm) from the top of the reservoir.

- When the fluid level in the master cylinder is low, you should inspect the brake system for leaks.

- When inspecting disc brakes, check the thickness of the brake pad linings. Pads should be replaced when the thinnest (most worn) part of the lining is no thicker than the metal shoe (approximately 1/8″ or 3 mm).

- A faulty master cylinder usually leaks fluid past the rear piston or leaks internally.

- A master cylinder is bench bled to remove air from inside the cylinder.

- Brake system bleeding is the use of fluid pressure to force air out of the brake line connections or wheel cylinder bleeder screws.

- Pressure bleeding a brake system is done using air pressure trapped inside a metal air tank.

- Brake disc runout is the amount of side-to-side movement measured on the outer friction surface of the rotor.

- Brake disc thickness is measured across the two friction surfaces in several locations.

- A wheel cylinder rebuild normally involves honing the cylinder and replacing the rubber cups and boots.

Important Terms

Symptoms	Brake fluid level
Brake vibration	Heat checked
Grabbing brakes	Bench bleeding
Excessive brake pedal effort	Brake system bleeding
	Manual bleeding
Pulling brakes	Pressure bleeding
Spongy brake pedal	Brake system flushing
Dropping brake pedal	Contaminated
Low brake pedal	Disc brake service
Dragging brakes	Brake disc runout
No brake pedal	Brake disc thickness
Brake warning light	Brake disc resurfacing
Braking noise	Vibration damper
Brake pedal check	Drum brake service
Brake pedal height	Wheel cylinder rebuild
Brake pedal free play	Brake drum resurfacing
Brake pedal reserve distance	Brake drum grinding

Review Questions—Chapter 72

Please do not write in this text. Place your answers on a separate sheet of paper.

1. What usually causes brake pedal vibration?

2. _____ brakes apply too quickly, even with light pedal application.

3. _____ brakes cause the vehicle to steer to the right or left when braking.

4. A spongy brake pedal is normally caused by air in the system. True or False?

5. A driver complains that the brake pedal slowly moves toward the floor when the vehicle is stopped at traffic lights. Pumping the pedal temporarily returns the pedal to the proper height. Also, the brake fluid has been checked and is at a normal height in the reservoir. Technician A says that a bad master cylinder could be the cause of the problem. Fluid could be leaking internally in the cylinder. Technician B says that a leaking wheel cylinder could be allowing pressure to drop. Who is correct?
   (A) A only.
   (B) B only.
   (C) Both A and B.
   (D) Neither A nor B.

6. List and explain three types of brake noise.

7. Which of the following is *not* a brake pedal measurement?
   (A) Brake pedal free play.
   (B) Brake pedal pressure.
   (C) Brake pedal height.
   (D) Brake pedal reserve distance.

8. How much fluid should typically be in a master cylinder?

9. How do you check brake pad lining wear and brake shoe lining wear?

10. How do you quickly test the basic operation of a vacuum brake booster?

11. Most technicians replace faulty master cylinders with new or factory rebuilt units. True or False?

12. Cleaning brake system parts in cold soak solvent is recommended. True or False?

13. A master cylinder should be _____ _____ before installation to remove air from the unit.

14. How do you pressure bleed a brake system?

15. List the four major operations done during disc brake service.

16. _____ _____ _____ is the amount of side-to-side movement measured near the outer friction surface of a disc.

17. What can happen if a brake disc is too thin?

18. List the seven major steps for drum brake service.

19. Typically, a brake drum should *not* be _____ oversize, or it can warp or break.

20. Summarize the preadjustment procedure for drum brakes.

21. A technician determines that an automobile's front brake pad linings are 0.025″ thick. Specs call for a minimum thickness of 0.020″. How much are the pads in or out of specifications?
   (A) 0.005″ out of specifications.
   (B) 0.005″ within specifications.
   (C) 0.045″ out of specifications.
   (D) 0.045″ within specifications.

⬢ ASE-Type Questions

1. A service writer says a customer's car has a brake system problem. Technician A road tests the car to detect any strange noises. Technician B checks the car for any unusual odors. Who is right?
   (A) *A only.*
   (B) *B only.*
   (C) *Both A and B.*
   (D) *Neither A nor B.*

2. A customer complains that his/her car vibrates when the brake pedal is applied. Technician A looks for a warped disc. Technician B checks for "hard spots" on the brake disc. Who is right?
   (A) *A only.*
   (B) *B only.*
   (C) *Both A and B.*
   (D) *Neither A nor B.*

3. A car is brought into the shop with grabbing brakes. Technician A looks for worn brake linings. Technician B looks for a faulty metering valve. Who is right?
   (A) *A only.*
   (B) *B only.*
   (C) *Both A and B.*
   (D) *Neither A nor B.*

4. A customer tells the shop service writer that there has been a noticeable increase in the amount of foot pressure needed to apply the brakes on her car. Technician A checks for a frozen wheel cylinder. Technician B checks the car's front end alignment. Who is right?
   (A) *A only.*
   (B) *B only.*
   (C) *Both A and B.*
   (D) *Neither A nor B.*

5. A car veers to the right when the brakes are applied. Technician A checks the vehicle's front end alignment. Technician B looks for a frozen caliper. Who is right?

(A) A only.
(B) B only.
(C) Both A and B.
(D) Neither A nor B.

6. An automobile's brake system has a soft, spongy pedal. Technician A checks the fluid level in the master cylinder. Technician B checks for a system leak. Who is right?
(A) A only.
(B) B only.
(C) Both A and B.
(D) Neither A nor B.

7. An automobile has a "spongy" brake pedal. Technician A checks to see if there is an excessive amount of fluid in the brake system. Technician B "bleeds" the brake system. Who is right?
(A) A only.
(B) B only.
(C) Both A and B.
(D) Neither A nor B.

8. An automobile has a dropping brake pedal. Technician A looks for an internal master cylinder leak. Technician B checks for a leaking wheel cylinder. Who is right?
(A) A only.
(B) B only.
(C) Both A and B.
(D) Neither A nor B.

9. A car has a low brake pedal problem. Technician A inspects the brake adjusters. Technician B looks for a maladjusted master cylinder push rod. Who is right?
(A) A only.
(B) B only.
(C) Both A and B.
(D) Neither A nor B.

10. An automobile's brakes are "dragging." Technician A checks the parking brake adjustment. Technician B checks for a master cylinder malfunction. Who is right?
(A) A only.
(B) B only.
(C) Both A and B.
(D) Neither A nor B.

11. A small pickup truck is brought into the shop with a no-brake-pedal condition. Technician A checks the hydraulic system for problems. Technician B checks the wheel cylinder return springs. Who is right?
(A) A only.
(B) B only.

(C) Both A and B.
(D) Neither A nor B.

12. An automobile has a metal-on-metal grinding sound when the brakes are applied. Technician A inspects the brake linings. Technician B looks for a leaking wheel cylinder. Who is right?
(A) A only.
(B) B only.
(C) Both A and B.
(D) Neither A nor B.

13. An automobile's brake system needs to be inspected. Technician A checks the system's brake pedal action. Technician B checks the system's master cylinder fluid level. Who is right?
(A) A only.
(B) B only.
(C) Both A and B.
(D) Neither A nor B.

14. Brake dust has accumulated on an automobile's wheel brake assemblies. Technician A uses a brake vacuum machine to remove this dust. Technician B uses compressed air to remove this dust. Who is right?
(A) A only.
(B) B only.
(C) Both A and B.
(D) Neither A nor B.

15. An automotive brake disc has 0.001 runout. Technician A wants to replace the brake disc. Technician B says this amount of runout is normally within specs. Who is right?
(A) A only.
(B) B only.
(C) Both A and B.
(D) Neither A nor B.

Activities—Chapter 72

1. Demonstrate the safe method of removing asbestos laden dust from a brake assembly.

2. Bleed a set of brakes and explain your procedure.

3. Using a flat rate manual and a parts catalog prepare a bill for a brake job.

4. Using a labor rate set by your instructor and a price list for parts, add up the cost of the brake job. Fill in the price on the bill prepared in Activity 3.

Brake System Diagnosis

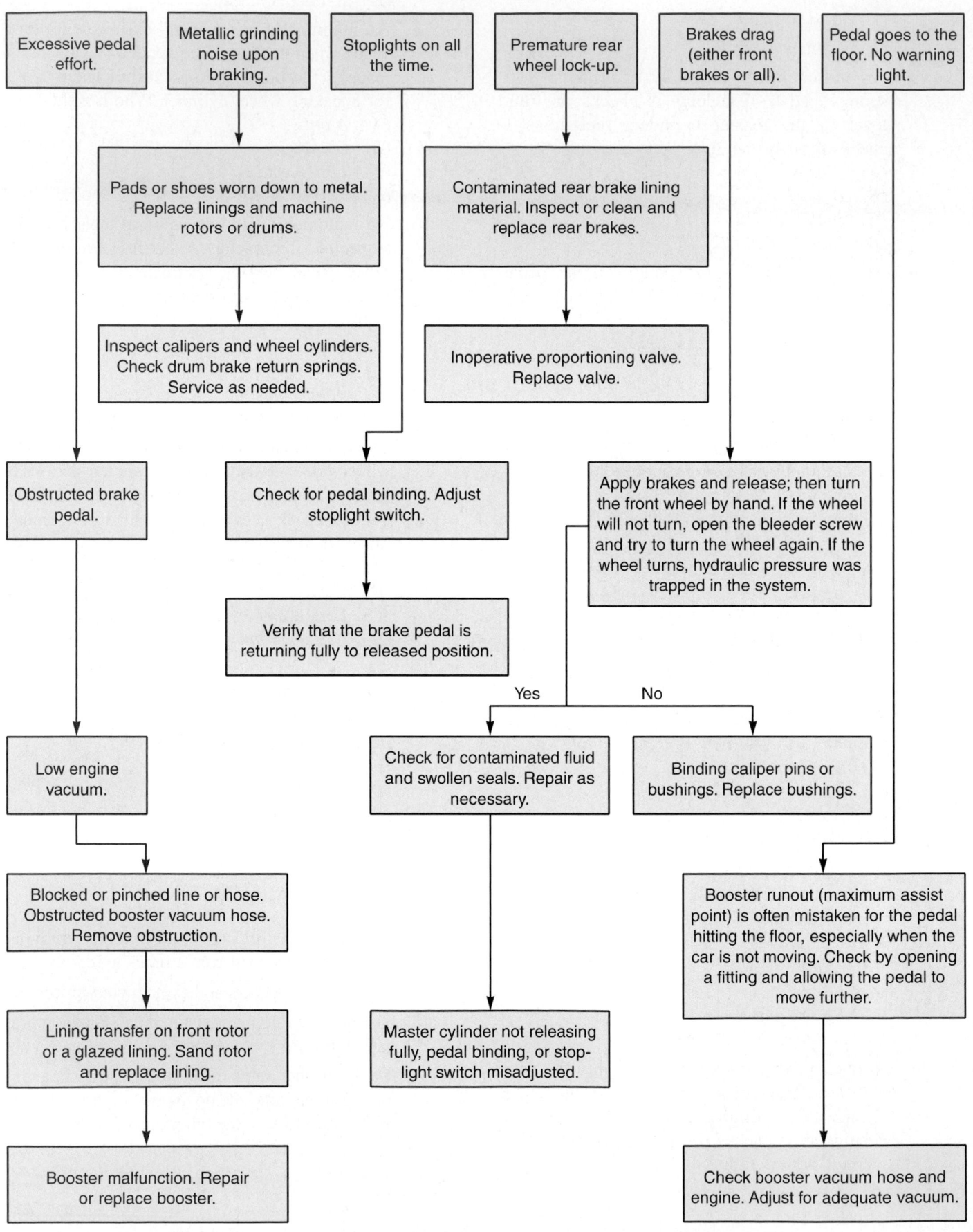

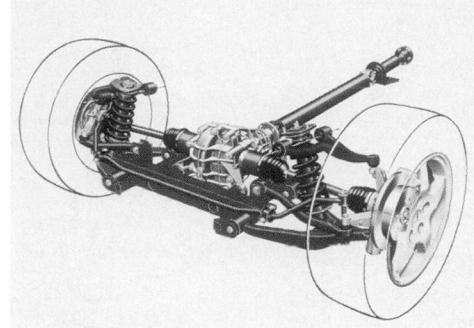

Chapter 73

Anti-Lock Brakes, Traction Control, and Stability Control

After studying this chapter, you will be able to:

- Identify the major parts of a typical anti-lock brake system.
- Describe the operation of anti-lock brake systems.
- Compare anti-lock brake design variations.
- Diagnose problems in anti-lock brake systems.
- Repair anti-lock brake systems.
- Describe the purpose and operation of traction control and stability control systems.
- Diagnose and repair traction control and stability control systems.
- Correctly answer ASE certification test questions requiring a knowledge of anti-lock brake systems traction control systems, and stability control systems.

This chapter will build on the information you learned in previous chapters, especially those on brake systems. It will introduce the additional parts needed to provide computer control of the brake system.

Anti-lock brake systems are now very common. They can be found on economy cars, luxury cars, mini vans, and even pickup trucks. This makes it essential that you understand the operation and service of these important systems.

The chapter also summarizes the operation and repair of traction control and stability control systems. These systems use parts of the anti-lock brake system to help prevent tire spin and skidding when accelerating or cornering.

Anti-Lock Brake Systems (ABS)

An *anti-lock brake system,* abbreviated *ABS,* uses wheel speed sensors, a computer (ECM), and a modulator unit to prevent loss of tire adhesion during hard braking. The major parts of a modern anti-lock brake system are illustrated in **Figure 73-1.**

If a tire locks up and skids on the road surface, braking distance can increase and steering control can be lost. In a panic stop situation, the driver will often press down hard on the brake pedal to try to avoid a collision. This will normally send too much hydraulic pressure to the wheel cylinders, locking the wheels and tires on vehicles without ABS. The tires will begin to squeal as they slide over the road surface. When the tires skid, they actually loose their friction with the road surface. This can increase stopping distance and reduce vehicle control in some situations.

On dry pavement, good drivers only apply enough pedal pressure to almost reach tire skid. They will release pedal pressure slightly when the squeal of tire skid is heard or felt. On slick pavement (water, snow, or ice on road surface), good drivers might pump the brake pedal manually to reduce skidding and stopping distance.

This is the principle of anti-lock brakes. For maximum stopping power, you want the tires to almost, but not quite, skid. When the tire skids, its friction with the road surface drops and stopping distance increases, which could cause an accident.

The anti-lock brake system improves driver and passenger safety by reducing stopping distances and increasing directional stability under panic stop conditions.

Figure 73-2 shows the advantages of ABS. In this example, one side of the road is very slippery and the other side is dry. This poses a problem if a panic stop is required.

Without ABS, the car would tend to skid to the right because of higher tire adhesion on the right. With ABS, the car would still travel straight ahead with hard braking. The brake units would be cycled to prevent tire skid and a loss of control. Also note that the car can be steered while braking with ABS.

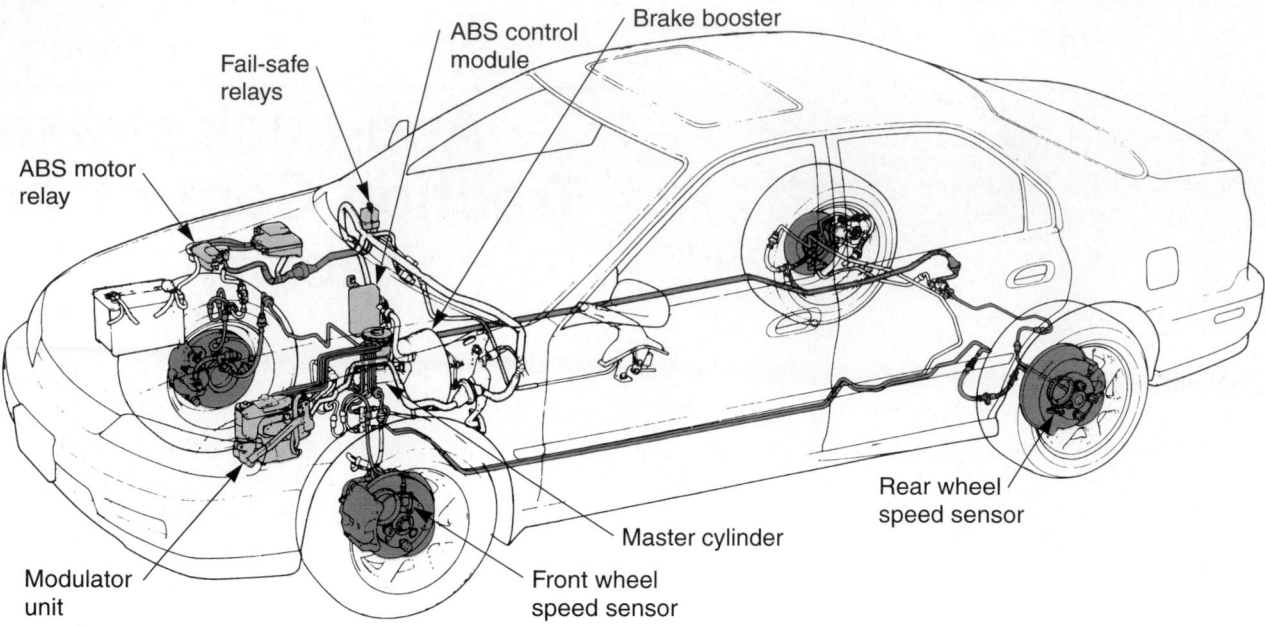

Figure 73-1. A modern anti-lock brake system uses a control module (computer), sensors, and a modulator to aid panic stops or stops on slick pavement. Also, note the locations of the relays. (Honda)

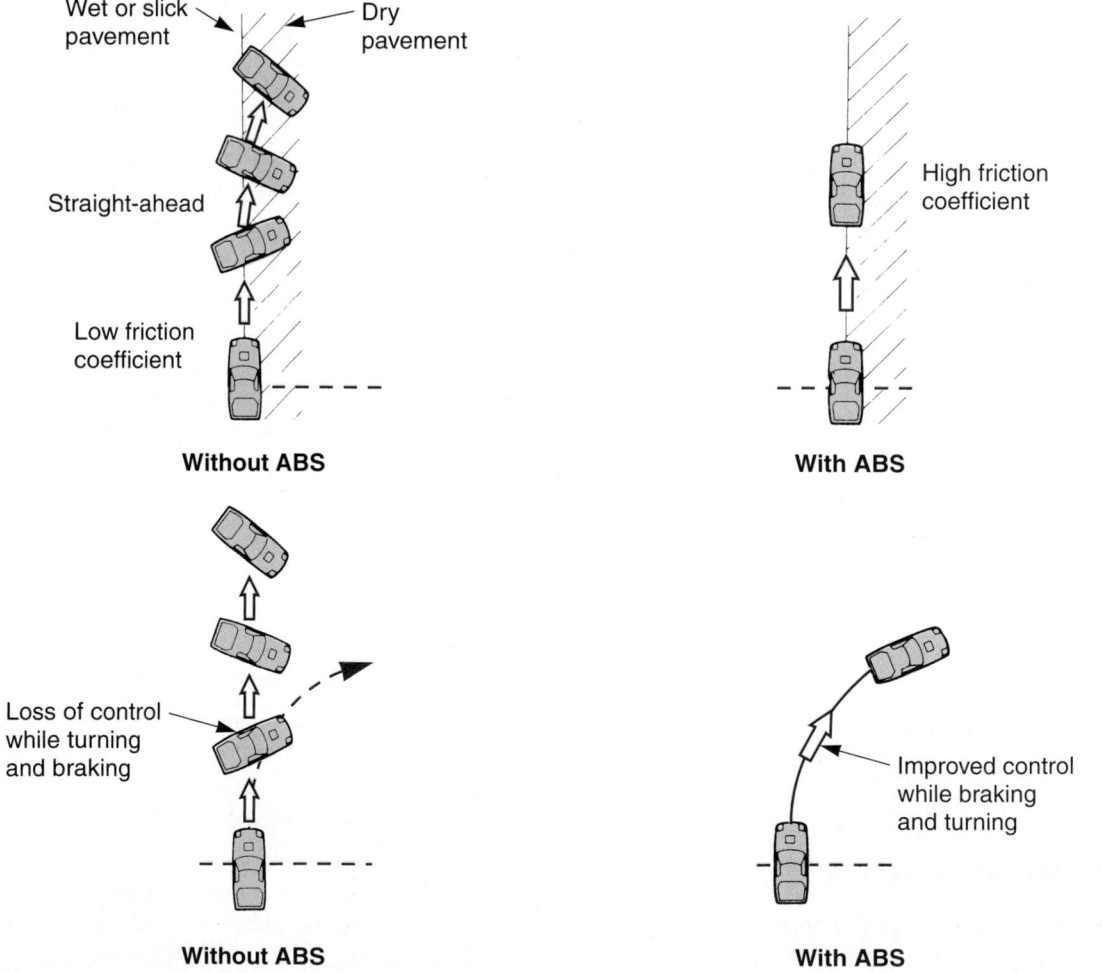

Figure 73-2. Without ABS, tires can lock up and lose adhesion with the road. This can cause a loss of control during hard braking or when braking and trying to steer. Even if one side of the road is slick, the brakes will be cycled so that none of the tires skid. Maximum braking results when tires are just about ready to skid; traction is reduced if the wheels lock up.

ABS Components

The major parts of an anti-lock braking system, **Figure 73-3,** include:

- *Wheel speed sensors*—magnetic pickups for detecting the rotating speed of each tire and wheel assembly.

- *Sensor rotors*—toothed wheels that rotate at the same rpm as the wheel and tire.

- *ABS control module*—small computer, or processor, that uses sensor inputs to control the hydraulic modulator.

- *Electro-hydraulic modulator*—solenoid-operated valve and electric pump mechanism for cycling pressure to each wheel brake cylinder.

- *Warning light*—dash light that informs the driver of problems in the anti-lock brake system.

Another illustration of the major parts of an anti-lock brake system is given in **Figure 73-4.**

Wheel Speed Sensors

The **wheel speed sensors** produce ac signals that correspond to wheel and tire speed. These signals are fed into the anti-lock brake system control module. With

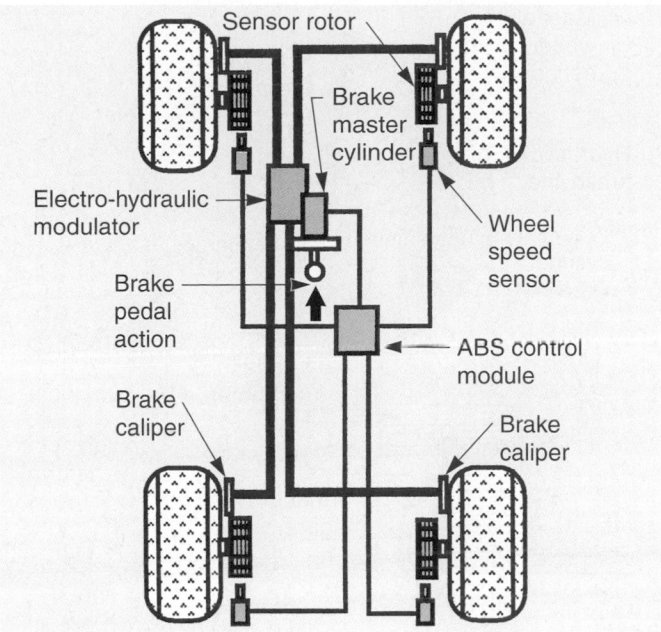

Figure 73-3. Note basic parts of an anti-lock brake system. Sensor rotors are mounted on each spindle. Magnetic sensor coils produce an ac signal that equals wheel speed. The ABS control module can detect if one or more wheels are skidding because the sensor signal will decrease in frequency. The control module can then trigger the electro-hydraulic modulator to cycle hydraulic pressure to keep the tires from skidding and losing traction.

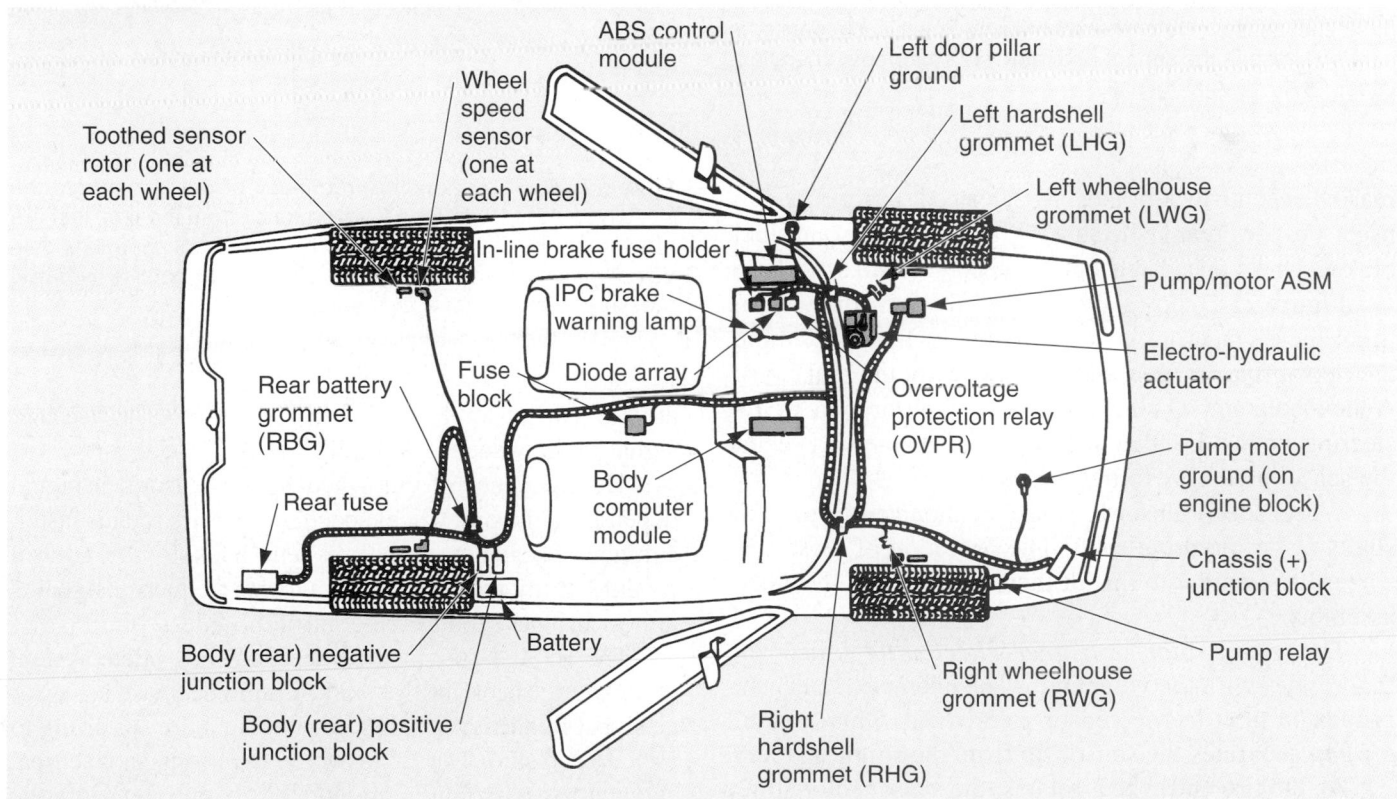

Figure 73-4. Study the parts of a typical anti-lock brake system. (General Motors)

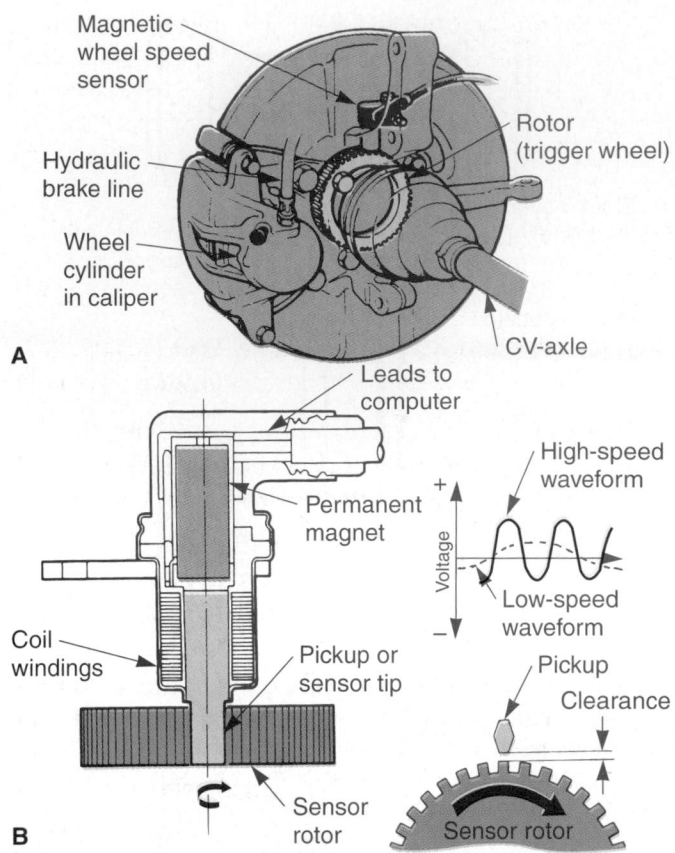

Figure 73-5. A—Wheel speed sensor is mounted next to the sensor rotor. As the tire rotates, teeth on the rotor make the magnetic sensor produce a weak ac signal for the ABS control module. B—Current is induced in the sensor windings as teeth pass the sensor. Note the waveform. Specific clearance is needed between the sensor and rotor teeth for proper operation. (Saab)

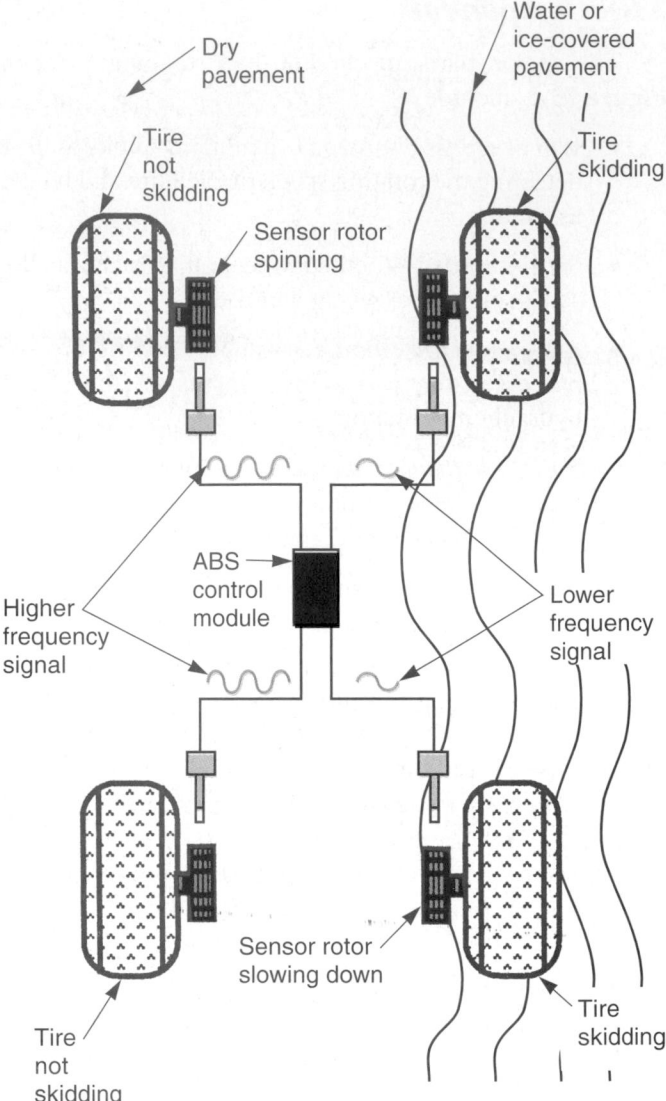

Figure 73-6. Study how the frequency of the signal from the skidding tires is lower than the frequency from the tires that are rotating on dry pavement. This is how the ABS control module detects when a tire or tires are skidding to control hydraulic pressure to those wheel cylinders.

rear-wheel anti-lock brakes, wheel speed sensors are only used on the rear wheels. With four-wheel anti-lock brakes, wheel speed sensors are needed on all wheels.

Figure 73-5 shows a wheel speed sensor mounted in the vehicle's steering knuckle assembly. In this example, the sensor rotor is mounted on the back of the brake disc. A hex bolt and O-rings secure the sensor coil in the steering knuckle. Other methods can be used to secure the sensor assembly to the vehicle.

Wheel speed sensors can also be mounted on the axle shafts or on the drive shaft. The operation of these sensors is very similar to that of those mounted on the wheel assembly.

The *sensor rotor,* or *trigger wheel,* is a toothed ring that rotates with the wheel hub. The magnetic sensor tip is located next to the sensor rotor teeth. Only a small *air gap* separates the sensor tip from the spinning teeth.

As the tire and wheel rotates, the sensor rotor spins. As each rotor tooth passes the sensor, the magnetic field

around the sensor is affected. This induces a weak ac signal in the wheel sensor coil, **Figure 73-5.**

The frequency of this wheel speed sensor's ac signal depends on tire and wheel speed. As the tire rotates faster, a higher frequency signal is sent back to the control module. If tire rotation slows, a lower frequency signal is almost instantly fed to the control module.

The ac signal is used by the anti-lock brake system computer to check for tire skid. A rapid decrease in sensor ac signal frequency would indicate that a tire is starting to lock up and skid. The ECM would then send an electrical signal to the hydraulic modulator to pulsate hydraulic pressure to the affected wheel cylinder, **Figure 73-6.**

Tech Tip
The operation of a wheel speed sensor is similar to that of a crankshaft sensor or a magnetic pickup coil in an ignition system. Refer to the discussion of these sensors for more information on coil and trigger wheel operating principles.

ABS Control Module

The *ABS control module* is a computer that uses wheel speed sensor inputs to control the operation of the electro-hydraulic modulator. It is constructed and operates like the other computers discussed in this book. See **Figure 73-7.** Refer to Chapter 17 for more information on computer operation.

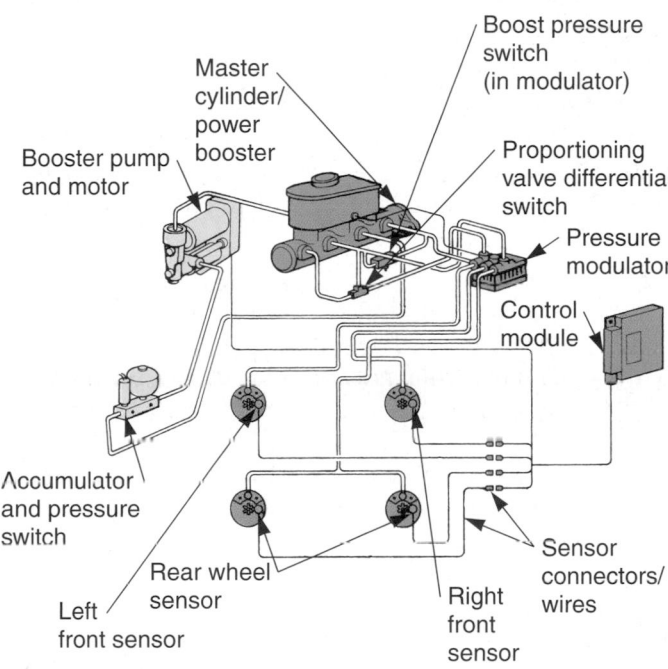

Figure 73-7. An accumulator is used to store small amount of pressure to reduce pressure cycling. An electric motor–driven pump produces pressure assist for this ABS system. (Chrysler)

ABS Electro-Hydraulic Modulator

The *ABS electro-hydraulic modulator,* also called a *ABS actuator,* regulates the fluid pressure applied to each wheel brake assembly during periods of hard braking. The modulator is controlled by the ABS computer. It is often mounted on the firewall in the engine compartment or somewhere in the passenger compartment, **Figure 73-8.**

The major parts of an ABS modulator include:

- *Fluid reservoir*—container for holding an extra supply of brake fluid, **Figure 73-9.**

- *Solenoid valve block*—valve block containing coil-operated valves that control brake fluid flow to the wheel brake cylinders.

- *Accumulator*—chamber for storing fluid under high pressure.

- *Hydraulic pump and motor*—high-pressure pump operated by small electric motor; they provide brake fluid pressure for system.

- *Pressure switch*—monitors system pressure and controls operation of electric motor for hydraulic pump.

- *Master cylinder-booster assembly*—conventional master cylinder with power assist for operating brakes under normal conditions.

ABS systems and hydraulic modulators vary with the specific make and model of car, van, or truck. However, their operating principles are very similar.

Note how the modulator assembly can control hydraulic brake line pressure to each wheel cylinder in **Figure 73-10.**

The primary purpose of the electro-hydraulic modulator is to produce hydraulic pressure pulses, or modulations. *Hydraulic pressure modulation* refers to the electro-hydraulic unit's ability to rapidly cycle pressure to the wheel cylinders or calipers on and off, preventing wheel lockup and skidding. Anti-lock systems can modulate the brakes about 15 to 20 times per second.

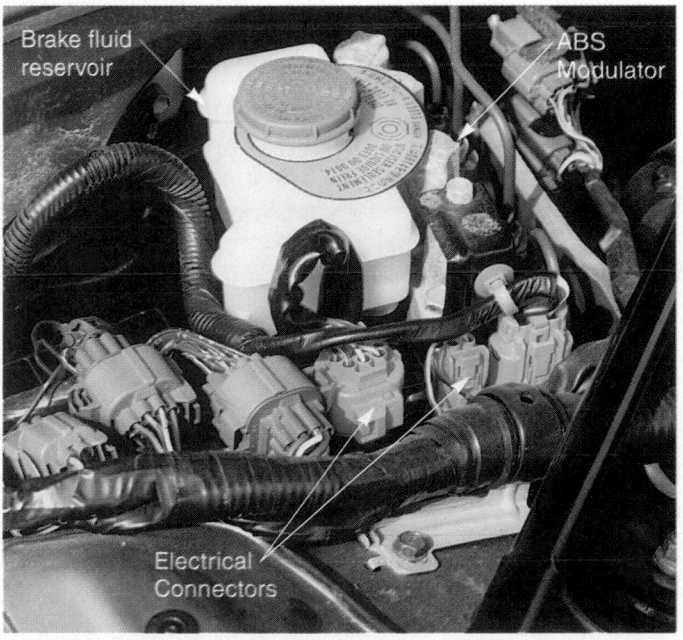

Figure 73-8. ABS modulator, or electro-hydraulic unit, in an engine compartment. Note the wire connections and the brake fluid reservoir. (Snap-on Tool Corp.)

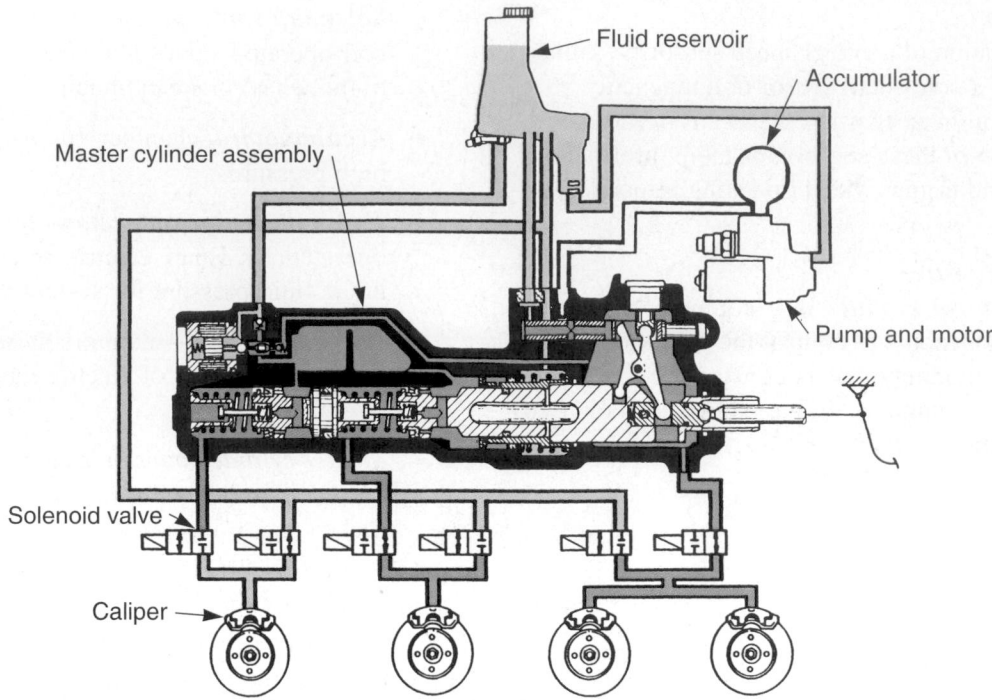

Figure 73-9. Study this variation of an ABS system. Note how the system is blocking pressure to one of the wheel cylinders to keep the brakes from locking and causing the tire to skid. Solenoid valves are positioned in the brake lines to the wheel cylinders. (Saab)

There are three modes of ABS operation: the isolation mode, the dump mode, and the reapply mode. In the *isolation mode,* the ABS control module detects that a tire is starting to skid and energizes the modulator to isolate that hydraulic circuit from the other brake lines. The nonskidding wheel circuits work normally, but the ABS takes over the one hydraulic circuit with the slowing wheel.

In the *dump mode,* the ABS releases hydraulic pressure to one or more of the brake wheel cylinders or calipers. This is done to stop tire skidding. Wheel circuits work normally, but the ABS takes over. The ECM will energize a solenoid that redirects brake fluid pressure away from the wheel brake assembly. This is illustrated in **Figure 73-11A.**

In the *reapply mode,* the ABS applies hydraulic pressure to one or more of the wheel brake assemblies. This is done to reapply the brakes during pressure modulation, **Figure 73-11B.**

These three control modes (isolation, dump, and reapply) happen in milliseconds (thousandths of a second). Several brake pumping cycles occur each second. This is much faster than a human could manually pump the brakes.

 Tech Tip
With hydraulic pressure modulation, the driver can actually feel and sometimes hear the brake pedal pulsing up and down or vibrating. This is caused by the modulator unit cycling pressure on and off, trying to reduce tire skid. This is a basic way to check to see if the system is functioning.

ABS Warning Light

An amber *ABS warning light* is mounted in the dash. This light is used to alert the driver to an ABS malfunction. It is important to note that the light generally glows during engine cranking and stays on for several seconds after the engine starts. This lets you know that the system is armed and functioning.

Integrated and Nonintegrated ABS

An *integrated ABS* combines the power booster, master cylinder, and modulator units into one assembly. All these units are mounted together.

In a *nonintegrated ABS,* the major units are separated. A conventional booster and master cylinder are used with a separate electro-hydraulic modulator.

ABS Channels

A *channel* is a separate hydraulic circuit that feeds out to one or more wheel cylinders or caliper pistons. Anti-lock brake systems can have from one to four channels.

A *one-channel ABS* normally operates the rear wheel brakes together. Only one channel is provided for both rear wheels. The front wheel brakes are not controlled by the ABS in a one-channel system. This was one of the earliest types of ABS.

A *two-channel ABS* has two separate hydraulic circuits controlled by the electro-hydraulic modulator. For example, with rear-wheel ABS, each rear brake has its own hydraulic control circuit. Pickup trucks sometimes use two-channel ABS, because the rear of the vehicle is

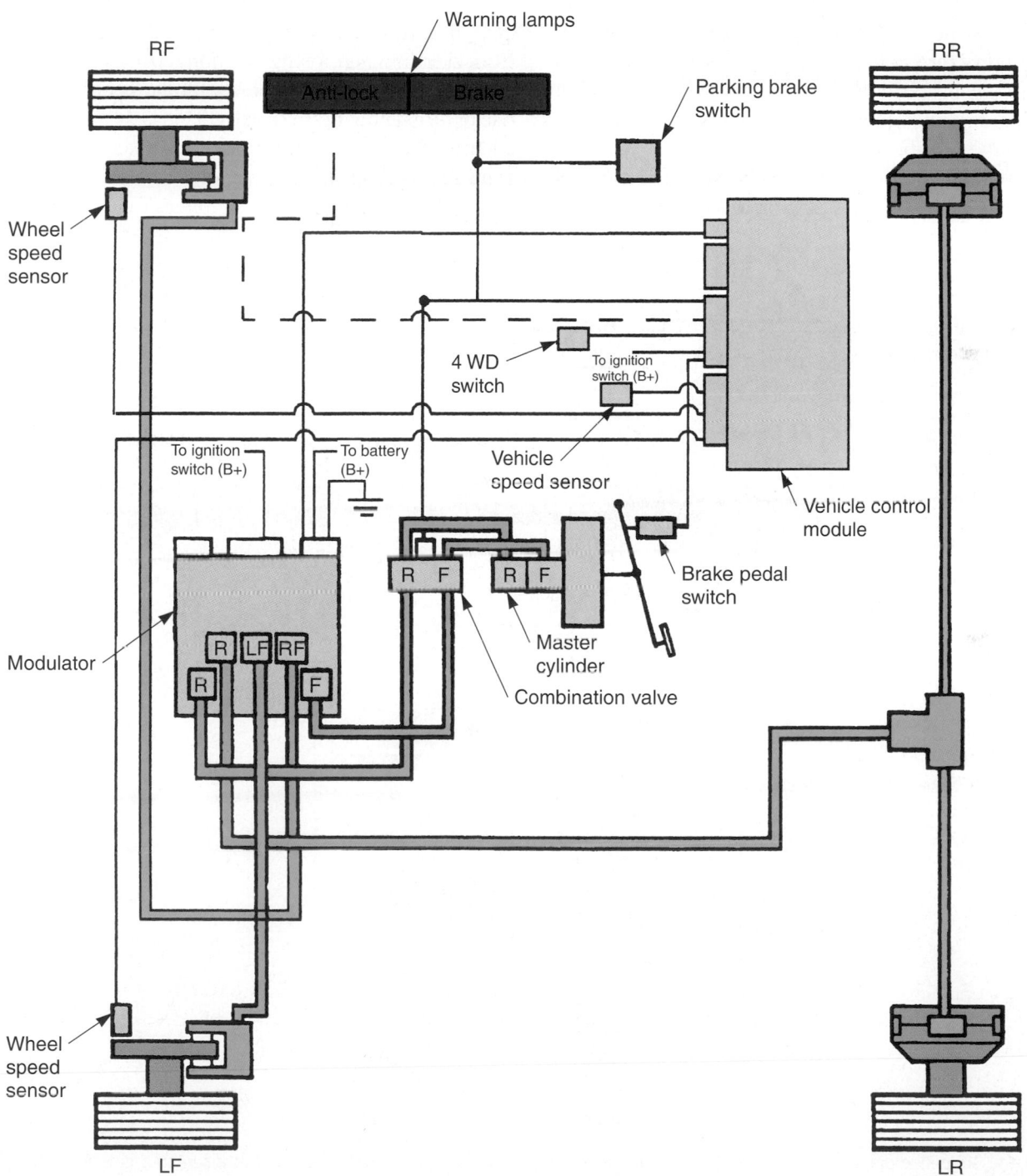

Figure 73-10. Diagram shows the major parts of another ABS. Study the electrical and hydraulic circuits closely. (General Motors)

considerably lighter than the front. Help is needed to keep the rear wheels from skidding because there is less weight pressing down on the rear tires.

A *four-channel ABS* has a separate hydraulic control circuit for each wheel. This is the most effective and common anti-lock brake system found on late-model vehicles. It can keep any one of the four tires from skidding.

ABS Operation

Under normal braking conditions, ABS is not used. The master cylinder reacts to brake pedal movement. It sends fluid pressure to each wheel cylinder without ABS intervention. A proportioning valve, like the one found in conventional braking systems, is commonly used to reduce pressure to the rear brakes.

If the brakes are applied in a panic stop and a wheel begins to lock up, the ABS is activated. The wheel sensor on the slowing wheel instantly reduces the frequency of the ac signal sent to the anti-lock control module. The control module detects that the wheel is slowing down more than the others and is getting ready to slide or skid. The module then sends an electrical current to the correct solenoid on the electro-hydraulic modulator assembly to bleed pressure from that channel.

With many designs, when current is sent to the solenoid, the solenoid opens a valve to reduce the pressure to the brake cylinder with the slowing wheel. The ABS control module quickly cycles the current to the solenoid on and off, artificially "pumping the brakes" to keep the tire from skidding and losing traction.

When ABS takes over, the brake pedal will usually rise or drop and vibrate slightly. This is due to the pressure entering the system from the accumulator or pump and from the cycling of the solenoid valves. This is normal and will stop when ABS is no longer functioning.

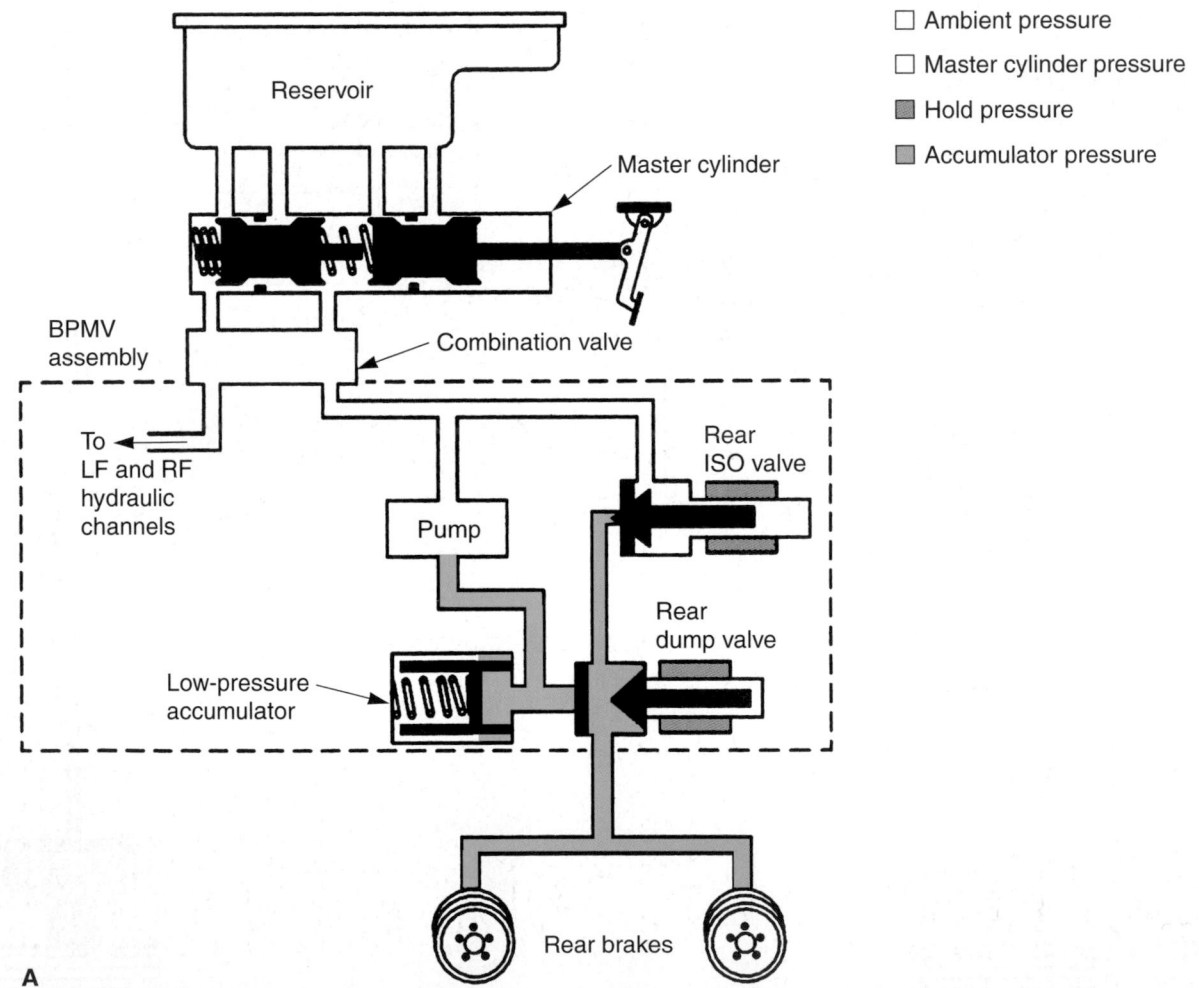

Figure 73-11. Simplified diagram shows how the modulator controls brake fluid pressure to the rear brakes in an anti-lock brake system. The pump maintains pressure in the accumulator. The accumulator stores and feeds fluid pressure to the wheel cylinders. A—The dump valve releases pressure to prevent tire skid. (General Motors) *(Continued)*

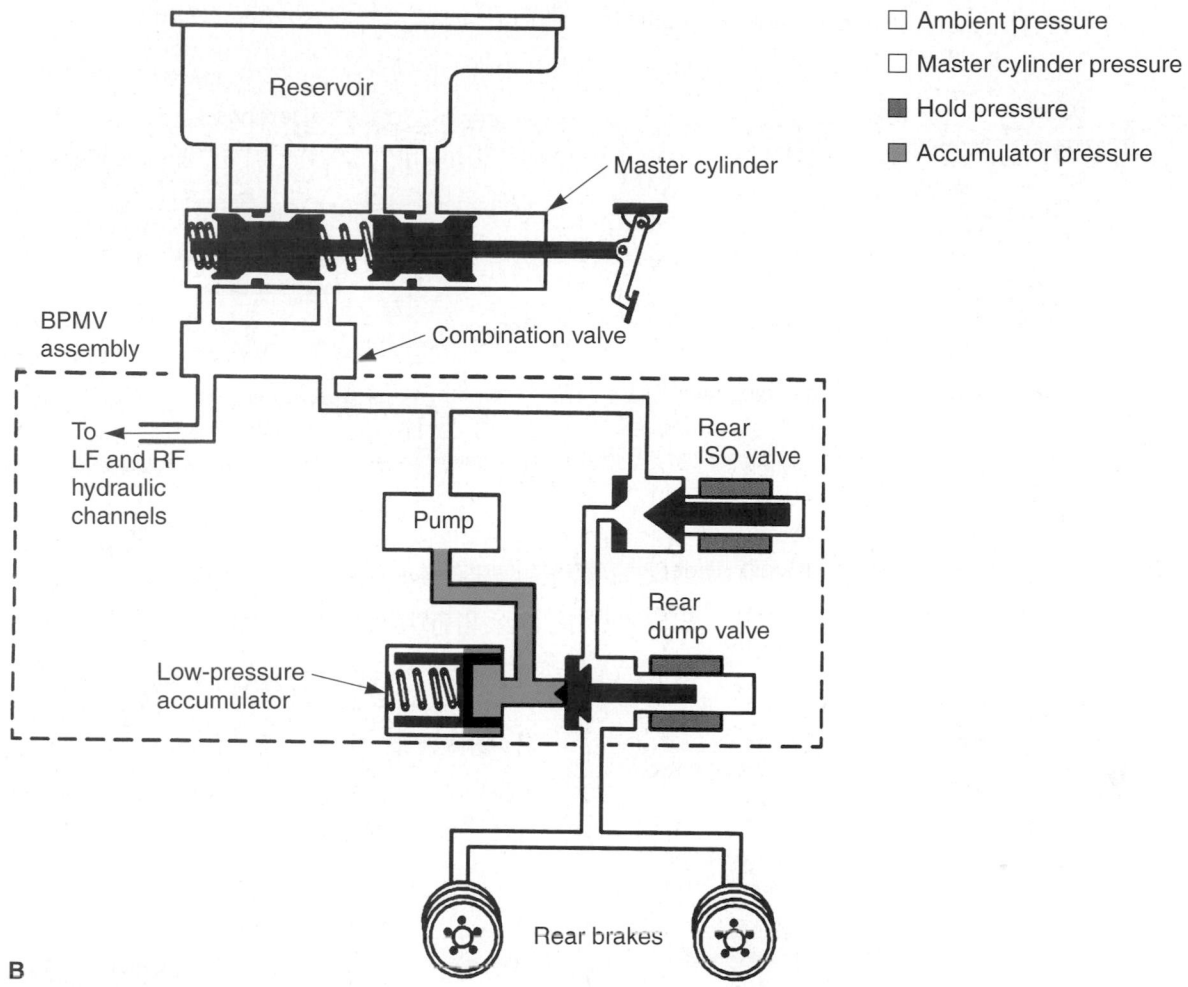

☐ Ambient pressure

☐ Master cylinder pressure

■ Hold pressure

■ Accumulator pressure

Figure 73-11. *(Continuod)* B In the reapply mode, pressure to the brakes is increased. (General Motors)

When hydraulic pressure in the system drops below a specific point, the pressure switch closes and energizes the electric motor. This drives the pump to build pressure back up. Once pressure is normal, the pressure switch opens and the motor and pump shut off.

If an ABS component malfunctions, the computer will detect an abnormal condition. It will then light the ABS warning light in the dash and deactivate the ABS. The brake system will still function normally, but without the anti-lock feature. The operation of another type of anti-lock brake system is shown in **Figure 73-12.**

Tech Tip

Warn new drivers that they should not pump the brake pedal on a vehicle equipped with ABS. This can prevent the system from functioning. Instruct them to apply hard, steady pedal pressure so that the ABS can modulate, or "pump," the brakes. For the experienced driver, this can take time to learn.

Traction and Stability Control Systems

Traction control systems are designed to prevent the vehicle's wheels from spinning and losing traction under hard acceleration. Most traction control systems work with the anti-lock brake system to cycle hydraulic pressure to the wheel spinning the fastest. The control module is capable of applying only one wheel brake at a time. Some systems also reduce engine power output to reduce skidding.

An indicator light comes on anytime the traction control system is activated. This warns the driver that the tires are losing traction and slipping on the road surface. He or she then knows that the traction control system is trying to limit or prevent this problem.

A *stability control system* is an advanced system that reduces tire spin upon acceleration *and* prevents tire skid when cornering too quickly. This system uses more input signals from various sensors to provide greater control over vehicle handling under severe cornering, braking, and

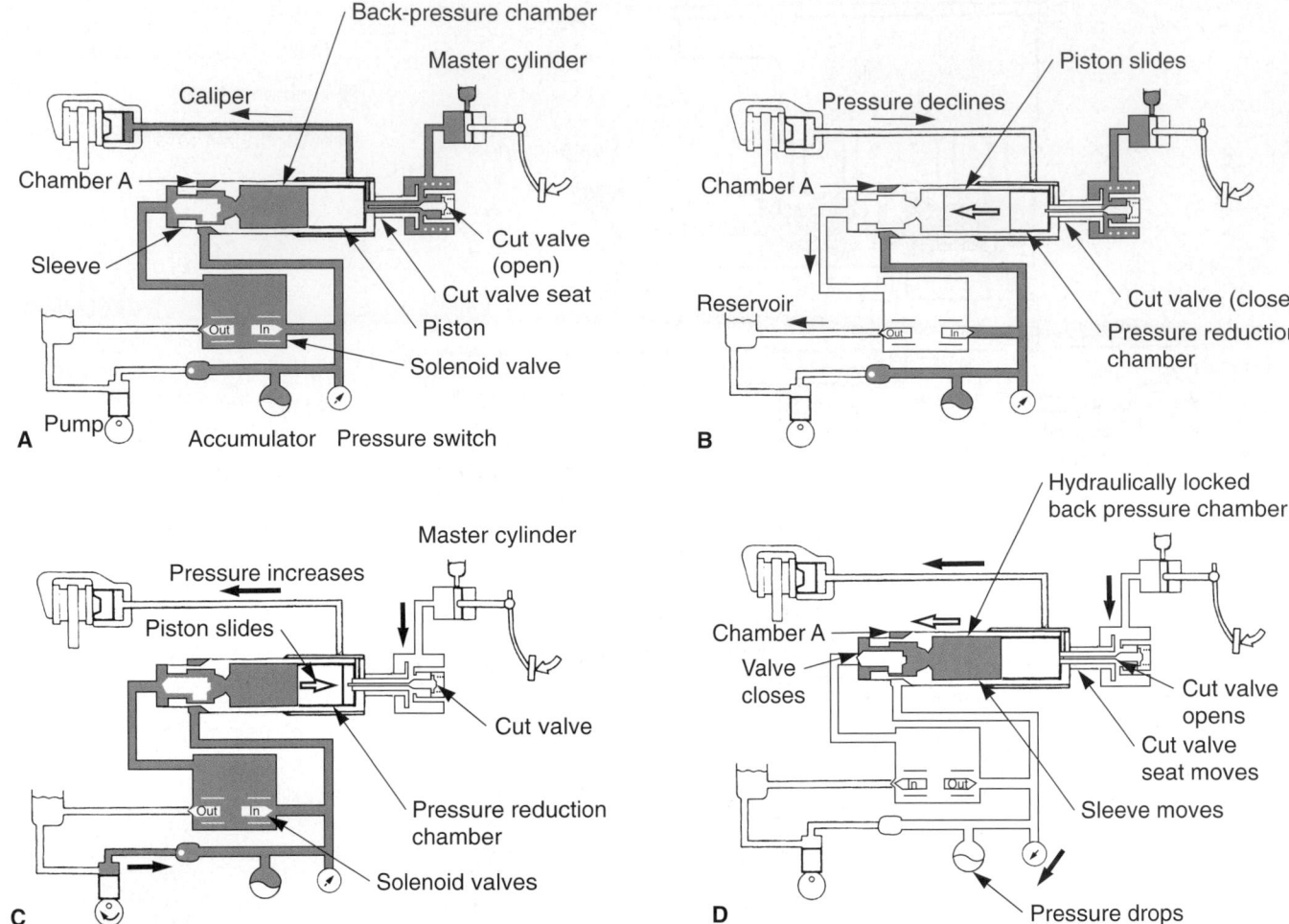

Figure 73-12. Study the operation of another type of ABS. A—ABS is not functioning. Master cylinder pressure flows to the caliper normally. B—ABS starts working to avoid a skid. The system operates the solenoid valve to close the inlet valve and open the outlet valve. High pressure is released to the reservoir. This reduces pressure to caliper, allowing that wheel to turn and not skid. C—ABS increases caliper pressure to maintain stopping action. The solenoid inlet valve is opened and the outlet valve is closed. High-pressure brake fluid is then sent to the back pressure chamber and the piston slides to increase pressure to brake caliper. D—The ABS control unit monitors pressure in the high-pressure passage by means of a pressure switch. The control unit turns on the ABS warning light. The sleeve moves and hydraulically locks the back pressure chamber. This connects the master cylinder to the caliper for normal braking again. (Honda)

acceleration conditions. It is designed to make the vehicle more stable under emergency handling situations, such as when trying to avoid a collision.

If *understeer* is detected (front tires lose adhesion with the road in a turn), the control module will apply braking force to the rear wheel on the opposite side of the vehicle. This will help bring the front of the vehicle back under control for making the turn. See **Figure 73-13.**

If the stability control system detects *oversteer* (rear tires lose adhesion with the road surface and slide sideways), additional braking force is automatically applied to the outside front wheel. This helps correct the oversteer so the rear of the vehicle does not spin out or slide sideways in a sharp turn. Refer to **Figure 73-14.**

In addition to the wheel speed sensors, a stability control system may use inputs from the following sensors:

- *Steering angle sensor*—measures how sharply the steering wheel is rotated to the right or left.

- *Lateral acceleration sensor*—measures how much side force is generated by a turn.

- *Yaw sensor*—measures the direction of the thrust generated by vehicle movement.

- *Throttle position sensor*—measures how far the driver has pressed down on the accelerator to control engine power output.

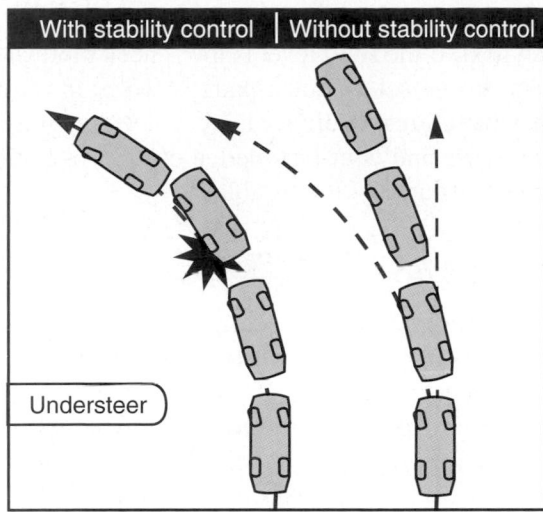

Figure 73-13. With an understeer problem, the stability control system might apply the left rear brake to keep the front of the vehicle from sliding or drifting to the right in a left-hand turn.

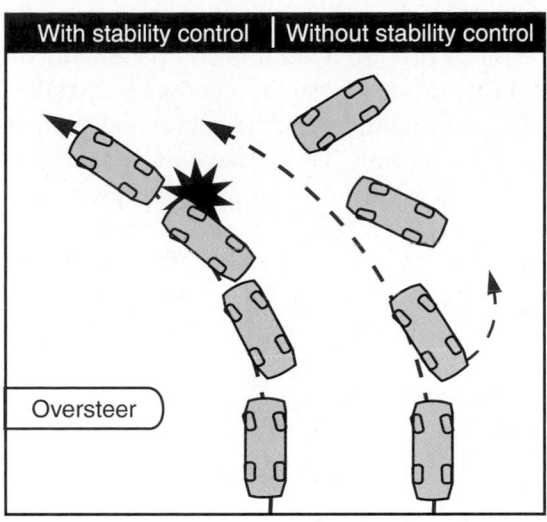

Figure 73-14. With an oversteer problem, the stability control system can apply the right front brake to keep the rear of the vehicle from sliding around in a high-speed turn.

- *Brake pressure sensor*—measures the amount of hydraulic pressure produced by the driver pressing on the brake pedal.

- *Other sensors*—give the control module information about the driver's actions and the resulting vehicle motion.

These sensors help the control module determine if the vehicle is skidding out of control, so corrective action can be taken. The control module is programmed to know when the vehicle is near its maximum rate of directional change. If this limit is reached, the system activates, **Figure 73-15.**

The control module is preprogrammed to apply one or more of the wheel brakes. It can also decrease or increase engine power to maintain vehicle control. If the driver rotates the steering wheel hard right for a sharp turn, the steering angle sensor signals the stability control module. The control module knows how fast each wheel should be turning. The inside wheels should be spinning more slowly than the outside wheels, because they are traveling less distance in the turn.

For example, the control module can use sensor signals to determine if the vehicle is following its intended direction of travel. The module knows what frequency each speed sensor should produce as the steering wheel is turned. If one or more of the wheels are not spinning at the right speed (skidding on road surface), the control module can react within 40 milliseconds to take corrective action. It might cut engine power slightly and apply one of the wheel brake calipers to reduce skidding.

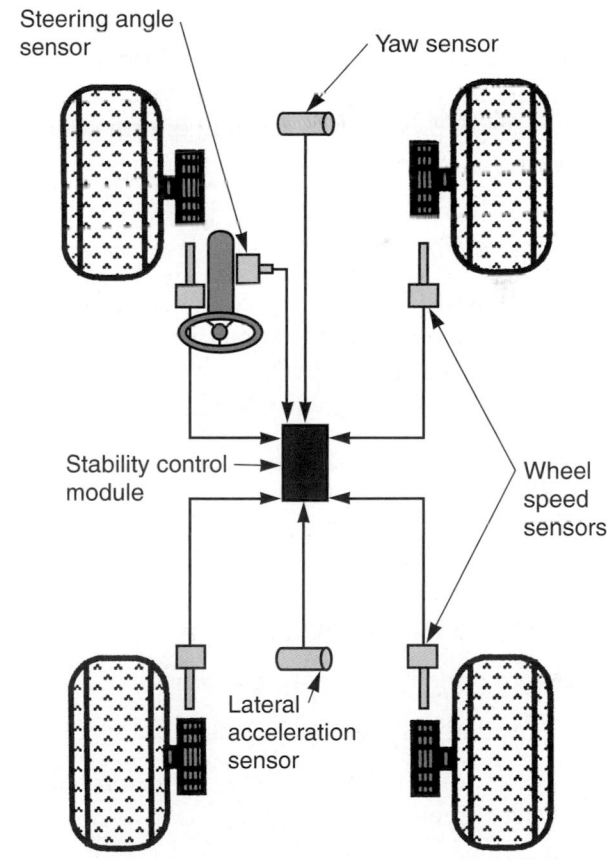

Figure 73-15. Note the extra sensors that are used in a stability control system.

Tech Tip
When driving a vehicle equipped with stability control system, a common mistake is to try to counter-steer the vehicle out of a skid when turning. This often causes overcorrection. It is best to continue steering in the desired direction of travel, allowing the system to correct the skid. This is similar to getting used to driving a vehicle with anti-lock brakes.

ABS Service

Before servicing an anti-lock brake system, you should be familiar with how a normally operating system works and feels. Become familiar with the pedal pulsations that occur when the ABS is activated. Test-drive known good vehicles with the same type of ABS. Memorize how the pedal bumps when ABS is first activated and how it vibrates as the system modulates. This will let you quickly detect some problems because you will know when the system in question does not "feel right."

You should also be familiar with the exact type ABS being repaired. Since designs and repair procedures vary, read the service manual for the vehicle at hand. This will give you the information needed (diagnosis charts, part locations, specs, procedures, etc.) to perform competent repair work.

ABS Inspection

To begin anti-lock brake system diagnosis, make a visual inspection of all the major parts. Look for obvious problems that could affect braking action. In addition to inspecting for conventional troubles, check for the following:

- ABS indicator light *on*.

- Low or contaminated brake fluid.

- Brake fluid leaks, especially at the modulator seals, gaskets, and lines.

- Worn or damaged brake pad linings and rotors.

- Tires that will not rotate freely.

- Loose or worn wheel bearings, which could affect wheel speed sensor operation.

- Badly worn tires or tires of unequal size, which could upset ABS operation.

- Damaged or improperly mounted speed sensors and trigger wheels.

Based on your findings, determine what step should be taken next. If the fluid level is low, check more closely for leaks. If one set of brake pads is worn down, make sure that the corresponding wheel is not dragging. Use common sense and your knowledge of ABS operation to help find the source of the trouble.

Scanning ABS

Late-model anti-lock brake systems have very sophisticated on-board diagnostics. Manufacturers have developed these elaborate self-test capabilities so the technician can troubleshoot and repair ABS with minimum difficulty. A typical ABS diagnostic system can produce over 50 trouble codes to help you find the cause of a malfunction.

When the ignition key is turned, an ABS warning light should glow for several seconds after the vehicle is started and then go off. This lets you know that the system is armed and functioning.

If the ABS light stays *on*, the ECM has detected an abnormal operating condition. This would signal you to use a scan tool to analyze the system and read trouble codes, **Figure 73-16.**

Some manufacturers have specialized testers, or scanners, for their anti-lock brake systems. Follow the operating and service manual instructions when using these tools. See **Figure 73-17.**

Tech Tip
The diagnostic connector for ABS can be in the trunk. If in doubt about connector locations, refer to the service manual.

First, use the scan tool to check for a ***code history*** (list of stored codes with the number of times each has occurred). This will let you know how often the problem occurs. It will also let you determine how many different codes have been set in the past.

Next, review all the trouble codes. This may help you quickly find the bad ABS component or circuit. Trouble codes are provided for all major electrical-electronic parts in the system. The scan tool will normally indicate problems with the following:

- Wheel speed sensor circuits.

- ABS warning light circuit.

- System relays.

- Low system voltage.

- Modulator motors.

- Solenoid circuits.

- Brake switch circuit.

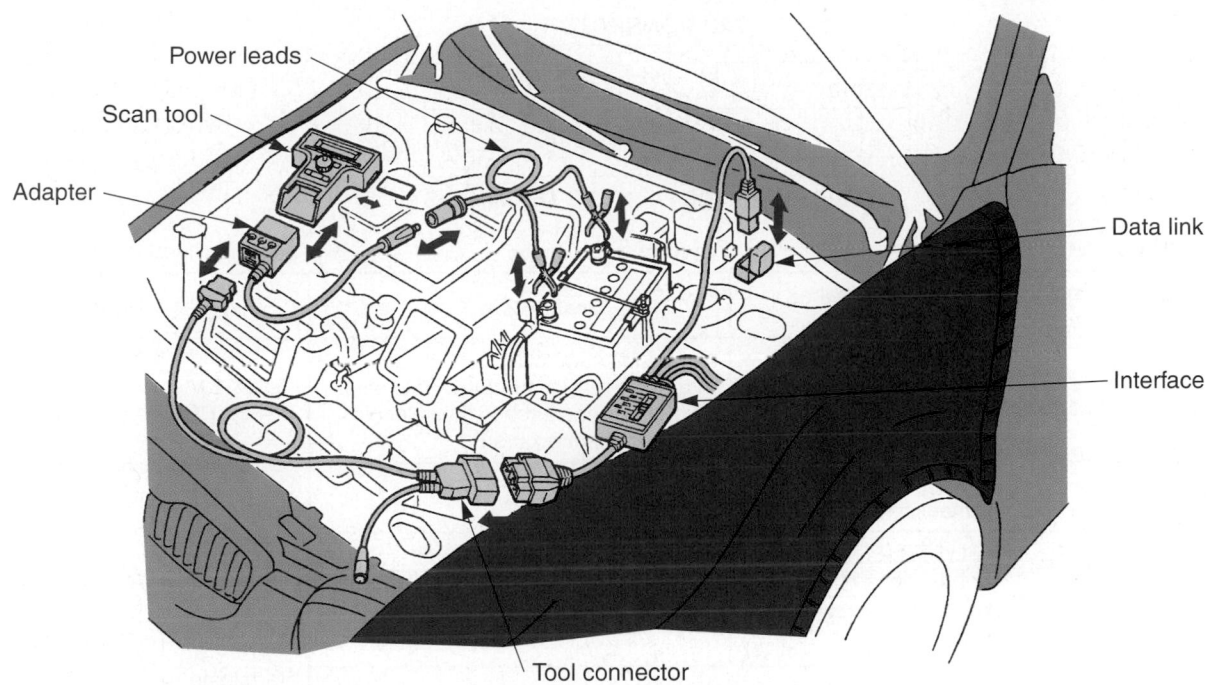

Figure 73-16. A scan tool is one of the first tools used when trying to find problems in anti-lock brake, traction control, and stability control systems. Connect the scan tool and any needed adapters to the vehicle's data link connector. (Mazda)

- Calibration memory in the control module.

- Other circuits and components.

With many late-model anti-lock brake systems, the scan tool can be used to perform several manual and automatic tests. You can use the scan tool to manually control motors, relays, solenoids, and other ABS parts. The automatic tests are used to check the operation of the modulator. This will let you make sure every part is functioning properly.

You can also use the scan tool to obtain system values (voltage, current, resistance). However, this is only done when a problem is difficult to locate. You can check to see if any value is almost out of spec, but not tripping a trouble code. This will narrow down your diagnosis on "boarder line" circuits and components that are causing intermittent problems.

ABS Pinpoint Tests

If your scan tool indicates a malfunction, you must still use pinpoint tests to determine the exact source of the problem. Use a multimeter to check the circuit or component. Compare your meter readings to factory specifications to verify the trouble.

The service manual will give electrical values for all major ABS circuits and components. One example is given in **Figure 73-18.**

ABS Speed Sensor Service

If your scan tool points to a problem with a wheel speed sensor, make sure the sensor is mounted securely. Check that all wire connectors are tight. Also check for chipped or broken trigger wheel teeth. Road debris and improper repair methods can damage the trigger wheel.

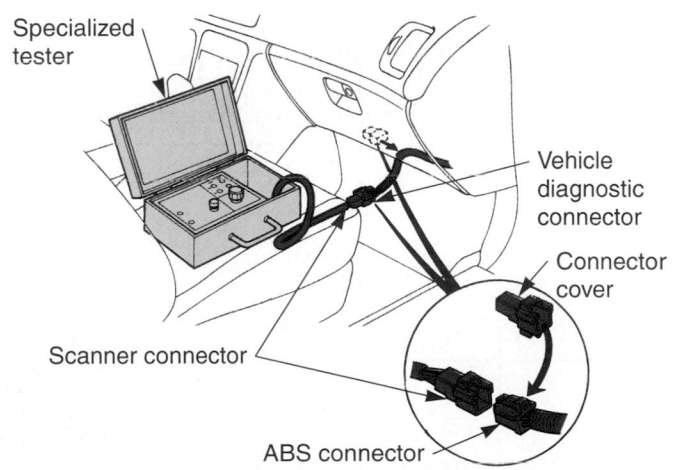

Figure 73-17. This scan tool can be placed on the seat so the vehicle can be test-driven while checking the operating conditions of the system. This is an advanced diagnostic procedure needed when a problem is hard to find. (Honda)

22P CONNECTOR

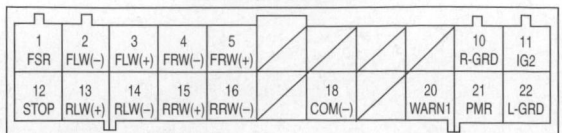

TERMINAL SIDE OF MALE TERMINALS

22P CONNECTOR   NOTE: Standard voltage is 12 V

Terminal number	Wire color	Terminal name	Description	Signal	
1	YEL/GRN	FSR (Fail-safe relay)	Drives fail-safe relay. (Fail-safe relay is turned OFF to shut off the power source to the solenoid when problem occurs.)	ON: 12 V OFF: 0 V	
2	BRN	FLW (–) (Front-left wheel sensor, negative)	Detects left-front wheel speed. (Ground level)	No. 2 - 3 terminals	When the wheel is turned at 1 turn/second: 70 mV or above on digital tester (AC range)
3	GRN/BLU	FLW (+) (Front-left wheel sensor, positive)	Detects left-front wheel speed.		
4	GRN	FRW (–) (Front-right wheel sensor, negative)	Detects right-front wheel speed. (Ground level)	No. 4 - 5 terminals	(Reference) 200 mVP-P or above on oscilloscope
5	GRN/BLK	FRW (+) (Front-right wheel sensor, positive)	Detects right-front wheel speed.		
10	BLK	R-GND (Rear solenoid valve ground)	Ground for rear inlet and outlet solenoid valves.		
11	YEL/BLK	IG2 (IG2 power source)	Detects ignition switch IG2 signal. (When IG2 is input, +B2 power source is switched to the power source for the ABS control unit (Vcc). Also IG2 monitors P-SW and MCK lines, and drives fail-safe relay.)	ON: 12 V OFF: 0 V	
12	GRN/WHT	STOP (Foot brake)	Detects brake switch signal. (Prevents unnecessary ABS operation when the brake pedal is not depressed.	ON: 12 V OFF: 0 V	
13	LT BLU	RLW (+) (Rear-left wheel sensor, positive)	Detects left-rear wheel speed.	No. 13 - 14 terminals	When the wheel is turned at 1 turn/second: 70 mV or above on digital tester (AC range)
14	GRY	RLW (–) (Rear-left wheel sensor, negative)	Detects left-rear wheel speed. (Ground level)		
15	GRN/YEL	RRW (+) (Rear-right wheel sensor, positive)	Detects right-rear wheel speed.	No. 15 - 16 terminals	(Reference) 200 mVP-P or above on oscilloscope
16	BLU/YEL	RRW (–) (Rear-right wheel sensor, negative)	Detects right-rear wheel speed. (Ground level)		
18	BLK/ORN	COM (–) (Common negative)	Ground for ALB checker when it is connected.		
20	BLU/WHT	WARN 1 (Warning lamp)	Drives ABS indicator light. (Shuts off the indicator light ground circuit inside the ABS control unit to turn off the light when the system is normal.)	Light ON: 0 V Light OFF: 12 V	
21	YEL/RED	PMR (Pump motor relay)	Drives pump motor relay. (Pump motor relay is turned ON to drive the pump motor when P-SW OFF signal is detected.)	ON: 0 V OFF: 12 V	
22	BLK	L-GND (Logic ground)	Ground for the ABS control unit control circuits.		

Figure 73-18. A service manual will normally give charts for doing pinpoint tests. Note how this chart gives electrical connector pin numbers, wire color codes, terminal names, circuit descriptions, and terminal specifications. (Honda)

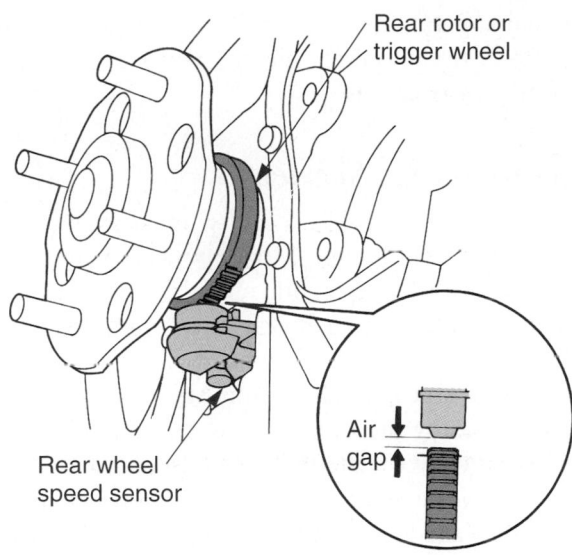

Figure 73-19. Some wheel speed sensors require adjustment of the air gap. The correct size non-metallic feeler gauge should fit snugly between the sensor tip and the trigger wheel tooth. (Honda)

If applicable, check the air gap between the speed sensor and its trigger wheel. Use a non-metal feeler gauge of the specified thickness. See **Figure 73-19.**

A hand-held scope will let you check the exact signal output from a suspicious wheel speed sensor. Unplug the wiring to the sensor and connect the scope to the sensor's output leads. To generate a waveform, spin the wheel by hand. Compare the actual sensor waveform to a known good waveform, **Figure 73-20.**

Tech Tip

If a pinpoint test of a wheel speed sensor shows no problem but the scan tool indicates a problem, check the wiring between the sensor and the anti-lock control module. If the wiring is not faulty, suspect the control module.

If the sensor is bad, replace it. Small bolts usually secure the sensor to the suspension knuckle. Clips may be used to secure the sensor wire to the body to prevent wiring damage, **Figure 73-21.**

After installing the new wheel speed sensor, adjust its air gap if needed using a brass or plastic feeler gauge.

Tech Tip

When working on vehicles with anti-lock brakes, be careful not to damage the wheel speed sensor trigger wheels. When pulling off calipers, axles, and related parts, it is easy to hit and chip one or more of the small teeth on the trigger wheel. When the vehicle is driven again, this would activate a trouble code and require trigger wheel replacement.

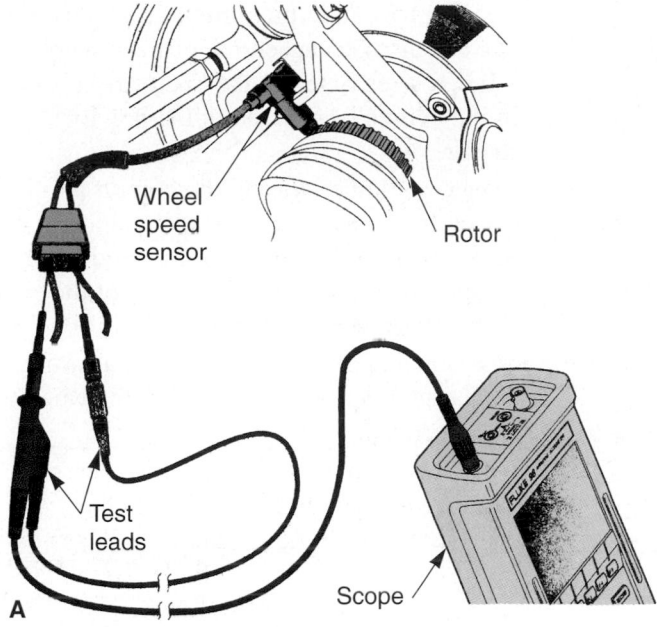

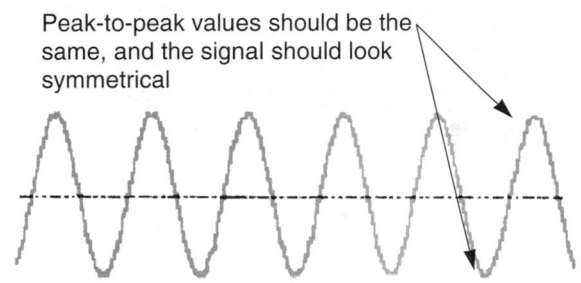

Peak-to-peak values should be the same, and the signal should look symmetrical

Improper air gap from a tone wheel or missing teeth on a wheel speed sensor will produce an erratic signal

Figure 73-20. A hand-held scope will measure the signal leaving a wheel speed sensor. A—Typical sensor hook up. B—Output signal from a speed sensor. The output should match the reading from the scan tool. If not, a wiring or control module problem is indicated. (Fluke)

Speed sensor trigger wheels often require major part removal for service. Refer to the service manual for instructions.

ABS Modulator Check

Most ABS modulator problems will show up when scanning. However, some automakers allow you to check the modulator in the following manner.

1. Place the vehicle on jack stands so all four tires are equally off the ground, **Figure 73-22A.**
2. Release the parking brake.
3. Rotate each wheel by hand. Make sure none of the wheels are dragging. A dragging wheel indicates that the system might not be

releasing hydraulic pressure normally. Brake drag can also be caused by a frozen wheel cylinder, a restricted brake line or hose, and similar sources discussed in the previous chapter.

4. Connect a jumper wire to the correct terminals on the vehicle's diagnostic connector or use the scan tool to energize the modulator to apply hydraulic pressure to one of the wheels, **Figure 73-22B.**

5. Press down and hold the brake pedal while you have a helper make sure that the wheel will not rotate.

6. Using service manual instructions, energize the modulator to dump pressure to the wheel. Make sure the modulator relieves pressure from the wheel cylinder so the tire will rotate freely, **Figure 73-22C.**

7. Repeat steps 4-6 on the other wheels.

This test will check the operation of the modulator, brake tubing, brake hoses, solenoids, relays, ECM, and other major parts. Remember, these are general procedures. Refer to the service manual and scan tool manual for specific instructions.

ABS Modulator Service

If the ABS modulator is faulty, most shops install a new or factory-rebuilt unit. Modulators are very complex and time consuming to rebuild.

If the vehicle is under warrantee, some dealerships repair or rebuild modulators in-house. Make sure the proper rebuild kit is available before starting. Also compare cost of rebuilding the modulator to the cost of a new or factory rebuilt unit to determine the best course of action.

To remove an ABS modulator, disconnect all wires between the vehicle and unit. See **Figure 73-23.** Then, use a tubing wrench to loosen the brake lines going to the modulator. Remove the bolts that hold the modulator on its brackets. Then, lift the unit out of the engine compartment.

If you are installing a new or rebuilt unit, transfer any necessary parts from the old unit to the new one. These parts might include the combination valve,

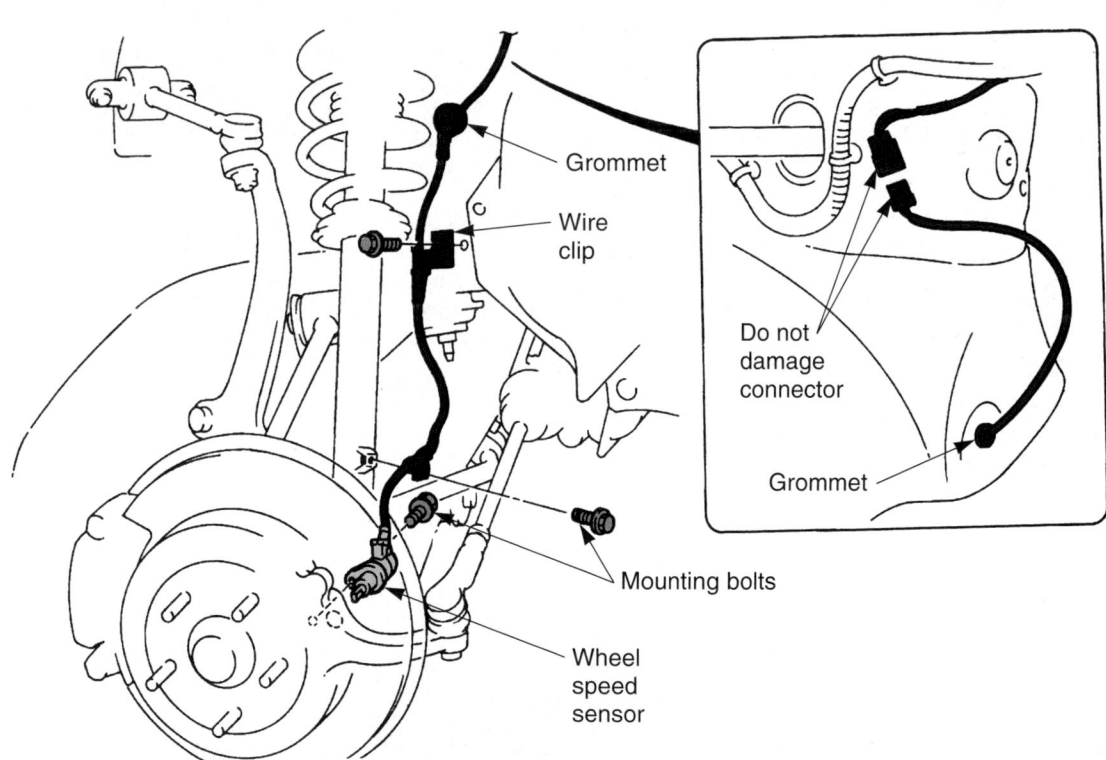

Figure 73-21. When replacing a wheel speed sensor, do not damage the electrical connectors and replace any clips that hold the wiring on the body. (Mazda)

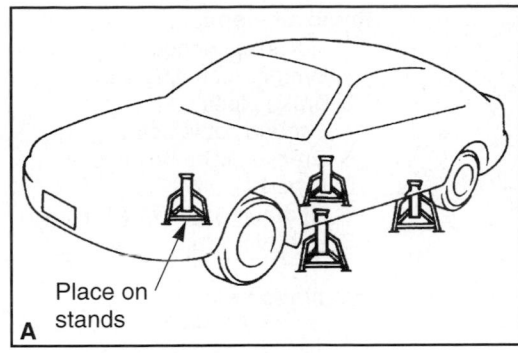

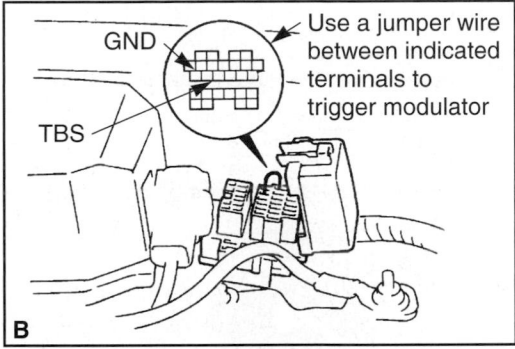

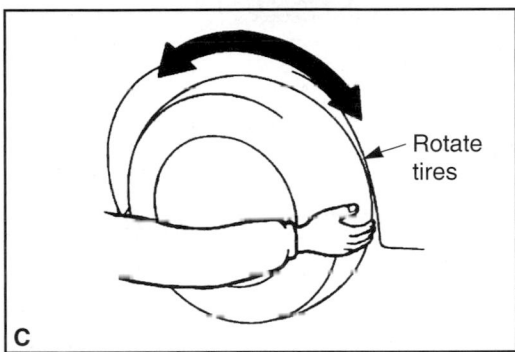

Figure 73-22. Some automakers recommend testing an ABS modulator by energizing the unit and rotating each tire by hand. A—Place the vehicle in level position on jack stands. B—Energize the modulator to free or lock up each wheel as directed in the service manual. C—Rotate each tire by hand to make sure the modulator frees and holds the brake as designed. (Mazda)

reservoir, master cylinder (integral units) and brackets. See **Figure 73-24.**

A few general rules to remember when repairing or replacing an ABS modulator include:

- Always install all parts included in the rebuild kit.
- If a new unit is being installed, make sure it is has the correct part number.

- Always lubricate rubber parts with clean DOT 3 brake fluid.
- Work on a clean, uncluttered workbench to avoid contamination.
- Avoid using compressed air on hydraulic parts. Oil from the compressor may deteriorate rubber parts.
- Prevent brake fluid from contacting electrical connections, painted surfaces, and electronic parts.
- Use correct torque values when tightening fasteners.
- Refer to the service manual for detailed procedures, specifications, and safety precautions.

ABS Bleeding

Bleeding an anti-lock brake system of air bubbles can be similar to bleeding conventional brakes or it might require special procedures and tools.

When servicing a master cylinder that is separate from the modulator, plug the brake lines to keep out as much air as possible. Bench bleed the new master cylinder. Then, re-bleed the master cylinder at its lines once it is mounted on the vehicle.

When the master cylinder is formed as an integral part of the modulator, you may need to use a very specific bleeding procedure. An electrical break-out box and an electronic unit might be required to properly energize the solenoids when bleeding some systems. Refer to the service manual when in doubt. See **Figure 73-25.**

ABS Test-Drive

When test-driving a vehicle that has a malfunctioning or newly repaired ABS, drive slowly and stay out of traffic. Make sure you have adequate braking action before driving on public streets.

A *wet pavement test* can be done to check your repair work or to help pinpoint a problem when all other methods of diagnosis have been exhausted. It verifies the operation of the "ABS electronics," as well as all mechanical parts associated with normal braking.

The wet pavement test is performed with two tires on wet pavement and the other two tires on dry pavement. Make sure the vehicle has enough stopping power before testing.

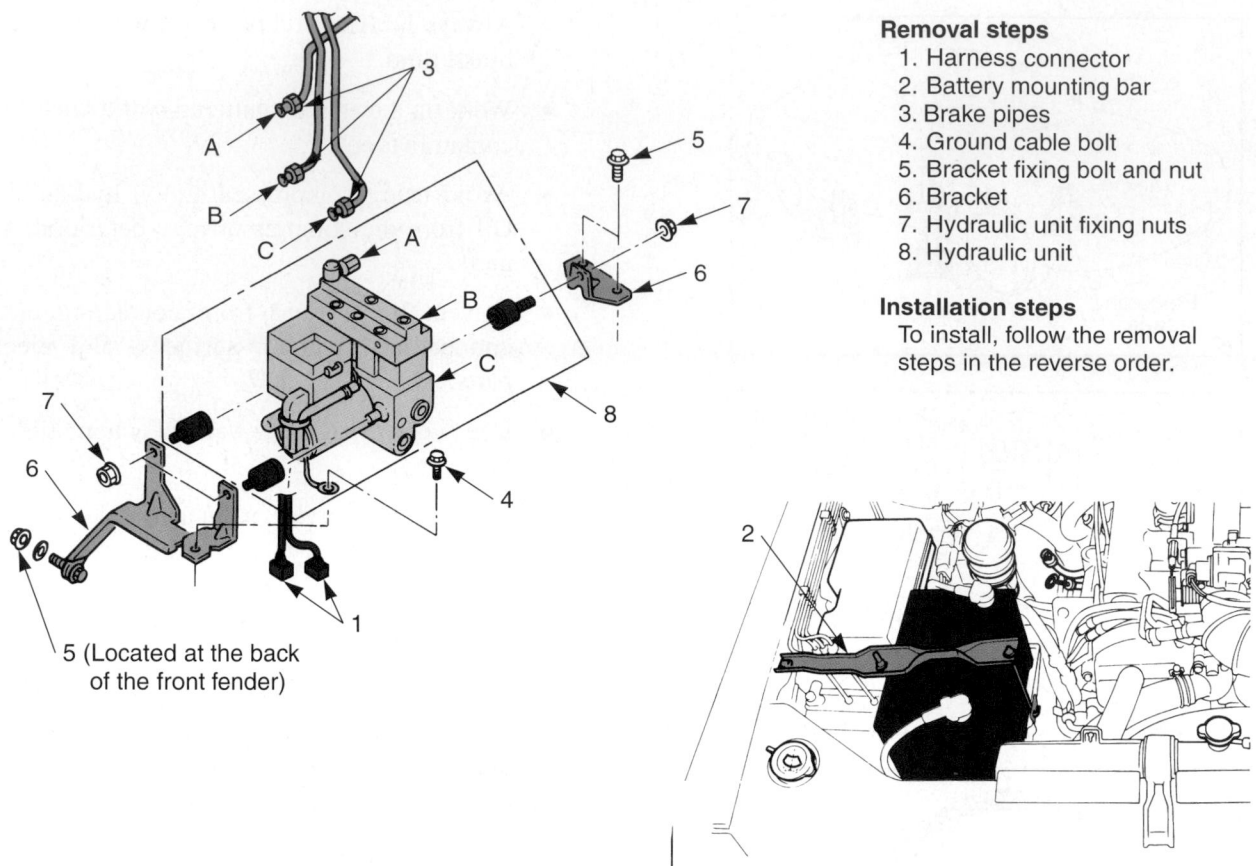

Removal steps
1. Harness connector
2. Battery mounting bar
3. Brake pipes
4. Ground cable bolt
5. Bracket fixing bolt and nut
6. Bracket
7. Hydraulic unit fixing nuts
8. Hydraulic unit

Installation steps
To install, follow the removal steps in the reverse order.

5 (Located at the back of the front fender)

Figure 73-23. Note the parts that must be disconnected to service a typical ABS modulator.

Use a garden hose to wet a small strip of pavement about a foot wide and 20 feet long in an open parking lot. Make sure the lot is large enough to provide ample stopping room if the brakes fail to work normally.

Position the vehicle so that only the right or left tires ride on the wet area. Drive the vehicle across the strip of wet pavement at about five miles per hour. Then, hit the brakes hard to try to make the tires skid on the pavement. Ask a helper to watch the tire contact on the wet pavement and to let you know if either tire is totally locking up or not modulating upon hard brake application. This will let you determine if any wheel is not being controlled by the ABS.

As you brake and two of the tires try to skid, you should feel the pedal modulate. The helper should not see the tires lock up or continue to rotate freely. The ABS system should modulate hydraulic pressure so the tires almost skid.

Repeat this test with the other set of tires on the wet strip of pavement. This will allow you to check modulation on all four wheels.

This test will give you useful information about the condition and operation of the ABS. Make sure the ABS

indicator light stays off before releasing the vehicle to the customer.

Traction and Stability Control System Service

The diagnosis and repair procedures for traction and stability control systems are similar to the methods described for ABS. You would use an initial inspection and your scan tool to find any obvious problems or trouble codes. If you have trouble codes, use pinpoint tests to verify the problem. After repairs are made, re-scan the system to check your work.

If needed, test-drive the vehicle while scanning to check the action of the traction control system. This should be done in your shop's parking lot or somewhere away from pedestrians and traffic. A gravel surface or wet pavement is best, because you can intentionally make the vehicle try to skid upon acceleration. If needed, use a garden hose to wet down an area in your parking lot.

When on the slick surface, accelerate the vehicle and try to make the drive tires spin. This should activate the traction control or stability control system. You can then

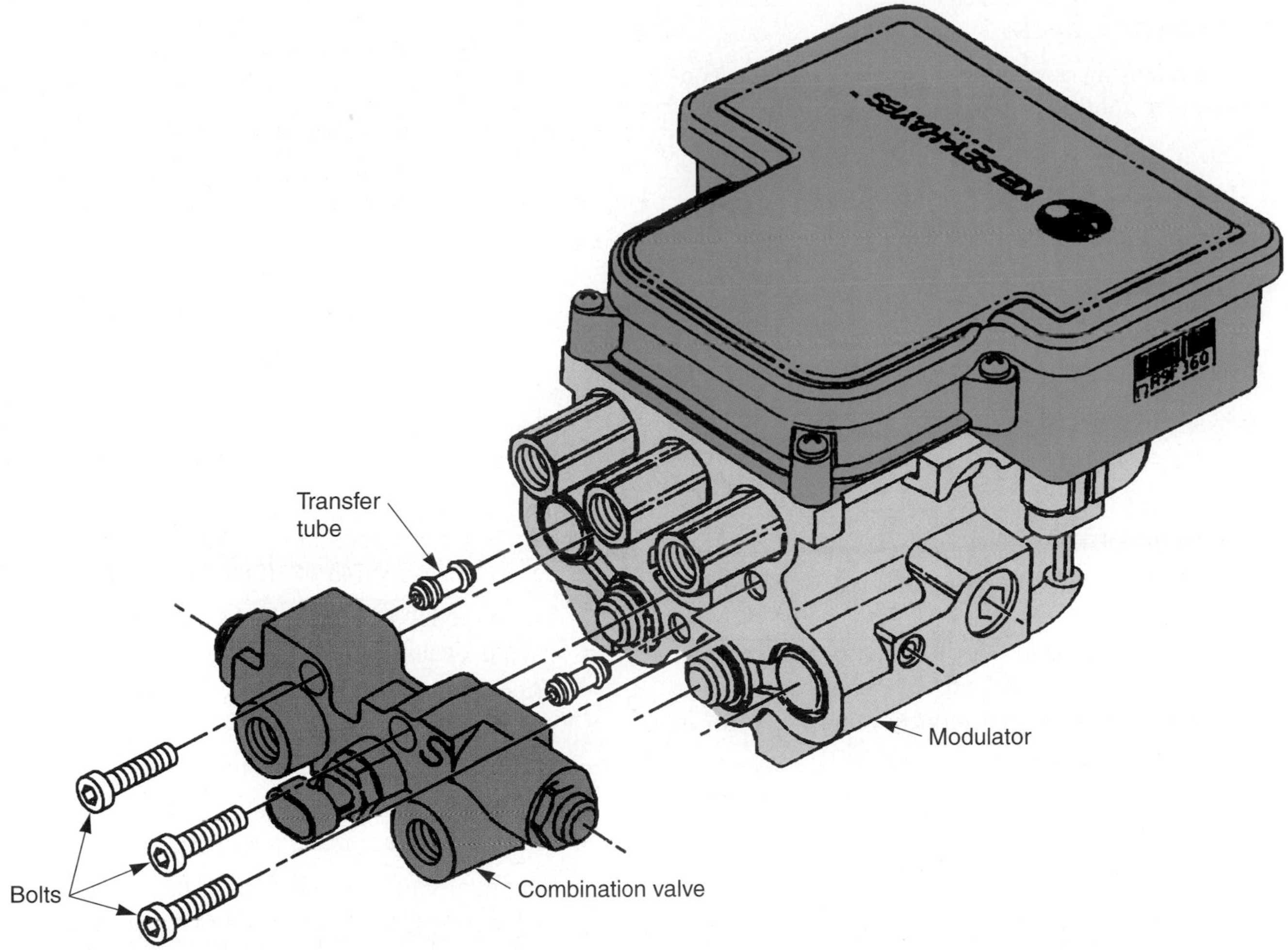

Transfer
tube

Modulator

Bolts

Combination valve

Figure 73-24. You may have to transfer some parts, like this combination valve, to the new modulator. Follow the manual instructions. Coat the O-ring seals with clean brake fluid to ensure a leakproof seal. (General Motors)

use scanner data about the wheel speed sensors, hydraulic unit, engine management computer, and other components to help isolate the problem.

Tech Tip
When the traction control or stability control system energizes, the indicator light should come on and you should usually hear a rapid buzzing or clicking sound from the modulator.

Basic pinpoint tests, as explained throughout this book, will generally be needed to find the exact part at fault. The service manual will give more detailed instructions for repairing a specific traction control or stability control system.

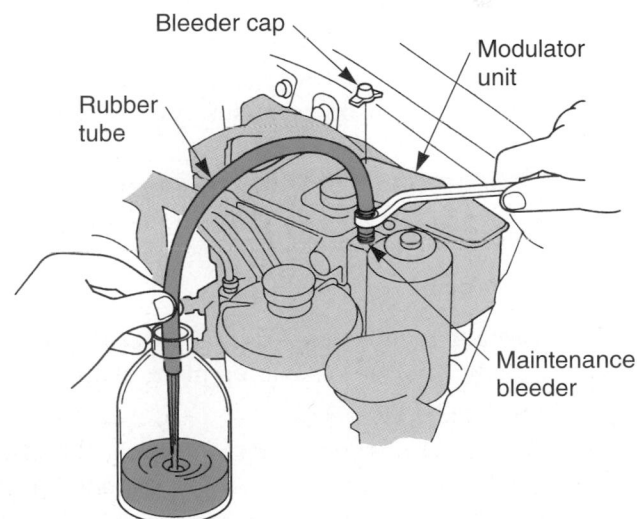

Bleeder cap

Modulator
unit

Rubber
tube

Maintenance
bleeder

Figure 73-25. Special procedures may be needed to bleed all the air from an anti-lock brake system. Follow the manufacturer's instructions. (Honda)

Final System Check

Before releasing a vehicle after ABS, traction control, or stability control system repairs, you should:

- Clear trouble codes and re-scan the system for problems.

- Check the brake fluid level in the system.

- Test-drive the vehicle in an area without traffic.

- Simulate a panic stop several times in a large parking lot.

- Make sure the ABS indicator light stays off.

Summary

- An anti-lock brake system, abbreviated ABS, uses wheel speed sensors, a computer, and a modulator unit to prevent tire skid during hard braking.

- When the tires skid, they actually lose traction or friction with the road surface. This can increase stopping distance and reduce vehicle control in some situations.

- The anti-lock brake system improves occupant safety by reducing stopping distances and increasing directional stability under panic stop conditions.

- The wheel speed sensors produce an ac type signal that corresponds to wheel and tire speed.

- The ABS control module uses wheel speed sensor inputs to control the operation of the electro-hydraulic modulator.

- The ABS electro-hydraulic modulator, also called a modulator, regulates the fluid pressure applied to each wheel brake assembly during hard braking.

- Hydraulic pressure modulation refers to the electro-hydraulic modulator's ability to rapidly cycle pressure on and off to prevent wheel lockup and skidding.

- The dump mode refers to when the ABS releases hydraulic pressure to one or more of the wheel brake assemblies.

- The pressure mode refers to when the ABS applies hydraulic pressure to one or more of the wheel brake assemblies.

- An integrated ABS system combines the power booster, master cylinder, and electro-hydraulic unit into one assembly.

- In a non-integrated ABS system, the major units are separated or mounted remotely.

- A four-channel ABS has a separate hydraulic control circuit for each wheel.

- Traction control systems are designed to prevent the wheels from spinning or skidding under hard acceleration.

- Stability control systems prevent the wheels from spinning under hard acceleration and also prevent skidding when cornering.

- You should also be familiar with the exact ABS being repaired. Since designs and repair procedures vary, read the service manual for the vehicle.

- To begin anti-lock brake system diagnosis, make a visual inspection of all the major parts.

- Late-model anti-lock brake systems have very sophisticated on-board diagnostics.

- Most anti-lock brake systems will trigger the indicator light if the ABS control module detects an abnormal operating condition.

- Many late-model anti-lock brake systems offer manual and automatic tests with the scan tool.

- If your scan tool indicates a problem, you still need to use pinpoint tests to determine its exact source.

- If applicable, check the air gap between the speed sensors and their trigger wheels. Use a non-metallic feeler gauge of the specified thickness.

- Most problems with the ABS modulator will show up when scanning. However, some automakers allow you to check the modulator by turning the wheels by hand while energizing the unit.

- If the ABS modulator tests bad, most shops simply install a new unit.

- Bleeding anti-lock brakes can be similar to conventional brakes or it might require special procedures and tools.

- A wet pavement test checks the ABS with two wheels on wet pavement and the other two on dry pavement.

- The diagnosis and repair procedures for traction control systems is similar to the methods described for ABS.

Important Terms

Anti-lock brake system (ABS)	Reapply mode
	ABS warning light
Wheel speed sensors	Integrated ABS
Sensor rotor	Non-integrated ABS
Trigger wheel	Channel
Air gap	One-channel ABS
ABS control module	Two-channel ABS
ABS electro-hydraulic modulator	Four-channel ABS
	Traction control systems
ABS modulator	Stability control system
Fluid reservoir	Understeer
Solenoid valve block	Oversteer
Accumulator	Steering angle sensor
Hydraulic pump and motor	Lateral acceleration sensor
Pressure switch	Yaw sensor
Master cylinder-booster assembly	Throttle position sensor
	Brake pressure sensor
Hydraulic pressure modulation	Other sensors
	Code history
Isolation mode	Wet pavement test
Dump mode	

Review Questions—Chapter 73

Please do not write in this text. Place your answers on a separate sheet of paper.

1. An anti-lock brake system prevents loss of tire _____ during hard braking.

2. In addition to reducing stopping distances, ABS improves driver and passenger safety by increasing _____ _____ under panic stop conditions.

3. What happens when a tire skids or slides during hard braking?

4. List and describe the five major parts of an anti-lock brake system.

5. The wheel speed sensors produce an ac type _____ that corresponds to wheel and tire speed.

6. The _____ _____, or _____ _____, is a toothed ring that rotates with the wheel hub.

7. List and summarize the six major parts of an ABS modulator.

8. Describe the three operating modes of an ABS modulator.

9. A scan tool is commonly used to troubleshoot a modern ABS. True or False?

10. Technician A says that a traction control system often uses parts common to the anti-lock brake system. Technician B says that stability control systems also use parts common to the anti-lock brake system. Who is right?
    (A) A only.
    (B) B only.
    (C) Both A and B.
    (D) Neither A nor B.

⬟ ASE-Type Questions

1. A car enters the shop with the ABS light glowing. Technician A says to inspect the system for obvious troubles. Technician B says to also use a scan tool to help find the problem. Who is right?
   (A) A only.
   (B) B only.
   (C) Both A and B.
   (D) Neither A nor B.

2. Which of the following could cause a trouble code for one of the wheel speed sensors?
   (A) Smaller diameter tire on one wheel.
   (B) Worn brake pad lining.
   (C) Leaking modulator.
   (D) Leaking wheel cylinder.

3. Which of the following *cannot* be checked with a scan tool?
   (A) Trouble code history.
   (B) Present trouble codes.
   (C) Brake lining condition.
   (D) Trigger wheel condition.

4. A scan tool shows a problem with the modulator pump motor. Technician A says to check for power to the motor from the ABS control module. Technician B says if no power is reaching the motor to check the wiring between the control module and the motor. Who is correct?
   (A) A only.
   (B) B only.
   (C) Both A and B.
   (D) Neither A nor B.

5. A wheel speed sensor has a broken tooth. Technician A says that you must set the sensor air gap on some vehicles. Technician B says that a steel feeler gauge should be used to set the sensor air gap. Who is correct?
   (A) A only.
   (B) B only.
   (C) Both A and B.
   (D) Neither A nor B.

6. A scan tool shows a problem with the right front wheel speed sensor circuit. Technician A says to replace the sensor. Technician B says to replace the sensor trigger wheel. Who is right?
   (A) A only.
   (B) B only.
   (C) Both A and B.
   (D) Neither A nor B.

7. Technician A says a scan tool will usually test the action of major ABS components by energizing them. Technician B says that a scan tool does not have this capability. Who is correct?
   (A) A only.
   (B) B only.
   (C) Both A and B.
   (D) Neither A nor B.

8. A new ABS modulator is being installed on a vehicle. Technician A says you may have to transfer the old combination valve to the new unit. Technician B says to coat all rubber parts with clean brake fluid. Who is correct?
   (A) A only.
   (B) B only.
   (C) Both A and B.
   (D) Neither A nor B.

9. A customer complains of a buzzing sound and pedal vibration when braking on slick pavement. This is:
   (A) caused by bad modulator.
   (B) caused by bad master cylinder.
   (C) caused by warped rotor.
   (D) a normal condition with most ABS.

10. After completing repairs on a stability control system, you should:
    (A) clear trouble codes and re-scan for problems.
    (B) recheck brake fluid level in system.
    (C) test-drive vehicle slowing in area without traffic.
    (D) All of the above.

Suggested Activities—Chapter 73

1. Inspect the ABS system on several different vehicles. Write down any problems you found on each. Also summarize any differences in part locations or part designs for a class discussion.

2. Scan a late-model ABS system. Unplug some of the wiring connectors in the system and note the scanner readout. Make a report of what you learned.

Anti-Lock Brake and Traction Control System Diagnosis

Note: Any possible base brake system defects must be ruled out before an ABS or TCS problem is suspected. If the red brake light is on, you must check the base brake system first.

Condition	Possible Causes	Correction
ABS or TCS light does not come on when the vehicle is started.	1. Defective or missing bulb. 2. Fuse blown. 3. Lamp circuit open. 4. Defective control module.	1. Replace bulb. 2. Replace fuse; check for shorts or grounds. 3. Reconnect or repair leads. 4. Replace control module.
ABS or TCS light comes on as soon as engine is started or when vehicle begins to move. DTC may or may not be stored.	1. Defect in base brake system. 2. Defect in vehicle electrical system. 3. Fuse missing or blown. 4. Short to vehicle positive wire. 5. Defective control module. 6. Defective hydraulic actuator. 7. Defective power relay. 8. Defective wheel speed sensor. 9. Defective pressure switches or sensors. 10. Inoperative hydraulic pump.	1. Correct base brake problem. 2. Repair electrical defect. 3. Replace fuse; check for shorts or grounds. 4. Repair short. 5. Replace control module. 6. Replace actuator. 7. Replace relay. 8. Replace wheel speed sensor. 9. Replace defective switch or sensor. 10. Replace hydraulic pump.
ABS or TCS light comes on intermittently. DTC may or may not be stored.	1. Defect in base brake systems. 2. Leaking accumulator. 3. Defective wheel speed sensor. 4. Hydraulic pump weak or leaking. 5. Electromagnetic interference (EMI). 6. Intermittent open connection.	1. Correct base brake problem. 2. Replace accumulator. 3. Replace wheel speed sensor. 4. Repair leak or replace pump. 5. Eliminate EMI source or reroute wiring. 6. Repair connection.
ABS or TCS does not operate when called for. Warning light may or may not be on. Self-energize test may not be performed.	1. Defect in base brake system. 2. Fuse blown. 3. Wheel speed sensor(s) defective or misadjusted. 4. Defective control module. 5. Defective hydraulic actuator solenoids. 6. Defective hydraulic actuator pump. 7. Defective accumulator. 8. Disconnected or defective wiring. 9. Electromagnetic interference (EMI).	1. Correct base brake problem. 2. Replace fuse. 3. Adjust or replace wheel speed sensor. 4. Replace module. 5. Replace actuator. 6. Replace pump or actuator. 7. Replace accumulator. 8. Reconnect or repair wiring. 9. Eliminate EMI source or reroute wiring.
ABS or TCS engages during light to moderate braking or acceleration. DTC may or may not be stored.	1. Defect in base brake system. 2. Misadjusted brake pedal/brake light switch. 3. Defective wheel speed sensor. 4. Hydraulic actuator leaking internally. 5. Defective control module.	1. Correct base brake problem. 2. Adjust or replace switch. 3. Replace wheel speed sensor. 4. Replace actuator. 5. Replace module.

Chapter 74

Wheel Alignment

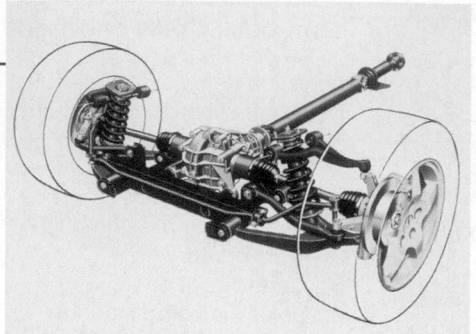

After studying this chapter, you will be able to:

- Explain the principles of wheel alignment.
- List the purpose of each wheel alignment setting.
- Perform a prealignment inspection of tires, steering, and suspension systems.
- Describe caster, camber, and toe adjustment.
- Explain toe-out on turns, steering axis inclination, and tracking.
- Describe the use of different types of wheel alignment equipment.
- Correctly answer ASE certification test questions requiring a knowledge of wheel alignment angles and procedures.

The term *alignment* means "to position in a straight line." Relating to vehicles, alignment means to position the four tires so that they roll freely and evenly over the road surface.

Correct wheel alignment is essential to automobile safety, sure handling, maximum fuel economy, and long tire life. This chapter introduces both the principles and the basic procedures for wheel alignment.

Note

For a complete understanding of the information in this chapter, you should have studied the previous chapters on wheel bearings, suspension systems, and steering systems. These systems must be in good condition before attempting wheel alignment.

Wheel Alignment Principles

The main purpose of *wheel alignment* is to make the tires roll without scuffing, slipping, or dragging under all operating conditions. Six fundamental angles or specifications are needed for proper wheel alignment:

- Caster.
- Camber.
- Toe.
- Steering axis inclination.
- Toe-out on turns (turning radius).
- Tracking (thrust line).

Caster

Caster is basically the forward or rearward tilt of the steering knuckle (spindle support) when viewed from the side of the vehicle. You are probably familiar with the term caster from furniture casters, **Figure 74-1.**

Caster controls where the tire touches the road in relation to an imaginary centerline drawn through the spindle support. Is it *not* a tire wearing angle.

The basic purposes of caster are:

- To aid directional control of the vehicle.
- To cause the wheels to return to the straight-ahead position.
- To offset *road crown pull* (steering wheel pull caused by hump in center of road).

Positive caster tilts the top of the steering knuckle toward the rear of the vehicle. Positive caster helps keep the vehicle's wheels traveling in a straight line. When you turn the wheels, it lifts the vehicle. Since this takes extra turning force, the wheels resist turning and try to return to the straight-ahead position. See **Figure 74-2A.**

Negative caster tilts the top of the steering knuckle toward the front of the vehicle. It is the opposite of positive caster. With negative caster, the wheels will be easier to turn. However, the wheels will tend to swivel and follow imperfections in the road. Look at **Figure 74-2B.**

Caster is measured in *degrees,* starting at the true vertical (plumb line). Auto manufacturer's give specs for caster as a specific number of degrees positive or negative. This is illustrated in **Figure 74-3.**

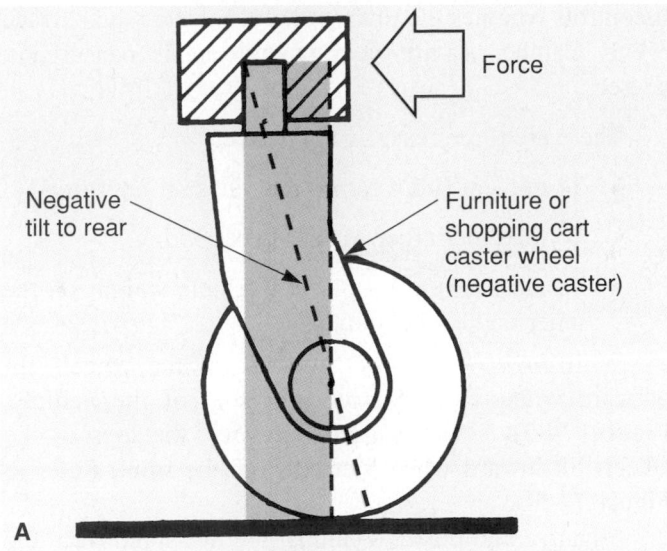

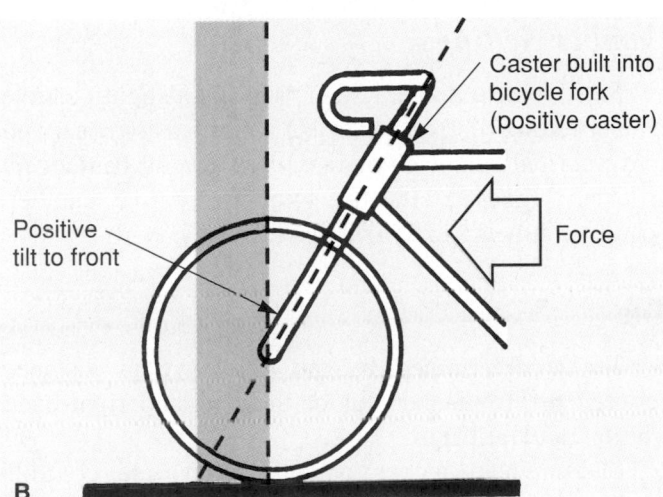

Figure 74-1. Caster is determined by the contact point of the tire and an imaginary centerline through the spindle support. A—Negative caster of ball joints is illustrated by a shopping cart wheel. The wheel follows the irregular surfaces in the floor. B—Positive caster, like that on a front bike tire, makes the front wheel travel straight ahead.

Typically, positive caster is recommended for vehicles with power steering. To ease steering effort, negative caster is recommended for vehicles with manual steering.

Caster-Road Crown Effect

Caster is a *directional control* angle. It determines whether the vehicle travels straight or pulls (steering wheel tries to turn) to the right or left.

Road crown is the normal slope toward the outer edge of the road surface. Most road surfaces angle

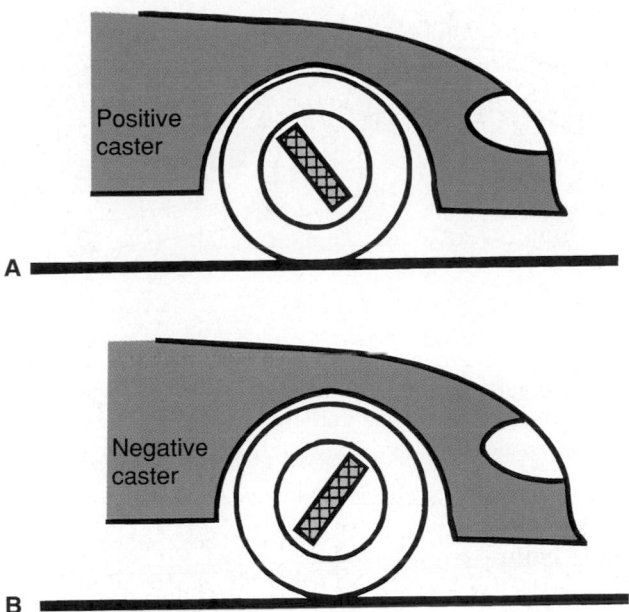

Figure 74-2. A—Positive caster tilts the steering knuckle toward the rear of the vehicle. B—Negative caster tilts the steering knuckle toward the front of the vehicle.

downward from the center. This helps keep rainwater from collecting on the pavement.

If the caster of both front wheels were the same, the road crown could make the vehicle pull or steer toward the outside (lower) edge of the road, **Figure 74-4.**

Since caster is a directional control angle, it is commonly used to offset the effect of road crown. The right front wheel may be set with slightly more positive caster than the left. This counteracts the forces caused by the road crown, and the vehicle will travel straight ahead.

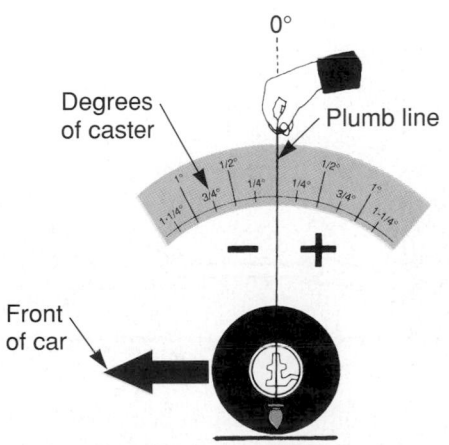

Figure 74-3. Caster is measured in degrees, as shown. (Bear)

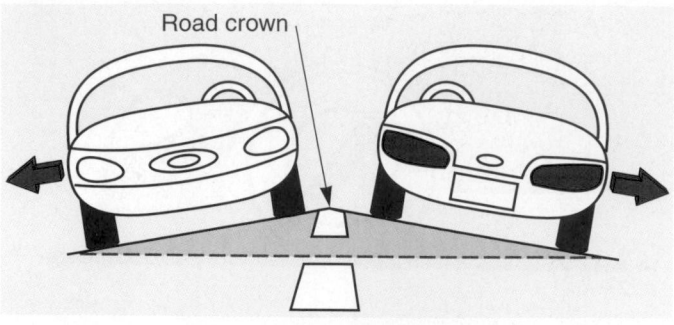

Figure 74-4. Most roads are crowned in the center to help water flow off the surface. Unequal caster can be used to offset road crown and keep the vehicle from trying to steer off the road.

Note
Always refer to the service manual for exact caster specs. They vary with vehicle design.

Camber

Camber is the inward or outward tilt of the wheel and tire assembly when viewed from the front of the vehicle.

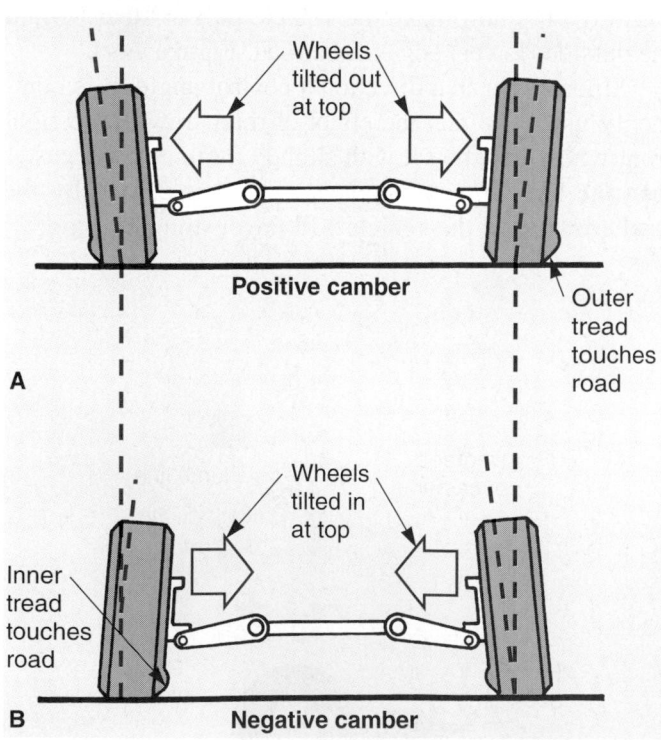

Figure 74-5. Camber is determined by the inward or outward tilt of the wheels when viewed from the front of a vehicle. Positive camber has the wheels tilted out at the top. Negative camber has the wheels tilted out at the bottom.

It controls whether the tire tread touches the road surface evenly. Camber is a tire-wearing angle and is measured in degrees.

There are three reasons for camber:

- To prevent tire wear on the outer or inner tread.

- To load the larger inner wheel bearing.

- To aid steering by placing vehicle weight on the inner end of the spindle.

With *positive camber*, the tops of the wheels tilt outward when viewed from the front of the vehicle, **Figure 74-5A**. With *negative camber*, the tops of the wheels tilt inward when viewed from the front. Refer to **Figure 74-5B**.

Negative and positive camber are measured from the true vertical (plumb line). If the wheel is aligned with the plumb line, camber is zero, **Figure 74-6**.

Camber Settings

Most vehicle manufacturers suggest a slight positive camber setting (about 1/4° to 1/2°). Suspension wear and above-normal curb weight caused by several passengers or extra luggage tend to increase negative camber. Positive camber counteracts this tendency.

Toe

Toe is determined by the difference in distance between the front and rear of the left- and right-hand wheels. Look at **Figure 74-7**.

Measured in inches or millimeters, toe controls whether the wheels roll in the direction of travel. Toe is very critical to *tire wear*. If the wheels do *not* have the correct toe setting, the tires will scuff, or skid sideways.

Toe-in is produced when the wheels are closer at the front than at the rear. Toe-in causes the wheels to point inward at the front, **Figure 74-7A**.

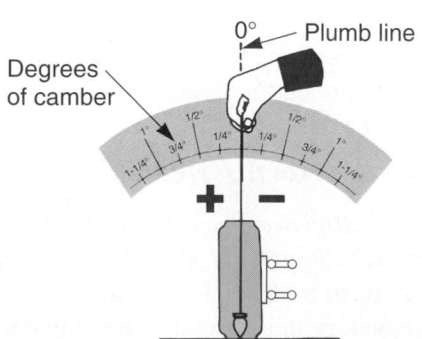

Figure 74-6. Camber is measured in degrees, as shown. (Bear)

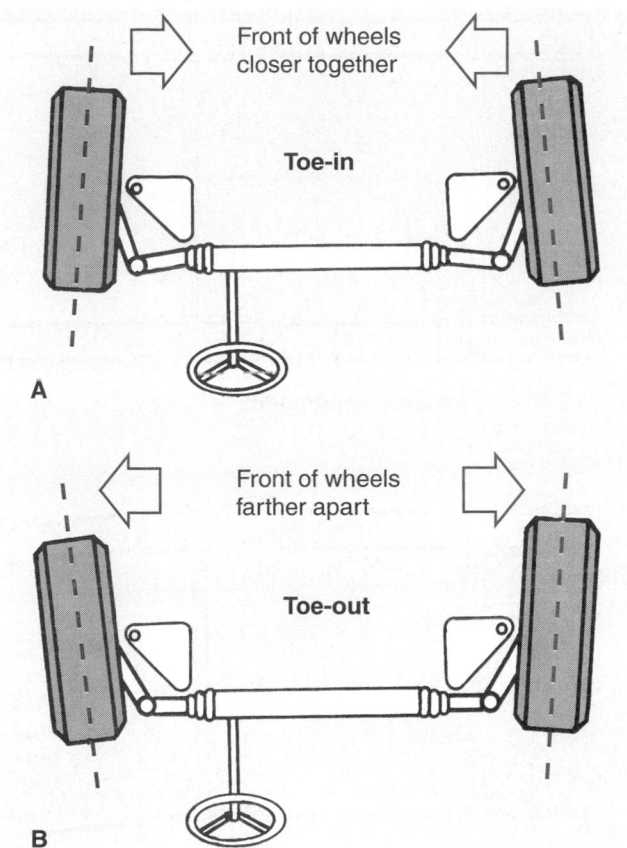

Figure 74-7. Toe is an inward or outward angle of the wheels. Toe-in is produced when the wheels are closer together at the front than at the rear. Toe-out results when the wheels are farther apart at the front than at the rear.

Toe-out results when the wheels are farther apart at the front than at the rear. Toe-out causes the front of the wheels to point away from each other. See **Figure 74-7B.**

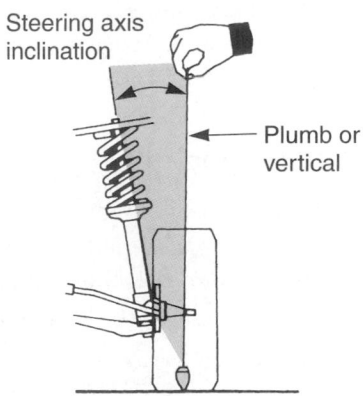

Figure 74-8. Steering axis inclination is determined by a vertical line and a projected line through upper ball joint, king pin, or MacPherson strut tube. (Bear)

Front-wheel-drive and rear-wheel-drive vehicles normally require much different toe settings. This is because the front-wheel drive tends to pull the front tires forward. With rear-wheel-drive vehicles, rolling resistance tends to push the front tires backward.

Rear-Wheel-Drive Toe Settings

Rear-wheel-drive vehicles are usually set to have *toe-in* at the front wheels. This is because the front wheels tend to toe-out while driving. Toe-in is needed to compensate for the action of tire rolling resistance, play in the steering system, and suspension system action.

As the tires roll over the road, they are pushed rearward. This turns the tires outward at the front, causing toe-out. By adjusting the wheels for a slight toe-in (approximately 1/16″- 1/4″ or 1.6 mm-6 mm), the wheels and tires roll straight ahead when driving.

Front-Wheel-Drive Toe Settings

Front-wheel-drive vehicles require different adjustment for toe. Since the front wheels propel the vehicle, they are pushed forward by engine torque. This makes the wheels toe in, or point inward when the vehicle is driven.

To compensate for this action, front-wheel-drive vehicles normally have front wheels adjusted for a slight toe-out (approximately 1/16″ or 1.5 mm). Theoretically, this will give the front end a zero toe setting when the vehicle moves down the road.

Steering Axis Inclination

Steering axis inclination is the angle, away from the vertical, formed by the inward tilt of the ball joints, king pin, or MacPherson strut tube. Steering axis inclination is always an inward tilt, regardless of whether the wheel tilts inward or outward. See **Figure 74-8.**

Steering axis inclination is *not* a tire wearing angle. Like caster, it aids directional stability by helping the steering wheel return to the straight-ahead position.

Steering axis inclination is *not* adjustable. It is designed into the suspension system by the vehicle manufacturer. If the angle is *not* correct, you must replace parts to correct the problem.

Toe-Out on Turns (Turning Radius)

Toe-out on turns, also termed *turning radius*, is the amount the front wheels toe-out when turning corners. As the vehicle goes around a turn, the inside tire must travel in a smaller radius circle than the outside tire. The

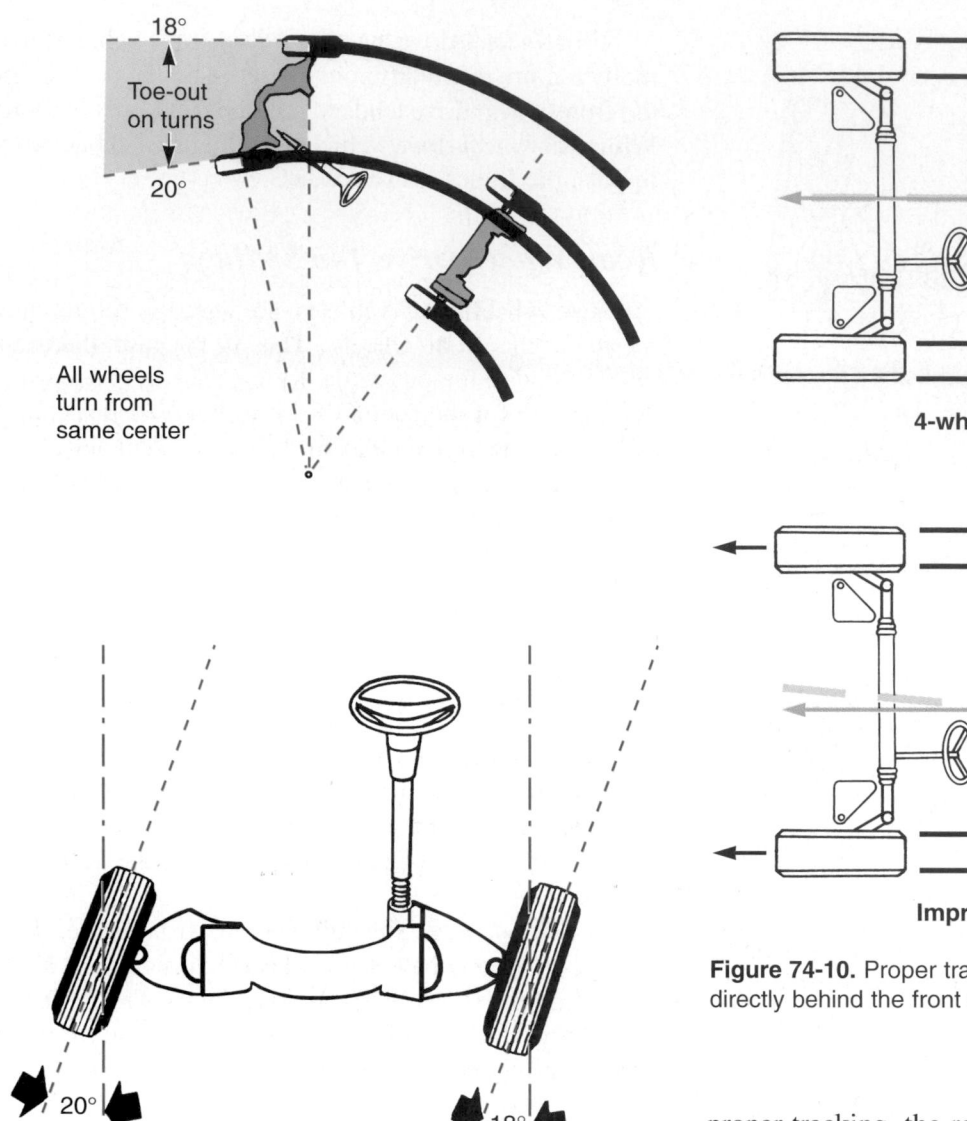

Figure 74-9. When rounding corners, the inside tire must turn more sharply than the outside tire. Angles built into the steering system produce proper toe-out on turns. (Hunter)

steering system is designed to turn the inside wheel sharper than the outside wheel.

Figure 74-9 illustrates toe-out on turns. Note how each front wheel turns at a different angle. This eliminates tire scrubbing and squealing by keeping the tires rolling in the right direction during turns.

Toe-out on turns is *not* an adjustable angle. It is controlled by the built-in angle of the steering arms. If this angle is incorrect, it indicates bent or damaged steering parts.

Tracking

Tracking refers to the position or direction of the two front wheels in relation to the two rear wheels. With

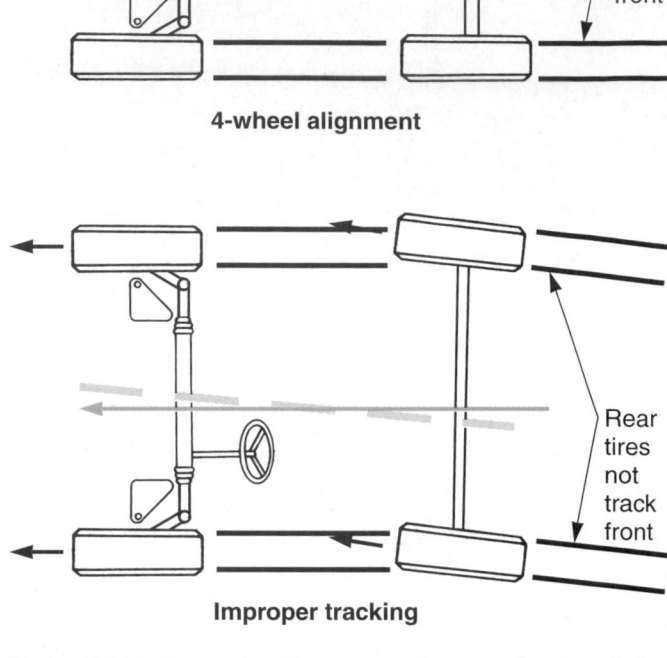

Figure 74-10. Proper tracking causes the rear wheels to follow directly behind the front wheels.

proper tracking, the rear tires follow in the tracks of the front tires, with the vehicle moving straight ahead. Look at **Figure 74-10.**

With improper tracking, or *"dog tracking,"* the rear tires do *not* follow the tracks of the front tires. This causes the back of the vehicle to actually shift sideways compared to the front when the vehicle is traveling down the road. Poor tracking will increase tire wear, lower fuel economy, and upset handling.

Prealignment Inspection

Before attempting wheel alignment, make sure all steering- and suspension-related parts are in good condition. It is impossible to properly align the wheels on a vehicle with worn or damaged parts. You should check for:

- Loose wheel bearings.
- Wheel or tire runout.
- Worn tires.
- Tires of different sizes and types.

- Incorrect tire inflation.
- Worn steering components.
- Worn suspension components.
- Incorrect curb height and weight.
- Incorrect cradle alignment.

Figure 74-11 shows several components that frequently cause problems during alignment. Adjust or replace worn or damaged parts before doing a wheel alignment. Refer to the text index for details on checking the above conditions.

Tech Tip

Many technicians like to road test the vehicle before doing a wheel alignment. This allows them to check for the kinds of problems just listed. The road test helps them detect steering wheel pull or play, abnormal noises from worn parts, and similar troubles that could affect wheel alignment.

Reading Tire Wear

Reading tires is done by inspecting tire tread wear and diagnosing the cause for any abnormal wear. Note that incorrect camber and toe show up as specific tread wear patterns. **Figure 74-12** shows various tire wear patterns.

Incorrect camber produces wear on one side of the tire tread. Too much negative camber would wear the *inside* of the tire tread. Too much positive camber would wear the *outer* tread only. With correct camber, the *full* tread area will wear evenly, **Figure 74-12.**

Incorrect toe will cause a feathered edge to form on the tire tread. A *feathered edge* is a tire wear pattern in which one side of each tread rib is sharp and raised and the other side of each rib is rounded or recessed. See **Figure 74-12.**

With too much toe-in, the sharp feathered edge points inward (toward the center of the vehicle). With too much toe-out, the sharp edge on the thread ribs points outward.

Covered in earlier chapters, tire wear patterns can also indicate incorrect wheel balance, incorrect tire inflation pressure, tire construction defects, and tire damage.

Cradle Alignment

The *vehicle cradle* is a strong metal structure mounted at the lower front and sometimes the lower rear of the unibody structure. The thick metal cradle is bolted to the frame rails on the body. The cradle often holds the lower control arms, steering rack, and engine in alignment in the body.

Loosening and moving the cradle can alter wheel alignment (camber, caster, setback, and steering axis inclination). This misalignment might happen after engine or transaxle and cradle removal. A major collision can also shift the cradle and upset wheel alignment.

Improper cradle location can reduce wheel alignment adjustment range. It can also cause premature tire wear, vehicle instability, and incorrect drive line angles.

If you ever remove a cradle, make sure it is reinstalled in the same location. This will help maintain proper wheel alignment. Alignment holes are often provided in the cradle and body. Make sure these holes align when reinstalling or adjusting a cradle. The service manual will give information on proper installation and location of the cradle.

Adjusting the position of the cradle is handy on front-wheel drive vehicles that do not provide a method of adjusting caster and camber. By shifting the cradle forward, rearward, or to one side, you can alter these angles. Keep this in mind if you run into problems when trying to perform a wheel alignment.

Adjusting Wheel Alignment

Caster, camber, and toe are the three commonly adjustable wheel alignment angles.

Before studying wheel alignment equipment, you should have a basic understanding of how alignment angles are changed. Then, you can relate this knowledge to the use of specific alignment equipment.

The basic sequence for wheel alignment is:

1. Inspect and correct tire, steering, and suspension problems.
2. Adjust caster.
3. Adjust camber and recheck caster.
4. Adjust toe.
5. Check toe-out in turns (needed if there is damage).
6. Check caster, camber, and toe on rear wheels (if needed).
7. Check tracking (if needed).

Caster Adjustment Methods

Caster is adjusted by moving the control arm so that the ball joint moves toward the front or rear of the vehicle. Depending on suspension system type, a control arm can be moved by adding or removing *shims,* adjusting the *strut rod,* turning an *eccentric bolt,* or shifting the control arm shaft bolts in *slotted holes.* See **Figure 74-13.**

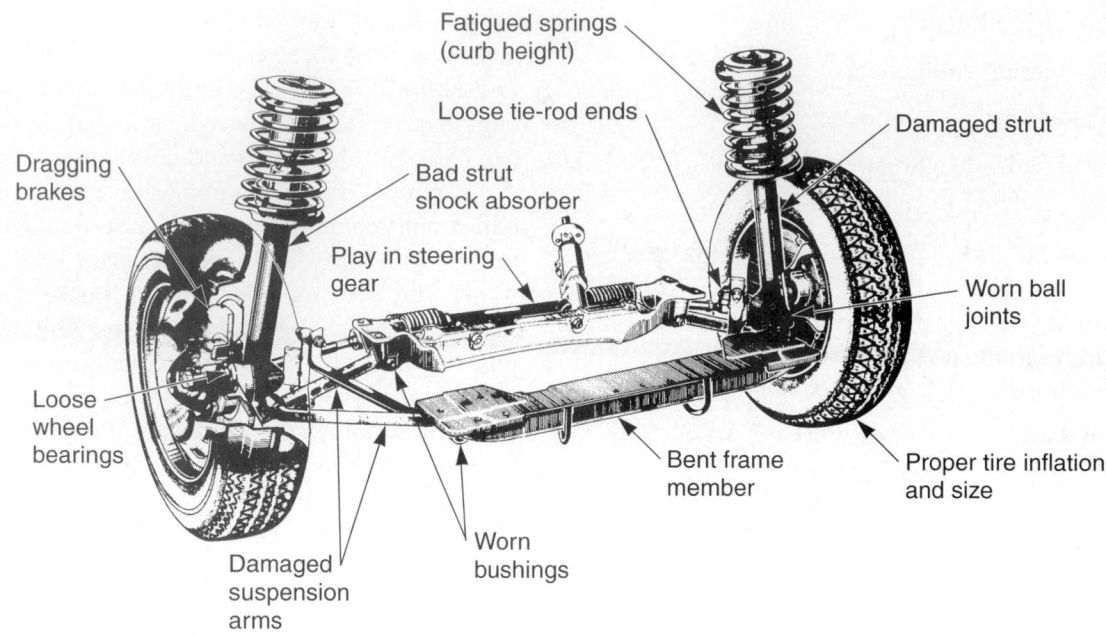

Alignment Inspection Report Form

Name_____ Date_____, 19_____

Address_____ Phone: Bus._____ Home_____

Make_____ Yr. and Model_____ License_____ Odometer _____

Tire and wheel checks

Tire condition-	Inspection	OK: LF__ RF__LR __ RR__	**Comments:**
Tire pressure		OK: LF__ RF__LR __ RR__	
Wheel bearings-	Adjustment	OK: LF__ RF__LR __ RR__	
	Roughness	OK: LF__ RF__LR __ RR__	
Runout-Lateral		OK: LF__ RF__LR __ RR__	
	Radial	OK: LF__ RF__LR __ RR__	
Wheel balance		OK: LF__ RF__LR __ RR__	
Shock absorbers-Operational		OK: LF__ RF__LR __ RR__	
	Leakage and	OK: LF__ RF__LR __ RR__	
	Bushings	OK: LF__ RF__LR __ RR__	
Riding height		OK: LF__ RF__LR __ RR__	

Suspension checks

Tracking	OK:__	**Comments:**
MacPherson-type struts	OK: LF__ RF__	
Ball joints	OK: LU__ LL__ RU __RL__	
Front control arm assembly	OK: LU__ LL__ RU __ RL__	
Strut rod and bushing assembly	OK: L__ R__	
Stabilizer (or sway) bar,		
mounting brackets,	Front: OK__	
and links	Rear: OK__	
Leaf spring assembly	OK: L__ R__	
Rear coil spring assembly	OK: L__ R__	
Rear control arm assembly	OK: LU__ RU__LL __ RL__	
Track bar and bushings	OK:__	

Steering linkage checks

Tie-rod end	OK: ____ L__ R__	**Comments:**
Tie-rods and inner ball/stud		
sockets	OK: ____ L__ R__	
Steering arms	OK: ____ L__ R__	
Tie-rod adjusting sleeves	OK: ____ L__ R__	**Comments:**
Relay rod	OK: ____	
Pitman arm	OK: ____	
Idler arm and bracket	OK: ____	
Steering shock absorber and		
bushings	OK: ____	
Steering gear mountings	OK: ____	

Manual steering gear – inspection

Lubricant leakage	OK: ____	**Comments:**
Operation	OK: ____	
Sector shaft and bearings	OK: ____	
Adjustment of gear	OK: ____	
Lubricant level	OK: ____	

Power steering gear – inspection

Fluid leakage	OK: ____	**Comments:**
Power steering hoses	OK: ____	
Power steering pump	OK: ____	
Fluid level	OK: ____	
Pump belt	OK: ____	
Power steering operation	OK: ____	
Steering gear adjustment	OK: ____	
Sector shaft and bushings	OK: ____	
Pinion shaft and bearings	OK: ____	
Control valve	OK: ____	

Front alignment check	Reading	Manufacturer's standard	OK
Caster	L__ ° R__ °	L__ ° R__ °	
Camber	L__ ° R__ °	L__ ° R__ °	
Steering axis inclination	L__ ° R__ °	L__ ° R__ °	
Turning radius	L__ ° R__ °	L__ ° R__ °	
Toe	In__ Out__	In__ Out__	

Figure 74-11. Study the problems that can affect front wheel alignment. All these must be corrected before wheels can be aligned.

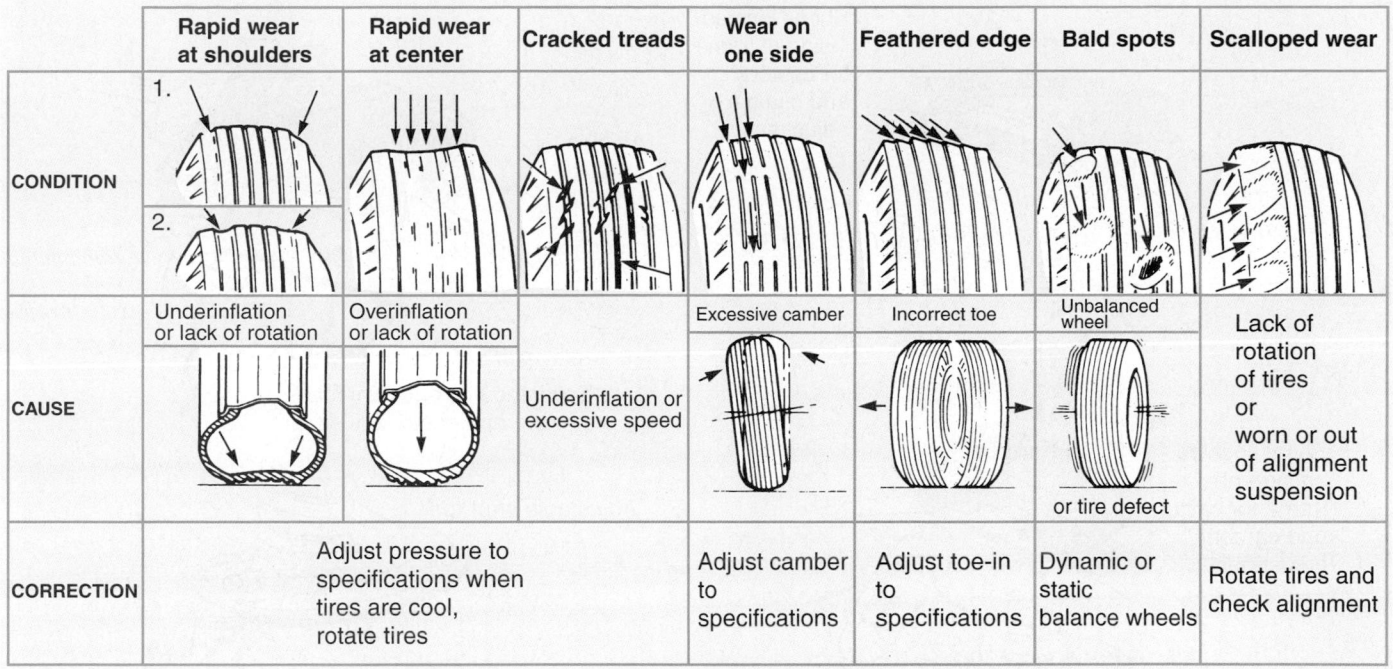

	Rapid wear at shoulders	Rapid wear at center	Cracked treads	Wear on one side	Feathered edge	Bald spots	Scalloped wear
CONDITION	1. 2.						
CAUSE	Underinflation or lack of rotation	Overinflation or lack of rotation	Underinflation or excessive speed	Excessive camber	Incorrect toe	Unbalanced wheel or tire defect	Lack of rotation of tires or worn or out of alignment suspension
CORRECTION	Adjust pressure to specifications when tires are cool, rotate tires			Adjust camber to specifications	Adjust toe-in to specifications	Dynamic or static balance wheels	Rotate tires and check alignment

Figure 74-12. Tire wear patterns should be read to help determine which steering or suspension parts are worn and to help with alignment checks. (Chrysler)

If the upper control arm ball joint is moved forward, negative caster is increased, **Figure 74-14.** If the upper control arm ball joint is moved rearward, positive caster is increased. The opposite is true for the lower control arm.

Camber Adjustment Methods

Camber is usually adjusted after setting caster. Camber is changed by moving the control arm in or out *without* moving the ball joint forward or rearward. *Shims* or *slots* in the control arm mount and *eccentric bolts* are the most common methods for adjustment. Again, refer to **Figure 74-14.**

Some MacPherson strut suspensions do not have provisions for caster and camber adjustments. However, other strut-type suspension systems have a camber adjustment at the connection between the steering knuckle and strut. See **Figure 74-13.**

The top of the steering knuckle and bottom of the strut can be pivoted in or out. The upper bolt on the steering knuckle may have an eccentric that moves the knuckle when turned.

Toe Adjustment

Toe is adjusted by lengthening or shortening the *tie-rods.* On most rack-and-pinion steering systems, the tie-rod is threaded into the outer ball socket, **Figure 74-15.**

Linkage type steering systems normally have a sleeve that is threaded on a two-piece tie-rod, **Figure 74-16.**

When the steering knuckle arms point to the rear of the vehicle, lengthen each tie-rod to increase toe-in and shorten each rod to increase toe-out. The opposite is true when the steering knuckle arms are pointed forward.

Centering Steering Wheel

To keep the steering wheel spokes centered, shorten or lengthen each tie-rod the same amount. Changing one tie-rod more than the other will rotate the steering wheel spokes. When the vehicle is traveling straight ahead, the wheel spokes should be positioned correctly. See **Figure 74-17.**

Adjusting Rear Wheel Alignment

Depending on vehicle make and model, a vehicle may or may *not* have provisions for adjusting rear wheel alignment. If the rear wheels fail to track properly, frame, unibody, or rear suspension damage may be the cause. The vehicle might have been in an accident that shifted the rear wheels out of place. Worn suspension system bushings can also upset tracking. **Figure 74-18** shows how shims can be used to align the rear wheels of one type of front-wheel drive vehicle. A shim of the correct thickness can be added to adjust camber and toe.

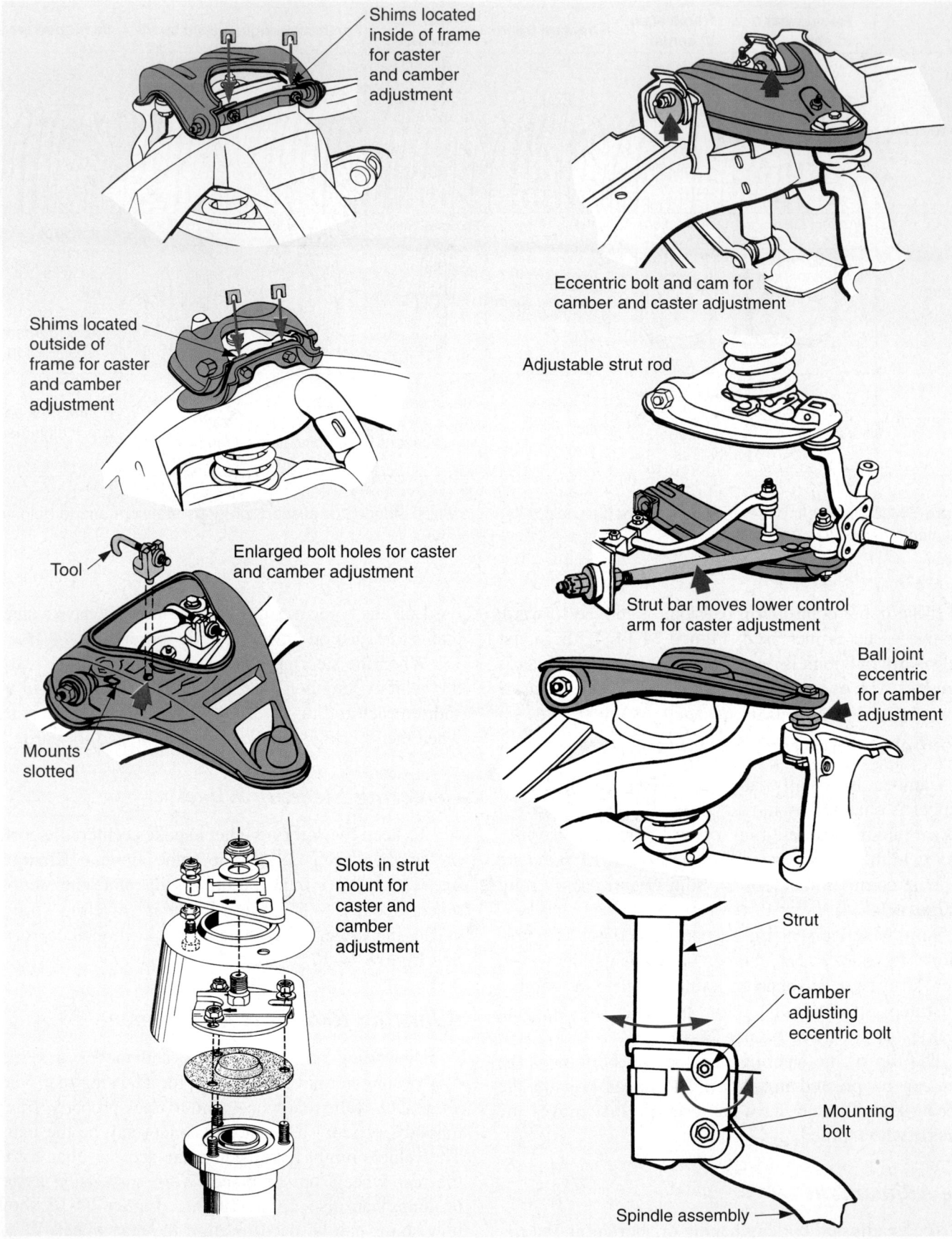

Figure 74-13. Study the various methods used to change caster and camber settings. (Hunter, Moog, Florida Dept. of Voc. Ed.)

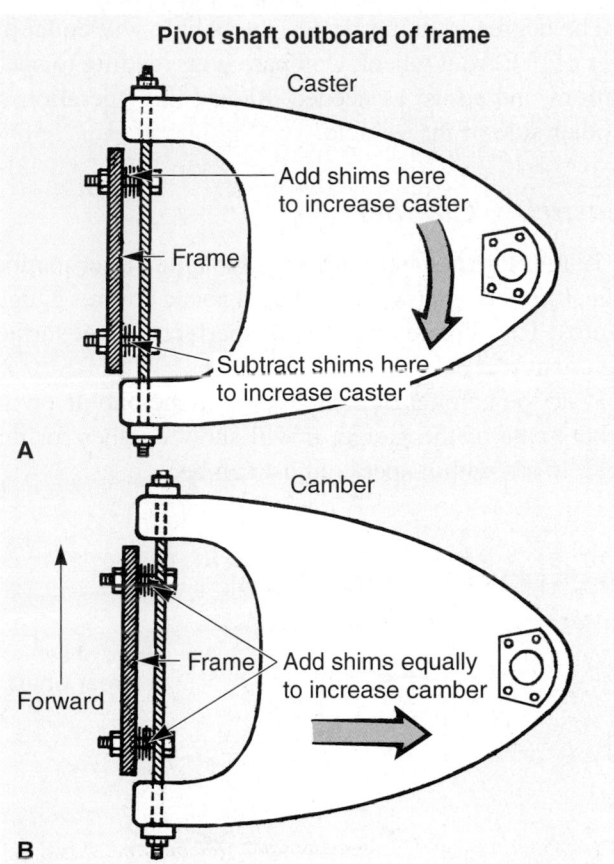

Figure 74-14. A—Caster is adjusted by moving either the upper or lower control arm to the front or rear of the vehicle. B—Camber is adjusted by moving the control arm in or out without moving the ball joint forward or rearward.

Other methods are sometimes used to align the rear wheels of a vehicle. They normally use the principles for aligning the front wheels covered earlier in this chapter.

Wheel Alignment Tools and Equipment

Various equipment and special tools are needed to adjust wheel alignment. Several special alignment tools are shown in **Figures 74-19** and **74-20.**

The most basic equipment for wheel alignment is the turning radius gauge, the caster-camber gauge, and the tram gauge. These are the least complicated of all alignment tools and easily illustrate the fundamentals of wheel alignment.

Covered shortly, these basic pieces of equipment are normally replaced with a large alignment rack. The rack will have special measuring instruments.

Turning Radius Gauges

Turning radius gauges measure how many degrees the front wheels are turned right or left. They are

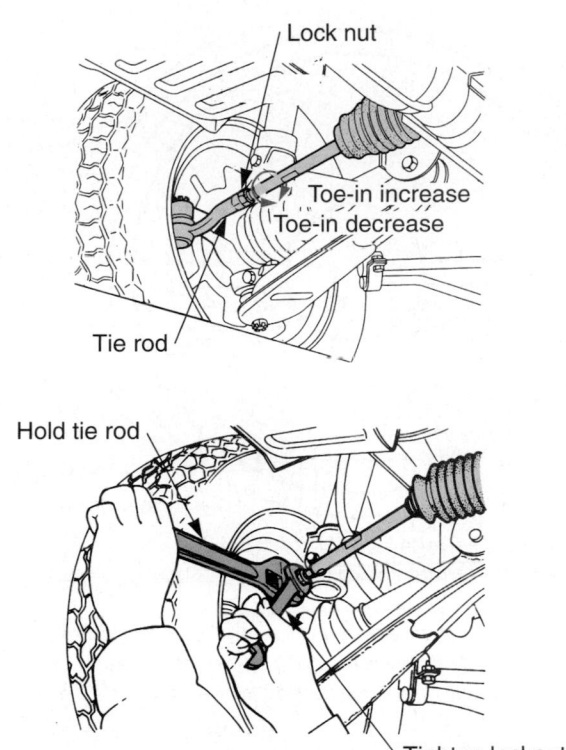

Figure 74-15. Toe is adjusted by lengthening or shortening the tie-rods. (Subaru)

commonly used when measuring caster, camber, and toe-out on turns. Look at **Figure 74-21.**

Turning radius gauges may be portable units. However, they are commonly mounted and an integral part of an alignment rack.

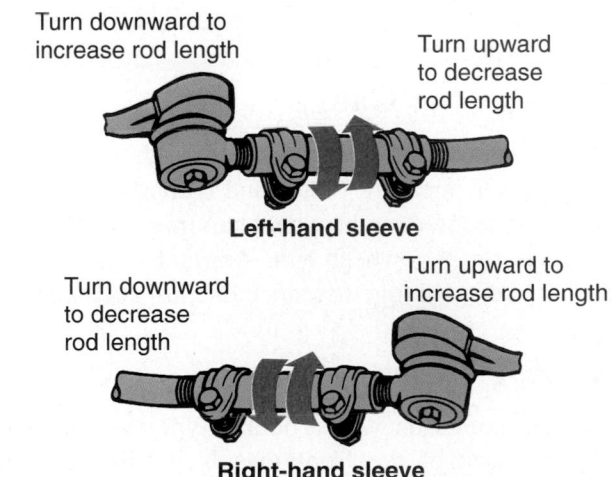

Figure 74-16. A linkage-type steering system uses a sleeve to lengthen or shorten the tie-rod. Note the adjustment method. (Ford)

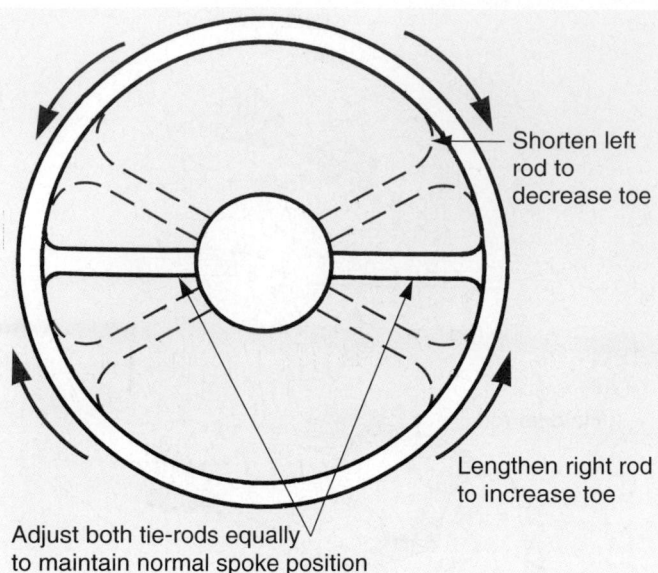

Shorten left rod to decrease toe

Lengthen right rod to increase toe

Adjust both tie-rods equally to maintain normal spoke position

Figure 74-17. When adjusting toe, the steering wheel must be kept in the center position. Study how turning each tie-rod end affects the position of the steering wheel spokes. (Ford)

The front wheels of the vehicle are centered on the turning radius gauges. Then, when the locking pins are pulled out, the gauge and tire turn together. The pointer on the gauge will indicate how many degrees the wheels have been turned.

Checking Toe-Out on Turns

To check toe-out on turns, center the front tires of the vehicle on the turning radius gauges. Turn one of the front wheels until the gauge reads 20°. Then, read the number of degrees showing on the other gauge. Check toe-out on turns on both the right and left sides. If toe-out is not within specs, check for bent or damaged parts.

Caster-Camber Gauge

A *caster-camber gauge* is used with the turning radius gauge to measure caster and camber in degrees. The gauge is secured on the wheel hub magnetically, or it may fasten on the wheel rim. Normally, caster and camber are adjusted together since one affects the other.

Measuring Caster

To measure caster with a bubble type caster-camber gauge, turn one of the front wheels inward until the radius gauge reads 20°. Turn the adjustment knob on the caster-camber gauge until the bubble is centered on zero. Then, turn the wheel out 20°.

The degree marking next to the bubble will equal the caster of that front wheel. Compare your reading to specifications and adjust as needed. Repeat this operation on the other side of the vehicle.

Measuring Camber

To measure camber with a bubble type caster-camber gauge, turn the front wheels straight ahead (radius gauges on zero). The vehicle must be on a perfectly level surface (alignment rack).

Read the number of degrees next to the bubble on the camber scale of the gauge. It will show camber for that wheel. If not within specs, adjust camber.

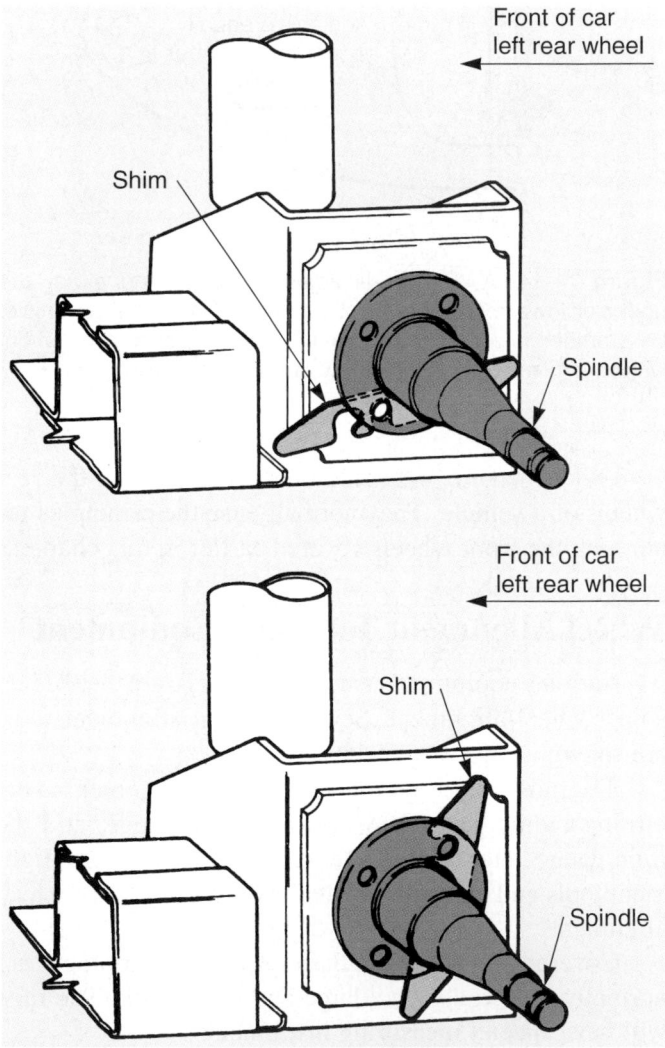

Front of car left rear wheel

Shim

Spindle

Front of car left rear wheel

Shim

Spindle

Figure 74-18. Note how a shim can be used to adjust alignment angles on the rear axle of this front-wheel-drive vehicle. Shims can be placed at the bottom, top, front, or rear to change any alignment angle. (Dodge)

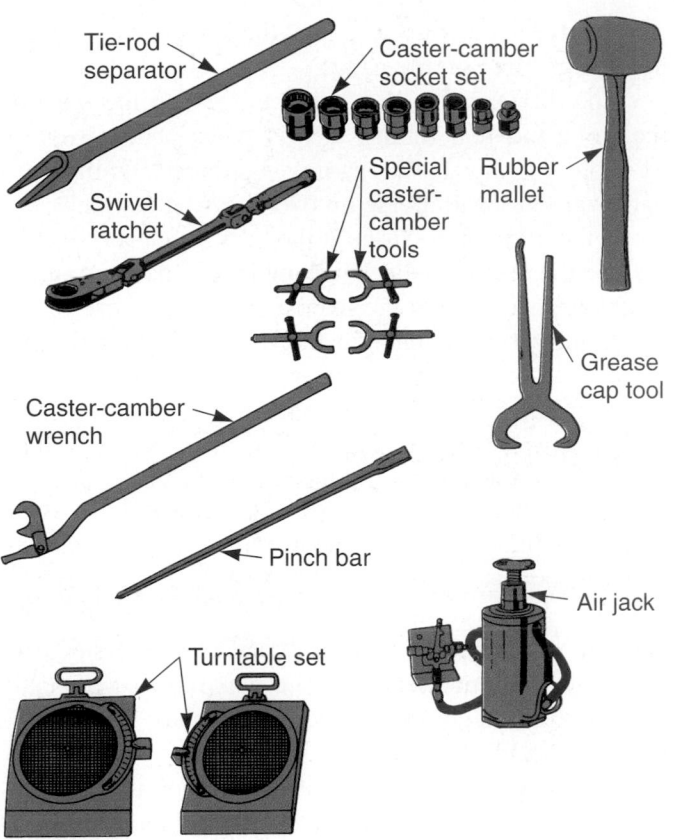

Figure 74-19. Common alignment tools. (Snap-on Tool Corp.)

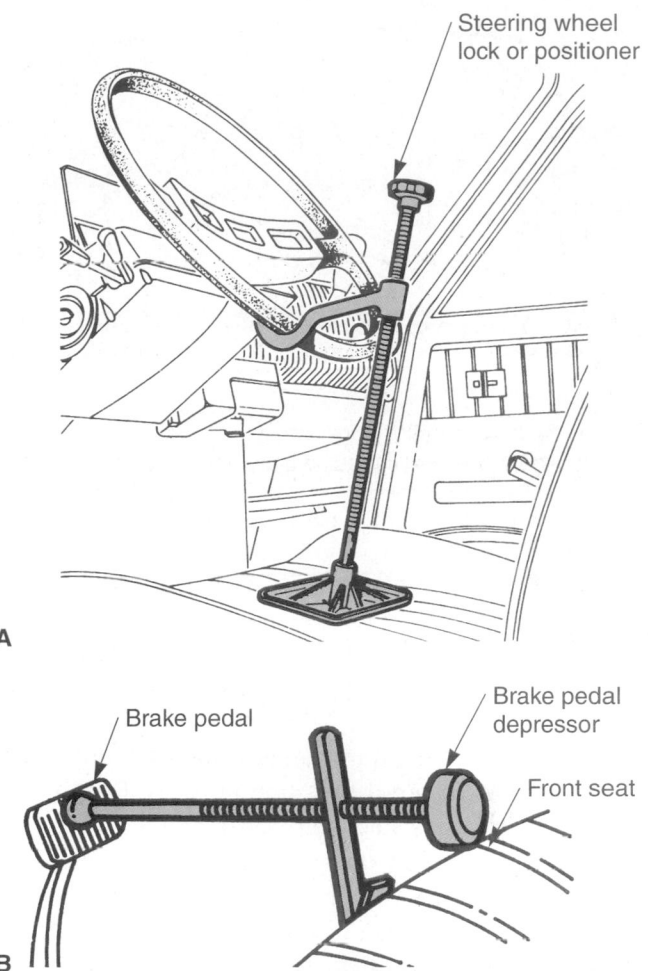

Figure 74-20. A—A steering wheel lock will hold the front wheels straight ahead. B—A brake pedal lock will keep the vehicle from rolling. (Florida Dept. of Voc. Ed. and Renault)

If shims are used, add or remove the same amount of shims from the front and rear of the control arm. This will keep the caster set correctly. Double check caster, especially when an excessive amount of camber adjustment is needed.

Tram Gauge

A *tram gauge* is used to compare the distance between the front and rear of a vehicle's tires for checking toe adjustment. It is an inexpensive tool that is sometimes used when an alignment machine is not available. Look at **Figure 74-22.**

A tram gauge is a metal rod or shaft with two pointers. The pointers slide on the gauge so that they can be set to measure the distance between the tires. The tram gauge will indicate toe-out or toe-in in either inches or millimeters.

Measuring Toe

To measure toe with a tram gauge, raise the wheels and rub a chalk line all the way around the center rib on each tire. Then, using a scribing tool, rotate each tire and scribe a fine line on the chalk line. This will give you a

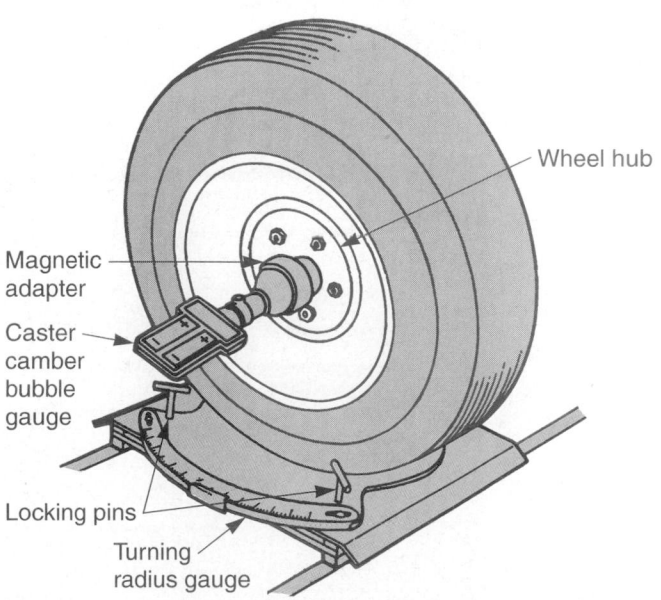

Figure 74-21. A turning radius gauge will measure the number of degrees the wheels are turned. Also note the caster-camber gauge mounted on hub. (Florida Dept. of Voc. Ed.)

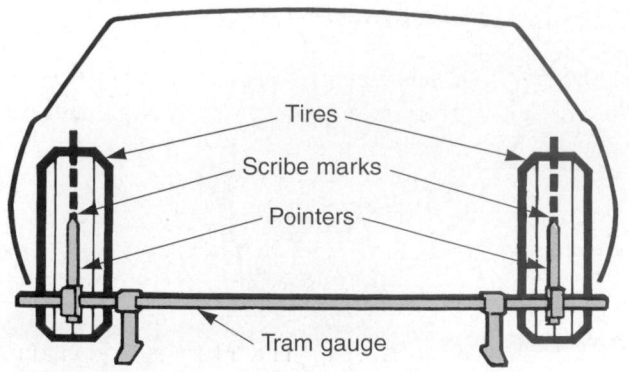

Figure 74-22. A tram gauge provides a simple method of adjusting toe. Lines are scribed on the chalked tire tread. The tram is then used to measure the distance between the lines at the front and rear of the tires. The difference equals toe. (Blackhawk)

very thin reference line that is used when measuring the distance between the tires. Lower the vehicle back on the radius gauges.

First, position the tram gauge at the back of the tires. Move the pointers until they line up with the thin lines scribed on the tires and note the reading on the tram gauge. Then, moving the pointers on the tram gauge, position the gauge on the lines at the front of the tires and note the reading on the gauge.

The difference in the distance between the lines on the front and rear of the tires is twice actual toe. For example, if the front reading is three eighths inch narrower than rear, divide by two to get actual toe.

If the lines on the front of the tires are closer together than on the rear, the wheels are toed in. If the lines are the same distance apart at the front and rear, toe is zero.

Using service manual instructions, adjust the tie-rods until the tram gauge readings are within specs.

Alignment Machines

Most medium and large service facilities have alignment machines, or alignment racks. The *alignment machine* consists of the rack, console, and related parts.

The *rack* is comprised of the ramps, turning radius gauges, and one of several types of equipment for measuring alignment angles. The rack must be perfectly level so that all equipment readings are accurate. Refer to **Figure 74-23.**

A modern *console* typically consists of a color monitor, keypad, and computer, all mounted in a roll-around cabinet. One is pictured in **Figure 74-24.**

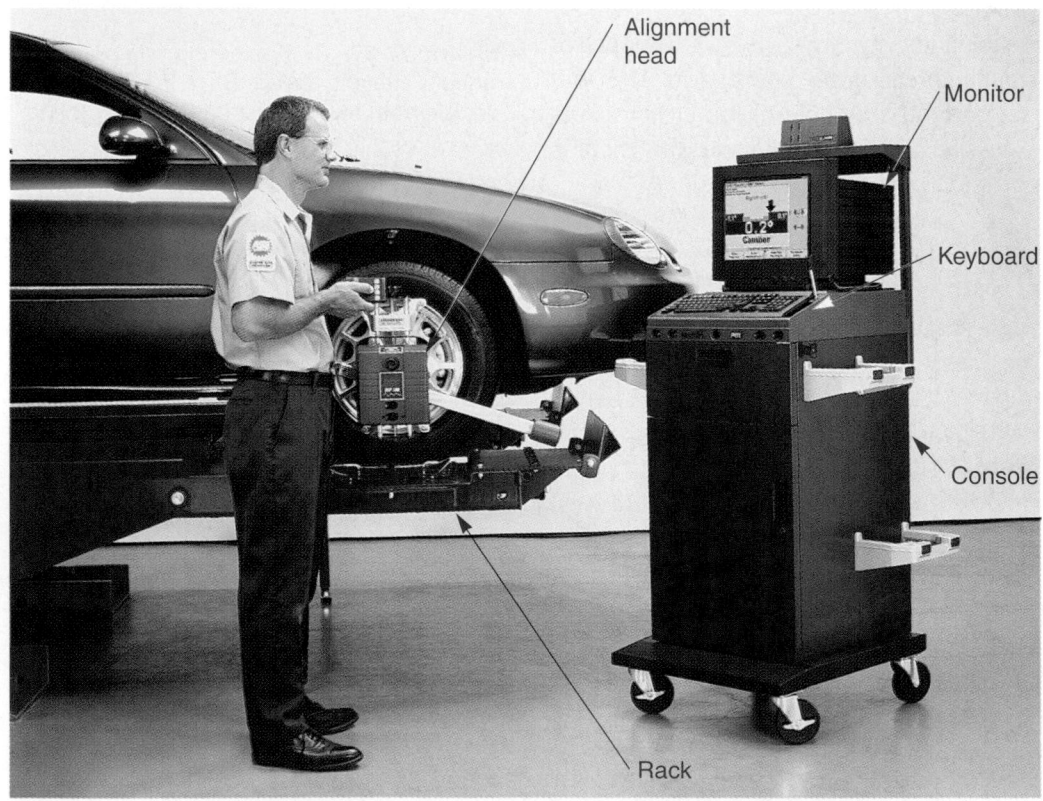

Figure 74-23. An alignment machine contains all the equipment needed to set alignment angles. Note the rack, alignment head, and console. (Hunter)

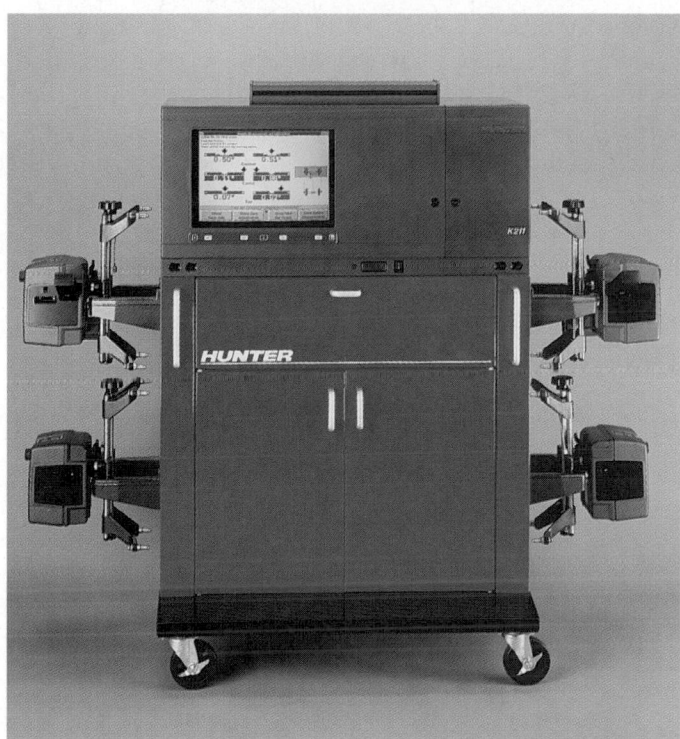

Figure 74-24. This modern console includes a color monitor that gives training, instructions, specifications, and feedback when doing wheel alignment. Note the alignment heads that are stored on the side of this console. (Hunter)

Alignment Equipment Software

Alignment equipment software contains computer instructions, information for using the alignment equipment, alignment specifications for various makes and models of vehicles, and other helpful information. When loaded into the console computer, the software will help you adjust all alignment angles quickly and easily. Most software will even give you step-by-step instructions.

The alignment machine software is usually stored on a CD-ROM (compact disc). The information contained in a complete set of manuals can be stored on one CD-ROM.

Depending on the type of the alignment machine, the monitor may prompt you to do a vehicle inspection before attempting wheel alignment. See **Figure 74-25.**

Alignment Heads

Alignment heads mount on the vehicle wheels and are used to check caster, camber, and toe. They often use lasers or proximity sensors to compare the alignment of each wheel. Refer to **Figure 74-26.**

Special brackets are used for mounting the alignment heads on the wheels. Sometimes the brackets are an integral part of the heads.

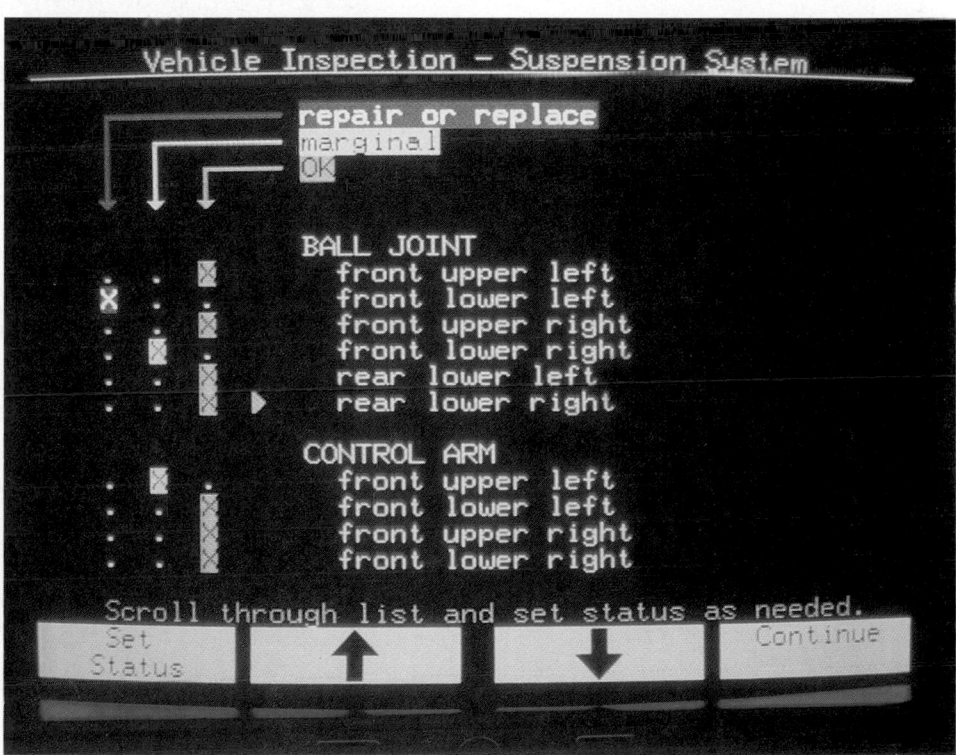

Figure 74-25. Some computerized alignment machines will even help you perform visual inspection for worn parts. This display shows how to check for bad suspension system parts.

Using Alignment Equipment

Since there are so many types of alignment equipment, always follow the operating instructions provided by the manufacturer. Remember that alignment principles are the same regardless of the equipment used. Apply your knowledge of wheel alignment to the specific type of equipment.

To use an alignment rack, drive the vehicle up on the ramps. Carefully center the front tires on the turning radius gauges. Once the vehicle is positioned on the rack, block the rear wheels to keep the vehicle from accidentally rolling off the ramps.

Warning

Use extreme care when positioning a vehicle on an alignment rack. Ask a friend to guide you up the ramps and onto the turning radius gauges. Always block the rear wheels!

Today's computerized alignment equipment is easy to use. Mount the alignment heads on the vehicle wheels. For a two-wheel alignment, only mount the heads on the front wheels. For an all-wheel alignment, mount heads on all four wheels. See **Figure 74-26.**

After turning on the alignment console, the computer might prompt you to enter the make, model, year, and other information about the car or truck. After you have keyed in the needed information, the computer will retrieve stored data about performing an alignment on the vehicle at hand from the compact disc or hard drive, **Figure 74-27.**

The monitor will then let you select different equipment functions, including:

- Training on equipment.
- Inspection of vehicle.
- Vehicle specifications.
- Vehicle measurements.
- Vehicle adjustments.
- Printing work order or measurements.
- Using help functions.

Figure 74-26. The alignment head must mount securely on the wheel. This will allow you to get accurate readouts of wheel angles. (Hunter)

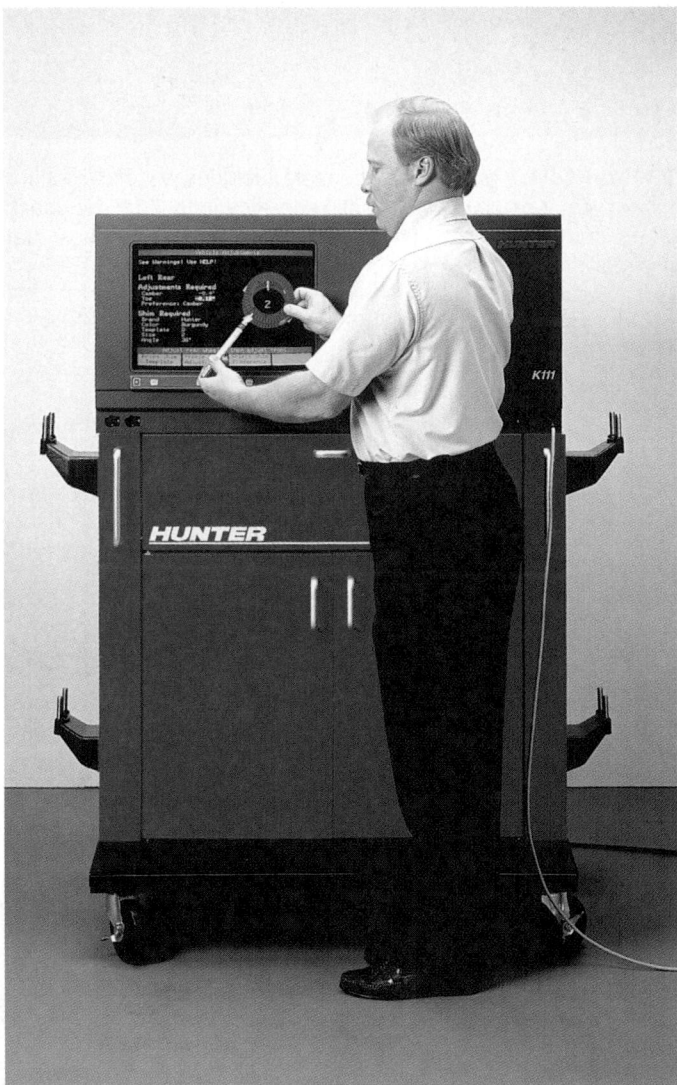

Figure 74-27. This technician is reading information on a color monitor during a four-wheel alignment. Today's alignment machines will guide you through the alignment process step-by-step. (Hunter)

Most modern alignment equipment is *"icon"* based, which means it uses small pictures to help you operate the equipment. Both words and pictures give you the feedback needed to successfully use the equipment.

As you make alignment adjustments, the equipment will monitor the changes in alignment angles. It will show you how much you have changed the wheel alignment and inform you when your adjustments are within specs. You can watch the monitor to check your progress. See **Figure 74-28.**

Warning
Always remember to use the operating manual provided with the specific alignment equipment. Procedures vary, and the slightest mistake could upset proper wheel alignment.

Road Test after Alignment

After completing your wheel alignment, you should road test the vehicle to check your work. As you test drive the car or truck on level pavement, check for a misaligned steering wheel, steering wheel pull, and similar troubles. If you detect problems, alter your alignment adjustments to correct any troubles.

A chart outlining noise and vibration problems is shown in **Figure 74-29.** Make sure the vehicle is not suffering from any of these problems before releasing it to the customer.

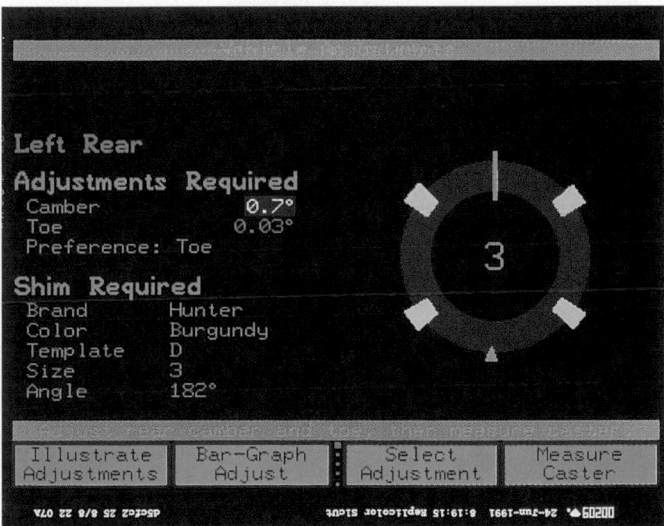

Figure 74-28. Close-up of an alignment machine monitor shows how the machine displays the various alignment angles. The alignment machine will tell you if the angles are in or out of specs and how much adjustment is needed to correct the alignment angles. Note the icon button along the bottom for selecting different functions. (Hunter)

Summary

- The main purpose of wheel alignment is to make the tires roll without scuffing, slipping, or dragging under all operating conditions.

- Caster is basically the forward or rearward tilt of the steering knuckle (spindle support) when viewed from the side of the vehicle.

- Road crown is the normal slope towards the outer edge of the road surface.

- Camber is the inward or outward tilt of the wheel and tire assembly when viewed from the front of the vehicle.

- Toe is determined by the difference in distance between the front and rear of the left- and right-hand wheels.

- Front-wheel-drive and rear-wheel-drive vehicles normally require much different toe settings.

- Steering axis inclination is the angle, away from the vertical, formed by the inward tilt of the ball joints, king pin, or MacPherson strut tube.

- Toe-out on turns, also termed turning radius, is the amount the front wheels toe-out when turning corners.

- Tracking, also called thrust line, refers to the position or direction of the two front wheels in relation to the two rear wheels.

- Reading tires is done by inspecting tire tread wear and diagnosing the cause for any abnormal wear.

- Camber is usually adjusted right after setting caster.

- Toe is adjusted by lengthening or shortening the tie-rods.

- Turning radius gauges measure how many degrees the front wheels are turned right or left.

- A caster-camber gauge is used with the turning radius gauge to measure caster and camber in degrees.

- The alignment rack consists of ramps, turning radius gauges, and one of several kinds of equipment for measuring alignment angles.

- A modern alignment console typically consists of a color monitor, keypad, and computer, all mounted in a roll-around cabinet.

- Alignment heads mount on the vehicle wheels to check caster, camber, and toe.

- After completing your wheel alignment, you should road test the vehicle to check your work.

Know These Terms

Wheel alignment
Caster
Camber
Toe

Steering axis inclination
Toe-out on turns
Tracking
Reading tires

Incorrect camber
Incorrect toe
Feathered edge
Turning radius gauges
Caster-camber gauge
Tram gauge

Alignment machine
Rack
Console
Alignment equipment
software
Alignment heads

Noise and Vibration Diagnosis Chart

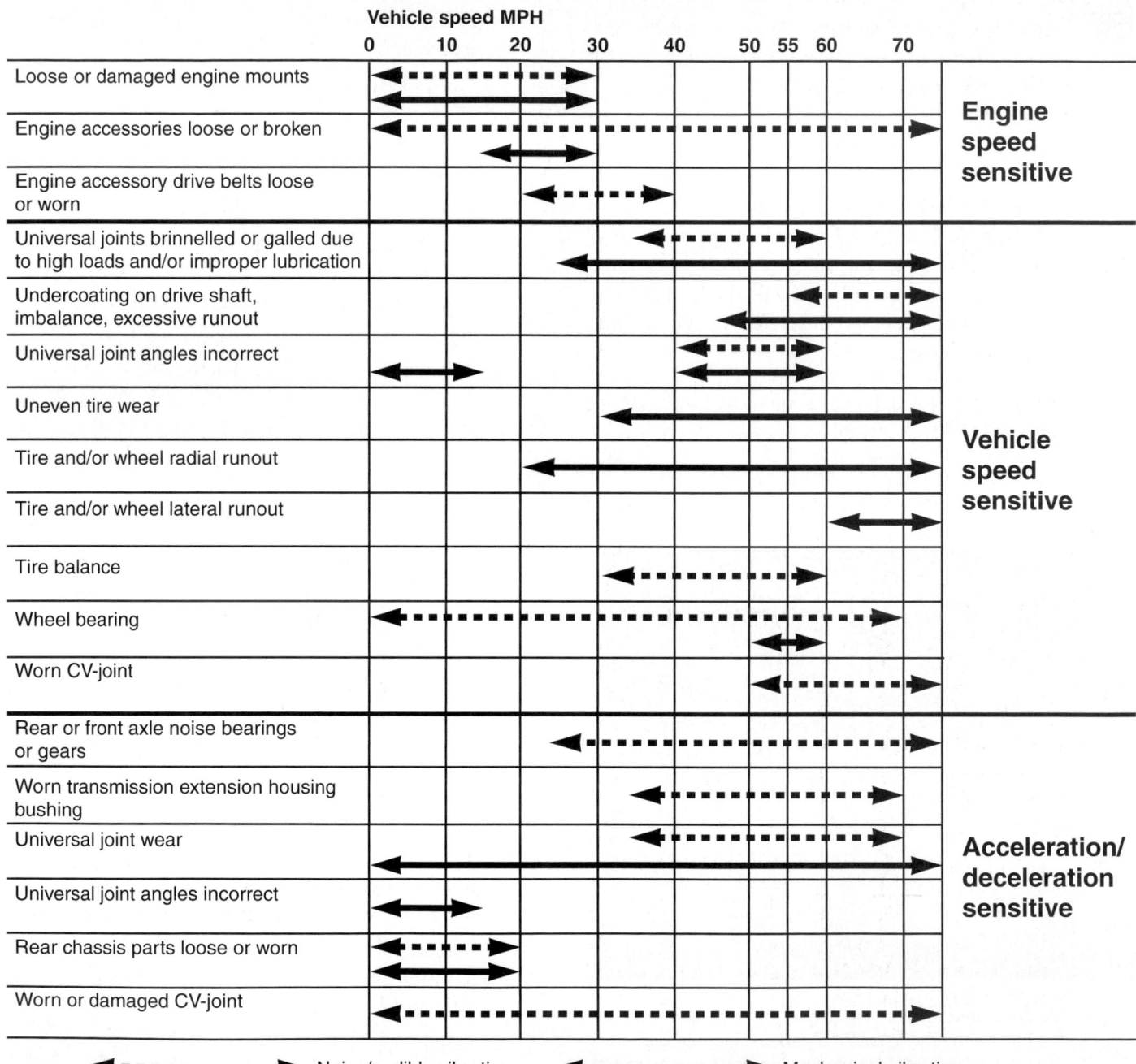

Figure 74-29. When doing a road test after alignment, you must make sure the vehicle drives and handles properly. There must not be any abnormal noises or vibration. (Dana-Perfect Circle)

Review Questions—Chapter 74

Please do not write in this text. Place your answers on a separate sheet of paper.

1. Define the term "alignment."

2. What is the main purpose of wheel alignment?

3. _____ is the forward or rearward tilt of the steering knuckle or steering support when viewed from the side of the vehicle.

4. List the three basic functions of caster.

5. Explain the difference between positive and negative caster.

6. Which of the following pertains to caster?
   (A) Measured in degrees.
   (B) A directional control angle.
   (C) Can be used to offset road crown.
   (D) All of the above.

7. _____ is the inward or outward tilt of the wheel and tire assembly when viewed from the front of the vehicle.

8. List three functions of camber.

9. Explain the difference between positive and negative camber.

10. Most vehicle manufacturers suggest a slight negative camber setting. True or False?

11. _____ is determined by the distance between the front and rear of the left- and right-hand wheels.

12. Explain the difference between toe-in and toe-out.

13. Rear-wheel-drive cars commonly use toe-in and front-wheel-drive cars commonly use toe-out. True or False?

14. Define the term "tracking."

15. Name seven possible problems you should check before attempting to align a vehicle's front end.

16. _____ _____ is done by inspecting tread wear and diagnosing the cause of abnormal wear.

17. List the seven basic steps for wheel alignment.

18. How do you change caster?

19. How do you change camber?

20. How do you adjust toe?

21. Turning radius gauges measure how many degrees the front wheels are turned right or left. True or False?

22. Summarize the use of a caster-camber gauge.

23. A _____ gauge is used to compare the distance between the front and rear of a vehicle's tires for toe adjustment.

24. What is an alignment console?

25. What is an alignment rack?

✹ ASE-Type Questions

1. Technician A says correct wheel alignment can improve an automobile's handling ability. Technician B says correct wheel alignment can improve an automobile's fuel economy. Who is right?
   (A) A only.
   (B) B only.
   (C) Both A and B.
   (D) Neither A nor B.

2. Technician A says one of the main functions of wheel alignment is to make the tires roll without slipping under all operating conditions. Technician B says one of the main functions of wheel alignment is to adjust for badly worn steering parts. Who is right?
   (A) A only.
   (B) B only.
   (C) Both A and B.
   (D) Neither A nor B.

3. Technician A says the term "caster" refers to the forward or rearward tilt of the steering knuckle when viewed from the side of the car. Technician B says the term "caster" refers to the forward or rearward tilt of the tie-rod when viewed from the side of the car. Who is right?
   (A) A only.
   (B) B only.
   (C) Both A and B.
   (D) Neither A nor B.

4. Technician A says one of the purposes of caster is to aid directional control of the vehicle. Technician B says one of the purposes of caster is to load the larger inner wheel bearing. Who is right?
   (A) A only.
   (B) B only.
   (C) Both A and B.
   (D) Neither A nor B.

5. A car's steering wheel does not return to the straight-ahead position after turning. Technician A adjusts the automobile's caster angle. Technician B says the car's caster angle has nothing to do with this problem. Who is right?
   (A) A only.
   (B) B only.
   (C) Both A and B.
   (D) Neither A nor B.

6. Technician A tilts the top of a car's steering knuckle toward the rear of the car to adjust for "positive" camber. Technician B tilts the top of the tie-rod toward the front of the car to adjust for "positive" caster. Who is right?
   (A) A only.
   (B) B only.
   (C) Both A and B.
   (D) Neither A nor B.

7. Technician A says cars with manual steering have more "negative" caster than cars with power steering. Technician B says cars with power steering have more "negative" caster than cars with manual steering. Who is right?
   (A) A only.
   (B) B only.
   (C) Both A and B.
   (D) Neither A nor B.

8. Technician A says camber is the inward or outward tilt of the wheel and tire assembly when viewed from the front of the car. Technician B says camber is the inward or outward tilt of the steering knuckle when viewed from the front of the car. Who is right?
   (A) A only.
   (B) B only.
   (C) Both A and B.
   (D) Neither A nor B.

9. Technician A adjusts an automobile's camber angle to help prevent tire wear on the inner or outer tread. Technician B adjusts an automobile's caster angle to help prevent tire wear on the inner or outer tread. Who is right?
   (A) A only.
   (B) B only.
   (C) Both A and B.
   (D) Neither A nor B.

10. Technician A says with "positive" camber, the tops of the wheels tilt outward when viewed from the front of the car. Technician B says with "positive" camber, the tops of the wheels tilt inward when viewed from the front of the car. Who is right?
    (A) A only.
    (B) B only.
    (C) Both A and B.
    (D) Neither A nor B.

11. Technician A normally adjusts a car's positive camber setting at about 1/4° to 1/2°. Technician B normally adjusts a car's positive camber setting at about 3° to 4°. Who is right?
    (A) A only.
    (B) B only.
    (C) Both A and B.
    (D) Neither A nor B.

12. Technician A says "toe-in" is produced when the front wheels are closer than the rear. Technician B says "toe-in" is produced when the front wheels are farther apart than the rear. Who is right?
    (A) A only.
    (B) B only.
    (C) Both A and B.
    (D) Neither A nor B.

13. Technician A adjusts a rear-wheel-drive car's toe-in setting at 1/16″ to 1/4″. Technician B adjusts a rear-wheel-drive car's toe-in setting at about 1.6 to 6 mm. Who is right?
    (A) A only.
    (B) B only.
    (C) Both A and B.
    (D) Neither A nor B.

14. An automobile needs a wheel alignment. Before performing this task, Technician A checks for loose wheel bearings. Before performing this task, Technician B checks for wheel or tire runout. Who is right?
    (A) A only.
    (B) B only.
    (C) Both A and B.
    (D) Neither A nor B.

15. Technician A adjusts a car's toe angle by lengthening or shortening the steering knuckle. Technician B adjusts a car's toe angle by lengthening or shortening the tie-rods. Who is right?
    (A) A only.
    (B) B only.
    (C) Both A and B.
    (D) Neither A nor B.

Activities—Chapter 74

1. Demonstrate for your instructor a prealignment inspection of tires, steering, and suspension of a vehicle in the shop for a front-end alignment.

2. Check and adjust the caster, camber, or toe-in of a front suspension.

3. Prepare a bill for an alignment. Use a flat labor rate established by your instructor and a $25/hour charge. Include the cost of any parts used. Add up all costs.

Wheel Alignment Diagnosis		
Condition	**Possible Causes**	**Correction**
Wheel shimmy.	1. Tire pressure uneven. 2. Improper or uneven caster. 3. Improper toe-in. 4. Defective stabilizer bar. 5. Front wheels misaligned. 6. Tire, wheel, or wheel bearing problems.	1. Inflate both front tires to same pressure. 2. Adjust caster angle. 3. Adjust to specifications. 4. Replace stabilizer. 5. Align front wheels. 6. Correct as necessary. *See* Chapter 66.
Poor recovery following turns and/or hard steering.	1. Lack of steering system lubrication. 2. Front wheels misaligned. 3. Bent spindle assembly.	1. Lubricate steering system. 2. Align front wheels properly. 3. Replace spindle assembly.
Vehicle pulls to one side.	1. Improper toe-in. 2. Incorrect or uneven caster. 3. Incorrect or uneven camber. 4. Improper rear wheel tracking. 5. Bent spindle assembly. 6. Dragging brakes. 7. Tire, wheel, or wheel bearing problems.	1. Adjust toe-in to specifications. 2. Adjust caster angle. 3. Adjust camber angle. 4. Align rear axle assembly. 5. Replace spindle. 6. Adjust brakes. 7. Correct as necessary. *See* Chapter 66.
Vehicle wanders from side to side.	1. Toe-in incorrect. 2. Improper caster. 3. Improper camber. 4. Vehicle overloaded or loaded too much on one side. 5. Bent spindle assembly.	1. Adjust toe-in. 2. Adjust caster angle. 3. Adjust camber angle. 4. Advise owner regarding vehicle load limits. 5. Replace spindle.
Tire squeal on corners.	1. Toe-out on turns incorrect. 2. Excessive cornering speed. 3. Bent spindle assembly. 4. Improper front wheel alignment. 5. Tire, wheel, or wheel bearing problems.	1. Replace bent steering arm. 2. Advise driver. 3. Replace spindle. 4. Align front wheels. 5. Correct as necessary. *See* Chapter 66.

(Continued)

Wheel Alignment Diagnosis		
Condition	**Possible Causes**	**Correction**
Improper wheel tracking.	1. Frame sprung. 2. Rear axle housing sprung. 3. Broken leaf spring. 4. Broken spring center bolt; spring shifted on axle housing. 5. Wheels misaligned.	1. Straighten frame. 2. Replace or straighten housing. 3. Replace spring. 4. Install new spring center bolt. 5. Align all wheels.
Tire wears in center.	1. Excessive pressure.	1. Reduce tire pressure to specifications.
Tire wears on one edge.	1. Improper camber. 2. High speed cornering.	1. Align camber angle. 2. Advise driver.
Tire wears on both sides.	1. Low pressure. 2. Overloading vehicle.	1. Inflate tires to specifications. 2. Advise driver.
Tire scuffing or feather edging.	1. Excessive toe-out (inside edges). 2. Excessive toe-in (outside edges). 3. Excessive cornering speed. 4. Improper tire pressure. 5. Improper toe-out on turns. 6. Improper camber. 7. Bent spindle assembly. 8. Tire, wheel, or wheel bearing problems.	1. Correct toe-out. 2. Correct toe-in. 3. Advise driver. 4. Inflate tires to specifications. 5. Replace bent steering arm. 6. Adjust camber angle. 7. Replace spindle. 8. Correct as necessary. *See* Chapter 66.
Tire cupping.	1. Uneven camber. 2. Bent spindle assembly. 3. Improper toe-in. 4. Improper tire pressure. 5. Tire, wheel, or wheel bearing problems. 6. Grabby brakes.	1. Correct camber angle. 2. Replace spindle. 3. Adjust toe-in. 4. Inflate tires to specifications. 5. Correct as necessary. *See* Chapter 66. 6. Repair brakes.

Heating and Air Conditioning

75. Heating and Air Conditioning Fundamentals
76. Heating and Air Conditioning Service

To the vehicle owner, the heating and air conditioning systems are vital. On a cold winter morning or a hot summer day, these systems keep the passenger compartment comfortable for the vehicle's occupants. If the heating or air conditioning system is not working properly, the passenger compartment can become very unpleasant and so can the customer.

This section explains the operation, service, and repair of heating and air conditioning systems. It details state-of-the-art methods for repairing these systems without damaging our environment. It will also help you pass the ASE Test A7, *Heating and Air Conditioning*.

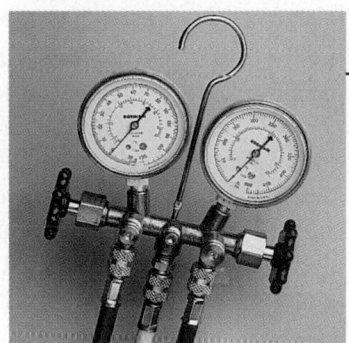

Chapter 75

Heating and Air Conditioning Fundamentals

After studying this chapter, you will be able to:

- Explain the principles of refrigeration.
- Describe the four cycles of refrigeration.
- Describe the high- and low-pressure sides of an air conditioning system.
- Explain the basic function and construction of each major part of a typical heating and air conditioning system.
- Summarize the operation and interaction of heating, ventilation, and air conditioning systems.
- Describe safety precautions to be observed when working on heating and air conditioning systems.
- Correctly answer ASE certification test questions requiring a knowledge of modern heating and air conditioning systems.

The term *"air conditioning"* generally means to control the temperature, humidity, and flow of air in the passenger compartment of a vehicle. An automotive air conditioning system uses a refrigerant system, a heating system, and a ventilation system to provide a comfortable environment in all weather conditions, **Figure 75-1.**

The *refrigerant system* uses a confined gas to provide cool, dehumidified (dried) air. The *heating system* uses heat from the engine cooling system to supply warm air. The *ventilation system* carries fresh outside air into the vehicle. These three systems work together to provide air conditioning.

 **Note**
In the trade, the term "air conditioning" is often used to refer to the refrigerant system. Actually, this term refers to all three systems that help control the passenger compartment environment.

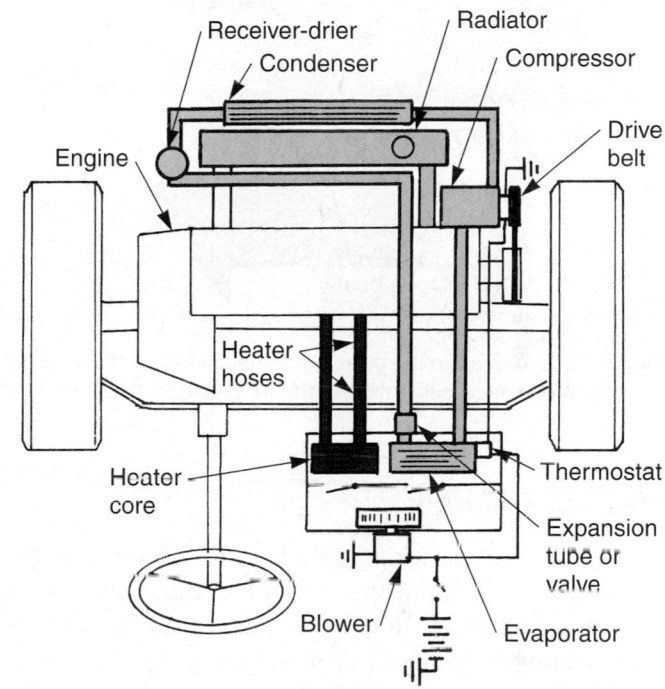

Figure 75-1. Top view shows the major parts of an air conditioning system. The heating, ventilation, and refrigerant systems work together to provide passenger comfort. Study the basic parts and their typical locations.

Principles of Refrigeration

An automotive refrigerant (air conditioning) system uses the same principles found in a home refrigerator. For this reason, you should understand a few rules that apply to refrigeration before studying actual automotive systems.

States of Matter

There are three basic *states of matter:* vapor, liquid, and solid. For example, water can exist as a *vapor* (steam), a *liquid* (tap water), or a *solid* (ice). The temperature of the water controls its state. Look at **Figure 75-2.**

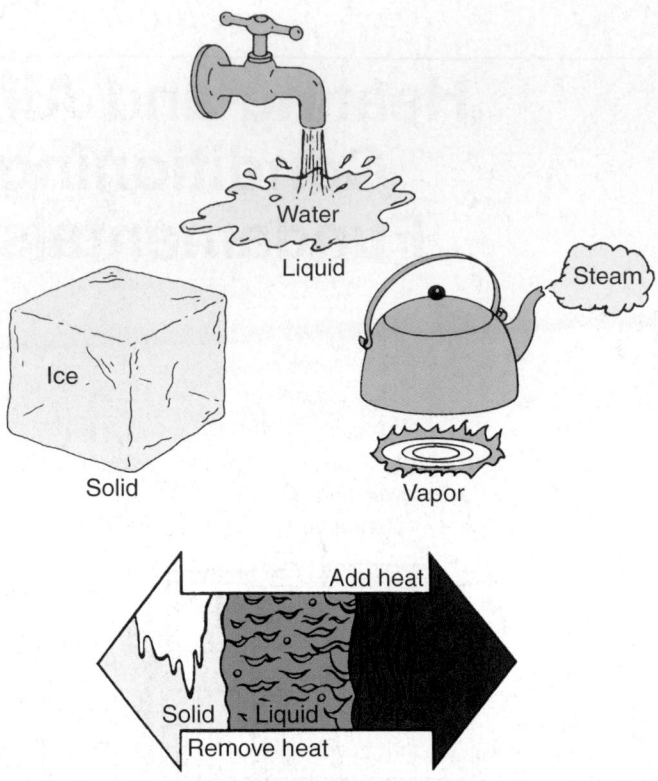

Figure 75-2. These are the three fundamental states of matter. As temperature changes, the state of matter changes. (Deere & Co.)

Radiation, the third method of heat transfer, is commonly caused by infrared waves (rays from the sun, for example). When the sun shines through the windows of a vehicle, the interior parts (seats, steering wheel) can become very hot because of heat radiation. Radiant energy moves through space in waves. When the waves strike a solid object, the energy is absorbed by the object. The internal energy in the object's molecules increases, and the temperature of the object rises. See **Figure 75-3C.**

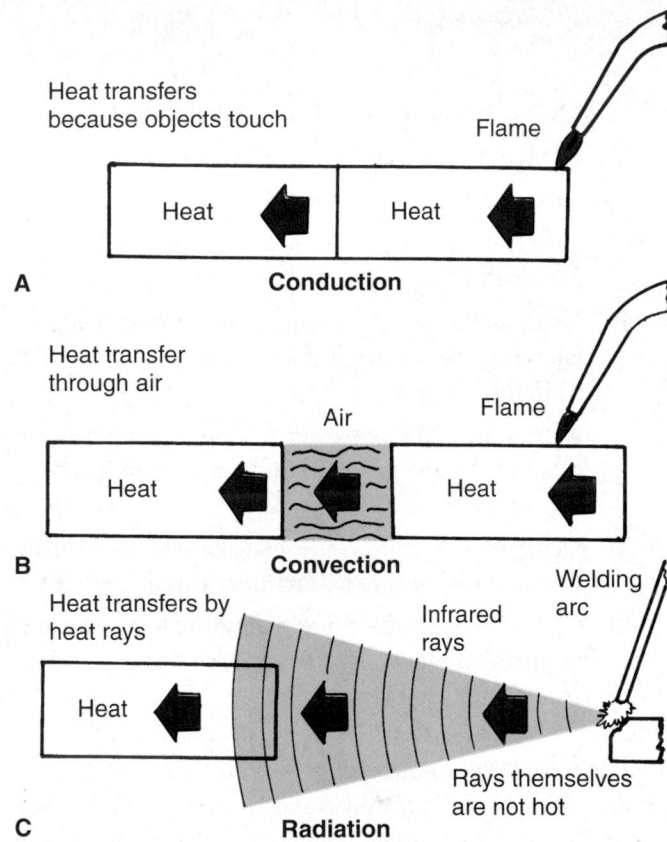

Figure 75-3. There are three means of heat transfer.

Heat and Matter

As you can see with the three states of water, heat is a controlling factor in the state of matter. Heat causes increased molecular motion inside a substance. The hotter an object, the faster its molecules move. The colder an object, the slower its molecules move.

At room temperatures, hydrogen and oxygen molecules are fluid and form a liquid. When water is hot enough, its molecules move fast enough to separate, forming a vapor. When water is cold enough, the molecules move slowly enough to join and form a solid chunk of ice.

Methods of Heat Transfer

There are three methods of heat transfer: conduction, convection, and radiation.

Conduction is heat transfer through objects that are touching each other. For example, engine combustion heat is conducted into the water jackets through the block and cylinder head. The metal parts of the engine would carry or *conduct* heat, **Figure 75-3A.**

Convection is heat transfer caused by the air surrounding objects. For instance, when air blows through a radiator, heat is convected into the air. Air would be the medium of heat transfer, **Figure 75-3B.**

Heat Movement

Heat always moves from a hotter object to a colder object. Technically, when you hold a piece of ice, the ice does not really cool your hand. Instead, it removes heat from your hand. This makes your hand feel cold.

Cold is actually an absence of heat. Hot and cold are relative (dependent on each other). When one object is colder than another, the colder object simply contains less heat.

Heat Measurement

The *British thermal unit,* abbreviated Btu, is the unit of measurement for heat transfer from one object to another. For example, when the temperature of one pound of water increases one degree Fahrenheit at sea level pressure, one Btu of heat has moved into the water. The cooling potential of home air conditioning units is normally rated in Btus.

Vaporization and Evaporation

Mentioned briefly, **vaporization** and **evaporation** generally refer to a liquid changing into a vapor (gas) state. For instance, boiling water changing into steam (vapor) demonstrates vaporization. Water slowly changing to a vapor while sitting in the hot sun would be evaporation. Heat is *absorbed* by the liquid as it changes into a vapor.

Condensation

Condensation, the opposite of vaporization, occurs when a vapor changes back into a liquid. For example, when water collects on the outside of a soft drink bottle on a warm day, moisture has condensed out of the air. Heat is *given off* when a vapor changes into a liquid.

Pressure and Temperature

When pressure is placed on a substance, the temperature of the substance increases. The substance gives off heat. When pressure is removed from a substance, the temperature of the substance drops. The substance absorbs heat (cools). Look at **Figure 75-4.**

Pressure also affects change of state. When a liquid is placed in a closed container and pressurized, its boiling (vaporizing) point increases.

For example, the boiling point of water at sea level pressure (14.7 psi [101 kPa]) is 212°F (100°C). However, when this pressure is increased, water's boiling point increases. If pressure is decreased, the boiling point decreases. This principle is true for other substances.

Refrigerant

A **refrigerant** is a substance with a very low boiling point. It is circulated through a refrigeration, or air conditioning, system.

A few years ago, **Refrigerant-12,** also called **R-12** or **Freon,** was the most efficient refrigerant. It has a boiling point of 22°F (−6°C) and vaporizes at room temperature and pressure. However, it has been found that R-12 is dangerous to the environment, because of a chemical reaction with the atmosphere. R-12 has been phased out and new vehicles no longer use this type refrigerant.

A new refrigerant, **R-134a,** is currently being phased in because it is less damaging to the ozone layer around the earth. New cars, vans, and trucks are equipped with refrigerant systems that allow the use of R-134a.

Discussed in the next chapter, several system parts, controls, sensors, tools, and procedures have changed with this new, environmentally safe refrigerant.

Figure 75-5 illustrates how a refrigerant can be used to reduce the temperature in an enclosed area.

Basic Refrigeration Cycle

In the **basic refrigeration cycle,** the refrigerant goes through four phases: compression, condensation, expansion, and vaporization. When the refrigerant condenses, heat transfers out of it and into the surrounding air. Then, when the refrigerant evaporates, it absorbs heat to cool the air in the surrounding area, **Figure 75-6.**

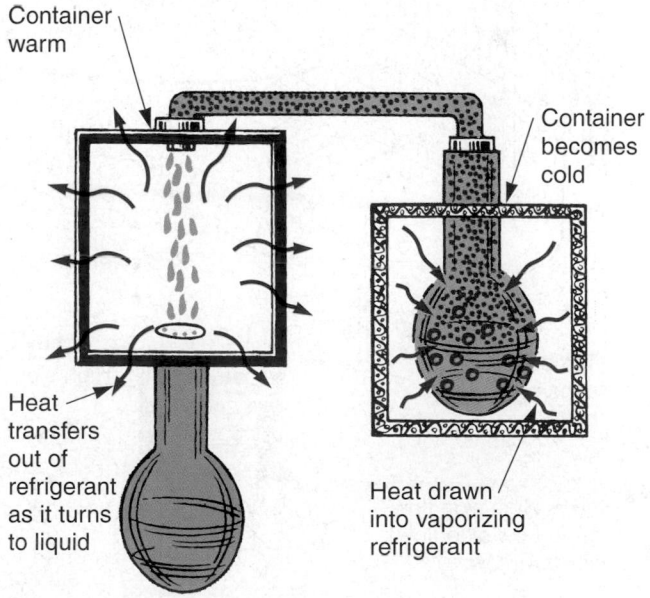

Figure 75-5. Refrigerant can be used to cool object when it vaporizes. It releases gathered heat when it condenses. (General Motors)

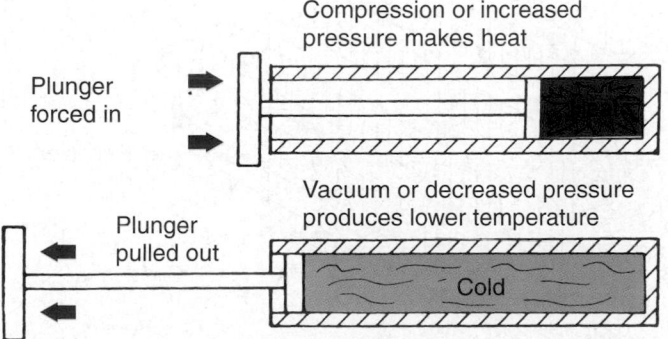

Figure 75-4. Study the relationship between pressure temperature.

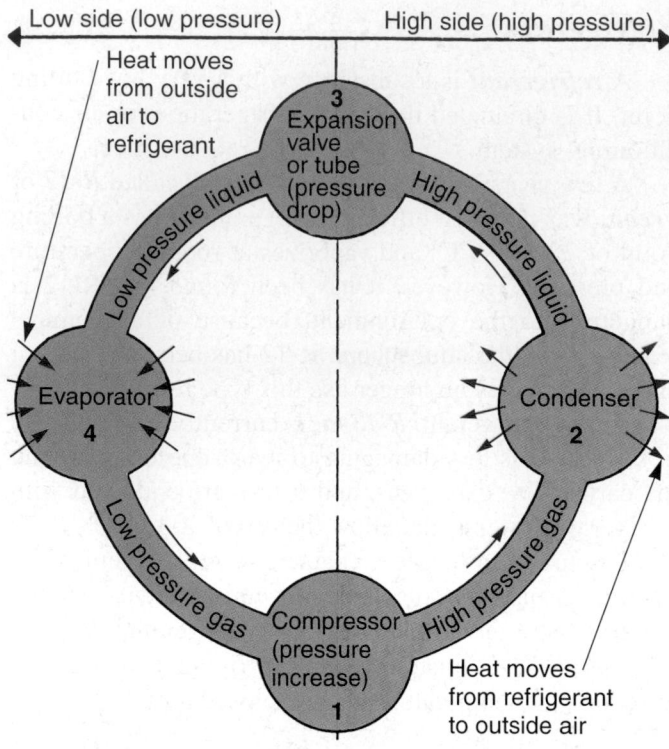

Low side (low pressure)   High side (high pressure)

Heat moves
from outside
air to
refrigerant

3
Expansion
valve
or tube
(pressure
drop)

High pressure liquid

Low pressure liquid

Evaporator
4

Condenser
2

Low pressure gas

High pressure gas

Compressor
(pressure
increase)
1

Heat moves
from refrigerant
to outside air

Figure 75-6. Study this diagram showing the basic refrigeration cycle. (Deere & Co.)

Figure 75-7 illustrates the operation of a refrigeration system. The eight basic parts of a typical refrigeration system include:

- *Refrigerant*—substance, usually R-134a, that carries heat through the system to lower the air temperature in the vehicle.

- *Compressor*—pump that pressurizes refrigerant and forces it through the system.

- *Condenser*—causes refrigerant to change from a gaseous state to a liquid state, causing it to give off its stored heat.

- *Flow-control device*—usually expansion valve or tube that causes refrigerant pressure and temperature to drop, cooling the evaporator.

- *Evaporator*—uses cooling action of vaporizing refrigerant to cool the air inside the vehicle.

- *Receiver-drier* or *accumulator*—removes moisture from and stores refrigerant.

- *Blower*—fan that forces air through the evaporator and into the passenger compartment.

- *Thermostatic switch*—shuts the compressor off when the evaporator temperature nears freezing.

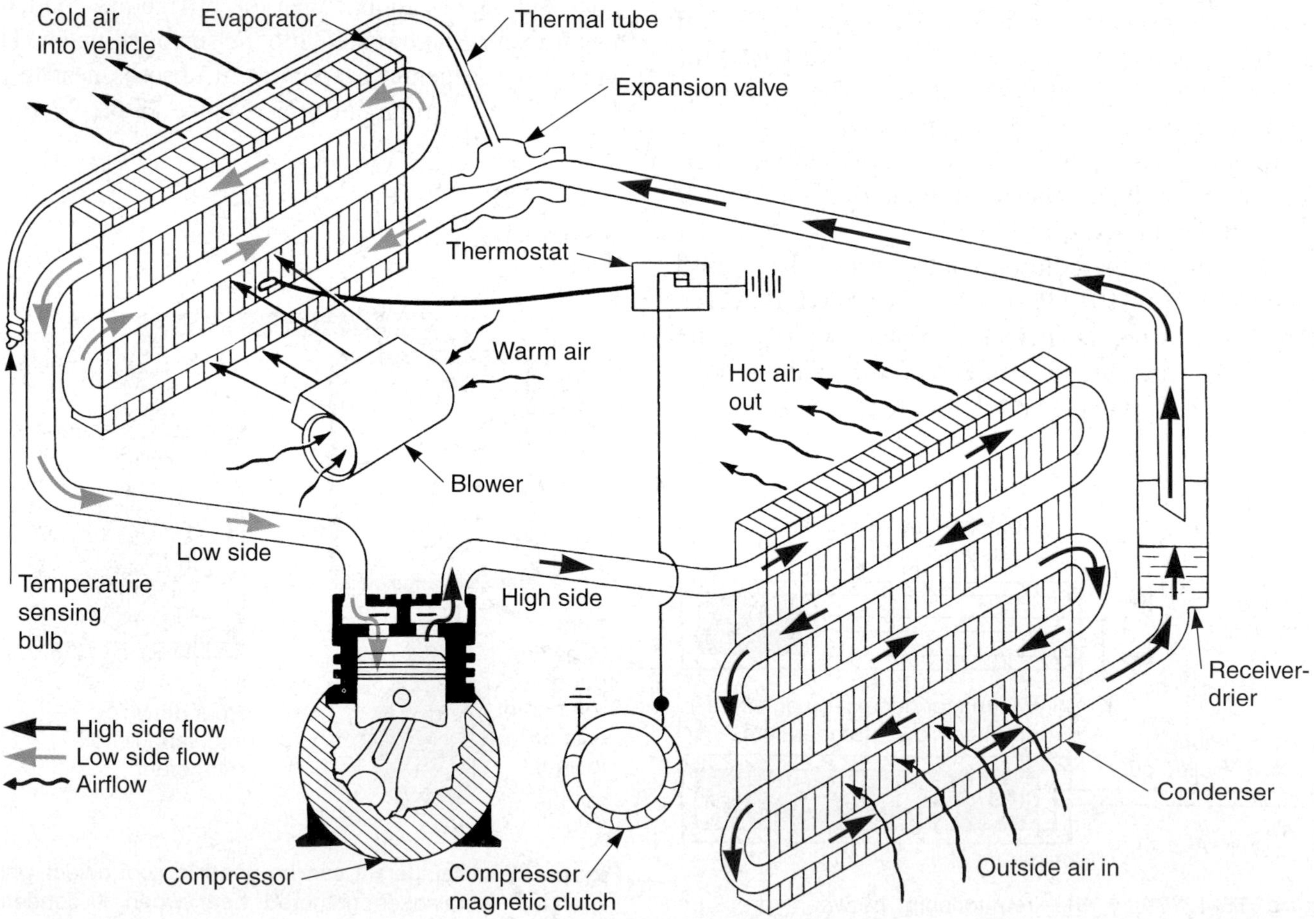

Cold air
into vehicle   Evaporator   Thermal tube

Expansion valve

Thermostat

Warm air

Hot air
out

Blower

Low side

Temperature
sensing
bulb

High side

Receiver-
drier

→ High side flow
→ Low side flow
↝ Airflow

Compressor

Compressor
magnetic clutch

Outside air in

Condenser

Figure 75-7. Fundamental parts and action of an automotive air conditioning system. (Deere & Co.)

Refrigerant circulates around inside the system. The compressor pressurizes the refrigerant and forces it into the condenser, where it condenses into a liquid. Heat is transferred into the condenser and then into the outside air.

Liquefied refrigerant then flows to a restriction (orifice tube or expansion valve). As it passes through the restriction, pressure suddenly drops. This makes the refrigerant turn into a vapor as it enters the evaporator. The refrigerant vapor absorbs heat from the evaporator, and the evaporator becomes very cold.

Since the blower is forcing air through the evaporator, cold air blows into the area to be cooled. The refrigerant vapor is then pulled back into the compressor for another cycle.

Automotive Air Conditioning System

An *automotive air conditioning system,* abbreviated A/C system, performs four basic functions—it cools, dehumidifies (dries), cleans, and circulates the air in the vehicle. All automotive air conditioning systems contain the basic components shown in **Figure 75-7.** Additional components (valves, ducts, flaps, and switches) are utilized to increase system efficiency and dependability, **Figure 75-8.**

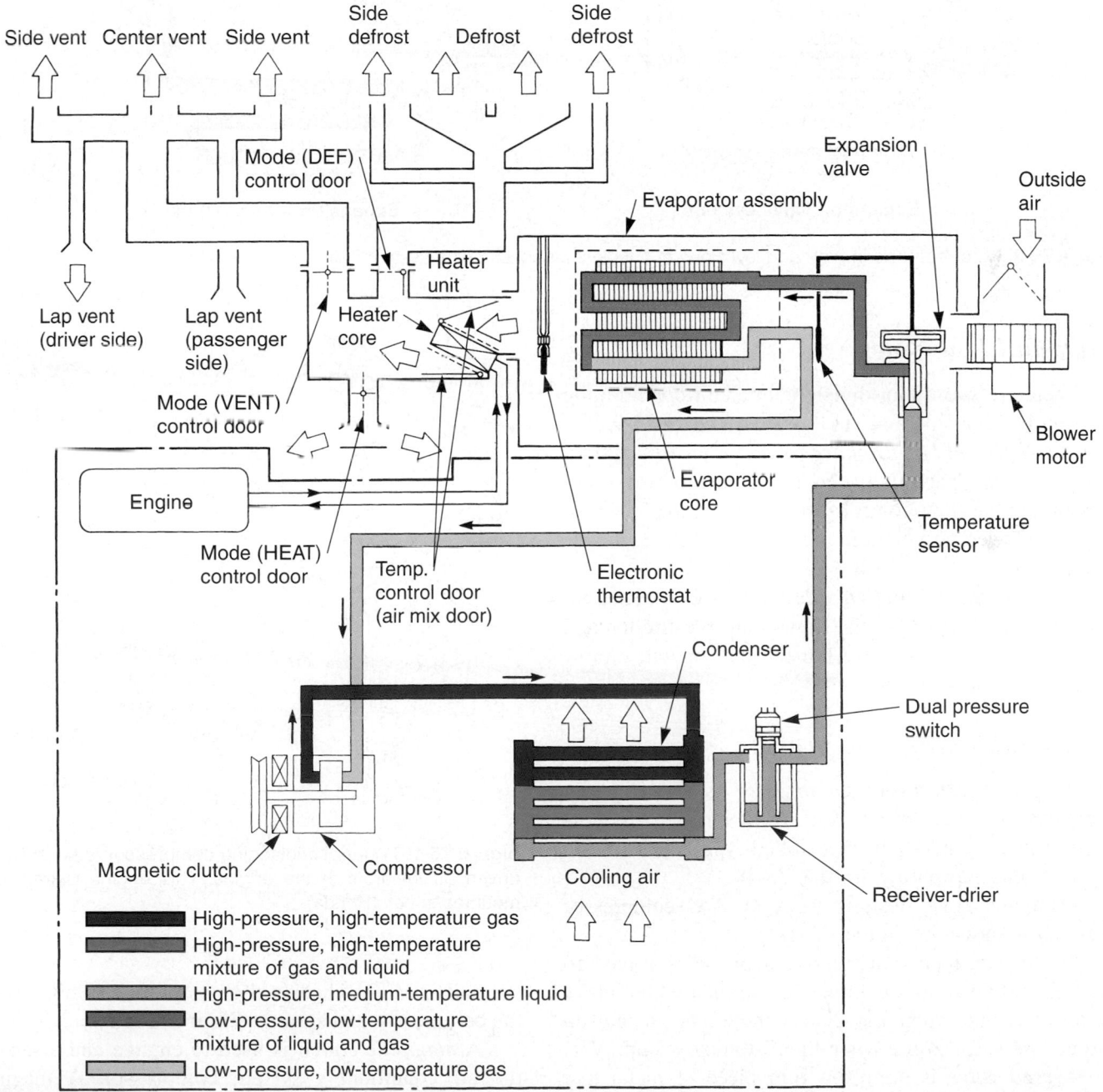

Figure 75-8. Trace the flow of refrigerant liquid and vapor through the system. Also study airflow through the vents. (General Motors)

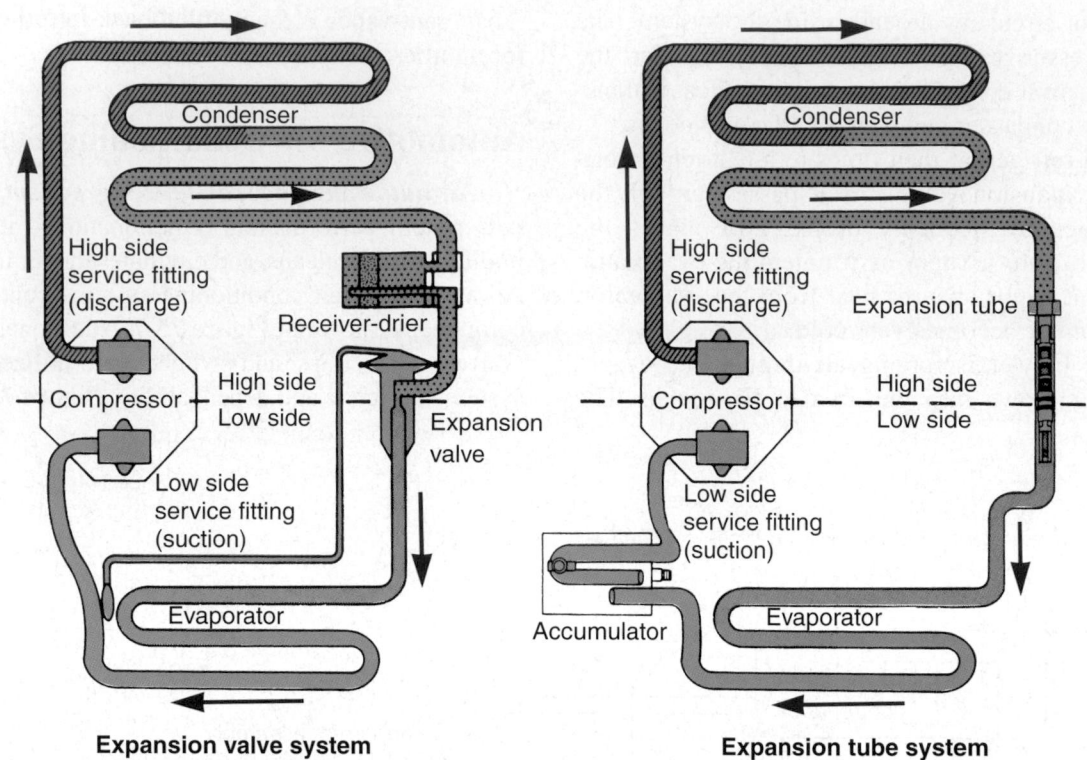

Expansion valve system   **Expansion tube system**

Figure 75-9. Note the parts in high and low sides of a typical air conditioning system. (Mitchell Manuals)

High and Low Sides

There are two major divisions of an air conditioning system: the high side and the low side, **Figure 75-9.**

The *high side,* or *discharge side,* of an A/C system consists of the parts between the output of the compressor and the flow-control restriction. All the lines and other parts operating on high pressure are in the high side of the system.

The *low side,* or *suction side,* of an A/C system consists of the parts between the flow-control restriction and the inlet of the compressor. All these parts operate on low pressure.

Air Conditioning Compressor

An *air conditioning compressor* is a refrigerant pump that is bolted to the front of the engine. It is normally driven by a V-belt or a ribbed-type belt from the engine crankshaft pulley, **Figure 75-10.**

Although exact designs vary, an A/C compressor operates as shown in **Figure 75-11.**

In a piston-type compressor, a piston is forced to slide up and down in a cylinder. This produces an intake (suction) stroke when the piston moves down and an output (pressure) stroke when the piston moves up.

A *reed valve* is simply a thin piece of metal that bends to open and close an opening. Reed valves are

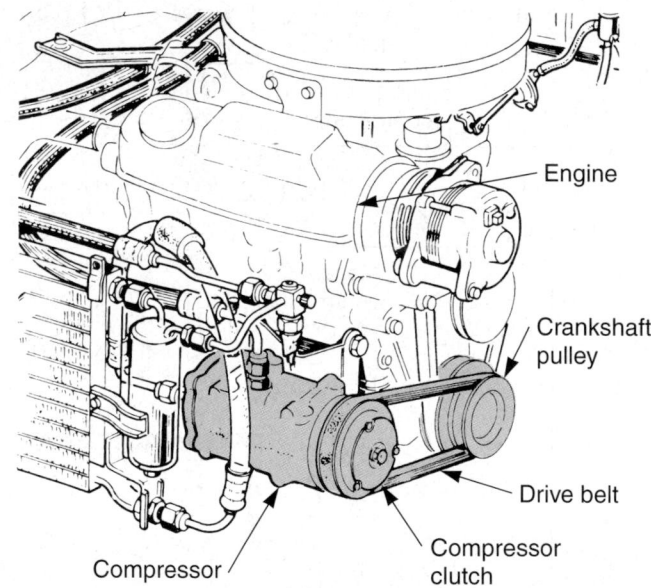

Figure 75-10. An air conditioning compressor is normally belt driven off the front of the engine. This engine is transverse mounted in car. (Honda)

normally used as check valves to produce flow through the compressor and system, **Figure 75-12.**

A *magnetic clutch* is used to engage and disengage the air conditioning system compressor. A magnetic clutch typically consists of an electric coil, a pulley, and

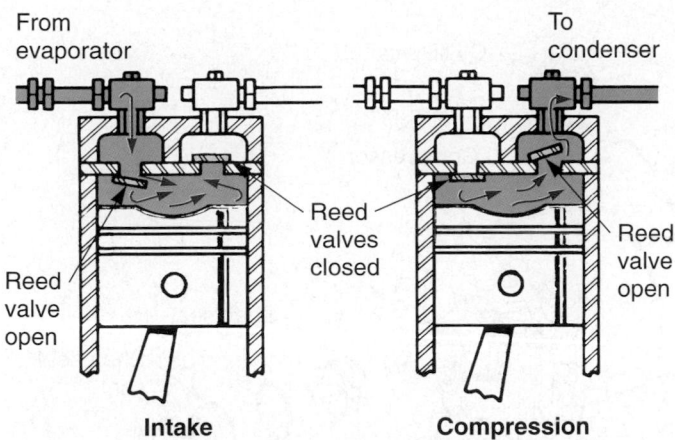

Figure 75-11. Study the basic operation of a compressor. During the intake stroke, the piston slides down and pulls vapor into the cylinder. On the compression stroke, the piston slides up and forces vapor through the system. (Ford)

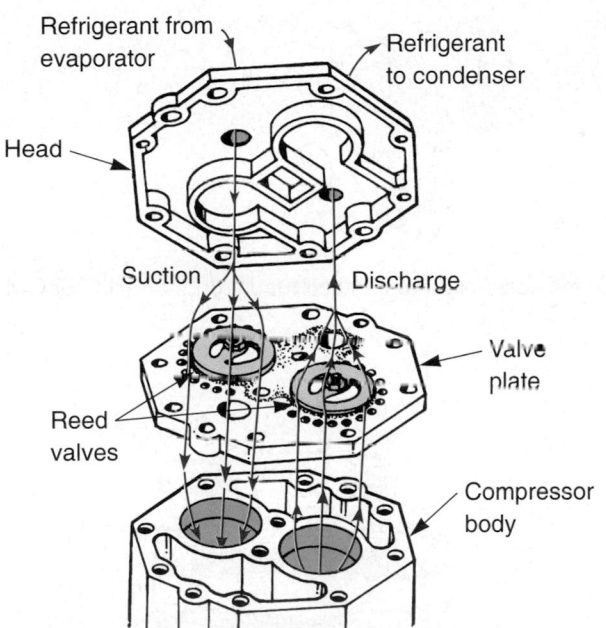

Figure 75-12. Compressor reed valves control flow in and out of the compressor. Note how they mount in the unit. (Florida Dept. of Voc. Ed.)

a front clutch plate. These parts bolt around or on the compressor shaft. See **Figure 75-13.**

When current is passed through the compressor coil, a strong magnetic field develops. The magnetic field pulls the front clutch plate into contact with the compressor pulley. This locks the clutch plate and pulley together. The compressor shaft then rotates to operate the internal parts of the compressor, **Figure 75-14.**

When current to the compressor coil is shut off, the clutch plate is released from the compressor pulley. This

lets the pulley freewheel, so the compressor doesn't function.

Compressor Types

There are five types of automotive air conditioning compressors:

- Crank compressor (inline or V-type), **Figure 75-15.**
- Axial compressor, **Figure 75-16.**
- Radial compressor, **Figure 75-17.**
- Rotary vane compressor, **Figure 75-18.**
- Scroll compressor, **Figure 75-19.**

Compare the shape and construction of each type. Note that three of the five types use pistons, cylinders, and reed valves to produce system pressure. The main difference is the position of these parts.

The crank compressor uses a *crankshaft* that is similar to the crankshaft used in an automotive engine. The axial compressor uses a *swash plate* to produce piston motion. A *shaft eccentric* (egg-shaped component) causes piston movement in a radial compressor.

Vanes (blades) similar to those found in oil or power steering pumps pressurize the refrigerant in a rotary vane compressor. A scroll compressor uses two scrolls—one movable and one stationary—to pressurize the refrigerant.

Refrigerant Oil

Refrigerant oil is high-viscosity, highly refined (pure) oil used to lubricate the parts of the air conditioning system, especially the compressor. Because the refrigerant oil and the refrigerant mix, oil droplets are carried throughout the system. The refrigerant oil also splashes around inside the compressor to reduce friction and wear.

Refrigerant Hoses

Refrigerant hoses carry refrigerant to each air conditioning system component and back to the compressor. These special high-pressure hoses are designed to withstand compressor pressures as high as 250 psi (1725 kPa). See **Figure 75-20.**

Special fittings are used on the ends of refrigerant hoses. O-ring seals, hose clamps, or flared tube ends are used on the fittings to prevent leakage. **Figure 75-21** shows the most common hose fittings.

Air Conditioning Condenser

An *air conditioning condenser* is a radiator-like device that transfers heat from the refrigerant to the outside

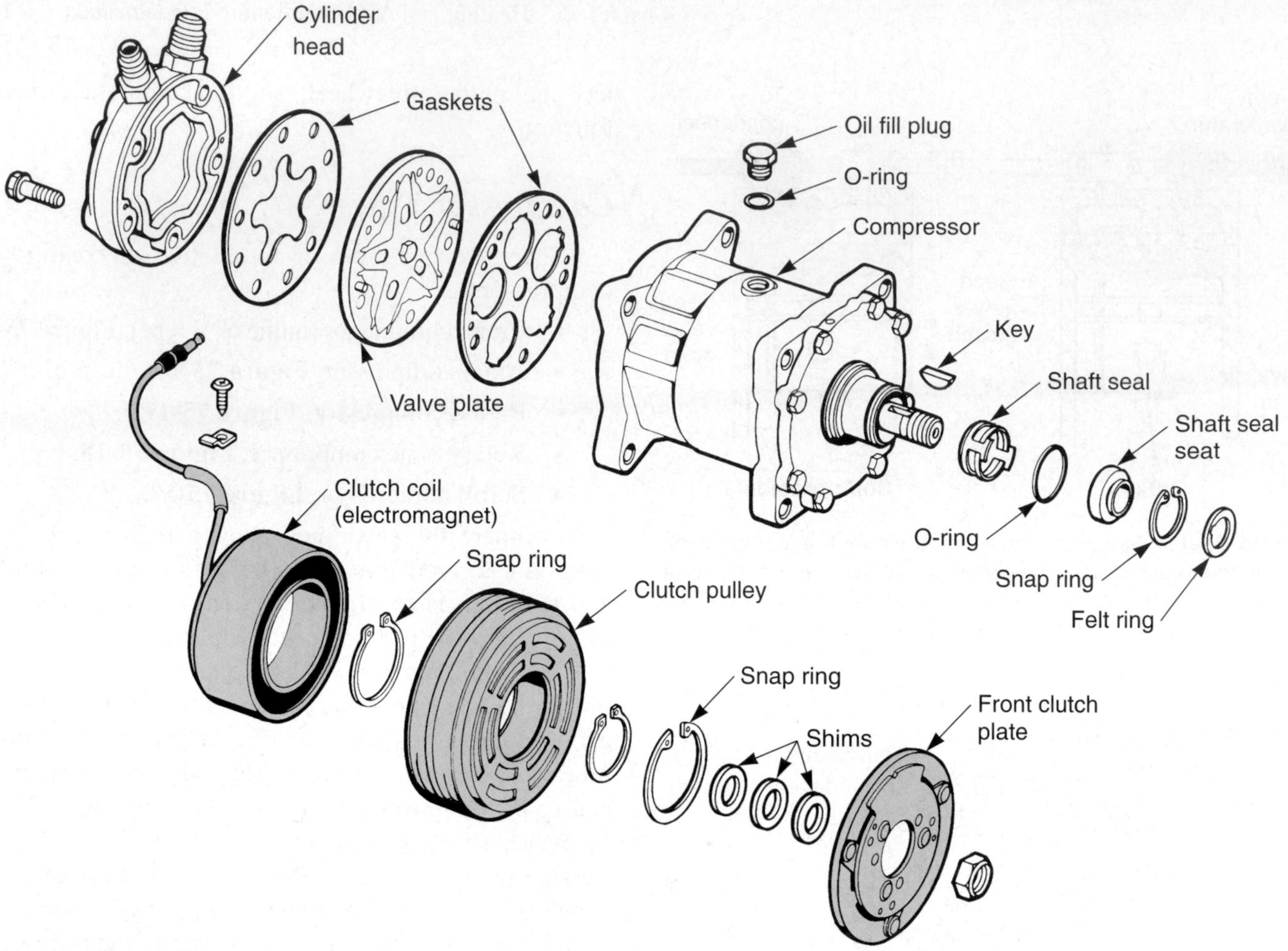

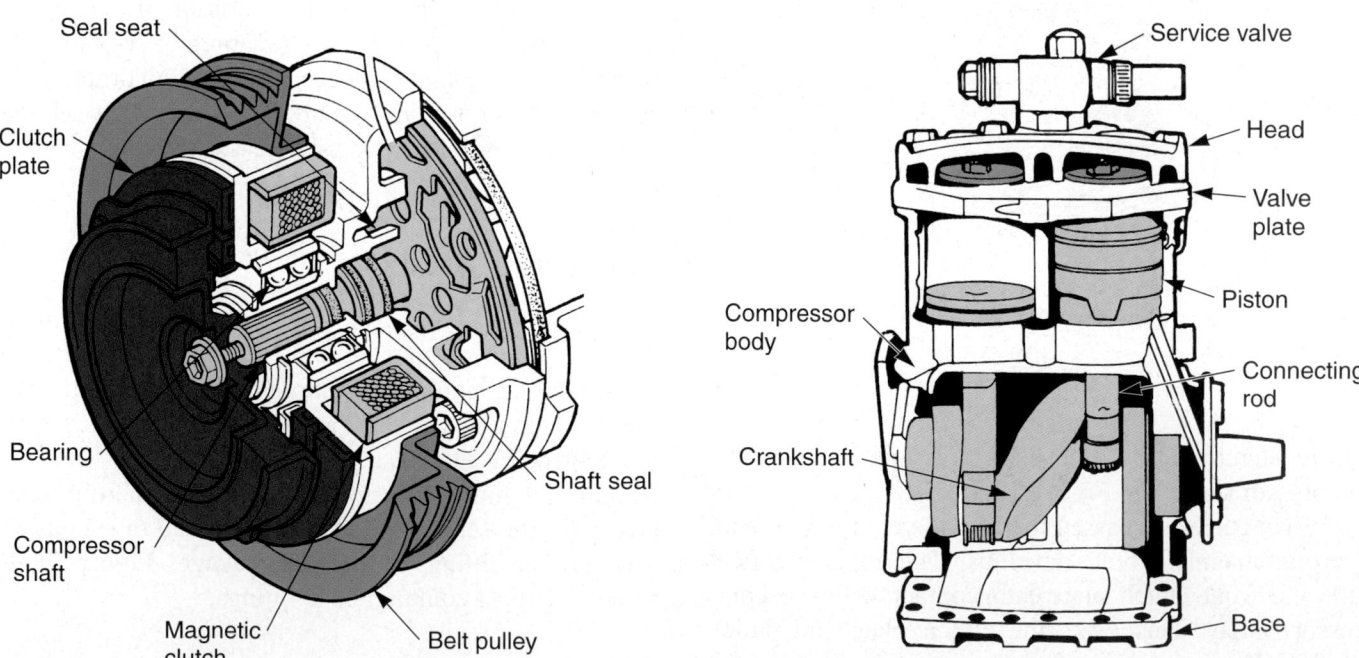

Figure 75-13. This exploded view of a modern compressor shows how the valve plate and clutch mount on the main body of the compressor. (Chrysler)

Figure 75-14. This cutaway of a compressor clutch shows the electromagnetic coil that pulls the clutch plate into the hub, locking the assembly together to rotate the input shaft. (Honda)

Figure 75-15. A crank-type compressor is similar to a small gasoline engine. The crankshaft, connecting rods, and pistons make up the reciprocating assembly. Note the other parts. (Florida Dept. of Voc. Ed.)

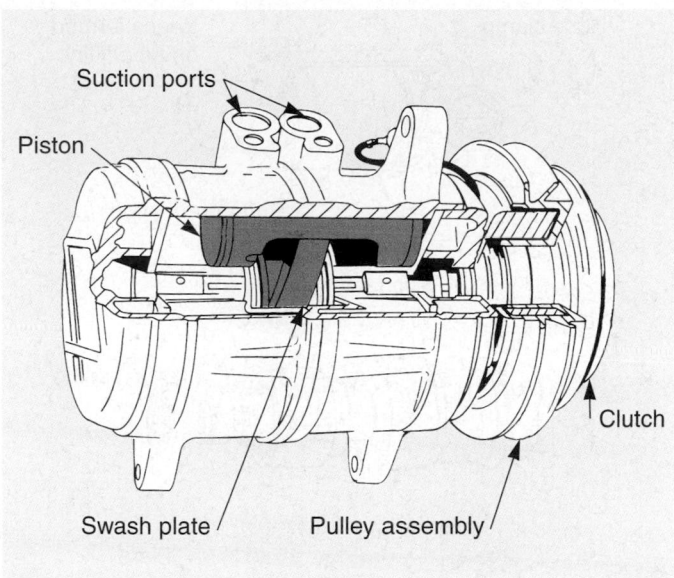

Figure 75-16. An axial compressor has pistons moving lengthwise in its body. The swash plate moves the pistons. (Mitchell Manuals)

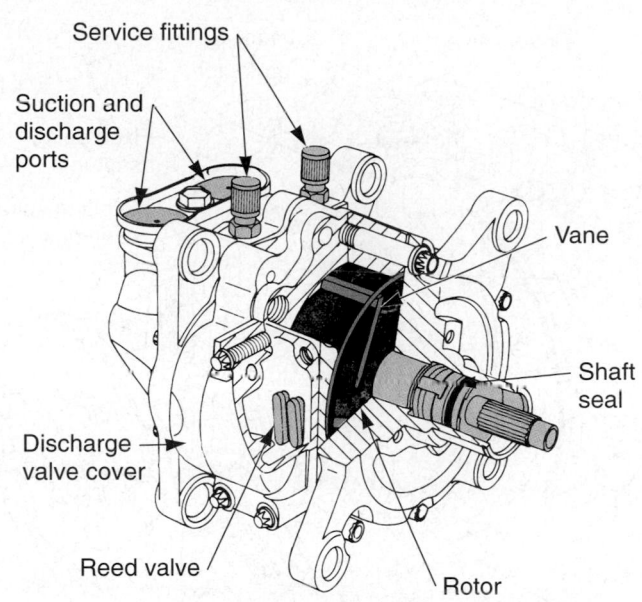

Figure 75-18. The modern rotary vane compressor works like an oil or air injection pump. Blades spin inside the housing to pressurize the refrigerant. (Mitchell Manuals)

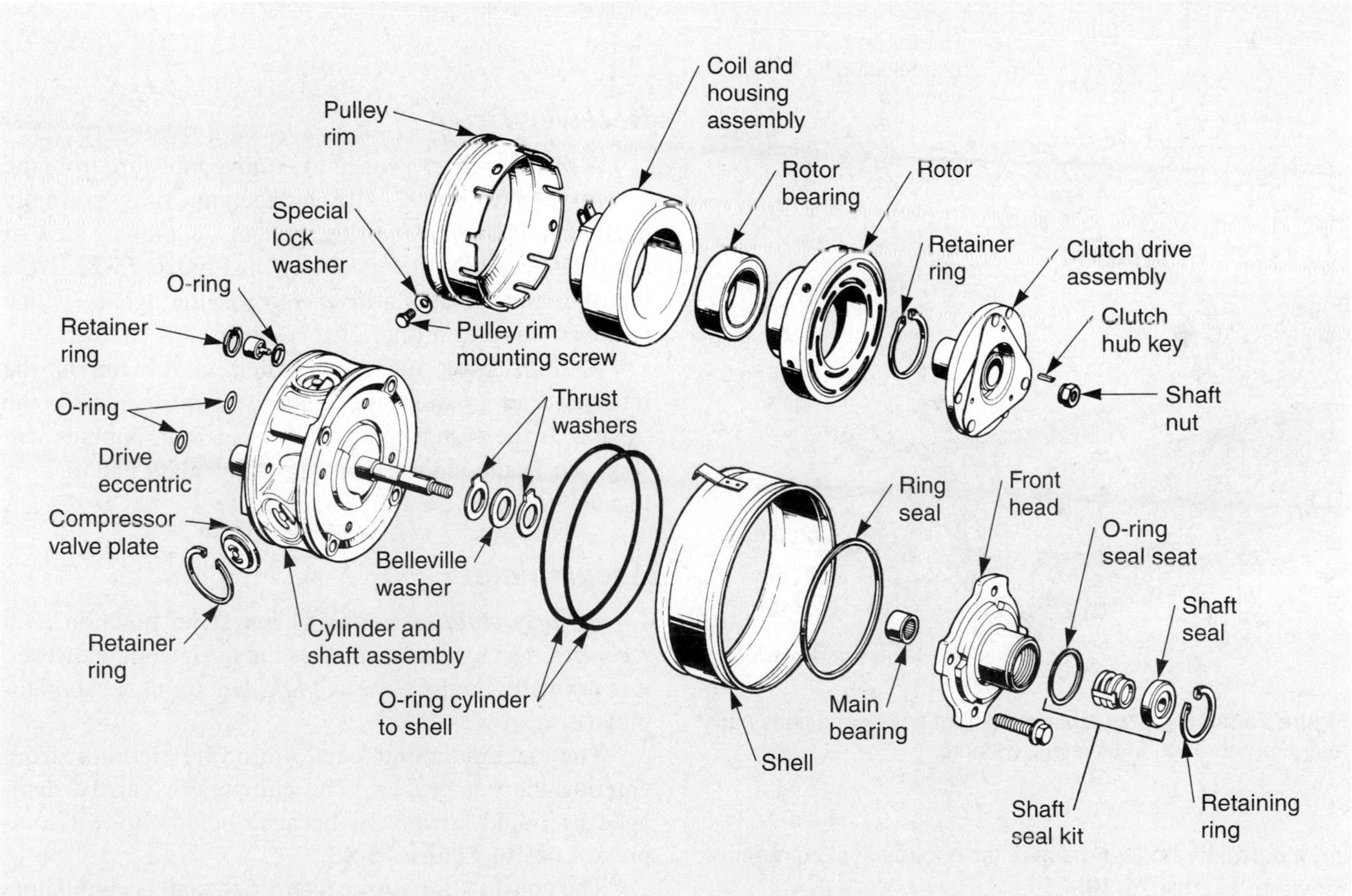

Figure 75-17. A radial compressor has a shaft-mounted eccentric that moves the pistons in and out. The pistons are arranged in a circle around the shaft. Study the parts. (Chrysler)

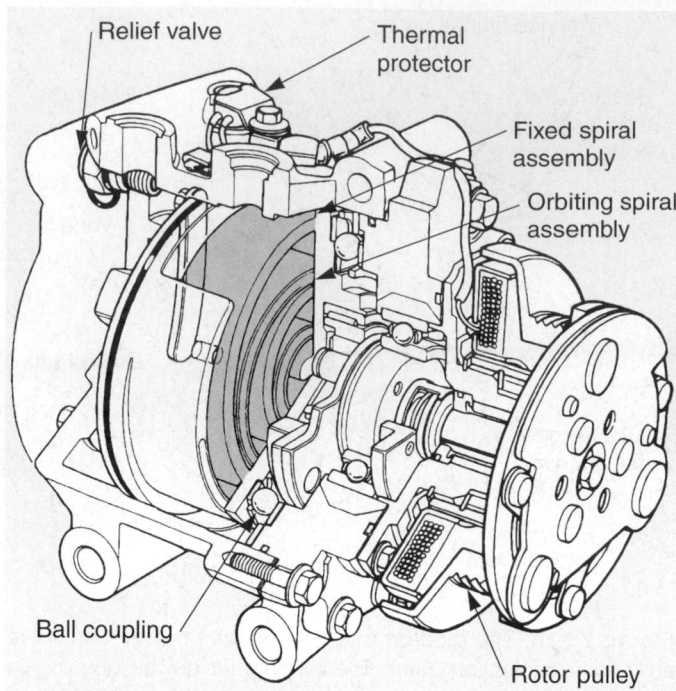

Figure 75-19. A cutaway view of a scroll compressor. (Honda)

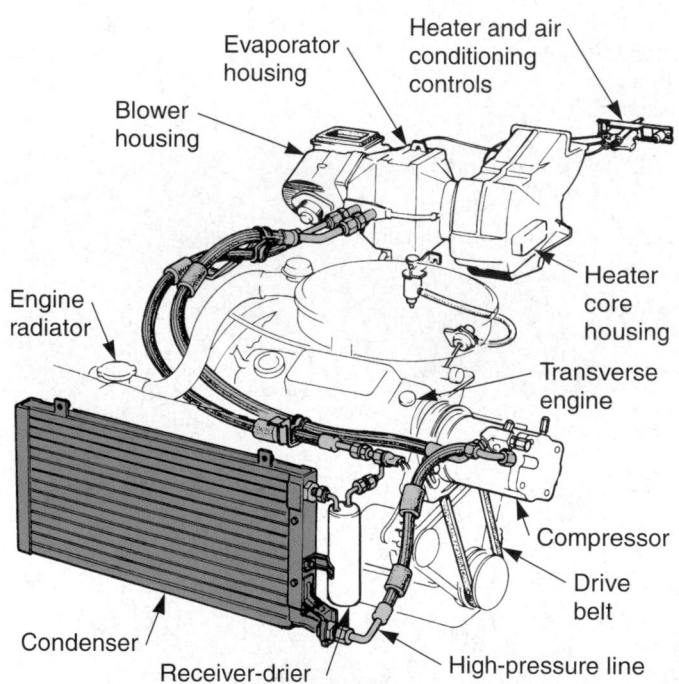

Figure 75-20. High-pressure refrigerant hoses and lines carry refrigerant through the system. (Honda)

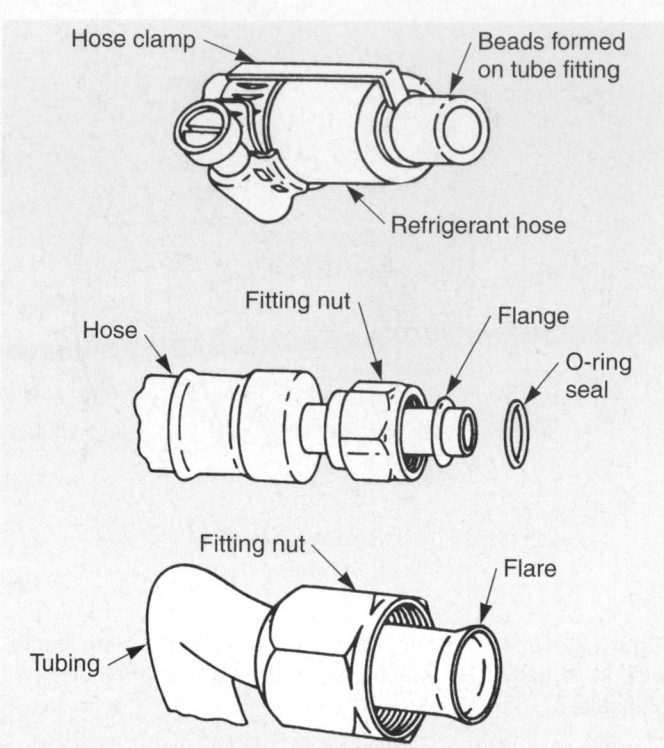

Figure 75-21. Note the common types of fittings found in automotive air conditioning systems. (GMC)

Receiver-Drier

A *receiver-drier* is used to remove moisture from the system and to store extra refrigerant. It is normally located in the *high side* of the system. A cutaway view of a typical receiver-drier is shown in **Figure 75-22.** Note that the unit contains a *desiccant* (drying agent), which removes water from the refrigerant.

A *sight glass* may be located in the top of the receiver-drier to show the amount of refrigerant in the system. If the system is low on refrigerant, bubbles will show up in the sight glass. The sight glass can also be located in one of the refrigerant lines.

Accumulator

An *accumulator* performs the same function as a receiver-drier: it stores and dries the refrigerant. However, it is normally on the system's *low side* (at the evaporator outlet).

The accumulator keeps liquid refrigerant from entering the compressor. The compressor can be damaged by liquid refrigerant because liquids do *not* compress. Look at **Figure 75-23.**

The construction of receiver-driers and accumulators varies. These examples are typical.

Figure 75-24 illustrates an air conditioning system that uses both a receiver-drier and an accumulator.

air. It normally bolts in front of the cooling system radiator, shown in **Figure 75-20.**

The condenser is simply a long metal tube wound back and forth inside hundreds of metal fins. The fins increase the amount of surface area, improving heat transfer.

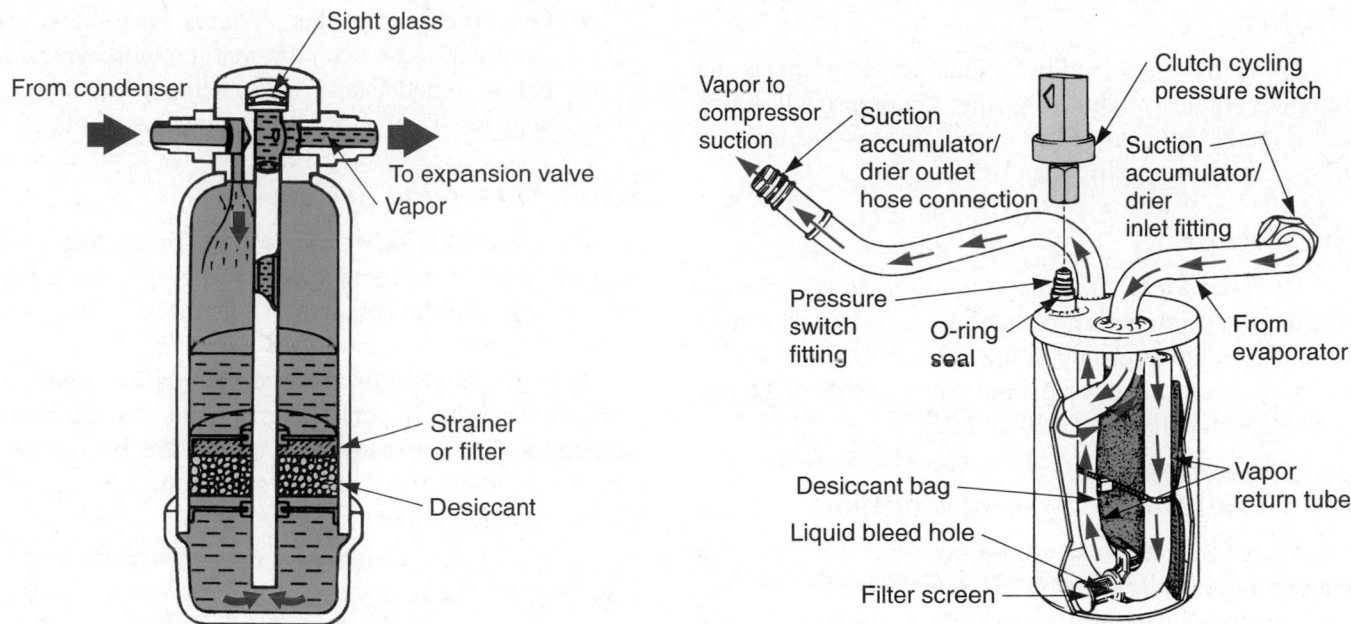

Figure 75-22. The receiver-drier, or dehydrator, removes moisture from the air conditioning system. Desiccant is a substance that absorbs and stores moisture. This unit has a sight glass in the top of housing. (Nissan)

Figure 75-23. An accumulator performs the same basic function as a receiver-drier. It serves as a reservoir and a moisture trap. Note how the clutch cycling switch mounts on the accumulator. (Ford)

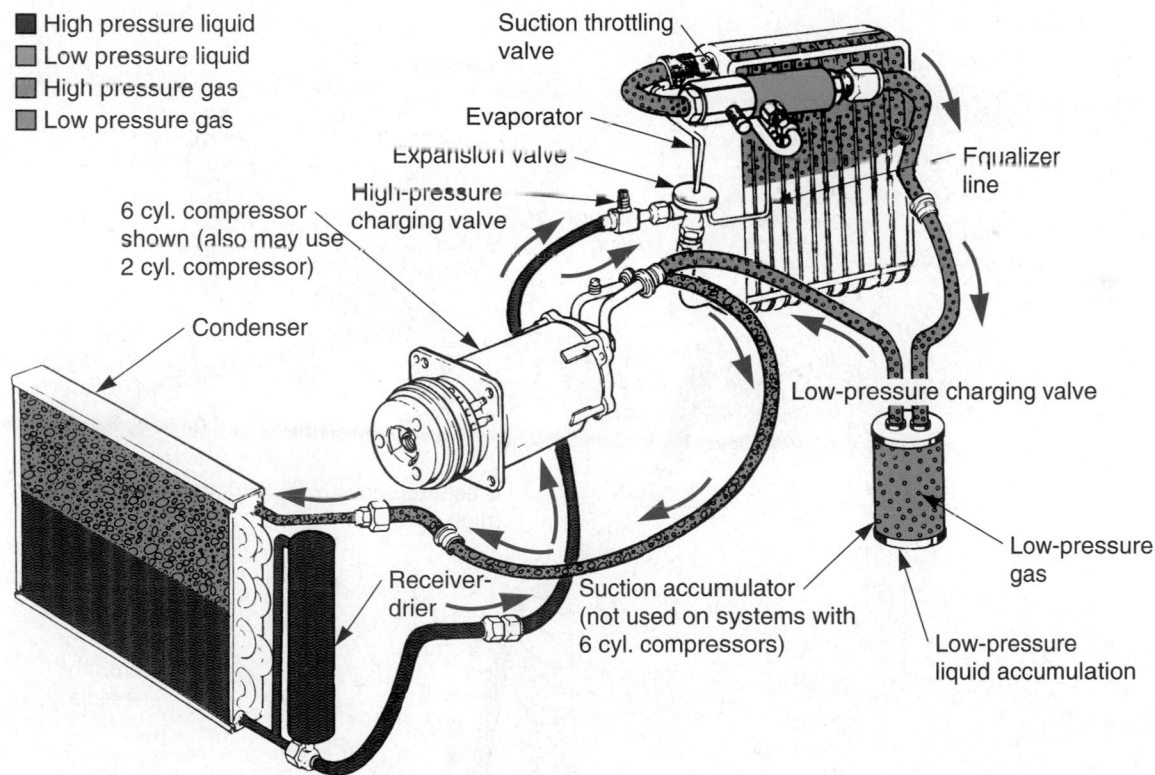

Figure 75-24. Trace the flow of liquid and vapor refrigerant through this system. Note the location of the accumulator. (Ford)

Muffler

A *muffler* is a hollow container that quiets the pumping sounds produced by the compressor. It is usually a metal can formed into the high-pressure, or compressor discharge, refrigerant line.

Evaporator

The *evaporator,* like the condenser, is a refrigerant coil mounted in cooling fins. The evaporator is usually mounted on the right side of the vehicle. It fits inside the blower housing, either on the outside or inside of the firewall. Refer to **Figures 75-20** and **75-24.**

Air Conditioning System Controls

All A/C systems use a compressor, a condenser, an evaporator, a drier, and/or an accumulator. However, many different devices are used to control the operation of the compressor and to meter the flow of refrigerant.

For example, the two most common types of air conditioning systems are:

- Expansion tube A/C system (cycling clutch system).
- Thermostatic expansion valve A/C system.

- Over the years, other systems have been used. However, these two represent the most typical and easiest to understand A/C systems found on today's vehicles. They will provide typical examples.

Expansion Tube System

An *expansion tube A/C system,* or *cycling clutch system,* controls refrigerant flow and evaporator temperature by cycling the compressor on and off. This system is one of the simplest. Look at **Figure 75-25.**

The *expansion tube,* also termed *orifice tube,* has a fixed opening that meters the refrigerant flowing into the evaporator. The expansion tube usually fits near or in the evaporator inlet. It is usually a straight tube made of plastic or sintered metal, **Figure 75-26.**

Cycling Switches

Two types of *compressor cycling switches* can be used in an expansion tube A/C system: thermostatic cycling switches and pressure cycling switches.

A *thermostatic cycling switch* turns the compressor on and off to maintain the correct evaporator temperature (approximately 32°F or 0°C).

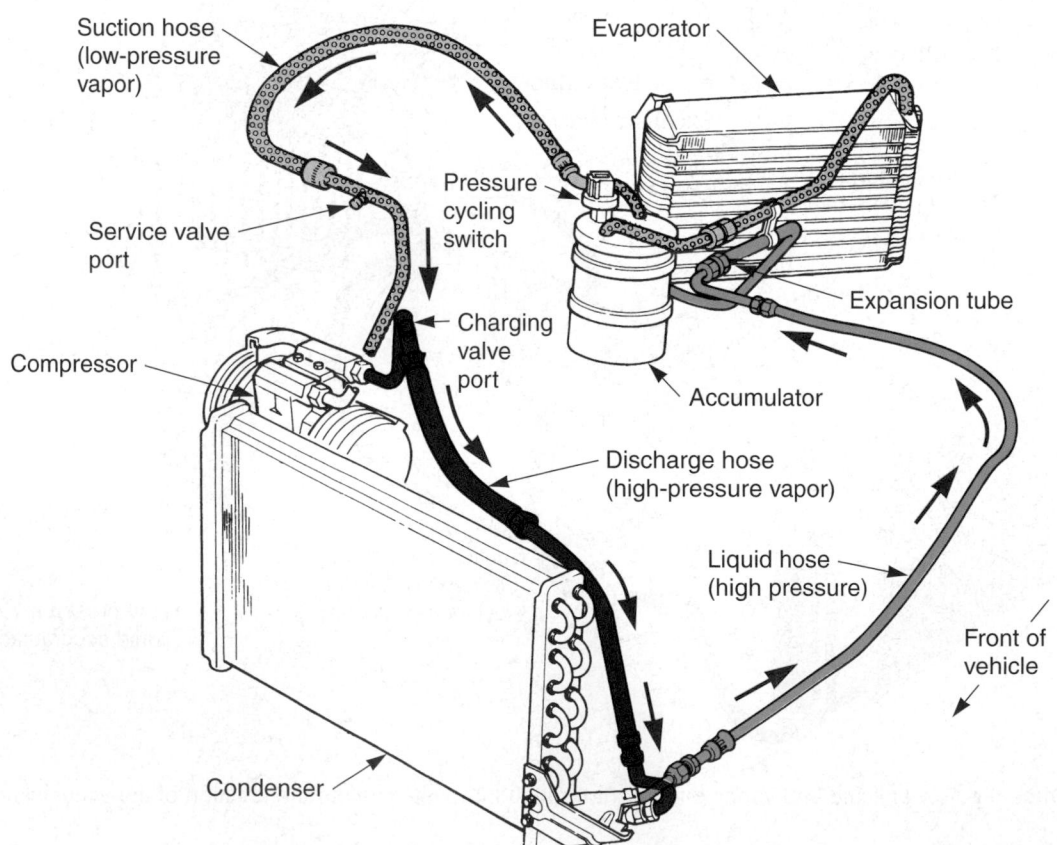

Figure 75-25. The expansion tube system, or clutch cycling system, turns the compressor on and off to control amount of refrigerant flowing through evaporator. This provides a simple method of maintaining temperature.

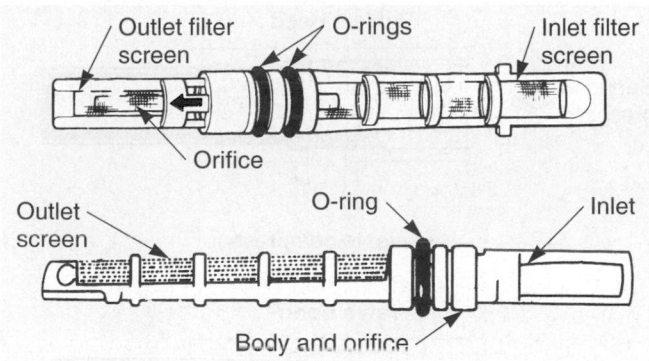

Figure 75-26. The expansion tube has a metering orifice that allows a specific amount of refrigerant to enter the evaporator. It is used in expansion tube A/C systems. (Mitchell Manuals)

When the evaporator is warm, the switch is closed and current flows to the compressor clutch coil. This causes the compressor to function, cooling the evaporator. Then, when the evaporator becomes cold enough

(near freezing point), the cycling switch opens the circuit to the compressor clutch. This allows the evaporator temperature to increase as needed.

Since evaporator pressure determines evaporator temperature, a ***pressure cycling switch*** can be used in place of or in conjunction with a thermostatic switch to maintain the correct evaporator temperature. See **Figure 75-25.**

Expansion Valve

An ***expansion valve*** is a temperature-sensitive valve that controls refrigerant flow and the evaporator temperature. See **Figure 75-27.**

Instead of constantly cycling the compressor on and off, the expansion valve can open and close to meter the right amount of refrigerant into the evaporator.

Basically, an expansion valve consists of a thermostatic bulb and tube, a diaphragm, actuating pins, a valve and seat, a metering orifice, and a spring. A cutaway view of an expansion valve is shown in **Figure 75-28.**

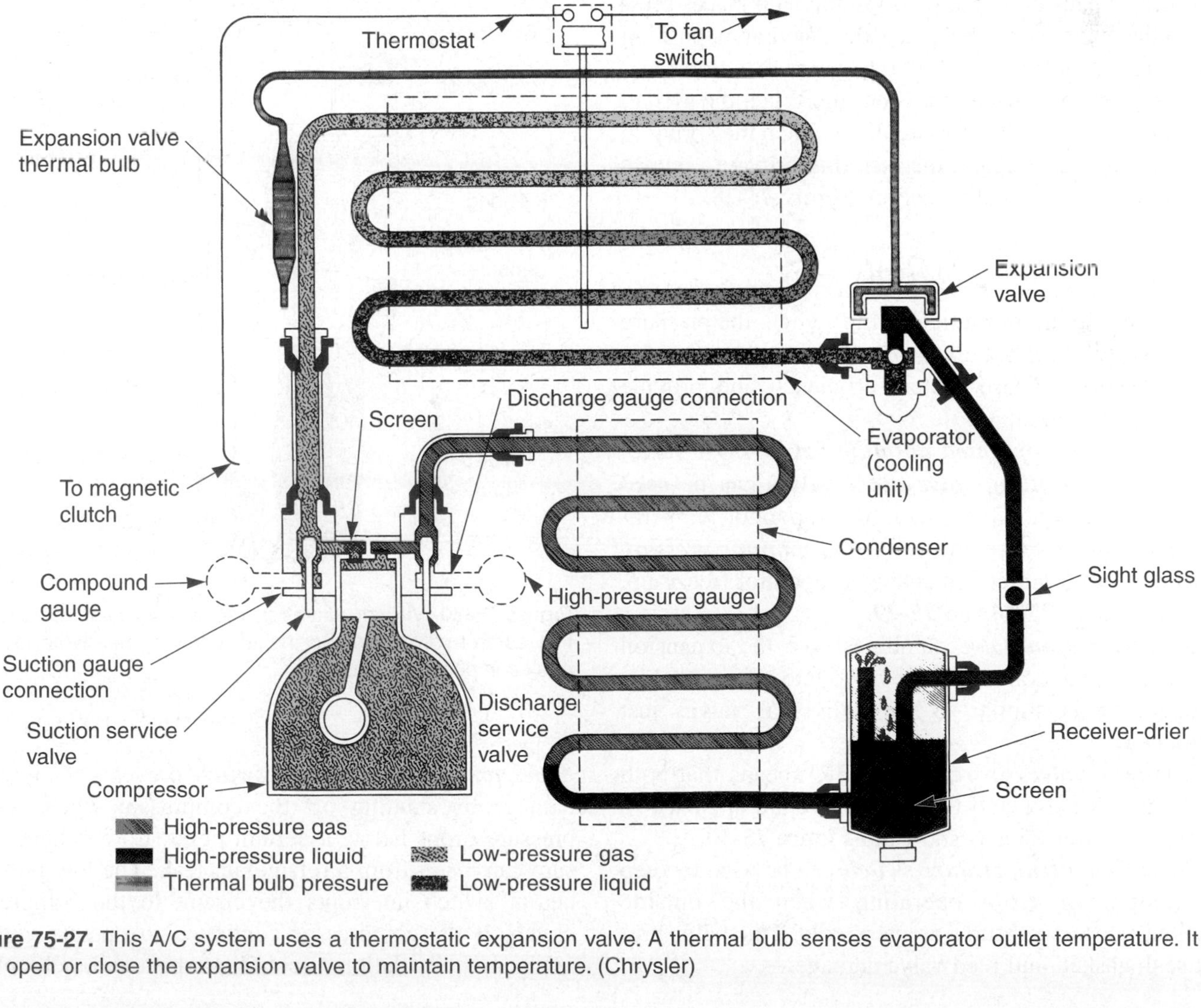

High-pressure gas
High-pressure liquid
Thermal bulb pressure
Low-pressure gas
Low-pressure liquid

Figure 75-27. This A/C system uses a thermostatic expansion valve. A thermal bulb senses evaporator outlet temperature. It can then open or close the expansion valve to maintain temperature. (Chrysler)

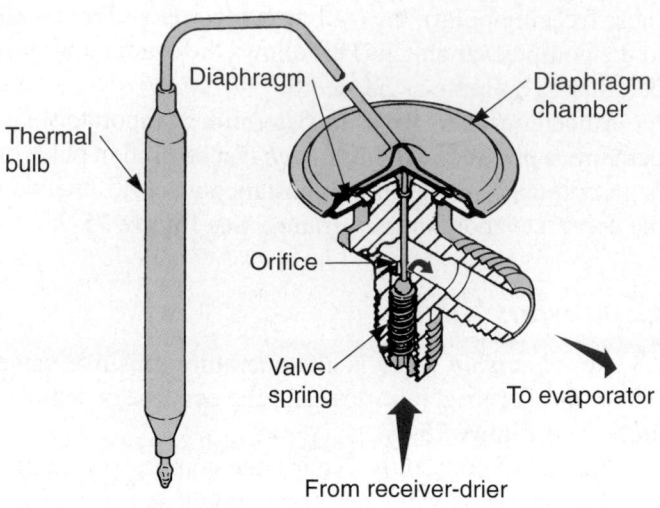

Figure 75-28. Expansion valve opens and closes with temperature changes to regulate the amount of refrigerant entering the evaporator. The compressor is not cycled on and off to control evaporator temperature. It runs all the time. (Nissan)

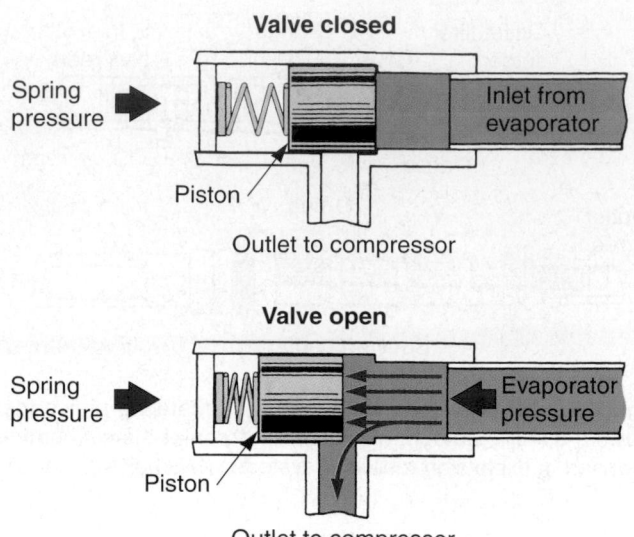

Figure 75-29. A suction throttling valve, like an expansion valve, allows the compressor to run constantly. Located in the evaporator outlet, it senses evaporator pressure to control refrigerant flow and temperature. (AC-Delco)

When the temperature of the evaporator and bulb is too warm, expansion in the thermostatic bulb places pressure on the diaphragm. This flexes the diaphragm, pushing the expansion valve open. More refrigerant then flows into the evaporator to provide more cooling. When the system becomes too cold, bulb pressure drops. Then the spring in the expansion valve closes the metering orifice to reduce refrigerant flow and cooling action, **Figure 75-28.**

Other A/C System Controls

Besides the thermostatic cycling switch, the pressure cycling switch, and the expansion valve, other controls are needed to increase system efficiency and provide additional protection.

Either a ***pilot-operated absolute valve*** (POA valve) or a ***suction throttling valve*** (STV valve) can be used with the expansion valve to control evaporator temperature. Both valves sense and control *evaporator pressure* to provide added protection against evaporator freeze-up. Look at **Figures 75-24** and **75-29.**

A ***combination valve*** combines both the expansion valve and the suction throttling valve into one assembly. Its operation is similar to the individual valves just introduced.

The term ***valves in receiver*** (VIR) means that both the expansion valve and the POA valve are enclosed in the receiver-drier. This is shown in **Figure 75-30.**

An ***ambient temperature switch*** can be used to keep the compressor from operating when the outside (ambient) air temperatures are very cold. This helps prevent seal, gasket, and reed valve damage.

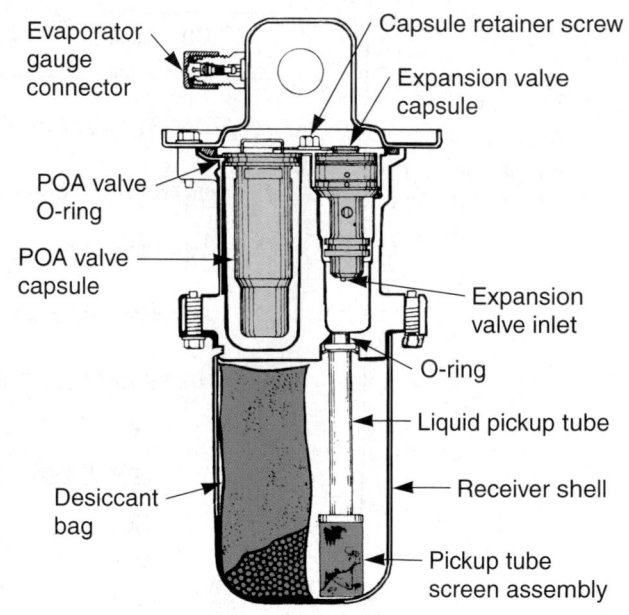

Figure 75-30. VIR, or valves in receiver, has air conditioning flow-control valves located inside receiver-drier. Note location of all components. (GMC)

A ***low-pressure cutout switch*** prevents compressor damage by shutting off the compressor when system pressure drops below a certain point. Low system pressure can result from a refrigerant leak. The low-pressure cutout switch interrupts the circuit to the compressor clutch. If the compressor continued to operate without refrigerant circulation, it could overheat and lock up.

A ***high-pressure cutout switch*** can be used to shut off the compressor if the discharge pressure is too high. Like the low-pressure cutout switch, it prevents compressor damage from abnormal operating conditions.

The high-pressure cutout switch is normally located in the service fitting of the discharge muffler. It opens the compressor coil circuit when discharge pressure reaches approximately 375 psi (2584 kPa).

A ***thermal limiter*** and a ***superheat switch*** can be used to prevent compressor damage when the refrigerant level or the oil level is too low. The superheat switch senses low system pressure and high compressor temperature. It can then send current to the thermal limiter. Current flow melts a fusible link in the thermal limiter and stops current to the compressor clutch. **Figure 75-31** shows these protection devices.

A ***pressure relief valve*** bleeds off excess pressure to prevent compressor damage. It is a pop-off, or spring-loaded, valve located on the receiver-drier, compressor, or other component in the high side of the system.

The ***wide open throttle (WOT) switch*** shuts off the compressor during rapid acceleration to save power. This switch is used on cars with small gasoline or diesel engines. It is normally located on the accelerator linkage, carburetor, or throttle body, **Figure 75-31.**

The ***A/C switch*** is the driver-operated switch in the car's dash that feeds current to the compressor clutch and other electric components. The A/C switch provides a manual means of controlling system operation.

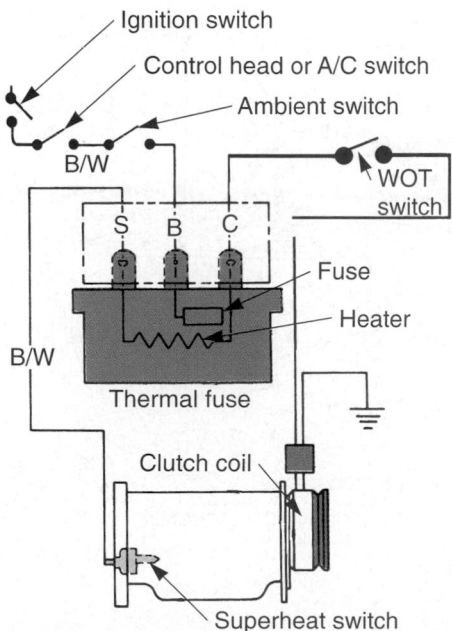

Figure 75-31. Study how switches are wired in an A/C system circuit. (AC-Delco)

Service Valves

Service valves are basically provided to allow refrigerant installation and removal, as well as pressure gauge tests of the A/C system.

The ***discharge service valve*** is located at the output from the compressor, in the system high side.

The ***suction service valve*** is located at the inlet to the compressor, in the system low side.

 Tech Tip
The service valves on older vehicles are different than those used on new vehicles. This helps prevent the accidental use of the wrong type of refrigerant (R-12 instead of R-134a). Service valves are explained fully in Chapter 76.

Heating System

The ***heating system*** uses engine cooling system heat to warm the passenger compartment in cold weather. A heating system typically consists of two heater hoses, a heater core, a blower (fan), and a control mechanism.

The ***heater hoses*** are small-diameter hoses that carry engine coolant to the heater core, **Figure 75-32.** Hot engine coolant flows through one heater hose, through the heater core, and back into the engine through another heater hose. A vacuum-operated valve may be used in one of the heater hoses to control coolant flow to the heater core.

The ***heater core*** is a small, radiator-like unit that provides a large surface area for heat dissipation into the passenger compartment. It is made of a series of tubes surrounded by fins. As air passes through the tubes and fins, heat is transferred into the air.

The ***blower*** consists of an electric motor and fan assembly. When current is fed to the electric motor, the fan spins to force air through the heater core or the air conditioning evaporator. The blower mounts in a housing that surrounds the heater core, the evaporator (if used), and the air doors. See **Figure 75-32.**

Heater controls can be set to move the duct flaps so that cool outside air is mixed with hot heater core air. This permits control of the passenger compartment temperature, **Figure 75-33.**

Heating and Air Conditioning Controls

The driver uses levers, buttons, or switches in the instrument panel to control the operation of the heating and air conditioning systems. When the driver activates one of the instrument panel controls, one of four methods may be used to operate the system components:

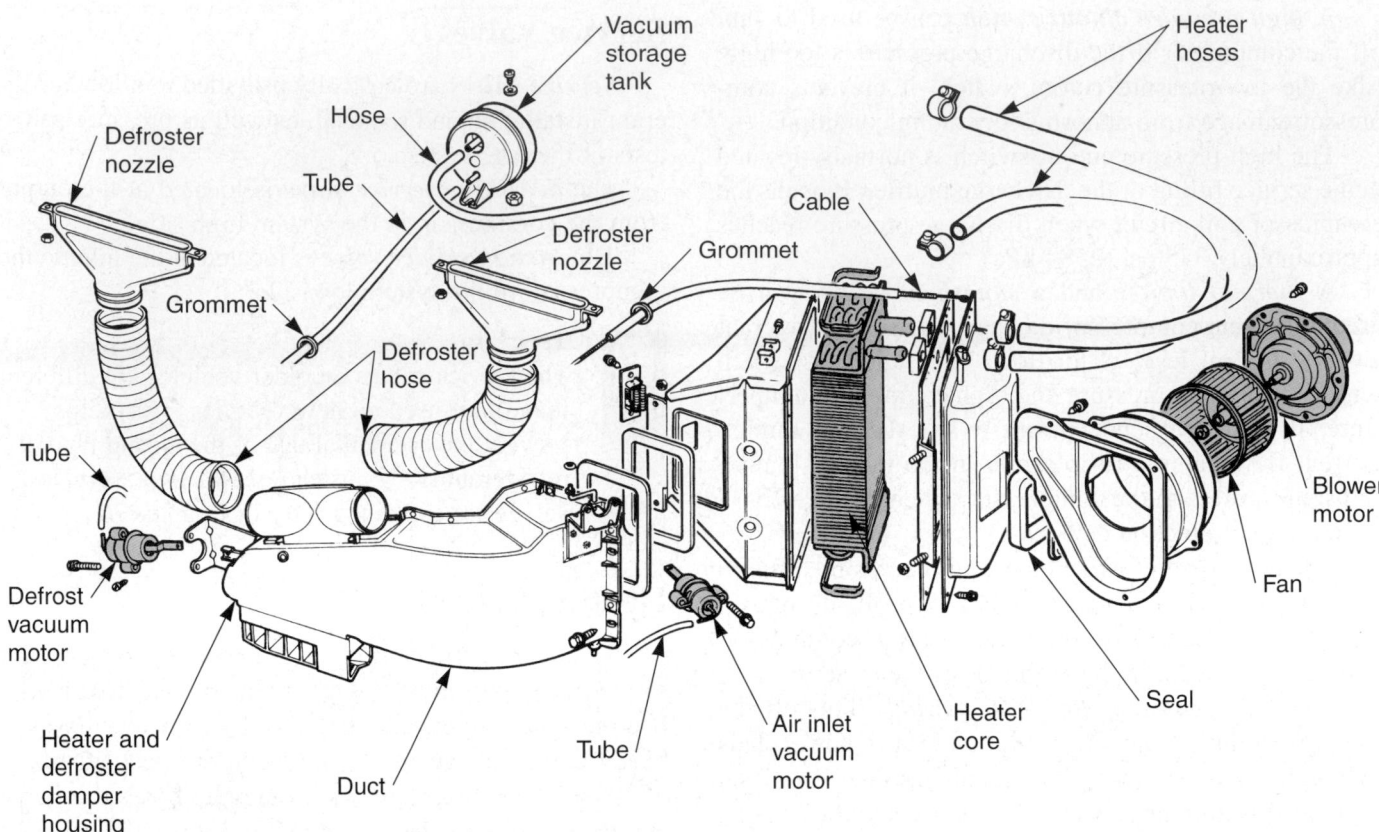

Figure 75-32. Exploded view of vent and heat components that mount under the dash of a car. Note the vacuum motors, the heater core, and the blower. (Chrysler)

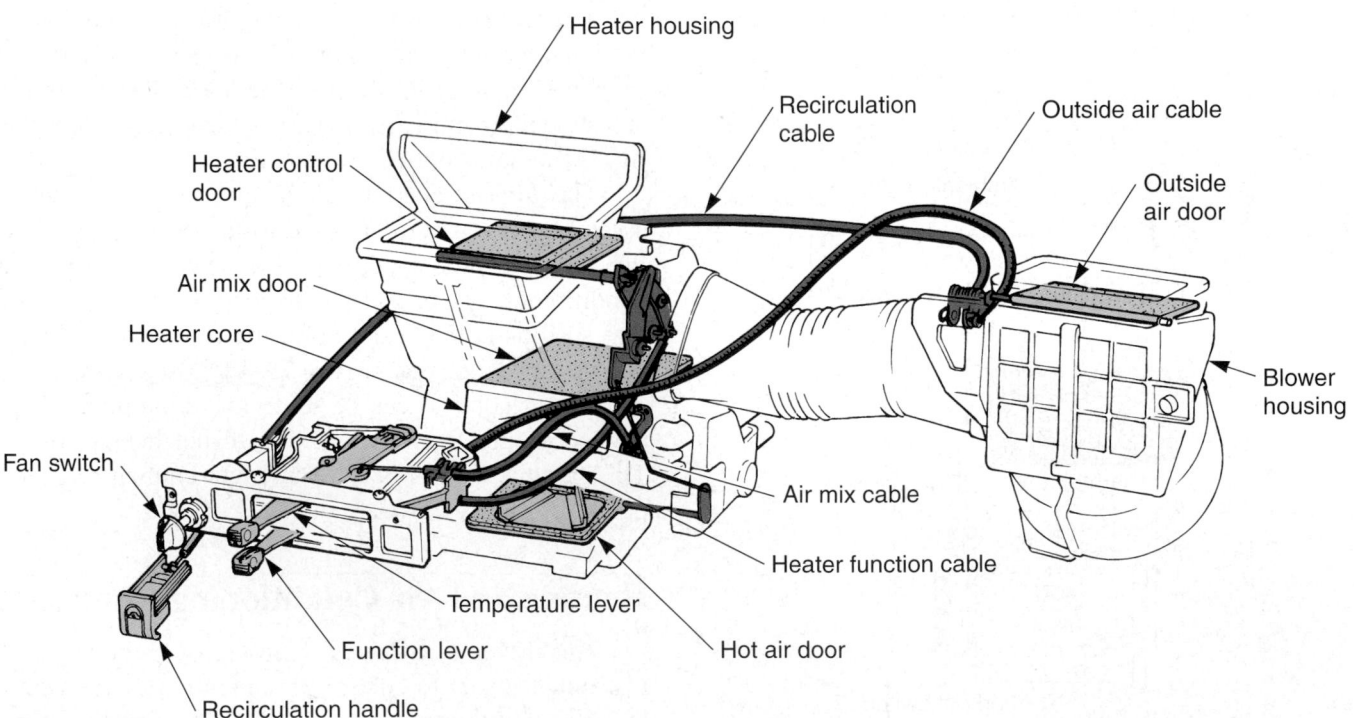

Figure 75-33. Driver's dash levers can be moved to operate cables to the air control doors. Doors then open or close the ducts to route warmed or cooled air to the passenger compartment. (Honda)

- An *electric switch* may operate, usually through relay(s), the blower motor, the compressor clutch, and other electric components.
- *Mechanical levers* and *cables* can operate air doors (air flaps) over the heater core, evaporator, and ducts.
- A *vacuum switch* can operate vacuum motors (vacuum diaphragms) that open and close the air doors.
- An *electronic control unit* (computer) and *sensors* can operate the heating and air conditioning systems automatically.

Manual Temperature Control

Manual controls for the heating and air conditioning systems are driver controlled. The driver must select the blower speed and air door positions to control the heating or cooling temperatures.

Push-pull cables use a stiff steel cable that slides back and forth in its housing. This design, after prolonged service, tends to bind when pushed. A ***pull-pull cable*** loops cable in its housing so that it always pulls when activated. Like a movable clothesline, turning the dash knob pulls the cable in both directions for smoother operation.

A typical cable-type manual control for a heater is given in **Figure 75-33.**

A manual control system can also use ***vacuum motors*** (sealed rubber diaphragms) to operate the air doors, **Figure 75-34.** The control panel will contain a ***vacuum switch,*** which routes engine vacuum to the correct vacuum motor. The vacuum motors then move the air doors.

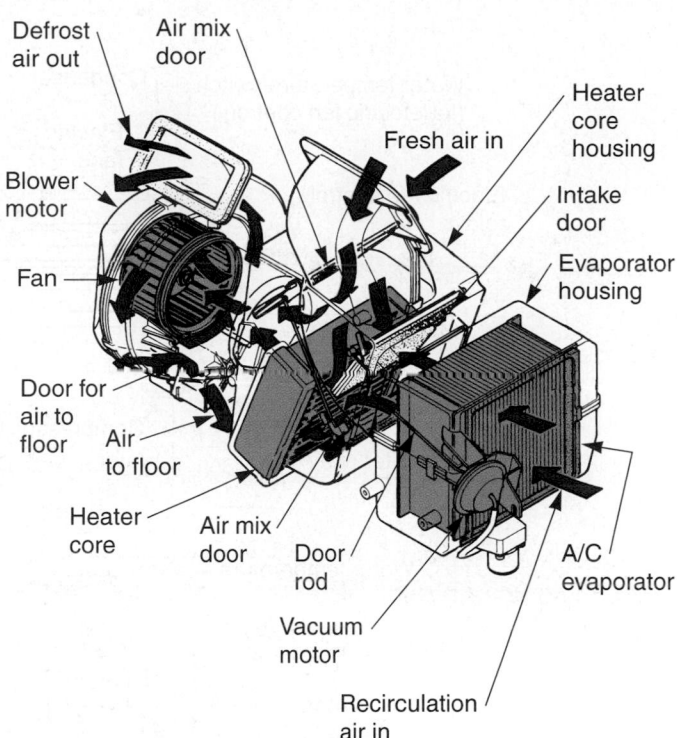

Figure 75-35. Trace the airflow through the heater core and air conditioning evaporator. Hot and cold air blend when needed to maintain a comfortable temperature output. (Nissan)

Figure 75-35 shows how air can be mixed to control the temperature of the air entering the passenger compartment.

Automatic Temperature Control

Automatic temperature control systems use various pressure and temperature sensors and a control module (computer) to maintain a preset passenger compartment temperature. The driver sets the instrument panel controls to a specified temperature; then the heating and air conditioning systems automatically keep the vehicle's passenger compartment at this temperature.

An automatic temperature control system can use three temperature sensors: an in-car sensor, a duct sensor, and an ambient (outside) air sensor. These temperature sensors may be thermistors. A ***thermistor*** is an electrical device that changes resistance with a change in temperature. For example, the in-car sensor's (thermistor's) internal resistance may drop when the passenger compartment temperature drops. The increased current flow could be used by the electronic control unit to turn on the heater or to turn off the air conditioning system. In this way, the system maintains the preset temperature.

Figure 75-36 illustrates the layout and major parts of one type of automatic temperature control system.

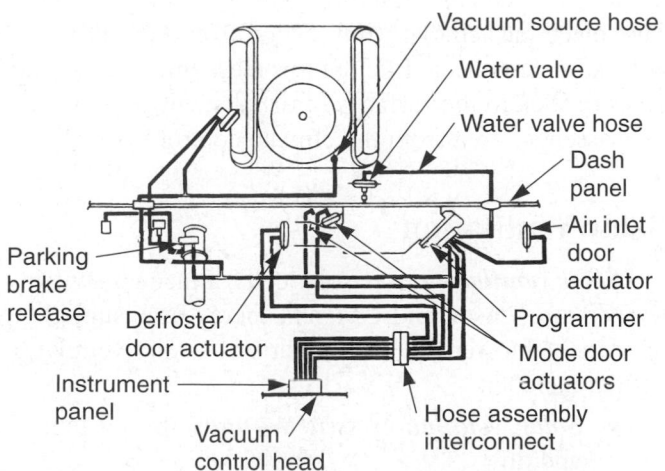

Figure 75-34. Vacuum diagram for a typical heating and air conditioning system. (Cadillac)

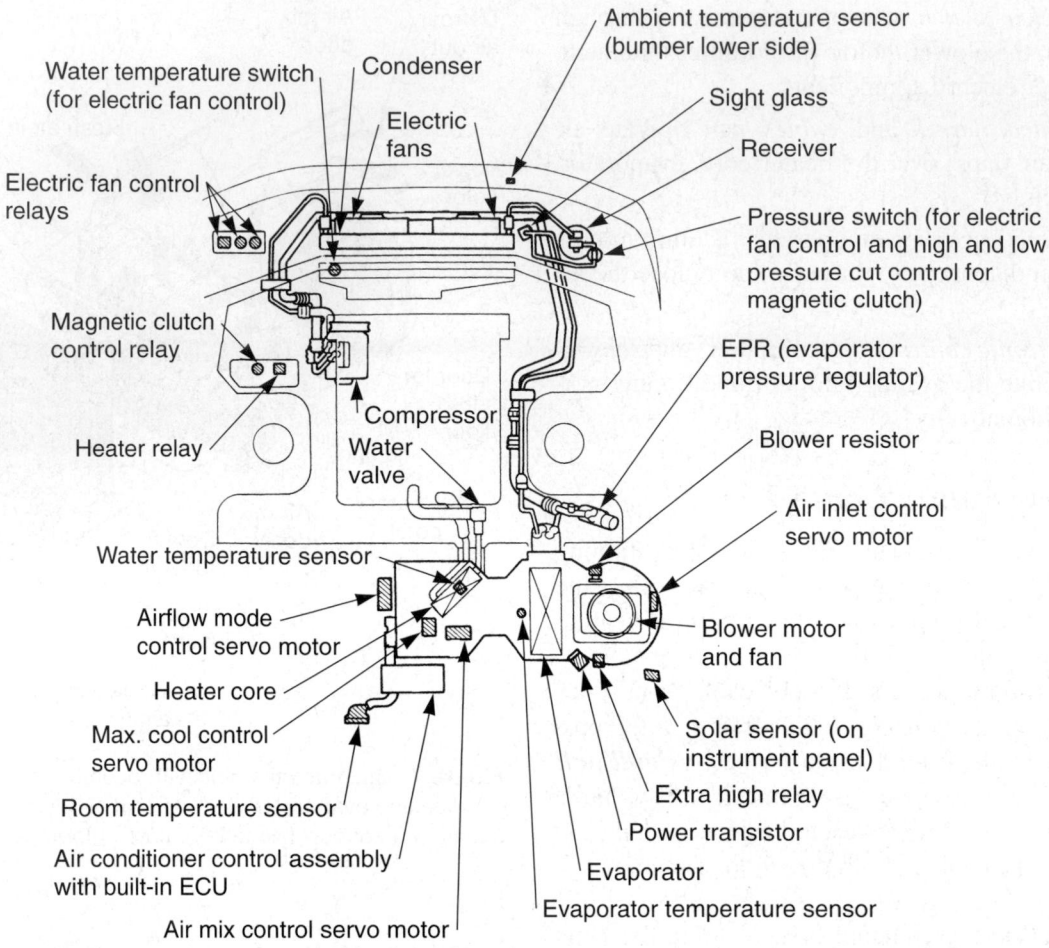

Water temperature switch
(for electric fan control)

Condenser

Electric
fans

Ambient temperature sensor
(bumper lower side)

Sight glass

Receiver

Electric fan control
relays

Pressure switch (for electric
fan control and high and low
pressure cut control for
magnetic clutch)

Magnetic clutch
control relay

EPR (evaporator
pressure regulator)

Heater relay

Compressor
Water
valve

Blower resistor

Air inlet control
servo motor

Water temperature sensor

Airflow mode
control servo motor

Heater core

Max. cool control
servo motor

Room temperature sensor

Air conditioner control assembly
with built-in ECU

Air mix control servo motor

Blower motor
and fan

Solar sensor (on
instrument panel)

Extra high relay

Power transistor

Evaporator

Evaporator temperature sensor

Figure 75-36. Study the general layout of this automatic temperature control system. The in-car temperature sensor is the primary controller of the system. (Lexus)

Figure 75-37 shows a wiring diagram for an automatic temperature control system. Study how the *control head* (driver-operated unit), the *programmer* (control module), and the other components are connected.

Ducts and Vents

Ducts are plastic tubes of various shapes that carry air over the evaporator and the heater core, as well as to the vents. See **Figure 75-38.**

Vents are the grilled openings where air blows into the passenger compartments. Adjustable grills are provided to open and close the duct and to change the direction of airflow.

Passenger Compartment Filters

Some climate control systems use a *high-efficiency air filter* to clean the air entering the passenger compartment. This filter helps people with allergies by trapping pollen and other particles in the air.

Electronic air filters use electrically produced ion action to help remove small particles, such as dust and pollen, from the air entering the passenger compartment. An electrically charged, plastic micron filter will trap particles as small as 5 microns (0.00005″ or 0.1 252 mm). The electrical charge (ion action) on the polyester-polycarbonate plastic filter causes these microscopic particles to stick to the surfaces of the element and not enter the passenger area from the climate control system.

Solar Ventilation

Solar ventilation uses solar cells, a manual switch, a temperature sensor, an ECM, and energy from sunlight to operate the blower. The major parts of a solar ventilation system include:

- *Solar ventilation switch*—turns the system on and off.

- *Ambient temperature switch*—turns the blower on if interior temperature is high enough.

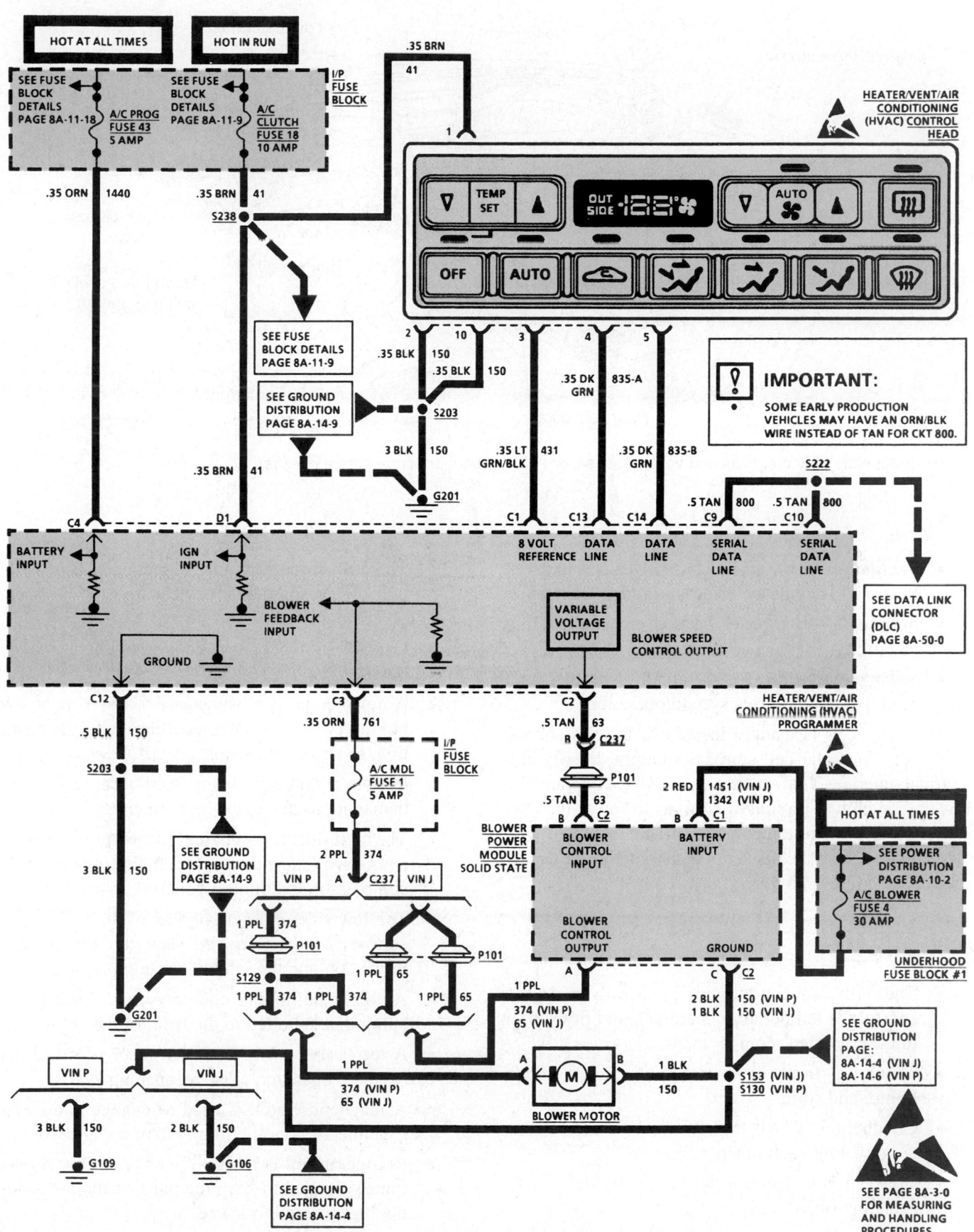

Figure 75-37. Study this wiring diagram for a typical automatic temperature control system. (Chevrolet)

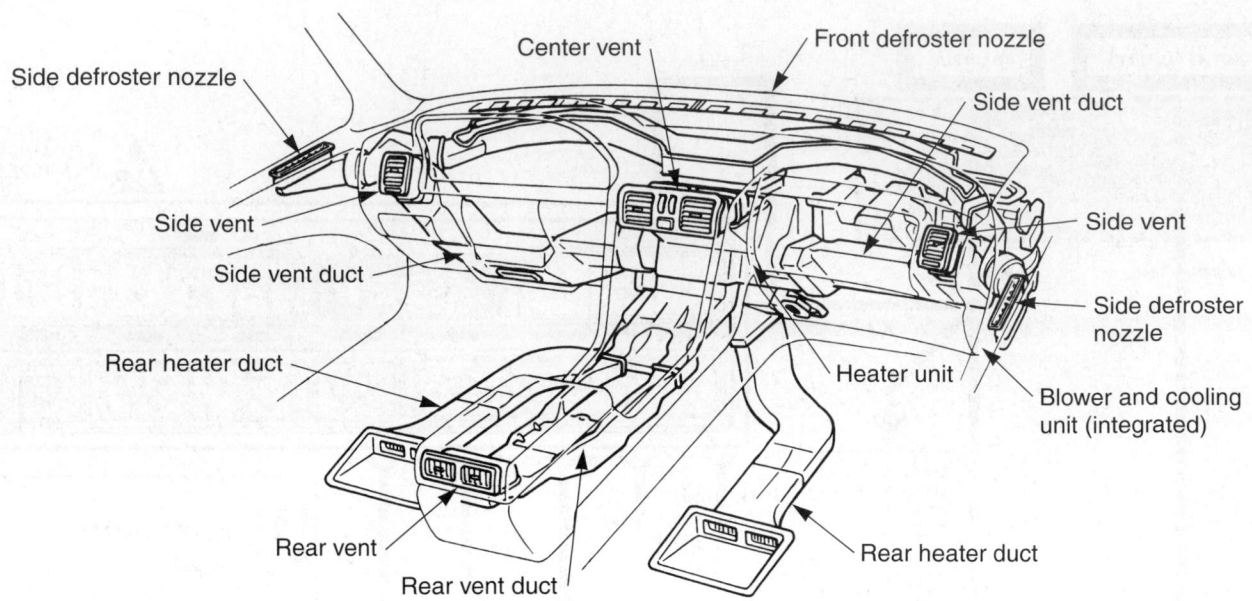

Figure 75-38. Note how the ducts and vents route air into the passenger compartment. (Lexus)

- *Ventilation fan*—small, low-power-consumption fan used to circulate outside air through the interior.
- *Solar cell*—solar panel that converts sunlight into electricity.
- *Solar ventilation ECM*—small computer that monitors and controls system operation.

The solar cell is usually located in the roof or the moon roof. The solar cell converts sunlight directly into electrical energy. This energy is used to operate the blower. The blower circulates cool air to lower the passenger compartment temperature in hot weather. The solar cell can also be used to recharge the battery if needed. See **Figure 75-39.**

Summary

- The term "air conditioning" generally means to control the temperature, humidity, and flow of air in the passenger compartment.
- There are three basic states of matter: vapor, liquid, and solid.
- Conduction is heat transfer through objects that are touching each other.
- Convection is heat transfer caused by the air surrounding objects.
- Radiation is commonly caused by infrared rays.
- Cold is actually an absence of heat.

- The British thermal unit, abbreviated Btu, is the unit of measurement for heat transfer from one object to another.
- Condensation occurs when a vapor changes into a liquid.
- A refrigerant is a substance with a very low boiling point. It is circulated through a refrigeration, or air conditioning, system.
- R-134a is currently being used because it is less damaging to the ozone layer than R-12.
- The high side, or discharge side, of an A/C system consists of the parts between the output of the compressor and the flow-control restriction.
- The low side, or suction side, of an A/C system consists of the parts between the flow-control restriction and the inlet of the compressor.
- An air conditioning compressor is a refrigerant pump that is bolted to the front of the engine.
- A reed valve is simply a thin piece of metal that bends to open and close an opening.
- A magnetic clutch is used to engage and disengage the air conditioning system compressor.
- Refrigerant oil is a high-viscosity, highly refined (pure) oil used to keep the parts of the air conditioning system lubricated.
- A condenser is a radiator-like device that transfers heat from the refrigerant to the outside air.

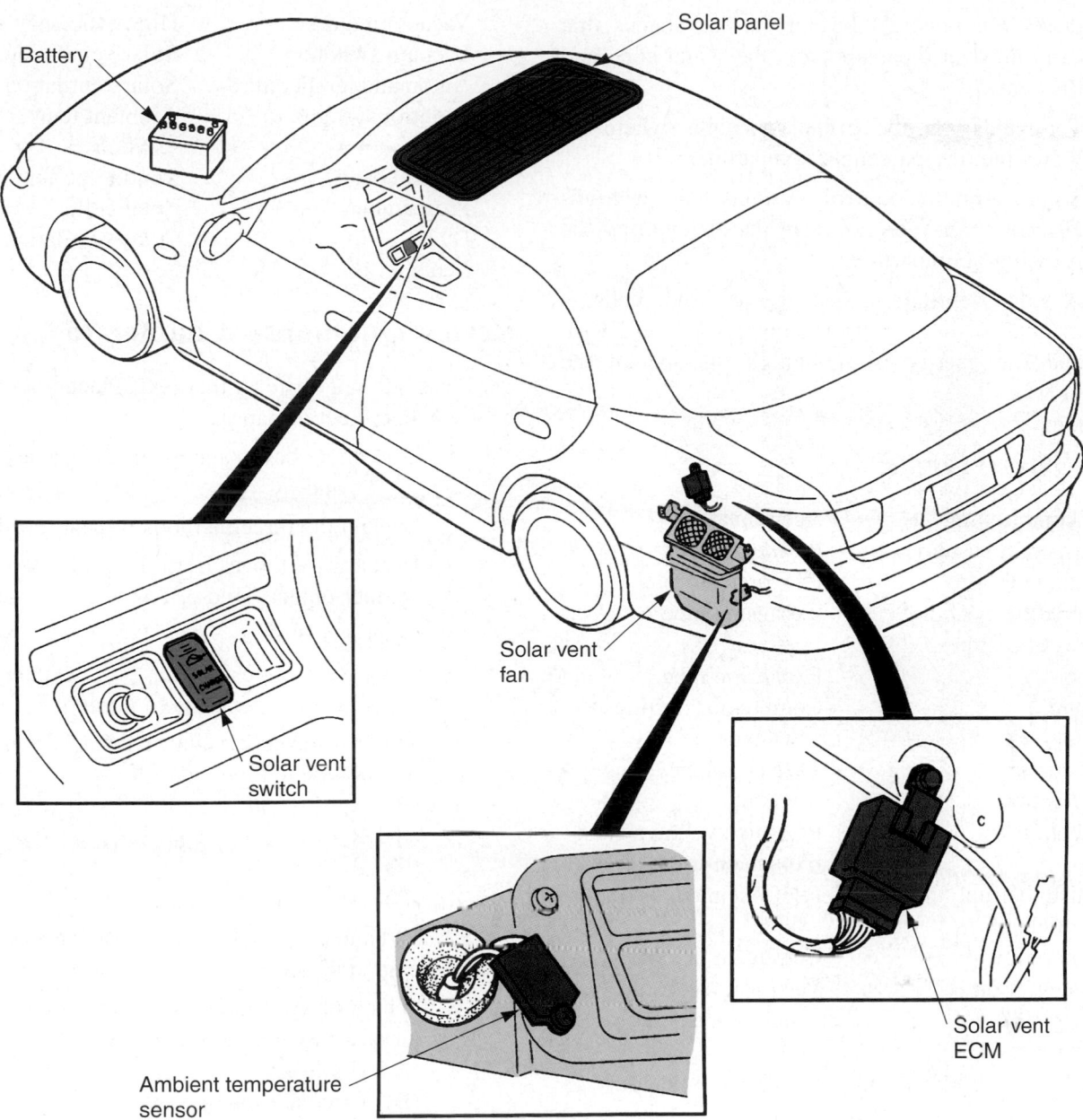

Figure 75-39. These are the basic parts of a solar ventilation system. This system helps keep the interior cool in very hot, sunny weather. (Mazda)

- A receiver-drier is used to remove moisture from the system and to store extra refrigerant.
- The evaporator, like the condenser, is a refrigerant coil mounted in cooling fins.
- The expansion tube, or orifice tube, has a fixed opening that meters the amount of refrigerant flowing into the evaporator.
- An expansion valve is a temperature-sensitive valve that controls refrigerant flow and the evaporator temperature.

- The heating system uses the engine cooling system heat to warm the passenger compartment in cold weather.
- The heater core is a small, radiator-like unit that provides a large surface area for heat dissipation into the passenger compartment.
- Automatic temperature control systems use temperature sensors and a control module (computer) to maintain a preset passenger compartment temperature.

- Ducts are plastic tubes of various shapes that carry air over the evaporator, the heater core, and the vents.

- The vents are the grilled openings where air blows into the passenger compartments.

- Some climate control systems use a high-efficiency air filter to clean the air entering the passenger compartment.

- A solar ventilation system uses solar cells, a manual switch, a temperature sensor, an ECM, and the energy from sunlight to operate the blower.

Know These Terms

Air conditioning	Accumulator
Refrigerant system	Muffler
Heating system	Evaporator
Ventilation system	Expansion tube A/C
States of matter	system
Vapor	Expansion tube
Liquid	Compressor cycling
Solid	switches
Conduction	Thermostatic cycling
Convection	switch
Radiation	Pressure cycling switch
Cold	Expansion valve
British thermal unit	Pilot-operated absolute
Vaporization	valve
Evaporation	Combination valve
Condensation	Valves in receiver
Refrigerant	Ambient temperature
Refrigerant-12	switch
R-12	Low-pressure cutout
Freon	switch
R-134a	High-pressure cutout
Basic refrigeration	switch
cycle	Thermal limiter
Automotive air	Pressure relief valve
conditioning system	Wide open. throttle
High side	(WOT) switch
Low side	A/C switch
Air conditioning	Service valves
compressor	Discharge service valve
Reed valve	Suction service valve
Magnetic clutch	Heating system
Refrigerant oil	Heater hoses
Refrigerant hoses	Heater core
Air conditioning	Blower
condenser	Manual controls
Receiver-drier	Push-pull cables
Sight glass	Pull-pull cable

Vacuum motors	High-efficiency air filter
Vacuum switch	Solar ventilation
Automatic temperature	Solar ventilation switch
control systems	Ambient temperature
Thermistor	switch
Control head	Ventilation fan
Programmer	Solar cell
Ducts	Solar ventilation ECM
Vents	

Review Questions—Chapter 75

Please do not write in this text. Place your answers on a separate sheet of paper.

1. The three basic states of matter are _____, _____, and _____.

2. Explain the three methods of heat transfer.

3. Heat always moves from the colder object to the warmer object. True or False?

4. What is a Btu?

5. _____, the opposite of _____, occurs when a vapor changes back into a liquid.

6. This substance circulates inside an air conditioning system.
   (A) Refrigerant.
   (B) R-12.
   (C) R-134a.
   (D) All of the above.

7. List and explain the eight basic parts of a refrigeration system.

8. Which of the following is *not* a function of an air conditioning system?
   (A) Cools air.
   (B) Circulates air.
   (C) Dehumidifies air.
   (D) Cleans air.
   (E) All of the above are correct.
   (F) None of the above are correct.

9. Describe the high and low side of an air conditioning system.

10. How does an air conditioning compressor turn on and off?

11. Name the five types of air conditioning compressors.

12. _____ _____ is used in the A/C system to keep the compressor lubricated.

13. How does the expansion tube A/C system control output temperature?

14. How does the thermostatic expansion valve A/C system control outlet temperature?

15. Why are service valves provided?

16. Most car heaters get heat from the engine exhaust. True or False?

17. The _____ _____ is a small radiator-like device that provides a large surface area for heat dissipation into the passenger compartment.

18. Explain four methods used to operate heater components.

19. Define the term "thermistor."

20. How does a solar ventilation system work?

🌀 ASE-Type Questions

1. Technician A says the operating principles of an automotive air conditioning system are completely different than those of a home refrigerator. Technician B says an automotive air conditioning system operates on the same principles as a home refrigerator. Who is right?
   (A) A only.
   (B) B only.
   (C) Both A and B.
   (D) Neither A nor B.

2. Technician A says vapor is one of the basic states of matter. Technician B says a molecule is one of the basic states of matter. Who is right?
   (A) A only.
   (B) B only.
   (C) Both A and B.
   (D) Neither A nor B.

3. Technician A says "conduction" is heat transfer through objects that are touching each other. Technician B says "conduction" is heat transfer caused by the air surrounding objects. Who is right?
   (A) A only.
   (B) B only.
   (C) Both A and B.
   (D) Neither A nor B.

4. Technician A says heat always transfers from a hotter object to a colder object. Technician B says heat always transfers from a colder object to a hotter object. Who is right?
   (A) A only.
   (B) B only.
   (C) Both A and B.
   (D) Neither A nor B.

5. Technician A says "evaporation" normally refers to a liquid changing into a vapor state. Technician B says "evaporation" normally refers to a vapor changing into a solid. Who is right?
   (A) A only.
   (B) B only.
   (C) Both A and B.
   (D) Neither A nor B.

6. Technician A says "condensation" occurs when liquid changes back into a vapor. Technician B says "condensation" occurs when a vapor changes back into a liquid. Who is right?
   (A) A only.
   (B) B only.
   (C) Both A and B.
   (D) Neither A nor B.

7. Technician A says when pressure is placed on a substance, the temperature of the substance increases. Technician B says when pressure is placed on a substance, the temperature of the substance decreases. Who is right?
   (A) A only.
   (B) B only.
   (C) Both A and B.
   (D) Neither A nor B.

8. Technician A says a "refrigerant" is a substance with a very low boiling point. Technician B says a "refrigerant" is a substance with a very high boiling point. Who is right?
   (A) A only.
   (B) B only.
   (C) Both A and B.
   (D) Neither A nor B.

9. Technician A says the refrigerant R-134a has been found dangerous to the environment and its production was ceased in 1995. Technician B says the refrigerant R-12 has been found dangerous to the environment and its production was ceased in 1995. Who is right?
   (A) A only.
   (B) B only.
   (C) Both A and B.
   (D) Neither A nor B.

10. Technician A says in a basic refrigeration cycle, the refrigerant goes through a compression phase. Technician B says in a basic refrigeration cycle, the refrigerant goes through a vaporization phase. Who is right?
    (A) A only.
    (B) B only.
    (C) Both A and B.
    (D) Neither A nor B.

11. Technician A says an automotive air conditioner's condenser causes refrigerant to change from a gaseous state into a liquid state. Technician B says an automotive air conditioner's condenser causes refrigerant to change from a liquid state into a gaseous state. Who is right?
    (A) A only.
    (B) B only.
    (C) Both A and B.
    (D) Neither A nor B.

12. Technician A says the "high side" of an automotive air conditioning system is known as the "discharge side." Technician B says the "high side" of an automotive air conditioning system is known as the "suction side." Who is right?
    (A) A only.
    (B) B only.
    (C) Both A and B.
    (D) Neither A nor B.

13. Technician A says a "crank type" compressor is used on certain automotive air conditioning systems. Technician B says a "rotary vane type" compressor is used on several automotive air conditioning systems. Who is right?
    (A) A only.
    (B) B only.
    (C) Both A and B.
    (D) Neither A nor B.

14. Technician A says an automotive air conditioner's accumulator forces air through the evaporator and into the passenger compartment. Technician B says an automotive air conditioner's accumulator keeps liquid refrigerant from entering the compressor. Who is right?
    (A) A only.
    (B) B only.
    (C) Both A and B.
    (D) Neither A nor B.

15. Which of the following is *not* a basic component of an automotive heating system?
    (A) *Blower.*
    (B) *Orifice tube.*
    (C) *Receiver-drier.*
    (D) *Both B and C.*

Activities—Chapter 75

1. Make a sketch (or use a Computer-Aided Drafting program, if one is available) of a basic refrigeration system. Create an overhead transparency from your sketch and use it to explain to the class the refrigeration cycle.

2. Obtain a junked automotive air conditioning compressor and disassemble it. Identify the type of compressor and the basic parts; then reassemble the components.

3. Visit the library and do research to find out why R-12 and other CFC refrigerants are dangerous to the environment. Write a short (1- to 2-page) report on the problem and what is being done to solve it.

Chapter 76

Heating and Air Conditioning Service

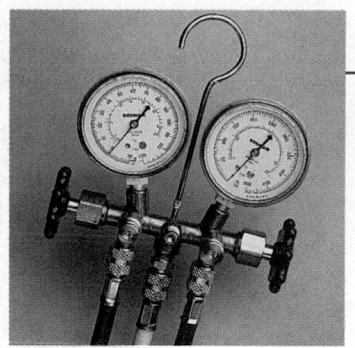

After studying this chapter, you will be able to:

- ■ Visually inspect a heating and air conditioning system and locate obvious troubles.
- ■ Diagnose common heating and air conditioning problems.
- ■ Describe the functions and uses of air conditioning test equipment.
- ■ Locate air conditioning and heating system leaks.
- ■ Explain how to replace major heating and air conditioning components.
- ■ Describe the general procedures for evacuating and charging an air conditioning system.
- ■ Demonstrate safe working practices when servicing heating and air conditioning equipment.
- ■ Correctly answer ASE certification test questions about the diagnosis and repair of heating and air conditioning systems.

Heating and air conditioning systems are prone to failure. Even the smallest leak (refrigerant or coolant) will eventually keep either system from working properly. As you will learn, correct service methods not only keep the vehicle operating normally, but they help protect our environment from damage caused by refrigerant gases and used antifreeze.

It is important that proper methods be used when servicing air conditioning systems. The refrigerant in older vehicles (R-12) has been found to be damaging to the upper atmosphere. It is up to us, as automobile technicians, to prevent refrigerant from leaking into the atmosphere.

This chapter will give you the basic information needed to repair the most common problems found in heating and air conditioning systems. This will increase your employability in almost any automotive repair facility.

Inspecting an Air Conditioning System

When an air conditioning system does not cool properly, visually inspect the system. Look for obvious signs of trouble, including:

- • Loose or missing compressor drive belt.
- • Inoperative compressor clutch.
- • Disconnected or damaged wiring or vacuum hoses.
- • Leaks (wetness) around lines or fittings.
- • Blockage (leaves, mud) in the condenser fins.
- • Inoperative air control doors.

Some common air conditioning system problems are illustrated in **Figure 76-1.**

Scanning the A/C System

Many late-model climate control systems will produce trouble codes if the on-board diagnostic system detects an abnormal circuit value or condition. If the indicator light in the dashboard glows, you would know to connect a scan tool to the vehicle. The scan tool will help you find the source of the problem. For example, the scan tool might indicate problems with the air conditioning system circuits and components shown in **Figure 79-2.**

You would then use pinpoint tests to verify where the problem is in each circuit. For more information on using a scan tool, refer to Chapter 18.

Checking Line Temperatures

To check the basic action of the A/C system, start the engine. Turn the air conditioning system on high and allow it to run for about ten minutes. Then, feel the refrigerant lines.

Internal compressor problems
Porous hose leaking
Condenser fins clogged with debris
Leaking condenser
Loose or missing drive belt
Inoperative compressor clutch
Inoperative blower motor
Leaking evaporator
Clogged or contaminated drier
Fittings leaking
Bad thermostat
Stuck expansion valve
Clogged water drain

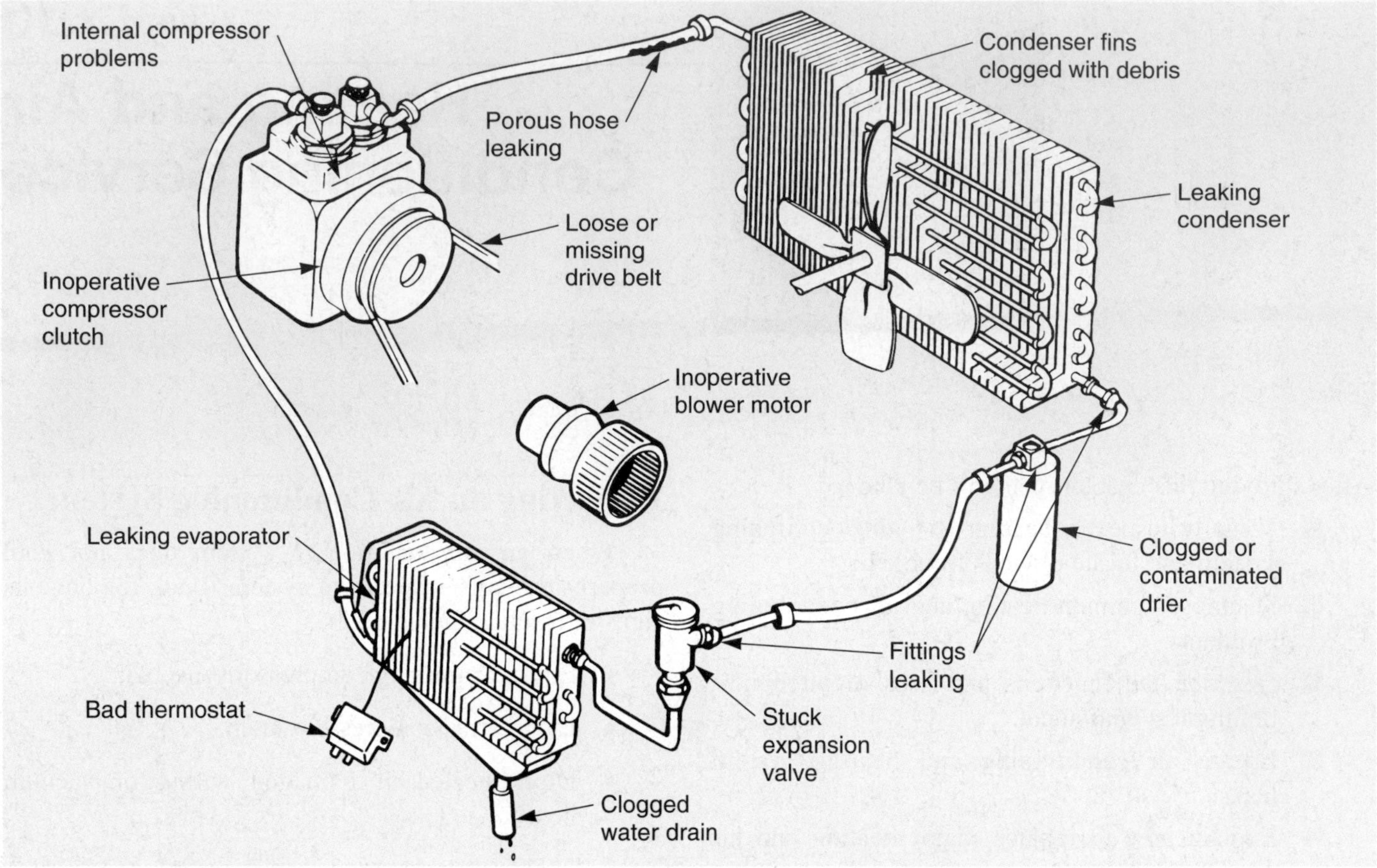

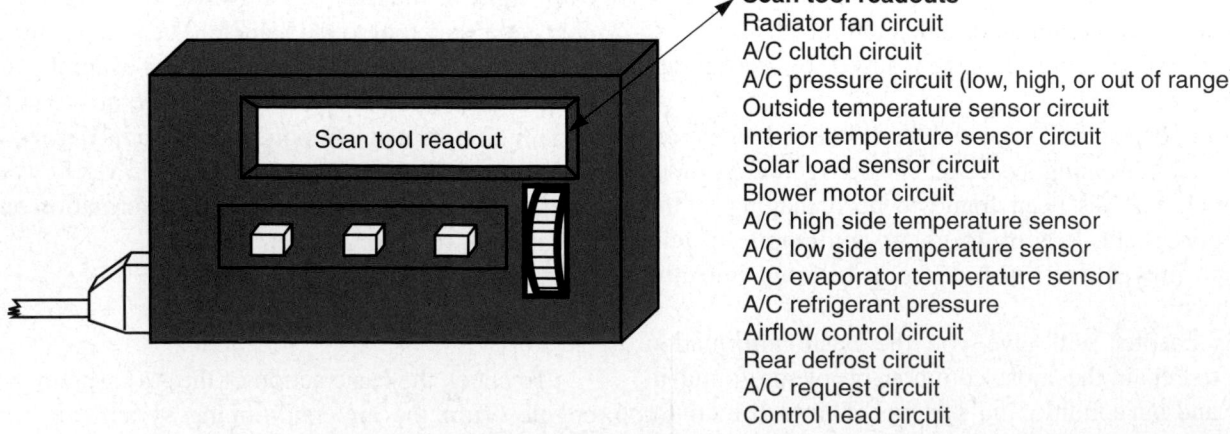

Figure 76-1. Note the types of problems that develop in a typical air conditioning system. (Florida Dept. of Voc. Ed.)

The *high-side* (discharge) line should be warm or hot. The *low-side* (suction) line should be cool. This will let you know that the refrigerant is moving through the system.

When the low-side line is cold but the system does *not* cool the passenger compartment, there might be a problem with the air control doors or the instrument panel controls.

If the high-side line is *not* warm and the low-side line is *not* cold, you know that a problem exists in the refrigeration section of the system. You must perform other tests to pinpoint the problem source.

Scan tool readout
Scan tool readouts
Radiator fan circuit
A/C clutch circuit
A/C pressure circuit (low, high, or out of range)
Outside temperature sensor circuit
Interior temperature sensor circuit
Solar load sensor circuit
Blower motor circuit
A/C high side temperature sensor
A/C low side temperature sensor
A/C evaporator temperature sensor
A/C refrigerant pressure
Airflow control circuit
Rear defrost circuit
A/C request circuit
Control head circuit

Figure 76-2. A scan tool will help find the source of trouble in most late-model climate control systems. It can be used to check the condition or operating values of the components listed.

Inspecting the Sight Glass

During your visual checks, also inspect the air conditioning system sight glass (if used). It can give you information on the condition of the system. The sight glass may be located in the top of the receiver-drier or in a refrigerant line. **Figure 76-3** shows how to read an A/C system sight glass.

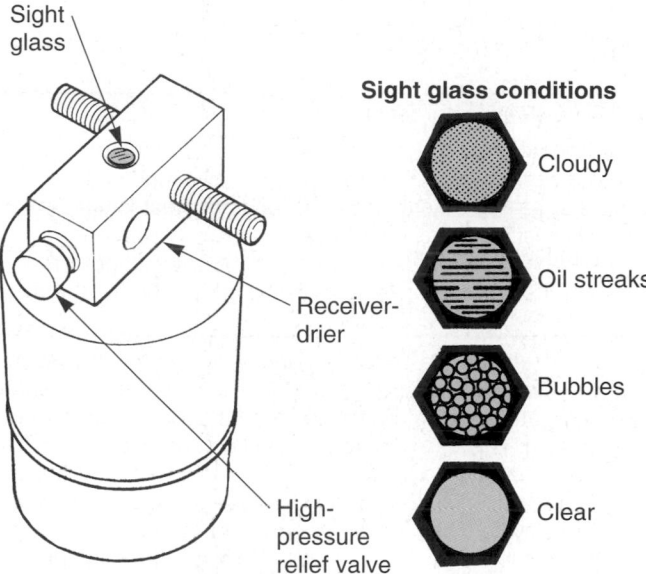

Figure 76-3. If used, inspect the sight glass. It will indicate the condition of the A/C system. The sight glass may be in the top of the receiver-dryer or in the line. (Chrysler)

A *clear sight glass* usually indicates that the A/C system has the correct charge of refrigerant; however, it can also indicate an empty system. If the low-side line is cool, there is refrigerant in the system. If the sight glass is clear and the low-side line is warm, the system could have a refrigerant leak.

A *foamy* or *bubbling sight glass* indicates that the A/C system is low on refrigerant and air is in the system. However, an occasional bubble during clutch cycling or system start-up may be normal.

An *oil-streaked sight glass* denotes a low refrigerant level, which is allowing excessive compressor oil to circulate through the system.

A *cloudy sight glass* may indicate that the desiccant (drying agent) in the receiver-drier or the accumulator has broken down and is circulating through the system. You would probably need to replace the unit or the desiccant bag.

Sight glass readings can be used only as indicators of system problems. They are not totally accurate. You must perform other tests to verify problems.

Refrigerant Safety Precautions

To avoid injury, observe the following safety precautions when working with refrigerant:

- Refrigerant can cause severe *frostbite* if it contacts your skin. Be careful when opening an air conditioning system line. Place a rag around the fitting before loosening.

- Always wear safety glasses when working with refrigerant. It can cause *blindness* if it sprays into your eyes. If refrigerant sprays into your eyes, flush them with water, without rubbing. Always consult a doctor.

- Keep refrigerants away from excessive heat. Pressure in a refrigerant container increases as temperature increases. High heat could make the tank explode, shooting metal fragments and refrigerant into your face or hands. Store refrigerant at temperatures below 104°F (40°C.).

- Keep refrigerant away from an open flame. When burned, refrigerant turns into phosgene gas, a highly *toxic poison*. When using a torch-type leak detector, carefully follow directions and make sure the work area is well ventilated.

- Never discharge refrigerant into the air. To prevent environmental damage, all refrigerant must be reclaimed or recycled. Also, discharging any vapor in a confined space, where it could displace air, could cause *suffocation*.

R-134a Service Differences

R-134a refrigerant systems require the use of special tools, fittings, and service procedures.

- Never use a flammable refrigerant, like OZ-12, in a system designed for R-134a or R-12. If a crash or collision occurs, this type refrigerant could catch fire and kill someone.

- The refrigerant tank for R-134a is normally blue, while R-12 is stored in a white tank.

- Tool fittings on R-12 systems use US Customary threads, while R-134a fittings have metric threads.

SERVICE TOOLS

Part/tool	R–12 system	R–134a system	Remarks
Tool joints	Inch threads	Metric threads	Threads standards for tool joints have been changed to avoid connecting R–12 system tools with R–134a system tools.
Charging valve joints	Screw-on type HI: 3/8–24 UNF LO: 7/16–20 UNF	Quick-connect type HI: 16 mm (0.6 in) dia. LO: 13 mm (0.5 in) dia.	The shape of the charging valve joints differ for each system to avoid confusion. The quick-connect type charging valve joint prevents refrigerant from leaking when the charging hose is connected to the valve.
Manifold gauge	High-pressure side maximum reading: 2.9 MPa (30 kgf/cm^2, 430 psi)	High-pressure side maximum reading: 3.5 MPa (35 kgf/cm^2, 500 psi)	R–134a requires a higher pressure to condense than R–12.
Gas leak tester	Gas type Electric type	Electric type	A gas leak tester reacts with chlorine in R–12 to indicate the location of a leak. This kind of tester does not work with an R–134a system, however, because R–134a has no chlorine. Two kinds of electric testers are available: those that work exclusively with one system or the other, and those that work with both. A tester built only for R–12 systems cannot be used with
Refrigerant	Chlorofluorocarbon–12 (CFC–12) (CCl$_2$F$_2$)	Hydrofluoro-carbon–134a (HFC–134a) (CH$_2$FCF$_3$)	If the refrigerants are mixed or one refrigerant is used in a system that requires the other, the compressor oil will separate from the refrigerant and not circulate within the system. This can damage the A/C compressor and cause abnormal A/C compressor vane noise. In addition, mixing R–134a with R–12 or using R–134a instead of R–12 in an R–12 system can lower the durability of the NBR O-ring and dissolve the fluorine rubber O-rings. If the fluorine rubber O-rings are dissolved, refrigerant may leak.
Compressor oil	Mineral oil	Polyalkylene glycol oil (PAG oil) (ATMOS GU10)	Special compressor oils for R–134a air conditioning systems are developed by each air conditioning vendor. Therefore, use only the specified oil for each model vehicle. If a PAG oil other than the specified type is used, the A/C compressor and refrigerant system can be damaged. If the compressor oils are mixed or one compressor oil is used in a system that requires the other, the refrigerant will separate from the compressor oil and not circulate within the system. This can damage the A/C compressor and cause abnormal A/C compressor vane noise. Mixing PAG oil with mineral oil or using PAG oil instead of mineral oil in an R–12 system can lower the durability of the NBR and fluorine rubber O-rings.
O-ring	Nitrile butadiene rubber (NBR) Fluorine rubber	High-circulated nitrile butadiene rubber (HNBR)	If an NBR O-ring is used in an R–134a system, the PAG oil and R–134a will lower the durability of the O-ring. If a fluorine rubber O-ring is used in an R–134a system, the R–134a will dissolve the O-ring and cause the refrigerant to leak.
Charging valve	Screw-on type HI: 3/8–24 UNF LO: 7/16–20 UNF	Quick-connect type HI: 16 mm (0.6 in) dia. LO: 13 mm (0.5 in) dia.	The shape of the charging valve differs for each system to avoid confusion. The quick-connect type charging valve prevents refrigerant from leaking when the charging hose is connected to the valve.

Figure 76-4. Chart compares R-12 and R-134a parts and tools. Study the differences closely. (Mazda)

- R-134a service fittings are male, while R-12 service fittings are female.

- R-12 systems use screw-on fittings on charging valve joints; R-134a systems use quick-disconnect fittings on charging valve joints.

- Pressure gauges must read higher values when testing R-134a systems.

- Never mix R-12 and R-134a. Mixing will make the refrigerant oil separate out of the refrigerant. Severe compressor damage will result. Mixing refrigerants can also deteriorate O-ring seals and cause leakage.

- Only use the refrigerant oil recommended by the manufacturer. Like refrigerants, mixing R-12 and R-134a refrigerant oils will cause separation and damage to the compressor and O-ring seals.

- Only use O-ring seals designed for the type of refrigerant and oil. R-134a will dissolve seals designed for R-12. See **Figure 76-4.**

Testing an Air Conditioning System

If your initial inspections and other checks do not find the source of the trouble, test the air conditioning system using pressure gauges. By comparing the high- and low-side pressures to specifications, you can determine the possible causes of the problem.

Pressure Gauge (Manifold) Assembly

A *pressure gauge assembly,* or *manifold gauge assembly,* typically consists of two pressure gauges, a manifold, two on-off valves, and three service hoses, **Figure 76-5.**

The high-pressure gauge is used to measure compressor discharge pressure, or high-side pressure. The low-pressure gauge measures suction, or low-side pressure.

The two outer service hoses connect to fittings on the air conditioning system. The center service hose is commonly connected to a recovery or recycling unit for cleaning or evacuating or to a refrigerant container for charging (filling) the system.

Service Valves

Service valves provide a means of connecting the pressure gauge assembly for testing, discharging, evacuating, and charging (filling) the air conditioning system. Most systems have two service valves. A few have three. The service valves may be located on the compressor fittings or in the refrigerant lines.

There are two basic types of service valves: Schrader valves and stem valves. These two types also vary for R-12 and R-134a systems.

A *Schrader service valve* is a spring-loaded valve, similar to the air valve in a tire. The service hoses on the pressure gauge set have depressors that open these valves. This is the most common type of service valve found on late-model vehicles. See **Figure 76-6.**

An *R-134a service valve* has a quick-disconnect fitting that accepts a corresponding fitting on the R-134a gauge set service hose. Before connecting these fittings, make sure the thumb screw on the hose fitting is turned all the way out. Pull back the sleeve on the hose fitting while pushing the hose fitting onto the service valve fitting. Release the sleeve to lock the fittings together. Finally, tighten down the thumb screw to depress the valve. An *R-12 service valve* has a threaded male fitting that accepts the female fitting on the R-12 service hose. This is shown in **Figure 76-6.**

Dust caps screw over the Schrader service valves to keep out debris and to prevent accidental opening of the spring-loaded valves. They should always be reinstalled after A/C system service, **Figure 76-7.**

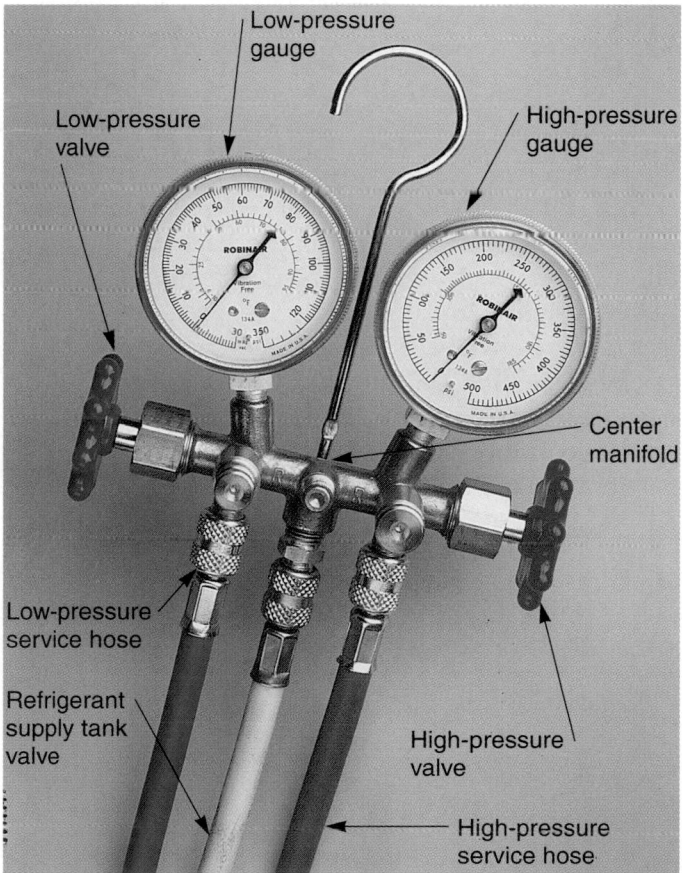

Figure 76-5. Study the parts of a pressure gauge assembly. It is commonly used when servicing air conditioning systems. (Imperial)

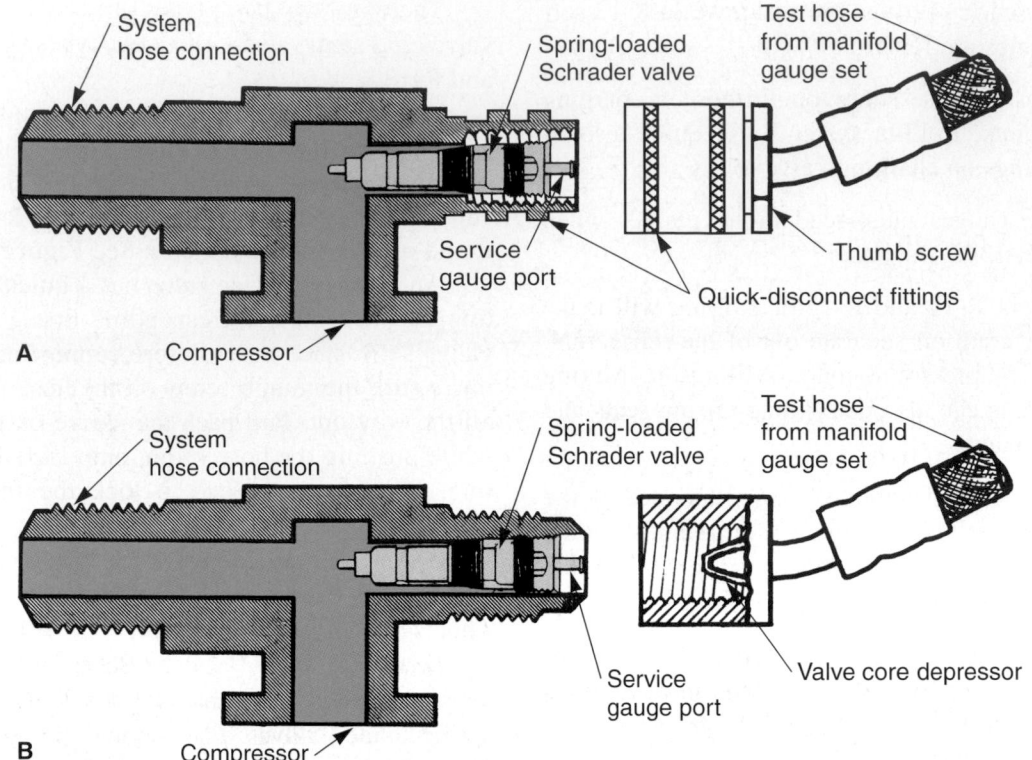

Figure 76-6. A Schrader service valve has a valve core that is similar to one used in a car tire. When the pressure gauge hose is connected to the valve, the valve is pushed open. A—Quick-disconnect fittings are used in R-134a systems. B—Service fittings for R-12 systems will have male threads. (Snap-on Tool Corp.)

A *stem-type service valve* is a manual valve that is opened and closed by screwing the valve stem in or out. Normally, when the valve stem is backseated (turned fully counterclockwise), the service port is blocked. Refer to **Figure 76-8.**

When the valve is turned midway between fully clockwise and fully counterclockwise, the service port is open to the pressure gauges. When the valve is frontseated (turned fully clockwise), the service port is only open to the compressor. The compressor is isolated

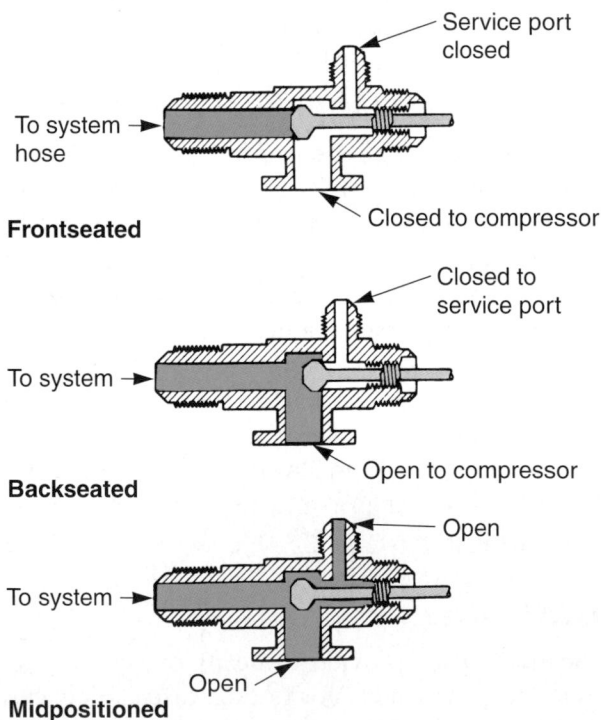

Figure 76-8. Stem-type service valve has a valve stem that must be turned. Note how the position of the stem opens different passages in the valve. (Chrysler)

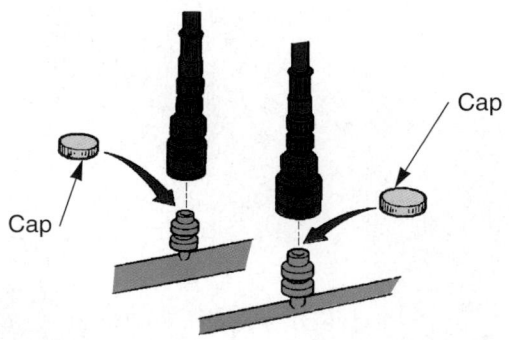

Figure 76-7. Dust caps protect Schrader valves from being filled with dirt and from accidental opening. Always reinstall the plastic caps on the service fittings. (Mazda)

from the rest of the system. This makes it possible to remove the compressor without losing the refrigerant in the other parts of the system.

Warning

Never operate an air conditioning system when stem-type valves are in the frontseated position. Excessive pressure could build in the compressor, causing part overheating and damage.

Connecting A/C Pressure Gauges

With the engine stopped and safety glasses on, remove the dust caps from the service valves. Make sure the pressure gauge valves are both closed.

Connect the high-side line first. Screw the *red* (high-side) hose fitting on the high-side service valve. Then screw the *blue* (low-side) hose fitting over the low-side service valve, **Figure 76-9.**

Note

With a few systems, a third pressure gauge is needed to check the action of the suction throttle valve. Refer to a service manual for details.

If you have stem-type service valves, turn them 1 1/2 turns clockwise. You can then measure air conditioning system pressures as a means of testing the system. This setup will also let you discharge or charge the system.

Static A/C Pressure Reading

A *static A/C pressure reading* will indicate how much refrigerant is in the system. With the engine *off*, read the high-side pressure gauge.

If the high pressure gauge shows approximately 50 psi (345 kPa), the system should have an adequate charge. If the pressure gauge reads below 50 psi (345 kPa), some of the refrigerant charge has leaked out and the system should *not* be operated. Correct any leaks and add refrigerant before making other tests.

Performance Testing A/C System

A *performance test* indicates air conditioning system condition by measuring system pressures with the engine running. Start the engine and operate it at fast idle (approximately 1500 rpm). Set the system for maximum cooling and operate it for about 10 minutes to allow pressures to stabilize. Close the vehicle's doors and windows. Leave the hood fully open.

Place a temperature gauge in one of the air outlets in the passenger compartment. Place another temperature gauge at the condenser to measure ambient (outside) air

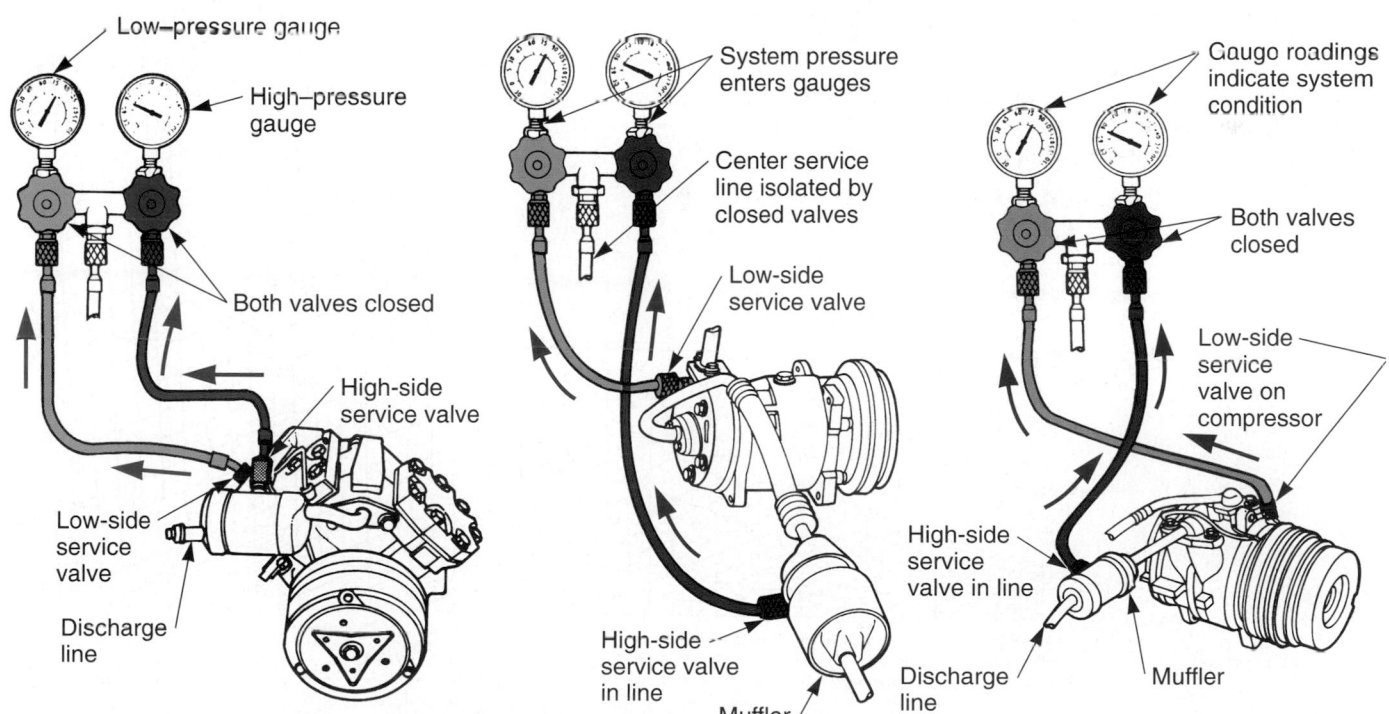

Figure 76-9. Study how a pressure gauge is connected. The manifold valves are kept closed, so that refrigerant cannot flow out the center hose on the manifold. (Florida Dept. of Voc. Ed.)

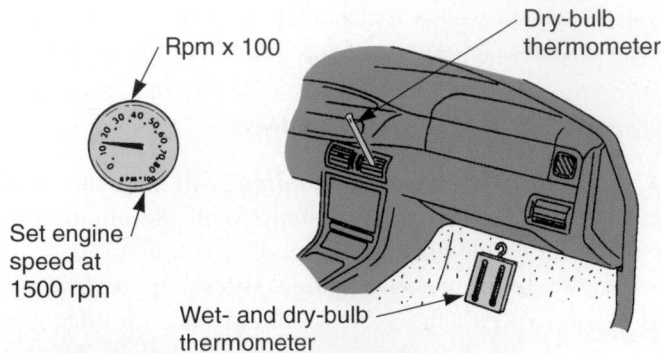

Figure 76-10. During a performance test of an A/C system, increase engine idle speed and check air temperature at the center vent outlet. With R-134a systems, you might also have to use wet- and dry-bulb thermometers at the blower outlet. Service manual specs will allow you to evaluate system operation. (Mazda)

APPROXIMATE TEST PRESSURE RANGES FOR NORMAL FUNCTIONING SYSTEMS

Outside temperatures	High-side pressures	Low-side pressures		
Ambient temperature in front of condenser	Psi at high pressure test fitting	Psi with STV, POA or VIR systems	Psi with expansion valve systems	Psi with orifice tube systems
60°F	120-170	28-31	7-15	–
70°F	150-250	28-31	7-15	24-31
80°F	180-275	28-31	7-15	24-31
90°F	200-310	28-31	7-15	24-32
100°F	230-330	28-35	10-30	24-32
110°F	270-360	28-38	10-35	24-32

Figure 76-12. Chart shows typical pressures for common A/C systems. Service manual values should be used during actual tests. (GMC)

temperature. Both temperatures are usually needed to analyze system performance. See **Figure 76-10.**

With R-134a, some manufacturers also require you to measure wet- and dry-bulb temperature at the blower inlet. This will let you use a chart to find the relative humidity, which affects system operation and pressures. See **Figure 76-11.**

Compare pressure and temperature readings, as well as relative humidity, to factory specs. **Figure 76-12** shows typical readings for several types of systems.

Note that pressure gauge readings vary with ambient air temperature, humidity, and system design. With an outside temperature of 70°F, for example, an orifice tube

HOW TO READ THE GRAPH

After measuring the temperatures of wet- and dry-bulb thermometer at the blower inlet, relative humidity (%) can be obtained.
Example: Supposing dry- and wet-bulb temperatures at the blower inlet are 77°F (25°C) and 67.1°F (19.5°C), respectively, the point of intersection of the dotted lines in the graph is 60%.

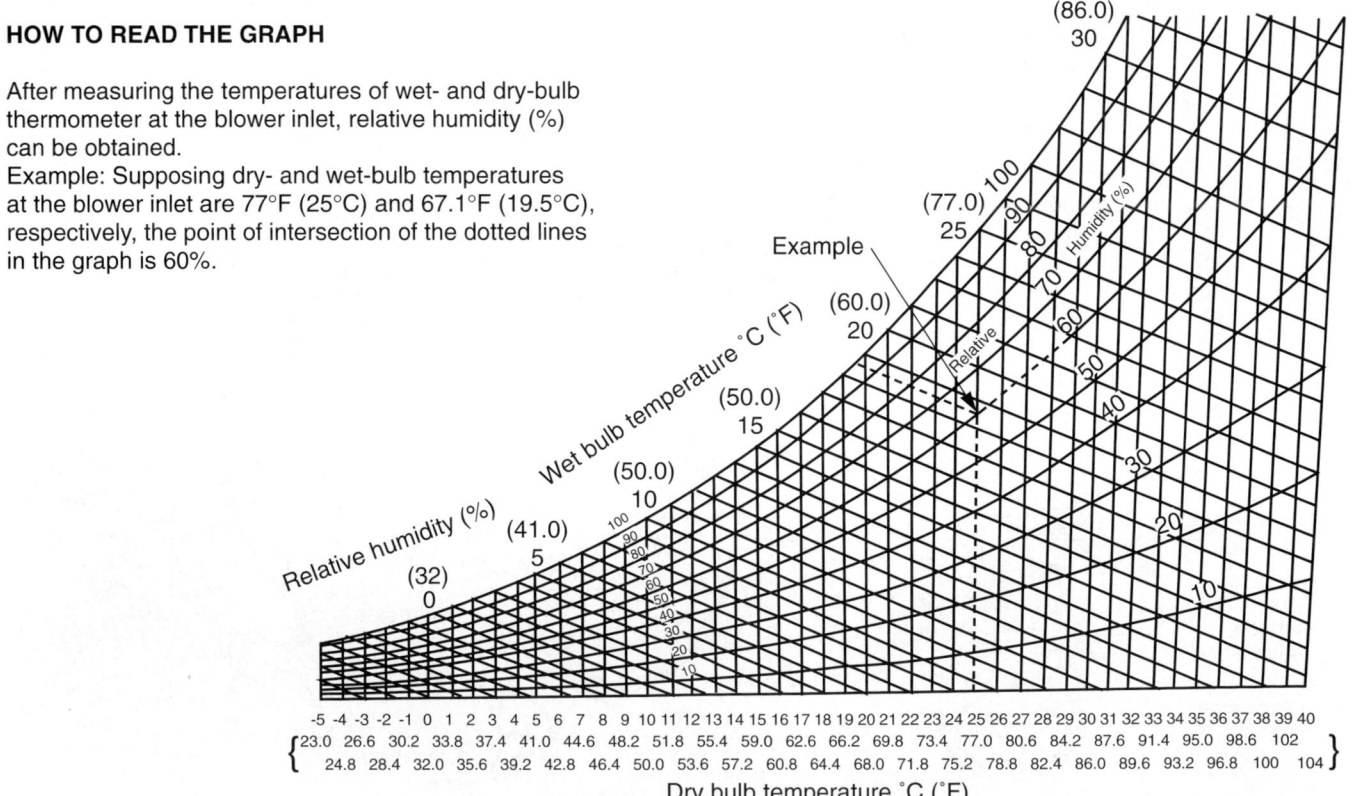

Figure 76-11. This chart will allow you to convert wet- and dry-bulb thermometer readings into humidity. Note the directions in the upper left corner. (Mazda)

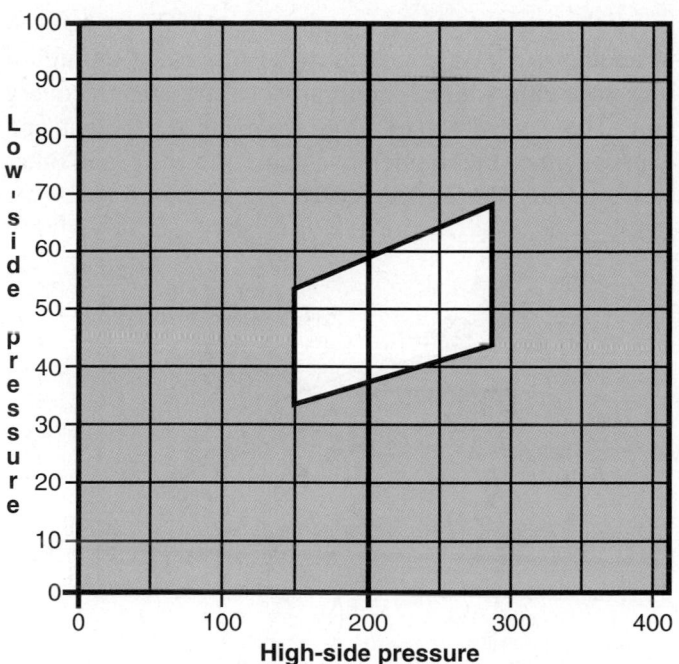

Figure 76-13. Graph shows normal pressures for one type of R-134a system. The white area indicates normal pressures. (General Motors)

system with a cycling switch should have approximately 150–250 psi high-side pressure and 24–31 psi low-side pressure.

A graph showing pressures for an R-134a system is given in **Figure 76-13.** The white area shows normal operating pressures for both the low and high sides. However, you must remember that variations in temperature and humidity can cause borderline pressure conditions.

If your pressure gauge readings are not within factory specs, there is a problem in the system. You must use the gauge readings, other symptoms, service manual diagnosis charts, and your knowledge of system operation to pinpoint the problem. A few examples of improper gauge readings are pictured in **Figure 76-14.**

Locating A/C System Leaks

An air conditioning system is considered leaking (needing repairs) when more than one-half pound of refrigerant must be added per year. There are several methods used to locate leaks.

An *internally charged detector* is a colored agent that can be charged into the system. Any leak will show up as a bright color spot (usually orange-red) at the point of leakage.

A *bubble detector* is a solution applied on the outside of possible leak points. A refrigerant leak will make bubbles or foam in the leak-detecting agent.

A *torch detector* uses a gas flame to indicate A/C leaks. Leaking refrigerant is drawn into and changes the color of the flame. If the flame does not change color, refrigerant is not present. This type of detector has been phased out for the safer electronic leak detector.

Warning
When refrigerant is burned, it turns into very toxic phosgene gas. Use a torch detector only in a ventilated area. Keep your face away from the detector to avoid inhaling fumes

An *electronic leak detector* uses a special sensor and an electronic amplifier to locate A/C

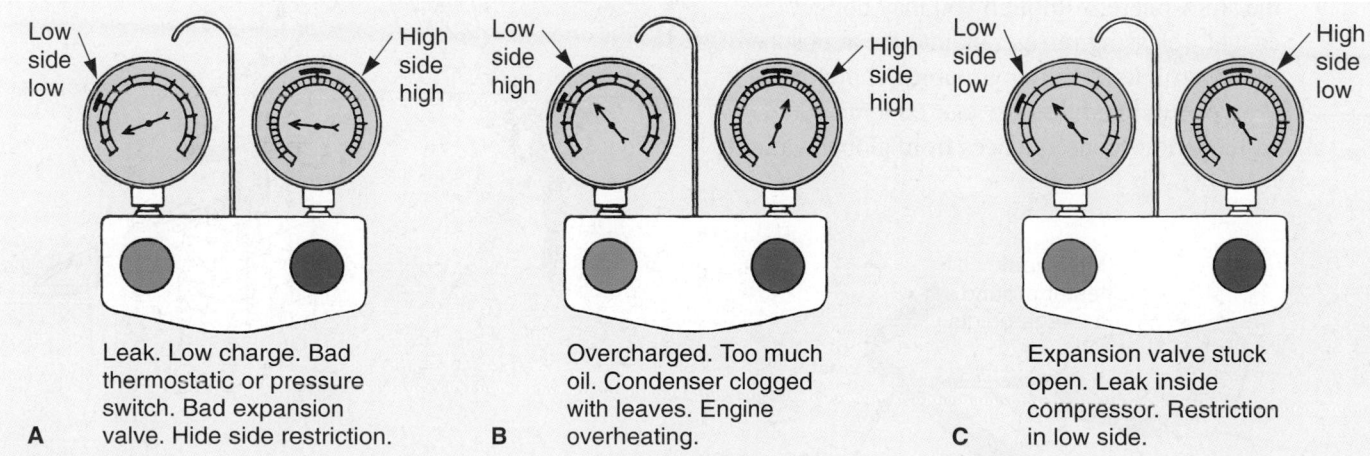

A Leak. Low charge. Bad thermostatic or pressure switch. Bad expansion valve. Hide side restriction.

B Overcharged. Too much oil. Condenser clogged with leaves. Engine overheating.

C Expansion valve stuck open. Leak inside compressor. Restriction in low side.

Figure 76-14. Note how basic gauge readings, whether too high or too low, can be used to troubleshoot possible problems. Gauge readings are only indicators. You must also use information on system design and your knowledge of A/C system operation to find the trouble. (Nissan)

leaks. It produces a sound or light signal when refrigerant is detected. See **Figure 76-15.**

An electronic leak detector is the fastest, safest method of locating system leaks. To use an electronic leak detector, move the tester's probe around possible leakage points (line fittings, condenser, compressor seals, valves, evaporator, etc.), **Figure 76-16.**

When the tester buzzes or lights up, you have found the leaking component. Check for leaks at the locations shown in **Figure 76-17.**

Tech Tip

Loose refrigerant line fittings are one of the most common sources of A/C system leaks. Tighten leaking fittings and retest them for leaks. If tightening does not correct the leak, replace the hose or line if needed. See **Figure 76-18.**

Recovering Refrigerant

Since refrigerant can no longer be discharged into the air when servicing air conditioning systems, technicians must use a *refrigerant recovery unit,* **Figure 76-19.**

The refrigerant that is drawn out of the system is stored in a refillable container and can be charged back into the system when repairs are completed.

Tech Tip

An automotive technician from St. Louis, Missouri, was the first person convicted of a crime for releasing automotive refrigerant into the atmosphere. The technician admitted to the offense and was prosecuted under a federal law written to protect the earth's atmosphere from further damage.

Never allow automotive refrigerant to escape into the atmosphere. Although you may not get caught, releasing refrigerant into the atmosphere will contribute to our environmental problems. If everyone ignored this law, our children could suffer serious consequences from global warming.

If the refrigerant is contaminated, or if it is to be used in another A/C system, it must be processed through a *recycling unit,* which is built into the refrigerant recovery unit. These units separate oil from the refrigerant and then use filter-driers to remove acids, moisture, and other contaminants. See **Figure 76-20.**

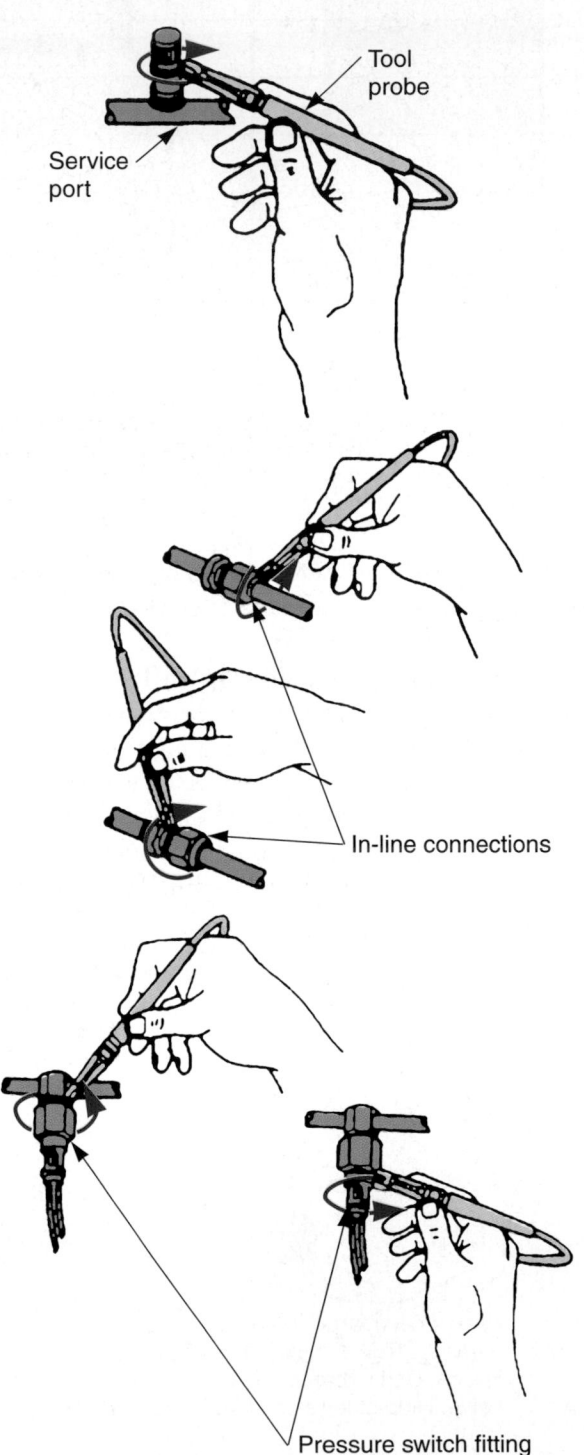

Figure 76-16. Note how a leak detector is used to "sniff" for leaking refrigerant gas. (Snap-on Tool Corp.)

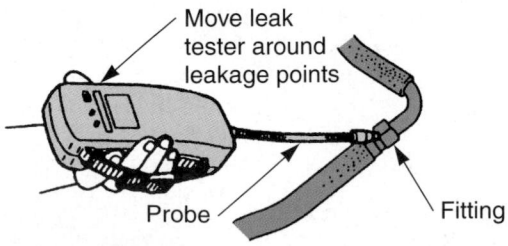

Figure 76-15. A leak detector will make an audible signal if refrigerant gases are present. (Snap-on Tool Corp.)

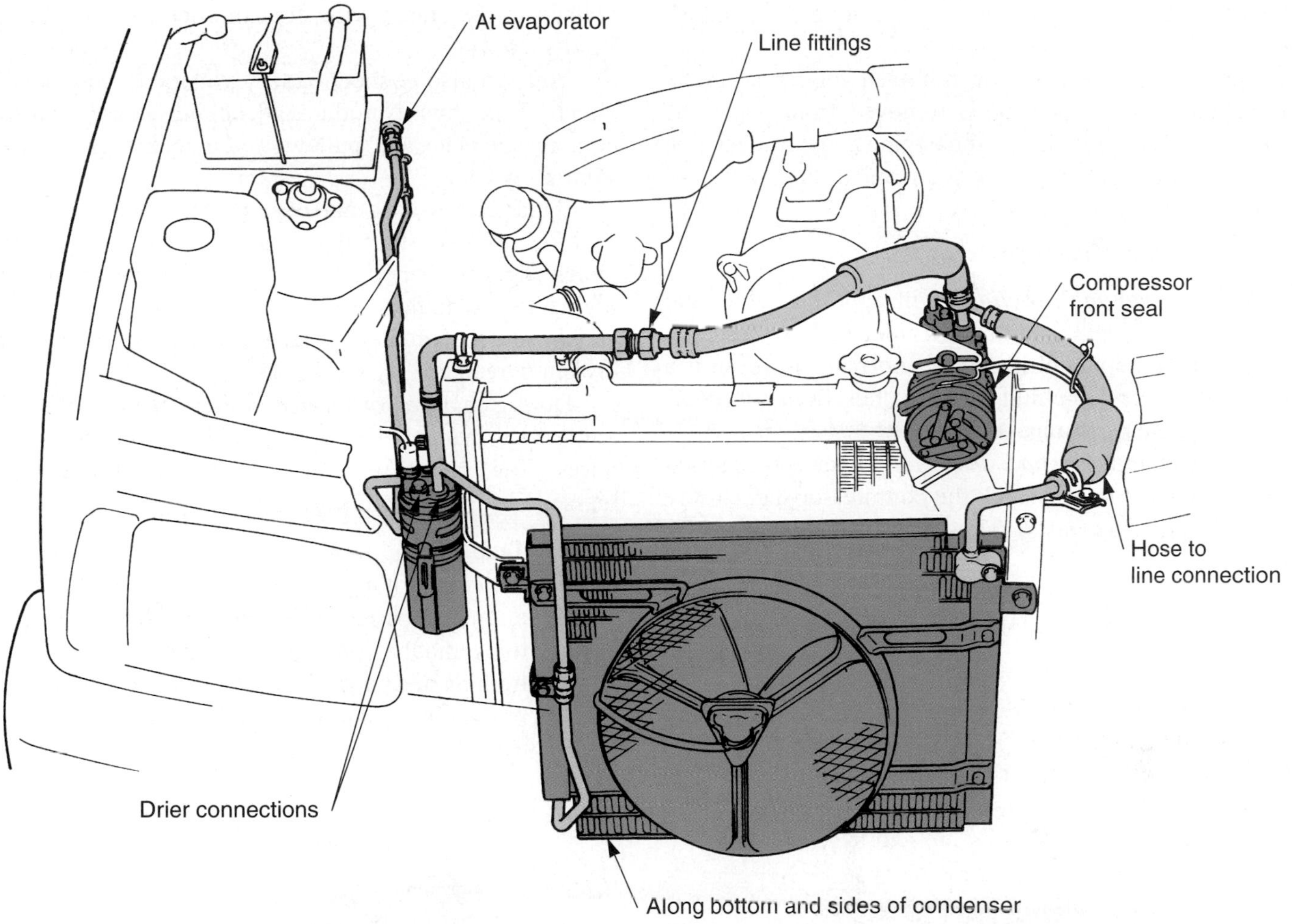

At evaporator

Line fittings

Compressor
front seal

Hose to
line connection

Drier connections

Along bottom and sides of condenser

Figure 76-17. The common leakage points on an air conditioning system are shown. (General Motors)

Recycling will become increasingly important as the reserve of R-12 and other CFC refrigerants is depleted. Environmentally safer refrigerants—namely, R-134a—are used in new vehicles, but the systems in many older cars and trucks still require R-12.

Tech Tip

Retrofit kits are available for converting A/C systems designed for R-12 to operate with R-134a refrigerant. The cost of these kits and the amount of labor involved in completing the retrofit can vary, so check with the vehicle manufacturer and your parts supplier.

Common A/C Component Problems

When your performance (pressure) tests indicate faulty components, remove and replace or repair the defective parts. The next section of the chapter outlines typical symptoms, problems, and service procedures for common air conditioning components. Refer to a service manual for details. It will outline special procedures and tools.

Evaporator Problems

With a **bad evaporator,** the trouble will normally show up as inadequate cooling. An evaporator can

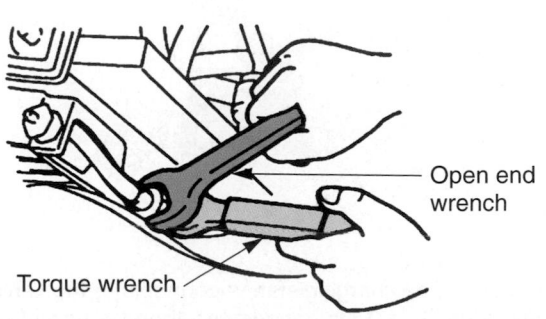

Open end
wrench

Torque wrench

Figure 76-18. If a fitting is leaking, try tightening it. Use two wrenches or one wrench and a torque wrench, if possible. (Mazda)

develop leaks that show up when testing with a leak detector around the evaporator case.

The evaporator must be removed for replacement. Sometimes, the evaporator is removed from under the dash or it may come out on the engine compartment side of the firewall. Refer to a service manual for details.

Compressor Problems

Compressor malfunctions will appear as abnormal noises, seizure, leakage, or high inlet or low discharge pressures. Some pumping noise is normal. However, if a loud rattling or knocking noise comes from the compressor, faulty parts may be indicated.

The *compressor shaft seal* is a common refrigerant leakage point. Always check the compressor seal closely during your leak tests. The seal can sometimes be replaced without compressor removal.

To check the *compressor clutch,* check for voltage being applied to the clutch magnet. If you are not getting

voltage to the clutch wire, find the open in the circuit, **Figure 76-21.**

You can also connect battery voltage directly to the clutch. This should produce a clicking sound, and the compressor plate and pulley should lock together. See **Figure 76-22.**

If the clutch does not engage, the compressor coil may be open or shorted and need replacement. If the clutch engages, there may be a problem in the circuit supplying voltage to the compressor clutch. Trace the circuit and locate the problem (bad switch, wire connection, broken wire, etc.), **Figure 76-23.**

During *compressor service,* follow the instructions in the service manual. The manual will outline the procedures for removal, disassembly, overhaul, and reassembly. Special tools are needed.

Tech Tip

Air conditioning compressors are complicated to the inexperienced technician. Never attempt to overhaul a compressor without proper training and supervision. Most technicians replace bad compressors with *new* or *rebuilt* units.

Figure 76-19. This typical charging station includes a recovery unit to withdraw refrigerant from an air conditioning system. The refrigerant is then stored in a refillable container for reuse or recycling. (Snap-on Tool Corp.)

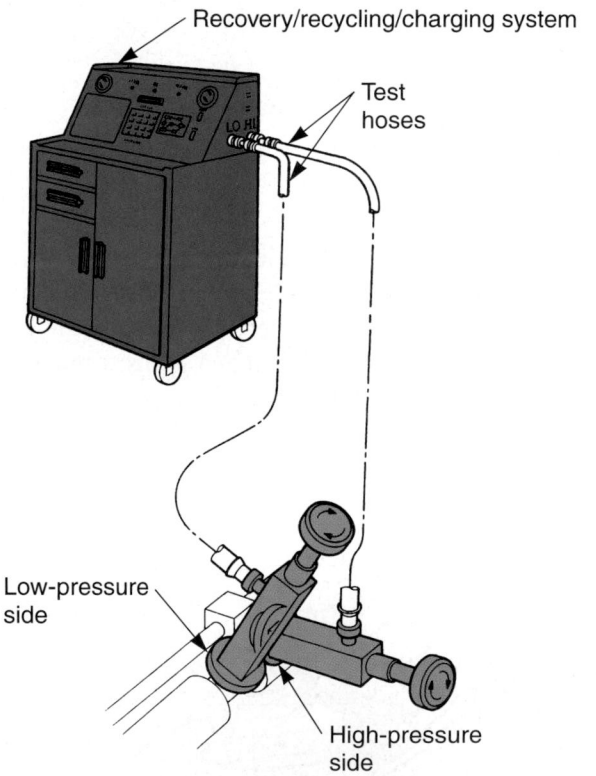

Figure 76-20. The charging station connects to the vehicle like a pressure gauge set. Its gauges can then be used to performance test the system. The charging station has a vacuum pump for evacuating the system and a refillable container for capturing old refrigerant. (Snap-on Tool Corp.)

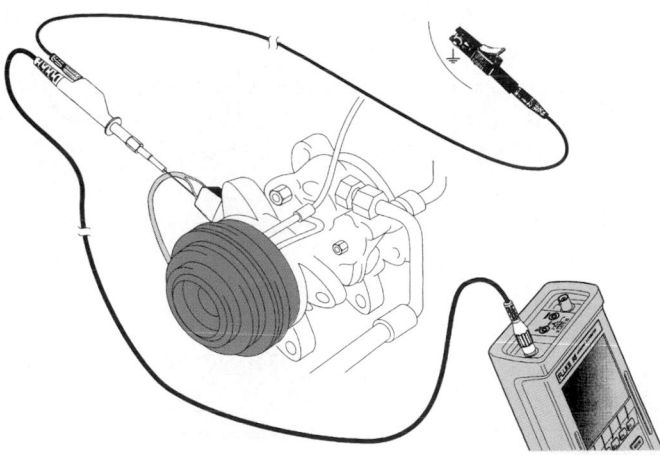

Figure 76-21. A multimeter can be used to check for current and voltage to the compressor clutch. If no voltage is reaching the magnetic clutch, check the wiring and the controller. (Fluke)

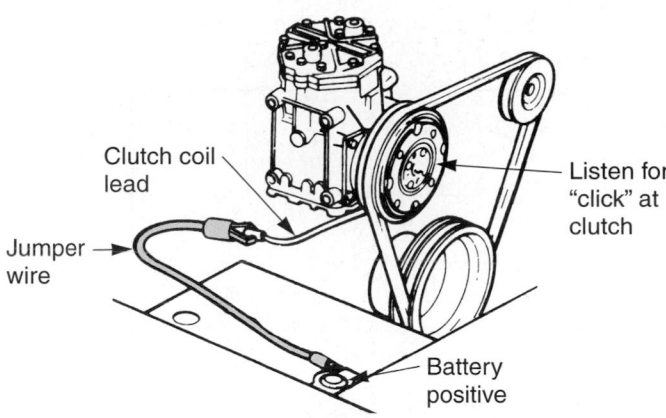

Figure 76-22. If the compressor fails to turn on, check the clutch by applying battery voltage to the clutch coil. This should make the clutch kick in. (Chrysler)

Condenser Problems

A *bad condenser* can leak refrigerant, or it can become restricted internally. A restricted condenser will cause a high pressure reading in the discharge (high) side of the system. A partial condenser restriction can cause ice or frost to form on the outside of the condenser. During normal condenser operation, the outlet tube will be slightly cooler than the inlet tube.

The condenser is fairly simple to replace. Its fittings must be loosened and removed. Use tubing wrenches to prevent fitting nut damage. Screws often hold the condenser to the vehicle's radiator core (metal framework across the front of the body). During installation, make sure you start the fitting nuts by hand to prevent crossthreading.

Receiver-Drier Problems

Receiver-drier problems are very common; the receiver-drier is one of the most frequently replaced air conditioning components. The desiccant (drying agent) can become contaminated, or the unit may become internally restricted. A restricted receiver-drier will alter system pressures, causing the outlet fitting on the unit to become abnormally cold.

Typically, replace the receiver-drier when:

- The sight glass is cloudy.
- System fittings have been left open for more than an hour.
- The system has been leaking and operated on a partial charge.

- The system has been opened and serviced for the third time.
- Moisture is found in the system.
- A restriction is indicated in the unit.

Expansion Valve or Tube Problems

A *faulty expansion valve* will usually show up as low pressure readings on both the low side (suction) and the high side (discharge) of the system. Usually, the failure is due to the power element (bulb) not operating the valve properly. A less common failure of the expansion valve is due to a clogged inlet screen (corrosion particles, loose desiccant beads, etc.).

A *clogged orifice tube* will produce symptoms similar to those caused by a restricted expansion valve. Contaminants can fill and clog the orifice tube or the tube screen, blocking refrigerant flow.

Refer to a service manual for detailed procedures for diagnosing and replacing these and other flow control valves. **Figure 76-24** shows the special tool needed to replace one type of orifice (expansion) tube.

Thermostat Problems

A *bad thermostat* will usually keep the compressor clutch from engaging. This will prevent the system from cooling. If the compressor clutch will not engage during normal system operation but engages when connected to battery voltage, you may need to test the thermostat. It may not be sending current to the clutch coil when needed.

Figure 76-25 shows how to test one type of thermostat. Replace the thermostat if faulty.

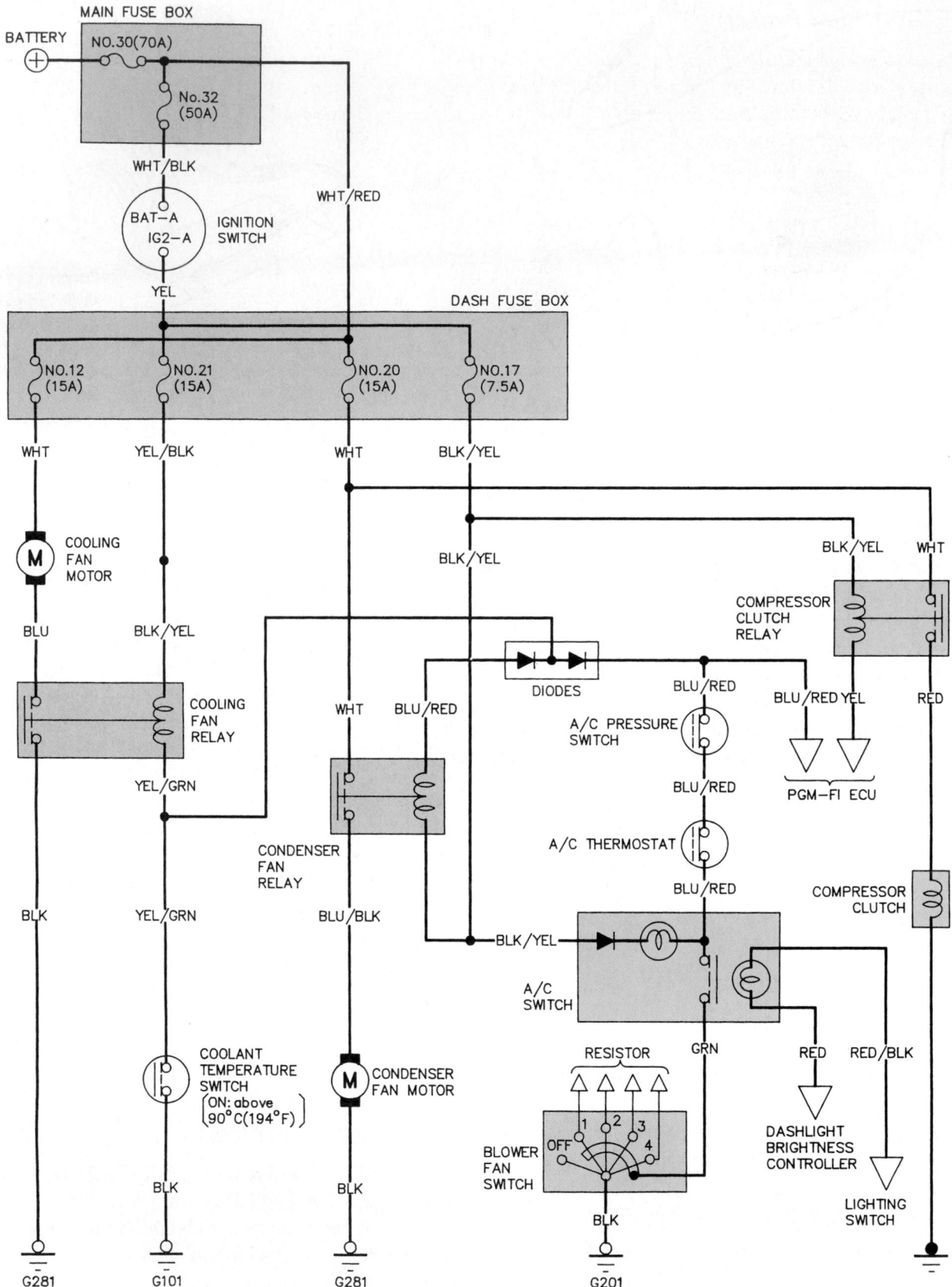

Figure 76-23. Study the basic compressor clutch circuit. Note the components that could prevent the clutch from functioning. (Honda)

Refrigerant Line Problems

Defective refrigerant lines may leak, harden, crack, and restrict refrigerant flow. Replace any refrigerant lines that are faulty. When installing a line using an O-ring seal, coat the seal with compressor oil. This will help prevent the seal from leaking, **Figure 76-26.**

Evacuating an Air Conditioning System

A/C system evacuation involves using a vacuum pump to remove air and moisture from the inside of the system. Evacuation is needed anytime the air conditioning system has been discharged or opened to the atmosphere.

To evacuate an A/C system, connect the pressure gauges to the system as shown in **Figure 76-27.** Connect the vacuum pump to the center service hose on the gauge set. If necessary, open the compressor valves (stem type). Open the valves on the pressure gauge set.

Plug in and turn on the vacuum pump. Operate the vacuum pump until the pressure gauges read approximately 26–28 in/hg at sea level. After reaching this vacuum, run the pump another 5 or 10 minutes to ensure complete removal of air and moisture.

If you cannot draw the service-manual-recommended vacuum, check the operation of the pump and examine the system for leaks. This is an easy way to detect major leaks before installing refrigerant.

After evacuating the system, shut off the vacuum pump and close both gauge valves. The system vacuum should not drop more than about 2–3 in/hg in a five-minute period. If it does, a small system leak is indicated. You must leak-test the system as described earlier.

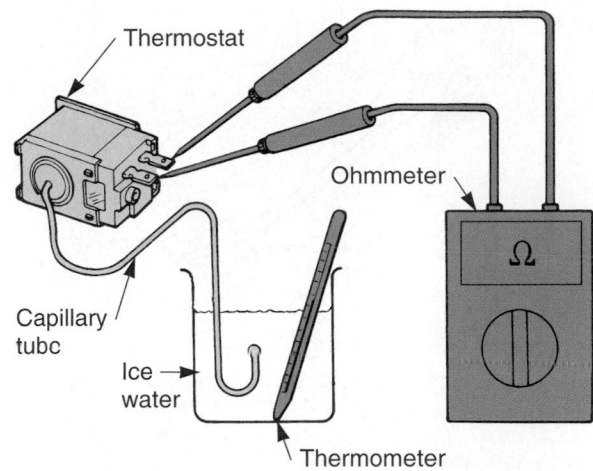

Figure 72-25. The thermostat should change resistance when a specific temperature is reached. This should make the ohmmeter reading change accordingly. Refer to the manual for temperature and resistance specs. (Honda)

Charging an Air Conditioning System

A/C system charging involves filling the system with the correct amount and type of refrigerant. The system should be charged only after it has been leak-tested and evacuated. Most shops use 10- to 30-pound containers of refrigerant. However, larger bulk containers are also used.

Figure 76-28 shows the basic connections for charging an air conditioning system. The pressure gauge set is left on the system service valves after evacuation. However, the center service hose is removed from the vacuum pump and connected to the refrigerant container.

With the gauge valves closed, connect the center hose to the refrigerant container. Open the center valve of

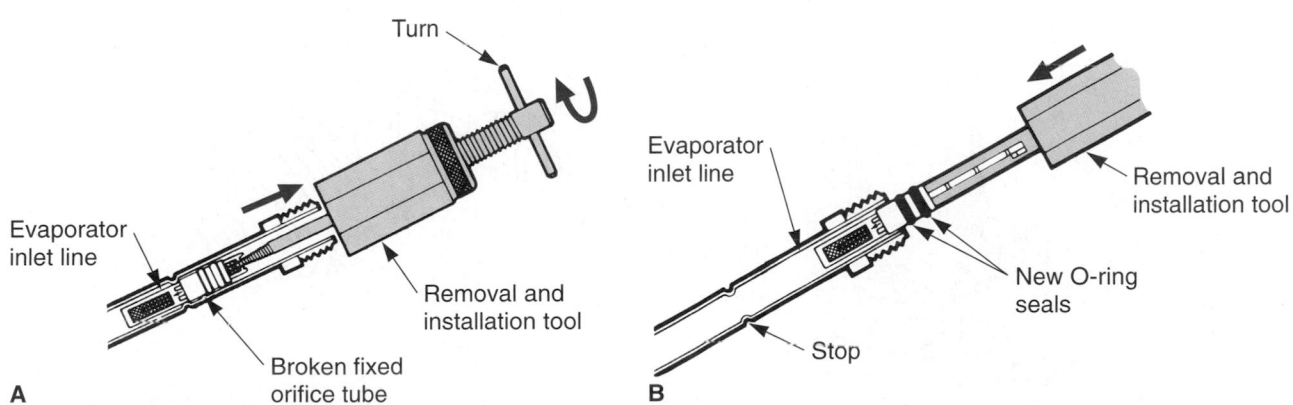

Figure 76-24. A special tool is being used to service a broken orifice tube. A—Thread the screw into the broken tube. Then tighten to pull out the unit. B—Forcing a new orifice tube into place. (Ford)

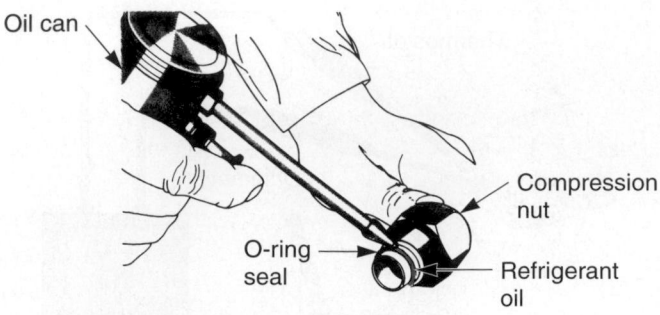

Figure 76-26. Lubricate all A/C system O-rings and seals with refrigerant oil. This will help seal them and prevent leaks. (Plymouth)

the manifold gauge set and the service valve on the refrigerant container. This will transfer control of refrigerant flow to the gauge valves.

To charge the system, open the gauge suction (low-pressure) valve but leave the gauge discharge (high-pressure) valve closed. Start the engine and turn on the air conditioning system. Compressor suction will then draw refrigerant into the system, **Figure 76-28.**

Adjust the gauge suction valve so that gauge pressure does *not* exceed 50 psi (345 kPa). This will ensure that

liquid refrigerant does not enter and damage the compressor. You may want to place the refrigerant container in warm water (not hotter than 125°F [52°C]). This will expand the refrigerant, helping it to flow into the system.

Note
Keep the refrigerant container right-side up when charging. This will ensure that refrigerant vapor, not liquid refrigerant, enters the system.

After charging the A/C system, double-check system pressures and temperatures. Make sure the high- and low-side pressures are satisfactory. Also, use a thermometer to measure the temperature of the air blowing from the passenger compartment vents.

If these checks are within specs, the air conditioning system is ready to be returned to service.

Charging Station

A *charging station* usually contains a vacuum pump, a pressure gauge set, an oil injection cylinder, and a charging tank of refrigerant. Refer back to **Figure 76-19.**

A charging station can be used to evacuate and charge the system without disconnecting the service

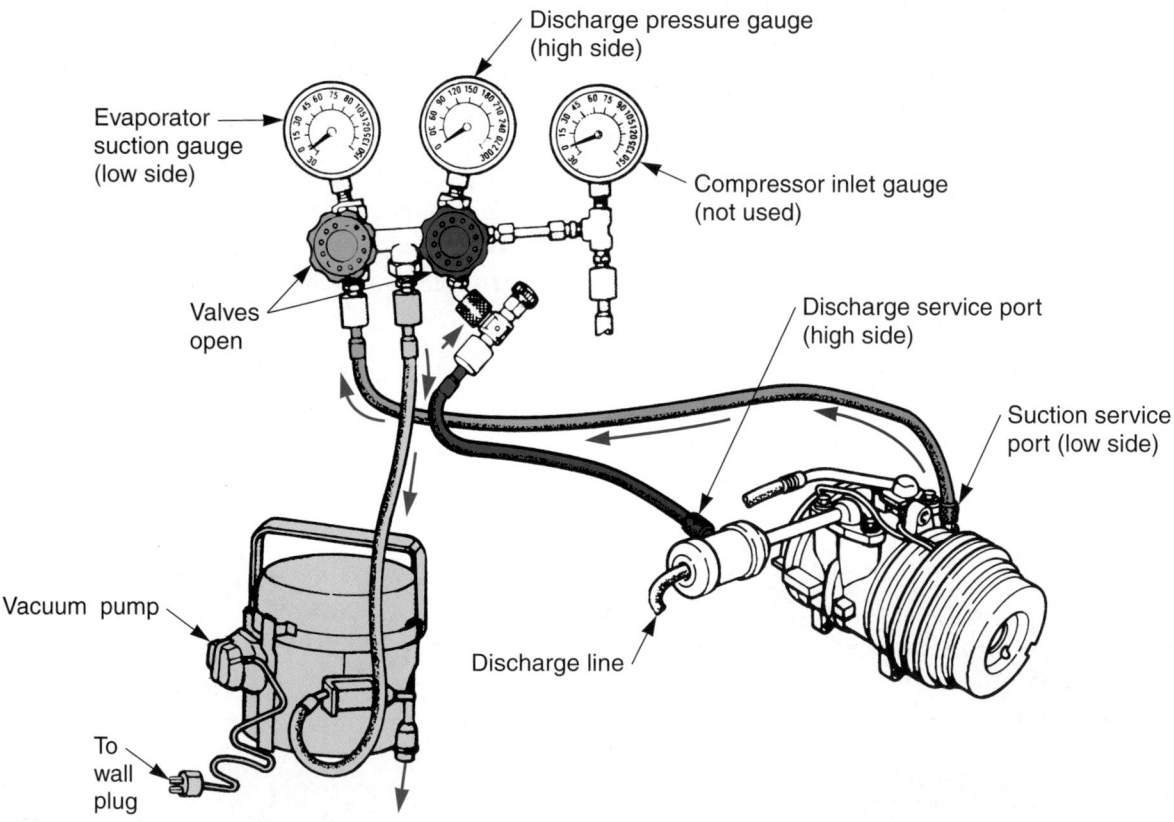

Figure 76-27. Note the connections for evacuating the system. Connect the vacuum pump as shown. Pump will draw moisture out. Operate the pump until the gauge reads 26-28 in/hg; then run the pump about five minutes more to ensure complete evacuation. (Chrysler)

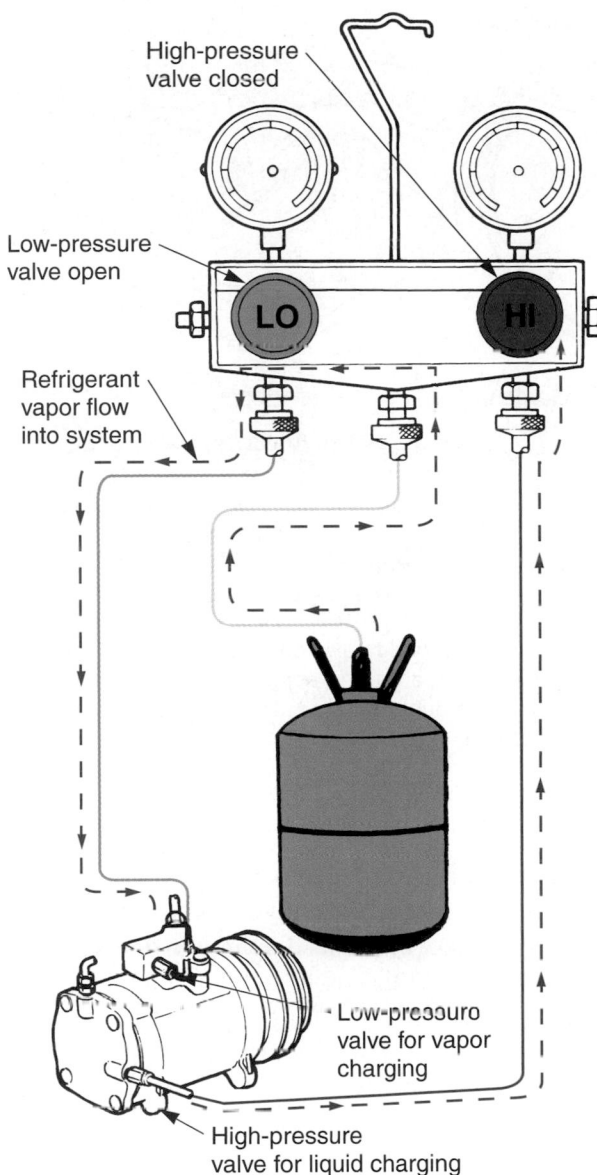

High-pressure
valve closed

Low-pressure
valve open

Refrigerant
vapor flow
into system

Low-pressure
valve for vapor
charging

High-pressure
valve for liquid charging

Figure 76-28. Adding refrigerant with the gauges attached to the service valves. Attach refrigerant container to the center hose. Open the lower pressure valve and allow refrigerant to flow into the low side of the system with the engine running. (Honda)

hose. The general procedures for using a separate vacuum pump, refrigerant container, and test gauge set should be followed when using a charging station. Remember to follow the directions for the specific type of equipment.

Adding Refrigerant Oil

The refrigerant oil level should be checked anytime you add or replace refrigerant. Before checking the refrigerant oil, operate the system for 10-15 minutes.

Then, isolate the compressor by closing the stem-type service valves. If Schrader valves are used, you must discharge the system to check the oil level.

Refer to a service manual for detailed procedures for checking the refrigerant oil level. Follow instructions for the type of compressor you are servicing.

If the oil is low, add the specified amount and type of refrigerant oil. Do not overfill the compressor, or system efficiency will be reduced.

 Tech Tip
Make sure you are using the right type of compressor oil. R-12 and R-134a oils differ. Mixing them will cause separation and lack of lubrication in the compressor.

A/C System Service Rules

When servicing an air conditioning system, remember to follow these basic rules:

- Inspect the system for obvious problems (loose or missing belt, clogged condenser fins, disconnected wires, inoperative compressor clutch, etc.).

- Check the sight glass to see if the system has a refrigerant charge.

- Feel hose temperatures to determine the general condition of the system.

- Leak test the system.

- Wear safety glasses when servicing a refrigerant system.

- Discharge the system into a recovery unit, never into the air.

- Make sure the system is fully discharged before attempting to remove any pressurized component.

- After disconnecting fittings, cap them to prevent moisture from entering the system.

- Coat all O-rings with refrigerant oil before installation. Add oil to the system as needed.

- Torque all fittings to specifications.

- Evacuate the system and check for leakage (vacuum drop) before charging.

- Charge the system with the correct amount of refrigerant.

- Add refrigerant to the suction side of the system with engine running. Do not let liquid refrigerant enter the compressor.

- Use service manual instructions and specifications.

- Recheck system operation after service.

Heater Service

Heater problems typically show up as coolant leaks or as insufficient warming of the passenger compartment. The heater hoses can harden, break, or leak, especially when they are routed near the hot engine. The heater core can rust and develop leaks, allowing coolant to drip on the car's floor or carpet.

When the system does not produce heat, the heater core may be clogged, a heater hose valve may be inoperative, or the air control doors may not be functioning properly.

Checking Heater Coolant Flow

To check heater coolant flow, let the engine and the heater hoses cool. Then, start the engine and operate it at a fast idle. Turn the heater on high.

As the engine coolant warms, feel both the inlet and outlet heater hoses. If used, feel the hose on both sides of the heater flow control valve.

If both heater hoses are about the same temperature, coolant is passing through the heater core and valve. If one hose is hot and the other is cool, there is *blockage* in the system. The heater core could be clogged, or the flow valve could be stuck closed.

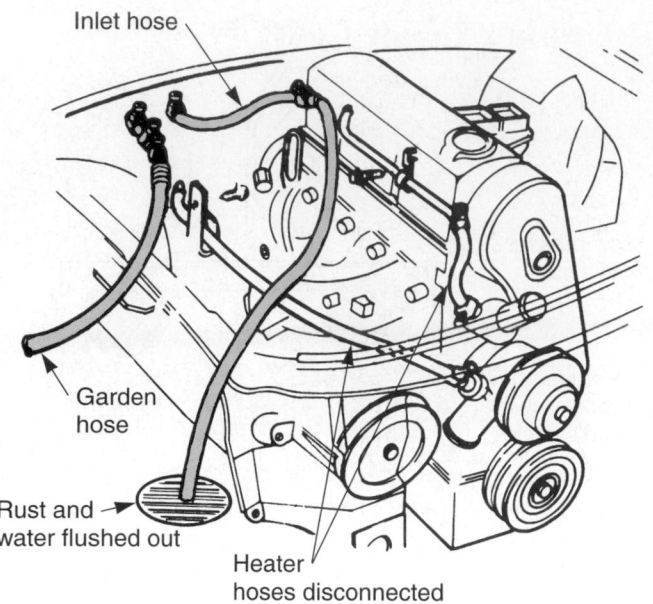

Figure 76-30. A clogged heater core is common. Remove the heater hoses. Attach a garden hose to the outlet heater hose and force water through the core. This may remove rust from the core. (Ford)

To check the operation of a heater flow valve, use a hand vacuum pump to apply suction to the diaphragm. When vacuum is applied, the valve should activate. If the valve does not move or the diaphragm does not hold vacuum, replace it. If the valve and diaphragm are good, check that vacuum is reaching the unit through the supply line, using a vacuum gauge or your finger. See **Figure 76-29.**

Flushing the Heater Core

A clogged heater core is a common problem that can reduce heating system efficiency. The inside of the core can become filled with rust from the engine cooling system.

To flush the heater core, remove both heater hoses. Connect a garden hose to the outlet heater hose. Turn the hose on and let water force rust out of the core for about five minutes, **Figure 76-30.**

Checking Air Control Doors

Inoperative air control doors will prevent either cool or warm air from leaving the desired vents. Use service manual diagrams of the doors to determine which ones are not working normally. You can sometimes lay under the dash and watch the air door shaft turn to check operation. If needed, adjust the air control door arm or cable until it closes and opens as needed.

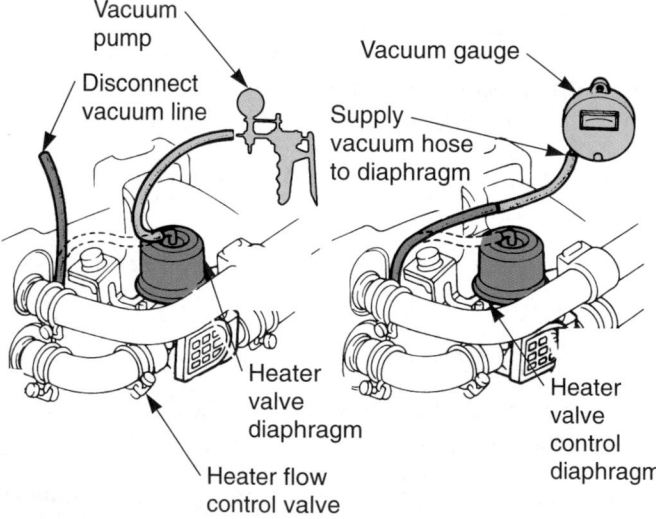

Figure 76-29. The heater flow valve can stick opened or closed and the vacuum diaphragm can rupture. A—Use a vacuum pump to check valve operation. B—Also check that vacuum is being fed to the diaphragm from the vacuum line. (Honda)

Electronic Climate Control Service

Electronic climate control service can be very challenging because there are so many design variations. Again, the service manual will give essential service information.

A *component location drawing* shows which parts affect heating and A/C operation and gives their general location on the vehicle. See **Figure 76-31.**

Wiring diagrams are also needed to troubleshoot electronic climate controls. These diagrams will let you trace wires from component to component. A blower control circuit is shown in **Figure 76-32.**

The control module is the brain of modern electronic climate control systems. To check its condition, you must basically determine if the unit is producing the right outputs for known inputs. For example, if you move the controls to the maximum heat position, the module should produce an output to move the correct air control door to route hot air into the passenger compartment.

The control module must also produce the correct readouts in the digital display. Various sensors provide feedback so that the module can control the climate control system and produce meaningful displays, **Figure 76-33.**

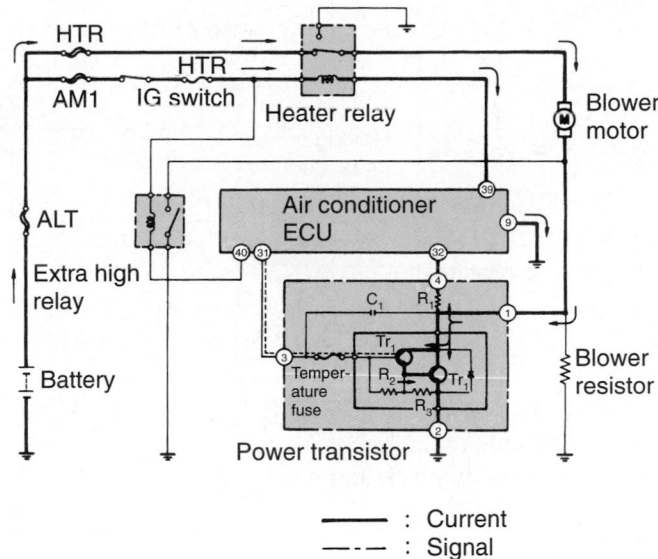

Figure 76-32. Circuit shows how the control module can control blower speed. This principle is used in other system designs. (Lexus)

A wiring diagram for a complete climate control system is shown in **Figure 76-34.** Refer to this type of diagram when diagnosing and repairing hard-to-find problems.

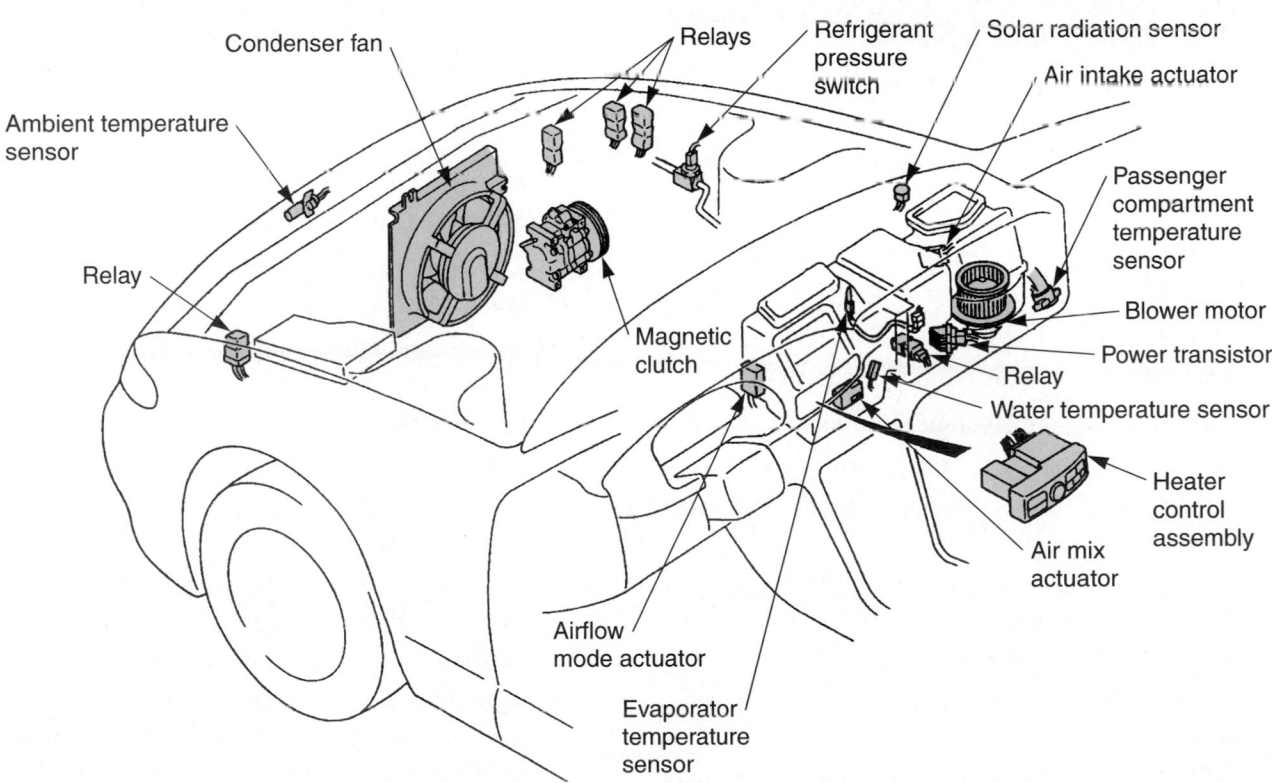

Figure 76-31. A component location drawing will show you the location of all major parts in climate control system. This can save time when troubleshooting. (Mazda)

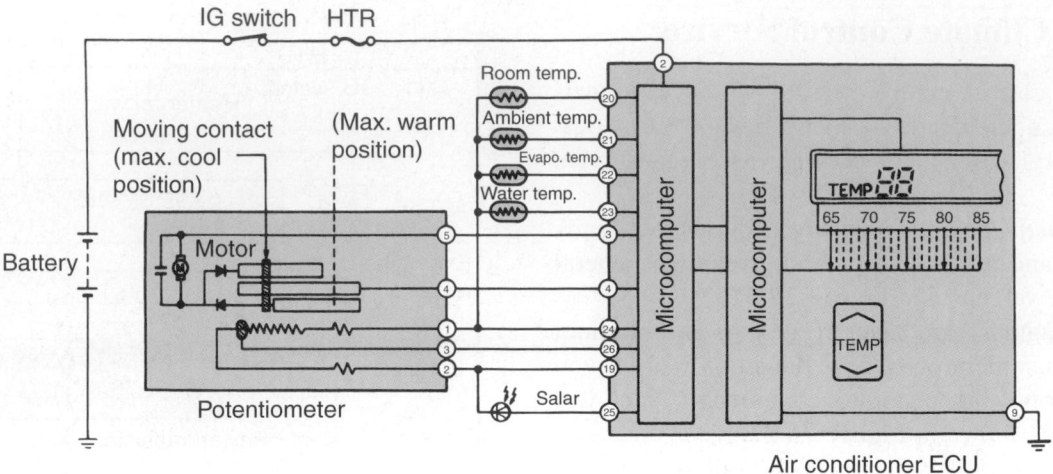

Figure 76-33. This circuit shows how two microprocessors are used to operate the digital display and to analyze inputs from various sensors and switches. (Lexus)

Summary

- Many late-model climate control systems produce trouble codes if the on-board diagnostic system detects an abnormal circuit value or condition.

- To check the basic action of the A/C system, start the engine. Turn the air conditioning system on high and allow it to run for about ten minutes. Then, feel the refrigerant lines.

- During visual checks of the A/C system, be sure to inspect the sight glass (if used).

- Refrigerant can cause severe *frostbite* if it contacts your skin.

- Always wear safety glasses when working with refrigerant.

- Keep refrigerants away from excessive heat.

- Keep refrigerant away from an open flame.

- Never discharge refrigerant into the air.

- Several different tools, fittings, and procedures must be used when servicing R-134a refrigerant systems.

- A pressure gauge assembly typically consists of two pressure gauges, a manifold, two on-off valves, and three service hoses.

- A Schrader service valve is a spring-loaded valve, similar to the air valve in a tire.

- An R-134a service valve has a threaded female fitting that accepts the male fitting on the service hose or gauge set.

- A performance test indicates air conditioning system condition by measuring system pressures with the engine running.

- With R-134a, some manufacturers require you to measure wet- and dry-bulb temperatures at the blower inlet. This will let you use a chart to find the relative humidity, which affects system operation and pressures.

- An electronic leak detector uses a special sensor and an electronic amplifier to locate refrigerant leaks.

- Since refrigerant can no longer be discharged into the air when servicing air conditioning systems, technicians must use a refrigerant recovery unit.

- The compressor shaft seal is a common refrigerant leakage point.

- A bad condenser can leak refrigerant or it can become restricted internally.

- Receiver-drier problems are very common.

- A faulty expansion valve will usually show up as low suction (low-side) and discharge (high-side) pressure readings.

- A clogged orifice tube will produce symptoms similar to those caused by a restricted expansion valve.

- A bad thermostat will usually keep the compressor clutch from engaging and prevent the A/C system from cooling.

- A/C system evacuation involves using a vacuum pump to remove air and moisture from the system.

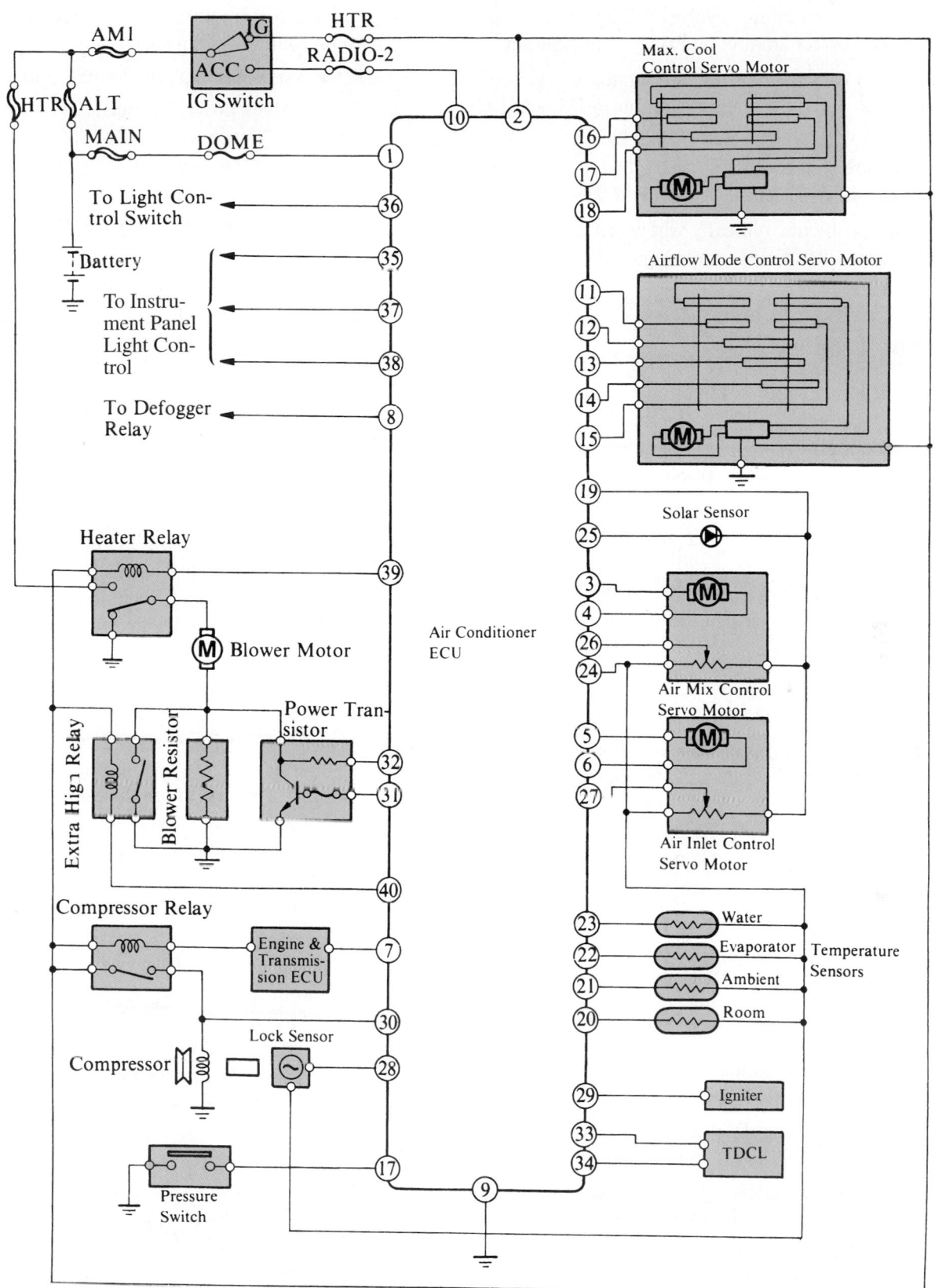

Figure 76-34. Study this wiring diagram for modern climate control system. Note the various inputs from sensors and outputs to motors and displays. (Lexus)

- A/C system charging involves filling the system with the correct amount and type of refrigerant.

- A charging station usually contains a vacuum pump, a pressure gauge set, an oil injection cylinder, and a refrigerant container.

- Make sure you are using the right type of compressor oil. R-12 and R-134a oils differ.

- Heater problems typically show up as coolant leaks or as insufficient warming of the passenger compartment.

Important Terms

Clear sight glass	Internally charged
Bubbling sight glass	detector
Oil-streaked sight	Bubble detector
glass	Torch detector
Cloudy sight glass	Electronic leak detector
Pressure gauge	Refrigerant recovery
assembly	unit
Manifold gauge	Recycling unit
assembly	Compressor shaft seal
Service valves	Compressor clutch
Schrader service valve	Compressor service
R-134a service valve	Clogged orifice tube
R-12 service valve	Defective refrigerant
Dust caps	lines
Stem-type service	A/C system evacuation
valve	A/C system charging
Static A/C pressure	Charging station
reading	Component location
Performance test	drawing

ReviewQuestions—Chapter 76

Please do not write in this text. Place your answers on a separate sheet of paper.

1. List six check points when inspecting an air conditioning system.

2. How can feeling line temperatures help you when inspecting an air conditioning system?

3. Which of the following sight glass conditions could indicate system troubles?
   (A) Clear sight glass.
   (B) Bubbling sight glass.
   (C) Cloudy sight glass.
   (D) All of the above.
   (E) None of the above.

4. Explain five important safety precautions for air conditioning system service.

5. What is a pressure gauge or manifold assembly?

6. _____ valves provide a means of connecting the pressure gauge assembly for testing, discharging, evacuating, and charging the system.

7. Describe the difference between Schrader valves and stem type service valves.

8. A static A/C pressure reading will indicate if the compressor is in good condition. True or False?

9. Why is an electronic leak detector safer than a torch type detector?

10. Give a reason why you should not discharge (empty) an A/C system into the air.

ASE-Type Questions

1. A car's air conditioning system is not operating properly. Technician A checks to see if the expansion valve is stuck. Technician B checks the system for a clogged water drain. Who is right?
   (A) A only.
   (B) B only.
   (C) Both A and B.
   (D) Neither A nor B.

2. An automotive air conditioning system is not cooling properly. Technician A checks for a loose compressor drive belt. Technician B looks for blockage in the condenser fins. Who is right?
   (A) A only.
   (B) B only.
   (C) Both A and B.
   (D) Neither A nor B.

3. An automobile's air conditioning is running on high. Technician A says the refrigerant lines on the "high side" of this system should be warm or hot. Technician B says the refrigerant lines on the "high side" of this system should be cool. Who is right?
   (A) A only.
   (B) B only.
   (C) Both A and B.
   (D) Neither A nor B.

4. An automobile air conditioning's low-side line is cold, but the system does not cool. Technician A says this problem may be caused by faulty air control doors. Technician B says this problem

may be caused by an instrument panel control malfunction. Who is right?

(A) A only.
(B) B only.
(C) Both A and B.
(D) Neither A nor B.

5. The condition of a car's air conditioning refrigerant needs to be checked through the system's sight glass. Technician A says the sight glass may be located on top of the receiver-drier. Technician B checks to see if the sight glass is located in a refrigerant line. Who is right?

(A) A only.
(B) B only.
(C) Both A and B.
(D) Neither A nor B.

6. An automotive air conditioning system's sight glass is clear. Technician A says a clear sight glass normally indicates that the system has the correct charge of refrigerant. Technician B says a clear sight glass may indicate an empty system. Who is right?

(A) A only.
(B) B only.
(C) Both A and B.
(D) Neither A nor B.

7. A pickup truck's air conditioning system sight glass is "foamy." Technician A adds refrigerant to the system. Technician B removes some of the refrigerant from the system. Who is right?

(A) A only.
(B) B only.
(C) Both A and B.
(D) Neither A nor B.

8. The sight glass on an auto air conditioning system appears to be "cloudy." Technician A checks the condition of the system's receiver-drier. Technician B says the system is operating properly. Who is right?

(A) A only.
(B) B only.
(C) Both A and B.
(D) Neither A nor B.

9. A car's A/C system is being serviced. Technician A places a shop rag around the fitting before removing a refrigerant line. Technician B wears safety glasses while removing a refrigerant line. Who is right?

(A) A only.
(B) B only.
(C) Both A and B.
(D) Neither A nor B.

10. Technician A uses the high-pressure gauge on a manifold assembly to measure the compressor discharge pressure. Technician B uses the high-pressure gauge on a manifold assembly to measure the A/C's suction-side pressure. Who is right?

(A) A only.
(B) B only.
(C) Both A and B.
(D) Neither A nor B.

11. During a static A/C pressure test, the manifold assembly's high-pressure gauge shows approximately 50 psi. Technician A says this reading indicates a very low refrigerant level in the system. Technician B says this reading indicates an adequate amount of refrigerant in the system. Who is right?

(A) A only.
(B) B only.
(C) Both A and B.
(D) Neither A nor B.

12. All of the following are used to detect automotive A/C system leaks *except:*

(A) Bubble detector.
(B) Torch detector.
(C) Electronic leak detector.
(D) Tube detector.

13. Moisture is found in an automotive air conditioning system. Technician A replaces the receiver-drier. Technician B removes refrigerant from the system. Who is right?

(A) A only.
(B) B only.
(C) Both A and B.
(D) Neither A nor B.

14. An automotive air conditioning system needs to be charged. Before charging this system, Technician A tests the system for leaks. Before charging this system, Technician B evacuates the A/C system. Who is right?

(A) A only.
(B) B only.
(C) Both A and B.
(D) Neither A nor B.

15. An automobile's heater system will not operate. Technician A checks for a clogged heater core. Technician B checks the operation of the heater hose valve. Who is right?

(A) A only.
(B) B only.
(C) Both A and B.
(D) Neither A nor B.

Activities—Chapter 76

1. Talk to the owners of several shops that service auto air conditioners. Ask if they have had to raise the price of servicing systems to cover the cost of adding recovery or recycling equipment. How have their customers reacted? Report to the class on your findings.

2. Demonstrate the proper connections and use of a gauge manifold set for making a static pressure reading on an A/C system.

3. Use a video camera to show proper use of an electronic refrigerant leak detector.

Climate Control System Diagnosis		
Condition	**Possible Causes**	**Correction**
No cooling.	1. Low refrigerant. 2. Refrigerant leak. 3. Bad compressor. 4. Compressor not engaging. 5. Missing drive belt. 6. Bad control circuit.	1. Add refrigerant. 2. Repair leak and recharge. 3. Replace compressor. 4. Repair clutch or sensors. 5. Install belt. 6. Test and repair circuit.
Poor oooling.	1. Low refrigerant 2. Blower will not work on high. 3. Bad air control system. 4. Debris in condenser. 5. Belt slippage.	1. Add refrigerant. 2. Repair blower circuit. 3. Repair air flap system. 4. Clean condenser fins. 5. Tighten compressor belt.
No heating.	1. Blown blower fuse. 2. Open blower switch. 3. Bad blower motor. 4. Fan loose on shaft. 5. Open wiring. 6. Inoperative flow valve.	1. Replace fuse. 2. Replace switch. 3. Replace blower. 4. Repair fan. 5. Repair wiring. 6. Repair or replace coolant valve at heater hose.
Poor heating.	1. Clogged heater core. 2. Inoperative flow valve. 3. Blower will not work on high. 4. Open thermostat. 5. Low engine coolant level. 6. Air in cooling system. 7. Airflow doors inoperative.	1. Flush or replace core. 2. Repair or replace coolant valve at heater hose. 3. Repair blower circuit. 4. Replace thermostat. 5. Add coolant. 6. Bleed out air. 7. Repair controls and doors.
Inadequate defrost.	1. Air flap problem. 2. Disconnected ducts. 3. Blockage in ducts.	1. Adjust air flaps or repair control circuit. 2. Reinstall ducts. 3. Remove debris.
Defrost fogs windshield.	1. AC not working. 2. Leaking heater core. 3. Leaking windshield seal.	1. Repair AC. 2. Replace heater core. 3. Replace or caulk glass.

(Continued)

Climate Control System Diagnosis		
Condition	**Possible Causes**	**Correction**
Abnormal noise.	1. Bad compressor. 2. Too much refrigerant. 3. Belt slippage. 4. Compressor cycling. 5. Loose parts. 6. Bad blower bearings. 7. Loose fan.	1. Replace compressor. 2. Install correct charge. 3. Adjust belt tension. 4. Add refrigerant or repair control circuit. 5. Tighten hold-downs. 6. Replace blower motor. 7. Install or replace blower fan.
Low-side pressure high.	1. Bad expansion valve. 2. Moisture in system. 3. Compressor problems. 4. Restricted suction line.	1. Replace expansion valve. 2. Replace desiccant and recharge system 3. Replace compressor. 4. Replace line.
Low-side pressure low.	1. Low refrigerant. 2. Low airflow. 3. Restricted liquid line. 4. Restricted suction line. 5. Compressor not disengaged.	1. Charge system. 2. Clean condenser fins and check blower operation. 3. Replace line. 4. Replace line. 5. Repair frozen clutch, bad sensor, or control circuit.
High-side pressure high.	1. Low refrigerant, air in system. 2. System overcharged. 3. Clogged condenser core. 4. Engine overheating. 5. Refrigerant system restriction.	1. Evacuate and charge. 2. Properly charge system. 3. Replace condenser. 4. Repair cooling system. 5. Replace parts as needed.
High-side pressure low.	1. Low refrigerant charge. 2. Expansion valve open. 3. Bad compressor.	1. Charge system. 2. Replace valve. 3. Replace compressor.
Water on floor.	1. Leaking heater core. 2. Loose heater hose clamps. 3. Evaporator drain clogged.	1. Replace core. 2. Tighten clamps. 3. Clean out drain tube.

A late-model V-6 engine. (Lexus)

Safety, Security, and Navigation Systems

77. Restraint Systems

78. Restraint System Service

79. Security, Navigation, and Future Systems

Every year, automakers are engineering more safety, security, and navigation devices into their vehicles. This makes it critical that you understand the operation and service of these systems.

Chapters 77 and 78 detail the operation and service of various restraint systems, including seat belts and air bags. Chapter 79 covers security and navigation systems, as well those systems that will be found on vehicles in the not-to-distant future.

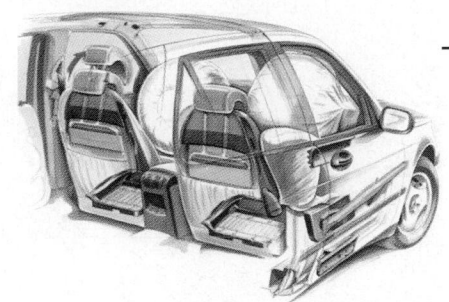

Chapter 77

Restraint Systems

After studying this chapter, you will be able to:

- Explain how vehicle body and frame construction works with restraint systems to protect a vehicle's occupants.

- Identify and locate the most important parts of vehicle restraint systems.

- Describe the purpose for restraint systems.

- Describe restraint system design variations.

- Summarize the operation of restraint system sensors, inflator modules, and electronic control modules.

- Correctly answer ASE certification test questions requiring a knowledge of restraint system operation and construction.

Restraint systems help hold a vehicle's occupants in their seats, protecting them from injury during an accident.

Restraint systems include the seat belts and the air bag system, as well as the vehicle's body, frame, steering column, and dash.

Seat belts and air bags are required on all cars and light trucks. Therefore, it is very important that you understand the operation and repair of these safety systems. This chapter covers the design and operation of common restraint systems. The next chapter summarizes their repair. Study both chapters carefully!

Vehicle Collisions

Vehicle collisions, or *accidents,* normally result from driver error that causes the car or truck to hit other objects (vehicles, trees, retaining walls, etc.). Tremendous force is generated when the vehicle, which can weigh almost two tons, crashes to a halt in a few feet (or meters) of travel, **Figure 77-1.** Automobile manufac-

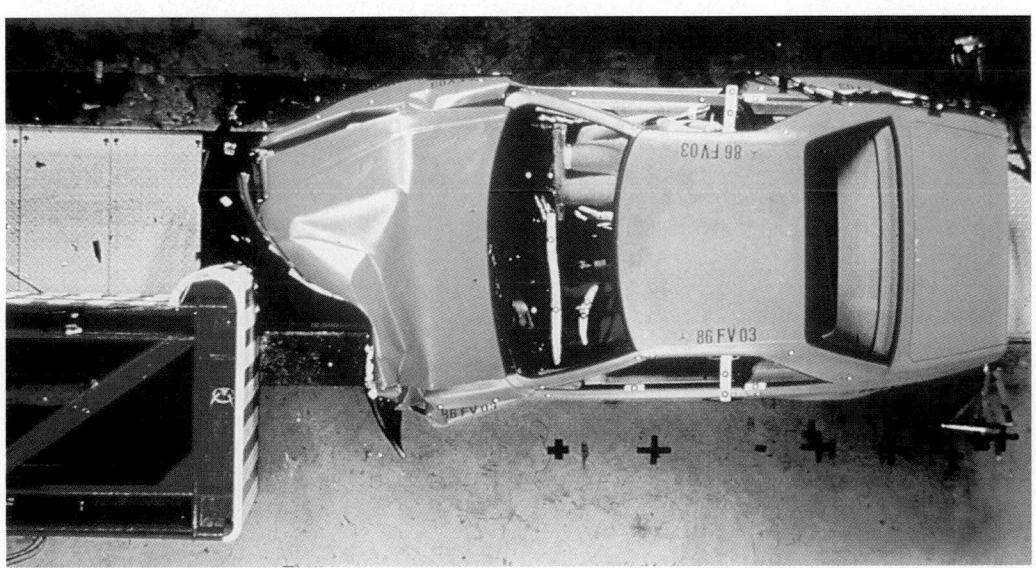

Figure 77-1. When a car or truck hits a large object, a tremendous amount of energy is absorbed to bring the vehicle to a halt. (Mercedes-Benz)

turers design their vehicles to absorb this force in a controlled manner.

Crush zones located at the front and rear of the body-frame assembly are designed to collapse during a severe impact. The passenger compartment is stiffer and stronger than these crush zones, so the occupants are protected from the forces of the accident. See **Figure 77-2.**

Side-impact beams, made of high-strength steel, are mounted in the vehicle's doors to help prevent intrusion into the passenger compartment. The *pillars* (body sections that extend up to the roof panel and are located in front of and behind the doors) are also strengthened to protect the passenger compartment, especially in the event of a rollover.

Crash tests are used by the auto manufacturer to measure how well the body-frame structure will protect the vehicle's occupants in the event of a major collision. The vehicle is crashed into a stationary wall and, sometimes, other vehicles to measure how well the vehicle withstands and reacts to the impact forces.

Crash test dummies are used to measure the impact forces acting upon the human body. Sensors in the test dummy record the forces acting upon vital parts of the body. This allows the manufacturer to estimate how badly people will be hurt during similar crash conditions, **Figure 77-3.**

Crash tests are performed from the front, side, and rear of the vehicle. This allows the manufacturer to study the effects of major impact forces from each direction. See **Figure 77-4.**

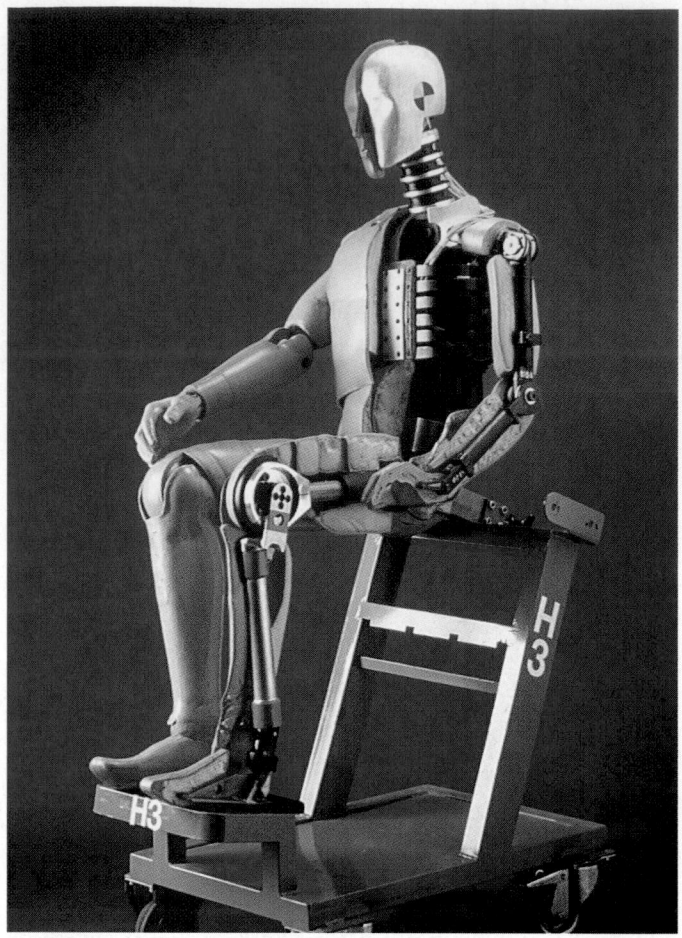

Figure 77-3. A crash test dummy is designed to measure forces that would act on the human body during an auto accident. Damage to the dummy indicates how badly people would be hurt in a similar accident. (Volvo)

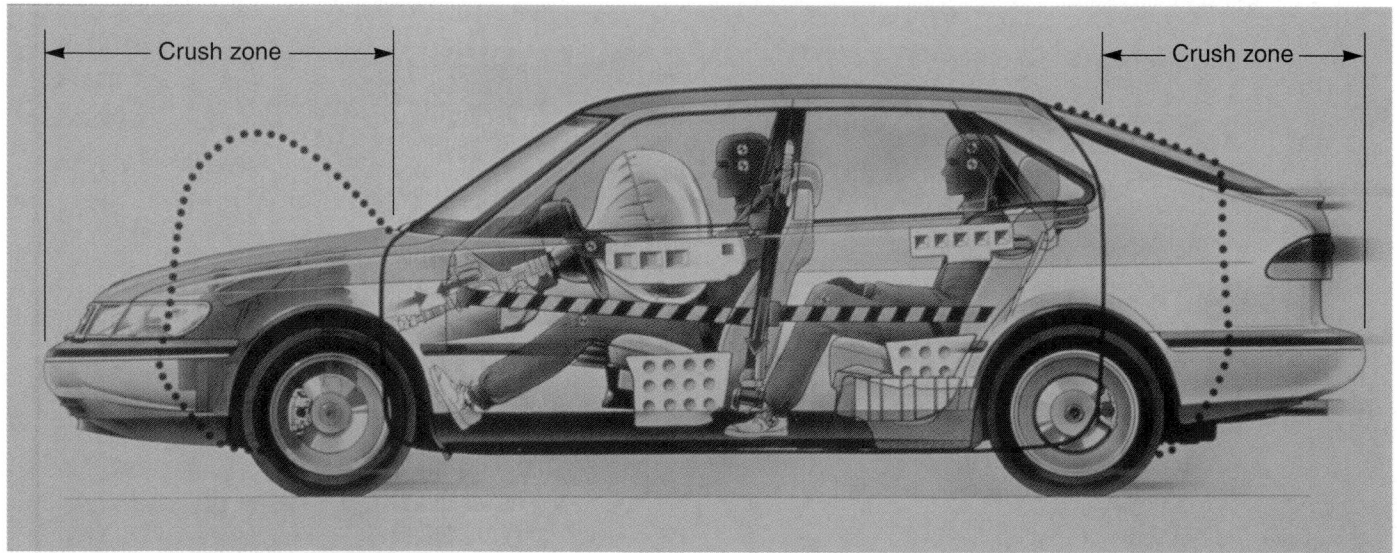

Figure 77-2. Crush zones are built into the front and rear areas of the vehicle's body and frame. These areas crush more easily than the passenger compartment, increasing occupant safety. Dotted lines show how the body deforms during a major collision. (Saab)

A

B

C

Figure 77-4. Auto makers perform extensive crash tests to analyze the safety of new car and truck designs. A—Frontal-impact crash test. B—Side-impact crash test. C—Rear-impact crash test. (Saab)

Active and Passive Restraints

Mentioned briefly, restraint systems hold the driver and passengers in their seats during an accident. These systems are designed to limit injury during a crash. Most injuries result when people are ejected from their seats or from the passenger compartment upon impact.

An *active restraint system* requires that the occupants ready the system. For example, seat belts that must be buckled manually are classified as an active system.

A *passive restraint system* operates without driver or passenger activation. Air bags and automatic seat belts are passive restraint systems. Refer to **Figure 77-5.**

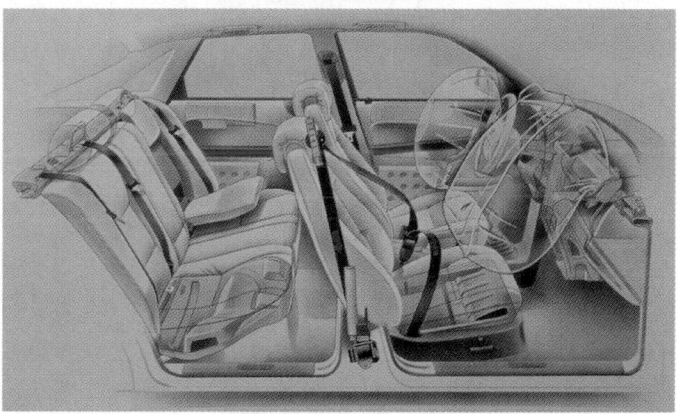

Figure 77-5. Modern vehicles are much safer than those produced in the past. Besides strong passenger compartments, seat belts and air bags help protect the driver and passengers from injury during collisions. (Saab)

Seat Belt Systems

Seat belts are strong nylon straps that hold people in their seats during a collision. *Lap belts* extend across a person's lap. *Shoulder belts* extend over a person's shoulder and chest.

The *seat belt buckle* allows you to engage and disengage the belt around your body. *Seat belt anchors* provide a means of bolting the seat belt to the car's body structure. Look at **Figure 77-6.**

A *belt retractor* takes the slack out of the seat belt so the belt fits snugly around the body. Designs vary. See **Figure 77-7.**

A *seat belt reminder system* lights a dash light and generates an audible tone to warn the vehicle's occupants to buckle their seat belts.

Knee Diverter

A *knee diverter,* or *knee bolster,* is formed into the lower part of the dash to protect the driver's and the front

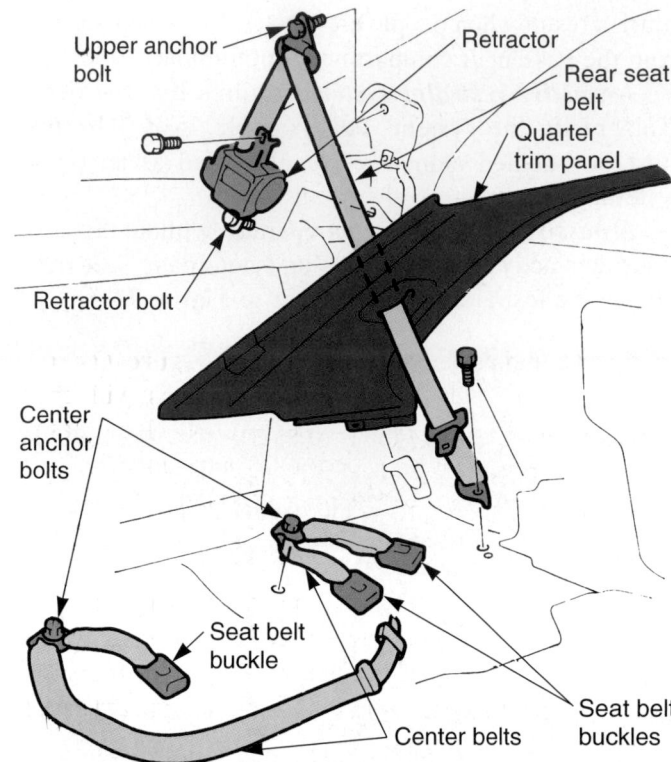

Figure 77-6. Note the basic parts of a seat belt assembly. (Honda)

passengers' knees from being injured on the metal frame of the dash. The diverter also prevents the driver and passengers from sliding under the air bag during a collision. It is usually a thick plastic panel that covers the metal frame of the dash, **Figure 77-8.**

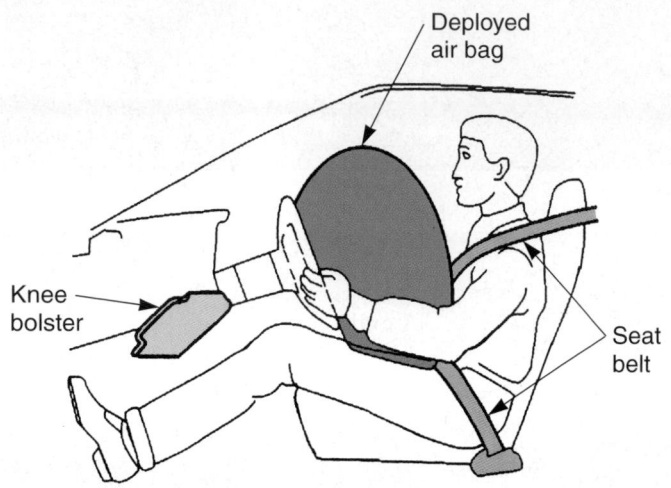

Figure 77-8. This drawing shows the three primary restraint devices: seat belts, air bag, and knee bolster. (General Motors)

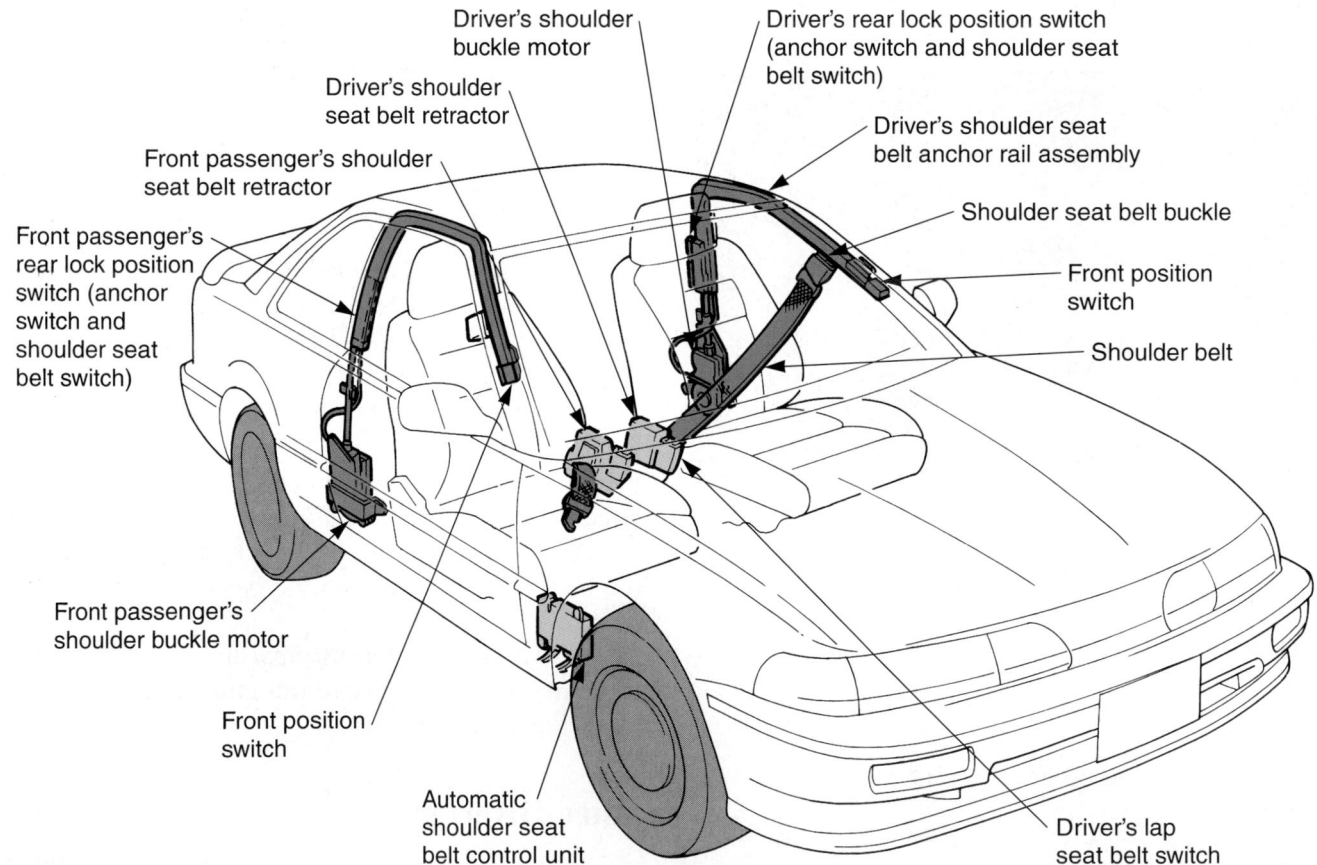

Figure 77-7. This vehicle has automatic shoulder belts that are tightened around the body by small electric motors. Note the location of the retractors. (Honda)

Air Bag Systems

An *air bag system* automatically inflates large nylon bags immediately after the start of a major collision. The air bag system is designed to supplement the protection provided by seat belts, **Figure 77-9.**

The major parts of an air bag system include:

- *Air bag sensors*—inertia sensors that signal the control module in the event of a collision.

- *Air bag module*—contains the inflator mechanism and the nylon air bag that expands to protect the driver and/or passengers during the collision.

- *Air bag controller*—computer that operates the system and detects faults.

- *Dash warning lamp*—Dash bulb that glows with system problem and goes out as system arms.

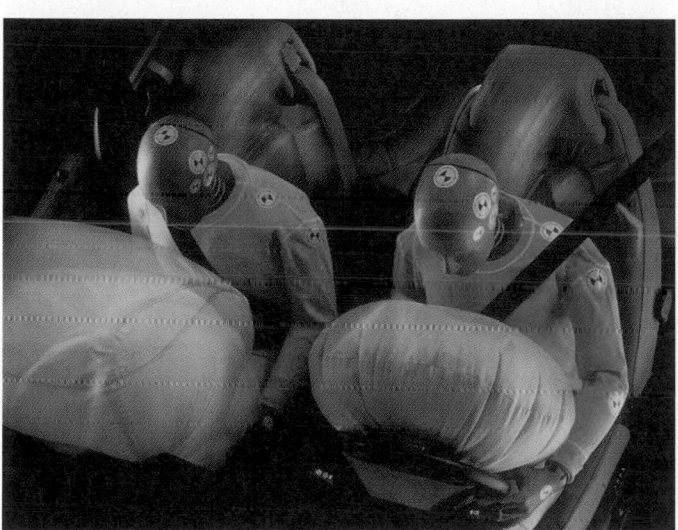

Figure 77-9. An air bag can shoot out at speeds up to 200 miles per hour (320 kilometers per hour). This is fast enough to inflate before the human body can fly forward in even the worst auto accidents. (Saab)

An electronic air bag system uses an inflatable nylon bag to help protect the driver during a collision. It uses impact sensors to detect a severe collision. The sensors feed their signals to the air bag controller. When at least two impact sensors are energized, the controller activates the air bag module.

The air bag inflates in about 1/20th of a second, well before the driver's body flies forward from the collision. The tough nylon bag can easily absorb the forward inertia of a human body. This helps protect both the driver and the passengers.

Air Bag Types

Some older cars use only one air bag, which is located in the steering wheel. This is termed a *driver-side air bag*.

Many new vehicles are equipped with dual air bags: a driver-side air bag and a *passenger-side air bag.* The passenger-side air bag deploys from the right side of the dash. It is much larger than a driver's-side air bag, since it is relatively far from the passengers and may have to protect two humans simultaneously. See **Figure 77-10.**

The driver- and passenger-side air bags will only deploy during frontal impacts. Steering wheel and dash-mounted air bags may *not* deploy in side impacts, rear impacts, or rollover situations. A collision must occur within about 30° of the vehicle's centerline for these air bags to inflate. This is illustrated in **Figure 77-11.**

Side-impact air bags can be located in the door panels or on the outside edge of each front seat. These small air bags deploy when the vehicle is hit from the side. They generally do not deploy during a frontal impact. Side-impact air bags are becoming more common and are used by several auto manufacturers, **Figure 77-12.**

When a vehicle is hit from the side, injury usually results when the occupant's shoulder and head fly through the side window glass. A side-impact air bag

Figure 77-10. The passenger-side air bag is much larger than other air bags. It must be able to protect two people from striking the dash and windshield during a frontal impact. (Volvo)

Direction of impact

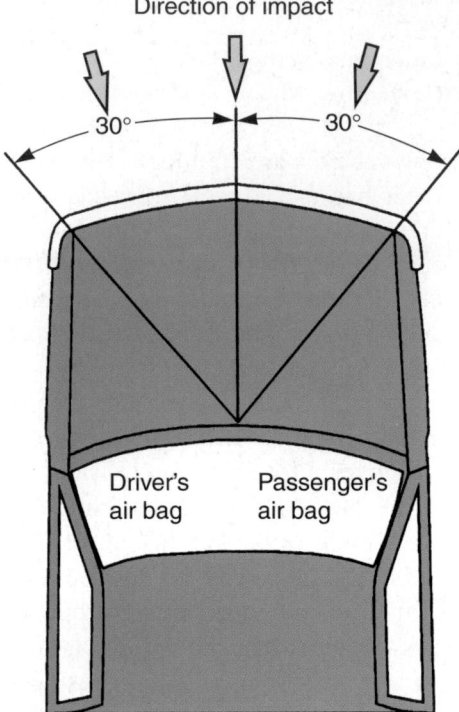

Figure 77-11. Driver's and passenger's air bags will normally deploy if impact is within 30° of a vehicle's centerline. (GM Trucks)

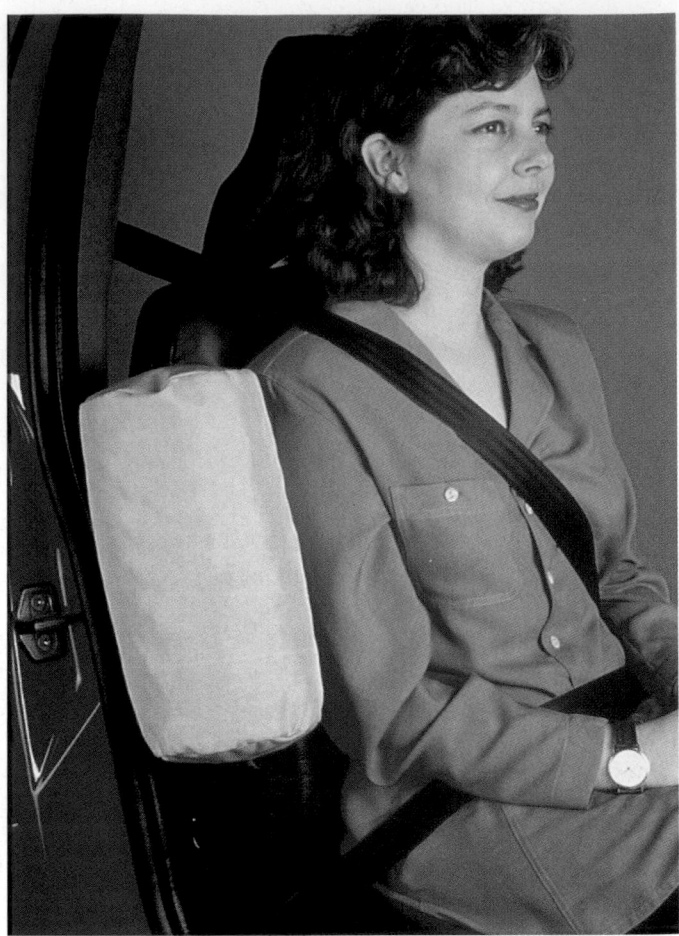

Figure 77-12. Side-impact air bags are becoming more common. They help hold your body in the seat when the vehicle is hit from the side. This unit deploys from the side of the seat cushion. (Volvo)

system senses the side thrust of the impact and deploys a small air bag to cushion the person as he or she is propelled sideways. See **Figure 77-13.**

Window air bags are designed to drop down like "curtains" over the side window glass. This helps protect the occupants from head and facial injuries caused by impact with the door glass.

Rear seat air bags fit into the rear cushion of the front seats. They inflate to protect the passengers in the rear seat from injury during a frontal collision. Although not very common, they can be found in a few expensive luxury cars.

Air Bag Module

The *air bag module* consists of a nylon bag and an igniter-inflator unit enclosed in a metal and plastic housing. The driver's-side air bag module is packaged in the center of the steering wheel pad. The passenger-side air bag module is mounted under a small door formed in the right side of the dash pad. Look at **Figure 77-14.**

The *air bag* itself is a strong, reinforced nylon sack attached to the metal frame of the module. It is tightly folded for storage in the steering wheel pad, the dash, the door panel, or the side of the seat, **Figure 77-15.**

Air bag vent holes allow for rapid deflation of the air bag after deployment. These small holes are formed

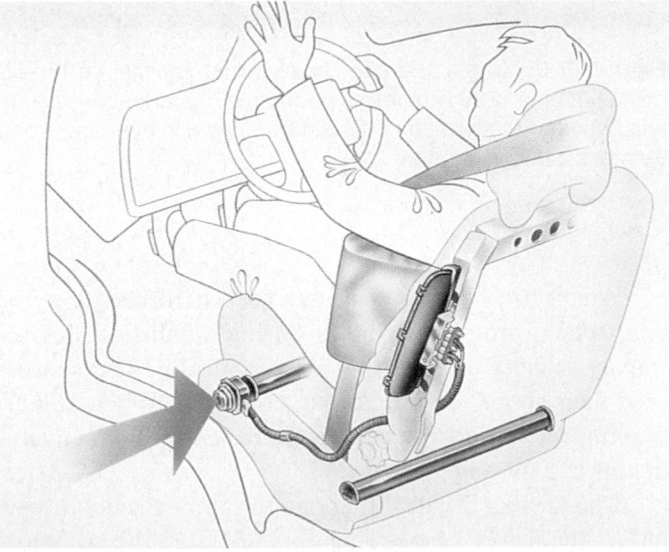

Figure 77-13. Side-impact air bags normally use their own sensors to detect side-impact forces. They only deploy when the impact force would tend to throw the body sideways and into the door. (Volvo)

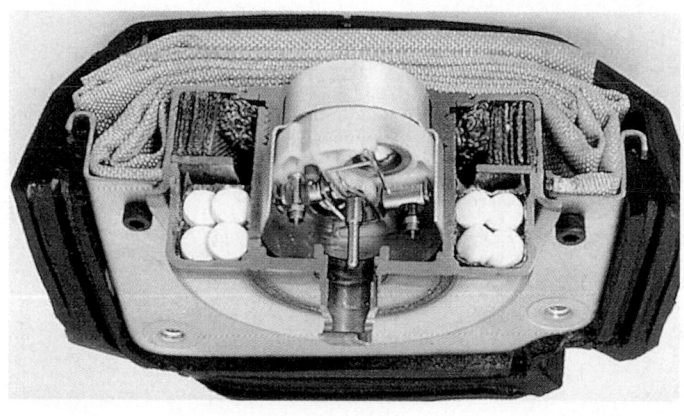

Figure 77-14. Cutaway shows inside of an air bag module. The air bag is folded neatly under the steering wheel cover. The air bag igniter generates a small electric arc across two metal pins when an electrical signal is sent to igniter from the controller. The arc fires an igniter charge, causing the gas-generating pellets to burn. The burning pellets generate a rapidly expanding gas that inflates the air bag. (Breed Automotive Corporation)

around the outer edge or back of the bag.

The *air bag igniter* generates a small electric arc when an electrical signal is sent to it from the air bag controller. The arc forms across two small pins in the igniter and fires an igniter charge. This flash of the igniter charge causes the gas-generating pellets to burn, generating a huge volume of expanding gas, **Figure 77-15.**

The air bag's *propellant charge,* or expanding gas, is usually produced by the burning of sodium azide pellets in the air bag module. The burning pellets form nitrogen gas, which inflates the air bag.

The large volume of nitrogen gas can inflate the air bag in a fraction of a second. This forces the steering wheel cover to split open, and the air bag shoots out at about 200 miles per hour (320 kilometers per hour). This is fast enough to cushion the forward thrust of the human body as it flies forward after the collision. The air bag also protects the driver's or passengers' heads from flying objects resulting from the accident.

Immediately after the air bag is inflated, the gas is vented out the small holes on the sides or rear of the bag. This prevents the occupants from being pinned in their seats. It also allows the driver to see out of the windshield right after deployment.

Passenger-side and side-impact air bags are very similar in design. The passenger-side air bag is much larger than a side-impact bag and, therefore, requires more gas for proper inflation. An exploded view of a passenger-side air bag is shown in **Figure 77-16.**

Hybrid Air Bags

A *hybrid air bag* uses a small explosive charge and a pressurized gas cartridge to cause deployment. The small, metal *gas cartridge* contains inert argon gas pressurized to 3000 psi (20 700 kPa).

When the air bag controller sends current to this type air bag module, it fires a tiny amount of pyrotechnic mate-

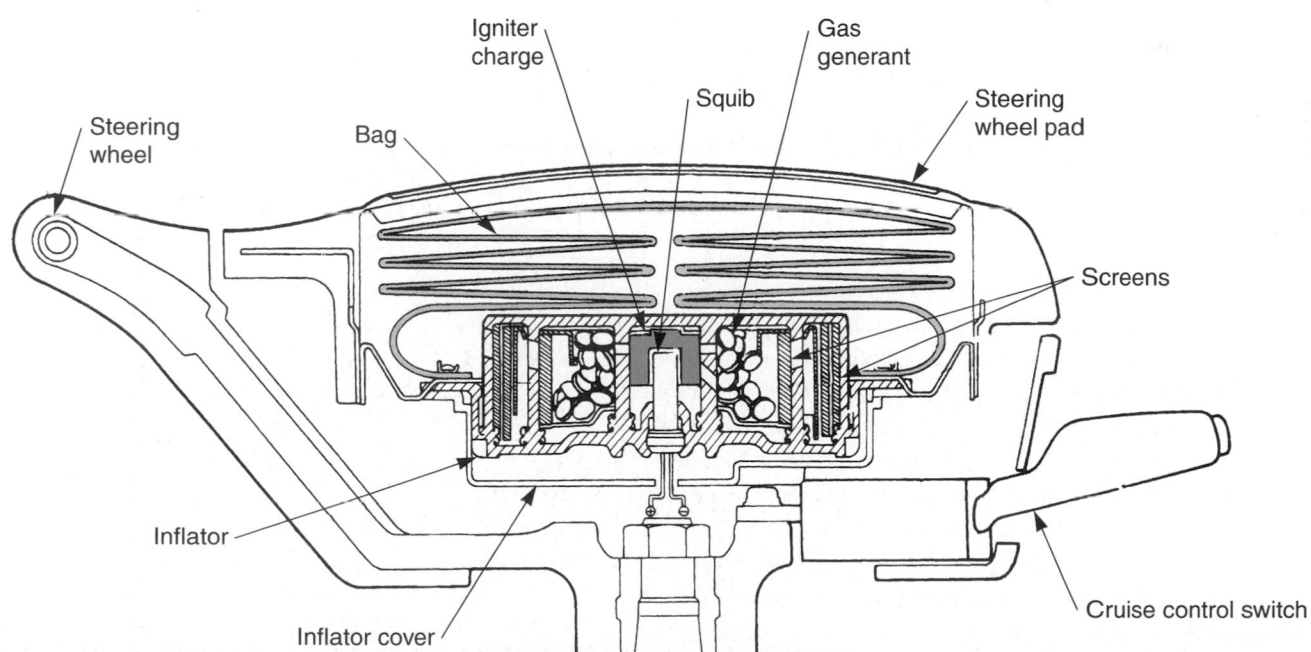

Figure 77-15. Line drawing shows the details of a common air bag module. Study the part names and locations carefully. (Toyota)

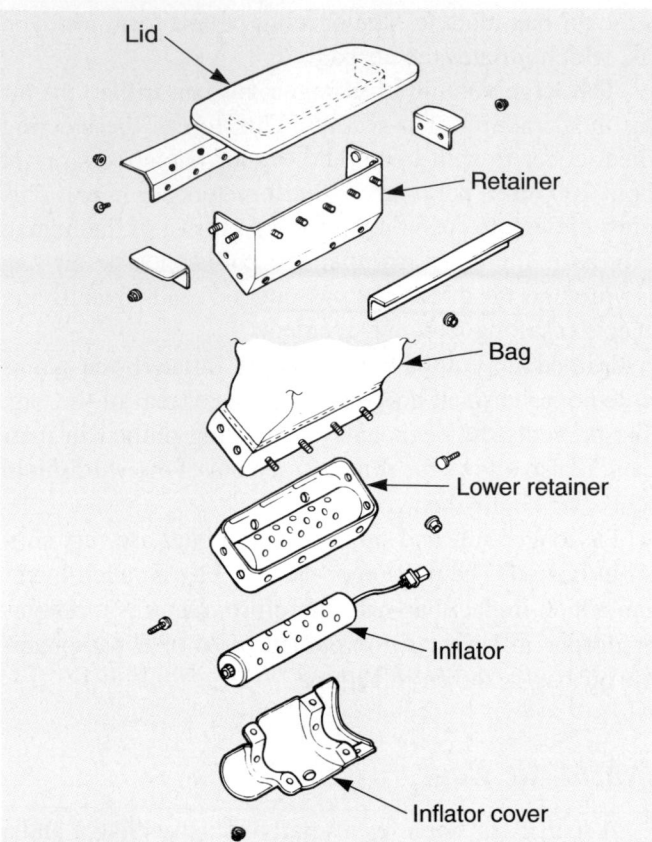

Figure 77-16. Exploded view shows the major parts of a passenger-side air bag, which fits in the right side of the dash. (Honda)

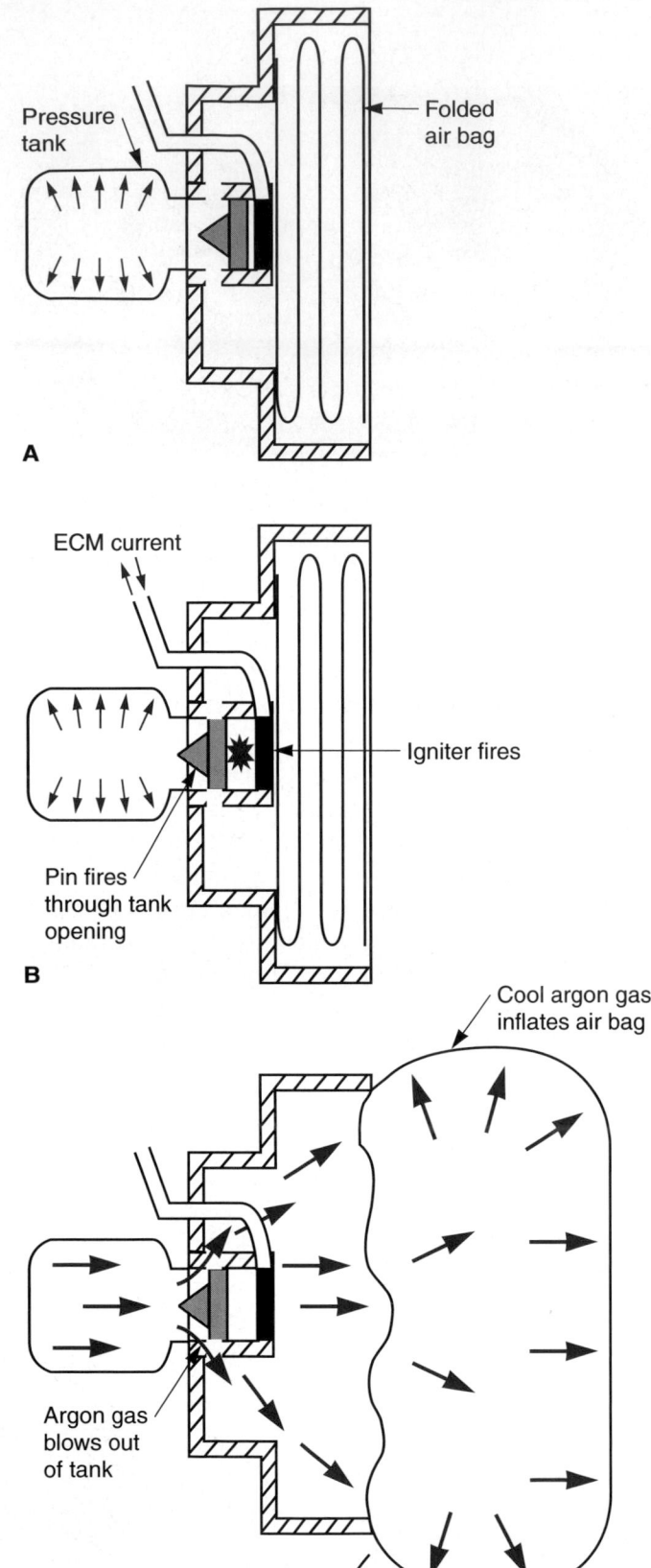

rial (rapidly burning substance) that forces a "plastic bullet" through the gas cartridge. The cool, pressurized argon gas then blows out to inflate the air bag, **Figure 77-17.**

A hybrid air bag is designed to help prevent minor skin burns that can result from the hot nitrogen gas generated by burning sodium azide pellets.

Mechanical Air Bags

All parts of a *mechanical air bag system* are contained in the steering wheel module. During an front-end collision, a metal ball in the module slams forward, striking a lever arm. The other end of the lever arm then pushes a firing pin into the igniter material. The igniter material burns, igniting the sodium azide pellets. The gas generated by the burning pellets quickly inflates the air bag.

Mechanical air bag systems are designed for aftermarket installations. Older vehicles, which were not originally equipped with air bags, can be retrofitted with the mechanical air bag system to increase driver protection during an accident.

Figure 77-17. A hybrid air bag uses both pyrotechnic material and a container of pressurized gas to cause air bag inflation. A—The metal tank is filled with argon gas. The igniter is mounted on one end of the tank. The air bag is folded inside the steering wheel cover. B—The controller detects a collision and sends current to the igniter. The igniter fires, forcing a plastic projectile through the end of the tank. C—A hole formed in the end of the tank allows cool argon gas to blow out and inflate the air bag.

Air Bag Sensors

Air bag sensors detect a collision by measuring vehicle deceleration during a collision. They are inertia sensors that detect a rapid change in speed or velocity. One or more sensors are commonly incorporated in air bag systems. The trend is to use fewer sensors than in the past.

Impact sensors are mounted in the front of the vehicle to first detect a collision. They are often located in the engine compartment, on or near the radiator support, **Figure 77-18.**

A *safing sensor,* or *arming sensor,* is a backup sensor designed to ensure that the vehicle is actually in a collision. It provides a fail-safe system to prevent accidental bag deployment. For the inflation of the air bags, the safing sensor and at least one impact sensor must be closed, **Figure 77-19.**

Magnet-and-Ball Sensors

A *magnet-and-ball sensor* is used in some air bag systems as the impact sensor. A small permanent magnet is used to hold a steel ball away from the electrical contacts in the sensor. During a severe collision, the rapid deceleration throws the steel ball forward, overcoming the force of the magnet. The ball rolls forward and strikes electrical contacts. This closes the sensor circuit to signal the controller of a possible collision requiring air bag deployment. See **Figure 77-20.**

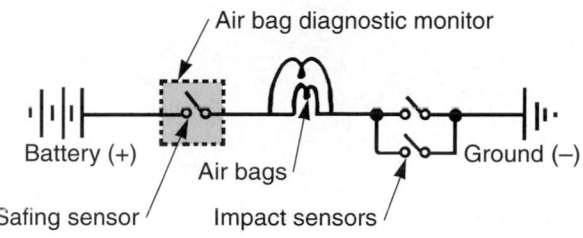

Air bag firing circuit diagram

Figure 77-19. Note the basic circuit for air bag sensors. A safing sensor and two primary impact sensors are wired in series. This requires that the safing sensor and at least one of the impact sensors be closed to fire air bag. (Ford)

Coil Spring Sensors

A *coil spring sensor* uses a small metal weight attached to a metal coil spring. During a severe frontal impact, the weight is thrown forward with enough force to overcome spring tension. This weight touches the sensor contacts and closes the circuit to the ECM.

Seat Cushion Sensors

Seat cushion sensors detect the weight of a person sitting in the passenger seat. If no one is sitting in the passenger seat, the air bag system may not deploy the passenger air bag. This saves the considerable cost of replacing an air bag without it protecting someone.

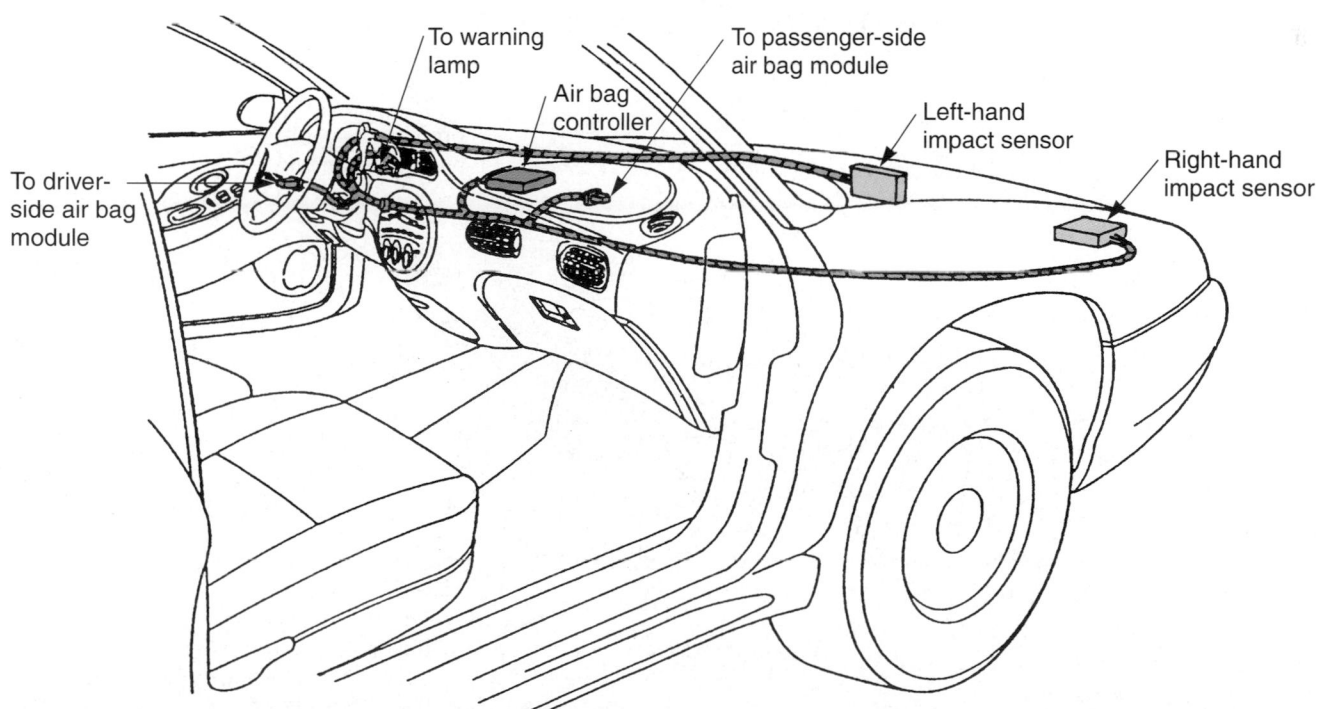

Figure 77-18. Typical location of the air bag impact sensors. They are often mounted near the radiator support. (Ford)

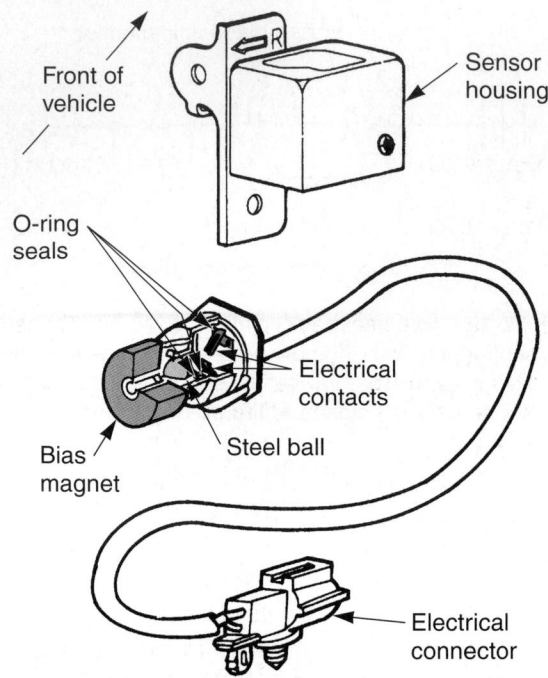

Figure 77-20. An air bag system sensor closes when exposed to rapid deceleration forces. This sensor uses a magnet to hold a steel ball. If impact is great enough, the steel ball is thrown away from the magnet and into the two metal contacts. This closes the circuit and signals the controller that the vehicle is in a collision. (Ford)

Accelerometer Sensors

Many late-model vehicles are equipped with one central *accelerometer* (inertia sensor) instead of separate impact and safing sensors. The accelerometer measures changes in motion or deceleration and is sometimes mounted in the air bag controller. Some accelerometers contain a thin wafer of semiconductor material that is deflected and warped by rapid deceleration. The bending of the semiconductor produces a piezo-electric, or pressure-generated, electrical signal that fires the air bag.

Air Bag Controller

The *air bag controller,* or *air bag control module,* uses inputs from the impact and safing sensors to determine if air bag deployment is needed. If at least one impact sensor and the safing sensor are closed, the controller sends a high current pulse to the air bag module. The pulse produces a small electric arc in the air bag module, igniting the pyrotechnic material to produce gas expansion and bag inflation, **Figure 77-21.**

The air bag controller also generates trouble codes and energizes a warning lamp if it detects something wrong with the system. Refer to **Figure 77-22.**

A *smart restraint system* uses additional inputs to affect the operation of the air bags. It uses conventional impact sensor inputs, as well as data from the seat sensors, side door impact sensors, yaw sensors, wheel speed sensors, seat belt sensors, and even collision-predicting sensors. This allows the smart system to adjust the speed and pressure applied to the air bags to better protect the vehicle's occupants from injury. For example, a small child would require less air bag inflation pressure than a very large adult.

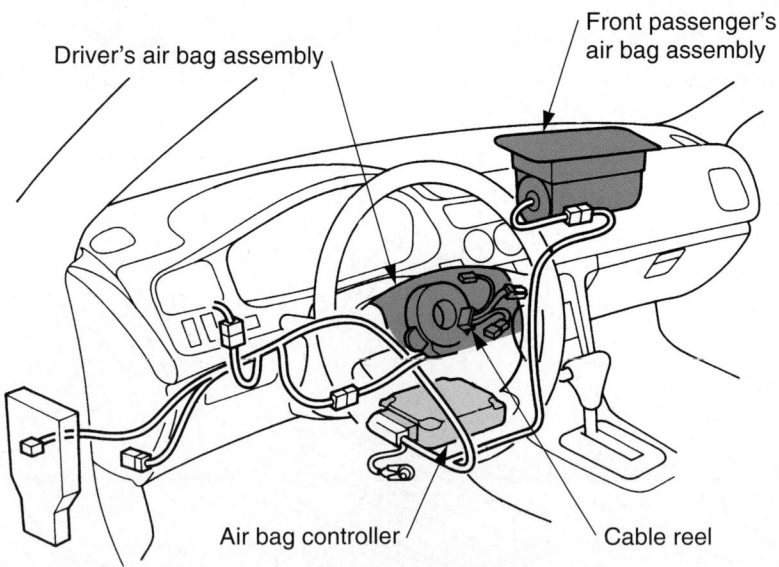

Figure 77-21. Study the general arrangement of the parts in this air bag system. Note the location of the controller. (Honda)

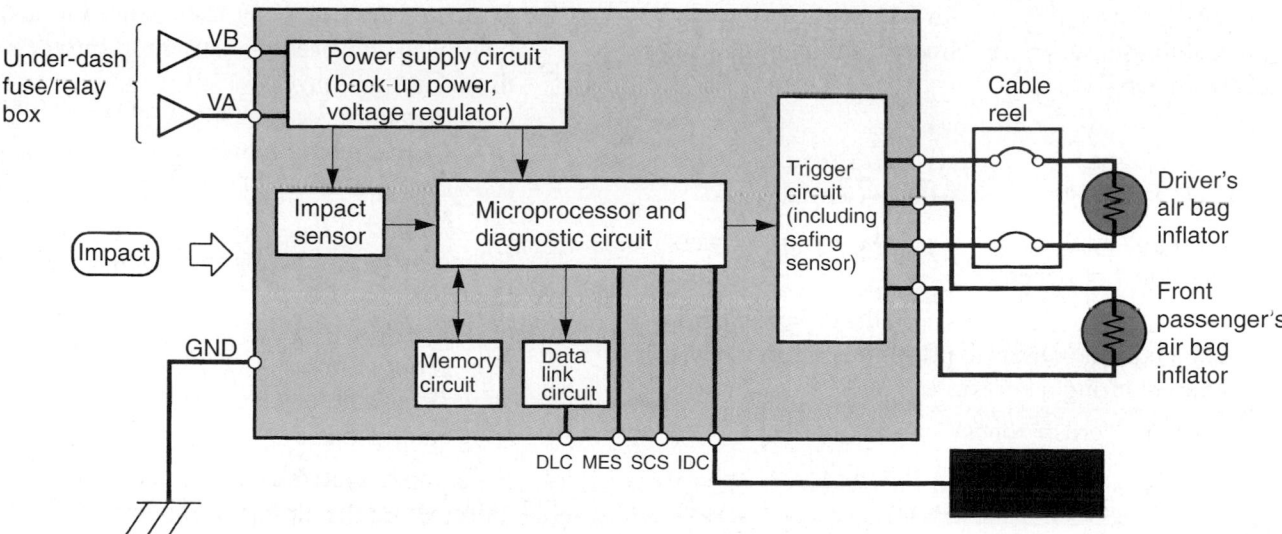

Figure 77-22. Diagram shows how the controller operates the air bags. Note that the controller has self-diagnostic capabilities and will generate a trouble code when an air bag system problem is detected. (Honda)

Summary

- Crush zones located at the front and rear of the body-frame assembly are designed to collapse during a severe impact.
- Crash tests are used by the auto manufacturer to measure how well the body-frame structure will protect the driver and passengers in a major collision.
- Crash test dummies are used to measure the forces acting upon the human body during a collision.
- Seat belts are strong nylon straps that hold people in their seats during a collision.
- A knee diverter is formed into the lower part of the dash to prevent the driver's and front passengers' knees from being injured on the metal frame of the dash.
- An air bag system automatically deploys a large nylon bag immediately after a major collision.
- Most vehicles are equipped with a passenger-side air bag, which deploys from the right side of the dash.
- Side-impact air bags can be located in the door panels or on the outside edges of the front seats.
- The air bag module comprises a nylon bag and an igniter-inflator unit enclosed in a metal-plastic housing.
- A hybrid air bag uses a small explosive charge and a pressurized container of gas to cause deployment.

- All parts of a mechanical air bag system are contained in the steering wheel module.
- Air bag sensors detect a collision by measuring vehicle deceleration.
- The trend is to replace several safing and impact sensors with one central accelerometer that measures changes in motion or deceleration.
- A smart restraint system uses additional inputs to affect the operation of the air bag system.

Important Terms

Restraint system	Knee diverter
Vehicle collisions	Air bag system
Accidents	Driver-side air bag
Crush zones	Passenger-side air bag
Side-impact beams	Side-impact air bags
Pillars	Rear seat air bags
Crash tests	Air bag module
Crash test dummies	Air bag
Active restraint system	Air bag vent holes
Passive restraint system	Air bag igniter
Seat belts	Propellant charge
Lap belts	Hybrid air bag
Shoulder belts	Gas cartridge
Seat belt buckle	Mechanical air bag system
Seat belt anchors	Air bag sensors
Belt retractor	Impact sensors
Seat belt reminder system	Safing sensor
	Magnet-and-ball sensor

Coil spring sensor   Air bag controller
Seat cushion sensors   Smart restraint system
Accelerometer

Review Questions—Chapter 77

Please do not write in this text. Place your answers on a separate sheet of paper.

1. A(n) _____ _____ helps hold the driver and passengers in their seats to prevent them from being injured during a collision.

2. What are "crush zones?"

3. The _____ takes the slack out of the seat belt, so it fits around the body snugly.
   (A) anchor.
   (B) buckle.
   (C) retractor.
   (D) propellant.

4. Name the major parts of an air bag system.

5. The largest air bag in a dual air bag system is the _____.
   (A) driver-side air bag.
   (B) passenger-side air bag.
   (C) side-impact air bag.
   (D) rear seat air bag.

6. Describe the air bag module.

7. How fast can an air bag travel during deployment?
   (A) 100 mph (160 km/h).
   (B) 150 mph (240 km/h).
   (C) 200 mph (320 km/h).
   (D) 300 mph (480 km/h).

8. Explain the operation of a hybrid air bag system.

9. Air bag sensors detect a collision by measuring _____ of the vehicle during a collision.

10. The trend is to replace several air bag sensors with a central _____.

ASE-Type Questions

1. Technician A says manual seat belts are passive restraints. Technician B says air bags are active restraints. Who is right?
   (A) A only.
   (B) B only.
   (C) Both A and B.
   (D) Neither A nor B.

2. Which of the following tests is used to analyze how well the vehicle structure protects the driver and passengers?
   (A) Beam test.
   (B) Crush zone test.
   (C) Crash test.
   (D) Dummy test.

3. Which of the following parts helps prevent leg injuries?
   (A) Knee diverter.
   (B) Pedal cushions.
   (C) Tilt wheel.
   (D) Padded dash.

4. With most systems, how many sensors must close before the air bag will deploy?
   (A) One.
   (B) Two.
   (C) Three.
   (D) Four.

5. Technician A says that the driver-side air bag may not deploy if the direction of impact is more than about 30° from the vehicle's centerline. Technician B says that side air bags may not deploy during a frontal impact. Who is right?
   (A) A only.
   (B) B only.
   (C) Both A and B.
   (D) Neither A nor B.

6. All of the following are found in motor vehicles *except:*
   (A) rear seat airbags.
   (B) knee air bags.
   (C) side air bags.
   (D) window air bags.

7. All parts of this system are contained in the steering wheel module.
   (A) Electronic air bag system.
   (B) Hybrid air bag system.
   (C) Pellet air bag system.
   (D) Mechanical air bag system.

8. Technician A says that the air bag controller is usually located on the vehicle's steering wheel. Technician B says that all major parts of a mechanical system are located in the steering wheel. Who is right?
   (A) A only.
   (B) B only.
   (C) Both A and B.
   (D) Neither A nor B.

9. Which of the following is *not* an air bag sensor?
   (A) *Arming sensor.*
   (B) *Safing sensor.*
   (C) *Accelerometer sensor.*
   (D) *Proximity sensor.*

10. Air bag sensors are usually located in the:
    (A) *engine compartment.*
    (B) *passenger compartment.*
    (C) *doors.*
    (D) *All of the above.*

Activities—Chapter 77

1. Visit a local salvage yard. Ask to inspect deployed air bags in wrecked vehicles. Write a report on the amount of damage to the vehicle.

2. Read service manual literature on air bags.

3. Visit the library and look up articles in periodicals on air bags. Give a presentation to the class about interesting facts found in these articles.

Chapter 78

Restraint System Service

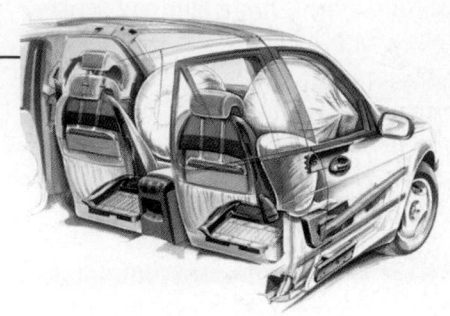

After studying this chapter, you will be able to:

- Explain how to inspect and repair seat belts.
- Summarize how to scan restraint systems for problems.
- Describe safety rules for working with air bags.
- Summarize the procedure for air bag replacement.
- Explain how to replace air bag sensors.
- Describe how to service an air bag controller.
- Correctly answer ASE certification test questions about the diagnosis and repair of restraint systems.

The proper operation of seat belts and air bags is critical to the safety of the driver and his or her passengers. If you make the slightest mistake when servicing these units, you will endanger the people who trust your work.

An air bag also has the potential to cause serious injury to you and others in the shop if handled improperly. Air bags "pack a powerful punch" when deployed. This chapter will summarize the most important procedures for correctly servicing air bags and seat belts. Study it closely.

Warning
Warn customers of the danger of placing children in front seats of vehicles equipped with dual air bags. Air bags have been blamed for the deaths of several small children. Children should be placed in the rear seats of the vehicles and, if necessary, buckled in an approved child seat.

Seat Belt Service

Seat belts can be damaged by the force of a collision and may deteriorate over time. Therefore, seat belt mechanisms should be inspected after an accident and after prolonged use.

Seat Belt Inspection

To inspect a seat belt for damage, pull the belt all the way out of its retractor. Check the webbing for cuts, pulled threads, broken thread loops, color fading, and fabric deterioration, **Figure 78-1.** Look for twisted webbing, which could make the belt stick in its retractor or guide plates. Untwist the belt so it moves freely in the retractor and guides.

Fit the two ends of the seat belt buckle together. Listen for a solid click. Then, try to pull the buckle apart by pulling sharply on the belt. The two ends must not disengage. Make sure the belt buckle will release with normal finger pressure. Finally, check the belt anchors for problems (bent, cracked, loose, etc.).

Replace the seat belt assembly if problems are found.

Seat Belt Service

Seat belt service generally involves removing the damaged belt and replacing it with a new seat belt assembly. When replacing a seat belt, remember the following precautions:

1. Sharp edges on sheet metal body parts and on tools can damage seat belts. Even small cuts can weaken the belt webbing, causing it to snap during a collision.

2. Never attempt to straighten any portion of the belt buckle, latch plate, or anchor plate. If bent, the components should be replaced.

3. Inspect seat belt anchors for bending, cracking, or looseness. When replacing anchors, use the correct case-hardened bolts.

4. Use a torque wrench to tighten all seat belt anchor bolts to factory specifications. See **Figure 78-2.**

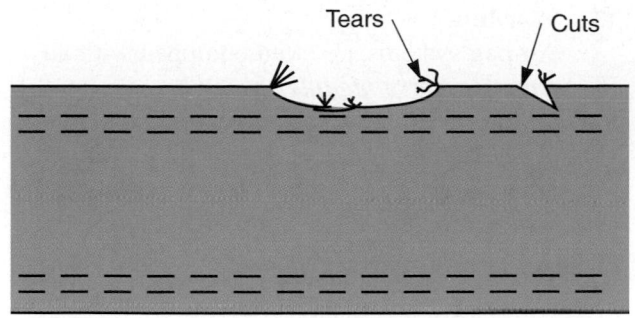

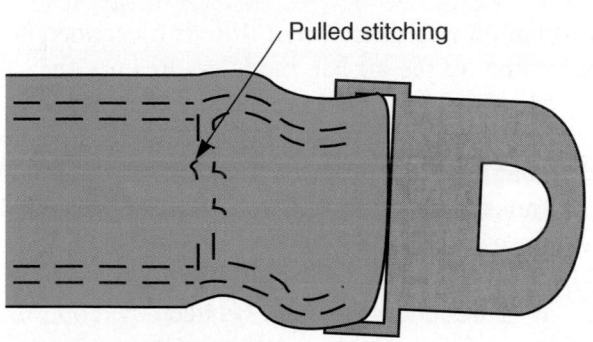

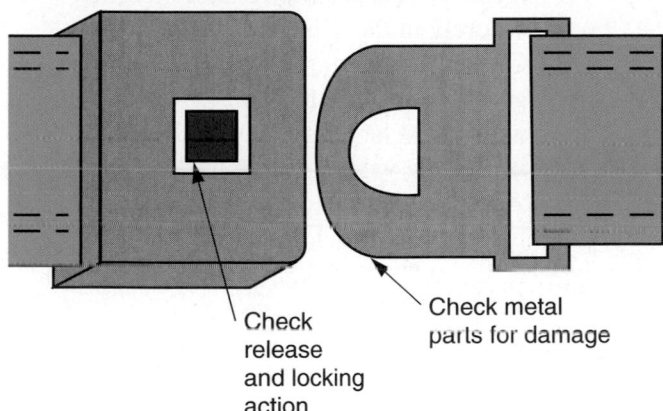

Figure 78-1. Inspect seat belt assemblies for these types of problems.

Belt Retractor Service

Seat belts commonly use one of two belt retractor mechanisms: a centrifugal retractor or a pendulum retractor.

A *centrifugal belt retractor* uses centrifugal force to keep the belt from extending. If you pull on the belt hard enough, the rapid centrifugal force engages the locking mechanism so the belt holds the driver or passenger in the seat.

To check the action of this type belt, simply pull out on the belt quickly. It should lock, and you should not be able to extend the belt further. Pulling slowly should allow free lengthening of the belt out of the retractor.

A *pendulum belt retractor* uses the inertia of vehicle deceleration to engage the belt locking mechanism. When the vehicle brakes or stops quickly, a small pendulum in the retractor is thrown forward, causing the belt locking mechanism to engage.

To test this type of belt, find an open road or an empty parking lot. Drive the vehicle at approximately 10 mph and then hit the brakes hard. The belt should lock up as the vehicle comes to a stop. If the belt still does not lock, the retractor should be replaced.

Tech Tip
Most manufacturers do not recommend the repair of seat belt parts. Defective parts should be replaced.

Air Bag System Service

Air bag service is needed after deployment or when the dash malfunction light indicates a problem. For your safety and the safety of the people relying on the air bags, proper repair methods must be followed.

Warning
Remember to be cautious to prevent accidental air bag deployment when working. An air bag has enough force to break bones and cause serious injury. Keep your head and arms to one side of the air bag when testing circuits, removing an undeployed bag, disconnecting sensors, and doing similar tasks. Never place your head close to the steering wheel or reach inside the wheel spokes during service.

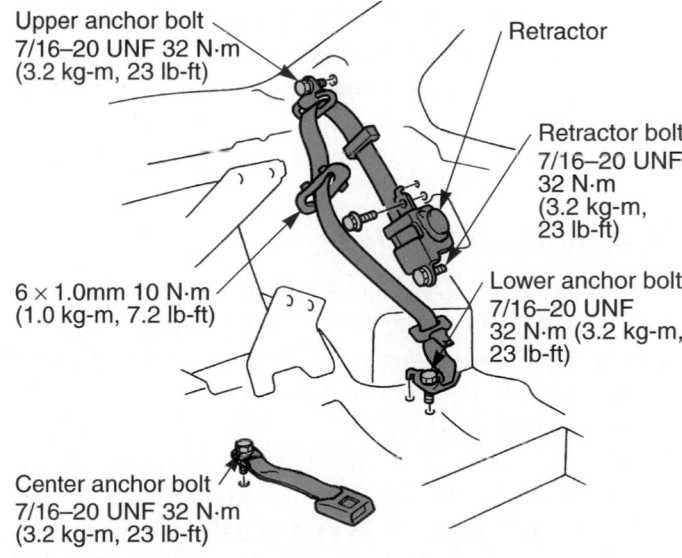

Figure 78-2. Always tighten the seat belt anchor bolts and retractor bolts to specifications.

Scanning an Air Bag System

If the air bag indicator light glows and the bag is not deployed, use a scan tool to analyze the system. Connect the scan tool to the vehicle's diagnostic connector and read any trouble codes. These codes will tell you where circuit problems might exist for further testing. Look at **Figure 78-3.**

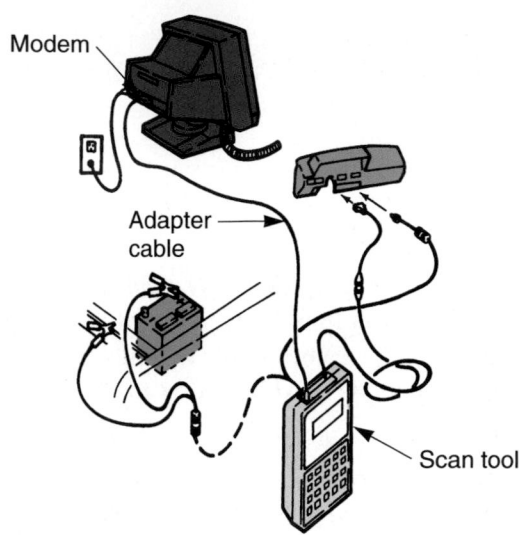

Figure 78-3. A scan tool will help you find most problems with air bag circuitry. This scanner is connected to a personal computer modem and communicates with the manufacturer to find difficult-to-locate problems. (OTC)

 Warning
Air bag systems are often equipped with an *energy reserve module,* which can deploy the air bag after a power failure. It must be removed or allowed to discharge for up to 30 minutes after disconnecting the battery. Refer to the service manual for details.

Manual Air Bag Deployment

Manual air bag deployment is done by connecting a voltage source to the air bag module wires to "fire" the air bag. Manual deployment is needed if the vehicle is to be scrapped at a salvage yard (total loss from accident) or if the air bag module is damaged or defective. Used air bags are not recommended for reuse in another vehicle for safety reasons.

 To manually deploy an air bag:
1. Disconnect the wires between the controller and air bag. Most manufacturers recommend that the air bag remain mounted securely in the vehicle.
2. Connect long jumper wires to the wires going into the air bag igniter. The jumper wires must be long enough to reach to the rear of the vehicle. Wrap the other ends of the jumper wires together to short them and prevent accidental deployment.

Air Bag Module Service

The air bag module must be serviced when it is deployed, defective, or damaged. You must often remove or disable the module when working on the steering column. It is important that specific procedures be used to ensure personal safety and proper air bag operation.

Disarming an Air Bag

To *disarm an airbag,* all sources of electricity for the igniter module must be disconnected from the air bag module. Before servicing a vehicle equipped with an air bag, the system must be disarmed. Procedures vary with the vehicle make and model.

Manufacturers may specify the removal of the system fuse, disconnection of the module, and disconnection of the battery. Always refer to the service manual for exact procedures for disarming the system. This will help prevent electrical system damage and accidental deployment, **Figure 78-4.**

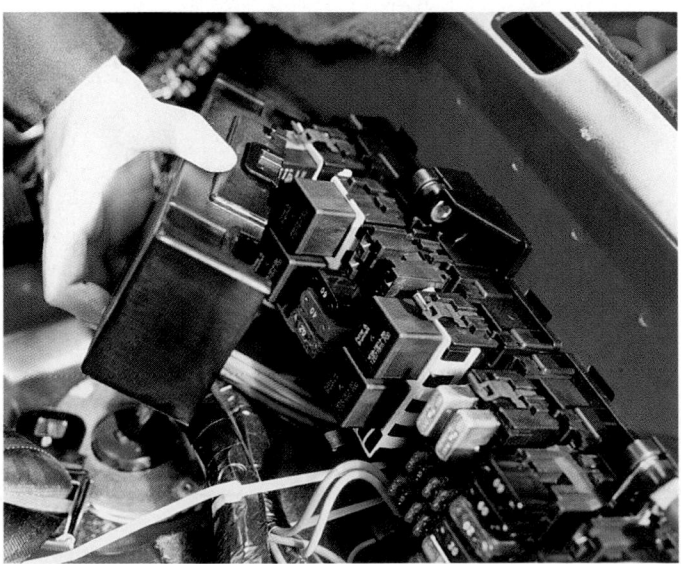

Figure 78-4. Follow the service manual instructions to disable the air bag. You may have to pull the fuse, disconnect the module, and then disconnect battery.

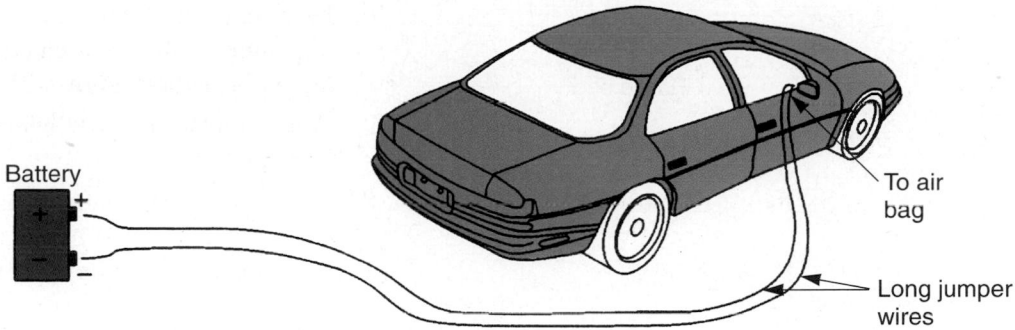

Figure 78-5. Manufacturers recommend that you deploy faulty or damaged air bags, as well as those in vehicles that are to be scrapped at a salvage yard. Connect long jumper wires to the air bag module wires. Then connect the jumper wires to a battery. The air bag will deploy with a loud bang. (Oldsmobile)

3. With everyone clear of the area, connect the other ends of the jumper wires to an automotive battery. The air bag should deploy with a loud pop, and the deployed bag should be visible. **See Figure 78-5.**

4. During deployment, the air bag becomes extremely hot and can cause serious burns. Wait at least thirty minutes before touching the deployed air bag assembly.

Warning
Never connect the jumper wires to the battery before connecting them to the air bag module. This will make the bag deploy as soon as the jumper wires are connected to the module, possibly causing serious injury.

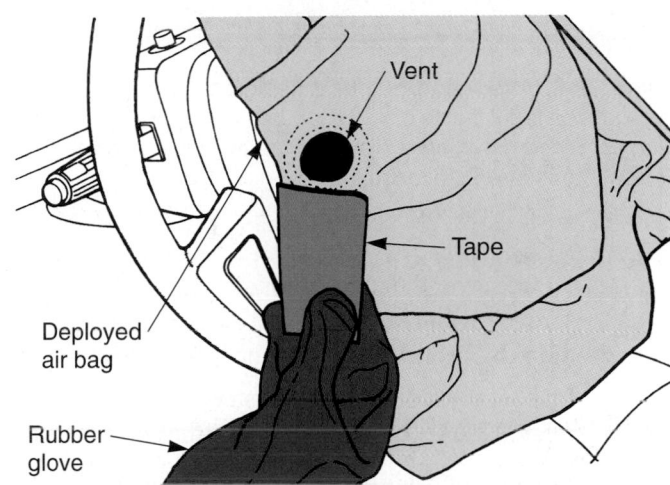

Figure 78-6. Place tape over the air bag vent holes to keep powder from leaking out during removal. (Chrysler)

Air Bag Disposal

After deployment, the air bag must be removed and disposed of properly. Follow all disposal information to avoid breaking federal law.

1. Wear rubber or plastic gloves, a respirator, and safety glasses when handling a deployed air bag.

2. *Air bag powder* is a potentially toxic substance that may blow out of the air bag upon deployment. Place pieces of tape over the vent holes in the air bag. This will help keep residual powder from leaking out the vents during bag removal, **Figure 78-6.**

3. Use a shop vacuum to clean up any air bag powder in the passenger compartment. The air bag powder can be an eye and skin irritant. Vacuum the dash, vents, seats, carpet, and other surfaces contaminated with this powder. Refer to **Figure 78-7.**

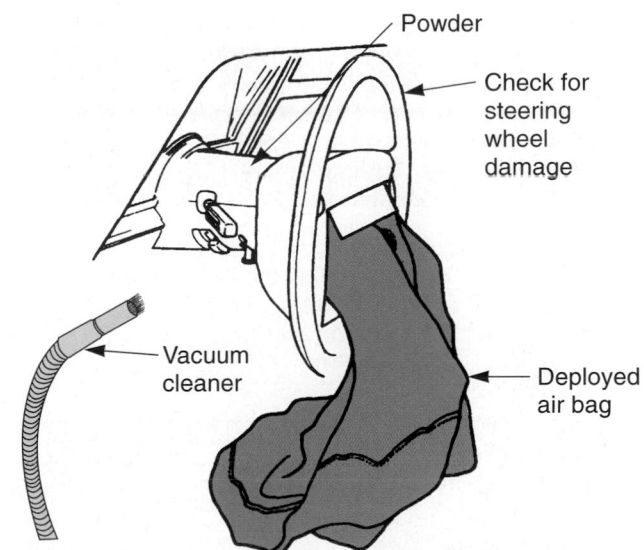

Figure 78-7. After air bag deployment, use a vacuum cleaner to pick up any powder. Wear safety goggles, rubber gloves, and a respirator. The powder can be a skin and eye irritant (Ford)

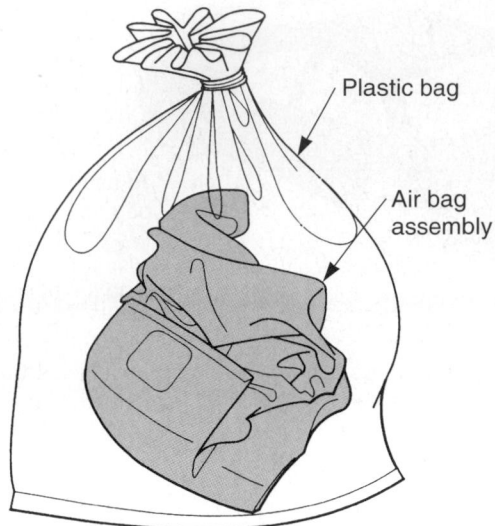

Figure 78-8. After removal, place deployed air bag in plastic bag. Use the label inside new air bag box to ship the old air bag to an approved recycling or disposal center. Never throw the old air bag into the trash. (Honda)

4. To remove the deployed air bag, remove the screws from the rear of the steering wheel. Lift the module out and disconnect the wires.

5. Inspect the steering wheel, steering column, and related parts for damage. Parts that have visible damage should be replaced. Damage to the wiring may require wiring harness replacement.

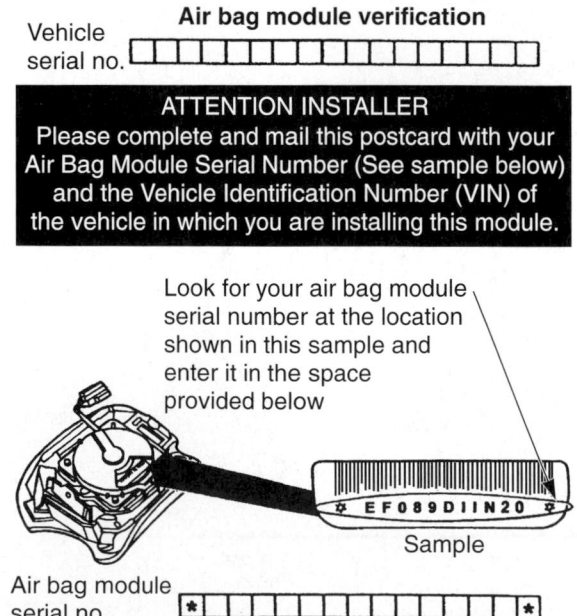

Figure 78-9. Compare part numbers to make sure you have the correct replacement air bag. (Ford)

6. Dispose of the air bag as instructed by the manufacturer. In all cases, seal the deployed bag in a large plastic bag, **Figure 78-8.**

7. Wash your hands after being exposed to the air bag powder.

8. A prepaid postage label is usually included with the new air bag. Use this label to ship the deployed air bag to its approved disposal station.

Air Bag Installation

Obtain the correct replacement air bag module. Make sure the part number on the new air bag is correct. Although air bag modules may look similar, they often have different internal designs. Look at **Figure 78-9.**

Warning

Keep your arms out of the steering wheel spokes and your head to one side of the steering wheel when working on an undeployed air bag. An air bag can shatter bones if accidentally deployed.

When carrying a live (undeployed) air bag module, aim the trim cover or bag away from your body. The metal housing should face your body. This will reduce the chance of injury if the bag accidentally deploys.

When placing the air bag module on a workbench, the bag and trim cover must face up. This will help prevent the module from flying violently through the shop if the bag suddenly inflates. See **Figure 78-10.**

If needed, service the clock spring that connects voltage to the air bag module through the steering wheel. See **Figure 78-11.**

Before installing the new air bag module, double-check that all sources of voltage are disconnected. You do not want the air bag to deploy when you reconnect its wires in the steering column.

You can then install the new air bag module. While keeping your head to one side of the bag, reconnect its wires. Fit the unit down into the steering wheel. Then, install the small screws that secure the module. Most manufacturers recommend using new screws, since the old ones might have been damaged by deployment. See **Figure 78-12.**

This procedure also applies to the passenger-side air bag. It must usually be secured from under the right side of the dash. Tighten the mounting bolts or nuts to specs. When reconnecting the wiring harness, keep your body away from the front of the air bag, **Figure 78-13.**

Air Bag Sensor Service

After deployment from a collision, some manufacturers recommend replacement of all air bag sensors.

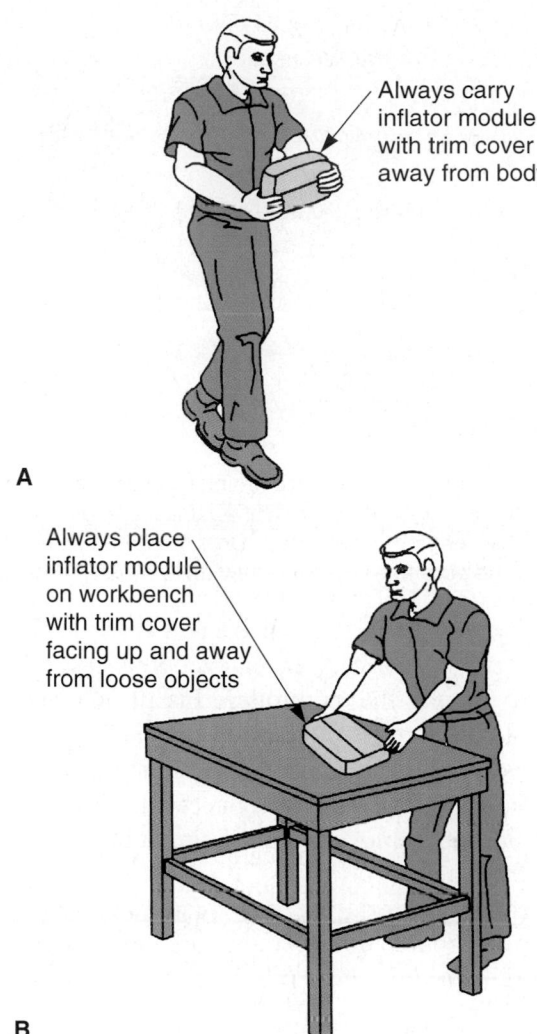

A

B

Figure 78-10. A—When carrying a "live" air bag, the trim cover should face away from your body. The metal housing should face your body. B—When placing an air bag on a workbench, the metal housing should face down and the trim cover should face upward. (Oldsmobile)

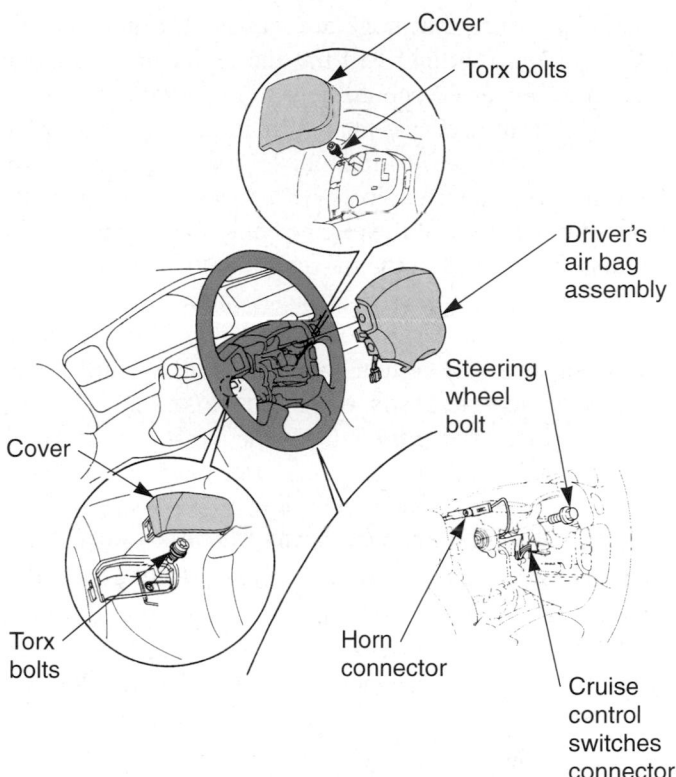

Figure 78-12. Before installing an air bag module, make sure the source of electricity is disconnected. Then, fit new air bag module into place. Use new fasteners to bolt unit into the steering wheel, if needed. Torque the fasteners to specs. (Honda)

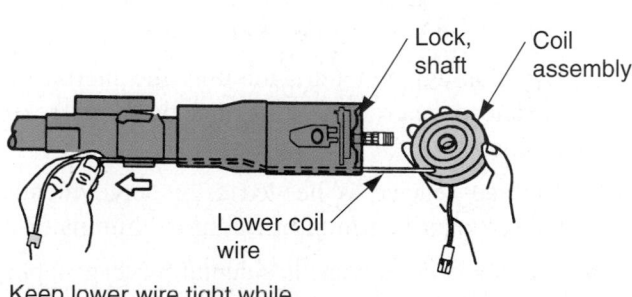

Keep lower wire tight while seating SIR coil ASM head in steering column.

Figure 78-11. During air bag service, you may have to replace the clock spring, or coil, that conducts current into the air bag module. Note how the wires must be routed down through the steering column. (General Motors)

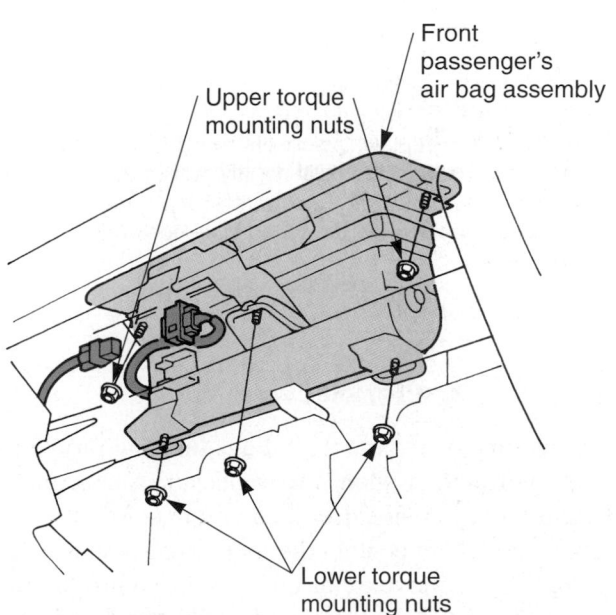

Figure 78-13. A passenger-side air bag must usually be installed from under the dash. Tighten the nuts to specifications before reconnecting the harness. (Honda)

This is to ensure that none are damaged. Some sensors will remain closed and will fire the replacement air bag as soon as power is reconnected.

Refer to the service manual if in doubt. It will give you critical service information: sensor locations, wiring diagrams, and specifications. Air bag sensors are usually located in the front of the engine compartment and under the dash or console in the passenger compartment.

Before replacing any air bag system sensors, make sure the system is disarmed. If you try to remove an air bag sensor with the system armed, the air bag could accidentally deploy—a costly, dangerous mistake.

Disconnect the wiring from the sensor first. Then, remove the screws holding the sensor to the vehicle. Make sure you have the correct replacement sensor.

Any arrow stamped or printed on the sensor must point toward the front of the vehicle. If you install a sensor backwards, the air bag will not deploy during a frontal impact. Tighten the sensor screws properly and reconnect the wiring connector. Refer to **Figure 78-14.**

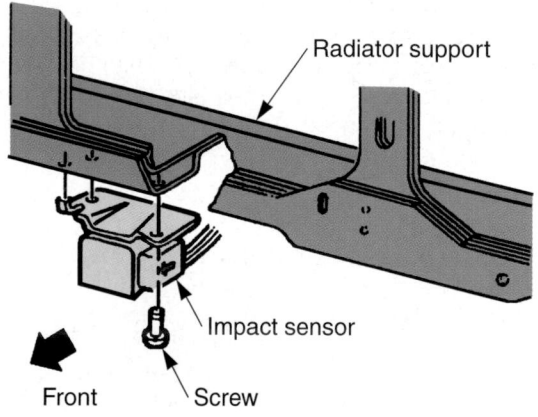

Figure 78-14. When replacing sensors, make sure air bag is disabled so it does not accidentally deploy. Also make sure any arrow on sensor points toward the front of the vehicle. If you install the sensor backwards, the air bag will not work. (General Motors)

Air Bag Controller Service

If symptoms or scan tool readings indicate problems with the controller, it should be replaced. Again, make sure the air bag system is disarmed before starting.

Some controllers contain the main impact sensor for the air bag. If you move or jar this type of controller, the air bag could accidentally deploy. See **Figure 78-15.**

First, disconnect the wiring harness from the controller. Do not force the connector from the controller. This can damage the connector or its pins. Remove the

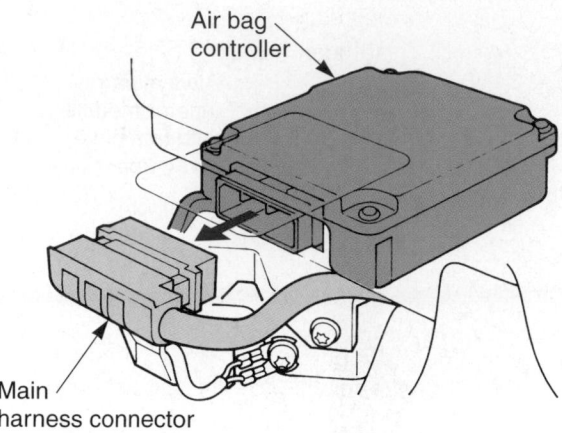

Figure 78-15. Follow the service manual instructions when replacing an air bag controller. Some units contain impact sensor that can deploy the air bag. Do not damage the connections and be sure to secure the new unit properly. (Honda)

screws that secure the controller. Install the correct replacement unit in reverse order of removal.

With everyone out of the vehicle and away from the air bags, rearm the system. Reconnect all harness connectors, install any removed fuses or circuit breakers, and reconnect the battery.

To check your work, use your scan tool to make a final sweep for trouble codes. Make sure the system is in perfect working order before releasing the vehicle to the customer.

Summary

- The proper service of seat belts and air bags is critical to the safety of the vehicle's occupants.

- Seat belt mechanisms should be inspected after an auto accident or after prolonged service.

- A centrifugal belt retractor uses centrifugal force to keep the belt from extending.

- A pendulum belt retractor uses the inertia of a vehicle deceleration to engage the belt locking mechanism.

- Air bag service is needed after deployment or when the air bag indicator light is illuminated.

- Take care to prevent accidental air bag deployment when working.

- If the indicator light glows but the bag is not deployed, use a scan tool to analyze the system.

- To disarm an airbag, all sources of electricity for the igniter module must be disconnected from the air bag module.

- Manual air bag deployment is done by connecting a source of voltage to the air bag module wires to "fire" the air bag.

- After deployment, the air bag must be serviced and disposed of properly.

- Keep your arms out of the steering wheel spokes and your head to one side of the module when working on an undeployed air bag.

- When carrying a live (undeployed) air bag module, aim the trim cover or bag away from your body.

- When placing the air bag module on a workbench, the bag and trim cover must face up.

- After deployment caused by a collision, some manufacturers recommend the replacement of all air bag sensors.

- Any air bag sensor arrow must point toward the front of the vehicle.

- If symptoms or scan tool readings indicate problems with the controller, it must usually be replaced.

Important Terms

Centrifugal belt retractor
Pendulum belt retractor
Disarm an airbag
Energy reserve module
Manual air bag deployment
Air bag powder

Review Questions—Chapter 78

Please do not write in this text. Place your answers on a separate sheet of paper.

1. Warn customers of the danger of placing _____ in front seats when air bags are used.
2. List five steps for inspecting seat belts.
3. List two of the four precautions that should be taken when replacing a seat belt.
4. How do you check the action of a centrifugal seat belt retractor?
5. How do you test the operation of a pendulum type seat belt retractor?
6. Air bag systems are often equipped with an energy reserve module which can deploy the air bag up to _____ minutes after power is disconnected.
7. How do you disable an air bag system?
8. How do you manually deploy an air bag?
9. After deployment from a collision or accident, some manufacturers recommend replacement of all air bag sensors. True or False?
10. What will happen if you install an air bag sensor backwards.

ASE-Type Questions

1. Technician A says that is acceptable to place your arms through the steering wheel spokes when installing a new air bag. Technician B says to place your face directly in front of the air bag during installation. Who is right?
   (A) A only.
   (B) B only.
   (C) Both A and B.
   (D) Neither A nor B.

2. An air bag indicator light shows a problem with the air bag system. Technician A says to use a scan tool to help determine the source of the trouble. Technician B says to replace the air bag sensors. Who is right?
   (A) A only.
   (B) B only.
   (C) Both A and B.
   (D) Neither A nor B.

3. After air bag deployment, which of the following should *not* be done?
   (A) Force excess powder out of air bag with hand pressure.
   (B) Wear a respirator, goggles, and rubber gloves.
   (C) Place tape over air bag vent holes.
   (D) Vacuum the passenger compartment.

4. When carrying a live (undeployed) air bag module, how should you aim the trim cover?
   (A) Upward.
   (B) Toward your body.
   (C) Toward the floor.
   (D) Away from your body.

5. Technician A says that you may have to replace all the sensors after air bag deployment. Technician B says that the air bag sensor arrow must point toward the front of the vehicle. Who is right?
   (A) A only.
   (B) B only.
   (C) Both A and B.
   (D) Neither A nor B.

6. Some controllers contain:
   (A) *azide pellets.*
   (B) *the malfunction indicator light.*
   (C) *the main impact sensor.*
   (D) *an igniter module.*

Activities—Chapter 78

1. Visit a salvage yard and locate several vehicles with front-end collision damage. While wearing a respirator, safety goggles, and gloves, look for steering wheel and steering column damage. Compare the steering wheel and column damage found on air bag–equipped vehicles to the damage found on vehicles without air bags.

2. Use a service manual to find the locations of impact sensors on different makes and models of vehicles.

Restraint System Diagnosis		
Condition	**Possible Causes**	**Correction**
Indicator does not work.	1. Burned out dash bulb. 2. Circuit problem. 3. Open in inflator module. 4. Controller failure.	1. Replace bulb. 2. Test and repair wiring. 3. Replace module. 4. Replace controller.
Air bag will not deploy during a collision.	1. Failed sensors. 2. Failed air bag module. 3. Bad controller. 4. Wiring problems.	1. Replace sensor. 2. Replace module. 3. Replace controller. 4. Test and repair wiring.
Air bag deploys accidentally.	1. Loose impact sensors. 2. Controller failure. 3. Shorted wires. 4. Moving or hitting sensors with key on.	1. Tighten sensors, test system. 2. Replace controller. 3. Test and repair wiring. 4. Use approved procedures.
Indicator light comes on.	1. Controller failure. 2. Circuit problem. 3. Bad air bag module.	1. Replace controller. 2. Repair wiring. 3. Replace air bag module.
Seat belt will not engage.	1. Broken spring in retractor. 2. Worn retractor. 3. Failed retractor motor. 4. Circuit problem.	1. Replace retractor. 2. Replace retractor. 3. Replace motor. 4. Repair wiring.
Seat belt will not retract.	1. Twisted belt. 2. Bad retractor. 3. Debris in retractor.	1. Straighten belt. 2. Replace retractor. 3. Remove debris.

Chapter 79

Security, Navigation, and Future Systems

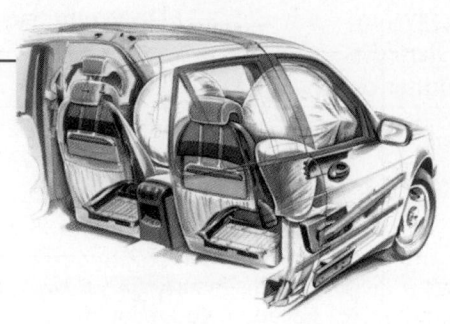

After studying this chapter, you will be able to:

- Explain the operation of vehicle security systems.
- Explain the operation of vehicle navigation systems.
- Compare security system design variations.
- Summarize the operation of alternate power sources for vehicles.
- Discuss how engineers might change vehicle designs in the future to increase safety, comfort, dependability, and environmental control of our planet.

The first half of this chapter will overview the operation of vehicle security and navigation systems. Security systems are now a very common option. Navigation systems are less common but should grow in popularity in the near future.

What will the future bring to automotive technology? What will happen if we run out of crude oil? What alternate forms of energy could we use? The second half of this chapter will try to answer these and other questions. Refer to **Figure 79-1.**

Security Systems

A *security system* is designed to help prevent the theft of a vehicle and its contents. A security system basically uses motion sensors and switches to feed signals to an electronic control unit. The control unit can then lock the vehicle and operate a warning siren or horn if someone tampers with the vehicle. Some systems also

Figure 79-1. Automotive technology has seen tremendous change over the years. We should see more dramatic improvements in the near future. (Chrysler)

disable the starting system and engine to prevent the vehicle from being started.

A remote transmitter or a dash-mounted keyboard is used to activate and deactivate the system. Much like a garage door opener, the transmitter produces a radio signal of a specific frequency. A receiver circuit responds to this signal and triggers the control unit to perform various tasks, such as turning the system on or off, unlocking the doors, sounding the siren, etc.

Security system designs vary tremendously. This is done so that thieves are not able to learn about the operation of any one system too well. For this reason, you should refer to the schematic and technical literature from the manufacturer of the security system. It will give the details for doing competent service and repair work.

Resistance Key

Modern theft deterrent systems often use a ***resistance key,*** which is an ignition key with a small electrical resistor mounted on its side. Several key resistance values are available. The resistor has small electrical connections on the sides, **Figure 79-2.** The key's metal teeth and the built-in resistor must match the vehicle to break security.

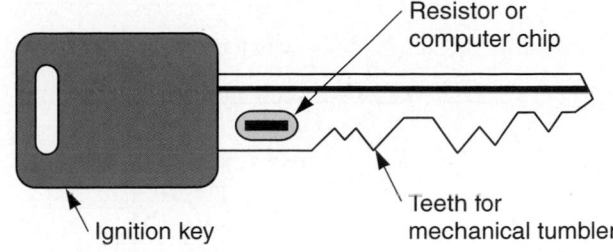

Figure 79-2. Resistor key has a small electrical resistance element built into a hole in the key. It engages contacts in the tumbler when inserted. This allows the vehicle computer to determine whether the correct key has been inserted.

The computer reads the resistance of the key to make sure it is not an unauthorized duplicate of only the metal teeth of the key. If the key resistance matches the one programmed into the control module, the vehicle starts normally.

However, it the key's resistance is not correct, the control module opens the starter relay circuit so the engine will not crank. The control module may also disable the engine support systems. Refer to **Figure 79-3.**

A basic wiring diagram for a theft deterrent system is given in **Figure 79-4.**

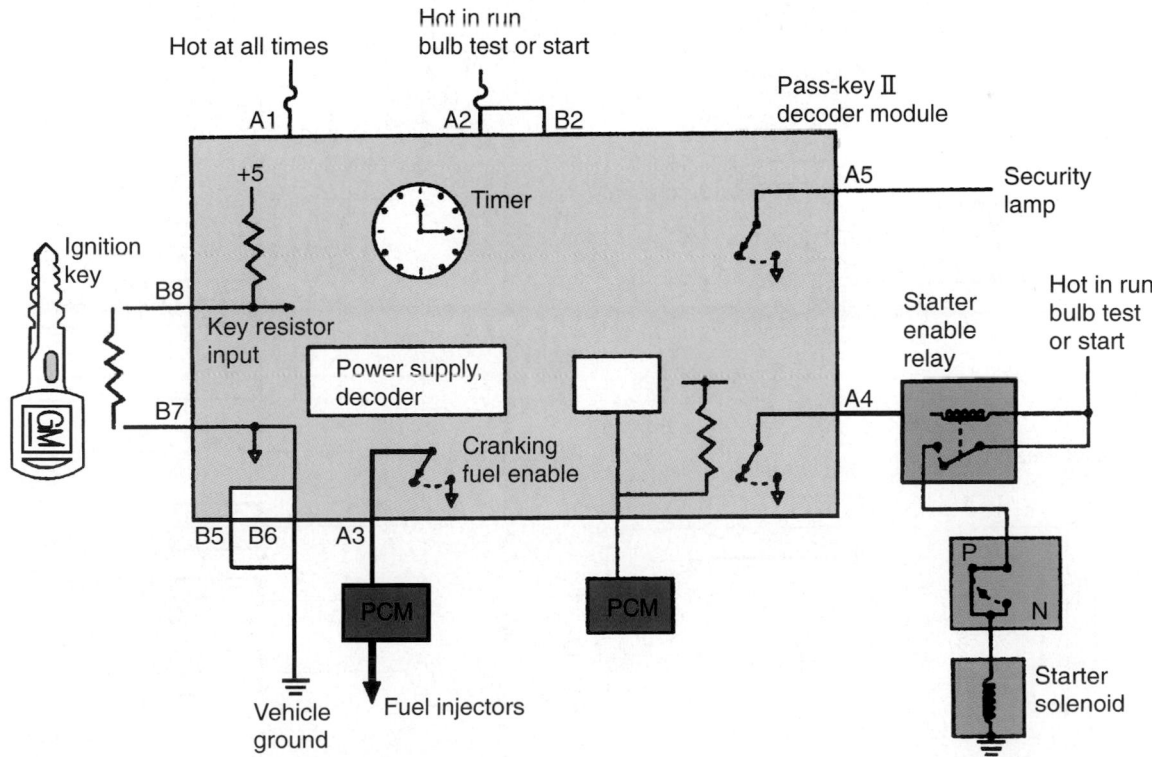

Figure 79-3. If someone breaks an ignition switch to try to steal a car, the ECM will not receive the correct signal from the resistor key and the ECM will disable the vehicle so it cannot be driven. (General Motors)

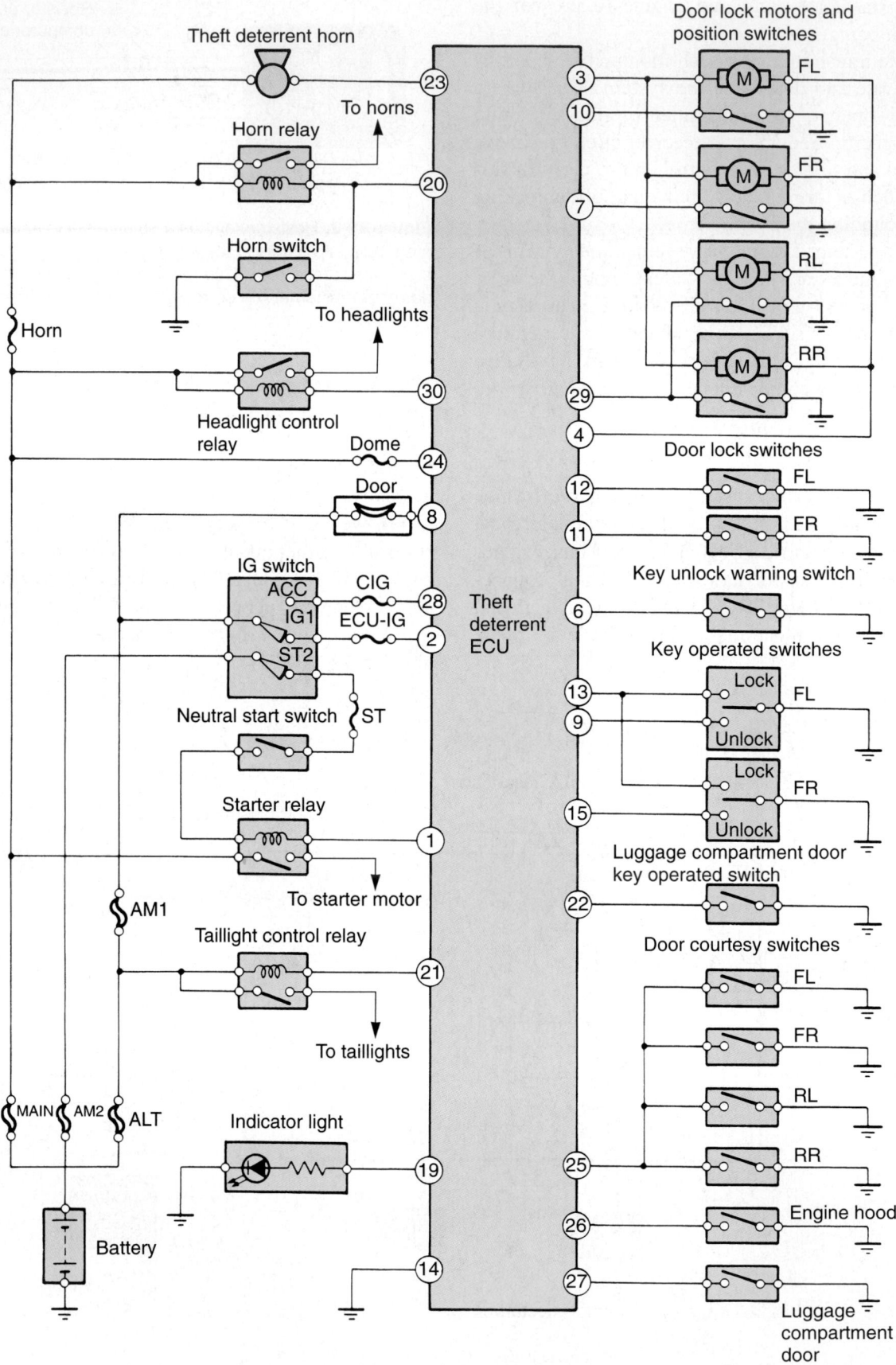

Figure 79-4. Wiring diagram shows typical components included in a theft deterrent system. (Lexus)

Transponder Key

A *transponder* is a device that receives a signal, alters it, and then retransmits it in an altered state to another device. Some late-model anti-theft systems now use a transponder (electronic chip) built into the ignition key. It is designed to prevent thieves with a knowledge of electronics from overriding the security system.

Every time the vehicle is started, the on-board computer sends a signal to the ignition key. The key's transponder alters this signal and sends it back to the computer. The transponder sends a new signal to the computer on every startup. This prevents a thief from duplicating the key as easily.

The computer knows what that key's circuitry should do with each signal variation. If the computer does not obtain the correct return signal from the key, it disables the starting and engine support systems so the vehicle cannot be driven.

Tech Tip
Resistance and transponder keys cannot be duplicated at your average hardware store or locksmith. You must usually contact the manufacturer to get service parts and codes.

Navigation Systems

Navigation systems use our nation's satellite global positioning system to display the geographic location of the vehicle. A dash display shows a road map with the position of the vehicle on it. Advanced systems can give other useful information: traffic information, road work, parking lot openings, miles to destination, turnoffs, etc. See **Figure 79-5.**

A

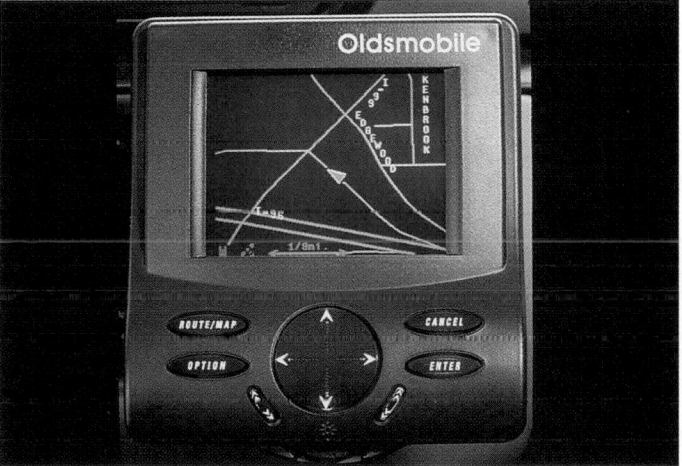

B

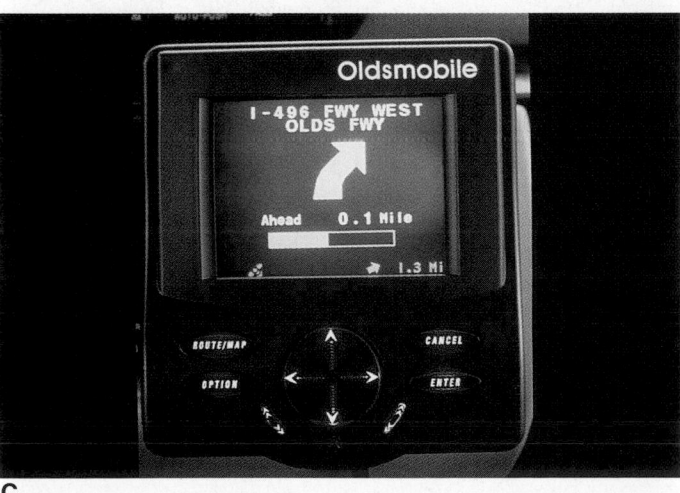

C

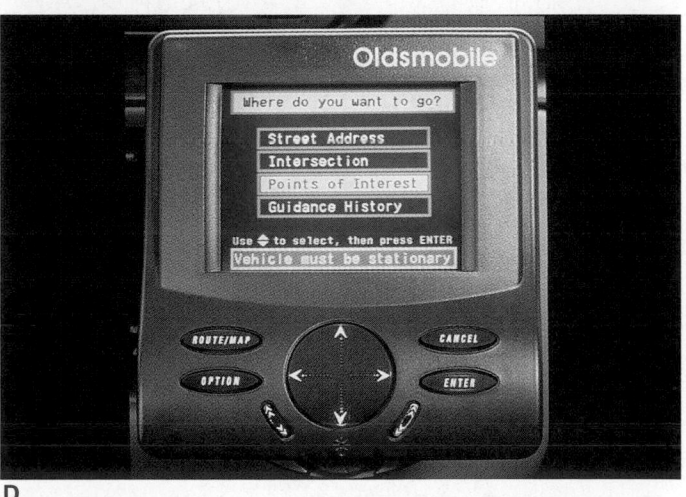

D

Figure 79-5. A navigation system improves convenience and safety. The system helps you find your destination with minimal confusion. A—Driver has pressed the keypad to enter the navigation system. B—System is showing the vehicle's location on a map. C—This navigation system tells you where to turn to reach your destination. D—This system is showing the street address and intersection for the destination. (Oldsmobile)

Basically, the navigation system communicates with the satellite system to determine vehicle direction and speed. The vehicle's on-board computer compares this data with a large database of roads and highways, usually from a compact disc. Look at **Figure 79-6.**

A road map of the area with a representation of the vehicle's location can then be shown on the dash display. This can help the driver navigate highways and roads more efficiently. Besides visually helping the driver navigate, some systems also use verbal output to help the driver.

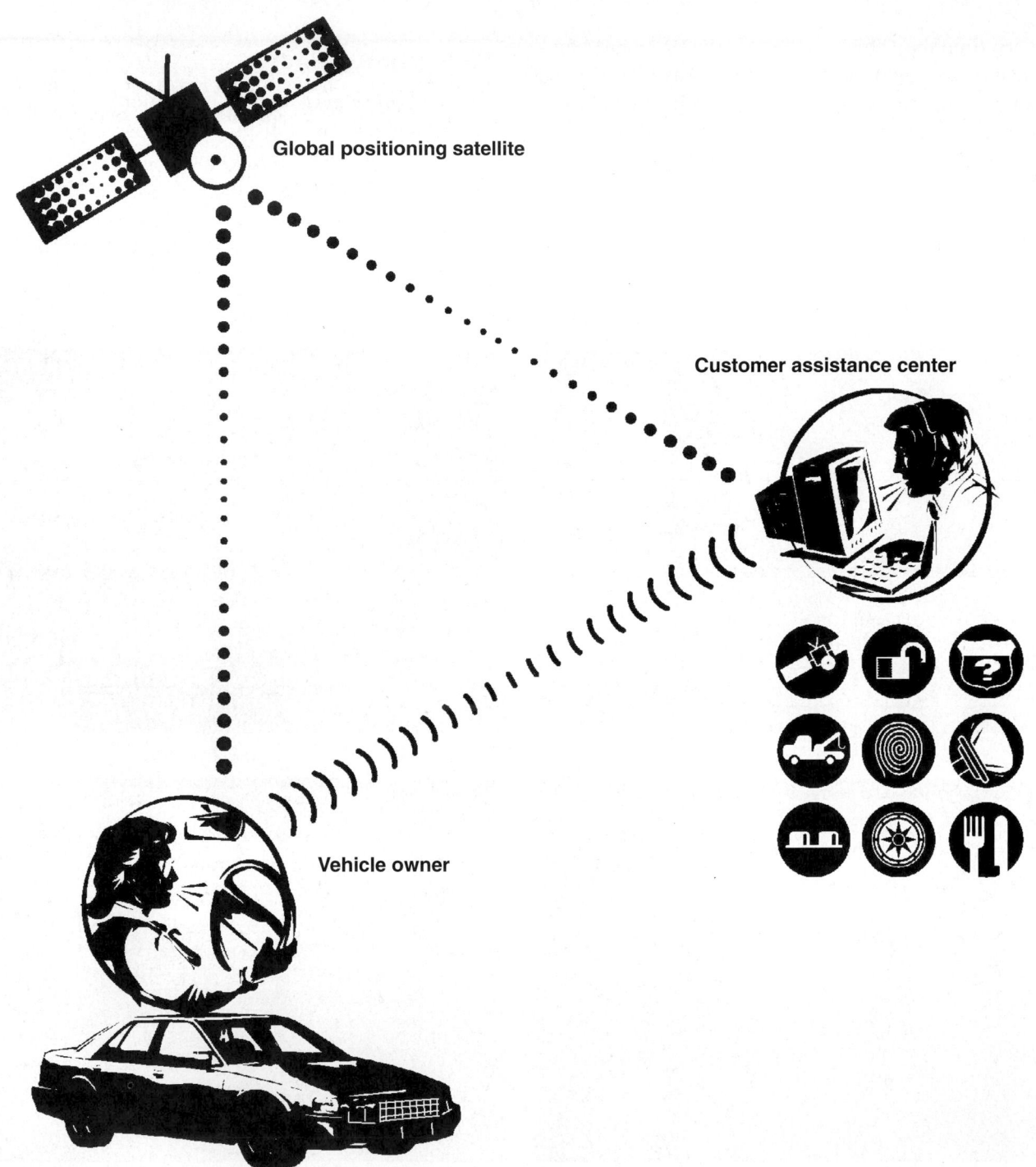

Global positioning satellite

Customer assistance center

Vehicle owner

Figure 79-6. Most navigation systems and tracking systems use global positioning system satellites to determine vehicle location. (Cadillac)

Vehicle Tracking Systems

A *vehicle tracking system* is a security system that can track the location of a car or truck if stolen. One system uses an on-board security control module, the vehicle's cellular telephone, the global positioning satellite system, and a 24-hour monitoring/response center, **Figure 79-7.**

Figure 79-7. A vehicle tracking system has trained operators at computer workstations 24 hours a day. If a vehicle is stolen or hijacked, an operator can phone police and give the exact location of the vehicle. (ATX Research)

The security control module interfaces with the cellular phone system. If the vehicle is stolen, the module and cellular phone are used by the satellite system to track the vehicle's geographic location. The tracking system is triggered if the perpetrator does not enter the pass code on the cellular phone before driving the vehicle.

Some tracking systems also provide added security for the driver during a car jacking or kidnapping. The driver can press an emergency call button on his or her key ring. This allows the people at the response center to listen to conversation in the vehicle and to take action as needed. They can call the police and tell them the exact location and description of the vehicle.

Future Systems

What automotive design changes will we see in the next century or two? Just look where the human race has taken us in the last 100 years! We have gone from the horseless carriage, to the motor car, to the moon in one lifetime.

If you sent a new car back in time to your great grandparents, they would think the car was created on another planet. Since human progress is increasing exponentially, we should see even more rapid progress in the future.

We have the technology to allow motor cars to clean up or even modify our atmosphere as desired. Just as a change from R-12 refrigerant to R-134a will have a desired affect on the earth's atmosphere, we can do many other things to help control the atmosphere and natural resources that support life on earth.

Recycling will become even more common in the automotive industry. It may expand to the point that we build whole car bodies out of recycled aluminum cans. A car body made out of "old soda pop cans" will producing a strong, lighter vehicle that will resist rust several hundreds times better than steel.

DUI Box

A *DUI box* is a computer that can test the driver's sobriety before the vehicle can be started. The computer, sometimes part of the security system, requires that the driver punch in keyboard responses in a specific time span.

If the driver cannot respond to the test quickly enough, the computer can determine whether the driver is "in any condition to drive" on public roads. If impaired through illness or a chemical, the computer will electronically "kill" the starting circuit and engine support circuits to prevent the vehicle from being driven unsafely.

Smog Scrubber Radiator

A *smog scrubber radiator* is coated with a catalytic platinum-based material or other chemicals to remove emissions from the outside air. As the polluted outside air flows over the hot radiator, the catalysts converts most of the smog-forming ozone and carbon monoxide into oxygen and harmless carbon dioxide. This experimental feature is being tested by several auto makers. It has the potential to make vehicles "large air cleaners." If all cars were equipped with smog scrubbers, we could theoretically help clean the earth's atmosphere as we drive.

Future vehicles could have radiators coated with other substances to further modify the atmosphere as desired. If a city is suffering from one particular type of pollution, its vehicles' radiators could be coated with a substance with a natural affinity for the particular chemical composition in the pollution.

Night Vision Systems

A *night vision system* could use an infra-red camera and monitor system to help the driver view through fog, snow, rain, and other vision impairing atmospheric conditions. The camera could be mounted in the front grill. When activated in bad weather or at night, the camera could produce an image on the dash-mounted monitor or on a heads-up display on the dash. This would let the driver see longer distances through fog to detect on-coming vehicles, people, curves in the road, or other dangers.

Automatic Tire Wear Adjustment

Future vehicles might have automatic tire wear adjustment systems to eliminate the need to rotate the tires. Automotive engineers could use the anti-lock brake sensors to monitor tire and wheel speed. If one tire was worn, it would be slightly smaller in diameter than the others.

The computer would detect a higher signal frequency from the wheel sensor monitoring the rotation of the worn tire. The computer could then adjust the differential solenoids to apply more pressure to the one side of the differential clutch pack upon acceleration. This would lower tread loss on the worn tire. More torque would be applied to the tire or tires with more tread to progressively even out tire wear.

Collision Avoidance System

A *collision avoidance system* uses a radar system to detect a vehicle or another object in front of the vehicle. The computer uses vehicle speed data and the distance to the leading vehicle to determine if the driver is too close for that road speed. If the control module determines that the driver is "tailgating," it will produce an audible warning, shut off the speed control, or even apply the anti-lock braking system to help avoid an accident.

Adaptive Cruise Control

An *adaptive cruise control system* can alter vehicle speed automatically when coming up on another vehicle too quickly. Front mounted radar works with the vehicle's on-board computer to reduce cruising speeds to prevent following too closely behind another vehicle.

Alternate Power Sources

Alternate power sources are other methods of producing energy (other than from gasoline and diesel oil) to propel a motor vehicle. The most promising include electric vehicles, hybrid vehicles, and multi-fuel vehicles.

Electric Vehicles

An *electric vehicle* uses a large, high efficiency electric motor and large storage batteries to power the automobile. Electric cars have been produced in limited numbers to help reduce air pollution from the burning of fossil fuels. They provide a viable, alternate means of vehicle propulsion for short trips, **Figure 79-8.**

Instead of going to a gas station to fill up with gasoline or diesel oil, the driver simply plugs the electric car into a recharging station. This re-energizes the storage batteries and readies the car for another commute. Look at **Figure 79-9.**

An all-electric vehicle replaces the conventional internal combustion engine with an electric motor, control module, and storage batteries. The electric motor-generator drives a transaxle that sends power out to the drive wheels. The electronic module monitors and controls the system so that little energy is wasted. See **Figure 79-10.**

Regenerative braking converts the inertia of the moving car into electrical energy to recharge the batteries. This is done by converting the drive motor into a large electric generator. When the brakes are applied, the ECM causes the transaxle to spin the motor-generator. The generator action produces a load that slows the vehicle while also converting this energy into electricity to recharge the batteries during braking.

Lead acid and lithium polymer batteries are sometimes used in electric cars because of their larger energy storage capability. Lead acid batteries are used for their dependability and availability.

Lithium polymer batteries can be formed into almost any shape for space savings during vehicle design. The lithium polymer cells are made from flexible laminate film as thin as 100 microns thick. The laminate is made up of metal foil, electrolyte, cathode, anode, and an insulator layers. This battery design offers high energy density and low manufacturing costs.

One drawback to electric vehicles is that we could have to build a large number of new power generating stations to recharge the millions of batteries every night. This would cause environmental pollution from burning much more coal, oil, or nuclear fuels. Battery recharging can also cause some pollution from the battery gases generated.

Solar Vehicles

A *solar vehicle* uses solar panels to generate electricity for propulsion. A large surface area of the vehicle body is covered with solar panels, which convert sunlight directly into electricity. This energy is used to power an electric motor that powers the vehicle. See **Figure 79-11.**

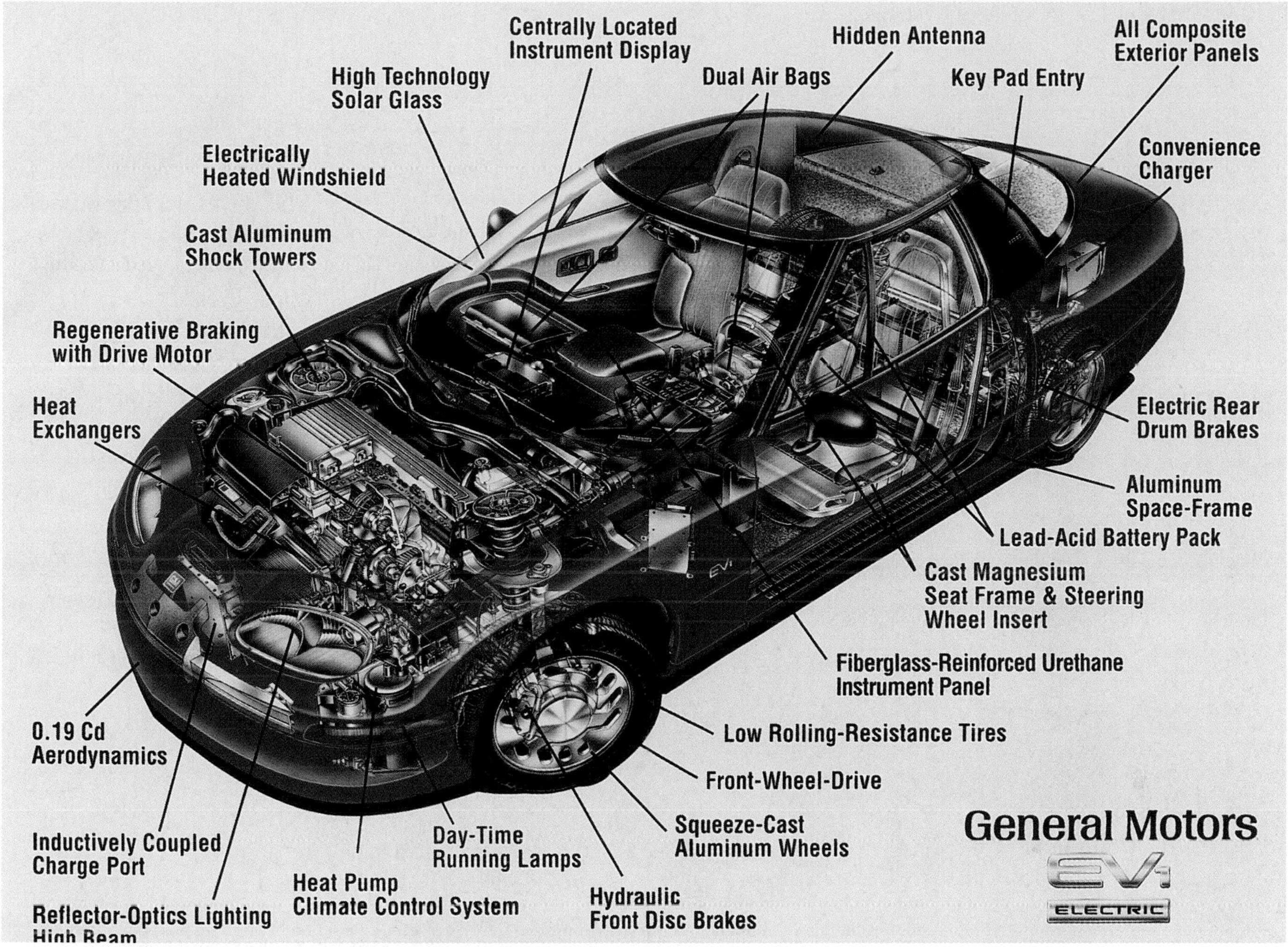

Centrally Located Instrument Display

High Technology Solar Glass

Hidden Antenna

Dual Air Bags

Key Pad Entry

All Composite Exterior Panels

Electrically Heated Windshield

Convenience Charger

Cast Aluminum Shock Towers

Regenerative Braking with Drive Motor

Heat Exchangers

Electric Rear Drum Brakes

Aluminum Space-Frame

Lead-Acid Battery Pack

Cast Magnesium Seat Frame & Steering Wheel Insert

Fiberglass-Reinforced Urethane Instrument Panel

0.19 Cd Aerodynamics

Low Rolling-Resistance Tires

Front-Wheel-Drive

Inductively Coupled Charge Port

Day-Time Running Lamps

Squeeze-Cast Aluminum Wheels

Heat Pump Climate Control System

Reflector-Optics Lighting High Beam

Hydraulic Front Disc Brakes

General Motors

EV1 ELECTRIC

Figure 79-8. Study the major parts of an electric car. (General Motors)

Figure 79-9. This car owner is plugging her electric vehicle into a charging station. (General Motors)

Hybrid Vehicles

A **hybrid vehicle** uses two different methods to generate power, usually a conventional internal combustion engine and an electric motor. This is a logical progression to a form of transportation that is dependable and also reduces the use of natural resources.

A tiny gasoline engine is used to power an electric generator-motor assembly. The gasoline engine is only used when the storage batteries become drained or to keep the vehicle rolling long distances over the highway. Look at **Figure 79-12.**

The batteries and electric motor supply power when the car first accelerates. This provides enough energy to accelerate the car quickly. Once cruising speeds are reached, the 15 to 20 hp gasoline or diesel engine takes over. It is a very small engine that can supply adequate power to keep the car moving on level ground. The gasoline engine also provides enough energy to recharge the batteries when needed. Hybrid cars also use regenerative braking to capture the energy of the moving vehicle for reuse upon initial acceleration.

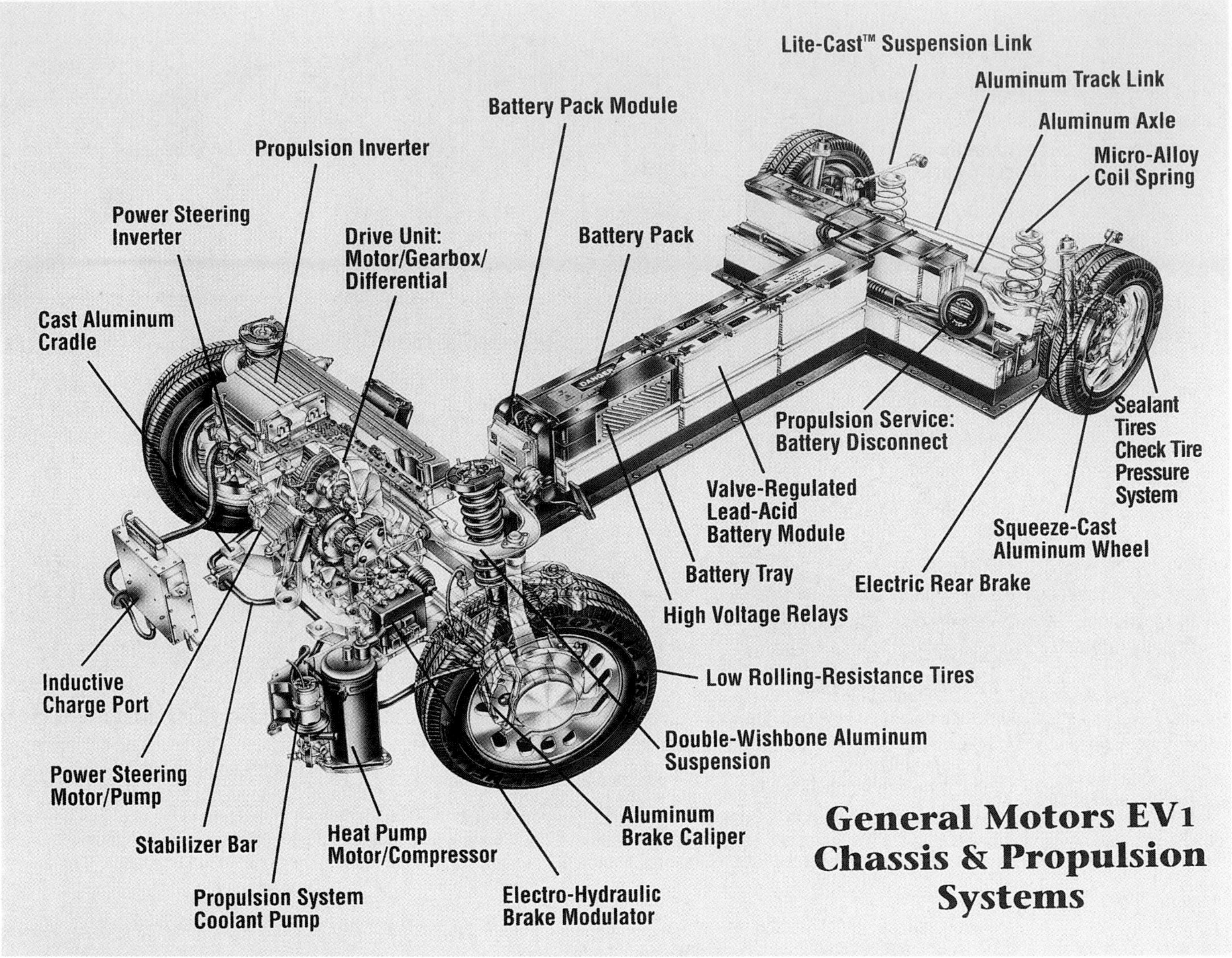

Lite-Cast™ Suspension Link

Aluminum Track Link

Aluminum Axle

Micro-Alloy Coil Spring

Battery Pack Module

Propulsion Inverter

Power Steering Inverter

Drive Unit: Motor/Gearbox/ Differential

Battery Pack

Cast Aluminum Cradle

Sealant Tires Check Tire Pressure System

Propulsion Service: Battery Disconnect

Valve-Regulated Lead-Acid Battery Module

Squeeze-Cast Aluminum Wheel

Battery Tray

Electric Rear Brake

High Voltage Relays

Inductive Charge Port

Low Rolling-Resistance Tires

Power Steering Motor/Pump

Double-Wishbone Aluminum Suspension

Stabilizer Bar

Heat Pump Motor/Compressor

Aluminum Brake Caliper

General Motors EV1 Chassis & Propulsion Systems

Propulsion System Coolant Pump

Electro-Hydraulic Brake Modulator

Figure 79-10. Study the drive train for a typical electric vehicle. (General Motors)

Figure 79-11. A solar vehicle is powered by sunlight. Its technology could someday be incorporated into a hybrid vehicle. (Honda)

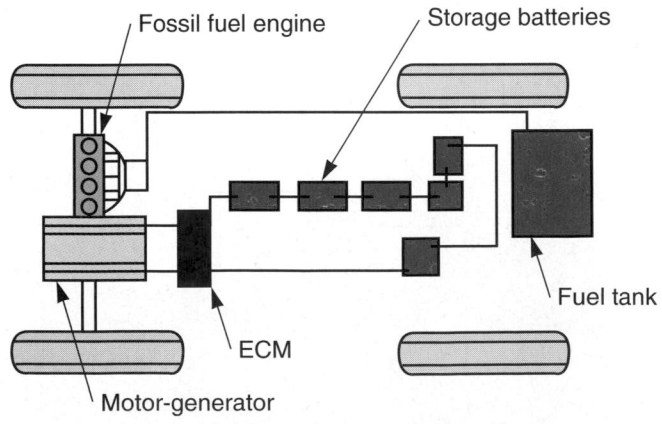

Fossil fuel engine

Storage batteries

Fuel tank

ECM

Motor-generator

Figure 79-12. A hybrid vehicle uses both a conventional internal combustion engine and an electric motor to reduce fuel consumption and emissions. It also eliminates the problem of ever having "dead batteries," which could strand you without power.

Inertia Storage

An *inertia energy storage system* uses the power of a high-speed flywheel to store electrical energy. The "heart" of the system is a special flywheel that spins on almost frictionless magnetic bearings, **Figure 79-13.**

A motor-generator works with the flywheel. On startup, the engine-driven generator feeds energy to the flywheel to bring it up to speed. When power is needed, the flywheel pumps electrical energy to the motor-generator for more rapid acceleration. The braking system can also use the motor-generator to use braking energy to feed power into the flywheel. This is presently an experimental system being developed for a race car.

Fuel Cell

A *fuel cell* combines hydrogen and water to produce electrical energy. It is electrolysis in reverse. A fuel cell is made of a plastic gas-permeable membrane coated with a catalytic platinum foil. This membrane is sandwiched between two electrode plates. Hydrogen gas flows through the anode plate and the membrane while compressed air flows the other way. The catalyst causes the hydrogen gas to separate into ions (protons) and electrons. This causes a potential difference of electrical energy across the anode and cathode. When current is drawn out of the fuel cell, oxygen molecules separate and combine with hydrogen protons to form water and heat.

Hydrogen Fuel

Could we someday see a shift to hydrogen as an automotive fuel? As we exhaust earth's supply of petroleum and as production costs go down for hydrogen, we could be burning a fuel that produces little or no air pollution. Hydrogen is the same fuel that powers the sun. It burns with little or no harmful emissions. We could greatly reduce air pollution in large cities by burning hydrogen gas.

Several automakers are experimenting with hydrogen-powered cars. There tests are very promising. We need more research in the production of hydrogen gas from salt water using solar-powered panels. This will allow us to produce hydrogen fuel with much lower production costs.

Multi-Fuel Vehicles

A *multi-fuel vehicle* is one designed to operate on many different types of fuel. Sensors are used to monitor the contents of the fuel through measurement of combustion efficiency and adjust engine functions for maximum efficiency for the type fuel being consumed.

A gas turbine is a very promising alternate type of engine that can burn a wide variety of fuels: gasoline, kerosene, or oil. A gas turbine can produce tremendous power for its size. Because of the spinning action, its power output is very smooth.

The gas turbine is not in use because of it's high manufacturing costs and poor fuel efficiency. It requires many special metals, ceramic parts, and precision machining and balancing. Some day, with a slight improvement in efficiency through engineering advances, the gas turbine may be a very common automotive engine.

Plant Fuels

Plant fuels are made from vegetation and other ingredients. One experimental fuel is made from a "burning bush." The leaves of the bush produces a flammable oil. The bush, even when alive, will burst into flames when ignited. When the plant's leaf oil is mixed with small amounts of gasoline, sugar, and other ingredients, tests have shown that it powers a car engine without modifications.

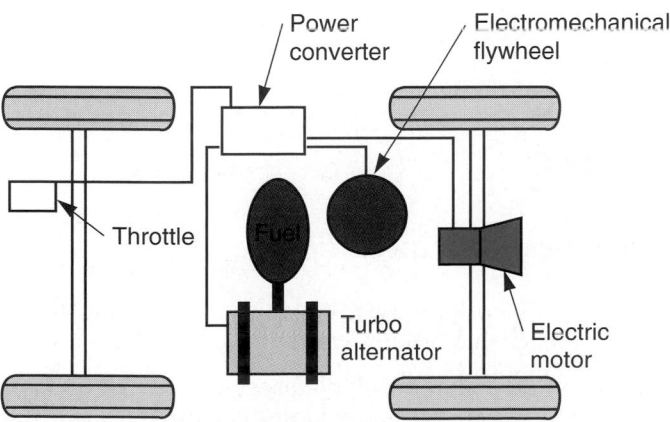

Figure 79-13. This hybrid car uses a turbine engine, an electric motor, and an energy storing flywheel to produce a tremendously powerful power plant.

Summary

- A security system is designed to help prevent theft of the vehicle and its contents.
- Security systems designs vary tremendously.
- Modern theft deterrent systems often use a resistor key which places a small electrical resistor in the side of the ignition key.
- A transponder is a device that receives a signal, alters the data, and then retransmits its information in an altered state back to another device.
- Navigation systems use our nation's satellite global positioning system to display the geographic location of the vehicle.

- A vehicle tracking system is a security system that can track or follow the location of a car or truck if stolen.
- A DUI box is an ECM that can test the driver's sobriety before the vehicle can be started.
- A collision avoidance system uses a radar system to detect a vehicle or other object in front of the vehicle.
- An adaptive cruise control system can alter vehicle speed automatically when coming up on another vehicle too quickly.
- Alternate power sources are other methods of producing energy (other than from gasoline and diesel oil) to propel a motor vehicle.
- A hybrid vehicle uses two different methods to generate power, usually a conventional internal combustion engine and electrical energy.
- An inertia energy storage system uses the power of a high speed flywheel to store electrical energy.
- A fuel cell combines hydrogen and water to produce electrical energy.
- A multi-fuel vehicle is one designed to operate on many different types of fuel.

Know These Terms

Security system	Collision avoidance
Resistance key	system
Transponder	Adaptive cruise control
Navigation system	system
Vehicle tracking	Alternate power sources
system	Electric vehicle
DUI box	Regenerative braking
Smog scrubber	Solar vehicle
radiator	Hybrid vehicle
Night vision system	Fuel cell
	Multi-fuel vehicle

Review Questions—Chapter 79

Please do not write in this text. Place your answers on a separate sheet of paper.

1. Describe the basic operation of a typical security system.
2. Why are security system designs so varied?
3. How does a resistance key work in a security system?
4. What is a transponder key?
5. What type of information can a navigation system provide?

● ASE-Type Questions

1. Technician A says vehicles will not change much in the next 30 years. Technician B says that this is not true. Who is correct?
   - (A) A only.
   - (B) B only.
   - (C) Both A and B.
   - (D) Neither A nor B.

2. Technician A says most collision avoidance systems use radar to measure the distance and speed of objects in front of the vehicle. Technician B says the ECM may also use signals from the vehicle speed sensor to detect a possible collision. Who is correct?
   - (A) A only.
   - (B) B only.
   - (C) Both A and B.
   - (D) Neither A nor B.

3. Technician A says lithium polymer batteries can be formed into almost any shape for space savings during vehicle design. Technician B says this is also true for lead-acid batteries. Who is correct?
   - (A) A only.
   - (B) B only.
   - (C) Both A and B.
   - (D) Neither A nor B.

4. In a(n) _____ vehicle, a tiny gasoline engine is used to power an electric generator-motor assembly.
   - (A) solar
   - (B) electric
   - (C) hybrid
   - (D) electronic

5. Which of the following burns with little or no air pollution.
   - (A) gasoline.
   - (B) diesel oil.
   - (C) gasohol.
   - (D) hydrogen.

Activities—Chapter 79

1. Visit your library and look up periodicals or magazine articles on future cars. Make a class report on your findings.
2. Research the production of hydrogen gas from salt water. Write a report on the current state of production and if solar energy can be used to produce hydrogen fuel.

Security System Diagnosis		
Condition	**Possible Causes**	**Corrections**
Alarm will not shut off.	1. Failed motion sensor. 2. Failed switch. 3. Circuit short or open. 4. Bad control module.	1. Replace sensor. 2. Replace switch. 3. Repair wiring. 4. Replace control module.
Alarm will not turn on.	1. Blown fuse. 2. Circuit short or open. 3. Bad control module. 4. Failed motion sensor. 5. Failed switch. 6. Bad siren or horn.	1. Replace fuse. 2. Repair wiring. 3. Replace control module. 4. Replace sensor. 5. Replace switch. 6. Replace parts.
Intermittent operation.	1. Loose connection. 2. Switch not adjusted properly. 3. Cracked circuit board.	1. Test and repair circuit. 2. Adjust switch to close fully. 3. Replace control module.

Fraction, Decimal, and Metric Equivalents

INCHES (FRACTIONS)	INCHES (DECIMALS)	MILLIMETERS	INCHES (FRACTIONS)	INCHES (DECIMALS)	MILLIMETERS
	.00394	.1	15/32	.46875	11.9063
	.00787	.2		.47244	12.00
	.01181	.3	31/64	.484375	12.3031
1/64	.015625	.3969	1/2	.5000	12.70
	.01575	.4		.51181	13.00
	.01969	.5	33/64	.515625	13.0969
	.02362	.6	17/32	.53125	13.4938
	.02756	.7	35/64	.546875	13.8907
1/32	.03125	.7938		.55118	14.00
	.0315	.8	9/16	.5625	14.2875
	.03543	.9	37/64	.578125	14.6844
	.03937	1.00		.59055	15.00
3/64	.046875	1.1906	19/32	.59375	15.0813
1/16	.0625	1.5875	39/64	.609375	15.4782
5/64	.078125	1.9844	5/8	.625	15.875
	.07874	2.00		.62992	16.00
3/32	.09375	2.3813	41/64	.640625	16.2719
7/64	.109375	2.7781	21/32	.65625	16.6688
	.11811	3.00		.66929	17.00
1/8	.125	3.175	43/64	.671875	17.0657
9/64	.140625	3.5719	11/16	.6875	17.4625
5/32	.15625	3.9688	45/64	.703125	17.8594
	.15748	4.00		.70866	18.00
11/64	.171875	4.3656	23/32	.71875	18.2563
3/16	.1875	4.7625	47/64	.734375	18.6532
	.19685	5.00		.74803	19.00
13/64	.203125	5.1594	3/4	.7500	19.05
7/32	.21875	5.5563	49/64	.765625	19.4469
15/64	.234375	5.9531	25/32	.78125	19.8438
	.23622	6.00		.7874	20.00
1/4	.2500	6.35	51/64	.796875	20.2407
17/64	.265625	6.7469	13/16	.8125	20.6375
	.27559	7.00		.82677	21.00
9/32	.28125	7.1438	53/64	.828125	21.0344
19/64	.296875	7.5406	27/32	.84375	21.4313
5/16	.3125	7.9375	55/64	.859375	21.8282
	.31496	8.00		.86614	22.00
21/64	.328125	8.3344	7/8	.875	22.225
11/32	.34375	8.7313	57/64	.890625	22.6219
	.35433	9.00		.90551	23.00
23/64	.359375	9.1281	29/32	.90625	23.0188
3/8	.375	9.525	59/64	.921875	23.4157
25/64	.390625	9.9219	15/16	.9375	23.8125
	.3937	10.00		.94488	24.00
13/32	.40625	10.3188	61/64	.953125	24.2094
27/64	.421875	10.7156	31/32	.96875	24.6063
	.43307	11.00		.98425	25.00
7/16	.4375	11.1125	63/64	.984375	25.0032
29/64	.453125	11.5094	1	1.0000	25.4001

Conversion Chart

METRIC/U.S. CUSTOMARY UNIT EQUIVALENTS

Multiply:	by:	to get:	Multiply:	by:	to get:
ACCELERATION					
feet/sec²	x 0.3048	= meters/sec² (m/s²)	x 3.281	– feet/sec²	
inches/sec²	x 0.0254	= meters/sec² (m/s²)	x 39.37	= inches/sec²	
ENERGY OR WORK (watt–second = joule = newton–meter)					
foot–pounds	x 1.3558	= joules (J)	x 0.7376	= foot–pounds	
calories	x 4.187	= joules (J)	x 0.2388	= calories	
Btu	x 1055	= joules (J)	x 0.000948	= Btu	
watt–hours	x 3600	= joules (J)	x 0.0002778	= watt–hours	
kilowatt–hrs.	x 3.600	= megajoules (MJ)	x 0.2778	= kilowatt–hrs	
FUEL ECONOMY AND FUEL CONSUMPTION					
miles/gal	x 0.42514	= kilometers/liter (km/L)	x 2.3522	= miles/gal	
Note:					
235.2/(mi/gal) = liters/100km					
235.2/(liters/100 km) = mi/gal					
LIGHT					
footcandles	x 10.76	= lumens/meter² (lm/m²)	x 0.0929	= footcandles	
PRESSURE OR STRESS (newton/sq meter = pascal)					
inches Hg(60˚F)	x 3.377	= kilopascals (kPa)	x 0.2961	= inches Hg	
pounds/sq in	x 6.895	= kilopascals (kPa)	x 0.145	= pounds/sq in	
inches H₂O(60˚F)	x 0.2488	= kilopascals (kPa)	x 4.0193	= inches H₂O	
bars	x 100	= kilopascals (kPa)	x 0.01	= bars	
pounds/sq ft	x 47.88	= pascals (Pa)	x 0.02088	= pounds/sq ft	
POWER					
horsepower	x 0.746	= kilowatts (kW)	x 1.34	= horsepower	
ft–lbf/min	x 0.0226	= watts (W)	x 44.25	= ft–lbf/min	
TORQUE					
pounds–inches	x 0.11298	= newton–meters (N·m)	x 8.851	= pound–inches	
pound–feet	x 1.3558	= newton–meters (N·m)	x 0.7376	= pound–feet	
VELOCITY					
miles/hour	x 1.6093	= kilometers/hour (km/h)	x 0.6214	= miles/hour	
feet/sec	x 0.3048	= meters/sec (m/s)	x 3.281	= feet/sec	
kilometers/hr	x 0.27778	= meters/sec (m/s)	x 3.600	= kilometers/hr	
miles/hour	x 0.4470	= meters/sec (m/s)	x 2.237	= miles/hour	

COMMON METRIC PREFIXES

mega	(M)	= 1 000 000	or 10⁶	centi	(c)	= 0.01	or 10⁻²
kilo	(k)	= 1 000	or 10³	milli	(m)	= 0.001	or 10⁻³
hecto	(h)	= 100	or 10²	micro	(μ)	= 0.000 001	or 10⁻⁶

METRIC/U.S. CUSTOMARY UNIT EQUIVALENTS

Multiply:	by:	to get:	Multiply:	by:	to get:
LINEAR					
inches	x 25.4	= millimeters (mm)	x 0.03937	= inches	
feet	x 0.3048	= meters (m)	x 3.281	= feet	
yards	x 0.9144	= meters (m)	x 1.0936	= yards	
miles	x 1.6093	= kilometers (km)	x 0.6214	= miles	
inches	x 2.54	= centimeters (cm)	x 0.3937	= inches	
microinches	x 0.0254	= micrometers (μm)	x 39.37	= microinches	
AREA					
inches²	x 645.16	= millimeters²(mm²)	x 0.00155	= inches²	
inches²	x 6.452	= centimeters²(cm²)	x 0.155	= inches²	
feet²	x 0.0929	= meters²(m²)	x 10.764	= feet²	
yards²	x 0.8361	= meters²(m²)	x 1.196	= yards²	
acres	x 0.4047	= hectares (10⁴m²)			
		ha	x 2.471	= acres	
miles²	x 2.590	= kilometers² (km²)	x 0.3861	= miles²	
VOLUME					
inches³	x 16387	= millimeters³ (mm³)	x 0.000061	= inches³	
inches³	x 16.387	= centimeters³ (cm³)	x 0.06102	= inches³	
inches³	x 0.01639	= liters (L)	x 61.024	= inches³	
quarts	x 0.94635	= liters (L)	x 1.0567	= quarts	
gallons	x 3.7854	= liters (L)	x 0.2642	= gallons	
feet³	x 28.317	= liters (L)	x 0.03531	= feet³	
feet³	x 0.02832	= meters³ (m³)	x 35.315	= feet³	
fluid oz	x 29.57	= milliliters (mL)	x 0.03381	= fluid oz	
yards³	x 0.7646	= meters³ (m³)	x 1.3080	= yards³	
teaspoons	x 4.929	= milliliters (mL)	x 0.2029	= teaspoons	
cups	x 0.2366	= liters (L)	x 4.227	= cups	
MASS					
ounces (av)	x 28.35	= grams (g)	x 0.03527	= ounces (av)	
pounds (av)	x 0.4536	= kilograms (kg)	x 2.2046	= pounds (av)	
tons (2000 lb)	x 907.18	= kilograms (kg)	x 0.001102	= tons (2000 lb)	
tons (2000 lb)	x 0.90718	= metric tons (t)	x 1.1023	= tons (2000 lb)	
FORCE					
ounces–f (av)	x 0.278	= newtons (N)	x 3.597	= ounces–f (av)	
pounds–f (av)	x 4.448	= newtons (N)	x 0.2248	= pounds f (av)	
kilograms–f	x 9.807	= newtons (N)	x 0.10197	= kilograms–f	
TEMPERATURE					

°F -40 0 32 40 80 98.6 120 160 200 212 240 280 320 °F

°C -40 -20 0 20 40 60 80 100 120 140 160 °C

°Celsius = 0.556 (°F – 32)   °F = (1.8 °C) + 32

Bolt Torquing Chart

SAE Standard/Foot-Pounds							**Metric Standard**						
Grade of Bolt	SAE 1 & 2	SAE 5	SAE 6	SAE 8			*Grade of Bolt*		5D	.8G	10K	12K	
Min. Ten. Strength	64,000 P.S.I.	105,000 P.S.I.	133,000 P.S.I.	150,000 P.S.I.			Min. Ten. Strength		71,160 P.S.I.	113,800 P.S.I.	142,200 P.S.I.	170,679 P.S.I.	
Markings on Head	⬢	⬡	⊕	✳	Size of Socket or Wrench Opening		Markings on Head		5D	.8G	10K	12K	Size of Socket or Wrench Opening
U.S. Standard	Foot Pounds				U.S. Regular		Metric						Metric
Bolt Dia.					Bolt Head	Nut	Bolt Dia.	U.S. Dec. Equiv.	Foot Pounds				Bolt Head
1/4	5	7	10	10.5	3/8	7/16	6mm	.2362	5	6	8	10	10mm
5/16	9	14	19	22	1/2	9/16	8mm	.3150	10	16	22	27	14mm
3/8	15	25	34	37	9/16	5/8	10mm	.3937	19	31	40	49	17mm
7/16	24	40	55	60	5/8	3/4	12mm	.4720	34	54	70	86	19mm
1/2	37	60	85	92	3/4	13/16	14mm	.5512	55	89	117	137	22mm
9/16	53	88	120	132	7/8	7/8	16mm	.6299	83	132	175	208	24mm
5/8	74	120	167	180	15/16	1	18mm	.7090	111	182	236	283	27mm
3/4	120	200	280	296	1-1/8	1-1/8	22mm	.8661	182	284	394	464	32mm

Wire Size Chart

The selection of the correct gage primary wire is very important for automotive and other low-voltage applications to ensure safe and reliable performance.

When too small a gage primary wire is used, a voltage drop occurs due to electrical resistance. On lighting equipment there is also a loss of candlepower (visible light measurement).

The two factors that should always be considered in selecting an adequate gage primary wire are (1) the total amperage the circuit will carry and (2) the total length of wire used in each circuit, including the return.

Select the correct gage wire from the wiring diagram and primary wiring guide below. Allowance for the return circuits, including grounded returns, has been computed on the recommendations below. The length cable should be determined by totaling both wires in a two-wire circuit.

Total Approx. Circuit Amperes	Total Circuit Watts	Total Candle Power	Wire Gage (for length in feet)											
12V	12V	12V	3'	5'	7'	10'	15'	20'	25'	30'	40'	50'	75'	100'
1.0	12	6	18	18	18	18	18	18	18	18	18	18	18	18
1.5		10	18	18	18	18	18	18	18	18	18	18	18	18
2	24	16	18	18	18	18	18	18	18	18	18	18	16	16
3		24	18	18	18	18	18	18	18	18	18	18	14	14
4	48	30	18	18	18	18	18	18	18	16	16	12	12	
5		40	18	18	18	18	18	18	18	16	14	12	12	
6	72	50	18	18	18	18	18	18	16	16	16	14	12	10
7		60	18	18	18	18	18	18	16	16	14	14	10	10
8	96	70	18	18	18	18	18	16	16	16	14	12	10	10
10	120	80	18	18	18	18	16	16	16	14	12	12	10	10
11		90	18	18	18	18	16	16	14	14	12	12	10	8
12	144	100	18	18	18	18	16	16	14	14	12	12	10	8
15		120	18	18	18	18	14	14	12	12	12	10	8	8
18	216	140	18	18	16	16	14	14	12	12	10	10	8	8
20	240	160	18	18	16	16	14	12	10	10	10	10	8	6
22	264	180	18	18	16	16	12	12	10	10	10	8	6	6
24	288	200	*18	18	16	16	12	12	10	10	10	8	6	6
30			18	16	16	14	10	10	10	10	10	6	4	4
40			18	16	14	12	10	10	8	8	6	6	4	2
50			16	14	12	12	10	10	8	8	6	6	2	2
100			12	12	10	10	6	6	4	4	4	2	1	1/0
150			10	10	8	8	4	4	2	2	2	1	2/0	2/0
200			10	8	8	6	4	4	2	2	1	1/0	4/0	4/0

*18 AWG indicated above this line could be 20 AWG electrically—18 AWG is recommended for mechanical strength.

How to use chart

1
Measure Length of Wire in Circuit—chart applies to ground return. Two-wire circuits will be total of both wire lengths. Be sure to indicate both vehicles on auto and trailer applications.

2
Find the total amperes, watts, or candlepower and choose nearest value in proper column.

3
Move horizontally to proper footage column and find nearest wire gage.

4
For 6 volt applications use two wire sizes larger. Example: If chart shows 16 ga. for 12 volt system use 14 ga. on 6 volt system.

Based on maximum of 10% voltage drop

Primary Wiring Amperage Guide

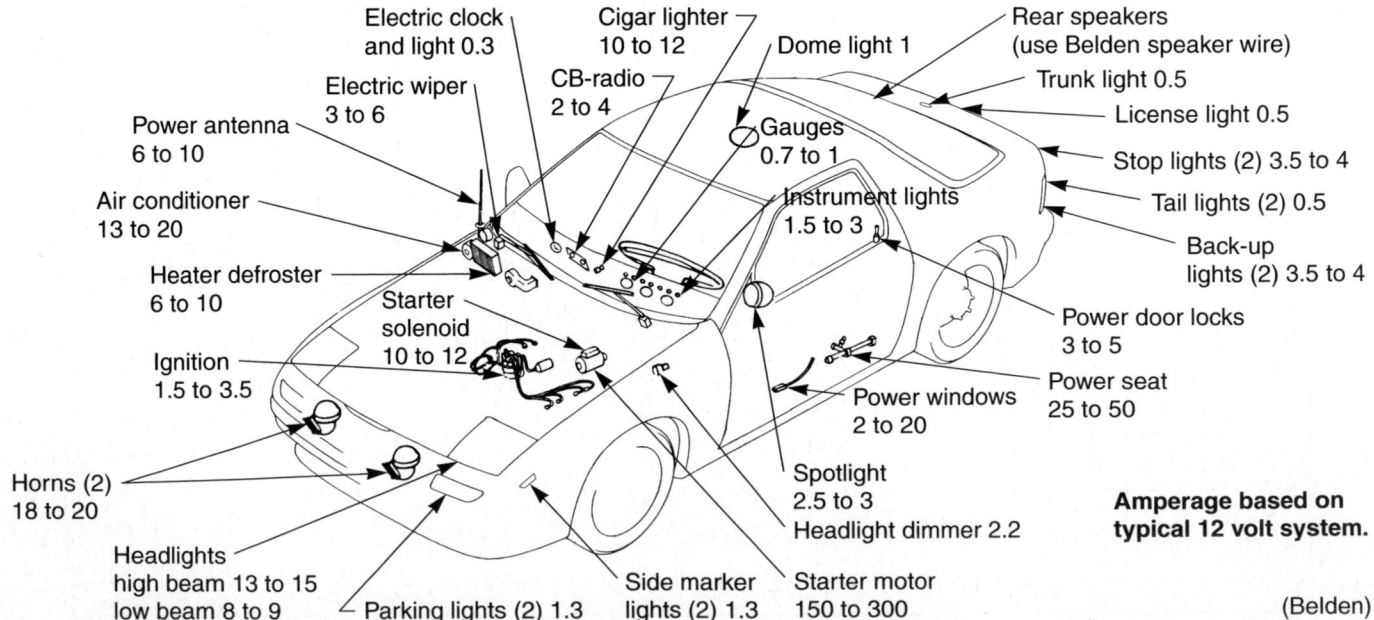

Electric clock and light 0.3
Electric wiper 3 to 6
Power antenna 6 to 10
Air conditioner 13 to 20
Heater defroster 6 to 10
Ignition 1.5 to 3.5
Horns (2) 18 to 20
Headlights high beam 13 to 15 low beam 8 to 9
Parking lights (2) 1.3
Starter solenoid 10 to 12
Cigar lighter 10 to 12
CB-radio 2 to 4
Dome light 1
Gauges 0.7 to 1
Instrument lights 1.5 to 3
Side marker lights (2) 1.3
Starter motor 150 to 300
Spotlight 2.5 to 3
Headlight dimmer 2.2
Power windows 2 to 20
Rear speakers (use Belden speaker wire)
Trunk light 0.5
License light 0.5
Stop lights (2) 3.5 to 4
Tail lights (2) 0.5
Back-up lights (2) 3.5 to 4
Power door locks 3 to 5
Power seat 25 to 50

Amperage based on typical 12 volt system.

(Belden)

SAE-Recommended Abbreviations/Acronyms

Term	Abbreviation/Acronym	Term	Abbreviation/Acronym
Accelerator Pedal	AP	Inertia Fuel Shutoff	IFS
Air Cleaner	AC	Intake Air	IA
Air Conditioning	A/C	Intake Air Temperature	IAT
Automatic Transmission	A/T		
Automatic Transaxle	A/T	Knock Sensor	KS
Barometric Pressure	BARO	Malfunction Indicator Lamp	MIL
Battery Positive Voltage	B+	Manifold Absolute Pressure	MAP
		Manifold Differential Pressure	MDP
Camshaft Position	CMP	Manifold Surface Temperature	MST
Carburetor	CARB	Manifold Vacuum Zone	MVZ
Charge Air Cooler	CAC	Mass Airflow	MAF
Closed Loop	CL	Mixture Control	MC
Closed Throttle Position	CTP	Multiport Fuel Injection	MFI
Clutch Pedal Position	CPP		
Continuous Fuel Injection CFI		Nonvolatile Random Access Memory	NVRAM
Continuous Trap Oxidizer	CTOX		
Crankshaft Position	CKP	On-Board Diagnostic	OBD
		Open Loop	OL
Data Link Connector	DLC	Oxidation Catalytic Converter	OC
Diagnostic Test Mode	DTM	Oxygen Sensor	O2S
Diagnostic Trouble Code	DTC		
Direct Fuel Injection	DFI	Park/Neutral Position	PNP
		Periodic Trap Oxidizer	PTOX
Early Fuel Evaporation	EFE	Positive Crankcase Ventilation	PCV
EGR Temperature	EGRT	Power Steering Pressure	PSP
Electronically Erasable Programmable		Powertrain Control Module	PCM
Read Only Memory	EEPROM	Programmable Read Only Memory	PROM
Electronic Ignition	EI	Pulsed Secondary Air Injection	PAIR
Engine Control	EC		
Engine Control Module	ECM	Random Access Memory	RAM
Engine Coolant Level	ECL	Read Only Memory	ROM
Engine Coolant Temperature	ECT	Relay Module	RM
Engine Modification	EM		
Engine Speed	RPM	Scan Tool	ST
Erasable Programmable		Secondary Air Injection	AIR
Read Only Memory	EPROM	Sequential Multiport Fuel Injection	SFI
Evaporative Emission	EVAP	Service Reminder Indicator	SRI
Exhaust Gas Recirculation	EGR	Smoke Puff Limiter	SPL
		Supercharger	SC
Fan Control	FC	Supercharger Bypass	SCB
Flash Electrically Erasable Programmable		System Readiness Test	SRT
Read Only Memory	FEEPROM		
Flash Erasable Programmable		Thermal Vacuum Valve	TVV
Read Only Memory	FEPROM	Third Gear	3GR
Flexible Fuel	FF	Three Way + Oxidation Catalytic Converter	TWC+OC
Fourth Gear	4GR	Three Way Catalytic Converter	TWC
Fuel Pump	FP	Throttle Body	TB
Fuel Trim	FT	Throttle Body Fuel Injection	TBI
		Throttle Position	TP
Generator	GEN	Torque Converter Clutch	TCC
Governor Control Module	GCM	Transmission Control Module	TCM
Ground	GND	Transmission Range	TR
		Turbocharger	TC
Heated Oxygen Sensor	HO2S		
		Vehicle Speed Sensor	VSS
Idle Air Control	IAC	Voltage Regulator	VR
Idle Speed Control	ISC	Volume Airflow	VAF
Ignition Control	IC		
Ignition Control Module	ICM	Warm Up Three Way Catalytic Converter	WU-TWC
Indirect Fuel Injection	IFI	Wide Open Throttle	WOT

Glossary of Terms

A

Acceleration sensor: Provides acceleration data to computer in electronic shock absorber system.

Accelerator: A driver-operated pedal used to control engine speed.

Accumulator: Receiver-drier combination used to remove moisture and store additional refrigerant. Gas-charged cylinder or chamber used to store brake fluid. Normally used in anti-lock brake systems.

Accumulator piston: Assembly used to cushion the application of a band or clutch in an automatic transmission.

Active sensor: Computer sensor that generates its own voltage signal in response to a change in a condition.

Active suspension system: A method of ride control that uses struts that are controlled electronically by the driver. A switch is used to adjust the vehicle's ride firmness to meet various driving conditions.

Actuator: Any output device controlled by a computer.

Adaptive strategy: Feature in an ECM that allows it to adjust its operation to compensate for sensor, actuator, or vehicle deterioration.

Add-on transmission cooler: Heat exchanger installed in front of the radiator to provide extra cooling for an automatic transmission.

Additives: Any material added to automotive oils to improve its performance under certain conditions, such as to decrease an engine oil's tendency to break down when heated.

Advance: Ignition timing setting so that a spark occurs earlier or more degrees before top dead center.

Air bag system: A protective system that inflates a protective bag in a frontal or side collision. Designed to act as a cushion between the front seat occupants and the vehicle's interior.

Air compressor: A pump that forces air, under pressure, into a storage tank.

Air conditioning: Process in which the air coming into the passenger compartment is cleaned, cooled, and dehumidified.

Air dam: A metal or plastic shroud placed beneath the front of a vehicle. Designed to channel air flow over the radiator and condenser.

Air filter: A pleated paper element installed inside the air cleaner. Used to filter dirt, dust, and foreign debris from the air stream entering the engine.

Air gap: The space between the electrodes of a spark plug, starting motor and alternator armatures, etc.

Air injection system: A method of reducing hydrocarbons and carbon monoxide emissions by forcing fresh air into the exhaust ports of the engine. Also called an Air Injection Reaction system or AIR.

Air pollution: Any release of harmful substances into the atmosphere by engine operation or other causes.

Air pump: Belt-driven pump that provides input for the air injection system.

Air shocks: Shock absorbers that use air pressure, rather than springs, to maintain vehicle height.

Air spring: Air-filled rubber cylinder that is lighter than an equivalent coil spring.

Air-cooled: An engine or other object that is cooled by passing a stream of air over its surface.

Air-fuel ratio: Ratio by weight between the air and fuel that makes up the engine's fuel charge.

Align: To bring various parts of a unit into their correct positions with respect to each other or to a predetermined location.

Alignment rack: Fixture onto which the vehicle is driven for alignment measurement and adjustment.

All-wheel drive: Type of drive system that transmits power to all wheels.

Alloy: A mixture of two or more metals. An example is solder, which is a mixture of lead and tin.

Alternating current (ac): An electrical current that moves in one direction and then the other.

Alternative engine: Engine types other than traditional internal combustion, four-stoke-cycle, piston engine.

Alternative fuel: Fuels other than gasoline and diesel fuel, such as compressed natural gas.

Alternator: An ac generator in which alternating current is first generated, then changed into direct (dc) current.

Aluminum: A metal known for its lightweight and good heat dissipation characteristics. It is often alloyed with other metals.

Ambient temperature switch: Temperature sensitive switch that is designed to prevent air conditioner operation when outdoor temperatures are below a set point.

Ammeter: Instrument used to measure the flow of electric current in a circuit in amperes. Normally connected in series in a circuit.

Ampere: The unit of measurement for the flow of electric current.

Amplifier: Electronic circuit that uses a small current to control a much larger current.

Anaerobic sealer: A chemical sealer that cures in the absence of air.

Analog signal: A signal that continually changes in strength.

Anti-lock brake system (ABS): A computer-controlled system that is part of the base brake system. The system cycles the brakes on and off to prevent wheel lockup and skidding.

Anti-rattle clip: One or more metal components designed to keep brake pads from vibrating and rattling.

Antifreeze: A liquid, usually ethylene or propylene glycol, that is added to water in a vehicle's cooling system. This mixture serves as the engine's coolant.

Antifriction bearing: A bearing that uses balls or rollers between a journal and bearing surface to decrease friction.

Antiknock additives: Chemical added to gasoline to increases its resistance to spark knock.

Armature: The revolving part in an alternator or motor. The moving part in a relay or horn.

Asbestos: A mineral material that has great heat resistant characteristics. Once widely used in brake and clutch linings. Asbestos is a known cancer causing agent.

ASE: Abbreviation for National Institute For Automotive Service Excellence.

Aspect ratio: The ratio between tire section height and section width.

Atmospheric pressure: The pressure exerted by the earth's atmosphere on all objects. Measured with reference to the pressure at sea level, which is around 14.7 psi (101 kPa).

Atomizes: The process of breaking a liquid, such as gasoline, into tiny drops and mixing with air.

Automatic transmission: A transmission that uses fluid pressure to shift itself. Shift points and range is determined by road speed, engine loading, altitude, and throttle position. Can also be computer-controlled.

Automatic transmission fluid (ATF): Mineral oil with special additives to make it compatible with clutch and band materials and to prevent foaming and breakdown.

Auxiliary oil cooler: A system designed to remove excess heat from engine oil by either passing it through a heat exchanger exposed to the air or by channeling engine coolant through an oil filter adapter mounted on the engine.

Axle: Shaft or shafts used to transmit driving force to the wheels.

Axle housing: A metal enclosure for the drive axles and differential. It is partially filled with differential fluid and other additives as needed.

B

Back pressure: The exhaust system's resistance to the flow of exhaust gases.

Backfiring: A condition where unburned fuel in the intake or exhaust manifold ignites.

Backing plate: A metal plate that holds the wheel cylinder, shoes, and other parts inside a drum brake.

Backlash: The clearance or play between two parts, such as the teeth of two gears.

Backup light: White lights in the rear of the vehicle. Lights turn on whenever the transmission is placed in reverse.

Baffle: An obstruction used to check of deflect the flow of gases, liquids, sounds, etc.

Balance shaft: One or more offset shafts geared to either the crankshaft or camshaft. Used to counterbalance crankshaft vibration.

Ball bearing (Antifriction): A bearing consisting of an inner and outer hardened steel race separated by a series of hardened steel balls.

Ball joint: A flexible joint which utilizes a ball and socket type of construction. Used primarily in steering linkage and steering knuckle pivot support.

Band: Holding member in an automatic transmission. Used to stop clutch rotation during certain transmission gear ranges.

Barometric pressure (BARO) sensor: A sensor that measures atmospheric pressure.

Base brakes: The hydraulic and friction members of the automotive brake system. Does not include any boosters or ABS components.

Battery: A device containing one or more cells that produce electricity through electrochemical action.

Battery cables: The heavy wires connecting the battery to the vehicle's electrical system.

Battery rating: Standardized measurement of a battery's ability to deliver an acceptable level of energy under specified conditions. Standards established by the Battery Council International (BCI).

Battery temperature sensor: Sensor installed in some batteries to control overcharging.

BDC: Abbreviation for bottom dead center.

Bead: The steel reinforced edge of a tire that engages the wheel rim.

Bearing: A part on which a shaft, journal, or pivot turns.

Bell housing: The metal shell surrounding the clutch assembly that bolts to the rear of the engine.

Belt: Fabric made of steel or other material that is placed between body plies and tread.

Belt tensioner: Spring loaded pulley which is used to keep tension and adjustment on serpentine belts.

Belt-driven oil pump: Type of pump used with dry sump lubrication systems. Oil pump is driven by the crankshaft or off another accessory pulley, which is driven by the crankshaft.

Belted bias tire: Tire construction method using crisscrossed plies with additional belts under the tread area.

Bendix drive: A self-engaging starter drive gear. It is mounted on a screw shaft attached to the starter motor armature. Also is used as a general term to describe the overrunning clutch on other starters.

Bezel: Crimped edge of metal used to secure a transparent instrument cover.

Bias ply tire: A type of tire construction in which the plies crisscross from bead to bead. No additional belts are used under the tread area.

Bleed: The process of removing air or fluid from a closed system, such as the brakes.

Blend door: Door installed in the air conditioning blower case. Controls temperature of the air going to the passenger compartment.

Block: The part of the engine containing the crankshaft, pistons, and cylinders. It is made of cast iron or aluminum and is the foundation for the entire engine.

Block heater: A heating device installed in place of one of the block's freeze plugs. Used to heat an engine during cold weather.

Blowby: Oil vapors and other emissions that leak past piston rings into the crankcase.

Blower motor: An electric motor used to force air to move through the evaporator in an air conditioning system.

Blueprinting: Process of assembling an engine to *exact* specifications.

Body control module (BCM): On-board computer responsible for controlling such functions as interior climate, radio, instrument cluster readings, and on some vehicles, cellular telephones. May also interact with the engine control module.

Body lubrication: Applying oil and grease to such friction points as hinges and latches.

Boost pressure: A measure of engine power provided by a supercharger or turbocharger.

Boot: Flexible pleated covers placed over the CV joints of a front-wheel-drive vehicle to keep road dirt out of them.

Bore: The diameter of a cylinder. Sometimes used to refer to the cylinder itself.

Boss: A rib or enlarged area designed to strengthen a portion or area of an object.

Brake anchor: Steel stud upon which one end of the brake shoes is attached to or firmly rests against. Anchor is normally cast as part of the backing plate.

Brake booster: Component operated by vacuum or power steering system to decrease braking effort needed.

Brake fluid: A special fluid compound used in hydraulic brake systems. It must meet exacting specifications, such as resistance to heating, freezing, and thickening.

Brake hoses: Flexible hose which connects solid brake lines to wheel brake assemblies.

Brake lines: Metal tubing and rubber hoses connecting master cylinder to wheel brake assemblies.

Brake pads: Replaceable friction surfaces mounted on caliper of disc brake system.

Brake pressure sensor: Provides brake fluid pressure data to computer in electronic shock absorber system.

Brake shoes: Curved, replaceable friction surfaces used with drum brakes.

Brake system: Components that are used to stop a vehicle.

Brake warning light: Dashboard indicator that warns of low brake system hydraulic pressure.

Braking ratio: A comparison of the front wheel to rear wheel braking effort. On most vehicles, the ratio is 55%-60% for the front brakes and 45%-50% for the rear brakes.

Break-away torque: In a limited-slip differential, the amount of torque needed to make one axle rotate the clutches.

Break-in: The process of wearing in two or more parts that have been replaced or reconditioned.

Breakout box: Diagnostic tool used to troubleshoot electronic systems. Usually used only as a last resort as it is time consuming to use.

Brushes: Bars of carbon, copper, or other conductive material that ride on the slip rings of a motor or alternator.

Burnish: To bring the surface of a metal to a high shine by rubbing it with a hard, smooth object.

Bushing: A smooth, removable liner used as a bearing for a shaft, piston pin, etc.

Butane: Petroleum gas that is liquefied under pressure. Sometimes referred to as liquefied petroleum gas and often combined with propane.

Bypass: To move around or detour from the normal route taken by a flowing substance, such as electricity, air, or fluid.

Bypass valve: Valve used to permit coolant flow when the thermostat is closed.

C

CAFE: Abbreviation for Corporate Average Fuel Economy.

Calibration gas: A test gas used to zero an exhaust gas analyzer's scales.

Caliper: A disc brake component which forms the cylinder and contains the piston and brake pads. It provides braking by producing a clamping action on a rotating disc.

Cam ground piston: Piston ground slightly egg-shaped. When heated, it becomes round.

Cam lobe: The offset portion of a shaft that, as it turns, will impart motion to another part.

Camber: The outward or inward tilt of a wheel from its centerline.

Camshaft: A shaft with offset lobes used to operate the valve train.

Camshaft gear: A gear that is used to drive the camshaft.

Camshaft position sensor (CPS): A sensor used by a computer control system to reference camshaft position in order to adjust spark timing, fuel, etc.

Capacitance: The property of a capacitor or condenser that permits it to receive and retain an electrical charge.

Carbon dioxide (CO_2): A dry chemical mixture that is an excellent fire retardant and is used to make dry ice. Also is produced as an end product of catalyzed exhaust emissions.

Carbon monoxide (CO): A toxic byproduct of partial burning of fuel. It is a deadly colorless, odorless gas.

Carbon trace: Lines of carbon found inside a distributor or on an ignition coil. Usually an indication of arcing.

Carburetor: A device used for many years to mix gasoline with air in correct proportions and deliver it to the engine.

Cardan universal joint: A type of universal joint that has yokes at right angles to each other.

Case-harden: A process of heating a piece of steel to harden its surface while the inside remains relatively soft.

Caster: The backward or forward tilt of a wheel away from its centerline.

Catalyst monitor: Name given to any post-converter oxygen sensor. Used in OBD II vehicles.

Catalytic converter: An emissions control device which uses the elements platinum, palladium, and rhodium as catalysts to reduce harmful exhaust emissions.

CD-ROM: Abbreviation for compact disc-read only memory. Used to store computer software.

Cell: The individual compartments in a battery which contain positive and negative plates suspended in electrolyte.

Center link: Also called a drag link, it is used to connect the pitman arm to the tie-rods in a parallelogram steering system.

Center of gravity: The point on an object on which it could be balanced.

Center support bearing: A bearing used to mount a two-piece Hotchkiss drive shaft.

Central processing unit (CPU): A microprocessor inside an electronic control unit responsible for controlling the ECM's operations.

Centrifugal advance mechanism: A series of weights used to advance engine timing as engine speed increases.

Centrifugal force: A force which tends to keep a rotating object away from the center of rotation.

Cetane number: A measurement of the performance characteristics of diesel fuel in cold weather.

Chamfer: To bevel an edge on an object at the edge of a hole.

Charge: A given or specified weight of refrigerant in an air conditioning system. Also the electrical rate of flow that passes through the battery to restore it to full power.

Charging station: A cart containing a manifold gauge, refrigerant tank, and pump. Used to recover and recharge air conditioning systems.

Chassis: The framework of a vehicle without its body panels.

Chassis dynamometer: A machine used to measure engine power at the vehicle's wheels.

Check engine light: See *Malfunction indicator light.*

Check valve: A valve that permits flow in only one direction.

CID: Abbreviation for cubic inch displacement.

Circuit: A source, resistance, and path to carry a flow from the source to the resistance and back to the source.

Circuit breaker: Device that interrupts current if a circuit is overloaded or a short occurs. Unlike a fuse, it can be reset.

Climate control computer: ECM that controls temperature levels in the vehicle's passenger compartment.

Closed cooling system: A type of cooling system that uses an overflow bottle.

Closed loop: Engine ECM operation mode used when an engine has reached its optimum operating temperature and conditions. When in closed loop, most ECM's rely on the oxygen sensor to determine if engine operating conditions need to be changed.

Cloud point: Temperature at which wax separates out of diesel fuel.

Cluster gear: A cluster of gears that are all cut on one long gear blank. The cluster gears ride in the bottom of a manual transmission, providing a connection between the transmission input and output shaft.

Clutch: Device that allows the driver to engage or disengage the engine and transmission.

Clutch fork: Lever that forces the throw-out (release) bearing into pressure plate of clutch.

Clutch housing: Cast iron or aluminum housing that surrounds the flywheel and clutch mechanism.

Clutch linkage: A cable or other mechanism which transfers movement from the clutch pedal to the throw-out fork.

Clutch master cylinder: The device that produces the hydraulic pressure needed to operate the clutch. Also known as a slave cylinder.

Clutch pilot bearing: A small bronze bushing or ball bearing placed in the end of the crankshaft or in the center of the flywheel. Used to support the outboard end of the transmission input shaft.

Clutch piston: A piston that moves back and forth inside the clutch drum to clamp driving and driven discs together.

Clutch start switch: Switch that prevents the engine from starting unless the clutch pedal is depressed.

Coast side: Concave side of gear tooth.

Coefficient of friction: The amount of friction developed when two objects are moved across each other. Coefficient of friction is calculated by dividing the force needed to push a load across a given surface.

Cogged belt: Rubber belt with built-in cogs (teeth) that engage the teeth of camshaft and crankshaft sprockets to form a positive drive.

Coil: Electrical transformer designed to increase voltage.

Coil spring: Length of spring steel rod wound into a spiral.

Coil wire: Conductor carrying high voltage from the coil to the distributor.

Cold cranking amps: Measurement of cranking amperes that a battery can deliver over a period of 30 seconds at 0°F (–18°C)

Cold patching: Repair method used to seal leaks in plastic fuel tanks.

Cold start injector: Valve that supplies extra fuel for cold starts.

Combination valve: One that includes both a suction valve and an expansion valve (used in air conditioning systems). Also braking system valve that can function as a metering or proportioning valve and a brake warning light switch.

Combustion chamber: Area at the top of the cylinder where a spark plug ignites the compressed air-fuel mixture.

Commutator: Sliding electrical connection between motor windings and brushes.

Compact spare: Small-diameter spare tire for use in emergencies.

Compensating port: A small hole placed in a master cylinder to permit fluid to return to the reservoir.

Composite headlamp: Reflector and lens assembly used with halogen bulbs. Designed to be used in a specific vehicle.

Compression: Opposite of tension. Reduction in volume, such as compressing a gas. Also applying pressure to a spring to reduce its length.

Compression ignition: Ignition of an air-fuel mixture by heat that results from high pressure (compression).

Compression ratio: The relationship of cylinder volumes with the piston at TDC and at BDC.

Compression rings: The top set of piston rings designed to seal between the piston and cylinder to prevent escape of gases from the combustion chamber.

Compression stroke: The portion of the piston's movement devoted to compressing the fuel mixture in the engine's cylinder.

Compressor: An engine-driven device used to compress refrigerant and causes it to flow in the air conditioning system.

Computer analyzer: A more complex testing instrument than a scanner.

Computer-controlled carburetor: One that calculates and sets air-fuel ratio based on input from sensors. Also referred to as a feedback carburetor.

Conduction: The transfer of heat from one object to another by being in direct contact.

Conductor: Any material that can form a path for electrical current.

Connecting rod: The connecting link between the crankshaft and the pistons.

Connecting rod bearing: Inserts which fit into the connecting rods and rides on the crankshaft journals.

Connecting rod cap: The removable lower part of the connecting rod.

Constant mesh gears: Gears that are always in mesh with each other.

Constant velocity axle: An axle that utilizes two constant velocity joints to effect torque transfer to the driving wheels.

Contact pattern: The area on a ring gear tooth that matches with the pinion gear tooth.

Continuous throttle body injection (CTBI): A type of fuel injection in which the fuel injectors are constantly open. The amount of fuel to the cylinders is controlled by the ECM.

Continuously variable transmission: One that has an infinite number of driving ratios and uses belts and pulleys, rather than planetary gearsets.

Contraction: Reduction in the size of an object when it cools.

Control arm: Movable lever arm that forms part of a vehicle's suspension system.

Control arm bushing: Sleeve that allows control arm to swing up and down.

Control rack: A toothed rod inside a mechanical diesel injector pump which rotates the pump plunger to control the quantity of injected fuel.

Convection: A transfer of heat from one object to another by heating the air surrounding the object.

Coolant: Ethylene glycol or other liquid used in a cooling system.

Coolant recovery system: A plastic bottle and hose that is used with a closed cooling system to recover and provide additional coolant when needed.

Coolant temperature sensor (CTS): A sensor that monitors engine coolant temperature. Used by the engine control ECM to determine how much fuel is needed and when to enter closed loop operation.

Cooling fins: A series of thin metal strips placed between cooling passages to help dissipate heat.

Cooling system: Radiator and other components that allow a coolant to circulate and maintain a constant engine operating temperature.

Corrosion inhibitors: Any chemical added to a substance that prevents the formation of oxidation or corrosion.

Countershaft: Shaft on which cluster gears are mounted. They transmit force from input gears to output gears.

Countersink: To make a counterbore so that the head of a screw may set flush or below the surface.

Cradle: Segmented frame on a front-wheel drive vehicle that supports the engine and front suspension.

Crankcase: The part of the engine that surrounds the crankshaft. Different from the pan, which simply covers the crankcase.

Crankshaft: The main shaft which supports the connecting rods and turns piston reciprocation into motion.

Crankshaft gear: A gear that is pressed on the crankshaft. Used to drive a chain or belt which drives the camshaft gear.

Crankshaft position (CKP) sensor: A sensor similar to a Hall-effect switch that is used to monitor crankshaft position and speed.

Crankshaft throw: The offset part of the crankshaft where the connecting rods fasten.

Cross and roller: A type of universal joint that uses a center cross (spider) mounted in needle bearings.

Crossmember: Part of the vehicle frame of unibody that runs crosswise in relation to the vehicle's length.

Curb weight: The weight of a unmanned vehicle with fuel, oil, coolant, and all standard equipment installed.

Current: The movement of free electrons in a conductor.

Cycle: A reoccurring period in which a series of actions take place in a definite order.

Cylinder: A hole or holes that have a set depth and contain pistons, diaphragm spring, rather than several coil springs, to help release the clutch disc.

Cylinder head: Metal section bolted on top of the engine block and forms part of the combustion chamber.

D

Damper: A unit or device used to reduce or eliminate vibration, or oscillation of a moving part, fluid, etc.

Dashpot: A cylinder with a piston or diaphragm with a small vent hole used to retard the movement of some other part.

Data bus: Pathway for data inside a computer. Can also be used to refer to a circuit used to connect two on-board computers.

Data link connector (DLC): Plug-in connector used to access vehicle computer data. Referred to by a variety of names.

Datastream values: Voltage, degree, and other values that come to the ECM from sensors and actuators in the form of analog and digital signals.

Daylight running lights: Type of headlight that turn on automatically whenever the ignition key is in the run position.

Dead axle: An axle that does not rotate, but merely forms a base on which to attach wheels.

Deflection rate: The measurement of the amount of force required to compress a leaf spring one inch.

Deglazer: An abrasive tool used to remove the glaze from cylinder walls before new piston rings are installed.

Degree: 1/360th of a circle.

Degree wheel: Wheel-like tool attached to the crankshaft. Used to time valves to a high degree of accuracy.

Deploy: Term used to describe the activation of an air bag system.

Desiccant: A material used to absorb and remove excess moisture.

Detonation: Condition where the fuel charge fires or burns too rapidly. Audible through the combustion chamber walls as a knocking noise.

Dexron III/Mercon: Type of automatic transmission fluid that replaces Dexron II, which was in use for many years.

Diagnosis: The process of analyzing symptoms, test results, etc., to determine the cause of a problem.

Diagnostic charts: Flowchart used in troubleshooting common vehicle system problems.

Diagnostic trouble code (DTC): A code that can be used to determine where a malfunction is located in a computer-controlled system.

Diagonally split brake system: A brake system that has the master cylinder pistons actuating diagonally opposed brake assemblies. Example: One piston will actuate the left front and right rear brakes.

Diaphragm: A flexible partition used to separate two different compartments.

Die: Tool for cutting threads on the outside of a rod or shaft.

Diesel engine: An engine that uses diesel oil for fuel. A diesel engine injects fuel oil directly into the cylinders. The compression is so great that the air itself is hot enough to ignite the diesel fuel without a spark.

Diesel injection pump: A mechanical pump that develops high pressure to force fuel out of the injectors and into the combustion chambers.

Dieseling: Condition in which the engine continues to run after the ignition key is turned off. Also referred to as run-on.

Differential: An assembly of gears used to provide power to the rear axles and allow them to rotate at different speeds as necessary.

Differential carrier: Component used to mount the differential assembly on the rear axle housing.

Differential yoke: Component that connects the rear universal to the pinion shaft.

Digital code: Trouble code displayed as actual digits (numbers), rather than flashes.

Digital display: An oscilloscope that displays a numerical reading (digital display) on a separate screen.

Digital signal: An electronic signal that uses on and off (or high-low) pulses.

Dimmer switch: A foot- or hand-operated switch that operates the high and low beam headlights.

Diode: A semiconductor device that allows current flow in one direction but resists it in the other.

Dipstick: A metal or plastic rod used to determine the quantity of oil or fluid in a system.

Direct current (dc): Electric current that flows steadily in one direction only.

Direct drive: Gear condition in an automatic transmission where the crankshaft and drive shaft turn at the same speed.

Direct fuel injection: Fuel injection system where fuel is sprayed directly into the combustion chamber.

Direct ignition system: A type of computer-controlled ignition system similar to the distributorless system, but does not use spark plug wires.

Disc brake assembly: A brake assembly that uses a hydraulic caliper to actuate brake pads against a metal rotor. Used for both front and rear brakes.

Discharge: Process of drawing electric current from a battery.

Discharging (air conditioning): Procedure for draining a vehicle's air conditioning system of its refrigerant charge.

Displacement: The volume displaced by the pistons in moving from BDC to TDC.

Display output: Actuator that provides readable characters on a small screen or liquid crystal display.

Distributor: A unit designed to open and close the ignition primary circuit, either mechanically or electronically. Also used to distribute secondary voltage to the proper cylinder at the correct time.

Distributor cap: A plastic, insulating cover that encloses the distributor rotor and other components.

Distributor injection pump: A pump that uses one or two cylinders to handle injection of diesel fuel for an engine, as compared to an inline pump with a plunger for each cylinder.

Distributor rotor: Part designed to transfer secondary current to the distributor cap outer terminals.

Distributorless ignition system: A type of computer-controlled ignition system that eliminates the distributor by using a sensor mounted on the crankshaft and/or camshaft to provide timing.

Diverter valve: Component that prevents air from entering the exhaust system during deceleration.

Documentation: Repair orders or other means used to record work performed on a vehicle.

Dog tracking: Condition where the rear wheels of a vehicle are not aligned with the front wheels.

DOHC: Abbreviation for dual overhead cam engine.

Door lock motors: Dc servo motors that lock and unlock the vehicle doors.

Double lap flare: A flare which, when made, utilizes two wall thicknesses. Used primarily in solid steel brake lines.

Draw: The amount of current required to operate an electrical device.

Drive chain: A chain used with some longitudinally mounted engines to transfer power from the engine crankshaft to the transaxle.

Drive housing: Case surrounding the pinion gear on a starter motor.

Drive shaft: Steel tube that transfers rotating motion from transmission to rear wheels of a vehicle.

Drive side: Convex side of gear tooth.

Drive train: All of the collective parts (engine, transmission, drive shaft, differential, drive axles, etc.) that generate power and transmit it to the wheels.

Driveability: The process of diagnosing, troubleshooting, isolating, and repairing a problem on a vehicle. Generalized term normally used when describing a repair of an engine performance problem.

Drive-by-wire system: Any system that uses servo motors, wiring, and sensors, rather than movable linkage, for control.

Drive-fit: A fit between two parts that is so tight that they must be driven together.

Driver information center (DIC): Small display inside the passenger compartment. Gives the driver information on vehicle and outside conditions.

Drivers: Power transistors in a computer that control current flow to actuators. Also known as quad drivers.

Driving hub: Mounting for wheel on end of axle.

Drop center: A type of rim in which the center section is lower than the two outer edges. This allows the bead of the tire to be pushed into the low area while the other side is pulled over and off the flange.

Drum: The housing that holds the parts of a clutch assembly for an automatic transmission.

Drum brake assembly: A brake system that uses a wheel cylinder to force two brake shoes against a rotating drum. Used primarily as rear brakes, but have been used in the front in older vehicles.

Dry charged battery: A battery that is charged, but lacks electrolyte. It is filled with electrolyte when it is placed into service.

Dry friction: Any resistance to movement between two unlubricated surfaces.

Dry sleeve: A thin cylinder liner that is not exposed to coolant.

Dual exhaust system: An exhaust system that uses separate pipes, mufflers, and catalytic converters for each bank of cylinders.

Dual master cylinder: Brake system pump with two pistons and fluid reservoirs for safety.

Dust cap: Metal cap installed on the end of an axle or spindle to keep dirt out and grease in.

Dwell: The amount of time distributor points remain closed between openings.

Dwell signal: Electronic signal output by carburetor that can be read on a special meter for troubleshooting.

Dye penetrant: A testing material that can be sprayed or painted onto iron and aluminum parts to locate cracks.

Dynamic imbalance: Unbalanced condition when the centerline of a revolving object is not in the same plane as the object itself.

Dynamometers: Instrument used to measure power output and performance of an engine. Also known as a dyno.

E

ECC: Abbreviation for electronic climate control.

Eccentric: A circle within a circle that has a different shape and center.

EGR sensor: Small sensor built into an electronic EGR valve. Used to determine EGR valve position and in some cases, gas flow.

ELC: Abbreviation for electronic level control.

Electrically erasable programmable read only memory (EEPROM): A type of microprocessor whose programming can be changed by scan tools and other electronic equipment that burns in the new programming.

Electrode: The insulated center and rod that is attached to the center of a spark plug. Also refers to a welding rod.

Electrolyte: A solution of sulfuric acid and water that surrounds the plates of a battery and allows a free flow of electrons.

Electromagnet: A magnet that is produced by placing a coil of wire around a steel or iron bar. When current flows through the wire, the bar becomes magnetized.

Electromagnetic interference (EMI): Electronic noise that is created when two or more wires carrying a strong voltage or signal are allowed to cross. Can also be created by radio waves.

Electron theory: The accepted theory of electronics that states that electricity flows from negative to positive.

Electronic advance: A system that uses sensor input and the vehicle's computer to control spark timing.

Electronic control module (ECM): General term used for any computer that controls a vehicle system. Abbreviation ECM also used for engine control module.

Electronic control unit (ECU): Another term for a computer used in a vehicle.

Electronic ignition system: Ignition system that uses an electronic control circuit and distributor pickup coil.

Element: One set of positive and negative plates with separators.

Emissions: Any release of harmful materials into the environment.

Emissions information label: Label normally located in the engine compartment that gives timing, idle speed, and vacuum hose routing information.

End play: Amount of lengthwise movement between two parts.

Engine: A device that converts energy into useful mechanical motion.

Engine displacement: The volume of space in which the piston moves in a full stroke multiplied by the number of cylinders in the engine.

Engine dynamometer: A device that tests engine output at the crankshaft.

Engine mounts: Pads made of metal, rubber, and plastic. Designed to hold engine to the frame. May also be liquid-filled.

Engine speed sensor: A sensor that sends information on engine speed and piston location to the ECM.

Equalizer pipe: A short pipe placed between the exhaust pipes on a dual exhaust system to equalize exhaust back pressure. May be used as location for oxygen sensor.

Erasable programmable read only memory (EPROM): A type of microprocessor whose programming can be altered only by erasing it with special equipment and reprogramming.

ESC: Abbreviation for electronic spark control.

EST: Abbreviation for electronic spark timing.

Ester oil: A type of refrigerant oil used in some R-134a air conditioning systems.

Ethanol: Grain alcohol, a gasoline additive.

Ethylene glycol: A liquid chemical used in engine coolant.

Evacuation (air conditioning): Process of removing oxygen from an air conditioning system by pumping air out of the system, creating a vacuum.

Evaporative emissions (EVAP) control: An emissions control system designed to prevent fuel vapor from escaping into the atmosphere.

Evaporator: A heat exchanger installed in the air conditioner blower case. Absorbs heat and humidity from the incoming air and transfers it to the circulating refrigerant.

Exhaust gas analyzer: Electronic calibrated device used to measure the amount of pollutants in exhaust emissions.

Exhaust gas recirculation (EGR) valve: A valve that allows a controlled amount of exhaust gas into the intake manifold during a certain period of engine operation. Used to lower exhaust emissions.

Exhaust manifold: Connecting pipes between the cylinder head exhaust ports and the exhaust pipes.

Exhaust stroke: The portion of the piston's travel that is devoted to expelling the burned fuel charge.

Exhaust system: Components which carry exhaust emissions to the rear of the vehicle. These components include the muffler, exhaust pipes, catalytic converter, and exhaust manifold.

Expansion tank: A plastic tank used to recover excess coolant as part of the coolant recovery system.

Expansion valve: A temperature sensitive device used to regulate the flow of refrigerant into the evaporator. Also called a thermostatic expansion valve (TXV).

Extension housing: A housing bolted to the transmission case. Contains the transmission output shaft and oil seal.

F

Failure record: ECM memory that shows the number of times (keystarts or warm-ups) that a computer system fault has occurred.

Fan: A mechanically or electrically operated device designed to create a moving stream of air, generally for cooling purposes.

Fan clutch: A temperature controlled device mounted to a engine-driven fan. It allows the fan to freewheel when the engine is cold or when the vehicle is at highway speed.

Feathering: Wear pattern that is raised on one side. Appears and feels like the edges of feathers on a bird.

Fiber optic: A path for electricity or data transmission in which light acts as the carrier.

Field windings: Stationary windings in a motor that creates a magnetic field to keep the armature rotating.

Filler neck restriction: Metal piece preventing introduction of the larger fuel nozzle used for leaded fuel.

Final drive ratio: Overall gear reduction at the front or rear wheels.

Firewall: Metal bulkhead between the engine and passenger compartment.

Firing line: The tall spike shown on an oscilloscope, representing the voltage needed to make the spark jump the plug gap.

Firing order: The order in which engine cylinders must be fired.

Five-gas exhaust analyzer: Exhaust gas analyzer that measures the amount of hydrocarbons, carbon monoxide, oxides of nitrogen, carbon dioxide, and oxygen in exhaust emissions.

Fixed caliper: Brake caliper rigidly mounted to steering knuckle.

Flank: Area on a gear tooth below the pitch line.

Flare: A flange applied to tubing in order to provide a seal and to keep a fitting in place.

Flash erasable programmable read only memory (FEPROM): A type of ECM memory which contains vehicle operation data. Is normally reprogrammed rather than replaced.

Flash programming: ECM programming that changes a portion or all the ECM operating parameters. Used with newer ECM containing fixed EEPROMS and FEPROMS.

Flashover: An arc or jump of ignition current across an open space.

Flash-to-pass feature: Turn signal lever position that allows the driver to flash the high beam lights, indicating to other drivers ahead of the desire to pass.

Flex fan: One with blades that alter airflow with engine speed.

Floating caliper: Brake caliper mounted on two rubber bushings, allowing some movement.

Flooding: Condition where the fuel mixture is overly rich or an excessive amount has reached the cylinders.

Flux: Lines of magnetic force moving through a magnetic field. Also, material used to join two pieces of metal being soldered or brazed.

Flywheel: A large heavy wheel that forms the base for the starter ring gear and provides a mounting surface for the torque converter or clutch assembly.

Forward bias: Arrangement in which diode acts as a conductor.

Four-gas analyzer: Exhaust gas analyzer that measures the amount of hydrocarbons, carbon monoxide, carbon dioxide, and oxygen in exhaust emissions.

Four-stroke cycle engine: An engine requiring two complete cycles of the crankshaft to fire each piston once.

Four-wheel alignment: Process in which all the wheels on a vehicle are aligned with each other.

Four-wheel drive: A vehicle in which the front wheels as well as the rear may be driven.

Four-wheel steering: System used to provide limited steering for the rear wheels. Operates in relation to the front wheels.

Frame: The steel structure that supports or is part of the vehicle's body.

Frame rails: Structural sections of the car frame.

Freeze frame: Portion of OBD II ECM memory that stores vehicle data whenever certain trouble codes are stored.

Freeze plugs: A stamped metal disc installed in the holes where the core was removed from the casting. Allows for expansion if coolant should freeze, preventing a cracked casting. Also called core plugs.

Frequency: The rate of change in direction, oscillation, cycles, etc., in a given time span.

Friction: Any resistance to movement between two objects placed in contact with each other.

Friction bearing: A bearing made of babbitt or bronze with a smooth surface.

Front drive axles: Shafts that transfer power from the transaxle differential to the vehicle's wheels.

Fuel: Any substance that will burn and release heat.

Fuel distributor: Hydraulically operated valve used to control fuel flow in a continuous injector system.

Fuel injection: A system that sprays fuel directly into or just ahead of the cylinders.

Fuel injector: Valve controlled by an electronic solenoid or spring pressure.

Fuel lines: The portion of the fuel system that carries fuel from the tank to the filter and on to the carburetor or fuel injectors. Can be made of metal, rubber, or plastic.

Fuel pressure regulator: System that controls pressure of fuel entering injector valves.

Fuel pulsation: Fuel pressure variations due to pump action.

Fuel pump: A mechanically or electrically driven vacuum device used to draw fuel from the tank and force it into the fuel system.

Fuel rail: Tubing that connects several injectors to the main fuel line.

Fuel return system: One that keeps cool fuel circulating to prevent vapor lock.

Fuel tank: A large tank made of steel or plastic used to store the vehicle's supply of fuel.

Fuel tank pressure sensor: Sensor similar to a MAP sensor that compares fuel tank pressure to ambient air pressure. Used to determine if there is a leak in the fuel system.

Full pressure lubrication system: A type of oiling system that draws oil from the sump and forces it through passages in the engine.

Full throttle cutout switch: A switch or sensor that is monitored by the ECM. Allows the ECM to disengage the air conditioning compressor when a wide open throttle condition is detected.

Full-floating axle: A rear drive axle that does not hold the wheel or support any weight, but merely drives the wheel.

Full-floating piston pin: One that is free to rotate. It is secured in place with snap rings.

Full-time transfer case: A four-wheel drive unit that drives all the wheels all the time.

Fuse: Device that interrupts current if a circuit is overloaded or a short occurs.

Fusible link: A special calibrated wire installed in a circuit. Will allow an overload condition for short periods. A constant overload will melt the wire and break the circuit.

G

Galvanometer: A device used for location, direction, and amount of an electric current.

Gas line freeze: Condition caused by water in fuel turning to ice in cold weather and blocking the fuel lines.

Gas-charged shock: Type that contains low-pressure gas to keep the oil from foaming, and thus improve performance.

Gasket: A soft, flexible material placed between two parts to prevent leaks.

Gasoline: A hydrocarbon fuel used in internal combustion engines.

Gassing: Bubbles that rise to the top of battery electrolyte during charging. The bubbles are caused by hydrogen gas produced as a by-product of the charging process.

Gear: A circular object, either flat or cone shaped, upon which a series of teeth are cut.

Gear clash: Noise that is heard when gears fail to mesh properly.

Gear oil pump: A type of pump that uses meshing gears to provide pressure and fluid movement.

Gear ratio: The relationship between the number of turns made by a driving gear to complete one full turn of the driven gear. If the driving gear makes four revolutions in order to turn the driven gear once, the gear ratio would be four to one (4:1).

Gear reduction: Setup in which a small gear is used to drive a larger gear. Produces an increase in torque.

Glaze: A highly smooth, glassy finish on a cylinder wall. Produced over a long period of time by piston ring friction.

Glow plugs: A heating element used to help diesel engines start initially or in cold weather.

Governor: A device designed to control and regulate speed.

Grid: Lead screen or plate to which the battery plate active material is affixed.

Gross axle weight rating (GAWR): The total load carrying capacity of a given axle. It can be expressed as a rating at the springs or at the ground.

Gross horsepower: The brake horsepower of an engine with optimum fuel and ignition settings and without allowing for power absorbed by the engine-driven accessories.

Gross torque: The maximum torque produced when measured at the engine's crankshaft. Does not allow for torque consumed by the engine-driven accessories.

Gross vehicle weight rating (GVWR): The total weight of the vehicle including passengers and load. It is used as an indicator of how much weight can be safely loaded in a certain vehicle.

Ground: The terminal of the battery connected to the vehicle's frame.

Gum: Any oxidized portions of petroleum products that accumulate in the engine and fuel system.

H

Hall-effect switch: A type of distributor pickup used in many ignition systems.

Halogen bulb: Quartz lightbulb in which a tungsten filament is surrounded by a halogen gas such as iodine, bromine, etc.

Hard failure: A computer system failure that sets a diagnostic code as soon as the ignition key is turned to run.

Harmonic balancer: A device that is mounted on one end of a crankshaft. Used to reduce torsional vibration.

Harmonic vibration: A high frequency vibration caused by the crankshaft.

Hazard flashers: Circuit used to flash the turn signal lamps as a warning to other drivers.

Hazardous waste: Any chemical or material that has one or more characteristics that makes it hazardous to health, life, and/or the environment.

Header: A steel pipe that connects the exhaust manifold to the catalytic converter.

Headlights: Main driving lights on the front of the vehicle.

Heat range: Rating given for the operating temperature of spark plugs.

Heat shield: A sheet of metal or fiberglass used to shelter heat sensitive components, such as wiring, from excessive engine and exhaust heat.

Heated oxygen sensor (HO2S): Oxygen sensor which contains a small heater element.

Heater core: A radiator-like unit in the blower case in which coolant circulates. Used to heat the vehicle interior.

Heat-shrink tubing: Plastic tube used to insulate electrical solder joints.

Heel: The outside or larger part of a gear tooth.

Height sensing proportioning valve: A brake valve that can change the braking ratio in relation to vehicle load.

Height sensor: An electronic switch used to measure changes in vehicle ride height. Normally used with electronic ride control systems.

Helical gear: A gear that has teeth cut at an angle to its centerline.

Helicoil: An insert used to repair damaged threads.

Hemispherical combustion chamber: A dome shaped combustion chamber that provides increased engine aspiration while suffering less heat loss than other designs. Often referred to as a hemi.

High side: Section of an air conditioning system in which the refrigerant is under high pressure.

Hone: Process of removing metal with a fine abrasive stone. Used to achieve close tolerances.

Horizontal-opposed engine: An engine that is placed perpendicular to the vehicle chassis.

Horn: A circuit that proves the driver with an audible warning signal.

Horsepower (hp): Measurement of an engine's ability to perform work. One horsepower is defined as the ability to move 33,000 pounds one foot in one minute.

Hose: A flexible rubber or neoprene tube for carrying water, oil, and other fluids.

Hose clamp: A device used to secure hoses to their fittings.

Hotchkiss drive: A drive axle in which the driving force is transmitted to the frame through rear springs or link arms that connect the rear housing to the frame.

Hub: Mounting plate for the wheels on the end of an axle, spindle, or bearing.

HUD: Abbreviation for heads-up display.

Hydraulic actuator: Solenoid-operated valve and electric pump mechanism.

Hydraulic modulator: An anti-lock brake system component that contains valves that are electronically controlled. Used to increase, decrease, or maintain brake system pressure to control wheel lockup (skidding). Also called a hydraulic actuator.

Hydraulic power booster: A brake booster that uses power steering fluid pressure to provide assistance in brake actuation. Also known as a Hydro-boost.

Hydraulic system: Arrangement of pistons and tubing that uses pressure to transmit force from one part to another.

Hydraulic valve lifters: A valve lifter that uses hydraulic pressure from the engine's oil system to keep it in constant contact with both the camshaft and pushrod/valve stem.

Hydrocarbon (HC): Chemical mixtures (12 percent hydrogen, 82 percent carbon) making up crude oil, or petroleum. Also an exhaust byproduct of burned fossil fuels.

Hydrogen (H): Light, flammable gas that is produced as part of a battery's chemical reaction.

Hypoid gearing: A system of gearing in which the pinion gear meshes with the ring gear below the centerline of the ring gear.

I

ID: Abbreviation for inside diameter.

Idle air control (IAC) solenoid: A computer-controlled valve used to supply air to a fuel injected engine at idle. Also regulates idle speed.

Idle speed: The crankshaft rotational speed in an engine with a closed throttle plate.

Idler arm: A steering system component that supports one end of the center link.

Ignition: Lighting or igniting a fuel charge by means of a spark or compression.

Ignition coil: Device used to produce the high voltage needed for ignition spark.

Ignition module: An electronic component that controls ignition spark sequence and fires the coil(s) when needed.

Ignition switch: A key operated switch mounted on the steering column for connecting and disconnecting power to the ignition and electrical system.

Ignition system: Components that produce a spark to ignite the air-fuel mixture in the engine.

Ignition timing: The relationship between the exact time a plug is fired and the position of the piston in terms of degrees of crankshaft rotation.

IM 240: Type of emissions test that is performed by state sponsored I/M programs. A vehicle is placed on a dyno and operated at real driving conditions while exhaust gas is sampled from the tailpipe.

Impact sensor: An open switch that is designed to close when an impact of sufficient force is encountered. Used in air bag systems to detect vehicle impact.

Impeller: A wheel-like device that has fins cast into it. It is mounted on a water pump shaft to turn for pumping coolant.

Inclined engine: An engine that is set at an angle. Allows engine to be installed in a smaller space.

Included angle: The angle formed by the steering knuckle and the center of the wheel if centerlines were drawn through it. Combines both camber and steering axis inclination angles.

Independent suspension: A suspension system that allows each wheel to move up and down without influence from the other wheel on that axle.

Induction: The imparting of electricity to an object by magnetic fields.

Inertia: Force which tends to keep stationary objects from moving and keeps moving objects in motion.

Inertia switch: A switch that is designed to operate only if a sudden movement occurs, such as a collision.

Infrared rays: Rays of light that are invisible to the human eye. Used with some dyes to trace fluid leaks.

Injection timing: The relationship between injection of fuel and the positions of the engine's pistons.

Injector: A valve that is controlled by a solenoid or spring pressure to inject fuel into the engine.

Inline engine: An engine in which all of its cylinders are in a straight row.

Inline pump: A diesel injection pump with one plunger (piston) for each cylinder.

Inner stub shaft: Section of front-drive axle that is splined to differential gears. It is connected to the interconnecting shaft through a universal joint.

Input sensors: Any sensor that provides information to an ECM.

Input shaft: A shaft that delivers power to a mechanism, such as the transmission input shaft delivering power to the transmission gears.

Inspection and maintenance (I/M) program: Government-operated or sponsored program to check vehicles periodically for excessive exhaust emissions. Some also perform safety checks.

Insulation: Any material that resists the flow of electrons, heat, or noise.

Intake air heaters: A wire grid that is placed as a spacer between the throttle plate and the intake manifold. Electricity is used to heat the air-fuel charge as it enters the intake manifold.

Intake air temperature (IAT) sensor: A thermistor installed in the air cleaner or air intake tube. Sensor used by the ECM to monitor the temperature of the air entering the engine.

Intake manifold: Series of connecting tubes or housing between the throttle plate and the openings to the intake valves.

Intake stroke: The portion of the piston's movement that is devoted to drawing the air-fuel mixture into the combustion chamber.

Intake valve: Valve through which air and fuel is admitted to the cylinder.

Integrated ABS: Type of anti-lock brake system that has the master cylinder, hydraulic actuator, pump, and accumulator as the heart of the hydraulic system.

Integrated circuit (IC): A single chip of semiconductor material which contains various electrical components in miniaturized form.

Intermediate gear: Any transmission gear between low and high.

Intermittent: An event that occurs at different intervals.

Intermittent codes: Computer diagnostic code that does not return immediately after it has been cleared.

Internal combustion engine: An engine that burns fuel within itself as a means of developing power.

J

Jet: An orifice used to control the flow of gasoline in various parts of a carburetor.

Jounce: An intentional push on the vehicle to test the shock absorbers or MacPherson struts. Also done to settle vehicle ride height.

Journal: The part of a shaft or axle that actually contacts a bearing.

Jump start: Method of starting a vehicle with a weak or dead battery through the use of jumper cables.

Jumper cables: A pair of electrical cables used to start a car with a weak or dead battery.

Jumper wire: A wire used to make a temporary electrical connection.

Junction block: Electrical block, usually located in the engine compartment. Contains fuses, circuit breakers, relays, normally used to power high amperage devices.

K

Keep alive memory (KAM): A type of computer memory that stores changes in sensor values. Provides information to the computer's CPU so it can properly adjust the signals to the various output devices. Sometimes referred to as adaptive strategy.

Key: A small metal piece that fits into a groove partially cut into two parts to allow them to turn together.

Keyway: A slot cut into a shaft, pulley, or hub that permits the insertion of a key.

Kickdown valve: Spool valve inside an automatic transmission's valve body that causes the transmission to shift into a lower gear during fast acceleration.

Kinetic energy: Any energy associated with motion.

Kingpin: A hardened steel shaft around which the steering knuckle pivots. Normally used on large trucks.

Knock sensor (KS): Engine sensor that detects preignition, detonation, and knocking.

Knurling: Technique in which metal is grooved, pushing it up a few thousandths of an inch to slightly increase a part's diameter.

L

Land: A portion of metal separating the grooves that rings ride against.

Lapping: The process of fitting two surfaces by rubbing them together with an abrasive between them.

Lateral runout: Side-to-side movement of a wheel, tire, or rotor.

Lathe: Machine on which a solid piece of material is spun and shaped by a fixed cutting tool.

Leaf spring: Suspension springs made of steel leaves bound together. A varying number of steel leaves are used depending on its intended use. Some leaf springs are solid units made of fiberglass.

Lean air-fuel mixture: An air-fuel mixture that is high in air concentration.

Lift: Maximum distance a valve head is raised off its seat.

Light emitting diode (LED): A special function diode that lights when forward biased.

Limited-slip differential: A differential unit designed to provide superior traction by transferring the driving torque to the wheel with the best traction. Extremely useful in wet or icy conditions.

Limp-in mode: ECM operation mode that engages in case of total failure. Allows limited vehicle operation so it can be driven to a shop.

Liner: A thin section, such as a cylinder liner, placed between two parts.

Linkage: Any moveable series of rods, levers, and cables used to transmit motion from one unit to another.

Liquid cooling system: Circulation of a heat-absorbing medium through engine passages, with the accumulated heat dissipated by further circulation through heat exchanger (radiator).

Live axle: An axle in which power travels from the differential to the wheels.

Load range: Tire designation which uses a letter system (A, B, C, etc.) to indicate specific tire load and inflation limits.

Locking hub: Components that transfer power from driving axles to driving wheels on a four-wheel-drive vehicle.

Lockup torque converter: A torque converter with an internal clutch that locks the turbine to the impeller when the automatic transmission is in direct drive or overdrive.

Longitudinal: Lengthwise; term used to identify an engine mounted with its centerline on or parallel to the centerline of the vehicle.

Low fluid warning switch: A sensor used in various systems to notify the driver of low or no fluid in a system, such as washer fluid or engine coolant.

Low pressure cutout switch: One that prevents air conditioning compressor clutch engagement if the system pressure falls below a specified point. May also be included as an input sensor to the engine's electronic control unit.

Low side: Section of an air conditioning system in which refrigerant is under low pressure.

LPG: Abbreviation for liquefied petroleum gas, an alternate fuel.

Lubrication system: Method of distributing lubricant (oil) to moving parts to minimize friction.

Lug nuts: Large steel bolts used to hold a wheel to the axle hub.

Lugs: Threaded bolts or studs that are press-fit into an axle, hub, or brake rotor and accept lug nuts to mount wheels.

M

MacPherson strut: Suspension system that uses one control arm and one strut for each wheel assembly.

Magnafluxing: Testing procedure that uses a magnet and metal powder to find cracks in cast iron parts.

Magnetic field: Field of force generated around an electrical conductor.

Magnetism: Invisible lines of force that attract ferrous metals.

Main bearing: Series of bearings that support the crankshaft in the engine.

Main bearing caps: Pieces that bolt to the bottom of the block to hold the crankshaft in place.

Main journals: Carefully machined surfaces on the ends of the crankshaft that fit into the block main bearings.

Main thrust bearing: Flanged version of main bearing, designed to limit crankshaft endplay.

Maintenance tune-up: Tune-up that includes replacing the spark plugs and air, fuel, and emissions filters.

Maintenance-free battery: A sealed battery that requires no additional water or electrolyte during its useful life.

Malfunction indicator light (MIL): Amber-colored light in the instrument cluster used to indicate that a problem exists in a vehicle's computer control system. Also called a Check Engine or Service Engine Soon light. Generalized term used for any instrument cluster light used to indicate a problem in a system.

Manifold: A pipe or series of pipes connecting a series of ports to a common opening.

Manifold absolute pressure sensor (MAP): Computer sensor used to measure the barometric pressure in relation to intake vacuum. Sometimes called a barometric pressure sensor.

Manifold vacuum: Vacuum that exists in the intake manifold under the carburetor or throttle body throttle plate.

Manual transmission/transaxle: A driver operated (shifted) transmission or transaxle.

Mass airflow (MAF) sensor: Computer sensor used to measure the amount of air entering the intake manifold. Also called an Airflow Meter.

Master cylinder: The part of the hydraulic brake system in which system pressure is generated.

Material Safety Data Sheet (MSDS): Information on a chemical or material that must be provided by the material's manufacturer. Lists potential health risks and proper handling procedures.

Mechanical lifter: Solid lifters used in some older vehicles that must be adjusted periodically.

Memory seats: Type of seat that is connected to a computer which remembers the position of the seat as dictated by the driver. Computer can usually store up to 2-3 driver settings.

Memory tilt wheel: Part of the memory seat system, adjusts the steering column position to the driver's preference.

Metering valve: A valve that limits hydraulic pressure to the front brakes until a predetermined line pressure is reached.

Methanol: Methyl alcohol or wood alcohol. A gasoline additive.

Microprocessor: A small silicon chip that contains elements in a computer. Often referred to as an integrated circuit or IC.

Mineral oil: A lightweight oil used as a refrigerant oil in R-12 air conditioning systems and as a base oil for power steering and automatic transmission fluid.

Misfire: Failure of one or more cylinders to fire.

Misfire data: Information stored in the computer indicating which cylinders are misfiring and how often.

Misfire failures: Number of times a cylinder has failed to fire.

Misfire history: Record of the number of times a cylinder has failed to fire.

Mixture control (MC) solenoid: An electronic solenoid located in the carburetor. Used to move the metering rods in and out.

Motor: A device which converts electrical or fluid energy to mechanical energy.

Muffler: A chambered unit attached to a pipe or hose to deaden noise.

Multimeter: An electrical test meter that can be used to test for voltage, current, or resistance.

Multiple disc clutch: A clutch assembly that contains several clutch discs in its construction.

Multiplexing: A method of using one communications path to carry two or more signals simultaneously.

Multiport fuel injection (MPFI) system: Fuel injection system in which there is one injector per cylinder. Also called Multi-Point, Tuned Port, or Sequential Fuel Injection.

Multiviscosity oil: An engine oil that can exhibit different viscosity characteristics when heated or cooled.

N

Natural gas: General term for any petroleum based gas, such as propane and liquefied petroleum gas.

Navigation systems: On-board electronic guidance system that acts as an interactive navigator.

Needle bearing: An antifriction roller bearing that uses many small diameter rollers in relation to their length.

Net horsepower: Maximum horsepower at the flywheel with all engine-driven accessories in use.

Net torque: Maximum torque at the flywheel with all engine-driven accessories in use.

Neutral safety switch: A switch that prevents starter engagement if the transmission is in gear.

Noid light: Small light used to test fuel injector wiring harnesses.

Nonintegrated ABS: Anti-lock brake system that is used with a conventional master cylinder and power booster.

Normally aspirated engine: Any engine that does not use a turbocharger or supercharger.

North pole: One of the poles of a magnet from which lines of force originate.

O

O-ring seal: A synthetic rubber ring that fits into a groove, and is compressed when parts are assembled.

OBD II drive cycle: Predetermined routine of acceleration, deceleration, and stops that allows the ECM to reset its emissions monitoring. Usually performed after ECM replacement or before some emissions tests.

Octane: Rating indicating a fuel's tendency to resist detonation. Does not have a bearing on the fuel's quality.

OD: Abbreviation for outside diameter.

OEM: Abbreviation for original equipment manufacturer.

Ohm: Unit of measurement for resistance to the flow of electric current in a given unit or circuit.

Ohmmeter: An electrical instrument used to measure the amount of resistance in a given unit or circuit.

Oil cooler: An air or liquid cooled device used to remove excess heat from the engine or transmission oil.

Oil filter: A filter element installed on the engine. Used to filter dirt and foreign debris from the oil entering the engine.

Oil gallery: Cast or drilled passageway or pipe in an engine. Used to carry oil from one part of the engine to another.

Oil pan: Metal or plastic pan located on the bottom of the engine or transmission. Acts as a reservoir for oil or fluid.

Oil pickup: Tube and screen that connects to the oil pump and extends to the bottom of the oil pan. Used by oil pump to pick up oil.

Oil pressure gauge: A dash mounted instrument that provides oil pressure readings in pounds per square inch or in kilopascals.

Oil pressure indicator: Warning light on control panel to alert driver to low pressure situation.

Oil pressure switch: Safety device that shuts off the fuel pump if engine oil pressure drops.

Oil pump: Device for forcing oil under pressure to the points where lubrication is needed.

Oil ring: The bottom piston ring which scrapes oil off the engine cylinder walls.

Oil seal: A device used to prevent oil leakage past certain areas, such as around a rotating shaft.

Oil service rating: Identification of type of service for which an oil is suited.

Oil slinger: Washer-shaped part mounted on crankshaft sprocket to throw oil onto timing chain during operation.

Oil spurt hole: Small hole drilled in connecting rod for improved cylinder lubrication.

Oil temperature sensor: Thermistor located in the oil pan or in an oil gallery. Allows the ECM or driver to monitor oil temperature.

On-board diagnostics generation one (OBD I): Classification referring to any computer control system installed on vehicles built prior to 1996.

On-board diagnostics generation two (OBD II): Enhanced diagnostic system required on 1996 and newer vehicles. Protocol also called for standardization of codes, data link connectors, and terminology.

One-wire circuit: One that uses the vehicle frame as a return wire to the power source.

Open circuit: Electrical circuit with a gap or break in continuity so that current cannot flow.

Open loop: Computer operation mode used when an engine has not reached its optimum operating temperature. The ECM operates the engine using a basic set of fixed variables provided by the ECM's PROM.

Open system: Cooling system that does not use a recovery tank.

Operating parameter: An acceptable maximum or minimum electrical value.

Opposed engine: One with cylinders lying flat on either side of the crankshaft.

Optical sensor: A light sensitive device used in some distributors to determine when to close the ignition circuit. Its operation is similar to a Hall-effect switch.

Orifice: A small hole or restricted opening used to control the flow of gasoline, oil, refrigerant, etc.

Orifice tube: A tube with a calibrated opening used to restrict the amount of liquid refrigerant. Used in place of an expansion valve in some systems.

Oscillate: Any back and forth swinging action like that of a pendulum.

Oscilloscope: Instrument that displays line patterns that relate voltages to time.

Out-of-round: Condition where a cylinder or other round object, such as a tire, has greater wear at one diameter than another.

Outer stub shaft: In a front-wheel drive vehicle, the short shaft connecting outer universal joint and the front wheel hub.

Output device: A computer-controlled device used to change an engine setting, such as fuel and spark timing.

Overdrive: An arrangement of transmission gears that results in the driven shaft turning more revolutions than the driving shaft.

Overflow tank: A plastic container used to hold extra coolant in a closed cooling system.

Overhead camshaft (OHC): A camshaft mounted above the cylinder head. Usually driven by a timing chain, belt, or a combination of the two.

Overhead valves (OHV): An engine design where the valves are located in the cylinder head.

Overrunning clutch: Device that locks a pinion gear in one direction and releases it in the other.

Oversteer: The tendency of a vehicle to turn more sharply in a corner than the driver intends.

Oxides of nitrogen (NO_x): An undesirable compound of nitrogen and oxygen in exhaust gases. Usually produced when combustion chamber temperatures are excessively high.

Oxidize: To combine an element with oxygen or a catalyst which converts it to its oxide form. This action can form rust in ferrous metals.

Oxygen sensor (O_2S): An exhaust sensor used to measure the amount of oxygen in the exhaust gases produced by the engine, the signal sent to the engine ECM is used to determine fuel mixture, spark timing, etc.

Oxygenates: Chemicals added to gasoline or other fluids to increase its ability to absorb or release oxygen.

P

Pad: Disc brake friction lining.

Pad wear sensor: Metal tab on brake pad that makes a squealing noise to signal the need for pad replacement.

Pan: A thin metal cover bolted to the bottom of side of an engine or transmission/transaxle to contain oil.

Parallel circuit: An electrical circuit that has two or more resistance units wired so that current can flow through them at the same time.

Parallelogram steering linkage: A steering system utilizing two short tie rods connected to steering arms and to a long center link. The link is supported on one end of an idler arm and the other end is attached to a pitman arm. The arrangement forms a parallelogram shape.

Parasitic load: Normal electrical load from the ECM, radio, and other electrical components placed on a vehicle's battery when the engine is not operating.

Parking brakes: A hand- or foot-operated brake which prevents vehicle movement while parked by actuating the rear brakes.

Parking lights: Small lights on the front and rear of the vehicle. Used to make vehicle more visible during nighttime driving.

Parking pawl: A latch that locks the transmission so that the vehicle will not roll when the selection lever is in the Park position.

Particulates: Solid particles of soot and other substances that result from combustion.

Passive sensor: One that changes an externally produced signal, but does not generate its own voltage.

PCV system: Positive crankcase ventilation, a system that decreases pollution by drawing toxic gases back through the combustion process.

Permanent magnet: A magnet capable of retaining its magnetic properties over a very long period of time.

Petcock: A valve placed in a tank or line for draining purposes.

Phosgene gas: A toxic, colorless, poisonous gas produced when refrigerant is burned.

Photosensitive diode: A semiconductor device that allows current to flow when exposed to light.

Pickup coil: Component that sends pulses to the control unit of an electronic ignition system as a result of trigger wheel rotation.

Pilot bearing: The bushing or bearing that supports the forward end of the transmission input shaft.

Pinging: A metallic rattling sound produced by the engine during acceleration. Sound associated with detonation or preignition.

Pinion carrier: Part of the rear axle that contains and supports the pinion gear shaft.

Pinion gear: A small gear that is either driven by or driving a larger gear.

Pinion shaft: Shaft holding the two differential idler (pinion) gears.

Piston: Component that rides up and down in the cylinder.

Piston boss: Built-up area around the piston pin hole.

Piston expansion: An increase in piston diameter due to normal heating.

Piston head: The portion of the piston above the top ring.

Piston lands: The portion of the piston between the ring grooves.

Piston pin: Fastening device that holds piston onto the connecting rod.

Piston ring: A split ring installed in a groove in the piston. Seals the compression chamber from the crankcase.

Piston ring gap: Clearance between ends of rings when installed on cylinder.

Piston ring grooves: Series of slots in the piston in which the rings are fitted.

Piston skirt: The portion of the piston below the rings.

Piston skirt expander: Spring device placed inside a piston skirt to increase its diameter.

Piston stroke: Distance the piston moves from BDC to TDC.

Pitch line: Imaginary line along the center of a gear tooth.

Pitman arm: Component that transfers gearbox motion to the steering linkage.

Planet carrier: Part of a planetary gearset upon which the planetary gears are affixed.

Planetary gears: The gears in a planetary gearset that mesh with the ring and sun gears. Referred to as planetary gears because they orbit or move around the center, or sun gear.

Plastigage: A measuring tool that is compressed between two tightly fitting surfaces, such as bearings, to measure clearance.

Plate: A grid, covered with porous lead, that will store electrical energy.

Plies: Layers of rubber impregnated with fabric or steel that make up the carcass of the tire.

Plug gap: The distance between the center and side electrodes of a spark plug.

Plug heat range: Numeric indicator of how hot a spark the plug will develop.

Plug reach: Length of the threaded portion of a spark plug.

Ply separation: Pulling apart of tire plies as a result of overheating due to underinflation, or other causes.

Polarity: The positive or negative terminals of a battery. Also the north and south poles of a magnet.

Pole piece: Magnetic component of motor that keeps the armature rotating.

Pole shoes: Metal pieces around which field coil winding are placed.

Polyalkylene glycol (PAG) oil: Oil used in air conditioning systems that have R-134a as a refrigerant.

Poppet valve: A valve used to open and close a circular port or hole.

Port fuel injection: A type of fuel injection system that uses one injector per cylinder. Also called multiport injection.

Positive crankcase ventilation (PCV) system: A system that uses a valve to clear the engine of blowby gases.

Potential horsepower: A measurement of the maximum amount of horsepower available.

Potentiometer: A variable resistor that can be used to adjust voltage in a circuit.

Power antenna: A radio antenna equipped with a small electric motor for raising and lowering.

Power booster: A vacuum or hydraulic operated device used to increase brake pedal force on the master cylinder during stops.

Power brakes: A brake system that has a vacuum or hydraulic powered booster as part of the overall system.

Power steering: A steering system that utilizes hydraulic pressure to increase the driver's turning effort.

Power stroke: The downward movement of the piston that occurs after the air-fuel charge has been ignited.

Powertrain: Group of components such as the engine, transmission, axles, joints, used to provide driving force to the wheels.

Powertrain control module (PCM): Name for computer that controls the engine and transmission.

Precombustion chamber: Used in diesel engines with a glow plug for easier winter starting.

Preheating: The application of heat in preparation for some further treatment, such as welding.

Preignition: Condition where the fuel charge is ignited before it is intended.

Preloading: Process of adjusting a bearing so that it has a mild pressure placed upon it.

Prelubricator: Pressure tank used to force oil through lubrication system without running the engine, as means of testing for worn engine bearings.

Press-fit: Condition of fit between two parts that requires pressure to force the two parts together.

Pressure differential switch: A hydraulic switch in the brake system that operates the brake warning light in the dashboard.

Pressure plate: A spring-loaded assembly that keeps pressure on the clutch disc against the flywheel.

Pressure regulator: A limiting device in an automatic transmission, regulating maximum hydraulic oil pressure.

Pressure relief valve: Spring-loaded bypass that operates when pressure reaches a preset point.

Pressure sensor: A sensor that is used to detect excessive high or low pressures in a system.

Pressure valve: Spring-loaded disc inside radiator cap that opens when system pressure increases past its setpoint.

Pressure-splash system: A system that uses an oil pump to supply oil to the camshaft and crankshaft bearings and the movement of the crankshaft to splash oil onto the cylinder walls and other nearby parts.

Primary brake shoe: Brake shoe installed facing the front of the vehicle.

Primary circuit: A low voltage circuit that is part of the ignition system.

Primary winding: Low voltage winding in a coil. It is made of heavy wire.

Primary wiring: Small insulated wires which serve the low voltage needs of the ignition and vehicle systems.

Printed circuit: An electrical circuit that is made by conductive strips printed on a board or panel.

Programmable read only memory (PROM): A semiconductor chip that contains instructions that are permanently encoded into the chip. Instructions contain base operating information on how components should operate under various conditions.

PROM carrier: A plastic case used to protect a PROM and make installation easier.

Propane: Hydrocarbon-based gas that is mixed with butane and is sometimes used as an engine fuel. Also known as LPG or CNG.

Proportioning valve: Valve designed to equalize pressure at wheel cylinders on vehicles with front disc and rear drum brakes.

Pulsation damper: A metal or plastic sleeve used to smooth out fuel pump pulsations or surges to the rest of the fuel system.

Pulse air injection system: An emission control system which feeds air into the exhaust system by using exhaust pulses.

Pulse fuel injection: Fuel system in which injectors are only open for a short period and remain closed the rest of the time. The amount of fuel delivered is controlled by how long the injector is open.

Pulse ring: Trigger wheel placed on the crankshaft damper in a crankshaft triggered ignition system.

Pulse width: The length of time a fuel injector is held open by the engine control computer.

Pump: A device that is designed to move coolant, oil, fuel, etc., from one area to another.

Purge line: Line connecting the charcoal canister and engine intake manifold.

Push rod: When camshaft is located in block, the long push-rod transmits motion from lifter to rocker arm.

Q

Quadrant: A gear position indicator often using a shift lever actuated pointer. Can be marked PRNDD21 (four-speed).

Quick charge test: A method of determining whether battery plates are sulfated (no longer able to hold a charge).

R

R-12 (CFC-12): Refrigerant used in older air conditioning systems. Gradually being replaced by R-134a in newer vehicles. Also called dichlorodifluoromethane (CCl_2F_2).

R-134a (HFC-134a): Refrigerant used in the air conditioning systems of most vehicles manufactured after 1992. Replaced R-12 due to environmental concerns. Also called tetrafluroethane (CF_3CH_2F).

Rack-and-pinion steering gear: A steering gear that utilizes a pinion gear on the end of the steering shaft. The pinion gear engages a long rack, which is connected to the steering arms via tie rods.

Radial: A line at right angles to a shaft, cylinder, etc., center line.

Radial compressor: An air conditioning compressor that uses reciprocating pistons set at right angles to the drive shaft and spaced around the shaft in a radial fashion.

Radial runout: Uneven rotation caused by differences in diameter.

Radial tire: One that has cord plies running straight across, from bead to bead. Additional stabilizer plies are placed beneath the tread.

Radial tire pull: A condition where a defect in a radial tire causes a vehicle to pull left or right.

Radiation: Method of heat transfer through infrared radiation.

Radiator: An arrangement of tubes and cooling fins that serves as a heat exchanger on a vehicle.

Radiator cap: Closure that seals and pressurizes the cooling system of a vehicle.

Radiator hoses: Flexible tubes that carry coolant between the engine and radiator.

Radius rods: Rods attached to the axle and pivoted on the frame. Used to keep an axle at a right angle to the frame.

Ram air: Air that is forced through a condenser or radiator or into the engine by vehicle movement.

Random access memory (RAM): A portion of computer memory that serves as temporary storage for data. This data is lost if power to the computer is lost. It is used to store sensor information and any diagnostic trouble codes.

Read only memory (ROM): A type of computer memory that cannot be changed and is not lost if power is cut off. Contains the general information to operate a computer.

Rear axle assembly: A combination of gears and axles converting rotary motion of the drive shaft to forward or backward motion of the vehicle.

Rear axle ratio: The relationship between the numbers of teeth on the pinion gear and ring gear. Ratio affects acceleration, pulling power, and fuel economy. Rear drive axle assembly: Differential, axles, and other components transferring power from driveline to rear wheels.

Rear main oil seal: Seal that fits around the rear of the crankshaft to prevent oil leakage.

Rear wheel bearing: Ball- or roller-type bearings that reduce friction between the axle and axle housing.

Receiver-drier: An air conditioning system component that is used to dry and store refrigerant.

Recirculating ball worm and nut: A steering gear that utilizes a series of ball bearings that feed through and around grooves in a worm and nut.

Rectified: A term used to describe alternating current (ac) that is changed to direct current (dc).

Reduction gear: A gear that increases torque by reducing the speed of a driven shaft in relation to the driving shaft.

Reduction starter: One that uses extra gears to increase the torque applied to the flywheel gear.

Reed switch: An electronic switch that consists of two metal strips or reeds. The reeds are influenced by a magnetic field, which causes them to open or close, depending on the application.

Reference voltage (Vref): A known voltage (usually 0-5V) fed to passive sensors by a computer. Changes in sensor resistance can then be read by the computer.

Refrigerant: A liquid used in refrigeration systems to remove heat from the evaporator and carry it to the condenser. Automotive systems use R-12 and R-134a.

Refrigerant oil: Lubricant used in the compressor of an air conditioning system.

Refrigerant recovery unit: Electronic station that combines a refrigerant storage tank, vacuum pump, gauges, and service valves. Used to recover, recycle, and recharge refrigerant in automotive air conditioning systems.

Relay: A magnetically operated switch used to make or break current flow in a circuit.

Relief valve: Valve that opens to protect steering or other hydraulic system when pressure becomes too high.

Reluctor: A component in an electronic ignition system distributor. It is affixed to the distributor shaft and triggers a magnetic pickup, which triggers the control module to fire the coil.

Remote keyless entry: Electronic system that is added to some vehicles to enable the vehicle's owner to lock and unlock the doors and open the trunk using a key fob transmitter.

Removable carrier: A rear axle assembly in which the differential housing, ring, and side gears can be removed as a single unit.

Reserve capacity: The amount of time a battery can produce an acceptable current when not charged by the alternator.

Reservoir: A tank or bottle used to hold a reserve of fluid, such as coolant or washer fluid.

Resistance: The measure of opposition to electrical flow in a circuit.

Resistor: A device placed in a circuit to lower voltage and current flow.

Resonator: A small muffler-like device that is placed in an exhaust system to further reduce exhaust noise.

Retard: To set the ignition timing so that the spark occurs later or less degrees before TDC.

Retread: A used tire that has new rubber bonded to it.

Retrofitting: Process of converting an air conditioning system that uses R-12 to handle R-134a refrigerant.

Return spring: A spring positioned to close a valve or return a brake shoe back to its resting position.

Reverse bias: Arrangement in which diode acts as an insulator.

Reverse flush: Method of cleaning by flushing a cleansing agent through a system in the reverse direction of normal fluid flow.

Reverse polarity: Accidental backward connection primary wires.

Ribbed belt: A V-type drive belt that has small ridges added along its length.

Rich air-fuel ratio: An air-fuel mixture that contains more fuel than a stoichiometric mixture.

Rim: The metal wheel upon which a tire is mounted.

Ring gear: Large gear in differential that is driven by the pinion gear and, in turn, drives the spider gears.

Ring spacers: Thin steel rings inserted next to tension rings to restore proper side clearance.

Ring-to-groove clearance: Also called ring side clearance, this is the space between a compression ring and the edges of the groove in the piston.

Road crown: Angle or slope of the roadway to allow water to run off.

Road feel: The feeling imparted to the steering wheel by the wheels of the vehicle in motion. This feeling is important in sensing and predetermining steering response.

Rocker arm: A lever arm used to direct downward motion on a valve stem.

Rocker shaft: A shaft upon which rocker arms are mounted in some engines.

Rod bolt covers: Temporary protective coverings, such as pieces of rubber hose, used when inserting piston and connecting rods in cylinders.

Rod cap: The lower removable half of a connecting rod.

Rod journals: Machined and polished surfaces on the crankshaft to which the connecting rods are attached.

Roller bearing: A bearing which contains hardened roller ball bearings between two races.

Roller clutch: A clutch that utilizes a series of rollers placed in ramps. The clutch will provide driving force in one direction, but will slip in the other.

Roller lifter: A valve lifter that has a roller bearing which rides on the cam lobe to reduce friction.

Roller vane pump: A pump which uses spring-loaded vanes that are shaped like rollers to provide the pumping action.

Rollover valve: A valve in the fuel tank or delivery lines that prevents the escape of raw fuel in the event of a vehicle rollover.

Room temperature vulcanizing (RTV): A type of sealant that cures at approximately 72°F.

Root cause of failure: The actual cause of a problem.

Rotary brush: A stiff brush used with an air tool for cleaning parts.

Rotary engine: An internal combustion engine that uses one or more triangular rotors instead of pistons to accomplish the intake, compression, power, and exhaust cycles. Also referred to as a Wankel engine.

Rotary pump: A pump that uses a star-shaped rotor

Rotary valve: A valve that exposes ported holes to allow the entrance and exit of gases.

Rotor: A rotating contact inside the distributor that routes electrical pulses from the coil to the spark plugs. Also, the metal disc against which brake pads are forced to stop vehicle.

Rotor pump: A type of pump that uses a central rotor with spring-loaded vanes.

Run-flat tires: Tire designed with a stiff sidewall to allow it to run while deflated.

Runout: The side-to-side distortion or play of a rotating part.

Rzeppa CV-joint: A type of constant velocity joint that uses a ball-and-cage assembly to provide joint motion.

S

SAE: Abbreviation for Society of Automotive Engineers.

Safety rim: A wheel designed with two safety ridges that prevent the tire from dropping into the center of the wheel in the event of a blowout.

Safing sensor: Sensor included in air bag systems to prevent accidental deployment.

Scale: Accumulation of corrosion and mineral deposits within a cooling system.

Scan tool: An electronic tool, usually hand held, that is used to read and interpret diagnostic codes and engine sensor information.

Scatter shield: A metal shield that surrounds a manual transmission bell or clutch housing. It is bolted or welded to the frame and is designed to protect the driver and spectator from flying debris in the event of a clutch explosion.

Scavenging: Refers to the cleaning or blowing out of exhaust gas.

Schrader valve: A spring-loaded valve, similar to a tire valve, used in air conditioning systems.

Scuffing: A roughening of the cylinder wall caused when there is no oil film separating the moving parts and metal-to-metal contact is made.

Seal: A formed device made of plastic, rope, neoprene, or Viton. Used to prevent oil leakage around a moving part, such as a shaft.

Sealant: A liquid or paste material applied to a surface along with or in place of a gasket to prevent oil leaks.

Sealed beam headlight: A headlight lens that has its lens, bulb, and reflector fused together into a single unit.

Sealed bearing: A bearing that has been sealed at the factory and cannot be serviced during its useful life.

Seat: A surface upon which another part rests or seats. An example would be a valve face resting on its valve seat.

Seat width: The area of the valve seat that is actually in contact with the valve face.

Secondary brake shoe: The brake shoe whose friction surface faces the rear of the vehicle.

Secondary circuit: In an ignition system, all components operating on coil (high) voltage. It carries high voltage from coil to spark plugs.

Secondary winding: The high voltage winding in a coil. It is made of very fine wire.

Section height: The overall height of a tire from the bottom of the bead to the top of the tread.

Section width: The overall width of the tire measured at the exterior sidewall's widest points.

Sector shaft: Output gear in a recirculating ball gearbox.

Sediment: An accumulation of matter or foreign debris, which settles to the bottom of a liquid.

Seize: Condition where two or more parts have forced the lubricant out from between them due to excessive heat and/or friction. When this happens, the parts stick together or freeze.

Self-adjusting brakes: A brake assembly that, through normal application, will maintain the friction members in close adjustment.

Self-diagnostics: The ability of a computer to not only check the operation of all of its sensors and output devices, but to check its own internal circuitry and indicate any problems via diagnostic trouble codes.

Self-energizing brakes: A drum brake assembly that, when applied, develops a wedging action that actually assists or boosts the braking force developed by the wheel cylinders.

Self-induction: The creation of voltage in a circuit by varying current in the circuit.

Self-sealing tire: One with a sealing compound applied to its liner to stop air leakage in case of puncture.

Semiconductor: A substance, such as silicon, that acts as a conductor or insulator, depending on its operating condition and application.

Semielliptical leaf spring: A spring commonly used on truck rear axles. It consists of one main leaf and a number of progressively shorter leaf springs.

Semi-floating axle: A type of axle used in most modern vehicles. The outer end turns the wheels and supports the weight of the vehicle. The splined inner end floats in the differential gear.

Semi-metallic: A friction material that has metal particles added to an organic compound to increase its useful life.

Sensible heat: Any additional heat that can be seen in a rise of temperature.

Sensor: A device that monitors a condition and reports on that particular condition to a computer.

Sensor rotor: A toothed wheel that operates at the same rpm as the vehicle wheel.

Separator: Plastic, rubber, or other insulating material placed between a battery's plates.

Series circuit: A circuit with only one path for current to flow.

Series-parallel circuit: A circuit in which a series and parallel circuits are combined.

Serpentine belt: A single belt used to drive all of the engine-driven accessories.

Service engine soon (SES) light: See *Malfunction indicator light.*

Service valves: Points at which pressures in an air conditioning system can be checked, and refrigerant removed or replaced.

Servo: Piston that operates a band in an automatic transmission. Also, a small dc motor that can turn or move parts.

Setback: A measurable condition in which one wheel spindle is positioned behind the spindle on the opposite side.

Shackle: A link for connecting a leaf spring to the frame.

Shell: A component of an automatic transmission clutch that connects the front clutch drum and the sun gear of a planetary gear set.

Shift forks: Devices in a manual transmission that straddle and move the synchronizers and gears back and forth on their shafts.

Shift lever: The handle operated by the vehicle driver to manually shift from gear to gear.

Shift linkage: A cable or mechanical linkage used to shift a transmission into its various gears.

Shift point: The points, at either engine rpm or vehicle speed, when a transmission should be shifted to the next gear.

Shim: A thin piece of brass, steel, or plastic inserted between two parts to adjust the distance between them.

Shimmy: A condition where a wheel shakes from side-to-side.

Shock absorber: An oil- or gas-filled device used to control spring oscillation in suspension systems.

Shock actuators: Solenoid-operated valves that control fluid flow inside shock absorbers in an electronic shock absorber system.

Short block: The bottom end of the engine, including the cylinders, pistons, and crankshaft.

Short circuit: An accidental grounding of an electrical circuit or electrical device.

Shrink fit: A fit so tight that one part must be cooled or heated to fit on another part.

Shroud: A plastic or metal enclosure around a fan to guide and facilitate airflow.

Shunt: An alternate or bypass portion of a circuit.

Side-impact beams: Steel beams installed in the doors of late-model vehicles. Meant to increase a vehicle's ability to resist a side impact.

Sidewall: The portion of a tire between the bead and tread.

Sight glass: A clear glass window in a receiver-dehydrator that can be used to check the refrigerant level.

Single exhaust system: An exhaust system that has only one pipe leading from the exhaust manifold to the catalytic converter, muffler, and out to the tailpipe.

Skid plate: A stout metal plate attached to the underside of a vehicle to protect the engine and transmission oil pans, drive shaft, etc., from damage from grounding out on rocks, curbs, and road surface.

Slant engine: An inline engine in which the cylinders have been tilted at an angle from the vertical plane.

Sleeve: A replaceable pipe-like section that is pressed or pushed into a block.

Sliding gear: A transmission gear that is splined to the shaft. This gear can be moved back and forth for shifting purposes.

Slip angle: The difference in the actual path taken by a vehicle and the path it would have taken if it had followed the direction the wheels were pointed.

Slip rings: Components mounted on the rotor shaft of a generator to provide current to rotor windings.

Slip yoke: Driveline component that allows back and forth drive shaft movement in response to road conditions.

Slow charger: One that feeds a small current into the battery over a long period of time.

Sludge: Black, mushy deposits found in the interior of the engine. Caused by oxidized petroleum products mixed with dirt and other contaminants.

Smog: Generalized term used to describe air pollution caused by chemical fumes and smoke.

Smoke meter: Device for testing the amount of smoke (ash or soot) in diesel exhaust.

Snap ring: Spring steel ring that snaps into a groove to act as a retainer on a shaft.

Sodium-filled valve: An engine valve that has metallic sodium added to its stem to speed up the transfer of heat from the valve head to the stem and from the guide and block.

Soft failure: One that is intermittent, such as the make/break connection from a loose terminal.

SOHC: Abbreviation for single overhead cam engine.

Solenoid: An electrically operated magnetic device used to operate some unit. An iron core is placed inside a coil. When electricity is applied to the coil, the iron core centers itself in the coil and, as a result, it will exert some force on anything it is connected to.

Solid axle: A single beam which runs between both wheels. Can be used on either the front or rear of the vehicle.

Solid state: Any electrical device that has no moving parts, such as a transistor, diode, or resistor.

South pole: One of the poles of a magnet from which lines of force originate.

Space saver spare tire: A spare tire that is smaller than a standard vehicle tire. Used for emergency purposes only.

Spark: A bridging or jumping of an air gap between two electrodes by electrical current.

Spark advance: To cause a spark plug to fire earlier by altering the ignition timing, by advancing the distributor, or by firing the coil earlier.

Spark gap: Air gap between the center and side spark plug electrodes.

Spark knock: Noise caused by one or more spark plugs firing too early.

Spark line: Oscilloscope line showing voltage needed to maintain an arc across the spark plug gap.

Spark plug: Devices that emit an electrical arc at the tip to ignite the air-fuel mixture in an engine cylinder.

Specific gravity: A relative weight of a given volume of a specific material as compared to an equal volume of water.

Speed control: Method of maintaining a set speed as determined by the driver. Usually referred to as cruise control.

Speed density: Method of determining the amount of air going into the intake manifold by monitoring sensor inputs and calculating the amount of airflow based on the sensor readings.

Speed rating: A letter designation indicating the maximum safe speed that a tire is designed to handle.

Speedometer: Instrument used to determine vehicle speed in miles or kilometers per hour.

Spider gears: Idler and axle gears in the differential that drive the rear axles of a vehicle.

Spindle: Stationary shaft used to support rotating wheel assembly on nondriving wheels.

Splash oiling: Oil distributed to needed areas by spraying or splashing.

Splines: Metal grooves cut into two mating parts.

Split hydraulic system: Brake system that is set up so that there are two separate hydraulic circuits. This provides some braking action if one section fails.

Spool valve: A hydraulic control valve shaped like a thread spool. Used primarily in automatic transmission valve body.

Sprag clutch: A clutch that will allow rotation in one direction, but not in the other. Commonly referred to as an overrunning clutch.

Spring: A suspension component that supports the vehicle chassis and compensates for uneven surfaces. Can use leaf, coil, or torsion bar construction. Other springs are used to close valves and provide tension in other components.

Spring oscillation: A process when a vehicle's spring is rapidly compressed or twisted and rebounds past its normal length and height.

Spring rate: The stiffness or tension; amount of weight needed to compress or bend a spring.

Spring tension: The stiffness of a valve spring.

Sprung weight: The weight of all the parts of the vehicle that is supported by the springs and suspension system.

Spur gear: A gear which has its teeth cut parallel to the shaft.

Squirrel cage: A type of circular fan blade attached to the blower motor in an air distribution system.

Squish area: The area between the piston head at TDC and the cylinder head, which makes up the combustion chamber.

Staking: Making a small dent in cylinder head metal next to a valve seat to hold it in place after replacement.

Stall speed: Highest speed of impeller rotation in a torque converter without rotation of the turbine.

Stall test: Method used to shop-test for transmission slippage.

Stalling: Condition in which the engine merely stops running.

Star wheel: Adjusting screw assembly for drum brakes.

Starter: An electric motor which uses a geardrive to crank (start) the engine.

Starter pinion gear: A small gear on the end of the starter shaft that engages and turns the flywheel ring gear.

Starter solenoid: A high current relay that energizes the starter motor.

Starting system: Electric motor and other components used to rotate the engine until it starts.

Static balance: Condition where a tire's weight mass is evenly distributed around the axis of rotation. If the wheel is raised from the floor and spun several times, it should always stop in a different place.

Static electricity: A charge of electricity generated by friction between two objects.

Static imbalance: Condition where a tire's weight mass is not evenly distributed. This condition is characterized by up and down vibration as the tire's weight mass tries to bring the tire back into balance.

Static pressure reading: A reading made with the engine off to determine whether a system has an adequate refrigerant charge.

Stator: A small hub in the torque converter that improves oil flow. Also, the stationary wire field in an alternator.

Steering arm: An arm bolted or forged to the steering knuckle. Used to transmit force from the tie rod to the knuckle.

Steering axis inclination (SAI): The angle formed by the ball joints, steering knuckle, kingpin, etc.

Steering column: Assembly consisting of the steering wheel, steering shaft, ignition key mechanism, and associated parts.

Steering gear: The assembly containing the gears, valves, and other components used to multiply turning force.

Steering geometry: Term sometimes used to describe the various angles formed by the components making up the vehicle suspension, such as caster, camber, toe, thrust angle, SAI, etc.

Steering knuckle: Component that provides support for wheel spindle or bearings surrounding an axle.

Steering linkage: Components connecting steering gearbox to steering knuckles.

Steering sensor: Provides wheel orientation and speed data to computer in electronic shock absorber system.

Steering shaft: A two-piece shaft that transfers turning motion from the steering wheel to the gearbox.

Steering system: The components that let the driver change direction of a vehicle.

Stethoscope: A device used by auto technicians to better hear internal engine noises.

Stoichiometric: A perfect (chemically correct) air-fuel mixture.

Stoplight: Red warning lights attached to the rear of a vehicle. Used to indicate that the vehicle is slowing down or stopping.

Straight roller bearing: A bearing which uses straight, non-tapered rollers in its construction.

Strap: Connector between cells of a battery.

Strategy-based diagnostics: Detailed step procedure used to locate problems in a logical manner.

Stratified charge: A combustion chamber design that first ignites the air-fuel mixture in a small chamber connected to the main chamber.

Stroke: The distance the piston moves between TDC and BDC.

Strut: A suspension component containing a shock damper cartridge and coil spring. It is used in many vehicles as a replacement for the shock absorber, front upper control arm, and some rear axle control arms.

Strut cartridge: Replaceable shock absorber unit on a MacPherson strut.

Strut rod: Rod that fastens to the control arm and frame to keep arm properly oriented.

Sump: The part of an oil pan that contains the oil.

Sun gear: The center gear in a planetary gear assembly around which the other gears revolve.

Supercharger: A belt-driven compressor which pumps air into an engine's intake manifold. Superchargers are dependent on engine speed and are most efficient at high engine speeds.

Supplemental restraint systems: See *Air bag system.*

Surging: Condition in which engine power fluctuates up and down.

Suspension leveling system: Suspension system designed to keep vehicle level and at proper height even when carrying a heavy load in the truck.

Suspension system: Components that let the wheels move up and down without body movement.

Suspension system computer: One that accepts sensor input and regulates the stiffness of the vehicle's suspension system.

Sway bar: A stabilizer that keeps the vehicle body from leaning excessively in turns.

Sweating: Process of joining two pieces of metal together by placing solder between them and, while clamping them tightly, applying sufficient heat to melt the solder.

Swing axle: Axle provided with U-joints to allow for up-and-down suspension movement. Used on vehicles with differential mounted solidly on frame.

Swirl chamber: Combustion chamber shape that causes the air-fuel mixture to spin as it enters, for better mixing.

Switch: A device to make or break the flow of current through a circuit.

Switching sensor: One that opens or closes a switch in response to a change in condition.

Synchronizer: Assembly of hub, sleeve, and other components that locks the selected output gear to the output shaft to transmit power. It permits meshing of gears without grinding.

T

Tachometer: Device used to measure engine speed in RPM.

Tailpipe: Tubing that carries exhaust from muffler to point at rear or side of the vehicle, where it can be dispersed.

Tandem booster: A brake booster with two internal diaphragms that increase vacuum boost pressure.

Tank gauge unit: A variable resistor float device placed inside the fuel tank to monitor fuel level.

Tank pickup-sending unit: Component that extends into fuel tank to withdraw fuel and send fuel-level information to the fuel gauge.

Tap: Tool for cutting threads inside a hole.

Tapered roller bearing: A bearing that utilizes a series of tapered steel rollers that operate between an outer and inner race.

Tappet clearance: The proper degree of tension (neither too tight nor too loose) for the valve train of an engine.

TDC: Abbreviation for top dead center.

Technical service bulletins (TSB): Information published by vehicle manufacturers in response to vehicle conditions, problems, etc., that may not be diagnosed by normal methods.

Temperature gauge: A dash mounted gauge that is used to indicate engine temperature.

Terminals: The connecting points in an electrical circuit.

Test light: A device that will show the presence of current by lighting a small light.

Thermal efficiency: A comparison of fuel burned to horsepower output.

Thermistor: A device that changes its resistance in relation to heat.

Thermostat: A temperature sensitive device used in cooling systems to control coolant flow in relation to temperature.

Thermostatic air cleaner: An emission control device used to control the temperature of the air entering the engine.

Thermostatic compressor switch: A switch that prevents air conditioning compressor engagement if the ambient air temperature is below a certain point.

Thermostatic switch: Electrical component that shuts off an air conditioning compressor when the evaporator temperature approaches the freezing point.

Throttle body: Throttle plate assembly that contains sensors and vacuum connectors. Used in place of a carburetor throttle plate on fuel injected vehicles.

Throttle body fuel injection (TBI): Fuel injection system that uses one or more fuel injectors mounted above or in the throttle body itself

Throttle plate: Movable valve inside a throttle body or at the base of a carburetor. Opens and closes to admit air to the engine.

Throttle position sensor (TPS): Input sensor to the engine control ECM. Used to monitor throttle position.

Throttle return dashpot: A carburetor solenoid that slows throttle closing to prevent stalling.

Throttle valve: A valve controlled by linkage from the engine throttle plates or by a vacuum modulator operated by engine manifold vacuum. The throttle valve linkage is often referred to as the throttle valve rod or cable, or just TV linkage. It controls the amount of fluid pressure to the shift valve depending on the distance the throttle is moved.

Throw: Offset portion of the crankshaft designed to accept a connecting rod.

Throw-out bearing: Bearing that is used to minimize pressure between the clutch surface and the throw-out fork.

Thrust angle: Imaginary lines of force that cross lengthwise through a vehicle's tires.

Thrust bearing: A bearing designed to resist side pressure.

Thrust load: A pushing or shoving force exerted against one body by another.

Thrust washer: A bronze or hardened steel washer placed between two moving parts to prevent longitudinal movement.

Tie rod: One or more rods used to connect steering arms and knuckles together.

Timed injection: System timed to inject fuel as the intake valves open.

Timing belt: A flexible toothed belt used to rotate the camshaft(s).

Timing chain: Sprocket-and-chain combination that performs same function as timing gears.

Timing gears: Meshing gears on crankshaft and camshaft that rotate camshaft at half crankshaft speed.

Timing light: A stroboscopic unit this is connected to the secondary circuit to produce flashes of light in unison with the firing of a specific spark plug.

Timing marks: Calibrating marks on timing gears or other timing devices. Used to set engine timing.

Timing sprocket: Chain or belt sprockets on the crankshaft and camshaft.

Tire: A rubber covered carcass made of steel and fiber cords.

Tire rating: Information shown on the sidewall to indicate inflation pressure, load carrying ability, size, and other data.

Tire rotation: Moving tires to different wheels periodically to even out wear.

Tire wear pattern: Areas of tread that are worn off, which can provide information on causes of the wear.

Toe: Degree to which opposing wheels are on converging or diverging lines (not parallel). Also, the narrow part of a gear tooth.

Toe-out on turns: The angle the front wheels assume when turning in relation to each other.

Torque: Any turning or twisting force.

Torque converter: Fluid coupling that acts as a clutch on an automatic transmission.

Torque multiplication: Increasing engine torque through the use of a torque converter or gears.

Torque steer: Condition where engine torque will cause the steering wheel to turn slightly to one side. Prevalent on front-wheel drive vehicles.

Torsion bar: A long bar made of spring steel attached in such a way that one end is anchored while the other is free to twist.

Torsion springs: Small coil springs that help absorb the shock and vibration that occur when the clutch engages.

Track rod: Metal rod used to prevent axle side-to-side movement when cornering.

Tracking: The position or direction of the front wheels in relation to the rear wheels.

Traction: The frictional force generated between the tires and the road.

Traction control system (TCS): A computer-controlled system that reduces idle speed and selectively applies the brakes to reduce excessive wheel spin.

Tramp: An up-and-down or hopping motion of the front wheels.

Transaxle: A combination of transmission and differential in one case, used on front-wheel-drive vehicles.

Transducer: A device that changes an action or signal from one medium to another (an electrical pulse into a physical movement, for example).

Transfer case: A transmission driven gearbox that provides driving force to both front and rear propeller shafts on a four-wheel drive vehicle.

Transistor: A semiconductor that is used as a switching device.

Transmission: A device that uses gearing and torque conversion to change the ratio between the engine RPM and driving wheel RPM.

Transmission case: Aluminum housing surrounding and supporting the transmission.

Transmission control solenoid: A computer-controlled solenoid used to control transmission shift patterns.

Transverse: Crosswise; term used to identify an engine rotated 90° from the traditional longitudinal mounting.

Tread: Outer surface of tire that contacts the road.

Trigger wheel: Rotating component with one tooth for each cylinder.

Tripod CV-joint: A constant velocity joint used on the inner part of a CV axle. It is triangular shaped and has three sets of needle bearings on a spider.

Trouble code chart: Diagnostic aid that lists the trouble for each code.

Troubleshooting chart: Diagnostic flow chart that provides step-by-step procedures to test automotive systems.

Tubeless tire: Tire that does not have a separate inner tube to hold air.

Tune-up: The process of checking and adjusting engine timing and replacing spark plugs, filters, etc.

Turbine: The driven fan assembly in a torque converter.

Turbocharger: A turbine device that utilizes exhaust pressure to increase the air pressure going into the cylinders.

Turn signal flasher: Bimetallic strip and heater unit that makes and breaks contact to cause on-off operation of the turn signals.

Turning radius: Diameter of a circle transcribed by the outer front wheel when making a turn.

Two-stroke cycle engine: An engine that requires one complete revolution of the crankshaft to fire each piston once.

Two-wheel alignment: A suspension alignment in which only the front wheel angles are checked and adjusted.

U

Umbrella valve seal: Rubber or plastic seals that fit over opening at top of valve guide to keep oil out of them.

Undercoating: A soft material sprayed on the underside of a vehicle to deaden noise and to resist corrosion.

Understeer: The tendency of a vehicle, when turning a corner, to turn less sharply than the driver intends.

Unibody construction: A vehicle design in which the frame and body is one unit.

Uniform tire quality grading system: A quality grading system that uses letters and numbers to grade a tire's temperature resistance, traction, and tread wear.

Universal joint: A flexible joint that permits changes in driving angles between a driving and driven shaft.

Unleaded gas: Automotive fuel that contains no tetraethyl lead.

Unsprung weight: The weight of all of the vehicle's parts not supported by the springs, such as wheels and tires.

Upshift: A shift into a higher transmission gear.

V

V-belt: A V-shaped belt that is used to turn engine driven accessories such as the alternator, water pump, and air conditioning compressor.

V-engine: An engine in which the cylinders are arranged in two separate banks set at a 60° or 90° angle to each other.

Vacuum: A pressure lower than atmospheric, in an enclosed area.

Vacuum advance: A mechanism installed inside a distributor that can advance and retard ignition timing in response to engine vacuum.

Vacuum booster: Braking system booster actuated by vacuum. Also known as a power booster.

Vacuum pump: A motor that creates extra vacuum to operate vehicle accessories.

Vacuum reservoir: A tank that is used to store vacuum for situations when the engine does not provide enough vacuum for vacuum-operated vehicle accessories.

Vacuum switch: A switch that opens or closes when vacuum is applied.

Vacuum valve: Valve inside radiator cap that allows flow of coolant from recovery tank back into radiator.

Valve: A metal device for opening and closing an aperture or port.

Valve body: Housing containing most of the valves used in operation of an automatic transmission.

Valve core: A threaded valve that screws into a tire's valve stem.

Valve cover: A metal or plastic cover over the top of the cylinder head.

Valve duration: The length of time in degrees of camshaft movement that the valve remains open.

Valve float: A tendency for valves to remain partly open, especially at high speeds. It usually results from a weak or broken valve spring.

Valve guide: A hole machined into the head to support the valves as they ride up and down. May be machined oversized to accept a removable guide.

Valve keeper: Device which snaps into a groove in the valve stem. Used to retain valve and spring assembly in the head.

Valve lash: Valve tappet clearance or total clearance in the valve train with the cam follower on the camshaft's base circle.

Valve lift: The distance that the valves move from fully closed to fully open.

Valve lifter: Component that is moved by the camshaft lobe and in turn moves the push rod or the rocker arm.

Valve overlap: Period in degrees of camshaft rotation in which both the intake and exhaust valves are partially open.

Valve ports: The openings through the head from the intake and exhaust manifolds to the combustion chamber.

Valve rotator: Device that turns the valve to prevent carbon buildup.

Valve seal: A seal that is placed over the valve stem to prevent oil leakage between the stem and the guide.

Valve seat: Machined or installed metal surface on the intake and exhaust ports, against which the valve rests and seals.

Valve spring: Assembly that closes the valve when rocker arm pressure is removed.

Valve stem: The portion of the valve that is inside the head. The stem rides in the valve guide.

Valve timing: Intervals at which valves open and close, determined by camshaft configuration.

Valve train: The parts that operate the engine valves: camshaft, lifters, push rods, rocker arms, and springs.

Vane pump: A type of pump that uses spring loaded vanes that throw off or are moved by liquid or air.

Vapor lock: Condition caused by bubbles in fuel due to overheating. Can cause stalling, hard starting, or failure to start.

Variable displacement compressor: An air conditioning compressor that is designed with internal valves that allow its pumping capacity to be varied. Used to control the system temperature to prevent evaporator icing (condition where the water on the surface of the evaporator freezes, which can block air movement).

Variable induction system: Intake system that uses a moving valve or plenum to allow increased airflow into the engine as needed. Operated by a vacuum switch, controlled by the ECM.

Variable resistance sensor: One with internal resistance that changes in response to changes in a condition (such as temperature).

Vehicle control module (VCM): Name given to computer that controls multiple vehicle systems, including the engine, transmission, and anti-lock brakes.

Vehicle identification number (VIN): Individual series of letters and numbers assigned to a vehicle by the manufacturer at the factory.

Vehicle speed sensor (VSS): Sensor placed in the transmission/transaxle or the rear axle assembly. Used by the engine's ECM to monitor vehicle speed.

Venturi: Restriction (narrowed area) in air horn.

Venturi vacuum: A vacuum that is created as the air entering the venturi suddenly speeds up, which causes it to be stretched. The more the air speeds up, the stronger the vacuum.

Vibration damper: A round metal or rubber weight attached to a crankshaft or other part to minimize vibration.

Viscosity: A measure of a fluid's ability to flow or its thickness.

Viscous coupling: A fluid-filled clutch used in some full-time four-wheel drive transfer cases. During periods of wheel slippage, it transfers torque to the axle that has the best traction.

Volt: Unit of measurement of electrical pressure or force that will move a current of one ampere through a resistance of one ohm.

Voltage: Electrical pressure that causes current flow.

Voltage drop: A lowering of circuit voltage due to excessive lengths of wire, undersize wire, or through a resistance.

Voltage regulator: Device used to control alternator output.

Voltmeter: Instrument used to measure voltage in a given circuit.

Volume: Unit of measurement of space in cubic inches or cubic centimeters.

Volumetric efficiency: A comparison between the actual and ideal efficiency of an internal combustion engine. The comparison is based on the actual volume of fuel mixture drawn in and what would be drawn in if the cylinder were completely filled on each stroke.

W

Waste gate: A valve that vents excess exhaust gas to limit the amount of boost delivered by a turbocharger.

Waste spark: A spark produced by a distributorless or direct ignition system in a cylinder during its exhaust stroke.

Water jacket: The area around the engine cylinders that is left hollow so that coolant may be admitted.

Water pump: The coolant pump; any pump used to circulate coolant through an engine.

Watt: Unit of measurement of electrical power. It is obtained by multiplying volts by amperes.

Waveform: The shape of a given voltage over a period of time.

Wear bar: Solid bars of rubber across the tread that appear when a tire has worn to the safe limit.

Wear sleeve: Bushing between the axle and seal on a front-wheel drive vehicle.

Wet charged: Battery that is filled with electrolyte and fully charged at the factory.

Wet friction: The resistance to movement between two lubricated surfaces.

Wet sleeve: A thick metal barrel inserted into an engine cylinder. Is constantly in contact with engine coolant.

Wheel alignment: Refers to checking and adjusting the various steering angles of both the front and rear wheels.

Wheel balancing machine: Device used to identify locations where weights must be placed to balance a tire.

Wheel bearing: Ball- or roller-bearing assemblies that reduce friction and support the wheels and axles as they rotate.

Wheel cover: A metal or plastic cover that fits over the center section of a steel wheel.

Wheel cylinder: A hydraulic cylinder used with drum brake systems to actuate the brake shoes.

Wheel shimmy: A side-to-side movement caused by dynamic imbalance.

Wheel speed sensor: Magnetic sensor used in an anti-lock brake system to measure wheel speed.

Wheel tracking: Ability of the rear wheels to follow directly behind the front wheels.

Wheel tramp: Hopping (up and down) vibration of a tire and wheel assembly. Different from wheel hop, as it is not torque related.

Wheel weight: A small lead weight used to balance a tire and wheel assembly. Can be clipped or taped onto the wheel.

Wheelbase: The distance between the center of the front tires and the center of the rear tires.

Wide open throttle (WOT) switch: A switch that signals the engine ECM to shut off the air conditioning compressor clutch during wide open throttle acceleration.

Wiggle test: Physically moving wires and connectors to locate broken wires or other causes of intermittent problems.

Winding: Loop of wire on a motor armature that generates a magnetic field.

Windup: A buildup of internal stresses between the front and rear axle parts, caused by differences in speeds between the front and rear axles.

Wiring diagram: Drawings that show relationships of components in an electrical circuit.

Wiring harness: A group of primary wire encased in a paper or plastic sleeve. Used to ease installation and to prevent wire damage.

Work: A force applied to a body, causing it to move. Measured in foot-pounds, watts, or joules.

Worm and sector gear: A type of steering gear that utilizes a worm gear engaging a portion of a gear on a cross shaft.

Wrist pin: See *Piston pin.*

Z

Zener diode: A silicon diode that serves as a rectifier. It will allow current to flow in one direction only until the applied voltage reaches a certain level. Once it reaches this point, the diode allows current to flow in the opposite direction.

Acknowledgments

For their assistance and contributions to this edition of *Modern Automotive Technology,* the author and publisher would like to thank the following:

3-M Company
AC-Delco
Airesearch Industrial Div.
Airtex Automotive Division
Alloy
Alfa Romeo, Inc.
American Bosch
American Bosch Diesel Products
American Hammered Automotive
    Replacement Division
American Honda Motor Co.
Ammco Tools, Inc.
AP Parts Co.
Applied Power, Inc.
Armstrong Bros. Tool Co.
Aston Martin Lagonda, Inc.
Automotive Control System Group
Beam Products Mfg. Co.
Bear Automotive
Belden Corp.
Bendix
Binks Mfg. Co.
Black & Decker, Inc.
Blackhawk Mfg. Co.
BMW of North America, Inc.
Bonney Tools
Borg-Warner Corp.
Bosch Power Tools
British Leyland Motors, Inc.
Buick Motor Car Division
C.A. Laboratories, Inc.
Cadillac Motor Car Division
Carter Div. of AFC Inc., Brodhead Garrett
Caterpillar Tractor Co.
Champion Spark Plug Co.
Chevrolet Motor Division
Chicago Rawhide Mfg. Co.
Chrysler Motor Corp.
Clayton Manufacturing Co.
Cleveland Motive Products
Clevite Corp.
Colt Industries
CRC Chemicals
Cummins Engine Co., In
Cy-lent Timing Gears Corp.
D. A. B. Industries, Inc.
Dake
Dana Corp.
Dayco Corp.
Debcor, Inc.
Deere & Co.
Delco-Remy Div. of GMC
Detroit Art Services, Inc.
Detroit Diesel Allison Div.
Duro-Chrome Hand Tools
E & L Instruments

Edu-Tech—A Division of Commercial
    Service Co.
Ethyl Corp.
Exxon Co. USA
Fairgate Fuel Co., Inc.
Federal Mogul
Fel-Pro Inc.
Fiat Motors of North America, Inc.
Firestone Tire and Rubber Co.
Florida Dept. of Vocational Education
Fluke Corporation
FMC Corporation
Ford Motor Company
Ford Parts and Service Division
Fram Corp.
Gates Rubber Co.
General Tire & Rubber Co.
GMC Truck & Coach Division
Goodall Manufacturing Co.
Gould Inc.
Gunk Chemical Div.
H.B. Egan Manufacturing Co.
Hartridge Equipment Corp.
Hastings Mfg. Co.
Heli-Coil Products
Helm Inc.
Hennessy Industries
Holley Carburetor Div.
Hunter Engineering Co.
Infinity
Ingersoll-Rand Co.
International Harvester Co.
Isuzu
J. McKelvey Co.
Jaguar
K-D Tool Manufacturing Co.
Kansas Jack, Inc.
Keller Crescent Co.
Kent-Moore
Kern Manufacturing Co., Inc.
Killian Corp.
Kline Diesel Acc.
Kwick-Way Mfg. Co.
Lexus
Lincoln St. Louis, Div. of McNeil Corp.
Lisle Corp.
Lister Diesels, Inc.
Lufkin Instrument Div.—Cooper
    Industries Inc.
Lyons, Marquette Corp.
Mac Tools Inc.
Maremont Corp.
Maserati Automobiles, Inc.
Mazda Motors of America, Inc.
McCord Replacement Products Division
Mercedes-Benz of North America, Inc.
Minnesota Curriculum Services Center
Mitsubishi Motor Sales of America
Mobile Oil Corp.
Moog Automotive Inc.
Motor Vehicle Manufacturer's Assn.
Motorola

NAPA
National Institute for Automotive Service
    Excellence (ASE)
Nissan Motor Corp.
Oldsmobile Division of GM
OTC Tools & Equipment
Owatonna Tool Co.
Parker Hannifin Corp.
Peugeot, Inc.
Pontiac Motor Division.
Precision Brand Products
Proto Tool Co.
Purolator Filter Division
Quaker State Corp.
Renault USA, Inc.
Rochester Div. of GM
Rolls-Royce, Inc.
Roto-Master
Saab-Scandia of America, Inc.
SATCO: Schwitzer Cooling Systems
Sealed Power Corp—Replacement Products
    Group
Sears, Roebuck and Co.
Sellstrom Mfg. Co.
Sern Products, Inc.
Shell Oil Co.
Simpson Electric Co.
Sioux Tools, Inc.
Snap-on Tools Corp.
Speed Clip Sales Co.
Stanadyne, Inc.
Stewart-Warner
Subaru of America, Inc.
Sun Electric Corp.
Sunnen Product Co.
Test Products Divison—The Allen Group, Inc.
Texaco Inc.
The B.F. Goodrich Co.
The DeVilbiss Co.
The Eastwood Co.
The Echlin Mfg. Co.
The Goodyear Tire & Rubber Co.
The L.S. Starrett Co.
TIF Instruments
Tomco (TI) Inc.
Toyota Motor Sales, USA, Inc.
TRW Inc.
TWECO Products, Inc.
U. S.A. Automobiles Citroen
Uniroyal, Inc.
Vaco Products Co.
Valvoline Oil Co.
Victor Gasket Co.
Volkswagen of America, Inc.-Porsche
Volvo of America.
Waukesha Engine Division, Dresser
    Industries, Inc.
Weatherhead Co.
White Diesel Div.
A special thanks goes to my wife (Jeanette)
    and children (Danielle, Jimmy, and DJ).

Index

A